# PLANT PHYSIOLOGY
*Fifth Edition*

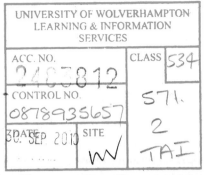

# PLANT PHYSIOLOGY

## Fifth Edition

## Lincoln Taiz
Professor Emeritus
*University of California, Santa Cruz*

## Eduardo Zeiger
Professor Emeritus
*University of California, Los Angeles*

Sinauer Associates Inc., Publishers
Sunderland, Massachusetts U.S.A.

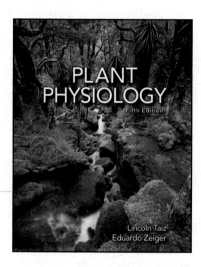

**Front Cover**

Southern beech trees (*Nothofagus* sp.) and pandani (also known as giant grass trees, *Richea pandanifolia*) grow among mosses in a Tasmanian rainforest. Walls of Jerusalem National Park, Tasmania, Australia. Photo © Rob Blakers/Photolibrary.com.

**Back Cover**

Torch ginger (*Etlingera elatior*) is native to the rainforests of Indonesia, Malaysia, and Thailand, where the flower buds and seed pods are used in cooking. Photo courtesy of Scott Zona.

**Plant Physiology, Fifth Edition**

Copyright © 2010 by Sinauer Associates, Inc.

For information, address
Sinauer Associates, Inc., 23 Plumtree Road, Sunderland, MA 01375 U.S.A.
FAX: 413-549-1118
E-mail: publish@sinauer.com
Internet: www.sinauer.com

**Library of Congress Cataloging-in-Publication Data**

Taiz, Lincoln.
 Plant physiology / Lincoln Taiz, Eduardo Zeiger. -- 5th ed.
    p. cm.
 Includes index.
 ISBN 978-0-87893-866-7 (casebound)
 1. Plant physiology. I. Zeiger, Eduardo. II. Title.
 QK711.2.T35 2010
 571.2--dc22

                            2010014391


Printed in U.S.A.
5 4 3 2 1

# Brief Table of Contents

# Authors

**Lincoln Taiz** is Professor Emeritus of Molecular, Cellular, and Developmental Biology at the University of California at Santa Cruz. He received his Ph.D. in Botany from the University of California at Berkeley in 1971. Dr. Taiz's main research focus has been on the structure, function, and evolution of vacuolar $H^+$-ATPases. He has also worked on gibberellins, cell wall mechanical properties, metal tolerance, auxin transport, and stomatal opening.

**Eduardo Zeiger** is Professor Emeritus of Biology at the University of California at Los Angeles. He received a Ph.D. in Plant Genetics at the University of California at Davis in 1970. His research interests include stomatal function, the sensory transduction of blue-light responses, and the study of stomatal acclimations associated with increases in crop yields.

## Principal Contributors

**Richard Amasino** is a Professor in the Department of Biochemistry at the University of Wisconsin-Madison. He received a Ph.D. in Biology from Indiana University in 1982 in the laboratory of Carlos Miller, where his interests in the induction of flowering were kindled. One of his research interests continues to be the mechanisms by which plants regulate the timing of flower initiation. (Chapter 25)

**Sarah M. Assmann** is a Professor in the Biology Department at the Pennsylvania State University. She received a Ph.D. in the Biological Sciences at Stanford University. Dr. Assmann's present research focuses on the systems biology of cellular signaling, particularly with regard to regulation of guard cell function, and on the roles of heterotrimeric G-proteins in plant growth, plasticity, and environmental response. (Chapter 6)

**Malcolm J. Bennett** is a Professor of Plant Sciences at the University of Nottingham, UK. He received his Ph.D in Biological Sciences from the University of Warwick in 1989. His research interests focus on hormone-regulated root development employing molecular genetic and systems biology approaches. (Chapter 14)

**Robert E. Blankenship** is a Professor of Biology and Chemistry at Washington University in St. Louis. He received his Ph.D. in Chemistry from the University of California at Berkeley in 1975. His professional interests include mechanisms of energy and electron transfer in photosynthetic organisms, and the origin and early evolution of photosynthesis. (Chapter 7)

**Arnold J. Bloom** is a Professor in the Department of Sciences at the University of California at Davis. He received a Ph.D. in Biological Sciences at Stanford University in 1979. His research focuses on plant-nitrogen relationships, especially the differences in plant responses to ammonium and nitrate as nitrogen sources. He is the co-author with Emanuel Epstein of the textbook, *Mineral Nutrition of Plants* and author of the textbook, *Global Climate Change: Convergence of Disciplines*. (Chapters 5 & 12)

**John Browse** is a Professor in the Institute of Biological Chemistry at Washington State University. He received his Ph.D. from the University of Aukland, New Zealand, in 1977. Dr. Browse's research interests include the biochemistry of lipid metabolism and the responses of plants to low temperatures. (Chapter 11)

**Thomas Brutnell** is an Associate Scientist at the Boyce Thompson Institute for Plant Research at Cornell University. He received his Ph.D. in Biology from Yale University in 1995. His research interests focus on $C_4$ photosynthesis in maize and dissecting the role of phytochromes in maize and other agronomically important grasses. (Chapter 17)

**Bob B. Buchanan** is a Professor of Plant and Microbial Biology at the University of California at Berkeley. He continues to work on thioredoxin-linked regulation in photosynthesis, seed germination, and related processes. His findings with cereals hold promise for societal application. (Chapter 8)

**Joanne Chory** is an Investigator with the Howard Hughes Medical Institute, Professor at The Salk Institute for Biological Studies, and Adjunct Professor of Biology at the University of California, San Diego. She received a Ph.D. in Microbiology from the University of Illinois at Urbana-Champaign. Dr. Chory's research focuses on plant responses to changes in the light environment. Her group's genetic studies have led to the identification of the plant steroid receptor and several components in the steroid signaling pathway. (Chapter 24)

**Daniel J. Cosgrove** is a Professor of Biology at the Pennsylvania State University at University Park. His Ph.D. in Biological Sciences was earned at Stanford University. Dr. Cosgrove's research interest is focused on plant growth, specifically the biochemical and molecular mechanisms governing cell enlargement and cell wall expansion. His research team discovered the cell wall loosening proteins called expansins and is currently studying the structure, function, and evolution of this gene family. (Chapter 15)

**Susan Dunford** is an Associate Professor of Biological Sciences at the University of Cincinnati. She received her Ph.D. from the University of Dayton in 1973 with a specialization in plant and cell physiology. Dr. Dunford's research interests include long-distance transport systems in plants, especially translocation in the phloem, and plant water relations. (Chapter 10)

**James Ehleringer** is at the University of Utah where he is a Distinguished Professor of Biology and serves as Director of the Stable Isotope Ratio Facility for Environmental Research (SIRFER). His research focuses on understanding terrestrial ecosystem processes through stable isotope analyses, gas exchange and biosphere–atmosphere interactions, water relations, and stable isotopes applied to homeland security issues. (Chapter 9)

**Jürgen Engelberth** is an Assistant Professor of Plant Biochemistry at the University of Texas at San Antonio. He received his Ph.D. in Plant Physiology at the Ruhr-University Bochum, Germany in 1995 and did postdoctoral work at the Max Planck Institute for Chemical Ecology, at USDA, ARS, CMAVE in Gainesville, and at Penn State University. His research focuses on signaling involved in plant–insect and plant–plant interaction. (Chapter 13)

**Ruth Finkelstein** is a Professor in the Department of Molecular, Cellular and Developmental Biology at the University of California at Santa Barbara. She received her Ph.D., also in Molecular, Cellular and Developmental Biology, from Indiana University in 1986. Her research interests include mechanisms of abscisic acid response, and their interactions with other hormonal, environmental, and nutrient signaling pathways. (Chapter 23)

**Lawrence Griffing** is an Associate Professor in the Biology Department at Texas A&M University. He received his Ph.D. in Biological Sciences at Stanford University. Dr. Griffing's research mainly focuses on plant cell biology, concentrating on the regulation of the dynamics of the endomembrane network. He also participates in several programs to incorporate authentic inquiry in genetics, biochemistry, cell biology, and behavioral ecology into undergraduate and pre-undergraduate programs. (Chapter 1)

**Paul M. Hasegawa** is a Professor of Plant Physiology at Purdue University. He earned a Ph.D. in Plant Physiology at the University of California at Riverside. His research has focused on plant morphogenesis and the genetic transformation of plants. He has used his expertise in these areas to study many aspects of stress tolerance in plants, especially ion homeostasis. (Chapter 26)

**N. Michele Holbrook** is a Professor in the Department of Organismic and Evolutionary Biology at Harvard University. She received her Ph.D. from Stanford University in 1995. Dr. Holbrook's research group focuses on water relations and water transport through the xylem. (Chapters 3 & 4)

**Joseph Kieber** is a Professor in the Biology Department at the University of North Carolina at Chapel Hill. He earned his Ph.D. in Biology from the Massachusetts Institute of Technology in 1990. Dr. Kieber's research interests include the role of hormones in plant development, with a focus on the signaling pathways for ethylene and ctyokinin, as well as circuitry regulating ethylene biosynthesis. (Chapters 21 & 22)

**Andreas Madlung** is a Professor in the Department of Biology at the University of Puget Sound. He received a Ph.D. in Molecular and Cellular Biology from Oregon State University in 2000. Research in his laboratory addresses fundamental questions concerning the influence of genome structure on plant physiology and evolution, especially with respect to polyploidy. (Chapter 2)

**Michael V. Mickelbart** is an Assistant Professor at Purdue University. He received his Ph.D. in Plant Physiology at Purdue. Dr. Mickelbart studies the genetic and physiological basis of plant water use and abiotic stress tolerance. (Chapter 26)

**Alistair Middleton** is a postdoctoral research fellow at the Centre for Plant Integrative Biology, Nottingham, UK. He received his Ph.D. in Mathematics from the University of Nottingham, UK in 2007. Dr. Middleton's research focuses on developing mathematical models of hormone response networks. (Chapter 14)

**Ian M. Møller** is Professor at Department of Genetics and Biotechnology at Aarhus University, Denmark. He received his Ph.D. in Plant Biochemistry from Imperial College, London, UK. He has worked at Lund University, Sweden and, more recently, at Risø National Laboratory and the Royal Veterinary and Agricultural University in Copenhagen, Denmark. Professor Møller has investigated plant respiration throughout his career. His current interests include turnover of reactive oxygen species and the role of protein oxidation in plant cells. (Chapter 11)

**Angus Murphy** is a Professor in the Department of Horticulture and Landscape Architecture at Purdue University. He earned his Ph.D. in Biology from the University of California, Santa Cruz in 1996. Dr. Murphy studies the regulation of auxin transport and the mechanisms by which transport proteins are regulated in plastic plant growth. (Chapter 19)

**Benjamin Péret** is a Marie Curie research fellow in Malcolm Bennett's lab at the Centre for Plant Integrative Biology, Nottingham, UK. He received his Ph.D. in Plant Physiology from the University of Montpellier, France. Dr. Péret's postdoctoral research interest focuses on building multiscale models of auxin response in the context of lateral root emergence. (Chapter 14)

**Allan G. Rasmusson** is a Professor of Plant Physiology at Lund University in Sweden. He received his Ph.D. in Plant Physiology at the same university in 1994. Dr. Rasmusson's current research centers on redox homeostasis and regulation in respiratory metabolism, especially regarding the roles of energy bypass enzymes. (Chapter 11)

**David E. Salt** is a Professor of Plant Biology at Purdue University. He received a Ph.D. in Plant Biochemistry from Liverpool University, UK, in 1989. He is interested in the understanding of gene networks regulating mineral ion homeostasis in plants, and the evolutionary forces that shape this regulation. A current interest is the coupling of high-throughput elemental analysis with bioinformatics and genomics, seeking to dissect the genetic architecture underlying natural variation in mineral ion uptake and accumulation in *Arabidopsis thaliana*. (Chapter 26)

**Darren Sandquist** is a Professor of Biological Science at California State University, Fullerton. He received his Ph.D. from the University of Utah. His research focuses on plant ecophysiological responses to disturbance, invasion, and climate change in arid and semi-arid ecosystems. (Chapter 9)

**Sigal Savaldi-Goldstein** is a Principal Investigator at the Technion, Israel. She received her Ph.D in Plant Sciences from the Weizmann Institute, Israel in 2003, and did her postdoctoral training at the Salk Institute. Dr. Savaldi-Goldstein's research group is specializing in the plant hormone field, primarily focusing on brassinosteroids. Her group's main interest is to understand how tissue identity and mechanical stimuli shape hormonal responses and how these processes are tuned by distinct environmental conditions to ensure coherent growth. (Chapter 24)

**Wendy Kuhn Silk** is Professor and Quantitative Plant Biologist at the University of California at Davis. She received her Ph.D. in Botany from the University of California at Berkeley in 1975. Her research is on plant-environment interactions including growth responses to environmental variation, nutrient cycling, and rhizosphere biology. Active in the Davis Art-Science fusion program, she often asks students to write songs and poetry as a way to learn and communicate science. (Appendix 2)

**Valerie Sponsel** is an Associate Professor in the Biology Department at the University of Texas at San Antonio. She received her Ph.D. from the University of Wales, U.K. in 1972 and a D.Sc from the University of Bristol, U.K. in 1984. Her research has focused on the biosynthesis and catabolism of gibberellins, and more recently, on the interaction of auxin with gibberellin biosynthesis and signaling pathways. (Chapter 20)

**Bruce Veit** is a senior scientist at AgResearch in Palmerston North, New Zealand. He received his Ph.D. in Genetics from University of Washington, Seattle in 1986 before undertaking postdoctoral research at the Plant Gene Expression Center in Albany, California. Dr. Veit's current research interests focus on mechanisms that influence the determination of cell fate. Dr. Veit wishes to acknowledge the contribution of Dr. Paul Dijkwel of the Institute of Molecular Biosciences, Massey University, for updating the senescence section. (Chapter 16)

**Philip A. Wigge** is a Principal Investigator at the John Innes Centre in Norwich, UK. He received his Ph.D in Cell Biology from the University of Cambridge, UK, in 2001. Dr. Wigge has studied how florigen controls plant development at the Salk Institute, CA, in the laboratory of Detlef Weigel. His research group is fascinated by how plants are able to sense and respond to climate change. (Chapter 25)

**Ricardo A. Wolosiuk** is Professor at the University of Buenos Aires and senior scientist at Instituto Leloir (Buenos Aires). He received his Ph.D. in Chemistry from the University of Buenos Aires in 1974. His current research centers on the modulation of photosynthetic $CO_2$ assimilation and the structure and function of plant proteins. (Chapter 8)

## Reviewers

Nick Battey
*University of Reading*

Magdalena Bezanilla
*University of Massachusetts, Amherst*

Ildefonso Bonilla
*Universidad Autònoma de Madrid*

Federica Brandizzi
*Michigan State University*

Thomas Buckley
*Sonoma State University*

Xumei Chen
*University of California, Riverside*

Asaph Cousins
*Washington State University*

Emmanuel Delhaize
*CSRIO*

Donald Geiger
*University of Dayton*

William Gray
*University of Minnesota*

Philip Harris
*University of Auckland*

Peter Hedden
*Rothamsted Research*

J. S. Heslop-Harrison
*University of Leicester*

John Hess
*Virginia Tech University*

Theodore Hsaio
*University of California, Davis*

Nick Kaplinsky
*Swarthmore College*

Eric Kramer
*Simon's Rock College of Bard*

Jianming Li
*University of Michigan*

David Macherel
*Universite d'Angers*

Massimo Maffei
*University of Turin*

Julin Maloof
*University of California, Davis*

Maureen McCann
*Purdue University*

Peter McCourt
*University of Toronto*

Sabeeha Merchant
*University of California, Los Angeles*

Jan Miernyk
*University of Missouri*

Don Ort
*University of Illinois at Urbana-Champaign*

Zhi Qi
*University of Connecticut*

Hitoshi Sakaibara
*RIKEN Plant Science Center*

George Schaller
*Dartmouth College*

Kathrin Schrick
*Kansas State University*

Julian Schroeder
*University of California, San Diego*

Johannes Stratmann
*University of South Carolina*

Tai-Ping Sun
*Duke University*

Sakis Theologis
*USDA Plant Gene Expression Center*

E. G. Robert Turgeon
*Cornell University*

John Ward
*University of Minnesota*

Philip A. Wigge
*John Innes Centre*

Yanhai Yin
*Iowa State University*

# Chapter and Appendix Histories

So many colleagues have given generously of their time and expertise to *Plant Physiology* over many editions of the text that it seems only fitting to record them all with our thanks. We have therefore added this important new section to the front matter, which recognizes the continuing intellectual legacy of *all* the chapter authors over the life time of the text, from the First Edition to the Fifth. Every one of these distinguished scholars has made important contributions to the worldwide success of *Plant Physiology*, which is now available in ten languages.

L.T.
E.Z.

**CHAPTER 1**   *Plant Cells*  Stephen M. Wolniak, Professor of Botany, University of Maryland (1E); Lincoln Taiz, Professor of Molecular, Cellular and Developmental Biology, University of California, Santa Cruz (1E–4E); Lawrence R. Griffing, Associate Professor of Biology, Texas A&M University (5E).

**CHAPTER 2**   *Genome Organization and Gene Expression* Lincoln Taiz, Professor of Molecular, Cellular and Developmental Biology, University of California, Santa Cruz (1E–4E); Andreas Madlung, Professor of Biology, University of Puget Sound (5E).

**CHAPTER 3**   *Water and Plant Cells*   Daniel Cosgrove, Professor of Biology, Pennsylvania State University (1E and 2E); Michele Holbrook, Professor of Organismic and Evolutionary Biology, Harvard University (3E–5E).

**CHAPTER 4**   *Water Balance of Plants*   Daniel Cosgrove, Professor of Biology, Pennsylvania State University (1E and 2E); Michele Holbrook, Professor of Organismic and Evolutionary Biology, Harvard University (3E–5E).

**CHAPTER 5**   *Mineral Nutrition*   Donald P. Briskin, Professor of Crop Sciences, University of Illinois at Urbana, Champaign (1E); Arnold Bloom, Professor, Department of Sciences, University of California at Davis (2E–5E).

**CHAPTER 6**   *Solute Transport*   George W. Bates, Professor of Biological Science, Florida State University (1E); Ronald J. Poole, Professor of Biology, McGill University (2E and 3E); Sarah M. Assmann, Professor of Biology, Pennsylvania State University (4E and 5E).

**CHAPTER 7**   *Photosynthesis: The Light Reactions*   Robert E. Blankenship, Professor of Biology and Chemistry, Washington University, St. Louis (1E–5E).

**CHAPTER 8**   *Photosynthesis: The Carbon Reactions* George H. Lorimer, professor, Department of Chemistry and Biochemistry, University of Maryland (1E); Bob B. Buchanan, Professor of Plant and Microbial Biology, University of California at Berkeley, and Ricardo A. Wolosiuk, Professor, Univ. de Buenos Aires, Senior Scientist, Instituto Leloir (Buenos Aires) (2E–5E).

**CHAPTER 9**   *Photosynthesis: Physiological and Ecological Considerations*   Thomas Sharkey, Professor, Department of Biochemistry and Molecular Biology, Michigan State University (1E); Thomas Vogelmann, Professor of Plant Biology, University of Vermont (2E and 3E); James Ehleringer, Professor of Biology, University of Utah (4E and 5E); and Darren Sandquist, Professor of Biological Science, California State University, Fullerton (5E).

**CHAPTER 10**   *Translocation in the Phloem*   Susan Dunford, Associate Professor of Biological Sciences, University of Cincinnati (1E–5E).

**CHAPTER 11**   *Respiration and Lipid Metabolism*   James Siedow, Professor of Biology, Duke University (1E); James Siedow and John Browse, Professor, Institute of Biological Chemistry, Washington State University (2E); Ian Max Møller, Head of Department of Genetics and Biotechnology at Aarhus University, Denmark; Allan G. Rasmusson, Professor of Plant Physiology, Lund University, Sweden, and John Browse (3E–5E).

**CHAPTER 12** *Assimilation of Mineral Nutrients* Donald P. Briskin, Professor of Crop Sciences, University of Illinois at Urbana, Champaign (1E); Arnold Bloom, Professor, Department of Sciences, University of California at Davis (2E–5E).

**CHAPTER 13** *Secondary Metabolites and Plant Defense* Jonathan Gershenzon, Director of Max-Planck Institute for Chemical Ecology (1E–3E); Jürgen E. Engelberth, Assistant Professor of Plant Biochemistry, University of Texas, San Antonio (4E and 5E).

**CHAPTER 14** *Signal Transduction* Lincoln Taiz, Professor of Molecular, Cellular and Developmental Biology, University of California, Santa Cruz (1E–4E); Malcolm Bennett, Professor of Plant Sciences, University of Nottingham, with Dr. Benjamin Peret and Dr. Alistair Middleton, University of Nottingham (5E).

**CHAPTER 15** *Cell Walls: Structure, Biogenesis, and Expansion* Daniel Cosgrove, Professor of Biology, Pennsylvania State University (2E–5E).

**CHAPTER 16** *Growth and Development* Donald E. Fosket, Professor of Developmental and Cell Biology, University of California at Irvine (1E–3E); Adrienne Hardham, Fellow of Plant Cell Biology Group, Australian National University at Canberra (1E); Wendy Kuhn Silk, Professor and Quantitative Plant Biologist, University of California at Davis (2E and 3E); Bruce Veit, Senior Scientist, AgResearch, Palmerston, New Zealand (4E and 5E).

**CHAPTER 17** *Phytochrome and Light Control of Plant Development* Stanley Roux, Professor of Botany, University of Texas, Austin (1E); Jane Silverthorne, Professor of Biology, University of California, Santa Cruz (2E and 3E); Thomas Brutnell, Associate Scientist, Boyce Thompson Institute at Cornell University (4E and 5E).

**CHAPTER 18** *Blue-Light Responses: Morphogenesis and Stomatal Movements* Eduardo Zeiger, Professor of Biology, University of California at Los Angeles (1E–5E).

**CHAPTER 19** *Auxin: The First Discovered Plant Growth Hormone* Richard G. Stout, Professor of Biology, Montana State University (1E); Paul Bernasconi, Director of Biochemistry for Syngenta Biotechnology, Research Triangle Park, North Carolina (2E); Angus Murphy, Professor of Horticulture and Landscape Architecture, Purdue University (3E– 5E).

**CHAPTER 20** *Gibberellins: Regulators of Plant Height and Seed Germination* Peter J. Davies, Professor of Plant Physiology, Cornell University (1E–3E); Valerie Sponsel, Associate Professor of Biology, University of Texas, San Antonio (4E and 5E).

**CHAPTER 21** *Cytokinins: Regulators of Cell Division* Donald E. Fosket, Professor of Developmental and Cell Biology, University of California, Irvine (1E and 2E); Joseph Kieber, Professor of Biology, University of North Carolina, Chapel Hill (3E–5E).

**CHAPTER 22** *Ethylene: The Gaseous Hormone* Shimon Gepstein, Professor of Biology, Israel Institute of Technology (1E); Joseph Kieber, Professor of Biology, University of North Carolina, Chapel Hill (2E–5E).

**CHAPTER 23** *Abscisic Acid: A Seed Maturation and Stress-Response Hormone* Shimon Gepstein, Professor of Biology, Israel Institute of Technology (1E); Joseph Kieber, Professor of Biology, University of North Carolina, Chapel Hill (2E); Ruth Finkelstein, Professor of Molecular, Cellular, and Developmental Biology, University of California, Santa Barbara (3E–5E).

**CHAPTER 24** *Brassinosteroids: Regulators of Cell Expansion and Development* Sigal Savaldi-Goldstein, Principal Investigator in the Faculty of Biology, Technion, Israel; Joanne Chory, Professor of Plant Biology, The Salk Institute, and Investigator with the Howard Hughes Medical Institute, La Jolla, California (4E and 5E).

**CHAPTER 25** *The Control of Flowering* Daphne Vince-Prue, Professor of Botany, University of Reading, England (1E); Donald E. Fosket, Professor of Developmental and Cell Biology, University of California, Irvine (2E); Richard Amasino, Professor of Biochemistry, University of Wisconsin, Madison (3E–5E).

**CHAPTER 26** *Responses and Adaptations to Abiotic Stress* John Radin, Research Leader, U.S. Department of Agriculture, deceased (1E); Ray Bressan, Professor of Plant Physiology, Purdue University, Malcom C. Drew, Professor of Plant Physiology, Texas A&M University, and Paul M. Hasegawa, Professor of Plant Physiology, Purdue University (2E); Robert Locy, Professor of Biological Science, Auburn University, Paul M. Hasegawa and Ray Bressan (3E and 4E); Michael V. Mickelbart, Assistant Professor, Purdue University, Paul M. Hasegawa, and David E. Salt, Professor of Plant Biology, Purdue University (5E).

**APPENDIX 1** *Energy and Enzymes* Frank Harold, Professor of Biochemistry, Colorado State University (1E–5E).

**APPENDIX 2** *The Analysis of Plant Growth* Wendy Kuhn Silk, Professor and Quantitative Plant Biologist, University of California, Davis (5E).

**APPENDIX 3** *Hormone Biosynthetic Pathways* Angus Murphy, Valerie Sponsel, Joseph Kieber, Ruth Finkelstein, Sigal Savaldi-Goldstein, Joanne Chory (5E).

# Preface

Once again we are privileged to present to the worldwide community of plant biologists a new edition of *Plant Physiology*. The Fifth Edition brackets the 20 years since the publication of the First Edition in 1991, and it embodies the dramatic progress experienced by the plant sciences during these two decades. The field of genomics launched by the publication of the Arabidopsis genome in 1999 has fueled major advances in bioinformatics, genomic analysis, molecular evolution, signal transduction, and systems biology. Exciting progress has also been achieved in the more established areas of plant physiology, such as bioenergetics, photosynthesis, environmental stress, and plant-microbial and plant-viral interactions, helping to propel the plant sciences into the forefront of biological research. As always, our challenge has been to keep pace with the new advances, while continuing to provide students with the vital background information they need for a better understanding of the principles and experimental techniques that characterize the science of plant biology today.

More than ever before, the Fifth Edition is the product of a remarkable collaboration of a truly outstanding team of authors and publishing professionals. All of our contributing authors are not only engaged in groundbreaking research in their respective fields, but can also draw on a thorough understanding of the pedagogical needs of students. Thanks to their efforts we have been able to incorporate into the text the most exciting new developments in the field, written in a lively style, and richly illustrated with drawings and micrographs of the highest scientific and artistic quality. As in previous editions, we have relied upon the advice of an impressive panel of reviewers, as well as numerous suggestions from instructors using the text, to ensure that each chapter is not only as authoritative as possible, but also readily accessible to students with varied backgrounds in biology.

Several new features of the Fifth Edition are noteworthy. Two new chapters have been added: Chapter 2, *Genome Organization and Gene Expression*, and Chapter 14, *Signal Transduction*. The chapters on plant hormones have been streamlined by simplifying the treatments of hormone biosynthesis and transferring the figures of the detailed pathways to a new Appendix 3. This Appendix also contains sections on Concepts in Bioenergetics and Plant Kinematics which were formerly available only on the text's website. Chapter 1, *Plant Cells*, has been extensively revised and updated, with stunning new micrographs and diagrams. Chapter 26, *Responses and Adaptations to Abiotic Stress*, has been completely rewritten and updated to reflect new paradigms in the field. Indeed, every chapter in the text has been updated to varying degrees to improve pedagogy and to reflect the current state-of-the-art of their respective fields. As in previous editions, the website (www.plantphys.net/) provides supplementary material referenced throughout the text—Web Topics and Web Essays. The website is continually upgraded and expanded, and provides links to relevant sites to enhance the student's learning experience.

As in previous editions, responsibility for the oversight of the chapters, and for integrating them into a coherent whole, has been divided between us, with one exception, Chapter 26, which was edited jointly. The division of labor was as follows: E.Z. was in charge of chapters 3–12, 18, and 26, while L.T. had oversight of chapters 1, 2, 13–17, 19–26, and Appendices 1–3.

We wish to express our gratitude to the outstanding team of professionals assembled by our publisher, Sinauer Associates, especially to our two production editors, Laura Green and Kathaleen Emerson. Laura shepherded the book nearly to the end, applying her deep understanding of plant biology, sharp eye for detail, tireless energy, and tactfulness to the task. Kathaleen, who has been with us since the Third Edition, saw the project through to completion with her usual diligence, patience, and exceptional editorial skills that we have come to expect from previous editions. Our long-time developmental editor, James Funston, deserves much of the credit for the pedagogical excellence of the text. James's unbending commitment to ensuring that the chapters are written in clear, well-organized, and accessible prose pitched at the appropriate level, and to an illustration program that is self-explanatory and makes no assumptions of prior knowledge, have been essential components of the overall success of the book. We are also fortunate to avail ourselves of the services of a truly outstanding scientific illustrator, Elizabeth Morales, whose beautiful drawings have graced the pages of *Plant Physiology* since the First Edition. Also we thank Carrie Crompton, our copy editor, David McIntyre, our extremely resourceful image specialist, Jason Dirks, our webmaster, Susan McGlew and Marie Scavotto, our marketing specialists, and Joanne Delphia and Chris Small for their expertise and a clean, contemporary design. And we offer our deepest acknowledgment to our publisher, Andy Sinauer, for his commitment, highest professional standards and for making Sinauer Associates our home. As always we thank Lee Taiz and Yael Zeiger-Fischman for their love and unconditional support.

Lincoln Taiz
Eduardo Zeiger
May 2010

# Media and Supplements

## to accompany Plant Physiology 5E

**eBook** (ISBN 978-0-87893-507-9)
*www.coursesmart.com*

New for the Fifth Edition, *Plant Physiology* is available as an eBook via CourseSmart, at a substantial discount off the price of the printed textbook. The CourseSmart eBook reproduces the look of the printed book exactly, and includes convenient tools for searching the text, highlighting, and note-taking.

### For the Student
**Companion Website**
*www.plantphys.net*

Available free of charge, this website supplements the coverage provided in the textbook with additional and more advanced material on selected topics of interest and current research. The site includes the following:

- *Web Topics:* Additional coverage of selected topics across all chapters
- *Web Essays:* Articles on cutting-edge research, written by the researchers themselves
- *Study Questions:* A set of short answer-style questions for each chapter
- *Suggested Readings:* Chapter-specific recommended readings for further study or research

References to specific Web Topics and Essays are included throughout each textbook chapter, as well as at the end of each chapter.

### For the Instructor
**Instructor's Resource Library**
(Available to qualified adopters)

The *Plant Physiology,* Fifth Edition Instructor's Resource Library includes the complete collection of visual resources from the textbook for use in preparing lectures and other course materials. The textbook figures have all been sized, formatted, and color-adjusted for optimal image quality and legibility when projected. The IRL includes:

- All textbook art, photos, and tables in JPEG format, in both high-resolution and low-resolution versions
- All textbook art, photos, and tables in PowerPoint® format

# Contents

## CHAPTER 4   Water Balance of Plants   85

## CHAPTER 5   Mineral Nutrition   107

## CHAPTER 6   Solute Transport   131

# UNIT II   Biochemistry and Metabolism   161

## CHAPTER 7   Photosynthesis: The Light Reactions   163

## CHAPTER 8  Photosynthesis: The Carbon Reactions  199

## CHAPTER 9  Photosynthesis: Physiological and Ecological Considerations  243

# UNIT III ■ Growth and Development   401

## CHAPTER 14  Signal Transduction   403

## CHAPTER 17 Phytochrome and Light Control of Plant Development  493

## CHAPTER 18 — Blue-Light Responses: Morphogenesis and Stomatal Movements 521

## CHAPTER 19 — Auxin: The First Discovered Plant Growth Hormone 545

## CHAPTER 20    Gibberellins: Regulators of Plant Height and Seed Germination    583

## CHAPTER 21 Cytokinins: Regulators of Cell Division 621

## CHAPTER 22   Ethylene: The Gaseous Hormone   649

## CHAPTER 23   Abscisic Acid: A Seed Maturation and Stress-Response Hormone   673

**CHAPTER 24   Brassinosteroids: Regulators of Cell Expansion and Development   699**

# CHAPTER 25   The Control of Flowering   719

## CHAPTER 26   Responses and Adaptations to Abiotic Stress   755

# Plant Cells

The term *cell* is derived from the Latin *cella*, meaning storeroom or chamber. It was first used in biology in 1665 by the English scientist Robert Hooke to describe the individual units of the honeycomb-like structure he observed in cork under a compound microscope. The "cells" Hooke observed were actually the empty lumens of dead cells surrounded by cell walls, but the term is an apt one, because cells are the basic building blocks that define plant structure.

Plant physiology is the study of plant *processes*—how plants function as they interact with their physical (abiotic) and living (biotic) environments. Although this book will emphasize the physiological and biochemical functions of plants, it is important to recognize that all such functions—whether gas exchange in the leaf, water conduction in the xylem, photosynthesis in the chloroplast, ion transport across the plasma membrane, or a signal transduction pathway that involves light and hormone action—depend on structures. At every level, structure and function represent different frames of reference of a biological unity.

This chapter provides an overview of the basic anatomy of plants, from the macroscopic structure of organs to the microscopic ultrastructure of cellular organelles. In subsequent chapters we will treat these structures in greater detail from the perspective of their physiological functions in the plant life cycle.

# Plant Life: Unifying Principles

The spectacular diversity of plant size and form is familiar to everyone. Plants range in height from less than 1 centimeter to more than 100 meters. Plant morphology (shape) is also surprisingly diverse. At first glance, the tiny plant duckweed (*Lemna*) seems to have little in common with a giant saguaro cactus or a redwood tree. No single plant shows the entire spectrum of adaptations to the range of environments plants occupy on Earth, so plant physiologists study **model organisms**, plants with short generation times and small **genomes** (the sum of their genetic information) (see WEB TOPIC 1.1). These models are useful because all plants, regardless of their specific adaptations, carry out fundamentally similar processes and are based on the same architectural plan.

We can summarize the major design elements of plants as follows:

- As Earth's primary producers, green plants are the ultimate solar collectors. They harvest the energy of sunlight by converting light energy to chemical energy, which they store in bonds formed when they synthesize carbohydrates from carbon dioxide and water.

- Other than certain reproductive cells, plants are nonmotile. As a substitute for motility, they have evolved the ability to grow toward essential resources, such as light, water, and mineral nutrients, throughout their life span.

- Terrestrial plants are structurally reinforced to support their mass as they grow toward sunlight against the pull of gravity.

- Terrestrial plants have mechanisms for moving water and minerals from the soil to the sites of photosynthesis and growth, as well as mechanisms for moving the products of photosynthesis to nonphotosynthetic organs and tissues.

- Terrestrial plants lose water continuously by evaporation and have evolved mechanisms for avoiding desiccation.

# Overview of Plant Structure

Despite their apparent diversity, all seed plants (see WEB TOPIC 1.2) have the same basic body plan (FIGURE 1.1). The vegetative body is composed of three organs: leaf, stem, and root. The primary function of the leaf is photosynthesis; that of the stem is support; and that of the root is anchorage and the absorption of water and minerals. Leaves are attached to the stem at nodes, and the region of the stem between two nodes is termed the internode. The stem, together with its leaves, is commonly referred to as the shoot.

There are two categories of seed plants: gymnosperms (from the Greek for "naked seed") and angiosperms (from the Greek "vessel seed," or seeds contained in a vessel). **Gymnosperms** are the less advanced type; about 700 species are known. The largest group of gymnosperms is the conifers ("cone-bearers"), which include such commercially important forest trees as pine, fir, spruce, and redwood.

**Angiosperms**, the more advanced type of seed plant, first became abundant during the Cretaceous period, about 100 million years ago. Today, the 250,000 known species of angiosperms easily outnumber the gymnosperms, but many more remain to be characterized. The major innovation of the angiosperms is the flower; hence they are referred to as *flowering plants* (see WEB TOPIC 1.3).

## Plant cells are surrounded by rigid cell walls

A fundamental difference between plants and animals is that each plant cell is surrounded by a rigid **cell wall**. In animals, embryonic cells can migrate from one location to another; developing tissues and organs may thus contain cells that originated in different parts of the organism. In plants, such cell migrations are prevented, because each walled cell is cemented to its neighbors by a **middle lamella**. As a consequence, plant development, unlike animal development, depends solely on patterns of cell division and cell enlargement.

Plant cells have two types of walls: primary and secondary (FIGURE 1.2). **Primary cell walls** are typically thin (less than 1 μm) and are characteristic of young, growing cells. **Secondary cell walls** are thicker and stronger than primary walls and are deposited when most cell enlargement has ended. Secondary cell walls owe their strength and toughness to **lignin**, a brittle, gluelike material (see Chapter 15). Circular gaps in the secondary wall give rise to **simple pits**, which always occur opposite simple pits in the neighboring secondary wall. Adjoining simple pits are called **pit-pairs**.

The evolution of lignified secondary cell walls provided plants with the structural reinforcement necessary to grow vertically above the soil and to colonize the land. Bryophytes, such as mosses and liverworts, which lack lignified cell walls, are unable to grow more than a few centimeters above the ground.

## New cells are produced by dividing tissues called meristems

Plant growth is concentrated in localized regions of cell division called **meristems**. Nearly all nuclear division (mitosis) and cell division (cytokinesis) occurs in these meristematic regions. In a young plant, the most active meristems are called **apical meristems**; they are located

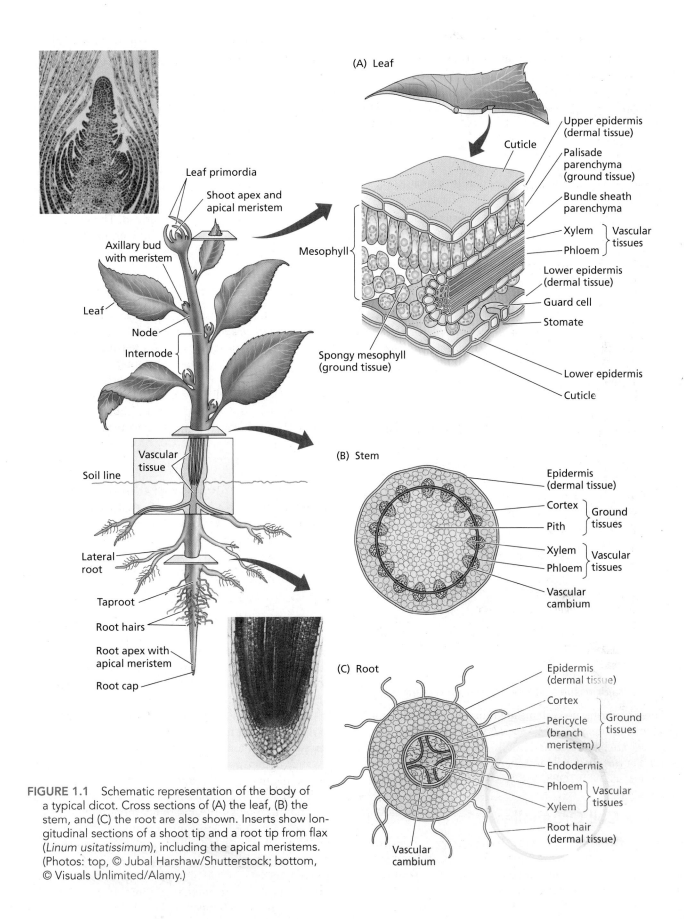

**FIGURE 1.1** Schematic representation of the body of a typical dicot. Cross sections of (A) the leaf, (B) the stem, and (C) the root are also shown. Inserts show longitudinal sections of a shoot tip and a root tip from flax (*Linum usitatissimum*), including the apical meristems. (Photos: top, © Jubal Harshaw/Shutterstock; bottom, © Visuals Unlimited/Alamy.)

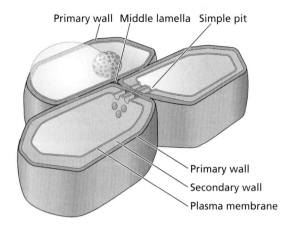

Primary wall  Middle lamella  Simple pit

Primary wall
Secondary wall
Plasma membrane

**FIGURE 1.2** Schematic representation of primary and secondary cell walls and their relationship to the rest of the cell.

at the tips of the stem and the root (see Figure 1.1). At the nodes, **axillary buds** contain the apical meristems for branch shoots. Lateral roots arise from the **pericycle**, an internal meristematic tissue (see Figure 1.1C). Proximal to (i.e., next to) and overlapping the meristematic regions are zones of cell elongation in which cells increase dramatically in length and width. Cells usually differentiate into specialized types after they have elongated.

The phase of plant development that gives rise to new organs and to the basic plant form is called **primary growth**. Primary growth results from the activity of apical meristems, in which cell division is followed by progressive cell enlargement, typically elongation, producing axial (top-to-bottom) polarity. After elongation in a given region is complete, **secondary growth** may occur, producing radial (inside-to-outside) polarity. Secondary growth involves two lateral meristems: the **vascular cambium** (plural *cambia*) and the **cork cambium**. The vascular cambium gives rise to secondary xylem (wood) and secondary phloem. The cork cambium produces the periderm, which consists mainly of cork cells.

### Three major tissue systems make up the plant body

Three major tissue systems are found in all plant organs: dermal tissue, ground tissue, and vascular tissue. These tissues are illustrated and briefly characterized in **FIGURE 1.3**. For further details and characterizations of these plant tissues, see **WEB TOPIC 1.4**.

## Plant Cell Organelles

All plant cells have the same basic eukaryotic organization: They contain a nucleus, a cytoplasm, and subcellular organelles; and they are enclosed in a membrane that defines their boundaries, as well as a cellulosic cell wall (**FIGURE 1.4**). Certain structures, including the nucleus,

**FIGURE 1.3** (A) The outer epidermis (dermal tissue) of a leaf of *Welwitschia mirabilis*. (120×) Diagrammatic representations of three types of ground tissue: (B) parenchyma, (C) collenchyma, (D) sclerenchyma, and (E) conducting cells of the xylem and phloem. (A © Steve Gschmeissner/Photo Researchers, Inc.)

can be lost during cell maturation, but all plant cells *begin* with a similar complement of organelles. These organelles fall into three main categories based on they ways in which they arise:

- *The endomembrane system*: the endoplasmic reticulum, the nuclear envelope, the Golgi apparatus, the vacuole, the endosomes, and the plasma membrane. The endomembrane system plays a central role in secretory processes, membrane recycling, and the cell cycle. The plasma membrane regulates transport into and out of the cell. Endosomes arise from vesicles derived from the plasma membrane and process or recycle the vesicle contents.

- *Independently dividing organelles derived from the endomembrane system*: the oil bodies, peroxisomes, and glyoxysomes, which function in lipid storage and carbon metabolism.

- *Independently dividing, semiautonomous organelles*: plastids and mitochondria, which function in energy metabolism and storage.

Because all of these cellular organelles are membranous compartments, we begin with a description of membrane structure and function.

### Biological membranes are phospholipid bilayers that contain proteins

All cells are enclosed in a membrane that serves as their outer boundary, separating the cytoplasm from the external environment. This **plasma membrane** (also called **plasmalemma**) allows the cell to take up and retain certain substances while excluding others. Various transport proteins embedded in the plasma membrane are responsible for this selective traffic of solutes, water-soluble ions, and small, uncharged molecules across the membrane. The accumulation of ions or molecules in the cytosol through the action of transport proteins consumes metabolic energy. Membranes also delimit the boundaries of the specialized internal organelles of the cell and regulate the fluxes of ions and metabolites into and out of these compartments.

According to the **fluid-mosaic model**, all biological membranes have the same basic molecular organization. They consist of a double layer (*bilayer*) of either phospholipids or, in the case of chloroplasts, glycosylglycerides, in which proteins are embedded (**FIGURE 1.5A AND C**). Each layer is called a *leaflet* of the bilayer. In most mem-

(A) Dermal tissue: epidermal cells

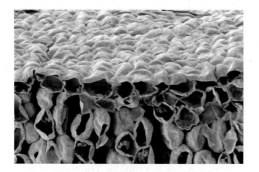

(B) Ground tissue: parenchyma cells

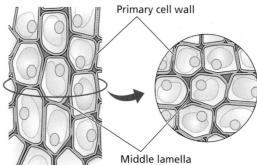

Primary cell wall

Middle lamella

(C) Ground tissue: collenchyma cells

Primary cell wall

Nucleus

(D) Ground tissue: sclerenchyma cells

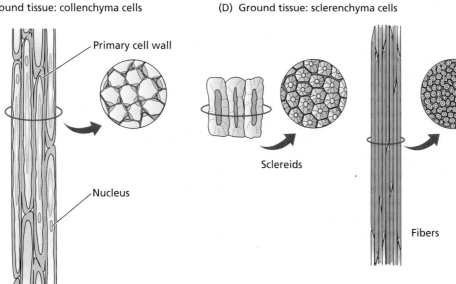

Sclereids

Fibers

(E) Vascular tisssue: xylem and phloem

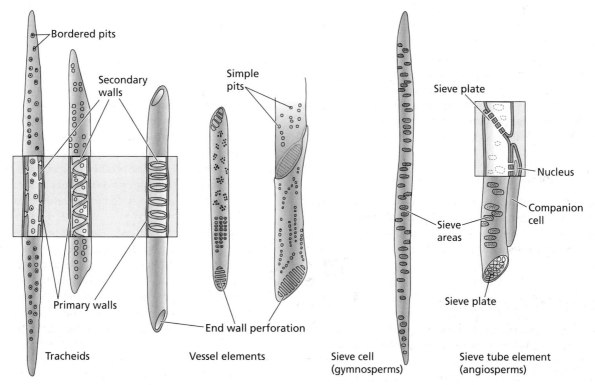

Bordered pits

Secondary walls

Simple pits

Sieve plate

Primary walls

End wall perforation

Nucleus

Companion cell

Sieve areas

Sieve plate

Tracheids

Vessel elements

Sieve cell (gymnosperms)

Sieve tube element (angiosperms)

**Xylem**

**Phloem**

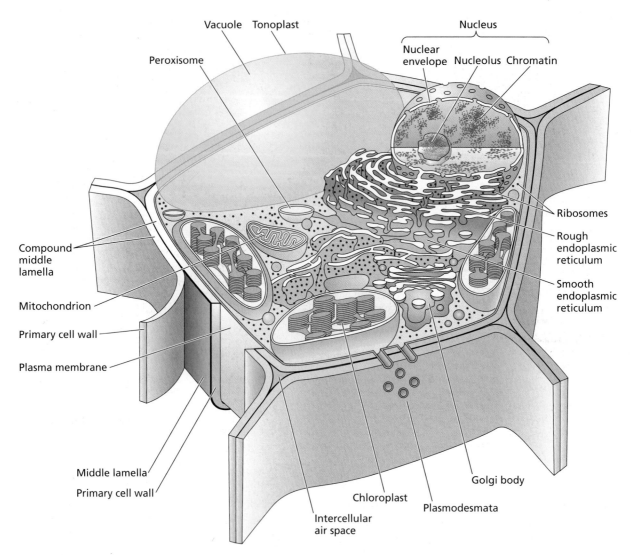

**FIGURE 1.4**  Diagrammatic representation of a plant cell. Various intracellular compartments are defined by their respective membranes, such as the tonoplast, the nuclear envelope, and the membranes of the other organelles. The two adjacent primary walls, along with the middle lamella, form a composite structure called the compound middle lamella.

branes, proteins make up about half of the membrane's mass. However, the composition of the lipid components and the properties of the proteins vary from membrane to membrane, conferring on each membrane its unique functional characteristics.

**PHOSPHOLIPIDS**  Phospholipids are a class of lipids in which two fatty acids are covalently linked to glycerol, which is covalently linked to a phosphate group. Also attached to this phosphate group is a variable component, called the *head group*, such as serine, choline, glycerol, or inositol (see Figure 1.5C). The nonpolar hydrocarbon chains of the fatty acids form a region that is exclusively

hydrophobic—that is, that excludes water. In contrast to the fatty acids, the head groups are highly polar; consequently, phospholipid molecules display both hydrophilic and hydrophobic properties (i.e., they are *amphipathic*). Various phospholipids are distributed asymmetrically across the plasma membrane, giving the membrane sidedness; in terms of phospholipid composition, the outside leaflet of the plasma membrane that faces the outside of the cell is different from inside leaflet that faces the cytosol.

**FIGURE 1.5**  (A) The plasma membrane, endoplasmic reticulum, and other endomembranes of plant cells consist of proteins embedded in a phospholipid bilayer. (B) Various anchored membrane proteins, attached to the membrane via GPI, fatty acids, and prenyl groups, enhance the sidedness of membranes. (C) Chemical structures and space-filling models of typical phospholipids: phosphatidylcholine and monogalactosyldiacylglycerol. (B after Buchanan et al. 2000.)

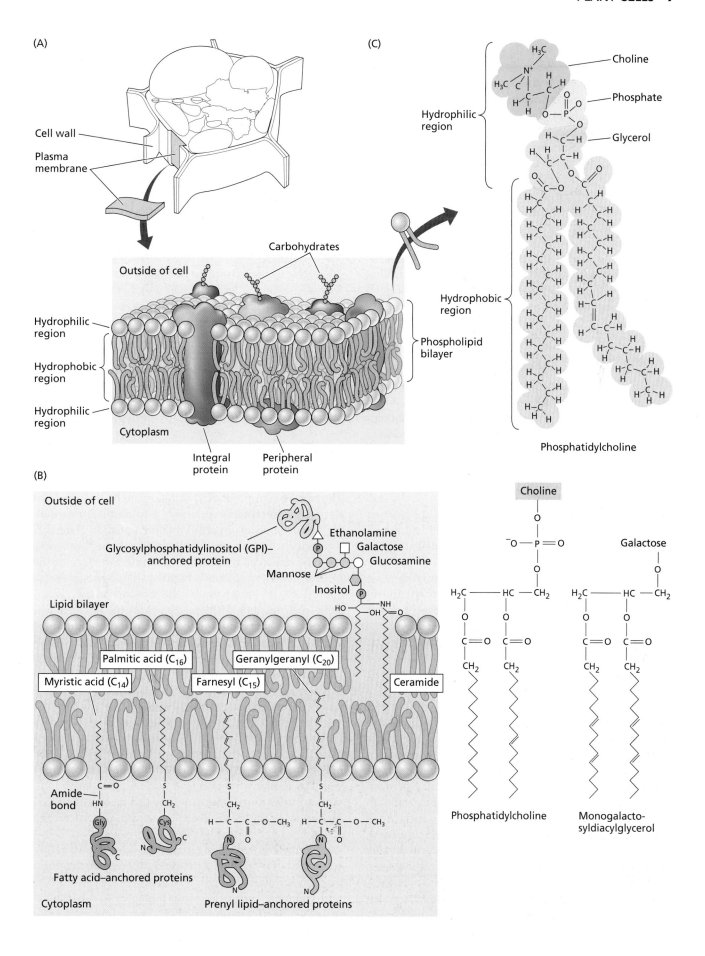

(A)

Cell wall

Plasma membrane

Carbohydrates

Outside of cell

Hydrophilic region

Hydrophobic region

Hydrophilic region

Cytoplasm

Phospholipid bilayer

Integral protein

Peripheral protein

(C)

Choline

Phosphate

Glycerol

Hydrophilic region

Hydrophobic region

Phosphatidylcholine

(B)

Outside of cell

Glycosylphosphatidylinositol (GPI)–anchored protein

Ethanolamine
Galactose
Mannose
Glucosamine
Inositol

Lipid bilayer

Palmitic acid (C₁₆)
Myristic acid (C₁₄)
Geranylgeranyl (C₂₀)
Farnesyl (C₁₅)
Ceramide

Amide bond

Fatty acid–anchored proteins

Cytoplasm

Prenyl lipid–anchored proteins

Choline

Galactose

Phosphatidylcholine

Monogalacto-syldiacylglycerol

The membranes of specialized plant organelles called **plastids**, the group of membrane-bound organelles to which chloroplasts belong, are unique in that their lipid component consists almost entirely of **glycosylglycerides** rather than phospholipids. In glycosylglycerides, the polar head group consists of galactose, digalactose, or sulfated galactose, without a phosphate group (see WEB TOPIC 1.5).

The fatty acid chains of phospholipids and glycosylglycerides are variable in length, but they usually consist of 14 to 24 carbons. If the carbons are linked by single bonds, the fatty acid chain is *saturated* (with hydrogen atoms), but if the chain includes one or more double bonds, it is *unsaturated*.

The double bonds within a fatty acid chain can rotate so as to create a kink in the chain that prevents tight packing of the phospholipids in the bilayer (i.e., the bonds adopt a kinked *cis* configuration, as opposed to an unkinked *trans* configuration). The kinks promote membrane fluidity, which is critical for many membrane functions. Membrane fluidity is also strongly influenced by temperature. Because plants generally cannot regulate their body temperatures, they are often faced with the problem of maintaining membrane fluidity under conditions of low temperature, which tends to decrease membrane fluidity. To maintain membrane fluidity at cold temperatures, plants can produce a higher percentage of unsaturated fatty acids, such as *oleic acid* (one double bond), *linoleic acid* (two double bonds), and *linolenic acid* (three double bonds) (see also Chapter 26).

PROTEINS  The proteins associated with the lipid bilayer are of three main types: integral, peripheral, and anchored. Proteins and lipids can also combine in transient aggregates in the membrane called *lipid rafts*.

**Integral proteins** are embedded in the lipid bilayer (see Figure 1.5A). Most integral proteins span the entire width of the phospholipid bilayer, so one part of the protein interacts with the outside of the cell, another part interacts with the hydrophobic core of the membrane, and a third part interacts with the interior of the cell, the cytosol. Proteins that serve as ion channels (see Chapter 6) are always integral membrane proteins, as are certain receptors that participate in signal transduction pathways (see Chapter 14). Some receptor-like proteins on the outer surface of the plasma membrane recognize and bind tightly to cell wall constituents, effectively cross-linking the membrane to the cell wall.

**Peripheral proteins** are bound to the membrane surface by noncovalent bonds, such as ionic bonds or hydrogen bonds, and can be dissociated from the membrane with high-salt solutions or chaotropic agents, which break ionic and hydrogen bonds, respectively (see Figure 1.5A). Peripheral proteins serve a variety of functions in the cell. For example, some are involved in interactions between the plasma membrane and components of the cytoskeleton, such as microtubules and actin microfilaments, which are discussed later in this chapter.

**Anchored proteins** are bound to the membrane surface via lipid molecules, to which they are covalently attached. These lipids include fatty acids (myristic acid and palmitic acid), prenyl groups derived from the isoprenoid pathway (farnesyl and geranylgeranyl groups), and glycosyl-phosphatidylinositol (GPI)-anchored proteins (FIGURE 1.5B). These lipid anchors make the two sides of the membrane even more different, with the fatty acid and prenyl anchors occurring on the cytoplasm-facing leaflet of the membrane and the GPI linkages occurring on the leaflet facing the outside of the cell.

## The Endomembrane System

The endomembrane system of eukaryotic cells is the collection of related internal membranes that divides the cell into functional and structural compartments, and distributes membranes and proteins via vesicular traffic among cellular organelles. Our discussion of the endomembrane system begins with the nucleus, where the genetic information for organelle biogenesis is mainly stored. This will be followed by a description of the independently dividing endomembrane organelles and the semiautonomous organelles.

### The nucleus contains the majority of the genetic material

The **nucleus** (plural *nuclei*) is the organelle that contains the genetic information primarily responsible for regulating the metabolism, growth, and differentiation of the cell. Collectively, these genes and their intervening sequences are referred to as the **nuclear genome**. The size of the nuclear genome in plants is highly variable, ranging from about $1.2 \times 10^8$ base pairs for the mustard relative *Arabidopsis thaliana*, to $1 \times 10^{11}$ base pairs for the lily *Fritillaria assyriaca*. The remainder of the genetic information of the cell is contained in the two semiautonomous organelles—chloroplasts and mitochondria—which we will discuss a little later in this chapter.

The nucleus is surrounded by a double membrane called the **nuclear envelope** (FIGURE 1.6A), which is a subdomain of the endoplasmic reticulum (ER, see below). **Nuclear pores** form selective channels across both membranes, connecting the nucleoplasm (the region inside the nucleus) with the cytoplasm (FIGURE 1.6B). There can be very few to many thousands of nuclear pores on an individual nuclear envelope and they can be arranged into higher-order aggregates (see Figure 1.6B) (Fiserova et al. 2009).

The nuclear "pore" is actually an elaborate structure composed of more than a hundred different **nucleoporin** proteins arranged octagonally to form a 105-nm **nuclear pore complex**. The nucleoporins lining the 40-nm channel of the NPC form a meshwork that acts as a supramolecular

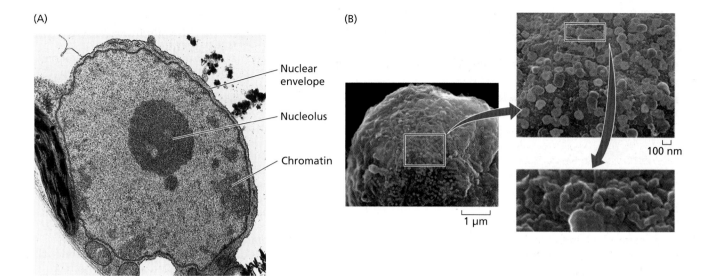

(A)

(B)

100 nm

1 μm

**FIGURE 1.6** (A) Transmission electron micrograph of a plant cell, showing the nucleolus and the nuclear envelope. (B) Organization of the nuclear pore complexes (NPCs) on the nuclear surface in tobacco culture cells. The NPCs that touch each other are colored brown; the rest are colored blue. The first inset (top right) shows that most of the NPCs are closely associated, forming rows of 5–30 NPCs. The second inset (bottom right) shows the tight associations of the NPCs. (A courtesy of R. Evert; B from Fiserova et al. 2009.)

sieve (Denning et al. 2003) (see **WEB TOPIC 1.6**). A specific amino acid sequence called the **nuclear localization signal** is required for a protein to gain entry into the nucleus. Several proteins required for nuclear import and export have been identified (see **WEB TOPICS 1.6 AND 1.7**).

The nucleus is the site of storage and replication of the **chromosomes**, composed of DNA and its associated proteins (**FIGURE 1.7**). Collectively, this DNA–protein complex is known as **chromatin**. The linear length of all the DNA within any plant genome is usually millions of times greater than the diameter of the nucleus in which it is found. To solve the problem of packaging this chromosomal DNA within the nucleus, segments of the linear double helix of DNA are coiled twice around a solid cylinder of eight **histone** protein molecules, forming a **nucleosome**. Nucleosomes are arranged like beads on a string along the length of each chromosome.

During mitosis, the chromatin condenses, first by coiling tightly into a **30-nm chromatin fiber**, with six nucleosomes per turn, followed by further folding and packing processes that depend on interactions between proteins and nucleic acids (see Figure 1.7). At interphase, two types of chromatin are distinguishable, based on their degree of condensation: heterochromatin and euchromatin. **Heterochromatin** is a highly compact and transcriptionally inactive form of chromatin and accounts for about 10% of the DNA. Most

of the heterochromatin is concentrated along the periphery of the nuclear membrane and is associated with regions of the chromosome containing few genes, such as telomeres and centromeres. The rest of the DNA consists of **euchromatin**, the dispersed, transcriptionally active form. Only about 10% of the euchromatin is transcriptionally active at any given time. The remainder exists in a state of condensation intermediate between that of the transcriptionally active euchromatin and that of heterochromatin. The chromosomes reside in specific regions of the nucleoplasm, each in its own separate space, giving rise to the possibility of separate regulation of each chromosome.

During the cell cycle chromatin undergoes dynamic structural changes. In addition to transient local changes required for transcription, heterochromatic regions can be converted to euchromatic regions, and vice versa, by the addition or removal of functional groups on the histone proteins (see Chapter 2). Such global changes in the genome can give rise to stable changes in gene expression. In general, stable changes in gene expression that occur without changes in the DNA sequence are referred to as *epigenetic regulation*.

Nuclei contain a densely granular region called the **nucleolus** (plural *nucleoli*), which is the site of ribosome synthesis (see Figure 1.6A). Typical cells have one nucleolus per nucleus; some cells have more. The nucleolus includes portions of one or more chromosomes where ribosomal RNA (rRNA) genes are clustered to form a structure called the **nucleolar organizer region**. The nucleolus assembles the proteins and rRNA of the **ribosome** into a large and a small subunit, each exiting the nucleus separately through the nuclear pores. They unite in the cytoplasm to form a complete ribosome (**FIGURE 1.8A**). Assembled ribosomes are protein-synthesizing machines. Those produced by the nucleus for cytoplasmic, "eukaryotic" protein synthesis, the 80S ribosomes, are larger than the ribosomes assem-

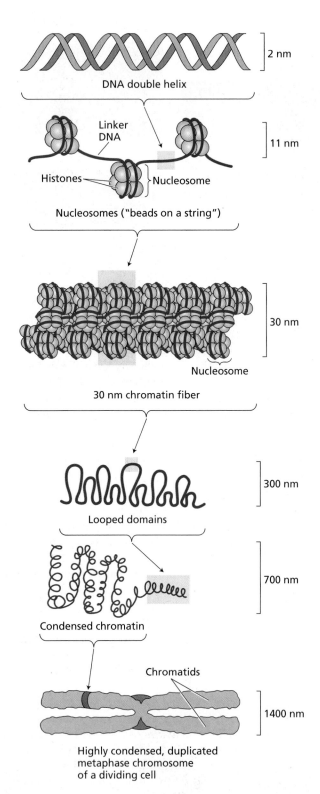

2 nm

DNA double helix

Linker
DNA

11 nm

Histones

Nucleosome

Nucleosomes ("beads on a string")

30 nm

Nucleosome

30 nm chromatin fiber

300 nm

Looped domains

700 nm

Condensed chromatin

Chromatids

1400 nm

Highly condensed, duplicated
metaphase chromosome
of a dividing cell

**FIGURE 1.7** Packaging of DNA in a metaphase chromosome. The DNA is first aggregated into nucleosomes and then wound to form the 30 nm chromatin fibers. Further coiling leads to the condensed metaphase chromosome. (After Alberts et al. 2002.)

**FIGURE 1.8** (A) Basic steps in gene expression, including transcription, processing, export to the cytoplasm, and translation. (1–2) Proteins may be synthesized on free or bound ribosomes. (3) Proteins destined for secretion are synthesized on the rough endoplasmic reticulum and contain a hydrophobic signal sequence. A signal recognition particle (SRP) binds the signal peptide to the ribosome, interrupting translation. (4) SRP receptors associate with protein-transporting channels called translocons. The ribosome-SRP complex binds to the SRP receptor on the ER membrane, and the ribosome docks with the translocon. (5) The translocon pore opens, the SRP particle is released, and the elongating polypeptide enters the lumen of the endoplasmic reticulum. (6) Translation resumes. Upon entering the lumen of the ER, the signal sequence is cleaved off by a signal peptidase on the membrane. (7, 8) After carbohydrate addition and chain folding, the newly synthesized polypeptide is shuttled to the Golgi apparatus via vesicles. (B) Amino acids are polymerized on the ribosome, with the help of tRNA, to form the elongating polypeptide chain.

bled in and remaining in mitochondria and plastids for their separate program of "prokaryotic" protein synthesis, the 70S ribosomes.

### Gene expression involves both transcription and translation

The complex process of protein synthesis starts with **transcription**—the synthesis of an RNA polymer bearing a base sequence that is complementary to a specific gene. The primary RNA transcript is processed to become messenger RNA (mRNA), which moves from the nucleus to the cytoplasm through the nuclear pore. In the cytoplasm, the mRNA attaches first to the small ribosomal subunit and then to the large subunit to initiate translation (see Figure 1.8A).

**Translation** is the process whereby a specific protein is synthesized from amino acids according to the sequence information encoded by the mRNA. A series of ribosomes (called a polysome) travel the entire length of the mRNA and serves as the site for the sequential bonding of amino acids as specified by the base sequence of the mRNA (**FIGURE 1.8B**). The process of translation on either cytosolic or membrane-bound polysomes produces the primary protein sequence of the protein, which includes not only the sequence involved in protein function, but also sequence information required to "ticket" (or "target") the protein to different destinations within the cell (see **WEB TOPIC 1.8**).

### The endoplasmic reticulum is a network of internal membranes

The ER is composed of tubules that join together to form a network of polygons and flattened saccules called **cister-**

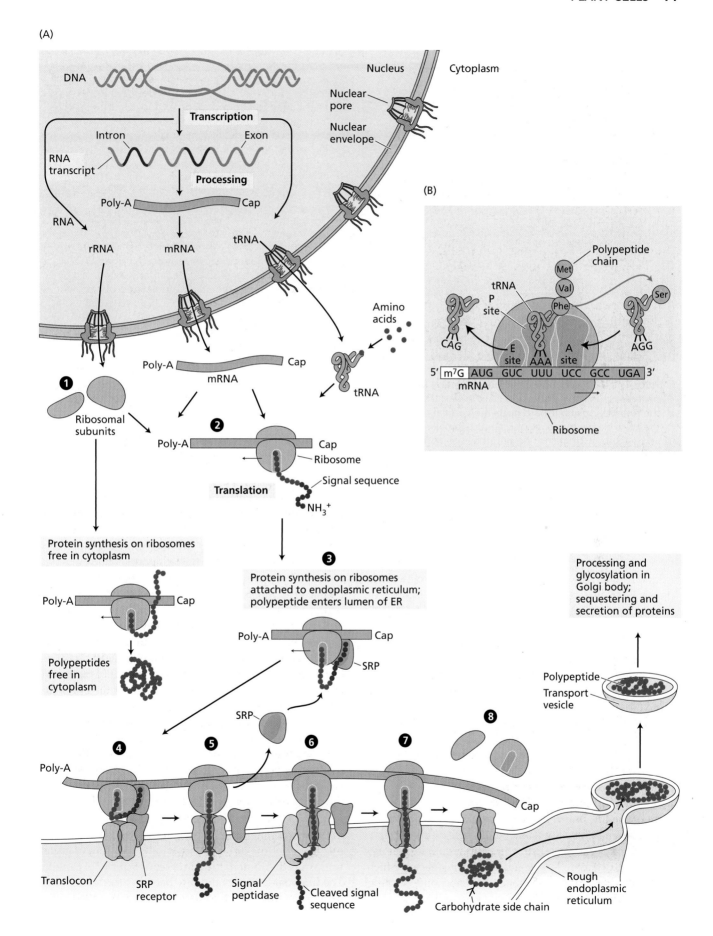

(A)

DNA

Nucleus          Cytoplasm

Nuclear pore

Nuclear envelope

**Transcription**

Intron          Exon

RNA transcript

**Processing**

Poly-A          Cap

RNA

rRNA          mRNA          tRNA

Poly-A          Cap
mRNA

Amino acids

tRNA

**1**
Ribosomal subunits

**2**
Poly-A          Cap
Ribosome
Signal sequence
$NH_3^+$
**Translation**

(B)

Polypeptide chain
Met
Val
tRNA          Phe          Ser
P site
CAG          A site          AGG
E site
AAA
5' m⁷G AUG GUC UUU UCC GCC UGA 3'
mRNA
Ribosome

Protein synthesis on ribosomes free in cytoplasm

Poly-A          Cap

Polypeptides free in cytoplasm

**3**
Protein synthesis on ribosomes attached to endoplasmic reticulum; polypeptide enters lumen of ER

Poly-A          Cap
SRP

SRP

Processing and glycosylation in Golgi body; sequestering and secretion of proteins

**8**

Polypeptide
Transport vesicle

**4**          **5**          **6**          **7**
Poly-A

Cap

Translocon          SRP receptor          Signal peptidase          Cleaved signal sequence          Carbohydrate side chain          Rough endoplasmic reticulum

(A)

(B)

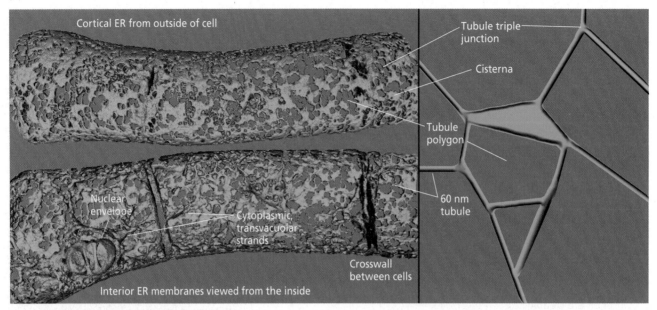

Cortical ER from outside of cell

Tubule triple junction

Cisterna

Tubule polygon

Nuclear envelope

Cytoplasmic, transvacuolar strands

60 nm tubule

Crosswall between cells

Interior ER membranes viewed from the inside

**FIGURE 1.9** Three-dimensional reconstruction of ER in tobacco suspension culture cells. (A) Viewing the cells from the outside looking inward (top), the cortical network of the ER is clearly made up of cisternal domains and polygonal tubule domains. Viewing the cells from the inside looking outward (bottom), transvacuolar strands containing ER tubules can be seen, as well as the nuclear envelope, a subdomain of the ER. The nuclei have transnuclear channels and invaginations in the nuclear envelope. (B) Diagrammatic representation of tubules and cisternae arranged in the network of polygons typical of cortical ER. (Courtesy of L. R. Griffing.)

nae (singular *cisterna*) (**FIGURE 1.9A**). A large part of the ER resides in the outer layer of cytoplasm called the *cell cortex*, but it can also form tubule bundles that traverse the cell through *transvacuolar strands*—strands of cytoplasm that extend through the central vacuole. The tubule network is physically connected to the plasma membrane at some of the points of intersection in the *polygonal network* (**FIGURE 1.9B**). Tubular and cisternal forms of the ER rapidly transition between each other. The transition may be controlled by a class of proteins called **reticulons**, which form tubules from membrane sheets. The actomyosin cytoskeleton, discussed later in the chapter, activates ER tubule rearrangement (Sparkes et al. 2009).

The region of the ER that has many membrane-bound ribosomes is called **rough ER** (**RER**) because the bound ribosomes give the ER a rough appearance in electron micrographs (**FIGURE 1.10**). ER without bound ribosomes is called **smooth ER** (**SER**). The distinction between rough and smooth ER is sometimes correlated with changes in form of the ER: rough ER being cisternal, that is, shaped like a flattened small sac, and smooth ER being tubular. However, this classical distinction applies only to certain cell types, such as in floral glands producing nectar, which contain much smooth ER. In general, ribosome binding sites are relatively uniform in both types of ER.

The ER provides the membrane building blocks and protein cargo for the other compartments in the endomembrane pathway: the nuclear envelope, the Golgi apparatus, vacuoles, the plasma membrane, and the endosomal system. It even transports some proteins to the chloroplast (Nanjo et al. 2006; Villarejo et al. 2005). The ER is the major source of membrane phospholipids used to construct the entire endomembrane system, in partnership with the chloroplast, with which the ER can exchange lipids (Xu et al. 2008). The portion of phospholipid biosynthesis carried out by the ER is called the *eukaryotic pathway* and the portion carried out by the chloroplast is called the *prokaryotic pathway* (see Chapter 11).

There is an intrinsic sidedness or asymmetry to membrane bilayers because the enzyme that initiates phospholipid synthesis on the ER adds new phospholipid precursors exclusively to the cytosolic leaflet of the bilayer. The enzymes involved in synthesizing the phospholipid head groups (serine, choline, glycerol or inositol) are also on the cytosolic leaflet. This causes intrinsic lipid asymmetry in the membranes of endomembranes, with the cytoplasmic leaflet of the organelles having a different composition from the lumenal (inside) leaflet of the organelles. The lumenal leaflet eventually becomes the leaflet of the membrane that faces the outside of the cell on the plasma

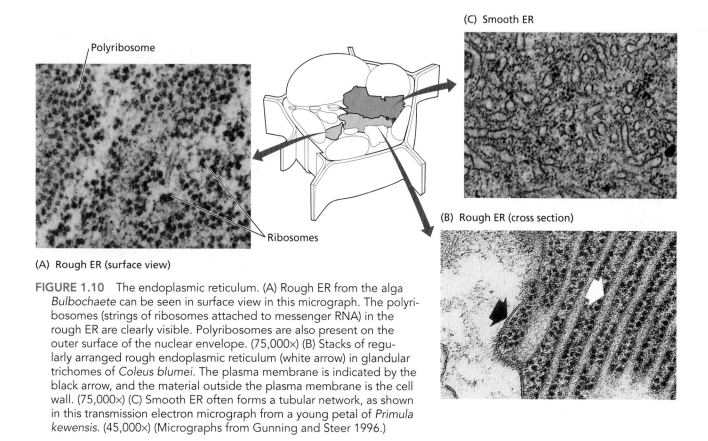

(C) Smooth ER

Polyribosome

Ribosomes

(A) Rough ER (surface view)

(B) Rough ER (cross section)

**FIGURE 1.10** The endoplasmic reticulum. (A) Rough ER from the alga *Bulbochaete* can be seen in surface view in this micrograph. The polyribosomes (strings of ribosomes attached to messenger RNA) in the rough ER are clearly visible. Polyribosomes are also present on the outer surface of the nuclear envelope. (75,000×) (B) Stacks of regularly arranged rough endoplasmic reticulum (white arrow) in glandular trichomes of *Coleus blumei*. The plasma membrane is indicated by the black arrow, and the material outside the plasma membrane is the cell wall. (75,000×) (C) Smooth ER often forms a tubular network, as shown in this transmission electron micrograph from a young petal of *Primula kewensis*. (45,000×) (Micrographs from Gunning and Steer 1996.)

membrane. Further asymmetrical modifications of lipid head groups and posttranslational modification of proteins by covalent addition of lipids and carbohydrates amplify the sidedness of membranes (see Figure 1.5).

In animal cells, membrane asymmetry can be counteracted by enzymes called **flippases**, which "flip" newly synthesized phospholipids across the bilayer to the inner leaflet. Recently, researchers have found evidence for flippases in plants (Sahu and Gummadi 2008). Furthermore, specialized regions of the ER can apparently exchange lipids with other organelles, such as the plasma membrane, chloroplasts, and mitochondria, when in close association (Larsson et al. 2007).

The ER and plastids are able to add new membrane directly through lipid and protein synthesis. For other organelles, however, the addition of new membrane occurs primarily through the process of **fusion** with other membranes. Because membranes are fluid, new membrane constituents can be transferred to the existing membrane even if the new membrane subsequently separates from the existing membrane by **fission**. These cycles of membrane fusion and fission are the basis for the growth and division of the endomembrane organelles that are derived directly or indirectly from the ER. These organelles include the Golgi apparatus, the vacuole, oil bodies, peroxisomes,

and the plasma membrane. These processes are carried out mainly by transport vesicles and tubules.

Selective fusion and fission of vesicles and tubules that serve as transporters between the compartments of the endomembrane system is achieved with a special class of targeting recognition proteins, called **SNAREs** and **Rabs** (see WEB TOPIC 1.9).

### Secretion of proteins from cells begins with the rough ER (RER)

Proteins destined for *secretion* take the secretory pathway, which passes through the ER and Golgi apparatus to the plasma membrane and the outside of the cell. Secretory proteins are synthesized while being inserted across the membrane into the ER lumen during the process of translation (*cotranslational insertion*). This process involves the ribosomes, the mRNA that codes for the secretory protein, and special receptors for ribosomes and partially synthesized proteins on the outer surface of the RER (see Figure 1.8). All secretory proteins and most integral membrane proteins of the secretory pathway have been shown to have a hydrophobic "leader" sequence of 18 to 30 amino acid residues, called the **signal peptide** sequence, at the amino-terminal end of the chain (see WEB TOPIC 1.8).

**FIGURE 1.11** Electron micrograph of a Golgi apparatus in a tobacco (*Nicotiana tabacum*) root cap cell. The *cis*, *medial*, and *trans* cisternae are indicated. The *trans*-Golgi network is associated with the *trans* cisternae. (60,000×) (From Gunning and Steer 1996.)

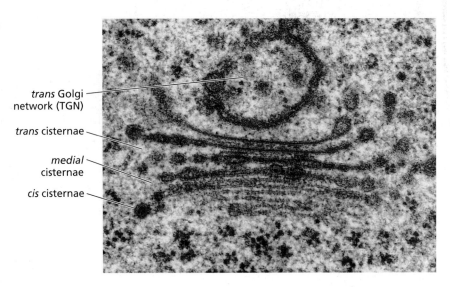

*trans* Golgi network (TGN)

*trans* cisternae

*medial* cisternae

*cis* cisternae

Early in translation, a **signal recognition particle (SRP)**, made up of protein and RNA, binds both to this hydrophobic leader and to the ribosome, interrupting translation. The ER membrane contains **SRP receptors**, which can associate with protein-lined channels called **translocons** (see Figure 1.8).

The ribosome–SRP complex in the cytosol binds to the SRP receptor on the ER membrane, and the ribosome docks with the translocon. Docking opens the translocon pore, the SRP particle is released, translation resumes, and the elongating polypeptide enters the lumen of the ER. For secretory proteins that enter the lumen of the ER, the signal sequence is cleaved off by a signal peptidase on the membrane (see Figure 1.8). For integral membrane proteins, some parts of the polypeptide chain are translocated across the membrane while others are not. The completed proteins are anchored to the membrane by one or more hydrophobic membrane-spanning domains.

Many of the proteins found in the lumen of the endomembrane system are **glycoproteins**—proteins with small sugar chains covalently attached—destined for secretion from the cell or delivery to the other endomembranes. In the vast majority of cases, a branched oligosaccharide chain made up of *N*-acetylglucosamine (GlcNAc), mannose (Man), and glucose (Glc) is attached to the free amino group of one or more specific asparagine residues of the secretory protein in the ER. This *N-linked glycan* is first assembled on a lipid molecule, **dolichol diphosphate**, which is embedded in the ER membrane (see Chapter 13). The completed 14-sugar glycan is then transferred to the nascent polypeptide as it enters the lumen. As in animal cells, these **N-linked glycoproteins** are then transported to the Golgi apparatus (discussed next) via small vesicles or tubules. However, in the Golgi, the glycans are further processed in a plant-specific way, making the proteins highly antigenic (recognized as foreign) in vertebrate immune systems.

## Glycoproteins and polysaccharides destined for secretion are processed in the Golgi apparatus

The Golgi body (in plants, also called a *dictyosome*) is a polarized stack of cisternae, with fatter cisternae occurring on the **cis** side or forming face, which accepts tubules and vesicles from the ER (**FIGURE 1.11**). The opposite, maturing face or **trans** side of the Golgi body has more flattened, thinner cisternae and includes a tubular network called the **trans-Golgi-network** or TGN. Secretory vesicles may bud off from the *trans* cisternae and TGN. There may be up to a hundred Golgi bodies in the Golgi apparatus (the entire collection of Golgi bodies) of a meristematic cell; other cell types differ in their Golgi content, but usually have fewer than a hundred. Golgi bodies can divide by splitting, enabling cells to regulate their capacity for secretion during growth and differentiation (Langhans et al. 2007).

Different cisternae within a single Golgi body have different enzymes, and have different biochemical functions depending on the type of polymer being processed—whether polysaccharides for the cell wall or glycoproteins for the cell wall or the vacuole. For example, as N-linked glycoproteins pass from the *cis* to the *trans* Golgi cisternae, they are successively modified by the specific sets of enzymes localized in the different cisternae. Certain sugars, such as mannose, are removed from the oligosaccharide chains, and other sugars are added. In addition to these modifications, glycosylation of the —OH groups of hydroxyproline, serine, threonine, and tyrosine residues (**O-linked oligosaccharides**) also occurs in the Golgi. The enzymes involved in polysaccharide biosynthesis in the Golgi bodies are substantially different, but occur side by side with glycoprotein modification enzymes.

Delivery of membrane and its contents to the Golgi body from the ER occurs at specialized **ER exit sites** (ERES). The ER exit site is determined by the presence of

(A)

1. COPII-coated vesicles bud from the ER and are transported to the *cis* face of the Golgi apparatus.

2. Cisternae progress through the Golgi stack in the anterograde direction, carrying their cargo with them.

3. Retrograde movement of COPI-coated vesicles maintains the correct distribution of enzymes in the *cis*-, *medial*-, and *trans*-cisternae of the stack.

4. Uncoated vesicles bud from the *trans*-Golgi membrane and fuse with the plasma membrane.

5. Endocytotic clathrin-coated vesicles fuse with the prevacuolar compartment.

6. Uncoated vesicles bud off from the prevacuolar compartment and carry their cargo to a lytic vacuole.

7. Proteins destined for lytic vacuoles are secreted from the *trans*-Golgi to the PVC via clathrin-coated vesicles, and then repackaged for delivery to the lytic vacuole.

8. Endocytic clathrin-coated vesicles can also uncoat and recycle via the early recycling endosome. Vesicles produced by the early recycling endosome either directly fuse with the plasma membrane or fuse with the *trans*-Golgi.

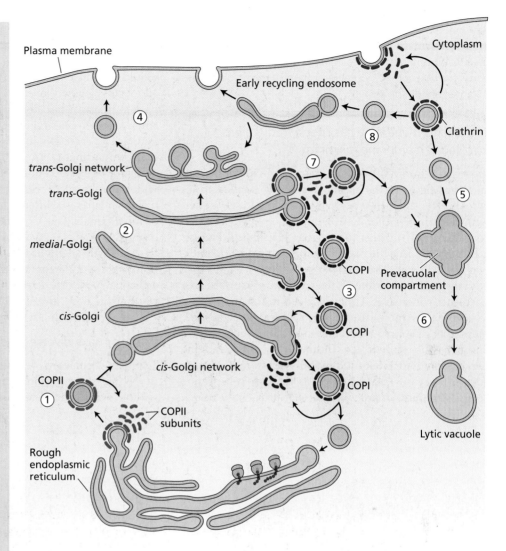

**FIGURE 1.12** Vesicular traffic along the secretory and endocytotic pathways. (A) Diagram of vesicular traffic mediated by three types of coat proteins. COPII is indicated in green, COPI in blue, and clathrin in red. (B) Electron micrograph of clathrin-coated vesicles isolated from bean leaves. (102,000×) (B courtesy of D. G. Robinson.)

(B)

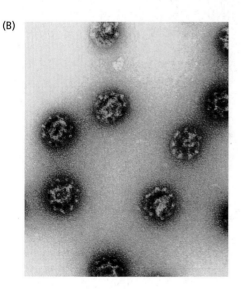

a **coat protein** called **COP II** (**FIGURE 1.12A**). This surface protein coat associates with the transmembrane receptors that bind the specific cargo destined for the Golgi. These membrane regions then bud off to form coated vesicles that lose their COP II coats prior to fusion with target membranes. By using fluorescent tags for both ERES and Golgi, it has been possible to show that the ERES move in concert with the Golgi bodies as the latter stream through the cell (see movie in **WEB TOPIC 1.10**). Movement out of the ER to the Golgi, within the Golgi from the *cis* to *trans* face, followed by transport to the plasma membrane or to prevacuolar structures via vesicles is called **anterograde** (forward) movement. In contrast, **recycling** of membrane vesicles, either from the Golgi to the ER or from the *trans* to *cis* face of the Golgi, is called **retrograde**, or backward, movement. **COP I**-coated vesicles are involved in retrograde movement within the Golgi and from the Golgi to the ER. Without retrograde membrane recycling, the Golgi body would soon be depleted of membrane through loss by anterograde movement.

### The plasma membrane has specialized regions involved in membrane recycling

Membrane internalization by retrograde movement of small vesicles from the plasma membrane is called **endocytosis**. The small (100 nm) vesicles are initially coated with **clathrin** (FIGURE 1.12B), but they quickly lose that coat and fuse with other tubules and vesicles (see Figures 1.12A and 1.13A); the organelles of this endocytic pathway are called **endosomes**. When secretory vesicles fuse with the plasma membrane, the membrane's surface area necessarily increases. Unless the cell is also expanding to keep pace with the added surface area, it needs some method of membrane recycling to keep the cell's surface area in balance with its size. The importance of membrane recycling is best illustrated by cells active in secretion, such as root cap cells (FIGURE 1.13). Root cap cells secrete copious amounts of mucopolysaccharides (slime), which lubricate the root tip as it grows through the soil. The increase in membrane surface area caused by the fusion of large slime-filled vesicles with the plasma membrane would become excessive if it were not for the process of endocytosis, which constantly recycles plasma membrane back to an organelle called the **early endosome**. The endosome can then be targeted either back to the *trans*-Golgi network for secretion or to a structure called the **prevacuolar compartment** for hydrolytic degradation.

Endocytosis and endocytotic recycling take place in a wide variety of plant cells. The control of endocytosis at the plasma membrane differentially regulates the abundance of ion channels (see Chapter 6), such as the potassium channel in stomatal guard cells and the boron transporter in roots. During gravitropism, the differential internalization of transporters for the growth hormone, auxin, causes a change in the concentration of hormone across the root, resulting in a bending of the root (see Chapter 19).

### Vacuoles have diverse functions in plant cells

The plant vacuole was originally defined by its appearance in the microscope—a membrane-enclosed compartment without cytoplasm. Instead of cytoplasm, it contains **vacu-**

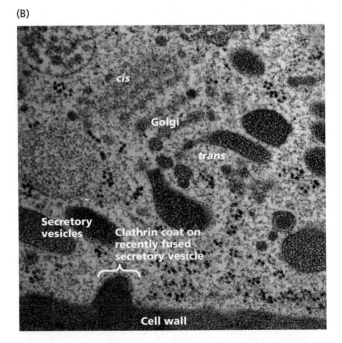

(A)

(B)

**FIGURE 1.13** Clathrin-coated pits are associated with secretion of slime in corn root cap. (A) Diagram of the recycling of membrane using clathrin-coated vesicles from recent secretion sites on the plasma membrane. (B) Recent secretion site showing a secretory vesicle that has just deposited its contents into the cell wall and a clathrin-coated invagination, which recycles membrane from the secretion site. There are 20 times more coated pits on secretion sites than on the membrane in general. (Mollenhauer et al. 1991; B micrograph by H. H. Mollenhauer, courtesy of L. R. Griffing.)

olar sap composed of water and solutes. The increase in volume of plant cells during growth takes place primarily through an increase in the volume of vacuolar sap. A large central vacuole occupies up to 95% of the total cell volume in many mature plant cells; sometimes there are two or more large central vacuoles, as in the case of certain flower petals with both pigmented and unpigmented vacuoles (see WEB TOPIC 1.9). The fact that vacuoles can differ in size and appearance suggests how diverse in form and function the vacuolar compartment can be. Some of the variations are probably due to differences in degree of vacuole maturation. For example, meristematic cells have no large central vacuole, but rather many small vacuoles. Some of these probably fuse together to form the large central vacuole as the cell matures.

The vacuolar membrane, the **tonoplast**, contains proteins and lipids that are synthesized initially in the ER. In addition to its role in cell expansion, the vacuole can also serve as a storage compartment for plant secondary metabolites involved in plant defense against herbivores (see Chapter 13). Inorganic ions, sugars, organic acids, and pigments are just some of the solutes that can accumulate in vacuoles, thanks to the presence of a variety of specific membrane transporters (see Chapter 6). Protein-storing vacuoles, called **protein bodies**, are abundant in seeds.

Vacuoles also play a role in protein turnover, analogous to the animal lysosome, as in the case of the **lytic vacuoles** that accumulate in senescing leaves, which release hydrolytic enzymes that degrade cellular constituents during senescence. Membrane sorting to plant vacuoles and animal lysosomes occurs by different mechanisms. Although in both cases the sorting to the prevacuolar compartment occurs in the Golgi, the recognition processes used in sorting receptors and processing proteins to the vacuole versus the lysosome are different. In mammalian cells, many lysosomal proteins are recognized by an ER enzyme that adds mannose 6-phosphate to the protein that is subsequently recognized by a sorting receptor in the Golgi, a pathway apparently missing in plants. On the other hand, some plant lytic vacuoles are derived directly from the ER, bypassing the Golgi entirely, a pathway apparently missing in mammals.

The delivery of Golgi-derived vesicles to the vacuole is indirect. As described above, there are multiple vacuolar compartments in the cell, not all of which serve as targets for Golgi vesicles. Those vacuoles that do receive Golgi-derived vesicles do so via an intermediate, a *prevacuolar compartment* that also serves as a sorting organelle for membrane endocytosed from the plasma membrane. This prevacuolar sorting compartment includes the **multivesicular body** (MVB), which, in some cases, is also a *postvacuolar compartment* that functions in the degradation of vacuoles and their membranes (Otegui et al. 2006).

# Independently Dividing Organelles Derived from the Endomembrane System

Several organelles are able to grow and proliferate independently even though they are derived from the endomembrane system. These organelles include oil bodies, peroxisomes, and glyoxysomes.

## Oil bodies are lipid-storing organelles

Many plants synthesize and store large quantities of oil during seed development. These oils accumulate in organelles called oil bodies (also known as the oleosomes, lipid bodies or spherosomes). Oil bodies are unique among the organelles in that they are surrounded by a "half–unit membrane"—that is, a phospholipid monolayer—derived from the ER. The phospholipids in the half–unit membrane are oriented with their polar head groups toward the aqueous phase of the cytosol and their hydrophobic fatty acid tails facing the lumen, dissolved in the stored lipid.

Oil bodies are initially formed as regions of differentiation within the ER. The nature of storage product, **triglycerides** (three fatty acids covalently linked to a glycerol backbone), dictates that this storage organelle will have a hydrophobic lumen. Consequently, as triglyceride is stored, it appears to be initially deposited in the hydrophobic region between the outer and inner leaflets of the ER membrane (FIGURE 1.14). Triglycerides do not have the polar head groups of membrane phospholipids, so are not exposed to the hydrophilic cytoplasm. Although the nature of the budding process that gives rise to the oil body is not yet fully understood, when the oil body separates from the ER it contains a single outer leaflet of phospholipid containing a special protein that coats oil bodies, **oleosin**. Oleosins are synthesized on cytosolic polysomes and are inserted into the oil body single leaflet membrane posttranslationally. The protein consists of a central, hairpin-like hydrophobic region, which inserts itself inside the oil-containing lumen, and two hydrophilic ends, which remain on the outside of the oil body (Alexander et al. 2002). When oil bodies break down during seed germination, they associate with other organelles that contain the enzymes for lipid oxidation, the glyoxysomes.

## Microbodies play specialized metabolic roles in leaves and seeds

**Microbodies** are a class of spherical organelles surrounded by a single membrane and specialized for one of several metabolic functions. **Peroxisomes** and **glyoxysomes** are microbodies specialized for the **β-oxidation** of fatty acids and the metabolism of **glyoxylate**, a two-carbon acid aldehyde (see Chapter 11). Microbodies lack DNA and are inti-

(A)

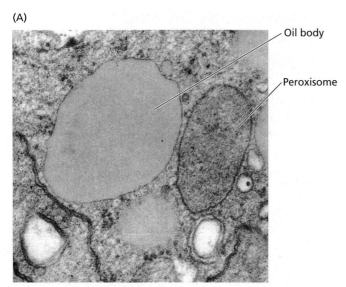

Oil body

Peroxisome

(B)

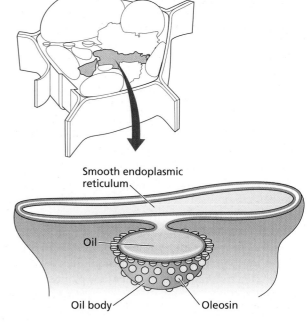

Smooth endoplasmic reticulum

Oil

Oil body

Oleosin

**FIGURE 1.14** (A) Electron micrograph of an oil body beside a peroxisome. (B) Diagram showing the formation of oil bodies by the synthesis and deposition of oil within the phospholipid bilayer of the ER. After budding off from the ER, the oil body is surrounded by a phospholipid monolayer containing the protein oleosin. (A from Huang 1987; B after Buchanan et al. 2000.)

mately associated with other organelles, with which they share intermediate metabolites. The glyoxysome is associated with mitochondria and oil bodies, while the peroxisome is associated with mitochondria and chloroplasts.

Initially, it was thought that peroxisomes and glyoxysomes were independent organelles, produced separately from the ER. However, experiments using antibodies specific for each type of organelle have supported a model in which peroxisomes develop directly from glyoxysomes, at least in greening cotyledons. For example, in cucumber seedlings, the nongreen cotyledon cells initially contain glyoxysomes, but after greening only peroxisomes are present. At intermediate stages of greening the microbodies react with both glyoxysomal and peroxisomal markers, demonstrating that the glyoxysomes are converted to peroxisomes during the greening process (Titus and Becker 1985).

In the peroxisome, glycolate, a two-carbon oxidation product of the photorespiratory cycle in an adjacent chloroplast, is oxidized to the acid aldehyde glyoxylate. During this conversion, hydrogen peroxide is generated, which can easily oxidize and destroy other compounds. However, the most abundant protein of the peroxisome is **catalase**, an enzyme that breaks hydrogen peroxide down into water. Catalase is often so abundant in peroxisomes that it forms crystalline arrays of the protein (**FIGURE 1.15**).

The observation that glyoxysomes can mature into peroxisomes explains the appearance of peroxisomes in developing cotyledons, but it does not explain how peroxisomes arise in other tissues. If they are inherited during cell divi-

sion, peroxisomes can grow and divide separately from other organelles, using proteins similar to those involved in division of mitochondria (Zhang and Hu 2008). Most proteins enter peroxisomes from the cytosol posttranslationally by means of a specific targeting signal, consisting of serine-lysine-leucine at the carboxyl terminus (see **WEB TOPIC 1.8**).

## Independently Dividing, Semiautonomous Organelles

A typical plant cell has two types of energy-producing organelles: mitochondria and chloroplasts. Both types are separated from the cytosol by a double membrane (an outer and an inner membrane) and contain their own DNA and ribosomes.

**Mitochondria** (singular *mitochondrion*) are the cellular sites of respiration, a process in which the energy released from sugar metabolism is used for the synthesis of ATP (adenosine triphosphate) from ADP (adenosine diphosphate) and inorganic phosphate ($P_i$) (see Chapter 11).

Mitochondria are highly dynamic structures that can undergo both fission and fusion. Mitochondrial fusion can result in long, tubelike structures that may branch to form mitochondrial networks. Regardless of shape, all mitochondria have a smooth outer membrane and a highly convoluted inner membrane (**FIGURE 1.16**). The inner membrane contains a proton-pumping **ATP synthase** that

FIGURE 1.15 Catalase crystal in a peroxisome of a mature leaf. Note the close association of the peroxisome with two chloroplasts and a mitochondrion, organelles that share metabolites with peroxisomes.

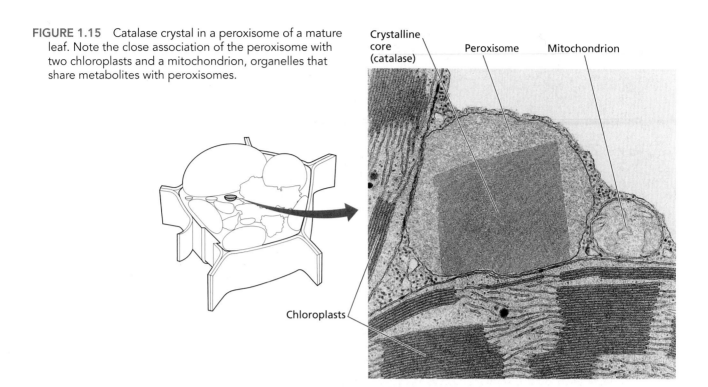

Crystalline core (catalase)  Peroxisome  Mitochondrion

Chloroplasts

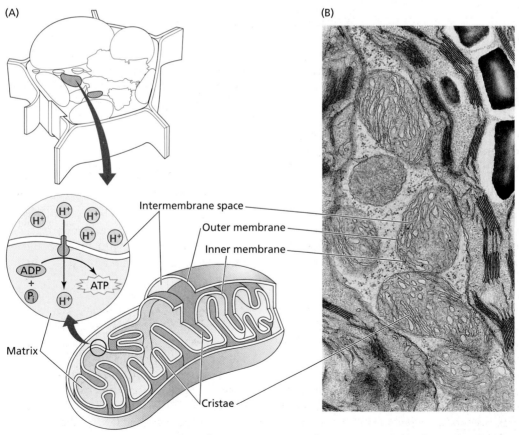

(A)

Intermembrane space

Outer membrane

Inner membrane

$H^+$  $H^+$  $H^+$  $H^+$  $H^+$  $H^+$

ADP + $P_i$  ATP  $H^+$

Matrix

Cristae

(B)

FIGURE 1.16 (A) Diagrammatic representation of a mitochondrion, including the location of the $H^+$-ATPases involved in ATP synthesis on the inner membrane. (B)

An electron micrograph of mitochondria from a leaf cell of Bermuda grass, *Cynodon dactylon*. (26,000×) (Micrograph by S. E. Frederick, courtesy of E. H. Newcomb.)

(A)

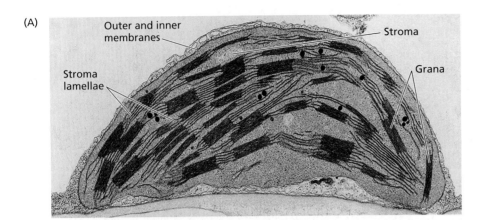

(B)

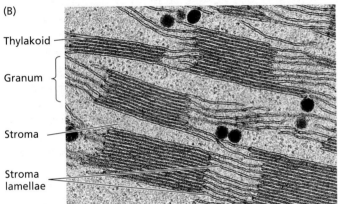

(C)

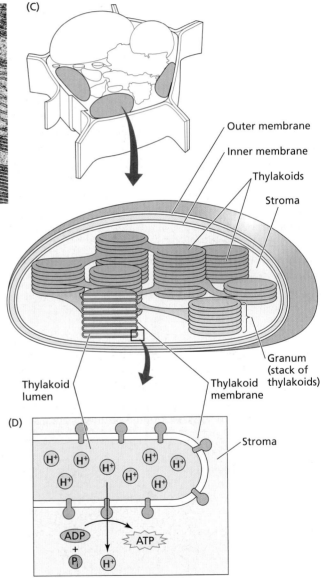

FIGURE 1.17 (A) Electron micrograph of a chloroplast from a leaf of timothy grass, *Phleum pratense*. (18,000×) (B) The same preparation at higher magnification. (52,000×) (C) A three-dimensional view of grana stacks and stroma lamellae, showing the complexity of the organization. (D) Diagrammatic representation of a chloroplast, showing the location of the H⁺-ATPases on the thylakoid membranes. (Micrographs by W. P. Wergin, courtesy of E. H. Newcomb.)

uses a proton gradient to synthesize ATP for the cell. The **proton gradient** is generated through the cooperation of electron transporters called the **electron transport chain**, which is embedded in, and peripheral to, the inner membrane (see Chapter 11).

The infoldings of the inner membrane are called **cristae** (singular *crista*). The compartment enclosed by the inner membrane, the mitochondrial **matrix**, contains the enzymes of the pathway of intermediary metabolism called the Krebs cycle.

**Chloroplasts** (FIGURE 1.17A) belong to another group of double-membrane–enclosed organelles called **plastids**. Chloroplast membranes are rich in glycosylglycerides (see WEB TOPIC 1.5). Chloroplast membranes contain chlorophyll and its associated proteins and are the sites of photosynthesis. In addition to their inner and outer envelope membranes, chloroplasts possess a third system of membranes called **thylakoids**. A stack of thylakoids forms a

PLANT CELLS **21**

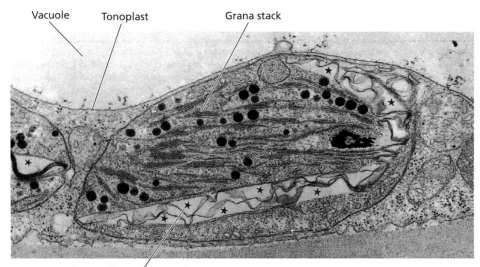

Vacuole  Tonoplast  Grana stack

Lycopene crystals

**FIGURE 1.18** Electron micrograph of a chromoplast from tomato (*Lycopersicon esculentum*) fruit at an early stage in the transition from chloroplast to chromoplast. Small grana stacks are still visible. Stars indicate crystals of the carotenoid lycopene. (27,000×) (From Gunning and Steer 1996.)

**granum** (plural *grana*) (FIGURE 1.17B). Proteins and pigments (chlorophylls and carotenoids) that function in the photochemical events of photosynthesis are embedded in the thylakoid membrane. The fluid compartment surrounding the thylakoids, called the **stroma**, is analogous to the matrix of the mitochondrion. Adjacent grana are connected by unstacked membranes called **stroma lamellae** (singular *lamella*).

The various components of the photosynthetic apparatus are localized in different areas of the grana and the stroma lamellae. The ATP synthases of the chloroplast are located on the thylakoid membranes (FIGURE 1.17C). During photosynthesis, light-driven electron transfer reactions result in a proton gradient across the thylakoid membrane (FIGURE 1.17D). As in the mitochondria, ATP is synthesized when the proton gradient is dissipated via ATP synthase. In the chloroplast, however, the ATP is not exported to the cytosol, but is used for many stromal reactions including the fixation of carbon from carbon dioxide in the atmosphere, as described in Chapter 8.

Plastids that contain high concentrations of carotenoid pigments, rather than chlorophyll, are called **chromoplasts**. Chromoplasts are responsible for the yellow, orange, or red colors of many fruits and flowers, as well as of autumn leaves (FIGURE 1.18).

Nonpigmented plastids are called **leucoplasts**. The most important type of leucoplast is the **amyloplast**, a starch-storing plastid. Amyloplasts are abundant in the storage tissues of shoots and roots, and in seeds. Specialized amyloplasts in the root cap also serve as gravity sensors that direct root growth downward into the soil (see Chapter 19).

### Proplastids mature into specialized plastids in different plant tissues

Meristem cells contain **proplastids**, which have few or no internal membranes, no chlorophyll, and an incomplete complement of the enzymes necessary to carry out photo-

synthesis (FIGURE 1.19A). In angiosperms and some gymnosperms, chloroplast development from proplastids is triggered by light. Upon illumination, enzymes are formed inside the proplastid or imported from the cytosol; light-absorbing pigments are produced; and membranes proliferate rapidly, giving rise to stroma lamellae and grana stacks (FIGURE 1.19B).

Seeds usually germinate in the soil in the dark, and their proplastids mature into chloroplasts only when the young shoot is exposed to light. If, instead, germinated seedlings are kept in the dark, the proplastids differentiate into **etioplasts**, which contain semicrystalline tubular arrays of membrane known as **prolamellar bodies** (FIGURE 1.19C). Instead of chlorophyll, etioplasts contain a pale yellow-green precursor pigment, protochlorophyllide.

Within minutes after exposure to light, an etioplast differentiates, converting the prolamellar body into thylakoids and stroma lamellae, and the protochlorophyllide into chlorophyll (for a discussion of chlorophyll biosynthesis, see WEB TOPIC 7.11). The maintenance of chloroplast structure depends on the presence of light; mature chloroplasts can revert to etioplasts during extended periods of darkness. Likewise, under different environmental conditions, chloroplasts can be converted to chromoplasts (see Figure 1.18), as in autumn leaves and ripening fruit.

### Chloroplast and mitochondrial division are independent of nuclear division

As noted earlier, plastids and mitochondria divide by the process of fission, consistent with their prokaryotic origins. Organellar DNA replication and fission are regulated independently of nuclear division. For example, the number of chloroplasts per cell volume depends on the cell's developmental history and its local environment. Accordingly, there are many more chloroplasts in the mesophyll cells in the interior of a leaf than in the epidermal cells forming the outer layer of the leaf.

(A)　　　　　　　　　　　(B)　　　　　　　　　　　(C)

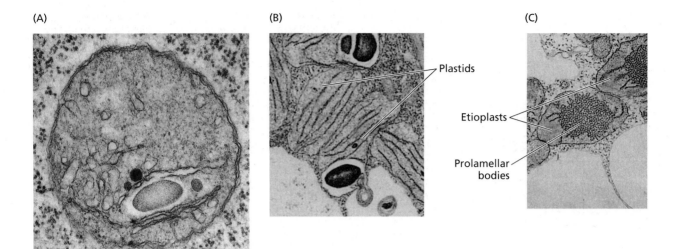

**FIGURE 1.19** Electron micrographs illustrating several stages of plastid development. (A) A high-magnification view of a proplastid from the root apical meristem of the broad bean (*Vicia faba*). The internal membrane system is rudimentary, and grana are absent. (47,000×) (B) A mesophyll cell of a young oat (*Avena sativa*) leaf grown in the light, at an early stage of differentiation. The plastids are developing grana stacks. (C) A cell of a young oat leaf from a seedling grown in the dark. The plastids have developed as etioplasts, with elaborate semicrystalline lattices of membrane tubules called prolamellar bodies. When exposed to light, the etioplast can convert to a chloroplast by the disassembly of the prolamellar body and the formation of grana stacks. (7,200×) (From Gunning and Steer 1996.)

Although the timing of the fission of chloroplasts and mitochondria is independent of the timing of cell division, these organelles require nuclear-encoded proteins to divide. In both bacteria and the semiautonomous organelles, fission is facilitated by proteins that form rings on the inner envelope at the site of the future division plane. In plant cells, the genes that encode these proteins are located in the nucleus. Mitochondria and chloroplasts can also increase in size without dividing in order to meet energy or photosynthetic demand. If, for example, the proteins involved in mitochondrial division are experimentally inactivated, the fewer mitochondria become larger, allowing the cell to meet its energy needs.

Protrusions of the outer and inner membrane occur in both mitochondria and chloroplasts. In chloroplasts these protrusions are called **stromules** because they contain stroma, but no thylakoids (see **WEB ESSAY 7.1**). In mitochondria, they are called **matrixules**. Stromules and matrixules may function by connecting the organelles from which they arise, thereby allowing them to exchange materials.

Both plastids and mitochondria can move around plant cells. In some plant cells the chloroplasts are anchored in the outer, cortical cytoplasm of the cell, but in others they are mobile. Like Golgi bodies and peroxisomes, mitochondria are motorized by plant myosins that move along actin microfilaments. Actin microfilament networks are among the main components of the plant cytoskeleton, which we will describe in the next section.

## The Plant Cytoskeleton

The cytosol is organized into a three-dimensional network of filamentous proteins called the **cytoskeleton**. This network provides the spatial organization for the organelles and serves as scaffolding for the movements of organelles and other cytoskeletal components. It also plays fundamental roles in mitosis, meiosis, cytokinesis, wall deposition, the maintenance of cell shape, and cell differentiation.

### The plant cytoskeleton consists of microtubules and microfilaments

Two major types of cytoskeletal elements have been demonstrated in plant cells: microtubules and microfilaments. Each type is filamentous, having a fixed diameter and a variable length, up to many micrometers. Microtubules and microfilaments are macromolecular assemblies of globular proteins.

**Microtubules** are hollow cylinders with an outer diameter of 25 nm; they are composed of polymers of the protein **tubulin**. The tubulin monomer of microtubules is a heterodimer composed of two similar polypeptide chains (**α-** and **β-tubulin**) (**FIGURE 1.20A**). A single microtubule consists of hundreds of thousands of tubulin monomers arranged in columns called **protofilaments**.

**Microfilaments** are solid, with a diameter of 7 nm. They are composed of a special form of the protein found in muscle, called globular actin, or **G-actin**. Monomers of G-actin polymerize to form a single chain of actin subunits, also called a *protofilament*. The actin in the polymerized protofilament is referred to as filamentous actin, or **F-actin**. A microfilament consists of two actin protofilaments intertwined in a double helical fashion (**FIGURE 1.20B**).

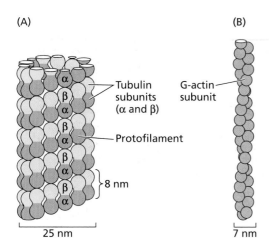

(A)

α
β
α
β
α
β
α
β
α

Tubulin
subunits
(α and β)

Protofilament

8 nm

25 nm

(B)

G-actin
subunit

7 nm

**FIGURE 1.20** (A) Drawing of a microtubule in longitudinal view. Each microtubule is composed of 13 protofilaments. The organization of the α and β subunits is shown. (B) Diagrammatic representation of a microfilament, showing two twisted F-actin strands (protofilaments), which are polymers of G-actin subunits.

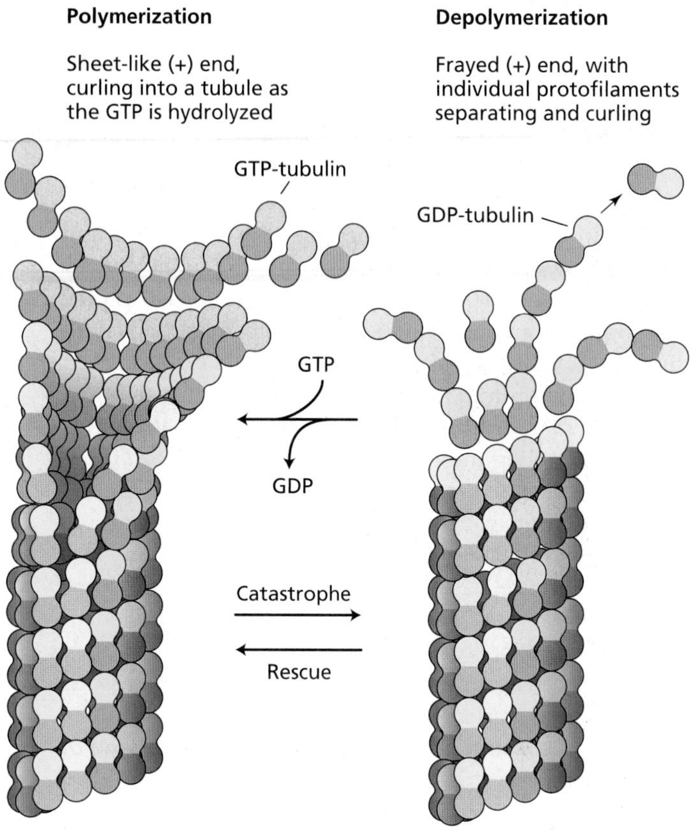

**Polymerization**

Sheet-like (+) end, curling into a tubule as the GTP is hydrolyzed

GTP-tubulin

**Depolymerization**

Frayed (+) end, with individual protofilaments separating and curling

GDP-tubulin

GTP

GDP

Catastrophe

Rescue

**FIGURE 1.21** Model for the dynamic equilibrium between polymerization and depolymerization of a microtubule. When the GTP bound to the β subunit of tubulin is hydrolyzed after the tubulin has been incorporated into the microtubule, the β subunit changes its orientation, causing separation and outward curling of the individual protofilaments, followed by catastrophic depolymerization. This leads to an increase in the local tubulin concentration. Rescue occurs when the exchange of GTP for GDP promotes the tubulin polymerization reaction that forms a sheet-GTP cap on the (+) end of the microtubule. As the sheet grows it curls around and fuses into a tubule.

## Microtubules and microfilaments can assemble and disassemble

In the cell, actin and tubulin subunits exist as pools of free proteins that are in dynamic equilibrium with the polymerized forms. Each of the monomers contains a bound nucleotide: ATP in the case of actin, GTP (guanosine triphosphate) in the case of tubulin. Both microtubules and microfilaments are polarized; that is, the two ends are different. In microfilaments, the polarity arises from the polarity of the actin monomer itself. In microtubules, the polarity arises from the polarity of the α- and β-tubulin heterodimer. The polarity is manifested by the different rates of growth of the two ends, with the more active end denoted the *plus* end and the less active end the *minus* end. In microtubules, the α-tubulin monomer is exposed on the minus end, and the β-tubulin is exposed on the plus end.

Microfilaments and microtubules have half-lives, usually counted in minutes, determined by accessory proteins: *actin-binding proteins (ABPs)* in microfilaments and *microtubule-associated proteins (MAPs)* in microtubules. A subset of these ABPs and MAPs stabilize microfilaments and microtubules, respectively, and prevent their depolymerization.

Purified actin polymerization proceeds by the direct association of G-actin at the growing, or plus, end of the filament to form F-actin in the absence of any accessory proteins. The attachments between the actin subunits in the polymers are noncovalent, and no energy is required for assembly. However, after a monomer is incorporated into a protofilament, the bound ATP is hydrolyzed to ADP. The energy released is stored in the microfilament itself,

making it more prone to dissociation. The roles of various actin-binding proteins in regulating microfilament growth are discussed in **WEB TOPIC 1.12**.

The assembly of microtubules follows a similar pattern, involving nucleation, elongation, and steady state phases. However, microtubule nucleation in vivo is not mediated by oligomerization of its constituent subunits, but by a small ring-shaped complex of a much less abundant type of tubulin called γ-*tubulin*. γ-Tubulin ring complexes (γ-TuRCs) are present at sites called *microtubule-organizing centers (MTOCs)*, where microtubules are nucleated. The principal sites of microtubule nucleation include the cortical cytoplasm (outer layer of cytoplasm) in cells at interphase, the periphery of the nuclear envelope, and the spindle poles in dividing cells. The function of γ-TuRCs is to prime the polymerization of α- and β-tubulin heterodimers into short longitudinal protofilaments. Next, the protofilaments (the number varies with species) associate laterally to form a flat sheet (**FIGURE 1.21**). Finally, the sheet curls into a cylindrical microtubule as GTP is hydrolyzed.

Each tubulin heterodimer contains two GTP molecules, one on the α-tubulin monomer and the other on the β-tubulin monomer. The GTP on the α-tubulin monomer is tightly bound and nonhydrolyzable, while the GTP on the β-tubulin site is readily exchangeable with the medium and is slowly hydrolyzed to GDP after the subunit assembles onto a microtubule. The hydrolysis of GTP to GDP on the β-tubulin subunit causes the dimer to bend slightly and, if not capped by a sheet of recently added GTP-charged tubulin, the protofilaments come apart from each other, initiating a "catastrophic" depolymerization that is much more rapid than the rate of polymerization (see Figure 1.21). Such depolymerization catastrophes can be rescued by the resulting increase in the local tubulin concentration, which (with the help of GTP) tends to favor polymerization (see Figure 1.21). Microtubules are thus said to be *dynamically unstable.* Moreover, the frequency and extent of catastrophic depolymerization that characterizes dynamic instability can be controlled by specific microtubule associated proteins, or MAPs, that can stabilize or destabilize the linkages between the tubulin heterodimers in the wall of the microtubule.

### Cortical microtubules can move around the cell by "treadmilling"

Once polymerized, microtubules do not necessarily remain anchored to the γ-tubulin primers of their microtubule organizing centers (MTOCs). In nonmitotic (interphase) cells, microtubules in the cortical cytoplasm can migrate laterally around the cell periphery by a process called **treadmilling**. During treadmilling tubulin heterodimers are added to the growing plus end at the same rate as tubulin is being lost from the minus end. As a consequence, the microtubule appears to "move" through the cytoplasm, although, in fact, it is not actually moving.

Newly formed microtubules typically treadmill around the cortical cytoplasm in the transverse direction relative to the cell axis (Paradez et al. 2006). As will be discussed in Chapter 15, the transverse orientation of the cortical microtubules determines the orientation of the newly synthesized cellulose microfibrils in the cell wall. The presence of transverse cellulose microfibrils in the cell wall reinforces the wall in the transverse direction, promoting growth along the longitudinal axis. In this way microtubules play an important role in the polarity of the plant.

### Cytoskeletal motor proteins mediate cytoplasmic streaming and organelle traffic

In general, the movements of organelles throughout the cytoplasm are the result of the actions of molecular motors, which have the ability to "walk" along cytoskeletal elements. Cytoskeletal molecular motors all have a similar structure. In their dimeric forms, they consist of a two

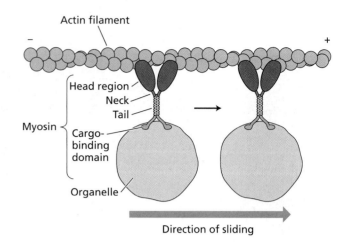

**FIGURE 1.22** Myosin-driven movement of organelles. In plant cells, most organelles move by means of myosin motors. Myosin moves towards the plus (+) end of the actin microfilament. Myosin is a homodimer with two heads and two tails. The two heads, shown in red, have ATPase and motor activity such that a change in the configuration of the neck region adjacent to the head produces a "walking," processive movement along the actin filament. The tails bind to the organelle via their cargo domains, but whether these domains interact directly with the organelle membrane is unknown.

large globular "heads" which actually serve as the "feet" of the motor. The "head" regions are attached to "necks," which function more like "legs." There is a "tail" of various lengths, which connects the heads to the cargo-binding domains, which function as two hands grasping the organelle (**FIGURE 1.22**). Each of the two heads of the motor has ATPase activity, and the energy released by ATP hydrolysis propels each head forward, allowing it to walk along the cytoskeletal element unidirectionally.

Different molecular motors move in different directions down the polar cytoskeletal polymers. The motors that move along microfilaments, the **myosins**, move toward the plus end of F-actin. There are two families of plant myosins, myosin VIII and myosin XI, each containing several members. The family of motor proteins that moves along microtubules is called the **kinesins**. Of the sixty-one members of the kinesin family, two-thirds move toward the plus end of microtubules and one-third move toward the minus end. Although kinesins function in organelle movements along microtubule pathways, kinesin cargo domains can also bind to chromatin or to other microtubules, and help to organize the spindle apparatus during mitosis (see **WEB TOPIC 1.13**). Dyneins, the predominantly minus end–directed microtubule motors in animals and protists, are absent in plants.

**Cytoplasmic streaming** is the coordinated movement of particles and organelles through the cytosol. In the giant

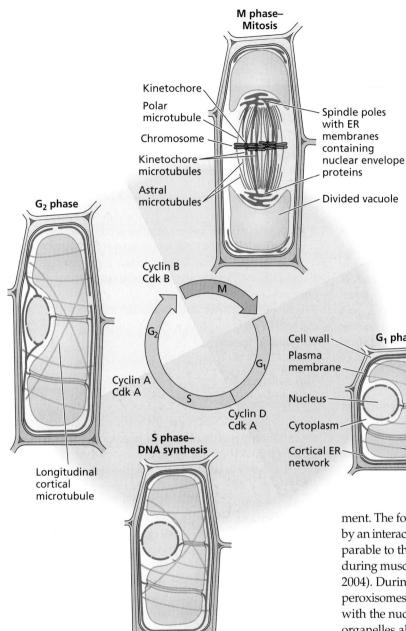

**M phase–Mitosis**

Kinetochore
Polar microtubule
Chromosome
Kinetochore microtubules
Astral microtubules

Spindle poles with ER membranes containing nuclear envelope proteins

Divided vacuole

**G₂ phase**

Cyclin B
Cdk B

M

G₂

Cyclin A
Cdk A

S

G₁

Cyclin D
Cdk A

Longitudinal cortical microtubule

**S phase–DNA synthesis**

Cell wall
Plasma membrane
Nucleus
Cytoplasm
Cortical ER network

**G₁ phase**

Transvacuolar strand
Tonoplast
ER
Transverse cortical microtubule

FIGURE 1.23 Cell cycle in a vacuolated cell type. The four phases of the cell cycle, $G_1$, S, $G_2$, and M are shown in relation to the elongation and division of a vacuolated cell. Various cyclins (CYC) and cyclin-dependent kinases (CDK) regulate the transitions from one phase to the next. Cyclin D and cyclin-dependent kinase A (Cdk A) are involved in the transition from $G_1$ to S. Cyclin A and Cdk A are involved in the transition from S into $G_2$. Cyclin B and cyclin-dependent kinase B (Cdk B) regulate the transition from $G_2$ into M. The kinases phosphorylate other proteins in the cell, causing major reorganization of the cytoskeleton and the membrane systems. The cyclin/cyclin-dependent kinase complexes have a finite lifetime, usually regulated by their own phosphorylation state; the decrease in their abundance toward the end of each phase allows progression to the next stage of the cell cycle.

ment. The forces necessary for movement may be generated by an interaction of the F-actin with myosin in a fashion comparable to that of the sliding protein interaction that occurs during muscle contraction in animals (Shimmen and Yokota 2004). During streaming, myosin independently motorizes peroxisomes, mitochondria, Golgi bodies, and the ER (along with the nuclear envelope and nucleus), transporting these organelles along actin microfilaments to specific regions of the cell (see Figure 1.22).

An example of developmentally regulated cytoplasmic streaming is the migration of the nucleus to the site where tip growth occurs during root hair formation in root epidermal cells. In contrast, the repositioning of leaf chloroplasts, which either maximizes or reduces exposure to light, represents an environmentally regulated form of movement (DeBlasio et al. 2005; see Chapter 9).

## Cell Cycle Regulation

The cell division cycle, or cell cycle, is the process by which cells reproduce themselves and their genetic material, the nuclear DNA (FIGURE 1.23). The cell cycle consists of four phases: $G_1$, S, $G_2$, and M. **$G_1$** is the phase when a newly formed daughter cell has not yet replicated its DNA. DNA is replicated during the **S phase**. **$G_2$** is the phase when a cell with replicated DNA has not yet proceeded to mitosis. Collectively, the $G_1$, S, and $G_2$ phases are referred to as

cells of the green algae *Chara* and *Nitella*, streaming occurs in a helical path down one side of a cell and up the other side, occurring at speeds of up to 75 µm s⁻¹. Cytoplasmic streaming occurs in most plant cells, but in diverse patterns and velocities, sometimes changing or stopping briefly in response to environmental insults (Hardham et al. 2008).

The term "cytoplasmic streaming" has been traditionally defined as bulk cytoplasmic flow of cytosol and organelles that is readily visible at low magnifications. This type of bulk flow results from the viscous drag that the motorized organelles exert on the cytosol. More recently, the term cytoplasmic streaming has also been applied to organellar movements in which neighboring organelles pass each other going in opposite directions. Such independent movements of organelles are clearly not due to bulk flow.

Cytoplasmic streaming involves bundles of actin microfilaments that are arranged along the axis of particle move-

interphase. The M phase is **mitosis**. In vacuolated cells, the vacuole enlarges throughout interphase, and the plane of cell division bisects the vacuole during mitosis (see Figure 1.23).

## Each phase of the cell cycle has a specific set of biochemical and cellular activities

Nuclear DNA is prepared for replication in $G_1$ by the assembly of a prereplication complex at the origins of replication along the chromatin. DNA is replicated during the S phase, and $G_2$ cells prepare for mitosis.

The whole architecture of the cell is altered as it enters mitosis. If the cell has a large central vacuole, this vacuole must first be divided in two by a coalescence of cytoplasmic transvacuolar strands that contain the nucleus; this becomes the region where nuclear division will occur (see Figure 1.23). Golgi bodies and other organelles partition themselves equally between the two halves of the cell. As described below, the endomembrane system and cytoskeleton are extensively rearranged.

As a cell enters mitosis, the chromosomes change from their interphase state of organization within the nucleus and begin to condense to form the metaphase chromosomes (**FIGURE 1.24**). The metaphase chromosomes are held together by special proteins called *cohesins*, which reside in the centromeric region of each chromosome pair.

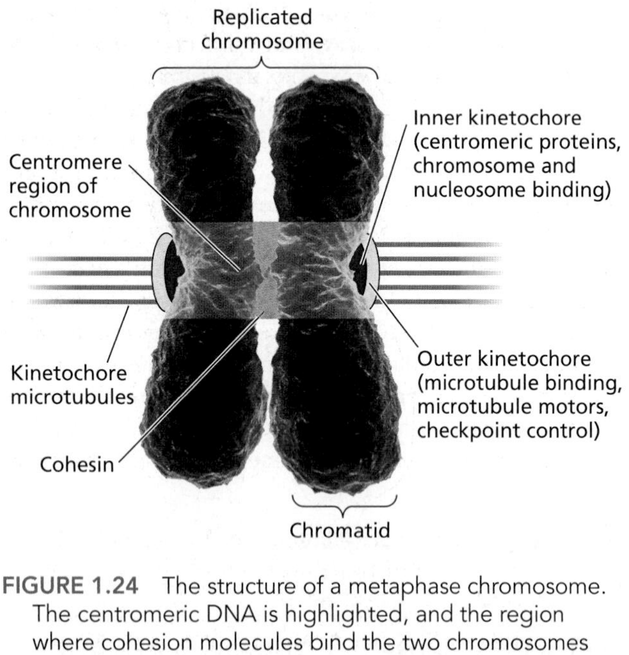

**FIGURE 1.24**  The structure of a metaphase chromosome. The centromeric DNA is highlighted, and the region where cohesion molecules bind the two chromosomes together is shown in orange. The kinetochore is a layered structure (inner layer purple, outer layer yellow) that contains microtubule-binding proteins, including kinesins that help depolymerize the microtubules during shortening of kinetochore microtubules in anaphase. (Chromosome model © Sebastian Kaulitzki/Shutterstock.)

In order for the chromosomes to separate, these proteins must be cleaved by the enzyme *separase*, which first has to be activated. This occurs when the kinetochore attaches to spindle microtubules (described in the next section).

At a key regulatory point, or **checkpoint**, early in $G_1$ of the cell cycle, the cell becomes committed to the initiation of DNA synthesis. In mammalian cells, DNA replication and mitosis are linked—the cell division cycle, once begun, is not interrupted until all phases of mitosis have been completed. In contrast, plant cells can leave the cell division cycle either before or after replicating their DNA (i.e., during $G_1$ or $G_2$). As a consequence, whereas most animal cells are diploid (having two sets of chromosomes), plant cells frequently are tetraploid (having four sets of chromosomes), or even polyploid (having many sets of chromosomes), after going through additional cycles of nuclear DNA replication without mitosis, a process called **endoreduplication**.

## The cell cycle is regulated by cyclins and cyclin-dependent kinases

The biochemical reactions governing the cell cycle are evolutionarily highly conserved in eukaryotes, and plants have retained the basic components of this mechanism. Progression through the cycle is regulated mainly at three checkpoints: during the late $G_1$ phase, late S phase, and at the $G_2$/M boundary.

The key enzymes that control the transitions between the different states of the cell cycle, and the entry of nondividing cells into the cell cycle, are the **cyclin-dependent protein kinases**, or **Cdks**. Protein kinases are enzymes that phosphorylate proteins using ATP. Most multicellular eukaryotes use several protein kinases that are active in different phases of the cell cycle. All depend on regulatory subunits called **cyclins** for their activities. Several classes of cyclins have been identified in plants, animals, and yeast. Three cyclins (A, B, and D) have been shown to regulate the tobacco cell cycle, as shown in Figure 1.23.

1. $G_1$/S cyclins: cyclin D, active late in $G_1$

2. S cyclins: cyclin A, active in late S phase.

3. M cyclins: cyclin B, active just prior to the mitotic phase

The critical restriction point late in $G_1$, which commits the cell to another round of cell division, is regulated primarily by the D-type cyclins. As we will see later in the book, plant hormones that promote cell division, including cytokinins (see Chapter 21) and brassinosteroids (see Chapter 24), appear to do so at least in part through an increase in cyclin D3, a plant D-type cyclin.

Cdk activity can be regulated in various ways, but two of the most important mechanisms are (1) cyclin synthesis and degradation and (2) the phosphorylation and dephosphorylation of key amino acid residues within the Cdk protein. In the first regulatory mechanism, Cdks are inactive unless they are associated with a cyclin. Most cyclins turn over rapidly; they are synthesized and then actively degraded (using ATP) at specific points in the cell cycle. Cyclins are degraded in the cytosol by a large proteolytic complex called the **26S proteasome** (see Chapter 2). Before being degraded by the proteasome, the cyclins are marked for destruction by the attachment of a small protein called *ubiquitin*, a process that requires ATP. Ubiquitination is a general mechanism for tagging cellular proteins destined for turnover (see Chapter 2).

The second mechanism regulating Cdk activity is phosphorylation and dephosphorylation. Cdks possess two tyrosine phosphorylation sites: One causes activation of the enzyme; the other causes inactivation. Specific kinases carry out both the stimulatory and the inhibitory phosphorylations. Similarly, protein phosphatases can remove phosphates from Cdks, either stimulating or inhibiting their activity depending on the position of the phosphate. The addition or removal of phosphate groups from Cdks is highly regulated and an important mechanism for the control of cell cycle progression. Further control of the pathway is exerted by the presence of Cdk inhibitors (ICKs) that can influence the $G_1/S$ transition.

### Mitosis and cytokinesis involve both microtubules and the endomembrane system

Mitosis is the process by which previously replicated chromosomes are aligned, separated, and distributed in an orderly fashion to daughter cells (**FIGURE 1.25**). Microtubules are an integral part of mitosis. The period immediately prior to prophase is called **preprophase**. During preprophase, the $G_2$ microtubules are completely reorganized into a **preprophase band** of microtubules around the nucleus at the site of the future cell plate—the precursor of the cross-wall (see Figure 1.25). The position of the preprophase band, the underlying *cortical division site* (Müller et al. 2009), and the partition of cytoplasm that divides central vacuoles determine the plane of cell division in plants, and thus play a crucial role in development (see Chapter 16).

At the start of prophase, microtubules, polymerizing on the surface of the nuclear envelope, begin to gather at two foci on opposite sides of the nucleus, initiating **spindle formation** (see Figure 1.25). Although not associated with the centrosomes, as they are in animal cells, these foci serve the same function in organizing microtubules. During **prophase**, the nuclear envelope remains intact, but it breaks down as the cell enters **metaphase** in a process that involves reorganization and re-assimilation of the nuclear envelope into the ER (see Figure 1.25). Throughout the division cycle, cell division kinases interact with microtubules to help reorganize the spindle by phosphorylating MAPs and kinesins.

As the chromosomes condense, the **nucleolar organizer** regions of the different chromosomes disassociate, causing the nucleolus to fragment. The nucleolus completely disappears during mitosis and gradually reassembles after mitosis, as the chromosomes decondense and re-establish their positions in the daughter nuclei.

In early metaphase, or **prometaphase**, the preprophase band is gone and new microtubules polymerize to complete the **mitotic spindle**. The mitotic spindles of plant cells, which lack centrosomes, are more boxlike in shape than those in animal cells. The spindle microtubules in a plant cell arise from a diffuse zone consisting of multiple foci at opposite ends of the cell and extend toward the middle in nearly parallel arrays.

**Metaphase** chromosomes are completely condensed through close packing of the histones into nucleosomes, which are further wound into condensed fibers (see Figure 1.7). The **centromere**, the region where the two chromatids are attached near the center of the chromosome, contains repetitive DNA, as does the **telomere**, which forms the chromosome ends that protect it from degradation. Some of the spindle microtubules bind to the chromosomes in the special region of the centromere called the **kinetochore**, and the condensed chromosomes align at the metaphase plate (see Figure 1.24). Some of the unattached microtubules overlap with microtubules from the opposite polar region in the spindle midzone.

Just as there are checkpoints that control the four phases of the cell cycle, there are also checkpoints that operate *within* mitosis. For example, a **spindle assembly checkpoint** stops cells from proceeding into anaphase if spindle microtubules have incorrectly interacted with the kinetochores. The cyclin B–Cdk B complex plays a central role in regulating this process. If the spindle microtubules have correctly attached to their kinetochores, the **anaphase promoting complex** brings about the proteolytic degradation of cyclin B. Without cyclin B, the cyclin B–Cdk B complex can no longer form; this allows the chromatids aligned at the metaphase plate to segregate to their respective poles. (Note that each chromatid has undergone a round of DNA replication and contains the diploid [2N] amount of DNA. Thus, as soon as separation occurs, the chromatids become chromosomes.)

The mechanism of chromosome separation during **anaphase** has two components:

- **anaphase A** or early anaphase, during which the sister chromatids separate and begin to move toward their poles; and

- **anaphase B** or late anaphase, during which the polar microtubules slide relative to each other and elongate to push the spindle poles farther apart. At the same time, the sister chromosomes are pushed to their respective poles.

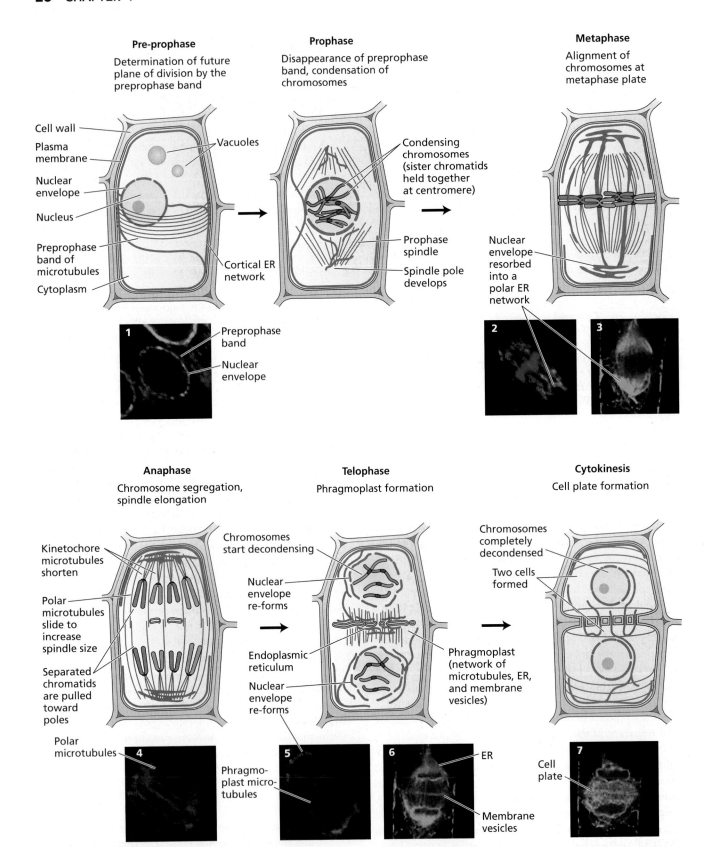

**FIGURE 1.25**  Changes in cellular organization that accompany mitosis in a meristematic (nonvacuolate) plant cell. (1, 2, 4, and 5) Red fluorescence is from anti-α-tubulin antibody (microtubules), green fluorescence is from WIP-GFP (green fluorescent protein fused to a nuclear en-velope protein), and blue fluorescence is from DAPI (a DNA-binding dye). (3, 6, and 7) The ER is labeled with green-fluorescing HDEL-GFP and the cell plate is labeled with red-fluorescing FM4-64. (1, 2, 4, and 5 from Xu et al. 2007; 3, 6, and 7 from Higaki et al. 2008.)

**Telophase**
Phragmoplast formation

(A)

(B)

**Cytokinesis**
Cell plate formation

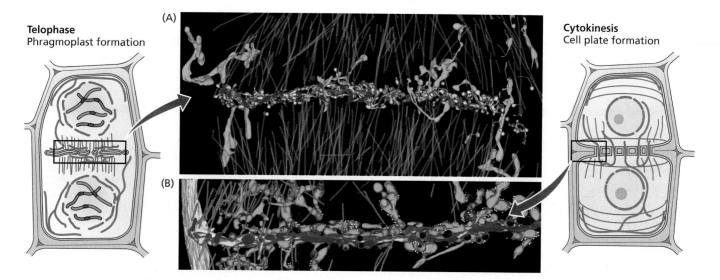

**FIGURE 1.26**  Changes in the organization of the phragmoplast and ER during cell plate formation. (A) The forming cell plate (yellow, seen from the side) in early telophase has only a few sites where it interacts with the tubulovesicular network of the ER (blue). The solid phragmoplast microtubules (purple) also have few ER cisternae among them. (B) Side view of the forming peripheral cell plate (yellow), showing that although many cytoplasmic ER tubules (blue) intermingle with microtubules (purple) in the peripheral growth zone, there is little direct contact between ER tubules and cell plate membranes. The small white dots are ER-bound ribosomes. (3D tomographic reconstruction of electron microscopy of the phragmoplast from Seguí-Simarro et al. 2004.)

In plants, the spindle microtubules are apparently not anchored to the cell cortex at the poles, and so the chromosomes cannot be pulled apart. Instead they are probably pushed apart by kinesins in the overlapping polar microtubules of the spindle (see WEB TOPIC 1.11).

At **telophase**, a new network of microtubules and F-actins called the **phragmoplast** appears. The phragmoplast organizes the region of the cytoplasm where cytokinesis takes place. The microtubules have now lost their spindle shape, but retain polarity, with their minus ends still pointed toward the separated, now decondensing, chromosomes, where the nuclear envelope is in the process of re-forming (see Figure 1.25, "Telophase"). The plus ends of the microtubules point toward the midzone of the phragmoplast, where small vesicles accumulate, partly derived from endocytotic vesicles from the parent cell plasma membrane. These vesicles have long tethers that may aid in the formation of the cell plate in the next cell cycle stage: *cytokinesis*.

**Cytokinesis** is the process that establishes the **cell plate**, a precursor of the cross-wall that will separate the two daughter cells (**FIGURE 1.26**). This cell plate, with its enclosing plasma membrane, forms as an island in the center of the cell that grows outward toward the parent wall by vesicle fusion. The targeting recognition protein **KNOLLE**, which belongs to the SNARE family of proteins involved in vesicle fusion (see WEB TOPIC 1.8), is present at the forming cell plate, as is **dynamin**, a noose-shaped GTPase that is involved in vesicle formation. The site at which the forming cell plate fuses with the parental plasma membrane is determined by the location of the preprophase band (which disappeared earlier) and specific microtubule-associated proteins (MAPS). As the cell plate assembles, it traps ER tubules in plasmalemma-lined membrane channels spanning the plate, thus connecting the two daughter cells (see Figure 1.26).

As discussed in the next section, the cell plate-spanning ER tubules establish the sites for the *primary plasmodesmata*. After cytokinesis, microtubules re-form in the cell cortex. The new cortical microtubules have a transverse orientation relative to the cell axis, and this orientation determines the polarity of future cell extension.

## Plasmodesmata

**Plasmodesmata** (singular *plasmodesma*) are tubular extensions of the plasma membrane, 40 to 50 nm in diameter, that traverse the cell wall and connect the cytoplasms of adjacent cells. Because most plant cells are interconnected in this way, their cytoplasms form a continuum referred to as the **symplast**. Intercellular transport of solutes through plasmodesmata is thus called **symplastic transport** (see Chapters 4 and 6).

### Primary and secondary plasmodesmata help to maintain tissue developmental gradients

**Primary plasmodesmata** form cytoplasmic connections between clonally derived cells (cells derived from each other by mitosis). The symplast allows water and solutes to be transported between cells without crossing a membrane. However, there is a restriction on the size of mol-

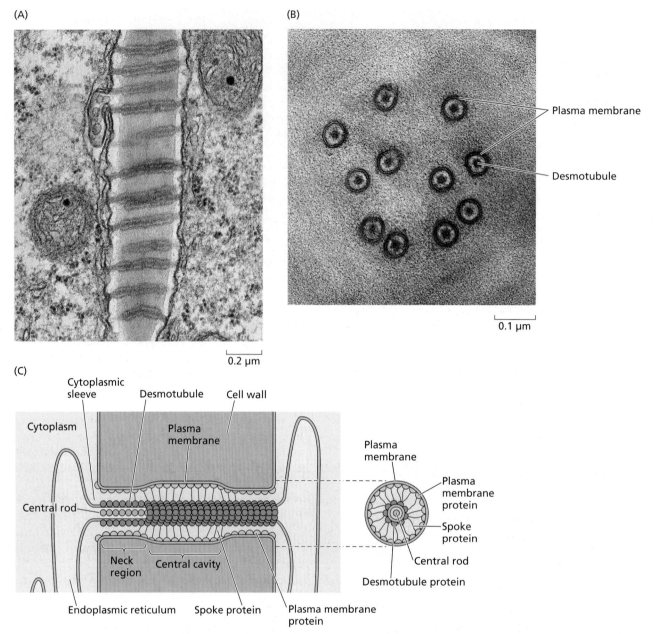

(A)

0.2 μm

(B)

Plasma membrane

Desmotubule

0.1 μm

(C)

Cytoplasmic sleeve

Desmotubule

Cell wall

Cytoplasm

Plasma membrane

Central rod

Neck region

Central cavity

Endoplasmic reticulum

Spoke protein

Plasma membrane protein

Plasma membrane

Plasma membrane protein

Spoke protein

Central rod

Desmotubule protein

**FIGURE 1.27** Plasmodesmata between cells. (A) Electron micrograph of a wall separating two adjacent cells, showing the plasmodesmata in longitudinal view. (B) Tangential section through a cell wall showing numerous plasmodesmata in cross section. (C) Schematic view of a cell wall with a plasmodesma. The pore consists of a central cavity between two narrow neck regions. The desmotubule is continuous with the ER of the adjoining cells. Proteins line the outer surface of the desmotubule and the inner surface of the plasma membrane; the two surfaces are thought to be connected by filamentous proteins, which divide the cytoplasmic sleeve into microchannels. The size of the gaps between the proteins controls the molecular sieving properties of plasmodesmata. (A and B courtesy of Ray Evert; from Robinson-Beers and Evert 1991.)

ecules that can be transported via the symplast; this restriction is called the **size exclusion limit** (**SEL**). This limit is imposed by the width of the cytoplasmic sleeve that surrounds the ER tubule, or **desmotubule**, which forms the center of the plasmodesma (**FIGURE 1.27**) (Roberts and Oparka 2003). In addition, globular proteins within the sleeve generate spiraling microchannels through the plasmodesma (Ding et al. 1992).

The limiting molecular mass can be determined by following the movement of fluorescent dye molecules of different sizes through plasmodesmata connecting leaf epidermal cells. The limit for transport among these cells appears to be about 700 to 1000 daltons, equivalent to a molecular size of about 1.5 to 2.0 nm. The mechanisms of solute transport through the cytoplasmic sleeve and the desmotubule itself are still unclear. Actin and myosin have been shown to be present in plasmodesmata (Roberts and Oparka 2003). Two possibilities have been suggested:

1. The actin and myosin may enable the pore to constrict, reducing the SEL of the pore.

2. The actin and myosin may themselves facilitate the movement of macromolecules and particles through the pore.

If the SEL of plasmodesmata restricts traffic between cells to molecules of 2.0 nm or less, how do particles as large as tobacco mosaic virus (a rod 18 nm wide and 300 nm long) pass through them? **Movement proteins** are nonstructural proteins encoded by the virus genome that facilitate viral movement through the symplast by one of two mechanisms. Some plant viruses encode movement proteins that can coat the surface of the viral genome (typically RNA), forming ribonucleoprotein complexes that can pass through the plasmodesmatal pore. The 30-kDa movement protein of tobacco mosaic virus acts in this way. Other viruses, such as cowpea mosaic virus and tomato spotted wilt virus, encode movement proteins that form a transport tubule within the plasmodesmatal pore that facilitates the movement of mature virus particles through plasmodesmata.

If the genetic information of pathogens can be transferred through plasmodesmata, could genetic information of clonally derived cells be shared as well? Polar gradients of morphogens (signaling molecules involved in development; see Chapter 16) that are sensed by the plant at a level above that of the individual cell—at the tissue level—probably contribute to the overall form of plants. For example, the number of cells in a leaf, above a minimum value, is independent of the size of the leaf. The signaling molecules involved in organogenesis involve not only small molecules, such as hormones, but proteins and RNA as well (Gallagher and Benfey 2005). In Chapter 16, we will examine this type of cell-to-cell communication in greater detail.

Symplastic transport can also occur between non-clonally related cells through the formation of **secondary plasmodesmata** (Lucas and Wolf 1993). These connections are made possible when areas of adjoining cell walls are locally digested; the plasma membranes of the adjacent cells then fuse. Finally, the ER network between the cells is connected. The fact that the symplast can expand in this way suggests that it plays an important role in nutrition as well as in developmental signaling.

## SUMMARY

Despite their great diversity in form and size, all plants carry out similar physiological processes. All plant tissues, organs, and whole organisms show a growth polarity, being derived from axial or radial polarity of cell division of meristems.

### Plant Life: Unifying Principles

- All plants convert solar energy to chemical energy; use growth instead of motility to obtain resources; have vascular systems; have rigid structures; and have mechanisms to avoid desiccation.

### Overview of Plant Structure

- All plants carry out fundamentally similar processes and have a common body plan (**Figure 1.1**).

- Because of their rigid cell walls, plant development depends solely on patterns of cell division and cell enlargement (**Figure 1.2**).

- Nearly all mitosis and cytokinesis occurs in meristems.

- Dermal tissue, ground tissue, and vascular tissue are the three major tissue systems present in all plant organs (**Figure 1.3**).

### Plant Cell Organelles

- In addition to cell walls and plasma membrane, plant cells contain compartments derived from the endomembrane system (**Figure 1.4**).

- Chloroplasts and mitochondria are not derived from the endomembrane system.

- The endomembrane system plays a central role in secretory processes, membrane recycling, and the cell cycle.

- The composition and fluid-mosaic structure of the plasma membrane permits regulation of transport into and out of the cell (**Figure 1.5**).

### The Endomembrane System

- The endomembrane system conveys both membrane and cargo proteins to diverse organelles.

- The specialized membranes of the nuclear envelope are derived from the endoplasmic reticulum (ER), a component of the endomembrane system (**Figures 1.6, 1.8**).

- The nucleus is the site of storage, replication, and transcription of the chromatin, as well as being the site for the synthesis of ribosomes (**Figures 1.7, 1.8**).

## SUMMARY continued

- The ER is a system of membrane-bound tubules that form a complex and dynamic structure (**Figure 1.9**).

- The rough ER (RER) is involved in synthesis of proteins that enter the lumen of the ER; the smooth ER is the site of lipid biosynthesis (**Figures 1.8, 1.10**).

- The ER provides the membrane and internal cargo for the other compartments of the endomembrane system.

- Secretion of proteins from cells begins with the RER (**Figure 1.8**).

- Glycoproteins and polysaccharides destined for secretion are processed in the Golgi apparatus (**Figure 1.11**).

- During endocytosis, membrane is removed from the plasma membrane by formation of small clathrin-coated vesicles (**Figures 1.12, 1.13**).

- Endocytosis from the plasma membrane provides membrane recycling (**Figure 1.13**).

### Independently Dividing Organelles Derived from the Endomembrane System

- Oil bodies, peroxisomes, and glyoxysomes grow and proliferate independently of the endomembrane system (**Figures 1.14, 1.15**).

### Independently Dividing, Semiautonomous Organelles

- Mitochondria and chloroplasts each have an inner and an outer membrane (**Figures 1.16, 1.17**).

- Plastids may contain high concentrations of pigments or starch (**Figure 1.18**).

- Proplastids pass through distinct developmental stages to form specialized plastids (**Figure 1.19**).

- In plastids and mitochondria, DNA replication and fission are regulated independently of nuclear division.

### The Plant Cytoskeleton

- A three-dimensional network of microtubules and microfilaments organizes the cytosol (**Figure 1.20**).

- Microtubules and microfilaments can assemble and disassemble (**Figure 1.21**).

- Molecular motors associated with components of the cytoskeleton move organelles throughout the cytoplasm (**Figure 1.22**).

- During cytoplasmic streaming, interaction of the F-actin with myosin provides for independent movement of organelles, including chloroplasts.

### Cell Cycle Regulation

- The cell cycle, during which cells replicate their DNA and reproduce themselves, consists of four phases (**Figures 1.23, 1.24**).

- Successful mitosis and cytokinesis require the participation of microtubules and the endomembrane system (**Figures 1.25, 1.26**).

### Plasmodesmata

- Tubular extensions of the plasma membrane traverse the cell wall and connect the cytoplasms of clonally derived cells, allowing water and small molecules to move between cells without crossing a membrane (**Figure 1.27**).

## WEB MATERIAL

### Web Topics

**1.1  Model Organisms**
Certain plant species are used extensively in the lab to study their physiology.

**1.2  The Plant Kingdom**
The major groups of the plant kingdom are surveyed and described.

**1.3  Flower Structure and the Angiosperm Life Cycle**
The steps in the reproductive style of angiosperms are discussed and illustrated.

**1.4  Plant Tissue Systems: Dermal, Ground, and Vascular**
A more detailed treatment of plant anatomy is given.

**1.5  The Structures of Chloroplast Glycosylglycerides**
The chemical structures of the chloroplast lipids are illustrated.

**1.6  A Model for the Structure of Nuclear Pores**
The nuclear pore is believed to be lined by a meshwork of unstructured nucleoporin proteins.

## WEB MATERIAL continued

## CHAPTER REFERENCES

Alberts, B., Johnson, A., Lewis, J., Raff, M., Roberts, K., and Walter, P. (2002) *Molecular Biology of the Cell*, 4th ed. Garland, New York.

Alexander, L., Sessions, R., Clarke, T., Tatham, A. S., Shewry, P. R., and Napier, J. A. (2002) Characterisation and modelling of the hydrophobic domain of a sunflower oleosin. *Planta* 214: 546–551.

Buchanan, B. B., Gruissem, W., and Jones, R. L., eds. (2000) *Biochemistry and Molecular Biology of Plants.* American Society of Plant Physiologists, Rockville, MD.

DeBlasio, S. L., Luesse, D. L., and Hangarter, R. P. (2005) A plant-specific protein essential for blue-light-induced chloroplast movements. *Plant Physiol.* 139: 101–114.

Denning, D. P., Patel, S. S., Uversky, V., Fink, A. L., and Rexach, M. (2003) Disorder in the nuclear pore complex: The FG repeat regions of nucleoporins are natively unfolded. *Proc. Natl. Acad. Sci. USA* 100: 2450–2455.

Ding, B., Turgeon, R., and Parthasarathy, M. V. (1992) Substructure of freeze substituted plasmodesmata. *Protoplasma* 169: 28–41.

Fiserova, J., Kiseleva, E., and Goldberg, M. W. (2009) Nuclear envelope and nuclear pore complex structure and organization in tobacco BY-2 cells. *Plant J.* 59: 243–255.

Gallagher, K. L., and Benfey, P. N. (2005) Not just another hole in the wall: Understanding intercellular protein trafficking. *Genes Dev.* 19: 189–195.

Gunning, B. E. S., and Steer, M. W. (1996) *Plant Cell Biology: Structure and Function of Plant Cells.* Jones and Bartlett, Boston.

Hardham, A. R., Takemoto, D., and White, R. G. (2008) Rapid and dynamic subcellular reorganization following mechanical stimulation of Arabidopsis epidermal cells mimics responses to fungal and oomycete attack. *BMC Plant Biol.* 8: 63.

Higaki, T., Kutsuna, N., Sano, T., and Hasezawa, S. (2008) Quantitative analysis of changes in actin microfilament contribution to cell plate development in plant cytokinesis. *BMC Plant Biol.* 8: 80.

Huang, A. H. C. (1987) Lipases. In *The Biochemistry of Plants: A Comprehensive Treatise*, Vol. 9, *Lipids: Structure and Function*, P. K. Stumpf, ed., Academic Press, New York, pp. 91–119.

Langhans, M., Hawes, C., Hillmer, S., Hummel, E., and Robinson, D. G. (2007) Golgi regeneration after brefeldin A treatment in BY-2 cells entails stack enlargement and cisternal growth followed by division. *Plant Physiol.* 145: 527–538.

Larsson, K. E., Kjellberg, J. M., Tjellström, H., and Sandelius, A. S. (2007) LysoPC acyltransferase/PC transacylase activities in plant plasma membrane and plasma membrane-associated endoplasmic reticulum. *BMC Plant Biol.* 7: 64.

Lucas, W. J., and Wolf, S. (1993) Plasmodesmata: The intercellular organelles of green plants. *Trends Cell Biol.* 3: 308–315.

Mollenhauer, H. H., Morre, D. J., and Griffing, L. R. (1991) Post Golgi apparatus structures and membrane removal in plants. *Protoplasma* 162: 55–60.

Müller, S., Wright, A. J., and Smith, L. G. (2009) Division plane control in plants: new players in the band. *Trends Cell Biol.* 19: 180–188.

Nanjo, Y., Oka, H., Ikarashi, N., Kaneko, K., Kitajima, A., Mitsui, T., Muñoz, F. J., Rodríguez-López, M., Baroja-Fernández, E., and Pozueta-Romero, J. (2006) Rice plastidial N-glycosylated nucleotide pyrophosphatase/phosphodiesterase is transported from the ER-golgi to the chloroplast through the secretory pathway. *Plant Cell* 18: 2582–2592.

Otegui, M., Herder, R., Schulze, J., Jung, R., and Staehelin, L. A. (2006) The proteolytic processing of seed storage proteins in Arabidopsis embryo cells starts in the multivesicular bodies. *Plant Cell* 18: 2567–2581.

Paradez, A., Wright, A., and Ehrhardt, D. W. (2006) Microtubule cortical array organization and plant cell morphogenesis. *Curr. Opin. Plant Biol.* 9: 571–578.

Roberts, A. G., and Oparka, K. J. (2003) Plasmodesmata and the control of symplastic transport. *Plant Cell Environ.* 26: 103–124.

Robinson-Beers, K., and Evert, R. F. (1991) Fine structure of plasmodesmata in mature leaves of sugar cane. *Planta* 184: 307–318.

Sahu, S. K., and Gummadi, S. N. (2008) Flippase activity in proteoliposomes reconstituted with Spinacea oleracea endoplasmic reticulum membrane proteins: Evidence of biogenic membrane flippase in plants. *Biochemistry* 47: 10481–10490.

Seguí-Simarro, J. M., Austin, J. R., White, E. A., and Staehelin, L. A. (2004) Electron tomographic analysis of somatic cell plate formation in meristematic cells of arabidopsis preserved by high-pressure freezing. *Plant Cell* 16: 836–856.

Shimmen, T., and Yokota, E. (2004) Cytoplasmic streaming in plants. *Curr. Opin. Cell Biol.* 16: 68–72.

Sparkes, I., Runions, J., Hawes, C., and Griffing, L. (2009) Movement and remodeling of the endoplasmic reticulum in non-dividing cells of tobacco leaves. *Plant Cell* 21: 3937–3949.

Titus, D. E., and Becker, W. M. (1985) Investigation of the glyoxysome-peroxisome transition in germinating cucumber cotyledons using double-label immunoelectron microscopy. *J. Cell Biol.* 101: 1288–1299.

Villarejo, A., Burén, S., Larsson, S., Déjardin, A., Monné, M., Rudhe, C., Karlsson, J., Jansson, S., Lerouge, P., Rolland, N., et al. (2005) Evidence for a protein transported through the secretory pathway en route to the higher plant chloroplast. *Nat. Cell Biol.* 7: 1224–1231.

Xu, C., Fan, J., Cornish, A. J., and Benning, C. (2008) Lipid trafficking between the endoplasmic reticulum and the plastid in Arabidopsis requires the extraplastidic TGD4 protein. *Plant Cell* 20: 2190–2204.

Xu, X. M., Meulia, T., and Meier, I. (2007) Anchorage of plant RanGAP to the nuclear envelope involves novel nuclear-pore-associated proteins. *Curr. Biol.* 17: 1157–1163.

Zhang, X., and Hu, J. (2008) Fission1A and fission1B proteins mediate the fission of peroxisomes and mitochondria in Arabidopsis. *Mol. Plant* 1: 1036–1047.

# 2

# Genome Organization and Gene Expression

A plant's phenotype is the result of three major factors: its genotype (all the genes, or alleles, that determine the plant's traits), the pattern of epigenetic modifications of its DNA, and the environment in which it lives. In Chapter 1 we reviewed the fundamental structure and function of DNA, its packaging into chromosomes, and the two major phases of gene expression: transcription and translation. In this chapter we will discuss how the composition of the genome beyond its genes influences the physiology and evolution of the organism. First we will look at the organization of the nuclear genome and the non-gene elements it contains. Then we will turn to the cytoplasmic genomes, which are contained inside the mitochondria and plastids. Next we will discuss the cellular machinery needed to transcribe and translate the genes into functional proteins, and we will see how gene expression is regulated both transcriptionally and posttranscriptionally. We will then introduce some of the tools used to study gene function, and we will finish up with a discussion of the use of genetic engineering in research and agriculture.

## Nuclear Genome Organization

As discussed in Chapter 1, the nuclear genome contains most of the genes required for the plant's physiological functions. The first plant genome to be fully sequenced, in 2000, was that of a small dicotyledonous angiosperm called thale cress, or *Arabidopsis thaliana*.

The genome of *A. thaliana* is made up of only about 157 million base pairs (157 Mbp), which are distributed over five chromosomes. (By contrast, the genome of the lily *Fritillaria assyriaca*, the plant species with the largest known genome, contains as many as 88,000 Mbp.) Within its nuclear genome, *A. thaliana* holds some 27,400 protein-coding genes and almost another 5000 genes that are either **pseudogenes** (nonfunctional genes) or parts of transposons (mobile DNA elements that we will discuss later in this chapter). In addition to these genes, more than 1300 genes that produce non-protein-coding RNAs (**ncRNAs**) are known in *A. thaliana*. Many of these RNAs are probably involved in gene regulation. Both transposons and ncRNAs are described in more detail later in this chapter.

The plant genome, however, consists of much more than genes. In this section we will take a look at the chemical makeup of the genome, then see how certain regions of the genome correspond to specific functions.

### The nuclear genome is packaged into chromatin

The nuclear genome consists of DNA molecules that are wrapped around histone proteins to form beadlike structures called **nucleosomes** (see Chapter 1). DNA and histones, together with other proteins that bind to the DNA, are referred to as **chromatin** (see Figure 1.10). Two types of chromatin can be distinguished: *euchromatin* and *heterochromatin*. Historically, these two types were differentiated based on their appearance in light microscopy when stained with specific dyes. **Heterochromatin** is usually more tightly packaged and thus appears darker than the less condensed **euchromatin**. Most genes that are actively transcribed in a plant are located within the euchromatic regions of a chromosome, while many genes located in heterochromatic regions are not transcribed—these genes are inactive, or silent. Compared with euchromatin, heterochromatin is relatively gene poor. Heterochromatic regions include the centromeres, several so-called knobs, and the regions immediately adjacent to the telomeres, or chromosome ends, known as the **subtelomeric regions** (Gill et al. 2008).

Heterochromatic structures often consist of highly repetitive DNA sequences, or **tandem repeats**: blocks of nucleotide motifs of about 150 to 180 bp that are repeated over and over. A second class of repeats is the **dispersed repeats**. One type of dispersed repeat is known as **simple sequence repeats (SSR)** or **microsatellites**. These repeats consist of sequence motifs as short as two nucle-

otides that are repeated hundreds or even thousands of times. Another dominant group of dispersed repeats found in heterochromatin is the "jumping genes," or *transposons* (Heslop-Harrison and Schmidt 2007).

### Centromeres, telomeres, and nucleolar organizers contain repetitive sequences

The most prominent structural landmarks on chromosomes are centromeres, telomeres, and nucleolar organizers. These regions contain repetitive DNA sequences that can be made visible by fluorescent in situ hybridization (FISH) (Kato et al. 2005), a technique that uses fluorescently labeled molecular probes—usually fragments of DNA—that bind specifically to the sequence to be identified (**FIGURE 2.1**). **Centromeres** are constrictions of the chromosomes where sister chromatids adhere to each other and where spindle fibers attach during cell division. The attachment of fibers to the centromere is mediated by the kinetochore, a protein complex surrounding the centromere (see Chapter 1). Centromeres consist of highly repetitive DNA regions and inactive transposable elements (Jiang et al. 2003; Ma et al. 2007). **Telomeres** are sequences located at the ends of each chromosome. Telomeres act as caps on the chromosome ends and prevent loss of DNA during DNA replication.

The RNA molecules that make up ribosomes (rRNA) are transcribed from **nucleolar organizer** (**NO**) regions. Because ribosomes are composed mostly of rRNA, and because many ribosomes are needed for translation, it is not surprising that NOs contain hundreds of repeated copies of each rRNA gene. Depending on the plant species, one or several nucleolar organizers are present within the

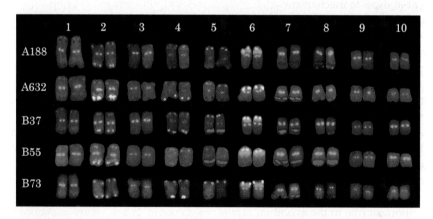

**FIGURE 2.1** Chromosomal landmarks, including centromeres, telomeres, and nucleolar organizers, can be used to identify individual chromosomes. Each row shows the ten chromosome pairs of a different inbred line of maize (5 common lines are shown here, from A188 to B73). DNA sequences (probes) complementary to certain chromosomal landmarks were labeled with fluorescent dyes and hybridized to chromosome preparations. The centromeres can be seen as green dots near the middle of the chromosomes, the nucleolar organizer as a larger green area on chromosome 6, and the telomeres as faint red dots, most clearly visible at the top of chromosomes 2–4. The larger blue areas are specific heterochromatic regions. (From Kato et al. 2004.)

genome (maize has one, on chromosome 6; see Figure 2.1). Due to their repetitive nature and their high GC content, NOs can be seen through a light microscope (after staining) and thus can serve as chromosome-specific markers. These markers can be used to map phenotypic traits to specific chromosomal regions. Despite its repetitive nature, rDNA (DNA that encodes rRNA) is actively transcribed. The rDNA of the nucleolar organizer, along with proteins that transcribe the rDNA and process the rRNA primary transcripts for assembly into ribosomes, forms a prominent nuclear structure called the nucleolus (see Figure 1.4).

## Transposons are mobile sequences within the genome

One dominant type of repetitive DNA within the heterochromatic regions of the genome is the transposon. **Transposons**, or **transposable elements**, are also known as "jumping genes" because some of them have the ability to insert a copy of themselves in a new location within the genome.

There are two large classes of transposons: the retroelements or retrotransposons (Class 1), and the DNA transposons (Class 2). These two classes are distinguished by their mode of replication and insertion into a new location (**FIGURE 2.2**). **Retrotransposons** make an RNA copy of themselves, which is then reverse-transcribed into DNA before it is inserted elsewhere in the genome (see Figure 2.2A). Because they do not normally leave their original location, but rather generate additional copies of themselves, active retrotransposons tend to multiply within the genome (Eickbush and Malik 2002). **DNA transposons**, by contrast, move from one position to another using a cut-and-paste mechanism catalyzed by an enzyme that is encoded within the transposon sequence. This enzyme, **transposase**, splices out the transposon and inserts it elsewhere in the genome, in most cases keeping the total transposon copy number the same (see Figure 2.2B).

Transposition into a gene can result in mutations. If a transposon lands within a coding region, the gene may be inactivated. Insertion of a transposon close to a gene can also alter that gene's expression pattern. For example, the transposon may disrupt the gene's normal regulatory elements, preventing transcription or, since transposons often carry promoters, increasing transcription of the gene.

The mutagenic capacity of transposons may play an important role in the evolution of the host genome. From an evolutionary perspective, a low level of mutagenesis could be advantageous to a species, since it could create the novel variation necessary to enable the species to adapt to a changing environment. If the transposition rate is high, however, resulting in individuals with many mutations, at least some of those mutations are likely to be deleterious and to decrease the overall fitness of the species.

Plants and other organisms seem to be able to regulate the activity of transposons through the methylation

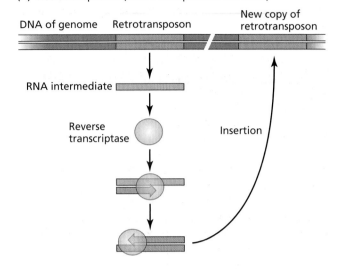

(A) Retrotransposons (class 1 transposable elements)

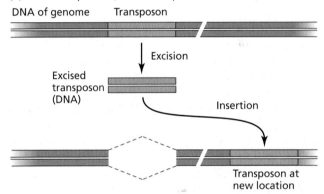

(B) DNA transposons (class 2 transposable elements)

**FIGURE 2.2** The two major classes of transposons differ in their mode of transposition. (A) Retrotransposons move via an RNA intermediate. (B) DNA transposons move using a cut-and-paste mechanism.

of DNA and histones. As we will see later in this chapter, these same processes are used to repress transcription in heterochromatic regions of the genome. As more genomic DNA sequences have become available, scientists have noticed large numbers of highly methylated transposons in heterochromatic regions. Did methylation of transposons cause heterochromatin to form there? Or did the transposons become methylated because they happened to land in heterochromatic regions?

Studies of mutants that are unable to maintain genome methylation have shown that slow loss of proper methylation over generations can activate dormant transposons and increase the frequency of transpositional mutations (Miura et al. 2001) (**FIGURE 2.3**). Such transposon activity may considerably decrease the fitness of offspring. Therefore, methylation and the formation of heterochromatin appear to play important roles in the stability of the genome.

(A)

(B)

Reverted
sector

**FIGURE 2.3** Loss of methylation can lead to mutations as unmethylated transposons become active. A mutation called *decrease in dna methylation* (*ddm1*) causes hypomethylation (decreased methylation) of endogenous transposons. The *clam* mutation, which arose in a *ddm1* mutant, is the result of insertion of a transposon into the *DWARF4* (*DWF4*) gene, which is required for the biosynthesis of the growth hormone brassinosteroid. (A) The transposon-induced *clam* mutant (left) next to wild-type Arabidopsis. (B) The *clam* mutant without (left) and with (right) a sector that has reverted to the wild-type phenotype after the transposon jumped back out of the *DWF4* gene. (From Miura et al. 2001.)

## Polyploids contain multiple copies of the entire genome

Ploidy level—the number of copies of the entire genome in a cell—is another important aspect of genome structure that may have implications for both physiology and evolution. In many organisms, but especially in plants, the entire diploid (2*N*) genome can undergo one or more additional rounds of replication without undergoing cytokinesis (see Chapter 1) to become **polyploid**. Polyploidy is not a rare phenomenon restricted to a few cells or organisms, nor is it normally associated with mutation or disease. In fact, polyploidy is the preferred way of life for many plant and

some animal species. Polyploidy in plants is so common that researchers now believe that the vast majority of all extant plants are of polyploid origin (Leitch and Leitch 2008). Essentially all plant species have had a whole-genome duplication event in their evolutionary history.

Two forms of polyploidy are distinguished: autopolyploidy and allopolyploidy. **Autopolyploids** contain multiple complete genomes of a single species, while **allopolyploids** contain multiple complete genomes derived from two or more separate species.

Both types of polyploidy can result from incomplete meiosis during gametogenesis. During meiosis, the chromosomes of a diploid germ cell undergo DNA replication followed by two rounds of division (meiosis I and meiosis II), producing four haploid cells (**FIGURE 2.4**). If chromosome duplication is not followed by cell division during meiosis, diploid gametes result. Within a species, or in a self-fertilizing individual, if a diploid egg is fertilized by a diploid sperm, the resulting zygote contains four copies of each chromosome, and is said to be *autotetraploid* (**FIGURE 2.5A**). Likewise, if cell division does not occur after chromosome duplication during mitosis, cells become autotetraploid. Both types of errors during meiosis or mitosis occur spontaneously in most plants, albeit at very low frequency.

Allopolyploids usually form in one of two ways:

1. A haploid sperm from one species and a haploid egg from another species may form a diploid interspecies hybrid. Meiosis in these plants generally fails but can lead to rare duplicated gametes, which can produce the allopolyploid. Additionally, if hybrid cells accidentally omit cell division these cells become allopolyploid (**FIGURE 2.5B**). This type of allopolyploidization can happen as early as in a hybrid zygote, or later in vegetative or reproductive tissues of the hybrid plant.

2. Diploid gametes from two different species may join to form a tetraploid zygote. The diploid gametes can come either from tetraploid parents that have undergone normal meiosis or from diploid parents in

**FIGURE 2.4** Generalized meiosis in plants. During anaphase I, homologous chromosomes separate and move to opposite poles. Chromatids separate during anaphase II. Meiosis results in four daughter cells, each of which has half the chromosome number of the parent. If a pair of homologous chromosomes accidentally fails to separate during anaphase I ("non-disjunction"), the resulting daughter cells contain extra or missing chromosomes and are either aneuploid or polyploid. Note that there is variation between plant species in telophase I: not all plants lay down new cell walls at this stage, and in some cases chromosomes decondense somewhat and can remain in an interphase-like stage for a period of time before entering meiosis II.

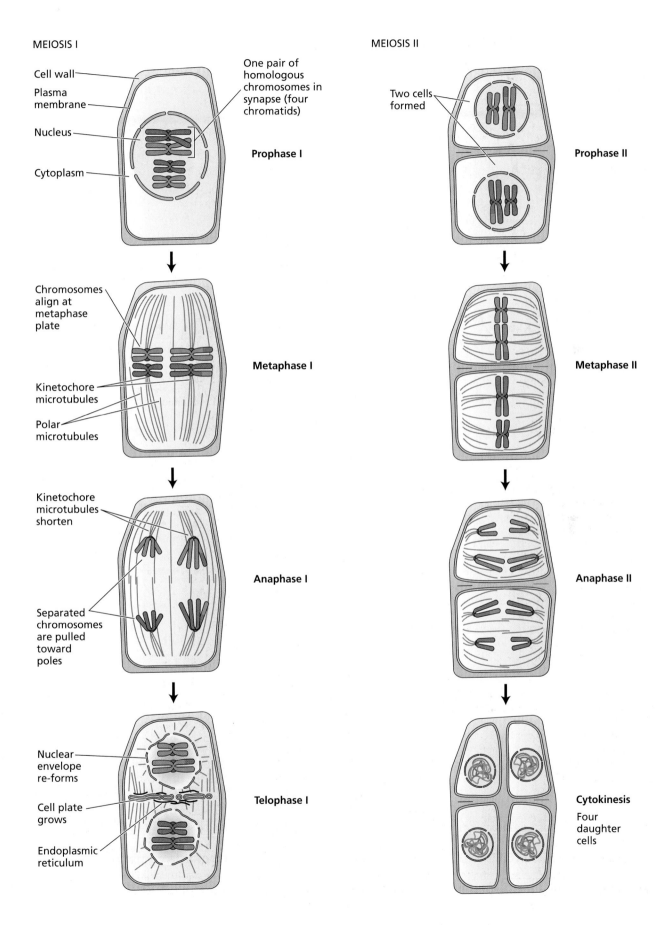

MEIOSIS I

Cell wall

Plasma membrane

Nucleus

Cytoplasm

One pair of homologous chromosomes in synapse (four chromatids)

**Prophase I**

Chromosomes align at metaphase plate

Kinetochore microtubules

Polar microtubules

**Metaphase I**

Kinetochore microtubules shorten

Separated chromosomes are pulled toward poles

**Anaphase I**

Nuclear envelope re-forms

Cell plate grows

Endoplasmic reticulum

**Telophase I**

MEIOSIS II

Two cells formed

**Prophase II**

**Metaphase II**

**Anaphase II**

**Cytokinesis**

Four daughter cells

(A) Autopolyploidy (by genome duplication)

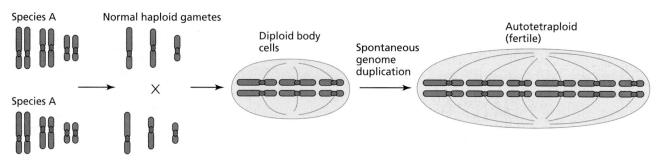

(B) Allopolyploidy (by genome duplication)

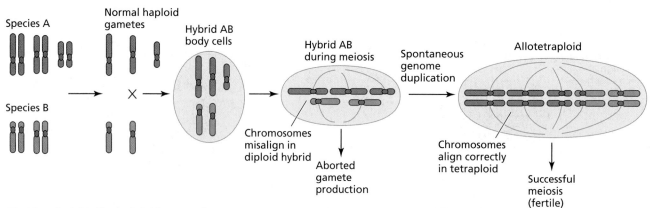

(C) Allopolyploidy (from diploid gametes)

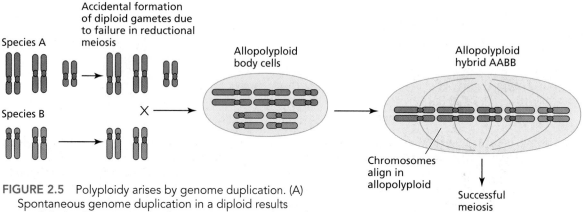

**FIGURE 2.5** Polyploidy arises by genome duplication. (A) Spontaneous genome duplication in a diploid results in a fully fertile autotetraploid. (B) Diploid interspecies hybrids are usually infertile due to the inability of chromosomes to align properly during meiosis. Spontaneous genome duplication allows each chromosome to have a pairing partner during meiosis, leading to restoration of fertility. (C) Fusion of diploid gametes from two different species results in a fertile allopolyploid.

which normal reduction meiosis has failed. (**FIGURE 2.5C**).

Diploid interspecies hybrids occur naturally, but they are frequently sterile because their chromosomes cannot pair properly during prophase I of meiosis (see Figure 2.5B). Spontaneous genome duplication in sterile inter-species hybrids can lead to the formation of a new, fertile allopolyploid—a phenomenon that has been observed in many species, such as the Brassicaceae (**FIGURE 2.6**). The lack of fertility in interspecies hybrids is in stark contrast to the phenomenon known as **hybrid vigor** or **heterosis**: the increased vigor often observed in the offspring of crosses between two inbred varieties of the same plant species. Heterosis can contribute to larger plants, greater biomass, and higher yields in agricultural crops.

Polyploidy can also be induced artificially by treatment with the natural cell toxin colchicine, which is derived from

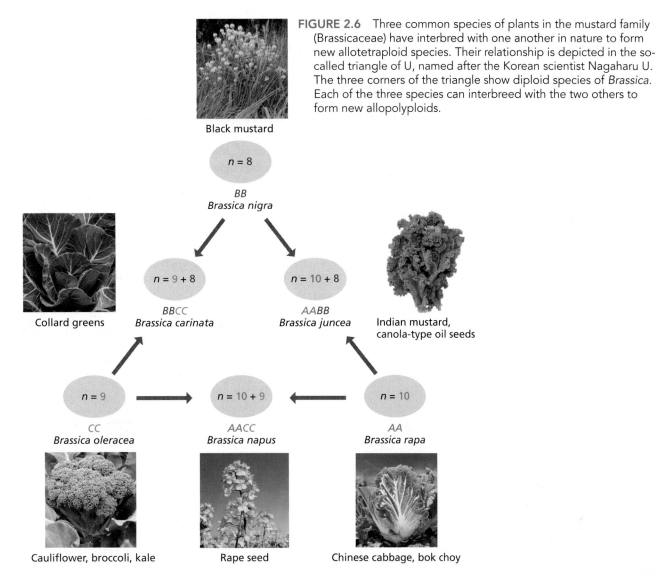

**FIGURE 2.6** Three common species of plants in the mustard family (Brassicaceae) have interbred with one another in nature to form new allotetraploid species. Their relationship is depicted in the so-called triangle of U, named after the Korean scientist Nagaharu U. The three corners of the triangle show diploid species of *Brassica*. Each of the three species can interbreed with the two others to form new allopolyploids.

the autumn crocus (*Colchicum autumnale*). Colchicine inhibits spindle fiber formation and prevents cell division, but does not interfere with DNA replication. Treatment with colchicine therefore results in an undivided nucleus containing multiple copies of the genome.

## Phenotypic and physiological responses to polyploidy are unpredictable

The widely held notion that autopolyploids are larger than their diploid progenitors does not always hold true. For example, when individual maize plants with the same genetic background, but differing in ploidy level, were compared, it was found that plant height increased from haploidy to diploidy, but *decreased* with further increases in ploidy level (**FIGURE 2.7A**). One hypothesis to explain the greater vigor of some autopolyploids compared with their diploid progenitors is that plant vigor increases with increasing ploidy only if hybridity (heterozygosity) also increases. If, instead, the level of *homozygosity* increases in

plants with increasing ploidy level (through inbreeding), their vigor actually decreases (**FIGURE 2.7B**) (Riddle et al. 2006; Riddle and Birchler 2008). Research to test this hypothesis is currently under way and could in the future help to guide breeding efforts in agriculturally important plants.

Allopolyploids differ from their parental diploid progenitor species in two major ways:

1. Their genomes, like those of autopolyploids, are duplicated.

2. They are hybrids between two different species.

When comparing allopolyploids with their progenitors, it is therefore difficult to determine whether the observed phenotypic differences are due to genome duplication or to hybridization. Current data indicate that hybridization makes a greater contribution than does genome duplication to the divergence of allopolyploid offspring from their parents. Allopolyploids are frequently more vigorous or higher yielding than their parent species and are very common among agriculturally important plants. Examples of

(A)

(B)

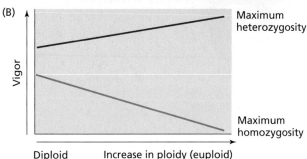

**FIGURE 2.7** (A) Ploidy series in maize. Plants of the same age are shown from left to right: haploid, diploid, triploid, and tetraploid. In inbred maize, autopolyploidy results in reduced vigor compared with the diploid. Each black or white section on the scale bar measures 20 cm. (B) A generalized view of the relationship between plant vigor and ploidy level. As ploidy increases, plant vigor increases only in plants whose overall level of heterozygosity also increases due to a greater number of different alleles per genome (red line). By contrast, increasing ploidy in homozygous or inbred plants is correlated with decreasing overall plant vigor (blue line). (A courtesy of E. Himelblau; B adapted from a diagram courtesy of J. Birchler.)

such allopolyploids include canola and collards, coffee, cotton, wheat, rye, oat, and sugarcane.

Regardless of how allopolyploids arise, the fusion of two divergent genomes has many consequences, although it is not yet clear if there is a common set of responses across all species during or immediately after allopolyploidization (Osborn et al. 2003; Chen 2007). Some of the genetic changes that have been observed in newly formed allopolyploids compared with their parent species are the following:

- Restructuring of the chromosomes, including loss of DNA sequences (Feldman et al. 1997; Pires et al. 2004)

- Changes in epigenetic modifications (Madlung et al. 2002; Salmon et al. 2005; Parisod et al. 2009)

- Changes in gene transcriptional activity (Adams et al. 2003; Chen 2007)

- Activation of previously dormant transposable elements through loss of gene silencing (Kashkush et al. 2003; Madlung et al. 2005)

Microarray analysis (which will be discussed later in this chapter) has also shown that the transcriptional activity of many genes throughout the entire genome is altered in allopolyploids as compared with the same genes in their progenitors (Wang et al. 2006; Rapp et al. 2009). It is likely that epigenetic modifications, including DNA and histone methylation and histone acetylation, are responsible for many of these changes.

Polyploidy leads to multiple, redundant copies of genes in the genome. As evolution acts on the duplicate genes, one copy may be either lost or changed in function, while the other retains its original function. This process is known as **subfunctionalization** (Lynch and Force 2000). Genome analysis shows that even in many diploid species there is clear evidence of genome duplication in the species' evolutionary history. In such cases, a subsequent gradual loss of DNA has led back to a diploid state (**FIGURE 2.8**). Species that show signs of ancient genome duplications followed by DNA loss are known as **paleopolyploids** and include maize and Arabidopsis (Wolfe 2001).

Polyploidy is in striking contrast to a condition called **aneuploidy**. Aneuploids are organisms whose genomes contain more or fewer individual chromosomes (not entire chromosome sets) than normal. Such states are known as **trisomies** if one type of chromosome is tripled or **monosomies** if only one chromosome of a given type is present. In humans and many animals, aneuploidy usually leads to death or to severe physiological problems, such as Down syndrome (trisomy 21). Aneuploid plants, although often phenotypically distinct from normal (euploid) plants, are generally viable. In polyploids, effects of aneuploidy can be masked by the additional chromosomes in the genome.

# Plant Cytoplasmic Genomes: Mitochondria and Chloroplasts

In addition to the nuclear genome, plant cells contain two additional genomes: the **mitochondrial genome**, which they share with animal cells, and the **chloroplast genome**. In this section we will see where these genomes come from and what roles they play. We will then look at their organization and discuss some important differences from the nuclear genome in the way their genetic information is transmitted.

## The endosymbiotic theory describes the origin of cytoplasmic genomes

Cytoplasmic genomes are probably the evolutionary remnants of the genomes of bacterial cells that were engulfed by another cell. The **endosymbiotic theory**, championed by Lynn Margulis in the 1980s, postulates that the original mitochondrion was an oxygen-using (aerobic) bacterium

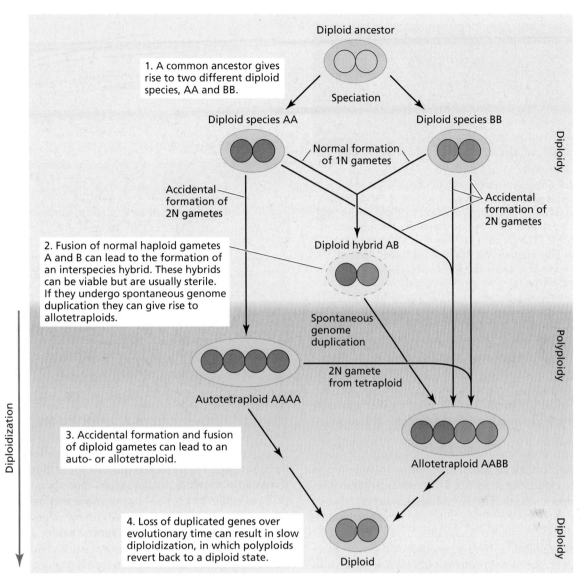

**FIGURE 2.8** Continuum in the evolution of polyploid species. Diploids can give rise to autopolyploids or allopolyploids by the mechanisms outlined in Figure 2.5. Polyploids can revert to diploidy via the gradual loss of duplicated genes over evolutionary time. Light purple underlying ovals represent the nuclei of a species; colored circles inside nuclei represent entire genomes. (After L. Comai 2005.)

that was absorbed by another prokaryotic organism. Over time, this original endosymbiont evolved into an organelle that was no longer able to live on its own. The host cell, along with its endosymbiont, gave rise to a lineage of cells that were able to use oxygen in aerobic metabolism; these cells, in turn, eventually gave rise to all animal cells. Plant cells, according to this theory, arose when a second endosymbiosis event took place. This time, a mitochondrion-containing cell engulfed a photosynthetic cyanobacterium, which over time evolved inside the cell into the chloroplast.

Two main lines of evidence are often cited in support of the endosymbiotic theory. First, both mitochondria and chloroplasts are enclosed by an outer and an inner membrane, and the inner membrane is continuous with additional membrane-bound compartments inside the

organelle. This observation is consistent with the idea that engulfment of the original aerobic or photosynthetic cell through invagination of the plasma membrane of the prokaryotic host cell left a double membrane around the new organelle. Second, both organellar genomes show sequence similarity to prokaryotic genomes. The organellar genomes, like those of prokaryotes, are not enclosed in a nuclear envelope and are called **nucleoids**.

## Organellar genomes consist mostly of linear chromosomes

Plastid genomes generally range in size from about 120 to 160 kilobases (kbp) and encode predominantly genes that are needed for photosynthesis and the expression of

(A)

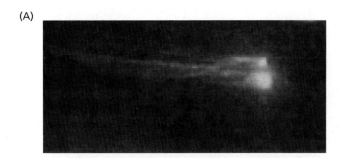

(B)

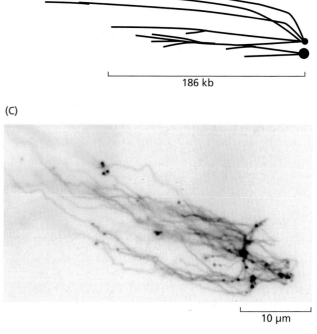

186 kb

(C)

10 μm

**FIGURE 2.9** The complex structure of mitochondrial and plastid genomes. Each photograph shows a single branched DNA molecule containing multiple connected organellar genomes. (A) Micrograph of ethidium-stained mitochondrial DNA from the liverwort *Marchantia polymorpha*. (B) Line drawing of the same molecule. The 186 kb scale bar shows the size of one genome. (C) Micrograph of multigenomic chloroplast DNA from maize stretched out in agarose. The 10 μm scale bar is equivalent to 30 kb of DNA sequence, which is roughly 20% of the total genome length (154 kb). (A, B from Oldenburg and Bendich 1998; C from Oldenburg and Bendich 2004.)

plastid genes. The mitochondrial genome is much more variable in size than the plastid genome. Plant mitochondrial genomes range between approximately 180 and nearly 3000 kbp—much larger than mitochondrial genomes of animals or fungi, many of which are only 15 to 50 kbp. Plant mitochondrial DNA contains genes that encode proteins needed in the electron transport chain, or that are involved in providing co-factors for electron transport. Additionally, plant mitochondrial DNA carries genes for proteins required for the organelle's own gene expression, such as ribosomal proteins, tRNAs, and rRNAs. In both organelles many genes required for proper chloroplast or mitochondrial function are no longer encoded in the organellar genome itself but over evolutionary time have been transferred to the nucleus of modern plants. These proteins are synthesized in the cytoplasm and then imported into the organelle.

For many years organellar chromosomes had been thought to contain a genome-sized DNA molecule in circular form, similar to the circular plasmids of bacteria. Recent data, however, show that most of the DNA in both plant mitochondria and chloroplasts is found in linear molecules that may contain more than one copy of the genome (**FIGURE 2.9**). These copies are connected to one another in head-to-tail orientation and the chromosomal DNA molecules can be highly branched, resembling a bush or tree (Oldenburg and Bendich 1998, 2004), unlike the simpler structures of linear nuclear chromosomes. While nuclear chromosomes are of constant size generation after generation, the chromosome size in mitochondria and chloroplasts can vary. Nonetheless, each organellar chromosome contains at least one complete genome.

## Organellar genetics do not obey Mendelian laws

The genetics of organellar genes are governed by two principles that distinguish them from Mendelian genetics. First, both mitochondria and plastids generally show **uniparental inheritance**, which means that sexual offspring (via pollen and eggs) only inherit organelles from one parent. Among the gymnosperms, the conifers normally inherit their plastids from the paternal parent. For angiosperms the general rule is that the plastids come from the maternal parent. However, there are a few angiosperms in which plastids are inherited either biparentally or paternally. Mitochondrial inheritance is usually maternal in the majority of plants but again a few exceptions can be found; for example some types of conifers, such as the cypresses, show paternal inheritance of mitochondria.

In plants in which plastids are inherited maternally, the paternal plastids are excluded as a consequence of the way the male gametophyte (pollen) is formed. Pollen formation involves two mitotic divisions of the microgametophyte. The first division results in a vegetative cell, which gives rise to the pollen tube, and a much smaller generative cell, which will later produce the two sperm cells. During the first division, plastids are excluded from the generative cell and thus the male germ line. Consequently, the zygote gets all its plastids from the egg in the female megagametophyte, resulting in maternal inheritance of plastid genes (Mogensen 1996).

The mechanism for paternal inheritance of plastids in some conifers, such as Douglas fir (*Pseudotsuga mensiesii*) or Chinese pine (*Pinus tabulaeformis*), is slightly more complicated and less well understood. In many gymnosperms the second mitotic division, which produces the two sperm

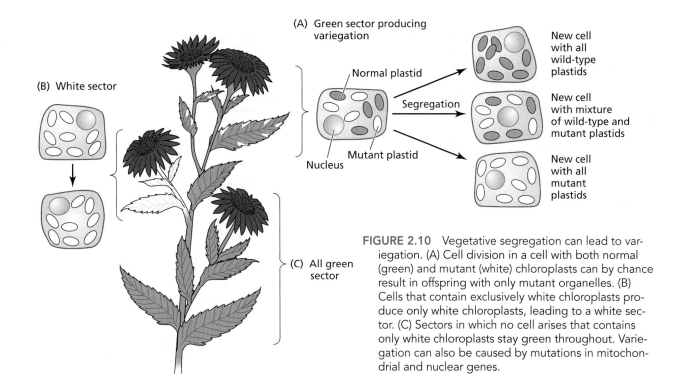

**FIGURE 2.10** Vegetative segregation can lead to variegation. (A) Cell division in a cell with both normal (green) and mutant (white) chloroplasts can by chance result in offspring with only mutant organelles. (B) Cells that contain exclusively white chloroplasts produce only white chloroplasts, leading to a white sector. (C) Sectors in which no cell arises that contains only white chloroplasts stay green throughout. Variegation can also be caused by mutations in mitochondrial and nuclear genes.

nuclei, occurs just as the pollen tube enters the female gametophyte. Paternal cytoplasm rich in plastids and some mitochondria clings to the sperm nucleus that enters the egg cell. At the time of fertilization, the egg nucleus is surrounded by maternal mitochondria but the maternal plastids are located in the periphery of the egg cell. After fusion of the sperm and egg nuclei and formation of the zygote, cellularization follows in such a fashion that the new cell walls essentially exclude all the maternal plastids and most of the paternal mitochondria from the embryo (Mogensen 1996; Guo et al. 2005).

The second major feature of organellar inheritance is the fact that both chloroplasts and mitochondria can **segregate vegetatively**. This means that a vegetative cell (as opposed to a gamete) can give rise to another vegetative cell via mitosis that is genetically different. For example, during mitosis, one daughter cell may receive plastids with one type of genome, while other plastids with different genetic information, perhaps containing one or more mutations, are by chance distributed into the other daughter cell. Vegetative segregation, which is also referred to as **sorting-out**, can result in the formation of phenotypically different sectors within a tissue (**FIGURE 2.10**). The presence of such sectors in leaves may result in what horticulturists often refer to as **variegation** (see Figure 16.31 for an example). Leaf variegation can be caused by mutations in nuclear, mitochondrial, or chloroplast genes.

Now that we have looked at the organization of nuclear and cytoplasmic genomes in plants, we will turn to the structure of the nuclear genome and how it influences the expression of the genes it contains. The basic mechanisms

of gene transcription will be reviewed first, followed by a description of the transcriptional regulation of gene expression.

## Transcriptional Regulation of Nuclear Gene Expression

The path from gene to protein is a multistep process catalyzed by many enzymes (**FIGURE 2.11**). Each step is subject to regulation by the plant to control the amount of protein that is produced by each gene. Regulation of the first step, transcription, determines when and whether an mRNA is made. This level of regulation, which is referred to as **transcriptional regulation**, includes the control of transcription initiation, maintenance, and termination. The next level in the regulation of gene expression, known as **posttranscriptional regulation**, occurs after transcription. This level, which will be covered later in this chapter, includes controls on mRNA stability, translation efficiency, and degradation. Finally, **protein stability** (posttranslational regulation) plays an important role in the overall activity of a gene or its product.

### RNA polymerase II binds to the promoter region of most protein-coding genes

Gene transcription is facilitated by an enzyme called **RNA polymerase**, which binds to the DNA to be transcribed and makes an mRNA transcript complementary to the DNA sequence. There are several types of RNA polymerase.

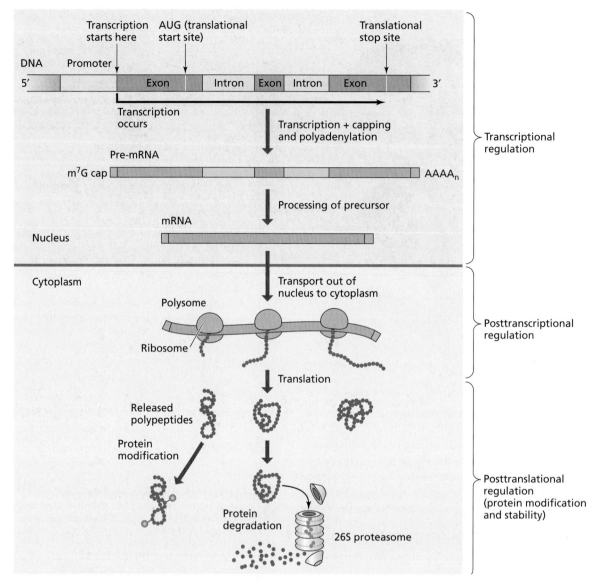

**FIGURE 2.11** Gene expression in eukaryotes. RNA polymerase II binds to the promoters of genes that encode proteins. Unlike prokaryotic genes, eukaryotic genes are not clustered in operons, and each is divided into introns and exons. Transcription from the template strand proceeds in the 3' to 5' direction at the transcription start site, and the growing RNA chain extends one nucleotide at a time in the 5' to 3' direction. Translation begins with the first AUG encoding methionine, as in prokaryotes, and ends with a stop codon. The pre-mRNA transcript is first "capped" by the addition of 7-methylguanylate (m7G) to the 5' end. The 3' end is shortened slightly by cleavage at a specific site, and a poly-A tail is added. The capped and polyadenylated pre-mRNA is then spliced by a spliceosome complex, and the introns are removed. The mature mRNA exits the nucleus through the nuclear pores and initiates translation on ribosomes in the cytosol. As each ribosome progresses toward the 3' end of the mRNA, new ribosomes attach at the 5' end and begin translating, leading to the formation of polysomes. After translation some proteins are modified by addition of chemical groups to the protein chain. The released polypeptides have characteristic half-lives, which are regulated by the ubiquitin pathway and a large proteolytic complex called the 26S proteasome (see Figure 2.15).

RNA polymerase II is the polymerase that transcribes most protein-coding genes.

The region of the gene that binds RNA polymerase is called the **promoter** (FIGURE 2.12A). We can divide the structure of the eukaryotic promoter into two parts: the **core promoter** or **minimum promoter**, consisting of the minimum upstream sequence required for gene expression, and the **regulatory sequences**, which control the activity of the core promoter.

Before transcription of a gene can start, several steps have to occur to allow the RNA polymerase to gain access to the gene's nucleotide sequence. Recall that nuclear DNA is wrapped around histones, forming beadlike nucleosomes. As will be discussed in more detail in the

(A)

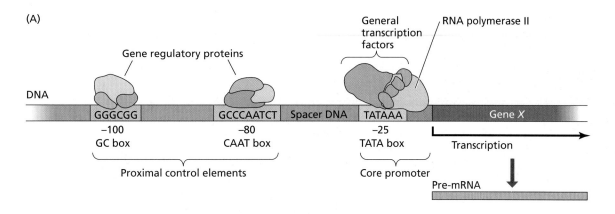

(B)

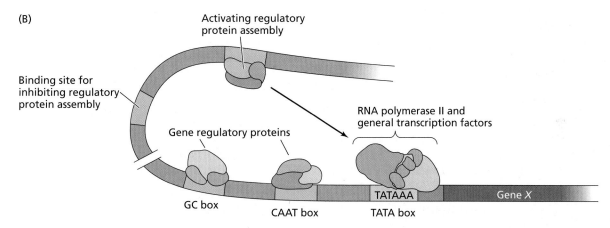

**FIGURE 2.12** Organization and regulation of a typical eukaryotic gene. (A) Features of a typical eukaryotic core promoter and the proteins that regulate gene expression. RNA polymerase II is situated at the TATA box in association with the general transcription factors about 25 bp upstream of the transcription start site. Two *cis*-acting regulatory sequences that enhance the activity of RNA polymerase II are the CAAT box and the GC box, located at about 80 and 100 bp upstream, respectively, of the transcription start site. The CAAT box and GC box bind gene regulatory proteins. (B) Regulation of transcription by distal regulatory sequences and *trans*-acting factors. The *trans*-acting factors can act in concert with the distal regulatory sequences to which they are bound to activate transcription by making direct physical contact with the transcription initiation complex. The details of this process are still not well understood.

next section, the histones are subject to modifications, and only if those modifications are favorable to transcription will RNA polymerase be able to bind to the DNA. Furthermore, to be functional, the RNA polymerases of eukaryotes require additional proteins called **general transcription factors** to position the polymerase at the transcription start site. These general transcription factors, together with the RNA polymerase, make up a large, multi-subunit **transcription initiation complex** (see Figure 2.12A). Transcription is initiated when the final transcription factor to join the complex phosphorylates the RNA polymerase. The RNA polymerase then separates from the initiation complex and proceeds along the antisense, or complementary, strand of the DNA in the 3′ to 5′ direction.

In addition to RNA polymerase and the general transcription factors, most genes, especially those that play important roles in development, require specific transcription factors (also often called gene regulatory proteins) for

RNA polymerase to become active. These regulatory proteins bind to the DNA and become part of the transcription initiation complex.

An example of a typical RNA polymerase II promoter is shown schematically in Figure 2.12A. The core promoter for genes transcribed by RNA polymerase II typically extends about 100 bp upstream of the transcription initiation site and includes several sequence elements referred to as **proximal promoter sequences**. About 25 to 35 bp upstream of the transcription initiation site is a short sequence called the **TATA box**, consisting of the sequence TATAAA(A). The TATA box plays a crucial role in transcription because it serves as the site of assembly of the transcription initiation complex discussed above.

In addition to the TATA box, the core promoters of eukaryotes also contain two additional regulatory sequences: the **CAAT box** and the **GC box** (see Figure 2.12A). These two sequences are binding sites for specific transcription factors. The DNA sequences themselves are termed *cis*-acting sequences, since they are adjacent (*cis*) to the transcription units they are regulating. The transcription factors that bind to the *cis*-acting sequences are also called *trans*-acting factors, since the genes that encode them are located elsewhere in the genome.

Numerous other *cis*-acting sequences located farther upstream of the proximal promoter sequences can exert either positive or negative control over eukaryotic promoters. These sequences, termed **distal regulatory sequences**, are usually located within 1000 bp of the transcription initiation site. The positively acting transcription factors that bind to these sites are called **activators**, while those that inhibit transcription are called **repressors**. In addition to having regulatory sequences within the promoter itself, eukaryotic genes can be regulated by control elements located tens of thousands of base pairs away from the transcription initiation site. Distantly located positive regulatory sequences are called **enhancers**. Enhancers may be located either upstream or downstream from the promoter.

How do all the transcription factors that bind to the *cis*-acting sequences regulate transcription? During the formation of the initiation complex, the DNA between the core promoter and the most distally located regulatory sequences loops out in such a way as to allow all of the transcription factors bound to that segment of DNA to make physical contact with the initiation complex (**FIGURE 2.12B**). Through this physical contact, each transcription factor exerts its control, either positive or negative, over transcription.

### Epigenetic modifications help determine gene activity

As mentioned in the previous section, transcription can be initiated only if the DNA is accessible to the RNA polymerase and other required binding proteins. In order to make the DNA accessible, its packaging has to be "loosened," a process mediated by covalent modifications of both DNA and histones. Because these modifications can change a gene's behavior without changing the DNA sequence of the gene itself, they are referred to as **epigenetic modifications** (from the Greek word *epi*, meaning "over" or "on top of").

One common type of DNA modification is the **methylation** of cytosine residues (**FIGURE 2.13A**). DNA sequences that are frequently methylated in plants are CG, CHG, and CHH (where H can be any nucleotide except guanine). In contrast, cytosine methylation in mammals occurs mostly on CG. Cytosine methylation is catalyzed by one of several methyltransferases, whereas DNA demethylation is

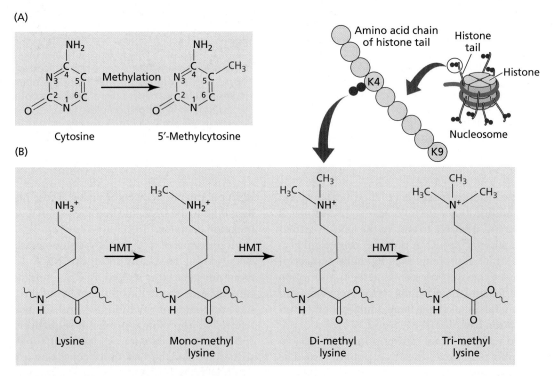

**FIGURE 2.13** (A) The addition of a methyl group to the C5 in cytosine is associated with transcriptional inactivity. (B) The amino acid lysine (K), which occurs at several positions in histones, can be mono-, di-, or trimethylated by histone methyltransferase (HMT). (C) Histones can be remodeled to activate gene transcription (top) or to repress it (bottom). Activation is associated with acetylation by histone acetyltransferases (HATs) and methylation by HMT at lysine residue H3K4. These modifications promote ATP-dependent chromatin remodeling and stimulate transcription. Repression of transcription is achieved by methylation of H3K9 and deacetylation by histone deacetylases.

catalyzed by glycosylases that replace methylcytosine with unmethylated cytosine.

Epigenetic modifications can also occur on histones, which, together with the DNA wrapped around them, make up the nucleosomes. Each histone has a "tail," which is made up of the first part of the histone's amino acid chain and protrudes outward from the nucleosome. Histone modifications occur on these tails, usually within the outermost 40 or so amino acids. These modifications can influence the conformation of the nucleosomes and thereby gene activity in the associated DNA.

One of the histone modifications that influences gene activity is methylation, especially at specific lysine residues (single-letter amino acid abbreviation K) in the tail of histone type H3. These residues are K4, K9, K27, and K36, counting from the outermost amino acid inward to the center of the histone coil. One, two, or three methyl groups can be added to a single lysine (**FIGURE 2.13B**). Histones dimethylated at position H3K4 are generally associated with active genes, whereas dimethylation at position H3K9 is often associated with inactive genes and elements, such as silent transposons. Methyl groups are removed by histone

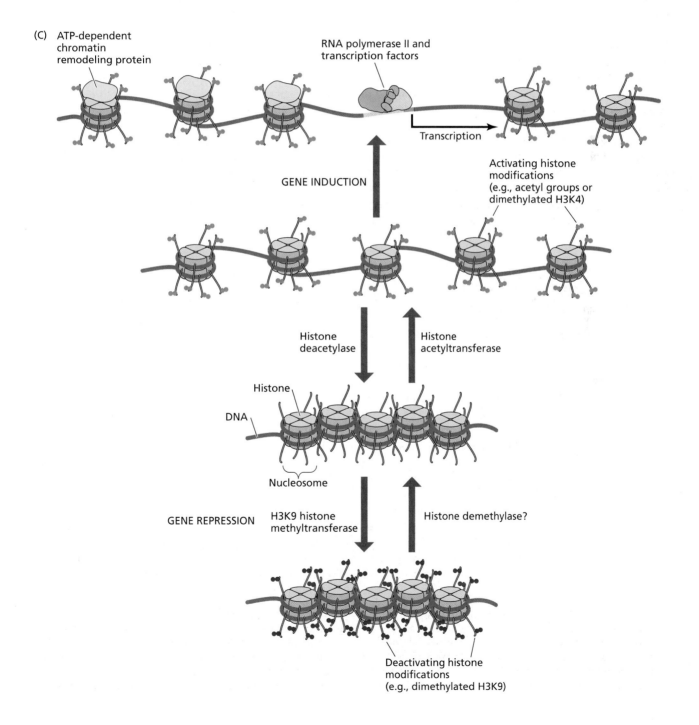

demethylases in mammals and yeast (Bannister and Kouzarides 2005), and candidate genes with a similar function have been identified in plants (Choi et al. 2007).

Another form of modification that occurs on the histone tail is **acetylation**, which is catalyzed by enzymes called histone acetyltransferases (HATs). Generally, acetylated histones are associated with genes that are actively transcribed. Histone deacetylases (HDACs) can reverse this activation by removing acetyl groups.

Both methylation and acetylation change the architecture of the chromatin complex, which may result in condensation or relaxation of the chromatin. These changes occur when multiprotein chromatin remodeling complexes bind to the modified histones. Using the energy released by ATP hydrolysis to drive the reaction, these complexes open up the chromatin by displacing the nucleosomes in the area slightly toward the 5′ or 3′ direction of the remodeling complex. The resulting gap between nucleosomes is now wide enough for RNA polymerase to bind and initiate transcription (**FIGURE 2.13C**). Alternatively, histone modifications may present novel binding sites for regulatory proteins that affect gene activity. Scientists are only beginning to understand the effects of specific chemical modifications on each of the first 40 or so amino acids of the histone tails. The entirety of histone modifications on a specific nucleosome is sometimes called the "histone code" to underline the strong link between nucleosome constitution and gene activity.

# Posttranscriptional Regulation of Nuclear Gene Expression

Immediately after transcription, the resulting mRNAs are processed: their introns are removed by splicing, and 5′ caps and 3′ poly-A tails are added. The transcripts are then exported to the cytoplasm for translation.

An organism often produces mRNA in response to a specific situation. In order to remain useful as a specific response to a specific situation, individual mRNAs must have a finite lifetime. For example, in order to cope with a transient environmental stress, a plant may need to briefly produce specific enzymes. After the stress subsides, it would be wasteful, maybe even detrimental, to continue to produce those enzymes. Thus mRNA production, activity, and stability are all regulated (Green 1993). We discussed the regulation of transcription (mRNA production) in the previous section. Now we turn to mechanisms of posttranscriptional regulation (regulation of activity and stability).

## RNA stability can be influenced by cis-elements

One mechanism by which mRNA stability is regulated depends on the presence of certain sequences within the mRNA molecule itself, called *cis*-**elements**—an unfortunate choice of terms, since the same term is used for the DNA regions that influence transcriptional activity. These

*cis*-elements can be bound by RNA-binding proteins, which may either stabilize the mRNA or promote its degradation by nucleases (Hollams et al. 2002). Depending on the types of *cis*-elements present, the stability of a given mRNA molecule can vary widely.

## Noncoding RNAs regulate mRNA activity via the RNA interference (RNAi) pathway

Another mechanism for regulating mRNA stability is the **RNA interference (RNAi) pathway**. This pathway involves several types of small RNA molecules that do not code for proteins and are thus called noncoding RNAs (ncRNAs). The RNAi pathway has important roles in gene regulation and genomic defense.

The RNAi pathway is a set of cellular reactions to the presence of double-stranded RNA (dsRNA) molecules. Recall that mRNA is usually a single-stranded molecule (ssRNA). In plant cells, dsRNAs usually occur as a result of one of three types of events:

1. The presence of **microRNAs (miRNAs)**, which are involved in normal developmental processes (**FIGURE 2.14A**)

2. The production of **short interfering RNAs (siRNAs)**, which silence certain genes (**FIGURE 2.14B**)

3. The introduction of foreign RNAs, either by viral infection or via transformation by a foreign gene (**FIGURE 2.14C**)

Regardless of how the dsRNAs are produced, the cell mounts the RNAi response. The dsRNAs are chopped up, or "diced," into small, 21- to 24-nucleotide RNAs, which bind to complementary single-stranded RNAs (e.g., mRNAs) from endogenous genes, viruses, or transgenes and promote their degradation or translational inhibition. In some cases, the RNAi pathway can also lead to gene silencing or **heterochromatization** of endogenous DNA or introduced foreign genes. To explore RNAi in more detail, we will first take a look at events leading to dsRNA accumulation in the cell. Next we will discuss the molecular components and downstream events of the RNAi process.

**MicroRNAs REGULATE MANY DEVELOPMENTAL GENES POSTTRANSCRIPTIONALLY** MicroRNAs (miRNAs) are involved in many developmental processes, such as leaf and flower development, cell division, and the orientation (polarity) of plant organs (Kidner and Martienssen 2005). All miRNAs arise from RNA polymerase II–mediated transcription of a specific locus that encodes a primary miRNA transcript (a pri-miRNA), which can vary in length from hundreds to thousands of nucleotides. This primary transcript is capped at the 5′ end and polyadenylated at the 3′ end and forms a double-stranded stem, whose base-paired arms border a single-stranded loop. Next, the pri-miRNAs are processed into pre-miRNAs, which are usually 70 to

(A) MicroRNA pathway

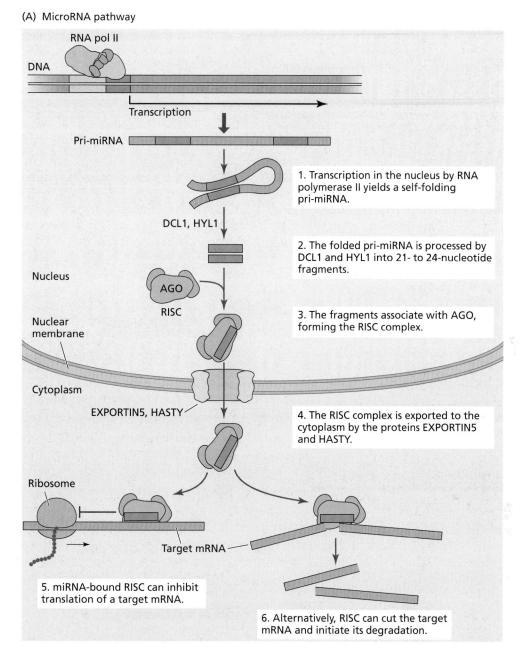

1. Transcription in the nucleus by RNA polymerase II yields a self-folding pri-miRNA.

2. The folded pri-miRNA is processed by DCL1 and HYL1 into 21- to 24-nucleotide fragments.

3. The fragments associate with AGO, forming the RISC complex.

4. The RISC complex is exported to the cytoplasm by the proteins EXPORTIN5 and HASTY.

5. miRNA-bound RISC can inhibit translation of a target mRNA.

6. Alternatively, RISC can cut the target mRNA and initiate its degradation.

**FIGURE 2.14** RNAi pathways in plants. (A) MicroRNAs (miRNAs) are part of many genetic pathways during plant development. (B) Short interfering RNAs (siRNAs) are required to maintain heterochromatin and to silence un-used genes. The siRNA pathway involving RdRP2 is illustrated here. (C) Plant cells can mount an RNAi response to infection by viruses.

80 nucleotides long in animals, but can be up to several hundred nucleotides long in plants. In plants, pri-miRNAs are converted into miRNAs inside the nucleus by the proteins **DICER-LIKE 1** (**DCL1**) and HYPONASTIC LEAVES 1 (HYL1), both of which are involved in processing the primary transcripts. After processing, the mature miRNAs are ready to be used in RNAi (see Figure 2.14A).

**SHORT INTERFERING RNAS ORIGINATE FROM REPETITIVE DNA** Mature short interfering RNAs (siRNAs) are structurally and functionally similar to miRNAs and also lead to the initiation of RNAi. However, siRNAs differ from miRNAs in the way they are generated. SiRNAs can be produced in two ways. First, they can arise from the transcription of opposing promoters that produce mRNA on opposite strands and can thereby generate two fully or partially complementary single-stranded RNA (ssRNA) molecules that can subsequently form one double-stranded molecule. The second way siRNAs can be generated is from ssRNAs that are converted into dsRNAs by a special class of **RNA-dependent RNA polymerases** (**RdRPs**) (see Figure 2.14B). Note that transcription of siRNAs is accom-

**(B) Short interfering RNA pathway**

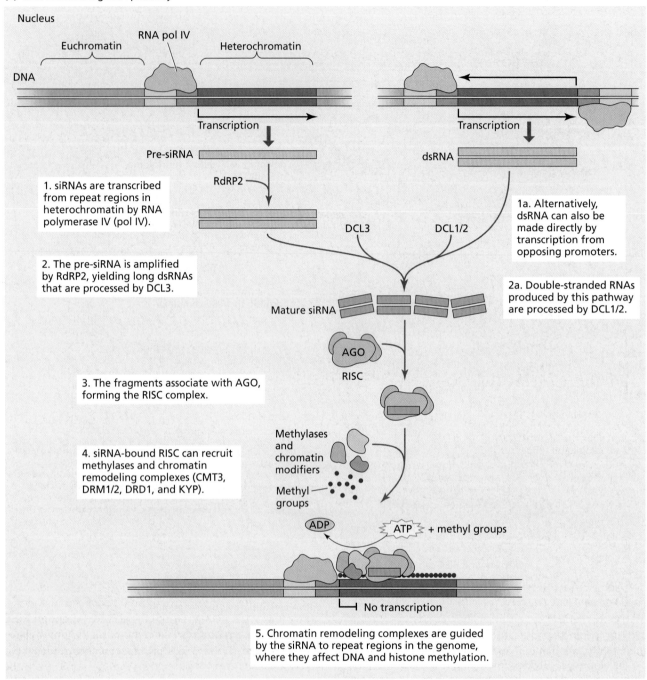

1. siRNAs are transcribed from repeat regions in heterochromatin by RNA polymerase IV (pol IV).

2. The pre-siRNA is amplified by RdRP2, yielding long dsRNAs that are processed by DCL3.

1a. Alternatively, dsRNA can also be made directly by transcription from opposing promoters.

2a. Double-stranded RNAs produced by this pathway are processed by DCL1/2.

3. The fragments associate with AGO, forming the RISC complex.

4. siRNA-bound RISC can recruit methylases and chromatin remodeling complexes (CMT3, DRM1/2, DRD1, and KYP).

5. Chromatin remodeling complexes are guided by the siRNA to repeat regions in the genome, where they affect DNA and histone methylation.

plished by RNA polymerase IV, not RNA polymerase II, as in miRNAs (Ramachandran and Chen 2008).

Endogenous siRNAs are transcribed from chromosomal regions that otherwise do not seem to support much transcription: repetitive DNA, transposons, and centromeric regions. Indeed, siRNAs originating from repeat regions are sometimes referred to as **repeat-associated silencing RNAs (rasiRNAs)**. As you will see in the next section, this may not be coincidental: it appears that the formation of siRNAs and the induction of RNAi actually cause these regions to become heterochromatic and largely transcriptionally silent. Once a double-stranded RNA is produced, either by transcription of inverted repeats or by RdRP2, it is cut into 21- to 24-nucleotide RNA duplexes by members of the DICER protein family (see Figure 2.14B).

Apart from these siRNAs of endogenous origin, exogenous RNAs can also trigger the formation of siRNAs. Sources for such exogenous RNAs include artificially introduced transgenes and viral RNA. In both cases, RdRPs and DICER-LIKE proteins are involved in producing mature siRNAs (see Figure 2.14C).

**(C)** RNAi response pathway to virus infection

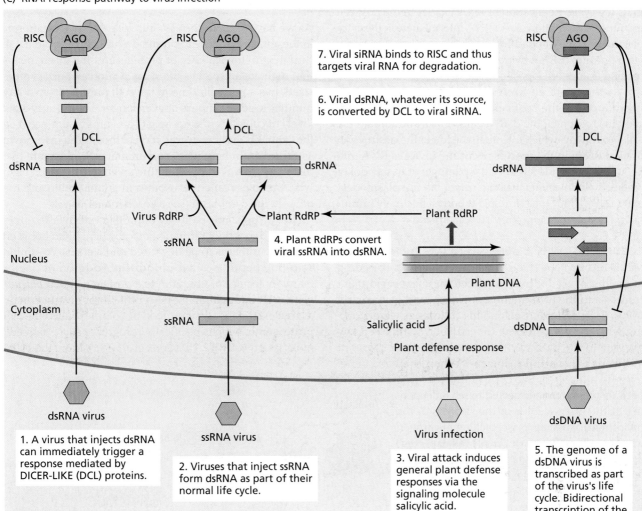

7. Viral siRNA binds to RISC and thus targets viral RNA for degradation.

6. Viral dsRNA, whatever its source, is converted by DCL to viral siRNA.

4. Plant RdRPs convert viral ssRNA into dsRNA.

1. A virus that injects dsRNA can immediately trigger a response mediated by DICER-LIKE (DCL) proteins.

2. Viruses that inject ssRNA form dsRNA as part of their normal life cycle.

3. Viral attack induces general plant defense responses via the signaling molecule salicylic acid.

5. The genome of a dsDNA virus is transcribed as part of the virus's life cycle. Bidirectional transcription of the viral DNA produces dsRNA.

**DOWNSTREAM EVENTS OF THE RNAi PATHWAY INVOLVE THE FORMATION OF AN RNA-INDUCED SILENCING COMPLEX** For miRNAs, siRNAs, and RNAs of exogenous origin, the end result of the RNAi process is similar: the inactivation or silencing of their complementary mRNAs or DNA sequences. After 21- to 24-nucleotide miRNAs or siRNAs have been formed by the DICER-LIKE proteins, one strand of the short RNA duplex associates with a ribonuclease complex called the **RNA-induced silencing complex** (**RISC**) (see Figure 2.14). In both animals and plants, RISC contains at least one catalytic **ARGONAUTE** (**AGO**) protein. In some cases, RISC can recruit additional proteins to the complex. In Arabidopsis, ten different members of the AGO gene family are known. The miRNA that binds to AGO guides RISC to a complementary mRNA. Upon binding to RISC, the target mRNA is cleaved by AGO, and the resulting fragments are released into the cytoplasm, where they are further degraded. RISC-bound mRNA molecules may also be translationally inhibited by RISC.

RISC-bound siRNAs have an additional function: they facilitate methylation of DNA and associated histones at sequences complementary to the siRNA. Although RISC probably doesn't interact directly with DNA methylases or histone methylases, the siRNA somehow guides those modification enzymes to the genomic sequence to be silenced. The chromatin structure is then "remodeled" in an ATP-requiring reaction and subsequently methylated, resulting in tighter condensation and heterochromatization of the DNA region involved (see Figure 2.13C) (Chapman and Carrington 2007).

**SMALL RNAS AND RNAi COMBAT VIRAL INFECTION** In addition to the processing of miRNAs and endogenous siRNAs, plants have also adopted the RNAi pathway as a type of molecular immune response against infection by viruses (Mlotshwa et al. 2008). The genomic structures of plant viruses are quite diverse. Some viruses inject double-stranded DNA, others use single- or double-stranded RNA.

Nevertheless, each virus produces dsRNA at some point in its life cycle. Replication of RNA viruses in the host cell requires the formation of a dsRNA intermediate in the cytoplasm. Double-stranded DNA viruses, on the other hand, often produce dsRNA by transcription of overlapping open reading frames on opposite strands of their DNA.

Whether from an RNA or a DNA virus, the dsRNA is produced in the host cell's nucleus. The plant's DCL proteins recognize the dsRNA molecules and initiate the RNAi pathway, which eventually leads to the destruction of viral RNA. In the process of cutting invasive RNA into 21- to 24-nucleotide siRNAs, the plant generates a pool of "memory" molecules that can travel via the plasmodesmata throughout the entire plant body, effectively immunizing the plant before the virus can spread.

**COSUPPRESSION IS A GENE SILENCING PHENOMENON MEDIATED BY RNA** One of the first experiments leading to the discovery of RNAi involved an unexpected plant response to the introduction of transgenes. In the early 1990s, Richard Jorgensen and his colleagues were working with the petunia gene for chalcone synthase, a key enzyme in the pathway that produces purple pigment molecules in petunia flowers. When they inserted a highly active copy of the gene into petunia plants, they expected to see a deepening of the purple color in the flowers of the offspring. To their surprise, the flowers ranged in petal color from dark purple (as expected) to completely white (as if chalcone synthase levels had gone *down* instead of up). This phenomenon—decreased expression of a gene when extra copies are introduced—was termed **cosuppression**. With our present understanding of RNAi, we now know that in some cells, overexpression of chalcone synthase triggered an RNA-dependent RNA polymerase to produce dsRNA molecules, which initiated the RNAi response. This response eventually led to posttranscriptional silencing and to methylation of both the introduced and the endogenous copies of the chalcone synthase gene. Interestingly, posttranscriptional silencing did not occur in all cells. The cells in which gene silencing took place gave rise to white sectors, explaining why some of the transgenic petunia plants had variegated purple and white flowers.

In summary, RNAi is a process in which dsRNA elicits a posttranscriptional response that leads to the silencing of specific transcripts. miRNAs aid in the regulation of genes—often developmental genes—while siRNAs help to keep heterochromatin transcriptionally inactive or act as a molecular immune response against viruses.

## Posttranslational regulation determines the life span of proteins

As we have seen, mRNA stability plays an important role in the ability of a gene to produce a functional protein. We turn next to the stability of proteins and the mechanisms that regulate a protein's life span. A protein, once synthesized, has a finite lifetime in the cell, ranging from a few minutes to several hours or even longer. Thus steady-state levels of cellular enzymes reflect an equilibrium between the synthesis and the degradation of those proteins, known as **turnover**. In both plant and animal cells, there are two distinct pathways of protein turnover, one in specialized lytic vacuoles (called lysosomes in animal cells) and the other in the cytoplasm (see also Chapter 1).

The cytoplasmic pathway of protein turnover involves the ATP-dependent formation of a covalent bond between the protein that is to be degraded and a small, 76–amino acid polypeptide called **ubiquitin**. Addition of one or many molecules of ubiquitin to a protein is called *ubiquitination*. Ubiquitination marks a protein for destruction by a large, ATP-dependent proteolytic complex called the **26S proteasome**, which specifically recognizes such "tagged" molecules (**FIGURE 2.15**) (Coux et al. 1996). More than 90%

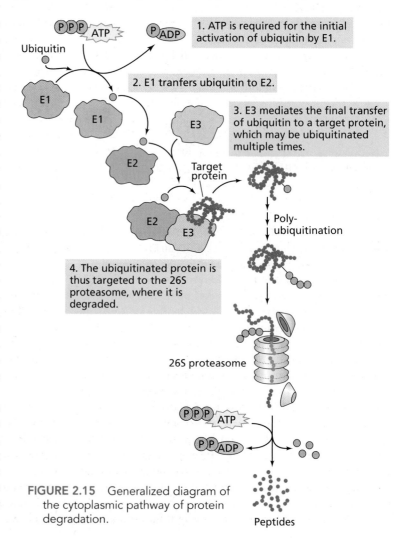

1. ATP is required for the initial activation of ubiquitin by E1.

2. E1 tranfers ubiquitin to E2.

3. E3 mediates the final transfer of ubiquitin to a target protein, which may be ubiquitinated multiple times.

4. The ubiquitinated protein is thus targeted to the 26S proteasome, where it is degraded.

Poly-ubiquitination

26S proteasome

Peptides

**FIGURE 2.15** Generalized diagram of the cytoplasmic pathway of protein degradation.

of the short-lived proteins in eukaryotic cells are degraded via the ubiquitin pathway (Lam 1997).

Ubiquitination is initiated when the **ubiquitin-activating enzyme (E1)** catalyzes the ATP-dependent adenylylation of the C terminus of ubiquitin. The adenylylated ubiquitin is then transferred to a cysteine residue on a second enzyme, the **ubiquitin-conjugating enzyme (E2)**. Proteins destined for degradation are bound by a third type of protein, a **ubiquitin ligase (E3)**. The E2–ubiquitin complex then transfers its ubiquitin to a lysine residue of the protein bound to E3. This process can occur multiple times to form a polymer of ubiquitin. The ubiquitinated protein is then targeted to the proteasome for degradation.

As we now know, there are a multitude of protein-specific ubiquitin ligases that regulate the turnover of specific target proteins (see Chapter 14). We will discuss an example of this pathway in more detail when we cover developmental regulation by the plant hormone auxin in Chapter 19.

## Tools for Studying Gene Function

Individuals that contain specific changes in their DNA sequence are called **mutants**. The analysis of mutants is an extremely powerful tool that can help scientists infer the function of a gene or map its location on the chromosomes. In this section we will discuss how mutants are generated and how they can be used in genetic analysis. We will also discuss some modern biotechnological tools that allow researchers to study or manipulate the expression of genes.

### Mutant analysis can help to elucidate gene function

Throughout this book we will discuss the genes and genetic pathways involved in physiological functions at length, often referring to certain types of mutants that allowed researchers to understand the genes and pathways under discussion. Why is a mutant gene a more powerful tool for elucidating gene function than the normal, wild-type gene on its own?

The use of mutants for gene identification relies on the ability to distinguish a mutant from a normal individual, so the change in the mutant's nucleotide sequence must result in an altered phenotype. If a mutant can be restored to the normal phenotype with a wild-type version of a candidate gene, the researcher knows that a mutation in that gene was responsible for conferring the originally observed mutant phenotype. This method is called **complementation**. For example, assume that a plant with a single-gene mutation shows a delay in producing flowers compared with the wild type. If we can determine the sequence and location of the gene responsible, we will presumably learn something about the mechanisms involved in floral development. Let's now assume that a researcher is able to find a gene in the mutant genome that differs from the wild-type gene in its DNA sequence. If the researcher is able to show that transferring the wild-type gene into the mutant restores the normal phenotype, we can be reasonably certain that the candidate gene plays a role in the initiation of flowering.

In the 1920s, H. J. Muller and L. J. Stadler independently experimented with the effects of X-rays on the stability of chromosomes in flies and in barley, respectively. Both researchers reported heritable changes in the treated organisms. In the following years, other techniques for inducing mutations were developed. These techniques include the use of ultraviolet or fast-neutron radiation and of mutagenic chemicals. For example, treatment with the chemical **ethylmethanesulfonate (EMS)** causes the addition of an ethyl group to a nucleotide, usually guanine. Ethylated guanine pairs with thymine instead of cytosine. The cell's DNA repair machinery then replaces the ethylated guanine with adenine, causing a permanent mutation from G/C to A/T at that site. Mutagenesis with either radiation or chemicals induces nucleotide changes randomly throughout the genome.

There are several ways to map a mutation to its chromosome and ultimately clone the affected gene. WEB TOPIC 2.1 explains a method called **map-based cloning**, which uses crosses between a mutant and a wild-type plant and genetic analysis of the offspring to narrow down the location of the mutation to a short segment of the chromosome, which is then sequenced.

Another method of mutagenesis is the random insertion of transposons into genes. This technique involves crossing the plant of interest with a plant carrying an active transposon and screening their offspring for mutant phenotypes caused by random insertion of the transposon into new locations. Because the sequence of the transposon is known, these mutations are "tagged"; thus the DNA sequences adjacent to the transposon can be readily found and analyzed to identify the mutated gene. This technique is called **transposon tagging** and is explained in detail in WEB TOPIC 2.2.

### Molecular techniques can measure the activity of genes

Once a gene of interest has been identified, scientists are usually interested in where and when the gene is expressed. For example, a gene may be expressed only in reproductive tissues, or only in vegetative ones. Likewise, a gene may encode general cell functions (so-called housekeeping functions) and be expressed continuously, or it may encode special functions and be expressed only in response to a certain stimulus, such as a hormone or an environmental cue. In the past, transcriptional analysis (the determination of the amount of mRNA produced from a gene at a given time) has been performed mainly on single genes. Tools developed for this type of analysis include Northern blotting, reverse transcription polymerase chain reaction (PCR), and in situ hybridization. You can find applications of each of these techniques throughout this book. A more recently developed technique, called **microarray** or gene chip technology, makes use of the information that has come out of massive genome sequencing projects.

All microarray techniques use a solid support, such as a glass slide, onto which DNA sequences are spotted that are representative of single genes of a given species. Such arrays can hold thousands of spots, which can all be investigated together in a single experiment, greatly increasing the through-put of gene analysis over the classic methods mentioned earlier. RNA extracted from a given tissue is first transformed by reverse transcription into a more stable DNA copy (a cDNA) of each RNA molecule represented in the extract (**FIGURE 2.16**). The cDNA mixture is then labeled with a fluorescent dye to allow its visualization later in the procedure. After labeling, the cDNA mixture is applied to the microarray.

Each single-stranded cDNA binds to (hybridizes with) its corresponding (complementary) DNA spot on the array, which represents the gene that produced the corresponding mRNA in the original tissue extract. For example, if mRNA from gene X is present in the tissue extract, its cDNA will bind to the spot on the array representing gene X. If gene Y, on the other hand, was not expressed at the time of tissue sampling, there will be no cDNA to bind to the DNA spot on the array representing gene Y, and that spot will stay blank. After hybridization, the microarray is scanned using a laser beam that can detect the fluorescent label applied to the cDNA. In our example, the spot for gene X will light up in the laser scan, whereas the spot for gene Y will not.

There are several types of microarray techniques. While some types analyze two mRNA samples—for example, from a treated and from a control plant—on a single array (using "two-color labeling"; see Figure 2.16), other types compare two samples using two separate arrays. Since its original application to the analysis of gene expression, microarray technology has been adapted to many other uses as well, ranging from diagnosing the genotypes of individuals in a population to determining the epigenetic status of genes or intergenic regions.

## Gene fusions can introduce reporter genes

The identification of a gene containing a mutation provides information about that gene's location in the genome and about the effect of its altered function on the plant's phenotype. From the sequence

**FIGURE 2.16** Two-color labeling is a microarray technique that can be used to compare gene expression in different individuals or under different conditions.

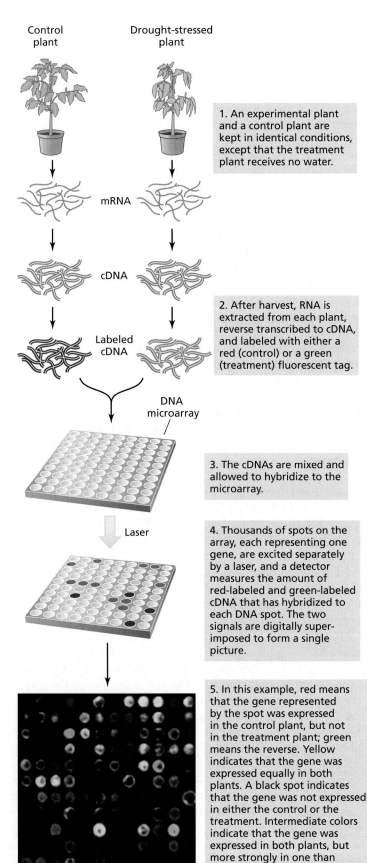

Control plant

Drought-stressed plant

mRNA

cDNA

Labeled cDNA

DNA microarray

Laser

1. An experimental plant and a control plant are kept in identical conditions, except that the treatment plant receives no water.

2. After harvest, RNA is extracted from each plant, reverse transcribed to cDNA, and labeled with either a red (control) or a green (treatment) fluorescent tag.

3. The cDNAs are mixed and allowed to hybridize to the microarray.

4. Thousands of spots on the array, each representing one gene, are excited separately by a laser, and a detector measures the amount of red-labeled and green-labeled cDNA that has hybridized to each DNA spot. The two signals are digitally super-imposed to form a single picture.

5. In this example, red means that the gene represented by the spot was expressed in the control plant, but not in the treatment plant; green means the reverse. Yellow indicates that the gene was expressed equally in both plants. A black spot indicates that the gene was not expressed in either the control or the treatment. Intermediate colors indicate that the gene was expressed in both plants, but more strongly in one than the other.

of a gene alone, scientists can make inferences about its cellular function by comparing the gene's structure with those of other known genes. For example, certain regions within the gene—so-called **domains**—might have similarity to domains found in certain families of genes, such as those encoding kinases, phosphatases, or membrane receptors. However, sequence information alone does not give direct evidence of the gene's cellular function, nor does it indicate where in the plant, or under what conditions, the gene is active.

One way to find out where a certain gene is expressed within a plant or cell is to make a gene fusion. A **gene fusion** is an artificial construct that combines part of the gene of interest—for example, the promoter—with another gene—referred to as a **reporter gene**—that produces a readily detectable protein. One such reporter gene is the green fluorescent protein gene (*GFP*), which produces a fluorescent protein that can be observed in an intact plant or cell by fluorescence microscopy (for example, see Figure 19.33). Recall that not all genes are transcribed in every cell of the plant at all times. A gene's expression is regulated by transcription factors that fine-tune its activity and allow it to be transcribed only where and when it is needed. If a plant carries a promoter–*GFP* gene fusion in all of its cells, *GFP* will be expressed only in those cells that would normally express the gene whose promoter was fused with *GFP*. In other words, green fluorescence will be visible wherever and whenever the gene under investigation is expressed.

To transform plants with gene fusions, scientists have harnessed the power of the microbial plant pathogen *Agrobacterium tumefaciens*. This bacterium causes infected plants to overproduce growth hormones, which induce the formation of a tumor called a **crown gall** (see Figure 21.4). Crown gall disease is a serious problem for certain agricultural crops, such as fruit trees, as it can reduce crop yield and decrease overall plant health.

*Agrobacterium* is sometimes referred to as the natural genetic engineer because it has the ability to transform plant cells with a small subset of its own genes. The genes transferred to the plant genome are part of a circular extrachromosomal piece of DNA called the tumor-inducing (Ti) plasmid (**FIGURE 2.17**). The Ti plasmid contains a number of virulence (*vir*) genes as well as a region called the transfer DNA (T-DNA). The *vir* genes are required for initiating and conducting the transfer of the T-DNA into the plant cell. Once transferred, the T-DNA inserts itself randomly into the plant nuclear genome. It carries genes for two general functions: first, the induction of the crown gall, which will provide housing for the bacterium, and second, the production of nonprotein amino acids called *opines*, which are used by the bacterium as a source of metabolic energy (see Figure 21.5). The steps involved in the transformation of plant cells by *Agrobacterium* are described in **FIGURE 2.18** (Citovsky et al. 2007; Dafny-Yelin et al. 2008).

Given that *Agrobacterium* is usually a plant pathogen, how can it be a useful biotechnological tool? When *Agrobacterium* is used in the laboratory, scientists use a strain that has a modified Ti plasmid. The hormone and opine genes are removed from the T-DNA, and a gene of interest is inserted in their place. Often a gene that confers resistance to an antibiotic is added as a selectable marker gene. The engineered Ti plasmid is then inserted into *Agrobacterium*. Any gene now contained within the T-DNA will be trans-

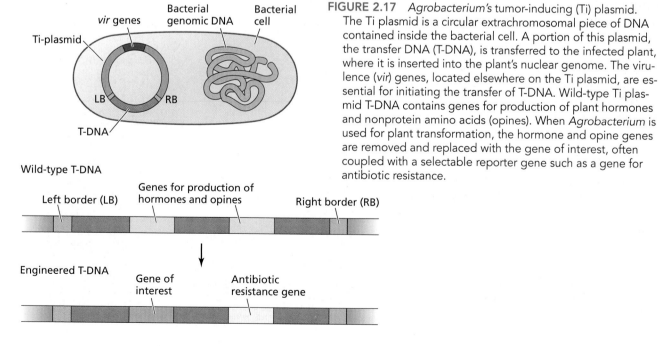

**FIGURE 2.17** *Agrobacterium's* tumor-inducing (Ti) plasmid. The Ti plasmid is a circular extrachromosomal piece of DNA contained inside the bacterial cell. A portion of this plasmid, the transfer DNA (T-DNA), is transferred to the infected plant, where it is inserted into the plant's nuclear genome. The virulence (*vir*) genes, located elsewhere on the Ti plasmid, are essential for initiating the transfer of T-DNA. Wild-type Ti plasmid T-DNA contains genes for production of plant hormones and nonprotein amino acids (opines). When *Agrobacterium* is used for plant transformation, the hormone and opine genes are removed and replaced with the gene of interest, often coupled with a selectable reporter gene such as a gene for antibiotic resistance.

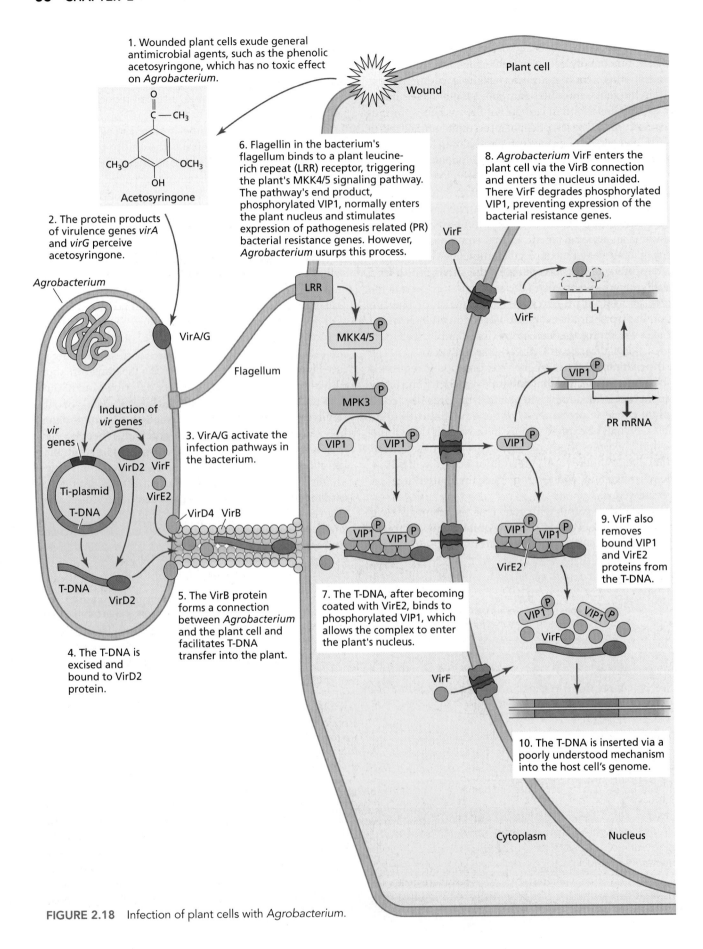

**FIGURE 2.18** Infection of plant cells with *Agrobacterium*.

ferred into a plant cell infected with the engineered bacterium. The antibiotic resistance gene allows the researcher to easily screen for transformed cells.

Plants can be infected with the engineered bacterium in several ways. Small leaf segments can be cut from a plant and co-cultivated with a solution of the bacterium, before culturing the washed off plant cells on tissue culture medium. The plant hormones auxin and cytokinin are then used to stimulate the generation of roots and shoots from the tissue. This technique eventually produces a transformed adult plant. Some plants, including Arabidopsis, are so easily transformed that just dipping the flowers in a suspension of the bacterium is sufficient to result in transformed embryos in the next generation.

Besides *Agrobacterium*-mediated transformation, several other techniques have been developed to incorporate foreign genes into plant genomes. One such technique is the fusion of two plant cells with different genomic information, called **protoplast fusion**. Another is **biolistics**, sometimes also referred to as the **gene gun** technique, in which tiny particles of gold coated with the genetic construct of interest are shot into cells growing in culture dishes. The genetic material is then randomly incorporated into the cells' genomes.

## Genetic Modification of Crop Plants

Humans have modified crop plants through selective breeding for many centuries to produce varieties that have higher yields, are better adapted to specific climates, or are resistant to plant pathogens. For example, modern maize varieties are the domesticated descendants of a subspecies of the genus *Zea*, known as teosinte (**FIGURE 2.19**). As is apparent from the photo, breeding and domestication have modified this crop plant substantially from its original form. Likewise, selective breeding has produced crop tomatoes that are much larger than the fruit from the original progenitor species. Breeding has even produced entirely new species, such as the common bread wheat *Triticum aestivum*, which arose from the cross-pollination of different progenitor species followed by allopolyploidization. While classic breeding techniques rely on random genetic recombination of traits in sexually compatible species, biotechnology allows the transfer of a controlled number of genes between species that cannot be crossed successfully. Let's discuss how classic breeding differs from breeding using biotechnological tools.

In classic breeding, desirable traits are introgressed into elite agricultural lines by cross-pollinating two varieties and selecting for those traits among the offspring. One disadvantage of this approach is that the genetic contributions of both parents are reshuffled in meiosis, so undesirable traits can be introduced into the recipient line along with the desirable ones. The undesirable traits must then be bred out again by repeated backcrosses with the elite line. Biotechnological

**FIGURE 2.19** Classic breeding and domestication of the wild grass teosinte (left) has led to the crop plant *Zea mays* (maize, right) over hundreds of years. (Courtesy of John Doebly.)

tools circumvent this problem by allowing insertion of only the desired genes into the recipient plant, most often either by *Agrobacterium*-mediated transformation or by biolistics. Plants produced in this way are commonly referred to as genetically modified organisms (GMOs).

There are three essential differences between GMOs and conventionally bred varieties of crops:

1. Gene transfer into GMOs occurs in the laboratory and does not require crossbreeding.

2. The donor genes of GMOs can be derived from any organism, not just those with which the recipient can be successfully crossed.

3. GMOs may carry gene constructs that are the product of splicing a variety of genetic components together to produce genes with altogether new functions (for example, the promoter–*GFP* fusion genes we described earlier).

We will turn next to some examples of genes commonly used to modify crops.

### Transgenes can confer resistance to herbicides or plant pests

Any gene artificially transferred into an organism is referred to as a **transgene**. Most often transgenes are introduced from one species into another species. Two of the types of transgenes most commonly used in commercial crops today are genes that allow plants to resist herbicide applications, or to withstand attack by certain insects. Weed invasion and insect infestation are two of the main causes of crop yield reductions in agriculture.

**FIGURE 2.20** Golden rice was produced by inserting two foreign genes involved in β-carotene synthesis into rice. (A) The β-carotene biosynthesis pathway in golden rice. (B) Normal white rice (left) compared with golden rice (right). (Photo courtesy of the Golden Rice Humanitarian Board, www.goldenrice.org)

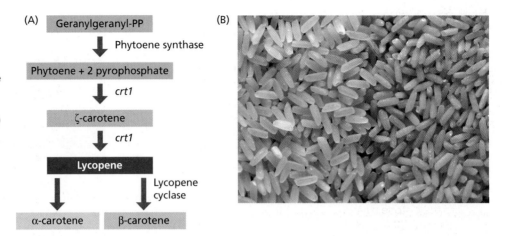

Plants carrying a transgene for **glyphosate resistance** will survive a field application of glyphosate (the commercial herbicide Roundup), which kills weeds but does not harm resistant crop plants. Glyphosate inhibits the enzyme enolpyruvalshikimate-3-phosphate synthase (EPSPS), which catalyzes a key reaction in the shikimic acid pathway, a plant-specific metabolic pathway necessary for the production of many secondary compounds, including auxin and aromatic amino acids (see Chapter 13). Glyphosate-resistant plants carry either a gene encoding a bacterial form of EPSPS that is insensitive to the herbicide, or transgene constructs that fuse high-activity promoters with the wild-type EPSPS gene to achieve herbicide resistance by overproduction of the enzyme.

Another commonly used transgene encodes an insecticidal toxin from the soil bacterium *Bacillus thuringiensis* (**Bt**). Bt toxin interferes with a receptor found only in the larval gut of certain insects, eventually killing them. Plants expressing Bt toxin are toxic to susceptible insects, but harmless to most other, nontarget organisms.

Transgenic plants are also being developed that have enhanced nutritional value. Every year, according to the World Health Organization, dietary vitamin A deficiency causes as many as 500,000 children in developing nations to go blind. Many of these children live in Southeast Asia, where rice is the main part of the diet. Although rice plants synthesize abundant levels of β-carotene (provitamin A) in their leaves, rice endosperm, which makes up the bulk of the grain, does not normally express the genes required for three of the steps in the β-carotene biosynthesis pathway (**FIGURE 2.20A**). Thus far, no naturally occurring rice mutant has arisen that accumulates β-carotene in its grain.

To overcome this block, Ingo Potrykus, Peter Beyer, and their colleagues inserted a bacterial gene, *crt1*, fused to an endosperm-specific promoter, into the rice genome. They found that the crt1 enzyme could catalyze two of the three blocked steps in the endosperm carotenoid pathway (see Figure 2.20A). When *crt1* was combined with

a plant gene for phytoene synthase, which catalyzed the remaining step, the resulting transgenic grain accumulated large amounts of β-carotene. The new variety was called "golden rice" (**FIGURE 2.20B**). This was not the first time that the β-carotene content of a crop plant had been altered by agriculturalists. Carrots, for example, were either red or yellow before the seventeenth century, when a Dutch horticulturist selected the first orange-colored varieties.

Other researchers are developing transgenic plants that express vaccines in their edible fruit as an alternative, more convenient way to vaccinate people in parts of the world where medical facilities are insufficient for the administration of conventional vaccines.

## Genetically modified organisms are controversial

The development of GMOs has not been greeted with universal enthusiasm and support. In spite of their enormous humanitarian potential, many individuals, as well as the governments of some countries, look on GMOs with suspicion and concern.

Opponents of the use of biotechnology in agriculture cite, for example, the possibility of inadvertently producing crops that express allergens introduced from another species. They also worry that windblown pollen from transgenic herbicide-resistant crops could cross-pollinate nearby wild species, thereby producing weeds with herbicide resistance (so-called superweeds). Another common concern is that overuse of genes for Bt toxin might select for insects that develop resistance to it.

While many such concerns have been addressed by proponents of plant biotechnology, research is ongoing to monitor the effects of the new technologies on human health and the environment. In the end, the controversy may come down to this question: How much risk is acceptable in an attempt to satisfy the needs of an ever-increasing world population for food and shelter?

## SUMMARY

Genotype, epigenetic modifications, and environmental interactions determine the phenotype of a plant. To understand thoroughly a plant's physiology requires understanding how genotype is translated into phenotype. Plant cells contain three genomes: the nuclear genome, the smaller genomes in the chloroplasts, and mitochondria.

### Nuclear Genome Organization

- The major protein components of the chromatin are the histones, around which the DNA is wound (**Figure 2.1**).
- Heterochromatin (often highly repetitive DNA sequences) is transcriptionally less active than euchromatin.
- Transposons are mobile DNA sequences within the nuclear genome. Some can insert themselves in new places along the chromosomes (**Figure 2.2**).
- Active transposons can significantly damage their host, but most mobile elements are inactivated by epigenetic modifications, such as methylation.
- Methylation of DNA and methylation and acetylation of histones play a role in determining whether chromatin is hetero- or euchromatic (**Figure 2.3**).
- Many plant species are polyploid, either due to exact genome duplication (autopolyploidy) or due to hybridization of two species followed by genome duplication (allopolyploidy) (**Figure 2.5**).
- Phenotypic and physiological responses to polyploidy are often unpredictable.
- Polyploids have multiple complete genomes; this altered genomic balance can phenotypically distinguish polyploids, especially allopolyploids, from their parents, and may result in reproductive isolation and speciation. Polyploidy is therefore an important evolutionary mechanism (**Figure 2.8**).

### Plant Cytoplasmic Genomes: Mitochondria and Chloroplasts

- Organellar genomes are complex in structure and they usually consist of multiple copies of the genome on the same DNA molecule (**Figure 2.9**).
- Organellar genetics do not obey Mendelian laws, but usually show uniparental inheritance and vegetative segregation (**Figure 2.10**).

### Transcriptional Regulation of Nuclear Gene Expression

- Gene activity is regulated on several levels: transcriptional, posttranscriptional, and posttranslational (**Figure 2.11**).

- For protein-coding genes, RNA Polymerase II binds to the promoter region and requires general transcription factors and other regulatory proteins to initiate gene transcription (**Figure 2.12**).
- Epigenetic modifications, such as methylation of DNA and methylation and acetylation of histone proteins, help determine gene activity (**Figure 2.13**).

### Posttranscriptional Regulation of Nuclear Gene Expression

- RNA-binding proteins may either stabilize mRNA or promote its degradation.
- The RNA interference (RNAi) pathway is a posttranscriptional response that leads to the silencing of specific transcripts. MicroRNAs (miRNAs) aid in gene regulation. Short interfering RNAs (siRNAs) help to keep heterochromatin transcriptionally inactive, or act as a molecular immune system against viruses (**Figure 2.14**).
- Proteins tagged with a small polypeptide called ubiquitin are targeted for destruction by the proteasome (**Figure 2.15**).

### Tools for Studying Gene Function

- Tools developed for transcriptional analysis of single genes include Northern blotting, reverse transcription PCR, and in situ hybridization.
- Microarray or gene-chip technology utilizes known genomic sequences for high-throughput analysis of gene expression or genotypes (**Figure 2.16**).
- Reporter gene fusions contain part of a gene of interest (e.g., the promoter) fused to a reporter gene that encodes a protein that can be readily detected when expressed. Such constructs can be used to monitor the time and place a particular gene is active.
- *Agrobacterium* can transform plant cells when targeted genes are transferred as part of a plasmid called the tumor-inducing (Ti-) plasmid (**Figures 2.17, 2.18**).

### Genetic Modification of Crop Plants

- In contrast to classical selective breeding, bioengineering allows the transfer of a specific gene or genes between species that cannot be crossed successfully.
- Artifically transferred genes can confer resistance to herbicides, plant pests, or provide enhanced nutrition (**Figure 2.20**).

## WEB MATERIAL

### Web Topics

**2.1  Recombination Mapping and Gene Cloning**
Mapped-based cloning can be used to isolate the gene(s) involved in a phenotype of interest.

**2.2  Transposon Tagging**
Mutagenesis using transposable elements is another approach to gene identification.

## CHAPTER REFERENCES

Adams, K. L., Cronn, R., Percifield, R., and Wendel, J. F. (2003) Genes duplicated by polyploidy show unequal contributions to the transcriptome and organ-specific reciprocal silencing. *Proc. Natl. Acad. Sci. USA* 100: 4649–4654.

Bannister, A. J., and Kouzarides, T. (2005) Reversing histone methylation. *Nature* 43: 1103–1106.

Chen, Z. J. (2007) Genetic and epigenetic mechanisms for gene expression and phenotypic variation in plant polyploids. *Ann. Rev. Plant Biol. Mol. Biol.* 58: 377–406.

Chapman, E., and Carrington, J. C. (2007) Specialization and evolution of endogenous small RNA pathways. *Nat. Rev. Genet*: 8: 884–896.

Choi, K., Park, C., Lee, J., Oh, M., Noh, B., and Lee, I. (2007) Arabidopsis homologs of components of the swr1 complex regulate flowering and plant development. *Development* 134: 1931–1941.

Citovsky, V., Kozlovsky, S. V., Lacroix, B., Zaltsman, A., Dafny-Yelin, M., Vyas, S., Tovkach, A., and Tzfira, T. (2007) Biological systems of the host cell involved in *Agrobacterium* infection. *Cell. Microbiol.* 9: 9–20.

Comai, L. (2005) The advantages and disadvantages of being polyploid. *Nat. Rev. Genet.* 6: 836–846.

Coux, O., Tanaka, K., and Goldberg, A. L. (1996) Structure and functions of the 20S and 26S proteasomes. *Annu. Rev. Biochem.* 65: 2069–2076.

Dafny-Yelin, M., Levy, A., and Tzfira, T. (2008) The ongoing saga of *Agrobacterium*–host interactions. *Trends Plant Sci.* 13: 102–105.

Eickbush, T. H., and Malik, H. S. (2002) Origins and evolution of retrotransposons. In *Mobile DNA II*, N. L. Craig et al., eds., ASM Press, Washington DC, pp. 1111–1144.

Feldman, M., Liu, B., Segal, G., Abbo, S., Levy, A. A., and Vega, J. M. (1997) Rapid elimination of low-copy DNA sequences in polyploid wheat: a possible mechanism for differentiation of homologous chromosomes. *Genetics* 147: 1381–1387.

Gill, N., Hans, C. S., and Jackson, S. (2008) An overview of plant chromosome structure. *Cytogenet. Genome Res.* 120: 194–201.

Green, P. J. (1993) Control of mRNA stability in higher plants. *Plant Physiol.* 102: 1065–1070.

Guo, F., Hu, S.-H., Yuan, Z., Zee, S.-Y., and Han, Y. (2005) Paternal cytoplasmic transmission in Chinese pine (*Pinis tabuleaformis*). *Protoplasma* 225: 5–14.

Heslop-Harrison, J. S., and Schmidt, T. (2007) *Plant nuclear genome composition*. In *Encyclopedia of Life Sciences*, John Wiley & Sons, Ltd., Chichester, UK, http://www.els. net/ [doi: 10.1002/9780470015902.a0002014].

Hollams, E. M., Giles, K. M., Thomson, A. M., and Leedman, P. J. (2002) mRNA stability and the control of gene expression: implications for human disease. *Neurochem. Res.* 27: 957–80.

Jiang, J., Birchler, J., Parrott, W. A., and Dawe, R. K. (2003) A molecular view of plant centromeres. *Trends Plant Sci.* 8: 570–575.

Kashkush, K., Feldman, M., and Levy, A. A. (2003) Transcriptional activation of retrotransposons alters the expression of adjacent genes in wheat. *Nat. Genet.* 33: 102–106.

Kato, A., Lamb, J. C., and Birchler, J. A. (2004) Chromosome painting using repetitive DNA sequences as probes for somatic chromosome identification in maize. *Proc. Natl. Acad. Sci. USA* 101: 13554–13559.

Kato, A., Vega, J. M., Han, F., Lamb, J. C., and Birchler, J. A. (2005) Advances in plant chromosome identification and cytogenetic techniques. *Curr. Opin. Plant Biol.* 8: 148–154.

Kidner, C. A., and Martienssen, R. A. (2005) The developmental role of microRNA in plants. *Curr. Opin. Plant Biol.* 8: 38–44.

Lam, E. (1997) Nucleic acids and proteins. In *Plant Biochemistry*, P. M. Dey and J. B. Harborne, eds., Academic Press, San Diego, pp. 316–352.

Leitch, A. R., and Leitch, I. J. (2008) Genomic plasticity and the diversity of polyploid plants. *Science* 320: 481–483.

Lynch, M., and Force, A. (2000) The probability of duplicate gene preservation by subfunctionalization. *Genetics* 154: 459–473.

Ma, J., Wing, R. A., Bennetzen, J. L., and Jackson, S. A. (2007) Plant centromere organization: a dynamic structure with conserved functions. *Trends Genet.* 23: 134–139.

Madlung, A., Masuelli, R., Watson, B., Reynolds, S., and Comai, L. (2002) Remodeling of DNA methylation and phenotypic and transcriptional changes in synthetic *Arabidopsis* allotetraploids. *Plant Physiol.* 129: 733–746.

Madlung, A., Tyagi, A. P., Watson, B., Jiang, H., Kagochi, T., Doerge, R. W., Martienssen, R., and Comai, L. (2005) Genomic changes in *Arabidopsis* polyploids. *Plant J.* 41: 221–230.

Miura, A., Yonebayashi, S., Watanabe, K., Toyama, T., Shimadak, H., and Kakutani, T. (2001) Mobilization of transposons by a mutation abolishing full DNA methylation in *Arabidopsis*. *Nature* 411: 212–214.

Mlotshwa, S., Pruss, G. J., and Vance, V. (2008) Small RNAs in viral infection and host defense. *Trends Plant Sci.* 13: 375–382.

Mogensen, L. (1996) The hows and whys of cytoplasmic inheritance in seed plants. *Am. J. Bot.* 83: 383–404.

Oldenburg, D. J., and Bendich, A. J. (1998) The structure of mitochondrial DNA from the liverwort, *Marchantia polymorpha*. *J. Mol. Biol.* 276: 745–758.

Oldenburg, D. J., and Bendich, A. J. (2004) Most chloroplast DNA of maize seedlings in linear molecules with defined ends and branched forms. *J. Mol. Biol.* 335: 953–970.

Osborn, T. C., Pires, J. C., Birchler, J. A., Auger, D., Chen, Z. J., Lee, H. S., Comai, L., Madlung, A. Doerge, R. W., Colot, V., and Martienssen, R. A. (2003) Understanding mechanisms of novel gene expression in polyploids. *Trends Genet.* 19: 141–147.

Parisod, C., Salmon, A., Zerjal, T., Tenaillon, M., Grandbastien, M. A., and Ainouche, M. (2009) Rapid structural and epigenetic reorganization near transposable elements in hybrid and allopolyploid genomes in *Spartina*. *New Phytologist* 184: 1003–1015.

Pires, J. C., Lim, K. Y., Kovarik, A., Matyasek, R., Boyd, A., Leitch, A. R., Leitch, I. J., Bennett, M. D., Soltis, P. S., and Soltis, D. E. (2004) Molecular cytogenetic analysis of recently evolved *Tragopogon* (Asteraceae) allopolyploids reveal a karyotype that is additive of the diploid progenitors. *Am. J. Bot.* 91: 1022–1035.

Ramachandran, V., and Chen, X. (2008) Small RNA metabolism in Arabidopsis. *Trends Plant Sci.* 13: 368–374.

Rapp, R., Udall, J., and Wendel, J. (2009) Genomic expression dominance in allopolyploids. *BMC Biol.* 7: 18.

Riddle, N. C., Kato, A., and Birchler, J. A. (2006) Genetic variation for the response to ploidy change in *Zea mays* L. *Theor. Appl. Genet.* 114: 101–111.

Riddle, N. C., and Birchler, J. A. (2008) Comparative analysis of inbred and hybrid maize at the diploid and tetraploid levels. *Theor. Appl. Genet.* 116: 563–576.

Salmon, A., Ainouche, M. L., and Wendel, J. F. (2005) Genetic and epigenetic consequences of recent hybridization and polyploidy in *Spartina* (Poaceae). *Mol. Ecol.* 14: 1163–1175.

Wang, J., Tian, L., Lee, H. S., Wei, N. E., Jiang, H., Watson, B., Madlung, A., Osborn, T. C., Doerge, R. W., Comai, L., and Chen, Z. J. (2006) Genomewide nonadditive gene regulation in Arabidopsis allotetraploids. *Genetics* 172: 507–517.

Wolfe, K. H. (2001) Yesterday's polyploidization and the mystery of diploidization. *Nat. Rev. Genet.* 2: 333–341.

# Unit One

# Transport and Translocation
of Water and Solutes

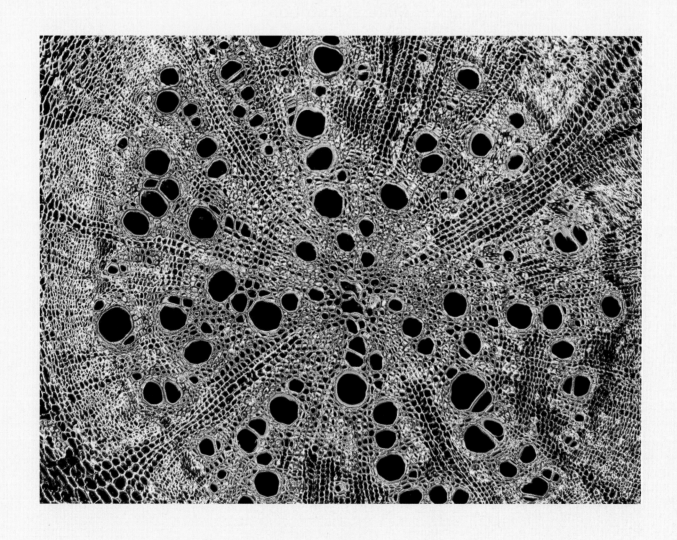

# UNIT ONE

*Previous page: This light micrograph of a soybean (Glycine max) root stele has been enhanced to accentuate the vascular tissue. Image © Garry DeLong/OSF/ Photolibrary.com.*

# 3

# Water and Plant Cells

Water plays a crucial role in the life of the plant. Photosynthesis requires that plants draw carbon dioxide from the atmosphere, and at the same time exposes them to water loss and the threat of dehydration. To prevent leaf desiccation, water must be absorbed by the roots, and transported through the plant body. Even slight imbalances between the uptake and transport of water and the loss of water to the atmosphere can cause water deficits and severe malfunctioning of many cellular processes. Thus, balancing the uptake, transport, and loss of water represents an important challenge for land plants.

A major difference between plant and animal cells, which has a large impact on their respective water relations, is that plant cells have cell walls and animal cells do not. Cell walls allow plant cells to build up large internal hydrostatic pressures, called **turgor pressure**. Turgor pressure is essential for many physiological processes, including cell enlargement, stomatal opening, transport in the phloem, and various transport processes across membranes. Turgor pressure also contributes to the rigidity and mechanical stability of nonlignified plant tissues. In this chapter we will consider how water moves into and out of plant cells, emphasizing the molecular properties of water and the physical forces that influence water movement at the cell level.

## Water in Plant Life

Of all the resources that plants need to grow and function, water is the most abundant and yet often the most limiting. The practice of crop irrigation reflects the fact that water is a key resource limiting agricultural productivity (**FIGURE 3.1**). Water availability likewise limits the productivity of natural ecosystems (**FIGURE 3.2**), leading to marked differences in vegetation type along precipitation gradients.

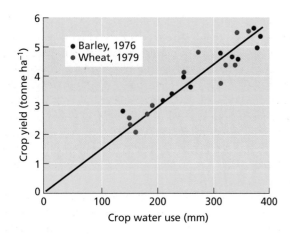

**FIGURE 3.1**  Grain yield as a function of water used under a range of irrigation treatments for barley in 1976 and wheat in 1979 in southeastern England. (After Jones 1992; data from Day et al. 1978 and Innes and Blackwell 1981.)

The reason that water is frequently a limiting resource for plants, but much less so for animals, is that plants use water in huge amounts. Most (~97%) of the water absorbed by a plant's roots is carried through the plant and evaporates from leaf surfaces. Such water loss is called **transpiration**. In contrast, only a small amount of the water absorbed by roots actually remains in the plant to supply growth (~2%) or to be consumed in the biochemical reactions of photosynthesis and other metabolic processes (~1%).

Water loss to the atmosphere appears to be an inevitable consequence of carrying out photosynthesis on land. The uptake of $CO_2$ is coupled to the loss of water through a

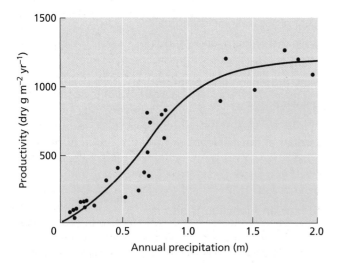

**FIGURE 3.2**  Productivity of various ecosystems as a function of annual precipitation. Productivity was estimated as net aboveground accumulation of organic matter through growth and reproduction. (After Whittaker 1970.)

common diffusional pathway: as $CO_2$ diffuses into leaves, water vapor diffuses out. Because the driving gradient for water loss from leaves is much larger than that for $CO_2$ uptake, as many as 400 water molecules are lost for every $CO_2$ molecule gained. This unfavorable exchange has had a major influence on the evolution of plant form and function and explains why water plays such a key role in the physiology of plants.

We will begin our study of water by considering how its structure gives rise to some of its unique physical properties. We will then examine the physical basis for water movement, the concept of water potential, and the application of this concept to cell–water relations.

## The Structure and Properties of Water

Water has special properties that enable it to act as a wide-ranging solvent and to be readily transported through the body of the plant. These properties derive primarily from the hydrogen bonding ability and polar structure of the water molecule. In this section we will examine how the formation of hydrogen bonds contributes to the high specific heat, surface tension, and tensile strength of water.

### Water is a polar molecule that forms hydrogen bonds

The water molecule consists of an oxygen atom covalently bonded to two hydrogen atoms (**FIGURE 3.3A**). Because the oxygen atom is more **electronegative** than hydrogen, it tends to attract the electrons of the covalent bond. This attraction results in a partial negative charge at the oxygen end of the molecule and a partial positive charge at each hydrogen, making water a **polar molecule**. These partial charges are equal, so the water molecule carries no *net* charge.

Water molecules are tetrahedral in shape. At two points of the tetrahedron are the hydrogen atoms, each with a partial positive charge. The other two points of the tetrahedron contain lone pairs of electrons, each with a partial negative charge. Thus each water molecule has two positive poles and two negative poles. These opposite partial charges create electrostatic attractions between water molecules, known as **hydrogen bonds** (**FIGURE 3.3B**).

Hydrogen bonds take their name from the fact that effective electrostatic bonds are formed only when highly electronegative atoms such as oxygen are covalently bonded to hydrogen. The reason for this is that the small size of the hydrogen atom allows the partial positive charges to be more concentrated, and thus more effective in bonding electrostatically.

Hydrogen bonds are responsible for many of the unusual physical properties of water. Water can form up to four hydrogen bonds with adjacent water molecules,

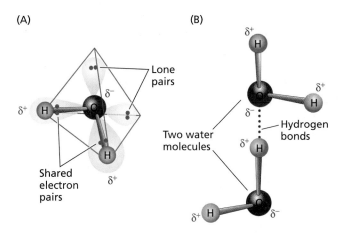

**FIGURE 3.3** Structure of the water molecule. (A) The strong electronegativity of the oxygen atom means that the two electrons that form the covalent bond with hydrogen are unequally shared, such that each hydrogen atom has a partial positive charge. The two lone pairs of electrons of the oxygen atom produce a pole with a partial negative charge. (B) The opposite partial charges ($\delta^-$ and $\delta^+$) on the water molecule lead to the formation of intermolecular hydrogen bonds with other water molecules. Oxygen has six electrons in the outer orbitals; each hydrogen has one.

resulting in very strong intermolecular interactions. Hydrogen bonds can also form between water and other molecules that contain electronegative atoms (O or N), especially when these are covalently bonded to H.

### Water is an excellent solvent

Water dissolves greater amounts of a wider variety of substances than do other related solvents. Its versatility as a solvent is due in part to the small size of the water molecule. However, it is the hydrogen bonding ability of water and its polar structure that make it a particularly good solvent for ionic substances and for molecules such as sugars and proteins that contain polar —OH or —NH$_2$ groups.

Hydrogen bonding between water molecules and ions, and between water and polar solutes, effectively decreases the electrostatic interaction between the charged substances and thereby increases their solubility. Similarly, hydrogen bonding between water and macromolecules such as proteins and nucleic acids reduces interactions between macromolecules, thus helping to draw them into solution.

### Water has distinctive thermal properties relative to its size

The extensive hydrogen bonding between water molecules results in water having both a high specific heat capacity and a high latent heat of vaporization.

**Specific heat** capacity is the heat energy required to raise the temperature of a substance by a set amount. Temperature is a measure of molecular kinetic energy (energy of motion). When the temperature of water is raised, the molecules vibrate faster and with greater amplitude. Hydrogen bonds act like rubber bands that absorb some of the energy from applied heat, leaving less energy available to increase motion. Thus, compared with other liquids, water requires a relatively large heat input to raise its temperature. This is important for plants, because it helps buffer temperature fluctuations.

**Latent heat of vaporization** is the energy needed to separate molecules from the liquid phase and move them into the gas phase—a process that occurs during transpiration. The latent heat of vaporization decreases as temperature increases, reaching a minimum at the boiling point (100°C). For water at 25°C, the heat of vaporization is 44 kJ mol$^{-1}$—the highest value known for any liquid. Most of this energy is used to break hydrogen bonds between water molecules.

Latent heat does not change the temperature of water molecules that have evaporated, but it does cool the surface from which the water has evaporated. Thus, the high latent heat of vaporization of water serves to moderate the temperature of transpiring leaves, which would otherwise increase due to the input of radiant energy from the sun.

### Water molecules are highly cohesive

Water molecules at an air–water interface are attracted to neighboring water molecules by hydrogen bonds, and this interaction is much stronger than any interaction with the adjacent gas phase. As a consequence, the lowest-energy (i.e., most stable) configuration is one that minimizes the surface area of the air–water interface. To increase the area of an air–water interface, hydrogen bonds must be broken, which requires an input of energy. The energy required to increase the surface area of a gas–liquid interface is known as **surface tension**.

Surface tension can be expressed in units of energy per area (J m$^{-2}$), but is generally expressed in the equivalent, but less intuitive, units of force per length (J m$^{-2}$ = N m$^{-1}$). A joule (J) is the SI unit of energy, with units of force × distance (N m); a newton (N) is the SI unit of force, with units of mass × acceleration (kg m s$^{-2}$). If the air–water interface is curved, surface tension produces a net force perpendicular to the interface (**FIGURE 3.4**). As we will see later, surface tension and adhesion (defined below) at the evaporative surfaces in leaves generate the physical forces that pull water through the plant's vascular system.

The extensive hydrogen bonding in water also gives rise to the property known as **cohesion**, the mutual attraction between molecules. A related property, called **adhesion**, is the attraction of water to a solid phase such as a cell wall or glass surface, again due primarily to the formation of hydro-

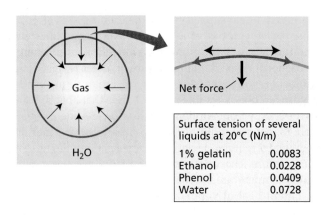

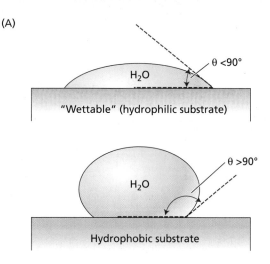

**FIGURE 3.4** A gas bubble suspended within a liquid assumes a spherical shape such that its surface area is minimized. Because surface tension acts along the tangent to the gas–liquid interface, the resultant (net) force will be inward, leading to compression of the bubble. The magnitude of the pressure (force/area) exerted by the interface is equal to $2T/r$, where $T$ is the surface tension of the liquid (N/m) and $r$ is the radius of the bubble (m). Water has an extremely high surface tension compared to other liquids at the same temperature.

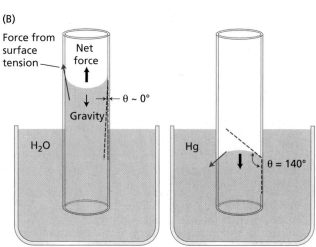

gen bonds. The degree to which water is attracted to the solid phase versus to itself can be quantified by measuring the **contact angle** (FIGURE 3.5A). The contact angle describes the shape of the air–water interface and thus the effect that the surface tension has on the pressure in the liquid.

Cohesion, adhesion, and surface tension give rise to a phenomenon known as **capillarity** (FIGURE 3.5B). Consider a vertically oriented glass capillary tube with wettable walls (contact angle < 90°). At equilibrium, the water level within the capillary tube will be higher than that of the water supply at its base. Water is drawn into the capillary tube due to (1) the attraction of water to the polar surface of the glass tube (adhesion) and (2) the surface tension of water. Together, adhesion and surface tension pull on the water molecules, causing them to move up the tube until this upward force is balanced by the weight of the water column. The narrower the tube, the higher the equilibrium water level. For calculations related to capillarity, see **WEB TOPIC 3.1**.

### Water has a high tensile strength

Hydrogen bonding gives water a high **tensile strength**, defined as the maximum force per unit area that a continuous column of water can withstand before breaking. We do not usually think of liquids as having tensile strength; however, such a property is evident in the rise of a water column in a capillary tube.

We can demonstrate the tensile strength of water by placing it in a clean glass syringe (FIGURE 3.6). When we

**FIGURE 3.5** (A) The shape of a droplet placed on a solid surface reflects the relative attraction of the liquid to the solid versus to itself. The contact angle (θ), defined as the angle from the solid surface through the liquid to the gas–liquid interface, is used to describe this interaction. "Wettable" surfaces have contact angles of less than 90°; a highly wettable (i.e., hydrophilic) surface (such as water on clean glass or primary cell walls) has a contact angle close to 0°. Water spreads out to form a thin film on highly wettable surfaces. In contrast, nonwettable (i.e., hydrophobic) surfaces have contact angles greater than 90°. Water "beads" up on such surfaces. (B) Capillarity can be observed when a liquid is supplied to the bottom of a vertically oriented capillary tube. If the walls are highly wettable (e.g., water on clean glass), the net force is upward. The water column will rise until this upward force is balanced by the weight of the water column. In contrast, if the liquid does not "wet" the walls (e.g., Hg on clean glass has a contact angle of approximately 140°), the meniscus will curve downward, and the force resulting from surface tension lowers the level of the liquid in the tube.

*push* on the plunger, the water is compressed, and a positive **hydrostatic pressure** builds up. Pressure is measured in units called *pascals* (Pa) or, more conveniently, *megapascals* (MPa). One MPa equals approximately 9.9 atmo-

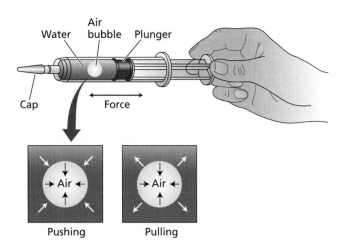

**FIGURE 3.6** A sealed syringe can be used to create positive and negative pressures in fluids such as water. Pushing on the plunger causes the fluid to develop a positive, hydrostatic pressure (white arrows) that acts in the same direction as the interfacial force resulting from surface tension (black arrows). Thus, a small air bubble trapped within the syringe will shrink as the pressure increases. Pulling on the plunger causes the fluid to develop a tension, or negative pressure. Air bubbles in the syringe will expand if the outward force on the bubble exerted by the fluid (white arrows) exceeds the inward force resulting from the surface tension of the gas–liquid interface (black arrows).

spheres. Pressure is equivalent to a force per unit area (1 Pa = 1 N m$^{-2}$) and to an energy per unit volume (1 Pa = 1 J m$^{-3}$). **TABLE 3.1** compares units of pressure.

If instead of pushing on the plunger we *pull* on it, a tension, or *negative hydrostatic pressure*, develops in the water to resist the pull. How hard must we pull on the plunger before the water molecules are torn away from each other and the water column breaks?

Careful studies have demonstrated that water can resist pressures more negative than –20 MPa (Wheeler

---

**TABLE 3.1**
**Comparison of units of pressure**

| | |
|---|---|
| 1 atmosphere | = 14.7 pounds per square inch |
| | = 760 mm Hg (at sea level, 45° latitude) |
| | = 1.013 bar |
| | = 0.1013 MPa |
| | = 1.013 × 10$^5$ Pa |

A car tire is typically inflated to about 0.2 MPa.

The water pressure in home plumbing is typically 0.2–0.3 MPa.

The water pressure under 15 feet (5 m) of water is about 0.05 MPa.

---

and Strook 2008), where the negative sign indicates tension, as opposed to compression. The water column in a syringe (see Figure 3.6), however, cannot sustain such large tensions due to the presence of microscopic gas bubbles. Because gas bubbles can expand when placed under tension, they interfere with the ability of the water in the syringe to resist the pull of exerted by the plunger. The expansion of gas bubbles due to tension is known as **cavitation**. As we will see in Chapter 4, cavitation can have a devastating effect on water transport through the xylem.

## Diffusion and Osmosis

Cellular processes depend upon the transport of molecules both to the cell and away from it. **Diffusion** is the spontaneous movement of substances from regions of higher to lower concentration. At the scale of a cell, diffusion is the dominant mode of transport. The diffusion of water across a selectively permeable barrier is referred to as **osmosis**.

In this section we will examine how the processes of diffusion and osmosis lead to the net movement of both water and solutes.

### Diffusion is the net movement of molecules by random thermal agitation

The molecules in a solution are not static; they are in continuous motion, colliding with one another and exchanging kinetic energy. The trajectory of any particular molecule after a collision is considered to be a random variable. Yet these random movements can result in the net movement of molecules.

Consider an imaginary plane dividing a solution into two equal volumes, A and B. As all the molecules undergo random motion, at each time step there is some probability that any particular solute molecule will cross our imaginary plane. The number we expect to cross from A to be B in any particular time step will be proportional to the number at the beginning of the time step on side A, and the number crossing from B to A will be proportional to the number starting on side B.

If the initial concentration on side A is higher than that on side B, more solute molecules will be expected to cross from A to B than B to A, and we will observe a net movement of solutes from A to B. Thus, diffusion results in the net movement of molecules from regions of high concentration to regions of low concentration, even though each molecule is moving in a random direction. The independent motion of each molecule explains why the system will evolve toward an equal number of A and B molecules on each side (**FIGURE 3.7**).

This tendency for a system to evolve toward an even distribution of molecules can be understood as a consequence of the second law of thermodynamics, which tells

Initial        Intermediate        Equilibrium

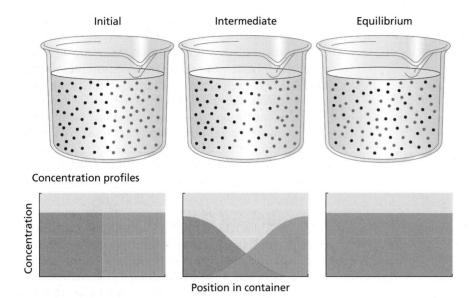

Concentration profiles

Concentration

Position in container

**FIGURE 3.7** Thermal motion of molecules leads to diffusion—the gradual mixing of molecules and eventual dissipation of concentration differences. Initially, two materials containing different molecules are brought into contact. The materials may be gas, liquid, or solid. Diffusion is fastest in gases, slower in liquids, and slowest in solids. The initial separation of the molecules is depicted graphically in the upper panels, and the corresponding concentration profiles are shown in the lower panels as a function of position. With time, the mixing and randomization of the molecules diminishes net movement. At equilibrium the two types of molecules are randomly (evenly) distributed.

us that spontaneous processes evolve in the direction of increasing entropy, or disorder. Increasing entropy is synonymous with a decrease in free energy. Thus, diffusion represents the natural tendency of systems to move toward the lowest possible energy state.

It was Adolf Fick who first noticed in the 1850s that the rate of diffusion is directly proportional to the concentration gradient ($\Delta c_s / \Delta x$)—that is, to the difference in concentration of substance $s$ ($\Delta c_s$) between two points separated by a very small distance $\Delta x$. In symbols, we write this relation as Fick's first law:

$$J_s = -D_s \frac{\Delta c_s}{\Delta x} \qquad (3.1)$$

The rate of transport, expressed as the **flux density** ($J_s$), is the amount of substance $s$ crossing a unit cross-sectional area per unit time (e.g., $J_s$ may have units of moles per square meter per second [$mol\ m^{-2}\ s^{-1}$]). The **diffusion coefficient** ($D_s$) is a proportionality constant that measures how easily substance $s$ moves through a particular medium. The diffusion coefficient is a characteristic of the substance (larger molecules have smaller diffusion coefficients) and depends on both the medium (diffusion in air is typically

10,000 times faster than diffusion in a liquid, for example) and the temperature (substances diffuse faster at higher temperatures). The negative sign in the equation indicates that the flux moves down a concentration gradient.

Fick's first law says that a substance will diffuse faster when the concentration gradient becomes steeper ($\Delta c_s$ is large) or when the diffusion coefficient is increased. Note that this equation accounts only for movement in response to a concentration gradient, and not for movement in response to other forces (e.g., pressure, electric fields, and so on).

### Diffusion is most effective over short distances

Consider a mass of solute molecules initially concentrated around a position $x = 0$ (**FIGURE 3.8A**). As the molecules undergo random motion, the concentration front moves away from the starting position, as shown for a later time point in **FIGURE 3.8B**.

Comparing the distribution of the solutes at the two times, we see that as the substance diffuses away from the starting point, the concentration gradient becomes less steep ($\Delta c_s$ decreases); that is, the number of solute molecules that happen to step "backward" (i.e., toward $x = 0$) relative to those that step "forward" (away from $x = 0$) increases, and thus net movement becomes slower. Note that the average position of the solute molecules remains at $x = 0$ for all time, but that the distribution slowly flattens out.

As a direct consequence of the fact that each molecule is undergoing its own random walk, and thus is as likely to step toward some point of interest as away from it, the average time needed for a particle to diffuse a distance $L$ grows as $L^2/D_s$. In other words, the average time required for a substance to diffuse a given distance increases as the *square* of that distance.

The diffusion coefficient for glucose in water is about $10^{-9}$ m$^2$ s$^{-1}$. Thus the average time required for a glucose molecule to diffuse across a cell with a diameter of 50 μm is 2.5 s. However, the average time needed for the same glucose molecule to diffuse a distance of 1 m in water is approximately 32 years. These values show that diffusion in solutions can be effective within cellular dimensions but is far too slow to be effective over long distances. For additional calculations on diffusion times, see **WEB TOPIC 3.2**.

### Osmosis describes the net movement of water across a selectively permeable barrier

Membranes of plant cells are **selectively permeable**; that is, they allow water and other small, uncharged substances to move across them more readily than larger solutes and charged substances (Stein 1986). If the concentration of solutes is greater within the cell than in the solution surrounding it, water will diffuse into the cell, but the solutes are unable to diffuse out of the cell. The net movement of water across a selectively permeable barrier is referred to as *osmosis*.

We saw earlier that the tendency of all systems toward increasing entropy results in solutes spreading out through the entire available volume. In osmosis, the volume available to solute movement is restricted by the membrane, and thus entropy maximization is realized by the volume of solvent diffusing through the membrane to dilute the solutes. Indeed, in the absence of any countervailing force, *all* the available water will flow to the solute side of the membrane.

Let's imagine what happens when we place a living cell in a beaker of pure water. The presence of a selectively permeable membrane means that the net movement of water will continue until one of two things happens: (1) the cell expands until the selectively permeable membrane ruptures, allowing solutes to diffuse freely, or (2) the expansion of the cell volume is mechanically constrained by the presence of a cell wall such that the driving force for water to enter the cell is balanced by a pressure exerted by the cell wall.

The first scenario describes what would happen to an animal cell, which lacks a cell wall. The second scenario is relevant to plant cells. The plant cell wall is very strong. The resistance of cells walls to deformation creates an inward force that raises the hydrostatic pressure within the cell. The word osmosis derives from the Greek word for *pushing*; it is an expression of the positive pressure generated when solutes are confined.

We will soon see how osmosis drives the movement of water into and out of plant cells. First, however, let's discuss the concept of a composite or total driving force, representing the free-energy gradient of water.

## Water Potential

All living things, including plants, require a continuous input of free energy to maintain and repair their highly organized structures, as well as to grow and reproduce.

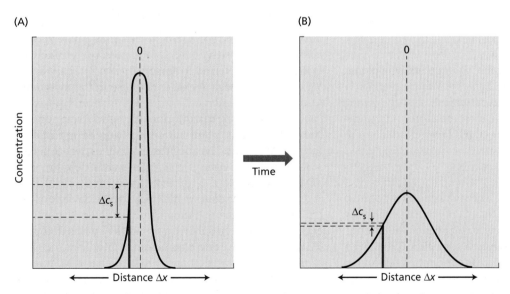

**FIGURE 3.8** Graphical representation of the concentration gradient of a solute that is diffusing according to Fick's law. The solute molecules were initially located in the plane indicated on the x-axis ("0"). (A) The distribution of solute molecules shortly after placement at the plane of origin. Note how sharply the concentration drops off as the distance, x, from the origin increases. (B) The solute distribution at a later time point. The average distance of the diffusing molecules from the origin has increased, and the slope of the gradient has flattened out. (After Nobel 1999.)

Processes such as biochemical reactions, solute accumulation, and long-distance transport are all driven by an input of free energy into the plant. (For a detailed discussion of the thermodynamic concept of free energy, see Appendix 1.) In this section we examine how concentration, pressure, and gravity influence free energy.

## The chemical potential of water represents the free-energy status of water

**Chemical potential** is a quantitative expression of the free energy associated with a substance. In thermodynamics, free energy represents the potential for performing work, force × distance. The unit of chemical potential is energy per mole of substance ($J\ mol^{-1}$). Note that chemical potential is a relative quantity: It represents the difference between the potential of a substance in a given state and the potential of the same substance in a standard state.

The chemical potential of water represents the free energy associated with water. Water flows spontaneously, i.e., without an input of energy, from regions of higher chemical potential to ones of lower chemical potential.

Historically, plant physiologists have most often used a related parameter called **water potential**, defined as the chemical potential of water divided by the partial molal volume of water (the volume of 1 mol of water): $18 \times 10^{-6}$ $m^3\ mol^{-1}$. Water potential is thus a measure of the free energy of water per unit volume ($J\ m^{-3}$). These units are equivalent to pressure units such as the pascal, which is the common measurement unit for water potential. Let's look more closely at the important concept of water potential.

## Three major factors contribute to cell water potential

The major factors influencing the water potential in plants are *concentration, pressure,* and *gravity.* Water potential is symbolized by $\Psi_w$ (the Greek letter psi), and the water potential of solutions may be dissected into individual components, usually written as the following sum:

$$\Psi_w = \Psi_s + \Psi_p + \Psi_g \qquad (3.2)$$

The terms $\Psi_s$ and $\Psi_p$ and $\Psi_g$ denote the effects of solutes, pressure, and gravity, respectively, on the free energy of water. (Alternative conventions for expressing the components of water potential are discussed in WEB TOPIC 3.3.) Energy levels must be defined in relation to a reference, analogous to how the contour lines on a map specify the distance above sea level. The reference state most often used to define water potential is pure water at ambient temperature and standard atmospheric pressure. The reference height is generally set either at the base of the plant (for whole plant studies) or at the level of the tissue under examination (for studies of water movement at the cellular

level). Let's consider each of the terms on the right-hand side of Equation 3.2.

**SOLUTES** The term $\Psi_s$, called the **solute potential** or the **osmotic potential**, represents the effect of dissolved solutes on water potential. Solutes reduce the free energy of water by diluting the water. This is primarily an entropy effect; that is, the mixing of solutes and water increases the disorder or entropy of the system and thereby lowers the free energy. This means that *the osmotic potential is independent of the specific nature of the solute.* For dilute solutions of nondissociating substances such as sucrose, the osmotic potential may be approximated by:

$$\Psi_s = -RTc_s \qquad (3.3)$$

where $R$ is the gas constant ($8.32\ J\ mol^{-1}\ K^{-1}$), $T$ is the absolute temperature (in degrees Kelvin, or K), and $c_s$ is the solute concentration of the solution, expressed as **osmolality** (moles of total dissolved solutes per liter of water [$mol\ L^{-1}$]). The minus sign indicates that dissolved solutes reduce the water potential of a solution relative to the reference state of pure water.

Equation 3.3 is valid for "ideal" solutions. Real solutions frequently deviate from the ideal, especially at high concentrations—for example, greater than $0.1\ mol\ L^{-1}$. Temperature also affects water potential (see WEB TOPIC 3.4). In our treatment of water potential, we will assume that we are dealing with ideal solutions (Friedman 1986; Nobel 1999).

**PRESSURE** The term $\Psi_p$ is the **hydrostatic pressure** of the solution. Positive pressures raise the water potential; negative pressures reduce it. Sometimes $\Psi_p$ is called *pressure potential.* The positive hydrostatic pressure within cells is the pressure referred to as *turgor pressure.* The value of $\Psi_p$ can also be negative, as is the case in the xylem and in the walls between cells, where a **tension**, or *negative hydrostatic pressure*, can develop. As we will see, negative pressures are important in moving water long distances through the plant. The question of whether negative pressures can occur within living cells is considered in WEB TOPIC 3.5.

Hydrostatic pressure is measured as the deviation from atmospheric pressure. Remember that water in the reference state is at atmospheric pressure, so by this definition $\Psi_p = 0$ MPa for water in the standard state. Thus the value of $\Psi_p$ for pure water in an open beaker is 0 MPa, even though its absolute pressure is approximately 0.1 MPa (1 atmosphere).

**GRAVITY** Gravity causes water to move downward unless the force of gravity is opposed by an equal and opposite force. The term $\Psi_g$ depends on the height ($h$) of the water above the reference-state water, the density of water ($\rho_w$), and the acceleration due to gravity ($g$). In symbols, we write the following:

$$\Psi_g = \rho_w g h \qquad (3.4)$$

where $\rho_w g$ has a value of 0.01 MPa m$^{-1}$. Thus a vertical distance of 10 m translates into a 0.1 MPa change in water potential.

The gravitational component ($\Psi_g$) is generally omitted in considerations of water transport at the cell level, because differences in this component among neighboring cells are negligible compared to differences in the osmotic potential and the hydrostatic pressure. Thus, in these cases Equation 3.2 can be simplified as follows:

$$\Psi_w = \Psi_s + \Psi_p \qquad (3.5)$$

### Water potentials can be measured

Cell growth, photosynthesis, and crop productivity are all strongly influenced by water potential and its components. Plant scientists have thus expended considerable effort in devising accurate and reliable methods for evaluating the water status of plants.

The principal approaches for determining $\Psi_w$ use psychrometers, of which there are two types, or the pressure chamber. Psychrometers take advantage of water's large latent heat of vaporization, which allows accurate measurements of (1) the vapor pressure of water in equilibrium with the sample, or (2) the transfer of water vapor between the sample and a solution of known $\Psi_s$. The pressure chamber measures $\Psi_w$ by applying external gas pressure to an excised leaf until water is forced out of the living cells.

In some cells, it is possible to measure $\Psi_p$ directly by inserting a liquid-filled microcapillary that is connected to a pressure sensor into the cell. In other cases, $\Psi_p$ is estimated as the difference between $\Psi_w$ and $\Psi_s$. Solute concentrations ($\Psi_s$) can be determined using a variety of methods, including psychrometers and instruments that measure freezing point depression. A detailed explanation of the instruments that have been used to measure $\Psi_w$, $\Psi_s$, and $\Psi_p$ can be found in WEB TOPIC 3.6.

In discussions of water in dry soils and plant tissues with very low water contents, such as seeds, one often finds reference to the **matric potential**, $\Psi_m$. Under these conditions, water exists as a very thin layer, perhaps one or two molecules deep, bound to solid surfaces by electrostatic interactions. These interactions are not easily separated into their effects on $\Psi_s$ and $\Psi_p$, and are thus sometimes combined into a single term, $\Psi_m$. The matric potential is discussed further in WEB TOPIC 3.7.

## Water Potential of Plant Cells

Plant cells typically have water potentials $\leq 0$ MPa. A negative value indicates that the free energy of water within the cell is less than that of pure water at ambient temperature, atmospheric pressure, and equal height. As the water potential of the solution surrounding the cell changes, water will enter or leave the cell via osmosis. In this section we will illustrate the osmotic behavior of water in plant cells with some numerical examples.

### Water enters the cell along a water potential gradient

First imagine an open beaker full of pure water at 20°C (**FIGURE 3.9A**). Because the water is open to the atmosphere, the hydrostatic pressure of the water is the same as atmospheric pressure ($\Psi_p = 0$ MPa). There are no solutes in the water, so $\Psi_s = 0$ MPa. Finally, because we focus here on transport processes that take place within the beaker, we define the reference height as equal to the level of the beaker, and thus $\Psi_g = 0$ MPa. Therefore, the water potential is 0 MPa ($\Psi_w = \Psi_s + \Psi_p$).

Now imagine dissolving sucrose in the water to a concentration of 0.1 $M$ (**FIGURE 3.9B**). This addition lowers the osmotic potential ($\Psi_s$) to –0.244 MPa and decreases the water potential ($\Psi_w$) to –0.244 MPa.

Next consider a flaccid plant cell (i.e., a cell with no turgor pressure) that has a total internal solute concentration of 0.3 $M$ (**FIGURE 3.9C**). This solute concentration gives an osmotic potential ($\Psi_s$) of –0.732 MPa. Because the cell is flaccid, the internal pressure is the same as atmospheric pressure, so the hydrostatic pressure ($\Psi_p$) is 0 MPa and the water potential of the cell is –0.732 MPa.

What happens if this cell is placed in the beaker containing 0.1 $M$ sucrose (see Figure 3.9C)? Because the water potential of the sucrose solution ($\Psi_w = -0.244$ MPa; see Figure 3.9B) is greater (less negative) than the water potential of the cell ($\Psi_w = -0.732$ MPa), water will move from the sucrose solution to the cell (from high to low water potential).

As water enters the cell, the cell wall is stretched by the contents of the enlarging protoplast. The wall resists deformation by pushing back on the cell. This increases the hydrostatic, or turgor, pressure ($\Psi_p$) of the cell. Consequently, the cell water potential ($\Psi_w$) increases, and the difference between inside and outside water potentials ($\Delta\Psi_w$) is reduced.

Eventually, cell $\Psi_p$ increases enough to raise the cell $\Psi_w$ to the same value as the $\Psi_w$ of the sucrose solution. At this point, equilibrium is reached ($\Delta\Psi_w = 0$ MPa), and net water transport ceases.

At equilibrium, water potential is equal everywhere: $\Psi_{w(cell)} = \Psi_{w(solution)}$. Because the volume of the beaker is much larger than that of the cell, the tiny amount of water taken up by the cell does not significantly affect the solute concentration of the sucrose solution. Hence $\Psi_s$, $\Psi_p$, and $\Psi_w$ of the sucrose solution are not altered. Therefore, at equilibrium, $\Psi_{w(cell)} = \Psi_{w(solution)} = -0.244$ MPa.

(A) **Pure water**

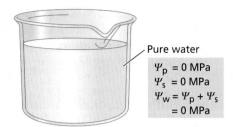

Pure water

$\Psi_p = 0$ MPa
$\Psi_s = 0$ MPa
$\Psi_w = \Psi_p + \Psi_s$
      $= 0$ MPa

(B) **Solution containing 0.1 *M* sucrose**

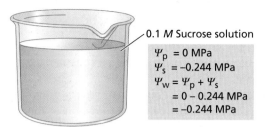

0.1 *M* Sucrose solution

$\Psi_p = 0$ MPa
$\Psi_s = -0.244$ MPa
$\Psi_w = \Psi_p + \Psi_s$
      $= 0 - 0.244$ MPa
      $= -0.244$ MPa

(C) **Flaccid cell dropped into sucrose solution**

Flaccid cell

$\Psi_p = 0$ MPa
$\Psi_s = -0.732$ MPa
$\Psi_w = -0.732$ MPa

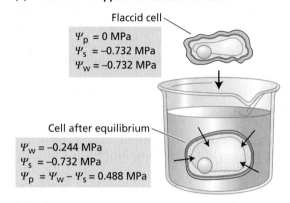

Cell after equilibrium

$\Psi_w = -0.244$ MPa
$\Psi_s = -0.732$ MPa
$\Psi_p = \Psi_w - \Psi_s = 0.488$ MPa

(D) **Concentration of sucrose increased**

Turgid cell

Cell wall
Plasma membrane
Vacuole
Cytosol
Nucleus

$\Psi_p = 0.488$ MPa
$\Psi_s = -0.732$ MPa
$\Psi_w = -0.244$ MPa

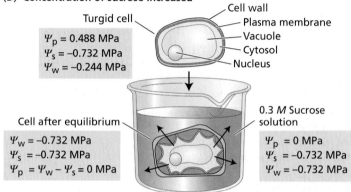

Cell after equilibrium

$\Psi_w = -0.732$ MPa
$\Psi_s = -0.732$ MPa
$\Psi_p = \Psi_w - \Psi_s = 0$ MPa

0.3 *M* Sucrose solution

$\Psi_p = 0$ MPa
$\Psi_s = -0.732$ MPa
$\Psi_w = -0.732$ MPa

(E) **Pressure applied to cell**

Applied pressure squeezes out half the water, thus doubling $\Psi_s$ from $-0.732$ to $-1.464$ MPa

0.1 *M* Sucrose solution

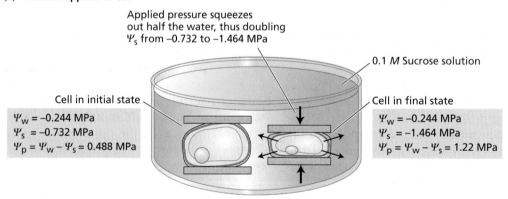

Cell in initial state

$\Psi_w = -0.244$ MPa
$\Psi_s = -0.732$ MPa
$\Psi_p = \Psi_w - \Psi_s = 0.488$ MPa

Cell in final state

$\Psi_w = -0.244$ MPa
$\Psi_s = -1.464$ MPa
$\Psi_p = \Psi_w - \Psi_s = 1.22$ MPa

**FIGURE 3.9** Five examples illustrating the concept of water potential and its components. (A) Pure water. (B) A solution containing 0.1 *M* sucrose. (C) A flaccid cell (in air) is dropped in a 0.1 *M* sucrose solution. Because the starting water potential of the cell is less than the water potential of the solution, the cell takes up water. After equilibration, the water potential of the cell rises to equal the water potential of the solution, and the result is a cell with a positive turgor pressure. (D) Increasing the concentration of sucrose in the solution makes the cell lose water. The increased sucrose concentration lowers the solution water potential, draws water out of the cell, and thereby reduces the cell's turgor pressure. In this case, the protoplast is able to pull away from the cell wall (i.e., the cell plasmolyzes), because sucrose molecules are able to pass through the relatively large pores of the cell walls. When this occurs, the difference in water potential between the cytoplasm and the solution is entirely across the plasma membrane, and thus the protoplast shrinks independently of the cell wall. In contrast, when a cell desiccates in air (e.g., the flaccid cell in C) plasmolysis does not occur, because the water held by capillary forces in the cell walls causes the plasma membrane to remain pressed against the cell wall even as the protoplast loses volume. Thus the cell (cytoplasm + wall) shrinks as a unit, resulting in the cell wall being mechanically deformed as the cell loses volume. (E) Another way to make the cell lose water is to press it slowly between two plates. In this case, half of the cell water is removed, so cell osmotic potential increases by a factor of 2.

The exact calculation of cell $\Psi_p$ and $\Psi_s$ requires knowledge of the change in cell volume. However, if we assume that the cell has a rigid cell wall, then the increase in cell volume will be small. Thus we can assume to a first approximation that $\Psi_{s(cell)}$ is unchanged during the equilibration process. This allows us to calculate the hydrostatic pressure of the cell by rearranging Equation 3.5 as follows: $\Psi_p = \Psi_w - \Psi_s = (-0.244) - (-0.732) = 0.488$ MPa.

### Water can also leave the cell in response to a water potential gradient

Water can also leave the cell by osmosis. If we now remove our plant cell from the 0.1 $M$ sucrose solution and place it in a 0.3 $M$ sucrose solution (**FIGURE 3.9D**), $\Psi_{w(solution)}$ (–0.732 MPa) is more negative than $\Psi_{w(cell)}$ (–0.244 MPa), and water will move from the turgid cell to the solution.

As water leaves the cell, the cell volume decreases. As the cell volume decreases, cell $\Psi_p$ and $\Psi_w$ decrease until $\Psi_{w(cell)} = \Psi_{w(solution)} = -0.732$ MPa. As before, we assume that the change in cell volume is small, so we can ignore the change in $\Psi_s$ due to the net outflow of water from the cell. This allows us to calculate that $\Psi_p = 0$ MPa using Equation 3.5.

If, instead of placing the turgid cell in the 0.3 M sucrose solution, we leave it in the 0.1 M solution and slowly squeeze it by pressing the cell between two plates (**FIGURE 3.9E**), we effectively raise the cell $\Psi_p$, consequently raising the cell $\Psi_w$ and creating a $\Delta\Psi_w$ such that water now flows *out* of the cell. This is analogous to the industrial process of reverse osmosis in which externally applied pressure is used to separate water from dissolved solutes by forcing it across a semipermeable barrier. If we continue squeezing until half the cell's water is removed and then hold the cell in this condition, the cell will reach a new equilibrium. As in the previous example, at equilibrium, $\Delta\Psi_w = 0$ MPa, and the amount of water added to the external solution is so small that it can be ignored. The cell will thus return to the $\Psi_w$ value that it had before the squeezing procedure. However, the components of the cell $\Psi_w$ will be quite different.

Because half of the water was squeezed out of the cell while the solutes remained inside the cell (the plasma membrane is selectively permeable), the cell solution is concentrated twofold, and thus $\Psi_s$ is lower (–0.732 MPa × 2 = –1.464 MPa). Knowing the final values for $\Psi_w$ and $\Psi_s$, we can calculate the turgor pressure, using Equation 3.5, as $\Psi_p = \Psi_w - \Psi_s = (-0.244$ MPa$) - (-1.464$ MPa$) = 1.22$ MPa.

In our example we used an external force to change cell volume without a change in water potential. In nature, it is typically the water potential of the cell's environment that changes, and the cell gains or loses water until its $\Psi_w$ matches that of its surroundings.

One point common to all these examples deserves emphasis: *Water flow across membranes is a passive process.* *That is, water moves in response to physical forces, toward regions of low water potential or low free energy.* There are no known metabolic "pumps" (e.g., reactions driven by ATP hydrolysis) that can be used to drive water across a semipermeable membrane against its free-energy gradient.

The only situation in which water can be said to move across a semipermeable membrane against its water potential gradient is when it is coupled to the movement of solutes. The transport of sugars, amino acids, or other small molecules by various membrane proteins can "drag" up to 260 water molecules across the membrane per molecule of solute transported (Loo et al. 1996).

Such transport of water can occur even when the movement is against the usual water potential gradient (i.e., toward a higher water potential), because the loss of free energy by the solute more than compensates for the gain of free energy by the water. The net change in free energy remains negative. The amount of water transported in this way will generally be quite small compared to the passive movement of water down its water potential gradient.

### Water potential and its components vary with growth conditions and location within the plant

In leaves of well-watered plants, $\Psi_w$ ranges from –0.2 to about –1.0 MPa in herbaceous plants and to –2.5 MPa in trees and shrubs. Leaves of plants in arid climates can have much lower $\Psi_w$, down to below –10 MPa under the most extreme conditions.

Just as $\Psi_w$ values depend on the growing conditions and the type of plant, so too, the values of $\Psi_s$ can vary considerably. Within cells of well-watered garden plants (examples include lettuce, cucumber seedlings, and bean leaves), $\Psi_s$ may be as high as –0.5 MPa (low cell solute concentration), although values of –0.8 to –1.2 MPa are more typical. In woody plants, $\Psi_s$ tends to be lower (higher cell solute concentration), allowing the more negative midday $\Psi_w$ typical of these plants to occur without a loss in turgor pressure.

Although $\Psi_s$ *within* cells may be quite negative, the apoplastic solution surrounding the cells—that is, in the cell walls and in the xylem—is generally quite dilute. The $\Psi_s$ of the apoplast is typically –0.1 to 0 MPa, although in certain tissues (e.g., developing fruits) and habitats (e.g., high salinity environments) the concentration of solutes in the apoplast can be large. In general, water potentials in the xylem and cell walls are dominated by $\Psi_p$, which is typically less than zero. In Chapter 4 we will examine what determines the $\Psi_p$ of the apoplast.

Values for $\Psi_p$ within cells of well-watered plants may range from 0.1 to as much as 3 MPa, depending on the value of $\Psi_s$ inside the cell. A plant **wilts** when the turgor pressure inside the cells of such tissues falls toward zero. As more water is lost from the cell, the walls become mechanically deformed, and the cell may be damaged as a result. WEB TOPIC 3.8 contrasts the situation in which a cell is dehydrated osmotically due to the presence of apoplastic solutes to that in which water is withdrawn from the cell due to lower (more negative) hydrostatic pressures in the apoplast.

## Cell Wall and Membrane Properties

Structural elements make important contributions to the water relations of plant cells. Cell wall elasticity defines the relation between turgor pressure and cell volume, while the permeability of the plasma and tonoplast membranes to water influence the rate at which cells exchange water with their surroundings. In this section, we examine how wall and membrane properties influence the water status of plant cells.

### Small changes in plant cell volume cause large changes in turgor pressure

Cell walls provide plant cells with a substantial degree of volume homeostasis relative to the large changes in water potential that they experience everyday as a consequence of the transpirational water losses associated with photosynthesis (see Chapter 4). Because plant cells have fairly rigid walls, a change in cell $\Psi_w$ is generally accompanied by a large change in $\Psi_p$, with relatively little change in cell (protoplast) volume, as long as $\Psi_p > 0$.

This phenomenon is illustrated by the *pressure-volume curve* shown in FIGURE 3.10. As $\Psi_w$ decreases from 0 to –1.2 MPa, the relative or percent water content is reduced by only slightly more than 5%. Most of this decrease is due to a reduction in $\Psi_p$ (by about 1.0 MPa); $\Psi_s$ decreases by less than 0.2 MPa as a result of increased concentration of cell solutes.

Measurements of cell water potential and cell volume can be used to quantify how wall properties influence the water status of plant cells. Turgor pressure in most cells approaches zero as the relative cell volume decreases by 10 to 15%. However, for cells with very rigid cell walls, the volume change associated with turgor loss can be much smaller. In cells with extremely elastic walls, such as the water-storing cells in the stems of many cacti, this volume change may be substantially larger.

The volumetric elastic modulus, symbolized by $\varepsilon$ (the Greek letter epsilon) can be determined by examining the relationship between $\Psi_p$ and cell volume: $\varepsilon$ is the change in $\Psi_p$ for a given change in relative volume ($\varepsilon = \Delta\Psi_p/\Delta[\text{relative volume}]$). Cells with a large $\varepsilon$ have stiff cell walls and thus experience larger changes in turgor pressure for the same change in cell volume than cells with a smaller $\varepsilon$ and more elastic walls. The mechanical properties of cell walls vary among species and cell types, resulting in significant differences in the extent to which water deficits affect cell volume.

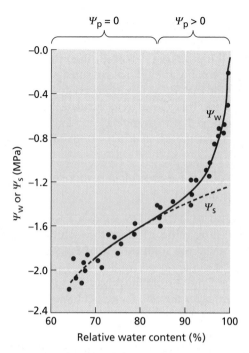

**FIGURE 3.10** The relation between water potential ($\Psi_w$), solute potential ($\Psi_s$), and relative water content ($\Delta V/V$) in cotton (*Gossypium hirsutum*) leaves. Note that water potential ($\Psi_w$) decreases steeply with the initial decrease in relative water content. In comparison, osmotic potential ($\Psi_s$) changes little. As cell volume decreases below 90% in this example, the situation reverses: Most of the change in water potential is due to a drop in cell $\Psi_s$, accompanied by relatively little change in turgor pressure. (After Hsiao and Xu 2000.)

A comparison of the cell water relations within stems of cacti illustrates the important role of cell wall properties. Cacti are stem succulent plants, typically found in arid regions. Their stems consist of an outer, photosynthetic layer that surrounds non-photosynthetic tissues that serve as a water storage reservoir (FIGURE 3.11). During drought, water is lost preferentially from these inner cells, despite the fact that the water potential of the two cell types remains in equilibrium (or very close to equilibrium) (Nobel 1988). How does this happen?

Detailed studies of *Opuntia ficus-indica* (Goldstein et al. 1991) demonstrate that the water storage cells are larger and have thinner walls than the photosynthetic cells, and are thus more flexible (have lower $\varepsilon$). For a given decrease in water potential, a water storage cell will lose a greater fraction of its water content than a photosynthetic cell. In addition, the solute concentration of the water storage cells decreases during drought, in part due to the polymerization of soluble sugars into insoluble starch granules. A more typical plant response to drought is to accumulate solutes, in part to prevent water loss from cells. However, in the case of cacti, the combination of more flexible cell walls and a decrease in solute concentration during drought allows water to be withdrawn preferentially from

FIGURE 3.11 Cross section of a cactus stem, showing an outer, photosynthetic layer, and an inner, nonphotosynthetic tissue that functions in water storage. During drought, water is lost preferentially from nonphotosynthetic cells, so the water status of the photosynthetic tissue is maintained. (Photograph by David McIntyre.)

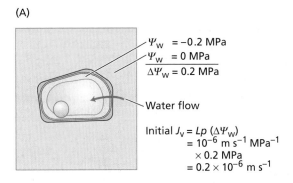

(A)

$$\Psi_w = -0.2 \text{ MPa}$$
$$\Psi_w = 0 \text{ MPa}$$
$$\overline{\Delta \Psi_w = 0.2 \text{ MPa}}$$

Water flow

Initial $J_v = Lp\,(\Delta \Psi_w)$
$= 10^{-6} \text{ m s}^{-1} \text{ MPa}^{-1}$
$\times\, 0.2 \text{ MPa}$
$= 0.2 \times 10^{-6} \text{ m s}^{-1}$

(B)

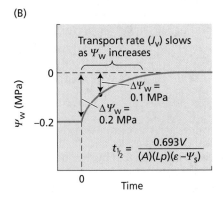

Transport rate ($J_v$) slows as $\Psi_w$ increases

$\Delta \Psi_w = 0.1 \text{ MPa}$

$\Delta \Psi_w = 0.2 \text{ MPa}$

$$t_{1/2} = \frac{0.693V}{(A)(Lp)(\varepsilon - \Psi_s)}$$

FIGURE 3.12 The rate of water transport into a cell depends on the water potential difference ($\Delta \Psi_w$) and the hydraulic conductivity of the cell membranes ($Lp$). (A) In this example, the initial water potential difference is 0.2 MPa and $Lp$ is $10^{-6}$ m s$^{-1}$ MPa$^{-1}$. These values give an initial transport rate ($J_v$) of $0.2 \times 10^{-6}$ m s$^{-1}$. (B) As water is taken up by the cell, the water potential difference decreases with time, leading to a slowing in the rate of water uptake. This effect follows an exponentially decaying time course with a half-time ($t_{1/2}$) that depends on the following cell parameters: volume ($V$), surface area ($A$), conductivity ($Lp$), volumetric elastic modulus ($\varepsilon$), and cell osmotic potential ($\Psi_s$).

the water storage cells, thus helping to maintain the hydration of the photosynthetic tissues.

### The rate at which cells gain or lose water is influenced by cell membrane hydraulic conductivity

So far, we have seen that water moves into and out of cells in response to a water potential gradient. The direction of flow is determined by the direction of the $\Psi_w$ gradient, and the rate of water movement is proportional to the magnitude of the driving gradient. However, for a cell that experiences a change in the water potential of its surroundings (e.g., see Figure 3.9), the movement of water across the cell membrane will decrease with time as the internal and external water potentials converge (**FIGURE 3.12**). The rate approaches zero in an exponential manner (Dainty 1976). The time it takes for the rate to decline by half—its half-time, or $t_{1/2}$—is given by the following equation:

$$t_{1/2} = \left(\frac{0.693}{(A)(Lp)}\right)\left(\frac{V}{\varepsilon - \Psi_s}\right) \tag{3.6}$$

where $V$ and $A$ are, respectively, the volume and surface of the cell, and $L_p$ is the **hydraulic conductivity** of the cell membrane. Hydraulic conductivity describes how readily water can move across a membrane; it is expressed in terms of volume of water per unit area of membrane per unit time per unit driving force (i.e., m$^3$ m$^{-2}$ s$^{-1}$ MPa$^{-1}$). For additional discussion on hydraulic conductivity, see WEB TOPIC 3.9.

A short half-time means fast equilibration. Thus, cells with large surface-to-volume ratios, high membrane hydraulic conductivity, and stiff cell walls (large $\varepsilon$) will come rapidly into equilibrium with their surroundings. Cell half-times typically range from 1 to 10 seconds, although some are much shorter (Steudle 1989). Because of their short half-times, single cells come to water potential equilibrium with their surroundings in less than 1 minute. For multicellular tissues, the half-times may be much longer.

### Aquaporins facilitate the movement of water across cell membranes

For many years, plant physiologists were uncertain about how water moves across plant membranes. Specifically,

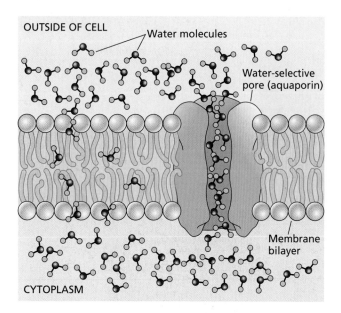

OUTSIDE OF CELL
Water molecules
Water-selective pore (aquaporin)
Membrane bilayer
CYTOPLASM

**FIGURE 3.13** Water can cross plant membranes by diffusion of individual water molecules through the membrane bilayer, as shown on the left, and by the linear diffusion of water molecules through water-selective pores formed by integral membrane proteins such as aquaporins.

it was unclear whether water movement into plant cells was limited to the diffusion of water molecules across the plasma membrane's lipid bilayer or if it also involved diffusion through protein-lined pores (**FIGURE 3.13**). Some studies suggested that diffusion directly across the lipid bilayer was not sufficient to account for observed rates of water movement across membranes, but the evidence in support of microscopic pores was not compelling.

This uncertainty was put to rest in 1991 with the discovery of **aquaporins** (see Figure 3.13). Aquaporins are integral membrane proteins that form water-selective channels across the membrane (Maurel et al. 2008). Because water diffuses much faster through such channels than through a lipid bilayer, aquaporins facilitate water movement into plant cells (Weig et al. 1997; Schäffner 1998; Tyerman et al. 1999).

Although aquaporins may alter the *rate* of water movement across the membrane, they do not change the *direction of transport* or the *driving force* for water movement. However, aquaporins can be reversibly "gated" (i.e., transferred between an open and a closed state) in response to physiological parameters such as intercellular pH and $Ca^{2+}$ levels (Tyerman et al. 2002). The discovery of "gating" opens the door to the possibility that plants actively regulate the permeability of their cell membranes to water.

# Plant Water Status

The concept of water potential has two principal uses: First, water potential governs transport across cell membranes, as we have described. Second, water potential is often used as a measure of the *water status* of a plant. In this section we discuss how the concept of water potential helps us evaluate the water status of a plant.

## Physiological processes are affected by plant water status

Because of transpirational water loss to the atmosphere, plants are seldom fully hydrated. During periods of drought, they suffer from water deficits that lead to inhibition of plant growth and photosynthesis. **FIGURE 3.14** lists some of the physiological changes that occur as plants experience increasingly drier conditions.

The sensitivity of any particular physiological process to water deficits is, to a large extent, a reflection of that plant's strategy for dealing with the range of water availability that it experiences in its environment. According to Figure 3.14, the process that is most affected by water deficit is cell expansion. In many plants reductions in water supply inhibit shoot growth and leaf expansion but *stimulate* root elongation. A relative increase in roots relative to leaves is an appropriate response to reductions in water availability, and thus the sensitivity of shoot growth to decreases in water availability can be seen as an adaptation to drought rather than a physiological constraint.

However, what plants cannot do is to alter the availability of water in the soil. (Figure 3.14 shows representative values for $\Psi_w$ at various stages of water stress.) Thus, drought does impose some absolute limitations on physiological processes, although the actual water potentials at which such limitations occur vary with species.

## Solute accumulation helps cells maintain turgor and volume

The ability to maintain physiological activity as water becomes less available typically incurs some costs. The plant may spend energy to accumulate solutes to maintain turgor pressure, invest in the growth of nonphotosynthetic organs such as roots to increase water uptake capacity, or build xylem conduits capable of withstanding large negative pressures. Thus, physiological responses to water availability reflect a tradeoff between the benefits accrued by being able to carry out physiological processes (e.g., growth) over a wider range of environmental conditions and the costs associated with such capability.

Plants that grow in saline environments, called **halophytes**, typically have very low values of $\Psi_s$. A low $\Psi_s$ low-

**Physiological changes due to dehydration:**

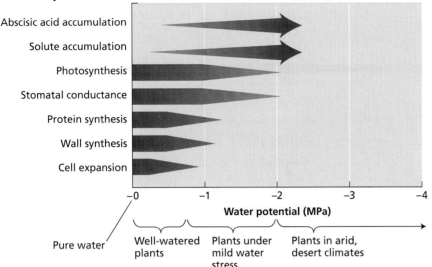

FIGURE 3.14  Sensitivity of various physiological processes to changes in water potential under various growing conditions. The intensity of the bar color corresponds to the magnitude of the process. For example, cell expansion decreases as water potential falls (becomes more negative). Abscisic acid is a hormone that induces stomatal closure during water stress (see Chapter 23). (After Hsiao and Acevedo 1974.)

ers cell $\Psi_w$ enough to allow root cells to extract water from saline water without allowing excessive levels of salts to enter at the same time. Plants may also exhibit quite negative $\Psi_s$ under drought conditions. Water stress typically leads to an accumulation of solutes in the cytoplasm and vacuole of plant cells, thus allowing the cells to maintain turgor pressure despite low water potentials.

A positive turgor pressure ($\Psi_p$) is important for several reasons. First, growth of plant cells requires turgor pressure to stretch the cell walls. The loss of $\Psi_p$ under water deficits can explain in part why cell growth is so sensitive to water stress, as well as why this sensitivity can be modified by varying the cell's osmotic potential (see Chapter 26). The second reason positive turgor is important is that turgor pressure increases the mechanical rigidity of cells and tissues.

Finally, some physiological processes appear to be influenced directly by turgor pressure. However, the existence of stretch-activated signaling molecules in the plasma membrane suggests that plant cells may sense changes in their water status via changes in volume, rather than by responding directly to turgor pressure.

## SUMMARY

Terrestrial life exposes plants to water loss and the threat of dehydration. To prevent desiccation, water must be absorbed by the roots and transported through the plant body.

### Water in Plant Life

- Cell walls allow plant cells to build up large internal hydrostatic pressures (turgor pressure). Turgor pressure is essential for many plant processes.
- Water limits both agricultural and natural ecosystem productivity (**Figures 3.1, 3.2**).
- About 97 percent of the water absorbed by roots is carried through the plant and is lost by transpiration from the leaf surfaces.
- The uptake of $CO_2$ is coupled to the loss of water through a common diffusional pathway.

### The Structure and Properties of Water

- The polarity and tetrahedral shape of water molecules permit them to form hydrogen bonds that give water its unusual physical properties: it is an excellent solvent and has a high specific heat, an unusually high latent heat of vaporization, and a high tensile strength (**Figures 3.3, 3.6**).
- Cohesion, adhesion, and surface tension give rise to capillarity (**Figures 3.4, 3.5**).

### Diffusion and Osmosis

- The random thermal motion of the molecules or ions results in diffusion (**Figures 3.7, 3.8**).
- Diffusion is important as small length scales. The average time for a substance to diffuse a given distance increases as the square of that distance.

## SUMMARY continued

- Osmosis is the net movement of water across a selectively permeable barrier and is driven by the sum of water's concentration gradient and the pressure gradient.

### Water Potential

- Water's chemical potential measures the free energy of water in a given state.
- Concentration, pressure, and gravity contribute to water potential ($\Psi_w$) in plants.
- $\Psi_s$, the solute potential (osmotic potential), represents solutes' dilution of water and the reduction of the free energy of water.
- $\Psi_p$, is the hydrostatic pressure of the solution. Positive pressure (turgor pressure) raises the water potential; negative pressure reduces it.
- The gravitational potential ($\Psi_g$) is generally omitted when calculating cell water potential. Thus, $\Psi_w = \Psi_s + \Psi_p$.

### Water Potential of Plant Cells

- Plant cells typically have negative water potentials.
- Water enters or leaves a cell according to the water potential gradient.
- When a flaccid cell is placed in a solution that has a water potential greater (less negative) than the cell's water potential, water will move from the solution into the cell (from high to low water potential) (**Figure 3.9**).
- As water enters, the cell wall resists being stretched, increasing the turgor pressure ($\Psi_p$) of the cell.
- At equilibrium ($\Psi_{w(cell)} = \Psi_{w(solution)}$; $\Delta\Psi_w = 0$), the cell $\Psi_p$ has increased sufficiently to raise the cell $\Psi_w$ to the same value as the $\Psi_w$ of the solution, and net water movement ceases.

- Water can also leave the cell by osmosis. When a turgid plant cell is placed in a sucrose solution that has a water potential more negative than the water potential of the cell, water will move from the turgid cell to the solution (**Figure 3.9**).
- If a cell is squeezed, its $\Psi_p$ is raised as is cell $\Psi_w$ resulting in a $\Delta\Psi_w$ such that water flows out of the cell (**Figure 3.9**).

### Cell Wall and Membrane Properties

- Cell wall elasticity relates turgor pressure and cell volume, while the water permeability of the plasma and tonoplast membranes determine how fast cells exchange water with their surroundings.
- Because plant cells have fairly rigid walls, small changes of plant cell volume cause large changes in turgor pressure (**Figure 3.10**).
- For $\Delta\Psi_w$, the movement of water across the membrane will decrease with time as the internal and external water potentials converge (**Figure 3.12**).
- Aquaporins are water-selective membrane channels (**Figure 3.13**).

### Plant Water Status

- During drought, photosynthesis and growth are inhibited, while ABA and solute accumulation increase (**Figure 3.14**).
- During drought, plants must use energy to maintain turgor pressure by accumulating solutes, as well as to support root and vascular growth.
- Stretch-activated signaling molecules in the plasma membrane may permit plant cells to sense changes in their water status via changes in volume.

## WEB MATERIAL
### Web Topics

**3.1  Calculating Capillary Rise**

Quantification of capillary rise allows us to assess its functional role in water movement of plants.

**3.2  Calculating Half-Times of Diffusion**

The assessment of the time needed for a molecule such as glucose to diffuse across cells, tissues, and organs shows that diffusion has physiological significance only over short distances.

**3.3  Alternative Conventions for Components of Water Potential**

Plant physiologists have developed several conventions to define water potential of plants. A comparison of key definitions in some of these convention systems provides us with a better understanding of the water relations literature.

**3.4  Temperature and Water Potential**

Variation in temperature between 0 and 30°C has a relatively minor effect on osmotic potential.

**3.5  Can Negative Turgor Pressures Exist in Living Cells?**

It is assumed that $\Psi_p$ is zero or greater in living cells; is this true for living cells with lignified walls?

**3.6  Measuring Water Potential**

Several methods are available to measure water potential in plant cells and tissues.

**3.7  The Matric Potential**

Matric potential is used to quantify the chemical potential of water in soils, seeds, and cell walls.

**3.8  Wilting and Plasmolysis**

Plasmolysis is a major structural change resulting from major water loss by osmosis.

**3.9  Understanding Hydraulic Conductivity**

Hydraulic conductivity, a measurement of the membrane permeability to water, is one of the factors determining the velocity of water movements in plants.

## CHAPTER REFERENCES

Dainty, J. (1976) Water relations of plant cells. In *Transport in Plants*, Vol. 2, Part A: *Cells* (Encyclopedia of Plant Physiology, New Series, Vol. 2.), U. Lüttge and M. G. Pitman, eds., Springer, Berlin, pp. 12–35.

Day, W., Legg, B. J., French, B. K., Johnston, A. E., Lawlor, D. W., and Jeffers, W. de C. (1978) A drought experiment using mobile shelters: the effect of drought on barley yield, water use and nutrient uptake. *J. Agricult. Sci., Cambridge* 91: 599–623.

Friedman, M. H. (1986) *Principles and Models of Biological Transport*. Springer Verlag, Berlin.

Goldstein, G., Andrade, J. L., and Nobel, P. S. (1991) Differences in water relations parameters for the chlorenchyma and parenchyma of *Opuntia ficus-indica* under wet versus dry conditions. *Aust. J. Plant Physiol.* 18: 95–107.

Hsiao, T. C., and Acevedo, E. (1974) Plant responses to water deficits, efficiency, and drought resistance. *Agricult. Meteorol.* 14: 59–84.

Hsiao, T. C., and Xu, L. K. (2000) Sensitivity of growth of roots versus leaves to water stress: Biophysical analysis and relation to water transport. *J. Exp. Bot.* 51: 1595–1616.

Innes, P., and Blackwell, R. D. (1981) The effect of drought on the water use and yield of two spring wheat genotypes. *J. Agricult. Sci., Cambridge* 102: 341–351.

Jones, H. G. (1992) *Plants and Microclimate*, 2nd ed., Cambridge University Press, Cambridge.

Loo, D. D. F., Zeuthen, T., Chandy, G., and Wright, E. M. (1996) Cotransport of water by the $Na^+$/glucose cotransporter. *Proc. Natl. Acad. Sci. USA* 93: 13367–13370.

Maurel, C., Verdoucq, L., Luu, D.-T., and Santoni, V. (2008) Plant aquaporins: Membrane channels with multiple integrated functions. *Annu. Rev. Plant Biol.* 59: 595–624.

Nobel, P. S. (1999) *Physicochemical and Environmental Plant Physiology*, 2nd ed. Academic Press, San Diego, CA.

Nobel, P. S. (1988) *Environmental Biology of Agaves and Cacti*. Cambridge University Press, New York.

Schäffner, A. R. (1998) Aquaporin function, structure, and expression: Are there more surprises to surface in water relations? *Planta* 204: 131–139.

Stein, W. D. (1986) *Transport and Diffusion across Cell Membranes*. Academic Press, Orlando, FL.

Steudle, E. (1989) Water flow in plants and its coupling to other processes: An overview. *Meth. Enzymol.* 174: 183–225.

Tyerman, S. D., Bohnert, H. J., Maurel, C., Steudle, E., and Smith, J. A. C. (1999) Plant aquaporins: Their molecular biology, biophysics and significance for plant–water relations. *J. Exp. Bot.* 50: 1055–1071.

Tyerman, S. D., Niemietz, C. M., and Bramley, H. (2002) Plant aquaporins: Multifunctional water and solute channels with expanding roles. *Plant Cell Environ.* 25: 173–194.

Weig, A., Deswarte, C., and Chrispeels, M. J. (1997) The major intrinsic protein family of Arabidopsis has 23 members that form three distinct groups with functional aquaporins in each group. *Plant Physiol.* 114: 1347–1357.

Wheeler, T. D., Strook, A. D. (2008) The transpiration of water at negative pressures in a synthetic tree. *Nature* 455: 208–212.

Whittaker, R. H. (1970) *Communities and Ecosystems.* Macmillan, New York.

# Water Balance of Plants

Life in Earth's atmosphere presents a formidable challenge to land plants. On the one hand, the atmosphere is the source of carbon dioxide, which is needed for photosynthesis. On the other hand, the atmosphere is usually quite dry, leading to a net loss of water due to evaporation. Because plants lack surfaces that can allow the inward diffusion of $CO_2$ while preventing water loss, $CO_2$ uptake exposes plants to the risk of dehydration. This problem is compounded because the concentration gradient for $CO_2$ uptake is much smaller than the concentration gradient that drives water loss. To meet the contradictory demands of maximizing carbon dioxide uptake while limiting water loss, plants have evolved adaptations to control water loss from leaves, and to replace the water lost to the atmosphere.

In this chapter we will examine the mechanisms and driving forces operating on water transport within the plant and between the plant and its environment. We will begin our examination of water transport by focusing on water in the soil.

## Water in the Soil

The water content and the rate of water movement in soils depend to a large extent on soil type and soil structure. At one extreme is sand, in which the soil particles may be 1 mm or more in diameter. Sandy soils have a relatively low surface area per gram of soil and have large spaces or channels between particles.

At the other extreme is clay, in which particles are smaller than 2 μm in diameter. Clay soils have much greater surface areas and smaller channels between particles. With the aid of organic substances such as humus (decomposing organic matter), clay particles may aggregate into "crumbs," allowing large channels to form that help improve soil aeration and infiltration of water.

When a soil is heavily watered by rain or by irrigation (see **WEB TOPIC 4.1**), the water percolates downward by gravity through the spaces between soil particles, partly displacing, and in some cases trapping, air in these channels. Because water is pulled into the spaces between soil particles by capillarity, the smaller channels become filled first. Depending on the amount of water available, water in the soil may exist as a film adhering to the surface of soil particles, it may fill the smaller but not the larger channels, or it may fill all of the spaces between particles.

In sandy soils, the spaces between particles are so large that water tends to drain from them and remain only on the particle surfaces and in the interstices between particles. In clay soils, the channels are so small that much water is retained against the force of gravity. A few days after a soaking rainfall, a clay soil might retain 40% water by volume. In contrast, sandy soils typically retain only ~15% water by volume after thorough wetting.

In the following sections we will examine how the physical structure of the soil influences soil water potential, how water moves in the soil, and how roots absorb the water needed by the plant.

### A negative hydrostatic pressure in soil water lowers soil water potential

Like the water potential of plant cells, the water potential of soils may be dissected into three components: the osmotic potential, the hydrostatic pressure, and the gravitational potential. The **osmotic potential** ($\Psi_s$; see Chapter 3) of soil water is generally negligible, because, except in saline soils, solute concentrations are low; a typical value might be –0.02 MPa. In soils that contain a substantial concentration of salts, however, $\Psi_s$ can be significant, perhaps –0.2 MPa or lower.

The second component of soil water potential is **hydrostatic pressure** ($\Psi_p$) (**FIGURE 4.1**). For wet soils, $\Psi_p$ is very close to zero. As soil dries out, $\Psi_p$ decreases and can become quite negative. Where does the negative pressure in soil water come from?

Recall from our discussion of capillarity in Chapter 3 that water has a high surface tension that tends to minimize air–water interfaces. However, because of adhesive forces, water also tends to cling to the surfaces of soil particles (**FIGURE 4.2**).

As the water content of the soil decreases, the water recedes into the interstices between soil particles, forming air–water surfaces whose curvature represents the balance

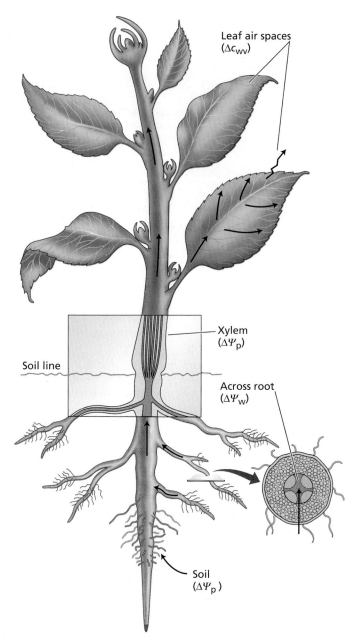

**FIGURE 4.1** Main driving forces for water flow from the soil through the plant to the atmosphere: differences in water vapor concentration ($\Delta c_{wv}$), hydrostatic pressure ($\Delta \Psi_p$), and water potential ($\Delta \Psi_w$).

between the tendency to minimize the surface area of the air–water interface and the attraction of the water for the soil particles. Water under a curved surface develops a negative pressure that may be estimated by the following formula:

$$\Psi_p = \frac{-2T}{r} \tag{4.1}$$

where $T$ is the surface tension of water ($7.28 \times 10^{-8}$ MPa m) and $r$ is the radius of curvature of the air–water interface. Note that this is the same capillarity equation discussed in

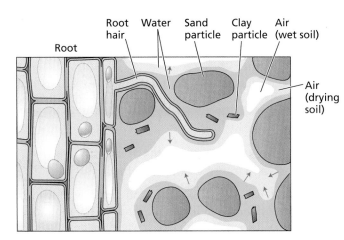

**FIGURE 4.2** Root hairs make intimate contact with soil particles and greatly amplify the surface area used for water absorption by the plant. The soil is a mixture of particles (sand, clay, silt, and organic material), water, dissolved solutes, and air. Water is adsorbed to the surface of the soil particles. As water is absorbed by the plant, the soil solution recedes into smaller pockets, channels, and crevices between the soil particles. At the air–water interfaces, this recession causes the surface of the soil solution to develop concave menisci (curved interfaces between air and water, marked in the figure by arrows), and brings the solution into tension (negative pressure) by surface tension. As more water is removed from the soil, the curvature of the air-water menisci increases, resulting in greater tensions (more negative pressures).

WEB TOPIC 3.1 (see also Figure 3.5), where here the soil particles are assumed to be fully wettable (contact angle $\theta = 0$; $\cos \theta = 1$).

As soil dries out, water is first removed from the largest spaces between soil particles and subsequently from successively smaller spaces between and within soil particles. In this process, the value of $\Psi_p$ in soil water can become quite negative due to the increasing curvature of air–water surfaces in pores of successively smaller diameter. For instance, a curvature of $r = 1$ µm (about the size of the largest clay particles) corresponds to a $\Psi_p$ value of –0.15 MPa. The value of $\Psi_p$ may easily reach –1 to –2 MPa as the air–water interface recedes into the smaller spaces between clay particles.

The third component is **gravitational potential** ($\Psi_g$). Gravity plays an important role in drainage. The downward movement of water is due to the fact that $\Psi_g$ is proportional to elevation: higher at higher elevations, and vice versa.

### Water moves through the soil by bulk flow

Bulk or mass flow is the concerted movement of molecules en masse, most often in response to a pressure gradient. Common examples of bulk flow are water moving through a garden hose or down a river. The movement of water through soils is predominantly by bulk flow.

Because the pressure in soil water is due to the existence of curved air–water interfaces, water flows from regions of higher soil water content, where the water-filled spaces are larger, to regions of lower soil water content, where the smaller size of the water-filled spaces is associated with more curved air–water interfaces. Diffusion of water vapor also accounts for some water movement, which can be important in dry soils.

As plants absorb water from the soil, they deplete the soil of water near the surface of the roots. This depletion reduces $\Psi_p$ in the water near the root surface and establishes a pressure gradient with respect to neighboring regions of soil that have higher $\Psi_p$ values. Because the water-filled pore spaces in the soil are interconnected, water moves down the pressure gradient to the root surface by bulk flow through these channels.

The rate of water flow in soils depends on two factors: the size of the pressure gradient through the soil, and the hydraulic conductivity of the soil. **Soil hydraulic conductivity** is a measure of the ease with which water moves through the soil, and it varies with the type of soil and its water content. Sandy soils, which have large spaces between particles, have a large hydraulic conductivity when saturated, whereas clay soils, with only minute spaces between their particles, have an appreciably smaller hydraulic conductivity.

As the water content (and hence the water potential) of a soil decreases, the hydraulic conductivity decreases dramatically. This decrease in soil hydraulic conductivity is due primarily to the replacement of water in the soil spaces by air. When air moves into a soil channel previously filled with water, water movement through that channel is restricted to the periphery of the channel. As more of the soil spaces become filled with air, water flow is limited to fewer and narrower channels, and the hydraulic conductivity falls. (WEB TOPIC 4.2 shows how soil texture influences both the water-holding capacity of soils and their hydraulic conductivity.)

## Water Absorption by Roots

Intimate contact between the surface of the root and the soil is essential for effective water absorption by the root. This contact provides the surface area needed for water uptake and is maximized by the growth of the root and of root hairs into the soil. **Root hairs** are filamentous outgrowths of root epidermal cells that greatly increase the surface area of the root, thus providing greater capacity for absorption of ions and water from the soil. When 3-month-old wheat plants were examined, their root hairs were found to constitute more than 60% of the surface area of the roots (see Figure 5.6).

Water enters the root most readily near the root tip. Mature regions of the root are less permeable to water, because they have developed an outer layer of protective

(A)

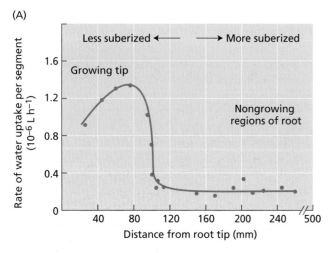

(B)

(C)

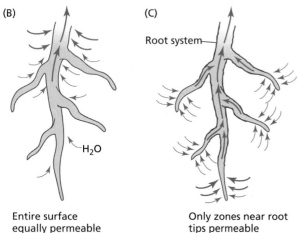

Entire surface
equally permeable

Only zones near root
tips permeable

**FIGURE 4.3** Rate of water uptake at various positions along a pumpkin (*Cucurbita pepo*) root (A). Diagram of water uptake in which the entire root surface is equally permeable (B) or is impermeable in older regions due to the deposition of suberin (C). When root surfaces are equally permeable, most of the water enters near the top of the root system, with more distal regions being hydraulically isolated as the suction in the xylem is relieved due to the inflow of water. Decreasing the permeability of older regions of the root allows xylem tensions to extend further into the root system, allowing water uptake from distal regions of the root system. (A after Kramer and Boyer 1995.)

tissue, called an *exodermis* or *hypodermis*, which contains hydrophobic materials in its walls. Although it might at first seem counterintuitive that any portion of the root system should be impermeable to water, the older regions of the root must be sealed off if there is to be water uptake (and thus bulk flow of nutrients) from the regions of the root system that are actively exploring new areas in the soil (**FIGURE 4.3**) (Zwieniecki et al. 2002).

The intimate contact between the soil and the root surface is easily ruptured when the soil is disturbed. It is for this reason that newly transplanted seedlings and plants

need to be protected from water loss for the first few days after transplantation. Thereafter, new root growth into the soil reestablishes soil–root contact, and the plant can better withstand water stress.

Let's consider how water moves within the root, and the factors that determine the rate of water uptake into the root.

### Water moves in the root via the apoplast, symplast, and transmembrane pathways

In the soil, water flows between soil particles. However, from the epidermis to the endodermis of the root, there are three pathways through which water can flow (**FIGURE 4.4**): the apoplast, the symplast, and the transmembrane pathway.

1. The apoplast is the continuous system of cell walls, intercellular air spaces, and the lumens of nonliving cells (e.g., xylem conduits and fibers). In this pathway, water moves through cell walls and extracellular spaces without crossing any membranes as it travels across the root cortex.

2. The symplast consists of the entire network of cell cytoplasm interconnected by plasmodesmata. In this pathway, water travels across the root cortex via the plasmodesmata (see Chapter 1).

3. The transmembrane pathway is the route by which water enters a cell on one side, exits the cell on the other side, enters the next in the series, and so on. In this pathway, water crosses the plasma membrane of each cell in its path twice (once on entering and once on exiting). Transport across the tonoplast may also be involved.

Although the relative importance of the apoplast, symplast, and transmembrane pathways has not yet been fully established, experiments with the pressure probe technique (see **WEB TOPIC 3.6**) indicate an important role for cell membranes in the movement of water across the root cortex (Frensch et al. 1996; Steudle and Frensch 1996; Bramley et al. 2009). And though there are three pathways, water moves not according to a single chosen path, but wherever the gradients and resistances direct it. A particular water molecule moving in the symplast may cross the membrane and move in the apoplast for a moment, and then move back into the symplast again.

At the endodermis, water movement through the apoplast pathway is obstructed by the Casparian strip (see Figure 4.4). The **Casparian strip** is a band within the radial cell walls of the endodermis that is impregnated with the waxlike, hydrophobic substance **suberin**. The endodermis becomes suberized in the nongrowing part of the root, several millimeters to several centimeters behind the root tip, at about the same time that the first protoxylem elements mature (Esau 1953). The Casparian strip breaks the conti-

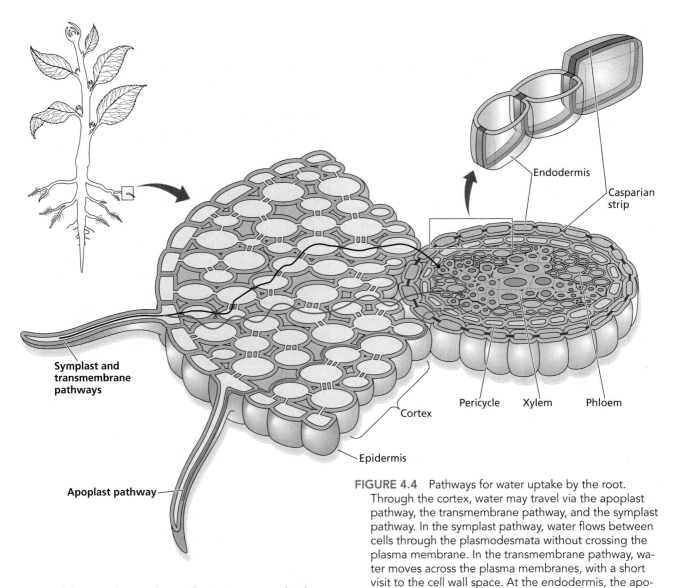

**Symplast and transmembrane pathways**

**Apoplast pathway**

Cortex

Epidermis

Endodermis

Casparian strip

Pericycle   Xylem   Phloem

**FIGURE 4.4** Pathways for water uptake by the root. Through the cortex, water may travel via the apoplast pathway, the transmembrane pathway, and the symplast pathway. In the symplast pathway, water flows between cells through the plasmodesmata without crossing the plasma membrane. In the transmembrane pathway, water moves across the plasma membranes, with a short visit to the cell wall space. At the endodermis, the apoplast pathway is blocked by the Casparian strip.

nuity of the apoplast pathway, forcing water and solutes to pass through the plasma membrane in order to cross the endodermis.

The requirement that water move symplastically across the endodermis helps explain why the permeability of roots to water depends strongly on the presence of aquaporins. Down-regulating the expression of aquaporin genes markedly reduces the hydraulic conductivity of roots and can result in plants that wilt easily (Siefritz et al. 2002) or that compensate by producing larger root systems (Martre et al. 2002).

Water uptake decreases when roots are subjected to low temperature or anaerobic conditions, or treated with respiratory inhibitors. Until recently, there was no explanation for the connection between root respiration and water uptake, or for the enigmatic wilting of flooded plants. We now know that the permeability of aquaporins can be regulated in response to intracellular pH (Tournaire-Roux et al. 2003). Decreased rates of respiration, in response to low temperature or anaerobic conditions, can lead to increases in intracellular pH. This increase in cytoplasmic pH alters

the conductance of aquaporins in root cells, resulting in roots that are markedly less permeable to water. The fact that aquaporins can be gated in response to pH provides a mechanism by which roots can actively alter their permeability to water in response to their local environment.

### Solute accumulation in the xylem can generate "root pressure"

Plants sometimes exhibit a phenomenon referred to as **root pressure**. For example, if the stem of a young seedling is cut off just above the soil, the stump will often exude sap from the cut xylem for many hours. If a manometer is sealed over the stump, positive pressures as high as 0.05 to 0.2 MPa can be measured.

When transpiration is low or absent, positive hydrostatic pressure builds up in the xylem because roots continue to absorb ions from the soil and transport them into the xylem. The buildup of solutes in the xylem sap leads to a decrease in the xylem osmotic potential ($\Psi_s$) and thus a decrease in the xylem water potential ($\Psi_w$). This lowering of the xylem $\Psi_w$ provides a driving force for water absorption, which in turn leads to a positive hydrostatic pressure in the xylem. In effect, the multicellular root tissue behaves as an osmotic membrane does, building up a positive hydrostatic pressure in the xylem in response to the accumulation of solutes.

Root pressure is most likely to occur when soil water potentials are high and transpiration rates are low. When transpiration rates are high, water is taken up into the leaves and lost to the atmosphere so rapidly that a positive pressure resulting from ion uptake never develops in the xylem.

Plants that develop root pressure frequently produce liquid droplets on the edges of their leaves, a phenomenon known as **guttation** (FIGURE 4.5). Positive xylem pressure causes exudation of xylem sap through specialized pores called *hydathodes* that are associated with vein endings at the leaf margin. The "dewdrops" that can be seen on the tips of grass leaves in the morning are actually guttation droplets exuded from hydathodes. Guttation is most noticeable when transpiration is suppressed and the relative humidity is high, such as at night. It is possible that root pressure reflects an unavoidable consequence of high rates of ion accumulation. However, the existence of positive pressures within the xylem at night can help to dissolve previously formed gas bubbles, and thus play a role in reversing the deleterious effects of cavitation described in the next section.

## Water Transport through the Xylem

In most plants, the xylem constitutes the longest part of the pathway of water transport. In a plant 1 m tall, more than 99.5% of the water transport pathway through the plant is within the xylem, and in tall trees the xylem represents an even greater fraction of the pathway. Compared with the movement of water through layers of living cells, the xylem is a simple pathway of low resistivity. In the following sections we will examine how the structure of the xylem contributes to the movement of water from the roots to the leaves, and how negative hydrostatic pressure generated by transpiration pulls water through the xylem.

### The xylem consists of two types of tracheary elements

The conducting cells in the xylem have a specialized anatomy that enables them to transport large quantities of water with great efficiency. There are two main types of **tracheary elements** in the xylem: tracheids and vessel elements (FIGURE 4.6). Vessel elements are found in angiosperms, a small group of gymnosperms called the Gnetales, and some ferns. Tracheids are present in both

**FIGURE 4.5**   Guttation in leaves from strawberry (*Fragaria grandiflora*). In the early morning, leaves secrete water droplets through the hydathodes, located at the margins of the leaves. Young flowers may also show guttation. (Courtesy of R. Aloni.)

angiosperms and gymnosperms, as well as in ferns and other groups of vascular plants.

The maturation of both tracheids and vessel elements involves the production of secondary cell walls and the subsequent death of the cell—the loss of the cytoplasm and all of its contents. What remain are the thick, lignified cell walls, which form hollow tubes through which water can flow with relatively little resistance.

**Tracheids** are elongated, spindle-shaped cells (see Figure 4.6A) that are arranged in overlapping vertical files (FIGURE 4.7). Water flows between tracheids by means of the numerous **pits** in their lateral walls (see Figure 4.6B). Pits are microscopic regions where the secondary wall is absent, and only the primary wall is present (see Figure 4.6C). Pits of one tracheid are typically located opposite pits of an adjoining tracheid, forming **pit pairs**. Pit pairs constitute a low-resistance path for water movement between tracheids. The water-permeable layer between pit pairs, consisting of two primary walls and a middle lamella, is called the **pit membrane**.

Pit membranes in tracheids of conifers have a central thickening, called a **torus** (plural *tori*) surrounded by a porous, and relatively flexible regions known as the **margo** (see Figure 4.6C). The torus acts like a valve: When it is centered in the pit cavity, the pit remains open; when it is lodged in the circular or oval wall thickenings bordering the pit, the pit is closed. Such lodging of the torus effec-

**FIGURE 4.6** Tracheary elements and their interconnections. (A) Structural comparison of tracheids and vessel elements, two classes of tracheary elements involved in xylem water transport. Tracheids are elongated, hollow, dead cells with highly lignified walls. The walls contain numerous pits—regions where secondary wall is absent but primary wall remains. The shapes of pits and the patterns of wall pitting vary with species and organ type. Tracheids are present in all vascular plants. Vessels consist of a stack of two or more vessel elements. Like tracheids, vessel elements are dead cells and are connected to one another by perforation plates—regions of the wall where pores or holes have developed. Vessels are connected to other vessels and to tracheids through pits. Vessels are found in most angiosperms and are lacking in most gymnosperms. (B) Scanning electron micrograph showing two vessels (running diagonally from lower left to upper right). Pits are visible on the side walls, as are the scalariform end walls between vessel elements. (200×) (C) Diagram of a coniferous bordered pit with the torus centered in the pit cavity (left) or lodged to one side of the cavity (right). When the pressure difference between two tracheids is small, the pit membrane will lie close to the center of the bordered pit, allowing water to flow through the porous margo region of the pit membrane; when the pressure difference between two tracheids is large, such as when one has cavitated and the other remains filled with water under tension, the pit membrane is displaced such that the torus becomes lodged against the overarching walls, thereby preventing the embolism from propagating between tracheids. (D) In contrast, the pit membranes of angiosperms and other nonconiferous vascular plants are relatively homogeneous in their structure. These pit membranes have very small pores compared with those of conifers, which prevents the spread of embolism but also imparts a significant hydraulic resistance. (B © Steve Gschmeissner/Photo Researchers, Inc.; C after Zimmermann 1983.)

(A)

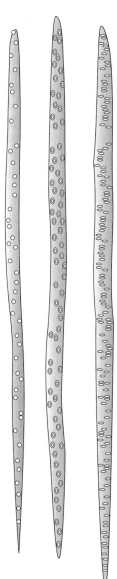

**Tracheids**

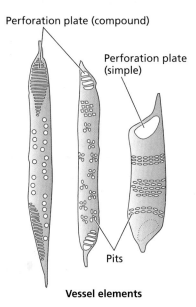

Perforation plate (compound)

Perforation plate (simple)

Pits

**Vessel elements**

(B)

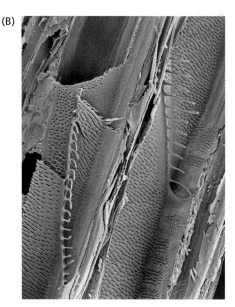

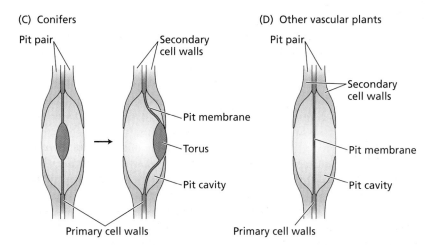

(C) Conifers

Pit pair

Secondary cell walls

Pit membrane

Torus

Pit cavity

Primary cell walls

(D) Other vascular plants

Pit pair

Secondary cell walls

Pit membrane

Pit cavity

Primary cell walls

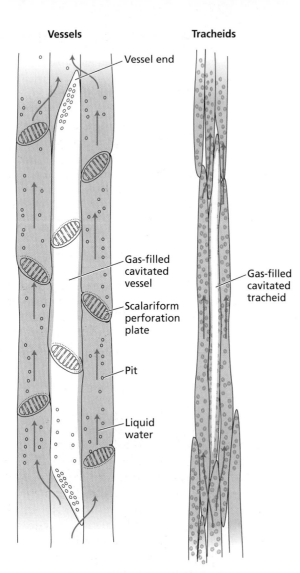

**Vessels**       **Tracheids**

Vessel end

Gas-filled
cavitated
vessel

Gas-filled
cavitated
tracheid

Scalariform
perforation
plate

Pit

Liquid
water

**FIGURE 4.7** Vessels (left) and tracheids (right) form a series of parallel, interconnected pathways for water movement. Cavitation blocks water movement because of the formation of gas-filled (embolized) conduits. Because xylem conduits are interconnected through openings ("bordered pits") in their thick secondary walls, water can detour around the blocked vessel by moving through adjacent tracheary elements. The very small pores in the pit membranes help prevent embolisms from spreading between xylem conduits. Thus, in the diagram on the right, the gas is contained within a single cavitated tracheid. In the diagram on the left, gas has filled the entire cavitated vessel, shown here as being made up of three vessel elements, each separated by scalariform (resembling the rungs of a ladder) perforation plates. In nature, vessels can be very long (up to several meters in length) and thus made up of many vessel elements.

tively prevents gas bubbles from spreading into neighboring tracheids (we will discuss this formation of bubbles, a process called cavitation, shortly). Pit membranes in all other plants, whether in tracheids or vessel elements, lack tori. But because the water-filled pores in the pit membranes of nonconifers are very small, they also serve as an effective barrier against the movement of gas bubbles.

Thus, pit membranes of both types play an important role in preventing the spread of gas bubbles, called *emboli*, within the xylem.

**Vessel elements** tend to be shorter and wider than tracheids and have perforated end walls that form a **perforation plate** at each end of the cell. Like tracheids, vessel elements have pits on their lateral walls (see Figure 4.6B). Unlike tracheids, the perforated end walls allow vessel elements to be stacked end to end to form a much longer conduit called a **vessel** (see Figure 4.7). Vessels are multicellular conduits that vary in length both within and between species. Vessels range from a few centimeters in length to many meters. The vessel elements found at the extreme ends of a vessel lack perforations in their end walls and are connected to neighboring vessels via pits.

### Water moves through the xylem by pressure-driven bulk flow

Pressure-driven bulk flow of water is responsible for long-distance transport of water in the xylem. It also accounts for much of the water flow through the soil and through the cell walls of plant tissues. In contrast to the diffusion of water across semipermeable membranes, pressure-driven bulk flow is independent of solute concentration gradients, as long as viscosity changes are negligible.

If we consider bulk flow through a tube, the rate of flow depends on the radius ($r$) of the tube, the viscosity ($\eta$) of the liquid, and the pressure gradient ($\Delta \Psi_p / \Delta x$) that drives the flow. Jean Léonard Marie Poiseuille (1797–1869) was a French physician and physiologist, and the relation just described is given by one form of Poiseuille's equation:

$$\text{Volume flow rate} = \left( \frac{\pi r^4}{8\eta} \right) \left( \frac{\Delta \Psi_p}{\Delta x} \right) \tag{4.2}$$

expressed in cubic meters per second ($m^3 \ s^{-1}$). This equation tells us that pressure-driven bulk flow is extremely sensitive to the radius of the tube. If the radius is doubled, the volume flow rate increases by a factor of 16 ($2^4$). Vessel elements up to 500 μm in diameter, nearly an order of magnitude greater than the largest tracheids, occur in the stems of climbing species. These large-diameter vessels permit vines to transport water long distances despite the slenderness of their stems.

Equation 4.2 describes water flow through a cylindrical tube and thus does not take into account the fact that xylem conduits are of finite length, such that water must cross many pit membranes as it flows from the soil to the leaves. All else being equal, pit membranes should impede water flow through single-celled (and thus shorter) tracheids to a greater extent than through multicellular (and thus longer) vessels. However, the pit membranes of conifers are much more permeable to water than those found in other plants (Pittermann et al. 2005), allowing conifers to grow into large trees despite producing only tracheids.

## Water movement through the xylem requires a smaller pressure gradient than movement through living cells

The xylem provides a pathway of low resistivity for water movement. Some numerical values will help us appreciate the extraordinary efficiency of the xylem. We will calculate the driving force required to move water through the xylem at a typical velocity and compare it with the driving force that would be needed to move water through a cell-to-cell pathway at the same rate.

For the purposes of this comparison, we will use a value of 4 mm s$^{-1}$ for the xylem transport velocity and 40 μm as the vessel radius. This is a high velocity for such a narrow vessel, so it will tend to exaggerate the pressure gradient required to support water flow in the xylem. Using a version of Poiseuille's equation (see Equation 4.2), we can calculate the pressure gradient needed to move water at a velocity of 4 mm s$^{-1}$ through an *ideal* tube with a uniform inner radius of 40 μm. The calculation gives a value of 0.02 MPa m$^{-1}$. Elaboration of the assumptions, equations, and calculations can be found in WEB TOPIC 4.3.

Of course, *real* xylem conduits have irregular inner wall surfaces, and water flow through perforation plates and pits adds resistance. Such deviations from the ideal increase the frictional drag: Measurements show that the actual resistance is greater by approximately a factor of 2 (Nobel 1999).

Let's now compare this value with the driving force that would be necessary to move water at the same velocity from cell to cell, crossing the plasma membrane each time. As calculated in WEB TOPIC 4.3, the driving force needed to move water through a layer of cells at 4 mm s$^{-1}$ is $2 \times 10^8$ MPa m$^{-1}$. This is ten orders of magnitude greater than the driving force needed to move water through our 40-μm-radius xylem vessel. Our calculation clearly shows that water flow through the xylem is vastly more efficient than water flow across living cells. Nevertheless, the xylem makes a significant contribution to the total resistance to water flow through the plant.

## What pressure difference is needed to lift water 100 meters to a treetop?

With the foregoing example in mind, let's see what pressure gradient is needed to move water up to the top of a very tall tree. The tallest trees in the world are the coast redwoods (*Sequoia sempervirens*) of North America and the mountain ash (*Eucalyptus regnans*) of Australia. Individuals of both species can exceed 100 m.

If we think of the stem of a tree as a long pipe, we can estimate the pressure difference that is needed to overcome the frictional drag of moving water from the soil to the top of the tree by multiplying the pressure gradient needed to move the water by the height of the tree. The pressure gradients needed to move water through the xylem of very tall trees are on the order of 0.01 MPa m$^{-1}$, smaller than in

our previous example. If we multiply this pressure gradient by the height of the tree (0.01 MPa m$^{-1}$ × 100 m), we find that the total pressure difference needed to overcome the frictional resistance to water movement through the stem is equal to 1 MPa.

In addition to frictional resistance, we must consider gravity. As described by Equation 3.5, for a height difference of 100 m, the difference in $\Psi_g$ is approximately 1 MPa. That is, $\Psi_g$ is 1 MPa higher at the top of the tree than at the ground level. So the other components of water potential must be 1 MPa more negative at the top of the tree to counter the effects of gravity.

To allow transpiration to occur, the pressure gradient due to gravity must be added to that required to cause water movement through the xylem. Thus we calculate that a pressure difference of roughly 2 MPa, from the base to the top branches, is needed to carry water up the tallest trees.

## The cohesion–tension theory explains water transport in the xylem

In theory, the pressure gradients needed to move water through the xylem could result from the generation of positive pressures at the base of the plant or negative pressures at the top of the plant. We mentioned previously that some roots can develop positive hydrostatic pressure in their xylem. However, root pressure is typically less than 0.1 MPa and disappears when the transpiration rate is high or when soils are dry, so it is clearly inadequate to move water up a tall tree. Furthermore, because root pressure is generated by the accumulation of ions in the xylem, relying on this for transporting water would require a mechanism for dealing with these solutes once the water evaporates from the leaves.

Instead, the water at the top of a tree develops a large tension (a negative hydrostatic pressure), and this tension *pulls* water through the xylem. This mechanism, first proposed toward the end of the nineteenth century, is called the *cohesion–tension theory of sap ascent* because it requires the cohesive properties of water to sustain large tensions in the xylem water columns. One can readily demonstrate xylem tension by puncturing intact xylem through a drop of ink on the surface of a stem from a transpiring plant. When the tension in the xylem is relieved, the ink is drawn instantly into the xylem, resulting in visible streaks along the stem.

What is the source of the negative pressure of water in leaves and how does it serve to pull water from the soil? The negative pressure that causes water to move up through the xylem develops at the surface of the cell walls in the leaf (FIGURE 4.8). The situation is thought to be analogous to that in the soil. Because water adheres to the cellulose microfibrils and other hydrophilic components of the cell wall, as water evaporates from cells within the leaf, the surface of the remaining water is drawn into the interstices of the cell wall (see Figure 4.8), where it forms

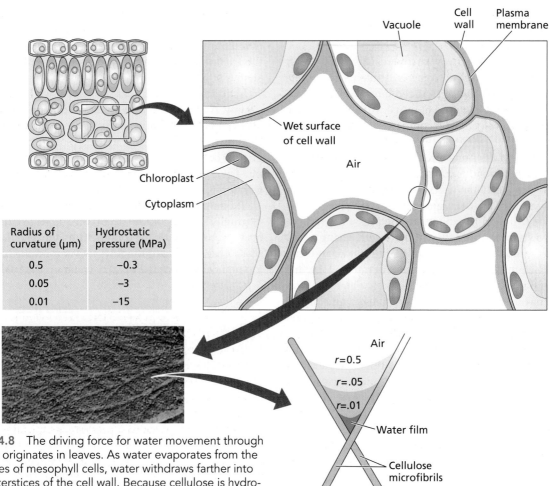

| Radius of curvature (μm) | Hydrostatic pressure (MPa) |
|---|---|
| 0.5 | −0.3 |
| 0.05 | −3 |
| 0.01 | −15 |

**FIGURE 4.8**   The driving force for water movement through plants originates in leaves. As water evaporates from the surfaces of mesophyll cells, water withdraws farther into the interstices of the cell wall. Because cellulose is hydrophilic (contact angle = 0°), the force resulting from surface tension causes a negative pressure in the liquid phase. As the radius of curvature of these air–water interfaces decreases, the hydrostatic pressure becomes more negative, as calculated from Equation 4.1. (Micrograph from Gunning and Steer 1996.)

curved air–water interfaces. Because of the high surface tension of water, the curvature of these interfaces induces a tension, or negative pressure, in the water. As more water is removed from the wall, the curvature of these air–water interfaces increases and the pressure of the water becomes more negative (see Equation 4.1).

The cohesion–tension theory explains how the substantial movement of water through plants can occur without the direct expenditure of metabolic energy: The energy that powers the movement water through plants comes from the sun, which, by increasing the temperature of both the leaf and the surrounding air, drives the evaporation of water.

The cohesion–tension theory has been a controversial subject for more than a century and continues to generate lively debate. The main controversy surrounds the question of whether water columns in the xylem can sustain the large tensions (negative pressures) necessary to pull water up tall trees. Most researchers believe that the cohesion–tension theory is sound (Steudle 2001). Recently, water transport through a microfluidic device designed to function as a synthetic "tree" demonstrated the stable flow of liquid water at negative pressures < –1.0 MPa (Wheeler and Strook 2008). For details on the history of research on water transport in the xylem, including the controversy surrounding the cohesion–tension theory, see **WEB ESSAYS 4.1 AND 4.2**.

## Xylem transport of water in trees faces physical challenges

The large tensions that develop in the xylem of trees (see **WEB ESSAY 4.3**) and other plants present significant physical challenges. First, the water under tension transmits an inward force to the walls of the xylem. If the cell walls were weak or pliant, they would collapse under this tension. The secondary wall thickenings and lignification of tracheids and vessels are adaptations that offset this tendency to collapse. Plants that experience large xylem tensions tend to have dense wood, reflecting the mechanical stresses imposed on the wood by water under tension (Hacke et al. 2001).

A second challenge is that water under such tensions is in a *physically metastable state*. Water is stable as a liquid when its hydrostatic pressure exceeds its saturated vapor pressure. When the hydrostatic pressure in liquid water becomes equal to its saturated vapor pressure, the water will undergo a phase change. We are all familiar with the idea of vaporizing water by increasing its temperature (raising its saturated vapor pressure). Less familiar, but still easily observed, is the fact that water can be made to boil at room temperature by placing it in a vacuum chamber (lowering the hydrostatic pressure of the liquid phase by reducing the pressure of the atmosphere).

In our earlier example, we estimated that a pressure gradient of 2 MPa would be needed to supply water to leaves to the top of a 100-meter tree. If we assume that the soil surrounding this tree is fully hydrated and lacks significant concentrations of solutes (i.e., $\Psi_w = 0$), the cohesion–tension theory predicts that the hydrostatic pressure of water in the xylem at the top of the tree will be −2 MPa. This value is substantially below the saturated vapor pressure (~0.002 MPa at 20°C), raising the question of what maintains the water column in its liquid state.

Water in the xylem is described as being in a metastable state because despite the possibility of its existing in a thermodynamically lower energy state—the vapor phase— it is in fact liquid. This situation occurs because (1) the cohesion and adhesion of water make the free-energy barrier for the liquid-to-vapor phase change very high, and (2) the structure of the xylem minimizes the presence of *nucleation sites*—sites that lower the energy barrier separating the liquid from the vapor phase.

The most important nucleation sites are gas bubbles. When a gas bubble grows to a sufficient size that the inward force resulting from surface tension is less than the outward force due to the negative pressure in the liquid phase, the bubble will expand. Furthermore, once a bubble starts to expand, the inward force due to surface tension decreases, because the air–water interface has less curvature. Thus, a bubble that exceeds the critical size for expansion will expand until it fills the entire conduit.

The absence of gas bubbles of sufficient size to destabilize the water column when under tension is, in part, due to the fact that in the roots, water must flow across pit membranes to enter the xylem. The pit membranes serve as filters, preventing gas bubbles from entering the xylem. However, when exposed to air on one side—due to injury, leaf abscission, or the existence of a neighboring gas-filled conduit—pit membranes can serve as sites of entry for air. Air enters when the pressure difference across the pit membrane is sufficient either to overcome the capillary forces of the air–water interfaces within the cellulose microfibriller matrix of structurally homogeneous pit membranes (see Figure 4.6D), or to dislodge the torus of a coniferous pit membrane (see Figure 4.6C). This phenomenon is called *air seeding*.

A second mode by which bubbles can form in xylem conduits is freezing of the xylem tissues (Davis et al. 1999). Because water in the xylem contains dissolved gases and the solubility of gases in ice is very low, freezing of xylem conduits can lead to bubble formation.

The phenomenon of bubble expansion is known as *cavitation* and the resulting gas-filled void is referred to as an *embolism*. Its effect is similar to that of a vapor lock in the fuel line of an automobile or an embolism in a blood vessel. Cavitation breaks the continuity of the water column and prevents the transport of water under tension (Tyree and Sperry 1989).

Such breaks in the water columns in plants are not unusual. When plants are deprived of water, sound pulses or clicks can be detected (Milburn and Johnson 1966, Jackson et al. 1999). The formation and rapid expansion of air bubbles in the xylem, such that the pressure in the water is suddenly increased by perhaps 1MPa or more, results in high-frequency acoustic shock waves through the rest of the plant. These breaks in xylem water continuity, if not repaired, would be disastrous to the plant. By blocking the main transport pathway of water, such embolisms would increase flow resistance and ultimately cause the dehydration and death of the leaves and other organs.

*Vulnerability curves* (**FIGURE 4.9**) provide a way of quantifying a species' susceptibility to cavitation and the impact of cavitation on flow through the xylem. A vulnerability curve plots the measured hydraulic conductivity (usually as a percent of maximum) of a branch, stem, or root segment versus the experimentally imposed level of xylem tension. Due to cavitation, xylem hydraulic conductivity decreases with

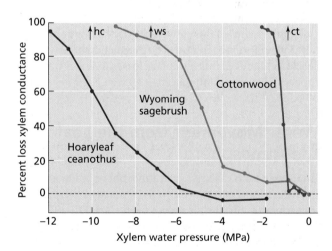

**FIGURE 4.9** Xylem vulnerability curves represent the percentage loss of hydraulic conductance in stem xylem versus xylem water pressure in three species of contrasting drought tolerance. Data were obtained from excised branches subjected experimentally to increasing levels of xylem tension using a centrifugal force technique (Alder et al. 1997). Arrows on the upper pressure axis indicate minimum xylem pressures measured in the field for each species. (After Sperry 2000.)

increasing tensions (more negative pressures) until flow ceases entirely. However, the decrease in xylem hydraulic conductivity occurs at much less negative xylem pressures in species in moist habitats, such as birch, than in species from more arid regions, such as sagebrush.

### Plants minimize the consequences of xylem cavitation

The impact of xylem cavitation on the plant can be minimized by several means. Because the tracheary elements in the xylem are interconnected, one gas bubble might, in principle, expand to fill the whole network. In practice, gas bubbles do not spread far, because the expanding gas bubble cannot easily pass through the small pores of the pit membranes. Because the capillaries in the xylem are interconnected, one gas bubble does not completely stop water flow. Instead, water can detour around the embolized conduit by traveling through neighboring, water-filled conduits (see Figure 4.7). Thus the finite length of the tracheid and vessel conduits of the xylem, while resulting in an increased resistance to water flow, also provides a way to restrict the impact of cavitation.

Gas bubbles can also be eliminated from the xylem. At night, when transpiration is low, xylem $\Psi_p$ increases, and water vapor and gases may simply dissolve back into the solution of the xylem. Moreover, as we have seen, some plants develop positive pressures (root pressures) in the xylem. Such pressures shrink bubbles and cause the gases to dissolve. Recent studies indicate that cavitation may be repaired even when the water in the xylem is under tension (Holbrook et al. 2001; Salleo et al. 2004). A mechanism for such repair is not yet known and remains the subject of active research (see WEB ESSAY 4.4).

Finally, many plants have secondary growth in which new xylem forms each year. The production of new xylem conduits allows plants to replace losses in water transport capacity due to cavitation.

# Water Movement from the Leaf to the Atmosphere

On its way from the leaf to the atmosphere, water is pulled from the xylem into the cell walls of the mesophyll, where it evaporates into the air spaces of the leaf (FIGURE 4.10). The water vapor then exits the leaf through the stomatal pore. The movement of liquid water through the living tissues of the leaf is controlled by gradients in water potential. However, transport in the vapor phase is by diffusion, so the final part of the transpiration stream is controlled by the *concentration gradient of water vapor.*

The waxy cuticle that covers the leaf surface is an effective barrier to water movement. It has been estimated that only about 5% of the water lost from leaves escapes through the cuticle. Almost all of the water lost from leaves is lost by diffusion of water vapor through the tiny stomatal pores. In most herbaceous species, stomata are present on both the upper and lower surfaces of the leaf, usually more abundant on the lower surface. In many tree species, stomata are located only on the lower surface of the leaf.

We will now examine the movement of liquid water through the leaf, the driving force for leaf transpiration, the main resistances in the diffusion pathway from the leaf to the atmosphere, and the anatomical features of the leaf that regulate transpiration.

### Leaves have a large hydraulic resistance

Although the distances that water must traverse within leaves are small relative to the entire soil-to-atmosphere pathway, the contribution of the leaf to the total hydraulic resistance is large. On average, leaves constitute 30% of the total liquid phase resistance, and in some plants their contribution is much larger (Sack and Holbrook 2006). This combination of short path length and large hydraulic resistance also occurs in roots, reflecting the fact that in both organs, water transport takes place across highly resistive living tissues as well as through the xylem.

Water enters into leaves and is distributed across the leaf lamina in xylem conduits. Water must exit through the xylem walls and pass through multiple layers of living cells before it evaporates. Leaf hydraulic resistance thus reflects the number, distribution, and size of xylem conduits, as well as the hydraulic properties of leaf mesophyll cells. The hydraulic resistance of leaves of diverse vein architectures varies as much as 40-fold (Brodribb et al. 2007). A large part of this variation appears to be due to the density of veins within the leaf and their distance from the evaporative leaf surface. Leaves with closely spaced veins tend to have lower hydraulic resistance and higher rates of photosynthesis, suggesting that the proximity of leaf veins to sites of evaporation exerts a significant impact on the rates of leaf gas exchange.

The hydraulic resistance of leaves is dynamic. For example, leaves of plants growing in shaded conditions exhibit greater resistance to water flow than leaves of plants grown in higher light (Sack et al. 2003). Leaf hydraulic resistance also typically increases with leaf age. Over shorter time scales, decreases in leaf water potential lead to marked increases in leaf hydraulic resistance. The increase in leaf hydraulic resistance may result from cavitation of xylem conduits in leaf veins or, in some cases, the physical collapse of xylem conduits under tension (Cochard et al. 2004; Brodribb and Holbrook 2005).

### The driving force for transpiration is the difference in water vapor concentration

Transpiration from the leaf depends on two major factors: (1) the **difference in water vapor concentration** between the leaf air spaces and the external bulk air and (2) the **diffusional resistance** (*r*) of this pathway. The difference in water vapor

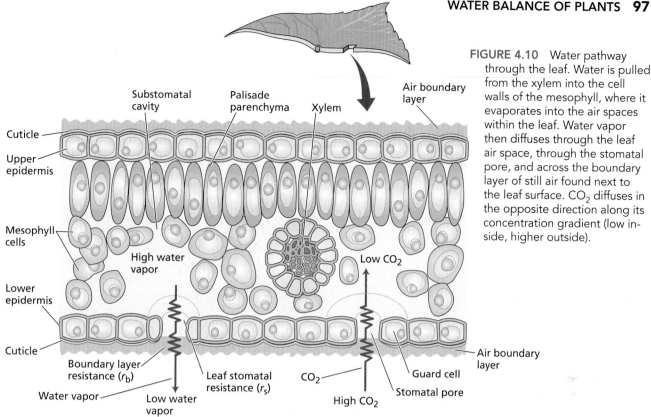

**FIGURE 4.10** Water pathway through the leaf. Water is pulled from the xylem into the cell walls of the mesophyll, where it evaporates into the air spaces within the leaf. Water vapor then diffuses through the leaf air space, through the stomatal pore, and across the boundary layer of still air found next to the leaf surface. $CO_2$ diffuses in the opposite direction along its concentration gradient (low inside, higher outside).

concentration is expressed as $c_{wv(leaf)} - c_{wv(air)}$. The water vapor concentration of air ($c_{wv(air)}$) can be readily measured, but that of the leaf ($c_{wv(leaf)}$) is more difficult to assess.

Whereas the volume of air space inside the leaf is small, the wet surface from which water evaporates is large. Air space volume is about 5% of the total leaf volume in pine needles, 10% in corn leaves, 30% in barley, and 40% in tobacco leaves. In contrast to the volume of the air space, the internal surface area from which water evaporates may be from 7 to 30 times the external leaf area. This high surface-to-volume ratio and the close distance within the air space (few to tens of micrometers) make for rapid vapor equilibration inside the leaf. Thus we can assume that the air space in the leaf is close to water potential equilibrium with the cell wall surfaces from which liquid water is evaporating.

Within the range of water potentials experienced by transpiring leaves (generally > –2.0 MPa), the equilibrium water vapor concentration is within two percentage points of the saturation water vapor concentration. This allows one to estimate the water vapor concentration within a leaf from its temperature, which is easy to measure. Because the saturated water vapor content of air increases exponentially with temperature, leaf temperature has a marked impact on transpiration rates. (WEB TOPIC 4.4 shows how we can calculate the water vapor concentration in the leaf air spaces and discusses other aspects of the water relations within a leaf.)

The concentration of water vapor, $c_{wv}$, changes at various points along the transpiration pathway. We see from TABLE 4.1 that $c_{wv}$ decreases at each step of the pathway

---

**TABLE 4.1**
**Representative values for relative humidity, absolute water vapor concentration, and water potential for four points in the pathway of water loss from a leaf**

| Location | Relative humidity | Water vapor Concentration (mol m⁻³) | Water vapor Potential (MPa)[a] |
|---|---|---|---|
| Inner air spaces (25°C) | 0.99 | 1.27 | –1.38 |
| Just inside stomatal pore (25°C) | 0.97 | 1.21 | –7.04 |
| Just outside stomatal pore (25°C) | 0.47 | 0.60 | –103.7 |
| Bulk air (20°C) | 0.50 | 0.50 | –93.6 |

*Source*: Adapted from Nobel 1999.

Note: See Figure 4.10.

[a]Calculated using Equation 4.5.2 in Web Topic 4.4, with values for $RT/\bar{V}_w$ of 135 MPa at 20°C and 137.3 MPa at 25°C.

from the cell wall surface to the bulk air outside the leaf. The important points to remember are that (1) the driving force for water loss from the leaf is the *absolute* concentration difference (difference in $c_{wv}$, in mol m$^{-3}$), and (2) this difference is markedly influenced by leaf temperature.

### Water loss is also regulated by the pathway resistances

The second important factor governing water loss from the leaf is the diffusional resistance of the transpiration pathway, which consists of two varying components (see Figure 4.10):

1. The resistance associated with diffusion through the stomatal pore, the **leaf stomatal resistance** ($r_s$).

2. The resistance due to the layer of unstirred air next to the leaf surface through which water vapor must diffuse to reach the turbulent air of the atmosphere. This second resistance, $r_b$, is called the leaf **boundary layer resistance**. We will discuss this type of resistance before considering stomatal resistance.

The thickness of the boundary layer is determined primarily by wind speed and leaf size. When the air surrounding the leaf is very still, the layer of unstirred air on the surface of the leaf may be so thick that it is the primary deterrent to water vapor loss from the leaf. Increases in stomatal apertures under such conditions have little effect on transpiration rate (**FIGURE 4.11**), although closing the stomata completely will still reduce transpiration.

When wind velocity is high, the moving air reduces the thickness of the boundary layer at the leaf surface, reducing the resistance of this layer. Under such conditions, stomatal resistance will largely control water loss from the leaf.

Various anatomical and morphological aspects of the leaf can influence the thickness of the boundary layer. Hairs on the surface of leaves can serve as microscopic windbreaks. Some plants have sunken stomata that provide a sheltered region outside the stomatal pore. The size and shape of leaves and their orientation relative to the wind direction also influence the way the wind sweeps across the leaf surface. Most of these factors, however, cannot be altered on an hour-to-hour or even a day-to-day basis. For short-term regulation of transpiration, control of stomatal apertures by the guard cells plays a crucial role in the regulation of leaf transpiration.

Some species are able to change the orientation of their leaves and thereby influence their transpiration rates. For example, by orienting leaves parallel to the sun's rays, leaf temperature is reduced and with it the driving force for transpiration $\Delta C_{wv}$ (Berg and Hsiao 1986). Many grass leaves roll up as they experience water deficits, in this way increasing their boundary layer resistance (Hsiao et al. 1984). Even wilting can help ameliorate high transpiration rates by reducing the amount of radiation intercepted,

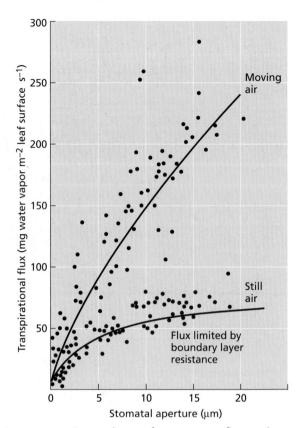

**FIGURE 4.11** Dependence of transpiration flux on the stomatal aperture of zebra plant (*Zebrina pendula*) in still air and in moving air. The boundary layer is larger and more rate limiting in still air than in moving air. As a result, the stomatal aperture has less control over transpiration in still air. (After Bange 1953.)

resulting in lower leaf temperatures and a decrease in $\Delta C_{wv}$ (Chiariello et al. 1987).

### Stomatal control couples leaf transpiration to leaf photosynthesis

Because the cuticle covering the leaf is nearly impermeable to water, most leaf transpiration results from the diffusion of water vapor through the stomatal pore (see Figure 4.10). The microscopic stomatal pores provide a *low-resistance pathway* for diffusional movement of gases across the epidermis and cuticle. Changes in stomatal resistance are important for the regulation of water loss by the plant and for controlling the rate of carbon dioxide uptake necessary for sustained $CO_2$ fixation during photosynthesis.

When water is abundant, the functional solution to the leaf's need to limit water loss while taking in $CO_2$ is the *temporal* regulation of stomatal apertures—open during the day, closed at night. At night, when there is no photosynthesis and thus no demand for $CO_2$ inside the leaf, stomatal apertures are kept small or closed, preventing unnecessary loss of water. On a sunny morning when the supply of water is abundant and the solar radiation incident on the leaf favors

(A)

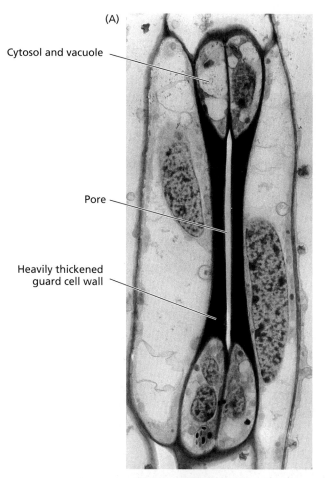

Cytosol and vacuole

Pore

Heavily thickened
guard cell wall

(B)

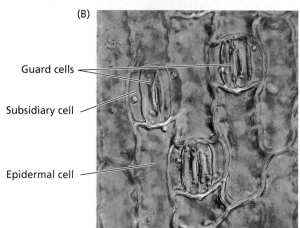

Guard cells

Subsidiary cell

Epidermal cell

(C)

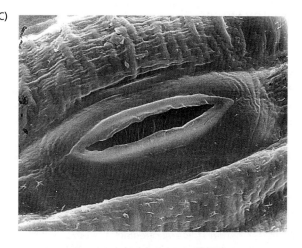

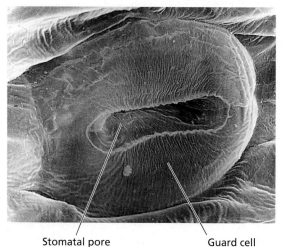

Stomatal pore          Guard cell

FIGURE 4.12  Electron micrographs of stomata. (A) A stoma from a grass. The bulbous ends of each guard cell show their cytosolic content and are joined by the heavily thickened walls. The stomatal pore separates the two midportions of the guard cells. (2560×) (B) Stomatal complexes of the sedge *Carex* viewed with differential interference contrast light microscopy. Each complex consists of two guard cells surrounding a pore and two flanking subsidiary cells. (550×) (C) Scanning electron micrographs of onion epidermis. The top panel shows the outside surface of the leaf, with a stomatal pore inserted in the cuticle. The bottom panel shows a pair of guard cells facing the stomatal cavity, toward the inside of the leaf. (1640×) (A from Palevitz 1981, B from Jarvis and Mansfield 1981, A and B courtesy of B. Palevitz; C from Zeiger and Hepler 1976 [top] and E. Zeiger and N. Burnstein [bottom].)

high photosynthetic activity, the demand for $CO_2$ inside the leaf is large, and the stomatal pores open wide, decreasing the stomatal resistance to $CO_2$ diffusion. Water loss by transpiration is substantial under these conditions, but since the water supply is plentiful, it is advantageous for the plant to trade water for the products of photosynthesis, which are essential for growth and reproduction.

On the other hand, when soil water is less abundant, the stomata will open less or even remain closed on a sunny morning. By keeping its stomata closed in dry conditions, the plant avoids dehydration. The leaf cannot control $(c_{wv(leaf)} - c_{wv(air)})$ or $r_b$. However, it can regulate its stomatal resistance ($r_s$) by opening and closing of the stomatal pore.

This biological control is exerted by a pair of specialized epidermal cells, the **guard cells**, which surround the stomatal pore (**FIGURE 4.12**).

## The cell walls of guard cells have specialized features

Guard cells are found in leaves of all vascular plants, and they are also present in more primitive plants, such as

**FIGURE 4.13** Electron micrograph showing a pair of guard cells from the dicot tobacco (*Nicotiana tabacum*). The section was made perpendicular to the main surface of the leaf. The pore faces the atmosphere; the bottom faces the substomatal cavity inside the leaf. Note the uneven thickening pattern of the walls, which determines the asymmetric deformation of the guard cells when their volume increases during stomatal opening. (Micrograph from Sack 1987, courtesy of F. Sack.)

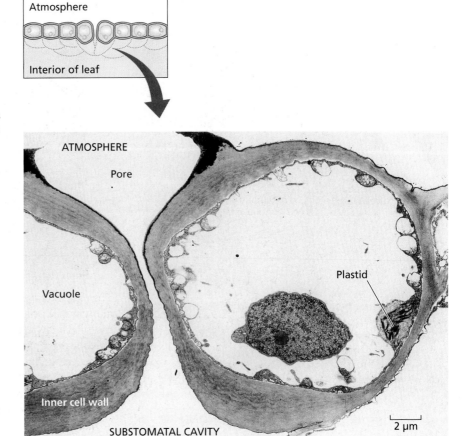

hornworts and mosses (Ziegler 1987). Guard cells show considerable morphological diversity, but we can distinguish two main types: One is typical of grasses and a few other monocots, such as palms; the other is found in all dicots, in many monocots, and in mosses, ferns, and gymnosperms.

In grasses (see Figure 4.12A), guard cells have a characteristic dumbbell shape, with bulbous ends. The pore proper is a long slit located between the two "handles" of the dumbbells. These guard cells are always flanked by a pair of differentiated epidermal cells called **subsidiary cells**, which help the guard cells control the stomatal pores (see Figure 4.12B). The guard cells, subsidiary cells, and pore are collectively called the **stomatal complex**.

In dicots and nongrass monocots, guard cells have an elliptical contour (often called "kidney-shaped") with the pore at their center (see Figure 4.12C). Although subsidiary cells are not uncommon in species with kidney-shaped stomata, they are often absent, in which case the guard cells are surrounded by ordinary epidermal cells.

A distinctive feature of guard cells is the specialized structure of their walls. Portions of these walls are substantially thickened (**FIGURE 4.13**) and may be up to 5 μm across, in contrast to the 1 to 2 μm typical of epidermal cells. In kidney-shaped guard cells, a differential thicken-

ing pattern results in very thick inner and outer (lateral) walls, a thin dorsal wall (the wall in contact with epidermal cells), and a somewhat thickened ventral (pore) wall. The portions of the wall that face the atmosphere extend into well-developed ledges, which form the pore proper.

The alignment of **cellulose microfibrils**, which reinforce all plant cell walls and are an important determinant of cell shape (see Chapter 15), plays an essential role in the opening and closing of the stomatal pore. In ordinary, cylindrically shaped cells, cellulose microfibrils are oriented transversely to the long axis of the cell. As a result, the cell expands in the direction of its long axis, because the cellulose reinforcement offers the least resistance at right angles to its orientation.

In guard cells the microfibril organization is different. Kidney-shaped guard cells have cellulose microfibrils fanning out radially from the pore (**FIGURE 4.14A**). The inner wall, facing the pore, is much thicker than the outer wall. Thus, as a guard cell increases in volume, the weaker outer wall bows outward, causing the pore to open (Sharpe et al. 1987). In grasses, the dumbbell-shaped guard cells function like beams with inflatable ends. As the bulbous ends of the cells increase in volume and swell, the beams are separated from each other and the slit between them widens (**FIGURE 4.14B**).

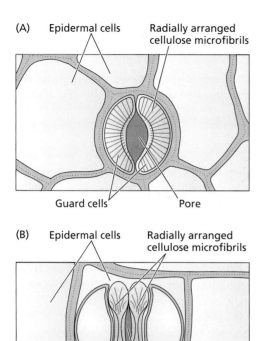

(A) Epidermal cells    Radially arranged cellulose microfibrils

Guard cells    Pore

(B) Epidermal cells    Radially arranged cellulose microfibrils

Guard cells    Pore    Subsidiary cell

Stomatal complex

**FIGURE 4.14** The radial alignment of the cellulose microfibrils in guard cells and epidermal cells of (A) a kidney-shaped stoma and (B) a grasslike stoma. (After Meidner and Mansfield 1968.)

### An increase in guard cell turgor pressure opens the stomata

Guard cells function as multisensory hydraulic valves. Environmental factors such as light intensity and quality, temperature, leaf water status, and intracellular $CO_2$ concentrations are sensed by guard cells, and these signals are integrated into well-defined stomatal responses. If leaves kept in the dark are illuminated, the light stimulus is perceived by the guard cells as an opening signal, triggering a series of responses that result in opening of the stomatal pore.

The early aspects of this process are ion uptake and other metabolic changes in the guard cells, which will be discussed in detail in Chapter 18. Here we will note the effect of decreases in osmotic potential ($\Psi_s$) resulting from ion uptake and from biosynthesis of organic molecules in the guard cells. Water relations in guard cells follow the same rules as in other cells. As $\Psi_s$ decreases, the water potential decreases, and water consequently moves into the guard cells. As water enters the cell, turgor pressure increases. Because of the elastic properties of their walls,

guard cells can reversibly increase their volume by 40 to 100%, depending on the species. Because of the differential thickening of guard cell walls, such changes in cell volume lead to opening or closing of the stomatal pore.

Subsidiary cells appear to play an important role in allowing stomata to open quickly and to achieve large apertures (Raschke and Fellows 1971; Franks and Farquhar 2007). A rapid transfer of solutes out of subsidiary cells and into the guard cells causes the former to decrease in both turgor pressure and size, facilitating the expansion of guard cells in the direction away from the stomatal pore. Conversely, the transfer of solutes from guard cells to the subsidiary cells increases the size and turgor pressure of the latter, thus pushing the guard cells together, and causing the stomates to close.

### The transpiration ratio measures the relationship between water loss and carbon gain

The effectiveness of plants in moderating water loss while allowing sufficient $CO_2$ uptake for photosynthesis can be assessed by a parameter called the **transpiration ratio**. This value is defined as the amount of water transpired by the plant divided by the amount of carbon dioxide assimilated by photosynthesis.

For plants in which the first stable product of carbon fixation is a 3-carbon compound ($C_3$ plants; see Chapter 8), as many as 400 molecules of water are lost for every molecule of $CO_2$ fixed by photosynthesis, giving a transpiration ratio of 400. (Sometimes the reciprocal of the transpiration ratio, called the *water use efficiency*, is cited. Plants with a transpiration ratio of 400 have a water use efficiency of 1/400, or 0.0025.)

The large ratio of $H_2O$ efflux to $CO_2$ influx results from three factors:

1. The concentration gradient driving water loss is about 50 times larger than that driving the influx of $CO_2$. In large part, this difference is due to the low concentration of $CO_2$ in air (about 0.038%) and the relatively high concentration of water vapor within the leaf.

2. $CO_2$ diffuses about 1.6 times more slowly through air than water does (the $CO_2$ molecule is larger than $H_2O$ and has a smaller diffusion coefficient).

3. $CO_2$ must cross the plasma membrane, the cytoplasm, and the chloroplast envelope before it is assimilated in the chloroplast. These membranes add to the resistance of the $CO_2$ diffusion pathway.

Some plants utilize variations in the usual photosynthetic pathway for fixation of carbon dioxide that substantially reduce their transpiration ratio. Plants in which a 4-carbon compound is the first stable product of photosynthesis ($C_4$ plants; see Chapter 8) generally transpire less water per molecule of $CO_2$ fixed than $C_3$ plants do; a typical transpiration ratio for $C_4$ plants is about 150. This is largely because $C_4$ photosynthesis results in a lower $CO_2$ concentration in the intercellular air space (see Chapter 8), thus creating a larger driving force for the uptake of $CO_2$ and allowing these plants to operate with smaller stomatal apertures and thus lower transpiration rates.

Desert-adapted plants with crassulacean acid metabolism (CAM) photosynthesis, in which $CO_2$ is initially fixed into 4-carbon organic acids at night, have even lower transpiration ratios; values of about 50 are not unusual. This is possible because their stomata have an inverted diurnal rhythm, opening at night and closing during the day. Transpiration is much lower at night, because the cool leaf temperature gives rise to only very small $\Delta C_{wv}$.

# Overview: The Soil–Plant–Atmosphere Continuum

We have seen that movement of water from the soil through the plant to the atmosphere involves different mechanisms of transport:

- In the soil and the xylem, liquid water moves by bulk flow in response to a pressure gradient ($\Delta \Psi_p$).
- When liquid water is transported across membranes, the driving force is the water potential difference across the membrane. Such osmotic flow occurs when cells absorb water and when roots transport water from the soil to the xylem.
- In the vapor phase, water moves primarily by diffusion, at least until it reaches the outside air, where convection (a form of bulk flow) becomes dominant.

However, the key element in the transport of water from the soil to the leaves is the generation of negative pressures within the xylem due to the capillary forces within the cell walls of transpiring leaves. At the other end of the plant, soil water is also held by capillary forces. This results in a "tug-of-war" on a rope of water by capillary forces at both ends. As a leaf loses water due to transpiration, water moves up the plant and out of the soil driven by physical forces, without the involvement of any metabolic pump. The energy for the movement of water is ultimately supplied by the sun.

This simple mechanism makes for tremendous efficiency energetically—which is critical when as many as 400 molecules of water are being transported for every $CO_2$ molecule being taken up in exchange. Crucial elements that allow this transport system to function are a low-resistivity xylem flow path that is protected from cavitation and a high-surface-area root system for extracting water from the soil.

## SUMMARY

There is an inherent conflict between a plant's need for $CO_2$ uptake and its need to conserve water, resulting from water being lost through the same pores that let $CO_2$ in. To manage this conflict, plants have evolved adaptations to control water loss from leaves, and to replace the water that is lost.

### Water in the Soil

- The water content and rate of movement in soils depends on soil type and structure, which influence the pressure gradient in the soil and its hydraulic conductivity.
- In soil, water may exist as a surface film on soil particles, or it may partially or completely fill the spaces between particles.
- Osmotic potential, hydrostatic pressure, and gravitational potential move water from the soil through the plant to the atmosphere (**Figure 4.1**).
- The intimate contact between root hairs and soil particles greatly increases the surface area for water absorption (**Figure 4.2**).

### Water Absorption by Roots

- Water uptake is mostly confined to regions near root tips (**Figure 4.3**).
- In the root, water may move via the apoplast, the symplast, or the transmembrane pathway (**Figure 4.4**).
- Water movement through the apoplast is obstructed by the Casparian strip in the endodermis (**Figure 4.4**).
- When transpiration is low, the continued transport of solutes into the xylem fluid leads to a decrease in $\Psi_s$ and a decrease in $\Psi_w$, providing the force for water absorption and a positive $\Psi_p$, which yields a positive hydrostatic pressure in the xylem (**Figure 4.5**).

### Water Transport through the Xylem

- Xylem conduits, which can be either single-celled tracheids or multicellular vessels, provide a low-resistance pathway for the transport of water (**Figure 4.6**).
- Elongated, spindle-shaped tracheids and stacked vessel elements have pits in lateral walls (**Figure 4.7**).

## SUMMARY continued

- Pressure-driven bulk flow moves water long distances through the xylem.
- The ascent of water through the xylem is due to negative pressure that develops at the surface of the cell walls in the leaf (**Figure 4.8**).
- Cavitation breaks the water column and prevents the transport of water under tension (**Figure 4.9**).

### Water Movement from the Leaf to the Atmosphere

- Water is pulled from the xylem into the cell walls of leaf mesophyll before evaporating into the leaf's air spaces (**Figure 4.10**).
- The hydraulic resistance of leaves is large and dynamic.
- Transpiration depends on the difference in water vapor concentration between the leaf air spaces and the external air and on the diffusional resistance of this pathway, which consists of leaf stomatal resistance and boundary layer resistance (**Figure 4.11**).
- Opening and closing of the stomatal pore is accomplished and controlled by guard cells (**Figures 4.12–4.14**).
- The effectiveness of plants in limiting water loss while allowing $CO_2$ uptake is given by the transpiration ratio.

### Overview: The Soil–Plant–Atmosphere Continuum

- Physical forces, without the involvement of any metabolic pump, drive the movement of water from soil, to plant, to atmosphere, with the sun being the ultimate source for the energy.

## WEB MATERIAL

### Web Topics

**4.1 Irrigation**

Irrigation has a dramatic impact on crop yield and soil salinity.

**4.2 Physical Properties of Soils**

The size distribution of soil particles influences its ability to hold and conduct water.

**4.3 Calculating Velocities of Water Movement in the Xylem and in Living Cells**

Water flows more easily through the xylem than across living cells.

**4.4 Leaf Transpiration and Water Vapor Gradients**

Leaf transpiration and stomatal conductance affect leaf and air water vapor concentrations.

### Web Essays

**4.1 A Brief History of the Study of Water Movement in the Xylem**

The history of our understanding of sap ascent in plants, especially in trees, is a beautiful example of how knowledge about plants is acquired.

**4.2 The Cohesion–Tension Theory at Work**

The cohesion–tension theory has withstood a number of challenges.

**4.3 How Water Climbs to the Top of a 112-Meter-Tall Tree**

Measurements of photosynthesis and transpiration in 112-meter-tall trees show that some of the conditions experienced by the top foliage are comparable to those of extreme deserts.

**4.4 Cavitation and Refilling**

A possible mechanism for cavitation repair is under active investigation.

# CHAPTER REFERENCES

Alder, N. N., Pockman, W. T., Sperry, J. S., and Nuismer, S. (1997) Use of centrifugal force in the study of xylem cavitation. *J. Exp. Bot.* 48: 293–301.

Bange, G. G. J. (1953) On the quantitative explanation of stomatal transpiration. *Acta Bot. Neerl.* 2: 255–296.

Berg, V. S., and Hsiao, T. C. (1986) Solar tracking: Light avoidance induced by water stress in leaves of kidney beans in the field. *Crop Sci.* 26: 980–986.

Bramley, H., Turner, N. C., Turner, D. W., and Tyerman, S. D. (2009) Roles of morphology, anatomy and aquaporins in determining contrasting hydraulic behavior of roots. *Plant Physiol.* 150: 348–364.

Brodribb, T. J., Field, T. S., Jordan, G. J. (2007) Leaf maximum photosynthetic rate and venation are linked by hydraulics. *Plant Physiol.* 144: 1890–1898.

Brodribb, T. J., and Holbrook, N. M. (2005) Water stress deforms tracheids peripheral to the leaf vein of a tropical conifer. *Plant Physiol.* 137: 1139–1146.

Chiariello, N. R., Field, C. B., and Mooney, H. A. (1987) Midday wilting in a tropical pioneer tree. *Funct. Ecol.* 1: 3–11.

Cochard, H., Froux, F., Mayr, F. F. S., and Courand, C. (2004) Xylem wall collapse in water-stressed pine needles. *Plant Physiol.* 134: 401–408.

Davis, S. D., Sperry, J. S., and Hacke, U. G. (1999) The relationship between xylem conduit diameter and cavitation caused by freezing. *Am. J. Bot.* 86: 1367–1372.

Esau, K. (1953) *Plant Anatomy*. John Wiley & Sons, Inc., New York.

Franks, P. J., and Farquhar, G. D. (2007) The mechanical diversity of stomata and its significance in gas-exchange control. *Plant Physiol.* 143: 78–87.

Frensch, J., Hsiao, T. C., and Steudle, E. (1996) Water and solute transport along developing maize roots. *Planta* 198: 348–355.

Gunning, B. S., and Steer, M. M. (1996) *Plant Cell Biology: Structure and Function*. Jones and Bartlett Publishers, Boston.

Hacke, U. G., Sperry, J. S., Pockman, W. T., Davis, S. D., and McCulloh, K. (2001) Trends in wood density and structure are linked to prevention of xylem implosion by negative pressure. *Oecologia* 126: 457–461.

Holbrook, N. M., Ahrens, E. T., Burns, M. J., and Zwieniecki, M. A. (2001) In vivo observation of cavitation and embolism repair using magnetic resonance imaging. *Plant Physiol.* 126: 27–31.

Hsiao, T. C., O'Toole, J. C., Yambao, E. B., and Turner, N. C. (1984) Influence of osmotic adjustment on leaf rolling and tissue death in rice (*Oryza sativa* L.). *Plant Physiol.* 75: 338–341.

Jackson, G. E., Irvine, J., and Grace, J. (1999) Xylem acoustic emissions and water relations of *Calluna vulgaris* L. at two climatological regions of Britain. *Plant Ecol.* 140: 3–14.

Jarvis, P. G., and Mansfield, T. A. (1981) *Stomatal Physiology*. Cambridge University Press, Cambridge.

Kramer, P. J., and Boyer, J. S. (1995) *Water Relations of Plants and Soils*. Academic Press, San Diego, CA.

Martre, P., Morillon, R., Barrieu, F., North, G. B., Nobel, P. S., and Chrispeels, M. J. (2002) Plasma membrane aquaporins play a significant role during recovery from water deficit. *Plant Physiol.* 130: 2101–2110.

Meidner, H., and Mansfield, D. (1968) *Stomatal Physiology*. McGraw-Hill, London.

Milburn, J. A., and Johnson, R. P. C. (1966) The conduction of sap. II Detection of vibrations produced by sap cavitation in *Ricinus* stems. *Planta* 69: 43–52.

Nobel, P. S. (1999) *Physicochemical and Environmental Plant Physiology*, 2nd ed., Academic Press, San Diego, CA.

Palevitz, B. A. (1981) The structure and development of guard cells. In *Stomatal Physiology*, P. G. Jarvis and T. A. Mansfield, eds., Cambridge University Press, Cambridge, pp. 1–23.

Pittermann, J., Sperry, J. S., Hacke, U. G., Wheeler, J. K., and Sikkema, E. H. (2005) Torus-margo pits help conifers compete with angiosperms. *Science* 310: 1924.

Raschke, K., and Fellows, M. (1971) Stomatal movement in *Zea mays*: Shuttle of potassium and chloride between guard cells and subsidiary cells. *Planta* 101: 296–316.

Sack, F. D. (1987) The development and structure of stomata. In *Stomatal Function*, E. Zeiger, G. Farquhar, and I. Cowan, eds., Stanford University Press, Stanford, CA, pp. 59–90.

Sack, L., Cowan, P. D., Jaikumar, N., and Holbrook, N. M. (2003) The 'hydrology' of leaves: coordination of structure and function in temperate woody species. *Plant Cell Environ.* 26: 1343–1356.

Sack, L. and Holbrook, N. M. (2006) Leaf hydraulics. *Annu. Rev. Plant Biol.* 57: 361–381.

Salleo, S., Lo Gullo, M. A., Trifilo, P., and Nardini, A. (2004) New evidence for a role of vessel-associated cells and phloem in the rapid xylem refilling of cavitated stems of *Laurus nobilis* L. *Plant Cell Environ.* 27: 1065–1076.

Sharpe, P. J. H., Wu, H.-I., and Spence, R. D. (1987) Stomatal mechanics. In *Stomatal Function*, E. Zeiger, G. Farquhar, and I. Cowan, eds., Stanford University Press, Stanford, CA, pp. 91–114.

Siefritz, F., Tyree, M. T., Lovisolo, C., Schubert, A., and Kaldenhoff, R. (2002) PIP1 plasma membrane aquaporins in tobacco: From cellular effects to function in plants. *Plant Cell* 14: 869–876.

Sperry, J. S. (2000) Hydraulic constraints on plant gas exchange. *Ag. For. Meteorol.* 104: 13–23.

Steudle, E. (2001) The cohesion-tension mechanism and the acquisition of water by plant roots. *Annu. Rev. Plant Physiol. Plant Mol. Biol.* 52: 847–875.

Steudle, E., and Frensch, J. (1996) Water transport in plants: Role of the apoplast. *Plant Soil* 187: 67–79.

Tournaire-Roux, C., Sutka, M., Javot, H., Gout, E., Gerbeau, P., Luu, D. T., Bligny, R., and Maurel, C. (2003) Cytosolic pH regulates root water transport during anoxic stress through gating of aquaporins. *Nature* 425: 393–397.

Tyree, M. T., and Sperry, J. S. (1989) Vulnerability of xylem to cavitation and embolism. *Annu. Rev. Plant Physiol. Plant Mol. Biol.* 40: 19–38.

Wheeler, T. D., and Strook, A. D. (2008) The transpiration of water at negative pressures in a synthetic tree. *Nature* 455: 208–212.

Zeiger, E., and Hepler, P. K. (1976) Production of guard cell protoplasts from onion and tobacco. *Plant Physiol.* 58: 492–498.

Ziegler, H. (1987) The evolution of stomata. In *Stomatal Function*, E. Zeiger, G. Farquhar, and I. Cowan, eds., Stanford University Press, Stanford, CA, pp. 29–58.

Zimmermann, M. H. (1983) *Xylem Structure and the Ascent of Sap.* Springer, Berlin.

Zwieniecki, M. A., Thompson, M. V., and Holbrook, N. M. (2002) Understanding the hydraulics of porous pipes: Tradeoffs between water uptake and root length utilization. *J. Plant Growth Regul.* 21: 315–323.

# Mineral Nutrition

Mineral nutrients are elements such as nitrogen, phosphorus, and potassium that plants acquire primarily in the form of inorganic ions from the soil. Although mineral nutrients continually cycle through all organisms, they enter the **biosphere** predominantly through the root systems of plants, so in a sense plants act as the "miners" of Earth's crust (Epstein 1999). The large surface area of roots and their ability to absorb inorganic ions at low concentrations from the soil solution increase the effectiveness of mineral absorption by plants. After being absorbed by the roots, the mineral elements are translocated to the various parts of the plant, where they are utilized in numerous biological functions. Other organisms, such as mycorrhizal fungi and nitrogen-fixing bacteria, often participate with roots in the acquisition of mineral nutrients.

The study of how plants obtain and use mineral nutrients is called **mineral nutrition**. This area of research is central to modern agriculture and environmental protection. High agricultural yields depend on fertilization with mineral nutrients. In fact, yields of most crop plants increase linearly with the amount of fertilizer they absorb (Loomis and Connor 1992). To meet increased demand for food, annual world consumption of the primary mineral elements used in fertilizer—nitrogen, phosphorus, and potassium—rose steadily from 30 million metric tons in 1960 to 143 million metric tons in 1990. For a decade after that, consumption remained relatively constant as fertilizer was used more judiciously in an attempt to balance rising

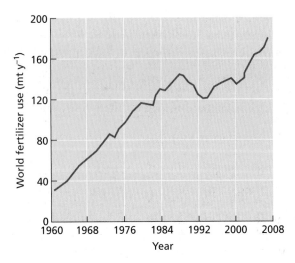

**FIGURE 5.1** Worldwide fertilizer consumption over the past five decades. (After www.faostat.fao.org/site/575/default.aspx#ancor.)

costs. During the past few years, however, annual consumption has climbed to 170 million metric tons (**FIGURE 5.1**).

Crop plants, however, typically use less than half of the fertilizer applied to the soils around them (Loomis and Connor 1992). The remaining minerals may leach into surface waters or groundwater, become attached to soil particles, or contribute to air pollution. As a consequence of fertilizer leaching, many water wells in the United States no longer meet federal standards for nitrate ($NO_3^-$) concentrations in drinking water (Nolan and Hitt 2006). Enhanced nitrogen availability through nitrate ($NO_3^-$) and ammonium ($NH_4^+$) released to the environment from human activities and deposited in the soil by rainwater, a process known as atmospheric nitrogen deposition, is altering ecosystems throughout the United States (Aber et al. 2003; Fenn et al. 2003).

On a brighter note, plants are the traditional means of recycling animal wastes and are proving useful for removing deleterious minerals, including heavy metals, from toxic-waste dumps (Ahluwalia and Goyal 2007). Because of the complex nature of plant–soil–atmosphere relationships, studies of mineral nutrition involve atmospheric chemists, soil scientists, hydrologists, microbiologists, and ecologists as well as plant physiologists.

In this chapter we will discuss the nutritional needs of plants, the symptoms of specific nutritional deficiencies, and the use of fertilizers to ensure proper plant nutrition. Then we will examine how soil structure (the arrangement of solid, liquid, and gaseous components) and root morphology influence the transfer of inorganic nutrients from the environment into a plant. Finally, we will introduce the topic of mycorrhizal symbiotic associations. Chapters 6 and 12 address additional aspects of solute transport and nutrient assimilation, respectively.

## Essential Nutrients, Deficiencies, and Plant Disorders

Only certain elements have been determined to be essential for plants. An **essential element** is defined as one that is an intrinsic component in the structure or metabolism of a plant or whose absence causes severe abnormalities in plant growth, development, or reproduction (Arnon and Stout 1939; Epstein and Bloom 2005). If plants are given these essential elements, as well as water and energy from sunlight, they can synthesize all the compounds they need for normal growth. **TABLE 5.1** lists the elements that are con-

---

**TABLE 5.1**
**Tissue levels of essential elements required by most plants**

| Element | Chemical symbol | Concentration in dry matter (% or ppm)[a] | Relative number of atoms with respect to molybdenum |
|---|---|---|---|
| **Obtained from water or carbon dioxide** | | | |
| Hydrogen | H | 6 | 60,000,000 |
| Carbon | C | 45 | 40,000,000 |
| Oxygen | O | 45 | 30,000,000 |
| **Obtained from the soil** | | | |
| **Macronutrients** | | | |
| Nitrogen | N | 1.5 | 1,000,000 |
| Potassium | K | 1.0 | 250,000 |
| Calcium | Ca | 0.5 | 125,000 |
| Magnesium | Mg | 0.2 | 80,000 |
| Phosphorus | P | 0.2 | 60,000 |
| Sulfur | S | 0.1 | 30,000 |
| Silicon | Si | 0.1 | 30,000 |
| **Micronutrients** | | | |
| Chlorine | Cl | 100 | 3,000 |
| Iron | Fe | 100 | 2,000 |
| Boron | B | 20 | 2,000 |
| Manganese | Mn | 50 | 1,000 |
| Sodium | Na | 10 | 400 |
| Zinc | Zn | 20 | 300 |
| Copper | Cu | 6 | 100 |
| Nickel | Ni | 0.1 | 2 |
| Molybdenum | Mo | 0.1 | 1 |

*Source*: Epstein 1972, 1999.

[a]The values for the nonmineral elements (H, C, O) and the macronutrients are percentages. The values for micronutrients are expressed in parts per million.

sidered to be essential for most, if not all, higher plants. The first three elements—hydrogen, carbon, and oxygen—are not considered mineral nutrients because they are obtained primarily from water or carbon dioxide.

Essential mineral elements are usually classified as *macronutrients* or *micronutrients* according to their relative concentrations in plant tissue. In some cases the differences in tissue content between macronutrients and micronutrients are not as great as those indicated in Table 5.1. For example, some plant tissues, such as leaf mesophyll, have almost as much iron or manganese as they do sulfur or magnesium. Often elements are present in concentrations greater than the plant's minimum requirements.

Some researchers have argued that a classification into macronutrients and micronutrients is difficult to justify physiologically. Mengel and Kirkby (2001) have proposed that the essential elements be classified instead according to their biochemical role and physiological function. TABLE 5.2 shows such a classification, in which plant nutrients have been divided into four basic groups:

1. Nitrogen and sulfur constitute the first group of essential elements. Plants assimilate these nutrients via biochemical reactions involving oxidation and reduction to form covalent bonds with carbon and create organic compounds.

2. The second group is important in energy storage reactions or in maintaining structural integrity. Elements in this group are often present in plant tissues as phosphate, borate, and silicate esters in which the elemental group is covalently bound to an organic molecule (e.g., sugar phosphate).

### TABLE 5.2
### Classification of plant mineral nutrients according to biochemical function

| Mineral nutrient | Functions |
|---|---|
| **Group 1** | **Nutrients that are part of carbon compounds** |
| N | Constituent of amino acids, amides, proteins, nucleic acids, nucleotides, coenzymes, hexosamines, etc. |
| S | Component of cysteine, cystine, methionine. Constituent of lipoic acid, coenzyme A, thiamine pyrophosphate, glutathione, biotin, 5'-adenylylsulfate, and 3'-phosphoadenosine. |
| **Group 2** | **Nutrients that are important in energy storage or structural integrity** |
| P | Component of sugar phosphates, nucleic acids, nucleotides, coenzymes, phospholipids, phytic acid, etc. Has a key role in reactions that involve ATP. |
| Si | Deposited as amorphous silica in cell walls. Contributes to cell wall mechanical properties, including rigidity and elasticity. |
| B | Complexes with mannitol, mannan, polymannuronic acid, and other constituents of cell walls. Involved in cell elongation and nucleic acid metabolism. |
| **Group 3** | **Nutrients that remain in ionic form** |
| K | Required as a cofactor for more than 40 enzymes. Principal cation in establishing cell turgor and maintaining cell electroneutrality. |
| Ca | Constituent of the middle lamella of cell walls. Required as a cofactor by some enzymes involved in the hydrolysis of ATP and phospholipids. Acts as a second messenger in metabolic regulation. |
| Mg | Required by many enzymes involved in phosphate transfer. Constituent of the chlorophyll molecule. |
| Cl | Required for the photosynthetic reactions involved in $O_2$ evolution. |
| Mn | Required for activity of some dehydrogenases, decarboxylases, kinases, oxidases, and peroxidases. Involved with other cation-activated enzymes and photosynthetic $O_2$ evolution. |
| Na | Involved with the regeneration of phosphoenolpyruvate in $C_4$ and CAM plants. Substitutes for potassium in some functions. |
| **Group 4** | **Nutrients that are involved in redox reactions** |
| Fe | Constituent of cytochromes and nonheme iron proteins involved in photosynthesis, $N_2$ fixation, and respiration. |
| Zn | Constituent of alcohol dehydrogenase, glutamic dehydrogenase, carbonic anhydrase, etc. |
| Cu | Component of ascorbic acid oxidase, tyrosinase, monoamine oxidase, uricase, cytochrome oxidase, phenolase, laccase, and plastocyanin. |
| Ni | Constituent of urease. In $N_2$-fixing bacteria, constituent of hydrogenases. |
| Mo | Constituent of nitrogenase, nitrate reductase, and xanthine dehydrogenase. |

*Source*: After Evans and Sorger 1966 and Mengel and Kirkby 2001.

3. The third group is present in plant tissue as either free ions dissolved in the plant water or ions electrostatically bound to substances such as the pectic acids present in the plant cell wall. Elements in this group have important roles as enzyme cofactors and in the regulation of osmotic potentials.

4. The fourth group, comprising metals such as iron, has important roles in reactions involving electron transfer.

Please keep in mind that this classification is somewhat arbitrary because many elements serve several functional roles. For example, manganese is listed in group 3 as a mineral element that remains in ionic form, yet it is involved in several key electron transfer reactions, which would place it in group 4.

Some naturally occurring elements, such as aluminum, selenium, and cobalt, that are not essential elements can also accumulate in plant tissues. Aluminum, for example, is not considered to be an essential element, but plants commonly contain from 0.1 to 500 ppm aluminum, and addition of low levels of aluminum to a nutrient solution may stimulate plant growth (Marschner 1995). Many species in the genera *Astragalus*, *Xylorhiza*, and *Stanleya* accumulate selenium, although plants have not been shown to have a specific requirement for this element. Cobalt is part of cobalamin (vitamin $B_{12}$ and its derivatives), a component of several enzymes in nitrogen-fixing microorganisms. Thus cobalt deficiency blocks the development and function of nitrogen-fixing nodules. Nonetheless, plants that do not fix nitrogen, as well as nitrogen-fixing plants that are supplied with ammonium or nitrate, do not require cobalt. Crop plants normally contain only relatively small amounts of such nonessential elements.

The following sections describe the methods used to examine the roles of nutrient elements within plants.

## Special techniques are used in nutritional studies

To demonstrate that an element is essential requires that plants be grown under experimental conditions in which only the element under investigation is absent. Such conditions are extremely difficult to achieve with plants grown in a complex medium such as soil. In the nineteenth century, several researchers, including Nicolas-Théodore de Saussure, Julius von Sachs, Jean-Baptiste-Joseph-Dieudonné Boussingault, and Wilhelm Knop, approached this problem by growing plants with their roots immersed in a **nutrient solution** containing only inorganic salts. Their demonstration that plants could grow normally with no soil or organic matter proved unequivocally that plants can fulfill all their needs from only inorganic elements, water, and sunlight.

The technique of growing plants with their roots immersed in a nutrient solution without soil is called **solution culture** or **hydroponics** (Gericke 1937). Successful hydroponic culture (**FIGURE 5.2A**) requires a large volume of nutrient solution or frequent adjustment of the nutrient solution to prevent nutrient uptake by roots from producing large changes in the nutrient concentrations and pH of the solution. A sufficient supply of oxygen to the root system—also critical—may be achieved by vigorous bubbling of air through the solution.

Hydroponics is used in the commercial production of many greenhouse crops, such as tomatoes (*Lycopersicon esculentum*). In one form of commercial hydroponic culture, plants are grown in a supporting material such as sand, gravel, vermiculite, or expanded clay (i.e., kitty litter). Nutrient solutions are then flushed through the supporting material, and old solutions are removed by leaching. In another form of hydroponic culture, plant roots lie on the surface of a trough, and nutrient solutions flow in a thin layer along the trough over the roots (Cooper 1979; Asher and Edwards 1983). This **nutrient film growth system** ensures that the roots receive an ample supply of oxygen (**FIGURE 5.2B**).

Another alternative, which has sometimes been heralded as the medium of the future for scientific investigations, is to grow the plants **aeroponically** (Weathers and Zobel 1992). In this technique plants are grown with their roots suspended in air while being sprayed continuously with a nutrient solution (**FIGURE 5.2C**). This approach provides easy manipulation of the gaseous environment around the roots, but it requires higher levels of nutrients than hydroponic culture does to sustain rapid plant growth. For this reason and other technical difficulties, the use of aeroponics is not widespread.

An ebb-and-flow system (**FIGURE 5.2D**) is yet another approach to solution culture. In such systems, the nutrient solution periodically rises to immerse plant roots and then recedes, exposing the roots to a moist atmosphere. Like aeroponics, ebb-and-flow systems require higher levels of nutrients than hydroponics or nutrient films.

## Nutrient solutions can sustain rapid plant growth

Over the years, many formulations have been used for nutrient solutions. Early formulations developed by Knop in Germany included only $KNO_3$, $Ca(NO_3)_2$, $KH_2PO_4$, $MgSO_4$, and an iron salt. At the time, this nutrient solution was believed to contain all the minerals required by plants, but these experiments were carried out with chemicals that were contaminated with other elements that are now known to be essential (such as boron or molybdenum). TABLE 5.3 shows a more modern formulation for a nutrient solution. This formulation is called a modified **Hoagland solution**, named after Dennis R. Hoagland, a researcher who was

**(A) Hydroponic growth system**

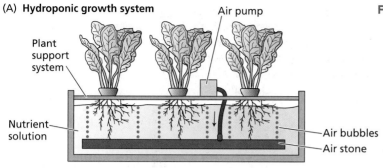

**(B) Nutrient film growth system**

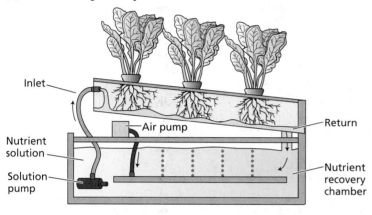

**(C) Aeroponic growth system**

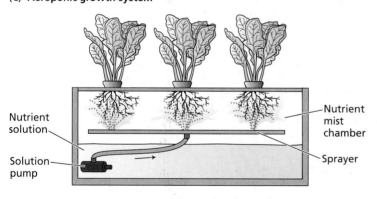

**(D) Ebb and flow system**

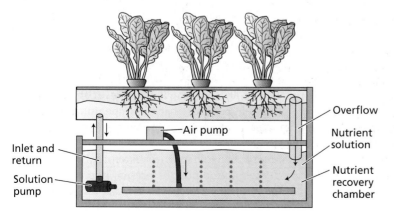

**FIGURE 5.2** Various types of solution culture systems. (A) In a standard hydroponic culture, plants are suspended by the base of the stem over a tank containing a nutrient solution. The pumping of air through an air stone, a porous solid that generates a stream of small bubbles, keeps the solution fully saturated with oxygen. (B) In the nutrient film technique, a pump drives nutrient solution from a main reservoir along the bottom of a tilted tank, and down a return tube back to the reservoir. (C) In one type of aeroponics, a high-pressure pump sprays nutrient solution on roots enclosed in a tank. (D) In an ebb-and-flow system, a pump periodically fills an upper chamber containing the plant roots with nutrient solution. When the pump is turned off, the solution drains back through the pump into a main reservoir. (From Epstein and Bloom 2005.)

prominent in the development of modern mineral nutrition research in the United States.

A modified Hoagland solution contains all the known mineral elements needed for rapid plant growth. The concentrations of these elements are set at the highest possible levels without producing toxicity symptoms or salinity stress, and thus may be several orders of magnitude higher than those found in the soil around plant roots. For example, whereas phosphorus is present in the soil solution at concentrations normally less than 0.06 ppm, here it is offered at 62 ppm (Epstein and Bloom 2005). Such high initial levels permit plants to be grown in a medium for extended periods without replenishment of the nutrient, but may injure young plants. Therefore many researchers dilute their nutrient solutions severalfold and replenish them frequently to minimize fluctuations of nutrient concentration in the medium and in plant tissue.

Another important property of the modified Hoagland formulation is that nitrogen is supplied as both ammonium ($NH_4^+$) and nitrate ($NO_3^-$). Supplying nitrogen in a balanced mixture of cations (positively charged ions) and anions (negatively charged ions) tends to reduce the rapid rise in the pH of the medium that is commonly observed when the nitrogen is supplied solely as nitrate anion (Asher and Edwards 1983). Even when the pH of the medium is kept neutral, most plants grow better if they have access to both $NH_4^+$ and $NO_3^-$ because absorption and assimilation of the two nitrogen forms promotes cation–anion balances within the plant (Raven and Smith 1976; Epstein and Bloom 2005).

## TABLE 5.3
## Composition of a modified Hoagland nutrient solution for growing plants

| Compound | Molecular weight | Concentration of stock solution | Concentration of stock solution | Volume of stock solution per liter of final solution | Element | Final concentration of element | |
|---|---|---|---|---|---|---|---|
| | g mol$^{-1}$ | m$M$ | g L$^{-1}$ | mL | | μ$M$ | ppm |
| **Macronutrients** | | | | | | | |
| KNO$_3$ | 101.10 | 1,000 | 101.10 | 6.0 | N | 16,000 | 224 |
| Ca(NO$_3$)$_2$·4H$_2$O | 236.16 | 1,000 | 236.16 | 4.0 | K | 6,000 | 235 |
| NH$_4$H$_2$PO$_4$ | 115.08 | 1,000 | 115.08 | 2.0 | Ca | 4,000 | 160 |
| MgSO$_4$·7H$_2$O | 246.48 | 1,000 | 246.49 | 1.0 | P | 2,000 | 62 |
| | | | | | S | 1,000 | 32 |
| | | | | | Mg | 1,000 | 24 |
| **Micronutrients** | | | | | | | |
| KCl | 74.55 | 25 | 1.864 | | Cl | 50 | 1.77 |
| H$_3$BO$_3$ | 61.83 | 12.5 | 0.773 | | B | 25 | 0.27 |
| MnSO$_4$·H$_2$O | 169.01 | 1.0 | 0.169 | 2.0 | Mn | 2.0 | 0.11 |
| ZnSO$_4$·7H$_2$O | 287.54 | 1.0 | 0.288 | | Zn | 2.0 | 0.13 |
| CuSO$_4$·5H$_2$O | 249.68 | 0.25 | 0.062 | | Cu | 0.5 | 0.03 |
| H$_2$MoO$_4$ (85% MoO$_3$) | 161.97 | 0.25 | 0.040 | | Mo | 0.5 | 0.05 |
| NaFeDTPA | 468.20 | 64 | 30.0 | 0.3–1.0 | Fe | 16.1–53.7 | 1.00–3.00 |
| **Optional**[a] | | | | | | | |
| NiSO$_4$·6H$_2$O | 262.86 | 0.25 | 0.066 | 2.0 | Ni | 0.5 | 0.03 |
| Na$_2$SiO$_3$·9H$_2$O | 284.20 | 1,000 | 284.20 | 1.0 | Si | 1,000 | 28 |

*Source:* After Epstein 1972.

*Note:* The macronutrients are added separately from stock solutions to prevent precipitation during preparation of the nutrient solution. A combined stock solution is made up containing all micronutrients except iron. Iron is added as sodium ferric diethylenetriaminepentaacetate (NaFeDTPA, trade name Ciba-Geigy Sequestrene 330 Fe; see Figure 5.3); some plants, such as maize, require the higher level of iron shown in the table.

[a]Nickel is usually present as a contaminant of the other chemicals, so it may not need to be added explicitly. Silicon, if included, should be added first and the pH adjusted with HCl to prevent precipitation of the other nutrients.

A significant problem with nutrient solutions is maintaining the availability of iron. When supplied as an inorganic salt such as FeSO$_4$ or Fe(NO$_3$)$_2$, iron can precipitate out of solution as iron hydroxide, particularly under alkaline conditions. If phosphate salts are present, insoluble iron phosphate will also form. Precipitation of the iron out of solution makes it physically unavailable to the plant, unless iron salts are added at frequent intervals. Earlier researchers solved this problem by adding iron together with citric acid or tartaric acid. Compounds such as these are called **chelators** because they form soluble complexes with cations such as iron and calcium in which the cation is held by ionic forces, rather than by covalent bonds. Chelated cations thus remain physically available to plants.

More modern nutrient solutions use the chemicals ethylenediaminetetraacetic acid (EDTA) or diethylenetriaminepentaacetic acid (DTPA, or pentetic acid) as chelating agents (Sievers and Bailar 1962). **FIGURE 5.3** shows the structure of DTPA. The fate of the chelation complex during iron uptake by the root cells is not clear; iron may be released from the chelator when it is reduced from ferric iron (Fe$^{3+}$) to ferrous iron (Fe$^{2+}$) at the root surface. The chelator may then diffuse back into the nutrient (or soil) solution and react with another Fe$^{3+}$ ion or other metal ions.

After uptake into the root, iron is kept soluble by chelation with organic compounds present in plant cells. Citric acid may play a major role as such an organic iron

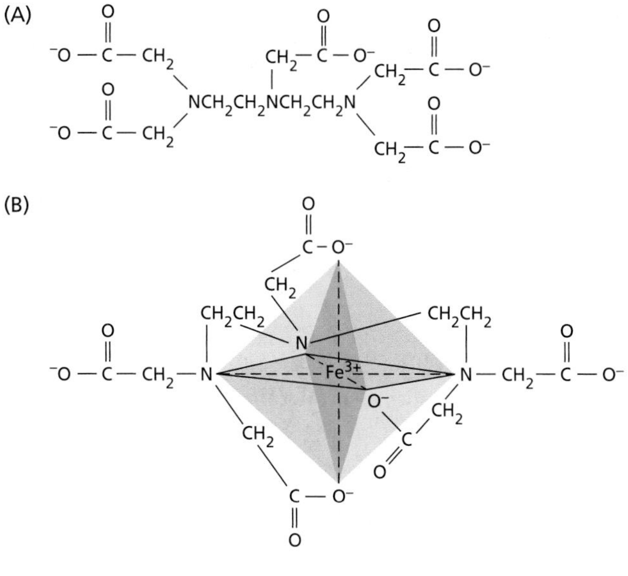

**FIGURE 5.3** Chelator and chelated cation. Chemical structure of the chelator diethylenetriaminepentaacetic acid (DTPA), by itself (A) and chelated to an $Fe^{3+}$ ion (B). Iron binds to DTPA through interactions with three nitrogen atoms and the three ionized oxygen atoms of the carboxylate groups (Sievers and Bailar 1962). The resulting ring structure clamps the metallic ion and effectively neutralizes its reactivity in solution. During the uptake of iron at the root surface, $Fe^{3+}$ appears to be reduced to $Fe^{2+}$, which is released from the DTPA–iron complex. The chelator can then bind to other available $Fe^{3+}$ ions.

chelator, and long-distance transport of iron in the xylem appears to involve an iron–citric acid complex.

## Mineral deficiencies disrupt plant metabolism and function

An inadequate supply of an essential element results in a nutritional disorder manifested by characteristic deficiency symptoms. In hydroponic culture, withholding of an essential element can be readily correlated with a given set of symptoms. For example, a particular deficiency might elicit a specific pattern of leaf discoloration. Diagnosis of soil-grown plants can be more complex for the following reasons:

• Deficiencies of several elements may occur simultaneously in different plant tissues.

• Deficiencies or excessive amounts of one element may induce deficiencies or excessive accumulations of another.

• Some virus-induced plant diseases may produce symptoms similar to those of nutrient deficiencies.

Nutrient deficiency symptoms in a plant are the expression of metabolic disorders resulting from the insufficient supply of an essential element. These disorders are related to the roles played by essential elements in normal plant metabolism and function (listed in Table 5.2).

Although each essential element participates in many different metabolic reactions, some general statements about the functions of essential elements in plant metabolism are possible. In general, the essential elements function in plant structure, metabolism, and cellular osmoregulation. More specific roles may be related to the ability of divalent cations such as calcium or magnesium to modify the permeability of plant membranes. In addition, research continues to reveal specific roles for these elements in plant metabolism; for example, calcium acts as a signal to regulate key enzymes in the cytosol (Hepler and Wayne 1985; Hetherington and Brownlee 2004). Thus most essential elements have multiple roles in plant metabolism.

An important clue in relating an acute deficiency symptom to a particular essential element is the extent to which an element can be recycled from older to younger leaves. Some elements, such as nitrogen, phosphorus, and potassium, can readily move from leaf to leaf; others, such as boron, iron, and calcium, are relatively immobile in most plant species (TABLE 5.4). If an essential element is mobile, deficiency symptoms tend to appear first in older leaves. Deficiency of an immobile essential element becomes evident first in younger leaves. Although the precise mechanisms of nutrient mobilization are not well understood, plant hormones such as cytokinins appear to be involved (see Chapter 21). In the discussion that follows, we will describe the specific deficiency symptoms and functional roles of the mineral essential elements as they are grouped in Table 5.2.

**TABLE 5.4**
**Mineral elements classified on the basis of their mobility within a plant and their tendency to retranslocate during deficiencies**

| Mobile | Immobile |
|---|---|
| Nitrogen | Calcium |
| Potassium | Sulfur |
| Magnesium | Iron |
| Phosphorus | Boron |
| Chlorine | Copper |
| Sodium | |
| Zinc | |
| Molybdenum | |

*Note:* Elements are listed in the order of their abundance in the plant.

*Group 1: Deficiencies in mineral nutrients that are part of carbon compounds.* This first group consists of nitrogen and sulfur. Nitrogen availability in soils limits plant productivity in most natural and agricultural ecosystems. By contrast, soils generally contain sulfur in excess. Despite this difference, nitrogen and sulfur are similar chemically in that their oxidation–reduction states range widely (see Chapter 12). Some of the most energy-intensive reactions in life convert the highly oxidized, inorganic forms, such as nitrate and sulfate, that roots absorb from the soil into the highly reduced forms found in organic compounds such as amino acids within plants.

**NITROGEN** Nitrogen is the mineral element that plants require in the greatest amounts. It serves as a constituent of many plant cell components, including amino acids, proteins, and nucleic acids. Therefore nitrogen deficiency rapidly inhibits plant growth. If such a deficiency persists, most species show **chlorosis** (yellowing of the leaves), especially in the older leaves near the base of the plant (for pictures of nitrogen deficiency and the other mineral deficiencies described in this chapter, see **WEB TOPIC 5.1**). Under severe nitrogen deficiency, these leaves become completely yellow (or tan) and fall off the plant. Younger leaves may not show these symptoms initially because nitrogen can be mobilized from older leaves. Thus a nitrogen-deficient plant may have light green upper leaves and yellow or tan lower leaves.

When nitrogen deficiency develops slowly, plants may have markedly slender and often woody stems. This woodiness may be due to a buildup of excess carbohydrates that cannot be used in the synthesis of amino acids or other nitrogen-containing compounds. Carbohydrates not used in nitrogen metabolism may also be used in anthocyanin synthesis, leading to accumulation of that pigment. This condition is revealed as a purple coloration in leaves, petioles, and stems of nitrogen-deficient plants of some species, such as tomato and certain varieties of corn (maize; *Zea mays*).

**SULFUR** Sulfur is found in amino acids (cystine, cysteine, and methionine) and is a constituent of several coenzymes and vitamins, such as coenzyme A, S-adenosylmethionine, biotin, vitamin $B_1$, and pantothenic acid, which are essential for metabolism.

Many of the symptoms of sulfur deficiency are similar to those of nitrogen deficiency, including chlorosis, stunting of growth, and anthocyanin accumulation. This similarity is not surprising, since sulfur and nitrogen are both constituents of proteins. The chlorosis caused by sulfur deficiency, however, generally arises initially in mature and young leaves, rather than in old leaves as in nitrogen deficiency, because sulfur, unlike nitrogen, is not easily remobilized to the younger leaves in most species. Nonetheless, in many plant species, sulfur chlorosis may occur simultaneously in all leaves, or even initially in older leaves.

*Group 2: Deficiencies in mineral nutrients that are important in energy storage or structural integrity.* This group consists of phosphorus, silicon, and boron. Phosphorus and silicon are found at concentrations within plant tissue that warrant their classification as macronutrients, whereas boron is much less abundant and is considered a micronutrient. These elements are usually present in plants as ester linkages to a carbon molecule (i.e., X–O–C–R, where the element X is attached to a carbon-containing molecule C–R via an oxygen atom O).

**PHOSPHORUS** Phosphorus (as phosphate, $PO_4^{3-}$) is an integral component of important compounds of plant cells, including the sugar–phosphate intermediates of respiration and photosynthesis as well as the phospholipids that make up plant membranes. It is also a component of nucleotides used in plant energy metabolism (such as ATP) and in DNA and RNA. Characteristic symptoms of phosphorus deficiency include stunted growth in young plants and a dark green coloration of the leaves, which may be malformed and contain small spots of dead tissue called **necrotic spots** (for a picture, see **WEB TOPIC 5.1**).

As in nitrogen deficiency, some species may produce excess anthocyanins, giving the leaves a slight purple coloration. In contrast to nitrogen deficiency, the purple coloration of phosphorus deficiency is not associated with chlorosis. In fact, the leaves may be a dark greenish purple. Additional symptoms of phosphorus deficiency include the production of slender (but not woody) stems and the death of older leaves. Maturation of the plant may also be delayed.

**SILICON** Only members of the family Equisetaceae—called *scouring rushes* because at one time their ash, rich in gritty silica, was used to scour pots—require silicon to complete their life cycle. Nonetheless, many other species accumulate substantial amounts of silicon within their tissues and show enhanced growth, fertility, and stress resistance when supplied with adequate amounts of silicon (Epstein 1999).

Plants deficient in silicon are more susceptible to lodging (falling over) and fungal infection. Silicon is deposited primarily in the endoplasmic reticulum, cell walls, and intercellular spaces as hydrated, amorphous silica ($SiO_2 \cdot nH_2O$). It also forms complexes with polyphenols and thus serves as an alternative to lignin in the reinforcement of cell walls. In addition, silicon can lessen the toxicity of many metals, including aluminum and manganese.

**BORON** Although the precise function of boron in plant metabolism is unclear, evidence suggests that it plays roles in cell elongation, nucleic acid synthesis, hormone responses, membrane function, and cell cycle regulation (Brown et al. 2002; Reguera et al. 2009). Boron-deficient plants may exhibit a wide variety of symptoms, depending on the species and the age of the plant.

A characteristic symptom is black necrosis of young leaves and terminal buds. The necrosis of the young leaves occurs primarily at the base of the leaf blade. Stems may be unusually stiff and brittle. Apical dominance may also be lost, causing the plant to become highly branched; however, the terminal apices of the branches soon become necrotic because of inhibition of cell division. Structures such as the fruits, fleshy roots, and tubers may exhibit necrosis or abnormalities related to the breakdown of internal tissues (see WEB ESSAY 5.1).

*Group 3: Deficiencies in mineral nutrients that remain in ionic form.* This group includes some of the most familiar mineral elements: the macronutrients potassium, calcium, and magnesium and the micronutrients chlorine, manganese, and sodium. These elements may be found as ions in solution in the cytosol or vacuoles, or they may be bound electrostatically or as ligands to larger, carbon-containing compounds.

**POTASSIUM**   Potassium, present within plants as the cation $K^+$, plays an important role in regulation of the osmotic potential of plant cells (see Chapters 3 and 6). It also activates many enzymes involved in respiration and photosynthesis.

The first observable symptom of potassium deficiency is mottled or marginal chlorosis, which then develops into necrosis primarily at the leaf tips, at the margins, and between veins. In many monocots, these necrotic lesions may initially form at the leaf tips and margins and then extend toward the leaf base. Because potassium can be mobilized to the younger leaves, these symptoms appear initially on the more mature leaves toward the base of the plant. The leaves may also curl and crinkle. The stems of potassium-deficient plants may be slender and weak, with abnormally short internodal regions. In potassium-deficient corn, the roots may have an increased susceptibility to root-rotting fungi present in the soil, and this susceptibility, together with effects on the stem, results in an increased tendency for the plant to be easily bent to the ground (lodging).

**CALCIUM**   Calcium ions ($Ca^{2+}$) are used in the synthesis of new cell walls, particularly the middle lamellae that separate newly divided cells. Calcium is also used in the mitotic spindle during cell division. It is required for the normal functioning of plant membranes and has been implicated as a second messenger for various plant responses to both environmental and hormonal signals (White and Broadley 2003, Hetherington and Brownlee 2004). In its function as a second messenger, calcium may bind to **calmodulin**, a protein found in the cytosol of plant cells. The calmodulin–calcium complex then binds to several different types of proteins, including kinases, phosphatases, second messenger signaling proteins, and cytoskeletal proteins, and

thereby regulates many cellular processes ranging from transcription control and cell survival to release of chemical signals (see Chapter 24).

Characteristic symptoms of calcium deficiency include necrosis of young meristematic regions such as the tips of roots or young leaves, where cell division and cell wall formation are most rapid. Necrosis in slowly growing plants may be preceded by a general chlorosis and downward hooking of young leaves. Young leaves may also appear deformed. The root system of a calcium-deficient plant may appear brownish, short, and highly branched. Severe stunting may result if the meristematic regions of the plant die prematurely.

**MAGNESIUM**   In plant cells, magnesium ions ($Mg^{2+}$) have a specific role in the activation of enzymes involved in respiration, photosynthesis, and the synthesis of DNA and RNA. Magnesium is also a part of the ring structure of the chlorophyll molecule (see Figure 7.6A). A characteristic symptom of magnesium deficiency is chlorosis between the leaf veins, occurring first in older leaves because of the mobility of this cation. This pattern of chlorosis results because the chlorophyll in the vascular bundles remains unaffected longer than that in the cells between the bundles. If the deficiency is extensive, the leaves may become yellow or white. An additional symptom of magnesium deficiency may be premature leaf abscission.

**CHLORINE**   The element chlorine is found in plants as the chloride ion ($Cl^-$). It is required for the water-splitting reaction of photosynthesis through which oxygen is produced (see Chapter 7) (Popelkova and Yocum 2007). In addition, chlorine may be required for cell division in both leaves and roots (Colmenero-Flores et al. 2007). Plants deficient in chlorine develop wilting of the leaf tips followed by general leaf chlorosis and necrosis. The leaves may also exhibit reduced growth. Eventually, the leaves may take on a bronzelike color ("bronzing"). Roots of chlorine-deficient plants may appear stunted and thickened near the root tips.

Chloride ions are highly soluble and are generally available in soils because seawater is swept into the air by wind and delivered to soil when it rains. Therefore chlorine deficiency is only rarely observed in plants grown in native or agricultural habitats (Engel et al. 2001). Most plants absorb chlorine at levels much higher than those required for normal functioning.

**MANGANESE**   Manganese ions ($Mn^{2+}$) activate several enzymes in plant cells. In particular, decarboxylases and dehydrogenases involved in the citric acid (Krebs) cycle are specifically activated by manganese. The best-defined function of manganese is in the photosynthetic reaction through which oxygen ($O_2$) is produced from water (Armstrong 2008) (see Chapter 7). The major symptom of manganese deficiency is intervenous chlorosis associated with the development of small necrotic spots. This chlorosis may occur on younger or older leaves, depending on plant species and growth rate.

**SODIUM** Most species utilizing the $C_4$ and crassulacean acid metabolism (CAM) pathways of carbon fixation (see Chapter 8) require sodium ions ($Na^+$). In these plants, sodium appears vital for regenerating phosphoenolpyruvate, the substrate for the first carboxylation in the $C_4$ and CAM pathways (Brownell and Bielig 1996). Under sodium deficiency, these plants exhibit chlorosis and necrosis, or even fail to form flowers. Many $C_3$ species also benefit from exposure to low levels of sodium ions. Sodium stimulates growth through enhanced cell expansion, and it can partly substitute for potassium as an osmotically active solute.

*Group 4: Deficiencies in mineral nutrients that are involved in redox reactions.* This group of five micronutrients consists of the metals iron, zinc, copper, nickel, and molybdenum. All of these can undergo reversible oxidations and reductions (e.g., $Fe^{2+} \leftrightarrow Fe^{3+}$) and have important roles in electron transfer and energy transformation. They are usually found in association with larger molecules such as cytochromes, chlorophyll, and proteins (usually enzymes).

**IRON** Iron has an important role as a component of enzymes involved in the transfer of electrons (redox reactions), such as cytochromes. In this role, it is reversibly oxidized from $Fe^{2+}$ to $Fe^{3+}$ during electron transfer.

As in magnesium deficiency, a characteristic symptom of iron deficiency is intervenous chlorosis. These symptoms, however, appear initially on younger leaves because iron, unlike magnesium, cannot be readily mobilized from older leaves. Under conditions of extreme or prolonged deficiency, the veins may also become chlorotic, causing the whole leaf to turn white. The leaves become chlorotic because iron is required for the synthesis of some of the chlorophyll–protein complexes in the chloroplast. The low mobility of iron is probably due to its precipitation in the older leaves as insoluble oxides or phosphates, or to the formation of complexes with phytoferritin, an iron-binding protein found in the leaf and other plant parts (Jeong and Guerinot 2009). The precipitation of iron diminishes subsequent mobilization of the metal into the phloem for long-distance translocation.

**ZINC** Many enzymes require zinc ions ($Zn^{2+}$) for their activity, and zinc may be required for chlorophyll biosynthesis in some plants (Broadley et al. 2007). Zinc deficiency is characterized by a reduction in internodal growth, and as a result plants display a rosette habit of growth in which the leaves form a circular cluster radiating at or close to the ground. The leaves may also be small and distorted, with leaf margins having a puckered appearance. These symptoms may result from loss of the capacity to produce sufficient amounts of the auxin indole-3-acetic acid (IAA). In some species (e.g., corn, sorghum, and beans), older leaves may become intervenously chlorotic and then develop white necrotic spots. This chlorosis may be an expression of a zinc requirement for chlorophyll biosynthesis.

**COPPER** Like iron, copper is associated with enzymes involved in redox reactions, through which it is reversibly oxidized from $Cu^+$ to $Cu^{2+}$. An example of such an enzyme is plastocyanin, which is involved in electron transfer during the light reactions of photosynthesis (Yruela 2009). The initial symptom of copper deficiency is the production of dark green leaves, which may contain necrotic spots. The necrotic spots appear first at the tips of young leaves and then extend toward the leaf base along the margins. The leaves may also be twisted or malformed. Under extreme copper deficiency, leaves may abscise prematurely.

**NICKEL** Urease is the only known nickel-containing ($Ni^{2+}$) enzyme in higher plants, although nitrogen-fixing microorganisms require nickel ($Ni^+$ through $Ni^{4+}$) for the enzyme that reprocesses some of the hydrogen gas generated during fixation (hydrogen uptake hydrogenase) (see Chapter 12). Nickel-deficient plants accumulate urea in their leaves and consequently show leaf tip necrosis. Nickel deficiency in the field has been found in only one crop, pecan trees in the southeastern United States (Wood et al. 2003), because the amounts of nickel required are minuscule.

**MOLYBDENUM** Molybdenum ions ($Mo^{4+}$ through $Mo^{6+}$) are components of several enzymes, including nitrate reductase and nitrogenase (Schwarz and Mendel 2006). Nitrate reductase catalyzes the reduction of nitrate to nitrite during its assimilation by the plant cell; nitrogenase converts nitrogen gas to ammonia in nitrogen-fixing microorganisms (see Chapter 12). The first indication of a molybdenum deficiency is general chlorosis between veins and necrosis of older leaves. In some plants, such as cauliflower or broccoli, the leaves may not become necrotic, but instead may appear twisted and subsequently die (whiptail disease). Flower formation may be prevented, or the flowers may abscise prematurely.

Because molybdenum is involved with both nitrate assimilation and nitrogen fixation, a molybdenum deficiency may bring about a nitrogen deficiency if the nitrogen source is primarily nitrate or if the plant depends on symbiotic nitrogen fixation. Although plants require only small amounts of molybdenum, some soils (for example, acidic soils in Australia) supply inadequate levels. Small additions of molybdenum to such soils can greatly enhance crop or forage growth at negligible cost.

## Analysis of plant tissues reveals mineral deficiencies

Requirements for mineral elements change during the growth and development of a plant. In crop plants, nutrient levels at certain stages of growth influence the yield of the economically important tissues (tuber, grain, and so on). To optimize yields, farmers use analyses of nutrient levels in soil and in plant tissue to determine fertilizer schedules.

**Soil analysis** is the chemical determination of the nutrient content in a soil sample from the root zone. As discussed later in the chapter, both the chemistry and the biology of soils are complex, and the results of soil analyses vary with sampling methods, storage conditions for the samples, and nutrient extraction techniques. Perhaps more important is that a particular soil analysis reflects the levels of nutrients *potentially* available to the plant roots from the soil, but soil analysis does not tell us how much of a particular mineral nutrient the plant actually needs or is able to absorb. This additional information is best determined by plant tissue analysis.

Proper use of **plant tissue analysis** requires an understanding of the relationship between plant growth (or yield) and the concentration of a nutrient in plant tissue samples (Bouma 1983). **FIGURE 5.4** identifies three zones (deficiency, adequate, and toxic) in the response of growth to increasing tissue concentrations of a nutrient. When the nutrient concentration in a tissue sample is low, growth is reduced. In this **deficiency zone** of the curve, an increase in nutrient availability is directly related to an increase in growth or yield. As nutrient availability continues to increase, a point is reached at which further addition of the nutrient is no longer related to increases in growth or yield, but is reflected in increased tissue concentrations. This region of the curve is called the **adequate zone**.

The point of transition between the deficiency and adequate zones of the curve reveals the **critical concentration** of the nutrient (see Figure 5.4), which may be defined as the minimum tissue content of the nutrient that is correlated with maximal growth or yield. As the nutrient concentration of the tissue increases beyond the adequate zone, growth or yield declines because of toxicity (this region of the curve is the **toxic zone**).

To evaluate the relationship between growth and tissue nutrient concentration, researchers grow plants in soil or a nutrient solution in which all the nutrients are present in adequate amounts except the nutrient under consideration. At the start of the experiment, the limiting nutrient is added in increasing concentrations to different sets of plants, and the concentrations of the nutrient in specific tissues are correlated with a particular measure of growth or yield. Several curves are established for each element, one for each tissue and tissue age.

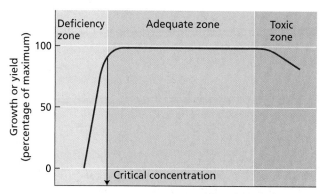

**FIGURE 5.4** Relationship between yield (or growth) and the nutrient content of the plant tissue defines zones of deficiency, adequacy, and toxicity. Yield or growth may be expressed in terms of shoot dry weight or height. To obtain data of this type, plants are grown under conditions in which the concentration of one essential nutrient is varied while all others are in adequate supply. The effect of varying the concentration of this nutrient during plant growth is reflected in the growth or yield. The critical concentration for that nutrient is the concentration below which yield or growth is reduced.

Because agricultural soils are often limited in the elements nitrogen, phosphorus, and potassium, many farmers routinely take into account, at a minimum, growth or yield responses for these elements. If a nutrient deficiency is suspected, steps are taken to correct the deficiency before it reduces growth or yield. Plant analysis has proved useful in establishing fertilizer schedules that sustain yields and ensure the food quality of many crops.

## Treating Nutritional Deficiencies

Many traditional and subsistence farming practices promote the recycling of mineral elements. Crop plants absorb nutrients from the soil, humans and animals consume locally grown crops, and crop residues and manure from humans and animals return the nutrients to the soil. The main losses of nutrients from such agricultural systems ensue from leaching that carries dissolved ions, especially nitrate, away with drainage water. In acidic soils, leaching may be decreased by the addition of lime—a mix of $CaO$, $CaCO_3$, and $Ca(OH)_2$—to make the soil more alkaline because many mineral elements form less soluble compounds when the pH is higher than 6 (**FIGURE 5.5**). This decrease in leaching, however, may be gained at the expense of decreased availability of some nutrients, especially iron.

In the high-production agricultural systems of industrialized countries, a large proportion of crop biomass leaves

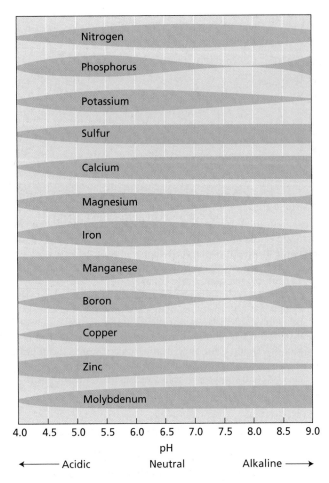

**FIGURE 5.5**  Influence of soil pH on the availability of nutrient elements in organic soils. The width of the shaded areas indicates the degree of nutrient availability to plant roots. All of these nutrients are available in the pH range of 5.5 to 6.5. (After Lucas and Davis 1961.)

be added to the soil as fertilizers. Adding micronutrients to the soil may also be necessary to correct a preexisting deficiency. For example, many acidic, sandy soils in humid regions are deficient in boron, copper, zinc, manganese, molybdenum, or iron (Mengel and Kirkby 2001) and can benefit from nutrient supplementation.

Chemicals may also be applied to the soil to modify soil pH. As Figure 5.5 shows, soil pH affects the availability of all mineral nutrients. Addition of lime, as mentioned previously, can raise the pH of acidic soils; addition of elemental sulfur can lower the pH of alkaline soils. In the latter case, microorganisms absorb the sulfur and subsequently release sulfate and hydrogen ions that acidify the soil.

**Organic fertilizers**, in contrast to chemical fertilizers, originate from the residues of plant or animal life or from natural rock deposits. Plant and animal residues contain many nutrient elements in the form of organic compounds. Before crop plants can acquire the nutrient elements from these residues, the organic compounds must be broken down, usually by the action of soil microorganisms through a process called **mineralization**. Mineralization depends on many factors, including temperature, water and oxygen availability, and the type and number of microorganisms present in the soil. As a consequence, rates of mineralization are highly variable, and nutrients from organic residues become available to plants over periods that range from days to months to years. This slow rate of mineralization hinders efficient fertilizer use, so farms that rely solely on organic fertilizers may require the addition of substantially more nitrogen or phosphorus, and suffer even higher nutrient losses, than farms that use chemical fertilizers. Residues from organic fertilizers do improve the physical structure of most soils, enhancing water retention during drought and increasing drainage in wet weather.

## Some mineral nutrients can be absorbed by leaves

In addition to absorbing nutrients added to the soil as fertilizers, most plants can absorb mineral nutrients applied to their leaves as sprays, a process known as **foliar application**. In some cases this method has agronomic advantages over the application of nutrients to the soil. Foliar application can reduce the lag time between application and uptake by the plant, which could be important during a phase of rapid growth. It can also circumvent the problem of restricted uptake of a nutrient from the soil. For example, foliar application of mineral nutrients such as iron, manganese, and copper may be more efficient than application through the soil, where these ions are adsorbed on soil particles and hence are less available to the root system.

Nutrient uptake by leaves is most effective when the nutrient solution remains on the leaf as a thin film (Mengel and Kirkby 2001). Production of a thin film often requires that the nutrient solutions be supplemented with surfactant chemicals, such as the detergent Tween 80, that reduce

the area of cultivation, and returning crop residues to the land where the crop was produced becomes difficult at best. This unidirectional removal of nutrients from agricultural soils make it important to restore the lost nutrients to these soil through the addition of fertilizers.

## Crop yields can be improved by addition of fertilizers

Most chemical fertilizers contain inorganic salts of the macronutrients nitrogen, phosphorus, and potassium (see Table 5.1). Fertilizers that contain only one of these three nutrients are termed *straight fertilizers*. Some examples of straight fertilizers are superphosphate, ammonium nitrate, and muriate of potash (a source of potassium). Fertilizers that contain two or more mineral nutrients are called *compound fertilizers* or *mixed fertilizers*, and the numbers on the package label, such as "10-14-10," refer to the percentages of N, P as $P_2O_5$, and K as $K_2O$, respectively, in the fertilizer.

With long-term agricultural production, consumption of micronutrients can reach a point at which they, too, must

surface tension. Nutrient movement into the plant seems to involve diffusion through the cuticle and uptake by leaf cells. Although uptake through the stomatal pore could provide a pathway into the leaf, the architecture of the pore (see Figures 4.12 and 4.13) largely prevents liquid penetration (Ziegler 1987).

For foliar nutrient application to be successful, damage to the leaves must be minimized. If foliar sprays are applied on a hot day, when evaporation is high, salts may accumulate on the leaf surface and cause burning or scorching. Spraying on cool days or in the evening helps to alleviate this problem. Addition of lime to the spray diminishes the solubility of many nutrients and limits toxicity. Foliar application has proved economically successful mainly with tree crops and vines such as grapes, but it is also used with cereals. Nutrients applied to the leaves could save an orchard or vineyard when soil-applied nutrients would be too slow to correct a deficiency. In wheat (*Triticum aestivum*), nitrogen applied to the leaves during the later stages of growth enhances the protein content of seeds.

# Soil, Roots, and Microbes

Soil is complex physically, chemically, and biologically. It is a heterogeneous substance containing solid, liquid, and gaseous phases (see Chapter 4). All of these phases interact with mineral elements. The inorganic particles of the solid phase provide a reservoir of potassium, calcium, magnesium, and iron. Also associated with this solid phase are organic compounds containing nitrogen, phosphorus, and sulfur, among other elements. The liquid phase of soil constitutes the soil solution, which contains dissolved mineral ions and serves as the medium for ion movement to the root surface. Gases such as oxygen, carbon dioxide, and nitrogen are dissolved in the soil solution, but roots exchange gases with soils predominantly through the air gaps between soil particles.

From a biological perspective, soil constitutes a diverse ecosystem in which plant roots and microorganisms compete strongly for mineral nutrients. Despite this competi-

tion, roots and microorganisms can form alliances for their mutual benefit (**symbioses**, singular *symbiosis*). In this section we will discuss the importance of soil properties, root structure, and mycorrhizal symbiotic relationships to plant mineral nutrition. Chapter 12 addresses symbiotic relationships with nitrogen-fixing bacteria.

## Negatively charged soil particles affect the adsorption of mineral nutrients

Soil particles, both inorganic and organic, have predominantly negative charges on their surfaces. Many inorganic soil particles are crystal lattices that are tetrahedral arrangements of the cationic forms of aluminum ($Al^{3+}$) and silicon ($Si^{4+}$) bound to oxygen atoms, thus forming aluminates and silicates. When cations of lesser charge replace $Al^{3+}$ and $Si^{4+}$ within the crystal lattice, these inorganic soil particles become negatively charged.

Organic soil particles originate from the products of microbial decomposition of dead plants, animals, and microorganisms. The negative surface charges of organic particles result from the dissociation of hydrogen ions from the carboxylic acid and phenolic groups present in this component of the soil. Most of the world's soil particles, however, are inorganic.

Inorganic soils are categorized by particle size:

- Gravel consists of particles larger than 2 mm.
- Coarse sand consists of particles between 0.2 and 2 mm.
- Fine sand consists of particles between 0.02 and 0.2 mm.
- Silt consists of particles between 0.002 and 0.02 mm.
- Clay consists of particles smaller than 0.002 mm.

The silicate-containing clay materials are further divided into three major groups—kaolinite, illite, and montmorillonite—based on differences in their structure and physical properties (**TABLE 5.5**). The kaolinite group is generally

**TABLE 5.5**
**Comparison of properties of three major types of silicate clays found in the soil**

| Property | Type of clay | | |
| --- | --- | --- | --- |
| | Montmorillonite | Illite | Kaolinite |
| Size (μm) | 0.01–1.0 | 0.1–2.0 | 0.1–5.0 |
| Shape | Irregular flakes | Irregular flakes | Hexagonal crystals |
| Cohesion | High | Medium | Low |
| Water-swelling capacity | High | Medium | Low |
| Cation exchange capacity (milliequivalents 100 $g^{-1}$) | 80–100 | 15–40 | 3–15 |

*Source*: After Brady 1974.

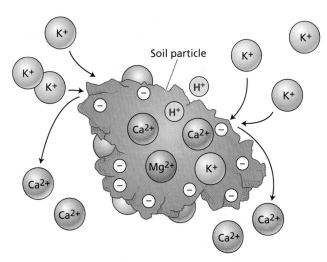

**FIGURE 5.6** The principle of cation exchange on the surface of a soil particle. Cations are adsorbed on the surface of a soil particle because that surface is negatively charged. Addition of one cation, such as potassium ($K^+$), to the soil can displace other cations, such as calcium ($Ca^{2+}$), from the surface of the soil particle and make it available for uptake by roots.

found in well-weathered soils; the montmorillonite and illite groups are found in less weathered soils.

Mineral cations such as ammonium ($NH_4^+$) and potassium ($K^+$) adsorb to the negative surface charges of inorganic and organic soil particles. This cation adsorption is an important factor in soil fertility. Mineral cations adsorbed on the surface of soil particles, which are not easily lost when the soil is leached by water, provide a nutrient reserve available to plant roots. Mineral nutrients adsorbed in this way can be replaced by other cations in a process known as **cation exchange** (FIGURE 5.6). The degree to which a soil can adsorb and exchange ions is termed its *cation exchange capacity* (*CEC*) and is highly dependent on the soil type. A soil with higher cation exchange capacity generally has a larger reserve of mineral nutrients.

Mineral anions such as nitrate ($NO_3^-$) and chloride ($Cl^-$) tend to be repelled by the negative charge on the surface of soil particles and remain dissolved in the soil solution. Thus the anion exchange capacity of most agricultural soils is small compared with the cation exchange capacity. Nitrate, in particular, remains mobile in the soil solution, where it is susceptible to leaching by water moving through the soil.

Phosphate ions ($H_2PO_2^-$) may bind to soil particles containing aluminum or iron because the positively charged iron and aluminum ions ($Fe^{2+}$, $Fe^{3+}$, and $Al^{3+}$) are associated with hydroxyl ($OH^-$) groups that are exchanged for phosphate. As a result, phosphate can be tightly bound, and its lack of mobility and availability in soil can limit plant growth.

Sulfate ($SO_4^{2-}$) in the presence of $Ca^{2+}$ forms gypsum ($CaSO_4$). Gypsum is only slightly soluble, but it releases sufficient sulfate to support plant growth. Most nonacidic soils contain substantial amounts of $Ca^{2+}$; consequently, sulfate mobility in these soils is low, so sulfate is not highly susceptible to leaching.

### Soil pH affects nutrient availability, soil microbes, and root growth

Hydrogen ion concentration (pH) is an important property of soils because it affects the growth of plant roots and soil microorganisms. Root growth is generally favored in slightly acidic soils, at pH values between 5.5 and 6.5. Fungi generally predominate in acidic (pH below 7) soils; bacteria become more prevalent in alkaline (pH above 7) soils. Soil pH determines the availability of soil nutrients (see Figure 5.5). Acidity promotes the weathering of rocks that releases $K^+$, $Mg^{2+}$, $Ca^{2+}$, and $Mn^{2+}$ and increases the solubility of carbonates, sulfates, and phosphates. Increasing the solubility of nutrients facilitates their availability to roots.

Major factors that lower soil pH are the decomposition of organic matter and the amount of rainfall. Carbon dioxide is produced as a result of the decomposition of organic matter and equilibrates with soil water in the following reaction:

$$CO_2 + H_2O \leftrightarrow H^+ + HCO_3^-$$

This reaction releases hydrogen ions ($H^+$), lowering the pH of the soil. Microbial decomposition of organic matter also produces ammonia ($NH_3$) and hydrogen sulfide ($H_2S$) that can be oxidized in the soil to form the strong acids nitric acid ($HNO_3$) and sulfuric acid ($H_2SO_4$), respectively. Hydrogen ions also displace $K^+$, $Mg^{2+}$, $Ca^{2+}$, and $Mn^{2+}$ from the surfaces of soil particles. Leaching may then remove these ions from the upper soil layers, leaving a more acidic soil. By contrast, the weathering of rock in arid regions releases $K^+$, $Mg^{2+}$, $Ca^{2+}$, and $Mn^{2+}$ into the soil, but because of the low rainfall, these ions do not leach from the upper soil layers, and the soil remains alkaline.

### Excess mineral ions in the soil limit plant growth

When excess mineral ions are present in soil, the soil is said to be *saline*, and plant growth may be restricted if these mineral ions reach levels that limit water availability or exceed the adequate zone for a particular nutrient (see Chapter 26). Sodium chloride and sodium sulfate are the most common salts in saline soils. Excess mineral ions in soils can be a major problem in arid and semiarid regions because rainfall is insufficient to leach them from the soil layers near the surface.

Irrigated agriculture fosters soil salinization if the amount of water applied is insufficient to leach the salt below the root zone. Irrigation water can contain 100 to 1000 g of mineral ions per cubic meter. An average crop requires about

4000 m³ of water per acre. Consequently, 400 to 4000 kg of mineral ions per acre may be added to the soil per crop (Marschner 1995), and over a number of growing seasons, high levels of mineral ions may accumulate in the soil.

In saline soil, plants encounter **salt stress**. Whereas many plants are affected adversely by the presence of relatively low levels of salt, other plants can survive (**salt-tolerant plants**) or even thrive (**halophytes**) at high salt levels. The mechanisms by which plants tolerate high salinity are complex (see Chapter 26), involving molecular synthesis, enzyme induction, and membrane transport. In some species, excess mineral ions are not taken up; in others, they are taken up but excreted from the plant by salt glands associated with the leaves. To prevent toxic buildup of mineral ions in the cytosol, many plants sequester them in the vacuole (Stewart and Ahmad 1983). Efforts are under way to bestow salt tolerance on salt-sensitive crop species using both classic plant breeding and biotechnology (Blumwald 2003), as detailed in Chapter 26.

Another important problem with excess mineral ions is the accumulation of heavy metals in the soil, which can cause severe toxicity in plants as well as humans (see WEB ESSAY 5.2). These heavy metals include zinc, copper, cobalt, nickel, mercury, lead, cadmium, silver, and chromium (Berry and Wallace 1981).

### Plants develop extensive root systems

The ability of plants to obtain both water and mineral nutrients from the soil is related to their capacity to develop an extensive root system. In the late 1930s, H. J. Dittmer examined the root system of a single winter rye plant after 16 weeks of growth. Dittmer estimated that the plant had 13 million primary and lateral root axes, extending more than 500 km in length and providing 200 m² of surface area (Dittmer 1937). This plant also had more than $10^{10}$ root hairs, providing another 300 m² of surface area. In total, the surface area of roots from a single rye plant equaled that of a professional basketball court.

In the desert, the roots of mesquite (genus *Prosopis*) may extend downward more than 50 m to reach groundwater. Annual crop plants have roots that usually grow between 0.1 and 2.0 m in depth and extend laterally to distances of 0.3 to 1.0 m. In orchards, the major root systems of trees planted 1 m apart reach a total length of 12 to 18 km per tree. The annual production of roots in natural ecosystems may easily surpass that of shoots, so in many respects, the aboveground portions of a plant represent only "the tip of the iceberg." Nonetheless, making observations on root systems is difficult and usually requires special techniques (see WEB TOPIC 5.2).

Plant roots may grow continuously throughout the year. Their proliferation, however, depends on the availability of water and minerals in the immediate microenvironment surrounding the root, the so-called **rhizosphere**. If the rhizosphere is poor in nutrients or too dry, root growth is slow. As rhizosphere conditions improve, root growth increases. If fertilization and irrigation provide abundant nutrients and water, root growth may not keep pace with shoot growth. Plant growth under such conditions becomes carbohydrate-limited, and a relatively small root system meets the nutrient needs of the whole plant (Bloom et al. 1993). Indeed, crops under fertilization and irrigation allocate more resources to the shoot and reproductive structures than to roots, and this shift in allocation patterns often results in higher yields.

### Root systems differ in form but are based on common structures

The *form* of the root system differs greatly among plant species. In monocots, root development starts with the emergence of three to six **primary** (or *seminal*) root axes from the germinating seed. With further growth, the plant extends new adventitious roots, called **nodal roots** or *brace roots*. Over time, the primary and nodal root axes grow and branch extensively to form a complex *fibrous root system* (FIGURE 5.7). In fibrous root systems, all the roots generally have the same diameter (except where environmental conditions or pathogenic interactions modify the root structure), so it is impossible to distinguish a main root axis.

In contrast to monocots, dicots develop root systems with a main single root axis, called a **taproot**, which may

(A) Dry soil    (B) Irrigated soil

30 cm

**FIGURE 5.7** Fibrous root systems of wheat (a monocot). (A) The root system of a mature (3-month-old) wheat plant growing in dry soil. (B) The root system of a mature wheat plant growing in irrigated soil. It is apparent that the morphology of the root system is affected by the amount of water present in the soil. In a mature fibrous root system, the primary root axes are indistinguishable. (After Weaver 1926.)

(A) Sugar beet  (B) Alfalfa

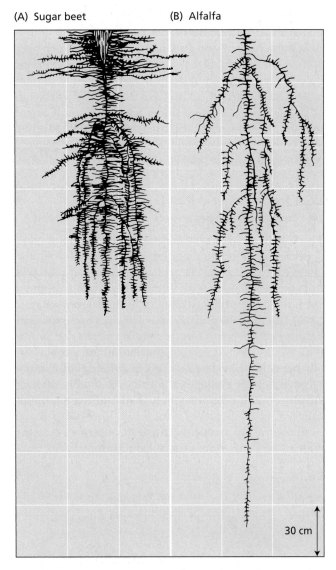

30 cm

**FIGURE 5.8** Taproot system of two adequately watered dicots: sugar beet (A) and alfalfa (B). The sugar beet root system is typical of 5 months of growth; the alfalfa root system is typical of 2 years of growth. In both dicots, the root system shows a major vertical root axis. In the case of sugar beet, the upper portion of the taproot system is thickened because of its function as storage tissue. (After Weaver 1926.)

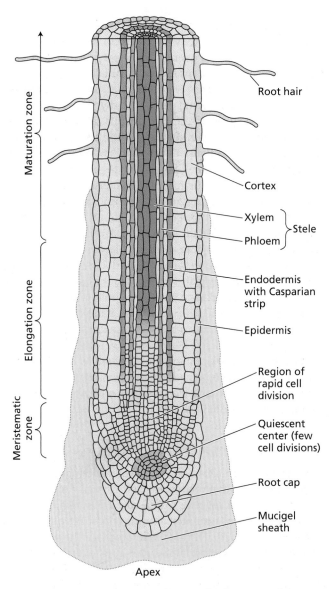

Root hair

Cortex

Xylem
Phloem
} Stele

Endodermis with Casparian strip

Epidermis

Region of rapid cell division

Quiescent center (few cell divisions)

Root cap

Mucigel sheath

Apex

Maturation zone

Elongation zone

Meristematic zone

**FIGURE 5.9** Diagrammatic longitudinal section of the apical region of the root. The meristematic cells are located near the tip of the root. These cells generate the root cap and the upper tissues of the root. In the elongation zone, cells differentiate to produce xylem, phloem, and cortex. Root hairs, formed in epidermal cells, first appear in the maturation zone.

thicken as a result of secondary cambial activity. From this main root axis, *lateral roots* develop to form an extensively branched root system (**FIGURE 5.8**).

The development of the root system in both monocots and dicots depends on the activity of the root apical meristem and the production of lateral root meristems. **FIGURE 5.9** is a generalized diagram of the apical region of a plant root and identifies three zones of activity: the meristematic, elongation, and maturation zones.

In the **meristematic zone**, cells divide both in the direction of the root base to form cells that will differentiate into the tissues of the functional root and in the direction

of the root apex to form the **root cap**. The root cap protects the delicate meristematic cells as the root expands into the soil. It commonly secretes a gelatinous material called *mucigel*, which surrounds the root tip. The precise function of mucigel is uncertain, but it may provide lubrication that eases the root's penetration of the soil, protect the root apex from desiccation, promote the transfer of nutrients to the root, or affect interactions between the root and soil microorganisms (Russell 1977). The root cap is central to the perception of gravity, the signal that directs the growth of roots downward. This process is termed the **gravitropic response** (see Chapter 19).

Cell division in the root apex proper is relatively slow; thus this region is called the **quiescent center**. After a few generations of slow cell divisions, root cells displaced from the apex by about 0.1 mm begin to divide more rapidly. Cell division again tapers off at about 0.4 mm from the apex, and the cells expand equally in all directions.

The **elongation zone** begins approximately 0.7 to 1.5 mm from the apex (see Figure 5.9). In this zone cells elongate rapidly and undergo a final round of divisions to produce a central ring of cells called the **endodermis**. The walls of this endodermal cell layer become thickened, and suberin (see Chapter 13) deposited on the radial walls forms the **Casparian strip**, a hydrophobic structure that prevents the apoplastic movement of water or solutes across the root (see Figure 4.4).

The endodermis divides the root into two regions: the **cortex** toward the outside and the **stele** toward the inside. The stele contains the vascular elements of the root: the **phloem**, which transports metabolites from the shoot to the root, and the **xylem**, which transports water and solutes to the shoot.

Phloem develops more rapidly than xylem, attesting to the fact that phloem function is critical near the root apex. Large quantities of carbohydrates must flow through the phloem to the growing apical zones in order to support cell division and elongation. Carbohydrates provide rapidly growing cells with an energy source and with the carbon skeletons required to synthesize organic compounds. Six-carbon sugars (hexoses) also function as osmotically active solutes in the root tissue. At the root apex, where the phloem is not yet developed, carbohydrate movement depends on symplastic diffusion and is relatively slow (Bret-Harte and Silk 1994). The low rates of cell division in the quiescent center may result from the fact that insufficient carbohydrates reach this centrally located region or that this area is kept in an oxidized state.

Root hairs, with their large surface area for absorption of water and solutes and for anchoring the root to the soil, first appear in the **maturation zone** (see Figure 5.9), and here the xylem develops the capacity to translocate substantial quantities of water and solutes to the shoot.

## Different areas of the root absorb different mineral ions

The precise point of entry of minerals into the root system has been a topic of considerable interest. Some researchers have claimed that nutrients are absorbed only at the apical regions of the root axes or branches (Bar-Yosef et al. 1972); others claim that nutrients are absorbed over the entire root surface (Nye and Tinker 1977). Experimental evidence supports both possibilities, depending on the plant species and the nutrient being investigated:

- Root absorption of calcium in barley (*Hordeum vulgare*) appears to be restricted to the apical region.

- Iron may be taken up either at the apical region, as in barley (Clarkson 1985), or over the entire root surface, as in corn (Kashirad et al. 1973).

- Potassium, nitrate, ammonium, and phosphate can be absorbed freely at all locations of the root surface (Clarkson 1985), but in corn the elongation zone has the maximum rates of potassium accumulation (Sharp et al. 1990) and nitrate absorption (Taylor and Bloom 1998).

- In corn and rice (*Oryza sativa*) (Colmer and Bloom 1998) and in wetland species (Fang et al. 2007), the root apex absorbs ammonium more rapidly than the elongation zone does. Ammonium and nitrate uptake by conifer roots vary significantly across different regions of the root and may be influenced by rates of root growth and maturation (Hawkins et al. 2008).

- In several species, root hairs are the most active in phosphate absorption (Föhse et al. 1991).

The high rates of nutrient absorption in the apical root zones result from the strong demand for nutrients in these tissues and the relatively high nutrient availability in the soil surrounding them. For example, cell elongation depends on the accumulation of solutes such as potassium, chloride, and nitrate to increase the osmotic pressure within the cell (see Chapter 15). Ammonium is the preferred nitrogen source to support cell division in the meristem because meristematic tissues are often carbohydrate-limited and because assimilation of ammonium into organic nitrogen compounds consumes less energy than assimilation of nitrate (see Chapter 12). The root apex and root hairs grow into fresh soil, where nutrients have not yet been depleted.

Within the soil, nutrients can move to the root surface both by bulk flow and by diffusion (see Chapter 3). In bulk flow, nutrients are carried by water moving through the soil toward the root. The amounts of nutrients provided to the root by bulk flow depend on the rate of water flow through the soil toward the plant, which depends on transpiration rates and on nutrient levels in the soil solution. When both the rate of water flow and the concentrations of nutrients in the soil solution are high, bulk flow can play an important role in nutrient supply.

In diffusion, mineral nutrients move from a region of higher concentration to a region of lower concentration. Nutrient uptake by roots lowers the concentrations of nutrients at the root surface, generating concentration gradients in the soil solution surrounding the root. Diffusion of nutrients down their concentration gradients along with bulk flow resulting from transpiration can increase nutrient availability at the root surface.

When the rate of absorption of a nutrient by roots is high and the nutrient concentration in the soil is low, bulk flow can supply only a small fraction of the total nutrient requirement (Mengel and Kirkby 2001). Under these conditions, diffusion rates limit the movement of the nutri-

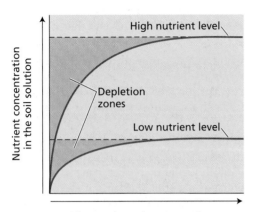

**FIGURE 5.10** Formation of a nutrient depletion zone in the region of the soil adjacent to the plant root. A nutrient depletion zone forms when the rate of nutrient uptake by the cells of the root exceeds the rate of replacement of the nutrient by bulk flow and diffusion in the soil solution. This depletion causes a localized decrease in the nutrient concentration in the area adjacent to the root surface. (After Mengel and Kirkby 2001.)

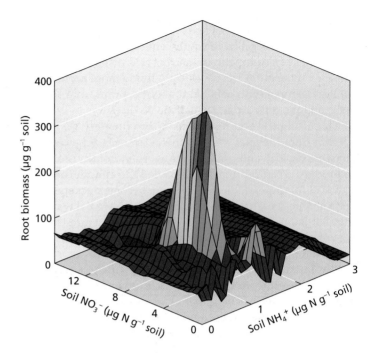

**FIGURE 5.11** Root biomass as a function of extractable soil $NH_4^+$ and $NO_3^-$. The root biomass is shown ($\mu g$ root dry weight $g^{-1}$ soil) plotted against extractable soil $NH_4^+$ and $NO_3^-$ ($\mu g$ extractable N $g^{-1}$ soil) for tomato, *Lycopersicon esculentum* cv T-5, growing in an irrigated field that had been fallow the previous 2 years. The colors emphasize the differences among biomasses, ranging from low (purple) to high (red). (After Bloom et al. 1993.)

ent to the root surface. When diffusion is too slow to maintain high nutrient concentrations near the root, a **nutrient depletion zone** forms adjacent to the root surface (**FIGURE 5.10**). This zone extends from about 0.2 to 2.0 mm from the root surface, depending on the mobility of the nutrient in the soil.

The formation of a depletion zone tells us something important about mineral nutrition: Because roots deplete the mineral supply in the rhizosphere, their effectiveness in mining minerals from the soil is determined not only by the rate at which they can remove nutrients from the soil solution, but by their continuous growth. Without growth, roots would rapidly deplete the soil adjacent to their surfaces. Optimal nutrient acquisition therefore depends both on the root system's capacity for nutrient uptake and on its ability to grow into fresh soil.

### Nutrient availability influences root growth

Plants, which have limited mobility for most of their lives, must deal with changes in their local environment because they cannot move away from unfavorable conditions. Above the ground, light level, temperature, and humidity may fluctuate substantially during the day and across the canopy, but $CO_2$ and $O_2$ concentrations remain relatively uniform. In contrast, soil buffers the roots from temperature extremes, but the belowground concentrations of $CO_2$ and $O_2$, water, and nutrients are extremely heterogeneous, both spatially and temporally. For example, inorganic nitrogen concentrations in soil may range a thousandfold over a dis-

tance of centimeters or the course of hours (Bloom 2005). Given such heterogeneity, plants seek the most favorable conditions within their reach.

Roots sense the belowground environment—through gravitropism, thigmotropism, chemotropism, and hydrotropism—to guide their growth toward soil resources. Some of these responses involve auxin (see Chapter 19). The extent to which roots proliferate within a soil patch varies with nutrient levels (Bloom et al. 1993) (**FIGURE 5.11**). Root growth is minimal in poor soils because the roots become nutrient-limited. As soil nutrient availability increases, roots proliferate (Robinson et al. 1999; Walch-Liu et al. 2005).

Where soil nutrients exceed an optimal level, root growth becomes carbohydrate-limited and eventually ceases (Durieux et al. 1994). With high soil nutrient levels, a few roots—3.5% of the root system in spring wheat (Robinson et al. 1991) and 12% in lettuce (Burns 1991)—are sufficient to supply all the nutrients required, so the plant may diminish the allocation of its resources to roots while increasing its allocation to the shoot and reproductive structures. This resource shifting is one mechanism through which fertilization stimulates crop yields.

## Mycorrhizal fungi facilitate nutrient uptake by roots

Our discussion so far has centered on the direct acquisition of mineral elements by roots, but this process may be modified by the association of mycorrhizal fungi with the root system. The host plant provides associated **mycorrhizae** (singular *mycorrhiza*) (from the Greek words for "fungus" and "root") with carbohydrates and in return receives nutrients or water from the mycorrhizae.

Mycorrhizae are not unusual; in fact, they are widespread under natural conditions. Much of the world's vegetation appears to have roots associated with mycorrhizal fungi: 83% of dicots, 79% of monocots, and all gymnosperms regularly form mycorrhizal associations (Wilcox 1991). In contrast, plants from the families Brassicaceae (e.g., cabbage [*Brassica oleracea*]), Chenopodiaceae (e.g., spinach [*Spinacea oleracea*]), and Proteaceae (e.g., macadamia nut [*Macadamia integrifolia*]), as well as aquatic plants, rarely, if ever, have mycorrhizae. Mycorrhizae are absent from roots in very dry, saline, or flooded soils, or where soil fertility is extreme, either high or low. In particular, plants grown in hydroponic culture and young, rapidly growing crop plants seldom have mycorrhizae.

Mycorrhizal fungi are composed of fine tubular filaments called *hyphae* (singular *hypha*). The mass of hyphae that forms the body of the fungus is called the *mycelium* (plural *mycelia*). There are two major classes of mycorrhizal fungi that are important in terms of mineral nutrient uptake by plants: ectotrophic mycorrhizae and arbuscular mycorrhizae (Brundrett 2004).

**Ectotrophic mycorrhizal fungi** typically form a thick sheath, or *mantle*, of mycelium around roots, and some of the mycelium penetrates between the cortical cells (**FIGURE 5.12**). The cortical cells themselves are not penetrated by the fungal hyphae, but instead are surrounded by a network of hyphae called the **Hartig net**. Often the amount of fungal mycelium is so extensive that its total mass is comparable to that of the roots themselves. The fungal mycelium also extends into the soil, away from the compact mantle, where it forms individual hyphae or strands containing fruiting bodies.

The capacity of the root system to absorb nutrients is improved by the presence of external fungal hyphae because they are much finer than plant roots and can reach beyond the nutrient depletion zone near the roots (Clarkson 1985). About 3% of higher plants, mainly forest trees in the Fagaceae, Betulaceae, and Pinaceae families and some woody legumes, form ectomycorrhiza. The fungi involved are mostly higher Basidiomycetes and Ascomycetes (Barea et al. 2005).

Unlike the ectotrophic mycorrhizal fungi, **arbuscular mycorrhizal fungi** (previously called vesicular–arbuscular mycorrhizae) do not produce a compact mantle of fungal mycelium around the root. Instead, the hyphae grow in a less dense arrangement, both within the root itself and

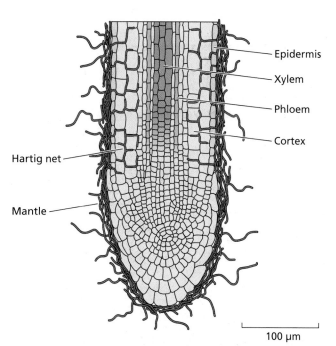

**FIGURE 5.12** Root infected with ectotrophic mycorrhizal fungi. The fungal hyphae surround the root to produce a dense fungal sheath, or mantle, and penetrate the intercellular spaces of the cortex to form the Hartig net. The total mass of fungal hyphae may be comparable to the root mass itself. (After Rovira et al. 1983.)

extending outward from the root into the surrounding soil (**FIGURE 5.13**). After entering the root through either the epidermis or a root hair via a mechanism that has commonalities with the entry of the bacteria responsible for the nitrogen-fixing symbiosis (see Chapter 12), the hyphae not only extend through the regions between cells, but also penetrate individual cells of the cortex. Within these cells, the hyphae can form oval structures called **vesicles** and branched structures called **arbuscules**. The arbuscules appear to be sites of nutrient transfer between the fungus and the host plant.

Outside the root, the external mycelium can extend several centimeters away from the root and may contain spore-bearing structures. Unlike the ectotrophic mycorrhizae, arbuscular mycorrhizae make up only a small mass of fungal material, which is unlikely to exceed 10% of the root weight. Arbuscular mycorrhizae are found in association with the roots of most species of herbaceous angiosperms, including most major agricultural crops (Smith and Read 2008). The fungi involved in this association are ubiquitous soil-borne microbes.

Fungal structures resembling those of extant arbuscular mycorrhizae have been found in 400-million-year-old fossils of plants. This observation indicates that these associations were present at the very early stages of vascular land plant evolution. Phylogenetic studies based on DNA

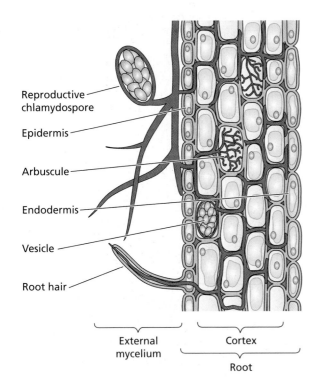

Reproductive chlamydospore

Epidermis

Arbuscule

Endodermis

Vesicle

Root hair

External mycelium

Cortex

Root

**FIGURE 5.13** Association of arbuscular mycorrhizal fungi with a section of a plant root. The fungal hyphae grow into the intercellular spaces of the cortex and penetrate individual cortical cells. As they extend into the cell, they do not break the plasma membrane or the tonoplast of the host cell. Instead, the hypha is surrounded by these membranes and forms structures known as arbuscules, which participate in nutrient ion exchange between the host plant and the fungus. (After Mauseth 1988.)

mycorrhizal associations often alter water movement into, through, and out of the host plant and can thus alleviate drought stress.

## Nutrients move from mycorrhizal fungi to root cells

Little is known about the mechanism by which the mineral nutrients absorbed by mycorrhizal fungi are transferred to the cells of plant roots. With ectotrophic mycorrhizae, inorganic phosphate may simply diffuse from the hyphae in the Hartig net and be absorbed by the root cortical cells. With arbuscular mycorrhizae, the situation may be more complex. Nutrients may diffuse from intact arbuscules to root cortical cells. Alternatively, because some root arbuscules are continually degenerating while new ones are forming, degenerating arbuscules may release their internal contents to the host root cells.

A key factor in the extent of mycorrhizal association with plant roots is the nutritional status of the host plant. Moderate deficiency of a nutrient such as phosphorus tends to promote mycorrhizal associations, whereas plants with abundant nutrients tend to suppress mycorrhizal fungi.

Mycorrhizal associations in well-fertilized soils may shift from symbiotic relationships to parasitic ones in that the fungus still obtains carbohydrates from the host plant, but the host plant no longer benefits from improved nutrient uptake efficiency. At high nutrient levels, the host plant may treat mycorrhizal fungi as it does other pathogens (Brundrett 1991), or it may still depend on the fungi for a substantial portion of its nutrient absorption from soil (Smith et al. 2009).

sequence data from living fungal taxa support this contention (Azcón-Aguilar et al. 2009).

The association of arbuscular mycorrhizae with plant roots facilitates the uptake of phosphorus, trace metals such as zinc and copper, and water. By extending beyond the depletion zone for phosphorus around the root, the external mycelium improves phosphorus absorption. Calculations show that a root associated with mycorrhizal fungi can transport phosphate at a rate more than four times faster than that of a root without mycorrhizae (Nye and Tinker 1977). The external mycelium of ectotrophic mycorrhizae can also absorb phosphate and make it available to plants. In addition, it has been suggested that ectotrophic mycorrhizae proliferate in the organic litter of the soil and hydrolyze organic phosphorus for transfer to the root (Smith and Read 2008). Finally, arbuscular

## SUMMARY

Plants are autotrophic organisms capable of using the energy from sunlight to synthesize all their components from carbon dioxide, water, and mineral elements. Although mineral nutrients continually cycle through all organisms, they enter the biosphere predominantly through the root systems of plants. After being absorbed by the roots, the mineral elements are translocated to the various parts of the plant, where they are utilized in numerous biological functions.

### Essential Nutrients, Deficiencies, and Plant Disorders

- Studies of plant nutrition have shown that specific mineral elements are essential for plant life (**Tables 5.1, 5.2**).

- These elements are classified as macronutrients or micronutrients, depending on the relative amounts found in plant tissue (**Table 5.1**).

## SUMMARY continued

- Certain visual symptoms are diagnostic for deficiencies in specific nutrients in higher plants. Nutritional disorders occur because nutrients have key roles within plant. They serve as components of organic compounds, in energy storage, in plant structures, as enzyme cofactors, and in electron transfer reactions.

- Mineral nutrition can be studied through the use of solution culture, which allow the characterization of specific nutrient requirements (**Figure 5.1, Table 5.3**).

- Soil and plant tissue analysis can provide information on the nutritional status of the plant–soil system and can suggest corrective actions to avoid deficiencies or toxicities (**Figure 5.3**).

- When crop plants are grown under modern high-production conditions, substantial amounts of nutrients, particularly nitrogen, phosphorus, or potassium, are removed from the soil.

### Treating Nutritional Deficiencies

- To prevent the development of deficiencies, nutrients can be added back to the soil in the form of fertilizers.

- Fertilizers that provide nutrients in inorganic forms are called chemical fertilizers; those that derive from plant or animal residues are considered organic fertilizers. In both cases, plants absorb the nutrients primarily as inorganic ions. Most fertilizers are applied to the soil, but some are sprayed on leaves.

### Soil, Roots, and Microbes

- The soil is a complex substrate—physically, chemically, and biologically. The size of soil particles and the cation exchange capacity of the soil determine the extent to which a soil provides a reservoir for water and nutrients (**Table 5.5, Figure 5.5**).

- Soil pH also has a large influence on the availability of mineral elements to plants (**Figure 5.4**).

- If mineral elements, especially sodium or heavy metals, are present in excess in the soil, plant growth may be adversely affected. Certain plants are able to tolerate excess mineral elements, and a few species—for example, halophytes in the case of sodium—may thrive under these extreme conditions.

- To obtain nutrients from the soil, plants develop extensive root systems (**Figures 5.6, 5.7**).

- Roots have a relatively simple structure with radial symmetry and few differentiated cell types (**Figure 5.8**).

- Roots continually deplete the nutrients from the immediate soil around them, and such a simple structure may permit rapid growth into fresh soil (**Figure 5.9**).

- Plant roots often form associations with mycorrhizal fungi (**Figure 5.10**).

- The fine hyphae of mycorrhizae extend the reach of roots into the surrounding soil and facilitate the acquisition of mineral elements, particularly those like phosphorus that are relatively immobile in the soil (**Figure 5.11**).

- In return, plants provide carbohydrates to the mycorrhizae. Plants tend to suppress mycorrhizal associations under conditions of high nutrient availability.

## WEB MATERIAL

### Web Topics

**5.1 Symptoms of Deficiency in Essential Minerals**
Deficiency symptoms are characteristic of each essential element and can be diagnostic for the deficiency. The color photographs in this topic illustrate deficiency symptoms for each essential element in tomato.

**5.2 Observing Roots below Ground**
The study of roots growing under natural conditions requires a means to observe roots below ground. State-of-the-art techniques are described in this topic.

## WEB MATERIAL continued

### Web Essays

**5.1  Boron Functions in Plants: Looking Beyond the Cell Wall**

Presents one long lists of "postulated roles of B essentiality" for microorganisms and for higher plant growth and development.

**5.2  From Meals to Metals and Back**

Heavy metal accumulation is toxic to plants. Understanding the molecular process involved in toxicity is helping researchers to develop better phytoremediation crops.

# CHAPTER REFERENCES

Aber, J. D., Goodale, C. L., Ollinger, S. V., Smith, M. L., Magill, A. H., Martin, M. E., Hallett, R. A., and Stoddard, J. L. (2003) Is nitrogen deposition altering the nitrogen status of northeastern forests? *BioScience* 53: 375–389.

Ahluwalia, S. S., and Goyal, D. (2007) Microbial and plant derived biomass for removal of heavy metals from wastewater. *Bioresour. Technol.* 98: 2243–2257.

Armstrong, F. A. (2008) Why did nature choose manganese to make oxygen? *Philos. Trans. R. Soc. Lond. B Biol. Sci.* 363: 1263–1270.

Arnon, D. I., and Stout, P. R. (1939) The essentiality of certain elements in minute quantity for plants with special reference to copper. *Plant Physiol.* 14: 371–375.

Asher, C. J., and Edwards, D. G. (1983) Modern solution culture techniques. In *Inorganic Plant Nutrition* (Encyclopedia of Plant Physiology, New Series, Vol. 15B), A. Läuchli and R. L. Bieleski, eds., Springer, Berlin, pp. 94–119.

Azcón-Aguilar, C., Barea, J. M., Gianinazzi, S., and Guaninazzi-Pearson, V. (2009) *Mycorrhizas: Functional Processes and Ecological Impacts*. Springer, Berlin.

Bar-Yosef, B., Kafkafi, U., and Bresler, E. (1972) Uptake of phosphorus by plants growing under field conditions. I. Theoretical model and experimental determination of its parameters. *Soil Sci.* 36: 783–800.

Barea, J. M., Azcón, R., and Azcón-Aguilar, C. 2005. Interactions between mycorrhizal fungi and bacteria to improve plant nutrient cycling and soil structure. In *Micro-organisms in Soils: Roles in Genesis and Functions*, F. Buscot and A. Varma, eds., Springer, Berlin, pp. 195–212.

Berry, W. L., and Wallace, A. (1981) Toxicity: The concept and relationship to the dose response curve. *J. Plant Nutr.* 3: 13–19.

Bloom, A. J. (2005) Coordination between roots and shoots. In *Long-Distance Transport in Plants*, N. M. Holbrook and M. A. Zwieniecki, eds., Academic Press, San Diego, CA, pp. 241–256.

Bloom, A. J., Jackson, L. E., and Smart, D. R. (1993) Root growth as a function of ammonium and nitrate in the root zone. *Plant Cell Environ.* 16: 199–206.

Blumwald, E. (2003) Engineering salt tolerance in plants. *Biotechnol. Genet. Eng. Rev.* 20: 261–275.

Bouma, D. (1983) Diagnosis of mineral deficiencies using plant tests. In *Inorganic Plant Nutrition* (Encyclopedia of Plant Physiology, New Series, Vol. 15B), A. Läuchli and R. L. Bieleski, eds., Springer, Berlin, pp. 120–146.

Brady, N. C. (1974) *The Nature and Properties of Soils*, 8th ed. Macmillan, New York.

Bret-Harte, M. S., and Silk, W. K. (1994) Nonvascular, symplasmic diffusion of sucrose cannot satisfy the carbon demands of growth in the primary root tip of *Zea mays* L. *Plant Physiol.* 105: 19–33.

Broadley, M. R., White, P. J., Hammond, J. P., Zelko, I., and Lux, A. (2007) Zinc in plants. *New Phytol.* 173: 677–702.

Brown, P. H., Bellaloui, N., Wimmer, M. A., Bassil, E. S., Ruiz, J., Hu, H., Pfeffer, H., Dannel, F., and Römheld, V. (2002) Boron in plant biology. *Plant Biol.* 4: 205–223.

Brownell, P. F., and Bielig, L. M. (1996) The role of sodium in the conversion of pyruvate to phosphoenolpyruvate in mesophyll chloroplasts of $C_4$ plants. *Aust. J. Plant Physiol.* 23: 171–177.

Brundrett, M. C. (1991) Mycorrhizas in natural ecosystems. *Adv. Ecol. Res.* 21: 171–313.

Brundrett, M. C. (2004) Diversity and classification of mycorrhizal associations. *Biol. Rev. Camb. Philos. Soc.* 79: 473–495.

Burns, I. G. (1991) Short- and long-term effects of a change in the spatial distribution of nitrate in the root zone on N uptake, growth and root development of young lettuce plants. *Plant Cell Environ.* 14: 21–33.

Clarkson, D. T. (1985) Factors affecting mineral nutrient acquisition by plants. *Annu. Rev. Plant Physiol.* 36: 77–116.

Colmenero-Flores, J. M., Martinez, G., Gamba, G., Vazquez, N., Iglesias, D. J., Brumos, J., and Talon, M. (2007) Identification and functional characterization of cation-chloride cotransporters in plants. *Plant J.* 50: 278–292.

Colmer, T. D., and Bloom, A. J. (1998) A comparison of net $NH_4^+$ and $NO_3^-$ fluxes along roots of rice and maize. *Plant Cell Environ.* 21: 240–246.

Cooper, A. (1979) *The ABC of NFT: Nutrient Film Technique: The World's First Method of Crop Production without a Solid Rooting Medium*. Grower Books, London.

Dittmer, H. J. (1937) A quantitative study of the roots and root hairs of a winter rye plant (*Secale cereale*). *Am. J. Bot.* 24: 417–420.

Durieux, R. P., Kamprath, E. J., Jackson, W. A., and Moll, R. H. (1994) Root distribution of corn: The effect of nitrogen fertilization. *Agron. J.* 86: 958–962.

Engel, R. E., Bruebaker, L., and Emborg, T. J. (2001) A chloride deficient leaf spot of durum wheat. *Soil Sci. Soc. Am. J.* 65: 1448–1454.

Epstein, E. (1972) *Mineral Nutrition of Plants: Principles and Perspectives*. John Wiley and Sons, New York.

Epstein, E. (1999) Silicon. *Annu. Rev. Plant Physiol. Plant Mol. Biol.* 50: 641–664.

Epstein, E., and Bloom, A. J. (2005) *Mineral Nutrition of Plants: Principles and Perspectives*, 2nd ed. Sinauer Associates, Sunderland, MA.

Evans, H. J., and Sorger, G. J. (1966) Role of mineral elements with emphasis on the univalent cations. *Annu. Rev. Plant Physiol.* 17: 47–76.

Fang, Y. Y., Babourina, O., Rengel, Z., Yang, X. E., and Pu, P. M. (2007) Spatial distribution of ammonium and nitrate fluxes along roots of wetland plants. *Plant Sci.* 173: 240–246.

Fenn, M. E., Baron, J. S., Allen, E. B., Rueth, H. M., Nydick, K. R., Geiser, L., Bowman, W. D., Sickman, J. O., Meixner, T., Johnson, D. W., et al. (2003) Ecological effects of nitrogen deposition in the western United States. *BioScience* 53: 404–420.

Föhse, D., Claassen, N., and Jungk, A. (1991) Phosphorus efficiency of plants. II. Significance of root radius, root hairs and cation–anion balance for phosphorus influx in seven plant species. *Plant Soil* 132: 261–272.

Gericke, W. F. (1937) Hydroponics—Crop production in liquid culture media. *Science* 85: 177–178.

Hawkins, B. J., Boukcim, H., and Plassard, C. (2008) A comparison of ammonium, nitrate and proton net fluxes along seedling roots of Douglas-fir and lodgepole pine grown and measured with different inorganic nitrogen sources. *Plant Cell Environ.* 31: 278–287.

Hepler, P. K., and Wayne, R. O. (1985) Calcium and plant development. *Annu. Rev. Plant Physiol.* 36: 397–440.

Hetherington, A. M., and Brownlee, C. (2004) The generation of $Ca^{2+}$ signals in plants. *Annu. Rev. Plant Biol.* 55: 401–427.

Jeong, J., and Guerinot, M. L. (2009) Homing in on iron homeostasis in plants. *Trends Plant Sci.* 14: 280–285.

Kashirad, A., Marschner, H., and Richter, C. H. (1973) Absorption and translocation of $^{59}Fe$ from various parts of the corn plant. *Z. Pflanzenernähr. Bodenk.* 134: 136–147.

Loomis, R. S., and Connor, D. J. (1992) *Crop Ecology: Productivity and Management in Agricultural Systems*. Cambridge University Press, Cambridge.

Lucas, R. E., and Davis, J. F. (1961) Relationships between pH values of organic soils and availabilities of 12 plant nutrients. *Soil Sci.* 92: 177–182.

Marschner, H. (1995) *Mineral Nutrition of Higher Plants*, 2nd ed. Academic Press, London.

Mauseth, J. D. (1988) *Plant Anatomy*. Benjamin/Cummings, Menlo Park, CA.

Mengel, K., and Kirkby, E. A. (2001) *Principles of Plant Nutrition*, 5th ed. Kluwer Academic Publishers, Dordrecht.

Nolan, B. T., and Hitt, K. J. (2006) Vulnerability of shallow groundwater and drinking-water wells to nitrate in the United States. *Environ. Sci. Technol.* 40: 7834–7840.

Nye, P. H., and Tinker, P. B. (1977) *Solute Movement in the Soil–Root System*. University of California Press, Berkeley.

Popelkova, H., and Yocum, C. F. (2007) Current status of the role of $Cl^-$ ion in the oxygen-evolving complex. *Photosyn. Res.* 93: 111–121.

Raven, J. A., and Smith, F. A. (1976) Nitrogen assimilation and transport in vascular land plants in relation to intracellular pH regulation. *New Phytol.* 76: 415–431.

Reguera, M., Espi, A., Bolaños, L., Bonilla, I., and Redondo-Nieto, M. (2009) Endoreduplication before cell differentiation fails in boron-deficient legume nodules. Is boron involved in signalling during cell cycle regulation? *New Phytol.* 183 (1): 8–12.

Robinson, D., Hodge, A., Griffiths, B. S., and Fitter, A. H. (1999) Plant root proliferation in nitrogen-rich patches confers competitive advantage. *Proc. R. Soc. Lond. B Biol. Sci.* 266: 431–435.

Robinson, D., Linehan, D. J., and Caul, S. (1991) What limits nitrate uptake from soil? *Plant Cell Environ.* 14: 77–85.

Rovira, A. D., Bowen, C. D., and Foster, R. C. (1983) The significance of rhizosphere microflora and mycorrhizas in plant nutrition. In *Inorganic Plant Nutrition* (Encyclopedia of Plant Physiology, New Series, Vol. 15B), A. Läuchli and R. L. Bieleski, eds., Springer, Berlin, pp. 61–93.

Russell, R. S. (1977) *Plant Root Systems: Their Function and Interaction with the Soil*. McGraw-Hill, London.

Sharp, R. E., Hsiao, T. C., and Silk, W. K. (1990) Growth of the maize primary root at low water potentials. 2. Role of growth and deposition of hexose and potassium in osmotic adjustment. *Plant Physiol.* 93: 1337–1346.

Sievers, R. E., and Bailar, J. C., Jr. (1962) Some metal chelates of ethylenediaminetetraacetic acid, diethylenetriaminepentaacetic acid, and triethylenetriaminehexaacetic acid. *Inorg. Chem.* 1: 174–182.

Smith, F. A., Grace, E. J., and Smith, S. E. (2009) More than a carbon economy: Nutrient trade and ecological sustainability in facultative arbuscular mycorrhizal symbioses. *New Phytol.* 182: 347–358.

Smith, S. E., and Read, D. J. (2008) *Mycorrhizal Symbiosis*, 3rd ed. Academic Press, Amsterdam, Boston.

Stewart, G. R., and Ahmad, I. (1983) Adaptation to salinity in angiosperm halophytes. In *Metals and Micronutrients: Uptake and Utilization by Plants*, D. A. Robb and W. S. Pierpoint, eds., Academic Press, New York, pp. 33–50.

Schwarz, G., and Mendel, R. R. (2006) Molybdenum cofactor biosynthesis and molybdenum enzymes. *Annu. Rev. Plant Biol.* 57: 623–647.

Taylor, A. R., and Bloom, A. J. (1998) Ammonium, nitrate and proton fluxes along the maize root. *Plant Cell Environ.* 21: 1255–1263.

Walch-Liu, P., Filleur, S., Gan, Y. B., and Forde, B. G. (2005) Signaling mechanisms integrating root and shoot responses to changes in the nitrogen supply. *Photosyn. Res.* 83: 239–250.

Weathers, P. J., and Zobel, R. W. (1992) Aeroponics for the culture of organisms, tissues, and cells. *Biotechnol. Adv.* 10: 93–115.

Weaver, J. E. (1926) *Root Development of Field Crops.* McGraw-Hill, New York.

White, P. J., and Broadley, M. R. (2003) Calcium in plants. *Ann. Bot.* 92: 487–511.

Wilcox, H. E. (1991) Mycorrhizae. In *Plant Roots: The Hidden Half*, Y. Waisel, A. Eshel, and U. Kafkafi, eds., Marcel Dekker, New York, pp. 731–765.

Wood, B. W., Reilly, C. C., and Nyczepir, A. P. (2003) Nickel corrects mouse-ear. *The Pecan Grower* 15: 3–5.

Yruela, I. (2009) Copper in plants: Acquisition, transport and interactions. *Functional Plant Biol.* 36: 409–430.

Ziegler, H. (1987) The evolution of stomata. In *Stomatal Function*, E. Zeiger, G. Farquhar, and I. Cowan, eds., Stanford University Press, Stanford, CA, pp. 29–57.

# 6

# Solute Transport

The interior of a plant cell is separated from the plant cell wall and the environment by a plasma membrane that is only two lipid molecules thick. This thin layer separates a relatively constant internal environment from variable external surroundings. In addition to forming a hydrophobic barrier to diffusion, the membrane must facilitate and continuously regulate the inward and outward traffic of selected molecules and ions as the cell takes up nutrients, exports solutes, and regulates its turgor pressure. The same is true of the internal membranes that separate the various compartments within each cell.

The plasma membrane also detects information about the physical environment, about molecular signals from other cells, and about the presence of invading pathogens. Often these signals are relayed by changes in ion fluxes across the membrane.

Molecular and ionic movement from one location to another is known as **transport**. Local transport of solutes into or within cells is regulated mainly by membranes. Larger-scale transport between plant organs, or between plant and environment, is also controlled by membrane transport at the cellular level. For example, the transport of sucrose from leaf to root through the phloem, referred to as **translocation**, is driven and regulated by membrane transport into the phloem cells of the leaf and from the phloem to the storage cells of the root (see Chapter 10).

In this chapter we will consider the physical and chemical principles that govern the movements of molecules in solution. Then we will show how these principles apply to membranes and biological systems. We will also discuss the molecular mechanisms of transport in living cells and the great variety of membrane transport proteins that are responsible for the particular transport properties of plant cells. Finally, we will examine the pathways that ions take when they enter the root as well as the mechanism of xylem loading, the process whereby ions are released into the tracheary elements of the stele.

## Passive and Active Transport

According to Fick's first law (see Equation 3.1), the movement of molecules by diffusion always proceeds spontaneously, down a gradient of free energy or chemical potential, until equilibrium is reached. The spontaneous "downhill" movement of molecules is termed **passive transport**. At equilibrium, no further net movements of solutes can occur without the application of a driving force.

The movement of substances against a gradient of chemical potential, or "uphill," is termed **active transport**. It is not spontaneous, and it requires that work be done on the system by the application of cellular energy. One common way (but not the only way) of accomplishing this task is to couple transport to the hydrolysis of ATP.

Recall from Chapter 3 that we can calculate the driving force for diffusion or, conversely, the energy input necessary to move substances against a gradient by measuring the potential-energy gradient. For uncharged solutes this gradient is often a simple function of the difference in concentration. Biological transport can be driven by four major forces: concentration, hydrostatic pressure, gravity, and electric fields. (However, recall from Chapter 3 that in small-scale biological systems, gravity seldom contributes substantially to the force that drives transport.)

The **chemical potential** for any solute is defined as the sum of the concentration, electric, and hydrostatic potentials (and the chemical potential under standard conditions). *The importance of the concept of chemical potential is that it sums all the forces that may act on a molecule to drive net transport* (Nobel 1991).

$$\tilde{\mu}_j \quad = \quad \mu_j^* \quad + \quad RT \ln C_j$$

| Chemical potential for a given solute, $j$ | Chemical potential of $j$ under standard conditions | Concentration (activity) component |
|---|---|---|

$$+ \quad z_j FE \quad + \quad \bar{V}_j P \qquad (6.1)$$

| Electric-potential component | Hydrostatic-pressure component |
|---|---|

Here $\tilde{\mu}_j$ is the chemical potential of the solute species $j$ in joules per mole (J mol$^{-1}$), $\mu_j^*$ is its chemical potential under standard conditions (a correction factor that will cancel out in future equations and so can be ignored), $R$ is the universal gas constant, $T$ is the absolute temperature, and $C_j$ is the concentration (more accurately the activity) of $j$.

The electrical term, $z_j FE$, applies only to ions; $z$ is the electrostatic charge of the ion (+1 for monovalent cations, –1 for monovalent anions, +2 for divalent cations, and so on), $F$ is Faraday's constant (equivalent to the electric charge on 1 mol of H$^+$), and $E$ is the overall electrical potential of the solution (with respect to ground). The final term, $\bar{V}_j P$, expresses the contribution of the partial molal volume of $j$ ($\bar{V}_j$) and pressure ($P$) to the chemical potential of $j$. (The partial molal volume of $j$ is the change in volume per mole of substance $j$ added to the system, for an infinitesimal addition.)

This final term, $\bar{V}_j P$, makes a much smaller contribution to $\tilde{\mu}_j$ than do the concentration and electrical terms, except in the very important case of osmotic water movements. As discussed in Chapter 3, the chemical potential of water (i.e., the water potential) depends on the concentration of dissolved solutes and the hydrostatic pressure on the system.

In general, diffusion (passive transport) always moves molecules energetically downhill from areas of higher chemical potential to areas of lower chemical potential. Movement against a chemical-potential gradient is indicative of active transport (**FIGURE 6.1**).

If we take the diffusion of sucrose across a cell membrane as an example, we can accurately approximate the chemical potential of sucrose in any compartment by the concentration term alone (unless a solution is concentrated, causing hydrostatic pressure to build up within the plant cell). From Equation 6.1, the chemical potential of sucrose inside a cell can be described as follows (in the next three equations, the subscript $s$ stands for sucrose and the superscripts $i$ and $o$ stand for inside and outside, respectively):

$$\tilde{\mu}_s^i \quad = \quad \mu_s^* \quad + \quad RT \ln C_s^i$$

| Chemical potential of sucrose solution inside the cell | Chemical potential of sucrose solution under standard conditions | Concentration component |
|---|---|---|

$$(6.2)$$

The chemical potential of sucrose outside the cell is calculated as follows:

$$\tilde{\mu}_s^o = \mu_s^* + RT \ln C_s^o \qquad (6.3)$$

We can calculate the difference in the chemical potential of sucrose between the solutions inside and outside the cell, $\Delta\tilde{\mu}_s$, regardless of the mechanism of transport. To get the signs right, remember that for inward transport, sucrose is being removed (–) from outside the cell and added (+) to the inside, so the change in free energy in joules per mole of sucrose transported will be as follows:

$$\Delta\tilde{\mu}_s = \tilde{\mu}_s^i - \tilde{\mu}_s^o \qquad (6.4)$$

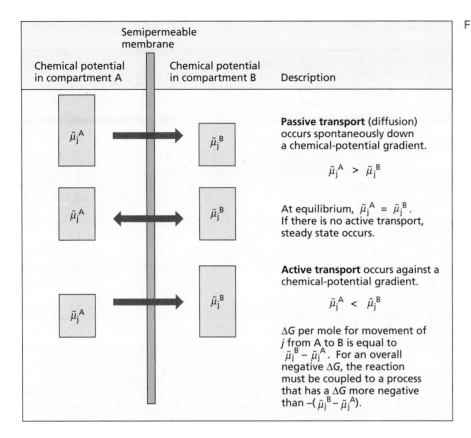

**FIGURE 6.1** Relationship between chemical potential, $\tilde{\mu}$, and the transport of molecules across a permeability barrier. The net movement of molecular species $j$ between compartments A and B depends on the relative magnitude of the chemical potential of $j$ in each compartment, represented here by the size of the boxes. Movement down a chemical gradient occurs spontaneously and is called passive transport; movement against or up a gradient requires energy and is called active transport.

Substituting the terms from Equations 6.2 and 6.3 into Equation 6.4, we get the following:

$$\Delta\tilde{\mu}_s = \left(\mu_s^* + RT \ln C_s^i\right) - \left(\mu_s^* + RT \ln C_s^o\right)$$
$$= RT \left(\ln C_s^i - \ln C_s^o\right)$$
$$= RT \ln \frac{C_s^i}{C_s^o} \tag{6.5}$$

If this difference in chemical potential is negative, sucrose can diffuse inward spontaneously (provided the membrane has a finite permeability to sucrose; see the next section). In other words, the driving force ($\Delta\tilde{\mu}_s$) for solute diffusion is related to the magnitude of the concentration gradient ($C_s^i/C_s^o$).

If the solute carries an electric charge (as does, for example, the potassium ion), the electrical component of the chemical potential must also be considered. Suppose the membrane is permeable to $K^+$ and $Cl^-$ rather than to sucrose. Because the ionic species ($K^+$ and $Cl^-$) diffuse independently, each has its own chemical potential. Thus for inward $K^+$ diffusion,

$$\Delta\tilde{\mu}_K = \tilde{\mu}_K^i - \tilde{\mu}_K^o \tag{6.6}$$

Substituting the appropriate terms from Equation 6.1 into Equation 6.6, we get

$$\Delta\tilde{\mu}_s = (RT \ln [K^+]^i + zFE^i) - (RT \ln [K^+]^o + zFE^o) \tag{6.7}$$

and because the electrostatic charge of $K^+$ is +1, $z = +1$, and

$$\Delta\tilde{\mu}_K = RT \ln \frac{[K^+]^i}{[K^+]^o} + F(E^i - E^o) \tag{6.8}$$

The magnitude and sign of this expression will indicate the driving force and direction for $K^+$ diffusion across the membrane. A similar expression can be written for $Cl^-$ (but remember that for $Cl^-$, $z = -1$).

Equation 6.8 shows that ions, such as $K^+$, diffuse in response to both their concentration gradients ($[K^+]^i/[K^+]^o$) and any electrical potential difference between the two compartments ($E^i - E^o$). One important implication of this equation is that ions can be driven passively against their concentration gradients if an appropriate voltage (electric field) is applied between the two compartments. Because of the importance of electric fields in the biological transport of any charged molecule, $\tilde{\mu}$ is often called the **electrochemical potential**, and $\Delta\tilde{\mu}$ is the difference in electrochemical potential between two compartments.

## Transport of Ions across Membrane Barriers

If the two KCl solutions in the previous example are separated by a biological membrane, diffusion is complicated by the fact that the ions must move through the membrane as well as across the open solutions. The extent to which a membrane permits the movement of a substance is called **membrane permeability**. As will be discussed later, permeability depends on the composition of the membrane as well as on the chemical nature of the solute. In a loose sense, permeability can be expressed in terms of a diffusion coefficient for the solute through the membrane. However, permeability is influenced by several additional factors, such as the ability of a substance to enter the membrane, that are difficult to measure.

Despite its theoretical complexity, we can readily measure permeability by determining the rate at which a solute passes through a membrane under a specific set of conditions. Generally the membrane will hinder diffusion and thus reduce the speed with which equilibrium is reached. For any particular solute, however, the permeability or resistance of the membrane itself cannot alter the final equilibrium conditions. Equilibrium occurs when $\Delta \tilde{\mu}_j = 0$.

In the sections that follow we will discuss the factors that influence the distribution of ions across a membrane. These parameters can be used to predict the relationship between the electrical gradient and the concentration gradient of an ion.

### Different diffusion rates for cations and anions produce diffusion potentials

When salts diffuse across a membrane, an electrical membrane potential (voltage) can develop. Consider the two KCl solutions separated by a membrane in **FIGURE 6.2**. The $K^+$ and $Cl^-$ ions will permeate the membrane independently as they diffuse down their respective gradients of electrochemical potential. And unless the membrane is very porous, its permeability to the two ions will differ.

As a consequence of these different permeabilities, $K^+$ and $Cl^-$ will initially diffuse across the membrane at different rates. The result is a slight separation of charge, which instantly creates an electrical potential across the membrane. In biological systems, membranes are usually more permeable to $K^+$ than to $Cl^-$. Therefore, $K^+$ will diffuse out of the cell (see compartment A in Figure 6.2) faster than $Cl^-$, causing the cell to develop a negative electric charge with respect to the extracellular medium. A potential that develops as a result of diffusion is called a **diffusion potential**.

The principle of electrical neutrality must always be kept in mind when the movement of ions across membranes is considered: bulk solutions always contain equal numbers of anions and cations. The existence of a membrane potential implies that the distribution of charges across the membrane is uneven; however, the actual number of unbalanced ions is negligible in chemical terms. For example, a membrane potential of –100 millivolts (mV), like that found across the plasma membranes of many plant cells, results from the presence of only one extra anion out of every 100,000 within the cell—a concentration

Compartment A    Compartment B

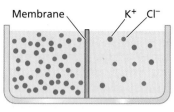

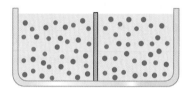

Initial conditions:
$[KCl]_A > [KCl]_B$

Diffusion potential exists until chemical equilibrium is reached.

Equilibrium conditions:
$[KCl]_A = [KCl]_B$

At chemical equilibrium, diffusion potential equals zero.

**FIGURE 6.2** Development of a diffusion potential and a charge separation between two compartments separated by a membrane that is preferentially permeable to potassium. If the concentration of potassium chloride is higher in compartment A ($[KCl]_A > [KCl]_B$), potassium and chloride ions will diffuse into compartment B. If the membrane is more permeable to potassium than to chloride, potassium ions will diffuse faster than chloride ions, and a charge separation (+ and –) will develop, resulting in establishment of a diffusion potential.

difference of only 0.001%! As Figure 6.2 shows, all of these extra anions are found immediately adjacent to the surface of the membrane; there is no charge imbalance throughout the bulk of the cell.

In our example of KCl diffusion across a membrane, electrical neutrality is preserved because as $K^+$ moves ahead of $Cl^-$ in the membrane, the resulting diffusion potential retards the movement of $K^+$ and speeds that of $Cl^-$. Ultimately, both ions diffuse at the same rate, but the diffusion potential persists and can be measured. As the system moves toward equilibrium and the concentration gradient collapses, the diffusion potential also collapses.

### How does membrane potential relate to ion distribution?

Because the membrane in the preceding example is permeable to both $K^+$ and $Cl^-$ ions, equilibrium will not be reached for either ion until the concentration gradients decrease to zero. However, if the membrane were permeable only to $K^+$, diffusion of $K^+$ would carry charges across the membrane until the membrane potential balanced the concentration gradient. Because a change in potential requires very few ions, this balance would be reached instantly. Potassium ions would then be at equilibrium, even though the change in the concentration gradient for $K^+$ would be negligible.

When the distribution of any solute across a membrane reaches equilibrium, the passive flux, $J$ (i.e., the amount of solute crossing a unit area of membrane per unit time), is the same in the two directions—outside to inside and inside to outside:

$$J_{o \to i} = J_{i \to o}$$

Fluxes are related to $\Delta \tilde{\mu}$ (for a discussion on fluxes and $\Delta \tilde{\mu}$, see Appendix 1); thus at equilibrium, the electrochemical potentials will be the same:

$$\tilde{\mu}_j^{\,o} = \tilde{\mu}_j^{\,i}$$

and for any given ion (the ion is symbolized here by the subscript $j$),

$$\mu_j^* + RT \ln C_j^o + z_j FE^o = \mu_j^* + RT \ln C_j^i + z_j FE^i \quad (6.9)$$

By rearranging Equation 6.9, we obtain the difference in electrical potential between the two compartments at equilibrium ($E^i - E^o$):

$$E^i - E^o = \frac{RT}{z_j F}\left( \ln \frac{C_j^o}{C_j^i} \right)$$

This electrical-potential difference is known as the **Nernst potential** ($\Delta E_j$) for that ion,

$$\Delta E_j = E^i - E^o$$

and

$$\Delta E_j = \frac{RT}{z_j F}\left( \ln \frac{C_j^o}{C_j^i} \right) \quad (6.10)$$

or

$$\Delta E_j = \frac{2.3 RT}{z_j F}\left( \log \frac{C_j^o}{C_j^i} \right)$$

This relationship, known as the *Nernst equation*, states that at equilibrium, the difference in concentration of an ion between two compartments is balanced by the voltage difference between the compartments. The Nernst equation can be further simplified for a univalent cation at 25°C:

$$\Delta E_j = 59 \text{mV} \log \frac{C_j^o}{C_j^i} \quad (6.11)$$

Note that a tenfold difference in concentration corresponds to a Nernst potential of 59 mV ($C_o/C_i = 10/1$; log 10 = 1). That is, a membrane potential of 59 mV would maintain a tenfold concentration gradient of an ion whose movement across the membrane is driven by passive diffusion. Similarly, if a tenfold concentration gradient of an ion existed across the membrane, passive diffusion of that

ion down its concentration gradient (if it were allowed to come to equilibrium) would result in a difference of 59 mV across the membrane.

All living cells exhibit a membrane potential that is due to the asymmetric ion distribution between the inside and outside of the cell. We can determine these membrane potentials by inserting a microelectrode into the cell and measuring the voltage difference between the inside of the cell and the extracellular medium (**FIGURE 6.3**).

The Nernst equation can be used at any time to determine whether a given ion is at equilibrium across a membrane. However, a distinction must be made between equilibrium and steady state. *Steady state* is the condition in which influx and efflux of a given solute are equal, and therefore the ion concentrations are constant over time. Steady state is not necessarily the same as equilibrium (see Figure 6.1); in steady state, the existence of active transport across the membrane prevents many diffusive fluxes from ever reaching equilibrium.

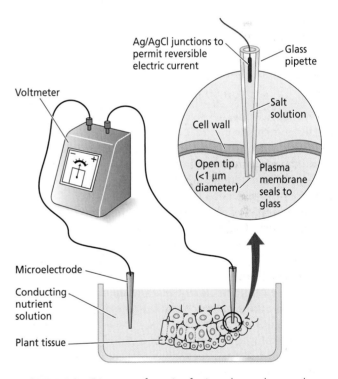

**FIGURE 6.3**  Diagram of a pair of microelectrodes used to measure membrane potentials across cell membranes. One of the glass micropipette electrodes is inserted into the cell compartment under study (usually the vacuole or the cytoplasm), while the other is kept in an electrolytic solution that serves as a reference. The microelectrodes are connected to a voltmeter, which records the electrical potential difference between the cell compartment and the solution. Typical membrane potentials across plant cell membranes range from –60 to –240 mV. The insert shows how electrical contact with the interior of the cell is made through the open tip of the glass micropipette, which contains an electrically conducting salt solution.

## TABLE 6.1
## Comparison of observed and predicted ion concentrations in pea root tissue

| Ion | Concentration in external medium (mmol L$^{-1}$) | Internal concentration[a] (mmol L$^{-1}$) | |
|---|---|---|---|
| | | Predicted | Observed |
| K$^+$ | 1 | 74 | 75 |
| Na$^+$ | 1 | 74 | 8 |
| Mg$^{2+}$ | 0.25 | 1340 | 3 |
| Ca$^{2+}$ | 1 | 5360 | 2 |
| NO$_3^-$ | 2 | 0.0272 | 28 |
| Cl$^-$ | 1 | 0.0136 | 7 |
| H$_2$PO$_4^-$ | 1 | 0.0136 | 21 |
| SO$_4^{2-}$ | 0.25 | 0.00005 | 19 |

*Source*: Data from Higinbotham et al. 1967.

*Note*: The membrane potential was measured as –110 mV.

[a]Internal concentration values were derived from ion content of hot water extracts of 1–2 cm intact root segments.

### The Nernst equation distinguishes between active and passive transport

TABLE 6.1 shows how experimental measurements of ion concentrations at steady state in pea root cells compare with predicted values calculated from the Nernst equation (Higinbotham et al. 1967). In this example, the concentration of each ion in the external solution bathing the tissue and the measured membrane potential were substituted into the Nernst equation, and a predicted internal concentration was calculated for that ion.

Notice that, of all the ions shown in Table 6.1, only K$^+$ is at or near equilibrium. The anions NO$_3^-$, Cl$^-$, H$_2$PO$_4^-$, and SO$_4^{2-}$ all have higher internal concentrations than predicted, indicating that their uptake is active. The cations Na$^+$, Mg$^{2+}$, and Ca$^{2+}$ have lower internal concentrations than predicted; therefore, these ions enter the cell by diffusion down their electrochemical-potential gradients and are then actively exported.

The example shown in Table 6.1 is an oversimplification: Plant cells have several internal compartments, each of which can differ in its ionic composition from the others. The cytosol and the vacuole are the most important intracellular compartments in determining the ionic relations of plant cells. In most mature plant cells, the central vacuole occupies 90% or more of the cell's volume, and the cytosol is restricted to a thin layer around the periphery of the cell.

Because of its small volume, the cytosol of most angiosperm cells is difficult to assay chemically. For this reason, much of the early work on the ionic relations of plants focused on certain green algae, such as *Chara* and *Nitella*, whose cells are several inches long and may contain an appreciable volume of cytosol. FIGURE 6.4 diagrams the

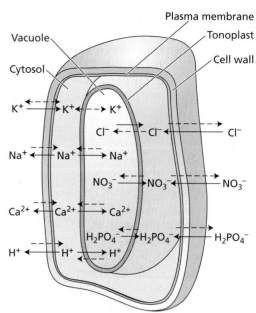

**FIGURE 6.4** Ion concentrations in the cytosol and the vacuole are controlled by passive (dashed arrows) and active (solid arrows) transport processes. In most plant cells the vacuole occupies up to 90% of the cell's volume and contains the bulk of the cell solutes. Control of the ion concentrations in the cytosol is important for the regulation of metabolic enzymes. The cell wall surrounding the plasma membrane does not represent a permeability barrier and hence is not a factor in solute transport.

conclusions from these studies and from related work with higher plants. In brief:

- Potassium ions are accumulated passively by both the cytosol and the vacuole. When extracellular K$^+$ concentrations are very low, K$^+$ may be taken up actively.

- Sodium ions are pumped actively out of the cytosol into the extracellular space and vacuole.

- Excess protons, generated by intermediary metabolism, are also actively extruded from the cytosol. This process helps maintain the cytosolic pH near neutrality, while the vacuole and the extracellular medium are generally more acidic by one or two pH units.

- Anions are taken up actively into the cytosol.

- Calcium ions are actively transported out of the cytosol at both the plasma membrane and the vacuolar membrane, which is called the tonoplast (see Figure 6.4).

Many different ions permeate the membranes of living cells simultaneously, but K$^+$ has the highest concentrations in plant cells, and it exhibits large permeabilities. A modified version of the Nernst equation, the **Goldman equation**, includes all permeant ions (all ions for which mechanisms of transmembrane movement exist) and therefore gives a more accurate value for the diffusion potential. When permeabilities and ion gradients are known, it is possible to calculate a diffusion potential across a biological membrane from the Goldman equation. The diffusion potential calculated from the Goldman equation is termed the *Goldman diffusion potential* (for a detailed discussion of the Goldman equation, see WEB TOPIC 6.1).

### Proton transport is a major determinant of the membrane potential

In most eukaryotic cells, $K^+$ has both the greatest internal concentration and the highest membrane permeability, so the diffusion potential may approach $E_K$, the Nernst potential for $K^+$. In some cells of some organisms—particularly in some mammalian cells such as neurons—the normal resting potential of the cell may also be close to $E_K$. This is not the case with plants and fungi, however, which often show experimentally measured membrane potentials (often –200 to –100 mV) that are much more negative than those calculated from the Goldman equation, which are usually only –80 to –50 mV. Thus, in addition to the diffusion potential, the membrane potential must have a second component. The excess voltage is provided by the plasma membrane electrogenic $H^+$-ATPase.

Whenever an ion moves into or out of a cell without being balanced by countermovement of an ion of opposite charge, a voltage is created across the membrane. Any active transport mechanism that results in the movement of a net electric charge will tend to move the membrane potential away from the value predicted by the Goldman equation. Such transport mechanisms are called *electrogenic pumps* and are common in living cells.

The energy required for active transport is often provided by the hydrolysis of ATP. We can study the dependence of the plasma membrane potential on ATP by observing the effect of cyanide ($CN^-$) on the membrane potential (FIGURE 6.5). Cyanide rapidly poisons the mitochondria, and the cell's ATP consequently becomes depleted. As ATP synthesis is inhibited, the membrane potential falls to the level of the Goldman diffusion potential (see WEB TOPIC 6.1).

Thus the membrane potentials of plant cells have two components: a diffusion potential and a component resulting from electrogenic ion transport (transport that results in the generation of a membrane potential) (Spanswick 1981). When cyanide inhibits electrogenic ion transport, the pH of the external medium increases while the cytosol becomes acidic because protons remain inside the cell. This observation is one piece of evidence that it is the active transport of protons out of the cell that is electrogenic.

As discussed earlier, a change in membrane potential caused by an electrogenic pump will change the driving forces for diffusion of all ions that cross the membrane. For example, the outward transport of $H^+$ can create an electrical driving force for the passive diffusion of $K^+$ into

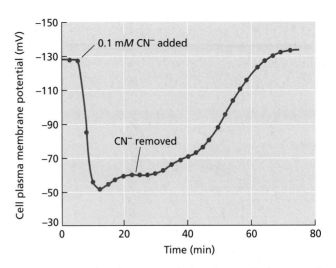

**FIGURE 6.5** The plasma membrane potential of a pea cell collapses when cyanide ($CN^-$) is added to the bathing solution. Cyanide blocks ATP production in the cell by poisoning the mitochondria. The collapse of the membrane potential upon addition of cyanide indicates that an ATP supply is necessary for maintenance of the potential. Washing the cyanide out of the tissue results in a slow recovery of ATP production and restoration of the membrane potential. (After Higinbotham et al. 1970.)

the cell. Protons are transported electrogenically across the plasma membrane not only in plants, but also in bacteria, algae, fungi, and some animal cells, such as those of the kidney epithelia.

ATP synthesis in mitochondria and chloroplasts also depends on a $H^+$-ATPase. In these organelles, this transport protein is sometimes called an *ATP synthase* because it forms ATP rather than hydrolyzing it (see Chapter 11). The structure and function of membrane proteins involved in active and passive transport in plant cells are discussed in detail later in this chapter.

## Membrane Transport Processes

Artificial membranes made of pure phospholipids have been used extensively to study membrane permeability. When the permeability of artificial phospholipid bilayers to ions and molecules is compared with that of biological membranes, important similarities and differences become evident (FIGURE 6.6).

Biological and artificial membranes have similar permeabilities to nonpolar molecules and many small polar molecules. On the other hand, biological membranes are much more permeable to ions, to some large polar molecules, such as sugars, and to water than artificial bilayers are. The reason is that, unlike artificial bilayers, biological membranes contain **transport proteins** that facilitate the passage of selected ions and other molecules. The general term *trans-*

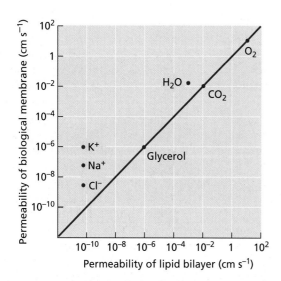

**FIGURE 6.6** Typical values for the permeability of a biological membrane to various substances compared with those for an artificial phospholipid bilayer. For nonpolar molecules such as $O_2$ and $CO_2$, and for some small uncharged molecules such as glycerol, permeability values are similar in both systems. For ions and selected polar molecules, including water, the permeability of biological membranes is increased by one or more orders of magnitude because of the presence of transport proteins. Note the logarithmic scale.

*port proteins* encompasses three main categories of proteins: **channels**, **carriers**, and **pumps** (**FIGURE 6.7**), each of which will be described in more detail later in this section.

Transport proteins exhibit specificity for the solutes they transport, hence their great diversity in cells. The simple prokaryote *Haemophilus influenzae*, the first organism for which the complete genome was sequenced, has only 1743 genes, yet more than 200 of those genes (more than 10% of the genome) encode various proteins involved in membrane transport. In Arabidopsis, out of a predicted 25,500 proteins, as many as 1800 (~7%) may execute transport functions (Schwacke et al. 2003).

Although a particular transport protein is usually highly specific for the kinds of substances it will transport, its specificity is often not absolute. In plants, for example, a $K^+$ transporter in the plasma membrane may transport $K^+$, $Rb^+$, and $Na^+$ with different preferences. In contrast, most $K^+$ transporters are completely ineffective in transporting anions such as $Cl^-$ or uncharged solutes such as sucrose. Similarly, a protein involved in the transport of neutral amino acids may move glycine, alanine, and valine with equal ease, but may not accept aspartic acid or lysine.

In the next several pages we will consider the structures, functions, and physiological roles of the various membrane transporters found in plant cells, especially in the plasma membrane and tonoplast. We begin with a discussion of the role of certain transporters (channels and carriers) in promoting the diffusion of solutes across membranes. We

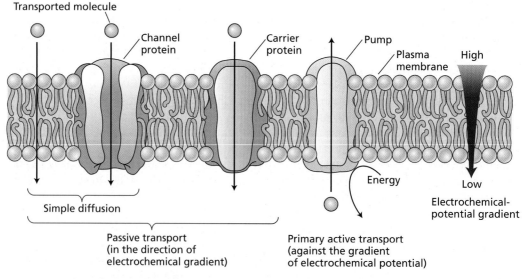

**FIGURE 6.7** Three classes of membrane transport proteins: channels, carriers, and pumps. Channels and carriers can mediate the passive transport of a solute across a membrane (by simple diffusion or facilitated diffusion) down the solute's gradient of electrochemical potential. Channel proteins act as membrane pores, and their specificity is determined primarily by the biophysical properties of the channel. Carrier proteins bind the transported molecule on one side of the membrane and release it on the other side. (The different types of carrier proteins are described in more detail in Figure 6.11.) Primary active transport is carried out by pumps and uses energy directly, usually from ATP hydrolysis, to pump solutes against their gradient of electrochemical potential.

will then distinguish between primary and secondary active transport and discuss the roles of the electrogenic $H^+$-ATPase and various symporters (proteins that transport two substances in the same direction simultaneously) in driving $H^+$-coupled secondary active transport.

## Channels enhance diffusion across membranes

**Channels** are transmembrane proteins that function as selective pores through which molecules or ions can diffuse across the membrane. The size of a pore and the density and nature of the surface charges on its interior lining determine its transport specificity. Transport through channels is always passive, and because the specificity of transport depends on pore size and electric charge more than on selective binding, channel transport is limited mainly to ions or water (**FIGURE 6.8**).

As long as the channel pore is open, solutes that can penetrate the pore diffuse through it extremely rapidly: about $10^8$ ions per second through each channel protein. Channel pores are not open all the time, however. Channel proteins have structures called **gates** that open and close the pore in response to external signals. Signals that can regulate channel activity include membrane potential changes, ligands, hormones, light, and posttranslational modifications such as phosphorylation. For example, voltage-gated channels open or close in response to changes in the membrane potential (see Figure 6.8B).

Individual ion channels can be studied in detail by an electrophysiological technique called patch clamping (see **WEB TOPIC 6.2**), which can detect the electrical current carried by ions diffusing through a single open channel or a collection of channels. Patch clamp studies reveal that for a given ion, such as $K^+$, a membrane has a variety of different channels. These channels may open over different voltage ranges, or in response to different signals, which may include $K^+$ or $Ca^{2+}$ concentrations, pH, reactive oxygen species, and so on. This specificity enables the transport of each ion to be fine-tuned to the prevailing conditions. Thus the ion permeability of a membrane is a variable that depends on the mix of ion channels that are open at a particular time.

As we saw in the experiment presented in Table 6.1, the distribution of most ions is not close to equilibrium across the membrane. Therefore, we know that channels for most ions are usually closed. Anion channels always function to allow anions to diffuse out of the cell, and other mechanisms are needed for anion uptake. Calcium ion channels are tightly regulated and essentially open only during signal transduction. Calcium ion channels function only to allow $Ca^{2+}$ release into the cytosol, and $Ca^{2+}$ must be expelled from the cytoplasm by active transport. In contrast, $K^+$ can diffuse either inward or outward, depending on whether the membrane potential is more negative or more positive than $E_K$, the potassium equilibrium potential.

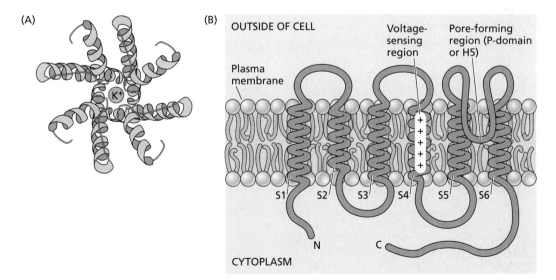

**FIGURE 6.8** Models of $K^+$ channels in plants. (A) Top view of a channel, looking through the pore of the protein. Membrane-spanning helices of four subunits come together in an inverted teepee with the pore at the center. The pore-forming regions of the four subunits dip into the membrane, forming a $K^+$ selectivity finger region at the outer part of the pore (more details on the structure of this channel can be found in **WEB ESSAY 6.1**). (B) Side view of an inwardly rectifying $K^+$ channel, showing a polypeptide chain of one subunit, with six membrane-spanning helices (S1–S6). The fourth helix contains positively charged amino acids and acts as a voltage sensor. The pore-forming region is a loop between helices 5 and 6. (A after Leng et al. 2002; B after Buchanan et al. 2000.)

K$^+$ channels that open only at potentials more negative than the prevailing Nernst potential for K$^+$ are specialized for inward diffusion of K$^+$ and are known as **inwardly rectifying**, or simply *inward*, K$^+$ channels. Conversely, K$^+$ channels that open only at potentials more positive than the Nernst potential for K$^+$ are **outwardly rectifying**, or *outward*, K$^+$ channels (**FIGURE 6.9**) (see **WEB ESSAY 6.1**). Inward K$^+$ channels function in the accumulation of K$^+$ from the apoplast, as occurs, for example, during K$^+$ uptake by guard cells in the process of stomatal opening (see Figure 6.9). Various outward K$^+$ channels function in the closing of stomata and in the release of K$^+$ into the xylem or apoplast.

### Carriers bind and transport specific substances

Unlike channels, carrier proteins do not have pores that extend completely across the membrane. In transport mediated by a carrier, the substance being transported is initially bound to a specific site on the carrier protein. This requirement for binding allows carriers to be highly selective for a particular substrate to be transported. Carriers therefore specialize in the transport of specific ions or organic metabolites. Binding causes a conformational change in the protein, which exposes the substance to the solution on the other side of the membrane. Transport is complete when the substance dissociates from the carrier's binding site.

Because a conformational change in the protein is required to transport an individual molecule or ion, the rate of transport by a carrier is many orders of magnitude slower than that through a channel. Typically, carriers may transport 100 to 1000 ions or molecules per second, while millions of ions can pass through an open ion channel in the same amount of time. The binding and release of molecules at a specific site on a carrier protein are similar to the binding and release of molecules by an enzyme in an enzyme-catalyzed reaction. As will be discussed later in this chapter, enzyme kinetics have been used to characterize transport carrier proteins.

Carrier-mediated transport (unlike transport through channels) can be either passive transport or secondary active transport (secondary active transport is discussed in a subsequent section). Passive transport via a carrier is sometimes called **facilitated diffusion**, although it resembles diffusion only in that it transports substances down their gradient of electrochemical potential, without an additional input of energy. (The term "facilitated diffusion" might seem more appropriately applied to transport through channels, but historically it has not been used in this way.)

### Primary active transport requires energy

To carry out active transport, a carrier must couple the energetically uphill transport of a solute with another, energy-releasing event so that the overall free-energy change is negative. **Primary active transport** is coupled directly to a source of energy other than $\Delta \tilde{\mu}_j$, such as ATP hydrolysis, an oxidation–reduction reaction (as in the electron transport chain of mitochondria and chloroplasts), or the absorption of light by the carrier protein (such as bacteriorhodopsin in halobacteria).

Membrane proteins that carry out primary active transport are called **pumps** (see Figure 6.7). Most pumps transport ions, such as H$^+$ or Ca$^{2+}$. However, as we will see later in this chapter, pumps belonging to the ATP-binding cassette (ABC) family of transporters can carry large organic molecules.

Ion pumps can be further characterized as either electrogenic or electroneutral. In general, **electrogenic transport** refers to ion transport involving the net movement of charge across the membrane. In contrast, **electroneutral transport**, as the name implies, involves no net movement of charge. For example, the Na$^+$/K$^+$-ATPase of animal cells pumps three Na$^+$ out for every two K$^+$ in, resulting in a net outward movement of one positive charge. The Na$^+$/K$^+$-ATPase is therefore an electrogenic ion pump. In contrast, the H$^+$/K$^+$-ATPase of the animal gastric mucosa pumps one H$^+$ out of the cell for every one K$^+$ in, so there is no net movement of charge across the membrane. Therefore, the H$^+$/K$^+$-ATPase is an electroneutral pump.

For the plasma membranes of plants, fungi, and bacteria, as well as for plant tonoplasts and other plant and animal

**FIGURE 6.9**  Current–voltage relationships. (A) A diagram showing the current that would result from K$^+$ flux through a set of hypothetical plasma membrane K$^+$ channels that were not voltage-regulated, given a K$^+$ concentration in the cytosol of 100 m$M$ and an extracellular K$^+$ concentration of 10 m$M$. Note that the current would be linear, and that there would be zero current at the equilibrium (Nernst) potential for K$^+$ ($E_K$). (B) Actual K$^+$ current data from an Arabidopsis guard cell protoplast, with the same intracellular and extracellular K$^+$ concentrations as in (A). These currents result from the activities of voltage-regulated K$^+$ channels. Note that, again, there is zero net current at the equilibrium potential for K$^+$. However, there is also zero net current over a broader voltage range because the channels are closed over this voltage range in these conditions. When the channels are closed, no K$^+$ can flow through them, hence zero current is observed over this voltage range. (C) The current–voltage relationship in (B) actually results from the activities of two sets of channels—the inwardly rectifying K$^+$ channels and the outwardly rectifying K$^+$ channels—which together produce the current–voltage relationship. (B after L. Perfus-Barbeoch and S. M. Assmann, unpublished data.)

(A)

Equilibrium or Nernst potential for K+: by definition, no net flux of K+, therefore, no current.

Current carried by the movement of K+ out of the cell. By convention, this **outward current** is given a **positive sign**.

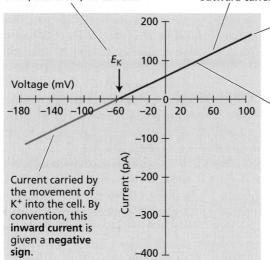

The opening and closing or "gating" of these channels is not regulated by voltage. Therefore, current through the channel is a linear function of voltage.

The slope of the line ($\Delta I/\Delta V$) gives the **conductance** of the channels mediating this K+ current.

Current carried by the movement of K+ into the cell. By convention, this **inward current** is given a **negative sign**.

$E_K = RT/ZF^* \ln \{[K_{out}]/[K_{in}]\}$
$E_K = 0.025^* \ln \{10/100\}$
$E_K = -59$ mV

(B)

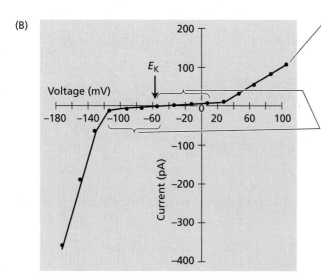

This current–voltage relationship is produced by K+ movement through channels that are regulated ("gated") by voltage. Note that the I/V relationship is nonlinear.

Little or no current over these voltage ranges because the channels are voltage-regulated and the effect of these voltages is to keep the channels in a closed state.

(C)

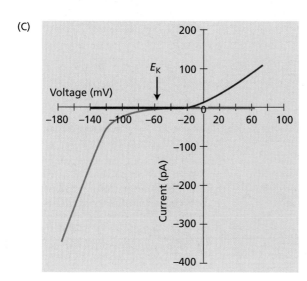

Current response illustrated in (B) is shown here to arise from the activity of two molecularly distinct types of K+ channels. The outward K+ channels (red) are voltage-gated such that they open only at membrane potentials >$E_K$; thus these channels mediate K+ efflux from the cell. The inward K+ channels (blue) are voltage-gated such that they open only at membrane potentials <$E_K$; thus these channels mediate K+ uptake into the cell.

endomembranes, H⁺ is the principal ion that is electrogenically pumped across the membrane. The **plasma membrane H⁺-ATPase** generates the gradient of electrochemical potential of H⁺ across the plasma membrane, while the **vacuolar H⁺-ATPase** and the **H⁺-pyrophosphatase** (H⁺-PPase) electrogenically pump protons into the lumen of the vacuole and the Golgi cisternae.

### Secondary active transport uses stored energy

In plant plasma membranes, the most prominent pumps are those for H⁺ and Ca²⁺, and the direction of pumping is outward from the cytosol to the extracellular space. Another mechanism is needed to drive the active uptake of mineral nutrients such as $NO_3^-$, $SO_4^{2-}$, and $PO_4^{3-}$; the uptake of amino acids, peptides, and sucrose; and the export of Na⁺, which at high concentrations is toxic to plant cells. The other important way that solutes are actively transported across a membrane against their gradient of electrochemical potential is by coupling the uphill transport of one solute to the downhill transport of another. This type of carrier-mediated cotransport is termed **secondary active transport** (FIGURE 6.10).

Secondary active transport is driven indirectly by pumps. In plant cells, protons are extruded from the cytosol by electrogenic H⁺-ATPases operating in the plasma membrane and at the vacuolar membrane. Consequently, a membrane potential and a pH gradient are created at the expense of ATP hydrolysis. This gradient of electrochemical potential for H⁺, referred to as $\Delta \tilde{\mu}_{jH^+}$, or (when expressed in other units) the **proton motive force (PMF)**, represents stored free energy in the form of the H⁺ gradient (see WEB TOPIC 6.3).

The proton motive force generated by electrogenic H⁺ transport is used in secondary active transport to drive the transport of many other substances against their gradients of electrochemical potential. Figure 6.10 shows how secondary active transport may involve the binding of a substrate (S) and an ion (usually H⁺) to a carrier protein and a conformational change in that protein.

There are two types of secondary active transport: symport and antiport. The example shown in Figure 6.10

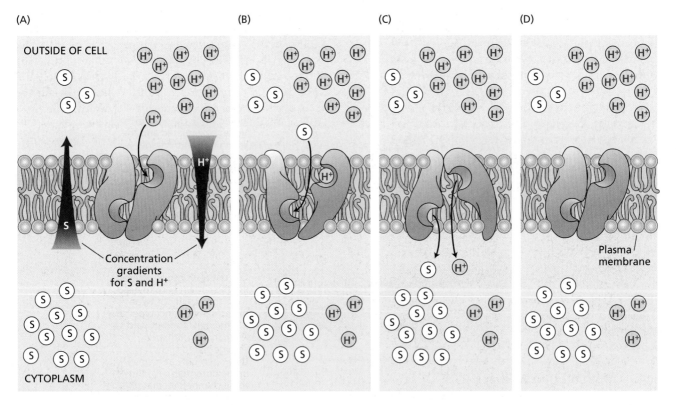

**FIGURE 6.10** Hypothetical model of secondary active transport. In secondary active transport, the energetically uphill transport of one solute is driven by the energetically downhill transport of another solute. In the illustrated example, energy that was stored as proton motive force ($\Delta\tilde{\mu}_{H^+}$, symbolized by the red arrow on the right in A) is being used to take up a substrate (S) against its concentration gradient (red arrow on the left). (A) In the initial conformation, the binding sites on the protein are exposed to the outside environment and can bind a proton. (B) This binding results in a conformational change that permits a molecule of S to be bound. (C) The binding of S causes another conformational change that exposes the binding sites and their substrates to the inside of the cell. (D) Release of a proton and a molecule of S to the cell's interior restores the original conformation of the carrier and allows a new pumping cycle to begin.

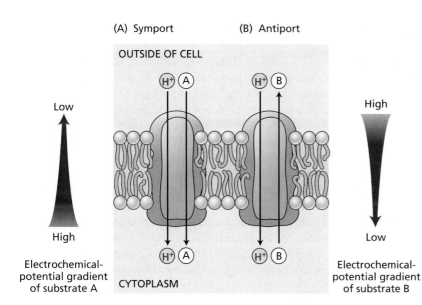

(A) Symport   (B) Antiport

OUTSIDE OF CELL

Low

High

Electrochemical-
potential gradient
of substrate A

High

Low

Electrochemical-
potential gradient
of substrate B

CYTOPLASM

**FIGURE 6.11** Two examples of secondary active transport coupled to a primary
proton gradient. (A) In symport, the energy dissipated by a proton moving back
into the cell is coupled to the uptake
of one molecule of a substrate (e.g., a
sugar) into the cell. (B) In antiport, the
energy dissipated by a proton moving
back into the cell is coupled to the active
transport of a substrate (for example, a
sodium ion) out of the cell. In both cases,
the substrate under consideration is moving against its gradient of electrochemical potential. Both neutral and charged
substrates can be transported by such
secondary active transport processes.

is called **symport** (and the proteins involved are called
*symporters*) because the two substances move in the same
direction through the membrane (see also **FIGURE 6.11A**).
**Antiport** (facilitated by proteins called *antiporters*) refers
to coupled transport in which the energetically downhill movement of protons drives the active (energetically
uphill) transport of a solute in the opposite direction (**FIG-
URE 6.11B**).

In both types of secondary transport, the ion or solute
being transported simultaneously with the protons is moving against its gradient of electrochemical potential, so its
transport is active. However, the energy driving this transport is provided by the proton motive force rather than
directly by ATP hydrolysis.

### Kinetic analyses can elucidate transport mechanisms

Thus far, we have described cellular transport in terms
of its energetics. However, cellular transport can also be
studied by use of enzyme kinetics because it involves
the binding and dissociation of molecules at active sites on
transport proteins (see **WEB TOPIC 6.4**). One advantage of
the kinetic approach is that it gives new insights into the
regulation of transport.

In kinetic experiments, the effects of external ion (or
other solute) concentrations on transport rates are measured. The kinetic characteristics of the transport rates can
then be used to distinguish between different transporters.
The maximum rate ($V_{max}$) of carrier-mediated transport,
and often channel transport as well, cannot be exceeded,
regardless of the concentration of substrate (**FIGURE 6.12**).
$V_{max}$ is approached when the substrate-binding site on the
carrier is always occupied or when flux through the channel is maximal. The concentration of transporter, not the
concentration of solute, becomes rate-limiting. Thus $V_{max}$

is an indicator of the number of molecules of the specific
transport protein that are functioning in the membrane.

The constant $K_m$ (which is numerically equal to the solute
concentration that yields half the maximal rate of transport)
tends to reflect the properties of the particular binding site.
Low $K_m$ values indicate high affinity of the transport site for
the transported substance. Such values usually imply the
operation of a carrier system. Higher values of $K_m$ indicate
a lower affinity of the transport site for the solute. The affinity is often so low that in practice $V_{max}$ is never reached. In
such cases, kinetics alone cannot distinguish between carriers and channels.

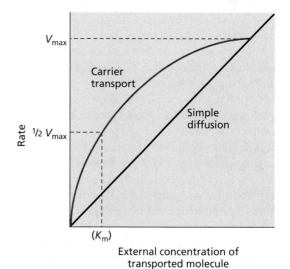

$V_{max}$

Carrier
transport

Simple
diffusion

Rate

$\frac{1}{2} V_{max}$

$(K_m)$

External concentration of
transported molecule

**FIGURE 6.12** Carrier transport often shows enzyme kinetics, including saturation ($V_{max}$) (see Appendix 1).
In contrast, simple diffusion through open channels
is ideally directly proportional to the concentration of
the transported solute, or for an ion, to the difference
in electrochemical potential across the membrane.

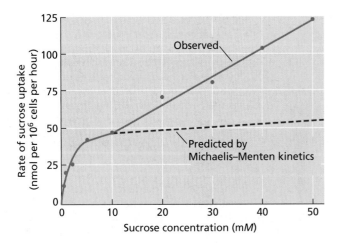

**FIGURE 6.13** The transport properties of a solute can change with solute concentrations. For example, at low concentrations (1–10 m$M$), the rate of uptake of sucrose by soybean cells shows saturation kinetics typical of carriers. A curve fitted to these data is predicted to approach a maximal rate ($V_{max}$) of 57 nmol per $10^6$ cells per hour. Instead, at higher sucrose concentrations, the uptake rate continues to increase linearly over a broad range of concentrations, consistent with the existence of facilitated transport mechanisms for sucrose uptake. (After Lin et al. 1984.)

Cells or tissues often display complex kinetics for the transport of a solute. Complex kinetics usually indicate the presence of more than one type of transport mechanism—for example, both high- and low-affinity transporters. FIGURE 6.13 shows the rate of sucrose uptake by soybean cotyledon protoplasts as a function of the external sucrose concentration (Lin et al. 1984). Uptake increases sharply with concentration and begins to saturate at about 10 m$M$. At concentrations above 10 m$M$, uptake becomes linear and nonsaturable within the concentration range tested. Inhibition of ATP synthesis with metabolic poisons blocks the saturable component, but not the linear one.

The interpretation of the pattern shown in Figure 6.13 is that sucrose uptake at low concentrations is an energy-dependent, carrier-mediated process (H⁺–sucrose symport). At higher concentrations, sucrose enters the cells by diffusion down its concentration gradient and is therefore insensitive to metabolic poisons. Consistent with these data, both H⁺–sucrose symporters (Lalonde et al. 2004) and sucrose facilitators (i.e., carrier proteins that mediate sucrose flux independent of energy supply or the proton gradient) (Zhou et al. 2007) have been identified at the molecular level.

## Membrane Transport Proteins

Numerous representative transport proteins located in the plasma membrane and the tonoplast are illustrated in FIGURE 6.14. Typically, transport across a biological membrane is energized by one primary active transport system coupled to ATP hydrolysis. The transport of one ionic species—for example, H⁺—generates an ion gradient and an electrochemical potential. Many other ions or organic substrates can then be transported by a variety of secondary active transport proteins, which energize the transport of their substrates by simultaneously carrying one or two H⁺ down their energy gradient. Thus protons circulate across the membrane, outward through the primary active transport proteins and back into the cell through the secondary active transport proteins. Most of the ion gradients across membranes of higher plants are generated and maintained by electrochemical-potential gradients of H⁺, which are generated by electrogenic H⁺ pumps.

Evidence suggests that in plants, Na⁺ is transported out of the cell by a Na⁺–H⁺ antiporter and that Cl⁻, $NO_3^-$, $H_2PO_4^-$, sucrose, amino acids, and other substances enter the cell via specific H⁺ symporters. In plants and fungi, sugars and amino acids are also taken up by symport with protons. What about potassium ions? Potassium ions can be taken up from the apoplast by symport with H⁺ (or, under some conditions, Na⁺). When the free-energy gradient favors passive K⁺ uptake, K⁺ can enter the cell by flux through specific K⁺ channels. However, even influx through channels is driven by the H⁺-ATPase, in the sense that K⁺ diffusion is driven by the membrane potential, which is maintained at a value more negative than the K⁺ equilibrium potential by the action of the electrogenic H⁺ pump. Conversely, K⁺ efflux requires the membrane potential to be maintained at a value more positive than $E_K$, which can be achieved by efflux of Cl⁻ or other anions through anion channels.

We have seen in preceding sections that some transmembrane proteins operate as channels for the controlled diffusion of ions. Other membrane proteins act as carriers for other substances (uncharged solutes and ions). Active transport utilizes carrier-type proteins that are energized directly by ATP hydrolysis or indirectly as symporters and antiporters. The latter systems use the energy of ion gradients (often a H⁺ gradient) to drive the energetically uphill transport of another ion or molecule. In the pages that follow we will examine in more detail the molecular properties, cellular locations, and genetic manipulations of some of the transport proteins that mediate the movement of organic and inorganic nutrients, as well as water, across plant cell membranes.

### The genes for many transporters have been identified

Transporter gene identification has aided greatly in the elucidation of the molecular properties of transporter proteins. One way to identify transporter genes is to screen plant complementary DNA (cDNA) libraries for genes that complement (i.e., compensate for) transport deficiencies in yeast. Many yeast transporter mutants are known and have

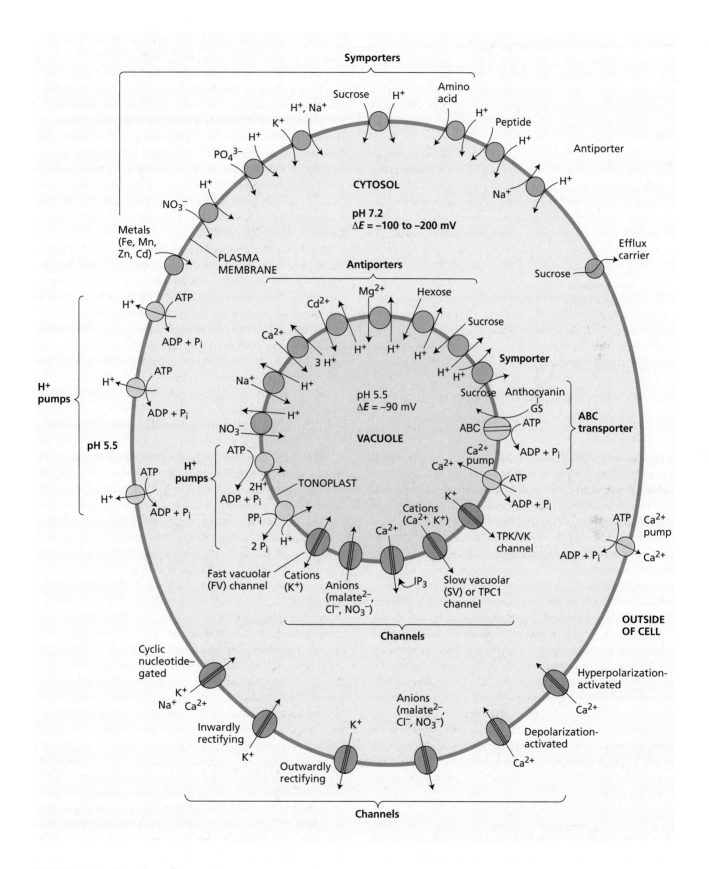

**FIGURE 6.14** Overview of the various transport proteins in the plasma membrane and tonoplast of plant cells.

been used to identify corresponding plant genes by complementation. In the case of genes for ion channels, researchers have also studied the behavior of the channel proteins by expressing the genes in oocytes of the frog *Xenopus*, which, because of their large size, are convenient for electrophysiological studies. Genes for both inwardly and outwardly rectifying $K^+$ channels have been cloned and characterized in this way.

As the number of sequenced genomes has increased, it has become increasingly common to identify putative transporter genes by phylogenetic analysis, in which sequence comparison with genes encoding transporters of known function in another organism allows one to predict function in the organism of interest. Computer-aided prediction of molecular structures is also becoming a useful tool in assigning putative function to a gene product.

The emerging picture of plant transporter genes shows that a family of genes, rather than an individual gene, exists in the plant genome for most transport functions. Within a gene family, variations in transport characteristics such as $K_m$, in mode of regulation, and in differential tissue expression give plants a remarkable plasticity to acclimate to a broad range of environmental conditions.

### Transporters exist for diverse nitrogen-containing compounds

Nitrogen, one of the macronutrients, can be present in the soil solution as nitrate ($NO_3^-$), ammonia ($NH_3$), or ammonium ($NH_4^+$). Plant $NH_4^+$ transporters are facilitators that promote $NH_4^+$ uptake down its free-energy gradient (Ludewig et al. 2007). Plant $NO_3^-$ transporters are of particular interest because of their complexity. Kinetic analysis shows that $NO_3^-$ transport, like the sucrose transport shown in Figure 6.13, has both high-affinity (low $K_m$) and low-affinity (high $K_m$) components. In contrast with sucrose, $NO_3^-$ is negatively charged, and such an electric charge imposes an energy requirement for uptake of nitrate. The energy is provided by symport with $H^+$. Nitrate transport is also strongly regulated according to $NO_3^-$ availability: The enzymes required for $NO_3^-$ transport, as well as for $NO_3^-$ assimilation (see Chapter 12), are induced in the presence of $NO_3^-$ in the environment, and uptake can be repressed if $NO_3^-$ accumulates in the cells.

Mutants with defects in $NO_3^-$ transport or $NO_3^-$ reduction can be selected by growth in the presence of chlorate ($ClO_3^-$). Chlorate is a $NO_3^-$ analog that is taken up and reduced in wild-type plants to the toxic product chlorite. If plants resistant to $ClO_3^-$ are selected, they are likely to show mutations that block $NO_3^-$ transport or reduction. Several such mutations have been identified in Arabidopsis. The first transporter gene identified in this way encodes an inducible $NO_3^-$–$H^+$ symporter that functions as a dual-affinity carrier, with its mode of action (high-affinity or low-affinity) being switched by its phosphorylation status (Liu and Tsay 2003). As more genes for $NO_3^-$

transport have been identified and characterized, it has become evident that both high- and low-affinity transport each involve more than one gene product.

Once nitrogen has been incorporated into organic molecules, there are a variety of mechanisms that distribute it throughout the plant. Peptide transporters provide one such mechanism. Peptide transporters are important for mobilizing nitrogen reserves during seed germination and for distribution of nitrogenous compounds via the vascular system. In the carnivorous pitcher plant (*Nepenthes alata*), high levels of expression of a peptide transporter are found in the pitcher, where the transporter presumably mediates uptake of peptides from digested insects into the internal tissues (Schulze et al. 1999).

Remarkably, the Arabidopsis genome encodes tenfold more peptide transporters than do the genomes of nonplant species, suggesting the importance of this group of transporters to plants (Stacey et al. 2002). One family of peptide transporters mediates di- and tripeptide uptake, and a second family of peptide transporters mediates uptake of tetra- and pentapeptides. Both of these families operate by coupling with the $H^+$ electrochemical gradient. A third group of peptide transporters differs from the other two in that the transporter directly utilizes energy from ATP hydrolysis for transport; thus transport does not depend on a primary electrochemical gradient (see **WEB TOPIC 6.5**). These transporters are members of the ATP-binding cassette (ABC) superfamily of proteins. ABC transporters are found both in the plasma membrane and in internal membranes.

The ABC superfamily is an extremely large protein family, and its members transport diverse substrates, ranging from ions to macromolecules. For example, large metabolites such as flavonoids, anthocyanins, and secondary products of metabolism are sequestered in the vacuole via the action of specific ABC transporters (Stacey et al. 2002; Rea 2007).

Amino acids constitute another important category of nitrogen-containing compounds. The plasma membrane amino acid transporters of eukaryotes are divided into five superfamilies, three of which rely on the proton gradient for coupled amino acid uptake and are present in plants (Wipf et al. 2002). In general, amino acid transporters can provide high- or low-affinity transport, and they have overlapping substrate specificities. Many amino acid transporters show distinct tissue-specific expression patterns, suggesting specialized functions in different cell types. Amino acids constitute an important form in which nitrogen is distributed long distances in plants, so it is not surprising that the expression patterns of many amino acid transporter genes include expression in vascular tissue.

Amino acid and peptide transporters have important roles in addition to their function as distributors of nitrogen resources. Because plant hormones are frequently found conjugated with amino acids and peptides, transporters for those molecules may also be involved in the distribution of hormone conjugates throughout the plant body. The hormone auxin is derived from tryptophan, and the genes encoding auxin transporters are related to those for some amino acid transporters. In another example, proline is an amino acid that accumulates under salt stress. This accumulation lowers cellular water potential, thereby promoting cellular water retention under stress conditions.

## Cation transporters are diverse

Cations are transported by both cation channels and cation carriers. The relative contributions of each type of transport mechanism differ depending on the membrane, cell type, and biological phenomenon under investigation.

**CATION CHANNELS** On the order of 50 genes in the Arabidopsis genome encode channels mediating cation uptake across the plasma membrane or intracellular membranes such as the tonoplast (Very and Sentenac 2002). Some of these channels are highly selective for specific ionic species, such as potassium ions. Others allow passage of a variety of cations, sometimes including $Na^+$, even though this ion is toxic when overaccumulated. As described in **FIGURE 6.15**, cation channels are categorized into six types based on their deduced structures and cation selectivity.

Of the six types of plant cation channels, the Shaker channels have been the most thoroughly characterized.

These channels are named after a *Drosophila* $K^+$ channel whose mutation causes the flies to shake or tremble. Plant Shaker channels are highly $K^+$-selective and can be either inwardly or outwardly rectifying or weakly rectifying. Some members of the Shaker family

- Mediate $K^+$ uptake or efflux across the guard cell membrane
- Provide a major conduit for $K^+$ uptake from the soil
- Participate in $K^+$ release to dead xylem vessels from the living stelar cells
- Play a role in $K^+$ uptake in pollen, a process that promotes water influx and pollen tube elongation

Some Shaker channels, such as those in roots, can function in high-affinity uptake, mediating $K^+$ uptake at micromolar external $K^+$ concentrations as long as the membrane potential is sufficiently hyperpolarized to drive this uptake (Hirsch et al. 1998).

Not all ion channels are as strongly regulated by membrane potential as the majority of Shaker channels. Some ion channels are not voltage-regulated, and the voltage sensitivity of others, such as the Kir-like channel, has not yet been determined (Lebaudy et al. 2007). Cyclic nucleotide–gated cation channels are only weakly voltage dependent.

### (A) K⁺ channels

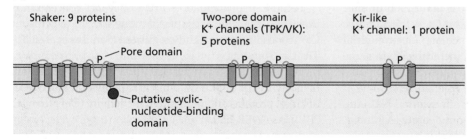

### (B) Poorly selective cation channels   (C) Ca²⁺-permeable channels (D) Cation selective, Ca²⁺ permeable

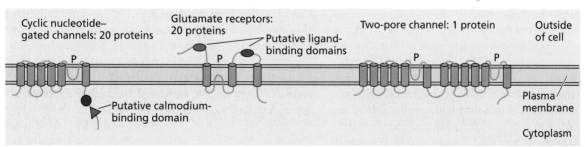

**FIGURE 6.15** Six families of Arabidopsis cation channels. Some channels have been identified solely from sequence homology with channels of animals, while others have been experimentally verified. (A) $K^+$-selective channels. (B) Weakly selective cation channels with activity regulated by binding of cyclic nucleotides. (C) Putative glutamate receptors; based on measurements of cytosolic $Ca^{2+}$ changes, these proteins probably function as $Ca^{2+}$-permeable channels. (D) TPC1 is the sole two-pore channel of this type encoded in the Arabidopsis genome. TPC1 is permeable to mono- and divalent cations, including $Ca^{2+}$. (After Very and Sentenac 2002; Lebaudy et al. 2007.)

Their activity is regulated by the binding of cyclic nucleotides such as cGMP. These channels can be permeable to $K^+$, $Ca^{2+}$, or $Na^+$ (Leng et al. 2002; Kaplan et al. 2007). Their functions are largely unknown, but mutational analyses have implicated several of these cyclic nucleotide–gated cation channels in disease-resistance responses to bacterial pathogens (Clough et al. 2000; Kaplan et al. 2007). Another interesting example is the glutamate receptor channels, which are homologous to a class of glutamate receptors in the mammalian nervous system that function as glutamate-gated cation channels. In plant cells, glutamate induces $Ca^{2+}$ uptake, suggesting that the glutamate receptors are $Ca^{2+}$-permeable cation channels (Qi et al. 2006; Tapken and Hollmann 2008).

Ion fluxes must also occur in and out of the vacuole, and both cation- and anion-permeable channels have been characterized in the vacuolar membrane (see Figure 6.14). Plant vacuolar channels include TPK/VK channels (see Figure 6.15A), which are highly selective $K^+$ channels that are activated by $Ca^{2+}$ (Gobert et al. 2007), and nonselective $Ca^{2+}$-permeable channels named TPC1/SV channels (see Figures 6.14, 6.15B). $Ca^{2+}$ efflux from internal storage sites such as the vacuole plays an important signaling role. $Ca^{2+}$ release from stores is triggered by several second messenger molecules, including cytosolic $Ca^{2+}$ itself and inositol trisphosphate ($InsP_3$). For a more detailed description of these signal transduction pathways, see Chapter 14.

**CATION CARRIERS** A variety of carriers also move cations into plant cells. There are two families of transporters that specialize in $K^+$ transport across plant membranes: the KUP/HAK/KT family and the HKTs. A third family, the cation–$H^+$ antiporters (CPAs), mediates electroneutral exchange of $H^+$ and other cations, including $K^+$ in some cases (Gierth and Mäser 2007). The KUP family contains both high-affinity and low-affinity transporters, some of which also mediate $Na^+$ influx at high external $Na^+$ concentrations. The HKT transporters can operate as $K^+$–$Na^+$ symporters, or as $Na^+$ channels under high external $Na^+$ concentrations. The functional roles of the HKT transporters within the plant remain incompletely elucidated, but one plausible role is retrieval of $Na^+$ from the transpiration stream (Davenport et al. 2007).

Irrigation increases soil salinity, and salinization of croplands is an increasing problem worldwide. Although halophytic plants, such as those found in salt marshes, are adapted to a high-salt environment, such environments are deleterious to other, glycophytic, plant species, including the majority of crop species. Plants have evolved mechanisms to sequester salt in the vacuole and to extrude it across the plasma membrane. Vacuolar $Na^+$ sequestration occurs by activity of $Na^+$–$H^+$ antiporters—a subset of CPA proteins—which couple the energetically downhill movement of $H^+$ into the cytoplasm with $Na^+$ uptake into the vacuole. When the Arabidopsis *AtNHX1* $Na^+$–$H^+$ anti-porter gene is overexpressed, it confers greatly increased salt tolerance to both Arabidopsis and crop species such as tomato (Apse et al. 1999; Zhang and Blumwald 2001).

At the plasma membrane, a $Na^+$–$H^+$ antiporter was uncovered in a screen to identify Arabidopsis mutants that showed enhanced sensitivity to salt, hence this antiporter was named Salt Overly Sensitive, or SOS1. SOS1 may function both to extrude $Na^+$ from the plant and to dilute root $Na^+$ concentrations via $Na^+$ extrusion to the transpiration stream, allowing eventual $Na^+$ sequestration in vacuoles of leaf mesophyll cells (Shi et al. 2002; Horie and Schroeder 2004).

Just as for $Na^+$, there is a large free-energy gradient for $Ca^{2+}$ that favors its entry into the cytosol from both the apoplast and intracellular stores. This entry is mediated by $Ca^{2+}$-permeable channels, which were described above. Calcium ion concentrations in the cell wall and in the apoplast are usually in the millimolar range; in contrast, free cytosolic $Ca^{2+}$ concentrations are kept in the hundreds of nanomolar ($10^{-9}$ $M$) to one micromolar ($10^{-6}$ $M$) range, against the large electrochemical-potential gradient for $Ca^{2+}$ diffusion into the cell. Calcium ion efflux from the cytosol is achieved by $Ca^{2+}$-ATPases found at the plasma membrane and in some endomembranes of plant cells, such as the tonoplast and endoplasmic reticulum (see Figure 6.14). Much of the $Ca^{2+}$ inside the cell is stored in the central vacuole, where it is sequestered via $Ca^{2+}$-ATPases and via $Ca^{2+}$–$H^+$ antiporters, which use the electrochemical potential of the proton gradient to energize the vacuolar accumulation of $Ca^{2+}$ (Hirschi et al. 1996).

Because small fluctuations in cytosolic $Ca^{2+}$ concentration drastically alter the activities of many enzymes, cytosolic $Ca^{2+}$ concentration is tightly regulated (Sanders et al. 2002). The $Ca^{2+}$-binding protein, calmodulin (CaM), participates in this regulation. Although CaM has no catalytic activity of its own, $Ca^{2+}$-bound CaM binds to many different classes of target proteins and regulates their activity (DeFalco et al. 2010) (see WEB ESSAY 6.2). $Ca^{2+}$-permeable cyclic nucleotide-gated channels are CaM-binding proteins, and there is evidence that this CaM binding results in downregulation of channel activity. One class of $Ca^{2+}$-ATPases also binds CaM. CaM binding releases these ATPases from autoinhibition, resulting in increased $Ca^{2+}$ extrusion into the apoplast, endoplasmic reticulum, and vacuole. Together, these two regulatory effects of CaM provide a mechanism whereby increases in cytosolic $Ca^{2+}$ concentration initiate a negative feedback loop, via activated CaM, that aids in restoration of resting cytosolic $Ca^{2+}$ levels.

## Anion transporters have been identified

Nitrate ($NO_3^-$), chloride ($Cl^-$), sulfate ($SO_4^{2-}$), and phosphate ($PO_4^{3-}$) are the major inorganic ions in plant cells, and malate$^{2-}$ is the major organic anion (Barbier-Brygoo et al. 2000). The free-energy gradient for all of these anions is in the direction of passive efflux. Several types of plant anion channels have been characterized by electrophysiological techniques, and most anion channels appear to be permeable to a variety of anions. In contrast to our detailed knowledge of $K^+$ transport mechanisms at the molecular level, it has proved more difficult to identify anion channel genes (Barbier-Brygoo et al. 2000). However, a protein, SLAC1, that likely mediates anion efflux from guard cells has recently been identified (Negi et al. 2008, Vahisalu et al. 2008).

In contrast to the relative lack of specificity of anion channels, anion carriers that mediate the energetically uphill transport of anions into plant cells exhibit selectivity for particular anions. In addition to the transporters for nitrate uptake described above, plants have transporters for various organic anions, such as malate and citrate. As discussed in Chapter 23, malate uptake is an important contributor to the increase in intracellular solute concentration that drives the water uptake into guard cells that leads to stomatal opening. One member of the ABC superfamily, AtABCB14, has been assigned this malate import function (Lee et al. 2008).

Phosphate availability in the soil solution often limits plant growth. In Arabidopsis, a family of about nine high-affinity plasma membrane phosphate transporters mediates phosphate uptake in symport with protons. These transporters are expressed primarily in root tissues, and their expression is induced upon phosphate starvation (Rausch and Bucher 2002). Phosphate–$H^+$ symporters with lower affinity for phosphate have also been identified in plants and have been localized to membranes of intracellular organelles such as plastids and mitochondria. Another group of phosphate transporters, the phosphate translocators, are located in the inner plastid membrane, where, in exchange for uptake of inorganic phosphate, they function to release phosphorylated carbon compounds derived from photosynthesis to the cytosol (Weber et al. 2004) (see **WEB TOPIC 8.15**).

## Metal transporters transport essential micronutrients

Several metals are essential nutrients for plants, although they are required in only trace amounts. One example is iron. Iron deficiency is the most common human nutritional disorder worldwide, so an increased understanding of how plants accumulate iron may also benefit efforts to improve the nutritional value of crops. Over 25 related ZIP transporters mediate the uptake of iron, manganese, and zinc ions into plants, and other transporter families that mediate the uptake of copper and molybdenum ions have been identified (Palmer and Guerinot 2009). Metal ions are usually present at low concentrations in the soil solution, so these transporters are typically high-affinity transporters. Some metal ion transporters mediate the uptake of cadmium ions, which are undesirable in crop species because cadmium is toxic to humans. However, this property may prove useful in the detoxifying of soils

by uptake of contaminants into plants (phytoremediation), which can then be removed and properly discarded (see **WEB ESSAY 26.2**).

Once in the plant, metal ions, usually chelated with other molecules, must be transported into the xylem for distribution throughout the plant body via the transpiration stream, and metals must also be sent to their appropriate subcellular destinations. For example, most of the iron in plants is found in chloroplasts, where it is incorporated into chlorophyll and components of the electron transport chain (see Chapter 7). The transporters that mediate the passage of iron across the chloroplast envelope await identification (Palmer and Guerinot 2009). Overaccumulation of metals such as iron and copper can lead to production of toxic reactive oxygen species (ROS). Compounds that chelate metal ions guard against this threat, and transporters that mediate metal uptake into the vacuole are also important in maintaining metal concentrations at nontoxic levels.

## Aquaporins have diverse functions

Water channels, or **aquaporins**, are a class of proteins that are relatively abundant in plant membranes (see Chapters 3 and 4). The existence of aquaporins was initially a surprise because it had been thought that the lipid bilayer was itself sufficiently permeable to water. Nevertheless, aquaporins are common in plant and animal membranes, and the Arabidopsis genome is predicted to encode approximately 35 aquaporins (Maurel et al. 2008).

Many aquaporins do not result in ion currents when expressed in oocytes, consistent with a lack of ion transport activity, but when the osmolarity of the external medium is reduced, expression of these proteins results in swelling and bursting of the oocytes. The bursting results from rapid influx of water across the oocyte plasma membrane, which normally has very low permeability to water. These results confirm that aquaporins form water channels in membranes (see Figure 3.13). Some aquaporin proteins also transport uncharged solutes (e.g., $NH_3$), and there is some evidence that aquaporins act as conduits for carbon dioxide uptake into plant cells. It has also been hypothesized that aquaporins function as sensors of gradients in osmotic potential and turgor pressure (Maurel et al. 2008).

Aquaporin activity is regulated by phosphorylation as well as by pH, calcium concentration, heteromerization, and reactive oxygen species (Tyerman et al. 2002; Maurel et al. 2008). Such regulation may account for the ability of plant cells to quickly alter their water permeability in response to circadian rhythm and to stresses such as salt, chilling, drought, and flooding (anoxia). Regulation also occurs at the level of gene expression. Aquaporins are highly expressed in epidermal and endodermal cells and in the xylem parenchyma, which may be critical points for control of water movement (Javot and Maurel 2002).

## Plasma membrane H⁺-ATPases are highly regulated P-type ATPases

As we have seen, the outward active transport of protons across the plasma membrane creates gradients of pH and electrical potential that drive the transport of many other substances (ions and uncharged solutes) through the various secondary active transport proteins. H⁺-ATPase activity is also important for the regulation of cytoplasmic pH and for the control of cell turgor, which drives organ (leaf and flower) movement, stomatal opening, and cell growth. **FIGURE 6.16** illustrates how a membrane H⁺-ATPase might work.

Plant and fungal plasma membrane H⁺-ATPases and $Ca^{2+}$-ATPases are members of a class known as P-type ATPases, which are phosphorylated as part of the catalytic cycle that hydrolyzes ATP. Because of this phosphorylation step, the plasma membrane ATPases are strongly inhibited by orthovanadate ($HVO_4^{2-}$), a phosphate ($HPO_4^{2-}$) analog that competes with phosphate from ATP for the aspartic acid phosphorylation site on the protein.

Plant plasma membrane H⁺-ATPases are encoded by a family of about a dozen genes (Sondergaard et al. 2004). The roles of each H⁺-ATPase isoform are starting to be understood, based on information from gene expression patterns and functional analysis of Arabidopsis plants harboring null mutations in individual H⁺-ATPase genes. Some H⁺-ATPases exhibit cell-specific patterns of expression. For example, several H⁺-ATPases are expressed in guard cells, where they energize the plasma membrane to drive solute uptake during stomatal opening (see Chapter 4).

In general, H⁺-ATPase expression is high in cells with key functions in nutrient movement, including root endodermal cells and cells involved in nutrient uptake from the apoplast that surrounds the developing seed (Sondergaard et al. 2004). In cells in which multiple H⁺-ATPases are coexpressed, they may be differentially regulated or may function redundantly, perhaps providing a "fail-safe" mechanism to this all-important transport function (Arango et al. 2003).

**FIGURE 6.17** shows a model of the functional domains of a yeast plasma membrane H⁺-ATPase, which is similar to those of plants. The protein has ten membrane-spanning domains that cause it to loop back and forth across the membrane. Some of the membrane-spanning domains make up the pathway through which protons are pumped. The catalytic domain, which catalyzes ATP hydrolysis, including the aspartic acid residue that becomes phosphorylated during the catalytic cycle, is on the cytosolic face of the membrane.

Like other enzymes, the plasma membrane H⁺-ATPase is regulated by the concentration of substrate (ATP), pH, temperature, and other factors. In addition, H⁺-ATPase molecules can be reversibly activated or deactivated by specific signals, such as light, hormones, or pathogen attack. This type of regulation is mediated by a specialized autoinhibitory domain at the C-terminal end of the polypeptide chain, which acts to regulate the activity of the H⁺-ATPase (see Figure 6.17). If the autoinhibitory domain is removed by a protease, the enzyme becomes irreversibly activated (Palmgren 2001).

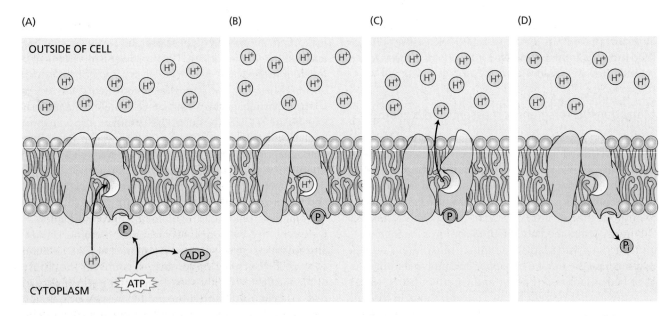

**FIGURE 6.16** Hypothetical steps in the transport of a proton against its chemical gradient by H⁺-ATPase. The pump, embedded in the membrane, binds the proton on the inside of the cell (A) and is phosphorylated by ATP (B). This phosphorylation leads to a conformational change that exposes the proton to the outside of the cell and makes it possible for the proton to diffuse away (C). Release of the phosphate ion (P) from the pump into the cytosol (D) restores the initial configuration of the H⁺-ATPase and allows a new pumping cycle to begin.

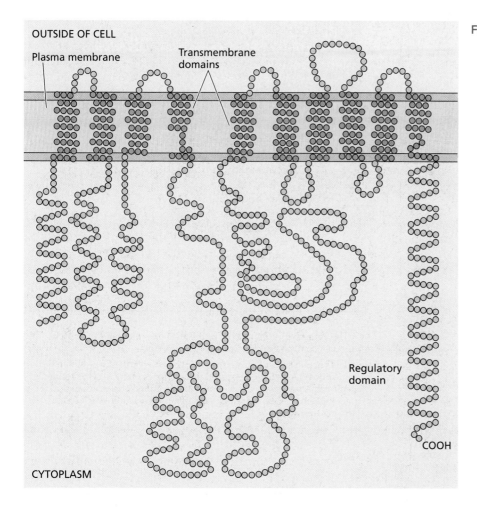

OUTSIDE OF CELL

Plasma membrane

Transmembrane domains

CYTOPLASM

Regulatory domain

COOH

**FIGURE 6.17** Two-dimensional representation of a plasma membrane H⁺-ATPase from yeast. The H⁺-ATPase protein has ten transmembrane segments. The regulatory domain is an autoinhibitory domain. Posttranslational modifications that lead to displacement of the autoinhibitory domain result in H⁺-ATPase activation. (After Palmgren 2001.)

The autoinhibitory effect of the C-terminal domain can also be regulated by protein kinases and phosphatases that add phosphate groups to or remove them from serine or threonine residues on this domain. Phosphorylation recruits ubiquitous enzyme-modulating proteins called 14-3-3 proteins, which bind to the phosphorylated region and are thought to thus displace the autoinhibitory domain, leading to H⁺-ATPase activation (Gaxiola et al. 2007). The fungal toxin **fusicoccin**, which is a strong activator of the H⁺-ATPase, activates this pump by increasing 14-3-3 binding affinity even in the absence of phosphorylation (Sondergaard et al. 2004). The effect of fusicoccin on the guard cell H⁺-ATPases is so strong that it can lead to irreversible stomatal opening, wilting, and even plant death.

## The tonoplast H⁺-ATPase drives solute accumulation in vacuoles

Plant cells increase their size primarily by taking up water into a large central vacuole. Therefore, the osmotic pressure of the vacuole must be kept sufficiently high for water to enter from the cytoplasm. The tonoplast regulates the traffic of ions and metabolites between the cytosol and

the vacuole, just as the plasma membrane regulates their uptake into the cell. Tonoplast transport became a vigorous area of research following the development of methods for the isolation of intact vacuoles and tonoplast vesicles (see **WEB TOPIC 6.6**). These studies led to the discovery of a new type of proton-pumping ATPase, which transports protons into the vacuole (see Figure 6.14).

The **vacuolar H⁺-ATPase** (also called **V-ATPase**) differs both structurally and functionally from the plasma membrane H⁺-ATPase (Kluge et al. 2003). The vacuolar ATPase is more closely related to the F-ATPases of mitochondria and chloroplasts (see Chapter 11). Because the hydrolysis of ATP by the vacuolar ATPase does not involve the formation of a phosphorylated intermediate, vacuolar ATPases are insensitive to vanadate, the inhibitor of plasma membrane ATPases discussed earlier. Vacuolar ATPases are specifically inhibited by the antibiotic bafilomycin as well as by high concentrations of nitrate, neither of which inhibit plasma membrane ATPases. Use of these selective inhibitors makes it possible to identify different types of ATPases and to assay their activity.

Vacuolar ATPases belong to a general class of ATPases that are present on the endomembrane systems of all

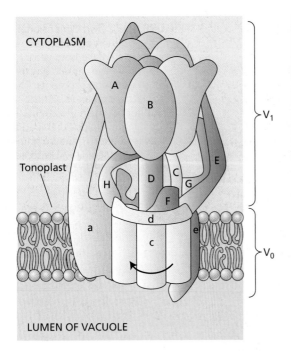

CYTOPLASM

Tonoplast

$V_1$

$V_0$

LUMEN OF VACUOLE

**FIGURE 6.18**   Model of the V-ATPase rotary motor. Many polypeptide subunits come together to make up this complex enzyme. The $V_1$ catalytic complex, which is easily dissociated from the membrane, contains the nucleotide-binding and catalytic sites. Components of $V_1$ are designated by uppercase letters. The integral membrane complex mediating $H^+$ transport is designated $V_0$, and its subunits are designated by lowercase letters. It is proposed that ATPase reactions catalyzed by each of the A subunits, acting in sequence, drive the rotation of the shaft (D) and the six c subunits. The rotation of the c subunits relative to the a subunit is thought to drive the transport of $H^+$ across the membrane. (After Kluge et al. 2003.)

**TABLE 6.2**
**The vacuolar pH of some hyperacidifying plant species**

| Tissue | Species | pH[a] |
|---|---|---|
| Fruits | | |
| | Lime (*Citrus aurantifolia*) | 1.7 |
| | Lemon (*Citrus limonia*) | 2.5 |
| | Cherry (*Prunus cerasus*) | 2.5 |
| | Grapefruit (*Citrus paradisi*) | 3.0 |
| Leaves | | |
| | Wax begonia (*Begonia semperflorens*) | 1.5 |
| | Begonia 'Lucerna' | 0.9–1.4 |
| | Oxalis sp. | 1.9–2.6 |
| | Sorrel (*Rumex* sp.) | 2.6 |
| | Prickly pear (*Opuntia phaeacantha*)[b] | 1.4 (6:45 A.M.) |
| | | 5.5 (4:00 P.M.) |

*Source*: Data from Small 1946.

[a] The values represent the pH of the juice or expressed sap of each tissue, usually a good indicator of vacuolar pH.

[b] The vacuolar pH of the cactus *Opuntia phaeacantha* varies with the time of day. As will be discussed in Chapter 8, many desert succulents have a specialized type of photosynthesis, called crassulacean acid metabolism (CAM), that causes the pH of the vacuoles to decrease during the night.

eukaryotes. They are large enzyme complexes, about 750 kDa, composed of multiple subunits (Kluge et al. 2003). These subunits are organized into a peripheral catalytic complex, $V_1$, and an integral membrane channel complex, $V_0$ (**FIGURE 6.18**). Because of their similarities to F-ATPases, vacuolar ATPases are assumed to operate like tiny rotary motors (see Chapter 11).

Vacuolar ATPases are electrogenic proton pumps that transport protons from the cytoplasm to the vacuole and generate a proton motive force across the tonoplast. Electrogenic proton pumping accounts for the fact that the vacuole is typically 20 to 30 mV more positive than the cytoplasm, although it is still negative relative to the external medium. To allow maintenance of bulk electrical neutrality, anions such as $Cl^-$ or $malate^{2-}$ are transported from the cytoplasm into the vacuole through channels in the membrane (Barkla and Pantoja 1996). Without this simultaneous movement of anions along with the pumped protons, the charge buildup across the tonoplast would make the pumping of additional protons energetically impossible.

The conservation of bulk electrical neutrality by anion transport makes it possible for the vacuolar $H^+$-ATPase to generate a large concentration gradient of protons (pH gradient) across the tonoplast. This gradient accounts for the fact that the pH of the vacuolar sap is typically about 5.5, while the cytoplasmic pH is typically 7.0 to 7.5. Whereas the electrical component of the proton motive force drives the uptake of anions into the vacuole, the electrochemical-potential gradient for $H^+$ ($\Delta \tilde{\mu}_{H^+}$) is harnessed to drive the uptake of cations and sugars into the vacuole via secondary transport (antiporter) systems (see Figure 6.14).

Although the pH of most plant vacuoles is mildly acidic (about 5.5), the pH of the vacuoles of some species is much lower—a phenomenon termed *hyperacidification*. Vacuolar hyperacidification is the cause of the sour taste of certain fruits (lemons) and vegetables (rhubarb). Some extreme examples are listed in **TABLE 6.2**. Biochemical studies have suggested that the low pH of lemon fruit vacuoles (specifically, those of the juice sac cells) is due to a combination of factors:

- The low permeability of the vacuolar membrane to protons permits a steeper pH gradient to build up.

- A specialized vacuolar ATPase is able to pump protons more efficiently (with less wasted energy) than normal vacuolar ATPases can (Müller et al. 1997).
- Organic acids such as citric, malic, and oxalic acids accumulate in the vacuole and help maintain its low pH by acting as buffers.

### $H^+$-pyrophosphatases also pump protons at the tonoplast

Another type of proton pump, a $H^+$-pyrophosphatase ($H^+$-PPase) (Gaxiola et al. 2007), works in parallel with the vacuolar ATPase to create a proton gradient across the tonoplast (see Figure 6.14). This enzyme consists of a single polypeptide that has a molecular mass of 80 kDa. The $H^+$-PPase harnesses energy from the hydrolysis of inorganic pyrophosphate ($PP_i$).

The free energy released by $PP_i$ hydrolysis is less than that from ATP hydrolysis. However, the $H^+$-PPase transports only one $H^+$ ion per $PP_i$ molecule hydrolyzed, whereas the vacuolar ATPase appears to transport two $H^+$ ions per ATP hydrolyzed. Thus the energy available per $H^+$ transported appears to be the same, and the two enzymes seem to be able to generate comparable proton gradients. Interestingly, the plant $H^+$-PPase is not found in animals or in yeast, although a similar enzyme is present in some bacteria and protists.

Both the V-ATPase and the $H^+$-PPase are found in other compartments of the endomembrane system in addition to the vacuole. Consistent with this distribution, evidence is emerging that these ATPases regulate not only $H^+$ gradients per se, but also vesicle trafficking and secretion. In addition, increased auxin transport and cell division in Arabidopsis plants overexpressing a $H^+$-PPase, and the opposite phenotypes in plants with reduced $H^+$-PPase activity, indicate connections between $H^+$-PPase activity and the synthesis, distribution, and regulation of auxin transporters (Gaxiola et al. 2007) (see Chapter 19).

## Ion Transport in Roots

Mineral nutrients absorbed by the root are carried to the shoot by the transpiration stream moving through the xylem (see Chapter 4). Both the initial uptake of nutrients and water and the subsequent movement of these substances from the root surface across the cortex and into the xylem are highly specific, well-regulated processes.

Ion transport across the root obeys the same biophysical laws that govern cellular transport. However, as we have seen in the case of water movement (see Chapter 4), the anatomy of roots imposes some special constraints on the pathway of ion movement. In this section we will discuss the pathways and mechanisms involved in the radial movement of ions from the root surface to the tracheary elements of the xylem.

### Solutes move through both apoplast and symplast

Thus far, our discussion of cellular ion transport has not included the cell wall. In terms of the transport of small molecules, the cell wall is an open lattice of polysaccharides through which mineral nutrients diffuse readily. Because all plant cells are separated by cell walls, ions can diffuse across a tissue (or be carried passively by water flow) entirely through the cell wall space without ever entering a living cell. This continuum of cell walls is called the **extracellular space**, or **apoplast** (see Figure 4.4). Typically, 5 to 20% of the plant tissue volume is occupied by cell walls.

Just as the cell walls form a continuous phase, so do the cytoplasms of neighboring cells, collectively referred to as the **symplast**. Plant cells are interconnected by cytoplasmic bridges called plasmodesmata (see Chapter 1), cylindrical pores 20 to 60 nm in diameter (**FIGURE 6.19**). Each plasmodesma is lined with plasma membrane and contains a narrow tubule, the *desmotubule*, that is a continuation of the endoplasmic reticulum.

In tissues where significant amounts of intercellular transport occur, neighboring cells contain large numbers of plasmodesmata, up to 15 per square micrometer of cell surface. Specialized secretory cells, such as floral nectaries and leaf salt glands, have high densities of plasmodesmata; so do the cells near root tips, where most nutrient absorption occurs.

By injecting dyes or by making electrical-resistance measurements on cells containing large numbers of plasmodesmata, investigators have shown that ions, water, and small solutes can move from cell to cell through these pores. Because each plasmodesma is partly occluded by the desmotubule and its associated proteins (see Chapter 1), the movement of large molecules such as proteins through plasmodesmata requires special mechanisms (Lucas and Lee 2004). Ions, on the other hand, appear to move symplastically through the plant by simple diffusion through plasmodesmata (see Chapter 4).

### Ions cross both symplast and apoplast

Ion absorption by the root (see Chapter 5) is more pronounced in the root hair zone than in the meristem and elongation zones. Cells in the root hair zone have completed their elongation but have not yet begun secondary growth. The root hairs are simply extensions of specific epidermal cells that greatly increase the surface area available for ion absorption.

An ion that enters a root may immediately enter the symplast by crossing the plasma membrane of an epider-

(A)

(B)

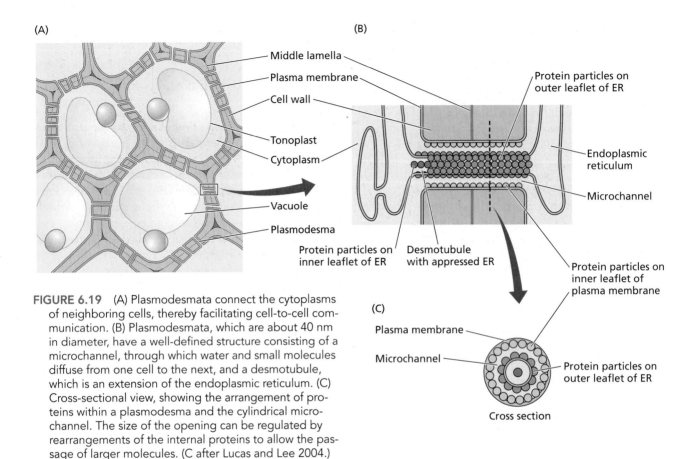

Middle lamella

Plasma membrane

Cell wall

Tonoplast

Cytoplasm

Vacuole

Plasmodesma

Protein particles on inner leaflet of ER

Desmotubule with appressed ER

Protein particles on outer leaflet of ER

Endoplasmic reticulum

Microchannel

Protein particles on inner leaflet of plasma membrane

(C)

Plasma membrane

Microchannel

Protein particles on outer leaflet of ER

Cross section

**FIGURE 6.19** (A) Plasmodesmata connect the cytoplasms of neighboring cells, thereby facilitating cell-to-cell communication. (B) Plasmodesmata, which are about 40 nm in diameter, have a well-defined structure consisting of a microchannel, through which water and small molecules diffuse from one cell to the next, and a desmotubule, which is an extension of the endoplasmic reticulum. (C) Cross-sectional view, showing the arrangement of proteins within a plasmodesma and the cylindrical microchannel. The size of the opening can be regulated by rearrangements of the internal proteins to allow the passage of larger molecules. (C after Lucas and Lee 2004.)

mal cell, or it may enter the apoplast and diffuse between the epidermal cells through the cell walls. From the apoplast of the cortex, an ion (or other solute) may either be transported across the plasma membrane of a cortical cell, thus entering the symplast, or diffuse radially all the way to the endodermis via the apoplast. The apoplast forms a continuous phase from the root surface through the cortex. However, in all cases, ions must enter the symplast before they can enter the stele, because of the presence of the Casparian strip. As discussed in Chapters 4 and 5, the Casparian strip is a suberized layer that forms rings around cells of the specialized endodermis (**FIGURE 6.20**) and effectively blocks the entry of water and solutes into the stele via the apoplast.

Once an ion has entered the stele through the symplastic connections across the endodermis, it continues to diffuse from cell to cell into the xylem. Finally, the ion is released into the apoplast and diffuses into a xylem tracheid or vessel element. Again, the Casparian strip prevents the ion from diffusing back out of the root through the apoplast. The presence of the Casparian strip allows the plant to maintain a higher ion concentration in the xylem than exists in the soil water surrounding the roots.

## Xylem parenchyma cells participate in xylem loading

Once ions have been taken up into the symplast of the root at the epidermis or cortex, they must be loaded into the tracheids or vessel elements of the stele to be translocated to the shoot. The stele consists of dead tracheary elements and living xylem parenchyma. Because the xylem tracheary elements are dead cells, they lack cytoplasmic continuity with the surrounding xylem parenchyma. To enter the tracheary elements, the ions must exit the symplast by crossing a plasma membrane a second time.

The process whereby ions exit the symplast and enter the conducting cells of the xylem is called **xylem loading**. Xylem loading is a highly regulated process. Xylem parenchyma cells, like other living plant cells, maintain plasma membrane $H^+$-ATPase activity and a negative membrane potential. Transporters that specifically function in the unloading of solutes to the tracheary elements have been identified by electrophysiological and genetic approaches. The plasma membranes of xylem parenchyma cells contain proton pumps, aquaporins, and a variety of ion channels and carriers specialized for influx or efflux (De Boer and Volkov 2003).

(A)

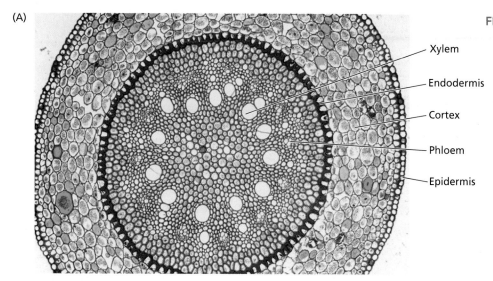

Xylem

Endodermis

Cortex

Phloem

Epidermis

**FIGURE 6.20** Tissue organization in roots. (A) Cross section through a root of carrion flower (genus *Smilax*), a monocot, showing the epidermis, cortex parenchyma, endodermis, xylem, and phloem. (B) Schematic diagram of a root cross section, illustrating the cell layers through which solutes pass from the soil solution to the xylem tracheary elements. (A ×30, ©Biodisc/Visuals Unlimited/Alamy; B after Dunlop and Bowling 1971.)

(B)

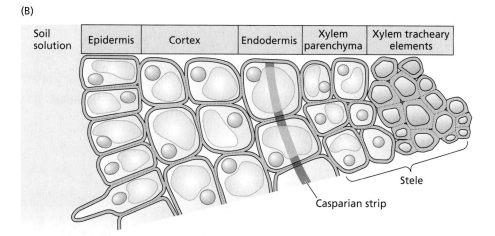

| Soil solution | Epidermis | Cortex | Endodermis | Xylem parenchyma | Xylem tracheary elements |
|---|---|---|---|---|---|

Stele

Casparian strip

In Arabidopsis xylem parenchyma, the stelar outwardly rectifying $K^+$ channel (SKOR) is expressed in cells of the pericycle and xylem parenchyma, where it functions as an efflux channel, transporting $K^+$ from the living cells out to the tracheary elements (Gaymard et al. 1998). In mutant Arabidopsis plants lacking the SKOR channel protein, or in plants in which SKOR has been pharmacologically inactivated, $K^+$ transport from the root to the shoot is severely reduced, confirming the function of this channel protein.

Several types of anion-selective channels have also been identified that participate in unloading of $Cl^-$ and $NO_3^-$ from the xylem parenchyma. Drought, abscisic acid (ABA) treatment, or elevation of cytosolic $Ca^{2+}$ concentrations (which often occurs as a response to ABA), all reduce the activity of SKOR and anion channels of the root xylem parenchyma, a response that could help maintain cellular hydration in the root under desiccating conditions.

Other, less selective ion channels found in the plasma membrane of xylem parenchyma cells are permeable to $K^+$, $Na^+$, and anions. In addition, the SOS1 $Na^+$–$H^+$ antiporter appears to load $Na^+$ into the xylem, possibly serving to ameliorate $Na^+$ levels in the root symplast. Other transport molecules have also been identified that mediate loading of boron, $Mg^{2+}$, and $PO_4^{2-}$. Thus the flux of ions from the xylem parenchyma cells into the xylem tracheary elements is under tight metabolic control through the regulation of plasma membrane $H^+$-ATPases, ion efflux channels, and carriers.

# SUMMARY

The biologically regulated movement of molecules and ions from one location to another is known as transport. Plants exchange solutes with their environment and among their tissues and organs. Both local and long-distance transport processes in plants are controlled largely by cellular membranes.

### Passive and Active Transport

- Concentration gradients and electrical-potential gradients, the main forces that drive transport across biological membranes, are integrated by a term called the electrochemical potential (**Equation 6.8**).

- Movement of solutes across membranes down their free-energy gradient is facilitated by passive transport mechanisms (**Figures 6.1, 6.2**).

- Movement of solutes against their free-energy gradient is known as active transport and requires energy input.

### Transport of Ions across Membrane Barriers

- The extent to which a membrane permits or restricts the movement of a substance is a property known as membrane permeability (**Figure 6.6**).

- Permeability depends on the chemical properties of the particular solute, on the lipid composition of the membrane, and particularly on the membrane proteins that facilitate the transport of specific substances.

- For each permeant ion, the distribution of that particular ionic species across a membrane that would occur at equilibrium is described by the Nernst equation (**Equation 6.10**).

- Transport of $H^+$ across the plant cell membrane by $H^+$-ATPases is a major determinant of the membrane potential (**Figures 6.16, 6.17**).

### Membrane Transport Processes

- Biological membranes contain specialized proteins—channels, carriers, and pumps—that facilitate solute transport (**Figure 6.7**).

- Channels are regulated protein pores that, when open, greatly enhance molecular fluxes across membranes (**Figures 6.7, 6.8**).

- Organisms have a great diversity of ion channel types. Depending on the channel type, channels can be nonselective or highly selective for just one ionic species. Channels can be regulated by

many parameters, including voltage, secondary messengers, and ligands (**Figures 6.9, 6.14, 6.15**).

- Carriers bind specific substances and transport them at a rate several orders of magnitude lower than that of channels (**Figures 6.7, 6.12**).

- Pumps require energy for transport. Active transport of $H^+$ and $Ca^{2+}$ across plant cell membranes is mediated by pumps (**Figure 6.7**).

- Secondary active transporters in plants harness energy from energetically downhill movement of protons to mediate energetically uphill transport of another solute (**Figure 6.10**).

- In symport, both transported solutes move in the same direction across the membrane, whereas in antiport, the two solutes move in opposite directions (**Figure 6.11**).

### Membrane Transport Proteins

- Many channels, carriers, and pumps of the plant plasma membrane and tonoplast have been identified at the molecular level (**Figure 6.14**) and characterized using electrophysiological (**Figure 6.9**) and biochemical techniques.

- Transporters exist for diverse nitrogenous compounds, including $NO_3^-$, amino acids, and peptides.

- Plants have a great variety of cation channels that can be classified according to their ionic selectivity and regulatory mechanisms (**Figure 6.15**).

- Several different classes of cation carriers mediate $K^+$ uptake into the cytosol (**Figure 6.14**).

- $Na^+$–$H^+$ antiporters on the tonoplast and plasma membrane extrude $Na^+$ into the vacuole and apoplast, respectively, thereby opposing accumulation of toxic levels of $Na^+$ in the cytosol (**Figure 6.14**).

- $Ca^{2+}$ is an important second messenger in signal transduction cascades, and its cytosolic concentration is tightly regulated. $Ca^{2+}$ enters the cytosol passively, via $Ca^{2+}$-permeable channels, and is actively removed from the cytosol by $Ca^+$ pumps and $Ca^{2+}$–$H^+$ antiporters (**Figure 6.14**).

- Selective carriers that mediate $NO_3^-$, $Cl^-$, $SO_4^-$, and $PO_4^{3-}$ uptake into the cytosol and anion channels that nonselectively mediate anion efflux from the cytosol regulate cellular concentrations of these macronutrients (**Figure 6.14**).

## SUMMARY continued

- Both essential and toxic metal ions are transported by high-affinity ZIP transport proteins (**Figure 6.14**).

- Aquaporins facilitate water flux across plant cell membranes, and their regulation allows for rapid changes in water permeability in response to environmental stimuli.

- Plasma membrane $H^+$-ATPases are encoded by a multigene family, and their activity is reversibly controlled by an autoinhibitory domain.

- Two types of $H^+$ pumps found on the vacuolar membrane, V-ATPases and $H^+$-pyrophosphatases, regulate the proton motive force across the tonoplast, which in turn drives the movement of other solutes across this membrane via antiport mechanisms (**Figures 6.14, 6.18**).

### Ion Transport in Roots

- Solutes such as mineral nutrients move between cells either through the extracellular space (the apoplast) or from cytoplasm to cytoplasm (via the symplast). The cytoplasm of neighboring cells is connected by plasmodesmata, which facilitate symplastic transport (**Figure 6.19**).

- When an ion enters the root, it may be taken up into the cytoplasm of an epidermal cell, or it may diffuse through the apoplast into the root cortex and then enter the symplast through a cortical or endodermal cell. From the root symplast, the ion is loaded into the xylem and moves to the shoot in the transpiration stream (**Figure 6.20**).

## WEB MATERIAL

### Web Topics

**6.1 Relating the Membrane Potential to the Distribution of Several Ions across the Membrane: The Goldman Equation**

The Goldman equation is used to calculate the membrane permeability to more than one ion.

**6.2 Patch Clamp Studies in Plant Cells**

Patch clamping is applied to plant cells for electrophysiological studies.

**6.3 Chemiosmosis in Action**

The chemiosmotic theory explains how electrical and concentration gradients are used to perform cellular work.

**6.4 Kinetic Analysis of Multiple Transporter Systems**

Application of principles of enzyme kinetics to transport systems provides an effective way to characterize different carriers.

**6.5 ABC Transporters in Plants**

ATP-binding cassette (ABC) transporters are a large family of active transport proteins energized directly by ATP.

**6.6 Transport Studies with Isolated Vacuoles and Membrane Vesicles**

Certain experimental techniques enable the isolation of tonoplast and plasma membrane vesicles for study.

### Web Essays

**6.1 Potassium Channels**

Several plant $K^+$ channels have been characterized.

**6.2 Calmodulin: A Simple but Multifaceted Signal Transducer**

This essay describes how CaM interacts with a broad array of cellular proteins and how these protein–protein interactions act to transduce changes in $Ca^{2+}$ concentration into a complex web of biochemical responses.

# CHAPTER REFERENCES

Apse, M. P., Aharon, G. S., Snedden, W. A., and Blumwald, E. (1999) Salt tolerance conferred by overexpression of a vacuolar Na⁺/H⁺ antiport in Arabidopsis. *Science* 285: 1256–1258.

Arango, M., Gevaudant, F., Oufattole, M., and Boutry, M. (2003) The plasma membrane proton pump ATPase: The significance of gene subfamilies. *Planta* 216: 355–365.

Barbier-Brygoo, H., Vinauger, M., Colcombet, J., Ephritikhine, G., Frachisse, J., and Maurel, C. (2000) Anion channels in higher plants: Functional characterization, molecular structure and physiological role. *Biochim. Biophys. Acta* 1465: 199–218.

Barkla, B. J., and Pantoja, O. (1996) Physiology of ion transport across the tonoplast of higher plants. *Annu. Rev. Plant Physiol. Plant Mol. Biol.* 47: 159–184.

Buchanan, B. B., Gruissem, W., and Jones, R. L., eds. (2000) *Biochemistry and Molecular Biology of Plants*. American Society of Plant Physiologists, Rockville, MD.

Clough, S. J., Fengler, K. A., Yu, I. C., Lippok, B., Smith, R. K., Jr., and Bent, A. F. (2000) The Arabidopsis *dnd1* "defense, no death" gene encodes a mutated cyclic nucleotide-gated ion channel. *Proc. Natl. Acad. Sci. USA* 97: 9323–9328.

Davenport, R. J., Munoz-Mayer, A., Jha, D., Essah, P. A., Rus, A., and Tester, M. (2007) The Na⁺ transporter AtHKT1;1 controls retrieval of Na⁺ from the xylem in Arabidopsis. *Plant Cell Environ.* 30: 497–507.

De Boer, A. H., and Volkov, V. (2003) Logistics of water and salt transport through the plant: Structure and functioning of the xylem. *Plant Cell Environ.* 26: 87–101.

DeFalco, T. A., Bender, K. W., and Snedden, W. A. (2010) Breaking the code: Ca²⁺ sensors in plant signaling. *Biochem. J.* 425: 27–40.

Dunlop, J., and Bowling, D. J. F. (1971) The movement of ions to the xylem exudate of maize roots. *J. Exp. Bot.* 22: 453–464.

Gaymard, F., Pilot, G., Lacombe, B., Bouchez, D., Bruneau, D., Boucherez, J., Michaux-Ferriere, N., Thibaud, J. B., and Sentenac, H. (1998) Identification and disruption of a plant shaker-like outward channel involved in K⁺ release into the xylem sap. *Cell* 94: 647–655.

Gaxiola, R. A., Palmgren, M. G., and Schumacher, K. (2007) Plant proton pumps. *FEBS Lett.* 581: 2204–2214.

Gierth, M., and Mäser, P. (2007) Potassium transporters in plants—involvement in K⁺ acquisition, redistribution and homeostasis. *FEBS Lett.* 581: 2348–2356.

Gobert, A., Isayenkov, S., Voelker, C., Czempinski, K., and Maathuis, F.J. (2007) The two-pore channel TPK1 gene encodes the vacuolar K⁺ conductance and plays a role in K⁺ homeostasis. *Proc. Natl. Acad. Sci. USA* 104: 10726–10731.

Higinbotham, N., Etherton, B., and Foster, R. J. (1967) Mineral ion contents and cell transmembrane electropotentials of pea and oat seedling tissue. *Plant Physiol.* 42: 37–46.

Higinbotham, N., Graves, J. S., and Davis, R. F. (1970) Evidence for an electrogenic ion transport pump in cells of higher plants. *J. Membr. Biol.* 3: 210–222.

Hirsch, R. E., Lewis, B. D., Spalding, E. P., and Sussman, M. R. (1998) A role for the AKT1 potassium channel in plant nutrition. *Science* 280: 918–921.

Hirschi, K. D., Zhen, R.-G., Rea, P. A., and Fink, G. R. (1996) CAX1, an H⁺/Ca²⁺ antiporter from *Arabidopsis*. *Proc. Natl. Acad. Sci. USA* 93: 8782–8786.

Horie, T., and Schroeder, J. I. (2004) Sodium transporters in plants. Diverse genes and physiological functions. *Plant Physiol.* 136: 2457–2462.

Javot, H., and Maurel, C. (2002) The role of aquaporins in root water uptake. *Ann. Bot. (Lond.)* 90: 301–313.

Kaplan, B., Sherma, T., and Fromm, H. (2007) Cyclic nucleotide-gated channels in plants. *FEBS Lett.* 581: 2237–2246.

Kluge, C., Lahr, J., Hanitzsch, M., Bolte, S., Golldack, D., and Dietz, K. J. (2003) New insight into the structure and regulation of the plant vacuolar H⁺-ATPase. *J. Bioenerg. Biomembr.* 35: 377–388.

Lalonde, S., Wipf, D., and Frommer, W. F. (2004) Transport mechanisms for organic forms of carbon and nitrogen between source and sink. *Annu. Rev. Plant Biol.* 55: 341–372.

Lebaudy, A., Véry, A., and Sentenac, H. (2007) K⁺ channel activity in plants: Genes, regulations and functions. *FEBS Lett.* 581: 2357–2366.

Lee, M., Choi, Y., Burla, B., Kim, Y. Y., Jeon, B., Maeshima, M., Yoo, J.Y., Martinoia, E., and Lee, Y. (2008) The ABC transporter AtABCB14 is a malate importer and modulates stomatal response to CO₂. *Nat. Cell Biol.* 10: 1217–1223.

Leng, Q., Mercier, R. W., Hua, B. G., Fromm, H., and Berkowitz, G. A. (2002) Electrophysiological analysis of cloned cyclic nucleotide-gated ion channels. *Plant Physiol.* 128: 400–410.

Lin, W., Schmitt, M. R., Hitz, W. D., and Giaquinta, R. T. (1984) Sugar transport into protoplasts isolated from developing soybean cotyledons. *Plant Physiol.* 75: 936–940.

Liu, K. H., and Tsay, Y. F. (2003) Switching between the two action modes of the dual-affinity nitrate transporter CHL1 by phosphorylation. *EMBO J.* 22: 1005–1013.

Lucas, W. J., and Lee, Y. J. (2004) Plasmodesmata as a supracellular control network in plants. *Nat. Rev. Mol. Cell Biol.* 5: 712–726.

Ludewig, U., Neuhäuser, B., and Bynowski, M. (2007) Molecular mechanisms of ammonium transport and accumulation in plants. *FEBS Lett.* 581: 2301–2308.

Maurel, C., Verdoucq, L., Luu, D., and Santoni, V. (2008) Plant aquaporins: Membrane channels with multiple integrated functions. *Annu. Rev. Plant Biol.* 59: 595–624.

Müller, M., Irkens-Kiesecker, U., Kramer, D., and Taiz, L. (1997) Purification and reconstitution of the vacuolar $H^+$-ATPases from lemon fruits and epicotyls. *J. Biol. Chem.* 272: 12762–12770.

Negi, J., Matsuda, O., Nagasawa, T., Oba, Y., Takahashi, H., Kawai-Yamada, M., Uchimiya, H., Hashimoto, M., and Iba, K. (2008). $CO_2$ regulator SLAC1 and its homologues are essential for anion homeostasis in plant cells. *Nature* 452: 483–486.

Nobel, P. (1991) *Physicochemical and Environmental Plant Physiology.* Academic Press, San Diego, CA.

Palmer, C. M., and Guerinot, M. L. (2009) Facing the challenges of Cu, Fe and Zn homeostasis in plants. *Nature Chem. Biol.* 5: 333–340.

Palmgren, M. G. (2001) Plant plasma membrane $H^+$-ATPases: Powerhouses for nutrient uptake. *Annu. Rev. Plant Physiol. Plant Mol. Biol.* 52: 817–845.

Qi, Z., Stephens, N. R., and Spalding, E. P. (2006) Calcium entry mediated by GLR3.3, an Arabidopsis glutamate receptor with a broad agonist profile. *Plant Physiol.* 142: 963–971.

Rausch, C., and Bucher, M. (2002) Molecular mechanisms of phosphate transport in plants. *Planta* 216: 23–37.

Rea, P. A. (2007) Plant ATP-binding cassette transporters. *Annu. Rev. Plant Biol.* 58: 347–375.

Sanders, D., Pelloux, J., Brownlee, C., and Harper, J. F. (2002) Calcium at the crossroads of signaling. *Plant Cell* 14 Suppl: S401–417.

Schulze, W., Frommer, W. B., and Ward, J. M. (1999) Transporters for ammonium, amino acids and peptides are expressed in pitchers of the carnivorous plant *Nepenthes*. *Plant J.* 17: 637–646.

Schwacke, R., Schneider, A., van der Graaff, E., Fischer, K., Catoni, E., Desimone, M., Frommer, W. B., Flugge, U. I., and Kunze R. (2003) ARAMEMNON, a novel database for Arabidopsis integral membrane proteins. *Plant Physiol.* 131: 16–26.

Shi, H., Quintero, F. J., Pardo, J. M., and Zhu, J. K. (2002) The putative plasma membrane $Na^+/H^+$ antiporter SOS1 controls long-distance $Na^+$ transport in plants. *Plant Cell.* 14: 465–477.

Small, J. (1946) *pH and Plants, an Introduction to Beginners.* D. Van Nostrand, New York.

Sondergaard, T. E., Schulz, A., and Palmgren, M. G. (2004) Energization of transport processes in plants: Roles of the plasma membrane $H^+$-ATPase. *Plant Physiol.* 136: 475–2482.

Spanswick, R. M. (1981) Electrogenic ion pumps. *Annu. Rev. Plant Physiol.* 32: 267–289.

Stacey, G., Koh, S., Granger, C., and Becker, J. M. (2002) Peptide transport in plants. *Trends Plant Sci.* 7: 257–263.

Tapken, D., and Hollmann, M. (2008) *Arabidopsis thaliana* glutamate receptor ion channel function demonstrated by ion pore transplantation. *J. Mol. Biol.* 383: 36–48.

Tyerman, S. D., Niemietz, C. M., and Bramley, H. (2002) Plant aquaporins: Multifunctional water and solute channels with expanding roles. *Plant Cell Environ.* 25: 173–194.

Vahisalu, T., Kollist, H., Wang, Y. F., Nishimura, N., Chan, W. Y., Valerio, G., Lamminmaki, A., Brosche, M. Moldau, H., Desikan, R., Schroeder, J. I., Kangasjarvi, J. (2008) SLAC1 is required for plant guard cell S-type anion channel function in stomatal signaling. *Nature* 452: 487–491.

Very, A. A., and Sentenac, H. (2002) Cation channels in the Arabidopsis plasma membrane. *Trends Plant Sci.* 7: 168–175.

Weber, A. P., Schneidereit, J., and Voll, L. M. (2004) Using mutants to probe the in vivo function of plastid envelope membrane metabolite transporters. *J. Exp. Bot.* 55: 1231–1244.

Wipf, D., Ludewig, U., Tegeder, M., Rentsch, D., Koch, W., and Frommer, W. B. (2002) Conservation of amino acid transporters in fungi, plants and animals. *Trends Biochem. Sci.* 27: 139–147.

Zhang, H. X., and Blumwald, E. (2001) Transgenic salt-tolerant tomato plants accumulate salt in foliage but not in fruit. *Nat. Biotechnol.* 19: 765–768.

Zhou, Y., Qu, H., Dibley, K. E., Offler, C. E., and Patrick, J. W. (2007) A suite of sucrose transporters expressed in coats of developing legume seeds includes novel pH-independent facilitators. *Plant J.* 49: 750–764.

# Unit Two

## Biochemistry and Metabolism

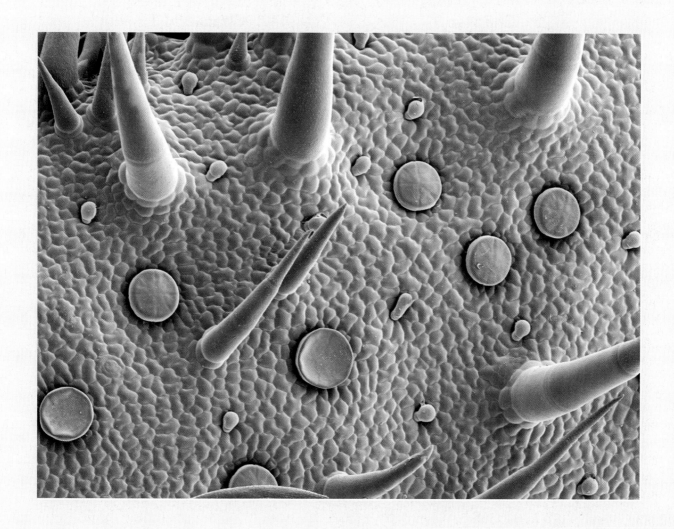

# UNIT TWO

*Previous page: The epidermal surface of a marjoram (Origanum majorana) leaf contains numerous glandular trichomes (yellow), as well as pointed non-glandular trichomes, as shown in this color-enhanced scanning electron micrograph. The glandular trichomes produce secondary metabolites that contribute to the plant's defenses against herbivores. Image © SPL/Photolibrary.com.*

# 7

# Photosynthesis: The Light Reactions

Life on Earth ultimately depends on energy derived from the sun. Photosynthesis is the only process of biological importance that can harvest this energy. A large fraction of the planet's energy resources results from photosynthetic activity in either recent or ancient times (fossil fuels). This chapter introduces the basic physical principles that underlie photosynthetic energy storage and the current understanding of the structure and function of the photosynthetic apparatus (Blankenship 2002).

The term *photosynthesis* means literally "synthesis using light." As we will see in this chapter, photosynthetic organisms use solar energy to synthesize complex carbon compounds. More specifically, light energy drives the synthesis of carbohydrates and generation of oxygen from carbon dioxide and water:

$$6\,CO_2 + 6\,H_2O \rightarrow C_6H_{12}O_6 + 6\,O_2$$

Carbon dioxide    Water    Carbohydrate    Oxygen

Energy stored in these molecules can be used later to power cellular processes in the plant and can serve as the energy source for all forms of life.

This chapter deals with the role of light in photosynthesis, the structure of the photosynthetic apparatus, and the processes that begin with the excitation of chlorophyll by light and culminate in the synthesis of ATP and NADPH.

# Photosynthesis in Higher Plants

The most active photosynthetic tissue in higher plants is the mesophyll of leaves. Mesophyll cells have many chloroplasts, which contain the specialized light-absorbing green pigments, the **chlorophylls**. In photosynthesis, the plant uses solar energy to oxidize water, thereby releasing oxygen, and to reduce carbon dioxide, thereby forming large carbon compounds, primarily sugars. The complex series of reactions that culminate in the reduction of $CO_2$ include the thylakoid reactions and the carbon fixation reactions.

The **thylakoid reactions** of photosynthesis take place in the specialized internal membranes of the chloroplast called thylakoids (see Chapter 1). The end products of these thylakoid reactions are the high-energy compounds ATP and NADPH, which are used for the synthesis of sugars in the **carbon fixation reactions**. These synthetic processes take place in the stroma of the chloroplast, the aqueous region that surrounds the thylakoids. The thylakoid reactions, also called the "light reactions" of photosynthesis, are the subject of this chapter; the carbon fixation reactions are discussed in Chapter 8.

In the chloroplast, light energy is converted into chemical energy by two different functional units called *photosystems*. The absorbed light energy is used to power the transfer of electrons through a series of compounds that act as electron donors and electron acceptors. The majority of electrons ultimately reduce NADP+ to NADPH and oxidize $H_2O$ to $O_2$. Light energy is also used to generate a proton motive force (see Chapter 6) across the thylakoid membrane; this proton motive force is used to synthesize ATP.

# General Concepts

In this section we will explore the essential concepts that provide a foundation for an understanding of photosynthesis. These concepts include the nature of light, the properties of pigments, and the various roles of pigments.

## Light has characteristics of both a particle and a wave

A triumph of physics in the early twentieth century was the realization that light has properties of both particles and waves. A wave (**FIGURE 7.1**) is characterized by a **wavelength**, denoted by the Greek letter lambda ($\lambda$), which is the distance between successive wave crests. The **frequency**, represented by the Greek letter nu ($v$), is the number of wave crests that pass an observer in a given time. A simple equation relates the wavelength, the frequency, and the speed of any wave:

$$c = \lambda v \qquad (7.1)$$

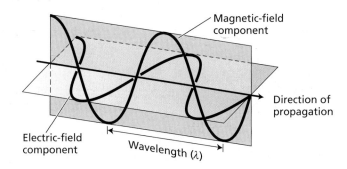

**FIGURE 7.1** Light is a transverse electromagnetic wave, consisting of oscillating electric and magnetic fields that are perpendicular to each other and to the direction of propagation of the light. Light moves at a speed of $3 \times 10^8$ m s$^{-1}$. The wavelength ($\lambda$) is the distance between successive crests of the wave.

where $c$ is the speed of the wave—in the present case, the speed of light ($3.0 \times 10^8$ m s$^{-1}$). The light wave is a transverse (side-to-side) electromagnetic wave, in which both electric and magnetic fields oscillate perpendicularly to the direction of propagation of the wave and at 90° with respect to each other.

Light is also a particle, which we call a **photon**. Each photon contains an amount of energy that is called a **quantum** (plural *quanta*). The energy content of light is not continuous but rather is delivered in discrete packets, the quanta. The energy ($E$) of a photon depends on the frequency of the light according to a relation known as Planck's law:

$$E = hv \qquad (7.2)$$

where $h$ is Planck's constant ($6.626 \times 10^{-34}$ J s).

Sunlight is like a rain of photons of different frequencies. Our eyes are sensitive to only a small range of frequencies—the visible-light region of the electromagnetic spectrum (**FIGURE 7.2**). Light of slightly higher frequencies (or shorter wavelengths) is in the ultraviolet region of the spectrum, and light of slightly lower frequencies (or longer wavelengths) is in the infrared region. The output of the sun is shown in **FIGURE 7.3**, along with the energy density that strikes the surface of Earth. The **absorption spectrum** of chlorophyll *a* (green curve in Figure 7.3) indicates approximately the portion of the solar output that is utilized by plants.

An absorption spectrum (plural *spectra*) provides information about the amount of **light energy** taken up or absorbed by a molecule or substance as a function of the wavelength of the light. The absorption spectrum for a particular substance in a nonabsorbing solvent can be determined by a spectrophotometer as illustrated in **FIGURE 7.4**. Spectrophotometry, the technique used to measure the absorption of light by a sample, is more completely discussed in **WEB TOPIC 7.1**.

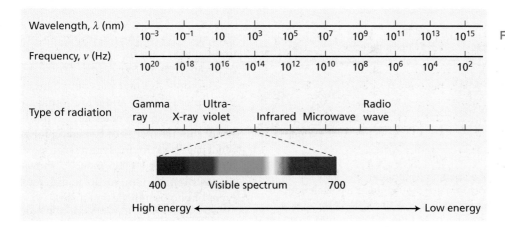

**FIGURE 7.2** Electromagnetic spectrum. Wavelength ($\lambda$) and frequency ($v$) are inversely related. Our eyes are sensitive to only a narrow range of wavelengths of radiation, the visible region, which extends from about 400 nm (violet) to about 700 nm (red). Short-wavelength (high-frequency) light has a high energy content; long-wavelength (low-frequency) light has a low energy content.

## When molecules absorb or emit light, they change their electronic state

Chlorophyll appears green to our eyes because it absorbs light mainly in the red and blue parts of the spectrum, so only some of the light enriched in green wavelengths (about 550 nm) is reflected into our eyes (see Figure 7.3).

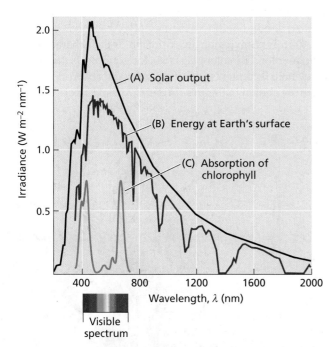

**FIGURE 7.3** The solar spectrum and its relation to the absorption spectrum of chlorophyll. Curve A is the energy output of the sun as a function of wavelength. Curve B is the energy that strikes the surface of Earth. The sharp valleys in the infrared region beyond 700 nm represent the absorption of solar energy by molecules in the atmosphere, chiefly water vapor. Curve C is the absorption spectrum of chlorophyll, which absorbs strongly in the blue (about 430 nm) and the red (about 660 nm) portions of the spectrum. Because the green light in the middle of the visible region is not efficiently absorbed, some of it is reflected into our eyes and gives plants their characteristic green color.

The absorption of light is represented by Equation 7.3, in which chlorophyll (Chl) in its lowest-energy, or ground, state absorbs a photon (represented by $hv$) and makes a transition to a higher-energy, or excited, state (Chl*):

$$\text{Chl} + hv \rightarrow \text{Chl*} \quad (7.3)$$

The distribution of electrons in the excited molecule is somewhat different from the distribution in the ground-state molecule (**FIGURE 7.5**). Absorption of blue light excites the chlorophyll to a higher energy state than absorption of red light, because the energy of photons is higher when their wavelength is shorter. In the higher excited state, chlorophyll is extremely unstable; it very rapidly gives up some of its energy to the surroundings as heat, and enters the **lowest excited state**, where it can be stable for a maximum of several nanoseconds ($10^{-9}$ s). Because of the inherent instability of the excited state, any process that captures its energy must be extremely rapid.

In the lowest excited state, the excited chlorophyll has four alternative pathways for disposing of its available energy:

1. Excited chlorophyll can re-emit a photon and thereby return to its ground state—a process known as **fluorescence**. When it does so, the wavelength of fluorescence is slightly longer (and of lower energy) than the wavelength of absorption, because a portion of the excitation energy is converted into heat before the fluorescent photon is emitted. Chlorophylls fluoresce in the red region of the spectrum.

2. The excited chlorophyll can return to its ground state by directly converting its excitation energy into heat, with no emission of a photon.

3. Chlorophyll may participate in **energy transfer**, during which an excited chlorophyll transfers its energy to another molecule.

4. A fourth process is **photochemistry**, in which the energy of the excited state causes chemical reactions to occur. The photochemical reactions of photosynthesis are among the fastest known chemical reactions. This

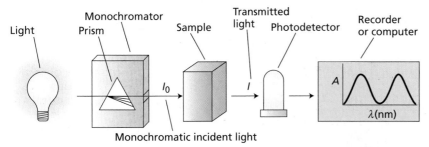

**FIGURE 7.4** Schematic diagram of a spectrophotometer. The instrument consists of a light source, a monochromator that contains a wavelength selection device such as a prism, a sample holder, a photodetector, and a recorder or computer. The output wavelength of the monochromator can be changed by rotation of the prism; the graph of absorbance ($A$) versus wavelength ($\lambda$) is called a spectrum.

extreme speed is necessary for photochemistry to compete with the three other possible reactions of the excited state just described.

## Photosynthetic pigments absorb the light that powers photosynthesis

The energy of sunlight is first absorbed by the pigments of the plant. All pigments active in photosynthesis are found in the chloroplast. Structures and absorption spectra of several photosynthetic pigments are shown in **FIGURES 7.6 AND 7.7**, respectively. The chlorophylls and **bacterio-**

chlorophylls (pigments found in certain bacteria) are the typical pigments of photosynthetic organisms.

Chlorophylls $a$ and $b$ are abundant in green plants, and $c$ and $d$ are found in some protists and cyanobacteria. A number of different types of bacteriochlorophyll have been found; type $a$ is the most widely distributed. **WEB TOPIC 7.2** shows the distribution of pigments in different types of photosynthetic organisms.

All chlorophylls have a complex ring structure that is chemically related to the porphyrin-like groups found in hemoglobin and cytochromes (see Figure 7.6A). A long hydrocarbon tail is almost always attached to the ring structure. The tail anchors the chlorophyll to the hydrophobic portion of its environment. The ring structure contains some loosely bound electrons and is the part of the molecule involved in electronic transitions and redox (reduction–oxidation) reactions.

The different types of **carotenoids** found in photosynthetic organisms are all linear molecules with multiple conjugated double bonds (see Figure 7.6B). Absorption bands in the 400 to 500 nm region give carotenoids their characteristic orange color. The color of carrots, for example, is due to the carotenoid β-carotene, whose structure and absorption spectrum are shown in Figures 7.6 and 7.7, respectively.

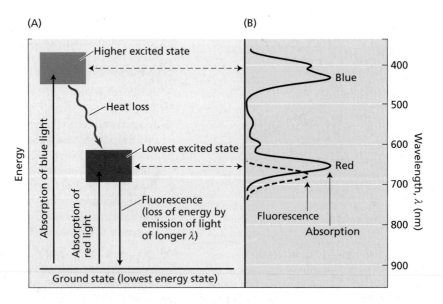

**FIGURE 7.5** Light absorption and emission by chlorophyll. (A) Energy level diagram. Absorption or emission of light is indicated by vertical lines that connect the ground state with excited electron states. The blue and red absorption bands of chlorophyll (which absorb blue and red photons, respectively) correspond to the upward vertical arrows, signifying that energy absorbed from light causes the molecule to change from the ground state to an excited state. The downward-pointing arrow indicates fluorescence, in which the molecule goes from the lowest excited state to the ground state while re-emitting energy as a photon. (B) Spectra of absorption and fluorescence. The long-wavelength (red) absorption band of chlorophyll corresponds to light that has the energy required to cause the transition from the ground state to the first excited state. The short-wavelength (blue) absorption band corresponds to a transition to a higher excited state.

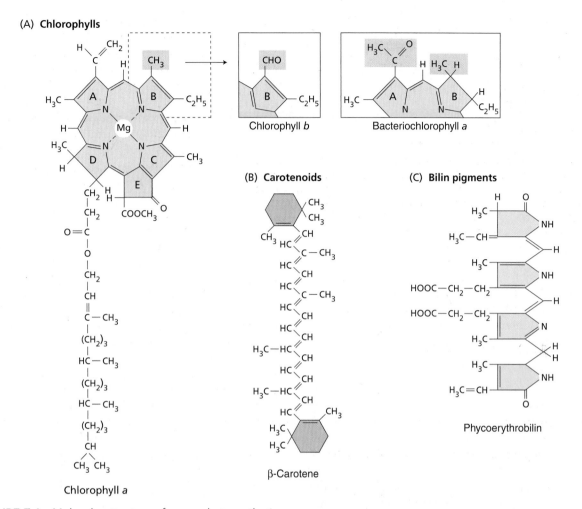

**(A) Chlorophylls**

Chlorophyll *b*

Bacteriochlorophyll *a*

Chlorophyll *a*

**(B) Carotenoids**

β-Carotene

**(C) Bilin pigments**

Phycoerythrobilin

**FIGURE 7.6** Molecular structure of some photosynthetic pigments. (A) The chlorophylls have a porphyrin-like ring structure with a magnesium atom (Mg) coordinated in the center and a long hydrophobic hydrocarbon tail that anchors them in the photosynthetic membrane. The porphyrin-like ring is the site of the electron rearrangements that occur when the chlorophyll is excited, and of the unpaired electrons when it is either oxidized or reduced. Various chlorophylls differ chiefly in the substituents around the rings and the pattern of double bonds. (B) Carotenoids are linear polyenes that serve as both antenna pigments and photoprotective agents. (C) Bilin pigments are open-chain tetrapyrroles found in antenna structures known as phycobilisomes that occur in cyanobacteria and red algae.

Carotenoids are found in all photosynthetic organisms, except for mutants incapable of living outside the laboratory. Carotenoids are integral constituents of the thylakoid membrane and are usually associated intimately with many of the proteins that make up the photosynthetic apparatus. The light energy absorbed by the carotenoids is transferred to chlorophyll for photosynthesis; because of this role they are called **accessory pigments**. Carotenoids also help protect the organism from damage caused by light (see p. 190 of this chapter and Chapter 13).

# Key Experiments in Understanding Photosynthesis

Establishing the overall chemical equation of photosynthesis required several hundred years and contributions by many scientists (literature references for historical developments can be found on the web site for this book). In 1771, Joseph Priestley observed that a sprig of mint growing in air in which a candle had burned out improved the air so that another candle could burn. He had discovered oxygen evolution by plants. A Dutch biologist, Jan Ingenhousz, documented the essential role of light in photosynthesis in 1779.

Other scientists established the roles of $CO_2$ and $H_2O$ and showed that organic matter, specifically carbohydrate, is a product of photosynthesis along with oxygen. By the end of the nineteenth century, the balanced overall chemical reaction for photosynthesis could be written as follows:

$$6\,CO_2 + 6\,H_2O \xrightarrow{\text{Light, plant}} C_6H_{12}O_6 + 6\,O_2 \qquad (7.4)$$

where $C_6H_{12}O_6$ represents a simple sugar such as glucose. As will be discussed in Chapter 8, glucose is not the actual product of the carbon fixation reactions, so this part of the equation should not be taken literally. However, the energetics for the actual reaction are approximately the same as represented here.

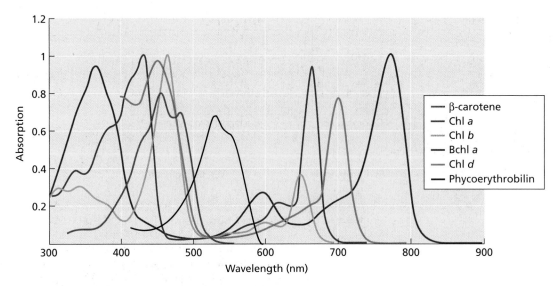

**FIGURE 7.7** Absorption spectra of some photosynthetic pigments, including β-carotene, chlorophyll *a* (Chl *a*), chlorophyll *b* (Chl *b*), bacteriochlorophyll *a* (Bchl *a*) chlorophyll *d* (Chl *d*), and phycoerythrobilin. The absorption spectra shown are for pure pigments dissolved in nonpolar solvents, except phycoerythrin, a protein from cyanobacteria that contains a phycoerythrobilin chromophore covalently attached to the peptide chain. In many cases the spectra of photosynthetic pigments in vivo are substantially affected by the environment of the pigments in the photosynthetic membrane.

The chemical reactions of photosynthesis are complex. In fact, at least 50 intermediate reaction steps have now been identified, and undoubtedly additional steps will be discovered. An early clue to the chemical nature of the essential chemical process of photosynthesis came in the 1920s from investigations of photosynthetic bacteria that did not produce oxygen as an end product. From his studies on these bacteria, C. B. van Niel concluded that photosynthesis is a redox process. This conclusion has served as a fundamental concept on which all subsequent research on photosynthesis has been based.

We now turn to the relationship between photosynthetic activity and the spectrum of absorbed light. We will discuss some of the critical experiments that have contributed to our present understanding of photosynthesis, and we will consider equations for the essential chemical reactions of photosynthesis.

### Action spectra relate light absorption to photosynthetic activity

The use of action spectra has been central to the development of our current understanding of photosynthesis. An **action spectrum** depicts the magnitude of a response of a biological system to light as a function of wavelength. For example, an action spectrum for photosynthesis can be constructed from measurements of oxygen evolution at different wavelengths (**FIGURE 7.8**). Often an action spectrum can identify the *chromophore* (pigment) responsible for a particular light-induced phenomenon.

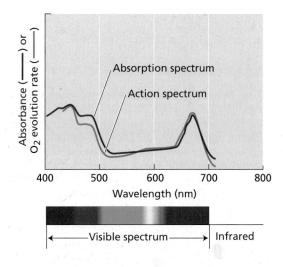

**FIGURE 7.8** Action spectrum compared with an absorption spectrum. The absorption spectrum is measured as shown in Figure 7.4. An action spectrum is measured by plotting a response to light, such as oxygen evolution, as a function of wavelength. If the pigment used to obtain the absorption spectrum is the same as that which causes the response, the absorption and action spectra will match. In the example shown here, the action spectrum for oxygen evolution matches the absorption spectrum of intact chloroplasts quite well, indicating that light absorption by the chlorophylls mediates oxygen evolution. Discrepancies are found in the region of carotenoid absorption, from 450 to 550 nm, indicating that energy transfer from carotenoids to chlorophylls is not as effective as energy transfer between chlorophylls.

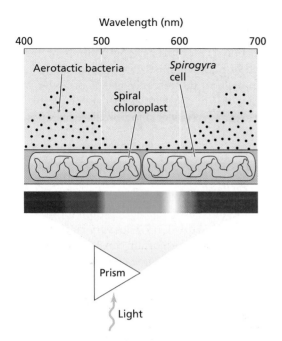

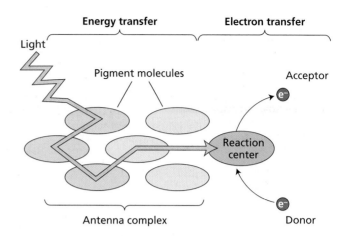

**FIGURE 7.9** Schematic diagram of the action spectrum measurements by T. W. Engelmann. Engelmann projected a spectrum of light onto the spiral chloroplast of the filamentous green alga *Spirogyra* and observed that oxygen-seeking bacteria introduced into the system collected in the region of the spectrum where chlorophyll pigments absorb. This action spectrum gave the first indication of the effectiveness of light absorbed by pigments in driving photosynthesis.

**FIGURE 7.10** Basic concept of energy transfer during photosynthesis. Many pigments together serve as an antenna, collecting light and transferring its energy to the reaction center, where chemical reactions store some of the energy by transferring electrons from a chlorophyll pigment to an electron acceptor molecule. An electron donor then reduces the chlorophyll again. The transfer of energy in the antenna is a purely physical phenomenon and involves no chemical changes.

Some of the first action spectra were measured by T. W. Engelmann in the late 1800s (**FIGURE 7.9**). Engelmann used a prism to disperse sunlight into a rainbow that was allowed to fall on an aquatic algal filament. A population of $O_2$-seeking bacteria was introduced into the system. The bacteria congregated in the regions of the filaments that evolved the most $O_2$. These were the regions illuminated by blue light and red light, which are strongly absorbed by chlorophyll. Today, action spectra can be measured in room-sized spectrographs in which a huge monochromator bathes the experimental samples in monochromatic light. The technology is more sophisticated, but the *principle* is the same as that of Engelmann's experiments.

Action spectra were very important for the discovery of two distinct photosystems operating in $O_2$-evolving photosynthetic organisms. Before we introduce the two photosystems, however, we need to describe the light-gathering antennas and the energy needs of photosynthesis.

## *Photosynthesis takes place in complexes containing light-harvesting antennas and photochemical reaction centers*

A portion of the light energy absorbed by chlorophylls and carotenoids is eventually stored as chemical energy via the formation of chemical bonds. This conversion of energy from one form to another is a complex process that depends on cooperation between many pigment molecules and a group of electron transfer proteins.

The majority of the pigments serve as an **antenna complex**, collecting light and transferring the energy to the **reaction center complex**, where the chemical oxidation and reduction reactions leading to long-term energy storage take place (**FIGURE 7.10**). Molecular structures of some of the antenna and reaction center complexes are discussed later in the chapter.

How does the plant benefit from this division of labor between antenna and reaction center pigments? Even in bright sunlight, a single chlorophyll molecule absorbs only a few photons each second. If there were a reaction center associated with each chlorophyll molecule, the reaction center enzymes would be idle most of the time, only occasionally being activated by photon absorption. However, if a reaction center receives energy from many pigments at once, the system is kept active a large fraction of the time.

In 1932, Robert Emerson and William Arnold performed a key experiment that provided the first evidence for the cooperation of many chlorophyll molecules in energy conversion during photosynthesis. They delivered very brief ($10^{-5}$ s) flashes of light to a suspension of the green alga *Chlorella pyrenoidosa* and measured the amount of oxygen produced. The flashes were spaced about 0.1 s apart, a time that Emerson and Arnold had determined in earlier work was long enough for the enzymatic steps of the process to be completed before the arrival of the next flash. The investigators varied the energy of the flashes and found that

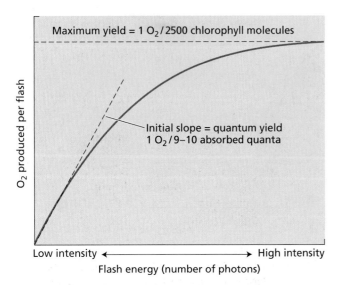

**FIGURE 7.11** Relationship of oxygen production to flash energy, the first evidence for the interaction between the antenna pigments and the reaction center. At saturating energies, the maximum amount of $O_2$ produced is one molecule per 2500 chlorophyll molecules.

at high energies the oxygen production did not increase when a more intense flash was given: The photosynthetic system was saturated with light (**FIGURE 7.11**).

In their measurement of the relationship of oxygen production to flash energy, Emerson and Arnold were surprised to find that under saturating conditions, only one molecule of oxygen was produced for each 2500 chlorophyll molecules in the sample. We know now that several hundred pigments are associated with each reaction center and that each reaction center must operate four times to produce one molecule of oxygen—hence the value of 2500 chlorophylls per $O_2$.

The reaction centers and most of the antenna complexes are integral components of the photosynthetic membrane. In eukaryotic photosynthetic organisms, these membranes are found within the chloroplast; in photosynthetic prokaryotes, the site of photosynthesis is the plasma membrane or membranes derived from it.

The graph shown in Figure 7.11 permits us to calculate another important parameter of the light reactions of photosynthesis, the quantum yield. The **quantum yield** of photosynthesis ($\Phi$) is defined as follows:

$$\Phi = \frac{\text{Number of photochemical products}}{\text{Total number of quanta absorbed}} \quad (7.5)$$

In the linear portion (low light intensity) of the curve, an increase in the number of photons stimulates a proportional increase in oxygen evolution. Thus the slope of the curve measures the quantum yield for oxygen production. The quantum yield for a particular process can range from 0 (if that process does not respond to light) to 1.0 (if every photon absorbed contributes to the process). A

more detailed discussion of quantum yields can be found in **WEB TOPIC 7.3**.

In functional chloroplasts kept in dim light, the quantum yield of photochemistry is approximately 0.95, the quantum yield of fluorescence is 0.05 or lower, and the quantum yields of other processes are negligible. Thus, the most common result of chlorophyll excitation is photochemistry.

### The chemical reaction of photosynthesis is driven by light

It is important to realize that equilibrium for the chemical reaction shown in Equation 7.4 lies very far in the direction of the reactants. The equilibrium constant for Equation 7.4, calculated from tabulated free energies of formation for each of the compounds involved, is about $10^{-500}$. This number is so close to zero that one can be quite confident that in the entire history of the universe no molecule of glucose has formed spontaneously from $H_2O$ and $CO_2$ without external energy being provided. The energy needed to drive the photosynthetic reaction comes from light. Here's a simpler form of Equation 7.4:

$$CO_2 + H_2O \xrightarrow{\text{Light, plant}} (CH_2O) + O_2 \quad (7.6)$$

where $(CH_2O)$ is one-sixth of a glucose molecule. About nine or ten photons of light are required to drive the reaction of Equation 7.6.

Although the photochemical quantum yield under optimum conditions is nearly 100%, the *efficiency* of the conversion of light into chemical energy is much less. If red light of wavelength 680 nm is absorbed, the total energy input (see Equation 7.2) is 1760 kJ per mole of oxygen formed. This amount of energy is more than enough to drive the reaction in Equation 7.6, which has a standard-state free-energy change of +467 kJ mol$^{-1}$. The efficiency of conversion of light energy at the optimal wavelength into chemical energy is therefore about 27%, which is remarkably high for an energy conversion system. Most of this stored energy is used for cellular maintenance processes; the amount diverted to the formation of biomass is much less (see Chapter 9).

There is no conflict between the fact that the photochemical quantum efficiency (quantum yield) is nearly 1.0 (100%) and the energy conversion efficiency is only 27%. The *quantum efficiency* is a measure of the fraction of absorbed photons that engage in photochemistry; the *energy efficiency* is a measure of how much energy in the absorbed photons is stored as chemical products. The numbers indicate that almost all the absorbed photons engage in photochemistry, but only about a fourth of the energy in each photon is stored, the remainder being converted to heat. The overall energy conversion efficiency into biomass, including all loss processes and considering the entire solar spectrum as energy source, is significantly lower still—approximately 4.3% for $C_3$ plants and 6% for $C_4$ plants (Zhu et al. 2008).

## Light drives the reduction of NADP and the formation of ATP

The overall process of photosynthesis is a redox chemical reaction, in which electrons are removed from one chemical species, thereby oxidizing it, and added to another species, thereby reducing it. In 1937, Robert Hill found that in the light, isolated chloroplast thylakoids reduce a variety of compounds, such as iron salts. These compounds serve as oxidants in place of $CO_2$, as the following equation shows:

$$4\ Fe^{3+} + 2\ H_2O \rightarrow 4\ Fe^{2+} + O_2 + 4\ H^+ \qquad (7.7)$$

Many compounds have since been shown to act as artificial electron acceptors in what has come to be known as the Hill reaction. The use of artificial electron acceptors has been invaluable in elucidating the reactions that precede carbon reduction. The demonstration of oxygen evolution linked to the reduction of artificial electron acceptors provided the first evidence that oxygen evolution could occur in the absence of carbon dioxide and led to the now accepted and proven idea that the oxygen in photosynthesis originates from water, not from carbon dioxide.

We now know that during the normal functioning of the photosynthetic system, light reduces nicotinamide adenine dinucleotide phosphate ($NADP^+$), which in turn serves as the reducing agent for carbon fixation in the Calvin–Benson cycle (see Chapter 8). ATP is also formed during the electron flow from water to $NADP^+$, and it, too, is used in carbon reduction.

The chemical reactions in which water is oxidized to oxygen, $NADP^+$ is reduced to NADPH, and ATP is formed are known as the *thylakoid reactions* because almost all the reactions up to $NADP^+$ reduction take place within the thylakoids. The carbon fixation and reduction reactions are called the *stroma reactions* because the carbon reduction reactions take place in the aqueous region of the chloroplast, the stroma. Although this division is somewhat arbitrary, it is conceptually useful.

## Oxygen-evolving organisms have two photosystems that operate in series

By the late 1950s, several experiments were puzzling the scientists who studied photosynthesis. One of these experiments, carried out by Emerson, measured the quantum yield of photosynthesis as a function of wavelength and revealed an effect known as the red drop (**FIGURE 7.12**). Note that while the quantum yield of photochemistry is nearly 1.0, as discussed above, the actions of about ten photons are required to produce each molecule of $O_2$, so the overall maximum quantum yield of $O_2$ production is about 0.1.

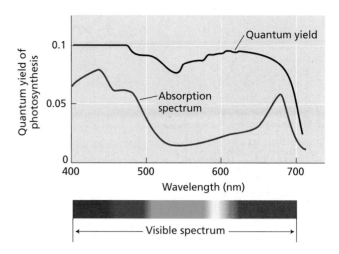

**FIGURE 7.12** Red drop effect. The quantum yield of oxygen evolution (upper, black curve) falls off drastically for far-red light of wavelengths greater than 680 nm, indicating that far-red light alone is inefficient in driving photosynthesis. The slight dip near 500 nm reflects the somewhat lower efficiency of photosynthesis using light absorbed by accessory pigments, carotenoids.

If the quantum yield is measured for the wavelengths at which chlorophyll absorbs light, the values found throughout most of the range are fairly constant, indicating that any photon absorbed by chlorophyll or other pigments is as effective as any other photon in driving photosynthesis. However, the yield drops dramatically in the far-red region of chlorophyll absorption (greater than 680 nm).

This drop cannot be caused by a decrease in chlorophyll absorption, because the quantum yield measures only light that has actually been absorbed. Thus, light with a wavelength greater than 680 nm is much less efficient than light of shorter wavelengths.

Another puzzling experimental result was the **enhancement effect**, also discovered by Emerson. He measured the rate of photosynthesis separately with light of two different wavelengths and then used the two beams simultaneously (**FIGURE 7.13**). When red and far-red light were given together, the rate of photosynthesis was greater than the sum of the individual rates, a startling and surprising observation. These and others observations were eventually explained by experiments performed in the 1960s (see **WEB TOPIC 7.4**) that led to the discovery that two photochemical complexes, now known as **photosystems I and II** (**PSI** and **PSII**), operate in series to carry out the early energy storage reactions of photosynthesis.

Photosystem I preferentially absorbs far-red light of wavelengths greater than 680 nm; photosystem II preferentially absorbs red light of 680 nm and is driven very poorly by far-red light. This wavelength dependence explains the enhancement effect and the red drop effect. Another difference between the photosystems is that:

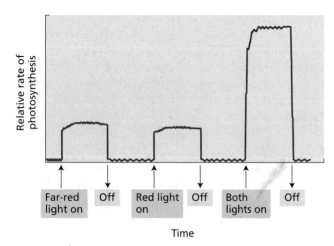

**FIGURE 7.13** Enhancement effect. The rate of photosynthesis when red and far-red light are given together is greater than the sum of the rates when they are given apart. When it was demonstrated in the 1950s, the enhancement effect provided essential evidence in favor of the concept that photosynthesis is carried out by two photochemical systems working in tandem but with slightly different wavelength optima.

- Photosystem I produces a strong reductant, capable of reducing $NADP^+$, and a weak oxidant.
- Photosystem II produces a very strong oxidant, capable of oxidizing water, and a weaker reductant than the one produced by photosystem I.

The reductant produced by photosystem II re-reduces the oxidant produced by photosystem I. These properties of the two photosystems are shown schematically in FIGURE 7.14.

The scheme of photosynthesis depicted in Figure 7.14, called the Z (for *zigzag*) *scheme*, has become the basis for understanding $O_2$-evolving (oxygenic) photosynthetic organisms. It accounts for the operation of two physically and chemically distinct photosystems (I and II), each with its own antenna pigments and photochemical reaction center. The two photosystems are linked by an electron transport chain.

# Organization of the Photosynthetic Apparatus

The previous section explained some of the physical principles underlying photosynthesis, some aspects of the functional roles of various pigments, and some of the chemical reactions carried out by photosynthetic organisms. We now turn to the architecture of the photosynthetic apparatus and the structure of its components, and learn how the molecular structure of the system leads to its functional characteristics.

## The chloroplast is the site of photosynthesis

In photosynthetic eukaryotes, photosynthesis takes place in the subcellular organelle known as the chloroplast. FIGURE 7.15 shows a transmission electron micrograph of a thin section from a pea chloroplast. The most striking aspect of the structure of the chloroplast is the extensive system of internal membranes known as **thylakoids**. All the chlorophyll is contained within this membrane system, which is the site of the light reactions of photosynthesis.

The carbon reduction reactions, which are catalyzed by water-soluble enzymes, take place in the **stroma**, the

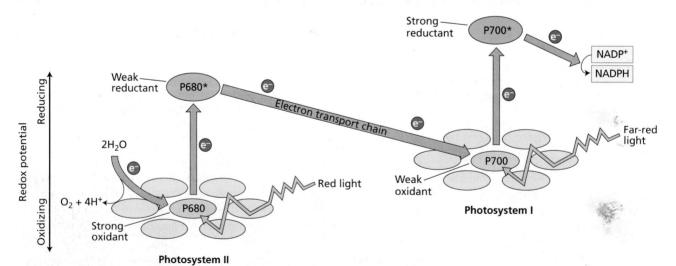

**FIGURE 7.14** Z scheme of photosynthesis. Red light absorbed by photosystem II (PSII) produces a strong oxidant and a weak reductant. Far-red light absorbed by photosystem I (PSI) produces a weak oxidant and a strong reductant. The strong oxidant generated by PSII oxidizes

water, while the strong reductant produced by PSI reduces $NADP^+$. This scheme is basic to an understanding of photosynthetic electron transport. P680 and P700 refer to the wavelengths of maximum absorption of the reaction center chlorophylls in PSII and PSI, respectively.

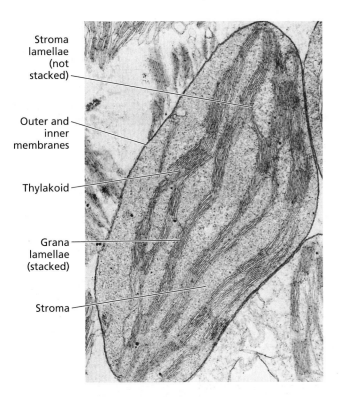

Stroma lamellae (not stacked)

Outer and inner membranes

Thylakoid

Grana lamellae (stacked)

Stroma

**FIGURE 7.15** Transmission electron micrograph of a chloroplast from pea (*Pisum sativum*) fixed in glutaraldehyde and OsO$_4$, embedded in plastic resin, and thin-sectioned with an ultramicrotome. (14,500×) (Courtesy of J. Swafford.)

region of the chloroplast outside the thylakoids. Most of the thylakoids appear to be very closely associated with each other. These stacked membranes are known as **grana lamellae** (singular *lamella*; each stack is called a *granum*), and the exposed membranes in which stacking is absent are known as **stroma lamellae**.

Two separate membranes, each composed of a lipid bilayer and together known as the **envelope**, surround most types of chloroplasts (**FIGURE 7.16**). This double-membrane system contains a variety of metabolite transport systems. The chloroplast also contains its own DNA, RNA, and ribosomes. Some of the chloroplast proteins are products of transcription and translation within the chloroplast itself, whereas most of the others are encoded by nuclear DNA, synthesized on cytoplasmic ribosomes, and then imported into the chloroplast. This remarkable division of labor, extending in many cases to different subunits of the same enzyme complex, will be discussed in more detail later in this chapter. For some dynamic structures of chloroplasts see **WEB ESSAY 7.1**.

## Thylakoids contain integral membrane proteins

A wide variety of proteins essential to photosynthesis are embedded in the thylakoid membranes. In many cases, portions of these proteins extend into the aqueous regions on both sides of the thylakoids. These **integral membrane proteins** contain a large proportion of hydrophobic amino acids and are therefore much more stable in a nonaqueous medium such as the hydrocarbon portion of the membrane (see Figure 1.5A).

The reaction centers, the antenna pigment–protein complexes, and most of the electron carrier proteins are all integral membrane proteins. In all known cases, integral membrane proteins of the chloroplast have a unique orientation within the membrane. Thylakoid membrane proteins have one region pointing toward the stromal side of the membrane and the other oriented toward the interior space of the thylakoid, known as the *lumen* (see Figure 7.16 and **FIGURE 7.17**).

The chlorophylls and accessory light-gathering pigments in the thylakoid membrane are always associated in a noncovalent but highly specific way with proteins, thereby forming pigment–protein complexes. Both

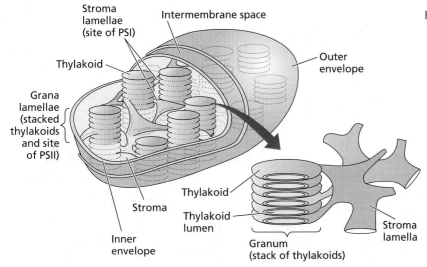

Stroma lamellae (site of PSI)

Intermembrane space

Thylakoid

Grana lamellae (stacked thylakoids and site of PSII)

Stroma

Inner envelope

Thylakoid

Thylakoid lumen

Granum (stack of thylakoids)

Outer envelope

Stroma lamella

**FIGURE 7.16** Schematic picture of the overall organization of the membranes in the chloroplast. The chloroplast of higher plants is surrounded by the inner and outer membranes (envelope). The region of the chloroplast that is inside the inner membrane and surrounds the thylakoid membranes is known as the *stroma*. It contains the enzymes that catalyze carbon fixation and other biosynthetic pathways. The thylakoid membranes are highly folded and appear in many pictures to be stacked like coins (the granum), although in reality they form one or a few large interconnected membrane systems, with a well-defined interior and exterior with respect to the stroma. (After Becker 1986.)

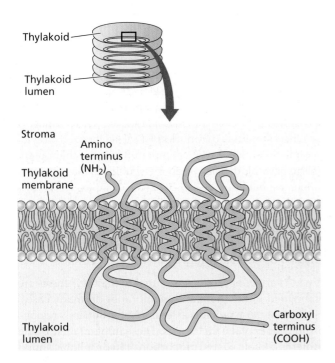

**FIGURE 7.17** Predicted folding pattern of the D1 protein of the PSII reaction center. The hydrophobic portion of the membrane is traversed five times by regions of the peptide chain rich in hydrophobic amino acid residues. The protein is asymmetrically arranged in the thylakoid membrane, with the amino ($NH_2$) terminus on the stromal side of the membrane and the carboxyl (COOH) terminus on the lumenal side. (After Trebst 1986.)

antenna and reaction center chlorophylls are associated with proteins that are organized within the membrane so as to optimize energy transfer in antenna complexes and electron transfer in reaction centers, while at the same time minimizing wasteful processes.

## Photosystems I and II are spatially separated in the thylakoid membrane

The PSII reaction center, along with its antenna chlorophylls and associated electron transport proteins, is located predominantly in the grana lamellae (**FIGURE 7.18A**) (Allen and Forsberg 2001). The PSI reaction center and its associated antenna pigments and electron transfer proteins, as well as the ATP synthase enzyme that catalyzes the formation of ATP, are found almost exclusively in the stroma lamellae and at the edges of the grana lamellae. The cytochrome $b_6f$ complex of the electron transport chain that connects the two photosystems (see Figure 7.21) is evenly distributed between stroma and grana lamellae. The structures of all these complexes are shown in **FIGURE 7.18B** (Nelson and Ben-Shem 2004).

Thus the two photochemical events that take place in $O_2$-evolving photosynthesis are spatially separated. This separation implies that one or more of the electron carriers that function between the photosystems diffuses from the grana region of the membrane to the stroma region, where electrons are delivered to photosystem I. These diffusible carriers are the blue-colored copper protein plastocyanin (PC) and the organic redox cofactor plastoquinone (PQ). These carriers are discussed in more detail on p. 184.

In PSII, the oxidation of two water molecules produces four electrons, four protons, and a single $O_2$ (see Equation 7.8 on p. 182). The protons produced by this oxidation of water must also be able to diffuse to the stroma region, where ATP is synthesized. The functional role of this large separation (many tens of nanometers) between photosystems I and II is not entirely clear, but is thought to improve the efficiency of energy distribution between the two photosystems (Allen and Forsberg 2001).

The spatial separation between photosystems I and II indicates that a strict one-to-one stoichiometry between the two photosystems is not required. Instead, PSII reaction centers feed reducing equivalents into a common intermediate pool of lipid-soluble electron carriers (plastoquinone). The PSI reaction centers remove the reducing equivalents from the common pool, rather than from any specific PSII reaction center complex.

Most measurements of the relative quantities of photosystems I and II have shown that there is an excess of photosystem II in chloroplasts. Most commonly, the ratio of PSII to PSI is about 1.5:1, but it can change when plants are grown in different light conditions. In contrast to the situation in chloroplasts of eukaryotic photosynthetic organisms, cyanobacteria usually have an excess of PSI over PSII.

## Anoxygenic photosynthetic bacteria have a single reaction center

Non-$O_2$-evolving (anoxygenic) organisms contain only a single photosystem similar to either photosystem I or II. These simpler organisms have been very useful for detailed structural and functional studies that have contributed to a better understanding of oxygenic photosynthesis. In most cases, these anoxygenic photosystems carry out cyclic electron transfer with no net reduction or oxidation. Part of the energy of the photon is conserved as a proton motive force (see below) and is used to make ATP.

Reaction centers from purple photosynthetic bacteria were the first integral membrane proteins to have structures determined to high resolution (Deisenhofer and Michel 1989) (see Figures 7.5.A and 7.5.B in **WEB TOPIC 7.5**). Detailed analysis of these structures, along with the characterization of numerous mutants, has revealed many of the principles involved in the energy storage processes carried out by all reaction centers.

The structure of the purple bacterial reaction center is thought to be similar in many ways to that found in photosystem II from oxygen-evolving organisms, especially in the electron acceptor portion of the chain. The proteins that make up the core of the bacterial reaction center are

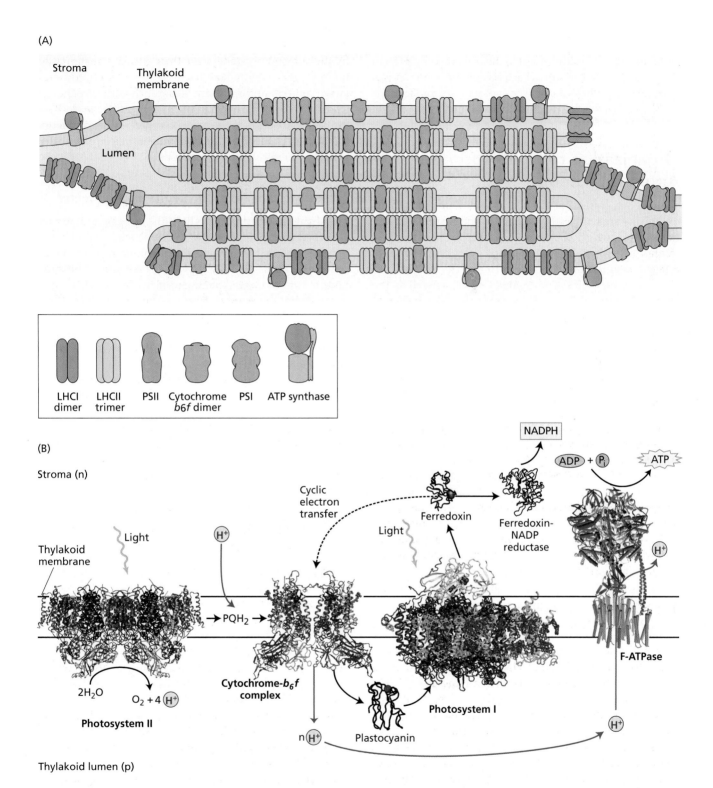

**FIGURE 7.18** Organization and structure of the four major protein complexes of the thylakoid membrane. (A) Photosystem II is located predominantly in the stacked regions of the thylakoid membrane; photosystem I and ATP synthase are found in the unstacked regions protruding into the stroma. Cytochrome $b_6f$ complexes are evenly distributed. This lateral separation of the two photosystems requires that electrons and protons produced by photosystem II be transported a considerable distance before they can be acted on by photosystem I and the ATP-coupling enzyme. (B) Structures of the four main protein complexes of the thylakoid membrane. Shown also are the two diffusible electron carriers—plastocyanin, which is located in the thylakoid lumen, and plastohydroquinone (PQH$_2$) in the membrane. The lumen has a positive electrical charge (p) with respect to the stroma (n). (A after Allen and Forsberg 2001; B after Nelson and Ben-Shem 2004.)

relatively similar in sequence to their photosystem II counterparts, implying an evolutionary relatedness. A similar situation is found with respect to the reaction centers from the anoxygenic green sulfur bacteria and the heliobacteria, compared to photosystem I. The evolutionary implications of this pattern are discussed later in this chapter.

# Organization of Light-Absorbing Antenna Systems

The antenna systems of different classes of photosynthetic organisms are remarkably varied, in contrast to the reaction centers, which appear to be similar in even distantly related organisms. The variety of antenna complexes reflects evolutionary adaptation to the diverse environments in which different organisms live, as well as the need in some organisms to balance energy input to the two photosystems (Grossman et al. 1995; Green and Durnford 1996; Green and Parson 2003). In this section we will learn how energy transfer processes absorb light and deliver energy to the reaction center.

## Antenna systems contain chlorophyll and are membrane associated

Antenna systems function to deliver energy efficiently to the reaction centers with which they are associated (Van Grondelle et al. 1994; Pullerits and Sundström 1996). The size of the antenna system varies considerably in different organisms, ranging from a low of 20 to 30 bacteriochlorophylls per reaction center in some photosynthetic bacteria, to generally 200 to 300 chlorophylls per reaction center in higher plants, to a few thousand pigments per reaction center in some types of algae and bacteria. The molecular structures of antenna pigments are also quite diverse, although all of them are associated in some way with the photosynthetic membrane. In almost all cases, the antenna pigments are associated with proteins to form **pigment–protein complexes**.

The physical mechanism by which excitation energy is conveyed from the chlorophyll that absorbs the light to the reaction center is thought to be **fluorescence resonance energy transfer**, often abbreviated as FRET. By this mechanism the excitation energy is transferred from one molecule to another by a nonradiative process.

A useful analogy for resonance transfer is the transfer of energy between two tuning forks. If one tuning fork is struck and properly placed near another, the second tuning fork receives some energy from the first and begins to vibrate. The efficiency of energy transfer between the two tuning forks depends on their distance from each other and their relative orientation, as well as on their vibrational frequencies, or pitches. Similar parameters affect the efficiency of energy transfer in antenna complexes, with energy substituted for pitch.

Energy transfer in antenna complexes is usually very efficient: Approximately 95 to 99% of the photons absorbed by the antenna pigments have their energy transferred to the reaction center, where it can be used for photochemistry. There is an important difference between energy transfer among pigments in the antenna and the electron transfer that occurs in the reaction center: Whereas energy transfer is a purely physical phenomenon, electron transfer involves chemical (redox) reactions.

## The antenna funnels energy to the reaction center

The sequence of pigments within the antenna that funnel absorbed energy toward the reaction center has absorption maxima that are progressively shifted toward longer red wavelengths (**FIGURE 7.19**). This red shift in absorption maximum means that the energy of the excited state is somewhat lower nearer the reaction center than in the more peripheral portions of the antenna system.

As a result of this arrangement, when excitation is transferred, for example, from a chlorophyll b molecule absorbing maximally at 650 nm to a chlorophyll a molecule absorbing maximally at 670 nm, the difference in energy between these two excited chlorophylls is lost to the environment as heat.

For the excitation to be transferred back to the chlorophyll b, the energy lost as heat would have to be resupplied. The probability of reverse transfer is therefore smaller simply because thermal energy is not sufficient to make up the deficit between the lower-energy and higher-energy pigments. This effect gives the energy-trapping process a degree of directionality or irreversibility and makes the delivery of excitation to the reaction center more efficient. In essence, the system sacrifices some energy from each quantum so that nearly all of the quanta can be trapped by the reaction center.

## Many antenna pigment–protein complexes have a common structural motif

In all eukaryotic photosynthetic organisms that contain both chlorophyll a and chlorophyll b, the most abundant antenna proteins are members of a large family of structurally related proteins. Some of these proteins are associated primarily with photosystem II and are called **light-harvesting complex II (LHCII)** proteins; others are associated with photosystem I and are called *LHCI* proteins. These antenna complexes are also known as **chlorophyll a/b antenna proteins** (Green and Durnford 1996; Green and Parson 2003).

The structure of one of the LHCII proteins has been determined (**FIGURE 7.20**) (Liu et al. 2004; Barros and Kühlbrandt 2009). The protein contains three α-helical regions and binds 14 chlorophyll a and b molecules, as well as four carotenoids. The structure of the LHCI proteins is

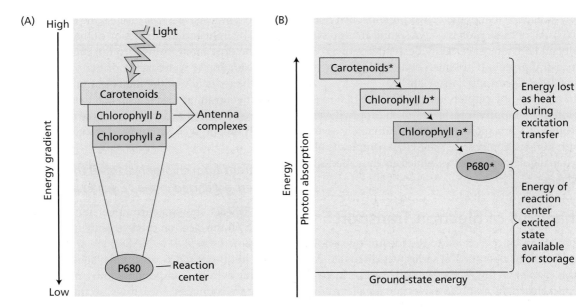

**FIGURE 7.19** Funneling of excitation from the antenna system toward the reaction center. (A) The excited-state energy of pigments increases with distance from the reaction center; that is, pigments closer to the reaction center are lower in energy than those farther from the reaction center. This energy gradient ensures that excitation transfer toward the reaction center is energetically favorable and that excitation transfer back out to the peripheral portions of the antenna is energetically unfavorable. (B) Some energy is lost as heat to the environment by this process, but under optimal conditions almost all the excitations absorbed in the antenna complexes can be delivered to the reaction center. The asterisks denote excited states.

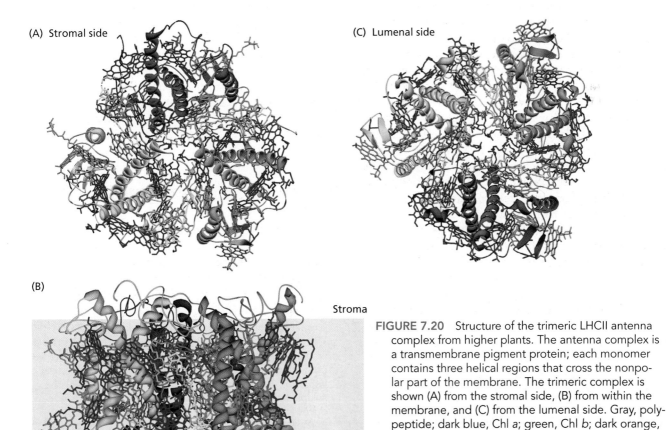

**FIGURE 7.20** Structure of the trimeric LHCII antenna complex from higher plants. The antenna complex is a transmembrane pigment protein; each monomer contains three helical regions that cross the nonpolar part of the membrane. The trimeric complex is shown (A) from the stromal side, (B) from within the membrane, and (C) from the lumenal side. Gray, polypeptide; dark blue, Chl a; green, Chl b; dark orange, lutein; light orange, neoxanthin; yellow, violaxanthin; pink, lipids. (After Barros and Kühlbrandt 2009.)

generally similar to that of the LHCII proteins (Ben-Shem et al. 2003). All of these proteins have significant sequence similarity and are almost certainly descendants of a common ancestral protein (Grossman et al. 1995; Green and Durnford 1996; Green and Parson 2003).

Light absorbed by carotenoids or chlorophyll *b* in the LHC proteins is rapidly transferred to chlorophyll *a* and then to other antenna pigments that are intimately associated with the reaction center. The LHCII complex is also involved in regulatory processes, which are discussed later in the chapter.

## Mechanisms of Electron Transport

Some of the evidence that led to the idea of two photochemical reactions operating in series was discussed earlier in this chapter. In this section we will consider in detail the chemical reactions involved in electron transfer during photosynthesis. We will discuss the excitation of chlorophyll by light and the reduction of the first electron acceptor, the flow of electrons through photosystems II and I, the oxidation of water as the primary source of electrons, and the reduction of the final electron acceptor ($NADP^+$). The chemiosmotic mechanism that mediates ATP synthesis will be discussed in detail later in the chapter (see "Proton Transport and ATP Synthesis in the Chloroplast").

### Electrons from chlorophyll travel through the carriers organized in the "Z scheme"

**FIGURE 7.21** shows a current version of the Z scheme, in which all the electron carriers known to function in electron flow from $H_2O$ to $NADP^+$ are arranged vertically at their midpoint redox potentials (see **WEB TOPIC 7.6** for further detail). Components known to react with each other are connected by arrows, so the Z scheme is really

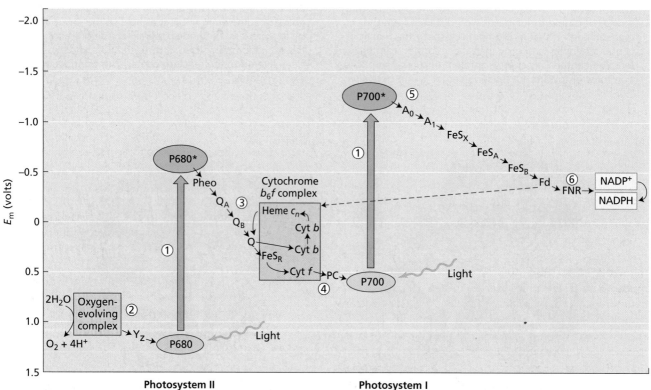

**FIGURE 7.21** Detailed Z scheme for $O_2$-evolving photosynthetic organisms. The redox carriers are placed at their midpoint redox potentials (at pH 7). (1) The vertical arrows represent photon absorption by the reaction center chlorophylls: P680 for photosystem II (PSII) and P700 for photosystem I (PSI). The excited PSII reaction center chlorophyll, P680*, transfers an electron to pheophytin (Pheo). (2) On the oxidizing side of PSII (to the left of the arrow joining P680 with P680*), P680 oxidized by light is re-reduced by $Y_z$, which has received electrons from oxidation of water. (3) On the reducing side of PSII (to the right of the arrow joining P680 with P680*), pheophytin transfers electrons to the acceptors $Q_A$ and $Q_B$, which are plastoquinones. (4) The cytochrome $b_6f$ complex transfers electrons to plastocyanin (PC), a soluble protein, which in turn reduces P700+ (oxidized P700). (5) The acceptor of electrons from P700* ($A_0$) is thought to be a chlorophyll, and the next acceptor ($A_1$) is a quinone. A series of membrane-bound iron–sulfur proteins ($FeS_X$, $FeS_A$, and $FeS_B$) transfers electrons to soluble ferredoxin (Fd). (6) The soluble flavoprotein ferredoxin–NADP reductase (FNR) reduces $NADP^+$ to NADPH, which is used in the Calvin–Benson cycle to reduce $CO_2$ (see Chapter 8). The dashed line indicates cyclic electron flow around PSI. (After Blankenship and Prince 1985.)

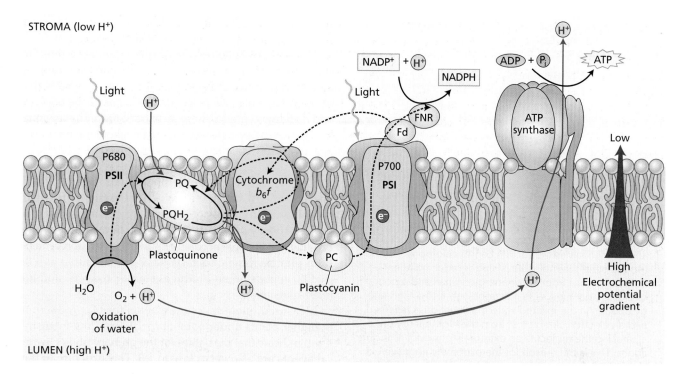

**FIGURE 7.22** The transfer of electrons and protons in the thylakoid membrane is carried out vectorially by four protein complexes (see Figure 7.18B for structures). Water is oxidized and protons are released in the lumen by PSII. PSI reduces NADP$^+$ to NADPH in the stroma, via the action of ferredoxin (Fd) and the flavoprotein ferredoxin–NADP reductase (FNR). Protons are also transported into the lumen by the action of the cytochrome $b_6f$ complex and contribute to the electrochemical proton gradient. These protons must then diffuse to the ATP synthase enzyme, where their diffusion down the electrochemical potential gradient is used to synthesize ATP in the stroma. Reduced plastoquinone (PQH$_2$) and plastocyanin transfer electrons to cytochrome $b_6f$ and to PSI, respectively. The dashed lines represent electron transfer; solid lines represent proton movement.

a synthesis of both kinetic and thermodynamic information. The large vertical arrows represent the input of light energy into the system.

Photons excite the specialized chlorophyll of the reaction centers (P680 for PSII; P700 for PSI), and an electron is ejected. The electron then passes through a series of electron carriers and eventually reduces P700 (for electrons from PSII) or NADP$^+$ (for electrons from PSI). Much of the following discussion describes the journeys of these electrons and the nature of their carriers.

Almost all the chemical processes that make up the light reactions of photosynthesis are carried out by four major protein complexes: photosystem II, the cytochrome $b_6f$ complex, photosystem I, and the ATP synthase. These four integral membrane complexes are vectorially oriented in the thylakoid membrane to function as follows (see Figure 7.18 and **FIGURE 7.22**):

- Photosystem II oxidizes water to O$_2$ in the thylakoid lumen and in the process releases protons into the lumen.

- Cytochrome $b_6f$ oxidizes plastohydroquinone (PQH$_2$) molecules that were reduced by PSII

and delivers electrons to PSI. The oxidation of plastohydroquinone is coupled to proton transfer into the lumen from the stroma, generating a proton motive force.

- Photosystem I reduces NADP$^+$ to NADPH in the stroma by the action of ferredoxin (Fd) and the flavoprotein ferredoxin–NADP reductase (FNR).

- ATP synthase produces ATP as protons diffuse back through it from the lumen into the stroma.

## Energy is captured when an excited chlorophyll reduces an electron acceptor molecule

As discussed earlier, the function of light is to excite a specialized chlorophyll in the reaction center, either by direct absorption or, more frequently, via energy transfer from an antenna pigment. This excitation process can be envisioned as the promotion of an electron from the highest-energy filled orbital of the chlorophyll to the lowest-energy unfilled orbital (**FIGURE 7.23**). The electron in the upper orbital is only loosely bound to the chlorophyll and is easily lost if a molecule that can accept the electron is nearby.

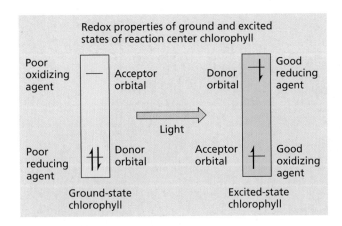

**Redox properties of ground and excited states of reaction center chlorophyll**

Poor oxidizing agent — Acceptor orbital

Donor orbital — Good reducing agent

Light →

Poor reducing agent — Donor orbital

Acceptor orbital — Good oxidizing agent

Ground-state chlorophyll

Excited-state chlorophyll

**FIGURE 7.23** Orbital occupation diagram for the ground and excited states of reaction center chlorophyll. In the ground state the molecule is a poor reducing agent (loses electrons from a low-energy orbital) and a poor oxidizing agent (accepts electrons only into a high-energy orbital). In the excited state the situation is reversed, and an electron can be lost from the high-energy orbital, making the molecule an extremely powerful reducing agent. This is the reason for the extremely negative excited-state redox potential shown by P680* and P700* in Figure 7.21. The excited state can also act as a strong oxidant by accepting an electron into the lower-energy orbital, although this pathway is not significant in reaction centers. (After Blankenship and Prince 1985.)

The first reaction that converts electron energy into chemical energy—that is, the primary photochemical event—is the transfer of an electron from the excited state of a chlorophyll in the reaction center to an acceptor molecule. An equivalent way to view this process is that the absorbed photon causes an electron rearrangement in the reaction center chlorophyll, followed by an electron transfer process in which part of the energy in the photon is captured in the form of redox energy.

Immediately after the photochemical event, the reaction center chlorophyll is in an oxidized state (electron deficient, or positively charged), and the nearby electron acceptor molecule is reduced (electron rich, or negatively charged). The system is now at a critical juncture. The lower-energy orbital of the positively charged oxidized reaction center chlorophyll shown in Figure 7.23 has a vacancy and can accept an electron. If the acceptor molecule donates its electron back to the reaction center chlorophyll, the system will be returned to the state that existed before the light excitation, and all the absorbed energy will be converted into heat.

This wasteful *recombination* process, however, does not appear to occur to any substantial degree in functioning reaction centers. Instead, the acceptor transfers its extra electron to a secondary acceptor and so on down the electron transport chain. The oxidized reaction center of the chlorophyll that had donated an electron is re-reduced by a secondary donor, which in turn is reduced by a tertiary donor. In plants, the ultimate electron donor is $H_2O$, and the ultimate electron acceptor is $NADP^+$ (see Figure 7.21).

The essence of photosynthetic energy storage is thus the initial transfer of an electron from an excited chlorophyll to an acceptor molecule, followed by a very rapid series of secondary chemical reactions that separate the positive and negative charges. These secondary reactions separate the charges to opposite sides of the thylakoid membrane in approximately 200 picoseconds (1 picosecond = $10^{-12}$ s).

With the charges thus separated, the reversal reaction is many orders of magnitude slower, and the energy has been captured. Each of the secondary electron transfers is accompanied by a loss of some energy, thus making the process effectively irreversible. The quantum yield for the production of stable products in purified reaction centers from photosynthetic bacteria has been measured as 1.0; that is, every photon produces stable products, and no reversal reactions occur.

Measured quantum requirements for $O_2$ production in higher plants under optimal conditions (low-intensity light) indicate that the values for the primary photochemical events are also very close to 1.0. The structure of the reaction center appears to be extremely fine-tuned for maximal rates of productive reactions and minimal rates of energy-wasting reactions.

### The reaction center chlorophylls of the two photosystems absorb at different wavelengths

As discussed earlier in the chapter, PSI and PSII have distinct absorption characteristics. Precise measurements of absorption maxima are made possible by optical changes in the reaction center chlorophylls in the reduced and oxidized states. The reaction center chlorophyll is transiently in an oxidized state after losing an electron and before being re-reduced by its electron donor.

In the oxidized state, chlorophylls lose their characteristic strong light absorbance in the red region of the spectrum; they become **bleached**. It is therefore possible to monitor the redox state of these chlorophylls by time-resolved optical absorbance measurements in which this bleaching is monitored directly (see **WEB TOPIC 7.1**).

Using such techniques, it was found that the reaction center chlorophyll of photosystem I absorbs maximally at 700 nm in its reduced state. Accordingly, this chlorophyll is named **P700** (the P stands for *pigment*). The analogous optical transient of photosystem II is at 680 nm, so its reaction center chlorophyll is known as **P680**. The reaction center bacteriochlorophyll from purple photosynthetic bacteria was similarly identified as **P870**.

The X-ray structure of the bacterial reaction center (see Figures 7.5.A and 7.5.B in **WEB TOPIC 7.5**) clearly indicates that P870 is a closely coupled pair or dimer of bacteriochlorophylls, rather than a single molecule. The primary donor of photosystem I, P700, is also a dimer of chlorophyll *a* molecules. Photosystem II also contains a dimer of

(A)　　　　　　　　　　　　　(B)　　　　　　　　　　　　　(C)

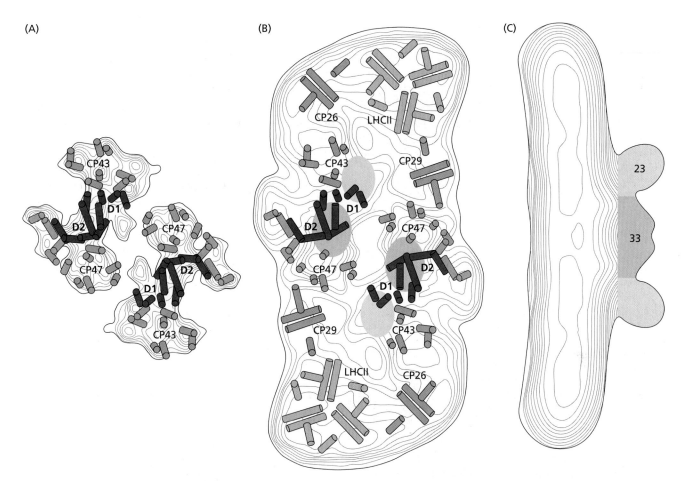

**FIGURE 7.24** Structure of the dimeric multisubunit protein supercomplex of photosystem II from higher plants, as determined by electron microscopy. The figure shows two complete reaction centers, each of which is a dimeric complex. (A) Helical arrangement of the D1 and D2 (red) and CP43 and CP47 (green) core subunits. (B) View from the lumenal side of the supercomplex, including addi-tional antenna complexes, LHCII, CP26, and CP29, and an extrinsic oxygen-evolving complex, shown as orange and yellow circles. Unassigned helices are shown in gray. (C) Side view of the complex illustrating the arrangement of the extrinsic proteins of the oxygen-evolving complex. (After Barber et al. 1999.)

chlorophylls, although the primary electron transfer event may not originate from these pigments. In the oxidized state, reaction center chlorophylls contain an unpaired electron.

Molecules with unpaired electrons often can be detected by a magnetic-resonance technique known as **electron spin resonance** (**ESR**). ESR studies, along with the spectroscopic measurements already described, have led to the discovery of many intermediate electron carriers in the photosynthetic electron transport system.

### The photosystem II reaction center is a multisubunit pigment–protein complex

Photosystem II is contained in a multisubunit protein supercomplex (**FIGURE 7.24**) (Barber et al. 1999). In higher plants, the multisubunit protein supercomplex has two complete reaction centers and some antenna complexes.

The core of the reaction center consists of two membrane proteins known as D1 and D2, as well as other proteins, as shown in **FIGURE 7.25** and **WEB TOPIC 7.7** (Ferreira et al. 2004; Guskov et al. 2009).

The primary donor chlorophyll, additional chlorophylls, carotenoids, pheophytins, and plastoquinones (two electron acceptors described in the following section) are bound to the membrane proteins D1 and D2. These proteins have some sequence similarity to the L and M peptides of purple bacteria. Other proteins serve as antenna complexes or are involved in oxygen evolution. Some, such as cytochrome $b_{559}$, have no known function but may be involved in a protective cycle around photosystem II.

### Water is oxidized to oxygen by photosystem II

Water is oxidized according to the following chemical reaction (Yano et al. 2006):

(A)

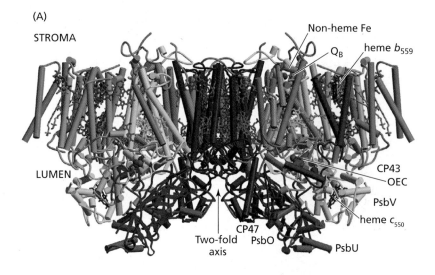

(B)

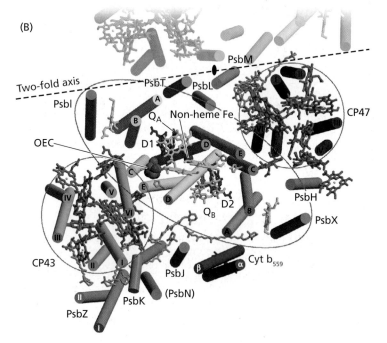

(C)

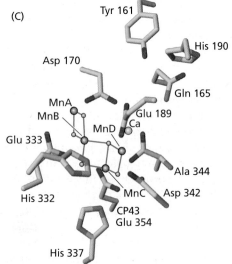

$$2\,H_2O \rightarrow O_2 + 4\,H^+ + 4\,e^- \qquad (7.8)$$

This equation indicates that four electrons are removed from two water molecules, generating an oxygen molecule and four hydrogen ions. (For more on oxidation–reduction reactions, see Appendix 1 and **WEB TOPIC 7.6**.)

Water is a very stable molecule. Oxidation of water to form molecular oxygen is very difficult: The photosynthetic oxygen-evolving complex is the only known biochemical system that carries out this reaction, and is the source of almost all the oxygen in Earth's atmosphere.

Many studies have provided a substantial amount of information about the process (see **WEB TOPIC 7.7**). The protons produced by water oxidation are released into the lumen of the thylakoid, not directly into the stromal compartment (see Figure 7.22). They are released into the lumen because of the vectorial nature of the membrane and the fact that the oxygen-evolving complex is localized near the interior surface of the thylakoid. These protons are eventually transferred from the lumen to the stroma by translocation through ATP synthase. In this way, the protons released during water oxidation contribute to the electrochemical potential driving ATP formation.

It has been known for many years that manganese (Mn) is an essential cofactor in the water-oxidizing process (see Chapter 5), and a classic hypothesis in photosynthesis research postulates that Mn ions undergo a series of oxidations—known as $S$ *states*, and labeled $S_0$, $S_1$, $S_2$, $S_3$, and $S_4$ (see **WEB TOPIC 7.7**)—that are perhaps linked to $H_2O$ oxidation and the generation of $O_2$. This hypothesis has received strong support from a variety of experiments, most notably X-ray absorption and ESR studies, both of which detect the manganese directly (Yano et al. 2006). Analytical experiments indicate that

**FIGURE 7.25** Structure of the photosystem II reaction center from the cyanobacterium *Thermosynechococcus elongatus*, resolved at 3.5 Å. The structure includes the $D_1$ (yellow) and $D_2$ (orange) core reaction center proteins, the CP43 (green) and CP47 (red) antenna proteins, cytochromes $b_{559}$ and $c_{550}$, the extrinsic 33-kDa oxygen evolution protein PsbO (dark blue), and the pigments and other cofactors. (A) Side view parallel to the membrane plane. (B) View from the lumenal surface, perpendicular to the plane of the membrane. (C) Detail of the Mn-containing water-splitting complex. (A, B after Ferreira et al. 2004; C after Yano et al. 2006.)

(A)

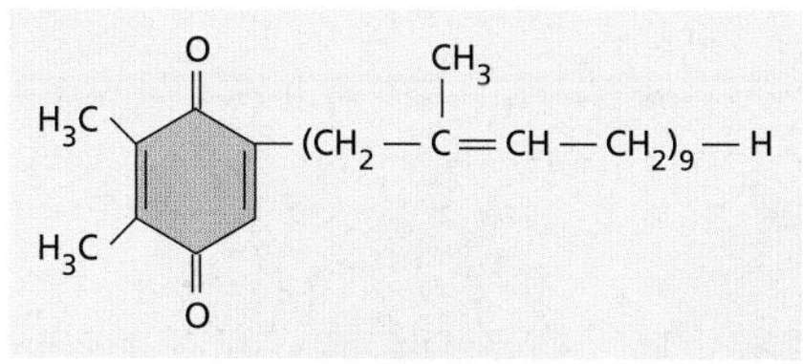

**Plastoquinone**

(B)

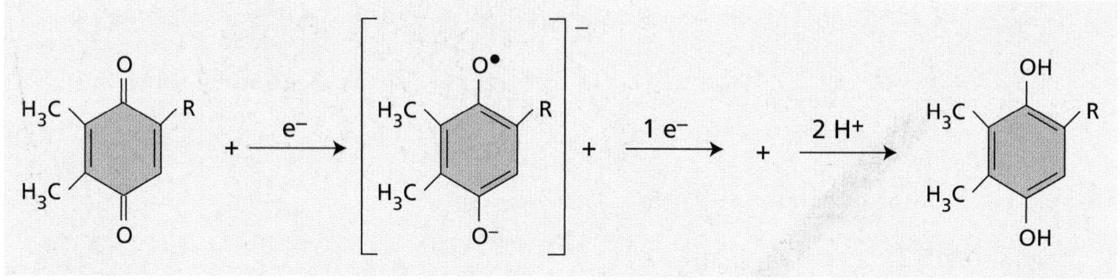

Quinone
(Q)

Plastosemiquinone
(Q•)

Plastohydroquinone
(QH₂)

**FIGURE 7.26** Structure and reactions of plastoquinones that operate in photosystem II. (A) The plastoquinone consists of a quinoid head and a long nonpolar tail that anchors it in the membrane. (B) Redox reactions of plastoquinone. The fully oxidized quinone (Q), anionic plastosemiquinone (Q•), and reduced plastohydroquinone ($PQH_2$) forms are shown; R represents the side chain.

four Mn atoms are associated with each oxygen-evolving complex. Other experiments have shown that $Cl^-$ and $Ca^{2+}$ ions are essential for $O_2$ evolution (see **WEB TOPIC 7.7**). The detailed chemical mechanism of the oxidation of water to $O_2$ is not yet well understood, but with structural information now available, rapid progress is being made in this area (Brudvig 2008).

One electron carrier, generally identified as $Y_z$, functions between the oxygen-evolving complex and P680 (see Figure 7.21). To function in this region, $Y_z$ needs to have a very strong tendency to retain its electrons. This species has been identified as a radical formed from a tyrosine residue in the D1 protein of the PSII reaction center.

### Pheophytin and two quinones accept electrons from photosystem II

Spectral and ESR studies have revealed the structural arrangement of the carriers in the electron acceptor complex. **Pheophytin**, a chlorophyll in which the central magnesium has been replaced by two hydrogens, acts as an early acceptor in photosystem II. The structural change gives pheophytin chemical and spectral properties that are slightly different from those of Mg-based chlorophylls. Pheophytin passes electrons to a complex of two plastoquinones in close proximity to an iron atom. The processes are very similar to those found in the reaction center of purple bacteria (for details, see Figure 7.5.B in **WEB TOPIC 7.5**).

The two plastoquinones, $PQ_A$ and $PQ_B$, are bound to the reaction center and receive electrons from pheophytin in a sequential fashion (Okamura et al. 2000). Transfer of the two electrons to $Q_B$ reduces it to $PQ_B^{2-}$, and the reduced $PQ_B^{2-}$ takes two protons from the stroma side of the medium, yielding a fully reduced **plastohydroquinone** ($PQH_2$) (**FIGURE 7.26**). The plastohydroquinone then dis-

sociates from the reaction center complex and enters the hydrocarbon portion of the membrane, where it in turn transfers its electrons to the cytochrome $b_6f$ complex. Unlike the large protein complexes of the thylakoid membrane, plastohydroquinone is a small, nonpolar molecule that diffuses readily in the nonpolar core of the membrane bilayer.

### Electron flow through the cytochrome $b_6f$ complex also transports protons

The **cytochrome $b_6f$ complex** is a large multisubunit protein with several prosthetic groups (**FIGURE 7.27**) (Kurisu et al. 2003; Stroebel et al. 2003; Baniulis et al. 2008). It contains two $b$-type hemes and one $c$-type heme (**cytochrome $f$**). In $c$-type cytochromes the heme is covalently attached to the peptide; in $b$-type cytochromes the chemically similar protoheme group is not covalently attached (see **WEB TOPIC 7.8**). In addition, the complex contains a **Rieske iron–sulfur protein** (named for the scientist who discovered it), in which two iron atoms are bridged by two sulfur atoms. The functional roles of all these cofactors are reasonably well understood, as described below. However, the cytochrome $b_6f$ complex also contains additional cofactors—including an additional heme group (called heme $c_n$), a chlorophyll, and a carotenoid—whose functions are yet to be resolved.

The structures of the cytochrome $b_6f$ complex and the related cytochrome $bc_1$ suggest a mechanism for electron and proton flow. The precise way by which electrons and protons flow through the cytochrome $b_6f$ complex is not yet fully understood, but a mechanism known as the **Q cycle** accounts for most of the observations. In this mechanism, plastohydroquinone (also called plastoquinol) ($PQH_2$) is oxidized, and one of the two electrons is passed along a linear electron transport chain toward photosystem I,

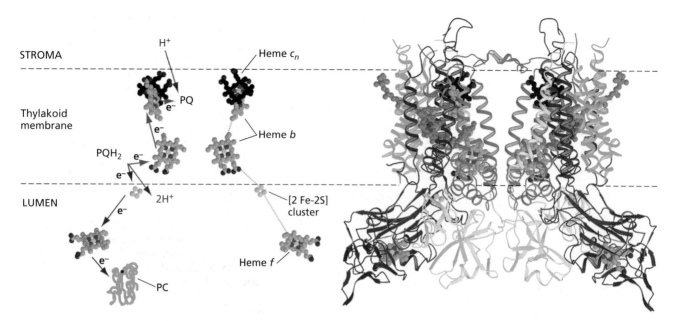

**FIGURE 7.27** Structure of the cytochrome $b_6f$ complex from cyanobacteria. The diagram on the right shows the arrangement of the proteins and cofactors in the complex. Cytochrome $b_6$ protein is shown in blue, cytochrome $f$ protein in red, Rieske iron-sulfur protein in yellow, and other smaller subunits in green and purple. On the left, the proteins have been omitted to more clearly show the positions of the cofactors. [2 Fe-2S] cluster, part of the Rieske iron–sulfur protein; PC, plastocyanin; PQ, plastoquinone; $PQH_2$; plastohydroquinone. (After Kurisu et al. 2003.)

while the other electron goes through a cyclic process that increases the number of protons pumped across the membrane (**FIGURE 7.28**).

In the linear electron transport chain, the oxidized Rieske protein (**FeSR**) accepts an electron from $PQH_2$ and transfers it to cytochrome $f$ (see Figure 7.28A). Cytochrome $f$ then transfers an electron to the blue-colored copper protein plastocyanin (PC), which in turn reduces oxidized P700 of PSI. In the cyclic part of the process (see Figure 7.28B), the plastosemiquinone (see Figure 7.26) transfers its other electron to one of the $b$-type hemes, releasing both of its protons to the lumenal side of the membrane.

The first $b$-type heme transfers its electron through the second $b$-type heme to an oxidized plastoquinone molecule, reducing it to the semiquinone form near the stromal surface of the complex. Another similar sequence of electron flow fully reduces the plastoquinone, which picks up protons from the stromal side of the membrane and is released from the $b_6f$ complex as plastohydroquinone.

The overall result of two turnovers of the complex is that two electrons are transferred to P700, two plastohydroquinones are oxidized to the plastoquinone form, and one oxidized plastoquinone is reduced to the plastohydroquinone form. In the process of oxidizing the plastoquinones, four protons are transferred from the stromal to the lumenal side of the membrane.

By this mechanism, electron flow connecting the acceptor side of the PSII reaction center to the donor side of the PSI reaction center also gives rise to an electrochemical potential across the membrane, due in part to $H^+$ concentration differences on the two sides of the membrane. This electrochemical potential is used to power the synthesis of ATP. The cyclic electron flow through the cytochrome $b$ and plastoquinone increases the number of protons pumped per electron beyond what could be achieved in a strictly linear sequence.

## Plastoquinone and plastocyanin carry electrons between photosystems II and I

The location of the two photosystems at different sites on the thylakoid membranes (see Figure 7.18) requires that at least one component be capable of moving along or within the membrane in order to deliver electrons produced by photosystem II to photosystem I. The cytochrome $b_6f$ complex is distributed equally between the grana and the stroma regions of the membranes, but its large size makes it unlikely that it is the mobile carrier. Instead, plastoquinone or plastocyanin or possibly both are thought to serve as mobile carriers to connect the two photosystems.

**Plastocyanin** (**PC**) is a small (10.5 kDa), water-soluble, copper-containing protein that transfers electrons between the cytochrome $b_6f$ complex and P700. This protein is found in the lumenal space (see Figure 7.28). In certain green algae and cyanobacteria, a $c$-type cytochrome is sometimes found instead of plastocyanin; which of these two proteins is synthesized depends on the amount of copper available to the organism.

(A) First QH$_2$ oxidized

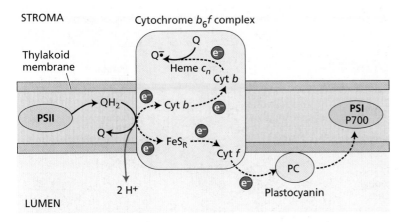

(B) Second QH$_2$ oxidized

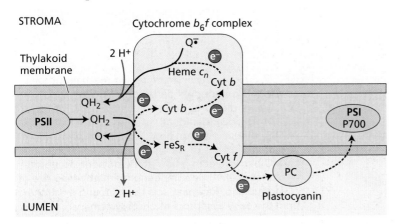

**FIGURE 7.28** Mechanism of electron and proton transfer in the cytochrome $b_6f$ complex. This complex contains two *b*-type cytochromes (Cyt *b*), a *c*-type cytochrome (Cyt *c*, historically called cytochrome *f*), a Rieske Fe–S protein (FeS$_R$), and two quinone oxidation–reduction sites. (A) The noncyclic or linear processes: A plastohydroquinone (PQH$_2$) molecule produced by the action of PSII (see Figure 7.26) is oxidized near the lumenal side of the complex, transferring its two electrons to the Rieske Fe–S protein and one of the *b*-type cytochromes and simultaneously expelling two protons to the lumen. The electron transferred to FeS$_R$ is passed to cytochrome *f* (Cyt *f*) and then to plastocyanin (PC), which reduces P700 of PSI. The reduced *b*-type cytochrome transfers an electron to the other *b*-type cytochrome, which reduces a plastoquinone (PQ) to the plastosemiquinone (PQ$^•$) state (see Figure 7.26). (B) The cyclic processes: A second PQH$_2$ is oxidized, with one electron going from FeS$_R$ to PC and finally to P700. The second electron goes through the two *b*-type cytochromes and reduces the plastosemiquinone to the plastohydroquinone, at the same time picking up two protons from the stroma. Overall, four protons are transported across the membrane for every two electrons delivered to P700.

## The photosystem I reaction center reduces NADP⁺

The PSI reaction center complex is a large multisubunit complex (**FIGURE 7.29**) (Jordan et al. 2001; Ben-Shem et al. 2003). In contrast to PSII, in which the antenna chlorophylls are associated with the reaction center, but present on separate pigment-proteins, a core antenna consisting of about 100 chlorophylls is an integral part of the PSI reaction center. The core antenna and P700 are bound to two proteins, PsaA and PsaB, with molecular masses in the range of 66 to 70 kDa (see **WEB TOPIC 7.8**) (Amunts and Nelson 2009). The PSI reaction center complex from pea contains four LHCI complexes in addition to the core structure similar to that found in cyanobacteria (see Figure 7.29). The total number of chlorophyll molecules in this complex is nearly 200.

The core antenna pigments form a bowl surrounding the electron transfer cofactors, which are in the center of the complex. In their reduced form, the electron carriers that function in the acceptor region of photosystem I are all extremely strong reducing agents. These reduced species are very unstable and thus difficult to identify. Evidence indicates that one of these early acceptors is a chlorophyll molecule, and another is a quinone species, phylloquinone, also known as vitamin K$_1$.

Additional electron acceptors include a series of three membrane-associated iron–sulfur proteins, also known as **Fe–S centers**: FeS$_X$, FeS$_A$, and FeS$_B$ (see Figure 7.29). Fe–S center X is part of the P700-binding protein; centers A and B reside on an 8-kDa protein that is part of the PSI reaction center complex. Electrons are transferred through centers A and B to **ferredoxin (Fd)**, a small, water-soluble iron–sulfur protein (see Figures 7.21 and 7.29). The membrane-associated flavoprotein **ferredoxin–NADP reductase (FNR)** reduces NADP⁺ to NADPH, thus completing the sequence of noncyclic electron transport that begins with the oxidation of water (Karplus et al. 1991).

In addition to the reduction of NADP⁺, reduced ferredoxin produced by photosystem I has several other functions in the chloroplast, such as the supply of reductants to reduce nitrate and the regulation of some of the carbon-fixation enzymes (see Chapter 8).

## Cyclic electron flow generates ATP but no NADPH

Some of the cytochrome $b_6f$ complexes are found in the stroma region of the membrane, where photosystem I is located. Under certain conditions, **cyclic electron flow** is

(A)

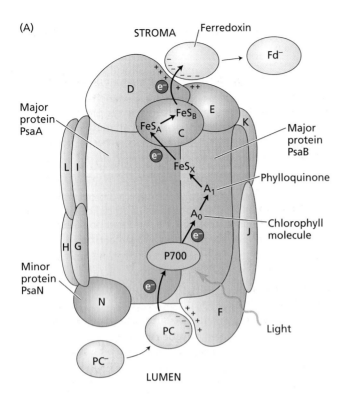

(B)

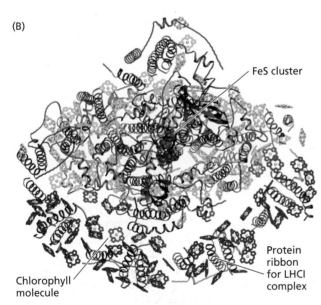

**FIGURE 7.29** Structure of photosystem I. (A) Structural model of the PSI reaction center from higher plants. Components of the PSI reaction center are organized around two major core proteins, PsaA and PsaB. Minor proteins PsaC to PsaN are labeled C to N. Electrons are transferred from plastocyanin (PC) to P700 (see Figures 7.21 and 7.22) and then to a chlorophyll molecule ($A_0$), to phylloquinone ($A_1$), to the Fe–S centers $FeS_X$, $FeS_A$, and $FeS_B$, and finally to the soluble iron–sulfur protein ferredoxin (Fd). (B) Structure of the photosystem I reaction center complex from pea at 4.4 Å resolution, including the LHCI antenna complexes. This is viewed from the stromal side of the membrane. (A after Buchanan et al. 2000; B after Nelson and Ben-Shem 2004.)

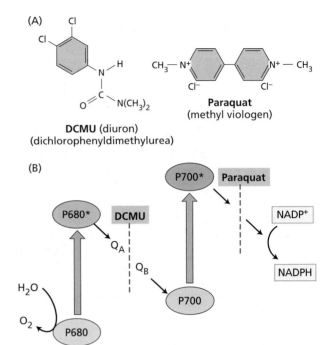

**FIGURE 7.30** Chemical structure and mechanism of action of two important herbicides. (A) Chemical structure of dichlorophenyldimethylurea (DCMU) and methyl viologen (paraquat), two herbicides that block photosynthetic electron flow. DCMU is also known as diuron. (B) Sites of action of the two herbicides. DCMU blocks electron flow at the quinone acceptors of photosystem II by competing for the binding site of plastoquinone. Paraquat acts by accepting electrons from the early acceptors of photosystem I.

known to occur from the reducing side of photosystem I via plastohydroquinone and the $b_6f$ complex and back to P700. This cyclic electron flow is coupled to proton pumping into the lumen, which can be utilized for ATP synthesis but does not oxidize water or reduce $NADP^+$. Cyclic electron flow is especially important as an ATP source in the bundle sheath chloroplasts of some plants that carry out $C_4$ carbon fixation (see Chapter 8). The molecular mechanism of cyclic electron flow is not well understood. Some proteins involved in regulating the process are just being discovered, and this remains an active area of research.

## Some herbicides block photosynthetic electron flow

The use of herbicides to kill unwanted plants is widespread in modern agriculture. Many different classes of herbicides have been developed. Some act by blocking amino acid, carotenoid, or lipid biosynthesis or by disrupting cell division. Other herbicides, such as dichlorophenyldimethylurea (DCMU, also known as diuron) and paraquat, block photosynthetic electron flow (**FIGURE 7.30**).

DCMU blocks electron flow at the quinone acceptors of photosystem II, by competing for the binding site of plastoquinone that is normally occupied by $PQ_B$. Paraquat accepts electrons from the early acceptors of photosystem I and then reacts with oxygen to form superoxide, $O_2^-$, a species that is very damaging to chloroplast components, especially lipids.

## Proton Transport and ATP Synthesis in the Chloroplast

In the preceding sections we learned how captured light energy is used to reduce NADP$^+$ to NADPH. Another fraction of the captured light energy is used for light-dependent ATP synthesis, which is known as **photophosphorylation**. This process was discovered by Daniel Arnon and his co-workers in the 1950s. In normal cellular conditions, photophosphorylation requires electron flow, although under some conditions electron flow and photophosphorylation can take place independently of each other. Electron flow without accompanying phosphorylation is said to be **uncoupled**.

It is now widely accepted that photophosphorylation works via the **chemiosmotic mechanism**. This mechanism was first proposed in the 1960s by Peter Mitchell. The same general mechanism drives phosphorylation during aerobic respiration in bacteria and mitochondria (see Chapter 11), as well as the transfer of many ions and metabolites across membranes (see Chapter 6). Chemiosmosis appears to be a unifying aspect of membrane processes in all forms of life.

In Chapter 6 we discussed the role of ATPases in chemiosmosis and ion transport at the cell's plasma membrane. The ATP used by the plasma membrane ATPase is synthesized by photophosphorylation in the chloroplast and oxidative phosphorylation in the mitochondrion. Here we are concerned with chemiosmosis and transmembrane proton concentration differences used to make ATP in the chloroplast.

The basic principle of chemiosmosis is that ion concentration differences and electric-potential differences across membranes are sources of free energy that can be utilized by the cell. As described by the second law of thermodynamics (see Appendix 1 for a detailed discussion), any nonuniform distribution of matter or energy represents a source of energy. Differences in **chemical potential** of any molecular species whose concentrations are not the same on opposite sides of a membrane provide such a source of energy.

The asymmetric nature of the photosynthetic membrane and the fact that proton flow from one side of the membrane to the other accompanies electron flow were discussed earlier. The direction of proton translocation is such that the stroma becomes more alkaline (fewer H$^+$ ions) and the lumen becomes more acidic (more H$^+$ ions) as a result of electron transport (see Figures 7.22 and 7.28).

Some of the early evidence supporting a chemiosmotic mechanism of photosynthetic ATP formation was provided by an elegant experiment carried out by André Jagendorf and co-workers (**FIGURE 7.31**). They suspended chloroplast thylakoids in a pH 4 buffer, and the buffer diffused across the membrane, causing the interior, as well as the exterior, of the thylakoid to equilibrate at this acidic pH. They then rapidly transferred the thylakoids to a pH 8 buffer, thereby creating a pH difference of 4 units across the thylakoid membrane, with the inside acidic relative to the outside.

They found that large amounts of ATP were formed from ADP and P$_i$ by this process, with no light input or electron transport. This result supports the predictions of the chemiosmotic hypothesis, described in the paragraphs that follow.

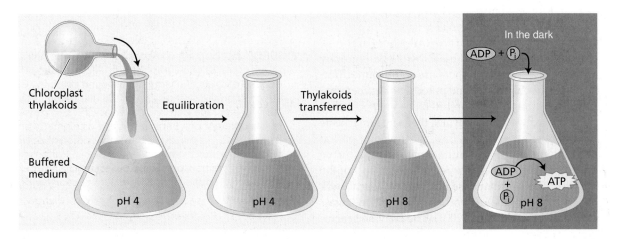

**FIGURE 7.31** Summary of the experiment carried out by Jagendorf and co-workers. Isolated chloroplast thylakoids kept previously at pH 8 were equilibrated in an acid medium at pH 4. The thylakoids were then transferred to a buffer at pH 8 that contained ADP and P$_i$. The proton gradient generated by this manipulation provided a driving force for ATP synthesis in the absence of light. This experiment verified a prediction of the chemiosmotic theory stating that a chemical potential across a membrane can provide energy for ATP synthesis. (After Jagendorf 1967.)

Mitchell proposed that the total energy available for ATP synthesis, which he called the **proton motive force** ($\Delta p$), is the sum of a proton chemical potential and a transmembrane electric potential. These two components of the proton motive force from the outside of the membrane to the inside are given by the following equation:

$$\Delta p = \Delta E - 59(pH_i - pH_o) \tag{7.9}$$

where $\Delta E$ is the transmembrane electric potential, and $pH_i - pH_o$ (or $\Delta pH$) is the pH difference across the membrane. The constant of proportionality (at 25°C) is 59 mV per pH unit, so a transmembrane pH difference of one pH unit is equivalent to a membrane potential of 59 mV.

In addition to the need for mobile electron carriers discussed earlier, the uneven distribution of photosystems II and I, and of ATP synthase at the thylakoid membrane (see Figure 7.18), poses some challenges for the formation of ATP. ATP synthase is found only in the stroma lamellae and at the edges of the grana stacks. Protons pumped across the membrane by the cytochrome $b_6 f$ complex or protons produced by water oxidation in the middle of the grana must move laterally up to several tens of nanometers to reach ATP synthase.

The ATP is synthesized by an enzyme complex (mass ~400 kDa) known by several names: **ATP synthase**, **ATPase** (after the reverse reaction of ATP hydrolysis), and **CF₀–CF₁** (Boyer 1997). This enzyme consists of two parts: a hydrophobic membrane-bound portion called $CF_o$ and a portion that sticks out into the stroma called $CF_1$ (**FIGURE 7.32**). $CF_o$ appears to form a channel across the membrane through which protons can pass. $CF_1$ is made up of several peptides, including three copies of each of the α and β peptides arranged alternately much like the sections of an orange. Whereas the catalytic sites are located largely on the β polypeptide, many of the other peptides are thought to have primarily regulatory functions. $CF_1$ is the portion of the complex that synthesizes ATP.

The molecular structure of the mitochondrial ATP synthase has been determined by X-ray crystallography (Stock et al. 1999). Although there are significant differences between the chloroplast and mitochondrial enzymes, they have the same overall architecture and probably nearly identical catalytic sites. In fact, there are remarkable similarities in the way electron flow is coupled to proton translocation in chloroplasts, mitochondria, and purple bacteria (**FIGURE 7.33**). Another remarkable aspect of the mechanism of the ATP synthase is that the internal stalk and probably much of the $CF_o$ portion of the enzyme rotate during catalysis (Yasuda et al. 2001). The enzyme is actually a tiny molecular motor (see **WEB TOPICS 7.9 AND 11.4**). Three molecules of ATP are synthesized for each rotation of the enzyme.

Direct microscopic imaging of the $CF_o$ part of the chloroplast ATP synthase indicates that it contains 14 copies of the integral membrane subunit (Seelert et al. 2000). Each

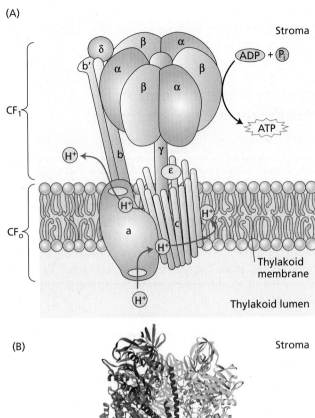

(A)

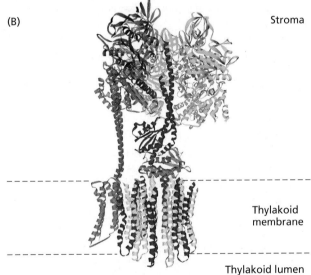

(B)

**FIGURE 7.32** Subunit composition (A) and compiled crystal structure (B) of chloroplast $F_1 F_o$ ATP synthase. This enzyme consists of a large multisubunit complex, $CF_1$, attached on the stromal side of the membrane to an integral membrane portion, known as $CF_o$. $CF_1$ consists of five different polypeptides, with stoichiometry of $\alpha_3$, $\beta_3$, $\gamma$, $\delta$, $\varepsilon$. $CF_o$ contains probably four different polypeptides, with a stoichiometry of a, b, b′, $c_{14}$. Protons from the lumen are transported by the rotating c polypeptide and ejected on the stroma side. (Figure courtesy of W. Frasch.)

subunit can translocate one proton across the membrane each time the complex rotates. This suggests that the stoichiometry of protons translocated to ATP formed is 14/3, or 4.67. Measured values of this parameter are usually somewhat lower than this value and the reasons for this discrepancy are not yet understood.

**(A) Purple bacteria**

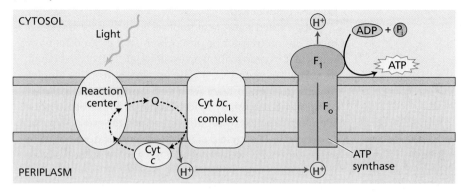

**(B) Chloroplasts**

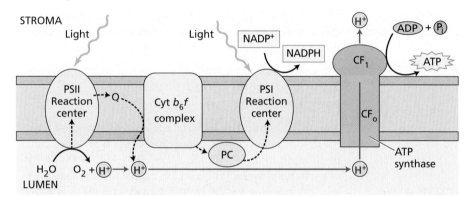

**(C) Mitochondria**

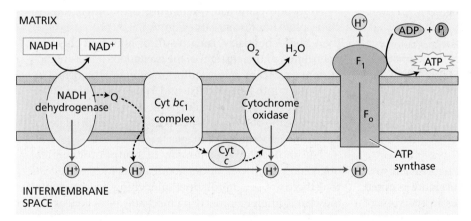

**FIGURE 7.33** Similarities of photosynthetic and respiratory electron flow in bacteria, chloroplasts, and mitochondria. In all three, electron flow is coupled to proton translocation, creating a transmembrane proton motive force ($\Delta p$). The energy in the proton motive force is then used for the synthesis of ATP by ATP synthase. (A) A reaction center in purple photosynthetic bacteria carries out cyclic electron flow, generating a proton potential by the action of the cytochrome $bc_1$ complex. (B) Chloroplasts carry out noncyclic electron flow, oxidizing water and reducing $NADP^+$. Protons are produced by the oxidation of water and by the oxidation of $PQH_2$ (labeled "Q" in the illustration) by the cytochrome $b_6f$ complex. (C) Mitochondria oxidize NADH to $NAD^+$ and reduce oxygen to water. Protons are pumped by the enzyme NADH dehydrogenase, the cytochrome $bc_1$ complex, and cytochrome oxidase. The ATP synthases in the three systems are very similar in structure.

# Repair and Regulation of the Photosynthetic Machinery

Photosynthetic systems face a special challenge. They are designed to absorb large amounts of light energy and process it into chemical energy. At the molecular level, the energy in a photon can be damaging, particularly under unfavorable conditions. In excess, light energy can lead to the production of toxic species such as superoxide, singlet oxygen, and peroxide, and damage can occur if the light energy is not dissipated safely (Asada 1999; Li et al. 2009).

Photosynthetic organisms therefore contain complex regulatory and repair mechanisms.

Some of these mechanisms regulate energy flow in the antenna system, to avoid excess excitation of the reaction centers and ensure that the two photosystems are equally driven. Although very effective, these processes are not entirely fail-safe, and sometimes toxic compounds are produced. Additional mechanisms are needed to dissipate these compounds—in particular, toxic oxygen species. In this section we will examine how some of these processes work to protect the system against photodamage.

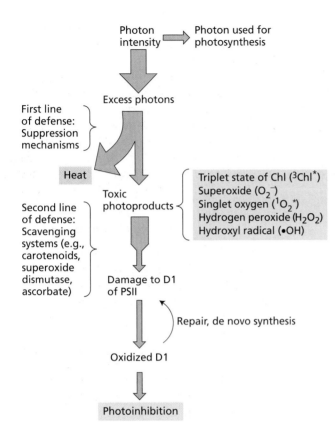

**FIGURE 7.34** Overall picture of the regulation of photon capture and the protection and repair of photodamage. Protection against photodamage is a multilevel process. The first line of defense is suppression of damage by quenching of excess excitation as heat. If this defense is not sufficient and toxic photoproducts form, a variety of scavenging systems eliminate the reactive photoproducts. If this second line of defense also fails, the photoproducts can damage the D1 protein of photosystem II. This damage leads to photoinhibition. The D1 protein is then excised from the PSII reaction center and degraded. A newly synthesized D1 is reinserted into the PSII reaction center to form a functional unit. (After Asada 1999.)

Despite these protective and scavenging mechanisms, damage can occur, and additional mechanisms are required to repair the system. **FIGURE 7.34** provides an overview of the several levels of the regulation and repair systems.

## Carotenoids serve as photoprotective agents

In addition to their role as accessory pigments, carotenoids play an essential role in **photoprotection**. The photosynthetic membrane can easily be damaged by the large amounts of energy absorbed by the pigments if this energy cannot be stored by photochemistry; this is why a protection mechanism is needed. The photoprotection mechanism can be thought of as a safety valve, venting excess energy before it can damage the organism. When the energy stored in chlorophylls in the excited state is rapidly dissipated by

excitation transfer or photochemistry, the excited state is said to be **quenched**.

If the excited state of chlorophyll is not rapidly quenched by excitation transfer or photochemistry, it can react with molecular oxygen to form an excited state of oxygen known as **singlet oxygen** ($^1O_2^*$). The extremely reactive singlet oxygen goes on to react with and damage many cellular components, especially lipids. Carotenoids exert their photoprotective action by rapidly quenching the excited state of chlorophyll. The excited state of carotenoids does not have sufficient energy to form singlet oxygen, so it decays back to its ground state while losing its energy as heat.

Mutant organisms that lack carotenoids cannot live in the presence of both light and molecular oxygen—a rather difficult situation for an $O_2$-evolving photosynthetic organism. Mutants of non-$O_2$-evolving photosynthetic bacteria that lack carotenoids can be maintained under laboratory conditions if oxygen is excluded from the growth medium.

## Some xanthophylls also participate in energy dissipation

Nonphotochemical quenching, a major process regulating the delivery of excitation energy to the reaction center, can be thought of as a "volume knob" that adjusts the flow of excitations to the PSII reaction center to a manageable level, depending on the light intensity and other conditions. The process appears to be an essential part of the regulation of antenna systems in most algae and plants.

**Nonphotochemical quenching** is the quenching of chlorophyll fluorescence (see Figure 7.5) by processes other than photochemistry. As a result of nonphotochemical quenching, a large fraction of the excitations in the antenna system caused by intense illumination are quenched by conversion into heat (Krause and Weis 1991; Baker 2008). Nonphotochemical quenching is thought to be involved in protecting the photosynthetic machinery against overexcitation and subsequent damage.

The molecular mechanism of nonphotochemical quenching is not well understood, and evidence suggests that there are several distinct quenching processes that may have different underlying mechanisms. It is clear that the pH of the thylakoid lumen and the state of aggregation of the antenna complexes are important factors. Three carotenoids, called **xanthophylls**, are involved in nonphotochemical quenching: violaxanthin, antheraxanthin, and zeaxanthin (**FIGURE 7.35**).

In high light, violaxanthin is converted into zeaxanthin, via the intermediate antheraxanthin, by the enzyme violaxanthin de-epoxidase. When light intensity decreases, the process is reversed. Binding of protons and zeaxanthin to light-harvesting antenna proteins is thought to cause conformational changes that lead to quenching and heat dissipation (Demmig-Adams and Adams 1992; Horton and Ruban 2004).

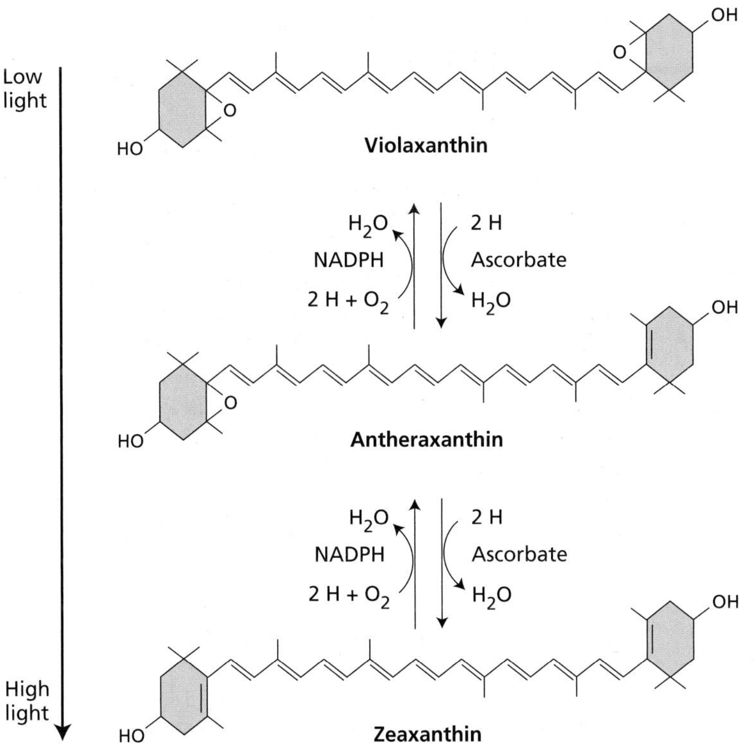

Low
light

**Violaxanthin**

HO

OH
O

H$_2$O  2 H
NADPH  Ascorbate
2 H + O$_2$  H$_2$O

HO

**Antheraxanthin**

OH
O

H$_2$O  2 H
NADPH  Ascorbate
2 H + O$_2$  H$_2$O

High
light

HO

**Zeaxanthin**

OH

**FIGURE 7.35** Chemical structure of violaxanthin, antheraxanthin, and zeaxanthin. The highly quenched state of photosystem II is associated with zeaxanthin, the unquenched state with violaxanthin. Enzymes interconvert these two carotenoids, with antheraxanthin as the intermediate, in response to changing conditions, especially changes in light intensity. Zeaxanthin formation uses ascorbate as a cofactor, and violaxanthin formation requires NADPH.

Nonphotochemical quenching appears to be preferentially associated with a peripheral antenna complex of photosystem II, the PsbS protein (Li et al. 2000). Recent evidence suggests that a transient electron transfer process may be an important part of the molecular quenching mechanism (Holt et al. 2005), although other molecular explanations have also been proposed (Johnson and Ruban 2009). This remains an active and controversial research area.

### The photosystem II reaction center is easily damaged

Another effect that appears to be a major factor in the stability of the photosynthetic apparatus is photoinhibition, which occurs when excess excitation arriving at the PSII reaction center leads to its inactivation and damage (Long et al. 1994). **Photoinhibition** is a complex set of molecular processes defined as the inhibition of photosynthesis by excess light.

As will be discussed in detail in Chapter 9, photoinhibition is reversible in early stages. Prolonged inhibition, however, results in damage to the system such that the PSII reaction center must be disassembled and repaired (Melis 1999). The main target of this damage is the D1 protein that makes up part of the PSII reaction center complex (see Figure 7.24). When D1 is damaged by excess light, it must be removed from the membrane and replaced with a newly synthesized molecule. The other components of the PSII reaction center are not damaged by excess excitation and are thought to be recycled, so the D1 protein is the only component that needs to be synthesized.

### Photosystem I is protected from active oxygen species

Photosystem I is particularly vulnerable to damage from active oxygen species. The ferredoxin acceptor of PSI is a very strong reductant that can easily reduce molecular oxygen to form superoxide ($O_2^-$). This reduction competes with the normal channeling of electrons to the reduction of NADP$^+$ and other processes. Superoxide is one of a series of active oxygen species that can be very damaging to biological membranes, but when formed in this way it can be eliminated by the action of a series of enzymes, including superoxide dismutase and ascorbate peroxidase (Asada 1999).

### Thylakoid stacking permits energy partitioning between the photosystems

The fact that photosynthesis in higher plants is driven by two photosystems with different light-absorbing properties poses a special problem. If the rate of delivery of energy to PSI and PSII is not precisely matched and conditions are such that the rate of photosynthesis is limited by the available light (low light intensity), the rate of electron flow will be limited by the photosystem that is receiving less energy. In the most efficient situation, the input of energy would be the same to both photosystems. However, no single arrangement of pigments would satisfy this requirement because at different times of day the light intensity and spectral distribution tend to favor one photosystem or the other (Allen and Forsberg 2001; Finazzi 2005).

This problem can be solved by a mechanism that shifts energy from one photosystem to the other in response to different conditions. Such a regulating mechanism has been shown to operate in different experimental conditions. The observation that the overall quantum yield of photosynthesis is nearly independent of wavelength (see Figure 7.12) strongly suggests that such a mechanism exists.

Thylakoid membranes contain a protein kinase that can phosphorylate a specific threonine residue on the surface of LHCII, one of the membrane-bound antenna pigment proteins described earlier in the chapter (see Figure 7.20). When LHCII is not phosphorylated, it delivers more energy to photosystem II, and when it is phosphorylated, it delivers more energy to photosystem I (Haldrup et al. 2001).

The kinase is activated when plastoquinone, one of the electron carriers between PSI and PSII, accumulates in the reduced state. Reduced plastoquinone accumulates when PSII is being activated more frequently than PSI. The phosphorylated LHCII then migrates out of the stacked regions of the membrane into the unstacked regions (see Figure 7.18), probably because of repulsive interactions with negative charges on adjacent membranes.

# Genetics, Assembly, and Evolution of Photosynthetic Systems

Chloroplasts have their own DNA, mRNA, and protein synthesis machinery, but most chloroplast proteins are encoded by nuclear genes and imported into the chloroplast. In this section we will consider the genetics, assembly, and evolution of the main chloroplast components.

## Chloroplast genes exhibit non-Mendelian patterns of inheritance

Chloroplasts and mitochondria reproduce by division rather than by de novo synthesis. This mode of reproduction is not surprising, since these organelles contain genetic information that is not present in the nucleus. During cell division, chloroplasts are divided between the two daughter cells. In most sexual plants, however, only the maternal plant contributes chloroplasts to the zygote. In these plants the normal Mendelian pattern of inheritance does not apply to chloroplast-encoded genes, because the offspring receive chloroplasts from only one parent. The result is **non-Mendelian**, or **maternal**, **inheritance**. Numerous traits are inherited in this way; one example is the herbicide-resistance trait discussed in WEB TOPIC 7.10.

## Most chloroplast proteins are imported from the cytoplasm

Chloroplast proteins can be encoded by either chloroplastic or nuclear DNA. The chloroplast-encoded proteins are synthesized on chloroplast ribosomes; the nucleus-encoded proteins are synthesized on cytoplasmic ribosomes and then transported into the chloroplast. Many nuclear genes contain introns—that is, base sequences that do not code for protein. The mRNA is processed to remove the introns, and the proteins are then synthesized in the cytoplasm.

The genes needed for chloroplast function are distributed in the nucleus and in the chloroplast genome with no evident pattern, but both sets are essential for the viability of the chloroplast. Some chloroplast genes are necessary for other cellular functions, such as heme and lipid synthe-

sis. Control of the expression of the nuclear genes that code for chloroplast proteins is complex and dynamic, involving light-dependent regulation mediated by both phytochrome (see Chapter 17) and blue light (see Chapter 18), as well as other factors (Eberhard et al. 2008).

The transport of chloroplast proteins that are synthesized in the cytoplasm is a tightly regulated process (Chen and Schnell 1999; Benz et al. 2009). For example, the enzyme rubisco (see Chapter 8), which functions in carbon fixation, has two types of subunits, a chloroplast-encoded large subunit and a nucleus-encoded small subunit. Small subunits of rubisco are synthesized in the cytoplasm and transported into the chloroplast, where the enzyme is assembled.

In this and other known cases, the nucleus-encoded chloroplast proteins are synthesized as precursor proteins containing an N-terminal amino acid sequence known as a **transit peptide**. This terminal sequence directs the precursor protein to the chloroplast, facilitates its passage through both the outer and the inner envelope membranes, and is then clipped off. The electron carrier plastocyanin is a water-soluble protein that is encoded in the nucleus but functions in the lumen of the chloroplast. It therefore must cross three membranes to reach its destination in the lumen. The transit peptide of plastocyanin is very large and is processed in more than one step as it directs the protein through two sequential translocations across the inner envelope membrane and the thylakoid membrane.

## The biosynthesis and breakdown of chlorophyll are complex pathways

Chlorophylls are complex molecules exquisitely suited to the light absorption, energy transfer, and electron transfer functions that they carry out in photosynthesis (see Figure 7.6). Like all other biomolecules, chlorophylls are made by a biosynthetic pathway in which simple molecules are used as building blocks to assemble more complex molecules (Beale 1999; Eckhardt et al. 2004; Tanaka and Tanaka 2007). Each step in the biosynthetic pathway is enzymatically catalyzed.

The chlorophyll biosynthetic pathway consists of more than a dozen steps (see WEB TOPIC 7.11). The process can be divided into several phases (FIGURE 7.36), each of which can be considered separately, but in the cell are highly coordinated and regulated. This regulation is essential because free chlorophyll and many of the biosynthetic intermediates are damaging to cellular components. The damage results largely because chlorophylls absorb light efficiently, but in the absence of accompanying proteins, they lack a pathway for disposing of the energy, with the result that toxic singlet oxygen is formed.

The breakdown pathway of chlorophyll in senescent leaves is quite different from the biosynthetic pathway (Takamiya et al. 2000; Eckhardt et al. 2004). The first step is removal of the phytol tail by an enzyme known as chlorophyllase, followed by removal of the magnesium by magnesium de-chelatase. Next the porphyrin structure

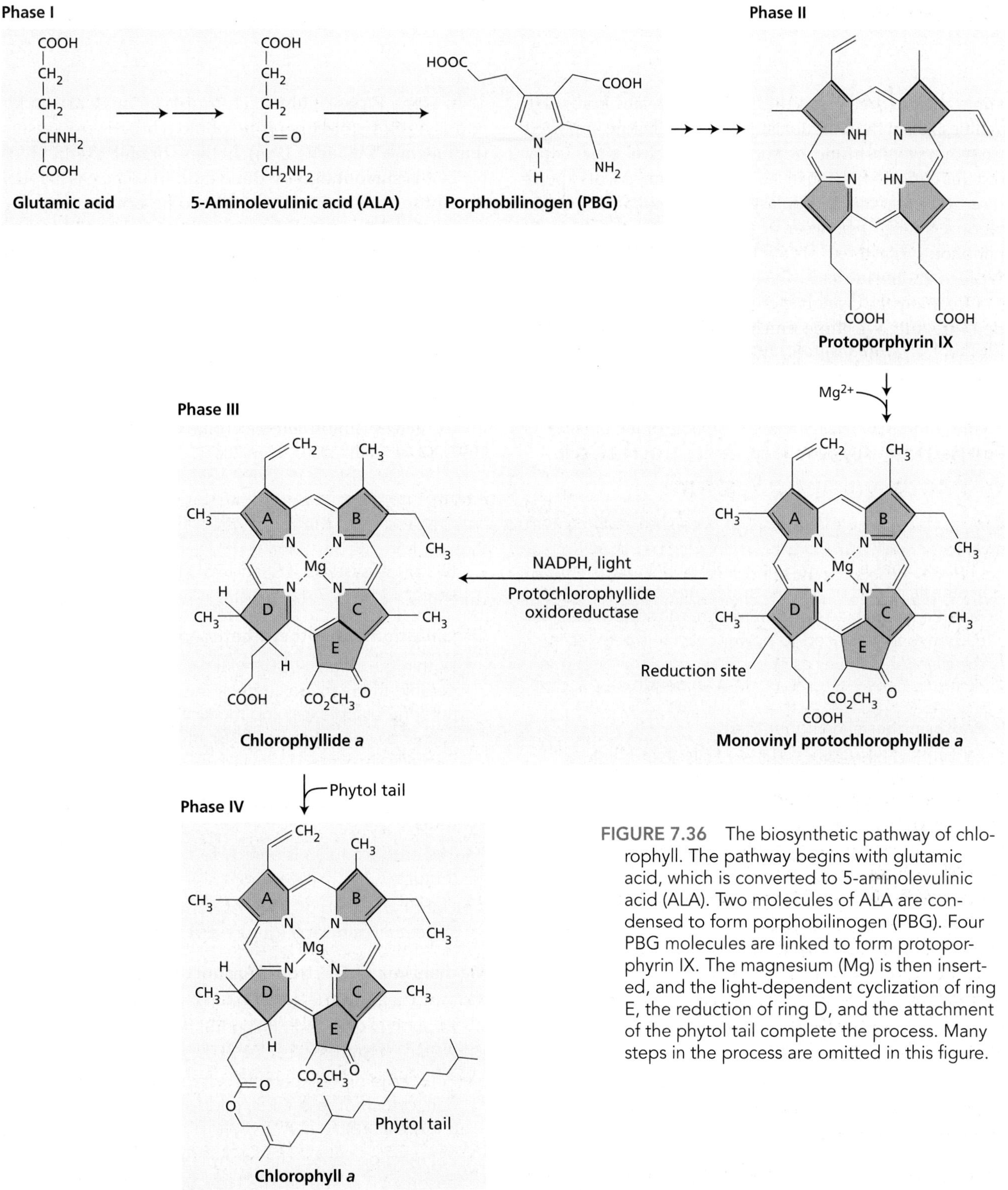

**FIGURE 7.36** The biosynthetic pathway of chlorophyll. The pathway begins with glutamic acid, which is converted to 5-aminolevulinic acid (ALA). Two molecules of ALA are condensed to form porphobilinogen (PBG). Four PBG molecules are linked to form protoporphyrin IX. The magnesium (Mg) is then inserted, and the light-dependent cyclization of ring E, the reduction of ring D, and the attachment of the phytol tail complete the process. Many steps in the process are omitted in this figure.

is opened by an oxygen-dependent oxygenase enzyme to form an open-chain tetrapyrrole.

The tetrapyrrole is further modified to form water-soluble, colorless products. These colorless metabolites are then exported from the senescent chloroplast and transported to the vacuole, where they are permanently stored. The chlorophyll metabolites are not further processed or recycled, although the proteins associated with them in the chloroplast are subsequently recycled into new proteins. The recycling of proteins is thought to be important for the nitrogen economy of the plant.

## Complex photosynthetic organisms have evolved from simpler forms

The complicated photosynthetic apparatus found in plants and algae is the end product of a long evolutionary sequence. Much can be learned about this evolutionary process from analysis of simpler prokaryotic photosynthetic organisms, including the anoxygenic photosynthetic bacteria and the cyanobacteria.

The chloroplast is a semiautonomous cell organelle, with its own DNA and a complete protein synthesis apparatus. Many of the proteins that make up the photosynthetic apparatus, as well as all the chlorophylls and lipids, are synthesized in the chloroplast. Other proteins are imported from the cytoplasm and are encoded by nuclear genes. How did this curious division of labor come about? Most experts now agree that the chloroplast is the descendant of a symbiotic relationship between a cyanobacterium and a simple nonphotosynthetic eukaryotic cell. This type of relationship is called **endosymbiosis** (Cavalier-Smith 2000).

Originally the cyanobacterium was capable of independent life, but over time much of its genetic information needed for normal cellular functions was lost, and a substantial amount of information needed to synthesize the photosynthetic apparatus was transferred to the nucleus. So the chloroplast was no longer capable of life outside its host and eventually became an integral part of the cell.

In some types of algae, chloroplasts have arisen by endosymbiosis of eukaryotic photosynthetic organisms (Palmer and Delwiche 1996). In these organisms the chloroplast is surrounded by three (and in some cases four) membranes, which are thought to be remnants of the plasma membranes of the earlier organisms. Mitochondria are also thought to have originated by endosymbiosis in a separate event much earlier than chloroplast formation.

The answers to other questions related to the evolution of photosynthesis are less clear. These include the nature of the earliest photosynthetic systems, how the two photosystems became linked, and the evolutionary origin of the oxygen evolution complex (Blankenship and Hartman 1998; Xiong et al. 2000; Allen 2005).

## SUMMARY

Photosynthesis in plants uses light energy for the synthesis of carbohydrates and generation of oxygen from carbon dioxide and water. Energy stored in carbohydrates is used to power cellular processes in the plant and serves as energy resources for all forms of life.

### Photosynthesis in Higher Plants

- Within chloroplasts, chlorophylls absorb light energy for the oxidation of water, releasing oxygen and generating NADPH and ATP (thylakoid reactions).
- NADPH and ATP are used to reduce carbon dioxide to form sugars (carbon fixation reactions).

### General Concepts

- Light behaves as both a particle and a wave, delivering energy as photons, some of which are absorbed and used by plants (**Figures 7.1–7.3**).
- Light-energized chlorophyll may fluoresce, transfer energy to another molecule, or use its energy to drive chemical reactions (**Figures 7.5, 7.10**).
- All photosynthesizers contain a mixture of pigments with distinct structures and light-absorbing properties (**Figures 7.6, 7.7**).

### Key Experiments in Understanding Photosynthesis

- An action spectrum for photosynthesis shows algal oxygen evolution at certain wavelengths (**Figures 7.8, 7.9**).
- Antenna pigment–protein complexes collect light energy and transfer it to the reaction center complexes (**Figure 7.10**).
- Light drives the reduction of NADP+ and the formation of ATP. Oxygen-evolving organisms have two photosystems (PSI and PSII) that operate in series (**Figures 7.12–7.14**).

### Organization of Photosynthetic Apparatus

- Within the chloroplast, thylakoid membranes contain the reaction centers, the light-harvesting antenna complexes, and most of the electron carrier proteins (**Figure 7.18**). PSI and PSII are spatially separated in thylakoids.

### Organization of Light Absorbing Antenna Systems

- The antenna funnels energy to the reaction center (**Figure 7.19**).
- Light-harvesting proteins of both photosystems are structurally similar (**Figure 7.20**).

### Mechanisms of Electron Transport

- The Z scheme identifies the flow of electrons through carriers in PSII and PSI from $H_2O$ to NADP+ (**Figures 7.14, 7.21**).
- Four large protein complexes transfer electrons: PSII, cytochrome $b_6f$, PSI, and ATP synthase (**Figure 7.22**).
- PSI reaction center chlorophyll has maximum absorption at 700 nm; PSII reaction center chlorophyll absorbs maximally at 680 nm.
- The PSII reaction center is a multisubunit protein–pigment complex (**Figures 7.24, 7.25**).
- Manganese ($Mn^{2+}$) ions are required to oxidize water.
- Two hydrophobic plastoquinones accept electrons from PSII (**Figures 7.22, 7.26**).
- Protons are transported into the thylakoid lumen when electrons pass through the cytochrome $b_6f$ complex (**Figure 7.22, 7.27**).

## SUMMARY continued

- Plastoquinone and plastocyanin carry electrons between PSII and PSI (**Figure 7.28**).
- $NADP^+$ is reduced by the PSI reaction center, using three Fe-S centers and ferredoxin as electron carriers (**Figure 7.29**).
- Cyclic electron flow generates ATP by proton pumping, but no NADPH.
- Herbicides may block photosynthetic electron flow (**Figure 7.30**).

### Proton Transport and ATP Synthesis in Chloroplasts

- In vitro transfer of pH 4–equilibrated chloroplast thylakoids to a pH 8 buffer resulted in the formation of ATP from ADP and $P_i$, supporting the chemiosmotic hypothesis (**Figure 7.31**).
- Protons move down an electrochemical gradient (proton motive force) passing through an ATP synthase and forming ATP (**Figure 7.32**).
- During catalysis, the $CF_o$ portion of the ATP synthase rotates like a miniature motor.
- Proton translocation in chloroplasts, mitochondria, and purple bacteria shows significant similarities (**Figure 7.33**).

### Regulation and Repair of Photosynthetic Machinery

- Protection and repair of photodamage consists of: quenching and heat dissipation, neutralizing toxic photoproducts, and synthetic repair of PSII (**Figure 7.34**).
- Xanthophylls (carotenoids) participate in nonphotochemical quenching (**Figure 7.35**).
- Kinase-mediated phosphorylation of LHCII causes its migration to stacked thylakoids and its delivery of energy to PSI (*state 2*). Upon dephosphorylation, LHCII migrates to unstacked thylakoids and delivers more energy to PSII (*state 1*).

### Genetics, Assembly, and Evolution of Photosynthetic Systems

- Chloroplasts have their own DNA, mRNA, and protein synthesis system but import many proteins encoded by nuclear genes.
- Chloroplasts show a non-Mendelian pattern of inheritance.
- Chlorophyll biosynthesis can be divided into four phases (**Figure 7.36**).
- The chloroplast is descended from a symbiotic relationship between a cyanobacterium and a simple nonphotosyntheic eukaryotic cell.

## WEB MATERIAL

### Web Topics

**7.1 Principles of Spectrophotometry**
Spectroscopy is a key technique for the study of light reactions.

**7.2 The Distribution of Chlorophylls and Other Photosynthetic Pigments**
The content of chlorophylls and other photosynthetic pigments varies among plant kingdoms.

**7.3 Quantum Yield**
Quantum yields measure how effectively light drives a photobiological process.

**7.4 Antagonistic Effects of Light on Cytochrome Oxidation**
Photosystems I and II were discovered in some ingenious experiments.

**7.5 Structures of Two Bacterial Reaction Centers**
X-ray diffraction studies resolved the atomic structure of the reaction center of photosystem II.

**7.6 Midpoint Potentials and Redox Reactions**
The measurement of midpoint potentials is useful for analyzing electron flow through photosystem II.

**7.7 Oxygen Evolution**
The S state mechanism is a valuable model for water splitting in PSII.

**7.8 Photosystem I**
The PSI reaction is a multiprotein complex.

**7.9 ATP Synthase**
The ATP synthase functions as a molecular motor.

## WEB MATERIAL continued

### Web Topics

**7.10 Mode of Action of Some Herbicides**
Some herbicides kill plants by blocking photosynthetic electron flow.

**7.11 Chlorophyll Biosynthesis**
Chlorophyll and heme share early steps of their biosynthetic pathways.

### Web Essay

**7.1 A Novel View of Chloroplast Structure**
Stromules extend the reach of the chloroplasts.

## CHAPTER REFERENCES

Allen, J. F. (2005) A redox switch hypothesis for the origin of two light reactions in photosynthesis. *FEBS Lett.* 579: 963–968.

Allen, J. F., and Forsberg, J. (2001) Molecular recognition in thylakoid structure and function. *Trends Plant Sci.* 6: 317–326.

Amunts, A., and Nelson, N. (2009) Plant photosystem I design in the light of evolution. *Structure* 17: 637–650.

Asada, K. (1999) The water–water cycle in chloroplasts: Scavenging of active oxygens and dissipation of excess photons. *Annu. Rev. Plant Physiol. Plant Mol. Biol.* 50: 601–639.

Avers, C. J. (1985) *Molecular Cell Biology.* Addison-Wesley, Reading, MA.

Baker, N. R. (2008) Chlorophyll fluorescence: a probe of photosynthesis in vivo. *Annu. Rev. Plant Biol.* 59: 89–113

Baniulis, D., Yamashita, E., Zhang, H., Hasan, S. S., and Cramer, W. A. (2008) Structure–function of the cytochrome $b_6f$ complex. *Photochem. Photobiol.* 84: 1349–1358.

Barber, J., Nield, N., Morris, E. P., and Hankamer, B. (1999) Subunit positioning in photosystem II revisited. *Trends Biochem. Sci.* 24: 43–45.

Barros, T., and Kühlbrandt, W. (2009) Crystallisation, structure and function of plant light-harvesting Complex II. *Biochim. Biophys. Acta* 1787: 753–772.

Beale, S. I. (1999) Enzymes of chlorophyll biosynthesis. *Photosyn. Res.* 60: 43–73.

Becker, W. M. (1986) *The World of the Cell.* Benjamin/Cummings, Menlo Park, CA.

Ben-Shem, A., Frolow, F., and Nelson, N. (2003) Crystal structure of plant photosystem I. *Nature* 426: 630–635.

Benz, J. P., Soll, J., and Bölter, B. (2009) Protein transport in organelles: The composition, function and regulation of the Tic complex in chloroplast protein import. *FEBS J.* 276: 1166–1176.

Blankenship, R. E. (2002) *Molecular Mechanisms of Photosynthesis.* Blackwell Science, Oxford.

Blankenship, R. E., and Hartman, H. (1998) The origin and evolution of oxygenic photosynthesis. *Trends Biochem. Sci.* 23: 94–97.

Blankenship, R. E., and Prince, R. C. (1985) Excited-state redox potentials and the Z scheme of photosynthesis. *Trends Biochem. Sci.* 10: 382–383.

Boyer, P. D. (1997) The ATP synthase: A splendid molecular machine. *Annu. Rev. Biochem.* 66: 717–749.

Brudvig, G. W. (2008) Water oxidation chemistry of photosystem II. *Philos. Trans. R. Soc. Lond. B, Biol. Sci.* 363: 1211–1219.

Buchanan, B. B., Gruissem, W., and Jones, R. L., eds. (2000) *Biochemistry and Molecular Biology of Plants.* American Society of Plant Physiologists, Rockville, MD.

Cavalier-Smith, T. (2000) Membrane heredity and early chloroplast evolution. *Trends Plant Sci.* 5: 174–182.

Chen, X., and Schnell, D. J. (1999) Protein import into chloroplasts. *Trends Cell Biol.* 9: 222–227.

Deisenhofer, J., and Michel, H. (1989) The photosynthetic reaction center from the purple bacterium *Rhodopseudomonas viridis. Science* 245: 1463–1473.

Demmig-Adams, B., and Adams, W. W., III. (1992) Photoprotection and other responses of plants to high light stress. *Annu. Rev. Plant Physiol. Plant Mol. Biol.* 43: 599–626.

Eberhard, S., Finazzi, G., and Wollman, F. A. (2008) The dynamics of photosynthesis. *Annu. Rev. Genet.* 17: 463–515.

Eckhardt, U., Grimm, B. and Hortensteiner, S. (2004) Recent advances in chlorophyll biosynthesis and breakdown in higher plants. *Photosyn. Res.* 56: 1–14.

Ferreira, K. N., Iverson, T. M., Maghlaoui, K., Barber, J., and Iwata, S. (2004) Architecture of the photosynthetic oxygen-evolving center. *Science* 303: 1831–1838.

Finazzi, G. (2005) The central role of the green alga *Chlamydomonas reinhardtii* in revealing the mechanism of state transitions. *J. Exp. Bot.* 56: 383–388.

Green, B. R., and Durnford, D. G. (1996) The chlorophyll-carotenoid proteins of oxygenic photosynthesis. *Annu. Rev. Plant Physiol. Plant Mol. Biol.* 47: 685–714.

Green, B. R., and Parson, W. W., eds. (2003) *Light-Harvesting Antennas in Photosynthesis*. Kluwer Academic Publishers, Dordrecht.

Grossman, A. R., Bhaya, D., Apt, K. E., and Kehoe, D. M. (1995) Light-harvesting complexes in oxygenic photosynthesis: Diversity, control, and evolution. *Annu. Rev. Genet.* 29: 231–288.

Guskov, A., Kern, J., Gabdulkhakov, A., Broser, M., Zouni, A., and Saenger, W. (2009) Cyanobacterial photosystem II at 2.9-Å resolution and the role of quinones, lipids, channels and chloride. *Nat. Struct. Mol. Biol.* 16: 334–342.

Haldrup, A., Jensen, P. E., Lunde, C., and Scheller, H. V. (2001) Balance of power: A view of the mechanism of photosynthetic state transitions. *Trends Plant Sci.* 6: 301–305.

Holt, N. E., Zigmantas, D., Valkunas, L., Li, X. P., Niyogi, K. K., and Fleming, G. R. (2005) Carotenoid cation formation and the regulation of photosynthetic light harvesting. *Science* 307: 433–436.

Horton, P., and Ruban, A. (2004) Molecular design of the photosystem II light harvesting antenna: photosynthesis and photoprotection. *J. Exp. Bot.* 56: 365–373.

Jagendorf, A. T. (1967) Acid-based transitions and phosphorylation by chloroplasts. *Fed. Proc. Am. Soc. Exp. Biol.* 26: 1361–1369.

Johnson, M. P., and Ruban, A. V. (2009) Photoprotective energy dissipation in higher plants involves alteration of the excited state energy of the emitting chlorophyll(s) in the light harvesting antenna II (LHCII). *J. Biol. Chem.* 284: 23592–23601.

Jordan, P., Fromme, P., Witt, H. T., Klukas, O., Saenger, W., and Krauss, N. (2001) Three-dimensional structure of cyanobacterial photosystem I at 2.5 Å resolution. *Nature* 411: 909–917.

Karplus, P. A., Daniels, M. J., and Herriott, J. R. (1991) Atomic structure of ferredoxin-NADP$^+$ reductase: Prototype for a structurally novel flavoenzyme family. *Science* 251: 60–66.

Krause, G. H., and Weis, E. (1991) Chlorophyll fluorescence and photosynthesis: The basics. *Annu. Rev. Plant Physiol. Plant Mol. Biol.* 42: 313–350.

Kurisu, G., Zhang, H. M., Smith, J. L., and Cramer, W. A. (2003) Structure of cytochrome $b_6 f$ complex of oxygenic photosynthesis: Tuning the cavity. *Science* 302: 1009–1014.

Li, X. P., Björkman, O., Shih, C., Grossman, A. R., Rosenquist, M., Jansson, S., and Niyogi, K. K. (2000) A pigment-binding protein essential for regulation of photosynthetic light harvesting. *Nature* 403: 391–395.

Li, Z., Wakao, S., Fischer, B. B., and Niyogi, K. K. (2009) Sensing and responding to excess light. *Annu. Rev. Plant Biol.* 60: 239–260.

Liu, Z. F., Yan, H. C., Wang, K. B., Kuang, T. Y., Zhang, J. P., Gui, L. L., An, X. M., and Chang, W. R. (2004) Crystal structure of spinach major light harvesting complex at 2.72 Å resolution. *Nature* 428: 287–292.

Long, S. P., Humphries, S., and Falkowski, P. G. (1994) Photoinhibition of photosynthesis in nature. *Annu. Rev. Plant Physiol. Plant Mol. Biol.* 45: 633–662.

Melis, A. (1999) Photosystem-II damage and repair cycle in chloroplasts: What modulates the rate of photodamage in vivo? *Trends Plant Sci.* 4: 130–135.

Nelson, N., and Ben-Shem, A. (2004) The complex architecture of oxygenic photosynthesis. *Nat. Rev. Mol. Cell Biol.* 5: 971–982.

Okamura, M. Y., Paddock, M. L., Graige, M. S., and Feher, G. (2000) Proton and electron transfer in bacterial reaction centers. *Biochim. Biophys. Acta* 1458: 148–163.

Palmer, J. D., and Delwiche, C. F. (1996) Second-hand chloroplasts and the case of the disappearing nucleus. *Proc. Natl. Acad. Sci. USA* 93: 7432–7435.

Pullerits, T., and Sundström, V. (1996) Photosynthetic light-harvesting pigment-protein complexes: Toward understanding how and why. *Acc. Chem. Res.* 29: 381–389.

Seelert, H., Poetsch, A., Dencher, N. A., Engel, A., Stahlberg, H., and Muller, D. J. (2000) Structural biology. Proton-powered turbine of a plant motor. *Nature* 405: 418–419.

Stock, D., Leslie, A. G. W., and Walker, J. E. (1999) Molecular architecture of the rotary motor in ATP synthase. *Science* 286: 1700–1705.

Stroebel, D., Choquet, Y., Popot, J. L., and Picot, D. (2003) An atypical heme in the cytochrome $b_6 f$ complex. *Nature* 426: 413–418.

Takamiya, K.-I., Tsuchiya, T., and Ohta, H. (2000) Degradation pathway(s) of chlorophyll: What has gene cloning revealed? *Trends Plant Sci.* 5: 426–431.

Tanaka, R., and Tanaka, A. (2007) Tetrapyrrole biosynthesis in higher plants. *Ann. Rev. Plant Biol.* 58: 321–346.

Trebst, A. (1986) The topology of the plastoquinone and herbicide binding peptides of photosystem II in the thylakoid membrane. *Z. Naturforsch. C.* 240–245.

Van Grondelle, R., Dekker, J. P., Gillbro, T., and Sundström, V. (1994) Energy transfer and trapping in photosynthesis. *Biochim. Biophys. Acta* 1187: 1–65.

Xiong, J., Fisher, W., Inoue, K., Nakahara, M., and Bauer, C. E. (2000) Molecular evidence for the early evolution of photosynthesis. *Science* 289: 1724–1730.

Yano, J., Kern, J., Sauer, K., Latimer, M., Pushkar, Y., Biesiadka, J., Loll, B., Saenger, W., Messinger, J., Zouni, A., and Yachandra, V. K. (2006) Where water is oxidized to dioxygen: structure of the photosynthetic $Mn_4 Ca$ cluster. *Science* 314: 821–825.

Yasuda, R., Noji, H., Yoshida, M., Kinosita, K., and Itoh, H. (2001) Resolution of distinct rotational substeps by submillisecond kinetic analysis of $F_1$-ATPase. *Nature* 410: 898–904.

Zhu, X.-G., Long, S. P., and Ort, D. R. (2008) What is the maximum efficiency with which photosynthesis can convert solar energy into biomass? *Curr. Opin. Biotechnol.* 19: 153–159.

# Photosynthesis: The Carbon Reactions

In Chapter 5 we discussed the requirements of plants for mineral nutrients and light in order to grow and complete their life cycle. Because the amount of matter in our planet remains constant, energy must be continually supplied to maintain the circulation of nutrients through the biosphere. Without energy input, entropy would increase and the flow of matter would ultimately stop. Solar radiant energy that strikes the Earth's surface is the ultimate source of energy for sustaining life in the biosphere (DOE 2008; Sinclair 2009). This energy (ca. $3 \times 10^{21}$ Joules/year) is converted via endergonic reactions in plants into carbohydrates (ca. $2 \times 10^{11}$ tonnes of carbon/year).

The capture of sunlight energy for transformation into various forms of chemical energy is one of the oldest biochemical reactions on Earth. One billion years ago, heterotrophic cells acquired the ability to convert sunlight into chemical energy through primary endosymbiosis with a cyanobacterium. Recent comparisons of the amino acid sequences of proteins from plastids, cyanobacteria, and eukaryotes have led us to group the progeny of this ancient event under the denomination of Archaeplastidae, which comprises three major lineages: Chloroplastidae (Viridiplantae: green algae, land plants), Rhodophyta (red algae), and Glaucophytae (Glaucocystophyceae: unicellular algae containing cyanobacteria-like plastids called cyanelles) (Rodríguez-Ezpeleta et al. 2005; Deschamps et al. 2008a).

The original endosymbiosis has given rise to an enormous variety of organelles. In general, the transition from endosymbiont to organelle involved both the loss of functions unnecessary in the protected milieu of the host cell and the gain of other metabolic pathways (Deschamps et al. 2008b). However, the details of this process vary

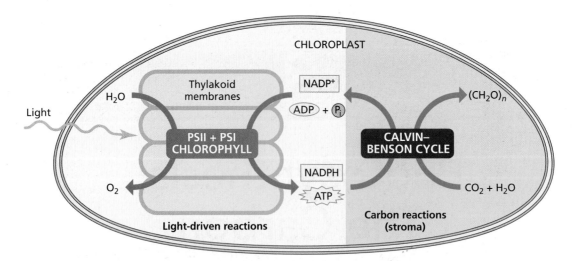

**FIGURE 8.1** The light and carbon reactions of photosynthesis in chloroplasts of land plants. In thylakoid membranes, the excitation of chlorophyll in the photosynthetic electron transport system [photosystem II (PSII) + photosystem I (PSI)] by light elicits the formation of ATP and NADPH (see Chapter 7). In the stroma, both ATP and NADPH are consumed by the Calvin–Benson cycle in a series of enzyme-catalyzed reactions that reduce atmospheric $CO_2$ to carbohydrates (triose phosphates).

among the different lineages of the Archaeplastidae. For example, the ancestral endosymbiont contributed the capacity not only to carry out oxygenic photosynthesis, but also to synthesize novel compounds, such as starch. All the descendants retain the ability to synthesize this polysaccharide, but Rhodophyta and Glaucophytae produce and store starch in the cytoplasm, while Chloroplastidae do so in the plastid.

In Chapter 7 we saw how the energy associated with the photochemical oxidation of water to molecular oxygen at the thylakoid membranes generates ATP and reduced ferredoxin and pyridine nucleotide (NADPH). Subsequently, the products of the light reactions, ATP and NADPH, flow from thylakoid membranes to the surrounding fluid phase (stroma) and drive the enzyme-catalyzed reduction of atmospheric $CO_2$ to carbohydrates and other cell components (**FIGURE 8.1**). The latter reactions in the stroma of chloroplasts were long thought to be independent of light and, as a consequence, were for many years referred to as the *dark reactions*. However, because these stroma-localized reactions depend on products of the photochemical processes and are also now known to be regulated directly by light, they are more properly referred to as the *carbon reactions of photosynthesis*.

In this chapter we will first examine the cyclic reactions that accomplish the incorporation of atmospheric $CO_2$ into organic compounds appropriate for life: the **Calvin–Benson cycle**. Subsequently we will consider how the unavoidable phenomenon of photorespiration releases part of the assimilated $CO_2$. Because photorespiration decreases the efficiency of photosynthetic $CO_2$ assimilation, we will also describe biochemical mechanisms for mitigating the loss of $CO_2$: $CO_2$ pumps (see **WEB TOPIC 8.1**), $C_4$ metabolism, and

crassulacean acid metabolism (CAM). We will close the chapter with a consideration of the two major products of the photosynthetic fixation of $CO_2$: starch, the reserve polysaccharide that accumulates transiently in chloroplasts; and sucrose, the disaccharide that is exported from leaves to developing and storage organs of the plant.

## The Calvin–Benson Cycle

Autotrophic organisms have the ability to use energy from physical and chemical sources to incorporate the carbon of atmospheric $CO_2$ into the skeletons of organic compounds that are compatible with the needs of the cell. The most important pathway of autotrophic $CO_2$ fixation is the Calvin–Benson cycle, which is found in many prokaryotes and in all photosynthetic eukaryotes, from the most primitive algae to the most advanced angiosperms. This pathway decreases the carbon oxidation state from the highest value, found in $CO_2$ (+4), to levels found in sugars (e.g., +2 in keto groups —CO—; 0 in secondary alcohols —CHOH—). In view of its notable capacity for lowering the carbon oxidation state, the Calvin–Benson cycle is also aptly named the *reductive pentose phosphate cycle*. In this section, we will examine how $CO_2$ is fixed via the Calvin–Benson cycle through the use of ATP and NADPH that are generated by the light reactions (see Figure 8.1), and how the cycle is regulated.

### The Calvin–Benson cycle has three stages: carboxylation, reduction, and regeneration

The Calvin–Benson cycle was elucidated in a series of elegant experiments by M. Calvin, A. Benson, J. A. Bassham,

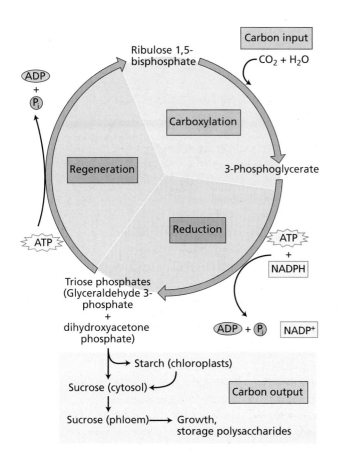

**FIGURE 8.2** The Calvin–Benson cycle proceeds in three stages: (1) *carboxylation*, which covalently links inorganic carbon ($CO_2$) to a carbon skeleton; (2) *reduction*, which forms a carbohydrate (triose phosphate) at the expense of photochemically generated ATP and reducing equivalents in the form of NADPH; and (3) *regeneration*, which restores the $CO_2$-acceptor ribulose 1,5-bisphosphate. At steady state, the input of $CO_2$ equals the output of triose phosphates. The latter either serve as precursors of starch biosynthesis in the chloroplast or flow to the cytosol for sucrose biosynthesis. Sucrose is loaded into the phloem sap and used for growth or polysaccharide biosynthesis in other parts of the plant.

## The carboxylation of ribulose 1,5-bisphosphate fixes $CO_2$ for the synthesis of triose phosphates

In the first step of the Calvin–Benson cycle, three molecules of $CO_2$ and three molecules of $H_2O$ react with three molecules of ribulose 1,5-bisphosphate to yield six molecules of 3-phosphoglycerate (**FIGURE 8.3** and **TABLE 8.1**, reaction 1). This reaction is catalyzed by the chloroplast enzyme ribulose-1,5-bisphosphate carboxylase/oxygenase, referred to as **rubisco** (see **WEB TOPIC 8.3**) (Cleland et al. 1998). In the first partial reaction, a $H^+$ is extracted from carbon 3 of ribulose 1,5-bisphosphate (**FIGURE 8.4**). The addition of gaseous $CO_2$ to the unstable rubisco-bound enediol intermediate drives the second partial reaction to the irreversible formation of 2-carboxy-3-ketoarabinitol 1,5-bisphosphate. Finally, the hydration of the resulting intermediate yields two molecules of 3-phosphoglycerate.

The reduction stage of the Calvin–Benson cycle reduces the carbon of the 3-phosphoglycerate coming from the carboxylation stage (see Figure 8.3, Table 8.1, reactions 2 and 3):

1. First, ATP formed in the light reactions phosphorylates 3-phosphoglycerate at the carboxyl group, yielding a mixed anhydride, 1,3-bisphosphoglycerate, in a reaction catalyzed by 3-phosphoglycerate kinase (Table 8.1, reaction 2).

2. Next, NADPH, also generated by the light reactions, reduces 1,3-bisphosphoglycerate to glyceraldehyde 3-phosphate, in a reaction catalyzed by the chloroplast enzyme NADP–glyceraldehyde-3-phosphate dehydrogenase (Table 8.1, reaction 3).

## Ribulose 1,5-bisphosphate is regenerated for the continuous assimilation of $CO_2$

To prevent depletion of Calvin–Benson cycle intermediates, the continuous uptake of atmospheric $CO_2$ requires constant regeneration of the $CO_2$ acceptor ribulose 1,5-bisphosphate. In the regeneration phase, three molecules of ribulose 1,5-bisphosphate (3 molecules × 5 carbons/

and their colleagues in the 1950s (Benson 1951) (see **WEB TOPIC 8.2**). The Calvin–Benson cycle proceeds in three stages that are highly coordinated in the chloroplast (**FIGURE 8.2**):

1. *Carboxylation* of the $CO_2$ acceptor molecule. The first committed enzymatic step of the cycle is the reaction of $CO_2$ and water with a 5-carbon acceptor molecule (ribulose 1,5-bisphosphate) to generate two molecules of a 3-carbon intermediate (3-phosphoglycerate).

2. *Reduction* of 3-phosphoglycerate. The 3-phosphoglycerate is reduced to 3-carbon carbohydrates (triose phosphates) by two enzymatic reactions driven by photochemically generated ATP and NADPH.

3. *Regeneration* of the $CO_2$ acceptor ribulose 1,5-bisphosphate. The cycle is completed by regeneration of ribulose 1,5-bisphosphate through a series of ten enzyme-catalyzed reactions, one requiring ATP.

The carbon output as triose phosphates balances the carbon input provided by atmospheric $CO_2$. Triose phosphates generated by the Calvin–Benson cycle in the chloroplasts are subsequently metabolized in other cellular compartments, yielding products such as sucrose and raffinose that are allocated to heterotrophic plant organs for sustaining growth or conversion into storage products.

## TABLE 8.1
### Reactions of the Calvin–Benson cycle

| Enzyme | Reaction |
|---|---|
| 1. Ribulose 1,5-bisphosphate carboxylase–oxygenase (rubisco) | 3 Ribulose 1,5-bisphosphate + 3 $CO_2$ + 3 $H_2O$ → 6 3-phosphoglycerate + 6 $H^+$ |
| 2. 3-Phosphoglycerate kinase | 6 3-Phosphoglycerate + 6 ATP → 6 1,3-bisphosphoglycerate + 6 ADP |
| 3. NADP–glyceraldehyde-3-phosphate dehydrogenase | 6 1,3-Bisphosphoglycerate + 6 NADPH + 6 $H^+$ → 6 glyceraldehyde 3-phosphate + 6 $NADP^+$ + 6 $P_i$ |
| 4. Triose phosphate isomerase | 2 Glyceraldehyde 3-phosphate ↔ 2 dihydroxyacetone phosphate |
| 5. Aldolase | Glyceraldehyde 3-phosphate + dihydroxyacetone phosphate → fructose 1,6-bisphosphate |
| 6. Fructose-1,6-bisphosphatase | Fructose 1,6-bisphosphate + $H_2O$ → fructose 6-phosphate + $P_i$ |
| 7. Transketolase | Fructose 6-phosphate + glyceraldehyde 3-phosphate → erythrose 4-phosphate + xylulose 5-phosphate |
| 8. Aldolase | Erythrose 4-phosphate + dihydroxyacetone phosphate → sedoheptulose 1,7-bisphosphate |
| 9. Sedoheptulose-1,7-bisphosphatase | Sedoheptulose 1,7-bisphosphate + $H_2O$ → sedoheptulose 7-phosphate + $P_i$ |
| 10. Transketolase | Sedoheptulose 7-phosphate + glyceraldehyde 3-phosphate → ribose 5-phosphate + xylulose 5-phosphate |
| 11a. Ribulose 5-phosphate epimerase | 2 Xylulose 5-phosphate → 2 ribulose 5-phosphate |
| 11b. Ribose 5-phosphate isomerase | Ribose 5-phosphate → ribulose 5-phosphate |
| 12. Phosphoribulokinase (Ribulose 5-phosphate kinase) | 3 Ribulose 5-phosphate + 3 ATP → 3 ribulose 1,5-bisphosphate + 3 ADP + 3 $H^+$ |

**Net**: 3 $CO_2$ + 5 $H_2O$ + 6 NADPH + 9 ATP → glyceraldehyde 3-phosphate + 6 $NADP^+$ + 3 $H^+$ + 9 ADP + 8 $P_i$

*Note*: $P_i$ stands for inorganic phosphate.

molecule = 15 carbons total) are formed by reactions that reshuffle the carbons from five molecules of glyceraldehyde 3-phosphate (5 molecules × 3 carbons/molecule = 15 carbons) (see Figure 8.3). The sixth molecule of glyceraldehyde 3-phosphate (1 molecule × 3 carbons/molecule = 3 carbons total) represents the net assimilation of three molecules of $CO_2$ and becomes available for the carbon metabolism of the plant. The reshuffling of the other five molecules of glyceraldehyde 3-phosphate to yield three molecules of ribulose 1,5-bisphosphate proceeds through reactions 4 to 12 in Table 8.1 and Figure 8.3:

1. *Two molecules* of glyceraldehyde 3-phosphate are converted to dihydroxyacetone phosphate in the reaction catalyzed by triose phosphate isomerase (Table 8.1, reaction 4).

2. One molecule of dihydroxyacetone phosphate undergoes aldol condensation with a *third molecule* of glyceraldehyde 3-phosphate, a reaction catalyzed by aldolase, to give fructose 1,6-bisphosphate (Table 8.1, reaction 5).

3. Fructose 1,6-bisphosphate is hydrolyzed to fructose 6-phosphate in a reaction catalyzed by a specific chloroplast fructose-1,6-bisphosphatase (Table 8.1, reaction 6).

4. A 2-carbon unit of the fructose 6-phosphate molecule (carbons 1 and 2) is transferred via the enzyme transketolase to a *fourth molecule* of glyceraldehyde 3-phosphate to form xylulose 5-phosphate. The other four carbons of the fructose 6-phosphate molecule (carbons 3, 4, 5, and 6) form erythrose 4-phosphate (Table 8.1, reaction 7).

5. The erythrose 4-phosphate then combines, via aldolase, with the remaining molecule of dihydroxyacetone phosphate to yield the 7-carbon sugar sedoheptulose 1,7-bisphosphate (Table 8.1, reaction 8).

6. Sedoheptulose 1,7-bisphosphate is then hydrolyzed by way of a specific phosphatase, sedoheptulose-1,7-bisphosphatase, to form sedoheptulose 7-phosphate (Table 8.1, reaction 9).

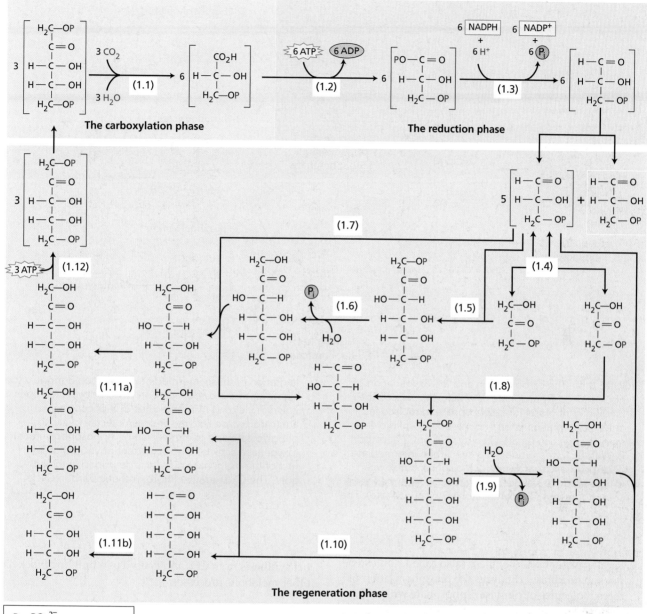

**FIGURE 8.3** The Calvin–Benson cycle. The carboxylation of three molecules of ribulose 1,5-bisphosphate yields six molecules of 3-phosphoglycerate (*carboxylation phase*). After phosphorylation of the carboxylic group, 1,3-bisphosphoglycerate is reduced to six molecules of glyceraldehyde 3-phosphate with the concurrent release of six molecules of inorganic phosphate (*reduction phase*). From the total of six molecules of glyceraldehyde 3-phosphate, one represents the net assimilation of three molecules of $CO_2$ while the other five undergo a series of reactions that finally regenerate the three starting molecules of ribulose 1,5-bisphosphate (*regeneration phase*). See Table 8.1 for a description of each numbered reaction.

7. Sedoheptulose 7-phosphate donates a 2-carbon unit (carbons 1 and 2) to the *fifth (and last) molecule* of glyceraldehyde 3-phosphate, via transketolase, producing xylulose 5-phosphate. The remaining five carbons (carbons 3–7) of sedoheptulose 7-phosphate molecule become ribose 5-phosphate (Table 8.1, reaction 10).

8. Two molecules of xylulose 5-phosphate are converted to two molecules of ribulose 5-phosphate by a ribulose 5-phosphate epimerase (Table 8.1, reaction 11a), while a third molecule of ribulose 5-phosphate originates from ribose 5-phosphate by the action of ribose 5-phosphate isomerase (Table 8.1, reaction 11b).

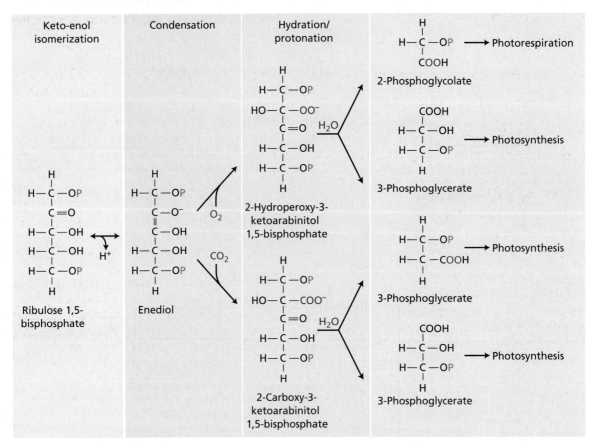

**FIGURE 8.4** The carboxylation and the oxygenation of ribulose 1,5-bisphosphate catalyzed by rubisco. The binding of ribulose 1,5-bisphosphate to rubisco facilitates the formation of an enzyme-bound enediol intermediate that can be attacked by $CO_2$ or $O_2$ at carbon 2. With $CO_2$, the product is a six-carbon intermediate (2-carboxy-3-ketoarabinitol 1,5-bisphosphate); with $O_2$ the product is a five-carbon reactive intermediate (2-hydroperoxy-3-ketoarabinitol 1,5-bisphosphate). The hydration of these intermediates at carbon 3 triggers the cleavage of the carbon–carbon bond between carbons 2 and 3, yielding two molecules of 3-phosphoglycerate (carboxylase activity) or one molecule of 2-phosphoglycolate and one molecule of 3-phosphoglyerate (oxygenase activity). The important physiological effect of the oxygenase activity is described in the section "The $C_2$ Oxidative Photosynthetic Carbon Cycle."

9. Finally, phosphoribulokinase (also called ribulose-5-phosphate kinase) catalyzes the phosphorylation of three molecules of ribulose 5-phosphate with ATP, thus regenerating the three needed molecules of the initial $CO_2$ acceptor, ribulose 1,5-bisphosphate (Table 8.1, reaction 12).

Hence, triose phosphates are formed in the carboxylation and reduction phases of the Calvin–Benson cycle at the expense of energy (ATP) and reducing equivalents (NADPH) generated in the thylakoid membranes of chloroplasts

$$3\ CO_2 + 3\ \text{ribulose 1,5-bisphosphate} + 3\ H_2O +$$
$$6\ NADPH + 6\ H^+ + 6\ ATP$$

$$\downarrow$$

$$6\ \text{Triose phosphates} + 6\ NADP^+ + 6\ ADP + 6\ P_i$$

From these six triose phosphates, five are used in the regeneration phase that restores ribulose 1,5-bisphosphate, the $CO_2$-acceptor, while the sixth triose phosphate represents net synthesis from $CO_2$ and is used as a building block for other metabolic processes.

$$5\ \text{Triose phosphates} + 3\ ATP \rightarrow 3\ \text{ribulose}$$
$$\text{1,5-bisphosphate} + 3\ ADP$$

In summary, the fixation of three $CO_2$ into one triose phosphate utilizes 9 ATP and 6 NADPH; that is, the ratio of ATP:NADPH required for the fixation of one $CO_2$ in the Calvin–Benson cycle is 3:2.

### An induction period precedes the steady state of photosynthetic $CO_2$ assimilation

When leaves are kept in darkness for long periods (e.g., at night), the stromal concentration of most biochemical intermediates of the Calvin–Benson cycle is low. Therefore, when leaves are transferred to the light, almost all stromal triose phosphates are committed to the production of the intermediates necessary to regenerate ribulose 1,5-bisphosphate. The illumination of previously darkened leaves or isolated chloroplasts illustrates clearly the relevance of metabolite buildup. In such experiments, the fixation of $CO_2$ starts after a lag, called the *induction period*, and the rate of photosynthesis increases with time in the first few minutes after the onset of illumination. This acceleration in the rate of photosynthesis during the induction period is due to both an increase in the concentration of intermediates of the Calvin–Benson cycle and the activation of enzymes by light (discussed later in this chapter). In summary, during the induction period, the six triose phosphates formed in the carboxylation and reduction phases of the Calvin–Benson cycle are used mainly for the regeneration of the $CO_2$ acceptor ribulose 1,5-bisphosphate.

By contrast, when photosynthesis reaches a *steady state*, the assimilation of atmospheric $CO_2$ increases the reserves of plant carbohydrates. At this stage, five-sixths of triose phosphates contribute to the regeneration of the $CO_2$ acceptor molecule ribulose 1,5-bisphosphate, while one-sixth is used to build up levels of carbohydrates (see Figure 8.2). This net production of triose phosphates is geared to both starch synthesis in the chloroplast and sucrose synthesis in the cytosol. Sucrose is generally the main product allocated to heterotrophic organs of the plant for further metabolism or conversion into storage products, such as starch and fructans. For a detailed analysis of the efficiency of the Calvin–Benson cycle in the use of energy see WEB TOPIC 8.4.

## Regulation of the Calvin–Benson Cycle

The efficient use of energy in the Calvin–Benson cycle requires the existence of specific regulatory mechanisms ensuring not only that all intermediates in the cycle are present at adequate concentrations in the light, but also that the cycle is turned off when not needed in the dark. To produce the necessary metabolites in response to environmental stimuli, chloroplasts achieve the appropriate rates of biochemical transformations through modification of enzyme levels (μmoles of enzyme/chloroplast) and catalytic capacities (μmoles of substrate converted/minute/μmole of enzyme).

The rates of gene expression and protein biosynthesis determine the amounts of enzymes in cell compartments. In particular, the amount of each enzyme present in the chloroplast stroma is regulated by mechanisms that control the concerted expression of nuclear and chloroplast genomes (Maier et al. 1995; Purton 1995). Nucleus-encoded enzymes are translated on 80S ribosomes in the cytosol and subsequently transported into the plastid. Plastid-encoded proteins are translated in the stroma on prokaryotic-like 70S ribosomes.

Light modulates the expression of stromal enzymes encoded by the nuclear genome via specific photorecep-

tors (e.g., phytochrome and blue-light receptors) (Neff et al. 2000). However, nuclear gene expression needs to be synchronized with the expression of other components of the photosynthetic apparatus in the organelle. Most of the regulatory signaling between nucleus and plastids is anterograde—that is, the products of nuclear genes control the transcription and translation of plastid genes. Such is the case in, for example, the assembly of stromal rubisco from eight nucleus-encoded small subunits (S) and eight plastid-encoded large subunits (L). However, in some cases (e.g., the synthesis of proteins associated with chlorophyll), regulation can be retrograde—that is, the signal flows from the plastid to the nucleus.

In contrast to the slow changes in catalytic rates caused by changes in the synthesis of enzymes, chloroplast enzymes respond rapidly to light through posttranslational modifications that change their specific activity (μmoles of substrate converted/minute/μmole of enzyme) (Wolosiuk et al. 1993). Two general mechanisms accomplish the modification of the kinetic properties of enzymes:

- Changes in covalent bonds that result in a chemically modified enzyme, such as the carbamylation of amino groups (Enz–$NH_2$ + $CO_2$ ↔ Enz–NH–$CO_2^-$ + $H^+$) or the reduction of disulfide bonds (Enz–$(S)_2$ + Prot–$(SH)_2$ ↔ Enz–$(SH)_2$ + Prot–$(S)_2$)

- Modifications of noncovalent interactions caused by the binding of metabolites or by changes in the ionic composition of the cellular milieu (e.g., pH, $Mg^{2+}$)

These two molecular mechanisms regulate the structure and the activity of an individual enzyme. However, every enzyme of the Calvin–Benson cycle is in close contact with proteins or thylakoid membranes that are inside or surround the stroma, respectively. Thus, enzymes of the Calvin–Benson cycle are regulated both by the mechanisms described above and by the reversible formation of supramolecular complexes with other components of chloroplasts. The close associations of enzymes with regulatory proteins in supramolecular complexes and the surface of thylakoid membranes enhance the efficiency of the Calvin–Benson cycle by facilitating the channeling of substrates and products inside the complexes.

In our further discussion of regulation, we will examine light-dependent modulation mechanisms that change the specific activity of five key enzymes within minutes of the light–dark transition:

- Rubisco
- Fructose-1,6-bisphosphatase
- Sedoheptulose-1,7-bisphosphatase
- Phosphoribulokinase
- NADP–glyceraldehyde-3-phosphate dehydrogenase

## The activity of rubisco increases in the light

Although rubisco plays a critical role in the carbon cycle of the biosphere, its catalytic rate is extremely slow (1–12 $CO_2$ fixations per second). This paradoxical feature was clarified when George Lorimer and colleagues found that rubisco must be activated before acting as a catalyst. Further studies revealed that the $CO_2$ molecule plays a dual role in the activity of rubisco: $CO_2$ participates in the transformation of the enzyme from an inactive to an active form (*modulation*) and is the substrate for the carboxylase reaction (*catalysis*) (Wolosiuk et al. 1993).

*As modulator*, $CO_2$ reacts slowly with the amino group of a specific lysine within the active site of rubisco ("Modulation" in **FIGURE 8.5**). The resulting carbamate derivative (a new anionic site) then rapidly binds $Mg^{2+}$ to yield the activated enzyme. Because two protons are released during the formation of the ternary rubisco–$CO_2$–$Mg^{2+}$ complex, the activation of rubisco in illuminated chloroplasts is promoted by the increase in stromal pH and $Mg^{2+}$ concentration (see below). At this stage, the molecule of $CO_2$ that reacts with ribulose 1,5-bisphosphate—*substrate* $CO_2$—binds to the active site of rubisco ("Catalysis" in Figure 8.5). Finally, rubisco releases two molecules of 3-phosphoglycerate ("Products" in Figure 8.5).

The tight binding of sugar phosphate–like molecules, such as ribulose 1,5-bisphosphate, to rubisco prevents carbamylation. However, the interaction of rubisco with an associated protein, rubisco activase, in a reaction that requires ATP, brings about a structural change of rubisco that releases sugar phosphate–like molecules and prepares the enzyme for activation via carbamylation and metal binding ("Rubisco activase" in Figure 8.5; also see **WEB TOPIC 8.5**) (Portis 2003). Rubisco activase is a member of a protein family that exhibits ATPase activity associated with chaperone-like functions. The binding of ATP to rubisco activase elicits the self-association of 14 to 16 rubisco activase polypeptides, which subsequently interact with rubisco. The stimulation of the carboxylase activity of rubisco is coupled to hydrolysis of the bound ATP, either for the oligomerization of rubisco activase or for rubisco conformational changes. Arabidopsis mutants that lack rubisco activase exhibit severely impaired photosynthesis at atmospheric levels of $CO_2$.

Many plant species contain two different rubisco activase polypeptides (ca. 42 and 47 kDa, respectively). These two isoforms come from alternative splicing of a unique pre-mRNA that produces two polypeptides that are identical except for the presence of extra amino acid residues at the C-terminal region of the longer form. The longer isoform of many plants contains two cysteines that are converted reversibly between the thiol and the disulphide forms [(rubisco activase)-(S)$_2$ + 2 H$^+$ + 2 e$^-$ ↔ (rubisco activase)-(SH)$_2$] (Portis 2003). The reversible thiol–disulphide exchange (which is catalyzed by a thioredoxin-dependent mechanism we will discuss later in the chap-

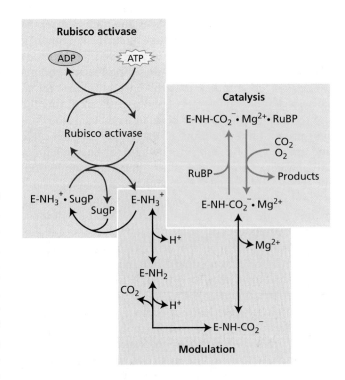

**FIGURE 8.5** $CO_2$ functions both as activator and as substrate in the reaction catalyzed by rubisco. *Modulation* (green panel): The reaction of *activator* $CO_2$ with rubisco (E) causes the formation of the E–carbamate adduct (E–NH–CO$_2^-$), whose stabilization by $Mg^{2+}$ yields the E–carbamate complex (E–NH–CO$_2^-$ • $Mg^{2+}$). In the stroma of illuminated chloroplasts, increases of both pH and $Mg^{2+}$ concentration facilitate the formation of the catalytically active form of rubisco, the E–NH–CO$_2^-$ • $Mg^{2+}$ complex. *Rubisco activase* (beige panel): Tight binding of sugar phosphates (SugP) impedes the production of the E–carbamate adduct. In the rubisco activase–mediated cycle, the hydrolysis of ATP by rubisco activase elicits a conformational change of rubisco that reduces its affinity for sugar phosphates. *Catalysis* (blue panel): Upon formation of the complex E–NH–CO$_2^-$ • $Mg^{2+}$ at the active site of the enzyme, rubisco combines with ribulose 1,5-bisphosphate (RuBP) and subsequently with the other *substrate*, $CO_2$ or $O_2$, initiating the carboxylase or oxygenase activities, respectively (see Figure 8.4). *Products* of catalysis are either two molecules of 3-phosphoglycerate (carboxylase activity) or one molecule each of 3-phosphoglycerate and 2-phosphoglycolate (oxygenase activity).

ter) modulates the sensitivity of the ATPase activity to the ATP/ADP ratio. Thus, the regulation of the ATPase activity of rubisco activase may communicate to rubisco the light-mediated stromal changes in the redox state of electron transport factors [e.g., (factor)-H$_2$ ↔ (factor)$_{oxidized}$ + 2 H$^+$ + 2 e$^-$] and in the phosphorylation potential (ATP:ADP ratio). However, other components still unknown may participate in this mechanism because both isoforms of rubisco activase are functional in the activation of rubisco.

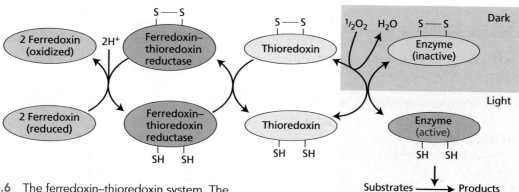

**FIGURE 8.6** The ferredoxin–thioredoxin system. The ferredoxin–thioredoxin system links the light signal sensed by thylakoid membranes to the activity of enzymes in the chloroplast stroma. Activation of enzymes of the reductive pentose phosphate cycle (Calvin–Benson cycle) starts in the light with the reduction of ferredoxin by the photosynthetic electron transport chain (PSII + PSI) (see Chapter 7). Reduced ferredoxin, together with two protons, is used to reduce a catalytically active disulfide bond (—S—S—) of the iron–sulfur enzyme ferredoxin–thioredoxin reductase, which, in turn, reduces the unique disulfide bond (—S—S—) of the regulatory protein thioredoxin (see **WEB TOPIC 8.6** for details). The reduced form (—SH HS—) of thioredoxin then reduces a critical (regulatory) disulfide bond of the target enzyme, triggering its conversion to a catalytically active state. In this state, the target enzyme catalyzes the transformation of substrates into products. In the dark, the formation of the disulfide bond in thioredoxin brings the reduced form (—SH HS—) of the enzyme back to the oxidized form (—S—S—) with a concurrent loss of activity. An enzyme of the oxidative pentose phosphate cycle, glucose-6-phosphate dehydrogenase, which is active in the oxidized (dark) state, is deactivated following reduction by thioredoxin.

Light stimulates the activity of rubisco in species that naturally produce only the shorter form (e.g., tobacco). On the other hand, light modulation of rubisco in transgenic lines of Arabidopsis that express only the longer (redox-linked) isoform is similar to that in the wild type.

In some plants, rubisco is also inhibited by the natural sugar phosphate 2-carboxyarabinitol 1-phosphate [$H_2O_3P$-O-$CH_2$-C($CO_2^-$)OH-CHOH-CHOH-$CH_2OH$], which closely resembles the 6-carbon intermediate of the carboxylation reaction (see 2-carboxy-3-ketoarabinitol 1,5-bisphosphate in Figure 8.4). Although present at low concentrations in leaves of many species, 2-carboxyarabinitol 1-phosphate is found at high concentrations in darkened leaves of legumes, such as soybean and bean. The inhibitor, which accumulates at night, binds to rubisco but is removed in the morning through the action of rubisco activase. The removal of 2-carboxyarabinitol 1-phosphate is enhanced by the action of a specific phosphatase that releases the phosphoryl group from the molecule.

## Light regulates the Calvin–Benson cycle via the ferredoxin–thioredoxin system

In addition to rubisco, light controls the activity of four other enzymes of the Calvin–Benson cycle via the ferredoxin–thioredoxin system, which consists of ferredoxin, ferredoxin–thioredoxin reductase, and thioredoxin (**FIGURE 8.6**). This oxidation–reduction mechanism, uncovered by B. B. Buchanan, P. Schürmann, R. A. Wolosiuk, and colleagues (Buchanan and Balmer 2005; Mora-Garcia et al. 2006), uses a product of the photosynthetic electron transport chain (reduced ferredoxin) to modulate the activity of four enzymes: fructose-1,6-bisphosphatase, sedoheptulose-1,7-bisphosphatase, phosphoribulokinase, and NADP–glyceraldehyde-3-phosphate dehydrogenase. Reduced ferredoxin transforms the unique disulfide bond of the ubiquitous regulatory protein thioredoxin (—S—S—) to the reduced state (—SH HS—) with the assistance of the iron–sulfur enzyme ferredoxin–thioredoxin reductase. Subsequently, the reduced thioredoxin converts the oxidized target enzyme to the reduced state, leading to enhancement of its catalytic capacity (see Figure 8.6 and **WEB TOPIC 8.6**).

Reduced thioredoxin is the common reductant for chloroplast enzymes, but the regulatory disulfide bonds targeted have unique structural and thermodynamic features. The amino acid sequences of enzymes regulated by thioredoxin do not exhibit a cysteine-containing consensus sequence, nor do the regulatory cysteines necessarily form part of the active site of the enzymes. In some cases, the activation of stromal enzymes with reduced thioredoxin is further coordinately regulated by other processes driven by light, such as the formation of metabolites and the transport of ions (Wolosiuk et al. 1993). Hence, the ferredoxin–thioredoxin system links the light absorbed by chlorophyll in thylakoid membranes to metabolic activities in the chloroplast stroma.

Initially associated with the Calvin–Benson cycle, the ferredoxin–thioredoxin system also reversibly reduces disulfide bonds of proteins that participate in a broad spectrum of chloroplast processes. Although the effect of reduction on many of the enzymes is unclear, the idea that thioredoxin modulates the structure and activity of many chloroplast proteins has been supported by proteomic approaches that have identified additional processes involving thioredoxin. For example, thioredoxin is the reductant that protects against damage caused by reactive oxygen species, such as hydrogen peroxide ($H_2O_2$), the superoxide anion ($O_2^{\bullet-}$), and the hydroxyl radical (OH•) (Buchanan and Balmer 2005; Mora-Garcia et al. 2006; Aran et al. 2009).

The deactivation of target enzymes in the dark appears to take place by reversal of the reduction (activation) pathway. Oxygen or reactive oxygen species transform reduced thioredoxin (—SH HS—) to the oxidized state (—S—S—), which in turn converts the reduced target enzyme to the oxidized state, leading to loss of catalytic activity. Thioredoxins have recently been found to alter enzyme activity by thiol–disulfide exchange in biological processes ranging from seed germination to cancer (Montrichard et al. 2009).

### Light-dependent ion movements modulate enzymes of the Calvin–Benson cycle

Concurrent with the posttranslational modification of chloroplast enzymes, light brings about reversible changes in the ionic composition of the stroma. Upon illumination, the flow of protons from the stroma into the thylakoid lumen is coupled to the release of $Mg^{2+}$ from the intrathylakoid space to the stroma. These ion fluxes decrease the stromal concentration of $H^+$ (the pH increases from 7 to 8) and increase that of $Mg^{2+}$ by 2–5 m$M$. These modifications in ionic composition of the chloroplast stroma are reversed upon darkening.

Several Calvin–Benson cycle enzymes that require $Mg^{2+}$ for catalysis are more active at pH 8 than at pH 7, including rubisco, fructose-1,6-bisphosphatase, sedoheptulose-1,7-bisphosphatase, and phosphoribulokinase. Hence, the light-mediated increase of $Mg^{2+}$ and $H^+$ enhances the activity of key enzymes of the Calvin–Benson cycle (Heldt 1979).

### Light controls the assembly of chloroplast enzymes into supramolecular complexes

The association of enzymes in supramolecular complexes has important effects on the catalytic capacity of constituent enzymes, thereby regulating the activity of metabolic pathways. The use of this strategy in the stroma was confirmed by the isolation of a supramolecular complex of the enzymes phosphoribulokinase and glyceraldehyde-3-phosphate dehydrogenase bound to CP12, a protein (ca. 8.5 kDa) with disulfide groups that hold the complex together. The formation of the supramolecular complex and the catalytic activities of the phosphoribulokinase and glyceraldehyde-3-phosphate dehydrogenase are tightly linked to the oxidation state of the –SH groups of specific cysteines (Marri et al. 2009).

Photosynthetic glyceraldehyde-3-phosphate dehydrogenases from cyanobacteria and most algae are homotetramers ($A_4$) that associate with phosphoribulokinase and CP12 when both the latter proteins contain intramolecular disulfide bonds (**FIGURE 8.7**). However, thioredoxin-mediated cleavage of the disulfide bonds in phosphoribulokinase and CP12 abolishes the ability to form a supramolecular complex with $A_4$-glyceraldehyde-3-phosphate dehydrogenase, thereby imparting full catalytic capacity to both glyceraldehyde-3-phosphate dehydrogenase and phosphoribulokinase. Under the less reducing conditions prevailing in the dark, the low ratio of reduced thioredoxin to oxidized thioredoxin restores disulfide bonds in both phosphoribulokinase and CP12. The oxidized forms of phosphoribulokinase and CP12 then recruit $A_4$-glyceraldehyde-3-phosphate dehydrogenase for the formation of the supramolecular complex, wherein both glyceraldehyde-3-phosphate dehydrogenase and phosphoribulokinase possess only marginal catalytic capacity.

Although land plants have $A_4$-glyceraldehyde-3-phosphate dehydrogenase and the associated regulatory system, their chloroplasts also contain a second form of the enzyme, the heterotetramer $A_2B_2$. The amino acid sequence of the B subunit is homologous to the A subunit but contains a C-terminal extension with two cysteines whose –SH groups are regulated by thioredoxin. When the cysteine residues in the C-terminal region of the B subunit are reduced (in the light), a catalytically functional glyceraldehyde-3-phosphate dehydrogenase is formed. In oxidizing conditions (in the dark), disulfide bonds are generated, thereby promoting the aggregation of $A_2B_2$ tetramers into enzymatically inactive $A_8B_8$ hexadecamers.

In summary, the formation of supramolecular complexes represents a photosynthetic mechanism for reversible modulation of the activity of phosphoribulokinase and glyceraldehyde-3-phosphate dehydrogenase in response to changes in light intensity. During the day, cloud cover and canopy shading change the intensity and quality of the sunlight that reach leaves. As a consequence, the dissociation and reassociation of the supramolecular complex have to occur in a time frame of minutes for the rapid regulation of the associated enzyme activities under natural light regimes. As this model would predict, mutants of cyanobacteria lacking CP12 exhibit impaired growth under light/dark cycles but grow normally under continuous light (Tamoi et al. 2005).

## The C$_2$ Oxidative Photosynthetic Carbon Cycle

As illustrated in Figure 8.4, rubisco has the capacity to catalyze both the carboxylation and oxygenation of ribulose 1,5-bisphosphate (Miziorko and Lorimer 1983). Carboxylation yields two molecules of 3-phosphoglycerate

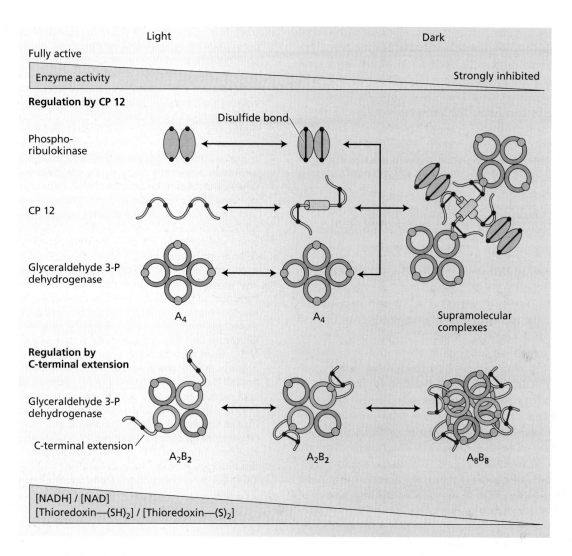

**FIGURE 8.7** Regulation of chloroplast phosphoribulokinase and glyceraldehyde-3-P dehydrogenase. Two forms of glyceraldehyde 3-phosphate dehydrogenase operate in the Calvin–Benson cycle of the chloroplast stroma. Under "dark conditions," phosphoribulokinase and glyceraldehyde 3-phosphate dehydrogenase are not active. The low ratio of reduced thioredoxin to oxidized thioredoxin [thioredoxin–$(SH)_2$] / [thioredoxin–$(S)_2$] promotes formation of disulfide bonds in CP12, phosphoribulokinase, and the C-terminal extension of the $A_2B_2$ form of glyceraldehyde 3-phosphate dehydrogenase. The oxidized states of CP12 and phosphoribulokinase recruit $A_4$-glyceraldehyde 3-phosphate dehydrogenase for the formation of the supramolecular complex CP12– phosphoribulokinase–$A_4$-glyceraldehyde 3-phosphate dehydrogenase, while the heterotetrameric $A_2B_2$-glyceraldehyde 3-phosphate dehydrogenase associates, forming the $A_8B_8$-glyceraldehyde-3-phosphate dehydrogenase. At this stage, the respective catalytic activities are strongly inhibited. Under "light conditions," phosphoribulokinase and glyceraldehyde 3-phosphate dehydrogenase exhibit maximum catalytic capacity. The increase in the ratio [thioredoxin–$(SH)_2$] / [thioredoxin–$(S)_2$] promotes the reduction of disulfide bonds in the enzyme assemblies, causing their dissociation. At this stage, the released (free) enzymes phosphoribulokinase and glyceraldehyde 3-phosphate dehydrogenase are fully active.

while oxygenation ("Catalysis" in Figure 8.5) produces one molecule each of 3-phosphoglycerate and 2-phosphoglycolate ("Products" in Figure 8.5). The oxygenation of ribulose 1,5-bisphosphate catalyzed by rubisco initiates a coordinated network of enzymatic reactions that are compartmentalized in chloroplasts, leaf peroxisomes, and mitochondria (see **WEB TOPIC 8.7**). This process, known as **photorespiration**, causes the partial loss of $CO_2$ fixed by the Calvin–Benson cycle and the concurrent uptake of oxygen in photosynthetically active leaves (Ogren 1984; Leegood et al. 1995; Foyer et al. 2009).

The negative impact of these competing reactions on plant growth has been demonstrated with a variety of photorespiratory mutants of Arabidopsis that exhibit retarded growth, precocious senescence, and cell death at the usual atmospheric $CO_2$ concentration (0.03%) but are normal in a high-$CO_2$ environment (0.3% or more). Moreover, several crops show a dramatic increase in yield when grown in greenhouses with elevated levels of $CO_2$. We will begin this section by describing the important features of the $C_2$ oxidative photosynthetic cycle—the reactions that result in partial recovery of the carbon diverted from the Calvin–Benson cycle by oxygen. Next, we will describe attempts to increase crop biomass through modification of leaf photorespiration.

### The carboxylation and the oxygenation of ribulose 1,5-bisphosphate are competing reactions

As alternative substrates for rubisco, $CO_2$ and $O_2$ compete for reaction with ribulose 1,5-bisphosphate because carboxylation and oxygenation occur within the same active site. The incorporation of one molecule of $O_2$ into the 2,3-enediol isomer of ribulose 1,5-bisphosphate generates an unstable intermediate that rapidly splits into 2-phosphoglycolate and 3-phosphoglycerate (see Figure 8.4, **FIGURE 8.8**, and **TABLE 8.2**, reaction 1). The ability to catalyze the oxygenation of ribulose 1,5-bisphosphate is a property of all rubiscos, regardless of taxonomic origin. Even the rubisco from anaerobic autotrophic bacteria catalyzes the oxygenase reaction in vitro. This notable property of anoxygenic bacteria suggests that the oxygenase reaction is an intrinsic feature of the enzyme active site and not an adaptive mechanism in response to the appearance of oxygen in the Earth's atmosphere.

The 2-phosphoglycolate formed in the chloroplast by oxygenation of ribulose 1,5-bisphosphate is rapidly hydrolyzed to glycolate by a specific chloroplast phosphatase (Table 8.2, reaction 2). The subsequent metabolism of glycolate involves the cooperation of two other organelles: peroxisomes and mitochondria (see Chapter 1) (Tolbert 1981).

Glycolate exits the chloroplast via a specific transporter protein in the inner envelope membrane and diffuses to the peroxisome. There, a flavin mononucleotide–dependent oxidase, glycolate oxidase (Table 8.2, reaction 3), catalyzes the oxidation of glycolate by producing $H_2O_2$ and glyoxylate. Catalase breaks down the $H_2O_2$, releasing $O_2$ (Table 8.2, reaction 4), while the glyoxylate undergoes transamination with glutamate, yielding the amino acid glycine (Table 8.2, reaction 5).

Glycine leaves the peroxisome and enters the mitochondrion (see Figure 8.8), where a multienzyme complex of glycine decarboxylase and serine hydroxymethyltransferase catalyzes the reaction of two molecules of glycine and one of NAD$^+$ to produce one molecule each of serine, NADH, NH$_4^+$, and $CO_2$ (Table 8.2, reactions 6 and 7). This reaction links the reduction of NAD$^+$ to the oxidation of carbon atoms because two molecules of gly-

**FIGURE 8.8** Operation of the $C_2$ oxidative photosynthetic cycle involves cooperation among three organelles: chloroplasts, peroxisomes, and mitochondria. *In chloroplasts*, the oxygenase activity of rubisco yields two molecules of 2-phosphoglycolate, which phosphoglycolate phosphatase converts into two molecules of glycolate and two molecules of inorganic phosphate. Two molecules of glycolate (four carbons) and one molecule of glutamate flow from chloroplasts to peroxisomes. *In peroxisomes*, the glycolate is oxidized by $O_2$ to glyoxylate in a reaction catalyzed by glycolate oxidase. Glyoxylate:glutamate aminotransferase catalyzes the conversion of glyoxylate and glutamate into glycine and 2-oxoglutarate. Glycine flows from peroxisomes to mitochondria, where two molecules of glycine (four carbons) yield a molecule of serine (three carbons) with the concurrent release of $CO_2$ (one carbon) and NH$_4^+$ by the successive action of the glycine decarboxylase complex and serine hydroxymethyl transferase. Serine is then transported back to the peroxisome and transformed into glycerate (three carbons) by the successive action of serine aminotransferase and hydroxypyruvate reductase. Glycerate and 2-oxoglutarate (from peroxisomes) and NH$_4^+$ (from mitochondria) return to chloroplasts in a process that recovers part of the carbon (three carbons) and all the nitrogen lost in photorespiration. Glycerate is phosphorylated to 3-phosphoglycerate and incorporated back into the Calvin–Benson cycle. In the chloroplast stroma, glutamine synthetase and ferredoxin-dependent glutamate synthase (GOGAT)—using the inorganic nitrogen (NH$_4^+$) and 2-oxoglutarate—recover the nitrogen initially lost in the exported glutamate. See Table 8.2 for a description of each numbered reaction.

cine (oxidation states, C1: +3; C2: –1) are converted to serine (oxidation states, C1: +3; C2: 0; C3: –1) and $CO_2$ (oxidation state, C: +4).

The newly formed serine diffuses from the mitochondrion back to the peroxisome (see Figure 8.8), where it is converted by transamination to hydroxypyruvate (Table 8.2, reaction 8), which, in turn, is reduced to glycerate (Table 8.2, reaction 9) via an NADH-dependent reductase. A malate–oxaloacetate shuttle transfers NADH from the cytoplasm to the peroxisome, thus maintaining an adequate concentration of NADH for this reaction. Finally, glycerate reenters the chloroplast, where it is phosphorylated to yield 3-phosphoglycerate (see Figure 8.8 and Table 8.2, reaction 10).

In parallel, the NH$_4^+$ released in the oxidation of glycine diffuses from the matrix of the mitochondrion to chloroplasts, where glutamine synthetase drives its ATP-dependent incorporation into glutamate, yielding glutamine (see Figure 8.8 and Table 8.2, reaction 11). Subsequently, a ferredoxin-dependent glutamate synthase catalyzes a reaction in which glutamine and 2-oxoglutarate react, leading to the production of two molecules of glutamate (Table 8.2, reaction 12).

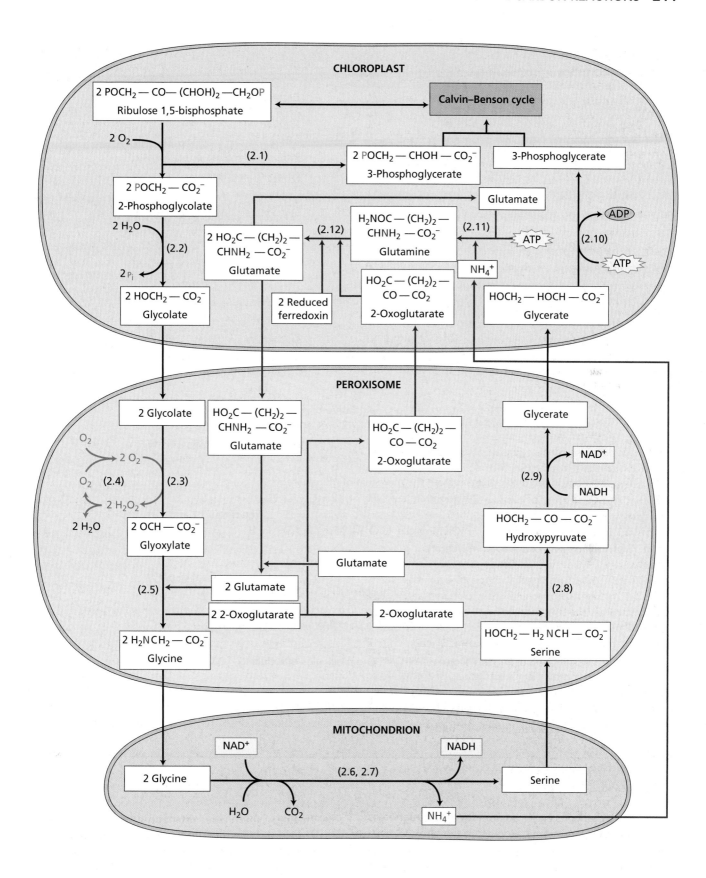

## TABLE 8.2
## Reactions of the $C_2$ oxidative photosynthetic cycle

| Reaction[a] | Enzyme |
|---|---|
| 1. 2 Ribulose 1,5-bisphosphate + 2 $O_2$ → 2 2-phosphoglycolate + 2 3-phosphoglycerate | Rubisco |
| 2. 2 2-Phosphoglycolate + 2 $H_2O$ → 2 glycolate + 2 $P_i$ | Phosphoglycolate phosphatase |
| 3. 2 Glycolate + 2 $O_2$ → 2 glyoxylate + 2 $H_2O_2$ | Glycolate oxidase |
| 4. 2 $H_2O_2$ → 2 $H_2O$ + $O_2$ | Catalase |
| 5. 2 Glyoxylate + 2 glutamate → 2 glycine + 2-oxoglutarate | Glyoxylate:glutamate aminotransferase |
| 6. Glycine + $NAD^+$ + [GDC] → $CO_2$ + $NH_4^+$ + NADH + methylene-[GDC] | Glycine decarboxylase complex (GDC) |
| 7. Methylene-[GDC] + glycine + $H_2O$ → serine + [GDC] | Serine hydroxymethyl transferase |
| 8. Serine + 2-oxoglutarate → hydroxypyruvate + glutamate | Serine aminotransferase |
| 9. Hydroxypyruvate + NADH + $H^+$ → glycerate + $NAD^+$ | Hydroxypyruvate reductase |
| 10. Glycerate + ATP → 3-phosphoglycerate + ADP | Glycerate kinase |
| 11. Glutamate + $NH_4^+$ + ATP → glutamine + ADP + Pi | Glutamine synthetase |
| 12. 2-Oxoglutarate + glutamine + 2 $Fd_{red}$ + 2 $H^+$ → 2 glutamate + 2 $Fd_{oxid}$ | Ferredoxin-dependent glutamate synthase (GOGAT) |

**Net reaction of the $C_2$ oxidative photosynthetic cycle**

2 Ribulose 1,5-bisphosphate + 3 $O_2$ + $H_2O$ + glutamate

↓          **(reactions 1 to 9)**

Glycerate + 2 3-phosphoglycerate + $NH_4^+$ + $CO_2$ + 2 $P_i$ + 2-oxoglutarate

Two reactions in the chloroplasts restore the molecule of glutamate:

2-Oxoglutarate + $NH_4^+$ + [(2 $Fd_{red}$ + 2 $H^+$), ATP]

↓          **(reactions 11 and 12)**

Glutamate + $H_2O$ + [(2 $Fd_{oxid}$), ADP + Pi]

and the molecule of 3-phosphoglycerate:

Glycerate + ATP

↓          **(reaction 10)**

3-Phosphoglycerate + ADP

Hence, the consumption of 3 molecules of atmospheric oxygen in the $C_2$ oxidative photosynthetic cycle (two in the oxygenase activity of Rubisco and one in peroxisomal oxidations) elicits

- the release of one molecule of $CO_2$ and
- the consumption of two molecules of ATP and two molecules of reducing equivalents (2 $Fd_{red}$ + 2 $H^+$)

for

- incorporating a 3-carbon skeleton back into the Calvin–Benson Cycle, and
- restoring glutamate from $NH_4^+$ and α-ketoglutarate.

[a]Locations: chloroplast; peroxisome; mitochondrion.

Three simultaneous cycles account for the circulation of carbon, nitrogen, and oxygen atoms through photorespiration (**FIGURE 8.9**).

- In the *first* cycle, carbon atoms exit the chloroplast as two molecules of glycolate and return as one molecule of glycerate, leaving a molecule of $CO_2$ in the mitochondrion.

- In the *second* cycle, nitrogen atoms leave the chloroplast as one molecule of glutamate and return as one molecule of $NH_4^+$, ultimately associated with one molecule of 2-oxoglutarate (see Figure 8.8). As a consequence, total nitrogen remains unchanged because the formation of inorganic nitrogen ($NH_4^+$) in the mitochondrion

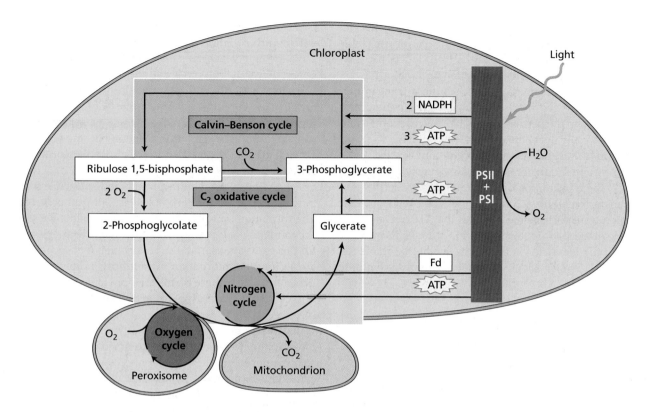

**FIGURE 8.9** Dependence of the $C_2$ oxidative photosynthetic cycle on chloroplast metabolism. ATP and reducing equivalents from light reactions in thylakoid membranes are needed for the functioning of the $C_2$ oxidative photosynthetic cycle in three compartments: chloroplasts, peroxisomes, and mitochondria. The "carbon cycle" uses (1) NADPH and ATP to maintain an adequate level of ribulose 1,5-bisphosphate in the Calvin–Benson cycle, and (2) ATP to convert glycerate to 3-phosphoglycerate in the $C_2$ oxidative photosynthetic cycle. The "nitrogen cycle" employs ATP and reducing equivalents to restore glutamate from $NH_4^+$ and 2-oxoglutarate coming from the photorespiratory cycle. In the peroxisome, the "oxygen cycle" contributes to the removal of $H_2O_2$ formed in the oxidation of glycolate by $O_2$.

is balanced by the synthesis of glutamine in the chloroplast.

- In the *third* ($O_2$-linked) cycle, the activities of rubisco and glycolate oxidase incorporate two molecules of $O_2$ each (Table 8.2, reactions 1 and 3), but catalase releases one molecule of $O_2$ from $H_2O_2$ (Table 8.2, reaction 4). Thus, a total of three molecules of $O_2$ are reduced when two molecules of ribulose 1,5-bisphosphate enter the $C_2$ oxidative photosynthetic cycle.

In vivo, the balance between the Calvin–Benson and the $C_2$ oxidative photosynthetic cycles is determined mainly by three factors: one inherent to the plant (the kinetic properties of rubisco) and two linked to the environment (tem-

perature and the concentration of substrates, $CO_2$ and $O_2$). An increase in the external temperature

- modifies the kinetic constants of rubisco, increasing the rate of oxygenation more than that of carboxylation (Ku and Edwards 1978), and
- lowers the concentration of $CO_2$ more than that of $O_2$ in a solution in equilibrium with air (see **WEB TOPIC 8.8**).

Hence, the increase in photorespiration (oxygenation) relative to photosynthesis (carboxylation) significantly limits the efficiency of photosynthetic carbon assimilation under warmer temperatures. Overall, a progressive increase in temperature tilts the balance away from the Calvin–Benson cycle and toward the $C_2$ oxidative photosynthetic cycle (see Chapter 9).

## Photorespiration depends on the photosynthetic electron transport system

Photosynthetic carbon metabolism in intact leaves reflects the integrated balance between two mutually opposed cycles, the Calvin–Benson cycle and the $C_2$ oxidative photosynthetic cycle. These are interlocked with the photosynthetic electron transport system for the supply of ATP and reducing equivalents (reduced ferredoxin) (see Figure 8.9). For salvaging 2-phosphoglycolate by conversion to 3-phosphoglycerate, photophosphorylation provides the ATP necessary for the transformation of glycerate to

3-phosphoglycerate (Table 8.2, reaction 10), while the consumption of NADH by hydroxypyruvate reductase (Table 8.2, reaction 9) is counterbalanced by its production by glycine decarboxylase (Table 8.2, reaction 6).

Nitrogen enters the $C_2$ cycle through a transamination step catalyzed by the glyoxylate:glutamate aminotransferase in the peroxisome (Table 8.2, reaction 5), and leaves the cycle in (1) the conversion of glycine to serine catalyzed by the glycine decarboxylase complex and serine hydroxymethyl transferase (Table 8.2, reactions 6 and 7), and (2) the transamination step catalyzed by serine aminotransferase (Table 8.2, reaction 8). The photosynthetic electron transport system supplies the ATP and reduced ferredoxin needed for salvaging $NH_4^+$ through its incorporation into glutamate via glutamine synthetase (Table 8.2, reaction 11) and ferredoxin-dependent glutamate synthase (GOGAT) (Table 8.2, reaction 12), respectively.

In summary,

$$2 \text{ Ribulose 1,5-bisphosphate} + 3 \text{ O}_2 + \text{H}_2\text{O} + \text{ATP} + \\ [2 \text{ Fd}_{red} + 2 \text{ H}^+ + \text{ATP}]$$

$$\downarrow$$

$$3 \text{ 3-Phosphoglycerate} + \text{CO}_2 + 2 \text{ Pi} + \text{ADP} + \\ [2 \text{ Fd}_{oxid} + \text{ADP} + \text{Pi}]$$

Overall, two molecules of phosphoglycolate (four carbons), which are lost from the Calvin–Benson cycle by the oxygenation of ribulose 1,5-bisphosphate, require the hydrolysis of one ATP molecule for their conversion into one molecule of 3-phosphoglycerate (three carbons) and one $CO_2$ (one carbon) (Lorimer 1981; Sharkey 1988). In addition, one molecule of ATP and two molecules of reduced ferredoxin are necessary for the total reassimilation of the ammonium molecule released during photorespiration. Because of this additional provision of ATP and reducing power, the quantum requirement for $CO_2$ fixation under photorespiratory conditions (air with high $O_2$ and low $CO_2$) is higher than it is under nonphotorespiratory conditions (low $O_2$ and high $CO_2$).

## Photorespiration protects the photosynthetic apparatus under stress conditions

Photorespiration functions alongside the Calvin–Benson cycle and contributes to a wide range of chloroplastic processes, from bioenergetics to carbon metabolism and nitrogen assimilation. However, the adapative value of this pathway is a matter of debate among plant physiologists. One line of reasoning is that it promotes the recovery of carbon that would otherwise be lost in the form of 2-phosphoglycolate. The $C_2$ oxidative photosynthetic cycle recovers as glycerate 75% of the carbon originally lost as 2-phosphoglycolate from the Calvin–Benson cycle.

The concentrations of $CO_2$ and $O_2$ at the active site of rubisco and the intrinsic affinity of the enzyme for these gases determine the relative rates of ribulose 1,5-bisphosphate carboxylation and oxygenation. The formation of 2-phosphoglycolate is the unavoidable consequence of the generation at the rubisco active site of an intermediate that is reactive not only with $CO_2$ but also with $O_2$ (see Figure 8.4). In early evolutionary times, when $CO_2$:$O_2$ ratios were higher, the oxygenation reaction would have had little relevance. But the oxygenation reaction became significant when levels of $O_2$ increased in the atmosphere. As a consequence, the low $CO_2$:$O_2$ ratio prevalent in the current atmosphere makes photorespiration indispensable for the partial recovery of the carbon lost as 2-phosphoglycolate. The formation of 3-phosphoglycerate prevents the lethal accumulation of glycerate in dead-end pathways. Support for this view comes from the observation that such glycerate accumulation occurs in Arabidopsis mutants lacking glycerate kinase; these plants are nonviable in normal air, but are viable in atmospheres with elevated levels of $CO_2$ (Boldt et al. 2005).

It can also be argued that evolution has acted on photorespiration to minimize not only the loss of carbon, but also the photoinhibition of the photosynthetic apparatus caused by excess reductant formed in chloroplasts under stress conditions—for example, high illumination, high temperature, and $CO_2$ or water deficit. When electron acceptors become limiting, the photosynthetic apparatus uses alternative substrates to remove the surplus of reducing power generated by electron flow. Unlike the other various mechanisms available to alleviate the damage caused by photoinhibition, the $C_2$ oxidative photosynthetic cycle dissipates excess reducing equivalents, thereby preventing over-reduction of the photosynthetic electron chain. Thus, Arabidopsis mutants devoid of photorespiratory enzymes grow normally at high levels of $CO_2$ (2%), but they die rapidly when transferred to normal air (0.03% $CO_2$).

An important development in this context was recognition that the major source of $H_2O_2$ in photosynthetic cells is photorespiration—specifically the glycolate oxidase reaction. Although this reactive oxygen species is considered toxic, $H_2O_2$ also acts as a signaling molecule linked to hormone and stress responses (Foyer et al. 2009). Many investigators have observed that the $H_2O_2$ produced by photorespiration not only triggers programmed cell death when the intracellular level exceeds the capacity of antioxidant enzymes such as catalases and peroxidases, but also regulates cellular redox homeostasis.

## Photorespiration may be engineered to increase the production of biomass

Solutions to the current food and energy shortages depend to a large extent on the degree to which land plants can be adapted to enhance photosynthetic $CO_2$ assimilation. Efficient ways to increase biomass, therefore, are being sought to feed and fuel the world. When $O_2$ outcompetes $CO_2$ at

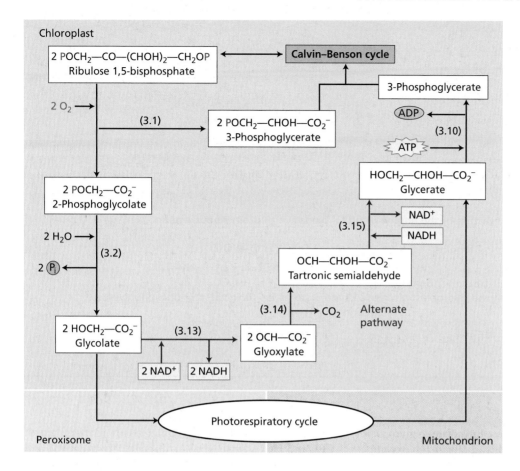

**FIGURE 8.10**  The recovery of photorespiratory glycolate by the *E. coli* glycolate catabolic pathway engineered into the chloroplasts of land plants. Upon illumination of transgenic plants, the oxygenase activity of rubisco yields 2-phosphoglycolate, whose phosphoryl moiety is hydrolyzed to generate glycolate. At this stage, the *E. coli* glycolate catabolic pathway introduced into the chloroplasts of transgenic *Arabidopsis* lines diverts glycolate from photorespiration to an alternate pathway (denoted by red boxes and arrows). First, glycolate is oxidized to glyoxylate by glycolate dehydrogenase. The bacterial glyoxylate carboxyligase then catalyzes the condensation of two molecules of glyoxylate, yielding one molecule of tartronic semialdehyde with the concurrent release of one molecule of $CO_2$. Finally, tartronic semialdehyde is reduced by tartronic semialdehyde reductase to generate glycerate, which proceeds to the Calvin–Benson cycle through the $C_2$ oxidative cycle. See Table 8.3 for a description of each numbered reaction.

the rubisco active site, the oxygenase activity of rubisco lowers the amount of carbon that enters the Calvin–Benson cycle. On this basis, an enhancement of the photosynthetic efficiency or a reduction of photorespiration should improve yields of crops such as wheat, rice, and cotton (Khan 2007). To learn how to genetically engineer leaf cells for improved production of biomass, scientists are attempting various alterations of the $C_2$ oxidative photosynthetic cycle, from modification of the active site of rubisco to the introduction of new, parallel photorespiratory pathways.

Despite considerable effort, attempts to reduce photorespiration by modifying the performance of rubisco have not met with success. Alternatively, mutants of the photorespiratory cycle have contributed to our understanding of the biochemical pathway, but these plants generally are stunted and exhibit conditionally lethal phenotypes when grown under atmospheric $CO_2$ (0.03%) (Reumann

and Weber 2006). Interestingly, some of these mutants, although stunted, are viable under conditions that suppress photorespiration, such as elevated concentrations of $CO_2$ (0.1–0.3 %).

Because the $C_2$ oxidative photosynthetic cycle appears to be essential for land plants, an alternative approach to improving photosynthetic efficiency is to introduce a complementary mechanism for retrieving the carbon atoms of 2-phosphoglycolate for use by the Calvin–Benson cycle. This promising approach would allow the plant to retain a fully functional photorespiratory pathway for other essential processes. The successful incorporation of a bacterial (*E. coli*) glycolate catabolic pathway into chloroplasts of land plants (Arabidopsis) supports this possibility (Kebeish et al. 2007) (**FIGURE 8.10**). Chloroplasts of these transgenic plants have an entirely functional $C_2$ oxidative photosynthetic cycle while additionally accommodating the bacterial

## TABLE 8.3
### Reactions of the engineered $C_2$ oxidative photosynthetic cycle

| Reaction[a] | Enzyme |
| --- | --- |
| 1. 2 Ribulose 1,5-bisphosphate + 2 $O_2$ → <br>　　2 2-phosphoglycolate + 2 3-phosphoglycerate | Rubisco |
| 2. 2 2-Phosphoglycolate + 2 $H_2O$ → 2 glycolate + 2 $P_i$ | Phosphoglycolate phosphatase |
| 13. 2 Glycolate + 2 $NAD^+$ → 2 glyoxylate + 2 NADH + 2 $H^+$ | Glycolate dehydrogenase |
| 14. 2 Glyoxylate + $H^+$ → tartronic semialdehyde + $CO_2$ | Glyoxylate carboligase |
| 15. Tartronic semialdehyde + NADH + $H^+$ → glycerate + $NAD^+$ | Tartronic semialdehyde reductase |
| 10. Glycerate + ATP → 3-phosphoglycerate + ADP | Glycerate kinase |

**Net reaction of the engineeered $C_2$ oxidative photosynthetic cycle**

2 Ribulose 1,5-bisphosphate + 2 $O_2$ + 2 $H_2O$ + $NAD^+$

↓　(reactions 1, 2, 13, 14, and 15)

Glycerate + 2 3-phosphoglycerate + NADH + $H^+$ + $CO_2$ + 2 Pi

The reaction catalyzed by the chloroplast glycerate kinase restores the molecule of 3-phosphoglycerate:

Glycerate + ATP

↓　(reaction 10)

3-Phosphoglycerate + ADP

[a]Locations: chloroplast

enzymes glycolate dehydrogenase [glycolate → glyoxylate] (Table 8.3, reaction 13), glyoxylate carboligase [glyoxylate → tartronic semialdehyde + $CO_2$] (TABLE 8.3, reaction 14), and tartronic semialdehyde reductase [tartronic semialdehyde → glycerate] (Table 8.3, reaction 15).

The integration of the bacterial glycolate pathway into chloroplasts of land plants lowers the flux of photorespiratory metabolites through peroxisomes and mitochondria. Although the engineered glycolate shunt is a photorespiratory pathway, in that it depends on the oxygenase activity of rubisco and evolves $CO_2$, it differs from the endogenous counterpart in sidestepping the mitochondrial and peroxisomal photorespiratory reactions. As a consequence, the shift of glycolate from the plant to the bacterial metabolic pathway avoids the use of the energy (one ATP and two reduced ferredoxins) required to recover ammonia via the successive action of glutamine synthetase (Table 8.2, reaction 11) and glutamate synthase (Table 8.2, reaction 12). Notably, consistent with the lower energy requirement and the increased recovery of carbon atoms in the chloroplast stroma, the engineered plants grow faster, have increased biomass, and contain higher levels of soluble sugars.

## Inorganic Carbon–Concentrating Mechanisms

The pronounced reduction in $CO_2$ and rise in $O_2$ levels that commenced about 350 million years ago triggered a series of adaptations to handle an environment that promoted photorespiration in photosynthetic organisms. These adaptations include various strategies for active uptake of $CO_2$ and $HCO_3^-$ from the surrounding environment and subsequent accumulation of inorganic carbon near rubisco (Giordano et al. 2005). The immediate consequence of $CO_2$-concentrating mechanisms that increase the levels of $CO_2$ around rubisco is a decrease in the oxygenation reaction. $CO_2$ and $HCO_3^-$ pumps at the plasma membrane have been studied extensively in prokaryotic cyanobacteria, eukaryotic algae, and aquatic plants (see WEB TOPIC 8.1).

The following sections focus on two different mechanisms developed by land plants for concentrating $CO_2$ at the rubisco carboxylation site:

- $C_4$ photosynthetic carbon fixation ($C_4$)
- Crassulacean acid metabolism (CAM)

These two mechanisms are found in some angiosperms and involve "add-ons" to the Calvin–Benson cycle that separate the uptake of atmospheric $CO_2$ from the supply of the substrate to rubisco.

## Inorganic Carbon–Concentrating Mechanisms: The $C_4$ Carbon Cycle

To minimize the oxygenase activity of rubisco and the concurrent loss of carbon through the photorespiratory cycle, $C_4$ photosynthesis appears to have evolved as one of the major carbon-concentrating mechanisms used by land plants to compensate for limitations associated with the low level of atmospheric $CO_2$. In this section we examine the biochemical features of $C_4$ photosynthesis, the

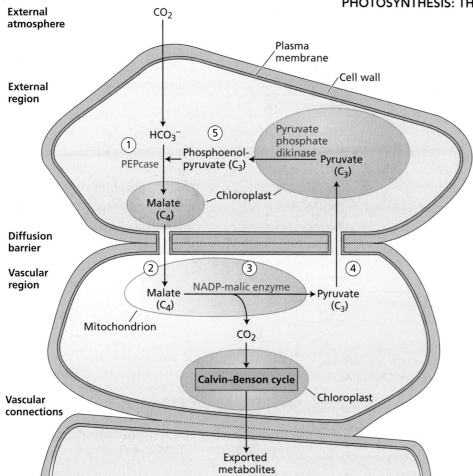

**External atmosphere**

**External region**

**Diffusion barrier**

**Vascular region**

**Vascular connections**

$CO_2$

Plasma membrane

Cell wall

$HCO_3^-$ (5)

(1)

PEPcase

Phosphoenol-pyruvate ($C_3$)

Pyruvate phosphate dikinase

Pyruvate ($C_3$)

Malate ($C_4$)

Chloroplast

(2) (3) (4)

Malate ($C_4$)

NADP-malic enzyme

Pyruvate ($C_3$)

Mitochondrion

$CO_2$

**Calvin–Benson cycle**

Chloroplast

Exported metabolites

**FIGURE 8.11** The $C_4$ photosynthetic carbon cycle involves five successive stages in two different compartments. (1) In the periphery of leaf cells (external region), PEPCase catalyzes the reaction of $HCO_3^-$ with phosphoenolpyruvate (PEP), a 3-carbon compound. The 4-carbon reaction product, oxaloacetate, is converted into malate or aspartate (depending on the species) by NADP-malate dehydrogenase or aspartate aminotransferase, respectively (see Table 8.4). For simplicity, malate is shown here. (2) The 4-carbon acid flows across a diffusion barrier to the vascular region. (3) A decarboxylating enzyme (e.g., NAD–malic enzyme) releases the $CO_2$ from the 4-carbon acid, yielding a 3-carbon acid (e.g., pyruvate). The uptake of the released $CO_2$ by the chloroplasts in the vascular region builds up a large excess of $CO_2$ relative to $O_2$ around rubisco, thereby facilitating the assimilation of $CO_2$ through the Calvin–Benson cycle. (4) The residual 3-carbon acid flows back to the region in contact with the external atmosphere. (5) Closing the $C_4$ cycle, the enzyme pyruvate–phosphate dikinase catalyzes the regeneration of PEP, the acceptor of the $HCO_3^-$, for another turn of the cycle. The consumption of two molecules of ATP per mole of fixed $CO_2$ (see Table 8.4, reactions 7 and 8) drives the $C_4$ cycle in the direction of the arrows, thus pumping $CO_2$ from the atmosphere to the Calvin–Benson cycle. The assimilated carbon leaves the chloroplast and, after being converted to sucrose in the cytoplasm, enters the phloem for translocation to other parts of the plant.

concerted action of different types of cells for the incorporation of inorganic carbon into carbon skeletons, the light-mediated modulation of enzyme activities, and the importance of this mechanism for sustaining plant growth in many tropical areas.

## Malate and aspartate are carboxylation products of the $C_4$ cycle

In the late 1950s, H. P. Kortschack and Y. Karpilov observed early labeling of 4-carbon acids when $^{14}CO_2$ was provided to sugarcane and maize. After leaves were exposed to $^{14}CO_2$ for a few seconds in the light, 70 to 80% of the label was found in the 4-carbon acids malate and aspartate—a pattern very different from the one observed in leaves that photosynthesize solely via the Calvin–Benson cycle.

In pursuing these initial observations, M. D. Hatch and C. R. Slack elucidated what is now named the $C_4$ *photosynthetic carbon cycle* (also known as the Hatch–Slack cycle or the $C_4$ cycle). They established that malate and aspartate are the first stable, detectable intermediates of photosynthesis in leaves of sugarcane, and that carbon 4 of malate subsequently becomes carbon 1 of 3-phosphoglycerate (Hatch and Slack 1966). The latter transformation became clear when they found that this novel metabolic pathway

takes place in two morphologically distinct cell types—the mesophyll and bundle sheath cells—that are separated by their respective membranes ("Diffusion barrier," **FIGURE 8.11**). In the $C_4$ cycle, the enzyme phosphoenolpyruvate carboxylase (PEPCase), rather than rubisco, catalyzes the primary carboxylation in a tissue that is close to the external atmosphere (**TABLE 8.4**, reaction 1) (see **WEB ESSAY 8.1**) (Sage 2004). The resulting 4-carbon acid flows across the diffusion barrier to the vascular region, where it is decarboxylated, releasing $CO_2$ that is refixed by rubisco via the Calvin–Benson cycle. Although all $C_4$ plants share primary carboxylation via PEPCase, the specific paths by

**TABLE 8.4**
**Reactions of the $C_2$ oxidative photosynthetic carbon cycle**

| Reaction | Enzyme |
|---|---|
| 1. PEPCase | Phosphoenolpyruvate + $HCO_3^- \rightarrow$ oxaloacetate + $P_i$ |
| 2. NADP–malate dehydrogenase | Oxaloacetate + NADPH + $H^+ \rightarrow$ malate + $NADP^+$ |
| 3. Aspartate aminotransferase | Oxaloacetate + glutamate $\leftrightarrow$ aspartate + 2-oxoglutarate |
| 4. NAD(P)–malic enzyme | Malate + $NAD(P)^+ \rightarrow$ pyruvate + $CO_2$ + $NAD(P)H + H^+$ |
| 5. Phosphoenolpyruvate carboxykinase | Oxaloacetate + ATP $\rightarrow$ phosphoenolpyruvate + $CO_2$ + ADP |
| 6. Alanine aminotransferase | Pyruvate + glutamate $\leftrightarrow$ alanine + 2-oxoglutarate |
| 7. Pyruvate–phosphate dikinase | Pyruvate + $P_i$ + ATP $\rightarrow$ phosphoenolpyruvate + AMP + $PP_i$ |
| 8. Adenylate kinase | AMP + ATP $\rightarrow$ 2 ADP |
| 9. Pyrophosphatase | $PP_i + H_2O \rightarrow 2 P_i$ |

*Note*: $P_i$ and $PP_i$ stand for inorganic phosphate and pyrophosphate, respectively.

which $CO_2$ is concentrated in the vicinity of rubisco vary substantially between different $C_4$ species (see **WEB TOPIC 8.9**).

Since the seminal studies of the 1950s and the 1960s, the $C_4$ cycle has been associated with a particular leaf structure, called **Kranz anatomy** (*Kranz*, German for "wreath"), that exhibits an inner ring of bundle sheath cells around the vascular tissues and an outer layer of mesophyll cells in close contact with the epidermis. This particular leaf anatomy ensures the compartmentalization of enzymes essential for the function of the $C_4$ pathway in two different types of cells. However, there are now clear examples of single-cell $C_4$ photosynthesis in a number of green algae, diatoms, and aquatic and land plants (Edwards et al. 2004; Muhaidat et al. 2007) (**FIGURE 8.12A**) (see **WEB TOPIC 8.10**). Thus, the shuttling of metabolites between the two compartments that is essential for the operation of the $C_4$ cycle is driven by diffusion gradients not only *between* but also *within* the cells.

## Two different types of cells participate in the $C_4$ cycle

The key features of the $C_4$ cycle were initially found in leaves of plants whose vascular tissues are surrounded by two distinctive photosynthetic cell types: an internal ring of **bundle sheath cells**, which is wrapped with an outer ring of **mesophyll cells**. The chloroplasts in bundle sheath cells are concentrically arranged and exhibit large starch granules and unstacked thylakoid membranes. On the other hand, mesophyll cells contain randomly arranged chloroplasts with stacked thylakoids and little or no starch. In this anatomical context, the transport of $CO_2$ from the external atmosphere to the bundle sheath cells proceeds through five successive stages (**FIGURE 8.12B** and Table 8.4):

1. Fixation of the $HCO_3^-$ through the carboxylation of phosphoenolpyruvate catalyzed by PEPCase in the mesophyll cells (Table 8.4, reaction 1). The reaction product, oxaloacetate, is subsequently reduced to malate by NADP–malate dehydrogenase (Table 8.4, reaction 2) or converted to aspartate by transamination with glutamate (Table 8.4, reaction 3).

2. Transport of the 4-carbon acids (malate or aspartate) to bundle sheath cells that surround the vascular bundles.

3. Decarboxylation of the 4-carbon acids and generation of $CO_2$, which is then reduced to carbohydrate via the Calvin–Benson cycle. Prior to this reaction, an aspartate aminotransferase catalyzes the conversion of aspartate back to oxaloacetate in some $C_4$ plants (Table 8.4, reaction 3). Different subtypes of $C_4$ plants make use of different decarboxylases to release $CO_2$ for the effective suppression of the oxygenase reaction of rubisco (Table 8.4, reactions 4 and 5) (see **WEB TOPIC 8.9**).

4. Transport of the 3-carbon backbone (pyruvate or alanine) formed by the decarboxylation step back to the mesophyll cells.

5. Regeneration of the $HCO_3^-$ acceptor. Pyruvate, the residual 3-carbon acid, is converted to phosphoenolpyruvate in the reaction catalyzed by pyruvate–phosphate dikinase (Table 8.4, reaction 7). At this stage, an additional molecule of ATP is required for the transformation of AMP to ADP catalyzed by adenylate kinase (see Table 8.4, reaction 8). When alanine is the 3-carbon compound exported by the bundle sheath cells, the formation of pyruvate by the action of alanine aminotransferase precedes phosphorylation by pyruvate–phosphate dikinase (Table 8.4, reaction 6).

(A)

Kranz anatomy

Single-cell $C_4$ cycle

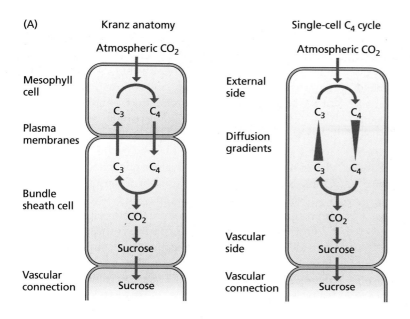

**FIGURE 8.12** The $C_4$ photosynthetic pathway in leaves. (A) In almost all known $C_4$ species, photosynthetic $CO_2$ assimilation requires the development of Kranz anatomy (left panel). This anatomical feature compartmentalizes photosynthetic reactions in two distinct types of cells—mesophyll cells and bundle sheath cells—that are arranged concentrically around the veins. Bundle sheath cells surround the vascular tissue, while an outer ring of mesophyll cells is peripheral to the bundle sheath and adjacent to intercellular spaces. Membranes that separate cells assigned to $CO_2$ fixation from cells destined to reduce carbon are essential for the efficient functioning of $C_4$ photosynthesis in land plants. A few land plants, typified by *Borszczowia aralocaspica* and *Bienertia cycloptera*, contain the equivalents of the $C_4$ compartmentalization in a single cell (right panel). Studies of the key photosynthetic enzymes of these plants also reveal two, dimorphic types of chloroplasts, which are located in different cytoplasmic compartments and have functions analogous to the mesophyll and bundle sheath cells in Kranz anatomy. (B) Kranz anatomy in a $C_4$ dicot, *Flaveria australasica* (family Asteraceae). (740×) (C) Single-cell $C_4$ photosynthesis. Diagrams of the $C_4$ cycle are superimposed on electron micrographs of *Borszczowia aralocaspica* (left) and *Bienertia cycloptera* (right). (B courtesy of Athena McKown; C from Edwards et al. 2004.)

(B) Kranz anatomy

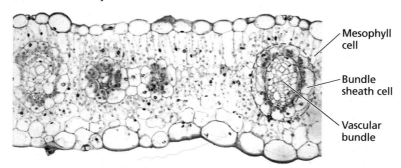

Mesophyll cell

Bundle sheath cell

Vascular bundle

(C) Single cell

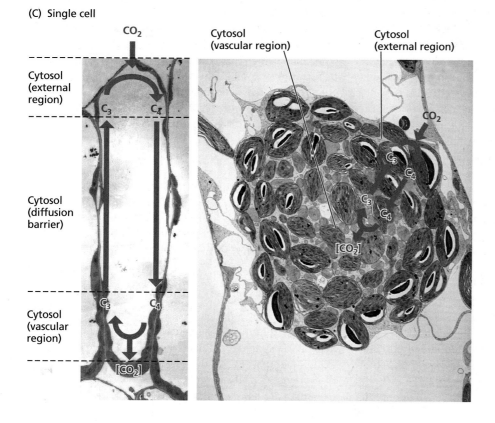

The compartmentalization of enzymes ensures that inorganic carbon from the surrounding atmosphere can be taken up initially by mesophyll cells, fixed subsequently by the Calvin–Benson cycle of bundle sheath cells, and finally exported to the phloem (see Figure 8.11). Because enzymes associated with the $C_4$ cycle are expressed abundantly and specifically in these cells, the lack of an effective mechanism for the fixation of $CO_2$ in mesophyll cells markedly enhances photorespiration in a $C_4$ plant, as revealed by a PEPCase-deficient mutant of *Amaranthus edulis* (Dever et al. 1996).

### The $C_4$ cycle concentrates $CO_2$ in the chloroplasts of bundle sheath cells

Originally described for tropical grasses, the $C_4$ cycle is now known to occur in 18 families of both monocots and dicots and is particularly prominent in the Poaceae (corn, millet, sorghum, sugarcane), Chenopodiaceae (*Atriplex* spp.), and Cyperaceae (many genera) (Edwards and Walker 1983). In all cases, the operation of the $C_4$ cycle requires the cooperative effort of the two distinct chloroplast-containing cell types discussed above: mesophyll and bundle sheath cells arranged in the distinctive Kranz anatomy (see Figure 8.12B). The transport process facilitated by plasmodesmata connecting the two cell types generates a much higher concentration of $CO_2$ in bundle sheath cells (the vascular region) than in mesophyll cells (the "external" region) (see Figure 8.11). This elevated concentration of $CO_2$ at the carboxylation site of rubisco results in the suppression of ribulose 1,5-bisphosphate oxygenation and hence of photorespiration.

In addition to their anatomical features, these cells show large differences in enzyme composition. PEPCase and rubisco are located in mesophyll and bundle sheath cells, respectively, while the decarboxylases are found in various intracellular compartments of bundle sheath cells: NADP–malic enzyme in chloroplasts (Table 8.4, reaction 4), NAD–malic enzyme in mitochondria (Table 8.4, reaction 4), and phosphoenolpyruvate carboxykinase in the cytosol (Table 8.4, reaction 5). Starch, the transitory storage product of photosynthesis (see below), accumulates predominantly in the chloroplasts of the bundle sheath cells.

The chloroplasts in mesophyll and bundle sheath cells of $C_4$ plants also exhibit other anatomical and biochemical differences. PSII activity and the accompanying linear electron flow to PSI, which yield both NADPH and ATP, are generally found in chloroplasts containing extensive grana, whereas PSI-mediated cyclic electron flow, which produces mostly ATP, is present in chloroplasts with few grana. These two types of chloroplasts correlate with the energy requirements of $C_4$ photosynthesis. $C_4$ species that are predominantly malate formers require more reducing power for the conversion of oxaloacetate than species that are aspartate formers (compare reactions 2 and 3 in Table 8.4). As a consequence, $C_4$ species of the NADP–malic

enzyme type, in which malate is shuttled from mesophyll to bundle sheath cells, exhibit well-developed grana in mesophyll chloroplasts, whereas the bundle sheath counterparts are deficient in grana. By contrast, in $C_4$ species of the NAD–malic enzyme type, the primary product of $CO_2$ fixation in mesophyll cells is aspartate, which requires less reducing power and therefore less linear electron flow to PSI. As the ATP produced by PSI cyclic electron flow drives the conversion of pyruvate to phosphoenolpyruvate (Table 8.4, reaction 7), chloroplasts of mesophyll cells in $C_4$ species of the NAD–malic enzyme type exhibit fewer grana than chloroplasts of bundle sheath cells.

Our understanding of $C_4$ photosynthesis advanced considerably with the application of recombinant DNA technology to plant metabolism. The use of transgenic plants with altered ratios of PEPCase to rubisco provided an opportunity to test whether the relative levels of these enzymes determine the efficiency with which $C_4$ photosynthesis operates. Antisense suppression of rubisco in *Flaveria* reduced its content in leaves without a concurrent effect on PEPCase (Siebke et al. 1997; von Caemmerer et al. 1997). Although these transgenic plants exhibited higher concentrations of $CO_2$ in the bundle sheath cells, the increased $CO_2$ was not retained, and photosynthetic efficiency declined accordingly. By contrast, high expression of tobacco carbonic anhydrase in bundle sheath cells caused leakage of $HCO_3^-$ to mesophyll cells through plasmodesmata and saturation of photosynthesis at low light, indicating that a high resistance to $CO_2$ diffusion out of the bundle sheath is essential for efficient $C_4$ photosynthesis (Ludwig et al. 1998).

As described in Chapter 1, chloroplasts of green plants are separated from the cytosol by the two membranes of the chloroplast envelope, which hold a large range of specific transporters for metabolites. The adaptation of chloroplast envelopes to the requirements of $C_4$ photosynthesis was revealed when chloroplast membranes from mesophyll cells of pea (a $C_3$ plant) and maize (a $C_4$ plant) were analyzed by liquid chromatography and tandem mass spectroscopy (Brautigam et al. 2008). Chloroplasts from mesophyll cells of $C_3$ and $C_4$ plants exhibit qualitatively similar but quantitatively different proteomes in their envelope membranes. In particular, translocators that participate in the transport of triose phosphates and phosphoenolpyruvate are more abundant in the envelopes of $C_4$ plants than in those of $C_3$ plants. This higher abundance ensures that fluxes of metabolic intermediates across the chloroplast envelope in $C_4$ plants are higher than in $C_3$ plants.

The existence of $C_4$-like photosynthesis in cells surrounding the vascular bundles of some $C_3$ plants suggests that only a few modifications are needed for $C_4$ photosynthesis to evolve from the $C_3$ pathway. Some $C_3$ plants (e.g., tobacco) exhibit features of $C_4$ photosynthesis in cells that surround the xylem and phloem, and use carbon supplied as malate from the vascular system (Hibberd and Quick 2002). Moreover, the presence of enzymes specific for the $C_4$ pathway (NAD[P]–malic enzyme, phosphoenolpyruvate carboxykinase) in these cells indicates that essential biochemical components for $C_4$ photosynthesis are already present in $C_3$ plants. These findings are consistent with the view that $C_4$ photosynthesis evolved from $C_3$ metabolism by changing the regulation and tissue specificity of key enzymes.

### The $C_4$ cycle also concentrates $CO_2$ in single cells

The finding of $C_4$ photosynthesis in organisms devoid of Kranz anatomy disclosed a much greater diversity in modes of $C_4$ carbon fixation than had previously been thought to exist (see WEB TOPIC 8.10). Two plants that grow in Asia—*Borszczowia aralocaspica* and *Bienertia cycloptera*—can perform complete $C_4$ photosynthesis within single chlorenchyma cells (Sage 2002; Edwards et al. 2004) (see Figure 8.12B). Diatoms, which are unicellular marine protists, also accomplish $C_4$ photosynthesis within a single cell.

### Light regulates the activity of key $C_4$ enzymes

In addition to supplying ATP and NADPH, light is essential for the regulation of several enzymes of the $C_4$ cycle. Thus, variations in photon flux density elicit changes in the activities of NADP–malate dehydrogenase, PEP-Case, and pyruvate–phosphate dikinase by two different mechanisms: thiol–disulfide exchange [Enz–(Cys-S)$_2$ ↔ Enz–(Cys-SH)$_2$] and phosphorylation–dephosphorylation of specific amino acid residues [e.g., serine, Enz–Ser-OH ↔ Enz–Ser-OP].

NADP–malate dehydrogenase is regulated via the classical ferredoxin–thioredoxin system (see Figure 8.6). The enzyme is reduced (activated) by thioredoxin when leaves are illuminated, but is oxidized (deactivated) in the dark. On the other hand, the diurnal phosphorylation of PEP-Case by a specific serine–threonine kinase, named PEP-Case kinase, increases the uptake of ambient $CO_2$ (see WEB ESSAY 8.1) (Izui et al. 2004). The nocturnal dephosphorylation of the phosphorylated serine by protein phosphatase 2A takes the catalytic capacity of PEPCase back to basal (low) activity. On the other hand, a highly unusual regulatory protein mediates the tight dark/light regulation of the activity of pyruvate–phosphate dikinase (Burnell and Chastain 2006). Pyruvate–phosphate dikinase (PPDK) is modified posttranslationally by a bifunctional threonine kinase–phosphatase that catalyzes both ADP-dependent phosphorylation [PPDK + ADP → PPDK-P + AMP] and $P_i$-dependent dephosphorylation [PPDK-P + $P_i$ → PPDK + PP$_i$] of pyruvate–phosphate dikinase. Dark induces the addition of a phosphoryl group to a specific threonine of pyruvate–phosphate dikinase by the regulatory kinase–phosphatase, causing the loss of enzyme activity. The phosphorolytic cleavage of this phosphoryl group restores the catalytic capacity of pyruvate–phosphate dikinase.

### In hot, dry climates, the $C_4$ cycle reduces photorespiration and water loss

As noted above, elevated temperatures decrease both the carboxylative capacity of rubisco and the solubility of $CO_2$, thus limiting the rate of photosynthetic $CO_2$ assimilation in $C_3$ plants. In $C_4$ plants, two features overcome the deleterious effects of high temperature:

- First, the affinity of PEPCase for its substrate, $HCO_3^-$, is sufficiently high to saturate the enzyme at the reduced $CO_2$ levels present in warm climates. Further, oxygenase activity is largely suppressed because $HCO_3^-$ does not compete with $O_2$ in the initial carboxylation. This high activity of PEPCase enables $C_4$ plants to reduce their stomatal aperture at high temperatures and thereby conserve water while fixing $CO_2$ at rates equal to or greater than those of $C_3$ plants.

- Second, the high concentration of $CO_2$ in bundle sheath cells minimizes the operation of the $C_2$ oxidative photosynthetic cycle (Maroco et al. 1998).

By enabling more efficient photosynthesis, these features give $C_4$ plants a competitive advantage where the costs of photorespiration are important—that is, in dry terrain and hot climates.

## Inorganic Carbon–Concentrating Mechanisms: Crassulacean Acid Metabolism (CAM)

Many plants that inhabit arid environments with seasonal water availability, including commercially important plants such as pineapple (*Ananas comosus*), agave (*Agave* spp.), cacti (Cactaceae), and orchids (Orchidaceae), exhibit another mechanism for concentrating $CO_2$ at the site of rubisco. This important variant of photosynthetic carbon fixation was historically named crassulacean acid metabolism (CAM) to recognize its initial observation in *Bryophyllum calycinum*, a succulent member of the Crassulaceae (Cushman 2001). Like the $C_4$ mechanism, CAM photosynthesis appears to have originated during the last 35 million years to capture atmospheric $CO_2$ and scavenge respiratory $CO_2$ in arid environments.

An important attribute of CAM plants is their capacity to attain high biomass in habitats where precipitation is inadequate, or where evaporation is so great that rainfall is insufficient for crop growth. CAM is generally associated with anatomical features that minimize water loss, such as thick cuticles, low surface-to-volume ratios, large vacuoles, and stomata with small apertures. In addition, tight packing of the mesophyll cells enhances CAM performance by restricting $CO_2$ loss during the day. Typically, a CAM

**Dark: Stomata opened**

**Light: Stomata closed**

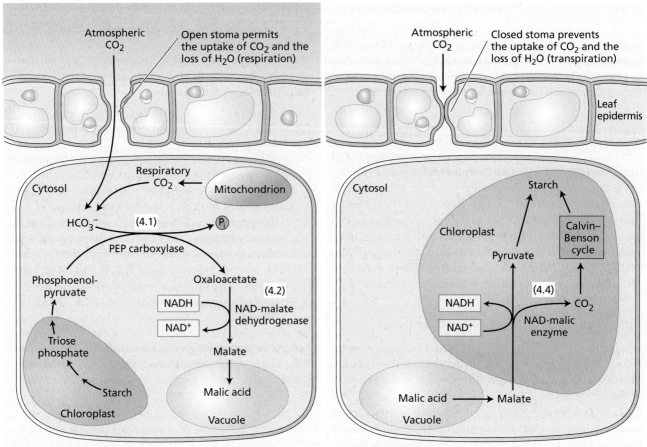

**FIGURE 8.13** Crassulacean acid metabolism (CAM). In CAM metabolism, $CO_2$ uptake is separated temporally from fixation via the Calvin–Benson cycle. The uptake of atmospheric $CO_2$ takes place at night when stomata are open. At this stage, gaseous $CO_2$ in the cytosol, coming from both the external atmosphere and mitochondrial respiration, increases levels of $HCO_3^-$ [$CO_2 + H_2O \leftrightarrow HCO_3^- + H^+$]. Then cytosolic PEPCase catalyzes a reaction between $HCO_3^-$ and phosphoenolpyruvate provided by the nocturnal breakdown of chloroplast starch. The resulting four-carbon acid, oxaloacetate, is reduced to malate which, in turn, proceeds to the acid milieu of the vacuole. During the day, the malic acid that was stored in the vacuole at night flows back to the cytosol. Malate decarboxylase (NAD-malic enzyme) acts on malate to release $CO_2$, which is refixed into carbon skeletons by the Calvin–Benson cycle. In essence, the diurnal accumulation of starch in the chloroplast constitutes the net gain of the nocturnal uptake of inorganic carbon. The adaptive advantage of stomatal closure during the day is that it prevents not only water loss by transpiration, but also the exchange of internal $CO_2$ with the external atmosphere.

plant loses 50 to 100 grams of water for every gram of $CO_2$ gained, compared with 250 to 300 grams for $C_4$ plants and 400 to 500 grams for $C_3$ plants (see Chapter 4). Thus, CAM plants have a competitive advantage in dry environments, such as deserts.

As described above, the formation of 4-carbon acids in $C_4$ plants proceeds in one compartment (e.g., mesophyll cells) within seconds before decarboxylation and refixation of the resulting $CO_2$ by the Calvin–Benson cycle takes place in another compartment (e.g., bundle sheath cells). In CAM plants, the initial capture of atmospheric $CO_2$ into $C_4$ acids and the final incorporation of $CO_2$ into carbon skeletons are spatially close but temporally out of phase—by almost 12 hours over the 24-hour light–dark cycle (**FIGURE 8.13**).

At night, cytosolic PEPCase fixes atmospheric (and respiratory) $CO_2$ into oxaloacetate using phosphoenolpyruvate formed via the glycolytic breakdown of stored carbohydrates (Table 8.4, reaction 1). A cytosolic NAD–malate dehydrogenase converts the oxaloacetate to malate, which is stored in the acid vacuole for the remainder of the night (Table 8.4, reaction 2). During the day, the stored malate is transported to the chloroplast and decarboxylated by mechanisms similar to those in $C_4$ plants—that is, by a cytosolic NADP–malic enzyme, mitochondrial NAD–malic enzyme or mitochondrial phosphoenolpyruvate carboxykinase (Table 8.4, reactions 4 and 5). The released $CO_2$ is made available to the chloroplast for processing via the Calvin–Benson cycle, while the complementary 3-carbon acids are converted to triose phosphates and subsequently to starch or sucrose via gluconeogenesis as in $C_4$ plants.

Changes in the rate of carbon uptake and in enzyme regulation throughout the day create a 24-hour CAM cycle that is divided into four distinct phases: phase I (night), phase II (early morning), phase III (daytime), and phase IV (late afternoon) (see WEB TOPIC 8.11). During the nocturnal phase I, when stomata are open and leaves are respiring, $CO_2$ is captured and stored as malate in the vacuole. $CO_2$ uptake by PEPCase dominates phase I. In the diurnal phase III, when stomata are closed and leaves are photosynthesizing, the stored malate is decarboxylated. This results in high concentrations of $CO_2$ around the active site of rubisco, thereby alleviating the adverse effects of photorespiration. The transient phases II and IV shift the metabolism in preparation for phases III and I, respectively. In phase II, rubisco activity increases, but it decreases in phase IV. In contrast, the activity of PEPCase increases in phase IV, but declines in phase II.

The duration and the contribution of each phase to the overall carbon balance vary considerably among different CAM plants. Constitutive CAM plants utilize the nocturnal uptake of $CO_2$ at all times, while their facultative counterparts resort to the CAM pathway only when induced by water stress.

Whether the triose phosphates produced by the Calvin–Benson cycle are stored as starch in chloroplasts or used for the synthesis of sucrose depends on the plant species. However, these carbohydrates ultimately ensure not only plant growth, but also the supply of substrates for the next nocturnal carboxylation phase. To sum up, the temporal separation of the virtually closed cycle of carboxylation and decarboxylation reduces the inevitable inefficiency of rubisco and optimizes the use of water, thereby enhancing photosynthetic performance in limiting environments.

### CAM is a versatile mechanism sensitive to environmental stimuli

The high efficiency of water use in CAM plants likely accounts for their extensive diversification and speciation in water-limited environments. CAM plants that grow in deserts, such as cacti, open their stomata during the cool nights and close them during the hot, dry days. Closing the stomata during the day minimizes the loss of water but, because $H_2O$ and $CO_2$ share the same diffusion pathway, $CO_2$ must then be taken up by the open stomata at night.

The availability of light mobilizes the reserves of vacuolar malic acid for the action of specific decarboxylating enzymes—NAD(P)–malic enzyme and phosphoenolpyruvate carboxykinase—and the assimilation of the resulting $CO_2$ via the Calvin–Benson cycle (Drincovich et al. 2001).

When the stomata are closed, neither the $CO_2$ released by decarboxylating enzymes nor the $CO_2$ released in mitochondrial respiration escape from the leaf. As a consequence, the internally generated $CO_2$ is fixed and converted to carbohydrates by the Calvin–Benson cycle. Thus, stomatal closure not only helps conserve water, but also assists in the buildup of the elevated internal concentration of $CO_2$ that enhances the photosynthetic carboxylation of ribulose 1,5-bisphosphate. The potential advantage of terrestrial CAM plants in arid environments is well illustrated by the inadvertent introduction of the African prickly pear (*Opuntia stricta*) into the Australian ecosystem (Osmond et al. 2008). From a few plants in 1840, the population of *O. stricta* progressively expanded to occupy 25 million hectares in less than a century.

Genotypic attributes and environmental factors modulate the extent to which the biochemical and physiological capacity of CAM plants is expressed. Although many species of succulent ornamental houseplants in the family Crassulaceae (e.g., *Kalanchoe*) are obligate CAM plants that exhibit circadian rhythmicity, others (e.g., *Clusia*) show $C_3$ photosynthesis and CAM simultaneously in opposing leaves (Lüttge 2008). The proportion of $CO_2$ taken up by PEPCase at night or by rubisco during the day (net $CO_2$ assimilation) is adjusted by (1) stomatal behavior, (2) fluctuations in organic acid and storage carbohydrate accumulation, (3) the activity of primary (PEPCase) and secondary (rubisco) carboxylating enzymes, (4) the activity of decarboxylating enzymes, and (5) synthesis and breakdown of $C_3$ carbon skeletons.

Many CAM representatives are able to adjust their pattern of $CO_2$ uptake in response to longer-term variations of environmental conditions. The ice plant (*Mesembryanthemum crystallinum* L.), agave, and *Clusia* are among the plants that use CAM when water is scarce but undergo a gradual transition to $C_3$ when water becomes abundant. Other environmental conditions, such as salinity, temperature, and light also contribute to the extent of CAM induction in these species. This form of regulation requires the expression of numerous CAM genes in response to stress signals (Adams et al. 1998; Cushman 2001; Lüttge 2008).

The water-conserving closure of stomata in arid lands may not be the unique basis of CAM evolution, because, paradoxically, CAM species are also found among aquatic plants (Mommer and Visser 2005; Lawlor 2009). Perhaps this mechanism also enhances the acquisition of inorganic carbon (as $HCO_3^-$) in aquatic habitats, where high resistance to gas diffusion restricts the availability of $CO_2$.

## Accumulation and Partitioning of Photosynthates—Starch and Sucrose

The intricate network of functions in different metabolic compartments is a remarkable feature of eukaryotic cells. Eukaryotic organisms have to mobilize sugars from the site of synthesis or absorption (source) to cells that use them for growth or energy (sinks). The arteries of animals transport glucose, whereas the conducting vessels of plants transport sucrose. The important feature of glucose is the aldehyde group on carbon 1 that enables the monosaccharide to reduce copper ($Cu^{2+} + e^- \rightarrow Cu^+$) under alkaline conditions (reducing end) (see **WEB TOPIC 8.12**). The aldehyde group makes glucose inherently reactive toward the amino groups of proteins (resulting in protein glycation); therefore, levels in blood vessels are kept within a narrow range (typically 6–8 m$M$ in healthy humans). In plants, however, a glycosidic linkage binds the hexose moieties of sucrose ($\beta$-D-Fruc-(2$\rightarrow$1)-$\alpha$-D-Gluc) and trehalose ($\alpha$-D-Gluc-(1$\rightarrow$1)-$\alpha$-D-Gluc). The formation of these disaccharides neutralizes the reactive groups of glucose (the aldehyde group on carbon 1) and fructose (the keto group on carbon 2), preventing the alkaline copper reduction. Given the lower reactivity of sucrose and trehalose relative to glucose, the contents of these two disaccharides can fluctuate widely in plant tissues. High extracellular levels of sucrose are associated with source to sink movement in the plant. Low levels of intracellular trehalose coordinate the synthesis of starch in the chloroplast with the carbon status of the cytosol (Kolbe et al. 2005).

A major challenge facing plant biochemists is to know how the compounds formed by photosynthesis in chloroplasts (photosynthates) are linked to the general metabolism of the leaf cell. The photosynthetic assimilation of $CO_2$ by most leaves yields sucrose and starch as end products, but the pathways that produce them are physically separated: sucrose is synthesized in the cytosol and starch in chloroplasts. During the day, sucrose flows continuously

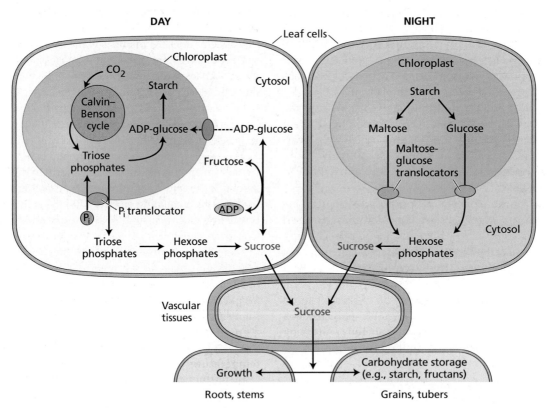

**FIGURE 8.14** Carbon mobilization in land plants. During the day, carbon assimilated photosynthetically is used either for the formation of starch in the chloroplast or is exported to the cytosol for the synthesis of sucrose, which then can be transported to nonphotosynthetic parts of the plant. External and internal stimuli control the partitioning between starch and sucrose. Triose phosphates from the Calvin–Benson cycle may be used for (1) the synthesis of chloroplast ADP–glucose, the glucosyl donor for starch synthesis, or (2) translocation to the cytosol for the synthesis of sucrose. At night, cleavage of the glycosidic linkages of starch releases both maltose and glucose, which flow across the chloroplast envelope to supplement the hexose phosphate pool and contribute to sucrose synthesis. On an ongoing basis, sucrose links the assimilation of inorganic carbon ($CO_2$) in leaves to the utilization of organic carbon for growth and storage in nonphotosynthetic parts of the plant. The indicated transport of ADP–glucose into the plastid is still controversial.

from the leaf cytosol to heterotrophic sink tissues, while starch accumulates as dense granules in chloroplasts (see WEB TOPIC 8.13). Sucrose is the principal carbohydrate exported from source leaves to sink tissues in most plants (see Figure 8.14). The retention of some photosynthate as starch in the chloroplast during the day ensures that there will be carbohydrate available for conversion to sucrose for export at night. Plants vary widely in the extent to which they accumulate starch and sucrose in leaves (FIGURE 8.14). In some species (e.g., soybean, sugar beet, Arabidopsis), the proportion of starch to sucrose in the leaf is almost constant throughout the day. In others (e.g., spinach, French beans), starch biosynthesis is stimulated when sucrose accumulation exceeds the storage capacity of the leaf or the demand by sink tissues. Environmental factors, especially day length, also influence the amount of fixed carbon allocated to sucrose and starch in the leaf; plants grown in short days divert relatively more of their photosynthates to starch than their counterparts grown in long days, thus ensuring an adequate supply of sugars during the longer nights.

Sugars produced by photosynthesis are transported from the source (leaf cells) to nonphotosynthetic sinks (stems, roots, tubers, grains) through the vascular tissues (phloem). Other carbohydrates are favored for export in some species: these include sucrose-galactosyl oligosaccharides (raffinose, stachyose, verbascose) in *Cucurbita* and Arabidopsis, sorbitol in apple, and mannitol in celery. Stems, roots, and young leaves utilize exported sugars as a source of energy for growth and as building blocks for storage polysaccharides (tubers and grains). Whereas starch is the principal reserve carbohydrate in most plants, other polysaccharides, mainly fructans, are also found as reserves in the vegetative tissues of many species (see WEB TOPIC 8.14).

The onset of darkness not only stops the assimilation of $CO_2$, but also starts the degradation of chloroplast starch. The content of starch in the chloroplast falls dramatically through the night, as it is converted to sucrose and exported. The large fluctuation in the amount of stromal starch in the light versus the dark is why the polysaccharide stored in chloroplasts is called *transitory starch*. Transitory starch functions as (1) an energy reserve to provide an adequate supply of carbohydrate at night when sugars are not formed by photosynthesis, and (2) a temporary storage form that allows photosynthesis to proceed faster than sucrose utilization during the day.

The carbon metabolism of the plant responds to the requirements of sink tissues for energy and growth (Rolland et al. 2006). Low levels of sugars in sink tissues stimulate the rate of photosynthesis and the mobilization of carbohydrates from reserve organs. On the other hand, an abundance of sugars in leaves promotes plant growth and carbohydrate storage in reserve organs. Sucrose trans-

port is critical to the allocation of the assimilated carbon because it links the availability of carbohydrates in source leaves to the formation of polysaccharide reserves in sink tissues (see Chapter 10). Carbon partitioning is an important aspect of plant physiology, and our understanding of it is essential to the task of crop improvement.

## Formation and Mobilization of Chloroplast Starch

Starch is the main storage carbohydrate in plants; it is surpassed only by cellulose in its abundance as a naturally occurring polysaccharide. Starch packs into regular semicrystalline arrays that give rise to insoluble granules (see WEB TOPIC 8.13). Electron micrograph studies showing prominent starch deposits and enzyme localization studies leave no doubt that the chloroplast is the site of starch synthesis in leaves. The architecture and the physicochemical characteristics of this macromolecular structure are critical for the location and the activity of many enzymes that partition between the soluble fraction of chloroplasts and the insoluble starch granules. In the sections that follow, we will consider the processes associated with the diurnal accumulation of photosynthetically assimilated carbon and the nocturnal degradation of the starch granule in the chloroplast.

### Starch is synthesized in the chloroplast during the day

Starch is a complex homopolymer made up of two components, amylose and amylopectin. The structure, size, and relative proportions of amylose and amylopectin vary among plant sources. The $\alpha$-D-glucosyl units associate in long linear chains linked through $\alpha$-D-1,4 glycosidic linkages wherein $\alpha$-D-1,6 glycosidic linkages are formed as branch points. The contribution of $\alpha$-D-1,6 glycosidic linkages to total bonds is extremely low in amylose (less than 1%) and moderately important in amylopectin (ca. 5–6%); thus, the former is essentially linear and the latter branched. Amylose, which is relatively extended, has a lower molecular weight (500–20,000 glucose units) than amylopectin (ca. $10^6$ glucose units), which has a compact shape.

The biosynthesis of the $\alpha$-D-1,4 glycosidic linkages of amylose proceeds through three successive steps: initiation, elongation, and termination of the polysaccharide chain. The sugar nucleotide ADP-glucose provides the glucosyl moiety for the biosynthesis of starch in photosynthetically active leaves. Although the origin of chloroplast ADP-glucose is a controversial issue (Baroja-Fernandez et al. 2005; Neuhaus et al. 2005), the chloroplast enzyme ADP-glucose pyrophosphorylase catalyzes the synthesis

**ADP-glucose biosynthesis**

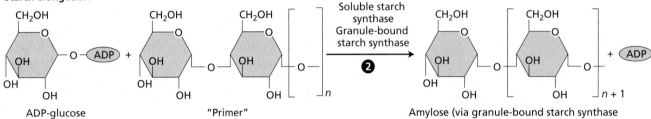

The enzyme ADP-glucose pyrophosphorylase catalyzes the formation of ADP-glucose from ATP and glucose 1-phosphate with the concurrent release of pyrophosphate.

**Starch elongation**

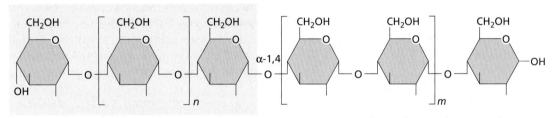

Starch synthases transfer the glucosyl moiety of ADP-glucose to the nonreducing end of a preexisting α-D-1,4 glucan primer, retaining the anomeric configuration of the glucose in the glycosidic bond.

**Starch branching**

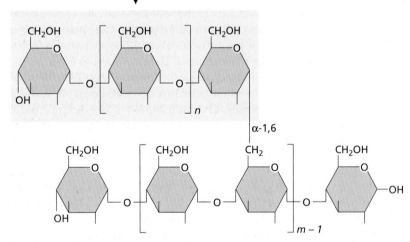

Branching enzymes cleave α-D-1,4 linkages and transfer the released oligosaccharide to a linear glucan, forming an α-D-1,6 linkage.

**FIGURE 8.15** The pathway of starch synthesis: Starch elongation and branching. Starch biosynthesis in plants is a complex process that includes biosynthesis of the sugar nucleotide ADP-glucose, formation of the "primer," elongation of the linear α-D-1,4 linked glucan, and various branching reactions. The first committed step of starch biosynthesis is the formation of ADP-glucose (reaction 1). The next step is the succes- sive addition by starch synthase of glucosyl moieties through α-D-1,4 linkages that elongate the polysac- charide (reaction 2). There are multiple isoforms of starch synthase (see text for details). The biosynthetic pathway for the formation of the primer remains elu- sive. Finally, branching enzymes catalyze the formation of branch points within the glucan chains (reaction 3).

of most of this starch precursor (**FIGURE 8.15**, reaction 1) (Zeeman et al. 2007). During the elongation process, the linear amylose chain periodically acquires novel $\alpha$-D-1,6 glycosidic linkages, leading to the formation of amylopectin in most plants. Subsequently, the elongation of starch proceeds via starch synthase, which catalyzes the transfer of the glucosyl moiety of ADP-glucose to the nonreducing end of a preexisting $\alpha$-D-1,4-glucan primer, retaining the anomeric configuration of the glycosidic bond; that is, the new glucosyl moiety in the growing glucan has an $\alpha$-configuration, as in the ADP-glucose donor (see Figure 8.15, reaction 2). Numerous studies have improved our understanding of the main steps of polysaccharide elongation and branching, but knowledge of initiation and termination remains limited.

Multiple isoforms of starch synthase have been found in oxygenic eukaryotes ranging from green algae to land plants. These isoforms can be broadly divided into two groups. The first group, primarily associated with the granule matrix, is involved in amylose synthesis and is responsible for the extension of long glucans within the amylopectin fraction. The second group of starch synthases, whose distribution between the stroma and starch granules varies with species, tissue, and developmental stage, is mainly confined to the biosynthesis of amylopectin.

The formation of amylopectin requires the participation of starch-branching enzymes that transfer a segment of an $\alpha$-D-1,4-glucan to carbon 6 of glucosyl moieties in the same glucan (see Figure 8.15, reaction 3). Like starch synthases, starch-branching enzymes are present in various isoforms that differ not only in the length of the glucan chain transferred, but also in their location in the soluble stroma or in particulate starch granules.

The construction of starch granules depends also on two debranching enzymes that process inappropriately positioned branches: the isoamylases and the **disproportionating enzyme (D-enzyme)**. As a randomly branched, water-soluble polysaccharide generally does not integrate into a semicrystalline, insoluble granule of starch, isoamylases trim the branches that impede the formation of the crystalline regions of amylopectin. The trimmed polysaccharide can be integrated into the starch granule (**FIGURE 8.16**, reaction 4). The D-enzyme recycles the residual (soluble) oligosaccharides back to the biosynthetic pathway of the starch granule via the glucan transferase reaction:

$$(\text{Glucose})_a + (\text{glucose})_b \rightarrow (\text{glucose})_{a+b-n} + (\text{glucose})_n$$

where $a$ and $b \geq 3$ and $n \leq 4$ (see Figure 8.16, reaction 5) (Zeeman et al. 2007). The products of this reaction become substrates for the action of starch synthases and branching enzymes (see Figure 8.16, reactions 2 and 3).

To function properly in the chloroplast environment, these basic enzyme reactions are regulated by homeostatic mechanisms. Thus, the biosynthesis of chloroplast starch is enhanced by activation of the chloroplast ADP-glucose pyrophosphorylase through thioredoxin-dependent mechanisms. Current evidence suggests that this enzyme can be regulated by free thioredoxin (Kolbe et al. 2005) or by the thioredoxin domain built into a special NADP–thioredoxin reductase C (Michalska et al. 2009). With free thioredoxin, ADP–glucose pyrophosphorylase appears to be regulated primarily by light through ferredoxin and ferredoxin–thioredoxin reductase. Alternatively, NADP–thioredoxin reductase C complements the ferredoxin–thioredoxin system in the redox activation of the enzyme in the light. However, in darkened leaves, the built-in thioredoxin acts independently via the NADPH formed during sugar oxidation in the oxidative pentose phosphate pathway (see Chapter 11). The response of amyloplast ADP–glucose pyrophosphorylase to dynamic changes of free thioredoxin or to NADP–thioredoxin reductase C also adjusts starch biosynthesis to the flux of metabolites in heterotrophic sink tissues. Recent work indicates that either free thioredoxin or the built-in thioredoxin of NADP–thioredoxin reductase C also regulates the amyloplast ADP–glucose pyrophosphorylase in response to phloem sucrose levels, which reflect the carbon status of the plant (Michalska et al. 2009).

Trehalose ($\alpha$-D-Gluc-(1→1)-$\alpha$-D-Gluc) is synthesized in all organisms except vertebrates, but this nonreducing disaccharide does not accumulate to any great extent in the vast majority of plants. However, Arabidopsis exhibits a surprising abundance of genes for trehalose synthesis (Paul et al. 2008). In this context, two experiments revealed that trehalose 6-phosphate may be another signal that links the carbon status of the cytosol to the synthesis of starch in the chloroplast (Kolbe et al. 2005). First, feeding of trehalose to Arabidopsis leaves leads to an enhancement of starch synthesis that is associated with activation of ADP–glucose pyrophosphorylase via thioredoxin. Second, the incubation of isolated chloroplasts with trehalose 6-phosphate increases significantly the reductive activation of ADP–glucose pyrophosphorylase.

Although the link between trehalose 6-phosphate and sucrose has not yet been fully resolved, it may involve the transcriptional or posttranscriptional regulation of enzymes involved in the metabolism of this sugar phosphate. In the cytosol, the concentrations of glucose 6-phosphate and UDP-glucose control the synthesis of trehalose 6-phosphate through the enzyme trehalose-phosphate synthase (UDP-glucose + glucose 6-phosphate → trehalose 6-phosphate + UDP). In turn, one of the precursors, UDP-glucose, is directly linked to the concentration of sucrose through the enzyme sucrose synthetase (sucrose + UDP ↔ UDP-glucose + fructose).

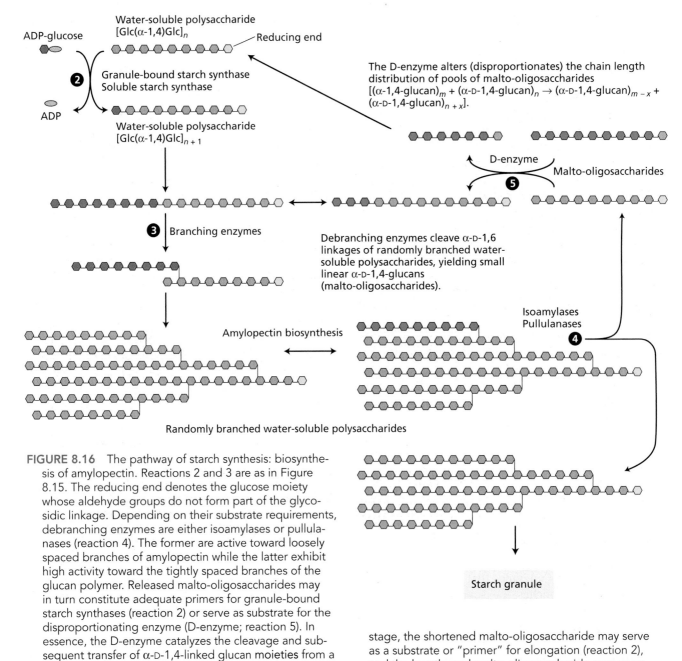

**FIGURE 8.16** The pathway of starch synthesis: biosynthesis of amylopectin. Reactions 2 and 3 are as in Figure 8.15. The reducing end denotes the glucose moiety whose aldehyde groups do not form part of the glycosidic linkage. Depending on their substrate requirements, debranching enzymes are either isoamylases or pullulanases (reaction 4). The former are active toward loosely spaced branches of amylopectin while the latter exhibit high activity toward the tightly spaced branches of the glucan polymer. Released malto-oligosaccharides may in turn constitute adequate primers for granule-bound starch synthases (reaction 2) or serve as substrate for the disproportionating enzyme (D-enzyme; reaction 5). In essence, the D-enzyme catalyzes the cleavage and subsequent transfer of α-D-1,4-linked glucan moieties from a malto-oligosaccharide donor [(α-D-1,4-glucan)$_m$] to a counterpart acceptor [(α-D-1,4-glucan)$_n$]. At this stage, the shortened malto-oligosaccharide may serve as a substrate or "primer" for elongation (reaction 2), and the lengthened malto-oligosaccharide may serve as a substrate for branching processes (reaction 3).

## Starch degradation at night requires the phosphorylation of amylopectin

Creative molecular biological approaches in constructing transgenic plants combined with biochemical analyses and genome sequence information have created a radically new picture of the pathway involved in the nocturnal degradation of transitory starch (**FIGURE 8.17**) (Lloyd et al. 2005; Smith et al. 2005). At night, chloroplast starch is broken down hydrolytically to maltose, the predominant form of carbon exported from the chloroplast. But for normal mobilization the starch must be phosphorylated prior to the hydrolysis of the α-D-1,4 and α-D-1,6 glycosidic linkages.

The release of soluble glucans from the dynamic pool of transitory starch requires the prior incorporation of phosphoryl groups into the polysaccharide, mainly amylopectin (see Figure 8.17). Unlike most kinases, the enzyme that catalyzes the phosphorylation reaction, glucan–water dikinase, transfers the β-*phosphate* of ATP (indicated by the red P in the equation below) to carbon 6 of glucosyl moieties of amylopectin (Ritte et al. 2006).

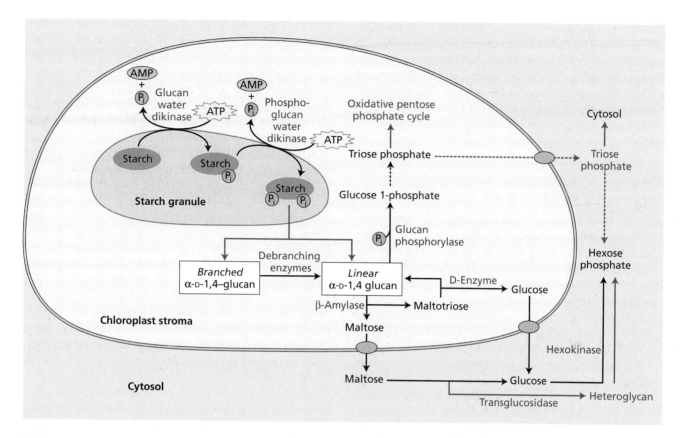

**FIGURE 8.17** Nocturnal starch degradation in *A. thaliana* leaves. The release of soluble glucans from the starch granule at night requires prior phosphorylation of the polysaccharide via the glucan–water dikinase and the phosphoglucan–water dikinase. At this stage, debranching enzymes transform the branched starch into linear glucans, which in turn can be converted into maltose via β-amylolysis catalyzed by chloroplast β-amylase. The residual maltotriose is transformed to maltopentaose and glucose via the D-enzyme (see Figure 8.16). The maltopentaose can be hydrolyzed via chloroplast β-amylase, while the glucose can be exported to the cytosol. In this context, the phosphorolytic cleavage of α-D-1,4-glucans catalyzed by chloroplast glucan phosphorylase would take place under stress conditions (see the text). Two transporters in the chloroplast envelope, one for maltose and another for glucose, facilitate the flow of the products of starch degradation to the cytosol. The utilization of maltose in the leaf cytosol proceeds via a transglucosidase that transfers a glucosyl moiety to a heteroglycan and concurrently releases a molecule of glucose. The cytosolic glucose can be phosphorylated via a hexokinase to glucose 6-phosphate for incorporation into the hexose phosphate pool.

$$\text{Adenosine-}\mathbf{P\text{-}P\text{-}P}\ (\text{ATP}) + (\text{glucan})\text{-}O\text{-}H + H_2O \rightarrow$$
$$\text{adenosine-}\mathbf{P}\ (\text{AMP}) + (\text{glucan})\text{-}O\text{-}P + P_i$$

Although phosphoryl groups occur infrequently in leaf starch (one phosphoryl group per 2000 glucosyl residues in Arabidopsis), diminished activities of glucan–water dikinase in transgenic plants causes a dramatic decrease not only in the incorporation of phosphoryl groups into amylopectin, but also in the rate of starch degradation. As a consequence, the content of starch in mature leaves of transgenic Arabidopsis lines (named *starch excess 1, sex1*) is up to seven times that of wild-type leaves (Yu et al. 2001). Glucan–water dikinase attaches to the starch granule in darkened leaves, but partly returns to the soluble state in illuminated leaves. The thioredoxin-dependent reduction of glucan–water dikinase modifies both the enzyme activity and the proportion of the enzyme bound to the starch granule (Mikkelsen et al. 2005). Hence, starch phosphorylation appears to be linked to light-driven redox processes through the regulation of the catalytic capacity and the stromal distribution of glucan–water dikinase.

Land plants contain a second enzyme, phosphoglucan–water dikinase, that catalyzes a reaction similar to glucan–water dikinase but strictly requires a phosphorylated glucan as substrate, adding the β-phosphate of ATP to carbon 3 of glucosyl moieties of amylopectin (see Figure 8.17) (Kötting et al. 2005).

$$\text{Adenosine-}\mathbf{P\text{-}P\text{-}P}\ (\text{ATP}) + (\mathbf{P}\text{-glucan})\text{-}O\text{-}H + H_2O \rightarrow$$
$$\text{adenosine-}\mathbf{P}\ (\text{AMP}) + (\mathbf{P}\text{-glucan})\text{-}O\text{-}P + P_i$$

Mutants lacking phosphoglucan–water dikinase also contain increased levels of starch but, unlike *sex1* mutants, do not exhibit altered content of phosphorylated amylopectin. Hence, the concerted action of glucan–water dikinase and phosphoglucan–water dikinase is essential for the transition of crystallized glucans to soluble maltodextrins in the nocturnal degradation of chloroplast starch (Hejazi et al. 2008).

## The export of maltose prevails in the nocturnal breakdown of transitory starch

At night, the formation of linear (soluble) oligosaccharides is indispensable for subsequent utilization of glucose moieties in the general metabolism of the plant. The enzymes α–amylase and β–amylase hydrolyze the internal and external α-D-1,4 glycosidic linkages of starch, respectively, but neither enzyme cleaves the α-D-1,6 bond. Thus, the endoglucanase activity of α–amylase and the exoglucanase activity of β–amylase inevitably yield different oligosaccharides, named collectively α– and β–limit dextrins.

Debranching enzymes are essential for the complete breakdown of starch granules to linear glucans, as α-D-1,6 branch points constitute 4–5% of the glycosidic linkages in amylopectin (see Figure 8.17). The genome of Arabidopsis encodes four proteins that may hydrolyze α-D-1,6 bonds in glucans: one pullulanase (limit-dextrinase) and three isoamylases. Although a mutation that eliminates pullulanase does not affect the cleavage of starch at night, the effects of mutations in each of the three isoamylases are difficult to interpret because the mutations also elicit abnormal starch synthesis. However, a double mutant devoid of both isoamylase-3 and pullulanase not only exhibits a severe "excess starch" phenotype, but also accumulates soluble branched oligosaccharides (limit dextrins) that are undetectable in single mutants and in the wild type (Delatte et al. 2006). This suggests that debranching of amylopectin may take place primarily at the surface of the granule starch via isoamylase-3, whereas debranching of soluble branched glucans is accomplished by the limit dextrinase.

Complementing the amylopectin debranching, two mechanisms accomplish the cleavage of the α-D-1,4 glycosidic bond of glucans (see Figure 8.17).

1. Hydrolysis, catalyzed by amylases:

$$[\text{Glucose}]_n + H_2O \rightarrow [\text{glucose}]_{n-m} + [\text{glucose}]_m$$
$$[\alpha\text{-}amylase]$$

$$[\text{Glucose}]_n + H_2O \rightarrow \text{linear } [\text{glucose}]_{n-2} + \text{maltose}$$
$$[\beta\text{-}amylase]$$

2. Phosphorolysis, catalyzed by α-glucan phosphorylases:

$$[\text{Glucose}]_n + P_i \rightarrow [\text{glucose}]_{n-1} + \text{glucose 1-phosphate}$$

Although chloroplast α-amylases were once thought to be important components of the starch-hydrolyzing pathway, nocturnal degradation proceeds at normal rates in transgenic Arabidopsis lacking all three isozymes of α-amylase. On the other hand, the genome of Arabidopsis codes for nine β-amylases, of which four are chloroplastic. The mutation of isoform-3 increases the nocturnal content of starch with a concurrent decrease of maltose (Kaplan et al. 2006). Thus, experimental evidence favors β-amylolysis as the source of maltose during starch breakdown.

On the other hand, mutation of the gene that encodes chloroplast α-glucan phosphorylase in Arabidopsis leaves fails to provoke significant changes in the diurnal accumulation and nocturnal remobilization of starch, indicating that the phosphorolytic cleavage of glycosidic linkages is not essential for starch degradation (Zeeman et al. 2004). However, under stress conditions such as drought or salinity, a minor portion of transitory starch may be degraded by the phosphorolytic pathway, yielding glucose 1-phosphate. The latter sugar phosphate can be converted subsequently by phosphoglucose isomerase into glucose 6-phosphate to be used in the oxidative pentose phosphate cycle (see Chapter 11).

The linear glucans provided by debranching enzymes are further degraded at night by chloroplast β-amylase. The production of appreciable amounts of the disaccharide maltose leads unavoidably to the formation of low amounts of maltotriose, because the exhaustive action of β-amylase cannot further process the trisaccharide (see Figure 8.17). The D-enzyme catalyzes the following transformation:

$$2\,[\text{Glucose}]_3 \rightarrow [\text{glucose}]_5 + \text{glucose}$$
$$\text{(Maltotriose)} \qquad \text{(maltopentaose)}$$

The formation of (1) maltopentaose, which is processed via β-amylases, and (2) glucose, which is exported to the cytosol via the glucose transporter in the inner chloroplast membrane, prevents the accumulation of maltotriose as starch breaks down during the night. Indeed, transgenic plants deficient in the D-enzyme exhibit lower rates of starch degradation and a concurrent increase in the level of maltotriose (Critchley et al. 2001).

Maltose is not metabolized in the stroma, because neither a maltose phosphorylase [maltose + $P_i$ → glucose 1-phosphate + glucose] nor an α-glucosidase [maltose + $H_2O$ → 2 glucose] appears to be functional in chloroplasts. More important, the inner chloroplast membrane contains a protein that facilitates the selective transport of maltose across the envelope (Niittyla et al. 2004). Transgenic plants in which the transporter locus *MEX1* is altered accumulate maltose at night to levels much higher than does the wild type. Because maltose export is a common feature of starch-storing plastids, angiosperms, gymnosperms, and mosses bear homologues of the *MEX1* gene.

The utilization of starch-derived maltose in the leaf cytosol follows a biochemical pathway unsuspected before the advent of transgenic plants. Transgenic lines devoid of a cytosolic transglucosidase degrade starch poorly and accumulate maltose to levels much higher than wild-type plants (Chia et al. 2004). The transglucosylation reaction catalyzed by this enzyme transfers a glucosyl moiety from maltose to cytosolic heteroglycans constituted of arabinose, galactose, and glucose [(heteroglycans) + maltose → (heteroglycans)-glucose + glucose] (Fettke et al. 2005; Fettke et al. 2006). Phosphorylation of the remaining glucose by hexokinase adds glucose 6-phosphate to the hexose phosphate pool for conversion to sucrose. In essence, the nocturnal export of maltose from starch breakdown in chloroplasts represents a critical process required for sustaining plant growth at night.

## Sucrose Biosynthesis and Signaling

Signaling mechanisms are essential to coordinate the development of multicellular organisms with physiological and environmental changes. In plants, carbohydrates act as signaling metabolites that are synthesized in leaves and used in compartments sustaining heterotrophic metabolism. The production of sucrose in the leaf cytosol, coupled to loading and translocation in the phloem, ensures the optimal development of the plant.

The tight regulation of these processes is invariably linked to the level of sucrose in each metabolic compartment. Sucrose thus not only provides carbon skeletons for growth and polysaccharide biosynthesis, but also constitutes a means of communication for the modulation of carbon partitioning between source leaves and sink tissues. This section describes the pathways that allocate products of photosynthetic $CO_2$ assimilation to the cytosol for the synthesis of sucrose.

### Triose phosphates supply the cytosolic pool of three important hexose phosphates in the light

The major factors that modulate the partitioning of assimilated carbon between the chloroplast and cytosol are the relative concentrations of triose phosphates and inorganic phosphate. Two products of the Calvin–Benson cycle—dihydroxyacetone phosphate and glyceraldehyde 3-phosphate (collectively named *triose phosphates*)—are rapidly interconverted by triose phosphate isomerases in the plastid and cytosol (**TABLE 8.5**, reaction 1) (Lüttge and Higinbotham 1979; Flügge and Heldt 1991). The exchange of chloroplast triose phosphates for cytosol phosphate is driven by a protein complex in the inner membrane of the chloroplast envelope—the triose phosphate translocator (see Table 8.5, reaction 2) (see **WEB TOPIC 8.15**) (Weber et al. 2005). The flux of triose phosphates across the chloro-

plast envelope conveys to the cytosol information on the status of photosynthesis in chloroplasts.

During active photosynthesis on a sunny day, the chloroplast exports part of the fixed carbon to the cytosol. In turn, the chloroplast imports the phosphate released from biosynthetic processes in the cytosol to replenish ATP and other phosphorylated metabolites, thereby sustaining photosynthetic electron transport and the Calvin–Benson cycle. This counter-exchange through the triose phosphate translocator balances the ratio of triose phosphates to inorganic phosphate in both the stroma and cytosol, reflecting the metabolic status of these two compartments. Transformed plants with impaired triose phosphate translocator activity fail to exhibit pronounced growth defects, but do show dramatic alterations in carbon metabolism. These transgenic lines accelerate the turnover of starch and the export of sugars from the chloroplast stroma, compensating for the lack of triose phosphates for sucrose biosynthesis in the cytosol.

The accumulation of triose phosphates in the cytosol increases the formation of fructose 1,6-bisphosphate from dihydroxyacetone phosphate and glyceraldehyde 3-phosphate catalyzed by cytosolic aldolase ($\Delta G^{0'} = -24$ kJ/mol) (**FIGURE 8.18**; Table 8.5, reaction 3). The $K_{eq}$ for this reaction is:

$$K_{eq} = [\text{dihydroxyacetone phosphate}] \times [\text{glyceraldehyde 3-phosphate}] / [\text{fructose 1,6-bisphosphate}]$$

Given that the action of the cytosolic aldolase yields two triose phosphates (which are rapidly interconverted by triose phosphate isomerase), the $K_{eq}$ may be transformed into

$$K_{eq} = [\text{triose phosphate}]^2 / [\text{fructose 1,6-bisphosphate}],$$

implying that the concentration of fructose 1,6-bisphosphate varies exponentially in response to changes in the concentration of triose phosphates. Hence, a constant input of triose phosphates from photosynthetically active chloroplasts biases the aldolase reaction in the cytosol of leaf cells toward the formation of fructose 1,6-bisphosphate. The reverse reaction—the aldol cleavage of fructose 1,6-bisphosphate to dihydroxyacetone phosphate and glyceraldehyde 3-phosphate—takes place when the proportion of fructose 1,6-bisphosphate is high in relation to triose phosphates, for example in glycolysis.

Cytosolic fructose-1,6-bisphosphatase subsequently catalyzes the hydrolysis of fructose 1,6-bisphosphate at the carbon 1 position, yielding fructose 6-phosphate and phosphate ($\Delta G^{0'} = -16.7$ kJ/mol) (see Figure 8.18; see Table 8.5, reaction 4). The cytosolic enzyme shows important differences from its chloroplastic counterpart both in primary structure and regulatory properties. Like the enzyme from heterotrophic organisms, the cytosolic fructose-1,6-bisphosphatase lacks a characteristic amino acid sequence that enables the chloroplastic enzyme to be regulated through

## TABLE 8.5
## Reactions in the conversion of photosynthetically formed triose phosphates to sucrose

1. *Triose phosphate isomerase*
   Dihydroxyacetone phosphate $\leftrightarrow$ glyceraldehyde 3-phosphate

$$
\begin{array}{c}
CH_2OH \\
| \\
C=O \\
| \\
CH_2OPO_3^{2-}
\end{array}
\qquad
\begin{array}{c}
CHO \\
| \\
CHOH \\
| \\
CH_2OPO_3^{2-}
\end{array}
$$

2. *Phosphate-triose phosphate translocator*
   Triose phosphate (*chloroplast*) + $P_i$ (*cytosol*) $\leftrightarrow$ triose phosphate (*cytosol*) + $P_i$ (*chloroplast*)

3. *Fructose-1,6-bisphosphate aldolase*
   Dihydroxyacetone phosphate + glyceraldehyde 3-phosphate $\rightarrow$ fructose 1,6-bisphosphate

4. *Fructose-1,6-bisphosphatase*
   Fructose 1,6-bisphosphate + $H_2O$ $\rightarrow$ fructose 6-phosphate + $P_i$

5a. *Fructose 6-phosphate 1-kinase (phosphofructokinase)*
   Fructose 6-phosphate + ATP $\rightarrow$ fructose 1,6-bisphosphate + ADP

5b. *PP$_i$-linked phosphofructokinase*
   Fructose 6-phosphate + $PP_i$ $\rightarrow$ fructose 1,6-bisphosphate + $P_i$

5c. *Fructose 6-phosphate 2-kinase*
   Fructose 6-phosphate + ATP $\rightarrow$ fructose 2,6-bisphosphate + ADP

6. *Fructose-2,6-bisphosphatase*
   Fructose 2,6-bisphosphate + $H_2O$ $\rightarrow$ fructose 6-phosphate + $P_i$

## TABLE 8.5 (continued)
## Reactions in the conversion of photosynthetically formed triose phosphates to sucrose

7. *Hexose phosphate isomerase*
Fructose 6-phosphate → glucose 6-phosphate

8. *Phosphoglucomutase*
Glucose 6-phosphate → glucose 1-phosphate

9. *UDP-glucose pyrophosporylase*
Glucose 1-phosphate  +  UDP  →  UDP-glucose + $PP_i$

10. *Sucrose $6^F$-phosphate synthase*
UDP-glucose  +  fructose 6-phosphate  →  UDP + sucrose $6^F$-phosphate

11. *Sucrose $6^F$-phosphate phosphatase*
Sucrose $6^F$-phosphate + $H_2O$ → sucrose + $P_i$

*Note:* The triose phosphate isomerase (reaction 1) catalyzes the equilibrium between dihydroxyacetone phosphate and glyceraldehyde 3-phosphate in the chloroplast stroma while the $P_i$ *translocator* (reaction 2) facilitates the exchange between triose phosphates and Pi across the chloroplast inner envelope membrane. All other enzymes catalyze cytosol reactions.

$P_i$ and $PP_i$ stand for inorganic phosphate and pyrophosphate, respectively.

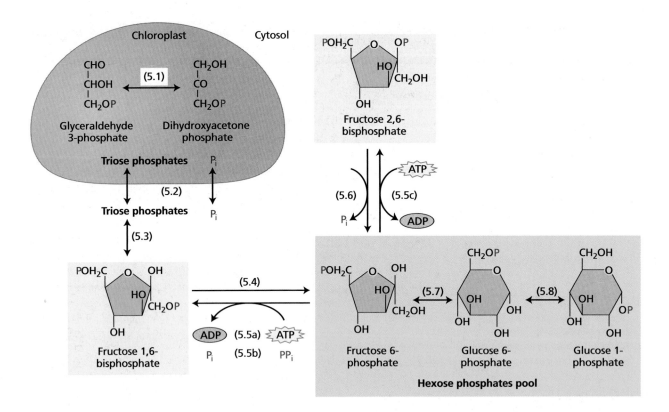

**FIGURE 8.18** Interconversion of hexose phosphates. Numbers in parentheses refer to reactions described in Table 8.5. Fructose 1,6-bisphosphate, formed from triose phosphates by aldolase, is cleaved at the carbon 1 position by cytosolic fructose-1,6-bisphosphatase, which differs structurally and functionally from the chloroplast counterpart. Fructose 6-phosphate constitutes the starting substrate for three transformations. *First*, land plants employ two different phosphorylation reactions of fructose 6-phosphate at the carbon 1 position of the furanose ring: (1) the classic ATP-dependent phosphofructokinase (see glycolysis in Chapter 11), and (2) a pyrophosphate-dependent phosphofructokinase (not shown) that catalyzes the readily reversible phospho-rylation of fructose 6-phosphate using pyrophosphate. *Second*, fructose-6-phosphate 2-kinase catalyzes the ATP-dependent phosphorylation of fructose 6-phosphate to fructose 2,6-bisphosphate and, in turn, fructose-2,6-bisphosphate phosphatase catalyzes the hydrolysis of fructose 2,6-bisphosphate, releasing the phosphoryl group and again yielding fructose 6-phosphate. *Third*, hexose phosphate isomerase and phosphogluco-mutase, respectively, favor the isomerization of fructose 6-phosphate to glucose 6-phosphate and of glucose 6-phosphate to glucose 1-phosphate. Collectively, fructose 6-phosphate, glucose 6-phosphate, and glucose 1-phosphate constitute the pool of hexose phosphates.

reduction by thioredoxin. Moreover, like the animal enzyme, cytosolic fructose-1,6-bisphosphatase is strongly inhibited by two important regulatory metabolites, fructose 2,6-bisphosphate and AMP.

Fructose 6-phosphate can proceed to different destinations through

1. Phosphorylation of carbon 1, which restores fructose 1,6-bisphosphate, catalyzed by two enzymes, phosphofructokinase and pyrophosphate-dependent phosphofructokinase (see Table 8.5, reactions 5a and b).

2. Phosphorylation of carbon 2, which yields fructose 2,6-bisphosphate, catalyzed by a unique bifunctional enzyme confined to the cytosol. Fructose-6-phosphate 2-kinase/fructose-2,6-bisphosphate phosphatase catalyzes both the incorporation and the hydrolysis of the phosphoryl group (Nielsen et al. 2004) (see Table 8.5, reaction 5c and 6).

3. Isomerization, which produces glucose 6-phosphate, catalyzed by hexose phosphate isomerase (see Table 8.5, reaction 7).

The cytosolic concentration of fructose 6-phosphate is kept close to equilibrium with glucose 6-phosphate and glucose 1-phosphate by readily reversible reactions catalyzed by hexose phosphate isomerase ($\Delta G^{0'} = 8.7$ kJ/mol) and phosphoglucomutase ($\Delta G^{0'} = 7.3$ kJ/mol) (see Table 8.5, reactions 7 and 8). These three sugar monophosphates collectively constitute the pool of hexose phosphates (see Figure 8.18).

Illuminated transgenic plants lacking either the triose phosphate translocator or cytosolic fructose-1,6-bisphosphatase sharply reduced export of carbon from the chloroplast to the cytosol, which prevents the incorporation of assimilated carbon into the hexose phosphate pool of the cytosol. Due to the impaired conversion of chloroplast triose phosphates to cytosolic hexose phosphates, the excess of chloroplast triose phosphates in transgenic lines drives the formation of unusually high levels of chloroplast starch (Zrenner et al. 1996; Hausler et al. 1998).

### Fructose 2,6-bisphosphate regulates the hexose phosphate pool in the light

Fructose 2,6-bisphosphate is an important signal that regulates the exchange of triose phosphates (chloroplast) and phosphate (cytosol) for the formation of the hexose phosphate pool (Huber 1986; Stitt 1990). A high ratio of triose phosphates to phosphate in the cytosol, typical of photosynthetically active leaves, suppresses the formation of fructose 2,6-bisphosphate, because triose phosphates strongly inhibit the kinase activity of the bifunctional enzyme fructose 6-phosphate 2-kinase/fructose 2,6-bisphosphate phosphatase. On the other hand, a low triose phosphates:phosphate ratio promotes the synthesis of fructose 2,6-bisphosphate because phosphate stimulates the fructose 6-phosphate 2-kinase activity and inhibits the fructose-2,6-bisphosphatase activity. Conversely, higher concentrations of fructose 2,6-bisphosphate inhibit the activity of fructose-1,6-bisphosphatase and, in so doing, deplete the level of hexose phosphates in the cytosol.

In turn, the hexose phosphate pool itself controls the intracellular content of fructose 2,6-bisphosphate, because fructose 6-phosphate inhibits the fructose-2,6-bisphosphatase activity and enhances the fructose 6-phosphate 2-kinase activity of the bifunctional enzyme fructose 6-phosphate 2-kinase/fructose 2,6-bisphosphate phosphatase. Thus, a high level of fructose 6-phosphate increases the concentration of fructose 2,6-bisphosphate, which inhibits fructose-1,6-bisphosphatase, restricting the formation of hexose phosphates. In summary, fructose 2,6-bisphosphate modulates the pool of hexose phosphates in response not only to photosynthates provided by the triose phosphate translocator, but also to the demands on the hexose phosphate pool itself.

At night, maltose and hexoses derived from the mobilization of chloroplast starch build up the pool of hexose phosphates in the cytosol. The hydrolysis of fructose bisphosphates does not contribute significantly to higher levels of hexose phosphates because the triose phosphates are not precursors of the hexose moieties. Therefore, the inhibition of the cytosolic fructose-1,6-bisphosphatase by fructose 2,6-bisphosphate does not affect the nocturnal synthesis of sucrose.

### The cytosolic interconversion of hexose phosphates governs the allocation of assimilated carbon

Cytosolic hexose-phosphate isomerase catalyzes the transformation of fructose 6-phosphate to glucose 6-phosphate, which, in turn, yields glucose 1-phosphate through the action of phosphoglucomutase (see Figure 8.18; see Table 8.5, reactions 7 and 8). Because *in vivo* concentrations of these three sugar phosphates are near equilibrium, the direction of carbon flow is dictated by the metabolic demands of the plant. Carbon enters the pool of hexose phosphates via phosphorylated products provided by diurnal $CO_2$ assimilation and nocturnal breakdown of chloroplast starch. On the other hand, carbon leaves the pool of hexose phosphates via glycolysis, the pentose phosphate pathway, and the synthesis of sucrose and polysaccharides.

### Sucrose is continuously synthesized in the cytosol

Photosynthates produced in leaves are transported, primarily as sucrose, to meristems and developing organs such as growing leaves, roots, flowers, fruits, and seeds (see Figure 8.14). The concentration of sucrose in the cytosol of leaves is largely dependent on the rates of two processes:

- Carbon import, which conveys diurnal triose phosphates and nocturnal maltose from chloroplasts to the leaf cytosol for sucrose synthesis
- Carbon export, which delivers sucrose from the leaf cytosol to other tissues for energy demands and polysaccharide synthesis

Cell fractionation studies, in which the organelles are physically separated from one another to analyze their intrinsic enzyme activities, have shown that sucrose is synthesized in the cytosol from the hexose phosphate pool as depicted in **FIGURE 8.19**, using the reactions described in Table 8.5. As was described for starch biosynthesis, the conversion of hexose phosphates to sugar nucleotides precedes the formation of sucrose. Members of the nucleoside diphosphate–sugar pyrophosphorylase family of enzymes catalyze the reaction between a nucleotide (NTP) and a hexose 1-phosphate, yielding the respective sugar-nucleotide and pyrophosphate:

$$NTP + \text{hexose 1-phosphate} \leftrightarrow NDP\text{–hexose} + PP_i$$

In the cytosol, a specific UDP-glucose pyrophosphorylase produces UDP-glucose from UTP and glucose 1-phosphate (Table 8.5, reaction 9). Because this reaction is not far from equilibrium ($\Delta G^{0'} = -2.88$ kJ/mol), reactions that utilize pyrophosphate in vivo drive the reaction to the

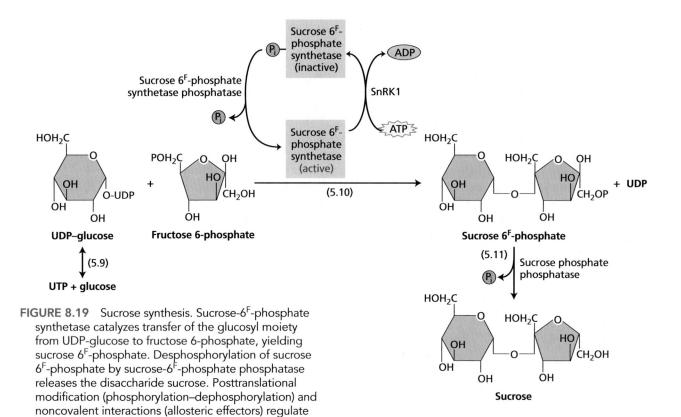

**FIGURE 8.19** Sucrose synthesis. Sucrose-6$^F$-phosphate synthetase catalyzes transfer of the glucosyl moiety from UDP-glucose to fructose 6-phosphate, yielding sucrose 6$^F$-phosphate. Desphosphorylation of sucrose 6$^F$-phosphate by sucrose-6$^F$-phosphate phosphatase releases the disaccharide sucrose. Posttranslational modification (phosphorylation–dephosphorylation) and noncovalent interactions (allosteric effectors) regulate the activity of sucrose-6$^F$-phosphate synthetase. Phosphorylation of a specific serine residue on the enzyme by the concerted action of ATP and a specific kinase, SnRK1, yields an inactive enzyme. Release of phosphate from phosphorylated sucrose-6$^F$-phosphate synthetase by a specific sucrose 6$^F$-phosphate synthetase–phosphatase restores the enzyme's basal activity. (The 6$^F$ notation in sucrose 6$^F$-phosphate indicates that sucrose is phosphorylated at carbon 6 of the fructose moiety.)

right. The absence of inorganic pyrophosphatases in the leaf cytosol promotes the use of pyrophosphate by other enzymes in transphosphorylation reactions (see Table 8.5, reaction 5b).

Two consecutive reactions complete the synthesis of sucrose from UDP-glucose. Sucrose 6$^F$-phosphate synthase (the superscript F indicates that sucrose is phosphorylated at carbon 6 of the fructose moiety) first catalyzes the formation of sucrose 6$^F$-phosphate from fructose 6-phosphate and UDP-glucose (see Table 8.5, reaction 10). Subsequently, sucrose 6$^F$-phosphate phosphatase releases inorganic phosphate from sucrose 6$^F$-phosphate, yielding sucrose (see Table 8.5, reaction 11) (Huber and Huber 1996; Lund et al. 2000).

The close association of the reversible formation of sucrose 6$^F$-phosphate ($\Delta G^{0'} = -5.7$ kJ/mol) with its irreversible hydrolysis ($\Delta G^{0'} = -16.5$ kJ/mol) renders the pathway of sucrose synthesis essentially irreversible in vivo. Sucrose 6$^F$-phosphate synthetase and sucrose 6$^F$-phosphate phosphatase may exist as a supramolecular complex. The

association of cooperating enzymes into macromolecular complexes facilitates more efficient metabolism than the isolated constituent enzymes due to substrate channeling: the direct transfer of one enzyme's product to the next enzyme in the pathway without complete mixing with the bulk phase (Salerno et al. 1996).

The sucrose 6$^F$-phosphate synthetase reaction has been identified as a key step in the control of sucrose synthesis (Huber and Huber 1996). Increased levels of sucrose 6$^F$-phosphate synthetase in transgenic plants have a dramatic impact on photosynthesis and leaf carbohydrate metabolism. For example, leaves overexpressing sucrose 6$^F$-phosphate synthetase have higher photosynthetic rates, higher ratios of sucrose to starch, and higher biomass than leaves of control plants.

Sucrose 6$^F$-phosphate synthetase is subject to a complex system of regulation involving direct control by posttranslational modifications (protein phosphorylation) and metabolites. In the dark, a specific kinase phosphorylates sucrose 6$^F$-phosphate synthetase, lowering its catalytic activity (Halford and Hey 2009). The kinase, SnRK1 (*Sucrose non-fermenting-1-Related* protein *Kinase*), is a hub within a network of interacting signaling pathways that phosphorylates and inactivates other enzymes (nitrate reductase, trehalose-phosphate synthase, fructose 6-phosphate 2-kinase–fructose 2,6-bisphosphate phosphatase). In the light, the inactive sucrose $6^F$-phosphate synthetase is activated by dephosphorylation via a protein phosphatase. The phosphorylation of sucrose $6^F$-phosphate synthetase is regulated by cytosolic metabolites: glucose 6-phosphate inhibits the kinase SnRK1 and phosphate inhibits the phosphatase.

In addition to its regulation by phosphorylation/dephosphorylation, the active form of sucrose $6^F$-phosphate synthetase is stimulated by glucose 6-phosphate and inhibited by phosphate. Thus, the increased levels of hexose phosphates and decreased levels of phosphate in the cytosol that are brought about by high rates of photosynthesis increase the synthesis of sucrose. Conversely, sucrose $6^F$-phosphate synthetase is inefficient when the level of hexose phosphates falls in response to lower rates of photosynthesis, a situation that is accompanied by an increase in the level of phosphate in the cytosol.

Sucrose plays a pivotal role in plant development because it not only provides the fuel for growth but also has a signaling function. In many plant tissues, sucrose regulates the expression of genes that encode enzymes, transporters, and storage proteins involved in developmental processes, such as flowering, tissue differentiation, seed development, and accumulation of storage products.

## SUMMARY

In chloroplasts, light energy is used by thylakoid-based photosystems to oxidize water, yielding $O_2$ and generating reduced ferredoxin, NADPH, and ATP. In the stroma, ATP and NADPH drive the fixation of atmospheric $CO_2$ and the production of carbon skeletons necessary for growth and development (**Figure 8.1**).

### The Calvin–Benson Cycle

- The Calvin–Benson cycle has three stages: (1) carboxylation of the $CO_2$ acceptor molecule; (2) reduction of the carboxylation product; and (3) regeneration of the $CO_2$ acceptor molecule (**Figures 8.2, 8.3; Table 8.1**).

- In the first stage, the rubisco catalyzes the carboxylation of ribulose 1,5-bisphosphate by $CO_2$, yielding 3-phosphoglycerate (**Figure 8.4**).

- In the second stage, 3-phosphoglycerate is transformed to triose phosphates using ATP and NADPH.

- In the third stage, five-sixths of the triose phosphates are used to regenerate the acceptor molecule ribulose 1,5-bisphosphate. The remaining one-sixth of the triose phosphates is used to supply carbon skeletons for biosynthesis, transport, and storage.

### Regulation of the Calvin–Benson Cycle

- $CO_2$ functions as both an activator and a substrate for rubisco (**Figure 8.5**). Rubisco activase complements the $CO_2$-dependent activation of rubisco by lowering the inhibition caused by sugar phosphates.

- Light controls the activity of rubisco activase and four enzymes of the Calvin–Benson cycle via the ferredoxin–thioredoxin system and by changes in $Mg^{2+}$ and pH (**Figures 8.5, 8.6**).

- Changes in light intensity regulate the formation of supramolecular complexes of enzymes controlling the activity of phosphoribulokinase and glyceraldehyde-3-phosphate dehydrogenase (**Figure 8.7**).

### The $C_2$ Oxidative Photosynthetic Carbon Cycle

- $CO_2$ and $O_2$ compete for reaction with ribulose 1,5-bisphosphate because carboxylation and oxygenation occur within the same active site of rubisco.

- The uptake of $O_2$ in the oxygenation of ribulose 1,5-bisphosphate causes the partial loss of fixed $CO_2$.

- The $C_2$ oxidative photosynthetic carbon cycle limits $CO_2$ losses due to photorespiration (**Table 8.2**).

- Operation of the $C_2$ oxidative photosynthetic cycle involves chloroplasts, peroxisomes, and mitochondria (**Figure 8.8**).

- Three simultaneous cycles move carbon, nitrogen, and oxygen atoms through photorespiration (**Figure 8.9**).

- The balance between the Calvin–Benson and the $C_2$ oxidative photosynthetic cycles is determined by three factors: the kinetic properties of rubisco, the temperature, and the concentrations of $CO_2$ and $O_2$.

- Improved photosynthetic efficiency may be engineered by introducing mechanisms for retrieving the carbon atoms of 2-phosphoglycolate for use by the Calvin–Benson cycle (**Figure 8.10, Table 8.3**).

### Inorganic Carbon-Concentrating Mechanisms

- Several mechanisms have evolved to increase the levels of $CO_2$ around the active site of rubisco and thereby decrease the rate of photorespiration.

### Inorganic Carbon-Concentrating Mechanisms: The $C_4$ Carbon Cycle

- The $C_4$ photosynthetic carbon cycle fixes the atmospheric $CO_2$ into carbon skeletons in one compartment and releases $CO_2$ in another compartment to increase $CO_2$ concentration for rubisco for refixing via the Calvin–Benson cycle (**Figure 8.11, Table 8.4**).

- The $C_4$ cycle involves five stages in two different compartments, with PEPCase, not rubisco, catalyzing the primary carboxylation (**Figure 8.11, Table 8.4**).

- The operation of the $C_4$ cycle may be driven by diffusion gradients within a single cell as well as by gradients between mesophyll and bundle sheath cells (Kranz anatomy) (**Figure 8.12**).

- Light regulates the activity of key $C_4$ cycle enzymes: NADP–malate dehydrogenase, PEPCase, and pyruvate–phosphate dikinase.

- The $C_4$ cycle reduces photorespiration and water loss in hot, dry climates.

### Inorganic Carbon-Concentrating Mechanisms: Crassulacean Acid Metabolism (CAM)

- CAM photosynthesis functions to capture atmospheric $CO_2$ and scavenge respiratory $CO_2$ in arid environments.

- CAM is generally associated with anatomical features that minimize water loss.

- In CAM plants, the initial capture of $CO_2$ and its final incorporation into carbon skeletons are temporally separated (**Figure 8.13**).

- Genetics and environmental factors determine CAM expression.

### The Accumulation and Partitioning of Photosynthates—Starch and Sucrose

- In most leaves, sucrose in the cytosol and starch in chloroplasts are the end products of the photosynthetic assimilation of $CO_2$ (**Figure 8.14, Table 8.5**).

- The glycosidic linkage in sucrose prevents glucose reactivity toward amino groups of proteins. This feature permits wide fluctuations of sucrose levels in the cytosol of leaves and in the phloem.

- During the day, sucrose flows from the leaf cytosol to sink tissues, while starch accumulates as granules in chloroplasts. At night, the starch content of chloroplasts falls to provide carbon skeletons for the synthesis of sucrose in the cytosol.

- Sugars produced by photosynthesis are transported from the source (leaf cells) to nonphotosynthetic sinks (stems, roots, tubers, grains) through the phloem. Sucrose sustains the metabolic and developmental requirements of sink tissues.

- An abundance of sugars in leaves promotes plant growth and carbohydrate storage.

- Low levels of sugars in sink tissues stimulate photosynthesis and the mobilization of carbohydrates from reserve organs.

### Formation and Mobilization of Chloroplast Starch

- Starch biosynthesis during the day proceeds through three steps: initiation, elongation, and termination (**Figure 8.15**).

- The biosynthesis of chloroplast starch is enhanced by activation of ADP-glucose pyrophosphorylase through thioredoxin-dependent mechanisms.

- Starch degradation at night requires prior phosphorylation of the polysaccharide (**Figure 8.17**). Glucan–water dikinase and phosphoglucan–water dikinase catalyze the transfer of the β–phosphate of ATP to the polysaccharide.

- Debranching enzymes transform the branched starch into linear glucans, which are converted to maltose by chloroplast β-amylase.

- Maltose from nocturnal starch breakdown is exported to the cytosol for the synthesis of sucrose.

### Sucrose Biosynthesis and Signaling

- During the day, the relative levels of triose phosphates and inorganic phosphate modulate the partitioning of carbon between chloroplasts and the cytosol. The accumulation of triose phosphates in the cytosol facilitates the hydrolysis of fructose 1,6-bisphosphate to fructose 6-phosphate, thereby building up the pool of hexose phosphates (**Figure 8.18**). Triose phosphates prevent the formation of fructose 2,6-bisphosphate, the inhibitor of fructose-1,6-bisphosphatase thus enhancing sucrose synthesis.

- At night, starch breakdown almost exclusively forms maltose, which, after transport to the cytosol, contributes to the hexose phosphate pool.

- Hexose phosphates are precursors for the cytosolic synthesis of sucrose catalyzed by sucrose $6^F$-phosphate synthetase and sucrose $6^F$-phosphate phosphatase (**Figure 8.19, Table 8.5**).

- A complex system of regulation involving posttranslational modification—phosphorylation—and noncovalent interactions with metabolites regulates the activity of sucrose $6^F$-phosphate synthetase.

- In addition to providing carbon for growth and polysaccharide biosynthesis, sucrose acts as a signal in the regulation of genes that encode enzymes, transporters, and storage proteins. Thus, sucrose also plays an important regulatory role in many developmental processes.

# WEB MATERIAL

## Web Topics

### 8.1 $CO_2$ Pumps

Cyanobacteria contain protein complexes ($CO_2$ pumps) and supramolecular complexes for the uptake and fixation of inorganic carbon.

### 8.2 How the Calvin–Benson Cycle Was Elucidated

Experiments carried out in the 1950s led to the discovery of the path of $CO_2$ fixation.

### 8.3 Rubisco: A Model Enzyme for Studying Structure and Function

As the most abundant enzyme on Earth, rubisco was obtained in quantities sufficient for elucidating its structure and catalytic properties.

### 8.4 Energy Demands for Photosynthesis in Land Plants

Evaluation of NADPH and ATP budget during the asimilation of $CO_2$.

### 8.5 Rubisco Activase

Rubisco is unique among Calvin–Benson Cycle enzymes in its regulation by a specific protein, rubisco activase.

### 8.6 Thioredoxins

First found to regulate chloroplast enzymes, thioredoxins are now known to play a regulatory role in all types of cells.

### 8.7 Operation of the $C_2$ Oxidative Photosynthetic Carbon Cycle

The enzymes of the $C_2$ oxidative photosynthetic carbon cycle are localized in three different organelles.

### 8.8 Carbon Dioxide: Some Important Physicochemical Properties

Plants have adapted to the properties of $CO_2$ by altering the reactions catalyzing its fixation.

### 8.9 Three Variations of $C_4$ Metabolism

Certain reactions of the $C_4$ photosynthetic pathway differ among plant species.

### 8.10 Single-Cell $C_4$ Photosynthesis

Some marine organisms and land plants accomplish $C_4$ photosynthesis in a single cell.

### 8.11 Photorespiration in CAM Plants

During the day, stomatal closing and photosynthesis in CAM leaves lead to very high intracellular concentrations of both oxygen and carbon dioxide. These unusual conditions pose interesting adaptive challenges to CAM leaves.

### 8.12 Glossary of Carbohydrate Biochemistry

An alphabetical list of definitions on the biochemistry of carbohydrates.

### 8.13 Starch Architecture

The morphology and composition of the starch granule influence the synthesis and degradation of the polysaccharide.

### 8.14 Fructans

Fructans are fructose polymers that serve as reserve carbohydrates in plants.

### 8.15 Chloroplast Phosphate Translocators

Chloroplast phosphate translocators are antiporters that catalyze a strict 1:1 exchange of phosphate with other metabolites between the chloroplast and the cytosol.

## Web Essay

### 8.1 Modulation of Phosphoenolpyruvate Carboxylase in $C_4$ and CAM Plants

The $CO_2$-fixing enzyme phosphoenolpyruvate carboxylase is regulated differently in $C_4$ and CAM species.

# Chapter References

Adams, P., Nelson, D. E., Yamada, S., Chmara, W., Jensen, R. G., Bohnert, H. J., and Griffiths, H. (1998) Tansley Review No. 97. Growth and development of *Mesembryanthemum crystallinum. New Phytol.* 138: 171–190.

Aran, M., Ferrero, D. S., Pagano, E., and Wolosiuk, R. A. (2009) Modulation of typical 2-Cys peroxiredoxins by covalent transformations and non-covalent interactions. *FEBS J.* 276: 2478–2493.

Baroja-Fernández, E., Muñoz, F. J., and Pozueta-Romero, J. (2005) Response to Neuhaus et al.: No need to shift the paradigm on the metabolic pathway to transitory starch in leaves. *Trends Plant Sci.* 10: 156–158.

Benson, A. A. (1951) Identification of ribulose in $C^{14}O_2$ photosynthesis products. *J. Am. Chem. Soc.* 73: 2971–2972.

Boldt, R., Edner, C., Kolukisaoglu, U., Hagemann, M., Weckwerth, W., Wienkoop, S., Morgenthal, K., and Bauwe, H. (2005) D-Gycerate 3-kinase, the last unknown enzyme in the photorespiratory cycle in Arabidopsis, belongs to a novel kinase family. *Plant Cell* 17: 2413–2420.

Brautigam, A., Hoffmann-Benning, S., and Weber, A. P. M. (2008) Comparative proteomics of chloroplast envelopes from C3 and C4 plants reveals specific adaptations of the plastid envelope to C4 photosynthesis and candidate proteins required for maintaining C4 metabolite fluxes. *Plant Physiol.* 148: 568–579.

Buchanan, B. B., and Balmer, Y. (2005) Redox regulation: A broadening horizon. *Annu. Rev. Plant Biol.* 56: 187–220.

Burnell, J. N., and Chastain, C. J. (2006) Cloning and expression of maize-leaf pyruvate, Pi dikinase regulatory protein gene. *Biochem. Biophys. Res. Commun.* 345: 675–680.

Chia, T., Thorneycroft, D., Chapple, A., Messerli, G., Chen, J., Zeeman, S. C., Smith, S. M., and Smith, A. M. (2004) A cytosolic glucosyl-transferase is required for conversion of starch to sucrose in *Arabidopsis* leaves at night. *Plant J.* 37: 853–863.

Cleland, W. W., Andrews, T. J., Gutteridge, S., Hartman, F. C., and Lorimer, G. H. (1998) Mechanism of Rubisco: The carbamate as general base. *Chem. Rev.* 98: 549–561.

Critchley, J. H., Zeeman, S. C., Takaha, T., Smith, A. M., and Smith, S. M. (2001) A critical role for disproportionating enzyme in starch breakdown is revealed by a knock-out mutant in *Arabidopsis. Plant J.* 26: 89–100.

Cushman, J. C. (2001) Crassulacean acid metabolism: A plastic photosynthetic adaptation to arid environments. *Plant Physiol.* 127: 1439–1448.

Delatte, T., Umhang, M., Trevisan, M., Eicke, S., Thorneycroft, D., Smith, S. M., and Zeeman, S. C. (2006) Evidence for distinct mechanisms of starch granule breakdown in plants. *J. Biol. Chem.* 281: 12050–12059.

Deschamps, P., Moreau, H., Worden, A. Z., Dauvillée, D., and Ball, S. G. (2008a) Early gene duplication within chloroplastida and its correspondence with relocation of starch metabolism to chloroplasts. *Genetics* 178: 2373–2387.

Deschamps, P., Haferkamp, I., d'Hulst, C., Neuhaus, H. E., and Ball, S. G. (2008b) The relocation of starch metabolism to chloroplasts: When, why and how. *Trends Plant Sci.* 13: 574–582.

Dever, L. V., Bailey, K. J., Lacuesta, M., Leegood, R. C., and Lea, P. J. (1996) The isolation and characterization of mutants of the $C_4$ plant *Amaranthus edulis. C. R. Acad. Sci. III. Sci. Vie* 319: 919–959.

DOE Solar Energy Workshop Report (2008) Basic research needs for solar energy utilization. http://www.sc.doe.gov/bes/reports/abstracts.html#SEU_rpt.pdf

Drincovich, M. F., Casati, P., and Andreo, C. S. (2001) NADP-malic enzyme from plants: A ubiquitous enzyme involved in different metabolic pathways. *FEBS Lett.* 490: 1–6.

Edwards, G. E., and Walker, D. (1983) *C3, C4: Mechanisms and Cellular and Environmental Regulation of Photosynthesis.* University of California Press, Berkeley.

Edwards, G. E., Franceschi, V. R., and Voznesenskaya, E. V. (2004) Single-cell $C_4$ photosynthesis versus the dual-cell (Kranz) paradigm. *Annu. Rev. Plant Biol.* 55: 173–196.

Fettke, J., Eckermann, N., Tiessen, A., Geigenberger, P., and Steup, M. (2005) Identification, subcellular localization and biochemical characterization of water-soluble heteroglycans (SHG) in leaves of Arabidopsis thaliana L.: Distinct SHG reside in the cytosol and in the apoplast. *Plant J.* 43: 568–586.

Fettke, J., Chia, T., Eckermann, N., Smith, A., and Steup, M. (2006) A transglucosidase necessary for starch degradation and maltose metabolism in leaves at night acts on cytosolic heteroglycans (SHG). *Plant J.* 46: 668–684.

Flügge, U. I., and Heldt, H. W. (1991) Metabolite translocators of the chloroplast envelope. *Annu. Rev. Plant Physiol. Plant Mol. Biol.* 42: 129–144.

Foyer, C. H., Bloom, A., Queval, G., and Noctor, G. (2009) Photorespiratory metabolism: Genes, mutants, energetics, and redox signaling. *Annu. Rev. Plant Biol.* 60: 455–484.

Giordano, M., Beardall, J., and Raven, J. A. (2005) CO2 concentrating mechanisms in algae: Mechanisms, environmental modulation, and evolution. *Annu. Rev. Plant Biol.* 56: 99–131.

Halford, N. G., and Hey, S. J. (2009) Snf1-related protein kinases (SnRKs) act within an intricate network that links metabolic and stress signaling in plants. *Biochem. J.* 419: 247–259.

Hatch, M. D., and Slack, C. R. (1966) Photosynthesis by sugarcane leaves. A new carboxylation reaction and the pathway of sugar formation. *Biochem. J.* 101: 103–111.

Häusler, R. E., Schlieben, N. H., Schulz, B., and Flügge, U. I. (1998) Compensation of decreased triose phosphate/phosphate translocator activity by accelerated starch turnover and glucose transport in transgenic tobacco. *Planta* 204: 366–376.

Hejazi, M., Fettke, J., Haebel, S., Edner, C., Paris, O., Frohberg, C., Steup, M., and Ritte, G. (2008) Glucan, water dikinase phosphorylates crystalline maltodextrins and thereby initiates solubilisation. *Plant J.* 55: 323–334.

Heldt, H. W. (1979) Light-dependent changes of stromal $H^+$ and $Mg^{2+}$ concentrations controlling $CO_2$ fixation. In *Photosynthesis II* (Encyclopedia of Plant Physiology, New Series, Vol. 6), M. Gibbs and E. Latzko, eds., Springer, Berlin, pp. 202–207.

Hibberd, J. M., and Quick, P. (2002) Characteristics of C4 photosynthesis in stems and petioles of C3 flowering plants. *Nature* 415: 451–454.

Huber, S. C. (1986) Fructose-2,6-bisphosphate as a regulatory metabolite in plants. *Annu. Rev. Plant Physiol.* 37: 233–246.

Huber, S. C., and Huber, J. L. (1996) Role and regulation of sucrose-phosphate synthase in higher plants. *Annu. Rev. Plant Physiol. Plant Mol. Biol.* 47: 431–444.

Izui, K., Matsumura, H., Furumoto, T., and Kai, Y. (2004) Phosphoenolpyruvate carboxylase: A new era of structural biology. *Annu. Rev. Plant Biol.* 55: 69–84.

Kaplan, F., Sung, D. Y., and Guy, C. L. (2006) Roles of β-amylase and starch breakdown during temperature stress. *Physiol. Plant.* 126: 120–128.

Kebeish, R., Niessen, M., Thiruveedhi, K., Bari, R., Hirsch, H. J., Rosenkranz, R., Stäbler, N., Schönfeld, B., Kreuzaler, F., and Peterhänsel, C. (2007) Chloroplastic photorespiratory bypass increases photosynthesis and biomass production in *Arabidopsis thaliana*. *Nat. Biotechnol.* 25: 593–599.

Khan, M. S. (2007) Engineering photorespiration in chloroplasts: A novel strategy for increasing biomass production. *Trends Biotechnol.* 25: 437–440.

Kolbe, A., Tiessen, A., Schluepmann, H., Paul, M., Ulrich, S., and Geigenberger, P. (2005) Trehalose-6-phosphate regulates starch synthesis via post-translational redox activation of ADP-glucose pyrophosphorylase. *Proc. Natl. Acad. Sci. USA* 102: 11118–11123.

Kötting, O., Pusch, K., Tiessen, A., Geigenberger, P., Steup, M., Ritte, G. (2005) Identification of a novel enzyme required for starch metabolism in *Arabidopsis* leaves. The phosphoglucan, water dikinase. *Plant Physiol.* 137: 242–252.

Ku, S. B., and Edwards, G. E. (1978) Oxygen inhibition of photosynthesis. III. Temperature dependence of quantum yield and its relation to $O_2/CO_2$ solubility ratio. *Planta* 140: 1–6.

Lawlor, D. W. (2009) Musings about the effects of environment on photosynthesis. *Ann. Bot.* 103: 543–549.

Leegood, R. C., Lea, P. J., Adcock, M. D., and Häusler, R. D. (1995) The regulation and control of photorespiration. *J. Exp. Bot.* 46: 1397–1414.

Lloyd, J. R., Kossmann, J., and Ritte, G. (2005) Leaf starch degradation comes out of the shadows. *Trends Plant Sci.* 10: 130–137.

Lorimer, G. H. (1981) The carboxylation and oxygenation of ribulose-1,5-bisphosphate: The primary events in photosynthesis and photorespiration. *Annu. Rev. Plant Physiol.* 32: 349–383.

Ludwig, M., von Caemmerer, S., Price, G. D., Badger, M. R., and Furbank, R. T. (1998) Expression of tobacco carbonic anhydrase in the $C_4$ dicot *Flaveria bidentis* leads to increased leakiness of the bundle sheath and a defective $CO_2$ concentrating mechanism. *Plant Physiol.* 117: 1071–1081.

Lund, J. E., Ashton, A. R., Hatch, M. D., and Heldt, H. W. (2000) Purification, molecular cloning, and sequence analysis of sucrose-6$^F$-phosphate phosphohydrolase from plants. *Proc. Natl. Acad. Sci. USA* 97: 12914–12919.

Lüttge, U. (2008) Clusia: Holy Grail and enigma. *J. Exp. Bot.* 59: 1503–1514.

Lüttge, U., and Higinbotham, N. (1979) *Transport in Plants.* Springer-Verlag, New York.

Maier, R. M., Neckermann, K., Igloi, G. L., and Kössel, H. (1995) Complete sequence of the maize chloroplast genome: Gene content, hotspots of divergence and fine tuning of genetic information by transcript editing. *J. Mol. Biol.* 251: 614–628.

Maroco, J. P., Ku, M. S. B., Lea, P. J., Dever, L. V., Leegood, R. C., Furbank, R. T., and Edwards, G. E. (1998) Oxygen requirement and inhibition of $C_4$ photosynthesis: An analysis of $C_4$ plants deficient in the $C_3$ and $C_4$ cycles. *Plant Physiol.* 116: 823–832.

Marri, L., Zaffagnini, M., Collin, V., Issakidis-Bourguet, E., Lemaire, S. D., Pupillo, P., Sparla, F., Miginiac-Maslow, M., and Trost, P. (2009) Prompt and easy activation by specific thioredoxins of calvin cycle enzymes of arabidopsis thaliana associated in the GAPDH/CP12/PRK supramolecular complex. *Mol. Plant* 2: 259–269.

Michalska, J., Zauber, H., Buchanan, B. B., Cejudo, F. J., and Geigenberger, P. (2009) NTRC links built-in thioredoxin to light and sucrose in regulating starch synthesis in chloroplasts and amyloplasts. *Proc. Natl. Acad. Sci. USA* 106: 9908–9913.

Mikkelsen, R., Mutenda, K. E., Mant, A., Schürmann, P., and Blennow, A. (2005) Alpha-glucan, water dikinase (GWD): A plastidic enzyme with redox-regulated and coordinated catalytic activity and binding affinity. *Proc. Natl. Acad. Sci. USA* 102: 1785–1790.

Miziorko, H. M., and Lorimer, G. H. (1983) Ribulose-1,5-bisphosphate oxygenase. *Annu. Rev. Biochem.* 52: 507–535.

Mommer, L., and Visser, E. J. W. (2005) Underwater photosynthesis in flooded terrestrial plants: A matter of leaf plasticity. *Ann. Bot.* 96: 581–589.

Montrichard, F., Alkhalfioui, F., Yano, H., Vensel, W. H., Hurkman, W. J., and Buchanan, B. B. (2009) Thioredoxin targets in plants: The first 30 years. *J. Proteomics* 72: 452–474.

Mora-Garcia, S., Stolowicz, F., and Wolosiuk, R. A. (2006) Redox signal transduction in plant metabolism. Control of primary metabolism in plants. In *Annual Plant*

*Reviews,* Vol. 22: *Control of Primary Metabolism in Plants.* Blackwell Publishing, Oxford, UK, pp. 50–186.

Muhaidat, R., Sage, R. F., and Dengler, N. G. (2007) Diversity of Kranz anatomy and biochemistry in C4 eudicots. *Am. J. Bot.* 94: 362–381.

Neff, M. M., Fankhauser, C., and Chory, J. (2000) Light: An indicator of time and place. *Genes Dev.* 14: 257–271.

Neuhaus, H. E., Häusler, R. E., and Sonnewald, U. (2005) No need to shift the paradigm on the metabolic pathway to transitory starch in leaves. *Trends Plant Sci.* 10: 154–156.

Nielsen, T. H., Rung, J. H., and Villadsen, D. (2004) Fructose-2,6-bisphosphate: A traffic signal in plant metabolism. *Trends Plant Sci.* 9: 556–563.

Niittylä, T., Messerli, G., Trevisan, M., Chen, J., Smith, A. M., and Zeeman, S. C. (2004) A previously unknown maltose transporter essential for starch degradation in leaves. *Science* 303: 87–89.

Ogren, W. L. (1984) Photorespiration: Pathways, regulation and modification. *Annu. Rev. Plant Physiol.* 35: 415–422.

Osmond, B., Neales, T. and Stange, G. (2008) Curiosity and context revisited: crassulacean acid metabolism in the Anthropocene. *J. Exp. Bot.* 59: 1489–1502.

Paul, M. J., Primavesi, L. F., Jhurreea, D., and Zhang, Y. (2008) Trehalose metabolism and signaling. *Annu. Rev. Plant Biol.* 59: 417–441.

Portis, A. R. (2003) Rubisco activase - Rubisco's catalytic chaperone. *Photosyn. Res.* 75: 11–27.

Purton, S. (1995) The chloroplast genome of *Chlamydomonas. Sci. Prog.* 78: 205–216.

Reumann, S., and Weber, A. P. M. (2006) Plant peroxisomes respire in the light: Some gaps of the photorespiratory C2 cycle have become filled - Others remain. *Biochim. Biophys. Acta* 1763: 1496–1510.

Ritte, G., Heydenreich, M., Mahlow, S., Haebel, S., Kötting, O., and Steup, M. (2006) Phosphorylation of C6- and C3-positions of glucosyl residues in starch is catalysed by distinct dikinases. *FEBS Lett.* 580: 4872–4876.

Rodríguez-Ezpeleta, N., Brinkmann, H., Burey, S. C., Roure, B., Burger, G., Löffelhardt, W., Bohnert, H. J., Philippe, H., and Lang, B. F. (2005) Monophyly of primary photosynthetic eukaryotes: Green plants, red algae, and glaucophytes. *Curr. Biol.* 15: 1325–1330.

Rolland, F., Baena-Gonzalez, E., and Sheen, J. (2006) Sugar sensing and signalling in plants: Conserved and novel mechanisms. *Annu. Rev. Plant Biol.* 57: 675–709.

Sage, R. F. (2002) C$_4$ photosynthesis in terrestrial plants does not require Kranz anatomy. *Trends Biochem. Sci.* 7: 283–285.

Sage, R. F. (2004) The evolution of C$_4$ photosynthesis. *New Phytol.* 161: 341–370.

Salerno, G. L., Echeverria, E., and Pontis, H. G. (1996) Activation of sucrose-phosphate synthase by a protein factor/sucrose-phosphate phosphatase. *Cell. Mol. Biol.* 42: 665–672.

Sharkey, T. D. (1988) Estimating the rate of photorespiration in leaves. *Physiol. Plant.* 73: 147–152.

Siebke, K., von Caemmerer, S., Badger, M., and Furbank, R. T. (1997) Expressing an *rbcS* antisense gene in transgenic *Flaveria bidentis* leads to an increased quantum requirement for $CO_2$ fixed in photosystems I and II. *Plant Physiol.* 115: 1163–1174.

Sinclair, T. R. (2009) Taking measure of biofuel limits. *Am. Sci.* 97: 400–407.

Smith, A. M., Zeeman, S. C., and Smith, S. M. (2005) Starch degradation. *Annu. Rev. Plant Biol.* 56: 73–97.

Stitt, M. (1990) Fructose-2,6-bisphosphate as a regulatory molecule in plants. *Annu. Rev. Plant Physiol. Plant Mol. Biol.* 41: 153–185.

Tamoi, M., Myazaki, T., Fukamizo, T., and Shigeoka, S. (2005) The Calvin cycle in cyanobacteria is regulated by CP12 via NAD(H)/NADP(H) ratio under light/dark conditions. *Plant J.* 42: 504–513.

Tolbert, N. E. (1981) Metabolic pathways in peroxisomes and glyoxysomes. *Annu. Rev. Biochem.* 50: 133–157.

von Caemmerer, S., Millgate, A., Farquhar, G. D., and Furbank, R. T. (1997) Reduction of ribulose-1,5-bisphosphate carboxylase/oxygenase by antisense RNA in the C$_4$ plant *Flaveria bidentis* leads to reduced assimilation rates and increased carbon isotope discrimination. *Plant Physiol.* 113: 469–477.

Weber, A. P. M., Schwacke, R., and Flügge, U. I. (2005) Solute transporters of the plastid envelope membrane. *Annu. Rev. Plant Biol.* 56: 133–164.

Wolosiuk, R. A., Ballicora, M. A., and Hagelin, K. (1993) The reductive pentose phosphate cycle for photosynthetic carbon dioxide assimilation: Enzyme modulation. *FASEB J.* 7: 622–637.

Yu, T. S., Kofler, H., Häusler, R. E., Hille, D., Flügge, U. I., Zeeman, S. C., Smith, A. M., Kossmann, J., Lloyd, J., Ritte, G., et al. (2001) The Arabidopsis sex1 mutant is defective in the R1 protein, a general regulator of starch degradation in plants, and not in the chloroplast hexose transporter. *Plant Cell* 13: 1907–1918.

Zeeman, S. C., Thorneycroft, D., Schupp, N., Chapple, A., Weck, M., Dunstan, H., Haldimann, P., Bechtold, N., Smith, A. M., and Smith, S. M. (2004) Plastidial α-glucan phosphorylase is not required for starch degradation in *Arabidopsis* leaves but has a role in the tolerance of abiotic stress. *Plant Physiol.* 135: 849–858.

Zeeman, S. C., Smith, S. M., and Smith, A. M. (2007) The diurnal metabolism of leaf starch. *Biochem. J.* 401: 13–28.

Zrenner, R., Krause, K. P., Apel, P., and Sonnewald, U. (1996) Reduction of the cytosolic fructose–1,6-bisphosphatase in transgenic potato plants limits photosynthetic sucrose biosynthesis with no impact on plant growth and tuber yield. *Plant J.* 9: 671–681.

CHAPTER

# 9

# Photosynthesis: Physiological and Ecological Considerations

The conversion of solar energy to the chemical energy of organic compounds is a complex process that includes electron transport and photosynthetic carbon metabolism (see Chapters 7 and 8). Earlier discussions of the photochemical and biochemical reactions of photosynthesis should not overshadow the fact that, under natural conditions, the photosynthetic process takes place in intact organisms that are continuously responding to internal and external changes. This chapter addresses some of the photosynthetic responses of the intact leaf to its environment. Additional photosynthetic responses to different types of stress are covered in Chapter 26.

The impact of the environment on photosynthesis is of interest to plant physiologists, ecologists, and agronomists. From a physiological standpoint, we wish to understand the direct responses of photosynthesis to environmental factors such as light, ambient $CO_2$ concentrations, and temperature, as well as the indirect responses (mediated through the effects of stomatal control) to environmental factors such as humidity and soil moisture. The dependence of photosynthetic processes on environmental conditions is also important to agronomists because plant productivity, and hence crop yield, depend strongly on prevailing photosynthetic rates in a dynamic environment. To the ecologist, the fact that photosynthetic rates and capacities vary among different environments is of great interest in terms of adaptation and evolution.

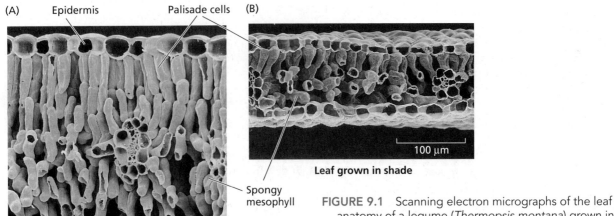

(A)  Epidermis        Palisade cells    (B)

100 μm

Leaf grown in shade

Spongy mesophyll

Epidermis

Leaf grown in sun

**FIGURE 9.1**  Scanning electron micrographs of the leaf anatomy of a legume (*Thermopsis montana*) grown in different light environments. Note that the sun leaf (A) is much thicker than the shade leaf (B) and that the palisade (columnlike) cells are much longer in the leaves grown in sunlight. Layers of spongy mesophyll cells can be seen below the palisade cells. (Courtesy of T. Vogelmann.)

In studying the environmental dependence of photosynthesis, a central question arises: How many environmental factors can limit photosynthesis at one time? The British plant physiologist F. F. Blackman hypothesized in 1905 that, under any particular conditions, the rate of photosynthesis is limited by the slowest step in the process, the so-called *limiting factor*.

The implication of this hypothesis is that at any given time, photosynthesis can be limited either by light or by $CO_2$ concentration, for instance, but not by both factors. This hypothesis has had a marked influence on the approach used by plant physiologists to study photosynthesis—that is, varying one factor and keeping all other environmental conditions constant. In the intact leaf, three major metabolic properties have been identified as important for optimal photosynthetic performance:

- Rubisco activity
- Regeneration of ribulose bisphosphate (RuBP)
- Metabolism of the triose phosphates

Farquhar and Sharkey (1982) added a fundamentally new perspective to our understanding of photosynthesis by pointing out that we should think of the controls on the overall rates of net photosynthetic rate in leaves in economic terms, considering "supply" and "demand" functions for carbon dioxide. Net photosynthesis is defined as net $CO_2$ uptake.

The biochemical activities referred to above take place in the palisade cells and spongy mesophyll of the leaf (**FIGURE 9.1**). These activities describe the "demand" by photosynthetic metabolism in the cells for $CO_2$ as a substrate. However, the actual rate of $CO_2$ "supply" to these cells is controlled by stomatal guard cells located on the epidermal portions of the leaf. These supply and demand functions associated with photosynthesis take place in different cells. It is the coordinated actions of "demand" by photosynthetic

cells and "supply" by guard cells that determine the leaf photosynthetic rate as measured by net $CO_2$ uptake.

In the following sections, we will focus on how naturally occurring variations in light and temperature influence photosynthesis in leaves and how leaves in turn adjust or acclimate to variations in light and temperature. In addition, we will explore how atmospheric carbon dioxide influences photosynthesis, an especially important consideration in a world where $CO_2$ concentrations are rapidly increasing as humans continue to burn fossil fuels for energy uses.

## Photosynthesis Is the Primary Function of Leaves

Scaling up from the chloroplast (the focus of Chapters 7 and 8) to the leaf adds new levels of complexity to photosynthesis. At the same time, the structural and functional properties of the leaf make possible other levels of regulation.

We will start by examining how leaf anatomy and leaf orientation control the absorption of light for photosynthesis. Then we will describe how chloroplasts and leaves acclimate to their light environment. We will see that the photosynthetic response of leaves grown under different light conditions also reflects the capacity of a plant to grow under different light environments. However, there are also limits in the extent to which photosynthesis in a species can acclimate to very different light environments.

It will become clear that under different environmental conditions, the rate of photosyntheis is limited by different factors. For example, in some situations photosynthesis is limited by an inadequate supply of light or $CO_2$. In other situations, absorption of too much light would cause severe problems if special mechanisms did not protect the photosynthetic system from excessive light. While plants have multiple levels of control over photosynthesis that allow them to grow successfully in constantly changing environments, there are ultimately limits to what is possible in terms of acclimation to sun and shade, high and low temperatures, and degrees of water stress.

Think of the different ways in which leaves are exposed to different spectra and quantities of light that result in photosynthesis. Plants grown outdoors are exposed to sunlight, and the spectrum of that sunlight will depend on whether it is measured in full sunlight or under the shade of a canopy. Plants grown indoors may receive either incandescent or fluorescent lighting, each of which is different from sunlight. To account for these differences in spectral quality and quantity, we need uniformity in how we measure and express the light that impacts photosynthesis.

The light reaching the plant is a flux and that flux can be measured in either energy or photon units. **Irradiance** is the amount of energy that falls on a flat sensor of known area per unit time, expressed in watts per square meter (W m$^{-2}$). (Recall that time [seconds] is contained within the term watt: 1 W = 1 joule [J] s$^{-1}$.) **Photon irradiance** is the number of incident **quanta** (singular *quantum*) striking the leaf, expressed in moles per square meter per second (mol m$^{-2}$ s$^{-1}$), where *moles* refers to the number of photons (1 mol of light = $6.02 \times 10^{23}$ photons, Avogadro's number). Quanta and energy units for sunlight can be interconverted relatively easily, provided that the wavelength of the light, $\lambda$, is known. The energy of a photon is related to its wavelength as follows:

$$E = \frac{hc}{\lambda}$$

where $c$ is the speed of light ($3 \times 10^8$ m s$^{-1}$), $h$ is Planck's constant ($6.63 \times 10^{-34}$ J s), and $\lambda$ is the wavelength of light, usually expressed in nm (1 nm = $10^{-9}$ m). From this equation it can be shown that a photon at 400 nm has twice the energy of a photon at 800 nm (see **WEB TOPIC 9.1**).

**Photosynthetically active radiation (PAR, 400–700 nm)** may also be expressed in terms of energy (W m$^{-2}$) but is more commonly expressed as quanta (mol m$^{-2}$ s$^{-1}$). Note that PAR is an irradiance-type measurement. In research on photosynthesis, PAR is expressed on a quantum basis.

Incoming sunlight can strike a flat leaf surface at a variety of angles depending on the time of day and the orientation of the leaf. When sunlight deviates from directly over the leaf (perpendicular), irradiance is proportional to the cosine of the angle at which the light rays hit the sensor or leaf (**FIGURE 9.2**).

How much light is there on a sunny day? Under direct sunlight, PAR irradiance is about 2000 μmol m$^{-2}$ s$^{-1}$ (900 W m$^{-2}$) at the top of a dense forest canopy, but may be only

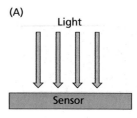

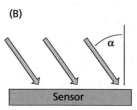

Irradiance = (A) × cosine α

**FIGURE 9.2** Incident sunlight relative to leaf angle. The maximum incident sunlight on a leaf will occur when the incoming sunlight is perpendicular to the leaf lamina (A). When the incoming sunlight is at any other angle (B), the incident light levels will be reduced by the cosine of the angle between the sunlight and the leaf lamina.

10 μmol m$^{-2}$ s$^{-1}$ (4.5 W m$^{-2}$) at the bottom of the canopy because of absorption of PAR by the leaves overhead.

## Leaf anatomy maximizes light absorption

While roughly 1.3 kW m$^{-2}$ of radiant energy from the sun reaches Earth, less than 5% of this energy is ultimately converted into carbohydrates by a photosynthesizing leaf. The reason this percentage is so low is that about half of the incident light is of a wavelength either too short or too long to be absorbed by the photosynthetic pigments (see Figure 7.3). Of the photosynthetically active radiation (PAR, 400–700 nm) that is absorbed, about 15% is reflected or transmitted through a green leaf. Because chlorophyll absorbs very strongly in the blue and the red regions of the spectrum (see Figure 7.3), the transmitted and reflected light are vastly enriched in green (**FIGURE 9.3**)—hence the green color of vegetation. Of the 85% the PAR absorbed by a green leaf, a significant fraction of the absorbed light is lost as heat and a smaller amount is lost as fluorescence (see Chapter 7), resulting in less than 5% of the incident energy being converted into the energy stored within a carbohydrate.

The anatomy of the leaf is highly specialized for light absorption (Terashima and Hikosaka 1995). The outermost cell layer, the epidermis, is typically transparent to visible light, and the individual cells are often convex. Convex epidermal cells can act as lenses and focus light so that the intensity reaching some of the chloroplasts can be many

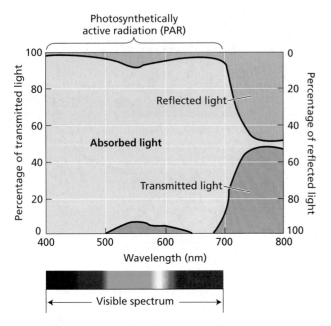

Photosynthetically active radiation (PAR)

**FIGURE 9.3** Optical properties of a bean leaf. Shown here are the percentages of light absorbed, reflected, and transmitted, as a function of wavelength. The transmitted and reflected green light in the wave band at 500 to 600 nm gives leaves their green color. Note that most of the light above 700 nm is not absorbed by the leaf. (After Smith 1986.)

times greater than the intensity of ambient light. Epidermal focusing is common among herbaceous plants and is especially prominent among tropical plants that grow in the forest understory, where light levels are very low.

Below the epidermis, the top layers of photosynthetic cells are called **palisade cells**; they are shaped like pillars that stand in parallel columns one to three layers deep (see Figure 9.1). Some leaves have several layers of columnar palisade cells, and we may wonder how efficient it is for a plant to invest energy in the development of multiple cell layers when the high chlorophyll content of the first layer would appear to allow little transmission of the incident light to the leaf interior. In fact, more light than might be expected penetrates the first layer of palisade cells because of the *sieve effect* and *light channeling*. To increase the efficiency of photosynthetic structures within palisade cells, chloroplasts have high surface-to-volume ratios (Evans et al. 2009).

The **sieve effect** is due to the fact that chlorophyll is not uniformly distributed throughout cells but instead is confined to the chloroplasts. This packaging of chlorophyll results in shading between the chlorophyll molecules and creates gaps between the chloroplasts where light is not absorbed—hence the reference to a sieve. Because of the sieve effect, the total absorption of light by a given amount of chlorophyll in a palisade cell is less than the light absorbed by the same amount of chlorophyll in a solution.

**Light channeling** occurs when some of the incident light is propagated through the central vacuoles of the palisade cells and through the air spaces between the cells, an arrangement that facilitates the transmission of light into the leaf interior (Vogelmann 1993).

Below the palisade layers is the **spongy mesophyll**, where the cells are very irregular in shape and are surrounded by large air spaces (see Figure 9.1). The large air spaces generate many interfaces between air and water that reflect and refract the light, thereby randomizing its direction of travel. This phenomenon is called **interface light scattering**.

Light scattering is especially important in leaves because the multiple reflections between cell–air interfaces greatly increase the length of the path over which photons travel, thereby increasing the probability for absorption. In fact, photon path lengths within leaves are commonly four times longer than the thickness of the leaf. Thus the palisade cell properties that allow light to pass through and the spongy mesophyll cell properties that are conducive to light scattering result in more uniform light absorption throughout the leaf.

Some environments, such as deserts, have so much light that it is potentially harmful to leaves. In these environments leaves often have special anatomical features, such as hairs, salt glands, and epicuticular wax, that increase the reflection of light from the leaf surface, thereby reducing light absorption (Ehleringer et al. 1976). Such adaptations can decrease light absorption by as much as 40%, minimizing heating and other problems associated with the absorption of too much solar energy.

## Plants compete for sunlight

Plants normally compete for sunlight. Held upright by stems and trunks, their leaves configure a canopy that absorbs light and influences photosynthetic rates and growth beneath them. Leaves that are shaded by other leaves experience lower light levels and different light quality than the leaves above them and have much lower photosynthetic rates.

Trees with their leaves high above the ground surface represent an outstanding adaptation for light interception. The elaborate branching structure of trees vastly increases the interception of sunlight. Very little PAR penetrates to the bottom of forest canopies; almost all of it is absorbed by leaves (**FIGURE 9.4**). At the other end of the growth spectrum are plants such as dandelion (*Taraxacum* sp.), which have a rosette growth habit in which leaves grow radially very close to each other on a very short stem, thus preventing the growth of any leaves below them.

In many shady habitats **sunflecks** are a common environmental feature. These are patches of sunlight that pass

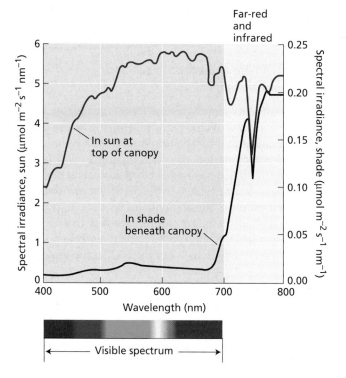

**FIGURE 9.4** The spectral distribution of sunlight at the top of a canopy and under the canopy. For unfiltered sunlight, the total irradiance was 1900 μmol m$^{-2}$ s$^{-1}$; for shade, 17.7 μmol m$^{-2}$ s$^{-1}$. Most of the photosynthetically active radiation was absorbed by leaves in the canopy. (After Smith 1994.)

the photons in sunflecks contain nearly 50% of the total light energy available during the day; such leaves often have mechanisms for taking advantage of sunflecks when they occur.

Sunflecks also play a role in the carbon metabolism of lower leaves in dense crops that are shaded by the upper leaves of the plant. Rapid responses by both the photosynthetic apparatus and the stomata to sunflecks have been of substantial interest to plant physiologists and ecologists (Pearcy et al. 2005), because they represent specialized physiological mechanisms for the capture of short bursts of sunlight.

## Leaf angle and leaf movement can control light absorption

How do leaves influence the light levels within a canopy? The angle of the leaf relative to the sun will determine the amount of sunlight incident upon it in a manner identical to that shown in Figure 9.2. If the sun is directly overhead, a horizontal leaf (such as the flat sensor in Figure 9.2A) will receive much more sunlight than a leaf at a steeper angle. Under natural conditions, leaves exposed to full sunlight at the top of the canopy tend to have steep leaf angles so that less than the maximum amount of sunlight is incident on the leaf blade; this allows more sunlight to penetrate into the canopy. It is common to see the angle of leaves within a canopy decrease (become more horizontal) with increasing depth in the canopy.

Leaves absorb the most light when the leaf blade, or lamina, is perpendicular to the incident light. Some plants control light absorption by **solar tracking** (Ehleringer and Forseth 1980); that is, their leaves continuously adjust the orientation of their laminae such that they remain perpendicular to the sun's rays (**FIGURE 9.5**). Many species, including alfalfa, cotton, soybean, bean, and lupine, have leaves capable of solar tracking.

through small gaps in the leaf canopy and move across shaded leaves as the sun moves. In a dense forest, sunflecks can change the photon flux incident on a leaf on the forest floor more than tenfold within seconds. This critical energy is available for only a few minutes now and then in a very high dose. For some leaves low in the canopy,

**FIGURE 9.5** Leaf movement in sun-tracking plants. (A) Initial leaf orientation in the lupine *Lupinus succulentus*. (B) Leaf orientation 4 hours after exposure to oblique light. The direction of the light beam is indicated by the arrows. Movement is generated by asymmetric swelling of a pulvinus, found at the junction between the lamina and the petiole. In natural conditions, the leaves track the sun's trajectory in the sky. (From Vogelmann and Björn 1983, courtesy of T. Vogelmann.)

Solar-tracking leaves present a nearly vertical position at sunrise, facing the eastern horizon. The leaf blades then begin to track the rising sun, following its movement across the sky with an accuracy of ±15° until sunset, when the laminae are nearly vertical, facing the west. During the night the leaf takes a horizontal position and reorients just before dawn so that it faces the eastern horizon in anticipation of another sunrise. Leaves track the sun only on clear days, and they stop moving when a cloud obscures the sun. In the case of intermittent cloud cover, some leaves can reorient as rapidly as 90° per hour and thus can catch up to the new solar position when the sun emerges from behind a cloud (Koller 2000).

Solar tracking is a blue-light response (see Chapter 18), and the sensing of blue light in solar-tracking leaves occurs in specialized regions of the leaf or stem. In species of *Lavatera* (Malvaceae), the photosensitive region is located in or near the major leaf veins (Koller 2000), but in many cases, leaf orientation is controlled by a specialized organ called the **pulvinus** (plural *pulvini*), found at the junction between the blade and the petiole. In lupines (*Lupinus*, Fabaceae), for example, leaves consist of five or more leaflets, and the photosensitive region is in a pulvinus located at the basal part of each leaflet lamina (see Figure 9.5). The pulvinus contains motor cells that change their osmotic potential and generate mechanical forces that determine laminar orientation. In other plants, leaf orientation is controlled by small mechanical changes along the length of the petiole and by movements of the younger parts of the stem (Ehleringer and Forseth 1980).

Building on the term **heliotropism** ("bending toward the sun"), used to describe sun-induced leaf movements, we call leaves that maximize light interception by solar tracking *diaheliotropic*. Some solar-tracking plants can also move their leaves so that they *avoid* full exposure to sunlight, thus minimizing heating and water loss. These sun-avoiding leaves are called *paraheliotropic*. Some plant species have leaves that can display diaheliotropic movements when they are well watered and paraheliotropic movements when they experience water stress.

By keeping leaves perpendicular to the sun, solar-tracking plants are able to maintain maximum photosynthetic rates throughout the day, including early morning and late afternoon. Air temperature is generally lower during the early morning and late afternoon, so water stress is lower at these times. Solar tracking therefore gives an advantage to rain-fed crop plants with short growing periods, such as pinto beans.

Diaheliotropic solar tracking appears to be a feature common to wild plants that are short-lived and must complete their life cycle before the onset of drought (Ehleringer and Forseth 1980). Paraheliotropic leaves are able to regulate the amount of sunlight incident on the leaf to a nearly constant value. Although the amount of incident sunlight is often only one-half to two-thirds of full sunlight, these levels may be advantageous under conditions of water stress or excessive solar radiation.

## Plants acclimate and adapt to sun and shade environments

Some plants have enough developmental plasticity to respond to a range of light regimes, growing as sun plants in sunny areas and as shade plants in shady habitats. We call this **acclimation**, a growth process in which each newly produced leaf has a set of biochemical and morphological characteristics suited to the particular environment in which it unfolds. The ability to acclimate is important, given that shady habitats can receive less than 20% of the PAR available in an exposed habitat, and deep shade habitats receive less than 1% of the PAR at the top of the canopy.

In some plant species, individual leaves that develop under very sunny or very shady environments are often unable to persist when transferred to the other type of habitat (see Figure 9.4). In such cases, the mature leaf will abscise and a new leaf will develop that is better suited for the new environment. You may notice this if you take a plant that developed indoors and transfer it outdoors; after some time, if it's the right type of plant, it develops a new set of leaves better suited to high sunlight. However, some species of plants are not able to acclimate when transferred from a sunny to a shady environment. The lack of acclimation suggests that these plants are **adapted** to either a sunny or a shady environment. When plants adapted to deep shade conditions are transferred into full sunlight, the leaves experience chronic photoinhibition and leaf bleaching, and the plants eventually die. Photoinhibition will be discussed later in this chapter.

Sun and shade leaves have contrasting biochemical characteristics:

- Shade leaves have more total chlorophyll per reaction center, have a higher ratio of chlorophyll *b* to chlorophyll *a*, and are usually thinner than sun leaves.
- Sun leaves have more rubisco and a larger pool of xanthophyll cycle components than shade leaves (see Chapter 7).

Contrasting anatomic characteristics can also be found in leaves of the same plant that are exposed to different light regimes. Figure 9.1 shows some anatomic differences between a leaf grown in the sun and a leaf grown in the shade. Most notably, sun-grown leaves are thicker and have longer palisade cells than leaves grown in the shade. Even different parts of a single leaf show adaptations to their light microenvironment (Terashima 1992).

Morphological and biochemical modifications are associated with specific functions found in response to variability in the amounts of sunlight in a plant's habitat. For example, far-red light, which is absorbed primarily by PSI, is proportionally more abundant in shady habitats than in sunny ones.

The adaptive response of some shade plants is to produce a 3:1 ratio of photosystem II to photosystem I reaction centers, compared with the 2:1 ratio found in sun plants (Anderson 1986). Other shade plants, rather than altering the ratio of PSII to PSI reaction centers, add more antenna chlorophyll to PSII to increase absorption by this photosystem and better balance the flow of energy through PSII and PSI. These changes appear to enhance light absorption and energy transfer in shady environments.

Sun and shade plants also differ in their dark respiration rates, and these differences alter the relationship between respiration and photosynthesis, as we'll see a little later in this chapter.

## Photosynthetic Responses to Light by the Intact Leaf

Light is a critical resource for plants that can limit growth and reproduction if too little or too much is received. The relationship between radiation and the photosynthetic properties of the leaf provides valuable information about plant adaptations to the light environment. In this section we describe typical photosynthetic responses to light as measured in light-response curves. We also consider how important features of a light-response curve can help explain contrasting physiological properties between sun and shade plants, and between $C_3$ and $C_4$ species. The section continues with descriptions of how leaves respond to excess light.

### Light-response curves reveal photosynthetic properties

Measuring net $CO_2$ fixation in intact leaves across varying levels of absorbed light allows us to construct light-response curves (FIGURE 9.6) that provide useful information about the photosynthetic properties of leaves. In the dark there is no photosynthetic carbon assimilation, but, because mitochondrial respiration continues, $CO_2$ is given off by the plant (see Chapter 11). $CO_2$ uptake is negative in this part of the light-response curve. At greater photon flux levels, photosynthetic $CO_2$ assimilation eventually reaches a point at which photosynthetic $CO_2$ uptake exactly balances $CO_2$ release. This is called the **light compensation point**.

The photon flux at which different leaves reach the light compensation point can vary among species and developmental conditions. One of the more interesting differences is found between plants that normally grow in full sunlight and those that grow in the shade (FIGURE 9.7). Light compensation points of sun plants range from 10 to

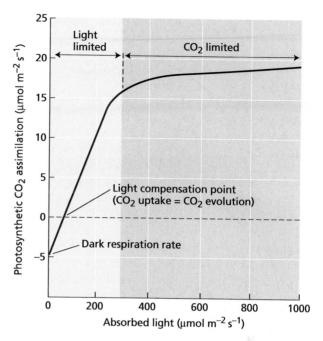

**FIGURE 9.6** Response of photosynthesis to light in a $C_3$ plant. In darkness, respiration causes a net efflux of $CO_2$ from the plant. The light compensation point is reached when photosynthetic $CO_2$ assimilation equals the amount of $CO_2$ evolved by respiration. Increasing light above the light compensation point proportionally increases photosynthesis, indicating that photosynthesis is limited by the rate of electron transport, which in turn is limited by the amount of available light. This portion of the curve is referred to as light-limited. Further increases in photosynthesis are eventually limited by the carboxylation capacity of rubisco or the metabolism of triose phosphates. This part of the curve is referred to as $CO_2$-limited.

$20$ μmol m$^{-2}$ s$^{-1}$, whereas corresponding values for shade plants are 1 to 5 μmol m$^{-2}$ s$^{-1}$.

Why are light compensation points lower for shade plants? For the most part, this is because respiration rates in shade plants are very low; therefore only a little photosynthesis is necessary to bring the net rates of $CO_2$ exchange to zero. Low respiratory rates allow shade plants to survive in light-limited environments through their ability to achieve positive $CO_2$ uptake rates at lower values of PAR than sun plants.

The linear relationship between photon flux and photosynthetic rate persists at light levels above the light compensation point (see Figure 9.6). Throughout this linear portion of the light response curve, photosynthesis is light-limited; more light stimulates proportionately more photosynthesis. The slope of this linear portion of the curve reveals the **maximum quantum yield** of photosynthesis for the leaf. Leaves of sun and shade plants show very similar quantum yields despite their different growth habi-

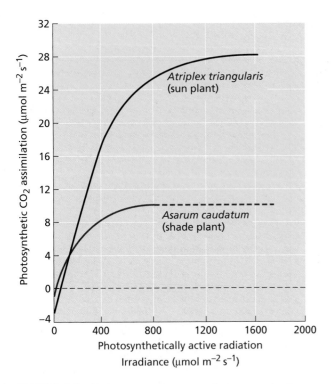

**FIGURE 9.7** Light-response curves of photosynthetic carbon fixation in sun and shade plants. *Atriplex triangularis* is a sun plant, and *Asarum caudatum* (a wild ginger) is a shade plant. Typically, shade plants have low light compensation points and have lower maximal photosynthetic rates than sun plants. The dashed red line has been extrapolated from the measured part of the curve. (After Harvey 1979.)

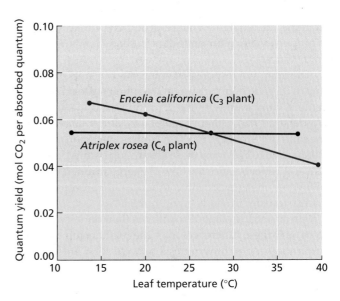

**FIGURE 9.8** The quantum yield of photosynthetic carbon fixation in a $C_3$ plant and a $C_4$ plant as a function of leaf temperature. In today's atmosphere, photorespiration increases with temperature in $C_3$ plants, and the energy cost of net $CO_2$ fixation increases accordingly. This higher energy cost is expressed in lower quantum yields at higher temperatures. Because of the $CO_2$-concentrating mechanisms of $C_4$ plants, photorespiration is low in these plants, and the quantum yield does not show temperature dependence. Note that at lower temperatures the quantum yield of $C_3$ plants is higher than that of $C_4$ plants, indicating that photosynthesis in $C_3$ plants is more efficient at lower temperatures. (After Ehleringer and Björkman 1977.)

tats. This is because the basic biochemical processes that determine quantum yield are the same for these two types of plants. But quantum yield can vary among plants with different photosynthetic pathways.

Recall that quantum yield is the ratio of a given light-dependent product to the number of absorbed photons (see Equation 7.5). Photosynthetic quantum yield can be expressed on either a $CO_2$ or an $O_2$ basis, and as explained in Chapter 7, the quantum yield of photochemistry is about 0.95. However, the photosynthetic quantum yield of an integrated process such as photosynthesis is lower than the theoretical yieldwhen measured in chloroplasts (organelles) or whole leaves. In fact, based on the biochemistry discussed in Chapter 8, we expect the maximum quantum yield for photosynthesis to be 0.125 for $C_3$ plants (one $CO_2$ molecule fixed per eight photons absorbed). But under today's atmospheric conditions (390 ppm $CO_2$, 21% $O_2$), the quantum yields for $CO_2$ of $C_3$ and $C_4$ leaves vary between 0.04 and 0.06 mole of $CO_2$ per mole of photons.

In $C_3$ plants the reduction from the theoretical maximum is caused primarily by energy loss through photorespiration. In $C_4$ plants the reduction is caused by the additional energy requirements of the $CO_2$-concentrating mechanism. If $C_3$ leaves are exposed to low $O_2$ concentrations, photorespiration is minimized and the quantum yield increases to about 0.09 mole of $CO_2$ per mole of photons. In contrast, if $C_4$ leaves are exposed to low $O_2$ concentrations, the quantum yields for $CO_2$ fixation remain constant at about 0.05 mole of $CO_2$ per mole of photons. This is because the carbon-concentrating mechanism in $C_4$ photosynthesis effectively eliminates $CO_2$ evolution via photorespiration.

Quantum yield also varies with temperature and $CO_2$ concentration because of their effect on the ratio of the carboxylase to oxygenase reactions of rubisco (see Chapter 8). Below 30°C in today's environment, quantum yields of $C_3$ plants are higher than those of $C_4$ plants; above 30°C, the situation is reversed (**FIGURE 9.8**).

At higher photon fluxes, the photosynthetic response to light starts to level off (**FIGURE 9.9**) and eventually reaches *saturation*. Light levels beyond the saturation point no longer affect photosynthetic rates, indicating that factors other than incident light, such as electron transport rate, rubisco activity, or the metabolism of triose phosphates, have become limiting to photosynthesis.

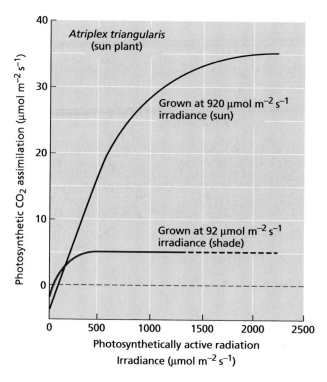

**FIGURE 9.9** Light-response of photosynthesis of a sun plant grown under sun or shade conditions. The upper curve represents an *A. triangularis* leaf grown at an irradiance ten times higher than that of the lower curve. In the leaf grown at the lower light levels, photosynthesis saturates at a substantially lower irradiance, indicating that the photosynthetic properties of a leaf depend on its growing conditions. The dashed red line has been extrapolated from the measured part of the curve. (After Björkman 1981.)

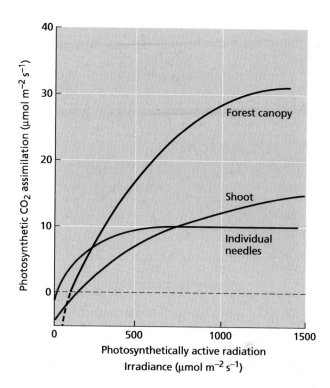

**FIGURE 9.10** Changes in photosynthesis (expressed on a per-square-meter basis) in individual needles, a complex shoot, and a forest canopy of Sitka spruce (*Picea sitchensis*) as a function of irradiance. Complex shoots consist of groupings of needles that often shade each other, similar to the situation in a canopy where branches often shade other branches. As a result of shading, much higher irradiance levels are needed to saturate photosynthesis. The dashed portion of the forest canopy trace has been extrapolated from the measured part of the curve. (After Jarvis and Leverenz 1983.)

Above the saturation point, photosynthesis is commonly referred to as *CO₂-limited* (see Figure 9.6), reflecting the inability of the Calvin–Benson cycle enzymes to keep pace with the production of ATP and NADPH from the light-dependent reactions. Light saturation levels for shade plants are substantially lower than those for sun plants. These levels usually reflect the maximum photon flux to which the leaf was exposed during growth.

The light-response curve of most leaves saturates between 500 and 1000 $\mu$mol m$^{-2}$ s$^{-1}$—well below full sunlight (which is about 2000 $\mu$mol m$^{-2}$ s$^{-1}$). Although individual leaves are rarely able to utilize full sunlight, whole plants usually consist of many leaves that shade each other, so only a small fraction of a plant's leaves are exposed to full sun at any given time of the day. The rest of the leaves receive subsaturating photon fluxes in the form of small patches of light that pass through gaps in the leaf canopy or in the form of light transmitted through other leaves.

Because the photosynthetic response of the intact plant is the sum of the photosynthetic activity of all the leaves, only rarely is photosynthesis light-saturated at the level

of the whole plant (**FIGURE 9.10**). Along these lines, crop productivity is related to the total amount of light received during the growing season, and given enough water and nutrients, the more light a crop receives, the higher the biomass (Ort and Baker 1988).

### Leaves must dissipate excess light energy

When exposed to excess light, leaves must dissipate the surplus absorbed light energy so that it does not harm the photosynthetic apparatus (**FIGURE 9.11**). There are several routes for energy dissipation that involve *nonphotochemical quenching* (see Chapter 7), the quenching of chlorophyll fluorescence by mechanisms other than photochemistry. The most important example involves the transfer of absorbed light energy away from electron transport toward heat production. Although the molecular mechanisms are not yet fully understood, the xanthophyll cycle appears to be an important avenue for dissipation of excess light energy (see **WEB ESSAY 9.1**).

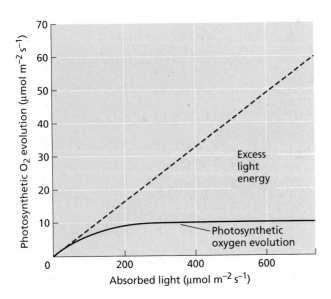

**FIGURE 9.11**  Excess light energy in relation to a light-response curve of photosynthetic oxygen evolution. The broken line shows theoretical oxygen evolution in the absence of any rate limitation to photosynthesis. At levels of photon flux up to 150 µmol m$^{-2}$ s$^{-1}$, a shade plant is able to utilize the absorbed light. Above 150 µmol m$^{-2}$ s$^{-1}$, however, photosynthesis saturates, and an increasingly larger amount of the absorbed light energy must be dissipated. At higher irradiances there is a large difference between the fraction of light used by photosynthesis versus that which must be dissipated (excess light energy). The differences are much greater in a shade plant than in a sun plant. (After Osmond 1994.)

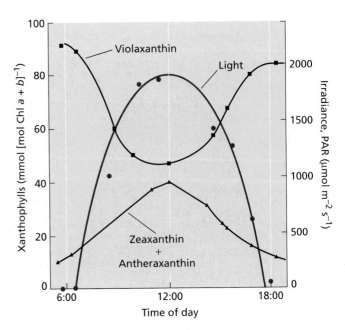

**FIGURE 9.12**  Diurnal changes in xanthophyll content as a function of irradiance in sunflower (*Helianthus annuus*). As the amount of light incident to a leaf increases, a greater proportion of violaxanthin is converted to antheraxanthin and zeaxanthin, thereby dissipating excess excitation energy and protecting the photosynthetic apparatus. (After Demmig-Adams and Adams 1996.)

**THE XANTHOPHYLL CYCLE**  Recall from Chapter 7 that the xanthophyll cycle, which comprises the three carotenoids violaxanthin, antheraxanthin, and zeaxanthin, is involved in the dissipation of excess light energy in the leaf (see Figure 7.35). Under high light, violaxanthin is converted to antheraxanthin and then to zeaxanthin. Note that in violaxanthin, both of the aromatic rings have a bound oxygen atom. In antheraxanthin only one of the two rings has a bound oxygen, and in zeaxanthin neither does. Experiments have shown that zeaxanthin is the most effective of the three xanthophylls in heat dissipation, and antheraxanthin is only half as effective. Whereas the levels of antheraxanthin remain relatively constant throughout the day, the zeaxanthin content increases at high irradiances and decreases at low irradiances.

In leaves growing under full sunlight, zeaxanthin and antheraxanthin can make up 60% of the total xanthophyll cycle pool at maximal irradiance levels attained at midday (**FIGURE 9.12**). In these conditions a substantial amount of excess light energy absorbed by the thylakoid membranes can be dissipated as heat, thus preventing damage to the photosynthetic machinery of the chloroplast (see Chapter 7). The fraction of light energy that is dissipated depends on irradiance, species, growth conditions, nutrient status, and ambient temperature (Demmig-Adams et al. 2006).

**THE XANTHOPHYLL CYCLE IN SUN AND SHADE**  Leaves that grow in full sunlight contain a substantially larger xanthophyll pool than do shade leaves, so they can dissipate higher amounts of excess light energy. Nevertheless, the xanthophyll cycle also operates in plants that grow in the low light of the forest understory, where they are only occasionally exposed to high light when sunlight passes through gaps in the overlying leaf canopy, forming sunflecks (described earlier in the chapter). Exposure to one sunfleck results in the conversion of much of the violaxanthin in the leaf to zeaxanthin. In contrast to typical leaves, in which violaxanthin levels increase again when irradiances drop, the zeaxanthin formed in shade leaves of the forest understory is retained and protects the leaf from damage due to subsequent sunflecks.

The xanthophyll cycle is also found in species such as conifers, the leaves of which remain green during winter, when photosynthetic rates are very low yet light absorption remains high. Contrary to the diurnal cycling of the xanthophyll pool observed in the summer, zeaxanthin levels remain high all day during the winter. Presumably this mechanism maximizes dissipation of light energy, thereby protecting the leaves against photooxidation during winter (Adams et al. 2001).

(A) Darkness     (B) Weak blue light     (C) Strong blue light

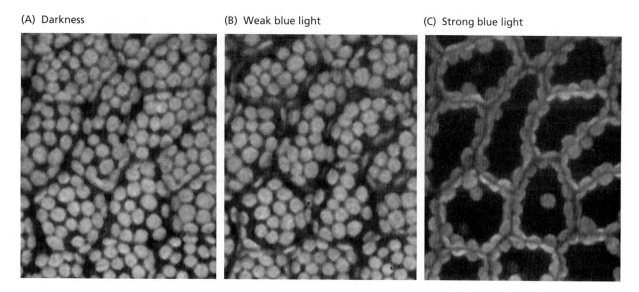

**FIGURE 9.13** Chloroplast distribution in photosynthesizing cells of the duckweed *Lemna*. These surface views show the same cells under three conditions: (A) darkness, (B) weak blue light, and (C) strong blue light. In A and B, chloroplasts are positioned near the upper surface of the cells, where they can absorb maximum amounts of light. When the cells are irradiated with strong blue light (C), the chloroplasts move to the side walls, where they shade each other, thus minimizing the absorption of excess light. (Courtesy of M. Tlalka and M. D. Fricker.)

**CHLOROPLAST MOVEMENTS** An alternative means of reducing excess light energy is to move the chloroplasts so that they are no longer exposed to high light. Chloroplast movement is widespread among algae, mosses, and leaves of higher plants (Haupt and Scheuerlein 1990; von Braun and Schleiff 2007). If chloroplast orientation and location are controlled, leaves can regulate how much of the incident light is absorbed. In the dark or under low light (**FIGURE 9.13A, B**), chloroplasts gather at the cell surfaces parallel to the plane of the leaf so that they are aligned perpendicular to the incident light—a position that maximizes absorption of light.

Under high light (**FIGURE 9.13C**), the chloroplasts move to the cell surfaces that are parallel to the incident light, thus avoiding excess absorption of light. Such chloroplast rearrangement can decrease the amount of light absorbed by the leaf by about 15% (Gorton et al. 1999). Chloroplast movement in leaves is a typical blue-light response (see Chapter 18). Blue light also controls chloroplast orientation in many of the lower plants, but in some algae, chloroplast movement is controlled by phytochrome (Haupt and Scheuerlein 1990; von Braun and Schleiff 2007). In leaves, chloroplasts move along actin microfilaments in the cytoplasm, and calcium regulates their movement (Tlalka and Fricker 1999).

**LEAF MOVEMENTS** Plants have evolved responses that reduce the excess light load on leaves during high sunlight periods, especially when transpiration and its cooling effects are reduced because of water stress. These responses often involve changes in the leaf orientation relative to the incoming sunlight. For example, paraheliotropic leaves of both alfalfa and lupine track the sun but at the same time can reduce incident light levels by folding leaflets together so that the leaf laminae become nearly parallel to the sun's rays. These movements are accomplished by changes in the turgor pressure of bulliform cells in the petiole. Another common response is wilting, as seen in many sunflowers, whereby a leaf droops to a vertical orientation, again effectively reducing the incident heat load and reducing transpiration and incident light levels.

### Absorption of too much light can lead to photoinhibition

Recall from Chapter 7 that when leaves are exposed to more light than they can utilize (see Figure 9.11), the reaction center of PSII is inactivated and often damaged in a phenomenon called **photoinhibition**. The characteristics of photoinhibition in the intact leaf depend on the amount of light to which the plant is exposed. The two types of photoinhibition are dynamic photoinhibition and chronic photoinhibition (Osmond 1994).

Under moderate excess light, **dynamic photoinhibition** is observed. Quantum efficiency decreases, but the maximum photosynthetic rate remains unchanged. Dynamic photoinhibition is caused by the diversion of absorbed light energy toward heat dissipation—hence the decrease in quantum efficiency. This decrease is often temporary, and quantum efficiency can return to its initial higher value when photon flux decreases below saturation lev-

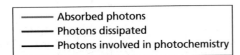

(A) Favorable environmental conditions

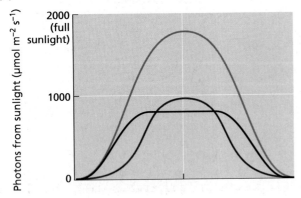

(B) Environmental stress conditions

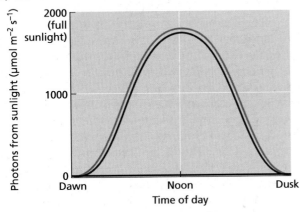

**FIGURE 9.14** Changes over the course of a day in the allocation of photons absorbed by sunlight. Shown here are contrasts in how the photons striking a leaf are either involved in photochemistry or thermally dissipated as excess energy by the leaves under favorable (upper panel) and stress (lower panel) conditions. (After Demmig-Adams and Adams 2000.)

els. **FIGURE 9.14** shows how the allocation of photons from sunlight is used for photosynthetic reactions versus being thermally dissipated as excess energy over the course of a day under favorable and stress environmental conditions.

**Chronic photoinhibition** results from exposure to high levels of excess light that damage the photosynthetic system and decrease both quantum efficiency and maximum photosynthetic rate. This would happen if the stress condition in Figure 9.14 persisted for an extended period of time. Chronic photoinhibition is associated with damage and replacement of the D1 protein from the reaction center of PSII (see Chapter 7). In contrast to dynamic photoinhibition, these effects are relatively long lasting, persisting for weeks or months.

Early researchers of photoinhibition interpreted all decreases in quantum efficiency as damage to the photosynthetic apparatus. It is now recognized that short-term decreases in quantum efficiency reflect protective mechanisms (see Chapter 7), whereas chronic photoinhibition represents actual damage to the chloroplast resulting from excess light or a failure of the protective mechanisms.

How significant is photoinhibition in nature? Dynamic photoinhibition appears to occur normally at midday, when leaves are exposed to maximum amounts of light and there is a corresponding reduction in carbon fixation. Photoinhibition is more pronounced at low temperatures, and it becomes chronic under more extreme climatic conditions.

# Photosynthetic Responses to Temperature

Photosynthesis ($CO_2$ uptake) and transpiration ($H_2O$ loss) share a common pathway. That is, $CO_2$ diffuses into the leaf, and $H_2O$ diffuses out, through the stomatal opening regulated by the guard cells. While these are independent processes, vast quantities of water are lost during photosynthetic periods, with the molar ratio of $H_2O$ loss to $CO_2$ uptake often reaching 250 to 500. This high water loss rate also removes heat from leaves through evaporative cooling, keeping them relatively cool under full sunlight conditions. Since photosynthesis is a temperature-dependent process, it is important to remember this linkage between two processes influenced by the degree of stomatal opening. As we will see, stomatal opening influences both leaf temperature and the extent of transpiration water loss.

## Leaves must dissipate vast quantities of heat

The heat load on a leaf exposed to full sunlight is very high. In fact, a leaf with an effective thickness of 300 μm of primarily water would warm up to a very high temperature if all available solar energy were absorbed and no heat were lost. However, this does not occur, because leaves absorb only about 50% of the total solar energy (300–3,000 nm), with most of the absorption occurring in the visible portion of the spectrum (see Figure 9.3). Yet the amount of the sun's energy absorbed by leaves is still enormous, and this heat load is dissipated by the emission of long-wave radiation (at about 10,000 nm), by sensible (i.e., perceptible) heat loss, and by evaporative (or latent) heat loss (**FIGURE 9.15**):

- Radiative heat loss: All objects emit radiation in proportion to their temperature. However, the maximum wavelength is inversely proportional to its temperature, and leaf temperatures are low enough that the wavelengths emitted are not visible to the human eye.

**Energy input**    **Heat dissipation**

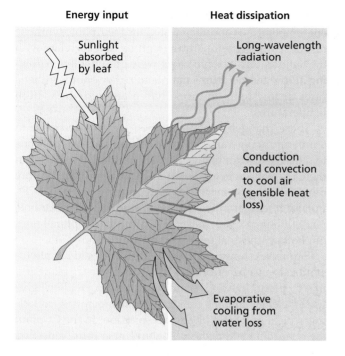

Sunlight absorbed by leaf

Long-wavelength radiation

Conduction and convection to cool air (sensible heat loss)

Evaporative cooling from water loss

**FIGURE 9.15** The absorption and dissipation of energy from sunlight by the leaf. The imposed heat load must be dissipated in order to avoid damage to the leaf. The heat load is dissipated by emission of long-wavelength radiation, by sensible heat loss to the air surrounding the leaf, and by the evaporative cooling caused by transpiration.

- Sensible heat loss: If the temperature of the leaf is higher than that of the air circulating around the leaf, the heat is convected (transferred) from the leaf to the air.

- Latent heat loss: Because the evaporation of water requires energy, when water evaporates from a leaf (transpiration), it withdraws large amounts of heat from the leaf and cools it. The human body is cooled by the same principle, through perspiration.

Sensible heat loss and evaporative heat loss are the most important processes in the regulation of leaf temperature, and the ratio of the two fluxes is called the **Bowen ratio** (Campbell and Norman 1996):

$$\text{Bowen ratio} = \frac{\text{Sensible heat loss}}{\text{Evaporative heat loss}}$$

In well-watered crops, transpiration (see Chapter 4), and hence water evaporation from the leaf, are high, so the Bowen ratio is low (see **WEB TOPIC 9.2**). Conversely, when evaporative cooling is limited, the Bowen ratio is large. For example, in a water-stressed crop, partial stomatal clo-

sure reduces evaporative cooling and the Bowen ratio is increased. The amount of evaporative heat loss (and thus the Bowen ratio) is influenced by the degree to which stomata remain open.

Plants with very high Bowen ratios conserve water, but also endure very high leaf temperatures. However, the high temperature difference between the leaf and the air does increase the amount of sensible heat loss. Reduced growth is usually correlated with high Bowen ratios, because a high Bowen ratio is indicative of at least partial stomatal closure.

### Photosynthesis is temperature sensitive

When photosynthetic rates are plotted as a function of temperature for either a $C_3$ leaf or a $C_4$ leaf under ambient $CO_2$ concentrations, the curve has a characteristic bell shape (**FIGURE 9.16**). Here we see two contrasting responses, in part reflecting the temperature optima expected when each species is grown under its natural temperature conditions. In this case, the $C_3$ species, *Atriplex glabriuscula*, commonly grows in cool coastal environments, while the $C_4$ plant, *Tidestromia oblongifolia*, was grown under its natural hot desert conditions. The ascending arm of the curve represents a temperature-dependent stimulation of enzymatic activities; the flat top portion of the curve represents a temperature range over which temperature is optimum for photosynthesis; the descending arm is associated with temperature-sensitive deleterious effects, some of which are reversible while others are not.

Temperature affects all biochemical reactions of photosynthesis as well as membrane integrity in chloroplasts,

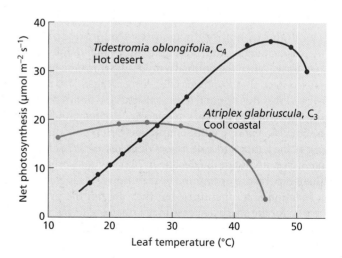

**FIGURE 9.16** Changes in photosynthesis as a function of temperature at normal atmospheric $CO_2$ concentrations for a $C_3$ plant grown in its natural cool habitat and a $C_4$ plant growing in its natural hot habitat under current ambient $CO_2$ concentrations. (After Berry and Björkman 1980.)

so it is not surprising that the responses to temperature are complex. At ambient $CO_2$ concentrations (see Figure 9.16), photosynthesis is limited by the activity of rubisco, and the response to temperature reflects two conflicting processes: an increase in carboxylation rate and a decrease in the affinity of rubisco for $CO_2$ as the temperature rises (see Chapter 8). There is evidence that rubisco activity decreases at high temperatures because of temperature effects on rubisco activase (see Chapter 8). These opposing effects dampen the temperature response of photosynthesis at ambient $CO_2$ concentrations.

By contrast, when photosynthetic rate is plotted as a function of temperature in a leaf with $C_4$ photosynthesis, the curves are bell-shaped in both cases (see Figure 9.16), since the leaf interior is $CO_2$-saturated (as was discussed in Chapter 8). This is one of the reasons that leaves of $C_4$ plants tend to have a higher photosynthetic temperature optimum than do leaves of $C_3$ plants when grown under common conditions.

At low temperatures, photosynthesis can also be limited by factors such as phosphate availability in the chloroplast (Sage and Sharkey 1987). When triose phosphates are exported from the chloroplast to the cytosol, an equimolar amount of inorganic phosphate is taken up via translocators in the chloroplast membrane. If the rate of triose phosphate utilization in the cytosol decreases, phosphate uptake into the chloroplast is inhibited and photosynthesis becomes phosphate limited (Geiger and Servaites 1994). Starch synthesis and sucrose synthesis decrease rapidly with decreasing temperature, reducing the demand for triose phosphates and causing the phosphate limitation observed at low temperatures.

### There is an optimal temperature for photosynthesis

The highest photosynthetic rates seen in response to increasing temperature represent the **optimal temperature response**. When the optimal temperature for a given plant is exceeded, photosynthetic rates decrease again. It has been argued that this optimal temperature is the point at which the capacities of the various steps of photosynthesis are optimally balanced, with some of the steps becoming limiting as the temperature decreases or increases. What factors are associated with the decline in photosynthesis beyond the temperature optimum? Respiration rates increase as a function of temperature, but they are not the primary reason for the sharp decrease in net photosynthesis at high temperatures. Rather, membrane-bound electron transport processes become unstable at high temperatures, cutting off the supply of reducing power and leading to a sharp overall decrease in photosynthesis.

Optimal temperatures have strong genetic (adaptation) and environmental (acclimation) components. Plants of different species growing in habitats with different temperatures have different optimal temperatures for photo-

synthesis, and plants of the same species, grown at different temperatures and then tested for their photosynthetic responses, show temperature optima that correlate with the temperature at which they were grown. Plants growing at low temperatures maintain higher photosynthetic rates at low temperatures than plants grown at high temperatures.

These changes in photosynthetic rates in response to temperature play an important role in plant adaptations to different environments. Plants are remarkably plastic in their adaptations to temperature. In the lower temperature range, plants growing in alpine areas are capable of net $CO_2$ uptake at temperatures close to 0°C; at the other extreme, plants living in Death Valley, California, have optimal rates of photosynthesis at temperatures approaching 50°C.

Figure 9.8 shows changes in quantum yield for photosynthesis as a function of temperature in a $C_3$ plant and in a $C_4$ plant. In the $C_4$ plant the quantum yield or light-use efficiency remains constant with temperature, reflecting typical low rates of photorespiration. In the $C_3$ plant the quantum yield decreases with temperature, reflecting a stimulation of photorespiration by temperature and an ensuing higher energy demand per net $CO_2$ fixed. While quantum yield effects are most expressed under light-limited conditions, a similar pattern is reflected in photorespiration rates under high light as a function of temperature.

The combination of reduced quantum yield and increased photorespiration leads to expected differences in the photosynthetic capacities of $C_3$ and $C_4$ plants in habitats with different temperatures. The predicted relative rates of primary productivity of $C_3$ and $C_4$ grasses along a latitudinal transect in the Great Plains of North America from southern Texas in the USA to Manitoba in Canada (Ehleringer 1978) are shown in **FIGURE 9.17**. This decline in $C_4$ relative to $C_3$ productivity moving northward very closely parallels the declining abundance of plants with these pathways in the Great Plains: $C_4$ species are more common below 40°N, and $C_3$ species dominate above 45°N (see Figure 9.17 and **WEB TOPIC 9.3**).

## Photosynthetic Responses to Carbon Dioxide

We have discussed how light and temperature influence plant growth and leaf anatomy. Now we turn our attention to how $CO_2$ concentration affects photosynthesis. $CO_2$ diffuses from the atmosphere into leaves—first through stomata, then through the intercellular air spaces, and ultimately into cells and chloroplasts. In the presence of adequate amounts of light, higher $CO_2$ concentrations support higher photosynthetic rates. The reverse is also true: Low $CO_2$ concentrations can limit the amount of photosynthesis in $C_3$ plants.

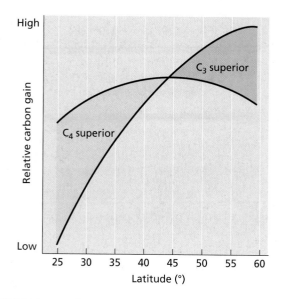

**FIGURE 9.17** The relative rates of photosynthetic carbon gain predicted for identical $C_3$ and $C_4$ grass canopies as a function of latitude across the Great Plains of North America. (After Ehleringer 1978.)

In this section we will discuss the concentration of atmospheric $CO_2$ in recent history, and its availability for carbon-fixing processes. Then we'll consider the limitations that $CO_2$ places on photosynthesis and the impact of the $CO_2$-concentrating mechanisms of $C_4$ plants.

## Atmospheric $CO_2$ concentration keeps rising

Carbon dioxide is a trace gas in the atmosphere, presently accounting for about 0.039%, or 390 parts per million (ppm), of air. The partial pressure of ambient $CO_2$ ($c_a$) varies with atmospheric pressure and is approximately 39 pascals (Pa) at sea level (see **WEB TOPIC 9.4**). Water vapor usually accounts for up to 2% of the atmosphere and $O_2$ for about 21%. The bulk of the atmosphere—77%—is nitrogen.

The current atmospheric concentration of $CO_2$ is almost twice the concentration that has prevailed during most of the last 420,000 years, as measured from air bubbles trapped in glacial ice in Antarctica (**FIGURE 9.18A, B**). Today's atmospheric $CO_2$ is likely higher than any that Earth has experienced in the last 2 million years. Except for the last 200 years, atmospheric $CO_2$ concentrations during the recent geologic past are thought to have been low; thus, the plants in the world today evolved in a low-$CO_2$ world.

The available evidence indicates that $CO_2$ concentrations greater than 1,000 ppm have not existed on Earth since the warm Cretaceous, over 70 million years ago. Thus, until the dawn of the Industrial Revolution, the geo-

**FIGURE 9.18** Concentration of atmospheric $CO_2$ from 420,000 years ago to the present. (A) Past atmospheric $CO_2$ concentrations, determined from bubbles trapped in glacial ice in Antarctica, were much lower than current levels. (B) In the last 1000 years, the rise in $CO_2$ concentration coincides with the Industrial Revolution and the increased burning of fossil fuels. (C) Current atmospheric concentrations of $CO_2$, measured at Mauna Loa, Hawaii, continue to rise. The wavy nature of the trace is caused by change in atmospheric $CO_2$ concentrations associated with seasonal changes in relative balance between photosynthesis and respiration rates. Each year the highest $CO_2$ concentration is observed in May, just before the Northern Hemisphere growing season, and the lowest concentration is observed in October. (After Barnola et al. 1994, Keeling and Whorf 1994, Neftel et al. 1994, and Keeling et al. 1995.)

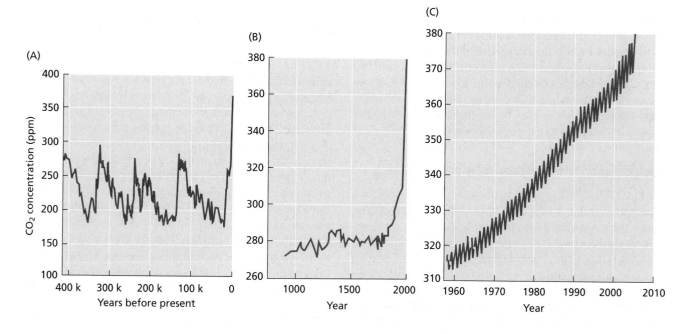

logic trend over the past 50 to 70 million years was one of decreasing atmospheric $CO_2$ concentrations (WEB TOPIC 9.5). What we would like to know is just how the recently elevated atmospheric $CO_2$ level affects photosynthesis and respiration processes, and how higher levels will affect these processes in the future.

Currently, the $CO_2$ concentration of the atmosphere is increasing by about 1 to 3 ppm each year, primarily because of the burning of fossil fuels such as coal, oil, and natural gas (FIGURE 9.18C). Since 1958, when C. David Keeling began systematic measurements of $CO_2$ in the clean air at Mauna Loa, Hawaii, atmospheric $CO_2$ concentrations have increased by more than 20% (Keeling et al. 2005). By 2100 the atmospheric $CO_2$ concentration could reach 600 to 750 ppm unless fossil fuel emissions are controlled (see WEB TOPIC 9.6).

**THE GREENHOUSE EFFECT**    The consequences of this increase in atmospheric $CO_2$ are under intense scrutiny by scientists and government agencies, particularly because of predictions that the *greenhouse effect* is altering the world's climate. The term **greenhouse effect** refers to the warming of Earth's climate that is caused by the trapping of long-wavelength radiation by the atmosphere.

A greenhouse roof transmits visible light, which is absorbed by plants and other surfaces inside the greenhouse. Some of the absorbed light energy is converted to heat, and some of it is re-emitted as long-wavelength radiation. Because glass transmits long-wavelength radiation very poorly, this radiation cannot leave the greenhouse through the glass roof, and the greenhouse heats up.

Certain gases in the atmosphere, particularly $CO_2$ and methane, play a role similar to that of the glass roof in a greenhouse. The increased $CO_2$ concentration and temperature associated with the greenhouse effect can influence photosynthesis. At current atmospheric $CO_2$ concentrations, photosynthesis in $C_3$ plants is $CO_2$ limited (as we will discuss later in the chapter), but this situation could change as atmospheric $CO_2$ concentrations continue to rise. Under laboratory conditions, most $C_3$ plants grow 30 to 60% faster when $CO_2$ concentration is doubled (to 600–750 ppm), and the growth rate becomes limited by the nutrients available to the plant (Bowes 1993).

### $CO_2$ diffusion to the chloroplast is essential to photosynthesis

For photosynthesis to occur, carbon dioxide must diffuse from the atmosphere into the leaf and into the carboxylation site of rubisco. Because diffusion rates depend on concentration gradients in leaves (see Chapters 3 and 6), appropriate gradients are needed to ensure adequate diffusion of $CO_2$ from the leaf surface to the chloroplast.

The cuticle that covers the leaf is nearly impermeable to $CO_2$, so the main port of entry of $CO_2$ into the leaf is the

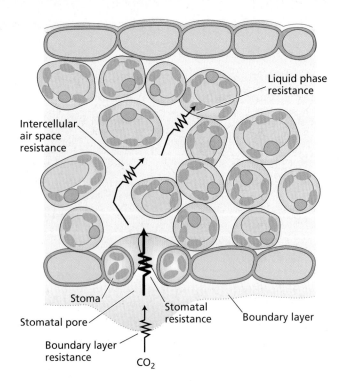

**FIGURE 9.19**    Points of resistance to the diffusion of $CO_2$ from outside the leaf to the chloroplasts. The stomatal pore is the major point of resistance to $CO_2$ diffusion.

stomatal pore. The same path is traveled in the reverse direction by $H_2O$. $CO_2$ diffuses through the pore into the substomatal cavity and into the intercellular air spaces between the mesophyll cells. This portion of the diffusion path of $CO_2$ into the chloroplast is a gaseous phase. The remainder of the diffusion path to the chloroplast is a liquid phase, which begins at the water layer that wets the walls of the mesophyll cells and continues through the plasma membrane, the cytosol, and the chloroplast. (For the properties of $CO_2$ in solution, see WEB TOPIC 9.6.)

The sharing of the stomatal entry pathway by $CO_2$ and water presents the plant with a functional dilemma. In air of high relative humidity, the diffusion gradient that drives water loss is about 50 times larger than the gradient that drives $CO_2$ uptake. In drier air, this difference can be even larger. Therefore, a decrease in stomatal resistance through the opening of stomata facilitates higher $CO_2$ uptake but is unavoidably accompanied by substantial water loss.

Each portion of this diffusion pathway imposes a resistance to $CO_2$ diffusion, so the supply of $CO_2$ for photosynthesis meets a series of different points of resistance. The gas phase of $CO_2$ diffusion into the leaf can be divided into three components—the boundary layer, the stomata, and the intercellular spaces of the leaf—each of which imposes a resistance to $CO_2$ diffusion (FIGURE 9.19). An evaluation of the magnitude of each point of resistance is helpful for understanding $CO_2$ limitations to photosynthesis.

The boundary layer consists of relatively unstirred air at the leaf surface, and its resistance to diffusion is called the **boundary layer resistance**. The magnitude of the boundary layer resistance decreases with leaf size and wind speed. The boundary layer resistance to water and $CO_2$ diffusion is physically related to the boundary layer resistance to sensible heat loss discussed earlier.

Smaller leaves have a lower boundary layer resistance to $CO_2$ and water diffusion, and to sensible heat loss. Leaves of desert plants are usually small, facilitating sensible heat loss. The large leaves often found in the shade of the humid tropics can have large boundary layer resistances, but these leaves can dissipate the radiation heat load by evaporative cooling made possible by the abundant water supply in these habitats.

After diffusing through the boundary layer, $CO_2$ enters the leaf through the stomatal pores, which impose the next type of resistance in the diffusion pathway, **stomatal resistance**. Under most conditions in nature, in which the air around a leaf is seldom completely still, the boundary layer resistance is much smaller than the stomatal resistance, and the main limitation to $CO_2$ diffusion is imposed by the stomatal resistance.

There is also a resistance to $CO_2$ diffusion in the air spaces that separate the substomatal cavity from the walls of the mesophyll cells, called the **intercellular air space resistance**. This resistance is usually small, causing a drop of 0.5 Pa or less in partial pressure of $CO_2$ from the 38 Pa outside the leaf.

The resistance to $CO_2$ diffusion of the liquid phase in $C_3$ leaves—the **liquid phase resistance**, also called **mesophyll resistance**—encompasses diffusion from the intercellular leaf spaces to the carboxylation sites in the chloroplast. Since localization of chloroplasts near the cell periphery minimizes the distance that $CO_2$ must diffuse through liquid to reach carboxylation sites within the chloroplast, the resistance to $CO_2$ diffusion is thought to be approximately one-tenth of the combined boundary layer resistance and stomatal resistance when the stomata are fully open. However, recent research has suggested that mesophyll resistance can be higher.

Because the stomatal pores usually impose the largest resistance to $CO_2$ uptake and water loss in the diffusion pathway, this single point of regulation provides the plant with an effective way to control gas exchange between the leaf and the atmosphere. In experimental measurements of gas exchange from leaves, the boundary layer resistance and the intercellular air space resistance are often ignored, and the stomatal resistance is used as the single parameter describing the gas phase resistance to $CO_2$ (see **WEB TOPIC 9.4**).

## Patterns of light absorption generate gradients of $CO_2$ fixation

We have discussed how leaf anatomy is specialized for capturing light and how it also facilitates the internal dif-

fusion of $CO_2$, but where inside an individual leaf do maximum rates of photosynthesis occur? In most leaves, light is preferentially absorbed at the upper surface, whereas $CO_2$ enters through the lower surface. Given that light and $CO_2$ enter from opposing sides of the leaf, does photosynthesis occur uniformly within the leaf tissues, or is there a gradient in photosynthesis across the leaf?

For most leaves, once $CO_2$ has diffused through the stomata, internal $CO_2$ diffusion is rapid, so limitations on photosynthetic performance within the leaf are imposed by factors other than internal $CO_2$ supply. When white light enters the upper surface of a leaf, blue and red photons are preferentially absorbed by chloroplasts near the irradiated surface (**FIGURE 9.20**), owing to the strong absorption bands of chlorophyll in the blue and red regions of the spectrum (see Figure 7.3). Green light, on the other hand, penetrates deeper into the leaf. Chlorophyll absorbs green

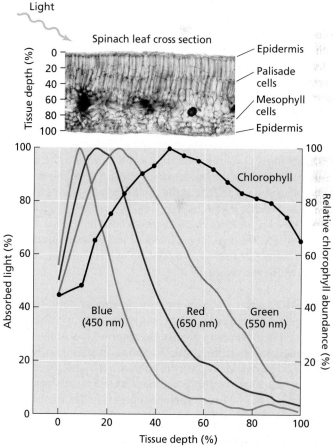

**FIGURE 9.20** Distribution of absorbed light in spinach sun leaves. Irradiation with blue, green, or red light results in different profiles of absorbed light in the leaf. The micrograph above the graph shows a cross section of a spinach leaf, with rows of palisade cells occupying nearly half of the leaf thickness. The shapes of the curves are in part a result of the unequal distribution of chlorophyll within the leaf tissues. (After Nishio et al. 1993 and Vogelmann and Han 2000; micrograph courtesy of T. Vogelmann.)

light poorly (again, see Figure 7.3), yet green light is very effective in supplying energy for photosynthesis in the tissues within the leaf depleted of blue and red photons.

The capacity of the leaf tissue for photosynthetic $CO_2$ assimilation depends to a large extent on its rubisco content. In spinach (*Spinacea oleracea*) and fava bean (*Vicia faba*), rubisco content is low at the top of the leaf, increases toward the middle, and decreases again toward the bottom, similar to the distribution of chlorophyll in a leaf, as shown in Figure 9.20. Like the distribution of chlorophyll, the distribution of photosynthetic carbon fixation within the leaf has a bell-shaped curve.

## $CO_2$ imposes limitations on photosynthesis

For many crops, such as tomatoes, lettuce, cucumbers, and roses growing in greenhouses under optimal water and nutrition, the carbon dioxide enrichment in the greenhouse environment above natural atmospheric levels results in increased productivity. Expressing photosynthetic rate as a function of the partial pressure of $CO_2$ in the intercellular air space ($c_i$) within the leaf (see **WEB TOPIC 9.4**) makes it possible to evaluate limitations to photosynthesis imposed by $CO_2$ supply. At very low intercellular $CO_2$ concentrations, photosynthesis is strongly limited by the low $CO_2$.

Increasing intercellular $CO_2$ to the concentration at which photosynthesis and respiration balance each other defines the **$CO_2$ compensation point**, at which the net efflux of $CO_2$ from the leaf is zero (**FIGURE 9.21**). This concept is analogous to that of the light compensation point discussed earlier in the chapter: *The $CO_2$ compensation point reflects the balance between photosynthesis and respiration as a function of $CO_2$ concentration, whereas the light compensation point reflects that balance as a function of photon flux under constant $O_2$ concentration.*

**$C_3$ PLANTS** In $C_3$ plants, increasing atmospheric $CO_2$ above the compensation point stimulates photosynthesis over a wide concentration range (see Figure 9.21). At low to intermediate $CO_2$ concentrations, photosynthesis is limited by the carboxylation capacity of rubisco. At high $CO_2$ concentrations, photosynthesis becomes limited by the capacity of the Calvin–Benson cycle to regenerate the acceptor molecule ribulose 1,5-bisphosphate, which depends on electron transport rates. However, photosynthesis continues to increase with increasing $CO_2$ because carboxylation replaces oxygenation on rubisco (see Chapter 8). By regulating stomatal conductance, most leaves appear to regulate their $c_i$ (internal partial pressure for $CO_2$) so that it is at an intermediate concentration between the limits imposed by carboxylation capacity and the capacity to regenerate ribulose 1,5-bisphosphate.

A plot of $CO_2$ assimilation as a function intercellular partial pressures of $CO_2$ tells us how photosynthesis is

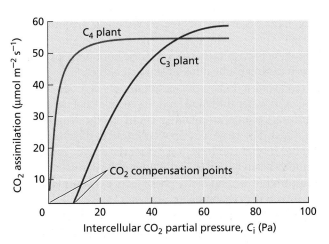

**FIGURE 9.21** Changes in photosynthesis as a function of intercellular $CO_2$ concentrations in Arizona honeysweet (*Tidestromia oblongifolia*), a $C_4$ plant, and creosote bush (*Larrea divaricata*), a $C_3$ plant. Photosynthetic rate is plotted against calculated intercellular partial pressure of $CO_2$ inside the leaf (see Equation 5 in **WEB TOPIC 8.4**). The partial pressure at which $CO_2$ assimilation is zero defines the $CO_2$ compensation point. (After Berry and Downton 1982.)

regulated by $CO_2$, independent of the functioning of stomata (see Figure 9.21). Inspection of such a plot for $C_3$ and $C_4$ plants reveals interesting differences between the two pathways of carbon metabolism:

- In $C_4$ plants, photosynthetic rates saturate at $c_i$ values of about 15 Pa, reflecting the effective $CO_2$-concentrating mechanisms operating in these plants (see Chapter 8).

- In $C_3$ plants, increasing $c_i$ levels continue to stimulate photosynthesis over a much broader $CO_2$ range.

- In $C_4$ plants, the $CO_2$ compensation point is zero or nearly zero, reflecting their very low levels of photorespiration (see Chapter 8).

- In $C_3$ plants, the $CO_2$ compensation point is about 10 Pa, reflecting $CO_2$ production because of photorespiration (see Chapter 8).

These responses indicate that $C_3$ plants may benefit more from ongoing increases in today's atmospheric $CO_2$ concentrations (see Figure 9.18). Because photosynthesis in $C_4$ plants is $CO_2$-saturated at low concentrations, $C_4$ plants do not benefit much from increases in atmospheric $CO_2$ concentrations.

In fact, the ancestral photosynthetic pathway is $C_3$ photosynthesis, and $C_4$ photosynthesis is a derived pathway. During geologic time periods when atmospheric $CO_2$ concentrations were very much higher than they are today, $CO_2$ diffusion through stomata into $C_3$ leaves would have resulted in higher $c_i$ values and therefore higher photosynthetic rates. While $C_3$ photosynthesis is typically $CO_2$-diffusion limited today, $C_3$ plants still account for nearly 70% of the world's primary productivity. The evolution of $C_4$ photosynthesis is one biochemical adaptation to a $CO_2$-limited atmosphere. Our current understanding is that $C_4$ photosynthesis may have evolved recently, some 10 to 15 million years ago.

**$C_4$ PLANTS**   If the ancient Earth of more than 50 million years ago had atmospheric $CO_2$ concentrations that were well above current atmospheric conditions, under what atmospheric conditions might we expect that $C_4$ photosynthesis should become a major photosynthetic pathway found in the Earth's ecosystems? Ehleringer et al. (1997) suggest that $C_4$ photosynthesis first became a prominent component of terrestrial ecosystems in the warmest growing regions of the Earth when global $CO_2$ concentrations decreased below some critical and as yet unknown threshold $CO_2$ concentration (**FIGURE 9.22**). That is, the negative impacts of high photorespiration and $CO_2$ limitation on $C_3$ photosynthesis would be greatest under warm to hot growing conditions, especially when atmospheric $CO_2$ is reduced. The $C_4$-favorable growing areas would have been located in those geographic regions with the warmest temperatures. $C_4$ plants would have been most favored during periods of Earth's history when $CO_2$ levels were lowest. In today's world, these regions are the subtropical grasslands and savannas. There are now extensive data to indicate that $C_4$ photosynthesis was more prominent during the glacial periods when atmospheric $CO_2$ levels were below 200 ppm than it is today (see Figure 9.18). Other factors may have contributed to the expansion of $C_4$ plants, but certainly low atmospheric $CO_2$ was one important factor favoring their geographic expansion.

Because of the $CO_2$-concentrating mechanisms in $C_4$ plants, $CO_2$ concentration at the carboxylation sites within $C_4$ chloroplasts is often saturating for rubisco activity. As a result, plants with $C_4$ metabolism need less rubisco than $C_3$ plants to achieve a given rate of photosynthesis, and require less nitrogen to grow (von Caemmerer 2000).

In addition, the $CO_2$-concentrating mechanism allows the leaf to maintain high photosynthetic rates at lower $c_i$ values, which require lower rates of stomatal conductance for a given rate of photosynthesis. Thus, $C_4$ plants can use water and nitrogen more efficiently than $C_3$ plants can. On the other hand, the additional energy cost of the concentrating mechanism (see Chapter 8) makes $C_4$ plants less efficient in their utilization of light. This is probably one of the reasons that most shade-adapted plants in temperate regions are $C_3$ plants.

**CAM PLANTS**   Plants with crassulacean acid metabolism (CAM), including many cacti, orchids, bromeliads, and other succulents, have stomatal activity patterns that contrast with those found in $C_3$ and $C_4$ plants. CAM plants open their stomata at night and close them during the day, exactly the opposite of the pattern observed in leaves of $C_3$ and $C_4$ plants (**FIGURE 9.23**). At night, atmospheric $CO_2$ diffuses into CAM plants where it is combined with phosphoenolpyruvate and fixed into malate (see Chapter 8).

The ratio of water loss to $CO_2$ uptake is much lower in CAM plants than it is in either $C_3$ or $C_4$ plants. This is because stomata are primarily open only at night, when lower temperatures and higher humidity contribute to a lower transpiration rate.

The main photosynthetic constraint on CAM metabolism is that the capacity to store malic acid is limited, and this limitation restricts the total amount of $CO_2$ uptake. However, some CAM plants are able to enhance total photosynthesis during wet conditions by fixing $CO_2$ via the Calvin–Benson cycle at the end of the day, when temperature gradients are less extreme. In water-limited conditions, stomata open only at night.

Cladodes (flattened stems) of cacti can survive after detachment from the plant for several months without water. Their stomata are closed all the time, and the $CO_2$ released by respiration is refixed into malate. This process, which has been called *CAM idling*, also allows the intact plant to survive for prolonged drought periods while losing remarkably little water.

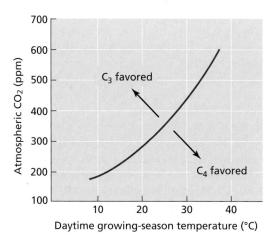

**FIGURE 9.22** The combination of atmospheric carbon dioxide levels and daytime growing season temperatures that are predicted to favor $C_3$ versus $C_4$ grasses. At any point in time, the Earth is at a single atmospheric carbon dioxide concentration, resulting in the expectation that $C_4$ plants would be most common in habitats with the warmest growing seasons. (After Ehleringer et al. 1997.)

## How will photosynthesis and respiration change in the future under elevated $CO_2$ conditions?

A central question in plant physiology today is: How are photosynthesis and respiration modified in an environment where $CO_2$ levels are 400 ppm, 500 ppm, or even

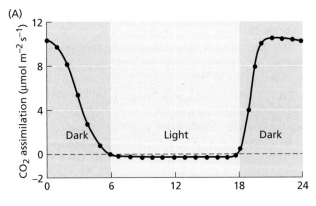

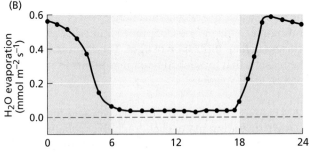

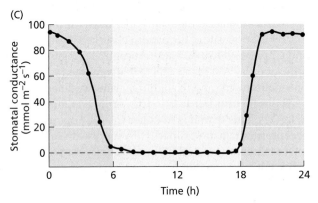

**FIGURE 9.23** Photosynthetic carbon assimilation, evaporation, and stomatal conductance of a CAM plant, the cactus *Opuntia ficus-indica*, during a 24-hour period. The whole plant was kept in a gas exchange chamber in the laboratory. The shaded areas indicate the dark periods. Three parameters were measured over the study period: (A) photosynthetic rate, (B) water loss, and (C) stomatal conductance. In contrast to plants with $C_3$ or $C_4$ metabolism, CAM plants open their stomata and fix $CO_2$ at night. (After Gibson and Nobel 1986.)

While these long-term FACE experiments are still underway, they are already providing key new insights into how plants will respond in the future. One key observation is that plants with the $C_3$ photosynthetic pathway are much more responsive than $C_4$ plants under well-watered conditions, with the net photosynthetic rate increasing 20% or more in $C_3$ plants and not at all in $C_4$ plants. Photosynthesis rises because intercellular $CO_2$ levels increase (recall Figure 9.21). At the same time, there is a down-regulation of photosynthetic capacity manifested in reduced activity of the enzymes associated with the dark reactions of photosynthesis (Ainsworth and Rogers 2007).

While $CO_2$ is indeed important for photosynthesis, other factors are important for growth under elevated $CO_2$ (Long et al. 2004, 2006). For example, a common FACE observation is that plant growth becomes quickly constrained by nutrient availability. A second, surprising observation is that soil moisture and the presence of trace gases, such as ozone, can reduce the net photosynthetic response below the maximum values predicted from initial greenhouse studies of a decade ago. Warmer and drier conditions are also predicted to occur as a result of increased $CO_2$ in the atmosphere. Important progress will be made in the near future through the study of how fertilized and irrigated crops compare to plants in natural ecosystems in their response to elevated $CO_2$. Understanding these responses is crucial as society looks for increased agricultural outputs to support rising human populations and to provide raw materials for biofuels.

Elevated $CO_2$ levels will affect many plant processes; for instance, leaves tend to keep their stomata more closed under elevated $CO_2$ levels. As a direct consequence of reduced transpiration, leaf temperatures are higher (see Figure 9.24C). Elevated temperatures will feed back on basic mitochondrial respiration and on the respiration of soil microbes and fungi. This is indeed an exciting and promising area of current research. From FACE studies, it is becoming increasingly clear that an acclimation process occurs under higher $CO_2$ levels in which respiration rates are different than they would be under today's atmospheric conditions, but not as high as would have been predicted without the down-regulation acclimation response (Long et al. 2004, 2006).

higher? This question is particularly relevant as humans continue to add $CO_2$ derived from fossil fuel combustion to the world's atmosphere. To study this question, scientists need to be able to create realistic models of future environments. A promising approach to the study of plant physiology and ecology in environments with elevated $CO_2$ levels has been the use of *Free Air CO₂ Enrichment* (FACE) experiments.

For FACE experiments, entire fields of plants or natural ecosystems are enclosed in rings of tubes that add $CO_2$ to the air to create the high-$CO_2$ environment we might expect 25–50 years from now. **FIGURE 9.24** shows FACE experiments in different major vegetation types. Figure 9.24A shows experiments being conducted in Wisconsin, where mixed and unmixed stands of aspen trees are growing in an elevated $CO_2$ environment. Figure 9.24B shows FACE experiments being conducted in a soybean field in Illinois.

(A)

**FIGURE 9.24**  Free Air $CO_2$ Enrichment (FACE) experiments are used to study how plants and ecosystems will respond to future $CO_2$ levels. Shown here are FACE experiments in stands of deciduous trees (A) and in a soybean field (B). (C) Under elevated $CO_2$ levels, leaf stomata are more closed, resulting in higher leaf temperatures as shown by the infrared image of a soybean canopy. (A courtesy of David F. Karnosky; B courtesy of USDA; C from Long et al. 2006.)

(B)

(C)

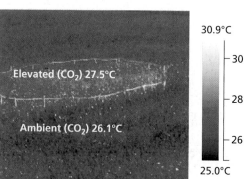

## Identifying Different Photosynthetic Pathways

We can learn more about the different photosynthetic pathways in plants by measuring the chemical composition of plant tissues. We do this using measurements of the abundances of stable isotopes in plants (Dawson et al. 2002). In particular, the stable isotopes of carbon atoms in a leaf contain useful information about photosynthesis. Recall that isotopes are simply different forms of an element. In the different isotopes of an element, the number of protons remains constant, since that defines the element, but the number of neutrons varies. Isotopes can be stable or radioactive.

Stable isotopes of an element remain constant in abundance, unchanged over time. In contrast, radioactive isotopes of an element decay to form different elements over time. The two stable isotopes of carbon are $^{12}C$ and $^{13}C$, differing in composition only by the addition of an additional neutron in $^{13}C$. $^{11}C$ and $^{14}C$ are radioactive isotopes of carbon that are frequently used in biological tracer experiments.

### How do we measure the stable carbon isotopes of plants?

Atmospheric $CO_2$ contains the naturally occurring stable carbon isotopes $^{12}C$ and $^{13}C$ in the proportions 98.9% and 1.1%, respectively. $^{14}CO_2$ is radioactive and is present in small quantities ($10^{-10}$%). The chemical properties of $^{13}CO_2$ are identical to those of $^{12}CO_2$, but plants assimilate less $^{13}CO_2$ than $^{12}CO_2$. In other words, leaves discriminate against the heavier isotope of carbon during photosynthesis, and therefore they have lower $^{13}C/^{12}C$ ratios than are found in atmospheric $CO_2$.

The $^{13}C/^{12}C$ isotope composition is measured by use of a mass spectrometer, which yields the following ratio:

$$R = \frac{^{13}CO_2}{^{12}CO_2} \tag{9.1}$$

The **carbon isotope ratio** of plants, $\delta^{13}C$, is quantified on a per mil (‰) basis:

$$\delta^{13}C \, ^{0}/_{00} = \left( \frac{R_{\text{sample}}}{R_{\text{standard}}} - 1 \right) \times 1000 \tag{9.2}$$

where the standard represents the carbon isotopes contained in a fossil belemnite from the Pee Dee limestone formation of South Carolina. The $\delta^{13}C$ of atmospheric $CO_2$ has a value of –8‰, meaning that there is less $^{13}C$ in atmospheric $CO_2$ than is found in the carbonate of the belemnite standard.

What are some typical values for carbon isotope ratios of plants? $C_3$ plants have a $\delta^{13}C$ value of about –28‰; $C_4$ plants have an average value of –14‰ (Farquhar et al. 1989). Both $C_3$ and $C_4$ plants have less $^{13}C$ than does $CO_2$ in the atmosphere, which means that leaf tissues discriminate against $^{13}C$ during the photosynthetic process. Cerling et al. (1997) provided $\delta^{13}C$ data for a large number of $C_3$ and $C_4$ plants from around the world (**FIGURE 9.25**).

What becomes clear from Figure 9.25 is that there is a wide spread of $\delta^{13}C$ values in $C_3$ and $C_4$ plants, with averages of –28‰ and –14‰, respectively. These $\delta^{13}C$ variations actually reflect the consequences of small variations in physiology associated with changes in stomatal conductance in different environmental conditions. Thus, $\delta^{13}C$ values can be used both to distinguish between $C_3$ and $C_4$ photosynthesis and to further reveal details about stomatal conditions for plants grown in different environments (such as the tropics versus deserts).

Differences in carbon isotope ratio are easily detectable with mass spectrometers that allow for very precise measurements of the abundance of $^{12}C$ and $^{13}C$ in either different molecules or different tissues. Many of our foods, such as wheat (*Triticum aestivum*), rice (*Oryza sativa*), potatoes (*Solanum tuberosum*), and beans (*Phaseolus* spp.) are products of $C_3$ plants. Yet many of our most productive crops are $C_4$ plants such as corn (maize; *Zea mays*), sugarcane (*Saccharum officinarum*), and sorghum (*Sorghum bicolor*). Carbohydrates extracted from all of these foods may be chemically identical, but they are $C_3$–$C_4$ distinguishable on the basis of their $\delta^{13}C$ values. For example, measuring the $\delta^{13}C$ values of table sugar (sucrose) makes it possible to determine if the sucrose came from sugar beet (*Beta vulgaris*; a $C_3$ plant) or sugarcane (a $C_4$ plant) (see **WEB TOPIC 9.7**).

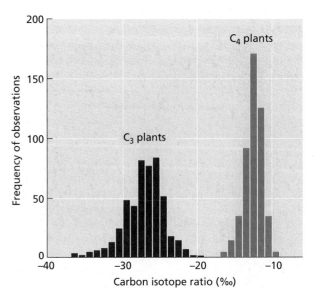

**FIGURE 9.25**  Frequency histograms for the observed carbon isotope ratios in $C_3$ and $C_4$ taxa from around the world. (After Cerling et al. 1997.)

## Why are there carbon isotope ratio variations in plants?

What is the physiological basis for $^{13}C$ depletion in plants relative to $CO_2$ in the atmosphere? It turns out that both the diffusion of $CO_2$ into the leaf and the carboxylation selectivity for $^{12}CO_2$ play a role.

We can predict the carbon isotope ratio of a $C_3$ leaf as

$$\delta^{13}C_L = \delta^{13}C_A - a - (b\text{-}a)(c_i/c_a) \tag{9.3}$$

where $\delta^{13}C_L$ and $\delta^{13}C_A$ are the carbon isotope ratios of the leaf and atmosphere, respectively; $a$ is the diffusion fraction; $b$ is the net carboxylase fraction in the leaf; and $c_i/c_a$ is the ratio of intercellular to ambient $CO_2$ concentrations.

$CO_2$ diffuses from air outside of the leaf to the carboxylation sites within leaves in both $C_3$ and $C_4$ plants. We express this diffusion using the term $a$. Because $^{12}CO_2$ is lighter than $^{13}CO_2$, it diffuses slightly faster toward the carboxylation site, creating an effective diffusion fractionation factor of –4.4‰. Thus, we would expect leaves to have a more negative $\delta^{13}C$ value simply because of this diffusion effect. Yet this factor alone is not sufficient to explain the $\delta^{13}C$ values of $C_3$ plants as shown in Figure 9.25.

The initial carboxylation event is a determining factor in the carbon isotope ratio of plants. Rubisco represents the first carboxylation reaction in $C_3$ photosynthesis and has an intrinsic discrimination value against $^{13}C$ of –30‰. By contrast, PEP carboxylase, the primary $CO_2$ fixation enzyme of $C_4$ plants, has a much smaller isotope discrimination effect—about 2‰. Thus, the inherent difference between the two carboxylating enzymes contributes to the different isotope ratio differences observed in $C_3$ and $C_4$ plants (Farquhar et al. 1989). We use $b$ to describe the net carboxylation effect.

Other physiological characteristics of plants affect its carbon isotope ratio. One primary factor is the partial pressure of $CO_2$ in the intercellular air spaces of leaves ($c_i$). In $C_3$ plants the potential isotope discrimination by rubisco of $-30‰$ is not fully expressed during photosynthesis because the availability of $CO_2$ at the carboxylation site becomes a limiting factor restricting the discrimination by rubisco. Greater discrimination against $^{13}CO_2$ occurs when $c_i$ is high, as when stomata are open. Yet open stomata also facilitate water loss. Thus, lower ratios of photosynthesis to transpiration are correlated with greater discrimination against $^{13}C$ (Ehleringer et al. 1993). When leaves are exposed to water stress, stomata tend to close, reducing $c_i$ values. As a consequence, $C_3$ plants grown under water stress conditions tend to have more positive carbon isotope ratios.

The application of carbon isotope ratios in plants has become very productive, because equation 9.3 provides a strong link between the carbon isotope ratio measurement and the intercellular $CO_2$ value in a leaf. Intercellular $CO_2$ levels are then directly linked with aspects of photosynthesis and stomatal constraints. As stomata close in $C_3$ plants or as water stress increases, we find that the leaf carbon isotope ratio increases. The carbon isotope ratio measurement then becomes a direct proxy to estimate several aspects of shorter-term water stress. These applications include using carbon isotopes to study plant performance in both agricultural and ecological studies (Ehleringer et al. 1993; Bowling et al. 2008).

One emergent environmental pattern is that, on average, leaf carbon isotope ratio values decrease as precipitation increases under natural conditions. **FIGURE 9.26** illustrates this pattern in a transect across Australia. Here we see that the $\delta^{13}C$ values are highest in the arid regions of Australia and become progressively lower in values along a precipitation gradient from desert to tropical rainforest ecosystems. Applying equation 9.3 to interpret these $\delta^{13}C$ data, we conclude that intercellular $CO_2$ levels of leaves of desert plants are lower than what we typically see in leaves of rainforest plants. Because of the sequential nature of tree ring formation, $\delta^{13}C$ observations in tree rings can help to separate the long-term effects of reduced water availability on plants (e.g., desert versus rainforest habitats) from short-term effects (e.g., seasonal drought cycles).

Carbon isotope ratio analyses are commonly used today to determine the dietary patterns of humans and other animals. The proportion of $C_3$ to $C_4$ foods in an animal's diet is recorded in its tissues—teeth, bones, muscles, and hair. Cerling and colleagues (2009) described an interesting application of carbon isotope ratio analysis to the eating habits of a family of wild African elephants. They

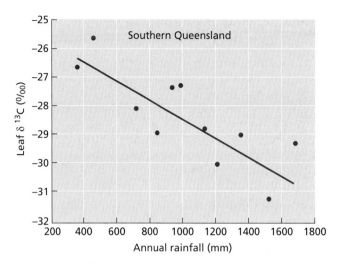

**FIGURE 9.26** Vegetation changes occur along rainfall gradients in Australia. Here we see that changes in carbon isotope ratios of vegetation appear to be strongly related to precipitation amounts in a region, suggesting that decreased moisture levels influence $c_i$ values and therefore carbon isotope ratios in $C_3$ species along a geographical gradient in Australia taxa. (After Stewart et al. 1995.)

examined sequential $\delta^{13}C$ values in segments of tail hair to reconstruct the daily diets of each animal. They observed very predictable seasonal shifts in shifts between trees ($C_3$) and grasses ($C_4$) as resource availability changes with rainfall patterns. Carbon isotope ratio analyses can be expanded to include consideration of human diets. One broad-scale observation is that the carbon isotope ratios of North Americans are higher than those observed in Europeans, indicating the prominent role that corn (a $C_4$ plant) plays in the diets of North Americans. Another application is measuring $\delta^{13}C$ in fossil, carbonate-containing soils and fossil teeth. From such observations it is possible to reconstruct the photosynthetic pathways of plants in the ancient past. These approaches have been used to determine that $C_4$ photosynthesis developed and became prevalent about 6 million years ago and to reconstruct the diets of ancient and modern animals (see **WEB TOPIC 9.8**).

CAM plants can have $\delta^{13}C$ values that are very close to those of $C_4$ plants. In CAM plants that fix $CO_2$ at night via PEP carboxylase, $\delta^{13}C$ is expected to be similar to that of $C_4$ plants. However, when some CAM plants are well watered, they can switch to $C_3$ mode by opening their stomata and fixing $CO_2$ during the day via rubisco. Under these conditions the isotope composition shifts toward that of $C_3$ plants. Thus the $\delta^{13}C$ values of CAM plants reflect how much carbon is fixed via the $C_3$ pathway versus the $C_4$ pathway.

# SUMMARY

In considering optimal photosynthetic performance, both the limiting factor hypothesis and an "economic perspective" emphasizing $CO_2$ "supply" and "demand" have guided research.

## Photosynthesis Is the Primary Function of Leaves

- Leaf anatomy is highly specialized for light absorption (**Figure 9.1**).
- Irradiance the amount of energy or photons that falls on a flat sensor of known area per unit time (**Figure 9.2**).
- About 5% of the solar energy reaching Earth is converted into carbohydrates by photosynthesis. Much absorbed light is lost as heat and fluorescence (**Figure 9.3**).
- In dense forests, almost all PAR is absorbed by leaves (**Figure 9.4**).
- Within a canopy, leaves maximize light absorption by solar tracking and chloroplast movements (**Figure 9.5**).
- Some plants respond to a range of light regimes. However, sun and shade leaves have contrasting biochemical characteristics.
- Some shade plants alter the ratios of photosystems I and II, while others add antenna chlorophyll to PSII.

## Photosynthetic Response to Light by the Intact Leaf

- Light response curves show the irradiance where photosynthesis is limited by light or by $CO_2$ (**Figure 9.6**). The slope of the linear portion of the light-response curve measures the quantum yield.
- Light compensation points for shade plants are lower than for sun plants because respiration rates in shade plants are very low (**Figure 9.7**).
- Below 30°C the quantum yield of $C_3$ plants is higher than that of $C_4$ plants; above 30°C, the situation is reversed (**Figure 9.8**).
- Beyond the saturation point, factors other than incident light, such as electron transport, rubisco activity, or triose metabolism, limit photosynthesis (**Figure 9.9**). Rarely is an entire plant light saturated (**Figure 9.10**).
- The xanthophyll cycle dissipates excess absorbed light energy to avoid damaging the photosynthetic apparatus (**Figures 9.11, 9.12**); chloroplast move-ments also limit excess light absorption (**Figure 9.13**).
- Dynamic photoinhibition temporarily diverts excess light absorption to heat but maintains maximal photosynthetic rate (**Figure 9.14**).

## Photosynthetic Responses to Temperature

- Plants are remarkably plastic in their adaptations to temperature. Optimal photosynthetic temperatures have strong genetic (adaptation) and environmental (acclimation) components.
- Leaf absorption of light energy generates a heat load that must be dissipated (**Figure 9.15**).
- The temperature sensitivity curves identify (*a*) a temperature range where enzymatic events are stimulated, (*b*) a range for optimal photosynthesis, and (*c*) a range where destructive events occur (**Figure 9.16**).
- Due to photorespiration, the quantum yield is strongly dependent on temperature in $C_3$ plants but is nearly independent of temperature in $C_4$ plants.
- Reduced quantum yield and increased photorespiration leads to differences in the photosynthetic capacities of $C_3$ and $C_4$ plants at different latitudes (**Figure 9.17**).

## Photosynthetic Responses to Carbon Dioxide

- Atmospheric $CO_2$ levels have been increasing since the Industrial Revolution due to human use of fossil fuels (**Figure 9.18**).
- Concentration gradients drive the diffusion of $CO_2$ from the atmosphere to rubisco, using both gaseous and liquid routes (**Figure 9.19**).
- Within the leaf depleted of blue and red photons, green light penetrates deeper into the leaf and effectively supplies energy for photosynthesis (**Figure 9.20**).
- In the greenhouse, enrichment of $CO_2$ above natural atmospheric levels results in increased productivity (**Figure 9.21**).
- $C_4$ photosynthesis may have become prominent in warmest regions when global $CO_2$ concentrations fell below a threshold value (**Figure 9.22**).
- Opening at night and closing during the day, the stomatal activity of CAM plants contrasts with those found in $C_3$ and $C_4$ plants (**Figure 9.23**).
- Free Air $CO_2$ Enrichment (FACE) experiments suggest that $C_3$ plants are more responsive to elevated $CO_2$ than are $C_4$ plants (**Figure 9.24**).

## Identifying Different Photosynthetic Pathways

- The carbon isotope ratios of leaves can be used to distinguish photosynthetic pathway differences among different plant species.

## SUMMARY continued

- Both $C_3$ and $C_4$ plants have less $^{13}C$ than does $CO_2$ in the atmosphere, indicating that leaf tissues discriminate against $^{13}C$ during photosynthesis (**Figure 9.25**).

- As stomata close in $C_3$ plants or as water stress increases, the leaf carbon isotope ratio increases and becomes a direct estimate of several aspects of shorter-term water stress (**Figure 9.26**).

## WEB MATERIAL

### Web Topics

**9.1 Working with Light**

Amount, direction, and spectral quality are important parameters for the measurement of light.

**9.2 Heat Dissipation from Leaves: The Bowen Ratio**

Sensible heat loss and evaporative heat loss are the most important processes in the regulation of leaf temperature.

**9.3 The Geographic Distributions of $C_3$ and $C_4$ Plants**

The geographic distribution of $C_3$ and $C_4$ plants corresponds closely with growing season temperature in today's world.

**9.4 Calculating Important Parameters in Leaf Gas Exchange**

Gas exchange methods allow us to measure photosynthesis and stomatal conductance in the intact leaf.

**9.5 Prehistoric Changes in Atmospheric $CO_2$**

Over the past 800,000 years, atmospheric $CO_2$ levels changed between 180 ppm (glacial periods) and 280 ppm (interglacial periods) as Earth moved between ice ages.

**9.6 Projected Future Increases in Atmospheric $CO_2$**

Atmospheric $CO_2$ reached 379 ppm in 2005 and is expected to reach 400 ppm by 2015.

**9.7 Using Carbon Isotopes to Detect Adulteration in Foods**

Carbon isotopes are frequently used to detect the substitution of $C_4$ sugars into $C_3$ food products, such as the introduction of sugar cane into honey to increase yield.

**9.8 Reconstruction of the Expansion of $C_4$ Taxa**

The $\delta^{13}C$ of animal teeth faithfully record the carbon isotope ratios of food sources and can be used to reconstruct the abundances of $C_3$ and $C_4$ plants eaten by mammalian grazers.

### Web Essay

**9.1 The Xanthophyll Cycle**

Molecular and biophysical studies are revealing the role of the xanthophyll cycle in the photoprotection of leaves.

# CHAPTER REFERENCES

Adams, W. W., Demmig-Adams, B., Rosenstiel, T. N., and Ebbert, V. (2001) Dependence of photosynthesis and energy dissipation activity upon growth form and light environment during the winter. *Photosyn. Res.* 67: 51–62.

Ainsworth, E. A., and Rogers, A. (2007) The response of photosynthesis and stomatal conductance to rising $[CO_2]$. *Plant Cell Environ.* 30: 258–270.

Anderson, J. M. (1986) Photoregulation of the composition, function, and structure of thylakoid membranes. *Annu. Rev. Plant Physiol.* 37: 93–136.

Barnola, J. M., Raynaud, D., Lorius, C., and Korothevich, Y. S. (1994) Historical $CO_2$ record from the Vostok ice core. In *Trends '93: A Compendium of Data on Global Change* (ORNL/CDIAC-65), T. A. Boden, D. P. Kaiser, R. J. Sepanski, and F. W. Stoss, eds., Carbon Dioxide Information Center, Oak Ridge National Laboratory, Oak Ridge, TN, pp. 7–10.

Berry, J., and Björkman, O. (1980) Photosynthetic response and adaptation to temperature in higher plants. *Annu. Rev. Plant Physiol.* 31: 491–543.

Berry, J. A., and Downton, J. S. (1982) Environmental regulation of photosynthesis. In *Photosynthesis: Development, Carbon Metabolism and Plant Productivity*, Vol. II, Govindjee, ed., Academic Press, New York, pp. 263–343.

Björkman, O. (1981) Responses to different quantum flux densities. In *Encyclopedia of Plant Physiology*, New Series, Vol. 12A, O. L. Lange, P. S. Nobel, C. B. Osmond, and H. Zeigler, eds., Springer, Berlin, pp. 57–107.

Bowes, G. (1993) Facing the inevitable: Plants and increasing atmospheric $CO_2$. *Annu. Rev. Plant Physiol. Plant Mol. Biol.* 44: 309–332.

Bowling, D. R., Pataki, D. E., and Randerson, J. T. (2008) Carbon isotopes in terrestrial ecosystem pools and CO2 fluxes. *New Phytol.* 178: 24–40.

Campbell, G. S., and Norman, J. H. (1996) *An Introduction to Environmental Biophysics*, 2nd ed. Springer-Verlag, New York.

Cerling, T. E., Wittemyer, G., Ehleringer, J. R., Remien, C. H., and Douglas-Hamilton, I. (2009) History of animals using isotope records (HAIR): A 6-year dietary history of one family of African elephants. *Proc. Natl. Acad. Sci. USA* 106: 8093–8100.

Cerling, T. E., Harris, J. M., MacFadden, B. J., Leakey, M. G., Quade, J., Eisenmann, V., and Ehleringer, J. R. (1997) Global vegetation change through the Miocene–Pliocene boundary. *Nature* 389: 153–158.

Dawson, T. E., Mambelli, S., Plamboeck, A. H., Templer, P. H., and Tu, K. P. (2002) Stable isotopes in plant ecology. *Annu. Rev. Ecol. Syst.* 33: 507–559.

Demmig-Adams, B., and Adams, W. (1996) The role of xanthophyll cycle carotenoids in the protection of photosynthesis. *Trends Plant Sci.* 1: 21–26.

Demming-Adams, B., and Adams, W. (2000) Harvesting sunlight safely. *Nature* 403: 371–372.

Demmig-Adams, B., Adams, W., and Matoo, A. (2006) *Photoprotection, Photoinhibition, Gene Regulation, and Environment.* Springer-Verlag, Dordrecht, Netherlands.

Ehleringer, J. R. (1978) Implications of quantum yield differences on the distributions of $C_3$ and $C_4$ grasses. *Oecologia* 31: 255–267.

Ehleringer, J. R., and Björkman, O. (1977) Quantum yields for $CO_2$ uptake in $C_3$ and $C_4$ plants. *Plant Physiol.* 59: 86–90.

Ehleringer, J. R., and Forseth, I. (1980) Solar tracking by plants. *Science* 210: 1094–1098.

Ehleringer, J. R., Björkman, O., and Mooney, H. A. (1976) Leaf pubescence: Effects on absorptance and photosynthesis in a desert shrub. *Science* 192: 376–377.

Ehleringer, J. R., Hall, A. E., and Farquhar, G. D., eds. (1993) *Stable Isotopes and Plant Carbon-Water Relations.* Academic Press, San Diego, CA.

Ehleringer, J. R., Cerling, T. E., and Helliker, B. R. (1997) $C_4$ photosynthesis, atmospheric $CO_2$, and climate. *Oecologia* 112: 285–299.

Evans, J. R., Kaldenhoff, R., Gentrry, B., and Terashima, I. (2009) Resistances along the $CO_2$ diffusion pathway inside leaves. *J. Exp. Bot.* 60: 2235–2248.

Farquhar, G. D., and Sharkey, T. D. (1982) Stomatal conductance and photosynthesis. *Annu. Rev. Plant Physiol.* 33: 317–345.

Farquhar, G. D., Ehleringer, J. R., and Hubick, K. T. (1989) Carbon isotope discrimination and photosynthesis. *Annu. Rev. Plant Physiol. Plant Mol. Biol.* 40: 503–538.

Geiger, D. R., and Servaites, J. C. (1994) Diurnal regulation of photosynthetic carbon metabolism in $C_3$ plants. *Annu. Rev. Plant Physiol. Plant Mol. Biol.* 45: 235–256.

Gibson, A. C., and Nobel, P. S. (1986) *The Cactus Primer.* Harvard University Press, Cambridge, MA.

Gorton, H. L., Williams, W. E., and Vogelmann, T. C. (1999) Chloroplast movement in *Alocasia macrorrhiza*. *Physiol. Plant.* 106: 421–428.

Harvey, G. W. (1979) Photosynthetic performance of isolated leaf cells from sun and shade plants. *Year B Carnegie Inst. Wash.* 79: 161–164.

Haupt, W., and Scheuerlein, R. (1990) Chloroplast movement. *Plant Cell Environ.* 13: 595–614.

Jarvis, P. G., and Leverenz, J. W. (1983) Productivity of temperate, deciduous and evergreen forests. In *Encyclopedia of Plant Physiology*, New Series, Vol. 12D, O. L. Lange, P. S. Nobel, C. B. Osmond, and H. Ziegler, eds., Springer, Berlin, pp. 233–280.

Keeling, C. D., Piper, S. C., Bacastow, R. B., Wahlen, M., Whorf, T. P., Heiman, M., and Meijer, H. A. (2005) Atmospheric $CO_2$ and $^{13}CO_2$ exchange with the terrestrial biosphere and oceans from 1978 to 2000: Observations and carbon cycle implications. In *A History of Atmospheric $CO_2$ and its Effects on Plants, Animals, and Ecosystems*, J. R. Ehleringer, T. E. Cerling, and M. D. Dearing, eds., Springer-Verlag, New York, pp 83–112.

Keeling, C. D., and Whorf, T. P. (1994) Atmospheric $CO_2$ records from sites in the SIO air sampling network. In *Trends '93: A Compendium of Data on Global Change* (ORNL/CDIAC-65), T. A. Boden, D. P. Kaiser, R. J. Sepanski, and F. W. Stoss, eds., Carbon Dioxide Information Center, Oak Ridge National Laboratory, Oak Ridge, TN, pp. 16–26.

Keeling, C. D., Whorf, T. P., Wahlen, M., and Van der Plicht, J. (1995) Interannual extremes in the rate of rise of atmospheric carbon dioxide since 1980. *Nature* 375: 666–670.

Koller, D. (2000) Plants in search of sunlight. *Adv. Bot. Res.* 33: 35–131.

Long, S. P., Ainsworth, E. A., Leakey, A. D., Nosberger, J., and Ort, D. R. (2006) Food for thought: Lower-than-expected crop stimulation with rising $CO_2$ concentrations. *Science* 312: 1918–1921.

Long, S. P., Ainsworth, E. A., Rogers, A., and Ort, D. R. (2004) Rising atmospheric carbon dioxide: Plants FACE the future. *Annu. Rev. Plant Biol.* 55: 591–628.

Neftel, A., Friedle, H., Moor, E., Lötscher, H., Oeschger, H., Siegenthaler, U., and Stauffer, B. (1994) Historical $CO_2$ record from the Siple Station ice core. In *Trends '93: A Compendium of Data on Global Change* (ORNL/CDIAC-65), T. A. Boden, D. P. Kaiser, R. J. Sepanski, and F. W. Stoss, eds., Carbon Dioxide Information Center, Oak Ridge National Laboratory, Oak Ridge, TN, pp. 11–15.

Nishio, J. N., Sun, J., and Vogelmann, T. C. (1993) Carbon fixation gradients across spinach leaves do not follow internal light gradient. *Plant Cell* 5: 953–961.

Ort, D. R., and Baker, N. R. (1988) Consideration of photosynthetic efficiency at low light as a major determinant of crop photosynthetic performance. *Plant Physiol. Biochem.* 26: 555–565.

Osmond, C. B. (1994) What is photoinhibition? Some insights from comparisons of shade and sun plants. In *Photoinhibition of Photosynthesis: From Molecular Mechanisms to the Field*, N. R. Baker and J. R. Bowyer, eds., BIOS Scientific, Oxford, pp. 1–24.

Pearcy, R. W., Muraoka, H., and Valladares, F. (2005) Crown architecture in sun and shade environments: Assessing function and trade-offs with a three-dimensional simulation model. *New Phytol.* 166: 791–800.

Sage, R. F., and Sharkey, T. D. (1987) The effect of temperature on the occurrence of $O_2$ and $CO_2$ insensitive photosynthesis in field grown plants. *Plant Physiol.* 84: 658–664.

Smith, H. (1986) The perception of light quality. In *Photomorphogenesis in Plants*, R. E. Kendrick and G. H. M. Kronenberg, eds., Nijhoff, Dordrecht, Netherlands, pp. 187–217.

Smith, H. (1994) Sensing the light environment: The functions of the phytochrome family. In *Photomorphogenesis in Plants*, 2nd ed., R. E. Kendrick and G. H. M. Kronenberg, eds., Nijhoff, Dordrecht, Netherlands, pp. 377–416.

Stewart, G. R., Turnbull, M. H., Schmidt, S., and Erskine, P. D. (1995) $^{13}C$ natural abundance in plant communities along a rainfall gradient: A biological integrator of water availability. *Aust. J. Plant Physiol.* 22: 51–55.

Terashima, I. (1992) Anatomy of non-uniform leaf photosynthesis. *Photosyn. Res.* 31: 195–212.

Terashima, I., and Hikosaka, K. (1995) Comparative ecophysiology of leaf and canopy photosynthesis. *Plant Cell Environ.* 18: 1111–1128.

Tlalka, M., and Fricker, M. (1999) The role of calcium in blue-light-dependent chloroplast movement in *Lemna trisulca* L. *Plant J.* 20: 461–473.

Vogelmann, T. C. (1993) Plant tissue optics. *Annu. Rev. Plant Physiol. Plant Mol. Biol.* 44: 231–251.

Vogelmann, T. C., and Björn, L. O. (1983) Response to directional light by leaves of a sun-tracking lupine (*Lupinus succulentus*). *Physiol. Plant.* 59: 533–538.

Vogelmann, T. C., and Han, T. (2000) Measurement of gradients of absorbed light in spinach leaves from chlorophyll fluorescence profiles. *Plant Cell Environ.* 23: 1303–1311.

von Braun, S. S., and Schleiff, E. (2007) Movement of endosymbiotic organelles. *Curr. Protein Pept. Sci.* 8: 426–438.

von Caemmerer, S. (2000) *Biochemical Models of Leaf Photosynthesis.* CSIRO, Melbourne, Australia.

# 10

# Translocation in the Phloem

Survival on land poses some serious challenges to terrestrial plants; foremost among these challenges is the need to acquire and retain water. In response to such environmental pressures, plants evolved roots and leaves. Roots anchor the plant and absorb water and nutrients; leaves absorb light and exchange gases. As plants increased in size, the roots and leaves became increasingly separated from each other in space. Thus, systems evolved for long-distance transport that allowed the shoot and the root to efficiently exchange products of absorption and assimilation.

You will recall from Chapters 4 and 6 that the xylem is the tissue that transports water and minerals from the root system to the aerial portions of the plant. The **phloem** is the tissue that translocates the products of photosynthesis—particularly sugars—from mature leaves to areas of growth and storage, including the roots.

The phloem also transmits signals between sources and sinks in the form of regulatory molecules, and redistributes water and various compounds throughout the plant body. All of these molecules appear to move with the transported sugars. The compounds to be redistributed, some of which initially arrive in the mature leaves via the xylem, can be either transferred out of the leaves without modification or metabolized before redistribution.

The discussion that follows emphasizes translocation in the phloem of angiosperms, because most of the research has been conducted on that group of plants. Gymnosperms will be compared briefly with angiosperms in terms of the anatomy of their conducting cells and possible differences in their mechanisms of translocation.

First we will examine some aspects of translocation in the phloem that have been researched extensively and are thought to be well understood, including the pathway and patterns of translocation, materials translocated in the phloem, and rates of movement. In the second part of the chapter we will explore aspects of translocation in the phloem that need further investigation. These include phloem loading and unloading and the allocation and partitioning of photosynthetic products. Finally, we will explore an area of intensive research at present: the phloem as a transport pathway for signaling molecules such as proteins and RNA.

## Pathways of Translocation

The two long-distance transport pathways—the phloem and the xylem—extend throughout the plant body. The phloem is generally found on the outer side of both primary and secondary vascular tissues (**FIGURES 10.1 AND 10.2**). In plants with secondary growth the phloem constitutes the inner bark. Although phloem is commonly found in a position external to the xylem, it is *also* found on the inner side in many eudicot families. In these families the phloem in the two positions is called external and internal phloem, respectively.

The cells of the phloem that conduct sugars and other organic materials throughout the plant are called **sieve elements**. *Sieve element* is a comprehensive term that includes both the highly differentiated **sieve tube elements** typical of the angiosperms and the relatively unspecialized **sieve cells** of gymnosperms. In addition to sieve elements, the phloem tissue contains companion cells (discussed below) and parenchyma cells (which store and release food molecules). In some cases the phloem tissue also includes fibers and sclereids (for protection and strengthening of the tissue) and laticifers (latex-containing cells). However, only the sieve elements are directly involved in translocation.

The small veins of leaves and the primary vascular bundles of stems are often surrounded by a **bundle sheath** (see Figure 10.1), which consists of one or more layers of compactly arranged cells. (You will recall the bundle sheath cells involved in C₄ metabolism discussed in Chapter 8.) In the vascular tissue of leaves, the bundle sheath surrounds the small veins all the way to their ends, isolating the veins from the intercellular spaces of the leaf.

We will begin our discussion of translocation pathways with the experimental evidence demonstrating that the sieve elements are the conducting cells in the phloem.

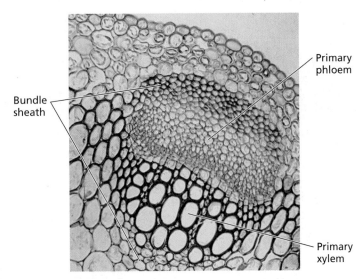

**FIGURE 10.1** Transverse section of a vascular bundle of trefoil, a clover (*Trifolium*). (130×) The primary phloem is toward the outside of the stem. Both the primary phloem and the primary xylem are surrounded by a bundle sheath of thick-walled sclerenchyma cells, which isolate the vascular tissue from the ground tissue. (© J. N. A. Lott/Biological Photo Service.)

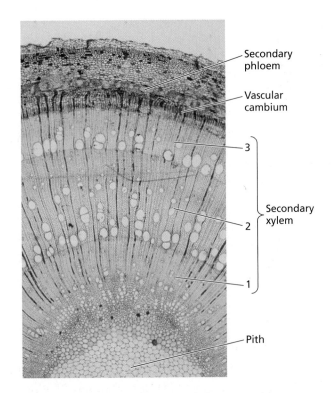

**FIGURE 10.2** Transverse section of a 3-year-old stem of an ash (*Fraxinus excelsior*) tree. (27×) The numbers 1, 2, and 3 indicate growth rings in the secondary xylem. The old secondary phloem has been crushed by expansion of the xylem. Only the most recent (innermost) layer of secondary phloem is functional. (© P. Gates/Biological Photo Service.)

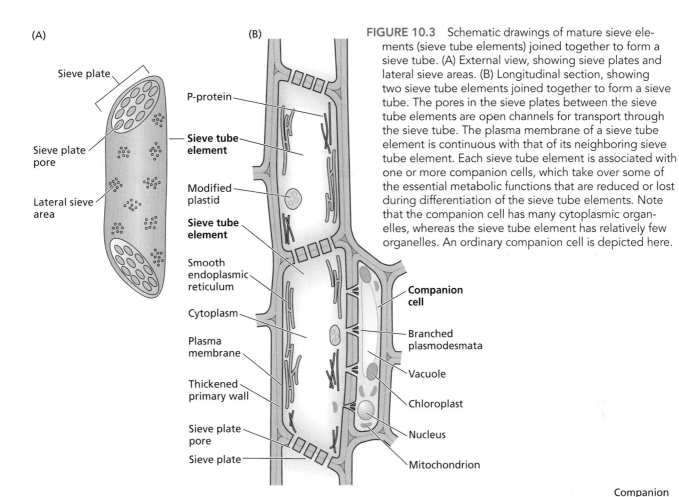

(A)

Sieve plate

Sieve plate pore

Lateral sieve area

(B)

P-protein

**Sieve tube element**

Modified plastid

**Sieve tube element**

Smooth endoplasmic reticulum

Cytoplasm

Plasma membrane

Thickened primary wall

Sieve plate pore

Sieve plate

**Companion cell**

Branched plasmodesmata

Vacuole

Chloroplast

Nucleus

Mitochondrion

**FIGURE 10.3** Schematic drawings of mature sieve elements (sieve tube elements) joined together to form a sieve tube. (A) External view, showing sieve plates and lateral sieve areas. (B) Longitudinal section, showing two sieve tube elements joined together to form a sieve tube. The pores in the sieve plates between the sieve tube elements are open channels for transport through the sieve tube. The plasma membrane of a sieve tube element is continuous with that of its neighboring sieve tube element. Each sieve tube element is associated with one or more companion cells, which take over some of the essential metabolic functions that are reduced or lost during differentiation of the sieve tube elements. Note that the companion cell has many cytoplasmic organelles, whereas the sieve tube element has relatively few organelles. An ordinary companion cell is depicted here.

Then we will examine the structure and physiology of these unusual plant cells.

## Sugar is translocated in phloem sieve elements

Early experiments on phloem transport date back to the nineteenth century, indicating the importance of long-distance transport in plants (see WEB TOPIC 10.1). These classical experiments demonstrated that removal of a ring of bark around the trunk of a tree, which removes the phloem, effectively stops sugar transport from the leaves to the roots without altering water transport through the xylem. When radioactive compounds became available, $^{14}CO_2$ was used to show that sugars made in the photosynthetic process are translocated through the phloem sieve elements (see WEB TOPIC 10.1).

## Mature sieve elements are living cells specialized for translocation

Detailed knowledge of the ultrastructure of sieve elements is critical to any discussion of the mechanism of translocation in the phloem. Mature sieve elements are unique among living plant cells (FIGURES 10.3 AND 10.4). They lack many structures normally found in living cells, even in the undifferentiated cells from which they are formed. For example, sieve elements lose their nuclei and tonoplasts (vacuolar membranes) during development. Micro-

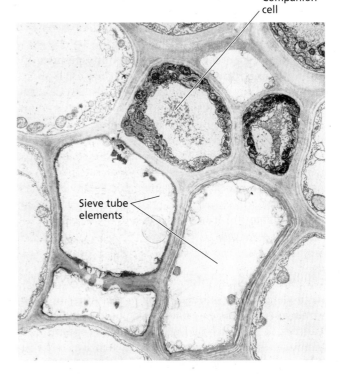

Companion cell

Sieve tube elements

**FIGURE 10.4** Electron micrograph of a transverse section of ordinary companion cells and mature sieve tube elements. (3600×) The cellular components are distributed along the walls of the sieve tube elements, where they offer less resistance to mass flow. (From Warmbrodt 1985.)

filaments, microtubules, Golgi bodies, and ribosomes are also generally absent from the mature cells. In addition to the plasma membrane, organelles that are retained include somewhat modified mitochondria, plastids, and smooth endoplasmic reticulum. The walls are nonlignified, though they are secondarily thickened in some cases.

Thus the cellular structure of sieve elements is different from that of tracheary elements of the xylem, which lack a plasma membrane, have lignified secondary walls, and are dead at maturity. As we will see, living cells are critical to the mechanism of translocation in the phloem.

### Large pores in cell walls are the prominent feature of sieve elements

Sieve elements (sieve cells and sieve tube elements) have characteristic sieve areas in their cell walls, where pores interconnect the conducting cells (FIGURE 10.5). The sieve area pores range in diameter from less than 1 µm to approximately 15 µm. Unlike sieve areas of gymnosperms, the sieve areas of angiosperms can differentiate into **sieve plates** (see Figure 10.5 and TABLE 10.1).

Sieve plates have larger pores than the other sieve areas in the cell and are generally found on the end walls of sieve tube elements, where the individual cells are joined together to form a longitudinal series called a **sieve tube** (see Figure 10.3). Furthermore, the sieve plate pores of sieve tube elements are open channels that allow transport between cells (see Figure 10.5A, B, and C).

In contrast, all of the sieve areas in gymnosperms such as conifers are structurally similar, though they can be more numerous on the overlapping end walls of sieve cells. The pores of gymnosperm sieve areas meet in large median cavities in the middle of the wall. Smooth endoplasmic reticulum (SER) covers the sieve areas (FIGURE 10.6) and is continuous through the sieve pores and median cavity, as indicated by ER-specific staining. Observation of living material with confocal laser scanning microscopy confirms that the observed distribution of SER is not an artifact of fixation. Table 10.1 lists characteristics of sieve tube elements and sieve cells.

### Damaged sieve elements are sealed off

Sieve-element sap is rich in sugars and other organic molecules. (*Sap* is a general term used to refer to the fluid contents of plant cells.) These molecules represent an energy investment for the plant, and their loss must be prevented when sieve elements are damaged. Short-term sealing mechanisms involve sap proteins, while the principal long-term mechanism for preventing sap loss entails closing sieve-plate pores with callose, a glucose polymer. (Another mechanism for blocking wounded sieve tubes over the short term occurs in plants in the legume family; see WEB TOPIC 10.2.)

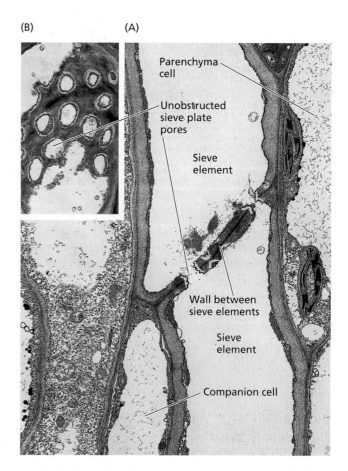

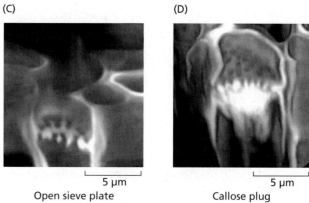

5 µm
Open sieve plate

5 µm
Callose plug

**FIGURE 10.5** Sieve elements and sieve plate pores. In images A, B, and C, the sieve plate pores are open—that is, unobstructed by P-protein or callose. Open pores provide a low-resistance pathway for transport between sieve elements. (A) Electron micrograph of a longitudinal section of two mature sieve elements (sieve tube elements), showing the wall between the sieve elements (called a sieve plate) in the hypocotyl of winter squash (*Cucurbita maxima*). (3685×) (B) The inset shows sieve plate pores in face view. (4280×) (C and D) Three-dimensional reconstructions of Arabidopsis sieve plates using a staining technique that can be used to image entire plant organs with confocal laser scanning microscopy. Open sieve pores are visible in C, while a callose plug, such as that formed in response to sieve tube damage, is visible in D. (A and B from Evert 1982; C and D from Truernit et al. 2008.)

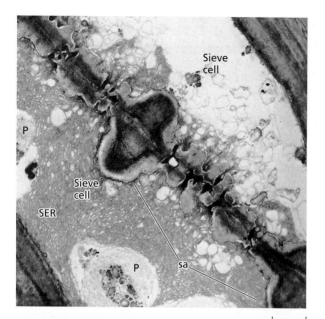

**FIGURE 10.6** Electron micrograph showing a sieve area (sa) linking two sieve cells of a conifer (*Pinus resinosa*). Smooth endoplasmic reticulum (SER) covers the sieve area on both sides and is also found within the pores and the extended median cavity. Such obstructed pores would result in a high resistance to solution flow between sieve cells. P, plastid. (From Schulz 1990.)

| **TABLE 10.1** |
|---|
| **Characteristics of the two types of sieve elements in seed plants** |
| **Sieve tube elements found in angiosperms** |
| 1. Some sieve areas are differentiated into sieve plates; individual sieve tube elements are joined together into a sieve tube. |
| 2. Sieve plate pores are open channels. |
| 3. P-protein is present in all dicots and many monocots. |
| 4. Companion cells are sources of ATP and perhaps other compounds. In some species, they serve as transfer cells or intermediary cells. |
| **Sieve cells found in gymnosperms** |
| 1. There are no sieve plates; all sieve areas are similar. |
| 2. Pores in sieve areas appear blocked with membranes. |
| 3. There is no P-protein. |
| 4. Albuminous cells sometimes function as companion cells. |

The main phloem proteins involved in sealing damaged sieve elements are structural proteins called **P-proteins** (see Figure 10.3B) (Clark et al. 1997). (In classical literature, P-protein was called *slime*.) The sieve tube elements of most angiosperms, including all dicots and many monocots, are rich in P-protein. However, P-protein is absent in gymnosperms. It occurs in several different forms (tubular, fibrillar, granular, and crystalline), depending on the species and maturity of the cell.

In immature cells, P-protein is most evident as discrete bodies in the cytosol known as **P-protein bodies**. P-protein bodies may be spheroidal, spindle-shaped, or twisted and coiled. They generally disperse into tubular or fibrillar forms during cell maturation.

P-proteins have been characterized at the molecular level (Dinant et al. 2003). For example, P-proteins from the genus *Cucurbita* consist of two major proteins: PP1, a phloem protein that forms filaments; and PP2, a phloem lectin associated with the filaments. (A lectin is a carbohydrate-binding protein associated with plant defenses.) Both PP1 and PP2 are synthesized in companion cells (discussed in the next section) and transported via the plasmodesmata to the sieve elements, where they associate to form P-protein filaments and P-protein bodies (Clark et al. 1997).

P-protein appears to function in sealing off damaged sieve elements by plugging up the sieve plate pores. Sieve tubes are under very high internal turgor pressure, and the sieve elements in a sieve tube are connected through open sieve plate pores. When a sieve tube is cut or punctured, the release of pressure causes the contents of the sieve elements to surge toward the cut end, from which the plant could lose much sugar-rich phloem sap if there were no sealing mechanism. When surging occurs, however, P-protein is trapped on the sieve plate pores, helping to seal the sieve element and prevent further loss of sap. Protein crystals released from ruptured plastids may play a similar role in some monocots (Paiva and Machado 2008). The sieve-element organelles and sometimes the ER, on the other hand, appear to be anchored to each other and to the sieve-element plasma membrane (Ehlers et al. 2000).

A longer-term solution to sieve tube damage is the production of the glucose polymer **callose** in the sieve pores (Figure 10.5D). Callose, a β-1,3-glucan, is synthesized by an enzyme in the plasma membrane (callose synthase) and is deposited between the plasma membrane and the cell wall. Callose is synthesized in functioning sieve elements in response to damage and other stresses, such as mechanical stimulation and high temperatures, or in preparation for normal developmental events, such as dormancy. The deposition of **wound callose** in the sieve pores efficiently seals off damaged sieve elements from surrounding intact tissue. In all cases, as sieve elements recover from damage or break dormancy, the callose disappears from the sieve pores; its dissolution is mediated by a callose-hydrolyzing enzyme.

Callose deposition is induced and callose synthase genes are up-regulated in rice (*Oryza sativa*) plants attacked by a phloem-feeding insect (brown planthopper); this occurs both in plants resistant to the insect and in susceptible plants. In the susceptible plants, however, feeding by the insects also activates genes for a callose-hydrolyzing enzyme. This unplugs the pores, allows continued feeding, and results in decreased sucrose and starch levels in the leaf sheath being attacked (Hao et al. 2008). Sealing off sieve elements that have been penetrated by insect mouthparts can thus play a key role in herbivore resistance.

### Companion cells aid the highly specialized sieve elements

Each sieve tube element is usually associated with one or more **companion cells** (see Figures 10.3B, 10.4, and 10.5). The division of a single mother cell forms the sieve tube element and the companion cell. Numerous plasmodesmata (see Chapter 1) penetrate the walls between sieve tube elements and their companion cells; the plasmodesmata are often complex and branched on the companion cell side. The presence of abundant plasmodesmata suggests a close functional relationship between a sieve element and its companion cell, an association that is demonstrated by the rapid exchange of solutes, such as fluorescent dyes, between the two cells.

Companion cells play a role in the transport of photosynthetic products from producing cells in mature leaves to the sieve elements in the minor (small) veins of the leaf. They also take over some of the critical metabolic functions, such as protein synthesis, that are reduced or lost during differentiation of the sieve elements. In addition, the numerous mitochondria in companion cells may supply energy as ATP to the sieve elements.

There are at least three different types of companion cells in the minor veins of mature, exporting leaves: "ordinary" companion cells, transfer cells, and intermediary cells. All three cell types have dense cytoplasm and abundant mitochondria.

**Ordinary companion cells** (FIGURE 10.7A) have chloroplasts with well-developed thylakoids and a cell wall with a smooth inner surface. The number of plasmodesmata connecting ordinary companion cells to surrounding cells is quite variable and apparently reflects the pathway taken by sugars as they move from the mesophyll to the minor veins (discussed in the section *Phloem Loading*, p. 285.)

**Transfer cells** are similar to ordinary companion cells, except for the development of fingerlike wall ingrowths, particularly on the cell walls that face away from the sieve element (FIGURE 10.7B). These wall ingrowths greatly increase the surface area of the plasma membrane, thus increasing the potential for solute transfer across the membrane. Relatively few plasmodesmata connect this type of companion cell to any of the surrounding cells except its own sieve element. As a result, the symplast of the sieve element and its transfer cell is relatively, if not entirely, symplastically isolated from that of surrounding cells.

Xylem parenchyma cells can also be modified as transfer cells, probably serving to retrieve and reroute solutes moving in the xylem, which is also part of the apoplast.

Although transfer cells (and some ordinary companion cells) are relatively isolated symplastically from surrounding cells, there are some plasmodesmata in the walls of these cells. The function of these plasmodesmata is not known. The fact that they are present indicates that they must have a function, and an important one, since the cost of having them is high: they are the avenues by which viruses become systemic in the plant. They are, however, difficult to study because they are so inaccessible.

In contrast to transfer cells, **intermediary cells** appear well suited for taking up solutes via cytoplasmic connections (FIGURE 10.7C). Intermediary cells have numerous plasmodesmata connecting them to bundle sheath cells. Although the presence of many plasmodesmatal connections to surrounding cells is their most characteristic feature, intermediary cells are also distinctive in having numerous small vacuoles, as well as poorly developed thylakoids and a lack of starch grains in the chloroplasts.

In general, transfer cells are found in plants where transport sugars enter the apoplast during the movement of sugars from mesophyll cells to sieve elements. Transfer cells transport sugars from the apoplast to the symplast of the sieve elements and companion cells in the source. Intermediary cells, on the other hand, function in symplastic transport of sugars from mesophyll cells to sieve elements. Ordinary companion cells can function in either symplastic or apoplastic short-distance transport in source leaves, depending in part on plasmodesmatal frequencies. (See section on *Phloem Loading*.)

## Patterns of Translocation: Source to Sink

Sap in the phloem is not translocated exclusively in either an upward or a downward direction, and translocation in the phloem is not defined with respect to gravity. Rather, sap is translocated from areas of supply, called *sources*, to areas of metabolism or storage, called *sinks*.

**Sources** include exporting organs, typically mature leaves, that are capable of producing photosynthate in excess of their own needs. The term *photosynthate* refers to products of photosynthesis. Another type of source is a storage organ during the exporting phase of its development. For example, the storage root of the biennial wild beet (*Beta maritima*) is a sink during the growing season of the first year, when it accumulates sugars received from the source leaves. During the second growing season the same root becomes a source; the sugars are remobilized and utilized to produce a new shoot, which ultimately becomes reproductive.

(A)

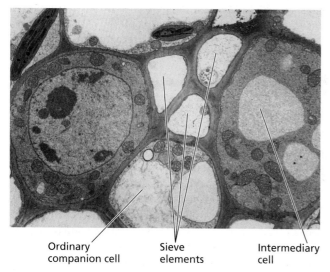

Ordinary companion cell    Sieve elements    Intermediary cell

(B)

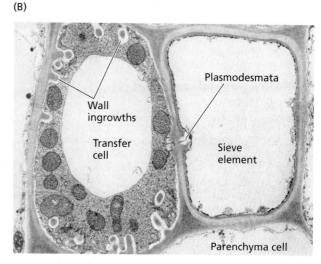

Wall ingrowths

Transfer cell

Plasmodesmata

Sieve element

Parenchyma cell

(C)

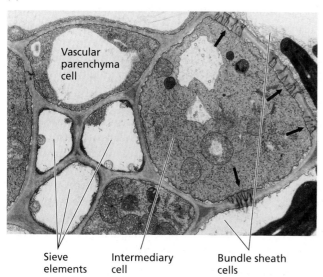

Vascular parenchyma cell

Sieve elements    Intermediary cell    Bundle sheath cells

**FIGURE 10.7**  Electron micrographs of companion cells in minor veins of mature leaves. (A) Three sieve elements abut two intermediary cells and a more lightly stained ordinary companion cell in a minor vein from scarlet monkey flower (*Mimulus cardinalis*). (6585×) (B) A sieve element adjacent to a transfer cell with numerous wall ingrowths in pea (*Pisum sativum*). (8020×) Such ingrowths greatly increase the surface area of the transfer cell's plasma membrane, thus increasing the transfer of materials from the mesophyll to the sieve elements. (C) A typical intermediary cell with numerous fields of plasmodesmata (arrows) connecting it to neighboring bundle sheath cells. These plasmodesmata are branched on both sides, but the branches are longer and narrower on the intermediary cell side. Minor-vein phloem was taken from heartleaf maskflower (*Alonsoa warscewiczii*). (4700×) (A and C from Turgeon et al. 1993, courtesy of R. Turgeon; B from Brentwood and Cronshaw 1978.)

**Sinks** include any nonphotosynthetic organs of the plant and organs that do not produce enough photosynthetic products to support their own growth or storage needs. Roots, tubers, developing fruits, and immature leaves, which must import carbohydrate for normal development, are all examples of sink tissues. Both girdling and labeling studies support the source-to-sink pattern of translocation in the phloem (**FIGURE 10.8A**).

Although the overall pattern of transport in the phloem can be stated simply as source-to-sink movement, the specific pathways involved are often more complex, depending on proximity, development, vascular connections (**FIGURE 10.8B**), and modification of translocation pathways. Not all sources supply all sinks on a plant; rather, certain sources preferentially supply specific sinks (see **WEB TOPIC 10.1**).

## Materials Translocated in the Phloem

Water is the most abundant substance in the phloem. Dissolved in the water are the translocated solutes, including carbohydrates, amino acids, hormones, some inorganic ions, RNAs and proteins, and some secondary compounds involved in defense and protection (Turgeon and Wolf 2009). Carbohydrates are the most significant and concentrated solutes in phloem sap (**TABLE 10.2**), with sucrose being the sugar most commonly transported in sieve elements. There is always some sucrose in sieve element sap, and it can reach concentrations of 0.3 to 0.9 *M*.

Complete identification of solutes that are mobile in the phloem and that have a significant function has been difficult; no one method of sampling phloem sap is free of artifacts or provides a complete picture of mobile solutes (Tur-

(A)

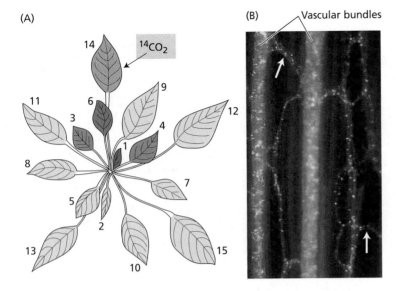

(B) Vascular bundles

**FIGURE 10.8** Source-to-sink patterns of phloem translocation. (A) Distribution of radioactivity from a single labeled source leaf in an intact plant. The distribution of radioactivity in leaves of a sugar beet plant (*Beta vulgaris*) was determined 1 week after $^{14}CO_2$ was supplied for 4 hours to a single source leaf (arrow). The degree of radioactive labeling is indicated by the intensity of shading of the leaves. Leaves are numbered according to their age; the youngest, newly emerged leaf is designated 1. The $^{14}C$ label was translocated mainly to the sink leaves directly above the source leaf (that is, sink leaves on the same orthostichy as the source; for example, leaves 1 and 6 are sink leaves directly above source leaf 14). (B) Longitudinal view of a typical three-dimensional structure of the phloem in a thick section (from an internode of dahlia [*Dahlia pinnata*]), viewed here after clearing, staining with aniline blue, and observing under an epifluorescence microscope. The sieve plates are seen as numerous small dots because of the yellow staining of callose in the sieve areas. Two large longitudinal vascular bundles are prominent. This staining reveals the delicate sieve tubes forming the phloem network; two phloem anastomoses (vascular interconnections) are marked by arrows. (A based on data from Joy 1964; B courtesy of R. Aloni.)

**TABLE 10.2**
**The composition of phloem sap from castor bean (*Ricinus communis*), collected as an exudate from cuts in the phloem**

| Component | Concentration (mg mL$^{-1}$) |
|---|---|
| Sugars | 80.0–106.0 |
| Amino acids | 5.2 |
| Organic acids | 2.0–3.2 |
| Protein | 1.45–2.20 |
| Potassium | 2.3–4.4 |
| Chloride | 0.355–0.675 |
| Phosphate | 0.350–0.550 |
| Magnesium | 0.109–0.122 |

*Source*: Hall and Baker 1972.

geon and Wolf 2009). We will begin this discussion with a brief examination of the available sampling methods, then continue with a description of the solutes that are currently accepted as significant mobile substances in the phloem.

### Phloem sap can be collected and analyzed

The collection of phloem sap is experimentally challenging because of the high turgor pressure in the sieve elements and the wound reactions described previously (see "Damaged sieve elements are sealed off" above and WEB TOPIC 10.3). In addition to plugging the sieve-plate pores, sudden pressure release in sieve elements can disrupt organelles and proteins and even pull substances from surrounding cells, especially the companion cells (Turgeon and Wolf 2009).

A few species exude phloem sap from wounds that sever sieve elements, making it possible to collect samples of the exuded sap from the incision. The initial samples may, however, be contaminated by the contents of surrounding damaged cells. Exudation of sap from cut petioles or stems, enhanced by the inclusion of EDTA in the collection fluid, has also been used in a number of studies. Chelating agents such as EDTA bind calcium, thus inhibiting callose synthesis (which requires calcium) and allowing exudation to occur for extended periods. However, exudation into EDTA is subject to a number of technical problems, such as the leakage of solutes, including carbohydrates, from the affected tissues.

A preferable approach is to use an aphid stylet as a "natural syringe." Aphids are small insects that feed by inserting their mouthparts, consisting of four tubular stylets, into a single sieve element of a leaf or stem. Sap can be collected from aphid stylets cut from the body of the insect, usually with a laser, after the aphid has been anesthetized with $CO_2$. The high turgor pressure in the sieve element forces the cell contents through the stylet to the cut end, where they can be collected. However, quantities of collected sap are small, and the method is technically difficult. Furthermore, exudation from severed stylets can continue for hours, suggesting that substances in aphid saliva prevent the plant's normal sealing mechanisms from operating and potentially altering the sap contents. Nonetheless, this method is thought to yield relatively pure sap from the sieve elements and companion cells (Doering-Saad et al. 2002; Gaupels et al. 2008) and to provide a fairly accurate picture of the composition of phloem sap (see WEB TOPIC 10.3).

**(A) Reducing sugars, which are not generally translocated in the phloem**

The reducing groups are aldehyde (glucose and mannose) and ketone (fructose) groups.

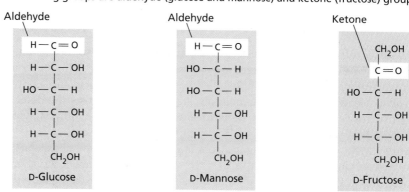

Aldehyde — D-Glucose

Aldehyde — D-Mannose

Ketone — D-Fructose

**(B) Compounds commonly translocated in the phloem**

Sucrose is a disaccharide made up of one glucose and one fructose molecule. Raffinose, stachyose, and verbascose contain sucrose bound to one, two, or three galactose molecules, respectively.

Mannitol is a sugar alcohol formed by the reduction of the aldehyde group of mannose.

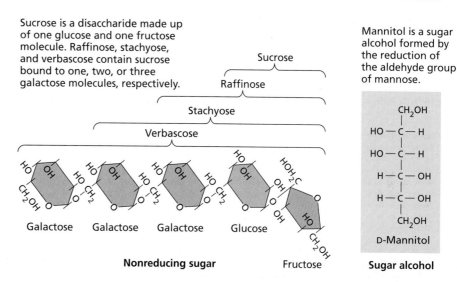

Sucrose
Raffinose
Stachyose
Verbascose

Galactose    Galactose    Galactose    Glucose                Fructose

**Nonreducing sugar**

D-Mannitol

**Sugar alcohol**

Glutamic acid, an amino acid, and glutamine, its amide, are important nitrogenous compounds in the phloem, in addition to aspartate and asparagine.

Glutamic acid

**Amino acid**

Glutamine

**Amide**

Species with nitrogen-fixing nodules also utilize ureides as transport forms of nitrogen.

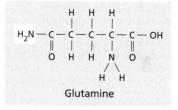

Allantoic acid              Allantoin              Citrulline

**Ureides**

**FIGURE 10.9** Structures of (A) compounds not normally translocated in the phloem and (B) compounds commonly translocated in the phloem.

## Sugars are translocated in nonreducing form

Results from many analyses of collected sap indicate that the translocated carbohydrates are nonreducing sugars. Reducing sugars, such as glucose and fructose, contain an exposed aldehyde or ketone group (**FIGURE 10.9A**). In a nonreducing sugar, such as sucrose, the ketone or aldehyde group is reduced to an alcohol or combined with a similar group on another sugar (**FIGURE 10.9B**). Most researchers believe that the nonreducing sugars are the major compounds translocated in the phloem because they are less reactive than their reducing counterparts.

Sucrose is the most commonly translocated sugar; many of the other mobile carbohydrates contain sucrose bound to varying numbers of galactose molecules. Raffinose consists of sucrose and one galactose molecule, stachyose consists of sucrose and two galactose molecules, and verbascose consists of sucrose and three galactose molecules (see Figure 10.9B). Translocated sugar alcohols include mannitol and sorbitol.

In science every concept can and should be questioned and further investigated, even those aspects of science that are thought to be "well understood." So it is with translocated sugars. While most studies report very low concentrations of reducing sugars in phloem sap, substantial quantities of hexoses are found in a few species. In these cases phloem exudates from cut petiole surfaces are collected in solutions containing chelating agents such as EDTA (van Bel and Hess 2008). Since high concentrations of hexoses in phloem sap are generally not detected by other types of studies and since exudation into EDTA can induce artifacts, further studies will be needed to clarify the significance of these findings.

### Other solutes are translocated in the phloem

Nitrogen is found in the phloem largely in amino acids—especially glutamate and aspartate—and their respective amides, glutamine and asparagine. Reported levels of amino acids and organic acids vary widely, even for the same species, but they are usually low compared with carbohydrates. (See WEB TOPIC 10.4 for more information on nitrogen transport in the phloem.) A variety of proteins and RNAs occur in phloem sap, in relatively low concentrations. RNAs found in the phloem include mRNAs (Doering-Saad et al. 2002), pathogenic RNAs, and small regulatory RNA molecules.

Almost all the endogenous plant hormones, including auxin, gibberellins, cytokinins, and abscisic acid (see Chapters 19, 20, 21, and 23), have been found in sieve elements. The long-distance transport of hormones, especially auxin, is thought to occur at least partly in the sieve elements. Nucleotide phosphates have also been found in phloem sap.

Some inorganic solutes move in the phloem, including potassium, magnesium, phosphate, and chloride (see Table 10.2). In contrast, nitrate, calcium, sulfur, and iron are relatively immobile in the phloem.

Proteins found in the phloem include structural P-proteins such as PP1 and PP2 (involved in the sealing of wounded sieve elements), as well as a number of water-soluble proteins. The functions of many of the proteins commonly found in phloem sap are related to stress and defense reactions (Walz et al. 2004; see the table in WEB TOPIC 10.12).

However, one large-scale proteomics analysis of pumpkin phloem sap revealed many unique proteins, in addition to those commonly found in phloem sap (Lin et al. 2009). This study identified 10 times as many proteins in phloem sap as previous studies. Perhaps most surprising was the detection of 100 proteins involved in protein synthesis, which is not thought to occur in the sieve elements.

Few ribosomal proteins were found, in keeping with the general observation that ribosomes are absent from mature sieve elements. Since EDTA was included in the collection fluid in this study, the possible artifacts mentioned earlier must be considered, and further studies will be necessary to clarify the significance of these observations. The possible roles of RNAs and proteins as signal molecules will be further discussed at the end of the chapter.

## Rates of Movement

The rate of movement of materials in the sieve elements can be expressed in two ways: as **velocity**, the linear distance traveled per unit time, or as **mass transfer rate**, the quantity of material passing through a given cross section of phloem or sieve elements per unit time. Mass transfer rates based on the cross-sectional area of the sieve elements are preferred because the sieve elements are the conducting cells of the phloem. Values for mass transfer rate range from 1 to 15 g $h^{-1}$ $cm^{-2}$ of sieve elements (see WEB TOPIC 10.5).

In early publications reporting on rates of transport in the phloem, the units of velocity were centimeters per hour (cm $h^{-1}$), and the units of mass transfer were grams per hour per square centimeter (g $h^{-1}$ $cm^{-2}$) of phloem or sieve elements. However, the currently preferred units (SI units) are meters (m) or millimeters (mm) for length, seconds (s) for time, and kilograms (kg) for mass.

Both velocities and mass transfer rates can be measured with radioactive tracers. (Methods of measuring mass transfer rates are described in WEB TOPIC 10.5.) In the simplest type of experiment for measuring velocity, [11]C- or [14]C-labeled $CO_2$ is applied for a brief period of time to a source leaf (pulse labeling), and the arrival of label at a sink tissue or at a particular point along the pathway is monitored with an appropriate detector.

In general, velocities measured by a variety of conventional techniques far exceed the rate of diffusion, averaging about 100 cm $h^{-1}$ and ranging from 30 to 150 cm $h^{-1}$. More recent measurements of velocity using NMR spectrometry and magnetic resonance imaging yielded an average velocity for castor bean of 0.25 mm $sec^{-1}$ (equivalent to 90 cm $h^{-1}$), which is remarkably close to the average obtained using older methods (Windt et al. 2006). Transport velocities in the phloem are clearly quite high and well in excess of the rate of diffusion over long distances. Any proposed mechanism of phloem translocation must account for these high velocities.

# The Pressure-Flow Model, a Passive Mechanism for Phloem Transport

The mechanism of phloem translocation in angiosperms is best explained by the pressure-flow model, which is consistent with most of the experimental and structural data currently available. The pressure-flow model explains phloem translocation as a flow of solution (mass flow or bulk flow) driven by an osmotically generated pressure gradient between source and sink. The pressure-flow model, predictions arising from mass flow, and supporting data will be described in this section. At the end of the section, the question of whether the model applies to gymnosperms will be briefly explored.

In early research on phloem translocation, both active and passive mechanisms were considered. All theories, both active and passive, assume an energy requirement in both sources and sinks. In sources, energy is necessary to synthesize the materials for transport and, in some cases, to move photosynthate into the sieve elements by active membrane transport. The movement of photosynthate into the sieve elements is called *phloem loading*, and it is discussed in detail later in the chapter. In sinks, energy is essential for some aspects of movement from sieve elements to sink cells, which store or metabolize the sugar. This movement of photosynthate from sieve elements to sink cells is called *phloem unloading* and will also be discussed later.

The passive mechanisms of phloem transport further assume that energy is required in the sieve elements of the path between sources and sinks simply to maintain structures such as the cell plasma membrane and to recover sugars lost from the phloem by leakage. The pressure-flow model is an example of a passive mechanism. The active theories, on the other hand, postulate an additional expenditure of energy by path sieve elements in order to drive translocation itself (Zimmermann and Milburn 1975). The active theories have largely been discounted and will not be discussed here.

## An osmotically-generated pressure gradient drives translocation in the pressure-flow model

Diffusion is far too slow to account for the velocities of solute movement observed in the phloem. Translocation velocities average $1 \text{ m h}^{-1}$; the rate of diffusion would be 1 m per 32 years! (See Chapter 3 for a discussion of diffusion velocities and the distances over which diffusion is an effective transport mechanism.)

The **pressure-flow model**, first proposed by Ernst Münch in 1930, states that a flow of solution in the sieve elements is driven by an osmotically generated *pressure gradient* between source and sink ($\Delta \Psi_p$). Phloem loading at the source and phloem unloading at the sink establish the pressure gradient.

As we will see later (see *Phloem Loading*), three different mechanisms exist to generate high concentrations of sugars in the sieve elements of the source: photosynthetic metabolism in the mesophyll, conversion of photoassimilate to transport sugars in intermediary cells (polymer trapping), and active membrane transport. Recall from Chapter 3 (Equation 3.5) that $\Psi_w = \Psi_s + \Psi_p$; that is, $\Psi_p = \Psi_w - \Psi_s$. In source tissues, an accumulation of sugars in the sieve elements generates a low (negative) solute potential ($\Delta \Psi_s$) and causes a steep drop in the water potential ($\Delta \Psi_w$). In response to the water potential gradient, water enters the sieve elements and causes the turgor pressure ($\Psi_p$) to increase.

At the receiving end of the translocation pathway, phloem unloading leads to a lower sugar concentration in the sieve elements, generating a higher (more positive) solute potential in the sieve elements of sink tissues. As the water potential of the phloem rises above that of the xylem, water tends to leave the phloem in response to the water potential gradient, causing a decrease in turgor pressure in the sieve elements of the sink. **FIGURE 10.10** illustrates the pressure-flow hypothesis; the figure specifically shows the case in which active membrane transport from the apoplast generates a high sugar concentration in the source sieve elements.

If no cross-walls were present in the translocation pathway—that is, if the entire pathway were a single membrane-enclosed compartment—the different pressures at the source and sink would rapidly equilibrate. Sieve plates present a sequence of resistances to the moving phloem sap; the resistance is thought to result in the generation and maintenance of a considerable pressure gradient in the sieve elements between source and sink.

The phloem sap moves by mass flow rather than by osmosis. That is, no membranes are crossed during transport from one sieve tube to another, and solutes move at the same rate as the water molecules. Since this is the case, mass flow can occur from a source organ with a lower water potential to a sink organ with a higher water potential, or vice versa, depending on the identities of the source and sink organs. In fact, Figure 10.10 illustrates an example in which the flow is against the water potential gradient. Such water movement does not violate the laws of thermodynamics, because it is an example of mass flow, which is driven by a pressure gradient, as opposed to osmosis, which is driven by a water potential gradient.

According to the pressure-flow model, movement in the translocation pathway is driven by transport of solutes and water into source sieve elements and out of sink sieve elements. The passive, pressure-driven, long-distance translocation in the sieve tubes ultimately depends on the mechanisms involved in phloem loading and unloading. These mechanisms are responsible for setting up the pressure gradient.

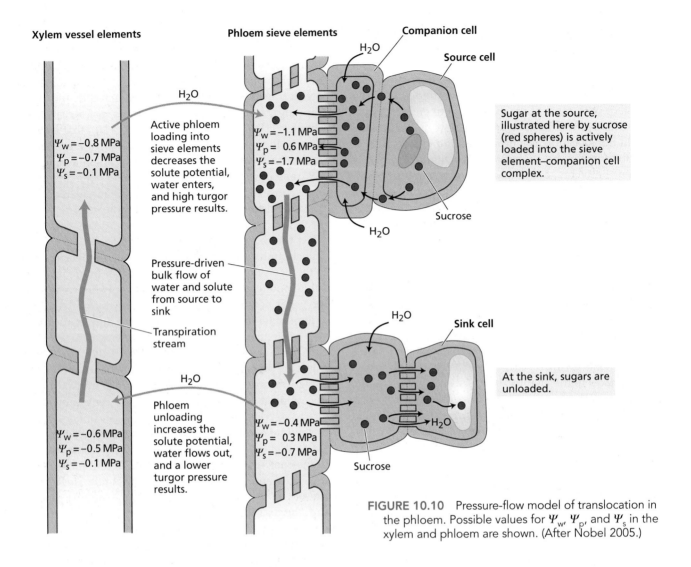

Xylem vessel elements

Phloem sieve elements

Companion cell

Source cell

$H_2O$

$\Psi_w = -0.8$ MPa
$\Psi_p = -0.7$ MPa
$\Psi_s = -0.1$ MPa

$H_2O$

Active phloem loading into sieve elements decreases the solute potential, water enters, and high turgor pressure results.

$\Psi_w = -1.1$ MPa
$\Psi_p = 0.6$ MPa
$\Psi_s = -1.7$ MPa

Sugar at the source, illustrated here by sucrose (red spheres) is actively loaded into the sieve element–companion cell complex.

Sucrose

$H_2O$

Pressure-driven bulk flow of water and solute from source to sink

Transpiration stream

$H_2O$

Sink cell

Phloem unloading increases the solute potential, water flows out, and a lower turgor pressure results.

$\Psi_w = -0.6$ MPa
$\Psi_p = -0.5$ MPa
$\Psi_s = -0.1$ MPa

$\Psi_w = -0.4$ MPa
$\Psi_p = 0.3$ MPa
$\Psi_s = -0.7$ MPa

At the sink, sugars are unloaded.

$H_2O$

Sucrose

**FIGURE 10.10**  Pressure-flow model of translocation in the phloem. Possible values for $\Psi_w$, $\Psi_p$, and $\Psi_s$ in the xylem and phloem are shown. (After Nobel 2005.)

## The predictions of mass flow have been confirmed

Some important predictions emerge from the model of phloem translocation as mass flow:

- The sieve plate pores must be unobstructed. If P-protein or other materials blocked the pores, the resistance to flow of the sieve element sap would be too great.

- No true bidirectional transport (i.e., simultaneous transport in both directions) in a single sieve element can occur. A mass flow of solution precludes such bidirectional movement because a solution can flow in only one direction in a pipe at any one time. Solutes within the phloem can move bidirectionally, but in different sieve elements or at different times.

- No great expenditures of energy are required in order to drive translocation in the tissues along the path. Therefore, treatments that restrict the supply of ATP in the path, such as low temperature,

anoxia, and metabolic inhibitors, should not stop translocation. However, energy *is* required to maintain the structure of the sieve elements, to reload any sugars lost to the apoplast by leakage, and perhaps to reload sugars at the termination of sieve tubes.

- The pressure-flow hypothesis demands the presence of a positive pressure gradient, with turgor pressure higher in sieve elements of sources than in those of sinks. According to the traditional picture of mass flow, the pressure difference must be large enough to overcome the resistance of the pathway and to maintain flow at the observed velocities. (See **WEB TOPIC 10.6** for an alternative view of sieve-element pressure gradients.)

The available evidence testing these predictions, presented below, is consistent with mass flow and with the pressure-flow hypothesis.

(A)

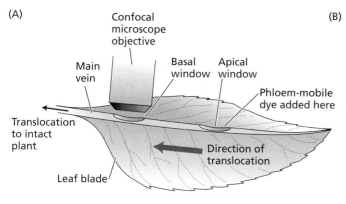

(B)

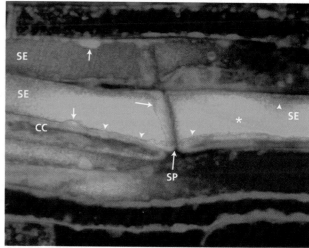

15 μm

(C)

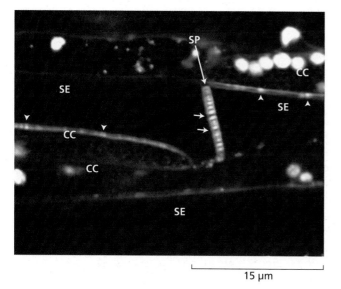

15 μm

**FIGURE 10.11** Sieve plate pores of living, translocating sieve elements are open. Sieve elements were observed in a leaf attached to an intact broad bean (*Vicia faba*) plant. (A) Two windows were sliced parallel to the epidermis on the lower side of the main vein of a mature leaf, exposing the phloem tissue. The objective of the laser confocal microscope was positioned over the basal window. A phloem-mobile fluorescent dye was added at the apical window. If translocation occurred, the dye would become visible in the microscope at the basal window of the leaf. In this way it could be demonstrated that the sieve elements being observed were alive and functional. (B) Phloem tissue of bean doubly stained with a locally applied fluorescent dye (red) that primarily stains membranes, and a different fluorescent dye (green) that is translocated. The presence of the green dye at the basal, observation window indicates that translocation has occurred from the apical, application window. Protein (arrows) deposited against the plasma membrane and the sieve plate does not impede translocation. A crystalline P-protein body (asterisk) is stained by the green dye. Plastids (arrowheads) are evenly distributed around the periphery of the sieve element. (C) Phloem tissue of bean stained only with the locally applied fluorescent dye that stains membranes. Arrows indicate sieve plate pores, which are not occluded. CC, companion cell; SP, sieve plate; SE, sieve element. (B and C from Knoblauch and van Bel 1998; courtesy of A. van Bel.)

## Sieve plate pores are open channels

Ultrastructural studies of sieve elements are challenging because of the high internal pressure in these cells. When the phloem is excised or killed slowly with chemical fixatives, the turgor pressure in the sieve elements is released. The contents of the cell, particularly P-protein, surge toward the point of pressure release and, in the case of sieve tube elements, accumulate on the sieve plates. This accumulation is probably the reason that many earlier electron micrographs show sieve plates that are obstructed.

Newer, rapid freezing and fixation techniques provide reliable pictures of undisturbed sieve elements. Electron micrographs of sieve tube elements prepared by such techniques show that P-protein is usually found along the periphery of the sieve tube elements (see Figures 10.3, 10.4, and 10.5), or it is evenly distributed throughout the lumen of the cell. Furthermore, the pores contain P-protein in similar positions, lining the pore or in a loose network. The open condition of the pores seen in many species, such as cucurbits, sugar beet (*Beta vulgaris*), and bean (*Phaseo-*

*lus vulgaris*) (e.g., see Figure 10.5), is consistent with mass flow.

In addition to obtaining the structural evidence provided by electron microscopy, it is important to determine whether the sieve plate pores are open in the intact tissue. The use of confocal laser scanning microscopy, which allows for the direct observation of translocation through living sieve elements, addresses this question (Knoblauch and van Bel 1998). Such experiments show that the sieve plate pores of living, translocating sieve elements are open (**FIGURE 10.11**).

### There is no bidirectional transport in single sieve elements

Researchers have investigated bidirectional transport by applying two different radiotracers to two source leaves, one above the other. Each leaf receives one of the tracers, and a point between the two sources is monitored for the presence of both tracers.

Transport in two directions has often been detected in sieve elements of different vascular bundles in stems. Transport in two directions has also been seen in adjacent sieve elements of the same bundle in petioles. Bidirectional transport in adjacent sieve elements can occur in the petiole of a leaf that is undergoing the transition from sink to source and simultaneously importing and exporting photosynthates through its petiole. However, simultaneous bidirectional transport in a single sieve element has never been demonstrated.

### The energy requirement for transport through the phloem pathway is small

In plants that can survive periods of low temperature, such as sugar beet, rapidly chilling a short segment of the petiole of a source leaf to approximately 1°C does not cause sustained inhibition of mass transport out of the leaf (FIGURE 10.12). Rather, there is a brief period of inhibition, after which transport slowly returns to the control rate. Chilling reduces respiration rate and both the synthesis and the consumption of ATP in the petiole by about 90%, at a time when translocation has recovered and is proceeding normally. These experiments show that the energy requirement for transport through the pathway of these herbaceous plants is small, consistent with mass flow.

Extreme treatments that inhibit all energy metabolism do inhibit translocation. For example, in bean, treating the petiole of a source leaf with a metabolic inhibitor (cyanide) inhibited translocation out of the leaf. However, examination of the treated tissue by electron microscopy revealed blockage of the sieve plate pores by cellular debris (Giaquinta and Geiger 1977). Clearly, these results do not bear on the question of whether energy is required for translocation along the pathway.

### Positive pressure gradients exist in the phloem sieve elements

Mass flow or bulk flow is the combined movement of all the molecules in a solution, driven by a pressure gradient. What are the pressure values in sieve elements, and how can they be determined?

Turgor pressure in sieve elements can be either calculated from the water potential and solute potential ($\Psi_p = \Psi_w - \Psi_s$) or measured directly. The most effective technique uses micromanometers or pressure transducers sealed over exuding aphid stylets (see Figure 10.2.A in WEB TOPIC 10.3) (Wright and Fisher 1980). The data obtained are accurate because aphids pierce only a single sieve element, and the plasma membrane apparently seals well around the aphid stylet. When the turgor pressure of sieve elements is measured by this technique, the pressure at the source is higher than that at the sink.

In soybean, the observed pressure difference between source and sink has been shown to be sufficient to drive a solution through the pathway by mass flow, taking into account the path resistance (caused mainly by the sieve plate pores), the path length, and the velocity of translocation (Fisher 1978). The actual pressure difference between source and sink was calculated from the water potential and solute potential to be 0.41 MPa, and the pressure difference required for translocation by pressure flow was calculated to be 0.12 to 0.46 MPa. Thus the observed pressure difference appears to be sufficient to drive mass flow through the phloem and supports traditional models of pressure flow, at least in this small herbaceous plant.

We can therefore conclude that all the experiments and data described here are consistent with the operation of mass flow in angiosperm phloem. The lack of an energy requirement in the pathway and the presence of open sieve plate pores provide definitive evidence for a mechanism in which the path phloem is relatively passive. The failure to detect bidirectional transport or motility proteins, as well as the positive data on pressure gradients, are in accord with the pressure-flow hypothesis.

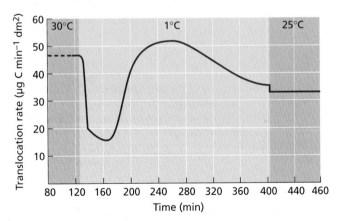

**FIGURE 10.12** The energy requirement for translocation in the path is small. Loss of metabolic energy resulting from the chilling of a source leaf petiole partially reduces the rate of translocation in sugar beet. However, translocation rates recover with time despite the fact that ATP production and utilization are still largely inhibited by chilling. $^{14}CO_2$ was supplied to a source leaf, and a 2-cm portion of its petiole was chilled to 1°C. Translocation was monitored by the arrival of $^{14}C$ at a sink leaf. (1 dm [decimeter] = 0.1 meter) (After Geiger and Sovonick 1975.)

## Does translocation in gymnosperms involve a different mechanism?

Although mass flow explains translocation in angiosperms, it may not be sufficient for gymnosperms. Very little physiological information on gymnosperm phloem is available, and speculation about translocation in these species is based almost entirely on interpretations of electron micrographs. As discussed previously, the sieve cells of gymnosperms are similar in many respects to sieve tube elements of angiosperms, but the sieve areas of sieve cells are relatively unspecialized and do not appear to consist of open pores (see Figure 10.6).

The pores in gymnosperms are filled with numerous membranes that are continuous with the smooth endoplasmic reticulum adjacent to the sieve areas. Such pores are clearly inconsistent with the requirements of mass flow. Although these electron micrographs might be artifactual and fail to show conditions in the intact tissue, translocation in gymnosperms might involve a different mechanism—a possibility that requires further investigation.

# Phloem Loading

Several transport steps are involved in the movement of photosynthate from the mesophyll chloroplasts to the sieve elements of mature leaves:

1. Triose phosphate formed by photosynthesis during the day (see Chapter 8) is transported from the chloroplast to the cytosol, where it is converted to sucrose. During the night, carbon from stored starch exits the chloroplast primarily in the form of maltose and is converted to sucrose. (Other transport sugars are later synthesized from sucrose in some species, while sugar alcohols are synthesized using hexose phosphate and in some cases hexose as the starting molecules.)

2. Sucrose moves from producing cells in the mesophyll to cells in the vicinity of the sieve elements in the smallest veins of the leaf (**FIGURE 10.13**). This **short-distance transport** pathway usually covers a distance of only a few cell diameters.

3. In the process called **phloem loading**, sugars are transported into the sieve elements and companion cells. Note that with respect to loading, the sieve elements and companion cells are often considered a functional unit, called the *sieve element–companion cell complex*. Once inside the sieve elements, sucrose and other solutes are translocated away from the source, a process known as **export**. Translocation through the vascular system to the sink is referred to as **long-distance transport**.

As discussed earlier, the processes of loading at the source and unloading at the sink provide the driving

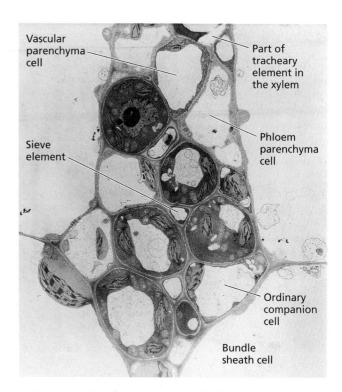

**FIGURE 10.13** Electron micrograph showing the relationship between the various cell types of a small vein in a source leaf of sugar beet (5000×). Photosynthetic cells (mesophyll cells) surround the compactly arranged cells of the bundle sheath layer. Photosynthate from the mesophyll must move a distance equivalent to only several cell diameters before being loaded into the sieve elements. Movement from the mesophyll to the sieve elements is thus known as short-distance transport. (From Evert and Mierzwa 1985, courtesy of R. Evert.)

force for long-distance transport and are thus of considerable basic, as well as agricultural, importance. A thorough understanding of these mechanisms should provide the basis of technology aimed at enhancing crop productivity by increasing the accumulation of photosynthate by edible sink tissues, such as cereal grains.

## Phloem loading can occur via the apoplast or symplast

We have seen that solutes (mainly sugars) in source leaves must move from the photosynthesizing cells in the mesophyll to the sieve elements. The initial short-distance pathway is probably symplastic (**FIGURE 10.14**). However, sugars might move entirely through the symplast (cytoplasm) to the sieve elements via the plasmodesmata (see Figure 10.14A), or they might enter the apoplast prior to phloem loading (see Figure 10.14B). (See Figure 4.4 for a general description of the symplast and apoplast.) The apoplastic and symplastic routes are used in different species.

**(A) Symplastic loading**

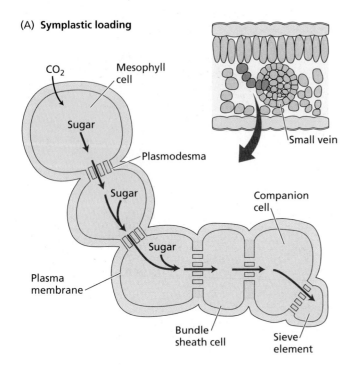

**(B) Apoplastic loading**

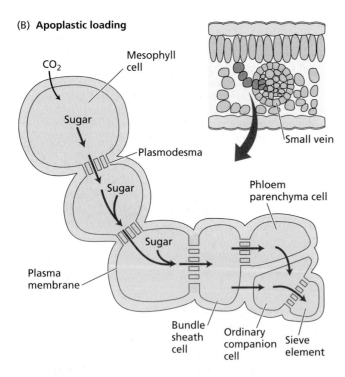

**FIGURE 10.14** Schematic diagram of pathways of phloem loading in source leaves. (A) In the totally symplastic pathway, sugars move from one cell to another in the plasmodesmata, all the way from the mesophyll to the sieve elements. (B) In the partly apoplastic pathway, sugars initially move through the symplast but enter the apoplast just prior to loading into the companion cells and sieve elements. Sugars loaded into the companion cells are thought to move through plasmodesmata into the sieve elements.

Several mechanisms for phloem loading are now recognized: apoplastic loading, symplastic loading with polymer trapping, and passive symplastic loading. Early research on phloem loading focused on the apoplastic pathway, apparently the most common mechanism, so this loading mechanism will be discussed first. The two types of symplastic loading will then be introduced in the order in which their importance was recognized.

## Abundant data support the existence of apoplastic loading in some species

In the case of apoplastic loading, the sugars enter the apoplast quite near the sieve element–companion cell complex. Sugars are then actively transported from the apoplast into the sieve elements and companion cells by an energy-driven, selective transporter located in the plasma membranes of these cells. Efflux into the apoplast is highly localized, probably into the walls of phloem parenchyma cells.

Apoplastic phloem loading leads to three basic predictions (Grusak et al. 1996):

1. Transported sugars should be found in the apoplast.

2. In experiments in which sugars are supplied to the apoplast, the exogenously supplied sugars should accumulate in sieve elements and companion cells.

3. Inhibition of sugar uptake from the apoplast should result in inhibition of export from the leaf.

Many studies devoted to testing these predictions have provided solid evidence for apoplastic loading in several species (see **WEB TOPIC 10.7**).

## Sucrose uptake in the apoplastic pathway requires metabolic energy

In many of the species initially studied, sugars become more concentrated in the sieve elements and companion cells than in the mesophyll. This difference in solute concentration can be demonstrated through measurement of the osmotic potential ($\Psi_s$) of the various cell types in the leaf (see Chapter 3).

In sugar beet, the osmotic potential of the mesophyll is approximately –1.3 MPa, and the osmotic potential of the sieve elements and companion cells is about –3.0 MPa (Geiger et al. 1973). Most of this difference in osmotic potential is thought to result from accumulated sugar, specifically sucrose, because sucrose is the major transport sugar in this species. Experimental studies have also demonstrated that both externally supplied sucrose and sucrose made from photosynthetic products accumulate in the sieve elements and companion cells of the minor veins of sugar beet source leaves (**FIGURE 10.15**) (see also **WEB TOPIC 10.7**).

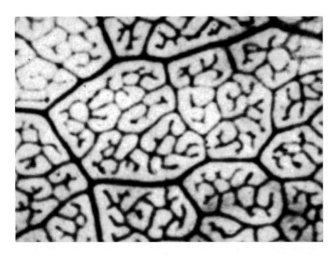

**FIGURE 10.15** This autoradiograph shows that labeled sugar moves from the apoplast into sieve elements and companion cells of sugar beet against its concentration gradient. A solution of $^{14}$C-labeled sucrose was applied for 30 minutes to the upper surface of a sugar beet leaf that had previously been kept in darkness for 3 hours. The leaf cuticle was removed to allow penetration of the solution to the interior of the leaf. Sucrose is actively transported against its concentration gradient into the sieve elements and companion cells of the small veins in the source leaf, as shown by the black accumulations. (From Fondy 1975, courtesy of D. Geiger.)

The fact that sucrose is at a higher concentration in the sieve element–companion cell complex than in surrounding cells indicates that sucrose is actively transported against its chemical-potential gradient. The dependence of sucrose accumulation on active transport is supported by the fact that treating source tissue with respiratory inhibitors both decreases ATP concentration and inhibits loading of exogenous sugar.

Plants that load sugars apoplastically into the phloem may also load amino acids and sugar alcohols (sorbitol and mannitol) actively. In contrast, other metabolites, such as organic acids and hormones, may enter sieve elements passively. (See **WEB TOPIC 10.7** for a discussion of these topics.)

## Phloem loading in the apoplastic pathway involves a sucrose–H⁺ symporter

A sucrose–H⁺ symporter is thought to mediate the transport of sucrose from the apoplast into the sieve element–companion cell complex. Recall from Chapter 6 that symport is a secondary transport process that uses the energy generated by the proton pump (see Figure 6.11A). The energy dissipated by protons moving back into the cell is coupled to the uptake of a substrate, in this case sucrose (**FIGURE 10.16**).

Data from a number of studies support the operation of a sucrose–H⁺ symporter in phloem loading:

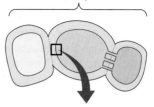
Sieve element–companion cell complex

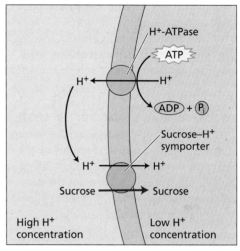

**FIGURE 10.16** ATP-dependent sucrose transport in apoplastic sieve-element loading. In the cotransport model of sucrose loading into the symplast of the sieve element–companion cell complex, the plasma membrane ATPase pumps protons out of the cell into the apoplast, establishing a high proton concentration there. The energy in this proton gradient is then used to drive the transport of sucrose into the symplast of the sieve element–companion cell complex through a sucrose–H⁺ symporter.

- Proton-pumping ATPases, localized by immunological techniques, have been found in the plasma membranes of companion cells (*Arabidopsis thaliana*) and in transfer cells (broad bean, *Vicia faba*). In transfer cells, the H⁺-ATPase molecules are most concentrated in the plasma membrane infoldings that face the bundle sheath and phloem parenchyma cells.

- The distribution of the H⁺-ATPases in companion cells appears to be correlated with the distribution of a sucrose–H⁺ symporter called SUC2 in Arabidopsis and broad-leaved plantain (*Plantago major*) (see **WEB TOPIC 10.7**). Thus, the H⁺-ATPase that supplies the driving force for sucrose uptake and the sucrose transporter that utilizes it have been shown to be located in the same cells in some species.

- Reducing transporter activity inhibits transport from source leaves. Tomato plants (*Lycopersicon esculentum*) transformed with antisense DNA to *SUT1*, an ortholog of the Arabidopsis SUC2, show reduced or absent fruit formation, high accumulation of soluble sugars in source leaves, and an inability to mobilize starch during prolonged dark periods. The changes in phenotype correlate with reduced transcript levels of *SUT1* (Hackel et al. 2006).

Several sucrose–H⁺ symporters have been cloned and localized in the phloem. SUT1 and SUC2 appear to be the major sucrose transporters in phloem loading into either companion cells or sieve elements. (See **WEB TOPIC 10.7** for more information about sucrose transporters in the phloem.)

**REGULATING SUCROSE LOADING IN THE APOPLASTIC PATHWAY** The mechanisms that regulate the loading of sucrose from the apoplast to the sieve elements by the sucrose–H⁺ symporter are not completely clear. Possible regulatory factors include the sieve element turgor pressure, the sucrose concentration in the apoplast, and the available number of symporter protein molecules:

- A decrease in sieve element turgor below a certain threshold would lead to a compensatory increase in loading. Although many investigations suggest that turgor pressure regulates membrane transport, very few have actually demonstrated such control (Daie and Wyse 1985).

- High sucrose concentrations in the apoplast would increase phloem loading. A number of investigators have reported a direct correlation between export and source-leaf sucrose concentration or photosynthetic rate, although none has measured the sucrose concentration in the apoplast at the site of loading. Measurement of this important parameter remains a significant challenge for future investigators.

- Data suggest that sucrose levels in source leaves can regulate the concentration of sucrose–H⁺ symporter molecules in the leaf, which in turn can regulate loading. The transcription of *SUT1* declines in sugar beet source leaves fed sucrose via the transpiration stream. Both SUT1 and its mRNA are degraded rapidly, so a decrease in transcription results in a decline in SUT1 protein. Uptake of sucrose into plasma membrane vesicles purified from the leaves declines at the same sucrose concentration as that which causes decreased transcription of *SUT1* (Vaughn et al. 2002), suggesting that the level of the symporter has declined. Protein phosphorylation has been demonstrated to play a role in these regulatory events (Ransom-Hodgkins et al. 2003). The levels of SUT1 transporter mRNA and protein are also regulated diurnally, being lower after 15 hours of darkness than after a light treatment.

Other studies have shown that sucrose efflux into the apoplast is enhanced by potassium availability in the apoplast, suggesting that a better nutrient supply increases translocation to sinks and enhances sink growth.

### Phloem loading is symplastic in some species

Many results point to apoplastic phloem loading in some species that transport only sucrose and that have few plasmodesmata leading into the minor vein phloem. On the other hand, a symplastic pathway has become evident in species that transport raffinose and stachyose, in addition to sucrose, in the phloem; that have intermediary cells in the minor veins; and that have abundant plasmodesmata leading into the minor veins. Some examples of such species are common coleus (*Coleus blumei*), pumpkin and squash (*Cucurbita pepo*), and melon (*Cucumis melo*).

The operation of a symplastic pathway requires the presence of open plasmodesmata between the different cells in the pathway. Many species have numerous plasmodesmata at the interface between the sieve element–companion cell complex and the surrounding cells (see Figure 10.7C).

### The polymer-trapping model explains symplastic loading in plants with intermediary cells

In many species the composition of sieve element sap is different from the solute composition in tissues surrounding the phloem. This difference indicates that certain sugars are specifically selected for transport in the source leaf. The involvement of symporters in apoplastic phloem loading provides a clear mechanism for selectivity, because symporters are specific for certain sugar molecules. Symplastic loading, in contrast, depends on the diffusion of sugars from the mesophyll to the sieve elements via the plasmodesmata. It is more difficult to envision how diffusion through plasmodesmata during symplastic loading could be selective for certain sugars.

Furthermore, data from several species showing symplastic loading indicate that sieve elements and companion cells have a higher osmotic content than the mesophyll. How could diffusion-dependent symplastic loading account for the observed selectivity for transported molecules and the accumulation of sugars against a concentration gradient?

The **polymer-trapping model** (**FIGURE 10.17**) has been developed to address these questions (Turgeon and Gowan 1990). This model states that the sucrose synthesized in the mesophyll diffuses from the bundle sheath cells into the intermediary cells through the abundant plasmodesmata that connect the two cell types. In the intermediary cells, raffinose and stachyose (polymers made of three and four hexose sugars, respectively; see Figure 10.9B) are synthesized from the transported sucrose and from galactinol (a metabolite of galactose). Because of the anatomy of the tis-

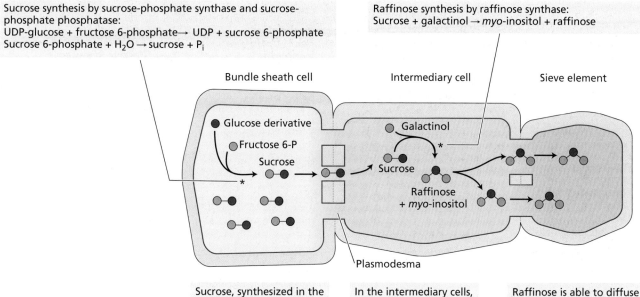

Sucrose synthesis by sucrose-phosphate synthase and sucrose-phosphate phosphatase:
UDP-glucose + fructose 6-phosphate→ UDP + sucrose 6-phosphate
Sucrose 6-phosphate + H₂O → sucrose + Pᵢ

Raffinose synthesis by raffinose synthase:
Sucrose + galactinol → *myo*-inositol + raffinose

Sucrose, synthesized in the mesophyll, diffuses from the bundle sheath cells into the intermediary cells through the abundant plasmodesmata.

In the intermediary cells, raffinose is synthesized from sucrose and galactinol, thus maintaining the diffusion gradient for sucrose. Because of its larger size, raffinose is not able to diffuse back into the mesophyll.

Raffinose is able to diffuse into the sieve elements. As a result, the concentration of transport sugar rises in the intermediary cells and the sieve elements. Note that stachyose is not shown here for clarity.

**FIGURE 10.17** Polymer-trapping model of phloem loading. For simplicity, the trisaccharide stachyose is omitted. (After van Bel 1992.)

sue and the relatively large size of raffinose and stachyose, the polymers cannot diffuse back into the bundle sheath cells, but they can diffuse into the sieve element. Sugar concentrations in the sieve elements of these plants can reach levels equivalent to those in plants that load apoplastically. Sucrose can continue to diffuse into the intermediary cells, because its synthesis in the mesophyll and its utilization in the intermediary cells maintain the concentration gradient (see Figure 10.17).

The polymer-trapping model makes three predictions:

1. Sucrose should be more concentrated in the mesophyll than in the intermediary cells.

2. The enzymes for raffinose and stachyose synthesis should be preferentially located in the intermediary cells.

3. The plasmodesmata linking the bundle sheath cells and the intermediary cells should exclude molecules larger than sucrose. Plasmodesmata between the intermediary cells and sieve elements must be wider to allow passage of raffinose and stachyose.

A number of studies support the polymer-trapping model of symplastic loading in some species (see **WEB TOPIC 10.7**).

## Phloem loading is passive in a number of tree species

Passive symplastic phloem loading has recently been recognized as a mechanism that is widespread among plant species. While the data supporting this mechanism are recent, passive symplastic loading was actually a part of Münch's original conception of pressure flow.

It has become apparent that a number of tree species possess abundant plasmodesmata between the sieve element–companion cell complex and surrounding cells but do not have intermediary cells and do not transport raffinose and stachyose. Willow (*Salix babylonica*) and apple (*Malus domestica*) trees are among the species that fall into this category (Rennie and Turgeon 2009). These plants have no concentrating step in the pathway from the mesophyll into the sieve element–companion cell complex. Since a concentration gradient from the mesophyll into the phloem drives diffusion along this short-distance pathway, the absolute levels of sugars in the source leaves of these species must be high in order to maintain the required high solute concentrations and the resulting high turgor pressures in the sieve elements. Although there is wide variation (over 50-fold) and considerable overlap between groups of plants with different loading mechanisms, source leaf sugar concentrations are generally higher in the tree species that load passively (Rennie and Turgeon 2009).

## TABLE 10.3
### Patterns in apoplastic and symplastic loading

| Feature | Apoplastic loading | Symplastic polymer trapping | Passive symplastic loading |
|---|---|---|---|
| Transport sugar | Sucrose | Raffinose and stachyose in addition to sucrose | Sucrose |
| Characteristic companion cells | Ordinary companion cells or transfer cells | Intermediary cells | Ordinary companion cells |
| Number and conductivity of plasmodesmata connecting the SE-CC complex to surrounding cells | Low | High | High |
| Dependence on active carriers in SE-CC complex | Transporter driven | Independent of transporters | Independent of transporters |
| Overall concentration of transport sugars in source leaves | Low | Low | High |
| Cell type in which driving force for long-distance transport is generated | Sieve element-companion cell complex | Intermediary cells | Mesophyll |
| Growth habit | Mainly herbaceous | Herbs and woody species | Mainly trees |

*Source:* Gamalei 1989; van Bel et al. 1992; Rennie and Turgeon 2009.

*Note:* Plants using all three mechanisms of phloem loading may also transport sugar alcohols. In addition, some species may load both apoplastically and symplastically, since different types of companion cells can be found within the veins of a single species. SE-CC complex, sieve element-companion cell complex.

### The type of phloem loading is correlated with a number of significant characteristics

As discussed above, the operation of apoplastic and symplastic phloem-loading pathways is correlated with a number of defining characteristics, listed in **TABLE 10.3** (Rennie and Turgeon 2009).

- Species that have apoplastic phloem loading translocate sucrose almost exclusively and have either ordinary companion cells or transfer cells in the minor veins. These species usually possess few connections between the sieve element–companion cell complex and the surrounding cells. Active carriers in the sieve element–companion cell complex concentrate sucrose in the phloem cells and generate the driving force for long distance transport.

- Species that use symplastic phloem loading with polymer trapping translocate oligosaccharides such as raffinose in addition to sucrose. They have intermediary-type companion cells in the minor veins, with abundant connections between the sieve element–companion cell complex and the surrounding cells. Polymer trapping concentrates transport sugars in the phloem cells and generates the driving force for long-distance transport.

- Species that have passive symplastic phloem loading translocate sucrose and have ordinary companion cells in the minor veins. These species also possess abundant connections between the sieve element–companion cell complex and the surrounding cells. Species with passive symplastic loading are characterized by high overall sugar concentrations in the source leaves, which maintains a concentration gradient between the mesophyll and the sieve element-companion cell complex. The high sugar concentrations give rise to the high turgor pressures in the sieve elements of source leaves, generating the driving force for long-distance transport. Many of the species with passive symplastic loading are trees.

**WEB TOPIC 10.7** discusses the relationships between loading characteristics (type of companion cell, transport sugars, and abundance of plasmodesmata) and the loading mechanisms in various species.

It has been suggested that active and passive loading may coexist in some species, simultaneously or at different times, or in different sieve elements in the same vein. Future research will likely reveal new combinations of loading pathways (Turgeon 2006). Certainly, the evolution of different loading types and the environmental pressures related to their evolution will continue to be important research areas in the future, as loading pathways are clarified in more species. At present, symplastic loading is thought to be the ancestral condition, while apoplastic loading is the derived condition. What adaptive advantage is conferred on the apoplastic loaders? Though cold temperatures and drought conditions are implicated in the evolution of apoplastic loading, only more research will provide complete answers to these questions.

# Phloem Unloading and Sink-to-Source Transition

Now that we have learned about the events leading up to the export of sugars from sources, let's take a look at **import** into sinks such as developing roots, tubers, and reproductive structures. In many ways the events in sink tissues are simply the reverse of the events in sources. The following steps are involved in the import of sugars into sink cells.

1. *Phloem unloading.* This is the process by which imported sugars leave the sieve elements of sink tissues.

2. *Short-distance transport.* After unloading, the sugars are transported to cells in the sink by means of a short-distance transport pathway. This pathway has also been called *post–sieve element transport.*

3. *Storage and metabolism.* In the final step, sugars are stored or metabolized in sink cells.

In this section we will discuss the following questions: Are phloem unloading and short-distance transport symplastic or apoplastic? Is sucrose hydrolyzed during the process? Do phloem unloading and subsequent steps require energy? Finally, we will examine the transition process by which a young, importing leaf becomes an exporting source leaf.

## *Phloem unloading and short-distance transport can occur via symplastic or apoplastic pathways*

In sink organs, sugars move from the sieve elements to the cells that store or metabolize them. Sinks vary from growing vegetative organs (root tips and young leaves) to storage tissues (roots and stems) to organs of reproduction and dispersal (fruits and seeds). Because sinks vary so greatly in structure and function, there is no single scheme of phloem unloading and short-distance transport. Differences in import pathways due to differences in sink types are emphasized in this section; however, the pathway often depends on the stage of sink development, as well.

As in sources, the sugars may move entirely through the symplast via the plasmodesmata in sinks, or they may enter the apoplast at some point. **FIGURE 10.18** diagrams the several possible pathways in sinks. Both unloading and the short-distance pathway appear to be completely

**(A) Symplastic phloem unloading and short-distance transport**

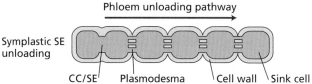

**(B) Apoplastic phloem unloading and short-distance transport**

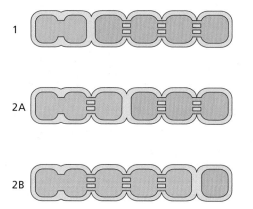

**FIGURE 10.18** Pathways for phloem unloading and short-distance transport. The sieve element–companion cell complex (CC/SE) is considered a single functional unit. The presence of plasmodesmata is assumed to provide functional symplastic continuity. An absence of plasmodesmata between cells indicates an apoplastic transport step. (A) Symplastic phloem unloading and short-distance transport. All steps are symplastic. (B) Apoplastic phloem unloading and short-distance transport.

**Type 1:** This short-distance pathway is designated apoplastic because one step, phloem unloading from the sieve element–companion cell complex, occurs in the apoplast. Once the sugars are taken back up into the symplast of adjoining cells, transport is symplastic.

**Type 2:** These pathways also have an apoplastic step. However, phloem unloading from the sieve element–companion cell complex is symplastic. The apoplastic step occurs later in the pathways. The upper figure (type 2A) shows an apoplastic step close to the sieve element–companion cell complex; the lower figure (type 2B), an apoplastic step that is further removed.

symplastic in some young dicot leaves, such as sugar beet (*Beta vulgaris*) and tobacco (*Nicotiana tabacum*) (see Figure 10.18A). Meristematic and elongating regions of primary root tips also appear to unload symplastically.

While symplastic import predominates in most sink tissues, part of the short-distance pathway is apoplastic in some sink organs—for example, in fruits, seeds and other storage organs that accumulate high concentrations of sugars (see Figure 10.18B). The apoplastic step could be located at the site of unloading itself (type 1 in Figure 10.18B) or farther removed from the sieve elements (type 2). This arrangement (type 2), typical of developing seeds, appears to be the most common in apoplastic pathways.

An apoplastic step is required in developing seeds because there are no symplastic connections between the maternal tissues and the tissues of the embryo. Sugars exit the sieve elements (phloem unloading) via a symplastic pathway and are transferred from the symplast to the apoplast at some point removed from the sieve element–companion cell complex (type 2 in Figure 10.18B). The apoplastic step permits membrane control over the substances that enter the embryo, because two membranes must be crossed in the process.

When an apoplastic step occurs in the import pathway, the transport sugar can be partly metabolized in the apoplast, or it can cross the apoplast unchanged (see **WEB TOPIC 10.8**). For example, sucrose can be hydrolyzed into glucose and fructose in the apoplast by invertase, a sucrose-splitting enzyme, and glucose and/or fructose would then enter the sink cells. As we will discuss later, such sucrose-cleaving enzymes play a role in the control of phloem transport by sink tissues.

### Transport into sink tissues requires metabolic energy

Inhibitor studies have shown that import into sink tissues is energy dependent. Growing leaves, roots, and storage sinks, in which carbon is stored as starch or in protein, appear to utilize symplastic phloem unloading and short-distance transport. Transport sugars are used as substrates for respiration and are metabolized into storage polymers and into compounds needed for growth. Sucrose metabolism thus results in a low sucrose concentration in the sink cells, maintaining a concentration gradient for sugar uptake. In this pathway, no membranes are crossed during sugar uptake into the sink cells, and transport is passive: Transport sugars move from a high concentration in the sieve elements to a low concentration in the sink cells. Metabolic energy is thus required in these sink organs mainly for respiration and for biosynthetic reactions.

In apoplastic import, sugars must cross at least two membranes: the plasma membrane of the cell that is releasing the sugar and the plasma membrane of the sink cell. When sugars are transported into the vacuole of the sink cell, they must also traverse the tonoplast. As discussed

earlier, transport across membranes in an apoplastic pathway may be energy dependent. While some evidence indicates that both efflux and uptake of sucrose can be active (see **WEB TOPIC 10.8**), the transporters have yet to be completely characterized.

Since these transporters have been shown to be bidirectional in some studies, some of the same sucrose transporters described above for sucrose loading could also be involved in sucrose unloading; the direction of transport would depend on the sucrose gradient, the pH gradient, and the membrane potential. Furthermore, symporters important in phloem loading have been found in some sink tissues— for example, SUT1 in potato tubers. The symporter may function in sucrose retrieval from the apoplast, in import into sink cells, or in both. Monosaccharide transporters must be involved in uptake into sink cells when sucrose is hydrolyzed in the apoplast (Hayes et al. 2007).

### The transition of a leaf from sink to source is gradual

Leaves of dicots such as tomato or bean begin their development as sink organs. A transition from sink to source status occurs later in development, when the leaf is approximately 25% expanded, and it is usually complete when the leaf is 40 to 50% expanded. Most of the species in which the sink-to-source transition has been investigated load the phloem from the apoplast, and this discussion is restricted to those species.

Export from the leaf begins at the tip or apex of the blade and progresses toward the base until the whole leaf becomes a sugar exporter. During the transition period, the tip exports sugar, while the base imports it from the other source leaves (**FIGURE 10.19**).

The maturation of leaves is accompanied by a large number of functional and anatomic changes, resulting in a reversal of transport direction from importing to exporting. In general, the cessation of import and the initiation of export are independent events (Turgeon 2006). In albino leaves of tobacco, which have no chlorophyll and therefore are incapable of photosynthesis, import stops at the same developmental stage as in green leaves, even though export is not possible. Therefore some change besides the initiation of export must occur in developing leaves of tobacco that causes them to cease importing sugars.

Sugars are unloaded and loaded almost entirely via different veins in tobacco (**FIGURE 10.20**) (Roberts et al. 1997), contributing to the conclusion that import cessation and export initiation are two separate events. The minor veins that are eventually responsible for most of the loading in tobacco and other *Nicotiana* species do not mature until about the time import ceases and cannot play a role in unloading.

The change that stops import must thus involve blockage of unloading from the large veins at some point in the development of mature leaves. Factors that could account

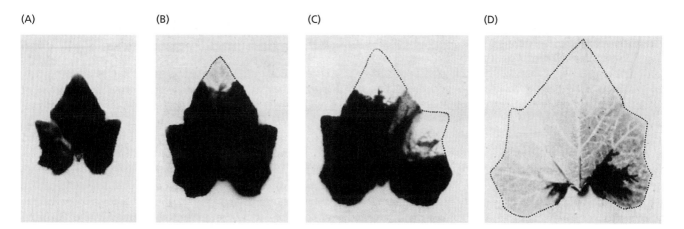

**FIGURE 10.19** Autoradiographs of a leaf of summer squash (*Cucurbita pepo*), showing the transition of the leaf from sink to source status. In each case, the leaf imported $^{14}C$ from the source leaf on the plant for 2 hours. Label is visible as black accumulations. (A) The entire leaf is a sink, importing sugar from the source leaf. (B–D) The base is still a sink. As the tip of the leaf loses the ability to unload and stops importing sugar (as shown by the loss of black accumulations), it gains the ability to load and to export sugar. (From Turgeon and Webb 1973.)

for the cessation of unloading include plasmodesmatal closure, a decrease in plasmodesmatal frequency, or another change in symplastic continuity. Experimental data have shown that both plasmodesmatal closure and elimination of plasmodesmata can occur.

Export of sugars begins when events have occurred that close the importing pathway and activate apoplastic loading and when loading has accumulated sufficient photosynthate in the sieve elements to drive translocation out of the leaf. The following conditions are necessary for export to begin:

- The leaf is synthesizing photosynthate in sufficient quantity that some is available for export. The sucrose-synthesizing genes are being expressed.

- Minor veins responsible for loading have matured. A regulatory element (enhancer) has been identified in the DNA of Arabidopsis that acts as part of a cascade of events leading to minor vein maturation. The enhancer can activate a reporter gene fused to a companion-cell specific promoter and does so in the same tip-to-base pattern as the sink-to-source transition (McGarry and Ayre 2008).

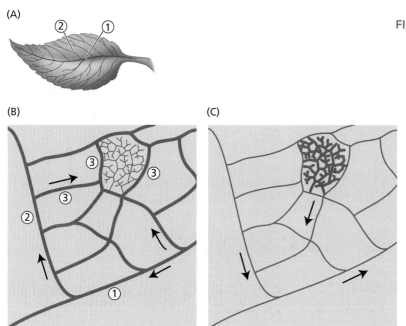

**FIGURE 10.20** Division of labor in the veins of a tobacco leaf shown in (A). When the leaf is immature and still in its sink phase (B), photosynthate is imported from mature leaves and distributed (arrows) throughout the blade (or lamina) via the larger, major veins (thicker lines). The major veins are numbered, with the midrib being the first-order vein. The imported photosynthate unloads from the same major veins into the mesophyll. The smallest, minor veins are shown within the areas enclosed by the third-order veins. The minor veins do not function in import and unloading because they are immature. In a source leaf (C), import has ceased, and export has begun. Photosynthate loads into the minor veins (thicker lines), while the larger veins serve only in export (arrows); they can no longer unload. Although (B) is drawn to scale from an autoradiograph, (C) is not to scale or in correct proportions, since the lamina grows considerably as the leaf matures. (From Turgeon 2006.)

- The sucrose–H$^+$ symporter is expressed and in place in the plasmalemma of the sieve element–companion cell complex. Regulation of these events is being investigated. For example, the promoter of the *SUC2* gene in Arabidopsis becomes active in companion cells in a pattern that corresponds to the sink-to-source transition (**FIGURE 10.21**). Binding sites for transcription factors have been identified within the *SUC2* promoter that mediate this source-specific and companion-cell specific gene expression (Schneidereit et al. 2008).

In leaves of plants such as sugar beet and tobacco, the ability to accumulate exogenous sucrose in the sieve element–companion cell complex is acquired as the leaves undergo the sink-to-source transition, suggesting that the symporter required for loading has become functional. In developing leaves of Arabidopsis, expression of the symporter that is thought to transport sugars during loading begins in the tip and proceeds to the base during a sink-to-source transition. The same basipetal pattern is seen in the development of export capacity.

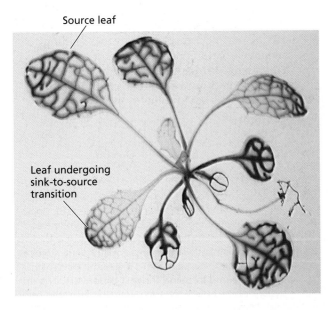

Source leaf

Leaf undergoing
sink-to-source
transition

**FIGURE 10.21** Export from source tissue depends on the placement and activity of active sucrose transporters. The figure shows an Arabidopsis rosette transformed with a construct consisting of a reporter gene under control of the *AtSUC2* promoter. SUC2, a sucrose–H$^+$ symporter, is one of the major sucrose transporters functioning in phloem loading. The GUS reporter system forms a visible product (blue) where the promoter is active. Staining is visible only in the vascular tissue of source leaves and in the tips of leaves undergoing the sink-to-source transition. (From Schneidereit et al. 2008)

# Photosynthate Distribution: Allocation and Partitioning

The photosynthetic rate determines the total amount of fixed carbon available to the leaf. However, the amount of fixed carbon available for translocation depends on subsequent metabolic events. The regulation of the distribution of fixed carbon into various metabolic pathways is termed **allocation** in this chapter.

The vascular bundles in a plant form a system of "pipes" that can direct the flow of photosynthates to various sinks: young leaves, stems, roots, fruits, or seeds. However, the vascular system is often highly interconnected, forming an open network that allows source leaves to communicate with multiple sinks. Under these conditions, what determines the volume of flow to any given sink? The differential distribution of photosynthates within the plant is termed **partitioning** in this chapter. (The terms "allocation" and "partitioning" are sometimes used interchangeably in current publications.)

After giving an overview of allocation and partitioning, we will examine the coordination of starch and sucrose synthesis. Throughout this section, keep in mind that a limited number of species has been studied, mainly those that load sucrose actively from the apoplast. It is likely that the mechanism of phloem loading affects the regulation of allocation. Studies of allocation will have to be extended to a wider range of species as more is learned about other loading mechanisms. We will conclude by discussing how sinks compete, how sink demand might regulate photosynthetic rate in the source leaf, and how sources and sinks communicate with each other.

## Allocation includes storage, utilization, and transport

The carbon fixed in a source cell can be used for storage, metabolism, and transport:

- *Synthesis of storage compounds.* Starch is synthesized and stored within chloroplasts and, in most species, is the primary storage form that is mobilized for translocation during the night. Plants that store carbon primarily as starch are called "starch storers."

- *Metabolic utilization.* Fixed carbon can be utilized within various compartments of the photosynthesizing cell to meet the energy needs of the cell or to provide carbon skeletons for the synthesis of other compounds required by the cell.

- *Synthesis of transport compounds.* Fixed carbon can be incorporated into transport sugars for export to various sink tissues. A portion of the transport sugar can also be stored temporarily in the vacuole.

Allocation is also a key process in sink tissues. Once the transport sugars have been unloaded and enter the sink cells, they can remain as such or can be transformed into various other compounds. In storage sinks, fixed carbon can be accumulated as sucrose or hexose in vacuoles or as starch in amyloplasts. In growing sinks, sugars can be utilized for respiration and for the synthesis of other molecules required for growth.

### Various sinks partition transport sugars

Sinks compete for the photosynthate being exported by the sources. Such competition determines the distribution of transport sugars among the various sink tissues of the plant (partitioning), at least in the short term. The ability of a sink to store or metabolize imported sugar (allocation) affects its ability to compete for available sugars. In this way, the processes of partitioning and allocation interact.

Of course, events in sources and sinks must be synchronized. Partitioning determines the patterns of growth, and growth must be balanced between shoot growth (photosynthetic productivity) and root growth (water and mineral uptake) in such a way that the plant can respond to the challenges of a variable environment. The goal is *not* a constant root-to-shoot ratio but one that secures a supply of carbon and mineral nutrients appropriate to the needs of the plant.

So an additional level of control lies in the interaction between areas of supply and demand. Turgor pressure in the sieve elements could be an important means of communication between sources and sinks, acting to coordinate rates of loading and unloading. Chemical messengers are also important in communicating the status of one organ to the other organs in the plant. Such chemical messengers include plant hormones and nutrients, such as potassium and phosphate, and even the transport sugars themselves. Recent findings suggest that macromolecules (RNA and protein) may also play a role in photosynthate partitioning, perhaps by influencing transport through plasmodesmata.

Attainment of higher yields of crop plants is one goal of research on photosynthate allocation and partitioning. Whereas grains and fruits are examples of edible yields, total yield includes inedible portions of the shoot. An understanding of partitioning should enable plant breeders to select and develop varieties that have improved transport to edible portions of the plant.

Allocation and partitioning in the whole plant must be coordinated such that increased transport to edible tissues does not occur at the expense of other essential processes and structures. Crop yield may also be improved if photosynthates that are normally "lost" by the plant are retained. For example, losses due to nonessential respiration or exudation from roots could be reduced. In the latter case, care must be taken not to disrupt essential processes outside the plant, such as growth of beneficial microbial species in the vicinity of the root that obtain nutrients from the root exudate.

### Source leaves regulate allocation

Increases in the rate of photosynthesis in a source leaf generally result in an increase in the rate of translocation from the source. Control points for the allocation of photosynthate (**FIGURE 10.22**) include the distribution of triose phosphates to the following processes:

- Regeneration of intermediates in the $C_3$ photosynthetic carbon reduction cycle (the Calvin–Benson cycle; see Chapter 8)
- Starch synthesis

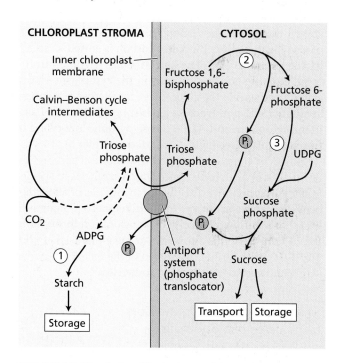

**FIGURE 10.22**   A simplified scheme for starch and sucrose synthesis during the day. Triose phosphate, formed in the Calvin–Benson cycle, can either be utilized in starch formation in the chloroplast or transported into the cytosol in exchange for inorganic phosphate ($P_i$) via the phosphate translocator in the inner chloroplast membrane. The outer chloroplast membrane (omitted here for clarity) is porous to small molecules. In the cytosol, triose phosphate can be converted to sucrose for either storage in the vacuole or transport. Key enzymes involved are starch synthetase (1), fructose-1,6-bisphosphatase (2), and sucrose-phosphate synthase (3). The second and third enzymes, along with ADP-glucose pyrophosphorylase, which forms adenosine diphosphate glucose (ADPG), are regulated enzymes in sucrose and starch synthesis (see Chapter 8). UDPG, uridine diphosphate glucose. (After Preiss 1982.)

- Sucrose synthesis, as well as distribution of sucrose between transport and temporary storage pools

Various enzymes operate in the pathways that process the photosynthate, and the control of these steps is complex. The research described below focuses on species that load sucrose actively from the apoplast, specifically during the daylight hours. Further studies will be needed to extend our knowledge to plants using other loading strategies and to the regulation of allocation in those species.

During the day the rate of starch synthesis in the chloroplast must be coordinated with sucrose synthesis in the cytosol. Triose phosphates (glyceraldehyde 3-phosphate and dihydroxyacetone phosphate) produced in the chloroplast by the Calvin–Benson cycle (see Chapter 8) can be used for either starch or sucrose synthesis. Sucrose synthesis in the cytoplasm diverts triose phosphate away from starch synthesis and storage. For example, it has been shown that when the demand for sucrose by other parts of a soybean plant is high, less carbon is stored as starch by the source leaves. The key enzymes involved in the regulation of sucrose synthesis in the cytoplasm and of starch synthesis in the chloroplast are sucrose-phosphate synthase (SPS) and fructose-1,6-bisphosphatase (FBPase) in the cytoplasm and ADP-glucose pyrophosphorylase in the chloroplast (see Figure 10.22) (see Chapter 8).

However, there is a limit to the amount of carbon that normally can be diverted from starch synthesis in species that store carbon primarily as starch. Studies of allocation between starch and sucrose under different conditions suggest that a fairly steady rate of translocation throughout the 24-hour period is a priority for most plants. See **WEB TOPIC 10.9** for further discussion of the balance between starch and sucrose synthesis in source leaves.

### Sink tissues compete for available translocated photosynthate

As discussed earlier, translocation to sink tissues depends on the position of the sink in relation to the source and on the vascular connections between source and sink. Another factor determining the pattern of transport is competition between sinks, for example, between terminal sinks or between terminal sinks and axial sinks along the transport pathway. For example, young leaves might compete with roots for photosynthates in the translocation stream. Competition has been shown by numerous experiments in which removal of a sink tissue from a plant generally results in increased translocation to alternative, and hence competing, sinks. Conversely, increased sink size, for example, increased fruit load, decreases translocation to other sinks, especially the roots.

In the reverse type of experiment, the source supply can be altered while the sink tissues are left intact. When the supply of photosynthates from sources to competing sinks is suddenly and drastically reduced by shading of all the source leaves but one, the sink tissues become dependent on a single source. In sugar beet and bean plants, the rates of photosynthesis and export from the single remaining source leaf usually do not change over the short term (approximately 8 hours; Fondy and Geiger 1980). However, the roots receive less sugar from the single source, while the young leaves receive relatively more. Shading in general decreases partitioning to roots. Presumably, the young leaves can deplete the sugar content of the sieve elements more readily and thus increase the pressure gradient and the rate of translocation toward themselves.

Treatments such as making the sink water potential more negative increase the pressure gradient and enhance transport to the sink. Treatment of the root tips of pea (*Pisum sativum*) seedlings with mannitol solutions increased the import of sucrose over the short term by more than 300%, possibly because of a turgor decrease in the sink cells (Schulz 1994). Longer-term experiments show the same trend. Moderate water stress induced with polyethylene glycol treatment of the roots increased the proportion of assimilates transported to the roots of apple plants over a period of 15 days but decreased the proportion transported to the shoot apex (Dai et al. 2007). This contrasts with shading treatments (above) in which source limitation diverts more sugar to the young leaves.

### Sink strength depends on sink size and activity

The ability of a sink to mobilize photosynthate toward itself is often described as **sink strength**. Sink strength depends on two factors—sink size and sink activity—as follows:

$$\text{Sink strength} = \text{sink size} \times \text{sink activity}$$

**Sink size** is the total biomass of the sink tissue, and **sink activity** is the rate of uptake of photosynthates per unit biomass of sink tissue. Altering either the size or the activity of the sink results in changes in translocation patterns. For example, the ability of a pea pod to import carbon depends on the dry weight of that pod as a proportion of the total number of pods (Jeuffroy and Warembourg 1991).

Changes in sink activity can be complex, because various activities in sink tissues can potentially limit the rate of uptake by the sink. These activities include unloading from the sieve elements, metabolism in the cell wall, uptake from the apoplast, and metabolic processes that use the photosynthate in either growth or storage.

Experimental treatments to manipulate sink strength have often been nonspecific. For example, cooling a sink tissue, which inhibits any activities that require metabolic energy, often results in a decrease in the speed of transport toward the sink. More recent experiments take advantage of our ability to specifically over- or underexpress enzymes related to sink activity—for example, those involved in sucrose metabolism in the sink. The two major enzymes that split sucrose are acid invertase and sucrose synthase, both of which can catalyze the first step in sucrose utilization. WEB TOPIC 10.10 discusses evidence for a correlation between the activity of sucrose-splitting enzymes, particularly invertase, and sink demand.

### The source adjusts over the long term to changes in the source-to-sink ratio

If all but one of the source leaves of a soybean plant are shaded for an extended period (e.g., 8 days), many changes occur in the single remaining source leaf. These changes include a decrease in starch concentration and increases in photosynthetic rate, rubisco activity, sucrose concentration, transport from the source, and orthophosphate concentration (Thorne and Koller 1974). Thus, in addition to the observed short-term changes in the distribution of photosynthate among different sinks, there are adjustments in the source leaf's metabolism in response to altered conditions over a longer term.

Photosynthetic rate (the net amount of carbon fixed per unit leaf area per unit time) often increases over several days when sink demand increases, and it decreases when sink demand decreases. An accumulation of photosynthate (sucrose or hexoses) in the source leaf can account for the linkage between sink demand and photosynthetic rate in starch-storing plants (see WEB TOPIC 10.11). Sugars act as signaling molecules that regulate many metabolic and developmental processes in plants. In general, carbohydrate depletion enhances the expression of genes for photosynthesis, reserve mobilization, and export processes, while abundant carbon resources favor genes for storage and utilization (Koch 1996).

Sucrose or hexoses that would accumulate as a result of decreased sink demand are well known to repress photosynthetic genes. Interestingly, the genes for invertase and sucrose synthase (discussed in the previous section) and genes for sucrose–H⁺ symporters (discussed in the earlier section *Regulating Sucrose Loading*) are also among those regulated by carbohydrate supply.

Such regulation of photosynthesis by sink demand suggests that sustained increases in photosynthesis in response to elevated $CO_2$ in the atmosphere may depend on increasing sink strength (increasing the strength of existing sinks or development of new sinks). See Chapter 12 and WEB ESSAY 12.1 for a discussion of the results of increased carbon dioxide levels in the atmosphere on photosynthesis and growth of plants.

## The Transport of Signaling Molecules

Besides its major function in the long-distance transport of photosynthate, the phloem is also a conduit for the transport of signaling molecules from one part of the organism to another. Such long-distance signals coordinate the activities of sources and sinks and regulate plant growth and development. As indicated earlier, the signals between

sources and sinks might be physical or chemical. Physical signals such as turgor change could be transmitted rapidly via the interconnecting system of sieve elements. Molecules traditionally considered to be chemical signals, such as proteins and plant hormones, are found in the phloem sap, as are mRNAs and small RNAs, which have more recently been added to the list of signal molecules. The translocated carbohydrates themselves may also act as signals.

### Turgor pressure and chemical signals coordinate source and sink activities

Turgor pressure may play a role in coordinating the activities of sources and sinks. For example, if phloem unloading were rapid under conditions of rapid sugar utilization at the sink tissue, turgor pressures in the sieve elements of sinks would be reduced, and this reduction would be transmitted to the sources. If loading were controlled in part by turgor in the sieve elements of the source, loading would increase in response to this signal from the sinks. The opposite response would be seen when unloading was slow in the sinks. Some data suggest that cell turgor can modify the activity of the proton-pumping ATPase at the plasma membrane and therefore alter transport rates.

Shoots produce growth regulators such as auxin (see Chapter 19), which can be rapidly transported to the roots via the phloem, and roots produce cytokinins (see Chapter 21), which move to the shoots through the xylem. Gibberellins (GA) and abscisic acid (ABA) (see Chapters 20 and 23) are also transported throughout the plant in the vascular system. Plant hormones play a role in regulating source–sink relationships. They affect photosynthate partitioning in part by controlling sink growth, leaf senescence, and other developmental processes. Plant defense responses against herbivores and pathogens can also change allocation and partitioning of photoassimilates, with plant defense hormones such as jasmonic acid mediating the responses.

Loading of sucrose has been shown to be stimulated by exogenous auxin but inhibited by ABA in some source tissues, while exogenous ABA enhances, and auxin inhibits, sucrose uptake by some sink tissues. Hormones might regulate apoplastic loading and unloading by influencing the levels of active transporters in plasma membranes. Other potential sites of hormone regulation of unloading include tonoplast transporters, enzymes for metabolism of incoming sucrose, wall extensibility, and plasmodesmatal permeability in the case of symplastic unloading (see the next section).

As indicated earlier, carbohydrate levels can influence the expression of genes encoding photosynthesis components, as well as genes involved in sucrose hydrolysis. Many genes have been shown to be responsive to sugar depletion and abundance (Koch 1996, Sivitz et al. 2008). Thus, not only is sucrose transported in the phloem, but sucrose or its metabolites can also act as signals that modify the activities of sources and sinks. As discussed earlier, sucrose–H$^+$ symporter mRNA declines in sugar-beet source leaves fed exogenous sucrose through the xylem. The decline in symporter mRNA is accompanied by a loss of symporter activity in plasma membrane vesicles isolated from the leaves. A working model includes the following steps:

1. Decreased sink demand leads to high sucrose levels in the vascular tissue.

2. High sucrose levels lead to down-regulation of the symporter in the source.

3. Decreased loading results in increased sucrose concentrations in the source.

Increased sucrose concentrations in the source can result in a lower photosynthetic rate (see WEB TOPIC 10.11). An increase of starch accumulation in source leaves of plants transformed with antisense DNA to the sucrose–H$^+$ symporter SUT1 also supports this model (Hackel et al. 2006).

Sugars and other metabolites have been shown to interact with hormonal signals to control and integrate many plant processes (Hammond and White 2008). Gene expression in some source–sink systems responds to both sugar and hormonal signals.

### Proteins and RNAs function as signal molecules in the phloem to regulate growth and development

It has long been known that viruses can move in the phloem, traveling as complexes of proteins and nucleic acids or as intact virus particles. More recently, endogenous RNA molecules and proteins have been found in phloem sap, and at least some of these can function as signal molecules or generate phloem-mobile signals.

To be assigned a signaling role in plants, a macromolecule must meet a number of significant criteria (Oparka and Santa Cruz 2000):

• The macromolecule must move from source to sink in the phloem.

• The macromolecule must be able to leave the sieve element–companion cell complex in sink tissues. Alternatively, the macromolecule might trigger the formation of a second signal that transmits information to the sink tissues surrounding the phloem; that is, it might initiate a signal cascade.

• Perhaps most important, the macromolecule must be able to modify the functions of specific cells in the sink.

How well do various macromolecules in the phloem meet these criteria?

Proteins synthesized in companion cells can clearly enter the sieve elements through the plasmodesmata that

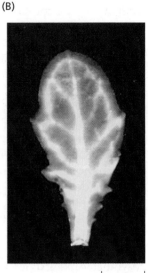

(A)　　　　　　　　(B)

|⊢ 2 mm ⊣|　　　|⊢ 0.1 mm ⊣|

**FIGURE 10.23**  GFP fluorescence in source and sink leaves from transgenic *Arabidopsis* plants expressing GFP under control of the *SUC2* promoter indicates that GFP moves through plasmodesmata from companion cells into sieve elements of source leaves and from sieve elements into the surrounding mesophyll of sink leaves. (A) GFP is synthesized in companion cells and moves into the sieve elements of the source, as indicated by the bright fluorescence in the veins. (B) Free GFP is imported into sink leaves and moves into the surrounding mesophyll. Because GFP has moved into surrounding tissues, the veins are no longer distinctly delineated, and GFP fluorescence is much more diffuse. Even though the source leaf in (A) appears to be the same size as the sink leaf in (B), the source leaf is actually much larger. The scales in (A) and (B) are different, as indicated by the bars on the photographs. (From Stadler et al. 2005.)

connect the two cell types and can move with the translocation stream to sink tissues. For example, passive movement of proteins from companion cells to sieve elements has been demonstrated in Arabidopsis and tobacco plants. These plants were transformed with the gene for green fluorescent protein (GFP) from jellyfish, under control of the *SUC2* promoter from Arabidopsis. The SUC2 sucrose–H$^+$ symporter is synthesized within the companion cells, so proteins expressed under the control of its promoter, including GFP, are also synthesized in the companion cells. GFP, which is localized by its fluorescence after excitation with blue light, moves through plasmodesmata from companion cells into sieve elements of source leaves (FIGURE 10.23A) and migrates within the phloem to sink tissues. However, only free GFP is able to move symplastically into sink tissues of the root (FIGURE 10.23B; Stadler et al. 2005). In fact, little evidence exists for movement of proteins from cells outside the sieve element–companion cell complex into source phloem or for movement of proteins from the phloem into sink tissues outside the sieve element–

companion cell complex. However, phloem transport of proteins that modify cellular functions has been demonstrated, implying that some signal, either the protein itself or some other signal molecule, moves between the sieve element–companion cell complexes and surrounding tissues of sources and sinks. For example, FLOWERING LOCUS T (FT) protein appears to be a significant component of the floral stimulus that moves from source leaf to apex and that induces flowering at the apex in response to inductive conditions (see Chapter 25).

RNAs transported in the phloem consist of endogenous mRNAs, pathogenic RNAs, and small RNAs associated with gene silencing (see Chapter 2). Most of these RNAs appear to travel in the phloem as complexes of RNA and protein (ribonucleoproteins [RNPs]) (Gomez et al. 2005; Kehr and Buhtz 2008). Like proteins in the phloem, little direct evidence exists for movement of these RNAs between sieve element–companion cell complexes and surrounding tissues. However, RNA transported in the phloem has been shown to cause visible changes in sinks. For example, mRNA for a regulator of gibberellic acid responses (called GAI) was localized to sieve elements and companion cells of pumpkin (*Cucurbita pepo*) and was found in pumpkin phloem sap. Transgenic tomato plants expressing a mutant version of the regulator gene were dwarf and dark green. The mRNA for the mutant regulator was localized to sieve elements, was able to be transported across graft unions into wild-type scions, and was unloaded into apical tissues. As a result, the mutant phenotype developed in new growth on the wild-type scion (Haywood et al. 2005).

See WEB TOPIC 10.12 for further discussion of these topics.

**PLASMODESMATA FUNCTION IN SIGNALING** Plasmodesmata have been implicated in nearly every aspect of phloem translocation, from loading to long-distance transport (pores in sieve areas and sieve plates are modified plasmodesmata) to allocation and partitioning. What role might plasmodesmata play in macromolecular signaling in the phloem?

The mechanism of plasmodesmatal transport (called trafficking) can be either passive (nontargeted) or selective and regulated. When a molecule moves passively, its size must be smaller than the *size exclusion limit* (*SEL*) of the plasmodesmata. As indicated earlier, green fluorescent protein moves passively through plasmodesmata. In contrast, when a molecule moves in a selective fashion, it must possess a trafficking signal or be targeted in some other way to the plasmodesmata. The transport of some developmental transcription factors and of viral movement proteins appears to occur by means of a selective mechanism. Viral movement proteins interact directly with plasmodesmata to allow the passage of viral nucleic acids between cells. They have been shown in a number of cases to use endogenous cellular proteins to "target" the plasmodesmata as their destination. Once at the plasmodesmata, movement proteins and/or the cellular proteins act to increase the SEL of the plasmodesmata to allow the viral genome to move between cells. The endogenous proteins commandeered by viruses to facilitate their specific movement to and through plasmodesmata are thought to carry out similar functions for endogenous macromolecules (see WEB TOPIC 10.12).

It is fitting to end this chapter with research topics that will continue to engage plant physiologists of the future: regulation of growth and development via the transport of endogenous RNA and protein signals, the nature of the proteins that facilitate the transport of signals through plasmodesmata, and the possibility of targeting signals to specific sinks in contrast to mass flow. Many other potential areas of inquiry have been indicated in this chapter as well, such as the mechanism of phloem transport in gymnosperms, the possibility of additional modes of phloem loading, and the nature of phloem unloading. As always in science, an answer to one question generates more questions!

## SUMMARY

Phloem translocation moves the products of photosynthesis from mature leaves to areas of growth and storage. It also transmits chemical signals and redistributes ions and other substances throughout the plant body.

### The Pathway of Translocation

- Sieve elements of the phloem conduct sugars and other organic materials throughout the plant (**Figures 10.1–10.3**).

- During development, sieve elements lose many organelles, retaining only the plasma membrane, and modified mitochondria, plastids, and smooth endoplasmic reticulum (**Figures 10.3, 10.4**).

- Sieve elements are interconnected through pores in their cell walls (**Figure 10.5**).

- In gymnosperms, smooth ER covers the sieve areas and is continuous through the sieve pores (**Figure 10.6, Table 10.1**).

- P proteins and callose seal off damaged phloem to limit loss of sap.

## SUMMARY continued

- Companion cells aid transport of photosynthetic products to the sieve elements. They also supply proteins and ATP to the sieve elements (**Figures 10.3, 10.4, 10.7**).

### Patterns of Translocation: Source to Sink

- Phloem translocation is not defined by gravity. Sap is translocated from sources to sinks. The pathways may be complex (**Figure 10.8**).

### Materials Translocated in the Phloem

- The composition of sap has been determined; non-reducing sugars are the main transported molecules (**Table 10.2, Figure 10.9**).
- Sap includes proteins, many of which may be related to stress and defense reactions.

### Rates of Movement

- Transport velocities in the phloem are high and exceed the rate of diffusion over long distances.

### The Pressure-Flow Model, a Passive Mechanism for Phloem Transport

- The pressure-flow model explains phloem translocation as a bulk flow of solution driven by an osmotically generated pressure gradient between source and sink.
- Phloem loading at the source and phloem unloading at the sink establish the pressure gradient for passive, long-distance bulk flow (**Figures 10.10, 10.11, 10.12**).
- The resistance offered by sieve plates results in the maintenance of a pressure gradient in the sieve elements between source and sink.

### Phloem Loading

- The export of sugars from sources involves: allocation of photosynthate to transport, short-distance transport, and phloem loading.
- Phloem loading can occur by way of the symplast or apoplast (**Figures 10.13, 10.14**).
- Sucrose is actively transported into the sieve element–companion cell complex in the apoplastic pathway (**Figures 10.15, 10.16**).
- The polymer-trapping model holds that polymers are synthesized from sucrose in the intermediate cells; the larger oligosaccharides can only diffuse into the sieve elements (**Figure 10.17**).
- Apoplastic and symplastic phloem-loading pathways have defining characteristics (**Table 10.3**).

### Phloem Unloading and Sink-to-Source Transition

- The import of sugars into sink cells involves: phloem unloading, short-distance transport, and storage/metabolism.
- Phloem unloading and short distance transport may operate by symplastic or apoplastic pathways in different sinks (**Figure 10.18**).
- Transport into sink tissues is energy dependent.
- Import cessation and export initiation are separate events, and there is a gradual transition from sink to source (**Figures 10.19, 10.20**).
- The transition from sink to source requires a number of conditions, including the expression and localization of the sucrose-H$^+$ symporter (**Figure 10.21**).

### Photosynthate Distribution: Allocation and Partitioning

- Allocation in source leaves includes storage, metabolic utilization, and synthesis of transport compounds.
- The regulation of allocation must thus control the distribution of fixed carbon to the Calvin–Benson cycle, starch synthesis, and sucrose synthesis (**Figure 10.22**).
- A variety of chemical and physical signals are involved in partitioning resources among the various sinks.
- In competing for photosynthate, sink strength depends on sink size and sink activity.
- In response to altered conditions, short-term changes alter the distribution of photosynthate among different sinks, while long-term changes take place in source metabolism and alter the amount of photosynthate available for transport.

### The Transport of Signaling Molecules

- Turgor pressure, cytokinins, gibberellins, and abscisic acid have signaling roles in coordinating source and sink.
- Proteins can move from companion cells into sieve elements of source leaves, and through the phloem to sink leaves (**Figure 10.23**).
- Proteins and RNAs transported in the phloem can alter cellular functions.
- Changes in the size exclusion limit may control what passes through plasmodesmata.

## WEB MATERIAL

### Web Topics

**10.1 Sieve Elements as the Transport Cells between Sources and Sinks**

Various methods demonstrate that sugar is tranported in the sieve elements of the phloem; anatomical and developmental factors affect the basic source to sink pattern of transport.

**10.2 An Additional Mechanism for Blocking Wounded Sieve Elements in the Legume Family**

Protein bodies rapidly disperse and block legume sieve tubes following wounding.

**10.3 Sampling Phloem Sap**

Exudation from wounds and from severed aphid stylets yield sufficient phloem sap for analysis.

**10.4 Nitrogen Transport in the Phloem**

Soybean is an economically important species widely studied in terms of nitrogen transport in the phloem.

**10.5 Monitoring Traffic on the Sugar Freeway: Sugar Transport Rates in the Phloem**

A variety of techniques measure mass transfer rate in the phloem, the dry weight moving through a cross-sectional area of sieve elements per unit time.

**10.6 Alternative Views of Pressure Gradient in Sieve Elements: Large or Small Gradients?**

Some mathematical models suggest that the pressure gradient in the sieve elements of angiosperms is small.

**10.7 Experiments on Phloem Loading**

Evidence exists for apoplastic loading of sieve elements in some species and for symplastic loading (polymer trapping) in others. While active carriers have been identified and characterized for some substances entering the phloem, other substances may enter sieve elements passively

**10.8 Experiments on Phloem Unloading**

Apoplastic unloading varies in its energy requirements and in the role of the cell-wall invertase.

**10.9 Allocation in Source Leaves: The Balance between Starch and Sucrose Synthesis**

Experiments with mutants and transgenic plants reveal flexibility in the regulation of starch and sucrose synthesis in source leaves.

**10.10 Partitioning: The Role of Sucrose-Metabolizing Enzymes in Sinks**

Increases in cell-wall invertase activity can enhance transport to a sink, while decreases in activity can inhibit transport to the sink.

**10.11 Possible Mechanisms Linking Sink Demand and Photosynthetic Rate in Starch Storers**

Photosynthate accumulation decreases the photosynthetic rate.

**10.12 Proteins and RNAs: Signal Molecules in the Phloem**

Proteins are transported between companion cells and sieve elements and travel in sieve elements between sources and sinks. Some proteins can modify cellular functions in the sinks. Little evidence exists for a movement of proteins outside the companion cells.

## CHAPTER REFERENCES

Brentwood, B., and Cronshaw, J. (1978) Cytochemical localization of adenosine triphosphatase in the phloem of *Pisum sativum* and its relation to the function of transfer cells. *Planta* 140: 111–120.

Clark, A. M., Jacobsen, K. R., Bostwick, D. E., Dannenhoffer, J. M., Skaggs, M. I., and Thompson, G. A. (1997) Molecular characterization of a phloem-specific gene encoding the filament protein, phloem protein 1 (PP1), from *Cucurbita maxima*. *Plant J.* 12: 49–61.

Dai, Z. W., Wang, L. J., Zhao, J. Y., Fan, P. G., and Li, S. H. (2007) Effect and after-effect of water stress on the distribution of newly-fixed $^{14}C$-photoassimilate in micropropagated apple plants. *Environ. Exp. Bot.* 60: 484–494.

Daie, J., and Wyse, R. E. (1985) Evidence on the mechanism of enhanced sucrose uptake at low cell turgor in leaf discs of *Phaseolus coccineus* cultivar scarlet. *Physiol. Plant.* 64: 547–552.

Dinant, S., Clark, A. M., Zhu, Y., Vilaine, F., Palauqui, J.-C., Kusiak, C., and Thompson, G. A. (2003) Diversity of the superfamily of phloem lectins (phloem protein 2) in angiosperms. *Plant Physiol.* 131: 114–128.

Doering-Saad, C., Newbury, H. J., Bale, J. S., and Pritchard, J. (2002) Use of aphid stylectomy and RT-PCR for the detection of transporter mRNAs in sieve elements. *J. Exp. Bot.* 53: 631–637.

Ehlers, K., Knoblauch, M., and van Bel, A. J. E. (2000) Ultrastructural features of well-preserved and injured sieve elements: Minute clamps keep the phloem transport conduits free for mass flow. *Protoplasma* 214: 80–92.

Evert, R. F. (1982) Sieve-tube structure in relation to function. *BioScience* 32: 789–795.

Evert, R. F., and Mierzwa, R. J. (1985) Pathway(s) of assimilate movement from mesophyll cells to sieve tubes in the *Beta vulgaris* leaf. In *Phloem Transport. Proceedings of an International Conference on Phloem Transport, Asilomar, CA*, J. Cronshaw, W. J. Lucas, and R. T. Giaquinta, eds. Liss, New York, pp. 419–432.

Fisher, D. B. (1978) An evaluation of the Münch hypothesis for phloem transport in soybean. *Planta* 139: 25–28.

Fondy, B. R. (1975) Sugar selectivity of phloem loading in *Beta vulgaris, vulgaris* L. and *Fraxinus americanus, americana* L. Thesis, University of Dayton, Dayton, OH.

Fondy, B. R., and Geiger, D. R. (1980) Effect of rapid changes in sink–source ratio on export and distribution of products of photosynthesis in leaves of *Beta vulgaris* L. and *Phaseolus vulgaris* L. *Plant Physiol.* 66: 945–949.

Gamalei, Y. V. (1985) Features of phloem loading in woody and herbaceous plants. *Fiziologiya Rastenii (Moscow)* 32: 866–875.

Gaupels, F., Buhtz, A., Knauer, T., Deshmukh, S., Waller, F., van Bel, A. J. E., Kogel, K.-H., and Kehr, J. (2008) Adaptation of aphid stylectomy for analyses of proteins and mRNAs in barley phloem sap. *J. Exp. Bot.* 59: 3297–3306.

Geiger, D. R., and Sovonick, S. A. (1975) Effects of temperature, anoxia and other metabolic inhibitors on translocation. In *Transport in Plants, 1: Phloem Transport* (Encyclopedia of Plant Physiology, New Series, Vol. 1), M. H. Zimmerman and J. A. Milburn, eds., Springer, New York, pp. 256–286.

Geiger, D. R., Giaquinta, R. T., Sovonick, S. A., and Fellows, R. J. (1973) Solute distribution in sugar beet leaves in relation to phloem loading and translocation. *Plant Physiol.* 52: 585–589.

Giaquinta, R. T., and Geiger, D. R. (1977) Mechanism of cyanide inhibition of phloem translocation. *Plant Physiol.* 59: 178–180.

Gomez, G., Torres, H., and Pallas, V. (2005) Identification of translocatable RNA-binding phloem proteins from melon, potential components of the long-distance RNA transport system. *Plant J.* 41: 107–116.

Grusak, M. A., Beebe, D. U., and Turgeon, R. (1996) Phloem loading. In *Photoassimilate Distribution in Plants and Crops: Source–Sink Relationships*, E. Zamski and A. A. Schaffer, eds., Dekker, New York, pp. 209–227.

Hackel, A., Schauer, N., Carrari, F., Fernie, A. R., Grimm, B., and Kuehn, C. (2006) Sucrose transporter LeSUT1 and LeSUT2 inhibition affects tomato fruit development in different ways. *Plant J.* 45: 180–192.

Hall, S. M., and Baker, D. A. (1972) The chemical composition of *Ricinus* phloem exudate. *Planta* 106: 131–140.

Hammond, J. P., and White, P. J. (2008) Sucrose transport in the phloem: Integrating root responses to phosphorus starvation. *J. Exp. Bot.* 59: 93–109.

Hao, P., Liu, C., Wang, Y., Chen, R., Tang, M., Du, B., Zhu, L., and He, G. (2008) Herbivore-induced callose deposition on the sieve plates of rice: An important mechansim for host resistance. *Plant Physiol.* 146: 1810–1820.

Hayes, M. A., Davies, C., and Dry, I. B. (2007) Isolation, functional characterization, and expression analysis of grapevine (*Vitis vinifera* L.) hexose transporters: Differential roles in sink and source tissues. *J. Exp. Bot.* 58: 1985–1997.

Haywood, V., Yu, T.-S., Huang, N.-C., and Lucas, W. J. (2005) Phloem long-distance trafficking of *GIBBERELLIC ACID-INSENSITIVE* RNA regulates leaf development. *Plant J.* 42: 49–68.

Jeuffroy, M.-H., and Warembourg, F. R. (1991) Carbon transfer and partitioning between vegetative and reproductive organs in *Pisum sativum* L. *Plant Physiol.* 97: 440–448.

Joy, K. W. (1964) Translocation in sugar beet. I. Assimilation of $^{14}CO_2$ and distribution of materials from leaves. *J. Exp. Bot.* 15: 485–494.

Kehr, J., and Buhtz, A. (2008) Long distance transport and movement of RNA through the phloem. *J. Exp. Bot.* 59: 85–92.

Knoblauch, M., and van Bel, A. J. E. (1998) Sieve tubes in action. *Plant Cell* 10: 35–50.

Koch, K. E. (1996) Carbohydrate-modulated gene expression in plants. *Annu. Rev. Plant Physiol. Plant Mol. Biol.* 47: 509–540.

Lin, M.-K., Lee, Y.-J., Lough, T. J., Phinney, B. S., and Lucas, W. J. (2009) Analysis of the pumpkin phloem proteome provides insights into angiosperm sieve tube function. *Mol. Cell Proteomics* 8: 343–356.

McGarry, R. C., and Ayre, B. G. (2008) A DNA element between At4g28630 and At4g28640 confers companion-cell specific expression following the sink-to-source transition in mature minor vein phloem. *Planta* 228: 839–849.

Münch, E. (1930) *Die Stoffbewegungen in der Pflanze.* Gustav Fischer, Jena, Germany.

Nobel, P. S. (2005) *Physicochemical and Environmental Plant Physiology*, 3rd ed., Academic Press, San Diego, CA.

Oparka, K. J., and Santa Cruz, S. (2000) The great escape: Phloem transport and unloading of macromolecules. *Annu. Rev. Plant Physiol. Plant Mol. Biol.* 51: 323–347.

Paiva, E. A. S., and Machado, S. R. (2008) Can sieve-element plastids in *Panicum maximum* (Poaceae) leaves act in the blockage of injured sieve-tube elements? *Flora* 203: 327–331.

Preiss, J. (1982) Regulation of the biosynthesis and degradation of starch. *Annu. Rev. Plant Physiol.* 33: 431–454.

Ransom-Hodgkins, W. D., Vaughn, M. W., and Bush, D. R. (2003) Protein phosphorylation plays a key role in sucrose-mediated transcriptional regulation of a phloem-specific proton-sucrose symporter. *Planta* 217: 483–489.

Rennie, E. A., and Turgeon, R. (2009) A comprehensive picture of phloem loading strategies. *Proc. Natl. Acad. Sci. USA* 106: 14163–14167.

Roberts, A. G., Santa Cruz, S., Roberts, I. M., Prior, D. A. M., Turgeon, R., and Oparka, K. J. (1997) Phloem unloading in sink leaves of *Nicotiana benthamiana*: Comparison of a fluorescent solute with a fluorescent virus. *Plant Cell* 9: 1381–1396.

Schneidereit, A., Imlau, A., and Sauer, N. (2008) Conserved *cis*-regulatory elements for DNA-binding-with-one-finger and homeo-domain-leucine-zipper transcription factors regulate companion cell-specific expression of the *Arabidopsis thaliana* SUCROSE TRANSPORTER 2 gene. *Planta* 228: 651–662.

Schulz, A. (1990) Conifers. In *Sieve Elements: Comparative Structure, Induction and Development*, H.-D. Behnke and R. D. Sjolund, eds., Springer-Verlag, Berlin.

Schulz, A. (1994) Phloem transport and differential unloading in pea seedlings after source and sink manipulations. *Planta* 192: 239–248.

Sivitz, A. B., Reinders, A., and Ward, J. M. (2008) Arabidopsis sucrose transporter AtSUC1 is important for pollen germination and sucrose-induced anthocyanin accumulation. *Plant Physiol.* 147: 92–100.

Stadler, R., Wright, K. M., Lauterbach, C., Amon, G., Gahrtz, M., Feuerstein, A., Oparka, K. J., and Sauer, N. (2005) Expression of GFP-fusions in Arabidopsis companion cells reveals non-specific protein trafficking into sieve elements and identifies a novel post-phloem domain in roots. *Plant J.* 41: 319–331.

Thorne, J. H., and Koller, H. R. (1974) Influence of assimilate demand on photosynthesis, diffusive resistances, translocation, and carbohydrate levels of soybean leaves. *Plant Physiol.* 54: 201–207.

Truernit, E., Bauby, H., Dubreucq, B., Grandjean, O., Runions, J., Barthelemy, J., and Palauqui, J.-C. (2008) High-resolution whole-mount imaging of three-dimensional tissue organization and gene expression enables the study of phloem development and structure in *Arabidopsis*. *Plant Cell* 20: 1494–1503.

Turgeon, R. (2006) Phloem loading: How leaves gain their independence. *BioScience* 56: 15–24.

Turgeon, R., and Gowan, E. (1990) Phloem loading in *Coleus blumei* in the absence of carrier-mediated uptake of export sugar from the apoplast. *Plant Physiol.* 94: 1244–1249.

Turgeon, R., and Webb, J. A. (1973) Leaf development and phloem transport in *Cucurbita pepo*: Transition from import to export. *Planta* 113: 179–191.

Turgeon, R., and Wolf, S. (2009) Phloem transport: Cellular pathways and molecular trafficking. *Annu. Rev. Plant Biol.* 60: 207–221.

Turgeon, R., Beebe, D. U., and Gowan, E. (1993) The intermediary cell: Minor-vein anatomy and raffinose oligosaccharide synthesis in the Scrophulariaceae. *Planta* 191: 446–456.

van Bel, A. J. E. (1992) Different phloem-loading machineries correlated with the climate. *Acta Botan. Neerl.* 41: 121–141.

van Bel, A. J. E., Gamalei, Y. V., Ammerlaan, A., and Bik, L. P. M. (1992) Dissimilar phloem loading in leaves with symplasmic or apoplasmic minor-vein configurations. *Planta* 186: 518–525.

van Bel, A. J. E., and Hess, P. H. (2008) Hexoses as phloem transport sugars: The end of a dogma? *J. Exp. Bot.* 59: 261–272.

Vaughn, M. W., Harrington, G. N., and Bush, D. R. (2002) Sucrose-mediated transcriptional regulation of sucrose symporter activity in the phloem. *Proc. Natl. Acad. Sci. USA* 99: 10876–10880.

Walz, C., Giavalisco, P., Schad, M., Juenger, M., Klose, J., and Kehr, J. (2004) Proteomics of cucurbit phloem exudate reveals a network of defence proteins. *Phytochemistry* 65: 1795–1804.

Warmbrodt, R. D. (1985) Studies on the root of *Hordeum vulgare* L.—Ultrastructure of the seminal root with special reference to the phloem. *Am. J. Bot.* 72: 414–432.

Windt, C. W., Vergeldt, F. J., De Jager, P. A., and Van As, H. (2006) MRI of long-distance water transport: A comparison of the phloem and xylem flow characteristics and dynamics in poplar, castor bean, tomato and tobacco. *Plant Cell Environ.* 29: 1715–1729.

Wright, J. P., and Fisher, D. B. (1980) Direct measurement of sieve tube turgor pressure using severed aphid stylets. *Plant Physiol.* 65: 1133–1135.

Zimmermann, M. H., and Milburn, J. A., eds. (1975) *Transport in Plants*, 1: *Phloem Transport* (Encyclopedia of Plant Physiology, New Series, Vol. 1). Springer, New York.

# Respiration and Lipid Metabolism

Photosynthesis provides the organic building blocks that plants (and nearly all other organisms) depend on. Respiration, with its associated carbon metabolism, releases the energy stored in carbon compounds in a controlled manner for cellular use. At the same time it generates many carbon precursors for biosynthesis.

We will begin this chapter by reviewing respiration in its metabolic context, emphasizing the interconnections among the processes involved and the special features that are peculiar to plants. We will also relate respiration to recent developments in our understanding of the biochemistry and molecular biology of plant mitochondria. Then we will describe the pathways of lipid biosynthesis that lead to the accumulation of fats and oils, which many plants use for energy and carbon storage. We will also examine lipid synthesis and the influence of lipids on membrane properties. Finally, we will discuss the catabolic pathways involved in the breakdown of lipids and the conversion of their degradation products into sugars that occurs during the germination of fat-storing seeds.

## Overview of Plant Respiration

Aerobic (oxygen-requiring) respiration is common to nearly all eukaryotic organisms, and in its broad outlines, the respiratory process in plants is similar to that found in animals and lower eukaryotes. However, some specific aspects of plant respiration distinguish it from its animal counterpart. **Aerobic respiration** is the biological process by which reduced organic compounds are mobilized and subsequently oxidized in a controlled manner. During respiration, energy is released and transiently stored in a compound, **adenosine triphosphate (ATP)**, that is readily utilized by the cell for maintenance and development.

Glucose is most commonly cited as the substrate for respiration. In a functioning plant cell, however, reduced carbon is derived mainly from sources such as the disaccharide sucrose, triose phosphates from photosynthesis, fructose-containing polymers (fructans), and other sugars, as well as from lipids (primarily triacylglycerols), organic acids, and on occasion, proteins (**FIGURE 11.1**).

From a chemical standpoint, plant respiration can be expressed as the oxidation of the 12-carbon molecule sucrose and the reduction of 12 molecules of $O_2$:

$$C_{12}H_{22}O_{11} + 13\ H_2O \rightarrow 12\ CO_2 + 48\ H^+ + 48\ e^-$$

$$12\ O_2 + 48\ H^+ + 48\ e^- \rightarrow 24\ H_2O$$

giving the following net reaction:

$$C_{12}H_{22}O_{11} + 12\ O_2 \rightarrow 12\ CO_2 + 11\ H_2O$$

This reaction is the reversal of the photosynthetic process; it represents a coupled redox reaction in which sucrose is completely oxidized to $CO_2$ while oxygen serves as the ultimate electron acceptor and is reduced to water in the process. The change in standard **Gibbs free energy** ($\Delta G^{0'}$) for the net reaction is –5760 kJ per mole (342 g) of sucrose oxidized. This large negative value is a consequence of the equilibrium point being strongly shifted to the right, and energy is therefore released by sucrose degradation. The controlled release of this free energy, along with its coupling to the synthesis of ATP, is the primary, although by no means the only, role of respiratory metabolism.

To prevent damage (incineration) of cellular structures, the cell mobilizes the large amount of free energy released in the oxidation of sucrose in a series of step-by-step reactions. These reactions can be grouped into four major processes: glycolysis, the oxidative pentose phosphate pathway, the citric acid cycle, and oxidative phosphorylation. These pathways do not function in isolation, but exchange metabolites at several levels. The substrates of respiration enter the respiratory process at different points in the pathways, as summarized in Figure 11.1:

- **Glycolysis** involves a series of reactions catalyzed by enzymes located in both the cytosol and the plastids. A sugar—for example, sucrose—is partly oxidized via six-carbon sugar phosphates (hexose phosphates) and three-carbon sugar phosphates (triose phosphates) to produce an organic acid—

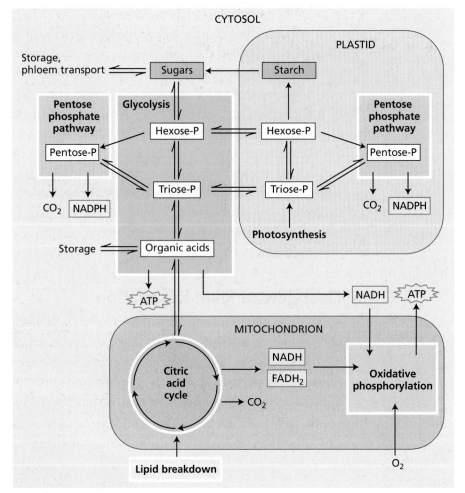

FIGURE 11.1 Overview of respiration. Substrates for respiration are generated by other cellular processes and enter the respiratory pathways. Glycolysis and the oxidative pentose phosphate pathways in the cytosol and plastids convert sugars into organic acids such as pyruvate, via hexose phosphates and triose phosphates, generating NADH or NADPH, and ATP. The organic acids are oxidized in the mitochondrial citric acid cycle, and the NADH and $FADH_2$ produced provide the energy for ATP synthesis by the electron transport chain and ATP synthase in oxidative phosphorylation. In gluconeogenesis, carbon from lipid breakdown is broken down in the glyoxysomes, metabolized in the citric acid cycle, and then used to synthesize sugars in the cytosol by reverse glycolysis.

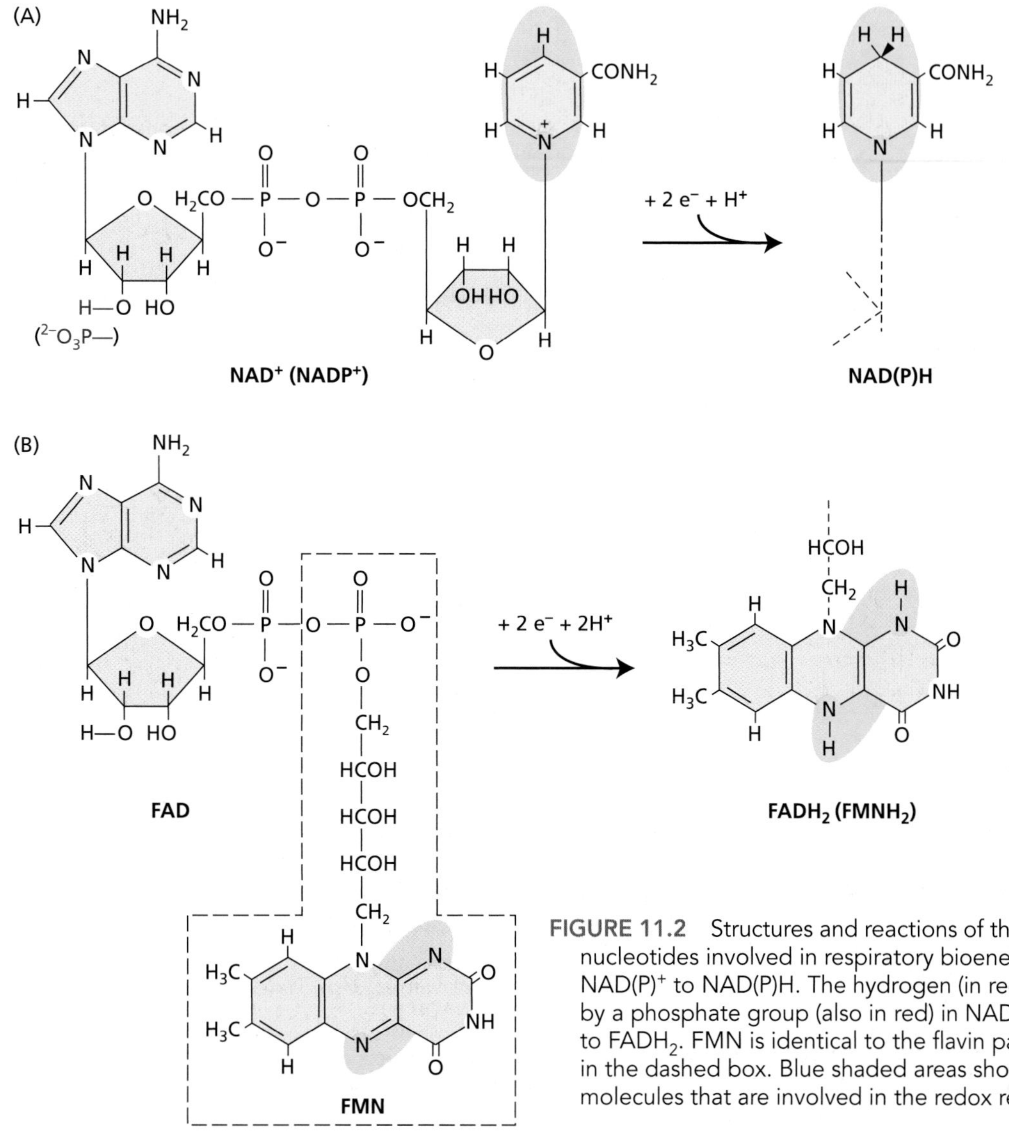

**FIGURE 11.2** Structures and reactions of the major electron-carrying nucleotides involved in respiratory bioenergetics. (A) Reduction of NAD(P)$^+$ to NAD(P)H. The hydrogen (in red) in NAD$^+$ is replaced by a phosphate group (also in red) in NADP$^+$. (B) Reduction of FAD to FADH$_2$. FMN is identical to the flavin part of FAD and is shown in the dashed box. Blue shaded areas show the portions of the molecules that are involved in the redox reaction.

for example, pyruvate. The process yields a small amount of energy as ATP and reducing power in the form of a reduced nicotinamide nucleotide, NADH.

- In the **oxidative pentose phosphate pathway**, also located in both the cytosol and the plastids, the six-carbon glucose 6-phosphate is initially oxidized to the five-carbon ribulose 5-phosphate. Carbon is lost as $CO_2$, and reducing power is conserved in the form of two molecules of another reduced nicotinamide nucleotide, NADPH. In subsequent near-equilibrium reactions of the pentose phosphate pathway, ribulose 5-phosphate is converted into sugars containing three to seven carbon atoms.

- In the **citric acid cycle**, pyruvate is oxidized completely to $CO_2$. This process generates the major amount of reducing power (16 NADH + 4 FADH$_2$ per sucrose) from the breakdown of sucrose. With one exception (succinate dehydrogenase), these reactions are carried out by enzymes located in the internal aqueous compartment, or matrix, of the mitochondrion. As we will discuss later, succinate dehydrogenase is localized in the inner of the two mitochondrial membranes.

- In **oxidative phosphorylation**, electrons are transferred along an electron transport chain consisting of a collection of electron transport proteins bound to the inner of the two mitochondrial membranes. This system transfers electrons from NADH (and related species)—produced by glycolysis, the oxidative pentose phosphate pathway, and the citric acid cycle—to oxygen. This electron transfer releases a large amount of free energy, much of which is conserved through the synthesis of ATP from ADP and P$_i$ (inorganic phosphate), catalyzed by the enzyme ATP synthase. Collectively, the redox reactions of the electron transport chain and the synthesis of ATP are called oxidative phosphorylation.

Nicotinamide adenine dinucleotide (NAD$^+$/NADH) is an organic cofactor (coenzyme) associated with many enzymes that catalyze cellular redox reactions. NAD$^+$ is the oxidized form of the cofactor, which undergoes a reversible two-electron reaction that yields NADH (**FIGURE 11.2**):

$$NAD^+ + 2\,e^- + H^+ \rightarrow NADH$$

(A)

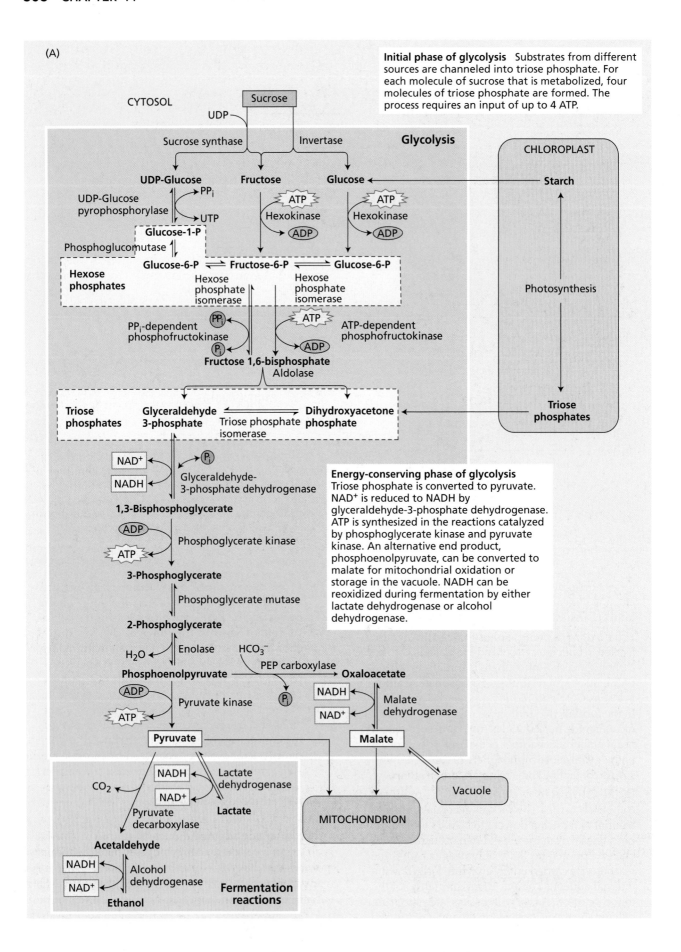

**Initial phase of glycolysis** Substrates from different sources are channeled into triose phosphate. For each molecule of sucrose that is metabolized, four molecules of triose phosphate are formed. The process requires an input of up to 4 ATP.

**Energy-conserving phase of glycolysis**
Triose phosphate is converted to pyruvate. $NAD^+$ is reduced to NADH by glyceraldehyde-3-phosphate dehydrogenase. ATP is synthesized in the reactions catalyzed by phosphoglycerate kinase and pyruvate kinase. An alternative end product, phosphoenolpyruvate, can be converted to malate for mitochondrial oxidation or storage in the vacuole. NADH can be reoxidized during fermentation by either lactate dehydrogenase or alcohol dehydrogenase.

(B)

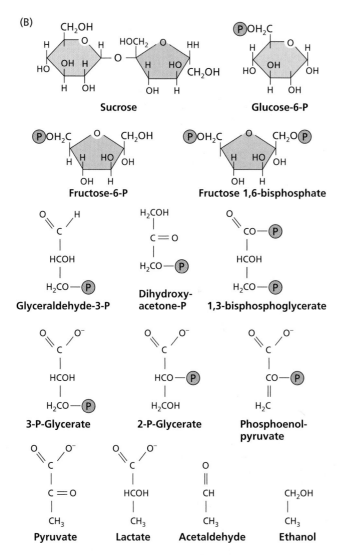

**FIGURE 11.3** Reactions of plant glycolysis and fermentation. (A) In the main glycolytic pathway, sucrose is oxidized via hexose phosphates and triose phosphates to the organic acid pyruvate, but plants also carry out alternative reactions. All the enzymes included in this figure have been measured at levels sufficient to support the respiration rates observed in intact plant tissues, and flux through the pathway has been observed in vivo. The double arrows denote reversible reactions; the single arrows, essentially irreversible reactions. (B) The structures of the carbon intermediates. P, phosphate group.

Keep in mind that not all the carbon that enters the respiratory pathway ends up as $CO_2$. Many respiratory intermediates are the starting points for pathways that assimilate nitrogen into organic form, pathways that synthesize nucleotides and lipids, and many others.

# Glycolysis

In the early steps of glycolysis (from the Greek words *glykos*, "sugar," and *lysis*, "splitting"), carbohydrates are converted into hexose phosphates, each of which is then split into two triose phosphates. In a subsequent energy-conserving phase, each triose phosphate is oxidized and rearranged to yield one molecule of pyruvate, an organic acid. Besides preparing the substrate for oxidation in the citric acid cycle, glycolysis yields a small amount of chemical energy in the form of ATP and NADH.

When molecular oxygen is unavailable—for example, in plant roots in flooded soils—glycolysis can be the main source of energy for cells. For this to work, the *fermentative pathways*, which are carried out in the cytosol, must reduce pyruvate to recycle the NADH produced by glycolysis. In this section we will describe the basic glycolytic and fermentative pathways, emphasizing features that are specific to plant cells. In the following section we will discuss the pentose phosphate pathway, another pathway for sugar oxidation in plants.

## Glycolysis metabolizes carbohydrates from several sources

Glycolysis occurs in all living organisms (prokaryotes and eukaryotes). The principal reactions associated with the classic glycolytic pathway in plants are almost identical to those in animal cells (**FIGURE 11.3**). However, plant glycolysis has unique regulatory features, alternative enzymatic routes for several steps, and a parallel partial glycolytic pathway in plastids.

In animals, the substrate of glycolysis is glucose, and the end product is pyruvate. Because sucrose is the major translocated sugar in most plants, and is therefore the form of carbon that most nonphotosynthetic tissues import, sucrose (not glucose) can be argued to be the true sugar

The standard reduction potential for this redox couple is about –320 mV, which makes it a relatively strong reductant (i.e., electron donor). NADH is thus a good molecule in which to conserve the free energy carried by the electrons released during the stepwise oxidations of glycolysis and the citric acid cycle. A related compound, nicotinamide adenine dinucleotide phosphate ($NADP^+/NADPH$), functions in the redox reactions of photosynthesis (see Chapters 7 and 8) and of the oxidative pentose phosphate pathway; it also takes part in mitochondrial metabolism (Møller and Rasmusson 1998). These roles will be discussed later in the chapter.

The oxidation of NADH by oxygen via the electron transport chain releases free energy (220 kJ mol$^{-1}$) that drives the synthesis of approximately 60 ATP (as we will see later). We can formulate a more complete picture of respiration as related to its role in cellular energy metabolism by coupling the following two reactions:

$$C_{12}H_{22}O_{11} + 12\ O_2 \rightarrow 12\ CO_2 + 11\ H_2O$$

$$60\ ADP + 60\ P_i \rightarrow 60\ ATP + 60\ H_2O$$

substrate for plant respiration. The end products of plant glycolysis include another organic acid, malate.

In the early steps of glycolysis, sucrose is split into its two monosaccharide units—glucose and fructose—which can readily enter the glycolytic pathway. Two pathways for the splitting of sucrose are known in plants, both of which also take part in the unloading of sucrose from the phloem (see Chapter 10): the invertase pathway and the sucrose synthase pathway.

*Invertases* present in the cell wall, vacuole, or cytosol hydrolyze sucrose into its two component hexoses (glucose and fructose). The hexoses are then phosphorylated by a hexokinase that uses ATP to form **hexose phosphates**. Alternatively, *sucrose synthase*, located in the cytosol, combines sucrose with UDP to produce fructose and UDP-glucose. UDP-glucose pyrophosphorylase then converts UDP-glucose and pyrophosphate ($PP_i$) into UTP and glucose 6-phosphate (see Figure 11.3). While the sucrose synthase reaction is close to equilibrium, the invertase reaction is essentially irreversible, driving the flux in the forward direction. In general, invertase predominates in tissues where carbohydrates are mainly catabolized for respiration, whereas sucrose synthase predominates in conversions providing monosaccharides for synthesis of carbohydrate polymers.

In plastids, a partial glycolysis occurs that produces metabolites for biosynthetic reactions there, but can also supply substrates for glycolysis in the cytoplasm. Starch is both synthesized and catabolized only in plastids, and carbon obtained from starch degradation (for example, in a chloroplast at night) enters the glycolytic pathway in the cytosol primarily as glucose (see Chapter 8). In the light, photosynthetic products can also enter the glycolytic pathway directly as triose phosphate (Hoefnagel et al. 1998). So glycolysis works like a funnel with an initial phase collecting carbon from different cellular sources, depending on physiological conditions.

In the initial phase of glycolysis, each hexose unit is phosphorylated twice and then split, eventually producing two molecules of **triose phosphate**. This series of reactions consumes two to four molecules of ATP per sucrose unit, depending on whether the sucrose is split by sucrose synthase or invertase. These reactions also include two of the three essentially irreversible reactions of the glycolytic pathway, which are catalyzed by hexokinase and phosphofructokinase (see Figure 11.3). As we will see later, the phosphofructokinase reaction is one of the control points of glycolysis in both plants and animals.

### The energy-conserving phase of glycolysis extracts usable energy

The reactions discussed thus far transfer carbon from the various substrate pools to triose phosphates. Once *glyceraldehyde 3-phosphate* is formed, the glycolytic pathway can begin to extract usable energy in the energy-conserving phase. The enzyme *glyceraldehyde-3-phosphate dehydrogenase*

catalyzes the oxidation of the aldehyde to a carboxylic acid, reducing $NAD^+$ to NADH. This reaction releases sufficient free energy to allow the phosphorylation (using inorganic phosphate) of glyceraldehyde 3-phosphate to produce 1,3-bisphosphoglycerate. The phosphorylated carboxylic acid on carbon 1 of 1,3-bisphosphoglycerate (see Figure 11.3) has a large standard free-energy change ($\Delta G^{0'}$) of hydrolysis ($-49.3$ kJ mol$^{-1}$). Thus 1,3-bisphosphoglycerate is a strong donor of phosphate groups.

In the next step of glycolysis, catalyzed by *phosphoglycerate kinase*, the phosphate on carbon 1 is transferred to a molecule of ADP, yielding ATP and 3-phosphoglycerate. For each sucrose entering the pathway, four ATPs are generated by this reaction—one for each molecule of 1,3-bisphosphoglycerate.

This type of ATP synthesis, traditionally referred to as **substrate-level phosphorylation**, involves the direct transfer of a phosphate group from a substrate molecule to ADP to form ATP. ATP synthesis by substrate-level phosphorylation is mechanistically distinct from ATP synthesis by the ATP synthases involved in oxidative phosphorylation in mitochondria (which will be described later in this chapter) or photophosphorylation in chloroplasts (see Chapter 7).

In the subsequent two reactions, the phosphate on 3-phosphoglycerate is transferred to carbon 2, and then a molecule of water is removed, yielding the compound *phosphoenolpyruvate* (*PEP*). The phosphate group on PEP has a high $\Delta G^{0'}$ of hydrolysis ($-61.9$ kJ mol$^{-1}$), which makes PEP an extremely good phosphate donor for ATP formation. Using PEP as substrate, the enzyme *pyruvate kinase* catalyzes a second substrate-level phosphorylation to yield ATP and pyruvate. This final step, which is the third essentially irreversible step in glycolysis, yields four additional molecules of ATP for each sucrose molecule that enters the pathway.

### Plants have alternative glycolytic reactions

The sequence of reactions leading to the formation of pyruvate from glucose occurs in all organisms that carry out glycolysis. In addition, organisms can operate this pathway in the opposite direction to synthesize sugars from organic acids. This process is known as **gluconeogenesis**.

Gluconeogenesis is particularly important in the seeds of plants (such as the castor oil plant *Ricinus communis* and sunflower) that store a significant quantity of their carbon reserves in the form of oils (triacylglycerols). After such a seed germinates, much of the oil is converted by gluconeogenesis into sucrose, which is then used to support the growing seedling. In the initial phase of glycolysis, gluconeogenesis overlaps with the pathway for synthesis of sucrose from photosynthetic triose phosphate described in Chapter 8, which is typical of plants.

Because the glycolytic reaction catalyzed by *ATP-dependent phosphofructokinase* is essentially irreversible (see Figure 11.3), an additional enzyme, *fructose-1,6-bisphosphate phosphatase*, converts fructose 1,6-bisphosphate into fructose

6-phosphate and $P_i$ during gluconeogenesis. ATP-dependent phosphofructokinase and fructose-1,6-bisphosphate phosphatase represent a major control point of carbon flux through the glycolytic/gluconeogenic pathways of both plants and animals as well as in sucrose synthesis in plants (see Chapter 8).

In plants, the interconversion of fructose 6-phosphate and fructose 1,6-bisphosphate is made more complex by the presence of an additional (cytosolic) enzyme, *PP$_i$-dependent phosphofructokinase* (pyrophosphate:fructose-6-phosphate 1-phosphotransferase), which catalyzes the following reversible reaction (see Figure 11.3):

Fructose-6-P + $PP_i$ ↔ fructose 1,6-bisphosphate + $P_i$

where -P represents bound phosphate. $PP_i$-dependent phosphofructokinase is found in the cytosol of most plant tissues at levels that are considerably higher than those of ATP-dependent phosphofructokinase (Kruger 1997). Suppression of $PP_i$-dependent phosphofructokinase in transgenic potato plants has shown that it contributes to glycolytic flux, but that it is not essential for plant survival, indicating that other enzymes can take over its function. The existence of different pathways that serve a similar function and can therefore replace each other without a clear loss in function is called **metabolic redundancy**; it is a common feature in plant metabolism.

The reaction catalyzed by $PP_i$-dependent phosphofructokinase is readily reversible, but it is unlikely to operate in sucrose synthesis (Dennis and Blakely 2000). Like ATP-dependent phosphofructokinase and fructose bisphosphate phosphatase, this enzyme appears to be regulated by fluctuations in cell metabolism (discussed later in the chapter), suggesting that under some circumstances operation of the glycolytic pathway in plants has some unique characteristics (see WEB ESSAY 11.1).

At the end of the glycolytic process, plants have alternative pathways for metabolizing PEP. In one pathway PEP is carboxylated by the ubiquitous cytosolic enzyme **PEP carboxylase** to form the organic acid oxaloacetate. The oxaloacetate is then reduced to malate by the action of *malate dehydrogenase*, which uses NADH as a source of electrons, and thus has a similar effect as the dehydrogenases during fermentation (see Figure 11.3). The resulting malate can be stored by export to the vacuole or transported to the mitochondrion, where it can enter the citric acid cycle. Thus the action of pyruvate kinase and PEP carboxylase can produce pyruvate or malate for mitochondrial respiration, although pyruvate dominates in most tissues.

## In the absence of oxygen, fermentation regenerates the NAD⁺ needed for glycolysis

Oxidative phosphorylation does not function in the absence of oxygen. Glycolysis thus cannot continue to operate because the cell's supply of NAD⁺ is limited and once all the NAD⁺ becomes tied up in the reduced state (NADH), the catalytic activity of glyveraldehyde-3-phosphate dehydrogenase comes to a halt. To overcome this limitation, plants and other organisms can further metabolize pyruvate by carrying out one or more forms of **fermentation** (see Figure 11.3).

Alcoholic fermentation is common in plants, although more widely known from brewer's yeast. Two enzymes, pyruvate decarboxylase and alcohol dehydrogenase, act on pyruvate, ultimately producing ethanol and $CO_2$ and oxidizing NADH in the process. In lactic acid fermentation (common in mammalian muscle, but also found in plants), the enzyme lactate dehydrogenase uses NADH to reduce pyruvate to lactate, thus regenerating NAD⁺.

Plant tissues may be subjected to low (hypoxic) or zero (anoxic) concentrations of ambient oxygen. These conditions force the tissues to carry out fermentative metabolism. The best-studied example involves flooded or waterlogged soils in which the diffusion of oxygen is sufficiently reduced to cause root tissues to become hypoxic.

In corn, the initial response to low oxygen concentrations is lactic acid fermentation, but the subsequent response is alcoholic fermentation. Ethanol is thought to be a less toxic end product of fermentation because it can diffuse out of the cell, whereas lactate accumulates and promotes acidification of the cytosol. In numerous other cases plants function under near-anoxic conditions by carrying out some form of fermentation.

It is important to consider the efficiency of fermentation. *Efficiency* is defined here as the energy conserved as ATP relative to the energy potentially available in a molecule of sucrose. The standard free-energy change ($\Delta G^{0'}$) for the complete oxidation of sucrose to $CO_2$ is –5760 kJ mol⁻¹. The $\Delta G^{0'}$ for the synthesis of ATP is 32 kJ mol⁻¹. However, under the nonstandard conditions that normally exist in both mammalian and plant cells, the synthesis of ATP requires an input of free energy of approximately 50 kJ mol⁻¹.

Normal glycolysis leads to a net synthesis of four ATP molecules for each sucrose molecule converted into pyruvate. With ethanol or lactate as the final product, the efficiency of fermentation is only about 4%. Most of the energy available in sucrose remains in the ethanol or lactate. Changes in the glycolytic pathway under oxygen deficiency can increase the ATP yield. This is the case when sucrose is degraded via sucrose synthase instead of invertase, avoiding ATP consumption by the hexokinase in the initial phase of glycolysis. Such modifications emphasize the importance of energetic efficiency for plant survival in the absence of oxygen (see WEB ESSAY 11.1).

Because of the low energy recovery of fermentation, an increased rate of carbohydrate breakdown is needed to sustain the ATP production necessary for cell survival. Glycolysis is up-regulated by changes in metabolite levels and by the induction of genes encoding the enzymes of glycolysis and fermentation. The increased glycolytic rate is called the *Pasteur effect* after the French microbiologist

Louis Pasteur, who first noted it when yeast switched from aerobic respiration to fermentation.

In contrast to the products of fermentation, the pyruvate produced by glycolysis during aerobic respiration is further oxidized by mitochondria, resulting in a much more efficient utilization of the free energy available in sucrose.

### Plant glycolysis is controlled by its products

In vivo, glycolysis appears to be regulated at the level of fructose 6-phosphate phosphorylation and PEP turnover. In contrast to animals, AMP and ATP are not major effectors of plant phosphofructokinase and pyruvate kinase. A more important regulator of plant glycolysis is the cytosolic concentration of PEP, which is a potent inhibitor of the plant ATP-dependent phosphofructokinase.

This inhibitory effect of PEP on phosphofructokinase is strongly decreased by inorganic phosphate, making the cytosolic ratio of PEP to $P_i$ a critical factor in the control of plant glycolytic activity. Pyruvate kinase and PEP carboxylase, the enzymes that metabolize PEP in the last steps of glycolysis (see Figure 11.3), are in turn sensitive to feedback inhibition by citric acid cycle intermediates and their derivatives, including malate, citrate, 2-oxoglutarate, and glutamate.

In plants, therefore, the control of glycolysis comes from the "bottom up" (as discussed later in the chapter), with primary regulation at the level of PEP metabolism by pyruvate kinase and PEP carboxylase. Secondary regulation is exerted by PEP at the conversion of fructose 6-phosphate into fructose 1,6-bisphosphate (see Figure 11.3). In contrast, regulation in animals operates from "top down," with primary activation occurring at the phosphofructokinase and secondary activation at the pyruvate kinase.

One possible benefit of bottom-up control of glycolysis is that it permits plants to regulate net glycolytic flux to pyruvate independently of related metabolic processes such as the Calvin–Benson cycle and sucrose–triose phosphate–starch interconversion (Plaxton 1996). Another benefit of this control mechanism is that glycolysis can adjust to the demand for biosynthetic precursors.

A consequence of bottom-up control of glycolysis is that its rate can influence cellular concentrations of sugars, in combination with sugar-supplying processes such as phloem transport. Glucose and sucrose are potent signaling molecules that make the plant adjust its growth and development to its sugar status. The glycolytic enzyme hexokinase not only functions as an enzyme in the cytosol, but also as a glucose receptor in the nucleus, where it modulates gene expression responses to several plant hormones (Rolland et al. 2006).

The presence of more than one enzyme metabolizing PEP in plant cells—pyruvate kinase and PEP carboxylase—may have consequences for the control of glycolysis. Although the two enzymes are inhibited by similar metabolites, PEP carboxylase can, under some conditions, catalyze a reaction that bypasses pyruvate kinase. The resulting malate can then enter the mitochondrial citric acid cycle.

Experimental support for multiple pathways of PEP metabolism comes from the study of transgenic tobacco plants with less than 5% of the normal level of cytosolic pyruvate kinase in their leaves (Plaxton 1996). In these plants, neither rates of leaf respiration nor rates of photosynthesis differed from those in controls with wild-type levels of pyruvate kinase. However, reduced root growth in the transgenic plants indicated that the pyruvate kinase reaction could not be circumvented without some detrimental effects.

Fructose 2,6-bisphosphate also affects the phosphofructokinase reaction, but unlike PEP, it affects the reaction in both the forward and reverse direction (see Chapter 8 for a detailed discussion). Therefore, fructose 2,6-bisphosphate mediates control of the partitioning of sugars between respiration and biosynthesis.

Another level of regulation may ensue from changes in the location of the glycolytic enzymes. These enzymes were believed to be dissolved in the cytosol; however, it is now clear that under high respiratory demand, there is a substantial pool of glycolytic enzymes bound to the mitochondrial outer surface. This positioning allows direct movement of intermediates from one enzyme to the next (called *substrate channeling*), which separates mitochondrially bound glycolysis from glycolysis in the cytosol. The latter can then contribute carbon intermediates for other processes without interfering with pyruvate production (Graham et al. 2007).

Understanding of the regulation of glycolysis requires the study of temporal changes in metabolite levels. Rapid extraction, separation, and analysis of many metabolites can be achieved by an approach called *metabolic profiling* (see WEB ESSAY 11.2).

## The Oxidative Pentose Phosphate Pathway

The glycolytic pathway is not the only route available for the oxidation of sugars in plant cells. The oxidative pentose phosphate pathway (also known as the *hexose monophosphate shunt*) can also accomplish this task (FIGURE 11.4). The reactions are carried out by soluble enzymes present in the cytosol and in plastids. Under most con-

**FIGURE 11.4** Reactions of the oxidative pentose phosphate pathway in plants. The first two reactions—which are oxidizing reactions—are essentially irreversible. They supply NADPH to the cytoplasm and to plastids in the absence of photosynthesis. The downstream part of the pathway is reversible (as denoted by double-headed arrows), so it can supply five-carbon substrates for biosynthesis even when the oxidizing reactions are inhibited; for example, in chloroplasts in the light.

NADPH is generated in the first two reactions of the pathway, where glucose 6-phosphate is oxidized to ribulose 5-phosphate. These reactions are essentially irreversible.

The ribulose 5-phosphate is converted to the glycolytic intermediates fructose 6-phosphate and glyceraldehyde 3-phosphate through a series of metabolic interconversions. These reactions are freely reversible.

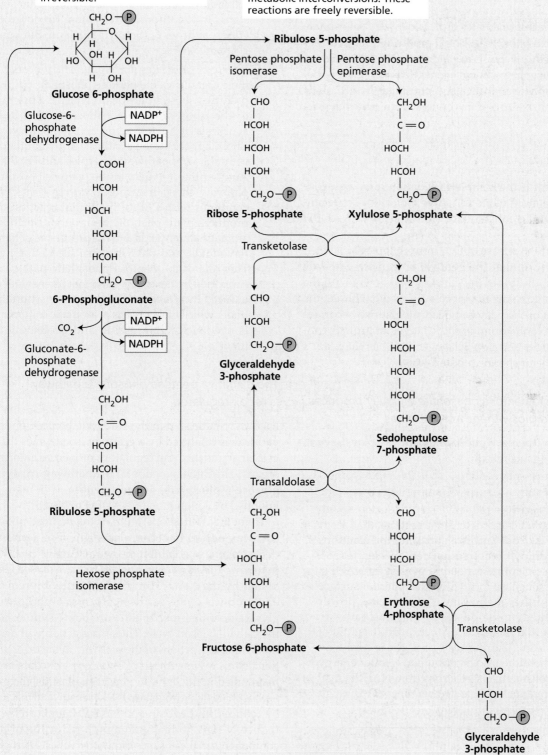

ditions, the pathway in plastids predominates over that in the cytosol (Dennis et al. 1997).

The first two reactions of this pathway involve the oxidative events that convert the six-carbon molecule glucose 6-phosphate into the five-carbon unit **ribulose 5-phosphate**, with loss of a $CO_2$ molecule and generation of two molecules of NADPH (not NADH). The remaining reactions of the pathway convert ribulose 5-phosphate into the glycolytic intermediates glyceraldehyde 3-phosphate and fructose 6-phosphate. These products can be further metabolized by glycolysis to yield pyruvate. Alternatively, glucose 6-phosphate can be regenerated from glyceraldehyde 3-phosphate and fructose 6-phosphate by glycolytic enzymes. For six turns of this cycle, we can write the reaction as follows:

$$6 \text{ Glucose-6-P} + 12 \text{ NADP}^+ + 7 \text{ H}_2\text{O}$$
$$5 \text{ glucose-6-P} + 6 \text{ CO}_2 + \text{P}_i + 12 \text{ NADPH} + 12 \text{ H}^+$$

The net result is the complete oxidation of one glucose 6-phosphate molecule to $CO_2$ (five molecules are regenerated) with the concomitant synthesis of 12 NADPH molecules.

Studies of the release of $CO_2$ from isotopically labeled glucose indicate that the pentose phosphate pathway accounts for 10–25% of the glucose breakdown, with the rest occurring mainly via glycolysis. As we will see, the contribution of the pentose phosphate pathway changes during development and with changes in growth conditions (Kruger and von Schaewen 2003) as the plant's requirements for specific products vary.

### The oxidative pentose phosphate pathway produces NADPH and biosynthetic intermediates

The oxidative pentose phosphate pathway plays several roles in plant metabolism:

- *NADPH supply in the cytosol.* The product of the two oxidative steps is NADPH. This NADPH drives reductive steps associated with biosynthetic and defensive reactions that occur in the cytosol and is a substrate for reactions that remove reactive oxygen species (ROS). Because plant mitochondria possess an NADPH dehydrogenase located on the external surface of the inner membrane, the reducing power generated by the pentose phosphate pathway can be balanced by mitochondrial NADPH oxidation. The pentose phosphate pathway may therefore also contribute to cellular energy metabolism; that is, electrons from NADPH may end up reducing $O_2$ and generating ATP through oxidative phosphorylation.

- *NADPH supply in plastids.* In nongreen plastids, such as amyloplasts in the root, and in chloroplasts functioning in the dark, the pentose phosphate pathway is a major supplier of NADPH. The NADPH is used for biosynthetic reactions such as lipid synthesis and nitrogen assimilation. The formation of NADPH by glucose 6-phosphate oxidation in amyloplasts may also signal sugar status to the thioredoxin system for control of starch synthesis (Schürmann and Buchanan 2008).

- *Supply of substrates for biosynthetic processes.* In most organisms, the pentose phosphate pathway produces ribose 5-phosphate, which is a precursor of the ribose and deoxyribose needed in the synthesis of nucleic acids. In plants, however, ribose appears to be synthesized by another, as yet unknown, pathway (Sharples and Fry 2007). Another intermediate in the pentose phosphate pathway, the four-carbon *erythrose 4-phosphate*, combines with PEP in the initial reaction that produces plant phenolic compounds, including aromatic amino acids and the precursors of lignin, flavonoids, and phytoalexins (see Chapter 13). This role of the pentose phosphate pathway is supported by the observation that its enzymes are induced by stress conditions such as wounding, under which biosynthesis of aromatic compounds is needed for reinforcing and protecting the tissue.

### The oxidative pentose phosphate pathway is redox-regulated

Each enzymatic step in the oxidative pentose phosphate pathway is catalyzed by a group of isoenzymes that vary in their abundance and regulatory properties among plant organs. The initial reaction of the pathway, catalyzed by **glucose-6-phosphate dehydrogenase**, is in many cases inhibited by a high ratio of NADPH to $NADP^+$.

In the light, little operation of the pentose phosphate pathway occurs in chloroplasts. Glucose-6-phosphate dehydrogenase is inhibited by a reductive inactivation involving the *ferredoxin–thioredoxin system* (see Chapter 8) and by the NADPH to $NADP^+$ ratio. Moreover, the end products of the pathway, fructose 6-phosphate and glyceraldehyde 3-phosphate, are being synthesized by the Calvin–Benson cycle. Thus mass action will drive the nonoxidative reactions of the pathway in the reverse direction. In this way, synthesis of erythrose 4-phosphate can be maintained in the light. In nongreen plastids, the glucose 6-phosphate dehydrogenase is less sensitive to inactivation by reduced thioredoxin and NADPH, and can therefore reduce $NADP^+$ to maintain a high reduction of plastid components in the absence of photosynthesis (Kruger and von Schaewen 2003).

# The Citric Acid Cycle

During the nineteenth century, biologists discovered that in the absence of air, cells produce ethanol or lactic acid, whereas in the presence of air, cells consume $O_2$ and produce $CO_2$ and $H_2O$. In 1937 the German-born British biochemist Hans A. Krebs reported the discovery of the citric acid cycle—also called the *tricarboxylic acid cycle* or *Krebs cycle*. The elucidation of the citric acid cycle not only explained how pyruvate is broken down into $CO_2$ and $H_2O$, but also highlighted the key concept of cycles in metabolic pathways. For his discovery, Hans Krebs was awarded the Nobel Prize in physiology or medicine in 1953.

Because the citric acid cycle occurs in the mitochondrial matrix, we will begin with a general description of mitochondrial structure and function, the knowledge of which was obtained mainly through experiments on isolated mitochondria (see WEB TOPIC 11.1). We will then review the steps of the citric acid cycle, emphasizing the features that are specific to plants and how they affect respiratory function.

## Mitochondria are semiautonomous organelles

The breakdown of sucrose into pyruvate releases less than 25% of the total energy in sucrose; the remaining energy is stored in the four molecules of pyruvate. The next two stages of respiration (the citric acid cycle and oxidative phosphorylation) take place within an organelle enclosed by a double membrane, the **mitochondrion** (plural *mitochondria*).

In electron micrographs, plant mitochondria usually look spherical or rodlike (FIGURE 11.5). They range from 0.5 to 1.0 μm in diameter and up to 3 μm in length (Douce 1985). With some exceptions, plant cells have substantially fewer mitochondria than are found in a typical animal cell. The number of mitochondria per plant cell varies; it is usually directly related to the metabolic activity of the tissue, reflecting the mitochondrial role in energy metabolism. Guard cells, for example, are unusually rich in mitochondria.

The ultrastructural features of plant mitochondria are similar to those of mitochondria in other organisms (see Figure 11.5). Plant mitochondria have two membranes: a smooth **outer membrane** completely surrounds a highly invaginated **inner membrane**. The

invaginations of the inner membrane are known as **cristae** (singular *crista*). As a consequence of its greatly enlarged surface area, the inner membrane can contain more than 50% of the total mitochondrial protein. The region between the two mitochondrial membranes is known as the **intermembrane space**. The compartment enclosed by the inner membrane is referred to as the mitochondrial **matrix**. It has a very high content of macromolecules, approximately 50% by weight. Because there is little water in the matrix, mobility is restricted, and it is likely that matrix proteins are organized into multienzyme complexes to facilitate substrate channeling.

Intact mitochondria are osmotically active; that is, they take up water and swell when placed in a hypo-osmotic medium. Most inorganic ions and charged organic molecules are not able to diffuse freely into the matrix. The inner membrane is the osmotic barrier; the outer membrane is permeable to solutes that have a molecular mass of less than approximately 10,000 Da—that is, most cellular metabolites and ions, but not proteins. The lipid fraction of

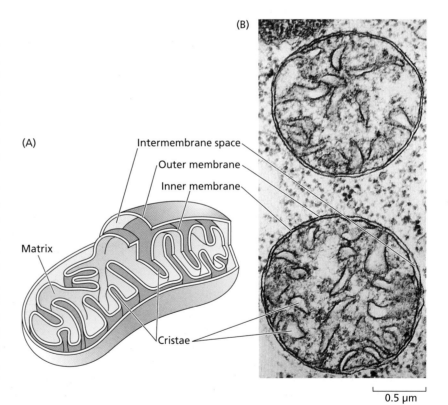

**FIGURE 11.5** Structure of plant mitochondria. (A) Three-dimensional representation of a mitochondrion, showing the invaginations of the inner membrane, called cristae, as well as the locations of the matrix and intermembrane space (see also Figure 11.9). (B) Electron micrograph of mitochondria in a mesophyll cell of broad bean (*Vicia faba*). Typically, individual mitochondria are 1 to 3 μm long in plant cells, which means that they are substantially smaller than nuclei and plastids. (B from Gunning and Steer 1996.)

both membranes is primarily made up of phospholipids, 80% of which are either phosphatidylcholine or phosphatidylethanolamine. About 15% is diphosphatidylglycerol (also called cardiolipin), which occurs in cells only in the inner mitochondrial membrane.

Like chloroplasts, mitochondria are semiautonomous organelles because they contain ribosomes, RNA, and DNA, which encodes a limited number of mitochondrial proteins. Plant mitochondria are thus able to carry out the various steps of protein synthesis and to transmit their genetic information. The number of mitochondria in a cell can vary dynamically due to divisions and fusions (see

WEB ESSAY 11.3) while keeping up with cell division. In most plants, mitochondria are maternally inherited during sexual reproduction.

## Pyruvate enters the mitochondrion and is oxidized via the citric acid cycle

The citric acid cycle is also known as the *tricarboxylic acid cycle* because of the importance of the tricarboxylic acids citric acid (citrate) and isocitric acid (isocitrate) as early intermediates (**FIGURE 11.6**). This cycle constitutes the second stage in respiration and takes place in the mitochondrial

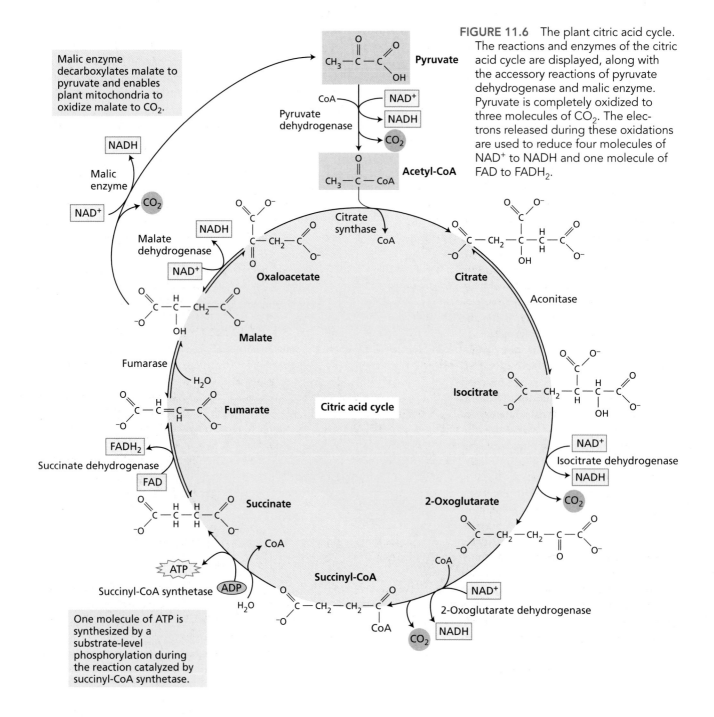

**FIGURE 11.6** The plant citric acid cycle. The reactions and enzymes of the citric acid cycle are displayed, along with the accessory reactions of pyruvate dehydrogenase and malic enzyme. Pyruvate is completely oxidized to three molecules of $CO_2$. The electrons released during these oxidations are used to reduce four molecules of $NAD^+$ to NADH and one molecule of FAD to $FADH_2$.

Malic enzyme decarboxylates malate to pyruvate and enables plant mitochondria to oxidize malate to $CO_2$.

One molecule of ATP is synthesized by a substrate-level phosphorylation during the reaction catalyzed by succinyl-CoA synthetase.

matrix. Its operation requires that the pyruvate generated in the cytosol during glycolysis be transported through the impermeable inner mitochondrial membrane via a specific transport protein (as will be described shortly).

Once inside the mitochondrial matrix, pyruvate is decarboxylated in an oxidation reaction catalyzed by **pyruvate dehydrogenase**, a large complex containing several enzymes. The products are NADH, $CO_2$, and acetyl-CoA, in which the acetyl group derived from pyruvate is linked by a thioester bond to a cofactor, coenzyme A (CoA) (see Figure 11.6).

In the next reaction, the enzyme citrate synthase, formally the first enzyme in the citric acid cycle, combines the acetyl group of acetyl-CoA with a four-carbon dicarboxylic acid (*oxaloacetate*) to give a six-carbon tricarboxylic acid (citrate). Citrate is then isomerized to isocitrate by the enzyme aconitase.

The following two reactions are successive oxidative decarboxylations, each of which produces one NADH and releases one molecule of $CO_2$, yielding a four-carbon product bound to CoA, succinyl-CoA. At this point, three molecules of $CO_2$ have been produced for each pyruvate that entered the mitochondrion, or 12 $CO_2$ for each molecule of sucrose oxidized.

In the remainder of the citric acid cycle, succinyl-CoA is oxidized to oxaloacetate, allowing the continued operation of the cycle. Initially the large amount of free energy available in the thioester bond of succinyl-CoA is conserved through the synthesis of ATP from ADP and $P_i$ via a substrate-level phosphorylation catalyzed by *succinyl-CoA synthetase*. (Recall that the free energy available in the thioester bond of acetyl-CoA was used to form a carbon–carbon bond in the step catalyzed by citrate synthase.) The resulting succinate is oxidized to fumarate by *succinate dehydrogenase*, which is the only membrane-associated enzyme of the citric acid cycle and also part of the electron transport chain.

The electrons and protons removed from succinate end up not on NAD$^+$, but on another cofactor involved in redox reactions: **flavin adenine dinucleotide (FAD)**. FAD is covalently bound to the active site of succinate dehydrogenase and undergoes a reversible two-electron reduction to produce FADH$_2$ (see Figure 11.2B).

In the final two reactions of the citric acid cycle, fumarate is hydrated to produce malate, which is subsequently oxidized by *malate dehydrogenase* to regenerate oxaloacetate and produce another molecule of NADH. The oxaloacetate produced is now able to react with another acetyl-CoA and continue the cycling.

The stepwise oxidation of one molecule of pyruvate in the mitochondrion gives rise to three molecules of $CO_2$, and much of the free energy released during these oxidations is conserved in the form of four NADH and one FADH$_2$. In addition, one molecule of ATP is produced by a substrate-level phosphorylation.

### The citric acid cycle of plants has unique features

The citric acid cycle reactions outlined in Figure 11.6 are not all identical to those carried out by animal mitochondria. For example, the step catalyzed by succinyl-CoA synthetase produces ATP in plants and GTP in animals. These nucleotides are energetically equivalent.

A feature of the plant citric acid cycle that is absent in many other organisms is the presence of **malic enzyme** in the mitochondrial matrix of plants. This enzyme catalyzes the oxidative decarboxylation of malate:

$$\text{Malate} + \text{NAD}^+ \rightarrow \text{pyruvate} + CO_2 + \text{NADH}$$

The activity of malic enzyme enables plant mitochondria to operate alternative pathways for the metabolism of PEP derived from glycolysis (see WEB ESSAY 11.1). As already described, malate can be synthesized from PEP in the cytosol via the enzymes PEP carboxylase and malate dehydrogenase (see Figure 11.3). For degradation, malate is transported into the mitochondrial matrix, where malic enzyme can oxidize it to pyruvate. This reaction makes possible the complete net oxidation of citric acid cycle intermediates such as malate (FIGURE 11.7A) or citrate (FIGURE 11.7B) (Oliver and McIntosh 1995). Many plant tissues, not only those that carry out crassulacean acid metabolism (see Chapter 8), store significant amounts of malate or other organic acids in their vacuoles. Degradation of malate via mitochondrial malic enzyme is important for regulating levels of organic acids in cells—for example, during fruit ripening.

Instead of being degraded, the malate produced via PEP carboxylase can replace citric acid cycle intermediates used in biosynthesis. Reactions that replenish intermediates in a metabolic cycle are known as *anaplerotic*. For example, export of 2-oxoglutarate for nitrogen assimilation in the chloroplast causes a shortage of malate for the citrate synthase reaction. This malate can be replaced through the PEP carboxylase pathway (FIGURE 11.7C).

Gamma-aminobutyric acid (GABA) is an amino acid that accumulates under several biotic and abiotic stress conditions in plants. GABA is synthesized from 2-oxoglutarate and degraded into succinate by a reaction that bypasses the citric acid cycle, called the **GABA shunt** (Bouché and Fromm 2004). The functional relationship between GABA accumulation and stress remains poorly understood.

## Mitochondrial Electron Transport and ATP Synthesis

ATP is the energy carrier used by cells to drive life processes, so chemical energy conserved during the citric acid cycle in the form of NADH and FADH$_2$ must be converted into ATP to perform useful work in the cell. This $O_2$-dependent process, called oxidative phosphorylation, occurs in the inner mitochondrial membrane.

(A)

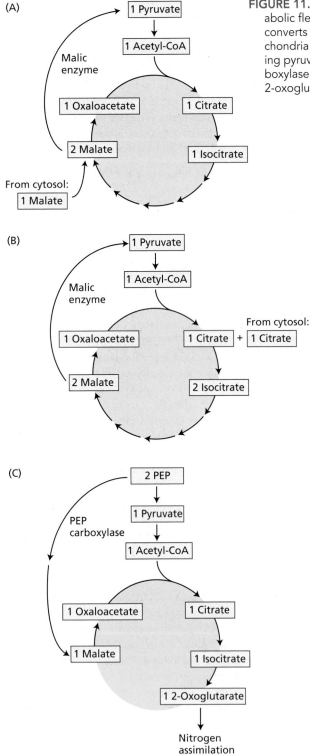

**FIGURE 11.7** Malic enzyme and PEP carboxylase provide plants with metabolic flexibility for the metabolism of PEP and pyruvate. Malic enzyme converts malate into pyruvate and thus makes it possible for plant mitochondria to oxidize both malate (A) and citrate (B) to $CO_2$ without involving pyruvate delivered by glycolysis. With the added action of PEP carboxylase to the standard pathway, glycolytic PEP can be converted into 2-oxoglutarate, which is used for nitrogen assimilation (C).

similar in all aerobic cells, the electron transport chain of plants (and fungi) contains multiple NAD(P)H dehydrogenases and an alternative oxidase not found in mammalian mitochondria.

We will also examine the enzyme that uses the energy of the proton gradient to synthesize ATP: the $F_oF_1$-ATP synthase. After examining the various stages in the production of ATP, we will summarize the energy conservation steps at each stage, as well as the regulatory mechanisms that coordinate the different pathways.

### The electron transport chain catalyzes a flow of electrons from NADH to $O_2$

For each molecule of sucrose oxidized through glycolysis and the citric acid cycle, four molecules of NADH are generated in the cytosol, and sixteen molecules of NADH plus four molecules of $FADH_2$ (associated with succinate dehydrogenase) are generated in the mitochondrial matrix. These reduced compounds must be reoxidized, or the entire respiratory process will come to a halt.

The electron transport chain catalyzes a transfer of two electrons from NADH (or $FADH_2$) to oxygen, the final electron acceptor of the respiratory process. For the oxidation of NADH, the reaction can be written as

$$NADH + H^+ + \tfrac{1}{2} O_2 \rightarrow NAD^+ + H_2O$$

From the reduction potentials for the NADH–$NAD^+$ pair (–320 mV) and the $H_2O$–$\tfrac{1}{2}$ $O_2$ pair (+810 mV), it can be calculated that the standard free energy released during this overall reaction ($-nF\Delta E0'$) is about 220 kJ per mole of NADH. Because the succinate–fumarate reduction potential is higher (+30 mV), only 152 kJ per mole of succinate is released. The role of the electron transport chain is to bring about the oxidation of NADH (and $FADH_2$) and, in the process, utilize some of the free energy released to generate an electrochemical proton gradient, $\Delta \tilde{\mu}_{H^+}$, across the inner mitochondrial membrane.

The electron transport chain of plants contains the same set of electron carriers found in the mitochondria of other organisms (**FIGURE 11.8**) (Siedow and Umbach 1995). The individual electron transport proteins are organized into four transmembrane multiprotein complexes (identified by roman numerals I through IV), all of which are localized in the inner mitochondrial membrane. Three of these complexes are engaged in proton pumping (I, III, and IV).

In this section we will describe the process by which the energy level of the electrons from NADH and $FADH_2$ is lowered in a stepwise fashion and conserved in the form of an electrochemical proton gradient across the inner mitochondrial membrane. Although fundamentally

INTERMEMBRANE SPACE

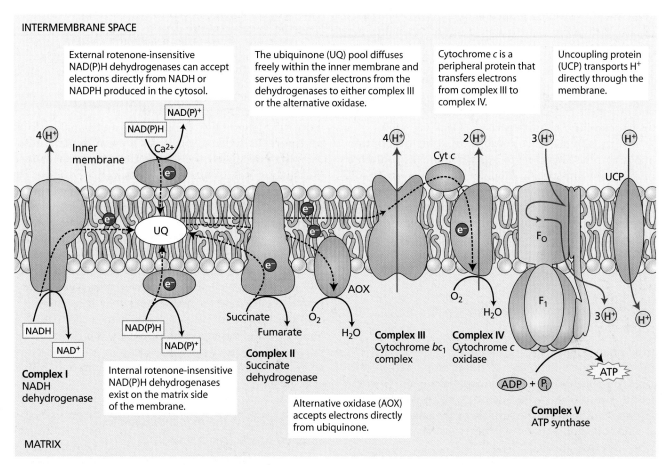

External rotenone-insensitive NAD(P)H dehydrogenases can accept electrons directly from NADH or NADPH produced in the cytosol.

The ubiquinone (UQ) pool diffuses freely within the inner membrane and serves to transfer electrons from the dehydrogenases to either complex III or the alternative oxidase.

Cytochrome *c* is a peripheral protein that transfers electrons from complex III to complex IV.

Uncoupling protein (UCP) transports $H^+$ directly through the membrane.

Internal rotenone-insensitive NAD(P)H dehydrogenases exist on the matrix side of the membrane.

Alternative oxidase (AOX) accepts electrons directly from ubiquinone.

**Complex I** NADH dehydrogenase

**Complex II** Succinate dehydrogenase

**Complex III** Cytochrome $bc_1$ complex

**Complex IV** Cytochrome *c* oxidase

**Complex V** ATP synthase

MATRIX

**FIGURE 11.8** Organization of the electron transport chain and ATP synthesis in the inner membrane of the plant mitochondrion. Mitochondria from nearly all eukaryotes contain the four standard protein complexes: I, II, III, and IV. The structures of most of these complexes have been determined, but they are shown here as simplified shapes. The electron transport chain of the plant mitochondrion contains additional enzymes (depicted in green) that do not pump protons. Additionally, uncoupling proteins directly bypass the ATP synthase by allowing passive proton influx. This multiplicity of bypasses in plants, whereas animals have only the uncoupling protein, gives a greater flexibility to plant energy coupling (see **WEB TOPIC 11.3**).

**COMPLEX I (NADH DEHYDROGENASE)** Electrons from NADH generated in the mitochondrial matrix during the citric acid cycle are oxidized by complex I (an **NADH dehydrogenase**). The electron carriers in complex I include a tightly bound cofactor (**flavin mononucleotide**, or **FMN**, which is chemically similar to FAD; see Figure 11.2B) and several iron–sulfur centers. Complex I then transfers these electrons to ubiquinone. Four protons are pumped from the matrix into the intermembrane space for every electron pair passing through the complex.

**Ubiquinone**, a small lipid-soluble electron and proton carrier, is localized within the inner membrane. It is not tightly associated with any protein, and it can diffuse within the hydrophobic core of the membrane bilayer.

**COMPLEX II (SUCCINATE DEHYDROGENASE)** Oxidation of succinate in the citric acid cycle is catalyzed by this complex, and the reducing equivalents are transferred via

$FADH_2$ and a group of iron–sulfur centers to ubiquinone. Complex II does not pump protons.

**COMPLEX III (CYTOCHROME $bc_1$ COMPLEX)** Complex III oxidizes reduced ubiquinone (ubiquinol) and transfers the electrons via an iron–sulfur center, two *b*-type cytochromes ($b_{565}$ and $b_{560}$), and a membrane-bound cytochrome $c_1$ to cytochrome *c*. Four protons per electron pair are pumped out of the matrix by complex III using a mechanism called the **Q-cycle** (see **WEB TOPIC 11.2**).

**Cytochrome *c*** is a small protein loosely attached to the outer surface of the inner membrane and serves as a mobile carrier to transfer electrons between complexes III and IV.

**COMPLEX IV (CYTOCHROME *c* OXIDASE)** Complex IV contains two copper centers ($Cu_A$ and $Cu_B$) and cytochromes *a* and $a_3$. This complex is the terminal oxidase and brings about the four-electron reduction of $O_2$ to two molecules of $H_2O$. Two protons are pumped out of the matrix per electron pair (see Figure 11.8).

Both structurally and functionally, ubiquinone and the cytochrome $bc_1$ complex are very similar to plastoquinone and the cytochrome $b_6f$ complex, respectively, in the photosynthetic electron transport chain (see Chapter 7).

Reality may be more complex than the description above implies. Plant respiratory complexes contain a number of plant-specific subunits whose function is still unknown. Several of the complexes contain subunits that participate in functions other than electron transport, such as protein import. Finally, several of the complexes appear to be present in supercomplexes, instead of freely mobile in the membrane, although the functional significance of these supercomplexes is not clear (Millar et al. 2005).

### The electron transport chain has supplementary branches

In addition to the set of protein complexes described in the previous section, the plant electron transport chain contains components not found in mammalian mitochondria (see Figure 11.8 and WEB TOPIC 11.3). These additional enzymes are bound to the surfaces of the inner membrane and do not pump protons, so energy conservation is lower whenever they are used.

- **NAD(P)H dehydrogenases**, mostly $Ca^{2+}$-dependent, are attached to the outer surface of the inner membrane facing the intermembrane space. They oxidize either NADH or NADPH from the cytosol. Electrons from these external NAD(P)H dehydrogenases—$ND_{ex}$(NADH) and $ND_{ex}$(NADPH)—enter the main electron transport chain at the level of the ubiquinone pool (Rasmusson et al. 2008).

- Plant mitochondria have two pathways for oxidizing matrix NADH. Electron flow through complex I, described in the previous section, is sensitive to inhibition by several compounds, including rotenone and piericidin. In addition, plant mitochondria have a rotenone-insensitive dehydrogenase, $ND_{in}$(NADH), on the matrix surface of the inner mitochondrial membrane. This enzyme oxidizes NADH derived from the citric acid cycle, and may also be a bypass engaged when complex I is overloaded, such as under photorespiratory conditions, as we will see shortly. An NADPH dehydrogenase, $ND_{in}$(NADPH), is also present on the matrix surface, but very little is known about this enzyme (Rasmusson et al. 2004).

- Most, if not all, plants have an "alternative" respiratory pathway for the oxidation of ubiquinol

and reduction of oxygen. This pathway involves the so-called **alternative oxidase**, which, unlike cytochrome $c$ oxidase, is insensitive to inhibition by cyanide, carbon monoxide, or the signal molecule nitric oxide (see WEB ESSAY 11.4).

The nature and physiological significance of these supplementary electron transport enzymes will be considered more fully later in the chapter. Some additional electron transport chain dehydrogenases present in plant mitochondria directly perform important carbon conversions (Rasmusson et al. 2008). A *proline dehydrogenase* oxidizes the amino acid proline. Proline accumulates during osmotic stress (see Chapter 26), and it is degraded by this mitochondrial pathway when water status returns to normal. An electron transfer flavoprotein:quinone oxidoreductase mediates the degradation of several amino acids that are used by plants as a reserve under carbon starvation conditions induced by light deprivation (Ishizaki et al. 2005). Finally, a galactono-gamma-lactone dehydrogenase, specific to plants, performs the last step in the major pathway for synthesis of the antioxidant *ascorbic acid* (also known as vitamin C). The enzyme uses cytochrome $c$ as its electron acceptor, in competition with normal respiration (Millar et al. 2003).

### ATP synthesis in the mitochondrion is coupled to electron transport

In oxidative phosphorylation, the transfer of electrons to oxygen via complexes I, III, and IV is coupled to the synthesis of ATP from ADP and $P_i$ via the $F_oF_1$-ATP synthase (complex V). The number of ATPs synthesized depends on the nature of the electron donor.

In experiments conducted on isolated mitochondria, electrons derived from matrix NADH (e.g., generated by malate oxidation) give ADP:O ratios (the number of ATPs synthesized per two electrons transferred to oxygen) of 2.4 to 2.7 (TABLE 11.1). Succinate and externally added NADH each give values in the range of 1.6 to 1.8, while ascorbate, which serves as an artificial electron donor to cytochrome $c$, gives values of 0.8 to 0.9. Results such as these (for both plant and animal mitochondria) have led to the general concept that there are three sites of energy conservation along the electron transport chain, at complexes I, III, and IV.

The experimental ADP:O ratios agree quite well with the values calculated on the basis of the number of $H^+$ pumped by complexes I, III, and IV and the cost of 4 $H^+$ for synthesizing one ATP (see next section and Table 11.1). For instance, electrons from external NADH pass only complexes III and IV, so a total of 6 $H^+$ are pumped, giving 1.5 ATP (when the alternative oxidase pathway is not used).

The mechanism of mitochondrial ATP synthesis is based on the **chemiosmotic hypothesis**, described in Chapter 7, which was first proposed in 1961 by Nobel laureate Peter

## TABLE 11.1
### Theoretical and experimental ADP:O ratios in isolated plant mitochondria

| Substrate | ADP:0 ratio | |
|---|---|---|
| | Theoretical[a] | Experimental |
| Malate | 2.5 | 2.4–2.7 |
| Succinate | 1.5 | 1.6–1.8 |
| NADH (external) | 1.5 | 1.6–1.8 |
| Ascorbate | 1.0[b] | 0.8–0.9 |

[a]It is assumed that complexes I, III, and IV pump 4, 4, and 2 $H^+$ per 2 electrons, respectively; that the cost of synthesizing one ATP and exporting it to the cytosol is 4 $H^+$ (Brand 1994); and that the nonphosphorylating pathways are not active.

[b]Cytochrome *c* oxidase pumps only two protons when it is measured with ascorbate as electron donor. However, two electrons move from the outer surface of the inner membrane (where the electrons are donated) across the inner membrane to the inner, matrix side. As a result, 2 $H^+$ are consumed on the matrix side. This means that the net movement of $H^+$ and charges is equivalent to the movement of a total of 4 $H^+$, giving an ADP:O ratio of 1.0.

Mitchell as a general mechanism of energy conservation across biological membranes (Nicholls and Ferguson 2002). According to the chemiosmotic hypothesis, the orientation of electron carriers within the inner mitochondrial membrane allows for the transfer of protons across the inner membrane during electron flow (see Figure 11.8).

Because the inner mitochondrial membrane is highly impermeable to protons, an **electrochemical proton gradient** can build up. As discussed in Chapters 6 and 7, the free energy associated with the formation of an electrochemical proton gradient ($\Delta \tilde{\mu}_{H^+}$, also referred to as a *proton motive force*, $\Delta p$, when expressed in units of volts) is made up of an electrical transmembrane potential component ($\Delta E$) and a chemical-potential component ($\Delta pH$) according to the following equation:

$$\Delta p = \Delta E - 59\Delta pH \text{ (at 25°C)}$$

where

$$\Delta E = E_{inside} - E_{outside}$$

and

$$\Delta pH = pH_{inside} - pH_{outside}$$

$\Delta E$ results from the asymmetric distribution of a charged species ($H^+$) across the membrane, and $\Delta pH$ is due to the proton concentration difference across the membrane. Because protons are translocated from the mitochondrial matrix to the intermembrane space, the resulting $\Delta E$ across the inner mitochondrial membrane has a negative value.

As this equation shows, both $\Delta E$ and $\Delta pH$ contribute to the proton motive force in plant mitochondria, although $\Delta E$ is consistently found to be of greater magnitude, prob-

ably because of the large buffering capacity of both cytosol and matrix, which prevents large pH changes. This situation contrasts with that in the chloroplast, where almost all of the proton motive force across the thylakoid membrane is due to $\Delta pH$ (see Chapter 7).

The free-energy input required to generate $\Delta \tilde{\mu}_{H^+}$ comes from the free energy released during electron transport. How electron transport is coupled to proton translocation is not well understood in all cases. Because of the low permeability (conductance) of the inner membrane to protons, the proton electrochemical gradient can be utilized to carry out chemical work (ATP synthesis). The $\Delta \tilde{\mu}_{H^+}$ is coupled to the synthesis of ATP by an additional protein complex associated with the inner membrane, the $F_oF_1$-ATP synthase.

The **$F_oF_1$-ATP synthase** (also called *complex V*) consists of two major components, $F_o$ and $F_1$ (see Figure 11.8). **$F_o$** (subscript "o" for oligomycin-sensitive) is an integral membrane protein complex of at least three different polypeptides. They form the channel through which protons cross the inner membrane. The other component, **$F_1$**, is a peripheral membrane protein complex that is composed of at least five different subunits and contains the catalytic site for converting ADP and $P_i$ into ATP. This complex is attached to the matrix side of $F_o$.

The passage of protons through the channel is coupled to the catalytic cycle of the $F_1$ component of the ATP synthase, allowing the ongoing synthesis of ATP and the simultaneous utilization of the $\Delta \tilde{\mu}_{H^+}$. For each ATP synthesized, 3 $H^+$ pass through the $F_o$ component from the intermembrane space to the matrix, down the electrochemical proton gradient.

A high-resolution structure for the $F_1$ component of the mammalian ATP synthase provided evidence for a model in which a part of $F_o$ rotates relative to $F_1$ to couple $H^+$ transport to ATP synthesis (Abrahams et al. 1994) (see WEB TOPIC 11.4). The structure and function of the mitochondrial ATP synthase is similar to that of the $CF_o$–$CF_1$ ATP synthase in chloroplasts (see Chapter 7).

The operation of a chemiosmotic mechanism of ATP synthesis has several implications. First, the true site of ATP formation on the inner mitochondrial membrane is the ATP synthase, not complex I, III, or IV. These complexes serve as sites of energy conservation whereby electron transport is coupled to the generation of a $\Delta \tilde{\mu}_{H^+}$. The synthesis of ATP decreases the $\Delta \tilde{\mu}_{H^+}$ and as a consequence, its restriction on the electron transport complexes. Electron transport is therefore stimulated by a large supply of ADP.

The chemiosmotic hypothesis also explains the action mechanism of **uncouplers**. These are a wide range of chemically unrelated, artificial compounds (including 2,4-dinitrophenol and *p*-trifluoromethoxycarbonylcyanide phenylhydrazone [FCCP]) that decrease mitochondrial

ATP synthesis but stimulate the rate of electron transport (see **WEB TOPIC 11.5**). All of these uncoupling compounds make the inner membrane leaky to protons, which prevents the buildup of a sufficiently large $\Delta\tilde{\mu}_{H^+}$ to drive ATP synthesis or restrict electron transport.

### Transporters exchange substrates and products

The electrochemical proton gradient also plays a role in the movement of the organic acids of the citric acid cycle, and of the substrates and products of ATP synthesis, into and out of mitochondria. Although ATP is synthesized in the mitochondrial matrix, most of it is used outside the mitochondrion, so an efficient mechanism is needed for moving ADP into and ATP out of the organelle.

The ADP/ATP (adenine nucleotide) transporter performs the active exchange of ADP and ATP across the inner membrane (**FIGURE 11.9**). The movement of the more negatively charged $ATP^{4-}$ out of the mitochondrion in exchange for $ADP^{3-}$—that is, one net negative charge out—is driven by the electrical-potential gradient ($\Delta E$, positive outside) generated by proton pumping.

The uptake of inorganic phosphate ($P_i$) involves an active phosphate transporter protein that uses the chemical-potential component ($\Delta pH$) of the proton motive force to drive the electroneutral exchange of $P_i^-$ (in) for $OH^-$ (out). As long as a $\Delta pH$ is maintained across the inner membrane, the $P_i$ content within the matrix remains high. Similar reasoning applies to the uptake of pyruvate, which is driven by the electroneutral exchange of pyruvate for $OH^-$, leading to continued uptake of pyruvate from the cytosol (see Figure 11.9).

The total energetic cost of taking up one phosphate and one ADP into the matrix and exporting one ATP is the movement of one $H^+$ from the intermembrane space into the matrix:

- Moving one $OH^-$ out in exchange for $P_i^-$ is equivalent to 1 $H^+$ in, so this electroneutral exchange consumes the chemical potential, but not the electrical potential.

- Moving one negative charge out ($ADP^{3-}$ entering the matrix in exchange for $ATP^{4-}$ leaving), is the same as moving one positive charge in, so this transport lowers only the electrical potential.

This proton, which drives the exchange of ATP for ADP and $P_i$, should also be included in our calculation of the cost of synthesizing one ATP. Thus the total cost is 3 $H^+$ used by the ATP synthase plus 1 $H^+$ for the exchange across the membrane, or a total of 4 $H^+$.

The inner membrane also contains transporters for dicarboxylic acids (malate or succinate) exchanged for $P_i^{2-}$ and for the tricarboxylic acid citrate exchanged for dicarboxylic acids (see Figure 11.9 and **WEB TOPIC 11.5**).

**FIGURE 11.9** Transmembrane transport in plant mitochondria. An electrochemical proton gradient, $\Delta\tilde{\mu}_{H^+}$, consisting of an electrical potential component ($\Delta E$, –200 mV, negative inside) and a chemical potential component ($\Delta pH$, alkaline inside), is established across the inner mitochondrial membrane during electron transport, as outlined in the text. Specific metabolites are moved across the inner membrane by specialized proteins called transporters or carriers. (After Douce 1985.)

### Aerobic respiration yields about 60 molecules of ATP per molecule of sucrose

The complete oxidation of a sucrose molecule leads to the net formation of

- Eight molecules of ATP by substrate-level phosphorylation (four from glycolysis and four from the citric acid cycle)

- Four molecules of NADH in the cytosol

- Sixteen molecules of NADH plus four molecules of $FADH_2$ (via succinate dehydrogenase) in the mitochondrial matrix

On the basis of theoretical ADP:O values (see Table 11.1), we can estimate that 52 molecules of ATP will be generated per molecule of sucrose by oxidative phosphorylation. The complete aerobic oxidation of sucrose (including substrate-level phosphorylation) results in a total of about 60 ATPs synthesized per sucrose molecule (**TABLE 11.2**).

Using 50 kJ $mol^{-1}$ as the actual free energy of formation of ATP in vivo, we find that about 3010 kJ $mol^{-1}$ of free energy is conserved in the form of ATP per mole of sucrose oxidized during aerobic respiration. This amount

**TABLE 11.2**
**The maximum yield of cytosolic ATP from the complete oxidation of sucrose to $CO_2$ via aerobic glycolysis and the citric acid cycle**

| Part reaction | ATP per sucrose[a] |
|---|---|
| Glycolysis | |
|     4 substrate-level phosphorylations | 4 |
|     4 NADH | $4 \times 1.5 = 6$ |
| Citric acid cycle | |
|     4 substrate-level phosphorylations | 4 |
|     4 $FADH_2$ | $4 \times 1.5 = 6$ |
|     16 NADH | $16 \times 2.5 = 40$ |
| Total | 60 |

*Source*: Adapted from Brand 1994.

*Note*: Cytosolic NADH is assumed oxidized by the external NADH dehydrogenase. The nonphosphorylating pathways are assumed not to be engaged.

[a]Calculated using the theoretical ADP/O values from Table 11.1.

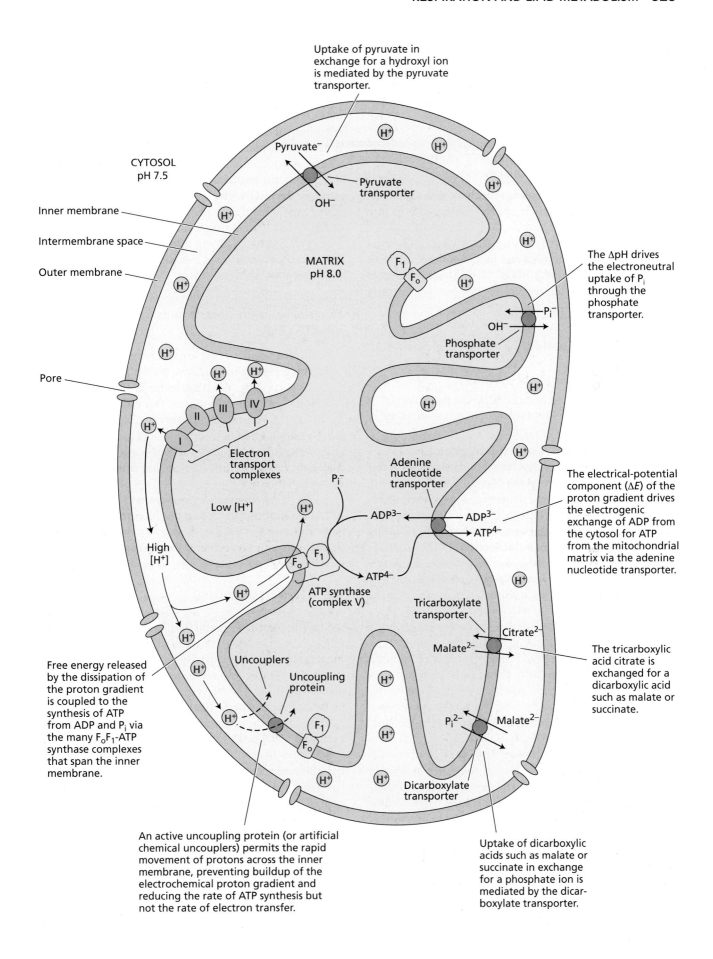

Uptake of pyruvate in exchange for a hydroxyl ion is mediated by the pyruvate transporter.

Pyruvate⁻

CYTOSOL
pH 7.5

Pyruvate transporter

OH⁻

Inner membrane

Intermembrane space

Outer membrane

MATRIX
pH 8.0

$F_1$

$F_o$

The ΔpH drives the electroneutral uptake of $P_i$ through the phosphate transporter.

$P_i^-$

OH⁻

Phosphate transporter

Pore

$H^+$

II    III    IV

I

Electron transport complexes

Adenine nucleotide transporter

$P_i^-$

ADP³⁻

ADP³⁻

ATP⁴⁻

The electrical-potential component (Δ$E$) of the proton gradient drives the electrogenic exchange of ADP from the cytosol for ATP from the mitochondrial matrix via the adenine nucleotide transporter.

Low [H⁺]

$H^+$

High [H⁺]

$F_o$    $F_1$

ATP⁴⁻

ATP synthase (complex V)

Tricarboxylate transporter

Citrate²⁻

Malate²⁻

The tricarboxylic acid citrate is exchanged for a dicarboxylic acid such as malate or succinate.

Free energy released by the dissipation of the proton gradient is coupled to the synthesis of ATP from ADP and $P_i$ via the many $F_oF_1$-ATP synthase complexes that span the inner membrane.

Uncouplers

Uncoupling protein

$H^+$

$F_1$

$F_o$

$P_i^{2-}$

Malate²⁻

Dicarboxylate transporter

An active uncoupling protein (or artificial chemical uncouplers) permits the rapid movement of protons across the inner membrane, preventing buildup of the electrochemical proton gradient and reducing the rate of ATP synthesis but not the rate of electron transfer.

Uptake of dicarboxylic acids such as malate or succinate in exchange for a phosphate ion is mediated by the dicarboxylate transporter.

represents about 52% of the standard free energy available from the complete oxidation of sucrose; the rest is lost as heat. It also represents a vast improvement over fermentative metabolism, in which only 4% of the energy available in sucrose is converted into ATP.

## Several subunits of respiratory complexes are encoded by the mitochondrial genome

The genetic system of the plant mitochondrion differs not only from that of the nucleus and the chloroplast, but also from those found in the mitochondria of animals, protists, or fungi. Most notably, processes involving RNA differ between plant mitochondria and mitochondria from most other organisms (see WEB TOPIC 11.6). Major differences are found in

- RNA splicing (for example, special introns are present)
- RNA editing (in which the nucleotide sequence is changed)
- Signals regulating RNA stability
- Translation (plant mitochondria use the universal genetic code, whereas mitochondria in other eukaryotes have deviant codons)

The size of the plant mitochondrial genome varies substantially even between closely related plant species, but at 180 to almost 3000 kilobase pairs (kbp), it is always much larger than the compact and uniform 16 kbp genome found in mammalian mitochondria. The size differences are due mainly to the presence of noncoding DNA including numerous introns, in plant **mitochondrial DNA (mtDNA)**. Mammalian mtDNA encodes only 13 proteins, in contrast to the 35 known proteins encoded by Arabidopsis mtDNA (Marienfeld et al. 1999). Both plant and mammalian mitochondria contain genes for rRNAs and tRNAs.

Plant mtDNA encodes several subunits of respiratory complexes I–V as well as proteins that take part in cytochrome biogenesis. The mitochondrially encoded subunits are essential for the activity of the respiratory complexes.

Except for the proteins encoded by mtDNA, all mitochondrial proteins (possibly more than 2000) are encoded by nuclear DNA—including all the proteins in the citric acid cycle (Millar et al. 2005). These nuclear-encoded mitochondrial proteins are synthesized by cytosolic ribosomes and imported via translocators in the outer and inner mitochondrial membranes. Therefore, oxidative phosphorylation is dependent on expression of genes located in two separate genomes. Any change in expression of these nuclear and mitochondrial genes must therefore be coordinated.

Whereas the expression of nuclear genes for mitochondrial proteins is regulated like that of other nuclear genes, less is known about the expression of mitochondrial genes. Genes can be down-regulated by a decreased copy number for the segment of mtDNA that contains the gene (Leon et al. 1998). Also, gene promoters in mtDNA are of several kinds and show different transcriptional activities. However, the biogenesis of respiratory complexes appears to be controlled by changes in the expression of the nuclear-encoded subunits; coordination with the mitochondrial genome takes place posttranslationally (Giegé et al. 2005).

The mitochondrial genome is important for pollen development. Naturally occurring rearrangements of genes in the mtDNA lead to so called *cytoplasmic male sterility* (cms). This trait leads to perturbed pollen development by inducing a premature **programmed cell death** (see WEB ESSAY 11.5), on otherwise unaffected plants. The cms traits are used in breeding of several crop plants for making hybrid seed stocks.

## Plants have several mechanisms that lower the ATP yield

As we have seen, a complex machinery is required for conserving energy in oxidative phosphorylation. So it is perhaps surprising that plant mitochondria have several functional proteins that reduce this efficiency (see WEB TOPIC 11.3). Plants are probably less limited by energy supply (sunlight) than by other factors in the environment (e.g., access to water and nutrients). As a consequence, metabolic flexibility may be more important to them than energetic efficiency.

In the following subsections we will discuss the role of three nonphosphorylating mechanisms and their possible usefulness in the life of the plant: the alternative oxidase, the uncoupling protein, and the rotenone-insensitive NADH dehydrogenases.

**THE ALTERNATIVE OXIDASE**  If cyanide in the mM range is added to actively respiring animal cells, cytochrome *c* oxidase is completely inhibited, and the respiration rate quickly drops to less than 1% of its initial level. However, most plants display a capacity for *cyanide-resistant respiration* that is comparable to the capacity of the cytochrome *c* oxidase pathway. The enzyme responsible for this cyanide-resistant oxygen uptake has been identified as a ubiquinol oxidase called the **alternative oxidase** (Vanlerberghe and McIntosh 1997) (see Figure 11.8 and WEB TOPIC 11.3).

Electrons feed off the main electron transport chain into this alternative pathway at the level of the ubiquinone pool (see Figure 11.8). The alternative oxidase, the only component of the alternative pathway, catalyzes a four-electron reduction of oxygen to water and is specifically inhibited by several compounds, most notably salicylhydroxamic acid (SHAM).

When electrons pass to the alternative pathway from the ubiquinone pool, two sites of proton pumping (at complexes III and IV) are bypassed. Because there is no energy conservation site in the alternative pathway between ubiquinone and oxygen, the free energy that would normally be conserved as ATP is lost as heat when electrons are shunted through this pathway.

How can a process as seemingly energetically wasteful as the alternative pathway contribute to plant metabolism? One example of the functional usefulness of the alternative oxidase is its activity during floral development in certain members of the Araceae (the arum family)—for example, the voodoo lily (*Sauromatum guttatum*). Just before pollination, parts of the inflorescence exhibit a dramatic increase in the rate of respiration via the alternative pathway. As a result, the temperature of the upper appendix increases by as much as 25°C over the ambient temperature. During this extraordinary burst of heat production, certain amines, indoles, and terpenes are volatilized, and the plant therefore gives off a putrid odor that attracts insect pollinators (see WEB ESSAY 11.6). Salicylic acid has been identified as the signal initiating this thermogenic event in the voodoo lily (Raskin et al. 1989) and was later found also to be involved in plant pathogen defense (see Chapter 13).

In most plants, the respiratory rates are too low to generate sufficient heat to raise the temperature significantly. What other role(s) does the alternative pathway play? To answer that question, we need to consider the regulation of the alternative oxidase: Its transcription is often specifically induced, for example, by various types of abiotic and biotic stress. The activity of the alternative oxidase, which functions as a dimer, is regulated by reversible oxidation–reduction of an intermolecular sulfhydryl bridge, by the

reduction level of the ubiquinone pool, and by pyruvate. The first two factors ensure that the enzyme is most active under reducing conditions, while the latter factor ensures that the enzyme has high activity when there is plenty of substrate for the citric acid cycle (see WEB TOPIC 11.3).

If the respiration rate exceeds the cell's demand for ATP (i.e., if ADP levels are very low), the reduction level in the mitochondrion will be high, and the alternative oxidase will be activated. Thus the alternative oxidase makes it possible for the mitochondrion to adjust the relative rates of ATP production and synthesis of carbon skeletons for use in biosynthetic reactions.

Another possible function of the alternative pathway is in the response of plants to a variety of stresses (phosphate deficiency, chilling, drought, osmotic stress, and so on), many of which can inhibit mitochondrial respiration (see Chapter 26). In response to stress, the electron transport chain generates increased amounts of reactive oxygen species, which act as a signal for the activation of alternative oxidase expression. By draining off electrons from the ubiquinone pool (see Figure 11.8), the alternative pathway prevents overreduction, which, if left unchecked, can lead to the generation of destructive reactive oxygen species such as hydroxyl radicals. In this way, the alternative pathway may lessen the detrimental effects of stress on respiration (Rhoads and Subbaiah 2007; Møller 2001) (see WEB ESSAY 11.7). This is an example of *retrograde regulation*, in which nuclear gene expression responds to changes in organellar status (FIGURE 11.10).

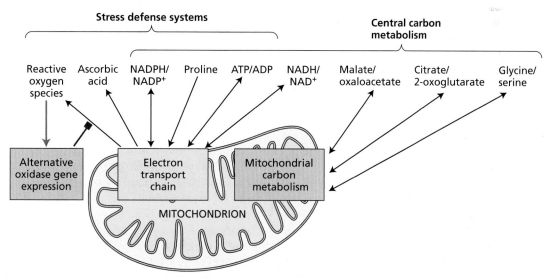

**FIGURE 11.10** Metabolic interactions between mitochondria and cytosol. Mitochondrial activities can influence the cytosolic levels of redox and energy molecules involved in stress defense and in central carbon metabolism (such as growth processes and photosynthesis). An exact distinction between stress defense and carbon metabolism cannot be made because they have components in common. Arrows denote influences caused by changes in mitochondrial synthesis (e.g., reactive oxygen species [ROS], ATP, or ascorbic acid) or degradation (e.g., NAD[P]H, proline, or glycine). The ROS-mediated activation of expression of nuclear genes for the alternative oxidase is an example of retrograde regulation.

**THE UNCOUPLING PROTEIN**  A protein found in the inner membrane of mammalian mitochondria, the **uncoupling protein**, can dramatically increase the proton permeability of the membrane and thus act as an uncoupler. As a result, less ATP and more heat are generated. Heat production appears to be one of the uncoupling protein's main functions in mammalian cells.

It had long been thought that the alternative oxidase in plants and the uncoupling protein in mammals were simply two different means of achieving the same end. It was therefore surprising when a protein similar to the uncoupling protein was discovered in plant mitochondria (Vercesi et al. 1995; Laloi et al. 1997). This protein is induced by stress and, like the alternative oxidase, may function to prevent overreduction of the electron transport chain and formation of reactive oxygen species (see WEB TOPIC 11.3 and WEB ESSAY 11.7). It remains unclear, however, why plant mitochondria require both mechanisms.

**ROTENONE-INSENSITIVE NADH DEHYDROGENASES**  The rotenone-insensitive NADH dehydrogenases are among the multiple NAD(P)H dehydrogenases found in plant mitochondria (see Figure 11.8 and WEB TOPIC 11.3). The internal, rotenone-insensitive NADH dehydrogenase ($ND_{in}$[NADH]) may work as a non-proton-pumping bypass when complex I is overloaded. Complex I has a higher affinity (ten times lower $K_m$) for NADH than $ND_{in}$(NADH). At lower NADH levels in the matrix, typically when ADP is available, complex I dominates, whereas when ADP is rate-limiting, NADH levels increase, and $ND_{in}$(NADH) is more active. For example, photorespiration leads to a massive generation of NADH from glycine oxidation in the matrix (see Chapter 8). $ND_{in}$(NADH) and the alternative oxidase probably recycle the NADH into $NAD^+$ to maintain pathway activity. Since reducing power can be shuttled from the matrix to the cytosol by the exchange of different organic acids, external NADH dehydrogenases can have bypass functions similar to those of $ND_{in}$(NADH). Taken together, these NADH dehydrogenases and the NADPH dehydrogenases are likely to make plant respiration more flexible and aid in the control of cellular redox homeostasis (see Figure 11.10).

## Short-term control of mitochondrial respiration occurs at different levels

The substrates of ATP synthesis—ADP and $P_i$—appear to be key short-term regulators of the rates of glycolysis in the cytosol and of the citric acid cycle and oxidative phosphorylation in the mitochondria. Control points exist at all three stages of respiration; here we will give just a brief overview of some major features of respiratory control.

The best-characterized site of posttranslational regulation of mitochondrial respiratory metabolism is the pyruvate dehydrogenase complex, which is phosphorylated by a *regulatory kinase* and dephosphorylated by a *phosphatase*.

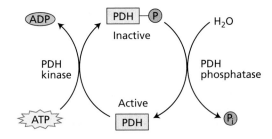

$$Pyruvate + CoA + NAD^+ \longrightarrow Acetyl\text{-}CoA + CO_2 + NADH$$

| Effect on PDH activity | Mechanism |
|---|---|
| **Activating** | |
| Pyruvate | Inhibits kinase |
| ADP | Inhibits kinase |
| $Mg^{2+}$ (or $Mn^{2+}$) | Stimulates phosphatase |
| **Inactivating** | |
| NADH | Inhibits PDH Stimulates kinase |
| Acetyl CoA | Inhibits PDH Stimulates kinase |
| $NH_4^+$ | Inhibits PDH Stimulates kinase |

**FIGURE 11.11**  Metabolic regulation of pyruvate dehydrogenase (PDH) activity, directly and by reversible phosphorylation. Upstream and downstream metabolites regulate PDH activity by direct actions on the enzyme itself and/or by regulating its kinase or phosphatase.

Pyruvate dehydrogenase is inactive in the phosphorylated state, and the regulatory kinase is inhibited by pyruvate, allowing the enzyme to be active when substrate is available (**FIGURE 11.11**). Pyruvate dehydrogenase forms the entry point to the citric acid cycle, so this regulation adjusts the activity of the cycle to the cellular demand.

Thioredoxins control many enzymes by reversible redox dimerization of cysteine residues (see Chapter 8). Numerous mitochondrial enzymes, representing virtually all pathways, are modified by thioredoxins (Buchanan and Balmer 2005). Although the detailed mechanisms have not been worked out yet, it is likely that mitochondrial redox status exerts an important control on respiratory processes.

The citric acid cycle oxidations, and subsequently respiration, are dynamically controlled by the cellular level of adenine nucleotides. As the cell's demand for ATP in the cytosol decreases relative to the rate of synthesis of ATP in the mitochondria, less ADP is available, and the electron transport chain operates at a reduced rate (see Figure 11.9). This slowdown could be signaled to citric acid cycle enzymes through an increase in matrix NADH, which inhibits the activity of several citric acid cycle dehydrogenases (Oliver and McIntosh 1995).

The buildup of citric acid cycle intermediates (such as citrate) and their derivates (such as glutamate) inhibits the action of cytosolic pyruvate kinase, increasing the cytosolic PEP concentration, which in turn reduces the rate of conversion of fructose 6-phosphate into fructose 1,6-bisphosphate, thus inhibiting glycolysis.

In summary, plant respiratory rates are *allosterically controlled* from the "bottom up" by the cellular level of ADP (**FIGURE 11.12**). ADP initially regulates the rate of electron transfer and ATP synthesis, which in turn regulates citric acid cycle activity, which, finally, regulates the rates of the glycolytic reactions. This bottom-up control allows the respiratory carbon pathways to adjust to the demand for biosynthetic building blocks, thereby increasing respiratory flexibility.

### Respiration is tightly coupled to other pathways

Glycolysis, the oxidative pentose phosphate pathway, and the citric acid cycle are linked to several other important metabolic pathways, some of which will be covered in greater detail in Chapter 12. The respiratory pathways produce the central building blocks for synthesis of a wide variety of plant metabolites, including amino acids, lipids and related compounds, isoprenoids, and porphyrins (**FIGURE 11.13**). Indeed, much of the reduced carbon that is metabolized by glycolysis and the citric acid cycle is diverted to biosynthetic purposes and not oxidized to $CO_2$.

Mitochondria are also integrated into the cellular redox network. Variations in consumption or production of redox and energy-carrying compounds such as NAD(P)H and organic acids are likely to affect metabolic pathways in the cytosol and in plastids. Of special importance is the synthesis of *ascorbic acid*, a central redox and stress defense molecule in plants, by the electron transport chain (see Figure 11.10) (Noctor et al. 2007). Mitochondria also carry out steps in the biosynthesis of *coenzymes* necessary for many metabolic enzymes in other cell compartments (see **WEB ESSAY 11.8**).

## Respiration in Intact Plants and Tissues

Many rewarding studies of plant respiration and its regulation have been carried out on isolated organelles and on cell-free extracts of plant tissues. But how does this knowledge relate to the function of the whole plant in a natural or agricultural setting?

In this section we will examine respiration and mitochondrial function in the context of the whole plant under a variety of conditions. First we will explore what happens when green organs are exposed to light: Respiration and photosynthesis operate simultaneously and are functionally integrated in the cell. Next we will discuss rates of respiration in different tissues, which may be under devel-

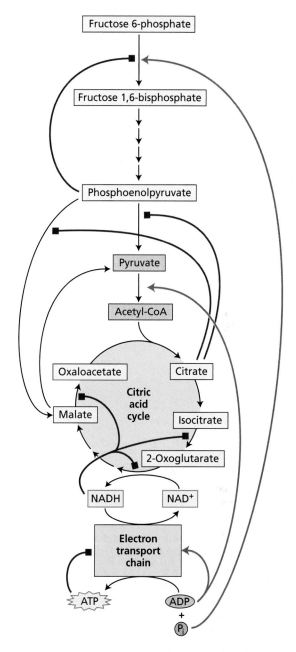

**FIGURE 11.12** Model of bottom-up regulation of plant respiration. Several substrates for respiration (e.g., ADP) stimulate enzymes in early steps of the pathways (green arrows). In contrast, accumulation of products (e.g., ATP) inhibits upstream reactions (red lines and squares) in a stepwise fashion. For instance, ATP inhibits the electron transport chain, leading to an accumulation of NADH. NADH inhibits citric acid cycle enzymes such as isocitrate dehydrogenase and 2-oxoglutarate dehydrogenase. Citric acid cycle intermediates such as citrate inhibit the PEP-metabolizing enzymes in the cytosol. Finally, PEP inhibits the conversion of fructose 6-phosphate into fructose 1,6-bisphosphate and restricts carbon flow into glycolysis. In this way, respiration can be up- or down-regulated in response to changing demands for either of its products: ATP and organic acids.

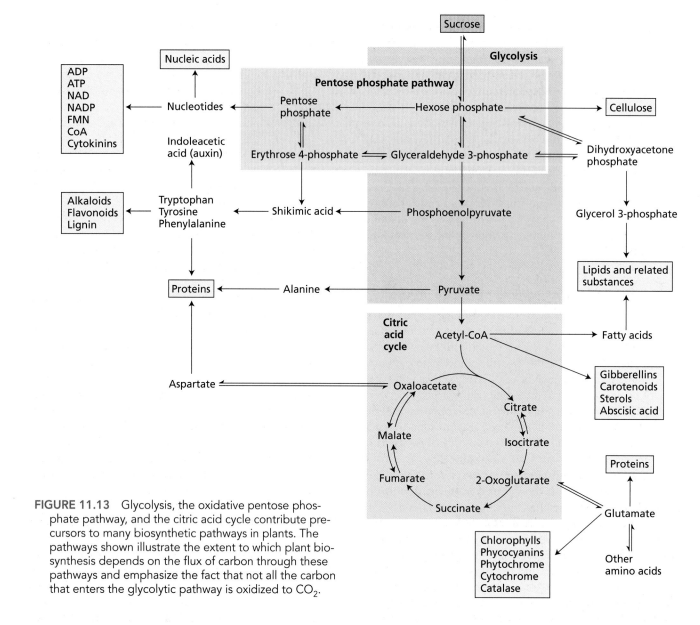

**FIGURE 11.13** Glycolysis, the oxidative pentose phosphate pathway, and the citric acid cycle contribute precursors to many biosynthetic pathways in plants. The pathways shown illustrate the extent to which plant biosynthesis depends on the flux of carbon through these pathways and emphasize the fact that not all the carbon that enters the glycolytic pathway is oxidized to $CO_2$.

opmental control. Finally, we will look at the influence of various environmental factors on respiration rates.

## Plants respire roughly half of the daily photosynthetic yield

Many factors can affect the respiration rate of an intact plant or of its individual organs. Relevant factors include the species and growth habit of the plant, the type and age of the specific organ, and environmental variables such as light, external $O_2$ and $CO_2$ concentrations, temperature, and nutrient and water supply (see Chapter 26). By measuring different oxygen isotopes, it is possible to measure the in vivo activities of the alternative oxidase and cytochrome $c$ oxidase simultaneously. Therefore we know that a significant part of respiration in most tissues takes place via the "energy-wasting" alternative pathway (see **WEB ESSAY 11.9**).

Whole-plant respiration rates, particularly when considered on a fresh-weight basis, are generally lower than respiration rates reported for animal tissues. This difference is mainly due to the presence in plant cells of a large vacuole and a cell wall, neither of which contains mitochondria. Nonetheless, respiration rates in some plant tissues are as high as those observed in actively respiring animal tissues, so the plant respiratory process is not inherently slower than in animals. In fact, isolated plant mitochondria respire as fast as or faster than mammalian mitochondria.

The contribution of respiration to the overall carbon economy of the plant can be substantial. Whereas only green tissues photosynthesize, all tissues respire, and they do so 24 hours a day. Even in photosynthetically active tissues, respiration, if integrated over the entire day, utilizes a substantial fraction of gross photosynthesis. A survey of several herbaceous species indicated that 30 to 60% of the daily gain in photosynthetic carbon is lost to respiration, although these values tend to decrease in older plants (Lambers 1985). Trees respire a similar fraction of their photosynthetic production, but their respiratory loss increases with age as the ratio of photosynthetic to non-photosynthetic tissue decreases.

## Respiration operates during photosynthesis

Mitochondria are involved in the metabolism of photosynthesizing leaves in several ways. The glycine generated by photorespiration is oxidized to serine in the mitochondrion in a reaction involving mitochondrial oxygen consumption (see Chapter 8). At the same time, mitochondria in photosynthesizing tissue carry out normal mitochondrial respiration (i.e., via the citric acid cycle). Relative to the maximum rate of photosynthesis, rates of mitochondrial respiration measured in green tissues in the light are far slower, generally by a factor of 6- to 20-fold. Given that rates of photorespiration can often reach 20 to 40% of the gross photosynthetic rate, daytime photorespiration is a larger provider of NADH for the respiratory chain than the normal respiratory pathways.

The activity of pyruvate dehydrogenase, one of the ports of entry into the citric acid cycle, decreases in the light to 25% of its activity in darkness (Budde and Randall 1990). Consistently, the overall rate of mitochondrial respiration decreases in the light, but the extent of the decrease remains uncertain at present. It is clear, however, that the mitochondrion is a major supplier of ATP to the cytosol (e.g., for driving biosynthetic pathways) even in illuminated leaves (Krömer 1995).

Another role of mitochondrial respiration during photosynthesis is to supply precursors for biosynthetic reactions, such as the 2-oxoglutarate needed for nitrogen assimilation (see Figures 11.7C and 11.13). The formation of 2-oxoglutarate also produces NADH in the matrix, linking the process to oxidative phosphorylation or to nonphosphorylating respiratory chain activities (Hoefnagel et al. 1998; Noctor and Foyer 1998).

Additional evidence for the involvement of mitochondrial respiration in photosynthesis has been obtained in studies with mitochondrial mutants defective in respiratory complexes. Compared with the wild type, these plants have slower leaf development and photosynthesis because changes in levels of redox-active metabolites are communicated between mitochondria and chloroplasts, negatively affecting photosynthetic function (Noctor et al. 2007).

## Different tissues and organs respire at different rates

Respiration is often considered to have two components of comparable magnitude. **Maintenance respiration** is needed to support the function and turnover of the tissues already present. **Growth respiration** provides the energy needed for converting sugars into the building blocks that make up new tissues. A useful rule of thumb is that the

greater the overall metabolic activity of a given tissue, the higher its respiration rate. Developing buds usually show very high rates of respiration, and respiration rates of vegetative organs usually decrease from the point of growth (e.g., the leaf tip in dicotyledons and the leaf base in monocotyledons) to more differentiated regions. A well-studied example is the growing barley leaf (Thompson et al. 1998).

In mature vegetative organs, stems generally have the lowest respiration rates, whereas leaf and root respiration varies with the plant species and the conditions under which the plants are growing. Low availability of soil nutrients, for example, increases the demand for respiratory ATP production in the root. This increase reflects increased active ion uptake and root growth in search of nutrients. (See **WEB TOPIC 11.7** for a discussion of how crop yield is affected by changes in respiration rates.)

When a plant organ has reached maturity, its respiration rate either remains roughly constant or decreases slowly as the tissue ages and ultimately senesces. An exception to this pattern is the marked rise in respiration, known as the *climacteric*, that accompanies the onset of ripening in many fruits (e.g., avocado, apple, and banana) and senescence in detached leaves and flowers. Both ripening and the climacteric respiratory rise are triggered by the endogenous production of ethylene, or may be induced by an exogenous application of ethylene (see Chapter 22). In general, *ethylene-induced respiration* is associated with the cyanide-resistant alternative pathway, but the role of this pathway in ripening is not clear (Tucker 1993).

Different tissues can use different substrates for respiration. Sugars dominate overall, but in specific organs other compounds, such as organic acids in maturing apples or lemons and lipids in germinating sunflower or canola seedlings, may provide the carbon for respiration. These compounds are built with different ratios of carbon to oxygen atoms. Therefore, the ratio of $CO_2$ release to $O_2$ consumption, which is called the **respiratory quotient**, or **RQ**, varies with the substrate oxidized. Lipids, sugars, and organic acids represent a series of rising RQ because lipids contain little oxygen per carbon and organic acids much. Alcoholic fermentation releases $CO_2$ without consuming $O_2$, so a high RQ is also a marker for fermentation. Since RQ can be determined in the field, it is an important parameter in analyses of carbon metabolism on a larger scale.

## Environmental factors alter respiration rates

Many environmental factors can alter the operation of metabolic pathways and change respiratory rates. Here we will examine the roles of environmental oxygen ($O_2$), temperature, and carbon dioxide ($CO_2$).

**OXYGEN** Oxygen can affect plant respiration because of its role as a substrate in the overall respiratory process. At 25°C, the equilibrium concentration of $O_2$ in an air-saturated (21%

$O_2$) aqueous solution is about 250 $\mu M$. The $K_m$ value for oxygen in the reaction catalyzed by cytochrome $c$ oxidase is well below 1 $\mu M$, so there should be no apparent dependence of the respiration rate on external $O_2$ concentrations. However, respiration rates decrease if the atmospheric oxygen concentration is below 5% for whole organs or below 2 to 3% for tissue slices. These findings show that oxygen supply can impose a limitation on plant respiration.

Oxygen diffuses slowly in aqueous solutions, so the network of intercellular air spaces (*aerenchyma*) found in plant tissues is needed to supply oxygen to the mitochondria. If this gaseous diffusion pathway throughout the plant did not exist, the respiration rates of many plants would be limited by an insufficient oxygen supply. Compact organs such as seeds and potato tubers have a noticeable $O_2$ concentration gradient from the surface to the center, which restricts the ATP/ADP ratio. Diffusion limitation is even more significant in seeds with a thick seed coat or in plant organs submerged in water. When plants are grown hydroponically, the solutions must be aerated to keep oxygen levels high in the vicinity of the roots (see Chapter 5). The problem of oxygen supply also arises with plants growing in very wet or flooded soils (see Chapter 26).

Some plants, particularly trees, have a restricted geographic distribution because of the need to maintain a supply of oxygen to their roots. For instance, dogwood (*Cornus florida*) and tulip tree poplar (*Liriodendron tulipifera*) can survive only in well-drained, aerated soils. On the other hand, many plant species are adapted to grow in flooded soils. For example, rice and sunflower rely on a network of aerenchyma running from the leaves to the roots to provide a continuous gaseous pathway for the movement of oxygen to the flooded roots.

Limitation in oxygen supply can be more severe for trees with very deep roots that grow in wet soils. Such roots must survive on anaerobic (fermentative) metabolism or develop structures that facilitate the movement of oxygen to the roots. Examples of such structures are outgrowths of the roots, called *pneumatophores*, that protrude out of the water and provide a gaseous pathway for oxygen diffusion into the roots. Pneumatophores are found in *Avicennia* and *Rhizophora*, both trees that grow in mangrove swamps under continuously flooded conditions.

**TEMPERATURE** Respiration operates over a wide temperature range (see WEB ESSAYS 11.4 and 11.6). It typically increases with temperatures between 0 and 30°C and reaches a plateau at 40 to 50°C. At higher temperatures, it again decreases because of inactivation of the respiratory machinery. The increase in respiration rate for every 10°C increase in temperature is commonly called the **temperature coefficient, $Q_{10}$**. This coefficient describes how respiration responds to short-term temperature changes, and it varies with plant development and external factors. On a longer time scale, plants acclimate to low temperatures by increasing their respiratory capacity so that ATP production can be continued (Atkin and Tjoelker 2003).

Low temperatures are utilized to retard postharvest respiration during the storage of fruits and vegetables, but those temperatures must be adjusted with care. For instance, when potato tubers are stored at temperatures above 10°C, respiration and ancillary metabolic activities are sufficient to allow sprouting. Below 5°C, respiration rates and sprouting are reduced, but the breakdown of stored starch and its conversion into sucrose impart an unwanted sweetness to the tubers. Therefore, potatoes are best stored at 7 to 9°C, which prevents the breakdown of starch while minimizing respiration and germination.

**CARBON DIOXIDE** It is common practice in commercial storage of fruits to take advantage of the effects of oxygen concentration and temperature on respiration by storing fruits at low temperatures under 2 to 3% $O_2$ and 3 to 5% $CO_2$ concentrations. The reduced temperature lowers the respiration rate, as does the reduced $O_2$ level. Low levels of oxygen, instead of anoxic conditions, are used to avoid lowering tissue oxygen tensions to the point at which fermentative metabolism sets in. Carbon dioxide has a limited direct inhibitory effect on respiration at the artificially high concentration of 3 to 5%.

The atmospheric $CO_2$ concentration is normally 360 ppm, but it is increasing as a result of human activities, and it is projected to double, to 700 ppm, before the end of the twenty-first century (see Chapter 9). The flux of $CO_2$ between plants and the atmosphere by photosynthesis and respiration is much larger than the flux of $CO_2$ to the atmosphere caused by the burning of fossil fuels. Therefore, the effects of elevated $CO_2$ concentrations on plant respiration will strongly influence future global atmospheric changes. Laboratory studies have shown that 700 ppm $CO_2$ does not directly inhibit plant respiration, but measurements on whole ecosystems indicate that respiration per biomass unit may decrease with increased $CO_2$ concentrations. The mechanism behind the latter effect is not yet clear, and it is at present not possible to fully predict the potential importance of plants as a sink for anthropogenic $CO_2$ (Gonzales-Meler et al. 2004).

## Lipid Metabolism

Whereas animals use fats for energy storage, plants use them for both energy and carbon storage. Fats and oils are important storage forms of reduced carbon in many seeds, including those of agriculturally important species such as soybean, sunflower, canola, peanut, and cotton. Oils serve a major storage function in many nondomesticated plants that produce small seeds. Some fruits, such as olives and avocados, also store fats and oils.

In this final part of the chapter we will describe the biosynthesis of two types of glycerolipids: the *triacylglycerols*

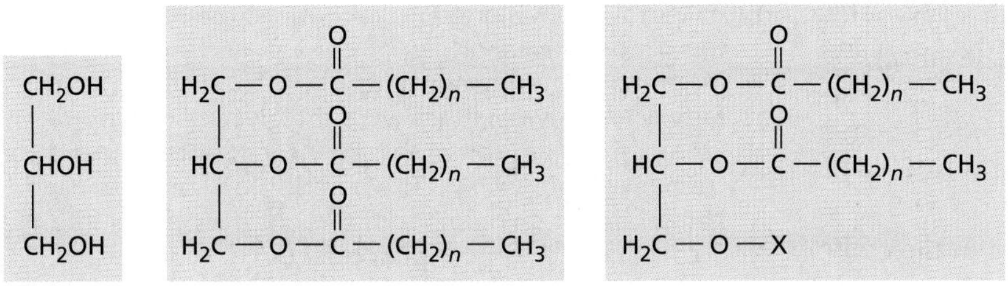

**FIGURE 11.14** Structural features of triacylglycerols and polar glycerolipids in higher plants. The carbon chain lengths of the fatty acids, which always have an even number of carbons, range from 12 to 20, but are typically 16 or 18. Thus the value of $n$ is usually 14 or 16.

| | |
|---|---|
| X = H | Diacylglycerol (DAG) |
| X = $HPO_3^-$ | Phosphatidic acid |
| X = $PO_3^-$—$CH_2$—$CH_2$—$\overset{+}{N}(CH_3)_3$ | Phosphatidylcholine |
| X = $PO_3^-$—$CH_2$—$CH_2$—$NH_2$ | Phosphatidylethanolamine |
| X = galactose | Galactolipids |

(the fats and oils stored in seeds) and the *polar glycerolipids* (which form the lipid bilayers of cellular membranes) (**FIGURE 11.14**). We will see that the biosynthesis of triacylglycerols and polar glycerolipids requires the cooperation of two organelles: the plastids and the endoplasmic reticulum. We will also examine the complex process by which germinating seeds obtain carbon skeletons and metabolic energy from the oxidation of fats and oils.

## Fats and oils store large amounts of energy

Fats and oils belong to the general class termed *lipids*, a structurally diverse group of hydrophobic compounds that are soluble in organic solvents and highly insoluble in water. Lipids represent a more reduced form of carbon than carbohydrates, so the complete oxidation of 1 g of fat or oil (which contains about 40 kJ of energy) can produce considerably more ATP than the oxidation of 1 g of starch (about 15.9 kJ). Conversely, the biosynthesis of lipids requires a correspondingly large investment of metabolic energy.

Other lipids are important for plant structure and function but are not used for energy storage. These lipids include the phospholipids that make up plant membranes, as well as sphingolipids, which are also important membrane components; waxes, which make up the protective cuticle that reduces water loss from exposed plant tissues, and terpenoids (also known as isoprenoids), which include carotenoids involved in photosynthesis and sterols present in many plant membranes (see Chapter 13).

## Triacylglycerols are stored in oil bodies

Fats and oils exist mainly in the form of **triacylglycerols** (*acyl* refers to the fatty acid portion), in which fatty acid molecules are linked by ester bonds to the three hydroxyl groups of glycerol (see Figure 11.14).

The fatty acids in plants are usually straight-chain carboxylic acids having an even number of carbon atoms. The carbon chains can be as short as 12 units and as long as 30 or more, but most commonly are 16 or 18 carbons long. *Oils* are liquid at room temperature, primarily because of the presence of unsaturated bonds in their component fatty acids; *fats*, which have a higher proportion of saturated fatty acids, are solid at room temperature. The major fatty acids in plant lipids are shown in **TABLE 11.3**.

### TABLE 11.3
### Common fatty acids in higher plant tissues

| Name[a] | Structure |
|---|---|
| **Saturated fatty acids** | |
| Lauric acid (12:0) | $CH_3(CH_2)_{10}CO_2H$ |
| Myristic acid (14:0) | $CH_3(CH_2)_{12}CO_2H$ |
| Palmitic acid (16:0) | $CH_3(CH_2)_{14}CO_2H$ |
| Stearic acid (18:0) | $CH_3(CH_2)_{16}CO_2H$ |
| **Unsaturated fatty acids** | |
| Oleic acid (18:1) | $CH_3(CH_2)_7CH{=}CH(CH_2)_7CO_2H$ |
| Linoleic acid (18:2) | $CH_3(CH_2)_4CH{=}CH{-}CH_2{-}CH{=}CH(CH_2)_7CO_2H$ |
| Linolenic acid (18:3) | $CH_3CH_2CH{=}CH{-}CH_2{-}CH{=}CH{-}CH_2{-}CH{=}CH{-}(CH_2)_7CO_2H$ |

[a]Each fatty acid has a numerical abbreviation. The number before the colon represents the total number of carbons; the number after the colon is the number of double bonds.

The proportions of fatty acids in plant lipids vary with the plant species. For example, peanut oil is about 9% palmitic acid, 59% oleic acid, and 21% linoleic acid, and cottonseed oil is 25% palmitic acid, 15% oleic acid, and 55% linoleic acid. The biosynthesis of these fatty acids will be discussed shortly.

In most seeds, triacylglycerols are stored in the cytoplasm of either cotyledon or endosperm cells in organelles known as **oil bodies** (also called *spherosomes* or *oleosomes*) (see Chapter 1). The oil-body membrane is a single layer of phospholipids (i.e., a half-bilayer) with the hydrophilic ends of the phospholipids exposed to the cytosol and the hydrophobic acyl hydrocarbon chains facing the triacylglycerol interior (see Chapter 1). The oil body is stabilized by the presence of specific proteins, called *oleosins*, that coat its outer surface and prevent the phospholipids of adjacent oil bodies from coming in contact and fusing with it.

The unique membrane structure of oil bodies results from the pattern of triacylglycerol biosynthesis. Triacylglycerol synthesis is completed by enzymes located in the membranes of the endoplasmic reticulum (ER), and the resulting fats accumulate between the two monolayers of the ER membrane bilayer. The bilayer swells apart as more fats are added to the growing structure, and ultimately a mature oil body buds off from the ER (Napier et al. 1996).

### Polar glycerolipids are the main structural lipids in membranes

As outlined in Chapter 1, each membrane in the cell is a bilayer of *amphipathic* (i.e., having both hydrophilic and hydrophobic regions) lipid molecules in which a polar head group interacts with the aqueous environment while hydrophobic fatty acid chains form the core of the membrane. This hydrophobic core prevents random diffusion of solutes between cell compartments and thereby allows the biochemistry of the cell to be organized.

**FIGURE 11.15** Major polar glycerolipid classes found in plant membranes: glyceroglycolipids and a sphingolipid (A) and glycerophospholipids (B). At least six different fatty acids may be attached to the glycerol backbone. One of the more common molecular species is shown for each glycerolipid class. The numbers given below each name refer to the number of carbons (number before the colon) and the number of double bonds (number after the colon). ▶

The main structural lipids in membranes are the **polar glycerolipids** (see Figure 11.14), in which the hydrophobic portion consists of two 16-carbon or 18-carbon fatty acid chains esterified to positions 1 and 2 of a glycerol backbone. The polar head group is attached to position 3 of the glycerol. There are two categories of polar glycerolipids:

1. **Glyceroglycolipids**, in which sugars form the head group (**FIGURE 11.15A**)

2. **Glycerophospholipids**, in which the head group contains phosphate (**FIGURE 11.15B**)

Plant membranes have additional structural lipids, including sphingolipids and sterols (see Chapter 13), but these are minor components. Other lipids perform specific roles in photosynthesis and other processes. Included among these lipids are chlorophylls, plastoquinone, carotenoids, and tocopherols, which together account for about one-third of the lipids in plant leaves.

Figure 11.15 shows the nine major glycerolipid classes in plants, each of which can be associated with many different fatty acid combinations. The structures shown in Figure 11.15 illustrate some of the more common molecular species.

Chloroplast membranes, which account for 70% of the membrane lipids in photosynthetic tissues, are dominated by glyceroglycolipids; other membranes of the cell contain glycerophospholipids (**TABLE 11.4**). In nonphotosynthetic tissues, glycerophospholipids are the major membrane glycerolipids.

**TABLE 11.4**
**Glycerolipid components of cellular membranes**

| | Lipid composition (percentage of total) | | |
|---|---|---|---|
| Lipid | Chloroplast | Endoplasmic reticulum | Mitochondrion |
| Phosphatidylcholine | 4 | 47 | 43 |
| Phosphatidylethanolamine | — | 34 | 35 |
| Phosphatidylinositol | 1 | 17 | 6 |
| Phosphatidylglycerol | 7 | 2 | 3 |
| Diphosphatidylglycerol | — | — | 13 |
| Monogalactosyldiacylglycerol | 55 | — | — |
| Digalactosyldiacylglycerol | 24 | — | — |
| Sulfolipid | 8 | — | — |

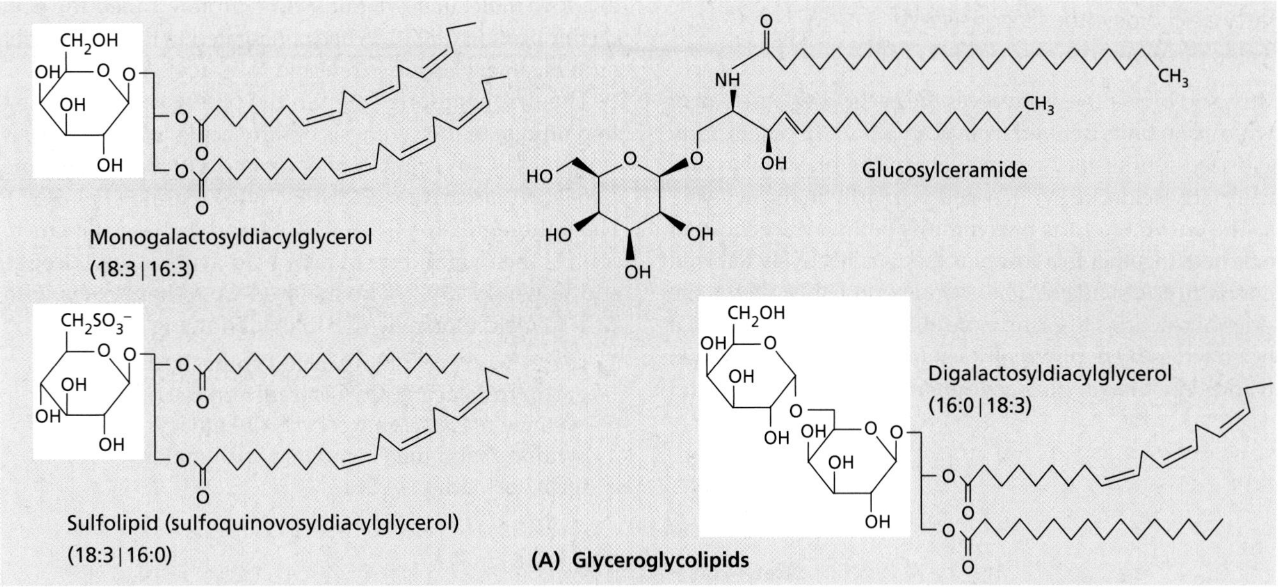

Monogalactosyldiacylglycerol
(18:3 | 16:3)

Glucosylceramide

Sulfolipid (sulfoquinovosyldiacylglycerol)
(18:3 | 16:0)

Digalactosyldiacylglycerol
(16:0 | 18:3)

**(A) Glyceroglycolipids**

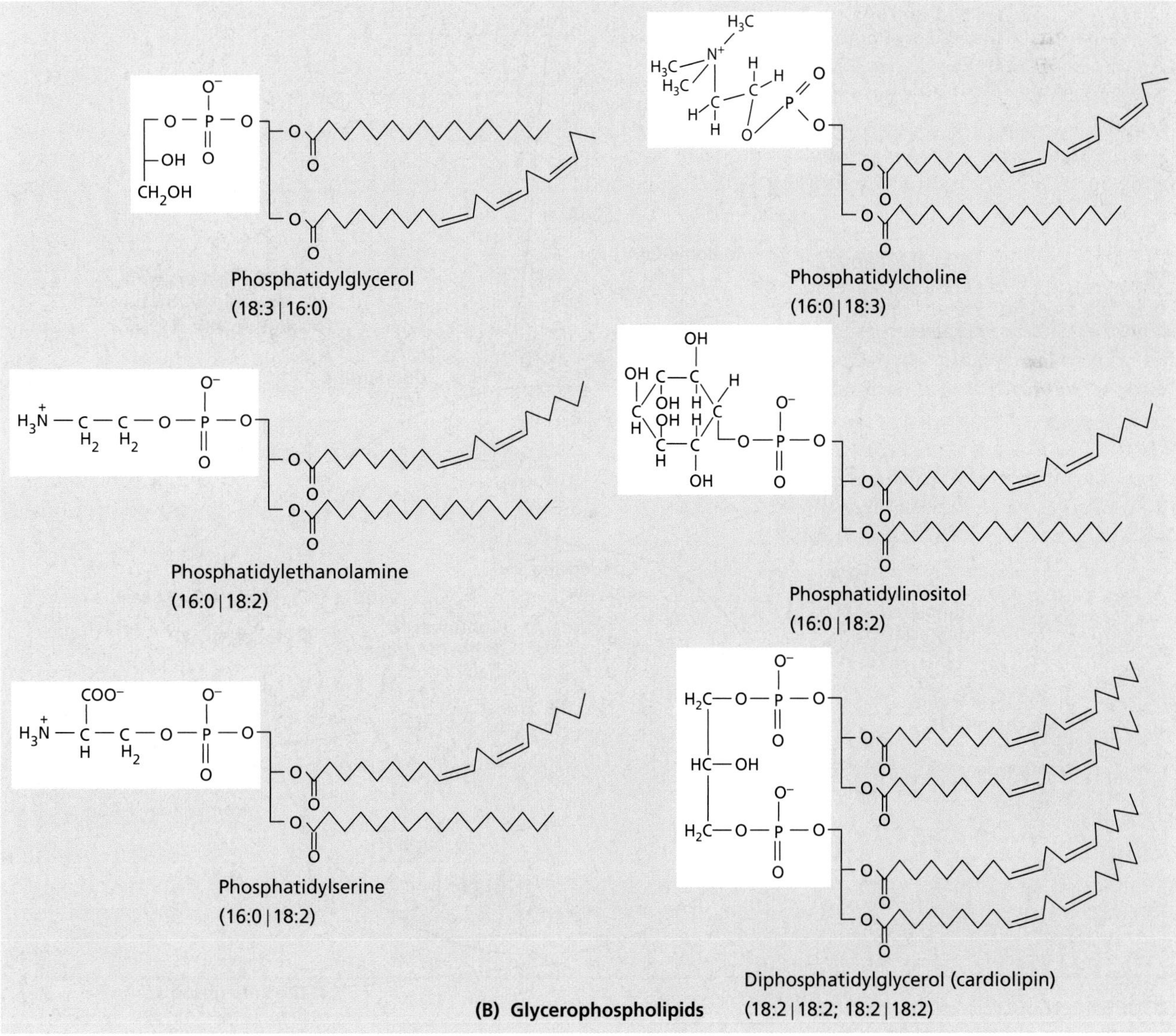

Phosphatidylglycerol
(18:3 | 16:0)

Phosphatidylcholine
(16:0 | 18:3)

Phosphatidylethanolamine
(16:0 | 18:2)

Phosphatidylinositol
(16:0 | 18:2)

Phosphatidylserine
(16:0 | 18:2)

Diphosphatidylglycerol (cardiolipin)
(18:2 | 18:2; 18:2 | 18:2)

**(B) Glycerophospholipids**

## Fatty acid biosynthesis consists of cycles of two-carbon addition

Fatty acid biosynthesis involves the cyclic condensation of two-carbon units derived from acetyl-CoA. In plants, fatty acids are synthesized exclusively in the plastids; in animals, fatty acids are synthesized primarily in the cytosol.

The enzymes of the biosynthesis pathway are thought to be held together in a complex that is collectively referred to as *fatty acid synthase*. The complex probably allows the series of reactions to occur more efficiently than it would if the enzymes were physically separated from one another. In addition, the growing acyl chains are covalently bound

to a low-molecular-weight, acidic protein called the **acyl carrier protein** (**ACP**). When conjugated to the acyl carrier protein, an acyl chain is referred to as **acyl-ACP**.

The first committed step in the pathway (i.e., the first step unique to the synthesis of fatty acids) is the synthesis of malonyl-CoA from acetyl-CoA and $CO_2$ by the enzyme *acetyl-CoA carboxylase* (**FIGURE 11.16**) (Sasaki et al. 1995). The tight regulation of acetyl-CoA carboxylase appears to control the overall rate of fatty acid synthesis (Ohlrogge and Jaworski 1997). The malonyl-CoA then reacts with ACP to yield malonyl-ACP in the following four steps:

1. In the first cycle of fatty acid synthesis, the acetate group from acetyl-CoA is transferred to a specific cysteine of *condensing enzyme* (3-ketoacyl-ACP synthase) and then combined with malonyl-ACP to form acetoacetyl-ACP.

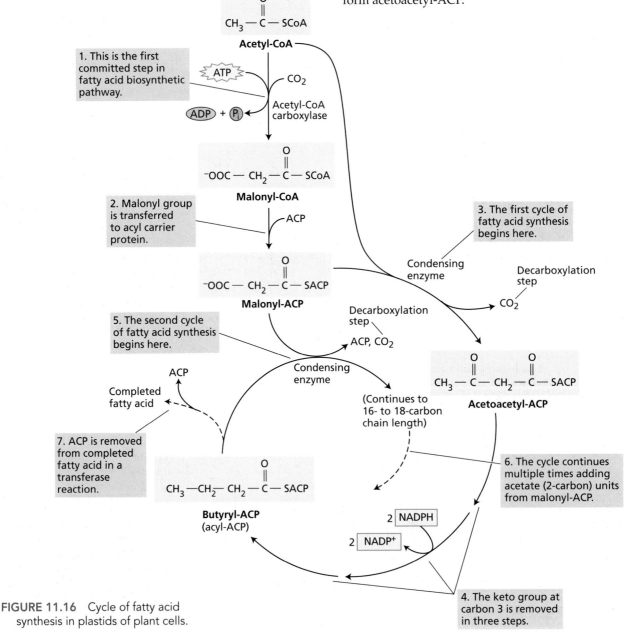

**FIGURE 11.16** Cycle of fatty acid synthesis in plastids of plant cells.

2. Next the keto group at carbon 3 is removed (reduced) by the action of three enzymes to form a new acyl chain (butyryl-ACP), which is now four carbons long (see Figure 11.16).

3. The four-carbon fatty acid and another molecule of malonyl-ACP then become the new substrates for condensing enzyme, resulting in the addition of another two-carbon unit to the growing chain. The cycle continues until 16 or 18 carbons have been added.

4. Some 16:0-ACP is released from the fatty acid synthase machinery, but most molecules that are elongated to 18:0-ACP are efficiently converted into 18:1-ACP by a desaturase enzyme. Thus 16:0-ACP and 18:1-ACP are the major products of fatty acid synthesis in plastids (**FIGURE 11.17**).

Fatty acids may undergo further modification after they are linked with glycerol to form glycerolipids. Additional double bonds are placed in the 16:0 and 18:1 fatty acids by a series of desaturase isozymes. *Desaturase isozymes* are integral membrane proteins found in the chloroplast and the endoplasmic reticulum (ER). Each desaturase inserts a double bond at a specific position in the fatty acid chain, and the enzymes act sequentially to produce the final 18:3 and 16:3 products (Ohlrogge and Browse 1995).

### Glycerolipids are synthesized in the plastids and the ER

The fatty acids synthesized in the chloroplast are next used to make the glycerolipids of membranes and oil bodies. The first steps of glycerolipid synthesis are two acylation reactions that transfer fatty acids from acyl-ACP or acyl-CoA to glycerol-3-phosphate to form **phosphatidic acid**.

The action of a specific phosphatase produces **diacylglycerol** (**DAG**) from phosphatidic acid. Phosphatidic acid can also be converted directly into phosphatidylinositol or phosphatidylglycerol; DAG can give rise to phosphatidylethanolamine or phosphatidylcholine (see Figure 11.17).

The localization of the enzymes of glycerolipid synthesis reveals a complex and highly regulated interaction between the chloroplast, where fatty acids are synthesized, and other membrane systems of the cell. In simple terms, the biochemistry involves two pathways referred to as the prokaryotic (or chloroplast) pathway and the eukaryotic (or ER) pathway (Ohlrogge and Browse 1995):

1. In chloroplasts, the **prokaryotic pathway** utilizes the 16:0-ACP and 18:1-ACP products of chloroplast fatty acid synthesis to synthesize phosphatidic acid and its derivatives. Alternatively, the fatty acids may be exported to the cytoplasm as CoA esters.

2. In the cytoplasm, the **eukaryotic pathway** uses a separate set of acyltransferases in the ER to incorporate the fatty acids into phosphatidic acid and its derivatives.

A simplified version of this two-pathway model is depicted in Figure 11.17.

In some higher plants, including Arabidopsis and spinach, the two pathways contribute almost equally to chloroplast lipid synthesis. In many other angiosperms, however, phosphatidylglycerol is the only product of the prokaryotic pathway, and the remaining chloroplast lipids are synthesized entirely by the eukaryotic pathway.

The biochemistry of *triacylglycerol synthesis* in oilseeds is generally the same as described for the glycerolipids. 16:0-ACP and 18:1-ACP are synthesized in the plastids of the cell and exported as CoA thioesters for incorporation into DAG in the endoplasmic reticulum (see Figure 11.17).

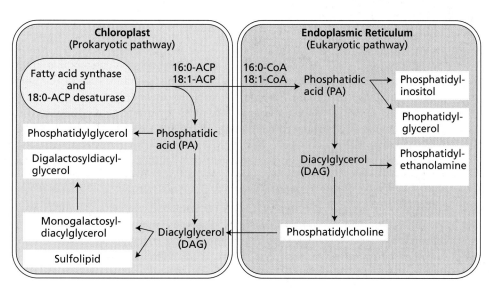

**FIGURE 11.17** The two pathways for glycerolipid synthesis in the chloroplast and endoplasmic reticulum of Arabidopsis leaf cells. The major membrane components are shown in boxes. Glycerolipid desaturases in the chloroplast, and enzymes in the ER convert 16:0 and 18:1 fatty acids into the more highly unsaturated fatty acids shown in Figure 11.15.

The key enzymes in oilseed metabolism (not shown in Figure 11.17) are *acyl-CoA:DAG acyltransferase* and *PC:DAG acyltransferase*, which catalyze triacylglycerol synthesis (Dahlqvist et al. 2000). As noted earlier, triacylglycerol molecules accumulate in specialized subcellular structures—the oil bodies—from which they can be mobilized during germination and converted into sugars.

### Lipid composition influences membrane function

A central question in membrane biology is the functional reason behind lipid diversity. Each membrane system of the cell has a characteristic and distinct complement of lipid types, and within a single membrane, each class of lipids has a distinct fatty acid composition (see Table 11.4).

Our understanding of a membrane is one in which lipids make up the fluid, semipermeable bilayer that is the matrix for the functional membrane proteins. Since this bulk lipid role could be satisfied by a single unsaturated species of phosphatidylcholine, such a simple model is obviously unsatisfactory. Why is lipid diversity needed? One aspect of membrane biology that might offer answers to this central question is the relationship between lipid composition and the ability of organisms to adjust to temperature changes (Iba 2002). For example, chill-sensitive plants experience sharp reductions in growth rate and development at temperatures between 0 and 12°C (see Chapter 26). Many economically important crops, such as cotton, soybean, maize, rice, and many tropical and subtropical fruits, are classified as chill-sensitive. In contrast, most plants that originate from temperate regions are able to grow and develop at chilling temperatures and are classified as chill-resistant plants.

It has been suggested that, because of the decrease in lipid fluidity at lower temperatures, the primary event of chilling injury is a *transition from a liquid-crystalline phase to a gel phase* in cellular membranes. According to this hypothesis, such a transition would result in alterations in the metabolism of chilled cells and would lead to injury and death of the chill-sensitive plants. The degree of unsaturation of the fatty acids would determine the temperature at which such damage occurred.

Recent research, however, suggests that the relationship between membrane unsaturation and plant responses to temperature is more subtle and complex (see WEB TOPIC 11.8). The responses of Arabidopsis mutants with increased saturation of fatty acids to low temperatures are not what is predicted by the chill-sensitivity hypothesis, suggesting that normal chilling injury may not be strictly related to the level of unsaturation of membrane lipids.

On the other hand, experiments with transgenic tobacco plants that are chill-sensitive show opposite results. The transgenic expression of exogenous genes in tobacco has been used specifically to decrease the level of saturated phosphatidylglycerol or to bring about a general increase in membrane unsaturation. In each case, damage caused by chilling was alleviated to some extent.

These new findings make it clear that either the extent of membrane unsaturation or the presence of particular lipids, such as desaturated phosphatidylglycerol, can affect the responses of plants to low temperatures. As discussed in WEB TOPIC 11.8, more work is required to fully understand the relationship between lipid composition and membrane function.

### Membrane lipids are precursors of important signaling compounds

Plants, animals, and microbes all use membrane lipids as precursors for compounds that are used for intracellular or long-range signaling. For example, jasmonate—derived from linolenic acid (18:3)—activates plant defenses against insects and many fungal pathogens (see Chapter 13). In addition, jasmonate regulates other aspects of plant growth, including the development of anthers and pollen (Browse 2009).

**Phosphatidylinositol-4,5-bisphosphate** (**PIP$_2$**) is the most important of several phosphorylated derivatives of phosphatidylinositol known as *phosphoinositides*. In animals, receptor-mediated activation of phospholipase C leads to the hydrolysis of PIP$_2$ into inositol trisphosphate (InsP$_3$) and diacylglycerol, both of which act as intracellular secondary messengers.

The action of InsP$_3$ in releasing Ca$^{2+}$ into the cytoplasm (through Ca$^{2+}$-sensitive channels in the tonoplast and other membranes), and thereby regulating cellular processes, has been demonstrated in several plant systems, including the stomatal guard cells (Schroeder et al. 2001). Information about other types of lipid signaling in plants is becoming available through biochemical and molecular genetic studies of phospholipases (Wang 2001) and other enzymes involved in the generation of these signals.

### Storage lipids are converted into carbohydrates in germinating seeds

After germinating, oil-containing seeds metabolize stored triacylglycerols by converting them into sucrose (Graham 2008). Plants are not able to transport fats from the cotyledons to other tissues of the germinating seedling, so they must convert stored lipids into a more mobile form of carbon, generally sucrose. This process involves several steps that are located in different cellular compartments: oil bodies, glyoxysomes, mitochondria, and the cytosol.

**OVERVIEW: LIPIDS TO SUCROSE** In oilseeds, the conversion of lipids into sucrose is triggered by germination. It begins with the hydrolysis of triacylglycerols stored in oil bodies into free fatty acids, followed by oxidation of those fatty acids to produce acetyl-CoA (**FIGURE 11.18**). The

(A)

Fatty acids are metabolized by β-oxidation to acetyl-CoA in the glyoxysome.

Triacylglycerols are hydrolyzed to yield fatty acids.

Lipase ● OIL BODY

Triacylglycerols

**Fatty acid**

Fatty-acyl-CoA synthetase

CoA

Acyl-CoA

$\frac{n}{2}O_2$

$n\,H_2O$

**β-oxidation**

$n$ NAD⁺

$n$ NADH

CoA

**Citrate**

**Oxaloacetate**

**Oxaloacetate**

**Citrate**

Aconitase

NADH

NAD⁺

Malate dehydrogenase

**Isocitrate**

$n$ **Acetyl-CoA**

**Glyoxylate cycle**

**Malate**

CoA

**Malate**

**Glyoxylate** ◄— **Isocitrate** ◄—

CHO
|
COOH

**Succinate**

Every two molecules of acetyl-CoA produced are metabolized by the glyoxylate cycle to generate one succinate.

CYTOSOL

GLYOXYSOME

**Phosphoenolpyruvate**

$CO_2$

PEP carboxykinase

ADP

ATP

**Oxaloacetate**

**Fructose-6-P**

**Succinate**

**Malate** ◄— **Fumarate**

MITOCHONDRION

Succinate moves into the mitochondrion and is converted to malate.

**Sucrose**

Malate dehydrogenase

NADH

NAD⁺

**Malate**

Malate is transported into the cytosol and oxidized to oxaloacetate, which is converted to phosphoenolpyruvate by the enzyme PEP carboxykinase. The resulting PEP is then metabolized to produce sucrose via the gluconeogenic pathway.

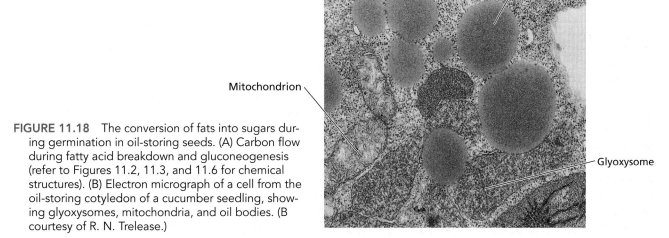

(B)

Oil body

Mitochondrion

Glyoxysome

**FIGURE 11.18** The conversion of fats into sugars during germination in oil-storing seeds. (A) Carbon flow during fatty acid breakdown and gluconeogenesis (refer to Figures 11.2, 11.3, and 11.6 for chemical structures). (B) Electron micrograph of a cell from the oil-storing cotyledon of a cucumber seedling, showing glyoxysomes, mitochondria, and oil bodies. (B courtesy of R. N. Trelease.)

fatty acids are oxidized in a type of peroxisome called a **glyoxysome**, an organelle enclosed by a single membrane bilayer that is found in the oil-rich storage tissues of seeds. Acetyl-CoA is metabolized in the glyoxysome and cytoplasm (see Figure 11.18A) to produce succinate, which is transported from the glyoxysome to the mitochondrion, where it is converted first into fumarate and then into malate. The process ends in the cytosol with the conversion of malate into glucose via gluconeogenesis, and then into sucrose. In most oilseeds, approximately 30% of the acetyl-CoA is used for energy production via respiration, and the rest is converted into sucrose.

**LIPASE-MEDIATED HYDROLYSIS**  The initial step in the conversion of lipids into carbohydrates is the breakdown of triacylglycerols stored in oil bodies by the enzyme lipase, which hydrolyzes triacylglycerols into three fatty acid molecules and one molecule of glycerol. During the breakdown of lipids, oil bodies and glyoxysomes are generally in close physical association (see Figure 11.18B).

**β–OXIDATION OF FATTY ACIDS**  The fatty acid molecules enter the glyoxysome, where they are activated by conversion into fatty-acyl-CoA by the enzyme *fatty-acyl-CoA synthetase*. Fatty-acyl-CoA is the initial substrate for the **β-oxidation** series of reactions, in which $C_n$ fatty acids (fatty acids composed of $n$ carbons) are sequentially broken down into $n/2$ molecules of acetyl-CoA (see Figure 11.18A). This reaction sequence involves the reduction of $\frac{1}{2} O_2$ to $H_2O$ and the formation of one NADH for each acetyl-CoA produced.

In mammalian tissues, the four enzymes associated with β-oxidation are present in the mitochondrion. In plant seed storage tissues, they are located exclusively in the glyoxysome or the equivalent organelle in vegetative tissues, the peroxisome (see Chapter 1).

**THE GLYOXYLATE CYCLE**  The function of the **glyoxylate cycle** is to convert two molecules of acetyl-CoA into succinate. The acetyl-CoA produced by β-oxidation is further metabolized in the glyoxysome through a series of reactions that make up the glyoxylate cycle (see Figure 11.18A). Initially, the acetyl-CoA reacts with oxaloacetate to give citrate, which is then transferred to the cytoplasm for isomerization to isocitrate by aconitase. Isocitrate is reimported into the glyoxysome and converted into malate by two reactions that are unique to the glyoxylate cycle:

1.  First, isocitrate ($C_6$) is cleaved by the enzyme isocitrate lyase to give succinate ($C_4$) and glyoxylate ($C_2$). The succinate is exported to the mitochondria.

2.  Next, malate synthase combines a second molecule of acetyl-CoA with glyoxylate to produce malate.

Malate is then transferred to the cytoplasm and converted into oxaloacetate by the cytoplasmic isozyme of malate dehydrogenase. Oxaloacetate is reimported into the glyoxysome and combines with another acetyl-CoA to continue the cycle (see Figure 11.18A). The glyoxylate produced keeps the cycle operating, but the succinate is exported to the mitochondria for further processing.

**THE MITOCHONDRIAL ROLE**  Moving from the glyoxysomes to the mitochondria, the succinate is converted into malate by the normal citric acid cycle reactions. The resulting malate can be exported from the mitochondria in exchange for succinate via the dicarboxylate transporter located in the inner mitochondrial membrane. Malate is then oxidized to oxaloacetate by malate dehydrogenase in the cytosol, and the resulting oxaloacetate is converted into carbohydrates by the reversal of glycolysis (gluconeogenesis). This conversion requires circumventing the irreversibility of the pyruvate kinase reaction (see Figure 11.3) and is facilitated by the enzyme PEP carboxykinase, which utilizes the phosphorylating ability of ATP to convert oxaloacetate into PEP and $CO_2$ (see Figure 11.18A).

From PEP, gluconeogenesis can proceed to the production of glucose, as described earlier. Sucrose is the final product of this process, and is the primary form of reduced carbon translocated from the cotyledons to the growing seedling tissues. Not all seeds quantitatively convert fat into sugar, however (see **WEB TOPIC 11.9**).

## SUMMARY

Using the building blocks provided by photosynthesis, respiration releases the energy stored in carbon compounds in a controlled manner for cellular use. At the same time it generates many carbon precursors for biosynthesis and cellular functions.

### Overview of Plant Respiration

- In plant respiration, reduced cellular carbon generated by photosynthesis is oxidized to $CO_2$ and water, and this oxidation is coupled to the synthesis of ATP.

- Respiration takes place by four main processes: glycolysis, the oxidative pentose phosphate pathway, the citric acid cycle, and oxidative phosphorylation (the electron transport chain and ATP synthesis) (**Figure 11.1**).

## SUMMARY continued

### Glycolysis

- In glycolysis, carbohydrates are converted into pyruvate in the cytosol, and a small amount of ATP is synthesized via substrate-level phosphorylation. NADH is also produced (**Figure 11.3**).

- Plant glycolysis has alternative enzymes for several steps. These allow differences in substrates used, products made, and the direction of the pathway.

- When $O_2$ is not available, fermentation regenerates $NAD^+$ for glycolysis. Only a minor fraction of the energy available in sugars is conserved by fermentation (**Figure 11.3**).

- Plant glycolysis is regulated from the "bottom up" by its products.

### The Oxidative Pentose Phosphate Pathway

- Carbohydrates can be oxidized via the oxidative pentose phosphate pathway, which provides biosynthetic building blocks and reducing power as NADPH (**Figure 11.4**).

### The Citric Acid Cycle

- Pyruvate is oxidized to $CO_2$ within the mitochondrial matrix through the citric acid cycle, generating a large number of reducing equivalents in the form of NADH and $FADH_2$ (**Figures 11.5, 11.6**).

- A unique feature of the plant citric acid cycle is malic enzyme, which participates in alternative pathways for the metabolism of malate derived from glycolysis (**Figures 11.6, 11.7**).

### Mitochondrial Electron Transport and ATP Synthesis

- Electron transport from NADH and $FADH_2$ to oxygen is coupled by enzyme complexes to proton transport across the inner mitochondrial membrane. This generates an electrochemical proton gradient used for powering synthesis and export of ATP (**Figure 11.8, 11.9**).

- During aerobic respiration, up to 60 molecules of ATP are produced per molecule of sucrose (**Table 11.2**).

- Typical for plant respiration is the presence of several proteins (alternative oxidase, NAD(P)H dehydrogenases, and uncoupling protein) that lower the energy recovery (**Figure 11.8**).

- The main products of the respiratory process are ATP and metabolic intermediates used in biosynthesis. The cellular demand for these compounds regulates respiration via control points in the electron transport chain, the citric acid cycle, and glycolysis (**Figures 11.10–11.13**).

### Respiration in Intact Plants and Tissues

- More than 50% of the daily photosynthetic yield may be respired by a plant.

- Many factors can affect the respiration rate observed at the whole-plant level. These factors include the nature and age of the plant tissue and environmental factors such as light, temperature, nutrient and water supply, and $O_2$ and $CO_2$ concentrations.

### Lipid Metabolism

- Triacylglycerols (fats and oils) are an efficient form for storage of reduced carbon, particularly in seeds. Polar glycerolipids are the primary structural components of membranes (**Figures 11.14, 11.15; Tables 11.3, 11.4**).

- Triacylglycerols are synthesized in the ER and accumulate within the phospholipid bilayer, forming oil bodies.

- Fatty acids are synthesized in plastids using acetyl-CoA, in cycles of two-carbon addition. Fatty acids from the plastids can be transported to the ER, where they are further modified (**Figures 11.16, 11.17**).

- The function of a membrane may be influenced by its lipid composition. The degree of unsaturation of the fatty acids influences the sensitivity of plants to cold, but does not seem to be involved in normal chilling injury.

- Certain membrane lipid breakdown products, such as jasmonic acid, can act as signaling agents in plant cells.

- During germination in oil-storing seeds, the stored lipids are metabolized to carbohydrates in a series of reactions that include the glyoxylate cycle. The glyoxylate cycle takes place in glyoxysomes, and subsequent steps occur in the mitochondria (**Figure 11.18**).

- The reduced carbon generated during lipid breakdown in the glyoxysomes is ultimately converted into carbohydrates in the cytosol by gluconeogenesis (**Figure 11.18**).

# WEB MATERIAL

## Web Topics

## Web Essays

# CHAPTER REFERENCES

Abrahams, J. P., Leslie, A. G. W., Lutter, R., and Walker, J. E. (1994) Structure at 2.8 Å resolution of $F_1$-ATPase from bovine heart mitochondria. *Nature* 370: 621–628.

Atkin, O. K., and Tjoelker, M. G. (2003) Thermal acclimation and the dynamic response of plant respiration to temperature. *Trends Plant Sci.* 8: 343–351.

Bouché, N., and Fromm, H. (2004) GABA in plants: Just a metabolite? *Trends Plant Sci.* 9: 110–115.

Brand, M. D. (1994) The stoichiometry of proton pumping and ATP synthesis in mitochondria. *Biochemist* 16: 20–24.

Browse, J. (2009) Jasmonate passes muster: A receptor and targets for the defense hormone. *Annu. Rev. Plant Biol.* 60: 183–205.

Buchanan, B. B., and Balmer, Y. (2005) Redox regulation: A broadening horizon. *Annu. Rev. Plant Biol.* 56: 187–220.

Budde, R. J. A., and Randall, D. D. (1990) Pea leaf mitochondrial pyruvate dehydrogenase complex is inactivated in vivo in a light-dependent manner. *Proc. Natl. Acad. Sci. USA* 87: 673–676.

Dahlqvist, A., Stahl, U., Lenman, M., Banas, A., Lee, M., Sandager, L., Ronne, H., and Stymne, S. (2000) Phospholipid:diacylglycerol acyltransferase: An enzyme that catalyzes the acyl-CoA-independent formation of triacylglycerol in yeast and plants. *Proc. Natl. Acad. Sci. USA* 97: 6487–6492.

Dennis, D. T., and Blakely, S. D. (2000) Carbohydrate metabolism. In *Biochemistry & Molecular Biology of Plants*, B. Buchanan, W. Gruissem, and R. Jones, eds., American Society of Plant Physiologists, Rockville, MD, pp. 630–674.

Dennis, D. T., Huang, Y., and Negm, F. B. (1997) Glycolysis, the pentose phosphate pathway and anaerobic respiration. In *Plant Metabolism*, 2nd ed., D. T. Dennis, D. H. Turpin, D. D. Lefebvre, and D. B. Layzell, eds., Longman, Singapore, pp. 105–123.

Douce, R. (1985) *Mitochondria in Higher Plants: Structure, Function, and Biogenesis.* Academic Press, Orlando, FL.

Giegé, P., Sweetlove, L. J., Cognat, V., and Leaver, C. J. (2005) Coordination of nuclear and mitochondrial genome expression during mitochondrial biogenesis in Arabidopsis. *Plant Cell* 17: 1497–1512.

Gonzalez-Meler, M. A., Taneva, L., and Trueman, R. J. (2004) Plant respiration and elevated atmospheric $CO_2$ concentration: Cellular responses and global significance. *Ann. Bot.* 94: 647–656.

Graham, I. A (2008) Seed storage oil mobilization. *Annu. Rev. Plant Biol.* 59: 115–42.

Graham, J. W. A., Williams, T. C. R., Morgan. M., Fernie, A. R., Ratcliffe, R. G., and Sweetlove, L. J. (2007) Glycolytic enzymes associate dynamically with mitochondria in response to respiratory demand and support substrate channeling. *Plant Cell* 19: 3723–3738.

Gunning, B. E. S., and Steer, M. W. (1996) *Plant Cell Biology: Structure and Function of Plant Cells.* Jones and Bartlett, Boston.

Hoefnagel, M. H. N., Atkin, O. K., and Wiskich, J. T. (1998) Interdependence between chloroplasts and mitochondria in the light and the dark. *Biochim. Biophys. Acta* 1366: 235–255.

Huang, J., Struck, F., Matzinger, D. F., and Levings, C. S. (1994) Flower-enhanced expression of a nuclear-encoded mitochondrial respiratory protein is associated with changes in mitochondrion number. *Plant Cell* 6: 439–448.

Iba, K. (2002) Acclimative response to temperature stress in higher plants: Approaches of Gene Engineering for Temperature Tolerance. *Annu. Rev. Plant Biol.* 53: 225–245.

Ishizaki, K., Larson, T. R., Schauer, N., Fernie, A. R., Graham, I. A., and Leaver, C. J. (2005) The critical role of Arabidopsis electron-transfer flavoprotein:ubiquinone oxidoreductase during dark-induced starvation. *Plant Cell* 17: 2587–2600.

Krömer, S. (1995) Respiration during photosynthesis. *Annu. Rev. Plant Physiol. Plant Mol. Biol.* 46: 45–70.

Kruger, N. J. (1997) Carbohydrate synthesis and degradation. In *Plant Metabolism*, 2nd ed., D. T. Dennis, D. H. Turpin, D. D. Lefebvre, and D. B. Layzell, eds., Longman, Singapore, pp. 83–104.

Kruger, N. J., and von Schaewen, A. (2003) The oxidative pentose phosphate pathway: Structure and organisation. *Curr. Opin. Plant Biol.* 6: 236–246.

Laloi, M., Klein, M., Riesmeier, J. W., Müller-Röber, B., Fleury, C., Bouillaud, F., and Ricquier, D. (1997) A plant cold-induced uncoupling protein. *Nature* 389: 135–136.

Lambers, H. (1985) Respiration in intact plants and tissues. Its regulation and dependence on environmental factors, metabolism and invaded organisms. In *Higher Plant Cell Respiration* (Encyclopedia of Plant Physiology, New Series, Vol. 18), R. Douce and D. A. Day, eds., Springer, Berlin, pp. 418–473.

Leon, P., Arroyo, A., and Mackenzie, S. (1998) Nuclear control of plastid and mitochondrial development in higher plants. *Annu. Rev. Plant Physiol. Plant Mol. Biol.* 49: 453–480.

Levings, C. S., III, and Siedow, J. N. (1992) Molecular basis of disease susceptibility in the Texas cytoplasm of maize. *Plant Mol. Biol.* 19: 135–147.

Marienfeld, J., Unseld, M., and Brennicke, A. (1999) The mitochondrial genome of Arabidopsis is composed of both native and immigrant information. *Trends Plant Sci.* 4: 495–502.

Millar, A. H., Heazlewood, J. L., Kristensen, B. K., Braun, H.-P., and Møller, I. M. (2005) The plant mitochondrial proteome. *Trends Plant Sci.* 10: 36–43.

Millar, A. H., Mittova, V., Kiddle, G., Heazlewood, J. L., Bartoli, C. G., Theodoulou, F. L., and Foyer, C. H. (2003) Control of ascorbate synthesis by respiration and its implications for stress responses. *Plant Physiol.* 133: 443–447.

Møller, I. M. (2001) Plant mitochondria and oxidative stress. Electron transport, NADPH turnover and metabolism of reactive oxygen species. *Annu. Rev. Plant Physiol. Plant Mol. Biol.* 52: 561–591.

Møller, I. M., and Rasmusson, A. G. (1998) The role of NADP in the mitochondrial matrix. *Trends Plant Sci.* 3: 21–27.

Napier, J. A., Stobart, A. K., and Shewry, P. R. (1996) The structure and biogenesis of plant oil bodies: The role of the ER membrane and the oleosin class of proteins. *Plant Mol. Biol.* 31: 945–956.

Nicholls, D. G., and Ferguson, S. J. (2002) *Bioenergetics 3*, 3rd ed. Academic Press, San Diego, CA.

Noctor, G., and Foyer, C. H. (1998) A re-evaluation of the ATP:NADPH budget during C3 photosynthesis: A contribution from nitrate assimilation and its associated respiratory activity? *J. Exp. Bot.* 49: 1895–1908.

Noctor, G., De Paepe, R., and Foyer, C. H. (2007) Mitochondrial redox biology and homeostasis in plants. *Trends Plant Sci.* 12: 125–134.

Ohlrogge, J. B., and Browse, J. A. (1995) Lipid biosynthesis. *Plant Cell* 7: 957–970.

Ohlrogge, J. B., and Jaworski, J. G. (1997) Regulation of fatty acid synthesis. *Annu. Rev. Plant Physiol. Plant Mol. Biol.* 48: 109–136.

Oliver, D. J., and McIntosh, C. A. (1995) The biochemistry of the mitochondrial matrix. In *The Molecular Biology of Plant Mitochondria*, C. S. Levings III and I. Vasil, eds., Kluwer, Dordrecht, Netherlands, pp. 237–280.

Plaxton, W. C. (1996) The organization and regulation of plant glycolysis. *Annu. Rev. Plant Physiol. Plant Mol. Biol.* 47: 185–214.

Raskin, I., Turner, I. M., and Melander, W. R. (1989) Regulation of heat production in the inflorescences of an *Arum* lily by endogenous salicylic acid. *Proc. Natl. Acad. Sci. USA* 86: 2214–2218.

Rasmusson, A. G., Soole, K. L., and Elthon, T. E. (2004) Alternative NAD(P)H dehydrogenases of plant mitochondria. *Annu. Rev. Plant Biol.* 55: 23–39.

Rasmusson, A. G., Geisler, D. A., and Møller, I. M. (2008) The multiplicity of dehydrogenases in the electron transport chain of plant mitochondria. *Mitochondrion* 8: 47–60.

Rhoads, D. M., and Subbaiah, C. C. (2007) Mitochondrial retrograde regulation in plants. *Mitochondrion* 7: 177–194.

Rolland, F., Baena-Gonzalez, E., and Sheen, J. (2006) Sugar sensing and signaling in plants: Conserved and novel mechanisms. *Annu. Rev. Plant Biol.* 57: 675–709.

Sasaki, Y., Konishi, T., and Nagano, Y. (1995) The compartmentation of acetyl-coenzyme A carboxylase in plants. *Plant Physiol.* 108: 445–449.

Schroeder, J. I., Allen, G. J., Hugouvieux, V., Kwak, J. M., and Waner, D. (2001) Guard cell signal transduction. *Annu. Rev. Plant Physiol. Plant Mol. Biol.* 52: 627–658.

Schürmann, P., and Buchanan, B. B. (2008) The ferredoxin/thioredoxin system of oxygenic photosynthesis. *Antioxid. Redox Signal.* 10: 1235–1273.

Sharples, S. C., and Fry, S. C. (2007) Radioisotope ratios discriminate between competing pathways of cell wall polysaccharide and RNA biosynthesis in living plant cells. *Plant J.* 52: 252–262.

Siedow, J. N., and Umbach, A. L. (1995) Plant mitochondrial electron transfer and molecular biology. *Plant Cell* 7: 821–831.

Thompson, P., Bowsher, C. G., and Tobin, A. K. (1998) Heterogeneity of mitochondrial protein biogenesis during primary leaf development in barley. *Plant Physiol.* 118: 1089–1099.

Tucker, G. A. (1993) Introduction. In *Biochemistry of Fruit Ripening*, G. Seymour, J. Taylor, and G. Tucker, eds., Chapman & Hall, London, pp. 1–51.

Vanlerberghe, G. C., and McIntosh, L. (1997) Alternative oxidase: From gene to function. *Annu. Rev. Plant Physiol. Plant Mol. Biol.* 48: 703–734.

Vercesi, A. E., Martins I. S., Silva, M. P., and Leite, H. M. F. (1995) PUMPing plants. *Nature* 375: 24.

Wang, X. (2001) Plant phospholipases. *Annu. Rev. Plant Physiol. Plant Mol. Biol.* 52: 211–231.

CHAPTER

# 12

# Assimilation of Mineral Nutrients

Higher plants are autotrophic organisms that can synthesize all of their organic molecular components out of inorganic nutrients obtained from their surroundings. For many mineral nutrients, this process involves absorption from the soil by the roots (see Chapter 5) and incorporation into the organic compounds that are essential for growth and development. This incorporation of mineral nutrients into organic substances such as pigments, enzyme cofactors, lipids, nucleic acids, and amino acids is termed **nutrient assimilation**.

Assimilation of some nutrients—particularly nitrogen and sulfur—involves a complex series of biochemical reactions that are among the most energy-consuming reactions in living organisms.

- In nitrate ($NO_3^-$) assimilation, the nitrogen in $NO_3^-$ is converted to a higher-energy form in nitrite ($NO_2^-$), then to a yet–higher-energy form in ammonium ($NH_4^+$), and finally into the amide nitrogen of the amino acid glutamine. This process consumes the equivalent of 12 ATPs per amide nitrogen (Bloom et al. 1992).

- Plants such as legumes form symbiotic relationships with nitrogen-fixing bacteria to convert molecular nitrogen ($N_2$) into ammonia ($NH_3$). Ammonia ($NH_3$) is the first stable product of natural fixation; at physiological pH, however, ammonia is protonated to form the ammonium ion ($NH_4^+$). The process of biological nitrogen fixation, together with the subsequent assimilation of $NH_3$ into an amino acid, consumes the equivalent of about 10 ATPs per amide nitrogen (Pate and Layzell 1990; Vande Broek and Vanderleyden 1995).

- The assimilation of sulfate ($SO_4^{2-}$) into the amino acid cysteine via the two pathways found in plants consumes about 14 ATPs (Hell 1997).

For some perspective on the enormous energies involved, consider that if these reactions run rapidly in reverse—say, from $NH_4NO_3$ (ammonium nitrate) to $N_2$—they become explosive, liberating vast amounts of energy as motion, heat, and light. Nearly all explosives (e.g., nitroglycerin, TNT, and gunpowder) are based on the rapid oxidation of nitrogen or sulfur compounds.

Assimilation of other nutrients, especially the macronutrient and micronutrient cations (see Chapter 5), involves the formation of complexes with organic compounds. For example, $Mg^{2+}$ associates with chlorophyll pigments, $Ca^{2+}$ associates with pectates within the cell wall, and $Mo^{6+}$ associates with enzymes such as nitrate reductase and nitrogenase. These complexes are highly stable, and removal of the nutrient from the complex may result in total loss of function.

This chapter outlines the primary reactions through which the major nutrients (nitrogen, sulfur, phosphate, cations such as $Mg^{2+}$ and $K^+$, and oxygen) are assimilated and discusses the organic products of these reactions. We emphasize the physiological implications of the required energy expenditures and introduce the topic of symbiotic nitrogen fixation. Plants serve as the major conduit through which nutrients pass from inert geophysical domains into dynamic biological ones; this chapter thus highlights the vital role of plant nutrient assimilation in the human diet.

# Nitrogen in the Environment

Many prominent biochemical compounds in plant cells contain nitrogen (see Chapter 5). For example, nitrogen is found in the nucleotides and amino acids that form the building blocks of nucleic acids and proteins, respectively. Only the elements oxygen, carbon, and hydrogen are more abundant in plants than nitrogen. Most natural and agricultural ecosystems show dramatic gains in productivity after fertilization with inorganic nitrogen, attesting to the importance of this element.

In this section we will discuss the biogeochemical cycle of nitrogen, the crucial role of nitrogen fixation in the conversion of molecular nitrogen into ammonium and nitrate, and the fate of ammonium and nitrate in plant tissues.

## Nitrogen passes through several forms in a biogeochemical cycle

Nitrogen is present in many forms in the biosphere. The atmosphere contains vast quantities (about 77% by volume) of molecular nitrogen ($N_2$) (see Chapter 9). For the most part, this large reservoir of nitrogen is not directly available to living organisms. Acquisition of nitrogen from the atmosphere requires the breaking of an exceptionally stable triple covalent bond between two nitrogen atoms ($N\equiv N$) to produce ammonia ($NH_3$) or nitrate ($NO_3^-$). These reactions, known as **nitrogen fixation**, can be accomplished by both industrial and natural processes.

Under elevated temperature (about 200°C) and high pressure (about 200 atmospheres) and in the presence of a metal catalyst (usually iron), $N_2$ combines with hydrogen to form ammonia. The extreme conditions are required to overcome the high activation energy of the reaction. This nitrogen fixation reaction, called the *Haber–Bosch process*, is a starting point for the manufacture of many industrial and agricultural products. Worldwide industrial production of nitrogen fertilizers amounts to more than $100 \times 10^{12}$ g $y^{-1}$ (FAOSTAT 2009).

Natural processes, which fix about $190 \times 10^{12}$ g $y^{-1}$ of nitrogen (**TABLE 12.1**), are the following (Schlesinger 1997):

- *Lightning.* Lightning is responsible for about 8% of the nitrogen fixed. Lightning converts water vapor and oxygen into highly reactive hydroxyl free radicals, free hydrogen atoms, and free oxygen atoms that attack molecular nitrogen ($N_2$) to form nitric acid ($HNO_3$). This nitric acid subsequently falls to Earth with rain.

- *Photochemical reactions.* Approximately 2% of the nitrogen fixed derives from photochemical reactions between gaseous nitric oxide (NO) and ozone ($O_3$) that produce nitric acid ($HNO_3$).

- *Biological nitrogen fixation.* The remaining 90% results from biological nitrogen fixation, in which bacteria or blue-green algae (cyanobacteria) fix $N_2$ into ammonia ($NH_3$). This ammonia dissolves in water to form ammonium ($NH_4^+$):

$$NH_3 + H_2O \rightarrow NH_4^+ + OH^- \qquad (12.1)$$

From an agricultural standpoint, biological nitrogen fixation is critical, because industrial production of nitrogen fertilizers seldom meets agricultural demand (FAOSTAT 2009).

Once fixed into ammonia or nitrate, nitrogen enters a biogeochemical cycle and passes through several organic or inorganic forms before it eventually returns to molecular nitrogen (**FIGURE 12.1**; see also Table 12.1). The ammonium ($NH_4^+$) and nitrate ($NO_3^-$) ions in the soil solution that are generated through fixation or released through decomposition of soil organic matter become the object of intense competition among plants and microorganisms. To remain competitive, plants have developed mechanisms for scavenging these ions rapidly from the soil solution (see Chapter 5). Under the elevated soil concentrations that occur after fertilization, the absorption of ammonium and nitrate by the roots may exceed the capacity of a plant to assimilate these ions, leading to their accumulation within the plant's tissues.

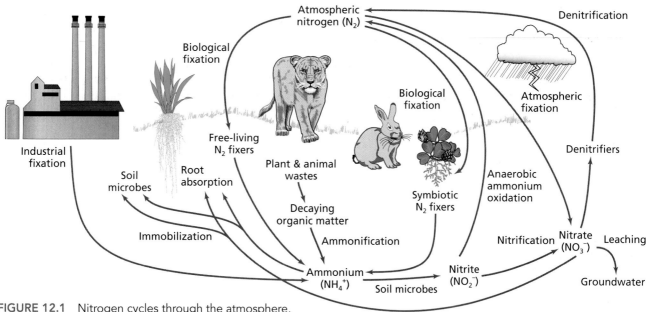

**FIGURE 12.1** Nitrogen cycles through the atmosphere, changing from a gas to reduced ions before it is incorporated into organic compounds in living organisms. Some of the steps involved in the nitrogen cycle are shown.

**TABLE 12.1**
**The major processes of the biogeochemical nitrogen cycle**

| Process | Definition | Rate ($10^{12}$ g y$^{-1}$)$^a$ |
|---|---|---|
| Industrial fixation | Industrial conversion of molecular nitrogen to ammonia | 100 |
| Atmospheric fixation | Lightning and photochemical conversion of molecular nitrogen to nitrate | 19 |
| Biological fixation | Prokaryotic conversion of molecular nitrogen to ammonia | 170 |
| Plant acquisition | Plant absorption and assimilation of ammonium or nitrate | 1200 |
| Immobilization | Microbial absorption and assimilation of ammonium or nitrate | N/C |
| Ammonification | Bacterial and fungal catabolism of soil organic matter to ammonium | N/C |
| Anammox | Anaerobic ammonium oxidation: bacterial conversion of ammonium and nitrate to molecular nitrogen | N/C |
| Nitrification | Bacterial (*Nitrosomonas* sp.) oxidation of ammonium to nitrite and subsequent bacterial (*Nitrobacter* sp.) oxidation of nitrite to nitrate | N/C |
| Mineralization | Bacterial and fungal catabolism of soil organic matter to mineral nitrogen through ammonification or nitrification | N/C |
| Volatilization | Physical loss of gaseous ammonia to the atmosphere | 100 |
| Ammonium fixation | Physical embedding of ammonium into soil particles | 10 |
| Denitrification | Bacterial conversion of nitrate to nitrous oxide and molecular nitrogen | 210 |
| Nitrate leaching | Physical flow of nitrate dissolved in groundwater out of the topsoil and eventually into the oceans | 36 |

*Note*: Terrestrial organisms, the soil, and the oceans contain about $5.2 \times 10^{15}$ g, $95 \times 10^{15}$ g, and $6.5 \times 10^{15}$ g, respectively, of organic nitrogen that is active in the cycle. Assuming that the amount of atmospheric $N_2$ remains constant (inputs = outputs), the *mean residence time* (the average time that a nitrogen molecule remains in organic forms) is about 370 years [(pool size)/(fixation input) = ($5.2 \times 10^{15}$ g + $95 \times 10^{15}$ g)/($80 \times 10^{12}$ g y$^{-1}$ + $19 \times 10^{12}$ g y$^{-1}$ + $170 \times 10^{12}$ g y$^{-1}$)] (Schlesinger 1997).

$^a$N/C, not calculated.

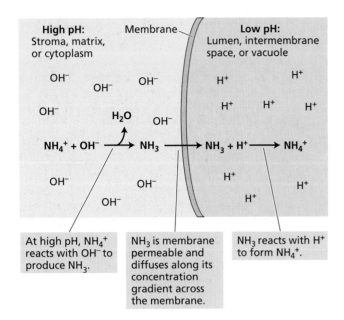

**High pH:**
Stroma, matrix, or cytoplasm

Membrane

**Low pH:**
Lumen, intermembrane space, or vacuole

At high pH, NH$_4^+$ reacts with OH$^-$ to produce NH$_3$.

NH$_3$ is membrane permeable and diffuses along its concentration gradient across the membrane.

NH$_3$ reacts with H$^+$ to form NH$_4^+$.

**FIGURE 12.2** NH$_4^+$ toxicity can dissipate pH gradients. The left side represents the stroma, matrix, or cytoplasm, where the pH is high; the right side represents the lumen, intermembrane space, or vacuole, where the pH is low; and the membrane represents the thylakoid, inner mitochondrial, or tonoplast membrane for a chloroplast, mitochondrion, or root cell, respectively. The net result of the reaction shown is that both the OH$^-$ concentration on the left side and the H$^+$ concentration on the right side have been diminished; that is, the pH gradient has been dissipated. (After Bloom 1997.)

### Unassimilated ammonium or nitrate may be dangerous

Ammonium, if it accumulates to high levels in living tissues, is toxic to both plants and animals. Ammonium dissipates transmembrane proton gradients (**FIGURE 12.2**) that are required for both photosynthetic and respiratory electron transport (see Chapters 7 and 11) and for sequestering metabolites in the vacuole (see Chapter 6). Because high levels of ammonium are dangerous, animals have developed a strong aversion to its smell. The active ingredient in smelling salts, a medicinal vapor released under the nose to revive a person who has fainted, is ammonium carbonate. Plants assimilate ammonium near the site of absorption or generation and rapidly store any excess in their vacuoles, thus avoiding toxic effects on membranes and the cytosol.

In contrast to ammonium, plants can store high levels of nitrate, or they can translocate it from tissue to tissue without deleterious effect. Yet if livestock or humans consume plant material that is high in nitrate, they may suffer methemoglobinemia, a disease in which the liver reduces nitrate to nitrite, which combines with hemoglobin and renders hemoglobin unable to bind oxygen. Humans and other animals may also convert nitrate into nitrosamines, which are potent carcinogens. Some countries limit the nitrate content in plant materials sold for human consumption.

In the next sections we will discuss the process by which plants assimilate nitrate into organic compounds via the enzymatic reduction of nitrate first into nitrite, next into ammonium, and then into amino acids.

## Nitrate Assimilation

Plant roots actively absorb nitrate from the soil solution via several low- and high-affinity nitrate–proton cotransporters (Crawford and Forde 2002; Miller et al. 2007). Plants eventually assimilate most of this nitrate into organic nitrogen compounds. The first step of this process is the reduction of nitrate to nitrite in the cytosol, a reaction that involves the transfer of two electrons (Oaks 1994). The enzyme **nitrate reductase** catalyzes this reaction:

$$NO_3^- + NAD(P)H + H^+ \rightarrow$$
$$NO_2^- + NAD(P)^+ + H_2O \qquad (12.2)$$

where NAD(P)H indicates NADH or NADPH. The most common form of nitrate reductase uses only NADH as an electron donor; another form of the enzyme that is found predominantly in nongreen tissues such as roots can use either NADH or NADPH (Warner and Kleinhofs 1992).

The nitrate reductases of higher plants are composed of two identical subunits, each containing three prosthetic groups: flavin adenine dinucleotide (FAD), heme, and a molybdenum atom complexed to an organic molecule called a *pterin* (Campbell 1999, 2001).

**A pterin (fully oxidized)**

Nitrate reductase is the main molybdenum-containing protein in vegetative tissues; one symptom of molybdenum deficiency is the accumulation of nitrate that results from diminished nitrate reductase activity (Mendel 2005).

X-ray crystallography (Fisher et al. 2005) and comparison of the amino acid sequences for nitrate reductase from several species with the sequences of other well-characterized proteins that bind FAD, heme, or molybdenum has led to a multiple-domain model for nitrate reductase; a simplified three-domain model is shown in **FIGURE 12.3**. The FAD-binding domain accepts two electrons from NADH or NADPH. The electrons then pass through the heme domain to the molybdenum complex, where they are transferred to nitrate.

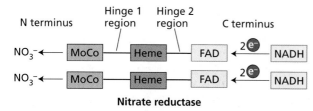

**Nitrate reductase**

**FIGURE 12.3** A model of the nitrate reductase dimer, illustrating the three binding domains whose polypeptide sequences are similar in eukaryotes: molybdenum complex (MoCo), heme, and FAD. The NADH binds at the FAD-binding region of each subunit and initiates a two-electron transfer from the carboxyl (C) terminus, through each of the electron transfer components, to the amino (N) terminus. Nitrate is reduced at the molybdenum complex near the amino terminus. The polypeptide sequences of the hinge regions are highly variable among species.

## Many factors regulate nitrate reductase

Nitrate, light, and carbohydrates influence nitrate reductase at the transcription and translation levels (Sivasankar and Oaks 1996). In barley seedlings, nitrate reductase mRNA was detected approximately 40 minutes after addition of nitrate, and maximum levels were attained within 3 hours (**FIGURE 12.4**). In contrast to the rapid mRNA accumulation, there was a gradual linear increase in nitrate reductase activity, reflecting the slower synthesis of the protein.

In addition, the protein is subject to posttranslational modification (involving a reversible phosphorylation) that is analogous to the regulation of sucrose phosphate synthase (see Chapters 8 and 10). Light, carbohydrate levels, and other environmental factors stimulate a protein phosphatase that dephosphorylates a key serine residue in the hinge 1 region of nitrate reductase (between the molybdenum complex and heme-binding domains; see Figure 12.3) and thereby activates the enzyme.

Operating in the reverse direction, darkness and $Mg^{2+}$ stimulate a protein kinase that phosphorylates the same serine residues, which then interact with a 14-3-3 inhibitor protein, and thereby inactivate nitrate reductase (Kaiser et al. 1999). *Regulation of nitrate reductase activity through phosphorylation and dephosphorylation provides more rapid control than can be achieved through synthesis or degradation of the enzyme (minutes versus hours).*

## Nitrite reductase converts nitrite to ammonium

Nitrite ($NO_2^-$) is a highly reactive, potentially toxic ion. Plant cells immediately transport the nitrite generated by nitrate reduction (see Equation 12.1) from the cytosol into chloroplasts in leaves and plastids in roots. In these organelles, the enzyme nitrite reductase reduces nitrite to ammonium, a reaction that involves the transfer of six electrons, according to the following overall reaction:

$$NO_2^- + 6\,Fd_{red} + 8\,H^+ \rightarrow$$
$$NH_4^+ + 6\,Fd_{ox} + 2\,H_2O \qquad (12.3)$$

where Fd is ferredoxin, and the subscripts *red* and *ox* stand for *reduced* and *oxidized*, respectively. Reduced ferredoxin is derived from photosynthetic electron transport in the chloroplasts (see Chapter 7) and from NADPH generated by the oxidative pentose phosphate pathway in nongreen tissues (see Chapter 11).

Chloroplasts and root plastids contain different forms of the enzyme, but both forms consist of a single polypeptide containing two prosthetic groups: an iron–sulfur cluster ($Fe_4S_4$) and a specialized heme (Siegel and Wilkerson 1989). These groups act together to bind nitrite and reduce it to ammonium. Although no nitrogen compounds of intermediate redox states accumulate, a small percentage (0.02–0.2%) of the nitrite reduced is released as nitrous oxide ($N_2O$), a greenhouse gas (Smart and Bloom 2001). The electron flow through ferredoxin, $Fe_4S_4$, and heme can be represented as in **FIGURE 12.5**.

Nitrite reductase is encoded in the nucleus and synthesized in the cytoplasm with an N-terminal transit peptide

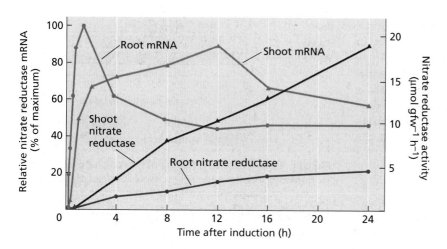

**FIGURE 12.4** Stimulation of nitrate reductase activity follows the induction of nitrate reductase mRNA in shoots and roots of barley; gfw, grams fresh weight. (After Kleinhofs et al. 1989.)

**FIGURE 12.5**  Model for coupling of photosynthetic electron flow, via ferredoxin, to the reduction of nitrite by nitrite reductase. The enzyme contains two prosthetic groups, $Fe_4S_4$ and heme, which participate in the reduction of nitrite to ammonium.

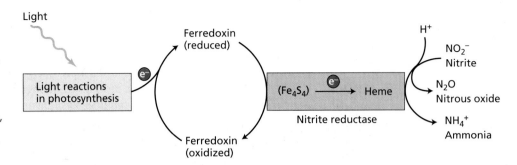

that targets it to the plastids (Wray 1993). Elevated concentrations of $NO_3^-$ or exposure to light induce the transcription of nitrite reductase mRNA. Accumulation of the end products in the process—asparagine and glutamine—repress this induction.

### Both roots and shoots assimilate nitrate

In many plants, when the roots receive small amounts of nitrate, nitrate is reduced primarily in the roots. As the supply of nitrate increases, a greater proportion of the absorbed nitrate is translocated to the shoot and assimilated there (Marschner 1995). Even under similar conditions of nitrate supply, the balance between root and shoot nitrate metabolism—as indicated by the proportion of nitrate reductase activity in each of the two organs or by the relative concentrations of nitrate and reduced nitrogen in the xylem sap—varies from species to species.

In plants such as the cocklebur (*Xanthium strumarium*), nitrate metabolism is restricted to the shoot; in other plants, such as white lupine (*Lupinus albus*), most nitrate is metabolized in the roots (**FIGURE 12.6**). Generally, species native to temperate regions rely more heavily on nitrate assimilation by the roots than do species of tropical or subtropical origins.

## Ammonium Assimilation

Plant cells avoid ammonium toxicity by rapidly converting the ammonium generated from nitrate assimilation or photorespiration (see Chapter 8) into amino acids. The primary pathway for this conversion involves the sequential actions of glutamine synthetase and glutamate synthase (Lea et al. 1992). In this section we will discuss the enzymatic processes that mediate the assimilation of ammonium into essential amino acids, and the role of amides in the regulation of nitrogen and carbon metabolism.

### Converting ammonium to amino acids requires two enzymes

**Glutamine synthetase** (GS) combines ammonium with glutamate to form glutamine (**FIGURE 12.7A**):

$$\text{Glutamate} + NH_4^+ + ATP \rightarrow \text{glutamine} + ADP + P_i \quad (12.4)$$

This reaction requires the hydrolysis of one ATP and involves a divalent cation such as $Mg^{2+}$, $Mn^{2+}$, or $Co^{2+}$ as a cofactor. Plants contain two classes of GS, one in the cytosol and the other in root plastids or shoot chloroplasts. The cytosolic forms are expressed in germinating seeds or in the vascular bundles of roots and shoots and produce glutamine for intracellular nitrogen transport. The GS in root plastids generates amide nitrogen for local consumption; the GS in shoot chloroplasts reassimilates photorespiratory $NH_4^+$ (Lam et al. 1996). Light and carbohydrate levels alter the expression of the plastid forms of the enzyme, but they have little effect on the cytosolic forms.

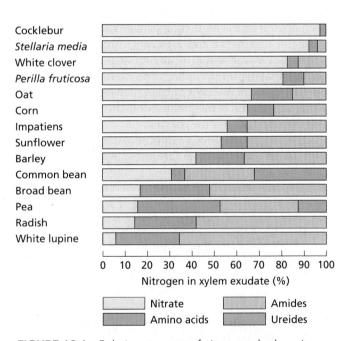

**FIGURE 12.6**  Relative amounts of nitrate and other nitrogen compounds in the xylem sap of various plant species. The plants were grown with their roots exposed to nitrate solutions, and xylem sap was collected by severing the stem. Note the presence of ureides in common bean and pea; only legumes of tropical origin export nitrogen in such compounds. (After Pate 1983.)

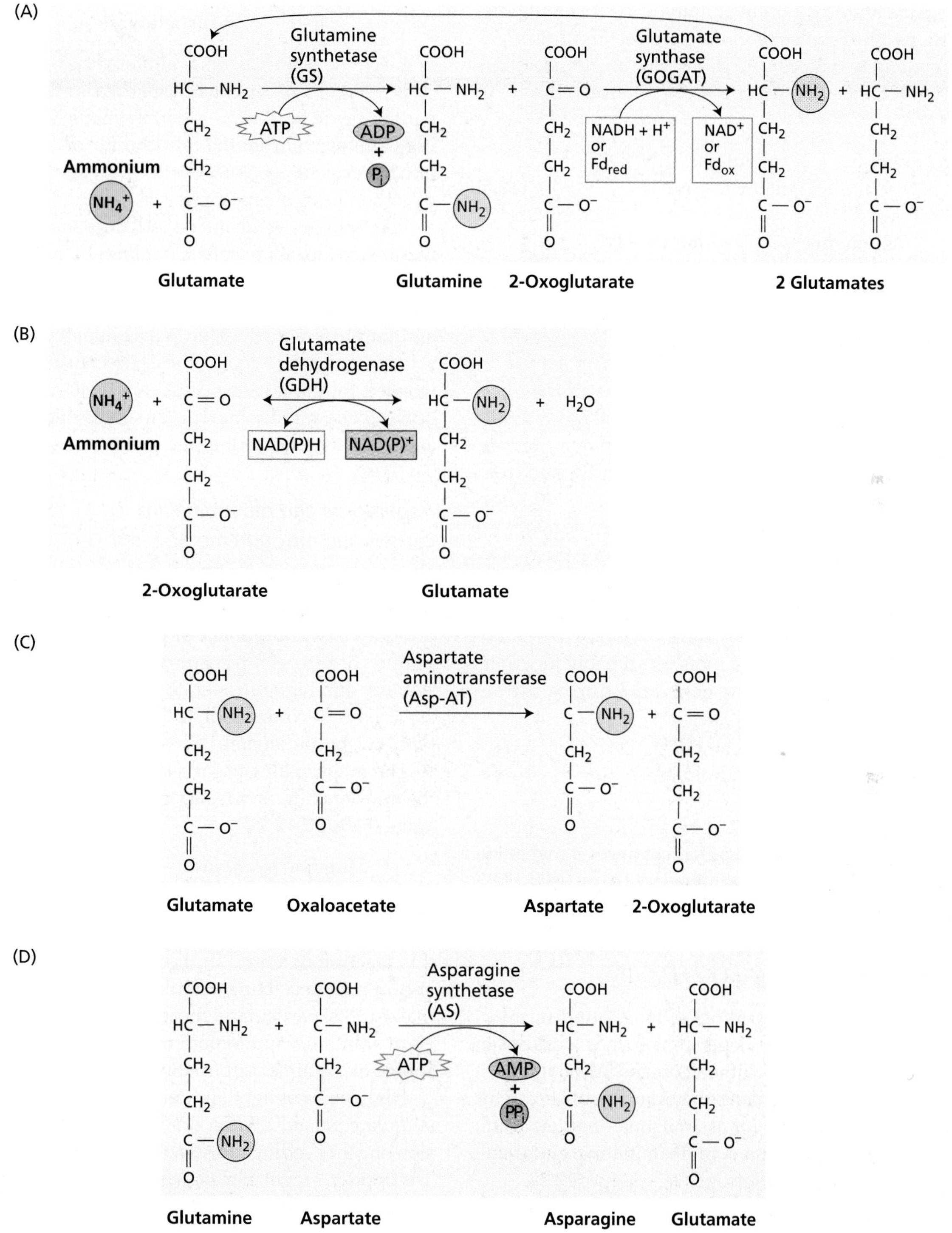

**FIGURE 12.7** Structure and pathways of compounds involved in ammonium metabolism. Ammonium can be assimilated by one of several processes. (A) The GS-GOGAT pathway that forms glutamine and glutamate. A reduced cofactor is required for the reaction: ferredoxin in green leaves and NADH in nonphotosynthetic tissue. (B) The GDH pathway that forms glutamate using NADH or NADPH as a reductant. (C) Transfer of the amino group from glutamate to oxaloacetate to form aspartate (catalyzed by aspartate aminotransferase). (D) Synthesis of asparagine by transfer of an amino acid group from glutamine to aspartate (catalyzed by asparagine synthetase).

Elevated plastid levels of glutamine stimulate the activity of **glutamate synthase** (also known as *glutamine:2-oxoglutarate aminotransferase*, or **GOGAT**). This enzyme transfers the amide group of glutamine to 2-oxoglutarate, yielding two molecules of glutamate (see Figure 12.7A). Plants contain two types of GOGAT: One accepts electrons from NADH; the other accepts electrons from ferredoxin (Fd):

$$\text{Glutamine} + \text{2-oxoglutarate} + \text{NADH} + \text{H}^+ \\ \rightarrow \text{2 glutamate} + \text{NAD}^+ \quad (12.5)$$

$$\text{Glutamine} + \text{2-oxoglutarate} + \text{Fd}_{red} \\ \rightarrow \text{2 glutamate} + \text{Fd}_{ox} \quad (12.6)$$

The NADH type of the enzyme (NADH-GOGAT) is located in plastids of nonphotosynthetic tissues such as roots or the vascular bundles of developing leaves. In roots, NADH-GOGAT is involved in the assimilation of $\text{NH}_4^+$ absorbed from the rhizosphere (the soil near the surface of the roots); in vascular bundles of developing leaves, NADH-GOGAT assimilates glutamine translocated from roots or senescing leaves.

The ferredoxin-dependent type of glutamate synthase (Fd-GOGAT) is found in chloroplasts and serves in photorespiratory nitrogen metabolism. Both the amount of protein and its activity increase with light levels. Roots, particularly those under nitrate nutrition, have Fd-GOGAT in plastids. Fd-GOGAT in the roots presumably functions to incorporate the glutamine generated during nitrate assimilation.

## Ammonium can be assimilated via an alternative pathway

**Glutamate dehydrogenase (GDH)** catalyzes a reversible reaction that synthesizes or deaminates glutamate (**FIGURE 12.7B**):

$$\text{2-Oxoglutarate} + \text{NH}_4^+ + \text{NAD(P)H} \leftrightarrow \\ \text{glutamate} + \text{H}_2\text{O} + \text{NAD(P)}^+ \quad (12.7)$$

An NADH-dependent form of GDH is found in mitochondria, and an NADPH-dependent form is localized in the chloroplasts of photosynthetic organs. Although both forms are relatively abundant, they cannot substitute for the GS–GOGAT pathway for assimilation of ammonium, and their primary function is in deaminating glutamate during the reallocation of nitrogen (see Figure 12.7B).

## Transamination reactions transfer nitrogen

Once assimilated into glutamine and glutamate, nitrogen is incorporated into other amino acids via transamination reactions. The enzymes that catalyze these reactions are known as aminotransferases. An example is **aspartate aminotransferase (Asp-AT)**, which catalyzes the following reaction (**FIGURE 12.7C**):

$$\text{Glutamate} + \text{oxaloacetate} \rightarrow \text{aspartate} + \\ \text{2-oxoglutarate} \quad (12.8)$$

in which the amino group of glutamate is transferred to the carboxyl group of aspartate. Aspartate is an amino acid that participates in the malate–aspartate shuttle to transfer reducing equivalents from the mitochondrion and chloroplast into the cytosol (see **WEB TOPIC 11.5**) and in the transport of carbon from the mesophyll to the bundle sheath for $\text{C}_4$ carbon fixation (see Chapter 8). All transamination reactions require pyridoxal phosphate (vitamin $\text{B}_6$) as a cofactor.

Aminotransferases are found in the cytoplasm, chloroplasts, mitochondria, glyoxysomes, and peroxisomes. The aminotransferases in chloroplasts may have a significant role in amino acid biosynthesis, because plant leaves or isolated chloroplasts exposed to radioactively labeled carbon dioxide rapidly incorporate the label into glutamate, aspartate, alanine, serine, and glycine.

## Asparagine and glutamine link carbon and nitrogen metabolism

Asparagine, isolated from asparagus as early as 1806, was the first amide to be identified (Lam et al. 1996). It serves not only as a component of proteins, but as a key compound for nitrogen transport and storage because of its stability and high nitrogen-to-carbon ratio (2 N to 4 C for asparagine in comparison to 2 N to 5 C for glutamine or 1 N to 5 C for glutamate).

The major pathway for asparagine synthesis involves the transfer of the amide nitrogen from glutamine to asparagine (**FIGURE 12.7D**):

$$\text{Glutamine} + \text{aspartate} + \text{ATP} \rightarrow \text{asparagine} \\ + \text{glutamate} + \text{AMP} + \text{PP}_i \quad (12.9)$$

**Asparagine synthetase (AS)**, the enzyme that catalyzes this reaction, is found in the cytosol of leaves and roots and in nitrogen-fixing nodules (see the next section). In maize roots, particularly those under potentially toxic levels of ammonia, ammonium may replace glutamine as the source of the amide group (Sivasankar and Oaks 1996).

High levels of light and carbohydrate—conditions that stimulate plastid GS and Fd-GOGAT—inhibit the expression of genes coding for AS and the activity of the enzyme. The opposing regulation of these competing pathways helps balance the metabolism of carbon and nitrogen in plants (Lam et al. 1996). Conditions of ample energy (i.e., high levels of light and carbohydrates) stimulate GS (see Equation 12.4) and GOGAT (see Equations 12.5 and 12.6), and inhibit AS; thus they favor nitrogen assimilation into glutamine and glutamate, compounds that are rich in carbon and participate in the synthesis of new plant materials.

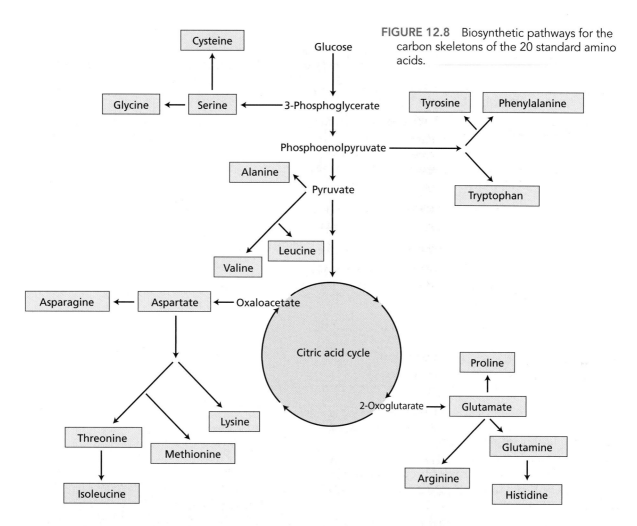

**FIGURE 12.8** Biosynthetic pathways for the carbon skeletons of the 20 standard amino acids.

In contrast, energy-limited conditions inhibit GS and GOGAT, stimulate AS, and thus favor nitrogen assimilation into asparagine, a compound that is rich in nitrogen and sufficiently stable for long-distance transport or long-term storage.

## Amino Acid Biosynthesis

Humans and most animals cannot synthesize certain amino acids—histidine, isoleucine, leucine, lysine, methionine, phenylalanine, threonine, tryptophan, and valine, and, in the case of young humans, arginine (human adults can synthesize arginine)—and thus must obtain these so-called essential amino acids from their diet. In contrast, plants synthesize all the 20 or so amino acids that are common in proteins. The nitrogen-containing amino group, as discussed in the previous section, derives from transamination reactions with glutamine or glutamate. The carbon skeleton for amino acids derives from 3-phosphoglycerate, phosphoenolpyruvate, or pyruvate generated during glycolysis, or from 2-oxoglutarate or oxaloacetate generated in the citric acid cycle (**FIGURE 12.8**). Parts of these pathways required for synthesis of the essential amino are appropri-

ate targets for herbicides (such as "Roundup," see Chapter 13), because they are missing from animals, so substances that block these pathways are lethal to plants but in low concentrations do not injure animals.

## Biological Nitrogen Fixation

Biological nitrogen fixation accounts for most of the conversion of atmospheric $N_2$ into ammonium, and thus serves as the key entry point of molecular nitrogen into the biogeochemical cycle of nitrogen (see Figure 12.1). In this section we will describe the symbiotic relations between nitrogen-fixing organisms and higher plants; nodules, the specialized structures that form in roots when infected by nitrogen-fixing bacteria; the genetic and signaling interactions that regulate nitrogen fixation by symbiotic prokaryotes and their hosts; and the properties of the nitrogenase enzymes that fix nitrogen.

### Free-living and symbiotic bacteria fix nitrogen

Some bacteria, as stated earlier, can convert atmospheric nitrogen into ammonium (**TABLE 12.2**). Most of these

**TABLE 12.2**
**Examples of organisms that can carry out nitrogen fixation**

| SYMBIOTIC NITROGEN FIXATION | |
| --- | --- |
| **Host plant** | **N-fixing symbionts** |
| Leguminous: legumes, *Parasponia* | *Azorhizobium, Bradyrhizobium, Photorhizobium, Rhizobium, Sinorhizobium* |
| Actinorhizal: alder (tree), *Ceanothus* (shrub), *Casuarina* (tree), *Datisca* (shrub) | *Frankia* |
| *Gunnera* | *Nostoc* |
| *Azolla* (water fern) | *Anabaena* |
| Sugarcane | *Acetobacter* |
| *Miscanthus* | *Azospirillum* |

| FREE-LIVING NITROGEN FIXATION | |
| --- | --- |
| **Type** | **N-fixing genera** |
| Cyanobacteria (blue-green algae) | *Anabaena, Calothrix, Nostoc* |
| Other bacteria | |
| Aerobic | *Azospirillum, Azotobacter, Beijerinckia, Derxia* |
| Facultative | *Bacillus, Klebsiella* |
| Anaerobic | |
| Nonphotosynthetic | *Clostridium, Methanococcus* (archaebacterium) |
| Photosynthetic | *Chromatium, Rhodospirillum* |

nitrogen-fixing prokaryotes live in the soil, generally independent of other organisms. A few form symbiotic associations with higher plants in which the prokaryote directly provides the host plant with fixed nitrogen in exchange for other nutrients and carbohydrates (Franche et al. 2009) (top portion of Table 12.2). Such symbioses occur in nodules that form on the roots of the plant and contain the nitrogen-fixing bacteria.

The most common type of symbiosis occurs between members of the plant family Fabaceae (Leguminosae) and soil bacteria of the genera *Azorhizobium, Bradyrhizobium, Photorhizobium, Rhizobium,* and *Sinorhizobium* (collectively called **rhizobia**; TABLE 12.3 and FIGURE 12.9). Another common type of symbiosis occurs between several woody plant species, such as alder trees, and soil bacteria of the genus *Frankia*; these plants are known as **actinorhizal** plants. Still other types of nitrogen-fixing symbioses involve the South American herb *Gunnera* and the tiny water fern *Azolla*, which form associations with the cyanobacteria *Nostoc* and *Anabaena*, respectively (see Table 12.2 and FIGURE 12.10). Finally, several types of nitrogen-fixing bacteria are associated with $C_4$ grasses such as sugarcane and *Miscanthus*.

## Nitrogen fixation requires anaerobic conditions

Because nitrogen fixation involves the expenditure of large amounts of energy, the **nitrogenase** enzymes that catalyze these reactions have sites that facilitate the high-energy exchange of electrons. Oxygen, being a strong electron acceptor, can damage these sites and irreversibly inactivate

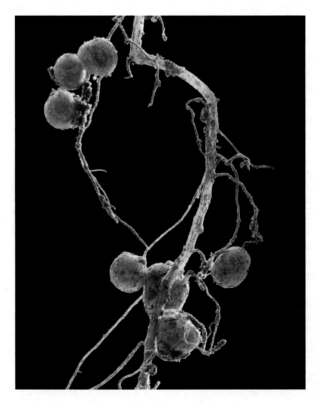

**FIGURE 12.9** Root nodules on a common bean (*Phaseolus vulgaris*). The nodules are a result of infection by *Rhizobium* sp. (Photo by David McIntyre.)

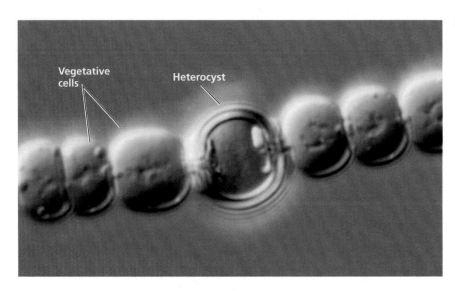

**FIGURE 12.10** A heterocyst in a filament of the nitrogen-fixing cyanobacterium *Anabaena*. The thick-walled heterocysts, interspersed among vegetative cells, have an anaerobic inner environment that allows cyanobacteria to fix nitrogen in aerobic conditions. (© Dr. Peter Siver/Visuals Unlimited/Alamy.)

Vegetative cells

Heterocyst

nitrogenase, so nitrogen must be fixed under anaerobic conditions. Each of the nitrogen-fixing organisms listed in Table 12.2 either functions under natural anaerobic conditions or creates an internal, local anaerobic environment in the presence of oxygen.

In cyanobacteria, anaerobic conditions are created in specialized cells called *heterocysts* (see Figure 12.10). Heterocysts are thick-walled cells that differentiate when filamentous cyanobacteria are deprived of $NH_4^+$. These cells lack photosystem II, the oxygen-producing photosystem of chloroplasts (see Chapter 7), so they do not generate oxygen (Burris 1976). Heterocysts appear to represent an adaptation for nitrogen fixation, in that they are widespread among aerobic cyanobacteria that fix nitrogen.

Cyanobacteria can fix nitrogen under anaerobic conditions such as those that occur in flooded fields. In Asian countries, nitrogen-fixing cyanobacteria of both the heterocyst and nonheterocyst types are a major means for maintaining an adequate nitrogen supply in the soil of rice fields. These microorganisms fix nitrogen when the fields are flooded and die as the fields dry, releasing the fixed nitrogen to the soil. Another important source of available nitrogen in flooded rice fields is the water fern *Azolla*, which associates with the cyanobacterium *Anabaena*. The *Azolla–Anabaena* association can fix as much as 0.5 kg of atmospheric nitrogen per hectare per day, a rate of fertilization that is sufficient to attain moderate rice yields.

Free-living bacteria that are capable of fixing nitrogen are aerobic, facultative, or anaerobic (see Table 12.2, bottom):

- *Aerobic* nitrogen-fixing bacteria such as *Azotobacter* are thought to maintain a low oxygen concentration (microaerobic conditions) through their high levels of respiration (Burris 1976). Others, such as

**TABLE 12.3**
**Associations between host plants and rhizobia**

| Plant host | Rhizobial symbiont |
| --- | --- |
| *Parasponia* (a nonlegume, formerly called *Trema*) | *Bradyrhizobium* spp. |
| Soybean (*Glycine max*) | *Bradyrhizobium japonicum* (slow-growing type); *Sinorhizobium fredii* (fast-growing type) |
| Alfalfa (*Medicago sativa*) | *Sinorhizobium meliloti* |
| *Sesbania* (aquatic) | *Azorhizobium* (forms both root and stem nodules; the stems have adventitious roots) |
| Bean (*Phaseolus*) | *Rhizobium leguminosarum* bv. *phaseoli*; *Rhizobium tropicii*; *Rhizobium etli* |
| Clover (*Trifolium*) | *Rhizobium leguminosarum* bv. *trifolii* |
| Pea (*Pisum sativum*) | *Rhizobium leguminosarum* bv. *viciae* |
| *Aeschenomene* (aquatic) | *Photorhizobium* (photosynthetically active rhizobia that form stem nodules, probably associated with adventitious roots) |

*Gloeothece*, evolve $O_2$ photosynthetically during the day and fix nitrogen during the night when respiration lowers oxygen levels.

- *Facultative* organisms, which are able to grow under both aerobic and anaerobic conditions, generally fix nitrogen only under anaerobic conditions.

- For *anaerobic* nitrogen-fixing bacteria, oxygen does not pose a problem, because it is absent in their habitat. These anaerobic organisms can be either photosynthetic (e.g., *Rhodospirillum*), or nonphotosynthetic (e.g., *Clostridium*).

### Symbiotic nitrogen fixation occurs in specialized structures

Symbiotic nitrogen-fixing prokaryotes dwell within **nodules**, the special organs of the plant host that enclose the nitrogen-fixing bacteria (see Figure 12.9). In the case of *Gunnera*, these organs are existing stem glands that develop independently of the symbiont. In the case of legumes and actinorhizal plants, the nitrogen-fixing bacteria induce the plant to form root nodules.

Grasses can also develop symbiotic relationships with nitrogen-fixing organisms, but in these associations root nodules are not produced. Instead, the nitrogen-fixing bacteria seem to colonize plant tissues or anchor to the root surfaces, mainly around the elongation zone and the root hairs (Reis et al. 2000). For example, the nitrogen-fixing bacterium *Acetobacter diazotrophicus* lives in the apoplast of stem tissues in sugarcane and may provide its host with sufficient nitrogen to lessen the need for fertilization (Dong et al. 1994; James and Olivares 1998). The potential for applying *Azospirillum* to corn and other grains has been explored, but *Azospirillum* seems to fix little nitrogen when associated with plants (Vande Broek and Vanderleyden 1995).

Legumes and actinorhizal plants regulate gas permeability in their nodules, maintaining a level of oxygen within the nodule that can support respiration but is sufficiently low to avoid inactivation of the nitrogenase (Kuzma et al. 1993). Gas permeability increases in the light and decreases under drought or upon exposure to nitrate. The mechanism for regulating gas permeability is not yet known, but may involve potassium fluxes into and out of infected cells (Wei and Layzell 2006).

Nodules contain an oxygen-binding heme protein called **leghemoglobin**. Leghemoglobin is present in the cytoplasm of infected nodule cells at high concentrations (700 $\mu M$ in soybean nodules) and gives the nodules a pink color. The host plant produces the globin portion of leghemoglobin in response to infection by the bacteria (Marschner 1995); the bacterial symbiont produces the heme portion. Leghemoglobin has a high affinity for oxygen (a $K_m$ of about 0.01 $\mu M$), about ten times higher than the β chain of human hemoglobin.

Although leghemoglobin was once thought to provide a buffer for nodule oxygen, more recent studies indicate that it stores only enough oxygen to support nodule respiration for a few seconds (Denison and Harter 1995). Its function is to help transport oxygen to the respiring symbiotic bacterial cells, analogous to the way hemoglobin transports oxygen to respiring tissues in animals (Ludwig and de Vries 1986). To continue aerobic respiration under such conditions, the bacteroid uses a specialized electron transport chain (see Chapter 11) in which the terminal oxidase has an affinity for oxygen even higher than that of leghemoglobin, a $K_m$ of about 0.007 $\mu M$ (Preisig et al. 1996).

### Establishing symbiosis requires an exchange of signals

The symbiosis between legumes and rhizobia is not obligatory. Legume seedlings germinate without any association with rhizobia, and they may remain unassociated throughout their life cycle. Rhizobia also occur as free-living organisms in the soil. Under nitrogen-limited conditions, however, the symbionts seek each other out through an elaborate exchange of signals. This signaling, the subsequent infection process, and the development of nitrogen-fixing nodules involve specific genes in both the host and the symbionts (Oldroyd and Downie 2008).

Plant genes specific to nodules are called **nodulin** (*Nod*) genes; rhizobial genes that participate in nodule formation are called **nodulation** (*nod*) genes (Heidstra and Bisseling 1996). The *nod* genes are classified as common *nod* genes or host-specific *nod* genes. The common *nod* genes—*nodA*, *nodB*, and *nodC*—are found in all rhizobial strains; the host-specific *nod* genes—such as *nodP*, *nodQ*, and *nodH*; or *nodF*, *nodE*, and *nodL*—differ among rhizobial species and determine the host range (the plants that can be infected). Only one of the *nod* genes, the regulatory *nodD*, is constitutively expressed, and as we will explain in detail, its protein product (NodD) regulates the transcription of the other *nod* genes.

The first stage in the formation of the symbiotic relationship between the nitrogen-fixing bacteria and their host is migration of the bacteria toward the roots of the host plant. This migration is a chemotactic response mediated by chemical attractants, especially (iso)flavonoids and betaines, secreted by the roots. These attractants activate the rhizobial NodD protein, which then induces transcription of the other *nod* genes (Phillips and Kapulnik 1995). The promoter region of all *nod* operons, except that of *nodD*, contains a highly conserved sequence called the *nod* box. Binding of the activated NodD to the *nod* box induces transcription of the other *nod* genes.

### Nod factors produced by bacteria act as signals for symbiosis

The *nod* genes, which NodD activates, code for nodulation proteins, most of which are involved in the biosynthesis

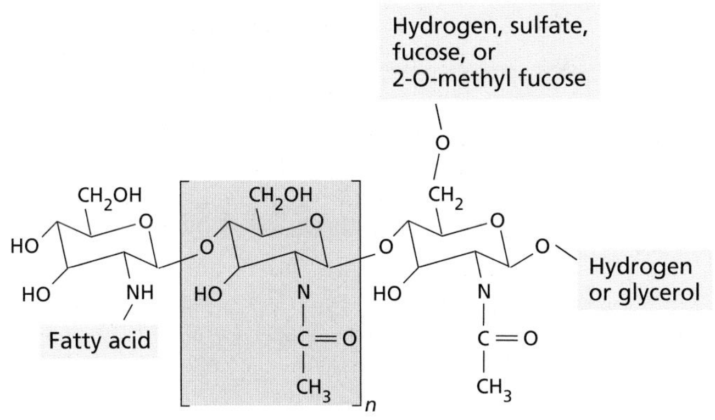

**FIGURE 12.11** Nod factors are lipochitin oligosaccharides. The fatty acid chain typically has 16 to 18 carbons. The number of repeated middle sections ($n$) is usually 2 or 3. (After Stokkermans et al. 1995.)

of Nod factors. **Nod factors** are lipochitin oligosaccharide signal molecules, all of which have a chitin β-1→4-linked $N$-acetyl-D-glucosamine backbone (varying in length from three to six sugar units) and a fatty acid chain on the C-2 position of the nonreducing sugar (**FIGURE 12.11**).

Three of the *nod* genes (*nodA*, *nodB*, and *nodC*) encode enzymes (NodA, NodB, and NodC, respectively) that are required for synthesizing this basic structure (Stokkermans et al. 1995):

1. NodA is an $N$-acyltransferase that catalyzes the addition of a fatty acyl chain.

2. NodB is a chitin-oligosaccharide deacetylase that removes the acetyl group from the terminal nonreducing sugar.

3. NodC is a chitin-oligosaccharide synthase that links $N$-acetyl-D-glucosamine monomers.

Host-specific *nod* genes that vary among rhizobial species are involved in the modification of the fatty acyl chain or the addition of groups important in determining host specificity (Carlson et al. 1995):

- NodE and NodF determine the length and degree of saturation of the fatty acyl chain; those of *Rhizobium leguminosarum* bv. *viciae* and *R. meliloti* result in the synthesis of an 18:4 and a 16:2 fatty acyl group, respectively. (Recall from Chapter 11 that the number before the colon gives the total number of carbons in the fatty acyl chain, and the number after the colon gives the number of double bonds.)

- Other enzymes, such as NodL, influence the host specificity of Nod factors through the addition of specific substitutions at the reducing or nonreducing sugar moieties of the chitin backbone.

A particular legume host responds to a specific Nod factor. The legume receptors for Nod factors appear to involve sugar-binding LysM domains (for lysine motif, a widespread protein module originally identified in enzymes that degrade bacterial cell walls but also present in many other proteins) in the root hairs (Radutoiu et al. 2007). Nod factors activate these domains, inducing calcium oscillations in root epidermal cells. Recognition of the calcium oscillations requires a calcium/calmodulin-dependent protein kinase (CaMK) that is associated with a protein of unknown function named CYCLOPS (Yano et al. 2008). Once the plant epidermal cell recognizes ongoing calcium oscillations, Nod factor–specific transcriptional regulators, including a complex of two GRAS proteins and ERF transcription factors, directly associate with the promoters of Nod factor–inducible genes (Hirsch et al. 2009). The overall process links Nod factor perception at the plasma membrane to gene expression changes in the nucleus and is called the symbiotic pathway because it has many similarities with the process through which arbuscular mycorrhizae initially interact with their hosts (see Chapter 5 and Oldroyd et al. 2009).

## Nodule formation involves phytohormones

Two processes—infection and nodule organogenesis—occur simultaneously during root nodule formation. During the infection process, rhizobia attached to the root hairs release Nod factors that induce a pronounced curling of the root hair cells (**FIGURE 12.12A AND B**). The rhizobia become enclosed in the small compartment formed by the curling. The cell wall of the root hair degrades in these regions, also in response to Nod factors, allowing the bacterial cells direct access to the outer surface of the plant plasma membrane (Geurts and Bissling 2002).

The next step is formation of the **infection thread** (**FIGURE 12.12C**), an internal tubular extension of the plasma membrane that is produced by the fusion of Golgi-derived membrane vesicles at the site of infection. The thread grows at its tip by the fusion of secretory vesicles to the end of the tube. Deeper into the root cortex, near the xylem, cortical cells dedifferentiate and start dividing, forming a distinct area within the cortex, called a *nodule primordium*, from which the nodule will develop. The nodule primordia form opposite the protoxylem poles of the root vascular bundle (Timmers et al. 1999) (see **WEB TOPIC 12.1**).

Different signaling compounds, acting either positively or negatively, control the position of nodule primordia. The nucleoside uridine diffuses from the stele into the cortex in the protoxylem zones of the root and stimulates cell division (Geurts and Bisseling 2002). Ethylene is synthesized in the region of the pericycle, diffuses into the cortex, and blocks cell division opposite the phloem poles of the root.

The infection thread filled with proliferating rhizobia elongates through the root hair and cortical cell layers, in the direction of the nodule primordium. When the infec-

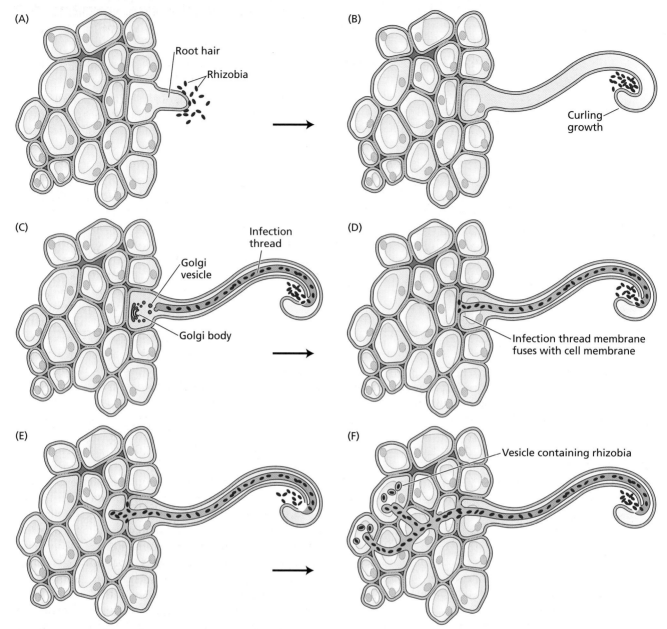

**FIGURE 12.12** The infection process during nodule organogenesis. (A) Rhizobia bind to an emerging root hair in response to chemical attractants sent by the plant. (B) In response to factors produced by the bacteria, the root hair exhibits abnormal curling growth, and rhizobia cells proliferate within the coils. (C) Localized degradation of the root hair wall leads to infection and formation of the infection thread from Golgi secretory vesicles of root cells. (D) The infection thread reaches the end of the cell, and its membrane fuses with the plasma membrane of the root hair cell. (E) Rhizobia are released into the apoplast and penetrate the compound middle lamella to the subepidermal cell plasma membrane, leading to the initiation of a new infection thread, which forms an open channel with the first. (F) The infection thread extends and branches until it reaches target cells, where vesicles composed of plant membrane that enclose bacterial cells are released into the cytosol.

tion thread reaches specialized cells within the nodule, its tip fuses with the plasma membrane of the host cell, releasing bacterial cells that are packaged in a membrane derived from the host cell plasma membrane (**FIGURE 12.12D**). Branching of the infection thread inside the nodule enables the bacteria to infect many cells (**FIGURE 12.12E AND F**) (Mylona et al. 1995).

At first the bacteria continue to divide, and the surrounding membrane increases in surface area to accom-modate this growth by fusing with smaller vesicles. Soon thereafter, upon an undetermined signal from the plant, the bacteria stop dividing and begin to enlarge and to differentiate into nitrogen-fixing endosymbiotic organelles called **bacteroids**. The membrane surrounding the bacteroids is called the *peribacteroid membrane*.

The nodule as a whole develops such features as a vascular system (which facilitates the exchange of fixed nitrogen produced by the bacteroids for nutrients contributed

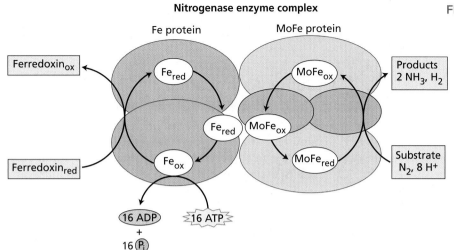

**Nitrogenase enzyme complex**

FIGURE 12.13 The reaction catalyzed by nitrogenase. Ferredoxin reduces the Fe protein. Binding and hydrolysis of ATP to the Fe protein is thought to cause a conformational change of the Fe protein that facilitates the redox reactions. The Fe protein reduces the MoFe protein, and the MoFe protein reduces the $N_2$. (After Dixon and Wheeler 1986; Buchanan et al. 2000.)

by the plant) and a layer of cells to exclude $O_2$ from the root nodule interior. In some temperate legumes (e.g., peas), the nodules are elongated and cylindrical because of the presence of a *nodule meristem*. The nodules of tropical legumes, such as soybeans and peanuts, lack a persistent meristem and are spherical (Rolfe and Gresshoff 1988).

### The nitrogenase enzyme complex fixes $N_2$

Biological nitrogen fixation, like industrial nitrogen fixation, produces ammonia from molecular nitrogen. The overall reaction is

$$N_2 + 8\ e^- + 8\ H^+ + 16\ ATP \rightarrow 2\ NH_3$$
$$+ H_2 + 16\ ADP + 16\ P_i \qquad (12.10)$$

Note that the reduction of $N_2$ to 2 $NH_3$, a six-electron transfer, is coupled to the reduction of two protons to evolve $H_2$. The **nitrogenase enzyme complex** catalyzes this reaction (Dixon and Kahn 2004; Seefeldt et al. 2009).

The nitrogenase enzyme complex can be separated into two components—the Fe protein and the MoFe protein—neither of which has catalytic activity by itself (FIGURE 12.13):

- The Fe protein is the smaller of the two components and has two identical subunits of 30 to 72 kDa each, depending on the organism. Each subunit contains an iron–sulfur cluster (4 Fe and 4 $S^{2-}$) that participates in the redox reactions involved in the conversion of $N_2$ to $NH_3$. The Fe protein is irreversibly inactivated by $O_2$ with typical half-decay times of 30 to 45 seconds (Dixon and Wheeler 1986).

- The MoFe protein has four subunits, with a total molecular mass of 180 to 235 kDa, depending on the species. Each subunit has two Mo–Fe–S clusters. The MoFe protein is also inactivated by oxygen, with a half-decay time in air of 10 minutes.

In the overall nitrogen reduction reaction (see Figure 12.13), ferredoxin serves as an electron donor to the Fe protein, which in turn hydrolyzes ATP and reduces the MoFe protein. The MoFe protein can then reduce numerous substrates (TABLE 12.4), although under natural conditions it reacts only with $N_2$ and $H^+$. One of the reactions catalyzed by nitrogenase, the reduction of acetylene to ethylene, is used in estimating nitrogenase activity (see WEB TOPIC 12.2).

The energetics of nitrogen fixation is complex. The production of $NH_3$ from $N_2$ and $H_2$ is an exergonic reaction (see Appendix 1 for a discussion of exergonic reactions), with a $\Delta G^{0'}$ (change in free energy) of –27 kJ mol$^{-1}$. However, industrial production of $NH_3$ from $N_2$ and $H_2$ is *endergonic*, requiring a very large energy input because of the activation energy needed to break the triple bond in $N_2$. For the same reason, the enzymatic reduction of $N_2$ by nitrogenase also requires a large investment of energy (see Equation 12.10), although the exact changes in free energy are not yet known.

Calculations based on the carbohydrate metabolism of legumes show that a plant consumes 12 g of organic carbon per gram of $N_2$ fixed (Heytler et al. 1984). On the basis of Equation 12.10, the $\Delta G^{0'}$ for the overall reaction of

**TABLE 12.4**
**Reactions catalyzed by nitrogenase**

| | |
|---|---|
| $N_2 \rightarrow NH_3$ | Molecular nitrogen fixation |
| $N_2O \rightarrow N_2 + H_2O$ | Nitrous oxide reduction |
| $N_3^- \rightarrow N_2 + NH_3$ | Azide reduction |
| $C_2H_2 \rightarrow C_2H_4$ | Acetylene reduction |
| $2\ H^+ \rightarrow H_2$ | $H_2$ production |
| $ATP \rightarrow ADP + P_i$ | ATP hydrolytic activity |

*Source:* After Burris 1976.

biological nitrogen fixation is about –200 kJ mol$^{-1}$. Because the overall reaction is highly exergonic, ammonium production is limited by the slow operation (number of N$_2$ molecules reduced per unit time is about 5 s$^{-1}$) of the nitrogenase complex (Ludwig and de Vries 1986). To compensate for this slow turnover time, the bacteroid synthesizes large amounts of nitrogenase (up to 20% of the total protein in the cell).

Under natural conditions, substantial amounts of H$^+$ are reduced to H$_2$ gas, and this process can compete with N$_2$ reduction for electrons from nitrogenase. In rhizobia, 30 to 60% of the energy supplied to nitrogenase may be lost as H$_2$, diminishing the efficiency of nitrogen fixation. Some rhizobia, however, contain hydrogenase, an enzyme that can split the H$_2$ formed and generate electrons for N$_2$ reduction, thus improving the efficiency of nitrogen fixation (Marschner 1995).

### Amides and ureides are the transported forms of nitrogen

The symbiotic nitrogen-fixing prokaryotes release ammonia that, to avoid toxicity, must be rapidly converted into organic forms in the root nodules before being transported to the shoot via the xylem. Nitrogen-fixing legumes can be classified as amide exporters or ureide exporters, depending on the composition of the xylem sap. Amides (principally the amino acids asparagine or glutamine) are exported by temperate-region legumes, such as pea (*Pisum*), clover (*Trifolium*), broad bean (*Vicia*), and lentil (*Lens*).

Ureides are exported by legumes of tropical origin, such as soybean (*Glycine*), common bean (*Phaseolus*), peanut (*Arachis*), and southern pea (*Vigna*). The three major ureides are allantoin, allantoic acid, and citrulline (**FIGURE 12.14**). Allantoin is synthesized in peroxisomes from uric acid, and allantoic acid is synthesized from allantoin in the endoplasmic reticulum. The site of citrulline synthesis from the amino acid ornithine has not yet been determined. All three compounds are ultimately released into

the xylem and transported to the shoot, where they are rapidly catabolized to ammonium. This ammonium enters the assimilation pathway described earlier.

## Sulfur Assimilation

Sulfur is among the most versatile elements in living organisms (Hell 1997). Disulfide bridges in proteins play structural and regulatory roles (see Chapter 8). Sulfur participates in electron transport through iron–sulfur clusters (see Chapters 7 and 11). The catalytic sites for several enzymes and coenzymes, such as urease and coenzyme A, contain sulfur. Secondary metabolites (compounds that are not involved in primary pathways of growth and development) that contain sulfur range from the rhizobial Nod factors discussed in the previous section to the antiseptic alliin in garlic and the anticarcinogen sulforaphane in broccoli.

The versatility of sulfur derives in part from the property that it shares with nitrogen: *multiple stable oxidation states*. In this section, we discuss the enzymatic steps that mediate sulfur assimilation, and the biochemical reactions that catalyze the reduction of sulfate into the two sulfur-containing amino acids, cysteine and methionine.

### Sulfate is the absorbed form of sulfur in plants

Most of the sulfur in higher-plant cells derives from sulfate (SO$_4$$^{2-}$) absorbed via an H$^+$–SO$_4$$^{2-}$ symporter (see Chapter 6) from the soil solution. Sulfate in the soil comes predominantly from the weathering of parent rock material. Industrialization, however, adds an additional source of sulfate: atmospheric pollution. The burning of fossil fuels releases several gaseous forms of sulfur, including sulfur dioxide (SO$_2$) and hydrogen sulfide (H$_2$S), which find their way to the soil in rain.

In the gas phase, sulfur dioxide reacts with a hydroxyl radical and oxygen to form sulfur trioxide (SO$_3$). SO$_3$ dissolves in water to become sulfuric acid (H$_2$SO$_4$), a strong acid, which is the major source of acid rain. Plants can

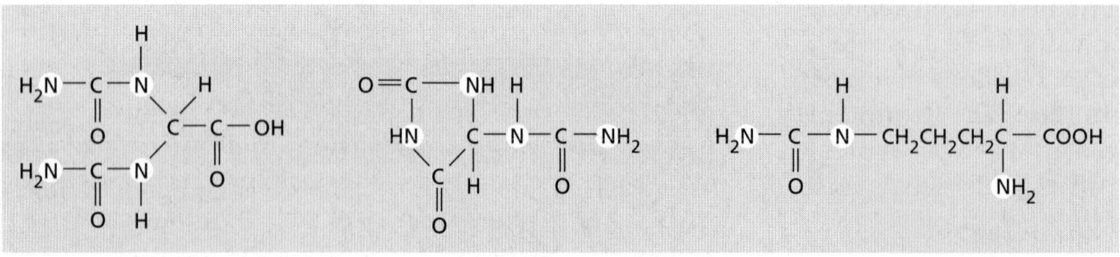

**Allantoic acid**          **Allantoin**          **Citrulline**

**FIGURE 12.14** The major ureide compounds that are used to transport nitrogen from sites of fixation to sites where their deamination will provide nitrogen for amino acid and nucleoside synthesis.

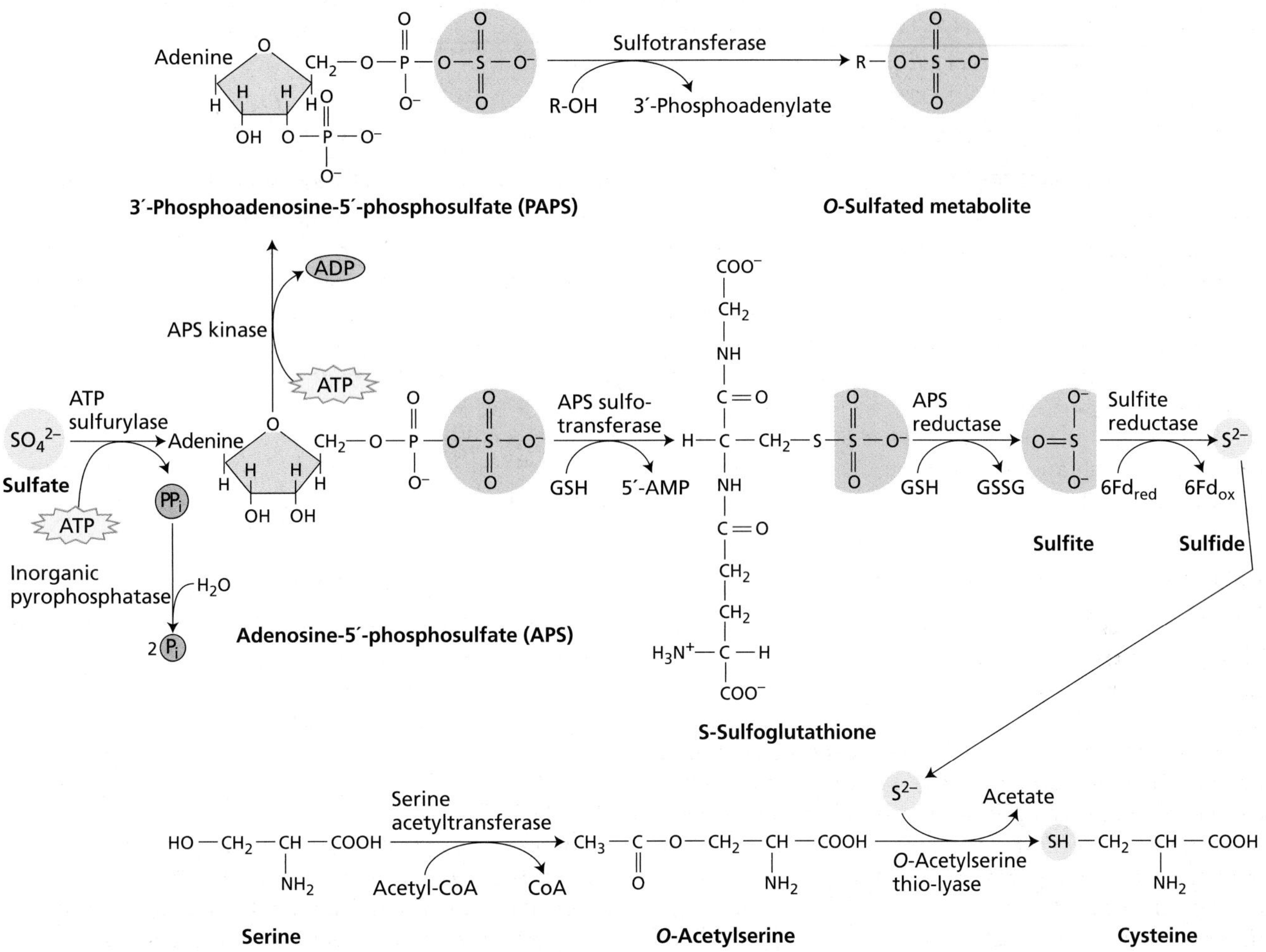

**FIGURE 12.15** Structure and pathways of compounds involved in sulfur assimilation. The enzyme ATP sulfurylase cleaves pyrophosphate from ATP and replaces it with sulfate. Sulfide is produced from APS through reactions involving reduction by glutathione and ferredoxin. The sulfide reacts with O-acetylserine to form cysteine. Fd, ferredoxin; GSH, glutathione, reduced; GSSG, glutathione, oxidized.

metabolize sulfur dioxide taken up in the gaseous form through their stomata. Nonetheless, prolonged exposure (more than 8 hours) to high atmospheric concentrations (greater than 0.3 ppm) of $SO_2$ causes extensive tissue damage because of the formation of sulfuric acid.

## Sulfate assimilation requires the reduction of sulfate to cysteine

The first step in the synthesis of sulfur-containing organic compounds is the reduction of sulfate to the amino acid cysteine (**FIGURE 12.15**). Sulfate is very stable and thus needs to be activated before any subsequent reactions may

proceed. Activation begins with the reaction between sulfate and ATP to form 5′-adenylylsulfate (which is sometimes referred to as adenosine-5′-phosphosulfate and thus is abbreviated APS) and pyrophosphate ($PP_i$) (see Figure 12.15):

$$SO_4^{2-} + \text{Mg-ATP} \rightarrow \text{APS} + PP_i \qquad (12.11)$$

The enzyme that catalyzes this reaction, ATP sulfurylase, has two forms: The major one is found in plastids, and a minor one is found in the cytoplasm (Leustek et al. 2000). The activation reaction is energetically unfavorable. To drive this reaction forward, the products APS and $PP_i$ must be converted immediately to other compounds. $PP_i$ is hydrolyzed to inorganic phosphate ($P_i$) by inorganic pyrophosphatase according to the following reaction:

$$PP_i + H_2O \rightarrow 2\,P_i \qquad (12.12)$$

The other product, APS, is rapidly reduced or phosphorylated. Reduction is the dominant pathway (Leustek et al. 2000).

The reduction of APS is a multistep process that occurs exclusively in the plastids. First, APS reductase transfers two electrons, apparently from reduced glutathione (GSH), to produce sulfite ($SO_3^{2-}$):

$$APS + 2\ GSH \rightarrow SO_3^{2-} + 2\ H^+$$
$$+ GSSG + AMP \qquad (12.13)$$

where GSSG stands for oxidized glutathione. (The *SH* in GSH and the *SS* in GSSG stand for S—H and S—S bonds, respectively.)

Second, sulfite reductase transfers six electrons from ferredoxin ($Fd_{red}$) to produce sulfide ($S^{2-}$):

$$SO_3^{2-} + 6\ Fd_{red} \rightarrow S^{2-} + 6\ Fd_{ox} \qquad (12.14)$$

The resultant sulfide then reacts with *O*-acetylserine (OAS) to form cysteine and acetate. The *O*-acetylserine that reacts with $S^{2-}$ is formed in a reaction catalyzed by serine acetyltransferase:

$$Serine + acetyl\text{-}CoA \rightarrow OAS + CoA \qquad (12.15)$$

The reaction that produces cysteine and acetate is catalyzed by OAS (thiol)lyase:

$$OAS + S^{2-} \rightarrow cysteine + acetate \qquad (12.16)$$

The phosphorylation of APS, localized in the cytosol, is the alternative pathway. First, APS kinase catalyzes a reaction of APS with ATP to form 3′-phosphoadenosine-5′-phosphosulfate (PAPS):

$$APS + ATP \rightarrow PAPS + ADP \qquad (12.17)$$

Sulfotransferases then may transfer the sulfate group from PAPS to various compounds, including choline, brassinosteroids, flavonol, gallic acid glucoside, glucosinolates, peptides, and polysaccharides (Leustek and Saito 1999).

### Sulfate assimilation occurs mostly in leaves

The reduction of sulfate to cysteine changes the oxidation number of sulfur from +6 to –2, thus entailing the transfer of 8 electrons. Glutathione, ferredoxin, NAD(P)H, or *O*-acetylserine may serve as electron donors at various steps of the pathway (see Figure 12.15). In *Arabidopsis thaliana*, all the enzymes of sulfate assimilation—with the exception of sulfite reductase and reduced glutathione synthesis—are encoded by small gene families. It is not yet clear whether this is a functional redundancy or if all genes have a specific function or location (Kopriva 2006).

Leaves are generally much more active than roots in sulfur assimilation, presumably because photosynthesis provides reduced ferredoxin, and photorespiration generates serine, that may stimulate the production of *O*-acetylserine (see Chapter 8). Sulfur assimilated in leaves is exported via the phloem to sites of protein synthesis (shoot and root apices, and fruits) mainly as glutathione (Bergmann and Rennenberg 1993):

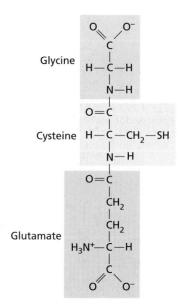

**Reduced glutathione**

Glutathione also acts as a signal that coordinates the absorption of sulfate by the roots and the assimilation of sulfate by the shoot.

### Methionine is synthesized from cysteine

Methionine, the other sulfur-containing amino acid found in proteins, is synthesized in plastids from cysteine (see **WEB TOPIC 12.3** for further detail). After cysteine and methionine are synthesized, sulfur can be incorporated into proteins and a variety of other compounds, such as acetyl-CoA and *S*-adenosylmethionine. The latter compound is important in the synthesis of ethylene (see Chapter 22) and in reactions involving the transfer of methyl groups, as in lignin synthesis (see Chapter 13).

## Phosphate Assimilation

Phosphate ($HPO_4^{2-}$) in the soil solution is readily absorbed by plant roots via an $H^+$–$HPO_4^{2-}$ symporter (see Chapter 6) and incorporated into a variety of organic compounds, including sugar phosphates, phospholipids, and nucleotides. The main entry point of phosphate into assimilatory pathways occurs during the formation of ATP, the energy "currency" of the cell. In the overall reaction for this process, inorganic phosphate is added to the second phosphate group in adenosine diphosphate to form a phosphate ester bond.

In mitochondria, the energy for ATP synthesis derives from the oxidation of NADH by oxidative phosphorylation (see Chapter 11). ATP synthesis is also driven by light-dependent photophosphorylation in the chloroplasts (see Chapter 7). In addition to these reactions in mitochondria and chloroplasts, reactions in the cytosol such as glycolysis also assimilate phosphate.

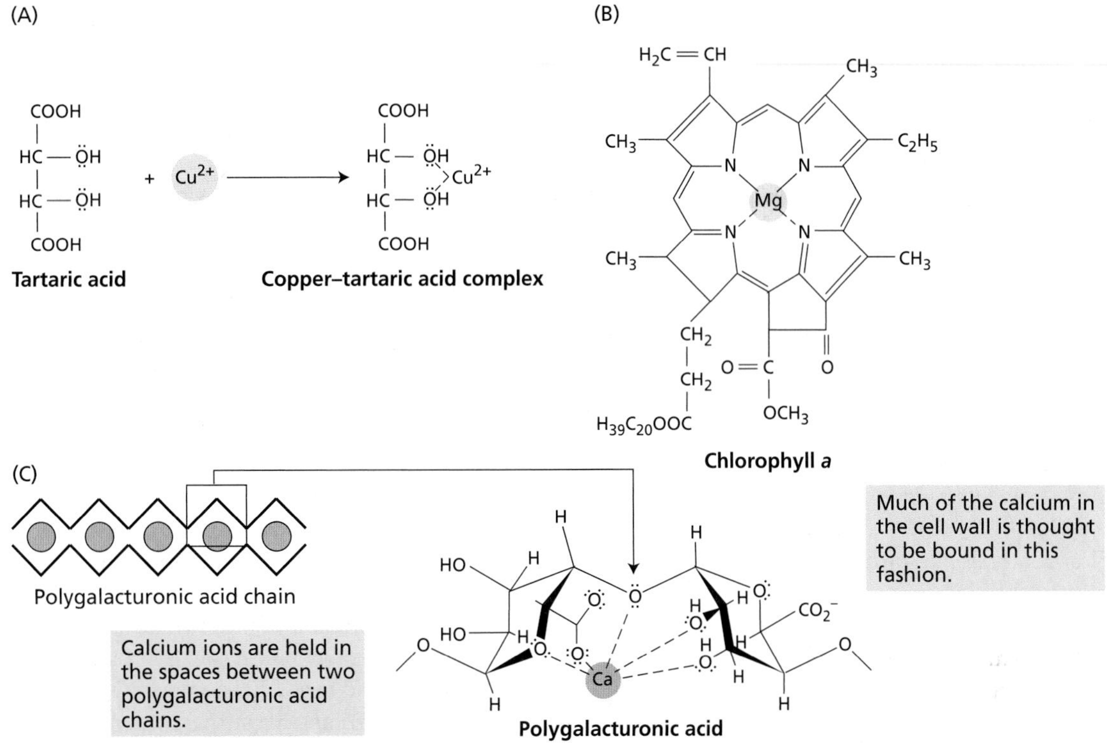

(A)

(B)

(C)

**FIGURE 12.16** Examples of coordination complexes. Coordination complexes form when oxygen or nitrogen atoms of a carbon compound donate unshared electron pairs (represented by dots) to form a bond with a cation. (A) Copper ions share electrons with the hydroxyl oxygens of tartaric acid. (B) Magnesium ions share electrons with nitrogen atoms in chlorophyll *a*. Dashed lines represent a coordination bond between unshared electrons from the nitrogen atoms and the magnesium cation. (C) The "egg box" model of the interaction of polygalacturonic acid, a major constituent of pectins in cell walls, and calcium ions. At right is an enlargement of a single calcium ion forming a coordination complex with the hydroxyl oxygens of the galacturonic acid residues. (After Rees 1977.)

Glycolysis incorporates inorganic phosphate into 1,3-bisphosphoglyceric acid, forming a high-energy acyl phosphate group. This phosphate can be donated to ADP to form ATP in a substrate-level phosphorylation reaction (see Chapter 11). Once incorporated into ATP, the phosphate group may be transferred via many different reactions to form the various phosphorylated compounds found in higher-plant cells.

## Cation Assimilation

Cations taken up by plant cells form complexes with organic compounds in which the cation becomes bound to the complex by noncovalent bonds (for a discussion of noncovalent bonds, see Appendix 1). Plants assimilate macronutrient cations such as potassium, magnesium, and calcium, as well as micronutrient cations such as copper, iron, manganese, cobalt, sodium, and zinc, in this manner. In this section we will describe coordination bonds and electrostatic bonds, which mediate the assimilation of several cations that plants require as nutrients, and the special requirements for the absorption of iron by roots and subsequent assimilation of iron within plants.

### Cations form noncovalent bonds with carbon compounds

The noncovalent bonds formed between cations and carbon compounds are of two types: coordination bonds and electrostatic bonds. In the formation of a coordination complex, several oxygen or nitrogen atoms of a carbon compound donate unshared electrons to form a bond with the cation nutrient. As a result, the positive charge on the cation is neutralized.

*Coordination bonds* typically form between polyvalent cations and carbon molecules—for example, complexes between copper and tartaric acid (**FIGURE 12.16A**) or magnesium and chlorophyll *a* (**FIGURE 12.16B**). The nutrients that are assimilated as coordination complexes include copper, zinc, iron, and magnesium. Calcium can also form coordination complexes with the polygalacturonic acid of cell walls (**FIGURE 12.16C**).

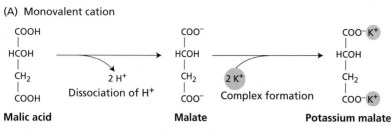

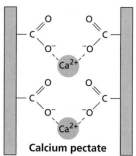

(A) Monovalent cation

**Malic acid**

**Malate**

**Potassium malate**

Dissociation of H⁺

Complex formation

(B) Divalent cation

**Calcium pectate**

**FIGURE 12.17** Examples of electrostatic (ionic) complexes. (A) The monovalent cation $K^+$ and malate form the complex potassium malate. (B) The divalent cation $Ca^{2+}$ and pectate form the complex calcium pectate. Divalent cations can form cross-links between parallel strands that contain negatively charged carboxyl groups. Calcium cross-links play a structural role in the cell walls.

*Electrostatic bonds* form because of the attraction of a positively charged cation for a negatively charged group such as carboxylate ($-COO^-$) on a carbon compound. Unlike the situation in coordination bonds, the cation in an electrostatic bond retains its positive charge. Monovalent cations such as potassium ($K^+$) can form electrostatic bonds with the carboxylic groups of many organic acids (**FIGURE 12.17A**). Nonetheless, much of the potassium that is accumulated by plant cells and functions in osmotic regulation and enzyme activation remains in the cytosol and the vacuole as the free ion. Divalent ions such as calcium form electrostatic bonds with pectates (**FIGURE 12.17B**) and the carboxylic groups of polygalacturonic acid (see Chapter 15).

In general, cations such as magnesium ($Mg^{2+}$) and calcium ($Ca^{2+}$) are assimilated by the formation of both coordination complexes and electrostatic bonds with amino acids, phospholipids, and other negatively charged molecules.

## Roots modify the rhizosphere to acquire iron

Iron is important in iron–sulfur proteins (see Chapter 7) and as a catalyst in enzyme-mediated redox reactions (see Chapter 5), such as those of nitrogen metabolism discussed earlier. Plants obtain iron from the soil, where it is present primarily as ferric iron ($Fe^{3+}$) in oxides such as $Fe(OH)^{2+}$, $Fe(OH)_3$, and $Fe(OH)_4^-$. At neutral pH, ferric iron is highly insoluble. To absorb sufficient amounts of iron from the soil solution, roots have developed several mechanisms that increase iron solubility and thus its availability (**FIGURE 12.18**). These mechanisms include:

- Soil acidification, which increases the solubility of ferric iron
- Reduction of ferric iron to the more soluble ferrous form ($Fe^{2+}$)
- Release of compounds that form stable, soluble complexes with iron (Marschner 1995). Recall from Chapter 5 that such compounds are called iron chelators (see Figure 5.3).

**FIGURE 12.18** Two processes through which plants roots absorb iron. (A) A process common to dicots such as pea, tomato, and soybean. The chelates include organic compounds such as malic acid, citric acid, phenolics, and piscidic acid. (B) A process common to grasses such as barley, maize, and oat. After the grass excretes the siderophore and it removes iron from soil particles, the complex may degrade and release the iron to the soil, exchange iron for another ligand, or be transported into the root. (After Guerinot and Yi 1994.)

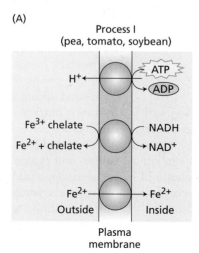

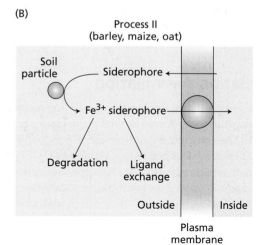

(A) Process I
(pea, tomato, soybean)

H⁺

ATP

ADP

Fe³⁺ chelate

NADH

Fe²⁺ + chelate

NAD⁺

Fe²⁺

Fe²⁺

Outside

Inside

Plasma membrane

(B) Process II
(barley, maize, oat)

Soil particle

Siderophore

Fe³⁺ siderophore

Degradation

Ligand exchange

Outside

Inside

Plasma membrane

Roots generally acidify the soil around them. They extrude protons during the absorption and assimilation of cations, particularly ammonium, and release organic acids, such as malic acid and citric acid, that enhance iron and phosphate availability (see Figure 5.5). Iron deficiencies stimulate the extrusion of protons by roots. In addition, plasma membranes in roots contain an enzyme, called *iron-chelate reductase*, that reduces ferric iron ($Fe^{3+}$) to the ferrous ($Fe^{2+}$) form, with cytosolic NADH or NADPH serving as the electron donor (see Figure 12.18A). The activity of this enzyme increases under iron deprivation.

Several compounds secreted by roots form stable chelates with iron. Examples include malic acid, citric acid, phenolics, and piscidic acid. Grasses produce a special class of iron chelators called *siderophores*. Siderophores are made of amino acids that are not found in proteins, such as mugineic acid, and form highly stable complexes with $Fe^{3+}$. Root cells of grasses have $Fe^{3+}$–siderophore transport systems in their plasma membranes that bring the chelate into the cytoplasm. Under iron deficiency, grass roots release more siderophores into the soil and increase the capacity of their $Fe^{3+}$–siderophore transport system (see Figure 12.18B).

### Iron forms complexes with carbon and phosphate

After the roots absorb iron or an iron chelate, they oxidize it to a ferric form and translocate much of it to the leaves as an electrostatic complex with citrate or nicotianamine (Jeong and Guerinot 2009).

Once in the leaves, iron undergoes an important assimilatory reaction through which it is inserted into the porphyrin precursor of heme groups found in the cytochromes located in chloroplasts and mitochondria (see Chapter 7). This reaction is catalyzed by the enzyme ferrochelatase (FIGURE 12.19) (Jones 1983). Most of the iron in the plant is found in heme groups. In addition, iron–sulfur proteins of the electron transport chain (see Chapter 7) contain nonheme iron covalently bound to the sulfur atoms of cysteine residues in the apoprotein. Iron is also found in $Fe_2S_2$ centers, which contain two irons (each complexed with the sulfur atoms of cysteine residues) and two inorganic sulfides.

Free iron (iron that is not complexed with carbon compounds) may interact with oxygen to form highly damaging hydroxyl radicals, OH• (Halliwell and Gutteridge 1992). Plant cells may limit such damage by storing surplus iron in an iron–protein complex called **phytoferritin** (Bienfait and Van der Mark 1983). Mutants of *Arabidopsis* show that, although ferritins are essential for protection against oxidative damage, they do not serve as a major iron pool for either seedling development or proper functioning of the photosynthetic apparatus (Ravet et al. 2009). Phytoferritin consists of a protein shell with 24 identical subunits forming a hollow sphere that has a molecular mass of about 480 kDa. Within this sphere is a core of 5400 to 6200 iron atoms present as a ferric oxide–phosphate complex.

How iron is released from phytoferritin is uncertain, but breakdown of the protein shell appears to be involved. The level of free iron in plant cells regulates the de novo biosynthesis of phytoferritin (Lobreaux et al. 1992). Interest in phytoferritin is high because iron in this protein-bound form may be highly available to humans, and foods rich in phytoferritin, such as soybean, may address dietary anemia problems (Welch and Graham 2004).

## Oxygen Assimilation

Respiration accounts for the bulk (about 90%) of the oxygen ($O_2$) assimilated by plant cells (see Chapter 11). Another major pathway for the assimilation of $O_2$ into organic compounds involves the incorporation of $O_2$ from water (see reaction 1 in Table 8.1). A small proportion of oxygen can be directly assimilated into organic compounds in the process of *oxygen fixation* via enzymes known as *oxygenases*. The most prominent oxygenase in plants is ribulose-1,5-bisphosphate carboxylase/oxygenase (rubisco) that during photorespiration incorporates oxygen into an organic compound and releases energy (see Chapter 8). Other oxygenases are discussed in WEB TOPIC 12.4.

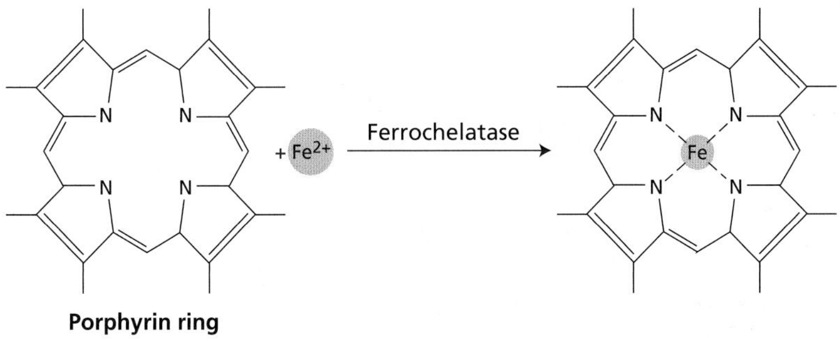

**Porphyrin ring**

FIGURE 12.19 The ferrochelatase reaction. The enzyme ferrochelatase catalyzes the insertion of iron into the porphyrin ring to form a coordination complex. See Figure 7.36 for illustration of the biosynthesis of the porphyrin ring.

# The Energetics of Nutrient Assimilation

Nutrient assimilation generally requires large amounts of energy to convert stable, low-energy inorganic compounds into high-energy organic compounds. For example, the reduction of nitrate to nitrite and then to ammonium requires the transfer of about ten electrons and accounts for about 25% of the total energy expenditures in both roots and shoots (Bloom 1997). Consequently, a plant may use one-fourth of its energy to assimilate nitrogen, a constituent that accounts for less than 2% of the total dry weight of the plant.

Many of these assimilatory reactions occur in the stroma of the chloroplast, where they have ready access to powerful reducing agents, such as NADPH, thioredoxin, and ferredoxin, generated during photosynthetic electron transport. This process—coupling nutrient assimilation to photosynthetic electron transport—is called **photoassimilation** (**FIGURE 12.20**).

Photoassimilation and the Calvin–Benson cycle occur in the same compartment, but only when photosynthetic electron transport generates reductant in excess of the needs of the Calvin–Benson cycle—for example, under conditions of high light and low $CO_2$—does photoassimilation proceed (Robinson 1988). High levels of $CO_2$ inhibit nitrate assimilation in the shoots of $C_3$ plants (**FIGURE 12.21**) (see **WEB ESSAY 12.1**) and in the bundle sheath cells (see Chapter 8) of $C_4$ plants (Becker et al. 1993).

The mechanisms that regulate the partitioning of reductant between the Calvin–Benson cycle and photoassimilation warrant investigation, because atmospheric levels of $CO_2$ are expected to double during this century (see Chapter 9), so this phenomenon will affect plant–nutrient relations.

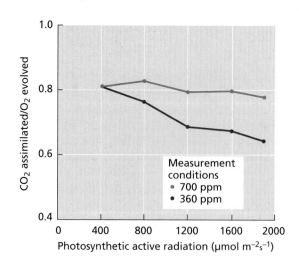

**FIGURE 12.21** The assimilatory quotient (AQ = $CO_2$ assimilated/$O_2$ evolved) of wheat seedlings as a function of light level (photosynthetically active radiation). Nitrate photoassimilation is directly related to assimilatory quotient, because transfer of electrons to nitrate and nitrite during photoassimilation increases $O_2$ evolution from the light-dependent reactions of photosynthesis, while $CO_2$ assimilation by the light-independent reactions continues at similar rates. Therefore, plants that are photoassimilating nitrate exhibit a lower AQ. In measurements carried out at ambient $CO_2$ concentrations (360 ppm, red trace), the AQ decreases as a function of incident radiation, indicating that nitrate photoassimilation rates increased. At elevated $CO_2$ concentrations (700 ppm, blue trace) the AQ remains constant at all light levels used, indicating that the $CO_2$-fixing reactions are competing for reductant and inhibiting nitrate photoassimilation. (After Bloom et al. 2002.)

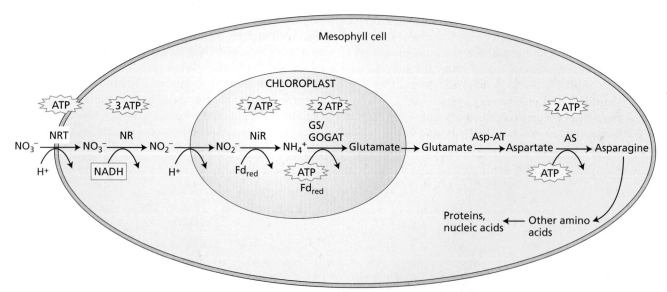

**FIGURE 12.20** Summary of the processes involved in the assimilation of mineral nitrogen in the leaf. Nitrate translocated from the roots through the xylem is absorbed by a mesophyll cell via one of the nitrate–proton symporters (NRT) into the cytoplasm. There it is reduced to nitrite via nitrate reductase (NR). Nitrite is translocated into the stroma of the chloroplast along with a proton. In the stroma, nitrite is reduced to ammonium via nitrite reductase (NiR) and this ammonium is converted into glutamate via the sequential action of glutamine synthetase (GS) and glutamate synthase (GOGAT). Once again in the cytoplasm, the glutamate is transaminated to aspartate via aspartate aminotransferase (Asp-AT). Finally, asparagine synthetase (AS) converts aspartate into asparagine. The approximate amounts of ATP equivalents are given above each reaction.

# SUMMARY

Nutrient assimilation is the often energy-requiring process by which plants incorporate inorganic nutrients into the carbon constituents necessary for growth and development.

## Nitrogen in the Environment

- When nitrogen is fixed into ammonia ($NH_3$) or nitrate ($NO_3^-$), it passes through several organic or inorganic forms before it eventually returns to molecular nitrogen ($N_2$) (**Figure 12.1**).
- At high concentrations, ammonium ($NH_4^+$) is toxic to living tissues, but nitrate can be safely stored and translocated in plant tissues (**Figure 12.2**).

## Nitrate Assimilation

- Plant roots actively absorb nitrate, then reduce it to nitrite ($NO_2^-$) in the cytosol (**Figure 12.3**).
- Nitrate, light, and carbohydrates affect the transcription and translation of nitrate reductase (**Figure 12.4**).
- Darkness and $Mg^{2+}$ can inactivate nitrate reductase. Such inactivation is faster than regulation by reduced synthesis or degradation of the enzyme.
- In chloroplasts or plastids, the enzyme nitrite reductase reduces nitrite to ammonium (**Figure 12.5**).
- Both roots and shoots assimilate nitrate (**Figure 12.6**).

## Ammonium Assimilation

- Plant cells avoid ammonium toxicity by rapidly converting ammonium into amino acids (**Figure 12.7**).
- Nitrogen is incorporated into other amino acids via transamination reactions involving glutamine and glutamate.
- The amino acid asparagine is a key compound for nitrogen transport and storage.

## Amino Acid Biosynthesis

- The carbon skeleton for amino acids derives from intermediates of glycolysis and the citric acid cycle (**Figure 12.8**).

## Biological Nitrogen Fixation

- Biological nitrogen fixation accounts for most of the ammonium formed from atmospheric $N_2$ (**Figure 12.1, Table 12.2**).
- Several types of nitrogen-fixing bacteria form symbiotic associations with higher plants (**Figures 12.9, 12.10; Table 12.3**).

- Nitrogen fixation requires anaerobic conditions.
- Symbiotic nitrogen-fixing prokaryotes function within specialized structures formed by the plant host (**Figures 12.9, 12.10**).
- The symbiotic relationship is initiated by the migration of the nitrogen-fixing bacteria toward the roots, which is mediated by chemical attractants secreted by the roots.
- Attractants activate the rhizobial NodD protein, which then induces the **biosynthesis of Nod factors that act as signals for symbiosis (Figure 12.11**).
- Nod factors induce root hair curling, sequestration of rhizobia, cell wall degradation, and bacterial access to the root hair plasma membrane, from which an infection thread forms (**Figure 12.12**).
- Filled with proliferating rhizobia, the infection thread elongates through root tissue in the direction of the developing nodule, which arises from cortical cells (**Figure 12.12**).
- In response to a signal from the plant, the bacteria in the nodule stop dividing and differentiate into nitrogen-fixing bacteroids.
- The reduction of $N_2$ to $NH_3$ is catalyzed by the nitrogenase enzyme complex (**Figure 12.13**).
- Fixed nitrogen is transported as amides or ureides (**Figure 12.14**).

## Sulfur Assimilation

- Most assimilated sulfur derives from sulfate ($SO_4^{2-}$) absorbed from the soil solution, but plants can also metabolize gaseous sulfur dioxide ($SO_2$) entering via stomata.
- Synthesis of sulfur-containing organic compounds begins with the reduction of sulfate to the amino acid cysteine (**Figure 12.15**).
- Sulfate is assimilated in leaves and exported as glutathione via the phloem to growing sites.

## Phosphate Assimilation

- Roots absorb phosphate ($HPO_4^{2-}$) from the soil solution, and its assimilation occurs with the formation of ATP.
- From ATP, the phosphate group may be transferred to many different carbon compounds in plant cells.

## Cation Assimilation

- Polyvalent cations form coordination bonds with carbon molecules (**Figure 12.16**).
- Monovalent cations form electrostatic bonds with the carboxylate groups (**Figure 12.17**).

## SUMMARY continued

- To absorb sufficient amounts of insoluble ferric iron ($Fe^{3+}$) from the soil solution, several mechanisms are used (**Figure 12.18**).

- Once in the leaves, iron undergoes an important assimilatory reaction (**Figure 12.19**).

- To limit the free radical damage that free iron can cause, plant cells may store surplus iron as phytoferritin.

### Oxygen Assimilation

- Respiration and the oxygenase activity of rubisco account for most of the oxygen ($O_2$) assimilated by plant cells, but direct oxygen fixation also is catalyzed by other oxygenases (**Figure 12.20**).

### The Energetics of Nutrient Assimilation

- Energy-requiring nutrient assimilation is coupled to photosynthetic electron transport, which generates powerful reducing agents (**Figure 12.21**).

- Photoassimilation operates only when photosynthetic electron transport generates reductant in excess of the needs of the Calvin–Benson cycle (**Figure 12.22**).

## WEB MATERIAL

### Web Topics

**12.1 Development of a Root Nodule**
Nodule primordia form opposite to the protoxylem poles of the root vascular bundles.

**12.2 Measurement of Nitrogen Fixation**
Acetylene reduction is used as an indirect measurement of nitrogen reduction.

**12.3 The Synthesis of Methionine**
Methionine is synthesized in plastids from cysteine.

**12.4 Oxygenases**
Oxygenases are enzymes that catalyze oxygen assimilation.

### Web Essay

**12.1 Elevated $CO_2$ and Nitrogen Photoassimilation**
In leaves grown under high $CO_2$ concentrations, $CO_2$ inhibits nitrogen photoassimilation because it competes for reductant.

## CHAPTER REFERENCES

Becker, T. W., Perrot-Rechenmann, C., Suzuki, A., and Hirel, B. (1993) Subcellular and immunocytochemical localization of the enzymes involved in ammonia assimilation in mesophyll and bundle-sheath cells of maize leaves. *Planta* 191: 129–136.

Bergmann, L., and Rennenberg, H. (1993) Glutathione metabolism in plants. In *Sulfur Nutrition and Assimilation in Higher Plants. Regulatory, Agricultural and Environmental Aspects*, L. J. De Kok, I. Stulen, H. Rennenberg, C. Brunold, and W. E. Rauser, eds., SPB Academic Publishing, The Hague, Netherlands, pp. 109–123.

Bienfait, H. F., and Van der Mark, F. (1983) Phytoferritin and its role in iron metabolism. In *Metals and Micronutrients: Uptake and Utilization by Plants*, D. A. Robb and W. S. Pierpoint, eds., Academic Press, New York, pp. 111–123.

Bloom, A. J. (1997) Nitrogen as a limiting factor: Crop acquisition of ammonium and nitrate. In *Ecology in Agriculture*, L. E. Jackson, ed., Academic Press, San Diego, CA, pp. 145–172.

Bloom, A. J., Smart, D. R., Nguyen, D. T., and Searles, P. S. (2002) Nitrogen assimilation and growth of wheat under elevated carbon dioxide. *Proc. Natl. Acad. Sci. USA* 99: 1730–1735.

Bloom, A. J., Sukrapanna, S. S., and Warner, R. L. (1992) Root respiration associated with ammonium and nitrate absorption and assimilation by barley. *Plant Physiol.* 99: 1294–1301.

Buchanan, B., Gruissem, W., and Jones, R., eds. (2000) *Biochemistry and Molecular Biology of Plants.* American Society of Plant Physiologists, Rockville, MD.

Burris, R. H. (1976) Nitrogen fixation. In *Plant Biochemistry*, 3rd ed., J. Bonner and J. Varner, eds., Academic Press, New York, pp. 887–908.

Campbell, W. H. (1999) Nitrate reductase structure, function and regulation: Bridging the gap between biochemistry and physiology. *Annu. Rev. Plant Physiol. Plant Mol. Biol.* 50: 277–303.

Campbell, W. H. (2001) Structure and function of eukaryotic NAD(P)H: Nitrate reductase. *Cell. Mol. Life Sci.* 58: 194–204.

Carlson, R. W., Forsberg, L. S., Price, N. P. J., Bhat, U. R., Kelly, T. M., and Raetz, C. R. H. (1995) The structure and biosynthesis of *Rhizobium leguminosarum* lipid A. In *Progress in Clinical and Biological Research*, Vol. 392: *Bacterial Endotoxins: Lipopolysaccharides from Genes to Therapy: Proceedings of the Third Conference of the International Endotoxin Society, held in Helsinki, Finland, on August 15–18, 1994*, J. Levin et al., eds., John Wiley and Sons, New York, pp. 25–31.

Crawford, N. M., and Forde, B. J. (2002) Molecular and developmental biology of inorganic nitrogen nutrition. In: *The Arabidopsis Book*, C. Somerville and E. Meyerowitz, eds., American Society of Plant Physiologists, Rockville, MD. doi:10.1199/tab.0011, http://www.aspb.org/publications/arabidopsis/

Denison, R. F., and Harter, B. L. (1995) Nitrate effects on nodule oxygen permeability and leghemoglobin. *Plant Physiol.* 107: 1355–1364.

Dixon, R. O. D., and Wheeler, C. T. (1986) *Nitrogen Fixation in Plants.* Chapman and Hall, New York.

Dixon, R., and Kahn, D. (2004) Genetic regulation of biological nitrogen fixation. *Nat. Rev. Microbiol.* 2: 621–631.

Dong, Z., Canny, M. J., McCully, M. E., Roboredo, M. R., Cabadilla, C. F., Ortega, E., and Rodes, R. (1994) A nitrogen-fixing endophyte of sugarcane stems: A new role for the apoplast. *Plant Physiol.* 105: 1139–1147.

FAOSTAT. (2009) *Agricultural Data.* Food and Agricultural Organization of the United Nations, Rome.

Fischer, K., Barbier, G. G., Hecht, H. J., Mendel, R. R., Campbell, W. H., and Schwarz, G. (2005) Structural basis of eukaryotic nitrate reduction: Crystal structures of the nitrate reductase active site. *Plant Cell* 17: 1167–1179.

Franche, C., Lindstrom, K., and Elmerich, C. (2009) Nitrogen-fixing bacteria associated with leguminous and non-leguminous plants. *Plant Soil* 321: 35–59.

Geurts, R., and Bisseling, T. (2002) Rhizobium nod factor perception and signalling. *Plant Cell* 14: S239–S249.

Guerinot, M. L., and Yi, Y. (1994) Iron: Nutritious, noxious, and not readily available. *Plant Physiol.* 104: 815–820.

Halliwell, B., and Gutteridge, J. M. C. (1992) Biologically relevant metal ion-dependent hydroxyl radical generation: An update. *FEBS Lett.* 307: 108–112.

Heidstra, R., and Bisseling, T. (1996) Nod factor-induced host responses and mechanisms of Nod factor perception. *New Phytol.* 133: 25–43.

Hell, R. (1997) Molecular physiology of plant sulfur metabolism. *Planta* 202: 138–148.

Heytler, P. G., Reddy, G. S., and Hardy, R. W. F. (1984) In vivo energetics of symbiotic nitrogen fixation in soybeans. In *Nitrogen Fixation and $CO_2$ Metabolism*, P. W. Ludden and I. E. Burris, eds., Elsevier, New York, pp. 283–292.

Hirsch, S., Kim, J., Munoz, A., Heckmann, A. B., Downie, J. A., and Oldroyd, G. E. D. (2009) GRAS proteins form a DNA binding complex to induce gene expression during nodulation signaling in medicago truncatula. *Plant Cell* 21: 545–557.

James, E. K., and Olivares, F. L. (1998) Infection and colonization of sugar cane and other graminaceous plants by endophytic diazotrophs. *CRC Crit. Rev. Plant Sci.* 17: 77–119.

Jeong, J., and Guerinot, M. L. (2009) Homing in on iron homeostasis in plants. *Trends Plant Sci.* 14: 280–285.

Jones, O. T. G. (1983) Ferrochelatase. In *Metals and Micronutrients: Uptake and Utilization by Plants*, D. A. Robb and W. S. Pierpoint, eds., Academic Press, New York, pp. 125–144.

Kaiser, W. M., Weiner, H., and Huber, S. C. (1999) Nitrate reductase in higher plants: A case study for transduction of environmental stimuli into control of catalytic activity. *Physiol. Plant.* 105: 385–390.

Kleinhofs, A., Warner, R. L., Lawrence, J. M., Melzer, J. M., Jeter, J. M., and Kudrna, D. A. (1989) Molecular genetics of nitrate reductase in barley. In *Molecular and Genetic Aspects of Nitrate Assimilation*, J. L. Wray and J. R. Kinghorn, eds., Oxford Science, New York, pp. 197–211.

Kopriva, S. (2006) Regulation of sulfate assimilation in Arabidopsis and beyond. *Ann. Bot.* 97: 479–495.

Kuzma, M. M., Hunt, S., and Layzell, D. B. (1993) Role of oxygen in the limitation and inhibition of nitrogenase activity and respiration rate in individual soybean nodules. *Plant Physiol.* 101: 161–169.

Lam, H.-M., Coschigano, K. T., Oliveira, I. C., Melo-Oliveira, R., and Coruzzi, G. M. (1996) The molecular-genetics of nitrogen assimilation into amino acids in higher plants. *Annu. Rev. Plant Physiol. Plant Mol. Biol.* 47: 569–593.

Lea, P. J., Blackwell, R. D., and Joy, K. W. (1992) Ammonia assimilation in higher plants. In *Nitrogen Metabolism of Plants* (Proceedings of the Phytochemical Society of Europe 33), K. Mengel and D. J. Pilbeam, eds., Clarendon, Oxford, pp. 153–186.

Leustek, T., and Saito, K. (1999) Sulfate transport and assimilation in plants. *Plant Physiol.* 120: 637–643.

Leustek, T., Martin, M. N., Bick, J.-A., and Davies, J. P. (2000) Pathways and regulation of sulfur metabolism revealed through molecular and genetic studies. *Annu. Rev. Plant Physiol. Plant Mol. Biol.* 51: 141–165.

Lobreaux, S., Massenet, O., and Briat, J.-F. (1992) Iron induces ferritin synthesis in maize plantlets. *Plant Mol. Biol.* 19: 563–575.

Ludwig, R. A., and de Vries, G. E. (1986) Biochemical physiology of *Rhizobium* dinitrogen fixation. In *Nitrogen Fixation*, Vol. 4: *Molecular Biology*, W. I. Broughton and S. Puhler, eds., Clarendon, Oxford, pp. 50–69.

Marschner, H. (1995) *Mineral Nutrition of Higher Plants*, 2nd ed. Academic Press, London.

Mendel, R. R. (2005) Molybdenum: Biological activity and metabolism. *Dalton Trans.* 2005: 3404–3409.

Miller, A. J., Fan, X. R., Orsel, M., Smith, S. J., and Wells, D. M. (2007) Nitrate transport and signalling. *J. Exp. Bot.* 58: 2297–2306.

Mylona, P., Pawlowski, K., and Bisseling, T. (1995) Symbiotic nitrogen fixation. *Plant Cell* 7: 869–885.

Oaks, A. (1994) Primary nitrogen assimilation in higher plants and its regulation. *Can. J. Bot.* 72: 739–750.

Oldroyd, G. E. D., and Downie, J. M. (2008) Coordinating nodule morphogenesis with rhizobial infection in legumes. *Annu. Rev. Plant Biol.* 59: 519–546.

Oldroyd, G. E. D., Harrison, M. J., and Paszkowski, U. (2009) Reprogramming plant cells for endosymbiosis. *Science* 324: 753–754.

Pate, J. S. (1983) Patterns of nitrogen metabolism in higher plants and their ecological significance. In *Nitrogen as an Ecological Factor: The 22nd Symposium of the British Ecological Society, Oxford 1981*, J. A. Lee, S. McNeill, and I. H. Rorison, eds., Blackwell, Boston, pp. 225–255.

Pate, J. S., and Layzell, D. B. (1990) Energetics and biological costs of nitrogen assimilation. In *The Biochemistry of Plants*, Vol. 16: *Intermediary Nitrogen Metabolism*, B. J. Miflin and P. J. Lea, eds., Academic Press, San Diego, CA, pp. 1–42.

Phillips, D. A., and Kapulnik, Y. (1995) Plant isoflavonoids, pathogens and symbionts. *Trends Microbiol.* 3: 58–64.

Preisig, O., Zufferey, R., Thony-Meyer, L., Appleby, C. A., and Hennecke, H. (1996) A high-affinity cbb3-type cytochrome oxidase terminates the symbiosis-specific respiratory chain of *Bradyrhizobium japonicum*. *J. Bacteriol.* 178: 1532–1538.

Rees, D. A. (1977) *Polysaccharide Shapes*. Chapman and Hall, London.

Radutoiu, S., Madsen, L. H., Madsen, E. B., Jurkiewicz, A., Fukai, E., Quistgaard, E. M. H., Albrektsen, A. S., James, E. K., Thirup, S., and Stougaard, J. (2007) LysM domains mediate lipochitin-oligosaccharide recognition and Nfr genes extend the symbiotic host range. *EMBO J.* 26: 3923–3935.

Ravet, K., Touraine, B., Boucherez, J., Briat, J. F., Gaymard, F., and Cellier, F. (2009) Ferritins control interaction between iron homeostasis and oxidative stress in Arabidopsis. *Plant J.* 57: 400–412.

Reis, V. M., Baldani, J. I., Baldani, V. L. D., and Dobereiner, J. (2000) Biological dinitrogen fixation in Gramineae and palm trees. *Crit. Rev. Plant Sci.* 19: 227–247.

Robinson, J. M. (1988) Spinach leaf chloroplast carbon dioxide and nitrite photoassimilations do not compete for photogenerated reductant: Manipulation of reductant levels by quantum flux density titrations. *Plant Physiol.* 88: 1373–1380.

Rolfe, B. G., and Gresshoff, P. M. (1988) Genetic analysis of legume nodule initiation. *Annu. Rev. Plant Physiol. Plant Mol. Biol.* 39: 297–320.

Schlesinger, W. H. (1997) *Biogeochemistry: An Analysis of Global Change*, 2nd ed. Academic Press, San Diego, CA.

Seefeldt, L. C., Hoffman, B. M., and Dean, D. R. (2009) Mechanism of Mo-dependent nitrogenase. *Annu. Rev. Biochem.* 78: 701–722.

Siegel, L. M., and Wilkerson, J. Q. (1989) Structure and function of spinach ferredoxin-nitrite reductase. In *Molecular and Genetic Aspects of Nitrate Assimilation*, J. L. Wray and J. R. Kinghorn, eds., Oxford Science, Oxford, pp. 263–283.

Sivasankar, S., and Oaks, A. (1996) Nitrate assimilation in higher plants—The effect of metabolites and light. *Plant Physiol. Biochem.* 34: 609–620.

Smart, D. R., and Bloom, A. J. (2001) Wheat leaves emit nitrous oxide during nitrate assimilation. *Proc. Natl. Acad. Sci. USA* 98: 7875–7878.

Stokkermans, T. J. W., Ikeshita, S., Cohn, J., Carlson, R. W., Stacey, G., Ogawa, T., and Peters, N. K. (1995) Structural requirements of synthetic and natural product lipo-chitin oligosaccharides for induction of nodule primordia on *Glycine soja*. *Plant Physiol.* 108: 1587–1595.

Timmers, A. C. J., Auriac, M.-C., and Truchet, G. (1999) Refined analysis of early symbiotic steps of the Rhizobium-Medicago: Interaction in relation with microtubular cytoskeleton rearrangements. *Development* 126: 3617–3628.

Vande Broek, A., and Vanderleyden, J. (1995) Review: Genetics of the *Azospirillum*-plant root association. *CRC Crit. Rev. Plant Sci.* 14: 445–466.

Warner, R. L., and Kleinhofs, A. (1992) Genetics and molecular biology of nitrate metabolism in higher plants. *Physiol. Plant.* 85: 245–252.

Wei, H. and Layzell, D. B. (2006) Adenylate-coupled ion movement. A mechanism for the control of nodule permeability to $O_2$ diffusion. *Plant Physiol.* 141: 280–287.

Welch, R. M., and Graham, R. D. (2004) Breeding for micronutrients in staple food crops from a human nutrition perspective. *J. Exp. Bot.* 55: 353–364.

Wray, J. L. (1993) Molecular biology, genetics and regulation of nitrite reduction in higher plants. *Physiol. Plant.* 89: 607–612.

Yano, K., Yoshida, S., Muller, J., Singh, S., Banba, M., Vickers, K., Markmann, K., White, C., Schuller, B., Sato, S., et al. (2008) CYCLOPS, a mediator of symbiotic intracellular accommodation. *Proc. Natl. Acad. Sci. USA* 105: 20540–20545.

CHAPTER

# 13

# Secondary Metabolites and Plant Defense

In natural habitats, plants are surrounded by an enormous number of potential enemies. Nearly all ecosystems contain a wide variety of bacteria, viruses, fungi, nematodes, mites, insects, mammals, and other herbivorous animals. By their nature, plants cannot avoid these herbivores and pathogens simply by moving away; they must protect themselves in other ways. Their first line of defense involves the plant surface. The cuticle (a waxy outer layer) and the periderm (secondary protective tissue), besides retarding water loss, provide passive barriers to bacterial and fungal entry. (For a discussion of these protective layers, see WEB TOPIC 13.1.)

A diverse group of plant compounds, commonly referred to as secondary metabolites, also defends plants against a variety of herbivores and pathogenic microbes. Some secondary metabolites serve other important functions as well, such as providing structural support, as in the case of lignin, or acting as pigments, as in the case of the anthocyanins.

In this chapter we will discuss some of the mechanisms by which plants protect themselves against both herbivores and pathogens. We will discuss the structures and biosynthetic pathways for the three major classes of secondary metabolites: terpenes, phenolics, and nitrogen-containing compounds. Induced plant defenses against insect herbivores will be discussed as an example of the important ecological functions of secondary metabolites. Finally, we will examine specific plant responses to pathogen attack, the genetic control of host–pathogen interactions, and cell signaling processes associated with infection.

# Secondary Metabolites

Plants produce a large, diverse array of organic compounds that appear to have no direct function in their growth and development. These compounds are known as **secondary metabolites**, *secondary products*, or *natural products*. Secondary metabolites have no generally recognized direct roles in the processes of photosynthesis, respiration, solute transport, translocation, protein synthesis, nutrient assimilation, or differentiation, or the formation of the *primary metabolites*—carbohydrates, proteins, nucleic acids, and lipids—discussed elsewhere in this book.

Secondary metabolites also differ from primary metabolites in having a restricted distribution within the plant kingdom. That is, certain secondary metabolites are only found in one plant species or related group of species, whereas primary metabolites are found throughout the plant kingdom.

## Secondary metabolites defend plants against herbivores and pathogens

For many years the adaptive significance of most secondary metabolites was unknown. These compounds were thought to be simply functionless end products of metabolism, or metabolic wastes. Study of these substances was pioneered by organic chemists of the nineteenth and early twentieth centuries who were interested in them because of their importance as medicinal drugs, poisons, flavors, and industrial materials.

Today we know that many secondary metabolites have important ecological functions in plants:

- They protect plants against being eaten by herbivores and against being infected by microbial pathogens.
- They serve as attractants (odor, color, taste) for pollinators and seed-dispersing animals.
- They function as agents of plant–plant competition and plant–microbe symbioses.

The ability of plants to compete and survive is therefore profoundly affected by the ecological functions of their secondary metabolites.

Secondary metabolism is also relevant to agriculture. The very defensive compounds that increase the reproductive fitness of plants by warding off fungi, bacteria, and herbivores may also make them undesirable as food for humans. Many important crop plants have been artificially selected to produce relatively low levels of these compounds (which, of course, can make them more susceptible to insects and disease).

In the remainder of this chapter we will discuss the major types of plant secondary metabolites, their biosynthesis, and what is known about their functions in the plant, particularly their defensive roles.

## Secondary metabolites are divided into three major groups

Plant secondary metabolites can be divided into three chemically distinct groups: terpenes, phenolics, and nitrogen-containing compounds. **FIGURE 13.1** shows in simplified form the pathways involved in the biosynthesis of secondary metabolites and their interconnections with primary metabolism. In the next three sections of this chapter, we will discuss each of these groups in turn.

# Terpenes

The **terpenes**, or *terpenoids*, constitute the largest class of secondary metabolites. Most of the diverse substances of this class are insoluble in water. They are synthesized from acetyl-CoA or its glycolytic intermediates. After discussing the biosynthesis of terpenes, we will examine how they act to repel herbivores.

## Terpenes are formed by the fusion of five-carbon isoprene units

All terpenes are derived from the union of 5-carbon elements (also called $C_5$ *units*) that have the branched carbon skeleton of isopentane:

$$\begin{array}{c} H_3C \\ \diagdown \\ \diagup \\ H_3C \end{array} CH - CH_2 - CH_3$$

The basic structural elements of terpenes are sometimes called **isoprene units** because terpenes can decompose at high temperatures to give isoprene:

$$\begin{array}{c} H_3C \\ \diagdown \\ \diagup \\ H_2C \end{array} CH - CH = CH_2$$

Thus, terpenes are occasionally also referred to as *isoprenoids*.

Terpenes are classified by the number of $C_5$ units they contain, although extensive metabolic modifications can sometimes make it difficult to pick out the original five-carbon residues. For example, 10-carbon terpenes, which contain two $C_5$ units, are called *monoterpenes*; 15-carbon terpenes (three $C_5$ units) are *sesquiterpenes*; and 20-carbon terpenes (four $C_5$ units) are *diterpenes*. Larger terpenes include *triterpenes* (30 carbons), *tetraterpenes* (40 carbons), and *polyterpenoids* ($[C_5]_n$ carbons, where $n > 8$).

## There are two pathways for terpene biosynthesis

Terpenes are synthesized from primary metabolites in at least two different ways. In the well-studied **mevalonic acid pathway**, three molecules of acetyl-CoA are joined together stepwise to form mevalonic acid (**FIGURE 13.2**). This key six-carbon intermediate is then pyrophosphorylated, decarboxylated, and dehydrated to yield **isopente-**

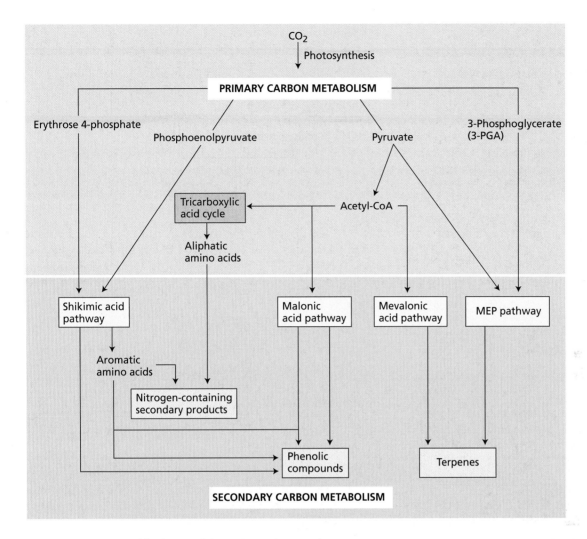

**FIGURE 13.1** A simplified view of the major pathways of secondary-metabolite biosynthesis and their interrelationships with primary metabolism.

nyl diphosphate (**IPP**)*. IPP is the activated five-carbon building block of terpenes.

IPP can also be formed from intermediates of glycolysis or the photosynthetic carbon reduction cycle via a separate set of reactions called the **methylerythritol phosphate (MEP) pathway** that operates in chloroplasts and other plastids (Lichtenthaler 1999). Glyceraldehyde 3-phosphate and two carbon atoms derived from pyruvate condense to form the five-carbon intermediate 1-deoxy-D-xylulose 5-phosphate. After this intermediate is rearranged and reduced to 2-C-methyl-D-erythritol 4-phosphate (MEP), it is eventually converted into IPP (see Figure 13.2).

---

*IPP is the abbreviation for isopentenyl *pyro*phosphate, an earlier name for this compound. The other pyrophosphorylated intermediates in this pathway are also now referred to as *di*phosphates.

## IPP and its isomer combine to form larger terpenes

IPP and its isomer, dimethylallyl diphosphate (DMAPP), are the activated 5-carbon building blocks of terpene biosynthesis that join together to form larger molecules. First IPP and DMAPP react to give geranyl diphosphate (GPP), the 10-carbon precursor of nearly all the monoterpenes (see Figure 13.2). GPP can then link to another molecule of IPP to give the 15-carbon compound farnesyl diphosphate (FPP), the precursor of nearly all the sesquiterpenes. Addition of yet another molecule of IPP gives the 20-carbon compound geranylgeranyl diphosphate (GGPP), the precursor of the diterpenes. Finally, FPP and GGPP can dimerize to give the triterpenes ($C_{30}$) and the tetraterpenes ($C_{40}$), respectively.

It is now generally accepted that sesquiterpenes and triterpenes are synthesized through the cytosolic mevalonic acid pathway, whereas mono-, di-, and tetraterpenes are

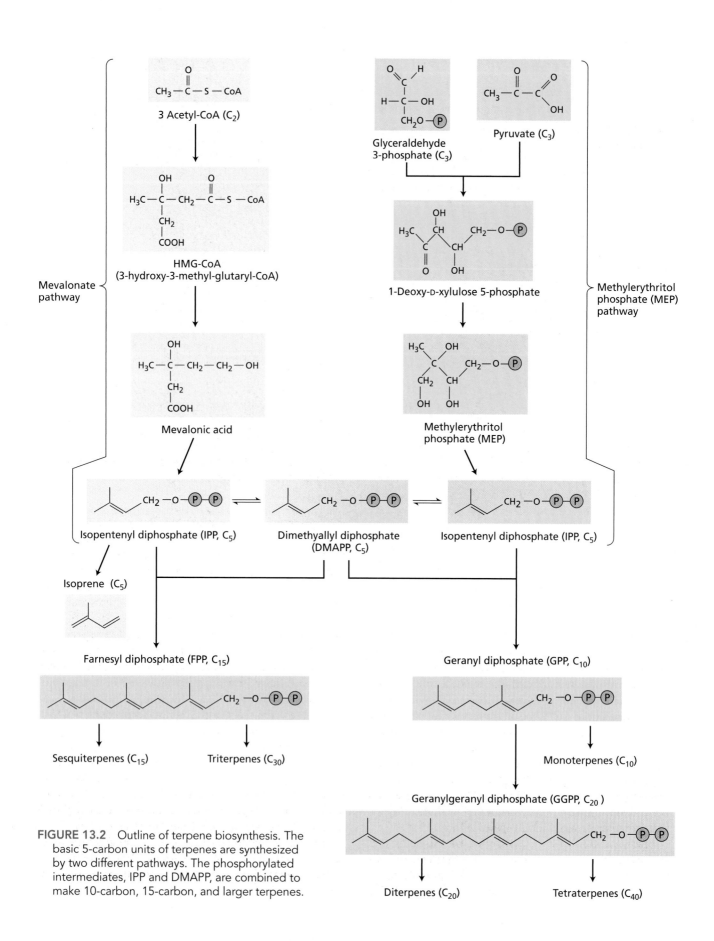

**FIGURE 13.2** Outline of terpene biosynthesis. The basic 5-carbon units of terpenes are synthesized by two different pathways. The phosphorylated intermediates, IPP and DMAPP, are combined to make 10-carbon, 15-carbon, and larger terpenes.

derived from the chloroplastic MEP pathway. However, cross talk between these two pathways does occasionally occur, leading to terpenes that are "mixed" with regard to their biosynthetic origin.

## Some terpenes have roles in growth and development

Certain terpenes have well-characterized functions in plant growth or development and so can be considered primary rather than secondary metabolites. For example, the *gibberellins* (see Chapter 20), an important group of plant hormones, are diterpenes. *Brassinosteroids* (see Chapter 24), another class of plant hormones with growth-regulating functions, originate from triterpenes.

Sterols are triterpene derivatives that are essential components of cell membranes, which they stabilize by interacting with phospholipids. The red, orange, and yellow carotenoids are tetraterpenes that function as accessory pigments in photosynthesis and protect photosynthetic tissues from photooxidation (see Chapter 7). The hormone abscisic acid (see Chapter 23) is a $C_{15}$ terpene produced by degradation of a carotenoid precursor.

Long-chain polyterpene alcohols known as *dolichols* function as carriers of sugars in cell wall and glycoprotein synthesis (see Chapter 15). Terpene-derived side chains, such as the phytol side chain of chlorophyll (see Chapter 7), help anchor certain molecules in membranes. The vast majority of terpenes, however, are secondary metabolites presumed to be involved in plant defenses.

## Terpenes defend many plants against herbivores

Terpenes are toxins and feeding deterrents to many herbivorous insects and mammals; thus they appear to play important defensive roles in the plant kingdom (Gershenzon and Croteau 1992). For example, monoterpene esters called **pyrethroids**, found in the leaves and flowers of *Chrysanthemum* species, show striking insecticidal activity. Both natural and synthetic pyrethroids are popular ingredients in commercial insecticides because of their low persistence in the environment and their negligible toxicity to mammals.

In conifers such as pine and fir, monoterpenes accumulate in resin ducts found in the needles, twigs, and trunk. These compounds are toxic to numerous insects, including bark beetles, which are serious pests of conifer species throughout the world. Many conifers respond to bark beetle infestation by producing additional quantities of monoterpenes (Trapp and Croteau 2001).

Many plants contain mixtures of volatile monoterpenes and sesquiterpenes, called **essential oils**, that lend a characteristic odor to their foliage. Peppermint, lemon, basil, and sage are examples of plants that contain essential oils. The

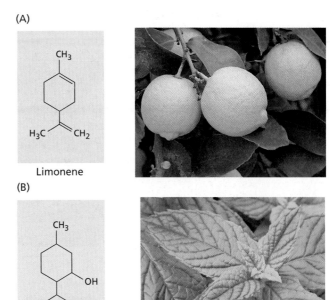

(A)

Limonene

(B)

Menthol

**FIGURE 13.3** Structures of limonene (A) and menthol (B). These two well-known monoterpenes serve as defenses against insects and other organisms that feed on the plants that produce them. (A, lemon tree [*Citrus × limon*], photo © Soren Pilman/istockphoto; B, peppermint [genus *Mentha*], photo © Jose Antonio Santiso Fernández/istockphoto.)

chief monoterpene constituent of lemon oil is limonene; that of peppermint oil is menthol (**FIGURE 13.3**).

Essential oils have well-known insect repellent properties. They are frequently found in glandular hairs that project outward from the epidermis and serve to "advertise" the toxicity of the plant, repelling potential herbivores even before they take a trial bite. Within the glandular hairs, the terpenes are stored in a modified extracellular space (**FIGURE 13.4**). Essential oils can be extracted from plants by steam distillation and are important commercially in flavoring foods and making perfumes (see **WEB ESSAY 13.1**).

Among the nonvolatile terpene antiherbivore compounds are the **limonoids**, a group of triterpenes ($C_{30}$) well known as bitter substances in citrus fruits. Perhaps the most powerful deterrent to insect feeding known is *azadirachtin* (**FIGURE 13.5A**), a complex limonoid from the neem tree (*Azadirachta indica*) of Africa and Asia. Azadirachtin is a feeding deterrent to some insects at doses as low as 50 parts per billion, and it exerts a variety of toxic effects (Aerts and Mordue 1997; Veitch et al. 2008). It has considerable potential as a commercial insect control agent because of its low toxicity to mammals, and several preparations containing azadirachtin are now being marketed in North America and India.

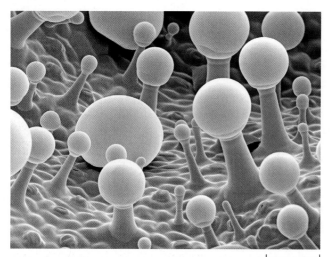

**FIGURE 13.4** Monoterpenes and sesquiterpenes are commonly found in glandular hairs on the surface of a plant. This false-colored scanning electron micrograph shows glandular trichomes (microscopic hairs, purple) on the calyx of a clary sage (*Salvia sclarea*) plant. The trichomes are secreting globules of essential oils (round, white). (© Andrew Syred/Photo Researchers, Inc.)

The **phytoecdysones**, first isolated from the common fern (*Polypodium vulgare*), are a group of plant steroids that have the same basic structure as insect molting hormones (**FIGURE 13.5B**). Ingestion of phytoecdysones by insects disrupts molting and other developmental processes, often with lethal consequences. In addition, phytoecdysones were recently found to have a defensive function against plant-parasitic nematodes (Soriano et al. 2004).

Triterpenes that defend plants against vertebrate herbivores include cardenolides and saponins. **Cardenolides** are glycosides (compounds containing an attached sugar or sugars) that taste bitter and are extremely toxic to higher animals. In humans they have dramatic effects on the heart muscle through their influence on $Na^+/K^+$-ATPases. In carefully regulated doses, they slow and strengthen the heartbeat. Cardenolides extracted from foxglove (*Digitalis*) are prescribed to millions of patients for the treatment of some types of heart disease.

**Saponins** are steroid and triterpene glycosides, so named because of their soaplike properties. The presence of both lipid-soluble (the steroid or triterpene) and water-soluble (the sugar) elements in one molecule gives saponins detergent properties, and they form a soapy lather when shaken with water. The toxicity of saponins is thought to be a result of their ability to form complexes with sterols. Saponins may interfere with sterol uptake from the digestive system or disrupt cell membranes after being absorbed into the bloodstream. (See WEB TOPIC 13.2 for more information about structures of triterpenes.)

## Phenolic Compounds

Plants produce a large variety of secondary compounds that contain a phenol group: a hydroxyl functional group on an aromatic ring:

(A) Azadirachtin, a limonoid

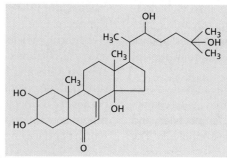

**FIGURE 13.5** Structure of two triterpenes, azadirachtin (A) and α-ecdysone (B), that serve as powerful insecticides. Azadirachtin affects more than 200 species of insects and can be considered a natural insecticide. α-Ecdysone, a plant-derived steroidal prohormone of the insect molting hormone 20-hydroxyecdysone, can cause irregular molting in insect herbivores. (A, photo of neem leaves © RN Photos/istockphoto; B, photo of *Polypodium vulgare* leaves, © blickwinkel/Alamy.)

(B) α-Ecdysone, an insect molting hormone

These substances are classified as *phenolic compounds*, or **phenolics**. Plant phenolics are a chemically heterogeneous group of nearly 10,000 individual compounds: Some are soluble only in organic solvents, some are water-soluble carboxylic acids and glycosides, and others are large, insoluble polymers.

In keeping with their chemical diversity, phenolics play a variety of roles in the plant. Many serve as defenses against herbivores and pathogens. Others function in mechanical support, in attracting pollinators and fruit dispersers, in absorbing harmful ultraviolet radiation, or in reducing the growth of nearby competing plants. After giving a brief account of phenolic biosynthesis, we will discuss three principal groups of phenolic compounds and what is known about their roles in the plant.

### Phenylalanine is an intermediate in the biosynthesis of most plant phenolics

Plant phenolics are synthesized by several different routes and thus constitute a heterogeneous group from a metabolic point of view. Two basic pathways are involved: the shikimic acid pathway and the malonic acid pathway (**FIGURE 13.6**). The shikimic acid pathway participates in the biosynthesis of most plant phenolics. The malonic acid pathway, although an important source of phenolic secondary products in fungi and bacteria, is of less significance in higher plants.

The **shikimic acid pathway** converts simple carbohydrate precursors derived from glycolysis and the pentose phosphate pathway into the three aromatic amino acids: phenylalanine, tyrosine, and tryptophan (see **WEB TOPIC 13.3**) (Herrmann and Weaver 1999). One of the pathway intermediates is shikimic acid, which has given its name to this whole sequence of reactions. The well-known broad-spectrum herbicide glyphosate (available commercially as Roundup) kills plants by blocking a step in this pathway (see Appendix 1). The shikimic acid pathway is present in plants, fungi, and bacteria but is not found in animals. Animals have no way to synthesize aromatic amino acids—phenylalanine, tyrosine, and tryptophan—which are therefore essential nutrients in animal diets.

The most abundant classes of phenolic secondary compounds in plants are derived from phenylalanine via the elimination of an ammonia molecule to form cinnamic acid (**FIGURE 13.7**). This reaction is catalyzed by **phenylalanine ammonia lyase** (**PAL**), perhaps the most studied enzyme in plant secondary metabolism. PAL is situated at a branch point between primary and secondary metabolism, so the reaction it catalyzes is an important regulatory step in the formation of many phenolic compounds.

The activity of PAL is increased by environmental factors such as low nutrient levels, light (through its effect on phytochromes), and fungal infection. The point of control appears to be the initiation of transcription. Fungal invasion, for example, triggers the transcription of messenger RNA that codes for PAL, thus increasing the amount of PAL in the plant, which then stimulates the synthesis of phenolic compounds. The regulation of PAL activity in many plant species is made more complex by the exis-

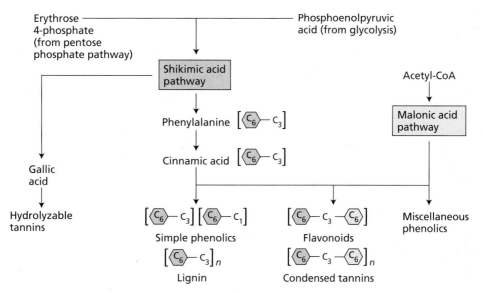

**FIGURE 13.6** Plant phenolics are synthesized in several different ways. In higher plants, most phenolics are derived at least in part from phenylalanine, a product of the shikimic acid pathway. Formulas in brackets indicate the basic arrangement of carbon skeletons: $C_6$ indicates a benzene ring, and $C_3$ is a three-carbon chain. More detail on the pathway from phenylalanine onward is given in Figure 13.7.

**FIGURE 13.7** Outline of phenolic biosynthesis from phenylalanine onward. The formation of many plant phenolics, including simple phenylpropanoids, coumarins, benzoic acid derivatives, lignin, anthocyanins, isoflavones, condensed tannins, and other flavonoids, begins with phenylalanine.

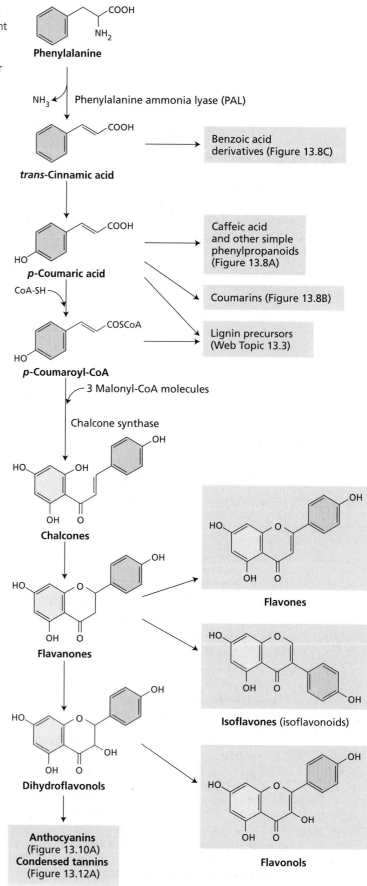

tence of multiple PAL-encoding genes, some of which are expressed only in specific tissues or only under certain environmental conditions (Logemann et al. 1995).

Reactions subsequent to that catalyzed by PAL lead to the addition of more hydroxyl groups and other substituents. The metabolites *trans*-cinnamic acid, *p*-coumaric acid, and their derivatives are simple phenolic compounds called **phenylpropanoids** because they contain a benzene ring:

and a three-carbon side chain.

Simple phenolic compounds are widespread in vascular plants and appear to function in various capacities. Their structures include the following:

- *Simple phenylpropanoids*, such as *trans*-cinnamic acid, *p*-coumaric acid, and their derivatives, such as caffeic acid, which have a basic phenylpropanoid carbon skeleton (**FIGURE 13.8A**)

$$C_6 - C_3$$

- Phenylpropanoid lactones (cyclic esters) called *coumarins*, which also have a phenylpropanoid carbon skeleton (**FIGURE 13.8B**)

- *Benzoic acid derivatives*, which have a carbon skeleton formed from phenylpropanoids by the cleavage of a two-carbon fragment from the side chain (**FIGURE 13.8C**) (see also Figure 13.7):

$$C_6 - C_1$$

As with many other secondary products, plants can elaborate on the basic carbon skeletons of these simple phenolic compounds to make more complex products.

Now that the biosynthetic pathways leading to most widespread phenolic compounds have been determined, researchers have turned

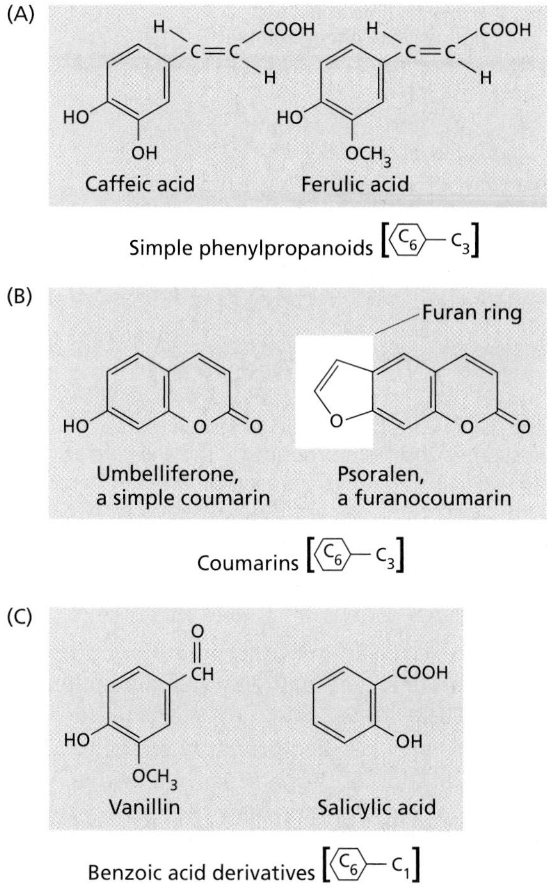

(A)

Caffeic acid     Ferulic acid

Simple phenylpropanoids [C$_6$ — C$_3$]

(B)

Furan ring

Umbelliferone,     Psoralen,
a simple coumarin     a furanocoumarin

Coumarins [C$_6$ — C$_3$]

(C)

Vanillin     Salicylic acid

Benzoic acid derivatives [C$_6$ — C$_1$]

**FIGURE 13.8** Simple phenolic compounds play a variety of roles in plants. (A) Caffeic acid and ferulic acid may be released into the soil and inhibit the growth of neighboring plants. (B) Psoralen is a furanocoumarin that exhibits phototoxicity to insect herbivores. (C) Salicylic acid is a plant hormone that is involved in systemic resistance to plant pathogens.

their attention to studying how those pathways are regulated. In some cases, specific enzymes, such as PAL, are important in controlling flux through the pathway. Several transcription factors have been shown to regulate phenolic metabolism by binding to the promoter regions of certain biosynthetic genes and activating transcription. Some of these factors activate the transcription of large groups of genes (Jin and Martin 1999).

### Ultraviolet light activates some simple phenolics

Many simple phenolic compounds have important roles in plants as defenses against insect herbivores and fungi. Of special interest is the phototoxicity of certain coumarins called **furanocoumarins**, which have an attached furan ring (see Figure 13.8B). These compounds are not toxic until they are activated by light. Sunlight in the ultravio-

let A (UV-A) region of the spectrum (320–400 nm) causes some furanocoumarins to become activated to a high-energy electron state. Activated furanocoumarins can insert themselves into the double helix of DNA and bind to the pyrimidine bases cytosine and thymine, thus blocking transcription and repair and leading eventually to cell death.

Phototoxic furanocoumarins are especially abundant in members of the Umbelliferae, including celery, parsnip, and parsley. In celery, the concentration of these compounds can increase about one-hundredfold if the plant is stressed or diseased. Celery pickers, and even some grocery shoppers, have been known to develop skin rashes from handling stressed or diseased celery. Some insects are adapted to survive on plants that contain furanocoumarins and other phototoxic compounds by living in silken webs or rolled-up leaves, which screen out the activating wavelengths (Sandberg and Berenbaum 1989).

### The release of phenolics into the soil may limit the growth of other plants

From leaves, roots, and decaying litter, plants release a variety of primary and secondary metabolites into the environment. The release of secondary compounds by one plant that have an effect on neighboring plants is referred to as **allelopathy**. If a plant can reduce the growth of nearby plants by releasing chemicals into the soil, it may increase its access to light, water, and nutrients and thus its evolutionary fitness. Generally speaking, the term *allelopathy* has come to be applied to the harmful effects of plants on their neighbors, although a precise definition also includes beneficial effects.

Simple phenylpropanoids and benzoic acid derivatives are frequently cited as having allelopathic activity. Compounds such as caffeic acid and ferulic acid (see Figure 13.8A) are found in soil in appreciable amounts and have been shown in laboratory experiments to inhibit the germination and growth of many plants (Inderjit et al. 1995).

Allelopathy is currently of great interest because of its potential agricultural applications (Kruse et al. 2000). Reductions in crop yields caused by weeds or residues from the previous crop may in some cases be a result of allelopathy. An exciting future prospect is the development of crop plants genetically engineered to be allelopathic to weeds (see **WEB ESSAY 13.2**).

### Lignin is a highly complex phenolic macromolecule

After cellulose, the most abundant organic substance in plants is **lignin**, a highly branched polymer of phenylpropanoid groups:

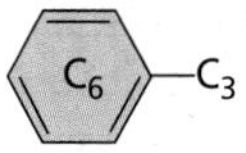

C$_6$ — C$_3$

Lignin plays both primary and secondary roles in plants. Its precise structure is not known because it is difficult to extract from plants, in which it is covalently bound to cellulose and other polysaccharides of the cell wall.

Lignin is generally formed from three different phenylpropanoid alcohols: coniferyl, coumaryl, and sinapyl, all of which are synthesized from phenylalanine via various cinnamic acid derivatives. The phenylpropanoid alcohols are joined into a polymer through the action of enzymes that generate free-radical intermediates. The proportions of the three phenylpropanoid alcohols in lignin vary among species, plant organs, and even layers of a single cell wall. In the polymer, there are often multiple C—C and C—O—C bonds in each phenylpropanoid alcohol unit, resulting in a complex structure that branches in three dimensions.

Unlike the monomeric units of polymers such as starch, rubber, or cellulose, those of lignin do not appear to be linked in a simple, repeating way. However, recent research suggests that a guiding protein may bind the individual units during lignin biosynthesis, giving rise to a scaffold that then directs the formation of a large, repeating unit (Davin and Lewis 2000; Hatfield and Vermerris 2001). (See **WEB TOPIC 13.4** for the partial structure of a hypothetical lignin molecule.)

Lignin is found in the cell walls of various cell types that make up supporting and conducting tissues, notably the tracheids and vessel elements of the xylem. It is deposited chiefly in the thickened secondary wall, but may also be present in the primary wall and middle lamella in close contact with the celluloses and hemicelluloses already present. The mechanical rigidity of lignin strengthens stems and vascular tissue, allowing upward growth and permitting water and minerals to be conducted through the xylem under negative pressure without collapse of the tissue. Because lignin is such a key component of water transport tissue, the ability to synthesize lignin must have been one of the most important adaptations permitting primitive plants to colonize dry land.

Besides providing mechanical support, lignin has significant protective functions in plants. Its physical toughness deters herbivory, and its chemical durability makes it relatively indigestible. By bonding to cellulose and protein, lignin also reduces the digestibility of those substances. Lignification blocks the growth of pathogens and is a common response to infection or wounding.

### There are four major groups of flavonoids

The **flavonoids** are one of the largest classes of plant phenolics. The basic carbon skeleton of a flavonoid contains 15 carbons arranged in two aromatic rings connected by a three-carbon bridge:

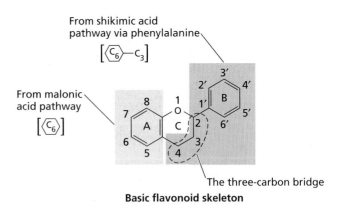

**FIGURE 13.9** Basic flavonoid carbon skeleton. Flavonoids are synthesized from products of the shikimic acid and malonic acid pathways. Flavonoids contain 15 carbons in the basic molecular skeleton provided by two aromatic rings and one 3-carbon bridge. Positions of carbons on the flavonoid ring system are numbered as shown.

This structure results from two separate biosynthetic pathways: the shikimic acid pathway and the malonic acid pathway (**FIGURE 13.9**).

Flavonoids are classified primarily on the basis of the degree of oxidation of the three-carbon bridge. We will discuss four of these groups here: the anthocyanins, the flavones, the flavonols, and the isoflavones (see Figure 13.7).

The basic flavonoid carbon skeleton may have numerous substituents. Hydroxyl groups are usually present at positions 3, 5, and 7, but they may also be found at other positions. Sugars are very common as well; in fact, the majority of flavonoids exist naturally as glycosides.

Whereas both hydroxyl groups and sugars increase the water solubility of flavonoids, other substituents, such as methyl ethers or modified isopentyl units, make flavonoids lipophilic (hydrophobic). Different types of flavonoids perform very different functions in the plant, including pigmentation and defense.

### Anthocyanins are colored flavonoids that attract animals

The colored pigments of plants provide visual cues that help to attract pollinators and seed dispersers. These pigments are of two principal types: carotenoids and flavonoids. *Carotenoids*, as we have already seen, are yellow, orange, and red terpenoid compounds that also serve as accessory pigments in photosynthesis (see Chapter 7). The flavonoids also include a wide range of colored substances. The most widespread group of pigmented flavonoids is the **anthocyanins**, which are responsible for most of the red, pink, purple, and blue colors observed in flowers and fruits.

Anthocyanins are glycosides that can have various sugars at position 3 (**FIGURE 13.10A**) and sometimes elsewhere. Without their sugars, anthocyanins are known as **anthocyanidins** (**FIGURE 13.10B**). Anthocyanin color is influenced by many factors, including the number of hydroxyl and methoxyl groups in ring B of the anthocyanidin (see Figure 13.10A), the presence of aromatic acids esterified to the main skeleton, and the pH of the cell vacuole in which the anthocyanins are stored. Anthocyanins may also exist in supramolecular complexes along with chelated metal ions and flavone copigments. The blue pigment of dayflower (*Commelina communis*) was found to consist of a large complex of six anthocyanin molecules, six flavones, and two associated magnesium ions (Kondo et al. 1992). The most common anthocyanidins and their colors are shown in Figure 13.10 and **TABLE 13.1**.

Considering the variety of factors affecting anthocyanin coloration and the possible presence of carotenoids as well, it is not surprising that so many different shades of flower and fruit color are found in nature. The evolution of flower color may have been governed by selection pressure to attract different types of pollinators, many of which have different color preferences.

## Flavones and flavonols may protect against damage by ultraviolet light

Two other groups of flavonoids found in flowers are **flavones** and **flavonols** (see Figure 13.7). These flavonoids

| TABLE 13.1 Effects of ring substituents on anthocyanidin color | | |
| --- | --- | --- |
| Anthocyanidin | Substituents | Color |
| Pelargonidin | 4'— OH | Orange red |
| Cyanidin | 3'— OH, 4'— OH | Purplish red |
| Delphinidin | 3'— OH, 4'— OH, 5'— OH | Bluish purple |
| Peonidin | 3'— $OCH_3$, 4'— OH | Rosy red |
| Petunidin | 3'— $OCH_3$, 4'— OH, 5'— $OCH_3$ | Purple |

generally absorb light at shorter wavelengths than do anthocyanins, so they are not visible to the human eye. However, insects such as bees, which see farther into the ultraviolet range of the spectrum than humans do, may respond to flavones and flavonols as visual attractant cues (**FIGURE 13.11**). Flavonols in a flower often form symmetric patterns of stripes, spots, or concentric circles called *nectar guides* (Lunau 1992). These patterns may be conspicuous to insects and are thought to help indicate the location of pollen and nectar.

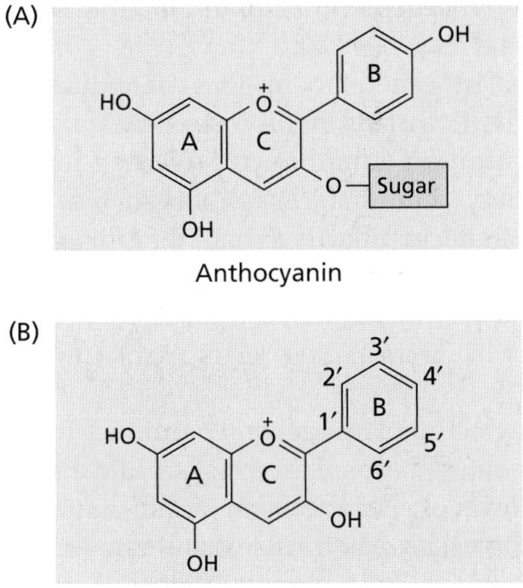

Flavones and flavonols are not restricted to flowers; they are also present in the leaves of all green plants. These two classes of flavonoids protect cells from excessive UV-B radiation (280–320 nm) because they accumulate in the epidermal layers of leaves and stems and absorb light strongly in the UV-B region while allowing the visible (photosynthetically active) wavelengths to pass through uninterrupted. In addition, exposure of plants to increased UV-B light has been demonstrated to increase the synthesis of flavones and flavonols. UV-B radiation is known to induce mutations in the DNA as well as oxidative stress, which has the potential to damage cellular macromolecules.

*Arabidopsis thaliana* mutants that lack the enzyme chalcone synthase produce no flavonoids. Lacking flavonoids, these plants are much more sensitive to UV-B radiation than wild-type individuals are, and they grow very poorly under normal conditions. When shielded from UV light, however, they grow normally (Li et al. 1993). A group of simple phenylpropanoid esters are also important in UV protection in Arabidopsis.

Other functions of flavones and flavonols have also been discovered. For example, these flavonoids, when secreted into the soil by legume roots, mediate the interaction of legumes and rhizobacteria, their nitrogen-fixing symbionts, as described in Chapter 12. As will be discussed in Chapter 19, recent work suggests that these flavonoids also play a regulatory role in plant development as modulators of polar auxin transport.

**FIGURE 13.10** The structures of anthocyanins (A) and anthocyanidins (B). The colors of anthocyanidins depend in part on the substituents attached to ring B (see Table 13.1). An increase in the number of hydroxyl groups shifts absorption to a longer wavelength and gives a bluer color. Replacement of a hydroxyl group with a methoxyl group (—$OCH_3$) shifts absorption to a slightly shorter wavelength, resulting in a redder color.

## Isoflavonoids have widespread pharmacological activity

The **isoflavones** (*isoflavonoids*) are a group of flavonoids in which the position of one aromatic ring (ring B) is shifted

(A)

(B)

**FIGURE 13.11** Black-eyed Susan (*Rudbeckia* sp.) as seen by humans (A) and as it might appear to honeybees (B). (A) To humans, the flowers have yellow rays and a brown central disc. (B) To bees, the tips of the rays appear "light yellow," the inner portion of the rays "dark yellow," and the central disc "black." UV-absorbing flavonols are found in the inner parts of the rays, but not in the tips. The distribution of flavonols in the rays creates a "bull's-eye" pattern visible to honeybees, which presumably helps them locate pollen and nectar. Special lighting was used to simulate the spectral sensitivity of the honeybee visual system. (Courtesy of Thomas Eisner.)

(see Figure 13.7). Isoflavonoids, which are found mostly in legumes, have several different biological activities. Some, such as rotenone, can be used effectively as insecticides, pesticides (e.g., as rat poison), and piscicides (fish poisons). Other isoflavones have anti-estrogenic effects; for example, sheep grazing on clover rich in isoflavonoids often suffer from infertility. The ring system of isoflavones has a three-dimensional structure similar to that of steroids (see Figure 13.5B), allowing these substances to bind to estrogen receptors. Isoflavones may also be responsible for the anticancer benefits of foods prepared from soybeans.

In the past few years, isoflavones have become best known for their role as *phytoalexins*, antimicrobial compounds synthesized in response to bacterial or fungal infection that help limit the spread of the invading pathogen. Phytoalexins are discussed in more detail later in this chapter.

### Tannins deter feeding by herbivores

A second category of plant phenolic polymers with defensive properties, besides lignin, is the **tannins**. The term *tannin* was first used to describe compounds that could convert raw animal hides into leather in the process known as *tanning*. Tannins bind the collagen proteins of animal hides, thereby increasing their resistance to heat, water, and microbes.

There are two categories of tannins: condensed and hydrolyzable. **Condensed tannins** are compounds formed by the polymerization of flavonoid units (**FIGURE 13.12A**). They are common constituents of woody plants. Because condensed tannins can often be hydrolyzed into anthocyanidins by treatment with strong acids, they are sometimes called *pro-anthocyanidins*.

**Hydrolyzable tannins** are heterogeneous polymers containing phenolic acids, especially gallic acid, and simple sugars (**FIGURE 13.12B**). They are smaller than condensed tannins and may be hydrolyzed more easily; only dilute acid is needed. Most tannins have molecular masses between 600 and 3000 Da.

Tannins are general toxins that can reduce the growth and survival of many herbivores when added to their diets. In addition, tannins act as feeding repellents to a great variety of animals. Mammals such as cattle, deer, and apes characteristically avoid plants or parts of plants with high tannin contents. Unripe fruits, for instance, frequently have very high tannin levels, which deter feeding on the fruits until their seeds are mature enough for dispersal.

Although crop plants generally produce fewer secondary metabolites, there are exceptions. Humans often prefer a certain level of astringency in tannin-containing foods, such as apples, blackberries, tea, and grapes. The tannins in red wine have been shown to block the formation of endothelin-1, a signaling molecule that makes blood vessels constrict (Corder et al. 2001). This effect of wine tannins may account for the often-touted health benefits of red wine, especially the reduction in the risk of heart disease associated with moderate red wine consumption. In recent years, however, another phenolic compound, the stilbene phenylpropanoid resveratrol, has also been identified as a health benefit factor in red wine.

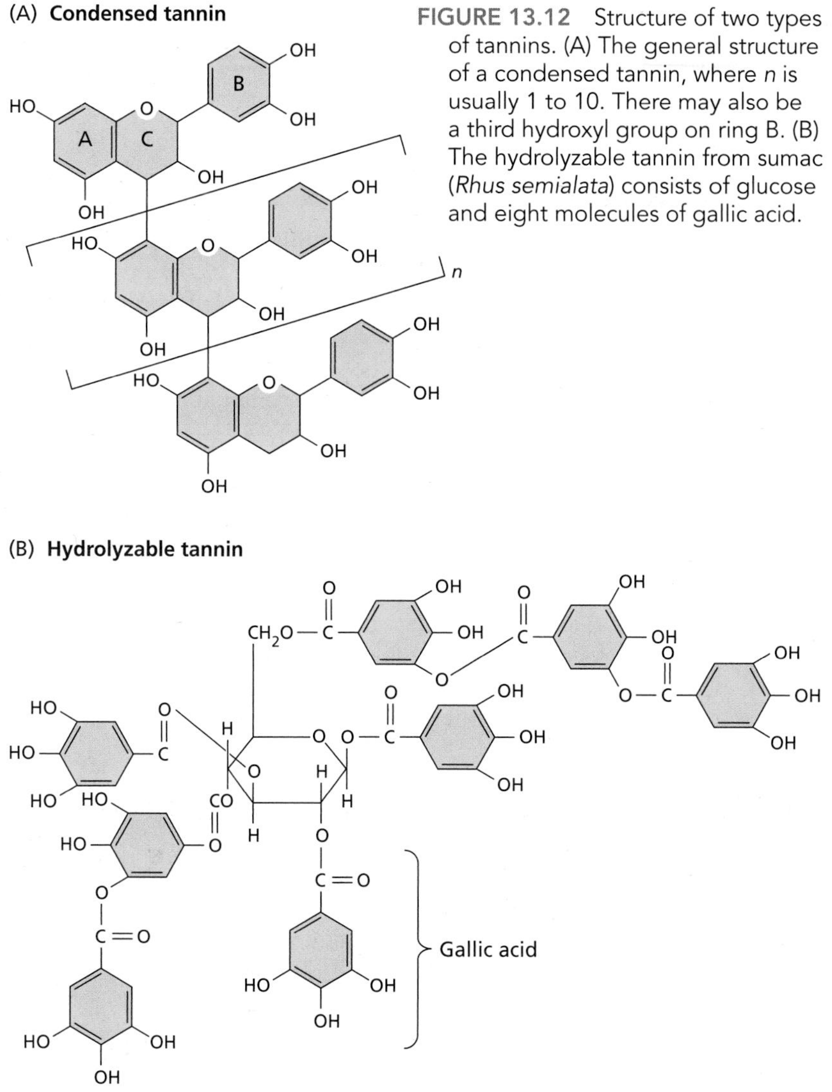

**(A) Condensed tannin**

FIGURE 13.12 Structure of two types of tannins. (A) The general structure of a condensed tannin, where *n* is usually 1 to 10. There may also be a third hydroxyl group on ring B. (B) The hydrolyzable tannin from sumac (*Rhus semialata*) consists of glucose and eight molecules of gallic acid.

**(B) Hydrolyzable tannin**

Gallic acid

Moderate amounts of specific tannins may have health benefits for humans, but the defensive properties of most tannins are due to their toxicity, which is generally attributed to their ability to bind proteins nonspecifically. It has long been thought that plant tannins bind proteins in the guts of herbivores by forming hydrogen bonds between their hydroxyl groups and electronegative sites on the proteins (**FIGURE 13.13A**). More recent evidence indicates that tannins and other phenolics can also bind to dietary protein in a covalent fashion (**FIGURE 13.13B**). The foliage of many plants contains enzymes that oxidize phenolics to their corresponding quinone forms in the guts of herbivores (Felton et al. 1989). Quinones are highly reactive electrophilic molecules that readily react with the nucleophilic —$NH_2$ and —SH groups of proteins (Appel 1993). By whatever mechanism protein–tannin binding occurs, this process has a negative effect on herbivore nutrition. Tannins can inactivate herbivore digestive enzymes and create complex aggregates of tannins and plant proteins that are difficult to digest.

Herbivores that habitually feed on tannin-rich plant material appear to possess some interesting adaptations to remove tannins from their digestive systems. For example, some mammals, such as rodents and rabbits, produce salivary proteins with a very high proline content (25–45%) that have a high affinity for tannins. Secretion of these proteins, which is induced by ingestion of food with a high tannin content, greatly diminishes the toxic effects of tannins (Butler 1989). The large number of proline residues gives these proteins a very flexible, open conformation and a high degree of hydrophobicity, which facilitate binding to tannins.

Plant tannins also serve as defenses against microorganisms. For example, the nonliving heartwood of many trees contains high concentrations of tannins, which help prevent fungal and bacterial decay.

# Nitrogen-Containing Compounds

A large variety of plant secondary metabolites have nitrogen as part of their structure. Included in this category are such well-known antiherbivore defenses as alkaloids and cyanogenic glycosides, which are of considerable interest because of their toxicity to humans as well as their medicinal properties. Most nitrogenous secondary metabolites are synthesized from common amino acids.

In this section we will examine the structures and biological properties of various nitrogen-containing secondary metabolites, including alkaloids, cyanogenic glycosides, glucosinolates, and nonprotein amino acids.

## Alkaloids have dramatic physiological effects on animals

The **alkaloids** are a large family of more than 15,000 nitrogen-containing secondary metabolites. They are found in approximately 20% of vascular plant species. The nitrogen atom in these compounds is usually part of a **heterocyclic ring**, a ring that contains both nitrogen and carbon atoms. As a group, alkaloids are best known for their striking pharmacological effects on vertebrate animals.

As their name would suggest, most alkaloids are alkaline. At the pH values commonly found in the cytosol (pH 7.2) or the vacuole (pH 5–6), the nitrogen atom is proto-

(A) Hydrogen bonding between tannin and protein

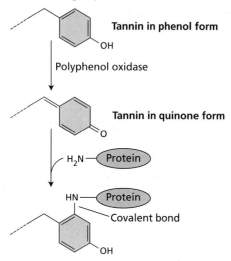

Tannin

$\delta^+$ $\delta^-$
O—H ••••• N—Protein
            H₂

**FIGURE 13.13** Proposed mechanisms for the interaction of tannins with proteins. (A) Hydrogen bonds may form between the phenolic hydroxyl groups of the tannin and electronegative sites on the protein. (B) Phenolic hydroxyl groups may bind covalently to proteins following activation by oxidative enzymes, such as polyphenol oxidase.

(B) Covalent bonding to protein after oxidation

**Tannin in phenol form**

OH

Polyphenol oxidase

**Tannin in quinone form**

O

H₂N—Protein

HN—Protein

Covalent bond

OH

**Tannin linked to protein**

nated; hence alkaloids are positively charged and are generally water soluble.

Alkaloids are usually synthesized from one of a few common amino acids—in particular, lysine, tyrosine, or tryptophan. However, the carbon skeleton of some alkaloids contains a component derived from the terpene pathway. TABLE 13.2 lists the major alkaloid types and their amino acid precursors. Several different types, including nicotine and its relatives (FIGURE 13.14), are derived from ornithine, an intermediate in arginine biosynthesis. The B vitamin nicotinic acid (niacin) is a precursor of the pyridine (six-membered) ring of this alkaloid; the pyrrolidine (five-membered) ring of nicotine arises from ornithine (FIGURE 13.15). Nicotinic acid is also a constituent of $NAD^+$ and $NADP^+$, which serve as electron carriers in metabolism.

The role of alkaloids in plants has been a subject of speculation for at least a century. Alkaloids were once thought to be nitrogenous wastes (analogous to urea and uric acid

## TABLE 13.2
## Major types of alkaloids, their amino acid precursors, and well-known examples of each type

| Alkaloid class | Structure | Biosynthetic precursor | Examples | Human uses |
|---|---|---|---|---|
| Pyrrolidine | | Ornithine (aspartate) | Nicotine | Stimulant, depressant, tranquilizer |
| Tropane | | Ornithine | Atropine | Prevention of intestinal spasms, antidote to other poisons, dilation of pupils for examination |
| | | | Cocaine | Stimulant of the central nervous system, local anesthetic |
| Piperidine | | Lysine (or acetate) | Coniine | Poison (paralyzes motor neurons) |
| Pyrrolizidine | | Ornithine | Retrorsine | None |
| Quinolizidine | | Lysine | Lupinine | Restoration of heart rhythm |
| Isoquinoline | | Tyrosine | Codeine | Analgesic (pain relief), treatment of coughs |
| | | | Morphine | Analgesic |
| Indole | | Tryptophan | Psilocybin | Hallucinogen |
| | | | Reserpine | Treatment of hypertension, treatment of psychoses |
| | | | Strychnine | Rat poison, treatment of eye disorders |

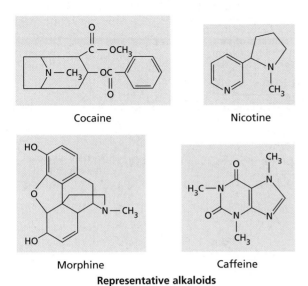

Cocaine      Nicotine

Morphine      Caffeine

**Representative alkaloids**

**FIGURE 13.14** Examples of alkaloids, a diverse group of secondary metabolites that contain nitrogen, usually as part of a heterocyclic ring. Caffeine is a purine-type alkaloid similar to the nucleotide bases adenine and guanine. The pyrrolidine (five-membered) ring of nicotine arises from ornithine; the pyridine (six-membered) ring is derived from nicotinic acid.

in animals), nitrogen storage compounds, or growth regulators, but there is little evidence to support any of these functions. Most alkaloids are now believed to function as defenses against herbivores, especially mammals, because of their general toxicity and deterrence capability (Hartmann 1992).

Large numbers of livestock deaths are caused by the ingestion of alkaloid-containing plants. In the United States, many grazing livestock animals are poisoned each year by consumption of large quantities of alkaloid-containing plants such as lupines (*Lupinus*), larkspur (*Delphinium*), and groundsel (*Senecio*). This phenomenon may be due to the fact that domestic animals, unlike wild animals, have not been subjected to natural selection for avoidance of toxic plants. Indeed, some livestock actually seem to prefer alkaloid-containing plants to less harmful forage.

Nearly all alkaloids are also toxic to humans when taken in sufficient quantities. For example, strychnine, atropine, and coniine (from poison hemlock, *Conium maculatum*) are classic alkaloid poisons. At lower doses, however, many are useful pharmacologically. Morphine, codeine, and scopolamine are just a few of the plant alkaloids currently used in medicine. Other alkaloids, including cocaine, nicotine, and caffeine (see Figure 13.14), have widespread nonmedical uses as stimulants or sedatives.

On a cellular level, the mode of action of alkaloids in animals is quite variable. Many alkaloids interfere with components of the nervous system, especially neurotrans-

mitters; others affect membrane transport, protein synthesis, or miscellaneous enzyme activities.

One group of alkaloids, the pyrrolizidine alkaloids, illustrates how herbivores can become adapted to tolerate plant defensive compounds and even use them in their own defense (Hartmann 1999). Within plants, pyrrolizidine alkaloids occur naturally as nontoxic N-oxides. In the alkaline digestive tracts of many insect herbivores, however, they are quickly reduced to uncharged, hydrophobic tertiary alkaloids (**FIGURE 13.16**), which easily pass through membranes and are toxic. Nevertheless, some insect herbivores, like the cinnabar moth (*Tyria jacobeae*), have developed the ability to reconvert tertiary pyrrolizidine alkaloids into the nontoxic N-oxide form immediately after their absorption from the digestive tract. These herbivores may then store the N-oxides in their bodies as defenses against their own predators.

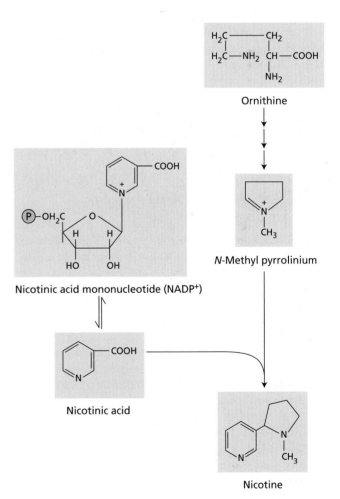

Ornithine

*N*-Methyl pyrrolinium

Nicotinic acid mononucleotide (NADP⁺)

Nicotinic acid

Nicotine

**FIGURE 13.15** Nicotine biosynthesis begins with the synthesis of nicotinic acid (niacin) from aspartate and glyceraldehyde 3-phosphate. Nicotinic acid is also a component of NAD⁺ and NADP⁺, important participants in biological oxidation–reduction reactions. The five-membered ring of nicotine is derived from ornithine, an intermediate in arginine biosynthesis.

**FIGURE 13.16** Two forms of pyrrolizidine alkaloids occur in nature: the N-oxide form and the tertiary alkaloid. The nontoxic N-oxide form found in plants is reduced to the toxic tertiary form in the digestive tracts of most insect herbivores. However, some adapted herbivores can convert the toxic tertiary form back into the nontoxic N-oxide. The two forms are illustrated here for senecionine, a pyrrolizidine alkaloid found in ragwort (*Senecio*).

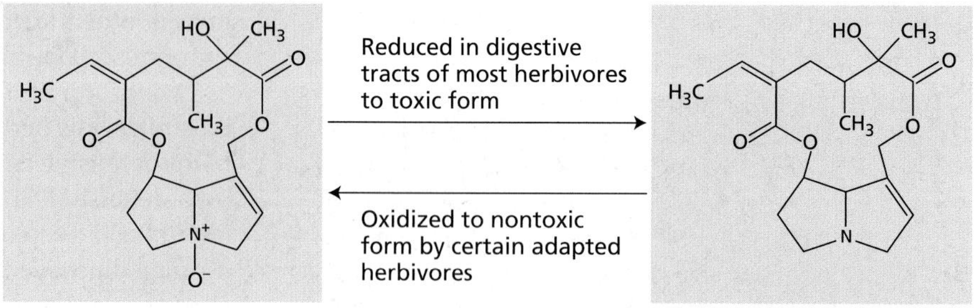

Reduced in digestive tracts of most herbivores to toxic form

Oxidized to nontoxic form by certain adapted herbivores

N-oxide (nontoxic form, stored in plants)

Tertiary alkaloid (toxic form)

Not all of the alkaloids that appear in plants are produced by the plant itself. Many grasses harbor endogenous fungal symbionts (endophytes) that grow in the apoplast and synthesize a variety of different alkaloids. Grasses with fungal symbionts often grow faster and are better defended against insect and mammalian herbivores than those without symbionts. Unfortunately, certain grasses with symbionts, such as tall fescue, are important pasture grasses that may become toxic to livestock when their alkaloid content is too high. Efforts are under way to breed tall fescue with alkaloid levels that are not poisonous to livestock but still provide protection against insects (see **WEB ESSAY 13.3**).

### Cyanogenic glycosides release the poison hydrogen cyanide

Various nitrogenous protective compounds other than alkaloids are found in plants. Two groups of these substances—cyanogenic glycosides and glucosinolates—are not themselves toxic but are readily broken down to give off poisons, some of which are volatile, when the plant is crushed. **Cyanogenic glycosides** release the well-known poisonous gas hydrogen cyanide (HCN).

The breakdown of cyanogenic glycosides in plants is a two-step enzymatic process. Species that make cyanogenic glycosides also make the enzymes necessary to hydrolyze the sugar and liberate HCN:

1. In the first step the sugar is cleaved by a glycosidase, a hydrolytic enzyme that separates sugars from other molecules to which they are linked (**FIGURE 13.17**).

2. In the second step the resulting hydrolysis product, called an α-hydroxynitrile or cyanohydrin, can decompose spontaneously at a low rate to liberate HCN. This second step can be accelerated by the lytic enzyme hydroxynitrile lyase.

Cyanogenic glycosides are not normally broken down in the intact plant because the glycoside and the degradative enzymes are spatially separated in different cellular compartments or in different tissues. In sorghum, for example, the cyanogenic glycoside dhurrin is present in the vacuoles of epidermal cells, while the hydrolytic and lytic enzymes are found in the mesophyll (Poulton 1990).

Under ordinary conditions this compartmentalization prevents decomposition of the glycoside. When the leaf is damaged, however, as during herbivore feeding, the cell contents of different tissues mix, and HCN forms. Cyanogenic glycosides are widely distributed in the plant kingdom and are found in many legumes, grasses, and species of the rose family.

Considerable evidence indicates that cyanogenic glycosides have a protective function in certain plants. HCN is a fast-acting toxin that inhibits metalloproteins such as the iron-containing cytochrome oxidase, a key enzyme of mitochondrial respiration. The presence of cyanogenic glycosides

**FIGURE 13.17** Enzyme-catalyzed hydrolysis of cyanogenic glycosides to release hydrogen cyanide. R and R′ represent various alkyl or aryl substituents. For example, if R is phenyl, R′ is hydrogen, and the sugar is the disaccharide β-gentiobiose, then the compound is amygdalin (the common cyanogenic glycoside found in the seeds of almonds, apricots, cherries, and peaches).

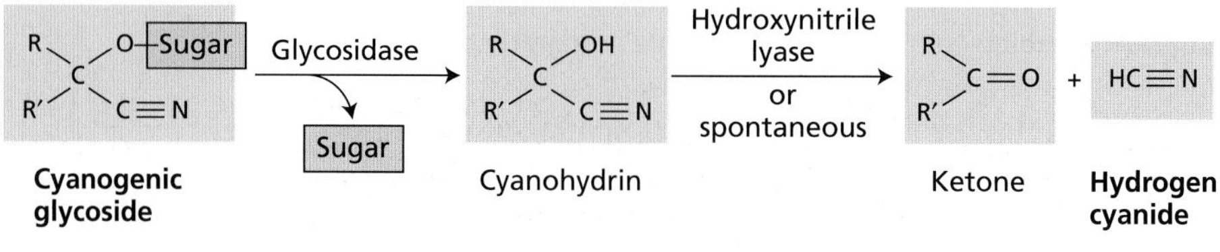

Cyanogenic glycoside → Glycosidase → Sugar → Cyanohydrin → Hydroxynitrile lyase or spontaneous → Ketone + Hydrogen cyanide

deters feeding by insects and other herbivores such as snails and slugs. As with other classes of secondary metabolites, however, some herbivores have adapted to feed on cyanogenic plants and can tolerate large doses of HCN.

The tubers of cassava (*Manihot esculenta*), a high-carbohydrate staple food in many tropical countries, contain high levels of cyanogenic glycosides. Traditional processing methods, such as grating, grinding, soaking, and drying, lead to the removal or degradation of a large fraction of the cyanogenic glycosides present in cassava tubers. However, chronic cyanide poisoning leading to partial paralysis of the limbs is still widespread in regions where cassava is a major food source because the traditional detoxification methods employed to remove cyanogenic glycosides from cassava are not completely effective. In addition, many populations that consume cassava have poor nutrition, which aggravates the effects of the cyanogenic glycosides.

Efforts are currently under way to reduce the cyanogenic glycoside content of cassava through both conventional breeding and genetic engineering approaches. However, the complete elimination of cyanogenic glycosides may not be desirable, because these substances are probably responsible for the pest resistance of stored cassava.

### Glucosinolates release volatile toxins

A second class of plant glycosides, called the **glucosinolates**, or mustard oil glycosides, break down to release defensive substances. Found principally in the Brassicaceae and related plant families, glucosinolates break down to produce the compounds responsible for the smell and taste of vegetables such as cabbage, broccoli, and radishes.

Glucosinolate breakdown is catalyzed by a hydrolytic enzyme, called a *thioglucosidase* or *myrosinase*, that cleaves glucose from its bond with the sulfur atom (**FIGURE 13.18**). The resulting aglycone—the nonsugar portion of the molecule—loses the sulfate and rearranges itself to give pungent and chemically reactive products, including isothiocyanates and nitriles, depending on the conditions of hydrolysis. These defensive products function as toxins and herbivore repellents. Like cyanogenic glycosides, glucosinolates are stored in the intact plant separately from the enzymes that

hydrolyze them, and they are brought into contact with these enzymes only when the plant is crushed.

As with other secondary metabolites, certain animals are adapted to feed on glucosinolate-containing plants without ill effects. For adapted herbivores such as the cabbage butterfly, glucosinolates serve as stimulants for adult feeding and egg laying, and the isothiocyanates produced after glucosinolate hydrolysis act as volatile attractants (Renwick et al. 1992). In addition, the caterpillars can redirect the glucosinolate hydrolysis reaction to the production of the less toxic nitriles (Wittstock et al. 2004).

Most of the recent research on glucosinolates in plant defense has concentrated on rape, or canola (*Brassica napus*), a major oilseed crop in both North America and Europe. Plant breeders have tried to lower the glucosinolate levels of rapeseed so that the high-protein seed meal remaining after oil extraction can be used as animal food. The first low-glucosinolate varieties tested in the field were unable to survive because of severe pest problems. However, more recently developed varieties with low glucosinolate levels in seeds but high glucosinolate levels in leaves are more resistant to pests and still provide a protein-rich seed residue for animal feeding.

### Nonprotein amino acids are toxic to herbivores

Plants and animals incorporate the same 20 amino acids into their proteins. However, many plants also contain unusual amino acids, called **nonprotein amino acids**, that are not incorporated into proteins. Instead, these amino acids are present in the free form and act as defensive substances. Many nonprotein amino acids are very similar to common protein amino acids. Canavanine, for example, is a close analog of arginine, and azetidine-2-carboxylic acid has a structure very much like that of proline (**FIGURE 13.19**).

Nonprotein amino acids exert their toxicity in various ways. Some block the synthesis or uptake of protein amino acids. Others, such as canavanine, can be mistakenly incorporated into proteins. After ingestion by an herbivore, canavanine is recognized by the enzyme that normally binds arginine to the arginine transfer RNA molecule, so

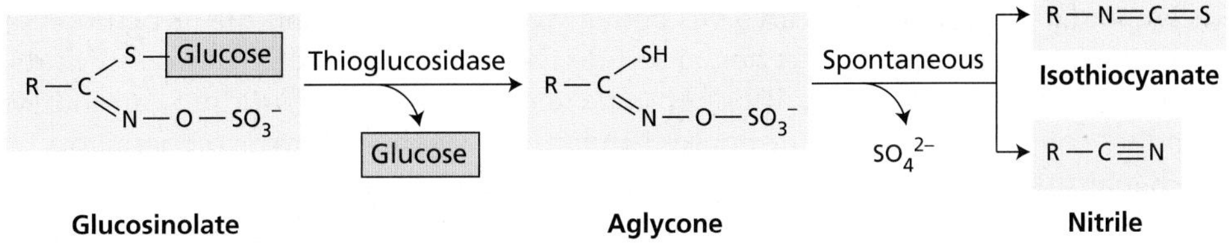

**FIGURE 13.18** Hydrolysis of glucosinolates into mustard-smelling volatiles. R represents various alkyl or aryl substituents. For example, if R is $CH_2{=}CH{-}CH_2^-$, the compound is sinigrin, a major glucosinolate of black mustard seeds and horseradish roots.

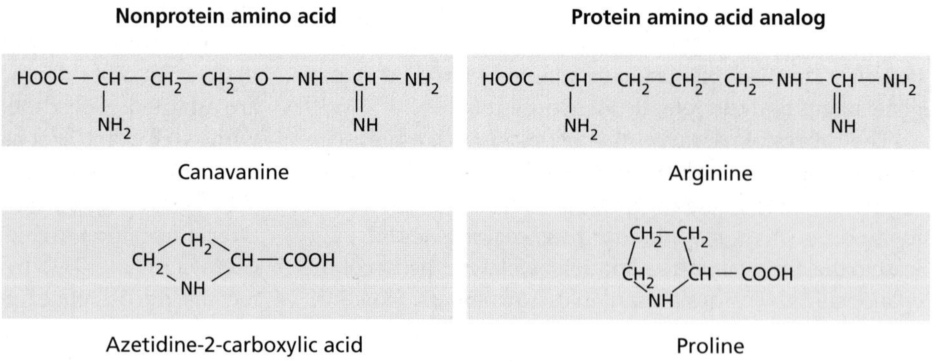

**FIGURE 13.19** Nonprotein amino acids and their protein amino acid analogs. The nonprotein amino acids are not incorporated into proteins, but are defensive compounds found in free form in plant cells. Their activity ranges from interference with the uptake of amino acids to the disruption of translation.

Nonprotein amino acid

Canavanine

Azetidine-2-carboxylic acid

Protein amino acid analog

Arginine

Proline

it becomes incorporated into herbivore proteins in place of arginine. Canavanine is less basic than arginine and its incorporation usually results in a nonfunctional protein because either its tertiary structure or its catalytic site is disrupted (Rosenthal 1991).

It has also been shown that several nonprotein amino acids, including 4-*N*-oxalyl-2,4-diaminobutyric acid, diaminobutyric acid (DABA), 2,3-diaminopropionic acid (DAPA), 3-*N*-oxalyl-2,3-diaminopropionic acid, and 2-amino-6-*N*-oxalylureidopropionic acid, as well as acetylated forms of some of these amino acids, occur in certain fodder legumes and have been linked to toxicity in ruminants (McSweeney et al. 2008).

Plants that synthesize nonprotein amino acids are not susceptible to the toxicity of these compounds. The jack bean (*Canavalia ensiformis*), which has high concentrations of canavanine in its seeds, has protein-synthesizing machinery that can discriminate between canavanine and arginine, and it does not incorporate canavanine into its own proteins. Some insects that are adapted to eat plants containing nonprotein amino acids have similar biochemical adaptations.

## Induced Plant Defenses against Insect Herbivores

Plants have developed a wide variety of defensive strategies against insect herbivory. These strategies can be divided into two categories: *constitutive defenses* and *induced defenses*. **Constitutive defenses** are defensive mechanisms that are always present in the plant. They are often species-specific and may exist as stored compounds, conjugated compounds (to reduce toxicity), or precursors of active compounds that can easily be activated if the plant is damaged. Most of the defensive secondary compounds that have been described so far in this chapter are constitutive defenses. In some cases, however, the same insecticidal defensive compounds are involved in both constitutive and induced defensive responses.

**Induced defenses** are initiated only after actual damage occurs. They include the production of defensive proteins such as lectins and protease inhibitors as well as the production of toxic secondary metabolites. In principle, induced defenses require a smaller investment of plant resources than constitutive defenses, but they must be activated quickly to be effective.

Three categories of insect herbivores can cause varying degrees of damage to plants:

1. *Phloem feeders*, such as aphids and whiteflies, cause little damage to the epidermis and mesophyll cells. The plant defensive response to phloem feeders more closely resembles the response to *pathogens* than the response to herbivores. (Although these insects cause little direct injury to the plant, they can cause much greater damage by serving as vectors for plant viruses.)

2. *Cell content feeders*, such as mites and thrips, are piercing–sucking insects that cause an intermediate amount of physical damage to plant cells.

3. *Chewing insects*, such as caterpillars (the larvae of moths and butterflies), grasshoppers, and beetles, cause the most significant damage to plants. In the discussion that follows, our definition of "insect herbivory" will be restricted to this third category.

In this section we will discuss the different strategies plants use to fend off insect herbivores and the signaling events that regulate the processes involved.

### Plants can recognize specific components of insect saliva

The plant response to damage by insect herbivores involves both a wound response and the recognition of certain insect-derived compounds referred to as **elicitors**. Although repeated mechanical wounding can induce responses similar to those caused by insect herbivory in some plants, certain molecules in insect saliva can serve as enhancers of this stimulus. In addition, such insect-derived elicitors can trigger signaling pathways systemically, thereby initiating defensive responses in distant regions of the plant in anticipation of further damage (see **WEB ESSAY 13.4**).

The first elicitors identified in insect saliva were *fatty acid–amino acid conjugates* (or *fatty acid amides*) (Alborn et al. 1997). The biosynthesis of these conjugates depends on the plant as the source of the fatty acids linolenic acid (18:3) and linoleic acid (18:2),* which are subsequently coupled to an amino acid in the insect gut. Since this initial discovery, other types of elicitors have been identified, but it was shown recently that their specific elicitor activity varies greatly among plant species (Schmelz et al. 2009).

After being regurgitated by an insect, elicitors become part of its saliva and are thus applied to the feeding site during herbivory. Plants then recognize these elicitors and activate a complex signal transduction pathway that induces their defenses.

## Jasmonic acid activates many defensive responses

A major signaling pathway involved in most plant defenses against insect herbivores is the *octadecanoid pathway*, which leads to the production of a plant hormone called **jasmonic acid** (**JA** or *jasmonate*) (FIGURE 13.20). Jasmonic acid levels rise steeply in response to insect herbivore damage and trigger the production of many proteins involved in plant defenses. Jasmonic acid is synthesized from linolenic acid, which is released from plant membrane lipids. Two organelles participate in jasmonic acid biosynthesis: the chloroplast and the peroxisome. An intermediate derived from linolenic acid is cyclized in the chloroplast and then transported to the peroxisome, where enzymes of the β-oxidation pathway (see Chapter 11) complete the conversion into jasmonic acid (see WEB ESSAY 13.5).

Jasmonic acid is known to induce the transcription of a host of genes involved in defensive metabolism. Among the genes it induces are those that encode key enzymes in all the major pathways for secondary metabolite biosynthesis. The mechanisms of this gene activation are slowly becoming clear. It has been shown that JA acts through a conserved signaling mechanism that bears close resemblance to those described for other plant hormones such as auxin (see Chapter 19) and gibberellins (see Chapter

20) (Katsir et al. 2008) (FIGURE 13.21). However, in contrast to those hormonal signals, JA first needs to be activated by conversion into an amino acid conjugate by *JAR proteins*, which belong to a family of carboxylic acid–conjugating enzymes. The enzyme JAR1, for example, exhibits a high substrate specificity for jasmonic acid and isoleucine (JA–Ile) and appears to be of particular importance for JA-dependent defensive signaling (Fonseca et al. 2009).

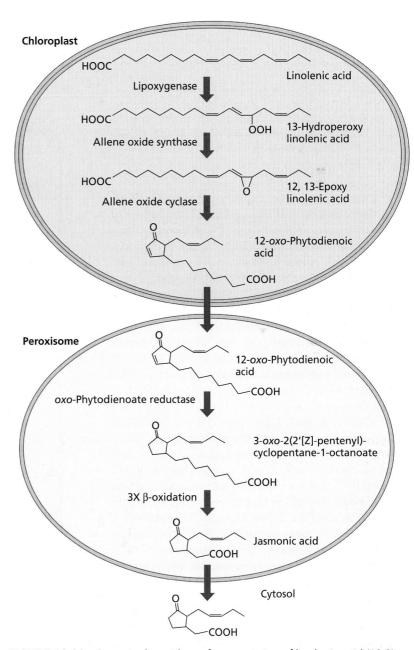

**FIGURE 13.20** Steps in the pathway for conversion of linolenic acid (18:3) into jasmonic acid. The first three enzymatic steps occur in the chloroplast, resulting in the cyclized product 12-*oxo*-phytodienoic acid. This intermediate is transported to the peroxisome, where it is first reduced and then converted into jasmonic acid by β-oxidation.

*Recall that the nomenclature for fatty acids is $X{:}Y$, where $X$ is the number of carbon atoms and $Y$ is the number of *cis* double bonds.

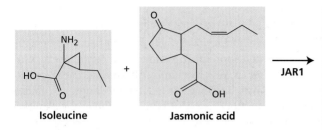

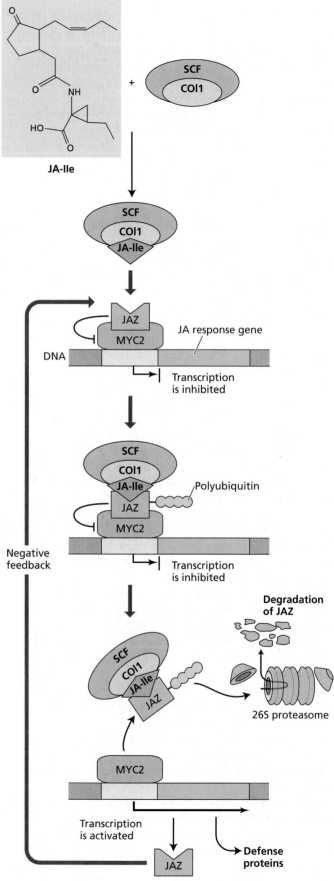

**FIGURE 13.21** A model for jasmonic acid signaling. In contrast to other hormonal signals, jasmononic acid needs to be conjugated first to an amino acid (here, isoleucine) by the enzyme JAR1. The resulting jasmonic acid–isoleucine (JA–Ile) conjugate then binds to COI1 as part of a SCF^COI1 protein complex. This complex targets JAZ, a repressor of transcription, leading to polyubiquitination and subsequent degradation of this protein in the 26S proteasome. Transcription factors such as MYC2 then initiate transcription of jasmonic acid–dependent genes encoding defensive proteins. In addition, *JAZ* genes are activated, thereby providing a negative feedback mechanism for this signaling pathway.

The bioactive JA–Ile conjugate then binds to COI1, an F-box protein that is an essential component of the SCF protein complex (SCF^COI1). SCF protein complexes are important regulators of protein degradation through polyubiquitination. The protein COI1 was first identified as an important signaling factor for coronatine, a phytotoxic virulence factor produced by certain strains of *Pseudomonas syringae*. Coronatine bears a structural resemblance to JA–Ile, and it has been shown that it binds to COI1 with an even higher affinity than does JA–Ile (see **WEB ESSAY 13.5**).

Recently two independent research groups simultaneously identified the target proteins for the SCF^COI1 protein complex as members of the JAZ protein family (Chini et al. 2007; Thines et al. 2007). It was clearly shown that the SCF^COI1 complex, after binding JA–Ile, targets a JAZ protein, which is a repressor of jasmonic acid–activated transcription factors. This binding of the SCF^COI1 JA–Ile complex to JAZ ultimately leads to the polyubiquitination and subsequent degradation of JAZ through the 26S proteasome. The removal of JAZ then activates transcription factors that initiate jasmonic acid–dependent gene expression.

This signaling system is regulated by an interesting negative feedback loop involving the primary target of the SCF^COI1 protein complex: Among the genes activated are those that encode the JAZ repressor protein. The ratio between degradation and synthesis of JAZ, together with the concentration of JA–Ile, appears to regulate the duration and intensity of the response.

An important transcription factor in this context is MYC2, which appears to be a major switch in the activation of JA-dependent genes. Another important JA-dependent transcription factor, ORCA3, was identified in Madagascar periwinkle (*Catharanthus roseus*), which makes some valuable anticancer alkaloids. This transcription factor not only activates the expression of several genes encoding alkaloid biosynthetic proteins (van der Fits and Memelink 2000), but also activates genes involved in certain primary metabolic pathways that provide precursors for alkaloid formation, so it appears to be a master regulator of metabolism in Madagascar periwinkle.

Direct demonstration of the role of jasmonic acid in resistance to insect herbivory has come from research with mutant lines of Arabidopsis that produce only low levels of JA (McConn et al. 1997). Such mutants are easily killed by insect pests, such as fungus gnats, that normally do not damage Arabidopsis. Application of exogenous JA restores resistance nearly to the levels of the wild-type plant.

Several other signaling compounds—including ethylene, salicylic acid, and methyl salicylate—are also induced by insect herbivory. In many cases, the concerted action of these signaling compounds is necessary for the full activation of induced defenses.

## Some plant proteins inhibit herbivore digestion

Among the diverse components of plant defensive arsenals induced by jasmonic acid are proteins that interfere with herbivore digestion. For example, some legumes synthesize **α-amylase inhibitors** that block the action of the starch-digesting enzyme α-amylase. Other plant species produce **lectins**, defensive proteins that bind to carbohydrates or carbohydrate-containing proteins. After ingestion by an herbivore, lectins bind to the epithelial cells lining the digestive tract and interfere with nutrient absorption (Peumans and Van Damme 1995).

The best-known antidigestive proteins in plants are the **protease inhibitors**. Found in legumes, tomatoes, and other plants, these substances block the action of herbivore proteolytic enzymes (proteases). After entering the herbivore's digestive tract, they hinder protein digestion by binding tightly and specifically to the active site of proteolytic enzymes such as trypsin and chymotrypsin. Insects that feed on plants containing protease inhibitors suffer reduced rates of growth and development.

The defensive role of protease inhibitors has been confirmed by experiments with transgenic tobacco. Plants that had been transformed to accumulate increased levels of protease inhibitors suffered less damage from insect herbivores than did untransformed control plants (Johnson et al. 1989). Some insect herbivores, however, have adapted to plant protease inhibitors by producing digestive proteases that are resistant to inhibition (Jongsma et al. 1995; Opperta et al. 2005).

## Damage by insect herbivores induces systemic defenses

In tomatoes, insect feeding leads to the rapid accumulation of protease inhibitors throughout the plant, even in undamaged areas far from the initial feeding site (Schilmiller and Howe 2005). The systemic production of protease inhibitors in young tomato plants is triggered by a complex sequence of events (FIGURE 13.22):

1. Wounded tomato leaves synthesize **prosystemin**, a large (200–amino acid) precursor protein.

2. Prosystemin is proteolytically processed to produce a short (18–amino acid) polypeptide called **systemin**.

3. Systemin is released from the damaged cells (phloem parenchyma) into the apoplast.

4. In adjacent intact tissue (companion cells), systemin binds to a cell surface receptor at the plasma membrane (see WEB ESSAY 13.6).

5. The binding of systemin to its receptor initiates intracellular signaling processes that result in the activation of jasmonic acid (JA) biosynthesis and accumulation.

6. JA is then transported through the phloem to other parts of the plant.

7. In target tissues, JA activates the expression of genes that encode protease inhibitors.

Since the initial discovery of systemin, many systemin-like signaling peptides have been identified in tomato that seem to play important roles in the regulation of defensive responses to insect herbivores and other pests and pathogens (Chen et al. 2008). These peptides have also been found in plants outside the Solanaceae, such as Arabidopsis and *Ipomoea*.

## Herbivore-induced volatiles have complex ecological functions

The induction and release of volatile organic compounds, also called **volatiles**, in response to insect herbivore damage provides an excellent example of the complex ecological functions of secondary metabolites in nature. The combination of molecules emitted is often specific for each insect herbivore species and typically includes representatives from the three major classes of secondary metabolites: terpenes, phenolics, and alkaloids. Additionally, in response to mechanical damage, all plants emit lipid-derived products such as **green-leaf volatiles**, a mixture of six-carbon aldehydes, alcohols, and esters.

The ecological functions of these volatiles are manifold (see WEB ESSAY 13.7). In many cases, they attract natural enemies—predators or parasites—of the attacking insect herbivore that utilize the volatiles as cues to find their prey

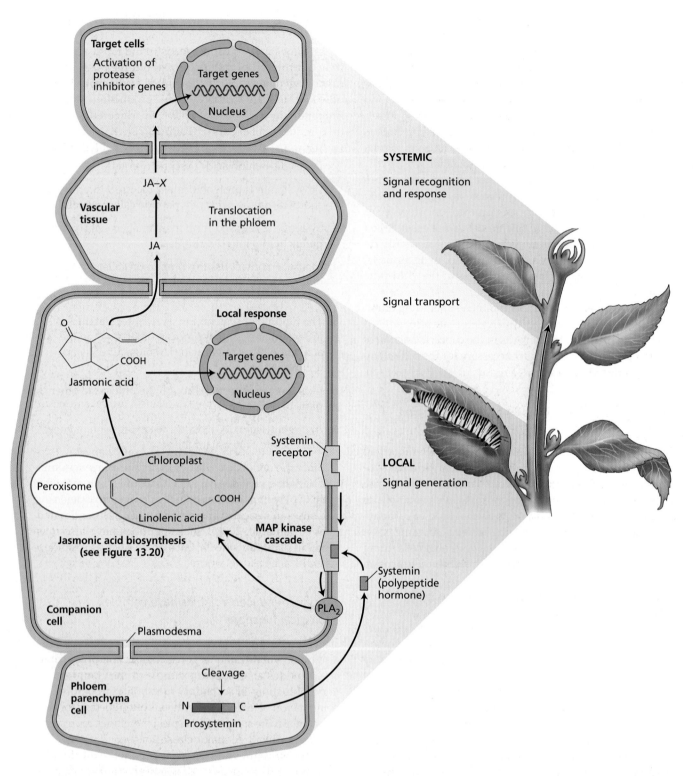

**FIGURE 13.22** Proposed systemin signaling pathway for the rapid induction of protease inhibitor biosynthesis in wounded tomato plants. Wounded tomato leaves (bottom of figure) synthesize prosystemin in phloem parenchyma cells, and the prosystemin is proteolytically processed to systemin. Systemin is released from phloem parenchyma cells and binds to receptors on the plasma membrane of adjacent companion cells. This binding activates a signaling cascade involving phospholipase A$_2$ (PLA$_2$) and mitogen-activated protein (MAP) kinases, which results in the biosynthesis of jasmonic acid (JA). JA is then transported via sieve elements, possibly in a conjugated form (JA–X), to unwounded leaves. There, JA initiates a signaling pathway in target mesophyll cells, resulting in the expresssion of genes that encode protease inhibitors. Plasmodesmata facilitate the spread of the signal at various steps in the pathway.

or hosts for their offspring. Volatiles released by the leaf during moth oviposition (egg laying) can act as repellents to other female moths, thereby preventing further egg deposition and herbivory. In addition, many of these compounds, although volatile, remain attached to the surface of the leaf and serve as feeding deterrents because of their taste.

Plants have the ability to distinguish among various insect herbivore species and to respond differentially. For example, *Nicotiana attenuata*, a wild tobacco that grows in the deserts of the Great Basin in the western United States, typically produces higher levels of nicotine following herbivory. However, when it is attacked by nicotine-tolerant caterpillars, there is no increase in nicotine production. Instead, the plant releases volatile terpenes that attract insect predators of the caterpillars (Karban and Baldwin 1997) (see **WEB ESSAY 13.8**). Clearly, wild tobacco and other plants must have ways of determining what type of insect herbivore is damaging their foliage. Herbivores might signal their presence by the type of damage they inflict or by the distinctive chemical compounds released in their oral secretions.

There is an interesting twist on the role of induced volatiles in plant defense. Besides attracting natural enemies of the attacking insect herbivore, certain volatiles emitted by infested plants can also serve as signals for neighboring plants to initiate expression of defense-related genes. In addition to several terpenoids, green-leaf volatiles act as potent signals in this process (Arimura et al. 2000). For example, when corn plants (*Z. mays*) were exposed to green-leaf volatiles, jasmonic acid production and JA-related gene expression were rapidly induced. More important, however, was the finding that exposure to green-leaf volatiles primed corn plant defenses to respond faster and more strongly to subsequent attacks by insect herbivores (Engelberth et al. 2004). Green-leaf volatiles have been shown to prime or sensitize the defensive mechanisms of a variety of plant species, including the induction of phytoalexins and other antimicrobial compounds (discussed in the next section) (see **WEB ESSAY 13.9**).

### Insects have developed strategies to cope with plant defenses

In spite of all the chemical mechanisms plants have evolved to protect themselves, herbivorous insects have acquired mechanisms for circumventing or overcoming those defenses by the process of *reciprocal evolution* (a type of coevolution). These adaptations, like plant defensive responses, can be either constitutive (always active) or induced (activated by the plant). Constitutive adaptations are more widely distributed among specialist insect herbivores, which feed on only one or a few plant species, whereas induced adaptations are more likely to be found among insects that are dietary generalists. Although it is not always obvious, in most natural environments, plant–insect interactions have led to a standoff in which each can develop and survive under suboptimal conditions.

## Plant Defenses against Pathogens

Even though they lack an immune system as complex as that in animals, plants are surprisingly resistant to diseases caused by the fungi, bacteria, viruses, and nematodes that are ever present in the environment. In this section we will examine the diverse array of mechanisms that plants have evolved to resist infection, including the production of antimicrobial agents and a type of programmed cell death (see Chapter 16) called the *hypersensitive response*. We will also discuss two special types of plant immunity referred to as *systemic acquired resistance* and *induced systemic resistance*.

### Pathogens have developed various strategies to invade host plants

Plants are continuously exposed to a diverse array of pathogens. To be successful, these pathogens have developed various strategies to invade their host plants. Some penetrate the cuticle and cell wall directly by secreting lytic enzymes, which digest these mechanical barriers. Others enter the plant through natural openings like stomata and lenticels. A third category invades the plant through wounding sites, for example those caused by insect herbivores. Additionally, many viruses, as well as other types of pathogens are transferred by insect herbivores, which serve as vectors, and invade the plant from the insect feeding site. Phloem feeders such as whiteflies and aphids deposit pathogens directly into the vascular system, from which they can easily spread throughout the plant.

Once inside the plant, pathogens generally use one of the three main attack strategies to utilize the host plant as a substrate for their own progression:

- **Necrotrophic** pathogens attack their host plant by secreting cell wall–degrading enzymes or toxins, which eventually kill the affected plant cells, leading to massive tissue laceration. This dead tissue is then colonized by the pathogen and serves as a food source.

- A different strategy is used by **biotrophic** pathogens. After infection, the plant tissue remains largely alive, and only minimal cell damage can be observed while the pathogen continues to feed on substrates provided by the host.

- **Hemibiotrophic** pathogens have an initial biotrophic stage, in which the host cells remain alive, as described for biotrophic pathogens. This stage is followed by a necrotrophic stage, in which the pathogen can cause extensive tissue damage.

Although these invasion and infection strategies are individually successful, plant disease epidemics are rare in natural ecosystems. This is because plants have developed effective defense strategies against this diverse array of pathogens. Later we will discuss some of these strategies in more detail.

## Some antimicrobial compounds are synthesized before pathogen attack

Several classes of secondary metabolites that we have already discussed have strong antimicrobial activity when tested in vitro; thus they have been proposed to function as defenses against pathogens in the intact plant. Among these are saponins, a group of triterpenes thought to disrupt fungal membranes by binding to sterols.

Experiments utilizing genetic approaches have demonstrated the role of saponins in defense against pathogens of oat (Papadopoulou et al. 1999). Mutant oat lines with reduced saponin levels had much less resistance to fungal pathogens than did wild-type oats. Interestingly, one fungal strain that normally grows on oats was able to detoxify one of the principal saponins in the plant. Mutants of this strain that could no longer detoxify saponins failed to infect oats, but could grow successfully on wheat that did not contain any saponins.

## Infection induces additional antipathogen defenses

After being infected by a pathogen, plants deploy a broad spectrum of defenses against the invading microbes. A common defense is the **hypersensitive response**, in which cells immediately surrounding the infection site die rapidly, depriving the pathogen of nutrients and preventing its spread. After a successful hypersensitive response, a small region of dead tissue is left at the site of the attempted invasion, but the rest of the plant is unaffected.

The hypersensitive response is often preceded by the rapid accumulation of reactive oxygen species and nitric oxide (NO). Cells in the vicinity of the infection synthesize a burst of toxic compounds formed by the reduction of molecular oxygen, including the superoxide anion ($O_2\bullet^-$), hydrogen peroxide ($H_2O_2$), and the hydroxyl radical (HO•). An NADPH-dependent oxidase located at the plasma membrane (**FIGURE 13.23**) is thought to produce $O_2\bullet^-$, which in turn is converted into HO• and $H_2O_2$.

The hydroxyl radical is the strongest oxidant of these active oxygen species and can initiate radical chain reactions with a range of organic molecules, leading to lipid peroxidation, enzyme inactivation, and nucleic acid degradation (Lamb and Dixon 1997). Active oxygen species may contribute to host cell death as part of the hypersensitive response or act to kill the pathogen directly.

A rapid spike of **nitric oxide** (**NO**) production accompanies the oxidative burst in infected leaves (Delledonne et al. 1998). NO, which acts as a second messenger in many signaling pathways in animals and plants, is synthesized from the amino acid arginine by the enzyme *NO synthase*. An increase in the cytosolic calcium concentration appears to be required for the activation of NO synthase during the hypersensitive response. An increase in *both* NO and reactive oxygen species is required for the activation of the hypersensitive response: Increasing only one of these signals has little effect on the induction of cell death.

Many plant species react to fungal or bacterial invasion by synthesizing lignin or callose (see Chapter 10). These polymers are thought to serve as barriers, walling off the pathogen from the rest of the plant and physically blocking its spread. A related response is the modification of cell wall proteins. Certain proline-rich proteins of the cell wall become oxidatively cross-linked after pathogen attack in

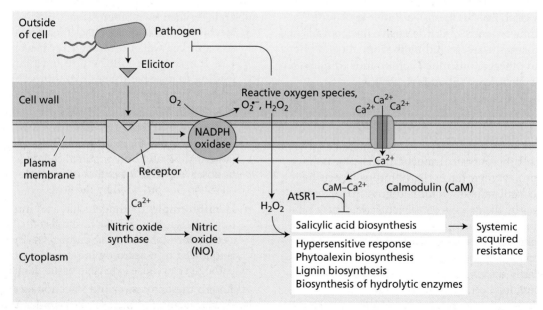

**FIGURE 13.23** Many types of antipathogen defenses are induced by infection. Fragments of pathogen molecules called elicitors initiate a complex signaling pathway leading to the activation of defensive responses. A burst of oxidation activity and nitric oxide production stimulates the hypersensitive response and other defense mechanisms. Note that $Ca^{2+}$ is necessary for the activation of some defenses, while it is also a negative regulator of salicylic acid biosynthesis (see text for details).

an $H_2O_2$-mediated reaction (see Figure 13.23) (Bradley et al. 1992). This process strengthens the walls of the cells in the vicinity of the infection site, increasing their resistance to microbial digestion.

Another defensive response to infection is the formation of hydrolytic enzymes that attack the cell wall of the pathogen. An assortment of glucanases, chitinases, and other hydrolases are induced by fungal invasion. (Chitin, a polymer of N-acetylglucosamine residues, is a principal component of fungal cell walls.) These hydrolytic enzymes belong to a group of proteins that are closely associated with pathogen infection and so are known as *pathogenesis-related (PR) proteins*.

## Phytoalexins often increase after pathogen attack

Perhaps the best-studied response of plants to bacterial or fungal invasion is the synthesis of **phytoalexins**. Phytoalexins are a chemically diverse group of secondary metabolites with strong antimicrobial activity that accumulate around the site of an infection.

Phytoalexin production appears to be a common mechanism of resistance to pathogenic microbes in a wide range of plants. However, different plant families employ different types of secondary products as phytoalexins. For example, in leguminous plants, such as alfalfa and soybean, isoflavonoids are common phytoalexins, whereas in solanaceous plants, such as potato, tobacco, and tomato, various sesquiterpenes are produced as phytoalexins (FIGURE 13.24).

Phytoalexins are generally undetectable in the plant before infection, but they are synthesized very rapidly after microbial attack. The point of control for the activation of these biosynthetic pathways is usually the initiation of gene transcription. Thus plants do not appear to store any of the enzymatic machinery required for phytoalexin synthesis. Instead, soon after microbial invasion, they begin transcribing and translating the appropriate mRNAs and synthesizing the enzymes de novo.

Although phytoalexins accumulate in concentrations that have been shown to be toxic to pathogens in bioassays, the defensive significance of these compounds in the intact plant is not fully known. Experiments on genetically modified plants and pathogens have provided the first direct proof of phytoalexin function in vivo. For example, tobacco transformed with a gene catalyzing the biosynthesis of the phenylpropanoid phytoalexin resveratrol become much more resistant to a fungal pathogen than nontransformed control plants (Hain et al. 1993). Similarly, resistance of Arabidopsis to a fungal pathogen depends on the tryptophan-derived phytoalexin camalexin because mutants deficient in camalexin production are more susceptible than the wild type. In other experiments, pathogens transformed with genes encoding phytoalexin-degrading enzymes were able to infect plants normally resistant to them (Kombrink and Somssich 1995).

## Some plants recognize specific pathogen-derived substances

Within a species, individual plants often differ greatly in their resistance to microbial pathogens. These differences often lie in the speed and intensity of a plant's reactions. Resistant plants respond more rapidly and more vigorously to pathogens than do susceptible plants. Hence it is important to learn how plants sense the presence of pathogens and initiate defensive responses.

A first line of resistance is provided by a system that recognizes broad categories of pathogens. Plants have a variety of receptors that recognize so-called **microbe-associated general molecular patterns (MAMPs)**. These elicitors are evolutionary conserved pathogen-derived molecules such as structural elements from the fungal cell wall or the bacterial flagellum (Hammond-Kosak and Kanyuka 2007). Among the best-studied elicitors in this category are pep13, a 13–amino acid sequence from a cell wall–localized transglutaminase of *Phytophthora*, and flg22, a 22–amino acid peptide derived from the bacterial flagellin protein.

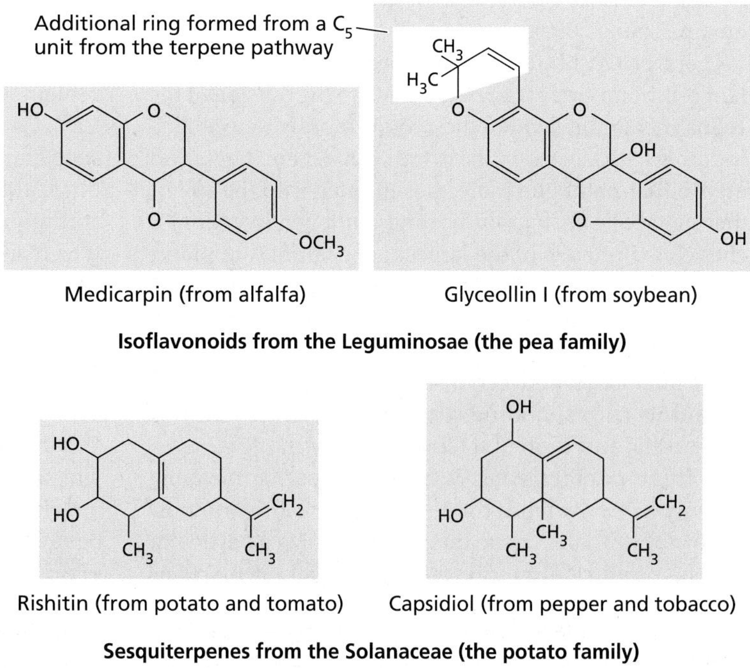

Medicarpin (from alfalfa)  Glyceollin I (from soybean)

**Isoflavonoids from the Leguminosae (the pea family)**

Rishitin (from potato and tomato)  Capsidiol (from pepper and tobacco)

**Sesquiterpenes from the Solanaceae (the potato family)**

**FIGURE 13.24** Structure of some phytoalexins found in two different plant families.

MAMPs are recognized by specific receptors, which then activate specific plant defensive responses, including massive phytoalexin production. The effectiveness of these MAMP receptors (or *pattern recognition receptors*) is amazing, considering the fact that with one receptor, a plant can recognize a complete taxonomic group that features a particular MAMP. For example, the flagellin (flg22) receptor FLS2 enables the plant to recognize all mobile (flagellated) bacteria. Similarly, the as yet uncharacterized receptor for pep13 enables plants to recognize all oomycete pathogens. Consequently, those pathogens cannot cause disease. This form of defensive strategy is also referred to as *innate immunity*.

A second system that provides specific resistance to pathogens is mediated through the interaction of plant *R* gene (or *r*esistance gene) products and pathogen-derived *a*virulent (*Avr*) gene products. Researchers have isolated more than 20 plant *R* genes that function in defense against fungi, bacteria, and nematodes. Most of the *R* genes are thought to encode protein receptors that recognize specific molecules originating from pathogens. Binding of these elicitors to the receptors alerts the plant to the pathogen's presence (see Figure 13.23). These pathogen-specific elicitors include proteins and peptides arising from the pathogen cell wall or outer membrane or from a secretion process (Boller 1995).

The *R* gene products themselves are nearly all proteins with a leucine-rich repeat domain that is repeated inexactly several times in the amino acid sequence. Such domains may be involved in elicitor binding and pathogen recognition. In addition, the *R* gene product is equipped to initiate signaling pathways that activate various antipathogen defenses. Some *R* genes encode a nucleotide-binding site that binds ATP or GTP; others encode a protein kinase domain (Young 2000).

*R* gene products are found in more than one place in the plant cell. Some appear to be attached to the outside of the plasma membrane, where they could rapidly detect elicitors; others are located in the cytoplasm, where they could detect either pathogen molecules injected into the cell or other metabolic changes indicating pathogen infection. *R* genes constitute one of the largest gene families in plants and are often clustered together in the genome. The structures of *R* gene clusters may help generate *R* gene diversity by promoting exchange between chromosomes.

The interaction between a *R* gene product (host receptor) and its corresponding *avr* gene product (elicitor) is very specific and is often referred to as *gene-for-gene resistance*. This type of resistance is genotype-specific, meaning that only a few varieties (or genotypes or cultivars) possess an appropriate *R* gene that can recognize a particular *avr* gene product. Similarly, only some pathogen strains (or *pathovars*) possess *avr* genes whose products are recognized by a particular R protein. This relationship also explains why certain plant species are susceptible to attack by certain pathogen strains but resistant to others. A single mutation in either gene can alter their interaction and result in disease, with severe consequences for the plant. Despite their name, *avr* genes appear to encode factors that promote infection.

## Exposure to elicitors induces a signal transduction cascade

Within a few minutes after pathogenic elicitors have been recognized by an *R* gene product or a MAMP receptor, complex signaling pathways are set in motion that eventually lead to defensive responses (see Figure 13.23). A common early element of these cascades is a transient change in the ion permeability of the plasma membrane. R-gene receptor activation stimulates an influx of $Ca^{2+}$ and $H^+$ ions into the cell and an efflux of $K^+$ and $Cl^-$ ions (Nürnberger and Scheel 2001). The influx of $Ca^{2+}$ activates the oxidative burst that may act directly in defense (as already described), as well as signaling other defensive reactions. Other components of pathogen-stimulated signal transduction pathways include nitric oxide, mitogen-activated protein (MAP) kinases, calcium-dependent protein kinases, and several hormones such as jasmonic acid, ethylene, and salicylic acid (SA). Besides being an important factor in the systemic activation of resistance (see next section), SA in particular also plays a significant regulatory role in the local hypersensitive response against various pathogens—for example, through the activation of PR (pathogenesis-related) proteins. Pathogenesis-related proteins are induced during pathogen attack and serve a protective function.

Recently it has been found that salicylic acid levels are regulated through a negative feedback loop involving $Ca^{2+}$—which is also necessary for the activation of the oxidative burst and nitrous oxide (NO) production (see Figure 13.23)—and calmodulin (CaM). CaM is a $Ca^{2+}$-binding protein with four binding sites for $Ca^{2+}$. Upon binding of $Ca^{2+}$, CaM undergoes conformational changes and in this state may activate other signaling elements. In Arabidopsis this CaM–$Ca^{2+}$ complex controls pathogen-induced SA-accumulation by activating a transcription factor (AtSR1), which acts as a suppressor of SA biosynthesis (see Figure 13.23) (Du et al. 2009).

## A single encounter with a pathogen may increase resistance to future attacks

When a plant survives infection by a pathogen at one site, it often develops increased resistance to subsequent attacks at sites throughout the plant and enjoys protection against a wide range of pathogenic species. This phenomenon, called **systemic acquired resistance (SAR)** (**FIGURE 13.25**), develops over several days following initial infection (Ryals et al. 1996). Systemic acquired resistance appears to result from increased levels of certain PR proteins that we have already mentioned, including chitinases and other hydrolytic enzymes.

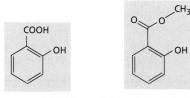

Salicylic acid (SA)    Methyl salicylate (MeSA)

**FIGURE 13.25** Initial pathogen infection may increase resistance to future pathogen attack through development of systemic acquired resistance (SAR). From the infection site, SAR is transmitted through the phloem to other parts of the plant, resulting in increased resistance throughout the plant. Salicylic acid and its methyl ester increase significantly in this process and cause the production of pathogenesis-related (PR) proteins. Methyl salicylate is often released during SAR and may serve as a SAR-inducing volatile signal in neigboring plants.

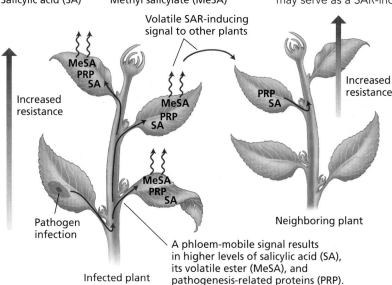

Volatile SAR-inducing signal to other plants

Increased resistance

MeSA PRP SA

MeSA PRP SA

PRP SA

Increased resistance

MeSA PRP SA

Pathogen infection

Neighboring plant

A phloem-mobile signal results in higher levels of salicylic acid (SA), its volatile ester (MeSA), and pathogenesis-related proteins (PRP).

Infected plant

ester of salicylic acid, methyl salicylate, might act as a volatile SAR-inducing signal transmitted to distant parts of the plant and even to neighboring plants (see Figure 13.25) (Shulaev et al. 1997).

## Interactions of plants with nonpathogenic bacteria can trigger induced systemic resistance

In contrast to SAR, which occurs as a consequence of actual pathogen infection, **induced systemic resistance (ISR)** is activated by nonpathogenic microbes (**FIGURE 13.26**). Colonialization of the root zone by rhizobacteria, for example, not only stimulates the formation of root nodules, but also initiates a signaling cascade throughout the plant. As a consequence of this signaling cascade, which involves JA and ethylene, pro-

Although the mechanism of SAR induction is still unknown, one of the endogenous signals involved is likely to be **salicylic acid**. This benzoic acid derivative accumulates dramatically in the zone of infection after the initial attack, and it is thought to establish SAR in other parts of the plant. Measurements of the rate of SAR transmission from the site of infection to the rest of the plant indicate that the movement is too rapid (3 cm h$^{-1}$) for simple diffusion and must therefore involve the vascular system (van Bel and Gaupels 2004). For tobacco, it has been shown that methyl salicylate acts as the mobile signal in the vasculature (Park et al. 2007); other plants, however, might use other means of SAR transmission. In Arabidopsis, mutations in the *DIR1* (*Defective in Induced Resistance 1*) gene, which is specifically expressed in the phloem, block the SAR response (Maldonado et al. 2002). The *DIR1* gene encodes a lipid transfer protein, suggesting that the long-distance signal in Arabidopsis could be derived from a lipid. Another compound that accumulates at the site of infection and may play a role in SAR is H$_2$O$_2$. However, like salicylic acid, H$_2$O$_2$ is unlikely to function as a long-distance signal (van Bel and Gaupels 2004).

In addition to phloem-mobile signals, airborne signaling via volatiles may induce SAR. For example, the methyl

**FIGURE 13.26** Exposure to nonpathogenic microorganisms may increase resistance to future pathogen attack through development of induced systemic resistance (ISR). Nonpathogenic microorganisms such as rhizobacteria activate signaling pathways involving jasmonic acid and ethylene that trigger ISR throughout the plant. Rather than activating immediate defensive measures, ISR is characterized by an increased level of preparedness against pathogen attack.

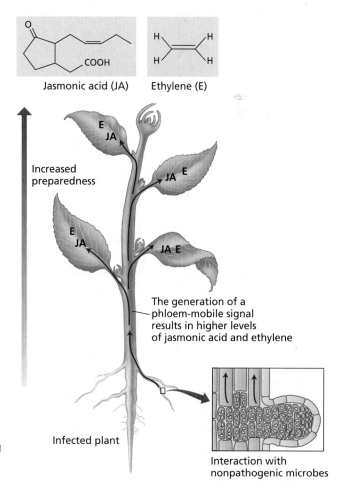

Jasmonic acid (JA)    Ethylene (E)

Increased preparedness

E JA

JA E

E JA

JA E

The generation of a phloem-mobile signal results in higher levels of jasmonic acid and ethylene

Infected plant

Interaction with nonpathogenic microbes

tective measures are activated throughout the plant, resulting in an enhanced mode of preparedness against pathogen attack. This form of systemic defense activation does not involve salicylic acid as a signaling compound and does not induce the accumulation of typical PR proteins.

While certain defensive measures are immediately put in place by ISR, other defensive responses are initiated only after actual pathogen infection, resulting in a faster and stronger response. The advantage of this defensive strategy lies in reducing the direct investment of resources in defensive measures, which would otherwise affect the performance of the plant, resulting, for example, in reduced growth and yield. In its mode of activation and responses, ISR bears a close resemblance to the processes involved in the priming of defensive responses by green-leaf volatiles as described previously.

# SUMMARY

Because they are sessile organisms, plants often use secondary metabolites in much the same way animals use behavior to interact with their environment. Among their many functions, secondary metabolites serve to attract beneficial organisms such as pollinators and seed-dispersing animals, to protect against environmental stresses, and to defend against herbivores and pathogens.

## Secondary Metabolites

- Plant secondary metabolites serve as defensive compounds, as attractants for pollinators and seed-dispersing animals, and as agents of plant–plant competition and plant–microbe symbioses.

- Secondary metabolites can be divided into three chemical groups: terpenes, phenolics, and nitrogen-containing compounds (**Figure 13.1**).

- Secondary metabolites exhibit very restricted distributions within the plant kingdom. A given secondary metabolite may be found in only one plant species or group of species.

## Terpenes

- Two pathways are involved in the biosynthesis of terpenes, the mevalonic acid pathway and the methylerythritol phosphate (MEP) pathway, both of which produce isopentenyl diphosphate (IPP) (**Figure 13.2**).

- IPP and its isomer, dimethylallyl diphosphate (DMAPP), combine to form geranyl diphosphate (GPP), the precursor of nearly all the monoterpenes. GPP reaction with another IPP yields farnesyl diphosphate (FPP), the precursor of nearly all the sesquiterpenes.

- Some terpenes have functions in plant growth or development. Other terpenes are toxins and herbivore deterrents (**Figures 13.3, 13.5**).

## Phenolic Compounds

- Phenolics are a heterogeneous group of compounds that are synthesized through the shikimic acid pathway (**Figures 13.6, 13.7**).

- Lignin is a highly branched polymer of phenyl-propanoid alcohols that is found in the cell walls of supporting and conducting tissue. In addition to providing mechanical support, it protects plants by deterring herbivory and blocking the growth of pathogens.

- Flavonoids comprise the anthocyanins, the flavones, the flavonols, and the isoflavones. Their molecular skeletons contain 15 carbons arranged in two aromatic rings linked by a three-carbon bridge. Flavonoids are synthesized through the shikimic acid and malonic acid pathways (**Figure 13.9**).

- The anthocyanins are pigmented flavonoids that are responsible for most of the red, pink, purple, and blue colors observed in plant parts (**Figure 13.10**).

- Flavones and flavonols are present in the leaves of all green plants and protect cells from excessive UV-B radiation. Isoflavonoids act as insecticides and poisons or as anti-estrogenic substances, since their ring structures are similar to those of steroids.

- Condensed tannins are compounds formed by the polymerization of flavonoid units. Hydrolyzable tannins are heterogeneous polymers containing phenolic acids. Because they bind to proteins, tannins are general toxins that reduce the growth and survival of many herbivores. They can also serve as defenses against microorganisms (**Figures 13.12, 13.13**).

## Nitrogen-Containing Compounds

- Nitrogen-containing secondary metabolites, which are synthesized from amino acids, include antiherbivore defenses such as alkaloids and cyanogenic glycosides. Many of these compounds are toxic to humans but also have medical uses.

- Alkaloids are alkaline, water-soluble compounds that are usually derived from an amino acid, most commonly lysine, tyrosine, or tryptophan. However, some alkaloids contain a derivative of the terpene pathway. Most alkaloids are toxic to animals and humans. Some alkaloids interfere with the nervous system, others affect membrane transport, protein synthesis, or miscellaneous enzyme activities (**Figures 13.14, 13.15**).

## SUMMARY continued

- The pyrrolizidine alkaloids can be tolerated and modified by some herbivores and used in their own defense (**Figure 13.16**). Fungal symbionts of grasses synthesize a variety of alkaloids, contributing to the host's faster growth and defenses against insect and mammalian herbivores.

- Cyanogenic glycosides and glucosinolates are not toxic per se, but are readily broken down to give off toxins when a plant containing them is crushed. Cyanogenic glycosides release the poisonous gas hydrogen cyanide (HCN) (**Figure 13.17**). The glucosinolates, or mustard oil glycosides, are found principally in the Brassicaceae. Their breakdown causes the release of isothiocyanate, a potent toxin (**Figure 13.18**).

- Nonprotein amino acids also act as defensive toxins. Some block the synthesis or uptake of protein amino acids; others can be mistakenly incorporated into proteins, altering their functions (**Figure 13.19**).

### Induced Plant Defenses against Insect Herbivory

- Plant defensive strategies against insect herbivores can be divided into constitutive (always present) and induced defenses.

- Induced defenses are initiated in response to mechanical damage inflicted by herbivores or by specific components of insect saliva, called elicitors.

- Jasmonic acid (JA) increases rapidly in response to insect damage and induces the transcription of key genes involved in plant defenses. JA is synthesized from linolenic acid, which is released from membrane lipids and then converted into JA through the octadecanoid pathway (**Figure 13.20**).

- JA is activated by conjugation with certain amino acids, such as isoleucine. Binding of JA–Ile to a receptor complex activates the degradation of a transcriptional suppressor, resulting in the initiation of JA-responsive gene expression (**Figure 13.21**).

- JA induces the production of defensive proteins, such as lectins and protease inhibitors, as well as toxic secondary metabolites.

- Herbivore damage can also induce systemic defenses by causing the synthesis of polypeptide signals such as systemin. Systemin is released to the apoplast and binds to cell surface receptors in the phloem, where it activates JA biosynthesis. Transport of JA throughout the plant leads to expression of genes that encode protease inhibitors in target cells (**Figure 13.22**).

- Volatile organic compounds, or volatiles, may be induced and released in response to insect damage. Volatiles attract natural enemies of the attacking herbivore, but may also serve as signals for neighboring plants to initiate expression of defense-related genes.

### Plant Defenses against Pathogens

- Pathogens have developed various strategies to invade their host plants, including secretion of lytic enzymes to penetrate the cuticle and cell walls, entry through natural openings such as stomata and lenticels, and invasion at wounds. Insect herbivores may serve as vectors for viruses and other pathogens.

- Pathogens generally use one of three main attack strategies: necrotrophism, biotrophism, or hemibiotrophism.

- A common defense against pathogens is the hypersensitive response, in which cells surrounding the site of infection die rapidly, thereby limiting the spread of the pathogen. The hypersensitive response is often preceded by rapid accumulation of reactive oxygen species and nitric oxide (NO) (**Figure 13.23**).

- Infected host plants may synthesize hydrolytic enzymes that attack the cell wall of fungal pathogens. Many plants produce phytoalexins, a diverse group of secondary metabolites with strong antimicrobial activity, in response to infection (**Figure 13.24**).

- Some plants have receptors that recognize evolutionarily conserved pathogen-derived substances, known as microbe-associated molecular patterns (MAMPs).

- Interaction of plant *R* gene products and pathogen-derived *Avr* gene products leads to recognition of specific pathogens. *R* genes encode protein receptors that recognize *Avr* gene products. Binding of an *Avr* gene product to its receptor initiates signaling pathways that activate antipathogen defenses.

- A plant that survives a pathogen infection at one site often develops increased resistance to subsequent attacks throughout the plant—a process termed systemic acquired resistance (SAR) (**Figure 13.25**).

- Interactions of plants with nonpathogenic bacteria can trigger induced systemic resistance (ISR) through a process that is mediated by JA and ethylene (**Figure 13.26**).

## WEB MATERIAL

### Web Topics

**13.1  Cutin, Waxes, and Suberin**

Plant surfaces are covered with layers of lipid material protecting them against water losses and blocking the entry of pathogenic microorganisms.

**13.2  Structure of Various Triterpenes**

The structures of several triterpenes are given.

**13.3  The Shikimic Acid Pathway**

The shikimic acid pathway converts simple carbohydrate precursors derived from glycolysis and the pentose phosphate pathway to the aromatic amino acids and salicylic acid.

**13.4  Detailed Chemical Structure of a Portion of a Lignin Molecule**

The partial structure of a hypothetical lignin molecule from European beech (*Fagus sylvatica*) is displayed, showing the complexity of this macromolecule.

### Web Essays

**13.1  Engineering Fruit Aromas**

The terpenoid pathway can be engineered to improve fruit aromas.

**13.2  Secondary Metabolites and Allelopathy in Plant Invasions: A Case Study of *Centaurea maculosa***

The invasive weed *Centaurea maculosa*, which is rapidly taking over pastureland in the western United States, secretes the polyphenol catechin into the rhizosphere, which suppresses the growth and germination of neighboring plants.

**13.3  Alkaloid-Making Fungal Symbionts**

Fungal endophytes can enhance plant growth, increase resistance to various stresses, and act as "defensive mutualists" against herbivores.

**13.4  Early Signaling Events in the Plant Wound Response**

A complex signaling network, which includes reactive oxygen species and rapid ion fluxes, is rapidly activated in wounded plants.

**13.5  Jasmonates and Other Fatty Acid–Derived Signaling Pathways in the Plant Defense Response**

The importance of fatty acid–derived signaling pathways as regulators of diverse plant defense strategies is becoming increasingly recognized. The complexity of the individual pathways and their mutual interactions are discussed in the context of direct and indirect defense strategies.

**13.6  The Systemin Receptor**

The systemin receptor from tomato is an LRR-receptor kinase.

**13.7  The Plant Volatilome**

The release of volatile organic compounds by plants provides an example of the diversity of secondary metabolites and the ecological implications thereof.

**13.8  Unraveling the Function of Secondary Metabolites**

Wild tobacco plants use alkaloids and terpenes to defend themselves against herbivores.

**13.9  Smelling the Danger and Getting Prepared: Volatile Signals as Priming Agents in the Defense Response**

By releasing volatiles, herbivore-damaged plants not only attract natural enemies of the attacking insect herbivore, but also signal this event to neighboring plants, allowing them to prepare their defenses against impending herbivory.

# CHAPTER REFERENCES

Aerts, R. J., and Mordue, A. J. (1997) Feeding deterrence and toxicity of neem triterpenoids. *J. Chem. Ecol.* 23: 2117–2132.

Alborn, H. T., Turlings, T. C., Jones, T. H., Stenhagen, G. S., Loughrin, J. H., and Tumlinson, J. H. (1997) An elicitor of plant volatiles identified from beet armyworm oral secretions. *Science* 276: 945–949.

Appel, H. M. (1993) The role of phenolics in ecological systems: The importance of oxidation. *J. Chem. Ecol.* 19: 1521–1552.

Arimura, G.-I., Ozawa, R., Shimoda, T., Nishioka, T., Boland, W., and Takabayashi, J. (2000) Herbivory-induced volatiles elicit defence genes in lima bean leaves. *Nature* 406: 512–515.

Boller, T. (1995) Chemoperception of microbial signals in plant cells. *Annu. Rev. Plant Physiol. Plant Mol. Biol.* 46: 189–214.

Bradley, D. J., Kjellbom, P., and Lamb, C. J. (1992) Elicitor- and wound-induced oxidative cross-linking of a proline-rich plant cell wall protein: A novel, rapid defense response. *Cell* 70: 21–30.

Butler, L. G. (1989) Effects of condensed tannin on animal nutrition. In *Chemistry and Significance of Condensed Tannins*, R. W. Hemingway and J. J. Karchesy, eds., Plenum, New York, pp. 391–402.

Chen, Y. C., Siems, W. E., Pearce, G., and Ryan, C. A. (2008) Six peptide wound signals derived from a single precursor protein in Ipomoea batatas leaves activate the expression of the defense gene sporamin. *J. Biol. Chem.* 283: 11469–11476.

Chini, A., Fonseca, S., Fernandez, G., Adie, B., Chico, J. M., Lorenzo, O., García-Casado, G., López-Vidriero, I., Lozano, F. M., Ponce, M. R., Micol, J. L., and Solano, R. (2007) The JAZ family of repressors is the missing link in jasmonate signalling. *Nature* 448: 666–671.

Corder, R., Douthwaite, J. A., Lees, D. M., Khan, N. Q., Viseu dos Santos, A. C., Wood, E. G., and Carrier, M. J. (2001) Endothelin-1 synthesis reduced by red wine. *Nature* 414: 863–864.

Davin, L. B., and Lewis, N. G. (2000) Dirigent proteins and dirigent sites explain the mystery of specificity of radical precursor coupling in lignan and lignin biosynthesis. *Plant Physiol.* 123: 453–461.

Delledonne, M., Zeier, J., Marocco, A., and Lamb, C. (2001) Signal interactions between nitric oxide and reactive oxygen intermediates in the plant hypersensitive disease resistance response. *Proc. Natl. Acad. Sci. USA* 98: 13454–13459.

Du, L., Ali, G. S., Simons, K. A., Hou, J., Yang, T., Reddy, A. S. N., and Poovaiah, B. W. (2009) $Ca^{2+}$/calmodulin regulates salicylic-acid-mediated plant immunity. *Nature* 457: 1154–1158.

Eigenbrode, S. D., Stoner, K. A., Shelton, A. M., and Kain, W. C. (1991) Characteristics of glossy leaf waxes associated with resistance to diamondback moth (Lepidoptera: Plutellidae) in *Brassica oleracea*. *J. Econ. Entomol.* 83: 1609–1618.

Engelberth, J., Alborn, H. T., Schmelz, E. A., and Tumlinson, J. H. (2004) Airborne signals prime plants against insect herbivore attack. *Proc. Natl. Acad. Sci. USA* 101: 1781–1785.

Felton, G. W., Donato, K., Del Vecchio, R. J., and Duffey, S. S. (1989) Activation of plant foliar oxidases by insect feeding reduces nutritive quality of foliage for noctuid herbivores. *J. Chem. Ecol.* 15: 2667–2694.

Fonseca, S., Chini, A., Hamberg, M., Adie, B., Porzel, A., Kramell, R., Miersch, O., Wasternack, C., and Solano, R. (2009) (+)-7-iso-Jasmonyl-L-isoleucine is the endogenous bioactive jasmonate. *Nat. Chem. Biol.* 5: 344–350.

Gershenzon, J., and Croteau, R. (1992) Terpenoids. In *Herbivores: Their Interactions with Secondary Plant Metabolites*, Vol. 1: *The Chemical Participants*, 2nd ed., G. A. Rosenthal and M. R. Berenbaum, eds., Academic Press, San Diego, CA, pp. 165–219.

Gunning, B. E. S., and Steer, M. W. (1996) *Plant Cell Biology: Structure and Function of Plant Cells.* Jones and Bartlett, Boston.

Hain, R., Reif, H.-J., Krause, E., Langebartels, R., Kindl, H., Vornam, B., Wiese, W., Schmelzer, E., Schreier, P. H., Stöcker, R. H., and Stenzel, K. (1993) Disease resistance results from foreign phytoalexin expression in a novel plant. *Nature* 361: 153–156.

Hammond-Kosack K. E., and Kanyuka, K. (2007) Resistance genes (*R* genes) in plants. In *Encyclopedia of Life Sciences*. John Wiley & Sons, Ltd: Chichester http://www.els.net/ [DOI: 10.1002/9780470015902.a0020119]

Hartmann, T. (1992) Alkaloids. In *Herbivores: Their Interactions with Secondary Plant Metabolites*, Vol. 1: *The Chemical Participants*, 2nd ed., G. A. Rosenthal and M. R. Berenbaum, eds., Academic Press, San Diego, CA, pp. 79–121.

Hartmann, T. (1999) Chemical ecology of pyrrolizidine alkaloids. *Planta* 207: 483–495.

Hatfield, R., and Vermerris, W. (2001) Lignin formation in plants. The dilemma of linkage specificity. *Plant Physiol.* 126: 1351–1357.

Herrmann, K. M., and Weaver, L. M. (1999) The shikimate pathway. *Annu. Rev. Plant Physiol. Plant Mol. Biol.* 50: 473–503.

Inderjit, Dakshini, K. M. M., and Einhellig, F. A., eds. (1995) *Allelopathy: Organisms, Processes, and Applications.* ACS Symposium series American Chemical Society, Washington, DC.

Jeffree, C. E. (1996) Structure and ontogeny of plant cuticles. In *Plant Cuticles: An Integrated Functional Approach*, G. Kerstiens, ed., BIOS Scientific, Oxford, pp. 33–85.

Jin, H., and Martin, C. (1999) Multifunctionality and diversity within the plant *MYB*-gene family. *Plant Mol. Biol.* 41: 577–585.

Johnson, R., Narvaez, J., An, G., and Ryan, C. (1989) Expression of proteinase inhibitors I and II in transgenic tobacco plants: Effects on natural defense against *Manduca sexta* larvae. *Proc. Natl. Acad. Sci. USA* 86: 9871–9875.

Jongsma, M. A., Bakker, P. L., Peters, J., Bosch, D., and Stiekema, W. J. (1995) Adaptation of *Spodoptera exigua* larvae to plant proteinase inhibitors by induction of gut proteinase activity insensitive to inhibition. *Proc. Natl. Acad. Sci. USA* 92: 8041–8045.

Karban, R., and Baldwin, I. T. (1997) *Induced Responses to Herbivory.* University of Chicago Press, Chicago.

Katsir, L., Chung, H. S., Koo, A. J., and Howe, G. A. (2008) Jasmonate signaling: A conserved mechanism of hormone sensing. *Curr. Opin. Plant Biol.* 11: 428–435.

Kombrink, E., and Somssich, I. E. (1995) Defense responses of plants to pathogens. *Adv. Bot. Res.* 21: 1–34.

Kondo, T., Yoshida, K., Nakagawa, A., Kawai, T., Tamura, H., and Goto, T. (1992) Structural basis of blue-color development in flower petals from *Commelina communis. Nature* 358: 515–518.

Kruse, M., Strandberg, M., and Strandberg, B. (2000) Ecological effects of allelopathic plants—a review. *NERI Technical Report* 315: 1–64.

Lamb, C., and Dixon, R. A. (1997) The oxidative burst in plant disease resistance. *Annu. Rev. Plant Physiol. Plant Mol. Biol.* 48: 251–275.

Li, J., Ou-Lee, T.-M., Raba, R., Amundson, R. G., and Last, R. L. (1993) *Arabidopsis* flavonoid mutants are hypersensitive to UV-B irradiation. *Plant Cell* 5: 171–179.

Lichtenthaler, H. K. (1999) The 1-deoxy-D-xylulose-5-phosphate pathway of isoprenoid biosynthesis in plants. *Annu. Rev. Plant Physiol. Plant Mol. Biol.* 50: 47–65.

Logemann, E., Parniske, M., and Hahlbrock, K. (1995) Modes of expression and common structural features of the complete phenylalanine ammonia-lyase gene family in parsley. *Proc. Natl. Acad. Sci. USA* 92: 5905–5909.

Lunau, K. (1992) A new interpretation of flower guide colouration: Absorption of ultraviolet light enhances colour saturation. *Plant Sys. Evol.* 183: 51–65.

McConn, M., Creelman, R. A., Bell, E., Mullet, J. E., and Browse, J. (1997) Jasmonate is essential for insect defense in *Arabidopsis. Proc. Natl. Acad. Sci. USA* 94: 5473–5477.

Maldonado, A. M., Doerner, P., Dixon, R. A., Lamb, C. J., and Cameron, R. K. (2002) A putative lipid transfer protein involved in systemic resistance signalling in *Arabidopsis. Nature* 419: 399–403.

McSweeney, C. S., Collins, E. M. C., Blackall, L. L., and Seawright A. A. (2008) A review of anti-nutritive factors limiting potential use of *Acacia angustissima* as a ruminant feed. *Animal Feed Sci. Tech.* 147: 158–171.

Nürnberger, T., and Scheel, D. (2001) Signal transmission in the plant immune response. *Trends Plant Sci.* 6: 372–379.

Opperta, T., Morgana, T. D., Hartzerb, K., and Kramer, K. J. (2005) Compensatory proteolytic responses to dietary proteinase inhibitors in the red flour beetle, *Tribolium castaneum* (Coleoptera: Tenebrionidae). *Comp. Biochem. Physiol. C Toxicol. Pharmacol.* 140: 53–58.

Papadopoulou, K., Melton, R. E., Legget, M., Daniels, M. J., and Osbourn, A. E. (1999) Compromised disease resistance in saponin-deficient plants. *Proc. Natl. Acad. Sci. USA* 96: 12923–12928.

Park, S. W., Kaimoyo, E., Kumar, D., Mosher, S., and Klessig, D. F. (2007) Methyl salicylate is a critical mobile signal for plant systemic acquired resistance. *Science* 318: 113–116.

Peumans, W. J., and Van Damme, E. J. M. (1995) Lectins as plant defense proteins. *Plant Physiol.* 109: 347–352.

Poulton, J. E. (1990) Cyanogenesis in plants. *Plant Physiol.* 94: 401–405.

Renwick, J. A. A., Radke, C. D., Sachdev-Gupta, K., and Städler, E. (1992) Leaf surface chemicals stimulating oviposition by *Pieris rapae* (Lepidoptera: Pieridae) on cabbage. *Chemoecology* 3: 33–38.

Rosenthal, G. A. (1991) The biochemical basis for the deleterious effects of L-canavanine. *Phytochemistry* 30: 1055–1058.

Ryals, J. A., Neuenschwander, U. H., Willits, M. G., Molina, A., Steiner, H.-Y., and Hunt, M. D. (1996) Systemic acquired resistance. *Plant Cell* 8: 1809–1819.

Sandberg, S. L., and Berenbaum, M. R. (1989) Leaf-tying by tortricid larvae as an adaptation for feeding on phototoxic *Hypericum perforatum. J. Chem. Ecol.* 15: 875–886.

Schilmiller, A. L., and Howe, G. A. (2005) Systemic signaling in the wound response. *Curr. Opin. Plant Biol.* 8: 369–377.

Schmelz, E. A., Engelberth, J., Alborn, H. T., Teal, P., and Tumlinson, J. (2009) Phytohormone-based activity mapping of arthropod-associated plant effectors. *Proc. Natl. Acad. Sci. USA* 106: 653–657.

Shulaev, V., Silverman, P., and Raskin, I. (1997) Airborne signalling by methyl salicylate in plant pathogen resistance. *Nature* 385: 718–721.

Soriano, I. R., Riley, I. T., Potter, M. J., and Bowers, W. S. (2004) Phytoecdysteroids: A novel defense against plant–parasitic nematodes. *J. Chem. Ecol.* 10: 1885–1899.

Thines, B., Katsir, L., Melotto, M., Niu, Y., Mandaokar, A., Liu, G. H., Nomura, K., He, S. Y., Howe, G. A., and Browse, J. (2007) JAZ repressor proteins are targets of the SCFCOI1 complex during jasmonate signalling. *Nature* 448: 661–662.

Trapp, S., and Croteau, R. (2001) Defensive resin biosynthesis in conifers. *Annu. Rev. Plant Physiol. Plant Mol. Biol.* 52: 689–724.

van Bel, A. J. E., and Gaupels, F. (2004) Pathogen-induced resistance and alarm signals in the phloem. *Mol. Plant Pathol.* 5: 495–504.

van der Fits, L., and Memelink, J. (2000) ORCA3, a jasmonate-responsive transcriptional regulator of plant primary and secondary metabolism. *Science* 289: 295–297.

Veitch, G. E., Boyer, A., and Ley, S. V. (2008) The azadirachtin story. *Angew. Chem. Int. Ed. Engl.* 47: 9402–9429.

Wittstock, U., Agerbirk, N., Stauber, E. J., Olsen, C. E., Hippler, M., Mitchell-Olds, T., Gershenzon, J., and Vogel, H. (2004) Successful herbivore attack due to metabolic diversion of a plant chemical defense. *Proc. Natl. Acad. Sci. USA* 101: 4859–4864.

Young, N. D. (2000) The genetic architecture of resistance. *Curr. Opin. Plant Biol.* 3: 285–290.

# Unit Three

## Growth and
## Development

# UNIT THREE

*Previous page: Stages in the germination and growth of a
rutabaga (Brassica napus var. napobrassica) seedlings are
shown in sharp contrast in this color-enhanced scanning
electron micrograph. Image © Power and Syred/SPL/
Photo Researchers, Inc.*

# Signal Transduction

Signal transduction in plants has long fascinated researchers. Charles Darwin performed pioneering studies on phototropic growth responses in canary grass and oat seedling tissues (Darwin 1881). He observed that a unidirectional light source was perceived at the coleoptile tip, yet the bending response took place further back along the shoot tissue. This led him to conclude that there must be a mobile signal that transferred information from one region of the coleoptile tissue to another and elicited the bending response. Went and co-workers later identified this signal as auxin (see Chapter 19).

Since these classic studies, researchers have demonstrated that plants perceive many environmental and physiological signals in order to fine-tune their growth and development. Plant cells perceive these external or internal signals by employing specialized sensor proteins termed *receptors*. Once receptors sense their specific signal, they must *transduce* the signal (i.e., convert it from one form to another) in order to amplify the signal and trigger the cellular response. Receptors often do this by modifying the activity of other proteins and/or employing intracellular signaling molecules called **secondary messengers**; these molecules then alter cellular processes such as gene transcription. Hence, all signal transduction pathways typically involve the following chain of events:

Signal → receptor → signal transduction → response

This framework has been central to our understanding of signal transduction in plants for the last 50 years. Many of the specific events and intermediate steps involved in plant signal transduction have now been identified, and these intermediates constitute the **signal transduction pathways**.

One of the milestones of plant signal transduction research in the mid-twentieth century involved the identification of the photoreceptor *phytochrome* (see Chapter 17). Earlier studies had demonstrated that the effects of red light on germination could be reversed by far-red light (Borthwick et al. 1952). The antagonistic effects of red and far-red light were later demonstrated to operate through phytochrome, a photo-interconvertible protein that functions in effect as a light switch (Butler et al. 1959).

The advent of molecular genetic studies in the model plant *Arabidopsis thaliana* in the late 1980s has led to an explosion of knowledge about signal transduction in plants. For example, genetic approaches have led to the identification of almost every class of hormone receptor described in plants to date (Chow and McCourt 2006). In addition, mutant screens in Arabidopsis have identified many new signal transduction components that act downstream of (that is, subsequent to) receptors for signals such as light and hormones, providing us with an entirely new level of understanding about the complexity of signaling in plants (McCourt 1999).

The completion of the Arabidopsis genome at the start of the twenty-first century has delivered an unparalleled new level of insight into the makeup of the plant signal transduction machinery. Genome sequencing has revealed that plants and animals share many signal transduction proteins. However, genome information also clearly demonstrates that plants contain a distinct repertoire of signaling proteins. Hence, key differences exist between animal and plant systems, reflecting their distinct evolutionary paths from single-cell progenitor to multicellular organisms.

In this chapter, we begin by comparing and contrasting signal transduction in plant and animal cells, highlighting the similarities and differences that exist between these systems at the gene sequence and functional levels. Next we discuss how the majority of plant signal transduction pathways function by inactivating, degrading or removing repressor proteins. This often results in activation of gene expression, which underpins any subsequent plant response. We then consider how plant responses are often integrated with other signaling pathways. Finally, we discuss how signal transduction processes in plants must operate across widely different physical distances and time scales, ranging from micrometers to meters and seconds to years.

# Signal Transduction in Plant and Animal Cells

Plants and animals employ distinct sets of signals to regulate their development. For example, many of the key developmental events in a plant's life cycle (like germination, leaf formation, and flowering) are regulated by environmental signals (such as temperature, light, and day length), whereas animal development is generally regulated by physiological (i.e., internal) signals. Such differences clearly reflect the sessile versus mobile life styles of plants and animals. In this section, we will consider both similarities and differences between signal transduction systems found in plants and animals.

## Plants and animals have similar transduction components

Despite obvious differences in their development, plants and animals share many similarities at the level of their signal transduction machinery. For example, plant and animal cells use common secondary messengers, like calcium, lipid signaling molecules, and pH changes, to trigger physiological responses (FIGURE 14.1).

Animals and plants also employ large numbers of kinase receptors and kinase signal transduction proteins. A **kinase** is an enzyme activity that catalyzes *phosphorylation*: the addition of a phosphate group from ATP to a substrate, such as a protein. Kinase receptors and kinase signal transduction proteins have kinase activity as well as func-

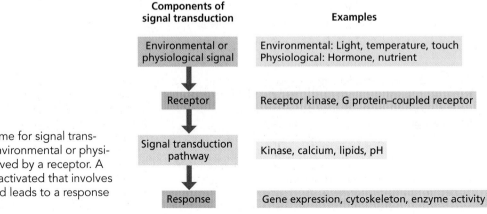

FIGURE 14.1 Generic scheme for signal transduction with examples. Environmental or physiological signals are perceived by a receptor. A signaling cascade is then activated that involves secondary messengers and leads to a response by the plant cell.

tioning as signal receptors or intermediates in transduction pathways. Phosphorylation often alters the biological activity of substrates, which, in the case of proteins, results in either their activation or their inactivation.

Plant and animal receptor kinases are often localized at the plasma membrane, where they are ideally positioned to directly detect extracellular signals. Binding of signaling molecules to a receptor on the extracellular face of a cell trig-

gers an intracellular signal. For example, the plant brassinosteroid receptor (BRI1, for *Br*assinosteroid *i*nsensitive 1) and the animal *f*ibroblast *g*rowth *f*actor *r*eceptor (FGFR) both contain kinase domains. In the case of the animal fibroblast growth factor receptor, ligand binding triggers receptor autophosphorylation and dimerization (**FIGURE 14.2A**). Similarly, binding of the plant hormone brassinosteroid to its receptor, BRI1, triggers receptor autophosphorylation and

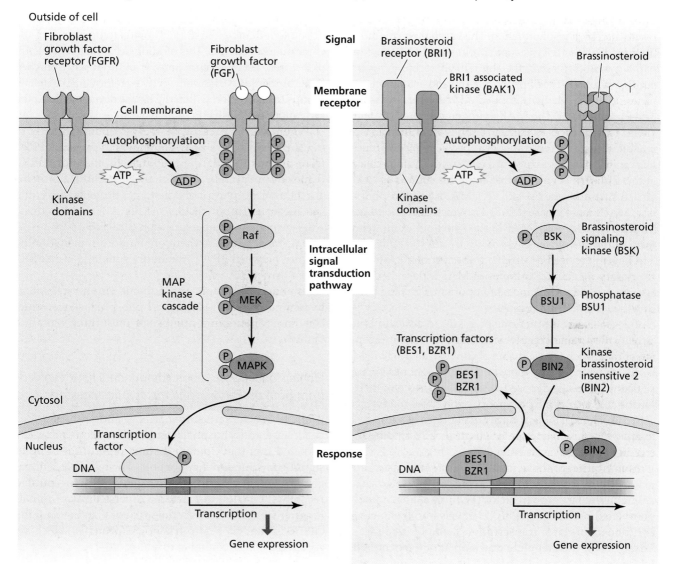

**FIGURE 14.2** Kinase-based signal transduction in animals and plants for communicating information from the plasma membrane to the nucleus. (A) Perception of the fibroblast growth factor (FGF) ligand by the tyrosine kinase receptor triggers the activation of the MAP kinase pathway (composed of the kinases Raf, MEK, and MAPK) to transduce the signal recognition event at the plasma membrane to the nucleus in an animal cell. (B) In plants, perception of brassinosteroids (BRs) causes the receptor BRI1 to phosphorylate itself and plasma membrane-

associated brassinosteroid signaling kinases (BSK). BSK phosphorylation promotes interaction with the protein phosphatase BSU1. BSU1 then dephosphorylates and inactivates the BIN2 kinase, a repressor that inhibits BR transcription factors by causing them to exit the nucleus. Inactivation of BIN2 allows BR-induced transcription to proceed. FGF, fibroblast growth factor; FGFR, FGF receptor; BRI1, brassinosteroid insensitive 1; BAK1, BRI1 associated kinase; BIN2, brassinosteroid insensitive 2.

dimerization with a second receptor termed *BRI1-a*ssociated receptor *k*inase (BAK1) (**FIGURE 14.2B**). This process is discussed in greater detail in Chapter 24.

## Receptor kinases can initiate a signal transduction cascade

Receptor kinases must modify the activity of other proteins in order to trigger an **intracellular signal transduction cascade**. A target protein can be phosphorylated at various amino acid residues (serine, threonine, tyrosine or histidine). The intracellular target proteins are often protein kinases themselves, and they may be activated or inactivated by phosphorylation. This mechanism for modulating kinase activity is common in signal transduction pathways. In the case of the animal FGF receptor, ligand binding results in the phosphorylation and activation of a series of kinase enzymes, termed a **MAP kinase cascade** (see Figure 14.2A). The MAP (*m*itogen-*a*ctivated *p*rotein) kinase cascade owes its name to a series of protein kinases that phosphorylate each other in a specific sequence, much like runners passing a baton in a relay race. The first kinase in the sequence is Raf, referred to in this context as MAP kinase kinase kinase (MAPKKK). MAPKKK phosphorylates MAP kinase kinase (MAPKK), which phosphorylates MAP kinase (MAPK). MAPK, the "anchor" of the relay team, enters the nucleus, where it activates still other protein kinases, as well as specific transcription factors and regulatory proteins. Examples of MAP kinase cascades also exist in plants and include important signaling pathways, such as defense responses.

One of the key functions of a kinase cascade is to amplify the original receptor signaling event at the plasma membrane. Each kinase that is phosphorylated will modify the activity of many more of its own target proteins. A signaling cascade composed of several kinases will therefore be able to alter the phosphorylation status (and hence activity) of tens of thousands of target proteins in response to relatively few ligand molecules originally binding the receptor at the plasma membrane.

Modification of the activities of protein kinases ultimately causes changes to gene expression. In the case of animal fibroblast growth factor (FGF) signaling, a phosphorylated protein termed ERK travels into the nucleus and phosphorylates transcription factors termed Et's. These Et's in turn regulate downstream targets of FGF. In the case of the plant hormone brassinosteroid, binding of the hormone to its receptor causes the inactivation of the repressor BIN2 (*BRASSINOSTEROID INSENSI-TIVE 2*). BIN2 is a protein kinase that normally blocks the nuclear accumulation of transcription factors involved in brassinosteroid-regulated gene expression. Inactivation of BIN2 allows nuclear accumulation of these transcription factors. The transcription factors in the nucleus are then able to bind their target promoters and trigger brassinosteroid-dependent changes in gene expression.

In addition to kinases, enzymes that remove phosphate groups from proteins, *protein phosphatases*, play important roles within signal transduction pathways. For example, the brassinosteroid signaling component BSU1 is a protein phosphatase that inactivates BIN2 by dephosphorylating a critical phospho-tyrosine residue. Phosphatases also play a key role in the signal transduction pathways of the plant hormone abscisic acid (ABA) (see also Chapter 23). A subgroup of phosphatases within the protein phosphatase 2C (PP2C) family have been shown to dephosphorylate a class of protein kinases (called SnRK2 kinases) in an ABA-dependent manner. First, PP2C binds to domain II of the C-terminus of SnRK2. This binding occurs independently of the presence or absence of ABA. In the absence of ABA, PP2C blocks SnRK2 kinase activity by removing phosphate groups from a region within the kinase domain termed the activation loop (**FIGURE 14.3A**).

When ABA is present, its receptor binds directly to PP2C proteins (Park et al. 2009), which prevents PP2C from dephosphorylating the kinase domain of SnRK2 (Umezawa et al. 2009; Vlad et al. 2009). SnRK2 proteins are then free to phosphorylate target proteins, including specific transcription factors (AREB/ABFs) that activate ABA-responsive gene expression (**FIGURE 14.3B**). ABA signal transduction is therefore based on reversing the balance between PP2C protein phosphatase and SnRK2 kinase activities.

In summary, plants and animals employ receptors in conjunction with kinase and phosphatase transduction components in a number of important signaling pathways.

## Plants signal transduction components have evolved from both prokaryotic and eukaryotic ancestors

While plants and animals share signaling proteins such as kinases and phosphatases, genome sequencing has revealed that they contain distinct repertoires of other signal transduction components. For example, animal genomes often contain hundreds of **G protein–coupled receptors (GPCRs)** that detect a diverse array of signals ranging from hormones, odors, flavors, and even light. GPCRs signal via hetero-trimeric G proteins, which are encoded by a large number of genes.

In contrast, no plant GPCRs have been functionally identified to date. Moreover, only single copies of hetero-trimeric G protein genes have been identified in the Arabidopsis genome (Jones and Assmann 2004). Such examples clearly illustrate that plants should not be considered to be "green animals" since key differences exist in the makeup of the signal transduction machinery between these eukaryotic systems.

(A) ABA absent

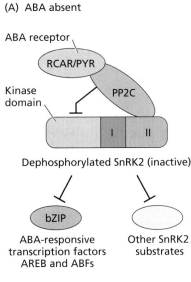

ABA receptor

Kinase domain

Dephosphorylated SnRK2 (inactive)

bZIP

ABA-responsive transcription factors AREB and ABFs

Other SnRK2 substrates

In the absence of ABA, the protein phosphatase PP2C keeps the protein kinase SnRK2 dephosphorylated and thereby inactivated.

(B) ABA present

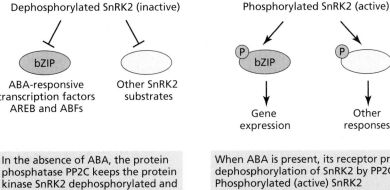

ABA

RCAR/PYR

PP2C

P  P

Phosphorylated SnRK2 (active)

P  bZIP

Gene expression

P

Other responses

When ABA is present, its receptor prevents dephosphorylation of SnRK2 by PP2C. Phosphorylated (active) SnRK2 phosphorylates downstream substrates, thereby inducing ABA responses.

**FIGURE 14.3** Abscisic acid (ABA) signaling involves kinase and phosphatase activities. (A) In the absence of ABA, the protein phosphatase PP2C dephosphorylates and inactivates the SnRK2 kinase. (B) In the presence of ABA, the ABA receptor protein RCAR/PYR interacts with PP2C, blocking phosphatase action and releasing SnRK2 from negative regulation. The activated SnRK2 phosphorylates ABA-responsive transcription factors (bZIP) and other unknown substrates to induce an ABA response. ABA, abscisic acid; SnRK2, SNF1-related protein kinase 2; PP2C, protein phosphatase 2C; KD, kinase domain.

One of the main reasons that higher plants have a distinct repertoire of signal transduction components is that they have evolved from both prokaryotic and eukaryotic ancestors. Plants have utilized signaling components from both their single-cell eukaryotic progenitor and the prokaryotic precursor of the chloroplast. For example, the Arabidopsis genome encodes two cryptochrome-related genes (*CRY1* and *CRY2*). **Cryptochromes** are bacterial flavoproteins that function as DNA photolyases that repair pyrimidine dimers produced by UV light. In Arabidopsis, cryptochromes lack the critical residues required for DNA repair. Instead, they mediate light control of stem elongation, leaf expansion, photoperiodic flowering, and the circadian clock (see Chapter 18).

Many important plant signaling components are derived from bacterial signaling proteins. Cytokinin and ethylene receptor sequences are related to two-component regulatory genes found in bacteria. Bacterial *two-component regulatory systems* are composed of a histidine kinase sensor protein and a response regulator protein (**FIGURE 14.4**). The function of the **sensor protein** is to receive the input signal and to pass the signal on to the response regulator, which brings about the cellular response, typically gene expression. Sensor proteins have two domains, an *input domain*, which receives the environmental signal, and a *transmitter domain*, which transmits the signal to the response regulator.

The **response regulator protein** also has two domains, a *receiver domain*, which receives the signal from the trans-

mitter domain of the sensor protein, and an *output domain*, such as a DNA-binding domain, which brings about the response. The signal is passed from transmitter domain to receiver domain via protein phosphorylation. Transmitter domains have the ability to phosphorylate themselves, using ATP, on a specific histidine residue near the amino terminus (see Figure 14.4). The phosphate is then transferred to a specific aspartate residue near the middle of the receiver domain of the response regulator protein. Phosphorylation of the aspartate residue causes the response regulator to undergo a conformational change that usually results in the activation of gene expression.

The hormone cytokinin is perceived via a phosphorelay that is organized in a similar way to the bacterial two-component systems (see Figure 14.4; Chapter 21). The cytokinin receptors designated CRE1, AHK2, and AHK3 are related in amino acid sequence to the histidine kinases in two-component systems. However, these cytokinin receptor sequences are described as *hybrid sensor histidine kinases* since they contain both the histidine kinase (transmitter) domain of a bacterial sensor protein and the receiver domain of a bacterial response regulator protein (see Figure 14.4).

As a result, cytokinin binding triggers autophosphorylation of the histidine residue on the transmitter domain, followed by phosphate transfer to the aspartate residue on the receiver domain. The phosphate is then transferred to proteins termed *histidine phosphotransfer proteins* (Hpt or AHP). The newly phosphorylated AHPs function as sig-

(A) Prokaryotic two-component system

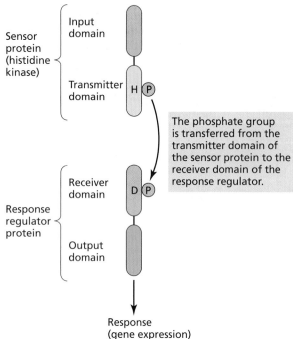

The phosphate group is transferred from the transmitter domain of the sensor protein to the receiver domain of the response regulator.

(B) Multistep version of the prokaryotic two-component system

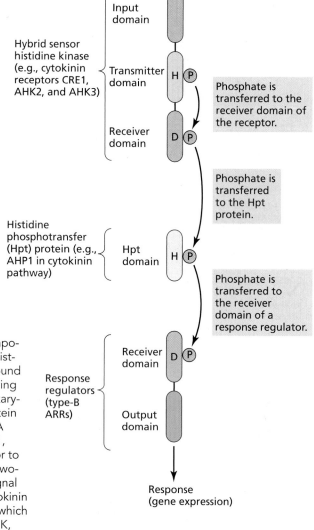

Phosphate is transferred to the receiver domain of the receptor.

Phosphate is transferred to the Hpt protein.

Phosphate is transferred to the receiver domain of a response regulator.

**FIGURE 14.4** Plants and bacteria employ similar signaling components and mechanisms. The two-component system (A), consisting of a sensor protein and a response regulator protein, is found only in prokaryotes, while a derived, three-step version involving a phosphorelay protein intermediate (B) is found in both prokaryotes and eukaryotes. The plant two-component receptor protein includes a receiver domain fused to the transmitter domain. A separate histidine phosphotransfer protein (Hpt), called AHP1, transfers phosphates from the receiver domain of the receptor to the receiver domain of the response regulator (ARR). Such a two-component system is involved in cytokinin perception and signal transduction. The genes *CRE1*, *AHK2*, and *AHK3* encode cytokinin receptors that phosphorylate the Hpt domain protein AHP1, which then transfers a phosphate group to type-B ARR proteins. AHK, Arabidopsis histidine kinase; AHP, Arabidopsis histidine phosphotransfer protein; ARR, Arabidopsis response regulator.

naling intermediates that transmit the plasma membrane perceived cytokinin signal to nuclear localized response regulators (termed ARRs), causing activation of gene expression.

## Signals are perceived at many locations within plant cells

The majority of examples we have provided so far have involved signal perception at the plasma membrane (see Figures 14.2 and 14.4). However, plants can perceive external signals at many different locations within their cells. For example, light is detected by distinct classes of photoreceptors at the plasma membrane and in the cytoplasm and nucleus (**FIGURE 14.5**). The genes *PHOT1* and *PHOT2* encode photoreceptors, called *phototropins*, that perceive

blue light at the plasma membrane and mediate phototropism, chloroplast movement, and stomatal opening. In contrast, cryptochrome genes *CRY1* and *CRY2* encode blue-light photoreceptors that are located in the nucleus.

Genes for phytochromes encode red/far-red light photoreceptors that partition between the cytoplasm and nucleus. Light induces a conformational change in the cytoplasmic protein, exposing a **nuclear localization signal** (**NLS**), which in turn causes the translocation of phytochrome to the nucleus. Once inside the nucleus, phytochrome interacts with transcription factors, triggering changes in gene expression (see Chapter 17).

Hormone signals are also perceived at many different locations in plant cells (see Figure 14.5). For example, brassinosteroids and cytokinins are perceived at the plasma membrane by their receptors BRI1 and CRE1/AHK2/

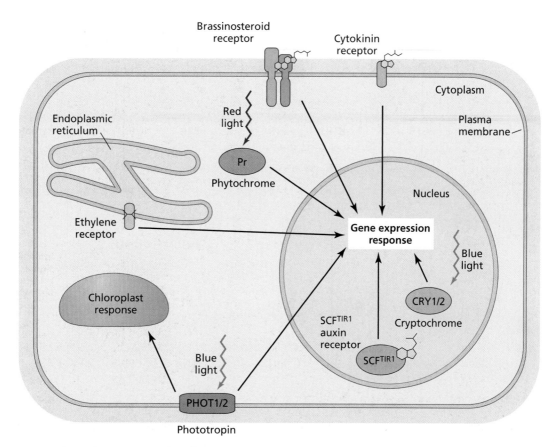

**FIGURE 14.5** Signals can be perceived at different locations within the plant cell. Membrane receptors can perceive signals at the plasma membrane (brassinosteroid, cytokinin, and blue light) or the endoplasmic reticulum (ethylene), while other signals are perceived in the cytoplasm (red light) or in the nucleus (auxin and blue light). Responses often result in the induction of gene expression, but can also involve modification of pre-existing biological structures such as proteins and organelles.

AHK3, respectively (Li and Chory 1997; Inoue et al. 2001). In contrast, the plant hormone ethylene is perceived by the receptor ETR1 in the endoplasmic reticulum (Chen et al. 2002). Ethylene ($C_2H_4$), being small and extremely lipophilic, can easily enter plant cells and their intracellular compartments. The auxin receptor, SCF$^{TIR1}$, is located in the nucleus, necessitating its electrostatically charged ligand, indole-3-acetic acid (IAA), to enter the plant cell either through diffusion or active transport (Kramer and Bennett 2006).

### Plant signal transduction often involves inactivation of repressor proteins

Most animal signal transduction pathways induce a response through the activation of a cascade of positive regulators. For example, in the case of fibroblast growth factor (FGF) signaling, receptor–ligand binding activates the receptor kinase. This induces changes in gene expression through the activation of regulatory kinases, in this case a MAP kinase cascade, which acts positively to stimulate gene expression (**FIGURE 14.6A**). In contrast, *the majority*

*of plant transduction pathways induce a response by inactivating repressor proteins.* For example, ethylene binding to its receptor, ETR1, results in the inactivation of the repressor CTR1 (**FIGURE 14.6B**). Inactivation of this negative regulator enables the activation of the transcription factor EIN3, which induces a transcriptional response (see also Chapter 22). Brassinosteroid binding to the protein BRI1 causes the repressor protein BIN2 to become inactivated, resulting in the activation of the transcription factors BES1 and BZR1 (**FIGURE 14.6C**; see Chapter 24).

Why have plant cells evolved signaling pathways based on negative regulation rather than positive regulation, as occurs in animal cells? Mathematical modeling of signal transduction pathways employing negative regulators suggests that negative regulators result in faster induction of downstream response genes (Rosenfeld et al. 2002). The speed of a response, particularly to an environmental stress such as drought, may be crucial to the survival of the sessile plant. Hence, the adoption of such a regulatory arrangement in the majority of plant signaling pathways is likely to have conferred a selective advantage during evolution.

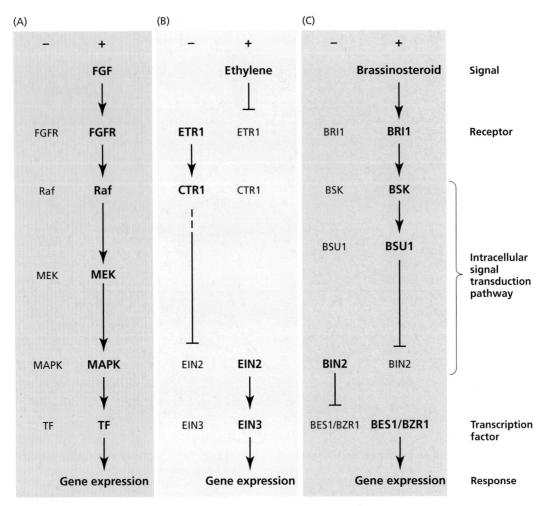

**FIGURE 14.6** Signal transduction pathways in animals and plants differ in their functional organization. The majority of animal signal transduction pathways feature a sequence of positive activation steps that ultimately activate transcription factors. In contrast, most plant signal transduction pathways function through the inactivation of a transcriptional repressor. Three examples are provided which illustrate this point. (A) In animal cells, the binding of fibroblast growth factor (FGF) to its receptor initiates the activation of a series of kinases (MAP kinase cascade). The last kinase in the sequence (MAPK) activates transcription factors, which initiate gene transcription. (B) The plant ethylene signal transduction pathway, unlike the FGF pathway, is active in the absence of a receptor–ligand interaction. In the absence of ethylene, the signaling intermediate CTR1 represses gene expression via a pathway—possibly involving a MAP kinase cascade—that inactivates the transcriptional regulator EIN2. In the presence of ethylene, the CTR pathway is blocked and EIN2 activates the appropriate transcription factors. (C) The brassinosteroid (BR) pathway also results in the inactivation of a transcriptional repressor, BIN2, which normally inhibits BR-induced gene expression. In the diagram light font indicates inactive intermediates and bold font indicates active components.

Several different molecular mechanisms have been described in plant cells to inactivate repressor proteins (**FIGURE 14.7**). As noted above, protein dephosphorylation is employed by the brassinosteroid pathway to inactivate the repressor protein BIN2 (see Chapter 24). Light triggers protein retargeting of the repressor COP1 out of the nucleus, enabling transcription factors like HY5 to accumulate and activate gene expression (see Chapter 17).

Protein degradation is another way to inactivate repressor proteins, a mechanism first described as part of the auxin signaling pathway (Gray et al. 2001). In this case, degradation is initiated when auxin binds to its receptor complex, causing the auxin/indole-3-acetic acid (AUX/IAA) repressor proteins to become tagged with multiple copies of a small protein called **ubiquitin** (see Figure 14.7C). Such ubiquitin tags act as a marker indicating that the targeted protein is to be degraded by what is called the **ubiquitination pathway** (see Chapter 2). In the presence of auxin, AUX/IAA repressor proteins are targeted for degradation, enabling transcription factors termed auxin response factors (ARFs) to activate gene expression (see Figure 14.7C).

(A) Kinase inactivation

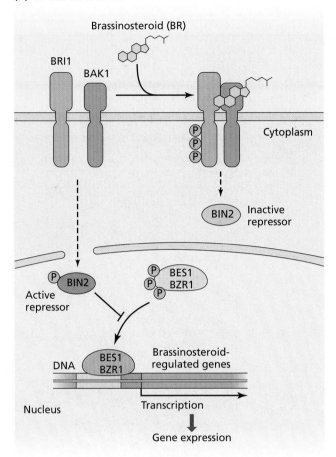

(B) Repressor retargeting

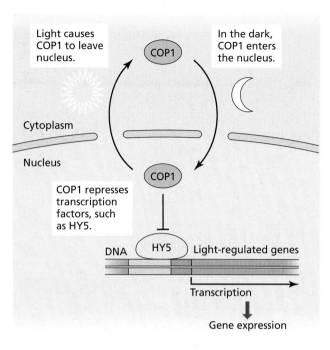

(C) Repressor degradation

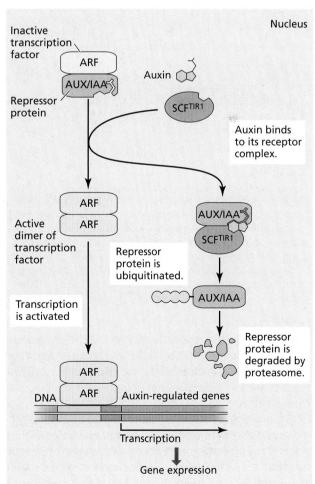

**FIGURE 14.7** Transduction pathways in plants often signal by inactivating repressor proteins. (A) In the signaling pathway for brassinosteroids, the negative regulator BIN2 is inactivated through inhibition of its kinase activity. (B) Absorption of red light by phytochrome causes the COP1 repressor protein to leave the nucleus. (C) The binding of auxin to its receptor complex initiates ubiquitin-dependent degradation of the AUX/IAA repressor protein by the 26S proteasome. In all three cases, inactivation of the repressor protein results in transcription factor activation.

## Protein degradation is a common feature in plant signaling pathways

There has been an explosion of interest in protein degradation as a signaling mechanism since the original observation linking auxin and ubiquitin (Leyser et al. 1993). Examples of ubiquitin-dependent protein degradation have been identified in almost every plant signal transduction pathway characterized to date. These include pathways for important plant signals such as jasmonate (JA) and bioactive gibberellins (GA). JA is associated with

(A)  Auxin response

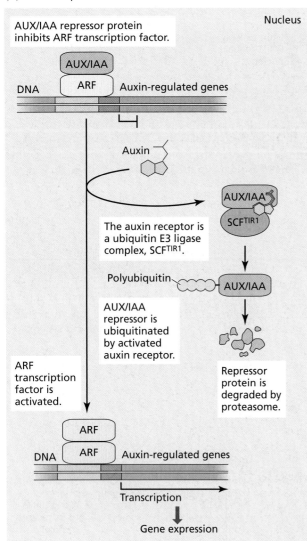

(B)  Jasmonate response

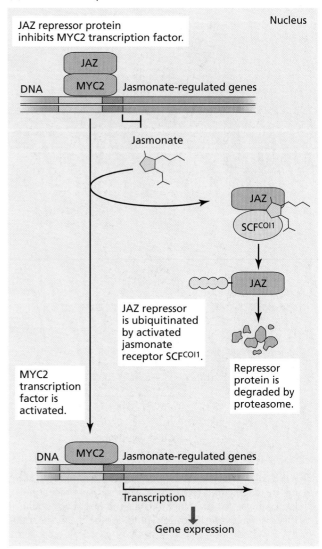

**FIGURE 14.8**   Several plant hormone receptors are part of SCF ubiquitination complexes. Auxin, jasmonate (JA), and gibberellins (GA) signal by promoting interaction between components of the SCF ubiquitination machinery and repressor proteins operating in each hormone's transduction pathway. Auxin (A) and JA (B) directly promote interaction between the SCF$^{TIR1}$ and SCF$^{COI1}$ complexes and the AUX/IAA and JAZ repressors, respectively. In contrast, GA (C) additionally requires an adaptor protein, GID1, to form the complex between SCF$^{SLY1}$ and DELLA proteins. Addition of multiple ubiquitins (polyubiquitin) marks these repressor proteins for degradation. This triggers the activation of ARF, MYC2, and PIF3/4 transcription factors, resulting in auxin-, JA-, and GA-induced changes in gene expression.

defense responses against herbivores and necrotrophic pathogens (see Chapter 13), while GA regulates important growth processes such as seed germination, stem and root elongation, and determination of leaf size and shape (see Chapter 20).

Jasmonate and GA employ similar signaling mechanisms to auxin (**FIGURE 14.8**). The pathway in each case involves one of a group of ubiquitin E3 ligase complexes called SCFs (see below) and results in degradation of a transcriptional repressor. The JA receptor, COI1, regulates JA responses by targeting JAZ-repressor proteins for degradation (Xie et al. 1998; Yan et al. 2007; Thines et al. 2007). Analogous to AUX/IAA proteins, JAZ proteins are thought to suppress JA-responsive genes. JA-induced ubiquitin-dependent degradation of JAZ repressor proteins results in the release and activation of the transcription factor MYC2, triggering the induction of JA-responsive gene expression.

Bioactive gibberellins act in large part by mediating the rapid turnover (through ubiquitination) of **DELLA repressor proteins** (Jiang and Fu 2007). DELLA proteins restrain plant growth (for example, root elongation and light-controlled hypocotyl elongation), which GA over-

**(C) Gibberellin response**

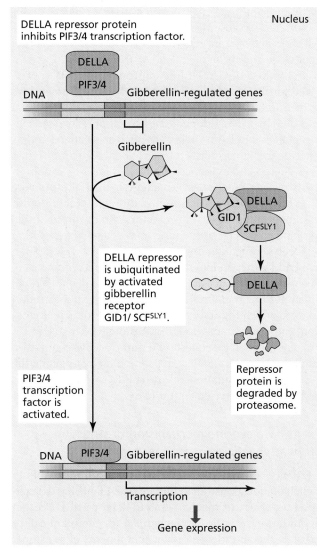

SCF, which is an **E3 ligase**. The superscript in the E3 ligase name (e.g., SCF$^{TIR1}$) indicates which F-box protein the complex contains. F-box proteins typically recruit target proteins to the SCF complex so that they can be ubiquitinated by the E3 ligase and subsequently degraded by the *26S proteasome*, a large multiprotein complex that degrades ubiquitin-tagged proteins.

The auxin receptor gene, *TIR1*, encodes an F-box component of the SCF complex (Ruegger et al. 1998). TIR1 functions as an auxin receptor by recruiting AUX/IAA proteins to the SCF complex (Dharmasiri et al. 2005; Kepinski and Leyser 2005). Often, modification of target proteins by phosphorylation (or other amino acid modification) is a prerequisite for binding to an F-box protein. However, binding of AUX/IAA proteins to TIR1 does not require prior protein modification. Instead, auxin acts as a "molecular glue" that promotes the interaction between TIR1 and AUX/IAA (Tan et al. 2007; see Figure 14.8).

Since the discovery of auxin's mechanism of action, several other important plant hormones have been found to promote the interaction between an F-box protein and its

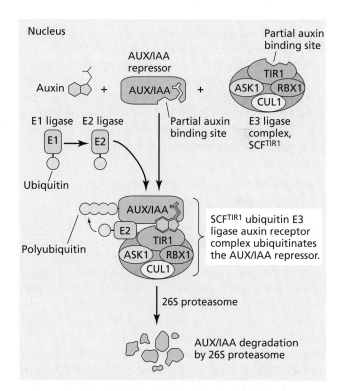

comes by promoting destruction of the DELLA proteins. GA does this by binding its receptor GID1, triggering the ubiquitin-dependent degradation of DELLA repressor proteins. DELLA degradation results in the release and activation of transcription factors, thus triggering changes in gene expression (de Lucas et al. 2008). To date, five DELLA genes have been identified in Arabidopsis: *REPRESSOR OF ga1-3* (*RGA*), **GA-INSENSITIVE** (*GAI*), **RGA-LIKE1** (*RGL1*), *RGL2*, and *RGL3* (Fleet and Sun 2005).

## Several plant hormone receptors encode components of the ubiquitination machinery

The ubiquitination pathway plays a key role in most, if not all, hormone perception pathways. In brief, the ubiquitin protein first gets conjugated to an activating enzyme termed an *E1 ligase* (**FIGURE 14.9**). The ubiquitin tag is then transferred to a second conjugating enzyme called an *E2 ligase*. This enzyme then associates with a large complex composed of *S*kp, *c*ullin, and *F*-box proteins, termed

**FIGURE 14.9** The auxin receptor is composed of two proteins: the SCF complex component TIR1 and the repressor protein AUX/IAA. Ubiquitin moieties are first activated by E1 ligase and added to target proteins by E2 ligase. TIR1 recruits AUX/IAA proteins to the SCF$^{TIR1}$ complex in an auxin-dependent manner. Once recruited by auxin, AUX/IAA proteins are ubiquitinated by the E3 ligase activity of the SCF$^{TIR1}$ complex, which marks the protein for destruction by the 26S proteasome.

target proteins. For example, the F-box protein COI1 has recently been demonstrated to be a JA receptor (Xie et al. 1998; Yan et al. 2007; Thines et al. 2007). Like auxin, JA promotes the interaction between COI1 and repressors of JA-induced gene expression called JAZ proteins (see Figure 14.8). GA signaling also involves components of the SCF complex. However, the GA receptor GID1 does not itself function as an F-box protein. Instead, GA promotes the binding of DELLA repressor proteins to the GA-receptor, GID1 (Griffiths et al. 2006; Willige et al. 2007; Nakajima et al. 2006). This in turn enhances the binding of DELLAs to SCF$^{SLY1}$, an E3 ubiquitin-ligase that contains the F-box protein SLY1 (McGinnis et al. 2003). In effect, the binding of GA receptor GID1 to DELLA repressor proteins triggers degradation of the latter via the F-box protein SLY1.

As the above discussion reveals, auxin, jasmonate, and gibberellins have bypassed the need for a membrane-bound receptor and a complex cytoplasmic signal transduction chain that ultimately results in a change in gene expression (see Figure 14.2). Instead, they signal by *directly* targeting the stability of nuclear-localized repressor proteins to induce a transcriptional response (see Figure 14.6). Such a short signal transduction pathway provides the means for a very rapid change in nuclear gene expression. However, there is no opportunity for signal amplification as in the case of a kinase cascade (see Figure 14.2). Instead, any resulting transcriptional response is directly related to the abundance of the signal molecule, since this will determine the number of repressor molecules that are degraded. These key differences in the organization of signal transduction pathways may help explain why much higher concentrations of signals like auxin and GA are required to elicit a biological response than is required for others such as brassinosteroids.

### Inactivation of repressor proteins results in a gene expression response

The majority of signal transduction pathways ultimately elicit a biological response by inducing changes in the expression of selected target genes. In plants, a change in gene expression almost always results from the inactivation of a repressor protein, thereby derepressing the activity of a transcription factor (see Figures 14.2, 14.7, and 14.8).

In the case of auxin, AUX/IAA repressor proteins are targeted for degradation, thereby activating auxin response factor (ARF)-dependent gene expression (see Figure 14.8A). Auxin responsive genes typically have auxin response element (AuxRE) binding sites located in their promoter regions (Hagen and Guilfoyle 2002). ARF transcription factors bind to these AuxRE motifs and thereby stimulate transcription (Hagen and Guilfoyle 2002; Ulmasov et al. 1999). AUX/IAA protein sequences contain four conserved motifs, termed domains I to IV (Dharmasiri and Estelle 2002). Domains III and IV are also found in ARF protein sequences, enabling them to heterodimerize with AUX/IAA proteins (Guilfoyle et al.

1998). This heterodimerization blocks ARF-mediated transcription (Guilfoyle et al. 1998; Tiwari et al. 2003). Auxin-stimulated degradation of AUX/IAA proteins releases ARF and allows transcription of AuxRE-containing target genes (see Figure 14.8).

In the case of GA, DELLA repressor proteins are targeted for degradation, which activates specific members of the PIF family of transcription factors (PIF3 and PIF4) and thereby PIF3/4-dependent gene expression (see Figure 14.8C). DELLA proteins are nuclear-localized and as such are thought to act as transcriptional regulators. However, there is no evidence of a DNA-binding domain in their coding sequence, which suggests that DELLAs may regulate gene expression indirectly by interacting with tissue-specific transcription factors. An example of this mechanism was recently found to mediate light control of hypocotyl elongation. Light inhibits hypocotyl elongation in a DELLA-dependent manner, whereas GA promotes it. The work of de Lucas et al. (2008) and Feng et al. (2008) illustrates that members of the PIF family of transcription factors (which regulate genes important for cell elongation) can bind directly to DELLA proteins. Thus, DELLA proteins function to sequester PIFs into inactive complexes. High levels of GA alleviate this inhibition by targeting the DELLAs for degradation, releasing PIFs to induce changes in gene expression (see Figure 14.8).

### Plants have evolved mechanisms for switching off or attenuating signaling responses

Arguably, it is equally important for a cell to be able to switch off a response as to initiate one. Plants have developed a variety of mechanisms to achieve this. Protein degradation provides a mechanism for the plant cell to regulate the abundance of key components of the signal transduction pathway, such as the receptor or a transcription factor. For example, the photoreceptor phytochrome is targeted for ubiquitin-dependent degradation by the E3 ligase COP1 in the nucleus (Chapter 17). Similarly, the transcription factor EIN3, which regulates ethylene-dependent gene expression (see Figure 14.6B), is targeted for ubiquitin-dependent degradation by the F-box proteins EBF1 and EBF2 (see Chapter 22).

Feedback regulation represents another key mechanism employed to attenuate a response. For example, the *AUX/IAA* genes, which encode the AUX/IAA auxin repressor proteins, have auxin response element binding sites located in their promoter regions (Hagen and Guilfoyle 2002). Thus, AUX/IAA proteins can bind to the promoters of their own genes and repress their own expression.

Hormone signaling pathways are often subject to several loops of negative feedback regulation. This is nicely illustrated by the GA pathway (**FIGURE 14.10**). Bioactive GA (GA$_4$ in this example) is synthesized through a complex biosynthetic pathway that involves multiple enzyme-catalyzed reactions (Hedden and Phillips 2000). The last two enzymes

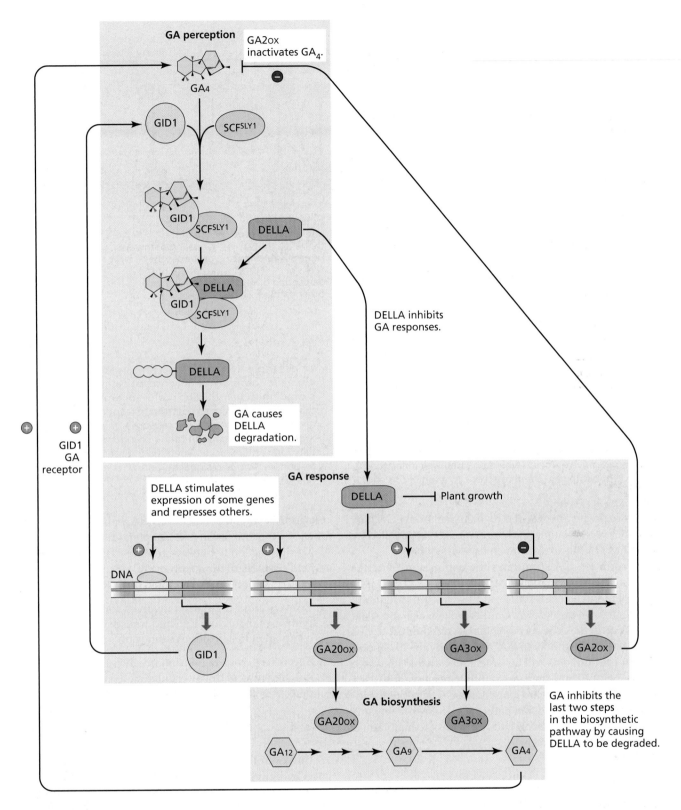

**FIGURE 14.10**  The GA response is regulated by a series of feedback mechanisms involving components of both GA signal transduction and GA biosynthesis. *GA20ox* and *GA3ox* genes encode enzymes that catalyze the last steps of the GA biosynthetic pathway, while GA2ox catalyzes the breakdown of the bioactive GA, GA₄. *GID1* encodes the GA receptor that, following ligand binding, recruits DELLA repressor proteins to the SCF^SLY1 complex for ubiquitination, triggering their degradation. In the absence of GA, the DELLA proteins positively regulate *GID1*, *GA20ox*, and *GA3ox* (plus signs), and negatively regulate *GA2ox* (minus sign). Conversely, both bioactive GA and the GID1 receptor enhance DELLA repressor degradation (plus signs), while GA2ox blocks DELLA repressor degradation (minus signs).

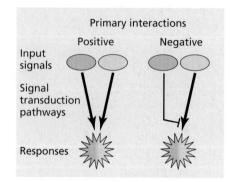

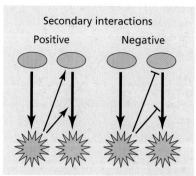

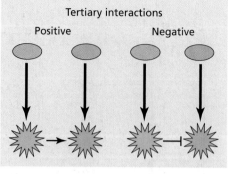

| | | | | | |
|---|---|---|---|---|---|
| Two input pathways regulate a single shared protein or multiple shared proteins controlling a response. Both pathways have the same effect on the response. | Two input pathways converge on shared protein(s), but one of the pathways inhibits the effect of the other. | Two input pathways regulate separate responses. In addition, one pathway enhances the input levels or perception of the other pathway. | As in the positive interaction, except that one pathway represses the input levels or perception of the other pathway. | The response of one of the signaling pathways promotes the response of the other pathway. | The response of one of the signaling pathways inhibits the response of the other pathway. |

**FIGURE 14.11** Signal transduction pathways operate as part of a complex web of signaling interactions. Three types of cross-regulation have been proposed: primary, secondary, and tertiary. Input signals are shown here as ovals, signal transduction pathways are indicated by heavy arrows, and responses (pathway outputs) are shown as stars. Light lines with arrowheads (positive) or bars (negative) indicate where one pathway influences the other pathway. The three types of interactions can be either positive or negative. Primary interactions involve the convergence of two input pathways on a single shared signal transduction protein or group of proteins that directly interact to control the response. In secondary interactions, the output response of one pathway regulates the perception of the second input signal. Tertiary interactions occur when the independent output responses from two signaling pathways influence the output responses of each other.

in this pathway are encoded by members of the GA20ox and GA3ox gene families, whose expression is inhibited by GA. Thus GA acts to slow down its own synthesis.

In contrast, GA stimulates the expression of genes encoding the GA catabolism enzyme GA2ox (Thomas et al. 1999; Zentella et al. 2007)—in this way GA essentially induces its own breakdown. DELLA proteins also stimulate expression of the gene encoding the GA receptor GID1; this increases the sensitivity to GA, and thereby the likelihood that DELLA proteins will be targeted for degradation (see Figure 14.10). Thus, there are multiple negative feedback and positive feed-forward loops in the GA perception, biosynthesis, and deactivation pathways. These mechanisms help ensure that GA levels and responses are maintained during plant development.

## Cross-regulation allows signal transduction pathways to be integrated

Within plant cells, signal transduction pathways never function in isolation, but operate as part of a complex web of signaling interactions. This realization helps us understand why plant hormones often exhibit *agonistic* (additive or positive) or *antagonistic* (inhibitory or negative) interactions with other signals. Classic examples include the antagonistic interaction between GA and ABA in the control of seed germination (see Chapters 20 and 23).

The interaction between signaling pathways has been termed **cross-regulation** and several categories have been proposed (Kappusamy et al. 2009) (**FIGURE 14.11**).

1. **Primary cross-regulation** involves distinct signaling pathways regulating a shared transduction component in a positive or a negative manner.

2. **Secondary cross-regulation** involves the output of one signaling pathway regulating the abundance or perception of a second signal.

3. **Tertiary cross-regulation** involves the outputs of two distinct pathways exerting influences on one another.

The control of cell elongation in the Arabidopsis hypocotyl by light and GA represents an example of negative primary cross-regulation (**FIGURE 14.12**; de Lucas et al. 2008; Feng et al. 2008; Sadeghi-aliabadi et al. 2008). In this case, the light and GA pathways share common downstream signaling components—the closely related transcription factors PIF3 and PIF4—that act to stimulate cell elongation. The accumulation of PIF3/4 is regulated in opposite ways by the two signals: positively by GA and negatively by light.

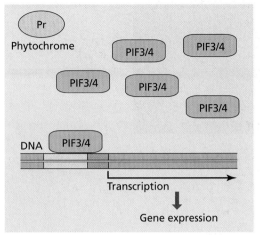

(A) PIF3/4 transcription factors accumulate in the dark.

Dark: Hypocotyl cells elongate

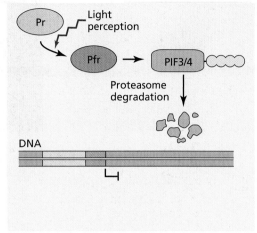

(B) PIF3/4 transcription factors are degraded in the light.

Light: Hypocotyl cells remain short

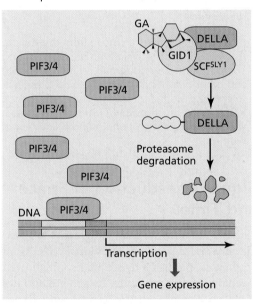

(C) GA causes DELLA degradation by the 26S proteosome and PIF3/4 accumulate.

+GA: Hypocotyl cells elongate

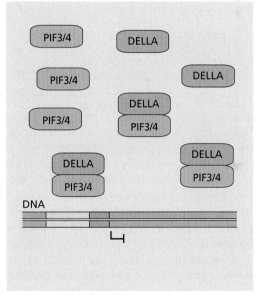

(D) DELLA binds PIF3/4 in the absence of GA.

−GA: Hypocotyl cells remain short

**FIGURE 14.12** GA and light positively and negatively regulate, respectively, the accumulation of their shared downstream signaling components, PIF3 and PIF4. (A) Dark-grown hypocotyl cells elongate due to the accumulation of the PIF3/4 transcription factors. (B) In the light, the red/far-red light photoreceptor phytochrome (Pr/Pfr) targets PIF3/4 for degradation, reducing cell elongation and thereby shortening hypocotyl length. (C) High levels of bioactive GAs target DELLAs for degradation, releasing PIF3/4 and promoting cell expansion. (D) In the absence of GA, PIF3/4 transcription factors bind directly to DELLA repressor proteins and are inactive.

Dark-grown hypocotyl cells elongate due to the accumulation of PIF3/4 (see Figure 14.12). However, in the light, the red-light photoreceptor PHYB targets PIF3/4 for degradation, which results in reduced cell elongation and shorter hypocotyls. In the absence of GA, PIF3/4 transcription factors bind directly to DELLA repressor proteins, and

are thus inactive. However, high levels of bioactive GA target DELLAs for degradation, releasing PIF3/4 to promote cell elongation (see Figure 14.8C).

Secondary cross-regulation involves the output of one signaling pathway regulating the abundance or perception of a second input signal (see Figure 14.12). An example

**(A)** Localization of *SHR* gene expression in Arabidopsis root

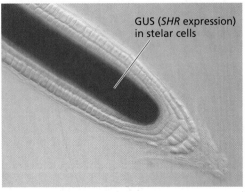

GUS (*SHR* expression) in stelar cells

**(B)** Localization of SHR protein in Arabidopsis root

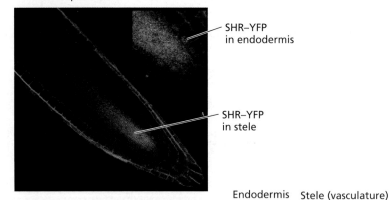

SHR–YFP in endodermis

SHR–YFP in stele

**(C)**

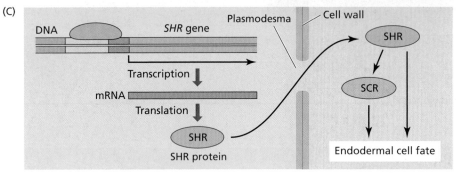

Endodermis   Stele (vasculature)

**FIGURE 14.13** Short distance movement of the SHORT-ROOT (SHR) protein between the stele and endodermal tissues regulates radial patterning. (A) The *SHR* gene is expressed in the vasculature, as visualized using a GUS reporter (blue) fused to the *SHR* promoter. (B) SHR protein is detected both in the vasculature and the endodermis (see inset), as visualized using a *SHR* promoter–driven SHR–YFP translational fusion (yellow-green fluorescence).

(C) The SHR protein can move from the stele to the adjacent layer of cells, which will become the endodermis. There it triggers endodermal cell differentiation via SCR and other output pathways. The SHR protein is necessary to produce a differentiated endodermis. SHR, SHORT-ROOT; SCR, SCARECROW; GUS, glucuronidase; YFP, yellow fluorescent protein. (A and B courtesy of Malcolm Bennett.)

of secondary cross-regulation involves the inhibition of Arabidopsis root cell elongation by auxin and ethylene. In this case, ethylene signaling inhibits root cell elongation indirectly by stimulating auxin biosynthesis (Stepanova et al. 2007; Swarup et al. 2007).

Lastly, in tertiary cross-regulation, the outputs of two distinct pathways exert influences on one another (see Figure 14.11). A recently discovered example is the interaction between auxin and cytokinin pathways to specify root stem cells (Muller and Sheen 2008). In this case, auxin negatively regulates the output of the cytokinin response pathway, the B-type **Arabidopsis response regulators** (**ARRs**) (see Figure 14.4). Auxin does this by increasing the expression of A-type ARRs (see Figure 21.8) that function as feedback inhibitors of B-type ARRs.

It is becoming clear that plant signaling is not based on a simple linear sequence of transduction events, but involves cross-regulation between many pathways. Understanding how such complex signaling pathways operate will demand a new scientific approach. This approach is often referred to as **systems biology** and employs mathematical and computational models to simulate these nonlinear biological networks and to better predict their outputs (Locke et al. 2006; Coruzzi et al. 2009; Middleton et al. 2010).

# Signal Transduction in Space and Time

Plant signal transduction pathways don't just operate at the network scale within individual cells. As multicellular organisms, plants have evolved sophisticated signaling mechanisms to coordinate the growth and development of each cell within a tissue, organ, and, ultimately, organism. In this section we will see that these signaling mechanisms often operate across a wide range of physical scales, ranging from nanometers to tens of meters, and of time, ranging from seconds to years.

## Plant signal transduction occurs over a wide range of distances

Signaling events can take place over very short distances (i.e., the next cell). For example, radial patterning in the Arabidopsis primary root is regulated by the transcription regulator SHORT-ROOT (SHR) (**FIGURE 14.13**). The gene *SHR* is transcribed and translated in cells of the stele, at the center of the root (Figure 14.13A and C). The SHR protein then moves via plasmadesmata to cells in the endodermis, where it activates the expression of cell fate regulators

(A) Arabidopsis root tip

(B) Localization of auxin influx carrier AUX1

(C) Localization of auxin efflux carrier PIN2

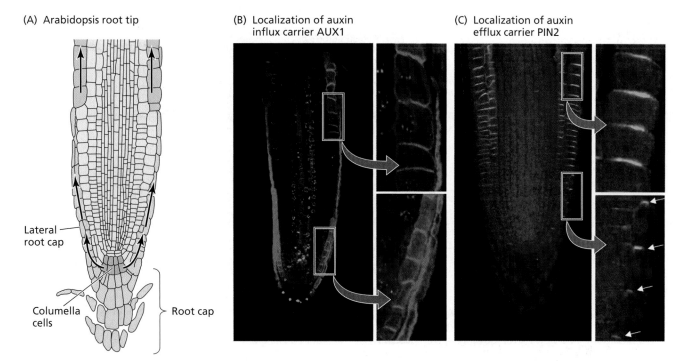

Lateral root cap

Columella cells

Root cap

**FIGURE 14.14** Auxin transport within the outer root tissue is mediated by specialized auxin influx and efflux proteins. (A) Auxin transporters move the hormone auxin (denoted by arrows) from gravity-sensing cells at the root tip via lateral root cap cells. This transport is fundamental to triggering growth responses in elongation zone cells.

(B, C) Confocal images illustrating the localization of auxin influx carrier protein AUX1 (B; in red), and auxin efflux carrier PIN2 (C; in green) within Arabidopsis root apical cells and tissues. AUX1, AUXIN RESISTANT 1; PIN2, PIN-FORMED 2. (B and C courtesy of Malcolm Bennett.)

such as SCARECROW (SCR) (Figure 14.12B and C). The transcription factor SHR can therefore be considered as a short-distance plant signal that operates between adjacent cells and tissues (see Chapter 16).

Complex growth responses within an organ often involve signaling between many cells. One example is the growth response that follows a change in root orientation relative to gravity (termed gravitropism; see Chapter 19). Briefly, columella cells at the root tip detect changes in the direction of gravity by using specialized starch-filled plastids termed *statoliths* (**FIGURE 14.14A**). Gravity-induced changes in statolith position result in rapid redistribution of the auxin efflux carrier PIN3 and the creation of an auxin gradient at the root apex (Friml et al. 2002). The gravity-induced auxin gradient is then transported from the columella cells to elongation zone tissues via the lateral root cap (Swarup et al. 2005).

Lateral root cap cells express auxin influx carrier AUX1 and auxin efflux carrier PIN2 to facilitate the rapid entry and exit of this hormone (**FIGURE 14.14B AND C**). In addition, the asymmetric localization of PIN2 at the upper face of lateral root cap cells directs the auxin towards the elongation zone tissues (see arrows in Figure 14.14C), while the evenly distributed AUX1 is essential for the efficient uptake of auxin by lateral root cap cells (Figure 14.14B). Hence, auxin can be considered a short- to long-distance plant signal, capable of operating between adjacent cells,

tissues, and organs, depending on the distribution of its specialized transport machinery (see Chapter 19).

The growth and development of one plant organ is often influenced by signals originating from another organ. For example, the transition of the shoot apical meristem from a vegetative to reproductive identity (termed floral induction) can be triggered by a signal originating from leaves (Imaizumi and Kay 2006). A long-day inductive signal can induce expression of the *FLOWERING TIME* (*FT*) gene, which encodes a transcription factor, in Arabidopsis leaf phloem companion cells (Takada and Goto 2003) (see Chapter 25). The resulting FT protein is then transported to the shoot apex (Corbesier et al. 2007; Jaeger and Wigge 2007; Matthieu et al. 2007; Notaguchi et al. 2008). In the shoot apex, FT interacts with another transcription factor, FLOWERING LOCUS D (FD), to coactivate target genes and trigger flowering (Abe et al. 2005; Wigge et al. 2005). Examples such as the FT–FD protein interaction elegantly illustrate that in plants (unlike in animals) transcriptional regulators are able to act over long distances to control developmental programs.

### The timescale of plant signal transduction ranges from seconds to years

How long does it take to induce a response in plants? The examples of signal transduction pathways described above operate over tens of minutes to hours because responses

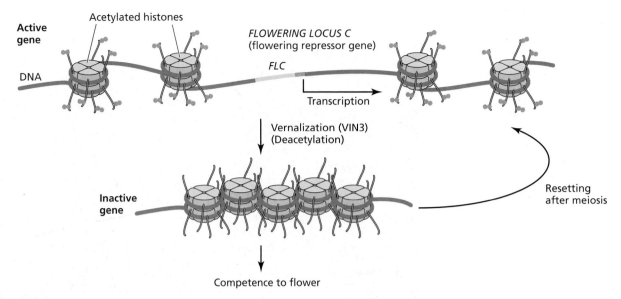

**FIGURE 14.15** Control of flowering by temperature is achieved through modification of chromatin proteins surrounding the *FLC* gene. Prolonged cold (vernalization) causes remodeling of *FLC* chromatin and accumulation of histone modifications characteristic of heterochromatin, which repress *FLC* transcription. After meiosis, the repressor FLC is expressed again to prevent early flowering.

that rely on changes in gene expression are relatively slow. Plant cells take up to 30 minutes to transcribe a gene, process its mRNA, and then export the mRNA to the cytoplasm. Additional time must also be factored in for protein synthesis and, possibly, intracellular trafficking. Such timescales are appropriate for plant growth and developmental processes.

However, many plant responses must occur much faster, some within seconds. For example, plants are often exposed to rapidly alternating periods of very high and low light as clouds pass in front of the sun. Exposure to very high light can be extremely damaging to a plant cell's photosynthetic machinery. One response of plant leaf cells to high-intensity light is to reorient chloroplasts rapidly so that their edges face the light source, thereby minimizing the surface area being exposed (see Chapter 9). *How can such responses to signals occur within seconds?*

Clearly such responses can't involve transcription and translation. Instead, a plant cell must employ signal transduction mechanisms that alter the activity of existing proteins. In the example described above, chloroplast reorientation is regulated by the PHOTOTROPIN (PHOT) class of blue-light photoreceptors, which are located in the plasma membrane (see Figure 14.5). Their light-activated kinase activities induce rapid changes in the organization of the plant cell's cytoskeleton, resulting in chloroplast reorientation within seconds.

Some plant responses operate over timescales of months, or even years. For example, a period of cold is required by many plant species for flowering to occur (Baurle and Dean 2006). Exposure to a relatively long period of cold (i.e.,

winter) must be sensed and transduced into a response (the "competence," or potential, to flower). This important process is termed **vernalization**. Although the cold-sensing mechanism is not yet known, the downstream part of the signal transduction pathway involves changes in chromatin organization. For example, vernalization depends on the gene *VERNALIZATION INDEPENDENT3* (*VIN3*), which is required for modification of chromatin proteins called histones around which DNA is wrapped (**FIGURE 14.15**) (see Chapter 2).

Expression of the *VIN3* gene is only induced after a prolonged period of cold treatment (Sung and Amasino 2004). VIN3 causes histones to become deacetylated at *FLOWERING LOCUS C* (*FLC*), which encodes a flowering repressor (see Chapter 25). The deacetylation results in *FLC* DNA being wrapped tightly in chromatin, which blocks *FLC* transcription and prevents it from repressing floral induction.

Finally, signaling response timescales that extend over years are not uncommon among plants. Some examples of multiyear signaling responses include vernalization in biennial species, flowering in trees, and seed dormancy, which in some cases can be maintained for centuries or even thousands of years.

In summary, signal transduction processes in plants must operate across timescales ranging from seconds to years. The speed of the response required will determine whether signal transduction will involve changes in protein activity (most rapid, perhaps seconds), gene expression (moderately fast), or chromatin organization (slowest, may take months).

## SUMMARY

Plant cells are able to sense and respond to a wide range of external and internal signals by a pathway that involves: signal → reception → transduction → response. To connect a signal to the appropriate response, plants employ a number of highly conserved transduction mechanisms.

### Signal Transduction in Plant and Animal Cells

- Signal transduction mechanisms include kinase receptors coupled to kinase cascades (**Figure 14.2**) and two-component systems (**Figure 14.4**), which are conserved between plants and their eukaryotic and prokaryotic progenitors, respectively.

- Plants have also developed a number of unique signaling mechanisms. These include employing components of the ubiquitination machinery to function as hormone receptors (**Figures 14.8, 14.9**).

- The majority of plant signal transduction pathways function by inactivating repressor proteins (**Figure 14.6**).

- Repressors can be posttranslationally modified (e.g., phosphorylated), retargeted, or subjected to ubiquitin-dependent degradation (**Figure 14.7**).

- Removal of repressor proteins often results in activation of transcription factors, inducing changes in gene expression that underpin any subsequent plant response.

- Mutant screens in Arabidopsis and rice have helped characterize plant signaling pathways. However, the power of genetic screening is sometimes limited by redundancy: the existence of multiple genes that encode functionally related sequences and can mask the loss other family members.

- New approaches such as chemical genetics can be used to overcome genetic redundancy; for example, these methods helped to identify a family of ABA receptors (**Figure 14.3**).

- Plant signaling pathways are not strictly linear sequences of transduction events. In reality, plant signaling pathways employ sophisticated feedback mechanisms to modulate or switch off a response (**Figure 14.10**).

- Transduction pathways cross-regulate one another, forming a complex web of signaling interactions (**Figure 14.11, 14.12**).

- Studying complex signaling pathways demands a new scientific approach termed systems biology, which employs mathematical and computational models to simulate these nonlinear biological networks to better predict their outputs.

### Signal Transduction in Space and Time

- Plant signaling transduction pathways do not just operate at the network scale within individual cells. They may operate across a wide range of physical scales, ranging from micrometers (within cells or tissues) to meters (organs, organ systems, and organisms) (**Figures 14.13, 14.14**).

- Intercellular signaling mechanisms involve a wide variety of molecules that include hormones, transcriptional regulators, and small RNAs.

- Signal transduction processes in plants must also operate across timescales spanning seconds to years. Very rapid plant responses (on the order of a few seconds or minutes) involve changes in protein activity rather than gene expression.

- Changes in gene expression are often required to sustain responses operating over timescales of hours and days.

- Longer-term plant responses (operating over months or years) frequently involve environmentally regulated posttranslational modifications to chromatin proteins near key regulatory genes, thereby influencing their expression (**Figure 14.15**).

## CHAPTER REFERENCES

Abe, K., Osakabe, K., Nakayama, S., Endo, M., Tagiri, A., Todoriki, S., Hiroaki, I., and Toki, S. (2005) Arabidopsis RAD51C gene is important for homologous recombination in meiosis and mitosis. *Plant Physiol.* 10: 896–908.

Bäurle, I., and Dean, C. (2006) The timing of developmental transitions in plants. *Cell* 125: 655–664.

Borthwick, H. A., Hendricks, S. B., Parker, M. W., Toole, E. H., and Toole, V. K. (1952) A reversible photoreaction controlling seed germination. *Proc. Natl. Acad. Sci. USA* 38: 662–666.

Butler, W. L., Norris, K. H., Siegelman, H.W., and Hendricks, S. B. (1959) Detection, assay, and preliminary purification of the pigment controlling photoresponsive development of plants. *Proc. Natl. Acad. Sci. USA* 45: 1703–1708.

Chen, Y. F., Randlett, M. D., Findell, J. L., and Schaller, G. E. (2002) Localization of the ethylene receptor ETR1 to the endoplasmic reticulum of Arabidopsis. *J. Biol. Chem.* 277: 19861–19866.

Chow, B., and McCourt, P. (2006) Plant hormone receptors: perception is everything. *Genes Devel.* 20: 1998–2008.

Corbesier, L., Vincent, C., Jang, S., Fornara, F., Fan, Q., Searle, I., Giakountis, A., Farrona, S., Gissot, L., Turnbull, C., et al. (2007) FT protein movement contributes to long distance signaling in floral induction of Arabidopsis. *Science* 316: 1030–1033.

Coruzzi, G. M., Rodrigo, A., and Guttiérrez, R. A., eds. (2009) *Plant Systems Biology. Annual Plant Reviews*, Vol. 35. Wiley-Blackwell, Oxford.

Darwin, C. (1881) *The Power of Movement in Plants*. D. Appleton and Company, New York.

de Lucas, M., Daviere, J. M., Rodriguez-Falcon, M., Pontin, M., Iglesias-Pedraz, J. M., Lorrain, S., Fankhauser, C., Blazquez, M. A., Titarenko, E., and Prat, S. (2008) A molecular framework for light and gibberellin control of cell elongation. *Nature* 451: 480–484.

Dharmasiri, S., Estelle, M. (2002) The role of regulated protein degradation in auxin reponse. *Plant Cell* 14: 2137–2144.

Dharmasiri, N., Dharmasiri, S., and Estelle, M. (2005) The F-box protein TIR1 is an auxin receptor. *Nature* 435: 441–445.

Feng, S., Martinez, C., Gusmaroli, G., Wang, Y., Zhou, J., Wang, F., Chen, L., Yu, L., Iglesias-Pedraz, J. M., and Kircher, S. (2008) Coordinated regulation of *Arabidopsis thaliana* development by light and gibberellins. *Nature* 451: 475–479.

Fleet, C. M., and Sun, T. (2005) A DELLAcate balance: the role of gibberellin in plant morphogenesis. *Curr. Opin. Plant Biol.* 8: 77–85.

Friml, J., Wisniewska, J., Benkova, E., Medgen, K., and Palme, K. (2002) Lateral relocation of auxin efflux regulator PIN3 mediates tropism in Arabidopsis. *Nature* 415: 806–809.

Gray, W. M., Kepinski, S., Rouse, D., Leyser, O., and Estelle, M. (2001) Auxin regulates SCF[TIR1]-dependent degradation of AUX/IAA proteins. *Nature* 414: 271–276.

Griffiths, J., Murase, K., Rieu, I., Zentella, R., Zhang, Z. L., Powers, S. J., Gong, F., Phillips, A. L., Hedden, P., and Sun, T. (2006) Genetic characterization and functional analysis of the GID1 gibberellin receptors in Arabidopsis. *Plant Cell* 18: 3399–3414.

Guilfoyle, T., Hagen, G., Ulmasov, T., and Murfett, J. (1998) How does auxin turn on genes? *Plant Physiol.* 118: 341–347.

Hagen, G., and Guilfoyle, T. (2002) Auxin-responsive gene expression: genes, promoters and regulatory factors. *Plant Mol. Biol.* 49: 373–385.

Hedden, P., and Phillips, A. L. (2000) Gibberellin metabolism: new insights revealed by the genes. *Trends Plant Sci.* 5: 523–530.

Imaizumi, T., and Kay, S. A. (2006) Photoperiodic control of flowering: not only by coincidence. *Trends Plant Sci.* 11: 550–558.

Inoue, T., Higuchi, M., Hashimoto, Y., Seki, M., Kobayashi, M., Kato, T., Tabata, S., Shinozaki, K., and Kakimoto, T. (2001) Identification of CRE1 as a cytokinin receptor from Arabidopsis. *Nature* 409: 1060–1063.

Jaeger, K. E., and Wigge, P. A. (2007) FT protein acts as a long-range signal in Arabidopsis. *Curr. Biol.* 17: 1050–1054.

Jiang, C., and Fu, X. (2007) GA action: turning on de-DELLA repressing signaling. *Curr. Opin. Plant Biol.* 10: 461–465.

Jones, A. M., and Assmann, S. M. (2004) Plants: the latest model system for G-protein research. *EMBO Rep.* 5: 572–578.

Kepinski, S., and Leyser, O. (2005) The *Arabidopsis* F-box protein TIR1 is an auxin receptor. *Nature* 435: 446–451.

Kuppusamy, K. T., Walcher, C. L., Nemhauser, J. L. (2009) Cross-regulatory mechanisms in hormone signaling. *Plant Mol. Biol.* 69: 375–381.

Leyser, H. M. O., Lincoln, C. A., Timpte, C., Lammer, D., Turner, J., and Estelle, M. (1993) Arabidopsis auxin-resistance gene *AXR1* encodes a protein related to ubiquitin-activating enzyme E1. *Nature* 364: 161–164.

Li, J., and Chory, J. (1997) A putative leucine-rich repeat receptor kinase involved in brassinosteroid signal transduction. *Cell* 90: 929–938.

Locke, J. C. W., Kozma-Bognar, L., Gould, P. D., Feher, B., Kevei, E., Nagy, F., Turner, M. S., Hall, A., and Millar, A. J. (2006) Experimental validation of a predicted feedback loop in the multi-oscillator clock of *Arabidopsis thaliana*. *Mol. Syst. Biol.* 2: 59, doi:10.1038/msb4100102.

Mathieu, J., Sung, H. H., Pugieux, C., Soetaert, J., Rorth, P. (2007) A sensitized PiggyBac-based screen for regulators of border cell migration in *Drosophila*. *Genetics* 176: 1579–1590.

McCourt, P. (1999) Genetic analysis of hormone signaling. *Annu. Rev. Plant Physiol. Plant Mol. Biol.* 50: 219–243.

McGinnis, K. M., Thomas, S. G., Soule, J. D., Strader, L. C., Zale, J. M., Sun, T., and Steber, C. M. (2003) The Arabidopsis *SLEEPY1* gene encodes a putative F-box subunit of an SCF E3 ubiquitin ligase. *Plant Cell* 15: 1120–1130.

Middleton, A. M., King, J. R., Bennett, M., and Owen, M. R. (2010) Mathematical modelling of the Aux/IAA negative feedback loop. *Bull. Math. Biol.*, doi:10.1007/s11538-009-9497-4.

Muller, B., and Sheen, J. (2008) Cytokinin and auxin interaction in root stem-cell specification during early embryogenesis. *Nature* 453: 1094–1097.

Nakajima, M., Shimada, A., Takashi, Y., Kim, Y. C., Park, S. H., Ueguchi-Tanaka, M., Suzuki, H., Katoh, E., Iuchi, S., and Kobayashi, M. (2006) Identification and characterization of Arabidopsis gibberellin receptors. *Plant J.* 46: 880–889.

Notaguchi, M., Abe, M., Kimura, T., Daimon, Y., Kobayashi, T., Yamaguchi, A., Tomita, Y., Dohi, K., Mori, M., and Araki, T. (2008) Long-distance, graft-transmissible action of *Arabidopsis* FLOWERING LOCUS T protein to promote flowering. *Plant Cell Physiol.* 49: 1645–1658.

Park, S. Y., Fung, P., Nishimura, N., Jensen, D.R., Fujii, H., Zhao, Y., Lumba, S., Santiago, J., Rodrigues, A., Chow, T.F., et al. (2009) Abscisic acid inhibits type 2C protein phosphatases via the PYR/PYL family of START proteins. *Science* 324: 1068–1071.

Rosenfeld, N., Elowitz, M. B., and Alon, U. (2002) Negative autoregulation speeds the response times of transcription networks. *J. Mol. Biol.* 323: 785–793.

Ruegger, M., Dewey, E., Gray, W. M., Hobbie, L., Turner, J., and Estelle, M. (1998) The TIR1 protein of Arabidopsis functions in auxin response and is related to human SKP2 and yeast grr1p. *Gene Develop.* 12: 198–207.

Sadeghi-aliabadi, H., Ghasemi, N., and Kohi, M. (2008) Cytotoxic effect of *Convolvulus arvensis* extracts on human cancerous cell line. *Res. Pharm. Sci.* 3: 31–34.

Stepanova, A., Yun, J., Likhacheva, A., and Alonson, J. (2007) Multilevel interactions between ethylene and auxin in *Arabidopsis* roots. *Plant Cell* 19: 2169–2185.

Sung, S. B., and Amasino, R. M. (2004) Vernalization and epigenetics: how plants remember winter. *Curr. Opin. Plant Biol.* 7: 4–10.

Swarup, R., Perry, P., Hagenbeek, D., Van Der Straeten, D., Beemster, G. T. S., Sandberg, G., Bhalerao, R., Ljung, K., and Bennett, M. J. (2007) Ethylene upregulates auxin biosynthesis in *Arabidopsis* seedlings to enhance inhibition of root cell elongation. *Plant Cell* 19: 2186–2196.

Takada, S., and Goto, K. (2003) TERMINAL FLOWER2, a HETEROCHROMATIN PROTEIN1-like protein of Arabidopsis, counteracts the activation of *FLOWERING LOCUS T* by CONSTANS in the vascular tissues of leaves to regulate flowering time. *Plant Cell* 15: 2856–2865.

Tan, X., Calderon-Villalobos, L. I. A., Sharon, M., Zheng, C., Robinson, C. V., Estelle, M., and Zheng, N. (2007) Mechanism of auxin perception by the TIR1 ubiquitin ligase. *Nature* 446: 640–645.

Thines, B., Katsir, L., Melotto, M., Niu, Y., Mandaokar, A., Liu, G., Nomura, K., He, S. Y., and Howe, G. A. (2007) JAZ repressor proteins are targets of the SCF[COI1] complex during jasmonate signaling. *Nature* 448: 661–665.

Thomas, S. G., Phillips, A. L., and and Hedden, P. (1999) Molecular cloning and functional expression of gibberellin 2-oxidases, multifunctional enzymes involved in gibberellin deactivation. *Proc. Natl. Acad. Sci. USA* 96: 4698–4703.

Tiwari, S. B., Hagen, G., and Guilfoyle, T. (2003) The roles of auxin response factor domains in auxin-responsive transcription. *Plant Cell* 15: 533–543.

Ulmasov, T., Hagen, G., and Guilfoyle, T.J. (1999) Dimerization and DNA binding of auxin response factors. *Plant J.* 19: 309–319.

Umezawa, T., Sugiyama, N., Mizoguchi, M., Hayashi, S., Myouga, F., Yamaguchi-Shinozaki, K., Ishihama, Y., Hirayama, T., and Shinozaki, K. (2009) Type 2C protein phosphatases directly regulate abscisic acid-activated protein kinases in Arabidopsis. *Proc. Natl. Acad. Sci. USA* 106: 17588–17593.

Vlad, F., Rubio, S., Rodrigues, A., Sirichandra, C., Belin, C., Robert, N., Leung, J., Rodriguez, P. L., Laurière, C., and Merlot, S. (2009) Protein phosphatases 2C regulate the activation of the Snf1-related kinase OST1 by abscisic acid in Arabidopsis. *Plant Cell* 21: 3170–3184.

Wigge, P., Kim, M. C., Jaeger, K. E., Busch, W., Schmid, M., Lohmann, J. U., and Weigel, D. (2005) Integration of spatial and temporal information during floral induction in *Arabidopsis*. *Science* 309: 1056–1059.

Willige, B. C., Ghosh, S., Nill, C., Zourelidou, M., Dohmann, E., Maier, A., and Schwechheimer, C. (2007) The DELLA domain of GA INSENSITIVE mediates the interaction with the GA INSENSITIVE DWARF1A gibberellin receptor of *Arabidopsis*. *Plant Cell* 19: 1209–1220.

Xie, D. X., Feys, B. F., James, S., Nieto-Rostro, M., and Turner, J. G. (1998) COI1: An Arabidopsis gene required for jasmonate-regulated defense and fertility. *Science* 280: 1091–1094.

Yan, Y., Stolz, S., Chételat, A., Reymond, P., Pagni, M., Dubugnon, L., and Farmer, E. E. (2007) A downstream mediator in the growth repression limb of the jasmonate pathway. *Plant Cell* 19: 2470–2483.

Zentella, R., Zhang, Z. L., Park, M., Thomas, S. G., Endo, A., Murase, K., Fleet, C. M., Jikumaru, Y., Nambara, E., Kamiya, Y., et al. (2007) Global analysis of DELLA direct targets in early gibberellin signaling in Arabidopsis. *Plant Cell* 19: 3037–3057.

# Cell Walls: Structure, Biogenesis, and Expansion

Plant cells, unlike animal cells, are surrounded by a mechanically strong cell wall. This thin layer consists of a scaffold of cellulose fibrils embedded in a matrix of polysaccharides that are secreted by the cell. The polysaccharide matrix and cellulose fibrils assemble into a strong network linked by a mixture of covalent and noncovalent bonds. The matrix may also contain structural proteins, enzymes, phenolic polymers, and other materials that modify the wall's physical and chemical characteristics.

The cell walls of prokaryotes, fungi, algae, and plants differ from each other in chemical composition and molecular structure, yet they all serve three common functions: regulating cell volume, determining cell shape, and protecting the delicate protoplast. As we will see, however, plant cell walls have acquired additional functions not evident in the cell walls of other organisms, and these diverse functions are reflected in the cell wall's structural complexity and diversity in composition and form.

In addition to these biological functions, the plant cell wall is important commercially. Plant cell walls are used in the production of paper, textiles (cotton, flax, linen, and others), lumber, and other wood products. Plant cell walls are also used to make synthetic fibers (such as rayon), plastics, films, coatings, adhesives, gels, and thickeners. Today, major efforts are underway worldwide to develop cost-effective methods for converting "cellulosic biomass" into biofuels to replace petroleum-based transportation fuels such as gasoline. According to some scenarios, a billion tons of cellulosic biomass will need to be harvested each year in the U.S. to replace nearly a third of the petroleum currently used for transportation. As the most abundant reservoir of organic carbon in nature, the plant cell wall also takes part in the processes of carbon flow through ecosystems (see **WEB TOPIC 15.1**).

We begin this chapter with a description of the general structure and composition of cell walls and the mechanisms of its biosynthesis and assembly. We then turn to the role of the primary cell wall in cell expansion. The mechanisms of tip growth, which occurs in a few specialized cell types, will be contrasted with those of diffuse growth, particularly with respect to the establishment of cell polarity and the control of the rate of cell expansion. Finally, we will describe the dynamic changes in the cell wall that often accompany cell differentiation, maturation and defense.

## The Structure and Synthesis of Plant Cell Walls

Without their cell walls, plants would be very different organisms from what we know. Instead of stately trees we would find formless blobs of amoeba-like cells. Indeed, the cell wall is essential for many critical processes in plant growth, development, maintenance, and reproduction:

- Cell walls determine the mechanical strength of plant structures, allowing plants to grow to great heights.

- Cell walls glue cells together, preventing them from sliding past one another. This constraint on cell movement contrasts markedly to the situation in animal cells, and it dictates the way in which plants develop (see Chapter 16).

- Plant morphogenesis ultimately depends on the control of cell wall properties, because the enlargement of plant cells is limited principally by the ability of the cell wall to expand.

- As a mechanically-strong layer encapsulating the cell, the wall acts as a cellular "exoskeleton" that controls cell shape and allows high turgor pressures to develop. Without a cell wall to resist the forces generated by turgor pressure, plant water relations would be profoundly different (see Chapter 3).

- Transpirational water flow in the xylem requires a mechanically strong wall that resists collapse in response to the negative pressure in the xylem.

- The wall acts as a diffusion barrier that limits the size of macromolecules that can reach the plasma membrane from outside, and it is a major structural barrier to pathogen invasion.

Much of the carbon assimilated in photosynthesis is channeled into polysaccharides that make up the wall. During specific phases of development or periods of sugar starvation, some of these polymers may be hydrolyzed into their constituent sugars to be scavenged by the cell and used to meet the cell's needs. This role is most notable in seeds with large storage reserves in the endosperm or cotyledons. Easily-digested polysaccharides are packed into these cell walls during seed development and are rapidly mobilized during germination to feed the growing embryo. Furthermore, oligosaccharide components of the cell wall may act as important signaling molecules during cell differentiation and during recognition of pathogens and symbionts.

The functional diversity and varying roles of the cell wall require diverse cell wall structures. In this section we begin with a brief description of the morphology and basic architecture of plant cell walls. Then we discuss the organization, composition, and synthesis of the wall in some of its diverse forms.

### Plant cell walls have varied architecture

In stained sections of plant organs, the most obvious visual objects are cell walls, which may vary greatly in appearance and composition in different cell types (**FIGURE 15.1**). For example, cell walls of parenchyma in the pith and cortex are generally thin (~100 nm) and have few distinguishing features. In contrast, epidermal cells, collenchyma, xylem vessels and tracheids, phloem fibers, and other forms of sclerenchyma cells have thicker walls (~1000 nm or more, sometimes multilayered). These walls may be intricately sculpted and impregnated with substances such as lignin, cutin, suberin, waxes, silica, or structural proteins, that alter the wall's chemical and physical properties.

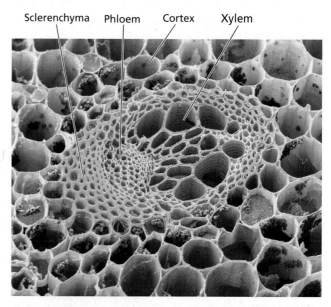

**FIGURE 15.1**   Cross section of a stem of a buttercup (*Ranuculus repens*), showing cells with varying wall morphology in different tissue types (see labels). Note the highly thickened walls of the sclerenchyma cells and the pitted walls of the xylem cells. (Photo © Andrew Syred/Photo Researchers, Inc.)

(A)

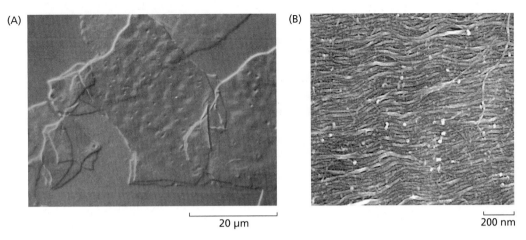

20 µm

200 nm

(C)

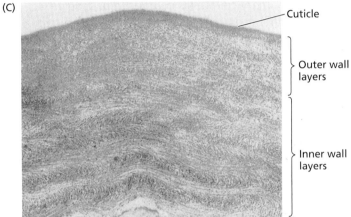

Cuticle

Outer wall layers

Inner wall layers

FIGURE 15.2 Three views of primary cell walls. (A) This surface view of cell wall fragments from onion parenchyma was taken with a light microscope using Nomarski optics. Note that at this scale the wall looks like a very thin sheet with small surface depressions; these depressions may be pit fields, places where plasmodesmatal connections between cells are concentrated. (B) This surface view of a cell wall from a growing cucumber hypocotyl was visualized by scanning electron microscopy. Note the fibrous texture of the wall and the more or less parallel orientation of the fibrils, which are oriented transverse to the long axis of the cell. The fibrils are cellulose microfibrils with matrix polymers bound to their surface. (C) Electron micrograph of the outer epidermal cell wall (cross section) from the growing region of a bean hypocotyl. Multiple layers are visible within the wall. The inner layers are thicker and more defined than the outer layers, because the outer layers are the older regions of the wall and have been stretched and thinned by cell expansion. (A from McCann et al. 1990; B from Marga et al. 2005; C from Roland et al. 1982.)

The walls on the individual sides of a cell may also vary in thickness, in amount and type of impregnating substances, in sculpting, and in frequency of pitting and plasmodesmata. For example, the outer wall of the epidermis lacks plasmodesmata, is impregnated with cutin and waxes, and is much thicker than the other walls of the cell. It is also distinguished by different polysaccharide components than those found in the other walls (Obel et al. 2009; Freshour et al. 1996). In guard cells, the wall adjacent to the stomatal pore is much thicker than the walls on the other sides of the cell. Such variations in wall architecture within a single cell reflect the cell's polarity and differentiated functions and arise from targeted secretion of wall components to the different cell surfaces.

Despite this morphological diversity, cell walls commonly are classified into two major types: primary walls and secondary walls. This classification is based not on structural or biochemical differences, but rather on the developmental state of the cell that is producing the cell wall. **Primary walls** are defined as walls formed by growing cells. Usually they are thin and architecturally simple (FIGURE 15.2A,B; FIGURE 15.3A), but some primary walls may be much thicker and multilayered, such as those found in collenchyma or in the epidermis (FIGURE 15.2C).

**Secondary walls** are formed after cell enlargement stops. They are formed between the plasma membrane and the primary cell wall. Secondary walls may become highly specialized in structure and composition, reflecting the differentiated state of the cell (FIGURE 15.3B,C). In the water-conducting tissue (xylem), fiber cells, tracheids, and vessels are notable for possessing thickened secondary walls that are strengthened and waterproofed by **lignin**. However, not all secondary walls are lignified or thickened. **Pits** and **pit fields** are thin areas where the primary wall is not covered by a secondary wall (see Figure 15.3B); they facilitate movement of water and other materials between cells.

A thin layer of material, called the **middle lamella**, is found at the interface where the walls of neighboring cells come into contact. The middle lamella is typically enriched in pectic polysaccharides (much less so in grasses) and may contain structural proteins different from those in the primary or secondary wall. Its origin can be traced to the cell plate formed during cell division.

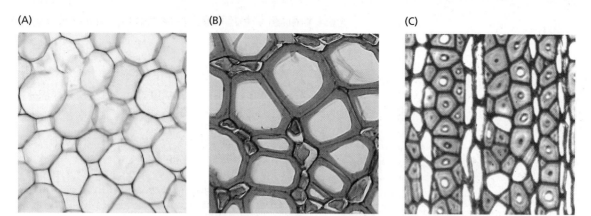

(A)        (B)        (C)

**FIGURE 15.3** Diversity of cell wall structure. The thin walls of rice stem (*Oryza sativa*) parenchyma (A) contrast with the thickened secondary cell walls of a vascular bundle of an underground fern stem (B) and the fibers of *Tetrameri-* *sta* xylem (C). (A, B © Garry DeLong/OSF/Photolibrary. com; C courtesy of the Bailey-Wetmore Wood Collection, Harvard University, Cambridge, MA.)

As we saw in Chapter 1, the cell wall is often penetrated by tiny membrane-lined channels, called **plasmodesmata** (singular *plasmodesma*), which connect neighboring cells. Plasmodesmata function in communication between cells by allowing passive transport of small molecules and active transport of proteins and nucleic acids between the cytoplasm of adjacent cells.

## The primary cell wall is composed of cellulose microfibrils embedded in a polysaccharide matrix

In primary cell walls, cellulose microfibrils are embedded in a hydrated matrix of noncellulosic polysaccharides and a small amount of structural protein (**FIGURE 15.4, TABLE 15.1**). This structure imparts an ideal combination of flexibility and strength to the growing cell wall, which must

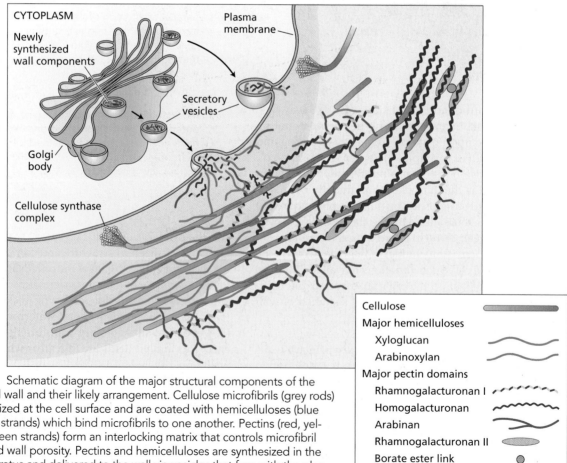

**FIGURE 15.4** Schematic diagram of the major structural components of the primary cell wall and their likely arrangement. Cellulose microfibrils (grey rods) are synthesized at the cell surface and are coated with hemicelluloses (blue and purple strands) which bind microfibrils to one another. Pectins (red, yellow, and green strands) form an interlocking matrix that controls microfibril spacing and wall porosity. Pectins and hemicelluloses are synthesized in the Golgi apparatus and delivered to the wall via vesicles that fuse with the plasma membrane and thus deposit these polymers to the cell surface. For clarity, the hemicellulose–cellulose network is emphasized on the left, and the pectin network is emphasized on the right. (After Cosgrove 2005.)

## TABLE 15.1
### Structural components of plant cell walls

| Class | Examples |
|---|---|
| Cellulose | Microfibrils of (1→4)-β-D-glucan |
| Pectins | Homogalacturonan |
| | Rhamnogalacturonan with arabinan, galactan, and arabinogalactan (Type 1) side chains |
| Hemicelluloses | Xyloglucan |
| | Xylan |
| | Glucomannan |
| | Arabinoxylan |
| | Callose (1→3)-β-D-glucan |
| | (1→3;1→4)-β-D-glucan [common in grasses] |
| Lignin | (see Figure 15.18) |
| Structural proteins | (see Table 15.2) |

be both extensible and strong at the same time. **Cellulose microfibrils** are nanometer scale (nm, $10^{-9}$ meter) crystalline ribbons that strengthen the cell wall, sometimes more in one direction than in another, depending on how the microfibrils are deposited in the wall (i.e., they give structural bias).

As a group, the **matrix polysaccharides** consist of several distinct polysaccharide structures, traditionally classified as hemicelluloses or pectins, a somewhat inexact classification based on extractability. The polysaccharides extracted from the cell wall with hot water or with calcium chelators are designated pectins, whereas the noncellulosic polysaccharides that are more tightly bound in the cell wall are defined as hemicelluloses and require stronger extraction conditions, such as 0.1–4 $M$ KOH.

**Hemicelluloses** are long, linear polysaccharides, often bearing short side branches, that characteristically bind to the surface of cellulose. They may form tethers that bind cellulose microfibrils together into a cohesive network (see the lower left part of Figure 15.4), or they may act as a slippery coating to prevent microfibril surfaces from sticking to each other. Another term for these molecules is *cross-linking glycan*, but in this chapter we'll use the more traditional term *hemicellulose*. As described later, the term *hemicellulose* includes several different polysaccharides.

**Pectins** form a hydrated gel phase in which the cellulose–hemicellulose network is embedded. They act as hydrophilic filler to prevent aggregation and collapse of the cellulose network, and they also determine the porosity of the cell wall to macromolecules. Like hemicelluloses, pectins include several different kinds of polysaccharides which are often called "pectin domains" because they are thought to be covalently linked to one another, forming massive macromolecular structures, unlike the distinct polymers of hemicellulose. The neutral side chains that make up some of these domains can bind to cellulose surfaces, although more weakly than do hemicelluloses (Zykwinska et al. 2007b).

Cell wall polysaccharides are made of different sugar monomers and are named after the principal sugars they contain (**FIGURE 15.5**). For example, a **galactan** is a polymer of galactose units linked together, a **glucan** is a polymer of

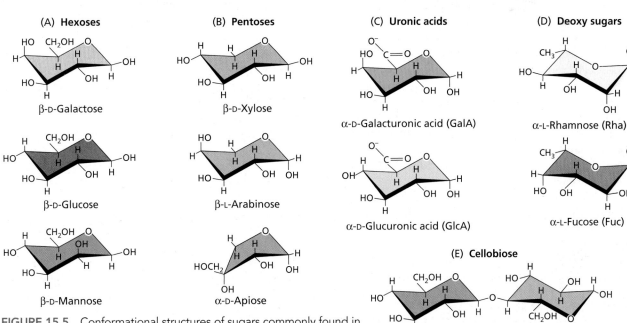

**FIGURE 15.5** Conformational structures of sugars commonly found in plant cell walls. (A) Hexoses (six-carbon sugars). (B) Pentoses (five-carbon sugars). (C) Uronic acids (acidic sugars). (D) Deoxy sugars. (E) Cellobiose, showing the (1→4)β-D-linkage between two glucose residues in inverted orientation. All sugars are shown in their pyranose forms except apiose which exists only in the furanose form. However, in cell-wall polysaccharides, L-arabinose most commonly occurs in the furanose form.

glucose, a **xylan** is a polymer of xylose, and so on. **Glycan** is the general term for a polymer made up of sugars and is synonymous with polysaccharide.

Polysaccharides may be linear polymers of sugar residues (units) linked end to end, or they may contain side branches that incorporate a different type of sugar. For branched polysaccharides, the backbone of the polysaccharide is usually indicated by the last part of the name. For example, **xyloglucan** has a glucan backbone (a linear chain of glucose residues) with xylose sugars attached as side chains. **Arabinoxylan** has a xylan backbone (a chain of xylose residues) with arabinose side chains. The names can get lengthy. For example, **glucuronoarabinoxylan** is an arabinoxylan decorated with glucuronic acid units. However, a compound name does not necessarily imply a branched structure. For example, **glucomannan** is the name given to a polymer containing both glucose and mannose in its backbone. Thus, the name is based on the major sugars in the polymer, but does not indicate its structural details.

Plant cell walls also contain **structural proteins** whose exact functions are uncertain; current research suggests that they may be involved in strengthening the cell wall and assisting in the proper assembly of other wall components, such as during cell plate formation (Cannon et al. 2008).

By dry mass, primary cell walls are typically composed of approximately 25% cellulose, 30% hemicelluloses, and 40% pectins, with perhaps 2 to 5% structural protein. However, large deviations from these values may be found among species (Harris and Stone 2008). For example, the walls of grass coleoptiles consist of 60 to 70% hemicelluloses, 20 to 25% cellulose, and only about 10% pectins. Cereal endosperm walls may have as little as ~2% cellulose, with arabinoxylan making up most of the wall. Parenchyma cell walls of celery and sugar beets contain mostly cellulose and pectin, with as little as 4% hemicellulose (Thimm et al. 2002). The wall at the tip of pollen tubes appears to be mostly pectin, with small amounts of cellulose to reinforce the tip structure. The cell walls of prickly pear (*Opuntia*) spines contain 50% cellulose and 50% galactan (a neutral polysaccharide typically classified as a pectin).

Secondary walls in fibers are at the other extreme, with high cellulose contents (>90% in cotton fibers) and negligible amounts of pectin. Wall compositions and polysaccharide structures are not static, but can change developmentally as a result of altered patterns of synthesis and the action of enzymes that can trim side branches and digest wall pectins and hemicelluloses (Gibeaut et al. 2005).

In living plant tissues, the primary wall contains a considerable amount of water, located mostly in the matrix, which is about 75 to 80% water. The hydration state of the matrix is a critical determinant of the physical properties of the wall; for example, removal of water makes the wall stiffer and less extensible—this is a contributing factor for plant growth inhibition by water deficits. Wall dehydration is also important in the strengthening of cell walls

during *lignification*, a process that drives water out of the cell wall and results in a more rigid wall that also resists enzymatic attack.

We will examine the structure of each of the major polymers of the cell wall in more detail in the sections that follow. We will also present a basic model of the primary wall, but be aware that the specific matrix polysaccharides may be quite different in different species and cell types, and that the relative proportions of the different kinds of wall polymers (cellulose, hemicellulose, pectin, and lignin in the case of some secondary walls) also vary considerably. The wide spectrum of cell wall compositions in nature make it clear that plant cells have wide latitude in constructing their cell walls to meet diverse needs.

### Cellulose microfibrils are synthesized at the plasma membrane

Cellulose is composed of numerous $(1{\rightarrow}4)\beta$-D-glucans —that is, linear chains of $(1{\rightarrow}4)$-linked $\beta$-D-glucose residues (for sugar structures, see Figure 15.5 and WEB TOPIC 15.2). Because each glucose is rotated 180° from its neighbors, the repeating unit in cellulose is considered to be cellobiose, a $(1{\rightarrow}4)$-linked $\beta$-D-glucose disaccharide (see Figure 15.5E).

In cellulose, many individual glucans are tightly packed to form a microfibril in which the glucans are closely aligned and bonded to each other by hydrogen bonds and van der Waals forces to make a highly ordered (**crystalline**) ribbon that excludes water and is relatively inaccessible to enzymatic attack. As a result, cellulose is insoluble, strong, stable, and very resistant to enzymatic degradation. The major barrier to enzymatic attack of cellulose is the energetic cost of stripping an individual glucan chain from this crystalline microfibril, a necessary step before an enzyme can attack the glycosidic bond that links the sugar residues together (Skopec et al. 2003). The highly ordered arrangement of glucans within a cellulose microfibril and the extensive noncovalent bonding between adjacent glucans within a cellulose microfibril gives this structure a very high tensile strength, equivalent to that of steel. Its chemical stability, insolubility, and resistance to enzymatic attack make cellulose a superb structural material for building a strong cell wall.

Cellulose microfibrils are many micrometers long (exact lengths are difficult to determine) and vary considerably in width and degree of order, depending on their biological source. For instance, cellulose microfibrils in land plants are 2 to 5 nm wide, whereas those formed by algae may be up to 20 nm wide and may be more highly ordered (more crystalline) than those found in land plants (Kennedy et al. 2007; Sturcova et al. 2004). This variation in microfibril width corresponds to the number of parallel chains that make up the cross section of a microfibril—estimated to be as few as six chains in the crystalline core of the thinnest microfibrils to as many as 30 to 50 chains in larger ones.

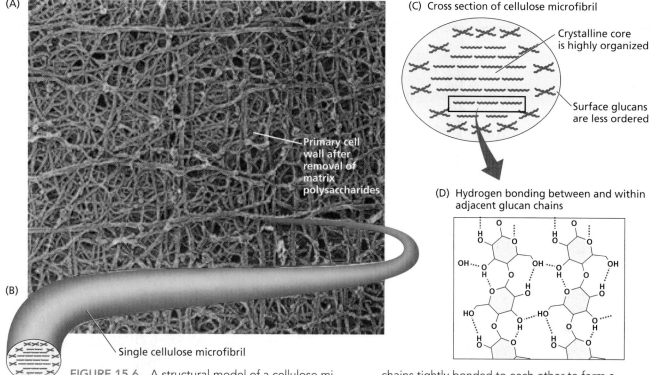

**FIGURE 15.6** A structural model of a cellulose microfibril. The microfibril has regions of high crystallinity intermixed with less ordered regions. Some hemicelluloses may also be trapped within the microfibril and bound to the surface. (A) Scanning EM of the primary cell wall from onion parenchyma, after pectic polysaccharides were extracted. Note its fibrillar texture, which arises from layers of cellulose microfibrils. (B) A single cellulose microfibril composed of two to four dozen (1→4)β-D-glucan chains tightly bonded to each other to form a crystalline ribbon. (C) Cross section of a cellulose microfibril, illustrating one model of cellulose structure, with a crystalline core of highly ordered (1→4)β-D-glucans surrounded by a less organized layer. (D) The crystalline regions of cellulose have precise alignment of glucans, with hydrogen bonding within, but not between, layers of (1→4)β-D-glucans. (After McCann et al. 1990 and Matthews et al. 2006.)

Individual microfibrils can also bundle together to form larger fibrils, as in the cell walls of woody tissues.

The individual glucan chains in cellulose microfibrils are composed of 2000 to more than 25,000 glucose residues, depending on biological source (Brown, Jr. et al. 1996). The microfibril may be longer than the individual glucans (1–12 μm) because of overlap and staggering of the glucans in the microfibril.

The precise molecular structure of the cellulose microfibril is not yet certain. Some models of microfibril organization suggest that it has a substructure consisting of highly crystalline domains linked together by less organized, relatively amorphous regions, while other models conceive of a solid crystalline core surrounded by a less organized surface layer (FIGURE 15.6). Within crystalline domains, adjacent glucans are highly ordered and firmly attached to each other by hydrogen bonds, van der Waals forces and hydrophobic interactions (all noncovalent interactions). Native cellulose in plants is found in two variant crystalline forms, called allomorphs Iα and Iβ, which differ slightly in the way the parallel glucan chains are packed. In vitro, these two forms may be interconverted by chemical and physical treatments. The biological significance of these two crystalline forms is unclear at present.

Evidence from electron microscopy of algae and land plants indicates that cellulose microfibrils are synthesized by large protein complexes, called particle rosettes or terminal complexes, that are embedded in the plasma membrane (FIGURE 15.7) (Kimura et al. 1999). These rosettes are made up of six subunits, each of which is believed to contain multiple units of **cellulose synthase**, the enzyme that synthesizes the individual (1→4)β-D-glucans that make up the microfibril (see WEB TOPIC 15.3). The cellulose synthase complexes likely contain additional proteins, but these have not yet been identified.

Cellulose synthases in higher plants are encoded by a gene family named *CesA* (*Cellulose synthase A*), which is a multigene family found in all land plants (Yin et al. 2009). The *CesA* family is part of a larger superfamily that also includes several families of *Csl* (*Cellulose synthase-like*) genes. *CslA* genes encode synthases for (1→4)β-D-mannan, *CslF* and *CslH* genes encode synthases for so-call "mixed-linkage" (1→3;1→4)β-D-glucan, and *CslC* genes likely encode synthases for the (1→4)β-D-glucan backbone of xyloglucan.

(A)

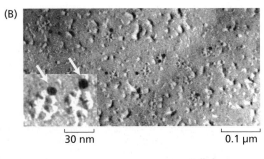

(B)

30 nm                              0.1 μm

(C)  (1→4)β-D-glucan chain ——→ Cellulose microfibril

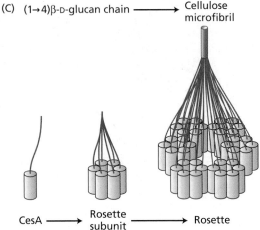

CesA ——→ Rosette subunit ——→ Rosette

**FIGURE 15.7** Cellulose synthesis by the cell. (A) Electron micrograph showing newly synthesized cellulose microfibrils immediately exterior to the plasma membrane. (B) Freeze-fracture labeled replicas showing reactions with antibodies against cellulose synthase. A field of labeled rosettes (arrows) with seven clearly labeled rosettes and one unlabeled rosette. The inset shows an enlarged view of two selected particle rosettes (terminal complexes) with immunogold labeling of CesA. The gold nanoparticles are the dark circles indicated with arrows. (C) Schematic model of the relationships between the particle rosettes (far right) and the single CesA proteins (far left). Six CesA proteins (encoded by three different genes) are thought to make one particle, which assembles to form a hexamer that is microscopically visible and that produces 36 glucan strands that form the ordered microfibril. The details of this model still await experimental verification. (A from Gunning and Steer 1996; B from Kimura et al. 1999; C after Doblin et al. 2002.)

The other *Csl* families likely encode enzymes that synthesize the backbones of other polysaccharides. However, the xylan backbone may be synthesized by a distinctly different group of synthases, named **GT43** (glycosyl transferase family 43) (Fincher 2009). All of these synthases are **sugar-nucleotide polysaccharide glycosyltransferases**, which transfer monosaccharides from sugar nucleotides to the growing end of the polysaccharide chain.

Cellulose synthase spans the plasma membrane but has its catalytic site on the cytoplasmic side of the plasma membrane. It transfers a glucose residue from a sugar nucleotide donor to the growing glucan chain. The sugar donor is uridine diphosphate D-glucose (UDP-glucose). There is some evidence that the glucose in the UDP-glucose used for cellulose synthesis is obtained from sucrose, a disaccharide composed of fructose and glucose (Amor et al. 1995; Salnikov et al. 2001). According to this idea, the enzyme **sucrose synthase** acts as a metabolic channel to transfer glucose taken from sucrose, via UDP-glucose,

to the growing cellulose chain (**FIGURE 15.8**). Genetic evidence also suggests that three different CesA proteins (encoded by three different *CesA* genes) are necessary to form a functional cellulose synthase complex.

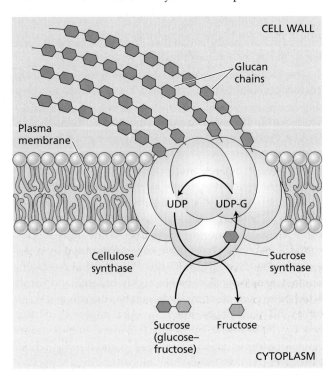

**FIGURE 15.8** Model of cellulose synthesis by a multisubunit complex containing cellulose synthase. Glucose residues are donated to the growing glucan chains by UDP-glucose (UDP-G). Sucrose synthase may act as a metabolic channel to transfer glucose taken from sucrose to UDP-glucose, or UDP-glucose may be obtained directly from the cytoplasm. (After Amor et al. 1995.)

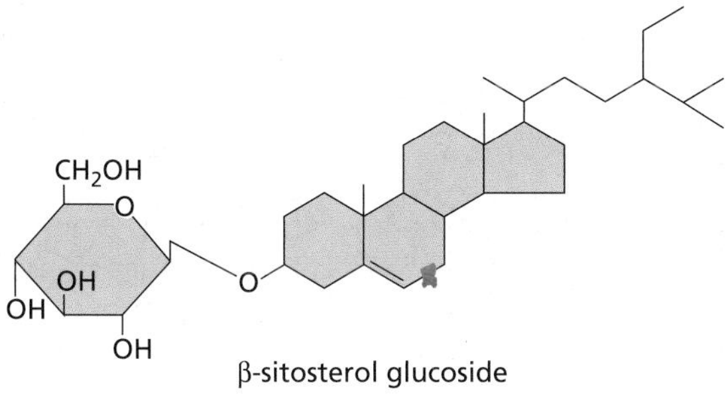

**FIGURE 15.9**  Structure of a sterol glucoside that may act as an initial primer (i.e., acceptor for glucan chain elongation) for cellulose synthesis. This β-sitosterol glucoside consists of a sterol (right) linked to a glucose residue (left). Additional glucose residues are added to the one shown here, forming a glucan used in the formation of a cellulose microfibril.

Sterol-glucosides (sterols linked to a chain of one or more glucose residues) (**FIGURE 15.9**) are hypothesized to serve as the primers, or initial acceptors, that start the elongation of the glucan chain (Peng et al. 2002). After the chain has elongated sufficiently, the sterol may be clipped from the glucan by an endoglucanase, allowing the growing glucan chain to be extruded through the membrane to the exterior of the cell, where, together with other glucan chains, it forms a crystalline ribbon and binds hemicelluloses to form a strong and resilient network (see Figure 15.4). It is also possible that some hemicellulose gets entrapped in the microfibril as it forms (Hayashi 1989); this may have the effect of creating disorder in the crystalline microfibril and also anchoring the microfibril to the matrix.

### Matrix polymers are synthesized in the Golgi apparatus and secreted via vesicles

The matrix of primary cell walls is a highly hydrated phase between the crystalline cellulose microfibrils. The matrix polysaccharides are synthesized by membrane-bound glycosyltransferases in the Golgi apparatus and are delivered to the cell wall via exocytosis of tiny vesicles (see Figure 15.4) (**WEB TOPIC 15.4**). As described above, genes in the *Csl* superfamily encode glycosyltransferases for synthesis of the backbone of some of these matrix polysaccharides. To make branched polysaccharides, additional sugar residues are added to the polysaccharide backbone by other sets of glycosyltransferases (Scheible and Pauly 2004).

The matrix polysaccharides are much less ordered than the microfibrils of cellulose, and most do not aggregate into highly ordered structures as cellulose does. This disordered character is a consequence of the structure of these polysaccharides—their branching and their nonlinear conformation. Nevertheless, studies using infrared spectroscopy and nuclear magnetic resonance (NMR) indicate partial order in the orientation of hemicelluloses and pectins

in the cell wall, probably as a result of a physical tendency of these polymers to align along the long axes of cellulose microfibrils (Wilson et al. 2000).

### Hemicelluloses are matrix polysaccharides that bind to cellulose

Hemicelluloses are a structurally heterogeneous group of polysaccharides (**FIGURE 15.10**) that are bound tightly in the wall. They can be solubilized from depectinated walls by the use of a strong alkali (2–4 $M$ NaOH), which disrupts hydrogen bonding and causes swelling, hydration, and disorder of cellulose. Plants can synthesize different kinds of hemicelluloses, varying with cell type, developmental state, and plant species.

In the primary cell wall of many land plants (eudicots and most monocots except grasses and closely related groups), the most abundant hemicellulose is **xyloglucan** (see Figure 15.10A), comprising ~20% of the cell wall. Like cellulose, this polysaccharide has a backbone of (1→4)-linked β-D-glucose residues. Unlike cellulose, however, xyloglucan has short side chains that contain xylose, often with a preterminal galactose and a terminal fucose.

Approximately one of every four glucose residues in the xyloglucan backbone is unsubstituted (does not carry a sugar side chain), but this fraction can be increased by the action of β-xylosidases, which remove the xylose side chain. With lower degrees of substitution, xyloglucans tend to bind more tightly to cellulose (Chambat et al. 2005). There is considerable taxonomic variation in the xyloglucan side branches; for example, solanaceous species have arabinose, but no fucose and sometimes no galactose in the side branches.

By interfering with the linear alignment of the glucan backbones with one another, these side chains prevent the assembly of xyloglucan into a crystalline microfibril. Because xyloglucans are longer (about 50–500 nm) than the spacing between cellulose microfibrils (20–40 nm), they have the potential to link adjacent microfibrils together. It is also possible that xyloglucans become trapped in cellulose microfibrils as they form; in such instances the crystallinity of the cellulose would be disrupted and the xyloglucan would be tightly anchored to the microfibril.

Compared with the walls of eudicots, the primary cell walls of grasses contain smaller amounts of xyloglucan and pectin (about 10% each), and more **glucuronoarabinoxylan** (see Figure 15.10B) and mixed-linkage **(1→3;1→4) β-D-glucan**. The mixed-linkage glucan is thought to bind tightly to the surface of cellulose, producing a less sticky surface, while the glucuronoarabinoxylan may serve a cross-linking function (Carpita et al. 2001). The degree of substitution of glucuronoarabinoxylan can vary significantly, with highly-substituted forms having weaker ability to bind to cellulose compared with less-substituted forms (Carpita 1983).

**FIGURE 15.10** Partial structures of common hemicelluloses. (For details on carbohydrate nomenclature, see **WEB TOPIC 15.1**.) (A) Xyloglucan has a backbone of (1→4)-linked β-D-glucose (Glc), with (1→6)-linked branches containing α-D-xylose (Xyl). In some cases galactose (Gal) and fucose (Fuc) are added to the xylose side chains. (B) Glucuronoarabinoxylans have a (1→4)-linked backbone of β-D-xylose (Xyl). They may also have side chains containing arabinose (Ara), 4-O-methyl-glucuronic acid (4-O-Me-α-D-GlcA). (After Carpita and McCann 2000.)

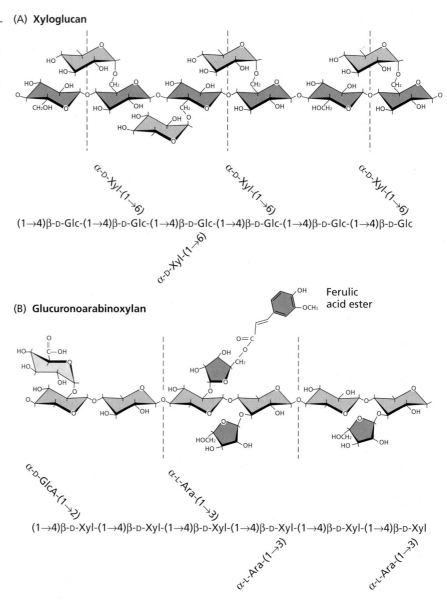

(A) **Xyloglucan**

α-D-Xyl-(1→6)

α-D-Xyl-(1→6)

α-D-Xyl-(1→6)

α-D-Xyl-(1→6)

(1→4)β-D-Glc-(1→4)β-D-Glc-(1→4)β-D-Glc-(1→4)β-D-Glc-(1→4)β-D-Glc-(1→4)β-D-Glc

(B) **Glucuronoarabinoxylan**

Ferulic acid ester

α-D-GlcA-(1→2)

α-L-Ara-(1→3)

α-L-Ara-(1→3)

(1→4)β-D-Xyl-(1→4)β-D-Xyl-(1→4)β-D-Xyl-(1→4)β-D-Xyl-(1→4)β-D-Xyl-(1→4)β-D-Xyl

α-L-Ara-(1→3)

Cross-linking of grass cell walls is aided by the presence of **ferulic acid**, a hydroxycinnamic acid, which is attached by an ester bond to the arabinose side chains of glucuronoarabinoxylan (see Figure 15.10B). Ferulate residues of neighboring molecules can be oxidatively cross-linked through the action of peroxidases, to form a covalent linkage between two glucuronoarabinoxylan chains (Burr and Fry 2009). Ferulic acid is also found esterified to the arabinan and galactan side chains of the pectin domain RG1 (see below) in certain plant families such as the *Amaranthaceae* (which includes sugar beet and spinach). These arabinans and galactans are able to bind cellulose surfaces (Zykwinska et al. 2007a); thus, binding to cellulose is not an exclusive property of hemicelluloses.

The secondary walls of woody tissues (secondary xylem) contain little xyloglucan or pectin; the matrix polysaccharides are relatively unbranched xylans and glucomannans, which bind tightly to cellulose. These walls are high in cellulose and lignin (~50% and 30%, respectively), and therefore both rigid and strong.

### Pectins are hydrophilic gel-forming components of the matrix

Like the hemicelluloses, pectins constitute a heterogeneous group of polysaccharides (**FIGURE 15.11**), characteristically containing the acidic sugar galacturonic acid and neutral sugars such as rhamnose, galactose, and arabinose. Pectins are the most soluble of the wall polysaccharides; most can be extracted with hot water or with calcium chelators. In the wall, pectins are very large and complex structures composed of different pectic polysaccharide

domains believed to be linked together by covalent and noncovalent bonds.

Some pectic polysaccharide domains, such as **homogalacturonan**, have a relatively simple primary structure (see Figure 15.11A). This polysaccharide, also called polygalacturonic acid, is a (1→4)-linked polymer of α-D-galacturonic acid residues. The fluorescence image in **FIGURE 15.12** shows that homogalacturonan is particularly concentrated at cell junctions where intercellular air spaces develop.

Another abundant pectin polysaccharide is **rhamnogalacturonan I (RG I)**, which has a long backbone of alternating rhamnose and galacturonic acid residues (see Figure 15.11B). This large and complex molecule carries both long and short side-chains of arabinans, galactans, and type-1 arabinogalactans attached to some of the rhamnose residues.

**(A) Homogalacturonan (HGA)**

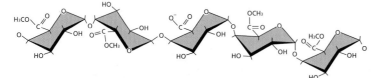

**(B) Rhamnogalacturonan I (RG I)**

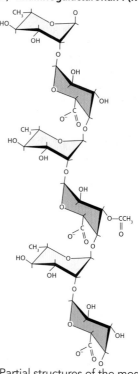

**(C) Arabinan**

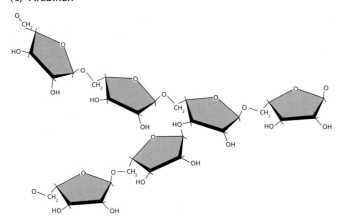

**(D) Arabinogalactan (Type I)**

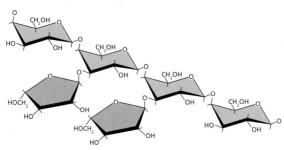

**FIGURE 15.11** Partial structures of the most common pectins. (A) Homogalacturonan, also known as polygalacturonic acid or pectic acid, is made up of (1→4)-linked α-D-galacturonic acid (GalA). The carboxyl residues are often methyl esterified. (B) Rhamnogalacturonan I (RG I) is a very large pectin domain, with a backbone of alternating (1→4)-α-D-galacturonic acid (GalA) and (1→2)-α-D-rhamnose (Rha). Side chains are attached to rhamnose and are composed principally of arabinans (C), galactans, and arabinogalactans (D). These side chains may be short or quite long. The galacturonic acid residues are often methyl esterified. (After Carpita and McCann 2000.)

Even further up the scale of molecular complexity is a highly branched pectic polysaccharide called **rhamnogalacturonan II (RG II)**, which contains a homogalacturonan backbone decorated with four different complex side chains comprising at least ten different sugars in a complicated pattern of linkages. Although RG I and RG II have similar names, they have very different structures. RG II units are cross-linked in the wall by borate diesters (Ishii et al. 1999) which are important for the structure and the mechanical strength of cell walls. For example, Ara-

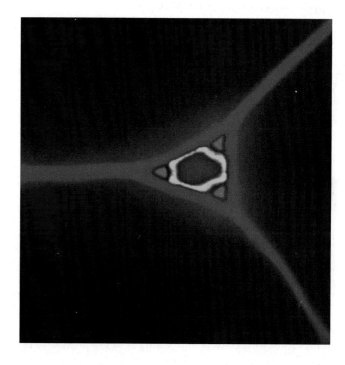

**FIGURE 15.12** Triple-fluorescence-labeled section of tobacco stem showing the primary cell walls of three adjacent parenchyma cells bordering an intercellular space. The blue color is calcofluor (indicating cellulose), and the red and green colors indicate the binding of two monoclonal antibodies to different epitopes (immunologically distinct regions) of pectic homogalacturonan. (Courtesy of W. Willats.)

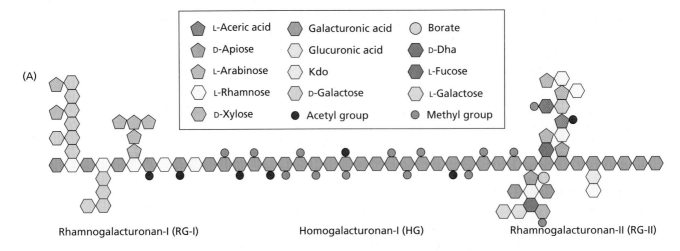

(A)

Rhamnogalacturonan-I (RG-I)          Homogalacturonan-I (HG)          Rhamnogalacturonan-II (RG-II)

(B)  Ionic bonding of pectin network by calcium

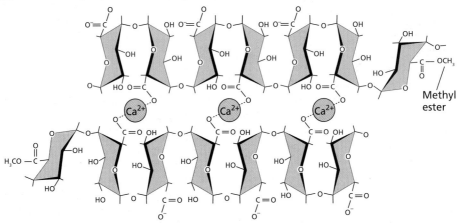

Methyl ester

**FIGURE 15.13**  (A) Schematic model illustrating the linear arrangement of the various pectin domains to each other, including rhamnogalacturonan I (RG I), homogalacturonan (HG) and rhamnogalacturonan II (RG II). The structure is not quantitatively accurate: HG should be about 10× more abundant and RG-I about 2× more abundant. Kdo = 3-Deoxy-D-manno-2-octulosonic acid; Dha = dihydroxy-acetone. (B) Formation of a pectin network involves ionic bridging of the nonesterified carboxyl groups ($COO^-$) by calcium ions. When blocked by methyl-esterified groups, the carboxyl groups cannot participate in this type of interchain network formation. Likewise, the presence of side chains on the backbone interferes with network formation. (A after Mohnen 2008; B after Carpita and Mc-Cann 2000.)

bidopsis mutants that synthesize an altered RG II show substantial growth abnormalities, apparently resulting from an unstable borate cross-link. Lack of RG II cross-linking by borate leads to radical swelling of the cell wall, increases in wall porosity, and mechanical weakening of the wall (O'Neill et al. 2004).

It is believed that in the wall all these pectic polysaccharides are covalently linked to one another, probably in a linear fashion (Coenen et al. 2007). **FIGURE 15.13A** illustrates a hypothetical scheme for how the various pectin domains may be linked to one another, but this model is not firmly established.

Pectins typically form gels—loose networks formed by highly hydrated polymers. Pectins are what make fruit jams and jellies solidify or "gel." In pectic gels, the charged carboxyl ($COO^-$) groups of neighboring pectin chains are linked via $Ca^{2+}$ to form a tight complex of calcium pectate.

A large calcium-bridged network may thus form, as illustrated in **FIGURE 15.13B**.

Pectins are subject to modifications that may alter their conformation and linkage in the wall. Many of the acidic residues are esterified with methyl, acetyl, and other unidentified groups during biosynthesis in the Golgi apparatus. Methyl esterification masks the charges of carboxyl groups and prevents calcium bridging between pectins, thereby reducing the ability of the pectin to form gels.

Once the pectin has been secreted into the wall, the ester groups may be removed by pectin esterases found in the wall, thus unmasking the charges of the carboxyl groups and increasing the ability of the pectin to form a rigid gel and making the cell wall less extensible (Derbyshire et al. 2007b). By creating free carboxyl groups, de-esterification also increases the electric-charge density in the wall, which in turn may influence the concentration of ions in the wall

**TABLE 15.2**
**Structural proteins of the cell wall**

| Class of cell wall proteins | Percentage carbohydrate | Tissue localization typically in: |
|---|---|---|
| HRGP (hydroxyproline-rich glycoprotein) | ~55 | Cambium, vascular parenchyma |
| PRP (proline-rich protein) | ~0–20 | Xylem, fibers, cortex |
| GRP (glycine-rich protein) | 0 | Primary xylem and phloem |

and the activities of wall enzymes. In addition to being connected by calcium bridging, pectins may be linked to each other by various covalent bonds, including those involving ferulate dimers (found only in some plant species), as described above.

### Structural proteins become cross-linked in the wall

In addition to the major polysaccharides described in the previous section, the cell wall contains several classes of structural proteins. These proteins usually are classified according to their predominant amino acid composition—for example, hydroxyproline-rich glycoprotein (HRGP), glycine-rich protein (GRP), proline-rich protein (PRP), and so on (**TABLE 15.2**). Some wall proteins have sequences that are characteristic of more than one class. Many structural proteins of the wall have highly repetitive primary structures that form simple helical rods, and some are highly glycosylated (**FIGURE 15.14**).

In vitro extraction studies have shown that newly secreted wall structural proteins are relatively soluble, but they become increasingly insoluble during cell maturation or in response to wounding. The biochemical nature of the insolubilization process may involve oxidative cross-linking of tyrosine residues in the protein.

Wall structural proteins vary greatly in their abundance, depending on cell type, maturation, and previous stimulation. Wounding, pathogen attack, and treatment with molecules that activate plant defense responses (elicitors) increase expression of the genes encoding many of these proteins (see Chapter 13). In histological studies, wall structural proteins are often localized to specific cell and tissue types. For example, HRGPs are associated mostly

with cambium, phloem parenchyma, and various types of sclerenchyma. GRPs and PRPs are most often localized to xylem vessels and fibers and thus are a distinctive feature of these cells with secondary walls.

In addition to structural proteins, cell walls contain **arabinogalactan proteins (AGPs)**, which usually amount to less than 1% of the dry mass of the wall (Seifert and Roberts 2007). These water-soluble proteins are very heavily glycosylated. More than 90% of the mass of AGPs may be sugar residues—primarily galactose and arabinose (**FIGURE 15.15**). Multiple AGP forms are found in plant tissues, either in the wall or associated with the plasma membrane (via a glycosylphosphatidylinositol group, known as a GPI anchor), and they display tissue- and cell-specific expression patterns.

AGPs may function in cell adhesion and in cell signaling during cell differentiation. As evidence for the latter idea, treatment of suspension cultures with exogenous AGPs or with agents that specifically bind AGPs affects cell proliferation and embryogenesis. AGPs are also implicated in the growth, nutrition, and guidance of pollen tubes through stylar tissues. Finally, AGPs may also function as a kind of polysaccharide chaperone within secretory vesicles to reduce spontaneous association of newly synthesized polysaccharides until they are secreted to the cell wall.

### New primary walls are assembled during cytokinesis

Primary walls originate *de novo* during the final stages of cell division, when the newly formed **cell plate** fuses with the mother cell walls to separate the two daughter cells and forms into a stable wall.

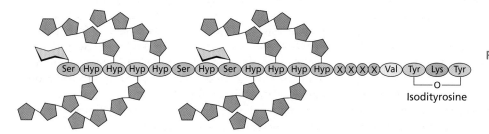

Tomato HRGP
(extensive glycosylation)

**FIGURE 15.14**  A repeated hydroxyproline-rich motif from a molecule of HRGP from tomato, showing extensive glycosylation and the formation of intramolecular isodityrosine bonds. (After Carpita and McCann 2000.)

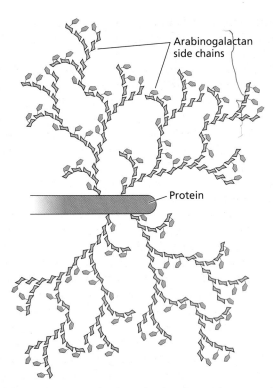

**FIGURE 15.15** A highly branched arabinogalactan molecule. (After Carpita and McCann 2000.)

The cell plate forms when Golgi vesicles and ER cisternae aggregate in the spindle midzone area of a dividing cell. This aggregation is organized by the **phragmoplast**, a complex assembly of microtubules, membranes, and vesicles that forms during late anaphase or early telophase (see Chapter 1). The membranes of the vesicles fuse with each other and with the lateral plasma membrane to form the new plasma membranes separating the daughter cells. The contents of the vesicles are the precursors from which the new middle lamella and the primary wall are assembled.

The "life" of an individual polymer may be outlined as follows:

Synthesis → deposition → assembly
→ modification (sometimes affecting wall extensibility)

At any given moment, wall polymers may populate any or all of these life stages. The synthesis and deposition of the major wall polymers were described earlier. Here we will consider their assembly into a cohesive network and later we will consider modifications that affect cell enlargement.

After their secretion into the extracellular space, the wall polymers must be assembled into a cohesive structure; that is, the individual polymers must attain the physical arrangement and bonding relationships that are characteristic of the primary (growing) cell wall and that confer it with both tensile strength and extensibility. Although the details of wall assembly are not fully understood, the prime candidates for this process are self-assembly and enzyme-mediated assembly.

**SELF-ASSEMBLY**   Self-assembly is an attractive model because it is mechanistically simple. Wall polysaccharides possess a marked tendency to aggregate spontaneously into organized structures. For example, isolated cellulose can be dissolved in strong solvents and then extruded to spontaneously form stable fibers, called rayon.

Similarly, hemicelluloses may be dissolved in strong alkali; when the alkali is removed, these polysaccharides may aggregate into ordered networks reminiscent of cell walls at the ultrastructural level. Aggregation can make the separation of hemicelluloses into component polymers technically difficult. In contrast, pectins are more soluble and tend to form dispersed, isotropic (randomly arranged) networks (gels).

These observations indicate that wall polymers have inherent tendencies to aggregate into somewhat ordered structures. However, this explanation may not be the whole story, because when hemicelluloses are bound to cellulose in vitro, their binding is much weaker than is the case in real cell walls. This discrepancy hints at the involvement of other processes needed to make strong networks in the wall.

**ENZYME-MEDIATED ASSEMBLY**   In addition to self-assembly, wall enzymes may take part in putting the wall together. A prime candidate for enzyme-mediated wall assembly is **xyloglucan endotransglucosylase (XET)**. This enzyme, which belongs to a large family of enzymes named **xyloglucan endotransglucosylase/hydrolases (XTHs)**, has the ability to cut the backbone of a xyloglucan and to join one end of the cut xyloglucan with the free end of an acceptor xyloglucan (**FIGURE 15.16**). Such a transfer reaction integrates newly synthesized xyloglucans into the wall (Thompson and Fry 2001; Rose et al. 2002), probably strengthening the cell wall.

Other wall enzymes that might aid in assembly of the wall include glycosidases, pectin methyl esterases, and various oxidases. Some glycosidases remove the side chains of hemicelluloses. This "debranching" activity increases the tendency of hemicelluloses to adhere to the surface of cellulose microfibrils. Pectin methyl esterases hydrolyze the methyl esters that block the carboxyl groups of pectins. By unblocking the carboxyl groups, these enzymes increase the concentration of acidic groups on the pectins and enhance the ability of pectins to form a $Ca^{2+}$-bridged gel network. Oxidases such as peroxidase catalyze cross-links between phenolic groups (tyrosine, phenylalanine, ferulic acid) in wall proteins, pectins, and other wall polymers. Such oxidative cross-links tie lignin subunits together in complex ways (**FIGURE 15.17**), and they may likewise link other wall components, such as feruloylated arabinoxylans, together.

### Secondary walls form in some cells after expansion ceases

After wall expansion ceases, cells sometimes continue to synthesize a **secondary wall**. Secondary walls may

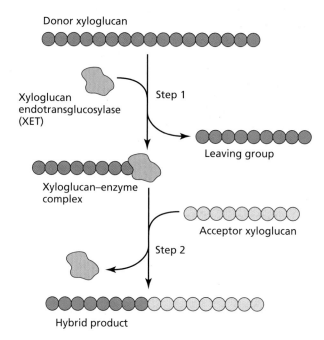

**FIGURE 15.16**   Action of xyloglucan endotransglucosylase (XET) to cut and stitch xyloglucan polymers into new configurations. Step 1: The enzyme cuts a xyloglucan molecule (the donor xyloglucan), forming a long-lived complex where the xyloglucan is covalently attached to the enzyme. Step 2: Subsequently the enzyme transfers the xyloglucan chain to the nonreducing end of a second xyloglucan (the acceptor xyloglucan), resulting in a hybrid product. (After Fry 2004.)

be quite thick, as in tracheids, fibers, and other cells that provide mechanical support to the plant (**FIGURE 15.18**). The rigid secondary wall in xylem is also important for preventing collapse of the water-conducting cells during periods of high water tension due to rapid transpiration. As mentioned above, the secondary walls in wood contain xylans rather than xyloglucans, as well as a higher proportion of cellulose. The orientation of the cellulose microfibrils may be more neatly aligned parallel to each other in secondary walls than in primary walls, and show

distinctive layering. In woody tissues, secondary walls are typically impregnated with lignin, which strengthens the wall and drives out water.

**Lignin** is a complex phenolic polymer built of phenylpropanoid subunits, called **monolignols**, which are derived from phenylalanine (see Chapter 13) and which contain 0, 1 or 2 methoxy groups on the phenyl ring (see Figure 15.18C). Lignin made from coniferyl alcohol is known as **G lignin**, whereas **S lignin** is made from sinapyl coniferyl alcohol and **H lignin** is made from *p*-coumaryl alcohol. The proportion of G:S:H lignin varies with plant species and cell type and influences the strength and pulping properties of woody tissues (Vanholme et al. 2008).

To form lignin, it is thought that cells secrete monolignol glucosides to the cell wall, where the glucose residue is enzymatically removed by a glucosidase. The details of the subsequent polymerization process are not fully understood, but it may start with monolignols being oxidatively coupled to ferulic acid or other phenolic residues that are covalently linked to xylans in the wall. The majority view at this time is that monolignols are randomly oxidized and coupled by free radicals formed by the enzymes peroxidase and laccase, resulting in a variable macromolecular structure (see Figure 15.18D). A second hypothesis (Davin and Lewis 2005) is that wall proteins control the polymerization process, producing a more regular structure.

As lignin forms in the wall, it displaces water from the matrix and forms a hydrophobic network that is physically intertwined with the polysaccharides in the matrix as well as covalently linked to them, primarily via ferulic acid or other phenolic residues linked to xylans. Lignin also binds to cellulose as well as encases it, thereby stiffening the whole cell wall (see Figure 15.17).

By strengthening the cell wall and dehydrating it, lignin reduces the susceptibility of walls to attack by hydrolytic enzymes from pathogens. Lignin also reduces the digestibility of plant material by animals and interferes with the pulping process (conversion of wood into free fibers) for paper making. Current efforts at genetic engineering of lignin content and structure may improve the digestibility

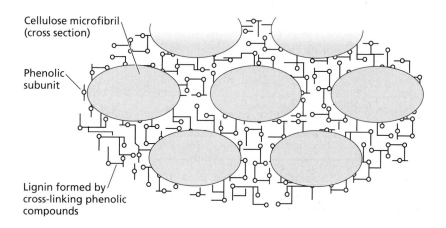

**FIGURE 15.17** Diagram illustrating how the phenolic subunits of lignin infiltrate the space between cellulose microfibrils, where they become cross-linked. (Other components of the matrix are omitted from this diagram.)

(A)

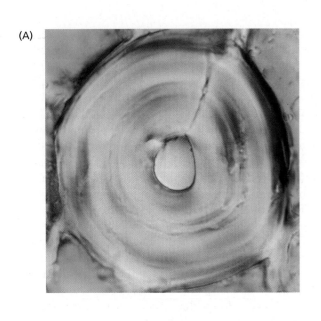

(B)

S₃
S₂ } Secondary wall
S₁

Primary wall

Middle lamella

S₁
S₂
S₃

(C) Monolignols

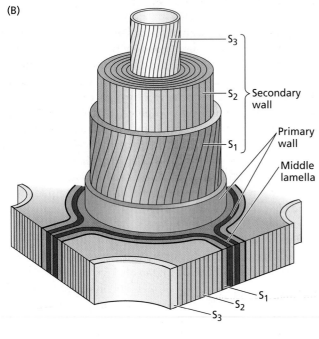

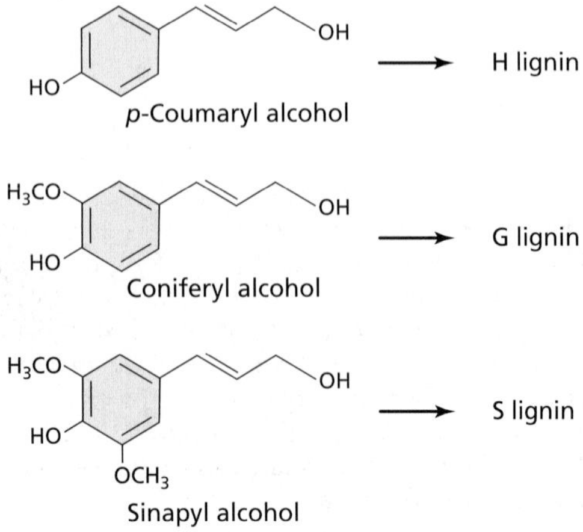

HO— [ring] —OH → H lignin

*p*-Coumaryl alcohol

H₃CO / HO— [ring] —OH → G lignin

Coniferyl alcohol

H₃CO / HO— [ring] / OCH₃ —OH → S lignin

Sinapyl alcohol

(D)

β-O-4, β-ether
β–β, resinol
β-5, phenylcoumaran
S-O-4, biphenyl ether
Cinnamyl alcohol endgroup
S Syringyl
G Guaiacyl

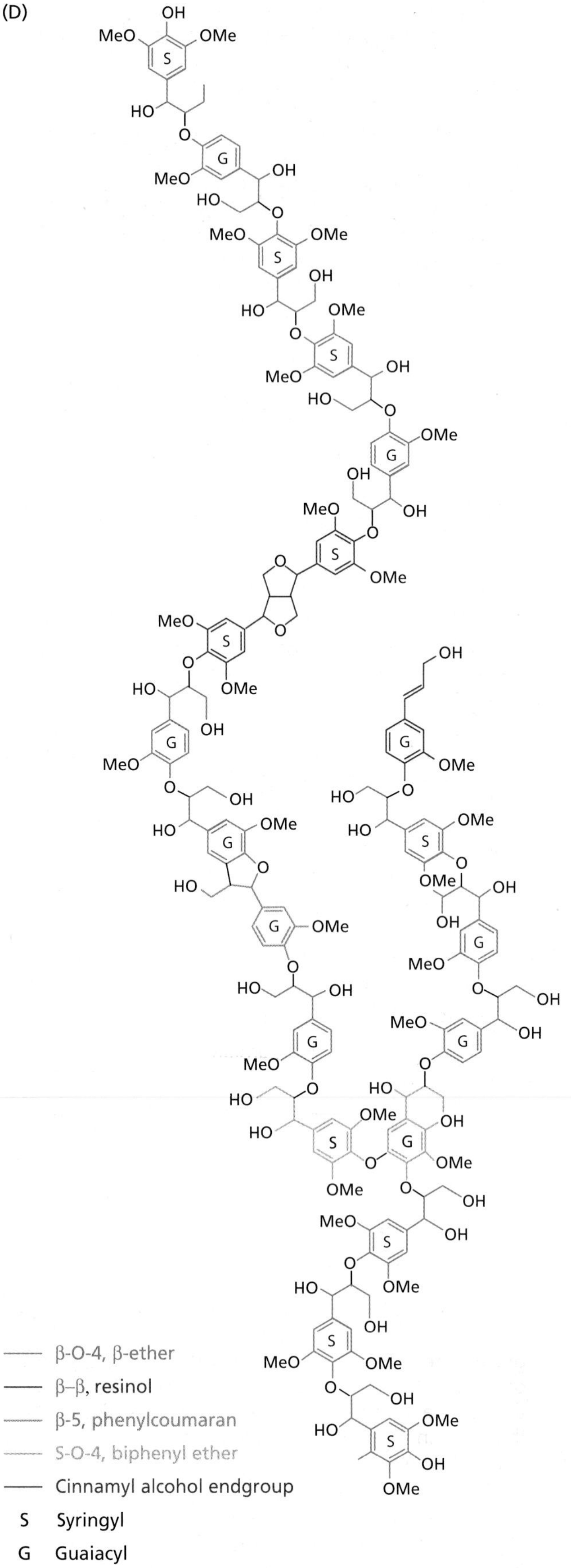

◀ **FIGURE 15.18** (A) Cross section of a *Podocarpus* sclereid, in which multiple layers in the secondary wall are visible. (B) Diagram of the cell wall organization often found in tracheids and other cells with thick secondary walls. Three distinct layers ($S_1$, $S_2$, and $S_3$) are formed interior to the primary wall. (C) Monolignols that make up so-called H, G, and S lignin differ in the number of methoxy substituents on the phenolic ring. (D) Current model of the structure of poplar lignin, composed of S and G monolignol subunits that are cross-linked by free radicals generated by peroxidase and laccase. Note that this is one of billions of possible isomers. (A © David Webb; D after Ralph et al. 2007.)

and nutritional content of plants used as animal fodder, as well as increase the value of cell walls for the production of paper and biofuel (ethanol used in autos).

# Patterns of Cell Expansion

During plant cell enlargement, new wall polymers are continuously synthesized and secreted at the same time that the preexisting wall is expanding. Wall expansion may be highly localized (as in the case of **tip growth**) or evenly distributed over the wall surface (**diffuse growth**) (FIGURE 15.19). Tip growth is characteristic of root hairs and pollen tubes; it is closely linked to cytoskeletal processes, especially actin microfilaments (see **WEB ESSAY 15.1**). Most of the other cells in the plant body exhibit diffuse growth, which is linked to the dynamics of both microtubules and actin microfilaments (Szymanski and Cosgrove 2009). Cells such as fibers, some sclereids, and trichomes grow in a pattern that is intermediate between diffuse growth and tip growth.

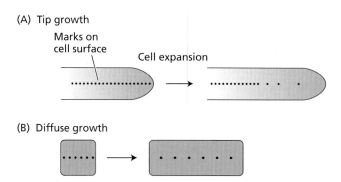

**FIGURE 15.19** The cell surface expands differently during tip growth and diffuse growth. (A) Expansion of a tip-growing cell is confined to an apical dome at one end of the cell. If marks are placed on the cell surface and the cell is allowed to continue to grow, only the marks that were initially within the apical dome grow farther apart. Root hairs and pollen tubes are examples of plant cells that exhibit tip growth. (B) If marks are placed on the surface of a diffuse-growing cell, the distance between all the marks increases as the cell grows. Most cells in multicellular plants grow by diffuse growth.

Even in cells with diffuse growth, however, different parts of the wall may enlarge at different rates or in different directions. For example, in cortical parenchyma cells of the stem, the end walls expand much less than side walls. This difference may be due to structural or enzymatic variations in specific walls or to variations in the stresses borne by different walls. As a consequence of this uneven pattern of wall expansion, plant cells morph into different shapes.

## *Microfibril orientation influences growth directionality of cells with diffuse growth*

During growth, the primary cell wall is biochemically loosened, permitting it to yield to the wall stresses generated by cell turgor pressure. Turgor pressure creates an outward-directed force, equal in all directions. The directionality of cell growth is determined largely by the structure of the cell wall—in particular, the orientation of cellulose microfibrils.

When cells first form in the meristem, they are isodiametric, having equal diameters in all directions. If the orientation of cellulose microfibrils in the primary cell wall were randomly arranged, or **isotropic**, the cell would grow equally in all directions, expanding radially to generate a sphere (FIGURE 15.20A). Cells in an enlarging apple grow in this way. In most plant cell walls, however, cellulose microfibrils are oriented in a particular direction—that is,

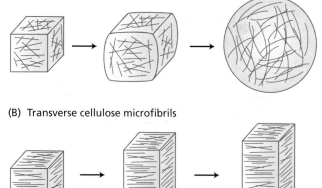

**FIGURE 15.20** The orientation of newly deposited cellulose microfibrils determines the direction of cell expansion. (A) If the cell wall is reinforced by randomly oriented cellulose microfibrils, the cell will expand equally in all directions, forming a sphere. (B) When most of the reinforcing cellulose microfibrils have the same orientation, the cell expands at right angles to the microfibril orientation and is constrained in the direction of the reinforcement. Here the microfibril orientation is transverse, so cell expansion is longitudinal.

FIGURE 15.21 Interdigitating cell growth of leaf pavement cells and its regulation by ROP GTPases. (A) Scanning electron micrograph of pavement cells from an Arabidopsis leaf. Note the jigsaw puzzle–like appearance. (B) Immunofluorescence image of pavement cells shows more clearly the lobes and indentations formed by interdigitated cells. (C) A model to explain the role of ROP GTPases and their effectors (RICs) in leaf morphogenesis. ROP2/4 GTPases, when activated by RIC4, promote actin microfilament formation in regions of growing lobes, whereas when activated by RIC1 they promote microtubule bundling at neck regions. These cytoskeletal changes somehow act as signals to direct the direction of wall growth. (A courtesy of Dan Szymanski; B from Settleman 2005, courtesy of J. Settleman; C after Fu et al. 2005.)

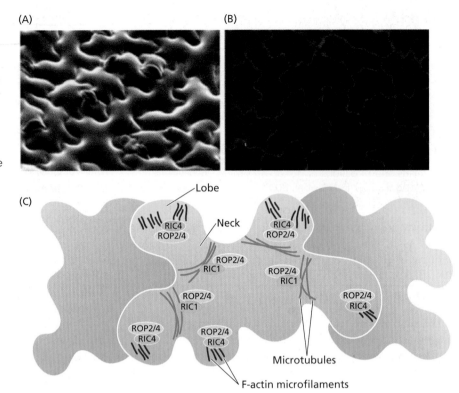

they are **anisotropic**, or aligned in a preferred direction. The alignment of cellulose reinforces the wall so that it is more extensible in the direction perpendicular to the alignment.

In the lateral walls of elongating cells such as cortical parenchyma and vascular cells of stems and roots, or the giant internode cells of the filamentous green alga *Nitella*, cellulose microfibrils are deposited circumferentially (transversely), at right angles to the long axis of the cell. The circumferentially arranged cellulose microfibrils have been likened to hoops in a barrel, restricting growth in girth and allowing growth in length (FIGURE 15.20B). However, because individual cellulose microfibrils do not actually form closed hoops around the cell, but form shallow helices, they separate from one another as the cell wall expands.

Cell wall deposition continues as cells enlarge. According to the **multinet growth hypothesis**, each successive wall layer is stretched and thinned during cell expansion, so the microfibrils would be expected to be passively reoriented in the longitudinal direction—that is, in the direction of growth. Consistent with the multinet growth hypothesis, electron micrographs of the cell wall indicate that the most recently deposited, inner wall layers have transversely oriented cellulose microfibrils, while the microfibrils of the older, outer wall layers have a more disordered arrangement. Likewise, with fluorescence imaging of elongating cells in Arabidopsis roots, bundles of cellulose fibrils were seen to rotate from the transverse to the longitudinal direction as the cells elongated (Anderson et al. 2009).

Other observations, however, cast doubt on the universality of multinet growth. Some cells have multilamellate walls, with the cellulose oriented in a criss-cross pattern that could not arise simply from passive realignment. In a study to test the ability of cell wall microfibrils to passively reorient in response to wall tension, isolated wall segments from growing hypocotyls were mechanically stretched under conditions that mimicked normal growth, and the effect of this extension on the orientation of the cellulose microfibrils of the wall was examined. Surprisingly, stretching the wall longitudinally by 20 to 30% failed to alter the transverse angle of the microfibrils of the inner wall surface, suggesting that the microfibrils at this surface separated from each other in a coordinate fashion, contrary to the prediction of the multinet growth hypothesis (Marga et al. 2005).

Thus, it seems that the most recently deposited (inner) layers of the cell wall behave differently than the older (outer) layers. This difference may result from thinning and fragmentation of the older wall layers as the wall extends, with the result that the outer layers have much less influence on the direction of cell expansion than do the newly deposited inner layers. Accordingly, the inner one-fourth of the wall is believed to bear nearly all the stress due to turgor pressure and determines the directionality of cell expansion (see WEB TOPIC 15.5).

So far we have considered only a simple pattern of diffuse growth. So-called "pavement cells" in the epidermis of many dicot leaves, however, present a more complicated situation. These cells are highly lobed, creating an interlocking pattern resembling a jigsaw puzzle (FIGURE 15.21A,B). This pattern of interdigitating cell wall expansion combines aspects of diffuse growth and tip growth

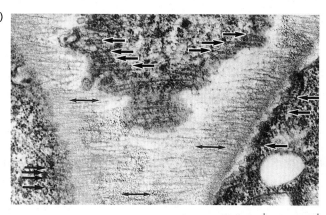

(A)

(B)

5 µm

**FIGURE 15.22** The orientation of microtubules in the cortical cytoplasm mirrors the orientation of newly deposited cellulose microfibrils in the walls of cells that are elongating. (A) The arrangement of microtubules can be revealed with fluorescently labeled antibodies to the microtubule protein tubulin. In this differentiating tracheary element from a zinnia cell suspension culture, the pattern of microtubules (green) mirrors the orientation of the cellulose microfibrils in the wall, as shown by calcofluor staining (blue). (B) The alignment of cellulose microfibrils in the cell wall can sometimes be seen in grazing sections prepared for electron microscopy, as in this micrograph of a developing sieve tube element in a root of *Azolla* (a water fern). The longitudinal axis of the root and the sieve tube element runs vertically. Both the wall microfibrils (double-headed arrows) and the cortical microtubules (single-headed arrows) are aligned transversely (see also Figure 15.7). (A courtesy of Robert W. Seagull; B courtesy of A. Hardham.)

and requires the action of GTP-binding proteins called ROP (*Rho*-related GTPase from *p*lants) GTPases, and their activating proteins called RICs (*ROP-i*nteracting *CRIB* motif-containing proteins) (**FIGURE 15.21C**). These proteins organize the cytoskeleton (actin microfilaments and tubulin microtubules), which delivers materials and catalysts for local control of cell wall growth (Szymanski and Cosgrove 2009). How does the cytoskeleton influence wall growth? This topic is covered in the next section.

### Cortical microtubules influence the orientation of newly deposited microfibrils

Newly deposited cellulose microfibrils usually are coaligned with microtubule arrays in the cytoplasm, close to the plasma membrane (**FIGURE 15.22**) (Baskin 2001; Baskin et al. 2004). A striking example occurs in xylem vessel elements, where bands of cortical microtubules mark the sites of secondary wall thickenings and also the sites of cellulose synthase A (CesA) localization (Gardiner et al. 2003). Moreover, experimental disruption of microtubule organization with drugs or by genetic defects often leads to disorganized wall structure and disorganized growth. For example, several drugs bind to tubulin, the subunit protein of microtubules, causing them to depolymerize. When growing roots are treated with a microtubule-depolymerizing drug such as oryzalin, the region of elongation expands laterally, becoming bulbous and tumorlike (**FIGURE 15.23A,B**).

The oryzalin-disrupted growth is due to the isotropic expansion of the cells; that is, they enlarge like a sphere instead of elongating. The drug-induced destruction of microtubules in the growing cells interferes with the transverse deposition of cellulose. Cellulose microfibrils continue to be synthesized in the absence of microtubules, but they are deposited randomly and consequently the cells expand equally in all directions.

These and related observations have led to the suggestion that microtubules serve as tracks that guide or direct the movement of complexes as they synthesize microfibrils (see **WEB ESSAY 15.2**). In recent studies, the movement of CesA in living cells was visualized by recombinant DNA methods to express a fusion of CesA with YFP, a fluorescent protein (Gutierrez et al. 2009; Paredez et al. 2006). The CesA units were observed to move within the plasma membrane along microtubule tracks (**FIGURE 15.23C**); they were also seen to be inserted into the plasma membrane from the Golgi apparatus at microtubule-tethered compartments. These results, obtained by confocal microscopy, reveal new details of how the cytoskeleton directs cell wall organization (Szymanski and Cosgrove 2009).

## The Rate of Cell Elongation

Plant cells typically expand 10- to 100-fold in volume before reaching maturity. In extreme cases, cells may enlarge more than 10,000-fold in volume (e.g., xylem vessel elements) compared with their meristematic initials. The cell wall undergoes this profound expansion without losing its mechanical integrity and without becoming thinner. Thus, newly synthesized polymers are integrated into the wall

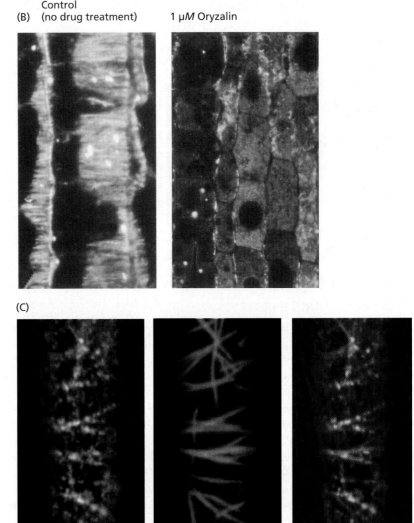

Control (no drug treatment) (A)   1 µM Oryzalin

Control (B) (no drug treatment)   1 µM Oryzalin

(C)

**FIGURE 15.23** The disruption of cortical microtubules results in a dramatic increase in radial cell expansion and a concomitant decrease in elongation. (A) Root of Arabidopsis seedling treated with the microtubule-depolymerizing drug oryzalin (1 µM) for 2 days before this photomicrograph was taken. The drug has altered the polarity of growth. (B) Microtubules were visualized by means of an indirect immunofluorescence technique and an antitubulin antibody. Whereas cortical microtubules in the control are oriented at right angles to the direction of cell elongation, very few microtubules remain in roots treated with 1 µM oryzalin. (C) Images of fluorescently tagged CesA proteins (left panel) and microtubules (middle panel) indicate that microtubules guide the trajectories of CesA movement in the plasma membrane, thus guiding the orientation of cellulose microfibrils. The right panel shows the superposition of the two images. (A,B from Baskin et al. 1994, courtesy of T. Baskin; C from Gutierrez et al. 2009.)

without destabilizing the existing structure. Exactly how this integration is accomplished is uncertain, although self-assembly and xyloglucan endotransglucosylase (XET) play important roles, as already described.

This integrating process may be particularly critical for rapidly growing root hairs, pollen tubes, and other specialized cells that exhibit tip growth, in which the region of wall deposition and surface expansion is localized to the hemispherical dome at the apex of the tube-like cell, and cell expansion and wall deposition must be closely coordinated.

In rapidly growing cells with tip growth, the wall doubles its surface area and is displaced to the nonexpanding part of the cell within minutes. This is a much greater

rate of wall expansion than is typically found in cells with diffuse growth, where growth rates are ~1–10% per hour. Because of their fast expansion rates, tip-growing cells are highly susceptible to wall thinning and bursting. Although diffuse growth and tip growth appear to be different growth mechanisms, both types of wall expansion must have analogous, if not identical, processes of polymer integration, wall stress relaxation, and wall polymer creep.

Many factors influence the rate of cell wall expansion. Cell type and age are important developmental factors. So, too, are the downstream consequences of the activities of hormones such as auxin and gibberellin. Environmental conditions such as light and water availability may likewise modulate cell expansion. These internal and external factors most likely modify cell expansion by altering the way in which the cell wall is loosened, so that it yields (stretches irreversibly) differently. In this context we speak of the **yielding properties** of the cell wall.

In this section we will first examine the biomechanical and biophysical parameters that characterize the yielding properties of the wall. For cells to expand at all, a mechanically strong cell wall must be loosened in some way. The type of wall loosening involved in plant cell expansion is termed stress relaxation, which is explained further in the next section.

According to the acid growth hypothesis for auxin action (see Chapter 19), one mechanism that causes wall stress relaxation and wall yielding is cell wall acidification, resulting from the activity of H$^+$ pumps in the plasma membrane. Cell wall loosening is enhanced at acidic pH. We will explore the biochemical basis for acid-induced wall loosening and stress relaxation, including the role of a special class of wall-loosening proteins called *expansins*.

As the cell approaches its maximum size, its growth rate diminishes and finally ceases altogether. At the end of this section we will consider the process of cell wall rigidification that leads to the cessation of growth.

## Stress relaxation of the cell wall drives water uptake and cell elongation

Because the cell wall is the major mechanical restraint that limits cell expansion, much attention has been given to its physical properties (Cosgrove 1993). As a hydrated polymeric material, the plant cell wall has physical properties that are intermediate between those of a solid and those of a liquid. We call these **viscoelastic**, or **rheological** (flow), **properties**. Walls of cells that are growing are generally less rigid than those of nongrowing cells, and under appropriate conditions they exhibit a long-term irreversible stretching, or **yielding**, that is lacking in nongrowing walls.

**Stress relaxation** is a crucial concept for understanding how cell walls enlarge (Cosgrove 1997). The term stress is used here in the mechanical sense, as force per unit area. Wall stresses arise as an inevitable consequence of cell turgor. The turgor pressure in growing plant cells is typically between 0.3 and 1.0 megapascals (MPa). Turgor pressure stretches the cell wall and generates a counterbalancing physical stress or tension in the wall. Because of cell geometry (a large pressurized volume contained by a thin wall), this wall tension is estimated to be 10 to 100 MPa of tensile stress—a very large stress indeed.

This simple fact has important consequences for the mechanics of cell enlargement. Whereas animal cells can change shape in response to cytoskeleton-generated forces, such forces are negligible compared with the turgor-generated forces that are resisted by the plant cell wall. To change shape, plant cells must thus control the direction and rate of wall expansion, which they do by depositing cellulose in a biased orientation (this determines the directionality of cell wall expansion) and by selectively loosening the bonding between cell wall polymers. This biochemical loosening enables cellulose microfibrils and their associated matrix polysaccharides to slip by each other, thereby increasing the wall surface area. At the same time, such loosening reduces the physical stress in the wall.

Wall stress relaxation is crucial because it allows growing plant cells to reduce their turgor and water potentials, which enables them to absorb water and to expand. Without stress relaxation, wall synthesis would only thicken the wall, not expand it; indeed, wall deposition and wall expansion are not closely linked in many cases (Derbyshire et al. 2007a). During secondary-wall deposition in nongrowing cells, for example, stress relaxation does not occur and consequently polysaccharide deposition results in a thickened cell wall.

When plant cells undergo expansive growth, the increase in volume is generated mostly by water uptake. This water ends up mainly in the vacuole, which takes up an increasing proportion of the cell volume as the cell enlarges. Here we will describe how growing cells regulate their water uptake and how this uptake is coordinated with wall yielding.

Water uptake in all plant cells is a passive process. There are no active water pumps; instead the growing cell uses the trick of wall stress relaxation to lower the water potential inside the cell so that water is taken up spontaneously in response to a water potential difference, without direct energy expenditure.

To put this in physical terms, we begin by defining the water potential difference, $\Delta\Psi_w$ (expressed in megapascals, MPa), as the water potential outside the cell ($\Psi_o$) minus the water potential inside ($\Psi_i$) (see Chapters 3 and 4). The rate of uptake depends on $\Delta\Psi_w$ times the surface area of the cell ($A$, in square meters) times the permeability of the plasma membrane to water ($Lp$, in meters per second per MPa). Membrane $Lp$ is a measure of how readily water crosses the membrane, and it is a function of the physical structure of the membrane and the activity of aquaporins (see Chapter 3). The rate of water uptake is defined as $\Delta V/\Delta t$, expressed in cubic meters per second. Assuming that a growing cell is in contact with pure water (with zero water potential), then we have:

$$\text{Rate of water uptake} = \Delta V/\Delta t = A \times Lp\,(\Psi_o - \Psi_i)$$
$$= A \times Lp\,(0 - \Psi_i) = -A \times Lp\,(\Psi_p + \Psi_s) \qquad (15.1)$$

This equation states that the rate of water uptake depends only on the cell area, membrane permeability to water, cell turgor ($\Psi_p$), and cell osmotic potential ($\Psi_s$).

Equation 15.1 is valid for both growing and nongrowing cells in pure water. But how can we account for the fact that growing cells can continue to take up water for a long time, whereas nongrowing cells soon cease water uptake?

In a nongrowing cell, water absorption would increase cell volume, causing the protoplast to push harder against the cell wall, thereby increasing cell turgor pressure, $\Psi_p$. This increase in $\Psi_p$ would increase cell water potential $\Psi_w$, quickly bringing $\Delta\Psi_w$ to zero. Water uptake would then cease, in a matter of a few seconds.

In a growing cell, $\Delta\Psi_w$ is prevented from reaching zero because the cell wall is "loosened": it yields irreversibly

**FIGURE 15.24** Graphic representation of the two equations that relate water uptake and cell expansion to cell turgor pressure and cell water potential. The values for the rates of cell expansion and water uptake are arbitrary. Steady-state growth is attained only at the point where the two equations intersect. Any imbalance between water uptake and wall expansion will result in changes in cell turgor and bring the cell back to this stable point of intersection between the two processes.

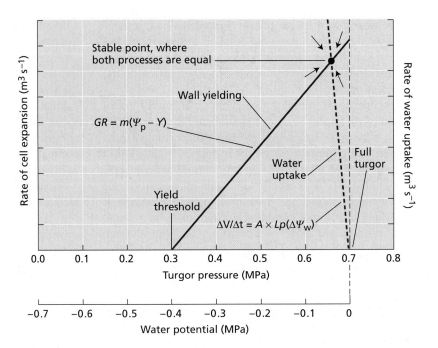

to the forces generated by turgor and thereby reduces simultaneously the wall stress and the cell turgor. This process is called **stress relaxation**, and it is *the* crucial physical difference between growing and nongrowing cells.

Stress relaxation can be understood as follows. In a turgid cell, the cell contents push against the wall, causing the wall to stretch elastically (i.e., reversibly) and giving rise to a counterforce, wall stress. In a growing cell, biochemical loosening causes the wall to yield inelastically (irreversibly) to the wall stress. Because water is nearly incompressible, only an infinitesimal expansion of the wall is needed to reduce cell turgor pressure and, simultaneously, wall stress. *Thus, stress relaxation is a decrease in wall stress with essentially no change in wall dimensions.*

As a consequence of wall stress relaxation, the cell water potential is reduced and water flows into the cell, extending the cell wall and increasing cell surface area and volume. Sustained growth of plant cells entails simultaneous stress relaxation of the wall (which tends to reduce turgor pressure) and water absorption (which tends to increase turgor pressure).

Many studies have shown that wall relaxation and expansion depend on turgor pressure. As turgor is reduced, wall relaxation and growth slow down. Growth usually ceases before turgor reaches zero. The turgor value at which growth ceases is called the **yield threshold** (usually represented by the symbol $Y$). This dependence of cell wall expansion on turgor pressure is expressed in the following equation:

$$GR = \Delta V/\Delta t = m(\Psi_p - Y) \qquad (15.2)$$

where $GR$ is the cell growth rate, and $m$ is the coefficient that relates growth rate to the turgor in excess of the yield threshold. The coefficient $m$ is usually called **wall extensibility**. Mathematically, it is defined as the slope of the line relating growth rate to turgor pressure.

Under conditions of steady-state growth, $GR$ in Equation 15.2 is the same as the rate of water uptake in Equation 15.1. That is, the increase in the volume of the cell equals the volume of water taken up. The two equations are plotted in

**FIGURE 15.24.** Note that the two processes of wall expansion and water uptake show opposing reactions to a change in turgor. For example, an increase in turgor increases wall extension but reduces water uptake. Under normal conditions, turgor is dynamically balanced in a growing cell exactly at the point where the two lines intersect. At this point both equations are satisfied, and water uptake is exactly matched by enlargement of the wall chamber.

This intersection point in Figure 15.24 is the steady-state condition, and any deviations from this point will cause transient imbalances between the processes of water uptake and wall expansion. The result of these imbalances is that turgor will return to the point of intersection, the point of dynamic stability for the growing cell.

The regulation of cell growth—for example, by hormones or by light—typically is accomplished by regulation of the biochemical processes that regulate wall loosening and stress relaxation. Such changes can be measured as a change in $m$ or in $Y$.

The water uptake that is induced by wall stress relaxation enlarges the cell and tends to restore wall stress and turgor pressure to their equilibrium values, as we have shown. However, if growing cells are physically prevented from taking up water, wall stress relaxation progressively reduces cell turgor. This situation may be detected, for example, by turgor measurements with a pressure probe or by water potential measurements with a psychrometer or a pressure chamber (see **WEB TOPIC 3.6**). **FIGURE 15.25** shows the results of such an experiment.

## Acid-induced growth and wall stress relaxation are mediated by expansins

A common characteristic of growing cell walls is that they extend much faster at acidic pH than at neutral pH (Cos-

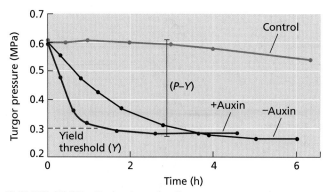

**FIGURE 15.25**   Reduction of cell turgor pressure (water potential) by stress relaxation. In this experiment, the excised stem segments from growing pea seedlings were incubated in solution with or without auxin, then blotted dry and sealed in a humid chamber. Cell turgor pressure (P) was measured at various time points. The segments treated with auxin rapidly reduced their turgor to the yield threshold (Y), as a result of rapid wall relaxation. The segments without auxin showed a slower rate of relaxation. The control segments were treated the same as the group treated with auxin, except that they remained in contact with a drop of water, which prevented wall relaxation. (After Cosgrove 1985.)

grove 1989; Rayle and Cleland 1992). This phenomenon is called **acid growth**. In living cells, acid growth is evident when growing cells are treated with acid buffers or with the drug fusicoccin, which induces acidification of the cell wall solution by activating an $H^+$-ATPase ($H^+$ pump) in the plasma membrane.

An example of acid-induced growth can be found in the initiation of a root hair, where the local wall pH drops to a value of 4.5 at the time when a bulge forms on the outer surface of the epidermal cell (Bibikova et al. 1998). Wall acidification is also associated with auxin-induced growth, but it is not sufficient to account for all of the growth mechanisms triggered by this hormone (see Chapter 19). Nevertheless, this pH-dependent mechanism of wall extension appears to be an evolutionarily conserved process common to all land plants (Cosgrove 2005) and is involved in a variety of growth processes.

Acid growth may also be observed in isolated cell walls which lack normal cellular, metabolic, and synthetic processes. Such an observation entails the use of an extensometer to put the walls under tension and to measure the pH-dependent wall extension, or "creep" (FIGURE 15.26).

The term **creep** refers to a time-dependent irreversible extension, typically the result of slippage of wall polymers relative to one another. When growing walls are incubated in neutral buffer (pH 7) and clamped in an extensometer, the walls extend briefly when tension is applied, but extension soon ceases. When transferred to an acidic buffer (pH 5 or less),

the wall begins to extend rapidly, in some instances continuing for many hours.

This acid-induced creep is characteristic of walls from growing cells, but it is not observed in mature (nongrowing) walls. When walls are pretreated with heat, proteases, or other agents that denature proteins, they lose their ability to respond to acidification. Such results indicate that acid growth is not due simply to the physical chemistry of the wall (e.g., a weakening of the pectin gel), but is catalyzed by one or more wall proteins.

The idea that proteins are required for acid growth was confirmed in reconstitution experiments in which heat-inactivated walls were restored to nearly full acid-growth responsiveness by addition of proteins extracted from growing walls (FIGURE 15.27). The active components proved to be a group of proteins that were named **expansins** (McQueen-Mason and Cosgrove 1995). These proteins catalyze the pH-dependent extension and stress relaxation of cell walls. They are effective in catalytic amounts (about 1 part protein per 5000 parts wall, by dry weight).

The molecular basis for expansin action on wall rheology is still uncertain, but most evidence indicates that expansins cause wall creep by loosening noncovalent adhesion between wall polysaccharides. Studies of protein structure and binding suggest that expansins may act at the interface between cellulose and one or more hemicelluloses (Yennawar et al. 2006).

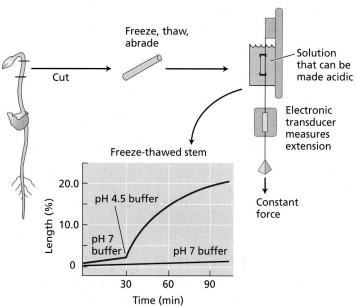

**FIGURE 15.26**   Acid-induced extension of isolated cell walls, measured in an extensometer. The wall sample from killed cells is clamped and put under tension in an extensometer that measures the length with an electronic transducer attached to a clamp. When the solution surrounding the wall is replaced with an acidic buffer (e.g., pH 4.5), the wall extends irreversibly in a time-dependent fashion (it creeps).

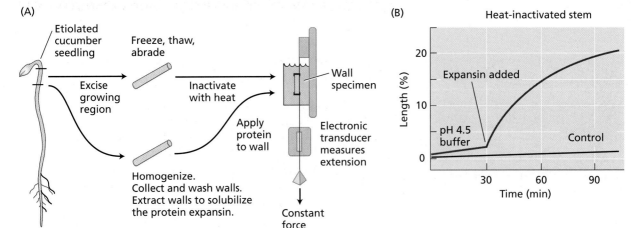

FIGURE 15.27 Scheme for the reconstitution of extensibility of isolated cell walls. (A) Cell walls are prepared as in Figure 15.26 and briefly heated to inactivate the endogenous acid extension response. To restore this response, proteins are extracted from growing walls and added to the solution surrounding the wall. (B) Addition of proteins containing expansins restores the acid extension properties of the wall. (After Cosgrove 1997.)

With the complete sequencing of several plant genomes, we now know that expansins belong to a large superfamily of genes, divided into two major expansin families, α-expansins (EXPA) and β-expansins (EXPB), plus two smaller families of unknown function (Sampedro and Cosgrove 2005). The two kinds of expansins act on different polymers of the cell wall and work coordinately during cell growth, fruit ripening, and other situations where wall loosening occurs (see WEB TOPICS 15.6 AND 15.7).

In addition to expansins, several types of experiments implicate **endo-(1→4)-β-D-glucanases** in cell wall loosening, especially during auxin-induced cell elongation (see WEB TOPIC 15.8).

## Many structural changes accompany the cessation of wall expansion

The growth cessation that occurs during cell maturation is generally irreversible and is typically accompanied by a reduction in wall extensibility, as measured by various biophysical methods. These physical changes in the wall might come about by (a) a reduction in wall-loosening processes, (b) an increase in wall cross-linking, or (c) an alteration in the composition of the wall, making for a more rigid structure or one less susceptible to wall loosening. There is some evidence for each of these ideas (Cosgrove 1997).

Several modifications of the maturing wall may contribute to wall rigidification:

- Newly secreted matrix polysaccharides may be altered in structure so as to form tighter complexes with cellulose or other wall polymers, or they may be resistant to wall-loosening activities.

- Removal of (1→3;1→4)β-D-glucan in grass cell walls is coincident with growth cessation in these walls and may cause wall rigidification.

- De-esterification of pectins, leading to more rigid pectin gels, is similarly associated with growth cessation in both grasses (where there is only small amounts of pectin) and eudicotyledons.

- Cross-linking of phenolic groups in the wall (such as tyrosine residues in HRGPs, ferulic acid residues attached to matrix polysaccharides, and lignin) generally coincides with wall maturation and is believed to be mediated by peroxidase, a putative wall rigidification enzyme.

Thus, many structural changes in the wall occur during and after cessation of growth, and it has not yet been possible to identify the significance of individual processes for cessation of wall expansion.

# SUMMARY

The architecture, mechanics, and function of plants depend on the structure of the cell wall. The wall is secreted and assembled as a complex structure that varies in form and composition as the cell differentiates.

## The Structure and Synthesis of Plant Cell Walls

- Cell walls vary greatly in form and composition, depending on cell type and species (**Figures 15.1–15.3**).

- Primary cell walls are synthesized in actively growing cells. In certain cells, secondary cell walls are deposited after cell expansion ceases.

- The basic model of the primary wall is of a network of cellulose microfibrils embedded in a matrix of hemicelluloses, pectins, and structural proteins (**Figures 15.4, 15.6; Table 15.1**).

- Cellulose microfibrils are highly ordered arrays of glucan chains synthesized at the surface of the cell

## SUMMARY continued

by particle rosettes containing multiple cellulose synthase (CesA) proteins (**Figures 15.7, 15.8**).

- Matrix polysaccharides are synthesized in the Golgi apparatus and secreted via vesicles (**Figure 15.4**).

- Hemicelluloses bind to the surface of cellulose microfibrils, and pectins form hydrophilic gels that can become cross-linked by calcium ions (**Figure 15.13**).

- Wall assembly is partly a physical process and partly mediated by enzymes such as xyloglucan endotransglucosylase (XET) (**Figure 15.16**).

- Lignin is a complex polymer made up of phenyl propanoid subunits that are oxidatively cross-linked in the cell wall and that lock the matrix and cellulose into a hydrophobic and indigestible material (**Figures 15.17, 15.18**).

- Secondary walls in woody tissues contain a higher percentage of cellulose and xylans than do primary walls and become highly impregnated with lignin (**Figure 15.18**).

### Patterns of Cell Expansion

- Wall expansion may be highly localized (tip growth) or evenly distributed over the wall surface (diffuse growth) (**Figure 15.19**).

- In diffuse-growing cells, the orientation of cell growth is determined by the orientation of cellulose microfibrils, which is determined by the orientation of microtubules in the cytoplasm (**Figure 15.20**).

- Complicated cell growth patterns, such found in the "jigsaw" pattern of the leaf epidermis of many plants, involve G protein signals that organize the cytoskeletal elements, thereby directing the local pattern of wall synthesis and growth (**Figure 15.21**).

### The Rate of Cell Elongation

- Biochemical loosening of the cell wall leads to wall stress relaxation, which dynamically links water uptake with cell wall expansion in the growing cell (**Figure 15.24**).

- The actions of hormones (such as auxin) and environmental conditions (light and water availability) modulate cell expansion by altering wall extensibility and/or yield threshold (**Figure 15.25**).

- Acid-induced cell wall extension is characteristic of primary walls and is mediated by the protein expansin, which loosens the linkages between microfibrils (**Figures 15.26, 15.27**).

- Cessation of cell growth during cell maturation involves multiple mechanisms of cell wall cross-linking and rigidification.

## WEB MATERIAL

### Web Topics

**15.1 Plant Cell Walls Play a Major Role in Carbon Flow Through Ecosystems**

Much of the carbon of ecosystems is tied up in plant cell walls.

**15.2 Terminology for Polysaccharide Chemistry**

A brief review of terms used to describe the structures, bonds, and polymers in polysaccharide chemistry is provided.

**15.3 Molecular Model for the Synthesis of Cellulose and Other Wall Polysaccharides That Consist of a Disaccharide Repeat**

A model is presented for the polymerization of cellobiose units into glucan chains by the enzyme cellulose synthase.

**15.4 Matrix Components of the Cell Wall**

The secretion of xyloglucan and glycosylated proteins by the Golgi can be demonstrated at the ultrastructural level.

**15.5 The Mechanical Properties of Cell Walls: Studies with *Nitella***

Experiments have demonstrated that the inner 25% of the cell wall determines the directionality of cell expansion.

**15.6 Wall Degradation and Plant Defense**

Cell wall degradation occurs during senescence, fruit ripening, and pathogen attack.

**15.7 Structure of Biologically Active Oligosaccharins**

Some cell wall fragments have been demonstrated to have biological activity.

**15.8 Glucanases and Other Hydrolytic Enzymes may Modify the Matrix**

Some evidence suggests that endo-$(1\rightarrow4)$-β-D-glucanases (EGases) may play a role in auxin-induced cell elongation.

# CHAPTER REFERENCES

Amor, Y., Haigler, C. H., Johnson, S., Wainscott, M., and Delmer, D. P. (1995) A membrane-associated form of sucrose synthase and its potential role in synthesis of cellulose and callose in plants. *Proc. Natl. Acad. Sci. USA* 92: 9353–9357.

Anderson, C. T., Carroll, A., Akhmetova, L., and Somerville, C. (2010) Real time imaging of cellulose reorientation during cell wall expansion in Arabidopsis roots. *Plant Physiology* 152: 787–796.

Baskin, T. I. (2001) On the alignment of cellulose microfibrils by cortical microtubules: a review and a model. *Protoplasma* 215: 150–171.

Baskin, T. I., Beemster, G. T., Judy-March, J. E., and Marga, F. (2004) Disorganization of cortical microtubules stimulates tangential expansion and reduces the uniformity of cellulose microfibril alignment among cells in the root of Arabidopsis. *Plant Physiol.* 135: 2279–2290.

Baskin, T. I., Wilson, J. E., Cork, A., and Williamson, R. E. (1994) Morphology and microtubule organization in Arabidopsis roots exposed to oryzalin or taxol. *Plant Cell Physiol.* 35: 935–942.

Bibikova, T. N., Jacob, T., Dahse, I., and Gilroy, S. (1998) Localized changes in apoplastic and cytoplasmic pH are associated with root hair development in *Arabidopsis thaliana*. *Development* 125: 2925–2934.

Brown, R. M., Jr., Saxena, I. M., and Kudlicka, K. (1996) Cellulose biosynthesis in higher plants. *Trends Plant Sci.* 1: 149–155.

Burr, S. J., and Fry, S. C. (2009) Extracellular cross-linking of maize arabinoxylans by oxidation of feruloyl esters to form oligoferuloyl esters and ether-like bonds. *Plant J.* 58: 544–567.

Cannon, M. C., Terneus, K., Hall, Q., Tan, L., Wang, Y., Wegenhart, B. L., Chen, L., Lamport, D. T., Chen, Y., and Kieliszewski, M. J. (2008) Self-assembly of the plant cell wall requires an extensin scaffold. *Proc. Natl. Acad. Sci. USA* 105: 2226–2231.

Carpita, N. C. (1983) Hemicellulosic polymers of cell walls of *Zea coleoptiles*. *Plant Physiol.* 72: 515–521.

Carpita, N. C., Defernez, M., Findlay, K., Wells, B., Shoue, D. A., Catchpole, G., Wilson, R. H., and McCann, M.

C. (2001) Cell wall architecture of the elongating maize coleoptile. *Plant Physiol.* 127: 551–565.

Carpita, N. C., and McCann, M. (2000) The cell wall. In *Biochemistry and Molecular Biology of Plants*, B. B. Buchanan, W. Gruissem, and R. L. Jones, eds., American Society of Plant Biologists, Rockville, MD, pp. 52–108.

Chambat, G., Karmous, M., Costes, M., Picard, M., and Joseleau, J. P. (2005) Variation of xyloglucan substitution pattern affects the sorption on celluloses with different degrees of crystallinity. *Cellulose* 12: 117–125.

Coenen, G. J., Bakx, E. J., Verhoef, R. P., Schols, H. A., and Voragen, A. G. J. (2007) Identification of the connecting linkage between homo- or xylogalacturonan and rhamnogalacturonan type I. *Carbohydr. Polym.* 70: 224–235.

Cosgrove, D. J. (1985) Cell wall yield properties of growing tissues. Evaluation by in vivo stress relaxation. *Plant Physiol.* 78: 347–356.

Cosgrove, D. J. (1989) Characterization of long-term extension of isolated cell walls from growing cucumber hypocotyls. *Planta* 177: 121–130.

Cosgrove, D. J. (1993) Wall extensibility: Its nature, measurement, and relationship to plant cell growth. *New Phytol.* 124: 1–23.

Cosgrove, D. J. (1997) Relaxation in a high-stress environment: The molecular bases of extensible cell walls and cell enlargement. *Plant Cell* 9: 1031–1041.

Cosgrove, D. J. (2005) Growth of the plant cell wall. *Nat. Rev. Mol. Cell Biol.* 6: 850–861.

Davin, L. B., and Lewis, N. G. (2005) Lignin primary structures and dirigent sites. *Curr. Opin. Biotechnol.* 16: 407–415.

Derbyshire, P., Findlay, K., McCann, M. C., and Roberts, K. (2007a) Cell elongation in Arabidopsis hypocotyls involves dynamic changes in cell wall thickness. *J. Exp. Bot.* 58: 2079–2089.

Derbyshire, P., McCann, M. C., and Roberts, K. (2007b) Restricted cell elongation in Arabidopsis hypocotyls is associated with a reduced average pectin esterification level. *BMC Plant Biol.* 7: 31.

Dickinson, W.G. (2000) *Integrative Plant Anatomy.* Harcourt/Academic Press, San Diego, CA, pp. 75 and 83.

Doblin, M. S., Kurek, I., Jacob-Wilk, D., and Delmer, D. P. (2002) Cellulose biosynthesis in plants: From genes to rosettes. *Plant Cell Physiol.* 43: 1407–1420.

Fincher, G. B. (2009) Revolutionary times in our understanding of cell wall biosynthesis and remodeling in the grasses. *Plant Physiol.* 149: 27–37.

Freshour, G., Clay, R. P., Fuller, M. S., Albersheim, P., Darvill, A. G., and Hahn, M. G. (1996) Developmental and tissue-specific structural alterations of the cell-wall polysaccharides of *Arabidopsis thaliana* roots. *Plant Physiol.* 110: 1413–1429.

Fry, S. C. (2004) Primary cell wall metabolism: tracking the careers of wall polymers in living plant cells. *New Phytol.* 161: 641–675.

Fu, Y., Gu, Y., Zheng, Z., Wasteneys, G., and Yang, Z. (2005) Arabidopsis interdigitating cell growth requires two antagonistic pathways with opposing action on cell morphogenesis. *Cell* 120: 687–700.

Gardiner, J. C., Taylor, N. G., and Turner, S. R. (2003) Control of cellulose synthase complex localization in developing xylem. *Plant Cell* 15: 1740–1748.

Gibeaut, D. M., Pauly, M., Bacic, A., and Fincher, G. B. (2005) Changes in cell wall polysaccharides in developing barley (*Hordeum vulgare*) coleoptiles. *Planta* 221: 729–738.

Gunning, B. E. S., and Steer, M. (1996) *Plant Cell Biology: Structure and Function.* Bartlet and Jones Publishers, Boston.

Gutierrez, R., Lindeboom, J. J., Paredez, A. R., Emons, A. M., and Ehrhardt, D. W. (2009) Arabidopsis cortical microtubules position cellulose synthase delivery to the plasma membrane and interact with cellulose synthase trafficking compartments. *Nat. Cell Biol.* 11: 797–806.

Harris, P. J., and Stone, B. A. (2008) Chemistry and molecular organization of plant cell walls. In *Biomass Recalcitrance*, M. Himmel, ed., Wiley-Blackwell, Hoboken, NJ, pp. 61–93.

Hayashi, T. (1989) Xyloglucans in the primary cell wall. *Annu. Rev. Plant Physiol. Plant Mol. Biol.* 40: 139–168.

Ishii, T., Matsunaga, T., Pellerin, P., O'Neill, M. A., Darvill, A., and Albersheim, P. (1999) The plant cell wall polysaccharide rhamnogalacturonan II self-assembles into a covalently cross-linked dimer. *J. Biol. Chem.* 274: 13098–13104.

Kennedy, C. J., Cameron, G. J., Sturcova, A., Apperley, D. C., Altaner, C., Wess, T. J., and Jarvis, M. C. (2007) Microfibril diameter in celery collenchyma cellulose: X-ray scattering and NMR evidence. *Cellulose* 14: 235–246.

Kimura, S., Laosinchai, W., Itoh, T., Cui, X. J., Linder, C. R., and Brown, R. M., Jr. (1999) Immunogold labeling of rosette terminal cellulose-synthesizing complexes in the vascular plant *Vigna angularis*. *Plant Cell* 11: 2075–2085.

Marga, F., Grandbois, M., Cosgrove, D. J., and Baskin, T. I. (2005) Cell wall extension results in the coordinate separation of parallel microfibrils: Evidence from scanning electron microscopy and atomic force microscopy. *Plant J.* 43: 181–190.

Matthews, J. F., Skopec, C. E., Mason, P. E., Zuccato, P., Torget, R. W., Sugiyama, J., Himmel, M. E., and Brady, J. W. (2006) Computer simulation studies of microcrystalline cellulose Ibeta. *Carbohydr. Res.* 341: 138–152.

McCann, M. C., Wells, B., and Roberts, K. (1990) Direct visualization of cross-links in the primary plant cell wall. *J. Cell Sci.* 96: 323–334.

McQueen-Mason, S. J., and Cosgrove, D. J. (1995) Expansin mode of action on cell walls. Analysis of wall hydrolysis, stress relaxation, and binding. *Plant Physiol.* 107: 87–100.

Mohnen, D. (2008) Pectin structure and biosynthesis. *Curr. Opin. Plant Biol.* 11: 266–277.

O'Neill, M. A., Ishii, T., Albersheim, P., and Darvill, A. G. (2004) Rhamnogalacturonan II: Structure and function of a borate cross-linked cell wall pectic polysaccharide. *Annu. Rev. Plant Biol.* 55: 109–139.

Obel, N., Erben, V., Schwarz, T., Kuehnel, S., Fodor, A., and Pauly, M. (2009) Microanalysis of plant cell wall polysaccharides. *Mol. Plant* 2: 922–932.

Paredez, A. R., Somerville, C. R., and Ehrhardt, D. W. (2006) Visualization of cellulose synthase demonstrates functional association with microtubules. *Science* 312: 1491–1495.

Peng, L., Kawagoe, Y., Hogan, P., and Delmer, D. (2002) Sitosterol-beta-glucoside as primer for cellulose synthesis in plants. *Science* 295: 147–150.

Ralph, J., Brunow, G., and Boerjan, W. (2007) Lignins. In *Encyclopedia of Plant Science*, Wiley, Hoboken, NJ, pp. 1123–1134.

Rayle, D. L., and Cleland, R. E. (1992) The acid growth theory of auxin-induced cell elongation is alive and well. *Plant Physiol.* 99: 1271–1274.

Roland, J. C., Reis, D., Mosiniak, M., and Vian, B. (1982) Cell wall texture along the growth gradient of the mung bean hypocotyl: Ordered assembly and dissipative processes. *J. Cell Sci.* 56: 303–318.

Rose, J. K., Braam, J., Fry, S. C., and Nishitani, K. (2002) The XTH family of enzymes involved in xyloglucan endotransglucosylation and endohydrolysis: Current perspectives and a new unifying nomenclature. *Plant Cell Physiol.* 43: 1421–1435.

Salnikov, V. V., Grimson, M. J., Delmer, D. P., and Haigler, C. H. (2001) Sucrose synthase localizes to cellulose synthesis sites in tracheary elements. *Phytochemistry.* 57: 823–833.

Sampedro, J., and Cosgrove, D. J. (2005) The expansin superfamily. *Genome Biol.* 6: 242.

Scheible, W. R., and Pauly, M. (2004) Glycosyltransferases and cell wall biosynthesis: Novel players and insights. *Curr. Opin. Plant Biol.* 7: 285–295.

Seifert, G. J., and Roberts, K. (2007) The biology of arabinogalactan proteins. *Annu. Rev. Plant Biol.* 58: 137–161.

Settleman, J. (2005) Intercalating Arabidopsis leaf cells: A jigsaw puzzle of lobes, necks, ROPs, and RICs. *Cell* 120: 570–572.

Skopec, C. E., Himmel, M. E., Matthews, J. F., and Brady, J. W. (2003) Energetics for displacing a single chain from the surface of microcrystalline cellulose into the active site of Acidothermus cellulolyticus Cel5A. *Protein Eng.* 16: 1005–1015.

Sturcova, A., His, I., Apperley, D. C., Sugiyama, J., and Jarvis, M. C. (2004) Structural details of crystalline cellulose from higher plants. *Biomacromolecules* 5: 1333–1339.

Szymanski, D. B., and Cosgrove, D. J. (2009) Dynamic coordination of cytoskeletal and cell wall systems during plant cell morphogenesis. *Curr. Biol.* 19: R800–R811.

Thimm, J. C., Burritt, D. J., Sims, I. M., Newman, R. H., Ducker, W. A., and Melton, L. D. (2002) Celery (*Apium graveolens*) parenchyma cell walls: cell walls with minimal xyloglucan. *Physiol. Plant* 116: 164–171.

Thompson, J. E., and Fry, S. C. (2001) Restructuring of wall-bound xyloglucan by transglycosylation in living plant cells. *Plant J.* 26: 23–34.

Vanholme, R., Morreel, K., Ralph, J., and Boerjan, W. (2008) Lignin engineering. *Curr. Opin. Plant Biol.* 11: 278–285.

Wilson, R. H., Smith, A. C., Kacurakova, M., Saunders, P. K., Wellner, N., and Waldron, K. W. (2000) The mechanical properties and molecular dynamics of plant cell wall polysaccharides studied by Fourier-transform infrared spectroscopy. *Plant Physiol.* 124: 397–405.

Yennawar, N. H., Li, L. C., Dudzinski, D. M., Tabuchi, A., and Cosgrove, D. J. (2006) Crystal structure and activities of EXPB1 (Zea m 1), a beta-expansin and group-1 pollen allergen from maize. *Proc. Natl. Acad. Sci. USA* 103: 14664–14671.

Yin, Y., Huang, J., and Xu, Y. (2009) The cellulose synthase superfamily in fully sequenced plants and algae. *BMC Plant Biol.* 9: 99.

Zykwinska, A., Gaillard, C., Buleon, A., Pontoire, B., Garnier, C., Thibault, J. F., and Ralet, M. C. (2007a) Assessment of in vitro binding of isolated pectic domains to cellulose by adsorption isotherms, electron microscopy, and X-ray diffraction methods. *Biomacromolecules* 8: 223–232.

Zykwinska, A., Thibault, J. F., and Ralet, M. C. (2007b) Organization of pectic arabinan and galactan side chains in association with cellulose microfibrils in primary cell walls and related models envisaged. *J. Exp. Bot.* 58: 1795–1802.

# 16

# Growth and Development

Plants offer intriguing developmental contrasts to animals, not only with respect to their diverse forms, but also in how those forms arise. A sequoia tree, for example, may grow for thousands of years before reaching a size big enough for an automobile to drive through its trunk. In contrast, an Arabidopsis plant can complete its life cycle in little more than a month, making hardly more than a handful of leaves (**FIGURE 16.1**). Dissimilar as they may be, both species employ growth mechanisms common to all multicellular plants, in which form is elaborated gradually through adaptive postembryonic growth processes. Animals, by contrast, have a more predictable pattern of development in which the basic body plan is largely determined during embryogenesis.

These differences between plants and animals can be understood partly in terms of contrasting survival strategies. Being photosynthetic, plants rely on flexible patterns of growth that allow them to adapt to fixed locations where conditions may be less than ideal, especially with respect to sunlight, and may vary over time. Animals, being heterotrophic, evolved mechanisms for mobility instead. In this chapter we consider the essential characteristics of plant development and the nature of the mechanisms that guide these flexible patterns of plant growth.

Biologists who wish to understand plant development are faced with two general issues. The first is the challenge of formulating clear and relevant descriptions of changes that occur over time. As an organism grows, are there corresponding increases in its complexity, and if so, how can this complexity be most simply described? To what extent is growth coupled to cell division, cell expansion, and specific differentiation processes? How do environmental factors influence growth processes?

With a detailed description of growth in place, biologists can begin to address a second set of questions that relate to the nature of the underlying mechanisms: How can characteristic patterns of growth be explained by genetically determined processes? How are these intrinsic programs of development coupled to external influences such as nutrient levels, energy inputs, and stress? What types of mechanisms mediate this coupling? What physical components are involved, how are they organized at the cellular and tissue levels, and how are their dynamic behaviors regulated in time and space?

To address these issues, this chapter begins with a brief overview of essential aspects of the organization and life cycle of plants and how they relate to basic growth processes. As background to this discussion, various approaches that can be used to provide a detailed and quantitative description of growth and development are reviewed in Appendix 2. Building on this foundation, we then consider how physiological, molecular, and genetic approaches can provide valuable insights into how these processes are regulated.

## Overview of Plant Growth and Development

An essential aspect of almost all land plants is their sedentary lifestyle. By virtue of their ability to photosynthesize, favorably positioned plants can readily obtain both the energy and the nutrients they need to grow and survive. Relieved of the need to move, plants have never evolved the sort of anatomical complexity that enables mobility in animals. In its place, one finds a relatively rigid anatomy adapted to the capture of light energy and nutrients. As a consequence, plant cells, unlike animal cells, are firmly attached to their neighbors in a relatively inflexible, often woody, matrix. This rigid anatomy imposes constraints on how the plant grows. Cells are added progressively to the body through the activity of localized structures called meristems. By contrast, many aspects of animal development, including the formation of primary tissue layers, are characterized by the migration of cells to new locations.

While the sedentary habit of plants allows a relatively simple organization, this lack of mobility presents significant challenges. Because plants are unable to relocate to optimal habitats, they must instead adapt to their local environments. While this adaptation can occur on a physi-

FIGURE 16.1   Two contrasting examples of plant form arising from indeterminate growth processes. (A) The Chandelier Tree, a famous *Sequoia sempervirens* that has adapted to many challenges during its roughly 2400-year existence. (B) The compact form and rapid life cycle of the much smaller *Arabidopsis thaliana* have made it a useful model for understanding mechanisms that guide plant growth and development. (A © David L. Moore/Alamy; B photo by David McIntyre.)

(A)

(B)

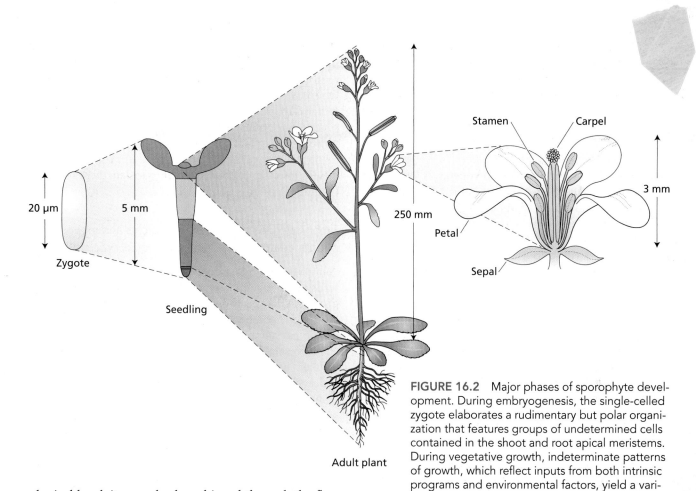

**FIGURE 16.2** Major phases of sporophyte development. During embryogenesis, the single-celled zygote elaborates a rudimentary but polar organization that features groups of undetermined cells contained in the shoot and root apical meristems. During vegetative growth, indeterminate patterns of growth, which reflect inputs from both intrinsic programs and environmental factors, yield a variable shoot and root architecture. During reproductive development, vegetative shoot apical meristems (SAMs) are reprogrammed to produce a characteristic series of floral organs, including carpels and stamens, in which the haploid gametophytic generation begins.

ological level, it may also be achieved through the flexible patterns of development that characterize vegetative growth. A key element of this adaptive growth is the presence of meristematic tissues, which contain reservoirs of cells whose fate remains undetermined. Through the regulated proliferation and differentiation of these cells, plants are able to produce a variety of complex forms adapted to the local environment.

## Sporophytic development can be divided into three major stages

The development of the seed plant sporophyte can be broken down into three major stages (**FIGURE 16.2**): embryogenesis, vegetative development, and reproductive development.

**EMBRYOGENESIS**  The term *embryogenesis* describes the process by which a single cell is transformed into a multicellular entity having a characteristic, but typically rudimentary, organization. In most seed plants, embryogenesis takes place within the confines of the ovule, a specialized structure formed within the carpels of the flower. The overall sequence of embryonic development is highly predictable, perhaps reflecting the need for the embryo to be effectively packaged within the maternally derived integuments to form the seed. Embryogenesis encompasses some

of the clearest examples of basic patterning processes in plants. Among these processes are those responsible for establishing polarity, by which the activities of cells differentiate according to their positions in the embryo.

Within the spatial framework of the embryo, groups of cells become functionally specialized to form epidermal, cortical, and vascular tissues. Certain groups of cells, known as apical meristems, are established at the growing points of the shoot and root and enable the elaboration of additional tissues and organs during subsequent vegetative growth. At the conclusion of embryogenesis, a number of physiological changes occur to enable the embryo to withstand long periods of dormancy and harsh environmental conditions (see **WEB TOPIC 16.1**).

**VEGETATIVE DEVELOPMENT**  With germination, the embryo breaks its dormant state and, by mobilizing stored reserves, commences a period of vegetative growth. Depending on the species, germination occurs in response to a variety of factors, including moisture levels or extended cold, heat, or light (see **WEB TOPIC 16.1**). Drawing initially on reserves stored in its cotyledons (e.g., beans) or in endosperm (e.g., grasses), the seedling builds on its rudimentary form through the activity of the root and shoot apical meristems (Weigel and Jürgens 2002). Through photomorphogenesis (see Chapter 17) and further development of the shoot, the seedling becomes photosynthetically competent, thus enabling further vegetative growth.

Unlike the growth of animals, vegetative growth is typically *indeterminate*—not predetermined, but subject to variation with no definite end point. This indeterminate growth is characterized by reiterated programs of lateral organ development that allow the plant to elaborate an architecture best suited to the local environment.

**REPRODUCTIVE DEVELOPMENT**  After a period of vegetative growth, plants respond to a combination of internal and external cues, including size, temperature, and photoperiod, to undergo the transition to reproductive development. In flowering plants, this transition involves the formation of specialized floral meristems that give rise to flowers. The processes by which floral meristems are specified and then develop to produce a stereotyped sequence of organ formation have provided some of the best-studied examples of plant development and are described in detail in Chapter 25.

In the following sections we examine several fundamental examples of plant development and consider how molecular and genetic methods have contributed to our understanding of how regional differences in growth are achieved.

# Embryogenesis: The Origins of Polarity

In seed plants, embryogenesis transforms a single-celled zygote into the considerably more complex individual contained in the mature seed. As such, embryogenesis provides many examples of developmental processes by which the basic architecture of the plant is established, including the elaboration of forms (**morphogenesis**), the associated formation of functionally organized structures (**organogenesis**), and the **differentiation** of cells to produce various tissues (**histogenesis**). An essential feature of this basic architecture is the presence of apical meristems at the tips of the shoot and root axes (see Figure 16.2), which are key to sustaining indeterminate patterns of vegetative growth. Finally, the development of the embryo features complex changes in physiology that enable the embryo to withstand prolonged periods of inactivity (**dormancy**) and to recognize and interpret environmental cues that signal the plant to resume growth (**germination**).

In the sections that follow we will examine from several perspectives how the complexity of the embryo arises. We begin with an anatomical comparison of embryogenesis in dicots and in monocots, which highlights both common themes and differences among seed plants. Next, we consider the nature of the signals that guide complex patterns of growth and differentiation in the embryo, with several lines of evidence highlighting the importance of position-dependent cues. Finally, we explore examples that illustrate how molecular and genetic approaches provide insight into the mechanisms that translate these cues into organized patterns of growth.

## Embryogenesis differs between dicots and monocots, but also features common fundamental processes

Anatomical comparisons highlight differences in the patterns of embryogenesis seen among different seed plant groups, such as those between monocots and dicots. Arabidopsis (a dicot) and rice (a monocot) provide two examples of embryogenesis that differ in detail, but which share certain fundamental features relating to the establishment of major growth axes. Arabidopsis embryogenesis is described in the next section. (For a description of rice embryogenesis see **WEB TOPIC 16.2**.)

**ARABIDOPSIS EMBRYOGENESIS**  By virtue of its small size, the patterns of cell division that produce the Arabidopsis embryo are relatively simple and easily followed (Mansfield and Briarty 1991). Five stages, each of which is linked to the shape of the embryo, are widely recognized:

1. **Zygotic** stage. The first stage of the diploid life cycle commences with the fusion of the haploid egg and sperm to form the single-celled zygote. Polarized growth of this cell, followed by an asymmetric transverse division, gives rise to a small apical cell and an elongated basal cell (**FIGURE 16.3A**).

2. **Globular** stage. The apical cell undergoes a series of divisions (**FIGURE 16.3B–D**) to generate a spherical, eight-cell (**octant**) globular embryo exhibiting radial symmetry (see Figure 16.3C). Additional cell divisions increase the number of cells in the globular embryo (see Figure 16.3D) and create the outer layer, termed the *protoderm*, which will become the epidermis.

3. **Heart** stage. Rapid cell division in two regions on either side of the future shoot apical meristem form the cotyledon primordia, giving the embryo bilateral symmetry (**FIGURE 16.3E AND F**).

4. **Torpedo** stage. Cell elongation throughout the embryonic axis and further development of the cotyledons occurs (**FIGURE 16.3G**).

5. **Mature** stage. Toward the end of embryogenesis, the embryo and seed lose water and become metabolically inactive as they enter dormancy (discussed in Chapter 23). Storage compounds accumulate in the cells at the mature stage (**FIGURE 16.3H**).

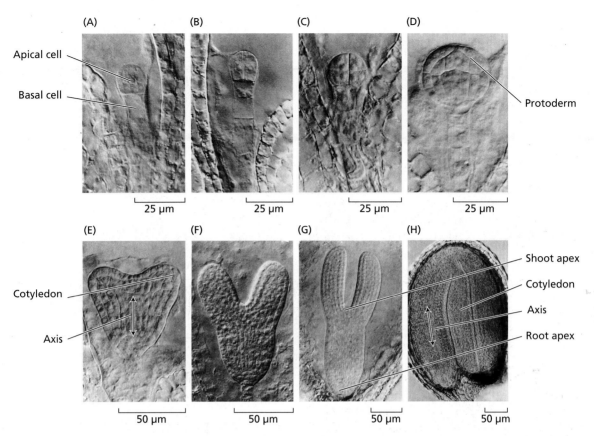

**FIGURE 16.3** The stages of Arabidopsis embryogenesis are characterized by precise patterns of cell division. (A) One-cell embryo after the first division of the zygote, which forms the apical and basal cells. (B) Two-cell embryo. (C) Eight-cell embryo. (D) Mid-globular stage, which has developed a distinct protoderm (surface layer). (E) Early heart stage; (F) late heart stage; (G) torpedo stage; and (H) mature embryo. (From West and Harada 1993; photographs taken by K. Matsudaira Yee; courtesy of John Harada, © American Society of Plant Biologists, reprinted with permission.)

A comparison of embryogenesis in Arabidopsis, a dicot, with that of rice, a monocot, illustrates several differences in embryo size, shape, cell number, and division patterns. Despite these differences, several common themes emerge that can be generalized to all seed plants. Perhaps the most fundamental of these relate to **polarity**. Beginning with the single-celled zygote, embryos become progressively more polarized throughout their development along two axes: an **apical–basal axis**, which runs between the tips of the embryonic shoot and root, and a **radial axis**, perpendicular to the apical–basal axis, which extends from the center of the plant outward.

In the following section we consider how these axes are established and discuss how specific molecular processes guide their development. Much of our discussion focuses on Arabidopsis, which is not only a powerful model for molecular and genetic studies, but also displays simple and highly stereotyped cell divisions during the early stages of its embryonic development. By observing changes in this simple pattern, we can more easily recognize both physiological and genetic factors that influence embryonic development. A graphic depiction of the earliest cell divisions in Arabidopsis, provided in **FIGURE 16.4**, offers a convenient guide for the discussion that follows. (For a discussion of the establishment of polarity in a simpler, algal zygote, see **WEB TOPIC 16.3**.)

### Apical–basal polarity is established early in embryogenesis

A characteristic feature of seed plants is a polarity in which tissues and organs are arrayed in a stereotyped order along an axis that extends from the shoot apical meristem to the root apical meristem. An early manifestation of this apical–basal axis is seen in the zygote itself, which elongates approximately threefold and becomes polarized with respect to its intracellular composition. The apical end of the zygote is densely cytoplasmic, in contrast to the basal end, which contains a large central vacuole. These differences in cytoplasmic density are captured when the zygote divides asymmetrically to give a short, cytoplasmically dense **apical cell** and a longer, vacuolated **basal cell** (see Figures 16.3A and 16.4).

The two cells produced by the division of the zygote are also distinguished by their subsequent developmental fates. Nearly the entire embryo, and ultimately the mature

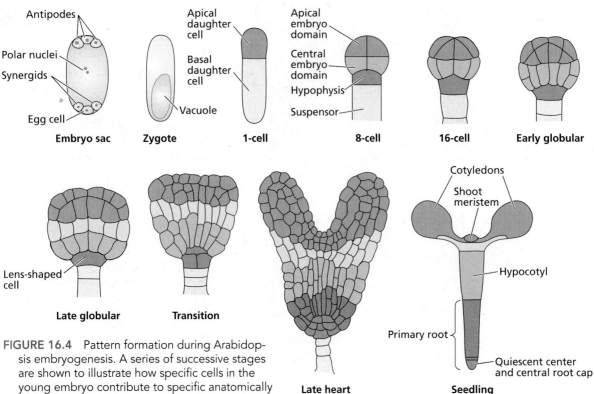

**FIGURE 16.4** Pattern formation during Arabidopsis embryogenesis. A series of successive stages are shown to illustrate how specific cells in the young embryo contribute to specific anatomically defined features of the seedling. Clonally related groups of cells (cells that can be traced back to a common progenitor) are indicated by distinct colors. Following the asymmetric division of the zygote, the smaller, apical daughter cell divides to form an eight-cell embryo consisting of two tiers of four cells each. The upper tier gives rise to the shoot apical meristem and most of the flanking cotyledon primordia. The lower tier produces the hypocotyl and some of the cotyledon, the embryonic root, and the upper cells of the root apical meristem. The basal daughter cell produces a series of nonembryonic cells that make up the suspensor, which attaches the embryo to the embryo sac. The uppermost cell of the suspensor becomes the hypophysis (blue), which is part of the embryo. The hypophysis divides to form the quiescent center and the stem cells (initials) that form the root cap.

plant, is derived from the smaller apical cell, which first undergoes two vertical divisions, then a set of horizontal divisions to generate the eight-cell (octant) globular embryo (see Figures 16.3C and 16.4).

The basal cell has a more limited developmental potential. A series of transverse divisions (producing new cell walls at right angles to the apical–basal axis) produces the filamentous **suspensor**, which attaches the embryo to the vascular system of the parent plant. Only the uppermost of the division products, known as the **hypophysis**, becomes incorporated into the mature embryo. Through further cell division, the hypophysis contributes to essential parts of the root apical meristem, including the columella and asso-

ciated root cap tissues and the quiescent center, which will be discussed later in this chapter (see Figure 16.4).

In the cells that make up the octant globular embryo, there is little, apart from position, to distinguish the appearance of the upper and lower tiers of cells. All eight cells divide **periclinally** (new cell walls form parallel to the tissue surface) to form a new cell layer called the **protoderm**. Cells of the protoderm subsequently divide **anticlinally** (new cell walls form perpendicular to the tissue surface) to increase the area of a one-cell-thick epidermis. By the early globular stage, broad distinctions between the fates of cells from the upper and lower tiers begin to emerge:

- The **apical region**, derived from the apical quartet of cells, gives rise to the cotyledons and shoot apical meristem.
- The **middle region**, derived from the basal quartet of cells, gives rise to the hypocotyl (embryonic stem), the root, and the apical domains of the root meristem.
- The **hypophysis**, derived from the uppermost cell of the suspensor, gives rise to the rest of the root meristem.

### Position-dependent signaling guides embryogenesis

The reproducible patterns of cell division during early embryogenesis in Arabidopsis might suggest that a fixed sequence of cell division is essential to this phase of devel-

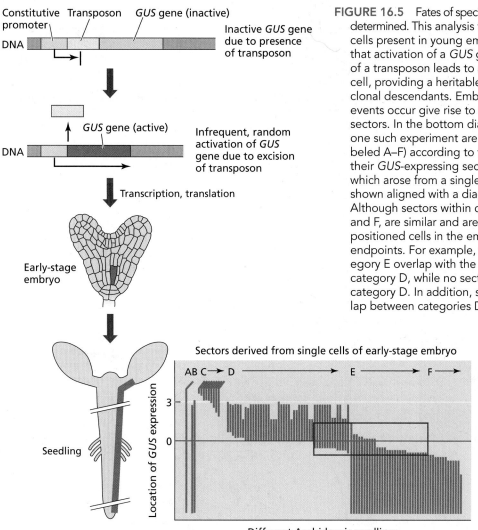

**FIGURE 16.5** Fates of specific embryonic cells are not rigidly determined. This analysis tracks the fates of individual cells present in young embryos. The top diagram shows that activation of a *GUS* gene by the random excision of a transposon leads to activation of *GUS* in a single cell, providing a heritable marker for that cell and its clonal descendants. Embryos in which these excision events occur give rise to seedlings with *GUS*-expressing sectors. In the bottom diagram, the seedlings from one such experiment are sorted into categories (labeled A–F) according to the positions and extents of their *GUS*-expressing sectors. These sectors, each of which arose from a single cell in the young embryo, are shown aligned with a diagram of a seedling to the left. Although sectors within certain categories, such as E and F, are similar and are likely to derive from similarly positioned cells in the embryo, there is variation in their endpoints. For example, the top ends of sectors in category E overlap with the bottom ends of some sectors in category D, while no sectors in category F overlap with category D. In addition, seedling cells within the overlap between categories D and E (at the root–hypocotyl boundary) provide examples of cells that differentiate according to their position rather than their clonal origin. This is direct evidence that the position of a cell—not its lineage—is an important determinant of cell fate. (From Scheres et al. 1994.)

opment. This consistency would be expected if the fates of individual cells within the embryo became fixed, or *determined*, early; once their fates were established, these cells would be committed to fixed programs of development. Such a *lineage-dependent* mechanism can be likened to assembling a structure from a standard set of parts according to self-contained instructions.

Although examples of lineage-dependent mechanisms have been documented in animal development, there are several difficulties in applying this type of explanation to plant embryogenesis. First, such lineage-dependent mechanisms are difficult to reconcile with the more variable patterns of cell division typically seen during embryogenesis in many other plants, including rice and even close relatives of Arabidopsis. Second, even for Arabidopsis, some limited variation in cell division behavior during normal embryogenesis can be seen by following the fates of individual cells with sensitive fate-mapping techniques (**FIGURE 16.5**). Finally, one can consider the extreme examples provided by certain Arabidopsis mutants that have mark-

edly different patterns of cell division, but still retain the capacity to form basic embryonic features (Torres-Ruiz and Jürgens 1994; **FIGURE 16.6**). From this perspective, it seems that the relatively predictable pattern of cell division seen in Arabidopsis may simply reflect the small size of its embryo, which places physical limits on the polarity and position of early cell divisions. Therefore some process other than a fixed sequence of cell division must guide further embryogenesis.

Given that embryogenesis can accommodate variable patterns of cell division, mechanisms that rely on **position-dependent** signaling are likely to play more significant roles than lineage-dependent mechanisms. Such mechanisms would operate by modulating the behavior of cells in a manner that depends on the position of these cells within the developing embryo. This type of mechanism would explain how equivalent forms can arise through different patterns of cell division. Such position-dependent signaling mechanisms could be expected to feature three general kinds of functional elements:

Wild-type Arabidopsis

(A)   (B)   (C)

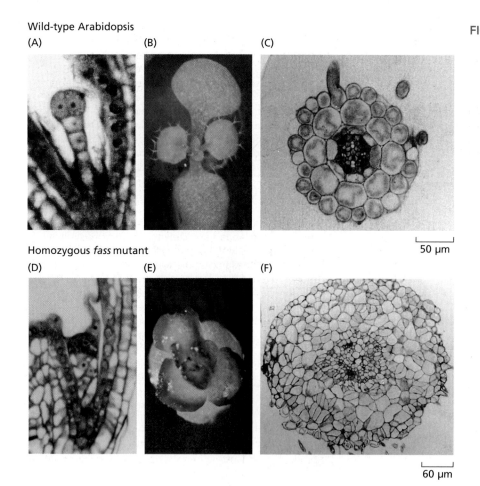

Homozygous *fass* mutant

(D)   (E)   (F)

50 μm

60 μm

**FIGURE 16.6** Extra cell divisions do not block the establishment of basic radial pattern elements. Arabidopsis plants with mutations in the *FASS* (alternatively, *TON2*) gene are unable to form a preprophase band of microtubules in cells at any stage of division. Plants carrying this mutation are highly irregular in their cell division and expansion planes, and as a result they are severely deformed. However, they continue to produce recognizable tissues and organs in their correct positions. Although the organs and tissues produced by these mutant plants are highly abnormal, the radial tissue pattern is not disturbed. (*Top*) Wild-type Arabidopsis: (A) early globular stage embryo; (B) seedling seen from the top; (C) cross section of a root. (*Bottom*) Comparable stages of Arabidopsis homozygous for the *fass* mutation: (D) early embryogenesis; (E) mutant seedling seen from the top; (F) cross section of a mutant root, showing the random orientation of the cells, but a nearly wild-type tissue order: an outer epidermal layer covers a multicellular cortex, which in turn surrounds the vascular cylinder. (From Traas et al. 1995.)

1. There must be cues that signify unique positions within the developing structure.

2. Individual cells must have the means to assess their location in relation to the positional cues.

3. Cells must have the capacity to respond in an appropriate way to the positional information.

In the following discussion, we consider the nature of mechanisms that guide axial patterning of the plant embryo, first by identifying critical elements of the process and second by considering how those elements might interact. We begin by discussing how the plant hormone auxin provides positional cues to the developing embryo and how this signaling contributes to embryonic development by regulating downstream auxin-dependent processes.

## Auxin may function as a mobile chemical signal during embryogenesis

In certain types of position-dependent development in animals, substances termed **morphogens** play key roles in providing positional cues. Through combinations of synthesis, transport, and turnover, morphogen molecules attain a graded distribution within the developing animal, which in turn evokes a range of concentration-dependent responses. Given the position-dependent nature of plant embryogenesis, it is possible that a similar morphogen-dependent mechanism operates in plants?

The varied levels and mobilities of certain plant hormones and the range of physiological responses they evoke suggest the potential of these molecules to act as morphogens, with auxin providing an especially noteworthy example. For many plant species, auxin (indole-3-acetic acid, IAA), or its synthetic analogs, can be used to induce the formation of embryos from somatic cells (see Chapter 19). This phenomenon not only highlights the potential involvement of auxin, but also indicates the intrinsic character of embryogenic programs that can proceed in the absence of information supplied by the maternal plant.

Later stages of embryogenesis are also sensitive to auxin levels, as demonstrated by the effects of auxin or its inhibitors on immature embryos propagated in vitro (**FIGURE 16.7A AND B**) (Liu et al. 1993). The cup-shaped apical regions induced by artificially perturbing auxin levels show a striking similarity to those of *PIN1* mutants, which are defective in auxin transport (**FIGURE 16.7C**), suggesting that auxin is essential for normal embryogenesis. If auxin were to function as a morphogen, it would not only

(A) Wild-type *Brassica juncea* plus *trans*-cinnamic acid

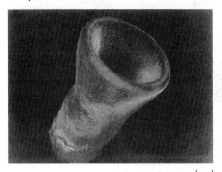

(B) Wild-type Arabidopsis

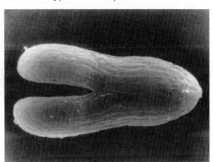

(C) Mutant *pin1-1* Arabidopsis

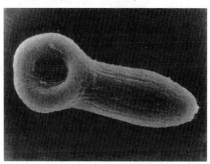

100 µm

50 µm

50 µm

**FIGURE 16.7** Evidence of a role for auxin in embryonic development. (A) Altered morphology of a *Brassica juncea* embryo, caused by culturing it in vitro for 10 days in the presence of *trans*-cinnamic acid (a flavonoid that has been shown to reduce auxin levels and inhibit auxin transport). (B) Wild-type Arabidopsis embryo. (C) A *pin1-1* mutant Arabidopsis embryo. Note the similar failure in cotyledon separation caused by chemical inhibition of auxin transport in vitro and by disruption of auxin transport by mutations in the *PIN* gene. (From Liu et al. 1993.)

possess the ability to elicit specific concentration-dependent responses in target tissues, but would also achieve a graded distribution throughout the developing embryo. By applying various measures of auxin (summarized in **TABLE 16.1**), provisional maps can be developed that suggest how the directed transport of auxin might contribute to a patterned distribution of auxin across the developing embryo (**FIGURE 16.8**).

### Mutant analysis has helped identify genes essential for embryo organization

Various types of mutants have been analyzed to gain insight into how the basic polarity of the embryo is established and how subsequent patterning occurs (reviewed by Laux et al. 2004). Mutations that block very early stages of embryogenesis are termed *embryo lethal* mutations. Given that these mutants are generally recessive, their presence

**TABLE 16.1**
**A summary of commonly used methods for estimating auxin levels in plants**

| Method | Sensitivity | Specificity | Resolution | Comments |
|---|---|---|---|---|
| **Mass spectroscopy** Molecules identified based on mass and charge | Medium | High | Tissue or organ level | Can discriminate between different forms of auxin |
| **Immunodection** Antibodies recognize molecules having a specific conformation | High* | Medium | Cellular | *Depends on the accessibility of auxin to antibody binding and specificity of antibody |
| **Reporters** Auxin activated promoter fused to gene that produces a visible readout | High | High | Cellular | Indicates location of auxin-dependent responses, but reporter activity may in some cases be limited by other factors |
| **PIN localization** Polarized distribution of auxin transporters used to infer auxin levels | Medium | Medium | Cellular | Provides a means to assess the likely direction of polar auxin transport mediated by a predominant auxin transport system, but other types of auxin are not directly measured |

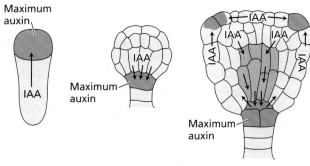

**2-cell stage**    **Globular embryo**    **Early heart stage**

**FIGURE 16.8** PIN-dependent movement of auxin (IAA) during early stages of embryogenesis. Auxin movement, as inferred from the asymmetric distribution of PIN proteins and the activity of *DR5* auxin response reporters (see Chapter 19), is depicted by arrows. Blue areas denote cells with maximum auxin concentrations.

meristem, the root apical meristem, or both were missing. The nature of the defects seen in these mutants suggests that the corresponding genes are required for establishing the normal apical–basal pattern (**FIGURE 16.9**). The cloning of several of these genes by map-based techniques has offered some insight into their molecular functions.

- *GURKE* (*GK*), named for the cucumber-like shape of the mutant, in which the cotyledons and shoot apical meristem are reduced or missing, encodes an acetyl-CoA carboxylase (Baud et al. 2004). Since acetyl-CoA carboxylase is required for the proper synthesis of very-long-chain fatty acids (VLCFA) and sphingolipids, these molecules or their derivatives appear to be crucial for proper patterning of the apical portion of the embryo.

- *FACKEL* (*FK*) was originally interpreted to be required for hypocotyl formation. Mutants exhibit

can be recognized in the partially filled seed pods, or siliques, of heterozygous plants that have been self-pollinated. While the arrested development of such mutants may indicate roles for the corresponding genes that are specific to embryogenesis, it be difficult to distinguish these mutants from those in which arrest reflects defects in more general metabolic activities.

To isolate mutations that specifically affect embryonic patterning processes, screens were performed for *seedling defective* mutants that were capable of developing into mature seeds, but which displayed an abnormal organization when germinated and examined as seedlings (Mayer et al. 1991). Among such mutants were those in which the normal apical–basal morphology was disrupted so that either the shoot apical

**(A) Wild type vs. *gnom* mutant**

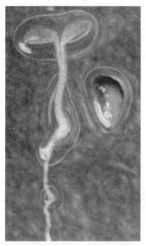

*GNOM* genes control apical–basal polarity

**(B) Wild type vs. *monopteros* mutant**

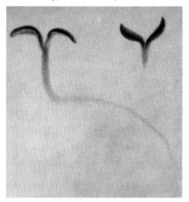

*MONOPTEROS* genes control formation of the primary root

**FIGURE 16.9** Genes essential for Arabidopsis embryogenesis have been identified from their mutant phenotypes. The development of mutant seedlings is contrasted here with that of wild-type seedlings at the same stage of development. (A) The *GNOM* gene helps establish apical–basal polarity. A plant homozygous for the *gnom* mutation is shown on the right. (B) The *MONOPTEROS* gene is necessary for basal patterning and formation of the primary root. A plant homozygous for the *monopteros* mutation (on the right) has a hypocotyl, a normal shoot apical meristem, and cotyledons, but lacks the primary root. (C) Schematic of four deletion mutant types. In each pair, the hatched regions of the wild-type plant on the left are missing from the mutant on the right. (A from Mayer et al. 1993; B from Berleth and Jürgens 1993; C from Mayer et al. 1991.)

**(C) Schematic of mutant types**

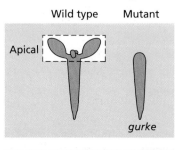

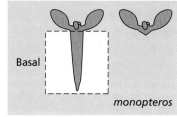

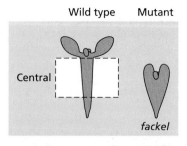

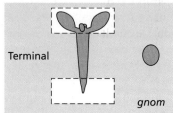

complex pattern formation defects that include malformed cotyledons, short hypocotyl and root, and often multiple shoot and root meristems. *FK* encodes a sterol C-14 reductase, suggesting that sterols are critical for pattern formation during embryogenesis.

- *GNOM* (*GN*) encodes a guanine nucleotide exchange factor (GEF), which enables the polar distribution of auxin by establishing a polar distribution of PIN auxin efflux carriers (Steinmann et al. 1999; Geldner et al. 2003).

- *MONOPTEROS* (*MP*), necessary for the normal formation of basal elements such as the root and hypocotyl, encodes an auxin response factor (ARF) (Berleth and Jürgens 1993; Hardtke and Berleth 1998).

Although cloning of these genes revealed the likely biochemical function of the proteins each encodes, further analyses are necessary to reveal how these activities contribute to the establishment of the normal apical–basal axis of the embryo. In the following section we focus on how the functions of *GN* and *MP* can be understood in terms of a model in which auxin acts as a morphogen.

### The GNOM protein establishes a polar distribution of auxin efflux proteins

By itself, the knowledge that the *GN* gene product is likely to function as a guanine nucleotide exchange factor (GEF) did not immediately suggest how it would contribute to the formation of the apical and basal regions of the embryo. It had been noted, however, that many aspects of the *gn* mutant phenotype can be mimicked, or *phenocopied*, by application of auxin transport inhibitors, suggesting that GN activity might be necessary for normal auxin transport. An explanation for how GN could enable this transport emerged through experiments demonstrating that the GEF activity of GN is required for the polarized localization of PIN proteins, which act as part of an auxin efflux transport system (Galweiler et al. 1998; see Chapter 19).

GN, like other related GEF proteins, promotes the intracellular movement of vesicles that deliver specific proteins to targeted sites within the cell, including PIN proteins. Mutations of *GN* disrupt the normal polarized distribution of PIN proteins. Further experiments demonstrated that the GEF activity of GN is responsible for PIN localization. Disruption of PIN localization is observed in cells treated with BFA, an inhibitor of GEF activity, but not in cells that contain an altered form of GN to which BFA is unable to bind. The notion that the altered pattern of embryonic development in *gn* mutants reflects a disruption of PIN activity is supported by the similar cell division defects that result from directly disrupting genes that encode PIN proteins.

These results suggest that apical–basal patterning of the embryo relies on auxin concentration differences across the embryo that are created, at least in part, through PIN-directed movement of auxin. In support of this model, the distribution of auxin inferred from auxin reporters at various stages of embryonic development is consistent with that inferred from the polarized distribution of PIN proteins. At the two-cell stage, the preferential accumulation of PIN proteins in the apical wall of the basal cell can be linked to the higher auxin levels in the apical cell. Later in the development of the embryo, the distribution of PIN proteins is reversed, with higher levels along the basal faces of apical cells, which in turn leads to higher auxin levels in basal regions (see Figure 16.8, globular stage). During the transition stage, the distribution of PIN proteins becomes more complex to facilitate a downward internal flow of auxin, which is balanced by an upward flow through superficial cell layers (see Figure 16.8, early heart stage) (Friml et al. 2003; Blilou et al. 2005).

### MONOPTEROS encodes a transcription factor that is activated by auxin

The cloning of the *MP* gene revealed that it encodes a member of a family of proteins called **auxin response factors** (**ARFs**), implicating it in auxin-dependent processes. In the presence of auxin, ARFs regulate the transcription of specific genes involved in auxin responses. In the absence of auxin, the activity of these proteins is inhibited through their association with specific repressors, termed IAA/AUX proteins. Auxin-dependent responses occur when auxin triggers the targeted degradation of these repressors, allowing ARFs to interact with their target genes (see Chapter 19).

Several lines of evidence support the view that *MP* mediates at least a subset of auxin responses. *mp* mutants not only lack the basal domains of the embryo (see Figure 16.9B and C), but also have defects in vascular patterning similar to those observed when auxin levels or movements are artificially blocked, suggesting that MP is likely to regulate genes that guide auxin-dependent vascular development.

Separate genetic studies have confirmed models of the regulation of MP activity by auxin. These studies focused on a mutant termed *bodenlos* (*bdl*), in which deletion of basal domains similar to that seen with *mp* suggested that the two genes might be functionally related (Hamann et al. 2002). Molecular cloning of *BDL* showed that it encodes one of many IAA/AUX repressor proteins. The normal form of BDL represses MP activity, but this repression can be relieved by auxin-promoted degradation of BDL. Biochemical studies demonstrated that the mutant form of BDL is resistant to auxin-induced degradation and would thus remain bound to MP, repressing its activity and producing a phenotype similar to *mp*. Taken together, GN and

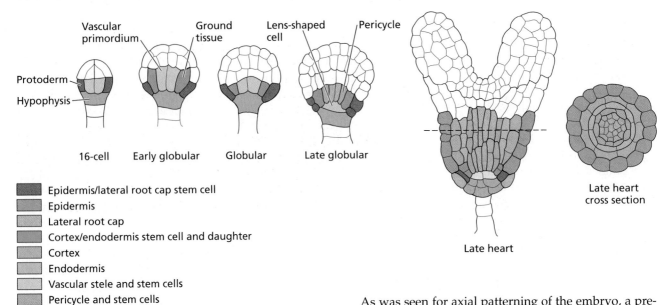

Epidermis/lateral root cap stem cell
Epidermis
Lateral root cap
Cortex/endodermis stem cell and daughter
Cortex
Endodermis
Vascular stele and stem cells
Pericycle and stem cells
Quiescent center
Columella and hypophysis.

**FIGURE 16.10** A summary of the sequence of radial patterning events during *Arabidopsis* embryogenesis. The five successive embryonic stages shown in longitudinal section illustrate the origin of distinct tissues, beginning with the delineation of the protoderm (*left*) and ending with the formation of the vascular tissues (*right*). Note how the number of tissues increases through the action of stem cells. A cross-sectional view of the basal portion of the late heart stage embryo is shown at the far right (the level of the cross section is shown by the line in the longitudinal section to its left).

MP form part of a mechanism by which auxin guides the establishment of a basic polar growth axis.

## Radial patterning guides formation of tissue layers

In addition to the distinctions among cells and tissues positioned along the apical–basal axis of the developing embryo, differences can also be seen along a radial axis extending from the interior to the surface. In Arabidopsis, differentiation of tissues along the radial axis is first observed in the globular embryo (**FIGURE 16.10**), in which periclinal divisions separate the embryo into three radially defined regions. The outermost cells form a one-cell-thick surface layer known as the **protoderm**, which eventually differentiates into the epidermis. Below this layer lie cells that will later become the **ground meristem**, which in turn gives rise to the cortex (the ground tissue between the vascular system and the epidermis) and, in the root and hypocotyl, the endodermis (the layer of suberized cells that restricts water and ion movements into and out of the stele via the apoplast; see Chapter 4). In the most central domain lies the **procambium**, which generates the vascular tissues, including the pericycle of the root.

As was seen for axial patterning of the embryo, a precisely defined sequence of cell division does not appear essential for the establishment of basic radial pattern elements, given observations of significant variation between species and of additional examples in which a radial pattern can still be established in mutants with disturbed patterns of cell division. However, compared with the apical–basal patterning of the embryo, in which the movement of auxin plays a key role in providing positional information, the molecular basis for radial patterning of the embryo is less clear. Nevertheless, some clues can be drawn from detailed anatomical and molecular analyses that highlight key aspects of the radial patterning process. (See **WEB ESSAY 16.1** and Chapter 1 for discussions of how cell division planes are established.)

One obvious and singular aspect of the radial axis is provided by the protoderm, a tissue that can be uniquely defined by its superficial position. Originating early in embryogenesis, protodermal cells have a set of exposed walls that could, in theory, facilitate the exchange of signals with the external environment or, alternatively, act as a boundary as signals move from cell to cell within the embryo. In either case, the protoderm would exhibit unique properties that would distinguish it from the internal cell layers. For example, studies in *Citrus* have shown the presence of a cuticle layer on the surface of the embryo from the earliest zygotic stages through maturity, suggesting that the walls of protodermal cells form a communication boundary (Bruck and Walker 1985).

These observations suggest that surface-associated factors might provide important cues for radial patterning. During postembryonic development, the protoderm-derived epidermis plays a key role by acting as an interface between the plant and surrounding environment. More recent studies suggest that the epidermis physically constrains the growth of more internal layers (see Chapter 24).

Genetic studies have pointed toward two genes, *Arabidopsis thaliana MERISTEM LAYER 1* (*ATML1*) and *PROTODERMAL FACTOR 2* (*PDF2*), as having essential roles

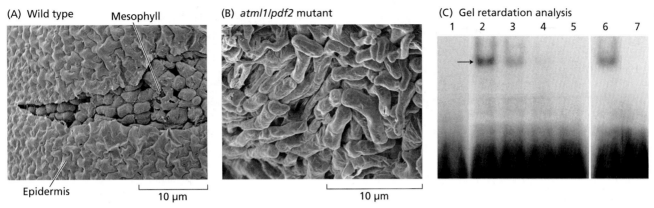

**(A) Wild type**    Mesophyll

Epidermis

10 µm

**(B) atml1/pdf2 mutant**

10 µm

**(C) Gel retardation analysis**

1    2    3    4    5    6    7

**FIGURE 16.11**  *ATML1* and *PDF2* are required for the establishment of a normal epidermis. Comparison of a wild-type plant (A) and a double *atml1/pdf2* mutant (B) shows the resemblance between the superficial layers of the mutant with the mesophyll of the wild-type plant (partially exposed in A). (C) Gel retardation analysis shows that the PDF2 protein binds specifically to a defined sequence found in promoters of genes regulated by PDF2. A labeled 21-nucleotide probe (L1) with the same sequence as the L1 box of the PDF1 pro-

moter was mixed with maltose-binding protein fused to PDF2 (MBP–PDF2). The DNA probe bound to the protein, producing a labeled complex that can be seen as a band in the gel (lane 2, arrow). No complex was produced if L1 was mixed with maltose-binding protein alone (lane 1) or if MBP–PDF2 was mixed with a mutated L1 probe (lane 7). Labeling of the complex diminished when unlabeled L1 probe (competitor) was added in increasing amounts (100-, 300-, or 1000-fold excess; lanes 3, 4, and 5). (From Abe et al. 2003.)

in establishing the epidermal identity of superficially positioned cells (Lu et al. 1996; Abe et al. 2003). Both genes encode homeodomain transcription factors and are expressed from early stages of embryogenesis in the outer cells of the embryo proper. This expression appears necessary for the establishment of normal epidermal identity, since mutant plants have an abnormal epidermis in which cells display characteristics normally associated with mesophyll cells (**FIGURE 16.11A AND B**).

The regulatory activities of the protein products of both genes appear to be linked to their recognition of a specific eight-base-pair recognition sequence shared by the promoters of epidermis-specific genes (**FIGURE 16.11C**). The binding of *ATML1* and *PDF2* to these promoter sequences promotes the transcription of the genes, whose active products lead to the differentiation of the epidermis. The *ATML1* and *PDF2* genes themselves contain this same recognition sequence, suggesting that their expression is maintained by a positive feedback loop. However, the nature of the signals that confine the expression of these genes to the epidermis remains unclear.

Genetic analysis has also revealed the identities of genes involved in the establishment of more internal tissues, including the vasculature and cortex. Arabidopsis mutants in which the *WOODEN LEG* (*WOL*) gene is disrupted fail to undergo a critical round of cell division that normally produces precursors for xylem and phloem (**FIGURE 16.12**). This defect leads to the development of a vascular system that contains xylem, but not phloem, elements. *WOL* (also known as *CYTOKININ RESPONSE 1* [*CRE1*]) encodes one of several related receptors for cytokinin, implicating this

hormone in the establishment of radial pattern elements (Mähönen et al. 2000; Inoue et al. 2001) (see Chapter 21). However, because these defects can be prevented by the *fass* mutation (that is, by making a *wol fass* double mutant), which causes extra rounds of cell division, it appears that the absence of phloem in *wol* may simply reflect the absence of an appropriately positioned precursor cell layer rather than the inability to specify phloem cell identity.

The mechanisms by which the apical–basal and radial axes are established offer examples in which a combination of hormone signaling and transcriptional control play key roles in patterning processes. In the following section we consider a well-studied example that illustrates how the movement of specific transcription factors between cell layers can contribute to the patterning process.

### The differentiation of cortical and endodermal cells involves the intercellular movement of a transcription factor

The development of endodermal and cortical tissues provides a classic example of how the radial patterning process can be regulated by gene activity communicated between adjacent layers. Two Arabidopsis genes, *SCARECROW* (*SCR*) and *SHORT-ROOT* (*SHR*), are both essential for the normal formation of cortical and endodermal cell layers (Di Laurenzio et al. 1996; Helariutta et al. 2000). The similar protein sequences encoded by these two genes place them in the GRAS family of transcription factors, whose name derives from the first known members, *GIBBERELLIN-INSENSITIVE* (*GAI*), *REPRESSOR OF GA1–3* (*RGA*), and *SCR*.

**(A) Wild type**

Protophloem sieve elements

Protoxylem

Pericycle

Protophloem sieve elements

Protoxylem

**(B) *wol* mutant**

Pericycle

Protoxylem

30 µm

**FIGURE 16.12** The cytokinin receptor encoded by the Arabidopsis *WOODEN LEG* (*WOL*) gene is required for normal phloem development. Comparison of wild-type (A) and *wol* mutant (B) roots shows an absence of phloem elements in *wol* that is accompanied by an apparent decrease in the number of cell layers. (From Mähönen et al. 2000.)

Mutants in which either *SCR* or *SHR* activity is reduced fail to undergo a round of cell division that produces the two layers that later differentiate as separate cortex and endodermis. Mutations in either gene block the round of cell division that creates these separate layers (**FIGURE 16.13**). In *scr* mutants, the single layer that remains exhibits characteristics of both endodermis and cortex, suggesting that the mutant is still able to express these characteristics, but is unable to separate them into discrete layers. This interpretation is supported by the ability of *fass* to restore more normal growth patterns. Much as it rescues *wol*, *fass* appears to compensate for the division defect of *scr*, and thus provides separate layers in which distinct endodermal and cortical traits can be expressed.

The mutant *shr*, by contrast, not only exhibits a cell division defect similar to *scr*, but is also unable to elaborate cellular characteristics typical of the endodermis. The single undivided layer in *shr* lacks endodermal traits, such as the Casparian strip, and instead displays gene activities that are normally limited

to the cortex. This apparent requirement for *SHR* gene activity to specify endodermal traits is puzzling, since the expression of *SHR* mRNA is normally restricted to more internal, provascular tissues.

More detailed analyses (Nakajima et al. 2001) have addressed this paradox, showing that although *SHR* mRNA is confined to the vascular cylinder, its translation product is not. The SHR protein is transported into the adjacent, more external layer, where it induces endodermal traits, presumably by regulating the transcription of specific genes (**FIGURE 16.14**). The contribution of the SHR protein to the differentiation of cortical and endodermal cells provides a clear example of how the functions of specific transcription factors may depend on their movement between cell layers.

**(A) Wild type**

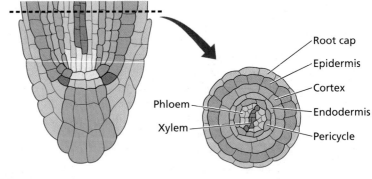

Root cap

Epidermis

Cortex

Phloem

Endodermis

Xylem

Pericycle

**(B) Mutants**

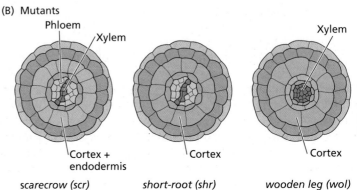

Phloem

Xylem

Cortex + endodermis

*scarecrow (scr)*

Cortex

*short-root (shr)*

Xylem

Cortex

*wooden leg (wol)*

**FIGURE 16.13** A comparison of normal and mutant radial root patterns shows the spatially defined functions of specific genes. (A) Wild-type root. (B) Defective radial root patterns of three Arabidopsis mutants: *scarecrow* (*scr*), *short-root* (*shr*), and *wooden leg* (*wol*). (After Nakajima and Benfey 2002.)

**SHR** expression

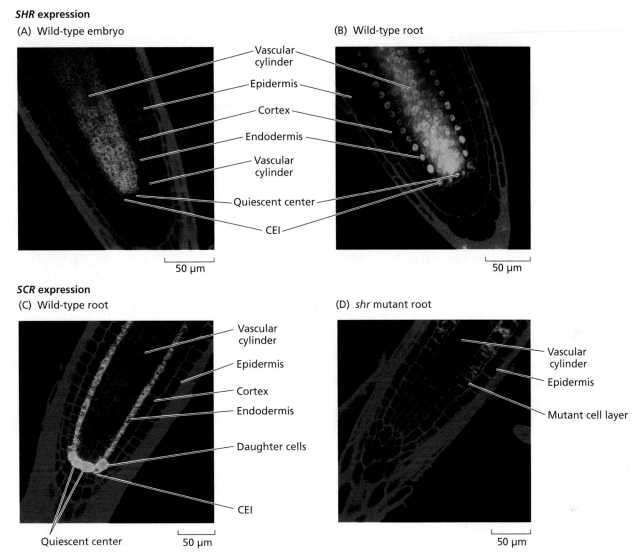

**FIGURE 16.14** The *SHORT-ROOT* (*SHR*) and *SCARE-CROW* (*SCR*) genes in Arabidopsis control tissue patterning during root development. Here, the SHR and SCR proteins have been localized by confocal laser scanning microscopy after being tagged with green fluorescent protein (GFP), which has a greenish yellow color. (A, B) SHR expression. (A) During embryogenesis in wild-type Arabidopsis, the SHR protein is localized in the provascular tissues. (B) The SHR protein continues to be localized in the vascular cylinder throughout growth of the primary root, but also moves into the adjacent endodermis. (C, D) SCR expression. (C) In wild-type roots, the SCR protein is localized in the quiescent center (QC), endodermis, and cortical–endodermal stem cell (CEI). It is not present in the cortex, vascular cylinder, or epidermis. (D) The expression of *SCR* is markedly reduced in the *shr* mutant root, and now appears only in the mutant cell layer that has characteristics of both endodermis and cortex. (From Helariutta et al. 2000.)

## Many developmental processes involve the intercellular movement of macromolecules

The interaction of adjacent tissues to trigger specific developmental processes, a phenomenon termed **induction**, is a strategy employed by many multicellular organisms. Induction ultimately depends on some type of intercellular communication. As in animals, many types of intercellular communication in plants are mediated by certain types of hormone molecules, whose movement may occur passively or, in some cases, via hormone-specific transport systems (e.g., PIN proteins). In addition to hormones, plants have the potential to transport much larger molecules, such as transcription factors, via plasmodesmata. These membrane-lined tubes, which penetrate the cell wall and connect the cytoplasm of adjacent cells (see Chapter 1), are dynamic structures that enable the selective passage of molecules based on their size and composition (Haywood et al. 2002; Maule 2008).

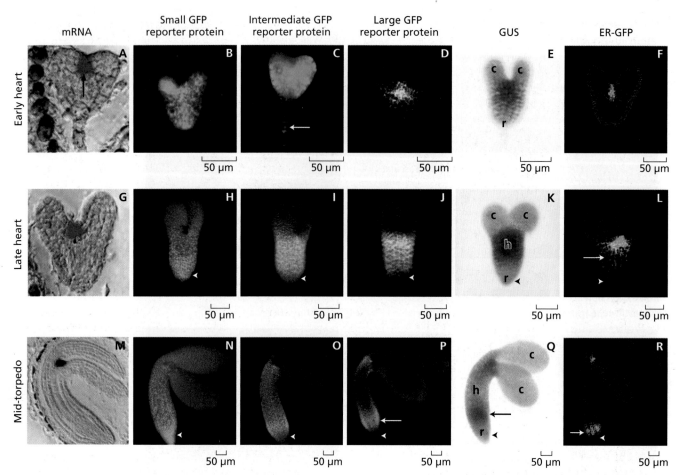

**FIGURE 16.15** Potential for intercellular protein movement changes during development. Figures show the distribution of small (B, H, N), intermediate (C, I, O), and large (D, J, P) GFP reporter proteins in embryos of different ages (early heart, A–F; late heart, G–L; and mid-torpedo, M–R). All constructs are transcribed from an *STM* promoter, which produces transcripts in relatively small regions of the embryos, as shown by in situ hybridization (A, G, and M), fusion to the nondiffusible GUS (E, K, and Q), or to ER–GFP reporters (F, L, and R). Small proteins appear to move readily in all stages of embryogenesis (B, H, and N), but the mobility of larger proteins is less and becomes more restricted in older embryos (C and D, I and J, O and P). Arrows indicate the nucleus in suspensor cells (C) and ectopic expression of the *STM* promoter in hypocotyls (L and P–R). Arrowheads indicate the root. Abbreviations: c, cotyledons; h, hypocotyl; r, root. (From Kim et al. 2005.)

The factors that govern the intercellular movement of macromolecules via plasmodesmata are complex. Although movement of molecules larger than 1 kDa is generally restricted, plasmodesmata can be regulated to permit the passage of much larger molecules or of specific proteins, such as the transcription factor SHR. Studies that tracked the movement of dye molecules or GFP-tagged proteins of various sizes showed that the behavior of plasmodesmata changes over the course of embryonic development: relatively large molecules move relatively freely during the early stages, but become restricted to more limited, tissue-defined domains later in development (**FIGURE 16.15**) (Kim et al. 2005). This regionalized communication is likely to explain in part why many developmental processes in plants seem to be regulated more in terms of tissue-defined domains than at the level of individual cells (Rinne and van der Schoot 1998).

## Meristematic Tissues: Foundations for Indeterminate Growth

The development of plants shows a remarkable degree of plasticity, which to a large extent can be attributed to specialized tissues called **meristems** (Esau 1965; Gifford and Foster 1987). A meristem can be broadly defined as a group of cells that retain the capacity to proliferate and whose ultimate fate remains undetermined. Several types of meristems that contribute to the vegetative development of plants can be distinguished based on their position in the plant.

The **root apical meristem (RAM)** and **shoot apical meristem (SAM)** are found at the tips of the root and shoot, respectively. **Intercalary meristems**, as their name suggests, represent proliferative tissues that are flanked by differentiated tissues, while **marginal meristems** function in a similar manner at the edges of developing organs. Small, superficial clusters of cells, known as **meristemoids**, give rise to structures such as trichomes or stomata (see WEB ESSAY 16.2 for a historical overview of plant meristems). In the following sections we consider the basic features of the root and shoot apical meristems that make them useful models for understanding the mechanisms that control the division of cells and determination of their fates.

### The root and shoot apical meristems use similar strategies to enable indeterminate growth

Although it might seem difficult to imagine two parts of a plant more different than a shoot and a root, certain features of the RAM and SAM and the roles they play in enabling indeterminate patterns of growth invite comparisons. Each of these structures features a spatially defined cluster of cells, termed **initials**, that are distinguished by their slow rate of division and undetermined fate. As the descendants of initials are displaced away by polarized patterns of cell division, they take on various differentiated fates that contribute to the radial and longitudinal organization of the root or shoot and to the development of lateral organs.

From this perspective, it is clear that both the RAM and the SAM must have mechanisms that balance the production of new cells with the ongoing recruitment of cells into differentiated tissues. Is it possible that common aspects of RAM and SAM behavior can be traced to similar underlying mechanisms? How are these mechanisms regulated to maintain the characteristic organizations of the shoot and root and to enable adaptive growth responses to a range of environments? Do the distinct patterns of growth and organogenesis in root and shoot impose special requirements on RAM and SAM function? To address these questions, we will discuss basic features of the RAM and the SAM as well as examples of genetically defined signaling pathways that contribute to their establishment and maintenance.

## The Root Apical Meristem

Many aspects of root growth reflect adaptations to a demanding environment. Roots, which anchor the plant and absorb water and mineral nutrients from the soil, display complex patterns of growth and tropisms that allow them to explore and exploit a heterogeneous environment laden with obstacles. To meet the challenges of their environment, roots initiate lateral organs differently from shoots. Although cells produced by the RAM divide, differentiate, and elongate as they are displaced away from the tip, like their counterparts in the shoot, lateral outgrowths such as root hairs or lateral branches form farther away from the root tip in regions where cell elongation is complete. In this manner, these outgrowths avoid the shearing damage that would result if they arose in more distal regions. Another difference between root and shoot is the presence of a root cap that covers the RAM.

In the section that follows we will consider the organization of the root in more detail, discussing regional differences in cellular behavior that contribute to the growth and functionality of the root. We will then review experimental evidence suggesting that the coordinated growth of the root depends on a combination of auxin-dependent and cytokinin-dependent programs of gene activity.

### The root tip has four developmental zones

The basic features of root development can best be described by first distinguishing zones within the root with distinct cellular behaviors. Although it is impossible to define their boundaries with absolute precision, the division of the root into the following zones provides a useful spatial framework that is relevant to our discussion of the underlying mechanisms (FIGURE 16.16).

- The **root cap** occupies the most distal part of the root. It represents a unique set of initial derivatives that are displaced distally away from the meristematic zone. The differentiated products of these divisions cover the apical meristem and protect it from mechanical injury as the root tip is pushed through the soil. Other functions of the root cap include perception of gravity, to enable gravitropism, and the secretion of compounds that help the root to penetrate the soil and to mobilize mineral nutrients.

- The **meristematic zone** lies just under the root cap. It contains a cluster of cells that act as initials, dividing with characteristic polarities to produce cells that divide further and differentiate into the various mature tissues that make up the root. Cells surrounding these initials have small vacuoles and expand and divide rapidly.

- The **elongation zone** is the site of rapid and extensive cell elongation. Although some cells continue to divide while they elongate within this zone, the rate of division decreases progressively to zero with increasing distance from the meristem.

- The **maturation zone** is the region in which cells acquire their differentiated characteristics. Cells enter the maturation zone after division and elongation have ceased, and in this region lateral organs such as lateral roots and root hairs may begin to form. Differentiation may begin much earlier, but cells do not achieve their mature state until they reach this zone.

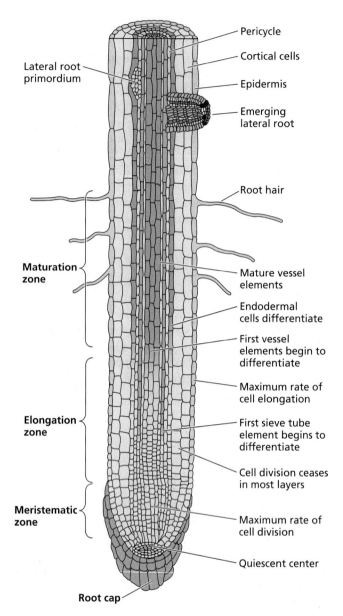

Pericycle

Cortical cells

Lateral root primordium

Epidermis

Emerging lateral root

Root hair

**Maturation zone**

Mature vessel elements

Endodermal cells differentiate

First vessel elements begin to differentiate

Maximum rate of cell elongation

**Elongation zone**

First sieve tube element begins to differentiate

Cell division ceases in most layers

**Meristematic zone**

Maximum rate of cell division

Quiescent center

**Root cap**

**FIGURE 16.16**   Simplified diagram of a primary root showing the root cap, the meristematic zone, the elongation zone, and the maturation zone.

In Arabidopsis these four developmental zones occupy little more than the first millimeter of the root tip. In many other species these zones extend over a longer distance, but growth is still confined to the distal regions of the root.

### The origin of different root tissues can be traced to specific initial cells

Given the progressive and linear development of the tissues that make up the root, it is relatively simple to trace their origin back to specific initial cells in the subapical region. In most plant roots, a medial longitudinal cross section reveals long cell files that converge in the subapical

region of the root (**FIGURE 16.17A**). At the center of this convergence zone is the so-called **quiescent center** (**QC**), named for its relatively low rate of cell division compared with those of surrounding tissues.

The close physical association between the initials that give rise to various tissues and the cells that make up the adjacent QC suggests a close functional interdependence between these cell types. Some have argued that the distinction between the quiescent center and adjacent meristematic cells is somewhat artificial because, in the roots of many higher plants, the cells that make up the QC occasionally divide to replace adjacent initials. In a similar line of reasoning, attention can be drawn to other plant species in which the relationship between the QC and initials is different. In some higher plants, the QC may include dozens or hundreds of cells, and this number may change during the plant's life cycle. By contrast, in some lower vascular plants, such as the water fern *Azolla*, a single, centrally positioned apical cell appears to fulfill the roles of both QC and initials by retaining low but consistent mitotic activity throughout vegetative development (see **WEB TOPIC 16.4**).

Like the patterns of cell division associated with embryogenesis, the behavior of the QC and the surrounding initials varies among plant species, suggesting that position-dependent mechanisms play an important role in specifying these cell types. As was the case for embryogenesis, considerable insight into the underlying mechanisms is afforded by models such as Arabidopsis, in which the behavior of individual cells can be monitored. The roots of Arabidopsis are well suited to this approach, given their small size and relatively transparent nature. Observations are also simplified by the relatively small number of cells of the Arabidopsis root.

The QC of Arabidopsis consists of only four to seven cells and, because division of cells in the QC is rare during postembryonic development, factors that perturb the activity of the QC or the surrounding initials are easily recognized. In Arabidopsis, four distinct sets of initials, all of which are adjacent to the QC, can be defined in terms of their position and the tissues they produce (**FIGURE 16.17B**):

1. **Columella initials**. Located directly below (distal to) the QC, these initials give rise to the central portion (columella) of the root cap.

2. **Epidermal–lateral root cap initials**. Located to the side of the QC, these initials first divide anticlinally to set off daughter cells, which then divide periclinally to form two files of cells that will mature into the lateral root cap and epidermis.

3. **Cortical–endodermal initials**. Located interior and adjacent to the epidermal–lateral root cap initials, the cortical–endodermal initials divide anticlinally to set off daughter cells, which then divide periclinally to form the cortical and endodermal cell layers.

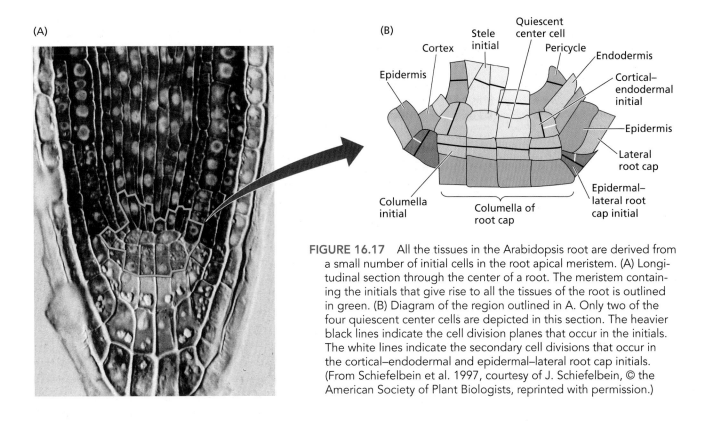

(A)

(B)

Quiescent
center cell

Stele
initial

Cortex

Pericycle

Epidermis

Endodermis

Cortical–
endodermal
initial

Epidermis

Lateral
root cap

Epidermal–
lateral root
cap initial

Columella
initial

Columella of
root cap

**FIGURE 16.17** All the tissues in the Arabidopsis root are derived from a small number of initial cells in the root apical meristem. (A) Longitudinal section through the center of a root. The meristem containing the initials that give rise to all the tissues of the root is outlined in green. (B) Diagram of the region outlined in A. Only two of the four quiescent center cells are depicted in this section. The heavier black lines indicate the cell division planes that occur in the initials. The white lines indicate the secondary cell divisions that occur in the cortical–endodermal and epidermal–lateral root cap initials. (From Schiefelbein et al. 1997, courtesy of J. Schiefelbein, © the American Society of Plant Biologists, reprinted with permission.)

4. **Stele initials**. Located directly above (proximal to) the QC, these initials give rise to the vascular system, including the pericycle.

### Cell ablation experiments implicate directional signaling processes in determination of cell identity

To test and refine the hypothesis that the behavior of the QC and surrounding initials is influenced by position-dependent signaling processes, a series of experiments was performed to assess the contributions of specific cells to the determination process. To evaluate these contributions, the highly stereotyped patterns of cell division in the RAM of normal Arabidopsis were compared with those in plants in which one or more specific cells had been destroyed (or ablated) using microscopically focused laser beams (Van den Berg 1995).

Ablation of the QC led to abnormal division and precocious differentiation of the adjacent initials (see Figure 16.17B), suggesting that the QC produces a mobile signal that acts on the adjacent initials to prevent their differentiation and thus maintain their capacity to divide. In a related experiment, ablation of differentiated cells that were proximal to initials caused the initials to take on abnormal identities, as revealed by the cell types they produced. These results suggest that the specification of the particular identities of initials relies on signals that emanate from more differentiated proximal tissues.

### Auxin contributes to the formation and maintenance of the RAM

Just as auxin seems to play a role in establishing apical–basal polarity in the embryo, a convincing case can be made for the involvement of auxin in positioning the RAM and guiding its complex behavior. In normal roots, the position of the QC coincides with an auxin concentration maximum. When the position of this maximum is shifted by chemical treatments, the position of the QC shows corresponding changes. By contrast, treatments that abolish this maximum lead to the loss of the QC (Sabatini et al. 1999; Jiang et al. 2003).

As in the embryo, relative levels of auxin across the root seem to be determined to a large extent by the polarized distribution of PIN proteins. Differences in the observed distribution of PIN proteins in various regions of the root can be used to predict flows and relative levels of auxin across the root. These models predict that auxin derived from the shoot and more superficial layers moves downward through central regions of the seedling before becoming concentrated in the QC and columella tissues (**FIGURE 16.18**). From these distal regions, auxin is transported outward into superficial layers, where it is transported upward toward the shoot. These flows have been likened to an inverted fountain.

Auxin levels predicted by such models agree remarkably well with experimentally observed values, both in normal roots and in roots whose behavior has been per-

**FIGURE 16.18** Patterns of PIN protein concentrations in the root can be used to predict auxin flows. (A) Asymmetric cellular localization of PIN proteins (red areas) in different root tissues. (B) The PIN protein patterns predict high auxin concentrations in the QC. (Regions predicted to have higher relative auxin concentrations are indicated by darker shading.) (C) Relative auxin concentrations inferred from DR5::GFP reporter activity are in agreement with these predictions. Abbreviations: EZ, elongation zone; MZ, meristematic (division) zone; QC, quiescent center. (Adapted from Grieneisen et al. 2007.)

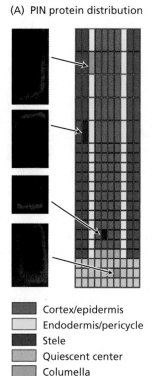

(A) PIN protein distribution

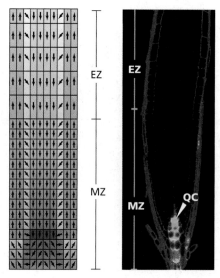

(B) Model of auxin flow    (C) DR5::GFP expression

EZ

MZ

QC

- Cortex/epidermis
- Endodermis/pericycle
- Stele
- Quiescent center
- Columella

turbed by chemical or physical treatments or by genetically altering auxin or PIN levels (Grieneisen et al. 2007). Together, these results suggest that a stable pattern of polarized auxin flow through the root is a key element in maintaining the activity of the RAM.

## Responses to auxin depend on specific transcription factors

With an understanding of how a graded distribution of auxin across the root is achieved, some explanation is still required of how these concentration differences evoke a

variety of downstream responses, including the localized zones of cell division, elongation, and differentiation (FIGURE 16.19). One part of the explanation involves auxin response factors (whose regulation by auxin is described in more detail in Chapter 19). Above some threshold concentration, auxin triggers the breakdown of IAA/AUX repressors, which would otherwise bind to ARFs such as MONOPTEROS and thus block their ability to regulate transcription. As in the establishment of the root during embryogenesis, MP and other ARFs play auxin-dependent roles to maintain the root during vegetative growth.

**FIGURE 16.19** Model for the specification of cell identity in the root. (A) Early auxin-dependent expression of the *MONOPTEROS* (*MP*) and *NONPHOTOTROPIC HYPOCOTYL 4* (*NPH4*) genes. *MP* and *NPH4* promote *PLETHORA* (*PLT*) expression in a basal domain. (B) *PLT* promotes the expression of *SCARECROW* (*SCR*) and *SHORT-ROOT* (*SHR*). (C) The combination of *PLT*, *SCR*, and *SHR* gene expression directs centrally positioned cells to become the quiescent center (QC), which signals surrounding cells (outlined in red) to maintain their meristematic activity. (From Aida et al. 2004.)

(A)

*MP* and *NPH4*

*PLT*

Auxin-dependent expression of *MP* and *NPH4*; *MP* and *NPH4* promote expression of *PLT*

(B)

*SCR* and *SHR*

*PLT*

*PLT* induces expression of *SCR* and *SHR*

(C)

Stem cells

Quiescent center

Combination of *PLT*, *SCR*, and *SHR* directs formation of the quiescent center

Genetic approaches have revealed additional types of transcription factors that act downstream of ARFs to coordinate specific aspects of root growth. Two of these transcription factors, belonging to the AP2 class, are encoded by the *PLETHORA 1* (*PLT1*) and *PLETHORA 2* (*PLT2*) genes (Aida et al. 2004). In the zone of high auxin concentration that includes the QC, the expression of these two *PLT* genes is activated. Mutants in which *PLT* genes have been disrupted are unable to form or maintain a functional QC, suggesting that these genes normally regulate programs of transcription that are essential for these processes. Conversely, the artificial expression of *PLT* genes in more proximal regions of the root led to the formation of an ectopic (in the wrong location) QC (Galinha et al. 2007). Together, these experiments support models in which auxin provides positional cues that lead to specific programs of transcription, which in turn mediate specific cellular behaviors that contribute to the formation and maintenance of the RAM.

*WOX* genes encode a third family of transcription factors that play key roles not only in the RAM, but also throughout the plant. These genes show sequence similarities to *WUSCHEL* (*WUS*), a gene essential for shoot apical meristem function, and also contain a distinctive class of the homeobox DNA binding motif (the name *WOX* is derived from *WUSCHEL* and homeob**ox**). Like that of *PLT* genes, the expression of *WOX* genes appears sensitive to auxin activity, as shown by changes in the distribution of their transcripts in *mp* or *bdl* mutants (which lack auxin activity) (Haecker et al. 2004). In the shoot, expression of *WUS* in subapical regions appears essential for maintaining undifferentiated initials (as discussed later in this chapter). In the root, *WOX5* appears to play an analogous role,

contributing to the stability of the initials that surround the QC (Sarkar et al. 2007).

## Cytokinin activity in the RAM is required for root development

Although much of our discussion of the growth and development of the root has focused on auxin, recent studies have drawn attention to cytokinin, suggesting that it acts in opposition to auxin. The contrasting activities of these hormones were first noted in physiological studies, which revealed that auxin is largely synthesized in the shoot and transported basipetally toward the root. By contrast, the opposite pattern, in which synthesis in the root is accompanied by acropetal transport toward the shoot, is seen for cytokinin. Similarly, whereas auxin promotes the development of roots, both in tissue culture and in vivo, cytokinin has the opposite effect, suppressing roots while promoting shoots.

Although the known elements that make up the cytokinin and auxin signal transduction pathways are largely distinct, similar experimental approaches have proved useful for the analysis of both pathways. Approaches analogous to the development of DR5 reporter fusions, which provide a measure of auxin activity (see Chapter 19), have been developed for cytokinin, in which promoter sequences that are activated by cytokinin are fused to GUS or GFP reporters (Müller and Sheen 2008).

The results of this reporter approach suggest that cytokinin signaling begins early in root development in the hypophysis of the globular embryo. Upon division of the hypophysis, cytokinin expression is lost in the basal cell, but is retained in the apical cell, which divides further to form the QC. At the same time, DR5-based reporters for auxin show an inverse pattern of expression, suggesting that auxin and cytokinin have opposing activities (**FIGURE 16.20**).

**FIGURE 16.20** Inverse correlation between cytokinin and auxin signaling in the embryo. (A) Expression of *TCS::GFP* (a reporter for cytokinin) in the hypophysis at the early globular stage. (B) Down-regulation of *TCS::GFP* expression in the basal cell lineage at the late globular stage. (C) Expression of *ARR7::GFP* is highest in the basal cell lineage. (D) Expression pattern of *ARR15::GFP*. (*ARR7* and *ARR15* are genes that suppress cytokinin responses.) (E) Expression of *DR5::GFP* (an auxin-responsive reporter) is highest in the basal cell lineage. The boxed sections in the upper panels are magnified in the middle panels; schematic interpretations are shown at the bottom. Abbreviations: hy, hypophysis; bc, basal cell; lsc, lens-shaped cell; s, suspensor. (From Müller and Sheen 2008.)

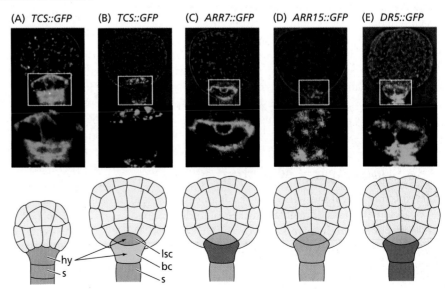

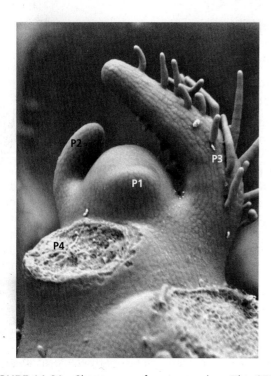

**FIGURE 16.21** Shoot apex of a tomato plant. This SEM micrograph shows the basic features of the shoot apex, including a central dome-shaped region, which maintains uncommitted initials, and a series of leaf primordia (P1, P2, P3), which have successively emerged at lateral positions on the flanks of the shoot apex. P4 indicates the base of an older leaf primordium that was removed to expose the younger primordia. (From Kuhlemeier and Reinhardt 2001; courtesy D. Reinhardt.)

Further molecular and genetic analyses indicate that the loss of cytokinin activity in the basal cell is a direct consequence of high auxin activity. Two genes that suppress cytokinin responses, *ARR7* and *ARR15*, have *a*uxin *r*esponse *e*lements (AuxRE) in their promoters, suggesting that, like DR5 reporters, they are regulated by auxin. Artificial deletion of these elements lowers *ARR7* and *ARR15* expression in basal cells, leading to ectopic cytokinin activity. Perturbing the expression of *ARR7* and *ARR15* results in abnormal phenotypes, confirming that suppression of cytokinin signaling in the basal cell is essential for normal development. Cytokinin activity in the apical cell is also essential, since suppressing this activity leads to major changes in root organization.

# The Shoot Apical Meristem

Like the root apical meristem, the shoot apical meristem is faced with the task of maintaining sets of undetermined cells that enable indeterminate growth (**FIGURE 16.21**). However, there are significant differences between the two meristem types in how the descendants of those cells become incorporated into organs. Whereas lateral root ini-

tiation occurs well back of the root tip, leaves and associated axillary branches form in close proximity to apical initials in the shoot. In place of the root cap that protects the apical initials of the root, young leaf primordia overlap and enclose the shoot tip.

Given the concentrated set of activities in the shoot tip, specific anatomical terminology has proved useful in their description. In this context, the term *shoot apical meristem* refers specifically to the initial cells and their undifferentiated derivatives, but excludes adjacent regions of the apex that contain cells that are fully committed to particular developmental fates. The more inclusive term **shoot apex** (plural *apices*) refers to the apical meristem plus the most recently formed leaf primordia.

As in the previously considered examples involving embryos and roots, the size, shape, and organization of the SAM vary according to a number of parameters, including plant species, developmental stage, and growth conditions (Steeves and Sussex 1989). Cycads have the largest SAM among vascular plants, which measures over 3 mm in diameter; at the other extreme, the SAM of Arabidopsis is less than 50 μm in diameter and contains only a few dozen cells. Within a given species, significant variations in SAM size can also occur over time, and SAM shapes may range from flat to mounded. Some of these variations are associated with successive rounds of leaf initiation, in which groups of cells on the flanks of the SAM become committed to a determinate fate. Further variations may be related to seasonal differences in growth rate, including the onset of dormancy or flowering.

In this section we will first consider the basic organization of the SAM, discussing in detail the regional differences in cellular behavior that contribute to its function. We will then discuss evidence suggesting that, like the RAM, the SAM relies on localized differences in hormone and transcription factor activities for its maintenance.

## The shoot apical meristem has distinct zones and layers

A discussion of the cellular organization of the SAM provides a useful framework for a more detailed description of its growth and development. Its organization is best appreciated by microscopic examination of shoot apices (reviewed by Bowman and Eshed 2000). Longitudinal sections of shoot apices reveal **zonation**, a term originally developed to describe regional cytological differences in the organization of the SAM of gymnosperms (Gifford and Foster 1987), but which has been extended to other seed plants to describe regional differences in cell division (**FIGURE 16.22**).

At the center of an active SAM lies the **central zone** (**CZ**), containing a cluster of infrequently dividing cells that can be compared to similar cells that make up the QC of roots. A flanking region, known as the **peripheral zone**,

(A)

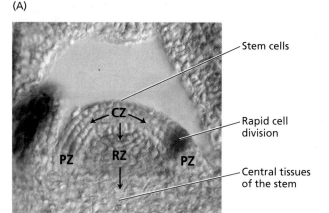

Stem cells

Rapid cell division

Central tissues of the stem

(B)

**FIGURE 16.22** The Arabidopsis shoot apical meristem can be analyzed in terms of cytological zones or cell layers. (A) The shoot apical meristem has cytological zones that represent regions with different identities and functions. The central zone (CZ) contains meristematic cells that divide slowly but are the ultimate source of the tissues that make up the plant body. The peripheral zone (PZ), in which cells divide rapidly, surrounds the central zone and produces the leaf primordia. A rib zone (RZ) lies to the interior of the central zone and generates the central tissues of the stem. (B) The shoot apical meristem also has cell layers that contribute to specific tissues of the shoot. Most cell divisions are anticlinal in the outer, L1 and L2 layers, while the planes of cell divisions are more randomly oriented in the L3 layer. The outermost (L1) layer generates the shoot epidermis; the L2 and L3 layers generate internal tissues. (From Bowman and Eshed 2000.)

consists of cytoplasmically dense cells that divide more frequently to produce cells that later become incorporated into lateral organs such as leaves. A centrally positioned **rib zone** more proximal to the CZ contains dividing cells that give rise to the internal tissues of the stem (see Figure 16.22A).

In addition to these regional differences in the frequency of division, patterns in the polarity of cell division are also observed. In most angiosperm species these differences are reflected in the layered organization of superficial cells, which are sometimes collectively referred to as the **tunica** (see Figure 16.22B). One or more adjacent layers that make up the tunica, including the most superficial protoderm, each derive from a separate set of apical initials, and the consistent thickness of each layer is maintained through predominantly anticlinal cell divisions (see discussion that follows). By contrast, the cells lying to the interior of the tunica, known as the **corpus**, also derive from a distinct set of initials, but display more variable division polarities.

### Shoot tissues are derived from several discrete sets of apical initials

Studies suggest that, like root tissues, shoot tissues are derived from a small number of apical initials (reviewed by Poethig 1987). In classic studies, the chemical colchicine was applied to shoot apices to induce the occasional formation of polyploid cells. These cells grow relatively normally, but can be easily recognized by their increased nuclear volume and cell size. Examination of sectioned shoot apices of plants that had been treated and allowed to grow for a time revealed large sectors of polyploid cells that were confined to specific layers and which extended into apical regions. Each sector could be explained by supposing that it originated from one of a small number of apical initials.

Analyses of a large number of such marked sectors, from both superficial layers and deeper tissues, indicate that several discrete sets of initials are typically maintained in the SAM. One set of superficial initials gives rise to a clonally distinct epidermal layer, termed **L1**, while more internal sets of initials give rise to the subepidermal **L2** layer and a centrally positioned **L3** layer (FIGURE 16.23). In many cases the marked sectors encompass only a portion of the shoot's circumference, suggesting that each layer derives from a small number of initials.

Analyses of cell lineages show that the identities of initial cells are determined by position-dependent mechanisms. Marked sectors that extend to apical regions of the shoot may exhibit abrupt changes in width or thickness over time. These changes can be explained by occasional divisions that lead to a marked initial displacing, or being displaced by, adjacent initials. This dynamic behavior indicates that the identities of apical initials, including their characteristic division patterns, reflect their relative position near the tip of the shoot apex, rather than a rigidly programmed identity. Similarly, the identities of cells derived from these initials also appear to be largely determined by position-dependent mechanisms. If a rare periclinal division leads to the derivative of an L2 cell adopting a superficial position, that cell

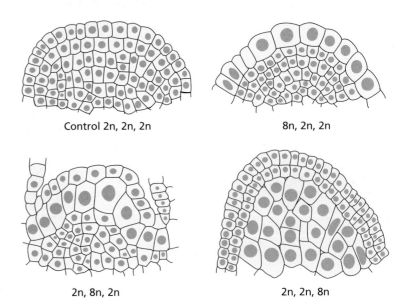

**FIGURE 16.23** In shoot apices treated with colchicine, one of the cell layers contains enlarged, polyploid nuclei, demonstrating the presence of clonally distinct layers in the shoot apical meristem. (From Steeves and Sussex 1989.)

Control 2n, 2n, 2n

8n, 2n, 2n

2n, 8n, 2n

2n, 2n, 8n

will typically adopt an epidermal identity that reflects its new location.

## The locations of PIN proteins influence SAM formation

The establishment of the SAM, like that of the RAM, is linked to complex patterns of intercellular auxin transport. During early stages of embryogenesis, the polar distribution of PIN proteins, especially PIN1, leads to the accumulation of auxin in apical regions, but by the transition stage, complex changes in the locations of PIN proteins lead to a basally directed redistribution of auxin. The factors that determine these changes are not completely understood, but changes in the phosphorylation state of PINs mediated by the kinase PINOID and the phosphatase PP2 can have significant effects on PIN localization (Michniewicz et al. 2007) (see Chapter 19).

Additional, but less direct, inputs into PIN localization are suggested by the mutant phenotypes that result from disrupting genes that encode several distinct classes of transcription factors, including members of the KANADI, DORN-

RÖSCHEN, and Class III HD-Zip families (Izhaki and Bowman 2007; Chandler et al. 2007). The altered patterns of embryonic development associated with these mutants have been interpreted to reflect abnormalities in the distribution of PIN proteins, which precede any overt changes in growth or cell division.

One aspect of the complex pattern of auxin movement in the embryo is the formation of a central apical region where auxin levels are relatively low (**FIGURE 16.24**).

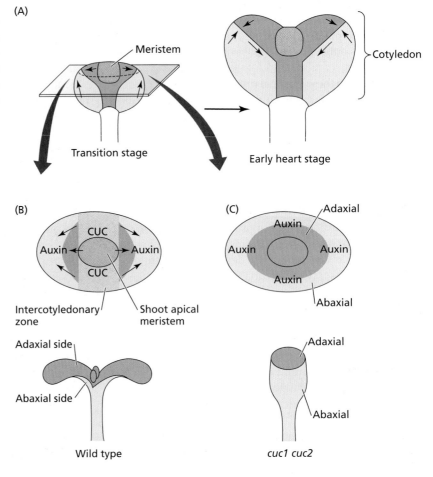

**FIGURE 16.24** A model for auxin-dependent patterning of the shoot apex. (A) The direction of auxin transport (arrows) during transition stage and early heart stage Arabidopsis embryos. (B, C) Cross sections (as shown in A) through the apical domain of a wild-type embryo (B) and a *CUP-SHAPED COTYLEDON* gene double mutant (*cuc1 cuc2*) (C), showing the region in the embryo that will develop into the shoot apical meristem, the intercotyledonary zones, and the adaxial and abaxial domains of the cotyledon. In the wild-type embryo, the SAM and intercotyledonary zones have low auxin levels and, consequently, high CUC levels, whereas the opposite pattern is seen in the flanking cotyledon primordia. In the *cuc1 cuc2* mutant, the cotyledons fail to separate, thus preventing the formation of a shoot apical meristem. (After Jenik and Barton 2005.)

Transport of auxin away from this region converges with upward superficial flows along the flanks of the embryo to create auxin maxima at the tips of the developing cotyledons. These pools of auxin feed into downward flows that converge in the hypocotyl, then continue to form the previously discussed auxin maximum in the QC. ARFs such as MP, which are activated to promote vascular development, reinforce this pattern of directional transport further.

Mutants that are defective for MP and the closely related ARF NON-PHOTOTROPIC HYPOCOTYL 4 (NPH4) are not only deficient in the basal structures, such as the root, but also lack cotyledons (reviewed by Jenik and Barton 2005). The similarities of these phenotypes to those associated with mutations affecting PIN-mediated auxin transport are consistent with models in which MP and NPH4 enable auxin-dependent responses.

## Embryonic SAM formation requires the coordinated expression of transcription factors

Although many types of genes are likely to be important in the formation and maintenance of the SAM, screens for mutants in which the SAM is unable to form, or is unstable, highlight the significance of three additional classes of transcription factors. One of these, a homeodomain transcription factor encoded by *WUS* (Mayer et al. 1998), is expressed in subapical regions as early as the 16-cell embryonic stage. Later, during the transition stage, transcription factors of the NAC class, encoded by *CUP-SHAPED COTYLEDON* (*CUC*) 1, 2, and 3, are expressed in an apically positioned stripe between the two developing cotyledons (Aida et al. 1997). Finally, during the heart stage, this sequence of gene activation concludes as another

class of homeodomain transcription factors, encoded by *SHOOT MERISTEMLESS* (*STM*), becomes expressed in a circular domain contained within the *CUC* expression domain (Long et al. 1996). Together, *WUS* and *STM* appear to help maintain cells in a state in which they can proliferate and thus ensure that the growth and differentiation of shoot tissues is balanced with the production of new undetermined cells (Lenhard et al. 2002).

The expression of *CUC* genes and the subsequent appearance of STM appear to reflect the low level of auxin activity in central apical regions (**FIGURE 16.25**; see Figure 16.24). For example, blocking auxin signaling in the flanking cotyledonary regions with mutations of *MP* and *NPH4* leads to ectopic expression of *CUC* genes in these regions. A role for auxin signaling is further supported by the observation that normal embryos treated with auxin transport inhibitors (see Figure 16.6) exhibit defects (cup-shaped cotyledons) similar to those observed among *cuc* mutants. The expression of *CUC*-like genes in the central apical domain of the embryo provides an environment that enables further patterning processes, including the localized expression of the *STM* gene, which initially coincides with the stripelike CUC expression domain, but later

**FIGURE 16.25** The formation of the apical domain involves a defined sequence of gene expression. The top row illustrates the early onset of *WUS* expression in an internal layer, which induces the expression of *CLV3* in adjacent external cell layers. The bottom row shows cross sections at the level indicated by the dashed line and emphasizes the gene expression patterns that demarcate the emerging cotyledonary and shoot apical domains. (After Laux et al. 2004.)

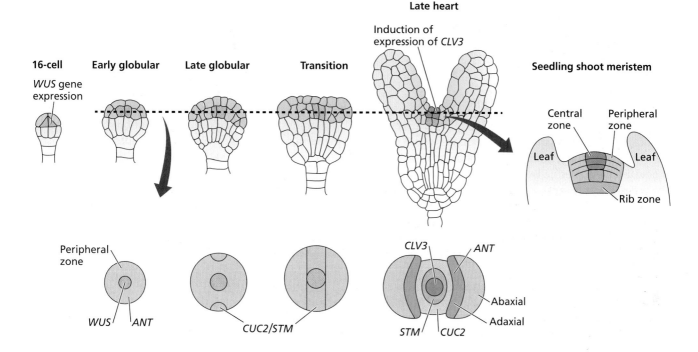

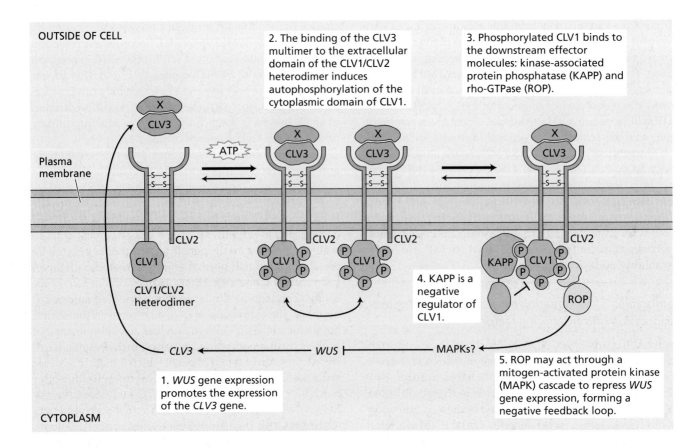

OUTSIDE OF CELL

2. The binding of the CLV3 multimer to the extracellular domain of the CLV1/CLV2 heterodimer induces autophosphorylation of the cytoplasmic domain of CLV1.

3. Phosphorylated CLV1 binds to the downstream effector molecules: kinase-associated protein phosphatase (KAPP) and rho-GTPase (ROP).

Plasma membrane

CLV1/CLV2 heterodimer

4. KAPP is a negative regulator of CLV1.

1. *WUS* gene expression promotes the expression of the *CLV3* gene.

5. ROP may act through a mitogen-activated protein kinase (MAPK) cascade to repress *WUS* gene expression, forming a negative feedback loop.

CYTOPLASM

FIGURE 16.26 Model of the CLAVATA1/CLAVATA2 (CLV1/CLV2) receptor kinase signaling cascade, which forms a negative feedback loop with the *WUS* gene. (After Clark 2001.)

becomes focused in a central circular domain. This expression depends on *CUC* gene activities, since *STM* expression does not occur in *cuc* mutant embryos.

The role that auxin might play in determining the early establishment of WUS is not completely clear; however, studies in which embryonic auxin signaling has been perturbed show changes in the expression levels of several related WOX genes, suggesting that the patterning of genes within this class is influenced by auxin (Haecker et al. 2004).

### Negative feedback limits apical meristem size

Given the ongoing recruitment of initial derivatives into various tissues and organs of the shoot, we would expect that some underlying mechanism balances the production of new undetermined cells with their recruitment into various types of tissues and organs. Analysis of the *CLAVATA* (*CLV*) 1, 2, and 3 genes has revealed at least part of such a mechanism and has suggested the involvement of a combination of positive and negative feedback regulation.

All three *CLV* genes were first described in Arabidopsis in terms of their mutant phenotypes, in which the SAM was grossly enlarged, leading to an easily recognizable increase in the number of lateral organs produced, especially floral organs (Sharma et al. 2003). The similar loss-of-function phenotypes for all three genes suggested that

their protein products normally function interdependently to limit meristem size, a prediction that has been supported by biochemical and molecular analyses. Molecular cloning revealed that all three *CLV* genes encode components of a membrane-spanning receptor kinase signaling complex (**FIGURE 16.26**; also see Chapter 14 for further information about receptor kinase signaling pathways).

*CLV1* encodes a leucine-rich repeat (LRR) kinase, which contains an extracellular ligand-binding domain, a membrane-spanning domain, and an intracellular kinase domain. Plants are known to contain hundreds of different genes that encode such proteins, which probably enable cellular responses to various types of extracellular signals. *CLV2* encodes a closely related protein that lacks the kinase domain. The leucine-rich domains of both proteins are thought to enable the formation of a heterodimer complex between *CLV1* and *CLV2*, which appears essential for the kinase activity of *CLV1* (see Figure 16.26).

The kinase activity of the CLV1/CLV2 complex also requires a small secreted protein encoded by the *CLV3* gene, which probably functions as a ligand to the CLV1/CLV2 complex (Rojo et al. 2002). Biochemical analyses have

**FIGURE 16.27** Model of the feedback loop that maintains initial cells in the SAM.

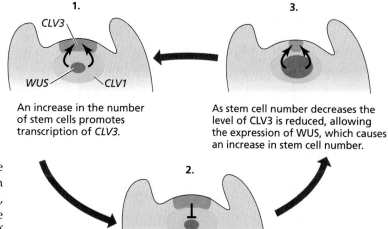

**1.** An increase in the number of stem cells promotes transcription of *CLV3*.

**2.** CLV3, a small peptide, binds to CLV1 and suppresses the expression of WUS. WUS is required for the maintenance of stem cell number.

**3.** As stem cell number decreases the level of CLV3 is reduced, allowing the expression of WUS, which causes an increase in stem cell number.

shown that, once synthesized, the water-soluble CLV3 protein is secreted into the apoplast. Given its small size (11 kDa) and its hydrophilic nature, CLV3 can diffuse relatively easily through the apoplast to activate surface-bound receptors of nearby cells, thus suppressing processes that would otherwise lead to abnormal increases in meristem size.

How does the kinase activity of the CLV complex suppress meristem growth? According to one well-supported model, CLV activity triggers a signaling cascade that ultimately represses *WUS*, a key gene required for ongoing cell division in the SAM (**FIGURE 16.27**) (Schoof et al. 2000). Consistent with this model, *WUS* transcription is increased in *clv* mutants, which leads to the increase in the size of the SAM. Conversely, *WUS* transcription can be completely suppressed by overexpression of *CLV3*, leading to a phenocopy of the *wus* mutant phenotype, in which the meristem is lost. Thus, although the complete set of elements of this signaling cascade has not been defined, it is evident that CLV kinase activity suppresses *WUS* transcription.

The foregoing observations suggest a mechanism that limits the size of the meristem, but leave open the question of how the meristem-limiting activity of CLV itself is regulated. Molecular analyses offer an explanation by showing that the transcription of the *CLV3* gene is promoted by WUS. Thus *WUS* expression not only stimulates the activity of the apical initials and promotes meristem growth, but also induces an increase in CLV activity, which in turn feeds back and limits *WUS* transcription. This system of feedback regulation provides a simple but elegant means of maintaining a dynamic equilibrium between the production of new undetermined cells and their recruitment into differentiated tissues and organs.

### Similar mechanisms maintain initials in the RAM and in the SAM

Parallels between the behavior of initial cells contained within the SAM and within the RAM raise the question of whether their regulation relies on similar mechanisms. Several lines of evidence support this hypothesis. First, as noted in our earlier discussion, the activities of initials in both the SAM and the RAM depend on the activity of

WOX class transcription factors: *WUS* promotes the activity of initials in the SAM, and *WOX5* contributes to the function of initials in the RAM. More recent experiments involving artificial expression of *WOX5* and *WUS* have demonstrated that these genes are functionally interchangeable if expressed in the appropriate tissue (Sarkar et al. 2007). Finally, experiments have shown that small proteins related to CLV3 have the potential to repress initial activity in the RAM when overexpressed (Fiers et al. 2005). The effect of these peptides can be likened to that of CLV3, whose binding to CLV1/CLV2 suppresses the SAM through its repression of *WUS*.

In addition to promoting the transcription of genes such as *CLV3*, which act as part of a self-limiting feedback mechanism, WUS is also expected to regulate genes that either help maintain initials in an undetermined state or promote the division of these cells. One strategy for identifying such genes involves overexpressing *WUS*, then using microarrays to reveal genes that show changes in transcript levels (Leibfried et al. 2005). Transcript levels for several type-A ARR cytokinin response regulators, which act to repress cytokinin responses, decreased significantly within 4 hours, suggesting that WUS represses the transcription of these genes directly. This repression is seen even when protein synthesis is inhibited by cycloheximide, suggesting that WUS represses *ARR* genes directly, rather than relying on the activation of an intermediate transcription factor. The direct interaction of WUS with the promoters of *ARR* genes is also supported by experiments in which complexes between the WUS protein and *ARR7* promoter sequences were detected with antibodies to WUS. Complementary genetic approaches, in which artificial overexpression of *ARR7* resulted in a *wus*-like phenotype, provided

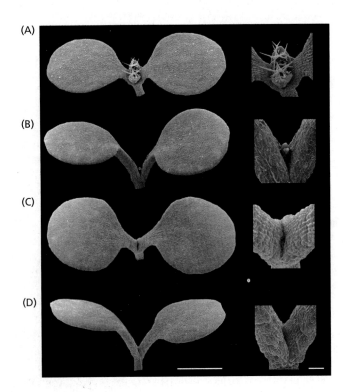

(A)

(B)

(C)

(D)

**FIGURE 16.28** Overexpression of a type-A ARR cytokinin response regulator (*ARR7*) in Arabidopsis phenocopies *wus*. (A) A weakly expressing line has a wild-type morphology. (B) A more highly expressing line has an intermediate phenotype. (C) A strongly expressing line has a phenotype very similar to that of *wus*. (D) A *wus* mutant seedling. (Scale bars 1 mm for seedlings and 100 μm for meristem insets.) (From Leibfried et al. 2005.)

further evidence for a role of cytokinin in maintaining the SAM (**FIGURE 16.28**).

A further link between cytokinin signaling and the SAM has been provided by analyses of STM. Although plants that are partly deficient for STM activity have difficulty in maintaining the SAM, this instability can be overcome by application of cytokinin, suggesting that STM might contribute to cytokinin-dependent signaling. This hypothesis is supported by molecular analyses, which show that STM promotes the transcription of genes encoding isopentenyl transferases that mediate a key step in the synthesis of cytokinins (Jasinski et al. 2005; Yanai et al. 2005).

In summary, several lines of evidence implicate cytokinin-dependent signaling processes in maintaining undetermined initials in both the SAM and the RAM. Moreover, the apparently interchangeable character of genes such as *WUS* and *WOX5* in the regulation of these processes suggests that both the root and shoot may have evolved from a common indeterminate structure.

## Vegetative Organogenesis

Although embryogenesis plays a critical role in establishing the basic polarity and growth axes of the plant, many other aspects of plant form reflect vegetative growth processes. For most plants, shoot architecture depends critically on the regulated production of determinate lateral organs, such as leaves, as well as the regulated formation and outgrowth of indeterminate branch systems. Root systems, though typically hidden from view, have com-

parable levels of complexity that result from the regulated formation and outgrowth of indeterminate lateral roots. In the following sections we consider the molecular nature of the mechanisms that underpin these growth patterns. Like embryogenesis, vegetative organogenesis relies on regional differences in hormone activities, which trigger complex transcription-based programs of gene expression that drive specific aspects of organ development.

### Localized zones of auxin accumulation promote leaf initiation

A long-standing question in plant biology is how the characteristic arrangement of leaves on the shoot, or **phyllotaxy**, is achieved. Three basic phyllotactic patterns, termed alternate, decussate (opposite), and spiral, can be directly linked to the pattern of initiation of leaf primordia on the shoot apical meristem (**FIGURE 16.29**). This pattern depends on a number of factors, including intrinsic factors that determine the characteristic phyllotaxy of a species. Environmental factors or mutations (e.g., *abphyl1* and *clv*) that lead to changes in meristem size or shape can also affect phyllotaxy, suggesting that position-dependent mechanisms play important roles. Position dependency is also supported by classic experiments in which surgical cuts to the shoot apex were shown to perturb the positioning of nearby leaf primordia (Steeves and Sussex 1989).

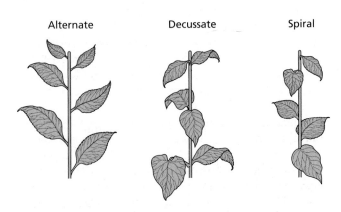

Alternate     Decussate     Spiral

**FIGURE 16.29** Three types of leaf arrangements (phyllotactic patterns) along the shoot axis. The same terms are also used for inflorescences and flowers.

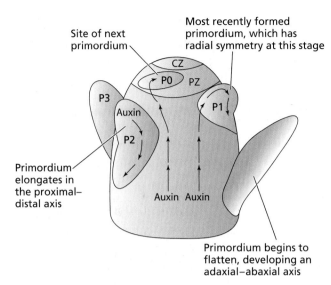

Site of next primordium

Most recently formed primordium, which has radial symmetry at this stage

CZ
P0
PZ
P3
P1
Auxin
P2
Auxin   Auxin

Primordium elongates in the proximal–distal axis

Primordium begins to flatten, developing an adaxial–abaxial axis

**FIGURE 16.30** Sites of leaf formation are related to patterns of polar auxin transport. Patterns of auxin movement (arrows) can be inferred from asymmetric localization of PIN proteins. P0, P1, P2, and P3 refer to the ages of leaf primordia; P0 corresponds to the stage at which the leaf begins its overt development, and P1, P2, and P3 represent increasingly older leaves. Leaf primordia are initiated where auxin accumulates. Acropetal (toward tip) movement of auxin is blocked at the boundary separating the central and peripheral zones (CZ and PZ, respectively), leading to increased auxin levels at this position and the initiation of a leaf (P0). The newly formed leaf primordium (P1) acts as an auxin sink, thus preventing initiation of new leaves directly above it. The displacement of a more mature leaf away from the PZ allows acropetal auxin movements to become reestablished, thus enabling the initiation of another leaf. (After Reinhardt et al. 2003.)

Studies involving experimental manipulation of the shoot apex in Arabidopsis and tomato have shown that auxin can influence the positions of leaves. For example, leaf primordia can be induced to form at abnormal positions on the shoot apex by applying small quantities of auxin directly to the shoot apical meristem, suggesting that auxin is a key factor in determining the position of leaf initiation (FIGURE 16.30). Further support for this hypothesis is provided by the changes in patterns of leaf initiation that result from experimental applications of auxin transport inhibitors.

Several complementary approaches have now provided compelling evidence that sites of leaf initiation correspond to localized zones of auxin accumulation. Although it is difficult to measure auxin levels directly in such small regions, localized concentration maxima can be inferred from DR5 reporters, whose activity shows a close correspondence to leaf initiation sites. The formation of these maxima can be explained by the asymmetric distribution of PIN proteins in cells, which would mediate the conver-

gence of superficial auxin flows from basal parts of the shoot with downward and lateral flows from the shoot apex (see Figure 16.30) (Reinhardt et al. 2003).

### Spatially regulated gene expression determines the planar form of the leaf

The emergence of a mound or ridge of cells on the flank of the SAM represents just one early step in the formation of a leaf. In addition to the formation of vascular tissues, subsequent development for most leaves features the elaboration of a characteristic planar form, which offers an efficient means to intercept light and promote gas exchange and enables highly regulated patterns of transpiration. How this planar architecture is established presents an intriguing puzzle that has been addressed by a variety of experimental approaches.

The most obvious question relates to how the planar geometry and tissue organization of the leaf arises. Much like the shoot, the leaf can be described by a *proximal–distal* axis—that is, one running from the base to the tip. However, in place of the single radial axis that can be seen in cross sections of the shoot, the leaf has two axes: the first, termed the *lateral* axis, runs from edge to edge across the breadth of the leaf; the second, termed the **adaxial–abaxial** axis (comparable to the "dorsal–ventral axis" applied to animals) runs from the upper to the lower surface of the leaf, with *adaxial* referring to the upper surface of the leaf and *abaxial* to the lower surface.

Developmental analyses reveal the complex patterns of cell division that accompany leaf planar development. The first indications of leaf initiation are periclinal divisions in subepidermal layers in the SAM, creating a bulge that defines the proximal–distal axis of the future leaf. Subsequently, divisions are observed throughout the primordium, but later are confined to more proximal regions as the leaf matures. Although the general distribution of cell division during leaf development is predictable, considerable variation is seen in the contributions that individual cells in the primordium make to the final leaf. The clearest examples of this variation are provided in plant material in which cell lineage relationships are evident (FIGURE 16.31). The variable shapes and sizes of clonal sectors seen in such material are difficult to explain by fixed programs of cell division. Instead, the development of leaves must be controlled by ongoing processes in which positional cues, including those relating to size and shape, fine-tune cell division and expansion to achieve predictable forms.

Although it is clear that the localized accumulation of auxin acts as a trigger to initiate the formation of a leaf, many questions remain regarding how subsequent patterns of polar growth are regulated. Considerable insight into these issues has been gained by analyzing mutants in which specific aspects of leaf growth are perturbed. One of these, the *Antirrhinum* mutant *phantastica* (*phan*), initiates

**FIGURE 16.31** Periclinal chimeras demonstrate that mesophyll tissue has more than a single clonal origin in English ivy (*Hedera helix*). These variegated leaves provide clues to the clonal origins of different leaf tissues. A mutation in a gene essential for chloroplast development occurred in some of the initial cells of the meristem. Cells derived from these mutated meristem cells are white, while cells derived from other, normal meristem cells appear green. (Courtesy of S. Poethig.)

leaves relatively normally; however, subsequent growth across the lateral axis fails to occur, leading to the formation of a cylindrical leaf. This abnormal pattern of lateral axis growth has been linked to abnormal abaxial–adaxial patterning in which the altered leaf displays only abaxial characters (Waites et al. 1998) (**FIGURE 16.32**).

By analogy to mechanisms that control wing growth in *Drosophila*, it has been proposed that the normal lateral growth of the lamina (leaf blade) is induced by the juxtaposition of distinct abaxial and adaxial tissue types. According to this model, the primary function of *PHAN* is to promote the development of tissues with an adaxial identity. Molecular analyses of *PHAN*, and the related Arabidopsis *ASYMMETRIC LEAVES 1* (*AS1*), reveal that

they encode MYB class transcription factors that function, at least in part, to down-regulate the expression of KNOX class genes that play key roles in the SAM.

The down-regulation of KNOX genes in leaf primordia appears essential for normal patterning of the leaf. Failure of this down-regulation disturbs processes that coordinate the normal polar growth of leaves, as exemplified by the abnormal leaves of *phan* and *as1* plants, as well as those of plants in which mutations of KNOX genes themselves prevents down-regulation of their expression in leaves.

In contrast to *PHAN* and *AS1*, which enable specification of adaxial cell types, a distinct class of genes, known as the YABBY gene family, is essential for specification of abaxial identity (Bowman and Eshed 2000). YABBY gene family members encode zinc finger proteins that probably func-

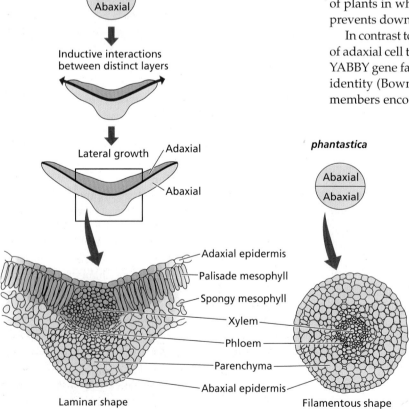

**FIGURE 16.32** A model to account for the lateral growth of the leaf that depends on the juxtaposition of adaxial (*top*) and abaxial (*bottom*) tissues. According to this model, interactions between these two tissue layers are required for lateral growth to occur. In the *phan* mutant, in which the specification of adaxial tissues fails, there is no juxtaposition of distinct tissue types, leading to a failure of lateral growth of the leaf and production of the mutant phenotype.

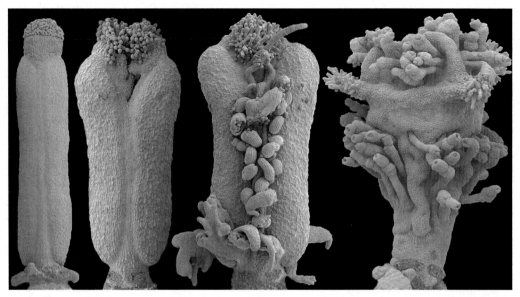

Wild type          crc-1          crc-1 kan1-2                    kan1-2 kan2-1

**FIGURE 16.33** The establishment of adaxial–abaxial po-larity requires YABBY and KANADI gene activities. A comparison of wild-type seed pods with those of single and double mutants of members of these two gene families illustrates how normal organ polarity depends on these gene activities. The mutant *crabs claw* (*crc*) has a defect in a YABBY gene. In mutants abaxial tis-sues (the outer surface of the seed pod, or silique) are replaced by placental tissues, including ovules, that are normally found on the inner face of the carpels that form the silique. (From Bowman et al. 2002.)

tion as transcription factors. The first member of this gene family identified, *CRABS CLAW* (*CRC*), was defined by the phenotype of its Arabidopsis loss-of-function mutant, in which the organization of the carpels is disturbed (**FIGURE 16.33**).

The more general activity of Arabidopsis YABBY genes is observed when mutations affecting several members of this family are combined. These multiple mutants show defects in both floral and vegetative leaflike organs in which abaxial characters have been replaced by adaxial characters, suggesting that there is functional redundancy among members of the YABBY gene family. The abaxial-promoting activity of YABBY genes is further supported by the phenotypes of plants in which YABBY genes are overexpressed, causing the ectopic formation of abaxial tissues.

A second class of genes required for specification of abaxial cell identity is represented by the KANADI genes (Eshed et al. 2004). Like the YABBY family, the genes of the KANADI family appear to have overlapping functions, with the most obvious loss of abaxial identity observed when loss-of-function mutations are combined (see Figure 16.33). Abnormal formation of abaxial tissues is observed when KANADI genes are overexpressed.

A third class of transcription factor genes involved in specifying abaxial/adaxial identity is the Class III HD-Zip proteins (see **WEB TOPIC 16.5**).

## Distinct mechanisms initiate roots and shoots

As we have seen, lateral roots begin their development some distance back from the root apical meristem (Malamy and Benfey 1997). Only after the derivatives of the RAM have ceased elongation do lateral root initials first appear. The first histological signs of root initiation are periclinal divisions among a small number of cells in the pericycle (see Chapter 1), leading to further divisions and growth in a plane perpendicular the parent root axis until the new root tip ruptures the endodermal and cortical layers of the parent root. Simultaneously, a new RAM becomes orga-nized at the tip of the lateral root, enabling it to grow in an indeterminate manner (**FIGURE 16.34**). The pattern of lateral root formation and growth can be influenced by intrinsic factors such as auxin concentration as well as by environmental factors such as soil nutrient availability.

Lateral shoots are initiated in a very different manner. Rather than drawing on cells from a single internal layer, lateral shoots incorporate cells from several clonally dis-tinct layers. In most species, the predominant means of shoot branching is the formation of axillary meristems. These are meristems that develop in the axils of leaf pri-mordia, so the pattern of branch formation is directly related to the SAM activities that regulate phyllotaxy. Like that of roots, the growth of lateral branches can be influenced by intrinsic factors such as hormones as well as environmental factors such as light. The phenomenon of

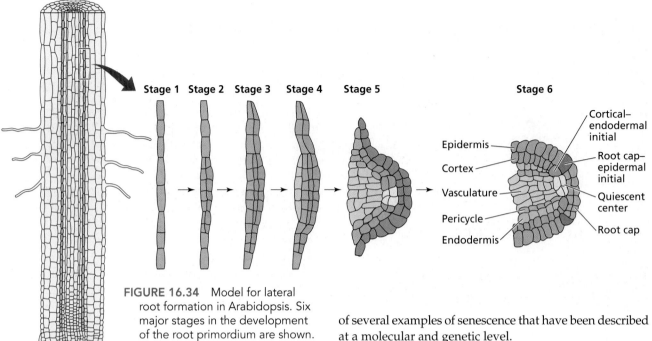

**FIGURE 16.34** Model for lateral root formation in Arabidopsis. Six major stages in the development of the root primordium are shown. The different tissue types are designated by colors. By stage 6, all tissues found in the primary root are present in the typical radial pattern of the lateral root.

**apical dominance**, in which the terminal bud suppresses the growth of the axillary buds, is primarily regulated by auxin from the terminal bud and a second signaling component, strigolactone. We will defer an in-depth discussion of the hormonal control of branching to Chapter 19.

# Senescence and Programmed Cell Death

Every autumn, people who live in temperate regions can enjoy the beautiful color changes that precede the loss of leaves from deciduous trees. So profound are these color changes in forests of such trees that they are apparent even to observers in orbiting spacecraft. These leaves change color because shorter day lengths and cooler temperatures trigger developmental processes that are collectively known as **senescence**. Although it leads to the death of targeted tissues, senescence is distinct from **necrosis**. Whereas necrosis is death brought about by physical damage, poisons, or other external injury, senescence is an energy-dependent developmental process that is controlled by the interaction of environmental factors with intrinsic, genetically regulated programs.

In this last section of the chapter we describe various types of senescence in more detail, and we present evidence that senescence represents a programmed and adaptive aspect of development. We conclude with a discussion of several examples of senescence that have been described at a molecular and genetic level.

## Leaf senescence is adaptive and strictly regulated

Leaf senescence involves the ordered degradation of cellular contents and results in the remobilization of nutrients. During senescence, hydrolytic enzymes break down many cellular proteins, carbohydrates, and nucleic acids. The component sugars, nucleosides, and amino acids are then transported back into the main body of the plant via the phloem, where they will be reused for synthetic processes. Many minerals are also transported out of senescing organs back into the plant. Since senescence redistributes nutrients to growing parts of the plant, it can serve as a survival strategy during environmentally adverse conditions such as drought or temperature stress.

Rather than being a response to stress, senescence may reflect part of a normal developmental program. As new leaves are initiated from the shoot apical meristem, older leaves may be shaded and lose the ability to function efficiently in photosynthesis. Leaf senescence is frequently associated with **abscission**, a process whereby specific cells in the petiole differentiate to form an abscission layer, allowing the senescent organ to separate from the plant. In Chapter 22 we will have more to say about the control of abscission by ethylene. Together, these programs of senescence (triggered by stress or normal development) recover a significant portion of the valuable resources that the plant invested in leaf formation.

The nutrient remobilization that occurs in early senescence is reversible, but advanced stages of senescence are irreversible. Although many materials in senescent organs can be reused in other parts of the plant, senescence means a considerable loss in energy potential. Given these costs, it can be expected that the initiation of senescence is a highly regulated process involving a complex interplay between

**FIGURE 16.35** Monocarpic senescence in soybean (*Glycine max*). The entire plant on the left underwent senescence after flowering and producing fruit (pods). The plant on the right remained green and vegetative because its flowers were continually removed. (Courtesy of L. Noodén.)

environmental and developmental cues. In this section we will examine the roles that senescence plays in plant development. We will see that there are many types of senescence, each with its own genetic program. In Chapters 21 and 22 we will describe how cytokinins and ethylene act as signaling agents that regulate plant senescence.

## Plants exhibit various types of senescence

Senescence occurs in a variety of organs and in response to many different cues. Many annual plants, including major crop plants such as wheat, maize (corn; *Zea mays*), and soybean, abruptly yellow and die following fruit production, even under optimal growing conditions. Senescence of the entire plant after a single reproductive cycle is called **monocarpic senescence** (FIGURE 16.35).

Other types of senescence include the following:

- Senescence of aerial shoots in herbaceous perennials
- Seasonal leaf senescence (as in deciduous trees)
- Sequential leaf senescence (in which leaves die when they reach a certain age)
- Senescence (ripening) of fleshy and of dry fruits
- Senescence of storage cotyledons and floral organs
- Senescence of specialized cell types (e.g., trichomes, tracheids, and vessel elements)

The triggers for the various types of senescence are different. They can be internal, such as reproductive processes in monocarpic senescence, or external, such as day length and temperature in the autumnal leaf senescence of deciduous trees. Regardless of the initial stimulus, the different senescence patterns may share common internal programs in which a regulatory senescence gene initiates a cascade of secondary gene expression that eventually brings about senescence.

Because of its effect on agriculture and storage of agricultural products, the regulation of plant senescence has received considerable attention in the past years. Here, the model plant Arabidopsis has served as an important model species. The genetic aspects of senescence regulation have been investigated using two complementary approaches: (1) the study of genes that are up-regulated during or prior to the onset of senescence, and (2) the analysis of genes that cause an altered senescence syndrome when mutated or ectopically expressed.

The expression of many genes changes dramatically upon the initiation of senescence. Not surprisingly, the expression of photosynthesis-related genes decreases, but at the same time the expression of thousands of other genes increases. Such genes, called senescence-associated genes (SAGs), include regulatory genes such as transcription factors and genes involved in degradative processes. SAGs include genes that encode hydrolytic enzymes, such as proteases, ribonucleases, and lipases, as well as enzymes involved in the biosynthesis of ethylene, such as l-aminocyclopropane-l-carboxylic acid (ACC) synthase and ACC oxidase (see Chapter 22). Other SAGs have secondary functions in senescence. These genes encode enzymes involved in the conversion or remobilization of breakdown products, such as glutamine synthetase, which catalyzes the conversion of ammonium into glutamine (see Chapter 12) and is responsible for nitrogen recycling from senescing tissues. Pectinases play important roles in cell wall breakdown during leaf abscission and fruit ripening.

Several key findings illustrate the complexity of senescence regulation:

- **Environmental cues** such as changing day length and temperature as a result of changing seasons affect senescence in deciduous trees, while stresses, including pathogen and temperature stress, induce leaf senescence in many plant species.
- **Hormones** control senescence. Ethylene has long been known to induce senescence, while cytokinin

is a potent inhibitor. Mutants that have a defect in ethylene perception generally show delayed senescence. Likewise, plants with increased cytokinin sensitivity show delayed senescence. In addition, most other plant hormones affect senescence to some extent.

- **Oxidative stress** can damage proteins and cell structures. The accumulation of cellular damage has been proposed to cause senescence in a number of species. However, plants have a large capacity for detoxifying reactive oxygen species, so oxidative stress more likely functions as a signal than a direct cause of deterioration and subsequent senescence.

- **Metabolic status** regulates leaf senescence. Sugars accumulate during senescence, and the sugar sensor hexokinase contributes to this process.

- **Macromolecule degradation** can affect senescence in multiple ways. Arabidopsis mutants with a defective ubiquitin-dependent proteolysis pathway show delayed senescence. This finding suggests that senescence may be induced when factors that inhibit the senescence syndrome have been degraded by targeted proteolysis. The senescence syndrome itself involves massive macromolecular degradation after senescence has been initiated.

- **Intrinsic developmental factors** prevent leaf senescence during early developmental stages and induce senescence during late development. Thus senescence depends on *age-related changes*. Even in environmentally adverse conditions or

after treatment with the senescence-inducing hormone ethylene, young leaves will not senesce. Conversely, no mutants have been isolated, and no environmental conditions have been found, in which plant leaves stay functionally green forever.

**FIGURE 16.36** summarizes these findings in a conceptual framework that helps explain the regulation of leaf senescence.

Environmental cues and stress, as well as metabolite and hormone levels, can affect the initiation of leaf senescence, but do not do so by default. Only a combination of factors, or a sufficiently high level of a single signal, will cause the induction of senescence. Thus plants must have a system in place that integrates relevant signals and transduces the senescence signal to the next level, where it is tested against intrinsic developmental factors and whether certain age-related changes have taken place (i.e., the leaf has reached a certain age). Similarly, senescence is induced after the plant has reached a certain age or, in the case of monocarpic senescence, once seed loading commences.

The model provided in Figure 16.36 clarifies how the leaf can integrate multiple cues and is therefore able to make an informed decision on whether it will be more beneficial to sacrifice itself and allow for the remobilization of nutrients, or to continue to be photosynthetically active and contribute to plant growth by producing assimilates. Major questions that need to be resolved are what the intrinsic developmental factors or age-related changes are, and how a leaf integrates the various environmental and developmental cues to regulate senescence.

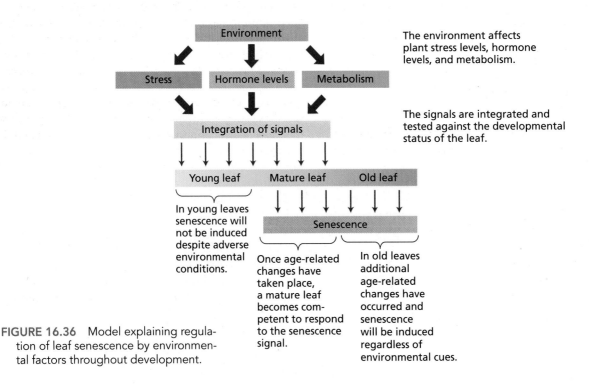

**FIGURE 16.36** Model explaining regulation of leaf senescence by environmental factors throughout development.

## Senescence involves the ordered degradation of potentially phototoxic chlorophyll

Senescence follows a predictable course of cellular events. On the cytological level, some organelles are destroyed while others remain active. The chloroplast is the first organelle to deteriorate during the onset of leaf senescence as its thylakoid protein components and stromal enzymes are destroyed. This ordered chloroplast degradation requires a functional cell, and nuclei remain structurally and functionally intact until the late stages of senescence. The degradation of the chloroplast is significant in terms of nutrient remobilization, since chlorophyll binding proteins contribute about 20% of total cellular nitrogen. However, the dismantling of the chloroplast also releases potentially phototoxic chlorophyll. Thus a mechanism that carefully disposes of the chlorophyll is required.

The elucidation of the chlorophyll degradation pathway has been facilitated by the availability of so-called *stay-green mutants*. Two fundamentally different types of stay-green mutants have been identified: Functional stay-green mutants retain their photosynthetic capacity and green color longer than the corresponding wild type. By contrast, nonfunctional stay-green mutants remain green for longer than the wild type, but dismantling of the cellular contents is carried out in the same way as in the wild type. It was soon recognized that this latter type of mutant may possess a dysfunctional chlorophyll degradation pathway. The maize *lls1* (for *lethal leaf spot 1*) and Arabidopsis *acd1* (for *accelerated cell death 1*) mutants were originally thought to have defects relating to disease resistance because they had lesions of a type usually induced only in response to pathogens. However, the mutant leaves have a stay-green phenotype when incubated in the dark. Molecular cloning revealed that the mutated genes encode an enzyme involved in chlorophyll breakdown (see WEB TOPIC 16.6).

## Programmed cell death is a specialized type of senescence

Senescence can occur at the level of the whole plant, as in monocarpic senescence; at the organ level, as in leaf senescence; and at the cellular level, as in tracheary element differentiation. The process whereby individual cells activate an intrinsic senescence program is called **programmed cell death** (**PCD**). PCD plays an important part in the development of animals, in which its molecular mechanism has been studied extensively. PCD can be initiated by specific developmental signals or by potentially lethal events, such as pathogen attack or errors in DNA replication during cell division. It involves the expression of a characteristic set of genes that orchestrate the dismantling of cellular components, ultimately resulting in cell death. PCD in animals is usually accompanied by a distinct set of morphological and biochemical changes called **apoptosis** (plural *apoptoses*) (from a Greek word meaning "falling off," as in autumn leaves). During apoptosis the cell nucleus condenses, and the nuclear DNA fragments in a specific pattern caused by degradation of the DNA between nucleosomes (see Chapter 1).

In contrast to animal systems, little is known about PCD in plants (Pennell and Lamb 1997). Some plant cells, particularly in senescing tissues, exhibit cytological changes similar to those observed in animals. PCD also appears to occur during the differentiation of xylem tracheary elements, during which the nuclei and chromatin degrade and the cytoplasm disappears. These changes result from the activation of genes that encode nucleases and proteases.

One of the important functions of PCD in plants is protection against pathogens. When a pathogen infects a plant, cells at the site of the infection quickly accumulate high concentrations of toxic phenolic compounds and die. The dead cells form a small circular island called a **necrotic lesion**, which isolates and prevents the infection from spreading to surrounding healthy tissues. This rapid, localized cell death due to pathogen attack is called the *hypersensitive response* (see Chapter 13). The existence of Arabidopsis mutants that can mimic the effect of infection and trigger the entire cascade of events leading to the formation of necrotic lesions, even in the absence of a pathogen, has demonstrated that the hypersensitive response is a genetically programmed process rather than simple necrosis.

## SUMMARY

Position-dependent mechanisms often direct plant development, such as the establishment of polarity in the developing embryo, maintenance of the root and stem meristems, and formation of leaves.

### Overview of Plant Growth and Development

- Meristematic cells are undetermined and are central to plant growth and development (**Figure 16.1**).

- There are three major stages of plant development: embryogenesis, vegetative development, and reproductive development (**Figure 16.2**).

### Embryogenesis: The Origins of Polarity

- Among seed plants, the apical–basal polarity is established early in embryogenesis (**Figures 16.3, 16.4**).

- Position-dependent signaling guides embryogenesis (**Figure 16.5**). Arabidopsis mutants demonstrate that some process other than a fixed sequence of cell division must guide radial pattern formation (**Figure 16.6**).

- Auxin (indole-3-acetic acid) may function as a mobile chemical signal during embryogenesis (**Figures 16.7, 16.8; Table 16.1**).

- Screens for *seedling defective* mutants reveal genes and functions required for establishing the normal apical–basal pattern (**Figure 16.9**).

- Radial patterning guides formation of tissue layers (**Figure 16.10**). Two Arabidopsis genes establish normal epidermal identity (**Figure 16.11**). Different genes establish internal tissues, including the vasculature and cortex (**Figures 16.12–16.14**).

### Meristematic Tissues: Foundations for Indeterminate Growth

- The root and shoot apical meristems use similar strategies to enable indeterminate growth.

- The organization and maintenance of apical meristems can be linked to regulated patterns of hormone activity.

### The Root Apical Meristem (RAM)

- The origin of different root tissues can be traced to four types of initial cells (**Figure 16.17**).

- Driven by PIN proteins, auxin moves down the shoot and becomes concentrated in the QC and columella tissues (**Figure 16.18**).

- High concentrations of auxin in the QC trigger the activation of a series of transcription factors (**Figure 16.19**).

- Cytokinin acting in opposition to auxin establishes the apical–basal identity of the two cells arising from the hypophysis (**Figure 16.20**).

### The Shoot Apical Meristem (SAM)

- The shoot apical meristem has a structure distinct from that of the root apical meristem (**Figures 16.16, 16.21, 16.22**).

- Shoot tissues are derived from several discrete sets of apical initials (**Figure 16.23**).

- PIN proteins determine auxin levels across the SAM, leading to auxin flows away from initials and triggering formation of leaf primordia (**Figure 16.24**). Embryonic SAM formation requires the coordinated expression of transcription factors and a feedback loop to promote cell proliferation (**Figure 16.25**).

- The CLV protein kinase acts to limit cell proliferation by repressing *WUS* (**Figures 16.26, 16.27**).

- In both RAM and SAM, cytokinin helps maintain a stable population of undetermined initial cells (**Figure 16.28**).

### Vegetative Organogenesis

- Phyllotactic patterns are directly linked to the pattern of leaf formation (**Figures 16.29, 16.30**).

- Spatially regulated gene expression of transcription factors determines the planar form of the leaf (**Figure 16.32**).

- Members of the YABBY and KANADI gene families are essential for specification of abaxial identity (**Figure 16.33**).

- Distinct mechanisms initiate roots and shoots (**Figure 16.34**).

### Senescence and Programmed Cell Death

- Senescence and programmed cell death are regulated and orderly processes in which cytological and biochemical events prepare to recycle critical nutrients for the benefit of future development.

- Plants exhibit a variety of senescence phenomena (**Figure 16.35**).

- Many signals, both external and internal, are integrated to regulate senescence (**Figure 16.36**).

# WEB MATERIAL

## Web Topics

**16.1 Embryonic Dormancy**
The ability of seeds to lie dormant for long periods and then germinate under favorable conditions reflects the activity of complex physiological programs.

**16.2 Rice Embryogenesis**
Embryogenesis in rice is typical of that found in most monocots, and is distinct from that of Arabidopsis.

**16.3 Polarity of *Fucus* Zygotes**
A wide variety of external gradients can polarize the growth of cells that are initially apolar.

**16.4 *Azolla* Root Development**
Anatomical studies of the root of the aquatic fern *Azolla* have provided insights into cell fate during root development.

**16.5 Class III HD-Zip Transcription Factors Promote Adaxial Development through a microRNA-Sensitive Mechanism**
Molecular genetic analyses of an important family of transcription factors have clarified their role in establishing adaxial identity.

**16.6 During Senescence Photoactive Chlorophyllide Is Converted into a Colorless Chlorophyll Catabolite**
The biochemistry of chlorophyll breakdown is described.

## Web Essays

**16.1 Division Plane Determination in Plant Cells**
Plant cells appear to utilize mechanisms different from those used by other eukaryotes to control their division planes.

**16.2 Plant Meristems: A Historical Overview**
Scientists have used many approaches to unraveling the secrets of plant meristems.

# CHAPTER REFERENCES

Abe, M., Katsumata, H., Komeda, Y., and Takahashi, T. (2003) Regulation of shoot epidermal cell differentiation by a pair of homeodomain proteins in Arabidopsis. *Development* 130: 635–643.

Aida, M., Ishida, T., Fukaki, H., Fujisawa, H., and Tasaka, M. (1997) Genes involved in organ separation in Arabidopsis: An analysis of the cup-shaped cotyledon mutant. *Plant Cell* 9: 841–857.

Aida, M., Beis, D., Heidstra, R., Willemsen, V., Blilou, I., Galinha, C., Nussaume, L., Noh, Y.-S., Amasino, R., and Scheres, B. (2004) The *PLETHORA* genes mediate patterning of the Arabidopsis root stem cell niche. *Cell* 119: 109–120.

Baud, S., Bellec, Y., Miquel, M., Bellini, C., Caboche, M., Lepiniec, L., Faure, J. D., and Rochat, C. (2004) *gurke* and *pasticcino3* mutants affected in embryo development are impaired in acetyl-CoA carboxylase. *EMBO Rep.* 5: 515–520.

Berleth, T., and Jürgens, G. (1993) The role of the *MONOPTEROS* gene in organising the basal body region of the Arabidopsis embryo. *Development* 118: 575–587.

Blilou, I., Xu, J., Wildwater, M., Willemsen, V., Paponov, I., Friml, J., Heidstra, R., Aida, M., Palme, K., and Scheres, B. (2005) The PIN auxin efflux facilitator network controls growth and patterning in Arabidopsis roots. *Nature* 433: 39–44.

Bowman, J. L., and Eshed, Y. (2000) Formation and maintenance of the shoot apical meristem. *Trends Plant Sci.* 5: 110–115.

Bowman, J. L., Eshed, Y., and Baum, S. F. (2002) Establishment of polarity in angiosperm lateral organs. *Trends Genet.* 18: 134–141.

Bruck, D. K., and Walker, D. B. (1985) Cell determination during embryogenesis in *Citrus jambhiri*. I. Ontogeny of the epidermis. *Bot. Gaz.* 146: 188–195.

Chandler, J. W., Cole, M., Flier, A., Grewe, B., and Werr, W. (2007) The AP2 transcription factors DORNROSCHEN and DORNROSCHEN-LIKE redundantly control Arabidopsis embryo patterning via interaction with PHAVOLUTA. *Development* 134: 1653–1662.

Clark, S. E. (2001) Cell signaling at the shoot meristem. *Nat. Rev. Mol. Cell Biol.* 2: 276–284.

Clouse, S. D. (2002) Arabidopsis mutants reveal multiple roles for sterols in plant development. *Plant Cell* 14: 1995–2000.

Di Laurenzio, L., Wysocka-Diller, J., Malamy, J. E., Pysh, L., Helariutta, Y., Freshour, G., Hahn, M. G., Feldmann, K. A., and Benfey, P. N. (1996) The *SCARECROW* gene regulates an asymmetric cell division that is essential for generating the radial organization of the Arabidopsis root. *Cell* 86: 423–433.

Esau, K. (1965) *Plant Anatomy*. John Wiley and Sons, New York.

Emery, J. F., Floyd, S. K., Alvarez, J., Eshed, Y., Hawker, N. P., Izhaki, A., Baum, S. F., and Bowman, J. L. (2003) Radial patterning of Arabidopsis shoots by class III HD-ZIP and KANADI genes. *Curr. Biol.* 13: 1768–1774.

Eshed, Y., Izhaki, A., Baum, S. F., Floyd, S. K., and Bowman, J. L. (2004) Asymmetric leaf development and blade expansion in Arabidopsis are mediated by KANADI and YABBY activities. *Development* 131: 2997–3006.

Fiers, M., Golemiec, E., Xu, J., van der Geest, L., Heidstra, R., Stiekema, W., and Liu, C.-M. (2005) The 14-amino acid CLV3, CLE19, and CLE40 peptides trigger consumption of the root meristem in Arabidopsis through a CLAVATA2-dependent pathway. *Plant Cell* 17: 2542–2553.

Friml, J., Vieten, A., Sauer, M., Weijers, D., Schwarz, H., Hamann, T., Offringa, R., and Jürgens, G. (2003) Efflux-dependent auxin gradients establish the apical-basal axis of Arabidopsis. *Nature* 426: 147–153.

Galinha, C., Hofhuis, H., Luijten, M., Willemsen, V., Blilou, I., Heidstra, R., and Scheres, B. (2007) PLETHORA proteins as dose-dependent master regulators of Arabidopsis root development. *Nature* 449: 1053–1057.

Gälweiler, L., Guan, C., Müller, A., Wisman, E., Mendgen, K., Yephremov, A., and Palme, K. (1998) Regulation of polar auxin transport by AtPIN1 in Arabidopsis vascular tissue. *Science* 282: 2226–2230.

Geldner, N., Anders, N., Wolters, H., Keicher, J., Kornberger, W., Muller, P., Delbarre, A., Ueda, T., Nakano, A., and Jürgens, G. (2003) The Arabidopsis GNOM ARF-GEF mediates endosomal recycling, auxin transport, and auxin-dependent plant growth. *Cell* 112: 219–230.

Gifford, E. M., and Foster, C. E. (1987) *Morphology and Evolution of Vascular Plants*. W. H. Freeman and Company, New York.

Grieneisen, V. A., Xu, J., Marée, A. F., Hogeweg, P., and Scheres, B. (2007) Auxin transport is sufficient to generate a maximum and gradient guiding root growth. *Nature* 449: 1008–1013.

Hardtke, C. S., and Berleth, T. (1998) The Arabidopsis gene *MONOPTEROS* encodes a transcription factor mediating embryo axis formation and vascular development. *EMBO J.* 17: 1405–1411.

Haecker, A., Gross-Hardt, R., Geiges, B., Sarkar, A., Breuninger, H., Herrmann, M., and Laux, T. (2004) Expression dynamics of *WOX* genes mark cell fate decisions during early embryonic patterning in *Arabidopsis thaliana*. *Development* 131: 657–668.

Hamann, T., Benkova, E., Bäurle, I., Kientz, M., and Jürgens, G. (2002) The Arabidopsis *BODENLOS* gene encodes an auxin response protein inhibiting MONOPTEROS-mediated embryo patterning. *Genes Dev.* 16: 1610–1615.

Haywood, V., Kragler, F., and Lucas, W. J. (2002) Plasmodesmata: Pathways for protein and ribonucleoprotein signaling. *Plant Cell* 14: S303–325.

Helariutta, Y., Fukaki, H., Wysocka-Diller, J., Nakajima, K., Jung, J., Sena, G., Hauser, M. T., and Benfey, P. N. (2000) The *SHORT-ROOT* gene controls radial patterning of the Arabidopsis root through radial signaling. *Cell* 101: 555–567.

Inoue, T., Higuchi, M., Hashimoto, Y., Seki, M., Kobayashi, M., Kato, T., Tabata, S., Shinozaki, K., and Kakimoto, T. (2001) Identification of CRE1 as a cytokinin receptor from Arabidopsis. *Nature* 409: 1060–1063.

Itoh, J., Nonomura, K., Ikeda, K., Yamaki, S., Inukai, Y., Yamagishi, H., Kitano, H., and Nagato, Y. (2005) Rice plant development: From zygote to spikelet. *Plant Cell Physiol.* 46: 23–47.

Izhaki, A., and Bowman, J. L. (2007) KANADI and Class III HD-Zip gene families regulate embryo patterning and modulate auxin flow during embryogenesis in Arabidopsis. *Plant Cell* 19: 495–508.

Jasinski, S., Piazza, P., Craft, J., Hay, A., Woolley, L., Rieu, I., Phillips, A., Hedden, P., and Tsiantis, M. (2005) *KNOX* action in Arabidopsis is mediated by coordinate regulation of cytokinin and gibberellin activities. *Curr Biol.* 15: 1560–1565.

Jenik, P. D., and Barton, M. K. (2005) Surge and destroy: The role of auxin in plant embryogenesis. *Development* 132: 3577–3585.

Jiang, K., Meng, Y. L., and Feldman, L. J. (2003) Quiescent center formation in maize roots is associated with an auxin-regulated oxidizing environment. *Development* 130: 1429–1438.

Kidner, C. A., and Martienssen, R. A. (2005) The developmental role of microRNA in plants. *Curr. Opin. Plant Biol.* 8: 38–44.

Kim, I., Kobayashi, K., Cho, E., and Zambryski, P. C. (2005) Subdomains for transport via plasmodesmata corresponding to the apical-basal axis are established during Arabidopsis embryogenesis. *Proc. Natl. Acad. Sci. USA* 102: 11945–11950.

Kuhlemeier, C., and Reinhardt, D. (2001) Auxin and phyllotaxis. *Trends Plant Sci.* 6: 187–189.

Laux, T., Würschum, T., and Breuninger, H. (2004) Genetic regulation of embryonic pattern formation. *Plant Cell* 16: S190–202.

Leibfried, A., To, J. P. C., Busch, W., Stehling, S., Kehle, A., Demar, M., Kieber, J. J., and Lohmann, J. U. (2005) WUSCHEL controls meristem function by direct regulation of cytokinin-inducible response regulators. *Nature* 438: 1172–1175.

Lenhard, M., Jürgens, G., and Laux, T. (2002) The *WUSCHEL* and *SHOOTMERISTEMLESS* genes fulfil complementary roles in Arabidopsis shoot meristem regulation. *Development* 129: 3195–3206.

Liu, C., Xu, Z., and Chua, N. H. (1993) Auxin polar transport is essential for the establishment of bilateral symmetry during early plant embryogenesis. *Plant Cell* 5: 621–630.

Long, J. A., Moan, E. I., Medford, J. I., and Barton, M. K. (1996) A member of the KNOTTED class of homeodomain proteins encoded by the *STM* gene of Arabidopsis. *Nature* 379: 66–69.

Lu, P., Porat, R., Nadeau, J. A., and O'Neill, D. (1996) Identification of a meristem L1 layer-specific gene in Arabidopsis that is expressed during embryonic pattern formation and defines a new class of homeobox genes. *Plant Cell* 8: 2155–2168.

Mähönen, A. P., Bonke, M., Kauppinen, L., Riikonen, M., Benfey, P. N., and Helariutta, Y. (2000) A novel two-component hybrid molecule regulates vascular morphogenesis of the Arabidopsis root. *Genes Dev.* 14: 2938–2943.

Malamy, J. E., and Benfey, P. N. (1997) Organization and cell differentiation in lateral roots of *Arabidopsis thaliana*. *Development* 124: 33–44.

Mansfield, S. G., and Briarty, L. G. (1991) Early embryogenesis in *Arabidopsis thaliana*: The developing embryo. *Can. J. Bot.* 69: 461–476.

Maule, A. J. (2008) Plasmodesmata: structure, function and biogenesis. *Curr. Opin. Plant Biol.* 11: 680–686.

Mayer, K. F., Schoof, H., Haecker, A., Lenhard, M., Jürgens, G., and Laux, T. (1998) Role of *WUSCHEL* in regulating stem cell fate in the Arabidopsis shoot meristem. *Cell* 95: 805–815.

Mayer, U., Torres-Ruiz, R. A., Berleth, T., Misera, S., and Jürgens, G. (1991) Mutations affecting body organisation in the Arabidopsis embryo. *Nature* 353: 402–407.

Mayer, U., Büttner, G., and Jürgens, G. (1993) Apical–basal pattern formation in the *Arabidopsis* embryo: studies on the role of the *gnom* gene. *Development* 117: 149–162.

Michniewicz, M., Zago, M. K., Abas, L., Weijers, D., Schweighofer, A., Meskiene, I., Heisler, M. G., Ohno, C., Zhang, J., Huang, F., et al. (2007) Antagonistic regulation of PIN phosphorylation by PP2A and PINOID directs auxin flux. *Cell* 130: 1044–1056.

Mukherjee, K., and Bürglin, T. R. (2006) MEKHLA, a novel domain with similarity to PAS domains, is fused to plant homeodomain-leucine zipper III proteins. *Plant Physiol* 140: 1142–1150.

Müller, B., and Sheen, J. (2008) Cytokinin and auxin interaction in root stem-cell specification during early embryogenesis. *Nature* 453: 1094–1097.

Nakajima, K., and Benfey, P. N. (2002) Signaling in and out: control of cell division and differentiation in the shoot and root. *Plant Cell* 14: S265–276.

Nakajima, K., Sena, G., Nawy, T., and Benfey, P. N. (2001) Intercellular movement of the putative transcription factor SHR in root patterning. *Nature* 413: 307–311.

Pennell, R. I., and Lamb, C. (1997) Programmed cell death in plants. *Plant Cell* 9: 1157–1168.

Poethig, R. S. (1987) Clonal analysis of cell lineage patterns in plant development. *Am. J. Bot.* 74: 581–594.

Pružinská, A., Tanner, G., Anders, I., Roca, M., and Hörtensteiner, S. (2003) Chlorophyll breakdown: Pheophorbide a oxygenase is a Rieske-type iron–sulfur protein, encoded by the accelerated cell death 1 gene. *Proc. Natl. Acad. Sci. USA* 100: 15259–15264.

Reinhardt, D., Pesce, E. R., Stieger, P., Mandel, T., Baltensperger, K., Bennett, M., Traas, J., Friml, J., and Kuhlemeier, C. (2003) Regulation of phyllotaxis by polar auxin transport. *Nature* 426: 255–260.

Rinne, P. L. H., and van der Schoot, C. (1998) Symplasmic fields in the tunica of the shoot apical meristem coordinate morphogenetic events. *Development* 125: 1477–1485.

Rojo, E., Sharma, V. K., Kovaleva, V., Raikhel, N. V., and Fletcher, J. C. (2002) CLV3 is localized to the extracellular space, where it activates the Arabidopsis CLAVATA stem cell signaling pathway. *Plant Cell* 14: 969–977.

Sabatini, S., Beis, D., Wolkenfelt, H., Murfett, J., Guilfoyle, T., Malamy, J., Benfey, P., Leyser, O., Bechtold, N., Weisbeek, P., et al. (1999) An auxin-dependent distal organizer of pattern and polarity in the Arabidopsis root. *Cell* 99: 463–472.

Sarkar, A. K., Luijten, M., Miyashima, S., Lenhard, M., Hashimoto, T., Nakajima, K., Scheres, B., Heidstra, R., and Laux, T. (2007) Conserved factors regulate signalling in *Arabidopsis thaliana* shoot and root stem cell organizers. *Nature* 446: 811–814.

Schiefelbein, J. W., Masucci, J. D., and Wang, H. (1997) Building a root: The control of patterning and morphogenesis during root development. *Plant Cell* 9: 1089–1098.

Scheres, B., Wolkenfelt, H., Willemsen, V., Terlouw, M., Lawson, E., Dean, C., and Weisbeek, P. (1994) Embryonic origin of the Arabidopsis primary root and root meristem initials. *Development* 120: 2475–2487.

Schoof, H., Lenhard, M., Haecker, A., Mayer, K. F., Jürgens, G., and Laux, T. (2000) The stem cell population of Arabidopsis shoot meristems in maintained by a regulatory loop between the *CLAVATA* and *WUSCHEL* genes. *Cell* 100: 635–644.

Schrick, K., Nguyen, D., Karlowski, W. M., and Mayer, K. F. (2004) START lipid/sterol-binding domains are amplified in plants and are predominantly associated with homeodomain transcription factors. *Genome Biol.* 5: R41.

Sharma, V. K., Carles, C., and Fletcher, J. C. (2003) Maintenance of stem cell populations in plants. *Proc. Natl. Acad. Sci. USA* 100: 11823–11829.

Steeves, T. A., and Sussex, I. M. (1989) *Patterns in Plant Development*. Cambridge University Press, Cambridge.

Steinmann, T., Geldner, N., Grebe, M., Mangold, S., Jackson, C. L., Paris, S., Gälweiler, L., Palme, K., and Jürgens, G. (1999) Coordinated polar localization of auxin efflux carrier PIN1 by GNOM ARF GEF. *Science* 286: 316–318.

Torres-Ruiz, R. A., and Jürgens, G. (1994) Mutations in the *FASS* gene uncouple pattern formation and morphogenesis in Arabidopsis development. *Development* 120: 2967–2978.

Traas, J., Bellini, C., Nacry, P., Kronenberger, J. Bouchez, D., and Caboche, M. (1995) Normal differentiation patterns in plants lacking microtubular preprophase bands. *Nature* 375: 676–677.

Van den Berg, C., Willemsen, V., Hage, W., Weisbeek, P., and Scheres, B. (1995) Cell fate in the Arabidopsis root meristem determined by directional signaling. *Nature* 378: 62–65.

Waites, R., Selvadurai, H. R., Oliver, I. R., and Hudson, A. (1998) The *PHANTASTICA* gene encodes a MYB transcription factor involved in growth and dorsoventrality of lateral organs in *Antirrhinum*. *Cell* 93: 779–789.

Weigel, D., and Jürgens, G. (2002) Stem cells that make stems. *Nature* 415: 751–754.

West, M. A. L., and Harada, J. J. (1993) Embryogenesis in higher plants: An overview. *Plant Cell* 5: 1361–1369.

Yanai, O., Shani, E., Dolezal, K., Tarkowski, P., Sablowski, R., Sandberg, G., Samach, A., and Ori, N. (2005) Arabidopsis KNOX proteins activate cytokinin biosynthesis. *Curr. Biol.* 15: 1566–1571.

# Phytochrome and Light Control of Plant Development

Have you ever dissected your sandwich and wondered why the bean sprouts looked the way they do? Most edible sprouts (e.g., alfalfa and mung bean) are germinated and grown in the dark, where they undergo a special kind of development called **skotomorphogenesis** (from *skotos*, the Greek word for darkness). Such **etiolated seedlings** have elongated stems and folded cotyledons, and they fail to accumulate chlorophyll (FIGURE 17.1). Now imagine these seedlings growing in soil rather than decorating your sandwich. Envision the elongating shoot pushing the delicate first leaves up through the soil, using the apical hook to clear a path (see Figure 17.1D). When the seedling emerges from the soil and the limited energy reserves from the cotyledons (dicots) or from the endosperm (monocots) are exhausted, the seedling must begin making food for itself.

This transition from skotomorphogenesis to **photomorphogenesis** is an extremely complex yet rapid process. Within minutes of applying a single flash of relatively dim light to a dark-grown bean seedling, several developmental changes occur:

- a decrease in the rate of stem elongation,

- the beginning of apical-hook straightening, and

- initiation of the synthesis of pigments that are characteristic of green plants.

(A) Light-grown corn

(B) Dark-grown corn

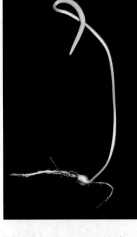

(C) Light-grown mustard

(D) Dark-grown mustard

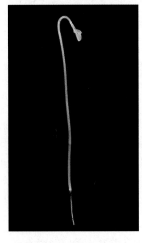

**FIGURE 17.1** Corn (maize; *Zea mays*) (A and B) and mustard (*Eruca* sp.) (C and D) seedlings grown either in the light (A and C) or the dark (B and D). Symptoms of etiolation in corn, a monocot, include the absence of greening, reduction in leaf width, failure of leaves to unroll, and elongation of the coleoptile and mesocotyl. In mustard, a dicot, etiolation symptoms include absence of greening, reduced leaf size, hypocotyl elongation, and maintenance of the apical hook. (A, B corn seedling photos courtesy of Patrice Dubois; C, D mustard seedling photos by David McIntyre.)

Thus, light acts as a signal to induce a change in the form of the seedling, from one that facilitates growth beneath the soil to one that will enable the plant to efficiently harvest light energy and convert it into the essential sugars, proteins, and lipids necessary for growth of the plant and for the people and animals that eat plants.

Among the different pigments that can promote photomorphogenic responses in plants, the most important are those that absorb red and blue light. The blue-light photoreceptors will be discussed in relation to guard cells and phototropism in Chapter 18. The focus of this chapter is **phytochrome**, a protein-pigment photoreceptor that absorbs red and far-red light most strongly, but that also absorbs blue light. As we will see in this chapter and in Chapter 25, phytochrome mediates several aspects of vegetative and reproductive development.

We begin with the discovery of phytochrome and the phenomenon of red/far-red photoreversibility. We then survey the various phytochrome responses in terms of the amount and quality of light required to produce the responses. Next we will discuss the structural properties of phytochrome and the conformational changes induced by light. Some of the physiological complexity of phytochrome response is explained by the fact that phytochromes are encoded by a multigene family, with members having both unique and redundant functions. Over the past few years, tremendous progress has been made in defining the molecular events that underlie phytochrome responses, including the identification of several interacting partners. Finally, we will examine ecological functions of phytochrome that enable plants to adapt to ever-changing environments.

## The Photochemical and Biochemical Properties of Phytochrome

Phytochrome, a blue protein pigment with a molecular mass of about 125 kilodaltons (kDa), was not identified as a unique chemical species until 1959, mainly because of technical difficulties in isolating and purifying the protein. However, many of the biological properties of phytochrome had been established earlier in studies of whole plants.

The first clues regarding the role of phytochrome in plant development came from studies that began in the 1930s on red light–induced morphogenic responses, especially seed germination. The list of such responses now includes one or more responses at almost every stage in the life history of a wide range of green plants (**TABLE 17.1**).

A key breakthrough in the history of phytochrome was the discovery that the effects of *red light* (650–680 nm) on morphogenesis could be reversed by a subsequent irradiation with light of longer wavelengths (710–740 nm), called *far-red light*. This phenomenon was first demonstrated in germinating seeds, but was also observed in stem and leaf growth, as well as in floral induction (see Chapter 25). The initial observation, made in 1935, was that the germination of lettuce seeds is stimulated by red light and inhibited by far-red light (Flint and McAlister 1935). But the real breakthrough came many years later, in 1952, when lettuce seeds were exposed to alternating treatments of red and far-red light. Nearly 100% of the seeds that received red light as the final treatment germinated; in seeds that received far-red light as the final treatment, however, germination was strongly inhibited (**FIGURE 17.2**) (Borthwick et al. 1952). This pivotal experiment demonstrated that the responses

**TABLE 17.1**
Typical photoreversible responses induced by phytochrome in a variety of higher and lower plants

| Group | Genus | Stage of development | Effect of red light |
|---|---|---|---|
| Angiosperms | *Lactuca* (lettuce) | Seed | Promotes germination |
| | *Avena* (oat) | Seedling (etiolated) | Promotes de-etiolation (e.g., leaf unrolling) |
| | *Sinapis* (mustard) | Seedling | Promotes formation of leaf primordia, development of primary leaves, and production of anthocyanin |
| | *Pisum* (pea) | Adult | Inhibits internode elongation |
| | *Xanthium* (cocklebur) | Adult | Inhibits flowering (photoperiodic response) |
| Gymnosperms | *Pinus* (pine) | Seedling | Enhances rate of chlorophyll accumulation |
| Pteridophytes | *Onoclea* (sensitive fern) | Young gametophyte | Promotes growth |
| Bryophytes | *Polytrichum* (moss) | Germling | Promotes replication of plastids |
| Chlorophytes | *Mougeotia* (alga) | Mature gametophyte | Promotes orientation of chloroplasts to directional dim light |

*Note*: The membrane potential was measured as –110 mV.

to red and far-red light are not merely opposite, they are also antagonistic.

Two interpretations of these results were possible. One is that there are two pigments, a red light–absorbing pigment and a far-red light–absorbing pigment, and the two pigments act antagonistically in the regulation of seed germination. Alternatively, there might be a single pigment that can exist in two interconvertible forms: a red light–absorbing form and a far-red light–absorbing form (Borthwick et al. 1952). The model chosen—the one-pigment model—was the more radical of the two because there was no precedent for such a photoreversible pigment. Several years later the presence of phytochrome was demonstrated in plant extracts for the first time, and its unique photoreversible properties were exhibited in vitro, confirming the prediction (Butler et al. 1959).

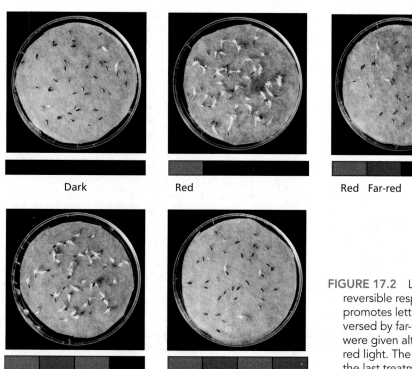

Dark     Red     Red   Far-red

Red   Far-red   Red     Red   Far-red   Red   Far-red

**FIGURE 17.2** Lettuce seed germination is a typical photoreversible response controlled by phytochrome. Red light promotes lettuce seed germination, but this effect is reversed by far-red light. Imbibed (water-moistened) seeds were given alternating treatments of red followed by far-red light. The effect of the light treatment depended on the last treatment given. Very few seeds germinated following the last far-red treatment (Photos by David McIntyre.)

## Phytochrome can interconvert between Pr and Pfr forms

In dark-grown or etiolated plants, phytochrome is present in a red light–absorbing form, referred to as **Pr**. This blue-colored inactive form is converted by red light to a far-red light–absorbing form called **Pfr**, which is blue-green. Pfr, in turn, can be converted back to Pr by far-red light.

Known as **photoreversibility**, this conversion/reconversion property is the most distinctive property of phytochrome, and it may be expressed in abbreviated form as follows:

$$\text{Pr} \underset{\text{Far-red light}}{\overset{\text{Red light}}{\rightleftarrows}} \text{Pfr}$$

The interconversion of the Pr and Pfr forms can be measured in vivo or in vitro. In fact, most of the spectral properties, such as absorption spectrum and photoreversibility, of carefully purified phytochrome measured in vitro are the same as those observed in vivo.

It is important to note that the phytochrome pool is never fully converted to the Pfr or Pr forms following red or far-red irradiation, because the absorption spectra of the Pfr and Pr forms overlap. Thus, when Pr molecules are exposed to red light, most of them absorb the photons and are converted to Pfr, but some of the Pfr made also absorbs the red light and is converted back to Pr (**FIGURE 17.3**). The proportion of phytochrome in the Pfr form after saturating irradiation by red light is only about 88%. Similarly, the very small amount of far-red light absorbed by Pr makes it impossible to convert Pfr entirely to Pr by broad-spectrum far-red light. Instead, an equilibrium of 98% Pr and 2% Pfr is achieved. This equilibrium is termed the **photostationary state**.

In addition to absorbing red light, both forms of phytochrome absorb light in the blue region of the spectrum (see Figure 17.3). Therefore, phytochrome effects can be elicited also by blue light, which can convert Pr to Pfr and vice versa. Blue-light responses can also result from the action of one or more specific blue-light photoreceptors (see Chapter 18). Whether phytochrome is involved in a response to blue light is often determined by a test of the ability of far-red light to reverse the response, since only phytochrome-induced responses are reversed by far-red light.

## Pfr is the physiologically active form of phytochrome

Because phytochrome responses are induced by red light, hypothetically they could result from either the appearance of Pfr or the disappearance of Pr. In most cases studied, a quantitative relationship holds between the magnitude of the physiological response and the amount of Pfr generated by light, but no such relationship holds between the physiological response and the loss of Pr. Evidence such as this has led to the conclusion that Pfr is the physiologically active form of phytochrome.

The use of narrow waveband red and far-red light was central to the discovery and eventual isolation of phytochrome. However, a plant growing outdoors is never exposed to strictly "red" or "far-red" light, as are plants used in laboratory-based photobiological experiments. In natural settings plants are exposed to a much broader spectrum of light, and it is under these conditions that phytochrome must work to regulate developmental responses to changes in the light environment. Indeed, as shown in Figure 17.3, the plant canopy itself can have a dramatic

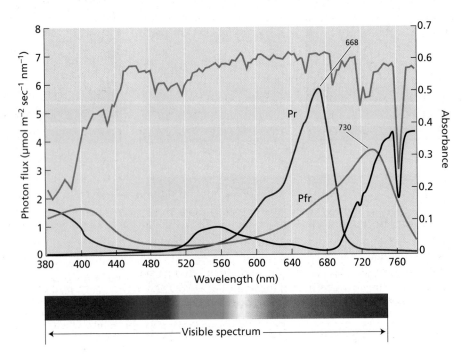

**FIGURE 17.3**  Absorption spectra of purified oat phytochrome in the Pr (red line) and Pfr (green line) forms overlap. At the top of the canopy, there is a relatively uniform distribution of visible-spectrum light (blue line), but under a dense canopy much of the red light is absorbed by plant pigments, resulting in transmittance of mostly far-red light. The black line shows the spectral properties of light that is filtered through a leaf. Thus, the relative proportions of Pr and Pfr are determined by the degree of vegetative shading in the canopy. (After Kelly and Lagarias 1985, courtesy of Patrice Dubois.)

effect on both the quality and quantity of incident light reaching individual plants and leaves.

# Characteristics of Phytochrome-Induced Responses

The variety of different phytochrome responses in intact plants is extensive, in terms of both the kinds of responses (see Table 17.1) and the quantity of light needed to induce the responses. A survey of this variety will show how diversely the effects of a single photoevent—the absorption of light by Pr—are manifested throughout the plant. For ease of discussion, phytochrome-induced responses may be logically grouped into two types:

- Rapid biochemical events
- Slower morphological changes, including movements and growth

Some of the early biochemical reactions affect later developmental responses. The nature of these early biochemical events, which comprise signal transduction pathways, will be treated in detail later in the chapter. Here we will focus on the effects of phytochrome on whole-plant responses. As we will see, such responses can be classified into various types depending on the amount and duration of light required and on their action spectra.

## Phytochrome responses vary in lag time and escape time

Morphological responses to the photoactivation of phytochrome may be observed visually after a *lag time*—the time between stimulation and observed response. The lag time may be as brief as a few minutes or as long as several weeks. The more rapid of these responses are usually reversible movements of organelles (see WEB TOPIC 17.1) or reversible volume changes (swelling, shrinking) in cells, but even some growth responses are remarkably fast.

Red-light inhibition of the stem elongation rate of light-grown pigweed (*Chenopodium album*) is observed within minutes after the proportion of Pfr to Pr in the stem is increased. Kinetic studies using Arabidopsis have confirmed this observation and further shown that phytochrome acts within 8 minutes after exposure to red light (Parks and Spalding 1999). But lag times of several weeks are observed for the induction of flowering (see Chapter 25).

Information about the lag time for a phytochrome response helps researchers evaluate the kinds of biochemical events that could precede and cause the induction of that response. The shorter the lag time, the more limited the range of biochemical events that could be involved.

Variety in phytochrome responses can also be seen in the phenomenon called **escape from photoreversibility**. Red light–induced events are reversible by far-red light for only a limited period of time, after which the response is said to have "escaped" from reversal control by light.

This escape phenomenon can be explained by a model based on the assumption that phytochrome-controlled morphological responses are the end result of a multi-step sequence of linked biochemical reactions in the responding cells. Early stages in the sequence may be fully reversible by removing Pfr, but at some point in the sequence a point of no return is reached, beyond which the reactions proceed irreversibly toward the response. *The escape time therefore represents the amount of time it takes before the overall sequence of reactions becomes irreversible*; essentially, the time it takes for Pfr to complete its primary action. The escape time for different responses ranges remarkably, from less than a minute to hours.

## Phytochrome responses can be distinguished by the amount of light required

Phytochrome responses can be distinguished by the amount of light required to induce them. The amount of light is referred to as the **fluence**, which is defined as the number of photons impinging on a unit surface area. The standard units for fluence are micromoles of quanta per square meter ($\mu mol\ m^{-2}$). Some phytochrome responses are sensitive not only to the fluence, but also to the **irradiance\***, or *fluence rate*, of light. The units of irradiance are micromoles of quanta per square meter per second ($\mu mol\ m^{-2}\ s^{-1}$). (For definitions of these and other terms used in light measurement see Chapter 9 and WEB TOPIC 9.1.)

Each phytochrome response has a characteristic range of light fluences within which the magnitude of the response is proportional to the fluence. As FIGURE 17.4 shows, phytochrome responses fall into three major categories based on the amount of light required: very low–fluence responses (VLFRs), low-fluence responses (LFRs), and high-irradiance responses (HIRs).

## Very low–fluence responses are nonphotoreversible

Some phytochrome responses can be initiated by fluences as low as 0.0001 $\mu mol\ m^{-2}$ (one-tenth of the amount of light emitted by a firefly in a single flash), and they saturate (i.e., reach a maximum) at about 0.05 $\mu mol\ m^{-2}$. For example, Arabidopsis seeds can be induced to germinate with red light in the range of 0.001 to 0.1 $\mu mol\ m^{-2}$. In dark-grown oat seedlings, red light can stimulate the growth of the coleoptile and inhibit the growth of the mesocotyl (the elongated axis between the coleoptile and the root) at similarly low fluences. These remarkable effects of van-

---

*Irradiance is sometimes loosely equated with light intensity. The term *intensity*, however, refers to light emitted by the source, whereas *irradiance* refers to light that is incident on the object.

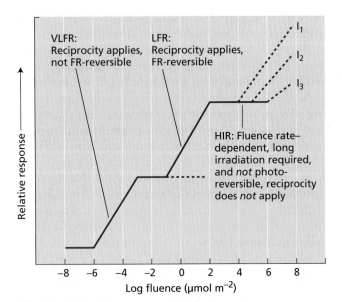

**FIGURE 17.4** Three types of phytochrome responses, based on their sensitivities to fluence. The relative magnitudes of representative responses are plotted against increasing fluences of red light. Short light pulses activate VLFRs and LFRs. Because HIRs are proportional to irradiance as well as to fluence, the effects of three different irradiances given continuously are illustrated ($I_1 > I_2 > I_3$). (After Briggs et al. 1984.)

ishingly low levels of illumination are called **very low–fluence responses** (**VLFRs**).

The minute amount of light needed to induce VLFRs converts less than 0.02% of the total phytochrome to Pfr. Because the far-red light that would normally reverse a red-light effect converts only 98% of the Pfr to Pr (as discussed earlier), about 2% of the phytochrome remains as Pfr—significantly more than the 0.02% needed to induce VLFRs (Mandoli and Briggs 1984). In other words, far-red light cannot lower the Pfr concentration below 0.02%, so it is unable to inhibit VLFRs. The VLFR action spectrum matches the absorption spectrum of Pr, supporting the view that Pfr is the active form for these responses (Shinomura et al. 1996).

Ecological implications of the VLFR in seed germination are discussed in **WEB ESSAY 17.1**.

## Low-fluence responses are photoreversible

Another set of phytochrome responses cannot be initiated until the fluence reaches 1.0 µmol m$^{-2}$, and are saturated at about 1000 µmol m$^{-2}$. These responses are referred to as **low-fluence responses** (**LFRs**); they include most of the red/far-red photoreversible responses, such as the promotion of lettuce seed germination and the regulation of leaf movements, that are mentioned in Table 17.1. The LFR action spectrum for Arabidopsis seed germination is shown in **FIGURE 17.5**. LFR spectra include a main peak for stimulation in the red region (660 nm), and a major peak for inhibition in the far-red region (720 nm).

Both VLFRs and LFRs can be induced by brief pulses of light, provided that the total amount of light energy adds up to the required fluence. The total fluence is a function of two factors: the fluence rate (µmol m$^{-2}$ s$^{-1}$) and the time of irradiation. Thus, a brief pulse of red light will induce a response, provided that the light is sufficiently bright, and conversely, very dim light will work if the irradiation

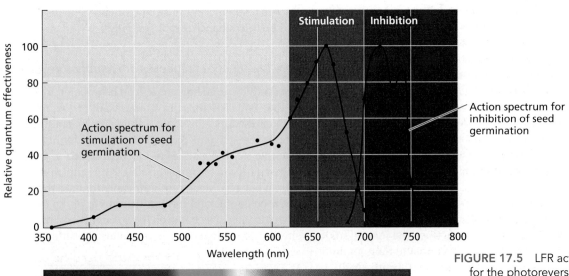

**FIGURE 17.5** LFR action spectra for the photoreversible stimulation and inhibition of seed germination in Arabidopsis. (After Shropshire et al. 1961.)

**TABLE 17.2**
## Some plant photomorphogenic responses induced by high irradiances

Synthesis of anthocyanin in various dicot seedlings and in apple skin segments

Inhibition of hypocotyl elongation in mustard, lettuce, and petunia seedlings

Induction of flowering in henbane (*Hyoscyamus*)

Plumular hook opening in lettuce

Enlargement of cotyledons in mustard

Production of ethylene in sorghum

time is long enough. This reciprocal relationship between fluence rate and time is known as the **law of reciprocity**, which was first formulated by R. W. Bunsen and H. E. Roscoe in 1850. VLFRs and LFRs both obey the law of reciprocity; that is, the magnitude of the response (for example, percent germination or degree of inhibition of hypocotyl elongation) is dependent on the product of the fluence rate and the time of irradiation.

### High-irradiance responses are proportional to the irradiance and the duration

Phytochrome responses of the third type are termed **high-irradiance responses** (**HIRs**), several of which are listed in **TABLE 17.2**. HIRs require prolonged or continuous exposure to light of relatively high irradiance. The response is proportional to the irradiance until the response saturates and additional light has no further effect (see **WEB TOPIC 17.2**).

The reason these responses are called high-irradiance responses rather than high-fluence responses is that they are proportional to fluence rate—the number of photons striking the plant tissue per second— rather than fluence— the total number of photons striking it in a given period of illumination. HIRs saturate at much higher fluences than LFRs—at least 100 times higher. Because neither continuous exposure to dim light nor transient exposure to bright light can induce HIRs, HIRs were thought not to obey the law of reciprocity. However, as elegantly shown by Shinomura and colleagues, inhibition of hypocotyl elongation by FR does obey the law of reciprocity when short pulses of FR light are provided, suggesting that photoperception by phytochrome is rate-limiting for this response (Shinomura et al. 2000).

Many of the photoreversible LFRs listed in Table 17.1, particularly those involved in de-etiolation, also qualify as HIRs. For example, at low fluences the action spectrum for anthocyanin production in seedlings of white mustard (*Sinapis alba*) shows a single peak in the red region of the spectrum. The effect is reversible with far-red light, and the response obeys the law of reciprocity. However, if the dark-grown seedlings are instead exposed to high-irradiance

light for several hours, the action spectrum now includes peaks in the far-red and blue regions, the effect is no longer photoreversible, and the response becomes proportional to the irradiance. Thus the same effect can be either an LFR or an HIR, depending on the history of a seedling's exposure to light.

As we will see later in the chapter, the various types of phytochrome responses (VLFR, LFR, and HIR) are mediated by different phytochrome molecules.

## Structure and Function of Phytochrome Proteins

Native phytochrome is a soluble protein with a molecular mass of about 250 kDa. It occurs as a dimer (a protein complex composed of two subunits). Each subunit consists of two components: a light-absorbing pigment molecule called the **chromophore**, and a polypeptide chain called the **apoprotein** (**FIGURE 17.6**). The apoprotein monomer has a molecular mass of about 125 kDa and is encoded in angiosperms by a small family of genes. Together, the apoprotein and its chromophore make up the **holoprotein**.

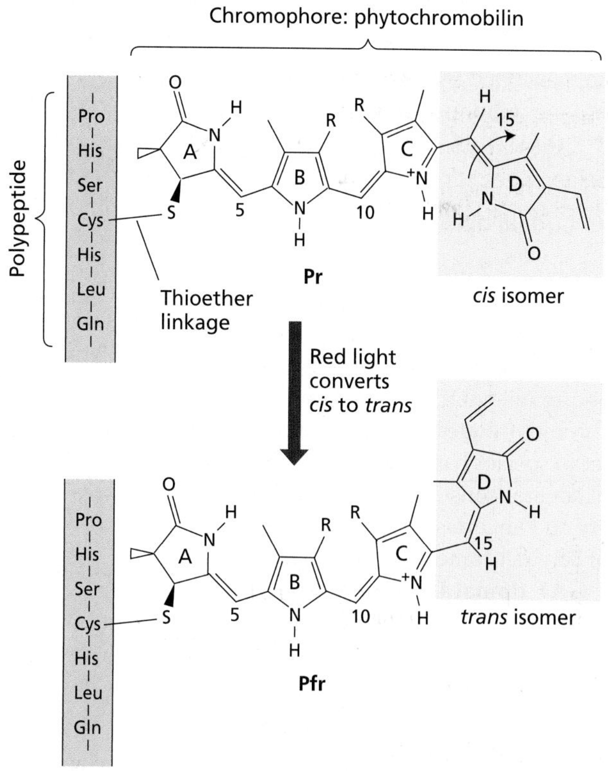

**FIGURE 17.6** Structure of the Pr and Pfr forms of the chromophore (phytochromobilin) and the peptide region bound to the chromophore through a thioether linkage. The chromophore undergoes a *cis–trans* isomerization at carbon 15 in response to red and far-red light. (After Andel et al. 1997.)

In higher plants the chromophore of phytochrome is a linear tetrapyrrole called **phytochromobilin**. The phytochrome apoprotein alone cannot absorb red or far-red light. Light can be absorbed only when the polypeptide is covalently linked with phytochromobilin to form the holoprotein. Phytochromobilin is synthesized inside plastids. It is derived from heme via a pathway that branches from the chlorophyll biosynthetic pathway. The phytochromobilin is exported to the cytosol where it attaches to the apoprotein through a thioether linkage to a cysteine residue (see Figure 17.6). Assembly of the phytochrome apoprotein with its chromophore is **autocatalytic**; that is, it occurs spontaneously when purified phytochrome polypeptide is mixed with purified chromophore in the test tube, with no additional proteins or cofactors (Li and Lagarias 1992).

In this section we will examine the functional domains of the phytochrome apoprotein, its activity as a protein kinase, and its localization within the cell. However, the situation is complicated by the fact that there is not one phytochrome but several, each encoded by a separate gene and each playing a unique role in development.

### Phytochrome has several important functional domains

**FIGURE 17.7** highlights several of the structural domains that have been identified in phytochrome and the diversity of cellular changes it mediates in response to light. Sequence comparisons between phytochromes isolated from plant and cyanobacterial species has enabled the identification of several conserved domains (Montgomery and Lagarias 2002). The N-terminal half of phytochrome contains a **PAS domain**\*, a **GAF domain** with bilin-lyase activity, which is necessary for autocatalytic assembly of the chromophore, and the **PHY**

\*The term PAS comes from the first letter of each of the three founding member proteins of the family: PER, ARNT, and SIM. In higher eukaryotes, the PAS domain functions in dimerization and protein–protein interactions. The GAF domain is related to the PAS domain and is found in c*GMP*-specific phosphodiesterases, cyanobacterial *a*denylate cyclase, and the transcriptional activator *FhlA*.

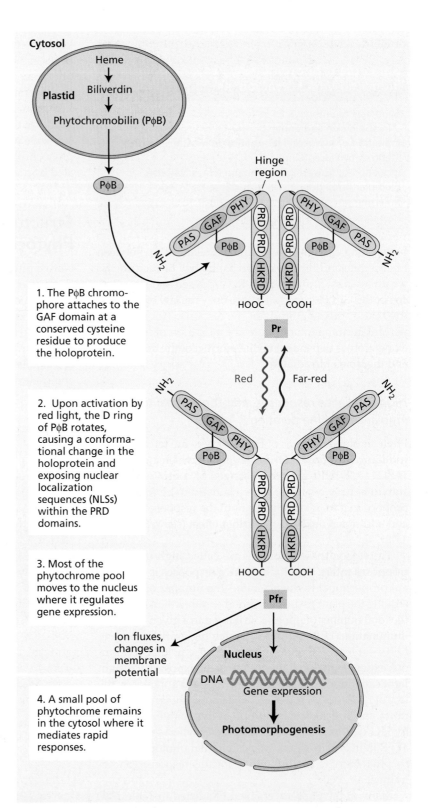

1. The PφB chromophore attaches to the GAF domain at a conserved cysteine residue to produce the holoprotein.

2. Upon activation by red light, the D ring of PφB rotates, causing a conformational change in the holoprotein and exposing nuclear localization sequences (NLSs) within the PRD domains.

3. Most of the phytochrome pool moves to the nucleus where it regulates gene expression.

4. A small pool of phytochrome remains in the cytosol where it mediates rapid responses.

**FIGURE 17.7**  After synthesis and assembly (1), phytochrome is activated by red light (2) and moves into the nucleus (3) to modulate gene expression. A small pool of phytochrome remains in the cytosol, where it may regulate rapid biochemical changes (4). Several conserved domains within the phytochrome are shown: PAS, GAF (contains bilin-lyase domain), PHY, PRD (PAS-related domain), and HKRD (HIS kinase–related domain). PφB, phytochromobilin. (After Montgomery and Lagarias 2002.)

domain, which stabilizes phytochrome in the Pfr form. A hinge region separates the N-terminal and C-terminal halves of the molecule and plays a critical role in conversion of the inactive, Pr form of phytochrome to the active, Pfr form.

Downstream of the hinge regions are two **PAS-related domain (PRD)** repeats that mediate phytochrome dimerization. Within the PRD are two **nuclear localization sequences (NLSs)** that when exposed, direct the active, Pfr form of phytochrome to the nucleus. The C-terminal histidine kinase–related domain (HKRD) is essential for autophosphorylation but can be deleted entirely if the phytochrome molecules are permitted to dimerize and activity is still retained (Matsushita et al. 2003). This was an unexpected result, as it was generally believed that the kinase domain would function in propagating the signal. Instead, it appears that it is involved in attenuating the response.

A major breakthrough in phytochrome research was achieved with the three-dimensional structure for the N-terminal region of a bacterial phytochrome bound to its chromophore, biliverdin, from the radiation-resistant, extremophilic bacterium *Deinococcus radiodurans* (Wagner et al. 2005). Unlike plant phytochromes, bacterial phytochromes utilize a range of tetrapyrrole chromophores and likely mediate very different downstream responses. Nevertheless, the structure of the chromophore-binding pocket is likely to be highly conserved. As shown in FIGURE 17.8, the chromophore is tightly associated with a pocket in the

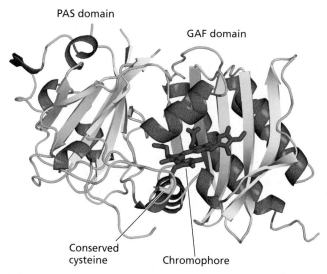

**FIGURE 17.8** A three-dimensional crystal structure is shown for the N-terminal portion of a bacterial phytochrome from *Deinococcus radiodurans*. The chromophore, biliverdin (shown in purple), is covalently attached to a conserved cysteine residue (shown in pink) and is closely associated with the protein backbone of the GAF domain. (After Wagner et al. 2005.)

GAF domain. This suggests a possible mechanism for how light can alter protein structure: isomerization (rotation) of the terminal D ring of the chromophore alters its association with certain amino acids in the GAF domain, thereby inducing a conformational change in the protein. (For a more detailed description of phytochrome chromophores see WEB ESSAY 17.2.)

### Phytochrome is a light-regulated protein kinase

The evidence for a potential role of phosphorylation in phytochrome action first came from red-light regulation of protein phosphorylation and phosphorylation-dependent binding of transcription factors to the promoters of phytochrome-regulated genes. Some highly purified preparations of phytochrome were also reported to have kinase activity. **Protein kinases** are enzymes that have the capacity to transfer phosphate groups from ATP to amino acids such as serine or tyrosine, either on themselves or on other proteins. Kinases are often found in signal transduction pathways in which the addition or removal of phosphate groups regulates enzyme activity (see Chapter 14). **Protein phosphatases** are enzymes that remove phosphate groups from proteins and thus work antagonistically to kinases in regulating protein activities. Thus, protein kinases and phosphatases play opposing roles in signaling pathways.

Phytochrome is now known to be a protein kinase that is capable of autophosphorylation (Yeh and Lagarias 1998). The evolutionary origin of phytochrome is very ancient, pre-dating the appearance of eukaryotes. Bacterial phytochromes are light-dependent histidine kinases that function as **sensor proteins** that phosphorylate corresponding **response regulator** proteins (FIGURE 17.9A) (see also Chapter 14 and WEB TOPIC 17.3). However, although higher-plant phytochromes have some homology with the histidine kinase domains, they do not function as histidine kinases. Instead, they are *serine/threonine kinases* (FIGURE 17.9B) and likely phosphorylate other proteins. Thus, although the sites of phosphorylation in plant phytochromes have diverged from those of bacterial phytochromes, the basic functional mechanism is likely conserved.

### Pfr is partitioned between the cytosol and the nucleus

In the cytosol, phytochrome holoproteins dimerize in the inactive Pr state. Upon absorption of light, the Pr chromophore undergoes a *cis–trans* isomerization of the double bond between carbons 15 and 16 and rotation of the C14–C15 single bond (see Figure 17.6). During the conversion of Pr to Pfr, the protein moiety of the phytochrome holoprotein also undergoes a conformational change in the hinge region that exposes a nuclear localization signal (NLS) in the C-terminal half of phytochrome resulting in the move-

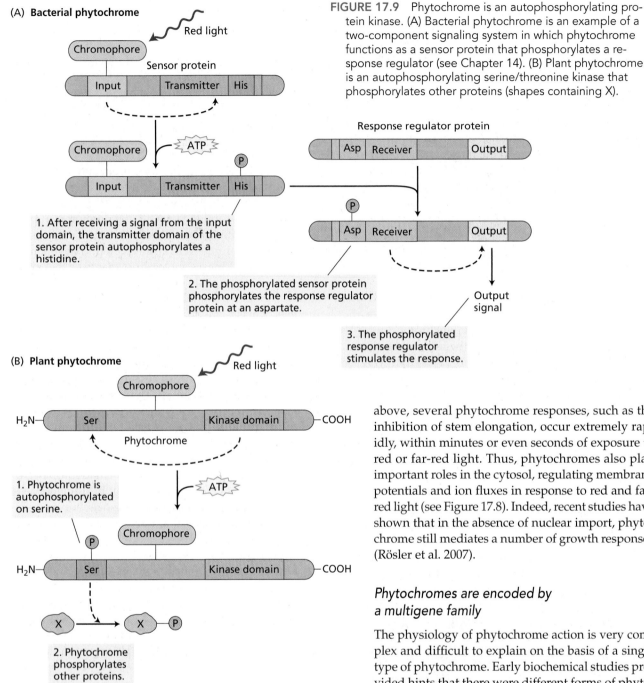

(A) **Bacterial phytochrome**

Red light

Chromophore

Sensor protein

| Input | Transmitter | His |

ATP

Chromophore

| Input | Transmitter | His | P

1. After receiving a signal from the input domain, the transmitter domain of the sensor protein autophosphorylates a histidine.

Response regulator protein

| | Asp | Receiver | | Output | |

P

| | Asp | Receiver | | Output | |

2. The phosphorylated sensor protein phosphorylates the response regulator protein at an aspartate.

Output signal

3. The phosphorylated response regulator stimulates the response.

(B) **Plant phytochrome**

Red light

Chromophore

H₂N— | Ser | Kinase domain | —COOH

Phytochrome

1. Phytochrome is autophosphorylated on serine.

ATP

P

Chromophore

H₂N— | Ser | Kinase domain | —COOH

X → X — P

2. Phytochrome phosphorylates other proteins.

**FIGURE 17.9** Phytochrome is an autophosphorylating protein kinase. (A) Bacterial phytochrome is an example of a two-component signaling system in which phytochrome functions as a sensor protein that phosphorylates a response regulator (see Chapter 14). (B) Plant phytochrome is an autophosphorylating serine/threonine kinase that phosphorylates other proteins (shapes containing X).

above, several phytochrome responses, such as the inhibition of stem elongation, occur extremely rapidly, within minutes or even seconds of exposure to red or far-red light. Thus, phytochromes also play important roles in the cytosol, regulating membrane potentials and ion fluxes in response to red and far-red light (see Figure 17.8). Indeed, recent studies have shown that in the absence of nuclear import, phytochrome still mediates a number of growth responses (Rösler et al. 2007).

### Phytochromes are encoded by a multigene family

The physiology of phytochrome action is very complex and difficult to explain on the basis of a single type of phytochrome. Early biochemical studies provided hints that there were different forms of phytochrome. For example, phytochrome was shown to be most abundant in etiolated seedlings; thus, most early biochemical studies were carried out on phytochrome purified from nongreen tissues, since very little phytochrome is extractable from green tissues.

It was soon apparent that there were two different classes of phytochrome with distinct properties, termed Type I and Type II phytochromes. Type I was shown to be about nine times more abundant than Type II in dark-grown pea seedlings; in light-grown pea seedlings the amounts of the two types were about equal, indicating that light causes Type I phytochromes to degrade. Thus phyto-

ment of phytochrome molecules from the cytosol to the nucleus (see Figure 17.7; Chen et al. 2005). This movement of phytochromes from the cytosol to the nucleus is light quality–dependent, in that the Pfr form of phytochrome is selectively imported into the nucleus (**FIGURE 17.10**).

Once in the nucleus, phytochromes interact with transcriptional regulators to mediate changes in gene transcription. Thus, one important function of phytochrome is to serve as a light-activated switch to bring about global changes in gene transcription. However, as mentioned

**FIGURE 17.10** Nuclear localization of phy–GFP fusion proteins in epidermal cells of Arabidopsis hypocotyls. Transgenic Arabidopsis plants expressing phyA–GFP (A) or phyB–GFP (B) were exposed to either continuous far-red light (A) or white light (B) and observed under a fluorescence microscope. Only nuclei are visible, demonstrating that the light treatments induced nuclear accumulation of the phy–GFP fusion proteins. These results indicate a role for nuclear–cytoplasmic partitioning in controlling phytochrome signaling. The smaller bright green dots inside the nucleus in B are called "speckles." The significance of speckles is unknown. (From Yamaguchi et al. 1999, courtesy of A. Nagatani.)

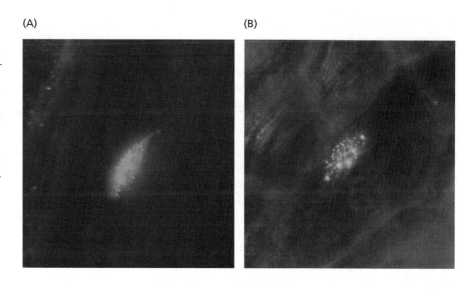

(A)   (B)

chrome was divided into two classes: the light-labile form (Type I) and the light-stable form (Type II). Actually, it is the Pfr form of Type I phytochromes that is unstable.

A major breakthrough came with the realization that the phytochromes are encoded by a family of genes and have different biochemical properties. In Arabidopsis the family is composed of five structurally related members: *PHYA*, *PHYB*, *PHYC*, *PHYD*, and *PHYE* (Sharrock and Quail 1989). In rice, a monocot, there are only three phytochrome-encoding genes: *PHYA*, *PHYB*, and *PHYC* (Mathews and Sharrock 1997). The apoprotein by itself (without the chromophore) is designated PHY; the holo-protein (with the chromophore) is designated phy. By convention, phytochrome sequences from other higher plants are named according to their homology with the Arabidopsis *PHY* genes. In both monocots and dicots, the light-labile (Type I) phytochrome is encoded by *PHYA*. In Arabidopsis, the most abundant light-stable (Type II) phytochrome is encoded by *PHYB*. As we will see, genetic studies have revealed that phyA and phyB have contrasting roles in development.

## Genetic Analysis of Phytochrome Function

Genetic screens have proven invaluable in defining light–signal transduction pathways in Arabidopsis. Marteen Koornneef was one of the first to use genetics to identify components of light signaling pathways (Koornneef et al. 1980). In a now classic study, he soaked Arabidopsis seeds in a solution of ethane methyl sulfonate (EMS), which creates point mutations in DNA. The plants derived from the mutagenized seed were allowed to self-pollinate, and large

pools of mutagenized seed were generated. The "families" were then grown under white light, and mutants with defects in light perception or response were identified as having a long hypocotyl (*hy*), similar to what would be observed if plants were grown in the dark. All of the mutations identified segregated as single recessive alleles.

Crosses made among the mutant plants defined five loci or complementation groups (*HY1–HY5*) that have served as the foundation for studies of light signaling in Arabidopsis. The cloning and sequence analysis of all five *HY* genes has led to the identification of several components of light-signal transduction, including genes necessary for phytochrome chromophore biosynthesis (*HY1* and *HY2*), the photoreceptor PHYB (*HY3*), the blue-light photoreceptor cryptochrome (*HY4*), and a light-induced transcription factor (*HY5*).

By combining genetic analysis, molecular biology, and detailed physiological studies, tremendous insights have been made into the mechanisms of light response over the past decade. Perhaps not surprisingly, these studies have revealed a complex interplay of light with plant signal transduction components that mediate developmental decisions of the plant throughout its life cycle, from seed germination to flowering.

In this section, we will examine the diverse physiological functions and complex interactions of the phytochrome gene family members as revealed by genetic analyses. In addition to their contrasting roles in VLFR and LFR responses, two types of high-irradiance responses have been identified, each dependent on a different phytochrome: the classic far-red light HIR of etiolated seedlings, which is mediated by phyA; and the red light (or white light) HIR of light-grown plants, which is mediated by phyB.

## Phytochrome A mediates responses to continuous far-red light

No phytochrome gene mutations other than for phyB were found in the original *hy* collection, so the identification of phyA mutants required the development of more ingenious screens. As discussed previously, because the far-red HIRs were known to require light-labile (Type I) phytochrome, it was suspected that phyA must be the photoreceptor involved in the perception of continuous far-red light. If this is true, then the phyA mutants should fail to respond to continuous far-red light and grow tall and spindly under these light conditions. However, mutants lacking chromophore would also look like this because phyA can detect far-red light only when assembled with the chromophore into holophytochrome.

To select for just the phyA mutants, the seedlings that grew tall in continuous far-red light were then grown under continuous red light. The phyA-deficient mutants can grow normally under this regimen, but a chromophore-deficient mutant, which also lacks functional phyB, does not respond. The *phyA* mutant seedlings selected in this screen had no obvious phenotypic abnormalities when grown in normal white light, suggesting that phyA has no discernible role in sensing white light (Whitelam et al. 1993). This also explains why *phyA* mutants were not detected in the original long-hypocotyl screen by Koornneef. Phytochrome A also appears to be involved in the germination VLFR of Arabidopsis seeds in response to broad-spectrum light. Thus, mutants lacking phyA cannot germinate in response to millisecond pulses of light, but they show a normal response to red light in the low-fluence range (Shinomura et al. 1996). This result demonstrates that phyA functions as the primary photoreceptor for this VLFR.

One extrapolation of these results is that phyA function is largely restricted to early stages of development such as seed germination and seedling de-etiolation. However, recent experiments suggest that the story is more complex. When *phyA phyB* double mutants are grown under high fluence red light (>100 umol/m/sec), they are more elongated than the *phyB* single mutant. Furthermore the *phyA* protein is detectable for a longer period of time following a shift of wild-type plants from dark to high light conditions (Franklin et al. 2007). PhyA has also been shown to play a role in the photoperiod control of flowering in Arabidopsis and rice (Valverde et al. 2004; Takano et al. 2005). Thus, it is likely that phyA functions throughout the life cycle of plant and mediates responses to both far-red and red light.

## Phytochrome B mediates responses to continuous red or white light

The characterization of the *hy3* mutant revealed an important role for phyB in de-etiolation, since mutant seedlings grown in continuous white light had long hypocotyls. The *phyB* mutant is deficient in chlorophyll and in some mRNAs that encode chloroplast proteins, and it is impaired in its ability to respond to plant hormones.

In addition to regulating white and red light–mediated HIR responses, phytochrome B also appears to regulate LFRs, such as photoreversible seed germination, the phenomenon that originally led to the discovery of phytochrome. Wild-type Arabidopsis seeds require light for germination, and the response shows red/far-red reversibility in the low-fluence range. Mutants that lack phyA respond normally to red light; mutants deficient in phyB are unable to respond to low-fluence red light (Shinomura et al. 1996). This experimental evidence strongly suggests that phyB mediates photoreversible seed germination.

As discussed later in this chapter, phyB plays an important role in regulating responses of plants to shade treatments. Plants that are deficient in phyB often look like wild-type plants that are grown under dense vegetative canopies. In fact, mediating responses to vegetative shade such as accelerated flowering and increased elongation growth may be one the most ecologically important roles of phytochromes (Smith 1982).

## Roles for phytochromes C, D, and E are emerging

Although phyA and phyB are the predominant forms of phytochrome in Arabidopsis, phyC, phyD, and phyE play unique roles in regulating response to red and far-red light. The creation of double and triple mutants has made it possible to assess the relative role of each phytochrome in a given response. PhyD and phyE are structurally similar to phyB, but are not functionally redundant. Responses mediated by phyD and phyE include petiole and internode elongation and the control of flowering time (see Chapter 25). The characterization of *phyC* mutants in Arabidopsis suggests a complex interplay between phyC, phyA, and phyB response pathways (Franklin et al. 2003; Monte et al. 2003). This specialization in *PHY* gene function is likely to be important in fine-tuning phytochrome responses to daily and seasonal changes in light regimes. Indeed, a study of natural variation has defined a role for phyC in regulating both plant growth and flowering time across a latitudinal cline (Balasubramanian et al. 2006).

## Phy gene family interactions are complex

Clearly, genetic analysis has been a powerful tool in dissecting *PHY* gene function in Arabidopsis. There are, however, several limitations to this approach. For one, *phy* mutant phenotypes are often interpreted under the assumption that the activity of the other family members is unchanged. However, detailed molecular studies have shown that phyC and to some extent phyD accumulation are dependent on active phyB, as are *PHYA* transcript levels. Furthermore,

(A)

(B)

Wild-type
pea

*phyA*-deficient mutants

Wild-type
tomato

Mutants lacking
*phyA* and both
copies of *phyB*

**FIGURE 17.11** Phytochrome deficiencies alter growth and development in pea and tomato. (A) Pea plants with lesions in *phyA* exhibit delayed flowering and shortened internodes. (B) In tomato, mutations in *phyA* and both copies of *phyB* prevent chlorophyll accumulation in the fruits and greatly extend the length of the fruit-bearing cluster, demonstrating the diversity of traits under phytochrome control. (From Weller et al. 1997, 2000.)

transgenic Arabidopsis plants overexpressing the gene encoding PHYB have increased levels of phyA, phyC, and phyD in Arabidopsis (Hirschfeld et al. 1998).

The interpretation of the genetic data is further complicated by the fact that the light-stable phy proteins form both homo- and heterodimers (e.g., phyB–phyC) (Clack et al. 2009). Thus, loss-of-function mutations in *PHYB* or plants overexpressing the PHYB apoprotein are likely to display phenotypes that are the consequence of an altered dynamic of many phytochrome interactions. Considering that there may be multiple alleles of each phytochrome in a natural population of a given species (Maloof et al. 2001), one can envision an incredibly complex molecular network. Unraveling the complexities of phytochrome interactions thus represents a challenging problem for the future.

### PHY gene functions have diversified during evolution

Although much of our understanding of the molecular events underlying phytochrome responses has been revealed through studies of Arabidopsis, the *PHY* gene family is rapidly evolving among the angiosperms (Mathews and Sharrock 1997). While most dicots have four subfamilies of phytochrome genes (*PHYA*, *PHYB/D*, *PHYC/F*, and *PHYE*), the monocots have only three (*PHYA*, *PHYB*, and *PHYC*). Through gene duplication/loss, genetic drift, and rapid diversification of *PHY* gene function, phytochrome signal transduction networks can be refashioned to suit the needs of plants in different habitats and under different selection pressures. Only when a detailed genetic

analysis is conducted in a given species can we begin to identify the similarities and differences in phytochrome-regulated pathways.

For example, a phyA deficiency has little phenotypic effect on white light–grown Arabidopsis and rice, but loss of phyA function in pea results in highly pleiotropic phenotypic effects, including shortened internodes, delayed senescence, and increased yield under long days. In tomato, a loss of phyA and both copies of the duplicated phyB genes prevents chlorophyll accumulation in the fruits and greatly extends the length of the fruit-bearing cluster or "truss" (**FIGURE 17.11**). However, chlorophyll accumulates in the leaves of these plants, indicating that the role of phyA and phyB in chlorophyll accumulation is fruit-specific in tomato.

The picture emerging from such cross-species comparisons indicates that while the mode of action of phy family members may be highly conserved (e.g., phyA mediates VLFR and far-red HIR), the downstream effectors and, ultimately, the physiological responses mediated by these photoreceptors, may be quite different across taxa (**FIGURE 17.12**).

## Phytochrome Signaling Pathways

All phytochrome-regulated changes in plants begin with absorption of light by the pigment. After light absorption, the molecular properties of phytochrome are altered, probably affecting the interaction of the phytochrome protein with other cellular components that ultimately bring about

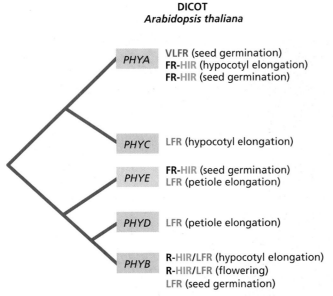
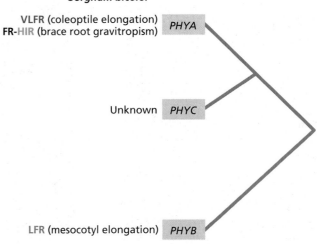

**DICOT**
*Arabidopsis thaliana*

PHYA — VLFR (seed germination)
FR-HIR (hypocotyl elongation)
FR-HIR (seed germination)

PHYC — LFR (hypocotyl elongation)

PHYE — FR-HIR (seed germination)
LFR (petiole elongation)

PHYD — LFR (petiole elongation)

PHYB — R-HIR/LFR (hypocotyl elongation)
R-HIR/LFR (flowering)
LFR (seed germination)

**MONOCOTS**
*Oryza sativa* (rice)
*Sorghum bicolor*

VLFR (coleoptile elongation)
FR-HIR (brace root gravitropism) — PHYA

Unknown — PHYC

LFR (mesocotyl elongation) — PHYB

**FIGURE 17.12** Differences in phytochrome gene family structure and function in the dicot *Arabidopsis thaliana* and the monocots *Oryza sativa* and *Sorghum bicolor*. *PHY* genes in monocots and dicots appear to utilize the same gene family members for VLFRs, LFRs, and HIRs (e.g., phyA is used for FR-HIR). However, the developmental phenomenon being regulated may be quite different (e.g., seed germination versus coleoptile elongation for the phyA-mediated VLFR). Also note that monocots do not contain *PHYD* or *PHYE*.

changes in the growth, development, or position of an organ (see Table 17.1).

Molecular and biochemical techniques are helping to unravel the early steps in phytochrome action and the signal transduction pathways that lead to physiological or developmental responses. These responses fall into two general categories:

- Ion fluxes, which cause relatively rapid turgor responses
- Altered gene expression, which result in slower, long-term processes

In this section we will examine the effects of phytochrome on both membrane permeability and gene expression, as well as the possible chain of events constituting the signal transduction pathways that bring about these effects.

### Phytochrome regulates membrane potentials and ion fluxes

Phytochrome can rapidly alter the properties of membranes, within seconds of a light pulse. Such rapid modulation has been measured in individual cells and has been inferred from the effects of red and far-red light on the

surface potential of roots and oat (*Avena*) coleoptiles, in which the lag between the production of Pfr and the onset of measurable hyperpolarization (membrane potential changes) is 4.5 seconds. Changes in the bioelectric potential of cells imply changes in the flux of ions across the plasma membrane and suggest that some of the cytosolic responses of phytochrome are initiated at or near the plasma membrane.

One longstanding conundrum has been how the alga *Mougeotia* uses red light to stimulate rapid chloroplast movement (see **WEB TOPIC 17.1**). In many species, including Arabidopsis, chloroplast movements are mediated by blue light through the action of phototropins. In *Mougeotia*, the chloroplast-movement photoreceptor is a fusion between a red-light and a blue-light photoreceptor (Suetsugu et al. 2005). Thus, *Mougeotia* appears to have evolved the ability to exploit red light as a signal and feed the response (chloroplast movement) through a more typical blue-light mediated signaling pathway (see Chapter 18). It is interesting to note that many of the cytosolic functions identified for phyA in Arabidopsis (Rösler et al. 2007) can also be mediated by blue-light photoreceptors, suggesting that these pathways may sometimes converge or act synergistically soon after signal perception in higher plants as well.

### Phytochrome regulates gene expression

As the term *photomorphogenesis* implies, plant development is profoundly influenced by light. Etiolation symptoms include spindly stems, small leaves (in dicots), and the absence of chlorophyll. Complete reversal of these symptoms by light involves major long-term alterations in metabolism that can be brought about only by changes in gene expression. Overall, the picture emerging for light-regulated plant promoters is similar to that for other eukaryotic genes: a collection of modular elements, the number, position, flanking sequences, and binding activities of which can lead to a wide range of transcriptional patterns. *No single DNA sequence or binding protein is common to all phytochrome-regulated genes.*

At first it may appear paradoxical that light-regulated genes have such a range of elements, any combination of which can confer light-regulated expression. However, this array of sequences allows for the differential light- and tissue-specific regulation of many genes through the action of multiple photoreceptors.

The stimulation and repression of transcription by light can be very rapid, with lag times as short as 5 minutes. Using **DNA microarray analysis**, global patterns of gene expression in response to changes in light can be monitored. (For a discussion of methods for transcriptional analysis see WEB TOPIC 17.4.) These studies have indicated that nuclear import triggers a transcriptional cascade involving thousands of genes that initiate photomorphogenic development. By monitoring gene expression profiles over time following a shift of plants from darkness to light, both early and late targets of *PHY* gene action were identified (Tepperman et al. 2001; Tepperman et al. 2004).

The nuclear import of phyA and phyB is highly correlated with the light quality that stimulates their activities. That is, nuclear import of phyA is mediated by far-red light and low fluence broad spectrum light, whereas phyB import is driven by red light exposure and is reversible by far-red light (Kircher et al. 1999; Yamaguchi et al. 1999). Thus, nuclear import of the phy proteins likely represents a major control point in phytochrome signaling. Unlike phyB, which appears to utilize a common nuclear import pathway, phyA is dependent on FHY1 for nuclear import (Genoud et al. 2008).

Some of the early gene products that are rapidly upregulated following a shift from darkness to light are themselves transcription factors that activate the expression of other genes. The genes encoding these rapidly up-regulated proteins are called **primary response genes**. Expression of the primary response genes depends on *signal transduction pathways* (discussed next) and is independent of protein synthesis. In contrast, the expression of the late genes, or **secondary response genes**, requires the synthesis of new proteins. DNA microarray analyses have thus revealed the global reprogramming of plant gene expression that accompanies the transition from *skotomorphogenic* to *photomorphogenic* development.

## Phytochrome interacting factors (PIFs) act early in phy signaling

Two techniques have been used extensively in recent years to identify interacting partners of plant proteins—yeast two-hybrid library screens and co-immunoprecipitation (see WEB TOPIC 17.5 for descriptions). Using these two methods, several **phytochrome-interacting factors** (**PIFs**) have been identified in Arabidopsis (Castillon et al. 2007). Proteins that interact with *either* phyA or phyB define branch points in the phy signaling networks, whereas proteins that interact with *both* phyA and phyB are likely to represent points of convergence.

One of the most extensively characterized of these factors is **PIF3**, a basic helix-loop-helix (bHLH) transcription factor that interacts with both phyA and phyB (Ni et al. 1998). PIF3 and several related PIF or **PIF-like proteins** (**PILs**) are particularly notable because at least five members of this gene family selectively interact with phytochromes in their active Pfr conformations. The fact that these proteins are localized to the nucleus and can bind to DNA suggests an intimate association between phytochrome and gene transcription.

Recent studies of PIF-family members have indicated that they act primarily as negative regulators of phytochrome response (Shin et al. 2009; Stephenson et al. 2009). A quadruple mutant, in which many PIF family members are disrupted, displays constitutive photomorphogenic development when plants are grown in the dark. Phytochromes appear to initiate the degradation of PIF proteins through phosphorylation, followed by degradation through the proteasome complex (Al-Sady et al. 2006). The phytochrome-induced rapid degradation of PIF proteins that act as negative regulators of phy responses may provide a mechanism for modulating light responses that is tightly coupled to the activities of phy proteins.

## Phytochrome associates with protein kinases and phosphatases

In addition to nucleus-localized transcription factors, two-hybrid screens also identified cytosolic proteins as potential partners for phy proteins. **Phytochrome kinase substrate 1** (**PKS1**) is capable of interacting with phyA and phyB in both the active Pfr and inactive Pr form (Fankhauser et al. 1999). This protein can accept a phosphate from phyA, further highlighting the importance of phosphorylation in phytochrome signaling. The PKS1 phosphorylation is regulated by phytochrome both in the test tube and in the plant, with Pfr having a twofold higher level of activity than Pr. Overexpression studies and loss-of-function mutants of PKS1 and the closely related PKS2 suggest that these two molecules maintain balanced levels through a negative feedback loop. Molecular and genetic analyses suggest that these proteins act selectively to promote phyA-mediated VLFR. More recent data suggest that these proteins are involved in cross talk between phyA and phototropin signaling pathways (Lariguet et al. 2006).

**Phytochrome-associated protein phosphatase 5** (**PAPP5**) is another factor that interacts with phyto-

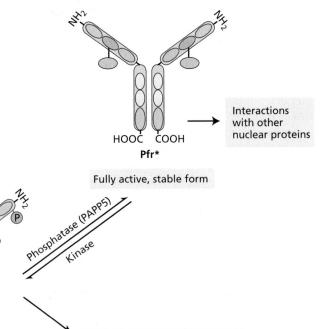

**FIGURE 17.13** Phytochrome activity is modulated by phosphorylation status. Following activation by red light, the phytochrome-associated phosphatase PAPP5 and as-yet unidentified kinases modulate phytochrome activity in response to the intensity or quality of light. (After Ryu et al. 2005.)

chromes and is probably involved in accentuating phytochrome response through dephosphorylation of the active phytochrome (Ryu et al. 2005). A possible model for the regulation of phy activity by phosphorylation is shown in **FIGURE 17.13**. In the dark, phytochrome in the Pr form is inactive and likely to be phosphorylated at serine residues in the N-terminal region. Absorption of red light induces a conformational change in phy that stimulates autophosphorylation of a serine residue in the hinge region and subsequent nuclear import; it may also target phy for degradation (discussed below).

Dephosphorylation of the serine residue in the hinge region enhances interaction of phy with downstream effectors and increases the stability of the protein in the light. The action of an unknown kinase and perhaps autophosphorylation serve to drive phy toward the less active, phosphorylated version that no longer interacts efficiently with its effector proteins.

### Phytochrome-induced gene expression involves protein degradation

We often equate the initiation of a signal transduction cascade to flipping a switch that activates a process, like a phone ringing when an incoming call arrives. However, what would happen if the phone never stopped ringing? It would lose its utility as a signaling mechanism after the first call! In much the same way, termination or resetting of a pathway is just as important as the initiation of the event. As suggested above, protein degradation is emerging as a ubiquitous mechanism regulating many cellular processes, including light and hormone signaling, circadian rhythms, and flowering time (for examples, see Chapters 19 and 25).

Genetic screens conducted independently by several groups identified mutants that exhibited light-grown phenotypes when grown in the dark, such as opened cotyledons, expanded leaves, and shortened hypocotyls. The genes identified in these screens were called *CONSTITUTIVE PHOTOMORPHOGENESIS* (*COP*), *DE-ETIOLATED* (*DET*), and *FUSCA* (*FUS*) (for the red color of the anthocyanins that accumulated in light-grown seedlings). Cloning and genetic complementation revealed that many of these genes were allelic or part of the same complex, and they are collectively known as *COP/DET/FUS*.

The cloning of several *COP/DET/FUS* genes has revealed an essential role for protein degradation in the regulation of the light response. **COP1** encodes an E3 ubiquitin ligase (Yi and Deng 2005) that is essential for placing a small peptide tag known as *ubiquitin* onto proteins (see Chapters 1 and 14). Once tagged by ubiquitin, the proteins are transported to the 26S proteasome, a cellular garbage disposal that chews up proteins into their constituent amino acids. COP9 and several other COP proteins compose the **COP9 signalosome** (**CSN**), which forms the lid of this garbage disposal, helping to determine which proteins enter the complex.

As shown in **FIGURE 17.14**, COP1 has been shown to interact with several proteins involved in the light response, including the transcription factors HFR1, HY5, and LAF1, targeting them for degradation in the dark. It is likely that COP1 functions with the SUPPRESSOR OF phyA-1 (SPA1) and related family members to form a multiprotein complex (Zhu et al. 2008) that mediates the **ubiquitination** (Vierstra 1994) of these transcription factors. In the light, COP1 is exported from the nucleus to the cytosol (see Figure 17.14), excluding it from interaction with many of the nucleus-localized transcription factors. These tran-

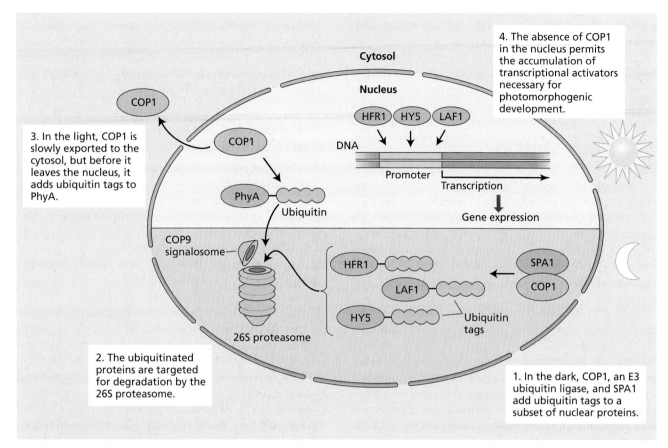

**FIGURE 17.14** COP proteins regulate the turnover of proteins required for photomorphogenic development. During the night, COP1 enters the nucleus, and the COP1/SPA1 complex adds ubiquitin to a subset of transcriptional activators. The transcription factors are then degraded by the COP9 signalosome–proteasome complex. During the day, COP1 exits the nucleus, allowing the transcriptional activators to accumulate. Blue tails represent ubiquitin tags on proteins destined for the COP9 signalosome complex (CSN) that serves as the gatekeeper of the 26S proteasome.

scription factors can then bind to promoter elements in genes that mediate photomorphogenic development.

As mentioned above, phyA protein is highly unstable in light, and its degradation has long been linked to the sequential attachment of multiple copies of the small protein ubiquitin to a specific site on the target protein. The discovery that COP1 can ubiquitinate phyA, thus targeting the protein for destruction, provides a satisfying link between COP1 function and the attenuation of phyA signaling in the light (Seo et al. 2004). As discussed in Chapter 25, COP1 is also responsible for the degradation of the flowering regulators CO and GI as well as proteins involved in auxin and GA response. Thus protein degradation is a central component to developmental signaling pathways triggered by light and hormones in plants.

## Circadian Rhythms

Various metabolic processes in plants, such as oxygen evolution and respiration, cycle alternately through high-activity and low-activity phases with a regular periodicity of about 24 hours. These rhythmic changes are referred to as **circadian rhythms** (from the Latin *circa diem*, meaning "approximately a day"). The **period** of a circadian rhythm is the time that elapses between successive peaks or troughs in the cycle. Because the rhythms persist in the absence of external controlling factors, they are considered to be **endogenous**. (For a more detailed description of circadian rhythms, see Chapter 25.)

The endogenous nature of circadian rhythms suggests that they are governed by an internal pacemaker; this mechanism is called an **oscillator**. The endogenous oscillator is coupled to a variety of physiological processes. An important feature of the oscillator is that it is unaffected by temperature, which enables the clock to function normally under a wide variety of seasonal and climatic conditions. The clock is said to exhibit **temperature compensation**.

Light is a strong modulator of rhythms in both plants and animals. Although circadian rhythms that persist under controlled laboratory conditions usually have periods one or more hours longer or shorter than 24 hours, in nature their periods tend to be uniformly closer to 24 hours because of the synchronizing effects of light at daybreak, referred to as **entrainment**.

Another important function of the clock is **gating**—regulating when physiological or molecular responses will occur within a period of the circadian rhythm. As discussed in Chapter 25 (control of flowering), gating is critical for regulating the transition from vegetative to reproductive growth. It also plays an important role in regulating plant growth in response to shade (Salter et al. 2003; Nozue et al. 2007).

The molecular basis of the circadian rhythms has long fascinated both plant and animal biologists, and the isolation of clock mutants has been an important tool for the identification of clock genes in other organisms. Isolating clock mutants in plants requires a convenient assay that allows monitoring of the circadian rhythms of many thousands of individual plants to detect the rare abnormal phenotype.

To allow screening for clock mutants in Arabidopsis, the promoter region of the *LHCB* (also called *CHLORO-PHYLL a/b BINDING* [*CAB*]) gene was fused to the gene that encodes luciferase, an enzyme that emits light in the presence of its substrate, luciferin. This reporter gene construct was then used to transform Arabidopsis with the Ti plasmid of *Agrobacterium* as a vector. Investigators were then able to monitor the temporal and spatial regulation of bioluminescence in individual seedlings in real time using a video camera (Millar et al. 1995). A total of 21 independent *toc* (*timing of CAB* [*LHCB*] *expression*) mutants were isolated, including both short-period and long-period lines. The *toc1* mutant in particular has been implicated in the core oscillator mechanism (Strayer et al. 2001).

Another important discovery came with the isolation and characterization of two MYB-related transcription factors, CIRCADIAN CLOCK–ASSOCIATED 1 (CCA1) and LATE ELONGATED HYPOCOTYLS (LHY). (For information on MYB, see Chapter 20.) Loss-of-function mutants or constitutive overexpression of *CCA1* or *LHY* abolishes the circadian and phytochrome regulation of several genes, and physiological responses such as leaf movements become arrhythmic. These observations strongly suggest that *CCA1* and *LHY* are components of the circadian clock.

### The circadian oscillator involves a transcriptional negative feedback loop

The circadian oscillators of a cyanobacterium (*Synechoc-occus*), a fungus (*Neurospora crassa*), an insect (*Drosophila melanogaster*), and mouse (*Mus musculus*) have now been elucidated. In these four organisms, the oscillator is composed of several "clock genes" involved in a transcriptional–translational negative feedback loop.

So far, three major clock genes have been identified in Arabidopsis: *TOC1*, *LHY*, and *CCA1*. The protein products of these genes are all transcriptional regulators. *TOC1* is not related to the clock genes of other organisms, suggesting that the plant oscillator is unique.

Models have been proposed that incorporate the findings of several genetic and molecular studies of circadian rhythms in Arabidopsis (Alabadi et al. 2001; Salome and McClung 2004). A simplified version of the Arabidopsis clock is shown in **FIGURE 17.15**. However, computational models have indicated that multiple loops are required for robust clock function (Locke et al. 2006).

In the simplified model, light and the TOC1 regulatory protein activate *LHY* and *CCA1* expression at dawn. The increase in LHY and CCA1 represses the expression of the *TOC1* gene. Because TOC1 is a positive regulator of the *LHY* and *CCA1* genes, the repression of *TOC1* expression causes a progressive reduction in the levels of LHY and CCA1, which reach their minimum levels at the end of the day. As LHY and CCA1 levels decline, *TOC1* gene expression is released from inhibition. TOC1 reaches its maximum at the end of the day, when LHY and CCA1

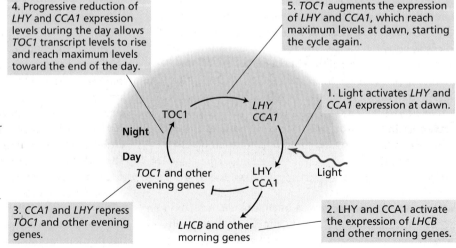

**FIGURE 17.15** Circadian oscillator model showing the hypothetical interactions between the *TOC1* and *MYB* genes *LHY* and *CCA1* in Arabidopsis. Light at dawn increases *LHY* and *CCA1* expression. *LHY* and *CCA1* regulate other daytime and evening genes.

4. Progressive reduction of *LHY* and *CCA1* expression levels during the day allows *TOC1* transcript levels to rise and reach maximum levels toward the end of the day.

5. *TOC1* augments the expression of *LHY* and *CCA1*, which reach maximum levels at dawn, starting the cycle again.

1. Light activates *LHY* and *CCA1* expression at dawn.

3. *CCA1* and *LHY* repress *TOC1* and other evening genes.

2. LHY and CCA1 activate the expression of *LHCB* and other morning genes.

TOC1

LHY CCA1

**Night**

**Day**

*TOC1* and other evening genes

LHY CCA1

Light

*LHCB* and other morning genes

are at their minimum. TOC1 then indirectly stimulates the expression of *LHY* and *CCA1* (Pruneda-Paz et al. 2009), and the cycle begins again.

A number of additional proteins help regulate this central oscillator. A protein kinase, CK2, can interact with and phosphorylate CCA1. The CK2 kinase is a multisubunit protein with serine/threonine kinase activity that, when mutated, changes the period of rhythmic expression of CCA1 (Sugano et al. 1999). The nuclear protein GIGANTEA (GI) is also required to maintain high levels of expression of LHY and CCA, in part through interaction of the F-box protein ZTL with TOC1 (Kim et al. 2007). **F-boxes** are protein motifs that promote protein–protein interactions. F-box proteins were discovered as components of ubiquitin E3 ligase complexes, which target proteins for degradation by the 26S proteasome (see Chapter 14). ZTL protein levels peak at dusk and are lowest at dawn. Interestingly, ZTL is also a blue-light photoreceptor, thus providing a mechanism for the tight control of the central oscillator through light signaling.

The two MYB regulator proteins—LHY and CCA1—have dual functions. In addition to serving as components of the oscillator, they regulate the expression of other genes, such as *LHCB* and other "morning genes," and they repress genes expressed at night. Light reinforces the effect of the *TOC1* gene in promoting *LHY* and *CCA1* expression. This reinforcement represents the underlying mechanism of *entrainment*.

**LIGHT REGULATION OF THE CIRCADIAN CLOCK**  To function properly, the oscillator must be entrained to the daily light/dark cycles of the external environment. In experiments designed to characterize the role of photoreceptors in this process, phytochrome-deficient mutants were crossed with lines carrying the luciferase reporter gene discussed above (Somers et al. 1998). The pace of the oscillator was slowed (i.e., period length increased) when *phyA* mutant plants were grown under dim red light, but not high-fluence red light. However, *phyB* mutants showed timing defects only under high-fluence red light. The **cryptochromes** (blue-light photoreceptors) CRY1 and CRY2 were required for blue light–mediated entrainment of the circadian clock.

These studies indicate that both phytochromes and cryptochromes entrain the circadian clock in Arabidopsis. This light input appears to be modulated by the genes *EARLY FLOWERING 3* (*ELF3*) and *TIME FOR COFFEE* (*TIC*). Mutations in *ELF3* stop the oscillations of the clock at dusk, whereas mutations in *TIC* stop the clock at dawn. The *elf3/ tic* double mutant is completely arrhythmic, suggesting that *TIC* and *ELF* interact with different components of the clock at different phases in the rhythm (Hall et al. 2003).

**GENE EXPRESSION AND CIRCADIAN RHYTHMS**  Phytochrome can also interact with circadian rhythms at the level of gene expression. The expression of genes in the *LHCB* family, encoding the light-harvesting chlorophyll *a/b*–binding proteins of photosystem II, is regulated at the transcriptional level by both circadian rhythms and phytochrome.

In leaves of pea and wheat, the level of *LHCB* mRNA has been found to oscillate during daily light–dark cycles, rising in the morning and falling in the evening. Since the rhythm persists even in continuous darkness, it appears to be a circadian rhythm. But phytochrome can perturb this cyclical pattern of expression.

When wheat plants are transferred from a cycle of 12 hours light and 12 hours dark to continuous darkness, the rhythm persists for a while, but it slowly *damps out* (i.e., reduces in amplitude until no peaks or troughs are discernible). If, however, the plants are given a pulse of red light before they are transferred to continuous darkness, no damping occurs (i.e., the levels of *LHCB* mRNA continue to oscillate as they do during the light–dark cycles).

In contrast, a far-red flash at the end of the day prevents the expression of *LHCB* in continuous darkness, and the effect of far-red is reversed by red light. Note that it is not the oscillator that damps out under constant conditions, but the coupling of the oscillator to the physiological event being monitored. Red light restores the coupling between the oscillator and the physiological process.

In Arabidopsis, it has been estimated that over 30% of expressed genes are under the control of the circadian clock (Michael and McClung 2003). Interestingly, many genes that participate in similar cellular activities display rhythms with a similar phase (Harmer et al. 2000). For instance, transcripts of many genes necessary for photosynthesis peak near the middle of the subjective day*, whereas transcripts required for cell wall biosynthesis peak near the middle of the subjective night. Careful examination of the sequence of promoter regions of these genes revealed a nine-nucleotide motif termed the **evening element** (AAAATATCT) that appears to mediate expression of many genes whose expression peaks at the end of the subjective day.

**CIRCADIAN RHYTHMS AND FITNESS**  Although circadian rhythms have long been known to play an essential role in photoperiodism during flowering (see Chapter 25), only recently has their importance in optimizing vegetative growth been tested experimentally (Dodd et al. 2005). Arabidopsis clock mutants with abnormally long or short periods were grown under artificial day–night cycles that either matched or were out of phase with oscillator periods. Plants whose circadian rhythms matched the light–dark

---

*According to standardized circadian time (CT), the *subjective day* begins at CT = 0 hour of the 24-hour day–night cycle, while the *subjective night* starts at CT = 12 hours. The middle of the subjective day would therefore correspond to CT = 6 hours, and the middle of the subjective night would occur at CT = 18 hours.

cycle of the environment (**circadian resonance**) contained more chlorophyll and had greater biomass than plants whose clocks were out of phase with the environment. Moreover, when grown together in competition experiments, plants with correctly matched circadian clocks out-competed plants with out-of-phase clocks. Thus, circadian resonance enhances evolutionary fitness by promoting vegetative growth (photosynthesis and biomass) and reproductive development at optimal times.

Clock function also plays an important role in promoting vigor in polyploids. In synthetic polyploids of Arabidopsis, altered clock function is correlated with increased expression of several genes required for photosynthesis and starch biosynthesis; this correlation suggests that circadian rhythms play an important role in regulating biomass (Ni et al. 2009).

## Ecological Functions

Thus far we have discussed phytochrome-regulated responses as studied in the laboratory. Phytochrome also plays important roles in plants growing in the natural environment. In the discussion that follows, we will learn how plants sense and respond to shading by other plants. We will also examine the specialized functions of the different phytochrome gene family members in these processes. For a discussion of the role of phytochrome in the sleep movements of leaves see WEB TOPIC 17.6.

### Phytochrome enables plant adaptation to changes in light quality

The presence of a red/far-red reversible pigment in all green plants, from algae to dicots, suggests that these wavelengths of light provide information that helps plants adjust to their environment. What environmental conditions change the relative levels of these two wavelengths in natural radiation?

The ratio of red light (R) to far-red light (FR) varies remarkably in different environments. This ratio can be defined as follows:

$$R/FR = \frac{\text{Photon fluence rate in 10 nm band centered on 660 nm}}{\text{Photon fluence rate in 10 nm band centered on 730 nm}}$$

TABLE 17.3 compares both the total light intensity in photons (400–800 nm) and the R:FR values in eight natural environments. Both parameters vary greatly in different environments.

Compared with direct daylight, there is proportionally more far-red in the light that reaches plants during sunset, under 5 mm of soil, or under the canopy of other plants (as on the floor of a forest). The canopy phenomenon results

### TABLE 17.3 Ecologically important light parameters

| | Fluence rate ($\mu$mol m$^{-2}$ s$^{-1}$) | R:FR[a] |
|---|---|---|
| Daylight | 1900 | 1.19 |
| Sunset | 26.5 | 0.96 |
| Moonlight | 0.005 | 0.94 |
| Ivy canopy | 17.7 | 0.13 |
| Lakes, at a depth of 1 m | | |
|   Black Loch | 680 | 17.2 |
|   Loch Leven | 300 | 3.1 |
|   Loch Borralie | 1200 | 1.2 |
| Soil, at a depth of 5 mm | 8.6 | 0.88 |

*Source*: Smith 1982, p. 493.

*Note*: The light intensity factor (400–800 nm) is given as the photon flux density, and phytochrome-active light is given as the R:FR ratio.

[a]Absolute values taken from spectroradiometer scans; the values should be taken to indicate the relationships between the various natural conditions and not as actual environmental means.

from the fact that green leaves absorb red light because of their high chlorophyll content, but are relatively transparent to far-red light.

### Decreasing the R:FR ratio causes elongation in sun plants

An important function of phytochrome is that it enables plants to sense shading by other plants. Plants that increase stem extension in response to shading are said to exhibit a **shade avoidance response**. As shading increases, the R:FR ratio decreases (see Figure 17.3). A higher proportion of far-red light converts more Pfr to Pr, and the ratio of Pfr to total phytochrome (Pfr/P$_{total}$) decreases.

When so-called sun plants (plants that normally grow in an open-field habitat) were grown in natural light under a system of shades so that F:FR was controlled, stem extension rates increased in response to a higher far-red content (i.e., a lower Pfr:P$_{total}$ ratio) (FIGURE 17.16). In other words, simulated canopy shading (high levels of far-red light; low Pfr:P$_{total}$ ratio) induced these plants to allocate more of their resources to growing taller. This correlation was not as strong for "shade plants," which normally grow in a shaded environment. Shade plants showed less reduction in their stem extension rate than did sun plants when they were exposed to higher R:FR values (see Figure 17.16). Thus there appears to be a systematic relationship between phytochrome-controlled growth and species habitat. Such results are taken as an indication of the involvement of phytochrome in shade perception.

For a "sun plant" or "shade-avoiding plant," there is a clear adaptive value in allocating its resources toward more

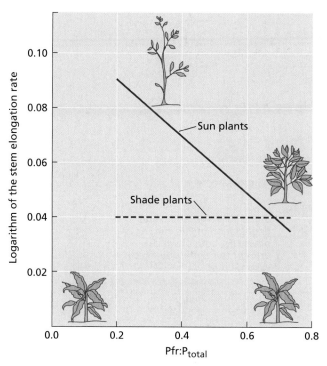

**FIGURE 17.16** Phytochromes appear to play a predominant role in controlling stem elongation rate in sun plants (solid line), but not in shade plants (dashed line). (After Morgan and Smith 1979.)

rapid extension growth when it is shaded by another plant. In this way it can enhance its chances of growing above the canopy and acquiring a greater share of unfiltered, photosynthetically active light. The price for increased internode elongation is usually reduced leaf area and reduced branching, but at least in the short run, this adaptation to canopy shade seems to work.

Genetic analyses of Arabidopsis have indicated that phyB plays the predominant role in mediating many shade avoidance responses, but phyD and phyE also contribute, particularly to petiole elongation. PhyA also plays a role by antagonizing the responses mediated by phyB, D, and E (discussed later in the chapter). Microarray analyses of shaded versus unshaded plants have confirmed and extended the genetic studies (see **WEB TOPIC 17.7**). For a discussion of how plants sense their neighbors using reflected light, see **WEB ESSAY 17.3**.

Evidence is also emerging for the integration of a number of hormonal pathways in the control of shade avoidance responses, including those of auxin, gibberellins, and ethylene. Several recent reports have suggested that the PIF proteins play important roles in mediating responses to shade and at least some of these responses are mediated through GA signaling pathways (**FIGURE 17.17**; see also Figures 14.12 and 20.20). As mentioned earlier, PIF proteins function in general as negative regulators of photomorphogenesis and are subject to negative regulation by

phytochromes. This antagonism between PIF function and PHY function allows the plants to fine-tune their responses to the light environment.

When plants are grown under high R:FR, as in an open canopy, phy proteins become nuclear localized and inactivate PIF proteins. In darkness or under low R:FR, a pool of phytochrome is excluded from the nucleus, enabling the accumulation of PIF proteins that promote elongation responses (Lorrain et al. 2008). In addition to PIF–PHY interactions, PIF proteins also are subject to negative regulation by DELLA proteins, which are components of the GA signaling pathway (de Lucas et al. 2008; Feng et al. 2008). Thus PIF proteins appear to integrate numerous light signals in the transition from skotomorphogenesis to photomorphogenesis (e.g., chlorophyll biosynthesis) as well as fine-tuning responses to changes in light quality (e.g., shade avoidance). As PIF proteins are transcription factors, these findings reinforce the notion that at least part of the phytochrome signal transduction pathway is relatively short.

## Small seeds typically require a high R:FR ratio for germination

Light quality also plays a role in regulating the germination of some seeds. As discussed earlier, phytochrome was discovered in studies of light-dependent lettuce seed germination.

In general, large seeds, which have ample food reserves to sustain prolonged seedling growth in darkness (e.g., underground), do not require light for germination. However, light is required by the small seeds of many herbaceous and grassland species, many of which remain dormant, even while hydrated, if they are buried below the depth to which light penetrates. Even when such seeds are on or near the soil surface, the level of shading by the vegetation canopy (i.e., the R:FR ratio they receive) is likely to affect their germination. For example, it is well documented that far-red enrichment imparted by a leaf canopy inhibits germination in a range of small-seeded species.

This inhibition can be reversed for the small seeds of the tropical species trumpet tree (*Cecropia obtusifolia*) and Veracruz pepper (*Piper auritum*) planted on the floor of a deeply shaded forest. If a light filter that blocks the far-red but permits the red component of the canopy-shaded light to pass through is placed immediately above the seeds, they germinate. Although the canopy transmits very little red light, it is enough to stimulate the seeds to germinate, probably because most of the inhibitory far-red light is excluded by the filter, and the R:FR ratio is very high. These seeds would also be more likely to germinate in spaces receiving sunlight through gaps in the canopy than in densely shaded spaces. The sunlight would help ensure that the seedlings became photosynthetically self-sustaining before their seed food reserves were exhausted.

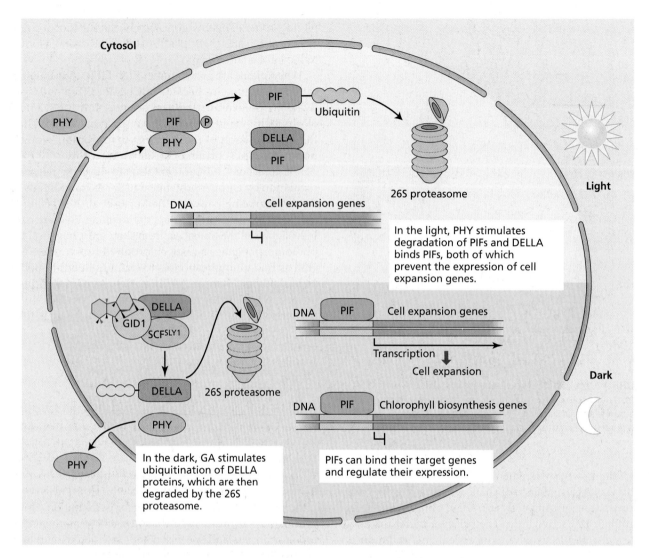

**Cytosol**

Ubiquitin

P

PHY

PIF

PHY

PIF

DELLA

PIF

26S proteasome

**Light**

DNA    Cell expansion genes

In the light, PHY stimulates degradation of PIFs and DELLA binds PIFs, both of which prevent the expression of cell expansion genes.

DELLA

GID1

SCF^SLY1

DELLA

26S proteasome

DELLA

PHY

PHY

In the dark, GA stimulates ubiquitination of DELLA proteins, which are then degraded by the 26S proteasome.

DNA    PIF    Cell expansion genes

Transcription

Cell expansion

**Dark**

DNA    PIF    Chlorophyll biosynthesis genes

PIFs can bind their target genes and regulate their expression.

**FIGURE 17.17** Convergence of light and hormone signaling. In the dark, the growth-promoting hormone gibberellic acid (GA) binds to its receptor and mediates the ubiquitination of DELLA proteins (see Chapters 14 and 20). The DELLA proteins are then targeted to the 26S proteasome for degradation. In the absence of the DELLA proteins, PIFs can act as both positive and negative regulators of gene expression, likely through interaction with different partners, perhaps mediated through different *cis*-regulatory elements upstream of target genes. In the light, DELLA proteins bind PIF proteins, preventing them from interacting with genes. PHY proteins also target PIF proteins, through phosphorylation, eventually leading to their ubiquitination and degradation. In the absence of PIF proteins, genes required for cell expansion are not expressed, and plant growth is retarded.

## Reducing shade avoidance responses can improve crop yields

Shade avoidance responses may be highly adaptive in a natural setting to help plants out-compete neighboring vegetation, but for many crop species, a reallocation of resources from reproductive to vegetative growth can reduce crop yield. In recent years, yield gains in crops such as maize have come largely through the breeding of new maize varieties with a higher tolerance to crowding (which induces shade avoidance responses) than through increases in basic yield per plant. As a consequence, today's maize crops can be grown at higher densities than older varieties without suffering decreases in plant yield (**FIGURE 17.18**).

As the mechanism of shade avoidance has become better understood, the prospect of engineering crop plants with decreased shade avoidance has improved. For example, Robson and colleagues created transgenic tobacco

**FIGURE 17.18** Modern maize varieties are planted at high density. Traditionally, Native Americans grew maize on small hills or mounds, which were separated by several feet. The plants were short and often produced multiple small ears. In contrast, modern hybrids are machine-planted in rows with little space between plants (typically 30,000 to 38,000 plants/ acre). Although yield per plant has not dramatically increased for many years in commercial hybrids, overall yields have continued to increase, largely because of better performance of plants at high planting density. As shown in this image of a typical upstate New York cornfield, modern varieties have upright leaves that help the plants to capture sunlight energy under crowded conditions. (Courtesy of T. Brutnell.)

plants that overexpressed phyA (Robson et al. 1996). When these plants were grown at high density, they failed to display a typical shade avoidance response and were actually shorter at high than at low density. Overexpression of phyA would lead to the persistence of phyA under far-red–rich shade beyond the early seedling stage, when phyB normally takes over. The persistence of phyA would cause an increase in the FR-HIR response (discussed earlier), which would counteract the phyB-mediated shade avoidance response.

Although much remains to be learned, such studies have shown that the manipulation of phytochromes or their downstream targets is a promising new approach to improving crop yield (Sawers et al. 2005).

## Phytochrome responses show ecotypic variation

To date, most of our understanding of light responses in any given model plant species has been derived from experiments performed on a limited number of varieties or accessions. For example, much of the genetic analysis of Arabidopsis has been performed using the Columbia or Landsberg ecotypes, and genome sequencing efforts in rice and maize have focused on two accessions for each species. As a result, plant research programs throughout the world have tended to focus on a small number of accessions.

When considering the role of phytochromes in an ecological context, however, it is essential to examine a much broader germplasm collection. Surveys of the light responses in Arabidopsis and maize have revealed tremendous ecotypic variation, both in the physiology of their light responses and in their phytochrome gene families. For instance, the Wassilewskija (Ws) accession of Arabidopsis contains a naturally occurring deletion of the phyD gene, whereas an accession from Le Mans, France (Lm-2) carries a light-stable form of phyA that fails to mediate responses to continuous far-red light (Aukerman et al. 1997; Maloof et al. 2001). These studies indicate that variations in phytochrome responses may have some adaptive value. Determining how such variations contribute to fitness in diverse habitats is a challenge for the future.

## Phytochrome action can be modulated

Can the actions of other photoreceptors modify the action of phytochrome? The isolation of the genes encoding the cryptochrome and phototropin photoreceptors (see Chapter 18) mediating blue light–regulated responses has made it possible to analyze whether these photoreceptors functionally overlap with phytochrome (Chory and Wu 2001). This possibility was suspected because mutations in the cryptochrome CRY2 gene led to delayed flowering under continuous white light, and flowering time was also known to be under phytochrome control.

In Arabidopsis, continuous blue or far-red light treatment leads to promotion of flowering, and red light inhibits flowering. Far-red light acts through phyA, and the antagonistic effect of red light is produced by of phyB. One might expect the cry2 mutant to be delayed in flowering, since blue light promotes flowering. However, cry2 mutants flower at the same time as the wild type under either continuous blue or continuous red light. Delay is observed only if both blue and red light are given together. Therefore, cry2 probably promotes flowering in blue light by repressing phyB function, most likely through a direct interaction with phyB (Mas et al. 2000).

Additional experiments have confirmed that the other cryptochrome, CRY1, also interacts with phytochromes. Both CRY1 and CRY2 interact with phyA in vitro and can be phosphorylated in a phyA-dependent manner. Phosphorylation of CRY1 has also been demonstrated to occur in vivo in a red light–dependent manner.

# SUMMARY

Light provides the signal for photomorphogenesis, and the pigmented phytochrome mediates several aspects of vegetative and reproductive development (**Figure 17.1, Table 17.1**).

### The Photochemical and Biochemical Properties of Phytochrome

- The effects of red light on morphogenesis can be reversed by a subsequent irradiation with far-red light, which can be reversed by red light (**Figure 17.2**).
- Phytochrome interconverts between Pr and Pfr forms, which exist in a photostationary state, with Pfr being the active form (**Figure 17.3**).

### Characteristics of Phytochrome-Induced Responses

- Photochemical responses are characterized by escape time and by the amount of light (fluence) required (**Figure 17.4**).
- A minimum amount of light is needed for photoreversible responses (LFRs) (**Figure 17.5**).
- VLFRs and LFRs can be induced by brief pulses of light, but for other responses (HIRs) the response is proportional to the irradiance (**Table 17.2**).

### Structure and Function of Phytochrome Proteins

- Phytochrome functions as a dimer. Each subunit consists of a light-absorbing chromophore covalently linked to a polypeptide chain, the apoprotein (**Figures 17.6, 17.8**). Only the holoprotein can absorb red and far-red light.
- After assembly, phytochrome is activated by red light, and Pfr moves to the nucleus for regulation of gene expression (**Figures 17.7, 17.10**).
- Phytochrome is a protein kinase that is capable of autophosphorylation (**Figure 17.9**). Evolutionarily it predates the origin of eukaryotes.
- In the cytosol, phytochrome acts within minutes or seconds to alter membrane potentials and ion fluxes in response to red and far-red light. In the nucleus, phytochrome acts on gene expression.
- Phytochromes with different properties are encoded by a family of genes; in Arabidopsis these are *PHYA, PHYB, PHYC, PHYD,* and *PHYE*.

### Genetic Analysis of Phytochrome Function

- Phytochrome A mediates responses to continuous far-red light but also mediates red-light responses. Phytochrome A controls flowering in Arabidopsis and rice.
- Phytochrome B mediates responses to continuous red or white light, and regulates responses to shade such as accelerated flowering and increased elongation.
- Phytochromes C, D, and E also have specific developmental roles that may be partially redundant with those of phytochrome A and B.
- The signaling pathway components downstream from phytochromes vary and can lead to different responses in different organs or taxa (**Figures 17.11, 17.12**).

### Phytochrome Signaling Pathways

- The early phytochrome-induced gene products are transcription factors that activate other genes.
- The signal transduction pathways for primary response genes are independent of protein synthesis; secondary response genes require the synthesis of new proteins.
- Acting early in the signaling pathway, PIF family members function primarily as negative regulators of phytochrome responses.
- Proteins that interact with either phytochrome A or phytochrome B define branch points in the signaling networks; proteins that interact with both A and B represent points of convergence.
- Phytochrome is associated with other kinases and phosphatases (**Figure 17.13**).
- Phytochrome-induced gene expression produces enzymes that add ubiquitin to proteins targeted for proteasome degradation (**Figure 17.14**).

### Circadian Rhythms

- Most organisms display endogenous circadian rhythms governed by an internal oscillator or clock whose operation is unaffected by temperature.
- In Arabidopsis, transcriptional and translational control are important regulatory elements of clock function, as are multiple feedback loops (**Figure 17.15**).
- Both phytochromes and cryptochromes entrain the circadian clock in Arabidopsis.
- Gene transcription for PSII light-harvesting proteins is regulated by both circadian rhythms and phytochrome.
- Circadian rhythms can optimize vegetative growth.

### Ecological Functions

- In the natural environment, phytochrome enables plants to respond to shading.
- Shading (high levels of far-red light; low $Pfr:P_{total}$ ratio) induces sun plants to grow taller (**Figure 17.16**).

## SUMMARY continued

- In Arabidopsis phytochrome B plays a major role in shade avoidance responses, with phytochromes D and E also contributing to petiole elongation.
- PIF proteins and GA signaling pathways interact in regulating responses to shade (**Figure 17.17**).

- Far-red enrichment imparted by shade inhibits germination in a range of small-seeded species.
- Reducing shade avoidance responses due to crop crowding can improve crop yields (**Figure 17.18**).
- Blue-light responses may functionally overlap with phytochrome signaling.

## WEB MATERIAL

### Web Topics

**17.1 *Mougeotia*: A Chloroplast with a Twist**
Microbeam irradiation experiments have been used to localize phytochrome in this filamentous green alga.

**17.2 Phytochrome and High-Irradiance Responses**
Dual-wavelength experiments helped demonstrate the role of phytochrome in HIRs.

**17.3 The Origins of Phytochrome as a Bacterial Two-Component Receptor**
The discovery of bacterial phytochrome led to the identification of phytochrome as a protein kinase.

**17.4 Profiling Gene Expression in Plants**
Progress in bioinformatic tool development and new sequencing technologies are changing the way we look at transcriptional networks.

**17.5 Two-Hybrid Screens and Co-immunoprecipitation**
Protein–protein interactions can be studied using both molecular-genetic and immunological techniques.

**17.6 Phytochrome Effects on Ion Fluxes**
Phytochrome regulates ion fluxes across membranes by altering the activities of ion channels and the plasma membrane proton pump.

**17.7 Microarray Studies on Shade Avoidance**
DNA microarray analyses have helped to characterize both global and specific effects of variations in the R:FR ratio on gene expression.

### Web Essays

**17.1 Awakened By a Flash of Sunlight**
When placed in the proper soil environment, seeds acquire extraordinary sensitivity to light, such that germination can be stimulated by less than 1 second of exposure to sunlight during soil cultivation.

**17.2 Diversity of Phytochrome Chromophores**
Bacterial and higher plant chromophores vary in their structure, attachment chemistries, and spectral properties. By replacing plant chromophores with bacterial chromophores, plants can be engineered to "see" different wavelengths of light.

**17.3 Know Thy Neighbor through Phytochrome**
Plants can detect the proximity of neighbors through phytochrome perception of the R:FR of reflected light and produce adaptive morphological changes before being shaded by potential competitors.

# CHAPTER REFERENCES

Alabadi, D., Oyama, T., Yanovsky, M. J., Harmon, F. G., Mas, P., and Kay, S. A. (2001) Reciprocal regulation between *TOC1* and *LHY/CCA1* within the *Arabidopsis* circadian clock. *Science* 293: 880–883.

Al-Sady, B., Ni, W., Kircher, S., Schafer, E., and Quail, P. H. (2006) Photoactivated phytochrome induces rapid PIF3 phosphorylation prior to proteasome-mediated degradation. *Mol. Cell* 23: 439–446.

Andel, F., Hasson, K. C., Gai, F., Anfinrud, P. A., and Mathies, R. A. (1997) Femtosecond time-resolved spectroscopy of the primary photochemistry of phytochrome. *Biospectroscopy* 3: 421–433.

Aukerman, M. J., Hirschfeld, M., Wester, L., Weaver, M., Clack, T., Amasino, R. M., and Sharrock, R. A. (1997) A deletion in the PHYD gene of the Arabidopsis Wassilewskija ecotype defines a role for phytochrome D in red/far-red light sensing. *Plant Cell* 9: 1317–1326.

Balasubramanian, S., Sureshkumar, S., Agrawal, M., Michael, T. P., Wessinger, C., Maloof, J. N., Clark, R., Warthmann, N., Chory, J., and Weigel, D. (2006) The PHYTOCHROME C photoreceptor gene mediates natural variation in flowering and growth responses of *Arabidopsis thaliana*. *Nat. Genet.* 38: 711–715.

Borthwick, H. A., Hendricks, S. B., Parker, M. W., Toole, E. H., and Toole, V. K. (1952) A reversible photoreaction controlling seed germination. *Proc. Natl. Acad. Sci. USA* 38: 662–666.

Briggs, W. R., Mandoli, D. F., Shinkle, J. R., Kaufman, L. S., Watson, J. C., and Thompson, W. F. (1984) Phytochrome regulation of plant development at the whole plant, physiological, and molecular levels. In *Sensory Perception and Transduction in Aneural Organisms*, G. Colombetti, F. Lenci, and P.-S. Song, eds., Plenum, New York, pp. 265–280.

Butler, W. L., Norris, K. H., Siegelman, H. W., and Hendricks, S. B. (1959) Detection, assay, and preliminary purification of the pigment controlling photosensitive development of plants. *Proc. Natl. Acad. Sci. USA* 45: 1703–1708.

Castillon, A., Shen, H., and Huq, E. (2007) Phytochrome interacting factors: central players in phytochrome-mediated light signaling networks. *Trends Plant Sci.* 12: 514–521.

Chen, M., Tao, Y., Lim, J., Shaw, A., and Chory, J. (2005) Regulation of phytochrome B nuclear localization through light-dependent unmasking of nuclear-localization signals. *Curr. Biol.* 15: 637–642.

Chory, J., and Wu, D. (2001) Weaving the complex web of signal transduction. *Plant Physiol.* 125: 77–80.

Clack, T., Shokry, A., Moffet, M., Liu, P., Faul, M., and Sharrock, R. A. (2009) Obligate heterodimerization of Arabidopsis phytochromes C and E and interaction with the PIF3 basic helix-loop-helix transcription factor. *Plant Cell* 21: 786–799.

de Lucas, M., Daviere, J. M., Rodriguez-Falcon, M., Pontin, M., Iglesias-Pedraz, J. M., Lorrain, S., Fankhauser, C., Blazquez, M. A., Titarenko, E., and Prat, S. (2008) A molecular framework for light and gibberellin control of cell elongation. *Nature* 451: 480–484.

Dodd, A. N., Salathia, N., Hall, A., Kevei, E., Toth, R., Nagy, F., Hibberd, J. M., Millar, A. J., and Webb, A. A. (2005) Plant circadian clocks increase photosynthesis, growth, survival, and competitive advantage. *Science* 309: 630–633.

Fankhauser, C., Yeh, K. C., Lagarias, J. C., Zhang, H., Elich, T. D., and Chory, J. (1999) PKS1, a substrate phosphorylated by phytochrome that modulates light signaling in Arabidopsis. *Science* 284: 1539–1541.

Feng, S., Martinez, C., Gusmaroli, G., Wang, Y., Zhou, J., Wang, F., Chen, L., Yu, L., Iglesias-Pedraz, J. M., Kircher, S., et al. (2008) Coordinated regulation of *Arabidopsis thaliana* development by light and gibberellins. *Nature* 451: 475–479.

Flint, L. H., and McAlister, E. D. (1935) Wavelengths of radiation in the visible spectrum inhibiting the germination of light-sensitive lettuce seed. *Smithsonian Misc. Pub.* 94: 5.

Franklin, K. A., Davis, S. J., Stoddart, W. M., Vierstra, R. D., and Whitelam, G. C. (2003) Mutant analyses define multiple roles for phytochrome C in *Arabidopsis* photomorphogenesis. *Plant Cell* 15: 1981–1989.

Franklin, K. A., Allen, T., and Whitelam, G. C. (2007) Phytochrome A is an irradiance-dependent red light sensor. *Plant J.* 50: 108–117.

Galston, A. (1994) *Life Processes of Plants*. Scientific American Library, New York.

Genoud, T., Schweizer, F., Tscheuschler, A., Debrieux, D., Casal, J. J., Schäfer, E., Hiltbrunner, A., and Fankhauser, C. (2008) FHY1 mediates nuclear import of the light-activated phytochrome A photoreceptor. *PLoS Genet.* 4: e1000143.

Hall, A., Bastow, R. M., Davis, S. J., Hanano, S., McWatters, H. G., Hibberd, V., Doyle, M. R., Sung, S., Halliday, K. J., Amasino, R. M., et al. (2003) The *TIME FOR COFFEE* gene maintains the amplitude and timing of *Arabidopsis* circadian clocks. *Plant Cell* 15: 2719–2729.

Harmer, S. L., Hogenesch, J. B., Straume, M., Chang, H. S., Han, B., Zhu, T., Wang, X., Kreps, J. A., and Kay, S. A. (2000) Orchestrated transcription of key pathways in *Arabidopsis* by the circadian clock. *Science* 290: 2110–2113.

Hirschfeld, M., Tepperman, J. M., Clack, T., Quail, P. H., and Sharrock, R. A. (1998) Coordination of phytochrome levels in phyB mutants of *Arabidopsis* as revealed by apoprotein-specific monoclonal antibodies. *Genetics* 149: 523–535.

Kelly, J. M., and Lagarias, J. C. (1985) Photochemistry of 124-kilodalton Avena phytochrome under constant illumination in vitro. *Biochemistry* 24: 6003–6010.

Kim, W. Y., Fujiwara, S., Suh, S. S., Kim, J., Kim, Y., Han, L., David, K., Putterill, J., Nam, H. G., and Somers, D. E. (2007) ZEITLUPE is a circadian photoreceptor stabilized by GIGANTEA in blue light. *Nature* 449: 356–360.

Kircher, S., Kozma-Bognar, L., Kim, L., Adam, E., Harter, K., Schafer, E., and Nagy, F. (1999) Light quality-dependent nuclear import of the plant photoreceptors phytochrome A and B. *Plant Cell* 11: 1445–1456.

Koornneef, M., Rolff, E., and Spruitt, C. J. P. (1980) Genetic control of light-induced hypocotyl elongation in *Arabidopsis thaliana* L. Z. *Pflanzenphysiol.* 100: 147–160.

Lariguet, P., Schepens, I., Hodgson, D., Pedmale, U. V., Trevisan, M., Kami, C., de Carbonnel, M., Alonso, J. M., Ecker, J. R., Liscum, E., et al. (2006) PHYTOCHROME KINASE SUBSTRATE 1 is a phototropin 1 binding protein required for phototropism. *Proc. Natl. Acad. Sci. USA* 103: 10134–10139.

Li, L., and Lagarias, J. C. (1992) Phytochrome assembly— Defining chromophore structural requirements for covalent attachment and photoreversibility. *J. Biol. Chem.* 267: 19204–19210.

Locke, J. C., Kozma-Bognár, L., Gould, P. D., Fehér, B., Kevei, E., Nagy, F., Turner, M. S., Hall, A., and Millar, A. J. (2006) Extension of a genetic network model by iterative experimentation and mathematical analysis. *Mol. Syst. Biol.* 2: 59.

Lorrain, S., Allen, T., Duek, P. D., Whitelam, G., and Fankhauser, C. (2008) Phytochrome-mediated inhibition of shade avoidance involves degradation of growth-promoting bHLH transcription factors. *Plant J.* 53: 312–313.

Maloof, J. N., Borevitz, J. O., Dabi, T., Lutes, J., Nehring, R. B., Redfern, J. L., Trainer, G. T., Wilson, J. M., Asami, T., Berry, C. C., et al. (2001) Natural variation in light sensitivity of *Arabidopsis*. *Nat. Genet.* 29: 441–446.

Mandoli, D. F., and Briggs, W. R. (1984) Fiber optics in plants. *Sci. Am.* 251: 90–98.

Mas, P., Devlin, P. F., Panda, S., and Kay, S. A. (2000) Functional interaction of phytochrome B and cryptochrome 2. *Nature* 408: 207–211.

Mathews, S., and Sharrock, R. A. (1997) Phytochrome gene diversity. *Plant Cell Environ.* 20: 666–671.

Matsushita, T., Mochizuki, N., and Nagatani, A. (2003) Dimers of the N-terminal domain of phytochrome B are functional in the nucleus. *Nature* 424: 571–574.

Michael, T., and McClung, R. (2003) Enhancer trapping reveals widespread circadian clock transcriptional control in Arabidopsis. *Plant Physiol.* 132: 629–639.

Millar, A. J., Carre, I. A., Strayer, C. A., Chua, N.-H., and Kay, S. A. (1995) Circadian clock mutants in *Arabidopsis* identified by luciferase imaging. *Science* 267: 1161–1163.

Monte, E., Alonso, J. M., Ecker, J. R., Zhang, Y., Li, X., Young, J., Austin-Phillips, S., and Quail, P. H. (2003) Isolation and characterization of phyC mutants in *Arabidopsis* reveals complex crosstalk between phytochrome signaling pathways. *Plant Cell* 15: 1962–1980.

Montgomery, B. L., and Lagarias, J. C. (2002) Phytochrome ancestry: Sensors of bilins and light. *Trends Plant Sci.* 7: 357–366.

Morgan, D. C., and Smith, H. (1979) A systematic relationship between phytochrome-controlled development and species habitat, for plants grown in simulated natural irradiation. *Planta* 145: 253–258.

Ni, M., Tepperman, J. M., and Quail, P. H. (1998) PIF3, a phytochrome-interacting factor necessary for normal photoinduced signal transduction, is a novel basic helix-loop-helix protein. *Cell* 95: 657–667.

Ni, Z., Kim, E. D., Ha, M., Lackey, E., Liu, J., Zhang, Y., Sun, Q., and Chen, Z. J. (2009) Altered circadian rhythms regulate growth vigour in hybrids and allopolyploids. *Nature* 457: 327–331.

Nozue, K., Covington, M. F., Duek, P. D., Lorrain, S., Fankhauser, C., Harmer, S. L., and Maloof, J. N. (2007) Rhythmic growth explained by coincidence between internal and external cues. *Nature* 448: 358–361.

Parks, B. M., and Spalding, E. P. (1999) Sequential and coordinated action of phytochromes A and B during *Arabidopsis* stem growth revealed by kinetic analysis. *Proc. Natl. Acad. Sci. USA* 96: 14142–14146.

Pruneda-Paz, J. L., Breton, G., Para, A., and Kay, S. A. (2009) A functional genomics approach reveals CHE as a component of the Arabidopsis circadian clock. *Science* 323: 1481–1485.

Robson, P. R., McCormac, A. C., Irvine, A. S., and Smith, H. (1996) Genetic engineering of harvest index in tobacco through overexpression of a phytochrome gene. *Nat. Biotechnol.* 14: 995–998.

Rösler, J., Klein, I., and Zeidler, M. (2007) Arabidopsis fhl/fhy1 double mutant reveals a distinct cytoplasmic action of phytochrome A. *Proc. Natl. Acad. Sci. USA* 104: 10737–10742.

Ryu, J. S., Kim, J. I., Kunkel, T., Kim, B. C., Cho, D. S., Hong, S. H., Kim, S. H., Fernandez, A. P., Kim, Y., Alonso, J. M., et al. (2005) Phytochrome-specific type 5 phosphatase controls light signal flux by enhancing phytochrome stability and affinity for a signal transducer. *Cell* 120: 395–406.

Salome, P. A., and McClung, C. R. (2004) The *Arabidopsis thaliana* clock. *J. Biol. Rhythms* 19: 425–435.

Salter, M. G., Franklin, K. A. and Whitelam, G. C. (2003) Gating of the rapid shade-avoidance response by the circadian clock in plants. *Nature* 426: 680–683.

Sawers, R. J., Sheehan, M. J., and Brutnell, T. P. (2005) Cereal phytochromes: Targets of selection, targets for manipulation? *Trends Plant Sci.* 10: 138–143.

Seo, H. S., Watanabe, E., Tokutomi, S., Nagatani, A., and Chua, N. H. (2004) Photoreceptor ubiquitination by COP1 E3 ligase desensitizes phytochrome A signaling. *Genes Dev.* 18: 617–622.

Sharrock, R. A., and Quail, P. H. (1989) Novel phytochrome sequences in *Arabidopsis thaliana*: Structure, evolution, and differential expression of a plant regulatory photoreceptor family. *Genes Dev.* 3: 1745–1757.

Shin, J., Kim, K., Kang, H., Zulfugarov, I. S., Bae, G., Lee, C. H., Lee, D., and Choi, G. (2009) Phytochromes promote seedling light responses by inhibiting four negatively-acting phytochrome-interacting factors. *Proc. Natl. Acad. Sci. USA* 106: 7660–7665.

Shinomura, T., Uchida, K., and Furuya, M. (2000) Elementary processes of photoperception by phytochrome A for high-irradiance response of hypocotyl elongation in Arabidopsis. *Plant Physiol.* 122: 147–156.

Shinomura, T., Nagatani, A., Hanzawa, H., Kubota, M., Watanabe, M., and Furuya, M. (1996) Action spectra for phytochrome A- and B-specific photoinduction of seed germination in *Arabidopsis thaliana. Proc. Natl. Acad. Sci. USA* 93: 8129–8133.

Shropshire, W., Jr., Klein, W. H., and Elstad, V. B. (1961) Action spectra of photomorphogenic induction and photoinactivation of germination in *Arabidopsis thaliana. Plant Cell Physiol.* 2: 63–69.

Smith, H. (1982) Light quality photoperception and plant strategy. *Annu. Rev. Plant Physiol.* 33: 481–518.

Somers, D. E., Devlin, P. F., and Kay, S. A. (1998) Phytochromes and cryptochromes in the entrainment of the *Arabidopsis* circadian clock. *Science* 282: 1488–1494.

Strayer, C., Oyama, T., Schultz, T. F., Raman, R., Somer, D. E., Mas, P., Panda, S., Kreps, J. A., and Kay, S. A. (2001) Cloning of the *Arabidopsis* clock gene *TOC1*, an autoregulatory response regulator homolog. *Science* 289: 768–771.

Stephenson, P. G., Fankhauser, C., and Terry, M. J. (2009) PIF3 is a repressor of chloroplast development. *Proc. Natl. Acad. Sci. USA* 106: 7654–7659.

Suetsugu, N., Mittmann, F., Wagner, G., Hughes, J., and Wada, M. (2005) A chimeric photoreceptor gene, NEOCHROME, has arisen twice during plant evolution. *Proc. Natl. Acad. Sci. USA* 102: 13705–13709.

Sugano, S., Andronis, C., Ong, M. S., Green, R. M., and Tobin, E. M. (1999) The protein kinase CK2 is involved in regulation of circadian rhythms in Arabidopsis. *Proc. Natl. Acad. Sci. USA* 96: 12362–12366.

Takano, M., Inagaki, N., Xie, X., Yuzurihara, N., Hihara, F., Ishizuka, T., Yano, M., Nishimura, M., Miyao, A., Hirochika, H., et al. (2005) Distinct and cooperative functions of phytochromes A, B, and C in the control of deetiolation and flowering in rice. *Plant Cell* 17: 3311–3325.

Tepperman, J. M., Hudson, M. E., Khanna, R., Zhu, T., Chang, S. H., Wang, X., and Quail, P. H. (2004) Expression profiling of phyB mutant demonstrates substantial contribution of other phytochromes to red-light-regulated gene expression during seedling de-etiolation. *Plant J.* 38: 725–739.

Tepperman, J. M., Zhu, T., Chang, H. S., Wang, X., and Quail, P. H. (2001) Multiple transcription factor genes are early targets of phytochrome A signaling. *Proc. Natl. Acad. Sci. USA* 98: 9437–9442.

Valverde, F., Mouradov, A., Soppe, W., Ravenscroft, D., Samach, A., and Coupland, G. (2004) Photoreceptor regulation of CONSTANS protein in photoperiodic flowering. *Science* 303: 1003–1006.

Vierstra, R. D. (1994) Phytochrome degradation. In *Photomorphogenesis in Plants*, 2nd ed., R. E. Kendrick and G. H. M. Kronenberg, eds., Martinus Nijhoff, Dordrecht, Netherlands, pp. 141–162.

Wagner, J. R., Brunzelle, J. S., Forest, K. T., and Vierstra, R. D. (2005) A light-sensing knot revealed by the structure of the chromophore binding domain on phytochrome. *Nature* 17: 325–321.

Weller, J. L., Murfet, I. C., and Reid, J. C. (1997) Pea mutants with reduced sensitivity to far-red light define an important role for phytochrome A in day-length detection. *Plant Physiol.* 114: 1225–1236.

Weller, J. L., Schreuder, M. E. L., Smith, H., Koornneef, M., and Kendrick, R. E. (2000) Physiological interactions of phytochromes A, B1 and B2 in the control of development in tomato. *Plant J.* 24: 345–356.

Whitelam, G. C., Johnson, E., Peng, J., Carol, P., Anderson, M. L., Cowl, J. S., and Harberd, N. P. (1993) Phytochrome A null mutants of Arabidopsis display a wild-type phenotype in white light. *Plant Cell* 5: 757–768.

Yamaguchi, R., Nakamura, M., Mochizuki, N., Kay, S. A., and Nagatani, A. (1999) Light-dependent translocation of a phytochrome B-GFP fusion protein to the nucleus in transgenic *Arabidopsis. J. Cell Biol.* 145: 437–445.

Yeh, K.-C., and Lagarias, J. C. (1998) Eukaryotic phytochromes: Light-regulated serine/threonine protein kinases with histidine kinase ancestry. *Proc. Natl. Acad. Sci. USA* 95: 13976–13981.

Yi, C., and Deng, X. W. (2005) COP1-from plant photomorphogenesis to mammalian tumorigenesis. *Trends Cell Biol.* 15: 618–625.

Zhu, D., Maier, A., Lee, J. H., Laubinger, S., Saijo, Y., Wang, H., Qu, L. J., Hoecker, U., and Deng, X. W. (2008) Biochemical characterization of Arabidopsis complexes containing CONSTITUTIVELY PHOTOMORPHOGENIC1 and SUPPRESSOR OF PHYA proteins in light control of plant development. *Plant Cell* 20: 2307–2323.

# 18

# Blue-Light Responses: Morphogenesis and Stomatal Movements

Most of us are familiar with the observation that the branches of houseplants placed near a window grow toward the incoming light. This phenomenon, called *phototropism*, is an example of how in nature plants alter their growth patterns in response to the direction of incident radiation. This response to light is intrinsically different from light trapping by photosynthesis. In photosynthesis, plants harness light and convert it into chemical energy (see Chapters 7 and 8). In contrast, phototropism is an example of the use of light as a *signal*. There are two major families of plant responses to light signals: the phytochrome responses, which were covered in Chapter 17, and the **blue-light responses**.

Some blue-light responses were introduced in Chapter 9—for example, chloroplast movement within cells in response to incident photon fluxes, and sun tracking by leaves. As with the phytochrome responses, there are numerous plant responses to blue light. Blue-light responses have been reported in higher plants, algae, ferns, fungi, and prokaryotes. Besides phototropism, they include anion uptake in algae, inhibition of seedling hypocotyl (stem) elongation, stimulation of chlorophyll and carotenoid synthesis, activation of gene expression, stomatal movements, and enhancement of respiration. Among motile unicellular organisms such as algae and bacteria, blue light mediates phototaxis, the movement of motile unicellular organisms toward or away from light (Senger 1984). Blue light also mediates the infection process in bacteria such as *Brucella* (Swartz et al 2007).

Some responses, such as electrical events at the plasma membrane, can be detected within seconds of irradiation by blue light. Other

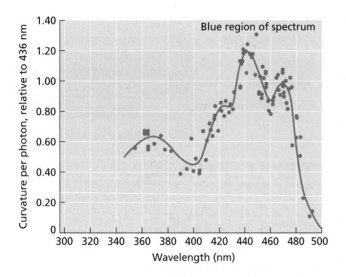

**FIGURE 18.1** Action spectrum for blue light–stimulated phototropism in oat coleoptiles. An action spectrum shows the relationship between a biological response and the wavelengths of light absorbed. The "three-finger" pattern in the 400–500 nm region is characteristic of specific blue-light responses. (After Thimann and Curry 1960.)

metabolic or morphogenetic responses may require minutes, hours, or even days, such as blue light–stimulated pigment biosynthesis in the fungus *Neurospora* or branching in the alga *Vaucheria* (Horwitz 1994).

Both chlorophylls and phytochrome absorb blue light (400–500 nm) from the visible spectrum. Other pigments and some amino acids, such as tryptophan, absorb light in the ultraviolet (250–400 nm) region. How, then, can we functionally distinguish intrinsic responses to blue light? Specific blue-light responses can be distinguished from photosynthetic responses by using red light, which stimulates photosynthetic responses but not blue-light responses. Blue-light responses can be distinguished from phytochrome responses by testing red/far-red reversibility, which is characteristic of phytochrome responses, but not of blue-light responses.

Another key distinction is that *many blue-light responses of higher plants share a characteristic action spectrum.* You will recall from Chapter 7 that an action spectrum is a graph that plots the magnitude of a light response as a function of wavelength (see **WEB TOPIC 7.1** for a detailed discussion of spectroscopy and action spectra). The action spectrum of the response can be compared with the *absorption spectra* of candidate photoreceptors. A close correspondence between action and absorption spectra provides a strong indication that the pigment under consideration is the photoreceptor mediating that particular light response (see Figure 7.8).

Action spectra for blue light–stimulated phototropism, stomatal movements, inhibition of hypocotyl elongation,

and other key blue-light responses share a characteristic "three-finger" fine structure in the 400–500 nm region (**FIGURE 18.1**) that is not observed in spectra for responses to light that are mediated by chlorophyll, phytochrome, or other photoreceptors (Cosgrove 1994).

In this chapter, we will describe representative blue-light responses in plants: phototropism, inhibition of stem elongation, and stomatal movements. The stomatal responses to blue light are discussed in detail because of the importance of stomata in leaf gas exchange and in plant acclimations and adaptations to their environment (see Chapter 9). We will also discuss blue-light photoreceptors and the signal transduction process that links light perception with the final expression of blue-light sensing in the organism.

## The Photophysiology of Blue-Light Responses

Blue-light signals are used in many responses, allowing the plant to sense the presence of light and its direction. This section describes the major morphological, physiological, and biochemical changes associated with some typical blue-light responses.

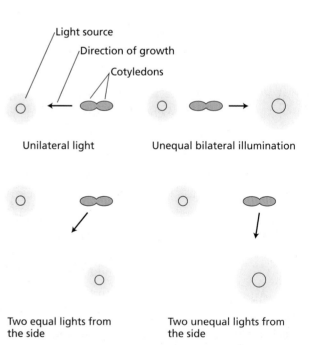

**FIGURE 18.2** Relationship between direction of growth and unequal incident light. Cotyledons from a young seedling are shown as viewed from the top. The arrows indicate the direction of phototropic curvature. The diagrams illustrate how the direction of growth varies with the location and the intensity of the light source, but growth is always toward light. (After Firn 1994.)

**FIGURE 18.3** Time-lapse photograph of a corn coleoptile growing toward unilateral blue light given from the right. In the first image on the left, the coleoptile is about 3 cm long. The consecutive exposures were made 30 minutes apart. Note the increasing angle of curvature as the coleoptile bends. (Courtesy of M. A. Quiñones.)

from the soil and the first true leaf has pierced the tip of the coleoptile. On the other hand, dark-grown, *etiolated* coleoptiles continue to elongate at high rates for several days and, depending on the species, can attain several centimeters in length. The dramatic phototropic response of these etiolated coleoptiles (see Figure 18.3) has made them a classic model for studies of phototropism (Firn 1994).

The action spectrum shown in Figure 18.1 was obtained by measuring the angles of curvature from oat coleoptiles that were irradiated with light of different wavelengths. The spectrum shows a peak at about 370 nm and the "three-finger" pattern in the 400–500 nm region discussed earlier. An action spectrum for phototropism in the dicot alfalfa (*Medicago sativa*) was found to be very similar to that of oat coleoptiles, suggesting that a common photoreceptor mediates phototropism in the two species.

Over the last three decades, phototropism of the stem of the small dicot Arabidopsis (**FIGURE 18.4**) has attracted much attention because of the ease with which advanced molecular techniques can be applied to Arabidopsis mutants. The genetics and the molecular biology of phototropism in Arabidopsis are discussed later in this chapter.

## Blue light rapidly inhibits stem elongation

In nature, the stems of seedlings growing in the dark elongate very rapidly, and the inhibition of stem elongation by light is a key morphogenetic response of the seedling

## Blue light stimulates asymmetric growth and bending

Directional growth toward (or in special circumstances away from) the light is called **phototropism**. It can be observed in fungi, ferns, and higher plants. Phototropism is a **photomorphogenetic** response that is particularly dramatic in dark-grown seedlings of both monocots and dicots. Unilateral light is commonly used in experimental studies, but phototropism can also be observed when a seedling is exposed to two unequally bright light sources from different directions (**FIGURE 18.2**), a condition that can occur in nature. *Light gradients* within the phototropic organs might play a role in the sensing of the direction of the light signal (see **WEB TOPIC 18.1**).

As it grows through the soil, the shoot of a grass seedling is protected by a modified leaf that covers it, called a **coleoptile** (**FIGURE 18.3**; see also Figure 19.1). As discussed in detail in Chapter 19, unequal light perception in the coleoptile results in unequal concentrations of auxin in the lighted and shaded sides of the coleoptile, which in turn cause unequal (asymmetric) growth and bending.

Keep in mind that phototropic bending occurs only in *growing* organs, and that coleoptiles and shoots that have stopped elongating will not bend when exposed to unilateral light. In grass seedlings growing in soil under sunlight, coleoptiles stop growing as soon as the shoot has emerged

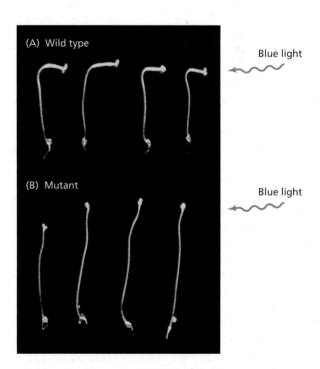

**FIGURE 18.4** Phototropism in wild-type (A) and mutant (B) Arabidopsis seedlings. Unilateral light was applied from the right. (Courtesy of Dr. Eva Huala.)

emerging from the soil surface (see Chapter 17). The conversion of Pr to Pfr (the red- and far red–absorbing forms of phytochrome, respectively) in etiolated seedlings causes a sharp decrease in elongation rates that is phytochrome-dependent (see Figure 17.1).

On the other hand, the action spectrum for the decrease in elongation rate shows strong activity in the blue region, which cannot be explained by the absorption properties of phytochrome (see WEB TOPIC 17.2). In fact, the 400–500-nm blue region of the action spectrum for the inhibition of stem elongation closely resembles that of phototropism (compare the action spectra in WEB TOPIC 17.2 and in Figure 18.1).

It is possible to experimentally separate a reduction in elongation rates mediated by phytochrome from a reduction mediated by a specific blue-light response. If lettuce seedlings are given low fluence rates of blue light under a strong background of yellow light, their hypocotyl elongation rate is reduced by more than 50%. The background yellow light establishes a well-defined Pr:Pfr ratio (see Chapter 17). Blue light added at low fluence rates does not significantly change this ratio, ruling out a phytochrome effect on the reduction in elongation rate observed upon the addition of blue light. These results indicate that the elongation rate of the hypocotyl is controlled by a specific blue-light response that is independent of the phytochrome-mediated response.

A specific blue light–mediated hypocotyl response can also be distinguished from one mediated by phytochrome by their contrasting time courses. Whereas phytochrome-mediated changes in elongation rates can be detected within 8 to 90 minutes, depending on the species, blue-light responses are rapid, and can be measured within 15 to 30 seconds (FIGURE 18.5A). Interactions between phytochrome and the blue light–dependent sensory transduction process in the regulation of elongation rates will be described later in the chapter.

Another rapid response elicited by blue light is a depolarization of the membrane of hypocotyl cells that precedes the inhibition of growth rate (FIGURE 18.5B). This membrane depolarization is caused by the activation of anion channels (see Chapter 6), which facilitates the efflux of anions such as chloride. Application of an anion channel blocker, NPPB (5-nitro-2-(4-phenylbutylamino)-benzoate), prevents the blue light–dependent membrane depolarization and decreases the inhibitory effect of blue light on hypocotyl elongation (Spalding 2000).

### Blue light stimulates stomatal opening

We now turn to the stomatal response to blue light. Stomata have a major regulatory role in gas exchange in leaves (see Chapter 9), and the efficiency of this regulation can impact yields of agricultural crops (see WEB TOPIC 26.1). Several characteristics of blue light–dependent stomatal move-

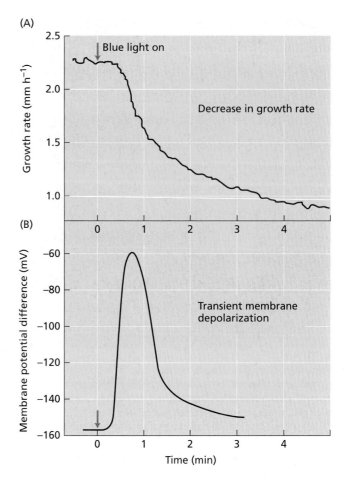

**FIGURE 18.5** Blue light–induced changes in elongation rates of etiolated cucumber seedlings (A) and transient membrane depolarization of hypocotyl cells (B). As the membrane depolarization (measured with intracellular electrodes) reaches its maximum, growth rate (measured with position transducers) declines sharply. Comparison of the two curves shows that the membrane starts to depolarize before the growth rate begins to decline, suggesting a cause–effect relation between the two phenomena. (After Spalding and Cosgrove 1989.)

ments make guard cells a valuable experimental system for the study of blue-light responses:

- The stomatal response to blue light is rapid and reversible, and it is localized in a single cell type, the **guard cell** (FIGURE 18.6).

- The stomatal response to blue light regulates stomatal movements *throughout the life of the plant*. This is unlike phototropism and hypocotyl elongation, which are functionally important only at early stages of development.

- The signal transduction process that links the perception of blue light with the opening of stomata is understood in considerable detail.

(A)

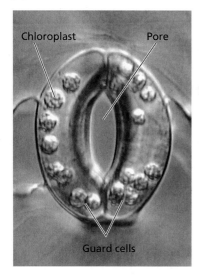

(B)

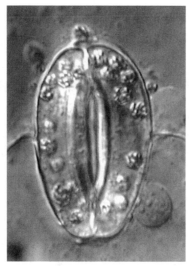

20 μm

**FIGURE 18.6** Light-stimulated stomatal opening in detached epidermis of *Vicia faba*. Open, light-treated stoma (A) is shown in the dark-treated, closed state in (B). Stomatal opening is quantified by microscopic measurement of the width of the stomatal pore. (Courtesy of E. Raveh.)

Since blue light stimulates both the specific blue-light response of stomata and guard cell photosynthesis (see the action spectrum for photosynthesis in Figure 7.8), blue light alone cannot be used to study the specific stomatal response to blue light. To effect a clear-cut separation between the two light responses, researchers use dual-beam experiments. First, high fluence rates of red light are used to *saturate* the photosynthetic response; such saturation prevents any further stomatal opening mediated by photosynthesis in response to further increases in red or blue light. Then, low photon fluxes of blue light are added after the response to the saturating red light has been established (**FIGURE 18.8**). The addition of blue light causes substantial additional stomatal opening that, as just explained, cannot be due to a further stimulation of guard cell photosynthesis, because the background red light has saturated photosynthesis.

An action spectrum for the stomatal response to blue light under saturating background red illumination shows

In the following sections we will discuss two central aspects of the stomatal response to light: the osmoregulatory mechanisms that drive stomatal movements, and the role of a blue light–activated H⁺-ATPase in ion uptake by guard cells.

Light is the dominant environmental signal controlling stomatal movements in leaves of well-watered plants growing in a natural environment. Stomata open as light levels reaching the leaf surface increase, and close as light decreases (see Figure 18.6). In greenhouse-grown leaves of broad bean (*Vicia faba*), stomatal movements closely track incident solar radiation at the leaf surface (**FIGURE 18.7**). This light dependence of stomatal movements has been documented in many species and conditions.

Early studies of the stomatal response to light showed that dichlorophenyldimethylurea (DCMU), an inhibitor of photosynthetic electron transport (see Figure 7.30), causes a partial inhibition of light-stimulated stomatal opening. These results indicated that photosynthesis in the guard cell chloroplast plays a role in light-dependent stomatal opening, but the partial response to DCMU suggested the involvement of a DCMU-insensitive, nonphotosynthetic component of the stomatal response to light. Detailed studies of the light responses of stomata, carried out under colored light, have shown that light activates two distinct responses of guard cells: photosynthesis in the guard cell chloroplast (Lawson 2009; see also **WEB ESSAY 18.1**), and a specific blue-light response.

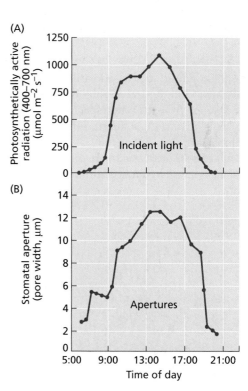

**FIGURE 18.7** Stomatal opening tracks photosynthetic active radiation at the leaf surface. Stomatal opening in the lower surface of leaves of *V. faba* grown in a greenhouse, measured as the width of the stomatal pore (A), closely follows the levels of photosynthetically active radiation (400–700 nm) incident to the leaf (B), indicating that the response to light is the dominant response regulating stomatal opening. (After Srivastava and Zeiger 1995a.)

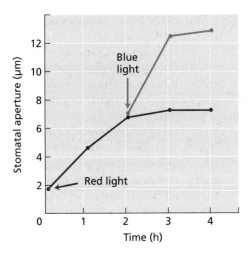

FIGURE 18.8 The response of stomata to blue light under a red-light background. Stomata from detached epidermis of common dayflower (*Commelina communis*) were treated with saturating photon fluxes of red light (red trace). In a parallel treatment, stomata illuminated with red light were also illuminated with blue light, as indicated by the arrow (blue trace). The increase in stomatal opening above the level reached in the presence of saturating red light indicates that a different photoreceptor system, stimulated by blue light, is mediating the additional increases in opening. (From Schwartz and Zeiger 1984.)

the three-finger pattern discussed earlier (**FIGURE 18.9**). This action spectrum, typical of blue-light responses and distinctly different from the action spectrum for photosynthesis, further indicates that guard cells respond specifically to blue light.

When guard cells are treated with cellulolytic enzymes that digest the cell walls, **guard cell protoplasts** are released

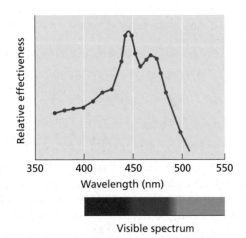

FIGURE 18.9 The action spectrum for blue light–stimulated stomatal opening (under a red-light background). (After Karlsson 1986.)

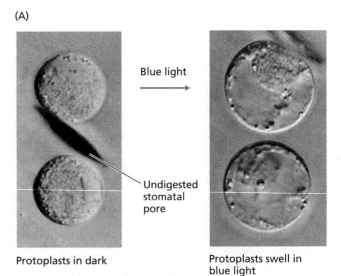

FIGURE 18.10 Blue light–stimulated swelling of guard cell protoplasts. (A) In the absence of a rigid cell wall, guard cell protoplasts of onion (*Allium cepa*) swell. (B) Blue light stimulates the swelling of guard cell protoplasts of broad bean (*V. faba*), and orthovanadate, an inhibitor of the $H^+$-ATPase, inhibits this swelling. Blue light stimulates ion and water uptake in the guard cell protoplasts, which in the intact guard cells provides a mechanical force working against the rigid cell wall that distorts the guard cells and drives increases in stomatal apertures. (A from Zeiger and Hepler 1977; B after Amodeo et al. 1992.)

and can be used for experimentation. In the laboratory, guard cell protoplasts swell when illuminated with blue light (**FIGURE 18.10**), indicating that blue light is sensed within the guard cells proper. The swelling of guard cell protoplasts also illustrates how intact guard cells function. Light stimulates the uptake of ions, and the accumulation of organic solutes in the guard cell protoplasts decreases the cell's osmotic potential (increases the osmotic pressure). As a result, water flows in, and the guard cell protoplasts swell. In guard cells with intact cell walls, this increase in turgor leads to deformation of the cell walls and an increase in the stomatal aperture (see Chapter 4).

### Blue light activates a proton pump at the guard cell plasma membrane

When guard cell protoplasts from broad bean (*V. faba*) are irradiated with blue light under saturating background red-light illumination, the pH of the suspension medium becomes more acidic (**FIGURE 18.11**). This blue light–induced acidification is blocked by inhibitors that dissipate pH gradients, such as CCCP (discussed shortly), and by inhibitors of the proton-pumping $H^+$-ATPase, such as orthovanadate, which was discussed in Chapter 6 (see Figure 18.10B). Such acidification studies have made it clear that *blue light activates a proton-pumping ATPase in the guard cell plasma membrane.*

In the intact leaf, this blue-light stimulation of proton pumping lowers the pH of the apoplastic space surrounding the guard cells, and generates the driving force needed for ion uptake and stomatal opening. The plasma membrane ATPase from guard cells has been isolated and extensively characterized (Kinoshita and Shimazaki 2001).

The activation of electrogenic pumps such as the proton-pumping ATPase can be measured in patch-clamping experiments as an outward electric current at the plasma membrane (see **WEB TOPIC 6.2** for a description of patch clamping). **FIGURE 18.12A** shows a patch clamp recording of a guard cell protoplast treated in the dark with the fungal toxin fusicoccin, a well-characterized activator of plasma membrane ATPases. Exposure to fusicoccin

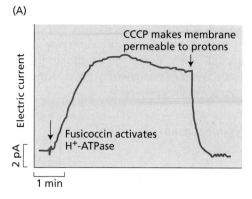

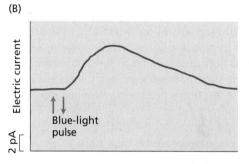

**FIGURE 18.12** Activation of the $H^+$-ATPase at the plasma membrane of guard cell protoplasts by fusicoccin and blue light can be measured as electric current in patch clamp experiments. (A) Outward electric current (measured in picoamps, pA) at the plasma membrane of a guard cell protoplast stimulated by the fungal toxin fusicoccin, an activator of the $H^+$-ATPase. The current is abolished by the proton ionophore carbonyl cyanide *m*-chlorophenylhydrazone (CCCP). (B) Outward electric current at the plasma membrane of a guard cell protoplast stimulated by a blue-light pulse. These results indicate that blue light stimulates the $H^+$-ATPase. (A after Serrano et al. 1988; B after Assmann et al. 1985.)

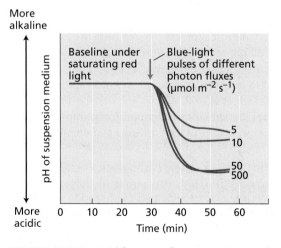

**FIGURE 18.11** Acidification of a suspension medium of guard cell protoplasts of *V. faba* stimulated by a 30-s pulse of blue light. The acidification results from the stimulation of an $H^+$-ATPase at the plasma membrane by blue light, and it is associated with protoplast swelling (see Figure 18.10). (After Shimazaki et al. 1986.)

stimulates an outward electric current, which generates a proton gradient. This proton gradient is abolished by carbonyl cyanide *m*-chlorophenylhydrazone (CCCP), a proton ionophore that makes the plasma membrane highly permeable to protons, thus precluding the formation of a proton gradient across the membrane and abolishing net proton efflux.

The relationship between proton pumping at the guard cell plasma membrane and stomatal opening is evident from the observations that (1) fusicoccin stimulates both proton extrusion from guard cell protoplasts and stomatal opening, and (2) CCCP inhibits the fusicoccin-stimulated opening. The increase in proton-pumping rates as a function of fluence rates of blue light further indicates that increasing the number of blue photons in the solar radiation reaching the leaf should cause a larger stomatal opening (see Figure 18.11) A pulse of blue light given under a saturating red-light background can also stimulate an outward electric current from guard cell protoplasts (**FIGURE 18.12B**). The relationship between these pulse-stimulated electrical currents and the acidification response to blue light pulses, shown in Figure 18.11, indicates that the measured electric current is carried by protons moving from the cell interior to the apoplast.

### Blue-light responses have characteristic kinetics and lag times

The temporal stomatal responses to blue-light pulses illustrate some important properties of blue-light responses: a persistence of the response after the light signal has been switched off, and a significant lag time separating the onset of the light signal and the beginning of the response. In contrast to typical photosynthetic responses, which are activated very quickly after a "light on" signal, and cease when the light goes off (see, for instance, Figure 7.13), blue-light responses proceed at maximal rates for several minutes after application of the pulse (see Figures 18.11 and 18.12B).

This persistence of the blue-light response after the "light off" signal can be explained by a physiologically inactive form of the blue-light photoreceptor that is converted to an active form by blue light, with the active form reverting slowly to the physiologically inactive form after the blue light is switched off (Iino et al. 1985). The rate of the response to a blue-light pulse would thus depend on the time course of the reversion of the active form to the inactive one.

Another property of the response to blue-light pulses is a lag time, which lasts about 25 seconds in both the acidification response and the outward electric currents stimulated by blue light (see Figures 18.11 and 18.12). This time interval is probably required for the signal transduction process to proceed from the photoreceptor site to the proton-pumping ATPase and for the proton gradient to form. Similar lag times have been measured for blue light–dependent inhibition of hypocotyl elongation, which was discussed earlier.

### Blue light regulates the osmotic balance of guard cells

Blue light modulates guard cell osmoregulation by means of its activation of proton pumping, solute uptake and stimulation of the synthesis of organic solutes. Before discussing these blue-light responses, let us briefly consider the major osmotically active solutes in guard cells.

The botanist Hugo von Mohl proposed in 1856 that turgor changes in guard cells provide the mechanical force for changes in stomatal apertures (see Chapter 4). The plant physiologist F. E. Lloyd hypothesized in 1908 that guard cell turgor is regulated by osmotic changes resulting from starch–sugar interconversions, a concept that led

to a starch–sugar hypothesis of stomatal movements. The discovery of potassium ion fluxes in guard cells in Japan in the 1940s and its re-discovery in the West in the 1960s, replaced the starch-sugar hypothesis with the modern theory of guard cell osmoregulation by potassium and its counterions, $Cl^-$ and malate$^{2-}$.

Potassium concentration in guard cells increases severalfold when stomata open, from 100 m$M$ in the closed state to 400 to 800 m$M$ in the open state, depending on the species and the experimental conditions. In most species, these large concentration changes in $K^+$ are electrically balanced by varying amounts of the anions $Cl^-$ and malate$^{2-}$ (**FIGURE 18.13A**) (Talbott et al. 1996; see also **WEB TOPIC 18.2**). On the other hand, in some species of the genus *Allium*, such as onion (*A. cepa*), $K^+$ ions are balanced solely by $Cl^-$ (Schnabl and Ziegler 1977).

Chloride ions are taken up into the guard cells from the apoplast during stomatal opening and extruded during stomatal closing. Malate, on the other hand, is synthesized in the guard cell cytosol, in a metabolic pathway that uses carbon skeletons generated by starch hydrolysis (**FIGURE 18.13B**). The malate content of guard cells decreases during stomatal closing, but it remains unclear whether malate is catabolized in mitochondrial respiration, extruded into the apoplast, or both.

Potassium and chloride are taken up into guard cells via secondary transport mechanisms driven by the electrochemical potential for $H^+$, $\Delta\mu_{H^+}$, generated by the proton pump discussed earlier in the chapter (see Figure 18.13; see also Chapter 6). Proton extrusion makes the electrical-potential difference across the guard cell plasma membrane more negative; light-dependent hyperpolarizations as high as 64 mV have been measured (Roelfsema et al. 2001). In addition, proton pumping generates a pH gradient of about 0.5 to 1 pH unit.

The electrical component of the proton gradient provides a driving force for the passive uptake of potassium ions via voltage-regulated potassium channels that were discussed in Chapter 6 (Schroeder et al. 2001). Chloride is thought to be taken up through a proton–chloride symporter (Pandey et al. 2007). Thus, the driving force generated by the blue light–dependent proton pumping plays a key role in the uptake of ions for stomatal opening.

**FIGURE 18.13** Three distinct osmoregulatory pathways ▶ in guard cells. The thick, dark arrows identify the major metabolic steps of each pathway that lead to the accumulation of osmotically active solutes in the guard cells. (A) Potassium and its counterions. Potassium and chloride are taken up in secondary transport processes driven by a proton gradient; malate is formed from the hydrolysis of starch. (B) Accumulation of sucrose from starch hydrolysis. (C) Accumulation of sucrose from photosynthetic carbon fixation. The possible uptake of apoplastic sucrose is also indicated. (From Talbott and Zeiger 1998.)

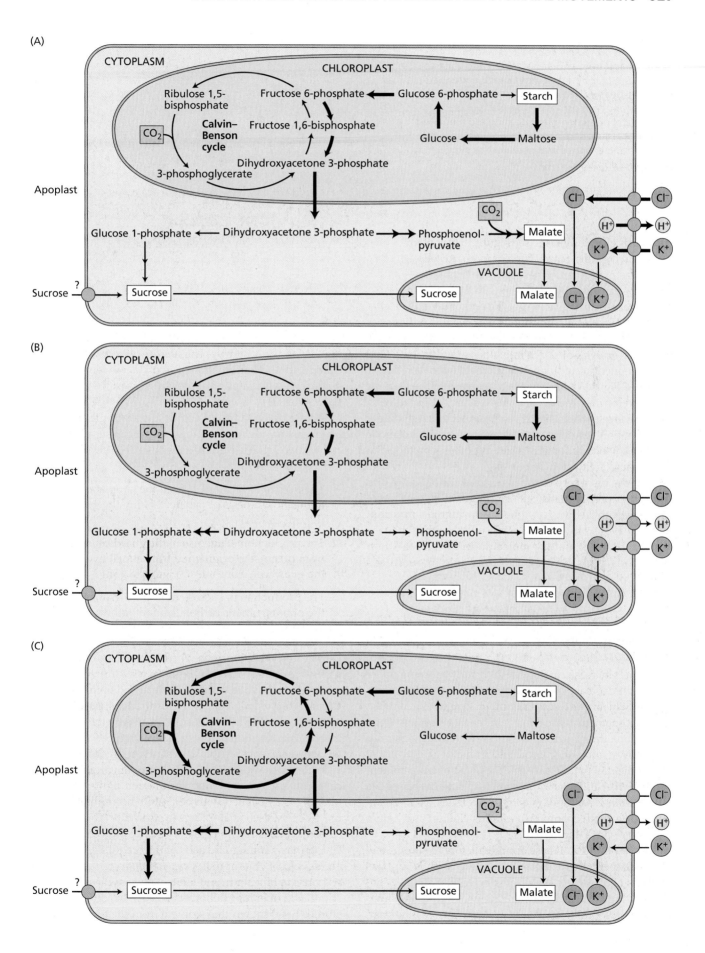

**FIGURE 18.14** Daily course of changes in stomatal aperture, and in potassium and sucrose content, of guard cells from intact leaves of broad bean (*V. faba*). These results indicate that the changes in osmotic potential required for stomatal opening in the morning are mediated by potassium and its counterions, whereas the afternoon changes are mediated by sucrose. (After Talbott and Zeiger 1998.)

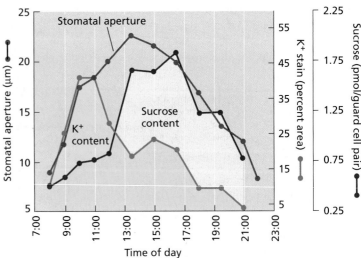

Guard cell chloroplasts (see Figure 18.6) contain large starch grains. Starch content in guard cells—in contrast with that in mesophyll chloroplasts—decreases during stomatal opening in the morning and increases during closing at the end of the day. Starch, an insoluble, high-molecular-weight polymer of glucose, does not contribute to the cell's osmotic potential, but the hydrolysis of starch into glucose and fructose and the subsequent accumulation of sucrose (Talbot and Zeiger 1998; see Figure 18.13B) cause a decrease in the osmotic potential (or increase in osmotic pressure) of guard cells. In the reverse process, starch synthesis decreases the sugar concentration, resulting in an increase of the cell's osmotic potential, which the starch–sugar hypothesis predicted to be associated with stomatal closing.

With the discovery of the major role of potassium and its counterions in guard cell osmoregulation, the starch-sugar hypothesis was no longer considered important (Outlaw 1983). However, we will now see that sucrose, a major osmoregulatory molecule in the starch-sugar hypothesis, does play an important role in guard cell osmoregulation.

### Sucrose is an osmotically active solute in guard cells

Recent studies of daily courses of stomatal movements in intact leaves have shown that the potassium content in guard cells increases in parallel with early-morning opening, but it decreases in the early afternoon under conditions in which apertures continue to increase. In contrast, sucrose content increases slowly in the morning, and upon potassium efflux, sucrose becomes the dominant osmotically active solute in guard cells. Stomatal closing at the end of the day parallels a decrease in the sucrose content (FIGURE 18.14). These osmoregulatory patterns have been seen in both *V. faba* and onion guard cells grown in greenhouse or growth chamber conditions (Talbott and Zeiger 1998).

One implication of these osmoregulatory features is that stomatal opening is associated primarily with K+ uptake, and closing is associated with a decrease in sucrose content (see Figure 18.14). The function of distinct potassium- and sucrose-dominated osmoregulatory phases is unclear; potassium might be the solute of choice for the daily opening that occurs at sunrise. The sucrose phase might be associated with the coordination of stomatal movements in the epidermis with rates of photosynthesis in the mesophyll (Lawson 2009).

Where do osmotically active solutes originate? Three major distinct metabolic pathways that can supply osmotically active solutes to guard cells have been clearly characterized (see Figure 18.13):

1. The uptake of K+ and Cl− from the apoplast, coupled to the biosynthesis of malate$^{2-}$ within the guard cells (Figure 18.13A)

2. The production of sucrose in the guard cell cytoplasm from precursors originating from starch hydrolysis in the guard cell chloroplast (Figure 18.13B)

3. The production of sucrose from precursors made in the photosynthetic carbon fixation pathway at the guard cell chloroplast (FIGURE 18.13C)

Depending on conditions, one or more osmoregulatory pathways can be activated. For instance, during red light–stimulated, stomatal opening in detached epidermis of *V. faba* kept under ambient $CO_2$ concentrations, the dominant solute in guard cells is sucrose generated by the photosynthetic carbon fixation pathway in the guard cell chloroplast, with no detectable K+ uptake or starch degradation (Figure 18.13C). On the other hand, in $CO_2$-free air, photosynthetic carbon fixation is inhibited and the red light-stimulated opening is associated with potassium accumulation (Figure 18.13A; Olsen et al. 2002; see also WEB TOPIC 18.2).

Some unusual osmoregulatory pathways occur in nature. Chloroplasts from guard cells of the orchid *Paphiopedilum* are devoid of chlorophyll. *Paphiopedilum* stomata open in response to blue light, and fail to show a typical red light–stimulated opening (Talbott et al. 2002). In contrast, recent studies have shown that stomata from the fern *Adiantum* lack a specific blue-light response and open in response to red light (Doi et al. 2008). *Adiantum* guard cells have an unusually large number of guard cell chloroplasts and the red light–dependent opening in *Adiantum* is blocked by DCMU, an inhibitor of photosynthetic electron transport. This implicates guard cell photosynthesis in the red light–stimulated opening. On the other hand, *Adiantum* stomata accumulate potassium under ambient $CO_2$ concentrations and are insensitive to $CO_2$ both in darkness and under red light. It is intriguing that these unusual osmoregulatory features are associated with an exceptionally high number of chloroplasts in these guard cells. Furthermore, *Adiantum* stomata are highly unusual in their lack of blue light sensitivity. In contrast, sensitivity to $CO_2$ and to blue light were shown to be linearly related to stomatal opening in *V. faba* (Zhu et al. 1998). These results suggest that the lack of a blue-light response and the $CO_2$ insensitivity in *Adiantum* could be associated with a defective blue-light sensory-transducing system (see WEB ESSAY 18.2).

The contrasting, unusual features of the *Paphiopedilum* and *Adiantum* illustrate the remarkable functional plasticity of guard cells, also shown in other studies in the intact leaf (Roelfsema et al. 2002; Outlaw 2003; Fan et al. 2004; Tallman 2004). These plastic features include acclimations of the responses to blue light and $CO_2$, and daily changes in guard cell photosynthetic rates (Zeiger et al. 2002).

## The Regulation of Blue Light–Stimulated Responses

Several key steps in the sensory transduction process of blue light–stimulated stomatal opening have been characterized. The proton-pumping $H^+$-ATPase has a central role in the regulation of stomatal movements. The C terminus of the $H^+$-ATPase has an autoinhibitory domain that regulates the activity of the enzyme (see Figure 6.17). If this autoinhibitory domain is experimentally removed by a protease, the $H^+$-ATPase becomes irreversibly activated. The autoinhibitory domain of the C terminus is thought to lower the activity of the enzyme by blocking its catalytic site. Conversely, fusicoccin appears to activate the enzyme by displacing the autoinhibitory domain away from the catalytic site (Kinoshita and Shimazaki 2001).

Upon blue-light irradiation, the $H^+$-ATPase shows a lower $K_m$ for ATP and a higher $V_{max}$, indicating that blue light activates the $H^+$-ATPase (see Chapter 6). Activation of the enzyme involves the phosphorylation of serine and threonine residues of the C-terminal domain of the $H^+$-ATPase. Blue light–stimulated proton pumping and stomatal opening are prevented by inhibitors of protein kinases, which might block phosphorylation of the $H^+$-ATPase. As with fusicoccin, phosphorylation of the C-terminal domain appears also to displace the autoinhibitory domain of the C terminus from the catalytic site of the enzyme.

A regulatory protein termed **14-3-3 protein** has been found to bind to the phosphorylated C terminus of the guard cell $H^+$-ATPase, but not to the nonphosphorylated one (FIGURE 18.15). The 14-3-3 proteins are ubiquitous regulatory proteins in eukaryotic organisms. In plants, 14-3-3 proteins regulate transcription by binding to activators in the nucleus, and they regulate metabolic enzymes such as nitrate reductase.

Only one of the four 14-3-3 isoforms found in guard cells binds to the $H^+$-ATPase, so the binding appears to be specific. The same 14-3-3 isoform binds to the guard cell $H^+$-ATPase in response to both fusicoccin and blue-light treatments. The 14-3-3 protein seems to dissociate from the $H^+$-ATPase upon dephosphorylation of the C-terminal domain (see Figure 18.15). Recent studies have shown that guard cell phototropins, a class of blue-light receptors discussed later in the chapter, are phosphorylated under blue light and that they bind the 14-3-3 protein upon phosphorylation (Kinoshita et al. 2003).

Proton-pumping rates of guard cells increase with fluence rates of blue light (see Figure 18.12), and the electrochemical gradient generated by the proton pump drives ion uptake into the guard cells, increasing turgor and turgor-mediated stomatal apertures. These processes define the major steps linking the activation of a serine/threonine protein kinase by blue light and blue light–stimulated stomatal opening (see Figure 18.15).

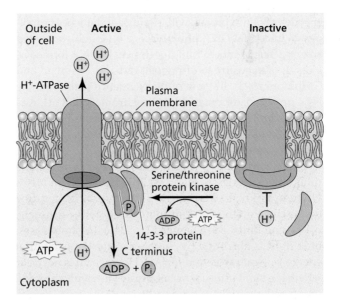

**FIGURE 18.15** The role of the proton-pumping ATPase in the regulation of stomatal movement. Blue light activates the $H^+$-ATPase. Activation of the enzyme involves the phosphorylation of serine and threonine residues of its C-terminal domain. A regulatory protein termed 14-3-3 protein binds to the phosphorylated C terminus of the guard cell $H^+$-ATPase.

# Blue-Light Photoreceptors

Experiments carried out by Charles Darwin and his son Francis in the nineteenth century determined that during phototropism, blue light is perceived in the coleoptile tip. Early hypotheses about blue-light photoreceptors focused on two pigments found in the coleoptile tip, carotenoids and flavins. Despite active research efforts, no significant advances toward the identification of blue-light photoreceptors were made until the early 1990s. In the case of phototropism and the inhibition of stem elongation, progress resulted from the identification of mutants for key blue-light responses, and the subsequent isolation of the relevant gene.

In the following section we will describe three photoreceptors associated with blue-light responses: cryptochromes, which function primarily in the inhibition of stem elongation and flowering; phototropins, which function primarily in phototropism; and zeaxanthin, which functions in the blue-light response of stomatal movement.

## Cryptochromes regulate plant development

Cryptochromes were first identified in Arabidopsis and later discovered in many organisms including cyanobacteria, ferns, algae, fruit flies, mice, and humans. The identification of cryptochrome was achieved in studies with the *hy4* mutant of Arabidopsis, which lacks the blue light–stimulated inhibition of hypocotyl elongation described earlier in the chapter. Isolation of the *HY4* gene showed that it encodes a 75-kDa protein with significant sequence homology to microbial DNA **photolyase**, a blue light–activated enzyme that repairs pyrimidine dimers in DNA formed as a result of exposure to ultraviolet radiation (Ahmad and Cashmore 1993). In view of this sequence similarity, the HY4 protein, later renamed **cryptochrome 1** (**CRY1**), was proposed to be a blue-light photoreceptor mediating the inhibition of stem elongation. Cryptochromes, however, show no photolyase activity in plants.

Like photolyases, cryptochromes bind a flavin adenine dinucleotide (FAD; see Figure 11.2B) and the pterin methyltetrahydrofolate. **Pterins** are light-absorbing pteridine derivatives often found in pigmented cells of insects, fishes, and birds (see Chapter 12, p. 346). In photolyases, blue light is absorbed by the pterin and the excitation energy is then transferred to FAD. However, it is not clear whether a similar mechanism operates in cryptochromes, or if blue light is absorbed directly by FAD.

Cryptochromes mediate several blue-light responses including suppression of hypocotyl elongation, promotion of cotyledon expansion, membrane depolarization, petiole elongation, anthocyanin production, and the regulation of circadian clocks. Overexpression of the CRY1 protein in transgenic tobacco or Arabidopsis plants results in a stronger blue light–stimulated inhibition of hypocotyl

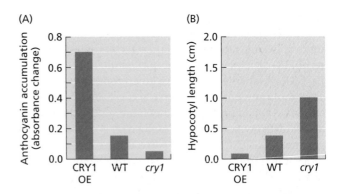

**FIGURE 18.16** Blue light stimulates the accumulation of anthocyanin (A) and the inhibition of stem elongation (B) in transgenic and mutant seedlings of Arabidopsis. These bar graphs show the phenotypes of a transgenic plant overexpressing the gene that encodes CRY1 (CRY1 OE), the wild type (WT), and *cry1* mutants. The enhanced blue-light response of the CRY1 overexpressor demonstrates the important role of this gene product in stimulating anthocyanin biosynthesis and inhibiting stem elongation. (After Ahmad et al. 1998.)

elongation, as well as increased production of anthocyanin (**FIGURE 18.16**).

A second gene product homologous to CRY1, named CRY2, has been isolated from Arabidopsis (Lin 2000). Both CRY1 and CRY2 appear to be ubiquitous throughout the plant kingdom. A major difference between them is that the CRY2 protein is preferentially degraded under blue light, whereas CRY1 is much more stable.

Transgenic plants overexpressing the *cry2* gene show only a small enhancement of the inhibition of hypocotyl elongation found in the wild type, indicating that unlike CRY1, CRY2 does not play a primary role in inhibiting stem elongation. On the other hand, transgenic plants overexpressing CRY2 show a large increase in blue light–stimulated cotyledon expansion. In addition, CRY1 is involved in the setting of the circadian clock in Arabidopsis (see Chapter 17), and both CRY1 and CRY2 play a major role in the induction of flowering (see Chapter 25). Cryptochrome homologs regulate the circadian clock in flies, mice, and humans.

Both CRY1 and CRY2 are detectable in the nucleus upon illumination, and they interact with COP1, the ubiquitin ligase discussed in Chapter 17, both in vivo and in vitro. In addition, the activity of cryptochrome is related to its phosphorylation state. CRY1 and CRY2 have been shown to interact with phytochrome A in vivo, and to be phosphorylated by phytochrome A in vitro (see Chapter 17 and **WEB ESSAY 18.3**). These studies indicate that both phosphorylation and protein degradation are important steps in the sensory transduction of cryptochrome-mediated blue-light signals (Spalding and Folta 2005).

Distinct nuclear and cytoplasmic CRY1 pools have been identified in Arabidopsis (Wu and Spalding 2007). Contrary to expectations, nuclear, rather than cytoplasmic, CRY1 molecules were found to mediate membrane depolarization. This response, showing a time course of a few seconds, is one of the fastest CRY1-mediated responses to blue light. The mechanism involved in this blue light–dependent, anion-channel activation is not yet known.

## Phototropins mediate blue light–dependent phototropism and chloroplast movements

Recall that plants growing under a directional light source bend toward the light to maximize light absorption, and that this phototropic response is mediated by blue light (see Figure 18.4) Arabidopsis mutants impaired in blue light–dependent phototropism of the hypocotyl have provided valuable information about cellular events preceding bending. One of these mutants, the *nph1* (nonphototropic *hypocotyl*) mutant, has been found to be genetically independent of the *hy4* (*cry1*) mutant discussed earlier. The *nph1* mutant lacks a phototropic response in the hypocotyl, but has normal blue light–stimulated inhibition of hypocotyl elongation, while *hy4* has the converse phenotype. The *nph1* gene was renamed *phot1*, and the protein it encodes was named **phototropin** (Christie 2007). Phototropins (phot1 and phot2) mediate phototropism, chloroplast movement, rapid inhibition of growth of etiolated seedlings, and leaf expansion.

The C-terminal half of phototropin is a serine/threonine kinase. The N-terminal half contains two similar domains, called LOV (light, oxygen, or voltage) domains, of about 100 amino acids each. The LOV domains bind flavins and have sequence similarity to proteins involved in signaling in bacteria and mammals. These proteins serve as oxygen sensors in *E. coli* and *Azotobacter* and as voltage sensors in potassium channels of *Drosophila* and vertebrates.

The N-terminal half of phototropin binds flavin mononucleotide (FMN) and undergoes a blue light–dependent autophosphorylation reaction. This reaction resembles the blue light–dependent phosphorylation of a 120-kDa membrane protein found in growing regions of etiolated seedlings. Despite intensive work from several research groups, the cellular events that follow the autophosphorylation of phototropin and its connection with blue-light responses remain unknown (Inoue et al. 2008).

Spectroscopic studies have shown that in the dark, a FMN molecule is noncovalently bound to each LOV domain. Upon blue light illumination, the FMN molecule becomes covalently bound to a cysteine residue in the phototropin molecule, forming a cysteine-flavin covalent adduct (**FIGURE 18.17**) (Schleicher et al. 2004). This reaction is reversed by a dark treatment.

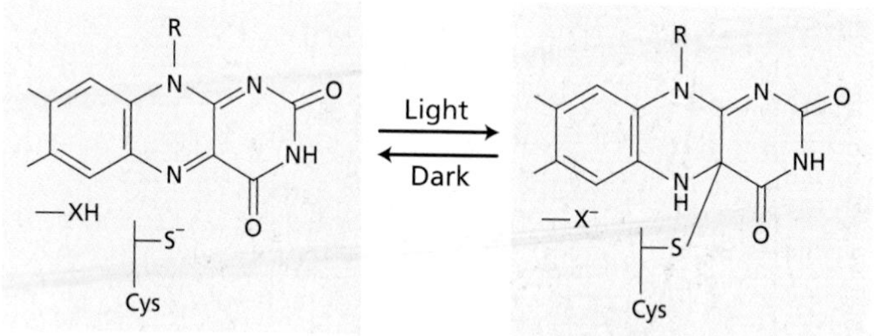

**FIGURE 18.17** Adduct formation of FMN and a cysteine residue of phototropin protein upon blue-light irradiation. XH and X⁻ represent an unidentified proton donor and acceptor, respectively. (After Briggs and Christie 2002.)

The Arabidopsis genome contains a second gene, *phot2*, which is related to *phot1*. The *phot1* mutant lacks hypocotyl phototropism in response to low-intensity blue light ($0.01–1$ µmol m$^{-2}$ s$^{-1}$) but retains a phototropic response at higher intensities ($1–10$ µmol m$^{-2}$ s$^{-1}$). The *phot2* mutant has a normal phototropic response, but the *phot1/phot2* double mutant is severely impaired at both low and high intensities. These data indicate that both *phot1* and *phot2* are involved in the phototropic response, with *phot2* functioning at high light fluence rates or over prolonged time courses.

**PHOTORECEPTOR INTERACTION IN THE INHIBITION OF STEM ELONGATION** High-resolution analysis of the changes in growth rate mediating the inhibition of hypocotyl elongation by blue light has provided valuable information about the interactions among phototropin, CRY1, CRY2, and the phytochrome PHYA (Parks et al. 2001). After a lag of 30 seconds, blue light–treated, wild-type Arabidopsis seedlings show a rapid decrease in elongation rates during the first 30 minutes, and then they grow very slowly for several days (**FIGURE 18.18**).

Analysis of the same response in *phot1*, *cry1*, *cry2*, and *phyA* mutants has shown that suppression of stem elongation by blue light during seedling de-etiolation is initiated by *phot1*, with *cry1*, and to a limited extent *cry2*, modulating the response after 30 minutes. The slow growth rate of stems in blue light–treated seedlings is primarily a result of the persistent action of CRY1, and this is the reason that *cry1* mutants of Arabidopsis show a long hypocotyl, compared to the short hypocotyl of the wild type. Phytochrome A appears to play a role in at least the early stages of blue light–regulated growth, because growth inhibition does not progress normally in *phyA* mutants.

**BLUE LIGHT–ACTIVATED CHLOROPLAST MOVEMENT** Leaves have an adaptive feature that can alter the intracellular distribution of their chloroplasts under changing light conditions. Redistribution of chloroplasts within the cells

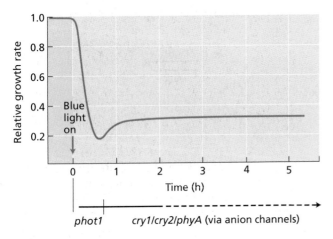

**FIGURE 18.18** Sensory transduction process of blue light–stimulated inhibition of stem elongation in Arabidopsis. Elongation rates in the dark (0.25 mm h$^{-1}$) were normalized to 1. Within 30 seconds of the onset of blue-light irradiation, growth rates decreased; they approached zero within 30 minutes, then continued at very reduced rates for several days. If blue light was applied to a *phot1* mutant, dark-growth rates remained unchanged for the first 30 minutes, indicating that the inhibition of elongation in the first 30 minutes is under phototropin control. Similar experiments with *cry1*, *cry2*, and *phyA* mutants indicated that the respective gene products control elongation rates at later times. (After Parks et al. 2001.)

modulates light absorption and prevents photodamage (see Figure 9.13). Under weak illumination, chloroplasts gather at the upper and lower surfaces of the mesophyll cells (the "accumulation" response; see Figure 9.13B), thus maximizing light absorption.

Under strong illumination, the chloroplasts move to the cell surfaces that are parallel to the incident light (the "avoidance" response; see Figure 9.13C), thus minimizing light absorption and avoiding photodamage. The action spectrum for the redistribution response shows the "three-finger" fine structure typical of specific blue-light responses. Recent studies have shown that mesophyll cells of the *phot1* mutant have a normal avoidance response and a poor accumulation response. Cells from the *phot2* mutant also show a poor accumulation response but lack the avoidance response. Cells from the *phot1/phot2* double mutant lack both the avoidance and accumulation responses (Kadota et al. 2009). These results indicate that *phot2* plays a key role in the avoidance response, and that both *phot1* and *phot2* contribute to the accumulation response.

### Zeaxanthin mediates blue-light photoreception in guard cells

Stomata from the Arabidopsis mutant *npq1* (nonphotochemical quenching) lack a specific blue light response. This

mutant has a lesion in the enzyme that converts the carotenoid violaxanthin to zeaxanthin (Niyogi et al. 1998). Recall from Chapters 7 and 9 that zeaxanthin is a component of the xanthophyll cycle of chloroplasts (see Figure 7.35), which protects photosynthetic pigments from excess excitation energy. In addition, zeaxanthin functions as a blue-light photoreceptor in guard cells, mediating blue light-stimulated stomatal opening. Compelling evidence for this role of zeaxanthin ensues from the observation that, in the absence of zeaxanthin, guard cells from *npq1* lack a specific blue-light response (**FIGURE 18.19**) (Frechilla et al. 1999).

(A)

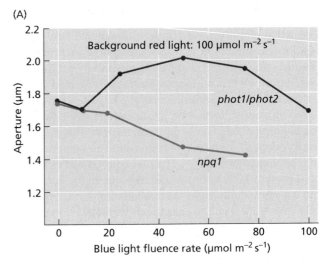

(B)

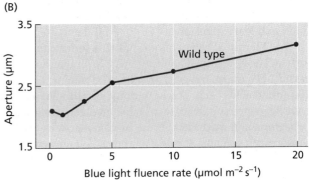

**FIGURE 18.19** (A) The blue-light sensitivity of the zeaxanthin-less mutant *npq1*, and of the phototropin-less double mutant *phot1/phot2*. The blue-light responses are assayed under 100 μmol m$^{-2}$ s$^{-1}$ red light to prevent stomatal opening resulting from the stimulation of photosynthesis by blue light. Darkness is shown as zero fluence rate. Neither mutant shows opening when illuminated with 10 μmol m$^{-2}$ s$^{-1}$ blue light. The *phot1/phot2* mutant opens at higher fluence rates of blue light, whereas the *npq1* mutant fails to show any blue light–stimulated opening. In fact, *npq1* stomata close, most likely because of an inhibitory effect of the additional blue light on photosynthesis-driven opening. (B) Blue light–stimulated opening in the wild type. Note the reduced scale of the y-axis, showing the reduced magnitude of the opening of *phot1/phot2* stomata, as compared with the wild type. (After Talbott et al. 2002.)

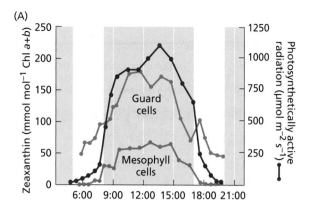

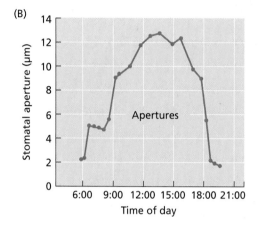

**FIGURE 18.20** The zeaxanthin content of guard cells closely tracks photosynthetically active radiation and stomatal apertures. (A) Daily course of photosynthetically active radiation reaching the leaf surface (red trace), and of zeaxanthin content of guard cells (blue trace) and mesophyll cells (green trace) of *V. faba* leaves grown in a greenhouse. The white areas within the graph highlight the contrasting sensitivity of the xanthophyll cycle in mesophyll and guard cell chloroplasts under the low irradiances prevailing early and late in the day. (B) Stomatal apertures in the same leaves used to measure guard cell zeaxanthin content. (After Srivastava and Zeiger 1995a.)

Additional evidence further indicates that zeaxanthin is a blue light photoreceptor in guard cells:

- In daily opening of stomata in intact leaves, incident radiation, zeaxanthin content of guard cells, and stomatal apertures are closely related (**FIGURE 18.20**).

- The absorption spectrum of zeaxanthin (**FIGURE 18.21**) closely matches the action spectrum for blue light–stimulated stomatal opening (see Figure 18.9).

- The blue-light sensitivity of guard cells increases as a function of their zeaxanthin concentration. The conversion of voilaxanthin to zeaxanthin depends on the pH of the thylakoid lumen.

Light-driven proton pumping at the thylakoid membrane alkalinizes the lumen compartment and increases the concentration of zeaxanthin (**FIGURE 18.22**). Because of this property of the xanthophyll cycle, guard cells illuminated with red light accumulate zeaxanthin. When guard cells from detached epidermis treated with increasing fluence rates of red light are exposed to a short blue-light pulse, the resulting blue light–stimulated stomatal opening is linearly related to the fluence rate of the red light pretreatment, and to the zeaxanthin content of the guard cells at the time of application of the blue-light pulse (Srivastava and Zeiger 1995b).

- Blue light–stimulated stomatal opening is inhibited by 3 m$M$ dithiothreitol (DTT), and the inhibition is concentration dependent. Zeaxanthin formation is blocked by DTT, a reducing agent that reduces S—S bonds to —SH groups and effectively inhibits the enzyme that converts violaxanthin into zeaxanthin. DTT does not block red light-stimulated opening (Srivastava and Zeiger 1995b).

- The facultative CAM species *Mesembryanthemum crystallinum* shifts its carbon metabolism from $C_3$ to CAM mode in response to salt stress (see Chapters 8 and 26). In the $C_3$ mode, stomata accumulate zeaxanthin and open in response to blue light. CAM induction inhibits both zeaxanthin accumulation and the ability of guard cells to open in response to blue light (Tallman et al. 1997).

- In corn coleoptiles bending in response to blue light, zeaxanthin content and the phototropic response are linearly related (see **WEB TOPIC 18.3**).

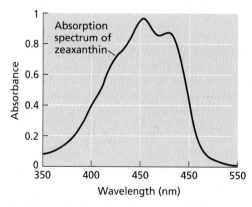

**FIGURE 18.21** The absorption spectrum of zeaxanthin in ethanol. (Courtesy of Professor Wieslaw Gruszecki.)

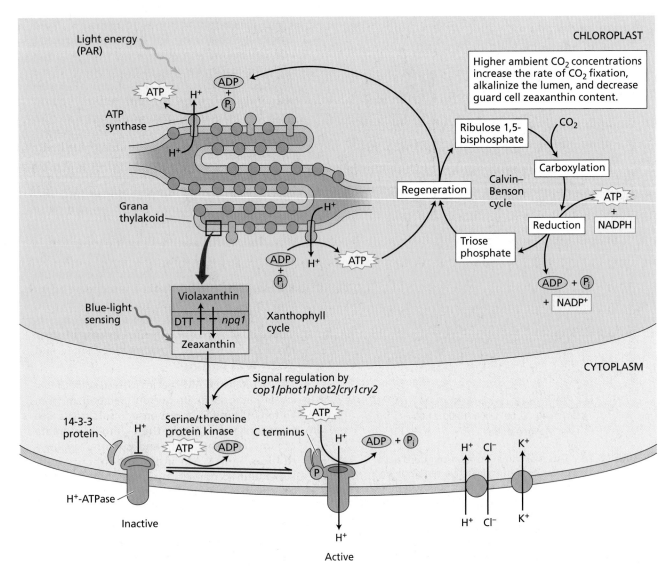

**FIGURE 18.22**  The role of zeaxanthin in blue light sensing in guard cells. Zeaxanthin concentration in guard cells varies with the activity of the xanthophyll cycle. The enzyme that converts violaxanthin to zeaxanthin is an integral thylakoid protein showing a pH optimum of 5.2 (Yamamoto 1979). Acidification of the lumen stimulates zeaxanthin formation, and alkalinization favors violaxanthin formation. Lumen pH depends on levels of incident photosynthetically active radiation (most effective at blue and red wavelengths; see Chapter 7), and on the rate of ATP synthesis, which consumes energy and dissipates the pH gradient across the thylakoid. Thus, photosynthetic activity in the guard cell chloroplast, lumen pH, zeaxanthin content, and blue-light sensitivity play an interactive role in the regulation of stomatal apertures. Compared with their mesophyll counterparts, guard cell chloroplasts are enriched in photosystem II, and they have unusually high rates of photosynthetic electron transport and low rates of photosynthetic carbon fixation (Zeiger et al. 2002). These properties favor lumen acidification at low photon fluxes, and they explain zeaxanthin formation in the guard cell chloroplast early in the day (see Figure 18.20). The regulation of zeaxanthin content by lumen pH, and the tight coupling between lumen pH and Calvin–Benson cycle activity in the guard cell chloroplast further suggest that rates of carbon dioxide fixation in the guard cell chloroplast can regulate zeaxanthin concentrations and integrate light and $CO_2$ sensing in guard cells (see **WEB ESSAY 18.2**). (Zeiger et al. 2002.)

## Green light reverses blue light–stimulated opening

Blue light–stimulated opening is abolished by green light in the 500–600 nm region of the spectrum. The blue-light response is inhibited when guard cells are simultaneously illuminated with both blue and green light (see **WEB ESSAY 18.4**). Green light also reverses blue light–stimulated stomatal opening in pulse experiments (**FIGURE 18.23**). Stomata in detached epidermis open in response to a 30-s blue-light pulse and the opening is abolished if the blue-light pulse is followed by a green-light pulse. The opening is restored if the green pulse is followed by a sec-

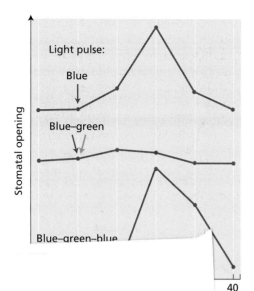

green light is turned off, but stomata from the zeaxanthin-less mutant *npq1* do not (see Figure 18.24). These results indicate that the green reversal of the blue light response requires zeaxanthin but not phototropin.

An action spectrum for the green reversal of blue light–stimulated opening shows a maximum at 540 nm and two

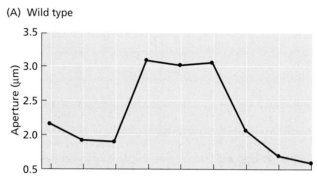

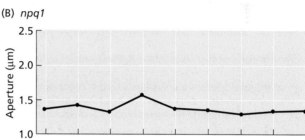

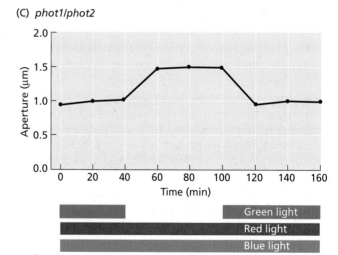

**FIGURE 18.24**   Green light regulates stomatal apertures in the intact leaf. Stomata from intact, attached leaves of Arabidopsis grown in a growth chamber under blue, red, and green light open when green light is removed and close when green light is restored. Blue light is required for the expression of this stomatal sensitivity to green light. Stomata from the zeaxanthin-less *npq1* mutant fail to respond to green light, whereas stomata from the *phot1/phot2* double mutant have a response similar to that of the wild type. (After Talbott et al. 2006.)

the presence of blue light, as observed in experiments with detached epidermis peels. An important ecophysiological implication of these stomatal responses to green light in the intact leaf is that green photons from solar radiation would be expected to down-regulate the stomatal response to blue light under natural conditions.

Stomata from the phototropin-less double mutant *phot1/phot2* respond to blue light and open further when

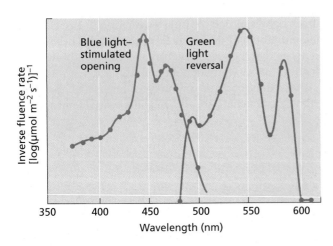

**FIGURE 18.25** Action spectrum for blue light–stimulated stomatal opening (left curve, from Karlsson 1986) and for its reversal by green light (Frechilla et al. 2000). The action spectrum for blue light–stimulated opening was obtained from measurements of transpiration as a function of wavelength in wheat leaves kept under a red light background. The action spectrum for the green light reversal of blue light–stimulated opening was calculated from measurements of aperture changes in stomata from detached epidermis of *Vicia faba* irradiated with a constant fluence rate of blue light and different wavelengths of green light. Note that the two spectra are similar, with the spectrum for the green reversal displaced by about 90 nm. Similar spectral red shifts have been observed upon the isomerization of carotenoids in a protein environment.

minor peaks at 490 and 580 nm (**FIGURE 18.25**). Such an action spectrum rules out the involvement of phytochrome or chlorophylls. Rather, the action spectrum is remarkably similar to the action spectrum for blue light–stimulated stomatal opening, red-shifted (displaced toward the longer, red wave-band of the spectrum) by about 90 nm. Similar spectral red shifts have been observed upon the isomerization of carotenoids in a protein environment. As discussed earlier, the action spectrum for blue light–stimulated stomatal opening matches the absorption spectrum of zeaxanthin (see Figure 18.21). Spectroscopic studies have shown that green light is very effective in the isomerization of zeaxanthin (Milanowska and Gruszecki 2005). Isomerization of zeaxanthin changes the orientation of the molecule within the membrane, a transition that would be very effective as a transduction signal.

**A CAROTENOID PROTEIN COMPLEX SENSES LIGHT INTENSITY** A carotenoid-protein complex which functions as a light intensity sensor provides a model system for the blue–green photocycle in guard cells (Wilson et al 2008). The **orange carotenoid protein** (**OCP**) is a soluble

protein associated with the phycobilisome antenna of photosystem II in cyanobacteria. Recall from Chapter 7 that cyanobacteria are photosynthetic bacteria common to freshwater and marine environments. The OCP is a 35 kDa protein that contains a single noncovalently bound carotenoid, 3'-hydroxyechinenone. Zeaxanthin and 3'-hydroxyechinenone have closely related chemical structures and both derive from β-carotene (Punginelli et. al. 2009).

Zeaxanthin

3'-Hydroxyechinenone

Blue light causes structural changes in both the carotenoid and the protein of the OCP and converts a dark form into a light, active form (**FIGURE 18.26**). Green light reverses the process. The active form of the OCP is essential for the induction of photoprotection in the photosynthetic cyanobacterium. In addition, the photocycle resulting from the interconversion of the blue- and green-absorbing forms of the OCP functions as an effective light intensity sensor.

These findings strongly suggest that the reversibility of stomatal opening in response to consecutive blue and green pulses results from the operation of a photocycle, most likely mediated by protein-bound zeaxanthin, converted by blue light into a physiologically active, green light–absorbing form and back-converted by green light into an inactive, blue light–absorbing form. It is also of interest that blue light–stimulated fluorescence quenching, likely to be associated with photoprotection, has been observed with guard cell and coleoptile chloroplasts of higher plants, paralleling the observation with the OCP in cyanobacteria (see **WEB ESSAY 18.5**).

In the 1940s, the discovery of carotenoids in the blue light–sensing coleoptile tip implicated the blue light–absorbing carotenoids as possible blue-light photoreceptors, but the hypothesis was ruled out because of the very short half-life of the excited carotenoid molecule. The OCP represents the first clearly documented case of a protein-bound carotenoid that functions as a photoreceptor, and of a protein–carotenoid complex sensing light intensity. The striking similarity between some of the photobiological properties of the OCP of cyanobacteria and blue light sensing by zeaxanthin in the guard cell chloroplast should stimulate further research in both systems.

**PHYTOCHROME-MEDIATED RESPONSES IN STOMATA** Many studies seeking to define a role for phytochrome in sto-

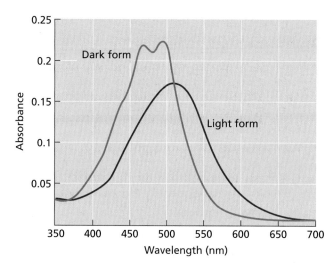

**FIGURE 18.26**  Absorption spectrum of blue–green reversal of the orange carotenoid protein.

matal movements have yielded mostly negative results. Recent findings, however, have shown that phytochrome modulates stomatal movements in the orchid *Paphiopedilum* and in the zeaxanthin-less mutant of Arabidopsis *npq1* (see **WEB TOPIC 18.4**).

It is therefore clear that guard cells can respond to blue light using three different sensory transduction pathways, mediated by

- a specific blue light photoreceptor,
- photosynthesis in the guard cell chloroplast, and
- phytochrome.

It is useful to identify the involved sensory transduction pathways in studies of stomatal responses, and experimental approaches to identify these distinct responses are available. The specific blue-light response can be reversed by green light; the blue component of the photosynthetic response is blocked under saturating red light; and the phytochrome response is far-red reversible. Applications of these approaches have been illustrated earlier. For example, blue light–stimulated stomatal opening observed in the Arabidopsis mutant *npq1* cannot be reversed by green light, but it is reversed by far red light, indicating that involved photoreceptor is phytochrome (see Figure 18.19). In contrast, the blue light–stimulated opening observed with *phot1/phot2* stomata is green light–reversible, indicating that a specific blue-light photoreceptor is mediating the response (Talbott et al. 2003).

## SUMMARY

Plants utilize light as a source of energy and as a signal that provides information about their environment. The responses to blue light signals are distinct from phytochrome responses.

### The Photophysiology of Blue-Light Responses

- Specific blue-light responses have a characteristic "three-finger" action spectrum in the 400 to 500 nm region (**Figure 18.1**).
- These blue-light signals are transduced into electrical, metabolic, and genetic processes that allow plants to alter growth, development, and function (**Figure 18.2**).
- Blue-light responses include phototropism by asymmetric growth, stomatal movements, inhibition of stem elongation involving membrane depolarization, gene activation, pigment biosynthesis, tracking of the sun by leaves, and chloroplast movements within cells (**Figures 18.3–18.9**).
- Blue light stimulates transcription and translation, yielding gene products that are required for the morphogenetic response to light.

- Blue light–stimulated stomatal movements are driven by changes in the osmoregulation of guard cells (**Figure 18.10**).
- Blue light stimulates an $H^+$-ATPase at the guard cell plasma membrane, generating an electrochemical-potential gradient that drives ion uptake (**Figures 18.11, 18.12**).
- Blue light also stimulates starch degradation and malate biosynthesis. Accumulation of sucrose or $K^+$ and its counter ions within guard cells leads to stomatal opening (**Figure 18.13**).
- Light quality can change the activity of different osmoregulatory pathways that modulate stomatal movements (**Figure 18.14**).

### The Regulation of Blue Light–Stimulated Responses

- In plants, 14-3-3 proteins regulate transcription as well as metabolic enzymes.
- A 14-3-3 protein binds to the guard cell $H^+$-ATPase only when the latter is phosphorylated.
- Blue light activates the $H^+$-ATPase in the regulation of stomatal movement (**Figure 18.15**).

## SUMMARY continued

### Blue-Light Photoreceptors

- In *Arabidopsis*, the genes *cry1* and *cry2* are involved in blue light–dependent inhibition of stem elongation, cotyledon expansion, anthocyanin synthesis, control of flowering, and setting of circadian rhythms (**Figure 18.16**).

- The CRY1 protein, and to a lesser extent CRY2, accumulate in the nucleus and interact with the ubiquitin ligase COP1, both in vivo and in vitro.

- The activity of cryptochrome is influenced by its phosphorylation state. The CRY1 protein also regulates anion channel activity at the plasma membrane.

- The protein phototropin aids in regulation of phototropism. Phototropin binds the flavin FMN and autophosphorylates in response to blue light (**Figure 18.17**).

- Phototropin-less mutants are defective in phototropism and in chloroplast movements.

- Blue light suppression of stem elongation by blue light is initiated by *phot1*, with *cry1*, and to a limited extent *cry2*, modulating the response (**Figure 18.18**).

- The chloroplast carotenoid zeaxanthin has been implicated in blue-light photoreception in guard cells (**Figure 18.19**).

- Daily stomatal opening, incident radiation, zeaxanthin content of guard cells, and stomatal apertures are closely related (**Figure 18.20**).

- The absorption spectrum of zeaxanthin matches the action spectrum for blue light–stimulated stomatal opening (**Figures 18.9, 18.21**).

- Blue light–stimulated stomatal opening is blocked if zeaxanthin accumulation in guard cells is blocked. Manipulation of zeaxanthin content in guard cells permits regulation of their response to blue light.

- Blue light signal transduction for guard cells crosses the chloroplast envelope, activates the $H^+$-ATPase, and produces turgor buildup to accomplish stomatal opening (**Figure 18.22**).

- The blue-light response of guard cells is reversed by green light (**Figure 18.23**).

- The reversibility of the stomatal response to blue light by green light can be observed in the intact leaf, indicating that this modulation of the stomatal response has functional implications under natural conditions (**Figure 18.24**).

- The zeaxanthin-less mutant *npq1* does not show the blue–green reversal, indicating that zeaxanthin is required for the response. Phototropin-less mutants have a normal blue-green reversal.

- The action spectrum of the green reversal is similar to the action spectrum of blue light–stimulated opening and to the absorption spectrum of zeaxanthin (**Figure 18.25**).

- A carotenoid–protein complex in cyanobateria, the orange carotenoid protein, shows blue–green reversibility and functions as a light sensor (**Figure 18.26**). The orange carotenoid protein provides a molecular model for blue light sensing by zeaxanthin in guard cells.

## WEB MATERIAL

### Web Topics

**18.1 Blue-Light Sensing and Light Gradients**
Light gradients within organs might serve as sensing mechanisms.

**18.2 Guard Cell Osmoregulation and a Blue Light–Activated Metabolic Switch**
Blue light controls major osmoregulatory pathways in guard cells and unicellular algae.

**18.3 The Coleoptile Chloroplast**
Both the coleoptile and the guard cell chloroplast specialize in sensory transduction.

**18.4 Phytochrome-Mediated Responses in Stomata**
Studies in the orchid *Paphiopedilum* and the zeaxanthin-less mutant of Arabidopsis, *npq1*, show that phytochrome regulates stomatal movements.

## WEB MATERIAL

### Web Essays

**18.1  Guard Cell Photosynthesis**
Photosynthesis in the guard cell chloroplast shows unique regulatory features.

**18.2  Predicted Involvement of the Zeaxanthin Sensory Transducing System in the Stomatal Response to CO$_2$ under Illumination**
Under constant light conditions, changes in ambient CO$_2$ concentrations modulate guard cell zeaxanthin concentrations and stomatal apertures.

**18.3  The Sensory Transduction of the Inhibition of Stem Elongation by Blue Light**
The regulation of stem elongation rates by

blue light has critical importance for plant development.

**18.4  The Blue–Green Reversibility of the Blue-Light Response of Stomata**
The responses of guard cells to blue and green light drive a unique photocycle.

**18.5  Blue-Light Stimulation of Fluorescence Quenching and Photoprotection in Cyanobacteria and Higher Plants**
Cyanobacteria and guard cells share some photobiological features involving carotenoids and photoprotection.

## CHAPTER REFERENCES

Ahmad, M., and Cashmore, A. R. (1993) *HY4* gene of *A. thaliana* encodes a protein with characteristics of a blue light photoreceptor. *Nature* 366: 162–166.

Ahmad, M., Jarillo, J. A., Smirnova, O., and Cashmore, A. R. (1998) Cryptochrome blue light photoreceptors of *Arabidopsis* implicated in phototropism. *Nature* 392: 720–723.

Amodeo, G., Srivastava, A., and Zeiger, E. (1992) Vanadate inhibits blue light–stimulated swelling of *Vicia* guard cell protoplasts. *Plant Physiol.* 100: 1567–1570.

Assmann, S. M., Simoncini, L., and Schroeder, J. I. (1985) Blue light activates electrogenic ion pumping in guard cell protoplasts of *Vicia faba*. *Nature* 318: 285–287.

Briggs, W. R., and Christie, J. M. (2002) Phototropins 1 and 2: Versatile plant blue-light receptors. *Trends Plant Sci.* 7: 204–210.

Christie, J. M. (2007) Phototropin blue-light receptors. *Annu. Rev. Plant Biol.* 58: 21–45.

Cosgrove, D. J. (1994) Photomodulation of growth. In *Photomorphogenesis in Plants*, 2nd ed., R. E. Kendrick and G. H. M. Kronenberg, eds., Kluwer, Dordrecht, Netherlands, pp. 631–658.

Doi, M., and Shimazaki, K. (2008) The Stomata of the fern *Adiantum capillus-veneris* do not respond to CO2 in the dark and open by photosynthesis in guard cells. *Plant Physiol.* 147: 922–930.

Fan, L. M., Zhao, Z., and Assmann, S. M. (2004) Guard cells: A dynamic signaling model. *Curr. Opin. Plant Biol.* 7: 537–546.

Firn, R. D. (1994) Phototropism. In *Photomorphogenesis in Plants*, 2nd ed., R. E. Kendrick and G. H. M. Kronenberg, eds., Kluwer, Dordrecht, Netherlands, pp. 659–681.

Frechilla, S., Zhu, J., Talbott, L. D., and Zeiger, E. (1999) Stomata from *npq1*, a zeaxanthin-less *Arabidopsis* mutant, lack a specific response to blue light. *Plant Cell Physiol.* 40: 949–954.

Frechilla, S., Talbott, L. D., Bogomolni, R. A., and Zeiger, E. (2000) Reversal of blue light-stimulated stomatal opening by green light. *Plant Cell Physiol.* 41: 171–176.

Horwitz, B. A. (1994) Properties and transduction chains of the UV and blue light photoreceptors. In *Photomorphogenesis in Plants*, 2nd ed., R. E. Kendrick and G. H. M. Kronenberg, eds., Kluwer, Dordrecht, Netherlands, pp. 327–350.

Iino, M., Ogawa, T., and Zeiger, E. (1985) Kinetic properties of the blue light response of stomata. *Proc. Natl. Acad. Sci. USA* 82: 8019–8023.

Inoue, S., Kinoshita, T., Matsumoto, M., Nakayama, K. I., Doi, M., and Shimazaki, K. (2008) Blue light-induced autophosphorylation of phototropin is a primary step for signaling. *Proc. Natl. Acad. Sci. USA* 105: 5626–5631.

Kadota, A., Yamada, N., Suetsugu, N., Hirose, M., Saito, C., Shoda, K., Ichikawa, S., Kagawa, T., Nakano, A., and Wada, M. (2009) Short actin-based mechanism for light-directed chloroplast movement in Arabidopsis. *Proc. Natl. Acad. Sci. USA* 106: 13106–13111.

Karlsson, P. E. (1986) Blue light regulation of stomata in wheat seedlings. II. Action spectrum and search for action dichroism. *Physiol. Plant.* 66: 207–210.

Kinoshita, T., and Shimazaki, K. (2001) Analysis of the phosphorylation level in guard-cell plasma membrane H+-ATPase in response to fusicoccin. *Plant Cell Physiol.* 42: 424–432.

Kinoshita, T., Doi, M., Suetsugu, N., Kagawa, T., Wada, M., and Shimazaki, K. (2001) *phot1* and *phot2* mediate blue light regulation of stomatal opening. *Nature* 414: 656–660.

Kinoshita, T., Emi, T., Tominaga, M., Sakamoto, K., Shigenaga, A., Doi, M., and Shimazaki, K. (2003) Blue-light- and phosphorylation-dependent binding of a 14-3-3 protein to phototropins in stomatal guard cells of broad bean. *Plant Physiol.* 133: 1453–1463.

Lawson, T. (2009) Guard cell photosynthesis and stomatal function. *New Phytol.* 181: 13–34.

Lin, C. (2000) Plant blue-light receptors. *Trends Plant Sci.* 5: 337–342.

Milanowska, J., and Gruszecki, W. I. (2005) Heat-induced and light-induced isomerization of the xanthophyll pigment zeaxanthin. *J. Photochem. Photobiol. B.* 80: 178–186.

Niyogi, K. K., Grossman, A. R., and Björkman, O. (1998) *Arabidopsis* mutants define a central role for the xanthophyll cycle in the regulation of photosynthetic energy conversion. *Plant Cell* 10: 1121–1134.

Olsen, R. L., Pratt, R. B., Gump, P., Kemper, A., and Tallman, G. (2002) Red light activates a chloroplast-dependent ion uptake mechanism for stomatal opening under reduced $CO_2$ concentrations in *Vicia* spp. *New Phytol.* 153: 497–508.

Outlaw, W. H., Jr. (1983) Current concepts on the role of potassium in stomatal movements. *Physiol. Plant.* 59: 302–311.

Outlaw, W. H., Jr. (2003) Integration of cellular and physiological functions of guard cells. *Crit. Rev. Plant Sci.* 22: 503–529.

Pandey, S., Zhang, W., and Assmann, S. (2007) Roles of ion channels and transporters in guard cell signal transduction. *FEBS Lett.* 581: 2325–2336.

Parks, B. M., Folta, K. M., and Spalding, E. P. (2001) Photocontrol of stem growth. *Curr. Opin. Plant Biol.* 4: 436–440.

Punginelli, C., Wilson, A., Routaboul, J. M., and Kirilovsky, D. (2009) Influence of zeaxanthin and echinenone binding on the activity of the Orange Carotenoid Protein. *Biochim. Biophys. Acta* 1787: 280–288.

Roelfsema, M. R. G., Steinmeyer, R., Staal, M., and Hedrich, R. (2001) Single guard cell recordings in intact plants: Light-induced hyperpolarization of the plasma membrane. *Plant J.* 26: 1–13.

Roelfsema, R. G., Hanstein, S., Felle, H. H., and Hedrich, R. (2002) $CO_2$ provides an intermediate link in the red light response of guard cells. *Plant J.* 32: 65–75.

Schleicher, E., Kowalczyk, R. M., Kay, C. W. M., Hegemann, P., Bacher, A., Fischer, M., Bittl, R., Richter, G., and Weber, S. (2004) On the reaction mechanism of adduct formation in LOV domains of the plant blue-light receptor phototropin. *J. Am. Chem. Soc.* 126: 11067–11076.

Schnabl, H., and Ziegler, H. (1977) The mechanism of stomatal movement in *Allium cepa* L. *Planta* 136: 37–43.

Schroeder, J. I., Allen, G. J., Hugouvieux, V., Kwak, J. M., and Waner, D. (2001) Guard cell signal transduction. *Annu. Rev. Plant Physiol. Plant Mol. Biol.* 52: 627–658.

Schwartz, A., and Zeiger, E. (1984) Metabolic energy for stomatal opening. Roles of photophosphorylation and oxidative phosphorylation. *Planta* 161: 129–136.

Senger, H. (1984) *Blue Light Effects in Biological Systems.* Springer, Berlin.

Serrano, E. E., Zeiger, E., and Hagiwara, S. (1988) Red light stimulates an electrogenic proton pump in *Vicia* guard cell protoplasts. *Proc. Natl. Acad. Sci. USA* 85: 436–440.

Shimazaki, K., Iino, M., and Zeiger, E. (1986) Blue light–dependent proton extrusion by guard cell protoplasts of *Vicia faba. Nature* 319: 324–326.

Spalding, E. P. (2000) Ion channels and the transduction of light signals. *Plant Cell Environ.* 23: 665–674.

Spalding, E. P., and Cosgrove, D. J. (1989) Large membrane depolarization precedes rapid blue-light induced growth inhibition in cucumber. *Planta* 178: 407–410.

Spalding, E. P., and Folta, K. M. (2005) Illuminating topics in plant photobiology. *Plant Cell Environ.* 28: 39–53.

Srivastava, A., and Zeiger, E. (1995a) Guard cell zeaxanthin tracks photosynthetic active radiation and stomatal apertures in *Vicia faba* leaves. *Plant Cell Environ.* 18: 813–817.

Srivastava, A., and Zeiger, E. (1995b) The inhibitor of zeaxanthin formation, dithiothreitol, inhibits blue-light-stimulated stomatal opening in *Vicia faba. Planta* 196: 445–449.

Swartz, T. E., Tseng, T.-S., Frederickson, M. A., Paris, G., Comerci, D. J., Rajashekara, G., Kim, J.-G., Mudgett, M. B., Splitter, G., A., Ugalde, R. A., et al. (2007) Blue-light-activated histidine kinases: Two-component sensors in bacteria. *Science* 317: 1090–1093.

Talbott, L. D., and Zeiger, E. (1998) The role of sucrose in guard cell osmoregulation. *J. Exp. Bot.* 49: 329–337.

Talbott, L. D., Srivastava, A., and Zeiger, E. (1996) Stomata from growth-chamber-grown *Vicia faba* have an enhanced sensitivity to $CO_2$. *Plant Cell Environ.* 19: 1188–1194.

Talbott, L. D., Zhu, J., Han, S. W., and Zeiger, E. (2002) Phytochrome and blue light-mediated stomatal opening in the orchid, *Paphiopedilum. Plant Cell Physiol.* 43: 639–646.

Talbott, L. D., Shmayevich, I. J., Chung, Y., Hammad, J. W., and Zeiger, E. (2003) Blue light and phytochrome-mediated stomatal opening in the *npq1* and *phot1 phot2* mutants of Arabidopsis. *Plant Physiol.* 133: 1522–1529.

Talbott, L. D., Hammad, J. W., Harn, L. C, Nguyen, V., Patel, J., and Zeiger, E. (2006) Reversal by green light of blue light-stimulated stomatal opening in intact, attached leaves of Arabidopsis operates only in the potassium dependent, morning phase of movement. *Plant Cell Physiol.* 47: 333–339.

Tallman, G. (2004) Are diurnal patterns of stomatal movement the result of alternating metabolism of endogenous guard cell ABA and accumulation of ABA delivered to the apoplast around guard cells by transpiration? *J. Exp. Bot.* 55: 1963–1976.

Tallman, G., Zhu, J., Mawson, B. T., Amodeo, G., Nouhi, Z., Levy, K., and Zeiger, E. (1997) Induction of CAM in *Mesembryanthemum crystallinum* abolishes the stomatal response to blue light and light-dependent zeaxanthin formation in guard cell chloroplasts. *Plant Cell Physiol.* 38: 236–242.

Thimann, K. V., and Curry, G. M. (1960) Phototropism and phototaxis. In *Comparative Biochemistry*, Vol. 1, M. Florkin and H. S. Mason, eds., Academic Press, New York, pp. 243–306.

Wu, G., and Spalding, E. P. (2007) Separate functions for nuclear and cytoplasmic chryptochrome 1 during photomorphogenesis of *Arabidopsis* seedlings. *Proc. Natl. Acad. Sci. USA* 104: 18813–18818.

Wilson, A., Punginelli, C., Gall, A., Bonetti, C., Alexandre, M., Routaboul, J. M., Kerfeld, C. A., van Grondelle, R., Robert, B., Kennis, J. T. et al. (2008) A photoactive carotenoid protein acting as light intensity sensor. *Proc. Natl. Acad. Sci. USA* 105: 12075–12080.

Yamamoto, H. Y. (1979) Biochemistry of the violaxanthin cycle in higher plants. *Pure Appl. Chem.* 51: 639–648.

Zeiger, E., and Hepler, P. K. (1977) Light and stomatal function: Blue light stimulates swelling of guard cell protoplasts. *Science* 196: 887–889.

Zeiger, E., Talbott, L. D., Frechilla, S., Srivastava, A., and Zhu, J. X. (2002) The guard cell chloroplast: A perspective for the twenty-first century. *New Phytol.* 153: 415–424.

Zhu, J., Talbott, L. D., Jin, X., and Zeiger, E. (1998) The stomatal response to $CO_2$ is linked to changes in guard cell zeaxanthin. *Plant Cell Environ.* 21: 813–820.

# Auxin: The First Discovered Plant Growth Hormone

The form and function of multicellular organisms would not be possible without efficient communication among cells, tissues, and organs. In higher plants, regulation and coordination of metabolism, growth, and morphogenesis often depend on chemical signals from one part of the plant to another. This idea originated in the nineteenth century with the German botanist Julius von Sachs (1832–1897).

Sachs proposed that chemical messengers are responsible for the formation and growth of different plant organs. He also suggested that external factors such as gravity could affect the distribution of these substances within a plant. Although Sachs did not know the identity of these chemical messengers, his ideas led to their eventual discovery.

**Hormones** are chemical messengers that are produced in one cell and modulate cellular processes in another cell by interacting with specific proteins that function as *receptors* linked to cellular transduction pathways. As is the case with animals, most plant hormones are synthesized in one tissue and act on specific target sites in another tissue at vanishingly low concentrations. Hormones that are transported to sites of action in tissues distant from their site of synthesis are referred to as *endocrine* hormones. Those that act on cells adjacent to the source of synthesis are referred to as *paracrine* hormones. Plant development is regulated by six major types of hormones: auxins, gibberellins, cytokinins, ethylene, abscisic acid, and brassinosteroids.

A variety of other signaling molecules that play roles in resistance to pathogens and defense against herbivores have also been identified in plants, including conjugated and unconjugated forms of jasmonic acid, salicylic acid, and small polypeptides (see Chapter 13). Another molecule, strigolactone, has recently been shown to be a transmissible signaling molecule that regulates the outgrowth of lateral buds (Brewer et al. 2009); this compound may also be a true plant hormone. Other classes of molecules, such as flavonoids (see Chapter 13), function as both intracellular and localized extracellular modulators of signal transduction pathways (Peer and Murphy 2007). Indeed, the list of signaling agents and growth regulators continues to expand.

The first signaling agent we will consider is the hormone **auxin**. Auxin was the first growth hormone to be studied in plants, and much of the early physiological work on the mechanism of plant cell expansion was carried out in relation to auxin action. Auxin signaling has been found to function in virtually every aspect of plant growth and development. Moreover, auxin and cytokinin differ from the other plant hormones and signaling agents in one important respect: they are required for viability of the plant embryo. Whereas other plant hormones seem to act as regulators of discrete developmental processes, auxin and cytokinin appear to be required at some level more or less continuously.

We begin our discussion of auxins with a brief history of their discovery, followed by a description of their chemical structures and the methods used to detect auxins in plant tissues. A brief look at the pathways of auxin biosynthesis and the polar nature of auxin transport follows (a more complete treatment of auxin biosynthesis can be found in Appendix 3). We will then describe our current understanding of auxin signal transduction pathways, from receptor binding to gene expression. Finally, we will briefly review the various developmental processes controlled by auxin: stem elongation, apical dominance, root initiation, fruit development, meristem development, and oriented, or *tropic*, growth. Auxin is involved in many physiological processes, many of which are also influenced by other hormones. A discussion of auxin's role in these growth processes is integrated into the descriptions of these processes in the relevant chapters.

## The Emergence of the Auxin Concept

During the latter part of the nineteenth century, Charles Darwin and his son Francis studied plant growth phenomena involving *tropisms* (from the Greek *tropos*, to turn). One of their interests was the bending of plants toward light. This phenomenon, which is caused by differential growth, is called **phototropism**. In some experiments the Darwins used seedlings of canary grass (*Phalaris canariensis*), in which the youngest leaves are sheathed in a protective organ called the **coleoptile** (FIGURE 19.1).

Coleoptiles and very young seedlings are highly sensitive to light, especially to blue light (see Chapter 18). If illu-

minated on one side with a short pulse of dim blue light, they will bend (grow) toward the source of the light pulse within an hour. The Darwins found that it was the tip of the coleoptile that perceived the light, for if they covered the tip with foil, the coleoptile would not bend. But the region of the coleoptile that is responsible for the bending toward the light, called the **elongation zone**, is several millimeters below the tip.

Thus they concluded that some sort of signal is produced in the tip, travels to the elongation zone, and causes the shaded side to grow faster than the illuminated side. The Darwins published the results of their experiments in 1881 in *The Power of Movement in Plants*.

There followed a long period of experimentation by many investigators on the nature of the growth stimulus in coleoptiles. It was known that if the tip of a coleoptile was removed, coleoptile growth ceased. Researchers attempted to isolate and identify the growth-promoting chemical by grinding up coleoptile tips and testing the activity of the extracts; but this approach failed, because grinding up the tissue destroyed the active compound when cellular compartments were disrupted.

This research culminated in 1926 with Frits Went's demonstration of a growth-promoting chemical in the tip of oat (*Avena sativa*) coleoptiles. Went's major breakthrough was to avoid grinding by allowing the material to diffuse out of excised coleoptile tips directly into gelatin blocks. Placed on one side or the other of the top of a decapitated coleoptile, these blocks could be tested for their ability to cause bending in the absence of a unilateral light source (see Figure 19.1). Because the substance that diffused from the coleoptile tips into the gelatin blocks promoted the elongation of the coleoptile sections (FIGURE 19.2), it was eventually named **auxin** from the Greek *auxein*, meaning "to increase" or "to grow." The next steps were to chemically identify this substance and to understand its production, destruction, and action.

## The Principal Auxin: Indole-3-Acetic Acid

Went's studies with gelatin (and, later, agar) blocks demonstrated unequivocally that the growth-promoting "influence" diffusing from the coleoptile tip was a chemical substance. The fact that it was produced at one location and transported in minute amounts to its site of action qualified it as an authentic plant hormone.

In the mid-1930s it was determined that the principal natural auxin is **indole-3-acetic acid (IAA)**. Several other auxins in higher plants were discovered later (FIGURE 19.3A), but IAA is by far the most abundant and physiologically important. Because the structure of IAA is relatively simple, academic and industrial laboratories were

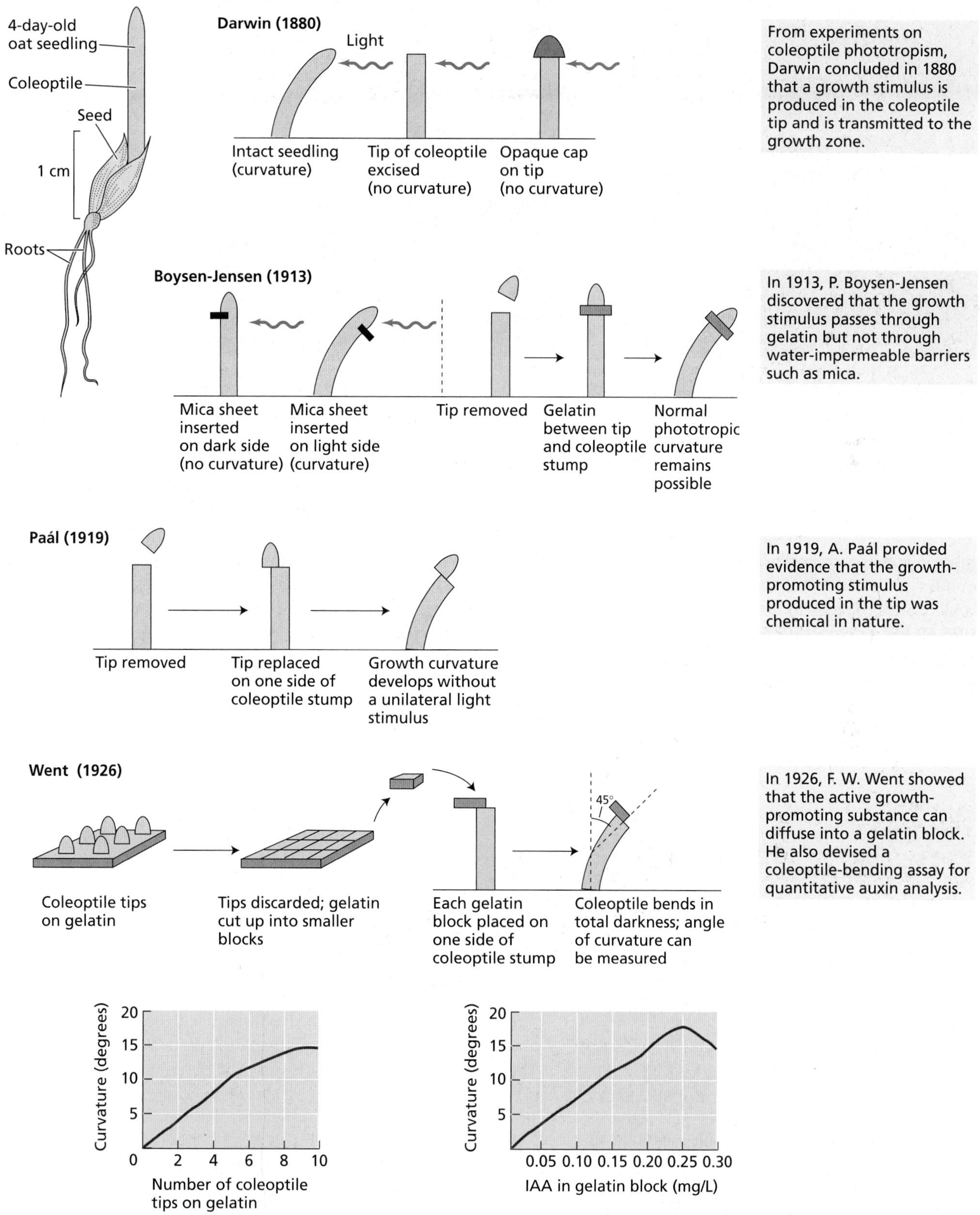
4-day-old oat seedling
Coleoptile
Seed
1 cm
Roots
Darwin (1880)
Light
Intact seedling (curvature)
Tip of coleoptile excised (no curvature)
Opaque cap on tip (no curvature)
From experiments on coleoptile phototropism, Darwin concluded in 1880 that a growth stimulus is produced in the coleoptile tip and is transmitted to the growth zone.
Boysen-Jensen (1913)
Mica sheet inserted on dark side (no curvature)
Mica sheet inserted on light side (curvature)
Tip removed
Gelatin between tip and coleoptile stump
Normal phototropic curvature remains possible
In 1913, P. Boysen-Jensen discovered that the growth stimulus passes through gelatin but not through water-impermeable barriers such as mica.
Paál (1919)
Tip removed
Tip replaced on one side of coleoptile stump
Growth curvature develops without a unilateral light stimulus
In 1919, A. Paál provided evidence that the growth-promoting stimulus produced in the tip was chemical in nature.
Went (1926)
Coleoptile tips on gelatin
Tips discarded; gelatin cut up into smaller blocks
Each gelatin block placed on one side of coleoptile stump
45°
Coleoptile bends in total darkness; angle of curvature can be measured
In 1926, F. W. Went showed that the active growth-promoting substance can diffuse into a gelatin block. He also devised a coleoptile-bending assay for quantitative auxin analysis.
Curvature (degrees)
20
15
10
5
0
2
4
6
8
10
Number of coleoptile tips on gelatin
Curvature (degrees)
20
15
10
5
0.05 0.10 0.15 0.20 0.25 0.30
IAA in gelatin block (mg/L)

FIGURE 19.1 Summary of early experiments in auxin research.

(A)

(B)

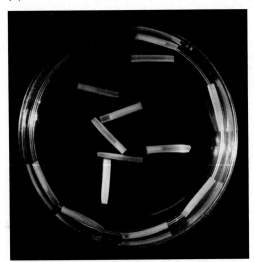

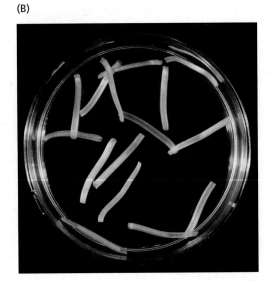

**FIGURE 19.2** Auxin stimulates the elongation of oat coleoptile sections. These coleoptile sections were incubated for 18 hours in either water (A) or auxin (B). The yellow tissue inside the translucent coleoptile is the primary leaves. (Photos © M. B. Wilkins.)

quickly able to synthesize a wide array of molecules with auxin activity. Some of these compounds are now used widely as herbicides in horticulture and agriculture (FIGURE 19.3B). Separate criteria were established for assessing the auxin activity and transport of these different compounds in plant tissues.

In general, auxins are defined as compounds with biological activities similar to those of IAA, including the ability to promote cell elongation in coleoptile and stem sections, cell division in callus cultures in the presence of cytokinins, formation of adventitious roots on detached leaves and stems, and other developmental phenomena associated with IAA action. Although they are chemically diverse, a common feature of all active auxins is a molecular distance of about 0.5 nm between a fractional positive charge on the aromatic ring and a negatively charged carboxyl group.

Our discussion begins with the chemical nature of auxin and continues with a brief description of its biosynthesis, transport, and metabolism. Increasingly powerful analytical methods and the application of molecular biological

(A)

**Indole-3-acetic acid (IAA)**

**4-Chloroindole-3-acetic acid (4-Cl-IAA)**

**Indole-3-butyric acid (IBA)**

(B)

**FIGURE 19.3** Structure of auxins. (A) Structures of naturally occurring auxins. Indole-3-acetic acid (IAA) occurs in all plants, but other related compounds in plants have auxin activity. Peas, for example, contain 4-chloroindole-3-acetic acid (4-Cl-IAA). Maize and legume contain indole-3-butyric acid (IBA). (B) Structures of two synthetic auxins. Most synthetic auxins are used as herbicides in horticulture and agriculture.

**2,4-Dichlorophenoxyacetic acid (2,4-D)**

**2-Methoxy-3, 6-dichlorobenzoic acid (dicamba)**

approaches have recently allowed scientists to identify auxin precursors and to study auxin turnover and distribution within the plant. Depending on the information that a researcher needs, the amounts and/or identity of auxins in biological samples can be determined by bioassay, mass spectrometry, or enzyme-linked immunoassays.

## IAA is synthesized in meristems and young dividing tissues

IAA biosynthesis is associated with rapidly dividing and growing tissues, especially in shoots (Ljung et al. 2005). Although virtually all plant tissues appear to be capable of producing low levels of IAA, shoot apical meristems and young leaves are the primary sites of auxin synthesis (Ljung et al. 2001). Root apical meristems are also important sites of auxin synthesis, especially as the roots elongate and mature, although the root remains dependent on the shoot for much of its auxin (Ljung et al. 2005). Young fruits and seeds contain high levels of auxin, but it is unclear whether this auxin is newly synthesized or transported from maternal tissues during development.

Auxin accumulates at the tips of very young leaf primordia of Arabidopsis. As the leaves develop, auxin accumulations are detectable at the leaf margins, but gradually shift toward the base of the leaf and, later, to the central region of the lamina. The basipetal shift in auxin production correlates closely with, and is probably causally related to, the basipetal maturation sequence of leaf development and vascular differentiation (Aloni 2001).

The *GUS* (β-glucuronidase) reporter gene is a useful analytical tool because its activity and location in the tissue can be visualized by treating the tissue with a substrate that produces a blue color when hydrolyzed by the GUS enzyme (Ulmasov et al. 1997). By genetically transforming plants with a *GUS* reporter gene fused to a DNA promoter sequence that responds to auxin it is possible to infer the distribution of free auxin in discrete groups of cells or whole organs (see Chapter 2). Wherever free auxin reaches a minimum threshold level, *GUS* expression occurs and can be detected as a blue color.

As shown in **FIGURE 19.4**, auxin appears to accumulate at specific sites at the margins of young leaves. These are the sites of future **hydathodes**, glandlike modifications of the ground and vascular tissues that allow the release of liquid water (guttation fluid) through pores in the epidermis in the presence of root pressure (see Chapter 4). During early stages of hydathode differentiation, a center of high auxin accumulation can be inferred from a concentrated dark blue GUS stain (arrow) in the lobes of young, serrated leaves of Arabidopsis. These lobes appear to be sites of auxin synthesis as well (Aloni et al. 2003). A trail of GUS activity, suggesting auxin movement, can be seen extending down to differentiating vessel elements in a developing vascular strand (see Figure 19.4). We will return to the

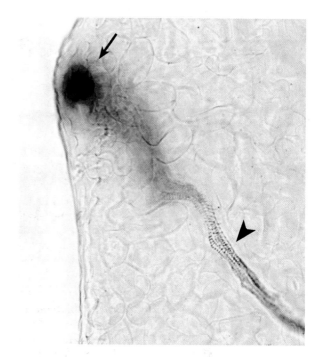

**FIGURE 19.4** Detection of apparent sites of auxin accumulation in a young leaf primordium of Arabidopsis using a synthetic auxin-responsive *DR5* promoter driving a *GUS* reporter gene. The *DR5* artificial promoter is derived from naturally occurring auxin response elements and reports the presence of auxin in tissues where auxin signaling mechanisms are intact. During the early stages of hydathode differentiation, auxin accumulation consistent with new synthesis is suggested by the presence of a concentrated dark blue *GUS* stain (arrow) in the lobes of the serrated leaf margin. A gradient of diluted GUS activity extends from the margin toward a differentiating vascular strand (arrowhead), which appears to function as a sink for auxin flow originating in the lobe. (Courtesy of R. Aloni and C. I. Ullrich.)

topic of auxin regulation of vascular differentiation later in the chapter.

## Multiple pathways exist for the biosynthesis of IAA

IAA is structurally related to the amino acid *tryptophan*, and to the tryptophan precursor *indole-3-glycerol phosphate*, both of which can serve as precursors for IAA biosynthesis. Molecular genetic and radioisotope labeling studies have been used to identify the enzymes and intermediate molecules involved in tryptophan-dependent IAA biosynthesis, and the order in which they function. Multiple biosynthetic pathways using tryptophan as a precursor have been shown to produce IAA in plants, and a bacterial pathway of tryptophan-dependent IAA biosynthesis has also been identified. These pathways are described in more detail in Appendix 3.

### Seeds and storage organs contain covalently bound auxin

Auxin can be covalently bound to both high and low molecular weight compounds, particularly in seeds and storage organs such as cotyledons. These conjugated, or "bound," auxins are found to a greater or lesser extent in all higher plants and are considered hormonally inactive. IAA can be conjugated to many different low molecular weight compounds like amino acids or sugars, or to high molecular weight molecules like peptides, complex glycans (multiple sugar units), or glycoproteins. IAA is rapidly released from many, but not all, conjugates by enzymatic processes. Those conjugates that can release free auxin serve as reversible storage forms of the hormone.

The extent and nature of IAA conjugation depends on the tissue involved and the specific conjugating enzymes present. The best-studied reaction is the conjugation of IAA to glucose in the endosperm of maize (corn; *Zea mays*), but recent molecular genetic studies in Arabidopsis have enhanced our understanding of the regulation of IAA conjugation/deconjugation processes in vegetative tissues.

Metabolism of conjugated auxin can be a major factor in the regulation of the levels of free auxin. For example, during the germination of *Z. mays* seeds, IAA-*myo*-inositol is translocated from the endosperm to the coleoptile via the phloem. Most of the free IAA produced in coleoptile tips

of *Z. mays* is believed to be derived from the hydrolysis of IAA-*myo*-inositol from the seed (**FIGURE 19.5A**).

In addition, environmental stimuli such as light and gravity have been shown to influence both the rate of auxin conjugation (removal of free auxin) and the rate of release of free auxin (hydrolysis of conjugated auxin). The formation of conjugated auxins may serve other functions as well, including storage and protection against oxidative degradation.

Indole-3-butyric acid (IBA) is another natural auxin. IBA and IBA conjugates also contribute to the pool of stored IAA (see Figure 19.5A). IBA-aminoacyl or glucosyl ester conjugates can be hydrolyzed to free IBA, which, in turn, can be enzymatically converted into IAA via β-oxidation in the peroxisome (see Chapter 11). A diagram of the storage forms of auxin can be found in Figure 19.5A.

### IAA is degraded by multiple pathways

To be effective developmental signals, hormones must be short-lived and should not accumulate over time. Auxin catabolism ensures the degradation of active hormone when the concentration exceeds the optimal level or when the response to the hormone is complete. Like IAA biosynthesis, the enzymatic breakdown (oxidation) of IAA involves more than one pathway (**FIGURE 19.5B**).

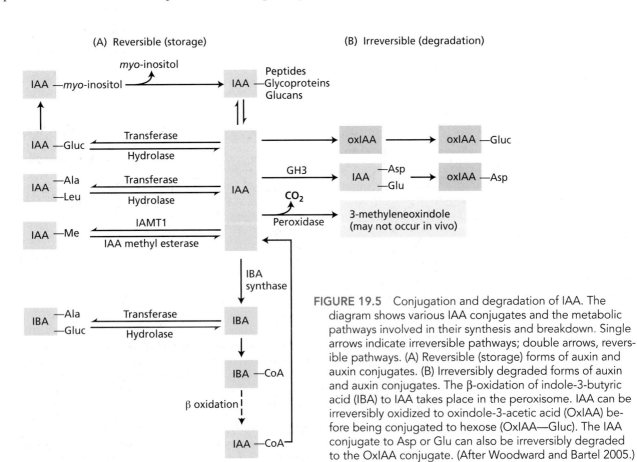

**FIGURE 19.5** Conjugation and degradation of IAA. The diagram shows various IAA conjugates and the metabolic pathways involved in their synthesis and breakdown. Single arrows indicate irreversible pathways; double arrows, reversible pathways. (A) Reversible (storage) forms of auxin and auxin conjugates. (B) Irreversibly degraded forms of auxin and auxin conjugates. The β-oxidation of indole-3-butyric acid (IBA) to IAA takes place in the peroxisome. IAA can be irreversibly oxidized to oxindole-3-acetic acid (OxIAA) before being conjugated to hexose (OxIAA—Gluc). The IAA conjugate to Asp or Glu can also be irreversibly degraded to the OxIAA conjugate. (After Woodward and Bartel 2005.)

On the basis of isotopic labeling and metabolite identification, two oxidative pathways are probably involved in the controlled degradation of IAA. In one pathway, the indole moiety of IAA is oxidized to form oxindole-3-acetic acid (OxIAA) and subsequently, OxIAA-glucose (OxIAA-Gluc). In another pathway, IAA-aspartate conjugates are oxidized to OxIAA.

IAA is also nonenzymatically oxidized when exposed to light, and plant pigments such as riboflavin can promote IAA's photodestruction in vitro. However, the physiological significance of auxin photooxidation is still not clear. Decarboxylation of IAA to 3-methyleneoxindole by plant peroxidases also can be demonstrated in vitro, but may not take place in plants (Normanly et al. 1995).

## Auxin Transport

The main axes of shoots and roots, along with their branches, exhibit apex–base structural polarity, and this structural polarity is dependent on the polarity of auxin transport. Soon after Went developed the coleoptile curvature test for auxin, it was discovered that IAA moves mainly from the apical to the basal end (*basipetally*) in excised oat coleoptile sections. (See **FIGURE 19.6A** for a diagrammatic explanation of the use of the terms "basipetal" and "acropetal.") This type of unidirectional transport is termed **polar transport**. Auxin is the only plant growth hormone that has been clearly shown to be transported polarly, and polar transport of this hormone is found in almost all plants, including bryophytes and ferns.

This transport process is ancient, as anatomical evidence of polar auxin flow has been reported in 375 million-year-old fossil wood. In living woody plants, auxin "whirlpools" arise wherever polar transport is disrupted by the presence of obstacles such as buds and branches (Kramer 2006). As a result, the tracheary elements that differentiate in these regions form circular patterns. Identical circular patterns at the same positions in the wood (indicative of polar auxin transport) have now been detected in the fossil wood of a primitive gymnosperm dating to the Upper Devonian period (Rothwell and Lev-Yadun 2005).

Because the shoot apex serves as the primary source of auxin in the plant, polar transport has long been believed to be the principal cause of an auxin gradient extending from the shoot tip to the root tip. The longitudinal gradient of auxin from the shoot to the root affects various devel-

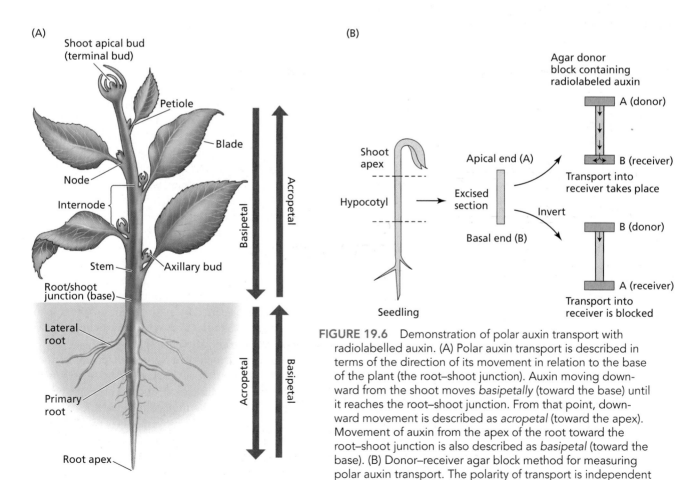

**FIGURE 19.6** Demonstration of polar auxin transport with radiolabelled auxin. (A) Polar auxin transport is described in terms of the direction of its movement in relation to the base of the plant (the root–shoot junction). Auxin moving downward from the shoot moves *basipetally* (toward the base) until it reaches the root–shoot junction. From that point, downward movement is described as *acropetal* (toward the apex). Movement of auxin from the apex of the root toward the root–shoot junction is also described as *basipetal* (toward the base). (B) Donor–receiver agar block method for measuring polar auxin transport. The polarity of transport is independent of the orientation of the plant tissue with respect to gravity.

opmental processes, including embryonic development, stem elongation, apical dominance, wound healing, and leaf senescence.

The major sites of polar auxin transport in the stems, leaves, and roots of most plants are the vascular parenchyma tissues, most likely those associated with the xylem. In grass coleoptiles, basipetal polar transport may also occur in nonvascular parenchyma tissues. In the vascular parenchyma, the overall direction of auxin transport is established in the embryo shortly after auxin accumulations are first detectable at the shoot apex (see Chapter 16). Embryonic polar auxin transport is initially described as entirely basipetal, as the embryo has no root. The downward direction of auxin transport in the embryonic vascular parenchyma is maintained in the root vascular cylinder throughout the life of the plant. However, in postgermination roots this downward stream, by convention, is described as *acropetal*, as auxin is then transported away from the root–shoot junction and toward the root tip.

Some auxin transport also occurs in phloem sieve tubes, and phloem-based movement, driven by "source–sink" translocation of sugars (see Chapter 10), contributes to the auxin transported acropetally (i.e., toward the tip) in the root.

Long-distance auxin transport in the phloem appears to also be important for controlling cambial cell divisions and callose accumulation or removal from sieve tube elements. Auxin transported in the phloem may also transfer to the xylem parenchymal system. Studies with radiolabeled IAA suggest that in pea (*Pisum sativum*), auxin can be transferred from the nonpolar phloem pathway in the immature tissues of the shoot apex.

Basipetal auxin transport (away from the apex) occurs in roots as well, but takes place in nonvascular tissues. In maize and Arabidopsis roots, for example, radiolabeled IAA applied to the root tip is transported basipetally for a distance of 2 to 8 mm. Basipetal auxin transport in the root plays a central role in gravitropism and contributes to lateral root elongation.

In the sections that follow we will discuss the cellular mechanisms underlying polar transport. Later in the chapter, we will see how such regulatory mechanisms enable the plant to adapt to various environmental signals.

## Polar transport requires energy and is gravity independent

Early studies of polar transport were carried out using the *donor–receiver agar block method* (FIGURE 19.6B): An agar block containing radioisotope-labeled auxin (donor block) is placed on one end of a tissue segment, and a receiver block is placed on the other end. The movement of auxin through the tissue into the receiver block can be determined over time by measurement of the radioactivity in the receiver block. This method has been refined to allow

FIGURE 19.7  Adventitious roots grow from the basal ends of grape hardwood cuttings, and shoots grow from the apical ends, whether the cuttings are maintained in the inverted (the two cuttings on the left) or upright orientation (the cuttings on the right). The roots always form at the basal ends because polar auxin transport is independent of gravity. (From Hartmann and Kester 1983.)

for the deposition of much smaller droplets of radiolabeled auxin onto discrete surfaces of plants, improving the accuracy of transport studies over short distances (Peer and Murphy 2007).

From a multitude of such studies, the general properties of polar IAA transport have emerged. Tissues differ in the degree of polarity of IAA transport. In coleoptiles, vegetative stems, leaf petioles, and the root epidermis, basipetal transport predominates, whereas in the stelar tissues of the root, auxin is transported acropetally. Polar transport is not affected by the orientation of the tissue (at least over short periods of time), so it is independent of gravity.

A demonstration of the lack of gravity effects on basipetal auxin transport is shown in FIGURE 19.7. When stem cuttings, in this case grape hardwood, are placed in a moist chamber, adventitious roots form at the basal ends of the cuttings, and shoots form at the apical ends, even when the cuttings are inverted. Roots form at the base because root differentiation is stimulated by auxin accumulation due to basipetal transport. Shoots tend to form at the apical end where the auxin concentration is lowest.

Polar transport proceeds in a cell-to-cell fashion, rather than via the symplast; that is, auxin exits the cell through the plasma membrane, diffuses across the compound middle lamella, and enters the next cell through its plasma membrane. The export of auxin from cells is termed *auxin efflux*; the entry of auxin into cells is called *auxin uptake* or *influx*. The overall process requires metabolic energy, as evidenced by the sensitivity of polar transport to $O_2$ deprivation, sucrose depletion, and metabolic inhibitors.

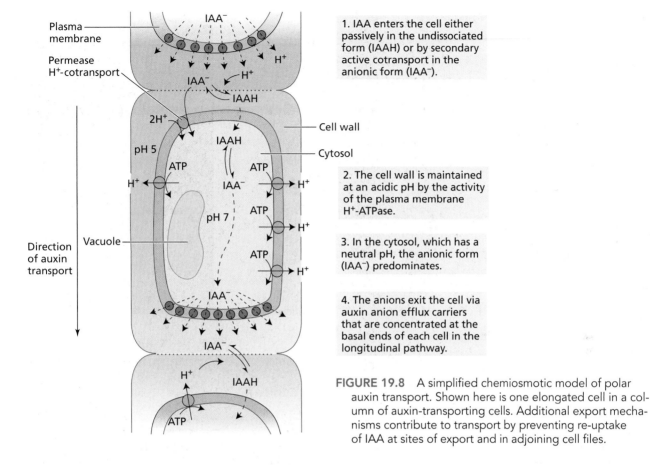

1. IAA enters the cell either passively in the undissociated form (IAAH) or by secondary active cotransport in the anionic form (IAA⁻).

2. The cell wall is maintained at an acidic pH by the activity of the plasma membrane H⁺-ATPase.

3. In the cytosol, which has a neutral pH, the anionic form (IAA⁻) predominates.

4. The anions exit the cell via auxin anion efflux carriers that are concentrated at the basal ends of each cell in the longitudinal pathway.

**FIGURE 19.8** A simplified chemiosmotic model of polar auxin transport. Shown here is one elongated cell in a column of auxin-transporting cells. Additional export mechanisms contribute to transport by preventing re-uptake of IAA at sites of export and in adjoining cell files.

The velocity of polar auxin transport can exceed 3 mm $h^{-1}$ in some tissues, which is faster than diffusion but slower than phloem translocation rates (see Chapter 10). Higher rates of polar transport are observed in tissues immediately adjacent to the shoot and root apical meristems. Polar transport is specific for active auxins, both natural and synthetic; other weak organic acids, inactive auxin analogs, and IAA conjugates are poorly transported. The specificity of polar transport indicates that it is mediated by protein carriers on the plasma membrane.

### Chemiosmotic potential drives polar transport

The discovery of the chemiosmotic mechanism of solute transport in the late 1960s (see Chapter 6) led to the application of this model to polar auxin transport. A **chemiosmotic model** for polar auxin transport proposes that auxin uptake is driven by the proton motive force ($\Delta E + \Delta pH$) across the plasma membrane, while auxin efflux is driven by the membrane potential, $\Delta E$. (Proton motive force is described in more detail in Chapter 7.) Polar flow of auxin is directed by polarly localized efflux carriers concentrated at the ends of the conducting cells (**FIGURE 19.8**). This model has been experimentally validated in whole plants (Li et al. 2005) and has been extended in recent years to

encompass the additional role of proton-driven uptake of auxin by symporters in polar transport processes. However, $\Delta E + \Delta pH$ are not the only energy source. ATP is also directly utilized by one class of auxin transporters.

**AUXIN INFLUX**  The first step in polar transport is auxin influx. Auxin enters plant cells nondirectionally via passive diffusion of the protonated form (IAAH) across the phospholipid bilayer or via secondary active transport of the dissociated form (IAA⁻) through a 2H⁺–IAA⁻ symporter.

The undissociated form of indole-3-acetic acid, in which the carboxyl group is protonated, is lipophilic and readily diffuses across lipid bilayer membranes. In contrast, the dissociated form of auxin is negatively charged and therefore does not cross membranes unaided. Because the plasma membrane H⁺-ATPase normally maintains the cell wall solution at pH 5 to 5.5, 15–25% of the auxin ($pK_a = 4.75$) in the apoplast will be in the undissociated form and will diffuse passively across the plasma membrane down a concentration gradient. Experimental support for pH-dependent, passive auxin uptake was first provided by the demonstration that IAA uptake by plant cells increases as the extracellular pH is lowered from a neutral to a more acidic value. More recently, the role of apoplastic acidification in driving auxin transport has been demonstrated in vivo (Li et al. 2005).

FIGURE 19.9 The auxin permease AUX1 is expressed in a subset of columella, lateral root cap, and stelar tissues. Another member of the AUX/LAX family, LAX3, functions in cortical and epidermal cells of the root. (A) Diagram of tissues in the Arabidopsis root tip. (B) Immunolocalization of AUX1 in protophloem cells of the stele, a central cluster of cells in the columella, and lateral root cap cells. (C) LAX3 mediates auxin uptake in epidermal cells surrounding emerging lateral roots to induce cell expansion and other changes that permit lateral root emergence. (B from Swarup et al. 2001; C from Swarup et al. 2008.)

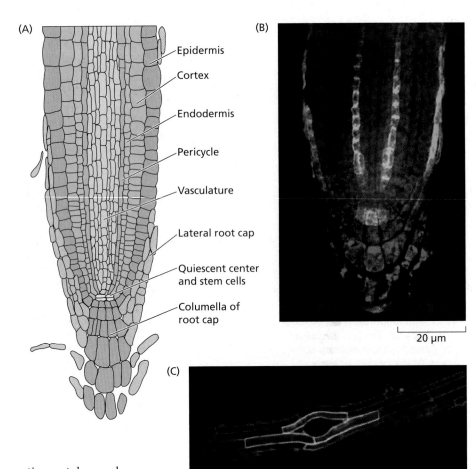

(A)

— Epidermis

— Cortex

— Endodermis

— Pericycle

— Vasculature

— Lateral root cap

— Quiescent center and stem cells

— Columella of root cap

(B)

20 μm

(C)

A carrier-mediated, secondary active uptake mechanism has been shown to be saturable and specific for active auxins. A small family of **AUX1/LAX** 2H$^+$–IAA$^-$ permeases are symporters that cotransport two protons along with the auxin anion (Yang et al. 2006; Swarup et al. 2008). This secondary active transport of auxin allows for greater auxin accumulation than does simple diffusion because it is driven across the membrane by the proton motive force (i.e., the high proton concentration in the apoplastic solution). Secondary active transport is particularly evident in cells where auxin sinks accelerate movement of polar streams, such as in the lateral root cap cells, where auxin uptake is the primary driver of auxin movement out of the lateral root cap (Kramer and Bennett 2005).

AUX1 functions in the uptake of auxin in the shoot and root apex. In the cells of the lateral root cap (**FIGURE 19.9A AND B**), AUX1 is required for mobilization of auxin away from the root apex into the basipetal auxin transport stream. The roots of Arabidopsis *aux1* mutants exhibit agravitropic growth, and this phenotype can be corrected by treatment with the synthetic auxin 1-naphthylene acetic acid (1-NAA), which readily crosses the lipid bilayer even in the absence of a carrier protein. The significance of this basipetal redirection at the root tip will become apparent when we examine the mechanism of root gravitropism.

Another auxin permease family member, LAX3, enhances auxin accumulation in some epidermal and cor-

tical cells, resulting in cell separation that facilitates the emergence of lateral roots from the central cylinder (**FIGURE 19.9C**). This is another illustration of the prominent role that AUX1/LAX proteins play in the maintenance of high auxin concentrations in some cell types.

**AUXIN EFFLUX** Once IAA enters the cytosol, which has a pH of approximately 7.2, nearly all of it dissociates to the anionic form. Because the membrane is impermeable to the anion, auxin accumulates inside the cell or along membrane surfaces unless it is exported by transport proteins on the plasma membrane. According to the chemiosmotic model, transport of IAA$^-$ out of the cell is driven by the negative membrane potential inside the cell.

As noted earlier, the central feature of the chemiosmotic model for polar transport is that IAA$^-$ efflux is directed by polar localization of some efflux carriers. Auxin uptake at one end of the cell and subsequent efflux from the other gives rise to a net polar transport. Members of a major subgroup of the **PIN** family of auxin carrier proteins are situated on the plasma membrane, transport auxin out of the cell, and are aligned with the direction of auxin transport (**FIGURE 19.10A**). (The PIN family of proteins is named

(A)

(B)

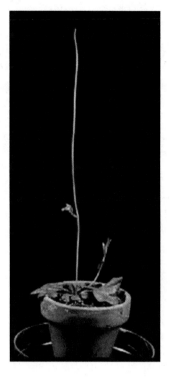

**FIGURE 19.10** PIN1 in Arabidopsis. (A) Localization of the PIN1 protein at the basal ends of conducting cells in Arabidopsis inflorescences as seen by immunofluorescence microscopy. (B) The *pin1* mutant of Arabidopsis. A normal wild-type Arabidopsis plant can be seen in Figure 16.1. (Courtesy of L. Gälweiler and K. Palme.)

sis, maize, and sorghum result in dwarf mutations of varying severity and in altered gravitropism and reduced auxin efflux (FIGURE 19.13).

There are 21 members of the ABCB family in Arabidopsis and 17 in rice. Three members of the ABCB family have been extensively characterized as participants in tissue-specific auxin transport (FIGURE 19.11B). It is still unknown how many other ABCB family members are auxin transporters, but some have been shown to transport malate or other small molecules rather than auxin. Unlike mammalian ABCB proteins, which are described as multiple drug resistance transporters and exhibit broad substrate specificity, plant auxin-transporting ABCBs appear to be more specific for the substrates they transport.

In general, ABCBs are uniformly, rather than polarly, distributed on the plasma membranes of cells in shoot and root apices. However, when specific ABCB and PIN proteins co-occur in the same location in the cell, the specificity of auxin transport is enhanced; PINs function synergistically with ABCBs to stimulate directional auxin transport (Blakeslee et al. 2005, 2007; Mravec et al. 2008; Titapiwatanakun et al. 2009). Further, in some cells, polarly localized ABCB transporters appear to make a more direct contribution to directional transport.

after the pin-shaped inflorescences formed by the *pin1* mutant of Arabidopsis; see FIGURE 19.10B.)

"Full-length" PIN proteins (in Arabidopsis, PIN1, 2, 3, 4, and 7) direct basal auxin efflux streams that are essential to normal plant development. Different PIN family members mediate auxin efflux in each tissue (FIGURE 19.11A). Of these, PIN1 is the most studied, as it essential to virtually every aspect of polar development and organogenesis in plants (see Chapter 16).

Modeling of auxin transport in intact tissues suggests that polar auxin transport also involves an energy-dependent mechanism, especially in small cells near the apical meristems. Auxin that has been directionally exported from one of these cells is likely to reenter the same cell unless actively excluded by efflux transporters on the plasma membrane (FIGURE 19.12). Mutational analyses also indicate that an additional efflux mechanism is required to prevent auxin that is moving in long distance transport streams from diffusing into adjacent tissues.

Plant cells contain ATP-dependent transporters belonging to the P-glycoprotein, or "B" subclass, of the large superfamily of plant ATP Binding Cassette (ABC) integral membrane transporters. A subgroup of these PGP/ABCB transporters are integral membrane proteins that function as ATP-dependent amphipathic anion carriers in cellular auxin efflux (see Figure 19.12). Defective *ABCB* genes in Arabidop-

## PIN and ABCB transporters regulate cellular auxin homeostasis

The subcellular distribution of IAA and its metabolites is poorly known, but IAA is thought to occupy both the cytoplasm and other subcellular compartments. As IAA is an *amphipathic* molecule (having both hydrophobic and hydrophilic regions), it is likely that most IAA within the cell is associated with endomembrane regions or proteins with hydrophobic regions. Further auxin conjugation and catabolism mechanisms appear to act rapidly on free IAA when concentrations increase in mature cells. A group of shorter PIN proteins (PIN5, 6, and 8) are localized to the membrane of the endoplasmic reticulum (ER). At least one, PIN5, appears to regulate IAA metabolite formation by transporting IAA into the ER (Mravec et al. 2009).

ABCB4 is a PGP-type transporter that is thought to regulate auxin levels in root hair cells. At low auxin concentrations, ABCB4 appears to function in cellular uptake, but rapidly reverses to efflux when auxin concentrations reach a threshold level (Yang and Murphy 2009). This suggests that free IAA levels can be regulated by transporters on both internal membranes and the plasma membrane (FIGURE 19.14). Not surprisingly, the abundance of all known auxin transporters is regulated to some extent by auxin levels.

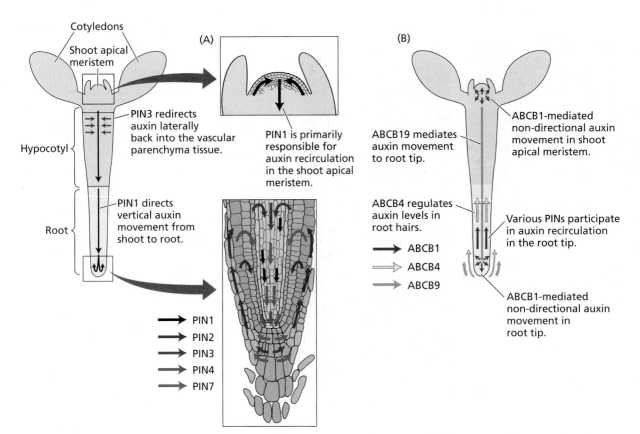

**FIGURE 19.11** In Arabidopsis, PIN and ABCB transport proteins direct the auxin efflux component of polar auxin transport throughout the plant. (A) PIN proteins determine the basal direction of auxin movement. Directional auxin movement is associated with the tissue-specific distribution of PIN efflux carrier proteins. PIN1 mediates vertical transport of IAA from the shoot to the root along the embryonic apical–basal axis (see Chapter 16). AUX1 (see Figure 19.9) creates an auxin sink that drives basipetal auxin transport upwards from the root apex via PIN2 efflux carriers. Since some lateral diffusion of auxin may occur, PIN2 and PIN3 are thought to redirect auxin back into vascular parenchymal tissue, where polar transport takes place. The two inserts show PIN1-mediated auxin movement in the shoot apical meristem (upper), and PIN-regulated auxin circulation in the root tip (lower). (B) Auxin flow associated with ATP-dependent ABCB transport proteins. The multidirectional arrows at the shoot and root apices indicate non-directional auxin transport. However, when combined with polarly localized PIN proteins, directional transport occurs. ABCB4 regulates auxin levels in elongating root hairs. (A, root model after Blilou et al. 2005.)

## Auxin influx and efflux can be chemically inhibited

Several compounds have been synthesized that can act as auxin transport inhibitors, including NPA (1-*N*-naphthylphthalamic acid), TIBA (2,3,5-triiodobenzoic acid), CPD (2-carboxyphenyl-3-phenylpropane-1,3-dione), NOA (1-napthoxyacetic acid), 2-[4-(diethylamino) -2-hydroxybenzoyl] benzoic acid, and gravacin (**FIGURE 19.15**). NPA, TIBA, CPD, and gravacin are **auxin efflux inhibitors (AEIs)**, while NOA is an **auxin influx inhibitor**.

Some AEIs, such as TIBA, have weak auxin activity and inhibit polar transport in part by competing with auxin at the efflux carrier site. Other AEIs, such as CPD, NPA, and gravacin interfere with auxin transport by binding to a regulatory site. Some inhibitors, such as gravacin, interfere more specifically with one type of transporter (Rojas-Pierce et al. 2007), while others, such as NPA, bind to and interfere with multiple proteins, some of which are only indirectly involved in auxin transport.

Some natural compounds, primarily flavonoids (see Figure 19.15B), also function as auxin efflux inhibitors. Flavonoids function as ROS scavengers and are inhibitors of some metalloenzymes, kinases, and phosphatases (Peer and Murphy 2007). Their effects on auxin transport appear to result primarily from these activities.

As is always the case with such pharmacological agents, the use of higher concentrations of the inhibitor results in less specific effects (Peer et al. 2009). When used at higher concentrations, AEIs can interfere with the trafficking of the plasma membrane proteins to which they bind, apparently by altering protein–protein interactions. The role of protein trafficking in regulating polar transport is discussed later in this section.

1. The plasma membrane H+-ATPase (purple) pumps protons into the apoplast. The acidity of apoplast affects the rate of auxin transport by altering the ratio of IAAH and IAA− present in the apoplast.

2. IAAH can enter the cell via proton symporters such as AUX1 (blue) or diffusion (dashed arrows). Once inside the cytosol, IAA is an anion, and may only exit the cell via active transport.

3. P-glycoproteins are localized nonpolarly on the plasma membrane and can drive active (ATP-dependent) auxin efflux.

4. Synergistically enhanced active polar transport occurs when polarly localized PIN proteins (yellow) associate with PGP proteins, overcoming the effects of back-diffusion.

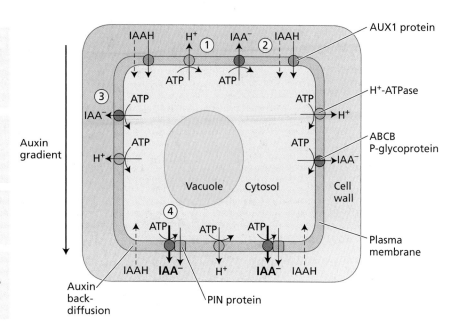

**FIGURE 19.12** Model for polar auxin transport in small cells with significant back-diffusion of auxin due to a high surface-to-volume ratio. ABCB proteins are thought to maintain polar streams by preventing reuptake of auxin exported at carrier sites. In larger cells, ABCB transporters appear to exclude movement of auxin out of polar streams into adjoining cell files.

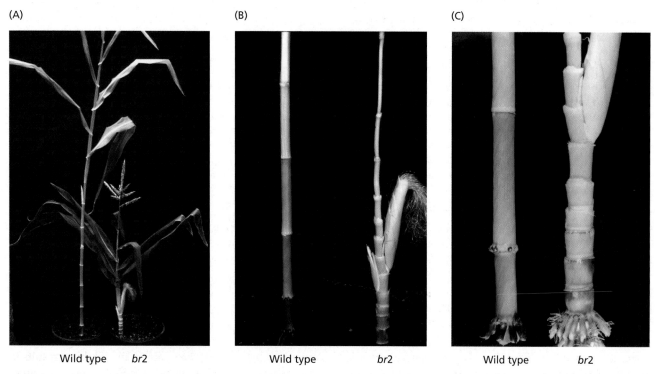

(A) Wild type  br2

(B) Wild type  br2

(C) Wild type  br2

**FIGURE 19.13** The *BR2* (*Brachytic 2*) gene encodes a P-glycoprotein required for normal auxin transport in corn, and *br2* mutants have short internodes. The mutant was created by insertional mutagenesis with the *Mutator* transposon. Unknown to the investigators, the *Mu8* transposon contained a fragment of the *BR2* gene. Expression of the *BR2* gene fragment produced interfering RNA (RNAi), which silenced BR2 expression (see Chapter 2). The *br2* mutants have compact lower stalks (B and C), but normal tassels and ear (A and B). (From Multani et al. 2003.)

AUX/LAX proteins regulate auxin uptake into the cell.

ABCB and PIN1-like PIN transporters control auxin efflux from the cell.

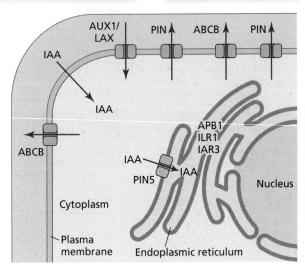

**FIGURE 19.14** Model of transporter-mediated auxin homeostasis. Polar, intercellular auxin transport is mediated by the PIN1-like and ABCB plasma membrane–localized auxin efflux carriers and AUX/LAX influx carriers. Intracellular auxin is compartmentalized into cytosolic and ER pools. PIN5 and PIN5-like (PIN6 and PIN8) transporters regulate auxin sequestration in the ER. Most of AUXIN BINDING PROTEIN1 (ABP1), which binds auxin with high affinity, localizes to the ER. ILR1 and IAR3 encode IAA-amino acid hydrolases that can release free IAA in the ER. (After Mravec et al. 2009.)

APB1 is an ER auxin-binding protein. Other ER proteins (ILR1 and IAR3) are hydrolases that deconjugate IAA.

"Short" PINs (e.g., PIN5) control auxin transport across the ER membrane.

## Auxin transport is regulated by multiple mechanisms

As would be expected for such an important mechanism, auxin transport is regulated by both gene transcription and posttranscriptional mechanisms. Developmental programs and environmental cues result in increased or decreased expression of auxin transporter genes and genes encoding proteins that regulate them in specific cells and tissues. Other factors such as phosphorylation state, interactions with regulatory proteins, cellular trafficking events, proteolytic processing, and membrane composition also deter-

mine rates and direction of auxin transport. Here we discuss some of the most important factors regulating auxin transport: hormones, protein trafficking, and flavonoids.

**INTERACTIONS WITH OTHER HORMONES** Almost all known plant signaling compounds have an effect on auxin transport and/or auxin-dependent gene expression. Auxin itself regulates expression of the genes encoding auxin transporters to increase or decrease their abundance and, thus, to regulate auxin levels. Auxin also appears to affect

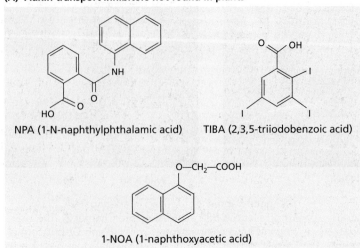

**(A) Auxin transport inhibitors not found in plants**

NPA (1-N-naphthylphthalamic acid)   TIBA (2,3,5-triiodobenzoic acid)

1-NOA (1-naphthoxyacetic acid)

**(B) Naturally occurring auxin transport inhibitors**

Quercetin (flavonol)

Genistein

**FIGURE 19.15** Structures of synthetic (A) and natural (B) auxin transport inhibitors.

(A)

(B)

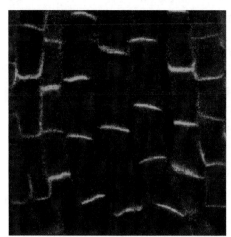
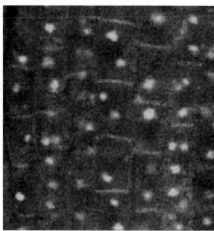

FIGURE 19.16 Auxin transport inhibitors block secretion of the auxin efflux carrier PIN1 to the plasma membrane. (A) Control, showing asymmetric localization of PIN1 in the root apex. (B) After treatment with brefeldin A (BFA). Following a washout with buffer, PIN1 localization is restored as in A. When BFA is washed out with the auxin transport inhibitor TIBA, PIN1 is retained in endomembrane compartments as in B. (Photos courtesy of Klaus Palme 1999.)

a variety of polar events in cells, and to *canalize* (create a channel for) its own directional transport streams. In other words, small directional flows of auxin are amplified and stabilized by establishment of transport proteins and vascular tissue in configurations that maintain directional flows to growing tissues. This effect is prominent in the establishment of embryonic polar growth and organogenesis. In particular, PIN1 has been shown to orient directionally with these developmentally important flows.

Ethylene influences auxin transport streams by changing the activity and abundance of AUX1 uptake and PIN efflux transporters. These effects are particularly pronounced in lateral root development, although ethylene appears to affect auxin transport primarily by altering auxin biosynthesis in the root (Swarup et al. 2007).

Brassinosteroids, cytokinins, jasmonic acid, gibberellins, strigolactone, and some flavonoids can also regulate auxin transport by altering the expression of auxin transporter genes, the activity of regulatory factors, and/or cellular trafficking mechanisms. We will discuss these interactions in the context of the physiological and developmental processes that they control in this and other chapters.

**ROLE OF PROTEIN TRAFFICKING** The plasma membrane localization of auxin transport proteins involves the movement of newly synthesized proteins through the endomembrane secretory system (protein trafficking). In the case of ABCB transporters, correctly folded and glycosylated proteins are transported via secretory vesicles to the plasma membrane where some, such as ABCB19, form extremely stable complexes that can include PIN1 in sterol-enriched portions of the membrane (Titapiwatanakun et al. 2009). However, the plasma membrane localization of both the full-length PIN efflux carriers and AUX1 influx carriers is regulated by trafficking mechanisms that involve endocytotic cycling between the plasma membrane and endosomal compartments (Geldner et al. 2001, 2003; Swarup

et al. 2001; Swarup and Bennett 2003). This mechanism appears to also regulate how polarly localized PIN transporters are mobilized from a symmetric distribution on the plasma membrane to a particular end of the cell.

This cycling mechanism was first demonstrated with PIN1, which is localized at the bottom of vascular cells in the Arabidopsis root tip (FIGURE 19.16A). Treatment of the roots with brefeldin A (BFA), which interferes with protein exocytosis and secretion regulated by the ADP ribosylation/guanine nucleotide exchange factor called GNOM (see Chapter 16), causes PIN1 to accumulate in abnormal intracellular compartments (FIGURE 19.16B). When the BFA is washed out with buffer, the normal localization on the plasma membrane at the base of the cell is restored. But when cytochalasin D, an inhibitor of actin polymerization, is included in the buffer washout solution, normal relocalization of PIN1 to the plasma membrane is prevented.

These results indicate that PIN1 is rapidly cycled between the plasma membrane at the base of the cell and an unidentified endosomal compartment by an actin-dependent mechanism. Several other proteins, including a plasma membrane $H^+$-ATPase, are targeted to the membrane by the same BFA-sensitive mechanism that regulates PIN1 cycling.

When high concentrations of the auxin efflux inhibitors TIBA and NPA were added to the washout buffer in the experiments above, they prevented the normal relocalization of PIN1 on the plasma membrane. Replacement of synthetic AEIs in washout buffers with natural flavonol inhibitors of auxin transport also prevented plasma membrane relocalization of PIN1, although only in flavonoid-deficient mutants (Peer et al. 2004). TIBA and NPA also altered the endocytotic cycling of the plasma membrane $H^+$-ATPase and other proteins, but, as was the case with PINs, were unaffected by AEI treatment unless a large scaffolding protein called BIG was absent. These results suggest that some AEIs can inhibit membrane trafficking

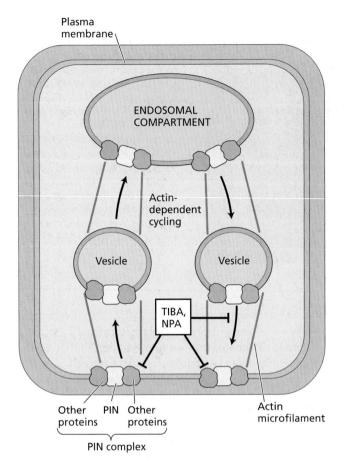

Plasma membrane

ENDOSOMAL COMPARTMENT

Actin-dependent cycling

Vesicle

Vesicle

TIBA, NPA

Other proteins    PIN    Other proteins

Actin microfilament

PIN complex

**FIGURE 19.17** PIN complex cycling as a result of actin-dependent vesicle trafficking between the plasma membrane and an endosomal compartment. Auxin transport inhibitors TIBA and NPA interfere with the relocalization of PIN1 proteins to basal plasma membranes after a washout with brefeldin A (see Figure 19.16). This suggests that auxin transport inhibitors may act by interfering with PIN1 protein cycling.

when used at high concentrations, but also indicate that NPA and TIBA must directly inhibit the transport activity of efflux carrier complexes on the plasma membrane as well. Similar experiments have shown that other full-length PIN proteins, such as Arabidopsis PIN2 and PIN3, are regulated by dynamic cellular trafficking mechanisms, although these differ slightly from isoform to isoform. A simplified model of the effects of TIBA and NPA on PIN1 cycling and auxin efflux is shown in **FIGURE 19.17**. A more complete model, which incorporates additional regulatory steps, is presented in **WEB ESSAY 19.1**.

High concentrations of auxin have also been shown to regulate full-length PIN trafficking and stability on the membrane (Paciorek et al. 2005). This is consistent with the concept of auxin canalization and provides a mechanism for this phenomenon. However, it is still not clear how auxin regulates PIN trafficking, as this phenomenon appears to be independent of the known auxin signaling mechanisms.

The polarized trafficking of PIN proteins is regulated by the protein kinase PINOID (named for the *pin1*-like phenotype of the *pinoid* mutant), which appears to act as a developmental switch controlling PIN polarity. A phosphatase 2A complex characterized by its RCN1 subunit (named for the phenotype of the *roots curling in NPA* mutant) appears to act antagonistically to PINOID activity in regulating PIN localization. However, these and other kinases have been shown to also directly regulate PIN and/or ABCB protein activity.

**ROLE OF FLAVONOIDS**  Arabidopsis mutants producing excess or decreased flavonoids exhibit altered auxin transport and growth, and some flavonoids, such as the flavonols quercetin (see Figure 19.15B) and kaempferol, can displace AEIs from plant membranes (Murphy et al. 2000). These naturally occurring plant compounds also inhibit the activity of certain kinases and phosphatases (see Chapter 14). PINOID, RCN1, and other phosphotransferase proteins that modify the activity or localization of auxin transporters are likely targets of flavonols, although flavonoids directly synthesized in vivo have not yet been shown to regulate the activity of these proteins.

Flavonoids are also inhibitors of ABCB transport activity, as they inhibit the ATP hydrolysis that powers these transporters. No direct inhibition of PIN transport activity by flavonoids has been detected, as would be expected for carrier proteins that have evolved in the presence of flavonoids. As auxin transport systems have largely evolved in the presence of these compounds, the effects of these compounds in vivo are limited (Peer and Murphy 2007).

## Auxin Signal Transduction Pathways

The ultimate goal of research on the molecular mechanism of hormone action is to reconstruct each step in the signal transduction pathway, from receptor binding to the physiological response. In the case of auxin, this would seem to be a particularly daunting task because auxin affects so many physiological and developmental processes. However, the initial steps in auxin signaling are surprisingly simple, and involve binding to a small group of receptor–enzyme complexes that regulate protein degradation via the ubiquitin-proteasome pathway (see Chapters 2 and 14).

Upon activation by auxin, the receptor–enzyme complex targets specific transcriptional repressors for proteolysis, resulting in the activation and derepression of auxin-responsive genes. While this mechanism appears to account for most auxin responses, a different type of auxin receptor protein may function in nontranscriptional activation and mobilization of plasma membrane $H^+$-ATPases to cause rapid cell wall acidification and cell elongation. In this last section of the chapter, we will examine the signaling pathways involved in both types of auxin responses.

**FIGURE 19.18** A model for auxin binding to the composite TIR1/ABF–AUX/IAA auxin receptors and subsequent transcriptional activation of auxin response genes. (A) In the absence of auxin, AUX/IAA repressors inhibit the transcription of auxin-induced genes by binding to ARF transcriptional activators, locking them into an inactive state. Auxin functions as "molecular glue" to initiate an interaction between an AUX/IAA and the TIR1/ABF component of an SCF$^{TIR1/ABF}$ complex. (B) Auxin-activated SCF$^{TIR1/ABF}$ complexes attach ubiquitin molecules to AUX/IAA proteins, promoting their destruction by the 26S proteasome. The removal and degradation of AUX/IAA proteins "unlocks" the ARF transcriptional activators. The ARF transcriptional activators bind to auxin response elements (AuxRE) and stimulate the transcription of auxin-induced genes.

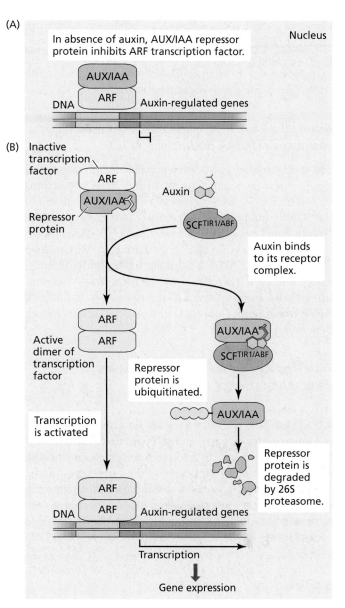

## The principal auxin receptors are soluble protein heterodimers

The principal auxin receptors have been identified as complexes made up of a soluble protein belonging to the **TIR1/AFB** family and a member of the larger family of AUX/IAA transcriptional repressor proteins (**FIGURE 19.18**) (Dharmasiri et al. 2005a, 2005b; Kepinski et al. 2005). TIR1/AFB proteins are named for the first member of the family identified in plants, *TRANSPORT INHIBITOR RESISTANT 1*, and their general classification as *Auxin F-box Binding* proteins. They are F-box protein components of an SCF ubiquitin E3 ligase complex that catalyzes the ATP-dependent, covalent addition of ubiquitin molecules to proteins targeted for proteolytic degradation (see Chapter 14 for a description of this signaling mechanism).

Mutant analyses in Arabidopsis led to the identification of TIR1 (*transport inhibitor response 1*), which functions in auxin-dependent hypocotyl elongation and lateral root formation. TIR1 is a component of a specific E3 ubiquitin ligase complex, designated SCF$^{TIR1}$, that is required for auxin signaling in cells (Dharmasiri et al. 2005a; Kepinski and Leyser 2005). Subsequently, researchers identified proteins called *auxin signaling F-box binding* (AFB) proteins that are structurally and functionally similar to TIR1.

Auxin acts as a "molecular glue" to bring together a heterodimer consisting of one TIR1/AFB and one AUX/IAA protein. Auxin stabilizes the formation of the TIR1/AFB–AUX/IAA heterodimers and, thus, promotes the association of the SCF$^{TIR1}$ complex with the AUX/IAA protein. This type of "combinatorial" receptor allows auxin to act as a signal that differentially regulates multiple processes in discrete tissues under different environmental conditions. In Arabidopsis, there could be more than 100 possible combinations of AUX/IAAs and TIR1/AFBs, thus providing the basis for the wide variety of auxin signaling responses seen throughout plant development.

## Auxin-induced genes are negatively regulated by AUX/IAA proteins

Two families of transcriptional regulators participate in the TIR1 auxin signaling pathway: auxin response factors and AUX/IAA proteins. **Auxin response factors (ARFs)** are short-lived nuclear proteins that bind to DNA **auxin response elements (AuxREs)** containing the consensus sequence TGTCTC in the promoters of primary, or early, auxin-response genes. There are 23 different ARF proteins in Arabidopsis. The binding of ARFs to AuxREs results in the activation or repression of gene transcription, depending on the particular ARF involved. ARFs appear to be present on the promoters of early auxin genes regardless of the auxin status of the tissue.

**AUX/IAA proteins** are important regulators of auxin-induced gene expression. There are 29 members of this family of short-lived small nuclear proteins in Arabidopsis. AUX/IAA proteins regulate gene transcription indirectly by binding to ARF protein bound to DNA. If the ARF in question is a transcriptional activator, the effect of the AUX/IAA protein is to repress transcription.

### Auxin binding to a TIR1/AFB-AUX/IAA heterodimer stimulates AUX/IAA destruction

The short signaling pathway responsible for auxin-induced gene expression begins with auxin-stabilized interactions between an AUX/IAA protein and a TIR1/AFB subunit of the SCF$^{TIR1}$ ubiquitin ligase complex (see Figure 19.18A). Unlike some other SCF complexes, no covalent modification is involved in SCF$^{TIR1}$ activation by auxin. As a consequence, the AUX/IAA proteins are rapidly ubiquitinated and then hydrolyzed via the proteasome (see Figure 19.18B). In the absence of their negative regulators, ARF proteins either stimulate or repress gene expression, depending on the ARF.

### Auxin-induced genes fall into two classes: early and late

Auxin-responsive genes that are directly activated by AUX/IAA-TIR/AFB signaling are described as **primary response genes** or **early genes**. These are distinguished from genes acting later in auxin-induced development. The time required for the expression of the early genes can be quite short, ranging from a few minutes to several hours. All of the primary response genes in auxin responses are induced via the SCF$^{TIR1}$ signaling pathway. Early response genes include the *AUX/IAA* genes, *SAUR* genes, and *GH3* genes.

In general, primary response genes have three main functions:

- *Transcription.* Some of the early genes encode proteins that regulate the transcription of **secondary response genes**, or **late genes**, which are required for the long-term responses to the hormone. Because late genes require de novo protein synthesis, their expression can be blocked by protein synthesis inhibitors.
- *Signaling.* Other early genes are involved in intercellular communication, or cell-to-cell signaling.
- *Auxin conjugation/catabolism.* Some rapidly induced genes encode proteins that are involved in elimination of active IAA by conjugation or degradation. Expression of these genes prevents excessive auxin accumulation that would limit the specificity of auxin responses.

**EARLY GENES FOR GROWTH AND DEVELOPMENT**  The expression of most *AUX/IAA* genes is stimulated by auxin within 5 to 60 minutes of hormone addition. AUX/IAA

proteins have short half-lives (about 7 minutes), which indicates that they turn over rapidly.

Auxin stimulates the expression of *SAUR* genes within 2 to 5 minutes of treatment, and the response is insensitive to the protein synthesis inhibitor cycloheximide, indicating that their expression does not require the synthesis of new transcription factors. The five *SAUR* genes of soybean are clustered together, contain no introns, and encode highly similar polypeptides of unknown function. Because of the rapidity of the response, expression of *SAUR* genes has proven to be a convenient probe for the lateral transport of auxin during photo- and gravitropism. The function of SAUR genes is still largely unknown.

*GH3* early-gene family members, identified in both soybean and Arabidopsis, are stimulated by auxin within 5 minutes. The GH3 protein functions in IAA conjugation (Staswick et al. 2005) (see Figure 19.5). Mutations in Arabidopsis *GH3*-like genes result in dwarfism (Nakazawa et al. 2001) and appear to function in light-regulated auxin responses (Hsieh et al. 2000). Because *GH3* expression is a good reflection of the presence of endogenous auxin, a synthetic *GH3*-based reporter gene promoter known as *DR5* is widely used in auxin bioassays (see Figure 19.5).

**LATE GENES FOR STRESS ADAPTATIONS**  Genes that act later in auxin-induced developmental processes—usually 2–4 hours after an auxin increase—have also been identified. For example, several genes encoding glutathione-S-transferases (GSTs), a class of proteins stimulated by various stress conditions, are induced by elevated auxin concentrations. Likewise, high levels of auxin induce ACC synthase, which is also induced by stress and is the rate-limiting step in ethylene biosynthesis (see Chapter 22). None of the late auxin response genes encode proteins shown to directly mediate primary auxin response mechanisms.

### Rapid, nontranscriptional auxin responses appear to involve a different receptor protein

As discussed earlier in the chapter, auxin induces cell elongation with a lag time of less than 15 minutes, and auxin-induced increases in plasma membrane H$^+$-ATPase activity begin even sooner. The earliest of these changes still take place in mutants that are missing components essential for expression of auxin primary response genes. Recent evidence suggests that auxin acts directly on cellular trafficking mechanisms, either within the secretory system or at the plasma membrane itself, to influence H$^+$-ATPase activity.

## Actions of Auxin: Cell Elongation

Auxin was discovered as the hormone involved in the bending of coleoptiles toward light. The coleoptile bends because of the unequal rates of cell elongation on its shaded versus its illuminated side (see Figure 19.1). The ability of

**FIGURE 19.19** Time course for auxin-induced growth of *Avena* (oat) coleoptile sections. Growth is plotted as the percent increase in length. Auxin was added at time zero. When sucrose (Suc) is included in the medium, the response can continue for as long as 20 hours. Sucrose prolongs the growth response to auxin mainly by providing osmotically active solute that can be taken up for the maintenance of turgor pressure during cell elongation. KCl can substitute for sucrose. The inset shows a short-term time course plotted with an electronic position-sensing transducer. In this graph, growth is plotted as the absolute length in millimeters versus time. The curve shows a lag time of about 15 minutes for auxin-stimulated growth to begin. (From Cleland 1995.)

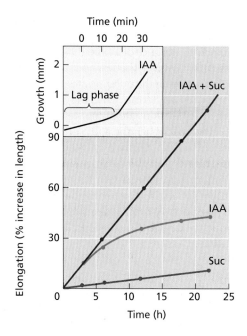

auxin to regulate the rate of cell elongation has long fascinated plant scientists. In this section we will review the physiology of auxin-induced cell elongation, some aspects of which were discussed in Chapter 15.

### Auxins promote growth in stems and coleoptiles, while inhibiting growth in roots

As we have seen, auxin synthesized in the shoot apex is transported basipetally to the tissues below. The steady supply of auxin arriving at the subapical region of the stem or coleoptile is required for the continued elongation of these cells. Because the level of endogenous auxin in the elongation region of a normal healthy plant is nearly optimal for growth, spraying the plant with exogenous auxin causes only a modest and short-lived stimulation in growth. Such spraying may even be inhibitory in the case of dark-grown seedlings, which are more sensitive to supraoptimal auxin concentrations than light-grown plants.

However, when the endogenous source of auxin is removed by excision of sections containing the elongation zones, the growth rate rapidly decreases to a low basal rate. Such excised sections will often respond dramatically to exogenous auxin by rapidly increasing their growth rate back to the level in the intact plant.

In long-term experiments, auxin treatment of excised sections of coleoptiles (see Figure 19.2) or dicot stems stimulates the rate of elongation of the section for up to 20 hours (**FIGURE 19.19**). The optimal auxin concentration for elongation growth in pea stems and oat coleoptiles is typically $10^{-6}$ to $10^{-5}$ M (**FIGURE 19.20**). In other species, such as Arabidopsis, the optimum concentration is slightly lower. The inhibition observed when auxin concentrations exceed optimal levels is attributed mainly to auxin-induced ethylene biosynthesis. As we will see in Chapter 22, the gaseous hormone ethylene inhibits stem elongation in many species.

Auxin control of root elongation has been more difficult to demonstrate, perhaps because auxin induces the production of ethylene, which also inhibits root growth. Recent evidence shows that these two hormones interact differentially in root tissue to control growth. However, even if ethylene biosynthesis is specifically blocked, low concentrations ($10^{-10}$ to $10^{-9}$ M) of auxin promote the growth of intact roots, whereas higher concentrations

($10^{-6}$ M) inhibit growth. Thus, while roots may require a minimum concentration of auxin to grow, root growth is strongly inhibited by auxin concentrations that promote elongation in stems and coleoptiles.

### The outer tissues of dicot stems are the targets of auxin action

Dicot stems are composed of many types of tissues and cells, only some of which may limit the growth rate. This point is illustrated by a simple experiment. When sections from growing regions of an etiolated dicot stem, such

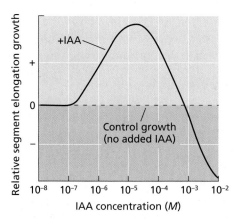

**FIGURE 19.20** Typical dose–response curve for IAA-induced growth in pea stem or oat coleoptile sections. Elongation growth of excised sections of coleoptiles or young stems is plotted versus increasing concentrations of exogenous IAA. At higher concentrations (above $10^{-5}$ M), IAA becomes less and less effective; above about $10^{-4}$ M it becomes inhibitory, as shown by the fact that the curve falls below the dashed line, which represents growth in the absence of added IAA.

as pea, are split lengthwise and incubated in buffer (no auxin), the two halves bend outward. This result indicates that in the absence of auxin, the central tissues—including the pith, vascular tissues, and inner cortex—elongate at a faster rate than the outer tissues, which consist of the outer cortex and epidermis. Thus the outer tissues must be limiting the extension rate of the stem in the absence of auxin.

However, when the split sections are incubated in buffer plus auxin, the two halves now curve inward, demonstrating that the outer tissues of dicot stems are the primary targets of auxin action during cell elongation. Studies of auxin movement and auxin-responsive growth in Arabidopsis roots suggest that epidermal cells are the principal targets of auxin action in the root elongation zone as well.

To reach its targets in the elongating regions of dicot shoots, auxin derived from the shoot apex must be diverted laterally from the polar transport stream in vascular parenchyma cells to the outer shoot tissues. In contrast, nonvascular tissues in monocot coleoptiles are capable of transporting auxin as well as responding to it.

This process of lateral auxin transport in dicots appears to be mediated by laterally oriented PIN proteins in vascular parenchyma cells, as well as evenly distributed ABCB transporters in the surrounding cell layer (Friml et al. 2002; Blakeslee et al. 2007). The positioning of these proteins is shown in FIGURE 19.21. The mechanism of lateral transport will be further discussed later in this chapter in relation to plant bending responses.

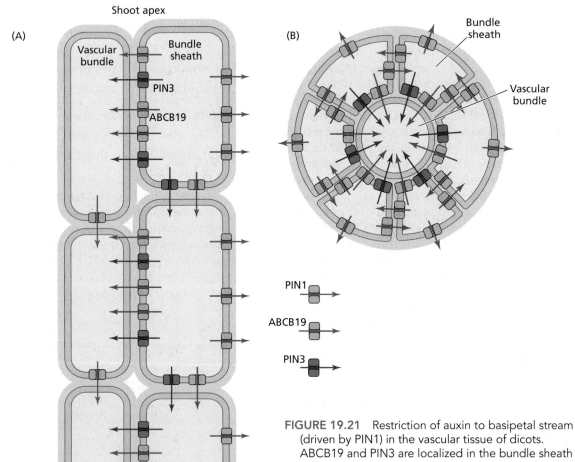

**FIGURE 19.21**  Restriction of auxin to basipetal stream (driven by PIN1) in the vascular tissue of dicots. ABCB19 and PIN3 are localized in the bundle sheath cells adjoining the vascular tissue. (A) PIN3 is localized to the inward lateral face of these cells and is thought to redirect auxin into the vascular stream. ABCB19 restricts uptake of auxin into these cells. The directions of the arrows indicate the directions of auxin flow. (B) A cross sectional view of this region indicates how ABCB19 export would contribute to redirection of auxin into the vascular cylinder. Mutational analyses indicate that both PIN3 and ABCB19 function in lateral redistribution of auxin in tropic bending.

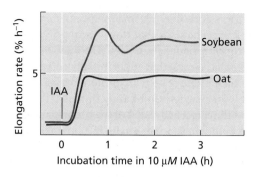

**FIGURE 19.22** Comparison of the growth kinetics of oat coleoptile and soybean hypocotyl sections incubated with 10 µM IAA and 2% sucrose. Growth is plotted as the rate of elongation, rather than the absolute length, at each time point. The growth rate of the soybean hypocotyl oscillates after 1 hour, whereas that of the oat coleoptile is constant. (After Cleland 1995.)

### The minimum lag time for auxin-induced elongation is ten minutes

When a stem or coleoptile section is excised and inserted into a sensitive growth-measuring device, the growth response to auxin can be monitored at very high resolution. Without auxin in the medium, the growth rate declines rapidly. Addition of auxin markedly stimulates the growth rate after a lag period of only 10 to 12 minutes (see the inset in Figure 19.19). As is shown in Figure 19.20, a threshold concentration of auxin must be reached to initiate this response. Beyond the optimum concentration, auxin becomes inhibitory. Both oat (*Avena sativa*) coleoptiles and soybean (*Glycine max*) hypocotyls (dicot stems) reach a maximum growth rate after 30 to 60 minutes of auxin treatment (**FIGURE 19.22**). This maximum represents a five- to tenfold increase over the basal rate. Oat coleoptile sections can maintain this maximum rate for up to 18 hours in the presence of osmotically active solutes such as sucrose or KCl.

The stimulation of growth by auxin requires energy, and metabolic inhibitors inhibit the response within minutes. Auxin-induced growth is also sensitive to inhibitors of protein synthesis such as cycloheximide, suggesting that protein synthesis is required for the response. Inhibitors of RNA synthesis also inhibit auxin-induced growth after a slightly longer delay.

Although the lag time for auxin-stimulated growth can be *lengthened* by lowering the temperature or by using low auxin concentrations so that it takes longer for auxin to diffuse into the tissue, the lag time cannot be *shortened* by raising the temperature, by using high auxin concentrations, or by abrading the waxy cuticle on the surface of the stem or coleoptile section to allow auxin to penetrate the tissue more rapidly. Thus the minimum lag time of 10 minutes is not determined by the time required for auxin to reach its site of action. Rather, the lag time reflects the time needed for the biochemical machinery of the cell to bring about the increase in the growth rate.

### Auxin rapidly increases the extensibility of the cell wall

How does auxin cause a five- to tenfold increase in the growth rate in only 10 minutes? To understand the mechanism, we must first review the process of cell enlargement in plants (see Chapter 15). Plant cells expand in three steps:

1. Osmotic uptake of water across the plasma membrane is driven by the gradient in water potential ($\Delta\Psi_w$).

2. Turgor pressure builds up because of the rigidity of the cell wall.

3. Biochemical wall loosening occurs, allowing the cell to expand in response to turgor pressure.

The effects of these parameters on the growth rate are encapsulated in the growth rate equation:

$$GR = m\,(\Psi_p - Y)$$

where $GR$ is the growth rate, $\Psi_p$ is the turgor pressure, $Y$ is the yield threshold, and $m$ is the coefficient (*wall extensibility*, see Chapter 15) that relates the growth rate to the difference between $\Psi_p$ and $Y$.

In principle, auxin could increase the growth rate by increasing $m$, increasing $_p$, or decreasing $Y$. Although extensive experiments have shown that auxin does not increase turgor pressure when it stimulates growth, conflicting results have been obtained regarding auxin-induced decreases in the yield threshold, $Y$. However, there is general agreement that auxin causes an increase in the wall extensibility parameter, $m$. This increase in $m$ is mediated by protons.

### Auxin-induced proton extrusion increases cell extension

According to the widely accepted **acid growth hypothesis**, hydrogen ions act as the intermediate between auxin and cell wall loosening. The source of the hydrogen ions is the plasma membrane $H^+$-ATPase, whose activity is thought to increase in response to auxin. The acid growth hypothesis allows five main predictions:

1. Acid buffers alone should promote short-term growth, provided the cuticle has been abraded to allow the protons access to the cell wall.

2. Auxin should increase the rate of proton extrusion (wall acidification), and the kinetics of proton extrusion should closely match those of auxin-induced growth.

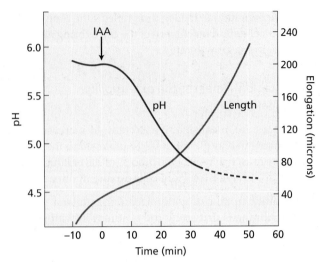

**FIGURE 19.23** Kinetics of auxin-induced elongation and cell wall acidification in maize coleoptiles. The pH of the cell wall was measured with a pH microelectrode. Note the similar lag times (10 to 15 minutes) for both cell wall acidification and the increase in the rate of elongation. (From Jacobs and Ray 1976.)

3. Neutral buffers should inhibit auxin-induced growth.

4. Compounds (other than auxin) that promote proton extrusion should stimulate growth.

5. Cell walls should contain a "wall-loosening factor" with an acidic pH optimum.

All five of these predictions have been confirmed. Acidic buffers cause a rapid and immediate increase in the growth rate, provided the cuticle has been abraded. Auxin stimulates proton extrusion into the cell wall after 10 to 15 minutes of lag time, consistent with the growth kinetics (**FIGURE 19.23**).

Auxin-induced growth is also inhibited by neutral buffers, as long as the cuticle has been abraded. Further, *fusicoccin*, a fungal phytotoxin that stimulates rapid proton extrusion, also induces rapid transient growth in stem and coleoptile sections. And finally, wall-loosening proteins called **expansins** loosen cell walls by weakening the hydrogen bonds between the polysaccharide components of the wall when the pH is acidic (see Chapter 15).

### Auxin-induced proton extrusion involves activation and protein mobilization

Recent studies in Arabidopsis have shown that the extrusion of protons by plasma membrane H+-ATPases is regulated by interactions with repressor proteins rather than by the abundance of the H+-ATPases on the membrane (see Chapter 6). Cell biology studies indicate that auxin directly influences the retention or endocytosis of PIN transporters and other plasma membrane proteins via an unknown mechanism that does not involve auxin-activated transcriptional responses (Dhonukshe et al. 2005). These results suggest that auxin initially acts directly at the plasma membrane to activate proton extrusion.

An ER-localized auxin-binding protein (ABP1) has been linked to the direct activation of the plasma membrane H+-ATPase in the presence of auxin, but how ABP1 mediates this response is unknown. Recent evidence that the short PINs (5, 6, and 8) are localized on the ER membrane suggests that secretion through the endomembrane system may play a role in acidification. Regardless of how acidification is stimulated, later auxin-induced growth events are dependent on well-characterized auxin receptors that activate gene transcription, as described in the previous section of this chapter.

## Actions of Auxin: Plant Tropisms

Although plant growth can be influenced by many environmental factors, three main guidance systems control the orientation of the plant axis:

- **Phototropism**, or growth with respect to light, is expressed in all shoots and some roots; it ensures that leaves will receive optimal sunlight for photosynthesis.

- **Gravitropism**, growth in response to gravity, enables roots to grow downward into the soil and shoots to grow upward away from the soil, responses that are especially critical during the early stages of germination.

- **Thigmotropism**, or growth with respect to touch, enables roots to grow around obstacles and is responsible for the ability of the shoots of climbing plants to wrap around other structures for support.

In this section we will examine the evidence that bending in response to light or gravity results from the lateral redistribution of auxin. We will also consider the cellular mechanisms involved in generating lateral auxin gradients during bending growth. Less is known about the mechanism of thigmotropism, although it, too, probably involves auxin gradients.

### Phototropism is mediated by the lateral redistribution of auxin

As we saw earlier, Charles and Francis Darwin provided the first clue concerning the mechanism of phototropism by demonstrating that the sites of perception and differential growth (bending) are separate: Light is perceived at the tip of a coleoptile, but bending occurs below the tip. The Darwins proposed that some "influence" that was transported from the tip to the growing region brought about the observed asymmetric growth response. This influence was later shown to be indole-3-acetic acid—auxin.

When a shoot is growing vertically, auxin is transported polarly from the growing tip to the elongation zone. The polarity of auxin transport from tip to base is developmentally determined and is independent of orientation with respect to gravity. However, auxin can also be transported laterally, and this lateral movement of auxin lies at the heart of a model for tropisms originally proposed independently in the 1920s by two plant physiologists: Nicolai Cholodny in Russia and Frits Went in the Netherlands.

According to the Cholodny–Went model of phototropism, the tips of grass coleoptiles are sites of high auxin concentration and have two other specialized functions:

- The perception of a unilateral light stimulus
- A decrease in basipetal IAA transport and diversion to lateral transport in response to the phototropic stimulus

Thus, in response to a directional light stimulus, the auxin produced at the tip, instead of being transported basipetally, is transported laterally toward the shaded side.

Although phototropic mechanisms appear to be highly conserved across plant species, the precise sites of auxin production, light perception, and lateral transport have been difficult to define. In maize coleoptiles, auxin accumulates in the upper 1 to 2 mm of the tip. The zones of photosensing and lateral transport extend farther, within the upper 5 mm of the tip. The response is also strongly dependent on the light *fluence* (the number of photons per unit area). Similar zones of auxin synthesis/accumulation, light perception, and lateral transport are seen in the true shoots of all monocots and dicots examined to date.

Two flavoproteins, **phototropins 1 and 2**, are the photoreceptors for the blue-light signaling pathway that induces phototropic bending in Arabidopsis hypocotyls and oat coleoptiles under both high- and low-fluence conditions. Phototropins are autophosphorylating protein kinases whose activity is stimulated by blue light. The action spectrum for blue-light activation of the kinase activity closely matches the action spectrum for phototropism, including multiple peaks in the blue region. Phototropin 1 displays a lateral gradient in phosphorylation during exposure to low-fluence unilateral blue light.

Phototropin phosphorylation results in dissociation of the protein from the plasma membrane and subsequent interactions with auxin transporters and/or proteins that regulate those transporters. In coleoptiles, the gradient in phototropin phosphorylation appears to induce the lateral movement of auxin to the shaded side of the coleoptiles, where the auxin activates cellular elongation. The resulting *differential growth*—the acceleration of growth on the shaded side and the slowing of growth on the illuminated side—produces a curvature toward the source of light (**FIGURE 19.24**).

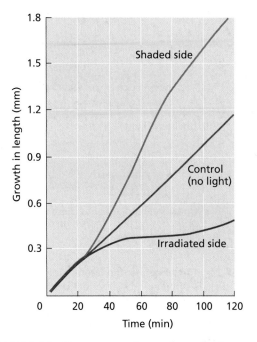

**FIGURE 19.24**  Time course of growth on the illuminated and shaded sides of a coleoptile responding to a 30-second pulse of unidirectional blue light. Control coleoptiles were not given a light treatment. (After Iino and Briggs 1984.)

Direct tests of the Cholodny–Went model using the agar block/coleoptile curvature bioassay have supported the model's prediction that auxin in coleoptile tips is transported laterally in response to unilateral light (**FIGURE 19.25**). The total amount of auxin diffusing out of the tip (here expressed as the angle of curvature) is the same in the presence of unilateral light as in darkness (compare Figure 19.25A and B). This result indicates that, in coleoptiles, light does not cause the photodestruction of auxin on the illuminated side, as had been proposed by some investigators.

Acidification of the apoplast appears to play a role in phototropic growth: The apoplastic pH on the shaded side of phototropically bending stems or coleoptiles is more acidic than on the side facing the light. Decreased pH increases auxin transport by increasing both the rate of IAA entry into the cell and the chemiosmotic proton potential–driven auxin efflux mechanisms. According to the acid growth hypothesis, this acidification would also be expected to enhance cellular elongation. Both processes—enhanced auxin uptake and increased cell elongation on the shaded side—would be expected to contribute to bending toward light.

Experiments in Arabidopsis suggest that phototropic auxin movement involves inhibition of ABCB19, destabilization of PIN1, and inhibition and/or relocalization of laterally localized PIN3 proteins (Noh et al. 2003; Friml

**Undivided agar block**

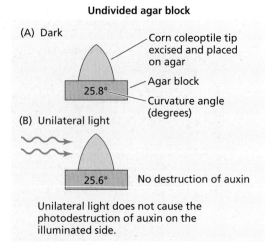

(A) Dark — Corn coleoptile tip excised and placed on agar

25.8° — Agar block

— Curvature angle (degrees)

(B) Unilateral light

25.6° No destruction of auxin

Unilateral light does not cause the photodestruction of auxin on the illuminated side.

**Divided agar block**

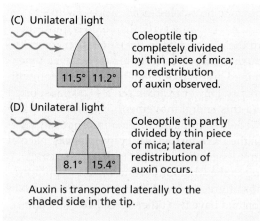

(C) Unilateral light

11.5° 11.2° — Coleoptile tip completely divided by thin piece of mica; no redistribution of auxin observed.

(D) Unilateral light

8.1° 15.4° — Coleoptile tip partly divided by thin piece of mica; lateral redistribution of auxin occurs.

Auxin is transported laterally to the shaded side in the tip.

**FIGURE 19.25** Evidence that the lateral redistribution of auxin is stimulated by unidirectional light in corn coleoptiles. The amount of auxin in the agar block is expressed as the angle of curvature the agar block induces when assayed by the coleoptile curvature bioassay (see Figure 19.1).

et al. 2003; Titapiwatanakun et al. 2009). These changes in auxin distribution can be visualized using the auxin-responsive reporter gene construct *DR5::GUS*, which utilizes the auxin-sensitive promoter, *DR5*, fused to the *GUS* reporter gene (**FIGURE 19.26**).

## Gravitropism involves lateral redistribution of auxin

When dark-grown *Avena* seedlings are oriented horizontally, the coleoptiles bend upward in response to gravity. According to the Cholodny–Went model, auxin in a horizontally oriented coleoptile tip is transported laterally to the lower side, causing the lower side of the coleoptile to grow faster than the upper side. Early experimental evidence indicated that the tip of the coleoptile could perceive gravity and redistribute auxin to the lower side. For example, if coleoptile tips are oriented horizontally, a greater amount of auxin diffuses into an agar block from the lower half than from the upper half (**FIGURE 19.27**).

Tissues below the tip are able to respond to gravity as well. For example, when vertically oriented maize coleoptiles are decapitated by removing the upper 2 mm of the tip and then oriented horizontally, gravitropic bending occurs at a slow rate for several hours even without the tip. Application of IAA to the cut surface restores the rate of bending to normal levels. This finding indicates that both the perception of the gravitational stimulus and the lateral redistribution of auxin can occur in the tissues below the tip, although the tip is still required for auxin production.

Lateral redistribution of auxin is more difficult to demonstrate in shoot apical meristems than in coleoptiles because of the presence of an apparent auxin recirculation ("fountain effect") in developing leaf and apical shoot pri-

(A)            (B)

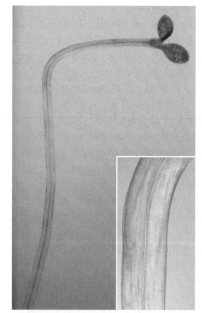

**FIGURE 19.26** Lateral redistribution of auxin during phototropism can be visualized by transforming plants with the *DR5::GUS* reporter gene construct. (A) Lateral auxin gradients are formed in Arabidopsis hypocotyls during the bending response to unidirectional light. Auxin accumulation on the shaded side of the hypocotyls is indicated by the blue staining (inset). (B) Treatment with the auxin efflux inhibitor NPA blocks both phototropic bending and auxin redistribution. A similar redistribution of auxin occurs during gravitropism. (Photos courtesy of Klaus Palme.)

(A)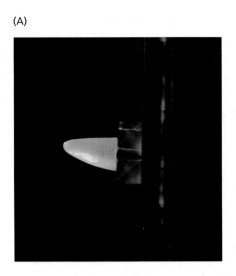

(B)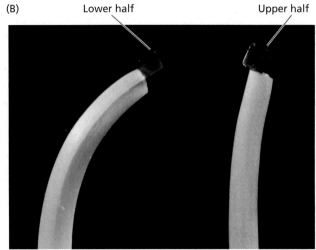

Lower half   Upper half

**FIGURE 19.27** Auxin is transported to the lower side of a horizontally oriented oat coleoptile tip. (A) Auxin from the upper and lower halves of a horizontal tip is allowed to diffuse into two agar blocks. (B) The agar block from the lower half (left) induces greater curvature in a decapitated coleoptile than the agar block from the upper half (right). (Photo © M. B. Wilkins.)

mordia that is similar to that observed in root tips (Benková et al. 2003). However, some of the same differential auxin transport mechanisms seen in phototropic bending are also involved in shoot gravitropic bending.

### Dense plastids serve as gravity sensors

Unlike unidirectional light, gravity does not form a gradient between the upper and lower sides of an organ. All parts of the plant experience the gravitational stimulus equally. How do plant cells detect gravity? The only way that gravity can be sensed is through the motion of a falling or sedimenting body.

Obvious candidates for intracellular gravity sensors in plants are the large, dense amyloplasts that are present in specialized gravity-sensing cells. These large amyloplasts (starch-containing plastids) are of sufficiently high density relative to the cytosol that they readily sediment to the bottom of the cell (**FIGURE 19.28**). Amyloplasts that function as gravity sensors are called **statoliths**, and the specialized gravity-sensing cells in which they occur are called **statocytes**. Whether the statocyte is able to detect the downward motion of the statolith as it passes through the cytoskeleton or whether the stimulus is perceived only when the statolith comes to rest at the bottom of the cell has not yet been resolved.

**GRAVITY PERCEPTION IN SHOOTS AND COLEOPTILES** In shoots and coleoptiles, gravity is perceived in the **starch sheath**, a layer of cells that surrounds the vascular tissues of the shoot. The starch sheath is continuous with the endodermis of the root, but unlike the endodermis it contains amyloplasts. Arabidopsis mutants lacking amyloplasts in the starch sheath display agravitropic shoot growth but normal gravitropic root growth.

As noted in Chapter 16, in the *scarecrow* (*scr*) mutant of Arabidopsis the cell layer from which the endodermis and the starch sheath are derived remains undifferentiated. As a result, the hypocotyl and inflorescence of the *scr* mutant are agravitropic, although the root exhibits a normal gravitropic response. On the basis of the phenotypes of these two mutants, we can conclude the following:

- The starch sheath is required for gravitropism in shoots. The starch sheath contains ABCB19 and PIN3, which function coordinately in the bundle sheath to restrict auxin streams to the vascular cylinder. Selective regulation of the downward auxin transport stream conducted by PIN1 inside the vascular cylinder and selective restriction of lateral auxin movement into starch sheath cells appear to play a fundamental role in tropic bending (Noh et al. 2003; Friml et al. 2003; Blakeslee et al. 2007).

- The root endodermis, which does not contain statoliths, is not required for gravitropism in roots; tropic bending in roots involves rerouting of auxin in a basipetal direction at the root tip rather than direct lateral movement from the central cylinder.

**GRAVITY PERCEPTION IN ROOTS** The site of gravity perception in primary roots is the root cap. Large, graviresponsive amyloplasts are located in the statocytes (**FIGURE 19.29A**) in the central cylinder, or **columella**, of the

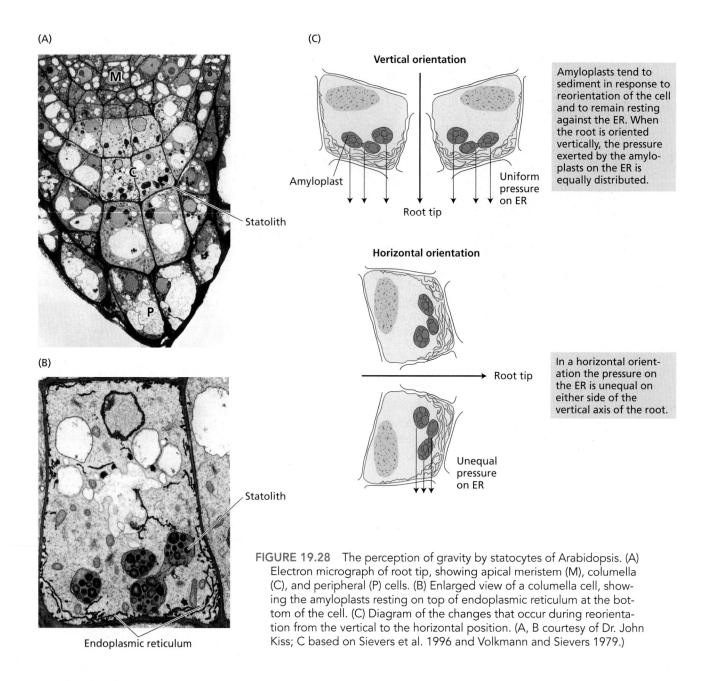

**FIGURE 19.28** The perception of gravity by statocytes of Arabidopsis. (A) Electron micrograph of root tip, showing apical meristem (M), columella (C), and peripheral (P) cells. (B) Enlarged view of a columella cell, showing the amyloplasts resting on top of endoplasmic reticulum at the bottom of the cell. (C) Diagram of the changes that occur during reorientation from the vertical to the horizontal position. (A, B courtesy of Dr. John Kiss; C based on Sievers et al. 1996 and Volkmann and Sievers 1979.)

root cap. Removal of the root cap from otherwise intact roots abolishes root gravitropism without inhibiting growth.

Precisely how the statocytes sense their falling statoliths is still poorly understood. According to one hypothesis, contact or pressure resulting from the amyloplast resting on the endoplasmic reticulum on the lower side of the cell triggers the response (see Figure 19.28C). The predominant form of endoplasmic reticulum in columella cells is of the tubular type, but an unusual form of ER, called "nodal ER," is also present. Nodal ER consists of five to seven rough ER sheets attached to a central nodal rod in a whorl, like petals on a flower. Nodal ER differs from the more

typical tubular cortical ER cisternae and may play a role in the gravity response.

This **starch–statolith hypothesis** of gravity perception in roots is supported by several lines of evidence. Amyloplasts are the only organelles that consistently sediment in the columella cells of different plant species, and the rate of sedimentation correlates closely with the time required to perceive the gravitational stimulus. The gravitropic responses of starch-deficient mutants are generally much slower than those of wild-type plants. Nevertheless, starchless mutants exhibit some residual gravitropism, suggesting that although starch is required for a normal gravitropic response,

(A)

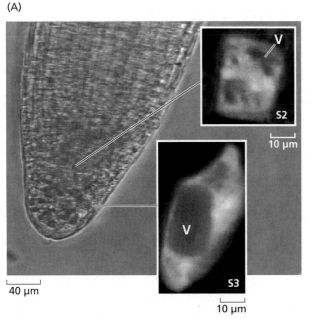

(B)

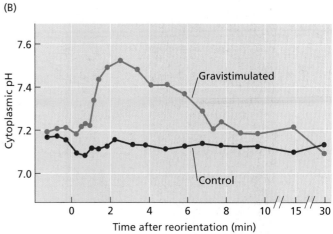

(C)

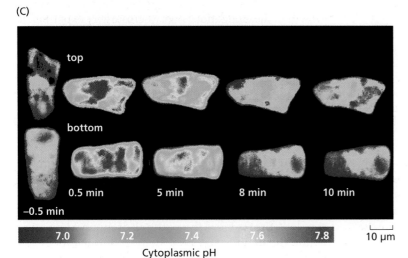

**FIGURE 19.29** Experiments with a pH-sensitive dye suggest that pH changes in columella cells of the root cap are involved in gravitropic signal transduction. (A) Micrograph showing a magnification of the root tip and two columella cells at different levels (stories) of the root cap, labeled S2 (story 2) and S3 (story 3) (insets). The cytosols of the two columella cells are fluorescing because the cells have been microinjected with a pH-sensitive fluorescent dye. The vacuoles (labeled V) contain no dye and therefore appear dark. (B) Cytoplasmic pH increases in less than one minute after gravistimulation. (C) Imaging of pH-sensitive dyes in the response of the two columella cells in (A) to gravitropic stimulus. The color scale below was used to generate the data in (B). (From Fasano et al. 2001.)

starch-independent gravity perception mechanisms may also exist.

Other organelles, such as nuclei, may be dense enough to act as statoliths. It also may not even be necessary for a statolith to come to rest at the bottom of the cell, as interactions with endomembranes and cytoskeletal components could transduce a gravitropic signal in an unknown manner.

## Gravity sensing may involve pH and calcium ions ($Ca^{2+}$) as second messengers

When gravity-sensing mechanisms detect that the root or shoot axis is out of alignment with the gravity vector, signal transduction mechanisms involving second messengers transmit this information to initiate corrective differential growth. This process is called **gravistimulation**. A variety of experiments suggest that localized changes in pH and $Ca^{2+}$ gradients are part of that signaling.

Changes in intracellular pH can be detected early in root columella cells responding to gravity. When pH-sensitive dyes were used to monitor both intracellular and extracellular pH in Arabidopsis roots, rapid changes were observed after roots were rotated to a horizontal position (Fasano et al. 2001). Within 2 minutes of gravistimulation, the cytoplasmic pH of the columella cells of the root cap increased from 7.2 to 7.6, and the apoplastic pH declined from 5.5 to 4.5 (**FIGURE 19.29B AND C**). These changes preceded any detectable tropic curvature by about 10 minutes.

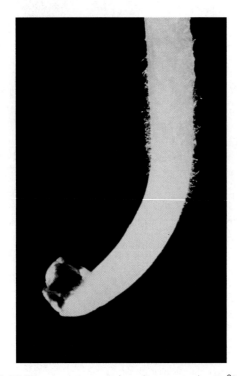

**FIGURE 19.30** A maize root bending toward a $Ca^{2+}$-containing agar block placed on one side of the cap. The result shows that an artificially imposed $Ca^{2+}$ gradient can override the normal gravitropic response. Although $Ca^{2+}$ does not appear to play a role in gravitropism itself, it may be part of a separate signaling pathway related to thigmotropism. (Courtesy of Michael L. Evans.)

The alkalinization of the cytosol combined with the acidification of the apoplast suggests that activation of the plasma membrane $H^+$-ATPase is one of the initial events that mediate root gravity perception or signal transduction. The chemiosmotic model of auxin transport predicts that differential acidification of the apoplast and alkalinization of the cytosol would result in increased directional uptake and efflux of IAA from the affected cells.

Early physiological studies suggested that $Ca^{2+}$ release from storage pools might be involved in root gravitropic signal transduction. For example, treatment of maize roots with EGTA [ethylene glycol-bis(β-aminoethyl ether)-N,N,N′,N′-tetraacetic acid], a compound that can chelate (form a complex with) $Ca^{2+}$, prevents $Ca^{2+}$ uptake by cells and inhibits root gravitropism. Placing a block of agar that contains $Ca^{2+}$ on the side of the cap of a vertically oriented maize root induces the root to grow toward the side with the agar block (**FIGURE 19.30**). As in the case of [³H]IAA, $^{45}Ca^{2+}$ is polarly transported to the lower half of a root cap that is stimulated by gravity. However, when sensitive $Ca^{2+}$-responsive dyes are used to monitor cellular $Ca^{2+}$ levels, no changes in localized $Ca^{2+}$ distribution are observed after gravistimulation. This suggests that $Ca^{2+}$ redistribution is not necessary for the gravitropic response.

A reconciliation of these apparently opposing observations is found in roots that temporarily bend away from the gravity vector after making contact with a solid surface (thigmotropism). In contrast to gravitropic bending, localized changes in internal $Ca^{2+}$ pools have been observed in root thigmotropic responses. In Arabidopsis, rapid changes in $Ca^{2+}$ levels have been shown to initiate independent increases in reactive oxygen species and decreases in plasma membrane pH gradients (Monshausen et al. 2009). It is therefore likely that the bending response to $Ca^{2+}$ shown in Figure 19.30 has more to do with thigmotropism than with gravitropism.

## Auxin is redistributed laterally in the root cap

In addition to protecting the sensitive cells of the apical meristem as the tip penetrates the soil, the root cap is the site of gravity perception. Because the cap is some distance away from the elongation zone where bending occurs, graviresponsive signaling events initiated in the root cap must induce production of a chemical messenger that modulates growth in the elongation zone. Microsurgery experiments in which half of the cap was removed showed that the cap supplies a root growth inhibitor to the lower side of the root during gravitropic bending (**FIGURE 19.31**).

Although root caps contain small amounts of IAA and abscisic acid (ABA) (see Chapter 23), IAA is more inhibitory to root growth than ABA is when applied directly to the elongation zone, suggesting that IAA is the root cap–derived inhibitor. Consistent with this conclusion, ABA-deficient Arabidopsis mutants have normal root gravitropism, whereas the roots of mutants defective in auxin transport, such as *aux1* and *pin2*, are agravitropic.

To understand how auxin is redistributed in the root cap, let's first review the patterns of auxin movement within the plant root. IAA is delivered to the root apex by the acropetal PIN1/ABCB19–directed stream (see Figure 19.11). IAA is also synthesized in the root meristem. However, the hormone is excluded from root cap apical cells by the combined activity of PIN3, PIN4, and ABCB1, while AUX1-mediated auxin uptake in lateral root cap cells drives a basipetal auxin stream out of the root apex (Swarup et al. 2003). PIN2, which is localized at the basal (top) end of root epidermal cells, conducts auxin away from the lateral root cap to the elongation zone, where it acts to stimulate or inhibit growth in a concentration-dependent manner (see Figure 19.11).

An *auxin reflux loop* mediated by PIN2 in root cortical cells is thought to redirect auxin back into the acropetal stelar transport stream at the boundary of the elongation zone (see Figure 19.11). Auxin circulation at the growing tip may allow root growth to continue for a time independent of auxin from the shoot (Blilou et al. 2005).

According to the current model for gravitropism, basipetal auxin transport in a vertically oriented root is equal

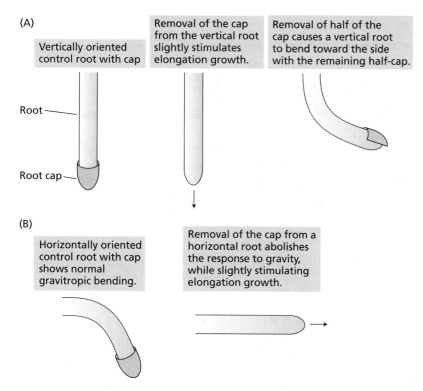

**FIGURE 19.31** Microsurgery experiments demonstrating that the root cap is required for redirection of auxin and subsequent differential inhibition of elongation in root gravitropic bending. Molecular genetic experiments have shown that AUX1 is the principal motivator of auxin streams out of the root cap in Arabidopsis. (After Shaw and Wilkins 1973.)

on all sides (**FIGURE 19.32A**). When the root is oriented horizontally, however, the cap redirects most of the auxin to the lower side, thus inhibiting the growth of that lower side (**FIGURE 19.32B**). Consistent with this model, the transport of [$^3$H]IAA across a horizontally oriented root cap is polar, with a preferential downward movement. The downward movement of auxin across a horizontal root cap has been confirmed using a reporter gene construct, *DR5::GFP*, consisting of green fluorescent protein (GFP) expressed under the control of the auxin-sensitive *DR5* promoter (see Chapter 2) (**FIGURE 19.33**). In transgenic plants containing this construct, the presence of elevated auxin levels in any given cell can be inferred from the green fluorescence observed. A fluorescent signal indicating apparent auxin accumulation is seen on the lower side of the root, where IAA is detected after rotation (Peer and Murphy 2007).

One of the members of the PIN protein family, PIN3, is thought to participate in the redirection of auxin in roots that are displaced from the vertical orientation. In a vertically oriented root, PIN3 is uniformly distributed around the columella cells, but when the root is placed on its side, PIN3 is preferentially targeted to the lower side of these cells. This redistribution of PIN3 is thought to accelerate

auxin transport to the lower side of the cap. As noted earlier, some PIN proteins are rapidly cycled between the plasma membrane and intracellular secretory compartments. PIN3 appears to be redistributed by such a mechanism. However, as *pin3* mutants are not completely agravitropic, other asymmetric events might act along with PIN3 localization to alter auxin flows. The most likely event would be an asymmetric change in apoplastic acidification, which would impose an asymmetric chemiosmotic potential to redirect auxin flow. This would cause PIN3 redistribution, which would amplify the flow of auxin in the new direction (canalization).

## Developmental Effects of Auxin

Although originally discovered in relation to growth, auxin influences nearly every stage of a plant's life cycle from germination to senescence. The morphology of a plant depends on the directed movement of auxin via the polar transport system, which maintains both basic shoot–root polarity and polarized outgrowth throughout development. Molecular genetic studies in Arabidopsis, rice, and maize have shown that polar auxin streams directed by PIN proteins function in every aspect of plant organogenesis, and that activity of AUX/LAX and ABCB transport proteins maintain these streams as tissues elongate and mature.

Loss of function of multiple PINs or of cellular components that regulate the polar localization of these carriers results in severe embryonic defects and loss of many plant organs. The polarity of auxin transport is established in the early stages of embryogenesis, and the cellular mechanisms that mediate polar auxin transport from the embryonic apical pole to the suspensor (point of maternal attachment) are maintained in mature tissues.

Because the effect that auxin produces depends on the identity of the target tissue, the response of a tissue to auxin is governed by its developmentally determined genetic program and is further influenced by the presence or absence of other signaling molecules such as hormones, $Ca^{2+}$, and reactive oxygen species. As we will see in this and subsequent chapters, interaction between two or more hormones is a recurring theme in plant development.

In this section we will go beyond cell elongation and examine some additional developmental processes regulated by auxin, including apical dominance, floral bud development, leaf arrangement (phyllotaxy), lateral root

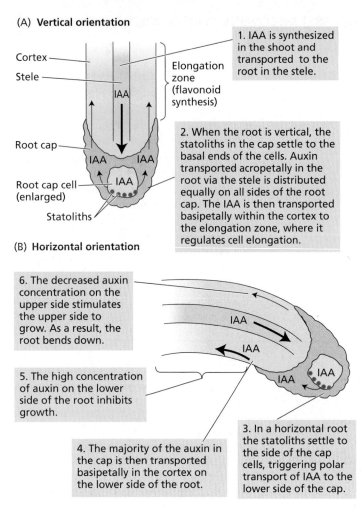

**(A) Vertical orientation**

Cortex

Stele

IAA

Elongation zone (flavonoid synthesis)

1. IAA is synthesized in the shoot and transported to the root in the stele.

Root cap

IAA ← IAA → IAA

Root cap cell (enlarged)

IAA

Statoliths

2. When the root is vertical, the statoliths in the cap settle to the basal ends of the cells. Auxin transported acropetally in the root via the stele is distributed equally on all sides of the root cap. The IAA is then transported basipetally within the cortex to the elongation zone, where it regulates cell elongation.

**(B) Horizontal orientation**

6. The decreased auxin concentration on the upper side stimulates the upper side to grow. As a result, the root bends down.

IAA

IAA

IAA

IAA ← IAA

5. The high concentration of auxin on the lower side of the root inhibits growth.

4. The majority of the auxin in the cap is then transported basipetally in the cortex on the lower side of the root.

3. In a horizontal root the statoliths settle to the side of the cap cells, triggering polar transport of IAA to the lower side of the cap.

**FIGURE 19.32** Current model for the redistribution of auxin during gravitropism in maize roots. (After Hasenstein and Evans 1988.)

formation, vascular development, leaf abscission, and fruit formation. Throughout this discussion we assume that the mechanisms of auxin action are comparable in all cases, involving similar receptors and signal transduction pathways.

## Auxin regulates apical dominance

In most higher plants, the growing apical bud inhibits the growth of lateral (axillary) buds—a phenomenon called **apical dominance**. Removal of the shoot apex (decapitation) results in the outgrowth of one or more of the lateral buds. Not long after the discovery of auxin, it was found that IAA could substitute for the apical bud in maintaining the inhibition of lateral buds. The classic experiment performed on kidney bean (*Phaseolus vulgaris*) plants by Kenneth Thimann and Folke Skoog in the 1920s is illustrated in **FIGURE 19.34**.

This result was soon confirmed for numerous other plant species, leading to the hypothesis that the outgrowth of the axillary bud is inhibited by auxin that is transported basipetally from the terminal bud. This hypothesis was supported by experiments that showed that a ring of the auxin transport inhibitor TIBA in lanolin paste (as a carrier) placed below the shoot apex releases the axillary buds from inhibition.

How does auxin from the shoot apex inhibit the outgrowth of lateral buds? Thimann and Skoog originally proposed that auxin synthesized in the shoot apex is transported basipetally to axillary buds, which they proposed were more sensitive to auxin than other tissues. If this is the case, the concentration of auxin in axillary buds should decrease following decapitation of the shoot apex. However,

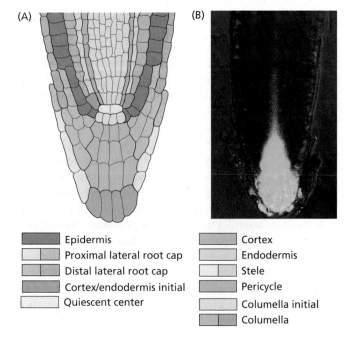

**(A)**

**(B)**

**(C)**

| | Epidermis |
| | Proximal lateral root cap |
| | Distal lateral root cap |
| | Cortex/endodermis initial |
| | Quiescent center |

| | Cortex |
| | Endodermis |
| | Stele |
| | Pericycle |
| | Columella initial |
| | Columella |

**FIGURE 19.33** Gravistimulation results in asymmetric auxin accumulation in lateral root cells on the non-elongating side. Auxin accumulation is indicated by green fluorescent protein (GFP) expressed under the control of the *DR5* promoter. (A) Diagram of the Arabidopsis root tip showing the tissues involved in auxin lateral redistribution. (B) Before gravistimulation. (C) Three hours after gravistimulation. Auxin has become redistributed to the lower side of the root cap. (Images courtesy of Jiří Friml.)

(A) Terminal bud intact

(B) Terminal bud removed

(C) Auxin added to decapitated stem

**FIGURE 19.34** Auxin suppresses the growth of axillary buds in bean (*Phaseolus vulgaris*) plants. (A) The axillary bud is suppressed in the intact plant because of apical dominance. (B) Removal of the terminal bud releases the axillary bud (arrow) from apical dominance. (C) Applying IAA in lanolin paste (contained in the gelatin capsule) to the cut surface prevents the outgrowth of the axillary bud. (Photos by David McIntyre.)

the reverse appears to be the case. Measurements of auxin levels in axillary buds have shown that following decapitation, the auxin content of the buds actually *increases*. In addition, application of auxin directly to the terminal bud raises the auxin concentration in the shoot but fails to inhibit normal axillary bud outgrowth. Finally, experiments with radiolabeled auxin have shown that the auxin applied at the terminal bud does not enter apical buds.

If auxin does not enter the bud, it must act remotely to suppress bud outgrowth, but where does auxin act? The answer to this question came from molecular studies using the *axr1* mutant of Arabidopsis (Booker et al. 2003). The AXR1 protein is a component of the primary auxin perception mechanism. Arabidopsis *axr1* mutants are unable to respond to auxin and, as a result, they exhibit increased lateral branching. However, normal apical dominance is completely restored in the mutant by expressing (using tissue-specific promoters) the wild-type AXR1 protein exclusively in the xylem and in the sclerenchyma cells between the vascular bundles (*interfascicular* sclerenchyma) of the stem, indicating that these tissues are the site of AXR1 action. Furthermore, grafting studies between *axr1* and the wild type demonstrated that auxin controls axillary bud growth by acting in the xylem and interfascicular sclerenchyma of the shoot. However, the mechanisms that regulate bud growth downstream of auxin signaling in these tissues have not been clearly elucidated. The role of auxin in regulating lateral branching is discussed further in WEB ESSAY 19.2.

**OTHER BRANCHING SIGNALS** Because AXR1 also regulates more than one type of SCF ubiquitin E3 ligase complex (see Chapters 2 and 14), other plant hormones are likely to be involved in the regulation of branching. Cytokinins were initially thought to interact with auxin to control axillary bud growth (see Chapter 21). Direct application of cytokinins to axillary buds stimulates bud growth in many species, overriding the inhibitory effect of the shoot apex. However, production of cytokinin in axillary buds is regulated by auxin levels (Tanaka et al. 2006). As such, cytokinins are unlikely to be primary or sole branching activation signals.

Analyses of pea, rice, and Arabidopsis mutants with increased branching patterns have led to the identification of *strigolactone* as the signal that interacts with auxin in regulating apical dominance (Hayward et al. 2009). Strigolactones (a group of terpenoid lactones derived from carotenoids) were first identified as an agent produced in host plant roots that stimulated germination of seeds of *Striga* Lour. (a parasitic plant commonly known as witchweed), and was subsequently found to enhance the formation of mycorrhizal structures by beneficial mycorrhizal fungi. Grafting studies between mutant and wild-type plants have shown that strigolactone is produced in both shoots and roots and is transported through the stem via the xylem. How this signal interacts with auxin to regulate apical dominance has not been resolved.

(A) (B)

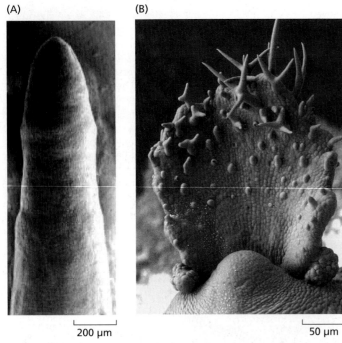

200 μm                    50 μm

**FIGURE 19.35** Scanning electron micrograph of leaf primordia on a vegetative *pin1* shoot apical meristem. (A) An untreated *pin1* inflorescence meristem fails to produce leaf primordia. (B) Leaf primordium induced on the inflorescence meristem of a *pin1* mutant by placing a microdrop of IAA in lanolin paste on the side of the meristem. (A from Vernoux et al. 2000; B from Reinhart et al. 2003)

## Auxin transport regulates floral bud development and phyllotaxy

The developing floral meristem depends on auxin being transported to it from subapical tissues. Auxin transport from these tissues also regulates leaf initiation and **phyllotaxy**, the pattern of leaf emergence from the shoot apex. In the absence of PIN1, auxin movement to the meristem is impaired and the initiation of both leaf and floral organ primordia is disrupted. This is the basis of the "pin-formed" phenotype of *pin1* mutants. Not surprisingly, therefore, leaf primordia can be induced to form on the meristem of a *pin1* mutant by applying a tiny spot of auxin in lanolin paste on the side of the apical meristem (**FIGURE 19.35**).

Microscopy of plants transformed with a reporter construct of PIN1 fused to green fluorescent protein demonstrates that PIN1 distribution correlates with the direction of auxin flow in leaf primordia of the shoot apical meristem. PIN1 localization within the cells of these tissues—and therefore the direction of auxin flow—is highly dynamic. Computer modeling of the direction of auxin transport (presumably through the combined action of AUX1, PIN1, and ABCB19) can accurately predict phyllotactic patterns, provided a basipetal transport sink is present (Smith 2008).

## Auxin promotes the formation of lateral and adventitious roots

Although elongation of the primary root is inhibited by auxin concentrations greater than $10^{-8}$ $M$, initiation of lateral (branch) roots and adventitious roots is stimulated by higher auxin levels. Lateral roots are commonly found above the elongation and root hair zones and originate from small groups of cells in the pericycle (see Chapter 16). Analyses of Arabidopsis mutants with aberrant patterns of root branching have shown that

- IAA transported acropetally (toward the tip) in the vascular parenchyma of the root is required to initiate cell division in the pericycle. Auxin uptake into the root is mediated by the concerted action of PIN2, AUX1, ABCB19, and ABCB1.

- LAX3 mediates uptake of auxin into cortical and epidermal cells of the root to promote cell expansion and cell wall modifications that facilitate lateral root emergence (see Figure 19.10 C).

- IAA derived from the root basipetal stream contributes to lateral root elongation and acts in combination with shoot-derived auxin to maintain cell division and growth.

Increased auxin levels or application of auxin can promote the formation of adventitious roots (roots originating from nonroot tissue). Adventitious roots arise from differentiated cells that begin to divide and develop into a root apical meristem in a manner somewhat analogous to the formation of a lateral root primordium. In horticulture, the stimulatory effect of auxin on the formation of adventitious roots has been very useful for the vegetative propagation of plants by cuttings.

## Auxin induces vascular differentiation

New vascular tissues differentiate directly below developing buds and young growing leaves (see Figure 19.4), and removal of the young leaves prevents vascular differentiation. The ability of an apical bud to stimulate vascular differentiation can be demonstrated in tissue culture; when the apical bud is grafted onto a clump of undifferentiated cells, or *callus*, xylem and phloem differentiate beneath the graft.

Studies in Arabidopsis using mutations in genes that encode auxin transport proteins, regulators of these transporters, and auxin signaling components, in conjunction with experiments using promoter::reporter fusions, have shown that auxin transport mediated by PIN1 is a primary determinant of vascular differentiation and patterning in leaves. ABCB19 plays an important role in the delivery of

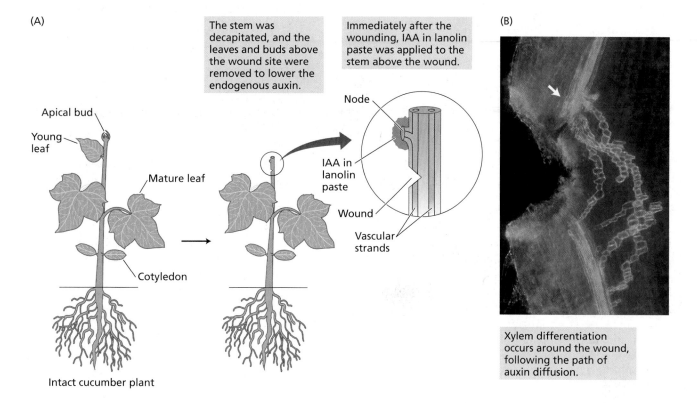

**FIGURE 19.36** IAA-induced xylem regeneration around a wound in cucumber (*Cucumis sativus*) stem tissue. (A) Method for carrying out the wound regeneration experiment. (B) Fluorescence micrograph showing regenerating vascular tissue around the wound. The arrow indicates the wound site where auxin accumulates and xylem differentiation begins. (B courtesy of R. Aloni.)

auxin to cotyledons (Lewis et al. 2009) and is thought to contribute to the auxin supply in leaves as well. Vascular differentiation in some tissues involves interactions with ethylene and other hormones.

Polar vascular differentiation is found in all plant organs. In woody perennials, auxin produced by growing buds in the spring stimulates activation of the cambium in a basipetal direction. The new round of secondary growth begins in the smallest twigs and progresses downward toward the root tip.

The regeneration of vascular tissue following wounding is also controlled by auxin produced by the young leaf directly above the wound site (**FIGURE 19.36**). Removal of the leaf prevents the regeneration of vascular tissue, and applied auxin can substitute for the leaf in stimulating regeneration.

### Auxin delays the onset of leaf abscission

The shedding of leaves, flowers, and fruits from the living plant is known as **abscission**. These parts abscise in a region called the **abscission zone**, which, in the case of leaves, is located near the base of the petiole. In most plants, leaf abscission is preceded by the differentiation of a distinct layer of cells, the **abscission layer**, within the abscission zone. During leaf senescence, the walls of the cells in the abscission layer are digested, which causes them to become soft and weak. The leaf eventually breaks off at the abscission layer as a result of stress on the weakened cell walls.

Auxin levels are high in young leaves, progressively decrease in maturing leaves, and are relatively low in senescing leaves when the abscission process begins. The role of auxin in leaf abscission can be readily demonstrated by excision of the blade from a mature leaf, leaving the petiole intact on the stem. Whereas removal of the leaf blade accelerates the formation of the abscission layer in the petiole, application of IAA in lanolin paste to the cut surface of the petiole prevents the formation of the abscission layer. (Lanolin paste alone does not prevent abscission.)

These results suggest the following:

- Auxin transported from the blade normally prevents abscission.
- Abscission is triggered during leaf senescence, when auxin is no longer being produced.

However, as will be discussed in Chapter 22, ethylene also plays a crucial role in the regulation of abscission.

### Auxin promotes fruit development

Much evidence suggests that auxin is involved in the regulation of fruit development. Auxin is produced or mobi-

CHAPTER 19

**FIGURE 19.37** The strawberry "fruit" is actually a swollen receptacle whose growth is regulated by auxin produced by the "seeds," which are actually achenes—true fruits. (A) When the achenes are present, the receptacle enlarges and develops its characteristic flavor, sweetness, and red color. (B) When the achenes are removed, the receptacle fails to develop normally. (C) Spraying the receptacle minus its achenes with IAA restores normal growth and development. (After Galston 1994.)

(A) Normal fruit  (B) Achenes removed  (C) Achenes removed; sprayed with auxin

Swollen receptacle

Achene

lized from storage in pollen, and the initial stimulus for fruit growth may result from pollination.

Successful pollination initiates ovule growth, which is known as **fruit set**. After fertilization, fruit growth may depend on auxin from developing seeds. The endosperm may contribute auxin during the initial stage of fruit growth, and the developing embryo may take over as the main auxin source during the later stages.

FIGURE 19.37 shows the influence of auxin produced by the achenes of strawberry (the actual fruits, each of which contains a single seed) on the growth of the receptacle (commonly known as the "berry").

### Synthetic auxins have a variety of commercial uses

Auxins have been used commercially in agriculture and horticulture for more than 50 years. Commercial uses include prevention of fruit and leaf drop, promotion of flowering in pineapple, induction of parthenocarpic fruit development, thinning of fruit, and rooting of cuttings for plant propagation. Rooting is enhanced if the excised leaf or stem cutting is dipped in an auxin solution, which increases the initiation of adventitious roots at the cut end. This is the basis of commercial rooting compounds, which consist mainly of a synthetic auxin mixed with talcum powder.

In some plant species, seedless fruits may be produced naturally, or they may be induced by treatment of the unpollinated flowers with auxin. The production of such seedless fruits is called **parthenocarpy**. In stimulating the formation of parthenocarpic fruits, auxin may act primarily to induce fruit set, which in turn may trigger the endogenous production of auxin by certain fruit tissues to complete the developmental process.

Ethylene is also involved in fruit development, and some of the effects of auxin on fruiting may result from the promotion of ethylene synthesis. The control of ethylene in the commercial handling of fruit is discussed in Chapter 22.

In addition to these applications, synthetic auxins, such as 2,4-D and dicamba (see Figure 19.3B) are widely used as herbicides that induce excessive cell expansion and subsequent plant death. Synthetic auxins are highly effective because they are not metabolized by the plant as quickly as is IAA. Some, like 2,4-D, are also transported more slowly than IAA, thus enhancing retention in shoots. Synthetic auxins are used by farmers for the control of dicot weeds (also called broad-leaved weeds) in commercial cereal fields, and by home gardeners for the control of weeds such as dandelions and daisies in lawns. Monocots such as maize and grasses are less sensitive to artificial auxins, because monocots are able to inactivate synthetic auxins rapidly by conjugation.

## SUMMARY

Auxin was the first plant hormone to be discovered and studied. It regulates many aspects of plant development, including stem elongation, apical dominance, root initiation, fruit development, meristem development, and tropic growth; auxin may function in association with other plant hormones.

### The Emergence of the Auxin Concept

- A diffusible chemical (auxin) in the tip of oat coleoptiles promotes cell elongation (**Figure 19.1**).

### The Principal Auxin: Indole-3-Acetic Acid

- The main natural auxin is indole-3-acetic acid (IAA), but other auxins have been identified (**Figure 19.3**).

- IAA is synthesized in meristems and young dividing tissues, and moves basipetally as leaves and vasculature mature (**Figure 19.4**).

- IAA may be stored or degraded. When conjugated to different substances, it is inactive, but enzyme action can release free (active) auxin. IAA can be degraded by several pathways involving irreversible oxidation (**Figure 19.5**).

## SUMMARY continued

### Auxin Transport

- IAA shows polar transport, moving from the apical to the basal regions of the shoot (**Figure 19.6**). Polar transport of IAA requires energy and is independent of gravity (**Figure 19.7**).

- Polar auxin flows are driven by a chemiosmotic potential. The flows are directed by polar distributions of efflux carriers, sinks generated by auxin uptake proteins, and nondirectional exporters that exclude auxin flows from adjoining cells (**Figures 19.8, 19.10**).

- Protonated auxin is taken up into cells by diffusion across the plasma membrane and by proton symporters of the AUX1/LAX family (**Figure 19.9**).

- PIN and ATP-dependent (ABCB) transport proteins direct auxin efflux throughout the plant (**Figures 19.11, 19.12**).

- Auxin compartmentalized in cytosolic and ER pools contributes to auxin homeostasis (**Figure 19.14**).

- Auxin influx and efflux can be inhibited (**Figures 19.15, 19.16**).

- Auxin transport is regulated by hormones and by trafficking of PIN proteins, which cycle between the plasma membrane and an endosomal compartment (**Figure 19.17**).

### Auxin Signal Transduction Pathways

- Upon binding to auxin, the receptor–enzyme complex targets specific transcriptional repressors for proteasome destruction, activating transcription of auxin response genes (**Figure 19.18**).

- Auxin-induced genes may be early or late.

### Actions of Auxin: Cell Elongation

- Auxin stimulates elongation of dicot stems and monocot coleoptiles but inhibits root growth (**Figures 19.19, 19.22**).

- Auxin-stimulated growth requires energy and is rapidly inhibited by metabolic inhibitors. Growth is also inhibited by protein and RNA synthesis inhibitors.

- Lateral auxin transport in dicots is mediated by laterally oriented PIN proteins in vascular cells, and ABCB transporters in the surrounding cells (**Figure 19.21**).

- Proton extrusion into the cell wall is stimulated by auxin, consistent with the acid growth hypothesis (**Figure 19.23**).

- Auxin influences the retention or the endocytosis of PIN transporters and other plasma membrane

proteins, suggesting auxin may act directly at the plasma membrane to stimulate proton extrusion.

### Actions of Auxin: Plant Tropisms

- The phototropins are autophosphorylating protein kinases that absorb blue-light and induce phototropic bending.

- Phosphorylation induces lateral movement of auxin to the shaded side of the coleoptiles, where it stimulates cell elongation and bending (**Figure 19.24**), consistent with the Cholodny–Went model (**Figure 19.25**).

- Phototropic auxin movement involves inhibition or redistribution of auxin efflux proteins resulting in lateral redistribution of auxin (**Figure 19.26**).

- The coleoptile tip perceives gravity and redistributes auxin to the lower side, inducing coleoptile bending (**Figure 19.27**).

- Large, dense amyloplasts sediment to the bottom of specialized root cells, where pressure on specialized ER signals orientation to gravity (**Figure 19.28**).

- In root cap columella cells, pH increase in the cytosol combined with acidification of the apoplast may activate the plasma membrane $H^+$-ATPase as the initial event in gravitropic signal transduction (**Figure 19.29**).

- The root cap supplies a growth inhibitor (IAA) to the lower side, of the root during gravitropic bending (**Figures 19.31–19.33**).

### Developmental Effects of Auxin

- Auxin influences many stages of a plant's life cycle—from germination to senescence—with its effects dependent on the developmentally active genetic program of specific tissues; auxin may act conjunction with other hormones, $Ca^{2+}$, and reactive oxygen.

- Auxin acts on the xylem and associated cells to suppress lateral bud development and to maintain apical dominance (**Figure 19.34**). Strigolactone interacts with auxin to regulate axillary branch development.

- For normal development, the floral meristem and leaf primordia require PIN1-mediated auxin transport from subapical tissues (**Figure 19.35**).

- PIN1-mediated auxin transport controls vascular differentiation and patterning in leaves. Vascular regeneration is also controlled by auxin (**Figure 19.36**).

- Auxin is involved in the regulation of fruit development (**Figure 19.37**).

## WEB MATERIAL

### Web Essays

**19.1 Exploring the Cellular Basis of Polar Auxin Transport**

Experimental evidence indicates that the polar transport of the plant hormone auxin is regulated at the cellular level. This implies that proteins involved in auxin transport must be asymmetrically distributed on the plasma membrane. How those transport proteins get to their destination is the focus of ongoing research.

**19.2 Strigolactones: The Unmasking of a New Branching Hormone**

A brief review of the discovery of a new plant hormone, strigolactone, how it is regulated and how it is integrated into the network of genes and signals controlling shoot branching.

## CHAPTER REFERENCES

Aloni, R. (2001) Foliar and axial aspects of vascular differentiation: Hypotheses and evidence. *J. Plant Growth Regul.* 20: 22–34.

Aloni, R., Schwalm, K., Langhans, M., and Ullrich, C. I. (2003) Gradual shifts in sites and levels of free-auxin production during leaf-primordium development and their role in vascular differentiation and leaf morphogenesis in *Arabidopsis*. *Planta* 216: 841–853.

Benková, E., Michniewicz, M., Sauer, M., Teichmann, T., Seifertová, D., Jurgens, G., and Friml, J. (2003) Local, efflux-dependent auxin gradients as a common module for plant organ formation. *Cell* 114: 591–602.

Blakeslee J. J., Bandyopadhyay, A, Lee, O. R., Sauer, M., Mravec, J., Titapiwatanakun, B., Geisler, M., Nagashima, A., Sakai, T., Martinoia, et al. (2007) Interactions among PINFORMED and P-glycoprotein auxin transporters in Arabidopsis. *Plant Cell* 19: 131–147.

Blakeslee, J. J., Peer, W. A., and Murphy, A. S. (2005) Auxin transport. *Curr. Opin. Plant Biol.* 8: 121–125.

Blilou, I., Xu, J., Wildwater, M., Willemsen, V., Paponov, I., Friml, J., Heidstra, R., Aida, M., Palme, K., and Scheres, B. (2005) The PIN auxin efflux facilitator network controls growth and patterning in *Arabidopsis* roots. *Nature* 433: 39–44.

Booker, J., Chatfield, S., and Leyser, O. (2003) Auxin acts in xylem-associated or medullary cells to mediate apical dominance. *Plant Cell* 15: 495–507.

Brewer, P. B., Dun, E. A., Ferguson, B. J., Rameau, C., and Beveridge, C. A. (2009) Strigolactone acts as a downstream of auxin to regulate bud outgrowth in pea and Arabidopsis. *Plant Physiol.* 150: 482-493.

Cleland, R. E. (1995) Auxin and cell elongation. In *Plant Hormones and Their Role in Plant Growth and Development*, 2nd ed., P. J. Davies, ed., Kluwer, Dordrecht, Netherlands, pp. 214–227.

Dharmasiri, N., Dharmasiri, S., and Estelle, M. (2005a) The F-box protein TIR1 is an auxin receptor. *Nature* 436: 441–451.

Dharmasiri, N., Dharmasiri, S., Weijers, D., Lechner, E., Yamada, M., Hobbie, L., Ehrismann, J. S., Jurgens, G., and Estelle, M. (2005b) Plant development is regulated by a family of auxin receptor F box proteins. *Dev. Cell* 9: 109–119.

Dhonukshe, P., Kleine-Vehn, J., and Friml, J. (2005) Cell polarity, auxin transport, and cytoskeleton-mediated division planes: Who comes first? *Protoplasma* 226: 67–73.

Fasano, J. M., Swanson, S. J., Blancaflor, E. B., Dowd, P. E., Kao, T. H., and Gilroy, S. (2001) Changes in root cap pH are required for the gravity response of the *Arabidopsis* root. *Plant Cell* 13: 907–921.

Friml, J., Wiśniewska, J., Benková, E. Mendgen, K., and Palme, K. (2002) Lateral relocation of auxin efflux regulator PIN3 mediates tropism in *Arabidopsis*. *Nature* 415: 806–809.

Galston, A. (1994) *Life Processes of Plants*. Scientific American Library, New York.

Geldner, N., Friml, J., Stierhof, Y. D., Jurgens, G., and Palme, K. (2001) Auxin transport inhibitors block PIN1 cycling and vesicle trafficking. *Nature* 413: 425–428.

Geldner, N., Anders, N., Wolters, H., Keicher, J., Kornberger, W., Muller, P., Delbarre, A., Ueda, T., Nakano, A., and Jurgens, G. (2003) The *Arabidopsis* GNOM ARF-GEF mediates endosomal recycling, auxin transport, and auxin-dependent plant growth. *Cell* 112: 219–230.

Hartmann, H. T., and Kester, D. E. (1983) *Plant Propagation: Principles and Practices*, 4th ed. Prentice-Hall, Inc., N.J.

Hasenstein, K. H., and Evans, M. L. (1988) Effects of cations on hormone transport in primary roots of *Zea mays*. *Plant Physiol.* 86: 890–894.

Hayward, A., Stirnberg, P., Beveridge, C., and Leyser, O. (2009) Interactions between auxin and strigolactone in shoot branching control. *Plant Physiol.* 151: 400-412.

Hsieh, H. L., Okamoto, H., Wang, M. L., Ang, L. H., Matsui, M., Goodman, H., and Deng, X. W. (2000) *FIN219*, an auxin-regulated gene, defines a link between phytochrome A and the downstream regulator COP1 in light control of *Arabidopsis* development. *Genes Dev.* 14: 1958–1970.

Iino, M., and Briggs, W. R. (1984) Growth distribution during first positive phototropic curvature of maize coleoptiles. *Plant Cell Environ.* 7: 97–104.

Jacobs, M., and Ray, P. M. (1976) Rapid auxin-induced decrease in free space pH and its relationship to auxin-induced growth in maize and pea. *Plant Physiol.* 58: 203–209.

Kepinski, S., and Leyser, O. (2005) The Arabidopsis F-box protein TIR1 is an auxin receptor. *Nature* 435: 446–451.

Kramer, E. M., and Bennett, M. (2006) Auxin transport: A field in flux. *Trends Plant Sci.* 11: 382–386.

Kramer, E. M. (2006) Wood grain pattern formation: A brief review. *J. Plant Growth Regul.* 25: 290–301.

Lewis, D. R., Wu, G., Ljung, K., and Spalding, E. P. (2009) Auxin transport into cotyledons and cotyledon growth depend similarly on the ABCB19 Multidrug Resistance-like Transporter. *Plant J.* 60: 91–101.

Li, J., Yang, H., Richter, G., Blakeslee, J. J., Bandyopadhyay, A., Peer, W. A., Titapiwantakun, B., Richards, E., Undurraga, S., Murphy, A. S., Gilroy, S., and Gaxiola, R. (2005) The H$^+$-PPase AVP1 is required for organ development in Arabidopsis. *Science* 310: 121–125.

Ljung, K., Bhalerao, R. P., and Sandberg, G. (2001) Sites and homeostatic control of auxin biosynthesis in Arabidopsis during vegetative growth. *Plant J.* 28: 465–474.

Ljung, K., Hull, A. K., Celenza, J., Yamada, M., Estelle, M., Normanly, J., and Sandberg, G. (2005) Sites and regulation of auxin biosynthesis in *Arabidopsis* roots. *Plant Cell* 17: 1090–1104.

Monshausen, G. B., Bibikova, T. N., Weisenseel, M. H., and Gilroy, S. (2009) Ca$^{2+}$ regulates reactive oxygen species production and pH during mechanosensing in Arabidopsis roots. *Plant Cell* 21: 2341–2356.

Mravec, J., Kubeš, M., Bielach, A., Gaykova, V., Petrášek, J., Skůpa, P., Chand, S., Benková, E., Zažímalová, E., and Friml, J. (2008) Interaction of PIN and PGP transport mechanisms in auxin distribution-dependent development. *Development* 135: 3345–3354.

Mravec, J., Skůpa, P., Bailly, A., Hoyerová, K., Křeček, P., Bielach, A., Petrášek, J., Zhang, J., Gaykova, V., Stierhof, Y.-D., et al. (2009) Subcellular homeostasis of phytohormone auxin is mediated by the ER-localized PIN5 transporter. *Nature* 459: 1136–1140.

Multani, D. S., Briggs, S. P., Chamberlin, M. A., Blakeslee, J. J., Murphy, A. S., and Johal, G. S. (2003) Loss of an MDR transporter in compact stalks of maize *br2* and sorghum *dw3* mutants. *Science* 302: 81–84.

Murphy, A. S., Peer, W. A., and Taiz, L. (2000) Regulation of auxin transport by aminopeptidases and endogenous flavonoids. *Planta* 211: 315–324.

Nakazawa, M., Yabe, N., Ishikawa, T., Yamamoto, Y. Y., Yoshizumi, T., Hasunuma, K., Matsui, M. (2001) *DFL1*, an auxin-responsive *GH3* gene homologue, negatively regulates shoot cell elongation and lateral root formation, and positively regulates the light response of hypocotyls length. *Plant J.* 25: 213–221.

Noh, B., Bandyopadhyay, A., Peer, W. A., Spalding, E. P., and Murphy, A. S. (2003) Enhanced gravi- and phototropism in plant mdr mutants mislocalizing the auxin efflux protein PIN1. *Nature* 423: 999–1002.

Normanly, J. P., Slovin, J., and Cohen, J. (1995) Rethinking auxin biosynthesis and metabolism. *Plant Physiol.* 107: 323–329.

Paciorek, T., Zažímalová, E., Ruthardt, N., Petrášek, J., Stierhof, Y.-D., Kleine-Vehn, J., David, A., Morris, N. E., Jürgens, G., Geldner, N., et al. (2005) Auxin inhibits endocytosis and promotes its own efflux from cells. *Nature* 435: 1251–1256.

Peer, W. A., Hosein, F. N., Bandhyopadhyay, A., Makam, S. N., Otegui, M., Lee, G. J., Blakeslee, J. J., Cheng, Y., Titapiwatanakun, B., Yakubov, B., et al. (2009) Mutation of the membrane-associated M1 protease APM1 results in embryonic and seedling developmental defects. *Plant Cell* 21:1693-1721.

Peer, W. A., Bandyopadhyay, A., Blakeslee, J. J., Srinivas, M. N., Chen, R., Masson, P., and Murphy, A. S. (2004) Variation in *PIN* gene expression and protein localization in flavonoid mutants with altered auxin transport. *Plant Cell* 16: 1898–1911.

Peer, W. A., and Murphy, A. S. (2007) Flavonoids and auxin transport: Regulators or modulators. *Trends Plant Sci.* 12: 556–563.

Reinhardt, D., Pesce, E. R., Stieger, P., Mandel, T., Baltensperger, K., Bennett, M., Traas, J., Friml, J., and Kuhlemeier, C. (2003) Regulation of phyllotaxis by polar auxin transport. *Nature* 426: 255–260.

Rojas-Pierce, M, Titapiwatanakun, B., Sohn, E. J., Fang, F., Larive, C. K., Blakeslee, J. J., Peer, W. A., Murphy, A. S., and Raikhel, N. V. (2007) Arabidopsis P-glycoprotein19 participates in the inhibition of gravitropism by gravacin. *Chem. Biol.* 14: 1366–1376.

Rothwell, G. W., and Lev-Yadun, S. (2005) Evidence of polar auxin flow in 375 million-year-old fossil wood. *Am. J. Bot.* 92: 903–906.

Shaw, S., and Wilkins, M. B. (1973) The source and lateral transport of growth inhibitors in geotropically stimulated roots of *Zea mays* and *Pisum sativum*. *Planta* 109: 11–26.

Sievers, A., Buchen, B., and Hodick, D. (1996) Gravity sensing in tip-growing cells. *Trends Plant Sci.* 1: 273–279.

Smith, R. S. (2008) The role of auxin transport in plant patterning mechanisms. *PLoS Biol.* 6: e323, 2631–2633.

Staswick, P. E., Serban, B., Rowe, M., Tiryaki, I., Maldonado, M. T., Maldonado, M. C., and Suza, W. (2005) Characterization of an *Arabidopsis* enzyme family that conjugates amino acids to indole-3-acetic acid. *Plant Cell* 17: 616–627.

Swarup, K., Benková, E., Swarup, R., Casimiro, I., Péret, B., Yang, Y., Parry, G., Nielsen, E., De Smet, I., Vanneste, S., et al. (2008) The auxin influx carrier LAX3 promotes lateral root emergence. *Nat. Cell Biol.* 10: 946–954.

Swarup, R., Perry, P., Hagenbeek, D., Van Der Straeten, D., Beemster, G. T., Sandberg, G., Bhalerao, R., Ljung, K., and Bennett, M. J. (2007) Ethylene upregulates auxin biosynthesis in Arabidopsis seedlings to enhance inhibition of root cell elongation. *Plant Cell* 19: 2186–2196.

Swarup, R., and Bennett, M. (2003) Auxin transport: The fountain of life in plants? *Dev. Cell* 5: 824–826.

Swarup, R., Friml, J., Marchant, A., Ljung, K., Sandberg, G., Palme, K., and Bennett, M. (2001) Localization of the auxin permease AUX1 suggests two functionally distinct hormone transport pathways operate in the *Arabidopsis* root apex. *Genes Dev.* 15: 2648–2653.

Tanaka, M., Takei, K., Kojima, M., Sakakibara, H., and Mori, H. (2006) Auxin controls local cytokinin biosynthesis in the nodal stem in apical dominance. *Plant J.* 45: 1028–1036.

Titapiwatanakun, B., Blakeslee, J. J., Bandyopadhyay, A., Sauer, M., Mravec, J., Cheng, Y., Adamec, J., Nagashima, A., Geisler, M., Sakai, T., et al. (2009) ABCB19/PGP19 stabilises PIN1 in membrane microdomain in Arabidopsis. *Plant J.* 57: 27–44.

Ulmasov, T., Murfett, J., Hagen, G., and Guilfoyle, T. J. (1997) Aux/IAA proteins repress expression of reporter genes containing natural and highly active synthetic auxin response elements. *Plant Cell* 9: 1963–1971.

Vernoux, T., Kronenberger, J., Grandjean, O., Laufs, P., and Traas, J. (2000) *PIN-FORMED1* regulates cell fate at the periphery of the shoot apical meristem. *Development* 127: 5157–5165.

Volkmann, D., and Sievers, A. (1979) Graviperception in multicellular organs. In *Encyclopedia of Plant Physiology*, New Series, Vol. 7, W. Haupt and M. E. Feinleib, eds., Springer, Berlin, pp. 573–600.

Woodward, W., and Bartel, B. (2005) Auxin: Regulation, action, and interaction. *Annals Botany* 95: 707–735.

Yang, Y., Hammes, U., Taylor, C., Schachtman, D., and Nielsen, E. (2006) High-affinity auxin transport by the AUX1 influx carrier protein. *Curr. Biol.* 16: 1123–1127.

Yang, H., and Murphy, A. S. (2009) Functional expression and characterization of Arabidopsis ABCB, AUX1 and PIN auxin transporters in *Schizosaccharomyces pombe*. *Plant J.* DOI: 10.1111/j.1365-313X.2009.03856.x

# Gibberellins: Regulators of Plant Height and Seed Germination

For nearly 30 years after the discovery of auxin in 1927, and more than 20 years after its structural elucidation as indole-3-acetic acid, plant scientists tried to ascribe the regulation of all developmental phenomena in plants to the hormone auxin. However, as we will see in this and subsequent chapters, plant growth and development are regulated by many hormones acting both individually and in concert.

The second group of plant hormones to be characterized was the gibberellins (GAs). At least 136 naturally occurring GAs have been identified (MacMillan 2002), and their structures can be viewed at http://www.plant-hormones.info/ga1info.htm. This website is frequently updated as naturally occurring GAs are newly characterized and named. Unlike the auxins, which are defined by their biological properties, the GAs all share a similar chemical structure but relatively few of them have intrinsic biological activity. Many of the GAs that do not have intrinsic biological activity are either metabolic precursors of the bioactive GAs or their deactivation products. There are often only a few bioactive GAs in any given plant, and their levels are generally correlated with stem length. Gibberellins also play important roles in a variety of other physiological phenomena, such as seed germination, transition to flowering, and pollen development.

The biosynthesis of GAs is under strict genetic, developmental, and environmental control. Gibberellins are best known for their promotion of stem elongation, and GA-deficient mutants that have dwarf phenotypes have been isolated. Mendel's tall/dwarf alleles in peas are a famous example of a single gene locus that can control the level of bioactive GA and hence stem length. Such mutants have been useful in elucidating the complex pathways of GA biosynthesis, and in determining which of the GAs in a plant has intrinsic biological activity.

We begin this chapter by describing the discovery of GAs in a fungal pathogen (*Gibberella fujikuroi*) of rice plants (*Oryza sativa*) and discussing their chemical structures. We then provide an overview of the many physiological processes that are regulated by GAs—seed germination, shoot growth, transition to flowering, anther development, pollen tube growth, floral development, fruit set and subsequent growth, and seed development. We then examine biosynthesis of GAs and the roles of factors that regulate the levels of bioactive GA in tissues or organs at specific developmental stages.

The identification of a GA receptor in rice in 2005 has paved the way for recent findings that include the crystal structures of the GA receptors for both rice and Arabidopsis. The binding of bioactive GA to its receptor initiates a chain of events that leads eventually to the responses observed at the whole plant level.

## Gibberellins: Their Discovery and Chemical Structure

A brief description of the groundbreaking discovery of gibberellins provides an explanation for the unusual terminology applied to GAs. After a consideration of their discovery, we will describe their chemical structures and numbering system. Finally, we will explore the relations of specific chemical structures to their biological activity.

### Gibberellins were discovered by studying a disease of rice

Although GAs first came to the attention of Western scientists in the 1950s, they had been discovered much earlier in Japan. Rice farmers had long known of a fungal disease (termed *bakanae* or "foolish seedling" disease) that caused rice plants to grow too tall and eliminated seed production. Plant pathologists found that these symptoms in rice were caused by a pathogenic fungus, **Gibberella fujikuroi**, that had infected the plants. Culturing this fungus in the laboratory and analyzing the culture filtrate enabled Japanese scientists in the 1930s to obtain impure crystals with plant growth–promoting activity. They named this mixture of compounds *gibberellin A*.

### Gibberellic acid was first purified from Gibberella culture filtrates

In the 1950s two research groups, one in Britain and one in the United States, elucidated the chemical structure of a compound that both had purified from *Gibberella* culture filtrates and which they named *gibberellic acid*. At about the same time, Japanese scientists separated and characterized three different gibberellins from the original gibberellin A sample, naming them gibberellin $A_1$ ($GA_1$), gibberellin $A_2$ ($GA_2$), and gibberellin $A_3$ ($GA_3$). The numbering system for gibberellins builds on this initial nomenclature. The Japanese scientists' $GA_3$ was later shown to be identical to the gibberellic acid isolated by the U.S. and British scientists. Thus the name 'gibberellic acid' refers specifically to $GA_3$, whereas 'gibberellin' is a general name that can refer to the entire class of hormones.

It soon became evident that many different GAs were present in *Gibberella* cultures, although $GA_3$ was usually the principal component. (For this reason $GA_3$ is produced commercially in industrial-scale fermentations of *Gibberella*, for agronomic, horticultural, and other scientific use.) As $GA_3$ became available, scientists began to test it on a wide variety of plants. Spectacular responses were obtained in the stem elongation of dwarf and rosette plants, particularly in genetically dwarf peas (*Pisum sativum*), dwarf maize (corn; *Zea mays*) (**FIGURE 20.1**), and many rosette plants (**FIGURE 20.2**).

Because applications of fungus-derived $GA_3$ could increase the height of dwarf mutants, scientists asked whether wild-type plants contain their own endogenous GA. Bioassays of extracts from a variety of plant species showed that GA-like substances* were indeed present. Higher concentrations were found in immature seeds (approximately 1 part per million) than in vegetative tissue (1–10 parts per billion). This made immature seeds the plant material of choice for GA extraction, but chemical characterization still required tens of kilograms of seeds. The first identification of a GA from a plant extract was the 1958 discovery of $GA_1$ from immature seeds of runner bean (*Phaseolus coccineus*). Nowadays, the availability of very-sensitive spectroscopic methods permits the identification and quantitation of known GAs with less than a gram of plant material.

As more GAs from *Gibberella* and different plant sources were characterized, a scheme was adopted to number them ($GA_1$–$GA_n$) in chronological order of their discovery. Only GAs that are naturally occurring and whose chemical structures have been conclusively determined are assigned **A numbers** using the procedure described at http://www.plant-hormones.info/gibberellin_nomenclature.htm. (The

---

*The term *GA-like substance* refers to a compound that shows activity in a GA bioassay, but has not been chemically characterized.

**FIGURE 20.1** The effect of exogenous $GA_1$ on wild-type (labeled as "normal" in the photograph) and dwarf mutant (*d1*) maize. Gibberellin stimulates dramatic stem elongation in the dwarf mutant, but has little or no effect on the tall, wild-type plant. (Courtesy of B. Phinney.)

**FIGURE 20.2** Cabbage, a long-day plant, remains a low-growing rosette in short days, but it can be induced to bolt (grow long internodes) and flower by applications of $GA_3$. In the case illustrated, giant flowering stalks were produced. (© Sylvan Wittwer/Visuals Unlimited.)

"number" of a GA is simply a cataloging convenience, and there is no implied metabolic relationship between GAs with adjacent numbers.)

## All gibberellins are based on an ent-*gibberellane* skeleton

The GAs are diterpenoids that are formed from *four* isoprenoid units each consisting of five carbons (see Chapter 13). They possess a tetracyclic (four-ringed) *ent*-gibberellane skeleton (containing 20 carbon atoms), or a 20-nor-*ent*-gibberellane skeleton (containing only 19 carbon atoms because carbon 20 is missing)*.

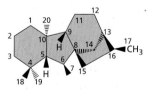

***ent*-gibberellane**

Gibberellins that have the full diterpenoid complement of 20 carbon atoms (e.g., $GA_{12}$) are referred to as **$C_{20}$-GAs**. Those that have only 19 carbons because carbon-20 (blue on next structure) has been lost by metabolism are referred to as **$C_{19}$-**

---

*The prefix *ent* refers to the fact that the skeleton is derived from *ent*-kaurene, a tetracyclic hydrocarbon that is enantiomeric to the naturally occurring compound, kaurene.

GAs (e.g., GA$_9$). In nearly all C$_{19}$-GAs, the carboxyl at C-4 forms a lactone at C-10 (red on structure below).

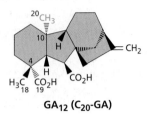

**GA$_{12}$ (C$_{20}$-GA)**

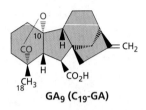

**GA$_9$ (C$_{19}$-GA)**

Other structural modifications include the insertion of additional features such as hydroxyl (—OH) groups or double bonds. The position and stereochemistry of these functional groups can have profound effects on biological activity. The GAs with highest intrinsic biological activity (GA$_1$, GA$_3$, GA$_4$, and GA$_7$) were among the first to be discovered: These GAs are all C$_{19}$-GAs. They all possess a 4,10-lactone (shown in red), a carboxylic acid (◄ COOH, shown in green on the structure below) at C-6, and a hydroxyl group at C-3 in β-orientation (◄ OH, shown in blue on the structure below). Elucidation of the tertiary structure of the GA receptor has identified the amino acid residues in the receptor that interact with each of these functional groups in a bioactive GA molecule, thereby holding the GA in the receptor pocket and allowing a "lid" to close in place (Murase et al. 2008; Shimada et al. 2008). This allosteric modification of the GA receptor upon binding an active GA facilitates subsequent interactions of the GA receptor with other proteins.

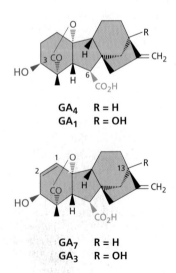

**GA$_4$  R = H**
**GA$_1$  R = OH**

**GA$_7$  R = H**
**GA$_3$  R = OH**

Although bioactive GAs *may* possess other features such as a double bond between C-1 and C-2 (as in GA$_7$ and GA$_3$) and/or an —OH group at C-13 (as in GA$_1$ and GA$_3$) these functional groups seem to neither help nor hinder GA binding in the receptor pocket. On the other hand, the presence of a 2β-OH group prevents the hydrophobic surface of the receptor lid from closing, and thus renders a GA inactive (Murase et al. 2008; Shimada et al. 2008). (A further discussion of GA structure can be found in **WEB TOPIC 20.1**.)

## Effects of Gibberellins on Growth and Development

Though they were originally identified as the cause of disease symptoms of rice that resulted in internode elongation, endogenous GAs can influence a large number of developmental processes in addition to stem elongation. Many of these properties of GAs have been exploited in agriculture for decades, and manipulation of the GA content of crop plants affects shoot size, fruit set, and fruit growth.

### Gibberellins promote seed germination

Many seeds, particularly those of wild plant species, do not germinate immediately after dispersal from the mother plant, and may experience a period of dormancy. Dormant seeds will not germinate even if provided with water. Abscisic acid (ABA) and bioactive GA act in an antagonistic manner, and the relative amounts of the two hormones within the seed can, in many species, determine the degree of dormancy. Light or cold treatments of dormant seeds have been shown to lower the amount of ABA and increase the concentration of bioactive GA, ending dormancy and promoting germination (Piskurewicz et al. 2008; Seo et al. 2006). Treatment of seeds with bioactive GA can often substitute for the light or cold treatment needed to break dormancy.

During germination, GAs induce the synthesis of hydrolytic enzymes, such as amylases and proteases in cereal grains. These enzymes degrade the stored food reserves accumulated in the endosperm or embryo as the seed matured. This degradation of carbohydrates and storage proteins provides nourishment and energy to support seedling growth. The GA-induced synthesis of α-amylase in germinating cereal grains has been studied extensively and is considered in detail later in the chapter.

### Gibberellins can stimulate stem and root growth

Applied GAs may not have dramatic effects on stem elongation in plants that are already "tall," since bioactive GA may *not* to be limiting in some tall plants. However, applied GAs can promote internode elongation very dramatically in genetically dwarf mutants, in "rosette" species, and in various members of the Poaceae (grass family). Exogenous GA causes such extreme stem elongation in dwarf maize plants that they resemble the tallest varieties of the same species (see Figure 20.1).

Rosette species are plants in which the first-formed internodes do not elongate under certain growing conditions. This results in a compact cluster or rosette of leaves, as seen in members of the Brassicaceae (cabbage family). Rosette formation is frequently observed when long-day plants are grown in short-day conditions. Bolting (stem growth) and flowering will result if plants are treated with a bioactive GA, or are transferred to long days (see Figure 20.2).

Gibberellins are also important for root growth. Extreme dwarf mutants of pea and Arabidopsis, in which GA biosynthesis is blocked, have shorter roots than wild-type plants, and GA application to the shoot enhances both shoot *and* root elongation (Yaxley et al. 2001; Fu and Harberd 2003).

## Gibberellins regulate the transition from juvenile to adult phases

Many woody perennials do not flower or produce cones until they reach a certain stage of maturity; up to that stage they are said to be juvenile (see Chapter 25). Applied GAs can regulate phase change, though whether GA hastens or retards the juvenile-to-adult transition will depend on the species. In many conifers, the juvenile phase, which may last up to 20 years, can be shortened by treatment with $GA_3$ or with a mixture of $GA_4$ and $GA_7$, and much younger plants can be induced to enter the reproductive, cone-producing phase precociously (**FIGURE 20.3**).

(A) White spruce

(B) White spruce

(C) Giant sequoia seedling

**FIGURE 20.3** Gibberellins induce conebud formation in juvenile conifers. (A, B) These photographs show female cones developing on sapling-size, grafted plants of white spruce (*Picea glauca*). The stems of these plants had been injected the previous summer with a mixture of $GA_4/GA_7$ in aqueous ethanol. (C) A 14-week-old seedling of giant sequoia (*Sequoiadendron giganteum*) that had been sprayed with an aqueous solution of $GA_3$ some 8 weeks earlier, showing the development of a female conebud. (Courtesy of S. D. Ross and R. P. Pharis.)

## Gibberellins influence floral initiation and sex determination

As already noted, GAs can substitute for the long-day requirement for flowering in many plants, especially rosette species. The interaction of photoperiod and GAs in flowering is complex, and this subject is discussed in Chapter 25.

In plants with imperfect (unisexual) rather than perfect (hermaphroditic) flowers, sex determination is genetically regulated. However, it is also influenced by environmental factors such as photoperiod and nutritional status, and these environmental effects may be mediated by GAs. Just as in the case of the juvenile-to-adult transition, the nature of the effect of GA on sex determination can vary with species. In dicots such as cucumber (*Cucumis sativus*), hemp (*Cannabis sativa*), and spinach, GAs promote the formation of staminate (male) flowers, and inhibitors of GA biosynthesis promote the formation of pistillate (female) flowers. In some other plants, such as maize, GAs suppress stamen formation and promote pistil formation.

## Gibberellins promote pollen development and tube growth

Gibberellin-deficient dwarf mutants (e.g., in Arabidopsis and rice) have impaired anther development and pollen formation, and both these defects, which lead to male sterility, can be reversed by treatment with bioactive GA. In other mutants in which GA response (rather than GA biosynthesis) is blocked, the defects in anther and pollen development cannot be reversed by GA treatment, so these mutants are male-sterile (Aya et al. 2009). In addition, reducing the level of bioactive GA in Arabidopsis by overexpressing a GA deactivating enzyme severely inhibits pollen tube growth (Swain and Singh 2005). Thus GAs seem to be required for both the development of the pollen grain and the formation of the pollen tube. The regulation of anther development in rice by bioactive GA will be discussed in more detail later in the chapter.

## Gibberellins promote fruit set and parthenocarpy

Gibberellin application can cause **fruit set** (the initiation of fruit growth following pollination) and growth of some fruits. For example, stimulation of fruit set by GA has been observed in pear (*Pyrus communis*). GA-induced fruit set may occur in the absence of pollination, resulting in parthenocarpic fruit (fruit without seeds). In grape (*Vitis vinifera*), the "Thompson Seedless" variety normally produces small fruits because of early seed abortion. Fruits can be stimulated to enlarge by treatment with $GA_3$ (FIGURE 20.4). This treatment also promotes growth of the pedicels (fruit-bearing stalks) and consequently reduces fungal infections that can be problematic in the compact clusters of untreated

**FIGURE 20.4** Gibberellin induces growth in "Thompson Seedless" grapes. Untreated grapes normally remain small because of natural seed abortion. The bunch on the left is untreated. The bunch on the right was sprayed with $GA_3$ during fruit development, leading to increased size of the fruits and elongation of the pedicels (fruit stalks). (© Sylvan Wittwer/Visuals Unlimited.)

grapes. Both these effects of GAs on grapes are exploited commercially to produce large, seedless fruits.

## Gibberellins promote early seed development

Some GA-deficient mutants, or transgenic plants with enhanced GA inactivation, have increased seed abortion. The failure of seeds to develop normally can be attributed to reduced levels of bioactive GAs in very young seeds. Treatment with GA will not restore normal seed development, because exogenous GA cannot enter the new seeds. However, the effect of GA deficiency on seed abortion can be negated by simultaneous expression of mutations that give a constitutive GA response (Swain and Singh 2005). Taken together, these results provide evidence for a role for GA in the early stages of seed development.

## Commercial uses of gibberellins and GA biosynthesis inhibitors

The major commercial uses of GAs (typically $GA_3$) are to promote the growth of fruit crops, to stimulate the barley malting process in the beer-brewing industry, and to increase sugar yield in sugarcane. A description of the malting process can be found in WEB TOPIC 20.2.

Inhibitors of GA biosynthesis have been useful for crops in which a *reduction* in plant height is desirable.

For example, tallness is a disadvantage for cereal crops grown in cool, damp climates, as occur in Europe, where lodging can be a problem. (*Lodging*—the bending of stems to the ground caused by the weight of water collecting on the ripened heads—makes it difficult to harvest the grain with a combine harvester.) Shorter internodes reduce the tendency of the plants to lodge, increasing the yield of the crop. Even genetically dwarf wheat cultivars grown in Europe are sprayed with inhibitors of GA biosynthesis such as Cycocel to further reduce stem length and lodging.

In the field or greenhouse, tall plants are often difficult to manage. For floral crops such as lilies, chrysanthemums, and poinsettias, short, sturdy stems are desirable. Applications of GA biosynthesis inhibitors are often used to control the size of container-grown ornamental plants in nurseries, greenhouses, and shade houses.

The chemical structures of several commercially available inhibitors of GA biosynthesis can be found in WEB TOPIC 20.1, and there is further discussion of the uses of these inhibitors in WEB TOPIC 20.2.

# Biosynthesis and Deactivation of Gibberellins

Gibberellins constitute a large family of tetracyclic diterpene acids synthesized via a terpenoid pathway. Early stages of that pathway are described in Chapter 13. In this section, we describe the later stages in the GA pathway. Additional details are discussed in WEB TOPIC 20.3 and in Yamaguchi (2008). We also discuss the regulatory enzymes in the pathway, as well as the genes that encode them. Knowledge of GA biosynthesis and deactivation is important, as it contributes to our understanding of *GA homeostasis*. By GA homeostasis, we mean the maintenance of appropriate levels of bioactive GA in plant cells and tissues throughout the life cycle. Homeostasis depends upon the regulation of GA biosynthesis, deactivation, and transport.

Fundamental to our understanding of how GAs control growth and development is the ability to identify and quantify the GAs present in our experimental plants. Highly sensitive physical techniques such as mass spectrometry are now used that allow precise identification and quantification of specific GAs from small amounts of tissue. GA identification and quantitation in plant extracts are discussed in WEB TOPIC 20.4.

The isolation of mutants with altered stem length has made it possible to determine which of the many GAs present in a plant have intrinsic biological activity. The use of mutants has also facilitated the identification and cloning of genes encoding enzymes in the GA biosynthetic pathway. Sequencing the Arabidopsis and rice genomes

has led to the development of comprehensive databases that facilitate the rapid identification of genes and proteins related to GA metabolism and its regulation.

## Gibberellins are synthesized via the terpenoid pathway

Terpenoids are compounds made up of five-carbon **isoprenoid** building blocks. The GAs are diterpenoids that are formed from *four* such isoprenoid units (see Chapter 13). The GA biosynthetic pathway can be divided into three stages, each residing in a different cellular compartment: plastid, ER, or cytosol. A simplified version of the pathway is shown in **FIGURE 20.5**. The complete pathway is presented in Appendix 3.

- In *stage 1*, which occurs in plastids, four isoprenoid units are assembled to give a 20-carbon linear molecule, geranylgeranyl diphosphate (GGPP). GGPP is then converted into a tetracyclic compound, *ent*-kaurene, in two steps, which are catalyzed by *ent*-copalyl-diphosphate synthase (CPS) and *ent*-kaurene synthase (KS).

- In *stage 2*, which occurs on the plastid envelope and in the endoplasmic reticulum, *ent*-kaurene is converted, in a stepwise manner, to the first-formed GA, which is $GA_{12}$. Two important enzymes in this part of the pathway are *ent*-kaurene oxidase (KO) and *ent*-kaurenoic acid oxidase (KAO). The pathway to $GA_{12}$ is essentially the same in all plant species studied so far.

- In *stage 3*, which occurs in the cytosol, $GA_{12}$ is converted, through a series of oxidative reactions, first into other $C_{20}$-GAs, and then into $C_{19}$-GAs, including the bioactive GA(s). Two major stage 3 pathways have been identified. They both comprise the same series of oxidative reactions, except that the intermediates in one pathway all have a —OH group at C-13 (and so it is called the 13-hydroxylation pathway), whereas the intermediates in the other pathway do not (and so it is referred to as the non-13-hydroxylation pathway). The series of oxidative reactions occur in the A-ring, and are the same in both stage 3 pathways.

The 13-hydroxylation pathway is the major pathway in many plants, although in Arabidopsis and in some crop plants in the Cucurbitaceae (pumpkin family) the non-13-hydroxylation pathway predominates. In the following discussion the reactions leading to bioactive GA (which is $GA_4$ in the non-13-hydroxylation pathway and $GA_1$ in the 13-hydroxylation pathway) are referred to as "biosynthesis." Further metabolism of the bioactive GA is referred to as "deactivation." GA biosynthesis and deactivation are described in more detail in WEB TOPIC 20.3.

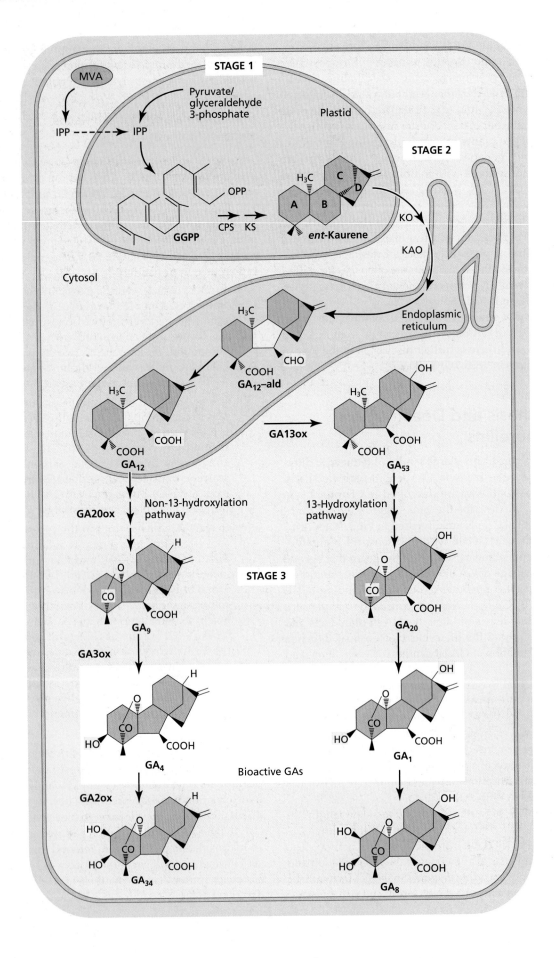

**FIGURE 20.5** The three stages of GA biosynthesis. In stage 1, geranylgeranyl diphosphate (GGPP) is converted to *ent*-kaurene. In stage 2 in the endoplasmic reticulum, *ent*-kaurene is converted to $GA_{12}$-aldehyde and $GA_{12}$. $GA_{12}$ is converted to $GA_{53}$ by hydroxylation at C-13. In stage 3 in the cytosol, $GA_{12}$ and $GA_{53}$ are converted, via parallel pathways, to other GAs. This conversion proceeds with a series of oxidations at C-20, resulting in the eventual loss of C-20 and the formation of $C_{19}$-GAs. 3β-hydroxylation then produces $GA_4$ and $GA_1$ as the bioactive GAs in each pathway. Hydroxylation at C-2 then converts $GA_4$ and $GA_1$ to the inactive forms $GA_{34}$ and $GA_8$, respectively. In most plants the 13-hydroxylation pathway predominates, although in Arabidopsis and some others, the non-13-OH-pathway is the main pathway. CPS, *ent*-copalyl diphosphate synthase; KS, *ent*-kaurene synthase; KO, *ent*-kaurene oxidase; KAO, *ent*-kaurenoic acid oxidase; GA20ox, GA 20-oxidase; GA3ox, GA 3-oxidase; GA2ox, GA 2-oxidase; GA13ox, GA 13-oxidase.

## Some enzymes in the GA pathway are highly regulated

Work with biosynthetic mutants of Arabidopsis, pea, and maize (**FIGURES 20.6 AND 20.7**) facilitated the cloning of

genes for many of the enzymes in GA biosynthesis and deactivation. Most notable from a regulatory standpoint are three enzymes in *stage 3* of the pathway. These are the GA 20-oxidase (GA20ox) and GA 3-oxidase (GA3ox) enzymes, which catalyze the steps prior to bioactive GA, and the GA 2-oxidase (GA2ox), which is involved in GA deactivation. All three of these enzymes are classified as **dioxygenases** and utilize 2-oxoglutarate as a co-substrate and $Fe^{2+}$ as a cofactor. For this reason they are referred to as 2-oxoglutarate-dependent dioxygenases (2ODDs). We shall discuss each of these enzymes, before describing some of what is known of their regulation.

- **GA 20-oxidase**. In Arabidopsis there is a small family of five GA 20-oxidases, named AtGA20ox1 through AtGA20ox5. These homologs (genes that have similar sequences to each other because they are derived from a single gene are said to be *homologous*) are expressed in different tissues and organs and at different developmental stages, although there may be some overlapping expression. The principal stem-expressed GA 20-oxidase is AtGA20ox1, which is encoded by a gene that was originally named *GA5* (Phillips et al. 1995; Xu et al. 1995). Mutation of the *GA5* gene results in a semidwarf, rather than extreme dwarf,

**FIGURE 20.6** Phenotypes of wild-type and GA-deficient mutants of Arabidopsis, showing the position in the GA biosynthetic pathway that is blocked in each mutant. All mutant alleles (denoted by lower case notation of the wild-type alleles) are homozygous. Plants were grown in continuous light and are 7 weeks old. Note that the *ga1*, *ga2*, and *ga3* seedlings are sterile and have not produced seed pods (siliques). See Figure 20.5 for abbreviations. (Courtesy of V. Sponsel.)

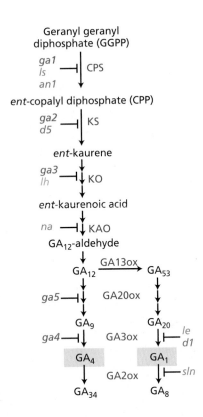

Geranyl geranyl
diphosphate (GGPP)

*ga1*
*ls*     CPS
*an1*

*ent*-copalyl diphosphate (CPP)

*ga2*
*d5*     KS

*ent*-kaurene

*ga3*
*lh*     KO

*ent*-kaurenoic acid

*na*     KAO

GA$_{12}$-aldehyde

GA13ox
GA$_{12}$ ——→ GA$_{53}$

*ga5*    GA20ox

GA$_9$     GA$_{20}$

*ga4*    GA3ox      *le*
                    *d1*

GA$_4$     GA$_1$

         GA2ox     *sln*

GA$_{34}$     GA$_8$

**FIGURE 20.7**  A portion of the GA biosynthetic pathway showing the metabolic steps that are blocked by known mutations (denoted by lower case notations of wild-type alleles) of Arabidopsis (green), pea (blue), and maize (orange). See Figure 20.5 for abbreviations. As a historical note, the names *GA1–GA5* for five nonallelic loci in Arabidopsis that encode enzymes in the GA biosynthetic pathway were assigned long before the nature of the enzymes or their genes were known (Koornneef and van der Veen 1980). After genetic tests, they were placed in sequence based on the anticipated order of action in the pathway.

phenotype (*ga5*) (see Figure 20.6). This is because of gene redundancy: one or more of the other GA 20-oxidases can partially compensate for the loss of AtGA20ox1. In contrast, the mutants *ga1*, *ga2*, and *ga3* are extreme dwarfs, since enzymes in stages 1 and 2 of the pathway (CPS, KS, and KO) are each encoded by only one gene (see Figure 20.6).

- **GA 3-oxidase.** The principal stem-expressed GA 3-oxidase in Arabidopsis (AtGA3ox1) is encoded by a gene originally named *GA4* (Chiang et al. 1995). Like *GA5*, *GA4* is one of a small gene family, and the *ga4* mutant, like *ga5*, is a semidwarf rather than an extreme dwarf (see Figure 20.6). In pea, this enzyme is encoded by the *LE* gene (see Figure 20.7 and later discussion). In maize this enzyme is encoded by *D1*, and a recessive mutation, *d1*, in this gene gives dwarf plants (see Figure 20.1).

- **GA 2-oxidase.** No mutant phenotype is known for the GA "deactivating" enzyme, GA 2-oxidase, in Arabidopsis. In contrast, a pea mutant identified because it grew taller than wild-type plants contains a mutation in a gene termed *SLENDER* (*SLN*), which encodes a GA 2-oxidase. In the *sln* mutant, GAs synthesized during seed maturation are not deactivated to the same extent as those in wild-type seeds, and some potentially bioactive GAs remain in the mature seed. During germination they are converted into bioactive GA$_1$, which enhances internode elongation in the young seedling and gives a "slender" phenotype (Reid et al. 1992).

## Gibberellin regulates its own metabolism

As explained in Chapter 19 for auxin, many factors are important for maintaining hormone homeostasis, including the relative balance between synthesis and deactivation. Part of a plant's response to bioactive GA is to depress GA biosynthesis and stimulate deactivation, to prevent excessive stem elongation. Depression of biosynthesis is achieved through down-regulation (inhibition of expression) of some of the *GA20ox* and *GA3ox* genes encoding the last two enzymes in the formation of bioactive GA. This effect of GA on its own biosynthesis is termed **negative feedback regulation.** Enhanced GA deactivation is also important for maintaining GA homeostasis. It is achieved by up-regulating (stimulating) the expression of some of the *GA2ox* genes encoding the enzyme that deactivates GA. The ability of GA to promote the expression of genes involved in its own deactivation is termed **positive feedforward regulation.**

The relative importance of each member of the gene family to feedback or feed-forward regulation varies with species and tissue (Hedden and Phillips 2000). The expression of one or more homologs that are not subject to feedback or feed-forward regulation may cause a "spike" in bioactive GA content, perhaps in response to a particular environmental signal; the subsequent expression of a regulated homolog provides a mechanism to reestablish GA concentrations to a normal level afterwards. In other circumstances, environmental regulation of expression can act on genes that are subject to feedback regulation by overriding homeostasis temporarily.

## GA biosynthesis occurs at multiple plant organs and cellular sites

All reactions in GA biosynthesis leading from GGPP to a C$_{19}$-GA can be demonstrated in cell-free systems from seed or seed-parts, providing definitive evidence that developing seeds are sites of GA biosynthesis. In pea seeds, a surge in GA biosynthesis occurs soon after fertilization and is

(A) 5-day-old seedling

(B) 3-week-old seedling

(C) Open flower

(D) Mature silique with immature seeds

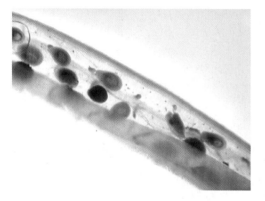

(E) Developing embryos

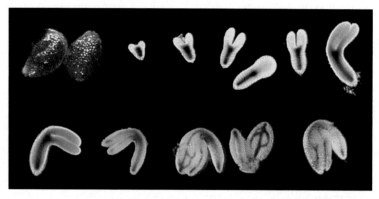

**FIGURE 20.8** Histochemical analysis of Arabidopsis plants containing the *GA1* promoter:*GUS* gene fusion, at five different stages of development. *GA1* encodes CPS, the first committed enzyme in GA biosynthesis. The blue staining shows where there is necessary for early seed and fruit growth. In later stages *GA1* promoter activity. These panels indicate that this early step in GA biosynthesis occurs in numerous tissues and organs, and at several stages in the life cycle. (From Silverstone et al. 1997a; courtesy of T-p. Sun.)

necessary for early seed and fruit growth. In later stages of seed maturation in many species, the level of GA accumulation is quite high.

Reporter gene studies have shown that *GA1*, which encodes CPS, the first committed enzyme in GA biosynthesis, is expressed in immature seeds, shoot apices, root tips, and anthers of wild-type Arabidopsis plants (**FIGURE 20.8**) (Silverstone et al. 1997a). Not unexpectedly, these organs are affected in *ga1* mutants, which exhibit seed dormancy, an extreme-dwarf growth habit, and male sterility. This correlation between CPS activity and phenotype indicates that the early stages of GA biosynthesis and GA action can occur in the same organs. Other elegant studies by Yamaguchi et al. (2001) and Ogawa et al. (2003) compared the cellular locations of GA 3-oxidase (the enzyme that converts inactive $GA_9$ to bioactive $GA_4$) and the products of genes known to be up-regulated by $GA_4$. Some cells did not make $GA_4$ but could respond to it.

Thus, within germinating Arabidopsis embryos we see $GA_4$ action both in the cells in which it is synthesized and in different cells, implying that $GA_4$, or a component downstream of $GA_4$ in the response pathway, must move from cell to cell.

### Environmental conditions can influence GA biosynthesis

Gibberellins play an important role in mediating the effects of environmental stimuli on plant development. Light and temperature can have profound effects on GA metabolism and GA response, as discussed in **WEB TOPIC 20.5**. In many cases the environment can alter the metabolism of, or response to, other hormones in addition to GAs. The ratio of bioactive GA to ABA is particularly important, and so is the relative responsiveness of different tissues to the two hormones.

## GA$_1$ and GA$_4$ have intrinsic bioactivity for stem growth

Seminal studies with GA biosynthetic mutants (also referred to as GA-deficient mutants) conducted in the 1980s achieved two important goals. Not only did they provide a way for the pathways of GA metabolism to be definitively established, but these studies also determined that GA$_1$ is the major bioactive GA for stem growth in pea and maize, and that its precursors have no intrinsic biological activity.

*LE* and *le* are two alleles of a gene that regulates tallness in peas, the genetic trait investigated by Gregor Mendel in his pioneering study published in 1866. If GA$_{20}$ is applied to the *le* mutant of pea, it is not bioactive, whereas GA$_1$ is bioactive, and rescues the mutant phenotype (the plants grow tall). Gibberellin A$_8$ is also inactive. We can infer from this information, and from knowledge of the GA metabolic pathway in pea (GA$_{20}$ → GA$_1$ → GA$_8$), that GA$_{20}$ is inactive unless it can be converted to GA$_1$ within the plant, and that *GA$_1$ has intrinsic bioactivity* (Ingram et al. 1984).

Metabolic studies using isotopically labeled GAs have demonstrated that the *LE* gene encodes an enzyme that 3β-hydroxylates GA$_{20}$ to produce GA$_1$. At the same time, it was confirmed that tall stems contain more GA$_1$ than dwarf stems (Reid and Howell 1995). Mendel's *LE* gene was eventually cloned, and the recessive *le* allele was shown to have a single base change leading to a defective enzyme (Lester et al. 1997; Martin et al. 1997). Less GA$_1$ is produced in plants homozygous for the recessive allele than in wild-type plants. However, since *le* plants still produce a partially active GA 3-oxidase, enough GA$_1$ is present for *le* seedlings to attain approximately 30% of the height of wild-type plants.

A study of other pea mutants has confirmed that the height of pea plants is directly correlated with the amount of endogenous GA$_1$ (Ross et al. 1989). For example, the *na* mutant of pea is deficient in KAO activity (see Figure 20.7), and mutant plants are almost completely devoid of GA$_1$. As a consequence they achieve a stature of only about 1 cm at maturity (**FIGURE 20.9**). In contrast, the seedlings of the *sln* mutant contain elevated levels of GA$_1$ because of impaired GA deactivation, and these mutant plants are actually taller than wild-type seedlings (see Figure 20.9). A personal account of this research on pea stem growth can be found in **WEB ESSAY 20.1**.

Work with dwarf mutants of other plants has confirmed that 3β-hydroxylated C$_{19}$-GAs have *intrinsic* biological activity for stem growth. Studies similar to those described above for pea determined that GA$_1$ is the major endogenous bioactive GA regulating stem growth in maize (Phinney 1984).

| Phenotype | Ultradwarf | Dwarf | Tall | Slender |
|---|---|---|---|---|
| Genotype | *na/LE/SLN* | *NA/le/SLN* | *NA/LE/SLN* | *NA/LE/sln* |
| GA$_1$ content | None | Lower than level in wild type | Wild-type level | Higher than level in wild type |

**FIGURE 20.9**  Phenotypes and genotypes of peas that differ in the GA$_1$ content of their vegetative tissue. (All alleles are homozygous.) (After Davies 1995.)

The situation is similar in rice, though a less abundant GA, $GA_4$ (non-13-hydroxylated $GA_1$), has higher affinity for the GA receptor than $GA_1$ when tested in vitro, and $GA_4$ may have an important role in, for example, reproductive growth of rice. In Arabidopsis and several members of the Cucurbitaceae (e.g., pumpkin and cucumber), applied $GA_1$ has less biological activity than $GA_4$. Therefore, in Arabidopsis and probably also in these cucurbits, $GA_4$ is assumed to be the main biologically active GA.

## Plant height can be genetically engineered

The identity of the bioactive GAs in crop plants, together with the characterization of key enzymes in GA biosynthesis and deactivation, has enabled genetic engineers to alter the levels of bioactive GA in crops, and thus affect plant height (Hedden and Phillips 2000). The conserved function of the GA20ox, GA3ox, and GA2ox enzymes between species means that genes encoding these enzymes in one species can be introduced and expressed in another species, thus extending the range of crop plants that can be manipulated. Several examples are given by Phillips (2004). For instance, a gene encoding an Arabidopsis GA 20-oxidase can be introduced into quaking aspen (*Populus tremuloides* × *P. tremula* hybrid) and expressed constitutively (i.e., expressed at high levels throughout the plant). The constitutive expression of the GA 20-oxidase is achieved using a promoter from cauliflower mosaic virus (CaMV 35S). As a consequence of this **overexpression** of GA 20-oxidase activity, the transgenic poplar seedlings contain higher levels of bioactive $GA_1$, are taller, and have enhanced xylem fiber quantity and length, which are desirable for paper manufacture (Eriksson et al. 2000).

In some other agricultural crops the desired effect is often to *decrease* growth. Reductions in $GA_1$ levels have been achieved by the transformation of crop plants with antisense constructs of *GA20ox* or *GA3ox*, thereby lowering the transcript levels (and hence expression) of these genes and reducing $GA_1$ biosynthesis. Alternatively, overexpressing the *GA2ox* gene, which encodes the enzyme catalyzing $GA_1$ deactivation, can also lead to a reduction in $GA_1$ levels. These approaches have been used to introduce more extreme dwarfing into wheat (Appleford et al. 2007) (**FIGURE 20.10**) and rice (Sakamoto et al. 2003).

## Dwarf mutants often show other phenotypic defects

We have learned that bioactive GAs control many aspects of plant growth and development in addition to stem length. What else happens in dwarf plants in which GA biosynthesis in stems is blocked by mutation? Clearly, many of these mutations give rise to pleiotropic phenotypes. The severely dwarfed *ga1*, *ga2*, and *ga3* mutants of Arabidopsis have dormant seeds that cannot germinate unless treated with GA. Moreover, the plants are male-

Wild type (untransformed)

Transformed plants overexpressing *GA 2-oxidase*

**FIGURE 20.10** Genetically engineered dwarf wheat plants. The wild-type (untransformed) wheat is shown on the extreme left. The three plants on the right were transformed with a *GA 2-oxidase* cDNA from bean under the control of a constitutive promoter. Consequently, the endogenous bioactive $GA_1$ is deactivated more in the transformed plants than in the wild-type plants. The varying degrees of dwarfing reflect varying degrees of overexpression of *GA 2-oxidase*, with the highest expression on the extreme right. (From Hedden and Phillips 2000; courtesy of A. Phillips.)

sterile unless treated with GA, because of a requirement for bioactive GA in anther and pollen development (see Figure 20.6).

In pea, too, one would expect that GA-deficient mutants would have short stems and reduced seed growth, as bioactive GA is required for both stem elongation and seed development. However, homozygous mutations in three genes encoding enzymes in the GA pathway all give dwarf plants and *normal* seed development because of gene redundancy. For instance, the *na* mutant of pea is blocked at the step catalyzed by the enzyme KAO (see Figure 20.7) and is an extreme dwarf, but these tiny plants are still able to produce pods containing viable seeds, because a second gene that encodes KAO is expressed in seeds (Davidson et al. 2003). In contrast, the *lh-2* mutation in the *LH* gene, which encodes the enzyme KO (see Figure 20.7), reduces $GA_1$ levels in both stems *and* seeds (Davidson and Reid 2004). Thus this *lh-2* mutant, in addition to being dwarf, has impaired seed development as well (**FIGURE 20.11**).

## Auxins can regulate GA biosynthesis

There is a considerable body of evidence, initially from studies in pea and then in other dicots and in monocots,

**FIGURE 20.11** Impaired seed development in a GA-deficient mutant of pea. Pods of the wild type (left) and the *lh-2* mutant (right), showing impaired seed development in the mutant. (Courtesy of J. B. Reid.)

that auxins can regulate GA biosynthesis. The "targets" for auxin regulation—that is, which genes are up- or down-regulated—are different in different species, and even in different organs or tissues of the same species. Some of these studies are described in **WEB TOPIC 20.6**.

# Gibberellin Signaling: Significance of Response Mutants

Single-gene mutants impaired in their response to GA have been valuable tools for identifying genes that encode possible GA receptors or components of their signal transduction pathways. In general terms, factors that affect signal transduction can be either *positive* or *negative* regulators. Three main classes of mutants can be distinguished:

1. A mutation that renders a **positive regulator** of GA signaling nonfunctional gives rise to a *dwarf* phenotype. These loss-of-function mutations are recessive, and the mutants do not respond to applied GA because of the deficiency in an essential component of the GA signal transduction pathway.

2. A mutation that renders a **negative regulator** of GA action nonfunctional gives rise to a *tall* phenotype. Again, these loss-of-function mutations are recessive.

3. A third class of mutants, in which a *negative regulator* is made *constitutively active*, also gives rise to GA-

nonresponsive *dwarf* plants, but in these cases the mutations are "gain-of-function," and thus semidominant.

What makes these GA response mutants different from those plants with mutations that block enzymes in the GA biosynthetic pathway? The difference is that the height of these GA response mutants is *not* proportional to the amount of endogenous bioactive GA. We know this because GA-insensitive dwarf plants will *not* grow tall when treated with bioactive GA. Nor do the constitutive extra-tall mutants (the phenotype referred to as "slender") exhibit reduced growth in the presence of inhibitors of GA biosynthesis.

The response mutants have been extensively studied in Arabidopsis, starting with the *GA insensitive (gai-1)* mutant that was isolated in the mid-1980s (Koornneef et al. 1985). Sequencing some of the positive and negative regulators in Arabidopsis paved the way for their characterization in a number of important crop plants, including rice, barley (*Hordeum vulgare*), wheat, maize, and pea. The discussion of GA signaling in the following sections is not treated historically, but the citations indicate the order in which the work was conducted.

## GID1 *encodes a soluble GA receptor*

A major breakthough in our understanding of GA signal transduction came with the characterization of a recessive dwarf mutant of rice, termed **GA-insensitive dwarf1 (gid1)** (Ueguchi-Tanaka et al. 2005). (This mutant fits the description for type 1 mutants listed earlier.) The wild-type allele codes for a globular protein, GID1, which is a **GA receptor**. Research to identify GA receptor proteins has been going on for at least two decades, and several putative GA-binding proteins have been investigated. So far, however, only GID1 fulfils all the criteria for a GA receptor. (Scientists doing this work used a GST–GID1 "fusion protein." The GST-tag allows the GID1 to be purified more easily, and it is known that GST–GID1 and GID1 have similar properties with respect to GA binding. For simplicity in the following discussion, we will refer to GST–GID1 as GID1. Likewise, we will refer to radiolabeled 16, 17 dihydro-GA$_4$ as radiolabeled GA$_4$.)

To be identified as a receptor, a protein must fulfill the following criteria:

- The binding of the ligand (in this case, GA) to the protein must be *specific*. This means you would expect bioactive GAs to bind to GID1 with higher affinity than less-active GAs. When the binding of ten different GAs (of varying relative biological activities) to GID1 was tested, bioactive GAs were shown to bind with higher affinity. Thus, GID1 can discriminate between bioactive GAs and those that have less activity.

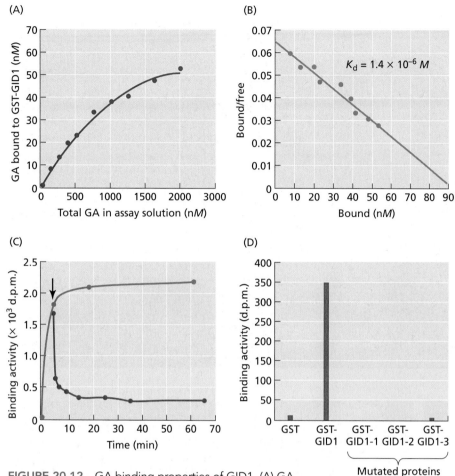

**FIGURE 20.12** GA binding properties of GID1. (A) GA-binding to GID1 protein (expressed as a GST–GID1 fusion protein) shows saturability when GID1 is incubated with a constant amount of radiolabeled $GA_4$ and increasing concentrations of unlabeled $GA_4$. (B) A Scatchard plot of the data from A gives the dissociation constant ($K_d$). (C) Association/dissociation rates of radiolabeled $GA_4$ and GID1. Total binding of radiolabeled $GA_4$ reached one-half of the maximum within 5 min (light green line). Adding an excess of unlabeled $GA_4$ (arrow) reduced binding to less than 10% within 5 min (dark green line). (D) Three different mutated GID1 proteins (expressed as the fusion proteins GST–GID1-1, GST–GID1-2, and GST–GID1-3) did not bind radiolabeled $GA_4$, providing strong evidence that $GA_4$ binding to the wild-type protein is specific. (After Ueguchi-Takana et al. 2005.)

- Ligand binding should be *saturable*; when all the receptor molecules have bound a GA molecule, no additional radiolabeled GA molecules will bind, and the curve flattens. This was shown to be the case for GID1 (see **FIGURE 20.12A**).

- Ligand binding should be of *high affinity*: the higher the affinity of GAs for the protein the more tightly bound they will be, and the less easily they will dissociate. The value for $K_d$ (the dissociation constant) can be obtained from a Scatchard analysis, in which the amount of bound radiolabeled GA (x-axis) is plotted against the amount of bound radiolabeled GA divided by the amount of unbound radiolabeled GA (y-axis) (**FIGURE 20.12B**). The slope is $-1/K_d$, and the low value for $K_d$ ($1.4 \times 10^{-6}$ $M$) indicates that GA is tightly bound to the protein.

- Ligand binding should be *rapid* and *reversible*. **FIGURE 20.12C** shows that total binding of radiolabeled GA to GID1 reached one-half of the maximal value in just 5 minutes. Also, the addition of unlabeled $GA_4$ showed that dissociation of radiolabeled GA from the protein is also very rapid. These features of a receptor allow it to accommodate rapid alterations in intracellular ligand concentration in a cell.

In addition to providing this compelling evidence that GID1 is a GA receptor, Ueguchi-Tanaka et al. (2005) also showed that single nucleotide substitutions in *gid1-1* and

(A)

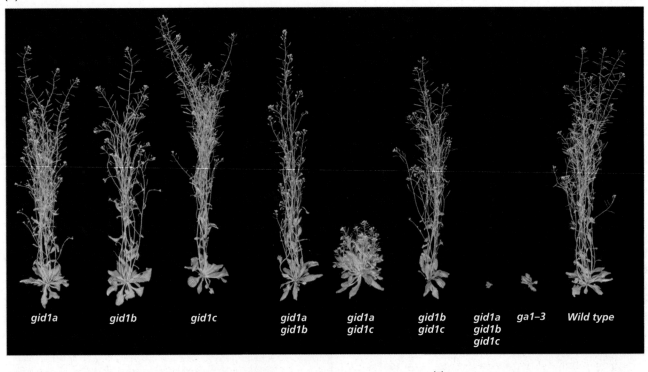

| gid1a | gid1b | gid1c | gid1a gid1b | gid1a gid1c | gid1b gid1c | gid1a gid1b gid1c | ga1–3 | Wild type |

(B)          (C)          (D)          (E)

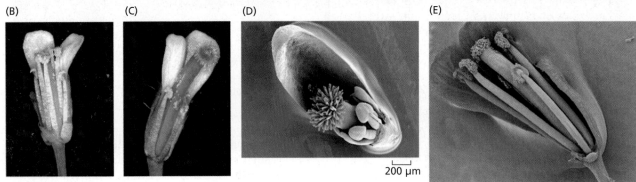

200 μm

500 μm

**FIGURE 20.13** Phenotypes of the *gid1a*, *gid1b*, and *gid1c* mutants of Arabidopsis. (A) Aerial portions of 37-day-old wild-type plant (Col-0) and homozygous mutant plants showing that single mutants do not have a noticeably altered stem length phenotype. The *gid1a/gid1c* double mutant is dwarf, whereas the other two double mutants show some functional redundancy and are tall. The triple mutant (*gid1a/gid1b/gid1c*) is an ex-treme dwarf. The GA-deficient mutant *ga1-3* is included for comparison. (B) Close-up view of a wild-type flower. (C) The *gid1a/gid1b* double mutant has defective anthers. (D, E) Scanning electron micrographs of flowers of *gid1a/gid1b/gid1c* (D), and wild type (E). In both panels, the sepals and petals were removed for clarity. (From Griffiths et al. 2006; courtesy of S. Thomas.)

*gid1-2* and a deletion in *gid1-3* all produced GA-insensitive dwarf plants and prevented the mutant proteins from binding radiolabeled GA$_4$ (**FIGURE 20.12D**).

Shortly after the identification of GID1 in rice, three orthologs were discovered in Arabidopsis, namely GID1a, GID1b, and GID1c (Griffiths et al. 2006; Nakajima et al. 2006). (Genes in different species that have similar sequences to each other because they are derived from a single gene in their common ancestor are said to be *ortholo-gous*.) Once the sequence of a gene, for example *GID1*, has been determined in one species (e.g., rice) it is relatively easy to search the genome databases of other species (e.g., Arabidopsis) in order to identify the orthologs. However in Arabidopsis there has been triplication of the original *GID1* gene, and each of the three genes, *GID1a*, *GID1b*, and *GID1c*, codes for a functional GA receptor. A mutation in only one of these Arabidopsis genes will *not* be expressed as a noticeable difference in stem length, but if all three are mutated then the so-called "triple mutant" is an extreme dwarf (**FIGURE 20.13**). In addition, severe anther defects lead to male sterility in the triple mutant.

The tertiary protein structures of both rice (GID1) and Arabidopsis (GID1a) proteins have recently been established, shedding light on just why a bioactive GA must have certain functional groups (Murase et al. 2008; Shimada et al. 2008). High-resolution analysis focused on the Arabidopsis protein bound to $GA_3$ or $GA_4$ and the rice protein bound to $GA_4$. The Arabidopsis protein was also cocrystallized with the amino-terminal part of a DELLA protein, GAI. As we shall see later, DELLA proteins are negative regulators of GA response. In both rice and Arabidopsis, the bioactive GA is anchored in a central pocket within the receptor protein by the carboxylic acid at C-6 in the GA molecule making several hydrogen bonds with two serine residues at the base of the pocket and one water molecule. In addition, the 3β-OH group of the GA is hydrogen bonded to a tyrosine residue and a bridging water molecule. Some modifications of the GA molecule can reduce binding: GAs in which the C-6 carboxylic acid is methylated or that do not possess a 3β-OH bind GID1 with an affinity that is several orders of magnitude less than that of a bioactive GA. Even if a GA has a 3β-OH and a C-6 COOH, the additional presence of a 2β-OH group, which has been known for a long time to reduce GA bioactivity, leads to steric interference and thus reduces binding to the receptor protein.

The "top" part of a bioactive GA molecule is hydrophobic and interacts with the N-terminal helical extension switch region of the GID1 receptor protein. The extension folds over the top of the GA, like a lid closing, and completely buries the GA inside the receptor (Murase et al. 2008; Shimada et al. 2008). Once the "lid" has closed, residues on the outer surface of the lid of GID1 are able to interact with specific residues on the amino-terminal end of GAI, which, as we will learn in the next section, is

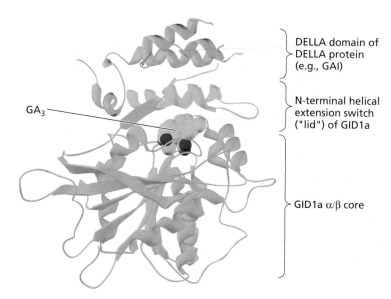

DELLA domain of DELLA protein (e.g., GAI)

N-terminal helical extension switch ("lid") of GID1a

$GA_3$

GID1a α/β core

**FIGURE 20.14** Structure of the $GA_3$-GID1a-DELLA complex. The bound $GA_3$ molecule is represented as a space-filling model with carbon in beige and oxygen in red. Binding of a bioactive GA (e.g., $GA_3$) within a pocket in the GID1a receptor allows an extension switch to close over the GA, rather like a "lid" closing. The lid closes over the hydrophobic "top" part of the GA, which has no exposed oxygen atoms. Once the extension switch has closed, the DELLA domain of a DELLA protein (e.g., GAI) can bind to the upper, outer surface of the switch. (After Murase et al. 2008.)

a DELLA protein (**FIGURE 20.14**). There is no *direct* interaction between the GA and the DELLA protein (the GA is buried within GID1), but bioactive GA is an **allosteric activator** of GID1, allowing a conformational change of shape in GID1 that facilitates its binding to a DELLA protein (**FIGURE 20.15**). In turn, the GA–GID1 complex appears to induce a coil-to-helix conformational change in the DELLA domain

**FIGURE 20.15** A model of the GA-induced change in the GID1 protein, and the changes in the DELLA protein induced by the GA–GID1 complex. (After Murase et al. 2008.)

1. GID1 is an allosteric protein. Binding GA causes a conformational change that leads to the extension switch closing like a lid.

2. Binding to the GA–GID1 complex causes a conformational change in the N-terminal DELLA domain of a DELLA protein.

3. Changes in the DELLA domain may also induce a conformational change in the GRAS domain.

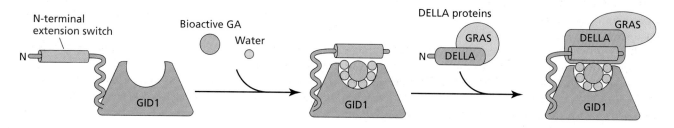

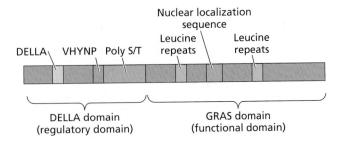

FIGURE 20.16 Domain structures of the RGA and GAI repressor proteins showing the regulatory DELLA domain and the functional GRAS domain.

of the DELLA protein, which may lead to further conformational changes in another domain (the so-called GRAS domain) of the DELLA protein.

### DELLA-domain proteins are negative regulators of GA response

DELLA-domain proteins are a subclass of the GRAS family of transcriptional regulators. All GRAS proteins have homology at the C-terminal (GRAS) domain. Those involved in GA responses also have a domain at the N-terminal end in which the first five amino acids are aspartic acid (D), glutamic acid (E), leucine (L), leucine (L), and alanine (A), hence the notation "DELLA domain" (FIGURE 20.16). Rice and barley each possess only one DELLA-domain protein, but it is now known that Arabidopsis has five. From extensive studies in cereals and Arabidopsis it was concluded that DELLA proteins are negative regulators of GA response. When they are degraded by proteolysis a GA response can occur. The history of the isolation of DELLA protein and the elucidation of their function is covered in WEB TOPIC 20.7. The work is summarized briefly here.

### Mutation of negative regulators of GA may produce slender or dwarf phenotypes

Mutations that prevent negative regulators of GA response from acting as negative regulators will produce tall plants. This is exactly what we see in "slender" mutants of rice (slender rice 1 [slr1]) and barley (slender 1 [sln1]). These mutants are excessively tall and, in fact, they look like plants that have been treated with high doses of bioactive GA (Ikeda et al. 2001; Chandler et al. 2002). But these plants do not contain high levels of endogenous GA, and even when they are grown in the presence of inhibitors of GA biosynthesis that can reduce the content of bioactive GA, *these plants are still tall*. Thus the tallness is *not* due to a high bioactive GA content, but to the GA response being constitutively expressed. (The *slender* mutations of cereals we are discussing here should not be confused with the

*slender* mutation of pea, which inhibits GA deactivation. The tall phenotype of *slender* pea mutants *is* due to high levels of endogenous GA.)

It was hypothesized that in the *slender* mutants of rice and barley a negative regulator of GA signaling is lost or nonfunctional. Cloning the cereal *slender* genes revealed that they are orthologs of the genes that encode DELLA proteins in Arabidopsis (Peng et al. 1997; Silverstone et al. 1997b, 1998). Remember, orthologs are genes in different species that are derived from a single ancestral gene. Just as we saw with *GID1* (for which there are three genes in Arabidopsis but only one in rice), there has been multiplication of *DELLA* genes in Arabidopsis but not in the cereals. In fact, there are five *DELLA* genes in Arabidopsis that encode five DELLA proteins: GA-INSENSITIVE (GAI), REPRESSOR of ga1-3 (RGA), RGL1, RGA2, and RGL3, all of which are negative regulators of GA response.

**RGA** was characterized from work with a homozygous recessive mutation (*rga*) in Arabidopsis plants that were also homozygous for *ga1*. Remember, the *ga1* mutation blocks an early step in the GA biosynthetic pathway, so these plants are GA-deficient. The *rga* mutation allowed enough stem elongation that it partially overcame the extreme dwarfism caused by the GA-deficient status of *ga1*, which is why the mutant was called *repressor of ga1*. When a recessive mutation causes growth, the inference is that the mutation must be making a negative regulator nonfunctional (type 2 of the mutants described on page 596). But the original mutant of *GAI* (gai-1), which was isolated by Koornneef et al. (1985) and is now known to also encode a DELLA protein, was a semi-dominant, gain-of-function *GA*-insensitive dwarf (i.e., fitting the description of a type 3 mutant). Further work revealed that both *RGA* and *GAI* encode negative regulators of GA response, but that different mutations in these negative regulators of GA signaling could produce opposite phenotypes, either slender (tall) or dwarf, depending on where within the genes the mutations occurred. The production of several mutant forms of each of these genes clarified that:

- If a mutation was in the C-terminal GRAS domain, the mutant plant was often taller than expected. In contrast,
- If the mutation was in the N-terminal part of the protein that contained the DELLA domain, the mutant protein was an irreversible repressor, thus giving rise to the insensitive dwarf phenotype of *gai-1*.

Since Arabidopsis contains five homologs, mutations that result in loss of function of several of the homologs are necessary before a GA-constitutive phenotype (i.e., tallness) is manifest.

The alternative phenotypes are seen in the *slender* mutations in cereals. For example, wild-type rice transformed with *SLR1*, which encodes a protein containing a 17–amino

*sln1c*      *WT*      *sln1d*

**FIGURE 20.17**   Demonstration of the opposite effects of two different mutations in the same *SLN1* repressor gene. Three shoots of 2-week-old barley seedlings are shown. Center: The wild-type seedling (*WT*). Left: The *sln1c* mutant has a GA-constitutive, slender phenotype associated with loss-of-function of the repressor protein due to a mutation in the GRAS repressor domain. Right: The *sln1d* mutant is a gain-of-function dominant dwarf, because a mutation in the DELLA domain prevents the repressor protein from being degraded. (From Chandler et al. 2002; courtesy of P. M. Chandler.)

acid deletion in the DELLA domain, gives rise to a GA-insensitive dwarf phenotype instead of the slender phenotype seen when mutations are in the GRAS domain (Ikeda et al. 2001). Similarly, as shown in **FIGURE 20.17**, the *sln1d* mutation in barley is a *d*ominant, insensitive dwarf, in contrast to the original slender *sln1c* mutant with a *c*onstitutive GA response (slender) phenotype (Chandler et al. 2002).

### Gibberellins signal the degradation of negative regulators of GA response

DELLA-domain proteins are nuclear-localized negative regulators of GA action that must be degraded in order for GA response to occur. To demonstrate the nuclear localization of RGA, Arabidopsis plants were transformed with the *RGA* gene fused to the gene encoding green fluorescent protein (GFP), so that the transgenic plants produce an RGA–GFP fusion protein. Expression of *GFP* was controlled by the promoter for *RGA* so that detection of the RGA–GFP fusion protein would mimic that of native RGA (Silverstone et al. 2001). The effect of bioactive GA on levels and localization of the fusion protein was determined by immunoblotting and fluorescence microscopy. Roots were used because the autofluorescence of chlorophyll makes GFP analysis in shoots more difficult.

RGA–GFP was localized in nuclei of Arabidopsis root tips, and when the plants were treated with bioactive GA, the GFP fluorescence disappeared. In contrast, if the GA content of roots was depleted by treatment with the GA biosynthesis inhibitor paclobutrazol, the nuclei

retained intense fluorescence (**FIGURE 20.18**) Since GFP reflects what is happening to RGA, we can conclude from these experiments that in the presence of bioactive GA the nuclear-localized RGA is degraded, whereas it is not degraded when GA levels are very low.

As we learned on page 599, bioactive GA is an allosteric activator of its receptor, GID1. In a GA-deficient plant, there will not be enough GA to bind GID1, and allow the conformational change in the GID1 protein that closes the "lid." Without this conformational change, a DELLA protein is unable to bind GID1, and as we shall now see, a DELLA protein needs to bind GID1 in order to be degraded. We can now also understand why a mutation in a DELLA protein that affects the N-terminal DELLA domain causes the mutated protein to be an irreversible repressor, since it is that part of the DELLA protein that binds the GA–GID1 receptor complex (see Figure 20.15)

### F-box proteins target DELLA domain proteins for degradation

Dwarf mutants of rice and Arabidopsis have been identified that are defective in their abilities to degrade wild-type DELLA proteins in the presence of bioactive GA. The *GA-insensitive dwarf 2* (*gid2*) and *sleepy 1* (*sly1*) recessive mutations are in orthologous proteins in rice and Arabidopsis respectively, and give rise to GA-insensitive dwarf phenotypes (Sasaki et al. 2003; McGinnis et al. 2003). These phenotypes suggest that the wild-type *GID2* and *SLY1* genes encode *positive* regulators of GA signaling (i.e., these are type 1 mutants, see page 596). In fact, both genes encode proteins with conserved F-box domains that are components of **ubiquitin E3 ligase complexes**, also known as SCF complexes (Dill et al. 2004; Itoh et al. 2003) (see also Chapter 14). For a protein (such as a DELLA protein) to be degraded by the 26S proteasome, it must be "tagged" for degradation by the attachment of a string of ubiquitin residues. This tagging occurs when F-box components recruit proteins into the SCF complex, leading to the addition of many ubiquitin molecules to the protein and its subsequent degradation by the 26S proteasome. (Further information on this topic can be found in **WEB ESSAY 20.2** and in Chapters 2 and 14.)

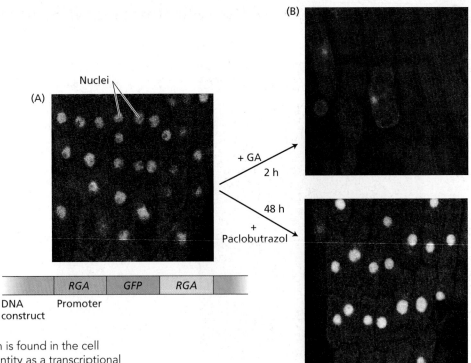

FIGURE 20.18  The RGA protein is found in the cell nucleus, consistent with its identity as a transcriptional regulator, and its level is affected by the level of GA. (A) Plant cells were transformed with the gene for RGA fused to the gene for green fluorescent protein (GFP), allowing detection of RGA in the nucleus by fluorescence microscopy. (B) A 2-hour pretreatment with GA causes the loss of RGA from the nucleus (top). When GA biosynthesis is inhibited by a 48-hour treatment with paclobutrazol (a GA biosynthesis inhibitor), the RGA content in the nucleus increases (bottom). These micrographs show that RGA is degraded in the presence of GA, but not in its absence. (From Silverstone et al. 2001.)

The orthologous proteins GID2 (in Arabidopsis) and SLY1 (in rice) target the DELLA proteins for degradation, and this targeting is enhanced when the DELLA protein is bound to the bioactive GA–GID1 complex (FIGURE 20.19). Just as binding bioactive GA causes a conformational change in GID1 so that it can bind the DELLA protein, the binding of the N-terminal DELLA domain to the GA–GID1 complex leads to a conformational (coil-to-helix) change in the DELLA domain. It has been hypothesized that this could, in turn, induce a conformational change in the GRAS domain that would favor binding to the F-box protein. For this reason the GA–GID1 complex has been termed a ubiquitinylation chaperone that stimulates recognition of the repressor protein by the SCF complex (Murase et al. 2008).

### Negative regulators with DELLA domains have agricultural importance

The experiments described above provide a considerable body of evidence that the DELLA-domain proteins are important negative regulators of GA response in both dicots and monocots. Additionally, in wheat the semidominant *Reduced height* mutations (*Rht-B1b* and *Rht-D1b*), and in maize the *d8* mutation, are all in the DELLA domains of the respective *GAI* orthologs, and these mutations give rise to GA-insensitive dwarf plants (Peng et al. 1999).

Classical breeding of wheat and rice in the early and middle of the twentieth century paved the way for the "Green Revolution" of the 1960s—the introduction of high-yielding dwarf varieties of wheat and rice into Latin America and Southeast Asia to keep abreast of human population growth (Hedden 2003). Normal cereals grow too tall when close together in a field, especially with high levels of fertilizer. The result is lodging (collapse of plants in wind and rain), with unacceptably large losses in yield. The dissemination of wheat cultivars expressing *Rht* mutations, with reduced GA response and shortened stems, was pivotal to the success of the Green Revolution. (More details on Dr. Norman Borlaug's Nobel Prize–winning work can be found in **WEB ESSAY 20.3**.) In barley, new mutant alleles in orthologs of the Green Revolution genes of wheat give potentially useful agronomic traits, as described in **WEB ESSAY 20.4**.

## Gibberellin Responses: Early Targets of DELLA Proteins

As you know from the earlier part of this chapter, GAs can affect many aspects of plant growth and development ranging from seed germination to flower and fruit development. The challenge, recently, has been to relate our newfound understanding of GA perception, at a molecular level, with downstream events that lead to changes in growth and

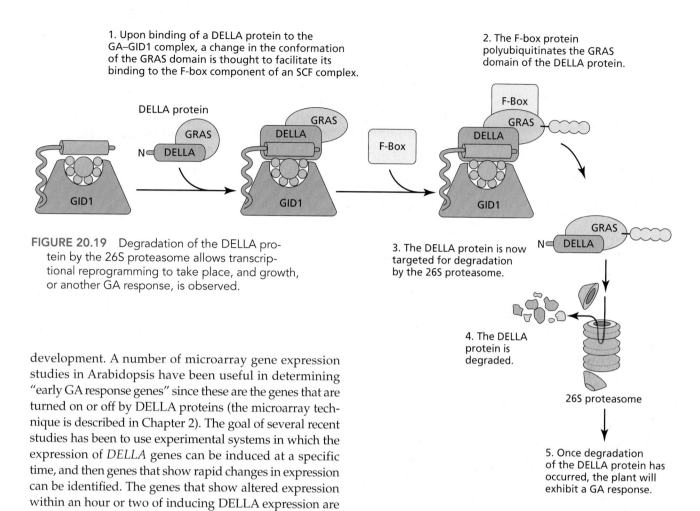

1. Upon binding of a DELLA protein to the GA–GID1 complex, a change in the conformation of the GRAS domain is thought to facilitate its binding to the F-box component of an SCF complex.

2. The F-box protein polyubiquitinates the GRAS domain of the DELLA protein.

3. The DELLA protein is now targeted for degradation by the 26S proteasome.

4. The DELLA protein is degraded.

26S proteasome

5. Once degradation of the DELLA protein has occurred, the plant will exhibit a GA response.

**FIGURE 20.19** Degradation of the DELLA protein by the 26S proteasome allows transcriptional reprogramming to take place, and growth, or another GA response, is observed.

development. A number of microarray gene expression studies in Arabidopsis have been useful in determining "early GA response genes" since these are the genes that are turned on or off by DELLA proteins (the microarray technique is described in Chapter 2). The goal of several recent studies has been to use experimental systems in which the expression of *DELLA* genes can be induced at a specific time, and then genes that show rapid changes in expression can be identified. The genes that show altered expression within an hour or two of inducing DELLA expression are likely to be early GA response genes.

### DELLA proteins can activate or suppress gene expression

Early GA response genes downstream of DELLA proteins have been identified in seedlings and inflorescences of Arabidopsis. In both experimental systems, several of the immediate targets of DELLA proteins are genes encoding GA biosynthetic enzymes or the GA receptor, implying strong homeostatic regulation to maintain GA levels and GA response within physiological limits. Other immediate targets of DELLA proteins are genes encoding transcription factors or transcriptional regulators. Although some of these DELLA targets in seedlings and in inflorescences belong to the same *classes* of transcriptional regulators—for example MYB and basic helix-loop-helix (bHLH) classes—what is remarkable is that there appears to be very little overlap in the individual genes regulated by DELLA proteins in seedlings and in inflorescences (Zentella et al. 2007; Hou et al. 2008). These observations may help to explain the specificity of GA action, and how a single hormone can lead to so many different types of responses depending on the developmental stage or target tissue.

### DELLA proteins regulate transcription by interacting with other proteins such as phytochrome-interacting factors

Since DELLA proteins do not have any recognizable DNA-binding domains, how can they activate and repress genes? It is possible that additional factors may be necessary to allow DELLA proteins to bind DNA, or that DELLA proteins regulate transcription through interaction with other transcription factors, rather than binding DNA themselves. There is recent evidence for the latter scenario; DELLA proteins interact with **Phytochrome-interacting factors** (**PIFs**), which are a type of bHLH transcription factor. When a DELLA protein binds to a PIF, it prevents the PIF from activating gene transcription; thus the target genes of a PIF are (indirectly) down-regulated by DELLA proteins.

There are at least five PIFs in Arabidopsis, which have some distinct and some overlapping functions. Several PIFs affect seedling growth, and are involved in the transition from skotomorphogenesis (growth and development in complete darkness) to photomorphogenesis (growth and development in light). In the dark, Arabidopsis seedlings are etiolated—the hypocotyl elongates and the cotyledons fail to open and expand. These effects are, in part, a conse-

(A) Dark/high GA ⟶ long hypocotyl

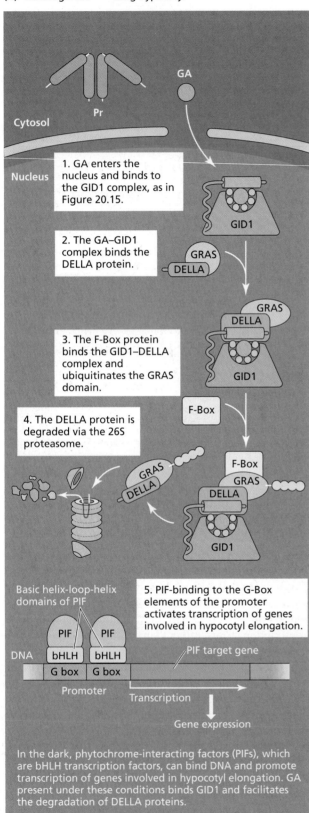

1. GA enters the nucleus and binds to the GID1 complex, as in Figure 20.15.

2. The GA–GID1 complex binds the DELLA protein.

3. The F-Box protein binds the GID1–DELLA complex and ubiquitinates the GRAS domain.

4. The DELLA protein is degraded via the 26S proteasome.

5. PIF-binding to the G-Box elements of the promoter activates transcription of genes involved in hypocotyl elongation.

Basic helix-loop-helix domains of PIF

In the dark, phytochrome-interacting factors (PIFs), which are bHLH transcription factors, can bind DNA and promote transcription of genes involved in hypocotyl elongation. GA present under these conditions binds GID1 and facilitates the degradation of DELLA proteins.

(B) Light/low GA ⟶ short hypocotyl

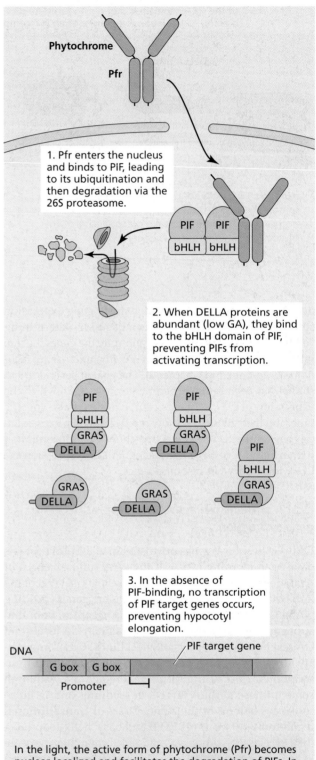

1. Pfr enters the nucleus and binds to PIF, leading to its ubiquitination and then degradation via the 26S proteasome.

2. When DELLA proteins are abundant (low GA), they bind to the bHLH domain of PIF, preventing PIFs from activating transcription.

3. In the absence of PIF-binding, no transcription of PIF target genes occurs, preventing hypocotyl elongation.

In the light, the active form of phytochrome (Pfr) becomes nuclear-localized and facilitates the degradation of PIFs. In addition, because of low GA under these conditions, DELLA proteins are not degraded. They bind to PIFs and prevent them from binding DNA. For both of these reasons, PIF-induced gene transcription is prevented and hypocotyls do not elongate.

**FIGURE 20.20** Integration of light and GA signaling in Arabidopsis seedlings controls hypocotyl length.

quence of PIFs activating the transcription of genes whose products lead to elongated hypocotyls (**FIGURE 20.20A**). For example, the bHLH domains of PIF3 and PIF4 bind to G-box elements in the promoters of target genes, such as the gene encoding expansin, which is a cell wall–loosening protein. In the light, photoactivated phytochrome moves into the nucleus and binds these PIFs, leading to their degradation. As a consequence their target genes (such as the one encoding expansin) are down-regulated and hypocotyl elongation is inhibited (**FIGURE 20.20B**).

Recently both PIF3 and PIF4 have been shown to bind DELLA proteins. This interaction is, in part, via the bHLH domain of the PIF protein, which is the domain needed for the PIF to bind DNA (de Lucas et al. 2008; Feng et al. 2008). When the PIF protein can no longer bind DNA, the PIF target genes are not transcribed (see Figure 20.20B), and hypocotyls do not elongate. This interaction of PIF and DELLA proteins occurs only when DELLA proteins are abundant. In the presence of bioactive GA, DELLA proteins are degraded so they cannot bind PIFs; the PIFs can activate transcription, and seedlings have long hypocotyls (see Figure 20.20A). Thus DELLA proteins have been shown to regulate gene transcription by binding directly to bHLH transcription factors, rather than to DNA, and the GA effect (long hypocotyls) is a consequence of PIF-regulated genes.

# Gibberellin Responses: The Cereal Aleurone Layer

In this section and the following two sections, we discuss examples of GA action at three different stages in the life cycle of a plant. First, we discuss the classic model system, the cereal aleurone layer in germinating cereal grain. We then turn to the development of anthers, also in cereals. In both the aleurone and the anther systems, MYB transcription factors up-regulate genes known to encode enzymes that implement the GA response. Finally we will discuss internode elongation and its suppression in deep-water rice.

Cereal grains have three parts: the embryo, the endosperm, and the fused testa–pericarp (seed coat–fruit wall) (**FIGURE 20.21**). The embryo, which will grow into the new seedling, has a specialized absorptive organ, the scutellum. The endosperm is composed of two tissues: the centrally located starchy endosperm and the **aleurone layer** (see Figure 20.21). The nonliving starchy endosperm consists of thin-walled cells filled with starch grains. Living cells of the aleurone layer, which surrounds the endosperm, synthesize and release hydrolytic enzymes into the endosperm during germination. As a consequence, the stored food reserves of the endosperm are broken down, and the solubilized sugars, amino acids, and other products are transported to the growing embryo. The isolated

aleurone layer, consisting of a homogeneous population of target cells for GA, provides a unique opportunity to study the molecular aspects of GA action in the absence of nonresponding cell types.

Two enzymes responsible for starch degradation are α- and β-amylase. α-Amylase (of which there are several isoforms) hydrolyzes starch chains internally to produce oligosaccharides consisting of α-1,4-linked glucose residues. β-Amylase degrades these oligosaccharides from the ends to produce maltose, a disaccharide. Maltase then converts maltose to glucose.

## GA is synthesized in the embryo

Experiments carried out in the 1960s confirmed Gottlieb Haberlandt's original 1890 observation that the secretion of starch-degrading enzymes by barley aleurone layers depends on the presence of the embryo. It was soon discovered that $GA_3$ could substitute for the embryo in stimulating starch degradation. The significance of the GA effect became clear when it was shown that the embryo synthesizes and releases GAs into the endosperm during germination.

## Aleurone cells may have two types of GA receptors

Rice mutants (*gid1*) that have a defective GA receptor are unable to synthesize α-amylase (Ueguchi-Tanaka et al. 2005), clearly implicating the soluble GA receptor in this classical GA response. Other evidence, which was obtained before the characterization of GID1, suggested that GA may bind to a protein in the plasma membrane of aleurone cells. Evidence for two types of hormone receptors has been obtained for auxin (see Chapter 19) and abscisic acid (see Chapter 23). Given the great diversity of GA responses, the existence of multiple receptors may not be too surprising, though at the present time there is no definitive identification of a plasma membrane–localized GA receptor. Within the aleurone cells there are both $Ca^{2+}$-independent and $Ca^{2+}$-dependent GA signaling pathways. The former leads to the production of α-amylase, while the latter regulates its secretion.

## Gibberellins enhance the transcription of α-amylase mRNA

Even before molecular biological approaches were developed, there was already physiological and biochemical evidence that GA enhanced α-amylase production at the level of gene transcription (Jacobsen et al. 1995). The two main lines of evidence were:

- $GA_3$-stimulated α-amylase production was shown to be blocked by inhibitors of transcription and translation.

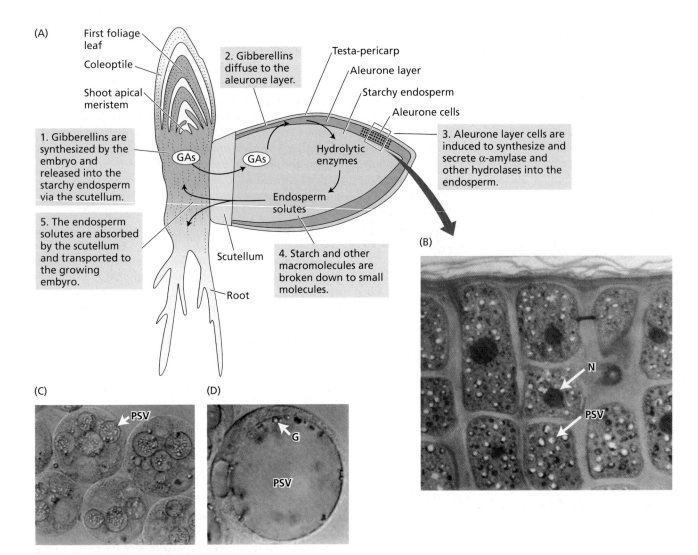

(A)

First foliage leaf

Coleoptile

Shoot apical meristem

2. Gibberellins diffuse to the aleurone layer.

Testa-pericarp

Aleurone layer

Starchy endosperm

Aleurone cells

1. Gibberellins are synthesized by the embryo and released into the starchy endosperm via the scutellum.

GAs

GAs

Hydrolytic enzymes

3. Aleurone layer cells are induced to synthesize and secrete α-amylase and other hydrolases into the endosperm.

5. The endosperm solutes are absorbed by the scutellum and transported to the growing embyro.

Endosperm solutes

Scutellum

4. Starch and other macromolecules are broken down to small molecules.

Root

(B)

N

PSV

(C)

PSV

(D)

G

PSV

**FIGURE 20.21** Structure of a barley grain and the functions of various tissues during germination. (A) Diagram of germination-initiated interactions. (B–D) Micrographs of the barley aleurone layer (B) and barley aleurone protoplasts at an early (C) and a late (D) stage of amylase production. Multiple protein storage vesicles (PSV) in (C) coalesce to form a large vesicle in (D), which will provide amino acids for α-amylase synthesis. G, phytin globoid that sequesters minerals; N, nucleus. (B–D from Bethke et al. 1997; courtesy of P. Bethke.)

- Isotope-labeling studies demonstrated that the stimulation of α-amylase activity by bioactive GA involved de novo synthesis of the enzyme from amino acids, rather than activation of preexisting enzyme.

Cereal grains can be cut in two, and half-grains that lack the embryo (the source of bioactive GA in intact grain) make a good experimental system for studying the action of applied GA. Microarray studies have confirmed the up-regulation of genes encoding several α-amylase isoforms in rice embryoless half-grains that have been treated for 8 hours with GA₃ (Bethke et al. 2006; Tsuji et al. 2006). In these half-grains the only living cells, and the only cells in which GA signaling occurs, are in the aleurone layer. Of all genes in the microarray analyses, those encoding α-amylase isoforms show the highest degree of up-regulation after GA treatment, followed closely by other hydrolases and proteases.

The purification of *α-amylase* mRNA, which is produced in relatively large amounts in aleurone cells, enabled the isolation of genomic clones containing both the structural gene for α-amylase and its upstream promoter sequences. Sequences conferring GA responsiveness, termed **GA response elements (GAREs)**, are located 200 to 300 base pairs upstream of the transcription start site (Gubler et al. 1995). Identical GAREs have been found in all cereal *α-amylase* promoters so far examined, and their presence has been shown to be necessary and sufficient for the induction of *α-amylase* gene transcription by GA.

## GAMYB is a positive regulator of α-amylase transcription

The sequence of the GARE in the α-*amylase* gene promoter (TAACAAA) is similar to the DNA sequence to which **MYB proteins** bind. MYB proteins are a class of transcription factors in eukaryotes. In plants there is a large MYB family that is divided into subgroups based on structural features of the proteins. In barley and rice, one MYB protein in the R2R3 subgroup has been implicated in GA signaling and so has been given the name **GAMYB**. The sequence of GAMYB in barley is quite similar to that of three MYB proteins in Arabidopsis (AtMYB33, AtMYB65, and AtMYB101). In fact, the structures of these AtMYBs are so similar to the cereal GAMYB that a barley mutant lacking GAMYB can be "rescued" by any of these AtMYBs (Gocal et al. 2001). In rice two additional GAMYB-like proteins have been identified, but they are not known to function in GA signaling in aleurone cells.

The hypothesis that GAMYB turns on α-*amylase* gene expression (i.e., that GAMYB is a *positive regulator* of α-*amylase*) is supported by the following findings:

- Synthesis of *GAMYB* mRNA begins to increase as early as 1 hour after GA treatment, preceding the increase in α-*amylase* mRNA by several hours (**FIGURE 20.22**).

- A mutation in the GARE that prevents MYB binding also prevents α-*amylase* expression.

- In the absence of GA, constitutive expression of GAMYB can induce the same responses that GA induces in aleurone cells, showing that GAMYB is necessary and sufficient for the enhancement of α-*amylase* expression.

Cycloheximide, an inhibitor of translation, has no effect on the production of *GAMYB* mRNA, indicating that protein synthesis is not required for *GAMYB* expression. *GAMYB* can therefore be defined as a **primary** or **early response gene**. In contrast, similar experiments show that the α-*amylase* gene is a **secondary** or **late response gene**.

## DELLA-domain proteins are rapidly degraded

How does GA application lead to the transcriptional activation of α-*amylase* by the GAMYB transcription factor? It is known that within 5 minutes of GA application to barley aleurone cells there is an effect on the DELLA protein, SLN1. Levels of a GFP-SLN1 fusion protein in nuclei decline, and indeed the protein is completely gone within 10 minutes of GA treatment (Gubler et al. 2002). However, GAMYB levels do not increase for 1–2 hours, suggesting that even though *GAMYB* is defined as an early response gene, it is probably not the *direct* target of the DELLA protein. We think that there are one or more unidentified steps between SLN1 degradation and *GAMYB* transcription.

Drawing together our information for the cereal aleurone system (**FIGURE 20.23**), we can hypothesize that the binding of bioactive GA to GID1 leads to degradation of the DELLA protein. As a consequence of DELLA degradation, and via some intermediary steps that have not yet been defined, the expression of *GAMYB* is up-regulated. Finally, the GAMYB protein binds to a highly conserved GARE in the promoter of the gene for α-amylase, activating its transcription. Alpha-amylase is secreted from aleurone cells by a pathway that requires $Ca^{2+}$ accumulation. Starch breakdown occurs in cells of the endosperm by the action of α-amylase and other hydrolases, and sugars are transported to the growing embryo.

Some of the genes encoding other hydrolytic enzymes whose synthesis is promoted by GA also have GAMYB binding motifs in their promoters, indicating that this is a common pathway for GA response in aleurone layers.

# Gibberellin Responses: Anther Development and Male Fertility

We have seen that the GAMYB transcription factor activates α-*amylase* gene expression during the GA response of cereal aleurone layers, leading to starch degradation in germinating grain. GAMYB is involved in other GA responses too. The effect of GA on flowering is discussed in **WEB TOPIC 20.8**, and, as we shall see here, the effect of bioactive GA on pollen development is also mediated by GAMYB.

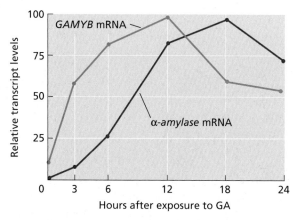

**FIGURE 20.22** Time course for the induction of *GAMYB* and α-*amylase* mRNA by $GA_3$. The production of *GAMYB* mRNA precedes that of α-*amylase* mRNA by about 3 hours. These and other results indicate that *GAMYB* is an early GA response gene that regulates the transcription of the gene for α-amylase. In the absence of GA, the levels of both *GAMYB* and α-*amylase* mRNAs are negligible. (After Gubler et al. 1995.)

1. $GA_1$ from the embryo enters an aleurone cell.

2. Once inside the cell, $GA_1$ may initiate a calcium-calmodulin–dependent pathway necessary for α-amylase secretion.

3. $GA_1$ binds to GID1 (a soluble GA receptor) in the nucleus.

4. Upon binding $GA_1$, the GID1 receptor undergoes an allosteric change that facilitates its binding to a DELLA protein.

5. Once the DELLA protein has bound the $GA_1$–GID complex, an F-box protein (part of an SCF complex) is now able to poly-ubiquitinate the GRAS domain of the DELLA protein.

6. The polyubiquitinated DELLA protein is degraded by the 26S proteasome.

7. Once the DELLA protein has been degraded, transcription of an early gene is activated. (GAMYB is shown, in this model, as an early gene, although there is evidence that transcriptional regulation of other early genes may occur first.) The mRNA for GAMYB is translated in the cytosol.

8. The newly synthesized GAMYB transcription factor enters the nucleus and binds the promoters of α- amylase and genes encoding other hydrolytic enzymes.

9. Transcription of these genes is activated.

10. α-amylase and other hydrolases are synthesized on the rough ER, processed, and packaged into secretion vesicles by the Golgi body.

11. Proteins are secreted by exocytosis.

12. The secretory pathway requires GA stimulation of the calcium-calmodulin–dependent pathway.

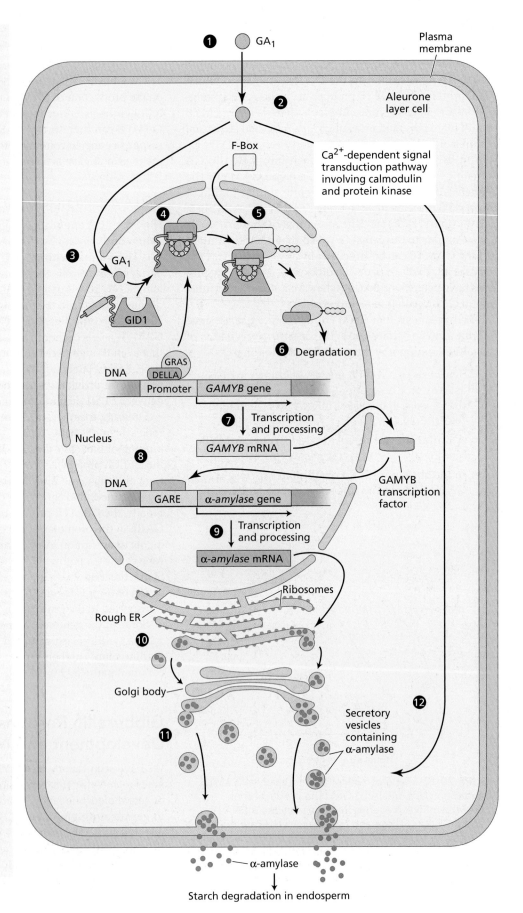

◀FIGURE 20.23   Composite model for the induction of α-amylase synthesis in barley aleurone layers by GA. A calcium-independent pathway induces *α-amylase* gene transcription; a calcium-dependent pathway is involved in α-amylase secretion.

## GAMYB regulates male fertility

Loss-of-function mutants of *GAMYB* in rice have revealed the role of this transcription factor as a positive regulator of floral organ development (Kaneko et al. 2004). Rice plants lacking a functional *GAMYB* gene (*gamyb*) develop flowers with shrunken, white (instead of yellow) anthers that contain no pollen. In some flowers, even on the same plant, phenotypic abnormalities are more pronounced, including grossly malformed pistils in the most severe cases (**FIGURE 20.24**).

Rice pollen grains develop from the innermost layer of the anther and are nourished by a secretory tissue called the tapetum. Tapetal cells form protrusions called orbicules (also known as Ubisch bodies) that are thought to function in the release of *sporopollenin*. **Sporopollenin** is a complex polymer consisting of fatty acid derivatives and phenylpropanoids that eventually forms the resilient outer wall of the pollen grains. As we shall see, bioactive GAs up-regulate the biosynthesis of some of these compounds. Wild-type tapetal cells normally undergo programmed cell death when microspores (which develop into pollen grains) are at the tetrad stage of development (**FIGURE 20.25**). If tapetal cell death occurs too soon, or if it does not occur at all, then viable pollen may not be produced. In *gamyb* mutants tapetal cells do not degenerate at the tetrad stage but continue to enlarge until they fill the anther cavity at the stage that is referred to in wild-type plants as the mature pollen stage. There are no mature pollen grains in the *gamyb* anthers because, in addition to this proliferation of tapetal cells, the microspores collapse between the tetrad and vacuolated pollen stages (see Figure 20.25) (Aya et al. 2009).

In wild-type plants, bioactive GA is involved in regulating programmed cell death of tapetal cells. In addition, GA regulates the biosynthesis of sporopollenin components that provide a tough outer coat for pollen grains. There are several lines of evidence:

- GA up-regulates the expression of genes that encode two enzymes in lipid metabolism during pollen development. The genes, *CYP703A3* and *KAR*, have GAMYB binding sites in their promoters, and alteration of these binding sites by mutation prevents their expression.

- Reporter line studies showed that GAMYB–GUS (a translational fusion to measure GAMYB protein levels) and CYP703A3:GUS (a promoter:GUS fusion to measure the transcription of *CYP703A3*) are co-expressed in wild-type anthers, but CYP703A3:GUS is not expressed in the *gamyb* mutant, in which there is no GAMYB protein (**FIGURE 20.26**).

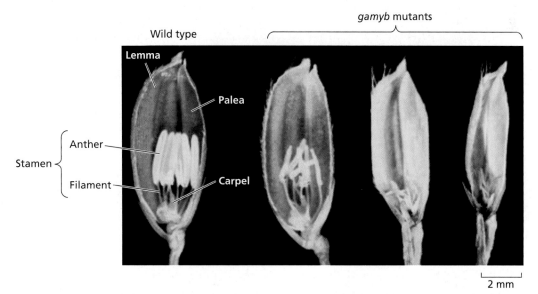

FIGURE 20.24   Phenotypes of floral organs of the *gamyb* mutants of rice, showing that normal pollen development requires the GAMYB transcription factor to be present and functional. A wild-type flower is shown on the far left, and progressing to the right are shown increasing severities of the *gamyb* mutant phenotype. In the absence of a functional *GAMYB* gene, the anthers are white instead of yellow, and lack pollen. More extreme phenotypes display white, shriveled bracts (lemma and palea) surrounding the flower and malformed carpels. (From Kaneko et al. 2004; courtesy of M. Matsuoka.)

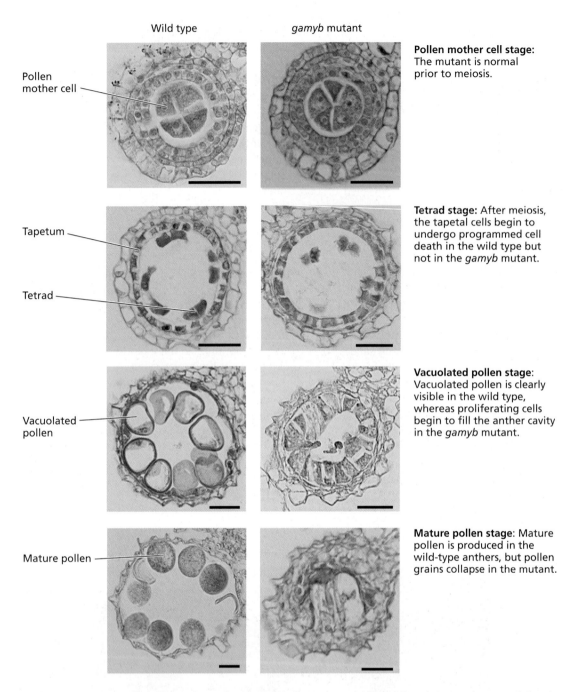

Wild type    *gamyb* mutant

Pollen mother cell

**Pollen mother cell stage:** The mutant is normal prior to meiosis.

Tapetum

Tetrad

**Tetrad stage:** After meiosis, the tapetal cells begin to undergo programmed cell death in the wild type but not in the *gamyb* mutant.

Vacuolated pollen

**Vacuolated pollen stage:** Vacuolated pollen is clearly visible in the wild type, whereas proliferating cells begin to fill the anther cavity in the *gamyb* mutant.

Mature pollen

**Mature pollen stage:** Mature pollen is produced in the wild-type anthers, but pollen grains collapse in the mutant.

**FIGURE 20.25** Histochemical analysis of anther development in wild-type and *gamyb* mutant rice plants at four different stages. Transverse sections of anthers show that tapetal cells in the mutant fail to undergo programmed cell death at the tetrad stage of development. In addition the resilient outer wall of pollen grains fails to develop in the mutant, and pollen grains collapse. Bars = 25 μm. (From Aya et al. 2009, courtesy of M. Matsuoka.)

These and other findings provide compelling evidence that in rice the formation of functional pollen grains requires GA-induced expression of GAMYB, which turns on transcription of genes involved in the biosynthesis of pollen wall components. In the absence of bioactive GA, or of any of the components of its signal transduction pathway, rice is male sterile.

In Arabidopsis, as in rice, GA-deficient mutants or those with defects in GA signaling are male sterile. Both *AtMYB33* and *AtMYB65* must be defective for male sterility to occur, because of functional redundancy between these two genes (Millar and Gubler 2005). The ortholog of *CYP703A3* in Arabidopsis also contains a MYB binding motif in its promoter, suggesting that in Arabidopsis as

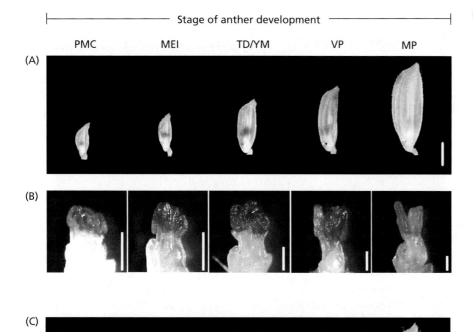

Stage of anther development

PMC MEI TD/YM VP MP

(A)

(B)

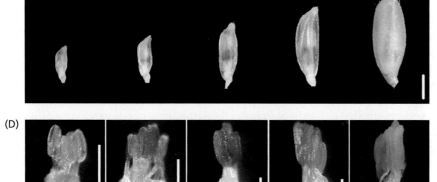

(C)

(D)

FIGURE 20.26 GUS expression under the control of the promoters of GA-inducible genes in rice anthers. (A and B) Expression of GAMYB–GUS in the wild type. GAMYB–GUS staining in flowers is localized in the anthers (B). (C and D) Expression of CYP703A3:GUS in the wild type. CYP703A3:GUS staining in the flowers is also localized to the anthers (D). (E) CYP703A3:GUS in the *gamyb* mutant at MEI stage of anther development. (A and C) Whole flowers at each developmental stage (bars = 1 mm). (B, D, and E) Close-up views of stamens. PMC, pollen mother cell stage; MEI, meiosis; TD/YM, tetrad/young microspore; VP, vacuolated pollen stage; MP, mature pollen stage. (From Aya et al. 2009; courtesy of M. Matsuoka.)

CYP703A3:GUS staining is shown to be in the same location as GAMYB–GUS. But, unlike GAMYB–GUS, it is not visible at the earliest developmental stage. This result indicates that expression of GAMYB may be required before CYP703A3 expression.

(E)

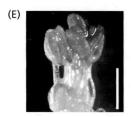

In a *gamyb* mutant there is no expression of CYP703A3:GUS, again indicating that GAMYB may be necessary for the expression of CYP703A4.

well as in rice GA regulates the synthesis of pollen wall components.

## Events downstream of GAMYB in rice aleurone and anthers are quite different

A comparison of GAMYB-induced gene expression as determined by microarray analyses of embryoless half-grains and anthers of rice has revealed some very surprising results: Although GAMYB regulates many genes in both aleurone layers and in anthers, the GAMYB-regulated genes in the two systems are quite different (Tsuji et al. 2006). Although many of the GAMYB-dependent genes in anthers do contain the conserved GA response element (GARE) first characterized in GAMYB-regulated genes in the aleurone, it appears that this GARE may not be sufficient for GAMYB-induced expression in anthers, and that additional regulatory factors are involved. In fact, the promoters of GAMYB-regulated genes in anthers lack some of the *cis*-acting elements present in the promoters of GAMYB-regulated genes in aleurone cells. One can postulate that other elements in the promoters of GAMYB-regulated genes in anthers may provide an additional level of specificity in GA-induced, GAMYB-regulated gene expression in the two systems.

## MicroRNAs regulate MYBs after transcription in anthers but not in aleurone

**MicroRNAs (miRNAs)** are small gene-encoded RNAs containing about 20 to 25 nucleotides that regulate the expression of other genes posttranscriptionally. MiRNAs bind to ribonuclease complexes and serve as guides to target the ribonuclease to specific mRNA molecules. The mRNAs that hybridize to these miRNA guides are quickly digested by the ribonuclease complex, thus effectively silencing the expression of the target genes. (For a more detailed discussion of miRNAs, see Chapter 2.)

Three miRNAs in Arabidopsis have been shown to be complementary to a conserved motif in the coding regions of *GAMYB-like* genes (Park et al. 2002; Rhoades et al. 2002). One of them, *miR159a*, can silence *AtMYB33* in Arabidopsis shoots, and overexpressing or mutating *miR159a* can cause striking defects in plants. *AtMYB33* is involved in floral organ development and flowering in Arabidopsis. Thus *miR159a* overexpressers, in which *AtMYB33* is silenced, are male-sterile and have delayed flowering (Achard et al. 2004). MicroRNAs can alter *GAMYB* expression in rice anthers, but not in aleurone (Tsuji et al. 2006). This shows another difference between GA-induced, GAMYB-regulated gene expression in the aleurone and in anthers.

Taken together, the work on both GA-induced α-amylase production and GA-induced floral organ development all support a role for MYB transcription factors as important intermediates of GA signaling in cereals and in Arabidopsis. Confirmation of their importance in other dicotyledonous plants in addition to Arabidopsis can be anticipated. There are intriguing differences in the signaling downstream from GAMYBs in different tissues and organs that may determine how it is that a single GA can induce different biological responses in different target tissues.

# Gibberellin Responses: Stem Growth

The effects of GAs on stem growth can be so dramatic that one might imagine it should be simple to determine how they act (see Figures 20.1 and 20.2). Unfortunately, this is not the case, because—as we have seen with auxin—much about plant cell growth is poorly understood. However, we do know some basic characteristics of GA-induced stem elongation. In the final section of this chapter we examine studies aimed at elucidating the physiological and biochemical mechanisms for GA-stimulated cell elongation and cell division in rice.

Rapid stem elongation occurs in rice in response to flash flooding, when the water level rises very quickly. However this tends to deplete carbohydrate reserves in the plants, and decreases their survival after the flooding subsides. Interestingly, submergence-tolerant lines of rice, which elongate less when underwater, have enhanced survival rates after flooding. This submergence-tolerance is directly linked to the persistence of DELLA proteins, which reduces GA sensitivity.

## Gibberellins stimulate cell elongation and cell division

Gibberellins stimulate both cell elongation and cell division, as evidenced by increases in cell length and cell number, in response to applications of bioactive GAs. Certain plants or plant organs associated with GA-induced growth display greater cell numbers than their respective control plants.

- Internodes of tall peas have more cells than those of dwarf peas, and the cells are longer.
- Mitosis increases markedly in the rib or ground meristem of rosette long-day plants growing in short days, after treatment with bioactive GA.
- The dramatic stimulation of internode elongation in deep-water rice when submerged, or when treated with GA, is due in part to increased cell division activity in the intercalary meristem found in specific monocots.

Because GA-induced cell elongation appears to precede GA-induced cell division, we begin our discussion with the role of GA in regulating cell elongation.

As discussed in Chapter 15, the rate of cell elongation can be influenced by both cell wall extensibility and the osmotically driven rate of water uptake. Gibberellin has no effect on the osmotic parameters but has consistently been observed to cause an increase in both the mechanical extensibility of cell walls and the stress relaxation of the walls of living cells. An analysis of pea genotypes differing in $GA_1$ content or sensitivity showed that GA decreases the *wall yield threshold* (the minimum force that will cause wall extension). Thus, both GA and auxin seem to exert their effects by modifying cell wall properties.

In the case of auxin, cell wall loosening appears to be mediated, in part, by cell wall acidification (see Chapter 19). However, this does not appear to be the mechanism of GA action, since GA-stimulated increase in proton extrusion has not been demonstrated. On the other hand, GA is never present in tissues in the complete absence of auxin, and the effects of GA on growth may depend on auxin-induced wall acidification.

The typical lag time before GA-stimulated growth begins is longer for GA than it is for auxin; in deep-water rice it is about 40 minutes, and in peas it is 2 to 3 hours (Yang et al. 1996). These longer lag times point to a growth-promoting mechanism distinct from that of auxin. Consistent with the existence of a separate GA-specific wall-loosening mechanism, the growth responses to applied GA and auxin are additive.

Various hypotheses have been explored regarding the mechanism of GA-stimulated stem elongation, and all have some experimental support; as yet, however, none provides a clear-cut answer. For example, there is evidence

**FIGURE 20.27**  Transgenic rice plants harboring sense and antisense constructs of *OsEXP4*. Shown are antisense (A), control (i.e., not transgenic) (B), and sense (C). Height of the plants is related to the amount of expansin 4. Although the gene is transcribed in the antisense line, it is not translated, diminishing the level of expansin 4 in the plants relative to control. In the sense line the *OsEXP4* gene is overexpressed leading to more expansin 4. (From Choi et al. 2003; courtesy of H. Kende.)

that the enzyme **xyloglucan endotransglucosylase/hydrolase** (**XTH**) is involved in GA-promoted wall extension (Xu et al. 1996). The function of XTH may be to facilitate the penetration of expansins into the cell wall. Expansins are cell wall proteins that cause wall loosening in acidic conditions by weakening hydrogen bonds between wall polysaccharides (see Chapter 15). Transcript levels of one particular expansin in rice, encoded by *OsEXP4*, increase in deep-water rice within 30 minutes of GA treatment, or in response to rapid submergence, both of which induce growth. Moreover, plants expressing an antisense version of *OsEXP4* are shorter and do not grow in submerged conditions; also, overexpression of *OsEXP4* leads to taller rice plants (**FIGURE 20.27**) (Choi et al. 2003). Taken together, these results indicate that GA-induced cell elongation is at least in part mediated by the expansins.

## GAs regulate the transcription of cell cycle kinases

The dramatic increase in growth rate of deep-water rice internodes with submergence is due partly to increased cell divisions in the intercalary meristem. To study the effect of GA on the cell cycle, researchers isolated nuclei from the intercalary meristem of deep-water rice and quantified the

amount of DNA per nucleus (Sauter and Kende 1992). In submergence-induced plants, GA activates the transition from $G_1$ to S phase, leading to an increase in mitotic activity. The stimulation of cell division results from a GA-induced expression of the genes for several **cyclin-dependent protein kinases** (**CDKs**) (see Chapter 1). The transcription of these genes—first, those regulating the transition from $G_1$ to S phase, then those regulating the transition from $G_2$ to M phase—is induced in the intercalary meristem by GA (Fabian et al. 2000).

## Reducing GA sensitivity may prevent crop losses

Deep-water rice plants respond to gradual flooding with accelerated internode elongation, which keeps some aerial parts of the shoot above the water. But flash flooding prompts very rapid growth, which causes severe depletion of nutrients and often the death of the plants. *Submergence-tolerant* cultivars of lowland rice do not respond to flash flooding with such extreme shoot elongation; consequently they are better able to conserve carbohydrate resources and recover once the flooding subsides. These cultivars possess a *Sub1A-1* gene that is highly induced during submergence and when introduced into intolerant cultivars gives flash flood tolerance. This is because *Sub1A* is associated with increased expression of *SLR1* leading to the accumulation of this DELLA protein, and of *SLR-like1* (*SLRL1*), which encodes another closely related GRAS protein. Both transgenic plants which overexpress *Sub1A* constitutively, or plants in which *Sub1A* is highly induced in submerged leaves, have reduced GA sensitivity and do not experience suicidal levels of growth in response to rapid submergence (Fukao and Bailey-Serres 2008) (**FIGURE 20.28**).

The introduction of cultivars expressing *Sub1A-1* is expected to reduce devastating flash flood–induced losses of rice in flash flood–prone areas of Laos, India, and Bangladesh. Like the Green Revolution cultivars of wheat that were developed to prevent crop losses from lodging, these flash flood–tolerant lines of rice also rely on DELLA proteins to modulate shoot growth.

Over the past few years it has become apparent that DELLA proteins are important in integrating a plant's

**FIGURE 20.28** Either constitutive or submergence-induced expression of *Sub1A* gives rice plants that are better able to survive very rapid flooding. Photographs show constitutive (*Sub1A*-OE#1 and *Sub1A*-OE#2), submergence-induced [M202(*Sub1A*)], and submergence intolerant (Nontransgenic and M202) plants before submergence (top), after 16 days of submergence (middle) and 7 days after the end of submergence (bottom). All plants were at the same developmental stage at the beginning of the experiment. The submergence-intolerant lines show very marked leaf and internode elongation during submergence (middle photograph, first and fourth plants), which depletes resources and prevents the plants from recovering afterwards (bottom photograph). The other plants show reduced GA sensitivity and survive the flooding. (From Fukao and Bailey-Serres 2008; courtesy of J. Bailey-Serres).

Before submergence

After submergence (16 d)

After recovery (7d)

response to several hormones in addition to GA, and certain environmental factors. Ethylene, auxins, ABA, light, and temperature are all implicated in the DELLA signaling pathway, though some affect the pathway upstream of the DELLA protein, and some downstream. This aspect of hormone cross talk is discussed further in **WEB TOPIC 20.9.**

## SUMMARY

Gibberellins play important roles in regulating seed germination, shoot growth, transition to flowering, anther development, pollen tube growth, floral development, fruit set and subsequent growth, and seed development. All GAs share a similar chemical structure but relatively few of them have intrinsic biological activity.

### Gibberellins: Their Discovery and Chemical Structure

• Gibberellins are best known for their dramatic effects on internode elongation in grasses and in dwarf and rosette species (**Figures 20.1, 20.2**).

• Gibberellins are tetracyclic diterpenoid acids made up of four 5-carbon isoprenoid units. In most plants $GA_1$ and/or $GA_4$ are the GAs with highest biological activity.

### Effects of Gibberellins on Growth and Development

• GAs are necessary throughout the entire plant life cycle to promote seed germination, floral initiation, sex determination, pollen development, pollen tube growth, fruit set, and promotion of fruit growth (**Figure 20.3**).

## SUMMARY continued

- Gibberellin A$_3$ (gibberellic acid), which is obtained from the fungus *Gibberella fujikuroi*, is used commercially on several different types of crops and in the brewing industry (**Figure 20.4**).

- Inhibitors of GA biosynthesis are important dwarfing agents used in cereal crop production.

- Mutations that affect GA biosynthesis and response to active GA also produce dwarf plants.

### Biosynthesis and Deactivation of Gibberellins

- GA biosynthesis occurs in multiple plant organs and at multiple cellular sites.

- GAs are mobile and may act either locally or distant from their sites of synthesis, which are in germinating embryos, young seedlings, shoot apices, and developing seeds.

- GA biosynthesis proceeds in plastids with the conversion of geranylgeranyl diphosphate to *ent*-kaurene, continues in the ER for the production of GA$_{12}$, and is completed in the cytosol where active GA$_1$ and GA$_4$ are produced (**Figure 20.5**).

- Research using GA-deficient mutants has helped to establish the biosynthetic pathway (**Figures 20.6, 20.7**).

- Endogenous bioactive GA regulates its own synthesis by inhibiting or enhancing the transcription of the genes for the enzymes of GA biosynthesis (GA 20-oxidases, and GA 3-oxidases) or GA deactivation (GA 2-oxidases).

- Photoperiod and temperature can modify gene transcription of GA biosynthetic enzymes.

- GA biosynthesis occurs at multiple plant organs and cellular sites (**Figure 20.8**).

- A study of GA-deficient pea mutants confirms that plant height is directly correlated with the amount of endogenous GA$_1$ (**Figure 20.9**).

- Genetic engineering to decrease GA$_1$ biosynthesis or increase its degradation can lead to decreases in stem growth that can enhance grain yield in cereal crops (**Figure 20.10**). However, some GA-deficient mutants may show defects in fruit or seed production (**Figure 20.11**).

### Gibberellin Signaling: Significance of Response Mutants

- The GA response for some mutants is *not* proportional to the amount of endogenous bioactive GA, indicating that the mutated genes encode GA receptors or components of the signal pathway.

- *GID1* encodes a soluble protein that meets the criteria to be a GA receptor in rice (**Figure 20.12**).

- In Arabidopsis, three genes each code for a functional GA receptor, and a "triple mutant" is required to produce an extreme dwarf phenotype (**Figure 20.13**).

- Active GA is an allosteric activator of the receptor protein GID1, producing a conformational change of shape in GID1 and facilitating binding to a DELLA-domain protein, which is a negative regulator of GA response (**Figures 20.14–20.16**).

- Binding of a DELLA protein to the GA–receptor complex results in polyubiquitination and degradation of the DELLA protein.

- Mutation of negative regulators of GA may produce either slender or dwarf phenotypes. The *sln1c* mutation in a barley DELLA protein (SLN) results in constitutive expression of the GA response (tallness). In contrast, the *sln1d* mutation blocks the response to GA (dwarf phenotype) (**Figure 20.17**).

- Gibberellins stimulate the degradation of DELLA-domain proteins located in the nucleus. An Arabidopsis DELLA protein, RGA, is degraded in the presence of GA, leading to a growth response (**Figures 20.18, 20.19**).

### Gibberellin Responses: Early targets of DELLA proteins

- DELLA proteins regulate transcription indirectly by interacting with other proteins.

- Phytochrome-interacting factors (PIFs) are transcription factors that in seedlings up-regulate gene expression in the dark and in the presence of bioactive GA, leading to hypocotyl elongation. Light signals PIF degradation and absence of GA response (**Figure 20.20**).

### Gibberellin Responses: The Cereal Aleurone Layer

- During germination, the GA produced by the embryo stimulates synthesis of hydrolytic enzymes and their release from the aleurone layer into the endosperm (**Figure 20.21**).

- Active GA binds to GID1 protein, leading to degradation of DELLA proteins; loss of DELLA up-regulates an early response gene that encodes a GAMYB transcription factor. Binding of this factor to a GARE sequence in promoters of α-*amylase* genes enhances their transcription (**Figures 20.22, 20.23**).

### Gibberellin Responses: Anther Development and Male Fertility

- The action of GA on pollen development and male fertility is also mediated by GAMYB transcription factors (**Figure 20.24**).

- GAMYB transcription factors are necessary for expression of genes involved in normal pollen development, and they are regulated in anthers by microRNAs (**Figures 20.25, 20.26**).

- Differences in GAMYB signaling in different tissues may help account for GA inducing different biological responses in different target tissues.

### Gibberellin Responses: Stem Growth

- Gibberellins stimulate both cell elongation and cell division, in some cases permitting rapid but unsustainable stem elongation in rice under stress (flooded) conditions.

- GA-induced cell elongation is partially mediated by a class of proteins named expansins. Transcription of one expansin gene, *OsEXP4*, increases in response to growth-stimulating GA treatment or rapid submergence. Seedlings expressing antisense *OsEXP4* do not grow in submerged conditions; overexpression of *OsEXP4* leads to taller rice plants (**Figure 20.27**).

- In rice, the *SUB1A-1* gene up-regulates DELLA proteins, reducing the sensitivity of stem tissues to GA, and preventing unsustainable stem elongation (**Figure 20.28**).

- DELLA proteins are important for integrating the effects of GAs, several other hormones, and environmental factors.

## WEB MATERIAL

### Web Topics

**20.1 Structures of Some Important Gibberellins and Their Precursors, Derivatives, and Biosynthetic Inhibitors**

The chemical structures of various gibberellins and the inhibitors of their biosynthesis are presented.

**20.2 Commercial Uses of Gibberellins**

Gibberellins have roles in agronomy, horticulture, and the brewing industry.

**20.3 Gibberellin Biosynthesis**

The GA biosynthetic pathways and GA conjugates in plants are described.

**20.4 Gas Chromatography–Mass Spectrometry of Gibberellins**

Identification and quantitation of individual GAs are accomplished by gas chromatography–mass spectrometry.

**20.5 Environmental Control of Gibberellin Biosynthesis**

The antagonistic relationship between gibberellins and abscisic acid is often mediated by environmental factors.

**20.6 Auxin Can Regulate Gibberellin Biosynthesis**

Studies in both monocot and dicot species have shown that auxin can up-regulate gibberellin biosynthesis and down-regulate gibberellin inactivation.

**20.7 Negative Regulators of GA Response**

The DELLA proteins are important regulators of GA response.

**20.8 Effects of GA on Flowering**

Bioactive GAs are involved in the transition to flowering, and interact with the floral meristem identity gene *Leafy*.

**20.9 DELLA Proteins as Integrators of Multiple Signals**

These proteins are involved in integrating signals from several hormones and multiple environmental factors.

## WEB MATERIAL continued

### Web Essays

**20.1 Gibberellins in Pea: From Mendel to Molecular Physiology**

A personal description of GA research using pea, that most recently focuses on the interaction of environmental factors and other hormones on the regulation of GA biosynthesis in this plant.

**20.2 Ubiquitin Becomes Ubiquitous in GA Signaling**

Homologs of SLY1/GID2 and DELLAs have been found in just about every plant species that has an EST sequence database.

**20.3 Green Revolution Genes**

High yielding dwarf varieties of wheat and rice introduced in the mid-1960s have altered GA biosynthesis or response.

**20.4 Overgrowth Mutants of Barley—from Model System to Potential Crops**

Self-fertile barley mutants with enhanced GA signaling have potentially useful agronomic traits.

## CHAPTER REFERENCES

Achard, P., Herr, A., Baulcombe, D. C., and Harberd, N. P. (2004) Modulation of floral development by a gibberellin-regulated microRNA. *Development* 131: 3357–3365.

Appleford, N. E., Wilkinson, M. D., Ma, Q., Evans, D. J., Stone, M. C., Pearce, S. P., Powers, S. J., Thomas, S. G., Jones, H. D., Phillips, A. L., et al. (2007) Decreased shoot stature and grain alpha-amylase activity following ectopic expression of a gibberellin 2-oxidase gene intransgenic wheat. *J. Exp. Bot.* 58: 3213–3226.

Aya, K., Ueguchi-Tanaka, M., Kondo, M., Hamada, K., Yano, K., Nishamura, M., and Matsuoka, M. (2009) Gibberellin modulates anther development in rice via the transcriptional regualtion of GAMYB. *Plant Cell* 21: 1453–1472.

Bethke, P. C., Hwang, Y.-S., Zhu, T., and Jones, R. L. (2006) Global patterns of gene expression in the aleurone of wild-type and *dwarf1* mutant rice. *Plant Physiol.* 140: 484–498.

Bethke, P. C., Schuurink, R., and Jones, R. L. (1997) Hormonal signalling in cereal aleurone. *J. Exp. Bot.* 48: 1337–1356.

Chandler, P. M., Marion-Poll, A., Ellis, M., and Gubler, G. (2002) Mutants at the *Slender1* locus of barley cv Himalaya. Molecular and physiological characterization. *Plant Physiol.* 129: 181–190.

Chiang, H. H., Hwang, I., and Goodman, H. M. (1995) Isolation of Arabidopsis *GA4* locus. *Plant Cell* 7: 195–201.

Choi, D., Lee, Y., Cho, H. T., and Kende, H. (2003) Regulation of expansin gene expression affects growth and development in transgenic rice plants. *Plant Cell* 15: 1386–1398.

Davidson, S. E., and Reid, J. B. (2004) The pea gene *LH* encodes *ent*-kaurene oxidase. *Plant Physiol.* 134: 1123–1134.

Davidson, S. E., Elliott, R. C., Helliwell, C. A., Poole, A. T., and Reid, J. B. (2003) The pea gene *NA* encodes *ent*-kaurenoic acid oxidase. *Plant Physiol.* 131: 335–344.

Davies, P. J. (1995) The plant hormones: Their nature, occurrence, and functions. In *Plant Hormones: Physiology, Biochemistry and Molecular Biology*, P. J. Davies, ed., Kluwer, Dordrecht, Netherlands, pp. 1–12.

De Lucas, M., Daviere, J.-M., Rodriguez-Falcon, M., Pontin, M., Iglesias-Pedraz, J. M., Lorrain, S., Frankhauser, C., Blazquez, M. A., Titarenko, E., and Prat, S. (2008) A molecular framework for light and gibberellin control of cell elongation. *Nature* 451: 480–484.

Dill, A., Thomas, S. G., Hu, J., Steber, C. M., and Sun, T.-P. (2004) The Arabidopsis F-box protein SLEEPY1 targets gibberellin repressors for gibberellin-induced degradation. *Plant Cell* 16: 1392–1405.

Eriksson, M. E., Israelsson, M., Olsson, O., and Moritz, T. (2000) Increased gibberellin biosynthesis in transgenic trees promotes growth, biomass production and xylem fiber length. *Nat. Biotechnol.* 18: 784–788.

Fabian, T., Lorbiecke, R., Umeda, M., and Sauter, M. (2000) The cell cycle genes *cycA1;1* and *cdc2Os-3* are coordinately regulated by gibberellin in plants. *Planta* 211: 376–383.

Feng, S., Martinez, C., Gusmaroli, G., Wang, Y., Zhou, Y., Zhou, J., Wang, F., Chen, L., Yu, L., Iglesias-Pedraz, J. M., et al. (2008) Coordinated regualtion of *Arabidopsis thaliana* development by light and gibberellins. *Nature* 451: 475–479.

Fu, X., and Harberd, N. P. (2003) Auxin promotes Arabidopsis root growth by modulating gibberellin response. *Nature* 421: 740–743.

Fukao, T., and Bailey-Serres, J. (2008) Submergence tolerance conferred by *Sub1A* is mediated by SLR1 and SLRL1 restriction of gibberellin responses in rice. *Proc. Natl. Acad. Sci. USA* 105: 16814–16819.

Griffiths, J., Murase, K., Rieu, I., Zentella, R., Zhang, Z.-L., Powers, S. J., Gong, F., Phillips, A. L., Hedden, P., Sun, T.-P., et al. (2006) Genetic characterization and functional analysis of the GID1 gibberellin receptors in *Arabidopsis. Plant Cell* 18: 3399–3414.

Gocal, G.F.W., Sheldon, C. C., Gubler, F., Moritz, T., Bagnall, D. J., MacMillan, C. P., Li, S. F., Parish, R. W., Dennis, E. S., Weigel, D., et al. (2001) *GAMYB*-like genes, flowering, and gibberellin signaling in Arabidopsis. *Plant Physiol.* 127: 1682–1693.

Gubler, F., Kalla, R., Roberts, J. K., and Jacobsen, J. V. (1995) Gibberellin-regulated expression of a *myb* gene in barley aleurone cells: Evidence of myb transactivation of a high-pI alpha-amylase gene promoter. *Plant Cell* 7: 1879–1891.

Gubler, F., Chandler, P. M., White, R. G., Llewellyn, D. J., and Jacobsen, J. V. (2002) Gibberellin signaling in barley aleurone cells. Control of SLN1 and GAMYB expression. *Plant Physiol.* 129: 191–200.

Hedden, P. (2003) The genes of the Green Revolution. *Trends Genet.* 19: 5–9.

Hedden, P., and Phillips, A. L. (2000) Gibberellin metabolism: New insights revealed by the genes. *Trends Plant Sci.* 5: 523–530.

Hou, X., Hu, W.-W., Shen, L., Lee, L.Y.C., Tao, Z., Han, J.-H., and Yu, H. (2008) Global identification of DELLA target genes during Arabidopsis flower development. *Plant Physiol.* 147: 1126–1142.

Ikeda, A., Ueguchi-Tanaka, M., Sonoda, Y., Kitano, H., Koshioka, M., Futsuhara, Y., Matsuoka, M., and Yamaguchi, J. (2001) *slender Rice*, a constitutive gibberellin response mutant, is caused by a null mutation of the *SLR1* gene, an ortholog of the height-regulating *GA1/RGA/RHT/D8. Plant Cell* 13: 999–1010.

Ingram, T. J., Reid, J. B., Murfet, I. C., Gaskin, P., Willis, C. L., and MacMillan, J. (1984) Internode length in *Pisum*: The *Le* gene controls the 3β-hydroxylation of gibberellin $A_{20}$ to gibberellin $A_1$. *Planta* 160: 455–463.

Itoh, H., Matsuoka, M., and Steber, C. (2003) A role for the ubiquitin-26S-proteasome pathway in gibberellin signaling. *Trends Plant Sci.* 8: 492–497.

Jacobsen, J. V., Gubler, F., and Chandler, P. M. (1995) Gibberellin action in germinated cereal grains. In *Plant Hormones: Physiology, Biochemistry and Molecular Biology*, P. J. Davies, ed., Kluwer, Dordrecht, Netherlands, pp. 246–271.

Kaneko, M., Inukai, Y., Ueguchi-Tanaka, M., Itoh, H., Izawa, T., Kobayashi, Y., Hattori, T., Miyao, A., Hirochika, H., Ashikari, M., et al. (2004) Loss-of-function mutations of the rice GAMYB gene impair alpha-amylase expression in aleurone and flower development. *Plant Cell* 16: 33–44.

Koornneef, M., and van der Veen, J. H. (1980) Induction and analysis of gibberellin sensitive mutants in *Arabidopsis thaliana* (L) Heynh. *Theor. Appl. Genet.* 58: 257–263.

Koornneef, M., Elgersma, A., Hanhart, C. J., van Loenen-Martinet, E. P., van Rijn, L., and Zeevaart, J.A.D. (1985) A gibberellin insensitive mutant of *Arabidopsis thaliana. Physiol. Plant.* 65: 33–39.

Lester, D. R., Ross, J. J., Davies, P. J., and Reid, J. B. (1997) Mendel's stem length gene (*Le*) encodes a gibberellin 3b-hydroxylase. *Plant Cell* 9: 1435–1443.

MacMillan, J. (2002) Occurrence of gibberellins in vascular plants, fungi and bacteria. *J. Plant Growth Regul.* 20: 387–442.

Martin, D. N., Proebsting, W. M., and Hedden, P. (1997) Mendel's dwarfing gene: cDNAs from the *Le* alleles and function of the expressed proteins. *Proc. Natl. Acad. Sci. USA* 94: 8907–8911.

McGinnis, K. M., Thomas, S. G., Soule, J. D., Strader, L. C., Zale, J. M., Sun, T.-P., and Steber, C. M. (2003) The Arabidopsis *SLEEPY1* gene encodes a putative F-box subunit of an SCF E3 ubiquitin ligase. *Plant Cell* 15: 1120–1130.

Millar, A. A., and Gubler, F. (2005) The Arabidopsis *GAMYB-like* genes, *MYB33* and *MYB 65*, are microRN-regulated genes that redundantly facilitate anther development. *Plant Cell* 17: 705–721.

Murase, K., Hirano, Y., Sun, T.-P., and Hakoshima, T. (2008) Gibberellin-induced DELLA recognition by the gibberellin receptor GID1. *Nature* 456: 459–464.

Nakajima, M., Shimada, A., Takashi, Y., Kim, Y. C., Park, S. H., Ueguchi-Tanaka, M., Suzuki, H., Katoh, E., Iuchi, S., Kobayashi, M., et al. (2006) Identification and characterization of Arabidopsis gibberellin receptors. *Plant J.* 46: 880–889.

Ogawa, M., Hanada, A., Yamauchi, Y., Kuwahara, A., Kamiya, Y., and Yamaguchi, S. (2003) Gibberellin biosynthesis and response during Arabidopsis seed germination. *Plant Cell* 15: 1591–1604.

Park, W. W., Li, J., Song, R., Messing, J., and Chen, X. (2002) CARPEL FACTORY, a Dicer homolog, and HEN1, a novel protein, act in microRNA metabolism in *Arabidopsis thaliana. Curr. Biol.* 2: 1484–1495.

Peng, J. R., Carol, P., Richards, D. E., King, K. E., Cowling, R. J., Murphy, G. P., and Harberd, N. P. (1997) The Arabidopsis *GAI* gene defines a signaling pathway that negatively regulates gibberellin responses. *Genes Dev.* 11: 3194–3205.

Peng, J. R., Richards, D. E., Hartley, N. M., Murphy, G. P., Flintham, J. E., Beales, J., Fish, L. J., Pelica, F., Sudhakar, D., Christou, P., et al. (1999) 'Green revolution' genes encode mutant gibberellin response modulators. *Nature* 400: 256–261.

Phillips, A. L. (2004) Genetic and transgenic approaches to improving crop performance. In *Plant Hormones:*

*Biosynthesis, signal transduction, action!* P. J. Davies, ed., Kluwer, Dordrecht, Netherlands, pp. 582–609.

Phillips, A. L., Ward, D. A., Uknes, S., Appleford, N.E.J., Lange, T., Huttley, A. K., Gaskin, P., Graebe, J. E., and Hedden, P. (1995) Isolation and expression of three gibberellin 20-oxidase cDNA clones from Arabidopsis. *Plant Physiol.* 108: 1049–1057.

Phinney, B. O. (1984) Gibberellin $A_1$, dwarfism and the control of shoot elongation in higher plants. In: *The Biosynthesis and Metabolism of Plant Hormones,* Vol. 23, A. Crozier and J. R. Hillman, eds., Cambridge University Press, London, pp. 17–41.

Piskurewicz, U., Jikumaru, Y., Kinoshita, N., Nambara, E., Kamiya, Y., Lopez-Molina, L. (2008) The gibberellic acid signaling repressor RGL2 inhibits *Arabidopsis* seed germination by stimulating abscisic acid synthesis and ABI5 activity. *Plant Cell* 20: 2729–2745.

Reid, J. B., and Howell, S. H. (1995) Hormone mutants and plant development. In *Plant Hormones: Physiology, Biochemistry and Molecular Biology*, P. J. Davies, ed., Kluwer, Dordrecht, Netherlands, pp. 448–485.

Reid, J. B., Ross, J. J., and Swain, S. W. (1992) Internode length in *Pisum*. A new *slender* mutant with elevated levels of $C_{19}$-gibberellins. *Planta* 188: 462–467.

Rhoades, M. W., Reinhart, B. J., Lim, L. P., Burges, C. B., Bartel, B., and Bartel, D. P. (2002) Prediction of plant microRNA targets. *Cell* 110: 513–520.

Ross, J. J., Reid, J. B., Gaskin, P., and Macmillan, J. (1989) Internode length in *Pisum*. Estimation of $GA_1$ levels in genotypes *Le, le* and *le^d*. *Physiol. Plant.* 76: 173–176.

Sakamoto, T., Morinaka, Y., Ishiyama, K., Kobayashi, M., Itoh., H., Kayano, T., Iwahori, S., Matsuoaka, M., and Tanaka, H. (2003) Genetic manipulation of gibberellin metabolism in transgenic rice. *Nature* 21: 909–913.

Sasaki, A., Itoh, H., Gomi, K., Ueguchi-Tanaka, M., Ishayama, K., Kobayashi, M., Jeong, D.-H., An, G., Kitano, H., Ashikari, M., et al. (2003) Accumulation of phosphorylated repressor for gibberellin signaling in an F box mutant. *Science* 299: 1896–1898.

Sauter, M., and Kende, H. (1992) Gibberellin-induced growth and regulation of the cell division cycle in deepwater rice. *Planta* 188: 362–368.

Seo, M., Hanada, A., Kuwahara, A., Endo, A., Okamato, M., Yamauchi, Y., North, H., Marion-Poll, A., Sun, T.-P., Koshiba, T., et al. (2006) Regulation of hormone metabolism in Arabidopsis seeds: phytochrome regulation of abscisic acid metabolism and abscisic acid regulation of gibberellin metabolism. *Plant J.* 48: 354–366.

Shimada, A., Ueguchi-Tanaka, M., Nakatsu, T., Nakajima, M., Naoe, Y., Ohmiya, H., Kato, H., and Matsuoka, M. (2008) Structural basis for gibberellin recognition by its receptor GID1. *Nature* 45: 520–524.

Silverstone, A. L., Chang, C.-W., Krol, E., and Sun, T.-P. (1997a) Developmental regulation of the gibberellin biosynthetic gene *GA1* in *Arabidopsis thaliana*. *Plant J.* 12: 9–19.

Silverstone, A. L., Mak, P. Y. A., Martinez, E. C., and Sun, T.-P. (1997b) The new *RGA* locus encodes a negative regulator of gibberellin responses in *Arabidopsis thaliana*. *Genetics* 146: 1087–1099.

Silverstone, A. L., Chang, C.-W., Krol, E., and Sun, T.-P. (1998) The Arabidopsis *RGA* encodes a transcriptional regulator repressing the gibberellin signal transduction pathway. *Plant Cell* 10: 155–169.

Silverstone, A. L., Jung, H. S., Dill, A., Kawaide, H., Kamiya, Y., and Sun, T.-P. (2001) Repressing a repressor: Gibberellin-induced rapid reduction of the RGA protein in Arabidopsis. *Plant Cell* 13: 1555–1565.

Swain, S. M., and Singh, D. P. (2005) Tall tales from sly dwarves: Novel functions of gibberellins in plant development. *Trends Plant Sci.* 10: 123–129.

Tsuji, H., Aya, K., Ueguchi-Tanaka, M., Shimada, Y., Nakazono, M., Watanabe, R., Nishizawa, N. K., Gomi, K., Shimada, A., Kitano, H., et al. (2006) GAMYB controls different sets of genes and is differentially regulated by microRNA in aleurone cells and anthers. *Plant J.* 47: 427–444.

Ueguchi-Tanaka, M., Ashikari, M., Nakajima, M., Itoh, H., Katoh, E., Kobayashi, M., Chow, T., Hsing, Y. C., Kitano, H., Yamaguchi, I., et al. (2005) *GIBBERELLIN INSENSITIVE DWARF1* encodes a soluble receptor for gibberellin. *Nature* 437: 693–698.

Xu, W., Campbell, P., Vargheese, A. V., and Braam, J. (1996) The Arabidopsis XET-related gene family: Environmental and hormonal regulation of expression. *Plant J.* 9: 879–889.

Xu, Y.-L., Li, L., Wu, K., Peeters, A. J. M., Gage, D. A., and Zeevaart, J. A. D. (1995) The *GA5* locus of *Arabidopsis thaliana* encodes a multi-functional gibberellin 20-oxidase: Molecular cloning and functional expression. *Proc. Natl. Acad. Sci. USA* 92: 6640–6644.

Yamaguchi, S. (2008) Gibberellin metabolism and its regulation. *Annu. Rev. Plant Biol.* 59: 225–251.

Yamaguchi, S., Kamiya, Y., and Sun, T.-P. (2001) Distinct cell-specific expression patterns of early and late gibberellin biosynthetic genes during Arabidopsis seed germination. *Plant J.* 28: 443–453.

Yang, T., Davies, P. J., and Reid, J. B. (1996) Genetic dissection of the relative roles of auxin and gibberellin in the regulation of stem elongation in intect light-grown peas. *Plant Physiol.* 110: 1029–1034.

Yaxley, J. R., Ross, J. J., Sherriff, L. J., and Reid, J. B. (2001) Gibberellin biosynthesis mutations and root development in pea. *Plant Physiol.* 125: 627–633.

Zentella, R., Zhang, Z. L., Park, M., Thomas, S. G., Endo, A., Murase, K., Fleet, C. M., Nambara, E., Kamiya, Y., and Sun, T.-P. (2007) Global analysis of DELLA direct targets in early gibberellin signaling in Arabidopsis. *Plant Cell* 19: 3037–3057.

CHAPTER

# 21

# Cytokinins: Regulators of Cell Division

The cytokinins were discovered in the search for factors that stimulate plant cells to divide (i.e., undergo cytokinesis). Since their discovery, cytokinins have been shown to have effects on many other physiological and developmental processes, including leaf senescence, nutrient mobilization, apical dominance, the formation and activity of apical meristems, vascular development, the breaking of bud dormancy, and interactions with other organisms. Cytokinins also appear to mediate many aspects of light-regulated development, including chloroplast differentiation, the development of autotrophic metabolism, and leaf and cotyledon expansion.

Although cytokinins regulate many cellular processes, the control of cell division is fundamental in plant growth and development and is a central function of this class of plant growth regulators. For these reasons we will preface our discussion of cytokinin function with a brief consideration of the roles of cell division in normal development, wounding, gall formation, and tissue culture.

We will discuss how cytokinins are synthesized and modified. We will then consider the molecular mechanisms underlying cytokinin perception and signaling. Finally, we will examine the biological roles of cytokinin and how the pathways for its biosynthesis, metabolism and signaling act to influence these growth and developmental processes.

# Cell Division and Plant Development

Plant cells form as the result of cell divisions in a primary or secondary meristem. Newly formed plant cells typically enlarge and differentiate, but once they have assumed their function—whether transport, photosynthesis, support, storage, or protection—usually they do not divide again during the life of the plant. In this respect they appear to be similar to animal cells, which are considered to be terminally differentiated.

However, this similarity to the behavior of animal cells is only superficial. Almost every type of plant cell that retains its nucleus at maturity has been shown to be capable of dividing. This property comes into play during such processes as wound healing and leaf abscission.

## Differentiated plant cells can resume division

Under some circumstances, mature, differentiated plant cells may resume cell division in the intact plant. In many species, mature cells of the cortex and/or phloem resume division to form secondary meristems, such as the vascular cambium or the cork cambium. The abscission zone at the base of a leaf petiole is a region where mature parenchyma cells begin to divide again after a period of mitotic inactivity, forming a layer of cells with relatively weak cell walls where abscission can occur (see Chapter 22).

Wounding of plant tissues induces cell divisions at the wound site. Even highly specialized cells, such as phloem fibers and guard cells, may be stimulated by wounding to divide at least once. Wound-induced mitotic activity typically is self-limiting; after a few divisions the derivative cells stop dividing and redifferentiate. However, when the soil-dwelling bacterium *Agrobacterium tumefaciens* invades a wound, it can cause the neoplastic (tumor-forming) disease known as **crown gall**. This phenomenon is dramatic natural evidence of the mitotic potential of mature plant cells.

Without *Agrobacterium* infection, the wound-induced cell division would subside after a few days and some of the new cells would differentiate as a protective layer of cork cells or vascular tissue. However, *Agrobacterium* changes the character of the cells that divide in response to the wound, making them tumorlike. They do not stop dividing; rather, they continue to divide throughout the life of the plant to produce an unorganized mass of tumorlike tissue called a **gall** (FIGURE 21.1). We will have more to say about this important disease later in this chapter.

## Diffusible factors control cell division

The considerations addressed in the previous section suggest that mature plant cells stop dividing because they no longer receive a particular signal—possibly a hormone—that is necessary for the initiation of cell division. The idea

**FIGURE 21.1** Tumor that formed on a tomato stem infected with the crown gall bacterium, *Agrobacterium tumefaciens*. Two months before this photo was taken the stem was wounded and inoculated with a virulent strain of the crown gall bacterium. (From Aloni et al. 1998, courtesy of R. Aloni.)

that cell division may be initiated by a diffusible factor originated with the Austrian plant physiologist Gottlieb Haberlandt, who, in 1913, demonstrated that vascular tissue contains a water-soluble substance or substances that will stimulate the division of wounded potato tuber tissue. The effort to determine the nature of this factor (or factors) led to the discovery of the cytokinins in the 1950s.

## Plant tissues and organs can be cultured

Biologists have long been intrigued by the possibility of growing organs, tissues, and cells in culture on a simple nutrient medium, in the same way that microorganisms can be cultured in test tubes or on petri dishes. In the 1930s, Philip White demonstrated that tomato roots can be grown indefinitely in a simple nutrient medium containing only sucrose, mineral salts, and a few vitamins, with no added hormones (White 1934).

In contrast to roots, isolated stem tissues exhibit very little growth in culture without added hormones in the medium. Even if auxin is added, only limited growth may occur, and usually this growth is not sustained. Frequently this auxin-induced growth is due to cell enlargement only. The shoots of most plants cannot grow on a simple medium lacking hormones, even if the cultured stem tissue contains apical or lateral meristems, until adventitious roots form. Once the stem tissue has rooted, shoot growth resumes, but now as part of an integrated, whole plant.

These observations indicate that there is a difference between the regulation of cell division in root and shoot meristems. They also suggest that some root-derived factor(s) may regulate growth in the shoot.

Crown gall stem tissue is an exception to these generalizations. After a gall has formed on a plant, heating the plant to 42°C will kill the bacterium that induced gall formation. The plant will survive the heat treatment, and its gall tissue will continue to grow as a bacteria-free tumor (Braun 1958).

Tissues removed from these bacteria-free tumors grow on simple, chemically defined culture media that would not support the proliferation of normal stem tissue of the same species. However, these stem-derived tissues are not organized. Instead they grow as a mass of disorganized, relatively undifferentiated cells called **callus tissue**.

Callus tissue sometimes forms naturally in response to wounding or in graft unions where stems of two different plants are joined. Crown gall tumors are a specific type of callus, whether they are growing attached to the plant or in culture. The finding that crown gall callus tissue can be cultured demonstrated that cells derived from stem tissues are capable of proliferating in culture and that contact with the bacteria may cause the stem cells to produce factors that stimulate cell division.

## The Discovery, Identification, and Properties of Cytokinins

A great many substances were tested in an effort to initiate and sustain the proliferation of normal stem tissues in culture. Materials ranging from yeast extract to tomato juice were found to have a positive effect, at least with some tissues. However, culture growth was stimulated most dramatically when the liquid endosperm of coconut (also known as coconut milk) was added to the culture medium.

Philip White's nutrient medium, supplemented with an auxin and 10 to 20% coconut milk, was found to support the continued cell division of mature, differentiated cells from a wide variety of tissues and species, leading to the formation of callus tissue (Caplin and Steward 1948). This finding indicated that coconut milk contains a substance or substances that stimulate mature cells to enter and remain in the cell division cycle.

Eventually coconut milk was shown to contain the cytokinin *zeatin*, but this finding was not obtained until several years after the discovery of the cytokinins (Letham 1974). The first cytokinin to be discovered was the synthetic analog *kinetin*.

### Kinetin was discovered as a breakdown product of DNA

In the 1940s and 1950s, Folke Skoog and co-workers at the University of Wisconsin tested many substances for their ability to initiate and sustain the proliferation of cultured tobacco pith tissue (see WEB ESSAY 21.1). They had observed that the nucleic acid base adenine had a slight promotive effect, so they tested the possibility that nucleic acids would stimulate division in this tissue. Surprisingly, aged or autoclaved herring sperm DNA had a powerful cell division–promoting effect.

After much work, a small molecule was identified from the autoclaved DNA and named **kinetin**. It was shown to be an adenine (6-aminopurine) derivative, 6-furfurylaminopurine (Miller et al. 1955):

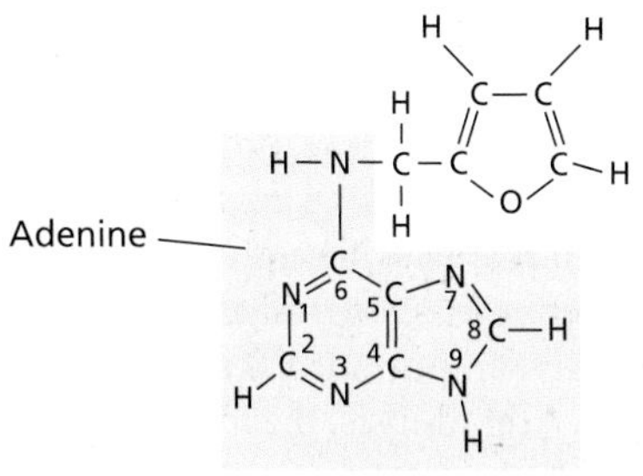

**Kinetin**

In the presence of an auxin, kinetin would stimulate tobacco pith parenchyma tissue to proliferate in culture. No kinetin-induced cell division occurs without auxin in the culture medium. (For more details, see WEB TOPIC 21.1.)

Kinetin is not a naturally occurring plant growth regulator, and it does not occur as a base in the DNA of any species. It is a by-product of the heat-induced degradation of DNA, in which the deoxyribose sugar of adenosine is converted to a furfuryl ring and shifted from the 9 position to the 6 position on the adenine ring.

The discovery of kinetin was important because it demonstrated that cell division could be induced by a simple chemical substance. Of greater importance, the discovery of kinetin suggested that naturally occurring molecules with structures similar to that of kinetin regulate cell division activity within the plant. This hypothesis proved to be correct.

### Zeatin was the first natural cytokinin discovered

Several years after the discovery of kinetin, extracts of the immature endosperm of maize (corn; *Zea mays*) were found to contain a substance that had the same biological effect as kinetin. This substance stimulated mature plant cells to divide when added to a culture medium along with an auxin. Letham (1973) isolated the molecule responsible for this activity and identified it as *trans*-6-(4-hydroxy-3-methylbut-2-enylamino)purine, which he called **zeatin**.

The molecular structure of zeatin is similar to that of kinetin. Both molecules are adenine (aminopurine) derivatives. Although they have different side chains, in both cases the side chain is linked to the nitrogen attached to C6 ($=N^6$) of adenine. Because the side chain of zeatin has a double bond, it can exist in either the *cis* or the *trans* configuration.

*trans*-Zeatin                     *cis*-Zeatin

**6-(4-Hydroxy-3-methylbut-2-enylamino)purine**

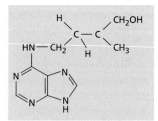

N⁶-(D²-Isopentenyl)-adenine (iP)

Dihydrozeatin (DZ)

**FIGURE 21.2** Structures of
other aminopurines that
are active as cytokinins.

Ribosylzeatin (zeatin riboside)

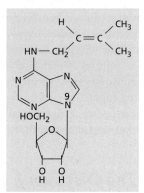

N⁶-(Δ²-Isopentenyl)adenosine
([9R]iP)

In higher plants, zeatin occurs in both the *cis* and the *trans* configurations, and these forms can be interconverted by an enzyme known as *zeatin isomerase*, which is found in some, but not all plants. Although the *trans* form of zeatin is much more active in biological assays, it is likely that the *cis* form also plays important roles in some plant species. For example, in Arabidopsis, the *trans* form predominates, but in maize and rice most zeatin is present in the *cis* form. Furthermore, *trans*-zeatin can bind to and activate cytokinin receptors from both Arabidopsis and maize, but the *cis*-form is much more active with the cytokinin receptors from maize.

Since its discovery in immature maize endosperm, zeatin has been found in many plants and in some bacteria. It is generally the most prevalent active cytokinin in higher plants, but other substituted aminopurines that are active as cytokinins have been isolated from many plant and bacterial species. These aminopurines differ from zeatin in the nature of the side chain attached to the $N^6$ position (**FIGURE 21.2**).

These cytokinins can be present in the plant as **ribosides** (in which a ribose sugar is attached to the 9 nitrogen of the purine ring), **ribotides** (in which the ribose sugar moiety contains a phosphate group), **glycosides** (in which a sugar molecule is attached to the 3, 7, or 9 nitrogen of the purine ring, or to the oxygen of the zeatin or dihydrozeatin side chain), or in other conjugated forms (see Appendix 3 and **WEB TOPIC 21.2**).

## Some synthetic compounds can mimic cytokinin action

Cytokinins are defined as compounds that have biological activities similar to those of *trans*-zeatin. These activities include:

- Inducing cell division in callus cells in the presence of an auxin
- Promoting bud or root formation from callus cultures when in the appropriate molar ratios to auxin
- Delaying senescence of leaves
- Promoting expansion of dicot cotyledons

Many chemical compounds have been synthesized and tested for cytokinin activity. Analysis of these compounds provides insight into the structural requirements for activity. Nearly all compounds active as cytokinins are $N^6$-substituted aminopurines, such as benzyladenine (BA):

**Benzyladenine
(benzylaminopurine)
(BA)**

and all the naturally occurring cytokinins are aminopurine derivatives. There are also synthetic cytokinin compounds that have not been identified in plants, most notably the diphenylurea-type cytokinins; one of these, thidiazuron, is used commercially as a defoliant and herbicide.

**Thidiazuron**

Naturally occurring molecules with cytokinin activity are detected and identified primarily by physical methods (see **WEB TOPIC 21.3**).

## Cytokinins occur in both free and bound forms

Biologically active cytokinins are present as free molecules (not covalently attached to any macromolecule) in plants and certain bacteria. Free cytokinins have been found in a wide spectrum of angiosperms and probably are universal in this group of plants. They have also been found in algae, diatoms, mosses, ferns, and conifers.

The regulatory role of cytokinins has been demonstrated in angiosperms, conifers, and mosses; it is likely that they function to regulate the growth, development, and metabolism of all plants. Usually zeatin is the most abundant naturally occurring free cytokinin, but *dihydrozeatin (DHZ)* and *isopentenyl adenine (iP)* also are commonly found in higher plants and bacteria. Numerous derivatives of these three cytokinins have been identified in plant extracts (see Figure 21.2 and **WEB TOPIC 21.2**). The most active forms of cytokinins are the free bases (see **WEB TOPIC 21.4**).

Transfer RNA (tRNA) contains not only the four nucleotides used to construct all other forms of RNA, but also some unusual nucleotides in which the base has been modified. Some of these modified bases act as cytokinins when the tRNA is hydrolyzed and tested in cytokinin bioassays. Some plant tRNAs contain *cis*-zeatin as a hypermodified base. However, cytokinins are not confined to plant tRNAs. They are part of certain tRNAs from all organisms, from bacteria to humans. (For details, see **WEB TOPIC 21.5**.)

## Some plant pathogenic bacteria, fungi, insects, and nematodes secrete free cytokinins

Some bacteria and fungi are intimately associated with higher plants. Many of these microorganisms produce and secrete substantial amounts of cytokinins and/or cause the plant cells to synthesize plant hormones, including cytokinins (Akiyoshi et al. 1987). The cytokinins produced by microorganisms include *trans*-zeatin, iP, *cis*-zeatin, and their ribosides (see Figure 21.2), as well as 2-methylthio-derivatives of zeatin (see Appendix 3). Infection of plant tissues with these microorganisms can induce the tissues to divide and, in some cases, to form special structures, such as mycorrhizal arbuscules, in which the microorganism can reside in a mutualistic relationship with the plant.

In addition to the crown gall bacterium, *A. tumefaciens*, other pathogenic bacteria may stimulate plant cells to divide. For example, increased cytokinin, supplied by interacting bacteria, fungi, viruses, or insects, can cause an increase in the proliferation of the shoot apical meristem and/or the growth of lateral buds, which normally remain dormant (Hamilton and Lowe 1972). This proliferation, known as fasciation, often manifests as a phenomenon known as a **witches' broom** (**FIGURE 21.3**), so-called because these growths can resemble an old-fashioned straw broom. One well-studied causative agent of fasciation is *Rhodococcus fascians* (Hamilton and Lowe 1972). *R. fascians* produces several different cytoki-

**FIGURE 21.3** Witches' broom on a fir tree (*Abies* sp.) caused by the fir broom rust fungus, *Melampsorella caryophyllacearum*. (Courtesy of Bob Erickson, Natural Resources Canada, Canadian Forest Service.)

nins, including both *cis*- and *trans*-zeatin as well as their 2-methylthio- derivatives (Pertry et al. 2009). This mixture of cytokinin species acts synergistically through the host's normal cytokinin signaling pathway (see below) to alter host development. *R. fascians* also secretes the auxin IAA, which contributes to the alteration in the growth of the host plant. Fasciation, which can also arise spontaneously by a mutation, is the basis for many of the horticultural dwarf conifers.

Certain insects secrete cytokinins, which play a role in the formation of the galls these insects use as feeding sites. Root-knot nematodes also produce cytokinins, which may be involved in manipulating host development to produce the giant cells from which the nematode feeds (Elzen 1983).

## Biosynthesis, Metabolism, and Transport of Cytokinins

The side chains of naturally occurring cytokinins are chemically related to rubber, carotenoid pigments, the plant hormones gibberellin and abscisic acid, and some of the plant defense compounds known as phytoalexins. All of these compounds are constructed, at least in part, from isoprene units (see Chapter 13).

Isoprene is similar in structure to the side chains of zeatin and iP (see the structures illustrated in Figures 21.2 and 21.5). These cytokinin side chains are synthesized from an isoprene derivative. Large molecules of rubber and the carotenoids are constructed by the polymerization of many

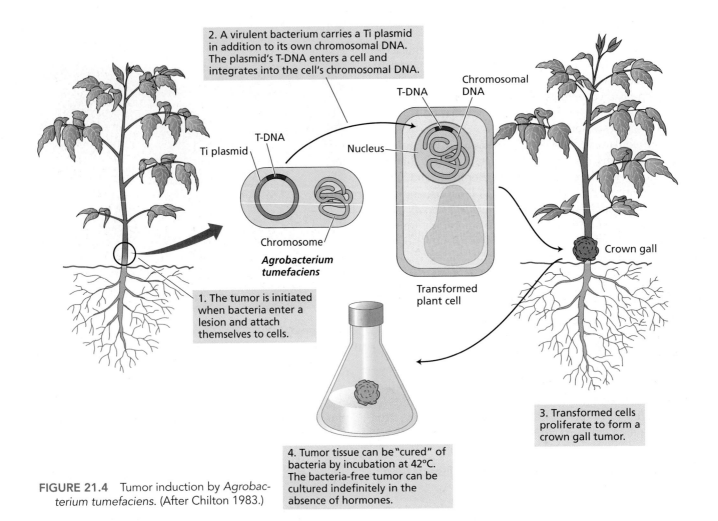

2. A virulent bacterium carries a Ti plasmid in addition to its own chromosomal DNA. The plasmid's T-DNA enters a cell and integrates into the cell's chromosomal DNA.

T-DNA

Chromosomal DNA

Ti plasmid

T-DNA

Nucleus

Chromosome

***Agrobacterium tumefaciens***

Transformed plant cell

Crown gall

1. The tumor is initiated when bacteria enter a lesion and attach themselves to cells.

3. Transformed cells proliferate to form a crown gall tumor.

4. Tumor tissue can be "cured" of bacteria by incubation at 42°C. The bacteria-free tumor can be cultured indefinitely in the absence of hormones.

**FIGURE 21.4** Tumor induction by *Agrobacterium tumefaciens*. (After Chilton 1983.)

isoprene units; cytokinins contain just one of these units. The precursor for the formation of these isoprene structures in cytokinins is dimethylallyl diphosphate (DMAPP), which is derived from either the mevalonate pathway (primarily for *cis*-zeatin) or the methylerythritol phosphate (MEP) pathway (primarily for DHZ, iP, and *trans*-zeatin).

### Crown gall cells have acquired a gene for cytokinin synthesis

Bacteria-free tissues from crown gall tumors proliferate in culture without the addition of any hormones to the culture medium. The crown gall tissues synthesize substantial amounts of both auxin and free cytokinins. During infection by *A. tumefaciens*, plant cells incorporate bacterial DNA into their chromosomes (the details of the infection process are described in Chapter 2). The virulent strains of *Agrobacterium* contain a large plasmid known as the **Ti plasmid**. Plasmids are circular pieces of extrachromosomal DNA that are not essential for the life of the bacterium. However, plasmids frequently contain genes that enhance the ability of the bacterium to survive in special environments.

A small portion of the Ti plasmid, known as the **T-DNA**, is incorporated into the nuclear DNA of the host plant cell (**FIGURE 21.4**) (Chilton et al. 1977). T-DNA carries genes necessary for the biosynthesis of cytokinins and auxin, as well as a member of a class of unusual carbon- and nitrogen-containing amino acid derivatives called *opines* (see **WEB TOPIC 21.6**). Opines are not synthesized by plants except after crown gall transformation.

The T-DNA gene involved in cytokinin biosynthesis, known as the *ipt* gene (the convention for bacterial genes is that they are written in lower case italics), encodes an **isopentenyl transferase** (**IPT**) enzyme that transfers the isopentenyl group from 1-hydroxy-2-methyl-2-(E)-butenyl 4-diphosphate (HMBDP) to adenosine monophosphate (AMP) to form tZRMP (*trans*-zeatin riboside 5′-monophosphate) (**FIGURE 21.5**) (Akiyoshi et al. 1984; Barry et al. 1984; Sakakibara et al. 2005). The *ipt* gene is also called the *tmr* locus because, when *inactivated* by mutation, it results in "rooty" tumors. This conversion route is somewhat similar to the pathway for cytokinin synthesis that has been postulated for normal tissue (see Figure 21.5).

The T-DNA also contains two genes encoding enzymes that convert tryptophan to the auxin indole-3-acetic acid

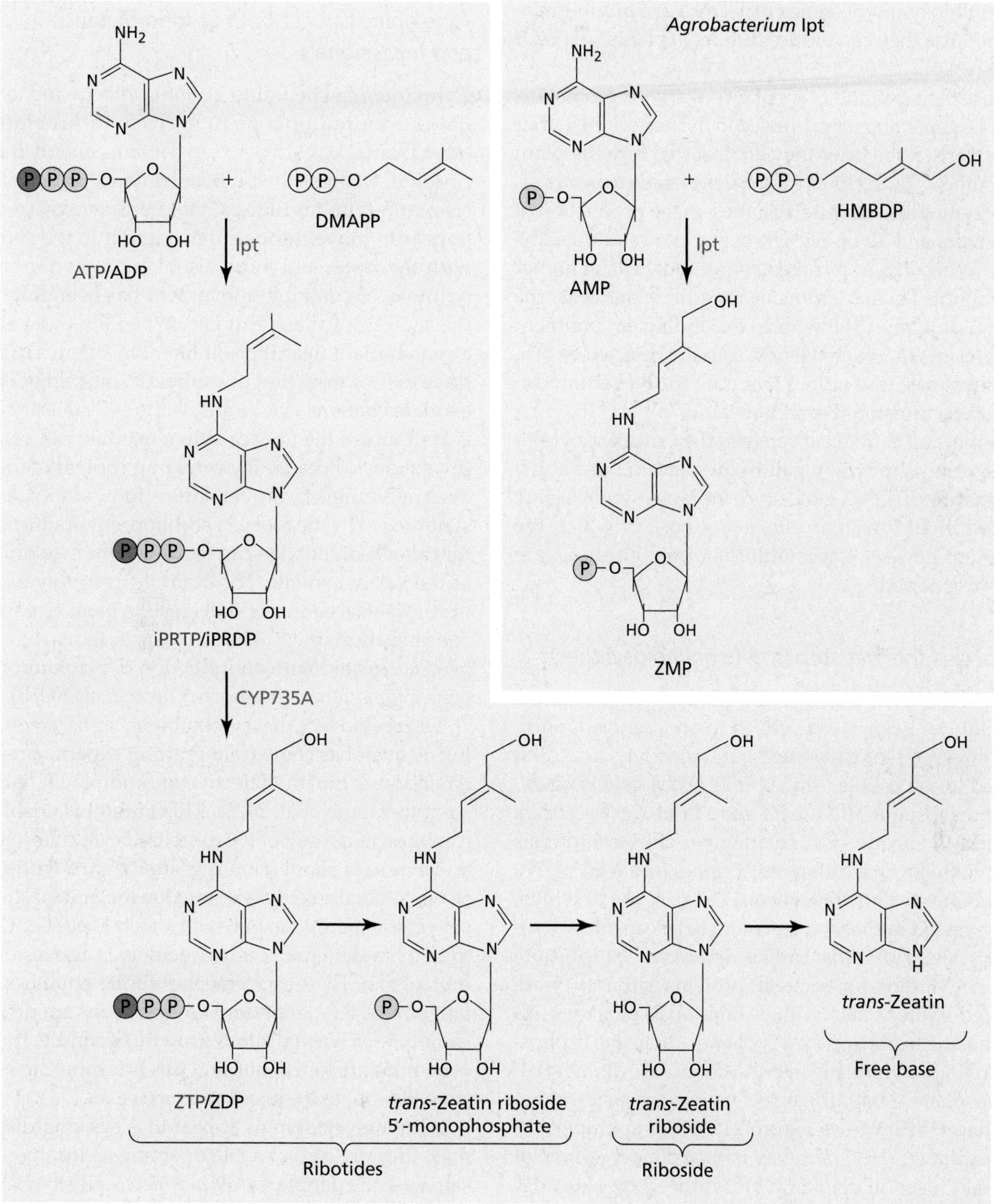

**FIGURE 21.5** Simplified biosynthetic pathway for cytokinin biosynthesis. The first committed step in cytokinin biosynthesis is the addition of the isopentenyl side chain from DMAPP (dimethylallyl diphosphate) to an adenosine moiety (ATP or ADP). The products of these reactions (iPRTP or iPRDP) are converted to zeatin (ZTP or ZDP) by a cytochrome P450 monooxygenase (CPY735A). Dihydrozeatin (DHZ) cytokinins are made from the various forms of *trans*-zeatin by an unknown enzyme (not shown). The ribotide and riboside forms of *trans*-zeatin can be interconverted and free *trans*-zeatin can be formed from the riboside by enzymes of general purine metabolism. *Inset*: The inset shows the pathway for cytokinin biosynthesis via *Agrobacterium* Ipt. The plant and bacterial Ipt enzymes differ in the adenosine substrate used and the side chain donor; the plant enzyme appears to utilize both ADP and ATP and couples this to DMAPP, and the bacterial enzyme utilizes AMP and couples this to HMBDP (1-hydroxy-2-methyl-2-(E)-butenyl 4-diphosphate). Note that the product of the *Agrobacterium* Ipt reaction is a zeatin ribotide.

(IAA). This pathway of auxin biosynthesis differs from the one in nontransformed cells and involves indoleacetamide as an intermediate (see Appendix 3). The *ipt* gene and the two auxin biosynthetic genes of T-DNA are **phyto-oncogenes**, because they can induce tumors in plants (see **WEB TOPIC 21.7**).

Because their promoters are plant eukaryotic promoters, the T-DNA genes are not expressed in the bacterium; rather they are transcribed after they are inserted into the plant chromosomes. Transcription of the genes leads to synthesis of the enzymes they encode, resulting in the production of zeatin, auxin, and an opine. The bacterium can utilize the opine as a carbon and nitrogen source, but cells of higher plants cannot. Thus, by transforming the plant cells, the bacterium provides itself with an expanding environment (the gall tissue) in which the host cells are directed to produce a substance (the opine) that only the bacterium can utilize for its nutrition (Bomhoff et al. 1976).

An important difference between the control of cytokinin biosynthesis in crown gall tissues and in normal tissues is that the T-DNA genes for cytokinin synthesis are expressed in all infected cells, even those in which the native plant genes for biosynthesis of the hormone are normally repressed.

### IPT catalyzes the first step in cytokinin biosynthesis

The first committed step in cytokinin biosynthesis is the transfer of the isopentenyl group to an adenosine moiety. An enzyme that catalyzes such an activity was first identified in the cellular slime mold *Dictyostelium discoideum*, and subsequently the *ipt* gene from *Agrobacterium* was found to encode such an enzyme. The Arabidopsis genome contains nine different *IPT* genes (see **WEB TOPIC 21.8**) (Kakimoto 2001; Takei et al. 2001a), seven of which, when expressed in *E. coli*, were found to be capable of synthesizing free cytokinins. Unlike *Agrobacterium* Ipt (note that the convention for bacterial proteins is that they are capitalized without italics), the Arabidopsis enzymes utilize adenosine triphosphate (ATP) and adenosine diphosphate (ADP) preferentially over AMP, and use dimethylallyl diphosphate (DMAPP) as the source of the side chain rather than HMBDP (see Figure 21.5). The methylerythritol phosphate (MEP) pathway is the primary source of the DMAPP used in cytokinin biosynthesis by plant IPT enzymes, and this pathway occurs in plastids, which is where the majority of IPT enzymes are localized in plants. Thus, cytokinin biosynthesis in plants occurs primarily in plastids. Measurements of the expression pattern of the *IPT* genes indicate there are multiple sources of cytokinin biosynthesis throughout the plant (Miyawaki et al. 2004).

The immediate products of the IPT reaction are iP-ribotides, which are next converted to zeatin ribotides (Takei et al. 2004). Cytokinin nucleotides can be converted to their most active free base forms via dephosphorylation

and deribosylation reactions (Kurakawa et al. 2007) (see Figure 21.5).

### Cytokinins can act both as long distance and local signals

Cytokinins can be found in both phloem and xylem exudates from multiple plant species. In Arabidopsis and other plant species, xylem sap contains mainly *trans*-zeatin riboside, whereas the phloem contains principally iP and *cis*-zeatin-type ribosides. Cytokinins synthesized in roots appear to move through the xylem into the shoot, along with the water and minerals taken up by the roots. This pathway of cytokinin movement has been inferred from the analysis of xylem exudate. When the shoot is cut from a rooted plant near the soil line, the xylem sap may continue to flow from the cut stump for some time. This xylem exudate contains cytokinins. If the soil covering the roots is kept moist, the flow of xylem exudate can continue for several days. Because the cytokinin content of the exudate does not diminish, the cytokinins found in it are probably synthesized by the roots. In addition, environmental factors that alter root function can modulate the cytokinin content of the xylem exudate. For example, resupply of nitrate to nitrogen-starved maize roots results in an elevation of the concentration of cytokinins in the xylem sap, which has been correlated with an induction of cytokinin-regulated gene expression in the shoots (Takei et al. 2001b).

Direct evidence that cytokinin can act as a mobile signaling element has come from grafting experiments using an Arabidopsis mutant defective in multiple *IPT* genes (Matsumoto-Kitano et al. 2008). This mutant is unable to form cambium as a result of the decreased cytokinin synthesis. If an *ipt* mutant shoot scion is grafted onto a wild-type rootstock, cambial activity is restored in the mutant shoot. In the converse experiment, in which a wild-type shoot is grafted to an *ipt* mutant root, cambial activity is also restored in the mutant root. These experiments indicate a number of important points. First, root-derived cytokinins are probably not essential for normal shoot growth. Secondly, transported cytokinins are functional, and this transport can occur both from the root to the shoot and vice versa.

Although cytokinins appear to act as long-distance signals, they are also capable of acting as local, or *paracrine* signals. For example, cytokinins act locally to release apical buds from dormancy (see below). They also act locally to promote the exit of cells from the root apical meristem.

### Cytokinins are rapidly metabolized by plant tissues

Many plant tissues contain the enzyme **cytokinin oxidase**, which cleaves the side chain from zeatin (both *cis* and *trans*), zeatin riboside, iP, and their *N*-glucosides (see Appendix 3). However, dihydrozeatin and its conjugates, as well as aromatic cytokinins such as benzyladenine, are resistant to cleavage. Cytokinin oxidase irreversibly inactivates cytokinins, and it is important in regulating and limiting cytokinin effects. The activity of the enzyme is induced by high cytokinin concentrations, due at least in part to an elevation of the RNA levels for a subset of the genes.

The lack of cleavage of aromatic cytokinins by cytokinin oxidase may explain why they have high activity in bioassays, despite their relatively poor binding to and activation of the cytokinin receptors.

A gene encoding cytokinin oxidase was first identified in maize (Houba-Herin et al. 1999; Morris et al. 1999). In Arabidopsis, cytokinin oxidase is encoded by a multigene family whose members show distinct patterns of expression. Interestingly, several of the genes contain putative secretory signals, suggesting that at least some of these enzymes may be extracellular.

As noted earlier, cytokinin levels can also be conjugated at various positions (see Appendix 3). The nitrogens at the 3, 7, and 9 positions of the adenine ring of cytokinins can be conjugated to glucose residues. Alanine can also be conjugated to the nitrogen at the 9 position, forming lupinic acid. The hydroxyl group of the side chain of cytokinins is also the target for conjugation to glucose residues, or in some cases xylose residues, yielding O-glucoside and O-xyloside cytokinins.

Enzymes that catalyze the conjugation of either glucose or xylose to zeatin have been purified, and their respective genes have been cloned (e.g., Martin et al. 1999). As noted above, both the N- and O-conjugates of cytokinin are functional only if the conjugate is removed. The N-conjugations are generally irreversible, and such conjugated forms of cytokinin are inactive in bioassays. The O-conjugations at the side chain can be removed by glucosidase enzymes to yield free cytokinins. Cytokinin glucosides may be a storage form of these compounds. A gene encoding a glucosidase that can release cytokinins from sugar conjugates has been cloned from maize (Brzobohaty et al. 1993), and ectopic overexpression of this gene in transgenic tobacco results in a perturbation of zeatin metabolism (Kiran et al. 2006).

Dormant seeds often have high levels of cytokinin glucosides but very low levels of hormonally active free cytokinins. Levels of free cytokinins increase rapidly, however, as germination is initiated, and this increase in free cytokinins is accompanied by a corresponding decrease in cytokinin glucosides.

The level of active cytokinin in a particular cell is the summation of the de novo biosynthesis, deconjugation, and transport into that cell, minus the conjugation, degradation, and transport out of cytokinin from that cell.

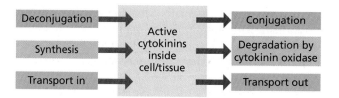

## Cellular and Molecular Modes of Cytokinin Action

The diversity of the effects of cytokinin on plant growth and development is consistent with the involvement of signal transduction pathways with branches leading to specific responses. Cytokinin perception and signaling is mediated by a two-component system, a response pathway remarkably similar to those utilized by bacteria to sense and respond to environmental changes.

### A cytokinin receptor related to bacterial two-component receptors has been identified

The first clue to the nature of the cytokinin receptor came from the discovery of the *CKI1* gene. *CKI1* was identified in a screen for genes that, when overexpressed, conferred cytokinin-independent growth on Arabidopsis cells in culture (Kakimoto 1996). As discussed previously, plant cells generally require cytokinin in order to divide in culture. However, a cell line that overexpresses *CKI1* is capable of growing in culture in the absence of added cytokinin.

*CKI1* encodes a protein similar in sequence to bacterial two-component sensor histidine kinases, which are ubiquitous receptors in prokaryotes (see Chapters 14 and 17). Bacterial two-component regulatory systems mediate a range of responses to environmental stimuli, such as osmoregulation and chemotaxis. Typically these systems are composed of two functional elements: a *sensor histidine kinase*, which senses signals, and a downstream *response regulator*, whose activity is regulated via phosphorylation by the sensor histidine kinase. The sensor histidine kinase is usually a membrane-bound protein that contains two distinct domains: the "input" domain and the "transmitter" domain (**FIGURE 21.6**).

Detection of a signal by the input domain alters the histidine kinase activity of the transmitter domain. Active sensor kinases are dimers that transphosphorylate a conserved histidine residue. This phosphate is then transferred to a conserved aspartate residue in the receiver domain of a cognate response regulator (see Figure 21.6), and this phosphorylation alters the activity of the response regulators. Most response regulators also contain *output* domains that act as transcription factors.

The phenotype resulting from *CKI1* overexpression, combined with its similarity to bacterial receptors, suggested that the CKI1 and/or similar histidine kinases are cytokinin receptors. Support for this model came from identification of the *CRE1* gene (Inoue et al. 2001; Yamada et al. 2001).

Like *CKI1*, *CRE1* encodes a protein similar to bacterial histidine kinases. Loss-of-function *cre1* mutations were identified in a genetic screen for mutants that failed to develop shoots from undifferentiated tissue culture cells in response to cytokinin. This is essentially the opposite screen from the one just described, from which the *CKI1*

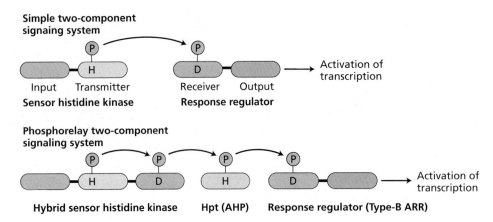

**FIGURE 21.6** Simple versus phosphorelay types of two-component signaling systems. (A) In simple two-component systems, the input domain is the site where the signal is sensed. This domain regulates the activity of the histidine kinase domain, which when activated autophosphorylates a conserved histidine residue. The phosphate is then transferred to an aspartate residue that resides within the receiver domain of a response regulator. Phosphorylation of this aspartate regulates the activity of the output domain of the response regulator, which in many cases is a transcription factor. (B) In the phosphorelay-type two-component signaling system, an extra set of phosphotransfers is mediated by a histidine phosphotransfer protein (Hpt) called AHPs in Arabidopsis. The Arabidopsis response regulators are called ARRs. H, histidine; D, aspartate.

gene was identified by a gain-of-function (ability to divide in the absence of cytokinin) mutation. The *cre1* mutants are also resistant to the inhibition of root elongation observed in response to cytokinin.

Convincing evidence that *CRE1* encodes a cytokinin receptor came from analysis of the expression of the protein in yeast and in *E. coli*. Yeast cells also contain a sensor histidine kinase, and deletion of the gene that encodes this kinase, *SLN1*, is lethal. Expression of *CRE1* in *SLN1*-deficient yeast can restore viability, *but only if cytokinins are present in the medium*. Similar experiments were done in an *E. coli* system in which CRE1 was found to complement a histidine kinase mutant in a cytokinin-dependent manner. Thus the activity of CRE1 (e.g., its ability to replace SLN1) is dependent on cytokinin, which, coupled with the cytokinin-insensitive phenotype of the *cre1* mutants in Arabidopsis, and the ability of purified CRE1 to bind cytokinin with high affinity, unequivocally demonstrates that CRE1 is a cytokinin receptor.

Two other genes in the Arabidopsis genome (*AHK2* [*ARABIDOPSIS HISTIDINE KINASE2*] and *AHK3*) are closely related to *CRE1*, suggesting that, like the ethylene receptors (see Chapter 22), cytokinin receptors are encoded by a multigene family. These transmembrane hybrid kinases contain an extracellular CHASE (Cyclase/*h*istidine kinase-*a*ssociated *s*ensing *e*xtracellular) domain, similar to the one found in CRE1. It has been demonstrated that cytokinins bind to the CHASE domains of CRE1, AHK2, and AHK3 with high affinity. Other members of the Arabidopsis histidine kinase family lack a CHASE domain (including CKI1), and probably do not act as cytokinin receptors.

Furthermore, the function of both AHK2 and AHK3, like that of CRE1, is dependent on cytokinin in the yeast and *E. coli* systems. Therefore, AHK2 and AHK3 are also cytokinin receptors. These receptors display distinct affinities for different cytokinin species, suggesting that different cytokinins may have unique signaling outputs, and therefore distinct functions in the plant. A triple mutant has been identified in which *CRE1*, *AHK2*, and *AHK3* are all disrupted (Higuchi et al. 2004; Nishimura et al. 2004). This triple mutant displays a variety of developmental abnormalities, including little or no floral development, greatly reduced root growth, and reduced rosette size (**FIGURE 21.7**). Nevertheless, the triple-receptor mutant is surprisingly viable, which could mean either that cytokinins are not essential for plant growth or that an alternative, second cytokinin response pathway is present in plants. Further research should help resolve this issue.

### Cytokinins increase expression of the type-A response regulator genes via activation of the type-B ARR genes

One of the primary effects of cytokinin is to alter the expression of various genes. Among the first genes to be up-regulated in response to cytokinin are the **ARR** (*ARABIDOPSIS RESPONSE REGULATOR*) genes. These genes are homologous to the receiver domain of bacterial two-component response regulators, the downstream target of sensor histidine kinases (see Figure 21.6).

In Arabidopsis, response regulators are encoded by a multigene family. They fall into two basic classes: the

Columbia          WS          *ahk2*          *ahk3*          *ahk2*          *ahk2*
                                              *cre1*          *cre1*          *ahk3*          *ahk3*
                                                                                             *cre1*

**FIGURE 21.7** Phenotypes of Arabidopsis plants harboring mutations in two or all three of the cytokinin receptors (*cre1*, *ahk2*, and *ahk3*). The parental wild-type ecotypes (Columbia and WS) are shown on the left. (From Nishimura et al. 2004.)

**type-A ARR** genes, the products of which are made up solely of a receiver domain, and the **type-B ARR** genes, which encode a transcription factor domain in addition to the receiver domain (**FIGURE 21.8**). The rate of transcription of the type-A genes, but not the type-B genes, increases very rapidly in response to applied cytokinin (D'Agostino et al. 2000). This rapid induction is specific for cytokinin and does not require new protein synthesis, features that are hallmarks of primary response genes (discussed in Chapters 17 and 19). The stability of a subset of the type-A ARR proteins is also increased by cytokinin (To et al. 2007), which acts synergistically with the transcriptional response to increase the level of type-A ARR proteins in response to cytokinin.

The rapid induction of the type-A genes, coupled with their similarity to signaling elements predicted to act downstream of sensor histidine kinases, suggests that these elements act downstream of the CRE1 cytokinin receptor family to mediate the primary cytokinin response. Analyses of single and multiple loss-of-function mutations in the type-A *ARR*s have confirmed that these elements are involved in cytokinin signaling with partially redundant functions (To et al. 2004). The type-A ARRs negatively regulate cytokinin signaling by interacting with other proteins in a manner dependent on their phosphorylation state (To et al. 2007).

The expression of a wide variety of other genes is altered in response to cytokinin. These include the gene that encodes nitrate reductase; light-regulated genes such as *LHCB* and *SSU*; defense-related genes such as *PR1*; and genes that encode rRNAs, cytochrome P450s, peroxidase, extensin (a cell wall protein rich in hydroxyproline), and various transcription factors. Cytokinin elevates the expression of these genes by increasing the rate of transcription (as in the case of the type-A *ARR*s) and/or by a stabilization of the RNA transcript.

The type-B *ARR*s, which contain a DNA-binding domain and a transcriptional activator domain in addition to the receiver domain, have been shown to be the direct upstream activators of type-A *ARR* transcription

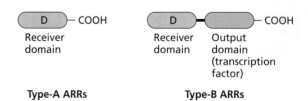

**Type-A ARRs**                    **Type-B ARRs**

**FIGURE 21.8** Comparison of the structures of the type-A and type-B ARRs. The type-A ARRs consist solely of an aspartate (D)-containing receiver domain, but the type-B proteins also contain a fused output domain at the carboxy terminus.

in response to cytokinin. Increased type-B ARR function leads to an increase in the transcription of the type-A *ARRs*, and disruption of multiple type-B *ARRs* compromises the induction of the type-A *ARRs* by cytokinin (Hwang and Sheen 2001; Sakai et al. 2001). Like the type-A ARRs, the type-B ARRs display partial functional redundancy, but in contrast to the type-A *ARRs*, loss-of-function mutations in the type-B *ARRs* lead to insensitivity to cytokinin. The current data suggest that phosphorylation of the receiver domain of the type-B ARRs enables them to increase the transcription of a set of genes including the type-A *ARRs*.

### Histidine phosphotransfer proteins are also involved in cytokinin signaling

From the preceding discussions we have seen that cytokinin binds to the CRE1/AHK receptors at the cell surface and initiates a phosphotransfer that ultimately leads to phosphorylation of the type-B ARRs in the nucleus. How then is the phosphate transferred from the receptors bound to the plasma membrane to the type-B ARRs in the nucleus? The answer is that another set of proteins, the AHP (*Arabidopsis histidine phosphotransfer*) proteins, acquire the phosphate from the activated receptors, and move into the nucleus, where they transfer the phosphate group to the type-B ARRs.

In two-component systems that involve a sensor kinase with a fused receiver domain (the structure of most eukaryotic sensor histidine kinases, including those of the CRE1 family), there is an additional set of phosphotransfers that are mediated by a **histidine phosphotransfer protein (Hpt)**. Phosphate is first transferred from ATP to a histidine within the histidine kinase domain of the sensor kinase, and then transferred to an aspartate residue on the fused receiver. From the aspartate residue, the phosphate group is then transferred to a histidine residue present in the Hpt protein, and then finally to an aspartate on the receiver domain of the response regulator (see Figure 21.6). This phosphorylation of the receiver domain of the response regulator alters its activity. Thus, Hpt proteins are predicted to mediate the phosphotransfer between sensor kinases and response regulators.

In Arabidopsis there are five *Hpt* genes, called *AHPs*. The AHP proteins have been shown to physically associate with receiver domains from both the cytokinin histidine kinase receptors and response regulators (Dortay et al. 2006), consistent with their role in mediating phosphotransfer among these signaling elements. In vitro, the AHPs are capable of both accepting a phosphate group from the cytokinin receptors and transferring it to the response regulators (Imamura et al. 1999). Finally, disruption of multiple *AHP* genes in Arabidopsis leads to cytokinin insensitivity (Hutchison et al. 2006). Together, these findings indicate that the AHPs are the immediate downstream targets of the activated cytokinin receptors,

and that these proteins transduce the cytokinin signal to the nucleus, where they phosphorylate and activate the type-B and type-A ARRs.

A model of cytokinin signaling is presented in **FIGURE 21.9**. Cytokinin binds to the CRE1, AHK2 and AHK3 receptors and initiates a phosphorelay that ultimately results in the phosphorylation and activation of the type-B ARR proteins. Activation of the type-B proteins (transcription factors) leads to the alteration of the transcription of various targets that mediate the changes in cellular function, such as an activation of the cell cycle. Phosphorylated type-A ARRs inhibit cytokinin responsiveness by interactions with unknown targets, providing a negative feedback loop (see Figure 21.9). These cytokinin two-component signaling elements are also found in monocots (rice and maize), as well as in the moss *Physcomitrella patens*, and may act throughout the plant kingdom to perceive and transduce cytokinin signals.

## The Biological Roles of Cytokinins

Although discovered as a cell division factor, cytokinins can stimulate or inhibit a variety of physiological, metabolic, biochemical, and developmental processes when they are applied to higher plants, and it is increasingly clear that endogenous cytokinins play an important role in the regulation of these events in the intact plant. In addition to its role in cell proliferation, cytokinin affects many other processes, including vascular development, apical dominance, nutrient acquisition, and leaf senescence.

In this section we will discuss the role of cytokinins in regulating cell division and survey some of their diverse effects on plant growth and development. WEB ESSAY 21.2 highlights an additional role of cytokinin in the formation of protonema in moss. These effects have been elucidated classically through the examination of the effects of exogenously applied cytokinin to various plant species and the expression of the *ipt* gene from *A. tumefaciens* (see page 626). More recently, insights into the roles of cytokinins in plants have been obtained by examining the phenotypes of mutants and transgenic lines with increased or decreased cytokinin function, such as lines altered in endogenous cytokinin levels or cytokinin signaling.

### Cytokinins promote shoot growth by increasing cell proliferation in the shoot apical meristem

As discussed earlier, cytokinins are generally required for cell division of plant cells in vitro. Several lines of evidence suggest that cytokinins also play key roles in the regulation of cell division in vivo.

Much of the cell division in an adult plant occurs in the meristems (see Chapter 16). Cytokinin plays a positive role in the proliferation of cells in the shoot apical meristem. Recall that elevated levels of cytokinins may

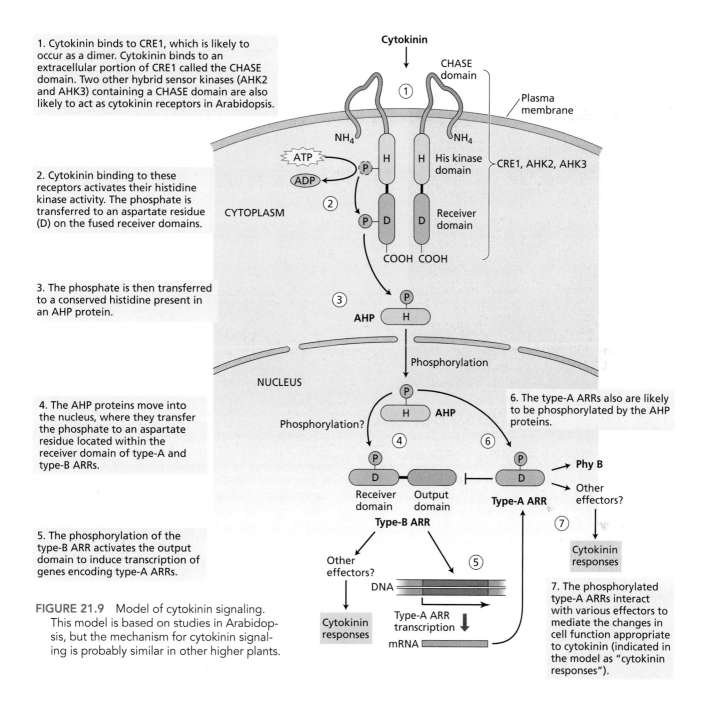

1. Cytokinin binds to CRE1, which is likely to occur as a dimer. Cytokinin binds to an extracellular portion of CRE1 called the CHASE domain. Two other hybrid sensor kinases (AHK2 and AHK3) containing a CHASE domain are also likely to act as cytokinin receptors in Arabidopsis.

2. Cytokinin binding to these receptors activates their histidine kinase activity. The phosphate is transferred to an aspartate residue (D) on the fused receiver domains.

3. The phosphate is then transferred to a conserved histidine present in an AHP protein.

4. The AHP proteins move into the nucleus, where they transfer the phosphate to an aspartate residue located within the receiver domain of type-A and type-B ARRs.

5. The phosphorylation of the type-B ARR activates the output domain to induce transcription of genes encoding type-A ARRs.

6. The type-A ARRs also are likely to be phosphorylated by the AHP proteins.

7. The phosphorylated type-A ARRs interact with various effectors to mediate the changes in cell function appropriate to cytokinin (indicated in the model as "cytokinin responses").

**FIGURE 21.9** Model of cytokinin signaling. This model is based on studies in Arabidopsis, but the mechanism for cytokinin signaling is probably similar in other higher plants.

result in fasciation of shoots (see Figure 21.3), a condition resulting from over-proliferation of the shoot apical meristem. Reduction of cytokinin function by reducing endogenous cytokinin levels via overexpression of cytokinin oxidase or by mutation of the *IPT* genes results in the opposite effect, a substantial retardation of shoot development (**FIGURE 21.10**) due to a reduction in the size of the shoot apical meristem (**FIGURE 21.11**) (Werner et al. 2001). Disruption of cytokinin perception (e.g., in a triple-receptor mutant) also results in a reduced shoot apical meristem, leading to a stunted shoot and little or no flower production (**FIGURE 21.12**) (Higuchi et al. 2004; Nishimura et al. 2004).

In contrast, disruption of certain *negative regulators* of cytokinin signaling (such as type-A ARRs, discussed earlier in the chapter) results in an *increase* in the size of the shoot apical meristem of maize (Giulini et al. 2004). Recall that disruption of a negative regulator would be expected to have a stimulatory effect on cytokinin signaling. Together, these findings strongly support the notion that endogenous cytokinins positively regulate cell division in the shoot apical meristem in vivo. Similarly, cytokinins promote cell

**FIGURE 21.10**  Tobacco plants overexpressing genes for cytokinin oxidase. The plant on the left is the wild type. The two plants on the right are each overexpressing one of two different Arabidopsis cytokinin oxidase genes: *AtCKX1* and *AtCKX2*. Shoot growth is strongly inhibited in the transgenic plants. (From Werner et al. 2001.)

(A)

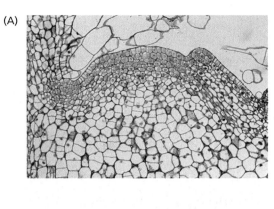

(B)

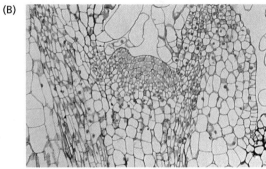

**FIGURE 21.11**  Cytokinin is required for normal growth of the shoot apical meristem. (A) Longitudinal section through the shoot apical meristem of a wild-type tobacco plant. (B) Longitudinal section through the shoot apical meristem of a transgenic tobacco overexpressing a gene that encodes cytokinin oxidase (*AtCKX1*). Note the reduction in the size of the apical meristem in the cytokinin-deficient plant. (From Werner et al. 2001.)

proliferation in the lateral meristems, also known as the vascular cambium (see below).

## Cytokinins interact with other hormones and with several key transcription factors

As discussed in Chapter 16, researchers have begun to elucidate the mechanism by which cytokinins interact with other hormones and with several key transcription factors involved in regulating shoot apical meristem function. Cytokinin levels positively regulate the expression of the KNOTTED1-like (KNOX) homeobox transcription factor homologs *KNAT1* and *STM*, genes that are important in the regulation of meristem function (see Chapter 16) (Rupp et al. 1999). Similarly, KNOX genes positively regulate cytokinin levels via the induction of a subset of *IPT* genes in Arabidopsis and rice. This suggests a positive feedback loop between KNOX transcription factors and cytokinin levels. Furthermore, a strong *stm* mutant was partially rescued by

(A)

Wild type ⊢——⊣ 10 mm

(B)

*ahk2 ahk3 cre1* ⊢——⊣ 10 mm

**FIGURE 21.12**  Comparison of the rosettes of wild-type Arabidopsis and the triple cytokinin receptor–knockout mutant, *ahk2 ahk3 cre1*. (From Nishimura et al. 2004.)

increasing cytokinin levels via the expression of an *IPT* gene from the STM promoter (Yanai et al. 2005; Jasinski et al. 2005), indicating that a major function of the KNOX genes is to elevate cytokinin biosynthesis in the shoot apical meristem.

Gibberellins (see Chapter 20) also play a role in regulating the shoot apical meristem. The KNOX proteins downregulate (reduce) the expression of *GA20 oxidase*, a gene encoding an enzyme in GA biosynthesis. Thus, the KNOX proteins act to establish a high cytokinin:GA ratio in the shoot apical meristem, which signals cells in the meristem to continue to proliferate, rather than differentiating into leaf primordia. Cytokinins in turn induce the expression of *GA2 oxidase* in the shoot apical meristem, which encodes an enzyme that degrades active GAs (Jasinski et al. 2005). This increase in the degradation of GA by cytokinin reinforces the effect of KNOX proteins on GA levels.

There is also a direct link between cytokinin signaling and the homeodomain protein WUSCHEL (WUS). WUS is expressed in the organizing center of the shoot apical meristem and induces a stem cell fate to the overlying cells (see Chapter 16). A subset of type-A *ARR* genes are directly repressed by WUS binding to their promoters (Leibfried et al. 2005). This down-regulation of type-A ARR gene expression results in a niche of cells within the shoot apical meristem that are hypersensitive to cytokinin. Thus, these WUS-expressing cells respond to the cytokinins synthesized within the shoot apical meristem in a unique manner, imparting a distinct fate to these cells.

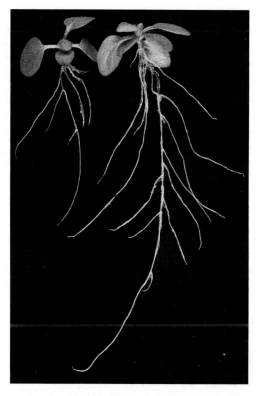

**FIGURE 21.13** Cytokinin suppresses the growth of roots. The cytokinin-deficient, *AtCKX1*-overexpressing roots (right) are larger than those of the wild-type tobacco plant (left). (From Werner et al. 2001.)

## Cytokinins inhibit root growth by promoting the exit of cells from the root apical meristem

Cytokinin plays a very different role in the root apical meristem than it does in the shoot apical meristem. In contrast to its effect on the shoot, overexpression of cytokinin oxidase in tobacco *increases* root growth (**FIGURE 21.13**), primarily by increasing the size of the root apical meristem (**FIGURE 21.14**). Similarly, mutations that partially disrupt cytokinin perception also cause enhanced root growth. However, disruption of all three cytokinin receptors in Arabidopsis results in reduced cell division in both root and shoot apical meristems. The roots of these triple-receptor mutants elongate for only a few days and then cease growing. The decreased root growth in the triple-receptor mutant may be the result of a disruption in the formation of vascular tissue (see below), in particular the lack of phloem development, which may result in dysfunction of the root apical meristem.

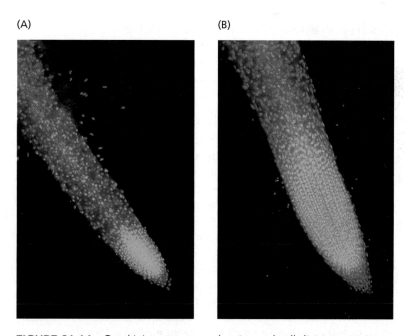

**FIGURE 21.14** Cytokinin suppresses the size and cell division activity of roots. (A) Wild type. (B) Plant overexpressing *AtCKX1*. These roots were stained with the fluorescent dye 4',6-diamidino-2-phenylindole (DAPI), which stains DNA in the nucleus. (From Werner et al. 2001.)

The mechanism by which cytokinins negatively regulate root apical meristems has recently been explored. The size of a meristem is determined by the rate at which cells divide minus the rate at which cells exit the meristem by growth and differentiation. Cytokinins accelerate the process of vascular differentiation in the root tip (Dello Ioio et al. 2008), causing a concomitant decrease in the size of the root apical meristem. That is, increased cytokinin function enhances the rate at which cells differentiate into vascular tissue; the result is fewer meristematic cells and, thus, less root growth. Conversely, decreased cytokinin function decreases the rate of vascular differentiation, and results in a greater number of apical meristem cells and more root growth (Werner et al. 2003).

Auxin also plays a role in the root apical meristem, acting as a positive regulator of cell division. Thus, auxin and cytokinin have distinct functions in the root apical meristem: auxin promoting cell division and cytokinin promoting cell differentiation.

Dello Ioio et al. have proposed a model for a simple regulatory circuit describing the interaction of auxin and cytokinin in the root apical meristem (Dello Ioio et al. 2008). These authors showed that cytokinin, acting through ARR1 (a type-B ARR; see Figure 21.8), stimulates the expression of the gene coding for an Aux/IAA repressor protein (*SHY2*), while auxin causes the degradation of the same Aux/IAA protein. Thus, auxin and cytokinin act antagonistically in the root by regulating in opposite ways the abundance of the Aux/IAA protein SHY2. SHY2 in turn positively regulates expression of the *IPT5* gene, involved in cytokinin biosynthesis, and negatively regulates the expression of a subset of *PIN* genes, which encode protein involved in polar auxin transport (see Chapter 19). These effects of SHY2 on cytokinin biosynthesis and auxin transport reinforce the regulatory loop regulating cell division and differentiation.

## Cytokinins regulate specific components of the cell cycle

Cytokinins regulate cell division by affecting the controls that govern the passage of the cell through the cell division cycle. Zeatin levels peak in synchronized culture tobacco cells at the end of S phase, the $G_2$/M phase transition, and in late $G_1$. Inhibition of cytokinin biosynthesis blocks cell division, and application of exogenous cytokinin allows cell division to proceed.

Cytokinins were discovered in relation to their ability to stimulate cell division in tissues supplied with an optimal level of auxin. Evidence suggests that both auxin and cytokinins participate in regulating the cell cycle and that they do so by controlling the activity of cyclin-dependent kinases. As discussed in Chapter 1, **cyclin-dependent protein kinases (CDKs)**, in concert with their regulatory subunits, the **cyclins**, are enzymes that regulate the eukaryotic cell cycle.

The expression of the gene that encodes the major CDK, Cdc2 (*cell division cycle 2*), is regulated by auxin. In pea root tissues, *CDC2* mRNA is induced within 10 minutes after treatment with auxin, and in tobacco pith, high levels of CDK are induced when the tissue is cultured on a medium containing auxin (John et al. 1993). However, the CDK induced by auxin is enzymatically inactive, and high levels of CDK alone are not sufficient to permit cells to divide.

Cytokinin has been linked to the activation of a Cdc25-like phosphatase, whose role is to remove an inhibitory phosphate group from the Cdc2 kinase (Zhang et al. 1996). This action of cytokinin provides one potential link between cytokinin and auxin in the cell cycle: regulating the passage from $G_2$ to M phase.

Cytokinins also elevate the expression of the *CYCD3* gene, which encodes a *D-type cyclin* (Soni et al. 1995; Riou-Khamlichi et al. 1999). In Arabidopsis, *CYCD3* is expressed in proliferating tissues such as shoot meristems and young leaf primordia. Overexpression of *CYCD3* can bypass the cytokinin requirement for cell proliferation in culture (**FIGURE 21.15**) (Riou-Khamlichi et al. 1999). These data suggest that a major mechanism for cytokinin's ability to stimulate cell division is its increase of *CYCD3* function. This is reminiscent of animal cell cycles, in which D-type cyclins are regulated by a wide variety of growth factors and play a key role in regulating the passage through the restriction point of the cell cycle in $G_1$.

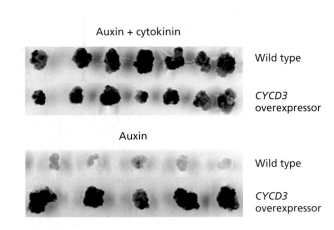

**FIGURE 21.15** *CYCD3*-expressing callus cells can divide in the absence of cytokinin. Leaf explants from transgenic Arabidopsis plants expressing *CYCD3* from a cauliflower mosaic virus 35S promoter were induced to form calluses through culturing in the presence of auxin plus cytokinin or auxin alone. The wild-type control calluses required cytokinin to grow. The *CYCD3*-expressing calluses grew well on medium containing auxin alone. The photographs were taken after 29 days. (From Riou-Khamlichi et al. 1999.)

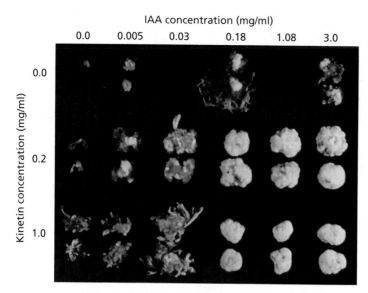

**FIGURE 21.16** The regulation of growth and organ formation in cultured tobacco callus at different concentrations of auxin and kinetin. At low auxin and high kinetin concentrations (lower left), buds developed. At high auxin and low kinetin concentrations (upper right), roots developed. At intermediate or high concentrations of both hormones (middle and lower right), undifferentiated callus developed. (Courtesy of Donald Armstrong.)

## The auxin:cytokinin ratio regulates morphogenesis in cultured tissues

Shortly after the discovery of kinetin, it was observed that the differentiation of cultured callus tissue derived from tobacco pith segments into either roots or shoots depends on the ratio of auxin to cytokinin in the culture medium. Whereas high auxin:cytokinin ratios stimulated the formation of roots, low auxin:cytokinin ratios led to the formation of shoots. At intermediate levels, the tissue grew as an undifferentiated callus (**FIGURE 21.16**) (Skoog and Miller 1965).

The effect of auxin:cytokinin ratios on morphogenesis can also be seen in crown gall tumors by mutation of the T-DNA of the *Agrobacterium* Ti plasmid (Garfinkel et al. 1981). Mutating the *ipt* gene (the *tmr* locus) of the Ti plasmid blocks zeatin biosynthesis in the infected cells. The resulting high auxin:cytokinin ratio in the tumor cells causes the proliferation of roots instead of undifferentiated callus tissue. In contrast, mutating either of the genes for auxin biosynthesis (*tms* locus) lowers the auxin:cytokinin ratio and stimulates the proliferation of shoots (**FIGURE 21.17**) (Akiyoshi et al. 1983). These partially differentiated tumors are known as teratomas.

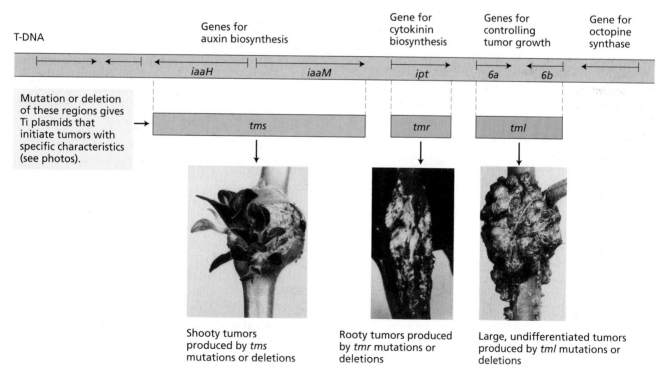

**FIGURE 21.17** Map of the T-DNA from an *Agrobacterium* Ti plasmid, showing the effects of T-DNA mutations on crown gall tumor morphology. The genes *iaaH* and *iaaM* encode the two enzymes involved in auxin biosynthesis, *ipt* encodes a cytokinin biosynthesis enzyme, and *6b* encodes a transcriptional regulator that controls tumor growth. Mutations in these genes produce the phenotypes illustrated. (From Morris 1986, courtesy of R. Morris.)

## Cytokinins modify apical dominance and promote lateral bud growth

One of the primary determinants of plant form is the degree of apical dominance (see Chapter 19). Plants with strong apical dominance, such as maize, have a single growing axis with few lateral branches. In contrast, many lateral buds initiate growth in shrubby plants. Branching patterns are normally determined by light, nutrients, and genotype. As discussed in Chapter 19, branching is also triggered by decapitation, a phenomenon that allows primary growth to continue after the loss of the terminal bud.

Physiologically, branching is regulated by a complex interplay of hormones, including auxin, cytokinin, and a recently identified root-derived signal. Auxin transported polarly from the apical bud suppresses the growth of axillary buds (see Chapters 16 and 19). In contrast, cytokinin stimulates cell division activity and outgrowth when applied directly to the axillary buds of many species, and cytokinin-overproducing mutants tend to be bushy. In the nodal region of pea stems, auxin was found to inhibit the expression of a subset of *IPT* genes, which encode the enzyme catalyzing the rate-limiting step in cytokinin biosynthesis, and to elevate the expression of cytokinin oxidase, which degrades cytokinins (**FIGURE 21.18**). The combined effect of the regulation of these genes by auxin is to keep cytokinin levels low in the apical buds. Removal of the shoot apex results in a decreased auxin flow, which allows IPT levels to rise and cytokinin oxidase levels to fall (see Figure 21.18). The net effect of terminal bud removal is thus an increased concentration of cytokinin in the nodal area of the stem. From there, the cytokinin moves into the adjacent bud and releases it from dormancy (Tanaka et al. 2006).

The finding that the cytokinins responsible for axillary bud growth are synthesized locally in the nodal tissue adjacent to the bud supersedes an earlier model in which the cytokinins were thought to be derived from the root. However, recent evidence from grafting studies in mutants with highly branched phenotypes suggests that a novel root-derived hormone called strigolactone acts, like auxin from the shoot apical meristem, as a *suppressor* of axillary bud growth (see Chapter 19 and **WEB TOPIC 21.9**). Thus, the dormancy of axillary buds may be regulated by the dual action of long-distance negative signals from both the shoot and root apical meristems.

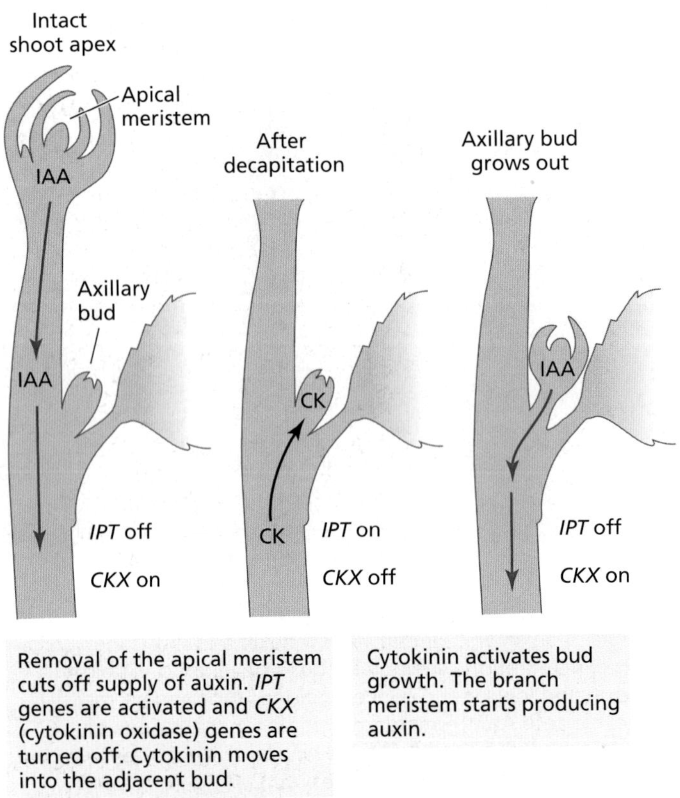

## Cytokinins delay leaf senescence

Leaves detached from the plant slowly lose chlorophyll, RNA, lipids, and protein, even if they are kept moist and provided with minerals. This programmed aging process leading to death is termed **senescence** (see Chapters 16 and 23). Leaf senescence is more rapid in the dark than

Removal of the apical meristem cuts off supply of auxin. *IPT* genes are activated and *CKX* (cytokinin oxidase) genes are turned off. Cytokinin moves into the adjacent bud.

Cytokinin activates bud growth. The branch meristem starts producing auxin.

**FIGURE 21.18**  Interaction of auxin and cytokinin in the regulation of shoot branching. In a plant with an intact shoot apex (left), auxin (IAA) derived from the shoot apex inhibits cytokinin levels in the apical bud by inhibiting *IPT* gene expression (cytokinin biosynthesis) and by elevating cytokinin oxidase expression (CKX) (cytokinin degradation), which prevents bud outgrowth. Removal of the shoot apex (middle) eliminates the flow of auxin, which results in elevated *IPT* expression and decreased *CKX* expression. This leads to elevated levels of cytokinins in the apical buds, causing them to grow. After the buds have grown for a period (right), they begin to make and export their own auxin. *IPT* is again shut down and *CKX* elevated, resulting in lower cytokinin levels once again. (Adapted from Shimizu-Sato et al. 2008.)

in the light. Treating isolated leaves of many species with cytokinins delays their senescence.

Although applied cytokinins do not prevent senescence completely, their effects can be dramatic, particularly when the cytokinin is sprayed directly on the intact plant. If only one leaf is treated, it remains green after other leaves of similar developmental age have yellowed and dropped off the plant. If a small spot on a leaf is treated with cytokinin, that spot will remain green after the surrounding tissues on the same leaf begin to senesce. This "green island" effect can also be observed in leaves infected by some fungal pathogens, as well as in those hosting galls produced by insects such as jumping plant lice (*Pachypsylla celtidis-mamma*). On hackberry (*Celtis occidentalis* L.) leaves, jumping plant lice galls produce green spots, correlated with a large increase in cytokinin content in the affected tissue, on an otherwise yellow, senescent leaf.

Unlike young leaves, mature leaves produce little, if any, cytokinin. Mature leaves may depend on root-derived cytokinins to postpone their senescence. Senescence is initiated in soybean leaves by seed maturation (a phenomenon known as *monocarpic senescence*) and can be delayed by seed removal. Although the seedpods control the onset of senescence, they do so by controlling the delivery of root-derived cytokinins to the leaves.

The cytokinins involved in delaying senescence are primarily zeatin riboside and dihydrozeatin riboside, which may be transported into the leaves from the roots through the xylem, along with the transpiration stream (Noodén et al. 1990).

To test the role of cytokinin in regulating the onset of leaf senescence, tobacco plants were transformed with a chimeric gene in which a senescence-specific promoter was used to drive the expression of the *Agrobacterium ipt* gene (Gan and Amasino 1995). The transformed plants had wild-type levels of cytokinins and developed normally until the onset of leaf senescence.

As the leaves aged, however, the senescence-specific promoter was activated, triggering the expression of the *ipt* gene within leaf cells just as senescence would have been initiated. The resulting elevated cytokinin levels not only blocked senescence, but also limited further expression of the *ipt* gene, preventing cytokinin overproduction (**FIGURE 21.19**). This result suggests that cytokinins are natural regulators of leaf senescence.

The AHK3 receptor appears to be the primary cytokinin receptor regulating leaf senescence in Arabidopsis. Elevated *AHK3* function results in a significant delay in leaf senescence. Conversely, disruption of *AHK3*, but not other cytokinin receptor genes, results in premature leaf senescence (Kim et al. 2006).

## Cytokinins promote movement of nutrients

Cytokinins influence the movement of nutrients into leaves from other parts of the plant, a phenomenon known as *cytokinin-induced nutrient mobilization*. This process can be observed when nutrients (sugars, amino acids, and so on) radiolabeled with $^{14}C$ or $^{3}H$ are fed to plants after one leaf or part of a leaf is treated with a cytokinin. Subsequent autoradiography of the whole plant reveals the pattern of movement and the sites at which the labeled nutrients accumulate.

Experiments of this nature have demonstrated that nutrients are preferentially transported to and accumulated in the cytokinin-treated tissues. It has been postulated that the hormone causes nutrient mobilization by creating a new source–sink relationship. As discussed in Chapter 10, nutrients translocated in the phloem move from a site of production or storage (the source) to a site of utilization (the sink). The hormone may stimulate the metabolism of

Plant expressing *ipt* gene remains green and photosynthetic.

Age-matched control shows advanced senescence.

**FIGURE 21.19** Leaf senescence is retarded in a transgenic tobacco plant containing a cytokinin biosynthesis gene, *ipt*, from *Agrobacterium tumefaciens* fused to a senescence-induced promoter. The *ipt* gene is expressed in response to signals that induce senescence. (From Gan and Amasino 1995, courtesy of R. Amasino.)

the treated area so that nutrients move toward it. However, it is not necessary for the nutrient itself to be metabolized in the sink cells because even nonmetabolizable substrate analogs are mobilized by cytokinins (**FIGURE 21.20**).

Cytokinin levels change in response to the concentration of nutrients to which plants are exposed. For example, application of nitrate to nitrogen-depleted maize seedlings results in a rapid rise in cytokinin levels in the roots, followed by mobilization of the cytokinins to the shoots via the xylem (Takei et al. 2001b). This increase is due, at least in part, to an induction of expression of one member of the IPT gene family, *IPT3*. Cytokinin levels are also influenced by the concentration of phosphate in the environment, and cytokinins alter the expression of phosphate- and sulfate-responsive genes, suggesting an interaction between these response pathways (Argueso et al. 2009).

Thus, *the nutrient status of the plant regulates cytokinin levels*, and in turn the ratio of cytokinin to auxin determines the relative growth rates of roots and shoots: High cytoki-

In seedling A, the left cotyledon was sprayed with water as a control. The left cotyledon of seedling B and the right cotyledon of seedling C were each sprayed with a solution containing 50 m*M* kinetin.

The dark stippling represents the distribution of the radioactive amino acid as revealed by autoradiography.

The results show that the cytokinin-treated cotyledon has become a nutrient sink. However, radioactivity is retained in the cotyledon to which the amino acid was applied when the labeled cotyledon is treated with kinetin (seedling C).

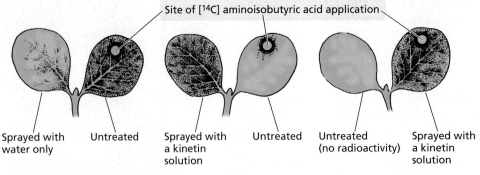

Site of [$^{14}$C] aminoisobutyric acid application

| Sprayed with water only | Untreated | Sprayed with a kinetin solution | Untreated | Untreated (no radioactivity) | Sprayed with a kinetin solution |

**Seedling A**          **Seedling B**          **Seedling C**

**FIGURE 21.20**  The effect of cytokinin on the movement of an amino acid in cucumber seedlings. A radioactively labeled amino acid that cannot be metabolized, such as aminoisobutyric acid, was applied as a discrete spot on the right cotyledon of each of these seedlings. The black color indicates the distribution of radioactivity. (Drawn from data obtained by K. Mothes.)

nin concentrations promote shoot growth, and, conversely, high auxin levels promote root growth. In the presence of low nutrient levels, cytokinin levels are also low, resulting in an increase in root growth and allowing the plant to more effectively acquire the nutrients present in the soil. In contrast, optimal levels of soil nutrients promote increased cytokinin levels, which favor shoot growth, thus maximizing photosynthetic capacity.

### Cytokinins affect light signaling via phytochrome

Although seeds can germinate in the dark, the morphology of dark-grown seedlings is very different from that of light-grown seedlings (see Chapter 17): Dark-grown seedlings are said to be **etiolated**. The hypocotyl and internodes of etiolated seedlings are more elongated, and cotyledons and leaves do not expand. Instead of maturing as chloroplasts, the proplastids of dark-grown seedlings develop into **etioplasts**, which do not synthesize chlorophyll or most of the enzymes and structural proteins required for the formation of the chloroplast thylakoid system and photosynthetic machinery. When seedlings germinate in the light, chloroplasts mature directly from the proplastids present in the embryo; etioplasts also can mature into chloroplasts when etiolated seedlings are illuminated.

If etiolated leaves are treated with cytokinin before being illuminated, they form chloroplasts with more extensive grana, and chlorophyll and photosynthetic enzymes are synthesized at a greater rate upon illumination (**FIGURE 21.21**). These results suggest that cytokinins—along

with other factors, such as light, nutrition, and development—regulate the synthesis of photosynthetic pigments and proteins. (For more on how cytokinins promote light-mediated development, see **WEB TOPIC 21.10**.)

A key mediator of many light responses is the red light photoreceptor phytochrome (see Chapter 17). The type-A response regulator ARR4 (see Figure 21.8) has been found to stabilize the active, Pfr form of one phytochrome, PhyB, by reducing the rate of dark reversion. Thus, ARR4 acts as a positive regulator of PhyB function (Sweere et al. 2001). Another point of convergence between light signaling and cytokinin occurs via the HY5 protein. This bZIP transcription factor is a positive regulator of photomorphogenesis, acting downstream of multiple families of photoreceptors, including the phytochromes and cryptochromes. Cytokinin increases the abundance of HY5 protein, most likely by increasing HY5 protein stability (Vandenbussche et al. 2007).

The regulation of cytokinin degradation by cytokinin oxidase has been linked to the response to shade in Arabidopsis. Plants grown under canopy shade receive a low ratio of red/far red light (R/FR), which induces a number of developmental changes, including increased hypocotyl elongation and a rapid arrest of leaf primordia growth (see Chapter 17; Carabelli et al. 2007). Exposure of Arabidopsis plants to low R/FR results in a rapid increase in auxin signaling in the leaf primordia, which in turn leads to elevated expression of a cytokinin oxidase gene (Carabelli et al. 2007). This elevated expression of cytokinin oxidase results in the reduction of cytokinin levels in the leaf primordia, and hence in a reduction in cell proliferation. Cytokinin has also been shown to promote the expansion and greening of several dicot cotyledons, such as those of radish (see **WEB TOPIC 21.11**).

Finally, elements of the two-component cytokinin response pathway have been found to interact with the circadian clock, which entrains plants to light/dark diurnal cycles (see **WEB TOPIC 21.12**).

(A) Etioplasts

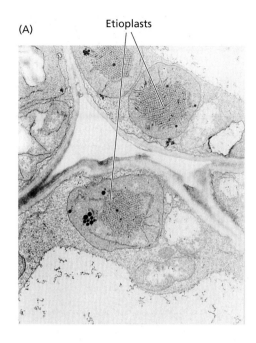

(B) Thylakoids

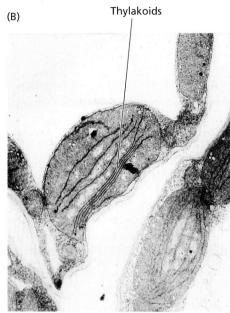

**FIGURE 21.21** Cytokinin influence on the development of the chloroplasts of wild-type Arabidopsis seedlings grown in darkness. (A) Plastids develop as etioplasts in the untreated, dark-grown control. (B) Cytokinin treatment resulted in thylakoid formation in the plastids of dark-grown seedlings. (From Chory et al. 1994, courtesy of J. Chory.)

## Cytokinins regulate vascular development

Cytokinins are involved in the development of vascular tissues. Arabidopsis mutants that are strongly insensitive to cytokinin (e.g., triple-receptor mutants) have reduced root vascular cell files that form only protoxylem, lacking both phloem and mature metaxylem. This defect is linked to a reduction in the number of embryonic root vascular initials. When the Hpt-like protein AHP6 is mutated so that it lacks phosphotransfer activity, this mutation rescues (reverses) these developmental defects (Mähönen et al. 2006). In vitro experiments suggest that AHP6 may interfere with phosphotransfer between two-component elements. These results suggest that cytokinin is necessary to regulate the balance of cell proliferation and differentiation during vascular development, with AHP6 acting as a negative regulator of cytokinin signaling in this context.

Cytokinins are also necessary for the development of the vascular cambium. In poplar, expression of a cytokinin oxidase gene in cambial cells in transgenic plants results in thinner stems as a result of reduced cambial function (Nieminen et al. 2008). Furthermore, disruption of multiple IPT genes in Arabidopsis also results in a plant with thinner stems and roots, due to the absence of cambium (Matsumoto-Kitano et al. 2008). Thus, cytokinins are required for both the initiation of vascular initials and the formation of the lateral meristems in the cambium.

## Manipulation of cytokinins to alter agriculturally important traits

Some of the consequences of altering cytokinin function could be highly beneficial for agriculture if synthesis of the hormone can be controlled. Because leaf senescence is delayed in the cytokinin-overproducing plants, it should be possible to extend their photosynthetic productivity.

Indeed, when an *ipt* gene is expressed in lettuce from a senescence-inducible promoter, leaf senescence is strongly retarded (**FIGURE 21.22**), similar to the results observed in tobacco (see Figure 21.19).

In addition, cytokinin production could be linked to damage caused by predators. For example, tobacco plants transformed with an *ipt* gene under the control of the promoter from a wound-inducible protease inhibitor II gene were more resistant to insect damage. The tobacco hornworm consumed up to 70% fewer tobacco leaves in plants that expressed the *ipt* gene driven by the protease inhibitor promoter (Smigocki et al. 1993).

Manipulation of cytokinin also has the potential to increase grain yield in rice. Humans have unwittingly taken advantage of the promotive effect of cytokinin on the shoot apical meristem in their breeding of cultivated rice varieties. The rice varieties *japonica* and *indica* differ dramatically in their yield, with the latter generally producing more grains in their main panicle and ultimately a higher yield (**FIGURE 21.23**). The increased grain number in *indica* varieties has recently been linked to a decrease in the function of a cytokinin oxidase gene (Ashikari et al. 2005). As a consequence of the reduced function of this cytokinin oxidase in the *indica* varieties, cytokinin levels are higher in the inflorescence, which alters the inflorescence meristem such that it produces more reproductive organs, more seeds per plant, and ultimately a higher yield.

## Cytokinins are involved in the formation of nitrogen-fixing nodules in legumes

Nitrogen-fixing rhizobacteria play an important role in the control of available nitrogen in the soil. Through the process of nitrogen fixation, these bacteria are able to convert $N_2$ into organic forms that can be assimilated by the plants.

**FIGURE 21.22** Leaf senescence is retarded in transgenic lettuce plants expressing the cytokinin biosynthesis gene *ipt* at the time of senescence. Control plants (upper five) lack the transgene; *SAG12::IPT* plants (lower five) utilize a "senescence-associated gene" promoter (*SAG12*) to drive the expression of *ipt* at the onset of senescence. (From McCabe et al. 2001.)

The bacteria obtain energy from the plant in exchange for ammonia produced during nitrogen fixation, an energetically costly process that is tightly regulated in legume plants.

The process of nitrogen fixation occurs in specialized structures, called *nodules*, in the root. In the course of nodule organogenesis, the bacteria induce morphogenic alterations in the roots of their legume plant hosts, a process that requires extensive modification of the root epidermis for infection initiation and subsequent activation of cell division leading to nodule formation.

Cytokinins have been implicated in nodulation based on various lines of evidence. Some nitrogen-fixing bacteria, such as *Rhizobium leguminosarum* and *Bradyrhizobium japonicum*, produce compounds with cytokinin-like activity. Application of exogenous cytokinin can cause the induction of cortical cell divisions and an up-regulation of early nodulin genes. Most compellingly, disruption of *MtCRE1* (an ortholog of Arabidopsis *CRE1*) in the legume *Medicago truncatula* results in a failure to initiate cortical cell divisions that are necessary for nodula-

tion (Gonzalez-Rizzo et al. 2006; Murray et al. 2007), while a gain-of-function mutation in the same receptor leads to the spontaneous formation of root nodules in the absence of rhizobia (Tirichine et al. 2007). Together, these studies demonstrate that cytokinin function is necessary and sufficient for nodulation.

(A)

Koshihikari (*japonica*)      Habataki (*indica*)

(B)

Koshihikari      Habataki

**FIGURE 21.23** Cytokinin regulates grain yield in rice. The grain number in the *indica* variety of rice is higher than that in the *japonica* variety as a result of a disruption in a cytokinin oxidase gene. (A) Whole plants of Koshihikari (a *japonica* strain) and Habataki (an *indica* strain). (B) Close-

ups of flower stalks from each variety. In the *indica* variety, cytokinin oxidase levels are decreased by a naturally occurring mutation that leads to higher levels of cytokinin in the developing panicles of the *indica* plants. This results in more flowers and ultimately a higher grain yield.

## SUMMARY

Cytokinins participate in the regulation of many plant processes, including cell proliferation, morphogenesis of shoots and roots, nutrient acquisition, vascular development, light responses, and senescence.

### Cell Division and Plant Development

- In the intact plant, cells can be stimulated to divide by wounding, infection, and plant hormones, including cytokinins (**Figure 21.1**).

- Cytokinins are $N^6$-substituted adenine derivatives that initiate cell proliferation in many plant cells in the presence of auxin.

### The Discovery, Identification and Properties of Cytokinins

- Zeatin is the most abundant naturally occurring free cytokinin, but dihydrozeatin (DHZ) and isopentenyl adenine (iP) also are found (**Figure 21.2**).

- Some plant pathogenic bacteria, fungi, insects, and nematodes secrete active cytokinins, which in some cases induce abnormal growths (**Figure 21.3**).

### Biosynthesis, Metabolism, and Transport of Cytokinins

- Cytokinins are synthesized in roots, developing embryos, young leaves, fruits, and crown gall tissues, as well as by plant-associated bacteria, fungi, insects, and nematodes.

- From infection with the Ti plasmid, crown gall cells have acquired T-DNA carrying genes for the biosynthesis of cytokinins, auxin, and opines (**Figure 21.4**).

- The product of the *IPT* gene catalyzes the first committed step in cytokinin biosynthesis (**Figure 21.5**).

- Cytokinins are passively transported through the xylem and phloem.

- Cytokinin oxidase irreversibly inactivates cytokinins, contributing to their regulation.

- The level of active cytokinin results from positive and negative influences on their synthesis, conjugation, and transport.

### Cellular and Molecular Modes of Cytokinin Action

- Cytokinin is perceived by a family of histidine kinase receptors.

- The cytokinin response pathway consists of a multistep phosphorelay in which the signal is transduced from plasma membrane receptors into the nucleus by alternating histidine and aspartate phosphorylation of signaling elements (**Figures 21.6, 21.9**).

- Disruption of the three genes encoding cytokinin receptors in Arabidopsis results in multiple cytokinin-related defects (**Figure 21.7**).

- Cytokinin binding to membrane receptors activates a set of transcription factors in the nucleus called type-B ARRs, which mediate the transcriptional response to cytokinin (**Figures 21.8, 21.9**).

- The type-A ARRs are cytokinin primary response genes that negatively regulate cytokinin responsiveness.

### The Biological Roles of Cytokinins

- Reducing endogenous cytokinin by overexpression of cytokinin oxidase or mutation of the *IPT* gene severely limits shoot growth and produces a smaller shoot apical meristem (**Figures 21.10, 21.11**).

- Mutation of all three cytokinin receptors in Arabidopsis eliminates perception of cytokinin and results in multiple developmental defects, including a reduced shoot apical meristem, a stunted shoot, and greatly reduced flowering (**Figure 21.12**).

- Cytokinins interact with other hormones and with key transcription factors involved in regulating shoot apical meristem function.

- In contrast to their action in shoots, cytokinins inhibit root growth (**Figures 21.13, 21.14**).

- In the root apical meristem, auxin promotes cell division and cytokinin promotes cell differentiation.

- Both cytokinin and auxin regulate the plant cell cycle and are needed for cell division (**Figures 21.15, 21.16**).

- The ratio of auxin to cytokinin determines the differentiation of cultured plant tissues into either roots or shoots (**Figure 21.17**).

- Auxin maintains apical dominance by repressing the local synthesis of cytokinin in lateral buds and increasing the expression of cytokinin oxidase (**Figure 21.18**).

- Cytokinins delay leaf senescence (**Figure 21.19**), promote nutrient mobilization (**Figure 21.20**), help regulate the synthesis of photosynthetic pigments and proteins (**Figure 21.21**), and regulate vascular development.

- Photosynthetic productivity may be increased by making cytokinin-overproducing plants that have delayed senescence or yield more grains (**Figures 21.22, 21.23**).

## WEB MATERIAL

### Web Topics

**21.1 Cultured Cells Can Acquire the Ability to Synthesize Cytokinins**

The phenomenon of habituation is described, whereby callus tissues become cytokinin independent.

**21.2 Structures of Some Naturally Occurring Cytokinins**

The structures of various naturally occurring cytokinins are presented.

**21.3 Various Methods Are Used to Detect and Identify Cytokinins**

Cytokinins can be qualified using immunological and sensitive physical methods.

**21.4 The Biologically Active Form of Cytokinin Is the Free Base**

The recent identification of the cytokinin receptors has allowed the question of the active form of cytokinin to be addressed directly.

**21.5 Cytokinins Are Also Present in Some tRNAs in Animal and Plant Cells**

Modified adenosines near the 3′ end of the anticodons of some tRNAs have cytokinin activity.

**21.6 The Structures of Opines**

Opines are amino acids that serve as substrates for *Agrobacterium* during crown gall formation.

**21.7 The Ti Plasmid and Plant Genetic Engineering**

Applications of the Ti plasmid of *Agro-bacterium* in bioengineering are described.

**21.8 Phylogenetic Tree of *IPT* Genes**

Arabidopsis contains nine different *IPT* genes, several of which form a distinct clade with other plant sequences.

**21.9 A Root-Derived Hormone, Strigolactone, Is Involved in the Suppression of Branching in Shoots**

Evidence for a root-derived signaling molecule that acts as a suppressor of branching in shoots comes from grafting studies in mutants with highly branched phenotypes.

**21.10 Cytokinin Can Promote Light-Mediated Development**

Cytokinins can mimic the effect of the *det* mutation on chloroplast development and de-etiolation.

**21.11 Cytokinins Promote Cell Expansion and Greening in Cotyledons**

Cytokinins increase the entensibilities of the cotyledon cell walls of several dicot species.

**21.12 Cytokinins Interact with Elements of the Circadian Clock**

An interdependent regulatory loop between the clock genes and cytokinin response genes is discussed.

### Web Essays

**21.1 1955: The Discovery of Kinetin**

An interesting description of the history of the discovery of cytokinins.

**21.2 Cytokinin-Induced Form and Structure in Moss**

The effects of cytokinins on the development of moss protonema are described.

# CHAPTER REFERENCES

Akiyoshi, D. E., Klee, H., Amasino, R. M., Nester, E. W., and Gordon, M. P. (1984) T-DNA of *Agrobacterium tumefaciens* encodes an enzyme of cytokinin biosynthesis. *Proc. Natl. Acad. Sci. USA* 81: 5994–5998.

Akiyoshi, D. E., Morris, R. O., Hinz, R., Mischke, B. S., Kosuge, T., Garfinkel, D. J., Gordon, M. P., and Nester, E. W. (1983) Cytokinin/auxin balance in crown gall tumors is regulated by specific loci in the T-DNA. *Proc. Natl. Acad. Sci. USA* 80: 407–411.

Akiyoshi, D. E., Regier, D. A., and Gordon, M. P. (1987) Cytokinin production by *Agrobacterium* and *Pseudomonas* spp. *J. Bacteriol.* 169: 4242–4248.

Aloni, R., Wolf, A., Feigenbaum, P., Avni, A., and Klee, H. J. (1998) The *Never ripe* mutant provides evidence that tumor-induced ethylene controls the morphogenesis of *Agrobacterium tumefaciens*-induced crown galls in tomato stems. *Plant Physiol.* 117: 841–849.

Argueso, C. T., Ferreira, F. J., and Kieber, J. J. (2009) Environmental perception avenues: The interaction of cytokinin and environmental response pathways. *Plant Cell Environ.* 32: 1147–1160.

Ashikari, M., Sakakibara, H., Lin, S., Yamamoto, T., Takashi, T., Nishimura, A., Angeles, E. R., Qian, Q., Kitano, H., and Matsuoka, M. (2005) Cytokinin oxidase regulates rice grain production. *Science* 309: 741–745.

Barry, G. F., Rogers, R. G., Fraley, R. T., and Brand, L. (1984) Identification of cloned biosynthesis gene. *Proc. Natl. Acad. Sci. USA* 81: 4776–4780.

Bomhoff, G., Klapwijk, P. M., Kester, H. C. M., and Schilperoort, R. A. (1976) Octopine and nopaline synthesis and breakdown genetically controlled by plasmid of *Agrobacterium tumefaciens*. *Mol. Gen. Genet.* 145: 177–181.

Braun, A. C. (1958) A physiological basis for autonomous growth of the crown-gall tumor cell. *Proc. Natl. Acad. Sci. USA* 44: 344–349.

Brzobohaty, B., Moore, I., Kristoffersen, P., Bako, L., Campos, N., Schell, J., and Palme, K. (1993) Release of active cytokinin by a β-glucosidase localized to the maize root meristem. *Science* 262: 1051–1054.

Caplin, S. M., and Steward, F. C. (1948) Effect of coconut milk on the growth of the explants from carrot root. *Science* 108: 655–657.

Carabelli, M., Possenti, M., Sessa, G., Ciolfi, A., Sassi, M., Morelli, G. and Ruberti I. (2007) Canopy shade causes a rapid and transient arrest in leaf development through auxin-induced cytokinin oxidase activity. *Genes Dev.* 21: 1863–1868.

Chilton, M.-D. (1983) A vector for introducing new genes into plants. *Sci. Am.* 248: 50–59.

Chilton, M.-D., Drummond, M. H., Merlo, D. J., Sciaky, D., Montoya, A. L., Gordon, M. P., and Nester, E. W. (1977) Stable incorporation of plasmid DNA into higher plant cells: The molecular basis of crown gall tumorigenesis. *Cell* 11: 263–271.

Chory, J., Reinecke, D., Sim, S., Washburn, T., and Brenner, M. (1994) A role for cytokinins in de-etiolation in *Arabidopsis. Det* mutants have an altered response to cytokinins. *Plant Physiol.* 104: 339–347.

D'Agostino, I. B., Deruère, J., and Kieber, J. J. (2000) Characterization of the response of the *Arabidopsis ARR* gene family to cytokinin. *Plant Physiol.* 124: 1706–1717.

Dello Ioio, R., Nakamura, K., Moubayidin, L., Perilli, S., Taniguchi, M., Morita, M. T., Aoyama, T., Costantino, P., and Sabatini, S. (2008) A genetic framework for the control of cell division and differentiation in the root meristem. *Science* 322: 1380–1384.

Dortay, H., Mehnert, N., Bürkle, L., Schmülling, T., and Heyl, A. (2006) Analysis of protein interactions within the cytokinin-signaling pathway of *Arabidopsis thaliana*. *FEBS J.* 273: 4631–4644.

Elzen, G. W. (1983) Cytokinins and insect galls. *Comp. Biochem. Physiol.* 76: 17–19.

Foo, E., Bullier, E., Goussot, M., Foucher, F., Rameau, C., and Beveridge, C. A. (2005) The branching gene *RAMOSUS1* mediates interactions among two novel signals and auxin in pea. *Plant Cell* 17: 464–474.

Gan, S., and Amasino, R. M. (1995) Inhibition of leaf senescence by autoregulated production of cytokinin. *Science* 270: 1986–1988.

Garfinkel, D. J., Simpson, R. B., Ream, L. W., White, F. F., Gordon, M. P., and Nester, E. W. (1981) Genetic analysis of crown gall: Fine structure map of the T-DNA by site-directed mutagenesis. *Cell* 27: 143–153.

Giulini, A., Wang, J., and Jackson, D. (2004) Control of phyllotaxy by the cytokinin-inducible response regulator homologue ABPHYL1. *Nature* 430: 1031–1034.

Gomez-Roldan, V., Fermas, S., Brewer, P. B., Puech-Pagès, V., Dun, E. A., Pillot, J. P., Letisse, F., Matusova, R., Danoun, S., Portais, et al. (2008) Strigolactone inhibition of shoot branching. *Nature* 455: 189–194.

Gonzalez-Rizzo, S., Crespi, M., and Frugier, F. (2006) The *Medicago truncatula* CRE1 cytokinin receptor regulates lateral root development and early symbiotic interaction with *Sinorhizobium meliloti*. *Plant Cell* 18: 2680–2693.

Hamilton, J. L., and Lowe, R. H. (1972) False broomrape: A physiological disorder caused by growth-regulator imbalance. *Plant Physiol.* 50: 303–304.

Higuchi, M., Pischke, M. S., Mahonen, A. P., Miyawaki, K., Hashimoto, Y., Seki, M., Kobayashi, M., Shinozaki, K., Kato, T., Tabata, S., et al. (2004) In planta functions of the *Arabidopsis* cytokinin receptor family. *Proc. Natl. Acad. Sci. USA* 101: 8821–8826.

Houba-Herin, N., Pethe, C., d'Alayer, J., and Laloue M. (1999) Cytokinin oxidase from *Zea mays*: Purification, cDNA cloning and expression in moss protoplasts. *Plant J.* 17: 615–626.

Huff, A. K., and Ross, C. W. (1975) Promotion of radish cotyledon enlargement and reducing sugar content by zeatin and red light. *Plant Physiol.* 56: 429–433.

Hutchison, C. E., Li, J., Argueso, C., Gonzalez, M., Lee, E., Lewis, M. W., Maxwell, B. B., Perdue, T. D., Schaller, G. E., Alonso, J. M., et al. (2006) The Arabidopsis histidine phosphotransfer proteins are redundant positive regulators of cytokinin signaling. *Plant Cell* 18: 3073–3087.

Hwang, I., and Sheen, J. (2001) Two-component circuitry in *Arabidopsis* signal transduction. *Nature* 413: 383–389.

Imamura, A., Hanaki, N., Nakamura, A., Suzuki, T., Taniguchi, M., Kiba, T., Ueguchi, C., Sugiyama, T., and Mizuno, T. (1999) Compilation and characterization of *Arabidopsis thaliana* response regulators implicated in His-Asp phosphorelay signal transduction. *Plant Cell Physiol.* 40: 733–742.

Inoue, T., Higuchi, M., Hashimoto, Y., Seki, M., Kobayashi, M., Kato, T., Tabata, S., Shinozaki, K., and Kakimoto, T. (2001) Identification of CRE1 as a cytokinin receptor from *Arabidopsis. Nature* 409: 1060–1063.

Ishida, K., Yamashino, T., and Mizuno, T. (2008) Expression of the cytokinin-induced type-A response regulator gene ARR9 is regulated by the circadian clock in *Arabidopsis thaliana. Biosci. Biotechnol. Biochem.* 72: 3025–3029.

Jasinski, S., Piazza, P., Craft, J., Hay, A., Woolley, L., Rieu, I., Phillips, A., Hedden, P., and Tsiantis, M. (2005) KNOX action in Arabidopsis is mediated by coordinate regulation of cytokinin and gibberellin activities. *Curr. Biol.* 15: 1560–1565.

John, P. C. L., Zhang, K., Don, C., Diederich, L., and Wightman, F. (1993) P34-cdc2 related proteins in control of cell cycle progression, the switch between division and differentiation in tissue development, and stimulation of division by auxin and cytokinin. *Aust. J. Plant Physiol.* 20: 503–526.

Kakimoto, T. (1996) CKI1, a histidine kinase homolog implicated in cytokinin signal transduction. *Science* 274: 982–985.

Kakimoto, T. (2001) Identification of plant cytokinin biosynthetic enzymes as dimethylallyl diphosphate: ATP/ADP isopentenyltransferases. *Plant Cell Physiol.* 42: 677–685.

Kim, H. J., Ryu, H., Hong, S. H., Woo, H. R., Lim, P. O., Lee, I. C., Sheen, J., Nam, H. G., Hwang, I. (2006) Cytokinin-mediated control of leaf longevity by AHK3 through phosphorylation of ARR2 in Arabidopsis. *Proc. Natl. Acad. Sci. USA* 103: 814–819.

Kiran, N. S., Polanská, L., Fohlerová, R., Mazura, P., Válková, M., Smeral, M., Zouhar, J., Malbeck, J., Dobrev, P. I., Macháčková, I., et al. (2006) Ectopic over-expression of the maize beta-glucosidase Zm-p60.1 perturbs cytokinin homeostasis in transgenic tobacco. *J. Exp. Bot.* 57: 985–996.

Kurakawa, T., Ueda, N., Maekawa, M., Kobayashi, K., Kojima, M., Nagato, Y., Sakakibara, H., and Kyozuka, J. (2007) Direct control of shoot meristem activity by a cytokinin-activating enzyme. *Nature* 445: 652–655.

Leibfried, A., To, J. P., Busch, W., Stehling, S., Kehle, A., Demar, M., Kieber, J. J., and Lohmann, J. U. (2005) WUSCHEL controls meristem function by direct regulation of cytokinin-inducible response regulators. *Nature* 438: 1172–1175.

Letham, D. S. (1973) Cytokinins from *Zea mays. Phytochemistry* 12: 2445–2455.

Letham, D. S. (1974) Regulators of cell division in plant tissues XX. The cytokinins of coconut milk. *Physiol. Plant.* 32: 66–70.

Martin, R. C., Mok, M. C., and Mok, D. W. S. (1999) Isolation of a cytokinin gene, ZOG1, encoding zeatin O-glucosyltransferase from *Phaseolus lunatus. Proc. Natl. Acad. Sci. USA* 96: 284–289.

Matsumoto-Kitano, M., Kusumoto, T., Tarkowski, P., Kinoshita-Tsujimura, K., Václavíková, K., Miyawaki, K., Kakimoto, T. (2008) Cytokinins are central regulators of cambial activity. *Proc. Natl. Acad. Sci. USA* 105: 20027–20031.

McCabe, M. S., Garratt, L. C., Schepers, F., Jordi, W. J., Stoopen, G. M. Davelaar, E., van Rhijn, J. H., Power, J. B., and Davey, M. R. (2001) Effects of P(SAG12)-IPT gene expression on development and senescence in transgenic lettuce. *Plant Physiol.* 127: 505–516.

Miller, C. O., Skoog, F., Von Saltza, M. H., and Strong, F. (1955) Kinetin, a cell division factor from deoxyribonucleic acid. *J. Am. Chem. Soc.* 77: 1392–1393.

Miyawaki, K., Matsumoto-Kitano, M., and Kakimoto, T. (2004) Expression of cytokinin biosynthetic isopentenyltransferase genes in *Arabidopsis*: Tissue specificity and regulation by auxin, cytokinin, and nitrate. *Plant J.* 37: 128–138.

Mähönen, A. P., Bishopp, A., Higuchi, M., Nieminen, K. M., Kinoshita, K., Törmäkangas, K., Ikeda, Y., Oka, A., Kakimoto, T., Helariutta, Y. (2006) Cytokinin signaling and its inhibitor AHP6 regulate cell fate during vascular development. *Science* 311: 94–98.

Morris, R. O. (1986) Genes specifying auxin and cytokinin biosynthesis in phytopathogens. *Annu. Rev. Plant Physiol.* 37: 509–538.

Morris, R., Bilyeu, K., Laskey, J., and Cheikh, N. (1999) Isolation of a gene encoding a glycosylated cytokinin oxidase from maize. *Biochem. Biophys. Res. Commun.* 225: 328–333.

Murray, J. D., Karas, B. J., Sato, S., Tabata, S., Amyot, L., and Szczyglowski, K. (2007) A cytokinin perception mutant colonized by *Rhizobium* in the absence of nodule organogenesis. *Science* 315: 101–104.

Nishimura, C., Ohashi, Y., Sato, S., Kato, T., Tabata, S., and Ueguchi, C. (2004) Histidine kinase homologs that act as cytokinin receptors possess overlapping functions in the regulation of shoot and root growth in *Arabidopsis. Plant Cell* 16: 1365–1377.

Nieminen, K., Immanen, J., Laxell, M., Kauppinen, L., Tarkowski, P., Dolezal, K., Tähtiharju, S., Elo, A., Decourteix, M., Ljung, K., et al. (2008) Cytokinin signaling regulates cambial development in poplar. *Proc. Natl. Acad. Sci. USA* 105: 20032–20037.

Noodén, L. D., Singh, S., and Letham, D. S. (1990) Correlation of xylem sap cytokinin levels with monocarpic senescence in soybean. *Plant Physiol.* 93: 33–39.

Pertry, I., Václavíková, K., Depuydt, S., Galuszka, P., Spíchal, L., Temmerman, W., Stes, E., Schmülling, T., Kakimoto, T., Van Montagu, M. C., et al. (2009) Identification of *Rhodococcus fascians* cytokinins and their modus operandi to reshape the plant. *Proc. Natl. Acad. Sci. USA* 106: 929–934.

Riou-Khamlichi, C., Huntley, R., Jacqmard, A., and Murray, J. A. (1999) Cytokinin activation of *Arabidopsis* cell division through a D-type cyclin. *Science* 283: 1541–1544.

Rivero, R. M., Kojima, M., Gepstein, A., Sakakibara, H., Mittler, R., Gepstein, S., and Blumwald, E. (2007) Delayed leaf senescence induces extreme drought tolerance in a flowering plant. *Proc. Natl. Acad. Sci. USA* 104: 19631–19636.

Romanov, G. A., Lomin, S. N., and Schmülling, T. (2006) Biochemical characteristics and differences in ligand-binding properties of *Arabidopsis* cytokinin receptors AHK3 and CRE1/AHK4 as revealed by a direct binding assay. *J. Exp. Bot.* 57: 4051–4058.

Rupp, H.-M., Frank, M., Werner, T., Strnad, M., and Schmülling, T. (1999) Increased steady state mRNA levels of the STM and KNATI homeobox genes in cytokinin overproducing *Arabidopsis thaliana* indicate a role for cytokinins in the shoot apical meristem. *Plant J.* 18: 557–563.

Sakai, H., Honma, T., Aoyama, T., Sato, S., Kato, T., Tabata, S., and Oka, A. (2001) *Arabidopsis ARR1* is a transcription factor for genes immediately responsive to cytokinins. *Science.* 294: 1519–1521.

Sakakibara, H., Kasahara, H., Ueda, N., Kojima, M., Takei, K., Hishiyama, S., Asami, T., Okada, K., Kamiya, Y., Yamaya, et al. (2005) Agrobacterium tumefaciens increases cytokinin production in plastids by modifying the biosynthetic pathway in the host plant. *Proc. Natl. Acad. Sci. USA* 102: 9972–9977.

Salomé, P. A., To, J. P. C., Kieber, J. J., and McClung, C. R. (2005) *Arabidopsis* response regulators ARR3 and ARR4 play cytokinin-Independent roles in the control of circadian period. *Plant Cell* 18: 55–69.

Shimizu-Sato, S., Tanaka, M., and Mori, H. (2008) Auxin-cytokinin interactions in the control of shoot branching. *Plant Mol. Biol.* 69: 429–435.

Skoog, F., and Miller, C. O. (1965) Chemical regulation of growth and organ formation in plant tissues cultured *in vitro.* In *Molecular and Cellular Aspects of Development*, E. Bell, ed., Harper and Row, New York, pp. 481–494.

Smigocki, A., Neal, J. W., Jr., McCanna, I., and Douglass, L. (1993) Cytokinin-mediated insect resistance in *Nicotiana* plants transformed with the *ipt* gene. *Plant Mol. Biol.* 23: 325–335.

Soni, R., Carmichael, J. P., Shah, Z. H., and Murray, J. A. H. (1995) A family of cyclin D homologs from plants differentially controlled by growth regulators and containing the conserved retinoblastoma protein interaction motif. *Plant Cell* 7: 85–103.

Sorefan, K., Booker, J., Haurogné, K., Goussot, M., Bainbridge, K., Foo, E., Chatfield, S., Ward, S., Beveridge, C., Rameau, C., et al. (2003) MAX4 and RMS1 are orthologous dioxygenase-like genes that regulate shoot branching in *Arabidopsis* and pea. *Genes Dev.* 17: 1469–1474.

Sweere, U., Eichenberg, K., Lohrmann, J., Mira-Rodado, V., Bäurle, I., Kudla, J., Nagy, F., Schäfer, E., and Harter, K. (2001) Interaction of the response regulator ARR4 with the photoreceptor phytochrome B in modulating red light signaling. *Science* 294: 1108–1111.

Tanaka, M., Takei, K., Kojima, M., Sakakibara, H., and Mori, H. (2006) Auxin controls local cytokinin biosynthesis in the nodal stem in apical dominance. *Plant J.* 45: 1028–1036.

Takei, K., Sakakibara, H., and Sugiyama, T. (2001a) Identification of genes encoding adenylate isopentyltransferase, a cytokinin biosynthetic enzyme, in *Arabidopsis thaliana.* *J. Biol. Chem.* 276: 26405–26410.

Takei, K., Sakakibara, H., Taniguchi, M., and Sugiyama, T. (2001b) Nitrogen-dependent accumulation of cytokinins in roots and the translocation to leaf: Implication of cytokinin species that induces gene expression of maize response regulator. *Plant Cell Physiol.* 42: 85–93.

Takei K., Yamaya, T., and Sakakibara, H. (2004) *Arabidopsis* CYP735A1 and CYP735A2 encode cytokinin hydroxylases that catalyze the biosynthesis of *trans*-zeatin. *J. Biol. Chem.* 279: 41866–41872.

Tirichine, L., Sandal, N., Madsen, L. H., Radutoiu, S., Albrektsen, A. S., Sato, S., Asamizu, E., Tabata, S., and Stougaard, J. (2007) A gain-of-function mutation in a cytokinin receptor triggers spontaneous root nodule organogenesis. *Science* 315: 104–107.

To, J. P., Haberer, G., Ferreira, F. J., Deruere, J., Mason, M. G., Schaller, G. E., Alonso, J. M., Ecker, J. R., and Kieber, J. J. (2004) Type-A Arabidopsis response regulators are partially redundant negative regulators of cytokinin signaling. *Plant Cell* 16: 658–671.

To, J. P., Deruère, J., Maxwell, B. B., Morris, V. F., Hutchison, C. E., Schaller, G. E., and Kieber, J. J. (2007) Cytokinin regulates type-A Arabidopsis response regulator activity and protein stability via two-component phosphorelay. *Plant Cell* 19: 3901–3914.

Umehara, M., Hanada, A., Yoshida, S., Akiyama, K., Arite, T., Takeda-Kamiya, N., Magome, H., Kamiya, Y., Shirasu, K., Yoneyama, K., et al. (2008) Inhibition of shoot branching by new terpenoid plant hormones. *Nature* 455: 195–200.

Vandenbussche, F., Habricot, Y., Condiff, A. S., Maldiney, R., Van Der Straeten, D., and Ahmad, M. (2007) HY5 is a point of convergence between cryptochrome and cytokinin signalling pathways in *Arabidopsis thaliana.* *Plant J.* 49: 428–441.

Werner, T., Motyka, V., Strnad, M., and Schmülling, T. (2001) Regulation of plant growth by cytokinin. *Proc. Natl. Acad. Sci. USA* 98: 10487–10492.

Werner, T., Motyka, V., Laucou, V., Smets, R., Van Onckelen, H., and Schmülling, T. (2003) Cytokinin-deficient transgenic Arabidopsis plants show multiple developmental alterations indicating opposite functions of cytokinins in regulating shoot and root meristem activity. *Plant Cell* 15: 2532–2550.

White, P. R. (1934) Potentially unlimited growth of excised tomato root tips in a liquid medium. *Plant Physiol.* 9: 585–600.

Yamada, H., Suzuki, T., Terada, K., Takei, K., Ishikawa, K., Miwa, K., Yamashino, T., and Mizuno, T. (2001) The *Arabidopsis* AHK4 histidine kinase is a cytokinin-binding receptor that transduces cytokinin signals across the membrane. *Plant Cell Physiol.* 42: 1017–1023.

Yanai, O., Shani, E., Dolezal, K., Tarkowski, P., Sablowski, R., Sandberg, G., Samach, A., and Ori, N. (2005) Arabidopsis KNOXI proteins activate cytokinin biosynthesis. *Curr. Biol.* 15: 1566–1571.

Zhang, K., Letham, D. S., and John, P. C. L. (1996) Cytokinin controls the cell cycle at mitosis by stimulating the tyrosine dephosphorylation and activation of p34cdc2-like H1 histone kinase. *Planta* 200: 2–12.

Zheng, B., Deng, Y., Mu, J., Ji, Z., Xiang, T., Niu, Q.-W., Chua, N.-H., and Zuo, J. (2006) Cytokinin affects circadian-clock oscillation in a phytochrome B- and Arabidopsis response regulator 4-dependent manner. *Physiol. Plant.* 127: 277–292.

# CHAPTER 22

# Ethylene: The Gaseous Hormone

During the nineteenth century, when coal gas was used for street illumination, it was observed that trees in the vicinity of streetlamps defoliated more extensively than other trees. Eventually it became apparent that coal gas and air pollutants affect plant growth and development, and ethylene was identified as the active component of coal gas (see WEB TOPIC 22.1).

In 1901, Dimitry Neljubov, a graduate student at the Botanical Institute of St. Petersburg in Russia, observed that dark-grown pea seedlings in the laboratory exhibited symptoms that were later termed the *triple response*: reduced stem elongation, increased lateral growth (swelling), and abnormal, horizontal growth (FIGURE 22.1). When the plants were allowed to grow in fresh air, they regained their normal morphology and rate of growth. Neljubov identified ethylene from coal gas, which was present in the laboratory air, as the molecule causing the response.

The first indication that ethylene is a natural product of plant tissues was published by H. H. Cousins in 1910. Cousins reported that "emanations" from oranges stored in a chamber caused the premature ripening of bananas when these gases were passed through a chamber containing the fruit. However, given that oranges synthesize relatively little ethylene compared to other fruits, such as apples, it is likely that the oranges used by Cousins were infected with the fungus *Penicillium*, which produces copious amounts of ethylene. In 1934, R. Gane and others identified ethylene chemically as a natural product of plant metabolism, and because of its dramatic effects on the plant it was classified as a hormone.

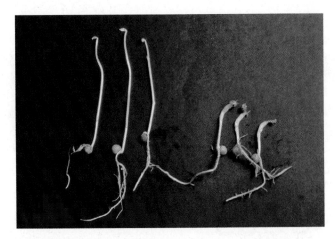

**FIGURE 22.1**   Triple response of etiolated pea seedlings. Six-day-old pea seedlings were grown in the presence of 10 ppm (parts per million) ethylene (right) or left untreated (left). The treated seedlings show radial swelling, inhibition of elongation of the epicotyl, and horizontal growth of the epicotyl (diagravitropism). (Courtesy of S. Gepstein.)

For 25 years ethylene was not recognized as an important plant hormone, mainly because many physiologists believed that the effects of ethylene were due to auxin, the first plant hormone to be discovered (see Chapter 19). Auxin was thought to be the main plant hormone, and ethylene was considered to play only an insignificant and indirect physiological role. Early work on ethylene was hampered by the lack of chemical techniques for its quantification. However, after gas chromatography was introduced in ethylene research in 1959, the importance of ethylene was rediscovered and its physiological significance as a plant growth regulator was recognized (Burg and Thimann 1959).

In this chapter we will describe the ethylene biosynthetic pathway and how ethylene acts at the cellular and molecular levels. At the end of the chapter we will outline some of the important effects of ethylene on plant growth and development.

## Structure, Biosynthesis, and Measurement of Ethylene

Ethylene is the simplest olefin (its molecular weight is 28):

$$\begin{array}{c} H \\ \diagdown \\ H \end{array} C = C \begin{array}{c} H \\ \diagup \\ H \end{array}$$

**Ethylene**

It is lighter than air under physiological conditions, and readily undergoes oxidation (see **WEB TOPIC 22.2**).

Ethylene can be produced by almost all parts of higher plants, although the rate of production depends on the type of tissue and the stage of development. It is usually measured by gas chromatography (see **WEB TOPIC 22.3**). Ethylene production increases during leaf abscission and flower senescence, as well as during fruit ripening. Any type of wounding can induce ethylene biosynthesis, as can physiological stresses such as flooding, disease, and temperature or drought stress. In additon, infection by various pathogens can also elevate ethylene biosynthesis.

The amino acid methionine is the precursor of ethylene, and 1-aminocyclopropane-1-carboxylic acid (ACC) serves as an intermediate in the conversion of methionine to ethylene. As we will see, the complete pathway is a cycle, taking its place among the many metabolic cycles that operate in plant cells.

### Regulated biosynthesis determines the physiological activity of ethylene

In vivo experiments showed that plant tissues convert [$^{14}$C]methionine to [$^{14}$C]ethylene, and that the ethylene is derived from carbons 3 and 4 of methionine. The $CH_3$—S group of methionine is recycled via the *Yang cycle* (**FIGURE 22.2**). The immediate precursor of ethylene is **1-aminocyclopropane-1-carboxylic acid** (**ACC**). In general, when ACC is supplied exogenously to plant tissues, ethylene production increases substantially. This observation indicates that the synthesis of ACC is usually the limiting biosynthetic step in ethylene production in plant tissues.

**ACC synthase** (**ACS**) is the enzyme that catalyzes the conversion of AdoMet to ACC (see Figure 22.2). Its level is regulated by environmental and internal factors, such as wounding, drought stress, flooding, and auxin. ACC synthase is encoded by members of a divergent multigene family that are differentially regulated by various inducers of ethylene biosynthesis. In tomato, for example, there are at least ten ACC synthase genes, different subsets of which are induced by auxin, wounding, and/or fruit ripening. (For more details, see **WEB TOPIC 22.4**.)

**ACC oxidase** catalyzes the last step in ethylene biosynthesis: the conversion of ACC to ethylene (see Figure 22.2). In tissues that show high rates of ethylene production, such as ripening fruit, ACC oxidase activity can be the rate-limiting step in ethylene biosynthesis. Like ACC synthase, ACC oxidase is encoded by a multigene family, the members of which are differentially regulated (see **WEB TOPIC 22.5**). For example, in ripening tomato fruits and senescing petunia flowers, the mRNA levels of a subset of ACC oxidase genes are highly elevated.

**CATABOLISM**   Researchers have studied the catabolism of ethylene by supplying $^{14}C_2H_4$ to plant tissues and tracing the radioactive compounds produced. Carbon dioxide, ethylene oxide, ethylene glycol, and the glucose conjugate

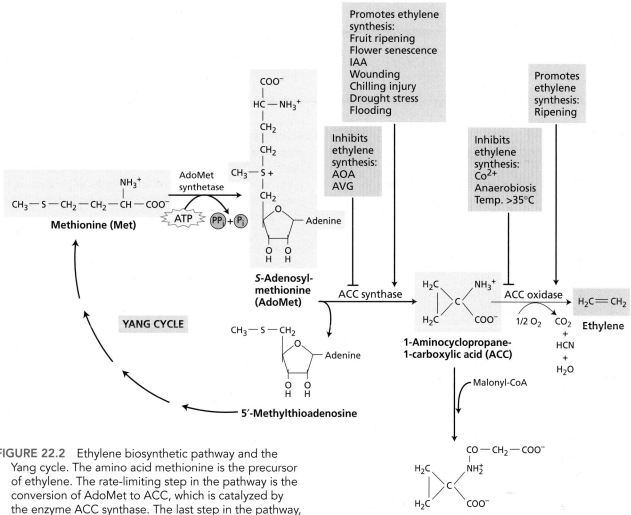

**FIGURE 22.2** Ethylene biosynthetic pathway and the Yang cycle. The amino acid methionine is the precursor of ethylene. The rate-limiting step in the pathway is the conversion of AdoMet to ACC, which is catalyzed by the enzyme ACC synthase. The last step in the pathway, the conversion of ACC to ethylene, requires oxygen and is catalyzed by the enzyme ACC oxidase. The $CH_3$—S group of methionine is recycled via the Yang cycle and thus conserved for continued synthesis. Besides being converted to ethylene, ACC can be conjugated to N-malonyl ACC. AOA = aminooxyacetic acid; AVG = aminoethoxy-vinylglycine. (After McKeon et al. 1995.)

of ethylene glycol have been identified as metabolic breakdown products. However, because certain cyclic olefin compounds, such as 1,4-cyclohexadiene, have been shown to block ethylene breakdown without inhibiting ethylene action, ethylene catabolism does not appear to play a significant role in regulating the level of the hormone (Raskin and Beyer 1989).

**CONJUGATION** Not all the ACC found in the tissue is converted to ethylene. ACC can also be converted to a conjugated form, N-malonyl ACC (see Figure 22.2), which does not appear to break down and accumulates in the tissue, primarily in the vacuole. A second, minor, conjugated form

of ACC, 1-(γ-L-glutamylamino) cyclopropane-1-carboxylic acid (GACC), has also been identified. The conjugation of ACC may play an important role in the control of ethylene biosynthesis, in a manner analogous to the conjugation of auxin and cytokinin.

**ACC DEAMINASE** A number of bacteria present in the soil express an enzyme called ACC deaminase that hydrolyzes ACC to ammonia and α-ketobutyrate (Glick 2005). These bacteria can promote plant growth by sequestering and cleaving ACC made and excreted by plants, thereby lowering the level of ethylene to which the plants are exposed. ACC deaminase has been expressed in transgenic plants to lower the level of ethylene produced. Recently, a gene encoding an ACC deaminase has been identified in Arabidopsis, suggesting that endogenous ACC deaminase plays a role in regulating ethylene biosynthesis.

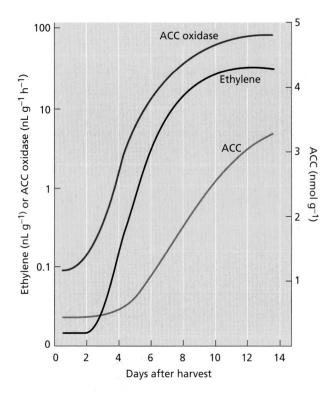

**FIGURE 22.3** Changes in the ACC concentrations, ACC oxidase activity, and ethylene during ripening of Golden Delicious apples. The data are plotted as a function of days after harvest. Increases in ethylene and ACC concentrations and in ACC oxidase activity are closely correlated with ripening. (After Yang 1987.)

## Ethylene biosynthesis is promoted by several factors

Ethylene biosynthesis is stimulated by several factors, including developmental state, environmental conditions, other plant hormones, and physical and chemical injury. Ethylene biosynthesis also varies in a circadian manner, peaking during the day and reaching a minimum at night.

**FRUIT RIPENING** As fruits mature, the rates of ACC and ethylene biosynthesis increase. Enzyme activities for both ACC oxidase (**FIGURE 22.3**) and ACC synthase increase, as do the mRNA levels for subsets of the genes encoding each enzyme. However, application of ACC to unripe fruits only slightly enhances ethylene production, indicating that an increase in the activity of ACC oxidase is the rate-limiting step in ripening (McKeon et al. 1995).

**STRESS-INDUCED ETHYLENE PRODUCTION** Ethylene biosynthesis is increased by stress conditions such as drought, flooding, chilling, exposure to ozone, and mechanical wounding. In all these cases ethylene is produced by the usual biosynthetic pathway, and the increased ethylene

production has been shown to result at least in part from an increase in transcription of ACC synthase mRNA. This "stress ethylene" is involved in the onset of stress responses such as abscission, senescence, wound healing, and increased disease resistance (see Chapter 26).

**CIRCADIAN REGULATION OF ETHYLENE PRODUCTION** The circadian clock regulates the biosynthesis of ethylene in a number of plant species. There is generally a peak of ethylene evolution at midday, with a trough in the middle of the night. This regulation likely results from the transcriptional control of a subset of ACC synthase genes, which is mediated by the TOC1/CCA1 clock in Arabidopsis (Thain et al. 2004).

**AUXIN-INDUCED ETHYLENE PRODUCTION** In some instances, auxins and ethylene can cause similar plant responses, such as induction of flowering in pineapple and inhibition of stem elongation. These responses might be due to the ability of auxins to promote ethylene synthesis by enhancing ACC synthase activity. These observations suggest that some responses previously attributed to auxin (indole-3-acetic acid, or IAA) are in fact mediated by the ethylene produced in response to auxin.

Following application of exogenous IAA, the transcription of multiple ACC synthase genes is elevated and increased ethylene production is observed (Nakagawa et al. 1991; Liang et al. 1992; Tsuchisaka and Theologis 2004). Inhibitors of protein synthesis block IAA-induced ethylene synthesis, indicating that the synthesis of new ACC synthase protein caused by auxins brings about the marked increase in ethylene production.

## Ethylene biosynthesis can be elevated through a stabilization of ACC synthase protein

In addition to auxin, brassinosteroids and cytokinins have been shown to elevate ethylene biosynthesis. As with auxin, these hormones act by increasing the activity of ACC synthase. However, in contrast to auxin, they act not by elevating ACC synthase transcription, but by increasing the stability of ACC synthase proteins (Chae and Kieber 2005).

For example, pathogen attack, cytokinin, and brassinosteroids all increase ethylene biosynthesis in part by stabilizing the ACC synthase protein so that it is broken down more slowly. The carboxy-terminal domain of ACC synthase plays a key role in this regulation (Vogel et al. 1998). This domain acts as a flag to target the protein for rapid degradation by the 26S proteasome (see Chapter 2). Phosphorylation of this domain by either a mitogen-activated protein (MAP) kinase (activated by pathogens) or a calcium-dependent kinase blocks its ability to target the protein for rapid turnover.

## Various inhibitors can block ethylene biosynthesis

Inhibitors of hormone synthesis or action are valuable for the study of the biosynthetic pathways and physiological roles of hormones. The use of inhibitors is particularly helpful when it is difficult to distinguish between different hormones that have identical effects in plant tissue or when a hormone affects the synthesis or the action of another hormone.

For example, ethylene mimics high concentrations of auxins by inhibiting stem growth and causing *epinasty* (a downward curvature of leaves). Use of specific inhibitors of ethylene biosynthesis and action made it possible to discriminate between the actions of auxin and ethylene. Studies using inhibitors showed that ethylene is the primary effector of epinasty and that auxin acts indirectly by causing a substantial increase in ethylene production.

**INHIBITORS OF ETHYLENE SYNTHESIS**  **Aminoethoxyvinylglycine (AVG)** and **aminooxyacetic acid (AOA)** block the conversion of AdoMet to ACC (see Figure 22.2). AVG and AOA are known to inhibit enzymes that use the cofactor pyridoxal phosphate, including ACC synthase. α-Aminoisobutyric acid (AIBA) and cobalt ions ($Co^{2+}$) also inhibit the ethylene biosynthetic pathway, blocking the conversion of ACC to ethylene by ACC oxidase, the last step in ethylene biosynthesis.

**INHIBITORS OF ETHYLENE ACTION**  Most of the effects of ethylene can be antagonized by specific ethylene inhibitors. Silver ions ($Ag^+$) applied as silver nitrate ($AgNO_3$) or as silver thiosulfate [$Ag(S_2O_3)_2^{3-}$] are potent inhibitors of ethylene action. Silver is very specific; the inhibition it causes cannot be induced by any other metal ion.

Carbon dioxide at high concentrations (in the range of 5 to 10%) also inhibits many effects of ethylene, such as the induction of fruit ripening, although $CO_2$ is less efficient than $Ag^+$. This effect of $CO_2$ has often been exploited in the storage of fruits, whose ripening is delayed at elevated $CO_2$ concentrations. The high concentrations of $CO_2$ required for inhibition make it unlikely that $CO_2$ acts as an ethylene antagonist under natural conditions. The volatile compound *trans*-cyclooctene, but not its isomer *cis*-cyclooctene, is a strong competitive inhibitor of ethylene binding (Sisler et al. 1990); *trans*-cyclooctene is thought to act by competing with ethylene for binding to the receptor. 1-Methylcyclopropene (MCP) binds almost irreversibly to the ethylene receptor (**FIGURE 22.4**), and effectively blocks multiple ethylene responses (Sisler and Serek 1997). This nearly odorless compound has been marketed under the trade name EthylBloc®, and is used to increase the shelf life of cut flowers and of some fruits, such as apples.

**ETHYLENE ABSORPTION**  Because ethylene gas is easily lost from its tissue of origin and may affect other tissues

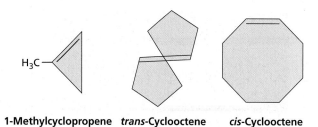

**1-Methylcyclopropene**    ***trans*-Cyclooctene**    ***cis*-Cyclooctene**
**(MCP)**

**FIGURE 22.4**  Two inhibitors that block ethylene binding to its receptor. The *cis* form of cyclooctene is not an effective inhibitor.

or organs, ethylene-trapping systems are used during the storage of fruits, vegetables, and flowers. Potassium permanganate ($KMnO_4$) is an effective absorbent of ethylene and can reduce the concentration of ethylene in apple storage areas from 250 to 10 µL $L^{-1}$, markedly extending the storage life of the fruit.

# Ethylene Signal Transduction Pathways

Despite the broad range of ethylene's effects on development, the primary steps in ethylene action are likely similar in all cases: They all involve binding to a receptor, followed by activation of one or more signal transduction pathways (see Chapter 14) leading to the cellular response. Ultimately, ethylene exerts its effects primarily by altering the pattern of gene expression. Molecular genetic studies of selected mutants of *Arabidopsis thaliana* have contributed greatly to the elucidation of ethylene signaling components.

The triple-response morphology of etiolated Arabidopsis seedlings has been used as a screen to isolate mutants affected in their response to ethylene (**FIGURE 22.5**) (Guzman and Ecker 1990). Two classes of mutants have been identified by experiments in which mutagenized Arabidopsis seeds were grown on an agar medium in the presence or absence of ethylene for 3 days in the dark:

1. Mutants that fail to respond to exogenous ethylene (ethylene-resistant or ethylene-insensitive mutants)

2. Mutants that display the response even in the absence of ethylene (constitutive mutants)

Ethylene-insensitive mutants are identified as tall seedlings extending above the lawn of short, triple-responding seedlings when grown in the presence of ethylene (see Figure 22.5). Conversely, constitutive ethylene response mutants are identified as seedlings displaying the triple response in the absence of exogenous ethylene (see Figure 22.9).

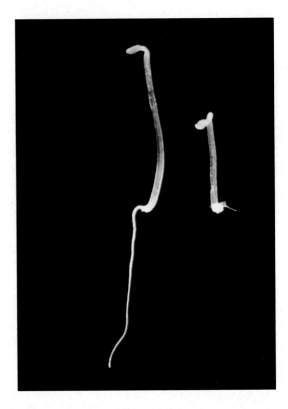

FIGURE 22.5 The triple response in Arabidopsis. Three-day-old etiolated seedlings grown in the presence (right) or absence (left) of 10 ppm ethylene. Note the shortened hypocotyl, reduced root elongation, and exaggeration of the curvature of the apical hook that results from the presence of ethylene. (Courtesy of J. Kieber.)

### Ethylene receptors are related to bacterial two-component system histidine kinases

The first ethylene-insensitive mutant isolated was *etr1* (*ethylene-response1*) (FIGURE 22.6). The *etr1* mutant was identified in a screen for mutations that block the response of Arabidopsis seedlings to ethylene. The amino acid sequence of the carboxy-terminal half of ETR1 is similar to bacterial two-component histidine kinases—receptors used by bacteria to perceive environmental cues such as chemosensory stimuli, phosphate availability, and osmolarity.

FIGURE 22.6 Screen for the *etr1* mutant of Arabidopsis. Seedlings were grown for 3 days in the dark in ethylene. Note that all but one of the seedlings exhibit the triple response: exaggeration in curvature of the apical hook, inhibition and radial swelling of the hypocotyl, and horizontal growth. The *etr1* mutant is completely insensitive to the hormone and grows like an untreated seedling. (Photograph by K. Stepnitz of the MSU/DOE Plant Research Laboratory.)

As previously discussed in Chapter 14, bacterial two-component systems consist of a sensor histidine kinase and a response regulator. ETR1 was the first example of a eukaryotic histidine kinase, but others have since been found in yeast, mammals, and plants. Phytochrome (see Chapter 17) and the cytokinin receptor (see Chapter 21) also share sequence similarity to bacterial two-component histidine kinases. The similarity to bacterial receptors and the ethylene insensitivity of the *etr1* mutants suggested that ETR1 might be an ethylene receptor. Binding studies confirmed this hypothesis (see WEB TOPIC 22.6).

The Arabidopsis genome encodes four additional proteins similar to ETR1 that also function as ethylene receptors: ETR2, ERS1 (*ETHYLENE-RESPONSE SENSOR 1*), ERS2, and EIN4 (FIGURE 22.7). Like ETR1, these receptors have been shown to bind ethylene, and missense mutations in the genes that encode these proteins, analogous to the original *etr1* mutation, prevent ethylene binding to the receptor while allowing the receptor to function normally as a regulator of the ethylene response pathway in the absence of ethylene.

All of these five receptor proteins share at least two domains:

1. The amino-terminal domain spans the membrane at least three times and contains the ethylene-binding site. Ethylene can readily access this site because of its hydrophobicity.

2. The carboxy-terminal half of the ethylene receptors contains a domain homologous to histidine kinase catalytic domains.

A subset of the ethylene receptors also have a carboxy-terminal domain that is similar to bacterial two-component receiver domains. In other two-component systems, binding of ligand regulates the activity of the histidine kinase domain, which autophosphorylates a conserved histidine residue. The phosphate is then transferred to an aspartic acid residue located within the fused receiver domain (see Figure 14.4). Histidine kinase activity has been demonstrated for the ethylene receptor ETR1. However, genetic studies have shown that, in contrast to bacterial two-component systems, the histidine kinase activity of ETR1 does *not* play a primary role in signaling by this ethylene receptor (Wang et al. 2003). Instead, the kinase activity of ETR1 appears to have more subtle effects on ethylene signaling that may modulate the ethylene response (Qu and Schaller 2004; Binder et al. 2004a). Several other ethylene receptors are missing amino acids critical for histidine kinase activity ("Subfamily 2" in Figure 22.7), making it unlikely that they possess histidine kinase activity.

The five Arabidopsis ethylene receptors have been shown to interact with each other in the plant, forming large multisubunit complexes (Gao et al. 2008). Furthermore, binding of ethylene appears to induce degradation of the receptors via the 26S proteasome (Kevany et al. 2007).

Unlike most receptors, which are associated with the plasma membrane, ETR1 and the other four ethylene receptors in Arabidopsis are located on the endoplasmic reticulum. However, ETR1 may also be localized to the Golgi apparatus, at least in roots. In either case, an intracellular location for the ethylene receptor is consistent with the hydrophobic nature of ethylene, which enables it to pass freely through the plasma membrane into the cell. In this respect ethylene is similar to the hydrophobic signaling molecules of animals, such as steroids and the gas nitric oxide, which also bind to intracellular receptors.

## High-affinity binding of ethylene to its receptor requires a copper cofactor

Even before the identification of its receptor, scientists had predicted that ethylene would bind to its receptor via a transition metal cofactor, most likely copper or zinc. This prediction was based on the high affinity of olefins, such as ethylene, for these transition metals. Recent genetic and biochemical studies have borne out these predictions.

Analysis of the ETR1 ethylene receptor expressed in yeast demonstrated that a copper ion was coordinated to the protein and that this copper was necessary for high-affinity ethylene binding (Rodriguez et al. 1999). Silver ion could substitute for copper to yield high-affinity binding, which indicates that silver blocks the action of ethylene not by interfering with ethylene binding, but by preventing the changes in the protein that normally occur when ethylene binds to the receptor.

Evidence that copper binding is required for ethylene receptor function in vivo came from identification of the **RAN1** (*RESPONSIVE-TO-ANTAGONIST1*) gene in Arabidopsis (Hirayama et al. 1999). Strong *ran1* mutations block the formation of functional ethylene receptors (Woeste and Kieber 2000). Cloning of *RAN1* revealed that it encodes a protein similar to a yeast protein required for the transfer of a copper ion to an iron transport protein. In an analogous manner, RAN1 is likely to be involved in the addition of a copper ion cofactor necessary for the function of the ethylene receptors.

## Unbound ethylene receptors are negative regulators of the response pathway

In Arabidopsis, tomato, and probably most other plant species, the ethylene receptors are encoded by multigene families. Targeted disruption (complete inactivation) of the five Arabidopsis ethylene receptors (ETR1, ETR2, ERS1, ERS2, and EIN4) has revealed that they are functionally redundant (Hua and Meyerowitz 1998). That is, disruption of any single gene encoding one of these proteins has no effect, but a plant with disruptions in multiple receptor

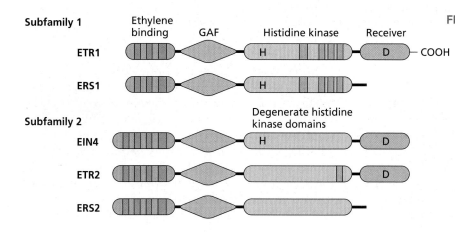

**FIGURE 22.7** Schematic diagram of five ethylene receptor proteins and their functional domains. The GAF domain is a conserved cGMP-binding domain, found in a diverse group of proteins, that generally acts as small molecule–binding regulatory domains. H and D are histidine and aspartate residues that participate in phosphorylation. Note that EIN4, ETR2, and ERS2 have degenerate histidine kinase domains, meaning that they are missing critical, highly conserved amino acids that are required for histidine kinase catalytic activity.

(A)

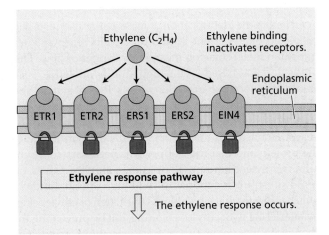

Ethylene (C₂H₄) — Ethylene binding inactivates receptors.

Endoplasmic reticulum

ETR1 ETR2 ERS1 ERS2 EIN4

**Ethylene response pathway**

⬇ The ethylene response occurs.

(B)

ETR1 ETR2 ERS1 ERS2 EIN4

**Ethylene response pathway**

In the absence of ethylene, the receptors are active and suppress the ethylene response.

(C)

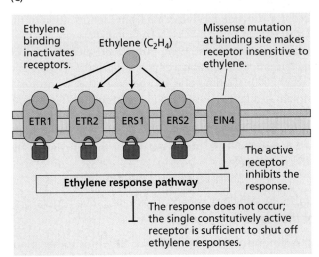

Ethylene binding inactivates receptors.

Ethylene (C₂H₄)

Missense mutation at binding site makes receptor insensitive to ethylene.

ETR1 ETR2 ERS1 ERS2 EIN4

The active receptor inhibits the response.

**Ethylene response pathway**

The response does not occur; the single constitutively active receptor is sufficient to shut off ethylene responses.

(D)

Disruptions in the regulatory domains of multiple ethylene receptors (at least three)

ETR1 ETR2 ERS1 ERS2 EIN4

Disrupted receptors are inactive in the presence or absence of ethylene.

**Ethylene response pathway**

⬇ Constitutive ethylene response

genes exhibits a constitutive ethylene response phenotype (**FIGURE 22.8D**).

The observation that ethylene responses, such as the triple response, become constitutive when the receptors are disrupted indicates that the receptors are normally "on" (i.e., in the active state) in the *absence* of ethylene, and that the function of the receptor *minus* its ligand (ethylene), is to *shut off* the signaling pathway that leads to the response (**FIGURE 22.8B**). Binding of ethylene "turns off" (inactivates) the receptors, thus allowing the response pathway to proceed (**FIGURE 22.8A**).

As discussed in Chapter 14, this somewhat counterintuitive model for ethylene receptors as negative regulators of a signaling pathway is unlike the mechanism of most animal receptors, which, after binding their ligands, serve as positive regulators of their respective signal transduction pathways. That is, animal receptors typically *activate* previously inactive signaling pathways, bringing about the response. Ethylene receptors, on the other hand, actively *repress* the hormone response in the absence of the hormone.

**FIGURE 22.8** Model for ethylene receptor action based on the phenotype of receptor mutants. (A) In the wild type, ethylene binding inactivates the receptors, allowing the response to occur. (B) In the absence of ethylene the receptors act as negative regulators of the response pathway. (C) A missense mutation that interferes with ethylene binding to its receptor, but leaves the regulatory site active, results in an ethylene-insensitive phenotype. (D) Disruption mutations in the regulatory sites result in a constitutive ethylene response.

In contrast to the disrupted receptors, receptors with missense mutations at the ethylene binding site (as occurs in the original *etr1* mutant) are unable to bind ethylene, but are still active as negative regulators of the ethylene response pathway. Such missense mutations result in a plant that expresses a subset of receptors that can no longer be turned off by ethylene, and thus confer a *dominant ethylene-insensitive phenotype* (**FIGURE 22.8C**). Even though the normal receptors can all be turned off by ethylene, the mutant receptors continue to signal the cell to suppress ethylene responses whether ethylene is present or not. In tomato, the *never-ripe* mutation, which, as the name suggests, sets fruit that fails to ripen, is such a dominant ethylene-insensitive mutation in a tomato ethylene receptor.

A consequence of this negative signaling is that a decrease in the level of ethylene receptors actually makes a tissue more sensitive to ethylene. Thus, the level of functional ethylene receptors is an important mechanism by which plants regulate their sensitivity to this hormone. For example, in tomato fruit, two of the ethylene receptors are rapidly degraded by the 26S proteasome in response to ethylene, resulting in an *increase* in ethylene sensitivity (Kevany et al. 2007). This is important in coordinating the timing of ripening throughout the large tomato fruit.

### A serine/threonine protein kinase is also involved in ethylene signaling

The recessive *ctr1* (constitutive *triple response* 1 = triple response in the absence of ethylene) mutation was identified in screens for mutations that constitutively activate ethylene responses (**FIGURE 22.9**). The fact that the recessive mutation caused an *activation* of the ethylene response suggests that the wild-type protein, like the ethylene receptors, acts as a *negative regulator* of the response pathway (Kieber et al. 1993).

CTR1 appears to be related to Raf, a MAPKKK (*mitogen-activated protein kinase kinase kinase*) type of serine/threonine protein kinase that is involved in the transduction of various external regulatory signals and developmental signaling pathways in organisms ranging from yeast to humans (see Chapter 14). In animal cells, the final product in the MAP kinase cascade is a phosphorylated transcription factor that regulates gene expression in the nucleus. There is some evidence that a MAP kinase cascade acts downstream of CTR1 in ethylene signaling (Yoo et al. 2008), but as yet there is no consensus on this point.

Various lines of evidence indicate that the CTR1 protein directly interacts with the ethylene receptors, forming part of a protein complex involved in perceiving ethylene. Genetic analysis has shown that the interaction of CTR1 with the ethylene receptors is necessary for its function, as mutations in CTR1 that block this interaction, but otherwise do not affect the protein, cause CTR1 to be inactive in the plant (Huang et al. 2003). The precise mechanism by which CTR1 is regulated by ETR1 and the other ethylene receptors is still not known.

### *EIN2* encodes a transmembrane protein

The *ein2* (*ethylene-insensitive 2*) mutation blocks all ethylene responses in both seedling and adult Arabidopsis plants. The *EIN2* gene encodes a protein containing 12 membrane-spanning domains that is most similar to the N-RAMP (*natural resistance–associated macrophage pro*-

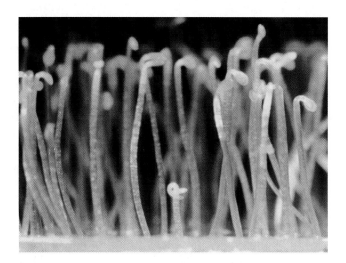

**FIGURE 22.9**  Screen for Arabidopsis mutants that constitutively display the triple response. Seedlings were grown for 3 days in the dark in air (no ethylene). A single *ctr1* mutant seedling is evident among the taller, wild-type seedlings. (Courtesy of J. Kieber.)

tein) family of cation transporters in animals (Alonso et al. 1999), suggesting that it may act as a channel or pore. To date, however, researchers have failed to demonstrate a transport activity for this protein, and the intracellular location of the protein is not known. The EIN2 protein is rapidly degraded by the 26S proteasome, and ethylene inhibits the degradation of EIN2 (Qiao et al. 2009). Because EIN2 alters the sensitivity to ethylene, the degradation of EIN2 provides a further mechanism for regulating the sensitivity of plant cells to ethylene.

## Ethylene Regulation of Gene Expression

One of the primary effects of ethylene signaling is an alteration in the expression of various target genes. Ethylene affects the mRNA transcript levels of numerous genes, including those that encode cellulase and genes related to ripening and ethylene biosynthesis. Regulatory sequences called **ethylene response elements**, or **ERES**, have been identified among the ethylene-regulated genes.

### Specific transcription factors are involved in ethylene-regulated gene expression

Key components mediating ethylene's effects on gene expression are the EIN3 family of transcription factors (Chao et al. 1997). There are at least four *EIN3*-like genes in Arabidopsis, and homologs have been identified in tomato and tobacco. In response to an ethylene signal, homodimers of EIN3 or closely related proteins bind to the promot-

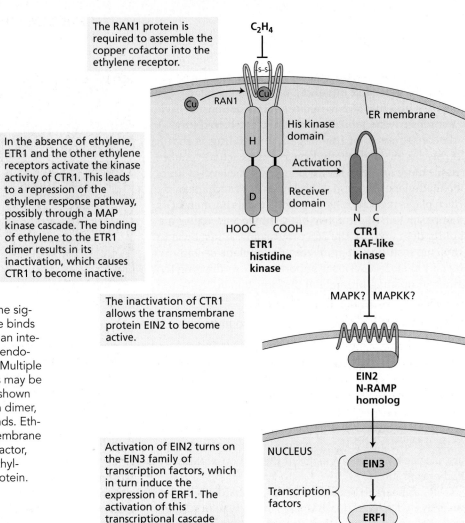

The RAN1 protein is required to assemble the copper cofactor into the ethylene receptor.

In the absence of ethylene, ETR1 and the other ethylene receptors activate the kinase activity of CTR1. This leads to a repression of the ethylene response pathway, possibly through a MAP kinase cascade. The binding of ethylene to the ETR1 dimer results in its inactivation, which causes CTR1 to become inactive.

The inactivation of CTR1 allows the transmembrane protein EIN2 to become active.

Activation of EIN2 turns on the EIN3 family of transcription factors, which in turn induce the expression of ERF1. The activation of this transcriptional cascade leads to large-scale changes in gene expression, which ultimately bring about alterations in cell functions.

**FIGURE 22.10** Model of ethylene signaling in Arabidopsis. Ethylene binds to the ETR1 receptor, which is an integral membrane protein of the endoplasmic reticulum membrane. Multiple isoforms of ethylene receptors may be present in a cell; only ETR1 is shown for simplicity. The receptor is a dimer, held together by disulfide bonds. Ethylene binds within the transmembrane domain, through a copper cofactor, which is assembled into the ethylene receptors by the RAN1 protein.

ers of genes that are rapidly induced by ethylene, including **ERF1** (*ETHYLENE RESPONSE FACTOR 1*), to activate their transcription (Solano et al. 1998).

*ERF1* encodes a protein that belongs to the **ERE-binding protein** (**EREBP**) family of transcription factors, which were first identified in tobacco as proteins that bind to ERE sequences (Ohme-Takagi and Shinshi 1995). Several EREBPs are rapidly up-regulated in response to ethylene. The *EREBP* genes exist in Arabidopsis as a very large gene family, but only a few of the genes are inducible by ethylene.

The regulation of EIN3 protein stability plays an important role in ethylene signaling, as well as in regulating the ethylene biosynthetic pathway. Two redundant F-box proteins, EBF1 and EBF2 (*EIN3-binding F-box 1 and 2*), promote ubiquitination and thus targeting of EIN3 for degradation by the 26S proteasome (see Chapter 14). Ethylene inhibits this EBF1/EBF2-dependent degradation of EIN3, possibly through phosphorylation of EIN3 by a MAP kinase (Yoo et al. 2008), resulting in the accumulation of EIN3 and the subsequent expression of ethylene-regulated genes. Thus, ethylene acts, at least in part, by regulating the level of the EIN3 and *EIN3-like* (EIL) proteins.

## Genetic epistasis reveals the order of the ethylene signaling components

The order of action of the genes *ETR1*, *EIN2*, *EIN3*, and *CTR1* has been determined by the analysis of how the mutations interact with each other (i.e., their epistatic order). Two mutants with opposite phenotypes are crossed, and a line harboring both mutations (the double mutant) is identified in the $F_2$ generation. In the case of the ethylene response mutants, researchers constructed a line doubly mutant for *ctr1* (a constitutive ethylene response mutant) and one of the ethylene-insensitive mutations.

The phenotype displayed by the double mutant reveals which of the mutations is epistatic to the other (Avery and Wasserman 1992). For example, if an *etr1 ctr1* double mutant displays a *ctr1* mutant phenotype, the *ctr1* mutation is said to be epistatic to *etr1*. From this it can be inferred that CTR1 acts downstream of ETR1. Similar genetic studies were used to determine the order of action of *ETR1*, *EIN2*, and *EIN3* relative to *CTR1*.

The ETR1 protein has been shown to interact physically with the predicted downstream protein, CTR1, suggesting that the ethylene receptors may directly regulate the kinase activity of CTR1 (Clark et al. 1998). The model in **FIGURE 22.10** summarizes these and other data. Genes similar to several of these Arabidopsis signaling genes have been found in other species (see **WEB TOPIC 22.7**).

This model is still incomplete and there are likely additional components of this pathway that have yet to be uncovered. In addition, we are only beginning to understand the biochemical properties of these proteins and how they interact. Further research will be needed to provide a more complete picture of the molecular basis for the perception and transduction of the ethylene signal.

## Developmental and Physiological Effects of Ethylene

As we have seen, ethylene was discovered in connection with its effects on seedling growth and fruit ripening. It has since been shown to regulate a wide range of responses in plants, including seed germination, cell expansion, cell differentiation, flowering, senescence, and abscission. In this section we will consider the phenotypic effects of ethylene in more detail.

### Ethylene promotes the ripening of some fruits

In everyday usage, the term *fruit ripening* refers to the changes in fruit that make it ready to eat. Such changes typically include softening due to the enzymatic breakdown of the cell walls, starch hydrolysis, sugar accumulation, and the disappearance of organic acids and phenolic compounds, including tannins.

From the perspective of the plant, fruit ripening means that the seeds are ready for dispersal. For seeds whose dispersal depends on animal ingestion, *ripeness* and *edibility* are synonymous. Brightly colored anthocyanins and carotenoids often accumulate in the epidermis of such fruits, enhancing their visibility. However, for seeds that rely on mechanical or other means for dispersal, *fruit ripening* may mean drying followed by splitting.

Because of their importance in agriculture, the vast majority of studies on fruit ripening have focused on edible fruits. Ethylene has long been recognized as the hormone that accelerates the ripening of edible fruits. Exposure of such fruits to ethylene hastens the processes associated with ripening, and a dramatic increase in ethylene production accompanies the initiation of ripening. However, surveys of a wide range of fruits have shown that not all of them respond to ethylene.

### Fruits that respond to ethylene exhibit a climacteric

All fruits that ripen in response to ethylene exhibit a characteristic respiratory rise called a **climacteric** before the

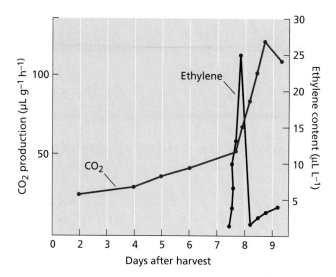

**FIGURE 22.11** Ethylene production and respiration. In banana, ripening is characterized by a climacteric rise in respiration rate, as evidenced by the increased $CO_2$ production. A climacteric rise in ethylene production precedes the increase in $CO_2$ production, suggesting that ethylene is the hormone that triggers the ripening process. (After Burg and Burg 1965.)

ripening phase.* Such fruits also show a spike of ethylene production immediately before the respiratory rise (**FIGURE 22.11**). Apples, bananas, avocados, and tomatoes are examples of climacteric fruits.

In contrast, fruits such as citrus fruits and grapes do not exhibit the respiration and ethylene production rise and are called **nonclimacteric** fruits. Other examples of climacteric and nonclimacteric fruits are given in **TABLE 22.1**.

In climacteric fruits, treatment with ethylene induces the fruit to produce additional ethylene, a response that can be described as **autocatalytic**. In climacteric plants, two systems of ethylene production operate:

- **System 1**, which acts in vegetative tissue, and in which ethylene inhibits its own biosynthesis
- **System 2**, which occurs in ripening climacteric fruit and in the senescing petals in some species, and in which ethylene stimulates its own biosynthesis—that is, it is autocatalytic

The positive feedback loop for ethylene biosynthesis in System 2 integrates ripening of the entire fruit once it has commenced. When unripe climacteric fruits are treated with ethylene, the onset of the climacteric rise is hastened.

---

*The term climacteric can be used either as a noun, as in "most fruits exhibit a climacteric during ripening" or as an adjective, as in "a climacteric rise in respiration." The term nonclimacteric, however, is used only as an adjective, as in "nonclimacteric fruit."

**TABLE 22.1**
**Climacteric and nonclimacteric fruits**

| Climacteric | | Nonclimacteric |
| --- | --- | --- |
| Apple | Olive | Bell pepper |
| Avocado | Peach | Cherry |
| Banana | Pear | Citrus |
| Cantaloupe | Persimmon | Grape |
| Cherimoya | Plum | Pineapple |
| Fig | Tomato | Snap bean |
| Mango | | Strawberry |
| | | Watermelon |

**FIGURE 22.12**  Leaf epinasty in tomato. Epinasty, or downward bending of the tomato leaves (right), is caused by ethylene treatment. Epinasty results when the cells on the upper side of the petiole grow faster than those on the bottom. (Courtesy of S. Gepstein.)

In contrast, when nonclimacteric fruits are treated with ethylene, the respiration rate increases as a function of the ethylene concentration, but the treatment does not trigger production of endogenous ethylene and does not accelerate ripening. Elucidation of the role of ethylene in the ripening of climacteric fruits has resulted in many practical applications aimed at either uniform ripening or the delay of ripening.

Although the effects of exogenous ethylene on fruit ripening are straightforward and clear, establishing a causal relation between the level of endogenous ethylene and fruit ripening is more difficult. Inhibitors of ethylene biosynthesis (such as AVG) or of ethylene action (such as $CO_2$, MCP, or $Ag^+$) have been shown to delay or even prevent ripening. However, the definitive demonstration that ethylene is required for fruit ripening was provided by experiments in which ethylene biosynthesis was blocked by expression of an antisense version of either ACC synthase or ACC oxidase in transgenic tomatoes (see **WEB TOPIC 22.8**). Elimination of ethylene biosynthesis in these transgenic tomatoes completely blocked fruit ripening, and application of exogenous ethylene restored ripening (Oeller et al. 1991).

### The receptors of never-ripe mutants of tomato fail to bind ethylene

Further demonstration of the requirement for ethylene in fruit ripening came from the analysis of the *never-ripe* mutation in tomato. As the name implies, this mutation completely blocks the ripening of tomato fruit. Molecular analysis revealed that *never-ripe* was due to a mutation in an ethylene receptor that rendered it unable to bind ethylene (Lanahan et al. 1994). This analysis, together with the demonstration that inhibiting ethylene biosynthesis via antisense technology blocked ripening, provided unequivocal proof of the role of ethylene in fruit ripening, and opened the door to the manipulation of fruit ripening through biotechnology.

In tomatoes many genes that are highly regulated during ripening have been identified using tomato complementary DNA (cDNA) microarrays.* During tomato fruit ripening, the fruit softens as the result of cell wall hydrolysis, and it changes from green to red as a consequence of chlorophyll loss and the synthesis of the carotenoid pigment lycopene. At the same time, aroma and flavor components are produced.

Analysis of mRNA from fruits of wild-type and transgenic tomato plants genetically engineered to lack ethylene has revealed that gene expression during ripening is regulated by at least two independent pathways:

1. *An ethylene-dependent pathway* includes genes involved in the biosynthesis of lycopene, volatile aromatic compounds, ACC synthase, and the enzymes of respiratory metabolism.

2. *A developmental, ethylene-independent pathway* includes genes that encode ACC oxidase and chlorophyllase.

Thus, not all of the processes associated with ripening in tomato are ethylene dependent.

### Leaf epinasty results when ACC from the root is transported to the shoot

The downward curvature of leaves that occurs when the upper (adaxial) side of the petiole grows faster than the lower (abaxial) side is termed *epinasty* (**FIGURE 22.12**). Eth-

---

*cDNA microarrays (also known as biochips, DNA chips, or gene arrays) are small chips on which cDNAs or oligonucleotides, each representing a given gene, have been immobilized. When probed with the appropriately tagged cDNAs, DNA chips allow one to measure the expression of thousands of genes in a tissue simultaneously. See Chapter 2 for a more detailed description of this technique.

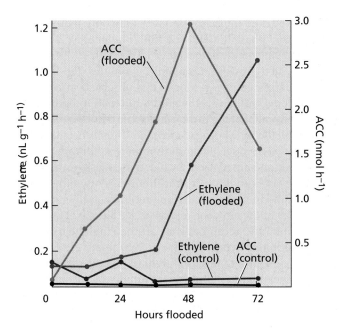

**FIGURE 22.13** Changes in the amounts of ACC in the xylem sap and ethylene production in the petiole following flooding of tomato plants. ACC is synthesized in roots, but it is converted to ethylene very slowly under the anaerobic conditions of flooding. ACC is transported via the xylem to the shoot, where it is converted to ethylene. Gaseous ethylene cannot be transported, so it usually affects the tissue nearest the site of its production. In contrast, the ethylene precursor ACC is transportable and can produce ethylene far from the site of ACC synthesis. (After Bradford and Yang 1980.)

ylene and high concentrations of auxin induce epinasty, and it has now been established that auxin acts indirectly by inducing ethylene production. As will be discussed later in the chapter, a variety of stress conditions, such as salt stress or pathogen infection, increase ethylene production and also induce epinasty.

In tomato and other dicots, flooding (waterlogging) or anaerobic conditions around the roots enhance the synthesis of ethylene in the shoot, leading to the epinastic response. Because these environmental stresses are sensed by the roots and the response is displayed by the shoots, a signal from the roots must be transported to the shoots. This signal is ACC, the immediate precursor of ethylene. ACC levels were found to be significantly higher in the xylem sap of tomato plants after the roots had been flooded for 1 to 2 days (**FIGURE 22.13**) (Bradford and Yang 1980).

Because water fills the air spaces in waterlogged soil, and $O_2$ diffuses slowly through water, the concentration of oxygen around flooded roots decreases dramatically. ACC accumulates in the anaerobic roots, and is then transported to shoots via the transpiration stream, where it is readily converted to ethylene in the presence of oxygen (see Figure 22.2).

### Ethylene induces lateral cell expansion

At concentrations above 0.1 $\mu$L $L^{-1}$, ethylene changes the growth pattern of seedlings by reducing the rate of elongation and increasing lateral expansion, leading to swelling of the hypocotyl or the epicotyl. In dicots, this swelling is part of the **triple response**, which, in Arabidopsis, consists of inhibition of hypocotyl elongation combined with hypo-

cotyl swelling, inhibition of root elongation, and exaggeration of the curvature of the apical hook (see Figure 22.5).

As discussed in Chapter 15, the directionality of plant cell expansion is determined by the orientation of the cellulose microfibrils in the cell wall. Transverse microfibrils reinforce the cell wall in the lateral direction, so that turgor pressure is channeled into cell elongation. The orientation of the microfibrils is in turn determined by the orientation of the cortical array of microtubules in the cortical (peripheral) cytoplasm. In typical elongating plant cells, the cortical microtubules are arranged transversely, giving rise to transversely arranged cellulose microfibrils.

During the seedling triple response to ethylene, in the hypocotyl, the transverse pattern of microtubule alignment is disrupted, and the microtubules switch over to a longitudinal orientation. This 90° shift in microtubule orientation leads to a parallel shift in cellulose microfibril deposition. The newly deposited wall is reinforced in the longitudinal direction rather than the transverse direction, which promotes lateral expansion instead of elongation.

How do microtubules shift from one orientation to another? To study this phenomenon, pea (*Pisum sativum*) epidermal cells were injected with the microtubule protein tubulin, to which a fluorescent dye was covalently attached. The fluorescent "tag" did not interfere with the assembly of microtubules. This procedure allowed researchers to monitor the assembly of microtubules in living cells using a confocal laser scanning microscope, which can focus in many planes throughout the cell.

It was found that microtubules do not reorient from the transverse to the longitudinal direction by complete depolymerization of the transverse microtubules followed by repolymerization of a new longitudinal array of microtubules. Instead, increasing numbers of nontransversely aligned microtubules appear in particular locations (**FIGURE 22.14**). Neighboring microtubules then adopt the new alignment, so at one stage different alignments coexist before all the microtubules adopt a uniformly longitudinal orientation (Yuan et al. 1994). Although the reorientations observed in this study were in response to wounding rather than induced by ethylene, it is presumed that

Transverse microtubules

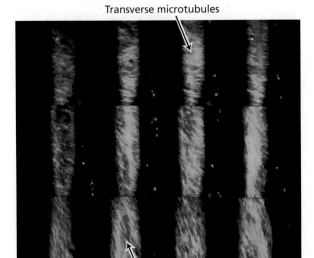

Longitudinal microtubules

**FIGURE 22.14** Reorientation of microtubules from transverse to vertical in pea stem epidermal cells in response to wounding. A living epidermal cell was microinjected with rhodamine-conjugated tubulin, which incorporates into the plant microtubules. A time series of approximately 6-minute intervals shows the cortical microtubules undergoing reorientation from net transverse to oblique/longitudinal. The reorientation seems to involve the appearance of patches of new, "discordant" microtubules in the new direction, concomitant with the disappearance of microtubules from the previous alignment. (From Yuan et al. 1994, photo courtesy of C. Lloyd.)

ethylene-induced microtubule reorientation operates by a similar mechanism.

### There are two distinct phases to growth inhibition by ethylene

As noted above, growth of seedlings in the presence of ethylene inhibits the elongation of hypocotyls in dark-grown seedlings. Careful kinetic analysis of this response indicates that it occurs in two distinct phases (**FIGURE 22.15A**) (Binder et al. 2004a).

- The first, rapid phase of inhibition occurs within 15 minutes of exposure to ethylene and lasts approximately 30 minutes.
- A second deceleration of growth then ensues, with the hypocotyl growth reaching a new steady-state level of growth that is slower than that of untreated seedlings.

These two phases are growth are mechanistically distinct: The first phase is more sensitive to ethylene as compared to the second phase; EIN2 is required for both phases, but the EIN3/EIL1 transcription factors are only required for the second phase (**FIGURE 22.15B**) (Binder et al. 2004b).

Following removal from ethylene, seedlings fully recover to untreated growth rates within 90 minutes (Binder et al. 2004a). How can we reconcile this observation with the fact that the half-life of ethylene bound to its receptors is 11 hours? Recall that the turnover of some ethylene receptors is stimulated by binding to ethylene. Thus, when ethylene is removed from seedlings, the bound receptors are rapidly degraded and are replaced by newly synthesized receptors that are not bound to ethylene. Remember that the newly synthesized unbound ethylene receptors act as *negative regulators* of the ethylene response (see Figure 22.8). Thus, these unbound receptors shut off the ethylene response (i.e., inhibited hypocotyl elongation) relatively rapidly, even though some receptors remain that are still bound to ethylene. The histidine kinase activity of the subfamily 1 receptors appears to play an important role during this recovery phase following removal of ethylene.

### The hooks of dark-grown seedlings are maintained by ethylene production

Etiolated dicot seedlings are usually characterized by a pronounced hook located just behind the shoot apex (see Figures 22.1 and 22.5). This hook shape facilitates penetration of the seedling through the soil, protecting the tender apical meristem.

Like epinasty, hook formation and maintenance result from ethylene-induced asymmetric growth. The closed shape of the hook is a consequence of the more rapid elongation of the outer side of the stem compared with the inner side. When the hook is exposed to white light it opens, because the elongation rate of the inner side increases, equalizing the growth rates on both sides (see Appendix 2).

Red light induces hook opening, and far-red light reverses the effect of red, indicating that phytochrome is the photoreceptor involved in this process (see Chapter 17). A close interaction between phytochrome and ethylene controls hook opening. As long as ethylene is produced by the hook tissue in the dark, elongation of the cells on the inner side is inhibited. Red light inhibits ethylene formation, promoting growth on the inner side, thereby causing the hook to open.

The auxin-insensitive mutation *axr1* does not develop an apical hook; and treatment of wild-type Arabidopsis seedlings with NPA (*N*-1-naphthylphthalamic acid), an inhibitor of polar auxin transport, blocks apical hook formation. These and other results indicate a role for auxin in maintaining hook structure. The more rapid growth rate of the outer tissues relative to the inner tissues could reflect an ethylene-dependent auxin gradient, analogous to the lateral auxin gradient that develops during phototropic curvature (see Chapter 19 and **WEB TOPIC 22.9**).

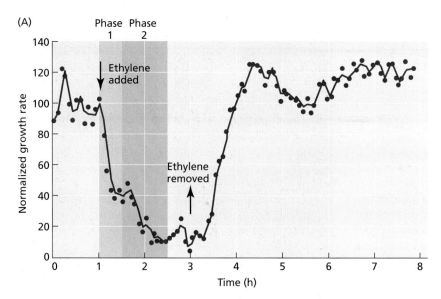

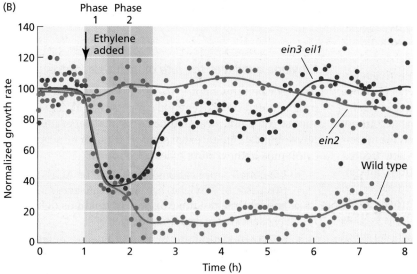

FIGURE 22.15 Kinetics of the effects of ethylene addition and removal on hypocotyl elongation in dark-grown Arabidopsis seedlings. (A) Growth rate of etiolated wild-type Arabidopsis after exposure to ethylene and subsequent removal of ethylene at the times indicated by the arrows. Note that the reduction in the growth rate following exposure to ethylene occurs in two distinct phases. (B) Growth rate of etiolated wild-type, *ein2*, and *ein3 eil1* Arabidopsis seedlings following exposure to ethylene at the time indicated by the arrow. Note that the phase 1 response of the *ein3 eil1* seedlings is identical to that of the wild type, but the phase 2 response is absent. (After Binder et al. 2004a, b.)

## Ethylene breaks seed and bud dormancy in some species

Seeds that fail to germinate under normal conditions (water, oxygen, temperature suitable for growth) are said to be dormant (see Chapter 23). Ethylene has the ability to break dormancy and initiate germination in certain seeds, such as cereals. In addition to its effect on dormancy, ethylene increases the rate of seed germination of several species. In peanuts (*Arachis hypogaea*), ethylene production and seed germination are closely correlated. Ethylene can also break bud dormancy, and ethylene treatment is sometimes used to promote bud sprouting in potato and other tubers.

## Ethylene promotes the elongation growth of submerged aquatic species

Although usually thought of as an inhibitor of stem elongation, ethylene is able to promote stem and petiole elongation in various submerged or partially submerged aquatic plants. These include the dicots *Ranunculus sceleratus*, *Nymphoides peltata*, and *Callitriche platycarpa*, and the fern *Regnellidium diphyllum*. Another agriculturally important example is deep-water rice (*Oryza sativa*), a cereal.

In these species, submergence induces rapid internode or petiole elongation, which allows the leaves or upper parts of the shoot to remain above water. Treatment with ethylene mimics the effects of submergence.

Growth is stimulated in the submerged plants because ethylene builds up in the tissues. In the absence of $O_2$, ethylene synthesis is diminished, but the loss of ethylene by diffusion is retarded under water. Sufficient oxygen for growth and ethylene synthesis in the underwater parts is usually provided by aerenchyma tissue (see Chapter 26).

Ethylene stimulates internode elongation in deep-water rice by increasing the amount of, and the sensitivity to, gibberellin in the cells of the intercalary meristem. The increased sensitivity to gibberellic acid (GA) in these cells in response to ethylene is brought about by a decrease in the level of abscisic acid (ABA), a potent antagonist of GA.

Two genes encoding transcription factors in the ethylene response factor family, SNORKEL1 and SNORKEL2, have recently been identified in deep-water rice that mediate this response (Hattori et al. 2009). In flooding conditions, ethylene accumulates and induces the expression of SNORKEL1 and SNORKEL2, which then triggers the dramatic internode elongation.

**FIGURE 22.16** Promotion of root hair formation by ethylene in lettuce seedlings. Two-day-old seedlings were treated with air (left) or 10 ppm ethylene (right) for 24 hours before the photo was taken. Note the profusion of root hairs on the ethylene-treated seedling. (From Abeles et al. 1992, courtesy of F. Abeles.)

Air

Ethylene

### Ethylene induces the formation of roots and root hairs

Ethylene is capable of inducing adventitious root formation in leaves, stems, flower stems, and even other roots. Vegetative stem cuttings from tomato and petunia make many adventitious roots in response to applied auxin, but in ethylene-insensitive mutants auxin has little or no effect, indicating that the promotive effect of auxin on adventitious rooting is mediated by ethylene (Clark et al. 1999). Ethylene also plays a role in the morphogenesis of crown gall tissue (see WEB ESSAY 22.1), and is a negative regulator of root nodule formation in legumes (see WEB TOPIC 22.10).

Ethylene has also been shown to act as a positive regulator of root hair formation in several species (FIGURE 22.16). This regulation has been best studied in Arabidopsis, in which root hairs normally are located in the epidermal cells that overlie a junction between the underlying cortical cells (Dolan et al. 1994). In ethylene-treated roots, cells not overlying a cortical cell junction differentiate into hair cells, and produce root hairs in abnormal locations (Tanimoto et al. 1995). Seedlings grown in the presence of ethylene inhibitors (such as Ag$^+$), as well as ethylene-insensitive mutants, display a reduction in root hair formation. These observations suggest that ethylene acts as a positive regulator in the differentiation of root hairs.

### Ethylene regulates flowering and sex determination in some species

Although ethylene inhibits flowering in many species, it induces flowering in pineapple and its relatives, and it is used commercially for synchronization of pineapple fruit set. Flowering of other species, such as mango, is also initiated by ethylene. On plants that have separate male and female flowers (monoecious species), ethylene may change the sex of developing flowers. The promotion of female flower formation in cucumber is one example of this effect. Recently, a gene responsible for andromonoecy (plants carrying both male and bisexual flowers) in melons was identified as encoding an ACC synthase (Boualem et al. 2008). A mutation that reduces the activity of this ACC synthase gene results in the formation of the bisexual flowers in these andromonoecious lines.

### Ethylene enhances the rate of leaf senescence

As described in Chapter 16, senescence is a genetically programmed developmental process that affects all tissues of the plant. Research has provided several lines of physiological evidence that support roles for ethylene and cytokinins in the control of leaf senescence:

- Exogenous applications of ethylene or ACC (the precursor of ethylene) accelerate leaf senescence, and treatment with exogenous cytokinins delays leaf senescence (see Chapter 21).

- Enhanced ethylene production is associated with chlorophyll loss and color fading, which are characteristic features of leaf and flower senescence; an inverse correlation has been found between cytokinin levels in leaves and the onset of senescence.

- Inhibitors of ethylene synthesis (e.g., AVG or Co$^{2+}$) and action (e.g., Ag$^+$ or CO$_2$) retard leaf and flower senescence (FIGURE 22.17).

Taken together, these physiological studies suggest that senescence is regulated by the balance of ethylene and cytokinin. In addition, abscisic acid has been implicated in the control of leaf senescence. The role of ABA in senescence will be discussed in Chapter 23.

Direct evidence for the involvement of ethylene in the regulation of leaf senescence has come from molecular genetic studies on Arabidopsis. As discussed above, ethylene-insensitive mutants, such as *etr1* (*e*thylene-*r*esistant *1*) and *ein2* (*e*thylene-*in*sensitive *2*), were identified by their failure to respond to ethylene. Consistent with a role for ethylene in leaf senescence, both *etr1* and *ein2* plants

**FIGURE 22.17** Inhibition of flower senescence by inhibition of ethylene action. Carnation flowers were held in deionized water for 14 days with (left) or without (right) silver thiosulfate (STS), a potent inhibitor of ethylene action. Blocking of ethylene action results in a marked inhibition of floral senescence. (From Reid 1995, courtesy of M. Reid.)

were found to be affected not only during the early stages of germination, but throughout the life cycle, including senescence (Zacarias and Reid 1990; Hensel et al. 1993; Grbić and Bleecker 1995). The ethylene mutants retained their chlorophyll and other chloroplast components for a longer period of time compared to the wild type. However, because the total life spans of these mutants were increased by only 30% over that of the wild type, ethylene appears to increase the *rate* of senescence, rather than acting as a developmental switch that initiates the senescence process (see **WEB TOPIC 22.11**).

### Ethylene mediates some defense responses

Pathogen infection and disease will occur only if the interactions between host and pathogen are genetically compatible. However, ethylene production generally increases in response to pathogen attack in both compatible (i.e., pathogenic) and noncompatible (nonpathogenic) interactions.

The discovery of ethylene-insensitive mutants has facilitated the assessment of the role of ethylene in the response to various pathogens. The involvement of ethylene in pathogenesis is complex and depends on the particular host–pathogen interaction. For example, blocking ethylene responsiveness does not affect the resistance responses of Arabidopsis to *Pseudomonas* bacteria or of tobacco to tobacco mosaic virus. In compatible interactions of these pathogens and hosts, however, elimination of ethylene responsiveness prevents the development of

disease symptoms, even though the growth of the pathogen appears to be unaffected.

On the other hand, ethylene, in combination with the plant hormone jasmonic acid (see Chapter 13), is required for the activation of several plant defense genes. In addition, ethylene-insensitive tobacco and Arabidopsis mutants are susceptible to several necrotrophic (growing on dead host tissue) soil fungi that are normally not pathogenic. Thus ethylene, in combination with jasmonic acid, plays an important role in plant defense against necrotrophic pathogens. On the other hand, ethylene does not appear to play a major role in the response of plants to biotrophic (growing on living tissue) pathogens.

### Ethylene acts on the abscission layer

The shedding of leaves, fruits, flowers, and other plant organs is termed **abscission** (see **WEB TOPIC 22.12**). Abscission takes place in specific layers of cells called **abscission layers**, which become morphologically and biochemically differentiated during organ development. Weakening of the cell walls at the abscission layer depends on cell wall–degrading enzymes such as cellulase and polygalacturonase (**FIGURE 22.18**).

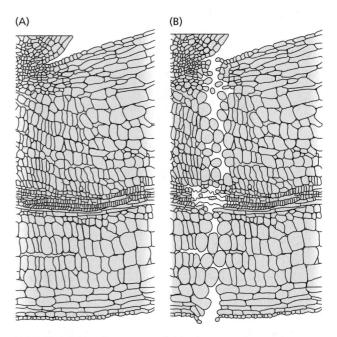

**FIGURE 22.18** Formation of the abscission layer of jewelweed (*Impatiens*). (A) During leaf abscission, two or three rows of cells in the abscission zone undergo cell wall breakdown because of an increase in cell wall–hydrolyzing enzymes. (B) The protoplasts, released from the restraint of their cell walls, expand and push apart the xylem tracheary cells, facilitating the separation of the leaf from the stem. (After Sexton et al. 1984.)

**FIGURE 22.19** Effect of ethylene on abscission in birch (*Betula pendula*). The plant on the left is the wild type; the plant on the right was transformed with a mutated version of the Arabidopsis ethylene receptor *etr1*. The expression of this gene was under the transcriptional control of its own promoter. One of the characteristics of these mutant trees is that they do not drop their leaves when fumigated for 3 days with 50 ppm ethylene. (From Vahala et al. 2003.)

The ability of ethylene gas to cause defoliation in birch trees is shown in **FIGURE 22.19**. Both trees have been subjected to 3 days of fumigation with 50 ppm ethylene. The wild-type tree on the left has lost most of its leaves. The tree on the right has been transformed with a gene for the Arabidopsis ethylene receptor, ETR1, which carries the dominant *etr1* mutation (discussed earlier). This tree is unable to respond to ethylene and therefore does not shed its leaves after ethylene treatment.

Ethylene appears to be the primary regulator of the abscission process, with auxin acting as a suppressor of the ethylene effect (see Chapter 19). However, supraoptimal auxin concentrations stimulate ethylene production, which has led to the use of auxin analogs as defoliants. For example, 2,4,5-T, the active ingredient in Agent Orange, was widely used as a defoliant during the Vietnam War. Its action is based on its ability to increase ethylene biosynthesis, thereby stimulating leaf abscission.

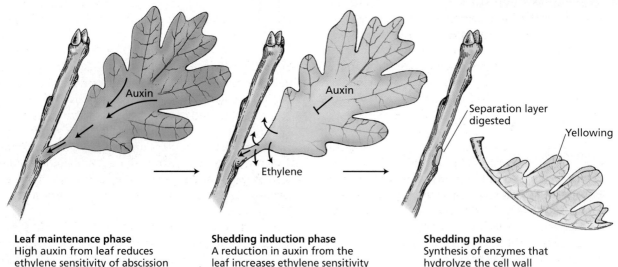

**Leaf maintenance phase**
High auxin from leaf reduces ethylene sensitivity of abscission zone and prevents leaf shedding.

**Shedding induction phase**
A reduction in auxin from the leaf increases ethylene sensitivity in the abscission zone, which triggers the shedding phase.

**Shedding phase**
Synthesis of enzymes that hydrolyze the cell wall polysaccharides results in cell separation and leaf abscission.

**FIGURE 22.20** Schematic view of the roles of auxin and ethylene during leaf abscission. In the shedding induction phase, the level of auxin decreases, and the level of ethylene increases. These changes in the hormonal balance increase the sensitivity of the target cells to ethylene. (After Morgan 1984.)

A model of the hormonal control of leaf abscission describes the process in three distinct sequential phases (**FIGURE 22.20**) (Reid 1995):

1. *Leaf maintenance phase*. Prior to the perception of any signal (internal or external) that initiates the abscission process, the leaf remains healthy and fully functional in the plant. A gradient of auxin from the blade to the stem maintains the abscission zone in a nonsensitive state.

2. *Shedding induction phase*. A reduction or reversal in the auxin gradient from the leaf, normally associated with leaf senescence, causes the abscission zone to become sensitive to ethylene. Treatments that enhance leaf senescence may promote abscission by interfering with auxin synthesis and/or transport in the leaf.

3. *Shedding phase*. The sensitized cells of the abscission zone respond to low concentrations of endogenous ethylene by synthesizing and secreting cellulase and other cell wall–degrading enzymes, resulting in shedding.

During the early phase of leaf maintenance, auxin from the leaf prevents abscission by maintaining the cells of the abscission zone in an ethylene-insensitive state. It has long been known that removal of the leaf blade (the site of auxin production) promotes petiole abscission. Application of exogenous auxin to petioles from which the leaf blade has been removed delays the abscission process. However, application of auxin to the proximal side of the abscission zone (i.e., the side closest to the stem) actually accelerates the abscission process. These results indicate that it is not the absolute amount of auxin at the abscission zone, but rather the auxin *gradient*, that controls the ethylene sensitivity of these cells.

In the shedding induction phase, the amount of auxin from the leaf decreases and the ethylene level rises. Ethylene appears to decrease the activity of auxin both by reducing its synthesis and transport and by increasing its destruction. The reduction in the concentration of free auxin increases the response of specific target cells to ethylene. The shedding phase is characterized by the induction of genes encoding specific hydrolytic enzymes of cell wall polysaccharides and proteins.

The target cells, located in the abscission zone, synthesize cellulase and other polysaccharide-degrading enzymes, and secrete them into the cell wall. The activities of these enzymes lead to cell wall loosening, cell separation, and abscission.

## Ethylene has important commercial uses

Because ethylene regulates so many physiological processes in plant development, it is one of the most widely used plant hormones in agriculture. Auxins and ACC can trigger the natural biosynthesis of ethylene and in several cases are used in agricultural practice. Because of its high diffusion rate, ethylene is very difficult to apply in the field as a gas, but this limitation can be overcome if an ethylene-releasing compound is used. The most widely used such compound is Ethephon, or 2-chloroethylphosphonic acid, which was discovered in the 1960s and is known by various trade names, such as Ethrel.

Ethephon is sprayed in aqueous solution and is readily absorbed and transported within the plant. It releases ethylene slowly by a chemical reaction, allowing the hormone to exert its effects:

$$Cl-CH_2-CH_2-\overset{\overset{O}{\|}}{\underset{\underset{O^-}{|}}{P}}-OH + OH^- \longrightarrow CH_2{=}CH_2 + H_2PO_4^- + Cl^-$$

**2-Chloroethylphosphonic acid
(Ethephon)**                **Ethylene**

Ethephon hastens fruit ripening of apple and tomato and degreening of citrus fruits, synchronizes flowering and fruit set in pineapple, and accelerates abscission of flowers and fruits. It can be used to induce fruit thinning or fruit drop in cotton, cherry, and walnut. It is also used to promote female sex expression in cucumber, to prevent self-pollination and increase yield, and to inhibit terminal growth of some plants in order to promote lateral growth and compact flowering stems.

Storage facilities developed to inhibit ethylene production and promote preservation of fruits have a controlled atmosphere of low $O_2$ concentration and low temperature for the inhibition of ethylene biosynthesis. A relatively high concentration of $CO_2$ (3 to 5%) prevents ethylene's action as a ripening promoter. Low pressure (vacuum) is used to remove ethylene and oxygen from the storage chambers, reducing the rate of ripening and preventing overripening. The ethylene binding inhibitor Ethylbloc® is increasingly being used to extend the shelf life of various climacteric fruits.

It is estimated that from 15 to 35% of the cut flowers harvested in the U.S. are lost because of postharvest spoilage. Specific inhibitors of ethylene biosynthesis and action have proven useful in postharvest preservation (see **WEB TOPIC 22.13**).

## SUMMARY

Ethylene regulates fruit ripening and processes associated with leaf and flower senescence and abscission, root hair development and nodulation, and seedling growth and hook opening, and does so at least in part by altering gene expression.

### Structure, Biosynthesis, and Measurement of Ethylene

- Ethylene gas induces the triple response in dicots (**Figures 22.1, 22.5**).

- The ethylene precursor is methionine, which is converted sequentially to *S*-adenosylmethionine, ACC, and ethylene (**Figure 22.2**). Ethylene acts near its site of synthesis. The immediate precursor of ethylene, ACC can be transported and thus can produce ethylene at a site distant from its synthesis.

- Ethylene biosynthesis is stimulated by several factors, including developmental state, environmental conditions, other plant hormones, and physical and chemical stimuli (**Figure 22.3**).

- The biosynthesis and perception of ethylene can be antagonized by inhibitors, some of which have commercial applications (**Figure 22.4**).

### Ethylene Signal Transduction Pathways

- The triple-response morphology of etiolated Arabidopsis seedlings has aided identification of functionally redundant ethylene receptor genes and other signaling elements (**Figures 22.5–22.7**).

- Ethylene receptors are located on the endoplasmic reticulum, and also on the Golgi apparatus.

- Ethylene binds to its receptor via a copper cofactor.

- Unbound receptors actively shut off the signaling pathway that leads to the response; binding of ethylene inactivates the receptors, allowing the response pathway to proceed (**Figures 22.8, 22.9**).

- ETR1 activates CTR1, a protein kinase that shuts off ethylene responses.

### Ethylene Regulation of Gene Expression

- Ethylene affects the transcription of numerous genes via specific transcription factors.

- Analysis of epistatic interactions revealed the sequence of action for the genes *ETR1*, *EIN2*, *EIN3*, and *CTR1* (**Figure 22.10**).

### Developmental and Physiological Effects of Ethylene

- Ethylene is involved in seedling growth and fruit ripening (**Figure 22.11, Table 22.1**), epinasty (**Figures 22.12, 22.13**), and seed germination.

- The hormone influences cell expansion and the orientation of the cellulose microfibrils in the cell wall (**Figure 22.14**).

- There are two mechanistically distinct phases to ethylene inhibition of hypocotyl elongation (**Figure 22.15**).

- Ethylene stimulates rapid internode or petiole elongation when some species are submerged. The hormone regulates flowering, sex determination, and defense responses in some species.

- Ethylene stimulates root hair formation (**Figure 22.16**).

- Ethylene is active in leaf and flower senescence and in leaf abscission (**Figures 22.17–22.20**).

## WEB MATERIAL

### Web Topics

**22.1 Ethylene in the Environment Arises Biotically and Abiotically**

Ethylene in the environment arises from a variety of sources, including pollution, photochemical reactions in the atmosphere, and production by microbes, algae, and plants.

**22.2 Ethylene Readily Undergoes Oxidation**

Ethylene can be oxidized to ethylene oxide, which can then be hydrolyzed to ethylene glycol.

**22.3 Ethylene Can Be Measured by Gas Chromatography**

Historically, bioassays based on the seedling triple response were used to measure ethylene levels, but they have been replaced by gas chromatography.

**22.4 Cloning of the Gene That Encodes ACC Synthase**

A brief description of the cloning of the gene for ACC synthase using antibodies raised against the partially purified protein.

## WEB MATERIAL continued

### Web Topics

**22.5 Cloning of the Gene That Encodes ACC Oxidase**

The ACC oxidase gene was cloned by a circuitous route using antisense DNA.

**22.6 Ethylene Binding to ETR1 and Seedling Response to Ethylene**

Ethylene binding to its receptor ETR1 was first demonstrated by expressing the gene in yeast.

**22.7 Conservation of Ethylene Signaling Components in Other Plant Species**

The evidence suggests that ethylene signaling is similar in all plant species.

**22.8 ACC Synthase Gene Expression and Biotechnology**

A discussion of the use of the ACC synthase gene in biotechnology.

**22.9 The *hookless* Mutation Alters the Pattern of Auxin Gene Expression**

The *hookless* mutation of Arabidopsis confirms the interaction between auxin and ethylene in hook formation.

**22.10 Ethylene Inhibits the Formation of Nitrogen-Fixing Root Nodules in Legumes**

Hyper-nodulating mutants are blocked in the ethylene signal transduction pathway.

**22.11 Ethylene Biosynthesis Can Be Blocked with Anti-Sense DNA**

Mutants expressing anti-sense DNA coding for ethylene biosynthesis enzymes have delayed leaf senescence and fruit ripening.

**22.12 Abscission and the Dawn of Agriculture**

A short essay on the domestication of modern cereals based on artificial selection for nonshattering rachises.

**22.13 Specific Inhibitors of Ethylene Biosynthesis Are Used Commercially to Preserve Cut Flowers**

Some inhibitors of ethylene biosynthesis are suitable for commercial use in flower preservation.

### Web Essay

**22.1 Tumor-Induced Ethylene Controls Crown Gall Morphogenesis**

*Agrobacterium tumefaciens*-induced galls produce very high ethylene concentrations, which reduce vessel diameter in the host stem adjacent to the tumor and enlarge the gall surface giving priority in water supply to the growing tumor over the host shoot.

## CHAPTER REFERENCES

Abeles, F. B., Morgan, P. W., and Saltveit, M. E., Jr. (1992) *Ethylene in Plant Biology*, 2nd ed. Academic Press, San Diego, CA.

Alonso, J. M., Hirayama, T., Roman, G., Nourizadeh, S., and Ecker, J. R. (1999) EIN2, a bifunctional transducer of ethylene and stress responses in *Arabidopsis*. *Science* 284: 2148–2152.

Avery, L., and Wasserman, S. (1992) Ordering gene function: The interpretation of epistasis in regulatory hierarchies. *Trends Genet.* 8: 312–316.

Binder, B. M., O'Malley, R. C., Moore, J. M., Parks, B. M., Spalding, E. P., and Bleecker, A. B. (2004a) Arabidopsis seedling growth response and recovery to ethylene: A kinetic analysis. *Plant Physiol.* 136: 2913–2920.

Binder, B. M., Mortimore, L. A., Stepanova, A. N., Ecker, J. R., and Bleecker, A. B. (2004b) Short term growth responses to ethylene in Arabidopsis seedlings are EIN3/EIL1 independent. *Plant Physiol.* 136: 2921–2927.

Boualem, A., Fergany, M., Fernandez, R., Troadec, C., Martin, A., Morin, H., Sari, M. A., Collin, F., Flowers, J. M., Pitrat, M., et al. (2008) A conserved mutation in an ethylene biosynthesis enzyme leads to andromonoecy in melons. *Science* 321: 836–838.

Bradford, K. J., and Yang, S. F. (1980) Xylem transport of 1-aminocyclopropane-1-carboxylic acid, an ethylene precursor, in waterlogged tomato plants. *Plant Physiol.* 65: 322–326.

Burg, S. P., and Burg, E. A. (1965) Relationship between ethylene production and ripening in bananas. *Bot. Gaz.* 126: 200–204.

Burg, S. P., and Thimann, K. V. (1959) The physiology of ethylene formation in apples. *Proc. Natl. Acad. Sci. USA* 45: 335–344.

Chae, H. S., and Kieber, J. J. (2005) Eto Brute? Role of ACS turnover in regulating ethylene biosynthesis. *Trends Plant Sci.* 10: 291–296.

Chao, Q., Rothenberg, M., Solano, R., Roman, G., Terzaghi, W., and Ecker, J. R. (1997) Activation of the ethylene gas response pathway in *Arabidopsis* by the nuclear protein ETHYLENE-INSENSITIVE3 and related proteins. *Cell* 89: 1133–1144.

Clark, D. G., Gubrium, E. K., Barrett, J. E., Nell, T. A., and Klee, H. J. (1999) Root formation in ethylene-insensitive plants. *Plant Physiol.* 121: 53–60.

Clark, K. L., Larsen, P. B., Wang, X., and Chang, C. (1998) Association of the *Arabidopsis* CTR1 Raf-like kinase with the ETR1 and ERS ethylene receptors. *Proc. Natl. Acad. Sci. USA* 95: 5401–5406.

Dolan, L., Duckett, C. M., Grierson, C., Linstead, P., Schneider, K., Lawson, E., Dean, C., Poethig, S., and Roberts, K. (1994) Clonal relationships and cell patterning in the root epidermis of *Arabidopsis*. *Development* 120: 2465–2474.

Gao, Z., Wen, C. K., Binder, B. M., Chen, Y. F., Chang, J., Chiang, Y. H., Kerris, R. J. III, Chang, C., and Schaller, G. E. (2008) Heteromeric interactions among ethylene receptors mediate signaling in Arabidopsis. *J. Biol. Chem.* 283: 23801–23810.

Glick, B. R. (2005) Modulation of plant ethylene levels by the bacterial enzyme ACC deaminase. *FEMS Microbiol. Lett.* 251: 1–7.

Grbi, V., and Bleecker, A. B. (1995) Ethylene regulates the timing of leaf senescence in *Arabidopsis*. *Plant J.* 8: 595–602.

Guzman, P., and Ecker, J. R. (1990) Exploiting the triple response of *Arabidopsis* to identify ethylene-related mutants. *Plant Cell* 2: 513–523.

Hai, L., Johnson, P., Stepanova, A., Alonso, J. M., and Ecker, J. M. (2004) Convergence of signaling pathways in the control of differential cell growth in *Arabidopsis*. *Dev. Cell* 7: 193–204.

Hattori, Y., Nagai, K., Furukawa, S., Song, X. J., Kawano, R., Sakakibara, H., Wu, J., Matsumoto, T., Yoshimura, A., Kitano, H., et al. (2009) The ethylene response factors SNORKEL1 and SNORKEL2 allow rice to adapt to deep water. *Nature* 460: 1026–1030.

Hensel, L. L., Grbi, V., Baumgarten, D. A., and Bleecker, A. B. (1993) Developmental and age-related processes that influence the longevity and senescence of photosynthetic tissues in *Arabidopsis*. *Plant Cell* 5: 553–564.

Hirayama, T., Kieber, J. J., Hirayama, N., Kogan, M., Guzman, P., Nourizadeh, S., Alonso, J. M., Dailey, W. P., Dancis, A., and Ecker, J. R. (1999) RESPONSIVE-TO-ANTAGONIST1, a Menkes/Wilson disease-related copper transporter, is required for ethylene signaling in *Arabidopsis*. *Cell* 97: 383–393.

Hua, J., and Meyerowitz, E. M. (1998) Ethylene responses are negatively regulated by a receptor gene family in *Arabidopsis thaliana*. *Cell* 94: 261–271.

Huang, Y., Li, H., Hutchison, C. E., Laskey, J., and Kieber, J. J. (2003) Biochemical and functional analysis of CTR1, a protein kinase that negatively regulates ethylene signaling in Arabidopsis. *Plant J.* 33: 221–233.

Kevany, B. M., Tieman, D. M., Taylor, M. G., Cin, V. D., and Klee, H. J. (2007) Ethylene receptor degradation controls the timing of ripening in tomato fruit. *Plant J.* 51: 458–467.

Kieber, J. J., Rothenburg, M., Roman, G., Feldmann, K. A., and Ecker, J. R. (1993) CTR1, a negative regulator of the ethylene response pathway in *Arabidopsis*, encodes a member of the Raf family of protein kinases. *Cell* 72: 427–441.

Lanahan, M., Yen, H.-C., Giovannoni, J., and Klee, H. (1994) The *Never-ripe* mutation blocks ethylene perception in tomato. *Plant Cell* 6: 427–441.

Lehman, A., Black, R., and Ecker, J. R. (1996) *Hookless1*, an ethylene response gene, is required for differential cell elongation in the *Arabidopsis* hook. *Cell* 85: 183–194.

Liang, X., Abel, S., Keller, J., Shen, N., and Theologis, A. (1992) The 1-aminocyclopropane-1-carboxylate synthase gene family of *Arabidopsis thaliana*. *Proc. Natl. Acad. Sci. USA* 89: 11046–11050.

McKeon, T. A., Fernández-Maculet, J. C., and Yang, S. F. (1995) Biosynthesis and metabolism of ethylene. In *Plant Hormones: Physiology, Biochemistry and Molecular Biology*, 2nd ed., P. J. Davies, ed., Kluwer, Dordrecht, Netherlands, pp. 118–139.

Morgan, P. W. (1984) Is ethylene the natural regulator of abscission? In *Ethylene: Biochemical, Physiological and Applied Aspects*, Y. Fuchs and E. Chalutz, eds., Martinus Nijhoff, The Hague, Netherlands, pp. 231–240.

Nakagawa, J. H., Mori, H., Yamazaki, K., and Imaseki, H. (1991) Cloning of the complementary DNA for auxin-induced 1-aminocyclopropane-1-carboxylate synthase and differential expression of the gene by auxin and wounding. *Plant Cell Physiol.* 32: 1153–1163.

Oeller, P., Min-Wong, L., Taylor, L., Pike, D., and Theologis, A. (1991) Reversible inhibition of tomato fruit senescence by antisense RNA. *Science* 254: 437–439.

Ohme-Takagi, M., and Shinshi, H. (1995) Ethylene-inducible DNA binding proteins that interact with an ethylene-responsive element. *Plant Cell* 7: 173–182.

Penmetsa, R. V., and Cook, D. R. (1997) A legume ethylene-insensitive mutant hyperinfected by its rhizobial symbiont. *Science* 275: 527–530.

Perovic, S., Seack, J., Gamulin, V., Müller, W. E. G., and Schröder, H. C. (2001) Modulation of intracellular calcium and proliferative activity of invertebrate and vertebrate cells by ethylene. *BMC Cell Biol.* 2: 7.

Qiao, H., Chang, K. N., Yazaki, J., and Ecker, J. R. (2009) Interplay between ethylene, ETP1/ETP2 F-box proteins, and degradation of EIN2 triggers ethylene responses in Arabidopsis. *Genes Dev.* 23: 512–521.

Qu, X., and Schaller, G. E. (2004) Requirement of the histidine kinase domain for signal transduction by the ethylene receptor ETR1. *Plant Physiol.* 136: 2961–2970.

Raskin, I., and Beyer, E. M., Jr. (1989) Role of ethylene metabolism in *Amaranthus retroflexus. Plant Physiol.* 90: 1–5.

Reid, M. S. (1995) Ethylene in plant growth, development and senescence. In *Plant Hormones: Physiology, Biochemistry and Molecular Biology*, 2nd ed., P. J. Davies, ed., Kluwer, Dordrecht, Netherlands, pp. 486–508.

Rodriguez, F. I., Esch, J. J., Hall, A. E., Binder, B. M., Schaller, E. G., and Bleecker, A. B. (1999) A copper cofactor for the ethylene receptor ETR1 from *Arabidopsis. Science* 283: 396–398.

Sexton, R., Burdon, J. N., Reid, J.S.G., Durbin, M. L., and Lewis, L. N. (1984) Cell wall breakdown and abscission. In *Structure, Function, and Biosynthesis of Plant Cell Walls*, W. M. Dugger and S. Bartnicki-Garcia, eds., American Society of Plant Physiologists, Rockville, MD, pp. 383–406.

Sisler, E. C., and Serek, M. (1997) Inhibitors of ethylene responses in plants at the receptor level: Recent developments. *Physiol. Plant.* 100: 577–582.

Sisler, E., Blankenship, S., and Guest, M. (1990) Competition of cyclooctenes and cyclooctadienes for ethylene binding and activity in plants. *Plant Growth Regul.* 9: 157–164.

Solano, R., Stepanova, A., Chao, Q., and Ecker, J. R. (1998) Nuclear events in ethylene signaling: A transcriptional cascade mediated by ETHYLENE-INSENSITIVE3 and ETHYLENE-RESPONSE-FACTOR1. *Genes Dev.* 12: 3703–3714.

Tanimoto, M., Roberts, K., and Dolan, L. (1995) Ethylene is a positive regulator of root hair development in *Arabidopsis thaliana. Plant J.* 8: 943–948.

Thain, S. C., Vandenbussche, F., Laarhoven, L. J., Dowson-Day, M. J., Wang, Z. Y., Tobin, E. M., Harren, F. J., Millar, A. J., and Van Der Straeten, D. (2004) Circadian rhythms of ethylene emission in Arabidopsis. *Plant Physiol.* 136: 3751–3761.

Tsuchisaka, A., and Theologis, A. (2004) Unique and overlapping expression patterns among the Arabidopsis 1-amino-cyclopropane-1-carboxylate synthase gene family members. *Plant Physiol.* 136: 2982–3000.

Vahala, J., Ruonala, R., Keinänen, M., Tuominen, H., and Kangasjärvi, J. (2003) Ethylene insensitivity modulates ozone-induced cell death in birch (*Betula pendula*). *Plant Physiol.* 132: 185–195.

Voesenek, L.A.C.J., Banga, M., Rijnders, J.H.G.M., Visser, E.J.W., Harren, F.J.M., Brailsford, R. W., Jackson, M. B., and Blom, C.W.P.M. (1997) Laser-driven photoacoustic spectroscopy: What we can do with it in flooding research. *Ann. Bot.* 79: 57–65.

Vogel, J. P., Woeste, K. E., Theologis, A., and Kieber, J. J. (1998) Recessive and dominant mutations in the ethylene biosynthetic gene ACS5 of Arabidopsis confer cytokinin insensitivity and ethylene overproduction, respectively. *Proc. Natl. Acad. Sci. USA* 95: 4766–4771.

Wang, W., Hall, A. E., O'Malley, R., and Bleecker, A. B. (2003) Canonical histidine kinase activity of the transmitter domain of the ETR1 ethylene receptor from *Arabidopsis* is not required for signal transmission *Proc. Natl. Acad. Sci. USA* 100: 352–357.

Woeste, K., and Kieber, J. J. (2000) A strong loss-of-function allele of *RAN1* results in constitutive activation of ethylene responses as well as a rosette-lethal phenotype. *Plant Cell* 12: 443–455.

Yang, S. F. (1987) The role of ethylene and ethylene synthesis in fruit ripening. In *Plant Senescence: Its Biochemistry and Physiology*, W. W. Thomson, E. A. Nothnagel, and R. C. Huffaker, eds., American Society of Plant Physiologists, Rockville, MD, pp. 156–166.

Yoo, S.-D., Cho, Y.-H., Tena, G., Xiong, Y., and Sheen, J. (2008) Dual control of nuclear EIN3 by bifurcate MAPK cascades in C2H4 signalling. *Nature* 451: 789–795.

Yuan, M., Shaw, P. J., Warn, R. M., and Lloyd, C. W. (1994) Dynamic reorientation of cortical microtubules, from transverse to longitudinal, in living plant cells. *Proc. Natl. Acad. Sci. USA* 91: 6050–6053.

Zacarias, L., and Reid, M. S. (1990) Role of growth regulators in the senescence of *Arabidopsis thaliana* leaves. *Physiol. Plant.* 80: 549–554.

# Abscisic Acid: A Seed Maturation and Stress-Response Hormone

The extent and timing of plant growth are controlled by the coordinated actions of positive and negative regulators. Some of the most obvious examples of regulated nongrowth are seed and bud dormancy, adaptive features that delay growth until environmental conditions are favorable. For years, plant physiologists suspected that the phenomena of seed and bud dormancy were caused by inhibitory compounds, so they attempted to extract and isolate such compounds from a variety of plant tissues, especially dormant buds.

Early experiments used paper chromatography for the separation of plant extracts, as well as bioassays based on oat coleoptile growth. These experiments led to the identification of a group of growth-inhibiting compounds, including a substance known as *dormin* purified from sycamore leaves collected in early autumn, when the trees were entering dormancy. Upon discovery that dormin was chemically identical to a substance that promotes the abscission of cotton fruits, *abscisin II*, the compound was renamed **abscisic acid** (**ABA**) (FIGURE 23.1) to reflect its supposed involvement in the abscission process.

ABA is now recognized as an important plant hormone that regulates growth and stomatal closure, particularly when the plant is under environmental stress. Another important function is its regulation of seed maturation and dormancy. Ironically, ABA's effects on abscission remain controversial: In many species, ABA appears to promote senescence (i.e., the events preceding abscission), but not abscission itself.

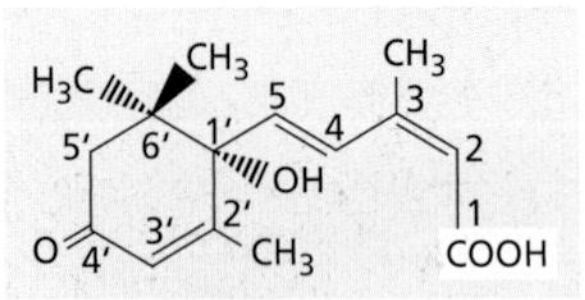

**(S)-cis-ABA**
**(naturally occurring**
**active form)**

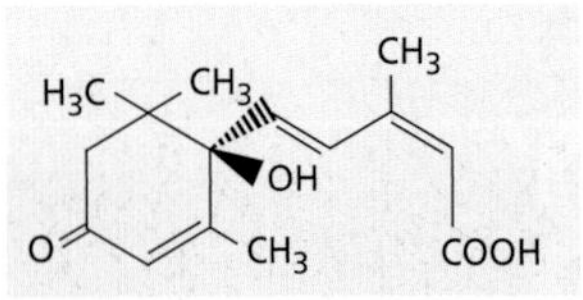

**(R)-cis-ABA**
**(inactive in stomatal closure)**

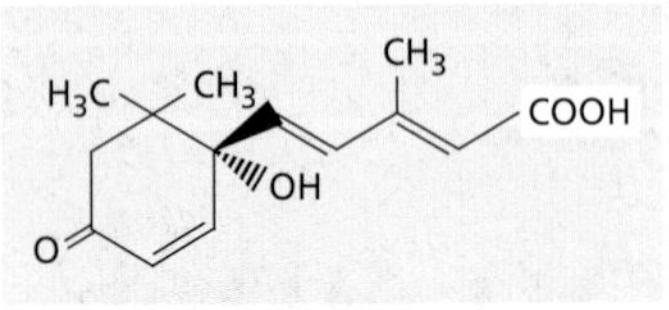

**(S)-2-trans-ABA (inactive, but**
**interconvertible with active**
**cis form)**

**FIGURE 23.1** Structures of active and inactive ABAs. The chemical structures of the S (+, or counterclockwise array) and R (–, or clockwise array) forms of cis-ABA, and the (S)-2-trans form of ABA. The numbers in the diagram of (S)-cis-ABA identify the carbon atoms.

In this chapter, we will survey the structure, synthesis, and movement of ABA and the general ABA signal transduction mechanisms. We will conclude with specific examples of ABA responses in growth and development.

# Occurrence, Chemical Structure, and Measurement of ABA

Abscisic acid is a ubiquitous plant hormone in vascular plants. It has been detected in mosses but appears to be absent in liverworts (see **WEB TOPIC 23.1**). Several genera of fungi make ABA as a secondary metabolite. ABA has also been found in metazoans ranging from sea sponges to humans, and some signaling mechanisms appear to be shared across kingdoms (see **WEB TOPIC 23.2**). Within the plant, ABA has been detected in every major organ or living tissue from the root cap to the apical bud. ABA is synthesized in almost all cells that contain chloroplasts or amyloplasts.

## The chemical structure of ABA determines its physiological activity

ABA is a 15-carbon compound that resembles the terminal portion of some carotenoid molecules (see Figure 23.1). The orientation of the carboxyl group at carbon 2 determines the cis and trans isomers of ABA. Nearly all naturally occurring ABA is in the cis form, and by convention the name *abscisic acid* refers to the cis isomer.

ABA also has an asymmetric carbon atom at position 1' in the ring, resulting in the S and R (or + and –, respectively) enantiomers. The S enantiomer is the natural form; commercially available synthetic ABA is a mixture of approximately equal amounts of the S and R forms. In long-term responses to ABA, such as seed maturation, both enantiomers are active. In other responses, such as stomatal closure, the S enantiomer is far more active.

Studies of the structural requirements for biological activity of ABA have shown that almost any change in the molecule results in loss of activity (see **WEB TOPIC 23.3**).

## ABA is assayed by biological, physical, and chemical methods

A variety of bioassays have been used for ABA, including inhibition of coleoptile growth, germination, and gibberellic acid (GA)–induced α-amylase synthesis. Rapid inductive responses such as promotion of stomatal closure and gene expression are also associated with ABA (see **WEB TOPIC 23.4**).

Physical methods of detection are much more reliable than bioassays because of their specificity and suitability for quantitative analysis. The most widely used techniques are those based on gas chromatography or high-performance liquid chromatography (HPLC). Gas chromatography allows detection of as little as $10^{-13}$ g of ABA, but it requires several preliminary purification steps, including thin-layer chromatography. Immunoassays are also highly sensitive and specific.

# Biosynthesis, Metabolism, and Transport of ABA

As with the other hormones, the response to ABA depends on its concentration within the tissue and on the sensitivity of the tissue to the hormone. The processes of biosynthesis, catabolism, compartmentation, and transport all contribute to the concentration of active hormone in plant tissues at any given stage of development.

## ABA is synthesized from a carotenoid intermediate

ABA biosynthesis begins in chloroplasts and other plastids via the simplfied pathway depicted in **FIGURE 23.2**. (For a more complete version see Appendix 3). The pathway begins with isopentenyl diphosphate (IPP)—the biological isoprene unit that is also a precursor of cytokinins, gibberellins, and brassinosteroids—and leads to the synthesis of the $C_{40}$ xanthophyll (i.e., oxygenated carotenoid) **violaxanthin** (see Figure 23.2). The discovery that the synthesis of violaxanthin is catalyzed by the enzyme encoded by the *ABA1* locus of Arabidopsis provided conclusive evidence that ABA synthesis occurs via the carotenoid pathway,

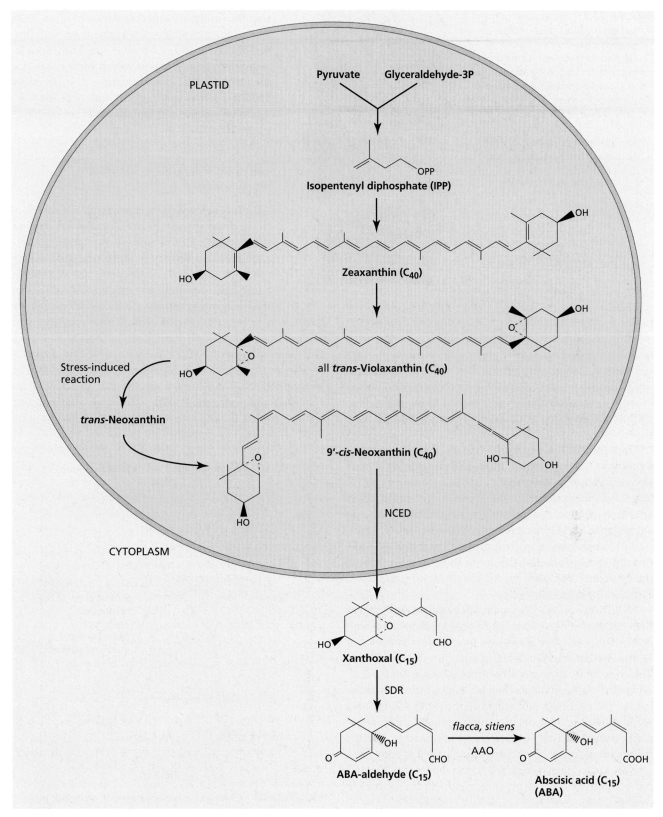

**FIGURE 23.2**  Simplified diagram of ABA biosynthesis via the terpenoid pathway. The initial stages occur in plastids, where isopentenyl diphosphate (IPP) is converted to the $C_{40}$ xanthophyll zeaxanthin. Zeaxanthin is further modified to 9-*cis*-neoxanthin, which is cleaved by the enzyme NCED to form the $C_{15}$ inhibitor, xanthoxin. Xanthoxin then is converted to ABA in the cytosol. ABA-deficient mutants that have been helpful in elucidating the pathway are shown in Appendix 3.

**FIGURE 23.3** Precocious germination involves germination of seeds in the fruit while still attached to the plant. Shown here is precocious germination in the ABA-deficient *vivipary 14* (*vp14*) mutant of maize. The VP14 protein catalyzes the cleavage of 9-*cis*-epoxycarotenoids to form xanthoxin, a precursor of ABA. (Courtesy of Bao Cai Tan and Don McCarty.)

rather than by modification of a $C_{15}$ isoprenoid, as occurs in some phytopathogenic fungi. Maize (corn; *Zea mays*) mutants (termed *viviparous*, *vp*) that are blocked at other steps in the carotenoid pathway also have reduced levels of ABA and exhibit **vivipary**—the precocious germination of seeds in the fruit still attached to the plant (**FIGURE 23.3**). Vivipary is a feature of many ABA-deficient seeds.

Under stress conditions, *trans*-violaxanthin is converted to another $C_{40}$ compound, **trans-neoxanthin**, by a reaction dependent on the product of the Arabidopsis *ABA4* locus. Following isomerization by an as-yet unidentified enzyme(s), 9-*cis*-neoxanthin can be cleaved by an enzyme abbreviated **NCED** (see Appendix 3) to form the $C_{15}$ compound **xanthoxin**, a neutral growth inhibitor that has physiological properties similar to those of ABA. This is the first committed step for ABA biosynthesis, and it is a rate-limiting regulatory step.

NCEDs are encoded by a family of genes that vary in their patterns of expression in response to stress and developmental signals; the genes thus vary in their importance to stress-induced versus organ-specific ABA synthesis. The enzyme functions on the stromal face of thylakoids, where the carotenoid substrate is located, but some members of the NCED family are present in both soluble and membrane-bound forms, providing a possibility for regulating enzyme activity via effects on localization. Finally, xanthoxin moves to the cytoplasm, where it is converted to ABA via oxidative steps involving the intermediate(s) **ABA-aldehyde** and/or possibly xanthoxic acid (abscisic alcohol).

Conversion of xanthoxin to abscisic aldehyde is catalyzed by a short-chain dehydrogenase/reductase-like (**SDR**) enzyme, encoded by the *ABA2* locus of Arabidopsis. The final step is catalyzed by a differentially regulated family of abscisic aldehyde oxidases (**AAOs**) that all require a molybdenum cofactor (see Figure 23.2). Mutations in a single AAO gene may have limited effects, however, because other, functionally redundant family members can compensate for the lost function. However, those AAO mutants lacking a functional molybdenum cofactor (e.g., Arabidopsis *aba3* and tomato *flacca*) are unable to synthesize ABA.

## ABA concentrations in tissues are highly variable

ABA biosynthesis and concentrations can fluctuate dramatically in specific tissues during development or in response to changing environmental conditions. In developing seeds, for example, ABA levels can increase 100-fold within a few days, reaching average concentrations in the micromolar range, and then decline to very low levels as maturation proceeds. Under conditions of water stress (i.e., dehydration stress), ABA in the leaves can increase 50-fold within 4 to 8 hours (**FIGURE 23.4**).

Part of this increase is due to increased expression of biosynthetic enzymes, but the specific enzymes depend on the tissue and the signal. For example, among biosynthetic

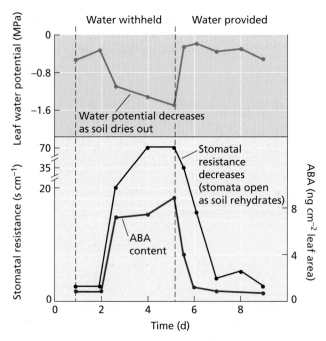

**FIGURE 23.4** Changes in water potential, stomatal resistance (the inverse of stomatal conductance), and ABA content in maize in response to water stress. As the soil dried out, the water potential of the leaf decreased, and the ABA content and stomatal resistance increased. Rewatering reversed the process. (After Beardsell and Cohen 1975.)

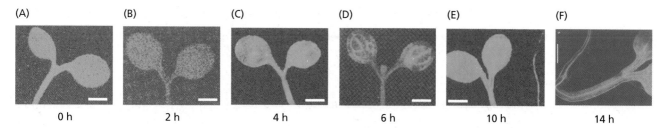

**FIGURE 23.5** Dynamics of ABA-dependent reporter response to root-sensed water stress. Seedlings of the reporter line pAtHB6::LUC were water-stressed via their roots and luciferase-dependent light emission (light blue and green patches) was recorded at the indicated times.

ABA-dependent expression was observed in hypocotyls by 2 hrs, cotyledon veins by 4 hrs, and spreading across the cotyledons by 6 hrs, but did not appear in roots before 10 hrs. Bars correspond to 1 mm. (From Christmann et al. 2005.)

enzymes, NCED is induced in all tissues; ZEP (zeaxanthin epoxidase) is induced in seeds and water-stressed roots; the AAOs are differentially induced in stressed tissues; and SDR is induced by sugar, but not by dehydration. The mechanism by which dehydration stress is perceived by cells has not yet been identified, but might involve sensors of cellular turgor or osmotic sensors. Upon rewatering, the ABA level declines to normal in the same amount of time.

As with other plant hormones, the concentration of free ABA in the cytosol is also regulated by degradation, compartmentation, conjugation, and transport. For example, cytosolic ABA increases during water stress as a result of synthesis in the leaf, redistribution within the mesophyll cell, import from the roots, and recirculation from other leaves. The concentration of ABA declines after rewatering because of degradation and export from the leaf, as well as a decrease in the rate of synthesis. Similarly, changing ABA levels in seeds reflect a shifting balance between synthesis and inactivation. ABA inactivation can occur by either oxidation to **phaseic acid** (**PA**) and **4-dihydrophaseic acid** (**DPA**) or by covalent conjugation to another molecule, such as a monosaccharide (see Appendix 3).

### ABA is translocated in vascular tissue

ABA is transported by both the xylem and the phloem, but it is normally much more abundant in the phloem sap. When radioactive ABA is applied to a leaf, it is transported both up the stem and down toward the roots. Most of the radioactive ABA is found in the roots within 24 hours. Destruction of the phloem by a stem girdle prevents ABA accumulation in the roots, indicating that the hormone is transported in the phloem sap.

ABA synthesized in the roots can also be transported to the shoot via the xylem. Whereas the concentration of ABA in the xylem sap of well-watered sunflower plants is between 1.0 and 15.0 n$M$, the ABA concentration in water-stressed sunflower plants rises to as much as 3000 n$M$ (3.0 μ$M$)—representing a 200- to 3000-fold increase (Schurr et

al. 1992). The magnitude of the stress-induced increase in xylem ABA content varies widely among species. ABA may also be transported in a conjugated form, and then released by hydrolysis in leaves.

Studies using ABA-dependent reporter gene activation to reflect localized ABA concentrations have shown that during water stress, ABA accumulates first in shoot vascular tissue, and only later appears in roots and guard cells (Christmann et al. 2005) (**FIGURE 23.5**). Furthermore, reciprocal grafting experiments with wild-type and ABA-deficient root stocks and scions demonstrated that ABA synthesis in shoots was essential to the plant's response to root drying (Christmann et al. 2007).

When stress was imposed rapidly, stomatal closure was correlated with a rapid drop in turgor of mesophyll cells and could be prevented by direct application of water to leaf upper surfaces. These studies suggest that the initial root-shoot communication is mediated by a change in the water potential gradient from roots to shoots. However, they do not rule out additional long-term signaling via transport of ABA or other chemical signal(s) from the root.

Although a concentration of 3.0 μ$M$ ABA in the apoplast is sufficient to close stomata, not all of the ABA in the xylem stream reaches the guard cells. Much of the ABA in the transpiration stream is taken up and metabolized by the mesophyll cells. At the cellular level, ABA distribution among plant cell compartments follows the "anion trap" mechanism: the dissociated (anion) form of this weak acid, ABA⁻, does not readily cross membranes, but ABA enters the cell in the protonated form and then accumulates in alkaline compartments. ABA is also transported into cells by specific uptake carriers, which help to maintain a low apoplastic ABA concentration in unstressed plants.

During the early stages of water stress, however, the pH of the xylem sap becomes more alkaline, increasing from about pH 6.3 to about pH 7.2 (Wilkinson and Davies 1997). Stress-induced alkalinization of the apoplast favors formation of the dissociated form of abscisic acid, ABA⁻. At the same time, dehydration also acidifies the cytosol,

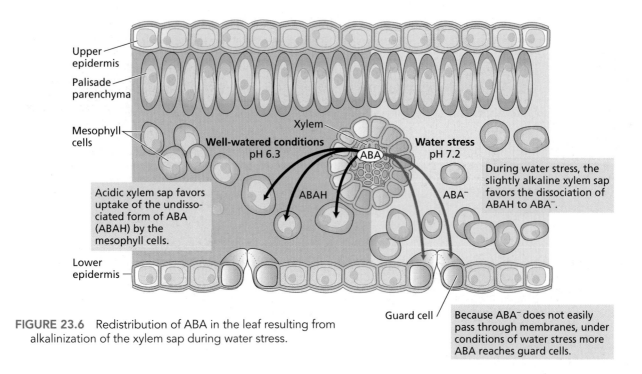

**Upper epidermis**

**Palisade parenchyma**

**Mesophyll cells**

Xylem

**Well-watered conditions** pH 6.3

ABA

**Water stress** pH 7.2

During water stress, the slightly alkaline xylem sap favors the dissociation of ABAH to ABA⁻.

ABAH

Acidic xylem sap favors uptake of the undissociated form of ABA (ABAH) by the mesophyll cells.

ABA⁻

**Lower epidermis**

Guard cell

Because ABA⁻ does not easily pass through membranes, under conditions of water stress more ABA reaches guard cells.

**FIGURE 23.6** Redistribution of ABA in the leaf resulting from alkalinization of the xylem sap during water stress.

contributing to ABA release from its sites of synthesis and decreasing uptake by mesophyll cells. Both of these pH changes increase the amount of ABA reaching the guard cells via the transpiration stream (**FIGURE 23.6**). In this way ABA can be redistributed in the leaf without any increase in the total ABA level. Therefore, the increase in xylem sap pH may function as an additional root signal that promotes early closure of the stomata.

## ABA Signal Transduction Pathways

ABA is involved in short-term physiological effects (e.g., stomatal closure), as well as long-term developmental processes (e.g., seed maturation).

- Rapid physiological responses frequently involve alterations in the fluxes of ions across membranes and usually involve regulation of certain genes as well, as evidenced by the fact that a variety of ABA-stimulated transcription factors that are expressed in guard cells regulate stomatal aperture.

- In contrast, long-term processes inevitably involve major changes in the pattern of gene expression.

Comparisons of total transcript populations have shown that at least 10% of the genes in both Arabidopsis and rice are regulated by ABA.

Signal transduction pathways, which amplify the primary signal generated when the hormone binds to its receptor, are required for both the short-term and the long-term effects of ABA (see Chapter 14). Genetic studies have

identified more than 100 loci involved in ABA signaling pathways (Wasilewska et al. 2008). Although many genes affect only a subset of ABA responses, some conserved signaling components regulate both short- and long-term responses, indicating that these responses share common signaling mechanisms. We will focus on mechanisms regulating stomatal aperture and gene expression, the two best-characterized ABA effects.

### Receptor candidates include diverse classes of proteins

Efforts to identify ABA receptors have employed biochemical, cellular, and genetic approaches. The results of such experiments have suggested the presence of both cell surface and intracellular types of ABA receptors (**FIGURE 23.7**) (see **WEB TOPIC 23.5**). Consistent with this inference, *no mutants have been identified with defects in all ABA responses.*

In recent years, several possible ABA receptor classes have been proposed, but most are controversial (McCourt and Creelman 2008; Risk et al. 2009; Müller and Hansson 2009). Three of the best candidates include:

- PYR/PYL/RCAR members of the START superfamily of ligand-binding proteins, which are present in the cytosol and nucleus (Park et al. 2009; Ma et al. 2009);

- CHLH, a plastid-localized subunit of the Mg-cheletase enzyme involved in chlorophyll synthesis and in signaling that coordinates plastid and nuclear gene expression (Shen et al. 2006);

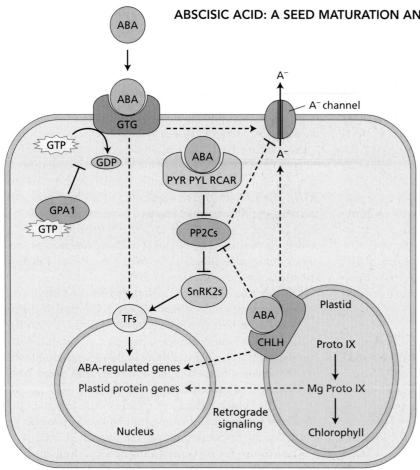

**FIGURE 23.7** A model of ABA interactions with three of its candidate receptor classes. ABA receptors identified to date are localized in plastid (CHLH), plasma membrane (GTG1 and GTG2), and cytosol and nucleus (PYR/PYL/RCAR). All mediate ABA effects on both gene expression and ion fluxes involved in stomatal regulation (represented by A⁻ channel for simplicity), but the intersections among downstream signaling events are not yet clear. PP2Cs, ABI-clade of protein phosphatases; SnRK2s, SNF-related kinases; TFs, transducing factors including ABRE-binding transcription factors. Solid lines, direct interactions; dotted lines, unknown interactions; arrows, positive regulation; bars, repression.

- *GPCR-type G* proteins (GTG1 and GTG2) are a pair of plasma membrane proteins that have intrinsic G protein activity and homology to G protein–coupled receptors (GPCRs) (Pandey et al. 2009). Although GPCRs appear to participate in ABA signaling, there is conflicting evidence for their role as receptors.

We discuss the PYR/PYL/RCAR proteins and CHLH in more detail below. For more information about GPCRs see **WEB TOPIC 23.6**.

**PYR/PYL/RCAR FAMILY**  The **PYR/PYL/RCAR** class of putative ABA receptors was identified by a chemical genetic screen for small molecules with ABA-like effects. In Arabidopsis seeds, this screen identified *pyrabactin* (*pyr*idyl-containing *ABA act*ivator) as an **agonist** of ABA; that is, it mimics the effects of ABA in plants (Park et al. 2009).

**Pyrabactin**

Subsequent screens for pyrabactin-resistant Arabidopsis mutants identified the **PYR1** locus, which encodes a member of the **START domain** (*St*eroidogenic *A*cute *R*egulatory

protein-*r*elated lipid-*t*ransfer, a predicted hydrophobic ligand-binding pocket) protein superfamily. Subsequent nuclear magnetic resonance experiments demonstrated direct binding between ABA and PYR1, suggesting that PYR1 could be a receptor.

Bioinformatic studies identified multiple closely related family members, designated **PYLs** for *PYR1-like*. This PYL family is conserved among plants ranging from dicots to mosses. Although *pyr1* mutants are pyrabactin-resistant, as are several previously characterized ABA-resistant mutants, the *pyr1* mutants do not display any significant resistance to ABA. However, triple or quadruple mutants combining defects in *PYR1* and several of its close homologs display strong ABA-insensitivity, indicating that these loci function redundantly and would therefore have been missed by a standard screen for ABA-resistance.

A yeast two-hybrid screen (see **WEB TOPIC 23.7**) for proteins with pyrabactin-dependent interactions with PYR1 identified a protein phosphatase of a PP2C class that had been previously implicated in ABA signaling. The PYR1–PP2C interaction was ligand-dependent in yeast and also occurred both in plants and in vitro, where it was found to suppress PP2C activity. Subsequent studies showed that distinct PYR/PYL family members showed differences in ligand selectivity and ligand-dependent binding to multiple members of a homologous set of PP2Cs.

Meanwhile, several research groups studying the PP2Cs made the reciprocal connection in their two-hybrid screens and this same family of PP2C-interacting proteins has also been named **Regulatory Components of ABA Receptors (RCARs)** (Ma et al. 2009). Recent crystallographic studies with two PYR/PYL family members show that ABA binds to an internal pocket in the receptor, changing its conformation to permit interaction with PP2Cs (Miyazono et al. 2009; Nishimura et al. 2009). These PP2Cs have previously been shown to interact specifically with multiple additional ABA signaling components, including transcription factors and protein kinases. Therefore, the PYR/PYL/RCARs appear to be a class of soluble receptors that directly link to many genetically well-characterized downstream signaling elements, effectively coalescing them into a network.

**CHLH** The **CHLH protein** was identified as the Arabidopsis homolog of an ABA-binding protein (ABAR) purified from broad bean leaf epidermis (Shen et al. 2006). Functional evidence of the role of CHLH in ABA signaling was provided by CHLH "knockdown" (partial loss-of-function) mutants, which display reduced ABA sensitivity for germination, dormancy, stomatal regulation, and ABA-induced gene expression. Consistent with this observation, CHLH overexpression lines are hypersensitive to ABA. CHLH "knockout" (complete loss-of-function) mutants fail to complete embryo maturation, producing inviable seeds that lack storage reserves and desiccation protectants. Mutants with severe defects in some ABA signaling components or very reduced seed ABA levels display the same phenotypic effects. However, for unknown reasons, conventional genetic ABA-insensitivity screens have not yet identified reduced-function alleles in *CHLH*.

Although CHLH also functions in chlorophyll synthesis and plastid-to-nucleus signaling, these functions appear to reside in distinct domains, because discrete mutations affect individual functions. The mechanism of CHLH function in ABA signaling is thus unknown, but its localization in plastids implies that some perception of ABA must occur within these organelles, presumably following transport of ABA back into the plastids.

In summary, the existence of multiple candidates for ABA receptors is consistent with the wide variety of signaling pathways known to operate (see Figure 23.7). Redundancy in signal transduction pathways permits cells to integrate a wide range of hormonal and environmental stimuli, and provide what biologists call "network robustness," which ensures that the response will occur under a variety of conditions. Although the immediate cellular events initiated by the CHLH and GTG proteins are unknown, all appear to be involved in ABA signaling elements, and future research will illuminate their individual roles.

## Secondary messengers function in ABA signaling

Numerous secondary messengers participate in ABA signaling, including $Ca^{2+}$, reactive oxygen species (ROS), cyclic nucleotides, and phospholipids. Cytosolic $Ca^{2+}$ levels are regulated by changes in the activity of channels permitting calcium influx through the plasma membrane and into the cytosol from internal compartments, such as the central vacuole or a variety of organelles (McAinsh and Pittman 2009). These channels are regulated in turn by additional messengers; for example, **reactive oxygen species (ROS)** such as hydrogen peroxide ($H_2O_2$) or superoxide ($O_2 \bullet^-$), which are produced by NADPH oxidase, function as secondary messengers leading to plasma membrane calcium channel activation (Kwak et al. 2003).

Calcium release from intracellular stores can be induced by a variety of second messengers, including inositol 1,4,5-trisphosphate ($InsP_3$), cyclic ADP-ribose (cADPR), and self-amplifying (calcium-induced) $Ca^{2+}$ release. ABA stimulates ADPR cyclase activity within 15 minutes, leading to increases in cADPR. In addition, ABA stimulates the synthesis of **nitric oxide (NO)** by nitrate reductase in guard cells, which induces stomatal closure in a cADPR-dependent manner, indicating that NO acts upstream of cADPR in the response pathway (Desikan et al. 2002).

The following sections examine signaling mechanisms involved in responding to some of these secondary messengers.

## $Ca^{2+}$-dependent and $Ca^{2+}$-independent pathways mediate ABA signaling

Intracellular free calcium has been measured by the use of microinjected calcium-sensitive ratiometric fluorescent dyes*, such as fura-2 or indo-1. Alternatively, transgenic plants expressing genes for calcium indicator proteins such as aequorin or **yellow cameleon** make it possible to monitor several fluorescing cells in parallel, without the need for invasive injections (Allen et al. 1999) (see WEB TOPIC 23.8). Studies with yellow cameleon have demonstrated that the cytosolic $Ca^{2+}$ concentration oscillates in guard cells, depending on the signals received (FIGURE 23.8).

Direct imaging of calcium fluctuations supports the hypothesis that an increase in cytosolic calcium, partly derived from intracellular stores, is responsible for ABA-induced stomatal closure. $Ca^{2+}$ elevations can then be interpreted by "conformational coupling," in which $Ca^{2+}$

---

*Ratiometric fluorescent dyes undergo a shift in their excitation and emission spectra when they bind calcium. On the basis of this property, one can determine the intracellular concentration of both forms of the dye (with and without bound calcium) by exciting them with the appropriate two wavelengths. The ratio of the two emissions provides a measure of the calcium concentration that is independent of dye concentration.

(A)

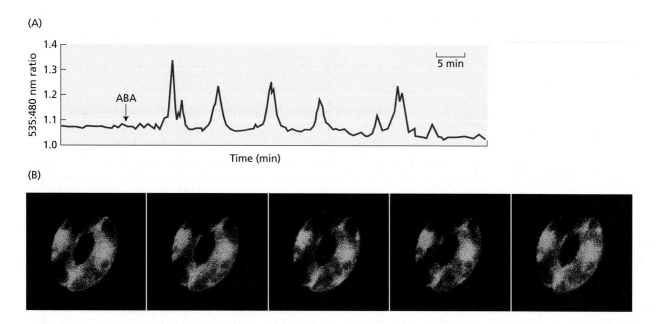

(B)

**FIGURE 23.8** ABA induces calcium oscillations in Arabidopsis guard cells, which are detected by transgenic expression of yellow cameleon, a calcium indicator protein dye. (A) Repetitive calcium transients elicited by ABA are indicated by increases in the ratio of fluorescence emission at 535 and 480 nm. (B) Pseudo- colored images of fluorescence in Arabidopsis guard cells; blue, green, yellow, and red represent increasing cytosolic calcium concentrations. (From Schroeder et al. 2001.)

binding to proteins such as calmodulin (CaM) or calcineurin B-like (CBL) proteins alters their conformations. The conformational changes allow these proteins to interact with and modify the function of a wide variety of cellular proteins. Mammalian calcineurin phosphatases are heterodimers of a catalytic A subunit and a $Ca^{2+}$-binding regulatory B subunit, which may be activated by interaction with calmodulin. Although calcineurin A-like subunits have not yet been found in plants, a family of plant **calcineurin B–like proteins** (**CBLs**) has been identified. The CBLs are differentially regulated by environmental stress and by ABA, and are predicted to have distinct organ and subcellular localizations (Luan et al. 2002).

In addition, $Ca^{2+}$ may either directly or indirectly regulate protein kinases or phosphatases that modify the phosphorylation state and therefore the activity of various proteins. Consequently, the overall response to ABA reflects the subcellular distribution, frequency, amplitude and duration of $Ca^{2+}$ elevations. It also reflects the availability of specific calcium binding proteins and their targets in a given cell. Some of these effects are mediated by changes in gene expression, while others affect ion channel activity directly.

Although an ABA-induced calcium response is a key feature of the current model for ABA-induced stomatal closure, ABA is able to induce stomatal closure even in guard cells that show no increase in cytosolic calcium (Allan et al. 1994). In other words, ABA seems to be able to act via both calcium-dependent and calcium-independent pathways. For example, ABA causes an alkalinization of the cytosol from about pH 7.7 to pH 7.9, which can lead to the activation of outward $K^+$ channels and stomatal closure.

An alternative explanation for apparently calcium-independent signaling is that ABA enhances the intracellular calcium *sensitivity* of stomatal closing mechanisms, including anion and $K^+$ channel regulation (Siegel et al. 2009). The hypersensitivity induced by ABA would thus allow cells to activate a calcium-dependent pathway in response to the resting calcium levels, in the absence of an intracellular calcium elevation. This mechanism of altering the threshold for calcium response has been referred to as the "calcium sensitivity priming" hypothesis and might also explain why, in the absence of ABA, spontaneously occurring intracellular calcium elevations in guard cells sometimes fail to induce stomatal closure (Young et al. 2006).

### ABA-induced lipid metabolism generates second messengers

In the classic model of G protein signaling in animal cells, the activated Gα subunit activates **phospholipase C**, resulting in the release of **inositol triphosphate** (**InsP₃**), along with **diacylglycerol** (**DAG**). Studies in broad bean (*Vicia faba*) guard cells have shown that ABA stimulates phosphoinositide metabolism, leading to the production of InsP₃ and **myo-inositol-hexaphosphate** (**InsP₆**). A 90%

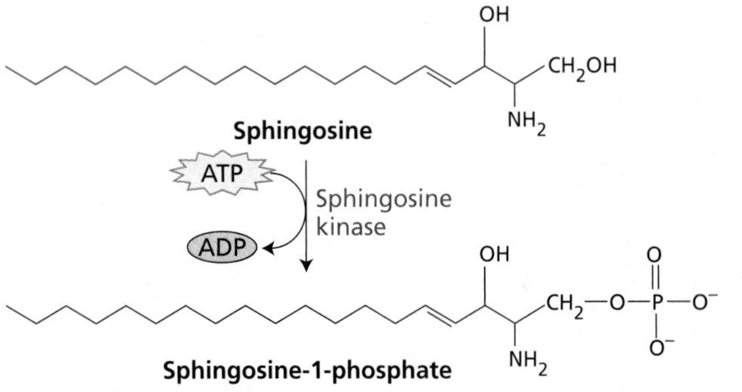

**FIGURE 23.9** Activated by ABA, sphingosine kinase catalyzes the phosphorylation of sphingosine to sphingosine-1-phosphate.

increase in the level of InsP$_3$ was measured within 10 seconds after application of ABA (Lee et al. 1996). Studies in Arabidopsis and tobacco using antisense DNA to block expression of an ABA-induced phospholipase C have shown that this enzyme is required for the effects of ABA on germination, growth, and gene expression (Sanchez and Chua 2001). Consistent with this observation, mutants with increased InsP$_3$ levels exhibit hypersensitivity to ABA (Xiong et al. 2001; Wilson et al. 2009).

ABA activation of the enzyme **sphingosine kinase** produces another class of phospholipids, **sphingosine-1-phosphate (S1P)** (FIGURE 23.9). S1P stimulates cytosolic Ca$^{2+}$ increases and stomatal closure in *Commelina communis* guard cells. In addition, S1P signaling inhibitors decrease the response to ABA. Although S1P stimulates stomatal closure in wild-type Arabidopsis, it does not cause stomatal closure in mutants lacking Gα subunits. This suggests that S1P acts via G proteins, and that G protein–mediated events are downstream of the S1P signal (Coursol et al. 2003).

Yet another potential second messenger mediating the ABA response is **phosphatidic acid**, which is released from phosphatidylcholine by the enzyme **phospholipase D (PLD)**. The most abundant isoform of PLD is activated by ABA, and PLD–mediated production of phosphatidic acid has been shown to promote ABA-induced stomatal closure and gene expression, as well as other stress responses, by multiple mechanisms (Zhang et al. 2005). The product of PLD activity, **phosphatidic acid (PA)**, binds a variety of targets including protein phosphatases, protein kinases, and metabolic enzymes. For at least one of these, the protein phosphatase ABI1 (discussed in the following section), PA binding alters its activity and localization such that ABA signaling is de-repressed, leading to stomatal closure, among other responses (see WEB TOPIC 23.9).

Thus, multiple lipid-derived messengers (the phosphoinositides, DAG, S1P, and phosphatidic acid) and the enzymes that produce and degrade them are involved in multiple feedback loops and pathways that enhance ABA signaling.

## Protein kinases and phosphatases regulate important steps in ABA signaling

Nearly all biological signaling systems involve protein phosphorylation and dephosphorylation reactions at some step in the pathway, and ABA signaling is no exception. Again, both biochemical and genetic studies have identified numerous families of protein kinases and phosphatases mediating specific aspects of ABA signaling.

**ABA-ACTIVATED PROTEIN KINASES** ABA- or stress-activated protein kinases (AAPK or SAPKs) have been identified in many species and are members of the **Sucrose Non-Fermenting Related Kinase2 (SnRK2)** family. The first SnRK2 to be identified is required for stomatal regulation and acts upstream of ROS and Ca$^{2+}$ signaling. Many of the Arabidopsis SnRK2 family members are induced by ABA, and others are stress-induced. A major aspect of induction is posttranslational activation of these kinases, but some are also transcriptionally induced. Genetic studies have shown that some of the SnRK2s function redundantly in regulating seedling ABA responses (Fujii et al. 2007).

All of the ABA- or stress-induced SnRKs tested can phosphorylate a conserved domain of a class of transcription factors known as **ABA-response element binding factors (AREBs or ABFs)** in vitro, and this phosphorylation is correlated with increased activity of these transcription factors. AAPK also modifies the activity of an RNA-binding protein that may mediate ABA effects on gene expression (Assmann 2004). Additional targets of stress-regulated SnRK2s include a variety of metabolic enzymes as well as several 14-3-3 family regulatory proteins expected to interact with and modify activity of additional signaling proteins.

The posttranslational control of these kinases permits rapid response to stress, and their ability to regulate numerous targets, including many regulators, rapidly amplifies this response.

Besides SnRK2 kinases, calcium-regulated kinases and Mitogen Activated Protein Kinases (MAPKs) also regulate key steps in the ABA signaling pathway (see WEB TOPIC 23.10).

**PROTEIN PHOSPHATASES** Pharmacological studies with inhibitors of **protein phosphatases (PPs)** have shown that several classes of serine/threonine phosphatases (PP1A, PP2A, calcineurin-like PP2Bs, and PP2C), and tyrosine phosphatases can all regulate aspects of guard cell signaling, but the effects may be positive or negative depending on numerous factors. Genetic studies permit analysis of individual protein phosphatases and have identified a specific PP2A and several PP2Cs as ABA signaling components with pleiotropic (multiple) effects on development.

### PP2Cs interact directly with the PYR/PYL/RCAR family of ABA receptors

The most dramatic effects on the ABA response are seen in mutants of the Arabidopsis **ABI1** and **ABI2** loci, which encode PP2Cs. The Arabidopsis *abi1-1* and *abi2-1* mutants display phenotypes consistent with a defect in ABA signaling, including reduced seed dormancy, a tendency to wilt (due to faulty regulation of stomatal aperture), and decreased expression of most ABA-inducible genes. The defects in the stomatal response include ABA insensitivity of S-type (slow) anion channels, both inward and outward K$^+$ channels, and actin reorganization. Although nonresponsive to ABA, the mutant stomata close when exposed to high external concentrations of Ca$^{2+}$, suggesting that they are defective in their ability to initiate Ca$^{2+}$ signaling. Consistent with this finding, ABA is less effective at inducing transient increases in cytosolic Ca$^{2+}$ in these mutants.

The Arabidopsis *abi1-1* and *abi2-1* mutations were identified in a screen for a *decreased* response to ABA, hence their designation as ABA-insensitive mutants. Based on their ABA-insensitive phenotypes, it was initially assumed that the wild-type *ABI1* genes must *promote* the ABA response. However, genetic studies showed that the original mutations were dominant: one defective copy of the gene was sufficient to disrupt the ABA response by blocking the activity of the functional gene products from the remaining wild-type allele. Subsequently, recessive mutants of *ABI1* were obtained that exhibited a simple loss of ABI1 activity. In contrast to the dominant *ABI1* mutants, these recessive *ABI1* mutants exhibited an *increased* sensitivity to ABA (Gosti et al. 1999). Thus, contrary to the initial assumption, the wild-type function of these protein phosphatases is to *inhibit* ABA response.

Genomic analyses have now shown that *ABI1* and *ABI2* are part of a subgroup of nine closely related genes belonging to a larger PP2C gene family in Arabidopsis (Schweighofer et al. 2004). As for *ABI1*, loss of function of many of the *ABI*-class PP2C genes results in *increased* sensitivity to ABA. Introducing these genes into plants under control of a highly expressed promoter results in *reduced* ABA sensitivity, confirming the role of these genes in repressing ABA responses. Many of these family members increase their expression in response to ABA, leading to feedback repression of continued ABA response. (Other negative regulators of ABA signaling are discussed in WEB TOPIC 23.11.)

The **ABI-class protein phosphatases** appear to interact with many other proteins in the cell, including protein kinases, Ca$^{2+}$-binding proteins, and transcription factors, presumably regulating their activity by dephosphorylat-

ing specific serine or threonine residues. The phosphatase activity of ABI1 is highly sensitive to pH and H$_2$O$_2$, such that ABA-induced alkalinization increases its activity, but H$_2$O$_2$ reduces it. ABI1 function is also controlled by subcellular localization: the wild-type protein interacts with phosphatidic acid, which may retain it at the plasma membrane (Mishra et al. 2006), but the ABA-resistant phenotype of the dominant negative *abi1-1* mutant depends on nuclear localization of the mutant protein (Moes et al. 2008).

As discussed in Chapter 14, and on pages 679–680 of this chapter these PP2Cs have now been shown to interact directly with the PYR/PYL/RCAR family of receptors (Park et al. 2009; Ma et al. 2009). This ABA-dependent interaction represses the PP2C phosphatase activity, which leads to de-repression of the events normally inhibited by these PP2Cs, including activation of the SnRKs. The dominant negative abi1-1 mutant protein does not interact with the receptors and therefore remains in a repressive state. From a signaling perspective, these PP2Cs can be viewed as "hubs" that integrate information from receptors and secondary messengers to modify the function of multiple downstream pathways.

In summary, protein kinases, such as SnRK2s, as well as others discussed in WEB TOPIC 23.10 (CPK/CDPK/CDKs, CIPKs, and the MAPK cascade system), and at least two classes of protein phosphatases (the ABI-related PP2Cs and a PP2A) regulate serine/threonine phosphorylation and therefore the activity of numerous downstream regulators in ABA signaling. Activity of many of these enzymes is regulated both pre- and posttranslationally, and some are subject to feedback regulation attenuating the response.

### ABA shares signaling intermediates with other hormonal pathways

*Enhanced response to ABA* (ERA3) was found to be allelic to a previously identified ethylene signaling locus, *ETHYLENE-INSENSITIVE 2* (EIN2) (Ghassemian et al. 2000) (see Chapter 22). In addition to displaying defects in ABA and ethylene responses, mutations in *ERA3* result in defects in the responses to auxin, jasmonic acid, and stress. This gene encodes a membrane-bound protein that appears to represent a common signaling intermediate for many different signals. As you will recall from Chapter 14, such an interaction is an example of *primary cross-regulation* between distinct signaling pathways.

## ABA Regulation of Gene Expression

Downstream of the early ABA signal transduction processes discussed above, ABA causes changes in gene expression. ABA regulates the expression of numerous genes during seed maturation and acclimation to stress conditions such as drought, low temperature, and salinity (salt stress). Transcriptional profiling studies in Arabidopsis and rice have shown that 5 to 10% of the genome is regulated by ABA and various stresses (e.g., drought, salinity, and cold). More than half of these changes are common to ABA, drought and salinity, but only about 10% of these stress-regulated genes are also cold-regulated (Shinozaki et al. 2003). The ABA- and stress-induced genes are presumed to contribute to adaptive aspects of induced tolerance (see Chapter 26).

### Gene activation by ABA is mediated by transcription factors

Four main classes of regulatory sequences conferring ABA inducibility have been identified, and proteins that bind to these sequences, including members of the basic leucine zipper (bZIP), B3, MYB, and MYC families, have been characterized (see WEB TOPIC 23.12). Under stress conditions, induction of gene expression may be ABA-dependent or ABA-independent, and additional transcription factors have been identified that specifically mediate responses to cold, drought, or salinity (see Chapter 26).

A few DNA elements that are involved in transcriptional repression by ABA have been identified. The best characterized of these are the gibberellin response elements (GAREs) that mediate the gibberellin-inducible, ABA-repressible expression of the barley α-amylase genes (see Chapter 20).

Four transcription factors involved in ABA-inducible gene activation in maturing seeds have been identified by genetic means: VIVIPAROUS 1 (VP1) in maize and ABA-INSENSITIVE (ABI) 3, ABI4, and ABI5 in Arabidopsis. A mutation in the gene encoding any of these four transcription factors reduces the ABA responsiveness of the seeds. The maize *VP1* and Arabidopsis *ABI3* genes encode highly similar proteins, while the *ABI4* and *ABI5* genes encode members of two other transcription factor families (Finkelstein et al. 2002).

The **ABA Response Element Binding Factors (AREBs/ABFs)** are additional members of the *ABI5* subfamily of bZIP transcription factors that are correlated with gene expression induced by ABA, embryo formation, drought, or salt stress.

Characterization of *vp1*, *abi3*, *abi4*, and *abi5* mutants has shown that each of these genes can either activate or repress transcription, depending on the target gene, but some of these effects may be indirect. Because the promoter of any given gene contains binding sites for a variety of regulators, it is likely that these transcription factors act in complexes made up of varying combinations of regulators whose composition is determined by the combination of available regulators and binding sites.

Availability and activity of these transcription factors is controlled by developmentally and environmentally regulated gene expression, regulation of their own expression or that of similar factors, posttranslational modification such as phosphorylation by specific SnRK2s and calcium-dependent protein kinases (CPKs), and in some cases, stabilization against proteasomal degradation (Finkelstein et al. 2002). As illustrated in FIGURE 23.10, many of these regulatory mechanisms require interactions among transcription factors or between transcription factors and kinases, phosphatases, or components of the degradation machinery. See WEB TOPIC 23.13 for additional regulatory factors.

Collectively, these studies demonstrate that many transcription factors can be present in a variety of regulatory complexes with distinct effects on ABA-induced gene expression depending on the combination of *cis*-acting sites in individual promoters and available factors in any given cell.

In addition to transcription factors, ABA-regulated gene expression also depends on proper mRNA processing and stabilization (see WEB TOPIC 23.14).

# Developmental and Physiological Effects of ABA

Abscisic acid plays primary regulatory roles in the initiation and maintenance of seed and bud dormancy and in the plant's response to stress, particularly water stress. In addition, ABA influences many other aspects of plant development by interacting, usually as an antagonist, with auxin, cytokinin, gibberellin, ethylene, and brassinosteroids. In this section we will explore the diverse physiological effects of ABA, including its roles in seed development, seed and bud dormancy, germination, vegetative growth, senescence, and stomatal regulation. For a discussion of ABA and pathogen responses see WEB TOPIC 23.15.

### ABA regulates seed maturation

Seed development can be divided into three phases of approximately equal duration:

1. During the first phase, which is characterized by cell divisions and tissue differentiation, the zygote undergoes embryogenesis and the endosperm tissue proliferates.

2. During the second phase, cell divisions cease and storage compounds accumulate.

3. In the final phase, embryos of "orthodox" seeds become tolerant to desiccation, and the seeds dehydrate, losing up to 90% of their water. As a consequence of dehydration, metabolism comes to a halt and the seed enters a **quiescent** ("resting") state. In some cases the seed becomes dormant as well. Unlike quiescent seeds, which will germinate upon rehydration, dormant seeds require additional

1. ABA and a variety of transcription factors are produced in response to developmental or environmental signals.

2. Phospholipase D (PLD)-dependent signaling, regulation by microRNA (miRNA), and activation of many kinases are implicated in the activation of transcription factors by ABA. The promoters of ABA-regulated genes have different combinations of recognition sites (MYBR, MYCR, etc.) that can be bound by various members of the corresponding transcription factor families. Multiple family members for each class of transcription factors shown participate in ABA signaling.

3. In addition to forming homo- and heterodimers within families, some of these factors interact with one another and additional components of the transcription machinery. The specific combinations determine the extent to which a given gene is activated or repressed.

4. Some of the transcription factor genes are cross- and autoregulated, in some cases enhancing the ABA response by positive feedback.

5. In the absence of ABA, these ABA-insensitive (ABI) transcription factors are degraded via the proteasome.

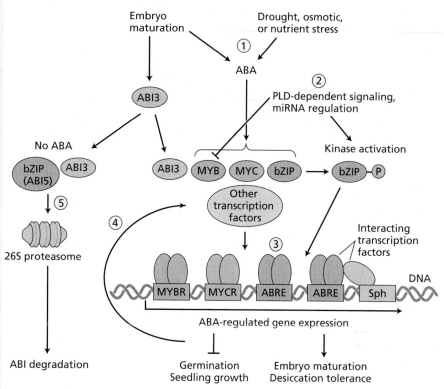

**FIGURE 23.10** Regulatory mechanisms and transcription factors that mediate ABA-regulated gene expression.

treatment or signals for germination to occur. In contrast to orthodox seeds, recalcitrant seeds do not complete this phase, so have a high moisture content at maturity and are not desiccation tolerant.

The latter two phases result in the production of viable seeds with adequate resources to support germination and the capacity to wait weeks to years before resuming growth. Typically, the ABA content of seeds is very low early in embryogenesis, reaches a maximum at about the halfway point, and then gradually falls to low levels as the seed reaches maturity. Thus there is a broad peak of ABA accumulation in the seed corresponding to mid- to late embryogenesis.

The hormonal balance of seeds is complicated by the fact that not all the tissues have the same genotype. The seed coat is derived from maternal tissues (see **WEB TOPIC 1.3**); the zygote and endosperm are derived from both parents. Genetic studies with ABA-deficient mutants of Arabidopsis have shown that the zygotic genotype controls ABA synthesis in the embryo and endosperm and is essential to dormancy induction, whereas the maternal genotype controls the major, early peak of ABA accumulation and helps suppress vivipary in mid-embryogenesis (Raz et al. 2001).

## ABA inhibits precocious germination and vivipary

When immature embryos are removed from their seeds and placed in culture midway through development before the onset of dormancy, they germinate precociously—that is, without passing through the normal quiescent and/or dormant stage of development. ABA added to the culture medium inhibits precocious germination. This result, in combination with the fact that the level of endogenous ABA is high during mid- to late seed development, suggests that ABA is the natural constraint that keeps developing embryos in their embryogenic state.

Further evidence for the role of ABA in preventing precocious germination has been provided by genetic studies of vivipary. The tendency toward vivipary, also known as *preharvest sprouting*, is characteristic of some grain crops when they mature in wet weather. In maize, several *viviparous* mutants have been selected in which the embryos germinate directly on the cob while still attached to the plant. Several of these mutants are ABA-deficient (*vp2, vp5, vp7, vp9,* and *vp14*) (see Figure 23.3); one is ABA-insensitive (*vp1*). Vivipary in the ABA-deficient mutants can be partially prevented by treatment with exogenous ABA. Vivipary in maize also requires synthesis of GA early in embryogenesis as a positive signal; double mutants deficient in both GA and ABA do not exhibit vivipary (White et al. 2000).

In contrast to the maize mutants, single-gene mutants of Arabidopsis with ABA deficiency or insensitivity fail to exhibit vivipary, although they are nondormant. The lack of vivipary might reflect a lack of moisture, because such seeds will germinate within the fruits under conditions of high relative humidity. However, other Arabidopsis mutants with a normal ABA response and only moderately reduced ABA levels exhibit some vivipary even at low humidities. One such mutant is *fusca3**, which belongs to a class of mutants defective in regulating the transition from embryogenesis to germination. Furthermore, double mutants combining either defects in ABA biosynthesis or ABA response with the *fusca3* mutation have a high frequency of vivipary, suggesting that redundant control mechanisms suppress vivipary in Arabidopsis (Finkelstein et al. 2002).

### ABA promotes seed storage reserve accumulation and desiccation tolerance

During mid- to late embryogenesis, when seed ABA levels are highest, seeds accumulate storage compounds that will support seedling growth at germination. Another important function of ABA in the developing seed is to promote the acquisition of **desiccation tolerance**. Desiccation can severely damage membranes and other cellular constituents (see Chapter 26). As maturing seeds begin to lose water, embryos accumulate sugars and so-called **late-embryogenesis–abundant** (**LEA**) proteins; these molecules are thought to interact to form a glassy state (a highly viscous liquid with very slow diffusion and therefore limited chemical reactions) involved in desiccation tolerance (Buitink and Leprince 2008).

Physiological and genetic studies have shown that ABA affects the synthesis of LEAs and of storage proteins and lipids. For example, exogenous ABA promotes accumulation of storage proteins and LEAs in cultured embryos of many species, but some ABA-deficient or ABA-insensitive mutants fail to accumulate these proteins in response to applied ABA. Synthesis of some LEA proteins, or related family members, can even be induced by ABA treatment of vegetative tissues. These results suggest that the synthesis of most LEA proteins is under ABA control (see WEB TOPIC 23.16). However, synthesis of both storage proteins and LEA proteins is also reduced in other seed development mutants with normal ABA levels and responses, indicating that ABA is only one of several signals controlling the expression of these genes during embryogenesis.

ABA not only regulates the accumulation of storage proteins and desiccation protectants during embryogenesis; it can also maintain the mature embryo in a dormant state until the environmental conditions are optimal for growth. Seed dormancy is an important factor in the adaptation of plants to unfavorable environments. As we will discuss in the next few sections, plants have evolved a variety of mechanisms, some of which involve ABA, that enable them to maintain their seeds in a dormant state.

### Seed dormancy can be regulated by ABA and environmental factors

During seed maturation, the embryo desiccates and enters a quiescent phase. Seed germination can be defined as the resumption of growth of the embryo of the mature seed. Germination depends on the same environmental conditions as vegetative growth does: Water and oxygen must be available, the temperature must be suitable, and there must be no inhibitory substances present.

In many cases a viable (living) seed will not germinate even if all the necessary environmental conditions for growth are satisfied. This phenomenon is termed **seed dormancy**. Seed dormancy introduces a temporal delay in the germination process that provides additional time for seed dispersal over greater geographic distances. It also maximizes seedling survival by preventing germination under unfavorable conditions.

Seed dormancy may result from coat-imposed dormancy, embryo dormancy, or both. Dormancy imposed on the embryo by the seed coat and other enclosing tissues, such as endosperm, pericarp, or extrafloral organs, is known as **coat-imposed dormancy**. The embryos of such seeds will germinate readily in the presence of water and oxygen once the seed coat and other surrounding tissues have been either removed or damaged. To learn more about the basic mechanisms of coat-imposed dormancy see Bewley and Black 1994 and WEB TOPIC 23.17.

Seed dormancy that is intrinsic to the embryo and is not due to any influence of the seed coat or other surrounding tissues is called **embryo dormancy**. In some cases, embryo dormancy can be relieved by amputation of the cotyledons. Species in which the cotyledons exert an inhibitory effect include European hazel (*Corylus avellana*) and European ash (*Fraxinus excelsior*).

A fascinating demonstration of the cotyledon's ability to inhibit growth is found in species (e.g., peach) in which dormant embryos that have been excised from the surrounding endosperm and seed coat germinate but grow extremely slowly to form a dwarf plant. If the cotyledons are removed at an early stage of development, however, the plant abruptly shifts to normal growth.

Embryo dormancy is thought to be due to the presence of inhibitors, especially ABA, as well as the absence of

---

*Named after the Latin term for the reddish-brown color of the embryos.

growth promoters, such as GA. Maintenance of dormancy in imbibed seeds requires de novo ABA biosynthesis, and the loss of embryo dormancy is often associated with a sharp decrease in the ratio of ABA to GA. The levels of ABA and GA are regulated by their synthesis and catabolism, which are catalyzed by specific isozymes whose expression is controlled by developmental and environmental factors.

Various external factors release the seed from embryo dormancy, and dormant seeds typically respond to more than one of three factors:

1. *After-ripening.* Many seeds lose their dormancy when their moisture content is reduced to a certain level by drying—a phenomenon known as **after-ripening**.

2. *Chilling.* Low temperature, or **chilling**, can release seeds from dormancy. Many seeds require a period of cold (0–10°C) while in a fully hydrated (imbibed) state in order to germinate.

3. *Light.* Many seeds have a light requirement for germination, which may involve only a brief exposure, as in the case of lettuce, an intermittent treatment (e.g., succulents of the genus *Kalanchoe*), or even a specific photoperiod involving short or long days.

For further information on environmental factors affecting seed dormancy, see WEB TOPIC 23.18. For a discussion of seed longevity, see WEB TOPIC 23.19.

## Seed dormancy is controlled by the ratio of ABA to GA

ABA mutants have been useful in demonstrating the role of ABA in seed dormancy. Dormancy of Arabidopsis seeds can be overcome with a period of after-ripening and/or cold treatment. ABA-deficient (*aba*) mutants of Arabidopsis have been shown to be nondormant at maturity. When reciprocal crosses between *aba* and wild-type plants were carried out, the seeds exhibited dormancy only when the embryo itself could produce ABA during seed maturation and following imbibition. Neither maternal nor exogenously applied ABA was effective in inducing dormancy in an *aba* embryo.

On the other hand, maternally derived ABA constitutes the majority of the ABA present in developing seeds and is required for other aspects of seed development—for example, helping suppress vivipary in mid-embryogenesis. Thus the two sources of ABA function in different developmental pathways. Dormancy is also greatly reduced in seeds from the ABA-insensitive mutants *ABA-insensitive 1* (*abi1*), *abi2*, and *abi3*, even though these seeds contain higher ABA concentrations than those of the wild type throughout development, possibly reflecting feedback regulation of ABA metabolism.

ABA-deficient tomato mutants seem to function in the same way, indicating that the phenomenon is probably a general one. However, other mutants with reduced dormancy, but normal ABA levels and sensitivity, point to additional regulators of dormancy, including transcriptional regulation by effects on chromatin structure or transcription elongation (Finkelstein et al. 2008). Additional factors are being identified through genetic studies of natural variation (see WEB TOPIC 23.20).

An alternate approach compares the entire transcriptomes or proteomes (all transcripts or proteins present in a given tissue) of seeds in different states of dormancy (e.g., primary, secondary, or nondormant) to identify genes whose expression correlates with dormancy. These can be additional candidates for dormancy regulators, but may also be effectors of dormancy (e.g., stress-induced genes) or factors that will eventually help the seed to escape from dormancy. These comparisons have revealed a surprising degree of activity for transcriptional and posttranscriptional processes in dry seeds.

Although the role of ABA in initiating and maintaining seed dormancy is well established, other hormones contribute to the overall effect. For example, in most plants the peak of ABA production in the seed coincides with a decline in the levels of indole-3-acetic acid (IAA, also called auxin, see Chapter 19) and GA.

An elegant demonstration of the importance of the ratio of ABA to GA in seeds was provided by the genetic screen that led to the isolation of the first ABA-deficient mutants of Arabidopsis (Koornneef et al. 1982). Seeds of a GA-deficient mutant that could not germinate in the absence of exogenous GA were mutagenized and then grown in the greenhouse. The seeds produced by these mutagenized plants were then screened for seeds that had regained their ability to germinate, called **revertants**.

Revertants were isolated, and they turned out to be mutants of abscisic acid synthesis. The revertants germinated because dormancy had not been induced, so subsequent synthesis of GA was no longer required to overcome it. Conversely, a screen for suppressors of ABA-resistant germination in an ABA insensitive mutant (*abi1-1*) background identified both GA-deficient and GA-resistant mutants (Steber et al. 1998). These studies elegantly illustrate the general principle that the balance of plant hormones, and the ability to respond to them, is often more critical than are their absolute concentrations in regulating development.

The ABA:GA balance is both adjusted and interpreted by the action of transcription factors. GA promotion of germination requires destruction of DELLA family proteins that repress germination in part by increasing expression of proteins that promote ABA synthesis. The increased ABA levels then promote expression of ABI transcription factor(s) and the germination-inhibiting DELLA protein, creating a positive feedback loop (Piskurewicz et al. 2008).

Recent genetic screens for suppressors of ABA insensitivity during seed germination have identified additional antagonistic interactions between ABA and ethylene, brassinosteroid, and auxin. In addition, many new alleles of ABA-deficient or *abi4* mutants have been identified in screens for altered sensitivity to sugar or salinity. These studies show that a complex regulatory web integrates hormonal, nutrient, and stress signaling controlling the commitment to growth of the next generation.

## ABA inhibits GA-induced enzyme production

In addition to the ABA–GA antagonism affecting seed dormancy, ABA inhibits the GA-induced synthesis of hydrolytic enzymes that are essential for the breakdown of storage reserves in germinating seeds. For example, GA stimulates the aleurone layer of cereal grains to produce α-amylase and other hydrolytic enzymes that break down stored resources in the endosperm during germination (see Chapter 20). ABA inhibits this GA-dependent enzyme synthesis by inhibiting the transcription of α-amylase mRNA. ABA exerts this inhibitory effect via at least two mechanisms, one direct and one indirect:

1. A protein originally identified as an activator of ABA-induced gene expression, VP1, acts as a

transcriptional repressor of some GA-regulated genes (Hoecker et al. 1995).

2. ABA represses the GA-induced expression of GAMYB, a transcription factor that mediates the GA induction of α-amylase expression (Gomez-Cadenas et al. 2001).

## ABA promotes root growth and inhibits shoot growth at low water potentials

ABA has different effects on the growth of roots and shoots, and the effects are strongly dependent on the water status of the plant. **FIGURE 23.11** compares the growth of shoots and roots of maize seedlings grown under either abundant water conditions (high water potential, $\Psi_w$) or dehydrating conditions (low $\Psi_w$). Two types of seedlings were used: (1) a wild type with normal ABA levels and (2) an ABA-deficient, *viviparous* mutant.

When the water supply is ample (high $\Psi_w$), shoot growth is greater in the wild-type plant with normal endogenous ABA levels than in the ABA-deficient mutant (see Figure 23.11A). Although the reduced shoot growth in the mutant could be due in part to excessive water loss from the leaves, the stunted shoot growth of ABA-deficient maize at high water potentials seems to be due

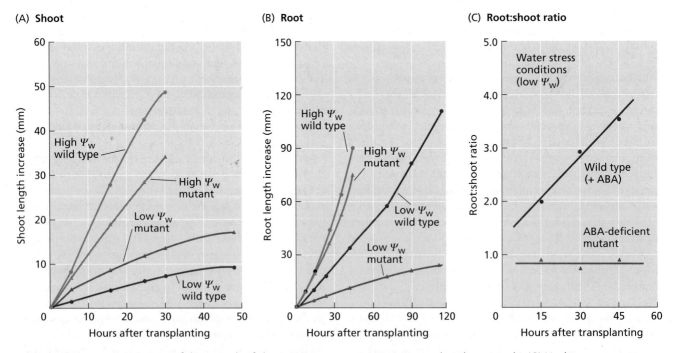

**FIGURE 23.11** Comparison of the growth of shoots (A) and roots (B) of normal versus ABA-deficient (*viviparous, vp*) maize plants growing in vermiculite maintained either at high water potential ($\Psi_w$ = –0.03 MPa) or at low water potential ($\Psi_w$ = –0.3 MPa in A and –1.6 MPa in B). Water stress (low $\Psi_w$) depresses the growth of both shoots and roots compared to the controls. (C) Under water stress conditions (low $\Psi_w$, defined slightly differently for shoot and root), the ratio of root growth to shoot growth is much higher when ABA is present (i.e., in the wild type) than when it is absent (in the mutant). (From Saab et al. 1990.)

to the overproduction of ethylene, which is normally inhibited by endogenous ABA. This finding suggests that endogenous ABA promotes shoot growth in well-watered plants by suppressing ethylene production. Similarly, root growth of well-watered plants is slightly greater in the wild type (normal endogenous ABA) than in the ABA-deficient mutant. Therefore, at high water potentials (when the total ABA levels are low), endogenous ABA exerts a slight positive effect on the growth of both roots and shoots.

In contrast, limiting water (i.e., low water potentials, low $\Psi_w$), has opposite effects on root and shoot growth. Under dehydrating conditions, growth is still inhibited relative to that of either genotype when water is abundant. However, whereas shoot growth is greater in the ABA-deficient mutant than in the wild type, the growth of the roots is much higher in the wild type than in the ABA-deficient mutant (see Figure 23.11B). Endogenous ABA appears to promote root growth by inhibiting ethylene production during water stress.

ABA also affects changes in the degree of root branching in response to auxin, nutrient availability, and stress. Auxin is well established as the primary signal promoting root branching, but some ABA response mutants show reduced sensitivity to auxin for initiation of lateral roots, implying that ABA signaling is part of this response. In contrast to auxin, high nitrate concentrations *inhibit* root branching, but this effect is also reduced in ABA-deficient and response mutants. A further specialization of ABA's role in controlling plant nutrient balance is seen in legumes, where it coordinates plant responses to microbial signals regulating root nodulation (Ding et al. 2008).

Another repressive response is seen under extended drought stress. In some species, roots may initiate many extra lateral roots but inhibit their growth until the stress is relieved, a phenomenon known as **drought rhizogenesis** (Vartanian et al. 1994). These repressive effects of stress and nutrient imbalance are at least partly dependent on ABA because ABA-deficient mutants and some ABA response mutants do not display this inhibition.

To summarize, despite the traditional view of ABA as a growth inhibitor, endogenous ABA restricts shoot growth only under water stress conditions. Moreover, under these conditions, when ABA levels are high, endogenous ABA exerts a strong positive effect on primary root growth by suppressing ethylene production. The overall effect is a dramatic increase in the root:shoot ratio at low water potentials (see Figure 23.11C), which, along with the effect of ABA on stomatal closure, helps the plant cope with water stress (Sharp 2002). Furthermore, the temporary inhibition of lateral root outgrowth promotes exploration of new areas of soil, and permits replacement of dehydrated laterals following rehydration. It is not clear how different ABA levels lead to opposite effects on growth, but these effects may reflect signaling through receptors

with different functional ranges of sensitivity or different downstream signaling elements in roots versus shoots.

For another example of the role of ABA in the response to dehydration, see WEB ESSAY 23.1.

## ABA promotes leaf senescence independently of ethylene

Abscisic acid was originally isolated as an abscission-causing factor. However, it has since become evident that ABA stimulates abscission of organs in only a few species and that the hormone primarily responsible for causing abscission is ethylene. On the other hand, ABA is clearly involved in leaf senescence, and through its promotion of senescence it might indirectly increase ethylene formation and stimulate abscission. (For more discussion on the relationship between ABA and ethylene, see WEB TOPIC 23.21.)

Leaf senescence has been studied extensively (the anatomical, physiological, and biochemical changes that take place during this process were described in Chapter 16). Leaf segments senesce faster in darkness than in light, and they turn yellow as a result of chlorophyll breakdown. In addition, the breakdown of proteins and nucleic acids is increased by the stimulation of several hydrolases. ABA greatly accelerates the senescence of both leaf segments and attached leaves.

## ABA accumulates in dormant buds

For woody species in cold climates, bud dormancy is an important adaptive feature. When a tree is exposed to very low temperatures in winter, bud scales protect its meristems and inhibit bud growth. This response to low temperatures requires a sensory mechanism that detects the environmental changes (sensory signals), and a control system that transduces the sensory signals and triggers the developmental processes leading to bud dormancy.

ABA was originally suggested as the dormancy-inducing hormone because it accumulates in dormant buds and decreases after the tissue is exposed to low temperatures. However, later studies showed that the ABA content of buds does not always correlate with the degree of dormancy. As we saw in the case of seed dormancy, this apparent discrepancy might reflect interactions between ABA and other hormones; perhaps bud dormancy and growth are regulated by the balance between bud growth inhibitors, such as ABA, and growth-inducing substances, such as cytokinins and gibberellins.

Much progress has been achieved in elucidating the role of ABA in seed dormancy by the use of ABA-deficient mutants. However, progress on the role of ABA in bud dormancy, a characteristic of woody perennials, has lagged because of the lack of a convenient genetic system. This discrepancy illustrates the tremendous contribution that genetics and molecular biology have made to plant physiology, and underscores the need for extending such approaches to woody species.

Analyses of traits such as dormancy are complicated by the fact that they are often controlled by the combined action of several genes, resulting in a gradation of phenotypes referred to as *quantitative traits*. Recent genetic mapping studies suggest that homologs of the protein phosphatase ABI1 may contribute to regulation of bud dormancy in poplar trees. For a description of such studies, see WEB TOPIC 23.20.

### ABA closes stomata in response to water stress

Elucidation of the roles of ABA in freezing, salt, and water stress (see Chapter 26) led to the characterization of ABA as a stress hormone. As noted earlier, ABA concentrations in leaves can increase up to 50 times under drought conditions—the most dramatic change in concentration reported for any hormone in response to an environmental signal. Redistribution or biosynthesis of ABA is very effective in causing stomatal closure, and its accumulation in stressed leaves plays an important role in the reduction of water loss by transpiration under water stress conditions (see Figure 23.4). Increases in humidity reduce ABA levels by increasing catabolism in both vascular tissue and guard cells, thereby permitting stomata to re-open.

Mutants that exhibit permanent wilting may be unable to close their stomata due to defects in either ABA synthesis or response. Application of exogenous ABA to ABA-deficient mutants causes stomatal closure and a restoration of turgor pressure. In contrast, *wilty* mutants blocked in their ability to respond to ABA are not rescued by ABA application.

### ABA regulates ion channels and the plasma membrane ATPase in guard cells

As discussed in Chapter 18, stomatal closure is driven by a reduction in guard cell turgor pressure caused by a massive long-term efflux of $K^+$ and anions from the cell. Opening of the $K^+$ efflux channels requires long-term membrane depolarization. This depolarization appears to be triggered by two factors: (1) an ABA-induced transient depolarization of the plasma membrane caused by the net influx of positive charge, coupled with (2) transient increases in cytosolic calcium (FIGURE 23.12). The combination of calcium influx and the release of calcium from internal stores raises the cytosolic calcium concentration from 50 to 350 n$M$ to as high as 1100 n$M$ (1.1 µ$M$) (FIGURE 23.13) (McAinsh et al. 1990). Furthermore, ABA enhances the sensitivity of stomatal closing mechanisms to intracellular calcium levels. These mechanisms cause ABA to open calcium-activated slow (S-type) anion channels on the plasma membrane (see Chapter 6). ABA has been shown to activate slow anion channels in guard cells (Schroeder et al. 2001). ABA also activates another class of anion channels in guard cells, the rapid (R-type) anion channels (Raschke et al. 2003).

The prolonged opening of these slow and rapid anion channels permits large quantities of $Cl^-$ and malate$^{2-}$ ions to escape from the cell, moving down their electrochemical gradients. (The inside of the cell is negatively charged, thus pushing $Cl^-$ and malate$^{2-}$ out of the cell, and the outside has lower $Cl^-$ and malate$^{2-}$ concentrations than the interior.) The outward flow of negatively charged $Cl^-$ and malate$^{2-}$ ions generated in this way strongly depolarizes the membrane, triggering the voltage-gated $K^+$ efflux channels to open.

In addition to increasing the cytosolic calcium concentration, ABA causes an alkalinization of the cytosol from

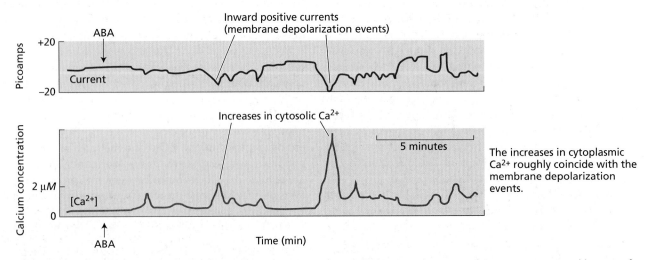

**FIGURE 23.12** Simultaneous measurements of ABA-induced inward positive currents and ABA-induced increases in cytosolic $Ca^{2+}$ concentrations in a guard cell of broad bean (*Vicia faba*). The current was measured by the patch clamp technique; calcium was measured by use of a fluorescent indicator dye. ABA was added to the system at the arrow in each case. (From Schroeder and Hagiwara 1990.)

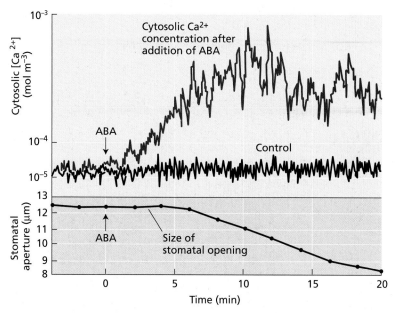

**FIGURE 23.13** Time course of the ABA-induced increase in guard cell cytosolic $Ca^{2+}$ concentration (upper panel) and ABA-induced stomatal aperture (lower panel). The rise in $Ca^{2+}$ begins within ~3 minutes, followed by decreased stomatal aperture within an additional 5 minutes. (From McAinsh et al. 1990.)

about pH 7.7 to pH 7.9. The increase in cytosolic pH has been shown to increase the activity of the $K^+$ efflux channels on the plasma membrane, apparently by increasing the number of channels available for activation (see Chapter 6). One effect of the *abi1* mutation is to render these $K^+$ channels insensitive to pH.

In addition to causing stomatal closure, ABA prevents light-induced stomatal opening. Another factor that can contribute to membrane depolarization is inhibition of the plasma membrane $H^+$-ATPase. ABA indirectly inhibits blue light–stimulated proton pumping by guard cell protoplasts (**FIGURE 23.14**). This inhibition is consistent with the model that the depolarization of the plasma

membrane by ABA is partially caused by a decrease in the activity of the plasma membrane $H^+$-ATPase.

In broad bean (*Vicia faba*), at least, the plasma membrane $H^+$-ATPase of the leaves is strongly inhibited by calcium. A calcium concentration of $0.3~\mu M$ blocks 50% of the activity of $H^+$-ATPase, and $1~\mu M$ calcium blocks the enzyme completely (Kinoshita et al. 1995). It appears that two factors contribute to ABA inhibition of the plasma membrane proton pump: an increase in the cytosolic $Ca^{2+}$ concentration and alkalinization of the cytosol.

ABA inhibition of stomatal closing also correlates with ABA inhibition of the inward $K^+$ channels, which are open when the membrane is hyperpolarized by the proton pump (see Chapters 6 and 18). Inhibition of the inward $K^+$ channels is mediated by the ABA-induced increase in cytosolic calcium concentration and appears to be mediated by G protein signaling. G protein activators, such as GTPγS, can inhibit the activity of the inward $K^+$ channels, but ABA does not inhibit inward $K^+$ channels or light-induced stomatal opening in an Arabidopsis mutant lacking a Gα subunit (Wang et al. 2001). Thus, calcium and pH affect guard cell plasma membrane channels in two ways:

1. They prevent stomatal opening by inhibiting inward $K^+$ channels and plasma membrane proton pumps.

2. They promote stomatal closing by activating outward anion channels, thus leading to activation of $K^+$ efflux channels.

Although ABA inhibition of stomatal opening is blocked in the Gα subunit mutant, ABA still promotes stomatal closure in this mutant, indicating that inhibition of opening and promotion of closing take two distinct paths to the

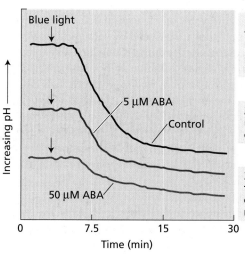

1. A pulse of blue light activates the plasma membrane $H^+$-ATPase, which pumps protons into the external medium and lowers the pH.

2. Addition of ABA to the medium inhibits the acidification by 40%.

3. These results demonstrate that ABA induces changes in the cell that inhibit the plasma membrane $H^+$-ATPase.

**FIGURE 23.14** ABA inhibition of blue light–stimulated proton pumping by guard cell protoplasts. A suspension of guard cell protoplasts was incubated under red-light irradiation, and the pH of the suspension medium was monitored with a pH electrode. The starting pH was the same in all cases (the curves are displaced for ease of viewing). (After Shimazaki et al. 1986.)

1. ABA binds to its receptors. (For clarity only the plasma membrane receptors are shown.)

2. ABA-binding induces the formation of reactive oxygen species (ROS) that activate $Ca^{2+}$ channels on the plasma membrane. Phospholipase D (PLD)-mediated phosphatidic acid (PA) production contributes to ROS production.

3. The influx of calcium initiates intracellular calcium transients and promotes the further release of calcium from vacuoles.

4. ABA stimulates NO production, and NO increases cADPR levels.

5. ABA increases $InsP_3$ levels via a signaling pathway that includes S1P, heterotrimeric G proteins, and phospholipase C and D (PLC and PLD).

6. The rise in cADPR and $InsP_3$ activates additional calcium channels on the tonoplast, releasing more $Ca^{2+}$ from vacuoles.

7. The rise in intracellular calcium blocks $K^+_{in}$ channels on the plasma membrane.

8. The rise in intracellular calcium promotes the opening of $Cl^-_{out}$ (anion) channels on the plasma membrane, causing depolarization.

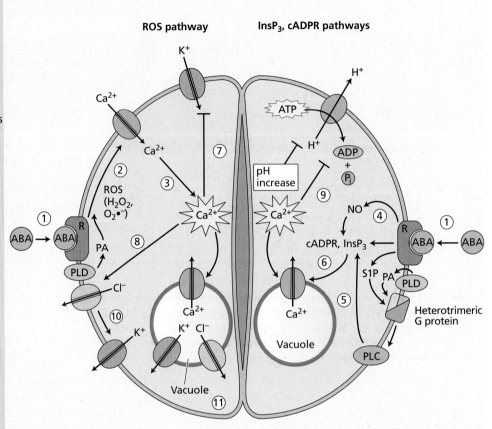

9. The plasma membrane proton pump is inhibited by the ABA-induced increase in cytosolic calcium and a rise in intracellular pH, further depolarizing the membrane.

10. Membrane depolarization activates $K^+_{out}$ channels on the plasma membrane.

11. $K^+$ and anions to be released across the plasma membrane are first released from vacuoles into the cytosol.

**FIGURE 23.15** Simplified model for ABA signaling in stomatal guard cells. The net effect is the loss of potassium ($K^+$) and its anion ($Cl^-$ or malate$^{2-}$) from the cell. cADPR, cyclic ADP-ribose; $InsP_3$, inositol 1,4,5-trisphosphate; NO, nitric oxide; PA, phosphatidic acid; PLC, phospholipase C; PLD, phospholipase D; R, receptor; ROS, reactive oxygen species; S1P, sphingosine-1-phosphate.

same end point—that is, closed stomata. As discussed previously, both phospholipase D (PLD) and its product phosphatidic acid (PA) affect G protein regulation of stomatal opening, and PA promotes stomatal closure by sequestering an inhibitory PP2C.

As a result of the large-scale (about 0.3 *M*) ion effluxes involved in stomatal closure, water is lost osmotically, and the surface area of the guard cell plasma membrane may contract by as much as 50%. Where does the extra membrane go? The answer seems to be that it is taken up as small vesicles by endocytosis—a process that also involves ABA-induced reorganization of the actin cytoskeleton

mediated by a family of plant Rho GTPases, or "Rops" (Assmann 2004).

Signal transduction in guard cells, with their multiple sensory inputs, involves protein kinases and phosphatases. The activities of the $H^+$-ATPases that drive the guard cell membrane potential are reduced by some protein kinases, SnRK, and calcium-dependent protein kinases. Consistent with this, protein kinase inhibitors block ABA-induced stomatal closing. Protein phosphatases of both the PP2C and PP2A class have also been implicated in modifying specific $H^+$-ATPase activities, leading to changes in the slow anion channel activities. In view of these results, it appears that protein phosphorylation and dephosphorylation play important roles in the ABA signal transduction pathway in guard cells.

A simplified general model for ABA action in stomatal guard cells is shown in **FIGURE 23.15**. For clarity, only the cell surface receptors are shown. For a more detailed model, see Li et al. 2006 and **WEB ESSAY 23.2**.

# SUMMARY

Abscisic acid (ABA) exerts both short-term (rapid and reversible) and long-term (lasting) controls over plant development. The hormone has major roles in plant responses to water stress, drought, low temperature, and salinity, as well as seed maturation and bud dormancy.

## Occurrence, Chemical Structure, and Measurement of ABA

- ABA resembles the terminal portion of some carotenoid molecules, and it exists in *cis* and *trans*, as well as S and R forms (**Figure 23.1**).
- ABA can be measured by bioassays, gas chromatography, high-pressure liquid chromatography, and immunoassays.

## Biosynthesis, Metabolism, and Transport of ABA

- ABA is produced from a carotenoid precursor that is synthesized from isopentenyl diphosphate (IPP) in plastids, with final steps completed in cytoplasm (**Figure 23.2**).
- Mutations that block carotenoid biosynthesis reduce ABA levels and cause precocious germination (**Figure 23.3**).
- ABA levels can change dramatically during development or in response to environmental changes including dehydration stress (**Figure 23.4**).
- ABA is inactivated by either oxidative degradation or conjugation.
- ABA is transported by both the xylem and the phloem, but it is normally more abundant in the phloem sap.
- During water stress, ABA accumulates first in shoot vascular tissue, and only later appears in roots and guard cells (**Figure 23.5**).
- During water stress, pH changes in apoplast and cytosol increase the amount of ABA reaching the guard cells to stimulate their closure (**Figure 23.6**).

## ABA Signal Transduction Pathways

- ABA short-term responses frequently involve alterations in the fluxes of ions across membranes but may also include regulation of certain genes (**Figure 23.15**).
- ABA-induced long-term developmental processes (e.g., seed maturation) involve major changes in the pattern of gene expression.
- ABA receptors have been identified in the cytosol and nucleus (**Figure 23.7**), as well as in plastids and the plasma membrane.

- Several second messengers operate in various ABA signaling responses, including $Ca^{2+}$ (**Figure 23.8**), reactive oxygen species, cyclic nucleotides, and phospholipids (**Figures 23.9, 23.15**).
- Protein kinases and phosphatases regulate important steps in ABA signaling (**Figure 23.7**).
- ABA shares signaling intermediates with other hormones and participates in primary cross-regulation between distinct signaling pathways.

## ABA Regulation of Gene Expression

- Gene activation by ABA is mediated by transcription factors.
- Transcription factor activity may be controlled by developmentally and environmentally regulated gene expression.
- Transcription can be controlled by interactions among transcription factors or between transcription factors and kinases, phosphatases, or components of the degradation machinery (**Figure 23.10**).

## Developmental and Physiological Effects of ABA

- ABA has roles in regulating seed development, seed and bud dormancy, germination, vegetative growth, senescence, stomatal regulation, as well as stress responses.
- In the developing seed, the genotype of the embryo and endosperm controls ABA synthesis that is essential to dormancy induction, whereas the maternal genotype of the seed coat controls ABA accumulation in mid-embryogenesis that suppresses vivipary.
- In seed development, ABA promotes the synthesis of storage proteins and lipids, as well as specific proteins involved in desiccation tolerance.
- Seed dormancy and germination are controlled by the ratio of ABA to gibberellic acid (GA).
- In germinating seeds, ABA inhibits the GA-induced synthesis of hydrolytic enzymes.
- ABA's effects on the growth of roots and shoots depend on the water status of the plant (**Figure 23.11**).
- ABA greatly accelerates the senescence of leaves, thereby increasing ethylene formation and stimulating abscission.
- ABA accumulates in dormant buds, inhibiting their growth; ABA may interact with growth-promoting hormones such as cytokinins and gibberellins.

## SUMMARY continued

- In response to water stress, ABA closes stomata by triggering a transient membrane depolarization due to influx of positive charge in guard cells (**Figure 23.12**). These fleeting changes cause a massive long-term efflux of K$^+$ and anions from the cell, reducing guard cell turgor pressure (**Figure 23.13**).

- ABA induced changes in the guard cells can inhibit the plasma membrane H$^+$-ATPase, thereby contributing to membrane depolarization (**Figure 23.14**).

## WEB MATERIAL

### Web Topics

**23.1   The Structure of Lunularic Acid from Liverworts**

Although inactive in higher plants, lunularic acid appears to have a function similar to ABA in liverworts.

**23.2   ABA May Be an Ancient Stress Signal**

Recent reports have documented a role for ABA in regulating stress responses in animals ranging from sea sponges to mammals.

**23.3   Structural Requirements for Biological Activity of Abscisic Acid**

To be active as a hormone, ABA requires certain functional groups.

**23.4   The Bioassay of ABA**

Several ABA-responding tissues have been used to detect and measure ABA.

**23.5   Evidence for Both Extracellular and Intracellular ABA Receptors**

Experiments supporting both types of ABA receptors in different cell types are described.

**23.6   The Existence of G Protein-Coupled ABA Receptors Is Still Unresolved**

Despite some experimental evidence for the participation of G-proteins in ABA responses, their role as receptors has not yet been demonstrated.

**23.7   The Yeast Two-Hybrid System**

The GAL4 transcription factor can be used to detect protein–protein interactions in yeast.

**23.8   Yellow Cameleon: A Noninvasive Tool for Measuring Intracellular Calcium**

The yellow cameleon protein has several features that enable it to act as a reporter for calcium concentration.

**23.9   Phosphatidic Acid May Stimulate Sphingosine-1-Phosphate Production**

The relationship between phosphatidic acid and the levels of sphingosine-1-phosphate is discussed.

**23.10   The ABA Signal Transduction Pathway Includes Several Protein Kinases**

Calcium-dependent kinases, calcineurin B-like interacting protein kinases, and MAP kinases have been implicated in the ABA signal transduction pathway.

**23.11   The *ERA1* and *ABH* Genes Code for Negative Regulators of the ABA Response**

The phenotypes of the *era1* and *abh* mutants are described.

**23.12   Promoter Elements That Regulate ABA Induction of Gene Expression**

ABA induction of gene expression is regulated by several different *cis*-acting sequences bound by distinct transcription factors.

**23.13   Regulatory Proteins Implicated in ABA-Stimulated Gene Transcription**

Techniques such as the yeast two-hybrid system have identified several additional ABA transcriptional regulators.

**23.14   ABA Gene Expression Can Also Be Regulated by mRNA Processing and Stability**

Some mutations affecting basic aspects of RNA metabolism have surprisingly specific effects on the ABA-mediated stress response.

## WEB MATERIAL continued

### Web Topics

**23.15  ABA May Play a Role in Plant Pathogen Responses**

Several studies link ABA with stimulation or inhibition of plant responses to bacterial and fungal pathogens.

**23.16  Proteins Required for Desiccation Tolerance**

ABA induces the synthesis of proteins that protect cells from damage due to desiccation.

**23.17  The Types of Coat-Imposed Seed Dormancy**

The seed coat can inhibit germination by five different mechanisms.

**23.18  Types of Seed Dormancy and the Roles of Environmental Factors**

Many types of seed dormancy exist; some are affected by various environmental factors.

**23.19  The Longevity of Seeds**

Under certain conditions, seeds can remain dormant for hundreds of years.

**23.20  Genetic Mapping of Dormancy: Quantitative Trait Locus (QTL) Scoring of Vegetative Dormancy Combined with a Candidate Gene Approach**

QTL analysis is a genetic method for determining the number and chromosomal locations of genes affecting a quantitative trait affected by many unlinked genes.

**23.21  ABA-Induced Senescence and Ethylene**

Hormone-insensitive mutants have made it possible to distinguish the effects of ethylene from those of ABA on senescence.

### Web Essays

**23.1  Heterophylly in Aquatic Plants**

Abscisic acid induces aerial-type leaf morphology in many aquatic plants.

**23.2  Hope for Humpty Dumpty: Systems Biology of Cellular Signaling**

ABA regulation of stomatal opening is used to illustrate the use of systems biology approaches for modelling cellular signaling processes.

# CHAPTER REFERENCES

Allan, A. C., Fricker, M. D., Ward, J. L., Beale, M. H., and Trewavas, A. J. (1994) Two transduction pathways mediate rapid effects of abscisic acid in *Commelina* guard cells. *Plant Cell* 6: 1319–1328.

Allen, G. J., Kwak, J. M., Chu, S. P., Llopis, J., Tsien, R. Y., Harper, J. F., and Schroeder, J. I. (1999) Cameleon calcium indicator reports cytoplasmic calcium dynamics in *Arabidopsis* guard cells. *Plant J.* 19: 735–747.

Anderson, B. E., Ward, J. M., and Schroeder, J. I. (1994) Evidence for an extracellular reception site for abscisic acid in *Commelina* guard cells. *Plant Physiol.* 104: 1177–1183.

Assmann, S. M. (2004) Abscisic acid signal transduction in stomatal responses. In *Hormones: Biosynthesis, Signal Transduction, Action!* P. J. Davies, ed., Springer, New York, pp. 391–412.

Beardsell, M. F., and Cohen, D. (1975) Relationships between leaf water status, abscisic acid levels, and stomatal resistance in maize and sorghum. *Plant Physiol.* 56: 207–212.

Bensmihen, S., Rippa, S., Lambert, G., Jublot, D., Pautot, V., Granier, F., Giraudat, J., and Parcy, F. (2002) The homologous ABI5 and EEL transcription factors function antagonistically to fine-tune gene expression during late embryogenesis. *Plant Cell* 14: 1391–1403.

Bewley, J. D., and Black, M. (1994) *Seeds: Physiology of Development and Germination*, 2nd ed. Plenum, New York.

Buitink, J., and Leprince, O. (2008) Intracellular glasses and seed survival in the dry state. *C. R. Biol.* 331: 788–795.

Christmann, A., Hoffmann, T., Teplova, I., Grill, E., and Müller, A. (2005) Generation of active pools of abscisic acid revealed by in vivo imaging of water-stressed *Arabidopsis*. *Plant Physiol.* 137: 209–219.

Christmann, A., Weiler, E. W., Steudle, E., and Grill, E. (2007) A hydraulic signal in root-to-shoot signalling of water shortage. *Plant J.* 52: 167–174.

Coursol, S., Fan, L., Le Stunff, H., Spiegel, S., Gilroy, S., and Assmann, S. M. (2003) Sphingolipid signaling in *Arabidopsis* guard cells involves heterotrimeric G proteins. *Nature* 423: 651–654.

Cutler, S., Ghassemian, M., Bonetta, D., Cooney, S., and McCourt, P. (1996) A protein farnesyl transferase involved in abscisic acid signal transduction in *Arabidopsis*. *Science* 273: 1239–1241.

Desikan, R., Griffiths, R., Hancock, J., and Neill, S. (2002) A new role for an old enzyme: Nitrate reductase-mediated nitric oxide generation is required for abscisic acid-induced stomatal closure in *Arabidopsis thaliana*. *Proc. Natl. Acad. Sci. USA* 99: 16314–16318.

Ding, Y. L., Kalo, P., Yendrek, C., Sun, J. H., Liang, Y., Marsh, J. F., Harris, J. M., Oldroyd, G. E. D. (2008) Abscisic acid coordinates nod factor and cytokinin signaling during the regulation of nodulation in *Medicago truncatula*. *Plant Cell* 20: 2681–2695.

Finkelstein, R. R, Gampala, S. S. L., and Rock, C. D. (2002) Abscisic acid signaling in seeds and seedlings. *Plant Cell* 14 Suppl.: S15–S45.

Finkelstein, R., Reeves, W., Ariizumi, T., and Steber, C. (2008) Molecular aspects of seed dormancy. *Annu. Rev. Plant Biol.* 59: 387–415.

Fujii, H., Verslues, P. E., and Zhu, J.-K. (2007) Identification of two protein kinases required for abscisic acid regulation of seed germination, root growth, and gene expression in *Arabidopsis*. *Plant Cell* 19: 485–494.

Ghassemian, M., Nambara, E., Cutler, S., Kawaide, H., Kamiya, Y., and McCourt, P. (2000) Regulation of abscisic acid signaling by the ethylene response pathway in *Arabidopsis*. *Plant Cell* 12: 1117–1126.

Gomez-Cadenas, A., Zentella, R., Walker-Simmons, M. K., and Ho, T.-H. D. (2001) Gibberellin/abscisic acid antagonism in barley aleurone cells: Site of action of the protein kinase PKABA1 in relation to gibberellin signaling molecules. *Plant Cell* 13: 667–679.

Gosti, F., Beaudoin, N., Serizet, C., Webb, A. A. R., Vartanian, N., and Giraudat, J. (1999) ABI1 protein phosphatase 2C is a negative regulator of abscisic acid signaling. *Plant Cell* 11: 1897–1909.

Hoecker, U., Vasil, I. K., and McCarty, D. R. (1995) Integrated control of seed maturation and germination programs by activator and repressor functions of Viviparous-1 of maize. *Genes Dev.* 9: 2459–2469.

Hugouvieux, V., Kwak, J. M., and Schroeder, J. I. (2001) A mRNA cap binding protein, ABH1, modulates early abscisic acid signal transduction in *Arabidopsis*. *Cell* 106: 477–487.

Jeannette, E., Rona, J.-P., Bardat, F., Cornel, D., Sotta, B., and Miginiac, E. (1999) Induction of *RAB18* gene expression and activation of K+ outward rectifying channels depend on an extracellular perception of ABA in *Arabidopsis thaliana* suspension cells. *Plant J.* 18: 13–22.

Kinoshita, T., Nishimura, M., and Shimazaki, K.-I. (1995) Cytosolic concentration of $Ca^{2+}$ regulates the plasma membrane H+-ATPase in guard cells of fava bean. *Plant Cell* 7: 1333–1342.

Koiwa, H., Barb, A. W., Xiong, L., Li, F., McCully, M. G., Lee, B. H., Sokolchik, I., Zhu, J., Gong, Z., Reddy, M., et al. (2002) C-terminal domain phosphatase-like family members (AtCPLs) differentially regulate *Arabidopsis thaliana* abiotic stress signaling, growth, and development. *Proc. Natl. Acad. Sci. USA* 99: 10893–10898.

Koornneef, M., Jorna, M. L., Brinkhorst-van der Swan, D. L. C., and Karssen, C. M. (1982) The isolation of abscisic acid (ABA) deficient mutants by selection of induced revertants in non-germinating gibberellin sensitive lines of *Arabidopsis thaliana* L. heynh. *Theor. Appl. Genet.* 61: 385–393.

Kwak, J. M., Mori, I. C., Pei, Z.-M., Leonhardt, N., Torres, M. A., Dangl, J. L., Bloom, R. E., Bodde, S., Jones, J. D. G., and Schroeder, J. I. (2003) NADPH oxidase *AtrbohD* and *AtrbohF* genes function in ROS-dependent ABA signaling in Arabidopsis. *EMBO J.* 22: 2623–2633.

Lee, Y., Choi, Y. B., Suh, S., Lee, J., Assmann, S. M., Joe, C. O., Kelleher, J. F., and Crain, R. C. (1996) Abscisic acid-induced phosphoinositide turnover in guard cell protoplasts of *Vicia faba*. *Plant Physiol.* 110: 987–996.

Li, S., Assmann, S. M., and Albert, R. (2006) Predicting essential components of signal transduction networks: A dynamic model of guard cell abscisic acid signaling. *PLoS Biol.* 4: e312.

Lu, C., Han, M. H., Guevara-Garcia, A., and Federow, N. V. (2002) Mitogen-activated protein kinase signaling in postgermination arrest of development by abscisic acid. *Proc. Natl. Acad. Sci. USA* 99: 15812–15817.

Luan, S., Kudla, J., Rodriguez-Concepcion, M., Yalovsky, S., and Gruissem, W. (2002) Calmodulins and calcineurin B–like proteins: Calcium sensors for specific signal response coupling in plants. *Plant Cell* 14: S389–S400.

Ma, Y., Szostkiewicz, I., Korte, A., Moes, D., Yang, Y., Christmann, A., and Grill, E. (2009) Regulators of PP2C phosphatase activity function as abscisic acid sensors. *Science* 324: 1064–1068.

McAinsh, M. R., Brownlee, C., and Hetherington, A. M. (1990) Abscisic acid-induced elevation of guard cell cytosolic $Ca^{2+}$ precedes stomatal closure. *Nature* 343: 186–188.

McAinsh, M. R., and Pittman, J. K. (2009) Shaping the calcium signature. *New Phytol.* 181: 275–294.

McCourt, P., and Creelman, R. (2008) The ABA receptors—we report you decide. *Curr. Opin. Plant Biol.* 11: 474–478.

Mishra, G., Zhang, W., Deng, F., Zhao, J., and Wang, X. (2006) A bifurcating pathway directs abscisic acid effects on stomatal closure and opening in Arabidopsis. *Science* 312: 264–266.

Miyazono, K.-I., Miyakawa, T., Sawano, Y., Kubota, K., Kang, H.-J., Asano, A., Miyauchi, Y., Takahashi, M., Zhi, Y., Fujita, Y., et al. (2009) Structural basis of abscisic acid signaling. *Nature* 462: 609–614.

Moes, D., Himmelbach, A., Korte, A., Haberer, G., and Grill, E. (2008) Nuclear localization of the mutant protein phosphatase abi1 is required for insensitivity towards ABA responses in Arabidopsis. *Plant J.* 54: 806–819.

Müller, A. H., and Hansson, M. (2009) The barley magnesium chelatase 150-kD subunit is not an abscisic acid receptor *Plant Physiol.* 150: 157–166.

Nishimura, N., Yoshida, T., Murayama, M., Asami, T., Shinozaki, K., and Hirayama, T. (2004) Isolation and characterization of novel mutants affecting the abscisic acid sensitivity of Arabidopsis germination and seedling growth. *Plant Cell Physiol.* 45: 1485–1499.

Nishimura, N., Hitomi, K., Arvai, A. S., Rambo, R. P., Hitomi, C., Cutler, S. R., Schroeder, J. I., and Getzoff, E. D. (2009) Structural mechanism of abscisic acid binding and signaling by dimeric PYR1. *Science* 326: 1373–1379.

Pandey, S., Nelson, D. C., and Assmann, S. M. (2009) Two novel GPCR-type G proteins are abscisic acid receptors in Arabidopsis. *Cell* 136: 136–148.

Park, S. Y., Fung, P., Nishimura, N., Jensen, D. R., Fujii, H., Zhao, Y., Lumba, S., Santiago, J., Rodrigues, A., Chow, T.-f. F., et al. (2009) Abscisic acid inhibits type 2C protein phosphatases via the PYR/PYL family of ABA binding START proteins. *Science* 324: 1068–1071.

Piskurewicz, U., Jikumaru, Y., Kinoshita, N., Nambara, E., Kamiya, Y., and Lopez-Molina, L. (2008) The gibberellic acid signaling repressor rgl2 inhibits *Arabidopsis* seed germination by stimulating abscisic acid synthesis and ABI5 activity. *Plant Cell* 20: 2729–2745.

Raschke, K., Shabahang, M., and Wolf, R. (2003) The slow and the quick anion conductance in whole guard cells: Their voltage-dependent alternation, and the modulation of their activities by abscisic acid and $CO_2$. *Planta* 217: 639–650.

Raz, V., Bergervoet, J. H. W., and Koornneef, M. (2001) Sequential steps for developmental arrest in *Arabidopsis* seeds. *Development* 128: 243–252.

Risk, J. M., Day, C. L., and Macknight, R. C. (2009) Reevaluation of abscisic acid-binding assays shows that G-protein-Coupled Receptor2 does not bind abscisic acid. *Plant Physiol.* 150: 6–11.

Saab, I. N., Sharp, R. E., Pritchard, J., and Voetberg, G. S. (1990) Increased endogenous abscisic acid maintains primary root growth and inhibits shoot growth of maize seedlings at low water potentials. *Plant Physiol.* 93: 1329–1336.

Sanchez, J.-P., and Chua, N.-H. (2001) *Arabidopsis* PLC1 is required for secondary responses to abscisic acid signals. *Plant Cell* 13: 1143–1154.

Schroeder, J. I., and Hagiwara, S. (1990) Repetetive increases in cytosolic $Ca^{2+}$ of guard cells by abscisic acid activation of nonselective $Ca^{2+}$ permeable channels. *Proc. Natl. Acad. Sci. USA* 87: 9305–9309.

Schroeder, J. I., Allen, G. J., Hugouvieux, V., Kwak, J. M., and Waner, D. (2001) Guard cell signal transduction. *Annu. Rev. Plant Phys. Plant Mol. Biol.* 52: 627–658.

Schultz, T. F., and Quatrano, R. S. (1997) Evidence for surface perception of abscisic acid by rice suspension cells as assayed by Em gene expression. *Plant Sci.* 130: 63–71.

Schurr, U., Gollan, T., and Schulze, E.-D. (1992) Stomatal response to drying soil in relation to changes in the xylem sap composition of *Helianthus annuus*. II. Stomatal sensitivity to abscisic acid imported from the xylem sap. *Plant Cell Environ.* 15: 561–567.

Schwartz, A., Wu, W.-H., Tucker, E. B., and Assmann, S. M. (1994) Inhibition of inward $K^+$ channels and stomatal response by abscisic acid: An intracellular locus of phytohormone action. *Proc. Natl. Acad. Sci. USA* 91: 4019–4023.

Schweighofer, A., Hirt, H., and Meskiene, I. (2004) Plant PP2C phosphatases: Emerging functions in stress signaling. *Trends Plant Sci.* 9: 236–243.

Sharp, R. E. (2002) Interaction with ethylene: Changing views on the role of abscisic acid in root and shoot growth responses to water stress. *Plant Cell Environ.* 25: 211–222.

Shen, Y. Y., Wang, X. F., Wu, F. Q., Du, S. Y., Cao, Z., Shang, Y., Wang, X. L., Peng, C. C., Yu, X. C., Zhu, S. Y., et al. (2006) The Mg-chelatase H subunit is an abscisic acid receptor. *Nature* 443: 823–826.

Shimazaki, K., Iino, M., and Zeiger, E. (1986) Blue light–dependent proton extrusion by guard cell protoplasts of *Vicia faba*. *Nature* 319: 324–326.

Shinozaki, K., Yamaguchi-Shinozaki, K., and Seki, M. (2003) Regulatory network of gene expression in the drought and cold stress responses. *Curr. Opin. Plant Biol.* 6: 410–417.

Siegel, R. S., Xue, S., Murata, Y., Yang, Y., Nishimura, N., Wang, A., and Schroeder, J. I. (2009) Calcium elevation-dependent and attenuated resting calcium-dependent abscisic acid induction of stomatal closure and abscisic acid-induced enhancement of calcium sensitivities of S-type anion and inward-rectifying K+ channels in Arabidopsis guard cells. *Plant J.* 59: 207–220.

Steber, C. M., Cooney, S. F., and McCourt, P. (1998) Isolation of the GA-response mutant *sly1* as a suppressor of *ABI1-1* in *Arabidopsis thaliana*. *Genetics* 149: 509–521.

Vartanian, N., Marcotte, L., and Giraudat, J. (1994) Drought rhizogenesis in *Arabidopsis thaliana*. *Plant Physiol.* 104: 761–767.

Vazquez, F., Gasciolli, V., Crete, P., and Vaucheret, H. (2004) The nuclear dsRNA binding protein HYL1 is required for micro RNA accumulation and plant development, but not posttranscriptional transgene silencing. *Curr. Biol.* 14: 346–351.

Wang, X.-Q., Ullah, H., Jones, A. M., and Assmann, S. M. (2001) G protein regulation of ion channels and abscisic acid signaling in *Arabidopsis* guard cells. *Science* 292: 2070–2072.

Wang, Y., Ying, J., Kuzma, M., Chalifoux, M., Sample, A., McArthur, C., Uchacz, T., Sarvas, C., Wan, J., Dennis, D. T., et al. (2005) Molecular tailoring of farnesylation for plant drought tolerance and yield protection. *Plant J.* 43: 413–424.

Wasilewska, A., Vlad, F., Sirichandra, C., Redko, Y., Jammes, F., Valon, C., Frei dit Frey, N., and Leung, J. (2008) An update on abscisic acid signaling in plants and more… *Mol. Plant* 1: 198–217.

White, C. N., Proebsting, W. M., Hedden, P., and Rivin, C. J. (2000) Gibberellins and seed development in maize. I. Evidence that gibberellin/abscisic acid balance governs germination versus maturation pathways. *Plant Physiol.* 122: 1081–1088.

Wilkinson, S., and Davies, W. J. (1997) Xylem sap pH increase: A drought signal received at the apoplastic face of the guard cell that involves the suppression of saturable abscisic acid uptake by the epidermal symplast. *Plant Physiol.* 113: 559–573.

Wilson, P. B., Estavillo, G. M., Field, K. J., Pornsiriwong, W., Carroll, A. J., Howell, K. A., Woo, N. S., Lake, J. A., Smith, S. M., Harvey Millar, A., et al. (2009) The nucleotidase/phosphatase SAL1 is a negative regulator of drought tolerance in Arabidopsis. *Plant J.* 58: 299–317.

Xiong, L., Lee, H., Ishitani, M., Zhang, C., and Zhu, J.-K. (2001) *FIERY1* encoding an inositol polyphosphate 1-phosphatase is a negative regulator of abscisic acid and stress signaling in *Arabidopsis*. *Genes Dev.* 15: 1971–1984.

Yamazaki, D., Yoshida, S., Asami, T., and Kuchitsu, K. (2003) Visualization of abscisic acid-perception sites on the plasma membrane of stomatal guard cells. *Plant J.* 35: 129–139.

Young, J. J., Mehta, S., Israelsson, M., Godoski, J., Grill, E., and Schroeder, J. I. (2006) $CO_2$ signaling in guard cells: Calcium sensitivity response modulation, a $Ca^{2+}$-independent phase, and $CO_2$ insensitivity of the *gca2* mutant. *Proc. Natl. Acad. Sci. USA* 103: 7506–7511.

Zhang, W., Yu, L., Zhang, Y., and Wang, X. (2005) Phospholipase D in the signaling networks of plant response to abscisic acid and reactive oxygen species. *Biochim. Biophys. Acta* 1736: 1–9.

Zheng, Z. L., Nafisi, M., Tam, A., Li, H., Crowell, D. N., Chary, S. N., Schroeder, J. I., Shen, J., and Yang, Z. (2002) Plasma membrane-associated ROP10 small GTPase is a specific negative regulator of abscisic acid responses in *Arabidopsis*. *Plant Cell* 14: 2787–2797.

Zhu, S.-Y., Yu, X.-C., Wang, X.-J., Zhao, R., Li, Y., Fan, R.-C., Shang, Y., Du, S.-Y., Wang, X.-F., Wu, F.-Q., et al. (2007) Two calcium-dependent protein kinases, CPK4 and CPK11, regulate abscisic acid signal transduction in Arabidopsis. *Plant Cell* 19: 3019–3036.

CHAPTER

# 24

# Brassinosteroids: Regulators of Cell Expansion and Development

Steroid hormones have long been known in animals, but they have only recently been discovered in plants. Animal steroid hormones include the sex hormones (estrogens, androgens, and progestins) and the adrenal cortex hormones (glucocorticoids and mineralocorticoids). The **brassinosteroids (BRs)** are a group of steroid hormones that play pivotal roles in a wide range of developmental phenomena in plants, including cell division and cell elongation in stems and roots, photomorphogenesis, reproductive development, leaf senescence, and stress responses (Clouse and Sasse 1998).

The identification of plant steroid hormones was the result of nearly 30 years of efforts to identify novel growth-promoting substances in pollen from many different plant species (Steffens 1991). Early studies by J. W. Mitchell and colleagues showed that the greatest growth-stimulating activity was found in the organic solvent extract of pollen from the rape plant (*Brassica napus L.*). The unidentified active compounds in rape pollen were named *brassins* (Mitchell et al. 1970).

The specific growth-promoting effects of the brassins were scored in several physiological tests, including the *bean second-internode bioassay*. In this assay, brassins behaved differently than other known phytohormones, causing both cell elongation and cell division, as well as bending, swelling, and splitting of the second internode (Mandava 1988).

Because of their ability to cause dramatic changes in growth and differentiation at low concentrations, Mitchell et al. (1970) proposed that brassins constituted a new family of plant hormones. Further work demonstrated that brassins not only induced stem elongation, they also increased total biomass and seed yield. Foreseeing likely practical applications, the U. S. Department of Agriculture provided funding for laboratories to purify and identify the active compounds in brassins.

Eventually, from 227 kg of bee-collected rape pollen, investigators were able to purify 10 mg of the most bioactive brassin compound, which they named *brassinolide* (Grove et al. 1979). X-ray analysis of the crystal structure and spectroscopic studies revealed brassinolide to be a polyhydroxylated steroid similar to animal steroid hormones.

Three years later, Japanese scientists purified another phytosteroid, castasterone—thought to be the precursor of brassinolide—from chestnut galls (Yokota et al. 1982). Shortly after, the same group identified a mixture of biologically active brassinolide-like substances in the broadleaf evergreen tree *Dystilium racemosum* (Abe and Marumo 1991).

Although brassinosteroids were known to be endogenous compounds that produced dramatic growth effects in bioassays such as the bean second-internode bioassay, they were not immediately accepted as plant hormones, because their role in normal plant growth and development remained elusive for many years. It was genetic studies in Arabidopsis in the mid-1990s that finally demonstrated that brassinosteroids are authentic plant hormones that participate with other plant hormones in the regulation of numerous aspects of plant development, including shoot growth, root growth, vascular differentiation, pollen tube growth, and seed germination (Clouse and Sasse 1998).

We will begin our discussion of brassinosteroids with a brief description of the chemical structure of the BRs and the genetic experiments that led to their identification as plant hormones. Next we will examine the signaling pathways of BRs, from their receptor to their target genes. We will next review the pathways for their biosynthesis, metabolism, and transport, and then describe some of the important physiological processes affected by this group of hormones. Finally, we will give a brief account of the potential applications of brassinosteroids in agriculture.

## Brassinosteroid Structure, Occurrence, and Genetic Analysis

Two main bioassays have been used to purify brassinosteroids: the bean second-internode bioassay (**FIGURE 24.1**) and the rice lamina (leaf) inclination bioassay (**FIGURE 24.2**). These bioassays distinguish between biologically active BRs and their inactive intermediates or metabolites, and allow quantitation of the amount of active compound present (**FIGURE 24.3**).

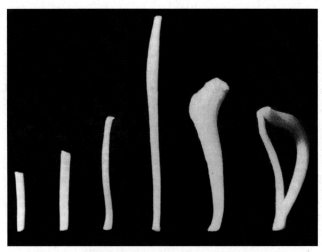

Increasing brassinosteroid concentration ⟶

**FIGURE 24.1**  Bean second-internode bioassay for brassinosteroids. Excised sections from the second internodes of bean plants were floated on solutions containing increasing concentrations of BRs for several days. The untreated control section is on the left. At low concentrations, BRs induce mainly elongation growth. Higher concentrations result in stem thickening, bending, and splitting. (From Mandava 1988.)

The basic chemical structure of a brassinosteroid was first resolved in 1979, when X-ray crystallographic analysis of the purified active substance determined that it was a steroidal lactone (**FIGURE 24.4**) (Grove et al. 1979). The compound was named **brassinolide** (**BL**), and its characterization ultimately led to the chemical identification of a group of about 60 related phytosteroids called brassinosteroids (Fujioka and Yokota 2003). The direct biosynthetic precursor of BL, **castasterone** (**CS**), has weak BR activity as well.

Brassinosteroids have now been identified in 27 families of seed plants (including both angiosperms and gymnosperms), one pteridophyte (*Equisetum arvense*), one bryophyte (*Marchantia polymorpha*), and one green alga (*Hydrodictyon reticulatum*). Thus, brassinosteroids appear to be ubiquitous plant hormones that predate the evolution of land plants. In the angiosperms, BRs are found at low levels in pollen, anthers, seeds, leaves, stems, roots, flowers, and young vegetative tissues.

Knowing the molecular structure of BL allowed investigators to synthesize both naturally occurring BRs and their related analogs. By testing these compounds with either the bean second internode or rice lamina inclination bioassay, the following key requirements for BR activity were determined (Mandava 1988):

- A *cis*-vicinal glycol function at C-2 and C-3 of ring A. The absence of a single hydroxyl group at either C-2 or C-3 or any change in their configuration results in a significant loss of function.

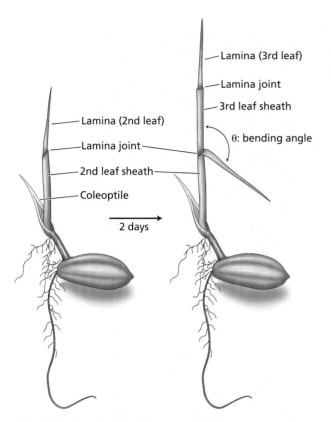

**FIGURE 24.2** Dwarf rice lamina inclination bioassay for brassinosteroids. A small droplet of sample dissolved in ethanol is applied to the joint between the lamina and the leaf sheath. After incubation for 2 days in high humidity, the bending angle *theta* (θ) between the lamina and the leaf sheath is measured. The angle is proportional to the amount of brassinosteroid in the sample.

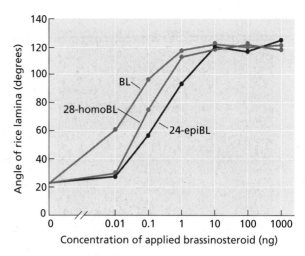

**FIGURE 24.3** Dose–response curves for three active BRs in the rice lamina inclination bioassay. The angle of inclination increases as a function of BR concentration, up to 10 ng. Note the relatively high activity of BL. 24-epiBL, 24-epibrassinolide; 28-homoBL, 28-homobrassinolide; BL, brassinolide.

- The seven-membered B-ring lactone. Although limited modifications of the B-ring don't eliminate activity (see, for example, the B-ring of castasterone, Figure 24.10), they significantly reduce activity.

- The steroid side chain with hydroxyl group at C-22 and C-23. The α orientation at C-22 and C-23 confers higher activity than the corresponding β orientation.

Variations in the alkyl side chain at C-24 are tolerated, although a reduction in activity is usually detected, as in the case of **24-epibrassinolide (24-epiBL)** and **28-homobrassinolide (28-homoBL)** (see Figures 24.3 and 24.4). The different compounds can be classified as $C_{27}$, $C_{28}$, or $C_{29}$ BRs, depending on the structures of their side chains. Since 24-epiBL can be synthesized more cheaply than brassinolide, 24-epiBL is often used in physiological experiments in preference to BL, although it is only 10% as active as brassinolide in most bioassays.

## BR-deficient mutants are impaired in photomorphogenesis

Definitive proof that brassinosteroids function as plant hormones came only during the past decade from genetic analyses in Arabidopsis, which led to the isolation and description of mutants defective in BR biosynthesis and perception. The abnormal phenotypes of these

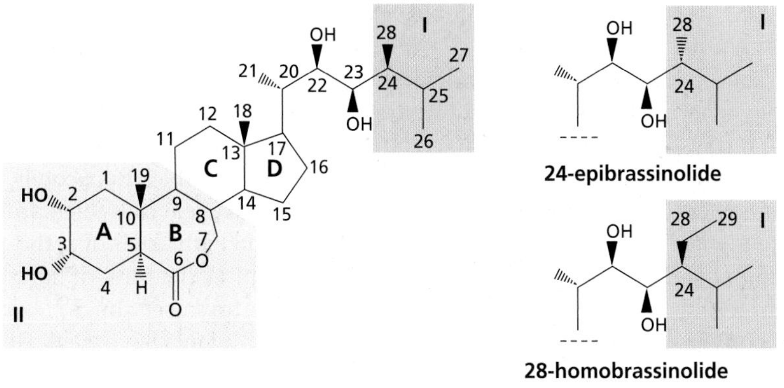

**FIGURE 24.4** The structures of brassinosteroids. Brassinolide (BL) is the most widespread and active BR in plants. The structure of BL is shown with its carbons numbered and the ring types indicated by letters. Regions where variations occur are indicated by roman numerals I and II. Variations in the side chain (region I) are present in 24-epibrassinolide and 28-homobrassinolide, but do not significantly affect activity. The hydroxyls on the side chain at carbons 22 and 23 are both essential for activity.

**24-epibrassinolide**

**28-homobrassinolide**

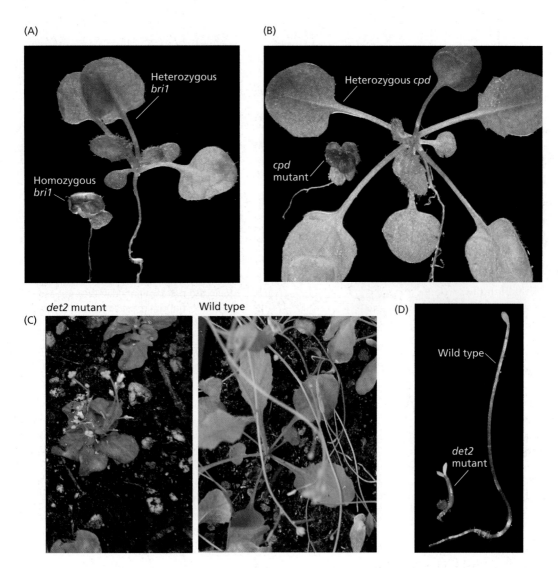

(A)

Heterozygous *bri1*

Homozygous *bri1*

(B)

Heterozygous *cpd*

*cpd* mutant

(C)   *det2* mutant          Wild type

(D)

Wild type

*det2* mutant

**FIGURE 24.5** Phenotypes of Arabidopsis BR mutants. (A) The 3-week-old light-grown homozygous *bri1* mutant (left) is a severe dwarf compared to the heterozygous *bri1* mutant (right), which exhibits wild-type morphology. (B) The 3-week-old light-grown homozygous *cpd* mutant (left) also exhibits a dwarf phenotype; the heterozygous mutant with a wild-type phenotype is on the right. (C) The light-grown adult *det2* mutant (left) is dwarfed compared to the wild-type plant (right). (D) The dark-grown *det2* on the left has a short, thick hypocotyls and expanded cotyledons; the dark-grown wild type is on the right. (Courtesy of S. Savaldi-Goldstein.)

mutants demonstrated that BRs were required for normal development.

The first characterized BR-deficient mutants, *det2* (*de-etiolated 2*) and *cpd* (*constitutive photomorphogenesis and dwarfism*), were identified in screens for Arabidopsis seedlings that have a light-grown morphology (that is, they are de-etiolated) after growing for several days in total darkness (**FIGURE 24.5**) (Li et al. 1996; Szekeres et al. 1996). Both mutants are impaired in brassinosteroid biosynthesis. The *DET2* locus encodes a protein with high amino acid sequence identity to that of mammalian steroid 5α-reductases (Li et al. 1996). Mammalian steroid 5α-reductases catalyze an NADPH-dependent conver-

sion of testosterone to dihydrotestosterone, a key step in steroid metabolism that is essential for normal embryonic development of male external genitalia and the prostate. Likewise, the *CPD* gene encodes a protein homologous to mammalian cytochrome P450 monooxygenase enzymes, including steroid hydroxylases.

Seedlings of *det2* and *cpd* mutants have an impaired photomorphogenic response, with short, thick hypocotyls, expanded cotyledons, young primary leaves (which are absent in dark-grown seedlings), and high levels of anthocyanins (all of which in the wild type are features of light-grown seedlings but not of dark-grown seedlings) (see Figure 24.5D). In addition, the two mutants have elevated levels of light-regulated mRNAs when grown in the dark. In contrast, dark-grown wild-type seedlings exhibit a typical etiolated phenotype (long hypocotyls, folded cotyledons, and an absence of anthocyanin pigment).

In addition to their atypical dark-grown phenotype, *det2* and *cpd* have an abnormal phenotype when grown in the light. Both mutants grow as dark-green dwarfs due to a reduction in cell size and intercellular air spaces (see Figure 23.5B and C), and they have reduced apical dominance (see Chapter 19) and male fertility. In addition, the *det2* and *cpd* mutants exhibit short roots, delayed flowering, and delayed leaf senescence even after flowering. In general, the *cpd* mutants have a more extreme phenotype than the *det2* mutants. The reason the *cpd* mutants have a more extreme phenotype than *det2* mutants is that *det2* mutants contain a residual amount of active BRs, while the levels of active BRs in *cpd* mutants are virtually undetectable (Szekeres et al. 1996). Treating the dwarf mutants with exogenous brassinolide restores the normal phenotype, providing further evidence that DET2 and CPD are required for both brassinosteroid production and normal photomorphogenesis (FIGURE 24.6).

## The Brassinosteroid Signaling Pathway

Genetic analyses of BR responses identified a plasma membrane-localized BR receptor protein and many other signaling elements in the BR signaling pathway. As a result, enormous progress has been achieved in understanding the molecular basis of BR action. Here we will discuss the main components of the BR signal transduction pathway, focusing on results obtained with Arabidopsis. (For comparison of the BR pathway to other signal transduction mechanisms see Chapter 14.)

### BR-insensitive mutants identified the BR cell surface receptor

To identify components of the BR signaling pathway in Arabidopsis, genetic screens were initially carried out to isolate mutants exhibiting normal root elongation in the presence of high BL concentrations. This resulted in the identification of a single *bri1* (*brassinosteroid-insensitive 1*) mutant (Clouse et al. 1996). Further screens for BL-insensitive mutations all yielded additional mutant alleles of *BRI1* (e.g., Li and Chory 1997), suggesting that BRI1 is an essential component of the BR signaling pathway. Subse-

(A) *cpd*

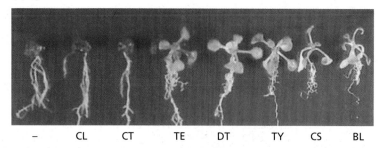

(B) Wild type

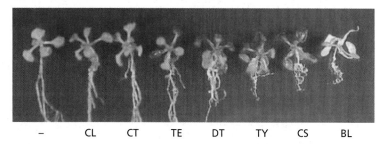

**FIGURE 24.6** BL and intermediates of the BL biosynthetic pathway (see Figure 24.10 and Appendix 3) restore normal growth to the *cpd* mutant. Wild-type and *cpd* mutant seedlings were grown for 14 days with no steroid (minus sign) or with 0.2 μM of BR intermediate compounds. Note that neither campesterol (CL) nor cathasterone (CT) has any effect on the *cpd* mutant phenotype, because these intermediates occur prior to the reaction catalyzed by CPD. In contrast, teasterone (TE), 3-dehydroteasterone (DT), typhasterol (TY), castasterone (CS), and brassinolide (BL) all rescue the phenotype because they occur after the CPD-catalyzed reaction. The wild type already contains optimal levels of these intermediates and is therefore slightly inhibited by BL and its immediate precursors. (From Szekeres et al. 1996.)

quent binding studies demonstrated that BL binds directly to BRI1 with high specificity, indicating that BRI1 is the BR receptor (Kinoshita et al. 2005).

Brassinolide binds to the extracellular domain of the BRI1 receptor, a plasma membrane leucine-rich repeat (LRR)-receptor kinase (FIGURE 24.7). BRI1 is also found in early endosomes (subcellular compartments) and can signal from there (Geldner et al. 2007). Interestingly, as opposed to the serine/threonine (S/T) specificity of plant kinases characterized thus far, BRI1 is a dual-specificity kinase. As such, BRI1 autophosphorylates both S/T and tyrosine residues; both of these modifications are necessary for BRI1's activation (Oh et al. 2009).

The LRR-receptor kinases constitute the largest receptor class predicted in the Arabidopsis genome, with over 230 family members. This family has a conserved domain structure, composed of an N-terminal extracellular domain with multiple tandem (adjacent) LRR motifs, a single transmembrane domain, and a cytoplasmic kinase domain with

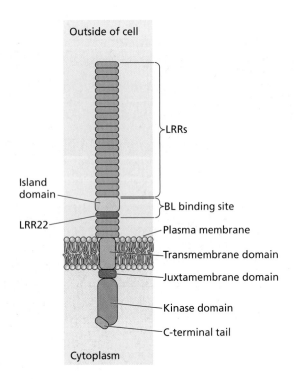

Outside of cell

LRRs

Island
domain

BL binding site

LRR22

Plasma membrane

Transmembrane domain

Juxtamembrane domain

Kinase domain

C-terminal tail

Cytoplasm

**FIGURE 24.7**   The domain structure of the BR receptor, BRI1. The receptor BRI1 is localized on the plasma membrane. The extracellular region consists of a stretch of leucine-rich repeat sequences (LRRs) containing an island domain that functions as part of the brassinolide (BL) binding site. The intracellular portion contains a juxtamembrane domain, a kinase domain, and the C-terminal tail.

specificity toward tyrosine, serine, and threonine residues (see Figure 24.7). In the case of BRI1, the number of LRRs is 25. BRI1 also has a unique feature that is required for BR binding: a stretch of amino acids called the *island domain* that interrupts the LRRs between LRRs 21 and 22 (Kinoshita et al. 2005). This domain plus the flanking LRR22 compose the minimum binding site for BRs.

### Phosphorylation activates the BRI1 receptor

Analysis of a large number of BR mutant alleles indicates that both the extracellular receptor domain and the internal kinase domain are necessary for transmitting the BR signal to the rest of the cell (Friedrichsen et al. 2000; Vert et al. 2005). BL binds BRI1 via a novel steroid-binding domain of about 100 amino acids that includes the island domain and its neighboring LRR sequence (see Figure 24.7). BL binding activates the receptor, as indicated by its increased autophosphorylating activity and by its increased association with a second LRR-receptor kinase, **BRI1-associated receptor kinase 1 (BAK1)** (FIGURE 24.8).

In the presence of brassinolide, BRI1 becomes phosphorylated in vivo at multiple intracellular domains, including the juxtamembrane* region (JM), the C-terminal tail (CT), and the kinase itself. These phosphorylation sites play regulatory roles in receptor activity and control the dissociation of BRI1's inhibitor **BKI1 (BRI1-kinase inhibitor 1)** from the plasma membrane and its interaction with other proteins, such as BAK1 and **BSK (BR-signaling kinase) proteins** (Wang et al. 2005a, b; Wang and Chory 2006; Tang et al. 2008).

As in the case of animal protein kinases, specific phosphorylation sites in the kinase domain of BRI1 are essential for its activation (see Figure 14.2). In addition, the CT of BRI1 negatively regulates the receptor. Upon ligand binding, this inhibitory effect is nullified, and BRI1 kinase activity increases (Wang et al. 2005b). However, the precise mechanism of this BL-induced activation will become clear only once a high-resolution structure for BRI1 has been determined (Wang et al. 2005b).

Receptor kinases in animal and plant cells often function as dimers in vivo. In vitro experiments have confirmed that BRI1 receptors normally function as homo-oligomers composed of identical monomers in the cell (Wang et al. 2005a).

Following BRI1's binding to and activation by its ligand, phosphorylated BRI1 forms a hetero-oligomer (i.e., composed of two different monomers) and phosphorylates BAK1, a second LRR kinase. Consequently, BAK1 is activated and transphosphorylates BRI1. The phosphorylated BRI1/BAK1 heterodimer appears to be the activated form of the receptor that, upon phosphorylation and activation of BSK proteins, induces the BR response by inactivating a repressor called BIN2.

### BIN2 is a repressor of BR-induced gene expression

The formation of the activated BRI1/BAK1 hetero-oligomer in the presence of BR initiates a signaling cascade that leads to BR-regulated gene transcription. The next step in the signal transduction pathway involves inactivation of the negative regulator **BIN2 (brassinosteroid insensitive-2)** (see Figure 24.8). *BIN2* encodes a protein kinase homologous to the glycogen synthase kinase 3 (GSK3) of yeast and animals (Li and Nam 2002). GSK3s function as constitutively active S/T kinases. They are involved in a wide range of signaling pathways in which they often act as repressors of gene expression.

In the absence of BRs, BIN2 phosphorylates two nuclear proteins, **BES1 (bri1-EMS-suppressor 1)** and **BZR1 (brassinazole-resistant 1)**, at multiple regulatory sites, thus inhibiting their activity. BES1 and BZR1 are closely related transcriptional regulators (sharing 90% identity in their amino acid sequence). They are short-lived proteins and are degraded by the 26S proteasome, a process that

---

*The side of the plasma membrane facing the cytoplasm.

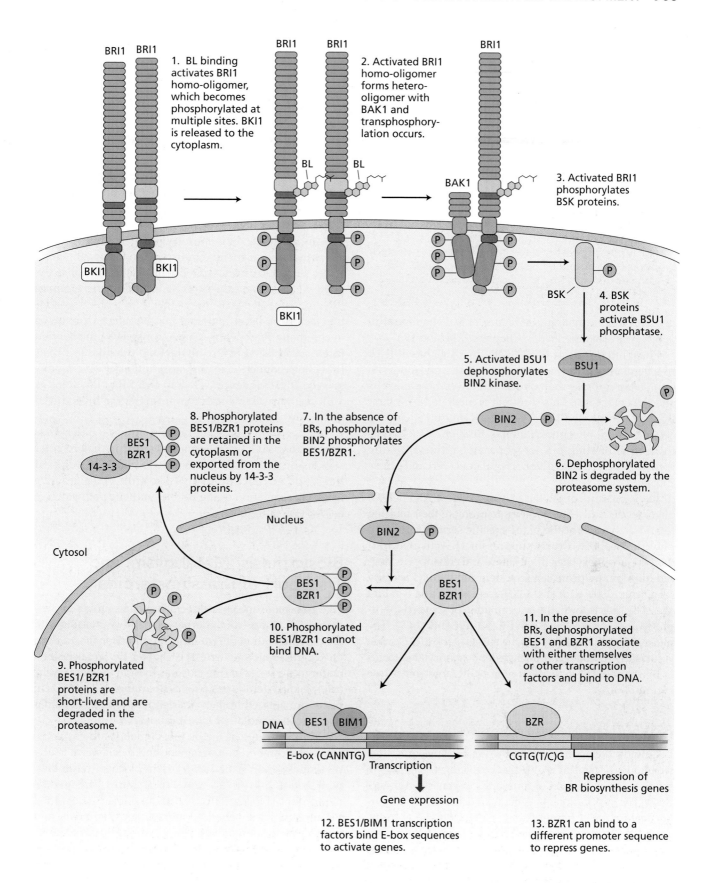

**FIGURE 24.8**  A model for BR signaling. Signal perception occurs at the cell surface.

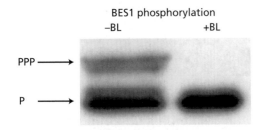

BES1 phosphorylation

−BL          +BL

PPP →

P →

**FIGURE 24.9** BL inhibits the phosphorylation of BES1. A Western blot was used to identify the two forms of BES1: the highly phosphorylated form (PPP) and the dephosphorylated form (P). In Arabidopsis plants not treated with brassinolide (−BL), much of the BES1 is in the highly phosphorylated state. In the brassinolide-treated plants (+BL) all of the BES1 is dephosphorylated. (Antibodies generated by Y. Yin; courtesy of G. Vert.)

involves ubiquitination (see Figure 24.8). (The ubiquitin pathway is also described in Chapters 2, 14, and 19.)

Phosphorylation of BES1/BZR1 by BIN2 has at least two regulatory roles. First, the BES1/BZR1 phosphorylation state regulates their shuttling between the nucleus and the cytoplasm. This shuttling is mediated by a family of proteins called 14-3-3 proteins (Gendron and Wang 2007). Second, phosphorylation of BES1/BZR1 prevents them from binding to a target promoter, thus blocking their activity as transcriptional regulators (Vert and Chory 2006).

In the presence of BR, the activated BRI1/BAK1 heterooligomer activates BSK proteins to promote their binding to and activation of the plant-specific serine/threonine phosphatase, **BSU1 (bri1 suppressor 1)**. Activated BSU then dephosphorylates BIN2 kinase and enhances its degradation by the proteasome system to block its activity (Peng et al. 2008; Kim et al. 2009) (see steps 4 and 5 in Figure 24.8). This leads to the accumulation of active, dephosphorylated forms of BES1 and BZR1 (**FIGURE 24.9**). The phosphatase that counteracts the effects of the BIN2 kinase is currently unknown. The active dephosphorylated forms of BES1 and BZR1 activate or repress BR target genes (see Figure 24.8).

### BES1/BZR1 regulate gene expression

Following the sequencing of the Arabidopsis genome, techniques became available that enabled investigators to monitor the expression of thousands of genes simultaneously. The application of techniques such as *DNA microarray analysis* (see Chapter 2 and WEB TOPIC 17.7) to the study of BR-regulated gene expression identified hundreds of BR-induced genes, many of which are predicted to play a role in growth processes. In addition, genes that are repressed by BRs were identified. Many of these downregulated genes are also controlled by the transcription factors BES1/BZR1 (see Figure 24.8) (Vert et al. 2005).

BES1 and BZR1 were identified by two independent genetic screens. When grown in the light, *bes1* mutants are larger than the wild type, similar to plants overexpressing DWF4 or BRI1. In contrast, light-grown *bzr1* mutants are semidwarf. When grown in the dark, *bes1* and *bzr1* mutants suppress the dwarfism of weak alleles of *bri1* and have normal hypocotyl lengths. This is because the *BES1* and *BZR1* mutations make the proteins they encode less susceptible to proteolysis. In genetic terms, *BES1* and *BZR1* are semidominant mutations resulting in a *gain* of function due to the accumulation of their corresponding protein products.

Despite their high sequence similarity, BES1 and BZR1 appear to regulate different subsets of genes in Arabidopsis, perhaps because these transcription factors are themselves subject to distinct spatio-temporal regulation in the plant. BES1 *enhances* the expression of a diverse group of BR-stimulated genes by interacting with target transcription factors of distinct families and with factors involved in chromatin remodeling. One example is a transcription factor called **BIM1 (BES1-interacting Myc-like 1)**. BES1/BIM1 heterodimers activate transcription by binding to a specific DNA sequence called an *E Box* that functions as a BR response element in the promoters of BR-induced genes (Yin et al. 2005).

BZR1 acts as a repressor of BR biosynthesis. BZR1 binds directly to CGTG(T/C)G elements found in the promoter regions of various BR biosynthetic genes, turning off transcription. BZR1 thus plays a key role in the negative feedback regulation of the BR biosynthetic pathway (see below) (He et al. 2005).

## Biosynthesis, Metabolism, and Transport of Brassinosteroids

Like gibberellin and abscisic acid, brassinosteroids are synthesized as a branch of the terpenoid pathway, starting with the polymerization of two farnesyl diphosphates to form the $C_{30}$ triterpene **squalene** (see Chapter 13). Squalene then undergoes a series of ring closures to form the pentacyclic triterpenoid (sterol) precursor **cycloartenol**. All steroids in plants are derived from cycloartenol by a series of oxidation reactions and other modifications.

Our knowledge of the BL biosynthetic pathway is the result of a combination of genetic and biochemical analyses (Fujioka and Yokota 2003). For the biochemical studies, periwinkle (*Catharanthus roseus*) cell cultures were used, as they produce BRs in relatively high amounts. Radiolabeled BR intermediates were used in feeding experiments, and their metabolic derivatives were identified by gas chromatography–mass spectroscopy analysis. Coupling this type of analysis to genetic studies with BR-deficient mutants in Arabidopsis, tomato, and other species has allowed the identification of the complete biosynthetic pathways. This section presents and discusses these discoveries.

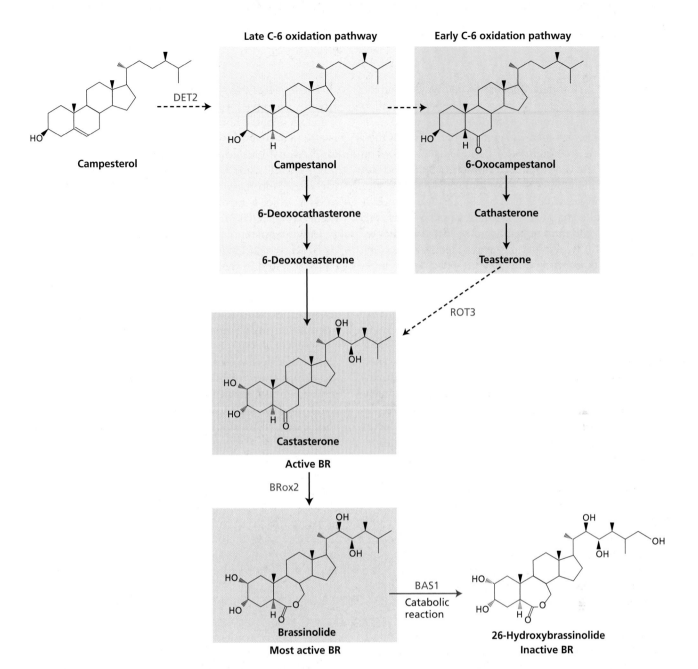

**FIGURE 24.10** Simplified pathways for brassinolide (BL) biosynthesis and catabolism. The precursor for BL biosynthesis is campesterol. Black arrows represent the sequence of biosynthetic events. Solid arrows indicate single reactions and dashed arrows represent multiple reactions. As shown, castasterone, the immediate precursor of BL, can be synthesized from two parallel pathways: the early and the late C-6 oxidation pathways. In the early C-6 oxidation pathway, oxidation at C-6 of the B ring occurs *before* the addition of vicinal hydroxyls at C-22 and C-23 of the side chain (refer to BL structure in Figure 24.3). In the late C-6 oxidation pathway, C-6 is oxidized *after* the introduction of hydroxyls at the side chain and C-2 of the A ring (further details may be found in Appendix 3). Both the early and the late pathways may be linked at various points, creating a biosynthetic network rather than a linear pathway. The Arabidopsis enzymes that catalyze the different steps are indicated. BL catabolism is indicated by a red arrow.

## Brassinolide is synthesized from campesterol

Differing by their C-24 alkyl substitutions, brassinosteroids are synthesized from campesterol, sitosterol, and cholesterol. While campesterol and sitosterol are abundant in plant membranes, cholesterol is present at relatively low levels. All three sterols are metabolized to a large number of intermediates in plant cells, but only a few of these metabolites have biological activity (Clouse and Sasse 1998; Sakurai 1999).

We will illustrate a simplified version of the BR biosynthetic pathway starting with the sterol progenitor **campesterol**, which is ultimately derived from cycloartenol (**FIGURE 24.10**). Campesterol is first converted to campestanol in steps involving DET2. Campestanol is then converted to castasterone (CS) through either of two pathways called the *early-* and *late C-6 oxidation pathways*. Additional information about these two pathways is provided in Appendix 3.

The two pathways merge at castasterone, which is then converted to BL (see Figures 24.4 and Figure 24.10). The early and the late C-6 oxidation pathways coexist and can be linked at different points in Arabidopsis, pea, and rice (Fujioka and Yokota 2003). The biological significance of having two linked pathways is currently unknown. In fact, the early C-6 oxidation branch is not detected in tomato. The presence of two linked pathways increases the complexity of BR biosynthesis and may provide an advantage under different physiological conditions, such as various types of stress.

All the mutants impaired in their ability to convert campestanol to BL have mutations in genes that encode **cytochrome P450 monooxygenases (CYPs)**. The Arabidopsis *DWARF4* (*DWF4*) and *CPD* genes encode two such monooxygenases, CYP90B1 and CYP90A1, which hydroxylate BR intermediates at the 22 and 23 positions, respectively (Fujioka and Yokota 2003) (see Figures 24.4 and 24.10).

CYP proteins involved in BR biosynthesis are located in the endoplasmic reticulum (ER), similar to the cytochrome P450 monooxygenases involved in gibberellin biosynthesis (see Chapter 20). Hence, it is likely that BL biosynthesis also takes place in the ER.

## Catabolism and negative feedback contribute to BR homeostasis

The level of active BRs is also regulated by metabolic processes that inactivate BL. Several types of reactions result in BL inactivation, including epimerization, oxidation, hydroxylation, sulfonation, and conjugation to glucose or lipids (Fujioka and Yokota 2003). Our limited knowledge in this area is based on experiments in which plants are fed radiolabeled BRs, and the resulting labeled products are identified and endogenous metabolites analyzed; however, their relevance for the BR pathway in the plant is still not clear.

The isolation of the Arabidopsis *BAS1* (*phyB a*ctivation-tagged *s*uppressor1-dominant) gene encoding a cytochrome P450 monooxygenase with steroid 26-hydroxylase activity (CYP72B1) has helped to elucidate the role of at least one metabolic enzyme in controlling BL concentrations. Overexpression of *BAS1* leads to decreased BL levels and an accumulation of an inactive 26-hydroxyBL (for 26-hydroxyBL structure, see Figure 24.6), resulting in a BR-deficient dwarf phenotype (Neff et al. 1999).

The levels of physiologically active BR are also regulated by negative feedback mechanisms. In other words, if an excess of the hormone accumulates, BR biosynthesis is attenuated and BR turnover is enhanced. Indeed, the mRNAs of all tested Arabidopsis BL biosynthetic genes (*DWF4*, *CPD*, *ROT3*, and *BR6ox1*) are down-regulated (decreased) in response to BL application, while *BAS1* mRNA, which is involved in BR turnover, accumulates to higher levels (**FIGURE 24.11**) (Tanaka et al. 2005).

Down-regulation of BR biosynthetic genes is mediated by BZR1, which binds directly to a conserved promoter element found in the above biosynthetic genes, thereby repressing their expression (see Figure 24.8) (He et al. 2005). Accordingly, Arabidopsis mutants impaired in their ability to respond to BL accumulate high levels of the active brassinosteroids CS and BL compared to wild-type plants (Noguchi et al. 1999).

A valuable tool for the genetic, physiological, and molecular study of BRs is the specific BR biosynthesis inhibitor, **brassinazole (Brz)** (**FIGURE 24.12**). Brz contains a triazole ring made up of two carbon and three nitrogen atoms. Various triazole compounds can act as inhibitors of cytochrome P450 monooxygenases. The triazole Brz spe-

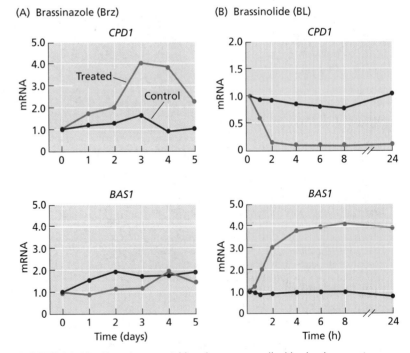

**FIGURE 24.11** Brassinosteroid levels are controlled by both negative and positive feedback. The mRNA levels of *CPD1* (BR synthesis) and *BAS1* (BR catabolism) were measured in Arabidopsis seedlings treated either (A) with 5 µM brassinazole (Brz); or (B) with 5 µM Brz (to deplete the endogenous BL levels), followed by 0.1 µM BL for 2 days. Expression of the BR biosynthetic gene *CPD1* is enhanced by Brz (A) and inhibited by BL (B). *CPD1* is thus negatively regulated by BL. In contrast, the expression of the BR-degrading enzyme BAS1 is stimulated by BL. *BAS1* is thus positively regulated by BL. (After Tanaka et al. 2005.)

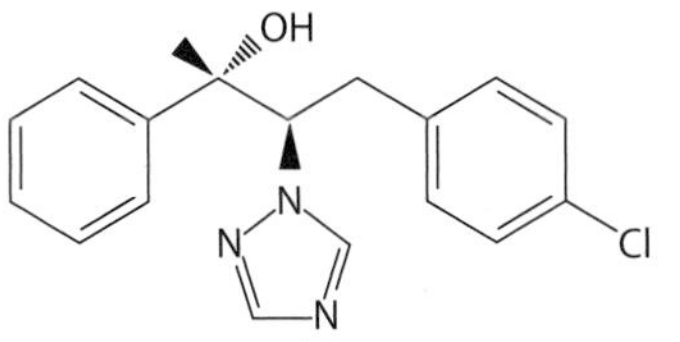

**FIGURE 24.12** The structure of brassinazole [4-(4-chloro-phenyl)-2-phenyl-3-(1,2,4-triazoyl)butan-2-ol], a triazole compound that inhibits brassinosteroid biosynthesis.

cifically inhibits the activity of the BL biosynthetic enzyme DWF4 (monooxygenase CYP90B1), which catalyses the hydroxylation reaction at C-22 (Asami et al. 2003). Plants grown on Brz show BR-deficient phenotypes, which can be reversed by the addition of BL to their growth medium (**FIGURE 24.13**). In this experiment both Brz and BL are taken up by the root system.

Numerous studies using Brz have yielded important information on BR homeostasis that complements the studies with BL described above. Thus, several BR biosynthetic genes were up-regulated in BR-depleted Arabidopsis plants grown in the presence of Brz (see Figure 24.11). Taken together, the results suggest that BR homeostasis is maintained by feedback regulation of multiple target genes (Tanaka et al. 2005).

BR homeostasis is also controlled by rate-limiting steps in the BL biosynthetic pathway. If an enzyme is rate-limiting, significant levels of its substrate should accumulate relative to its immediate product. Measurements of endogenous BRs in Arabidopsis have shown that CPD, DWF4, and BR6ox1/2 could be rate-limiting steps in BR biosynthesis and thus contribute to BR homeostasis. Indeed, overexpression of DWF4 and BR6ox2 results in increased vegetative growth of the plant (**FIGURE 24.14**) (Choe et al. 2001; Kim et al. 2005).

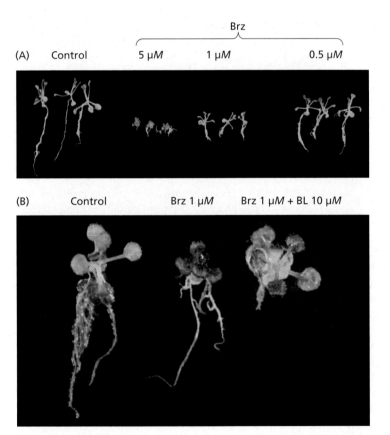

**FIGURE 24.13** The effect of brassinazole (Brz) on light-grown 14-day-old Arabidopsis seedlings. (A) Control seedlings (left) and 5 µM, 1 µM, and 0.5 µM brassinazole-treated Arabidopsis seedlings (right). The Brz-treated seedlings exhibit dwarfism in a concentration-dependent manner. (B) Control light-grown 14-day-old Arabidopsis seedling (left) and seedlings treated with 1 µM brassinazole (middle) or 1 µM brassinazole plus 10 nM brassinolide (BL) (right). Brz inhibits the BL biosynthetic enzyme DWF4. The addition of BL, which occurs downstream of the DWF4 catalytic reaction, alleviates the inhibitory effect of Brz on growth. (From Asami et al. 2000.)

**FIGURE 24.14** Overexpression of the BR biosynthetic gene, *DWF4*, in Arabidopsis results in a dramatic increase in plant size. Plants were grown for 25 days. (From Choe et al. 2001.)

## Brassinosteroids act locally near their sites of synthesis

An important determinant of hormone responses in general is the extent and rate of hormone transport from the site of synthesis to the site of action. Exogenously applied 24-epi-brassinolide (24-epiBL) undergoes long-distance transport from the root to the shoot. For example, when roots of cucumber, tomato, or wheat plants were treated with [14]C-24-epiBL, the radioactivity was readily translocated to the shoot (Schlagnhaufer and Arteca 1991; Nishikawa et al. 1994). Moreover, the dwarf phenotype of the BR-deficient Arabidopsis mutants could be restored (rescued) back to wild-type size, when grown on agar media supplemented with BL. In addition, the leaf petioles of wild-type plants elongated upon BL application to their root systems (Clouse and Sasse 1998).

In contrast, when [14]C-24-epiBL was applied to the upper surface of a young cucumber leaf, it was readily taken up, but was only slowly transported out of the leaf. In all, only about 6% of the applied [14]C-24-epiBL was transported to the younger leaves (Nishikawa et al. 1994). These results suggest that *exogenous* BRs are readily translocated from the root to the shoot, but are poorly translocated out of leaves. Presumably, 24-epiBL that is taken up by the roots moves to the shoot via the xylem transpiration stream. Since the xylem stream is unidirectional, however, exogenous 24-epiBL applied to the leaf can only exit the leaf via the phloem. The lack of 24-epiBL movement out of the leaf indicates that it is poorly translocated in the phloem.

Moreover, despite the evidence for root-to-shoot transport of 24-epiBL, *endogenous* BRs do not seem to undergo root-to-shoot translocation. For example, experiments in pea and tomato indicate that reciprocal stock/scion graftings of wild type to BR-deficient mutants do not rescue the phenotype of the latter in either the acropetal or the basipetal direction (**FIGURE 24.15**) (Symons and Reid 2004; Montoya et al. 2005). Furthermore, comparisons of the temporal and spatial distribution of BR intermediates indicate that they are present in all plant organs, although different intermediates predominate in different organs (Shimada et al. 2003).

In parallel with the ubiquity of the BR biosynthetic pathway, components of the BR signaling pathway (see above) also appear to be expressed throughout the plant, especially in young growing tissues (Friedrichsen et al. 2000). Within the shoot, BR activity in the epidermis plays a dominant role in regulating organ growth: Epidermal expression of CPD and BRI1 is sufficient to drive growth of the inner tissues, while epidermal expression of the BR catabolic enzyme BAS1 restricts organ growth (Savaldi-Goldstein et al. 2007). In summary, these results suggest that (1) endogenous BRs act locally, at or near their sites of synthesis; and (2) each organ synthesizes and responds to its own active BRs.

Wild type grafted to wild type gives normal growth

The BR-deficient mutant shoot is not rescued by the wild-type root

The BR-deficient mutant root does not inhibit the wild-type shoot

BR-deficient dwarf mutant shoot grafted to dwarf grows as a dwarf

**FIGURE 24.15** Effects of reciprocal grafting between the wild type and a BR-deficient mutant of pea (*lkb*) on the phenotype of the shoot in 45-day-old plants. The grafts were made epicotyl-to-epicotyl using 7-day-old seedlings. The BR-deficient dwarf shoot is not rescued by a wild-type root. Conversely, the growth of the wild-type shoot is unaffected by a BR-deficient root. Both results show an absence of long-distance BR signaling. (Adapted from Symons and Reid 2004.)

## Brassinosteroids: Effects on Growth and Development

BRs were originally discovered as growth-promoting substances isolated from pollen, and their role as plant hormones was confirmed in studies of photomorphogenesis. Since their discovery, it has been shown that BRs are involved in a wide range of developmental processes that include fiber development in cotton, development of lateral roots, maintenance of apical dominance, vascular differentiation, and pollen tube growth. BRs are also involved in plant defense, seed germination, and leaf senescence.

(For a discussion of BRs and apical hook maintenance see WEB ESSAY 24.1.) Much remains to be learned about the physiological role of BRs in development. In this section we will discuss some of the better-understood BR responses, including shoot and root growth, vascular differentiation, pollen tube elongation, and seed germination.

## BRs promote both cell expansion and cell division in shoots

The growth-promoting effects of BRs are reflected in acceleration of both cell elongation and cell division. These were first characterized using the bean second-internode bioassay, as discussed above (see Figure 24.1) (Mandava 1988).

The rice leaf lamina inclination bioassay (see Figure 24.2) is dependent on BR-induced cell expansion. Lamina inclination resembles the epinasty caused by ethylene (see Chapter 22). In response to BR, the cells on the adaxial (upper) surface of the leaf near the joint region expand more than the cells on the abaxial (lower) surface, causing the vertically oriented leaf to bend outward. An increase in cell wall loosening is required for BL-induced cell expansion on the adaxial side of the leaf.

In genetic studies, the dwarf phenotype of BR mutants strongly demonstrated the requirement of BRs for normal plant growth (see Figure 24.5). Microscopic examination of leaves from BR-deficient mutants showed not only smaller cell size, but also fewer cells in the leaf blade when compared to wild-type leaves, indicating that BRs are an important class of growth hormones in shoots (Nakaya et al. 2002). Therefore it is not surprising that overexpression of the BR biosynthetic gene *DWF4* results in elevated levels of endogenous BRs and causes an increase in plant size (see Figure 24.14) (Choe et al. 2001). Indeed, one of the prominent characteristics of BRs (as was first observed in the bean second-internode bioassay) is its promotion of both cell elongation and cell proliferation.

The stimulatory effect of BRs on growth is most pronounced in young, growing shoot tissues. The kinetics of cell expansion in response to nanomolar concentrations of BL differ from those of auxin-induced cell expansion. In soybean epicotyl sections, for example, BL begins to enhance the elongation rate after a 45-minute lag period, and reaches a maximum rate only after several hours of treatment. In contrast, auxin stimulates elongation after a 15-minute lag time and reaches a maximum rate within 45 minutes (**FIGURE 24.16**) (Zurek et al. 1994) (see Chapter 19). These results suggest that the growth response to BRs may involve a slower pathway involving gene transcription, whereas the rapid response to auxin may not require gene transcription.

Another possibility is that the stimulation of gene expression by auxin is much greater than the gene expression induced by BL (see Chapter 19). In fact, auxin and BRs

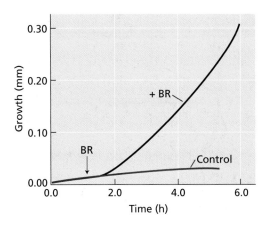

**FIGURE 24.16** The kinetics of BR stimulation of soybean epicotyl elongation. A 1.5-cm soybean epicotyl section was treated with 0.1 μM BR in a sensitive growth measuring system. BR-induced growth was observed after a 45-minute lag period. Five or more hours are required to achieve the maximum steady-state growth rate (not shown). (After Zurek et al. 1994.)

have been shown to enhance the growth of shoot tissues synergistically and in an interdependent manner, indicating that each hormone requires the presence of the other for optimal activity. A synergism between IAA and BR has also been demonstrated in the rice lamina inclination bioassay. At the molecular level, the combined activity of auxin and BRs leads to an elevated and prolonged expression of common downstream targets.

While the exact molecular mechanisms by which the transcriptional enhancement occurs are largely unknown, distinct cross-talk points between the BR and auxin pathways have been identified. For example, BIN2, a negative regulator of BR signaling, may phosphorylate transcription factors involved in both auxin and BR-regulated gene expression (Vert et al. 2008). BRs also enhance auxin transport and stimulate lateral root growth and differential growth in response to gravitropism (see below). Finally, the expression of the BR biosynthetic gene CPD is positively regulated by BRX1, a nuclear protein, whose expression is stimulated by auxin (Hardtke 2007). Thus, the cross talk between the two hormones occurs at various levels.

The process of cell expansion involves cell wall relaxation followed by the osmotic transport of water into the cell to maintain turgor pressure, and cell wall synthesis to maintain wall thickness (see Chapter 15). Each of these steps is likely to be modulated by BRs. BRs are thought to increase the uptake of water through aquaporins (Morillon et al. 2001), enhance cell wall loosening (**FIGURE 24.17**), and induce the expression of wall-modifying enzymes such as xyloglucan endotransglycosylase/hydrolases (XTHs) and expansins (see Chapter 15).

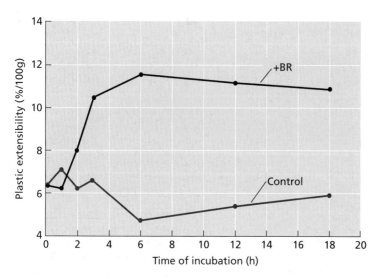

**FIGURE 24.17** BRs increase the plastic wall extensibility of soybean epicotyls. Soybean epicotyl sections (1.5 cm) were incubated with or without 0.1 µM BR for the times indicated, after which their plastic extensibility was determined using an extensometer (see Chapter 15). BR had a significant effect on plastic extensibility at 2 h. Plastic extensibility continued to increase up to about 6 h and then remained constant. The increase in wall plasticity in the presence of BR indicates that BR has induced cell wall loosening, which is required for cell expansion. (After Zurek et al. 1994.)

An additional requirement for normal elongation is the control of microtubule organization. As discussed in Chapter 15, microtubule orientation helps to align cellulose microfibrils during synthesis, and transversely arranged microfibrils in the wall are required for normal cell elonga-

tion. Microscopy of microtubules in a BR-deficient mutant of Arabidopsis has shown that mutant cells contain very few microtubules, and that those present lack organization. Treatment of the mutant with BR restored normal microtubule abundance and organization (**FIGURE 24.18**). Since BR did not increase the total amount of tubulin protein in the cell, BR must act by promoting microtubule nucleation and organization (Catterou et al. 2001).

In addition to cell elongation, BR also stimulates cell proliferation. As we saw in Chapter 21, cytokinin-induced cell division has been linked to the expression of a D-type cyclin, CYCD3. 24-epiBL has also been shown to increase *CYCD3* gene expression, and 24-epiBL can substitute for zeatin in the growth of Arabidopsis callus and cell suspension cultures (Hu et al. 2000). Thus the BRs and cytokinins appear to regulate the cell cycle via similar mechanisms.

## BRs both promote and inhibit root growth

The phenotypes of BR-deficient mutants, which typically exhibit reduced root growth, suggest that BRs are required for normal root elongation. However, like auxin, exogenously applied BRs may have positive or negative effects on root growth, depending on the concentration (Mussig 2005). When applied exogenously, BRs promote root growth at low concentrations and inhibit root growth at high concentrations. The threshold concentration for inhibition depends on the activity of the BR analog used. Thus, the threshold concen-

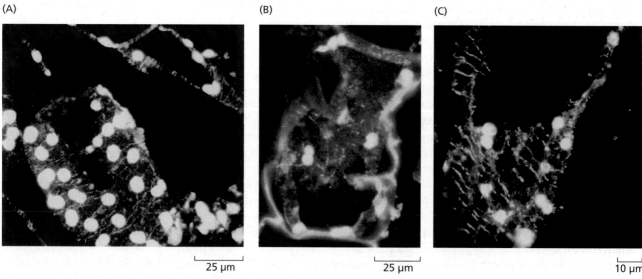

**FIGURE 24.18** Effect of BR on microtubule organization in Arabidopsis seedlings. Immunofluorescence study shows microtubules in green (note their arrangement as "stripes"). Yellow spots correspond to the autofluorescence signal of the chloroplasts. (A) Wild-type pa-

renchyma cell, showing normal transverse microtubule arrangement. (B) BR-deficient mutant parenchyma cell with few, nonaligned microtubules. (C) BR-deficient mutant treated with BR. The normal microtubule organization has been restored. (From Catterou et al. 2001.)

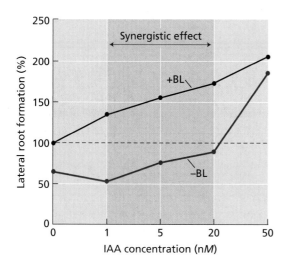

**FIGURE 24.19** BL and IAA act synergistically to promote lateral root development. Arabidopsis seedlings were grown vertically on agar plates containing 0, 1, 5, 20, or 50 nM IAA with or without 1 nM BL for 8 days, and the numbers of lateral roots and visible lateral root primordia per centimeter of primary root were counted. The number of lateral roots per centimeter in each treatment was plotted as the percentage of the number of lateral roots per centimeter in the 1 nM BL, zero auxin treatment (100%, horizontal dashed line). The synergistic effect of BL and auxin occurs in the 1–20 nM IAA range. (After Bao et al. 2004.)

tration is lower for the relatively active analog 24-epiBL than for the less active analog 24-epicastasterone.

The effects of BR on root growth are independent of both auxin and gibberellin action. An inhibitor of polar auxin transport, 2,3,5-triiodobenzoic acid (TIBA) (see Chapter 19), does not prevent BR-induced growth (Mussig 2005). When BR and auxin are applied simultaneously, both the promotive and inhibitory effects on root growth are additive. Moreover, the reduced root growth phenotype of BR-deficient mutants is not reversed by gibberellin application. Taken together, these observations indicate that BR inhibition of root growth does not involve interactions with either auxin or GA. On the other hand, high concentrations of BR, like auxin, stimulate ethylene production, so it is possible that at least some of BR's inhibitory effects on root growth are due to ethylene.

At low concentrations, BRs can also induce the formation of lateral roots (**FIGURE 24.19**) (Bao et al. 2004). In these conditions, however, BRs and auxin act synergistically. The current model suggests that BRs promote lateral root development partially by influencing polar auxin transport (see Chapter 19). BR treatment promotes acropetal auxin transport, which is required for the development of lateral roots, while 1-N-naphthylphthalamic acid (NPA), an auxin transport inhibitor, eliminates the promotive effect of BR (Bao et al. 2004). Finally, BR promotes gravitropism responses,

and this effect is associated with enhanced expression of the auxin efflux transporter PIN2 in the root elongation zone (see Chapter 19) (Li et al. 2005). Thus, BR exerts strong effects on overall root morphology, influencing both the elongation rate and the branching habit.

### BRs promote xylem differentiation during vascular development

BRs play an important role in vascular development, by promoting differentiation of the xylem and suppressing that of the phloem. This is evident in the impaired vasculature systems of BR mutants, which have a higher phloem-to-xylem ratio than the wild type (**FIGURE 24.20**) (reviewed in Fukuda 2004). BR-deficient mutants also have a reduced number of vascular bundles with irregular spacing between the bundles. In contrast, mutants overexpressing the BR receptor protein (discussed later in the chapter) produce more xylem than the wild type.

Cell cultures of zinnia (*Zinnia elegans*) serve as an in vitro system to study the sequential stages of xylem differentiation. When single cells are mechanically isolated from young zinnia leaves and cultured in liquid medium in the dark, they differentiate into tracheary elements between days 2 and 3 after culture (**FIGURE 24.21**). Measurements of BRs during xylem differentiation in this system have shown that BRs are actively synthesized in procambial-like cells and are essential for the differentiation of these cells into tracheary elements. BRs are likely to mediate the dif-

(A) Wild-type Arabidopsis stem cross section

(B) Arabidopsis *det2* mutant stem cross section

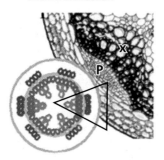
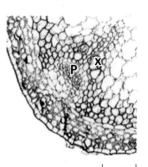

20 μm

**FIGURE 24.20** BR is required for a normal vascular development. The left panel insert shows a schematic representation of the Arabidopsis vascular system at the basal part of the inflorescence stem of a mature plant. The procambial cells (yellow) give rise to phloem tissue (red) to the outside and xylem tissue (blue) to the inside. The black triangle encloses a single vascular bundle. The vascular bundle of the BR-deficient *det2* mutant (right) has a lower xylem-to-phloem ratio than that of the wild type (left). P, phloem; x, xylem. (From Caño-Delgado et al. 2004.)

Zinnia leaf mesophyll cells

Tracheary cells that differentiated from zinnia leaf mesophyll cells

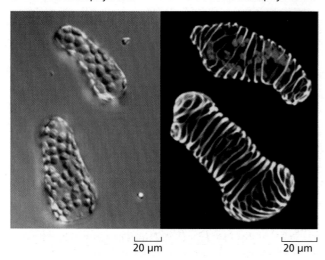

20 μm          20 μm

**FIGURE 24.21**   In vitro culture of zinnia leaf mesophyll cell before (left) and after (right) differentiation into a tracheary element. Brassinosteroids are essential for this differentiation process. (From Fukuda 2004.)

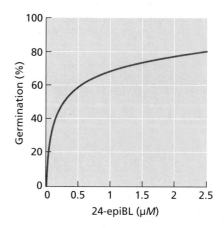

**FIGURE 24.22**   BR stimulates germination of Arabidopsis seeds. Seeds of a gibberellin-insensitive mutant were treated with increasing concentrations of 24-epiBL, and the percent germination was determined. The results show that the stimulation of germination by BR is independent of gibberellin. (After Steber and McCourt 2001.)

ferentiation from procambium to xylem by regulating the expression of homeobox genes (see Chapter 16) that often play crucial roles in development (Fukuda 2004).

### BRs are required for the growth of pollen tubes

Pollen is a rich source of BRs, so it is not surprising that BRs are important for male fertility. BR has been shown to promote the growth of the pollen tube from the stigma through the style to the embryo sac (Mussig 2005). For example, in the BR-deficient Arabidopsis mutant *cpd*, pollen tubes failed to elongate after germination on the stigma, and pollen tube elongation was shown to be partially dependent on BR application (Szekeres et al. 1996; Clouse and Sasse 1998).

Similarly, when a BR-insensitive mutant with a defective receptor gene was self-pollinated, the pollen tube failed to develop, resulting in sterile seeds. However, when the mutant was hand-pollinated with wild-type pollen, fertile seeds were produced (Clouse et al. 1996). Thus, for normal pollen tube growth, both BR and the BR signaling pathway are needed.

Reduced male fertility has also been attributed to a discrepancy in the heights of the stamens versus the pistil. The Arabidopsis *dwf4* mutant is BR-deficient, and its cells fail to elongate. The stamens of the flower are also shorter than those of the wild type. Because Arabidopsis is self-pollinating, the shorter filaments of *dwf4* stamens result in less pollen being deposited on the stigmatic surface. Since

the pollen grains are viable, hand pollination of the mutant flowers results in normal seed production.

### BRs promote seed germination

Seeds, like pollen grains, contain very high levels of BRs, and BRs promote seed germination as well (Mussig 2005). BRs promote seed germination by interacting with other plant hormones, although the molecular basis for these interactions is not known. It is well established that GA and abscisic acid (ABA) play positive and negative roles, respectively, in stimulating seed germination. BRs can enhance germination of tobacco seeds, independent of GA signaling (Leubner-Metzger 2001). Moreover, BRs can rescue the delayed germination phenotype of both GA-deficient and GA-perception mutants (**FIGURE 24.22**), and BR mutants are more sensitive to the inhibition by ABA than the wild type (Steber and McCourt 2001). Thus, BRs can stimulate germination and are needed to overcome the inhibitory effect of ABA. As BRs are known to stimulate cell expansion and division, it is likely that BRs facilitate germination by stimulating the growth of the embryo.

## Prospective Uses of Brassinosteroids in Agriculture

Brassinosteroids were discovered as a class of growth-promoting hormones, and researchers immediately recognized their potential applications to agriculture. For the past 20 years, numerous small-scale studies have been conducted to test the ability of BRs to increase yields of crop plants. BL has been found to increase bean crop yield (based on the weight of seeds per plant) by about 45%, and to enhance the leaf weight of various lettuce varieties by 25%. Similar increases in the yields of rice, barley, wheat, and lentils have been observed. BL also promoted potato tuber growth and increased its resistance to infections. Tomato fruit set was also enhanced by BL.

In addition to such small-scale studies, large-scale field trials using brassinosteroid derivatives have now been conducted in Japan, China, Korea, and Russia. The results of the field trials have been highly variable and appear to reflect the degree of stress under which the crop was grown. A crop grown under optimal conditions shows little effect of applied BR, while a crop grown under conditions of stress shows dramatic effects of BR application on yield. Thus it appears that BR application is most ben-

eficial to growth under stress conditions (Ikekawa and Zhao 1991).

BRs have also proven to be a useful aid for plant propagation. Pretreatment of woody cuttings of plants such as Norway spruce and apple trees with BR enhanced the rooting response. Micropropagation of cassava and pineapple by tissue culture has also been improved by BR treatment. The expression of a maize *DWARF4* (*DWF4*) gene in transgenic rice increased grain yield by 15–44% (Wu et al. 2008).

Reduced BR function can contribute to agriculture as well. For example, decreased BR synthesis or signaling in rice results in dwarfed plants with an erect leaf habit, which allows higher planting densities, leading to increased biomass and final seed yields (Sakamoto et al. 2006). As researchers continue to explore BR's effects on plant development, additional applications of brassinosteroids to agriculture are bound to emerge.

## SUMMARY

The brassinosteroids (BRs) are steroid hormones that regulate plant developmental processes, including cell division and cell elongation in stems and roots, photomorphogenesis, reproductive development, leaf senescence, and stress responses.

### Brassinosteroid Structure, Occurrence, and Genetic Analysis

- Bioassays distinguish active BRs from inactive intermediates and permit quantitation (**Figures 24.1–24.3**).

- BRs are a group of polyhydroxylated steroid hormones, with brassinolide (BL) being the most widespread and active BR in plants (**Figure 24.4**).

- BRs have been detected in all tissues examined with greatest activity in the apical shoot.

- BRs are ubiquitous plant hormones that predate the evolution of land plants.

- BR-deficient mutants show abnormal photomorphogenesis, which can be prevented with exogenous application of BL (**Figures 24.5, 24.6**).

### The Brassinosteroid Signaling Pathway

- BL binds to the BRI1 receptor found in the plasma membrane and endosomal membranes (**Figure 24.7**).

- BL binding activates BRI1, which becomes phosphorylated at multiple sites (**Figure 24.8**).

- The activation of BRI1/BAK1 initiates a signaling

cascade that leads to BR-regulated gene transcription.

- The dephosphorylated forms of BES1 and BZR1 activate or repress BR target genes (**Figure 24.9**).

### Biosynthesis, Metabolism, and Transport of Brassinosteroids

- Brassinosteroids are synthesized from campesterol, which is derived from the plant sterol precursor, cycloartenol (**Figure 24.10**).

- All the enzymes converting campestanol to BL are cytochrome P450 monooxygenases located on the ER.

- BR levels are regulated through multiple control mechanisms, including catabolism, conjugation, and negative feedback from the signaling pathway (**Figures 24.11–24.14**).

- BRs act near their sites of synthesis and do not undergo long-distance transport (**Figure 24.15**).

### Brassinosteroids: Effects on Growth and Development

- BRs are involved in development of fiber, lateral roots, and vasculature, as well as maintenance of apical dominance, pollen tube growth, seed germination, leaf senescence, and plant defenses.

- BRs promote both cell proliferation and cell elongation (**Figures 24.16, 24.17**).

- BR maintains the normal microtubule abundance and organization needed for cell wall growth (**Figure 24.18**).

## SUMMARY continued

- BRs promote root growth at low concentrations and inhibit root growth at high concentrations.
- BRs promote lateral root development by altering polar auxin transport.
- BRs promote differentiation of the xylem and suppress that of the phloem (**Figures 24.20, 24.21**).
- BRs promote seed germination by interacting with other hormones, such as GA and ABA (**Figure 24.22**).

**Prospective Uses of Brassinosteroids in Agriculture**

- BR application to crop plants is most effective under stress conditions.
- BRs are useful in plant propagation.

## WEB MATERIAL

### Web Essay

**24.1  Brassinosteroids and the Apical Hook—An Ongoing Story in Plant Architecture**
A model is proposed for the interactions between ethylene, auxin, and brassinosteroids in the formation of the hook of etiolated seedlings.

## CHAPTER REFERENCES

Abe, H., and Marumo, S. (1991) Brassinosteroids in leaves of *Distylium recemosum* Sieb. et Zucc. The beginning of brassinosteroid research in Japan. In *Brassinosteroids: Chemistry, Bioactivity, and Applications*, H. G. Cutler, T. Yokota, and G. Adam, eds., American Chemical Society, Washington, D. C., pp. 18–24.

Asami, T., Nakano, T., Nakashita, H., Sekimata, K., Shimada, Y., and Yoshida, S. (2003) The influence of chemical genetics on plant science: Shedding light on functions and mechanism of action of brassinosteroids using biosynthesis inhibitors. *J. Plant Growth Regul.* 22: 336–349.

Asami, T., Min, Y. K., Nagata, N., Yamagishi, K., Takatsuto, S., Fujioka, S., Murofushi, N., Yamaguchi, I., and Yoshida, S. (2000) Characterization of brassinazole, a triazole-type brassinosteroid biosynthesis inhibitor. *Plant Physiol.* 123: 93–100.

Bao, F., Shen, J., Brady, S. R., Muday, G. K., Asami, T., and Yang, Z. (2004) Brassinosteroids interact with auxin to promote lateral root development in *Arabidopsis*. *Plant Physiol.* 134: 1624–1631.

Caño-Delgado, A., Yin, Y., Yu, C., Vafeados, D., Mora-Garcia, S., Cheng, J. C., Nam, K. H., Li, J., and Chory, J. (2004) BRL1 and BRL3 are novel brassinosteroid receptors that function in vascular differentiation in *Arabidopsis*. *Development* 131: 5341–5351.

Catterou, M., Dubois, F., Schaller, H., Aubanelle, L., Vilcot, B., Sangwan-Norreel, B. S., and Sangwan, R. S. (2001) Brassinosteroids, microtubules and cell elongation in *Arabidopsis thaliana*. II. Effects of brassinosteroids on microtubules and cell elongation in the bul1 mutant. *Planta* 212: 673–683.

Choe, S., Fujioka, S., Noguchi, T., Takatsuto, S., Yoshida, S., and Feldmann, K. A. (2001) Overexpression of DWARF4 in the brassinosteroid biosynthetic pathway results in increased vegetative growth and seed yield in *Arabidopsis*. *Plant J.* 26: 573–582.

Clouse, S. D., and Sasse, J. M. (1998) Brassinosteroids: Essential regulators of plant growth and development. *Annu. Rev. Plant Physiol. Plant Mol. Biol.* 49: 427–451.

Clouse, S. D., Langford, M., and McMorris, T. C. (1996) A brassinosteroid-insensitive mutant in *Arabidopsis thaliana* exhibits multiple defects in growth and development. *Plant Physiol.* 111: 671–678.

Friedrichsen, D. M., Joazeiro, C. A., Li, J., Hunter, T., and Chory, J. (2000) Brassinosteroid-insensitive-1 is a ubiquitously expressed leucine-rich repeat receptor serine/threonine kinase. *Plant Physiol.* 123: 1247–1256.

Fujioka, S., and Yokota, T. (2003) Biosynthesis and metabolism of brassinosteroids. *Annu. Rev. Plant Biol.* 54: 137–164.

Fukuda, H. (2004) Signals that control plant vascular cell differentiation. *Nat. Rev. Mol. Cell Biol.* 5: 379–391.

Geldner, N., Hyman, D. L., Wang, X., Schumacher, K., and Chory, J. (2007) Endosomal signaling of plant steroid receptor kinase BRI1. *Genes Dev.* 21: 1598–1602.

Gendron, J. M., and Wang, Z. Y. (2007) Multiple mechanisms modulate brassinosteroid signaling. *Curr. Opin. Plant Biol.* 10: 436–441.

Grove, M. D., Spencer, G. F., Rohwedder, W. K., Mandava, N., Worley, J. F., Warthen, J. D., Steffens, G. L., Flippenanderson, J. L., and Cook, J. C. (1979) Brassinolide, a plant growth-promoting steroid isolated from *Brassica napus* pollen. *Nature* 281: 216–217.

Hardtke, C. S. (2007) Transcriptional auxin-brassinosteroid crosstalk: who's talking? *Bioessays* 29: 1115–1123.

He, J. X., Gendron, J. M., Sun, Y., Gampala, S. S., Gendron, N., Sun, C. Q., and Wang, Z. Y. (2005) BZR1 is a transcriptional repressor with dual roles in brassinosteroid homeostasis and growth responses. *Science* 307: 1634–1638.

Hu, Y., Bao, F., and Li, J. (2000) Promotive effect of brassinosteroids on cell division involves a distinct CycD3-induction pathway in *Arabidopsis*. *Plant J.* 24: 693–701.

Ikekawa, N., and Zhao, Y. (1991) Application of 24-epibrassinolide in agriculture. In *Brassinosteroids: Chemistry, Bioactivity, and Applications*, H. G. Cutler, T. Yokota, and G. Adam, eds., American Chemical Society, Washington, D. C., pp. 280–291.

Kim, T. W., Hwang, J. Y., Kim, Y. S., Joo, S. H., Chang, S. C., Lee, J. S., Takatsuto, S., and Kim, S. K. (2005) *Arabidopsis* CYP85A2, a cytochrome P450, mediates the Baeyer-Villiger oxidation of castasterone to brassinolide in brassinosteroid biosynthesis. *Plant Cell* 17: 2397–2412.

Kim, T. W., Guan, S., Sun, Y., Deng, Z., Tang, W., Shang, J. X., Sun, Y., Burlingame, A. L., and Wang, Z. Y. (2009) Brassinosteroid signal transduction from cell-surface receptor kinases to nuclear transcription factors. *Nat. Cell Biol.* 11: 1254–1260.

Kinoshita, T., Caño-Delgado, A., Seto, H., Hiranuma, S., Fujioka, S., Yoshida, S., and Chory, J. (2005) Binding of brassinosteroids to the extracellular domain of plant receptor kinase BRI1. *Nature* 433: 167–171.

Leubner-Metzger, G. (2001) Brassinosteroids and gibberellins promote tobacco seed germination by distinct pathways. *Planta* 213: 758–763.

Li, J., and Chory, J. (1997) A putative leucine-rich repeat receptor kinase involved in brassinosteroid signal transduction. *Cell* 90: 929–938.

Li, J., Nagpal, P., Vitart, V., McMorris, T. C., and Chory, J. (1996) A role for brassinosteroids in light-dependent development of *Arabidopsis*. *Science* 272: 398–401.

Li, J., and Nam, K. H. (2002) Regulation of brassinosteroid signaling by a GSK3/SHAGGY-like kinase. *Science* 295: 1299–1301.

Li, L., Xu, J., Xu, Z. H., and Xue, H. W. (2005) Brassinosteroids stimulate plant tropisms through

modulation of polar auxin transport in *Brassica* and *Arabidopsis*. *Plant Cell* 17: 2738–2753.

Mandava, N. B. (1988) Plant growth-promoting brassinosteroids. *Annu. Rev. Plant Physiol. Plant Mol. Biol.* 39: 23–52.

Mitchell, J. W., Mandava, N. B., Worley, J. F., Plimmer, J. R., and Smith, M. V. (1970) Brassins: A new family of plant hormones from rape pollen. *Nature* 225: 1065–1066.

Montoya, T., Nomura, T., Yokota, T., Farrar, K., Harrison, K., Jones, J. G., Kaneta, T., Kamiya, Y., Szekeres, M., and Bishop, G. J. (2005) Patterns of Dwarf expression and brassinosteroid accumulation in tomato reveal the importance of brassinosteroid synthesis during fruit development. *Plant J.* 42: 262–269.

Morillon, R., Catterou, M., Sangwan, R. S., Sangwan, B. S., and Lassalles, J. P. (2001) Brassinolide may control aquaporin activities in *Arabidopsis thaliana*. *Planta* 212: 199–204.

Mussig, C. (2005) Brassinosteroid-promoted growth. *Plant Biol. (Stuttg.)* 7: 110–117.

Nakaya, M., Tsukaya, H., Murakami, N., and Kato, M. (2002) Brassinosteroids control the proliferation of leaf cells of *Arabidopsis thaliana*. *Plant Cell Physiol.* 43: 239–244.

Neff, M. M., Nguyen, S. M., Malancharuvil, E. J., Fujioka, S., Noguchi, T., Seto, H., Tsubuki, M., Honda, T., Takatsuto, S., Yoshida, S., et al. (1999) *BAS1*: A gene regulating brassino-steroid levels and light responsiveness in *Arabidopsis*. *Proc. Natl. Acad. Sci. USA* 96: 15316–15323.

Nishikawa, N., Toyama, S., Shida, A., and Futatsuya, F. (1994) The uptake and the transport of $^{14}$C-labeled epibrassinolide in intact seedlings of cucumber and wheat. *J. Plant Res.* 107: 125–130.

Noguchi, T., Fujioka, S., Choe, S., Takatsuto, S., Yoshida, S., Yuan, H., Feldmann, K. A., and Tax, F. E. (1999) Brassinosteroid-insensitive dwarf mutants of *Arabidopsis* accumulate brassinosteroids. *Plant Physiol.* 121: 743–752.

Oh, M. H., Wang, X., Kota, U., Goshe, M. B., Clouse, S. D., and Huber, S. C. (2009) Tyrosine phosphorylation of the BRI1 receptor kinase emerges as a component of brassinosteroid signaling in Arabidopsis. *Proc. Natl. Acad. Sci. USA* 106: 658–663.

Peng, P., Yan, Z., Zhu, Y., and Li, J. (2008) Regulation of the Arabidopsis GSK3-like kinase BRASSINOSTEROID-INSENSITIVE 2 through proteasome-mediated protein degradation. *Mol. Plant* 1: 338–346.

Sakamoto, T., Morinaka, Y., Ohnishi, T., Sunohara, H., Fujioka, S., Ueguchi-Tanaka, M., Mizutani, M., Sakata, K., Takatsuto, S., Yoshida, S., et al. (2006) Erect leaves caused by brassinosteroid deficiency increase biomass production and grain yield in rice. *Nat. Biotechnol.* 24: 105–109.

Sakurai, A. (1999) Biosynthesis. In *Brassinosteroids: Steroidal Plant Hormones*, A. Sakurai, T. Yokota, and S. D. Clouse, eds., Springer, Tokyo, pp. 91–112.

Savaldi-Goldstein, S., Peto, C., and Chory, J. (2007) The epidermis both drives and restricts plant shoot growth. *Nature* 446: 199–202.

Schlagnhaufer, C. D., and Arteca, R. N. (1991) The uptake and metabolism of brassinosteroid by tomato (*Lycopersicon esculentum*) plants. *J. Plant Physiol.* 138: 191–194.

Shimada, Y., Goda, H., Nakamura, A., Takatsuto, S., Fujioka, S., and Yoshida, S. (2003) Organ-specific expression of brassino-steroid-biosynthetic genes and distribution of endogenous brassinosteroids in *Arabidopsis. Plant Physiol.* 131: 287–297.

Steber, C. M., and McCourt, P. (2001) A role for brassinosteroids in germination in *Arabidopsis. Plant Physiol.* 125: 763–769.

Steffens, G. L. (1991) U.S. Department of Agriculture brassins project: 1970–1980. In *Brassinosteroids: Chemistry, Bioactivity, and Applications*, H. G. Cutler, T. Yokota, and G. Adam, eds., American Chemical Society, Washington, D.C., pp. 2–17.

Symons, G. M., and Reid, J. B. (2004) Brassinosteroids do not undergo long-distance transport in pea. Implications for the regulation of endogenous brassinosteroid levels. *Plant Physiol.* 135: 2196–2206.

Szekeres, M., Nemeth, K., Koncz-Kalman, Z., Mathur, J., Kauschmann, A., Altmann, T., Redei, G. P., Nagy, F., Schell, J., and Koncz, C. (1996) Brassinosteroids rescue the deficiency of CYP90, a cytochrome P450, controlling cell elongation and de-etiolation in *Arabidopsis. Cell* 85: 171–182.

Tanaka, K., Asami, T., Yoshida, S., Nakamura, Y., Matsuo, T., and Okamoto, S. (2005) Brassinosteroid homeostasis in *Arabidopsis* is ensured by feedback expressions of multiple genes involved in its metabolism. *Plant Physiol.* 138: 1117–1125.

Tang, W., Kim, T. W., Oses-Prieto, J. A., Sun, Y., Deng, Z., Zhu, S., Wang, R., Burlingame, A. L., and Wang, Z. Y. (2008) BSKs mediate signal transduction from the receptor kinase BRI1 in Arabidopsis. *Science* 321: 557–560.

Vert, G., and Chory, J. (2006) Downstream nuclear events in brassinosteroid signaling. *Nature* 441: 96–100.

Vert, G., Nemhauser, J. L., Geldner, N., Hong, F., and Chory, J. (2005) Molecular mechanisms of steroid hormone signaling in plants. *Annu. Rev. Cell Dev. Biol.* 21: 177–201.

Vert, G., Walcher, C. L., Chory, J. and Nemhauser, J. L. (2008) Integration of auxin and brassinosteroid pathways by Auxin Response Factor 2. *Proc. Natl. Acad. Sci. USA* 105: 9829–9834.

Wang, X., Li, X., Meisenhelder, J., Hunter, T., Yoshida, S., Asami, T., and Chory, J. (2005a) Autoregulation and homodimerization are involved in the activation of the plant steroid receptor BRI1. *Dev. Cell* 8: 855–865.

Wang, X., Goshe, M. B., Soderblom, E. J., Phinney, B. S., Kuchar, J. A., Li, J., Asami, T., Yoshida, S., Huber, S. C., and Clouse, S. D. (2005b) Identification and functional analysis of in vivo phosphorylation sites of the *Arabidopsis* BRASSINOSTEROID-INSENSITIVE1 receptor kinase. *Plant Cell* 17: 1685–1703.

Wang, X., and Chory, J. (2006) Brassinosteroids regulate dissociation of BKI1, a negative regulator of BRI1 signaling, from the plasma membrane. *Science* 313: 1118–1122.

Wu, C. Y., Trieu, A., Radhakrishnan, P., Kwok, S. F., Harris, S., Zhang, K., Wang, J., Wan, J., Zhai, H., Takatsuto, S., et al. (2008) Brassinosteroids regulate grain filling in rice. *Plant Cell* 20: 2130–2145.

Yin, Y., Vafeados, D., Tao, Y., Yoshida, S., Asami, T., and Chory, J. (2005) A new class of transcription factors mediates brassinosteroid-regulated gene expression in *Arabidopsis. Cell* 120: 249–259.

Yokota, T., Arima, M., and Takahashi, N. (1982) Castasterone, a new phytosterol with plant-hormone potency, from chestnut insect. *Tetrahedron Lett.* 23: 1275–1278.

Zurek, D. M., Rayle, D. L., McMorris, T. C., and Clouse, S. D. (1994) Investigation of gene expression, growth kinetics, and wall extensibility during brassinosteroid-regulated stem elongation. *Plant Physiol.* 104: 505–513.

# 25

# The Control of Flowering

Most people look forward to the spring season and the profusion of flowers it brings. Many vacationers carefully time their travels to coincide with specific blooming seasons: *Citrus* along Blossom Trail in southern California, tulips in Holland. In Washington, DC, and throughout Japan, the cherry blossoms are received with spirited ceremonies. As spring progresses into summer, summer into fall, and fall into winter, wildflowers bloom at their appointed times. Plants have evolved a remarkable diversity of complex floral structures to attract different pollinators. Flowering at the correct time of year is crucial for the reproductive fitness of the plant; plants that are cross-pollinated must flower in synchrony with themselves as well as with their pollinators at the time of year that is optimal for seed set.

Although the strong correlation between flowering and seasons is common knowledge, the phenomenon poses fundamental questions that will be addressed in this chapter:

- How do plants keep track of the seasons of the year and the time of day?
- Which environmental signals influence flowering, and how are those signals perceived?
- How are environmental signals transduced to bring about the developmental changes associated with flowering?

In Chapter 16 we discussed the role of the root and shoot apical meristems in vegetative growth and development. The transition to flowering involves major changes in the pattern of morphogenesis and cell differentiation at the shoot apical meristem. Ultimately this process leads to the production of the floral organs—sepals, petals, stamens, and carpels (see Figure 1.2A in **WEB TOPIC 1.3**).

Specialized cells in the anther undergo meiosis to produce four haploid microspores that develop into pollen grains. Similarly, a cell within the ovule divides meiotically to produce four haploid megaspores, one of which undergoes three mitotic divisions to produce the cells of the embryo sac (see Figure 1.2B in **WEB TOPIC 1.3**). The embryo sac represents the mature female gametophyte. The pollen grain, with its germinating pollen tube, is the mature male gametophyte generation. The two gametophytic structures produce the gametes (egg and sperm cells), which fuse to form the diploid zygote, the first stage of the new sporophyte generation.

Clearly, flowers represent a complex array of functionally specialized structures that differ substantially from the vegetative plant body in form and cell types. The transition to flowering therefore entails radical changes in cell fate within the shoot apical meristem. In the first part of this chapter we will discuss these changes, which are manifested as *floral development*. Recently genes have been identified that play crucial roles in the formation of the floral organs. Such studies demonstrate how plants are able to achieve complex developmental outcomes with a relatively small number of master regulators.

The events occurring in the shoot apex that specifically commit the apical meristem to produce *flowers* are collectively referred to as **floral evocation**. In the second part of this chapter we will discuss the events leading to floral evocation. The developmental signals that stimulate floral evocation include endogenous factors, such as *circadian rhythms*, *phase change*, and *hormones*, and external factors, such as *photoperiod* (day length) and temperature.

In the case of photoperiodism, a transmissible signal from the leaves, referred to as the **floral stimulus**, is translocated to the shoot apical meristem. The plant apex must be competent to respond to this positive signal. Certain plants go through a *juvenile phase* during which the meristem may not be competent to flower. Another factor can be a requirement for *vernalization* (exposure to a prolonged period of cold before flowering can occur). These are some examples of plants integrating endogenous and environmental signals to flower at the correct time.

## Floral Meristems and Floral Organ Development

Floral meristems can usually be distinguished from vegetative meristems by their larger size. In the vegetative meristem, the cells of the central zone complete their division cycles slowly. The transition from vegetative to reproductive development is marked by an increase in the frequency of cell divisions within the central zone of the shoot apical meristem (see Chapter 16). The increase in the size of the meristem is largely a result of the increased division rate of these central cells (**FIGURE 25.1**). Genetic and molecular studies have now identified a network of genes that control floral morphogenesis in Arabidopsis, snapdragon (*Antirrhinum*), and other species.

In this section we will focus on floral development in Arabidopsis, which has been studied extensively (**FIGURE 25.2**). First we will outline the basic morphological changes that occur during the transition from the vegetative to the reproductive phase. Next we will consider the arrangement of the floral organs in four whorls on the meristem, and the types of genes that govern the normal pattern of floral development. According to the widely accepted ABC model (see Figure 25.6), the specific locations of floral organs in the flower are regulated by the overlapping expression of three types of floral organ identity genes.

(A)

(B)

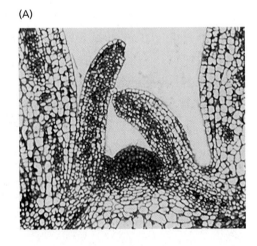
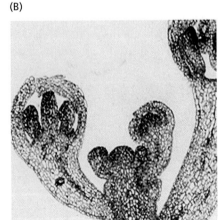

**FIGURE 25.1** Longitudinal sections through a vegetative (A) and a reproductive (B) shoot apical region of Arabidopsis. (Courtesy of V. Grbic and M. Nelson.)

(A)

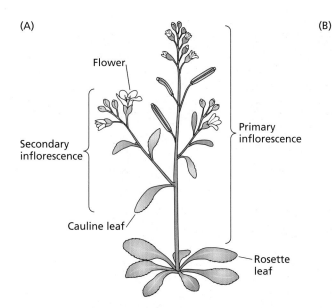

Flower

Secondary inflorescence

Cauline leaf

Primary inflorescence

Rosette leaf

(B)

<br>
**FIGURE 25.2**  (A) The shoot apical meristem in *Arabidopsis thaliana* generates different organs at different stages of development. Early in development the shoot apical meristem forms a rosette of basal leaves. When the plant makes the transition to flowering, the shoot apical meristem is transformed into a primary inflorescence meristem that ultimately produces an elongated stem bearing flowers. Leaf primordia initiated prior to the floral transition develop on the stem (cauline leaves), and secondary inflorescences develop in the axils of these stem-borne leaves. (B) Photograph of a flowering Arabidopsis plant. (Courtesy of Richard Amasino.)

## The shoot apical meristem in Arabidopsis changes with development

During the vegetative phase of growth, the Arabidopsis apical meristem produces leaves with very short internodes, resulting in a basal rosette of leaves (see Figure 25.2A). When reproductive development is initiated, the vegetative meristem is transformed into the primary inflorescence meristem. The **primary inflorescence meristem** produces an elongated inflorescence axis bearing two types of lateral organs: stem-borne (or inflorescence) leaves and flowers (see Figure 25.2).

The axillary buds of the stem-borne leaves develop into **secondary inflorescence meristems**, and their activity repeats the pattern of development of the primary inflorescence meristem, as shown in Figure 25.2A. The Arabi-

dopsis inflorescence meristem has the potential to grow indefinitely and thus exhibits *indeterminate* growth. Flowers arise from **floral meristems** that form on the flanks of the inflorescence meristem. In contrast to the inflorescence meristem, the floral meristem is determinate.

## The four different types of floral organs are initiated as separate whorls

Floral meristems initiate four different types of floral organs: sepals, petals, stamens, and carpels (Coen and Carpenter 1993). These sets of organs are initiated in concentric rings, called **whorls**, around the flanks of the meristem (**FIGURE 25.3**). The initiation of the innermost

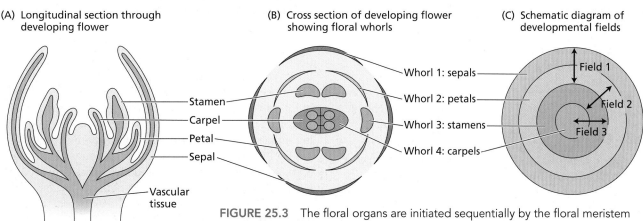

**FIGURE 25.3**  The floral organs are initiated sequentially by the floral meristem of Arabidopsis. (A and B) The floral organs are produced as successive whorls (concentric circles), starting with the sepals and progressing inward. (C) According to the combinatorial model, the functions of each whorl are determined by three overlapping developmental fields. These fields correspond to the expression patterns of specific floral organ identity genes. (After Bewley et al. 2000.)

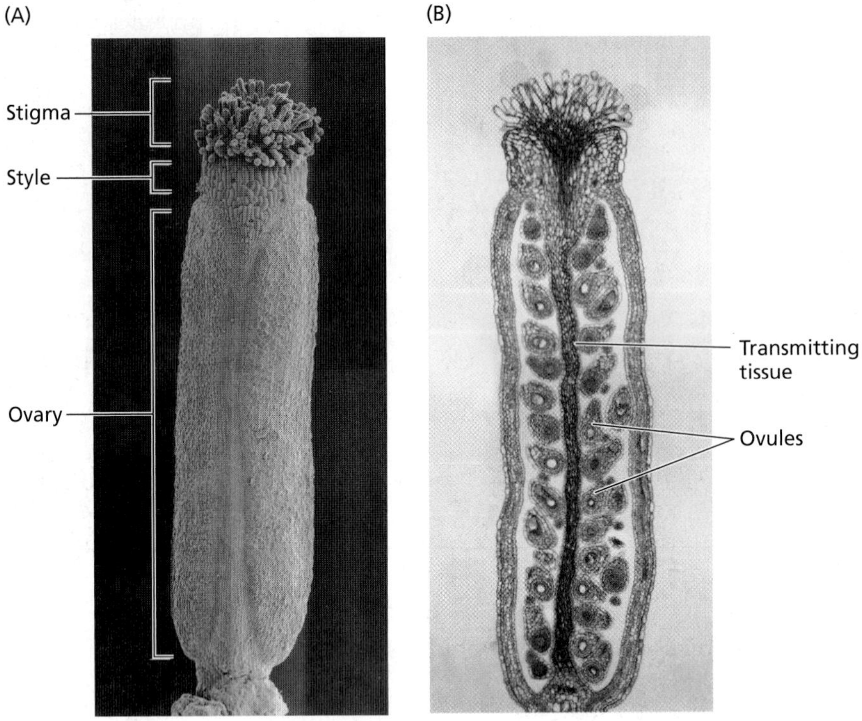

**FIGURE 25.4** The Arabidopsis pistil consists of two fused carpels, each containing many ovules. (A) Scanning electron micrograph of a pistil, showing the stigma, a short style, and the ovary. (B) Longitudinal section through the pistil, showing the many ovules. (From Gasser and Robinson-Beers 1993, courtesy of C. S. Gasser, © American Society of Plant Biologists, reprinted with permission.)

(A) Stigma — Style — Ovary

(B) Transmitting tissue — Ovules

organs, the carpels, consumes all of the meristematic cells in the apical dome, and only the floral organ primordia (localized regions of cell division) are present as the floral bud develops. In the wild-type Arabidopsis flower, the whorls are arranged as follows:

- The first (outermost) whorl consists of four sepals, which are green at maturity.

- The second whorl is composed of four petals, which are white at maturity.

- The third whorl contains six stamens (the male reproductive structures), two of which are shorter than the other four.

- The fourth (innermost) whorl is a single complex organ, the gynoecium or pistil (the female reproductive structure), which is composed of an ovary with two fused carpels, each containing numerous ovules, and a short style capped with stigma (**FIGURE 25.4**).

## Two major types of genes regulate floral development

Studies of mutations have enabled identification of two key classes of genes that regulate floral development: meristem identity genes and floral organ identity genes.

1. **Meristem identity genes** encode transcription factors that are necessary for the initial induction of organ identity genes. They are the positive regulators of floral organ identity in the developing floral meristem.

2. **Floral organ identity genes** directly control floral identity. The proteins encoded by these genes are transcription factors that likely control the expression of other genes whose products are involved in the formation and/or function of *floral* organs.

## Meristem identity genes regulate meristem function

Meristem identity genes must be active for the immature primordia formed at the flanks of the shoot or *inflorescence apical meristem* to become floral meristems. (Recall that an apical meristem that is forming floral meristems on its flanks is known as an inflorescence meristem.) For example, mutants of snapdragon (*Antirrhinum*) that have a defect in the meristem identity gene *FLORICAULA* (*FLO*) develop an inflorescence that does not produce flowers. Instead of developing floral meristems in the axils of the bracts, *flo* mutants develop additional inflorescence meristems in the bract axils. Thus the wild-type *FLO* gene controls the determination step in which floral meristem identity is established.

In Arabidopsis, *SUPPRESSOR OF CONSTANS 1* (*SOC1*)*, *APETALA1* (*AP1*), and *LEAFY* (*LFY*) are among the critical genes in the genetic pathway that must be activated to establish floral meristem identity. *LFY* is the Arabidopsis version of the snapdragon *FLO* gene. Both *LFY* and *SOC1* play a central role in floral evocation by integrating signals from several different pathways involving both environmental and internal cues (Blázquez and Weigel 2000; Borner et al. 2000). *LFY* and *SOC1* therefore serve as master regulators for the initiation of floral development.

Once activated, *SOC1* triggers the expression of *LFY*, and *LFY* turns on the expression of *AP1* (Simon et al. 1996). In Arabidopsis, *LFY* and *AP1* are involved in a positive feedback loop; that is, *AP1* expression also stimulates the

---

*Also known as *AGAMOUS-LIKE 20* (*AGL20*).

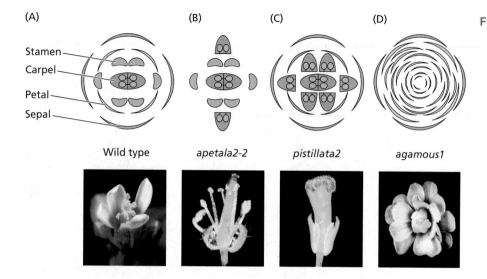

**FIGURE 25.5** Mutations in the floral organ identity genes dramatically alter the structure of the flower. (A) Wild-type Arabidopsis shows normal structure in all four floral components. (B) *apetala2-2* mutants lack sepals and petals. (C) *pistillata2* mutants lack petals and stamens. (D) *agamous1* mutants lack both stamens and carpels. (Photos from Meyerowitz 2002; courtesy of J. L. Riechmann.)

expression of *LFY*. Once initiated, this positive feedback loop is irreversible, and the meristem becomes committed to flowering.

## Homeotic mutations led to the identification of floral organ identity genes

The genes that determine floral organ identity were discovered as **floral homeotic mutants**. Mutations in the fruit fly, *Drosophila*, led to the identification of a set of homeotic genes encoding transcription factors that determine the locations at which specific structures develop. Homeotic genes act as major developmental switches that activate the entire genetic program for a particular structure. The expression of homeotic genes thus gives organs their identity.

The floral organ identity genes were identified through homeotic mutations that alter floral organ identity so that some of the floral organs appear in the wrong places. For example, Arabidopsis plants with mutations in the *APETALA2* (*AP2*) gene produce flowers with carpels where sepals should be, and stamens where petals normally appear.

The homeotic genes that have been identified so far encode transcription factors—proteins that control the expression of other genes. Most plant homeotic genes belong to a class of related sequences known as **MADS box genes**, whereas animal homeotic genes contain sequences called homeoboxes.

Many of the genes that determine floral organ identity are MADS box genes, including the *DEFICIENS* gene of snapdragon and the *AGAMOUS* (*AG*), *PISTILLATA* (*PI*), and *APETALA3* (*AP3*) genes of Arabidopsis. The MADS box genes share a characteristic, conserved nucleotide sequence known as a *MADS box*, which encodes a protein structure known as the *MADS domain*. The **MADS domain** enables these transcription factors to bind to DNA that has a specific nucleotide sequence.

Not all genes containing the MADS box domain are homeotic genes. For example, *SOC1* is a MADS box gene, but it functions as a meristem identity gene.

## Three types of homeotic genes control floral organ identity

Five of the key genes that specify floral organ identity in Arabidopsis are: *AP1*, *AP2*, *AP3*, *PI*, and *AG* (Bowman et al. 1989; Weigel and Meyerowitz 1994). The organ identity genes initially were identified through mutations that dramatically alter the structure and thus the identity of the floral organs produced in two adjacent whorls (**FIGURE 25.5**). For example, plants with the *ap2* mutation lack sepals and petals (see Figure 25.5B). Plants bearing *ap3* or *pi* mutations produce sepals instead of petals in the second whorl, and carpels instead of stamens in the third whorl (see Figure 25.5C). Plants homozygous for the *ag* mutation lack both stamens and carpels (see Figure 25.5D).

Because mutations in these genes change floral organ identity without affecting the initiation of flowers, they are homeotic genes. These homeotic genes fall into three classes—types A, B, and C—defining three different kinds of activities encoded by three distinct types of genes, *AP1* and *AP2*, *AP3* and *PI*, and *AG*. The control of organ identity by type A, B, and C homeotic genes is referred to as the ABC model (**FIGURE 25.6**) and is described in more detail in the next section.

- Type A activity, encoded by *AP1* and *AP2*, controls organ identity in the first and second whorls. Loss of type A activity results in the formation of carpels instead of sepals in the first whorl, and of stamens instead of petals in the second whorl.

- Type B activity, encoded by *AP3* and *PI*, controls organ determination in the second and third whorls. Loss of type B activity results in the formation of sepals instead of petals in the second whorl, and of carpels instead of stamens in the third whorl.

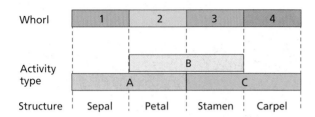

**FIGURE 25.7** A quadruple mutant of Arabidopsis (*ap1, ap2, ap3/pi, ag*) produces leaf-like structures in place of floral organs. (Courtesy of John Bowman.)

**FIGURE 25.6** The ABC model for the acquisition of floral organ identity is based on the interactions of three different types of activities of floral homeotic genes: A, B, and C. In the first whorl, expression of type A (*AP1* and *AP2*) alone results in the formation of sepals. In the second whorl, expression of both type A (*AP1/AP2*) and type B (*AP3/PI*) results in the formation of petals. In the third whorl, the expression of type B (*AP3/PI*) and C (*AG*) causes the formation of stamens. In the fourth whorl, activity type C (*AG*) alone specifies carpels. In addition, activity type A (*AP1* and *AP2*) represses activity of C (*AG*) in whorls 1 and 2, while type C represses A in whorls 3 and 4.

- Type C activity, encoded by *AG*, controls events in the third and fourth whorls. Loss of type C activity results in the formation of petals instead of stamens in the third whorl. Moreover, in the absence of type C activity, the fourth whorl (normally a carpel) is replaced by a *new flower*. As a result, the fourth whorl of an *ag* mutant flower is occupied by sepals. The floral meristem is no longer determinate. Flowers continue to form *within* flowers, and the pattern of organs (from outside to inside) is: sepal, petal, petal; sepal, petal, petal; and so on.

The role of organ identity genes in floral development is dramatically illustrated by experiments in which two or three activities are eliminated by loss-of-function mutations. Quadruple-mutant Arabidopsis plants (*ap1, ap2, ap3/pi,* and *ag*) produce floral meristems that develop as pseudoflowers; all the floral organs are replaced with green leaf-like structures, although these organs are produced with a whorled phyllotaxy typical of normal flowers (**FIGURE 25.7**). This experiment supports the ideas of the eighteenth-century German poet and natural scientist Johann Wolfgang von Goethe (1749–1832), who speculated that floral organs are highly modified leaves.

Since the A, B, and C genes were identified, another class, the D genes, has been discovered. D activity is specified by the *SEPALLATAs* (*SEP1-3*), which are also MADS box transcription factors. Remarkably, by expressing D class genes in combination with A and B genes, it is pos-

sible to convert leaves into petals (Pelaz et al. 2001; Honma and Goto 2001).

### The ABC model explains the determination of floral organ identity

In 1991 the **ABC model** was proposed to explain how homeotic genes control organ identity (Coen and Meyerowitz 1991). The strength of the model is that it immediately accounts for many observations in two distantly related species (snapdragon and Arabidopsis), and it provides a way of understanding how relatively few key regulators can combinatorially provide a complex outcome. The ABC model postulates that organ identity in each whorl is determined by a unique combination of the three organ identity gene activities (see Figure 25.6):

- Activity of type A alone specifies sepals.
- Activities of both A and B are required for the formation of petals.
- Activities of B and C form stamens.
- Activity of C alone specifies carpels.

The model further proposes that activities A and C mutually repress each other (see Figure 25.6); that is, both A- and C-type genes exclude each other from their expression domains, in addition to their function in determining organ identity.

The patterns of organ formation in wild-type flowers and most of the mutants are predicted and explained by this model (**FIGURE 25.8**). The challenges now are to understand (1) how the precise expression pattern of these organ identity genes is achieved; (2) how organ identity genes, which encode transcription factors, alter the pattern

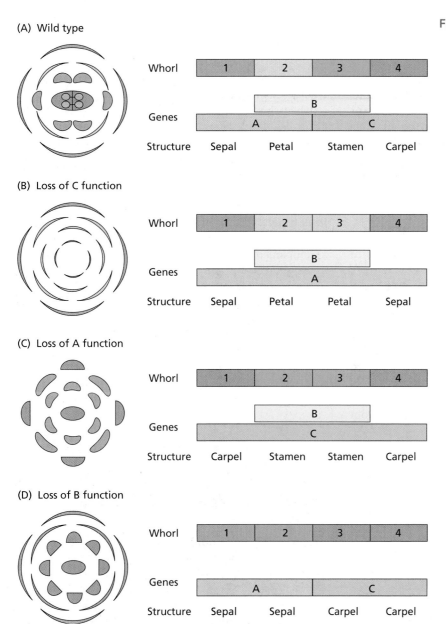

**(A) Wild type**

Whorl: 1 | 2 | 3 | 4

Genes: B (over petal–stamen); A (sepal–petal) | C (stamen–carpel)

Structure: Sepal | Petal | Stamen | Carpel

**(B) Loss of C function**

Whorl: 1 | 2 | 3 | 4

Genes: B; A

Structure: Sepal | Petal | Petal | Sepal

**(C) Loss of A function**

Whorl: 1 | 2 | 3 | 4

Genes: B; C

Structure: Carpel | Stamen | Stamen | Carpel

**(D) Loss of B function**

Whorl: 1 | 2 | 3 | 4

Genes: A | C

Structure: Sepal | Sepal | Carpel | Carpel

**FIGURE 25.8** Interpretation of the phenotypes of floral homeotic mutants based on the ABC model. (A) All three activity types are functional in the wild type. (B) Loss of type C function results in expansion of the A function throughout the floral meristem. (C) Loss of type A function results in the spread of C function throughout the meristem. (D) Loss of type B function results in the expression of only A and C functions.

information from the environment and integrate this information to regulate the genetic programs encoded in their genomes.

A particularly important developmental decision during the plant life cycle is when to flower. Delaying this decision will increase the carbohydrate reserves that will be available through photosynthesis, allowing more seeds to mature, while potentially increasing the danger that the plant will be eaten, killed by abiotic stress, or out-competed by other plants. Reflecting this, plants have evolved an extraordinary range of reproductive adaptations—for example, annual versus perennial life cycles.

Annual plants such as groundsel (*Senecio vulgaris*) may flower within a few weeks after germinating. But trees may grow for 20 or more years before they begin to produce flowers. Across the plant kingdom, different species flower at a wide range of ages, indicating that the age, or perhaps the size, of the plant is an *internal* factor controlling the switch to reproductive development.

of expression of other genes in the developing organ; and (3) how this altered pattern of gene expression results in the development of a specific floral organ.

# Floral Evocation: Integrating Environmental Cues

Plants are sessile (immobile) organisms, and therefore they must continually adapt their growth and development to the external environment. Unlike animals, in which most development is completed during embryogenesis, plant development continues throughout the life cycle. Indeed, there is no particular developmental outcome that is predetermined, since development will unfold in response to a whole host of environmental factors such as day length, temperature, competition from other plants, nutrient availability, and interactions with animals. Plants can therefore be thought of as ideal systems for understanding how organisms perceive

The case in which flowering occurs strictly in response to internal developmental factors, independently of any particular environmental condition is referred to as *autonomous regulation*. In species which exhibit an absolute requirement for a specific set of environmental cues in order to flower, flowering is considered to be an *obligate* or *qualitative* response. If flowering is promoted by certain environmental cues but will eventually occur in the absence of such cues, the flowering response is *facultative* or *quantitative*. A species with a facultative flowering response, such as Arabidopsis, relies on both environmental and autonomous signals to promote reproductive growth.

Photoperiodism and vernalization are two of the most important mechanisms underlying seasonal responses. *Photoperiodism* (see Chapter 17) is a response to the length of day or night; *vernalization* is the promotion of flowering by prolonged cold temperature. Other signals, such as light quality, ambient temperature and abiotic stress are also important external cues for plant development.

The evolution of both internal (autonomous) and external (environment-sensing) control systems enables plants to precisely regulate flowering at the optimal time for reproductive success. For example, in many populations of a particular species, flowering is synchronized. This synchrony favors crossbreeding and allows seeds to be produced in favorable environments, particularly with respect to water and temperature.

## The Shoot Apex and Phase Changes

All multicellular organisms pass through a series of more or less defined developmental stages, each with its characteristic features. In humans, infancy, childhood, adolescence, and adulthood represent four general stages of development, with puberty as the dividing line between the nonreproductive and the reproductive phases. Similarly, plants pass through distinct developmental phases. The timing of these transitions often depends on environmental conditions, allowing plants to adapt to a changing environment. This is possible because plants continuously produce new organs from the shoot apical meristem.

Primordia start as small bumps of cells on the flanks of the shoot apex, and depending on the interactions between the signals they receive from the environment and the plant's genetic program, these primordia may give rise to leaves (during vegetative growth) or flowers (reproductive). In this way plant growth can be thought of as modular process, whereby the architecture and form of the plant will depend on how the successive primordia are specified. Along the stem each primordium leads to the production of a **phytomer**, a basic repeating unit consisting of a vegetative section and a meristem section. In the case of the leaf, the vegetative section accounts for nearly all the cells, whereas in many flowers, the meristem section is large and often subsumes the vegetative region. This simple repeating unit is enormously flexible, and is common to plants as diverse as daisies and oaks.

The transitions between different phases are tightly developmentally regulated, since the plant must integrate information from the environment as well as endogenous signals to maximize its reproductive fitness. The following sections describe the major pathways that control these decisions.

### Plant development has three phases

Postembryonic development in plants can be divided into three phases:

1. The juvenile phase
2. The adult vegetative phase
3. The adult reproductive phase

The transition from one phase to another is called **phase change** (Poethig 2003).

The primary distinction between the juvenile and the adult vegetative phases is that the latter has the ability to form reproductive structures: flowers in angiosperms, cones in gymnosperms. However, flowering, which represents the expression of the reproductive competence of the adult phase, often depends on specific environmental and developmental signals. Thus the absence of flowering itself is not a reliable indicator of juvenility.

The transition from juvenile to adult is frequently accompanied by changes in vegetative characteristics, such as leaf morphology, phyllotaxy (the arrangement of leaves on the stem), thorniness, rooting capacity, and leaf retention in deciduous plants such as English ivy (*Hedera helix*) (**FIGURE 25.9**; see also **WEB TOPIC 25.1**). Such changes are most evident in woody perennials, but they are apparent in many herbaceous species as well. Unlike the abrupt transition from the adult vegetative phase to the reproductive phase, the transition from juvenile to vegetative adult is usually gradual, involving intermediate forms.

Ovate adult leaves

Fruit

Lobed juvenile leaves

**FIGURE 25.9** Juvenile and adult forms of English ivy (*Hedera helix*). The juvenile form has lobed palmate leaves arranged alternately, a climbing growth habit, and no flowers. The adult form (projecting out to the right) has entire ovate leaves arranged in spirals, an upright growth habit, and flowers that develop into fruits. (Courtesy of L. Rignanese.)

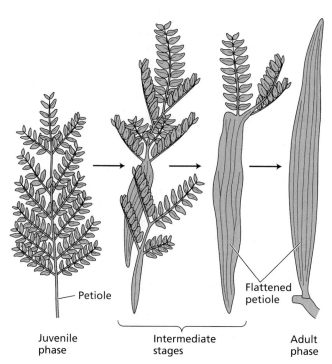

**FIGURE 25.10** Leaves of *Acacia heterophylla*, showing transitions from pinnately compound leaves (juvenile phase) to phyllodes (adult phase). Note that the previous phase is retained at the top of the leaf in the intermediate forms.

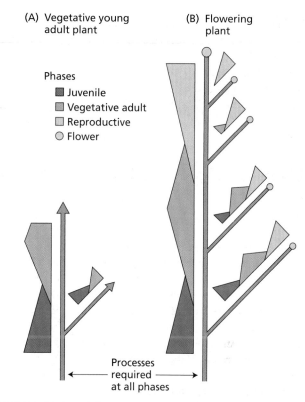

**FIGURE 25.11** Schematic representation of the combinatorial model of shoot development in maize. Overlapping gradients of expression of the juvenile, vegetative adult, and reproductive phases are indicated along the length of the main axis and branches. The continuous brown lines represent processes that are required during all phases of development. Each of the three phases may be regulated by separate developmental programs, with intermediate phases arising when the programs overlap. (A) Vegetative young adult plant. (B) Flowering plant. (After Poethig 1990.)

Sometimes the transition from juvenile to vegetative adult can be observed in a single leaf. A dramatic example of this is the progressive transformation of juvenile leaves of the leguminous tree *Acacia heterophylla* into phyllodes (petioles with the form and functions of leaves), a phenomenon first noted in the eighteenth century by Goethe. Whereas the juvenile pinnately compound leaves consist of rachis (stalk) and leaflets, adult phyllodes are specialized structures representing flattened petioles (**FIGURE 25.10**).

Intermediate structures also form during the transition from aquatic to aerial leaf types of aquatic plants such as common marestail (*Hippuris vulgaris*). As in the case of *A. heterophylla*, these intermediate forms possess distinct regions with different developmental patterns. To account for intermediate forms during the transition from juvenile to adult in maize (corn; *Zea mays*), a **combinatorial model** has been proposed (**FIGURE 25.11 and WEB TOPIC 25.2**). According to this model, shoot development can be described as a series of independently regulated, *overlapping* programs (juvenile, adult, and reproductive) that modulate the expression of a common set of developmental processes.

In the transition from juvenile to adult leaves, the intermediate forms indicate that different regions of the same leaf can express different developmental programs. Thus the cells at the tip of the leaf remain committed to the juvenile program, while the cells at the base of the leaf become committed to the adult program. The developmental fates of the two sets of cells in the same leaf are quite different.

### Juvenile tissues are produced first and are located at the base of the shoot

The time sequence of the three developmental phases results in a spatial gradient of juvenility along the shoot axis. Because growth in height is restricted to the apical meristem, the juvenile tissues and organs, which form first, are located at the base of the shoot. In rapidly flowering herbaceous species, the juvenile phase may last only a few days, and few juvenile structures are produced. In contrast, woody species have a more prolonged juvenile phase, in some cases lasting 30 to 40 years (**TABLE 25.1**). In these cases the juvenile structures can account for a significant portion of the mature plant.

Once the meristem has switched to the adult phase, only adult vegetative structures are produced, culminat-

**TABLE 25.1**
**Length of juvenile period in some woody plants**

| Species | Length of juvenile period |
| --- | --- |
| Rose (*Rosa* [hybrid tea]) | 20–30 days |
| Grape (*Vitis* spp.) | 1 year |
| Apple (*Malus* spp.) | 4–8 years |
| *Citrus* spp. | 5–8 years |
| English ivy (*Hedera helix*) | 5–10 years |
| Redwood (*Sequoia sempervirens*) | 5–15 years |
| Sycamore maple (*Acer pseudoplatanus*) | 15–20 years |
| English oak (*Quercus robur*) | 25–30 years |
| European beech (*Fagus sylvatica*) | 30–40 years |

*Source*: Clark 1983.

ing in floral evocation. The adult and reproductive phases are therefore located in the upper and peripheral regions of the shoot.

Attainment of a sufficiently large size appears to be more important than the plant's chronological age in determining the transition to the adult phase. Conditions that retard growth, such as mineral deficiencies, low light, water stress, defoliation, and low temperature tend to prolong the juvenile phase or even cause **rejuvenation** (reversion to juvenility) of adult shoots. In contrast, conditions that promote vigorous growth accelerate the transition to the adult phase. When growth is accelerated, exposure to the correct flower-inducing treatment can result in flowering.

Although plant size seems to be the most important factor, it is not always clear which specific component associated with size is critical. In some *Nicotiana* species, it appears that plants must produce a certain number of leaves to transmit a sufficient amount of the floral stimulus to the apex.

Once the adult phase has been attained, it is relatively stable and is maintained during vegetative propagation or grafting. For example, cuttings taken from the basal region of mature plants of English ivy (*Hedera helix*) develop into juvenile plants, while those taken from the tip develop into adult plants. When scions were taken from the base of a flowering silver birch (*Betula verrucosa*) and grafted onto seedling rootstocks, there were no flowers on the grafts for the first 2 years. In contrast, grafts taken from the top of the mature tree flowered freely.

### Phase changes can be influenced by nutrients, gibberellins, and other signals

The transition at the shoot apex from the juvenile to the adult phase can be affected by transmissible factors from the rest of the plant. In many plants, exposure to low-light conditions prolongs juvenility or causes reversion to juvenility. A major consequence of a low-light regime is a reduction in the supply of carbohydrates to the apex; thus carbo-

hydrate supply, especially sucrose, may play a role in the transition between juvenility and maturity. Carbohydrate supply as a source of energy and raw material can affect the size of the apex. For example, in the florist's chrysanthemum (*Chrysanthemum × morifolium*), flower primordia are not initiated until a minimum apex size has been reached.

The apex receives a variety of hormonal and other factors from the rest of the plant in addition to carbohydrates and other nutrients. Experimental evidence shows that the application of gibberellins (GA) causes reproductive structures to form in young, juvenile plants of several conifer families. The involvement of *endogenous* GAs in the control of reproduction is also indicated by the fact that other treatments that accelerate cone production in pines (e.g., root removal, water stress, and nitrogen starvation) often also result in a buildup of GAs in the plant.

### Competence and determination are two stages in floral evocation

The term *juvenility* has different meanings for herbaceous and woody species. Whereas juvenile herbaceous meristems flower readily when grafted onto flowering adult plants (see WEB TOPIC 25.3), juvenile woody meristems generally do not. What is the difference between the two?

Extensive studies in tobacco (*Nicotiana tabacum*, an herbaceous plant) have demonstrated that floral evocation requires the apical bud to pass through two developmental stages (FIGURE 25.12) (McDaniel et al. 1992). One stage is the acquisition of competence. A bud is said to be **competent** if it is able to flower when given the appropriate developmental signal.

For example, if a vegetative shoot (scion) is grafted onto a flowering stock and the scion flowers immediately, it is demonstrably capable of responding to the level of floral stimulus present in the stock and is therefore competent. Failure of the scion to flower would indicate that the shoot apical meristem has not yet attained competence.

The next stage that a competent vegetative bud goes through is determination. A bud is said to be *determined* if it progresses to the next developmental stage (flowering) even after being removed from its normal context. Thus a florally determined bud will produce flowers even if it is grafted onto a vegetative plant that is not producing any floral stimulus.

In a strain of day-neutral tobacco, plants typically flower after producing about 41 leaves or nodes. In an experiment to measure the floral determination of the axillary buds, flowering tobacco plants were decapitated just above the thirty-fourth leaf (from the bottom). Released from apical dominance, the axillary bud of the thirty-fourth leaf grew out, and after producing about seven more leaves (for a total of 41), it flowered (FIGURE 25.13A) (McDaniel

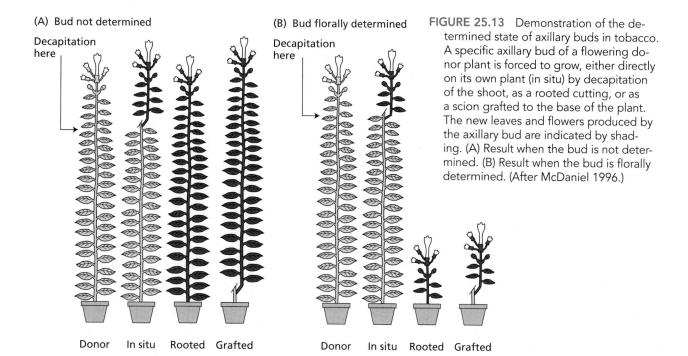

Induction

**Competent:**

Able to respond in expected manner when given the appropriate developmental signals.

Photoperiod

**Determined:**

Able to follow same developmental program even after removal from its normal position in plant.

Signal

Hormones ?

**Expressed:**

The apical meristem undergoes morphogenesis.

Vegetative growth

Flowers

**FIGURE 25.12** A simplified model for floral evocation at the shoot apex in which the cells of the vegetative meristem acquire new developmental fates. To initiate floral development, the cells of the meristem must first become competent. A competent vegetative meristem is one that can respond to a floral stimulus (induction) by becoming florally determined (committed to producing a flower). The determined state is usually expressed, but this may require an additional signal. (After McDaniel et al. 1992.)

1996). However, if the thirty-fourth bud was excised from the plant and either rooted or grafted onto a stock without leaves near the base, it produced nearly a complete set of leaves before flowering. These results show that the thirty-fourth bud was not yet florally determined.

In another experiment, the donor plant was decapitated above the thirty-seventh leaf. This time the thirty-seventh axillary bud flowered after producing only about four leaves *in all three situations* (**FIGURE 25.13B**). This result demonstrates that the terminal bud became florally determined after initiating 37 leaves.

Extensive grafting of shoot tips among tobacco varieties has established that the number of nodes a meristem produces before flowering is a function of two factors: the strength of the floral stimulus from the leaves and the competence of the meristem to respond to the signal (McDaniel et al. 1996).

In some cases the *expression* of flowering may be delayed or arrested even after the apex becomes determined, unless it receives a second developmental signal that stimulates expression (see Figure 25.12). For example, intact darnel ryegrass (*Lolium temulentum*) plants become committed to flowering after a single exposure to a long day. If the *Lolium* shoot apical meristem is excised 28 hours after the beginning of the long day and cultured in vitro, it will produce normal inflorescences in culture, but only if the hormone gibberellic acid is present in the medium.

Because apices cultured from plants grown exclusively in short days never flower, even in the presence of GA, we can conclude that long days are required for determination in *Lolium*, whereas GA is required for *expression* of the determined state.

(A) Bud not determined

Decapitation here

Donor    In situ    Rooted    Grafted

(B) Bud florally determined

Decapitation here

Donor    In situ    Rooted    Grafted

**FIGURE 25.13** Demonstration of the determined state of axillary buds in tobacco. A specific axillary bud of a flowering donor plant is forced to grow, either directly on its own plant (in situ) by decapitation of the shoot, as a rooted cutting, or as a scion grafted to the base of the plant. The new leaves and flowers produced by the axillary bud are indicated by shading. (A) Result when the bud is not determined. (B) Result when the bud is florally determined. (After McDaniel 1996.)

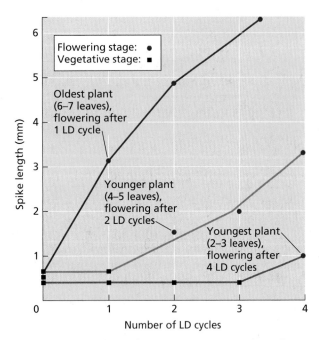

**FIGURE 25.14**   Effect of plant age on the number of long-day (LD) inductive cycles required for flowering in the long-day plant darnel ryegrass (*Lolium temulentum*). An inductive long-day cycle consisted of 8 hours of sunlight followed by 16 hours of low-intensity incandescent light. The older the plant is, the fewer photoinductive cycles are needed to produce flowering.

In general, once a meristem has become competent, it exhibits an increasing tendency to flower with age (leaf number). For example, in plants controlled by day length, the number of short-day or long-day cycles necessary to achieve flowering is often fewer in older plants (**FIGURE 25.14**). As will be discussed later in the chapter, this increasing tendency to flower with age has its physiological basis in the greater capacity of the leaves to produce a floral stimulus.

Before discussing how plants perceive day length, however, we will lay the foundation by examining how organisms measure time in general. This topic is known as **chronobiology**, or the study of **biological clocks**. The best-understood biological clock is the circadian rhythm.

# Circadian Rhythms: The Clock Within

Organisms are normally subjected to daily cycles of light and darkness, and both plants and animals often exhibit rhythmic behavior in association with these changes. Examples of such rhythms include leaf and petal movements (day and night positions), stomatal opening and closing, growth and sporulation patterns in fungi (e.g., *Pilobolus* and *Neurospora*), time of day of pupal emergence (the fruit fly *Drosophila*), and activity cycles in rodents, as well as daily changes in the rates of metabolic processes such as photosynthesis and respiration.

When organisms are transferred from daily light–dark cycles to continuous darkness or continuous light, many of these rhythms continue to be expressed, at least for several days. Under such uniform conditions the period of the rhythm is close to 24 hours, and consequently the term circadian rhythm (from Latin *circa*, about; *diem*, day) is applied (see Chapter 17). Because they continue in a constant light or dark environment, these circadian rhythms cannot be direct responses to the presence or absence of light but must be based on an internal pacemaker, often called an endogenous oscillator. A molecular model for a plant endogenous oscillator was described in Chapter 17.

The endogenous oscillator is coupled to a variety of physiological processes, such as leaf movement or photosynthesis, and it maintains the rhythm. For this reason the endogenous oscillator can be considered the clock mechanism, and the physiological functions that are being regulated, such as leaf movements or photosynthesis, are sometimes referred to as the hands of the clock.

## Circadian rhythms exhibit characteristic features

Circadian rhythms arise from cyclic phenomena that are defined by three parameters:

1. **Period** is the time between comparable points in the repeating cycle. Typically the period is measured as the time between consecutive maxima (peaks) or minima (troughs) (**FIGURE 25.15A**).

2. **Phase*** is any point in the cycle that is recognizable by its relationship to the rest of the cycle. The most obvious phase points are the peak and trough positions.

3. **Amplitude** is usually considered to be the distance between peak and trough. The amplitude of a biological rhythm can often vary while the period remains unchanged (as, for example, in **FIGURE 25.15C**).

In constant light or darkness, rhythms depart from an exact 24-hour period. The rhythms then drift in relation to solar time, either gaining or losing time depending on whether the period is shorter or longer than 24 hours. Under natural conditions, the endogenous oscillator is **entrained** (synchronized) to a true 24-hour period by environmental signals, the most important of which are the light-to-dark transition at dusk and the dark-to-light transition at dawn (**FIGURE 25.15B**).

Such environmental signals are termed **zeitgebers** (German for "time givers"). When such signals are removed—for example, by transfer to continuous darkness—the rhythm is said to be **free-running**, and it reverts to the circadian period that is characteristic of the particular organism (see Figure 25.15B).

---

*The term *phase* in this context should not be confused with the term *phase change* in meristem development discussed earlier.

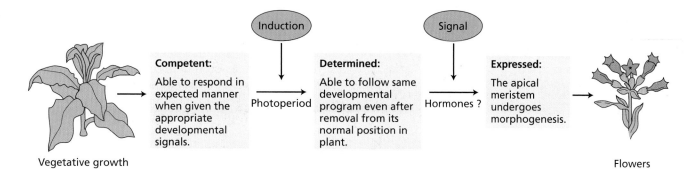

Vegetative growth

Flowers

**FIGURE 25.12** A simplified model for floral evocation at the shoot apex in which the cells of the vegetative meristem acquire new developmental fates. To initiate floral development, the cells of the meristem must first become competent. A competent vegetative meristem is one that can respond to a floral stimulus (induction) by becoming florally determined (committed to producing a flower). The determined state is usually expressed, but this may require an additional signal. (After McDaniel et al. 1992.)

1996). However, if the thirty-fourth bud was excised from the plant and either rooted or grafted onto a stock without leaves near the base, it produced nearly a complete set of leaves before flowering. These results show that the thirty-fourth bud was not yet florally determined.

In another experiment, the donor plant was decapitated above the thirty-seventh leaf. This time the thirty-seventh axillary bud flowered after producing only about four leaves *in all three situations* (**FIGURE 25.13B**). This result demonstrates that the terminal bud became florally determined after initiating 37 leaves.

Extensive grafting of shoot tips among tobacco varieties has established that the number of nodes a meristem produces before flowering is a function of two factors: the strength of the floral stimulus from the leaves and the competence of the meristem to respond to the signal (McDaniel et al. 1996).

In some cases the *expression* of flowering may be delayed or arrested even after the apex becomes determined, unless it receives a second developmental signal that stimulates expression (see Figure 25.12). For example, intact darnel ryegrass (*Lolium temulentum*) plants become committed to flowering after a single exposure to a long day. If the *Lolium* shoot apical meristem is excised 28 hours after the beginning of the long day and cultured in vitro, it will produce normal inflorescences in culture, but only if the hormone gibberellic acid is present in the medium.

Because apices cultured from plants grown exclusively in short days never flower, even in the presence of GA, we can conclude that long days are required for determination in *Lolium*, whereas GA is required for *expression* of the determined state.

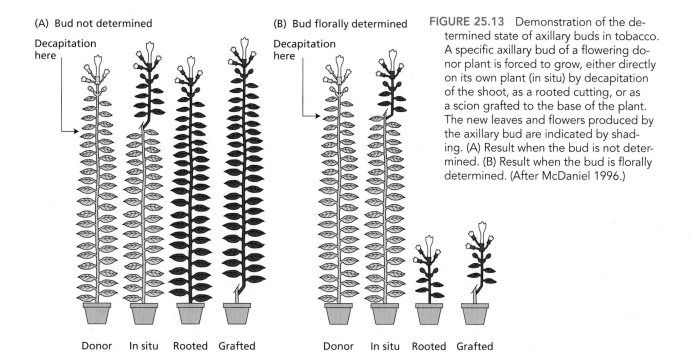

(A)  Bud not determined

Decapitation here

Donor    In situ    Rooted    Grafted

(B)  Bud florally determined

Decapitation here

Donor    In situ    Rooted    Grafted

**FIGURE 25.13** Demonstration of the determined state of axillary buds in tobacco. A specific axillary bud of a flowering donor plant is forced to grow, either directly on its own plant (in situ) by decapitation of the shoot, as a rooted cutting, or as a scion grafted to the base of the plant. The new leaves and flowers produced by the axillary bud are indicated by shading. (A) Result when the bud is not determined. (B) Result when the bud is florally determined. (After McDaniel 1996.)

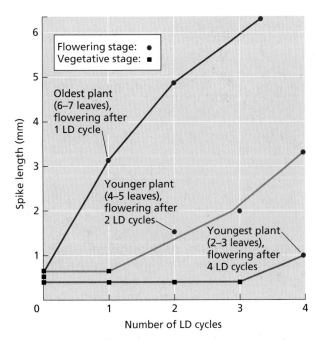

**FIGURE 25.14** Effect of plant age on the number of long-day (LD) inductive cycles required for flowering in the long-day plant darnel ryegrass (*Lolium temulentum*). An inductive long-day cycle consisted of 8 hours of sunlight followed by 16 hours of low-intensity incandescent light. The older the plant is, the fewer photoinductive cycles are needed to produce flowering.

In general, once a meristem has become competent, it exhibits an increasing tendency to flower with age (leaf number). For example, in plants controlled by day length, the number of short-day or long-day cycles necessary to achieve flowering is often fewer in older plants (**FIGURE 25.14**). As will be discussed later in the chapter, this increasing tendency to flower with age has its physiological basis in the greater capacity of the leaves to produce a floral stimulus.

Before discussing how plants perceive day length, however, we will lay the foundation by examining how organisms measure time in general. This topic is known as **chronobiology**, or the study of **biological clocks**. The best-understood biological clock is the circadian rhythm.

## Circadian Rhythms: The Clock Within

Organisms are normally subjected to daily cycles of light and darkness, and both plants and animals often exhibit rhythmic behavior in association with these changes. Examples of such rhythms include leaf and petal movements (day and night positions), stomatal opening and closing, growth and sporulation patterns in fungi (e.g., *Pilobolus* and *Neurospora*), time of day of pupal emergence (the fruit fly *Drosophila*), and activity cycles in rodents, as well as daily changes in the rates of metabolic processes such as photosynthesis and respiration.

When organisms are transferred from daily light–dark cycles to continuous darkness or continuous light, many of these rhythms continue to be expressed, at least for several days. Under such uniform conditions the period of the rhythm is close to 24 hours, and consequently the term circadian rhythm (from Latin *circa*, about; *diem*, day) is applied (see Chapter 17). Because they continue in a constant light or dark environment, these circadian rhythms cannot be direct responses to the presence or absence of light but must be based on an internal pacemaker, often called an endogenous oscillator. A molecular model for a plant endogenous oscillator was described in Chapter 17.

The endogenous oscillator is coupled to a variety of physiological processes, such as leaf movement or photosynthesis, and it maintains the rhythm. For this reason the endogenous oscillator can be considered the clock mechanism, and the physiological functions that are being regulated, such as leaf movements or photosynthesis, are sometimes referred to as the hands of the clock.

### Circadian rhythms exhibit characteristic features

Circadian rhythms arise from cyclic phenomena that are defined by three parameters:

1. **Period** is the time between comparable points in the repeating cycle. Typically the period is measured as the time between consecutive maxima (peaks) or minima (troughs) (**FIGURE 25.15A**).

2. **Phase*** is any point in the cycle that is recognizable by its relationship to the rest of the cycle. The most obvious phase points are the peak and trough positions.

3. **Amplitude** is usually considered to be the distance between peak and trough. The amplitude of a biological rhythm can often vary while the period remains unchanged (as, for example, in **FIGURE 25.15C**).

In constant light or darkness, rhythms depart from an exact 24-hour period. The rhythms then drift in relation to solar time, either gaining or losing time depending on whether the period is shorter or longer than 24 hours. Under natural conditions, the endogenous oscillator is **entrained** (synchronized) to a true 24-hour period by environmental signals, the most important of which are the light-to-dark transition at dusk and the dark-to-light transition at dawn (**FIGURE 25.15B**).

Such environmental signals are termed **zeitgebers** (German for "time givers"). When such signals are removed—for example, by transfer to continuous darkness—the rhythm is said to be **free-running**, and it reverts to the circadian period that is characteristic of the particular organism (see Figure 25.15B).

---

*The term *phase* in this context should not be confused with the term *phase change* in meristem development discussed earlier.

(A)

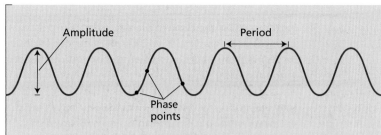

A typical circadian rhythm. The **period** is the time between comparable points in the repeating cycle; the **phase** is any point in the repeating cycle recognizable by its relationship with the rest of the cycle; the **amplitude** is the distance between peak and trough.

(B)

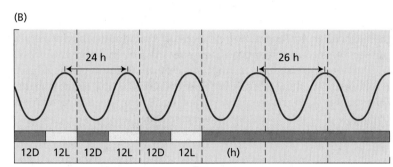

A circadian rhythm entrained to a 24 h light–dark (L–D) cycle and its reversion to the free-running period (26 h in this example) following transfer to continuous darkness.

(C)

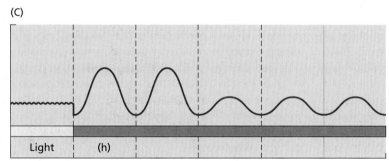

Suspension of a circadian rhythm in continuous bright light and the release or restarting of the rhythm following transfer to darkness.

(D)

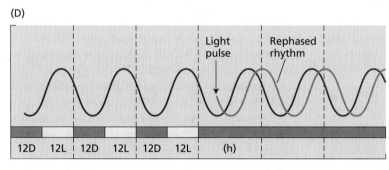

Typical phase-shifting response to a light pulse given shortly after transfer to darkness. The rhythm is rephased (delayed) without its period being changed.

**FIGURE 25.15** Some characteristics of circadian rhythms.

Although the rhythms are generated internally, they normally require an environmental signal, such as exposure to light or a change in temperature, to initiate their expression. In addition, many rhythms damp out (i.e., the amplitude decreases) when the organism is subjected to a constant environment for several cycles. When this occurs an environmental zeitgeber, such as a transfer from light to dark or a change in temperature, is required to restart the rhythm (see Figure 25.15C). Note that *the clock itself does not damp out; only the coupling between the molecular clock (endogenous oscillator) and the physiological function is affected.*

The circadian clock would be of no value to the organism if it could not keep accurate time under the fluctuating temperatures experienced in natural conditions. Indeed, temperature has little or no effect on the period of the free-running rhythm. The feature that enables the clock to keep time at different temperatures is called **temperature compensation.** Although all of the biochemical steps in the pathway are temperature-sensitive, their temperature responses probably cancel each other. For example, changes in the rates of synthesis of intermediates could be compensated for by parallel changes in their rates of degradation. In this way, the steady-state levels of clock regulators would remain constant at different temperatures.

### Phase shifting adjusts circadian rhythms to different day–night cycles

In circadian rhythms, physiological responses are coupled to a specific time point of the endogenous oscillator so that the response occurs at a particular time of day. A single oscillator can be coupled to multiple circadian rhythms, which may even be out of phase with each other.

How do such responses remain on time when the daily durations of light and darkness change with the seasons? Investigators typically test the response of the endogenous oscillator by placing the organism in continuous darkness and examining the response to a short pulse of light (usually less than 1 hour) given at different phase points in the free-running rhythm. When an organism is entrained to a cycle of 12 hours light and 12 hours dark and then allowed to free-run in darkness, the phase of the rhythm that coincides with the light period of the previous entraining cycle is called the **subjective day**, and the phase that coincides with the dark period is called the **subjective night**.

If a light pulse is given during the first few hours of the subjective night, the rhythm is delayed; the organism interprets the light pulse as the end of the previous day (**FIGURE 25.15D**). In contrast, a light pulse given toward the end of the subjective night advances the phase of the rhythm; now the organism interprets the light pulse as the beginning of the following day.

This is precisely the response that would be expected if the rhythm were able to stay on local time even when the seasons change. These phase-shifting responses enable the rhythm to be entrained to approximately 24-hour cycles with different durations of light and darkness, and they demonstrate that the rhythm can adjust to seasonal variations in day length.

### Phytochromes and cryptochromes entrain the clock

The molecular mechanism whereby a light signal causes phase shifting is not yet known, but studies in Arabidopsis have identified some of the key elements of the circadian oscillator and its inputs and outputs (see Chapter 17). The low levels and specific wavelengths of light that can induce phase shifting indicate that the light response must be mediated by specific photoreceptors rather than by photosynthetic rate. For example, the red-light entrainment of rhythmic day/night leaf movements in *Samanea*, a semitropical leguminous tree, is a low-fluence response mediated by phytochrome (see Chapter 17).

Arabidopsis has five **phytochromes**, and all but one of them (phytochrome C) have been implicated in clock entrainment. Each phytochrome acts as a specific photoreceptor for red, far-red, or blue light. As well as phyto-

chromes, plants sense light through **cryptochromes** (CRY), and the CRY1 and CRY2 proteins participate in blue-light entrainment of the clock in plants, as they do in insects and mammals (Devlin and Kay 2000; see Chapter 18). Surprisingly, CRY proteins also appear to be required for normal entrainment by red light. Since these proteins do not absorb red light, this requirement suggests that CRY1 and CRY2 may act as intermediates in phytochrome signaling during entrainment of the clock (Yanovsky and Kay 2001).

In *Drosophila*, CRY proteins interact physically with clock components and thus constitute part of the oscillator mechanism (Devlin and Kay 2000). However, this does not appear to be the case in Arabidopsis, in which *cry1/cry2* double mutants are impaired in entrainment but otherwise have normal circadian rhythms. In plants it has been shown that photoactivated CRY2 is able to activate flowering in response to blue light by up-regulating expression of *FT* directly (Liu et al. 2008).

## Photoperiodism: Monitoring Day Length

As we have seen, the circadian clock enables organisms to repeat particular molecular or biochemical events at specific times of day or night. **Photoperiodism**, or the ability of an organism to detect day length, makes it possible for an event to occur at a particular time of *year*, thus allowing for a *seasonal* response. Circadian rhythms and photoperiodism have the common property of responding to cycles of light and darkness.

Precisely at the equator, day length and night length are equal and constant throughout the year. As one moves away from the equator toward the poles, the days become longer in summer and shorter in winter (**FIGURE 25.16**). Plant species have evolved the ability to detect these seasonal changes in day length, and their specific photoperiodic responses are strongly influenced by the latitude in which they originated.

Photoperiodic phenomena are found in both animals and plants. In the animal kingdom, day length controls such seasonal activities as hibernation, development of summer or winter coats, and reproductive activity. Plant responses controlled by day length are numerous; they include the initiation of flowering, asexual reproduction, the formation of storage organs, and the onset of dormancy.

### Plants can be classified according to their photoperiodic responses

Numerous plant species flower during the long days of summer, and for many years plant physiologists believed that the correlation between long days and flowering was a consequence of the accumulation of photosynthetic products synthesized during long days.

(A)

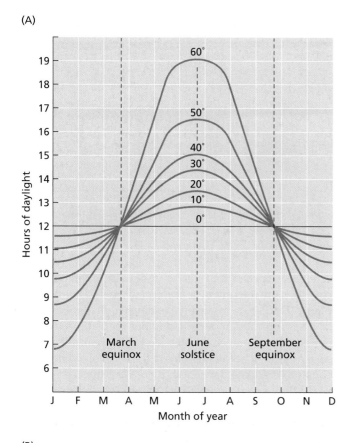

(B)

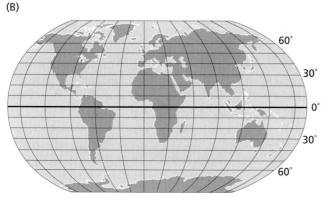

**FIGURE 25.16**   (A) The effect of latitude on day length at different times of the year in the northern hemisphere. Day length was measured on the twentieth of each month. (B) Global map showing longitudes and latitudes.

**FIGURE 25.17**   'Maryland Mammoth' mutant of tobacco (right) compared to wild-type tobacco (left). Both plants were grown during summer in the greenhouse. (University of Wisconsin graduate students used for scale.) (Courtesy of R. Amasino.)

This hypothesis was shown to be incorrect by the work of Wightman Garner and Henry Allard, conducted in the 1920s at the U.S. Department of Agriculture laboratories in Beltsville, Maryland. Garner and Allard found that a mutant variety of tobacco, 'Maryland Mammoth,' grew profusely to about 5 m in height but failed to flower in the prevailing conditions of summer (**FIGURE 25.17**). However, the plants flowered in the greenhouse during the winter under natural light conditions.

These results ultimately led Garner and Allard to test the effect of artificially shortened days by covering plants grown during the long days of summer with a light-tight tent from late in the afternoon until the following morning. These artificial short days also caused the plants to flower. Garner and Allard concluded that day length, rather than the accumulation of photosynthate, was the determining factor in flowering. They were able to confirm their hypothesis in many different species and conditions. This work laid the foundations for the extensive subsequent research on photoperiodic responses.

Although many other aspects of plants' development may also be affected by day length, flowering is the response that has been studied the most. Flowering species tend to fall into one of two main photoperiodic response categories: short-day plants and long-day plants.

- **Short-day plants (SDPs)** flower only in short days (*qualitative* SDPs), or their flowering is accelerated by short days (*quantitative* SDPs).

- **Long-day plants (LDPs)** flower only in long days (*qualitative* LDPs), or their flowering is accelerated by long days (*quantitative* LDPs).

The essential distinction between long-day and short-day plants is that flowering in LDPs is promoted only when the day length *exceeds* a certain duration, called the **critical day**

**FIGURE 25.18** The photoperiodic response in long- and short-day plants. The critical duration varies between species: In this example, both the SDPs and the LDPs would flower in photoperiods between 12 and 14 h long.

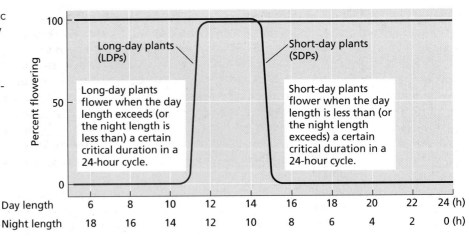

length, in every 24-hour cycle, whereas promotion of flowering in SDPs requires a day length that is *less than* the critical day length. The absolute value of the critical day length varies widely among species, and only when flowering is examined for a range of day lengths can the correct photoperiodic classification be established (**FIGURE 25.18**).

Long-day plants can effectively measure the lengthening days of spring or early summer and delay flowering until the critical day length is reached. Many varieties of wheat (*Triticum aestivum*) behave in this way. SDPs often flower in the fall when the days shorten below the critical day length, as in many varieties of *Chrysanthemum* × *morifolium*. However, day length alone is an ambiguous signal, because it cannot distinguish between spring and fall.

Plants exhibit several adaptations for avoiding the ambiguity of the day length signal. One is the presence of a juvenile phase that prevents the plant from responding to day length during the spring. Another mechanism for avoiding the ambiguity of day length is the coupling of a temperature requirement to a photoperiodic response. Certain plant species, such as winter wheat, do not respond to photoperiod until after a cold period (vernalization or overwintering) has occurred. (We will discuss vernalization later in the chapter.)

Other plants avoid seasonal ambiguity by distinguishing between *shortening* and *lengthening* days. Such "dual day length plants" fall into two categories:

- **Long–short-day plants (LSDPs)** flower only after a sequence of long days followed by short days. LSDPs, such as *Bryophyllum*, *Kalanchoe*, and night-blooming jasmine (*Cestrum nocturnum*), flower in the late summer and fall, when the days are shortening.

- **Short–long-day plants (SLDPs)** flower only after a sequence of short days followed by long days. SLDPs, such as white clover (*Trifolium repens*), Canterbury bells (*Campanula medium*), and echeveria (*Echeveria harmsii*), flower in the early spring in response to lengthening days.

Finally, species that flower under any photoperiodic condition are referred to as *day-neutral plants*. **Day-neutral plants (DNPs)** are insensitive to day length. Flowering in DNPs is typically under autonomous regulation—that is, internal developmental control. Some day-neutral species, such as kidney bean (*Phaseolus vulgaris*), evolved near the equator where the day length is constant throughout the year. Many desert annuals, such as desert paintbrush (*Castilleja chromosa*) and desert sand verbena (*Abronia villosa*), evolved to germinate, grow, and flower quickly whenever sufficient water is available. These are also DNPs.

### The leaf is the site of perception of the photoperiodic signal

The photoperiodic stimulus in both LDPs and SDPs is perceived by the leaves. For example, treatment of a single leaf of the SDP *Xanthium* with short photoperiods is sufficient to cause the formation of flowers, even when the rest of the plant is exposed to long days. Thus, in response to photoperiod the leaf transmits a signal that regulates the transition to flowering at the shoot apex. The photoperiod-regulated processes that occur in the leaves resulting in the transmission of a floral stimulus to the shoot apex are referred to collectively as **photoperiodic induction**.

Photoperiodic induction can take place in a leaf that has been separated from the plant. For example, in the SDP *Perilla crispa*, an excised leaf exposed to short days can cause flowering when subsequently grafted to a noninduced plant maintained in long days (Zeevaart and Boyer 1987). This result indicates that photoperiodic induction depends on events that take place exclusively in the leaf.

### Plants monitor day length by measuring the length of the night

Under natural conditions, day and night lengths configure a 24-hour cycle of light and darkness. In principle, a plant could perceive a critical day length by measuring the duration of

(A)

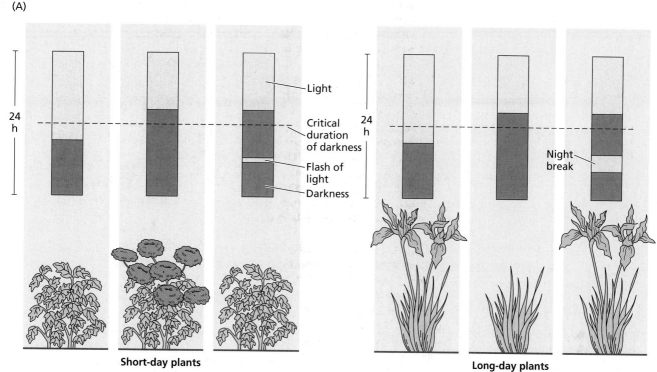

**Short-day plants**

Short-day (long-night) plants flower when night length exceeds a critical dark period. Interruption of the dark period by a brief light treatment (a night break) prevents flowering.

**Long-day plants**

Long-day (short-night) plants flower if the night length is shorter than a critical period. In some long-day plants, shortening the night with a night break induces flowering.

(B)

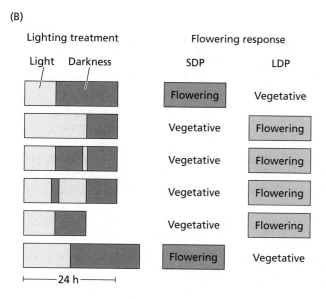

**FIGURE 25.19** The photoperiodic regulation of flowering. (A) Effects on SDPs and LDPs. (B) Effects of the duration of the dark period on flowering. Treating short- and long-day plants with different photoperiods clearly shows that the critical variable is the length of the dark period.

SDPs did not flower when short days were followed by short nights.

More detailed experiments demonstrated that photoperiodic timekeeping in SDPs is a matter of measuring the duration of darkness. For example, flowering occurred only when the dark period exceeded 8.5 hours in cocklebur (*Xanthium strumarium*) or 10 hours in soybean (*Glycine max*). The duration of darkness was also shown to be important in LDPs (see Figure 25.19). These plants were found to flower in short days, provided that the accompanying night length was also short; however, a regime of long days followed by long nights was ineffective.

## Night breaks can cancel the effect of the dark period

A feature that underscores the importance of the dark period is that it can be made ineffective by interruption with a short exposure to light, called a **night break** (see Figure 25.19A). In contrast, interrupting a long day with a brief dark period does not cancel the effect of the long day (see Figure 25.19B). Night-break treatments of only a few minutes are effective in *preventing* flowering in many SDPs,

either light or darkness. Much experimental work in the early studies of photoperiodism was devoted to establishing which part of the light–dark cycle is the controlling factor in flowering. Results showed that flowering of SDPs is determined primarily by the duration of darkness (**FIGURE 25.19A**). It was possible to induce flowering in SDPs with light periods longer than the critical value, provided that these were followed by sufficiently long nights (**FIGURE 25.19B**). Similarly,

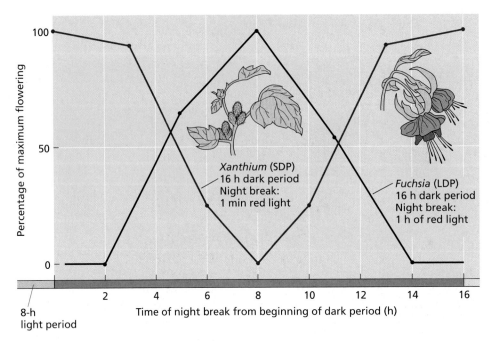

**FIGURE 25.20** The time at which a night break is given determines the flowering response. When given during a long dark period, a night break promotes flowering in LDPs and inhibits flowering in SDPs. In both cases, the greatest effect on flowering occurs when the night break is given near the middle of the 16-hour dark period. The LDP *Fuchsia* was given a 1-hour exposure to red light in a 16-hour dark period. *Xanthium* was exposed to red light for 1 minute in a 16-hour dark period. (Data for *Fuchsia* from Vince-Prue 1975; data for *Xanthium* from Salisbury 1963 and Papenfuss and Salisbury 1967.)

including *Xanthium* and *Pharbitis*, but much longer exposures are often required to *promote* flowering in LDPs.

In addition, the effect of a night break varies greatly according to the time when it is given. For both LDPs and SDPs, a night break was found to be most effective when given near the middle of a dark period of 16 hours (**FIGURE 25.20**).

The discovery of the night-break effect, and its time dependence, had several important consequences. It established the central role of the dark period and provided a valuable probe for studying photoperiodic timekeeping. Because only small amounts of light are needed, it became possible to study the action and identity of the photoreceptor without the interfering effects of photosynthesis and other nonphotoperiodic phenomena. This discovery has also led to the development of commercial methods for regulating the time of flowering in horticultural species, such as *Kalanchoe*, chrysanthemum, and poinsettia (*Euphorbia pulcherrima*).

### The circadian clock and photoperiodic timekeeping

The decisive effect of night length on flowering indicates that measuring the passage of time in darkness is central to photoperiodic timekeeping. Most of the available evidence favors a mechanism based on a circadian rhythm (Bünning 1960). According to the **clock hypothesis**, photoperiodic timekeeping depends on an endogenous circadian oscillator of the type discussed earlier in the chapter (see also Chapter 17). The central oscillator is coupled to various physiological processes that involve gene expression, including flowering in photoperiodic species.

Measurements of the effect of a night break on flowering can be used to investigate the role of circadian rhythms in photoperiodic timekeeping. For example, when soybean plants, which are SDPs, are transferred from an 8-hour light period to an extended 64-hour dark period, the flowering response to night breaks shows a circadian rhythm (**FIGURE 25.21**).

This type of experiment provides strong support for the clock hypothesis. If this SDP were simply measuring the length of night by the accumulation of a particular intermediate in the dark, any dark period greater than the critical night length should cause flowering. Yet long dark periods are not inductive for flowering if the light break is given at a time that does not properly coincide with a certain phase of the endogenous circadian oscillator. This finding demonstrates that flowering in SDPs requires both a dark period of sufficient duration and a dawn signal at an appropriate time in the circadian cycle (see Figure 25.15).

Further evidence for the role of a circadian oscillator in photoperiod measurement is the observation that the photoperiodic response can be phase-shifted by light treatments (see **WEB TOPIC 25.4**).

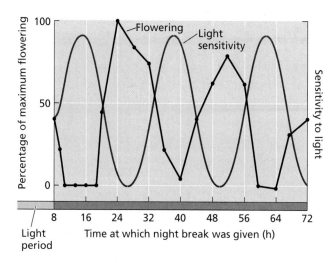

**FIGURE 25.21** Rhythmic flowering in response to night breaks. In this experiment, the SDP soybean (*Glycine max*) received cycles of an 8-hour light period followed by a 64-hour dark period. A 4-hour night break was given at various times during the long inductive dark period. The flowering response, plotted as the percentage of the maximum, was then plotted for each night break given. Note that a night break given at 26 hours induced maximum flowering, while no flowering was obtained when the night break was given at 40 hours. Moreover, this experiment demonstrates that the sensitivity to the effect of the night break shows a circadian rhythm. These data support a model in which flowering in SDPs is induced only when dawn (or a night break) occurs after the completion of the light-sensitive phase. In LDPs the light break must coincide with the light-sensitive phase for flowering to occur. (Data from Coulter and Hamner 1964.)

## The coincidence model is based on oscillating light sensitivity

How does an oscillation with a 24-hour period measure a critical duration of darkness of, say, 8 to 9 hours, as in the SDP *Xanthium*? In 1936 Erwin Bünning proposed that the control of flowering by photoperiodism is achieved by an oscillation of phases with different sensitivities to light. This proposal has evolved into the **coincidence model** (Bünning 1960), in which the circadian oscillator controls the timing of light-sensitive and light-insensitive phases.

The ability of light either to promote or to inhibit flowering depends on the phase in which the light is given. When a light signal is administered during the light-sensitive phase of the rhythm, the effect is either to *promote* flowering in LDPs or to *prevent* flowering in SDPs. As shown in Figure 25.21, the phases of sensitivity and insensitivity to light continue to oscillate in darkness. Flowering in SDPs is induced only when exposure to light from a night break or from dawn occurs after completion of the light-sensitive phase of the rhythm.

If a similar experiment is performed with an LDP, flowering is induced only when the night break occurs *during* the light-sensitive phase of the rhythm. In other words, *flowering in both SDPs and LDPs is induced when the light exposure is coincident with the appropriate phase of the rhythm*. This continued oscillation of sensitive and insensitive phases in the absence of dawn and dusk light signals is characteristic of a variety of processes controlled by the circadian oscillator.

## The coincidence of CONSTANS expression and light promotes flowering in LDPs

According to the coincidence model, plant flowering responses are sensitive to light only at certain times of the day–night cycle. A key component of a regulatory pathway that promotes flowering of Arabidopsis in long days is a gene called **CONSTANS** (*CO*), which encodes a zinc finger protein that regulates the transcription of other genes. *CO* was first identified in an Arabidopsis mutant, *co*, that was incapable of a photoperiodic flowering response. The expression of *CO* is controlled by the circadian clock, with the peak of activity occurring 12 hours after dawn (**FIGURE 25.22A**). Genetic and molecular studies have shown that in Arabidopsis CO protein accumulates in response to long days, and this accelerates flowering (**FIGURE 25.22B**).

As indicated in Figure 25.22B, a critical feature of the coincidence mechanism in the LDP Arabidopsis is that flowering is promoted when the *CO* gene is expressed in the leaf (the site of perception of the photoperiodic stimulus) during the light period. The increase in *CO* mRNA that occurs during short days does not lead to an increase in CO protein, because *CO* expression occurs entirely in the dark. In contrast, during long days *CO* gene expression is accompanied by a sharp increase in CO protein level, because at least some of the expression overlaps with the light period (see Figure 25.22B) (Suarez-Lopez et al. 2001).

As a result, long days are inductive for Arabidopsis flowering because CO protein increases. Short days are noninductive because CO protein level does not increase in the absence of light. Thus, an important feature of the coincidence model is that there must be overlap (coincidence) between *CO* mRNA synthesis and daylight so that light can permit active CO protein to accumulate to a level that promotes flowering. The circadian oscillation of *CO* mRNA provides an explanation for the link between photoperiod perception and the circadian clock. But how does daylight bring about the accumulation of CO protein?

A clue to the function of light was provided by experiments in which *CO* was expressed from a constitutive promoter. Under these conditions, *CO* mRNA was expressed continuously, and its level remained constant throughout the day–night cycle. Nevertheless, the abundance of CO protein continued to cycle, suggesting that CO protein abundance is regulated by a posttranscriptional mechanism.

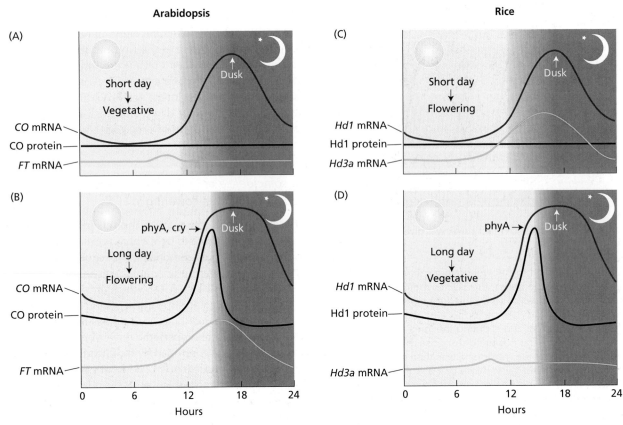

**FIGURE 25.22** Molecular basis of the coincidence model in Arabidopsis (A and B) and rice (C and D). (A) In Arabidopsis under short days, there is little overlap between *CO* mRNA expression and daylight. CO protein does not accumulate to sufficient levels in the phloem to promote the expression of the transmissible floral stimulus, FT protein, and the plant remains vegetative. (B) Under long days, the peak of *CO* mRNA abundance (at hours 12 through 16) overlaps with daylight (sensed by phyA and cryptochrome), allowing CO protein to accumulate. CO activates *FT* mRNA expression in the phloem, which causes flowering when the FT protein is translocated to the apical meristem. (C) In rice under short days, the lack of coincidence between *Hd1* mRNA expression and daylight prevents the accumulation of the Hd1 protein, which acts as a repressor of the gene encoding the rice transmissible floral stimulus and FT relative, Hd3a. In the absence of the Hd1 protein repressor, *Hd3a* mRNA is expressed and the protein it encodes is translocated to the apical meristem where it causes flowering. (D) Under long days (sensed by phytochrome), the peak of *Hd1* mRNA expression overlaps with the day, allowing the accumulation of the Hd1 repressor protein. As a result, *Hd3a* mRNA is not expressed, and the plant remains vegetative. (After Hayama and Coupland 2004.)

The posttranscriptional mechanism is based in part on differences in the rates of CO degradation in the light versus the dark. During the dark, CO is tagged with ubiquitin and rapidly degraded by the 26S proteasome (see Chapter 2). Light appears to enhance the *stability* of the CO protein, allowing it to accumulate during the day. This explains why *CO* promotes flowering only when its mRNA expression coincides with the light period (Valverde et al. 2004). In the dark, CO protein does not accumulate, because it is rapidly degraded.

However, the situation is more complicated than a simple light–dark switch regulating CO turnover. The effect of light on CO stability depends on the photoreceptor involved. Different photoreceptors not only contribute to setting the phase of the circadian rhythm, but, more directly, they also affect CO protein accumulation and flowering. In the morning, phyB signaling appears to enhance CO degradation, whereas in the evening (when CO protein accumulates in long days), cryptochromes and phyA antagonize this degradation and allow the CO protein to build up (see Figure 25.22) (Valverde et al. 2004).

How does the CO protein stimulate flowering in long day plants? CO, a transcriptional regulator, promotes flowering by stimulating the expression of a key floral signal, *FLOWERING LOCUS T* (*FT*). As will be described later in the chapter, there is now evidence that FT protein is the phloem-mobile signal that stimulates flower evocation in the meristem. A similar pathway is used to promote flowering in SDPs, as is discussed next.

## SDPs use a coincidence mechanism to inhibit flowering in long days

Studies of flowering in the SDP rice have shown that the basic coincidence mechanism for photoperiod-sensing is conserved in rice and Arabidopsis. In the long history of rice cultivation, breeders have identified variant alleles of several genes that modify flowering behavior. The rice genes **Heading-date 1** (**Hd1**) and **Heading-date 3a** (**Hd3a**) encode proteins homologous to Arabidopsis CO and FT, respectively. In transgenic plants, overexpression of *FT* in Arabidopsis, and of *Hd3a* in rice, result in rapid flowering regardless of photoperiod, demonstrating that both *FT* and *Hd3a* are strong promoters of flowering. Moreover, expression of both the native *FT* and *Hd3a* genes is substantially elevated during inductive photoperiods (long days in Arabidopsis and short days in rice) (**FIGURE 25.22C**). In addition, rice *Hd1* and Arabidopsis *CO* exhibit similar patterns of circadian mRNA accumulation.

The difference between rice and Arabidopsis is that in the SDP rice, Hd1 acts as an *inhibitor* of *Hd3a* expression. That is, in rice the coincidence of *Hd1* expression and light signaling through phytochrome *suppresses* flowering by inhibiting the expression of *Hd3a* (**FIGURE 25.22D**) (Izawa et al. 2003; Hayama and Coupland 2004). In contrast, CO *promotes* the expression of its downstream gene, *FT*, in the LDP Arabidopsis. Flowering in the SDP rice thus occurs only when *Hd1* is expressed exclusively in the dark. Remarkably, the different responses to photoperiod of

SDPs versus LDPs are due in part to the opposite effects of this one component, CO/Hd1, of the photoperiodic sensing system.

However, it is important to note that photoperiodism is highly complex, and other regulatory mechanisms that fine-tune the responses of SDPs and LDPs to changing day length are certain to be present.

## Phytochrome is the primary photoreceptor in photoperiodism

Night-break experiments are well suited for studying the nature of the photoreceptors involved in the reception of light signals during the photoperiodic response. The inhibition of flowering in SDPs by night breaks was one of the first physiological processes shown to be under the control of phytochrome (**FIGURE 25.23**).

In many SDPs, a night break becomes effective only when the supplied dose of light is sufficient to saturate the photoconversion of **Pr** (phytochrome that absorbs red light) to **Pfr** (phytochrome that absorbs far-red light) (see Chapter 17). A subsequent exposure to far-red light, which photoconverts the pigment back to the physiologically inactive Pr form, restores the flowering response.

Action spectra for the inhibition and restoration of the flowering response in SDPs are shown in **FIGURE 25.24**. A peak at 660 nm, the absorption maximum of Pr, is obtained when dark-grown *Pharbitis* seedlings are used

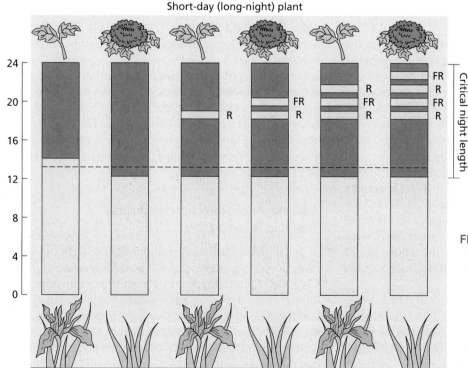

**FIGURE 25.23** Phytochrome control of flowering by red (R) and far-red (FR) light. A flash of red light during the dark period induces flowering in an LDP, and the effect is reversed by a flash of far-red light. This response indicates the involvement of phytochrome. In SDPs, a flash of red light prevents flowering, and the effect is reversed by a flash of far-red light.

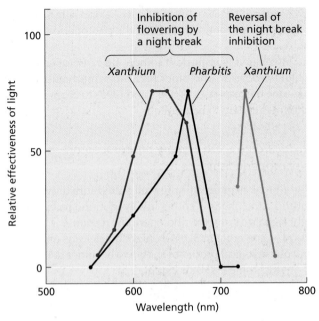

**FIGURE 25.24** Action spectra for the control of flowering by night breaks implicates phytochrome. Flowering in SDPs is inhibited by a short light treatment (night break) given in an otherwise inductive period. In the SDP *Xanthium strumarium*, red-light night breaks of 620 to 640 nm are the most effective. Reversal of the red-light effect is maximal at 725 nm. In the dark-grown SDP *Pharbitis nil*, which is devoid of chlorophyll and its interference with light absorption, night breaks of 660 nm are the most effective. This 660 nm maximum coincides with the absorption maximum of phytochrome. (Data for *Xanthium* from Hendricks and Siegelman 1967; data for *Pharbitis* from Saji et al. 1983.)

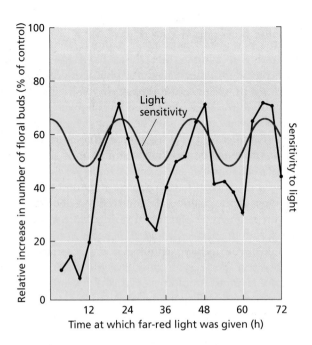

**FIGURE 25.25** Effect of far-red light on floral induction in Arabidopsis. At the indicated times during a continuous 72-hour daylight period, 4 hours of far-red light was added. Data points in the graph are plotted at the centers of the 6-hour treatments. The data show a circadian rhythm of sensitivity to the far-red promotion of flowering (red line). This supports a model in which flowering in LDPs is promoted when the light treatment (in this case far-red light) coincides with the peak of light sensitivity. (After Deitzer 1984.)

to avoid interference from chlorophyll. In contrast, the spectra for *Xanthium* provide an example of the response in green plants, in which the presence of chlorophyll can cause some discrepancy between the action spectrum and the absorption spectrum of Pr. These action spectra plus the red/far-red reversibility of the night break responses confirm the role of phytochrome as the photoreceptor that is involved in photoperiod measurement in SDPs.

Another demonstration of the critical role of phytochrome in photoperiodism in SDPs comes from genetic analyses. In rice, the gene *PHOTOPERIOD SENSITIVITY 5* (*Se5*) encodes a protein similar to Arabidopsis HY1. Se5 and HY1 are enzymes that catalyze a step in the biosynthesis of phytochrome chromophore. Mutations in Se5 cause rice to flower extremely rapidly regardless of the day length (Izawa et al. 2000).

Night break experiments with LDPs have also implicated phytochrome. Thus, in some LDPs, a night break of red light promotes flowering, and a subsequent exposure to far-red light prevents this response (see Figure 25.23).

A circadian rhythm in the promotion of flowering by far-red light has been observed in the LDPs barley (*Hor-

deum vulgare*), darnel ryegrass (*Lolium temulentum*), and Arabidopsis (**FIGURE 25.25**) (Deitzer 1984). The response is proportional to the irradiance and duration of far-red light and is therefore a high-irradiance response (HIR). As in other HIRs, phyA is the phytochrome that mediates the response to far-red light. Consistent with a role of phyA in the flowering of LDPs, mutations in the *PHYA* gene delay flowering in Arabidopsis (Johnson et al. 1994). However, in some LDPs the role of phytochrome is more complex than in SDPs, because a blue light photoreceptor also participates in the response.

## A blue-light photoreceptor regulates flowering in some LDPs

In some LDPs, such as Arabidopsis, blue light can promote flowering, suggesting the possible participation of a blue-light photoreceptor in the control of flowering. As discussed in Chapter 18, the *cryptochromes*, encoded by the *CRY1* and *CRY2* genes, are blue-light photoreceptors that control seedling growth in Arabidopsis.

As noted earlier, the CRY protein has also been implicated in the entrainment of the circadian oscillator. The role of blue light in flowering and its relationship to circadian rhythms have been investigated by use of the luciferase reporter gene construct mentioned in WEB TOPIC 25.5. In continuous white light, the cyclic luminescence has a period of 24.7 hours, but in constant darkness the period lengthens to 30 to 36 hours. Either red or blue light, given individually, shortens the period to 25 hours.

To distinguish between the effects of phytochrome and a blue-light photoreceptor, researchers transformed phytochrome-deficient *hy1* mutants, which are defective in chromophore synthesis and are therefore deficient in *all* phytochromes, with the luciferase construct to determine the effect of the mutation on the period length (Millar et al. 1995). Under continuous white light, the *hy1* plants had a period similar to that of the wild type, indicating that little or no phytochrome is required for white light to affect the period. Furthermore, under continuous red light, which would be perceived only by phyB, the period of *hy1* was significantly lengthened (i.e., it became more like constant darkness), whereas the period was not lengthened by continuous blue light. These results indicate that *both phytochrome and a blue-light photoreceptor are involved in period control.*

The role of blue light in regulating both circadian rhythmicity and flowering is also supported by studies with an Arabidopsis flowering-time mutant, *elf3* (*early flowering 3*) (see WEB TOPICS 25.5 AND 25.6). Confirmation that a blue-light photoreceptor is involved in sensing inductive photoperiods in Arabidopsis was provided by experiments demonstrating that mutations in one of the cryptochrome genes, *CRY2* (see Chapter 18), caused a delay in flowering

and an inability to perceive inductive photoperiods (Guo et al. 1998).

In contrast, plants carrying a gain-of-function allele of *CRY2* flower much earlier than the wild type (El-Din El-Assal 2003). In addition, the *cry1/cry2* double mutants flowered slightly later than *cry2* in long days, indicating some functional redundancy of CRY1 and CRY2 in promoting flowering time in Arabidopsis.

In addition to their role in entraining the circadian clock, it is likely that the cryptochromes, like phyA, also regulate flowering directly by stabilizing the CO protein, allowing it to accumulate under long day conditions. As noted above, the CO protein acts as a promoter of flowering in LDPs.

# Vernalization: Promoting Flowering with Cold

**Vernalization** is the process whereby repression of flowering is alleviated by a cold treatment given to a hydrated seed (i.e., a seed that has imbibed water) or to a growing plant (dry seeds do not respond to the cold treatment because vernalization is an active metabolic process). Without the cold treatment, plants that require vernalization show delayed flowering or remain vegetative, and they are not competent to respond to floral signals such as inductive photoperiods. In many cases these plants grow as rosettes with no elongation of the stem (FIGURE 25.26).

**Winter-annual Arabidopsis without vernalization**

**Winter-annual Arabidopsis with vernalization**

FIGURE 25.26 Vernalization induces flowering in the winter-annual types of *Arabidopsis thaliana*. The plant on the left is a winter-annual type that has not been exposed to cold. The plant on the right is a genetically identical winter-annual type that was exposed to 40 days of temperatures slightly above freezing (4°C) as a seedling. It flowered 3 weeks after the end of the cold treatment with about nine leaves on the primary stem. (Courtesy of Colleen Bizzell.)

In this section we will examine some of the characteristics of the cold requirement for flowering, including the range and duration of the inductive temperatures, the sites of perception, the relationship to photoperiodism, and a possible molecular mechanism.

### Vernalization results in competence to flower at the shoot apical meristem

Plants differ considerably in the age at which they become sensitive to vernalization. Winter annuals, such as the winter forms of cereals (which are sown in the fall and flower in the following summer), respond to low temperature very early in their life cycle. In fact, many winter annuals can be vernalized before germination (i.e., radicle emergence from the seed) if the seeds have imbibed water and become metabolically active. Other plants, including most biennials (which grow as rosettes during the first season after sowing and flower in the following summer), must reach a minimal size before they become sensitive to low temperature for vernalization.

The effective temperature range for vernalization is from just below freezing to about 10°C, with a broad optimum usually between about 1 and 7°C (Lang 1965). The effect of cold increases with the duration of the cold treatment until the response is saturated. The response usually requires several weeks of exposure to low temperature, but the precise duration varies widely with species and variety.

Vernalization can be lost as a result of exposure to devernalizing conditions, such as high temperature (**FIGURE 25.27**), but the longer the exposure to low temperature, the more permanent the vernalization effect.

Vernalization appears to take place primarily in the shoot apical meristem. Localized cooling causes flowering when only the stem apex is chilled, and this effect appears to be largely independent of the temperature experienced by the rest of the plant. Excised shoot tips have been successfully vernalized, and where seed vernalization is possible, fragments of embryos consisting essentially of the shoot tip are sensitive to low temperature.

In developmental terms, vernalization results in the acquisition of competence of the meristem to undergo the floral transition. Yet, as discussed earlier in the chapter, competence to flower does not guarantee that flowering will occur. A vernalization requirement is often linked with a requirement for a particular photoperiod (Lang 1965). The most common combination is a requirement for cold treatment *followed* by a requirement for long days—a combination that leads to flowering in early summer at high latitudes (see **WEB TOPIC 25.7**).

### Vernalization can involve epigenetic changes in gene expression

For vernalization to occur, active metabolism is required during the cold treatment. Sources of energy (sugars) and

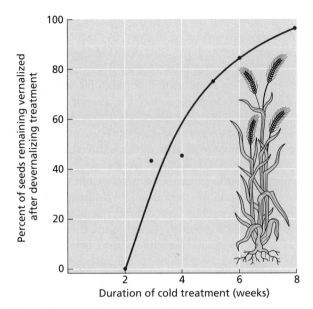

**FIGURE 25.27**   The duration of exposure to low temperature increases the stability of the vernalization effect. The longer that winter rye (*Secale cereale*) is exposed to a cold treatment, the greater the number of plants that remain vernalized when the cold treatment is followed by a devernalizing treatment. In this experiment, seeds of rye that had imbibed water were exposed to 5°C for different lengths of time, then immediately given a devernalizing treatment of 3 days at 35°C. (Data from Purvis and Gregory 1952.)

oxygen are required, and temperatures below freezing at which metabolic activity is suppressed are not effective for vernalization. Furthermore, cell division and DNA replication also appear to be required. In some plant species, vernalization causes a stable change in the competence of the meristem to form an inflorescence (Amasino 2004).

One model for how vernalization stably affects competence is that there are changes in the pattern of gene expression in the meristem after cold treatment that persist into the spring and throughout the remainder of the life cycle. Stable changes in gene expression that do not involve alterations in the DNA sequence and which can be passed on to descendant cells through mitosis or meiosis are known as **epigenetic changes**. As such, epigenetic changes in gene expression are stable even after the signal (in this case cold) that induced the change is no longer present. Epigenetic changes of gene expression occur in many organisms, from yeast to mammals, and often require cell division and DNA replication, as is the case for vernalization.

The involvement of epigenetic regulation of a specific target gene in the vernalization process has been confirmed in the LDP Arabidopsis. In winter-annual types of Arabidopsis that require both vernalization and long days for flowering to be accelerated, a gene that acts as a repressor of flowering has been identified: ***FLOWERING LOCUS C (FLC)***. *FLC* is highly expressed in nonvernalized shoot apical regions (Michaels and Amasino 2000).

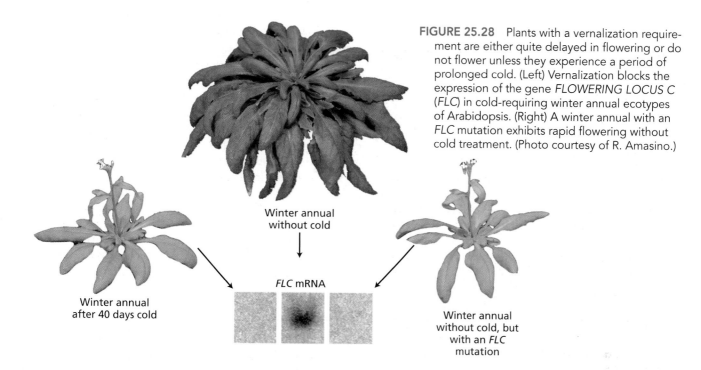

**FIGURE 25.28** Plants with a vernalization requirement are either quite delayed in flowering or do not flower unless they experience a period of prolonged cold. (Left) Vernalization blocks the expression of the gene *FLOWERING LOCUS C* (*FLC*) in cold-requiring winter annual ecotypes of Arabidopsis. (Right) A winter annual with an *FLC* mutation exhibits rapid flowering without cold treatment. (Photo courtesy of R. Amasino.)

Winter annual
without cold

*FLC* mRNA

Winter annual
after 40 days cold

Winter annual
without cold, but
with an *FLC*
mutation

After vernalization, this gene is epigenetically switched off for the remainder of the plant's life cycle, permitting flowering in response to long days to occur (**FIGURE 25.28**). In the next generation, however, the gene is switched on again, restoring the requirement for cold. Thus in Arabidopsis, the state of expression of the *FLC* gene represents a major determinant of meristem competence (Amasino 2004). In Arabidopsis it has been shown that FLC works by directly repressing the expression of the key floral signal *FT* in the leaves as well as the transcription factors *SOC1* and *FD* at the shoot apical meristem (Searle et al. 2006).

The epigenetic regulation of *FLC* involves stable changes in chromatin structure resulting from **chromatin remodeling**. As discussed in Chapter 2, in eukaryotes nuclear DNA can exist in a transcriptionally inactive, compact form called heterochromatin or a more dispersed, transcriptionally active form called euchromatin. Heterochromatin and euchromatin are characterized by distinct sets of covalent modifications of certain histones in the nucleosome (short stretch of DNA wrapped around a histone core) (see Chapter 2), and these modifications are thought to favor the formation of either heterochromatin or euchromatin. The set of covalent modifications that favors a particular chromatin structure is often called the **histone code** (Jenuwein and Allis 2001).

Vernalization causes the chromatin of the *FLC* to lose histone modifications characteristic of euchromatin and to acquire modifications, such as the methylation of specific lysine residues, characteristic of heterochromatin (Bastow et al. 2004; Sung and Amasino 2004). The cold-induced conversion of *FLC* from euchromatin to heterochromatin effectively silences the gene.

## A range of vernalization pathways may have evolved

Many vernalization-requiring plants germinate in the fall, taking advantage of the cool and moist conditions optimal for their growth. The vernalization requirement of such plants ensures that flowering does not occur until spring, allowing the plants to survive winter vegetatively (flowers are especially sensitive to frost). A vernalizing plant must not only sense cold exposure but also have a mechanism to measure the duration of cold exposure. For example, if a plant is exposed to a short period of cold early in the fall followed by a return of warm temperatures later that fall, it is important for the plant not to perceive the brief exposure to cold as winter and the following warm weather as spring. Accordingly, vernalization occurs only after an exposure to a duration of cold sufficient to indicate that a complete winter season has passed.

A similar system of measuring the duration of cold before buds can be released from dormancy operates in many perennials that grow in temperate climates. The mechanism that plants have evolved to measure the duration of cold is not known, but in Arabidopsis there are genes that are induced only after exposure to a long period of cold, and these genes are critical to the vernalization process (Amasino 2004; Sung and Amasino 2004).

There does not appear to be a particular vernalization pathway that is conserved among all flowering plants. As discussed above, *FLC* is the flowering repressor that is responsible for the vernalization requirement in Arabidopsis. *FLC* encodes a MADS box protein that is related to regulatory proteins discussed earlier in the chapter, such as DEFICIENS and AGAMOUS, that are involved in floral development. In cereals, a gene encoding a different type of protein, a zinc finger-containing protein called VRN2 (*vernalization 2*), acts as the flowering repressor that creates a vernalization requirement (Greenup et al. 2009; Distelfeld and Dubcovsky 2009).

It appears that the major groups of flowering plants evolved in warm climates and therefore did not evolve a mechanism to measure the duration of winter. Over geological time, regions of the earth gradually developed a temperate climate due to continental drift and other factors. Members of many groups of plants adapted to these new temperate niches with the development of responses like vernalization and bud dormancy, and thus these responses are likely to have evolved independently in different groups of plants.

# Long-Distance Signaling Involved in Flowering

Although floral evocation occurs at the apical meristems of shoots, in photoperiodic plants inductive photoperiods are sensed by the leaves. This suggests that a long-range signal must be transmitted from the leaves to the apex, which has been shown experimentally through extensive grafting experiments in many different plant species. The biochemical nature of this signal had long baffled physiologists. The problem was finally solved using molecular genetic approaches, and the floral stimulus was identified as a protein. In this section we will review the background for the discovery of the floral stimulus, historically referred to as *florigen*, which serves as the long-distance signal during flowering. We will also describe various other biochemical signals that can serve either as activators or as inhibitors of flowering.

## The floral stimulus is transported in the phloem

The leaf-derived photoperiodic floral stimulus is translocated via the phloem to the shoot apical meristem, where it promotes floral evocation. Treatments that block phloem translocation, such as girdling or localized heat-killing, block flowering by preventing the movement of the floral stimulus out of the leaf.

It is possible to measure rates of movement of the floral stimulus by removing a leaf at different times after induction, and comparing the time it takes for the signal to reach two buds located at different distances from the induced leaf. The rationale for this type of measurement is that a threshold amount of the signaling compound has reached the bud when flowering takes place, despite the removal of the leaf. In this way, the time for a sufficient amount of signal to exit the leaf can be determined. Furthermore, comparing the times of induction for two differently positioned buds provides a measure of the rate of movement of the signal along the stem.

Studies using this method have shown that the rate of movement of the flowering signal is comparable to, or somewhat slower than, the rate of translocation of sugars in the phloem (see Chapter 10). For example, export of the floral stimulus from adult leaves of the SDP *Chenopodium* is complete within 22.5 hours from the beginning of the long night period. In the LDP *Sinapis*, movement of the floral stimulus out of the leaf is complete as early as 16 hours after the start of the long-day treatment (Zeevaart 1976).

Because the floral stimulus is translocated along with sugars in the phloem, it is subject to source–sink relations. An induced leaf positioned close to the shoot apex is more likely to cause flowering than an induced leaf at the base of a stem, which normally feeds the roots. Similarly, noninduced leaves positioned between the induced leaf and the apical bud will tend to inhibit flowering by serving as the preferred source leaves for the bud, thus preventing the floral stimulus from the more distal induced leaf from reaching its target.

## Grafting studies have provided evidence for a transmissible floral stimulus

The production in photoperiodically induced leaves of a biochemical signal that is transported to a distant target tissue (the shoot apex) where it stimulates a response (flowering) satisfies an important criterion for a hormonal effect. In the 1930s, Mikhail Chailakhyan, working in Russia, postulated the existence of a universal flowering hormone, which he named **florigen**.

The evidence in support of florigen comes mainly from experiments in which noninduced receptor plants were stimulated to flower by having a leaf or shoot from a photoperiodically induced donor plant grafted to them. For example, in the SDP *Perilla crispa*, a member of the mint family, grafting a leaf from a plant grown under inductive short days onto a plant grown under noninductive long days causes the latter to flower (**FIGURE 25.29**). Moreover, the floral stimulus seems to be the same in plants with different photoperiodic requirements. Thus, grafting an induced shoot from the LDP *Nicotiana sylvestris*, grown under long days, onto the SDP 'Maryland Mammoth' tobacco caused the latter to flower under noninductive (long day) conditions.

The leaves of DNPs have also been shown to produce a graft-transmissible floral stimulus (**TABLE 25.2**). For example, grafting a single leaf of a day-neutral variety of soybean, 'Agate,' onto the short-day variety, 'Biloxi,' caused flowering in 'Biloxi' even when the latter was maintained in noninductive long days. Similarly, a shoot from a day-neutral variety of tobacco (*Nicotiana tabacum*, cv. Trapezond) grafted onto the LDP *Nicotiana sylvestris* induced the latter to flower under noninductive short days.

Grafting studies also showed that in some species, such as *Xanthium* (SDP), *Bryophyllum* (SLDP), and *Silene* (LDP),

**FIGURE 25.29** Demonstration by grafting of a leaf-generated floral stimulus in the SDP *Perilla*. (Left) Grafting an induced leaf from a plant grown under short days onto a noninduced shoot causes the axillary shoots to produce flowers. The donor leaf has been trimmed to facilitate grafting, and the upper leaves have been removed from the stock to promote phloem translocation from the scion to the receptor shoots. (Right) Grafting a noninduced leaf from a plant grown under LDs results in the formation of vegetative branches only. (Courtesy of J. A. D. Zeevaart.)

not only can flowering be induced by grafting, but the induced state itself appears to be self-propagating (Zeevaart 1976) (see **WEB TOPIC 25.8**). In a few cases, flowering has even been induced by grafts between different genera. The SDP *Xanthium strumarium* flowered under long-day

conditions when shoots of flowering *Calendula officinalis* were grafted onto a vegetative *Xanthium* stock. Similarly, grafting a shoot from the LDP *Petunia hybrida* onto a stock of the cold-requiring biennial henbane (*Hyoscyamus niger*) caused the latter to flower under long days, even though it was nonvernalized (**FIGURE 25.30**).

In *Perilla* (see Figure 25.29), the movement of the floral stimulus from a donor leaf to the stock across the graft union correlated closely with the translocation of $^{14}$C-labeled assimilates from the donor, and this movement was dependent on the establishment of vascular continuity across the graft union (Zeevaart 1976). These results confirmed earlier girdling studies showing that the floral stimulus is translocated along with photoassimilates in the phloem.

## The Discovery of Florigen

Pioneering grafting experiments of the kind described above established the importance of a long-range signal from the leaf to the apical meristem to stimulate flowering. Since the 1930s, there have been many unsuccessful attempts to isolate and characterize both florigen and "antiflorigen." Based on the structures of the known plant hormones, nearly all of these early studies focused on small molecules, which are relatively easy to bioassay. Typically, extracts from induced and uninduced leaf tissue were prepared and tested for their ability either to promote or inhibit flowering. In one notable case, an investigator laboriously collected and tested phloem sap obtained from hundreds of aphids stationed on induced plants (see Chapter 10). Although positive results were occasionally reported, none of them has stood the test of time. According to one hypothesis, the reason for the failure to detect florigen in extracts was that "florigen" was not a single substance, but a combination of factors, which were experimentally difficult to reconstitute. The alternative hypoth-

---

**TABLE 25.2**
**Transmission of the flowering signal occurs through a graft junction**

| Donor plants maintained under flower-inducing conditions | Photoperiod type[a,b] | Vegetative receptor plant induced to flower | Photoperiod type[a,b] |
|---|---|---|---|
| *Helianthus annus* | DNP in LD | *H. tuberosus* | SDP in LD |
| *Nicotiana tabacum* Delcrest | DNP in SD | *N. sylvestris* | LDP in SD |
| *Nicotiana sylvestris* | LDP in LD | *N. tabacum* Maryland Mammoth | SDP in LD |
| *Nicotiana tabacum* Maryland Mammoth | SDP in SD | *N. sylvestris* | LDP in SD |

Note: The successful transfer of a flowering induction signal by grafting between plants of different photoperiodic response groups shows the existence of a transmissible floral hormone that is effective.

[a]LDPs = Long-day plants; SDPs = Short-day plants; DNPs = Day-neutral plants.

[b]LD, long days; SD, short days.

**FIGURE 25.30** Successful transfer of the floral stimulus between different genera: The scion (right branch) is the LDP *Petunia hybrida*, and the stock is nonvernalized henbane (*Hyoscyamus niger*). The graft combination was maintained under long days. (Courtesy of J. A. D. Zeevaart.)

esis was that florigen might not be a small hormonelike molecule after all, but a macromolecule, such as protein or RNA. Because such macromolecules are not taken up readily by plant cells, the latter hypothesis would explain why it was so difficult to detect florigen activity in extracts. Researchers had reached a dead end and a new approach to the problem was needed.

A major breakthrough was the identification of *FLOWERING LOCUS T* (*FT*) in Arabidopsis through genetic screens.

### The Arabidopsis protein FLOWERING LOCUS T is florigen

According to the coincidence model, flowering in LDPs such as Arabidopsis occurs when the *CONSTANS* gene is expressed during the light period. *CO* gene expression appears to be highest in the phloem of leaves and

stems (An et al. 2004). The downstream target gene of *CO*, *FLOWERING LOCUS T* (*FT*), is also specifically expressed in the phloem.

Consistent with a phloem localization of CO, *co* mutants with a defective photoperiodic response could be rescued by expressing *CO* specifically in the phloem of the minor veins of mature leaves using a promoter construct specific for companion cells (An et al. 2004; Ayre and Turgeon 2004). In contrast, expressing *CO* in the apical meristems of *co* mutants did not restore the photoperiodic response. Thus, *CO* seems to act specifically in the phloem of leaves to stimulate flowering in response to long days. In addition, flowering could be induced in the *co* mutant by grafting transgenic shoots expressing *CO* in the phloem of their leaves onto the mutant. This observation suggests that *CO* expression gives rise to a graft-transmissible floral stimulus that can cause flowering at the apical meristem (Ayre and Turgeon 2004).

The signaling output of CO activity is mediated by the expression of *FT*. In Arabidopsis, *CO* expression during long days results in an increase in *FT* mRNA. However, unlike *CO*, *FT* stimulates flowering when expressed in either the phloem or the apical meristem (An et al. 2004; Corbesier and Coupland 2005).

Biochemically, FT is a small globular protein that is related to a family of regulatory proteins conserved between budding yeast and vertebrates. Could either *FT* mRNA or FT protein be the long-sought phloem-mobile floral stimulus, or "florigen," involved in photoperiodic signaling to the apical meristem? Expression of the *FT* gene (or its relatives, such as *Hd3a* in rice, discussed earlier) is induced in a range of species during their floral-inductive photoperiods. When the *FT* gene is introduced into a range of plant species whose flowering is not influenced by photoperiod, it causes photoperiod-independent flowering. Finally, evidence has now been obtained that FT protein can move from the leaves to the apical meristem (Corbesier et al. 2007; Jaeger and Wigge 2007; Mathieu et al. 2007; Lin et al. 2007; Tamaki et al. 2007; summarized in Zeevaart 2008). Thus, *the FT protein exhibits the properties that would be expected of florigen.*

According to the current model, FT protein moves via the phloem from the leaves to the meristem under inductive photoperiods (**FIGURE 25.31**). Once in the meristem, the FT protein forms a complex with FD, a basic leucine zipper (bZIP) transcription factor that is expressed in the meristem (Abe et al. 2005; Wigge et al. 2005). The complex of FT and FD then activates floral identity genes such as *APETALA 1* (see Figure 25.31). In Arabidopsis, this sets in motion positive feedback loops that keep the meristem in a flowering state. However, the meristems of some species lack such positive feedback loops and, as a consequence, revert to producing leaves in the absence of a continuous inductive photoperiod (Tooke et al. 2005).

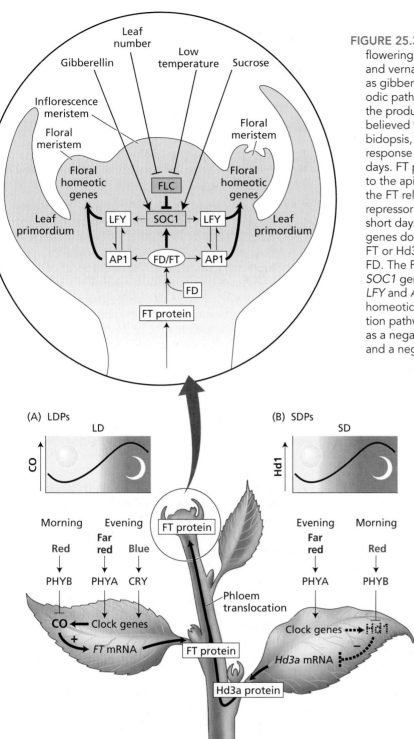

FIGURE 25.31 Multiple developmental pathways for flowering in Arabidopsis: photoperiod, autonomous, and vernalization (low temperature) pathways as well as gibberellins can affect flowering. The photoperiodic pathway is located in the leaves and involves the production of a transmissible floral stimulus, believed to be FT protein. (A) In LDPs such as Arabidopsis, *FT* mRNA is produced in the phloem in response to CO protein accumulation under long days. FT protein is then translocated via sieve tubes to the apical meristem. (B) In SDPs such as rice, the FT relative called Hd3a is expressed when the repressor protein, Hd1, fails to accumulate under short days (dotted lines indicate that in SD the clock genes do not activate *Hd1*). In the meristem (inset), FT or Hd3a protein interact with another protein, FD. The FT/FD complex then activates the *AP1* and *SOC1* genes, which trigger *LFY* gene expression. *LFY* and *AP1* then trigger the expression of the floral homeotic genes. The autonomous and vernalization pathways negatively regulate *FLC*, which acts as a negative regulator of *SOC1* in the meristem and a negative regulator of *FT* in the leaves.

GA appears to promote flowering in Arabidopsis by activating expression of the *LEAFY* gene (Blázquez and Weigel 2000). The activation of *LFY* by GA is mediated by the transcription factor GAMYB, which is negatively regulated by DELLA proteins (see Chapter 20). In addition, GAMYB levels are also modulated by a microRNA that promotes the degradation of the GAMYB transcript (Achard et al. 2004) (see Chapter 20).

Exogenously applied GAs can also evoke flowering in a few SDPs in noninductive conditions and in cold-requiring plants that have not been vernalized. As previously discussed, cone formation can also be promoted in juvenile plants of several gymnosperm families by the addition of GAs. Thus, in some plants exogenous GAs can bypass the endogenous trigger of age in autonomous flowering, as well as the primary environmental signals of day length and temperature. As discussed in Chapter 20, plants contain many GA-like compounds. Most of these compounds are either precursors to, or inactive metabolites of, the active forms of GA.

In some species, such as in the LDP darnel ryegrass (*Lolium temulentum*), different GAs have markedly different effects on flowering and stem elongation (see **WEB TOPIC 25.10** and **WEB ESSAY 25.1**). In the quantitative

## Gibberellins and ethylene can induce flowering

Among the naturally occurring growth hormones, gibberellins (GAs) (see Chapter 20) can have a strong influence on flowering (see **WEB TOPIC 25.9**). Exogenous GA can evoke flowering when applied either to rosette LDPs like Arabidopsis, or to dual–day length plants such as *Bryophyllum*, when grown under short days (Lang 1965; Zeevaart 1985).

LDP Arabidopsis, a mutation in GA biosynthesis prevents flowering altogether in noninductive short days but has little effect on flowering in long days, demonstrating that endogenous GA is required for flowering in specific situations (Wilson et al. 1992). Thus, in Arabidopsis GA appears to be an alternative stimulus to flowering when the photoperiodic stimulus is low or lacking. A certain level of GA is probably required for flowering in many species.

Considerable attention has been given to the effects of day length on GA metabolism in the plant (see Chapter 20). For example, in the LDP spinach (*Spinacia oleracea*), the levels of GAs are relatively low in short days, and the plants maintain a rosette form. After the plants are transferred to long days, the levels of all the GAs of the 13-hydroxylated pathway ($GA_{53} \rightarrow GA_{44} \rightarrow GA_{19} \rightarrow GA_{20} \rightarrow GA_1$; see Chapter 20) increase. However, the fivefold increase in the physiologically active GA, $GA_1$, is what causes the marked stem elongation that accompanies flowering.

In addition to GAs, other growth hormones can either inhibit or promote flowering. One commercially important example is the striking promotion of flowering in pineapple (*Ananas comosus*) by ethylene and ethylene-releasing compounds—a response that appears to be restricted to members of the pineapple family (Bromeliaceae).

### Climate change has already caused measurable changes in flowering time of wild plants

Plants are highly sensitive to temperature, and appear to use temperature signals to control development in several ways. We have seen that some plants require a long period of cold (vernalization) in order to become competent to flower. Another influence of temperature is the ambient growing temperature. Plants are able to sense as little as a 1°C difference in temperature. Among many of the plants studied, increasing ambient temperature accelerates flowering. Remarkably, climate change has already caused a significant change in the flowering times of wild plants (Fitter and Fitter 2002). The pathway by which plants sense temperature has not yet been identified, but it is dependent on expression of *FT*, because in the absence of *FT*, ambient temperature increases do not accelerate flowering (Balasubramanian et al. 2006).

It appears that different species have different sensitivities to ambient temperature, some plants responding very little and others much more. This has important implications for how species will adapt to climate change, since it has been shown that those plants which respond to climate change appear to have been able to extend their range, while those plants that do not use ambient temperature as an important developmental cue are much more likely to go extinct (Willis et al. 2008).

### The transition to flowering involves multiple factors and pathways

It is clear that the transition to flowering involves a complex system of interacting factors. Leaf-generated transmissible signals are required for determination of the shoot apex in both autonomously regulated and photoperiodic species.

Genetic studies have established that there are four distinct developmental pathways that control flowering in the LDP Arabidopsis (see Figure 25.31):

- The *photoperiodic pathway* begins in the leaf and involves phytochromes and cryptochromes. (Note that PHYA and PHYB have contrasting effects on flowering; see WEB TOPIC 25.11.) In LDPs under long-day conditions, the interaction of these photoreceptors with the circadian clock initiates a pathway that results in the expression of *CO* in the phloem companion cells of the leaf. *CO* activates the expression of its downstream target gene, *FT*, in the phloem. FT protein ("florigen") a phloem-mobile signal that stimulates flowering in the apical meristem. As shown in the enlargement of the meristem, FT protein forms a complex with the transcription factor, FD. The FD/FT complex then activates downstream target genes such as *SOC1*, *AP1*, and *LFY*, which turn on floral homeotic genes on the flanks of the inflorescence meristem.

- In rice, a SDP, the *CO* homolog *Heading-date 1* (*Hd1*) acts as an inhibitor of flowering. During inductive short-day conditions, however, Hd1 protein is not produced. The absence of Hd1 stimulates the expression of the *Hd3a* gene in the phloem companion cells (Hd3a is a relative of FT). Hd3a protein is then translocated via the sieve tubes to the apical meristem, where it is believed to stimulate flowering by a pathway similar to that in Arabidopsis.

- In the *autonomous* and *vernalization pathways*, flowering occurs either in response to internal signals—the production of a fixed number of leaves—or to low temperatures. In the autonomous pathway of Arabidopsis, all of the genes associated with the pathway are expressed in the meristem. The autonomous pathway acts by reducing the expression of the flowering repressor gene *FLOWERING LOCUS C* (*FLC*), an inhibitor of *SOC1* expression (Michaels and Amasino 2000). Vernalization also represses *FLC*, but perhaps by a different mechanism (an epigenetic switch). Because the *FLC* gene is a common target, the autonomous and vernalization pathways are grouped together.

- The *gibberellin pathway* is required for early flowering and for flowering under noninductive

Short days to long days at time 0

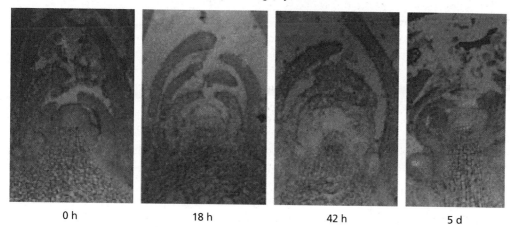

0 h          18 h          42 h          5 d

**FIGURE 25.32** Increase in expression of *SOC1* (indicated by red staining) during floral evocation in the shoot apical meristem of Arabidopsis. The times after shifting the plants from SDs to LDs are indicated. (From Borner et al. 2000.)

short days. The gibberellin pathway involves GAMYB as an intermediary, which promotes the *LFY*; GA may also interact with *SOC1* by a separate pathway.

All four pathways converge by increasing the expression of key floral regulators: *FT* in the vasculature and *SOC1*, *LFY* and *AP1* in the meristem (see Figure 25.31). **FIGURE 25.32** shows the level of *SOC1* gene expression in the shoot apical meristem of an Arabidopsis plant after shifting from noninductive short (8-hour) days to inductive long (16-hour) days. Note that an increase in *SOC1* expression can be detected as early as 18 hours after the start of the long-day treatment (Borner et al. 2000). Thus it takes only 10 hours beyond an 8-hour short day for the meristem to begin responding to the floral stimulus from

the leaves. This timing is consistent with previous measurements of the rates of export of the floral stimulus from induced leaves (discussed earlier in the chapter).

Expression of genes like *SOC1*, *LFY*, and *AP1* in turn activate downstream genes required for floral organ development such as *AP3*, *PISTILLATA* (*PI*), and *AGAMOUS* (*AG*) (see Figure 25.6).

The existence of multiple flowering pathways provides angiosperms with maximum reproductive flexibility, enabling them to produce seeds under a wide variety of conditions. Redundancy within the pathways ensures that reproduction, the most crucial of all physiological functions, will be relatively insensitive to mutations and evolutionarily robust.

The details of the pathways undoubtedly vary among different species. In maize, for example, at least one of the genes involved in an autonomous pathway is expressed in leaves (see WEB TOPIC 25.12). The presence of multiple flowering pathways is a theme with many variations among angiosperms.

## SUMMARY

Formation of the floral organs (sepals, petals, stamens, and carpels) occurs at the shoot apical meristem and is linked to both endogenous timekeeping and environmental signals. A network of genes that control floral morphogenesis has been identified in several species (**Figures 25.1, 25.2**).

### Floral Meristems and Floral Organ Development

- The four different types of floral organs are initiated sequentially in separate, concentric whorls (**Figure 25.3**).

- Formation of floral meristems requires active meristem identity genes such as *SOC1*, *AP1*, and *LFY* in Arabidopsis.

- Mutations in homeotic floral identity genes alter the types of organs produced in each of the whorls (**Figure 25.5**).

- The ABC model suggests that organ identity in each whorl is determined by the combined activity of three organ identity genes (**Figures 25.6–25.8**).

### Floral Evocation: Investigating Environmental Cues

- Internal (autonomous) and external (environment-sensing) control systems enable plants to precisely regulate and time flowering for reproductive success.

## SUMMARY continued

- Floral development can be a response to changes in day length (photoperiodism) or to prolonged cold (vernalization).

- Synchronous flowering favors crossbreeding and allows seed production under favorable conditions.

### The Shoot Apex and Phase Changes

- In plants, the change from juvenile to adult is usually accompanied by changes in vegetative characteristics (**Figures 25.9, 25.10**).

- The combinatorial model accounts for intermediate forms during the transition from juvenile to adult (**Figure 25.11**).

- Floral evocation requires the apical bud to pass through two developmental stages, competence and determination (**Figure 25.12**).

- Competence is demonstrated when a vegetative shoot (scion) is grafted onto a flowering stock and the scion flowers.

- Determination is demonstrated when a competent vegetative bud progresses to flowering (**Figure 25.13**).

- With competence, a meristem exhibits an increasing tendency to flower with age (leaf number) (**Figure 25.14**).

### Circadian Rhythms: The Clock Within

- Circadian rhythms are based on an endogenous oscillator, not the presence or absence of light; they are defined by three parameters: period, phase, and amplitude (**Figure 25.15**).

- Temperature compensation processes prevent temperature changes from affecting the period of the clock.

- Phytochromes and cryptochromes have been implicated in clock entrainment.

### Photoperiodism: Monitoring Day Length

- Plants can detect seasonal changes in day length at latitudes away from the equator (**Figure 25.16**).

- Flowering in LDPs requires a critical day length. Flowering in SDPs requires a day length that is less than the critical day length (**Figure 25.18**).

- Leaves perceive the photoperiodic stimulus in both LDPs and SDPs.

- Day length is monitored by measuring the length of the night; SDP flowering is determined primarily by the duration of darkness (**Figure 25.19**).

- For both LDPs and SDPs, the dark period can be made ineffective by interruption with a short exposure to light (a night break) (**Figure 25.20**).

- The flowering response to night breaks shows a circadian rhythm, supporting the clock hypothesis (**Figure 25.21**).

- In the coincidence model, flowering is induced in both SDPs and LDPs when light exposure is coincident with the appropriate phase of the oscillator.

- CO (in Arabidopsis) and Hd1 (in rice) regulate flowering by controlling the transcription of floral stimulus genes (**Figure 25.22**).

- CO protein is degraded at different rates in the light versus the dark. Light enhances the *stability* of CO, allowing it to accumulate during the day; in the dark it is rapidly degraded.

- The effects of red and far-red night breaks implicate phytochrome in the control of flowering in SDPs and LDPs (**Figures 25.23, 25.24**).

- Flowering in LDPs is promoted when the inductive light treatment coincides with a peak in light sensitivity, which follows a circadian rhythm (**Figure 25.25**).

### Vernalization: Promoting Flowering with Cold

- In sensitive plants, a cold treatment is required for plants to respond to floral signals such as inductive photoperiods (**Figures 25.26, 25.27**).

- For vernalization to occur, active metabolism is required during the cold treatment.

- After vernalization, the *FLC* gene is epigenetically switched off for the remainder of the plant's life cycle, permitting flowering in response to long days to occur in Arabidopsis (**Figure 25.28**).

- The epigenetic regulation of *FLC* involves stable changes in chromatin structure.

- A variety of vernalization pathways have evolved in flowering plants.

### Long-Distance Signaling Involved in Flowering

- In photoperiodic plants, a long-range signal is transmitted in the phloem from the leaves to the apex, permitting floral evocation (**Figures 25.29, 25.30**).

### The Discovery of Florigen

- FT is a small, globular protein that exhibits the properties that would be expected of florigen.

- FT protein moves via the phloem from the leaves to the shoot apical meristem under inductive photoperiods. In the meristem, FT forms a complex with the transcription factor FD to activate floral identity genes (**Figure 25.31**).

## SUMMARY continued

- The four distinct pathways that control flowering converge to increase the expression of key floral regulators: *FT* in the vasculature and *SOC1*, *LFY*, and *AP1* in the meristem (**Figure 25.31**).

## WEB MATERIAL

### Web Topics

**25.1 Contrasting the Characteristics of Juvenile and Adult Phases of English Ivy (*Hedera helix*) and Maize (*Zea mays*)**

A table of juvenile vs. adult morphological characteristics is presented.

**25.2 Regulation of Juvenility by the *TEOPOD* (*TP*) Genes in Maize**

The genetic control of juvenility in maize is discussed.

**25.3 Flowering of Juvenile Meristems Grafted to Adult Plants**

The competence of juvenile meristems to flower can be tested in grafting experiments.

**25.4 Characteristics of the Phase-Shifting Response in Circadian Rhythms**

Petal movements in *Kalanchoe* have been used to study circadian rhythms.

**25.5 Support for the Role of Blue-Light Regulation of Circadian Rhythms**

ELF3 plays a role in mediating the effects of blue light on flowering time.

**25.6 Genes That Control Flowering Time**

A discussion of genes that control different aspects of flowering time is presented.

**25.7 Regulation of Flowering in Canterbury Bells by Both Photoperiod and Vernalization**

Short days acting on the leaf can substitute for vernalization at the shoot apex in Canterbury bell.

**25.8 The Self-Propagating Nature of the Floral Stimulus**

In certain species, the induced state can be transferred by grafting almost indefinitely.

**25.9 Examples of Floral Induction by Gibberellins in Plants with Different Environmental Requirements for Flowering**

A table of the effects of gibberellins on plants with different photoperiodic requirements is presented.

**25.10 The Effects of Two Different Gibberellins on Flowering (Spike Length) and Elongation (Stem Length)**

$GA_1$ and $GA_{32}$ have different effects on flowering in *Lolium*.

**25.11 The Contrasting Effects of Phytochromes A and B on Flowering**

PhyA and phyB affect flowering in Arabidopsis and other species.

**25.12 A Gene That Regulates the Floral Stimulus in Maize**

The *INDETERMINATE 1* gene of maize regulates the transition to flowering and is expressed in young leaves.

### Web Essay

**25.1 The Role of Gibberellins in Floral Evocation of the Grass *Lolium temulentum***

Evidence that GA functions as a leaf-derived floral stimulus in *Lolium* is presented.

# CHAPTER REFERENCES

Abe, M., Kobayashi, Y., Yamamoto, S., Daimon, Y., Yamaguchi, A., Ikeda, Y., Ichinoki, H., Notaguchi, M., Goto, K., and Araki, T. (2005) FD, a bZIP protein mediating signals from the floral pathway integrator FT at the shoot apex. *Science* 309: 1052–1056.

Achard, P., Herr, A., Baulcombe, D. C., and Harberd, N. P. (2004) Modulation of floral development by a gibberellin-regulated microRNA. *Development* 131: 3357–3365.

Amasino, R. M. (2004) Vernalization, competence, and the epigenetic memory of winter. *Plant Cell* 16: 2553–2559.

An, H., Roussot, C., Suárez-López, P., Corbesier, L., Vincent, C., Piñeiro, M., Hepworth, S., Mouradov, A., Justin, S., Turnbull, C., et al. (2004) CONSTANS acts in the phloem to regulate a systemic signal that induces photoperiodic flowering of *Arabidopsis*. *Development* 131: 3615–3626.

Ayre, B. G., and Turgeon, R. (2004) Graft transmission of a floral stimulant derived from CONSTANS. *Plant Physiol.* 135: 2271–2278.

Balasubramanian, S., Sureshkumar, S., Lempe, J., and Weigel, D. (2006) Potent induction of *Arabidopsis thaliana* flowering by elevated growth temperature. *PLoS Genet.* 2: 106.

Bastow, R., Mylne, J. S., Lister, C., Lippman, Z., Martienssen, R. A., and Dean, C. (2004) Vernalization requires epigenetic silencing of FLC by histone methylation. *Nature* 427: 164–167.

Bewley, J. D., Hempel, F. D., McCormick, S., and Zambryski, P. (2000) Reproductive Development. In: *Biochemistry and Molecular Biology of Plants*, B. B. Buchanan, W. Gruissem, and R. L. Jones, eds., American Society of Plant Biologists, Rockville, MD, pp. 988–1034.

Blázquez, M. A., and Weigel, D. (2000) Integration of floral inductive signals in *Arabidopsis*. *Nature* 404: 889–892.

Borner, R., Kampmann, G., Chandler, J., Gleissner, R., Wisman, E., Apel, K., and Melzer, S. (2000) A *MADS* domain gene involved in the transition to flowering in *Arabidopsis*. *Plant J.* 24: 591–599.

Bowman, J. L., Smyth, D. R., and Meyerowitz, E. M. (1989) Genes directing flower development in *Arabidopsis*. *Plant Cell* 1: 37–52.

Bünning, E. (1960) Biological clocks. *Cold Spring Harb. Symp. Quant. Biol.* 15: 1–9.

Clark, J. R. (1983) Age-related changes in trees. *J. Arboriculture* 9: 201–205.

Coen, E. S., and Meyerowitz, E. M. (1991) The war of the whorls: Genetic interactions controlling flower development. *Nature* 353: 31–37.

Coen, E. S., and Carpenter, R. (1993) The metamorphosis of flowers. *Plant Cell* 5: 1175–1181.

Corbesier, L., and Coupland, G. (2005) Photoperiodic flowering in *Arabidopsis*: Integrating genetic and physiological approaches to characterization of the floral stimulus. *Plant Cell Environ.* 28: 54–66.

Corbesier, L., Vincent, C., Jang, S., Fornara, F., Fan, Q., Searle, I., Giakountis, A., Farrona, S., Gissot, L., Turnbull, C., et al. (2007) FT protein movement contributes to long-distance signaling in floral induction of Arabidopsis. *Science* 316: 1030–1033.

Coulter, M. W., and Hamner, K. C. (1964) Photoperiodic flowering response of Biloxi soybean in 72 hour cycles. *Plant Physiol.* 39: 848–856.

Deitzer, G. (1984) Photoperiodic induction in long-day plants. In *Light and the Flowering Process*, D. Vince-Prue, B. Thomas, and K. E. Cockshull, eds., Academic Press, New York, pp. 51–63.

Devlin, P. F., and Kay, S. A. (2000) Cryptochromes are required for phytochrome signaling to the circadian clock but not for rhythmicity. *Plant Cell* 12: 2499–2509.

Distelfeld, A., Li, C., and Dubcovsky, J. (2009) Regulation of flowering in temperate cereals. *Curr. Opin. Plant Biol.* 12: 178–184.

El-Din El-Assal, S., Alonso-Blanco, C., Peeters, A. J., Wagemaker, C., Weller, J. L., and Koornneef, M. (2003) The role of cryptochrome 2 in flowering in *Arabidopsis*. *Plant Physiol.* 133: 1504–1516.

Fitter, A. H., and Fitter, R. S. (2002) Rapid changes in flowering time in British plants. *Science* 296: 1968–1991.

Gasser, C. S., and Robinson-Beers, K. (1993) Pistil development. *Plant Cell* 5: 1231–1239.

Greenup, A., Peacock, W. J., Dennis, E. S., and Trevaskis, B. (2009) The molecular biology of seasonal flowering-responses in Arabidopsis and the cereals. *Ann. Bot.* 103: 1165–1172.

Guo, H., Yang, H., Mockler, T. C., and Lin, C. (1998) Regulation of flowering time by *Arabidopsis* photoreceptors. *Science* 279: 1360–1363.

Hayama, R., and Coupland, G. (2004) The molecular basis of diversity in the photoperiodic flowering responses of *Arabidopsis* and rice. *Plant Physiol.* 135: 677–684.

Hendricks, S. B., and Siegelman, H. W. (1967) Phytochrome and photoperiodism in plants. *Comp. Biochem.* 27: 211–235.

Honma, T., and Goto, K. (2001) Complexes of MADS-box proteins are sufficient to convert leaves into floral organs. *Nature* 409: 525–529.

Izawa, T., Oikawa, T., Tokutomi, S., Okuno, K., and Shimamoto, K. (2000) Phytochromes confer the photoperiodic control of flowering in rice (a short-day plant). *Plant J.* 22: 391–399.

Izawa, T., Takahashi, Y., and Yano, M. (2003) Comparative biology comes into bloom: Genomic and genetic comparison of flowering pathways in rice and *Arabidopsis*. *Curr. Opin. Plant Biol.* 6: 113–120.

Jaeger, K. E., and Wigge, P. A. (2007) FT protein acts as a long-range signal in Arabidopsis. *Curr. Biol.* 17: 1050–1054.

Jenuwein, T., and Allis, C. D. (2001) Translating the histone code. *Science* 293: 1074–1080.

Johnson, E., Bradley, M., Harberd, N., and Whitelam, G. C. (1994) Photoresponses of light grown phyA mutants of *Arabidopsis*. *Plant Physiol.* 105: 141–149.

Lang, A. (1965) Physiology of flower initiation. In *Encyclopedia of Plant Physiology,* Old Series, Vol. 15, W. Ruhland, ed., Springer, Berlin, pp. 1380–1535.

Lin, M.-K., Belanger, H., Lee, Y.-L., Varkonyi-Gasic, E., Taoka, K.-I., Miura, E., Xoconostie-Cázares, B., Gendler, K., Jorgensen, R. A., Phinney, B., et al. (2007) FLOWERING LOCUS T-protein may act as the long-distance florigenic signal in the cucurbits. *Plant Cell* 19: 1488–1506.

Liu, H., Yu, X., Li, K., Klejnot, J., Yang, H., Lisiero, D., and Lin, C. (2008) Photoexcited CRY2 interacts with CIB1 to regulate transcription and floral initiation in Arabidopsis. *Science* 322: 1535–1539.

Mathieu, J., Warthmann, N., Kuttner, F., and Schmid, M. (2007) The export of FT protein from Phloem companion cells is sufficient for floral induction in Arabidopsis *Curr. Biol.* 17: 1055–1060.

McDaniel, C. N. (1996) Developmental physiology of floral initiation in *Nicotiana tabacum* L. *J. Exp. Bot.* 47: 465–475.

McDaniel, C. N., Hartnett, L. K., and Sangrey, K. A. (1996) Regulation of node number in day-neutral *Nicotiana tabacum*: A factor in plant size. *Plant J.* 9: 56–61.

McDaniel, C. N., Singer, S. R., and Smith, S. M. E. (1992) Developmental states associated with the floral transition. *Dev. Biol.* 153: 59–69.

Meyerowitz, E. M. (2002) Plants compared to animals: The broadest comparative study of development. *Science* 295: 1482–1485.

Michaels, S. D., and Amasino, R. M. (2000) Memories of winter: Vernalization and the competence to flower. *Plant Cell Environ.* 23: 1145–1154.

Millar, A. J., Carre, I. A., Strayer, C. A., Chua, N.-H., and Kay, S. A. (1995) Circadian clock mutants in *Arabidopsis* identified by luciferase imaging. *Science* 267: 1161–1163.

Papenfuss, H. D., and Salisbury, F. B. (1967) Aspects of clock resetting in flowering of *Xanthium*. *Plant Physiol.* 42: 1562–1568.

Pelaz, S., Gustafson-Brown, C., Kohalmi, S. E., Crosby, W. L., and Yanofsky, M. F. (2001) APETALA1 and SEPALLATA3 interact to promote flower development. *Plant J.* 26: 385–394.

Poethig, R. S. (1990) Phase change and the regulation of shoot morphogenesis in plants. *Science* 250: 923–930.

Poethig, R. S. (2003) Phase change and the regulation of developmental timing in plants. *Science* 301: 334–336.

Purvis, O. N., and Gregory, F. G. (1952) Studies in vernalization of cereals. XII. The reversibility by high temperature of the vernalized condition in Petkus winter rye. *Ann. Bot.* 1: 569–592.

Saji, H., Vince-Prue, D., and Furuya, M. (1983) Studies on the photoreceptors for the promotion and inhibition of flowering in dark-grown seedlings of *Pharbitis nil* choisy. *Plant Cell Physiol.* 67: 1183–1189.

Salisbury, F. B. (1963) Biological timing and hormone synthesis in flowering of *Xanthium*. *Planta* 49: 518–524.

Searle, I., He, Y., Turck, F., Vincent, C., Fornara, F., Kröber, S., Amasino, R. A., and Coupland, G. (2006) The transcription factor FLC confers a flowering response to vernalization by repressing meristem competence and systemic signaling in Arabidopsis. *Genes Dev.* 20: 898–912.

Simon, R., Igeno, M. I., and Coupland, G. (1996) Activation of floral meristem identity genes in *Arabidopsis*. *Nature* 384: 59–62.

Suarez-Lopez, P., Wheatley, K., Robson, F., Onouchi, H., Valverde, F., and Coupland, G. (2001) CONSTANS mediates between the circadian clock and the control of flowering in *Arabidopsis*. *Nature* 410: 1116–1120.

Sung, S., and Amasino, R. M. (2004) Vernalization in *Arabidopsis thaliana* is mediated by the PHD finger protein VIN3. *Nature* 427: 159–164.

Tamaki, S., Matsuo, S., Wong, H. L., Yokoi, S., and Shimamoto, K. (2007) Hd3a protein is a mobile flowering signal in rice. *Science* 316: 1033–1036.

Tooke, F., Ordridge, M., Chiurugwi, T., and Battey, N. (2005) Mechanisms and function of flower and inflorescence reversion. *J. Exp. Bot.* 56: 2587–2599.

Valverde, F., Mouradov, A., Soppe, W., Ravenscroft, D., Samach, A., and Coupland, G. (2004) Photoreceptor regulation of CONSTANS protein in photoperiodic flowering. *Science* 303: 1003–1006.

Vince-Prue, D. (1975) *Photoperiodism in Plants*. McGraw-Hill, London.

Weigel, D., and Meyerowitz, E. M. (1994) The ABCs of floral homeotic genes. *Cell* 78: 203–209.

Wigge, P. A., Kim, M. C., Jaeger, K. E., Busch, W., Schmid, M., Lohmann, J. U., and Weigel, D. (2005) Integration of spatial and temporal information during floral induction in *Arabidopsis*. *Science* 309: 1056–1059.

Willis, C. G., Ruhfel, B., Primack, R. B., Miller-Rushing, A. J., Davis, C. C. (2008) Phylogenetic patterns of species loss in Thoreau's woods are driven by climate change. *Proc. Natl. Acad. Sci.* 105: 17029–10733.

Wilson, R. A., Heckman, J. W., and Sommerville, C. R. (1992) Gibberellin is required for flowering in *Arabidopsis thaliana* under short days. *Plant Physiol.* 100: 403–408.

Yanovsky, M. J., and Kay, S. A. (2001) Signaling networks in the plant circadian rhythm. *Curr. Opin. Plant Biol.* 4: 429–435.

Zeevaart, J.A.D. (1976) Physiology of flower formation. *Annu. Rev. Plant Physiol.* 27: 321–348.

Zeevaart, J.A.D. (1985) Bryophyllum. In *Handbook of Flowering*, Vol. 2, A. H. Halevy, ed., CRC Press, Boca Raton, FL, pp. 89–100.

Zeevaart, J.A.D. (2008) Leaf-produced floral signals. *Curr. Opin. Plant Biol.* 11: 541–547.

Zeevaart, J.A.D., and Boyer, G. L. (1987) Photoperiodic induction and the floral stimulus in *Perilla*. In *Manipulation of Flowering*, J. G. Atherton, ed., Butterworths, London, pp. 269–277.

# Responses and Adaptations to Abiotic Stress

Plants grow and reproduce in a complex environment composed of a multitude of abiotic (non-living) chemical and physical factors, which vary both in time and with geographic location. These abiotic factors include air quality and air flow (wind), light intensity and quality, temperature, water availability, mineral nutrient and trace element concentrations, salinity, and the soil chemical environment (pH and redox potential).

Fluctuations of these environmental factors outside of their normal ranges usually have negative biochemical and physiological consequences for plants. In this chapter we will assess these consequences, which include the generation of reactive oxygen species (ROS), reduction in membrane stability, increased protein denaturation, altered ion balance, perturbed metabolism, and physical injury. We will also provide an integrated view of how plants adapt and respond to the abiotic environment.

We will begin with a general discussion of the contributions of genetic changes and phenotypic responses to the plant's overall reproductive fitness. We will then describe the components of the abiotic environment, the potential biological effects of these environmental factors on the plant, and the physiological, biochemical, and molecular processes available to the plant to avoid or mitigate damage that could result from abiotic stress.

# Adaptation and Phenotypic Plasticity

Plants have various mechanisms that allow them to survive and often prosper in the complex environments in which they live. **Adaptation** to the environment is characterized by genetic changes in the entire population that have been fixed by natural selection over many generations. In contrast, individual plants can also *respond* to changes in the environment, by directly altering their physiology or morphology to allow them to better survive the new environment. These responses require no new genetic modifications, and if the response of an individual improves with repeated exposure to the new environmental condition then the response is one of **acclimation**. Such responses are often referred to as **phenotypic plasticity** (Debat and David 2001), and represent nonpermanent changes in the physiology or morphology of the individual that can be reversed if the prevailing environmental conditions change

## Adaptations involve genetic modification

A remarkable example of adaptation to an extreme abiotic environment is the growth of plants in *serpentine soils* (Brady et al. 2005). Serpentine soils are characterized by low moisture, low concentrations of macronutrients, and elevated levels of heavy metals. These conditions would result in severe stress conditions for most plants. However, it is not unusual to find populations of plants genetically adapted to serpentine soils growing not far from closely related nonadapted plants growing on "normal" soils. Simple transplant experiments have shown that only the adapted populations can grow and reproduce on the serpentine soil, and genetic crosses reveal the stable genetic basis of this adaptation.

The evolution of adaptive mechanisms in plants to a particular set of environmental conditions generally involves processes that allow *avoidance* of the potentially damaging effects of these conditions. For example, in southwestern England, populations of Yorkshire fog grass (*Holcus lanatus*) contain a specific genetic modification that reduces the uptake of arsenate, allowing the plants to avoid arsenic toxicity and thrive on contaminated mine sites (Meharg and MacNair 1992; Meharg et al. 1992). In contrast, populations growing on uncontaminated soils are less likely to contain this genetic modification (Meharg et al. 1993).

## Phenotypic plasticity allows plants to respond to environmental fluctuations

In addition to genetic changes in entire populations, individual plants may also show *phenotypic plasticity*; they may respond to fluctuations in the environment by directly altering their morphology and physiology. The changes associated with phenotypic plasticity require no new genetic modifications, and many are reversible.

Both genetic adaptation and phenotypic plasticity can contribute to the plant's overall tolerance of extremes in their abiotic environment. In the previous example, genetic adaptation in the tolerant grass population only *reduces* arsenic uptake—it does not *stop* it. To mitigate the toxic effects of the arsenic that does accumulate, the adapted plants use the same biochemical mechanism that nonadapted plants use to respond to the toxic effects of arsenic accumulation in tissues. This mechanism involves the biosynthesis of low-molecular-weight, arsenic-binding molecules called *phytochelatins* (these will be discussed in more detail later), which reduce arsenic toxicity (Cobbett and Goldsbrough 2002). Thus, the ability of Yorkshire fog grass to prosper on arsenic-contaminated mine waste depends on both a genetic adaptation specific to the tolerant population (arsenic exclusion) and on phenotypic plasticity, which is common to all plants that respond to arsenic by producing phytochelatins (Hartley-Whitaker et al. 2001).

Another example of phenotypic plasticity is the response of salt-sensitive plants, termed *glycophytic* plants, to salinity. Although glycophytic plants are not genetically adapted to growth in saline environments, when exposed to elevated salinity they can activate a number of stress responses that allow the plants to cope with the physiological perturbations imposed by elevated salinity in their environment. For example, the SOS pathway (a signaling pathway based on a serine/threonine protein kinase discovered in the *Salt Overly Sensitive* mutants) leads to enhanced efflux of $Na^+$ and a reduction in salinity-induced toxicity (Zhu 2002). This type of response is usually called a **stress response**.

Stress responses can be elicited by a variety of abiotic environmental factors, such as flooding, drought, elevated UV radiation, salinity, heavy metals, and high and low temperatures. The commonly used terms **stress resistance** and **stress tolerance** are best understood as different expressions of phenotypic plasticity—how a given plant (genotype) responds to a change in the abiotic environment. The ability of a plant to survive and prosper in any given environment is thus related to a balance between genetic adaptation and phenotypic plasticity. Overall, such adaptations and responses enhance the reproductive fitness of a plant in an ecological setting, and translate into stable yields in agricultural settings.

In the sections that follow, we will look first at the ways the abiotic environment can harm plants, and then at the mechanisms used by plants to prevent or minimize the damage.

# The Abiotic Environment and its Biological Impact on Plants

The principal abiotic factors that influence plant growth and development are water, the mineral elements in the soil solution, temperature, and light. As we discuss these factors in this section, we will also consider the primary and secondary physiological and biochemical changes that occur in the plant in response to environmental extremes.

## Climate and soil influence plant fitness

Climatic and edaphic (soil) factors greatly impact plant fitness—that is, their growth, development, reproduction, and survival. **Climatic factors** affecting physiological homeostasis include atmospheric gases, light, temperature, humidity, precipitation, and wind. Humans influence the climate in various ways: by reducing water availability, by increasing the levels of greenhouse gases in the atmosphere, and through the release of air pollutants (see **WEB ESSAY 26.1**). Abiotic factors can also influence one another. For example, high light increases air temperature, wind modulates temperature through evaporative cooling, and ocean currents regulate atmospheric temperature and rainfall. Climatic factors may vary seasonally, or on the scale of a decade or longer, in any given ecosystem. The variations may be gradual and predictable or abrupt and intermittent.

Globally, light intensity variation is exemplified by the high light of deserts and the low light under the canopy of temperate rainforests. Photoperiod in the polar regions changes during the year from continuous light to constant darkness. In contrast, day length at the equator is relatively constant (approximately 12 hours) throughout the year (see Chapter 25). Temperatures also may be relatively constant or may vary substantially on a daily and seasonal basis. For example, the seasonal temperature range is about 10°C in humid rainforests and about 80°C in dry grasslands (from as high as 40°C in the summer down to −40°C in the winter). Monthly temperatures in the tropics may remain above 18°C and in polar regions below 0°C. Yearly precipitation also varies from an average of 15 mm in deserts to more than 11,500 mm in monsoon areas, such as Mawsynram, India. **Edaphic factors** are those relating to soil conditions that affect plant growth, development, and survival; they include air, moisture, and mineral element composition. Soil mineral and organic composition, hydraulic conductivity, ion exchange capacity, pH, and microfauna and microflora, together with climate, determine the availability to plants of air, water, and mineral nutrients in the soil.

Soils are classified according to particle size, ranging from the largest (sand) to the smallest (clay) (see Chapter 5). Soils with large particles and higher porosity have less water-holding capacity than soils with smaller particles and lower porosity. Plant roots must have access to $O_2$, and aeration is greater in soils that are highly porous. Soil organic material derives from the decomposition of animals, plants, and soil microfauna and microflora. Physi-

cally, soils anchor plants to the substratum and govern root development.

## Imbalances in abiotic factors have primary and secondary effects on plants

Plants may experience physiological stress when an abiotic factor is deficient or in excess (referred to as an *imbalance*). The deficiency or excess may be chronic or intermittent. As explained previously, abiotic conditions to which native plants are adapted may cause physiological stress to non-native plants. Most agricultural crops, for example, are cultivated in regions to which they are not highly adapted. Field crops are estimated to produce only 22% of their genetic potential for yield because of suboptimal climatic and soil conditions (Boyer 1982).

Imbalances of abiotic factors in the environment cause primary and secondary effects in plants (**TABLE 26.1**). *Primary effects* such as reduced water potential and cellular dehydration directly alter the physical and biochemical properties of cells, which then lead to secondary effects. These *secondary effects*, such as reduced metabolic activity, ion cytotoxicity, and the production of reactive oxygen species, initiate and accelerate the disruption of cellular integrity, and may lead ultimately to cell death. Different abiotic factors may cause similar primary physiological effects because they affect the same cellular processes. This is the case for water deficit, salinity, and freezing, all of which cause reduction in hydrostatic pressure (turgor pressure, $\Psi_p$) (see Chapter 4) and cellular dehydration. Secondary physiological effects caused by different abiotic imbalances may overlap substantially. From Table 26.1 it is evident that imbalances in many abiotic factors reduce cell proliferation, photosynthesis, membrane integrity, and protein stability, and induce production of reactive oxygen species (ROS), oxidative damage, and cell death.

In the following sections we discuss the effects of imbalances in the following abiotic factors on the physiology and fitness of plants: water, minerals, temperature, and light.

# Water Deficit and Flooding

Water in plants, as in most other organisms, makes up the largest proportion of the cellular volume and is the most limiting resource. About 97% of water taken up by plants is lost to the atmosphere (mostly by transpiration). About 2% is used for volume increase or cell expansion, and 1% for metabolic processes, predominantly photosynthesis (see Chapters 3 and 4). Plant growth can be limited both by water deficit and by excess water. Water deficit (insufficient water availability) occurs in most natural and agricultural habitats and is caused mainly by intermittent to continuous periods without precipitation. **Drought** is the meteorological term for a period of insufficient precipitation that results in plant water deficit. Excess water occurs

**TABLE 26.1**
Physiological and biochemical perturbations in plants caused by fluctuations in the abiotic environment

| Environmental factor | Primary effects | Secondary effects |
|---|---|---|
| Water deficit | Water potential ($\Psi_p$) reduction<br>Cellular dehydration<br>Hydraulic resistance | Reduced cell/leaf expansion<br>Reduced cellular and metabolic activities<br>Stomatal closure<br>Photosynthetic inhibition<br>Leaf abscission<br>Altered carbon partitioning<br>Cytorrhysis<br>Cavitation<br>Membrane and protein destabilization<br>ROS production<br>Ion cytotoxicity<br>Cell death |
| Salinity | Water potential ($\Psi_p$) reduction<br>Cellular dehydration<br>Ion cytotoxicity | Same as for water deficit (see above) |
| Flooding and soil compaction | Hypoxia<br>Anoxia | Reduced respiration<br>Fermentative metabolism<br>Inadequate ATP production<br>Production of toxins by anaerobic microbes<br>ROS production<br>Stomatal closure |
| High Temperature | Membrane and protein destabilization | Photosynthetic and respiratory inhibition<br>ROS production<br>Cell death |
| Chilling | Membrane destabilization | Membrane dysfunction |
| Freezing | Water potential ($\Psi_p$) reduction<br>Cellular dehydration<br>Symplastic ice crystal formation | Same as for water deficit (see above)<br>Physical destruction |
| Trace element toxicity | Disturbed cofactor binding to proteins and DNA<br>ROS production | Disruption of metabolism |
| High light intensity | Photoinhibition<br>ROS production | Inhibition of PSII repair<br>Reduced $CO_2$ fixation |

as the result of flooding or soil compaction. The deleterious effects of excess water are a consequence of the displacement of oxygen from the soil.

*Soil water content and the relative humidity of the atmosphere determine the water status of the plant*

Most of the water utilized by plants is absorbed by roots from the soil. When soil is water-saturated (i.e., when it is at field capacity), the water potential ($\Psi_w$) of the soil solution may approach zero, but drying can reduce the soil $\Psi_w$ to below –1.5 MPa, the point at which permanent wilting can occur (see Chapter 4). Soil dehydration also increases

the salt concentration in the soil solution, resulting in a further lowering of the soil water potential, causing both *osmotic stress* and *specific ion effects* (to be discussed later in the chapter).

The relative humidity of the air determines the vapor pressure gradient between the leaf stomatal cavity and the atmosphere, and this vapor pressure gradient is the driving force for transpirational water loss. Very low relative humidity, which causes a large vapor pressure gradient, can lead to a water deficit in the plant, even in the presence of adequate soil water.

When a soil dries, its hydraulic conductivity decreases very sharply, particularly near the permanent wilting point (that is, the soil water content at which plants cannot regain turgor upon rehydration). Redistribution of water within the roots often occurs at night, when evaporative demand from leaves is low. But at the permanent wilting point (usually about –1.5 MPa), water delivery to the roots is too slow to allow the overnight rehydration of plants that have wilted during the day. Thus, decreasing soil water conductivity hinders rehydration after wilting. (For more details on the relationship between soil hydraulic conductivity and soil water potential, see WEB TOPIC 4.2.)

Rehydration is further hindered by the resistance within the wilted plant, which is larger than the resistance within the soil over a wide range of water deficits (Blizzard and Boyer 1980). Several factors may contribute to the increased plant resistance to water flow during drying. As plant cells lose water, they shrink, primarily because of the collapse of the cell wall (called **cytorrhysis**). When roots shrink, the root surface can move away from the soil particles that hold the water, and the delicate root hairs through which water is absorbed may be damaged. Root hair damage severely impairs water uptake. In addition, as root growth slows during soil drying, the outer layer of the root cortex (the hypodermis) often becomes more extensively covered with suberin, a water-impermeable lipid, further increasing the resistance to water flow into the root.

Another important factor that increases resistance to water flow within a plant is *cavitation*, or the breakage of water columns under tension within the xylem. As discussed in Chapter 4, transpiration from leaves "pulls" water through the plant by creating a tension on the water column. The cohesive forces that are required to support large tensions are present only in very narrow columns in which the water adheres to the walls.

In most plants, cavitation begins at moderate water potentials (–1 to –2 MPa), and the largest vessels cavitate first. For example, in trees such as oak (*Quercus spp.*), the large-diameter vessels that are laid down in the spring function as a low-resistance pathway early in the growing season, when ample water is available. As soil water content decreases during the summer, these large vessels cease functioning, leaving the small-diameter vessels produced during the period of decreasing water supply to carry the transpiration stream. Even if water becomes available later in the season, the original low-resistance pathway may not function efficiently again.

## Water deficits cause cell dehydration and an inhibition of cell expansion

When plant cells experience water deficit, **cellular dehydration** occurs. Cellular dehydration adversely affects basic physiological processes (see Table 26.1). Water deficit causes reductions in cell turgor ($\Psi_p$) and cell volume, which are associated with cellular dehydration and with

the water potential ($\Psi_w$) of the apoplast becoming more negative than that of the symplast. A secondary effect of cellular dehydration is that ions become more concentrated and may become cytotoxic.

Plant water status (cellular water potential, $\Psi_w$) and relative water content (RWC)—defined as the water content of a plant as a percentage of its water content at full turgor or "saturation"—are dependent on the soil moisture content, the capacity for water absorption by roots, and the hydraulic conductivity of root and shoot tissues. Even when water absorption by roots equals water loss by transpiration, the RWC of cells is usually less than 100%. Cells are close to 100% RWC only at night when the vapor pressure deficit is low, the leaf transpiration rate is very low, and the soil is near field capacity.

Cell expansion is a turgor-driven process and is extremely sensitive to water deficit (see Figure 3.14). As discussed in Chapter 15, cell expansion is described by the relationship:

$$GR = m(\Psi_p - Y) \qquad (26.1)$$

where $GR$ is growth rate, $\Psi_p$ is turgor pressure, $Y$ is the yield threshold (the turgor pressure below which the cell wall resists plastic, or nonreversible, deformation), and $m$ is wall extensibility (the responsiveness of the wall to pressure). This equation shows that a decrease in turgor pressure causes a decrease in growth rate. Note also that $\Psi_p$ need decrease only to the value of $Y$, not to zero, to eliminate expansion. In normal conditions, $Y$ is usually only 0.1 to 0.2 MPa less than $\Psi_p$; therefore, small decreases in water content and turgor can result in a decrease in or cessation of growth.

Water deficit not only decreases turgor pressure, but also decreases wall extensibility ($m$) and increases $Y$. Wall extensibility is normally greatest when the cell wall solution is slightly acidic (see Chapter 15). In part, $m$ decreases because cell wall pH typically increases during water deficit. The effects of water deficit on $Y$ are not well understood, but presumably they involve complex structural changes of the cell wall that may not be readily reversed after relief of water deficit. Water-deficient plants tend to become rehydrated at night, allowing leaf growth to occur then. Nonetheless, because of changes in $m$ and $Y$, the growth rate is still lower than that of plants with sufficient water having the same turgor (FIGURE 26.1).

## Flooding, soil compaction, and $O_2$ deficiency are related stresses

Water deficit is stressful, but too much water can also have several potentially negative consequences for a plant. Both flooding and soil compaction result in poor drainage, leading to reduced $O_2$ availability to cells (Bailey-Serres and Voesenek 2008). Roots in well-drained, well-structured soils usually obtain sufficient $O_2$ for aerobic respiration

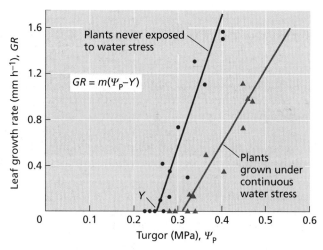

**FIGURE 26.1** Dependence of leaf expansion on leaf turgor. Sunflower (*Helianthus annuus*) plants were grown either with ample water or with limited soil water to produce mild water stress. After rewatering, plants of both treatment groups were stressed by the withholding of water, and leaf growth rates (*GR*) and turgor ($\Psi_p$) were periodically measured. Both decreased extensibility (*m*) and an increased threshold turgor for growth (*Y*) in the previously stressed plants limited the capacity of their leaves to grow after exposure to stress. (After Matthews et al. 1984.)

directly from the gaseous spaces in the soil (see Chapter 4). These gas-filled spaces in soil permit the diffusion of gaseous $O_2$ to a depth of several meters. Consequently, the $O_2$ concentration deep in the soil can be similar to that in humid air.

Flooding, however, fills soil pores with water, reducing $O_2$ availability. Dissolved oxygen diffuses so slowly in stagnant water that only a few cm of soil near the surface remain oxygenated. When temperatures are low and plants are dormant, $O_2$ depletion is very slow and the consequences are relatively harmless. However, when temperatures are higher (greater than 20°C), $O_2$ consumption by plant roots, soil fauna, and microorganisms can totally deplete $O_2$ from the soil in as little as 24 hours.

Flooding-sensitive plants are severely damaged by 24 hours of *anoxia* (the absence of $O_2$). Anoxia causes significant reductions in yield in crop plants. For example, 24 hours of flooding can reduce the yield of flooding-sensitive garden pea (*Pisum sativum*) by fifty percent. Other plants, particularly crop plants such as corn and other species not adapted to growth in continuously wet conditions, are affected by flooding in a milder way, and are more resistant to flooding. These plants can withstand anoxia temporarily, but not for periods of more than a few days. Short periods of flooding may result in *hypoxia* (abnormally low $O_2$).

Soil hypoxia and anoxia damage plant roots directly by inhibiting cellular respiration (see Chapter 11). The **critical oxygen pressure (COP)** is the oxygen pressure below which respiration rates decrease as a result of $O_2$ deficiency. The COP for a corn root tip growing in a well-stirred nutrient solution at 25°C is about 20 kilopascals (kPa), or 20% $O_2$ by volume, close to the oxygen concentration in ambient air. When $O_2$ concentrations are below the critical oxygen pressure, the center of the root becomes hypoxic or anoxic. Diffusion of $O_2$ in an aqueous solution is very slow over long distances. Further, there are substantial resistances to $O_2$ diffusion as the gas moves from the root surface to the cells in the center of the root. As a result, there is insufficient $O_2$ for respiration in these cells below the COP, leading to reduced ATP production, an inability to perform biochemical work, and ultimately cell death.

In healthy cells, the vacuole is more acidic (pH 5.4–5.8) than the cytoplasm (pH 7.0–7.2). However, under extreme $O_2$ deficiency, active transport of $H^+$ into the vacuole by the tonoplast $H^+$-pumping ATPase is slowed by a lack of ATP, and without this activity the normal pH gradient between cytosol and vacuole cannot be maintained. Protons gradually leak from the vacuole into the cytoplasm, adding to the acidity generated by the cell's switch from aerobic respiration to lactic acid fermentation in the cytosol under low $O_2$ conditions (which will be discussed later in the chapter). Acidosis irreversibly disrupts metabolism in the cytoplasm and results in cell death.

Soil $O_2$ deficiency can also have indirect effects on plants by promoting the growth of anaerobic bacteria. Some soil-borne anaerobic microorganisms reduce $Fe^{3+}$ to $Fe^{2+}$. Because of the greater solubility of $Fe^{2+}$, the concentrations of $Fe^{2+}$ can rise to toxic levels in hypoxic soils. Other anaerobes may reduce sulfate ($SO_4^{2-}$) to hydrogen sulfide ($H_2S$), which is a respiratory poison. When anaerobes have an abundant supply of organic substrates, they release metabolites such as acetic acid and butyric acid into the soil water. These substances are toxic to plants at high concentrations.

## Imbalances in Soil Minerals

Imbalances in the mineral content of soils can affect plant fitness either indirectly, by affecting plant nutritional status or water uptake, or directly, through toxic effects on plant cells. In this section we will discuss the effects of high salt concentrations (salinity) and toxic levels of trace elements on plant growth and fitness.

### Soil mineral content can result in plant stress in various ways

Several anomalies associated with the elemental composition of soils can result in plant stress, including high concentrations of salts (e.g., $Na^+$ and $Cl^-$) and toxic ions (e.g., As and Cd), and low concentrations of essential mineral nutrients, such as $Ca^{2+}$, $Mg^{2+}$, N, and P (Epstein and Bloom 2005). In this chapter, the term **salinity** is used to describe

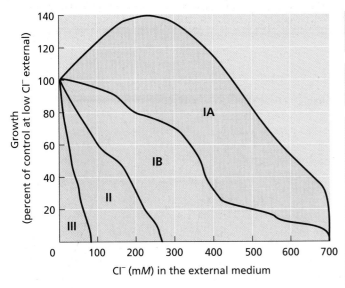

**Group IA (halophytes)** includes sea blite (*Suaeda maritima*) and salt bush (*Atriplex nummularia*). These species show growth stimulation with Cl⁻ levels below 400 m*M*.

**Group IB (halophytes)** includes Townsend's cordgrass (*Spartina x townsendii* ) and sugar beet (*Beta vulgaris*).These plants tolerate salt, but their growth is retarded.

**Group II (halophytes and nonhalophytes)** includes salt-tolerant halophytic grasses that lack salt glands, such as red fescue (*Festuca rubra* subsp. *littoralis*) and *Puccinellia peisonis*, and nonhalophytes, such as cotton (*Gossypium* spp.) and barley (*Hordeum vulgare*). All are inhibited by high salt concentrations. Within this group, tomato (*Lycopersicon esculentum*) is intermediate, and common bean (*Phaseolus vulgaris*) and soybean (*Glycine max*) are sensitive.

The species in **Group III (very salt-sensitive nonhalophytes)** are severely inhibited or killed by even low salt concentrations. Included are many fruit trees, such as citrus, avocado, and stone fruits.

**FIGURE 26.2**   The growth of different species subjected to salinity relative to that of unsalinized controls. The curves dividing the regions are based on data for different species. Plants were grown for 1 to 6 months. (After Greenway and Munns 1980.)

excessive accumulation of salt in the soil solution. Salinity stress has two components: nonspecific **osmotic stress** that causes water deficits, and **specific ion effects** resulting from the accumulation of toxic ions, which disturb nutrient acquisition and result in cytotoxicity (Munns and Tester 2008). Salt-tolerant plants genetically adapted to salinity are termed **halophytes** (from *halo*, the Greek word for "salty"), while less salt-tolerant plants that are not adapted to salinity are termed **glycophytes** (from *glyco*, the Greek word for "sweet").

### Soil salinity occurs naturally and as the result of improper water management practices

In natural environments, there are many causes of salinity. Terrestrial plants encounter high salinity close to the seashore and in estuaries where seawater and freshwater mix or replace each other with the tides. The movement of seawater upstream into rivers can be substantial, depending on the strength of the tidal surge. Far inland, natural seepage from geologic marine deposits can wash salt into adjoining areas. Evaporation and transpiration remove pure water (as vapor) from the soil, concentrating the salts in the soil solution. Soil salinity is also increased when water droplets from the ocean disperse over land and evaporate.

Human activities also contribute to soil salinization. Improper water management practices associated with intensive agriculture can cause substantial salinization of croplands. In many areas of the world, salinity threatens the production of staple foods. Irrigation water in semiarid and arid regions is often saline. In the United States, the salt content of the headwaters of the Colorado River is only 50 mg L⁻¹, but 2000 km downstream, in southern California, the salt content of the same river reaches about 900 mg L⁻¹. This high salinity can preclude growth of some salt-sensitive crops, such as corn, beans, and strawberries. Water from some wells used for irrigation in Texas may contain as much as 2000 to 3000 mg salt L⁻¹. An annual application of 1 m of irrigation water from such wells would add 20 to 30 tons of salts per hectare (8–12 tons per acre) to the soil. Only halophytes, the most salt-tolerant plants, can tolerate such high levels of salts (**FIGURE 26.2**). Glycophytic crops cannot be grown with saline irrigation water.

Saline soils are often associated with high concentrations of NaCl, but in some areas Ca²⁺, Mg²⁺, and SO₄⁻ are also present in high concentrations in saline soils (Epstein and Bloom 2005). High Na⁺ concentrations that occur in *sodic* soils (soils in which Na⁺ occupies ≥10% of the cation exchange capacity) not only injure plants but also degrade the soil structure, decreasing porosity and water permeability. Salt incursion into the soil solution causes water deficits in leaves and inhibits plant growth and metabolism.

### The toxicity of high Na⁺ and Cl⁻ in the cytosol is due to their specific ion effects

Outside the cell, high salt concentrations can result in osmotic stress. Once in the cytosol, however, certain ions act specifically, either singly or in combination, to disturb the nutrient status of the plant. Such specific ion effects can affect the whole plant because ions move to the shoot in the transpirational stream (Munns and Tester 2008).

The most widespread example of a specific ion effect is the cytotoxic accumulation of $Na^+$ and $Cl^-$ ions under saline conditions. Under nonsaline conditions, the cytosol of higher plant cells contains about 100 mM $K^+$ and less than 10 mM $Na^+$, an ionic environment in which enzymes are optimally functional. In saline environments, cytosolic $Na^+$ and $Cl^-$ increase to more than 100 mM, and these ions become cytotoxic. High concentrations of salt cause protein denaturation and membrane destabilization by reducing the hydration of these macromolecules. However, $Na^+$ is a more potent denaturant than $K^+$.

At high concentrations, apoplastic $Na^+$ also competes for sites on transport proteins that are necessary for high-affinity uptake of $K^+$ (see Chapter 6), an essential macronutrient (see Chapter 5). Further, $Na^+$ displaces $Ca^{2+}$ from sites on the cell wall, reducing $Ca^{2+}$ activity in the apoplast and resulting in greater $Na^+$ influx, presumably through nonselective cation channels (Epstein and Bloom 2005; Apse and Blumwald 2007). Reduced apoplastic $Ca^{2+}$ concentrations caused by excess $Na^+$ may also restrict the availability of $Ca^{2+}$ in the cytosol. Since cytosolic $Ca^{2+}$ is necessary to activate $Na^+$ detoxification via efflux across the plasma membrane, elevated external $Na^+$ has the ability to block its own detoxification.

Potentially toxic trace elements such as iron (Fe), zinc (Zn), copper (Cu), cadmium (Cd), nickel (Ni), and arsenic (As) can accumulate to toxic levels in plants, depending on their concentration in the soil (see WEB ESSAY 5.2). Plants require some of these trace elements (e.g., Fe, Zn, Ni, and Cu) for essential biochemical processes, and take up others due to their chemical similarities to essential elements; thus, cadmium can be taken up in place of Zn, and As in place of phosphate.

## Temperature Stress

**Mesophytic plants** (terrestrial plants adapted to temperate environments that are neither excessively wet nor dry) have a relatively narrow temperature range of about 10°C for optimal growth and development. Outside of this range, varying amounts of damage occur, depending on the magnitude and duration of the temperature fluctuation. In this section we will discuss three types of temperature stress: high temperatures, low temperatures above freezing, and temperatures below freezing.

### High temperatures are most damaging to growing, hydrated tissues

Most actively growing tissues of higher plants are unable to survive extended exposure to temperatures above 45°C (113°F) or even short exposure to temperatures of 55°C (131°F) or above. However, nongrowing cells or dehy-

| TABLE 26.2 Heat-killing temperatures for plants | | |
|---|---|---|
| **Plant** | **Heat-killing temperature (°C)** | **Time of exposure** |
| *Nicotiana rustica* (wild tobacco) | 49–51 | 10 min |
| *Cucurbita pepo* (squash) | 49–51 | 10 min |
| *Zea mays* (corn) | 49–51 | 10 min |
| *Brassica napus* (rape) | 49–51 | 10 min |
| *Citrus aurantium* (sour orange) | 50.5 | 15–30 min |
| *Opuntia* (cactus) | >65 | — |
| *Sempervivum arachnoideum* (succulent) | 57–61 | — |
| Potato leaves | 42.5 | 1 hour |
| Pine and spruce seedlings | 54–55 | 5 min |
| *Medicago* seeds (alfalfa) | 120 | 30 min |
| Grape (ripe fruit) | 63 | — |
| Tomato fruit | 45 | — |
| Red pine pollen | 70 | 1 hour |
| Various mosses | | |
| Hydrated | 42–51 | — |
| Dehydrated | 85–110 | — |

*Source:* After Table 11.2 in Levitt 1980.

drated tissues (e.g., seeds and pollen) remain viable at much higher temperatures (TABLE 26.2). Pollen grains of some species can survive 70°C (158°F) and some dry seeds can tolerate temperatures as high as 120°C (248°F).

Most plants with access to abundant water are able to maintain leaf temperatures below 45°C by evaporative cooling, even at elevated ambient temperatures (see WEB TOPIC 26.1 and Chapter 9). However, high leaf temperatures combined with minimal evaporative cooling causes heat stress. Leaf temperatures can rise to 4 to 5°C above ambient air temperature in bright sunlight near midday, when soil water deficit causes partial stomatal closure or when high relative humidity reduces the gradient driving evaporative cooling. Increases in leaf temperature during the day can be more pronounced in plants experiencing drought and high irradiance from direct sunlight. Poor air circulation also decreases the rate of leaf evaporative cooling. Seedlings emerging in moist soil may suffer from heat stress, because wet, bare soil is typically darker and absorbs more solar radiation than drier, lighter soil.

### Temperature stress can result in damaged membranes and enzymes

Plant membranes consist of a lipid bilayer interspersed with proteins and sterols (see Chapters 1 and 11), and any abiotic factor that alters membrane properties can disrupt

(A)

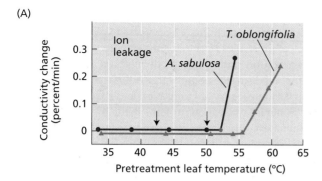

(B)

(C)

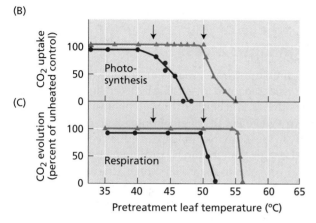

**FIGURE 26.3** Effect of heat stress on frosted orache (*Atriplex sabulosa*) and Arizona honeysweet (*Tidestromia oblongifolia*). (A) Ion leakage was measured in leaf slices submerged in water. Photosynthesis (B) and respiration (C) were measured on attached leaves. At the beginning of the experiment, control rates were measured at a noninjurious 30°C. Attached leaves were then exposed to the indicated temperatures for 15 minutes and returned to the initial control conditions before the rates were recorded. Arrows indicate the temperature thresholds for inhibition of photosynthesis in each of the two species. Membrane permeability was more sensitive to heat damage in *A. sabulosa* than in *T. oblongifolia*, but in both species was less sensitive to heat stress than was photosynthesis. Photosynthesis and respiration were more sensitive to heat damage in *A. sabulosa* than in *T. oblongifolia*. In both species, however, photosynthesis was more sensitive to heat stress than respiration, and photosynthesis was completely inhibited at temperatures that were noninjurious to respiration. (After Björkman et al. 1980.)

cellular processes. The physical properties of the lipids greatly influence the activities of the integral membrane proteins, including H⁺-pumping ATPases, carriers, and channel-forming proteins that regulate the transport of ions and other solutes (see Chapter 6). High temperatures cause an increase in the fluidity of membrane lipids and a decrease in the strength of hydrogen bonds and electrostatic interactions between polar groups of proteins within the aqueous phase of the membrane. High temperatures thus modify membrane composition and structure, and can cause leakage of ions (**FIGURE 26.3A**).

High tempeatures can also lead to a loss of the three-dimensional structure required for correct function of enzymes or structural cellular components, thereby leading to loss of proper enzyme structure and activity. Misfolded proteins often aggregate and precipitate, creating serious problems within the cell.

## Temperature stress can inhibit photosynthesis

Photosynthesis and respiration are both inhibited by temperature stress. Typically, photosynthetic rates are inhibited by high temperatures to a greater extent than respiratory rates (**FIGURE 26.3B AND C**). Although chloroplast enzymes such as rubisco, rubisco activase, NADP–G3P dehydrogenase, and PEP carboxylase become unstable at high temperatures, the temperatures at which these enzymes began to denature and lose activity are distinctly higher than the temperatures at which photosynthetic rates begin to decline (see Chapter 9). This would indicate that the early stages of heat injury to photosynthesis are more directly related to changes in membrane properties and to uncoupling of the energy transfer mechanisms in chloroplasts.

The temperature at which the amount of $CO_2$ fixed by photosynthesis equals the amount of $CO_2$ released by respiration in a given time interval is called the **temperature compensation point**. At temperatures above the temperature compensation point, photosynthesis cannot replace the carbon used as a substrate for respiration. As a result, carbohydrate reserves decline, and fruits and vegetables lose sweetness.

This imbalance between photosynthesis and respiration is one of the main reasons for the deleterious effects of high temperatures. On an individual plant, leaves growing in the shade have a lower temperature compensation point than leaves that are exposed to the sun (and heat). Reduced photosynthate production may also result from stress-induced stomatal closure, reduction in leaf canopy area, and regulation of assimilate partitioning. Enhanced respiration rates relative to photosynthesis at high temperatures are more detrimental in $C_3$ plants than in $C_4$ or CAM plants, because the rates of both dark respiration and photorespiration increase in $C_3$ plants at higher temperatures (see Chapter 9).

## Low temperatures above freezing can cause chilling injury

*Chilling injury* occurs at temperatures that are suboptimal for growth and development but not low enough to result in ice formation. Tropical and subtropical species are typically susceptible to chilling injury. Corn, bean, rice, tomato, cucumber, sweet potato, and cotton are examples of chilling-sensitive food and fiber crops, while *Passiflora*, *Coleus*, and *Gloxinia* are examples of susceptible ornamental plants. Plants grown in relatively warm temperatures (25 to 35°C, 77 to 95°F) experience chilling injury if they are cooled rapidly to 10 to 15°C (50 to 59°F). Chilling injury causes reduced growth, discoloration and lesions of leaves, and vitrified (translucent) foliage. If roots are chilled, the plants may wilt because of disturbances to basic physiological functions such as water absorption. Storage at refrigeration temperatures (5 to 10°C) can also cause chilling injury.

As in the case of high temperatures, membranes can be destabilized by chilling, but in this case it is because of *reduced* membrane fluidity rather than excessive membrane fluidity. As the membrane lipids become less fluid at low temperatures, their protein components can no longer function normally. The result is inhibition of many biochemical processes, including H⁺-pumping ATPase activity, solute transport into and out of cells, energy transduction (see WEB TOPIC 26.2 and Chapters 7 and 11), and enzyme-dependent metabolism.

## Freezing temperatures cause ice crystal formation and dehydration

Freezing temperatures result in intra- and extracellular ice crystal formation. Intracellular ice formation physically shears membranes and organelles. Extracellular ice crystals, which usually form before the cell contents freeze, may not cause immediate physical damage to cells, but they do cause cellular dehydration. This is because ice formation substantially lowers the water potential ($\Psi_w$) in the apoplast, resulting in a gradient from high $\Psi_w$ in the symplast to low $\Psi_w$ in the apoplast. Consequently, water moves from the symplast to the apoplast, resulting in cellular dehydration (for a detailed description of this process, see WEB TOPIC 26.3). Dehydration caused by freezing causes irreversible membrane damage (Uemura et al. 2006). Cells that are already dehydrated, such as those in seeds and pollen, are relatively less affected by ice crystal formation.

Ice usually forms first within the intercellular spaces and in the xylem vessels, along which the ice can quickly propagate. This ice formation is not lethal to hardy plants, and the tissue recovers fully if warmed. However, when plants are exposed to freezing temperatures for an extended period, the growth of extracellular ice crystals leads to physical destruction of membranes and excessive dehydration.

Several hundred water molecules are needed for an ice crystal to begin forming. The process by which these hundreds of molecules start to form a stable ice crystal is called **ice nucleation**, and it strongly depends on the properties of the involved surfaces. Some large polysaccharides and proteins facilitate ice crystal formation and are called *ice nucleators*. Some ice nucleation proteins made by bacteria appear to facilitate ice nucleation by aligning water molecules along repeated amino acid domains within the protein. In plant cells, ice crystals begin to grow from endogenous ice nucleators; the resulting relatively large intracellular ice crystals cause extensive damage to the cell and are usually lethal.

# High Light Stress

As photoautotrophs, plants are dependent upon—and exquisitely adapted to—visible light for the maintenance of a positive carbon balance through photosynthesis. Higher energy wavelengths of electromagnetic radiation, especially in the ultraviolet range, can inhibit cellular processes by damaging membranes, proteins, and nucleic acids. However, even in the visible range, irradiances far above the light saturation point of photosynthesis cause *high light stress*, which can disrupt chloroplast structure and reduce photosynthetic rates, a process known as *photoinhibition*.

## Photoinhibition by high light leads to the production of destructive forms of oxygen

The inhibition of photosynthesis by high light is called **photoinhibition** (see Chapters 7 and 9). Excess light excitation arriving at the PSII reaction center can lead to its inactivation by the direct damage of the D1 protein (see Chapter 7). Excess absorption of light energy by photosynthetic pigments also produces excess electrons outpacing the availability of NADP⁺ to act as an electron sink at PSI. The excess electrons produced by PSI lead to the production of **reactive oxygen species** (**ROS**), notably superoxide ($O_2^{\bullet-}$) (see Chapter 7). Superoxide and other ROS are low-molecular-weight molecules that function in signaling and, in excess, cause oxidative damage to proteins, lipids, RNA, and DNA. The oxidative stress generated by excessive ROS destroys cellular and metabolic functions and leads to cell death.

The extent of photoinhibition depends on the balance between photodamage to the photosystem II (PSII) complex and its repair (Takahashi and Murata 2008) (see Chapters 7 and 9). Repair of PSII after photodamage is critical for the maintenance of photosynthesis. Under oxidative, saline, or low temperature stresses, increased photoinhibition seems to result more from impaired repair capacity than from increased photodamage to PSII. For example, excess ROS inhibit the translation of mRNAs that encode proteins in the PSII complex, thereby limiting PSII repair (Takahashi and Murata 2008). Conversely, fatty acid unsaturation of thylakoid membrane lipids, which allows for membrane fluidity, is associated with an enhanced capacity for photosystem repair and recovery of PSII activity (Vijayan and Browse 2002).

Excess production of ROS occurs during many stresses, not just high light. ROS are byproducts of respiration and photosynthesis. Accumulation of these products from oxygen metabolism in different cellular compartments (predominantly in chloroplasts, mitochondria, and peroxisomes) is regulated by antioxidative mechanisms, which normally maintain a balance between ROS production and ROS scavenging (Apel and Hirt 2004). Temperature extremes, dehydration, high light, salinity, and ion toxicity all lead to an imbalance between ROS production and scavenging that causes increased oxidative damage of macromolecules and signaling dysfunction.

# Developmental and Physiological Mechanisms that Protect Plants against Environmental Extremes

In this section, we describe the plant responses that mitigate the impact of the environmental stresses we have just discussed. These responses range from temporary metabolic adjustments to changes in organ shape, plant architecture, and life cycle. Some responses allow plants to avoid stresses, while others allow plants to tolerate them.

## Plants can modify their life cycles to avoid abiotic stress

One way plants can adapt to extreme environmental conditions is through modification of their life cycles. For example, annual desert plants have short life cycles: they complete them during the periods when water is available, and are dormant (as seeds) during dry periods. Deciduous trees of the temperate zone shed their leaves before the winter so that sensitive leaf tissue is not damaged by cold temperatures.

During less predictable stressful events (e.g., a summer of significant but erratic rainfall) the growth habits of some species may confer a degree of tolerance to these conditions. For example, plants that can grow and flower over an extended period (*indeterminate growth*) are often more tolerant to erratic environmental extremes than plants that develop preset numbers of leaves and flower over only very short periods (*determinate growth*).

## Phenotypic changes in leaf structure and behavior are important stress responses

Because of their roles in photosynthesis, leaves (or their equivalent) are crucial to the survival of a plant. To func-

tion, leaves must be exposed to sunlight and air, but this also makes them particularly vulnerable to environmental extremes. Plants have thus evolved various mechanisms that enable them to avoid or mitigate the effects of abiotic extremes to leaves. Such mechanisms include changes in leaf area, leaf orientation, trichomes, and the cuticle.

**LEAF AREA** Total leaf area is a major component of biomass accumulation and yield. Large leaf areas provide significant surfaces for the production of photosynthate, but can be detrimental to crop growth and survival under stressful conditions. Large individual leaves or large total leaf areas provide a large surface for evaporation of water that is advantageous for leaf cooling, but can lead to quick depletion of soil water or excessive, damaging absorption of solar energy. Plants can reduce their leaf area by

- reducing leaf cell division and expansion,
- altering leaf shapes, and/or
- initiating senescence and abscission of leaves.

Both cell division and expansion are reduced in plants growing under water deficit and salinity. Specific signaling cascades can slow or arrest the cell cycle and slow DNA replication, thereby limiting plant growth (Anami et al. 2009).

Turgor reduction is the earliest significant biophysical effect of water deficit. As a result, turgor-dependent processes such as leaf expansion (**FIGURE 26.4**) and root elongation are the most sensitive to water deficits. When water deficit develops slowly enough to allow changes in developmental processes, it has several effects on growth, one of which is a limitation of leaf expansion. As discussed earlier in the chapter, a decrease in the water content of the plant leads to a reduction of the turgor pressure acting on the cell walls, and cell expansion decreases.

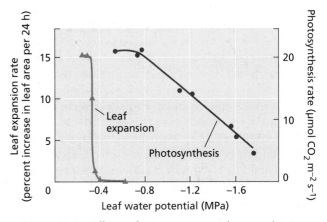

**FIGURE 26.4** Effects of water stress on photosynthesis and leaf expansion of sunflower (*Helianthus annuus*). In this species, leaf expansion is completely inhibited under mild stress levels that hardly affect photosynthetic rates. (After Boyer 1970.)

Because leaf expansion depends mostly on cell expansion, the principles that underlie the two processes are similar. Inhibition of cell expansion results in a slowing of leaf expansion early in the development of water deficits. The resulting smaller leaf area transpires less water, effectively conserving a limited water supply in the soil over a longer period. The reduction in leaf area is often considered a physiological response to water deficit.

Many arid-zone plants have very small leaves, which minimizes the thin film of still air at the surface of the leaf (the *boundary layer*). This very thin boundary layer (low boundary layer resistance) permits the transfer of heat from the leaf to the air (see Figure 9.15). Because of their low boundary layer resistance, small leaves are able to maintain surface temperatures that are close to that of the air even when transpiration is greatly reduced, and avoid overheating. In contrast, large leaves have higher boundary layer resistance and dissipate less thermal energy (per unit leaf area) by direct transfer of heat to the air.

In plants with indeterminate growth, water limitation not only limits leaf size, but also leaf number, because it decreases both the number and the growth rate of branches. Stem growth has been studied less than leaf expansion, but stem growth is probably affected by the same conditions that limit leaf growth during periods of low water availability. Cell and leaf expansion also depend on biochemical and molecular factors beyond those that control water flux. Plants change their growth rates in response to water limitation by coordinately controlling many other important processes such as cell wall and membrane biosynthesis, cell division, and protein synthesis (Burssens et al. 2000).

Altering leaf shape is another way that plants can reduce leaf area. Under conditions of water, heat, or salinity extremes, leaves may be narrower or may develop deeper lobes during development (**FIGURE 26.5**). The result is a reduced leaf surface area and therefore, reduced water loss and *heat load* (defined as amount of heat loss [cooling] required to maintain a leaf temperature close to air temperature) (see Chapter 9). Furthermore, the total leaf area of a plant (number of leaves × surface area of each leaf) does not remain constant after all the leaves have matured. In some plants, leaf abscission may occur under mild water deficit, effectively reducing the leaf area losing water.

Leaves will senesce and eventually abscise if plants experience water deficit after a substantial leaf area has developed (**FIGURE 26.6**). Many drought-adapted, deciduous desert plants abscise all leaves during a drought and develop new leaves after a rain. Abscission during periods of water deficit results largely from enhanced synthesis of and responsiveness to the endogenous plant hormone ethylene (see Chapter 22). The negative consequence of leaf abscission is that it also reduces total canopy area, limiting total photosynthate production of the plant.

In a leaf senescing under well-watered conditions, the leaf cells undergo a form of programmed cell death

**FIGURE 26.5** Altered leaf shape can occur in response to environmental changes. The oak (*Quercus sp.*) leaf on the left is from the outside of a tree canopy, where temperatures are higher than they are inside the canopy. The leaf on the right is from the inside of the canopy. The deeper sinuses of the leaf on the left result in a lower boundary layer, which allows for better evaporative cooling. (Photograph by David McIntyre.)

**FIGURE 26.6** The leaves of young cotton (*Gossypium hirsutum*) plants abscise in response to water stress. The plants at left were watered throughout the experiment; those in the middle and at right were subjected to moderate stress and severe stress, respectively, before being watered again. Only a tuft of leaves at the top of the stem is left on the severely stressed plants. (Courtesy of B. L. McMichael.)

that involves controlled degradation and reallocation of their molecular building blocks to other cells within the plant. Leaf senescence is often a normal part of a plant's life cycle—for example, when it switches from vegetative to reproductive growth. Abiotic extremes, in contrast, can damage leaf and root tissues and eventually lead to premature, uncontrolled senescence; in a crop plant, this can dramatically reduce yield by reducing grain size and quality. In many crops, tolerance to drought during grain maturation is correlated with enhanced crop productivity, grain quality, and lodging resistance.

Under extreme environmental conditions, some varieties of crop plants have the ability to maintain green leaf area during grain maturation, when the leaves of many other genotypes of the same crop would begin to senesce. In grass crops such as corn and sorghum, the retention of photosynthetically active leaves is a trait known as *stay-green* (**FIGURE 26.7**). Stay-green varieties maintain green stems and upper leaves under drought conditions during grain fill and have a significant yield advantage over senescent hybrids under late-season drought conditions. The grain yield of stay-green sorghum cultivars may be increased by 0.35 Mg ha$^{-1}$ for every day that leaf senescence is delayed (Borrell et al. 2000). Transgenic tobacco plants able to delay leaf senescence (through the activation of cytokinin biosynthesis) display enhanced drought tolerance (Rivero et al. 2007) (**FIGURE 26.8**).

**LEAF ORIENTATION** One of the major contributors to increased corn yields in North American hybrids over the past sixty years has been an increase in leaf angle that

**FIGURE 26.7** Certain corn genotypes exhibit the trait known as stay-green. The corn inbred line B73 on the left is displaying the typical senescence that occurs during grain maturation, while the inbred Mo20W on the right is exhibiting stay-green. (Courtesy of M. R. Tuinstra.)

(A) Wild type

(B) P$_{SARK}$::IPT

**FIGURE 26.8** Effects of drought on wild-type and transgenic tobacco plants expressing isopentenyl transferase (a key enzyme in the production of cytokinin) under the control of P$_{SARK}$ (promoter region of *Senescence-Associ-* ated *Receptor Kinase*), a maturation and stress-induced promoter. Shown are wild-type (A) and transgenic (B) plants after 15 days of drought following by 7 days re-watering. (Courtesy of E. Blumwald).

allows greater light absorption by total plant leaf area. However, in warm, sunny environments, a transpiring leaf may be near its upper limit of temperature tolerance. If transpiration is decreased, increases in leaf temperature or additional energy absorption can damage the leaf. In environments with intense solar radiation, high temperatures, and/or soil water deficits, plants that can alter the orientation of their leaves have the advantage of being able to avoid excessive heating of the leaves.

For protection against overheating during water deficit, the leaves of some plants may orient themselves away from the sun; such leaves are said to be *paraheliotropic*. Leaves that gain energy by orienting themselves perpendicular to the sunlight are referred to as *diaheliotropic* (see Chapter 9). FIGURE 26.9 shows the strong effect of water deficit on leaf position in soybean. Other factors that can alter the interception of radiation include wilting and leaf rolling. Wilting changes the angle of the leaf, and leaf rolling minimizes the profile of tissue exposed to the sun.

Leaf orientation may also change in response to low oxygen availability. Hypoxia accelerates production of the ethylene precursor ACC (1-aminocyclopropane-1-carboxylic acid) in roots (see Chapter 22). In tomato, ACC travels via the xylem sap to the shoot, where, in contact with oxygen, it is converted by ACC oxidase to ethylene. The upper (adaxial) surfaces of the leaf petioles of tomato and sunflower have ethylene-responsive cells that expand more rapidly when ethylene concentrations are high. This expansion results in **epinasty**, downward growth of the leaves such that they appear to droop (see Figure 22.12). Unlike wilting, epinasty does not involve loss of turgor.

**TRICHOMES** Many leaves have hair-like epidermal cells known as trichomes. Densely packed trichomes on a leaf surface (also referred to as pubescence) keep leaves cooler by reflecting radiation. Leaves of some plants have a silvery-white appearance because the densely packed hairs reflect a large amount of light. Dry nonglandular trichomes tend to reflect the most light, especially when they are very dense. The desert shrub white brittlebush (*Encelia farinosa*) can produce two types of leaves at different times of the year: green nearly hairless leaves in the winter and silvery-white pubescent leaves in the summer. The silvery leaves produced in the summer are several degrees cooler than they would be if they lacked the thick layer trichomes, which reflect infrared radiation that causes excessive heating. However, pubescent leaves are a disadvantage in the cooler spring months because the trichomes also reflect the visible light needed for photosynthesis. Thus, brittlebrush has adapted to the desert environment by producing non-reflectant leaves lacking trichomes during the spring.

**CUTICLE** The cuticle is a multilayered structure of waxes and related hydrocarbons deposited on the outer cell walls of the leaf epidermis. The cuticle, like trichomes, can reflect light, thereby reducing heat load. The cuticle appears to

(A) Well-watered

(B) Mild water stress

(C) Severe water stress

FIGURE 26.9 Leaf movements in soybean in response to osmotic stress. Leaflet orientation of field-grown soybean (*Glycine max*) plants in the normal, unstressed position (A); during mild water stress (B); and during severe water stress (C). The large leaf movements induced by mild stress are quite different from wilting, which occurs during severe stress. Note that during mild stress (B), the terminal leaflet has been raised, whereas the two lateral leaflets have been lowered; each is almost vertical. (Courtesy of D. M. Oosterhuis.)

also restrict the diffusion of water and gases, as well as the entrance of pathogens. A developmental response to water deficit in some plants is the production of a thicker cuticle, which decreases transpiration. It remains unclear whether the altered thickness results from changes that are quantitative (more wax) or qualitative (different wax composition or altered structure of the inner layer of the cuticle). As cuticular transpiration accounts for only 5 to 10% of total leaf transpiration, this response becomes significant only under extreme conditions.

### The ratio of root-to-shoot growth increases in response to water deficit

Mild water deficits also affect the development of the root system. Root-to-shoot biomass ratio appears to be governed by a functional balance between water uptake by the root and photosynthesis by the shoot. Within the limits set by the plant's genetic potential, a shoot will tend to grow until water uptake by the roots becomes limiting to further growth; conversely, roots will tend to grow until their demand for photosynthate from the shoot exceeds the supply. This functional balance is shifted if the water supply decreases.

When water to the shoot becomes limiting, leaf expansion is reduced before photosynthetic activity is affected (see Figure 26.4). Inhibition of leaf expansion reduces the consumption of carbon and energy, and a greater proportion of the plant's assimilates can be allocated to the root system, where they can support further root growth. This root growth is sensitive to the water status of the soil microenvironment; the root apices in dry soil lose turgor, while roots in the soil zones that remain moist continue to grow.

As water deficits progress, the upper layers of the soil usually dry first. Plants usually have relatively shallow root systems when soil moisture is abundant, but the roots proliferate into deeper, moister soil when water is depleted from the upper soil layers (Koike et al. 2003). This change in root architecture can be considered another line of defense against drought.

Enhanced root growth into deeper soil during water deficit requires allocation of photosynthates to the growing root tips. In reproductive plants, enhanced water uptake resulting from root growth is less pronounced than in vegetative plants, because during water deficit, photosynthates are often directed to the fruits and away from the roots (see Chapter 10). Competition for photosynthates between roots and fruits is one explanation for the fact that plants are generally more sensitive to water deficit during reproduction.

### Plants can regulate stomatal aperture in response to dehydration stress

The ability to control stomatal aperture allows plants to respond quickly to a changing environment, for example to avoid excessive water loss or limit uptake of liquid or gaseous pollutants through stomata. Stomatal opening and closing is modulated by uptake and loss of water in guard cells, which changes their turgor pressure (see Chapters 4 and 18).

Although guard cells can lose turgor as a result of a direct loss of water by evaporation to the atmosphere, stomatal closure in response to dehydration is almost always an active, energy-dependent process rather than a passive one (Buckley 2005). Abscisic acid (ABA) mediates the solute loss from guard cells that is triggered by a decrease in the water content of the leaf (see Chapter 23 and **WEB TOPIC 26.4**). Plants constantly modulate the concentration and cellular localization of ABA, and this allows them to respond quickly to environmental changes, such as fluctuations in water availability. Abscisic acid is synthesized in a diurnal pattern in both leaf and root cells. Modulation of ABA biosynthesis is one way that the plant regulates ABA metabolism in response to dehydration in both below- and above-ground parts of the plant.

The redistribution of ABA to the guard cell depends on pH gradients within the leaf, on the weak-acid properties of the ABA molecule, and on the permeability of cell membranes (see Figure 23.6). ABA is synthesized at a higher rate when leaves are dehydrated, and more ABA accumulates in the leaf apoplast. The higher ABA concentrations resulting from the higher rates of ABA synthesis appear to enhance or prolong the initial stomatal closing effect of the ABA stored in the leaf. The mechanism of ABA-induced stomatal closure is discussed in Chapter 23.

Increased stomatal conductance has also been found to confer heat resistance to plants grown under ample irrigation in a hot climate (see **WEB TOPIC 26.1**). Studies have shown that intensive selection for higher yields in Pima cotton and wheat has increased stomatal conductance, and that these increases are linearly related to yield. Higher stomatal conductance enhances leaf cooling and reduces the gap between air temperature, which might exceed $40°C$, and optimal temperatures for leaf photosynthesis, which are usually below $30°C$.

### Plants adjust osmotically to drying soil by accumulating solutes

Water can move through the soil-plant-atmosphere continuum only if water potential decreases along that path (see Chapters 3 and 4). Recall from Chapter 3 that $\Psi_w = \Psi_s + \Psi_p$, where $\Psi_w$ = water potential, $\Psi_s$ = osmotic potential, and $\Psi_p$ = hydrostatic pressure. When the water potential of the rhizosphere (the microenvironment surrounding the root) decreases due to water deficit (see **WEB TOPIC 3.7**) or salinity, plants can continue to absorb water only as long as $\Psi_w$ is lower (more negative) than that of the soil water. Consider that a decrease in $\Psi_p$ would lower $\Psi_w$ but would also result in loss of turgor and growth cessation. On the other hand, a decrease in $\Psi_s$ can maintain a water potential gradient between cells, the soil and the plant, or

(A) External $\Psi_w$ –0.6 MPa

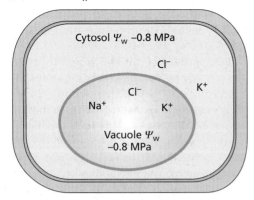

(B) External $\Psi_w$ –0.8 MPa

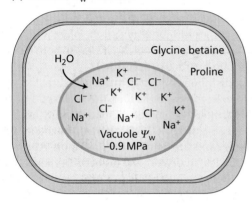

**FIGURE 26.10** Solute adjustments during osmotic stress. The water potential of the cytosol and vacuole of cells must be slightly lower than the surrounding environment to maintain a water potential gradient that allows for water uptake. (A) Cell with an external water potential of –0.6 MPa. Equilibrium is maintained inside the cell by ion accumulation in the vacuole and cytosol. (B) Cell with an external water potential of –0.8 MPa because of salinity, drought, or other dehydration stresses. The cell can adjust osmotically by increasing the cellular concentration of solutes in the vacuole and cytosol. If inorganic ions are utilized for osmotic adjustment, they are typically stored in the vacuole where they cannot affect metabolic processes in the cytosol. Equilibrium in the cytosol is maintained with compatible solutes (typically uncharged) such as proline and glycine betaine.

between the plant and the atmosphere without a decrease in turgor or growth. **Osmotic adjustment** is the capacity of plant cells to accumulate solutes and use them to lower $\Psi_w$ during periods of osmotic stress. The adjustment involves a net increase in solute content per cell that is independent of the volume changes that result from loss of water (**FIGURE 26.10**). The decrease in $\Psi_s$ is typically limited to about 0.2 to 0.8 MPa, except in plants adapted to extremely dry conditions.

There are two main ways by which osmotic adjustment can take place. A plant may take up ions from the soil, or transport ions from other plant organs to the root, so

that the solute concentration of the root cells increases. For example, increased uptake and accumulation of $K^+$ will lead to decreases in $\Psi_s$, due to the effect of the potassium ions on the osmotic pressure within the cell. This is a common event in saline areas, where ions such as potassium and calcium are readily available to the plant.

There is a potential problem, however, when ions are used to decrease $\Psi_s$. Some ions, such as sodium or chloride, are essential to plant growth in low concentrations, but higher concentrations can have a detrimental effect on cellular metabolism. Other ions, such as potassium, are required in larger quantities, but at high concentrations can still have a detrimental effect on the plant, mostly through disruption of cell membranes or proteins.

The accumulation of ions during osmotic adjustment is predominantly restricted to the vacuoles, where the ions are kept out of contact with cytosolic enzymes or organelles. Many halophytes utilize vacuolar compartmentalization of $Na^+$ and $Cl^-$ to facilitate osmotic adjustment that sustains or enhances growth in saline environments. Since an increase in vacuolar volume is the driving force for cell expansion, a high vacuolar concentration of ions provides such a driving force.

When ions are compartmentalized in the vacuole, other solutes must accumulate in the cytoplasm to maintain water potential equilibrium within the cell. These solutes are called compatible solutes (or compatible osmolytes). **Compatible solutes** are organic compounds that are osmotically active in the cell, but do not destabilize the membrane or interfere with enzyme function, as high concentrations of ions can. Plant cells can hold large concentrations of these compounds without detrimental effects on metabolism. Common compatible solutes include amino acids such as proline, sugar alcohols such as mannitol, and quaternary ammonium compounds such as glycine betaine (**FIGURE 26.11**). Some of these solutes, such as proline, also seem to have an osmoprotectant function whereby they protect plants from toxic byproducts produced during periods of water shortage, and provide a source of carbon and nitrogen to the cell when conditions return to normal. Each plant family tends to use one or two compatible solutes in preference to others. Because the synthesis of compatible solutes is an active metabolic process, energy is required. The amount of carbon used for the synthesis of these organic solutes can be rather large, and for this reason, the synthesis of these compounds tends to reduce crop yield.

## Submerged organs develop aerenchyma tissue in response to hypoxia

We now turn to the mechanisms used by plants to cope with too *much* water. In most wetland plants, such as rice, and in many plants that acclimate well to wet conditions, the stem and roots develop longitudinally interconnected,

**Amino acids**

Proline

**Sugar alcohols**

$HOCH_2$—C—C—C—C—$CH_2OH$

Sorbitol

**Quaternary ammonium compunds (QACs)**

$CH_3$—$N^+$—$CH_2$—$COO^-$

Glycine betaine

**Tertiary sulfonium compounds (TSCs)**

$CH_3$—$S$—$CH_2$—$CH_2$—$COO^-$

3-Dimethylsulfoniopropionate (DMSP)

**FIGURE 26.11** Four groups of molecules frequently serve as compatible solutes: amino acids, sugar alcohols, quaternary ammonium compounds, and tertiary sulfonium compounds. Note that these compounds are small and have no net charge.

gas-filled channels that provide a low-resistance pathway for the movement of oxygen and other gases. The gases (air) enter through stomata, or through **lenticels** on woody stems and roots, and travel by molecular diffusion or by convection driven by small pressure gradients. In many plants adapted to wetland growth, root cells are separated by prominent, gas-filled spaces that form a tissue called **aerenchyma**. These cells develop in the roots of wetland plants independently of environmental stimuli. In some nonwetland monocots and dicots, however, oxygen deficiency induces the formation of aerenchyma in the stem base and newly developing roots.

An example of induced aerenchyma occurs in corn (**FIGURE 26.12**). Hypoxia stimulates the activity of ACC synthase and ACC oxidase in the root tips of corn, and causes ACC and ethylene to be produced faster (see Chapter 22). Ethylene triggers programmed cell death and disintegration of cells in the root cortex. The spaces formerly occupied by these cells provide gas-filled voids that facilitate movement of $O_2$. Ethylene-signaled cell death is highly selective; only some cells have the potential to initiate the developmental program that creates the aerenchyma (Drew et al. 2000).

When aerenchyma formation is induced, a rise in cytosolic $Ca^{2+}$ concentration is thought to be part of the ethylene signal transduction pathway leading to cell death. Signals that elevate cytosolic $Ca^{2+}$ concentration can promote cell death in the absence of hypoxia. Conversely, signals that lower cytosolic $Ca^{2+}$ concentration block cell death in hypoxic roots that would normally form aerenchyma.

Some tissues can tolerate anaerobic conditions in flooded soils for an extended period (weeks or months) before developing aerenchyma. These include the embryo and coleoptile of rice (*Oryza sativa*) and rice grass (*Echi-*

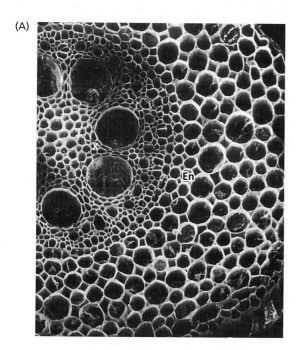

(A)

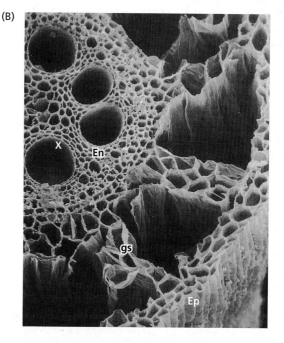

(B)

**FIGURE 26.12** Scanning electron micrographs of transverse sections through roots of corn, showing changes in structure with oxygen supply. (150×) (A) Control root, supplied with air, with intact cortical cells. (B) Oxygen-deficient root growing in a nonaerated nutrient solution.

Note the prominent gas-filled spaces (gs) in the cortex (cx), formed by degeneration of cells. The stele (all cells interior to the endodermis, En) and the epidermis (Ep) remain intact. X, xylem. (Courtesy of J. L. Basq and M. C. Drew.)

*nochloa crus-galli* var. *oryzicola*) and the rhizomes (underground horizontal stems) of giant bulrush (*Schoenoplectus lacustris*), salt marsh bulrush (*Scirpus maritimus*), and narrow-leafed cattail (*Typha angustifolia*). These rhizomes can survive for several months and expand their leaves under anaerobic conditions.

In nature, these rhizomes overwinter in anaerobic mud at the edges of lakes. In spring, once the leaves have expanded above the mud or water surface, $O_2$ diffuses down through aerenchyma into the rhizome. Metabolism then switches from an anaerobic (fermentative) to an aerobic mode, and roots begin to grow using the available oxygen. Likewise, during germination of paddy (wetland) rice and of rice grass, the coleoptile breaks through the water surface and becomes a diffusion pathway for $O_2$ into submerged parts of the plant, including the roots. Although rice is a wetland species, its roots are as intolerant of anoxia as corn roots are. As the root extends into oxygen-deficient soil, the continuous formation of aerenchyma just behind the tip allows oxygen movement within the root to supply the apical zone.

In roots of rice and other typical wetland plants, structural barriers composed of suberized and lignified cell walls prevent $O_2$ diffusion outward to the soil. The $O_2$ thus retained supplies the apical meristem and allows growth to extend 50 cm or more into anaerobic soil. In contrast, roots of nonwetland species, such as corn, leak $O_2$. Internal $O_2$ becomes insufficient for aerobic respiration in the root apex of these nonwetland plants, and this lack of $O_2$ severely limits the depth to which such roots can extend into anaerobic soil.

## Plants have evolved two different strategies to protect themselves from toxic ions: exclusion and internal tolerance

Two basic strategies are employed by plants to tolerate the presence of elevated concentrations of various toxic trace elements in the environment, including sodium (Na), arsenic (As), cadmium (Cd), copper (Cu), nickel (Ni), zinc (Zn), and selenium (Se):

- *Exclusion*, by which the concentration of these elements inside the plant is maintained below a toxic threshold value, and

- *Internal tolerance*, through which various biochemical adaptations allow the plant to tolerate, compartmentalize, or chelate elevated concentrations of these elements (see WEB ESSAY 26.2).

**EXCLUSION**  Salt-sensitive plants generally tolerate moderate salinity mainly because of root mechanisms that reduce uptake of potentially harmful ions to the shoots. Horticulturists utilize this whole-plant ion exclusion mechanism when they graft salt-sensitive scions onto salt-tolerant rootstocks. Calcium ions play a key role in mini-

mizing the uptake of $Na^+$ ions from the external medium. The electrochemical potential normally maintained by cells at the plasma membrane (see Chapter 6) can facilitate passive $Na^+$ accumulation into the cytosol 100- to 1000-fold above the external concentrations. As a charged ion, $Na^+$ has very low permeability to the plasma membrane, but it is transported by both low- and high-affinity transport systems, many of which are normally used for $K^+$ uptake (Epstein and Bloom 2005) (see also Chapter 6).

External $Ca^{2+}$ at millimolar concentrations (a typical physiological concentration of $Ca^{2+}$ in the apoplast) increases the selectivity of the $K^+$ transporters and minimizes $Na^+$ uptake through mechanisms that reduce uptake and facilitate efflux of $Na^+$ to the apoplast. Influx of $Na^+$ occurs through nonselective, voltage-insensitive gated channels that passively transport cytotoxic cations into cells. When such channels are open, $Na^+$ diffuses into the cytosol, where it can be substantially concentrated. Physiological levels of $Ca^{2+}$ cause closure of these channels, restricting $Na^+$ uptake (Tracy et al. 2008).

**INTERNAL TOLERANCE**  In contrast to glycophytes, halophytes can accumulate ions in the shoot because they have a greater capacity for *vacuolar sequestration* of ions in leaf cells. Furthermore, halophytes seem to have a greater ability to restrict net intracellular $Na^+$ uptake in leaf cells. As a result of this increased vacuolar compartmentalization and reduced cellular uptake of $Na^+$ in the shoots, these plants have an enhanced capacity to tolerate an increased flux of $Na^+$ from roots in the transpirational stream (Apse and Blumwald 2007).

A restriction of the transport of $Na^+$ from root to shoot by a reduction of the amount of $Na^+$ loaded into the xylem in the roots appears to be more critical for glycophytes than for halophytes. However, it is clear that salt tolerance of both halophytes and glycophytes is dependent on the ion transport processes that control net ion uptake across the plasma membrane and compartmentalization into the vacuole (see WEB TOPIC 26.6).

Some halophytes, such as saltcedar (*Tamarix* spp.) and saltbush (*Atriplex* spp.) have evolved specialized salt glands on the surfaces of their leaves that excrete salt. These glands are unique structures with a highly specialized function in salt tolerance. Other plants accumulate toxic ions in older leaves and allow those leaves to senesce and abscise to allow younger, more photosynthetically productive leaves to survive.

Abundant evidence makes it clear that both halophytes and glycophytes accumulate ions intracellularly and use them for the osmotic adjustment necessary for cell expansion (Hasegawa et al. 2000). Since high ion concentrations are toxic to the cellular metabolism of all plants, both halophytes and glycophytes compartmentalize cytotoxic ions into the vacuole or actively pump them outside the cell to the apoplast (see WEB TOPIC 26.6).

An extreme example of internal tolerance to toxic ions is the *hyperaccumulation* of certain trace elements that occurs in certain species. Hyperaccumulating plants can tolerate foliar concentrations of various trace elements, including As, Cd, Ni, Zn, and Se, of up to 1% (10,000 µg g$^{-1}$) of their shoot dry weight.

Hyperaccumulation is a genetic adaptation that has been identified in over 400 plant taxa, and it can occur in plants growing on soils with low concentrations of the hyperaccumulated element. The primary function of hyperaccumulation is to protect the plants against pathogens and insect herbivores. Hyperaccumulating plants not only resist the cytotoxic burden of the accumulated trace elements, but also have a powerful scavenging mechanism for the efficient uptake of these potentially toxic elements from the soil. Hyperaccumulation involves heritable genetic changes that drive the elevated expression of the ion transporters involved in uptake and vacuolar compartmentalization of these elements from the soil (see WEB TOPIC 26.5).

## Chelation and active transport contribute to internal tolerance

**Chelation** is the binding of an ion with at least two ligating atoms within a chelating molecule. Chelating molecules can have different atoms available for ligation, such as sulfur (S), nitrogen (N), or oxygen (O), and these different atoms have different affinities for the ions they chelate. By wrapping itself around the ion it binds to form a complex, the chelating molecule renders the ion less chemically active, thereby reducing its potential toxicity. The complex is then usually translocated to other parts of the plant, or stored away from the cytoplasm (typically in the vacuole). Long-distance transport of chelated ions from roots to shoots is also a critical process for hyperaccumulation of metals in shoot tissues. Both the iron chelator nicotianamine and the free amino acid histidine have been implicated in chelation of metals during this transport process (Ingle et al. 2005). In addition, plants also synthesize other ligands for ion chelation, such as phytochelatins.

**Phytochelatins** are low-molecular-weight thiols consisting of the amino acids glutamate, cysteine, and glycine, with the general form of ($\gamma$-Glu-Cys)$_n$Gly. Phytochelatins are synthesized by the enzyme phytochelatin synthase (Corbett and Goldsbrough 2002). The thiol groups act as ligands for ions of trace elements such as Cd and As (FIGURE 26.13). Once formed, the phytochelatin-metal complex is transported into the vacuole for storage. Synthesis of phytochelatins has been shown to be necessary for resistance to Cd and As. In addition to chelation, active transport into the vacuole and out of the cell also contribute to internal metal tolerance.

As discussed in Chapter 6, two H$^+$ pumps in the tonoplast generate the electrochemical gradient for secondary

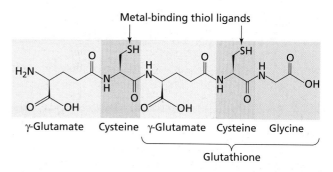

**FIGURE 26.13** Molecular structure of the metal chelate phytochelatin. Phytochelatin utilizes the sulfur in cysteine to bind metals such as cadmium (Cd), zinc (Zn), and arsenic (As).

active transport of ions from the cytosol into the vacuole: a V-type H$^+$-ATPase and a H$^+$-pyrophosphatase (see Chapter 6). The H$^+$–Na$^+$ antiporter AtNHX1 (FIGURE 26.14) is responsible for influx of Na$^+$ into the vacuole. In the plasma membrane, a P-type H$^+$-pumping ATPase provides the driving force (H$^+$ electrochemical potential) for secondary active transport of ions (see Figure 26.14), and is required for cellular extrusion of excess Na$^+$ associated with plant responses to elevated salinity. Na$^+$ efflux across the plasma membrane occurs via the SOS1 Na$^+$–H$^+$ antiporter. The SOS1 Na$^+$–H$^+$ antiporter at the plasma membrane is activated by high intracellular concentrations of NaCl, a response that is mediated by the Ca$^{2+}$-signaling SOS pathway (see WEB TOPIC 26.6). In the presence of elevated NaCl, SOS3 binds Ca$^{2+}$, which allows SOS3 to activate a serine/threonine kinase, SOS2. SOS2 phosphorylates SOS1, which activates the latter's Na$^+$–H$^+$ antiporter function. By this mechanism, plants have the capacity to control net Na$^+$ fluxes across the plasma membrane through the regulation of Na$^+$ efflux. Overexpression of SOS1 or AtNHX1 in transgenic plants enhances salt tolerance (Apse et al. 1999; Shi et al. 2003) (see WEB TOPIC 26.6).

## Many plants have the capacity to acclimate to cold temperatures

The ability to tolerate freezing temperatures under natural conditions varies greatly among tissues. Seeds and other partially dehydrated tissues, as well as fungal spores, can be kept indefinitely at temperatures near absolute zero (0 K, or –273°C), indicating that these very low temperatures are not intrinsically harmful. Hydrated, vegetative cells can also retain viability at freezing temperatures, provided that ice crystal formation can be restricted to the intercellular spaces and cellular dehydration is not too extreme.

Temperate plants have the capacity for **cold acclimation**—a process whereby exposure to low but nonlethal temperatures (typically above freezing) increases the capac-

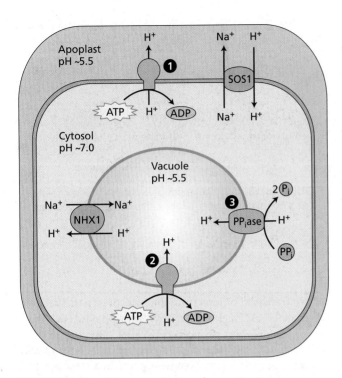

**FIGURE 26.14** Primary and secondary active transport. The plasma membrane–localized $H^+$-pumping ATPase (P-type ATPase) (1), and the tonoplast–localized $H^+$-pumping ATPase (V-type ATPase) (2), and pyrophosphatase ($PP_i$ase) (3) are primary active transport systems that energize the plasma membrane and the tonoplast, respectively. By coupling the energy released by hydrolysis of ATP or pyrophosphate, these $H^+$-pumps are able to transport $H^+$ across the plasma membrane and the tonoplast against an electrochemical gradient. The $H^+$–$Na^+$ antiporters SOS1 and NHX1 are secondary active transport systems that couple the transport of $Na^+$ against its electrochemical gradient with that of $H^+$ down its electrochemical gradient. SOS1 transports $Na^+$ out of the cell, whereas NHX1 transports $Na^+$ into the vacuole.

ity for low temperature survival. Cold acclimation in nature is induced in the early autumn by exposure to short days and nonfreezing, chilling temperatures, which combine to stop growth. A diffusible factor that promotes acclimation, most likely ABA, moves from leaves via the phloem to overwintering stems. ABA accumulates during cold acclimation and is necessary for this process (Survila et al. 2009).

During cold acclimation, temperate woody species withdraw water from the xylem vessels, thereby preventing the stem from splitting in response to the expansion of water during subsequent freezing. Cold acclimation also involves the ability to prevent cellular damage caused by ice crystal formation during freezing. If warmed again, acclimated plants lose their freezing tolerance rapidly, and they can become susceptible to freezing in 24 hours.

How do plants detect the temperature during cold acclimation? A current hypothesis is that low temperatures alter the physical properties of lipids, resulting in increased membrane rigidity and leading to mechanical activation of an as-yet-unknown cold sensor, which initiates signal transduction pathways (Survila et al. 2009). These pathways activate the expression of a suite of genes that encode signaling and cryoprotective proteins and enzymes that mediate lipid and sugar metabolism, compatible solute biosynthesis, ROS scavenging, and metabolic reprogramming. The expression of these genes is regulated by low temperature (Guy 1990; Thomashow 1999; Yamaguchi-Shinozaki and Shinozaki 2006a, 2006b; Chinnusamy et al. 2006; Survila et al. 2009).

Cold acclimation signaling pathways are integrated into a network of interacting signaling pathways that includes those for acclimation to other abiotic conditions. Thus, some of the genes that are induced by low temperatures are also induced by water deficit and salinity (Shinozaki and Yamaguchi-Shinozaki 2006; Survila et al. 2009).

Low temperature signaling has recently been shown to be regulated by calcium ions, abscisic acid (ABA), specific $Ca^{2+}$ signal transduction proteins, and other kinases and phosphatases (Survila et al. 2009). Mutations of *Arabidopsis thaliana* that cause ABA deficiency (e.g., *aba*) or insensitivity (e.g., *abi1*) result in plants that are incapable of cold acclimation. However, ABA-dependent signaling and transcriptional regulation are necessary but not sufficient for full cold acclimation, and not all genes induced by low temperature are ABA-dependent (Yamaguchi-Shinozaki and Shinozaki 2006a, 2006b; Survila et al. 2009) (see **WEB TOPIC 26.7**).

Cold acclimation involves $Ca^{2+}$ influx (mediated by the hypothetical cold sensor described above) to the cytosol from the apoplast, the endoplasmic reticulum, and vacuolar pools, resulting in cytosolic transients necessary for induction of some low temperature–responsive genes required for cold acclimation (Survila et al. 2009). $Ca^{2+}$ signal transduction proteins implicated in cold acclimation include calmodulin and calmodulin-like proteins, $Ca^{2+}$-dependent protein kinases, and calcineurin B-like proteins (see Chapters 6 and 23). (For a discussion of additional signal transduction pathways involved in cold acclimation see **WEB TOPIC 26.7**)

## Plants survive freezing temperatures by limiting ice formation

During rapid freezing, the protoplast, including the vacuole, may **supercool**; that is, the cellular water remains liquid because of its solute content, even at temperatures several degrees below its theoretical freezing point. Supercooling is common to many species of the hardwood forests of southeastern Canada and the eastern United States (see **WEB TOPIC 26.3**). Cells can supercool to only about –40°C, the temperature at which ice forms spontaneously. Spontaneous ice formation sets the *low-temperature limit*

**TABLE 26.3**
Fatty acid composition of mitochondria isolated from chilling-resistant and chilling-sensitive species

| Major fatty acids[a] | Percent weight of total fatty acid content | | | | | |
|---|---|---|---|---|---|---|
| | Chilling-resistant species | | | Chilling-sensitive species | | |
| | Cauliflower bud | Turnip root | Pea shoot | Bean shoot | Sweet potato | Maize shoot |
| Palmitic (16:0) | 21.3 | 19.0 | 17.8 | 24.0 | 24.9 | 28.3 |
| Stearic (18:0) | 1.9 | 1.1 | 2.9 | 2.2 | 2.6 | 1.6 |
| Oleic (18:0) | 7.0 | 12.2 | 3.1 | 3.8 | 0.6 | 4.6 |
| Linoleic (18:2) | 16.1 | 20.6 | 61.9 | 43.6 | 50.8 | 54.6 |
| Linolenic (18:3) | 49.4 | 44.9 | 13.2 | 24.3 | 10.6 | 6.8 |
| Ratio of unsaturated to saturated fatty acids | 3.2 | 3.9 | 3.8 | 2.8 | 1.7 | 2.1 |

*Source*: After Lyons et al. 1964.

[a]Shown in parentheses are the number of carbon atoms in the fatty acid chain and the number of double bonds.

at which many alpine and subarctic species that undergo deep supercooling can survive. It may also explain why the altitude of the timberline in mountain ranges is at or near the –40°C minimum isotherm.

Several specialized plant proteins, termed **antifreeze proteins**, limit the growth of ice crystals through a mechanism independent of lowering of the freezing point of water. Synthesis of these antifreeze proteins is induced by cold temperatures. The proteins bind to the surfaces of ice crystals to prevent or slow further crystal growth. Sugars, polysaccharides, osmoprotectant solutes, and some cold-induced proteins also have cryoprotective effects (Survila et al. 2009).

### The lipid composition of membranes affects their response to temperature

Plant cell membranes are often damaged under extreme environmental conditions. Even slight changes in temperature or other environmental stresses can result in changes in membrane structure and function that alter normal metabolism.

As temperatures drop, membranes may go through a phase transition from a flexible liquid-crystalline structure to a solid gel structure. The phase transition temperature varies with species (tropical species: 10–12°C; apples: 3–10°C) and the actual lipid composition of the membranes. Chilling-resistant plants tend to have membranes with more unsaturated fatty acids (**TABLE 26.3**). Chilling-sensitive plants, on the other hand, have a high percentage of saturated fatty acid chains, and membranes with this composition tend to solidify into a semicrystalline state at a temperature well above 0°C. In general, saturated fatty acids that have no double bonds and lipids containing *trans*-monounsaturated fatty acids solidify at higher tem-

peratures than do lipids that contain polyunsaturated fatty acids, because the latter have kinks in their hydrocarbon chains and do not pack as closely as saturated fatty acids (see Table 26.3 and **WEB TOPIC 26.2**).

Prolonged exposure to extreme temperatures may result in an altered composition of membrane lipids, a form of acclimation. Certain transmembrane enzymes can alter lipid saturation, by introducing one or more double bonds into fatty acids. For example, during acclimation to cold temperatures, the activities of desaturase enzymes increase and the proportion of unsaturated lipids rises (Williams et al. 1988; Palta et al. 1993). This modification lowers the temperature at which the membrane lipids begin a gradual phase change from fluid to semicrystalline form and allows membranes to remain fluid at lower temperatures, thus protecting the plant against damage from chilling. Conversely, a greater degree of saturation of the fatty acids in membrane lipids makes the membranes less fluid (Raison et al. 1982). Certain mutants of *A. thaliana* have reduced activity of omega-3 fatty acid desaturases. These mutants show increased thermotolerance of photosynthesis, presumably because the degree of saturation of chloroplast lipids is increased (Falcone et al. 2004).

The importance of membrane lipids in chilling tolerance has been demonstrated by work with mutant and transgenic plants in which the activity of particular enzymes led to a specific change in membrane lipid composition independent of acclimation to low temperature. For example, *A. thaliana* was transformed with a gene from *Escherichia coli* that raised the proportion of high-melting-point (saturated) membrane lipids. The expression of this gene greatly increased the chilling sensitivity of the transformed plants. Similarly, *fab1* mutants of *A. thaliana* have increased levels of saturated fatty acids, particularly 16:0 fatty acids (see Tables 26.3 and 11.3). When this mutant was grown at chilling temperatures, photosynthesis and growth were gradually inhibited over the course of 3 to 4 weeks, and the chloroplasts were eventually destroyed. At nonchilling temperatures, the mutant grew as well as wild-type controls did (Wu et al. 1997). (For additional experiments with transgenic plants, see WEB TOPIC 26.2.)

### Plant cells have mechanisms that maintain protein structure during temperature stress

Under environmental extremes, protein structure is sensitive to disruption. Plants have several mechanisms to limit or avoid such problems, including osmotic adjustment for maintenance of hydration and chaperone proteins that physically interact with other proteins to facilitate protein folding, reduce misfolding and aggregation, and stabilize protein tertiary structure. In response to sudden 5 to 10°C increases in temperature, plants produce a unique set of chaperone proteins referred to as **heat shock proteins** (**HSPs**). Cells that have been induced to synthesize HSPs show improved thermal tolerance and can tolerate subsequent exposure to temperatures that otherwise would be lethal.

Heat shock proteins are also induced by widely different environmental conditions, including water deficit, ABA treatment, wounding, low temperature, and salinity. Thus, cells that have previously experienced one condition may gain cross-protection against another. Such is the case with tomato fruits, in which heat shock (48 hours at 38°C) has been observed to promote HSP accumulation and to protect cells for 21 days from chilling at 2°C.

Heat shock proteins were discovered in the fruit fly (*Drosophila melanogaster*) and appear to be ubiquitous, having since been identified in other animals, as well as in plants, fungi, and microorganisms. Some HSPs are found in normal, unstressed cells, and some essential cellular proteins are homologous to HSPs but do not increase in response to increased temperatures (Vierling 1991). The heat shock response appears to be mediated by one or more signal transduction pathways, one of which involves a specific transcription factor (heat shock factor, or HSF) that regulates the transcription of HSP mRNAs.

Numerous other proteins have been identified that act in a similar fashion to stabilize proteins during dehydration, temperature extremes, and ion imbalance. These include the protein family RAB/LEA/DHN (RESPONSIVE TO ABA, LATE EMBRYO ABUNDANT, and DEHYDRIN, respectively) (see Chapter 23). Compatible solutes have also been implicated in the stabilization of proteins, membranes, and organelles.

### Scavenging mechanisms detoxify reactive oxygen species

Reactive oxygen species (ROS) are highly reactive forms of oxygen, having at least one unpaired electron, that are produced during the normal operation of photosynthetic and respiratory electron transport. Under normal aerobic conditions, production of ROS is balanced by cellular antioxidative mechanisms that prevent cellular damage. Many types of environmental extremes, particularly exposure to high light intensity under low temperature (conditions that also cause photoinhibition), generate high ROS levels because of an imbalance between the light-driven excitation of the photosynthetic reaction centers and decreased energy dissipation by carbon fixation. High levels of ROS production also occur during exposure to ionizing radiation, UV light, and environmental pollutants. Strong oxidants such as ozone can generate ROS directly.

The most common ROS forms in plant cells are superoxide ($O_2^{\bullet-}$), singlet oxygen ($^1O_2$) hydrogen peroxide ($H_2O_2$), and hydroxyl radicals (OH•) (see Figure 7.34). Certain enzymes or antioxidants can reduce ROS and limit the amount of damage within plant cells. Superoxide dismutase (SOD) catalyzes the dismutation of the superoxide anion: $2\ O_2^{\bullet-} + 2\ H^+ \rightarrow O_2 + H_2O_2$. There are three identified forms of SOD, which differ in their metal cofactors: Cu/Zn-SOD (cytoplasm), Mn-SOD (mitochondria), and Fe-SOD (plastid). In each form, the metal ion within the protein acts to accept electrons from the superoxide radical and then donates them for the production of $O_2$ and $H_2O_2$. Catalase is a heme protein that catalyzes the detoxification of hydrogen peroxide. There are two forms of catalase: the more predominant peroxisome form and a cytoplasmic form. Glutathione peroxidase also catalyzes detoxification of hydrogen peroxide (discussed in greater detail below).

Antioxidants and antioxidant systems also act to reduce ROS. Common antioxidants in plants include the water-soluble ascorbate (vitamin C) and reduced glutathione, and the lipid-soluble α-tocopherol (vitamin E) and β-carotene (vitamin A). Ozone is an environmental gas that acts as an oxidizing agent that can generate ROS in plants. Ascorbate plays an important role in detoxifying ROS generated by ozone exposure.

Many antioxidants act together as systems to detoxify free radicals by eliminating the highly reactive unpaired electron associated with the radical. For example, α-tocopherol works in conjunction with ascorbate and glutathione as a lipid peroxide (LOO) scavenger (in the equations below, the dot represents the unpaired electron):

1. Vitamin E accepts a free radical from ROS (LOO in this case)

$$\text{Vit E} + \text{LOO}^{\bullet} \Rightarrow \text{Vit E}^{\bullet} + \text{LOOH}$$

2. The vitamin E radical is reduced by vitamin C

$$\text{Vit E}^{\bullet} + \text{Vit C} \Rightarrow \text{Vit C}^{\bullet} + \text{Vit E}$$

3. The vitamin C radical is reduced by GSH or vitamin C reductase

$$\text{Vit C}^{\bullet} + \text{GSH} \Rightarrow \text{Vit C} + \text{GSSG}$$

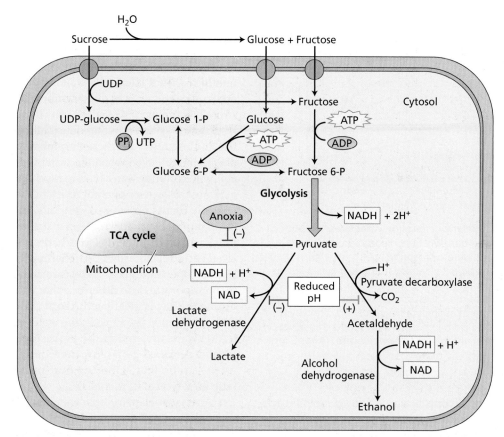

**FIGURE 26.15**  During episodes of anoxia, pyruvate produced by glycolysis is initially fermented to lactate. Proton production by glycolysis and other metabolic pathways, and decreased proton translocation out of the cytosol across the plasma membrane (out of cell) and tonoplast (into vacuole) lead to a lowering of cytosolic pH. At lower pH, lactate dehydrogenase activity is inhibited, and pyru-vate decarboxylase is activated. This leads to an increase in the fermentation to ethanol and a decrease in the fermentation to lactate at lower pH. The pathway of ethanol fermentation consumes more protons than does the pathway of lactate fermentation. This increases the cytosolic pH and enhances the ability of the plant to survive the episode of anoxia.

Glutathione (GSH) is a tripeptide derived from glycine, glutamate, and cysteine, with the cysteine supplying the SH group that neutralizes ROS. Glutathione is highly water-soluble and can be accumulated in large concentrations. The unusual peptide bond of glutathione between the gamma carboxyl group of glutamate and the alpha amino group of cysteine prevents nonspecific destruction by hydrolytic enzymes. Glutathione reductase uses NADPH to maintain a pool of reduced glutathione that can accept an electron from superoxide or $H_2O_2$. In plants, the glutathione antioxidant system actively neutralizes ROS, which may include $H_2O_2$, $O_2^{•-}$, and $H_2O_2^{•-}$:

1. Reduced glutathione can accept $e^-$ from $O_2^{•}$ and $H_2O_2^{•-}$:

$$O_2^{•-} + H^+ + GSH \Rightarrow GS^• + H_2O_2$$

$$H_2O_2^{•-} + 2\,GSH \Rightarrow GSSG + H_2O$$

2. Glutathione peroxidase oxidizes reduced glutathione:

$$H_2O_2 + 2\,GSH \Rightarrow GSSG + 2\,H_2O$$

3. Reduced glutathione is regenerated by glutathione reductase:

$$GSSG + NADPH \Rightarrow 2\,GSH + NADP^+$$

Hyperaccumulation of various toxic trace elements has the potential to cause extensive oxidative damage to plant tissues. To help protect against such oxidative damage, nickel (Ni) hyperaccumulators in the *Thlaspi* genus overaccumulate the antioxidant glutathione (Freeman et al. 2004).

## Metabolic shifts enable plants to cope with a variety of abiotic stresses

Changes in the environment may stimulate shifts in metabolic pathways. When the supply of $O_2$ is insufficient for aerobic respiration, roots first begin to ferment pyruvate to lactate through the action of lactate dehydrogenase; this recycles NADH to NAD$^+$, allowing the maintenance of ATP production through glycolysis (**FIGURE 26.15**) (see Chapter 11). Production of lactate (lactic acid) lowers the

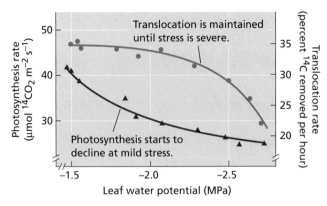

**FIGURE 26.16** Relative effects of water stress on photosynthesis and translocation in sorghum (*Sorghum bicolor*). Plants were exposed to $^{14}CO_2$ for a short time interval. The radioactivity fixed in the leaf was taken as a measure of photosynthesis, and the loss of radioactivity after removal of the $^{14}CO_2$ source was taken as a measure of the rate of assimilate translocation. Photosynthesis was affected by mild stress, whereas translocation was unaffected until stress became severe. (After Sung and Krieg 1979.)

intracellular pH, inhibiting lactate dehydrogenase and activating pyruvate decarboxylase. These changes in enzyme activity quickly lead to a switch from lactate to ethanol production. The net yield of ATP in fermentation is only 2 moles of ATP per mole of hexose sugar catabolized (compared with 36 moles of ATP per mole of hexose respired in aerobic respiration). Thus, injury to root metabolism by $O_2$ deficiency originates in part from a lack of ATP to drive essential metabolic processes such as root absorption of essential nutrients (Drew 1997).

Water shortage decreases both photosynthesis and the consumption of assimilates in the expanding leaves. As a consequence, water shortage indirectly decreases the amount of photosynthate exported from leaves. Because phloem transport depends on pressure gradients (see

Chapter 10), decreased water potential in the phloem during water deficit may inhibit the movement of assimilates. However, experiments have shown that translocation is unaffected until late in the water shortage period, when other processes, such as photosynthesis, have already been strongly inhibited (**FIGURE 26.16**). This relative insensitivity of translocation to water deficit allows plants to mobilize and use reserves where they are needed (e.g., in seed development), even when the water deficit is severe. The ability to continue translocating assimilates is a key factor in almost all aspects of plant resistance to drought.

Some plants possess adaptations, such as the $C_4$ and CAM modes of photosynthesis, that allow them to exploit more arid environments (see Chapters 8 and 9). Crassulacean acid metabolism (CAM) is an adaptation in which stomata open at night and close during the day. The leaf-to-air vapor pressure difference that drives transpiration during the day is greatly reduced at night, when both leaf and air are cool. As a result, the water-use efficiencies of CAM plants are among the highest measured. A CAM plant may gain 1 g of dry matter for only 125 g of water used—a ratio that is three to five times greater than the ratio for a typical $C_3$ plant.

CAM is prevalent in succulent plants such as cacti. Some succulent species display facultative CAM, switching to CAM when subjected to water deficits or saline conditions (see Chapter 8). This switch in metabolism is a remarkable adaptation that allows the plant to acclimate to these conditions, involving the synthesis of the enzymes phosphoenolpyruvate (PEP) carboxylase, pyruvate–orthophosphate dikinase, and NADP–malic enzyme, among others.

As discussed in Chapters 8 and 9, CAM metabolism involves many structural, physiological, and biochemical features, including changes in carboxylation and decarboxylation patterns, the transport of large quantities of malate into and out of the vacuoles, and the reversal of the periodicity of stomatal movements. Thus, the ability to induce CAM is a remarkable response to water deficit.

# SUMMARY

Plants have well-defined responses to life-threatening environmental extremes that help them tolerate and even thrive in a variety of habitats.

### Adaptation and Phenotypic Plasticity

- Adaptations to environmental conditions involve genetic changes.
- Phenotypic plasticity involves changes in physiology or development that are temporary, often reversible, and non-heritable.

### The Abiotic Environment and Its Biological Impact on Plants

- Atmospheric gases, light, temperature, humidity, precipitation, and wind are climatic factors that affect plant homeostasis.
- Edaphic factors—soil composition, hydraulic conductivity, ion exchange capacity, and pH—influence availability of air, water, and mineral nutrients to plant roots.
- Imbalances of abiotic factors have primary and secondary effects on plants (**Table 25.1**).

# SUMMARY continued

## Water Deficit and Flooding

- Water deficit may cause root hair damage, which impairs water uptake, and cavitation, which reduces water flow in vascular tissues.
- Water deficit reduces cell turgor ($\Psi_p$) and cell volume.
- Water deficit decreases wall extensibility ($m$) and increases $Y$ (**Figure 26.1**).
- Flooding and soil compaction reduce $O_2$ availability to root cells.
- $O_2$ concentrations below the critical oxygen pressure (COP) reduce respiration rates and ATP production in roots.

## Imbalances in Soil Minerals

- Salinity stress has two components: nonspecific osmotic stress, which causes water deficits; and specific ion effects due to toxic ions.
- Improper water management practices cause increased soil salinity (**Figure 26.2**).
- High cytosolic $Na^+$ and $Cl^-$ denature proteins and destabilize membranes.

## Temperature Stress

- High leaf temperatures combined with minimal evaporative cooling cause heat stress (**Table 26.2**).
- High temperatures inhibit photosynthesis and respiration, as well as alter membrane fluidity (**Figure 26.3**).
- Above the temperature compensation point, photosynthesis cannot replace the fixed carbon consumed by respiration.
- Chilling reduces membrane fluidity, impairing $H^+$-pumping ATPase activity, solute transport, and general energy transduction pathways.
- Freezing temperatures result in ice crystal formation, which dehydrates cells and disrupts membranes and organelles.

## High Light Stress

- Above the light saturation point for photosynthesis, light causes photoinhibition by disrupting the photosystems.
- Excess electrons produced under high light generate excess ROS, damaging proteins, lipids, RNA, and DNA.
- Excess light, temperature extremes, dehydration, salinity, and ion toxicity all lead to an imbalance between ROS production and scavenging.

## Developmental and Physiological Mechanisms that Protect Plants against Environmental Extremes

- Altered life cycles as well as changes in leaf area, orientation, trichome density, cuticle thickness, and premature senescence are all responses to various environmental fluctuations (**Figures 26.4–24.9**).
- In response to water deficit, the root-to-shoot growth ratio increases.
- To respond to dehydration, plants can regulate their stomatal opening via ABA.
- When soil water potential decreases due to water deficit or salinity, roots can accumulate solutes and continue to absorb water (**Figure 26.10**).
- Root osmotic adjustment is accomplished by taking up soil ions or by transport of ions to the roots from other plants organs.
- Ion accumulation for osmotic adjustment takes place in vacuoles but requires cytosolic accumulation of counter-ions (**Figure 26.11**).
- Aerenchyma tissue can develop in some submerged organs, allowing for continued gas exchange (**Figure 26.12**).
- Phytochelatins chelate certain ions, reducing their reactivity and toxicity (**Figure 26.13**).
- Salt tolerance depends on control of net ion uptake across the plasma membrane, translocation to the shoot and vacuolar compartmentalization.
- A $H^+$–$Na^+$ antiporter is responsible for influx of $Na^+$ into the vacuole (**Figure 26.14**).
- A diffusible factor that promotes cold acclimation moves from leaves to stems via the phloem.
- Low temperatures may alter membrane lipids and thereby activate signal pathways for gene expression (**Table 26.3**).
- $Ca^{2+}$ signal transduction proteins are implicated in cold acclimation.
- Cold temperatures induce the synthesis of antifreeze proteins that limit ice crystal growth.
- Cold-resistant plants tend to have membranes with more unsaturated fatty acids (**Table 26.3**).
- A large variety of heat shock proteins can be induced by different environmental conditions.
- Normal operation of photosynthesis and respiration generates destructive ROS and is balanced by antioxidative mechanisms to prevent cellular damage.
- When $O_2$ is insufficient for aerobic respiration, roots ferment pyruvate to lactate and then to ethanol (**Figure 26.15**).
- During mild or short-term water shortage, photosynthesis is strongly inhibited, but phloem translocation is unaffected until the shortage becomes severe (**Figure 26.16**).

# WEB MATERIAL

## Web Topics

## Web Essays

# CHAPTER REFERENCES

Anami, S., De Block, M., Machuka, J., and Van Lijsebettens, M. (2009) Molecular improvement of tropical corn for drought stress tolerance in Sub-Saharan Africa. *Crit. Rev. Plant Sci.* 28: 16–35.

Apel, K., and Hirt, H. (2004) Reactive oxygen species: Metabolism, oxidative stress, and signal transduction. *Annu. Rev. Plant. Biol.* 55: 373–399.

Apse, M. P., Aharon, G. S., Snedden, W. A., and Blumwald, E. (1999) Salt tolerance conferred by overexpression of a vacuolar Na$^+$/H$^+$ antiport in Arabidopsis. *Science* 285: 1256–1258.

Apse, M. P., and Blumwald, E. (2007) Na$^+$ transport in plants. *FEBS Lett.* 581: 2247–2254.

Bailey-Serres, J., and Voesenek, L. A. (2008) Flooding stress: Acclimations and genetic diversity. *Annu. Rev. Plant. Biol.* 59: 313–339.

Björkman, O., Badger, M. R., and Armond, P. A. (1980) Response and adaptation of photosynthesis to high temperatures. In *Adaptation of Plants to Water and High Temperatures Stress*, N. C. Turner and P. J. Kramer, eds., Wiley, New York, pp. 233–249.

Blizzard, W. E., and Boyer, J. S. (1980) Comparative resistance of the soil and the plant to water transport. *Plant Physiol.* 66: 809–814.

Borrell, A. K., Hammer, G. L., and Henzell, R. O. (2000) Does maintaining green leaf area in sorghum improve yield under drought? II. Dry matter production and yield. *Crop Sci.* 40: 1037–1048.

Boyer, J. S. (1970) Leaf enlargement and metabolic rates in corn, soybean, and sunflower at various leaf water potentials. *Plant Physiol.* 46: 233–235.

Boyer, J. S. (1982) Plant productivity and environment. *Science* 218: 443–448.

Brady, K. U., Kruckeberg, A. R., and Bradshaw, H. D. (2005) Evolutionary ecology of plant adaptation to serpentine soils. *Annu. Rev. Ecol. Evol. Syst.* 36: 243–266.

Buckley, T. N. (2005) The control of stomata by water balance. *New Phytol.* 168: 275–292.

Burssens, S., Himanen, K., van de Cotte, B., Beeckman, T., Van Montagu, M., Inze, D., and Verbruggen, N. (2000) Expression of cell cycle regulatory genes and morphological alterations in response to salt stress in *Arabidopsis thaliana*. *Planta* 211: 632–640.

## SUMMARY continued

### Water Deficit and Flooding

- Water deficit may cause root hair damage, which impairs water uptake, and cavitation, which reduces water flow in vascular tissues.

- Water deficit reduces cell turgor ($\Psi_p$) and cell volume.

- Water deficit decreases wall extensibility ($m$) and increases Y (**Figure 26.1**).

- Flooding and soil compaction reduce $O_2$ availability to root cells.

- $O_2$ concentrations below the critical oxygen pressure (COP) reduce respiration rates and ATP production in roots.

### Imbalances in Soil Minerals

- Salinity stress has two components: nonspecific osmotic stress, which causes water deficits; and specific ion effects due to toxic ions.

- Improper water management practices cause increased soil salinity (**Figure 26.2**).

- High cytosolic $Na^+$ and $Cl^-$ denature proteins and destabilize membranes.

### Temperature Stress

- High leaf temperatures combined with minimal evaporative cooling cause heat stress (**Table 26.2**).

- High temperatures inhibit photosynthesis and respiration, as well as alter membrane fluidity (**Figure 26.3**).

- Above the temperature compensation point, photosynthesis cannot replace the fixed carbon consumed by respiration.

- Chilling reduces membrane fluidity, impairing $H^+$-pumping ATPase activity, solute transport, and general energy transduction pathways.

- Freezing temperatures result in ice crystal formation, which dehydrates cells and disrupts membranes and organelles.

### High Light Stress

- Above the light saturation point for photosynthesis, light causes photoinhibition by disrupting the photosystems.

- Excess electrons produced under high light generate excess ROS, damaging proteins, lipids, RNA, and DNA.

- Excess light, temperature extremes, dehydration, salinity, and ion toxicity all lead to an imbalance between ROS production and scavenging.

### Developmental and Physiological Mechanisms that Protect Plants against Environmental Extremes

- Altered life cycles as well as changes in leaf area, orientation, trichome density, cuticle thickness, and premature senescence are all responses to various environmental fluctuations (**Figures 26.4–24.9**).

- In response to water deficit, the root-to-shoot growth ratio increases.

- To respond to dehydration, plants can regulate their stomatal opening via ABA.

- When soil water potential decreases due to water deficit or salinity, roots can accumulate solutes and continue to absorb water (**Figure 26.10**).

- Root osmotic adjustment is accomplished by taking up soil ions or by transport of ions to the roots from other plants organs.

- Ion accumulation for osmotic adjustment takes place in vacuoles but requires cytosolic accumulation of counter-ions (**Figure 26.11**).

- Aerenchyma tissue can develop in some submerged organs, allowing for continued gas exchange (**Figure 26.12**).

- Phytochelatins chelate certain ions, reducing their reactivity and toxicity (**Figure 26.13**).

- Salt tolerance depends on control of net ion uptake across the plasma membrane, translocation to the shoot and vacuolar compartmentalization.

- A $H^+$–$Na^+$ antiporter is responsible for influx of $Na^+$ into the vacuole (**Figure 26.14**).

- A diffusible factor that promotes cold acclimation moves from leaves to stems via the phloem.

- Low temperatures may alter membrane lipids and thereby activate signal pathways for gene expression (**Table 26.3**).

- $Ca^{2+}$ signal transduction proteins are implicated in cold acclimation.

- Cold temperatures induce the synthesis of antifreeze proteins that limit ice crystal growth.

- Cold-resistant plants tend to have membranes with more unsaturated fatty acids (**Table 26.3**).

- A large variety of heat shock proteins can be induced by different environmental conditions.

- Normal operation of photosynthesis and respiration generates destructive ROS and is balanced by antioxidative mechanisms to prevent cellular damage.

- When $O_2$ is insufficient for aerobic respiration, roots ferment pyruvate to lactate and then to ethanol (**Figure 26.15**).

- During mild or short-term water shortage, photosynthesis is strongly inhibited, but phloem translocation is unaffected until the shortage becomes severe (**Figure 26.16**).

## WEB MATERIAL

### Web Topics

**26.1 Stomatal Conductance and Yields of Irrigated Crops**

Stomatal conductance predicts yields of irrigated crops grown in hot environments.

**26.2 Membrane Lipids and Low Temperatures**

Lipid enzymes from mutant and transgenic plants mimic the effects of low-temperature acclimation.

**26.3 Ice Formation in Higher-Plant Cells**

Heat is released when ice forms in intercellular spaces.

**26.4 Water-Deficit-Regulated ABA Signaling and Stomatal Closure**

Plant response to drought includes the accumulation of ABA in leaves, which reduces transpiration by inducing stomatal closure.

**26.5 Genetic and Physiological Adaptations Required for Zinc Hyperaccumulation**

Zinc hyperaccumulation is driven by enhancements in zinc uptake, transport to the shoot, and storage in leaf vacuoles.

**26.6 Cellular and Whole Plant Responses to Salinity Stress**

Control of sodium efflux from cells, translocation to the shoot, and vacuolar compartmentalization allow plants to regulate whole plant sodium level.

**26.7 Signaling during Cold Acclimation Regulates Genes That Are Expressed in Response to Low Temperature and Enhances Freezing Tolerance**

Cold acclimation is the process by which temperate plants are able to survive over winter at freezing temperatures.

### Web Essays

**26.1 The Effect of Air Pollution on Plants**

Polluting gases inhibit stomatal conductance, photosynthesis, and growth.

**26.2 An Extreme Plant Lifestyle: Metal Hyperaccumulation**

Plants can overaccumulate highly toxic metals.

## CHAPTER REFERENCES

Anami, S., De Block, M., Machuka, J., and Van Lijsebettens, M. (2009) Molecular improvement of tropical corn for drought stress tolerance in Sub-Saharan Africa. *Crit. Rev. Plant Sci.* 28: 16–35.

Apel, K., and Hirt, H. (2004) Reactive oxygen species: Metabolism, oxidative stress, and signal transduction. *Annu. Rev. Plant. Biol.* 55: 373–399.

Apse, M. P., Aharon, G. S., Snedden, W. A., and Blumwald, E. (1999) Salt tolerance conferred by overexpression of a vacuolar Na$^+$/H$^+$ antiport in Arabidopsis. *Science* 285: 1256–1258.

Apse, M. P., and Blumwald, E. (2007) Na$^+$ transport in plants. *FEBS Lett.* 581: 2247–2254.

Bailey-Serres, J., and Voesenek, L. A. (2008) Flooding stress: Acclimations and genetic diversity. *Annu. Rev. Plant. Biol.* 59: 313–339.

Björkman, O., Badger, M. R., and Armond, P. A. (1980) Response and adaptation of photosynthesis to high temperatures. In *Adaptation of Plants to Water and High Temperatures Stress*, N. C. Turner and P. J. Kramer, eds., Wiley, New York, pp. 233–249.

Blizzard, W. E., and Boyer, J. S. (1980) Comparative resistance of the soil and the plant to water transport. *Plant Physiol.* 66: 809–814.

Borrell, A. K., Hammer, G. L., and Henzell, R. O. (2000) Does maintaining green leaf area in sorghum improve yield under drought? II. Dry matter production and yield. *Crop Sci.* 40: 1037–1048.

Boyer, J. S. (1970) Leaf enlargement and metabolic rates in corn, soybean, and sunflower at various leaf water potentials. *Plant Physiol.* 46: 233–235.

Boyer, J. S. (1982) Plant productivity and environment. *Science* 218: 443–448.

Brady, K. U., Kruckeberg, A. R., and Bradshaw, H. D. (2005) Evolutionary ecology of plant adaptation to serpentine soils. *Annu. Rev. Ecol. Evol. Syst.* 36: 243–266.

Buckley, T. N. (2005) The control of stomata by water balance. *New Phytol.* 168: 275–292.

Burssens, S., Himanen, K., van de Cotte, B., Beeckman, T., Van Montagu, M., Inze, D., and Verbruggen, N. (2000) Expression of cell cycle regulatory genes and morphological alterations in response to salt stress in *Arabidopsis thaliana*. *Planta* 211: 632–640.

Chinnusamy, V., Zhu, J., and Zhu, J.-K. (2007) Cold stress regulation of gene expression in plants. *Trends Plant Sci.* 12: 444–451.

Cobbett, C., and Goldsbrough, P. (2002) Phytochelatins and metallothioneins: Roles in heavy metal detoxification and homeostasis. *Annu. Rev. Plant Biol.* 53: 159–182.

Debat, V., and David, P. (2001) Mapping phenotypes: canalization, plasticity and developmental stability. *Trends Ecol. Evol.* 16: 555–561.

Drew, M. C. (1997) Oxygen deficiency and root metabolism: Injury and acclimation under hypoxia and anoxia. *Annu. Rev. Plant Physiol. Plant Mol. Biol.* 48: 223–250.

Drew, M. C., He, C. J., and Morgan, P. W. (2000) Programmed cell death and aerenchyma formation in roots. *Trends Plant Sci.* 5: 123–127.

Epstein, E., and Bloom, A. J. (2005) *Mineral Nutrition of Plants. Principles and Perspectives,* 2nd ed. Sinauer Associates, Inc., Sunderland, MA.

Falcone, D. L., Ogas, J. P., and Somerville, C. R. (2004) Regulation of membrane fatty acid composition by temperature in mutants of Arabidopsis with alterations in membrane lipid composition. *BMC Plant Biol.* 4:17.

Freeman, J. L., Persans, M. W., Nieman, K., Albrecht, C., Peer, W., Pickering, I. J., and Salt, D. E. (2004) Increased glutathione biosynthesis plays a role in nickel tolerance in Thlaspi nickel hyperaccumulators. *Plant Cell* 16: 2176–2191.

Greenway, H., and Munns, R. (1980) Mechanisms of salt tolerance in nonhalophytes. *Annu. Rev. Plant Physiol. Plant Mol. Biol.* 31: 149–190.

Guy, C. L. (1990) Cold acclimation and freezing stress tolerance: Role of protein metabolism. *Annu. Rev. Plant Physiol. Plant Mol. Biol.* 41: 187–223.

Hartley-Whitaker, J., Ainsworth, G., Vooijs, R., Bookum, W. T., Schat, H., and Meharg, A. A. (2001) Phytochelatins are involved in differential arsenate tolerance in *Holcus lanatus*. *Plant Physiol.* 126: 299–306.

Hasegawa, P. M., Bressan, R. A., Zhu, J. K., and Bohnert, H. J. (2000) Plant cellular and molecular responses to high salinity. *Annu. Rev. Plant Physiol. Plant Mol. Biol.* 51: 463–499.

Ingle, R. A., Mugford, S. T., Rees, J. D, Campbell, M. M., and Smith, J. A. (2005) Constitutively high expression of the histidine biosynthetic pathway contributes to nickel tolerance in hyperaccumulator plants. *Plant Cell* 17: 2089–2106.

Koike, T., Kitao, M., Quoreshi, A. M., and Matsuura, Y. (2003) Growth characteristics of root-shoot relations of three birch seedlings raised under different water regimes. *Plant Soil* 255: 303–310.

Levitt, J. (1980) *Responses of Plants to Environmental Stresses,* Vol. 1, 2nd ed. Academic Press, New York.

Lyons, J. M., Wheaton, T. A., and Pratt, H. K. (1964) Relationship between the physical nature of mitochondrial membranes and chilling sensitivity in plants. *Plant Physiol.* 39: 262–268.

MacNair, M. R., Cumbes, Q. J., and Meharg, A. A. (1992) The genetics of arsenate tolerance in Yorkshire fog, *Holcus lanatus* L. *Heredity* 69: 325–335.

Matthews, M. A., Van Volkenburgh, E., and Boyer, J. S. (1984) Acclimation of leaf growth to low water potentials in sunflower. *Plant Cell Environ.* 7: 199–206.

Meharg, A. A., and MacNair, M. R. (1992) Genetic correlation between arsenate tolerance and the rate of influx of arsenate and phosphate in *Holcus lanatus* L. *Heredity* 69: 336–341.

Meharg, A. A., Cumbes, Q. J., and MacNair, M. R. (1993) Pre-adaptation of Yorkshire fog, *Holcus lanatus* L. (Poaceae) to arsenate tolerance. *Evolution* 47: 313–316.

Munns, R., and Tester, M. (2008) Mechanisms of salinity tolerance. *Annu. Rev. Plant. Biol.* 59: 651–681.

Palta, J. P., Whitaker, B. D., and Weiss, L. S. (1993) Plasma membrane lipids associated with genetic variability in freezing tolerance and cold acclimation of Solanum species. *Plant Physiol.* 103: 793–803.

Raison, J. K., Pike, C. S., and Berry, J. A. (1982) Growth temperature-induced alterations in the thermotropic properties of *Nerium oleander* membrane lipids. *Plant Physiol.* 70: 215–218.

Rivero, R. M., Kojima, M., Gepstein, A., Sakakibara, H., Mittler, R., Gepstein, S., and Blumwald, E. (2007) Delayed leaf senescence induces extreme drought tolerance in a flowering plant. *Proc. Natl. Acad. Sci. USA* 104: 19631–19636.

Shi, H., Lee, B. H., Wu, S. J., and Zhu, J. K. (2003) Overexpression of a plasma membrane Na+/H+ antiporter gene improves salt tolerance in *Arabidopsis thaliana. Nat Biotechnol.* 21: 81–85.

Shinozaki, K., and Yamaguchi-Shinozaki, K. (2006) Transcriptional regulatory networks in cellular responses and tolerance to dehydration and cold stresses. *Annu. Rev. Plant Biol.* 57: 781–803.

Sung, F.J.M., and Krieg, D. R. (1979) Relative sensitivity of photosynthetic assimilation and translocation of [14]carbon to water stress. *Plant Physiol.* 64: 852–856.

Survila, M., Heino, P., and Palva, E. T. (2009) Genes and gene regulation for low temperature tolerance. In *Genes for Plant Abiotic Stress,* M. A. Jenks and A. J. Wood, eds., Blackwell Publishing, pp. 185–219.

Takahashi, S., and Murata, N. (2008) How do environmental stresses accelerate photoinhibition. *Trends Plant Sci.* 13: 178–182.

Thomashow, M. F. (1999) Plant cold acclimation: Freezing tolerance genes and regulatory mechanisms. *Annu. Rev. Plant Physiol. Plant Mol. Biol.* 50: 571–599.

Tracy, F. E., Gilliham, W., Dodd, A. N., Webb, A.A.R., and Tester, M. (2008) NaCl-induced changes in cytosolic free calcium in Arabidopsis thaliana are heterogenous and modified by external ionic composition. *Plant Cell Environ.* 31: 1063–1073.

Uemura, M., Tominaga, Y., Nakagawara, C., Shigematsu, S., and Minami, A. (2006) Responses of the plasma membrane to low temperatures. *Physiol. Plant.* 126: 81–89.

Valenzuela-Estrada, L. R., Richards, J. H., Diaz, A., and Eissensat, D. M. (2009) Patterns of nocturnal rehydration in root tissues of Vaccinium corymbosum L. under severe drought conditions. *J. Exp. Bot.* 60: 1241–1247.

Vierling, E. (1991) The roles of heat shock proteins in plants. *Annu. Rev. Plant Physiol. Plant Mol. Biol.* 42: 579–620.

Vijayan, P., and Browse, J. (2002) Photoinhibition in mutants of Arabidopsis deficient in thylakoid unsaturation. *Plant Physiol.* 129: 867–885.

Williams, J. P., Khan, M. U., Mitchell, K., and Johnson, G. (1988) The effect of temperature on the level and biosynthesis of unsaturated fatty acids in diacylglycerols of Brassica napus leaves. *Plant Physiol.* 87: 904–910.

Wu, J., Lightner, J., Warwick, N., and Browse, J. (1997) Low-temperature damage and subsequent recovery of *fab1* mutant *Arabidopsis* exposed to 2°C. *Plant Physiol.* 113: 347–356.

Yamaguchi-Shinozaki, K., and Shinozaki, K. (2006a) Transcriptional regulatory networks in cellular responses and tolerance to dehydration and cold stress. *Annu. Rev. Plant Biol.* 57: 781–803.

Yamaguchi-Shinozaki, K., and Shinozaki, K. (2006b) Transcriptional regulatory networks in an Arabidopsis gene is involved in responsiveness to drought, low-temperature, or high-salt stress. *Plant Cell* 6: 251–264.

Zhang, X., Fowler, S. G., Cheng, H., Lou, Y., Rhee, S. Y., Stockinger, E. J., and Thomashow, M. F. (2004) Freezing-sensitive tomato has a functional CBF cold response pathway, but a CBF regulon that differs from that of freezing-tolerant Arabidopsis. *Plant J.* 29: 905–919.

Zhang, J., and Zhang, X. (1994) Can early wilting of old leaves account for much of the ABA accumulation in flooded pea plants? *J. Exp. Bot.* 45: 1335–1342.

Zhu, J., Dong, C. H., and Zhu, J.-K. (2007) Interplay between cold-responsive gene regulation, metabolism and RNA processing during plant cold acclimation. *Curr. Opin. Plant Biol.* 10: 290–295.

Zhu, J. K. (2002) Salt and drought stress signal transduction in plants. *Annu. Rev. Plant Biol.* 53: 247–273.

Zhu, J. K., Liu, J., and Xiong, L. (1998) Genetic analysis of salt tolerance in Arabidopsis. Evidence for a critical role of potassium nutrition. *Plant Cell* 10: 1181–1191.

# APPENDIX 1

# Energy and Enzymes

*The force that through the green fuse drives the flower*
*Drives my green age; that blasts the roots of trees*
*Is my destroyer.*
*And I am dumb to tell the crooked rose*
*My youth is bent by the same wintry fever.*

*The force that drives the water through the rocks*
*Drives my red blood; that dries the mouthing streams*
*Turns mine to wax.*
*And I am dumb to mouth unto my veins*
*How at the mountain spring the same mouth sucks.*

Dylan Thomas, Collected Poems (1952)

In these opening stanzas from Dylan Thomas's famous poem, the poet proclaims the essential unity of the forces that propel animate and inanimate objects alike, from their beginnings to their ultimate decay. Scientists call this force energy. Energy transformations play a key role in all the physical and chemical processes that occur in living systems. But energy alone is insufficient to drive the growth and development of organisms. Protein catalysts called enzymes are required to ensure that the rates of biochemical reactions are rapid enough to support life. In this chapter we will examine basic concepts about energy, the way in which cells transform energy to perform useful work (bioenergetics), and the structure and function of enzymes.

## Energy Flow through Living Systems

The flow of matter through individual organisms and biological communities is part of everyday experience; the flow of energy is not, even though it is central to the very existence of living things. What makes concepts such as energy, work, and order so elusive is their insubstantial nature: We find it far easier to visualize the dance of atoms and molecules than the forces and fluxes that determine the direction and extent of natural processes. The branch of physical science that deals with such matters is thermodynamics, an abstract and demanding discipline that most biologists are content to skim over lightly. Yet bioenergetics is so shot through with concepts and quantitative relationships derived from thermodynamics that it is scarcely possible to discuss the subject without frequent reference to free energy, potential, entropy, and the second law.

The purpose of this chapter is to collect and explain, as simply as possible, the fundamental thermodynamic concepts and relationships that recur throughout this book. Readers who prefer a more extensive treatment of the subject should consult either the introductory texts by Klotz (1967) and by Nicholls and Ferguson (1992) or the advanced texts by Morowitz (1978) and by Edsall and Gutfreund (1983).

Thermodynamics evolved during the nineteenth century out of efforts to understand how a steam engine works and why heat is produced when one bores a cannon. The very name "thermodynamics," and much of the language of this science, recall these historical roots, but it would be more appropriate to speak of energetics, for the principles involved are universal. Living plants, like all other natural phenomena, are constrained by the laws of thermodynamics. By the same token, thermodynamics supplies an indispensable framework for the quantitative description of biological vitality.

## Energy and Work

Let us begin with the meanings of "energy" and "work." **Energy** is defined in elementary physics, as in daily life, as the capacity to do work. The meaning of work is harder to come by and more narrow. **Work**, in the mechanical sense, is the displacement of any body against an opposing force. The work done is the product of the force and the distance displaced, as expressed in the following equation:*

$$W = f \Delta \tag{A1.1}$$

Mechanical work appears in chemistry because whenever the final volume of a reaction mixture exceeds the initial volume, work must be done against the pressure of the atmosphere; conversely, the atmosphere performs work when a system contracts. This work is calculated by the expression $P\Delta V$ (where $P$ stands for pressure and $V$ for volume), a term that appears frequently in thermodynamic formulas. *In biology, work is employed in a broader sense to describe displacement against any of the forces that living things encounter or generate: mechanical, electric, osmotic, or even chemical potential.*

A familiar mechanical illustration may help clarify the relationship of energy to work. The spring in FIGURE A1.1 can be extended if force is applied to it over a particular distance—that is, if work is done on the spring. This work can be recovered by an appropriate arrangement of pulleys and used to lift a weight onto the table. The extended spring can thus be said to possess energy that is numerically equal to the work it can do on the weight (neglecting friction). The weight on the table, in turn, can be said to possess energy by virtue of its position in Earth's gravitational field, which can be utilized to do other work, such as turning a crank. The weight thus illustrates the concept of **potential energy**, a capacity to do work that arises from the position of an object in a field of force, and the sequence as a whole illustrates the conversion of one kind of energy into another, or **energy transduction**.

### The First Law: The total energy is always conserved

It is common experience that mechanical devices involve both the performance of work and the production or absorption of heat. We are at liberty to vary the amount of

---

* We may note in passing that the dimensions of work are complex—$ml^2t^{-2}$—where $m$ denotes mass, $l$ distance, and $t$ time, and that work is a scalar quantity, that is, the product of two vectorial terms.

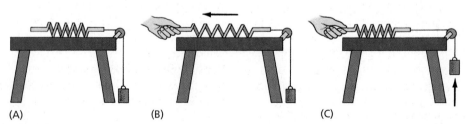

FIGURE A1.1 Energy and work in a mechanical system. (A) A weight resting on the floor is attached to a spring via a string. (B) Pulling on the spring places the spring under tension. (C) The potential energy stored in the extended spring performs the work of raising the weight when the spring contracts.

work done by the spring, up to a particular maximum, by using different weights, and the amount of heat produced will also vary. But much experimental work has shown that, under ideal circumstances, the sum of the work done and of the heat evolved is constant and depends only on the initial and final extensions of the spring. We can thus envisage a property, the internal energy of the spring, with the characteristic described by the following equation:

$$\Delta U = \Delta Q + \Delta W \qquad (A1.2)$$

Here $Q$ is the amount of heat absorbed by the system, and $W$ is the amount of work done on the system.* In Figure A1.1 the work is mechanical, but it could just as well be electrical, chemical, or any other kind of work. Thus $\Delta U$ is the net amount of energy put into the system, either as heat or as work; conversely, both the performance of work and the evolution of heat entail a decrease in the internal energy. We cannot specify an absolute value for the energy content; only changes in internal energy can be measured. Note that Equation A1.2 assumes that heat and work are equivalent; its purpose is to stress that, under ideal circumstances, $\Delta U$ depends only on the initial and final states of the system, not on how heat and work are partitioned.

Equation A1.2 is a statement of the first law of thermodynamics, which is the principle of energy conservation. If a particular system exchanges no energy with its surroundings, its energy content remains constant; if energy is exchanged, the change in internal energy will be given by the difference between the energy gained from the surroundings and that lost to the surroundings. The change in internal energy depends only on the initial and final states of the system, not on the pathway or mechanism of energy exchange. Energy and work are interconvertible; even heat is a measure of the kinetic energy of the molecular constituents of the system. To put it as simply as possible, Equation A1.2 states that no machine, including the chemical machines that we recognize as living, can do work without an energy source.

An example of the application of the first law to a biological phenomenon is the energy budget of a leaf. Leaves absorb energy from their surroundings in two ways: as direct incident irradiation from the sun and as infrared irradiation from the surroundings. Some of the energy absorbed by the leaf is radiated back to the surroundings as infrared irradiation and heat, while a fraction of the absorbed energy is stored, as either photosynthetic products or leaf temperature changes. Thus we can write the following equation:

Total energy absorbed by leaf =
energy emitted from leaf + energy stored by leaf

Note that although the energy absorbed by the leaf has been transformed, the total energy remains the same, in accordance with the first law.

### The change in the internal energy of a system represents the maximum work it can do

We must qualify the equivalence of energy and work by invoking "ideal conditions"—that is, by requiring that the process be carried out reversibly. The meaning of "reversible" in thermodynamics is a special one: The term describes conditions under which the opposing forces are so nearly balanced that an infinitesimal change in one or the other would reverse the direction of the process.* Under these circumstances the process yields the maximum possible amount of work. Reversibility in this sense does not often hold in nature, as in the example of the leaf. Ideal conditions differ so little from a state of equilibrium that any process or reaction would require infinite time and would therefore not take place at all. Nonetheless, the concept of thermodynamic reversibility is useful: If we measure the change in internal energy that a process entails, we have an upper limit to the work that it can do; for any real process the maximum work will be less.

In the study of plant biology we encounter several sources of energy—notably light and chemical transformations—as well as a variety of work functions, including mechanical, osmotic, electrical, and chemical work. The meaning of the first law in biology stems from the certainty, painstakingly achieved by nineteenth-century physicists, that the various kinds of energy and work are measurable, equivalent, and, within limits, interconvertible. Energy is to biology what money is to economics: the means by which living things purchase useful goods and services.

### Each type of energy is characterized by a capacity factor and a potential factor

The amount of work that can be done by a system, whether mechanical or chemical, is a function of the size of the system. Work can always be defined as the product of two factors—force and distance, for example. One is a potential or intensity factor, which is independent of the size of the system; the other is a capacity factor and is directly proportional to the size (Table A1.1).

In biochemistry, energy and work have traditionally been expressed in calories; 1 calorie is the amount of heat required to raise the temperature of 1 g of water by 1°C,

---

\* Equation A1.2 is more commonly encountered in the form $\Delta U = \Delta Q - \Delta W$, which results from the convention that $Q$ is the amount of heat absorbed by the system from the surroundings and $W$ is the amount of work done by the system on the surroundings. This convention affects the sign of $W$ but does not alter the meaning of the equation.

---

*In biochemistry, reversibility has a different meaning: Usually the term refers to a reaction whose pathway can be reversed, often with an input of energy.

specifically, from 15.0 to 16.0°C . In principle, one can carry out the same process by doing the work mechanically with a paddle; such experiments led to the establishment of the mechanical equivalent of heat as 4.186 joules per calorie (J cal$^{-1}$).* We will also have occasion to use the equivalent electrical units, based on the volt: A volt is the potential difference between two points when 1 J of work is involved in the transfer of a coulomb of charge from one point to another. (A coulomb is the amount of charge carried by a current of 1 ampere [A] flowing for 1 s. Transfer of 1 mole [mol] of charge across a potential of 1 volt [V] involves 96,500 J of energy or work.) The difference between energy and work is often a matter of the sign. Work must be done to bring a positive charge closer to another positive charge, but the charges thereby acquire potential energy, which in turn can do work.

## The Direction of Spontaneous Processes

Left to themselves, events in the real world take a predictable course. The apple falls from the branch. A mixture of hydrogen and oxygen gases is converted into water. The fly trapped in a bottle is doomed to perish, the pyramids to crumble into sand; things fall apart. But there is nothing in the principle of energy conservation that forbids the apple to return to its branch with absorption of heat from the surroundings or that prevents water from dissociating into its constituent elements in a like manner. The search for the reason that neither of these things ever happens led to profound philosophical insights and generated useful quantitative statements about the energetics of chemical reactions and the amount of work that can be done by them. Since living things are in many respects chemical machines, we must examine these matters in some detail.

### The Second Law: The total entropy always increases

From daily experience with weights falling and warm bodies growing cold, one might expect spontaneous processes to proceed in the direction that lowers the internal energy—that is, the direction in which $\Delta U$ is negative. But there are too many exceptions for this to be a general rule. The melting of ice is one exception: An ice cube placed in water at 1°C will melt, yet measurements show that liquid water (at any temperature above 0°C) is in a state of higher energy than ice; evidently, some spontaneous processes are accompanied by an increase in internal energy. Our melting ice cube does not violate the first law, for heat is

---

* In current standard usage based on the meter, kilogram, and second, the fundamental unit of energy is the joule (1 J = 0.24 cal) or the kilojoule (1 kJ = 1000 J).

### TABLE A1.1
### Potential and capacity factors in energetics

| Type of energy | Potential factor | Capacity factor |
| --- | --- | --- |
| Mechanical | Pressure | Volume |
| Electrical | Electric potential | Charge |
| Chemical | Chemical potential | Mass |
| Osmotic | Concentration | Mass |
| Thermal | Temperature | Entrop |

absorbed as it melts. This suggests that there is a relationship between the capacity for spontaneous heat absorption and the criterion determining the direction of spontaneous processes, and that is the case. The thermodynamic function we seek is called **entropy**, the amount of energy in a system not available for doing work, corresponding to the degree of randomness of a system. Mathematically, entropy is the capacity factor corresponding to temperature, $Q/T$. We may state the answer to our question, as well as the second law of thermodynamics, thus: The direction of all spontaneous processes is to increase the entropy of a system plus its surroundings.

Few concepts are so basic to a comprehension of the world we live in, yet so opaque, as entropy—presumably because entropy is not intuitively related to our sense perceptions, as mass and temperature are. The explanation given here follows the particularly lucid exposition by Atkinson (1977), who states the second law in a form bearing, at first sight, little resemblance to that given above:

> We shall take [the second law] as the concept that any system not at absolute zero has an irreducible minimum amount of energy that is an inevitable property of that system at that temperature. That is, a system requires a certain amount of energy just to be at any specified temperature.

The molecular constitution of matter supplies a ready explanation: Some energy is stored in the thermal motions of the molecules and in the vibrations and oscillations of their constituent atoms. We can speak of it as isothermally unavailable energy, since the system cannot give up any of it without a drop in temperature (assuming that there is no physical or chemical change). The isothermally unavailable energy of any system increases with temperature, since the energy of molecular and atomic motions increases with temperature. Quantitatively, the isothermally unavailable energy for a particular system is given by $ST$, where $T$ is the absolute temperature and $S$ is the entropy.

But what is this thing, entropy? Reflection on the nature of the isothermally unavailable energy suggests that, for any particular temperature, the amount of such energy will be greater the more atoms and molecules are free to move and to vibrate—that is, the more chaotic is the system. By contrast, the orderly array of atoms in a crystal, with a place for each and each in its place, corresponds to a state of low entropy. At absolute zero, when all motion ceases, the entropy of a pure substance is likewise zero; this statement is sometimes called the third law of thermodynamics.

A large molecule, a protein for example, within which many kinds of motion can take place, will have considerable amounts of energy stored in this fashion—more than would, say, an amino acid molecule. But the entropy of the protein molecule will be less than that of the constituent amino acids into which it can dissociate, because of the constraints placed on the motions of those amino acids as long as they are part of the larger structure. Any process leading to the release of these constraints increases freedom of movement, and hence entropy.

This is the universal tendency of spontaneous processes as expressed in the second law; it is why the costly enzymes stored in the refrigerator tend to decay and why ice melts into water. The increase in entropy as ice melts into water is "paid for" by the absorption of heat from the surroundings. As long as the net change in entropy of the system plus its surroundings is positive, the process can take place spontaneously. That does not necessarily mean that the process will take place: The rate is usually determined by kinetic factors separate from the entropy change. All the second law mandates is that the fate of the pyramids is to crumble into sand, while the sand will never reassemble itself into a pyramid; the law does not tell how quickly this must come about.

### A process is spontaneous if ΔS for the system and its surroundings is positive

There is nothing mystical about entropy; it is a thermodynamic quantity like any other, measurable by experiment and expressed in entropy units. One method of quantifying it is through the heat capacity of a system, the amount of energy required to raise the temperature by 1°C. In some cases the entropy can even be calculated from theoretical principles, though only for simple molecules. For our purposes, what matters is the sign of the entropy change, $\Delta S$: A process can take place spontaneously when $\Delta S$ for the system and its surroundings is positive; a process for which $\Delta S$ is negative cannot take place spontaneously, but the opposite process can; and for a system at equilibrium, the entropy of the system plus its surroundings is maximal and $\Delta S$ is zero.

"Equilibrium" is another of those familiar words that is easier to use than to define. Its everyday meaning implies that the forces acting on a system are equally balanced, such that there is no net tendency to change; this is the sense in which the term "equilibrium" will be used here. A mixture of chemicals may be in the midst of rapid interconversion, but if the rates of the forward reaction and the backward reaction are equal, there will be no net change in composition, and equilibrium will prevail.

The second law has been stated in many versions. One version forbids perpetual-motion machines: Because energy is, by the second law, perpetually degraded into heat and rendered isothermally unavailable ($\Delta S > 0$), continued motion requires an input of energy from the outside. The most celebrated yet perplexing version of the second law was provided by R. J. Clausius (1879): "The energy of the universe is constant; the entropy of the universe tends towards a maximum."

How can entropy increase forever, created out of nothing? The root of the difficulty is verbal, as Klotz (1967) neatly explains. Had Clausius defined entropy with the opposite sign (corresponding to order rather than to chaos), its universal tendency would be to diminish; it would then be obvious that spontaneous changes proceed in the direction that decreases the capacity for further spontaneous change. Solutes diffuse from a region of higher concentration to one of lower concentration; heat flows from a warm body to a cold one. Sometimes these changes can be reversed by an outside agent to reduce the entropy of the system under consideration, but then that external agent must change in such a way as to reduce its own capacity for further change. In sum, "entropy is an index of exhaustion; the more a system has lost its capacity for spontaneous change, the more this capacity has been exhausted, the greater is the entropy" (Klotz 1967). Conversely, the farther a system is from equilibrium, the greater is its capacity for change and the less its entropy. Living things fall into the latter category: *A cell is the epitome of a state that is remote from equilibrium.*

## Free Energy and Chemical Potential

Many energy transactions that take place in living organisms are chemical; we therefore need a quantitative expression for the amount of work a chemical reaction can do. For this purpose, relationships that involve the entropy change in the system plus its surroundings are unsuitable. We need a function that does not depend on the surroundings but that, like $\Delta S$, attains a minimum under conditions of equilibrium and so can serve both as a criterion of the feasibility of a reaction and as a measure of the energy available from it for the performance of work. The function universally employed for this purpose is free energy, abbreviated $G$ in honor of the nineteenth-century physical chemist J. Willard Gibbs, who first introduced it.

## ΔG is negative for a spontaneous process at constant temperature and pressure

Earlier we spoke of the isothermally unavailable energy, $ST$. **Free energy** is defined as the energy that is available under isothermal conditions, and by the following relationship:

$$\Delta H = \Delta G + T\Delta S \qquad (A1.3)$$

The term $H$, **enthalpy** or heat content, is not quite equivalent to $U$, the internal energy (see Equation A1.2). To be exact, $\Delta H$ is a measure of the total energy change, including work that may result from changes in volume during the reaction, whereas $\Delta U$ excludes this work. (We will return to the concept of enthalpy a little later.) However, in the biological context we are usually concerned with reactions in solution, for which volume changes are negligible. For most purposes, then,

$$\Delta U \cong \Delta G + T\Delta S \qquad (A1.4)$$

and

$$\Delta G \cong \Delta U - T\Delta S \qquad (A1.5)$$

What makes this a useful relationship is the demonstration that *for all spontaneous processes at constant temperature and pressure, $\Delta G$ is negative*. The change in free energy is thus a criterion of feasibility. Any chemical reaction that proceeds with a negative $\Delta G$ can take place spontaneously; a process for which $\Delta G$ is positive cannot take place, but the reaction can go in the opposite direction; and a reaction for which $\Delta G$ is zero is at equilibrium, and no net change will occur. For a given temperature and pressure, $\Delta G$ depends only on the composition of the reaction mixture; hence the alternative term "chemical potential" is particularly apt. Again, nothing is said about rate, only about direction. Whether a reaction having a given $\Delta G$ will proceed, and at what rate, is determined by kinetic rather than thermodynamic factors.

There is a close and simple relationship between the change in free energy of a chemical reaction and the work that the reaction can do. Provided the reaction is carried out reversibly,

$$\Delta G = \Delta W_{max} \qquad (A1.6)$$

That is, for a reaction taking place at constant temperature and pressure, $-\Delta G$ is a measure of the maximum work the process can perform. More precisely, $-\Delta G$ is the maximum work possible, exclusive of pressure–volume work, and thus is a quantity of great importance in bioenergetics. Any process going toward equilibrium can, in principle, do work. We can therefore describe processes for which $\Delta G$ is negative as "energy-releasing," or **exergonic**. Conversely, for any process moving away from equilibrium, $\Delta G$ is positive, and we speak of an "energy-consuming," or **endergonic**, reaction. Of course, an endergonic reaction cannot occur: All real processes go toward equilibrium, with a negative $\Delta G$. The concept of endergonic reactions

is nevertheless a useful abstraction, for many biological reactions appear to move away from equilibrium. A prime example is the synthesis of ATP during oxidative phosphorylation, whose apparent $\Delta G$ is as high as 67 kJ mol$^{-1}$ (16 kcal mol$^{-1}$). Clearly, the cell must do work to render the reaction exergonic overall. The occurrence of an endergonic process in nature thus implies that it is coupled to a second, exergonic process. Much of cellular and molecular bioenergetics is concerned with the mechanisms by which energy coupling is effected.

## The standard free-energy change, ΔG⁰, is the change in free energy when the concentration of reactants and products is 1 M

Changes in free energy can be measured experimentally by calorimetric methods. They have been tabulated in two forms: as the free energy of formation of a compound from its elements, and as $\Delta G$ for a particular reaction. It is of the utmost importance to remember that, by convention, the numerical values refer to a particular set of conditions. *The standard free-energy change, $\Delta G^0$, refers to conditions such that all reactants and products are present at a concentration of 1 M*; in biochemistry it is more convenient to employ $\Delta G^{0'}$, which is defined in the same way except that the pH is taken to be 7. The conditions obtained in the real world are likely to be very different from these, particularly with respect to the concentrations of the participants. To take a familiar example, $\Delta G^{0'}$ for the hydrolysis of ATP is about –33 kJ mol$^{-1}$ (–8 kcal mol$^{-1}$). In the cytoplasm, however, the actual nucleotide concentrations are approximately 3 m$M$ ATP, 1 m$M$ ADP, and 10 m$M$ P$_i$. As we will see, changes in free energy depend strongly on concentrations, and $\Delta G$ for ATP hydrolysis under physiological conditions thus is much more negative than $\Delta G^{0'}$, about –50 to –65 kJ mol$^{-1}$ (–12 to –15 kcal mol$^{-1}$). *Thus, whereas values of $\Delta G^{0'}$ for many reactions are easily accessible, they must not be used uncritically as guides to what happens in cells.*

## The value of ΔG is a function of the displacement of the reaction from equilibrium

The preceding discussion of free energy shows that there must be a relationship between $\Delta G$ and the equilibrium constant of a reaction: At equilibrium, $\Delta G$ is zero, and the farther a reaction is from equilibrium, the larger $\Delta G$ is and the more work the reaction can do. The quantitative statement of this relationship is

$$\Delta G^0 = -RT \ln K = -2.3RT \log K \qquad (A1.7)$$

where $R$ is the gas constant, $T$ the absolute temperature, and $K$ the equilibrium constant of the reaction. This equation is one of the most useful links between thermodynamics and biochemistry and has a host of applications. For example, the equation is easily modified to allow computation of the change in free energy for concentrations other than the standard ones. For the reactions shown in the equation

$$A + B \leftrightarrow C + D \qquad (A1.8)$$

the actual change in free energy, $\Delta G$, is given by the equation

$$\Delta G = \Delta G^0 + RT \ln \frac{[C][D]}{[A][B]} \qquad (A1.9)$$

where the terms in brackets refer to the concentrations at the time of the reaction. Strictly speaking, one should use activities, but these are usually not known for cellular conditions, so concentrations must do.

Equation A1.9 can be rewritten to make its import a little plainer. Let $q$ stand for the mass:action ratio, $[C][D]/[A][B]$. Substitution of Equation A1.7 into Equation A1.9, followed by rearrangement, then yields the following equation:

$$\Delta G = -2.3 \, RT \, \log \frac{K}{q} \qquad (A1.10)$$

In other words, the value of $\Delta G$ is a function of the displacement of the reaction from equilibrium. In order to displace a system from equilibrium, work must be done on it and $\Delta G$ must be positive. Conversely, a system displaced from equilibrium can do work on another system, provided that the kinetic parameters allow the reaction to proceed and a mechanism exists that couples the two systems. Quantitatively, a reaction mixture at 25°C whose composition is one order of magnitude away from equilibrium ($\log K/q = 1$) corresponds to a free-energy change of 5.7 kJ mol$^{-1}$ (1.36 kcal mol$^{-1}$). The value of $\Delta G$ is negative if the actual mass:action ratio is less than the equilibrium ratio and positive if the mass:action ratio is greater.

The point that $\Delta G$ is a function of the displacement of a reaction (indeed, of any thermodynamic system) from equilibrium is central to an understanding of bioenergetics. FIGURE A1.2 illustrates this relationship diagrammatically for the chemical interconversion of substances A and B, and the relationship will reappear shortly in other guises.

### The enthalpy change measures the energy transferred as heat

Chemical and physical processes are almost invariably accompanied by the generation or absorption of heat, which reflects the change in the internal energy of the system. The amount of heat transferred and the sign of the reaction are related to the change in free energy, as set out in Equation A1.3. The energy absorbed or evolved as heat under conditions of constant pressure is designated as the change in heat content or enthalpy, $\Delta H$. Processes that generate heat, such as combustion, are said to be **exothermic**; those in which heat is absorbed, such as melting or evaporation, are referred to as **endothermic.** The oxidation of glucose to $CO_2$ and water is an exergonic reaction ($\Delta G^0$ = –2858 kJ mol$^{-1}$ [–686 kcal mol$^{-1}$] ); when this reaction

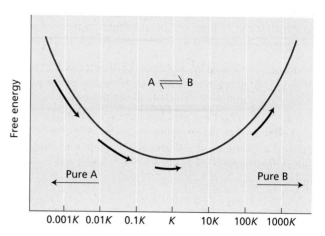

**FIGURE A1.2** Free energy of a chemical reaction as a function of displacement from equilibrium. Imagine a closed system containing components A and B at concentrations [A] and [B]. The two components can be interconverted by the reaction A × B, which is at equilibrium when the mass:action ratio, [B]/[A], equals unity. The curve shows qualitatively how the free energy, G, of the system varies when the total [A] + [B] is held constant but the mass:action ratio is displaced from equilibrium. The arrows represent schematically the change in free energy, $\Delta G$, for a small conversion of [A] into [B] occurring at different mass:action ratios. (After Nicholls and Ferguson 1992.)

takes place during respiration, part of the free energy is conserved through coupled reactions that generate ATP. The combustion of glucose dissipates the free energy of reaction, releasing most of it as heat ($\Delta H$ = –2804 kJ mol$^{-1}$ [–673 kcal mol$^{-1}$]).

Bioenergetics is preoccupied with energy transduction and therefore gives pride of place to free-energy transactions, but at times heat transfer may also carry biological significance. For example, water has a high heat of vaporization, 44 kJ mol$^{-1}$ (10.5 kcal mol$^{-1}$) at 25°C, which plays an important role in the regulation of leaf temperature. During the day, the evaporation of water from the leaf surface (transpiration) dissipates heat to the surroundings and helps cool the leaf. Conversely, the condensation of water vapor as dew heats the leaf, since water condensation is the reverse of evaporation, is exothermic. The abstract enthalpy function is a direct measure of the energy exchanged in the form of heat.

## Redox Reactions

Oxidation and reduction refer to the transfer of one or more electrons from a donor to an acceptor, usually to another chemical species; an example is the oxidation of ferrous iron by oxygen, which forms ferric iron and water. Reactions of this kind require special consideration, for they play a central role in both respiration and photosynthesis.

## The free-energy change of an oxidation–reduction reaction is expressed as the standard redox potential in electrochemical units

Redox reactions can be quite properly described in terms of their change in free energy. However, the participation of electrons makes it convenient to follow the course of the reaction with electrical instrumentation and encourages the use of an electrochemical notation. It also permits dissection of the chemical process into separate oxidative and reductive half-reactions. For the oxidation of iron, we can write

$$2Fe^{2+} \leftrightarrow 2Fe^{3+} + 2e^- \qquad (A1.11)$$

$$\tfrac{1}{2}O_2 + 2H^+ + 2e^- \leftrightarrow H_2O \qquad (A1.12)$$

$$2Fe^{2+} + \tfrac{1}{2}O_2 + 2H^+ \leftrightarrow 2Fe^{3+} + H_2O \qquad (A1.13)$$

The tendency of a substance to donate electrons, its "electron pressure," is measured by its standard reduction (or redox) potential, $E_0$, with all components present at a concentration of 1 $M$. In biochemistry, it is more convenient to employ $E'_0$, which is defined in the same way except that the pH is 7. By definition, then, $E'_0$ is the electromotive force given by a half cell in which the reduced and oxidized species are both present at 1 $M$, 25°C, and pH 7, in equilibrium with an electrode that can reversibly accept electrons from the reduced species. By convention, the reaction is written as a reduction. The standard reduction potential of the hydrogen electrode* serves as reference: at pH 7, it equals –0.42 V. The standard redox potential as defined here is often referred to in the bioenergetics literature as the **midpoint** potential, $E_m$. A negative midpoint potential marks a good reducing agent; oxidants have positive midpoint potentials.

The redox potential for the reduction of oxygen to water is +0.82 V; for the reduction of $Fe^{3+}$ to $Fe^{2+}$ (the direction opposite to that of Equation A1.11), +0.77 V. We can therefore predict that, under standard conditions, the $Fe^{2+}$–$Fe^{3+}$ couple will tend to reduce oxygen to water rather than the reverse. A mixture containing $Fe^{2+}$, $Fe^{3+}$, and oxygen will probably not be at equilibrium, and the extent of its displacement from equilibrium can be expressed in terms of either the change in free energy for Equation A1.13 or the difference in redox potential, $\Delta E'_0$, between the oxidant and the reductant couples (+0.05 V in the case of iron oxidation). In general,

$$\Delta G^{0'} = -nF \, \Delta E'_0 \qquad (A1.14)$$

where $n$ is the number of electrons transferred and $F$ is Faraday's constant (23.06 kcal V$^{-1}$ mol$^{-1}$). In other words, the standard redox potential is a measure, in electrochemical units, of the change in free energy of an oxidation–reduction process.

As with free-energy changes, the redox potential measured under conditions other than the standard ones depends on the concentrations of the oxidized and reduced species, according to the following equation (note the similarity in form to Equation A1.9):

$$E_h = E'_0 + \frac{2.3RT}{nF} \, \log \frac{[\text{oxidant}]}{[\text{reductant}]} \qquad (A1.15)$$

Here $E_h$ is the measured potential in volts, and the other symbols have their usual meanings. It follows that the redox potential under biological conditions may differ substantially from the standard reduction potential.

# The Electrochemical Potential

In the preceding section we introduced the concept that a mixture of substances whose composition diverges from the equilibrium state represents a potential source of free energy (see Figure A1.2). Conversely, a similar amount of work must be done on an equilibrium mixture in order to displace its composition from equilibrium. In this section, we will examine the free-energy changes associated with another kind of displacement from equilibrium—namely, gradients of concentration and of electric potential.

## Transport of an uncharged solute against its concentration gradient decreases the entropy of the system

Consider a vessel divided by a membrane into two compartments that contain solutions of an uncharged solute at concentrations $C_1$ and $C_2$, respectively. The work required to transfer 1 mol of solute from the first compartment to the second is given by the following equation:

$$\Delta G = 2.3RT \, \log \frac{C_2}{C_1} \qquad (A1.16)$$

This expression is analogous to the expression for a chemical reaction (Equation A1.10) and has the same meaning. If $C_2$ is greater than $C_1$, $\Delta G$ is positive, and work must be done to transfer the solute. Again, the free-energy change for the transport of 1 mol of solute against a tenfold gradient of concentration is 5.7 kJ, or 1.36 kcal.

The reason that work must be done to move a substance from a region of lower concentration to one of higher concentration is that the process entails a change to a less probable state and therefore a decrease in the entropy of the system. Conversely, diffusion of the solute from the region of higher concentration to that of lower concentration takes place in the direction of greater probability; it results in

---

* The standard hydrogen electrode consists of platinum, over which hydrogen gas is bubbled at a pressure of 1 atm. The electrode is immersed in a solution containing hydrogen ions. When the activity of hydrogen ions is 1, approximately 1 $M$ H$^+$, the potential of the electrode is taken to be 0.

an increase in the entropy of the system and can proceed spontaneously. The sign of $\Delta G$ becomes negative, and the process can do the amount of work specified by Equation A1.16, provided a mechanism exists that couples the exergonic diffusion process to the work function.

### The membrane potential is the work that must be done to move an ion from one side of the membrane to the other

Matters become a little more complex if the solute in question bears an electric charge. Transfer of positively charged solute from compartment 1 to compartment 2 will then cause a difference in charge to develop across the membrane, the second compartment becoming electropositive relative to the first. Since like charges repel one another, the work done by the agent that moves the solute from compartment 1 to compartment 2 is a function of the charge difference; more precisely, it depends on the difference in electric potential across the membrane. This difference, called membrane potential for short, will appear again in later pages.

The **membrane potential**, $\Delta E$,* is defined as the work that must be done by an agent to move a test charge from one side of the membrane to the other. When 1 J of work must be done to move 1 coulomb of charge, the potential difference is said to be 1 V. The absolute electric potential of any single phase cannot be measured, but the potential difference between two phases can be. By convention, the membrane potential is always given in reference to the movement of a positive charge. It states the intracellular potential relative to the extracellular one, which is defined as zero.

The work that must be done to move 1 mol of an ion against a membrane potential of $\Delta E$ volts is given by the following equation:

$$\Delta G = zF\,\Delta E \qquad (A1.17)$$

where $z$ is the valence of the ion and $F$ is Faraday's constant. The value of $\Delta G$ for the transfer of cations into a positive compartment is positive and so calls for work. Conversely, the value of $\Delta G$ is negative when cations move into the negative compartment, so work can be done. *The electric potential is negative across the plasma membrane of the great majority of cells; therefore cations tend to leak in but have to be "pumped" out.*

### The electrochemical-potential difference, $\Delta\tilde{\mu}$, includes both concentration and electric potentials

In general, ions moving across a membrane are subject to gradients of both concentration and electric potential. Con-

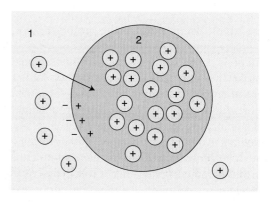

**FIGURE A1.3**  Transport against an electrochemical-potential gradient. The agent that moves the charged solute (from compartment 1 to compartment 2) must do work to overcome both the electrochemical-potential gradient and the concentration gradient. As a result, cations in compartment 2 have been raised to a higher electrochemical potential than those in compartment 1. Neutralizing anions have been omitted.

sider, for example, the situation depicted in **FIGURE A1.3**, which corresponds to a major event in energy transduction during photosynthesis. A cation of valence $z$ moves from compartment 1 to compartment 2, against both a concentration gradient ($C_2 > C_1$) and a gradient of membrane electric potential (compartment 2 is electropositive relative to compartment 1). The free-energy change involved in this transfer is given by the following equation:

$$\Delta G = zF\Delta E + 2.3RT \, \log \frac{C_2}{C_1} \qquad (A1.18)$$

$\Delta G$ is positive, and the transfer can proceed only if coupled to a source of energy, in this instance the absorption of light. As a result of this transfer, cations in compartment 2 can be said to be at a higher electrochemical potential than the same ions in compartment 1.

The electrochemical potential for a particular ion is designated $\tilde{\mu}_{ion}$. Ions tend to flow from a region of high electrochemical potential to one of low potential and in so doing can in principle do work. The maximum amount of this work, neglecting friction, is given by the change in free energy of the ions that flow from compartment 2 to compartment 1 (see Equation A1.6) and is numerically equal to the electrochemical-potential difference, $\Delta\tilde{\mu}_{ion}$. This principle underlies much of biological energy transduction.

The electrochemical-potential difference, $\Delta\tilde{\mu}_{ion}$, is properly expressed in kilojoules per mole or kilocalories per mole. However, it is frequently convenient to express the driving force for ion movement in electrical terms, with the dimensions of volts or millivolts. To convert $\Delta\tilde{\mu}_{ion}$ into millivolts (mV), divide all the terms in Equation A1.18 by $F$:

$$\frac{\Delta\tilde{\mu}_{ion}}{F} = z\Delta E + \frac{2.3RT}{F} \, \log \frac{C_2}{C_1} \qquad (A1.19)$$

---

* Many texts use the term $\Delta\Psi$ for the membrane potential difference. However, to avoid confusion with the use of $\Delta\Psi$ to indicate water potential (see Chapter 3), the term $\Delta E$ is used here and throughout the text.

An important case in point is the proton motive force, which is considered at length in Chapter 6.

Equations A1.18 and A1.19 have proved to be of central importance in bioenergetics. First, they measure the amount of energy that must be expended on the active transport of ions and metabolites, a major function of biological membranes. Second, since the free energy of chemical reactions is often transduced into other forms via the intermediate generation of electrochemical-potential gradients, these gradients play a major role in descriptions of biological energy coupling. It should be emphasized that the electrical and concentration terms may be either added, as in Equation A1.18, or subtracted, and that the application of the equations to particular cases requires careful attention to the sign of the gradients. We should also note that free-energy changes in chemical reactions (see Equation A1.10) are scalar, whereas transport reactions have direction; this is a subtle but critical aspect of the biological role of ion gradients.

Ion distribution at equilibrium is an important special case of the general electrochemical equation (Equation A1.18). FIGURE A1.4 shows a membrane-bound vesicle (compartment 2) that contains a high concentration of the salt $K_2SO_4$, surrounded by a medium (compartment 1) containing a lower concentration of the same salt; the membrane is impermeable to anions but allows the free passage of cations. Potassium ions will therefore tend to diffuse out of the vesicle into the solution, whereas the sulfate anions are retained. Diffusion of the cations generates a membrane potential, with the vesicle interior negative, which restrains further diffusion. At equilibrium, $\Delta G$ and $\Delta\tilde{\mu}_{K^+}$ equal zero (by definition). Equation A1.18 can then be arranged to give the following equation:

$$\Delta E = \frac{-2.3RT}{zF} \log \frac{C_2}{C_1} \qquad (A1.20)$$

where $C_2$ and $C_1$ are the concentrations of $K^+$ ions in the two compartments; $z$, the valence, is unity; and $\Delta E$ is the membrane potential in equilibrium with the potassium concentration gradient.

This is one form of the celebrated **Nernst equation**. It states that at equilibrium, a permeant ion will be so distributed across the membrane that the chemical driving force (outward in this instance) will be balanced by the electric driving force (inward). For a univalent cation at 25°C, each tenfold increase in concentration factor corresponds to a membrane potential of 59 mV; for a divalent ion the value is 29.5 mV.

The preceding discussion of the energetic and electrical consequences of ion translocation illustrates a point that must be clearly understood—namely, that an electric potential across a membrane may arise by two distinct mechanisms. The first mechanism, illustrated in Figure A1.4, is the diffusion of charged particles down a preexisting concentration gradient, an exergonic process. A potential generated

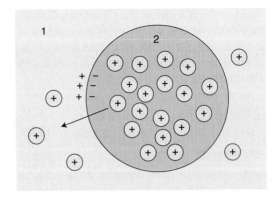

**FIGURE A1.4**   Generation of an electric potential by ion diffusion. Compartment 2 has a higher salt concentration than compartment 1 (anions are not shown). If the membrane is permeable to the cations but not to the anions, the cations will tend to diffuse out of compartment 2 into compartment 1, generating a membrane potential in which compartment 2 is negative.

by such a process is described as a **diffusion potential** or as a Donnan potential. (**Donnan potential** is defined as the diffusion potential that occurs in the limiting case where the counterion is completely impermeant or fixed, as in Figure A1.4.) Many ions are unequally distributed across biological membranes and differ widely in their rates of diffusion across the barrier; therefore diffusion potentials always contribute to the observed membrane potential. But in most biological systems the measured electric potential differs from the value that would be expected on the basis of passive ion diffusion. In these cases one must invoke electrogenic ion pumps, transport systems that carry out the exergonic process indicated in Figure A1.3 at the expense of an external energy source. Transport systems of this kind transduce the free energy of a chemical reaction into the electrochemical potential of an ion gradient and play a leading role in biological energy coupling.

## Enzymes: The Catalysts of Life

Proteins constitute about 30% of the total dry weight of typical plant cells. If we exclude inert materials, such as the cell wall and starch, which can account for up to 90% of the dry weight of some cells, proteins and amino acids represent about 60 to 70% of the dry weight of the living cell. As we saw in Chapter 1, cytoskeletal structures such as microtubules and microfilaments are composed of protein. Proteins can also occur as storage forms, particularly in seeds. But the major function of proteins in metabolism is to serve as enzymes, biological catalysts that greatly increase the rates of biochemical reactions, making life possible. Enzymes participate in these reactions but are not themselves fundamentally changed in the process (Mathews and Van Holde 1996).

Enzymes have been called the "agents of life"—a very apt term, since they control almost all life processes. A typical cell has several thousand different enzymes, which carry out a wide variety of actions. The most important features of enzymes are their *specificity*, which permits them to distinguish among very similar molecules, and their *catalytic efficiency*, which is far greater than that of ordinary catalysts. The stereospecificity of enzymes is remarkable, allowing them to distinguish not only between enantiomers (mirror-image stereoisomers), for example, but between apparently identical atoms or groups of atoms (Creighton 1983).

This ability to discriminate between similar molecules results from the fact that the first step in enzyme catalysis is the formation of a tightly bound, noncovalent complex between the enzyme and the substrate(s): the **enzyme–substrate complex**. Enzyme-catalyzed reactions exhibit unusual kinetic properties that are also related to the formation of these very specific complexes. Another distinguishing feature of enzymes is that they are subject to various kinds of regulatory control, ranging from subtle effects on the catalytic activity by effector molecules (inhibitors or activators) to regulation of enzyme synthesis and destruction by the control of gene expression and protein turnover.

Enzymes are unique in the large rate enhancements they bring about, orders of magnitude greater than those effected by other catalysts. Typical orders of rate enhancements of enzyme-catalyzed reactions over the corresponding uncatalyzed reactions are $10^8$ to $10^{12}$. Many enzymes will convert about a thousand molecules of substrate to product in 1 s. Some will convert as many as a million!

Unlike most other catalysts, enzymes function at ambient temperature and atmospheric pressure and usually in a narrow pH range near neutrality (there are exceptions; for instance, vacuolar proteases and ribonucleases are most active at pH 4 to 5). A few enzymes are able to function under extremely harsh conditions; examples are pepsin, the protein-degrading enzyme of the stomach, which has a pH optimum around 2.0, and the hydrogenase of the hyperthermophilic ("extreme heat–loving") archaebacterium *Pyrococcus furiosus*, which oxidizes $H_2$ at a temperature optimum greater than 95°C (Bryant and Adams 1989). The presence of such remarkably heat-stable enzymes enables *Pyrococcus* to grow optimally at 100°C.

Enzymes are usually named after their substrates by the addition of the suffix "-ase"—for example, α-amylase, malate dehydrogenase, β-glucosidase, phosphoenolpyruvate carboxylase, horseradish peroxidase. Many thousands of enzymes have already been discovered, and new ones are being found all the time. Each enzyme has been named in a systematic fashion, on the basis of the reaction it catalyzes, by the International Union of Biochemistry. In addition, many enzymes have common, or trivial, names.

Thus the common name *rubisco* refers to D-ribulose-1,5-bisphosphate carboxylase/oxygenase (EC 4.1.1.39*).

The versatility of enzymes reflects their properties as proteins. The nature of proteins permits both the exquisite recognition by an enzyme of its substrate and the catalytic apparatus necessary to carry out diverse and rapid chemical reactions (Stryer 1995).

## Proteins are chains of amino acids joined by peptide bonds

Proteins are composed of long chains of amino acids (FIGURE A1.5) linked by amide bonds, known as **peptide bonds** (FIGURE A1.6). The 20 different amino acid side chains endow proteins with a large variety of groups that have different chemical and physical properties, including hydrophilic (polar, water-loving) and hydrophobic (nonpolar, water-avoiding) groups, charged and neutral polar groups, and acidic and basic groups. This diversity, in conjunction with the relative flexibility of the peptide bond, allows for the tremendous variation in protein properties, ranging from the rigidity and inertness of structural proteins to the reactivity of hormones, catalysts, and receptors. The three-dimensional aspect of protein structure provides for precise discrimination in the recognition of **ligands**, the molecules that interact with proteins, as shown by the ability of enzymes to recognize their substrates and of antibodies to recognize antigens, for example.

All molecules of a particular protein have the same sequence of amino acid residues, determined by the sequence of nucleotides in the gene that codes for that protein. Although the protein is synthesized as a linear chain on the ribosome, upon release it folds spontaneously into a specific three-dimensional shape, the **native** state. The chain of amino acids is called a polypeptide. The three-dimensional arrangement of the atoms in the molecule is referred to as the **conformation**. Changes in conformation do not involve breaking of covalent bonds. *Denaturation* involves the loss of this unique three-dimensional shape and results in the loss of catalytic activity.

The forces that are responsible for the shape of a protein molecule are noncovalent (FIGURE A1.7). These noncovalent interactions include hydrogen bonds; electrostatic interactions (also known as ionic bonds or salt bridges); van der Waals interactions (dispersion forces), which are transient dipoles between spatially close atoms; and hydrophobic "bonds"—the tendency of nonpolar groups to avoid contact with water and thus to associate with themselves. In addition, covalent disulfide bonds are found in many proteins. Although each of these types of noncovalent interaction is weak, there are so many

---

* The Enzyme Commission (EC) number indicates the class (4 = lyase) and subclasses (4.1 = carbon–carbon cleavage; 4.1.1 = cleavage of C—COO⁻ bond).

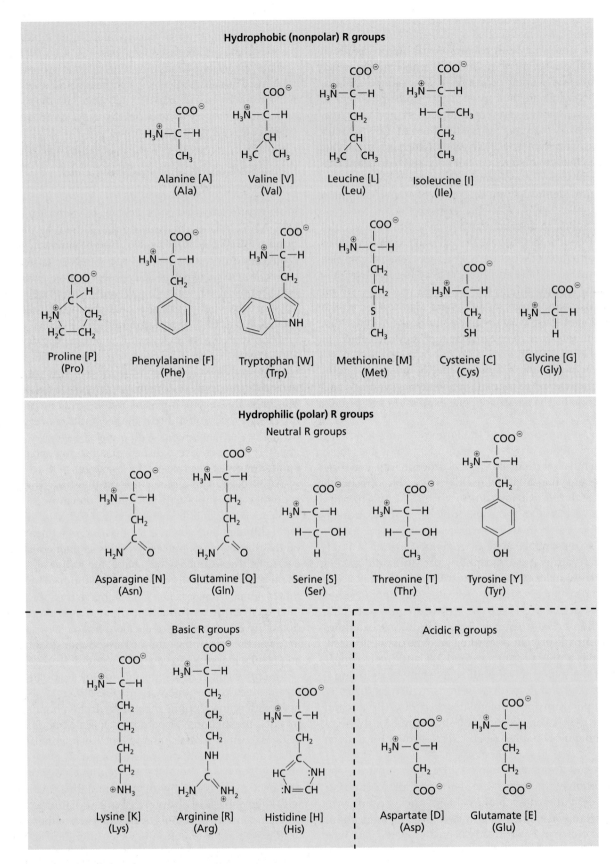

**FIGURE A1.5**   The structures, names, single-letter codes (in square brackets), three-letter abbreviations, and classification of the amino acids.

(A)

Peptide bond

(B)

Rigid unit

**FIGURE A1.6** (A) The peptide (amide) bond links two amino acids. (B) Sites of free rotation, within the limits of steric hindrance, about the N—C$_\alpha$ and C$_\alpha$—C bonds ($\psi$ and $\phi$); there is no rotation about the peptide bond, because of its double-bond character.

noncovalent interactions in proteins that in total they contribute a large amount of free energy to stabilizing the native structure.

## Protein structure is hierarchical

Proteins are built up with increasingly complex organizational units. The **primary structure** of a protein refers to the sequence of amino acid residues. The **secondary structure** refers to regular, local structural units, usually held together by hydrogen bonding. The most common of these units are the α helix and β strands forming parallel and antiparallel β pleated sheets and turns (**FIGURE A1.8**). The **tertiary structure**—the final three-dimensional structure of the polypeptide—results from the packing together of the secondary structure units and the exclusion of solvent. The **quaternary structure** refers to the association of two or more separate three-dimensional polypeptides to form complexes. When associated in this manner, the individual polypeptides are called **subunits**.

A protein molecule consisting of a large single polypeptide chain is composed of several independently folding units known as **domains**. Typically, domains have a molecular mass of about $10^4$ daltons. The active site of an enzyme—that is, the region where the substrate binds and the catalytic reaction occurs—is often located at the interface between two domains. For example, in the enzyme papain (a vacuolar protease that is found in papaya and

HYDROGEN BONDS

Between elements of peptide linkage

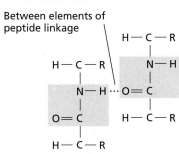

Between side chains

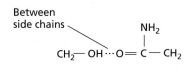

Serine    Asparagine

ELECTROSTATIC ATTRACTIONS

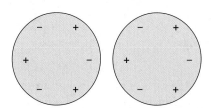

VAN DER WAALS INTERACTIONS

**FIGURE A1.7** Examples of noncovalent interactions in proteins. Hydrogen bonds are weak electrostatic interactions involving a hydrogen atom between two electronegative atoms. In proteins the most important hydrogen bonds are those between the peptide bonds. Electrostatic interactions are ionic bonds between positively and negatively charged groups. The van der Waals interactions are short-range transient dipole interactions. Hydrophobic interactions (not shown) involve restructuring of the solvent water around nonpolar groups, minimizing the exposure of nonpolar surface area to polar solvent; these interactions are driven by entropy.

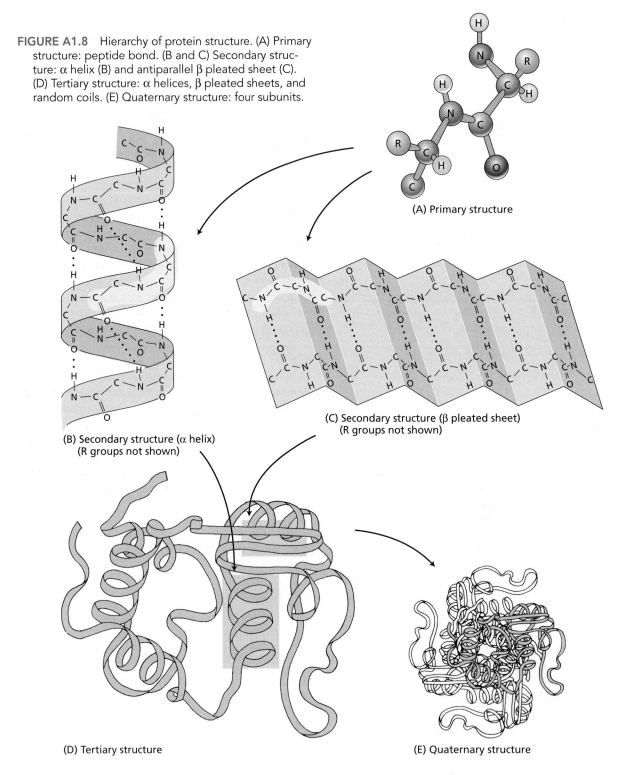

**FIGURE A1.8** Hierarchy of protein structure. (A) Primary structure: peptide bond. (B and C) Secondary structure: α helix (B) and antiparallel β pleated sheet (C). (D) Tertiary structure: α helices, β pleated sheets, and random coils. (E) Quaternary structure: four subunits.

(A) Primary structure

(B) Secondary structure (α helix)
(R groups not shown)

(C) Secondary structure (β pleated sheet)
(R groups not shown)

(D) Tertiary structure

(E) Quaternary structure

is representative of a large class of plant thiol proteases), the active site lies at the junction of two domains (**FIGURE A1.9**). Helices, turns, and β sheets contribute to the unique three-dimensional shape of this enzyme.

Determinations of the conformation of proteins have revealed that there are families of proteins that have common three-dimensional folds, as well as common patterns of supersecondary structure, such as β-α-β.

## Enzymes are highly specific protein catalysts

All enzymes are proteins, although recently some small ribonucleic acids and protein–RNA complexes have been found to exhibit enzymelike behavior in the processing of RNA. Proteins have molecular masses ranging from $10^4$ to $10^6$ daltons, and they may be a single folded polypeptide chain (subunit, or protomer) or oligomers of several subunits (oligomers are usually dimers or tetram-

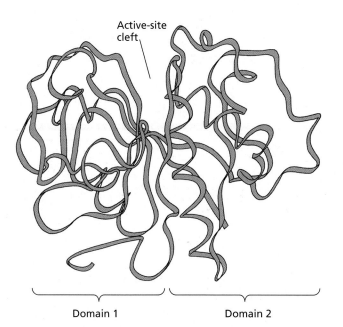

Active-site cleft

Domain 1          Domain 2

**FIGURE A1.9** The backbone structure of papain, showing the two domains and the active-site cleft between them.

ers). Normally, enzymes have only one type of catalytic activity associated with the same protein; **isoenzymes**, or **isozymes**, are enzymes with similar catalytic function that have different structures and catalytic parameters and are encoded by different genes. For example, various different isozymes have been found for peroxidase, an enzyme in plant cell walls that is involved in the synthesis of lignin. An isozyme of peroxidase has also been localized in vacuoles. Isozymes may exhibit tissue specificity and show developmental regulation.

Enzymes frequently contain a nonprotein **prosthetic group** or **cofactor** that is necessary for biological activity. The association of a cofactor with an enzyme depends on the three-dimensional structure of the protein. Once bound to the enzyme, the cofactor contributes to the specificity of catalysis. Typical examples of cofactors are metal ions (e.g., zinc, iron, molybdenum), heme groups or iron–sulfur clusters (especially in oxidation–reduction enzymes), and coenzymes (e.g., nicotinamide adenine dinucleotide [$NAD^+$/NADH], flavin adenine dinucleotide [$FAD$/$FADH_2$], flavin mononucleotide [FMN], and pyridoxal phosphate [PLP]). Coenzymes are usually vitamins or are derived from vitamins and act as carriers. For example, $NAD^+$ and FAD carry hydrogens and electrons in redox reactions, biotin carries $CO_2$, and tetrahydrofolate carries one-carbon fragments. Peroxidase has both heme and $Ca^{2+}$ prosthetic groups and is glycosylated; that is, it contains carbohydrates covalently added to asparagine, serine, or threonine side chains. Such proteins are called **glycoproteins**.

A particular enzyme will catalyze only one type of chemical reaction for only one class of molecule—in some cases, for only one particular compound. Enzymes are also very stereospecific and produce no by-products. For example, β-glucosidase catalyzes the hydrolysis of β-glucosides, compounds formed by a glycosidic bond to D-glucose. The substrate must have the correct anomeric configuration: it must be β-, not α-. Furthermore, it must have the glucose structure; no other carbohydrates, such as xylose or mannose, can act as substrates for β-glucosidase. Finally, the substrate must have the correct stereochemistry, in this case the D absolute configuration. Rubisco (D-ribulose-1,5-bisphosphate carboxylase/oxygenase) catalyzes the addition of carbon dioxide to D-ribulose 1,5-bisphosphate to form two molecules of 3-phospho-D-glycerate, the initial step in the $C_3$ photosynthetic carbon reduction cycle, and is the world's most abundant enzyme. Rubisco has very strict specificity for the carbohydrate substrate, but it also catalyzes an oxygenase reaction in which $O_2$ replaces $CO_2$, as is discussed further in Chapter 8.

## Enzymes lower the free-energy barrier between substrates and products

Catalysts speed the rate of a reaction by lowering the energy barrier between substrates (reactants) and products and are not themselves used up in the reaction, but are regenerated. Thus a catalyst increases the rate of a reaction but does not affect the equilibrium ratio of reactants and products, because the rates of the reaction in both directions are increased to the same extent. It is important to realize that enzymes cannot make a nonspontaneous (energetically uphill) reaction occur. However, many energetically unfavorable reactions in cells proceed because they are coupled to an energetically more favorable reaction usually involving ATP hydrolysis (**FIGURE A1.10**).

$$A + B \rightarrow C \qquad \Delta G = +4.0 \text{ kcal mol}^{-1}$$
$$ATP + H_2O \rightarrow ADP + P_i + H^+ \qquad \Delta G = -7.3 \text{ kcal mol}^{-1}$$

$$A + B + ATP + H_2O \rightarrow C + ADP + P_i + H^+ \quad \Delta G = -3.3 \text{ kcal mol}^{-1}$$

$$A + ATP \rightarrow A - P + ADP$$
$$A - P + B + H_2O \rightarrow C + H^+ + P_i$$

$$A + B + ATP + H_2O \rightarrow C + ADP + P_i + H^+$$

**FIGURE A1.10** Coupling of the hydrolysis of ATP to drive an energetically unfavorable reaction. The reaction A + B → C is thermodynamically unfavorable, whereas the hydrolysis of ATP to form ADP and inorganic phosphate ($P_p$) is thermodynamically very favorable (it has a large negative $\Delta G$). Through appropriate intermediates, such as A–P, the two reactions are coupled, yielding an overall reaction that is the sum of the individual reactions and has a favorable free-energy change.

FIGURE A1.11   Free-energy curves for the same reaction, either uncatalyzed or enzyme catalyzed. As a catalyst, an enzyme lowers the free energy of activation of the transition state between substrates and products compared with the uncatalyzed reaction. It does this by forming various complexes and intermediates, such as enzyme–substrate and enzyme–product complexes. The ground state free energy of the enzyme–substrate complex in the enzyme-catalyzed reaction may be higher than that of the substrate in the uncatalyzed reaction, and the transition state free energy of the enzyme-bound substrate will be signficantly less than that in the corresponding uncatalyzed reaction.

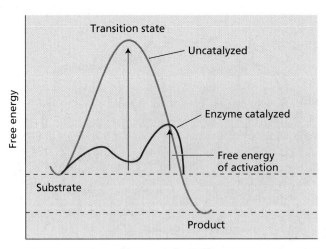

Enzymes act as catalysts because they lower the free energy of activation for a reaction. They do this by a combination of raising the **ground state** $\Delta G$ of the substrate and lowering the $\Delta G$ of the **transition state** of the reaction, thereby decreasing the barrier against the reaction (**FIGURE A1.11**). The presence of the enzyme leads to a new reaction pathway that is different from that of the uncatalyzed reaction.

## Catalysis occurs at the active site

The **active site** of an enzyme molecule is usually a cleft or pocket on or near the surface of the enzyme that takes up only a small fraction of the enzyme surface. It is convenient to consider the active site as consisting of two components: the **binding site** for the substrate (which attracts and positions the substrate) and the **catalytic groups** (the reactive side chains of amino acids or cofactors, which carry out the bond-breaking and bond-forming reactions involved).

Binding of substrate at the active site initially involves noncovalent interactions between the substrate and either side chains or peptide bonds of the protein. The rest of the protein structure provides a means of positioning the substrate and catalytic groups, flexibility for conformational changes, and regulatory control. The shape and polarity of the binding site account for much of the specificity of enzymes, and there is complementarity between the shape and the polarity of the substrate and those of the active site. In some cases, binding of the substrate induces a conformational change in the active site of the enzyme. Conformational change is particularly common where there are two substrates. Binding of the first substrate sets up a conformational change of the enzyme that results in formation of the binding site for the second substrate. Hexokinase is a good example of an enzyme that exhibits this type of conformational change (**FIGURE A1.12**).

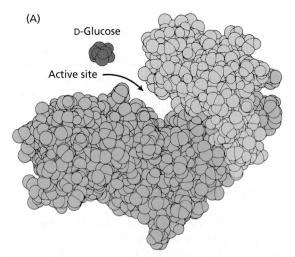

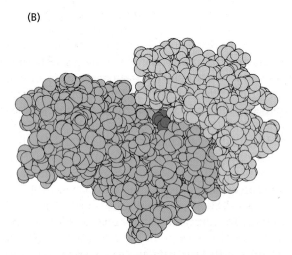

FIGURE A1.12   Conformational change in hexokinase, induced by the first substrate of the enzyme, D-glucose. (A) Before glucose binding. (B) After glucose binding. The binding of glucose to hexokinase induces a conformational change in which the two major domains come together to close the cleft that contains the active site.

This change sets up the binding site for the second substrate, ATP. In this manner the enzyme prevents the unproductive hydrolysis of ATP by shielding the substrates from the aqueous solvent. The overall reaction is the phosphorylation of glucose and the formation of ADP.

The catalytic groups are usually the amino acid side chains and/or cofactors that can function as catalysts. Common examples of catalytic groups are acids (—COOH from the side chains of aspartic acid or glutamic acid, imidazole from the side chain of histidine), bases (—NH$_2$ from lysine, imidazole from histidine, —S$^-$ from cysteine), nucleophiles (imidazole from histidine, —S$^-$ from cysteine, —OH from serine), and electrophiles (often metal ions, such as Zn$^{2+}$). The acidic catalytic groups function by donating a proton, the basic ones by accepting a proton. Nucleophilic catalytic groups form a transient covalent bond to the substrate.

The decisive factor in catalysis is the direct interaction between the enzyme and the substrate. In many cases, there is an intermediate that contains a covalent bond between the enzyme and the substrate. Although the details of the catalytic mechanism differ from one type of enzyme to another, a limited number of features are involved in all enzyme catalysis. These features include acid–base catalysis, electrophilic or nucleophilic catalysis, and ground state distortion through electrostatic or mechanical strains on the substrate.

### A simple kinetic equation describes an enzyme-catalyzed reaction

Enzyme-catalyzed systems often exhibit a special form of kinetics, called Michaelis–Menten kinetics, which are characterized by a hyperbolic relationship between reaction velocity, $v$, and substrate concentration, [S] (**FIGURE A1.13**). This type of plot is known as a saturation plot because when the enzyme becomes saturated with substrate (i.e., each enzyme molecule has a substrate molecule associated with it), the rate becomes independent of substrate concentration. Saturation kinetics implies that an equilibrium process precedes the rate-limiting step:

$$E + S \underset{}{\overset{fast}{\longleftrightarrow}} ES \overset{slow}{\longrightarrow} E + P$$

where E represents the enzyme, S the substrate, P the product, and ES the enzyme–substrate complex. Thus, as the substrate concentration is increased, a point will be reached at which all the enzyme molecules are in the form of the ES complex, and the enzyme is saturated with substrate. Since the rate of the reaction depends on the concentration of ES, the rate will not increase further, because there can be no higher concentration of ES.

When an enzyme is mixed with a large excess of substrate, there will be an initial very short time period (usually milliseconds) during which the concentrations of enzyme-substrate complexes and intermediates build up to certain levels; this is known as the pre–steady-state period. Once the intermediate levels have been built up, they remain relatively constant until the substrate is depleted; this period is known as the steady state.

Normally enzyme kinetic values are measured under steady-state conditions, and such conditions usually pre-

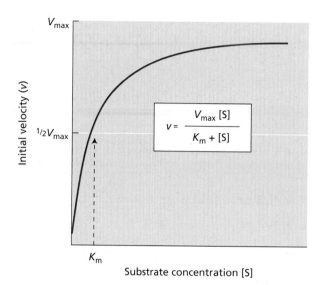

**FIGURE A1.13**  Plot of initial velocity, $v$, versus substrate concentration, [S], for an enzyme-catalyzed reaction. The curve is hyperbolic. The maximal rate, $V_{max}$, occurs when all the enzyme molecules are fully occupied by substrate. The value of $K_m$, defined as the substrate concentration at $1/2V_{max}$, is a reflection of the affinity of the enzyme for the substrate. The smaller the value of $K_m$, the tighter the binding.

vail in the cell. For many enzyme-catalyzed reactions the kinetics under steady-state conditions can be described by a simple expression known as the Michaelis–Menten equation:

$$v = \frac{V_{max}[S]}{K_m + [S]} \tag{A1.21}$$

where $v$ is the observed rate or velocity (in units such as moles per liter per second), $V_{max}$ is the maximum velocity (at infinite substrate concentration), and $K_m$ (usually measured in units of molarity) is a constant that is characteristic of the particular enzyme–substrate system and is related to the association constant of the enzyme for the substrate (see Figure A1.13). $K_m$ represents the concentration of substrate required to half-saturate the enzyme and thus is the substrate concentration at $V_{max}/2$. In many cellular systems the usual substrate concentration is in the vicinity of $K_m$. The smaller the value of $K_m$, the more strongly the enzyme binds the substrate. Typical values for $K_m$ are in the range of $10^{-6}$ to $10^{-3}$ M.

We can readily obtain the parameters $V_{max}$ and $K_m$ by fitting experimental data to the Michaelis–Menten equation, either by computerized curve fitting or by a linearized form of the equation. An example of a linearized form of the equation is the Lineweaver–Burk double-reciprocal plot shown in **FIGURE A1.14A**. When divided by the concentration of enzyme, the value of $V_{max}$ gives the **turnover number**, the number of molecules of substrate converted

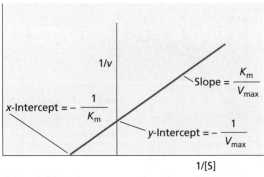
(A) Uninhibited enzyme-catalyzed reaction

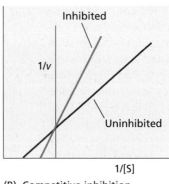
(B) Competitive inhibition

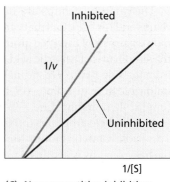
(C) Noncompetitive inhibition

**FIGURE A1.14** Lineweaver–Burk double-reciprocal plots. A plot of $1/v$ versus $1/[S]$ yields a straight line. (A) Uninhibited enzyme-catalyzed reaction showing the calculation of $K_m$ from the x-intercept and of $V_{max}$ from the y-intercept. (B) The effect of a competitive inhibitor on the parameters $K_m$ and $V_{max}$. The apparent $K_m$ is increased, but the $V_{max}$ is unchanged. (C) A noncompetitive inhibitor reduces $V_{max}$ but has no effect on $K_m$.

to product per unit of time per molecule of enzyme. Typical turnover number values range from $10^2$ to $10^3$ s$^{-1}$.

### Enzymes are subject to various kinds of inhibition

Any agent that decreases the velocity of an enzyme-catalyzed reaction is called an inhibitor. Inhibitors may exert their effects in many different ways. Generally, if inhibition is irreversible the compound is called an **inactivator**. Other agents can increase the efficiency of an enzyme; they are called **activators**. Inhibitors and activators are very important in the cellular regulation of enzymes. Many agriculturally important insecticides and herbicides are enzyme inhibitors. The study of enzyme inhibition can provide useful information about kinetic mechanisms, the nature of enzyme–substrate intermediates and complexes, the chemical mechanism of catalytic action, and the regulation and control of metabolic enzymes. In addition, the study of inhibitors of potential target enzymes is essential to the rational design of herbicides.

Inhibitors can be classified as reversible or irreversible. **Irreversible inhibitors** form covalent bonds with an enzyme or they denature it. For example, iodoacetate ($ICH_2COOH$) irreversibly inhibits thiol proteases such as papain by alkylating the active-site —SH group. One class of irreversible inhibitors is called affinity labels, or active site–directed modifying agents, because their structure directs them to the active site. An example is tosyl-lysine chloromethyl ketone (TLCK), which irreversibly inactivates papain. The tosyl-lysine part of the inhibitor resembles the substrate structure and so binds in the active site. The chloromethyl ketone part of the bound inhibitor reacts with the active-site histidine side chain. Such compounds

are very useful in mechanistic studies of enzymes, but they have limited practical use as herbicides because of their chemical reactivity, which can be harmful to the plant.

**Reversible inhibitors** form weak, noncovalent bonds with the enzyme, and their effects may be competitive, noncompetitive, or mixed. For example, the widely used broad-spectrum herbicide glyphosate (Roundup®) works by competitively inhibiting a key enzyme in the biosynthesis of aromatic amino acids, 5-*enol*pyruvylshikimate-3-*p*hosphate (EPSP) synthase (see Chapter 13). Resistance to glyphosate has recently been achieved by genetic engineering of plants so that they are capable of overproducing EPSP synthase (Donahue et al. 1995).

**COMPETITIVE INHIBITION** Competitive inhibition is the simplest and most common form of reversible inhibition. It usually arises from binding of the inhibitor to the active site with an affinity similar to or stronger than that of the substrate. Thus the effective concentration of the enzyme is decreased by the presence of the inhibitor, and the catalytic reaction will be slower than if the inhibitor were absent. Competitive inhibition is usually based on the fact that the structure of the inhibitor resembles that of the substrate; hence the strong affinity of the inhibitor for the active site. Competitive inhibition may also occur in **allosteric enzymes**, where the inhibitor binds to a distant site on the enzyme, causing a conformational change that alters the active site and prevents normal substrate binding. Such a binding site is called an **allosteric site**. In this case, the competition between substrate and inhibitor is indirect.

Competitive inhibition results in an apparent increase in $K_m$ and has no effect on $V_{max}$ (**FIGURE A1.14B**). By measuring the apparent $K_m$ as a function of inhibitor concentration, one can calculate $K_i$, the inhibitor constant, which reflects the affinity of the enzyme for the inhibitor.

**NONCOMPETITIVE INHIBITION** In noncompetitive inhibition, the inhibitor does not compete with the substrate for binding to the active site. Instead, it may bind to another site on the protein and obstruct the substrate's access to the active site, thereby changing the catalytic properties of the

(A)

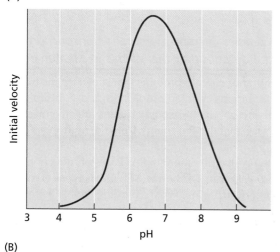

(B)

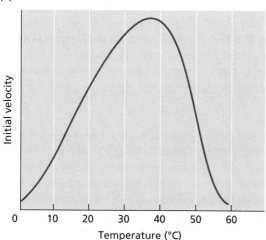

**FIGURE A1.15** pH and temperature curves for typical enzyme reactions. (A) Many enzyme-catalyzed reactions show bell-shaped profiles of rate versus pH. The inflection point on each shoulder corresponds to the $pK_a$ of an ionizing group (that is, the pH at which the ionizing group is 50% dissociated) in the active site. (B) Temperature causes an exponential increase in the reaction rate until the optimum is reached. Beyond the optimum, thermal denaturation dramatically decreases the rate.

curves, one for an ionizable group acting as an acid and the other for the group acting as a base (**FIGURE A1.15A**). Although the effects of pH on enzyme catalysis usually reflect the ionization of the catalytic group, they may also reflect a pH-dependent conformational change in the protein that leads to loss of activity as a result of disruption of the active site.

The temperature dependence of most chemical reactions also applies to enzyme-catalyzed reactions. Thus, most enzyme-catalyzed reactions show an exponential increase in rate with increasing temperature. However, because the enzymes are proteins, another major factor comes in to play—namely, denaturation. After a certain temperature is reached, enzymes show a very rapid decrease in activity as a result of the onset of denaturation (**FIGURE A1.15B**). The temperature at which denaturation begins, and hence at which catalytic activity is lost, varies with the particular protein as well as the environmental conditions, such as pH. Frequently, denaturation begins at about 40 to 50°C and is complete over a range of about 10°C.

## Cooperative systems increase the sensitivity to substrates and are usually allosteric

Cells control the concentrations of most metabolites very closely. To keep such tight control, the enzymes that control metabolite interconversion must be very sensitive. From the plot of velocity versus substrate concentration (see Figure A1.13), we can see that the velocity of an enzyme-catalyzed reaction increases with increasing substrate concentration up to $V_{max}$. However, we can calculate from the Michaelis–Menten equation (Equation A1.21) that raising the velocity of an enzyme-catalyzed reaction from $0.1\,V_{max}$ to $0.9\,V_{max}$ requires an enormous (81-fold) increase in the substrate concentration:

$$0.1V_{max} = \frac{V_{max}[S]}{K_m+[S]}, \quad 0.9V_{max} = \frac{V_{max}[S]'}{K_m+[S]'}$$

$$0.1K_m = 0.9[S], \quad 0.9K_m = 0.1[S]'$$

$$\frac{0.1}{0.9} = \frac{0.9}{0.1} \times \frac{[S]}{[S]'}$$

$$\frac{[S]}{[S]'} = \left(\frac{0.1}{0.9}\right)^2 = \frac{0.01}{0.81}$$

enzyme, or it may bind to the enzyme–substrate complex and thus alter catalysis. Noncompetitive inhibition is frequently observed in the regulation of metabolic enzymes. The diagnostic property of this type of inhibition is that $K_m$ is unaffected, whereas $V_{max}$ decreases in the presence of increasing amounts of inhibitor (**FIGURE A1.14C**).

**MIXED INHIBITION** Mixed inhibition is characterized by effects on both $V_{max}$ (which decreases) and $K_m$ (which increases). Mixed inhibition is very common and results from the formation of a complex consisting of the enzyme, the substrate, and the inhibitor that does not break down to products.

## pH and temperature affect the rate of enzyme-catalyzed reactions

Enzyme catalysis is very sensitive to pH. This sensitivity is easily understood when one considers that the essential catalytic groups are usually ionizable ones (imidazole, carboxyl, amino) and that they are catalytically active in only one of their ionization states. For example, imidazole acting as a base will be functional only at pH values above 7. Plots of the rates of enzyme-catalyzed reactions versus pH are usually bell-shaped, corresponding to two sigmoidal

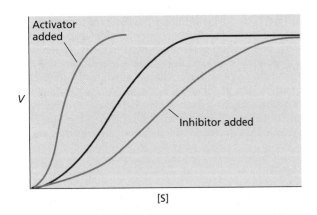

**FIGURE A1.16** Allosteric systems exhibit sigmoidal plots of rate versus substrate concentration. The addition of an activator shifts the curve to the left; the addition of an inhibitor shifts it to the right.

This calculation shows that reaction velocity is insensitive to small changes in substrate concentration. The same factor applies in the case of inhibitors and inhibition. In **cooperative systems**, on the other hand, a small change in one parameter, such as inhibitor concentration, brings about a *large change* in velocity. A consequence of a cooperative system is that the plot of $v$ versus [S] is no longer hyperbolic, but becomes *sigmoidal* (**FIGURE A1.16**). The advantage of cooperative systems is that a small change in the concentration of the critical effector (substrate, inhibitor, or activator) will bring about a large change in the rate. In other words, the system behaves like a switch.

Cooperativity is typically observed in allosteric enzymes that contain multiple active sites located on multiple subunits. Such oligomeric enzymes usually exist in two major conformational states, one active and one inactive (or relatively inactive). Binding of ligands (substrates, activators, or inhibitors) to the enzyme perturbs the position of the equilibrium between the two conformations. For example, an inhibitor will favor the inactive form; an activator will favor the active form. The cooperative aspect comes in as follows: A positive cooperative event is one in which binding of the first ligand makes binding of the next one easier. Similarly, negative cooperativity means that the second ligand will bind less readily than the first.

Cooperativity in substrate binding (homoallostery) occurs when the binding of substrate to a catalytic site on one subunit increases the substrate affinity of an identical catalytic site located on a different subunit. Effector ligands (inhibitors or activators), in contrast, bind to sites other than the catalytic site (heteroallostery). This relationship fits nicely with the fact that the end products of metabolic pathways, which frequently serve as feedback inhibitors, usually bear no structural resemblance to the substrates of the first step.

## The kinetics of some membrane transport processes can be described by the Michaelis–Menten equation

Membranes contain proteins that speed up the movement of specific ions or organic molecules across the lipid bilayer. Some membrane transport proteins are enzymes, such as ATPases, that use the energy from the hydrolysis of ATP to pump ions across the membrane. When these reactions run in the reverse direction, the ATPases of mitochondria and chloroplasts can synthesize ATP. Other types of membrane proteins function as carriers, binding their substrate on one side of the membrane and releasing it on the other side.

The kinetics of carrier-mediated transport can be described by the Michaelis–Menten equation in the same manner as the kinetics of enzyme-catalyzed reactions are (see Chapter 6). Instead of a biochemical reaction with a substrate and product, however, the carrier binds to the solute and transfers it from one side of a membrane to the other. Letting X be the solute, we can write the following equation:

$$X_{out} + carrier \rightarrow [X\text{-}carrier] \rightarrow X_{in} + carrier$$

Since the carrier can bind to the solute more rapidly than it can transport the solute to the other side of the membrane, solute transport exhibits saturation kinetics. That is, a concentration is reached beyond which adding more solute does not result in a more rapid rate of transport (**FIGURE A1.17**). $V_{max}$ is the maximum rate of transport of X across the membrane; $K_m$ is equivalent to the binding constant of the solute for the carrier. Like enzyme-catalyzed reactions, carrier-mediated transport requires a high degree of structural specificity of the protein. The actual transport of the solute across the membrane apparently involves conformational changes, also similar to those in enzyme-catalyzed reactions.

## Enzyme activity is often regulated

Cells can control the flux of metabolites by regulating the concentration of enzymes and their catalytic activity. By using allosteric activators or inhibitors, cells can modulate enzymatic activity and obtain very carefully controlled expression of catalysis.

**CONTROL OF ENZYME CONCENTRATION** The amount of enzyme in a cell is determined by the relative rates of synthesis and degradation of the enzyme. The rate of synthesis is regulated at the genetic level by a variety of mechanisms, which are discussed in greater detail in the last section of this chapter.

**COMPARTMENTALIZATION** Different enzymes or isozymes with different catalytic properties (e.g., substrate affinity) may be localized in different regions of the cell, such as mitochondria and cytosol. Similarly, enzymes associated

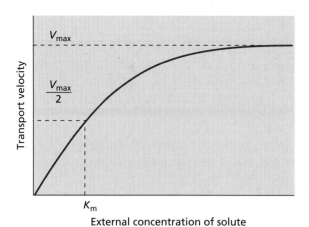

**FIGURE A1.17** The kinetics of carrier-mediated transport of a solute across a membrane are analogous to those of enzyme-catalyzed reactions. Thus, plots of transport velocity versus solute concentration are hyperbolic, becoming asymptotic to the maximal velocity at high solute concentration.

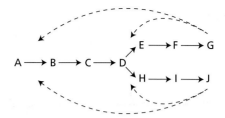

**FIGURE A1.18** Feedback inhibition in a hypothetical metabolic pathway. The letters (A–J) represent metabolites, and each arrow represents an enzyme-catalyzed reaction. The boldface arrow for the first reaction indicates that two different enzymes with different inhibitor susceptibilities are involved. Dashed lines indicate metabolites that inhibit particular enzymes. The first step in the metabolic pathway and the branch points are particularly important sites for feedback control.

with special tasks are often compartmentalized; for example, the enzymes involved in photosynthesis are found in chloroplasts. Vacuoles contain many hydrolytic enzymes, such as proteases, ribonucleases, glycosidases, and phosphatases, as well as peroxidases. The cell walls contain glycosidases and peroxidases. The mitochondria are the main location of the enzymes involved in oxidative phosphorylation and energy metabolism, including the enzymes of the tricarboxylic acid (TCA) cycle.

**COVALENT MODIFICATION** Control by covalent modification of enzymes is common and usually involves their phosphorylation or adenylylation*, such that the phosphorylated form, for example, is active and the nonphosphorylated form is inactive. These control mechanisms are normally energy dependent and usually involve ATP.

Proteases are normally synthesized as inactive precursors known as zymogens or proenzymes. For example, papain is synthesized as an inactive precursor called propapain and becomes activated later by cleavage (hydrolysis) of a peptide bond. This type of covalent modification avoids premature proteolytic degradation of cellular constituents by the newly synthesized enzyme.

---

* Although some texts refer to the conjugation of a compound with adenylic acid (AMP) as "adenylation," the chemically correct term is *"adenylylation."*

**FEEDBACK INHIBITION** Consider a typical metabolic pathway with two or more end products such as that shown in **FIGURE A1.18**. Control of the system requires that if the end products build up too much, their rate of formation is decreased. Similarly, if too much reactant A builds up, the rate of conversion of A to products should be increased. The process is usually regulated by control of the flux at the first step of the pathway and at each branch point. The final products, G and J, which might bear no resemblance to the substrate A, inhibit the enzymes at A $\varnothing$ B and at the branch point.

By having two enzymes at A $\varnothing$ B, each inhibited by one of the end metabolites but not by the other, it is possible to exert finer control than with just one enzyme. The first step in a metabolic pathway is usually called the *committed step*. At this step enzymes are subject to major control.

Fructose 2,6-bisphosphate plays a central role in the regulation of carbon metabolism in plants. It functions as an activator in glycolysis (the breakdown of sugars to generate energy) and an inhibitor in gluconeogenesis (the synthesis of sugars). Fructose 2,6-bisphosphate is synthesized from fructose 6-phosphate in a reaction requiring ATP and catalyzed by the enzyme fructose-6-phosphate 2-kinase. It is degraded in the reverse reaction catalyzed by fructose-2,6-bisphosphatase, which releases inorganic phosphate ($P_i$). Both of these enzymes are subject to metabolic control by fructose 2,6-bisphosphate, as well as ATP, $P_i$, fructose 6-phosphate, dihydroxyacetone phosphate, and 3-phosphoglycerate. The role of fructose 2,6-bisphosphate in plant metabolism is discussed further in Chapters 8 and 11.

# REFERENCES

Atkinson, D. E. (1977) *Cellular Energy Metabolism and Its Regulation.* Academic Press, New York.

Bryant, F. O., and Adams, M.W.W. (1989) Characterization of hydrogenase from the hyperthermophilic archaebacterium, *Pyrococcus furiosus. J. Biol. Chem.* 264: 5070–5079.

Clausius, R. (1879) *The Mechanical Theory of Heat.* Tr. by Walter R. Browne. Macmillan, London.

Creighton, T. E. (1983) *Proteins: Structures and Molecular Principles.* W. H. Freeman, New York.

Donahue, R. A., Davis, T. D., Michler, C. H., Riemenschneider, D. E., Carter, D. R., Marquardt, P. E., Sankhla, N., Sahkhla, D., Haissig, B. E., and Isebrands, J. G. (1995) Growth, photosynthesis, and herbicide tolerance of genetically modified hybrid poplar. *Can. J. For. Res.* 24: 2377–2383.

Edsall, J. T., and Gutfreund, H. (1983) *Biothermodynamics: The Study of Biochemical Processes at Equilibrium.* Wiley, New York.

Klotz, I. M. (1967) *Energy Changes in Biochemical Reactions.* Academic Press, New York.

Mathews, C. K., and Van Holde, K. E. (1996) *Biochemistry,* 2nd ed. Benjamin/Cummings, Menlo Park, CA.

Morowitz, H. J. (1978) *Foundations of Bioenergetics.* Academic Press, New York.

Nicholls, D. G., and Ferguson, S. J. (1992) *Bioenergetics 2.* Academic Press, San Diego.

Stryer, L. (1995) *Biochemistry,* 4th ed. W. H. Freeman, New York.

# APPENDIX 2

# The Analysis of Plant Growth

To gain a more complete understanding of plant growth and development, it is first essential to obtain clear and concise descriptions of changes that occur over time. Classically, plant growth has been analyzed in terms of cell number or overall size (i.e., mass). However, these measures tell only part of the story.

Growth in plants is defined as an irreversible increase in volume. The largest component of plant growth is cell expansion driven by turgor pressure. During this process, cells increase in volume many fold and become highly vacuolate. However, size is only one criterion that may be used to measure growth.

Growth also can be measured in terms of change in fresh weight—that is, the weight of the living tissue—over a particular period of time. However, the fresh weight of plants growing in soil fluctuates in response to changes in water status, so this criterion may be a poor indicator of actual growth. In these situations, measurements of dry weight are often more appropriate.

Cell number is a common and convenient parameter by which to measure the growth of unicellular organisms, such as the green alga *Chlamydomonas* (FIGURE A2.1). In multicellular plants, however, cell number can be a misleading growth measurement because cells can divide without increasing in volume. For example, during the early stages of embryogenesis, the zygote subdivides into progressively smaller cells with no net increase in the size of the embryo. Only after it reaches the eight-cell stage does the increase in volume begin to parallel the increase in cell number. Another example of the lack

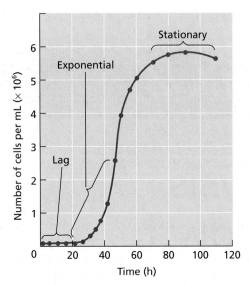

**FIGURE A2.1** Growth of the unicellular green alga *Chlamydomonas*. Growth is assessed by a count of the number of cells per milliliter at increasing times after the cells are placed in fresh growth medium. Temperature, light, and nutrients provided are optimal for growth. An initial lag period during which cells may synthesize enzymes required for rapid growth is followed by a period in which cell number increases exponentially. This period of rapid growth is followed by a period of slowing growth in which the cell number increases linearly. Then comes the stationary phase, in which the cell number remains constant or even declines as nutrients are exhausted from the medium.

of correlation between cell number and growth occurs under the influence of environmental stress, which typically affects cell division and cell elongation differentially.

In this brief overview, we will discuss both the classical definitions of growth and a more recent approach, termed **kinematics**, which views growth and the related problem of cell expansion in terms of motions of cells or "tissue elements." As we will see, the advantage of the kinematic approach is that it allows one to describe the growth patterns of organs mathematically in terms of the expansion patterns of their component cells. These quantitative and localized descriptions of growth provide a useful perspective from which to develop models for underlying mechanisms that control growth.

## Growth Rates and Growth Curves

As we have seen, growth can be defined as an increase in volume or mass or cell number. Several kinds of growth rates are used in plant physiology. The absolute growth rate ($g$) is the time ($t$) rate of change in size ($s$)

$$g = \frac{\Delta s}{\Delta t}$$

The relative growth rate ($r$) is the absolute growth rate divided by the size

$$r = \frac{1}{s}\frac{\Delta s}{\Delta t}$$

If $\Delta s$ is small relative to s,

$$r \approx \frac{\Delta \ln s}{\Delta t}$$

"Growth curves" are data on size or weight or dry weight ("biomass") versus time. The slope of the growth curve represents the growth rate at an instant in time. For building models, or characterizing seasonal patterns, simple global functions such as a straight line, an exponential curve, or a logistic (S-shaped) curve are often fit to growth curves. Growth rates can be obtained by analytic differentiation of the global fits. However, physiologists are usually concerned with mechanistic explanations. Local fits and local derivatives evaluated over short distances along the growth curve are often valuable in physiological work.

### Cell division and cell expansion are independent processes that are often synchronized during development

For multicellular plants, many aspects of growth relate to newly formed cells produced by meristematic tissues. Growth associated with these cells is neither uniform nor random. For example, newly formed cells in apical meristems first enlarge slowly, but later expand more rapidly as they are displaced into sub-apical regions. The resulting increase in cell volume can range from several-fold to one hundred-fold, depending on the species and environmental conditions. The predictable and site-specific ways these derivatives expand is closely linked to the final size and shape of the primary plant body. In essence, the total growth of the plant can be thought of as the sum of the local patterns of cell division and expansion.

In many plant axes, progressively formed cells go through similar patterns of expansion during their displacement through the meristem into adjacent growth zones. In such cases, the final axis length is related to the number of cells produced. In some cases, however, cell division is not closely correlated to organ growth or growth rate. Cell division, the partitioning of an existing cell into two initially smaller cells, must be considered as physically independent of cell expansion. The two processes may be either synchronized or uncoupled. Also the two processes can be affected differently by environmental stress. Kinematic analysis (see below) allows us to see the cell size profile as the result of the concurrent processes of cell division and expansion. Recent studies have combined kinematic analysis, flow cytometry, and microarray analysis to characterize cell cycle regulation during the growth process of leaves and roots of Arabidopsis (Beemster et al. 2005). Genome-wide microarray analysis allowed cell cycle genes to be categorized into three major classes: constitutively expressed, proliferative, and inhibitory. Comparison with published expression data corresponding to similar developmental stages in other growth zones and from synchronized cell cultures supported this categorization and enabled identification of over 100 proliferation genes. Other genes independently regulate the expansion process (e.g., Zhu et al. 2007).

### The production and fate of meristematic cells is comparable to a fountain

Moving fluids such as waterfalls, fountains, and the wakes of boats can generate specific forms. The study of the motion of fluid particles and the shape changes that the fluids undergo is called **kinematics**. The ideas and numerical methods used to study these fluid forms are useful for characterizing plant growth. In both cases, an unchanging form is produced even though it is composed of moving and changing elements.

An example of an unchanging form composed of changing and displaced elements in plants is the hypocotyl hook of a dicot, such as the common bean (**FIGURE A2.2**). As the bean seedling emerges from the seed coat, the apical end of the hypocotyl grows in a curved form, a hook. The hook is thought to protect the seedling apex from damage during growth through the soil. During seedling growth (in soil or dim light) the hook migrates up the stem, from the hypocotyl into the epicotyl and then to the first and second internodes, but the form of the hook remains constant.

If we mark a specific epidermal cell on the seedling stem located close to the seedling apex, we can watch it as it flows into the hook summit, then down into the straight region below the hook (see Figure A2.2). The mark is not crawling over the plant surface, of course; plant cells are

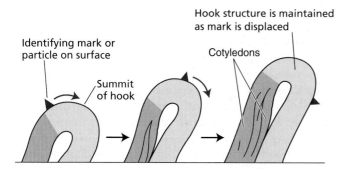

**FIGURE A2.2** The dicot hypocotyl hook is an example of a constant form composed of changing elements. The hooked form is maintained over time, while different tissues first curve and then straighten as they are displaced from the seedling apex during growth. If a mark is placed at a fixed point on the surface, it will be displaced (indicated by the arrow), appearing to flow through the hook over time. (After Silk 1984.)

cemented together and do not experience much relative motion during development. The change in position of the mark relative to the hook implies that the hook is composed of a procession of tissue elements, each of which first curves and then straightens as it is displaced from the plant apex during growth. The steady form is produced by a parade of changing cells.

A root tip is another example of a steady form composed of changing tissue elements. Here, too, the form is observed to be steady only when distance is measured from the root tip. A region of cell division occupies perhaps 2 mm of the root tip. The elongation zone extends for about 10 mm behind the root tip. Phloem differentiation is first observed beginning at 3 mm from the tip, and functional xylem elements may be seen at about 12 mm from the tip. A marked cell near the tip will seem to flow first through the region of cell division, then through the elongation zone and into the region of xylem differentiation, and so on. This shifting implies that developing tissue elements first divide and elongate, and then differentiate.

In an analogous fashion, the shoot bears a succession of leaves of different developmental stages. During a period of 24 hours, a leaf may grow to the same size, shape, and biochemical composition that its neighbor had a day earlier. Thus, shoot form is also produced by a parade of changing elements that can be analyzed with kinematics. Such an analysis is not merely descriptive; it permits calculations of the growth and biosynthetic rates of individual tissue elements (cells) within a dynamic structure.

### Tissue elements are displaced during expansion

As we have seen, growth in shoots and roots is localized in regions at the tips of these organs. Regions with expanding tissue are called **growth zones**. With time, the organ tips, bearing meristems, move away from the plant base by the growth of the cells in the growth zone.

If successive marks are placed on the stem or root, the distance between the marks will change depending on where they are within the growth zone. In addition, all of these marks will move away from the tip of the root or shoot, and their rate of movement will differ depending on their distance from the tip.

From another perspective, if you were to stand at the tip of a root that had marks placed at intervals along the axis, you would see that all marks would move farther away from you with time. The reason is that discrete tissue elements on the plant axis experience displacement as well as expansion during growth and development.

### The growth trajectory is a cell-specific growth description that relates cell position to time

One way to characterize the growth pattern along an axis mathematically is to plot the distance of a tissue element

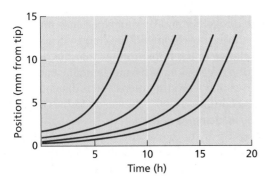

**FIGURE A2.3** Growth trajectories of *Zea mays* (maize):Growth trajectories plot the positions of particles spaced along the root axis as a function of time. Note that distance is measured from the root tip. This is termed a material specification of growth, or a cell- particle- specific description, because real or material particles are followed.

from the apex versus time. The resulting curve is called the growth trajectory. With the apex as the reference point, the growth trajectory is concave. Marked tissue elements at different locations may be followed simultaneously to generate a family of growth trajectories (**FIGURE A2.3**).

The growth trajectory provides a way to convert spatial information into a developmental time course. For example, if the amount of suberin in the root increases with position, the growth trajectory will show us the time required for the observed accumulation of suberin in cells moving between any two spatial points.

## As cells move away from the apex, their displacement rate increases

Looking at the growth trajectories of Figure A2.3, we see that as a cell of the plant axis moves away from the apex, its growth velocity increases (i.e., the cell accelerates) until a constant limiting velocity is reached equal to the overall organ extension rate. The reason for this increase in growth velocity is that with time, progressively more tissue is located between the moving particle and the apex, and progressively more cells are expanding, so the particle is displaced more and more rapidly. In a rapidly growing maize root, a tissue element takes about 8 hours to move from 2 mm (the end of the meristem) to 12 mm (the end of the elongation zone). The root tissue element accelerates through this region (see Figure A2.3 and **FIGURE A2.4A**).

Beyond the growth zone, elements do not separate; neighboring elements have the same velocity, and the rate at which particles are displaced from the tip is the same as the rate at which the tip is pushed away from the soil surface. The root tip of maize is pushed through the soil at 3 mm h$^{-1}$. This is also the rate at which the non-growing region recedes from the apex, and it is equal to the final slope of the growth trajectory.

## The growth velocity and the relative elemental growth rate profiles are spatial descriptions of growth

The slope of the growth trajectory (see Figure A2.3) at any given point is equivalent to the growth velocity at that region of the axis. The velocities of different tissue elements are plotted against their distance from the apex to give the spatial pattern of growth velocity, or **growth velocity profile** (see Figure A2.4A). Figure A2.4A confirms that growth velocity increases with position in the growth zone. A constant value is obtained at the base of the growth zone. The final growth velocity is the final, constant slope of the growth trajectory equal to the elongation rate of the organ, as discussed in the previous section. In the rapidly growing maize root, the growth velocity is 1 mm h$^{-1}$ at 4 mm, and it reaches its final value of nearly 3 mm h$^{-1}$ at 12 mm.

To characterize the local expansion rate, we must consider the growth velocities at the apical and basal ends of a small segment of tissue. If the apical end of the segment is moving faster than the basal end, the segment is elongating. The velocity difference, divided by the segment length, gives the relative growth rate of the segment. Now, imagine that the segment shrinks to a point at location $x$. To find the relative growth rate, $r$, at point $x$, we can use calculus: The value of $r$ is given by the velocity gradient, $dv/dx$, where $v$ is the growth velocity. This measure of the local growth rate is called the relative elemental growth rate (Erickson and Sax 1956) or REGR. It represents the fractional change in length per unit time and has units of h$^{-1}$. If the growth velocity is known, the relative elemental growth rate can be evaluated by differentiation of the growth velocity with respect to position (**FIGURE A2.4B**). Or we can approximate $r$ by measuring the relative growth rates of closely spaced segments of initial length $L$. For each segment, $r = (1/L)(dL/dt)$. The relative elemental growth rate shows the location and magnitude of the extension rate and can be used to quantify the effects of environmental variation on the growth pattern.

The relative elemental growth rate profile is similar to growth localizations obtained from discrete marking experiments used to find percent size change per hour over the surface of an expanding organ. However, because marked tissue elements are constantly moving, care must be used to find appropriate time and space intervals for a marking experiment to localize growth. In contemporary research computer-assisted methods, including automated particle tracking, are the basis for growth analyses with high resolution in space and time. Recent work includes quantitative descriptions of effects of environmental variation on growth rate patterns, and molecular studies to find the genes and proteins responsible for the responses to environmental stress (Zhu et al. 2007; Walter et al. 2009).

(A) Growth velocity profile

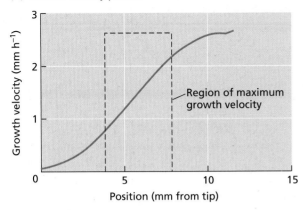

(B) Relative elemental growth rate

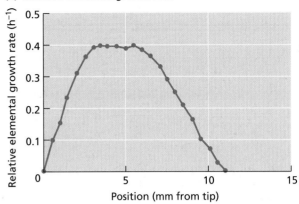

**FIGURE A2.4**   The growth of the primary root of *Zea mays* (maize) can be represented kinematically by two related growth curves. (A) The growth velocity profile plots the velocity of movement away from the tip of points at different distances from the tip. This tells us that growth velocity increases with distance from the tip until it reaches a uniform velocity equal to the rate of elongation of the root. (B) The relative elemental growth rate tells us the rate of expansion of any particular point on the root. It is the most useful measure for the physiologist because it tells us where the most rapidly expanding regions are located. (Data from W. Silk.)

# REFERENCES

Beemster, G.T.S., De Veylder, L., Vercruysse, S., West, G., Rombaut, D., Van Hummelen, P., Galichet, A., Gruissem, W., Inze, D., and Vuylsteke, M. (2005) Genome-wide analysis of gene expression profiles associated with cell cycle transitions in growing organs of Arabidopsis. *Plant Physiol.* 138: 734–743.

Erickson, R. O., and Sax, K. B. (1956) Elementary growth rate of the primary root of *Zea mays*. *Proc. Am. Philos. Soc.* 100: 487–498.

Silk, W. K. (1984) Quantitative descriptions of development. *Annu. Rev. Plant Physiol.* 35: 479–518.

Walter, A., Silk, W. K., and Schurr, U. (2009) Environmental effects on spatial and temporal patterns of root and leaf growth. *Annu. Rev. Plant Biol.* 60: 279–304.

Zhu, J. M., Alvarez, S., Marsh, E. L., LeNoble, M. E., Cho, I. J., Sivaguru, M., Chen, S. X., Nguyen, H. T., Wu, Y. J., Schachtman, D. P., and Sharp, R. E. (2007) Cell wall proteome in the maize primary root elongation zone. II. Region-specific changes in water soluble and lightly ionically bound proteins under water deficit. *Plant Physiol.* 145: 1533–154.

# APPENDIX 3

# Hormone Biosynthetic Pathways

Despite their diverse chemical structures, most of the known plant hormones are all derived from three main types of metabolic precursors: amino acids, isoprenoid compounds, and lipids (**FIGURE A3.1**). The amino acids tryptophan and methionine serve as precursors for IAA (indole-3-acetic acid) and ethylene, respectively. The isoprenoid pathway gives rise to five classes of plant hormones: cytokinins, brassinosteroids, gibberellins, abscisic acid, and strigolactones. And finally, jasmonic acid is synthesized from a lipid precursor.

The major intermediates and subcellular locations of most of the biosynthetic pathways have now been elucidated, although new details continue to emerge. Of particular interest to researchers is how the individual hormone biosynthetic pathways are regulated and how they interact with one another; such regulation and interactions have profound effects on plant development.

Here we provide some additional details about the biosynthetic pathways beyond the treatments in the main chapters, for those students who wish to further their understanding of plant hormone biochemistry. For some of the hormones, such as auxin and ethylene, the biosynthetic pathways and their regulation are discussed in greater depth than in the main chapters, while in others, such as gibberellin, ABA, and brassinolide, diagrams of more complete versions of the pathways are presented.

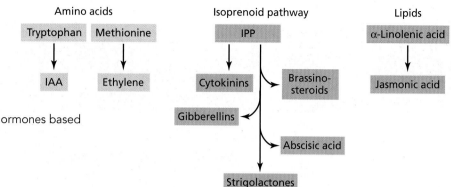

**FIGURE A3.1** Three categories of hormones based on their biosynthetic precursors.

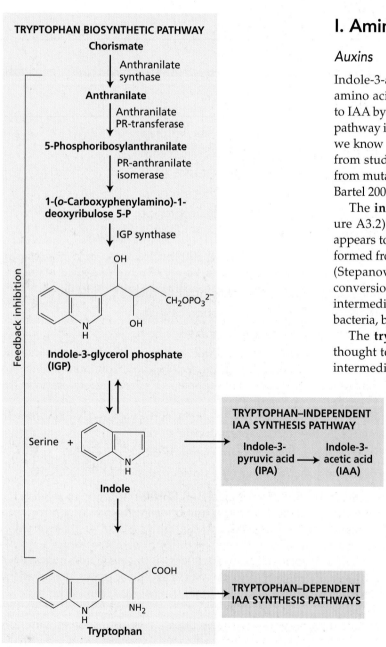

# I. Amino Acid Pathways

## Auxins

Indole-3-acetic acid (IAA) is structurally related to the amino acid tryptophan, and plants convert tryptophan to IAA by several pathways. The tryptophan biosynthetic pathway in plants is shown in **FIGURE A3.2**. Much of what we know about the biosynthesis of IAA has been learned from studies of microbial pathways and, more recently, from mutational analyses in Arabidopsis (Woodward and Bartel 2005; Normanly 2010).

The **indole-3-pyruvic acid** (**IPA**) pathway (see Figure A3.2) was initially characterized in microbes, and appears to function in most plants. In Arabidopsis, IPA is formed from tryptophan by tryptophan aminotransferase (Stepanova et al. 2008; Tao et al. 2008). The enzymatic conversion of IPA via an indole-3-acetaldehyde (IAAld) intermediate is assumed to be similar to what is seen in bacteria, but has not been directly demonstrated.

The **tryptamine** (**TRM**) **pathway** (**FIGURE A3.3**) is thought to produce indole-3-acetaldoxime (IAOx) as an intermediate in IAA biosynthesis. IAOx is then used to produce IAA by two pathways that generate indole-3-acetonitrile (IAN) as an intermediate and one pathway that generates indole-3-acetamide (IAM) as an intermediate. Species that do not utilize the IPA pathway possess the TRM pathway. There is evidence that both the IPA and the TRM

**FIGURE A3.2** The tryptophan biosynthetic pathway provides precursors for IAA biosynthesis. In most plants, tryptophan synthesis takes place in the chloroplast. The branchpoint precursor for tryptophan-independent IAA biosynthesis is indole. Indole pyruvic acid is thought to be an intermediate in the pathway.

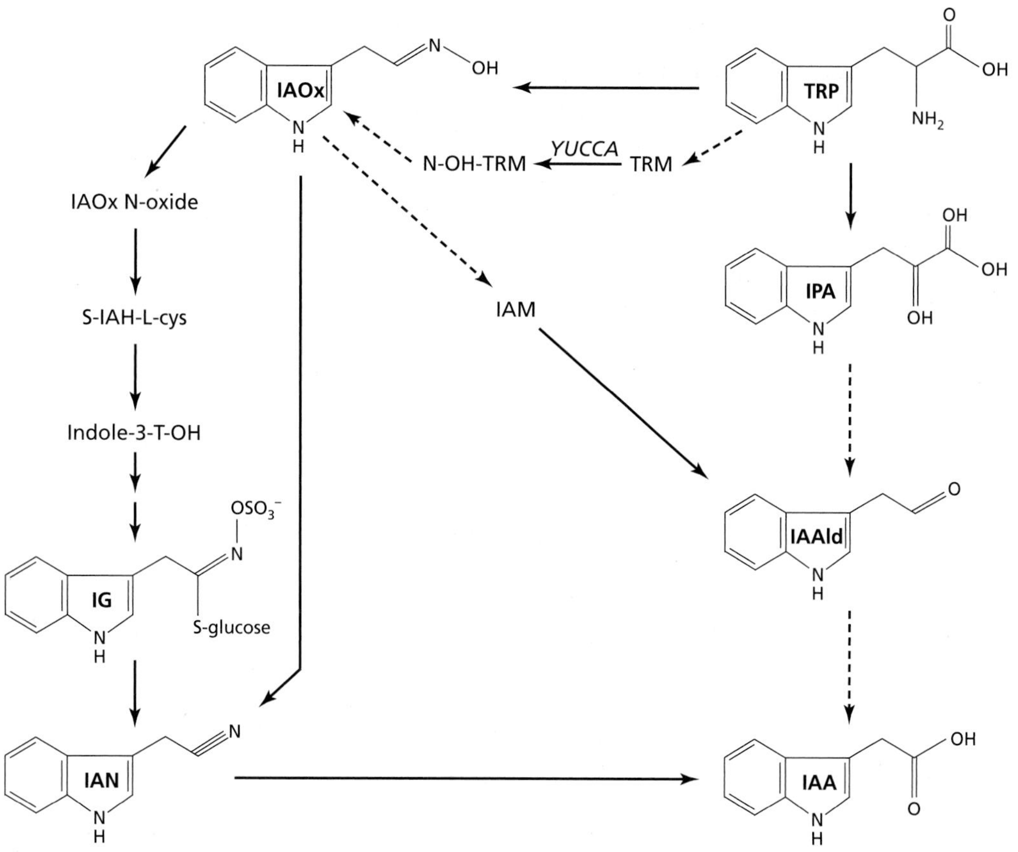

**FIGURE A3.3** Tryptophan-dependent pathways of IAA biosynthesis in Arabidopsis. Dashed arrows indicate that neither a gene nor enzyme activity has been identified in Arabidopsis. TRP, tryptophan; IAM, indole-3-acetamide; IPA, indole-3-pyruvic acid; IAAld, indole-3-acetaldehyde; IAOx, indole-3-acetaldoxime; S-IAH-L-cys, S-(indolylacetohydroximoyl)-L-cysteine; indole-3-T-OH, indole-3-thiohydroximate; IG, indole-3-methylglucosinolate; TRM, tryptamine; IAN, indole-3-acetonitrile. (After Normanly 2010.)

pathways exist in tomato (Nonhebel et al. 1993). However, all recent evidence suggests that the TRM/YUCCA pathway is specific to plants that produce indole glucosinolate defense compounds (Sugawara et al. 2009).

When IAM is produced as an intermediate, it is converted to IAAld which, in turn, is used to synthesize IAA as in the IPA pathway. In plants that produce indole glucosinolates, a well-characterized sequence of biosynthetic events converts IAOx to IAN before its conversion by nitrilase to IAA. If a second IAN-dependent pathway is present in plant species that do not produce indole glucosinolates, it would involve other enzymes.

A tryptophan-independent pathway of IAA synthesis using indole as a precursor has been described in multiple plant species using precursors labeled with stable isotopes. IAN and IPA are possible intermediates. However, little is known about the enzymatic processes involved.

## Ethylene

In vivo experiments have shown that plant tissues convert l-[$^{14}$C]methionine to [$^{14}$C]ethylene, and that the ethylene is derived from carbons 3 and 4 of methionine (FIGURE A3.4). The $CH_3$—S group of methionine is recycled via the Yang cycle. Without this recycling, the amount of reduced sulfur present would limit the available methionine and the synthesis of ethylene. S-adenosylmethionine (AdoMet), which is synthesized from methionine and adenosine triphosphate (ATP), is an intermediate in the ethylene biosynthetic pathway, and the immediate precursor of ethylene is **1-aminocyclopropane-1-carboxylic acid** (**ACC**) (see Figure A3.4).

The role of ACC became evident in experiments in which plants were treated with [$^{14}$C]methionine. Under anaerobic conditions, ethylene was not produced from the [$^{14}$C]methionine, and labeled ACC accumulated in the tissue. On exposure to oxygen, however, ethylene production surged. The labeled ACC was rapidly converted to ethylene in the presence of oxygen by various plant tissues, suggesting that ACC is the immediate precursor of ethylene in higher plants and that oxygen is required for the conversion.

In general, when ACC is supplied exogenously to plant tissues, ethylene production increases substantially. This observation indicates that the synthesis of ACC is usually

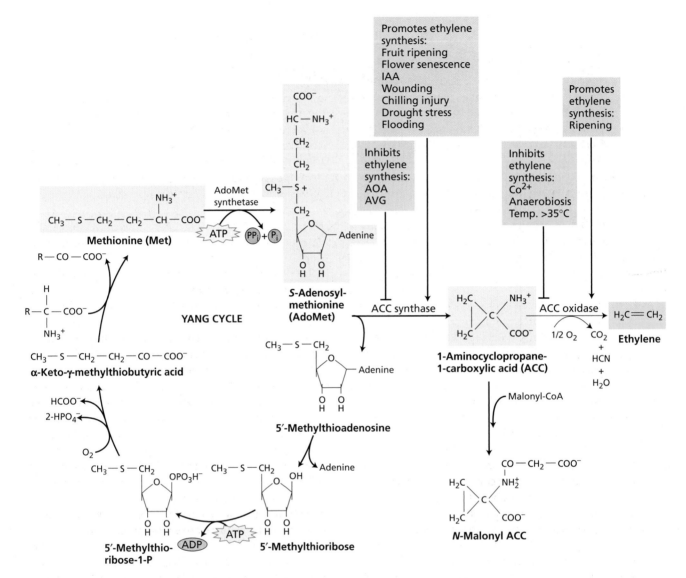

**FIGURE A3.4** Ethylene biosynthetic pathway and the Yang cycle. The amino acid methionine is the precursor of ethylene. The rate-limiting step in the pathway is the conversion of AdoMet to ACC, which is catalyzed by the enzyme ACC synthase. The last step in the pathway, the conversion of ACC to ethylene, requires oxygen and is catalyzed by the enzyme ACC oxidase. The CH₃—S group of methionine is recycled via the Yang cycle and thus conserved for continued synthesis. Besides being converted to ethylene, ACC can be conjugated to N-malonyl ACC. AOA, aminooxyacetic acid; AVG, aminoethoxy-vinylglycine. (After McKeon et al. 1995.)

the biosynthetic step that limits ethylene production in plant tissues.

**ACC synthase (ACS)**, the enzyme that catalyzes the conversion of AdoMet to ACC (see Figure A3.4), has been characterized in many types of tissues of various plants. ACC synthase is an unstable, cytosolic enzyme. Its level is regulated by environmental and internal factors, such as wounding, drought stress, flooding, and auxin. Because ACC synthase is present in such low amounts in plant tissues (0.0001% of the total protein of ripe tomato) and is

very unstable, it is difficult to purify the enzyme for biochemical analysis (see **WEB TOPIC 22.4**).

ACC synthase is encoded by members of a divergent multigene family that are differentially regulated by various inducers of ethylene biosynthesis. In tomato, for example, there are at least ten ACC synthase genes, different subsets of which are induced by auxin, wounding, and/or fruit ripening; the Arabidopsis genome contains nine ACC synthase genes. An analysis of the purified Arabidopsis ACC synthase proteins revealed a diversity of kinetic properties (for example, various affinities for the substrate AdoMet), suggesting that these eight isoforms might be optimized for different roles in various tissues and cell types (Yamagani et al. 2003). The deduced crystal structure of both apple and tomato ACC synthase reveals a dimeric protein with shared active sites, similar to aminotransferases (Capitani et al. 1999; Huai et al. 2001).

**ACC oxidase** catalyzes the last step in ethylene biosynthesis: the conversion of ACC to ethylene (see Figure A3.4). In tissues that show high rates of ethylene production, such as ripening fruit, ACC oxidase activity can be the rate-limiting step in ethylene biosynthesis. The gene that encodes ACC oxidase has been cloned (see **WEB TOPIC 22.5**). Like ACC synthase, ACC oxidase is encoded by a multigene family that is differentially regulated. For example, in ripening tomato fruits and senescing petunia flowers, the mRNA levels of a subset of ACC oxidase genes are highly elevated.

The deduced amino acid sequences of ACC oxidases revealed that these enzymes belong to the $Fe^{2+}$/ascorbate oxidase superfamily. Their similarity suggested that ACC oxidase might require $Fe^{2+}$ and ascorbate for activity—a requirement that has been confirmed by biochemical analysis of the protein. The requirement of ACC oxidase for cofactors presumably explains why the purification of this enzyme eluded researchers for so many years.

Researchers have studied the catabolism of ethylene by supplying $^{14}C_2H_4$ to plant tissues and tracing the radioactive compounds produced. Carbon dioxide, ethylene oxide, ethylene glycol, and the glucose conjugate of ethylene glycol have been identified as metabolic breakdown products. However, because certain cyclic olefin compounds, such as 1,4-cyclohexadiene, have been shown to block ethylene breakdown without inhibiting ethylene action, ethylene catabolism does not appear to play a significant role in regulating the level of the hormone (Raskin and Beyer 1989).

Not all the ACC found in the tissue is converted to ethylene. ACC can also be converted to a conjugated form, *N*-malonyl ACC (see Figure A3.4), which does not appear to break down and accumulates in the tissue, primarily in the vacuole. A second, minor conjugated form of ACC, 1-(γ-L-glutamylamino)cyclopropane-1-carboxylic acid (GACC), has also been identified. The conjugation of ACC may play an important role in the control of ethylene biosynthesis, in a manner analogous to the conjugation of auxin and cytokinin.

A number of bacteria present in the soil express an enzyme called ACC deaminase that hydrolyzes ACC to ammonia and α-ketobutyrate (Glick, 2005). These bacteria can promote plant growth by sequestering and cleaving ACC made and excreted by plants, thereby lowering the level of ethylene to which the plants are exposed. ACC deaminase has also been expressed in transgenic plants to lower the level of ethylene produced.

## II. Isoprenoid Pathways

### Cytokinins

As described earlier in Chapter 21, the first committed step in cytokinin biosynthesis is the transfer of the isopentenyl group to an adenosine moiety (**FIGURE A3.5**).

Cytokinins are present in the tRNAs of most cells, including plant and animal cells. The tRNA cytokinins are synthesized by modification of specific adenine residues within the fully transcribed tRNA. As with the free cytokinins, isopentenyl groups are transferred to the adenine molecules from DMAPP by an enzyme called tRNA-IPT. The genes for tRNA-IPT have been cloned from many species.

The possibility that free cytokinins are derived from tRNA has been explored extensively. Although the tRNA-bound cytokinins can act as hormonal signals for plant cells if the tRNA is degraded and fed back to the cells, it is unclear if a significant amount of the free hormonal cytokinin in plants is derived from the turnover of tRNA.

When the Arabidopsis genome was searched for potential *ipt*-like sequences, nine different *IPT* genes were identified—more than are present in animal genomes, which usually contain only one or two *ipt*-like genes used in tRNA modification (Kakimoto 2001; Takei et al. 2001a).

Phylogenetic analysis revealed that two of the Arabidopsis *IPT* genes resemble tRNA-IPTs found in animals and bacteria, and the other seven form a distinct group or clade. The grouping of the seven Arabidopsis *IPT* genes in this unique plant clade provided a clue that these genes may encode the cytokinin biosynthetic enzyme.

The proteins encoded by these genes were expressed in *E. coli* and analyzed. It was found that with the exception of the genes most closely related to the tRNA-*IPT* genes, these genes encoded proteins capable of synthesizing free cytokinins. Unlike *Agrobacterium* Ipt (note that the names for bacterial proteins are capitalized without italics), however, the Arabidopsis enzymes utilize adenosine triphosphate (ATP) and adenosine diphosphate (ADP) preferentially over AMP, and use DMAPP as the source of the side chain rather than HMBDP (see Figure A3.5).

The MEP pathway is the primary source of the DMAPP used in cytokinin biosynthesis by plant IPT enzymes. This pathway occurs in plastids, which is where the majority of IPT enzymes are localized in plants. Thus, the primary site of cytokinin biosynthesis in plants is in the plastids. However, in Arabidopsis, at least one IPT protein, IPT3, can be modified by the addition of a farnesyl moiety (Galichet et al. 2008). Farnesylation is the addition of a hydrophobic farnesyl group (a long chain lipid molecule made from DMAPP subunits) to C-terminal cysteine(s) of the target protein, which can alter its subcellular localization, often

**FIGURE A3.5**   Simplified biosynthetic pathway for cytokinin biosynthesis. The first committed step in cytokinin biosynthesis is the addition of the isopentenyl side chain from DMAPP (dimethylallyl diphosphate) to an adenosine moiety. The products of this reaction (iPRDP or iPRTP) are converted to zeatin by a cytochrome P450 monooxygenase (CPY735A). Dihydrozeatin (DHZ) cytokinins are made from the various forms of *trans*-zeatin by an unknown enzyme (not shown). The ribotide and riboside forms of *trans*-zeatin can be interconverted, and free *trans*-zeatin can be formed from the riboside by enzymes of the general purine metabolism. In addition, the LONELY GUY (LOG) enzyme can convert zeatin ribotide (but only the monophosphate) directly to the free base form. Note that iP and DHZ ribotides can also be converted to the corresponding ribosides and free base forms in a similar manner (not shown). *Inset:* The pathway for cytokinin biosynthesis via *Agrobacterium* Ipt. The plant and bacterial Ipt enzymes differ in the adenosine substrate used and the side chain donor; the plant enzyme appears to utilize both ADP and ATP and couples this to DMPP, and the bacterial enzyme utilizes AMP and couples this to HMBDP (1-hydroxy-2-methyl-2-(E)-butenyl 4-diphosphate). Note that the product of the *Agrobacterium* Ipt reaction is a zeatin ribotide.

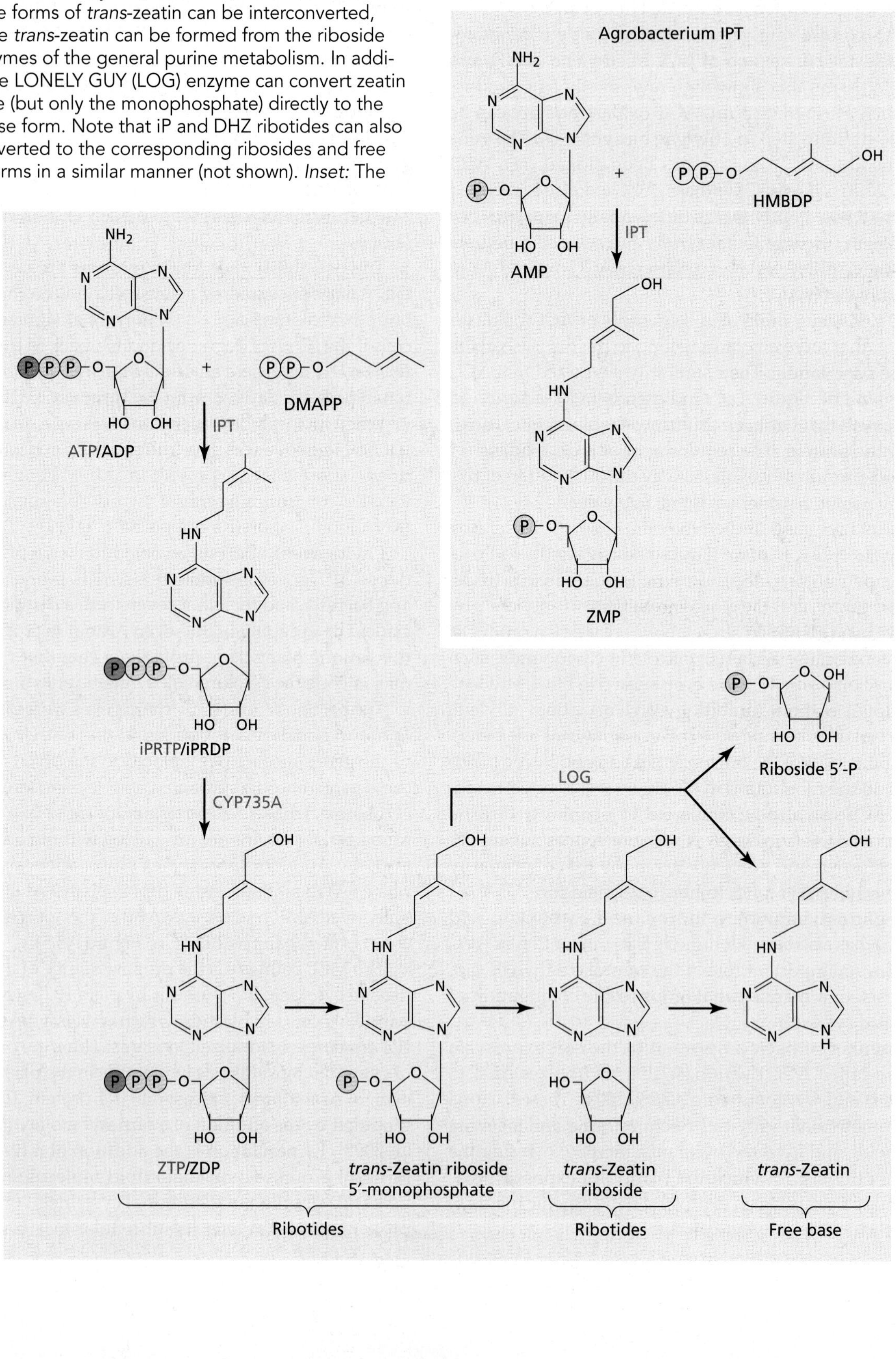

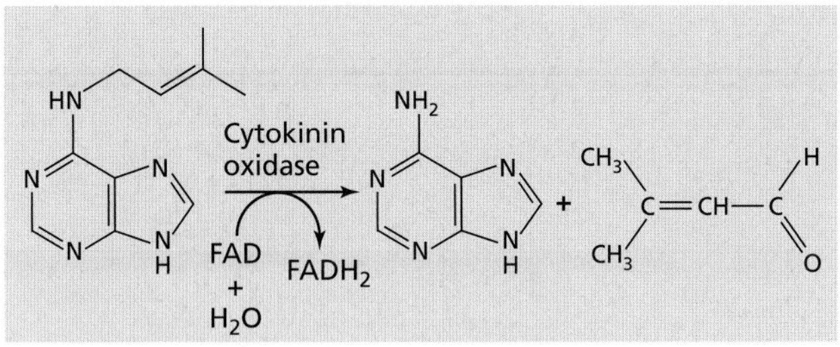

**FIGURE A3.6** Cytokinin oxidase irreversibly degrades some cytokinins.

targeting the protein to a membrane. The farnesylation of IPT3 directs it to the nucleus, rather than to the plastids, where the non-farnesylated IPT3 protein is localized. Furthermore, the IPT4 protein is found in the cytoplasm, and the IPT7 protein is found in mitochondria. Thus, although the plastids are the primary sites of cytokinin biosynthesis, they are not the only sites.

The expression pattern of the *IPT* genes indicates multiple sources of cytokinin biosynthesis throughout the plant (Miyawaki et al. 2004). The expression of a subset of *IPT* genes is down-regulated by cytokinin, indicating that cytokinin exerts a negative feedback control on its own biosynthesis. *IPT* gene expression is also affected by other regulatory inputs, including auxin, nitrate, and the meristem identity gene *SHOOT-MERISTEMLESS* (*STM*).

The immediate products of the IPT reaction are iP-ribotides. The isoprene side chain of iP ribotides is subsequently trans-hydroxylated by the P450 monooxygenases CYP735A1 and CYP735A2 to yield zeatin ribotides (Takei et al. 2004). Cytokinin nucleotides can be converted to their most active free base forms via dephosphorylation and deribosylation. Such interconversions may involve enzymes common to purine metabolism. In addition to these enzymes, the monophosphate forms of cytokinin ribotides can be directly converted to the free base forms by the phosphoribohydro-

lase LONELY GUY (LOG) (see Figure A3.5), which was identified in a rice mutant that displayed shoot meristem defects (Kurakawa et al. 2007). LOG expression is localized to the tip of shoot meristems and it likely fine tunes the spatial distribution of bioactive cytokinins to regulate meristem activity.

As discussed in Chapter 21, zeatin can be oxidized; it can also be conjugated to form *O*-glucosides or *N*-glucosides. These reactions are shown in **FIGURES A3.6 and A3.7**, respectively.

### Brassinolide

The BR biosynthetic pathway, starting with the sterol progenitor campesterol, which is ultimately derived from cycloartenol, is illustrated in **FIGURE A3.8**. As discussed in Chapter 24, campesterol is first converted to campestanol in steps involving DET2. Campestanol is then converted to castasterone (CS) through either of two pathways, called the *early* and *late C-6 oxidation pathways*.

The two pathways merge at castasterone, which is then converted to brassinolide (BL) (see Figure 24.10). The early and the late C-6 oxidation pathways coexist and can be

(A)

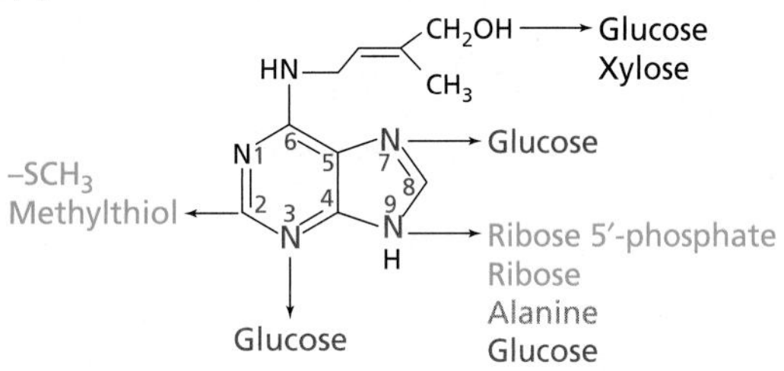

(B)

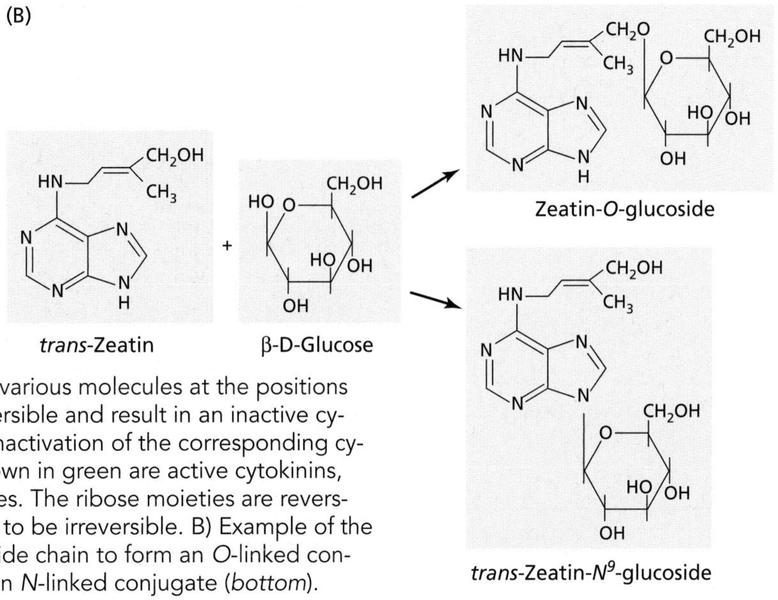

**FIGURE A3.7** Cytokinins can be conjugated to various molecules at the positions shown. The conjugates shown in red are irreversible and result in an inactive cytokinin. The conjugates shown in blue cause inactivation of the corresponding cytokinin, but are reversible. The conjugates shown in green are active cytokinins, albeit less so than the corresponding free bases. The ribose moieties are reversible, but the methylthiol modification appears to be irreversible. B) Example of the conjugation of *trans*-zeatin to glucose at the side chain to form an *O*-linked conjugate (*top*) and on the adenine ring to form an *N*-linked conjugate (*bottom*).

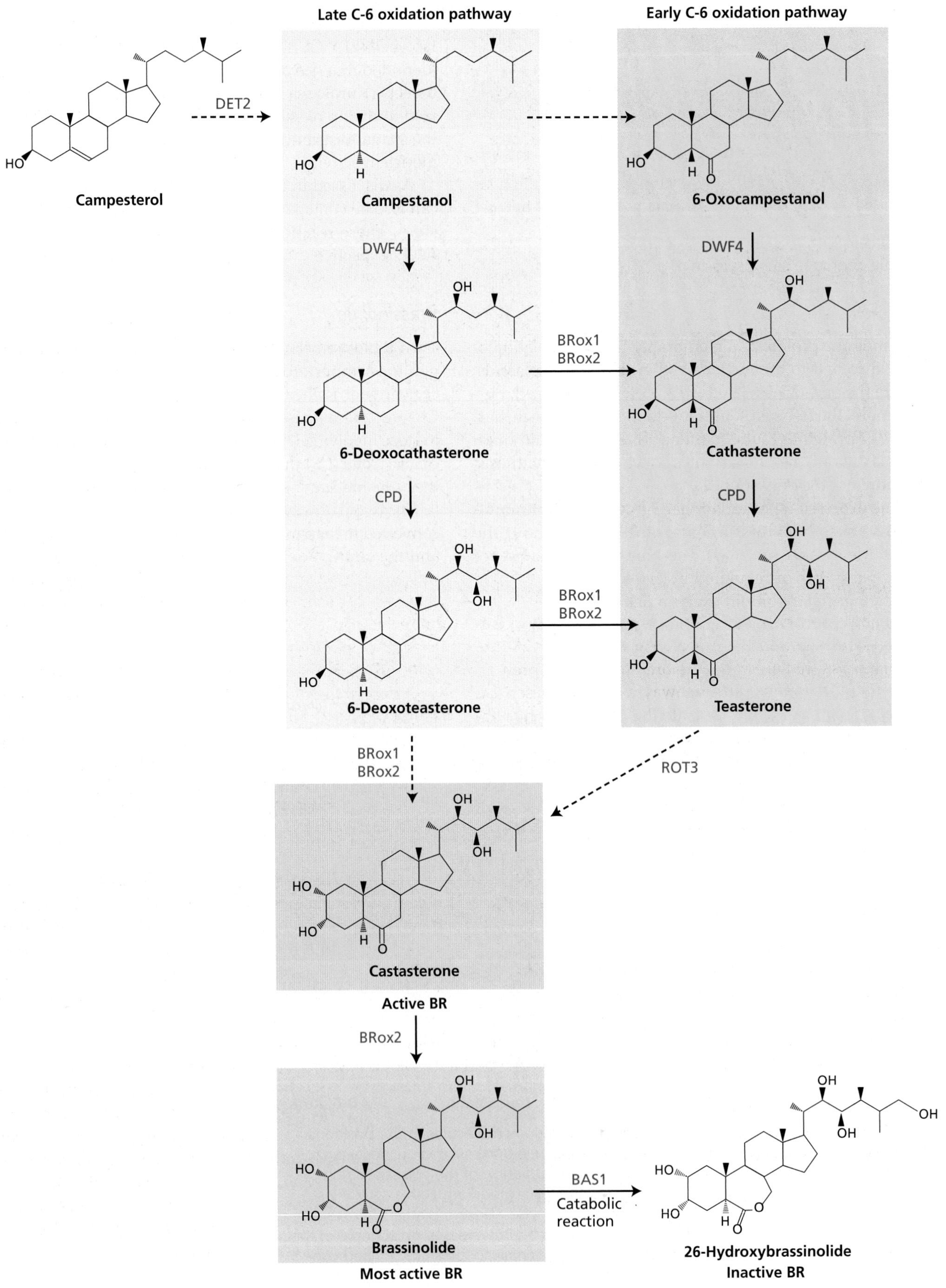

Late C-6 oxidation pathway

Early C-6 oxidation pathway

Campesterol

Campestanol

6-Oxocampestanol

6-Deoxocathasterone

Cathasterone

6-Deoxoteasterone

Teasterone

Castasterone

Active BR

Brassinolide

Most active BR

26-Hydroxybrassinolide

Inactive BR

DET2

DWF4

DWF4

BRox1
BRox2

CPD

CPD

BRox1
BRox2

BRox1
BRox2

ROT3

BRox2

BAS1
Catabolic
reaction

◀ **FIGURE A3.8** Simplified pathways for BL biosynthesis and catabolism. The precursor for BL biosynthesis is campesterol. The sequence of biosynthetic events is represented by black arrowheads. Solid arrows indicate single reactions; dashed arrows represent multiple reactions. As shown, castasterone, the immediate precursor of BL, can be synthesized from two parallel pathways: the early and the late C-6 oxidation pathways. In the early C-6 oxidation pathway, oxidation at C-6 of the B ring occurs *before* the addition of vicinal hydroxyls at C-22 and C-23 of the side chain (refer to BL structure in Figure 24.4). In the late C-6 oxidation pathway, C-6 is oxidized *after* the introduction of hydroxyls at the side chain and C-2 of the A ring. Both the early and the late pathways may be linked at various points, creating a biosynthetic network rather than a linear pathway. The Arabidopsis enzymes that catalyze the different steps are indicated.

linked at different points in Arabidopsis, pea, and rice (Fujioka and Yokota 2003). The presence of two linked pathways increases the complexity of BR biosynthesis and may provide an advantage under different physiological conditions, such as various types of stress.

## Gibberellins

Terpenoids are compounds made up of five-carbon **isoprenoid** building blocks, joined head to tail. The GAs are diterpenoids that are formed from *four* such isoprenoid units. The GA biosynthetic pathway, as already described in Chapter 20, can be divided into three stages, each residing in a different cellular compartment (**FIGURE A3.9**) (Hedden and Phillips 2000): *Stage 1*, which occurs in plastids ; *Stage 2*, which occurs on the plastid envelope and in the endoplasmic reticulum; and Stage 3, which occurs in the cytosol, as shown in Figure A3.9. The entire pathway is described in WEB TOPIC 20.3.

## Abscisic Acid

As discussed in Chapter 23, ABA biosynthesis begins in chloroplasts and other plastids. The complete pathway is depicted in **FIGURE A3.10**. Several ABA-deficient mutants have been identified with lesions at specific steps of the pathway. These mutants exhibit abnormal phenotypes that can be corrected by the application of exogenous ABA. For example, *flacca* (*flc*) and *sitiens* (*sit*) are "wilty" mutants of tomato in which the tendency of the leaves to wilt (due to an inability to close their stomata) can be prevented by the application of exogenous ABA. The *aba* mutants of Arabidopsis also exhibit a "wilty" phenotype. These and other mutants have been useful in elucidating the details of the pathway and in cloning the genes encoding ABA biosynthetic enzymes (Wasilewska et al. 2008).

The pathway begins with isopentenyl diphosphate (IPP)—the biological isoprene unit that is also a precursor of cytokinins, gibberellins, and brassinosteroids—and leads to the synthesis of the $C_{40}$ xanthophyll (i.e., oxygenated carotenoid) **violaxanthin** (see Figure A3.10). Synthesis of violaxanthin is catalyzed by **zeaxanthin epoxidase** (**ZEP**), the enzyme encoded by the *ABA1* locus of Arabidopsis. This discovery provided conclusive evidence that ABA synthesis occurs via the "indirect" or carotenoid pathway, rather than by modification of a $C_{15}$ isoprenoid, as in the "direct pathway" of some phytopathogenic fungi. Maize (corn; *Zea mays*) mutants (termed *viviparous, vp*) that are blocked at other steps in the carotenoid pathway also have reduced levels of ABA and exhibit **vivipary**—the precocious germination of seeds in the fruit while still attached to the plant. Vivipary is a feature of many ABA-deficient seeds.

As noted earlier, trans-violaxanthin is converted to another $C_{40}$ compound, **trans-neoxanthin**, under stress conditions, by a reaction dependent on the product of the Arabidopsis *ABA4* locus. Isomerization by as-yet unidentified enzyme(s), is followed by the cleavage of both 9-*cis*-violaxanthin and 9-*cis*-neoxanthin by **9-*cis*-epoxycarotenoid dioxygenase** (**NCED**) to form the $C_{15}$ compound **xanthoxin**, a neutral growth inhibitor that has physiological properties similar to those of ABA. This is the first committed step for ABA biosynthesis, and it is a rate-limiting regulatory step.

A major cause of the inactivation of free ABA is oxidation to **phaseic acid** (**PA**) and **4′-dihydrophaseic acid** (**DPA**) (see Figure A3.10). ABA increases the expression of the oxidative enzymes in some tissues, resulting in negative feedback regulation of ABA levels. This constitutes a major part of the regulation of ABA levels, such that mutants lacking these oxidases accumulate far more ABA than lines overexpressing any of the biosynthetic enzymes.

The degradation product phaseic acid is usually inactive, or it exhibits greatly reduced activity, in bioassays. However, phaseic acid can induce stomatal closure in some species, and it is as active as ABA in inhibiting gibberellic acid–induced α-amylase production in barley aleurone layers. These effects suggest that phaseic acid may be able to bind to some ABA receptors. In contrast to phaseic acid, the other degradation product, DPA, has no detectable activity in any of the bioassays tested.

Free ABA is also inactivated by covalent conjugation to another molecule, such as a monosaccharide. A common example of an ABA conjugate is **ABA-β-D-glucosyl ester** (**ABA-GE**). Conjugation not only renders ABA inactive as a hormone; it also alters its polarity and cellular distribution. Whereas free ABA is localized in the cytosol, ABA-GE accumulates in vacuoles and apoplastic space and might serve as a storage form of the hormone.

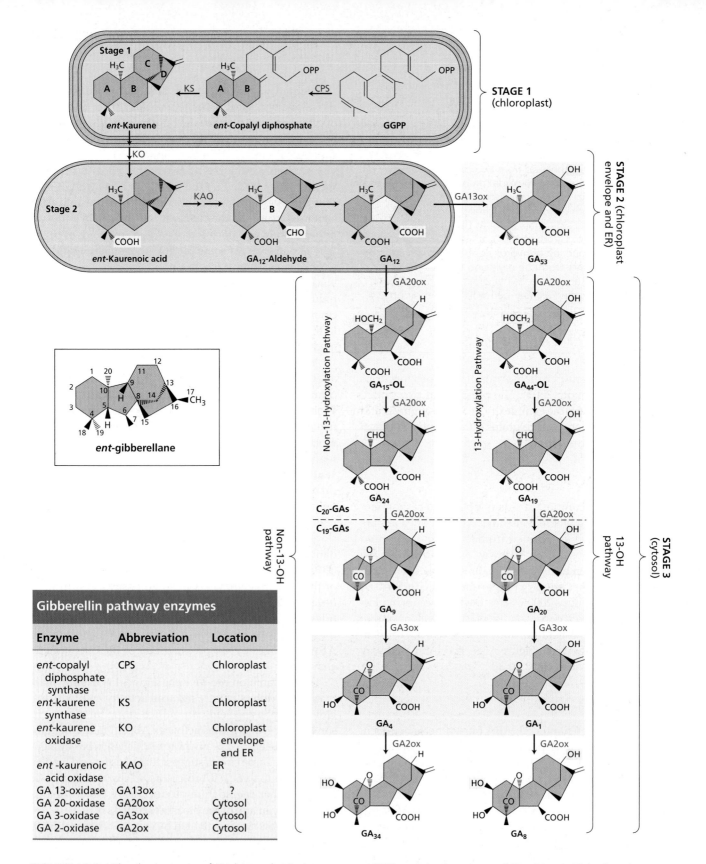

**FIGURE A3.9** The three stages of GA biosynthesis. In stage 1, geranylgeranyl diphosphate (GGPP) is converted to *ent*-kaurene. In stage 2, *ent*-kaurene is converted to GA$_{12}$. In many plants, GA$_{12}$ is converted to GA$_{53}$ by hydroxylation at C-13. In stage 3 in the cytosol, GA$_{12}$ or GA$_{53}$ is converted, via parallel pathways, to other GAs. This conversion proceeds with a series of oxidations at C-20, resulting in the eventual loss of C-20 and the formation of C$_{19}$-GAs. A 3β-hydroxylation reaction then produces GA$_4$ and GA$_1$ as the bioactive GAs in each pathway. In most plants the 13-hydroxylation pathway predominates, although in Arabidopsis and some others, the non-13-OH-pathway is the main pathway. OL, open lactone. See the table for full names and subcellular locations of the enzymes.

---

## Gibberellin pathway enzymes

| Enzyme | Abbreviation | Location |
|---|---|---|
| *ent*-copalyl diphosphate synthase | CPS | Chloroplast |
| *ent*-kaurene synthase | KS | Chloroplast |
| *ent*-kaurene oxidase | KO | Chloroplast envelope and ER |
| *ent*-kaurenoic acid oxidase | KAO | ER |
| GA 13-oxidase | GA13ox | ? |
| GA 20-oxidase | GA20ox | Cytosol |
| GA 3-oxidase | GA3ox | Cytosol |
| GA 2-oxidase | GA2ox | Cytosol |

Dehydration stress results in relocalization of ABA-GE from vacuoles to the endoplasmic reticulum, where it may be cleaved by β-glucosidases. Dehydration also rapidly activates the glucosidases, apparently by inducing polymerization of these enzymes into larger complexes. Consistent with the importance of ABA-GE as a source of ABA, mutants lacking the β-glucosidase have lower ABA levels and impaired stomatal regulation, germination, and stress responses.

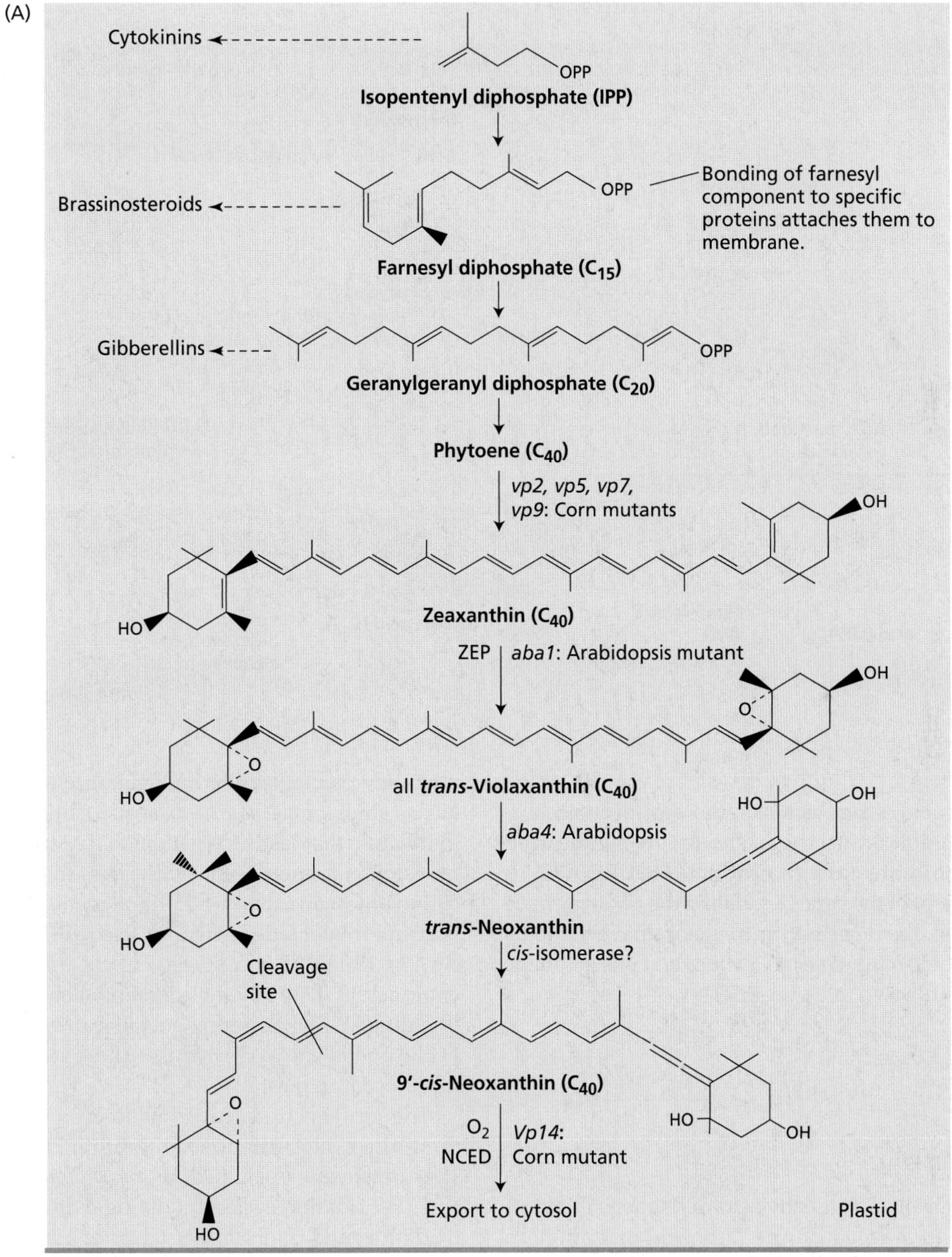

FIGURE A3.10   ABA biosynthesis and metabolism. In higher plants, ABA is synthesized via the terpenoid pathway (see Chapter 13), as are cytokinins, brassinosteroids, and gibberellins (see Chapters 21, 24, and 20 respectively). Some ABA-deficient mutants that have been helpful in elucidating the pathway are shown at the steps at which they are blocked. The pathways for ABA catabolism include conjugation to form ABA-β-D-glucosyl ester or oxidation to form phaseic acid and then dihydrophaseic acid. NCED, 9-cis-epoxycarotenoid dioxygenase; ZEP, zeaxanthin epoxidase. The pathway occurs in two compartments, (A) plastid; (B) cytosol (see next page).

(B)

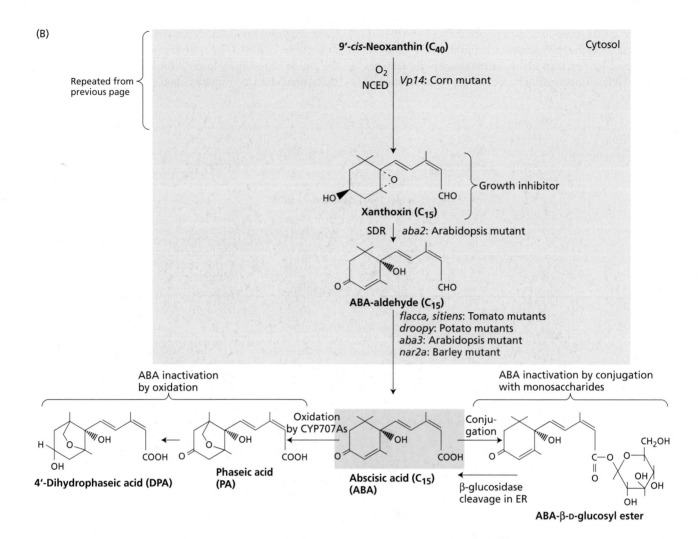

### Strigolactone

Like abscisic acid, strigolactones, the root-derived hormones that inhibit branching in shoots, are synthesized via a carotenoid breakdown product (**FIGURE A3.11**). Although several of the genes regulating the pathway have been identified, the details of the reactions have yet to be elucidated. Three natural strigolactones and one synthetic analog are also shown in Figure A3.11.

## III. Lipids

### Jasmonic Acid (JA)

Several biologically active fatty acid derivatives in plants and animals are formed during fatty acid oxidation. In animals, the *arachidonic acid cascade* generates numerous important metabolic mediators known as eicosanoids,

including prostaglandins and leukotrienes. Higher plants have a similar *linolenate cascade* (also called "octadecanoid pathway") that leads to jasmonate (JA) biosynthesis. The first step in jasmonate biosynthesis is the peroxidation of α-linolenic acid (18:3) by 13-lipoxygenase to form (13S)-hydroperoxyoctadecatrienoic acid (13-HPOT) (**FIGURE A3.12**). 13-HPOT is converted to cis-(+)-12-oxophytodienoic acid (OPDA) by the action of allene oxide synthase [yielding (13S)-12,13-epoxy-octadecatrienoic acid (12,13-EOT)] and allene oxide cyclase. These steps in JA biosynthesis occur in plastids.

The subsequent reactions all occur in the peroxisomes. First, the cyclopentenone ring of OPDA is reduced to 12-oxophytoenoic acid (OPC-8) by OPDA reductase. Next, three β-oxidation cycles, involving oxidation, hydration, oxidation, and thiolysis (see Chapter 11), have been proposed to shorten the carboxylic side chain of OPC-8 to produce the 12-carbon JA. (Acosta et al. 2009)

(A)

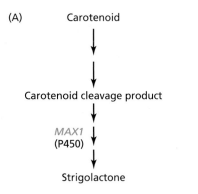

Carotenoid

↓

↓

Carotenoid cleavage product

*MAX1*
(P450) ↓

↓

Strigolactone

(B)

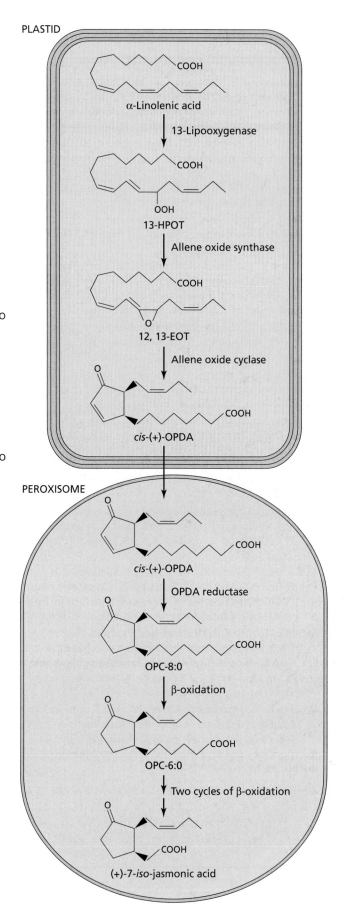

(+)-5-Deoxystrigol

(+)-Orbanchol

(+)-Strigol

GR24
(synthetic analog)

**FIGURE A3.11** (A) Strigolactone is synthesized via a carotenoid breakdown product. (B) Some strigolactone structures.

**FIGURE A3.12** Jasmonic acid is synthesized in two different organelles, the pastid and peroxisome, via the octadecanoid pathway.

PLASTID

COOH

α-Linolenic acid

↓ 13-Lipooxygenase

COOH

OOH
13-HPOT

↓ Allene oxide synthase

COOH

12, 13-EOT

↓ Allene oxide cyclase

COOH
*cis*-(+)-OPDA

PEROXISOME

COOH
*cis*-(+)-OPDA

↓ OPDA reductase

COOH
OPC-8:0

↓ β-oxidation

COOH
OPC-6:0

↓ Two cycles of β-oxidation

COOH
(+)-7-*iso*-jasmonic acid

# REFERENCES

Acosta, I. F., Laparra, H., Romero, S. P., Schmelz, E., Hamberg, M., Mottinger, J. P., Moreno, M. A., and Dellaporta, S. L. (2009) *tasselseed1* is a lipoxygenase affecting jasmonic acid signaling in sex determination of maize. *Science* 323: 262–265.

Capitani, G., Hohenester, E., Feng, L., Storici, P., Kirsch, J. F., and Jansonius, J. N. (1999) Structure of 1-aminocyclopropane-1-carboxylate synthase, a key enzyme in the biosynthesis of the plant hormone ethylene. *J. Mol. Biol.* 294: 745–756.

Fujioka, S., and Yokota, T. (2003) Biosynthesis and metabolism of brassinosteroids. *Annu. Rev. Plant Biol.* 54: 137–164.

Galichet, A., Hoyerová, K., Kamínek, M., and Gruissem, W. (2008) Farnesylation directs AtIPT3 subcellular localization and modulates cytokinin biosynthesis in Arabidopsis. *Plant Physiol.* 146: 1155–1164.

Glick, B. R. (2005) Modulation of plant ethylene levels by the bacterial enzyme ACC deaminase. *FEMS Microbiol. Lett.* 251: 1–7.

Hedden, P., and Phillips, A. L. (2000) Gibberellin metabolism: New insights revealed by the genes. *Trends Plant Sci.* 5: 523–530.

Huai, Q., Xia, Y., Chen, Y., Callahan, B., Li, N., and Ke, H. (2001) Crystal structures of 1-aminocyclopropane-1-carboxylate (ACC) synthase in complex with aminoethoxyvinylglycine and pyridoxal-5′-phosphate provide new insight into catalytic mechanisms. *J. Biol. Chem.* 276: 38210–38216.

Kakimoto, T. (2001) Identification of plant cytokinin biosynthetic enzymes as dimethylallyl diphosphate: ATP/ADP isopentenyltransferases. *Plant Cell Physiol.* 42: 677–685.

Kurakawa, T., Ueda, N., Maekawa, M., Kobayashi, K., Kojima, M., Nagato, Y., Sakakibara, H., and Kyozuka, J. (2007) Direct control of shoot meristem activity by a cytokinin-activating enzyme. *Nature* 445: 652–655.

McKeon, T. A., Fernández-Maculet, J. C., and Yang, S. F. (1995) Biosynthesis and metabolism of ethylene. In *Plant Hormones: Physiology, Biochemistry and Molecular Biology*, 2nd ed., P. J. Davies, ed., Kluwer, Dordrecht, Netherlands, pp. 118–139.

Miyawaki, K., Matsumoto-Kitano, M., and Kakimoto, T. (2004) Expression of cytokinin biosynthetic isopentenyltransferase genes in *Arabidopsis*: Tissue specificity and regulation by auxin, cytokinin, and nitrate. *Plant J.* 37: 128–138.

Nonhebel, H. M., Cooney, T. P., and Simpson, R. (1993) The route, control and compartmentation of auxin synthesis. *Aust J. Plant Physiol.* 20: 527–539.

Normanly, J. (2010) Approaching cellular and molecular resolution of auxin biosynthesis and metabolism. *Cold Spring Harb. Perspect. Biol.* 2: a001594.

Raskin, I., and Beyer, E. M., Jr. (1989) Role of ethylene metabolism in *Amaranthus retroflexus*. *Plant Physiol.* 90: 1–5.

Stepanova, A. N., Robertson-Hoyt, J., Yun, J., Benavente, L. M., Xie, D. Y., Dolezal, K., Schlereth, A., Jurgens, G., and Alonso, J. M. (2008) TAA1-mediated auxin biosynthesis is essential for hormone crosstalk and plant development. *Cell* 133: 177–191.

Sugawara, S., Hishiyama, S., Jikumaru, Y., Hanada, A., Nishimura, T., Koshiba, T., Zhao, Y., Kamiya, Y., and Kasahara, H. (2009) Biochemical analyses of indole-3-acetaldoxime-dependent auxin biosynthesis in Arabidopsis. *Proc. Natl. Acad. Sci. USA* 106: 5430–5435.

Takei, K., Sakakibara, H., and Sugiyama, T. (2001a) Identification of genes encoding adenylate isopentyltransferase, a cytokinin biosynthetic enzyme, in *Arabidopsis thaliana*. *J. Biol. Chem.* 276: 26405–26410.

Takei K., Yamaya, T., and Sakakibara, H. (2004) *Arabidopsis CYP735A1* and *CYP735A2* encode cytokinin hydroxylases that catalyze the biosynthesis of *trans*-zeatin. *J. Biol. Chem.* 279: 41866–41872.

Tao, Y., Ferrer, J. L., Ljung, K., Pojer, F., Hong, F., Long, J. A., Li, L., Moreno, J. E., Bowman, M. E., Ivans, L. J., et al. (2008) Rapid synthesis of auxin via a new tryptophan-dependent pathway is required for shade avoidance in plants. *Cell* 133: 164–176.

Yamagami, T., Tsuchisaka, A., Yamada, K., Haddon, W. F., Harden, L. A., and Theologis, A. (2003) Biochemical diversity among the 1-amino-cyclopropane-1-carboxylate synthase isozymes encoded by the *Arabidopsis* gene family. *J. Biol. Chem.* 278: 49102–49112.

Wasilewska, A., Vlad, F., Sirichandra, C., Redko, Y., Jammes, F., Valon, C., Frei dit Frey, N., and Leung, J. (2008) An update on abscisic acid signaling in plants and more… *Molecular Plant* 1: 198–217.

Woodward, A. W., and Bartel, B. (2005) Auxin: Regulation, action, and interaction. *Ann. Bot.* 95: 707–735.

# Glossary

**A number** An A number is assigned to a newly characterized, naturally-occurring gibberellin for which the structure has been fully determined. Numbers are assigned gibberellins in the order of their discovery.

**abscisic aldehyde oxidases (AAOs)** Enzymes that catalyze the final step in abscisic acid biosynthesis.

**ABA response element binding factors (AREBs/ABFs)** A class of transcription factors that are activated by phosphorylation and mediate ABA effects on gene expression.

**ABA-aldehyde** Precursor of abscisic acid.

**ABA-response element (ABRE)** A six-nucleotide sequence element found in the promoters of genes regulated by abscisic acid (ABA), a plant hormone.

**abaxial** Refers to the lower surface of the leaf.

**ABC model** Proposal for the way in which floral homeotic genes control organ formation in flowers. According to the model, organ identity in each whorl is determined by a unique combination of the three organ identity gene activities.

**ABC transporters** *See* ATP-binding cassette transporters.

**ABFs** *See* ABA Response Element Binding Factors (AREBs/ABFs).

**ABI-class protein phosphatases** Enzymes that interact with many other proteins in the cell, including protein kinases, Ca$^{2+}$-binding proteins, and transcription factors, by binding and regulating their activity through dephosphorylating specific serine or threonine residues.

**abiotic** Referring to what is non-living. In the context of plant stress refers to stresses caused by environmental factors.

**ABP1 gene** Gene that encodes the ABP1 protein.

**ABP1 protein** A glycosylated *a*uxin-*b*inding *p*rotein found in the endoplasmic reticulum (ER).

**abscisic acid (ABA)** An important plant hormone that regulates growth and stomatal closure (particularly when the plant is under environmental stress), as well as regulation of seed maturation and dormancy.

**abscission** The shedding of leaves, flowers, and fruits from a living plant. The process whereby specific cells in the leaf petiole (stalk) differentiate to form an abscission layer, allowing a dying/dead organ to separate from the plant. *See* abscission layer.

**abscission layer** Located within the abscission zone, a distinct layer of cells with weakened cell walls that permit abscission, usually of a leaf or fruit.

**abscission zone** A region that contains the abscission layer and is located near the base of the petiole of leaves.

**absorption spectrum** A graphic representation of the amount of light energy absorbed by a substance plotted against the wavelength of the light.

**ACC** The immediate precursor of ethylene; 1-*a*minocyclopropane-1-*c*arboxylic acid.

**ACC oxidase** Catalyzes the conversion of ACC to ethylene, the last step in ethylene biosynthesis.

**ACC synthase** The enzyme that catalyzes the synthesis of ACC from *S*-adenosylmethionine.

**accessory pigments** Light-absorbing molecules in photosynthetic organisms that work with chlorophyll *a* in the absorption of light used for photosynthesis. They include carotenoids, other chlorophylls, and phycobiliproteins.

**acclimation (hardening)** The increase in plant stress tolerance due to exposure to prior stress. May involve gene expression. Contrast with adaptation.

**acetogenic bacteria** Obligatory anaerobes that use CO$_2$ and H$_2$ as electron acceptor and donor, respectively, in anaerobic respiration for the production of acetyl-CoA. The latter compound serves as building blocks for biosynthetic processes while the excess is excreted as acetate.

**acetylation** The catalyzed chemical addition of an acetate group to another molecule.

**acid growth** A characteristic of growing cell walls in which they extend more rapidly at acidic pH than at neutral pH.

**acid growth hypothesis** Cell wall acidification resulting from proton extrusion across the plasma membrane causing cell wall stress relaxation and extension.

**acropetal** From the base to the tip of an organ, such as a stem, root, or leaf.

**ACS (ACC synthase)** The enzyme that catalyzes the conversion of

S-adenosylmethionine to 1-amino-cyclopropane-1-carboxylic acid (ACC), the immediate precursor of ethylene.

**actinorhizal** Pertaining to several woody plant species, such as alder trees, in which symbiosis occurs with soil bacteria of the nitrogen-fixing genus *Frankia*.

**action spectrum** A graphic representation of the magnitude of a biological response to light as a function of wavelength.

**activators** In the control of transcription, positively acting transcription factors that bind to distal regulatory sequences usually located within 1000 bp of the transcription initiation site.

**active transport** The use of energy to move a solute across a membrane against a concentration gradient, a potential gradient, or both (electrochemical potential). Uphill transport.

**acyl** Refers to the fatty acid residue linked to another compound, often glycerol (e.g., triacylglycerol).

**acyl-ACP** A fatty acid chain bonded to the acyl carrier protein.

**acyl carrier protein (ACP)** A low-molecular-weight, acidic protein to which growing acyl chains are covalently bonded on fatty acid synthetase.

**adaptation (to stress)** An inherited level of stress resistance acquired by a process of selection over many generations. Contrast with acclimation.

**adaxial** Refers to the upper surface of a leaf.

**adaxial–abaxial axis** The leaf axis that runs from the upper to the lower surface of the leaf.

**adenosine-5′-phosphosulfate (APS)** A short-lived, activated form of sulfate, formed by the reaction between sulfate and ATP. Product of first reaction in the several pathways from sulfate to the amino acid cysteine. *See* 3′-phosphoadenosine-5′-phosphosulfate and thiosulfonate.

**adequate zone** Range of mineral nutrient concentrations beyond which further addition of nutrients no longer increases growth or yield.

**adhesion** The attraction of water to a solid phase such as a cell wall or glass surface, due primarily to the formation of hydrogen bonds.

**AdoMet** S-adenosylmethionine.

**ADP/ATP transporter** A protein that catalyzes an antiport exchange of ADP and ATP across the inner mitochondrial membrane.

**ADPG pyrophosphorylase** Catalyzes ATP + α-D-glucose 1-phosphate → pyrophosphate + ADP-glucose.

**adventitious roots** Roots that arise from structures other than roots, such as stems or leaves.

**aerenchyma** Anatomical feature of roots found in hypoxic conditions, showing large, gas-filled intercellular spaces in the root cortex.

**aerobic respiration** *See* respiration.

**aeroponically** The technique by which plants are grown without soil with their roots suspended in air while being sprayed continuously with a nutrient solution.

**after-ripening** Technique for breaking seed dormancy by storage at room temperature under dry conditions, usually for several months.

**agar block** A semi-solid cube of agar used to donate or receive auxin when in contact with plant tissue. Used in studies of polar auxin transport.

**agonist** A substance that binds to a cellular receptor and stimulates a response, in some cases mimicking the action of another substance.

**agravitropic mutants** Mutants that do not respond to gravity. Useful in understanding the mechanism of gravitropism.

**AHP genes** *Arabidopsis Histidine Phosphotransfer* genes involved in cytokinin signal propagation from the plasma membrane receptor to the nucleus.

**alcohol dehydrogenase (ADH)** The enzyme that catalyzes the conversion of acetaldehyde to ethanol.

**alcoholic fermentation** Metabolism of pyruvate from glycolysis that produces ethanol and $CO_2$, while reoxidizing NADH to NAD$^+$.

**aldose** A sugar with a terminal aldehyde.

**aleurone cells** Cells of the aleurone layer enclosed in thick primary cell walls and containing large numbers of protein-storing organelles called protein bodies.

**aleurone layer** Layer of aleurone cells surrounding and distinct from starchy endosperm of cereal grains.

**alkaloids** A large family of nitrogen-containing secondary metabolites found in many vascular plants. Defend against predators, especially mammals. Includes toxins such as strychnine and atropine, and medicinals such as morphine, codeine, atropine, and ephedrine. Others are stimulants and sedatives (e.g., cocaine, nicotine, and caffeine).

**allelopathy** Release by plants of substances into the environment that have harmful effects on neighboring plants.

**allocation** The regulated diversion of photosynthate into storage, utilization, and/or transport.

**allopolyploidy** A form of polyploidy in which multiple complete genomes are derived from two separate species.

**allosteric activation** Activation of an effector binding site by means of a slight change in protein conformation initiated by an allosteric activator.

**α-(1,4)-glucan branching enzyme** Catalyzes the transfer of a segment of an α-D-1,4-linked glucan chain to a primary hydroxyl group in a similar glucan chain.

**α-amylase** Catalyzes the endohydrolysis of α-D-1,4 glucosidic linkages in polysaccharides containing three or more 1,4-α-linked D-glucose units. The term α indicates the initial anomeric configuration of the free sugar group released and not the configuration of the linkage hydrolysed. In the aleurone layer of cereal grains, α-amylase is synthesized de novo in response to gibberellin.

**α-amylase inhibitors** Substances synthesized by some legumes that interfere with herbivore digestion by blocking the action of the starch-digesting enzyme α-amylase.

**α-expansins (EXPA)** One of two major families of expansin proteins that catalyze the pH-dependent extension and stress relaxation of cell walls.

**α-glucosidase** Catalyzes the hydrolysis of terminal, nonreducing 1,4-linked α-D-glucose residues with the release of α-D-glucose.

**α-tubulin** Along with β–tubulin, a component of the heterodimer monomer that polymerizes to form microtubules.

**alternative oxidase** An enzyme in the mitochondrial electron transport chain that reduces oxygen and oxidizes ubihydroquinone.

**alternative pathway** The pathway comprising the oxidation of ubiquinol and the reduction of oxygen by the alternative oxidase. Also known as cyanide-resistant pathway.

**ambiphotoperiodic** Describes plants that flower in long days or short days but not at intermediate day lengths.

**amide** A nitrogen-containing compound formed by the reaction of an amine and a carboxylic acid to form a –$CONR_2$ group.

**amide exporters** Legumes from temperate region that convert toxic ammonia to amides, such as the amino acids asparagine or glutamine for transport to the shoot via the xylem. Contrast with ureide exporters.

**amine** A nitrogen-containing compound derived from ammonia by replacing hydrogens with carbon-containing groups.

**1-aminocyclopropane-1-carboxylic acid (ACC)** An intermediate in the conversion of methionine to ethylene.

**aminoethoxy-vinylglycine (AVG)** A substance that blocks the physiological effects of ethylene by inhibiting the synthesis of ethylene.

**aminooxyacetic acid (AOA)** A substance that blocks the physi-ological effects of ethylene by inhibiting the synthesis of ethylene.

**aminotransferases** Enzymes that carry out transaminations.

**amphipathic** In a molecule, the quality of having both hydrophilic and hydrophobic regions.

**amplitude** In a biological rhythm, the distance between peak and trough; it can often vary while the period remains unchanged.

**amylopectin** The constituent of starch having a polymeric, branched structure in which an α-D-1,6 bond occurs every 20–30 glucose units linked via α-D-1,4 bonds.

**amyloplast** A starch-storing plastid found abundantly in storage tissues of shoots and roots, and in seeds. Specialized amyloplasts in the root cap also serve as gravity sensors.

**amylose** The constituent of starch in which α-D-1,4 glucosidic bonds form linear chains of 200–2000 glucose units.

**anaphase** The stage of mitosis during which the two chromatids of each replicated chromosome are separated and move toward opposite poles.

**anaphase A** Early anaphase, during which the sister chromatids separate and begin to move toward opposite poles.

**anaphase B** Late anaphase, during which the polar microtubules slide passed each other and elongate to push the spindle poles farther apart. Simultaneously, the sister chromosomes are pushed to their respective poles.

**anaphase-promoting complex** During mitosis, this protein complex controls the proteasomal destruction of cyclin proteins, allowing the metaphase-aligned chromatids to segregate to their respective poles.

**anaplerotic** Quality of a reaction to provide a reactant that is limiting in another reaction or pathway. For example, the PEP carboxylase reaction replenishes oxaloacetate to the citric acid cycle that has been used for biosynthesis.

**anastomoses** Vascular interconnections that provide a pathway between source and sink tissues that are not directly connected.

**anchored proteins** Proteins that are bound to the membrane surface via lipid molecules, to which they are covalently attached.

**ancymidol (A-Rest)** A commercial inhibitor of gibberellin biosynthesis. Like paclobutrazol, ancymidol blocks the oxidation reaction between kaurene and kaurenoic acid on the endoplasmic reticulum.

**aneuploidy** A condition in which genomes contain additional or fewer individual chromosomes (not entire chromosome sets) than normal.

**angiosperms** The flowering plants. With their innovative reproductive organ, the flower, they are more advanced type of seed plant and dominate the landscape with at least 250,000 known species. Distinguished from gymnosperms by the presence of a carpel that encloses the seeds.

**anisotropic** Characteristic of materials showing different mechanical properties in different directions. Usually due to a bias in the orientation or linkage of constituent molecules. Contrast with isotropic.

**anomer** Epimers that differ only in the configuration at the carbonyl carbon (C=O) of aldoses or ketoses.

**anoxic** Refers to the absence of oxygen. Compare to hypoxic.

**anoxygenic organism** Photosynthetic organism that does not produce molecular oxygen.

**antenna complex** A group of pigment molecules that cooperate to absorb light energy and transfer it to a reaction center complex.

**antenna pigments** Chlorophylls and carotenoids are the pigments found in the antenna complex.

**anterograde transport** The movement of membrane cisternae or vesicles in the forward direction.

**anther** The apical structure of stamen in which pollen forms and from which it is released.

**anthocyanidins** Pigmented flavonoids derived from anthocyanin by removal of attached sugars.

**anthocyanins** Pigmented glycosylated flavonoids responsible for most of the red, pink, purple, and blue colors in plants.

**anticlinal** Pertaining to the orientation of the cell plate at right angles to the longitudinal axis during cytokinesis.

**antiflorigen** Hypothetical hormone produced by uninduced leaves and translocated to the shoot apical meristem. Proposed to inhibit the formation of flowers in certain long-day plants under noninductive conditions.

**antifreeze proteins** Proteins that confer to aqueous solutions the property of thermal hysteresis. When induced by cold temperatures, these plant proteins bind to the surfaces of ice crystals to prevent or slow further crystal growth, thereby limiting or preventing freeze damage. Some antifreeze proteins may be identical to pathogenesis-related proteins.

**antiport** A type of secondary active transport in which the passive (downhill) movement of protons or other ions drives the active (uphill) transport of a solute in the opposite direction.

**antiporter** A protein involved in antiport.

**antisense DNA** DNA of a gene whose transcription produces antisense mRNA, which hybridizes with sense mRNA and inhibits its translation.

**AOA** Aminooxyacetic acid, an inhibitor of ethylene biosynthesis, blocking the conversion of *AdoMet* to ACC.

**AP1 gene** In Arabidopsis, *APETALA1* (*AP1*) is a gene involved in establishing floral meristem identity.

**ap2 mutants** In Arabidopsis, mutations in the *APETALA2* (*AP2*) homeotic gene produce flowers with carpels where sepals should be and stamens where petals normally appear.

**apical** Relating to the apex or tip. Distinguishing one end of an axis. Contrast with *basal*.

**apical cell** In ferns and other primitive vascular plants, the single initial or stem cell of roots and shoots that gives rise to all the other cells of the organ. In angiosperm embryogenesis, the smaller, cytoplasm-rich cell formed by the first division of the zygote.

**apical dominance** In most higher plants, the growing apical bud's inhibition of the growth of lateral buds (axillary buds).

**apical meristems** Localized regions of cell division located at the tips of shoots and roots.

**apical region** In the globular embryo of seed plants, the cells that are derived from the apical quartet of cells; gives rise to the cotyledons and shoot apical meristem.

**apical–basal axis** An axis that extends from the shoot apical meristem to the root apical meristem.

**apoplast** The mostly continuous system of cell walls, intercellular air spaces, and xylem vessels in a plant.

**apoplastic pathway** The route by which water and water-soluble solutes move exclusively through the cell walls without crossing any membranes.

**apoprotein** The polypeptide component which may be bound to a chromophore or other cofactor or prosthetic group to form an active protein, the holoprotein.

**apoptosis** A type of programmed cell death found in animals showing characteristic morphological and biochemical changes, including fragmentation of nuclear DNA between the nucleosomes. Apoptosis-like changes also occur in some senescing plant tissues, differentiating xylem tracheary elements, and in the hypersensitive response against pathogens.

**APS** *See* adenosine-5′-phosphosulfate.

**aquaporins** Integral membrane proteins that form water-selective channels across the membrane. Such channels facilitate water movement across the membrane.

**Arabidopsis Response Regulator (ARR)** Arabidopsis genes that are similar to bacterial two-component signaling proteins called response regulators. There are two classes: Type-A ARRs, whose transcription is up-regulated by cytokinin; type-B ARRs, whose expression is not affected by cytokinin.

**arabinogalactan proteins (AGPs)** A family of extensively glycosylated (mostly galactose and arabinose), water soluble cell wall proteins that usually amount to less than 1% of the wall dry mass. Some may associate with the plasma membrane via a glycosylphosphatidylinositol anchor. They often display tissue- and cell-specific expression.

**arabinoxylan** A branched cell wall polysaccharide consisting of a backbone of xylose residues with arabinose side chains.

**arbuscular mycorrhizal fungi** Symbiotic fungi that form hyphae growing in a loose arrangement, both within the root itself and extending outward from the root into the surrounding soil. The hyphae form unique structures, such as the arbuscules that enhance the exchange of nutrients between the fungus and its host.

**arbuscules** Branched structures of mycorrhizal fungi that form within penetrated cells; the sites of nutrient transfer between the fungus and the host plant.

**ARGONAUTE (AGO)** A catalytic protein that is part of the RNA-induced silencing complex.

**ARR** *See* Arabidopsis Response Regulator.

**asparagine synthetase (AS)** Enzyme transferring nitrogen as an amino group from glutamine to aspartate, forming asparagine.

**aspartate aminotransferase (AspAT)** An aminotransferase that transfers the amino group from glutamate to the carboxyl atom of oxaloacetate to form asparatate.

**assimilation** *See* nutrient assimilation.

**assimilatory power** The combined energy available in NADPH and ATP that can be used to drive the photosynthetic fixation of atmospheric $CO_2$ into organic molecules.

**ATI** *See* auxin transport inhibitor.

***AtNHX1* gene** In Arabidopsis, gene encoding a $Na^+/H^+$ antiporter that is active in vacuolar compartmentalization of $Na^+$.

**ATP (adenosine triphosphate)** The major carrier of chemical energy in the cell, which by hydrolysis is converted to adenosine diphosphate (ADP) or adenosine monophosphate (AMP).

**ATP synthase (ATPase or $CF_o$–$CF_1$)** The enzyme that synthesizes ATP from ADP and phosphate (P). $F_oF_1$ and $CF_o$-$CF_1$ types are present in mitochondria and chloroplasts, respectively.

**ATP-binding cassette transporters (ABC transporters)** A group of active transport proteins, of distinctive structure, energized by ATP hydrolysis and involved in moving organic molecules across a membrane.

**ATPase** An enzyme that hydrolyses ATP into ADP and inorganic phosphate (Pi).

**autocatalysis** The capacity of the Calvin–Benson cycle to use most of the fixed carbon (triose phosphate) to build up the concentrations of intermediates to adequately function at steady state. As a result, the concentration of the acceptor (ribulose-1,5-bisphosphate) doubles every five revolutions of the cycle.

**autocatalytic** An action or reaction that promotes the action or reaction. For example, ethylene production by fruit is stimulated by ethylene. Relating to a process that occurs spontaneously when purified components of a system are mixed in vitro, requiring no additional proteins or cofactors.

**autoinhibitory domain** A domain at the C-terminal end of the plasma membrane $H^+$-ATPase and $Ca2^+$-ATPase that inhibits the activity of the enzyme. Regulated by phosphorylation and, in the case of the PM $H^+$-ATPase, by agents such as fusicoccin.

**autopolyploid** A form of polyploidy containing multiple complete genomes of a single species.

**autoradiography** A technique used to determine the location of a radioactive isotope that has been supplied to living cells, using photographic film (overall localization) or photographic emulsion (precise cellular localization).

**autotrophic** Cells or organisms that synthesize cellular components from structural elements found in the most oxidized states (for example, carbohydrates from $CO_2$). As the reduction is usually endothermic, the energy is supplied either by sunlight (photoautotrophs) or by chemical energy (chemoautotrophs).

***Aux/IAA* genes** Family of primary response genes encoding short-lived transcription factors that function as repressors or activators of the expression of late auxin-inducible genes.

**Aux/IAA proteins** A family of short-lived small proteins that combine with the TIR1/AFB proteins to form the primary auxin receptor. This family of short-lived small proteins in Arabidopsis regulates auxin-induced gene expression by binding to ARF protein that is bound to DNA. If the specific ARF is a transcriptional activator, the Aux/IAA binding represses transcription.

**AUX1/LAX** A small family of $H^+$–$IAA^-$ symporters that cotransport two protons along with the auxin anion across the plasma membrane.

**auxin** A compound with biological activities similar to, but not necessarily identical with, those of IAA. Induces cell elongation in isolated coleoptile or stem sections, cell division in callus tissues in the presence of a cytokinin, lateral root formation at the cut surfaces of stems, parthenocarpic fruit growth, and ethylene formation.

**auxin efflux inhibitors (AEIs)** Any of several synthetic substances that act in different ways to inhibit auxin transport out of a cell.

**auxin influx inhibitor** A synthetic substance that inhibits the import of auxin into a cell.

**auxin response domains (AuxRDs)** Auxin responsive promoter sequences composed of multiple *AuxREs*.

**auxin response elements (AuxREs)** DNA promoter sequences that modulate gene expression when bound by auxin-responsive transcription factors.

**auxin response factors (ARFs)** A family of proteins that regulate the transcription of specific genes involved in auxin responses; they are inhibited by association with specific Aux/IAA repressor proteins, which are degraded in the presence of auxin.

**auxin transport** The energy-requiring, polar movement of IAA from the shoot apex to the root tip, and subsequent redistribution of auxin from the root tip to the basal portion of the root. Consists of protein-mediated auxin efflux from cells driven by the membrane potential, $\Delta E$, and auxin uptake into cells driven by the total proton motive force ($\Delta E + \Delta pH$).

**auxin transport inhibitor (ATI)** Any compound that prevents the polar transport of auxin, usually by preventing auxin exit (efflux) from cells. *See* NPA and TIBA.

**auxin-induced proton extrusion** Increased rate of proton extrusion by both activation of preexisting plasma membrane $H^+$-ATPases and synthesis of new $H^+$-ATPases.

**AVG** Aminoethoxy-vinylglycine, an inhibitor of ethylene biosynthesis, blocking the conversion of AdoMet to ACC.

***avr* genes** Pathogen avirulence genes that encode for specific elicitors of plant defense responses.

**axil** The angled juncture between the upper side of a leaf and the stem to which it is attached. Usual site for insertion of an axillary bud.

**axillary buds** Secondary meristems that are formed in the axils of leaves. If they are also vegetative meristems, they will have a structure and developmental potential similar to that of the vegetative apical meristem. Axillary buds can also form flowers, as in inflorescences. *See* lateral bud.

**axis (plant axis)**   The hypothetical central line of a body around which parts are arranged. *See* linear axis, radial axis.

**axr1 mutant**   Arabidopsis auxin-resistant mutant deficient in many auxin response deficiencies, including gravitropism.

**Bacillus thuringiensis (Bt)**   A soil bacterium that is the source for a commonly used transgene encoding an insecticidal toxin.

**bacteriochlorophyll**   Light-absorbing pigments active in photosynthesis in anoxygenic photosynthetic organisms.

**bacteroids**   Nitrogen-fixing organelles that develop from endosymbiotic bacteria upon a signal from the host plant.

**BAK1**   A second LRR-receptor kinase which associates with the BR receptor, BRI1, forming an active complex.

**bark**   Collective term for all the tissues outside the cambium of a woody stem or root, and composed of phloem and periderm.

**basal**   Relating to the base. Distinguishing one end of an axis. Contrast with apical.

**basal cell**   In embryogenesis, the larger, vacuolated cell formed by the first division of the zygote. Gives rise to the suspensor.

**basipetal**   From the growing tip of a shoot or root toward the base (junction of the root and shoot).

**basipetal transport**   Transport away from the apical meristems, in both root and shoot.

**BES1-interacting Myc-like 1 (BIM1)**   A transcription factor that activates transcription by binding to a specific DNA sequence called an *E Box* that functions as a brassinosteroid response element in the promoters of brassinosteroid-induced genes.

**β-amylase**   Catalyzes the hydrolysis of β-D-1,4 glucosidic linkages in polysaccharides so as to remove successive maltose units from the nonreducing ends of the chains. The term β denotes the initial anomeric configuration of the free sugar group released and not the

configuration of the linkage hydrolyzed.

**β-amylolysis**   The stepwise hydrolysis of alternate glucosidic bonds in starch-type polysaccharides with the liberation of maltose.

**beta-(β-) oxidation**   Oxidation of fatty acids into fatty acyl-CoA, and the sequential breakdown of the fatty acids into acetyl-CoA units. NADH is also produced.

**β-tubulin**   Along with α–tubulin, a component of the heterodimer monomer that polymerizes to form microtubules.

**bidirectional transport**   Simultaneous transport in two directions in a single sieve element.

**biennial**   Plant that requires two growing seasons to flower and produce seed.

**bilayer**   *See* lipid bilayer.

**BIN2**   A repressor of brassinosteroid-induced gene expression.

**bioassay**   Quantitation of a known or suspected biologically active substance by measuring the effect the substance has on living material.

**biolistics**   A procedure, also called the "gene gun" technique, in which tiny gold particles coated with the genes of interest are mechanically shot into cultured cells. Some of the DNA is randomly incorporated into the genome of the targeted cells.

**biological clocks**   In most if not all biological systems, an endogenous oscillator that can maintain a variety of rhythmic physiological processes independent of input from environmental rhythms.

**biological nitrogen fixation**   Nitrogen fixation carried out by bacteria or blue-green algae (cyanobacteria); about 90% of all nitrogen fixation on Earth.

**biosphere**   Parts of the surface and atmosphere of the Earth that support life as well as the organisms living there.

**biotic**   Referring to what is living.

**biotrophic**   Refers to pathogens that leave infected tissue alive and only minimally damaged while the

pathogen continues to feed on host resources.

**bleached**   The loss of chlorophyll's characteristic absorbance due to its conversion into another structural state, often by oxidation.

**blue-light responses**   Responses of plant cells and organs to blue light (400 to 500 nm). Includes phototropism, chloroplast movement within cells, sun tracking by leaves, inhibition of hypocotyl elongation, stimulation of chlorophyll and carotenoid synthesis, activation of gene expression, and stomatal movements.

**bolting**   Premature elongation of the stem of a rosette plant, usually associated with flowering.

**boundary layer**   A thin film of still air at the leaf surface. Its resistance to water vapor diffusion is proportional to its thickness.

**boundary layer resistance ($r_b$)**   The resistance to the diffusion of water vapor due to the layer of unstirred air next to the leaf surface. A component of diffusional resistance.

**Bowen ratio**   The ratio of sensible heat loss to evaporative heat loss, the two most important processes in the regulation of leaf temperature.

**BR-signaling kinase (BSK)**   Enzymes that transmit the brassinolide signal through the cytoplasm by activating the phosphatase BSU1, which dephosphorylates and inactivates the repressor BIN2.

**bract**   Small leaf-like structure with underdeveloped blade.

**brassinazole (Brz)**   A triazole inhibitor of cytochrome p450 monooxygenases which acts as a specific inhibitor of brassinolide biosynthesis.

**BRASSINAZOLE-RESISTANT 1 (BZR1)**   Transcriptional regulator that is inhibited by phosphorylation by BIN2 in the absence of brassinosteroids.

**brassinolide (BL) [24]**   A plant steroidal hormone with growth-promoting activity, first isolated from *Brassica napus* pollen. One of a group of plant steroidal hormones with similar activities called brassinosteroids.

**BRASSINOSTEROID INSENSITIVE-2 (BIN2)** A repressor of gene expression whose inactivation induces the response to brassinosteroids.

**brassinosteroids (BRs)** A group of plant steroid hormones that play important roles in many developmental processes, including cell division and cell elongation in stems and roots, photomorphogenesis, reproductive development, leaf senescence, and stress responses.

**bri1 suppressor 1 (BSU1)** A plant-specific serine/threonine phosphatase, which is activated by brassinosteroid-signaling kinases in the presence of brassinosteroids.

**BRI1** The plasma membrane receptor protein for brassinosteroids. BRI1 belongs to the general class of leucine-rich repeat (LRR)-receptor serine/threonine (S/T) kinases.

**BRI1-associated receptor kinase 1 (BAK1)** A second LRR-receptor kinase which is associated with the brassinolide-activated receptor.

**bri1-EMS-suppressor 1 (BES1)** A nuclear protein phosphorylated and inhibited by BIN2.

**BRI1-kinase inhibitor 1 (BKI1)** A negative regulator of BRI1 signaling, whose dissociation from the plasma membrane is regulated by brassinosteroids.

**bulk flow (or mass flow)** The concerted movement of molecules en masse, most often in response to a pressure gradient.

**bundle sheath** One or more layers of closely packed cells surrounding the small veins of leaves and the primary vascular bundles of stems.

**bundle sheath cells** Chloroplast-containing cells found in the leaves of $C_4$ plants but not in $C_3$ plants.

**Bünning hypothesis** Hypothesis that photoperiodic control of flowering is achieved by a coincidence of the light or dark phase of an inductive photoperiod with a particular phase of oscillation of the circadian rhythm that has a different sensitivity to light.

**$C_{19}$-GAs** *See* $C_{19}$-gibberellins.

**$C_{19}$-gibberellins** Gibberellins that have only 19 carbons because they have lost carbon-20 by metabolism.

**$C_{20}$-GAs** *See* $C_{20}$-gibberellins.

**$C_{20}$-gibberellins** Gibberellins that possess the full diterpenoid complement of 20 carbon atoms, and which are precursors of $C_{19}$-GAs.

**$C_3$ metabolism** *See* Calvin–Benson cycle.

**$C_3$ plants** Plants in which the first stable product of photosynthetic $CO_2$ fixation is a three-carbon compound (i.e., 3-phosphoglycerate). *See* Calvin–Benson cycle.

**$C_4$ cycle** The photosynthetic carbon metabolism of certain plants in which the initial fixation of $CO_2$ and its subsequent reduction take place in different cells, the mesophyll and bundle sheath cells respectively. The initial carboxylation is catalyzed by phosphoenylpyruvate carboxylase, (not by rubisco as in $C_3$ plants), producing a four-carbon compound (oxaloacetate), which is immediately converted to malate or aspartate.

**$C_4$ metabolism** *See* $C_4$ cycle.

**$C_4$ plants** Plants in which the first stable product of $CO_2$ assimilation in mesophyll cells is a four-carbon compound that is immediately transported to bundle sheath cells and decarboxylated. The $CO_2$ released enters the Calvin–Benson cycle. *See* $C_4$ cycle.

**CAAT box** A sequence of nucleotides involved in the initiation of transcription in eukaryotes.

**cadastral genes** Genes that act as spatial regulators of the floral organ identity genes by setting boundaries for their expression.

**calcineurin B–like proteins (CBLs)** A family of plant $Ca^{2+}$ binding proteins that are differentially regulated by environmental stress and by ABA. Upon binding $Ca^{2+}$ the conformation changes, allowing these proteins to interact with and modify the function of a wide variety of cellular proteins.

**callose** A β-1,3-glucan synthesized in the plasma membrane and deposited between the plasma membrane and the cell wall. Syn-

thesized by sieve elements in response to damage, stress, or as part of a normal developmental process.

**callus tissue** The product of the disorganized growth of undifferentiated plant cells in tissue culture.

**calmodulin** A conserved calcium-binding protein found in all eukaryotes that regulates many calcium-driven, metabolic reactions.

**Calvin–Benson cycle** The biochemical pathway for the reduction of $CO_2$ to carbohydrate. The cycle involves three phases: the carboxylation of ribulose 1,5-bisphosphate with atmospheric $CO_2$, catalyzed by rubisco, the reduction of the formed 3-phosphoglycerate to trioses phosphate by 3-phosphoglycerate kinase and NADP-glyceraldehyde-3-phosphate dehydrogenase, and the regeneration of ribulose 1,5-bisphospate through the concerted action of ten enzymatic reactions.

**CAM plants** Plants that fix $CO_2$ during the night into a four-carbon compound (malate) that, after storage in the vacuole, is transported out of the vacuole and decarboxylated during the day. The $CO_2$ released is assimilated by the Calvin–Benson cycle in the chloroplast stroma.

**cambium (vascular cambium)** Layer of meristematic cells between the xylem and phloem that produces cells of these tissues and results in the lateral (secondary) growth of the stem or root.

**campesterol** The sterol progenitor of brassinosteroids, ultimately derived from cycloartenol.

**canavanine** Toxic, nonprotein amino acid that is a close analog of the protein amino acid arginine.

**capillarity** The movement of water for small distances up a glass capillary tube or within the cell wall, due to water's cohesion, adhesion, and surface tension.

**carbon dioxide compensation point** *See* $CO_2$ compensation point.

**carbon fixation reactions** The synthetic reactions occurring in the stroma of the chloroplast that use the high-energy compounds ATP and NADPH for the incorporation of $CO_2$ into carbon compounds.

**carbon isotope ratio** The ratio $^{13}C/^{12}C$ isotope composition of carbon compounds as measured by use of a mass spectrometer.

**carboxylation** The reaction in which a carboxylase catalyzes the formation of a carbon–carbon bond between $CO_2$ and the carbon atom of an organic molecule.

**cardenolides** Steroidal glycosides that taste bitter and are extremely toxic to higher animals through their action on $Na^+K^+$-activated ATPases. Extracted from foxglove (*Digitalis*) for treatment of human heart disorders.

**carotenoid protein** A protein with a carotenoid molecule bound to it. For example, the orange carotenoid protein.

**carotenoids** Linear polyenes arranged as a planar zigzag chain, with the repeating conjugated double-bond system —CH=CH—CCH$_3$=CH—. These orange pigments serve both as antenna pigments and photoprotective agents.

**carpels** The structure that contains the ovules of flowering plants, consisting of ovary, style, and stigma. The carpel develops into the fruit and the ovule develops into the seed. *See also* ovule, sepals, petals, and stamens.

**carrier-mediated transport** Active or passive transport of a solute accomplished by a carrier.

**carriers** Membrane-transport proteins that bind to a solute, undergo conformational change, and release the solute on the other side of the membrane.

**cascade** A succession of interactions so that each interaction derives from or acts upon the product of the preceding. A series of interactions that involves an amplification of the initial action.

**Casparian strip** A band in the cell walls of the endodermis that is impregnated with the waxlike, hydrophobic substance suberin. Prevents water and solutes from entering the xylem by moving between the endodermal cells.

**castasterone (CS)** The bioactive, immediate precursor of brassinolide.

**catalase** An enzyme that breaks hydrogen peroxide down into water. When it is abundant in peroxisomes it may form crystalline arrays.

**cation exchange** The replacement of mineral cations adsorbed to the surface of soil particles by other cations.

**cavitation** The collapse of tension in a column of water resulting from the indefinite expansion of a tiny gas bubble.

**CCCP (Carbonyl cyanide m-chlorophenylhydrazone)** An ionophore that makes membranes highly permeable to protons, thus abolishing proton gradients.

**Cdc2 protein** The major cyclin-dependent protein kinase, *cell division cycle 2*. Biosynthesis stimulated by auxin.

**CDK** *See* cyclin-dependent protein kinase.

**cell number** A convenient means to measure the growth of unicellular organisms. In multicellular plants, increasing numbers of cells are usually associated with cell expansion, particularly in meristems.

**cell plate** Wall-like structure that separates newly divided cells. Formed by the phragmoplast and later becomes the cell wall.

**cell wall** The rigid cell surface structure external to the plasma membrane that supports, binds, and protects the cell. Composed of cellulose and other polysaccharides and proteins. *See also* primary cell walls and secondary cell walls.

**cell wall matrix** Plant cell wall material consisting of hemicelluloses and pectins plus a small amount of structural protein.

**cellobiose** A 1→4-linked β-D-glucose disaccharide that makes up cellulose.

**cellular dehydration** Water loss from cells, resulting in a loss of turgor pressure and a reduction in cell volume.

**cellulose** Linear chains of 1→4-linked β-D-glucose. The repeating unit is cellobiose.

**cellulose microfibril** Thin, ribbon-like structure of indeterminate length and variable width composed of 1→4-linked β-D-glucan chains tightly packed in crystalline arrays alternating with less organized amorphous regions. Provides structural integrity to the cell walls of plants and determines the directionality of cell expansion.

**cellulose synthase** Enzyme that catalyzes the synthesis of individual 1→4-linked β-D-glucans that make up the cellulose microfibril.

**central zone** A central cluster of relatively large, highly vacuolate, slow-dividing cells in shoot apical meristems, comparable to the quiescent center of root meristems.

**centromere** The constricted region on the mitotic chromosome where the kinetochore forms and to which spindle fibers attach.

**cereal grains** Seeds of grasses consisting of the diploid embryo, the triploid endosperm, and the fused seed coat–fruit wall.

**CesA** *Cellulose synthase A*. A multigene family of cellulose synthases found in all land plants.

**CF$_o$-CF$_1$ ATPase** A multi-protein complex associated with the thylakoid membrane that couples the passage of protons across the membrane to the synthesis of ATP from ADP and phosphate. Similar to $F_oF_1$ ATP synthase in oxidative phosphorylation but much less sensitive to oligomycin.

**channels** Transmembrane proteins that function as selective pores for passive transport of ions or water across the membrane.

**checkpoint** A key regulatory point early in $G_1$ of the cell cycle that determines if the cell is committed to the initiation of DNA synthesis.

**chelates** Substances such as EGTA that form a complex with divalent cations eliminating their biological activity.

**chelator** A carbon compound that can form a noncovalent complex with certain cations facilitating their uptake (e.g., malic acid, citric acid).

**chemical potential** The free energy associated with a substance that is available to perform work.

**chemical potential of water** *See* water potential.

**chemical-potential gradient** A change in the free energy per mole of a substance, measured over a given distance. A substance moving spontaneously does so down its chemical potential gradient.

**chemiosmotic mechanism (model)** The mechanism whereby the electrochemical gradient of protons established across a membrane by an electron transport process is used to drive energy-requiring ATP synthesis. It operates in mitochondria and chloroplasts.

**chilling** Exposure to low temperatures (0–10°C), sometimes causing chilling injury; can be used to release certain seeds from dormancy.

**chilling injury** Changes that occur when plants growing at 25 to 35°C are cooled to 10 to 15°C. Includes slowed growth, leaf discoloration, and/or lesions. Contrast with freezing injury.

**chilling-sensitive plants** Plants that experience a sharp reduction in growth rate at temperatures between 0 and 12°C.

**CHLH protein** Group of ABA receptor proteins localized to plastids.

**2-chloroethylphosphonic acid** *See* ethephon.

**chlorophyll** A group of light absorbing green pigments active in photosynthesis.

**chlorophyll *a/b* antenna proteins** *See* light harvesting complex proteins.

**chlorophyllase** An enzyme that removes the phytol from chlorophyll as part of the chlorophyll breakdown process.

**chlorophyte** Unicellular photosynthetic eukaryote whose chloroplasts contain chlorophyll *a* and *b* (green algae).

**chloroplast** The organelle that is the site of photosynthesis in eukaryotic photosynthetic organisms.

**chloroplast genome** About 5 to 10 percent of the cellular DNA is found in a cell's chloroplasts. Chloroplast DNA has recently been shown to consist of linear molecules that may contain more than one copy of the genome connected to one another in head-to-tail orientation. Chloroplast DNA can be highly branched, resembling a bush or tree.

**chlorosis** The yellowing of older, lower plant leaves characteristic of prolonged nitrogen deficiency.

**Cholodny–Went model** Early mechanism proposed for tropisms involving stimulation of the bending of the plant axis by lateral transport of auxin in response to a stimulus, such as light, gravity, or touch. The original model has been supported and expanded by recent experimental evidence.

**chromatin** The DNA–protein complex found in the interphase nucleus. Condensation of chromatin forms the mitotic and meiotic chromosomes.

**30-nm chromatin fiber** The irregular helical structure formed by nucleosome wrapped with DNA.

**chromatin remodeling** Stable changes in chromatin structure accomplished by epigenetic factors.

**chromophore** A light-absorbing pigment molecule that is usually bound to a protein (an apoprotein).

**chromoplasts** Plastids that contain high concentrations of carotenoid pigments, rather than chlorophyll. Chromoplasts are responsible for the yellow, orange, or red colors of many fruits, flowers, and autumn leaves.

**chromosomes** The condensed form of chromatin that forms early in mitosis and meiosis.

**chronic photoinhibition** Photoinhibition of photosynthentic activity in which both quantum efficiency and the maximum rate of photosynthesis are decreased. Occurs under excess light.

**chronobiology** The study of biological clocks, such as the circadian rhythm.

**cinnamic acid** A phenylpropanoid derived from the amino acid phenylalanine that is a key intermediate in the biosynthesis of many phenolic compounds.

**circadian resonance** Applies to a match between a plant's circadian rhythms and the light–dark cycle of the environment.

**circadian rhythm** A biological activity that shows a cycle of high-activity and low-activity independent of external stimuli, with a regular periodicity of about 24 hours (L. *circa diem*: about a day).

***cis*-Golgi** The face of the Golgi where vesicles and tubules from the ER are accepted.

***cis*-acting sequences** DNA sequences that bind transcription factors and are adjacent (*cis*) to the transcription units they regulate. Not to be confused with *cis*-elements.

***cis*-elements** Certain nucleotide sequences within the mRNA molecule by which mRNA stability is regulated. Not to be confused with *cis*-acting sequences in DNA that influence transcriptional activity.

**cisternae** (singular *cisterna*) A network of flattened saccules and tubules that compose the endoplasmic reticulum, the ER.

**cisternal maturation/progression model** Model for Golgi membrane development in which the Golgi stack is not fixed, but is a dynamic structure in which cisternae progress through *cis*, *medial*, and *trans* faces, carrying their cargo with them.

**citric acid cycle (Krebs cycle, tricarboxylic acid cyle)** A cycle of reactions localized in the mitochondrial matrix that catalyzes the oxidation of pyruvate to $CO_2$. ATP and NADH are generated in the process.

***CKI1* gene** Gene whose overexpression confers cytokinin-independent growth on Arabidopsis cells in culture. Encodes a protein similar to bacterial histidine kinases functioning in signal transduction.

**clathrin** Proteins that have a unique *triskelion* structure that spontaneously assemble into 100-nm cages that coat vesicles associated with endocytosis at the plasma membrane and other cellular trafficking events.

**climacteric**   Marked rise in respiration at the onset of ripening that occurs in all fruits that ripen in response to ethylene, and in the senescence process of detached leaves and flowers.

**climatic factors**   Atmospheric gases, light, temperature, humidity, precipitation, and wind that affect physiological processes.

**clock hypothesis**   Currently accepted hypothesis of how plants measure night length. Proposes that photoperiodic timekeeping depends on the endogenous oscillator of circadian rhythms. *See* hourglass hypothesis.

**$CO_2$ compensation point**   The $CO_2$ concentration at which the rate of respiration balances the photosynthetic rate

**CoA**   *See* coenzyme A.

**coat proteins**   Specific proteins on the surface of vesicles that determine the delivery of vesicle membrane and contents to the Golgi or the ER. COP1, COP2, and clathrin are coat proteins.

**coat-imposed dormancy**   Dormancy imposed on the embryo by the seed coat and other enclosing tissues, such as endosperm, pericarp, or extrafloral organs.

**coconut milk**   The liquid endosperm of coconut seeds that contains cytokinins and other nutritional factors. Stimulates the growth of normal stem tissues when added to liquid culture media.

**coenzyme A**   A coenzyme with an —SH group that serves as an acyl group carrier for many enzymatic reactions.

**cohesion**   The mutual attraction between water molecules due to extensive hydrogen bonding.

**cohesion–tension theory**   A model for sap ascent in the xylem up the stem of the plant, stating that evaporation of water from the leaves at the top of the stem causes a tension (negative hydrostatic pressure) that pulls water up the long water columns in the xylem.

**coincidence model**   A model for flowering in photoperiodic plants in which the circadian oscillator controls the timing of light-sensitive and light-insensitive phases during the twenty-four hour cycle.

**colchicine**   A drug that destroys microtubules and blocks cell division.

**cold acclimation**   A process whereby exposure to low but non-lethal temperatures (usually above freezing) increases the capacity for low temperature survival.

**coleoptile**   A modified ensheathing leaf that covers and protects the young primary leaves of a grass seedling as it grows through the soil. Unilateral light perception, especially blue light, by the tip results in asymmetric growth and bending due to unequal auxin distribution in the lighted and shaded sides.

**collenchyma**   A specialized parenchyma with irregularly thickened, pectin-rich, primary cell walls that function in support in growing parts of a stem or leaf.

**colligative properties**   Properties of solutions that depend on the number of dissolved particles and not on their chemical characteristics.

**columella**   The central cylinder of the root cap.

**columella initials**   Located directly below (distal to) the quiescent center, these cells give rise to the central portion of the root cap.

**columella root cap**   The central region of the root cap that contains the statocytes—cells containing large, dense amyloplasts that function in gravity perception during root gravitropism.

**combinatorial model**   Proposal that during the transition from juvenile to adult shoot in maize, a series of independently regulated, overlapping programs (juvenile, adult, and reproductive) modulate the expression of a common set of developmental processes.

**companion cell**   In angiosperms, a metabolically active cell that is connected to its sieve element by large, branched plasmodesmata and takes over many of the metabolic activities of the sieve element.

In source leaves, it functions in the transport of photosynthate into the sieve elements.

**compatible solutes (compatible osmolytes)**   Organic compounds that are accumulated in the cytosol during osmotic adjustment. Compatible solutes do not inhibit cytosolic enzymes as do high concentrations of ions. Examples of compatible solutes include proline, sorbitol, mannitol, glycine betaine.

**competence**   The capacity of a particular cell or group of cells to respond in the expected manner when given the appropriate developmental signal.

**complementation**   Genetic procedure by which two recessive mutations are introduced into the same cell to discover whether they effect the same genetic function and are therefore alleles. If the *trans* configuration ($m +/+ m_1$) exhibits a mutant phenotype, the mutations are allelic, but if they show a wild-type phenotype, they are nonallelic.

**Complex I**   A protein complex in the mitochondrial electron transport chain that oxidizes NADH and reduces ubiquinone.

**Complex II**   A protein complex in the mitochondrial electron transport chain that oxidizes succinate and reduces ubiquinone.

**Complex III**   A protein complex in the mitochondrial electron transport chain that oxidizes reduced ubiquinone (ubiquinol) and reduces cytochrome *c*.

**Complex IV**   A protein complex in the mitochondrial electron transport chain that oxidizes reduced cytochrome *c* and reduces $O_2$ to $H_2O$.

**Complex V**   *See* $F_oF_1$-ATP synthase.

**compound (or mixed) fertilizers**   Contain two or more mineral nutrients; numbers such as 10–14–10 refer to the effective percentages of nitrogen, phosphorus, and potassium.

**condensed tannins**   Tannins that are polymers of flavonoid units. Require use of strong acid for hydrolysis.

**CONSTANS (CO)**  Gene for a key component of a regulatory pathway that promotes flowering of Arabidopsis in long days; it encodes a protein that regulates the transcription of other genes.

**constitutive**  Constantly present or expressed, whether there is demand or not. Refers to the ongoing synthesis of a particular protein. Contrast with inducible.

**constitutive defenses**  Plant defenses that are always immediately available or operational; that is, defenses that are not induced.

**constitutive ethylene response mutants**  Mutants that show the ethylene triple response in the absence of exogenous ethylene. *See ctr* mutant.

**contact angle**  A quantitative measure of the degree to which a water molecule is attracted to a solid phase versus to itself.

**COP**  *See* critical oxygen pressure.

**COP I protein**  A vesicle-coating protein that directs vesicles involved in the retrograde movement within the Golgi and from the Golgi to the ER.

**COP II protein**  A vesicle-coating protein that directs delivery of vesicle membrane and contents to the Golgi from the ER.

**COP1 gene**  A gene that encodes an E3 ubiquitin ligase that is essential for placing the peptide ubiquitin onto proteins targeted for destruction.

**COP9 signalosome**  A protein that forms a lid-like structure on the proteasome and helps to determine which proteins enter the complex.

**core promoter**  One part of the two part eukaryotic promoter consisting of the minimum upstream sequence required for gene expression.

**cork cambium**  A layer of lateral meristem that develops within mature cells of the cortex and the secondary phloem. Produces the secondary protective layer, the periderm.

**corpus**  The internal region of the shoot apical meristem in which the planes of cell division are not strongly polarized, leading to increases in the volume of the shoot.

**cortex**  The outer layer of the root delimited on the outside by the epidermis and on the inside by the endodermis.

**cortical cytoplasm**  The outer region or layer of cytoplasm adjacent to the plasma membrane.

**cortical–endodermal stem cells**  A ring of stem cells that surround the quiescent center and generate the cortical and endodermal layers in roots.

**cosuppression**  Decreased expression of a gene when extra copies are introduced.

**cotransport**  The simultaneous transport of two solutes by the same carrier. Usually one solute is moving down its chemical-potential gradient, while the other is moving against its chemical-potential gradient. *See* symport and antiport.

**cotyledons**  The one or more seed leaves contained in the seed of seed plants. In some seeds they are storage organs supporting early non-photosynthetic growth of the seedling. In other seeds, they absorb and transmit to the seedling resources stored in the endosperm. *See also* monocot and dicot.

**coumarins**  A group of phenylpropanoid compounds including the phototoxic furanocoumarins, and other substances responsible for the odor of fresh hay.

**coupled reactions**  *See* coupling.

**coupling**  A process by which a chemical reaction releasing free energy is linked to a reaction requiring free energy.

**crassulacean acid metabolism (CAM)**  A biochemical process for concentrating $CO_2$ at the carboxylation site of rubisco. Found in the family Crassulaceae (*Crassula, Kalanchoe, Sedum*) and numerous other families of angiosperms. In CAM, $CO_2$ uptake and fixation take place at night, and decarboxylation and reduction of the internally released $CO_2$ occur during the day.

**CRE1 gene**  Arabidopsis gene that encodes a cytokinin receptor protein, similar to bacterial two-component histidine kinases.

**creep**  pH-dependent cell wall extension. Contributes to cell wall expansion along with polymer integration and wall stress relaxation.

**cristae**  Folds in the inner mitochondrial membrane that project into the mitochondrial matrix.

**critical concentration (of a nutrient)**  The minimum tissue content of a mineral nutrient that is correlated with maximal growth or yield.

**critical day length**  The minimum length of the day required for flowering of a long-day plant; the maximum length of day that will allow short-day plants to flower. However, studies have shown that it is the length of the night, not the length of the day, that is important. *See* critical night length.

**critical night length**  The night length that must be exceeded for flowering of short-day plants, or for inhibition of flowering in long-day plants.

**critical oxygen pressure (COP)**  The oxygen pressure at which the respiration rate is first decreased by $O_2$ deficiency.

**cross-regulation**  Refers to the interaction of two or more signaling pathways.

**crown gall**  A tumor-forming plant disease resulting from wound infection of the stem or trunk by the soil-dwelling bacterium *Agrobacterium tumefaciens*. A tumor resulting from the disease.

**CRY1 gene**  The gene encoding cryptochrome, a flavoprotein implicated in many blue light responses that has homology with photolyase. Formerly *HY4*, *see also hy4* mutant.

**cry1 mutant**  Arabidopsis mutant that lacks the blue light-stimulated inhibition of hypocotyl elongation.

**cry^s mutant**  *See la* mutant.

**cryptochrome 1 (CRY1)**  A flavoprotein involved in the inhibition of stem elongation and other blue-light responses.

**crystalline**  Pertaining to solids having a highly ordered and repetitive geometric form.

*Csl*  A family of *cellulose synthase-like* genes.

*CslA*  A family of cellulose synthase-like genes that encode synthases for (1,4)- β-D-mannan.

*CslC*  A family of cellulose synthase-like genes that encode synthases for the (1,4)- β-D-glucan backbone of xyloglucan.

*CslF*  A family of cellulose synthase-like genes that encode synthases for 'mixed-linkage' (1,3;1,4)-β-D-glucans.

*CslH*  A family of cellulose synthase-like genes that encode synthases for 'mixed-linkage' (1,3;1,4)-β-D-glucans.

**CSN**  The COP9 signalosome complex of proteins.

*CTR1* **gene**  Encodes for a negative regulator of ethylene responses.

*ctr1* **mutant**  In Arabidopsis, a recessive mutation causing the constitutive expression of the ethylene triple responses (*constitutive triple response 1* = triple response in the absence of ethylene).

**CTR1 protein**  Regulator of the ethylene triple response; resembles serine/threonine protein kinases involved in signal transduction.

**curvature test**  A bioassay for auxin using the curvature of the *Avena* coleoptile in response to asymmetrically applied auxin in an agar block.

**cutan**  A lipid polymer made up of long-chain hydrocarbons that is a constituent of the cuticle. *See* cutin.

**cuticle**  A multilayered structure that coats the outer cell walls of the epidermis and restricts the passage of water and gases into and out of the plant. Includes cutin, cutan, and waxes.

**cutin**  A rigid three-dimensional polymer of hydroxyl-bearing fatty acids that are attached to each other by ester linkages. The principal constituent of the cuticle.

**cyanide**  An inhibitor of complex IV and other heme-containing enzymes.

**cyanide-resistant pathway**  *See* alternative pathway.

**cyanogenic glycosides**  Nonalkaloid, nitrogenous protective compounds that break down to give off the poisonous gas hydrogen cyanide when the plant is crushed.

**cyclic electron flow**  In photosystem I, flow of electrons from the electron acceptors through the cytochrome $b_6 f$ complex and back to P700, coupled to proton pumping into the lumen. This electron flow energizes ATP synthesis but does not oxidize water or reduce $NADP^+$.

**cyclin-dependent protein kinase (CDK)**  Protein kinases that regulate the transitions from $G_1$ to S, and from $G_2$ to mitosis, during the cell cycle.

**cyclins**  Regulatory proteins associated with CDKs that play a crucial role in regulating the cell cycle.

**cycloartenol**  The direct biochemical precursor of all plant steroid compounds, formed from squalene by a series of ring closures.

**cycloheximide**  Inhibits eukaryotic protein synthesis on 80S ribosomes, but does not block protein synthesis in prokaryotes, mitochondria, or chloroplasts.

**cytochalasin B**  A drug that destroys actin filaments.

**cytochrome $b_6 f$ complex**  A large multi-subunit protein containing two b-type hemes, one c-type heme (cytochrome $f$ ), and a Rieske iron–sulfur protein. A nonmobile protein distributed equally between the grana and the stroma regions of the membranes.

**cytochrome c**  A peripheral, mobile component of the mitochondrial electron transport chain that oxidizes complex III and reduces complex IV.

**cytochrome $f$**  A member of the multi-protein cytochrome $b_6 f$ complex that plays a role in electron transport between Photosystem I and II.

**cytochrome P450 monooxygenases (CYPs)**  A generic term for a large number of related, but distinct, mixed-function oxidative enzymes localized on the endoplasmic reticulum. CYPs participate in a variety of oxidative processes, including steps in the biosynthesis of gibberellins and brassinosteroids.

**cytohistological zones**  Regions of the shoot apical meristem showing differences in morphological appearance and rates of mitosis.

**cytokinesis**  In plant cells, following nuclear division, the separation of daughter nuclei by the formation of new cell wall.

**cytokinin oxidase**  Enzyme that inactivates cytokinins by removing the iosprene moiety from adenine or its derivatives.

**cytokinin synthase**  Plant enzyme that transfers the isopentenyl group from isopentenyl diphosphate to AMP to form isopentenyl adenine ribotide, the first unique intermediate in the synthesis of cytokinin. An isopentenyl transferase.

**cytokinins**  Compounds with many developmental effects on plants, including leaf senescence, nutrient mobilization, apical dominance, the formation and activity of shoot apical meristems, floral development, the breaking of bud dormancy, and seed germination. Cytokinins mediate many light-regulated processes, including chloroplast development and metabolism, and expansion of leaves and cotyledons. May exist as free and bound forms. Operationally defined as compounds with biological activities similar to those of *trans*-zeatin.

**cytoplasm**  The cellular matter enclosed by the plasma membrane exclusive of the nucleus, that contains the cytosol, ribosomes, and the cytoskeleton which, in eukaryotes, surrounds intracellular and membrane-limited organelles (chloroplasts, mitochondria, endoplasmic reticulum, etc.).

**cytoplasmic male sterility (*cms*)**  A plant trait caused by mutations in mtDNA in which viable pollen is not formed.

**cytoplasmic streaming**  The coordinated movement of particles and organelles through the cytosol.

**cytorrhysis**  The collapse of the plant cell wall and consequent cell shrinkage due to water loss.

**cytoskeleton**  Composed of polarized microfilaments of actin or microtubules of tubulin, the cytoskeleton helps control the organization and polarity of organelles and cells during growth.

**cytosol**  The colloidal-aqueous phase of the cytoplasm containing dissolved solutes but excluding supramolecular structures, such as ribosomes and components of the cytoskeleton.

*d8* **mutant**  In maize, phenotypically dwarf mutant that is insensitive to added GA.

**day-neutral plant**  A plant whose flowering is not regulated by day length.

**DCMU (dichlorophenyldimethylurea, diuron)**  An herbicide that blocks photosynthetic electron flow by displacing $Q_B$ from the photosystem II reaction center complex. It also causes a partial inhibition of light-stimulated stomatal opening.

**de novo synthesis**  Synthesis and/or assembly from simple molecular species.

**de-differentiation**  Process by which cells lose their differentiated characteristics, reinitiate cell division, and may, under appropriate conditions, regenerate whole plants.

**de-embryonate**  To remove the embryo from a cereal seed, leaving only the starchy endosperm, aleurone layer, and fused seed coat-fruit wall. De-embryonated half-seeds are dependent on exogenous GA to produce α-amylase.

**de-etiolation**  Rapid developmental changes associated with loss of the etiolated form due to the action of light. *See* photomorphogenesis.

**deciduous trees**  Trees that shed their leaves seasonally.

**decussate phyllotaxy**  Phyllotactic pattern in which the two opposite leaves at one node are at right angles to the two leaves at the next node.

**deficiency zone**  Concentrations of a mineral nutrient in plant tissue below the critical concentration that reduces plant growth.

**dehydration-response element (DRE)**  A nine-nucleotide regulatory sequence element found in the promoters of genes regulated by abscisic acid (ABA), a plant hormone.

**DELLA domain**  N-terminal domain consisting of 17 amino acids in GRAS transcriptional regulators that is necessary for their proteolysis.

**DELLA proteins**  A sub-class of GRAS transcriptional regulators that are negative regulators of gibberellin response.

**denitrification**  Process by which soil anaerobic microbes convert nitrate ($NO_3^-$) or nitrite ($NO_2^-$) to the gases, nitrous oxide ($N_2O$) and molecular nitrogen ($N_2$), which are then lost to the atmosphere.

**depolarization**  Refers to a decrease in the usually negative membrane potential difference across the plasma membrane of plant cells. May be caused by the activation of anion channels and loss of anions, such as chloride from the cell interior, which is negative with respect to the outside.

**desiccation postponement**  A plant's ability to maintain tissue hydration when environmental water is limited.

**desiccation tolerance (drought avoidance)**  Plant's ability to function while dehydrated.

**desmotubule**  A narrow tubule of the ER that passes through plasmodesmata and connects the ER in adjacent cells.

**determinate growth**  Inability to grow beyond the mature, reproductive stage due to the loss of meristematic activity. For example, plants that form flowers after the development of a preset numbers of leaves. Flowering times are generally short and all meristems of the plant are used up in the production of flowers.

**determination**  The process by which a cell acquires a fixed identity or fate.

**dextranase**  Catalyzes the endohydrolysis of α-D-1,6 glucosidic linkages in dextran.

**diacylglycerol (DAG)**  A molecule consisting of the three-carbon glycerol molecule to which two fatty acids are covalently attached by ester linkages.

**diaheliotropism**  Leaf movements that maximize light interception by solar tracking and minimize overexposure to light.

**DICER-LIKE 1 (DCL1)**  One of the plant nuclear proteins that convert pri-miRNAs into miRNAs.

**dicot**  One of the two classes of flowering plants characterized by two seed leaves (cotyledons) in the embryo. Contrast with monocot.

**difference in water vapor concentration**  Referring to the difference between the water vapor concentration of the air spaces inside the leaf and that of the air outside the leaf. One of the two major factors that drive transpiration from the leaf.

**differentiation**  Process by which a cell acquires metabolic, structural, and functional properties that are distinct from those of its progenitor cell. In plants, differentiation is frequently reversible, when excised differentiated cells are placed in tissue culture.

**diffuse growth**  A type of cell growth in plants in which expansion occurs more or less uniformly over the entire surface. Contrast with tip growth.

**diffusion**  The movement of substances due to random thermal agitation from regions of high free energy (high concentration) to regions of low free energy (low concentration).

**diffusion coefficient ($D_s$)**  The proportionality constant that measures how easily a specific substance *s* moves through a particular medium. The diffusion coefficient is a characteristic of the substance and depends on the medium.

**diffusion potential**  The potential (voltage) difference that develops across a semipermeable membrane as a result of the differential permeability of solutes with opposite charges (for example $K^+$ and $Cl^-$).

**diffusional resistance**  Restriction posed by the boundary layer and the stomata to the free diffusion of gases from and into the leaf.

**4'-dihydrophaseic acid (DPA)** An inactive product of abscisic acid oxidation.

**dioecious** Refers to plants in which male and female flowers are found on different individuals (e.g., spinach [*Spinacia*] and hemp [*Cannabis sativa*]). Contrast with monoecious.

**dioxygenases** Group of oxygenase enzymes that incorporate the two oxygen atoms from $O_2$ into one or two-carbon compounds.

**direct-inhibition model** Hypothesis for *apical dominance* stating that the growth of lateral buds is directly inhibited by the relatively high auxin concentration in the stem produced by the shoot apex.

**dispersed repeats** A type of repeated sequence that is not restricted to a single location in the genome. May occur as microsatellites or transposons.

**disproportionating enzyme** One of two debranching enzymes that process inappropriately positioned oligosaccharide branches in the construction of starch granules. Catalyzes the transfer of a segment of a α-D-1,4 glucan to a new position in an acceptor, which may be glucose or a 1,4-linked α-D-glucan.

**distal regulatory sequences** Located upstream of the proximal promoter sequences, these *cis*-acting sequences can exert either positive or negative control over eukaryotic promoters.

**diterpenes** Terpenes having twenty carbons, four five-carbon isoprene units.

**DNA microarray analysis** *See* microarray.

**DNA transposons** Dominant group of dispersed repeats found in heterochromatin that can move or be copied from one location to another within the genome of the same cell.

**dolichol diphosphate** Embedded in the ER membrane, this lipid is the assembly site for a branched oligosaccharide (*N*-acetylglucosamine, mannose, and glucose) that will be transferred to the free amino group of one or more asparagine residues of a protein in the ER that is destined for secretion.

**domains** (1) Regions (nucleotide sequences) within the gene that are similar to regions found in other genes. (2) Region of a protein (amino acid sequence) with a particular structure or function. (3) The three major taxonomic groups of living organisms.

**dormancy** A living condition in which growth does not occur under conditions that are normally favorable to growth.

**drought** A period of insufficient water availability for the plant (from precipitation or irrigation) that results in plant water deficit.

**drought avoidance** *See* desiccation tolerance.

**drought escape** Capacity of a plant to grow and complete its life cycle during the wet season, before the onset of drought.

**drought resistance** Plant's capacity to limit and control the consequences of water deficit. Mechanisms include desiccation postponement and desiccation tolerance.

**drought rhizogenesis** During extended drought stress, the roots of some plant species may initiate many additional lateral roots but inhibit their growth until the stress is relieved.

**dry weight** Weight of desiccated (dried) tissue. Often used to measure growth. Avoids variation in water content when fresh weight measurements are made.

**DTT (dithiothreitol)** An inhibitor of the enzyme that converts violaxanthin to zeaxanthin. It inhibits blue light-stimulated stomatal opening. A reducing agent that reduces S—S bonds to –SH groups.

**dynamic photoinhibition** Photoinhibition of photosynthesis in which quantum efficiency decreases but the maximum photosynthetic rate remains unchanged. Occurs in moderate, not high, excess light.

**dynamin** A large GTPase that is involved in the formation of a many vesicles and organelles, including the cell plate.

**E3 ligase** A protein that attaches ubiquitin tags to proteins targeted for destruction by the 26S proteasome.

**early endosome** The small (100 nm) vesicles first formed in endocytosis. Initially they are coated with clathrin, but it is quickly lost. Part of the endomembrane system.

**early genes** *See* primary response genes.

**early response gene** A gene that does not require protein synthesis for its expression and is therefore expressed rapidly in response to a stimulus.

**ectotrophic mycorrhizal fungi** A dense sheath of fungal mycelium around the roots and extending into the surrounding soil; some hyphae may penetrate between, but not into, the root cortical cells.

**edaphic factors** The physical and chemical characteristics of soil that influences plants.

**EGTA (ethylene glycol-bis[β-aminoethyl ether]-N,N,N′,N′-tetraacetic acid)** A compound that chelates calcium ions, preventing their uptake by cells and inhibiting both root gravitropism and the asymmetric distribution of auxin in response to gravity. *See* plasmalemma central control model.

**eicosanoids** A group of substances involved in the mammalian inflammatory response and similar to jasmonic acid found in plants.

***ein* mutants** In Arabidopsis, mutants blocked in their ethylene responses (ethylene-*in*sensitive).

***EIN2* gene** In Arabidopsis, ethylene response gene encoding a protein that may act as a membrane channel protein. *See ein* mutants.

***EIN3* gene** In Arabidopsis, ethylene response gene encoding a transcription factor. *See ein* mutants.

**electrochemical potential** The chemical potential of an electrically charged solute.

**electrochemical proton gradient** The sum of the electrical charge gradient and the pH gradient across the membrane, resulting from a concentration gradient of protons.

**electrogenic transport** Active ion transport involving the net movement of charge across a membrane.

**electron spin resonance (ESR)** A magnetic resonance technique that detects unpaired electrons in molecules. Instrumental measurements that identify intermediate electron carriers in the photosynthetic electron transport system.

**electron transport chain (in the mitochondrion)** A series of protein complexes in the inner mitochondrial membrane linked by the mobile electron carriers ubiquinone and cytochrome *c*, that catalyze the transfer of electrons from NADH to $O_2$. In the process a large amount of free energy is released. Some of that energy is conserved as an electrochemical proton gradient.

**electronegative** Having the capacity to attract electrons and thus producing a slightly negative electric charge.

**electroneutral transport** Active ion transport that involves no net movement of charge across a membrane.

***elf3* mutant** In Arabidopsis, a flowering-time mutant (*early flowering 3*).

**elicitors** Specific pathogen molecules or cell wall fragments that bind to plant proteins and thereby signal for plant defense against a pathogen. *See avr* genes.

**ELISA** Enzyme-linked immunosorbent assay. A very sensitive detection method that uses radioisotopes or chemiluminescent compounds linked to antibodies to detect compounds (like IAA) or proteins in tissue extracts.

**elongation** Growth of the plant axis or cell primarily in the longitudinal direction.

**elongation zone (of a root)** The region of rapid and extensive root cell elongation showing few, if any, cell divisions.

**embryo** Immature plant formed after sexual or asexual reproduction. Found in seed (of seed plants) and consists of an embryonic axis bearing a terminal bud (plumule), root (radicle), and one or more seed leaves (cotyledons).

**embryo dormancy** Seed dormancy that is caused directly by the embryo and is not due to any influence of the seed coat or other surrounding tissues.

**embryo sac** In flowering plants, large oval cell (arising from the megaspore) in the ovule that develops into the female gametophyte and in which fertilization of the egg and development of the embryo take place.

**embryogenesis** The development of a zygote into a multicellular embryo. In plants, the processes of cell division and differentiation that take place in the ovule and immature seed and establishes the basic developmental patterns of the adult plant: the radial pattern of tissues; the apical–basal axis; and the primary meristems.

**embryonic axis** The hypothetical central line of an embryo around which lateral organs are arranged.

***emf* mutant** In Arabidopsis, a mutant of *EMBRYONIC FLOWER* (*EMF*) that produces flowers soon after germination; the shoot apical meristem initiates no vegetative structures, but immediately initiates floral organs during germination.

**endo-(1→4)-β-D-glucanases (EGases)** Enzymes involved in hydrolyzing cell wall β-D-glucans; may help to loosen the wall during auxin-induced cell elongation.

**endocytosis** The formation of small vesicles from the plasma membrane, which detach and move into the cytosol, where they fuse with elements of the endomembrane system.

**endodermis** A specialized layer of cells with a Casparian strip surrounding the vascular tissue in roots and some stems.

**endogenous** Relating to the interior of a living system. Concerning what is found within or originates within such a system.

**endogenous oscillator** An internal, molecular pacemaker that maintains circadian rhythms in a constant light or dark environment.

**endogenous rhythm** A rhythm that persists in the absence of external controlling factors such as light.

**endoreduplication** Cycles of nuclear DNA replication without mitosis resulting in polyploidization.

**endosomes** Early in endocytosis, vesicles that have lost their clathrin coats and moved away from the plasma membrane into the cell interior.

**endosymbiosis** A theory that explains the evolutionary origin of the chloroplast and mitochondrion through formation of a symbiotic relationship between a prokaryotic cell and a simple nonphotosynthetic eukaryotic cell, followed by extensive gene transfer to the nucleus.

**endosymbiotic origin** A theory in which the engulfment of a free-living ancestral prokaryotic cell by a eukaryotic host evolved into a cell with semi-autonomous chloroplasts and mitochondria.

**endosymbiotic theory** The widely accepted proposal that the mitochondrion and the chloroplast each arose when an early prokaryotic cell was engulfed, but not destroyed, by another cell and established a mutually beneficial and enduring relationship. Over time, the original endosymbionts evolved into mitochondria and chloroplasts that were no longer able to live on their own.

**energy transfer** In the light reactions of photosynthesis, the direct transfer of energy from an excited molecule, such as carotene, to another molecule, such as chlorophyll. Energy transfer can also take place between chemically identical molecules, such as chlorophyll-to-chlorophyll transfer.

**enhancement effect** The synergistic (higher) effect of red and far-red light on the rate of photosynthesis, as compared with the sum of the rates when the two different wavelengths are delivered separately.

**enhancers** Positive regulatory sequences located tens of thousands of base pairs away from a gene's start site. Enhancers may be located either upstream or downstream from the promoter.

**entrainment** The synchronizing effects of external controlling factors, such as light and darkness, on the period of biological rhythms.

**envelope** The double-membrane system surrounding the chloroplast or the nucleus. The outer membrane of the nuclear envelope is continuous with the endoplasmic reticulum.

**environmental cues** Environmental conditions, such as changing day length or temperature, that can influence plant development.

**24-epibrassinolide (24-epiBL)** A brassinosteroid hormone differing from brassinolide (BL) in the alkyl side chain at C-24, and possessing slightly lower activity than BL.

**epicotyl** The region of the seedling stem above the cotyledons.

**epidermal–lateral root cap initials** Cells located to the side of the quiescent center. In Arabidopsis, these initials first divide anticlinally to set off daughter cells, which then divide periclinally to form two files of cells that will mature into the lateral root cap and epidermis.

**epidermis (epidermal cells)** The outermost layer of plant cells, typically one cell thick.

**epigenetic modifications** Chemical modifications to DNA and histones that cause heritable changes in gene activity without altering the underlying DNA sequence.

**epimers** A pair of isomers that differ from each other only in their configuration at a single asymmetric center.

**epinasty** A downward curvature of leaves due to asymmetric growth of the petiole. A response to ethylene production during flooding.

**equilibrium** For a particular solute, a state or condition in which there is no gradient of (electro) chemical potential, and therefore no net passive transport of the solute.

**ER exit sites (ERES)** On the ER, specialized sites characterized by the coat protein COP II from which delivery of vesicles to the Golgi occurs.

**ERE** *See* ethylene response element.

**ERE-binding proteins (EREBPs)** Proteins that bind to *ERE* sequences.

***ERF1 (ETHYLENE RESPONSE FACTOR 1)*** A gene that encodes a protein belonging to the ERE-binding protein family of transcription factors.

**escape from photoreversibility** The loss of photoreversibility by far-red light of phytochrome-mediated red light–induced events after a short period of time.

**ESR** *See* electron spin resonance.

**essential element** A chemical element that is part of a molecule that is an intrinsic component of the structure or metabolism of a plant. When the element is in limited supply, a plant suffers abnormal growth, development, or production.

**essential oils** Mixtures of volatile terpenes and other secondary metabolites that give characteristic odors to some plants, e.g., peppermint, lemon, basil, and sage.

**ethephon** An ethylene-releasing compound, 2-chloroethylphosphonic acid, that makes practical the application of the plant hormone ethylene gas under field conditions. Trade name, Ethrel.

**ethylene** Ethylene gas ($CH_2{=}CH_2$) functions as a plant hormone that is synthesized from the amino acid methionine via ACC. Has important effects on plant growth and development, including stimulating or inhibiting elongation of stems, roots, depending on conditions and species; enhances fruit development; suppresses flowering in most species; increases abscission of flowers and fruits. Increases RNA transcription of numerous genes. *See also* ethylene triple response.

**ethylene response element** Key regulatory sequence found in genes regulated by ethylene.

**ethylene triple response** Common responses to ethylene of growing etiolated seedlings of most dicots and of coleoptiles and mesocotyls of grass seedlings (e.g., wheat and oats). Reduced rate of elongation, increased lateral expansion, and swelling in the region below the hook.

**ethylene-insensitive mutants** Mutants that do not respond to ethylene. In Arabidopsis, tall seedlings protruding above short seedlings showing ethylene triple response when grown in the presence of ethylene.

**ethylmethanesulfonate (EMS)** A chemical mutagen that causes the addition of an ethyl group to a nucleotide, and results in a permanent mutation from G/C to A/T at that site.

**etiolated seedlings** Dark-grown seedlings in which the hypocotyl and stem are more elongated, cotyledons and leaves do not expand, and chloroplasts do not mature.

**etiolation** The form and growth of seedlings grown in darkness. A pale, unusually tall and slender appearance, dramatically different from the stockier, green appearance of seedlings grown in the light.

**etioplast** Photosynthetically inactive form of chloroplast found in etiolated seedlings. Does not synthesize chlorophyll or most of the enzymes and structural proteins required for the formation of thylakoids and operation of photosynthesis. Contains an elaborate system of interconnected membrane tubules called the prolamellar body.

***etr1*** In Arabidopsis, a dominant mutation that blocks the response to ethylene, (*et*hylene-*r*esistant *1*).

**euchromatin** The dispersed, transcriptionally active form of chromatin, in contrast to heterochromatin.

**eukaryotic pathway (of lipid synthesis)** In the cytoplasm, the series of reactions for the synthesis of glycerolipids. *See also* prokaryotic pathway.

**evaporative heat loss** Loss of heat due to the cooling resulting from the evaporation of water.

**exodermis (or hypodermis)** In mature roots, an outer layer of cells of the cortex that is relatively impermeable to water.

**exogenous**   Relating to or coming from outside a living system. Originating outside such a system.

**expansins**   Class of wall-loosening proteins that accelerate wall stress relaxation and cell expansion, typically with an optimum at acidic pH. Appears to mediate acid growth.

**expansion**   Cell and tissue growth due to increase in cell size, not cell number.

**export**   The movement of photosynthate in sieve elements away from the source tissue.

**extracellular space (apoplast)**   In plants, the space continuum outside the plasma membrane made up of interconnecting cell walls through which water and mineral nutrients readily diffuse.

**$F_1$**   The ATP-binding part of the $F_oF_1$-ATP synthase.

**F-actin**   Filamentous actin, the form of actin in the polymerized protofilament, which is formed from G-actin monomers.

**F-boxes**   Protein motifs that promote protein–protein interactions. F-box proteins are components of ubiquitin E3 ligase complexes.

**facilitated diffusion**   Passive transport across a membrane using a carrier.

**FACKEL (FK)**   Gene that encodes a sterol C-14 reductase that seems to be critical for pattern formation during embryogenesis. Mutants exhibit pattern formation defects: malformed cotyledons, short hypocotyl and root, and often multiple shoot and root meristems.

**FAD**   *See* flavin adenine dinucleotide.

**feed-forward regulation**   A type of homeostatic regulation in which an early step in a pathway (e.g., gene expression) anticipates a later step in the pathway. Feed-forward regulation can be either positive or negative. In contrast, feed-back regulation occurs when the output of a pathway regulates an earlier step in the pathway. Gibberellin action involves both positive and negative feed-forward regulation.

**Fe–S centers**   Prosthetic groups consisting of inorganic iron and sulfur that are abundant in proteins in respiratory and photosynthetic electron transport.

**fermentation**   *See* fermentative metabolism.

**fermentative metabolism**   The metabolism of pyruvate in the absence of oxygen, leading to the oxidation of the NADH generated in glycolysis to $NAD^+$. Allows glycolytic ATP production to function in the absence of oxygen. *See also* alcoholic fermentation and lactic acid fermentation.

**ferredoxin (Fd)**   A small, water-soluble, iron–sulfur protein involved in electron transport in photosystem I.

**ferredoxin–NADP reductase (FNR)**   The membrane-associated flavoprotein that reduces $NADP^+$ to NADPH, completing the sequence of noncyclic electron transport that begins with the oxidation of water.

**ferredoxin–thioredoxin system**   Three chloroplast proteins (ferredoxin, ferredoxin-thioredoxin reductase, thioredoxin). The concerted action of the three proteins uses reducing power from the photosynthetic electron transport system to reduce protein disulfide bonds by a cascade of thiol/disulfide exchanges. As a result, light controls the activity of several enzymes of the Calvin cycle.

**ferulic acid**   A hydroxycinnamic acid that is attached by an ester bond to the arabinose side chains of glucuronoarabinoxylan and aids in cross linking grass cell walls

**$FeS_A$**   *See* Fe–S centers.

**$FeS_B$**   *See* Fe–S centers.

**$FeS_R$**   An iron and sulfer containing protein of the cytochrome $b_6f$ complex, isolated by John Rieske, involved in electron and proton transfer.

**$FeS_X$**   *See* Fe–S centers.

**fiber**   An elongated, tapered sclerenchyma cell that provides support in vascular plants.

**fibrous root system**   The complex root system of monocots, which lack a main root axis; all roots have about the same diameter.

**Fick's first law**   The rate of diffusion is directly proportional to the concentration gradient, defined as the difference in concentration of a substance between two points separated by the distance $\Delta x$.

**field capacity**   The water content of a soil after it has been saturated with water and excess water has been allowed to drain away. The moisture-holding capacity of soils.

**fission**   The process by which portions of a membrane separate from the remaining membrane, forming vesicles.

**fixed carbon**   *See* photosynthate.

**flavin adenine dinucleotide (FAD)**   A riboflavin-containing cofactor that undergoes a reversible two electron reduction to produce $FADH_2$.

**flavin hypothesis**   Discredited suggestion that riboflavin activated by blue light could participate in the in vivo photodestruction of an auxin. Although blue light can reduce a flavin such as riboflavin in vitro, and the reduced flavin can in turn reduce cytochrome *c*, no in vivo role for these photoreactions has been shown.

**flavin mononucleotide (FMN)**   A riboflavin-containing cofactor that undergoes a reversible one or two electron reduction to produce FMNH or $FMNH_2$.

**flavones**   Group of ultraviolet light-absorbing, protective flavonoids that may also attract pollinating insects to flowers. Secreted along with flavonoids into the soil by legume roots, they mediate interaction with nitrogen-fixing symbionts.

**flavonoids**   A large group of plant phenolics with the basic carbon structure of two aromatic rings connected by three carbons. Includes the anthocyanins, the flavones, the flavonols, and the isoflavones. Functions in plant pigmentation, protection against UV irradiation, and defense against herbivores and pathogens.

**flavonols**   Group of ultraviolet light-absorbing, protective flavonoids that may also attract pollinating insects to flowers. Secreted into the soil by legume roots, they mediate interactions with nitrogen-fixing symbionts.

**flavoprotein ferredoxin-NADP reductase (Fp)** In photosystem I, a membrane-associated flavoprotein that reduces $NADP^+$, using electrons from ferredoxin. This reaction completes the noncyclic electron transport that begins with the oxidation of water.

**flippases** Enzymes that "flip" newly synthesized phospholipids across the bilayer from the outer (cytoplasmic) face of the membrane to the inner leaflet, thereby assuring symmetrical lipid composition of the membrane.

**flooding-sensitive** Refers to plants that are severely damaged by 24 hours of anoxia due to flooding of roots.

**flooding-tolerant** Refers to plants that can withstand anoxia temporarily, but not for more than a few days.

**floral evocation** The events occurring in the shoot apex that specifically commit the apical meristem to produce flowers.

**floral homeotic genes** Key regulatory genes that determine the positions and identities of floral organs in flowers.

**floral meristem** Forms floral (reproductive) organs: sepals, petals, stamens, and carpels. May form directly from vegetative meristems or indirectly via an inflorescence meristem.

**floral organ identity genes** Three types of genes that control the specific locations of floral organs in the flower.

**floral organs** Angiosperm organs involved directly or indirectly in sexual reproduction; sepals, petals, stamens, and carpels.

**floral primordia** A primordium is the earliest recognizable stage when a group of cells begins to form a structure. In the developing Arabidopsis flower, four discrete whorls of floral primordia are formed from a floral meristem, and these primordia give rise to sepals, petals, stamens, ovules, style, and stigma.

**floral reversion** The conversion of a floral meristem to a vegetative or inflorescence meristem, causing a shoot or inflorescence to grow directly out of the developing flower.

**floral stimulus** In photoperiodism, signals that are translocated from the leaves to the shoot apical meristem. *See* florigen.

**florigen** The hypothetical, universal flowering hormone synthesized by leaves and translocated to the shoot apical meristem via the phloem. So far, it has not been isolated or characterized.

**flower** Specialized reproductive shoot structure of angiosperms. Consists of nonreproductive organs (sepals and petals) and reproductive organs (stamens and carpels).

**FLOWERING LOCUS C (FLC)** In Arabidopsis, a gene that represses flowering.

**FLOWERING LOCUS T (FT)** The gene coding for the protein that acts as a florigen in Arabidopsis and other species.

**fluence** The number of photons absorbed per unit surface area.

**fluence rate** A unit for the measurement of light falling on a spherical sensor from many directions expressed as watts per square meter ($W\ m^{-2}$) or moles of photons per square meter per second (mol $m^{-2}\ s^{-1}$). *See* irradiance.

**fluid-mosaic model** The common molecular lipid–protein structure for all biological membranes. A double layer (*bilayer*) of polar lipids (phospholipids or, in chloroplasts, glycosylglycerides) has a hydrophobic, fluid-like interior. Membrane proteins are embedded in the bilayer and may move laterally due to its fluid-like properties.

**fluorescence** Following light absorption, the emission of light at a slightly longer wavelength (lower energy) than the wavelength of the absorbed light.

**fluorescence resonance energy transfer** The physical mechanism by which excitation energy is conveyed from the pigment that absorbs the light to the reaction center.

**flux density ($J_s$)** The rate of transport of a substance $s$ across a unit area per unit time. $J_s$ may have units of moles per square meter per second [mol $m^{-2}\ s^{-1}$].

**FMN** *See* flavin mononucleotide.

**$F_oF_1$-ATP synthase** A multiprotein complex associated with the inner mitochondrial membrane that couples the passage of protons across the membrane to the synthesis of ATP from ADP and phosphate. The subscript 'o' in $F_o$ refers to the binding of the inhibitor oligomycin. Similar to $CF_o$–$CF_1$ ATPsynthase in photophosphorylation.

**foliar application** The application and absorption of some mineral nutrients to leaves as sprays.

**formative divisions** Cell divisions that form new cell files in an axis, as in a root apical meristem, usually longitudinally oriented.

**free energy** *See* Gibbs free energy.

**free-running** Designation of the biological rhythm that is characteristic for a particular organism when environmental signals (zeitgebers) are removed, as in total darkness.

**freezing injury** Injury that occurs when plants are cooled below the freezing point of water. Contrast with chilling injury.

**frequency (ν)** A unit of measurement that characterizes waves, in particular light energy. The number of wave crests that pass an observer in a given time.

**fresh weight** Weight of living tissue.

**fruit** In angiosperms, one or more mature ovaries containing seeds and sometimes adjacent attached parts.

**fruit set** The start of fruit growth following pollination.

**furanocoumarins** Group of coumarins with attached furan rings whose toxicity may result from exposure to light.

**fusicoccin** A fungal toxin that induces acidification of plant cell walls by activating an $H^+$-ATPase in the plasma membrane. Fusicoccin stimulates rapid acid growth in stem and coleoptile sections. It also stimulates stomatal opening by stimulating proton pumping at the guard cell plasma membrane.

**fusiform stem cells** Elongated, vacuolate stem cells of the vascular cambium that divide longitudinally and whose derivatives form the conducting cells of the secondary xylem and phloem.

**G$_1$** The phase of the cell cycle preceding the synthesis of DNA.

**G$_2$** The phase of the cell cycle following the synthesis of DNA.

**G lignin** A form of lignin made from coniferyl alcohol; it is distinct from H lignin and S lignin.

**G protein** GTP-binding protein involved in signal transduction.

**G protein–coupled receptors (GPCRs)** In animals, a large diverse group of receptors that detect a diverse array of signals ranging from hormones, odors, flavors, and even light. GPCRs signal via hetero-trimeric G proteins, which are encoded by a large number of genes.

**GA** *See* gibberellins.

**GA1 gene** Gene encoding enzyme early in the gibberellin biosynthetic pathway.

*ga1* **mutant** In Arabidopsis, dwarf mutants deficient in biologically active gibberellins.

**GA$_1$** *See* gibberellin A$_1$.

**GA$_3$** *See* gibberellin A$_3$.

**GA$_4$** *See* gibberellin A$_4$.

**GA 2-oxidase** An enzyme that deactivates gibberellins.

**GA 20-oxidase** An enzyme in stage 3 of the gibberellin biosynthetic pathway.

**GA 3-oxidase** An enzyme in stage 3 of the gibberellin biosynthetic pathway.

**GA$_{12}$-aldehyde** The end-product of the second stage of gibberellin biosynthesis and the precursor for all other gibberellins.

**G-actin** The globular, monomeric form of actin from which F-actin is formed.

**GA-INSENSITIVE DWARF1 (GID1)** Gibberellin receptor protein in rice.

*GA-insensitive dwarf (gid)1a, b, or c* Arabidopsis mutants that do not grow when treated with bioactive GA. Loss-of-function mutations in GA receptor proteins that produce dwarf phenotypes

**GA receptor** Protein to which bioactive GA binds, initiating a signal transduction pathway that leads to GA-associated responses. The GID1 protein has been identified as a soluble GA receptor.

**GA response elements (GAREs)** Promoter sequences conferring GA responsiveness are located 200 to 300 base pairs upstream of the transcription start site.

**GABA shunt** A pathway supplementing the citric acid cycle with the ability to form and degrade GABA.

**GAF domain** Part of the N-terminal half of phytochrome with bilin-lyase activity.

*GAI gene* In Arabidopsis, encodes a Della protein.

**galactan** A cell wall polysaccharide composed of galactose residues.

**gall** An unorganized mass of tumorlike plant tissue resulting from uncontrolled cell division.

**GAMYB** A MYB eukaryotic transcription factor implicated in GA signaling. Barley GAMYB is similar to that of three MYB proteins in Arabidopsis.

**gas chromatography (GC)** Method that separates components of a mixture according to their affinity for inert column material versus their tendency to volatilize and move through the column on a current of inert gas.

**gate** A structural domain of the channel protein that opens or closes the channel in response to external signals such as voltage changes, hormone binding, or light.

**gating** The function of the biological clock that regulates when physiological or molecular responses will occur within a period of the circadian rhythm.

**G-box** A specific sequence of DNA nucleotides that bind G-box

type *cis*-acting transcription factors, leading to the transcriptional activation of genes.

**GC box** A sequence of nucleotides involved in the initiation of transcription in eukaryotes.

**GDHs** *See* glutamate dehydrogenase.

**gel** A highly-hydrated, dispersed network of long polymers, typically with elastic properties intermediate between that of a liquid and that of a solid.

**gene activators** Proteins that, either alone or in concert with other proteins, increase gene expression. *See* transcription factor.

**gene fusion** An artificial construct that links a promoter for one gene with the coding sequence of another gene. Often includes a reporter gene, such as green fluorescent protein gene, *GFP*, that produces a readily detected protein.

**gene gun** *See* biolistics.

**general transcription factors** Proteins that are required by RNA polymerases of eukaryotes for proper positioning at the transcription start site.

**genetic tumors** Spontaneous tumors produced by certain genotypes. Form in about ten percent of interspecific crosses of the genus *Nicotina* due to overproduction of cytokinin.

**genome** Refers to all the genes in a haploid complement of eukaryotic chromosomes, in an organelle, a microbe, or the DNA or RNA content of a virus.

**geranylgeranyl diphosphate (GGPP)** A precursor in the synthesis of gibberellins.

**germinaton** The beginning or resumption of growth by a spore, seed, or bud.

**GGPP** *See* geranylgeranyl diphosphate.

*GH3* **genes** In soybean and Arabidopsis, family of primary response genes that are stimulated by auxin within 5 minutes of treatment.

*Gibberella fujikuroi* The fungal pathogen of rice in which gibberellins were originally identified.

**gibberellic acid** Identical to gibberellin A$_3$.

**gibberellin A$_1$ (GA$_1$)** A chemically distinct form of gibberellin. The primary active GA in stem growth for most species.

**gibberellin A$_3$ (GA$_3$)** The main gibberellin found in fungal cultures; commonly available and used in the management of fruit crops, the malting of barley, and the extension of sugarcane (to increase sugar yield). Occurs rarely in plants.

**gibberellin A$_4$ (GA$_4$)** Thought to be the primary active gibberellin for stem growth in Arabidopsis.

**gibberellin glycosides** Inactive or stored forms of gibberellin in which the hormone is covalently linked to a sugar (usually glucose).

**gibberellin A$_{12}$-aldehyde** *See* GA$_{12}$- aldehyde.

**gibberellins (GAs)** A large group of chemically related plant hormones synthesized by a branch of the terpenoid pathway and associated with the promotion of stem growth (especially in dwarf and rosette plants), seed germination, and many other functions. *See* gibberellic acid, Gibberellin A$_1$.

**Gibbs free energy** The energy that is available to do work; in biological systems the work of synthesis, transport, and movement.

**GID1 gene** *Gibberellin insensitive dwarf 1* encodes a soluble gibberellin receptor in rice.

**GID2 gene** *Gibberellin insensitive dwarf 2* encodes an F-box protein in rice that targets SLR repressor protein for proteolytic degradation.

**girdling** Removal of a ring of bark from a woody stem that severs the vascular system.

**globular stage embryo** The first stage of embryogenesis. A radially symmetrical, but not developmentally uniform, sphere of cells produced by initially synchronous cell divisions of the zygote. *See* heart stage, torpedo stage.

**glucan** A polysaccharide made from glucose units.

**(1→3; 1→4)β-D-glucan** Mixed-linkage glucan found in cell walls of grasses. It may bind tightly to cellulose surface, producing a less sticky network.

**glucomannan** A polysaccharide made from both glucose and mannose units.

**gluconeogenesis** The synthesis of carbohydrates through the reversal of glycolysis.

**glucose-6-phosphate dehydrogenase** Cytosolic and plastidic enzyme catalyzing the initial reaction of the oxidative pentose phosphate pathway.

**glucosinolates (mustard oil glycosides)** Plant glycosides that break down to release volatile substances that defend against herbivores and are responsible for the odor and flavor of broccoli, cabbage, and other cruciferous vegetables.

**glucosinolates** A class of plant glycosides that break down to release defense compounds. Found principally in the Brassicaceae and related plant families.

**glucuronoarabinoxylan** A hemicellulose with a (1,4)-linked backbone of β-D-xylose (Xyl) and side chains containing arabinose (Ara) and 4-O-methyl-glucuronic acid (4-O-Me-α-D-GlcA).

**glutamate dehydrogenase (GDH)** Catalyzes a reversible reaction that synthesizes or deaminates glutamate as part of the nitrogen assimilation process.

**glutamate synthase** Enzyme that transfers the amide group of glutamine to 2-oxoglutarate, yielding two molecules of glutamate. Also known as glutamine:2-oxoglutarate aminotransferase (GOGAT).

**glutamine synthetase GS** Catalyzes the condensation of ammonium and glutamate to form glutamine. Reaction is critical for the assimilation of ammonium into essential amino acids. Two forms of GS exist—one in the cytosol and one in chloroplasts.

**glutathione** A short (3 amino acids) peptide that is one of the major cellular antioxidants. It is also involved in long-distance signaling between organs.

**glycan** A general term for a polymer made up of sugar units; it is synonymous with polysaccharide.

**glyceroglycolipids** Glycerolipids in which sugars form the polar head group. Glyceroglycolipids are the most abundant glycerolipid in chloroplast membranes.

**glycerolipids** Polar lipids that form the lipid bilayer of cellular membranes.

**glycerophospholipids** Polar glycerolipids in which the hydrophobic portion consists of two 16-carbon or 18-carbon fatty acid chains esterified to positions 1 and 2 of a glycerol backbone. The phosphate-containing polar head group is attached to position 3 of the glycerol.

**glycine oxidation** The part of the photorespiratory carbon cycle in which glycine is converted into serine in the mitochondrial matrix, producing NADH, CO$_2$, and NH$_4^+$.

**glycolysis** A series of reactions in which a sugar is oxidized to produce two molecules of pyruvate. A small amount of ATP and NADH is produced.

**glycophytes** Plants that are not able to resist salts to the same degree as halophytes. Show growth inhibition, leaf discoloration, and loss of dry weight at soil salt concentrations above a threshold. Contrast with halophytes.

**glycoside** Compound containing an attached sugar or sugars.

**glycosidic bond** A bond between the C-1 of glucose and the oxygen atom of another hexose moeity sugar can be linked to each other by O-glycosidic bonds to form oligo- and polysaccharides.

**glycosyl transferase family 43** Group of synthases that synthesize the backbone of xylan polysaccharides.

**glycosylglycerides** Polar lipid molecules that are found in the chloroplast membrane. In glycosylglycerides, there is no phosphate group, and the polar head group is galactose, digalactose, or a sulfated galactose.

**glyoxylate** A two-carbon acid aldehyde that is an intermediate of the glyoxylate cycle.

**glyoxylate cycle** The sequence of reactions that convert two molecules of acetyl-CoA to succinate in the glyoxysome.

**glyoxysome** An organelle found in the oil-rich storage tissues of seeds in which fatty acids are oxidized. A type of microbody.

**glyphosate resistance** Genetic capacity to survive a field application of the commercial herbicide Roundup, which kills weeds but does not harm resistant crop plants.

*GNOM* **gene** Arabidopsis gene for the development of roots and cotyledons. Homozygous *GNOM* mutant produces seedlings lacking both roots and cotyledons.

*GNOM (GN)* Gene that encodes a guanine nucleotide exchange factor, which establishes a polar distribution of PIN auxin efflux carriers, enabling the polar distribution of auxin.

**GOGAT** *See* glutamate synthases.

**Goldman diffusion potential** The diffusion potential calculated from the Goldman equation.

**Goldman equation** An equation that predicts the diffusion potential across a membrane, as a function of the concentrations and permeabilities of all ions (e.g., $K^+$, $Na^+$ and $Cl^-$) that permeate the membrane.

**gossypol** An aromatic 30-carbon sesquiterpene dimer that defends against insects, fungi, and bacterial pathogens in cotton.

**grana lamellae** Stacked thylakoid membranes within the chloroplast. Each stack is called a granum, while the exposed membranes in which stacking is absent are known as stroma lamellae.

**granum (plural grana)** In the chloroplast, a stack of thylakoids.

**GRAS domain** The C-terminal domain in GRAS proteins that functions as a transcriptional repressor.

**GRAS proteins** A family of transcriptional regulators named after the first three to be characterized (GAI, RGA and SCR).

**gravistimulation** A multicomponent process whereby gravity-sensing mechanisms detect that the root or shoot axis is out of alignment with the vertical orientation established by gravity. Signal transduc-

tion mechanisms initiate corrective growth.

**gravitropic response** The growth initiated by the root cap's perception of gravity and the signal that directs the roots to grow downward.

**gravitropism** Plant growth in response to gravity, enabling roots to grow downward into the soil and shoots to grow upward.

**gravity potential, $\Psi_g$** A component of water potential determined by the effect of gravity on the free energy of water.

**Green Revolution** The introduction of high-yielding dwarf varieties of wheat and rice into Latin America and Southeast Asia in the 1960s to keep abreast of human population growth.

**green-leaf volatiles** A mixture of lipid-derived 6-carbon aldehydes, alcohols, and esters released by plants in response to mechanical damage.

**greenhouse effect** The warming of Earth's climate, caused by the trapping of long-wavelength radiation by $CO_2$ and other gases in the atmosphere. Term derived from the heating of a greenhouse resulting from the penetration of long-wavelength radiation through the glass roof, the conversion of the long-wave radition to heat, and the blocking of the heat by the glass roof.

**ground meristem** In the plant meristems, cells that will give rise to the cortical and pith tissues and, in the root and hypocotyl, will produce the endodermis.

**growth respiration** The respiration that provides the energy needed for converting sugars into the building blocks that make up new tissue. Contrast to maintenance respiration.

*GSA* **gene** Gene of unicellular alga *Chlamydomonas reinhardtii* whose expression is mediated solely by a blue light-sensing system. Encodes the enzyme glutamate-1-semialdehyde aminotransferase, a key enzyme in the chlorophyll biosynthesis pathway.

**GT43** *See* glycosyl transferase family 43.

**guard cell protoplasts** Protoplasts prepared from guard cells by removing their walls through application of enzymes that degrade cell wall components.

**guard cells** A pair of specialized epidermal cells which surround the stomatal pore and regulate its opening and closing.

*GURKE (GK)* Gene involved in pattern formation. Encodes an acetyl-CoA carboxylase that is required for the proper synthesis of very-long-chain fatty acids and sphingolipids, which are involved in the proper patterning of the apical portion of the embryo.

**GUS** β-Glucuronidase from bacteria used as a reporter when fused to the promoter of a gene of interest in transgenic plants. Activity shows as blue staining or fluorescence depending on the substrate used to assay its activity.

**guttation** An exudation of liquid from the leaves due to root pressure.

**gymnosperm** An early type of seed plant. Distinguished from angiosperm by having seeds borne unprotected (naked) in cones.

**H lignin** A lignin made from *p*-coumaryl alcohol; it is distinct from G lignin and S lignin.

**halophytes** Plants that are native to saline soils and complete their life cycles in that environment. Contrast with glycophytes.

**hartig net** A fungal network of hyphae that surround but do not penetrate the cortical cells of roots.

*Heading-date 1( Hd1)* A gene for a *CO* homolog that acts as an inhibitor of flowering in rice.

*Heading-date 3a (Hd3a)* The gene for the FT-like protein in rice that is translocated via the sieve tubes to the apical meristem, where it stimulates flowering.

**heart stage embryo** The second stage of embryogenesis. A bilaterally symmetrical structure produced by rapid cell divisions in two regions on either side of the future shoot apex. *See* globular stage, torpedo stage.

**heat shock cognate proteins**
Molecular chaperon proteins that are constitutively expressed and function like HSPs.

**heat shock proteins (HSPs)**  A specific set of proteins that are induced by a rapid rise in temperature, and by other factors that lead to protein denaturation. Most act as molecular chaperones.

**heat shock response**  The increased synthesis of HSPs and the reduced synthesis of other proteins following a stressful but nonlethal heat episode.

**heliotropism**  Movements of leaves toward or away from the sun.

**hemibiotrophic**  Plant pathogens that show an initial biotrophic stage, which is followed by a necrotrophic stage, in which the pathogen causes extensive tissue damage.

**hemicelluloses**  Heterogeneous group of polysaccharides that bind to the surface of cellulose, linking cellulose microfibrils together into a network. Typically solubilized by strong alkali solutions.

**herb**  A plant with no enduring above-ground parts, as distinct from trees and shrubs.

**herbaceous**  *See* herb.

**herbivores**  Plant-feeding animals, including many species of insects and mammals.

**hermaphroditic flowers**  Containing both carpels and stamens. *See* perfect flowers.

**heterochromatin**  Chromatin that is densely packed, darkly staining, and transcriptionally inactive; it accounts for about 10% of the nuclear DNA.

**heterochromatization**  The condensation of euchromatin into heterochromatin, resulting in gene silencing.

**heterocyclic ring**  A ring structure that contains both carbon and non-carbon (nitrogen or oxygen) atoms.

**heterocysts**  Specialized, thick-walled structures formed by cyanobacteria within plant cells to create an anaerobic environment for nitrogen fixation.

**heterosis**  The increased vigor often observed in the offspring of crosses between two inbred varieties of the same plant species.

**heterotrimeric G protein**  A membrane-bound GTP-binding protein composed of three subunits, α, β, and γ; mediates the signal transduction pathways of G-protein-linked membrane receptors.

**heterotrophic cells**  Organisms that rely upon reduced forms of carbon for the synthesis of structural elements and the supply of metabolic energy.

**hexose monophosphate shunt**  *See* pentose phosphate pathway.

**hexose phosphates**  Six-carbon sugars with phosphate groups attached.

**HIR (high irradiance response)**
Phytochrome responses whose magnitude is proportional to the irradiance (rather than fluence). HIRs saturate at fluences 100 times higher than LFRs and are not photoreversible. HIRs do not obey the law of reciprocity.

**histidine phosphotransfer protein (Hpt)**  Proteins active in the signal transduction pathway for cytokinin. Hpts act in the primary cytokinin signal transduction pathway to shuttle phosphates from the AHK receptors to the ARR proteins.

**histogenesis**  The differentiation of cells to produce various tissues.

**histogenic layers**  The tissue-producing regions of the shoot apical meristem: tunica layers and corpus.

**histone code**  The set of covalent modifications that favor the formation of either heterochromatin or euchromatin.

**histones**  A family of proteins that interact with DNA and around which DNA is wound to form a nucleosome.

**Hoagland solution**  A nutrient solution for plant growth, originally formulated by Dennis R. Hoagland.

**holoprotein**  An apoprotein bound to a smaller, nonprotein molecule such as a chromophore.

**homeobox**  A sequence coding for the homeodomain, a 60-amino acid domain in transcription factors that binds to specific regions of DNA.

**homeotic genes**  First discovered in *Drosophila*, homeotic genes regulate the locations of body parts in flies. These transcription factors are typically characterized by homeodomains.

**homeotic mutations**  In *Drosophila*, homeotic mutations in homeobox genes cause body segments and other structures to form at inappropriate locations. In plants, mutations with similar phenotypic effect on flower development, but do not involve homeobox genes.

**28-homobrassinolide (28-homo-BL)**  A brassinosteroid hormone differing from brassinolide (BL) in the alkyl side chain at C-24, and possessing slightly lower activity than BL.

**homogalacturonan**  This pectin polysaccharide is a (1,4)-linked polymer of β-D-galacturonic acid residues; also called polygalacturonic acid.

**hormone**  Organic molecule often (but not always) synthesized in one location of the organism and transported to another, where it produces dramatic effects on growth or development at vanishingly low concentrations. *See* plant hormone.

**hourglass hypothesis**  One hypothesis of how plants measure night length. Proposes that time is measured by a unidirectional series of biochemical reactions that start at the beginning of the dark period. *See* clock hypothesis.

**hpt protein**  In Arabidopsis, *hi*stidine *p*hospho*t*ransfer protein.

**H$^+$-pyrophosphatase (H$^+$-PPase)**
An electrogenic pump that moves protons into the vacuole, energized by the hydrolysis of pyrophosphate.

**HSF**  A specific transcription factor that acts during the transcription of heat shock proteins.

**hy mutants**  Arabidopsis mutants in which white light does not inhibit hypocotyl growth as it does in wild-type. Some *hy* mutants cannot synthesize phytochrome.

*hy4* **mutant**   *See cry1* mutant.

**hybrid vigor**   *See* heterosis.

**hydathodes**   Specialized pores associated with vein endings at the leaf margin from which xylem sap may exude when there is positive hydrostatic pressure in the xylem. Also a site of auxin synthesis in immature leaves of Arabidopsis.

**hydraulic conductivity**   Describes how readily water can move across a membrane; it is expressed in terms of volume of water per unit area of membrane per unit time per unit driving force (i.e., $m^3$ $m^{-2}$ $s^{-1}$ $MPa^{-1}$).

**hydroactive stomatal closure**   Closing of stomata in response to a closing signal which includes water stress. Depends on metabolic processes that reduce guard cell solute content resulting in water loss. Reversal of the mechanism of stomatal opening.

**hydrogen bond**   A weak chemical bond formed between a hydrogen atom and an oxygen or nitrogen atom.

**hydrolyzable tannins**   Tannins that are polymers of phenolic acids, especially gallic acid, and simple sugars. May be hydrolyzed by dilute acid.

**hydropassive closure**   Closing of stomata due to loss of water directly from guard cells and the consequent loss of cell turgor. Operates at low external humidity, when direct water loss is not rapidly compensated.

**hydrophilic**   Ability of an atom or a molecule to attract water molecules. For example, substances that can engage in hydrogen bonding are hydrophilic.

**hydrophobic**   Substances, molecules, or functional groups that repel water molecules.

**hydroponics**   A technique for growing plants with their roots immersed in nutrient solution without soil.

**hydroquinone ($QH_2$, quinol)**   A fully reduced form of quinone.

**hydrostatic pressure**   Pressure generated by compression of water into a confined space. Measured in units called pascals (Pa) or, more conveniently, megapascals (MPa).

**hydroxyproline-rich proteins (HRGPs)**   A class of wall structural proteins, rich in hydroxyproline, with possible roles in protection against pathogens and desiccation.

**hyperpolarization**   An increase in the normally negative inside, electrical potential (mV) that exists across the plasma membrane of a cell.

**hypersensitive response**   A common plant defense following microbial infection, in which cells immediately surrounding the infection site die rapidly, depriving the pathogen of nutrients and preventing its spread.

**hyphae (singular hypha)**   The small, tubular filaments of fungi.

**hypocotyl**   The region of the seedling stem below the cotyledons and above the root.

**hypocotyl hook**   An inverted "J" formed when the apical end of the dicot hypocotyl bends back on itself as dicot seedlings emerge from the seed coat. Protects the shoot apex from damage during growth through the soil.

**hypophysis**   In seed plant embryogenesis, the apical-most progeny of the basal cell which contributes to the embryo and will form part of the root apical meristem.

**hypoxic**   Refers to oxygen concentrations (pressures) lower than normal. *02.* also anoxic.

**IAA**   *See* indole-3-acetic acid.

**IAM**   *See* indole-3-acetamide.

**IAN**   *See* indole-3-acetonitrile.

**ice nucleation**   Process by which water molecules start to form a stable ice crystal. The surface properties of some polysaccharides and proteins facilitate ice crystal formation.

**immunocytochemistry**   The use of specific antibodies attached to identifying molecules in order to indicate the presence and perhaps location of molecules with antigenic properties.

**imperfect flowers**   Flowers that lack either the male (stamens) or female (pistils) structures. Unisexual flowers.

**import**   The movement of photosynthate in sieve elements into sink organs.

**in vitro**   Biological experiments performed in "the test tube" isolated from a whole organism. Contrast with in vivo.

**in vivo**   Within the intact organism. Contrast with in vitro.

**indeterminate growth**   Capacity for both vegetative growth and flowering over an extended period of time. Growth is not genetically limited, and will continue as long as environmental conditions and resources permit.

**indole**   A precursor of IAA biosynthesis by a tryptophan-independent pathway.

**indole-3-acetamide (IAM)**   A biosynthetic intermediate for the synthesis of IAA in various pathogenic bacteria, such as *Pseudomonas savastanoi* and *Agrobacterium tumefaciens*.

**indole-3-acetic acid (IAA)**   The most common naturally occurring auxin. When protonated, IAA is lipophilic and diffuses across lipid bilayer membranes, but when dissociated, the negatively charged IAA cannot cross membranes unaided. Free IAA is biologically active; IAA covalently bound to other molecules is inactive.

**indole-3-acetonitrile (IAN)**   An intermediate in one of three biosynthetic pathways for the synthesis of IAA in plants. IAN converted to IAA by action of nitrilase.

**indole-3-glycerol phosphate**   A precursor of IAA biosynthesis by a tryptophan-independent pathway.

**indole-3-pyruvic acid (IPA)**   An intermediate in one of three biosynthetic pathways for the synthesis of IAA in plants. Formed by deamination of the amino acid tryptophan.

**induced defenses**   Plant defenses that are initiated only after actual damage occurs.

**induced systemic resistance (ISR)**   Plant defenses that are activated by non-pathogenic microbes such as rhizobacteria. A defense response elicited by a local infection is mediated by JA and ethylene and that leads to systemic and long-lasting disease resistance, effective against fungi, bacteria, and viruses.

**induced thermotolerance** Tolerance to lethal, high temperatures accomplished by periodic brief exposure to sublethal heat stress.

**inducible** The capacity for increased synthesis of a particular protein or proteins in response to a particular external signal, such as a hormone.

**induction period** The period of time (time lag) elapsed between the perception of a signal and the activation of the response. In the Calvin–Benson cycle, the time elapsed between the onset of illumination and the full activitation of the cycle.

**infection thread** An internal tubular extension of the plasma membrane of root hairs through which rhizobia enter root cortical cells.

**inflorescence meristem** Produces cauline leaves and inflorescence meristems in the axils of the leaves, as well as bracts and floral meristems in the axils of the bracts, and does not directly produce floral organs.

**initials** In the root and shoot meristems, a cluster of slowly dividing and undetermined cells. Their descendants are displaced away by polarized patterns of cell division and take on various differentiated fates, contributing to the radial and longitudinal organization of the root or shoot and to the development of lateral organs.

**inner membrane** The inner of a mitochondrion's or chloroplast's two membranes.

**inositol triphosphate (InsP$_3$)** One of several second messengers that trigger the release of calcium from intracellular stores.

**integral membrane proteins** Proteins that are embedded in the lipid bilayer. Most span the entire width of the bilayer, so one part of the protein interacts with the outside of the cell, another part interacts with the hydrophobic core of the membrane, and a third part interacts with the cytosol.

**integument** The outer tissue layers surrounding the nucellus of an ovule; develops into the seed coat.

**intercalary meristem** Meristem located near the base, rather than the tip, of a stem or leaf, as in grasses.

**intercellular air space resistance** The resistance or hindrance that slows down the diffusion of $CO_2$ inside leaf, from the substomatal cavity to the walls of the mesophyll cells.

**intermediary cell** A type of companion cell with numerous plasmodesmatal connections to surrounding cells, particularly to the bundle sheath cells.

**intermediate-day plant** A plant that flowers only between narrow day length limits (for example between 12 and 14 hours).

**intermembrane space** The fluid-filled space between the two mitochondrial or chloroplast membranes.

**internode** Portion of a stem between nodes.

**interphase** Collectively the G$_1$, S, and G$_2$ phases of the cell life cycle.

**invertase** An enzyme that catalyzes the hydrolysis of sucrose to glucose and fructose.

**inwardly rectifying** Ion channels that open only at potentials more negative than the prevailing Nernst potential for a cation, or more positive than the prevailing Nernst potential for an anion, and thus mediate inward current.

**ionophore** A molecule that allows ions to cross lipid bilayers. There are two classes: carriers and channels. Carriers, like valinomycin, form cage-like structures around specific ions, diffusing freely through the hydrophobic regions of the bilayer. Channels, like gramicidin, form continuous aqueous pores through the bilayer, allowing ions to diffuse through.

**IPA** *See* indole-3-pyruvic acid.

***ipt* gene** The T-DNA gene involved in cytokinin biosynthesis; encodes cytokinin synthase, an isopentenyl transferase enzyme. Also called the *tmr* locus.

**irradiance** The amount of energy that falls on a flat sensor of known area per unit time. Expressed as watts per square meter (W m$^{-2}$). Note, time (seconds) is contained within the term watt: 1 W = 1 joule (J) s$^{-1}$, or as moles of quanta per square meter per second (mol m$^{-2}$ s$^{-1}$), also referred to as fluence rate.

**isoamylase** Catalyzes the hydrolysis of α-D-1,6 glucosidic branch linkages in glycogen, amylopectin, and their β–limit dextrins.

**isoflavonoids (isoflavones)** Group of flavonoids with antimicrobial activity in which the position of one aromatic ring is shifted.

**isopentenyl diphosphate (IPP)** The activated five-carbon building block of terpenes. Formerly named isopentenyl *pyro*phosphate.

**isopentenyl transferase (IPT)** The enzyme catalyzing the rate-limiting step in cytokinin biosynthesis.

**isoprene (2-methyl-1,3-butadiene)** A gaseous, branched five-carbon molecule emitted by many plants. Isoprene emission protects leaves against damage from high temperatures. Provides the structure of the basic five-carbon unit in terpene formation.

**isoprene units** The five-carbon units out of which terpenes, gibberellins, and other natural plant products are constructed.

**isoprenoid** *See* isoprene units.

**isotropic** Characteristic of materials showing uniform structural and mechanical properties in all directions. Contrast with anisotropic.

**isozymes** Proteins that are structurally similar, but not identical, and share the same catalytic activity. Isozymes might be regulated by different mechanisms.

**jasmonate** *See* jasmonic acid.

**jasmonic acid** A plant signaling molecule derived from linolenic acid (18:3) found in membrane lipids. Activates plant defenses against insects, fungal pathogens, and regulates plant growth including the development of anthers and pollen. Activates the expression of genes involved in response to various biotic and abiotic stresses, such as proteinase inhibitors.

**ketose**  Sugar with a terminal ketone group.

**kinases**  Enzymes that have the capacity to transfer phosphate groups from ATP to other molecules. *See* protein kinase.

**kinematics**  The concepts and numerical methods applicable to the motion of fluid particles and the shape changes that the fluids undergo. Useful in analyzing meristematic growth.

**kinesins**  Motor proteins that interact with microtubules and are responsible for movements of organelles during cytoplasmic streaming. Kinesins also bind to the spindle microtubules during cell division.

**kinetin**  Substance originally isolated from autoclaved herring sperm DNA that greatly promotes cell division. Not a naturally occurring cytokinin. Chemically, a derivative of adenine (or aminopurine), 6-furfurylaminopurine. *See* zeatin.

**kinetochore**  The site of spindle fiber attachment to the chromosome in anaphase. A layered structure associated with the centromere that contains microtubule-binding proteins and kinesins that help depolymerize and shorten the kinetochore microtubules.

**KNOLLE**  A target recognition protein involved in vesicle fusion during cell plate formation. Belongs to the SNARE family of proteins.

**Kranz anatomy**  (G: *kranz*: wreath or halo.) The wreathlike arrangment of mesophyll cells around a layer of large bundle-sheath cells. The two concentric layers of photosynthetic tissue surrounds the vascular bundle. This anatomical feature is typical of leaves of $C_4$ plants.

**Krebs cycle**  *See* citric acid cycle.

**L1**  A clonally distinct epidermal layer derived from one set of initials in the shoot apical meristem.

**L2**  A subepidermal layer of cells derived from an internal set of initials in the shoot apical meristem.

**L3**  A centrally positioned layer of cells derived from an internal set of initials in the shoot apical meristem.

***la* mutant**  Occurs in peas, when the *cry^s* mutation is present. Results in an ultra tall constitutive response mutant caused by the loss of function of a negative regulator. *See also cry^s* mutant.

**lactate dehydrogenase (LDH)**  The enzyme that catalyzes the reversible conversion of pyruvate to lactate using the coenzyme NAD.

**lactic acid fermentation**  Reaction in which pyruvate from glycolysis is reduced to lactate using NADH and thus regenerating $NAD^+$.

**late-embryogenesis–abundant (LEA) proteins**  These proteins are involved in desiccation tolerance. They interact to form a highly viscous liquid with very slow diffusion and therefore limited chemical reactions. Encoded by a group of genes that are regulated by osmotic stress first characterized in desiccating embryos during seed maturation.

**late genes**  *See* secondary response genes.

**late response gene**  *See* secondary response genes.

**latent heat of vaporization**  The energy needed to separate molecules from the liquid phase and move them into the gas phase at constant temperature.

**lateral bud**  Undeveloped shoot consisting of an axillary meristem, a short stem and immature leaves, often covered with bud scales and located above the point of attachment of a leaf to the stem.

**lateral meristems**  Secondary meristems found in mature woody stems and roots as cylinders of meristematic cells. Their activity increases stem or root circumference. *See* vascular cambium, cork cambium.

**lateral roots (branch roots)**  Arise from the pericycle in mature regions of the root through establishment of secondary meristems

that grow out through the cortex and epidermis, establishing a new growth axis.

**latex**  A complex, often milky solution exuded from cut surfaces of some plants that represents the cytoplasm of laticifers and may contain defensive substances.

**laticifers**  In many plants, elongated, often interconnected phloem cells that contain rubber, latex, and other secondary metabolites.

**law of reciprocity**  The reciprocal relationship between fluence rate (mol m$^{-2}$ s$^{-1}$) and duration of light exposure characteristic of many photochemical reactions as well as some developmental responses of plants to light. Total fluence depends on two factors: the fluence rate and the irradiation time. A brief light exposure can be effective with bright light; conversely, dim light requires a long exposure time. Also referred to as Bunsen–Roscoe Law.

***LE* gene**  Dominant allele for tall stems in peas, first studied by Mendel; codes for an enzyme, 3β-hydroxylase, that hydroxylates $GA_{20}$ to produce $GA_1$.

***le* mutant**  In peas, a recessive mutant allele that causes a dwarf phenotype by interfering with the synthesis of active gibberellin. Mendel's dwarf gene.

**LEA proteins**  *See* late-embryogenesis-abundant (LEA) proteins.

**leaf laminae**  The blade of a leaf.

**leaf primordia**  Region of the shoot apical meristem that will form a leaf during the normal course of development.

**leaf stomatal resistance**  Resistance to $CO_2$ diffusion imposed by the stomatal pores.

**lectins**  Defensive plant proteins that bind to carbohydrates; or carbohydrate-containing proteins inhibiting their digestion by a herbivore.

**leghemoglobin**  An oxygen-binding heme protein found in the cytoplasm of infected nodule cells that facilitates the diffusion of oxygen to the respiring symbiotic bacteria.

**legume**  A member of the family Leguminosae often associated with rhizobia. Legumes include the pea (*Pisum*), clover (*Trifolium*), broad bean (*Vicia*), lentil (*Lens*), soybean (*Glycine*), kidney bean (*Phaseolus*), peanut (*Arachis*), and southern pea (*Vigna*).

**lenticels**  Porous aggregations of cells on the outer surfaces of woody stems and roots through which gases can enter the plant.

**leucoplasts**  Nonpigmented plastids, the most important of which is the amyloplast.

**LFR (low fluence response)**  Phytochrome responses whose magnitude is proportional to low fluence (1.0 to 1000 $\mu$mol m$^{-2}$). These include the classic red/far-red photoreversible responses.

**LFY gene**  In Arabidopsis, *LEAFY* (*LFY*) is a gene involved in establishing floral meristem identity.

**LHCB gene family**  Genes encoding the chlorophyll *a/b*–binding proteins (also called CAB proteins) of photosystem II. Regulated at the transcriptional level by both circadian rhythms and phytochrome.

**light**  A form of radiant energy with properties of both particles and waves.

**light channeling**  In photosynthetic cells, the propagation of some of the incident light through the central vacuole of the palisade cells and through the air spaces between the cells.

**light compensation point**  The amount of light reaching a photosynthesizing leaf at which photosynthetic $CO_2$ uptake exactly balances respiratory $CO_2$ release.

**light energy**  The energy associated with photons.

**light harvesting complex proteins (LHC proteins)**  Chlorophyll-containing proteins associated with one or the other of the two photosystems in eukaryotic organisms. Also known as chlorophyll *a/b* antenna proteins.

**light scattering**  The randomization of the direction of photon movement within plant tissues due to the reflecting and refracting of light from the many air–water interfaces. Greatly increases the probability of photon absorption within a leaf.

**lignin**  Highly branched phenolic polymer with a complex structure made up of phenylpropanoid alcohols that may be associated with celluloses and proteins. Deposited in secondary walls, it adds strength allowing upward growth and permitting conduction through the xylem under negative pressure. Lignin has significant defensive functions.

**limit dextrinase**  Catalyzes the hydrolysis of $\alpha$-D-1,6 glucosidic linkages in $\alpha$- and $\beta$-limit dextrins of amylopectin and glycogen, and in amylopectin and pullulan.

**limiting factor**  In any physiological process, the factor whose operation limits the rate at which the entire process can operate. For example, leaf photosynthesis can be limited by light or by ambient $CO_2$ concentrations.

**limonoids**  Antiherbivore triterpenoids (30 carbons) which give citrus fruits their bitter taste.

**linear axis**  In most plants, the body pattern in which the root and the shoot are at opposite ends. *See* radial axis.

**lipid bilayer**  The core of cellular membranes formed by two layers of phospholipid molecules facing each other through their nonpolar tails.

**lipid rafts**  Detergent-resistant aggregates of lipids and proteins that may represent transient microdomains of tightly packed fatty acid chains enriched in sphingolipids and sterols. Lipid rafts may play a role in membrane signaling.

**liquid phase resistance**  The resistance or hindrance that slows down the diffusion of $CO_2$ inside a leaf, from the walls of the mesophyll cells to the carboxylation sites in the chloroplast.

**lodging**  The bending of cereal stalks to the ground because of the weight of moisture collecting on the ripened heads. Makes mechanical harvesting ineffective.

**long-day plant (LDP)**  A plant that flowers only in long days (qualitative LDP) or whose flowering is accelerated by long days (quantitative LDP).

**long-distance phloem transport**  Transport via the vascular tissue from source to sink.

**long–short-day plants (LSDPs)**  Plants that flower only after a sequence of long days followed by short days.

**lowest excited state**  The lowest energy electronically excited state attained when chlorophyll that is in a higher energy state gives up some of its energy to the surroundings as heat.

**low-fluence responses (LFRs)**  Phytochrome responses that require the fluence to reach 1.0 $\mu$mol m$^{-2}$, and are saturated at about 1000 $\mu$mol m$^{-2}$. They include most of the red/far-red photoreversible responses, such as the promotion of lettuce seed germination and the regulation of leaf movements.

**lumen**  The cavity or space within a tube or sac, especially the space inside thylakoid membranes.

**LUX genes**  Bacterial genes for luciferase that catalyze a light emitting reaction. Used as a reporter gene in transgenic plants in order to visibly indicate the activity of another gene sharing the same promoter.

**lysophosphatidylcholine**  A product of PLA$_2$ phospholipid degradation that activates protein kinases; present on plant membranes in vitro and may be a second messenger in the auxin signal transduction pathway.

**lytic vacuoles**  Analogous to lysosomes in animal cells, they release hydrolytic enzymes that degrade cellular constituents during senescence.

**macromolecule degradation**  The process of ubiquitin-dependent, proteasome proteolysis; may have a role in senescence.

**macronutrient**  Minerals obtained from the soil and present in plant tissues at concentrations usually greater than 30 $\mu$mol g$^{-1}$ dry matter. Nitrogen, potassium, calcium, magnesium, phosphorus, sulfur, silicon.

**MADS box genes**  Genes encoding a family of transcription factors containing a conserved sequence called the MADS box. It is the family that includes most floral homeotic genes and some of the genes involved in regulating flowering time.

**MADS domain**  A conserved DNA-binding/dimerization region present in MADS box family transcription factors.

**magnesium dechelatase**  The enzyme that removes magnesium from chlorophyll as part of the chlorophyll breakdown process.

**maintenance respiration**  The respiration needed to support the function and turnover of existing tissue. Contrast to growth respiration.

**malic enzyme**  Catalyzes the oxidation of malate to pyruvate, permitting plant mitochondria to oxidize malate or citrate to $CO_2$ without involving pyruvate generated by glycolysis.

**malting**  The first step in the brewing process. Germination of barley seeds at temperatures that maximize the production of hydrolytic enzymes.

**malto-oligosaccharides (maltose, malto*triose*, malto*tetraose*, malto*pentaose*, malto*hexaose*)**  The series of linear oligosaccharides composed of two, three, four, five and six, respectively, units of glucose all linked via an α-D-1,4 bond.

**maltose phosphorylase**  Catalyzes the phosphorolysis of maltose yielding D-glucose and β-D-glucose 1-phosphate.

**map-based cloning**  A technique that uses genetic analysis of the offspring of crosses between a mutant and a wild-type plant to narrow the location of the mutation to a short segment of the chromosome, which can then be sequenced.

**MAP kinase cascade (*mitogen-activated protein kinase cascade*)**  The binding of a ligand signal that results in the phosphorylation and activation of a series of kinase enzymes.

**marginal meristems**  Proliferative tissues that are flanked by differentiated tissues at the edges of developing organs.

**mass spectrometry (MS)**  A method that identifies chemical compounds by their molecular mass to charge (m/z) ratios and fragmentation patterns. MS can detect as little as $10^{-12}$ g (1 picogram, or pg) of IAA in plant extracts.

**mass transfer rate**  The quantity of material passing through a given cross section of phloem or sieve elements per unit time.

**material**  Pertaining to matter, the amount and kind of matter under consideration.

**maternal inheritance**  A non-Mendelian pattern of inheritance in which offspring receive genes from only the female parent.

**matric potential, $\Psi_m$ (or matric pressure)**  The sum of osmotic potential ($\Psi_s$) + hydrostatic pressure ($\Psi_p$). Useful in situations (dry soils, seeds, and cell walls) where the separate measurement of $\Psi_s$ and $\Psi_p$ is difficult or impossible.

**matrix polysaccharides**  Polysaccharides comprising the matrix of plant cell walls. In primary cell walls they consist of pectins, hemicelluloses, and proteins.

**matrix**  The colloidal-aqueous phase contained within the inner membrane of a mitochondrion.

**matrixules**  Protrusions of the outer and inner membrane in mitochondria.

**maturation zone**  The region of the root that has completed its differentiation and shows root hairs for the absorption of water and solutes; and competent vascular tissue.

**maximum quantum yield**  Ratio between photosynthetic product and the number of photons absorbed by a photosynthetic tissue. In a graphic plot of photon flux and photosynthetic rate, the quantum yield is given by the slope of the linear portion of the curve.

**MCP**  A strong competitive inhibitor of ethylene binding 1-methylcyclopropene.

**megapascals (MPa)**  $10^6$ Pa.

**membrane depolarization**  *See* depolarization.

**membrane permeability**  The extent to which a membrane permits or restricts the movement of a substance.

**meristem identity genes**  Genes necessary for the initial induction of the floral organs.

**meristematic zone**  A region at the tip of the root containing the meristem that generates the body of the root. Located just above the root cap.

**meristemoids**  Small, superficial clusters of dividing cells that give rise to structures such as trichomes or stomata.

**meristems**  Localized regions of ongoing cell division that enable growth during post-embryonic development.

**mesophyll**  Leaf tissue found between the upper and lower epidermal layers, consisting of palisade parenchyma and spongy mesophyll.

**mesophyll cells**  Photosynthetic cells found in the mesophyll of leaves of both $C_4$ and $C_3$ plants.

**mesophyll resistance (liquid phase resistance)**  The resistance to $CO_2$ diffusion imposed by the liquid phase inside leaves. The liquid phase includes diffusion from the intercellular leaf spaces to the carboxylation sites in the chloroplast.

**mesophytic plants**  Plants that grow in areas with adequate moisture for normal plant growth and development.

**metabolic redundancy**  Refers to a common feature of plant metabolism in which different pathways serve a similar function and can therefore replace each other without apparent loss in function.

**metaphase**  A stage of mitosis during which the nuclear envelope breaks down and the condensed chromosomes align in the middle of the cell.

**methanogenic bacteria**  Obligate anaerobes that use $CO_2$ and $H_2$ as electron acceptor and donor, respectively, in anaerobic respiration for the production of methane.

**methemoglobinemia**  A disease of humans and livestock caused by the consumption of plant material high in nitrate. The liver reduces nitrate to nitrite, which combines with hemoglobin blocking its ability to bind oxygen.

**methylation**  The chemical addition of methyl groups to alter structure or function. A common modification of cytosine residues in DNA.

**methylerythritol phosphate (MEP) pathway**  Operating in chloroplast and other plastids, the pathway that synthesizes isopentenyl diphosphate (IPP) from intermediates of glycolysis or the photosynthetic carbon reduction cycle.

**mevalonic acid pathway**  The reactions by which three molecules of acetyl CoA are joined together stepwise to form mevalonic acid, an intermediate in one pathway for terpene biosynthesis.

**Michaelis–Menten constant ($K_m$)**  A constant in the Michaelis-Menten equation for enzyme kinetics. The constant reflects the binding affinity of the substrate for the enzyme or a solute for its carrier and corresponds to the concentration of substrate that gives ½ the maximum velocity that the enzyme can catalyze.

**microaerobic conditions**  Reduced oxygen conditions maintained by some aerobic nitrogen-fixing bacteria such as *Azotobacter* through high rates of cellular respiration

**microarray**  Techniques that use a solid support onto which are spotted thousands of DNA sequences that are representative of single genes of a given species. The genes on an array can be investigated all in a single experiment, increasing the throughput of gene analysis many-fold over the classical methods.

**microbe-associated general molecular patterns (MAMPs)**  Evolutionarily conserved pathogen-derived molecules that serve as elicitors of plant defenses.

**microbodies**  A class of spherical organelles surrounded by a single membrane and specialized for one of several metabolic functions, such as the β-oxidation of fatty acids and the metabolism of glyoxylate (in

peroxisomes and glyoxysomes, respectively).

**microfilament**  A component of the cell cytoskeleton made of actin; it is involved in organelle motility within cells.

**micronutrient**  Minerals obtained from the soil and present in plant tissues at concentrations usually less than 3 µmol $g^{-1}$ dry matter. Chloride, iron, boron, manganese, sodium, zinc, copper, nickel, molybdenum.

**microRNAs (miRNAs)**  Short (21–24nt) RNAs that have double stranded stem-loop structures and mediate RNA interference.

**microsatellites**  One group of heterochromatic dispersed repeats that consist of sequences as short as two nucleotides repeated hundreds or even thousands of times. Also known as simple sequence repeats.

**microtubule**  Component of the cell cytoskeleton made of tubulin, a component of the mitotic spindle, and a player in the orientation of cellulose microfibrils in the cell wall.

**middle lamella**  A thin layer of pectin-rich material at the junction where the primary walls of neighboring cells come into contact. Originates as the cell plate during cell division.

**middle region**  In the globular embryo, the cells derived from the basal quartet of cells that give rise to the hypocotyl (embryonic stem), the root, and the apical domains of the root meristem.

**mineral nutrients**  Inorganic ions absorbed from soil. *See also* macronutrients, micronutrients.

**mineralization**  Process of breaking down organic compounds by soil microorganisms that releases mineral nutrients in forms that can be assimilated by plants.

**minimum promoter**  *See* core promoter.

**mitochondrion (plural mitochondria)**  The organelle that is the site for most reactions in the respiratory process in eukaryotes.

**mitochondrial DNA (mtDNA)**  *See* mitochondrial genome.

**mitochondrial genome**  The DNA found in plant mitochondria which consists of between approximately 200 and 2000 kb and is much larger than mitochondrial genomes of animals or fungi. Mitochondrial genes encode a variety of proteins necessary for cellular respiration.

**mitosis**  The ordered cellular process by which replicated chromosomes are distributed to daughter cells formed by cytokinesis.

**mitotic spindle**  The mitotic structure involved in chromosome movement. Polymerized from α- and β-tubulin monomers formed by the disassembly of the preprophase band in early metaphase.

**mixed-function oxidases**  *See* monooxygenases.

**model organisms**  Organisms that are particularly accessible and convenient for research and which provide information for hypothesis testing in other organisms.

**moiety**  A part of a larger molecule or structure.

**molecular motors**  Large protein complexes associated with microtubules that move flagella, vesicles, chromosomes, and the cellulose synthase complex.

**monocarpic senescence**  Senescence of the entire plant after a single reproductive cycle, initiated by fruit and seed development. Can be prevented by continual removal of flowers.

**monocot**  One of the two classes of flowering plants characterized by a single seed leaf (cotyledon) in the embryo. Contrast with dicot.

**monoecious**  Refers to plants in which male and female flowers are found on the same individuals, e.g., cucumber (*Cucumis sativus*) and maize (*Zea mays*). Contrast with dioecious.

**monolignols**  The phenylpropanoid subunits that make up lignin.

**monooxygenases**  Group of oxygenase enzymes that incorporate only one of the oxygen atoms from $O_2$ into a carbon compound; the other oxygen atom is reduced to water using NADH or NADPH as an electron donor.

***MONOPTEROS (MP)***  Gene involved in embryonic patterning. Encodes an auxin response factor that is necessary for the normal formation of basal elements such as the root and hypocotyls.

**monosomies**  A type of aneuploidy in which only one chromosome of a given kind is present.

**monoterpenes**  Terpenes having ten carbons, two five-carbon isoprene units.

**morphogenesis**  The developmental processes that give rise to biological form.

**morphogens**  In animals, substances that play key roles in providing positional cues in certain types of position-dependent development.

**movement proteins**  Nonstructural proteins encoded by the virus genome that facilitate viral movement through the symplast.

**mtDNA**  Mitochondrial DNA.

**multidrug resistance/P-glycoproteins (MDR/PGPs)**  MDR/PGP proteins are integral membrane proteins that function as ATP-dependent hydrophobic anion carriers that facilitate cellular auxin efflux during polar transport.

**multinet growth hypothesis**  Concerns cell wall deposition during cell expansion. Holds that each successive wall layer is stretched and thinned during cell expansion, so the microfibrils would be expected to be passively reoriented in the direction of growth.

**multivesicular body (MVB)**  Part of the prevacuolar sorting compartment that functions in the degradation of vacuoles and their membranes.

**mutants**  Individuals that contain specific changes in their DNA sequence and may show an altered phenotype.

**mutualism**  A symbiotic relationship in which both organisms benefit.

**MYB proteins**  A class of transcription factors in eukaryotes. In plants, one subgroup of a large MYB family that has been implicated in GA signaling (GAMYB).

**mycelium**  The mass of hyphae that forms the body of a fungus.

**mycorrhizae**  (singular mycorrhiza, from the Greek words for "fungus" and "root"). The symbiotic (mutualistic) association of certain fungi and plant roots. Facilitate the uptake of mineral nutrients by roots. *See also* arbuscular mycorrhizal fungi.

***myo*-inositol-hexaphosphate (InsP$_6$)**  In broad bean (*Vicia faba*) guard cells, one of the phosphoinositides whose production ABA stimulates, and which functions as a signaling intermediate during ABA-induced stomatal closure.

**myosins**  Myosins are a large family of motor proteins found in eukaryotic cells. In plants, myosin motors are responsible for the movements of organelles along microtubules.

**N-linked glycoproteins**  Glycan linked via a nitrogen atom to a protein. Formed by transfer of a 14-sugar glycan from the ER membrane-embedded dolichol diphosphate to the nascent polypeptide as it enters the lumen of the ER.

***na* mutant**  In peas, a recessive mutant allele causing an extreme dwarf phenotype by completely blocking gibberellin biosynthesis—no GA$_{12}$-aldehyde is produced.

**NAD**  A major carrier of energy-rich electrons in redox processes, especially in respiration, by conversion between oxidized NAD$^+$ and reduced NADH.

**NADH dehydrogenase**  A membrane-bound enzyme oxidizing NADH and reducing a quinone. Several are present in the electron transport chain of mitochondria, for example complex I.

**NADP**  A major carrier of energy-rich electrons in redox processes, especially in biosynthesis and photosynthesis. It is converted between NADP$^+$ and NADPH, as for NAD.

**NAD(P)H dehydrogenases**  A collective term for dehydrogenases oxidizing NADH or NADPH, or both.

**natural products**  *See* secondary metabolites.

**NCED**  An enzyme (9-*cis*-epoxy-carotenoid dioxygenase) catalyzing the first committed step for ABA biosynthesis, forming an intermediate that is a neutral growth inhibitor and has physiological properties similar to those of ABA.

**ncRNAs**  Non-protein-coding RNAs that may be involved in gene regulation and may be active in the RNA interference (RNAi) pathway.

**necrosis**  A type of cell death.

**necrotic lesions**  Regions within an organ in which cells die while surrounding regions remain alive.

**necrotic spots**  Small spots of dead leaf tissue. For example, a characteristic of phosphorus deficiency.

**necrotrophic**  Refers to cell and tissue killing by pathogens that attack their host plant first by secreting cell wall-degrading enzymes and/or toxins, which will lead to massive tissue laceration and plant death.

**nectar**  Sugary fluid produced in specialized glands of some flowers to attract insects as pollinators. Sometimes contains amino acids and other nitrogenous substance.

**nectar guides**  On a flower, symmetric patterns of stripes, spots, or concentric circles formed by flavonols and other compounds to attract and orient pollinating insects.

**negative feedback regulation**  A regulatory system in which the end product of a pathway inhibits the activity of one or more key enzymes of that pathway, or depresses the synthesis of that enzyme. For example, gibberellin (GA) depresses GA biosynthesis through inhibition of the expression of *GA20ox* and *GA3ox* genes which encode the last two enzymes in the formation of bioactive GA.

**negative regulator**  A factors that suppresses a cellular process, such as transcription, and whose absence allows the process to occur.

**neoplasm**  *See* tumor.

**Nernst equation**  An equation that predicts the electrical potential at which a charged ion will be in equilibrium across a membrane, as a function of the relative concentrations of the ion on each side.

**Nernst potential**  The electrical potential described by the Nernst equation.

**night break**  An interruption of the dark period with a short exposure to light that makes the entire dark period ineffective.

**NIT gene**  One of the genes (*NIT1* through *NIT4*) encoding nitrilase enzymes that convert IAN to IAA.

**nitrate reductase**  Enzyme that reduces nitrate ($NO_3^-$) to nitrite ($NO_2^-$). Catalyzes the first step by which nitrate absorbed by roots is assimilated into organic form.

**nitric oxide (NO)**  A gas that acts as a second messenger in many signaling pathways in animals and plants. It is synthesized from the amino acid arginine by the enzyme *NO synthase.*

**nitrite reductase**  The enzyme that reduces nitrite ($NO_2^-$) to ammonium ($NH_4^+$).

**nitrogen fixation**  The natural or industrial processes by which atmospheric nitrogen $N_2$ is converted to ammonia ($NH_3$) or nitrate ($NO_3^-$).

**nitrogenase enzyme complex**  The two-component protein complex that conducts biological nitrogen fixation in which ammonia is produced from molecular nitrogen.

**Nod factors**  Lipochitin oligosaccharide signal molecules active in gene expression during nitrogen-fixing nodule formation. All nod factors have a chitin β-1→4-linked *N*-acetyl-D-glucosamine backbone (varying in length from three to six sugar units) and a fatty acid chain on the C-2 position of the nonreducing sugar.

**nod genes**  *See* nodulation genes.

**Nod genes**  *See* nodulin genes.

**nodal roots**  Adventitious roots that form after the emergence of primary roots.

**node**  Position on the stem where leaves are attached.

**nodulation genes (nod)**  Rhizobial genes, the products of which participate in nodule formation.

**nodule primordium**  Rhizobia-infected root cortical cells that dedifferentiate and start dividing.

**nodules**  Specialized organs of a plant host containing symbiotic nitrogen-fixing prokaryotes.

**nodulin (Nod) genes**  Plant genes specific to nodule formation.

**nonclimacteric fruits**  Fruits such as citrus fruits and grapes that do not exhibit the increased cellular respiration and ethylene production rise seen in climacteric fruits.

**non-Mendelian**  A pattern of inheritance that does not conform to Mendelian expectations. *See* maternal inheritance.

**nonphotochemical quenching**  The quenching of chlorophyll fluorescence by processes other than photochemistry—the converting of excess excitation into heat.

**nonprotein amino acids**  Amino acids that are not found in proteins, but are free in plant fluids and can act as defense compounds.

**nonreducing end**  The terminal glucose moiety of the starch molecule whose C-1 is β-linked to the C-4 of the preceding moiety.

**nonreducing sugar**  A sugar in which the aldehyde or ketone group is reduced to an alcohol or combined with a similar group on another sugar, e.g., sucrose. Less reactive than a reducing sugar.

**northern-blot analysis**  A method for detecting and quantifying specific RNAs (after electrophoresis and transfer of the RNA onto nitrocellulose paper or a nylon membrane filter) by hybridization with complementary strands of RNA or DNA.

**NPA (1-N-naphthylphthalamic acid)**  A noncompetitive inhibitor of auxin anion efflux carriers that blocks the polar transport of auxin.

**npq1 mutant**  An Arabidopsis mutant (*non*photochemical *q*uenching) defective in the enzyme that converts violaxanthin into zeaxanthin. Lacks a specific stomatal response to blue light.

**nuclear division**  Mitosis or meiosis. Distinguished from cytokinesis.

**nuclear envelope**  The double membrane surrounding the nucleus.

**nuclear genome**  The entire complement of DNA found in the nucleus.

**nuclear localization signal**  A specific amino acid sequence required for a protein to gain entry into the nucleus.

**nuclear pore complex (NPC)**  An elaborate structure, 120 nm wide, composed of more than a hundred different nucleoporin proteins arranged octagonally. The NPC forms a large, protein-lined pore in the nuclear membrane.

**nuclear pores**  Sites where the two membranes of the nuclear envelope join, forming a partial opening between the interior of nucleus and the cytosol. Contains an elaborate structure of more than a hundred different nucleoporin proteins that form a nuclear pore complex (NPC).

**nucleoids**  Organellar and prokaryotic genomes that are not enclosed in a nuclear envelope.

**nucleolar organizer region**  Associated with the nucleolus in the interphase nucleus. Sites where portions of one or more chromosomes containing ribosomal RNA genes are clustered and are transcribed.

**nucleolus (plural nucleoli)**  A densely granular region in the nucleus that is the site of ribosome synthesis.

**nucleoporins**  Proteins that form the nuclear pore complex in the nuclear envelope.

**nucleosome**  A structure consisting of eight histone proteins around which DNA is coiled.

**nucleus (plural nuclei)**  The organelle that contains the genetic information primarily responsible for regulating cellular metabolism, growth, and differentiation.

**nutrient assimilation**  The incorporation of mineral nutrients into carbon compounds such as pigments, enzyme cofactors, lipids, nucleic acids, or amino acids.

**nutrient depletion zone**  The region surrounding the root surface showing diminished nutrient concentrations due to slow diffusion.

**nutrient film growth system**  A form of hydroponic culture in which the plant roots lie on the surface of a trough, and nutrient solutions flow over the roots in a thin layer along the trough.

**nutrient solution**  A solution containing only inorganic salts that supports the growth of plants in sunlight without soil or organic matter.

**nyctinasty**  Sleep movements of leaves. Leaves extend horizontally to face the light during the day and fold together vertically at night.

**O-linked oligosaccharides**  Oligosaccharides that are linked to proteins via the OH groups of hydroxyproline, serine, threonine, and tyrosine residues.

**OCP (orange carotenoid protein)**  *See* carotenoid protein.

**octant embryo**  The spherical, eight-cell, globular embryo exhibiting radial symmetry.

**oil bodies**  Also known as the oleosomes or spherosomes, these organelles accumulate and store triacylglycerols. They are bounded by a single phospholipid leaflet ("half–unit membrane" or phospholipid monolayer) derived from the ER.

**oleosin**  A specific protein that coats oil bodies.

**oleosomes**  Also known as oil bodies or spherosomes, these organelles bounded by an unusual single layer phospholipid membrane that store triacylglycerols.

**oligogalacturonans**  Pectin fragments (10 to 13 residues) resulting from plant cell wall degradation that elicit multiple defense responses. May also function during the normal control of cell growth and differentiation.

**oligosaccharins**  Fragments resulting from plant cell wall degradation that affect plant development or defenses.

**opines**  Nitrogen-containing compounds found only in crown gall tumors. Genes required for their synthesis found only in Ti plasmid

T-DNA of *Agrobacterium tumefaciens*, which can utilize the opines, not the plant. The bacterium can utilize the opine as a nitrogen source, but cells of higher plants cannot.

**optimal temperature response**  The environmental temperature that affords the highest rate of photosynthesis.

**ordinary companion cell**  A type of companion cell with relatively few plasmodesmata connecting it to any of the surrounding cells other than its associated sieve element.

**organ identity genes**  Genes that directly control floral development. Genes for transcription factors that likely control the expression of other genes whose products are involved in the formation and/or function of floral organs.

**organic fertilizers**  Contain nutrient elements derived from natural sources without any synthetic additions.

**organogenesis**  The formation of functionally organized structures during embryogenesis.

***orp* mutant**  In maize, a mutation that inactivates the enzyme for the final step in tryptophan biosynthesis. Prevents conversion of tryptophan to IAA. Named for *orange pericarp* mutant of maize.

**orthostichy**  A set of leaves inserted in the stem directly above or below one another.

**osmolality**  A unit of concentration expressed as moles of total dissolved solutes per liter of water [$mol\ L^{-1}$].

**osmoregulation**  Control of the osmotic potential of a cell or organism.

**osmosis**  The movement of water across a selectively permeable membrane toward the region of more negative water potential, $\Psi_w$ (lower concentration of water).

**osmotic adjustment**  The ability of the cell to accumulate compatible solutes and lower water potential during periods of osmotic stress.

**osmotic pressure (π)**  The negative of the osmotic potential, ($\Psi_s$).

Because $\Psi_s$ has negative values, π has positive values

**osmotic solutes**  The major osmotically active solutes in guard cells. Include potassium, chloride, malate, and sucrose.

**osmotic stress**  Stress imposed on cells or whole plants when the osmotic potential of external solutions is more negative than that of the solution inside the plant.

**outer membrane**  The outer membrane of the mitochondrion or chloroplast.

**outwardly rectifying**  Refers to ion channels that open only at potentials more positive than the prevailing Nernst potential for a cation, or more negative than the prevailing Nernst potential for an anion, and thus mediate outward current.

**overexpression**  Genetic manipulation such that a specific target gene is expressed at high levels throughout the plant with a consequent change in phenotype.

**ovule**  In seed plants, the structure that contains the embryo sac and develops into the seed after fertilization of the egg contained within it.

**oxIAA (oxindole-3-acetic acid)**  End product of oxidative degradation of IAA.

**oxidation**  The chemical reaction whereby electrons or hydrogen atoms are removed from a substance.

**oxidation number**  Number used to keep track of the redistribution of electrons during chemical reactions. The sum of the oxidation numbers of all atoms in a neutral compound is zero. A useful method to balance redox equations.

**oxidative pentose phosphate pathway**  *See* pentose phosphate pathway.

**oxidative phosphorylation**  Transfer of electrons to oxygen in the mitochondrial electron transport chain that is coupled to ATP synthesis from ADP and phosphate by the ATP synthase.

**oxidative stress**  A signal or a cause of senescence brought on by reactive oxygen species.

**oxigenic organism** Photosynthetic organisms that produce molecular oxygen as an end product of photosynthesis.

**oxygenases** Group of enzymes that add oxygen from $O_2$ directly to organic compounds. *See* dioxygenases, monooxygenases.

**oxygenation** The oxygenase-catalyzed reaction in which the carbon of an organic molecule binds the oxygen atoms of the $O_2$ molecule.

**P protein** Phloem proteins which act to seal damaged sieve elements by plugging the sieve-element pores. Abundant in the sieve elements of most angiosperms, but absent from gymnosperms. Formerly called "slime."

**P-ATPase** An ion pump in which the enzyme protein is phosphorylated by ATP during the catalytic cycle, e.g., the plasma membrane $H^+$-ATPase.

**P-protein bodies** Discrete spheroidal, spindle-shaped, or twisted/coiled structures of P-proteins present in the cytosol of immature sieve tube elements. Generally disperse into tubular or fibrillar forms during cell maturation.

**P680** The chlorophyll of the photosystem II reaction center that absorbs maximally at 680 nm in its neutral state. The P stands for pigment.

**P700** The chlorophyll of the photosystem I reaction center that absorbs maximally at 700 nm in its neutral state. The P stands for pigment.

**P870** The reaction center bacteriochlorophyll from purple photosynthetic bacteria that absorbs maximally at 870 nm in its neutral state. The P stands for pigment.

**paclobutrazol (aka: Bonzi)** A commercial inhibitor of gibberellin biosynthesis. Paclobutrazol inhibits P450 mono-oxygenases, blocking the synthesis $GA_{12}$-aldehyde on the endoplasmic reticulum.

**paleopolyploids** Species that show signs of ancient genome duplications followed by DNA loss.

**palisade cells** Below the leaf upper epidermis, the top one to three layers of pillar-shaped photosynthetic cells.

**PAPS** *See* 3'-phosphoadenosine-5'-phosphosulfate.

**PAR** *See* photosynthetically active radiation.

**paraheliotropism** Movement of leaves away from incident sunlight.

**paraquat** An herbicide that blocks photosynthetic electron flow by accepting electrons from photosystem I.

**parenchyma** Metabolically active plant tissue consisting of thin-walled cells, with air-filled spaces at the cell corners.

**parthenocarpy** The production of fruits without fertilization, consequently fruits lacking mature functioning seeds. Occurs naturally in banana and pineapple.

**particle rosettes (terminal complexes)** Large, ordered protein complexes that are embedded in the plasma membrane and contain cellulose synthase.

**partitioning** The differential distribution of photosynthate to multiple sinks within the plant.

**PAS domain** A domain of phytochrome, which is necessary for chromophore attachment to the protein.

**PAS-related domain (PRD)** On the phytochrome protein, two domains that mediate phytochrome dimerization.

**pascal (Pa)** SI unit of pressure: $1\ Pa = 1\ kg\ m^{-1}\ s^{-2}$, or $1\ Pa = 1\ J\ m^{-3}$.

**passive transport** Diffusion across a membrane. The spontaneous movement of a solute across a membrane in the direction of a gradient of (electro)chemical potential (from higher to lower potential). Downhill transport.

**patch-clamp** Electrophysiological method used to study ion pumps and single ion channels.

**pathogenesis-related proteins (PR proteins)** Group of proteins produced by pathogen attack; includes hydrolytic enzymes that attack the cell wall of the pathogen, particularly fungi, e.g., glucanases, chitinases.

**pathogens** Microorganisms capable of infecting other organisms and causing disease.

**PCD** *See* programmed cell death.

**PCMBS (*p*-chloromercuribenzene-sulfonic acid)** A reagent that inhibits the transport of sucrose across plasma membranes but does not permeate the cell membrane.

**pectins** A heterogeneous group of complex cell wall polysaccharides that form a gel in which the cellulose–hemicellulose network is embedded. Typically contain acidic sugars such as galacturonic acid, and neutral sugars such as rhamnose, galactose, and arabinose. Often include calcium as a structural component, allowing extractions from the wall with chelators or dilute acids.

**pentose phosphate pathway (hexose monophosphate shunt)** A cytosolic and plastidic pathway that oxididizes glucose and produces NADPH and a number of sugar phosphates.

**PEP carboxylase** A cytosolic enzyme that forms oxaloacetate by the carboxylation of phosphoenolpyruvate.

**perfect flowers** Flowers with both male (stamens) and female (pistils) structures present. Hermaphroditic flowers.

**perforation plate** The perforated end walls of vessel elements.

**periclinal division** Cell divisions parallel to the surface or longitudinal axis.

**periclinal** Pertaining to the orientation of cell division such that the new cell walls form parallel to the tissue surface.

**pericycle** Meristematic cells forming the outermost layer of the vascular cylinder in the stem or root, interior to the endodermis. An internal meristematic tissue from which lateral roots arise.

**periderm** Tissue produced by the cork cambium that contributes to the outer bark of stems and roots during secondary growth of woody plants, replacing the epidermis. Also forms over wounds and abscission layers after the shedding of plant parts.

**period**  In cyclic (rhythmic) phenomena, the time between comparable points in the repeating cycle, such as peaks or troughs.

**peripheral proteins**  Proteins that are bound to the membrane surface by noncovalent bonds, such as ionic bonds or hydrogen bonds.

**peripheral zone**  A doughnut-shaped region surrounding the central zone in shoot apical meristems consisting of small, actively dividing cells with inconspicuous vacuoles. Leaf primordia are formed in the peripheral zone.

**permanent wilting point**  Water content of the soil (or soil water potential) from which plants cannot regain turgor and therefore remain wilted even if all water loss through transpiration stops.

**peroxisome**  Organelle in which organic substrates are oxidized by $O_2$. These reactions generate $H_2O_2$ that is broken down to water by the peroxysomal enzyme catalase.

**petals**  Conspicuous, often brightly colored flower structures surrounded at their base by sepals. *See also* sepals stamens, and carpels.

**petiole**  The leaf stalk that joins the leaf blade to the stem.

**Pfr**  The far-red light–absorbing form of phytochrome converted from Pr by the action of red light. The blue-green colored Pfr is converted back to Pr by far-red light. Pfr is the physiologically active form of phytochrome.

**phase**  In cyclic (rhythmic) phenomena, any point in the cycle recognizable by its relationship to the rest of the cycle, for example, the maximum and minimum positions.

**phase change**  The phenomenon in which the fates of the meristematic cells become altered in ways that cause them to produce new types of structures.

**phaseic acid (PA)**  One of the products formed by oxidative inactivation of abscisic acid.

**phenolics**  Plant secondary metabolites that contain a hydroxyl group on an aromatic ring, a phenol. Many plant phenolic compounds function as defenses against insect herbivores and pathogens, UV protectants, pollinator attractants and allelopathic agents.

**phenotypic plasticity**  Physiological or developmental responses of a plant to its environment that do not involve genetic changes.

**phenylalanine ammonia lyase (PAL)**  Catalyzes the conversion of phenylalanine into cinnamic acid with loss of an ammonia molecule. Situated at a branch point between primary and secondary metabolism, PAL catalyzes an important regulatory step in the formation of many phenolic compounds from phenylalanine.

**phenylpropanoids**  Phenolic derivatives of cinnamic acid containing a benzene ring and a three-carbon side chain.

**pheophytin**  A chlorophyll in which the central magnesium atom has been replaced by two hydrogen atoms.

**phloem**  The tissue that transports the products of photosynthesis from mature leaves to areas of growth and storage, including the roots.

**phloem loading**  The movement of photosynthetic products from the mesophyll chloroplasts to the sieve elements of mature leaves. Includes short-distance transport steps and sieve-element loading. *See also* phloem unloading.

**phloem unloading**  The movement of photosynthate from the sieve elements to the sink cells that store or metabolize them. Includes sieve-element unloading and short-distance transport. *See also* phloem loading.

**phosphate**  Orthophosphate; inorganic phosphate, $HPO_4^{2-}$, $P_i$.

**phosphate translocator (phosphate/triose phosphate translocator)**  An integral membrane protein that catalyzes the reversible transport of inorganic phosphate in exchange for trioses phosphates (or 3-phosphoglycerate) across the chloroplast envelope. Under illumination, this protein facilitates the increase of trioses phosphates in the cytoplasm while concurrently replenishing inorganic phosphate in the chloroplast for further photosynthetic $CO_2$ fixation.

**phosphatidic acid (PA)**  A diacylglycerol that has a phosphate on the third carbon of the glycerol backbone.

**phosphatidylinositol-4,5-bisphosphate (PIP$_2$)**  A group of phosphorylated derivatives of phosphatidylinositol.

**3′-phosphoadenosine-5′-phospho-sulfate (PAPS)**  The first stable intermediate in the bacterial and fungal reduction of sulfate ($SO_4^{2-}$) to sulfite ($SO_3^{2-}$) and then to sulfide ($S^{2-}$). Probably formed by reaction of adenosine-5′-phosphosulfate (APS) with ATP. A few plants have more specialized day length requirements or no requirement.

**phospholipase C**  An enzyme whose action on phosphoinositides releases inositol triphosphate ($InsP_3$), along with diacylglycerol (DAG).

**phospholipase D (PLD)**  An enzyme active in ABA signaling; it releases phosphatidic acid from phosphatidylcholine.

**phosphorolysis**  The cleavage of a glycosidic linkage by the addition of a phosphoryl group to one of the products.

***phot1* mutant**  Arabidopsis mutant (phototropin 1) whose blue light–dependent phototropism of the hypocotyl is blocked. Also defective in a blue light–stimulated phosphorylation. The mutant *phot1* is genetically independent of the *hy4* (*cry1*) mutant. Formerly known as *nph1* (nonphototropic hypocotyls).

**photoassimilation**  The coupling of nutrient assimilation to photosynthetic electron transport.

**photochemistry**  Very rapid chemical reactions in which light energy absorbed by a molecule causes a chemical reaction to occur.

**photoinhibition**  The inhibition of photosynthesis by excess light.

**photolyase**  A blue light–activated enzyme that repairs pyrimidine dimers in DNA that have been damaged by ultraviolet radiation. Contains an FAD and a pterin.

**photomorphogenesis**  The influence and specific roles of light on plant development. In the seedling, light-induced changes in gene expression to support above-ground growth in the light rather than below-ground growth in the dark.

**photon**   A discrete physical unit of radiant energy.

**photon irradiance**   Measurement of light energy expressed in moles per square meter per second (mol $m^{-2} s^{-1}$), where 1 mol of light = 6.02 × $10^{23}$ photons, Avogadro's number.

**photoperiod**   The amount of time per day that a plant is exposed to light or darkness. May control various aspects of sexual or vegetative reproductive development, including flowering and tuberization.

**photoperiodic induction**   The photoperiod-regulated processes that occur in leaves resulting in the transmission of a floral stimulus to the shoot apex.

**photoperiodism**   A biological response to the length and timing of day and night, making it possible for an event to occur at a particular time of year.

**photophosphorylation**   The formation of ATP from ADP and inorganic phosphate ($P_i$) using light energy stored in the proton gradient across the thylakoid membrane.

**photoprotection**   A carotenoid-based system for dissipating excess energy absorbed by chlorophyll in order to avoid forming singlet oxygen and damaging pigments. Involves quenching.

**photorespiration**   Uptake of atmospheric $O_2$ with a concomitant release of $CO_2$ by illuminated leaves. Molecular oxygen serves as substrate for rubisco and the formed 2-phosphoglycolate enters the photorespiratory carbon oxidation cycle. The activity of the cycle recovers some of the carbon found in 2-phosphoglycolate, but some is lost to the atmosphere.

**photorespiratory carbon oxidation (PCO) cycle**   *See* photorespiration.

**photoreversibility**   Relating to phytochrome, the interconversion of the Pr and Pfr forms.

**photostationary state**   Relating to phytochrome under natural light conditions, the equilibrium of 97% Pr and 3% Pfr.

**photosynthate**   The carbon-containing products of photosynthesis.

**photosynthesis**   The conversion of light energy to chemical energy by photosynthetic pigments using water and $CO_2$, and producing carbohydrates.

**photosynthetic electron transport**   Electron flow from light-excited chlorophyll and the oxidation of water, through PSII and PSI, to the final electron acceptor $NADP^+$.

**photosynthetic photon flux density (PPFD)**   Photosynthetically active radiation expressed on a quantum basis, quanta (mol $m^{-2} s^{-1}$).

**photosynthetically active radiation (PAR)**   Light of wavelengths between 400 to 700 nm range that corresponds to the wave band absorbed by photosynthetic pigments.

**photosystem**   A functional unit in the chloroplast that harvests light energy to power electron transfer and to generate a proton motive force used to synthesize ATP.

**photosystem I (PSI)**   A system of photoreactions that absorbs maximally far-red light (700 nm), oxidizes plastocyanin and reduces ferredoxin.

**photosystem II (PSII)**   A system of photoreactions that absorbs maximally red light (680 nm), oxidizes water and reduces plastoquinone. Operates very poorly under far-red light.

**phototropins 1 and 2**   Two flavoproteins that are the photoreceptors for the blue-light signaling pathway that induces phototropic bending in Arabidopsis hypocotyls and in oat coleoptiles. They also mediate chloroplast movements and participate in stomatal opening in response to blue light. Phototropins are autophosphorylating protein kinases whose activity is stimulated by blue light.

**phototropism**   The alteration of plant growth patterns in response to the direction of incident radiation.

**phragmoplast**   An assembly of microtubules, membranes, and vesicles that forms during late anaphase or early telophase and precedes fusion of vesicles to form cell plate.

**phy**   The designation for the phytochrome holoprotein (with the chromophore).

**PHY**   The designation for the phytochrome apoprotein (without the chromophore).

**PHY domain**   In phytochrome, the domain necessary for stabilizing the active form.

***PHY* genes**   The genes for the apoproteins of phytochrome; phytochrome gene family members. In Arabidopsis these are *PHYA, PHYB, PHYC, PHYD,* and *PHYE.*

**phyllotaxy**   The arrangement of leaves on the stem.

**phyto-oncogenes**   Genes that can induce tumors in plants.

**phytoalexins**   Chemically diverse group of secondary metabolites with strong antimicrobial activity that are synthesized following infection and accumulate at the site of infection.

**phytochelatins**   Low molecular-weight peptides synthesized from glutathione by the enzyme phytochelatin synthase. These peptides can bind a variety of metal(oids) and play an important role in tolerance of plants to As, Cd and Zn.

**phytochrome**   A plant growth-regulating photoreceptor protein that absorbs primarily red light and far-red light, but also absorbs blue light. The holoprotein that contains the chromophore phytochromobilin.

**phytochrome-associated protein phosphatase 5 (PAPP5)**   A factor that interacts with phytochromes through dephosphorylation of the active phytochrome.

**phytochrome-interacting factors (PIFs)**   Families of phytochrome-interacting proteins that may activate and repress gene transcription; some are targets for phytochrome-mediated degradation.

**phytochrome kinase substrate 1 (PKS1)**   A protein capable of interacting with phytochrome A and B in both the active (Pfr) and inactive (Pr) form.

**phytochromobilin** The linear tetrapyrrole chromophore of phytochrome.

**phytoecdysones** Group of plant steroids that are toxic to insects because of chemical similarity to the insect molting hormone.

**phytoferritin** An iron–protein complex which stores surplus iron in plant cells.

**phytoglycogen** A minor constituent of starch having a branched structure in which an α-D-1,6 bond occurs every 10–15 glucose units linked via α-D-1,4 bonds.

**phytohormone** *See* plant hormone.

**phytol** The long hydrocarbon chain found on the chlorophyll molecule that anchors it to the thylakoid membrane.

**phytomere** A developmental unit consisting of one or more leaves, the node to which the leaves are attached, the internode below the node, and one or more axillary buds.

**phytosiderophores** A special class of iron chelators produced by grasses and made of amino acids that are not found in proteins.

**P$_i$** *See* phosphate.

**PIF3** A basic helix-loop-helix transcription factor that interacts with both phytA and phyB.

**PIF-like proteins (PILs)** Nuclear, DNA-binding proteins that selectively interact with phytochromes in their active Pfr conformations.

**PIN1** Membrane protein that forms part of the complex that transports auxin out of the basal ends of conducting cells during polar auxin transport.

**pin1-1 mutant** An *Arabidopsis* mutant with flowers similar to those of plants treated with auxin transport inhibitors. The inflorescence is impaired in polar auxin transport. Useful in investigating the role of polar auxin transport in floral development.

**PIN family of auxin carrier proteins** A family of membrane carrier proteins that transport auxin. A group of short PIN proteins appear to be ancestral and regulate the movement of auxin in and out of the ER. A second group of PINs are situated on the plasma membrane and regulate polar auxin flows. Different tissues have different PIN proteins.

**pit** Microscopic regions where the secondary wall of tracheary elements is absent and the primary wall is thin and porous, facilitating sap movement between one tracheid and the adjacent one.

**pit fields** Depressions in the primary cell walls where numerous plasmodesmata make connections with adjacent cells. When present, secondary walls are not deposited in the locations of pit fields, giving rise to pits.

**pit membrane** The porous layer between pit pairs, consisting of two thinned primary walls and a middle lamella.

**pit pairs** The adjacent pits of adjoined tracheid cells. A low-resistance path for water movement between tracheids.

**plant growth substance** *See* plant hormone.

**plant hormones** Substances that influence plant growth and development at low concentrations. Major classes are abscisic acid, auxin, brassinosteroid, cytokinin, ethylene, and gibberellin.

**plasma membrane** A fluid mosaic structure composed of a bilayer of polar lipids (phospholipids or glycosylglycerides) and embedded proteins that together confer selective permeability on the structure.

**plasma membrane H$^+$-ATPase** A P-ATPase that transports H$^+$ across the plasma membrane.

**plasmalemma central control model (PCC model)** Proposed mechanism for gravitropism in which calcium channels in the plasma membrane open in response to gravity-induced changes in the distribution of forces exerted by the protoplast, the cyto-skeleton, or the cell wall. Contrast with starch–statolith hypothesis.

**plasmalemma** *See* plasma membrane.

**plasmid** Circular pieces of extra-chromosomal bacterial DNA that are not essential for the life of the bacterium. Plasmids frequently contain genes that enhance the ability of the bacterium to survive in special environments.

**plasmodesmata (singular plasmodesma)** Microscopic membrane-lined channel connecting adjacent cells through the cell wall and filled with cytoplasm and a central rod derived from the ER called the desmotubule. Allows the movement of molecules from cell to cell through the symplast. The pore size can apparently be regulated by globular proteins lining the channel inner surface and the desmotubule to allow particles as large as viruses to pass through.

**plasmolysis** Shrinking of the protoplasm of a cell placed in an hypertonic (low osmotic potential) solution, away from the cell wall, due to loss of water.

**plastids** Cellular organelles found in eukaryotes, bounded by a double membrane, and sometimes containing extensive membrane systems. They perform many different functions: photosynthesis, starch storage, pigment storage, and energy transformations.

**plastocyanin** A small (10.5 kDa), water-soluble, copper-containing protein that transfers electrons between the cytochrome $b_6f$ complex and P700. This protein is found in the lumenal space.

**plastohydroquinone (PQH$_2$)** The fully reduced form of plastoquinone.

**plastoquinone** A small, nonpolar molecule that diffuses readily in the nonpolar core of the thylakoid membrane and is capable of undergoing reduction to plastohydroquinone. A mobile electron carrier connecting PSII and PSI. Chemically and functionally, similar to ubiquinone in the mitochondrial electron transport chain.

**pleiotropic** Referring to a gene having more than one (perhaps many) phenotypic effects.

**plumule** *See* embryo.

**pneumatophores** Plant structures that protrude out of the water and provide a gaseous pathway for oxygen diffusion to the roots growing in water or in water-saturated soils.

**polar glycerolipids**  The main structural lipids in membranes, in which the hydrophobic portion consists of two 16-carbon or 18-carbon fatty acid chains esterified to positions 1 and 2 of a glycerol.

**polar molecule**  Property of some molecules, such as water, in which differences in the electronegativity of some atoms results in a partial negative charge at one end of a molecule and a partial positive charge at the other end.

**polar transport**  Refers to directional movement of a small molecule such as auxin through a plant tissue or organ.

**polarity**  Refers to the distinct ends and intermediate regions along an axis. Beginning with the single-celled zygote, the progressive development of distinctions along two axes: an apical–basal axis and a radial axis.

**pollen**  Small structures (microspores) produced by anthers of seed plants. Contain haploid male nuclei that will fertilize egg in ovule.

**pollination**  Transfer of pollen from anther to stigma.

**polymer-trapping model**  A model that explains the specific accumulation of sugars in the sieve elements of symplastically loading species.

**polyploidy**  The condition of being polyploid, that is, having one or more extra sets of chromosomes.

**polysaccharide**  A polymer made of many sugar residues.

**polyterpenes ($[C_5]_n$)**  High-molecular-weight terpenes containing 1500 to 15,000 isoprene units; e.g., rubber.

**polyterpenoids**  Terpenes having more than fifty carbons.

**position-dependent**  Mechanisms that operate by modulating the behavior of cells in a manner that depends on the position of the cells within the developing embryo.

**positional information**  The concept that cell position, cell relationships, and associations rather than cell lineage is the important determinant of cell differentiation and tissue formation.

**positive regulator**  Factor whose presence causes a change in function.

**posttranscriptional regulation**  Following transcription, the control of gene expression by altering mRNA stability or translation efficiency.

**PPFD**  *See* photosynthetic photon flux density.

**$PP_i$**  *See* pyrophosphate.

**Pr**  Red light–absorbing phytochrome form. This is the form in which phytochrome is assembled. The blue-colored Pr is converted by red light to the far-red light–absorbing form, Pfr.

**preprophase**  In mitosis, the stage just before prophase during which the $G_2$ microtubules are completely reorganized into a preprophase band.

**preprophase band**  A circular array of microtubules and microfilaments formed in the cortical cytoplasm just prior to cell division that encircles the nucleus and predicts the plane of cytokinesis following mitosis.

**pressure-flow model**  A widely accepted model of phloem translocation in angiosperms. It states that transport in the sieve elements is driven by a pressure gradient between source and sink. The pressure gradient is osmotically generated and results from the loading at the source and unloading at the sink.

**pressure potential ($\Psi_p$)**  The hydrostatic pressure of a solution in excess of ambient atmospheric pressure.

**prevacuolar compartment (PVC)**  A membrane compartment equivalent to the late endosome in animal cells where sorting occurs before cargo is delivered to the lytic vacuole.

**primary active transport**  The direct coupling of a metabolic energy source such as ATP hydrolysis, oxidation-reduction reaction, or light absorption to active transport by a carrier protein.

**primary cell walls**  The thin (less than 1 μm) cell walls that are characteristic of young, growing cells.

**primary cross-regulation**  Involves distinct signaling pathways regulating a shared transduction component in a positive or a negative manner.

**primary growth**  The phase of plant development that gives rise to new organs and to the basic plant form. It results from cell proliferation in apical meristems, followed by cell elongation and differentiation.

**primary inflorescence meristem**  The meristem that produces stem-bearing flowers; it is formed from the shoot apical meristem.

**primary meristem**  Meristems that are formed during embryogenesis and are found at the tip of the root and the shoot. *See* secondary meristem.

**primary phloem**  Phloem derived from the procambium during growth and development of a vascular plant.

**primary plasmodesmata**  Tubular extensions of the plasma membrane, 40 to 50 nm in diameter, that traverse the cell wall and form cytoplasmic connections between cells derived from each other by mitosis.

**primary response genes ("early genes")**  Genes whose expression is necessary for plant morphogenesis and that are expressed rapidly following exposure to a light signal. Often regulated by phytochrome-linked activation of transcription factors. Genes whose expression does not require protein synthesis. *See* secondary response genes.

**primary root**  Root generated directly by growth of the embryonic root or radicle.

**primary walls**  The first-formed, unspecialized cell walls with similarities in molecular architecture in diverse types of growing plant cells. About 85% polysaccharide and 10% protein by dry weight.

**primary xylem**  Xylem derived from the procambium during growth and development of a vascular plant.

**primordia**  Localized region of the shoot apical meristem characterized by higher cell division, leading to the formation or growth of cells with identifiable future, such as leaf.

**pro mutant**  In tomato, an ultra-tall constitutive response mutant resulting from the loss of function of a negative regulator.

**procambium**  Primary meristematic cells that differentiate into xylem, phloem, and cambium.

**programmed cell death (PCD)**  Process whereby individual cells activate an intrinsic senescence program accompanied by a distinct set of morphological and biochemical changes similar to mammalian apoptosis.

**prohexadione (BX-112)**  Inhibitor of GA biosynthesis. Inhibits gibberellin 3β-hydroxylase that converts inactive $GA_{20}$ to growth-active $GA_1$.

**prokaryotic pathway**  In the chloroplast, the series of reactions for the synthesis of glycerolipids.

**prolamellar bodies**  Elaborate semicrystalline lattices of membrane tubules that develop in plastids that have not been exposed to light (etioplasts).

**promeristem**  That part of the shoot or root apical meristem containing the stem cells and their immediate derivatives which have not yet begun to differentiate.

**prometaphase**  Early metaphase stage in which the preprophase band disassembles and new microtubules polymerize to form the mitotic spindle.

**promoter**  The region of the gene that binds RNA polymerase.

**prophase**  The first stage of mitosis (and meiosis) prior to disassembly of the nuclear envelope, during which the chromatin condenses to form distinct chromosomes.

**proplastid**  Type of immature, undeveloped plastid found in meristematic tissue that can convert to various specialized plastid types, such as chloroplasts, amyloplasts, and chromoplasts, during development.

**prosthetic group**  A metal ion or small carbon compound (other than an amino acid) that is covalently bound to a protein and is essential for its function.

**prosystemin**  A large precursor protein whose cleavage initiates a series of reactions that lead to the activation of jasmonic acid (JA) biosynthesis and accumulation.

**protease inhibitors**  Protein inhibitors encoded by genes whose expression is activated by release of jasmonic acid in target tissues.

**proteasome**  *See* 26S proteasome.

**26S proteasome**  A large proteolytic complex that degrades intracellular proteins marked for destruction by the attachment of one or more copies of the small protein, *ubiquitin*.

**14-3-3 proteins**  Phosphoserine-binding proteins that regulate the activities of a wide array of targets via direct protein–protein interactions. The 14-3-3 proteins were first identified as abundant brain proteins and named after their elution position on ion exchange chromatography and mobility in starch gel electrophoresis.

**protein bodies**  Protein storage organelles enclosed by a single unit membrane; found mainly in seed tissues.

**protein kinases**  Enzymes that have the capacity to transfer phosphate groups from ATP to specific amino acid, such as histidine, serine, threonine, or tyrosine, located either within themselves or on other proteins. Play important roles in enzyme regulation, gene expression, and signal transduction.

**protein phosphatases**  Enzymes that remove regulatory phosphate groups from proteins. Have important roles in enzyme regulation, gene expression, and signal transduction.

**protein stability**  The rate of protein destruction or inactivation; can contribute to posttranslational regulation and also plays an important role in the overall activity of a gene or its product.

**protoderm**  In the plant embryo, the surface layer one cell thick that covers both halves of the embryo and will generate the epidermis.

**protofilaments**  Polymerized α- and β-tubulin heterodimers.

**protomeristem**  An embryonic structure that will become the root or shoot meristem upon germination.

**proton gradient**  *See* proton motive force.

**proton motive force (PMF, $\Delta p$)**  The energetic effect of the electrochemical $H^+$ gradient across a membrane, expressed in units of electrical potential.

**protonema (plural protonemata)**  An algal-like filament of moss cells generated by germination of moss spores.

**protoplasm**  Classically, the living substance of all cells. The plasma membrane and all the active organelles, substances, and processes required for life and contained within the plasma membrane.

**protoplast**  The living contents of a cell enclosed by the plasma membrane: the cytosol, organelles, and nucleus. What remains after removal of the cell wall.

**protoplast fusion**  A technique for incorporating foreign genes into plant genomes by fusion of two genetically different cells from which the cell walls have been removed.

**proximal promoter sequences**  Sequence elements that are part of the core promoter.

**pseudogenes**  Stable but nonfunctional genes; apparently derived from active genes by mutation.

**$\Psi_s = -RTc_s$**  Where $R$ is the gas constant, $T$ is the absolute temperature (in degrees Kelvin, or K), and $c_s$ is the solute concentration of the solution, expressed as osmolality.

**pteridine**  Nitrogen-containing compound composed of two six-membered rings. Component of riboflavin and parent compound of pterins.

**pterin**  A carbon compound to which molybdenum is complexed, forming a prosthetic group in the nitrate reductases of higher plants. Also a blue light-absorbing chromophore in the DNA-repair enzyme, photolyase. Chemically derived from pteridines.

**pullulanase** Catalyzes the hydrolysis of α-D-1,6 glucosidic linkages in pullulan (a linear polymer of α-1,6-linked maltotriose units) and in amylopectin and glycogen, and the α- and β-limit dextrins of amylopectin and glycogen.

**pulvinus (plural pulvini)** A turgor-driven organ found at the junction between the blade and the petiole of the leaf that provides a mechanical force for leaf movements.

**pumps** Membrane proteins that carry out primary active transport across a biological membrane. Most pumps transport ions, such as $H^+$ or $Ca^{2+}$.

**PYLs** Proteins implicated in the plant response to abscisic acid.

**PYR/PYL/RCAR** A group of soluble ABA receptors that directly link to many genetically well-characterized downstream signaling elements, effectively coalescing them into a network.

**pyrethroids** Monoterpene esters with high insect toxicity. Both natural and synthetic forms are popular ingredients in commercial insecticides.

**pyridoxal phosphate (vitamin B6)** A cofactor required by all transamination reactions.

**pyrophosphate** Two phosphate groups linked by a phosphate ester bond, $PP_i$.

**pyruvate dehydrogenase** An enzyme in the mitochondrial matrix that decarboxylates pyruvate, producing NADH (from $NAD^+$), $CO_2$, and acetic acid in the form of acetyl CoA (acetic acid bound to coenzyme A).

**Q cycle** A mechanism for oxidation of plastohydroquinone in chloroplasts and ubihydroquinone in mitochondria.

**$Q_{10}$ (temperature coefficient)** The increase in rate for a process (e.g., respiration) for every 10°C increase in temperature.

**quantum (plural quanta)** A discrete packet of energy contained in a photon.

**quantum efficiency** Photosynthetic yield per photon flux of absorbed light.

**quantum yield** The ratio of the yield of a particular product of a photochemical process to the total number of quanta absorbed.

**quenching** The process by which energy stored in light excited chlorophylls is rapidly dissipated mainly by excitation transfer or photochemistry.

**quercetin** An endogenous flavonoid compound that may inhibit and regulate auxin transport by blocking auxin efflux.

**quiescent center (QC)** Central region of root meristem where cells divide more slowly than surrounding cells, or do not divide at all.

**quinone** A small, nonpolar molecule that diffuses readily in the nonpolar core of the membrane and is capable of undergoing reduction to hydroquinone.

*R* **genes** Resistance genes that function in plant defense against fungi, bacteria, and nematodes in some cases by encoding protein receptors that bind to specific pathogen molecules, elicitors.

**Rabs** A class of targeting recognition proteins for the selective fusion and fission of vesicles and tubules within the endomembrane system.

**radial axis** The pattern of concentric tissues extending from the outside of a root or stem into its center. *See* linear axis.

**radicle** The embryonic root. Usually the first organ to emerge on germination.

***RAN1* (RESPONSIVE-TO-ANTAGONIST1)** Gene involved in the addition of a copper ion cofactor necessary for the function of ethylene receptors.

**ray stem cells** Small cells whose derivatives include the radially oriented files of parenchyma cells known as rays.

***rcn1* mutant** Arabidopsis (roots curl in NPA) mutant that shows high sensitivity to NPA, inhibition of hypocotyl elongation, and auxin efflux. *RCN1* gene is closely related to the regulatory subunit of protein phosphatase 2A, a serine/threonine phosphatase.

**reaction center complex** A group of electron transfer proteins that receive energy from the antenna complex and convert it to chemical energy using oxidation-reduction reactions.

**reactive oxygen species (ROS)** These include the superoxide anion ($O_2^{\cdot-}$), hydrogen peroxide ($H_2O_2$), and the hydroxyl radical ($HO\bullet$) and singlet oxygen. They are generated in several cell compartments and can act as signals or cause damage to cellular components.

**redox reactions** Chemical reactions involving simultaneous oxidation and reduction of molecular species.

**reducing sugar** A sugar with an aldehyde or ketone group available for oxidation, e.g, glucose, fructose.

**reduction** The chemical process by which electrons or hydrogen atoms are added to a substance.

**reductive pentose phosphate (RPP) cycle** *See* Calvin–Benson cycle.

**regulatory components of ABA receptors (RCARs)** A family of soluble ABA receptors identified as proteins that interact with the PP2C protein phosphatases.

**regulatory sequences** Parts of the eukaryotic promoter that control the activity of the core promoter (minimum promoter).

**rejuvenation** The reversion of adult shoots to juvenile shoots. May be promoted by hormones, mineral deficiencies, low light, water stress, defoliation, and low temperature.

**repeat-associated silencing RNAs (rasiRNAs)** Repeat regions from which short interfering RNAs originate.

**reporter gene** A gene whose expression conspicuously reveals the activity of another gene. Gene engineered to share the same promoter as another gene.

**repressors**   Proteins that either alone or in concert with other proteins repress expression of a gene. *See* transcription factor.

**resistivity (specific electrical resistance)**   A measure indicating how strongly a material opposes the flow of electric current. A low resistivity indicates a material that readily allows the movement of electrons. The SI unit for electrical resistivity is the ohm meter.

**resonance transfer**   The nonradiative, molecule-to-molecule transfer of energy from one excited molecule to another, such as the transfer on energy from an antenna complex to the reaction center.

**respiration**   The complete oxidation of carbon compounds to $CO_2$ and $H_2O$, using oxygen as the final electron acceptor. Energy is released and conserved as ATP.

**respiratory quotient (RQ)**   The ratio of $CO_2$ evolution to $O_2$ consumption.

**response regulator protein**   One component of the two-component regulatory systems that are composed of a histidine kinase sensor protein and a response regulator protein.

**reticulons**   A class of proteins that control the transition between tubular and cisternal forms of the ER by forming tubules from membrane sheets.

**retrograde**   A backward movement in transport or signaling.

**retrograde vesicular transport**   Movement of secretory vesicles in a backward fashion, from the *trans* to the *cis* face of the Golgi. Maintains the spatial distribution of enzymes and other functional proteins within the Golgi stack by acting as a countercurrent to cisternal progression.

**retrotransposons**   In contrast to DNA transposons, these make an RNA copy of themselves, which is then reverse transcribed into DNA before it is inserted elsewhere in the genome.

**revertants**   Mutants that by some process or treatment have regained their former or wild-type characteristic.

**RGA (REPRESSOR of *ga1-3*)**   One of five DELLA-domain GRAS proteins in Arabidopsis that function as repressors of the gibberellin response.

**rhamnogalacturonan I (RG I)**   An abundant pectin polysaccharide that has a long backbone of alternating rhamnose and galacturonic acid residues.

**rheological properties**   Referring to the flow properties of a material. *See* viscoelastic properties.

**rhizobia**   Collective term for the genera of soil bacteria that form symbiotic (mutualistic) relationships with members of the plant family Leguminosae.

**rhizosphere**   The immediate microenvironment surrounding the root.

**rib meristem zone**   Meristematic cells beneath the central zone that give rise to the internal tissues of the stem in shoot apical meristems.

**rib zone**   *See* rib meristem zone.

**riboflavin**   A vitamin that is part of FAD and FMN.

**ribosome**   The site of cellular protein synthesis and consisting of RNA and protein.

**ribulose 5-phosphate**   In the pentose phosphate pathway, the initial five-carbon product of the oxidation of glucose 6-phosphate; in subsequent reactions, it is converted into sugars containing three to seven carbon atoms.

**Rieske iron–sulfur protein**   A protein in which two iron atoms are bridged by two sulfur atoms, with two histidine and two cysteine ligands.

**RNA-dependent RNA polymerases (RdRPs)**   A special class of RNA polymerases that convert single-stranded RNA into double-stranded RNA.

**RNA-induced silencing complex (RISC)**   A multiprotein complex that incorporates one strand of a small interfering RNA (siRNA) or micro RNA (miRNA). RISC complexes bind to and cleave mRNA, thereby preventing translation.

**RNA interference pathway**   An RNA-dependent gene silencing process that is controlled by the RNA-induced silencing complex (RISC) and is initiated by short double-stranded RNA molecules in a cell's cytoplasm.

**RNA polymerase**   A class of enzymes that bind to a gene and transcribe it into an RNA complementary to the DNA sequence.

**RNAi pathway**   *See* RNA interference pathway.

**root apical meristem (RAM)**   At the tip of the root, a group of cells that retain the capacity to proliferate and whose ultimate fate remains undetermined.

**root cap**   Cells at the root apex that cover and protect the meristematic cells from mechanical injury as the root moves through the soil. Site for the perception of gravity and signaling for the gravitropic response.

**root cap stem cells**   Meristematic cells that give rise to the root cap.

**root cap–epidermal stem cells**   Generate the epidermis of the root cap by anticlinal cell divisions and generate the lateral root cap by periclinal divisions followed by anticlinal divisions.

**root hairs**   Microscopic extensions of root epidermal cells that greatly increase the surface area of the root, thus providing greater capacity for absorption of soil ions and, to a lesser extent, soil water.

**root pressure**   A positive hydrostatic pressure in the xylem of roots.

**rotenone**   Specific inhibitor of complex I.

**rough ER**   The endoplasmic reticulum to which ribsomes are attached.

**R-type channels**   A type of gated channel for anions that opens or closes very rapidly in response a voltage change.

**rubisco**   The acronym for the chloroplast enzyme *ribulose bisphosphate carboxylase/oxygenase*. In a carboxylase reaction, rubisco uses atmospheric $CO_2$ and ribulose 1,5-bisphosphate to form two molecules of 3-phosphoglycerate. It also functions as an oxygenase that incorporates $O_2$ to ribulose 1,5-bis-phosphate to yield one molecule of 3-phosphoglycerate and another of 2-phosphoglycolate. The competition between $CO_2$ and $O_2$ for ribulose 1,5-bisphosphate limits net $CO_2$ fixation.

**rubisco activase**   An enzyme that facilitates the dissociation of sugar bisphosphates-rubisco complexes and, in so doing, activates rubisco.

**RUBs**   A family of small, ubiquitin-related proteins. Proteins linked to RUB are usually activated rather than degraded.

**S lignin**   A form of lignin made from sinapyl alcohol; it is distinct from G lignin and H lignin.

**S phase**   In the cell life cycle, the stage during which DNA is replicated; it follows $G_1$ and precedes $G_2$.

**S-type channels**   A type of gated channel for anions that remains open for the duration of the stimulus.

**SAG genes**   Senescence-associated genes whose expression is induced during senescence.

**salicylic acid**   A benzoic acid derivative believed to be an endogenous signal for *SAR*.

**salicylhydroxamic acid**   Specific inhibitor of the alternative oxidase.

**salinity**   Refers to high concentrations of total salts in the soil. Contrast with sodicity.

**salinization**   The accumulation of mineral ions, particularly sodium chloride and sodium sulfate, in soil often due to irrigation.

**salt stress**   The adverse effects of excess minerals on plants.

**salt-tolerant plants**   Plants that can survive or even thrive in high-salt soils. *See also* halophytes.

**sap**   Fluid content of the xylem, sieve elements of the phloem and the cell vacuole.

**saponins**   Toxic glycosides of steroids and triterpenes with detergent properties. They may interfere with sterol uptake from the digestive system or disrupt cell membranes.

**SAR**   *See* systemic acquired resistance.

**saturation**   Refers to a condition under which an increase in a stimulus does not elicit a further increase in a response. A maximum state; not capable of further increase, movement, or inclusion.

**SAUR genes**   In soybean, a group of primary response genes stimulated by auxin within 2 to 5 minutes of treatment.

**sclereid**   A type of nonelongated sclerenchyma cell commonly found in hard structures such as seed coats.

**sclerenchyma**   Plant tissue composed of cells, often dead at maturity, with thick, lignified secondary cell walls. It functions in support of nongrowing regions of the plant.

**SCR gene**   Arabidopsis *SCARECROW* gene that controls tissue organization and cell differentiation in the embryo, hypocotyl, primary roots, and secondary roots.

**scr mutant**   Arabidopsis mutant in which hypocotyl and inflorescence are agravitropic and lack both endodermis and starch sheath.

**scutellum**   The single cotyledon of the grass embryo, specialized for nutrient absorption from the endosperm.

**second messenger**   Intracellular molecule (e.g., cyclic AMP, cyclic GMP, calcium, $InsP_3$, or diacylglycerol) whose production has been elicited by a systemic hormone (the primary messenger) binding to a receptor (often on the plasma membrane). Diffuses intracellularly to the target enzymes or intracellular receptor to produce and amplify the response.

**secondary active transport**   Active transport that uses energy stored in the proton motive force or other ion gradient, and operates by symport or antiport.

**secondary cell walls**   *See* secondary wall.

**secondary cross-regulation**   Involves the output of one signal pathway regulating the abundance or perception of a second signal.

**secondary growth**   The tissue growth that occurs after elongation

is complete. It involves the vascular cambium (producing the secondary xylem and phloem) and the cork cambium (producing the periderm).

**secondary inflorescence meristems**   The inflorescence meristems that develop from the axillary buds of stem-borne leaves.

**secondary meristems**   Meristems that are formed after seed germination and include axillary meristems and lateral meristems. Their activity may be suppressed by active primary meristem.

**secondary messengers**   Activated in response to the binding of an initial signal, an intracellular, mobile signal that directly or indirectly alters cell function.

**secondary metabolites (secondary products)**   Plant compounds that have no direct role in plant growth and development, but function as defenses against herbivores and microbial infection by microbial pathogens, attractants for pollinators and seed-dispersing animals, and as agents of plant–plant competition.

**secondary plasmodesmata**   Plasmodesmata that form and permit symplastic transport between nonclonally related cells.

**secondary products (natural products)**   *See* secondary metabolites.

**secondary response genes ("late genes")**   Genes whose expression requires protein synthesis and follows that of primary response genes.

**secondary transport**   Active transport driven by the proton gradient established by the proton pump.

**secondary wall**   Cell wall synthesized by nongrowing cells. Often multilayered and containing lignin, it differs in composition and structure from the primary wall. Forms during cell differentiation after cell expansion ceases.

**seed**   Develops from the ovule after fertilization of the egg, consisting of protective layers enclosing embryo of seed plants. May contain nutritive endosperm tissue separate from the embryo.

**seed coat** The outer layer of the seed, derived from the integument of the ovule.

**seed dormancy** The state in which a living seed will not germinate even if all the necessary environmental conditions for growth are met. Seed dormancy introduces a temporal delay in the germination process, providing additional time for seed dispersal.

**seed plants** Plants in which the embryo is protected and nourished within a seed. The gymnosperms and angiosperms.

**segregate vegetatively** *See* vegetative segregation.

**selectively permeability (of a membrane)** Membrane property that allows diffusion of some molecules across the membrane to a different extent than other molecules.

**selectivity filter** The domain of a channel protein that determines its specificity of transport.

**self-assembly** Tendency for large molecules under appropriate conditions to aggregate spontaneously into organized structures.

**senescence** An active, genetically controlled, developmental process in which cellular structures and macromolecules are broken down and translocated away from the senescing organ (typically leaves) to actively growing regions that serve as nutrient sinks. Initiated by environmental cues, regulated by hormones.

**sense RNA** RNA capable of translation into a functional protein. *See* antisense RNA.

**sensible heat loss** Loss of heat from leaf surfaces to the air circulating around the leaf, under conditions in which leaf surface temperature is higher than that of the air.

**sensor protein** Specialized plant cellular receptor proteins that perceive external or internal signals. They consist of two domains, an *input domain*, which receives the environmental signal, and a *transmitter domain*, which transmits the signal to the response regulator.

**sepals** Green, leaf-like structures that form the outermost part of a flower. In bud, they enclose and protect other flower parts. *See also* petals, stamens, and carpels.

**sesquiterpene lactones** Bitter, anti-herbivore, 15-carbon terpenes found in members of the composite family, such as sunflower and sagebrush.

**sesquiterpenes** Terpenes having fifteen carbons, three five-carbon isoprene units.

**sex determination** The process whereby unisexual flowers are produced by the early selective abortion of either the stamen or the pistil primordia. Genetically regulated, but also influenced by photoperiod and nutritional status. Mediated by GA.

**shade avoidance response** A response to shading; includes lengthening of the stem.

**shikimic acid pathway** Reactions that convert simple carbohydrate precursors to the aromatic amino acids—phenylalanine, tyrosine, and tryptophan.

**shoot apex** Consists of the shoot apical meristem plus the most recently formed leaf primordia (organs derived from the apical meristem).

**shoot apical meristem** Meristem at the tip of a shoot. Consists of the terminal Central Zone (CZ), which contains slowly dividing, undetermined initials, and the flanking Peripheral Zone (PZ) and Rib Meristem (RM), in which derivatives of the CZ divide more rapidly and then differentiate.

**short-day plant (SDP)** A plant that flowers only in short days (qualitative SDP) or whose flowering is accelerated by short days (quantitative SDP).

**short-distance transport** Transport over a distance of only two or three cell diameters. Involved in phloem loading, when sugars move from the mesophyll to the vicinity of the smallest veins of the source leaf, and in phloem unloading, when sugars move from the veins to the sink cells.

**short interfering RNAs (siRNAs)** RNAs that are structurally and functionally quite similar to miRNAs and also lead to the initiation of the RNA interference pathway.

**short–long-day plants (SLDPs)** Plants that flower only after a sequence of short days followed by long days.

**sieve cells** The relatively unspecialized sieve elements of gymnosperms. Contrast with sieve tube elements.

**sieve effect** The penetration of photosynthetically active light through several layers of cells due to the gaps between chloroplasts permitting the passage of light.

**sieve element–companion cell complex** A functional unit consisting of a sieve element and its companion cell.

**sieve element loading** The movement of sugars into the sieve elements and companion cells of source leaves, where they become more concentrated than in the mesophyll.

**sieve element unloading** The process by which imported sugars leave the sieve elements of sinks.

**sieve elements** Cells of the phloem that conduct sugars and other organic materials throughout the plant. Refers to both sieve tube elements (angiosperms) and sieve cells (gymnosperms).

**sieve plates** Sieve areas found in angiosperm sieve-tube elements; they have larger pores (sieve-plate pores) than other sieve areas and are generally found in end walls of sieve tube elements.

**sieve tube** Tube formed by the joining together of individual sieve tube elements at their end walls.

**sieve tube elements** The highly differentiated sieve elements typical of the angiosperms. Contrast with sieve cells.

**signal peptide** A hydrophobic sequence of 18 to 30 amino acid residues at the amino-terminal end of a chain; it is found on all secretory proteins and most integral membrane proteins and permits their transit across the membrane of the rough ER.

**signal recognition particle** The signal recognition particle (SRP) is a ribonucleoprotein (protein-RNA complex) that recognizes and targets specific proteins to the endoplasmic reticulum in eukaryotes.

**signal transduction** A sequence of processes by which an extracellular signal (typically light, a hormone or neurotransmitter) interacts with a receptor at the cell surface, causing a change in the level of a second messenger and ultimately a change in cell functioning.

**signal transduction cascade** The molecular process by which an initial binding of a signal molecule sequentially alters the activity of a series of proteins, amplifying the initial signal and altering cellular function or gene activity.

**simple pits** Regions where the secondary wall is absent and the primary wall is thin and porous, facilitating water movement between cells. Simple pits always occur opposite simple pits in the neighboring wall, forming pit pairs and a low resistance path for water movement.

**simple sequence repeats (SSR)** *See* microsatellites.

**singlet oxygen ($^1O_2^*$)** An extremely reactive and damaging form of oxygen formed by reaction of excited chlorophyll with molecular oxygen. Causes damage to cellular components especially lipids.

**sink** Any organ that imports photosynthate, including nonphotosynthetic organs and organs that do not produce enough photosynthetic products to support their own growth or storage needs, e.g., roots, tubers, developing fruits, and immature leaves. Contrast with source.

**sink activity** The rate of uptake of photosynthate per unit weight of sink tissue.

**sink size** The total weight of the sink.

**sink strength** The ability of a sink organ to mobilize assimilates toward itself. Depends on two factors: sink size and sink activity.

**size exclusion limit of plasmodesmata** The restriction on the size of molecules that can be transported via the symplast. It is imposed by the width of the cytoplasmic sleeve that surrounds the desmotubule in the center of the plasmodesma.

**skotomorphogenesis** The developmental program plants follow when seeds are germinated and grown in the dark.

**slender mutants** Plants that are phenotypically very tall, due either to constitutive gibberellin response (as in *slr1* and *sln1* mutants of rice and barley respectively) or to elevated levels of bioactive gibberellins (as in the *sln* mutant of pea).

**sln mutant** In peas, a mutant having abnormally high levels of $GA_{20}$ in the seed due to the impairment of a hydroxylation step in GA deactivation. In barley, an ultra-tall mutant resulting from a recessive allele causing negation of a negative signal transduction factor.

**SLY gene** *Sleepy* gene encodes an F-box protein in Arabidopsis that targets GAI and RGA for proteolytic degradation.

**smooth ER** The endoplasmic reticulum lacking attached ribosomes and usually consisting of tubules.

**SNAREs** A class of targeting recognition proteins for the selective fusion and fission of vesicles and tubules within the endomembrane system.

**SnRK1** The plant protein SnRK1 is homologous to the product of the *SNF1* (sucrose *non-fermenting-1*) gene that was identified genetically in screenings of budding yeast mutants. These variants fail to express the invertase gene, *SUC2*, in response to glucose deprivation. SnRK1 phosphorylates and, in so doing, inactivates *in vitro* sucrose 6F-phosphate phosphatase, 3-hydroxy-3-methylglutaryl-Coenzyme A reductase (linked to sterol/isoprenoid synthesis), and nitrate reductase (associated to nitrogen assimilation).

**sodicity** Refers to the high concentration of $Na^+$ in the soil. Contrast with salinity.

**soil analysis** The chemical determination of the nutrient content in a soil sample from the root zone.

**soil hydraulic conductivity** A measure of the ease with which water moves through the soil.

**solar tracking** The movement of leaf blades throughout the day so that the planar surface of the blade remains perpendicular to the sun's rays.

**solute potential (or osmotic potential) $\Psi_s$** The effect of dissolved solutes on water potential.

**solution culture (hydroponics)** The technique of growing plants without soil whereby their roots are immersed in a nutrient solution.

**sorting-out** *See* vegetative segregation.

**source** Any exporting organ that is capable of producing photosynthetic products in excess of its own needs, e.g., a mature leaf or a storage organ. Contrast with sink.

**spatial** Pertaining to space, the form and distribution of matter in space.

**specific heat** Ratio of the heat capacity of a substance to the heat capacity of a reference substance, usually water. Heat capacity is the amount of heat needed to change the temperature of a unit mass by 1° C. The heat capacity of water is 1 calorie per gram per degree Celsius.

**spectrophotometer** Instrument that measures the amount of light absorbed at different wavelengths by a substance.

**spectrophotometry** The technique used to measure the absorption of light by a sample.

**sphingosine kinase** When activated by ABA, the enzyme that catalyzes the phosphorylation of sphingosine to sphingosine-1-phosphate.

**sphingosine-1-phosphate (S1P)** Formed by active sphingosine kinase, stimulates an increase in cytosolic $Ca^{2+}$ and stomatal closure.

**spindle assembly checkpoint** A control point in the cell cycle that stops cells from proceeding into anaphase if spindle microtubules have incorrectly interacted with the kinetochores.

**spiral phyllotaxy** One leaf is produced at each node, with each subsequent leaf produced at a 137° angle to the previously formed leaf.

**spongy mesophyll** Mesophyll cells of very irregular shape located below the palisade cells and surrounded by large air spaces.

**sporopollenin** A complex polymer consisting of fatty acid derivatives and phenylpropanoids that forms the resilient outer wall of the pollen grains.

*spy* **mutant** In Arabidopsis, an ultra-tall constitutive response mutant resulting in the loss of function of a negative regulator.

**squalene** A triterpene ($C_{30}H_{50}$) which serves as the starting point for the synthesis of the whole family of steroids in both plants and animals.

**SRP receptors** A receptor protein on the ER membrane that binds to the ribosome-SRP complex, permitting the ribosome to dock with the translocon pore through which the elongating polypeptide will enter the lumen of the ER.

**stamen** Male reproductive organ of the flower that produces pollen. Consists of stalk (filament) and an anther. *See also* sepals, petals, and carpels.

**starch phosphorylase** Catalyzes the phosphorolysis of ($\alpha$-D-1,4 glucosyl)$_n$ yielding ($\alpha$-D-1,4 glucosyl)$_{n-1}$ and $\alpha$-D-glucose 1-phosphate.

**Starch sheath** A layer of cells that surrounds the vascular tissues of the shoot and coleoptile and is continuous with the root endodermis. Required for gravitropism in Arabidopsis shoots.

**starch–statolith hypothesis** Proposed mechanism for gravitropism involving sedimentation of statoliths in statocytes.

**starch synthase** Catalyzes the reaction: ADP-glucose + ($\alpha$-D-1,4 glucosyl)$_n$ → ADP + ($\alpha$-D-1,4 glucosyl)$_{n+1}$.

**statocytes** Specialized gravity-sensing plant cells that contain statoliths.

**statoliths** Cellular inclusions such as amyloplasts that act as gravity sensors by having a high density relative to the cytosol and sedimenting to the bottom of the cell.

**steady state** A condition where part of a system is constant; for example, the concentrations of intermediates in a reaction or the concentration gradient of a solute across a membrane.

**stele** In the root, the tissues located interior to the endodermis. The stele contains the vascular elements of the root: the phloem and the xylem.

**stele initials** *See* stele stem cells.

**stele stem cells** In the root, cells immediately above (promimal to) the quiescent center that give rise to the pericycle and vascular tissue.

**stem cells** Uncommitted, slowly dividing initial cells that produce all the cells in the meristem and thereby all the cells in the entire plant body (stem and root). Upon cell division, one daughter cell remains a stem cell, while the other is committed to a developmental pathway.

**stigma** The receptive surface for pollen atop the style. *See* carpel.

*STM* **gene** Arabidopsis *SHOOTMERISTEMLESS* gene that suppresses cell differentiation, ensuring that shoot meristem cells remain undifferentiated. Function required for formation of shoot protomeristem.

**stoichiometry** The quantitative relationship between the amounts of reactants consumed and the products produced during a chemical reaction.

**stoma (plural stomata)** A microscopic pore in the leaf epidermis surrounded by a pair of guard cells and in some species, also including subsidiary cells. Stomata regulate the gas exchange (water and $CO_2$) of leaves by controlling the dimension of stomatal pore.

**stomata** *See* stoma.

**stomatal apertures** The opening in the leaf epidermis through which gases enter and leave the interior spaces of the leaf. Changes in stomatal apertures are controlled by guard cells.

**stomatal complex** The guard cells, subsidiary cells, and stomatal pore, which together regulate leaf transpiration.

**stomatal conductance** A measurement of the flux of water and carbon dioxide through the stomata, in and out of the leaf. The inverse of stomatal resistance.

**stomatal movements** Opening and closing of the stomata due to tugor changes in the guard cells.

**stomatal pore** An opening through the epidermis to the leaf interior. Surrounded by a pair of guard cells.

**stomatal resistance ($r_s$)** A measurement of the limitation to the free diffusion of gases from and into the leaf posed by the stomatal pores. The inverse of stomatal conductance.

**straight fertilizers** Chemical fertilizers that contain inorganic salts of only one of the three macronutrients: nitrogen, phosphorus, and potassium (e.g., superphosphate, ammonium nitrate, or muriate of potash, a source of potassium).

**straight-growth test** A bioassay based on the ability of auxin to stimulate the elongation of *Avena* coleoptile sections.

**stratification** In some plants, a cold temperature requirement for seed germination. The term is derived from the former practice of breaking dormancy by allowing seeds to overwinter in small mounds of alternating layers of seeds and soil.

**stress** Disadvantageous influences exerted on a plant by external abiotic or biotic factor(s), such as infection, or heat, water, and anoxia. Measured in relation to plant survival, crop yield, biomass accumulation, or $CO_2$ uptake.

**stress relaxation** Selective loosening of bonds between primary cell wall polymers, allowing the polymers to slip by each other, simultaneously increasing the wall surface area and reducing the physical stress in the wall.

**stress resistance** *See* stress tolerance.

**stress response** A plant response that limits or reduces the damage or potential damage due to a variety of abiotic environmental factors, such as flooding, drought, elevated UV radiation, and high and low temperatures.

**stress tolerance (stress resistance)** A plant's ability to cope with an unfavorable environment.

**stroma** The fluid component surrounding the thylakoid membranes of a chloroplast.

**stroma lamellae** Unstacked thylakoid membranes within the chloroplast.

**stroma reactions** The carbon fixation and reduction reactions of photosynthesis that take place in the stroma of the chloroplast.

**stromules** Protrusions of the inner and outer membranes of the chloroplast.

**style** A stalk-like extension of the carpel. *See* carpel.

**suberin** A wax-like lipid polymer similar to cutin that acts as a barrier to water and solute movement through the Casparian strips of the endodermis, the outer cell walls of underground organs, cork cells of the periderm, and sites of leaf abscission or wounding.

**subfunctionalization** The process by which evolution acts on duplicate genes, causing one copy to be either lost or changed in function, while the other retains its original function.

**subjective day** When an organism is placed in total darkness, the phase of the rhythm that coincides with the light period of a preceding light/dark cycle. *See* subjective night.

**subjective night** When an organism is placed in total darkness, the phase of the rhythm that coincides with the dark period of a preceding light/dark cycle. *See* subjective day.

**subsidiary cells** Specialized epidermal cells that flank the guard cells and work with the guard cells in the control of stomatal apertures.

**substomatal cavity** The air space within the leaf to which the stomatal pore opens and which is bounded by mesophyll cells.

**substrate-level phosphorylation** Involves the direct transfer of a phosphate group from a substrate molecule to ADP to form ATP.

**subunit** A polypeptide that is part of a protein complex.

**SUC2 protein** A sucrose-$H^+$ symporter found in the plasma membrane of companion cells.

**sucrose non-fermenting related kinase2 (SnRK2)** A family of kinases that include ABA-activated protein kinases or stress-activated protein kinases.

**sucrose synthase** One of the enzymes interconverting sucrose and its hexose units. May participate in cellulose synthesis by transferring glucose units from sucrose to UDP-glucose.

**sucrose-$H^+$ symporter** An active carrier that couples the energy dissipated by a proton, diffusing back into the cell to the uptake of a sucrose molecule. Mediates the transport of sucrose from the apoplast into the sieve element-companion cell complex in source leaves of apoplastically loading species.

**sugar-nucleotide polysaccharide glycosyltransferases** A group of enzymes that synthesize the backbones of cell wall polysaccharides.

**sunflecks** Patches of sunlight that pass through openings in a forest canopy to the forest floor. Major source of incident radiation for plants growing under the forest canopy.

**supercool** The condition in which cellular water remains liquid because of its solute content, even at temperatures several degrees below its theoretical freezing point.

**superoxide ($O_2^-$)** A reduced form of oxygen that is damaging to biological membranes.

**superoxide dismutase (SOD)** This enzyme converts superoxide radicals to hydrogen peroxide.

**surface tension** A force exerted by water molecules at the air–water interface, resulting from the cohesion and adhesion properties of water molecules. This force minimizes the surface area of the air–water interface.

**suspensor** In seed plant embryogenesis, the structure that develops from the basal cell following the first division of the zygote. Supports, but is not part of, the embryo that develops from the apical cell and hypophysis.

**SUT1 protein** One of several sucrose-$H^+$ symporters found in the plasma membrane of sieve elements.

**symbionts** Either one of the two organisms associated in a symbiotic relationship that may or may not be mutually beneficial. *See* mutualistic.

**symbiosis** The close association of two organisms in a relationship that may or may not be mutually beneficial. Often applied to beneficial (mutualistic) relationships. *See* mutualism.

**symplast** The continuous system of cell protoplasts interconnected by plasmodesmata.

**symplast pathway** The transport route by which xylem and phloem sap travel from one cell to the next via the plasmodesmata.

**symplastic transport** The intercellular transport of water and solutes through plasmodesmata.

**symport** A type of secondary active transport in which two substances are moved in the same direction across a membrane.

**symporter** A protein carrier involved in symport.

**syntaxins** Proteins that integrate into membranes, permitting the membranes to fuse.

**synthetic auxin** A substance with auxin activity that is not produced by plants, for example, 2,4-D and dicamba. Often used as herbicides because they are not degraded by the plant as quickly as IAA.

**System 1** The regulatory system in which ethylene production in vegetative tissue of climacteric plants inhibits its own biosynthesis.

**System 2** The regulatory system in which ethylene production in ripening climacteric fruit and in the senescing petals in some species stimulates its own biosynthesis.

**systemic acquired resistance (SAR)** The increased resistance throughout a plant to a range of pathogens following the infection of a pathogen at one site.

**systemin** A polypeptide plant hormone that signals production of some defenses not continuously present in plants; initiates the biosynthesis of jasmonic acid.

**systems biology**   An approach to examining complex living processes that employs mathematical and computational models to simulate nonlinear biological networks and to better predict their operation.

**TAM**   *See* tryptamine.

**tandem repeats**   Heterochromatic structures that consist of highly repetitive DNA sequences.

**tannins**   Plant phenolic polymers that often bind proteins and function as defenses against microorganisms, insects, and many mammals, such as cattle, deer, and apes.

**taproot**   The main single root axis from which lateral roots develop.

**TATA box**   Located about 25 to 35 bp upstream of the transcriptional start site, this short sequence TATAAA(A) serves as the assembly site for the transcription initiation complex.

**T-DNA**   A small portion of the *Ti* plasmid that is incorporated into the nuclear DNA of the host plant cell carries genes necessary for the biosynthesis of *trans*-zeatin and auxin, and opines. Because their promoters are plant eukaryotic promoters, none of these T-DNA genes are expressed in the bacterium, but are transcribed after they are inserted into the plant chromosomes.

**telomere**   Region of repetitive DNA that forms the chromosome ends and protects it from degradation.

**telophase**   Prior to cytokinesis the final stage of mitosis (or meiosis) during which the chromatin decondenses, the nuclear envelope reforms, and the cell plate extends.

**temperature coefficient**   *See* $Q_{10}$.

**temperature compensation point**   The temperature at which the amount of $CO_2$ fixed by photosynthesis equals the amount of $CO_2$ released by respiration in a given time.

**tensile strength**   The ability to resist a pulling force. Water has a high tensile strength.

**tension**   Often used to refer to the partial pressure of a gas.

**teratomas**   Tumors that contain partially developed structures. Mutated *tmr* locus of T-DNA produced teratomas with abnormal proliferation of roots.

**terpenes (terpenoids or isoprenoids)**   A large class of plant compounds formed from five-carbon isoprene units, many of which are secondary metabolites with anti-herbivore activity.

**terpenoids (isoprenoids)**   Class of plant lipids that includes carotenoids and sterols.

**tertiary cross-regulation**   Involves the outputs of two distinct signaling pathways exerting influences on one another.

**tetraterpenes**   Terpenes having forty carbons, eight five-carbon isoprene units. The latter compound serves as building blocks for anabolic processes while the excess is excreted as acetate.

**thermal hysteresis**   Phenomenon in which the liquid to solid transition is promoted at a lower temperature than the solid to liquid transition.

**thermal hysteresis proteins**   *See* antifreeze proteins.

**thigmotropism**   Plant growth in response to touch, enabling roots to grow around rocks and shoots of climbing plants to wrap around structures for support.

**thioredoxin**   A small, ubiquitous protein (ca. 12 kDa) whose active site cysteines participate in thiol-disulfide exchange reactions.

**thiosulfide (R–S–)**   An intermediate in the reduction of sulfate. Formed from thiosulfonate. Reacts with O-acetylserine to form cysteine.

**thiosulfonate (R–SO₃–)**   An enzyme-bound intermediate in the reduction of sulfate. Formed from APS. *See* thiosulfide.

**threshold**   The magnitude of a stimulus that needs to be exceeded to elicit a response.

**thylakoid reactions**   The chemical reactions of photosynthesis that occur in the specialized internal membranes of the chloroplast (called thylakoids). Include photosynthetic electron transport and ATP synthesis.

**thylakoids**   The specialized, internal, chlorophyll containing membranes of the chloroplast where light absorption and the chemical reactions of photosynthesis take place.

**Ti plasmid**   A large tumor-inducing plasmid found in virulent strains of *Agrobacterium tumefaciens*.

**TIBA (2,3,5-triiodobenzoic acid)**   A competitive inhibitor of polar auxin transport.

**tip growth**   Localized growth at the tip of a plant cell caused by localized secretion of new wall polymers. Occurs in pollen tubes, root hairs, some sclerenchyma fibers, and cotton fibers, as well as moss protonema and fungal hyphae.

**TIR1/AFB proteins**   A family of F box soluble proteins, subunits of the ubiquitin E3 ligase complex, which are components of the principal auxin receptors.

**TIR1/AFB–AUX/IAA**   The composite auxin receptor located in the nucleus. TIR1/AFBs function as components of a complex of proteins involved in ubiquitination of proteins targeted for destruction by the 26S proteasome.

***tir3* mutant**   Arabidopsis (*transport inhibitor response 3*) mutant that exhibits reduced polar auxin transport and NPA binding. *tir3* contains a mutation in the BIG calossin-like protein, which is also defective in the *doc1* mutant.

**tissue culture**   Growth of isolated plant cells, tissues, or organs in the laboratory on an artificial medium consisting of various essential minerals, vitamins, hormones, and a carbon source.

***tmr* locus**   Mutations at this locus of *T-DNA* produced teratomas with abnormal proliferation of roots. "Rooty" mutations. *See ipt* gene.

***tms* loci**   Mutations in these sites of T-DNA resulted in an abnormal proliferation of shoots. "Shooty" mutations. *See tmr* locus.

**toc** Group of clock mutants in Arabidopsis.

**tonoplast** The vacuolar membrane.

**torpedo stage embryo** The third stage of embryogenesis. The structure produced by elongation of the axis of the heart stage embryo and further development of the cotyledons. *See* globular stage.

**torus** A central thickening found in the pit membranes of tracheids of most gymnosperms.

**totipotency** In differentiated cells, retention of full genetic capacity for the development into a complete plant.

**toxic zone** The range of nutrient concentrations in excess of the adequate zone and where growth or yield declines.

**toxicity effects** Injuries caused by high concentrations of ions (particularly $Na^+$, $Cl^-$, or $SO_4^{2-}$) that accumulate in cells at low water potentials.

**tracheary elements** Water transporting cells of the xylem.

**tracheids** Spindle-shaped, water-conducting cells with tapered ends and pitted walls without perforations found in the xylem of both angiosperms and gymnosperms.

*trans* **Golgi** The face of the Golgi where vesicles and tubules are produced.

*trans*-**acting factors** The transcription factors that bind to the *cis*-acting sequences.

*trans*-**cyclooctene** A strong competitive inhibitor of ethylene binding.

*trans*-**Golgi-network** The tubular network produced at the maturing face or *trans* side of the Golgi.

*trans*-**neoxanthin** Under stress conditions, a $C_{40}$ product formed from *trans*-violaxanthin in the biosynthetic pathways for abscisic acid.

*trans*-**zeatin** The principal free cytokinin; chemically similar to kinetin. Exogenously applied in presence of auxin, induces cell division in callus cells and promotes bud or root formation from callus cultures. Endogenous *trans*-zeatin delays senescence of leaves and promotes expansion of dicot cotyledons. *See* zeatin.

**transamination** Reversible reactions catalyzed by transaminases in which nitrogen as an amino group is transferred from an α-amino acid to an α-keto acid.

**transcellular current** A positive current, due largely to calcium ions, that flows into the cell on one side and loops back through the external medium. Establishes polarity prior to morphogenesis in *Fucus* zygotes, root hairs and germinating pollen grains.

**transcription** The process by which the base sequence information in DNA is copied into an RNA molecule.

**transcription factors** Proteins that interact with DNA promoter elements to modulate gene expression (RNA transcription).

**transcription initiation complex** A multiprotein complex of transcription factors required for binding RNA polymerase and initiating transcription.

**transcriptional regulation** The level of regulation that determines if and when RNA is transcribed from DNA.

**transfer cells** A type of companion cell similar to an ordinary companion cell, but with finger-like projections of the cell wall that greatly increase the surface area of the plasma membrane and increase the capacity for solute transport across the membrane from the apoplast.

**transgene** A foreign or altered gene that has been inserted into a cell or organism.

**transgenic plants** A plant expressing a foreign gene introduced by genetic engineering techniques.

**transit peptide** An N-terminal amino acid sequence that facilitates the passage of a precursor protein through both the outer and the inner membranes of an organelle such as the chloroplast. The transit peptide is then clipped off.

**translation** The process whereby a specific protein is synthesized according to the sequence information encoded by the mRNA.

**translocation** The movement of photosynthate from sources to sinks in the phloem.

**translocator** A membrane-based protein that functions as a carrier in the transport of one or more substances across the membrane.

**translocons** Protein-lined channels in the rough endoplasmic reticulum that form associations with SRP receptors and enable proteins synthesized on ribosomes to enter the ER lumen.

**transmembrane pathway** The route followed by the xylem sap that sequentially passes from cell to cell, crossing the cell plasma membrane both on entering and exiting the cell. Transport across the tonoplast may also be involved.

**transpiration ratio** The ratio of water loss to photosynthetic carbon gain. Measures the effectiveness of plants in moderating water loss while allowing sufficient $CO_2$ uptake for photosynthesis.

**transpiration** The evaporation of water from the surface of leaves and stems.

**transport** Molecular or ionic movement from one location to another; may involve crossing a diffusion barrier such as one or more membranes.

**transport proteins** Transmembrane proteins that are involved in the movement of molecules or ions from one side of a membrane to the other side.

**transposable elements (transposons)** DNA elements that can move or be copied from one site in the genome to another site.

**transposase** An enzyme that catalyzes the movement of a DNA sequence from one site to a different site in the DNA molecule.

**transposon tagging** The technique of inserting a transposon into a gene and thereby marking that gene with a known DNA sequence.

**transposons** *See* transposable elements.

**treadmilling** During interphase, a process by which microtubules in the cortical cytoplasm appear to migrate around the cell periphery due to addition of tubulin heterodimers to the plus end at the same rate as their removal from the minus end.

**triacylglycerols (triglycerides)** Three fatty acyl groups in ester linkage to three hydroxyl groups of glycerol. Fats and oils.

**tricarboxylic acid cycle** *See* citric acid cycle.

**triglycerides** *See* triacylglycerols.

**triose phosphate** A three-carbon sugar phosphate.

**triple response** In Arabidopsis three associated responses to ethylene: inhibition of hypocotyl elongation combined with hypocotyl swelling; inhibition of root elongation; and exaggeration of the curvature of the apical hook.

**trisomies** One kind of aneuploid in which there are three copies of one type of chromosome rather than the normal two.

**triterpenes** Terpenes having thirty carbons, six five-carbon isoprene units.

**tropism** Oriented plant growth in response to a perceived directional stimulus from light, gravity, or touch.

**tryptamine (TAM)** An intermediate in one of three biosynthetic pathways for the synthesis of IAA in plants. Formed by decarboxylation of the amino acid tryptophan.

**tumor** A mass of rapidly dividing, undifferentiated, and disorganized cells.

**tunica layers** The outer cell layers of the shoot apical meristem. Outermost tunica layer generates the shoot epidermis.

**turgor** The firmness of a cell resulting from its hydrostatic or turgor pressure.

**turgor pressure (or hydrostatic pressure)** Force per unit area in a liquid. In a plant cell, the turgor pressure pushes the plasma membrane against the rigid cell wall and provides a force for cell expansion.

**turnover** The balance between the rate of synthesis and the rate of degradation, usually applied to protein or RNA. An increase in

*turnover* typically refers to an increase in degradation.

**type-A ARR** In Arabidopsis, genes that encode response regulators that are made up solely of a receiver domain.

**type-B ARR** In Arabidopsis, genes that encode response regulators that have an output domain in addition to a receiver domain.

**ubiquinone** A mobile electron carrier of the mitochondrial electron transport chain. Chemically and functionally similar to plastoquinone in the photosynthetic electron transport chain.

**ubiquitin** A small polypeptide that is covalently attached to proteins by the enzyme ubiquitin ligase using energy from ATP, and that serves as a recognition site for a large proteolytic complex, the proteasome.

**ubiquitin-activating enzyme (E1)** Part of the ubiquitination pathway. Initiates ubiquitination by catalyzing the ATP-dependent adenylylation of the C terminus of ubiquitin.

**ubiquitin-conjugating enzyme (E2)** Part of the ubiquitination pathway. A cysteine residue on E2 receives the adenylylated ubiquitin produced by ubiquitin-activating enzyme, E1.

**ubiquitin ligase (E3)** SCF complexes that are part of the ubiquitination pathway. Bind to proteins destined for degradation. Lysine residues on E3 receive ubiquitin from the conjugate of ubiquitin-activating enzyme (E2) and ubiquitin.

**ubiquitination pathway** A sequence of reactions that add a small protein, ubiquitin, to proteins targeted for destruction by the proteasome.

**uncoupled** *See* uncoupling.

**uncoupler** A chemical compound that increases the proton permeability of membranes and thus uncouples the formation of the proton gradient from ATP synthesis.

**uncoupling** A process by which coupled reactions are separated in such a way that the free energy released by one reaction is not available to drive the other reaction.

**uncoupling protein** A protein that increases the proton permeability of the inner mitochondrial membrane and thereby decreases energy conservation.

**uniparental inheritance** Form of inheritance shown by both mitochondria and plastids, meaning that offspring of sexual reproduction (via pollen and eggs) inherit organelles from only one parent.

**unisexual** *See* imperfect flowers.

**ureide exporters** Legumes of tropical origin that convert toxic ammonia to ureides such as allantoin, allantoic acid, and citrulline for transport to the shoot via the xylem. Contrast with amide exporters.

**V-ATPase** A vacuolar $H^+$-ATPase.

**vacuolar $H^+$-ATPase** A large, multi-subunit enzyme complex, related to $F_oF_1$-ATPases, present in endomembranes (tonoplast, Golgi). Acidifies the vacuole and provides the proton motive force for the secondary transport of a variety of solutes into the lumen. V-ATPases also function in the regulation of intracellular protein trafficking.

**vacuolar sap** The fluid contents of a vacuole, which may include water, inorganic ions, sugars, organic acids, and pigments.

**vanadate** Inhibitor of the $H^+$-ATPase. The $H^+$-ATPase is phosphorylated as part of the catalytic cycle that hydrolyze ATP. Because of this phosphorylation step, the plasma membrane ATPases are strongly inhibited by orthovanadate ($HVO_4^{2-}$), a phosphate ($HPO_4^{2-}$) analog that competes with phosphate from ATP for the aspartic acid phosphorylation site on the enzyme.

**van't Hoff equation** Relates the solute potential, $\Psi_s$, to the solute concentration.

**variegation** A condition in which leaves show patterns of white and green. Produced by vegetative segregation and may be due to mutations in nuclear, mitochondrial, or chloroplast genes.

**vascular bundles** Strands of primary phloem and xylem, separated by the vascular cambium and often surrounded by a bundle sheath, found in shoots, but continuous with the vascular cylinder of the root.

**vascular cambium** A lateral meristem consisting of fusiform and ray stem cells, giving rise to secondary xylem and phloem elements, as well as ray parenchyma.

**vascular tissue** Plant tissues specialized for the transport of water (xylem) and photosynthetic products (phloem).

**vegetative segregation** A major consequence of organellar inheritance (choroplasts and mitochondria) is that a vegetative cell (as opposed to a gamete) can give rise to another vegetative cell via mitosis that is genetically different because one daughter cell may receive organelles with one type of genome, while the other receives organelles with different genetic information.

**veins** The fine branches and intricate network of leaf vascular bundles.

**velocity** The rate of movement of materials in the sieve elements expressed as the linear distance traveled per unit time.

**vernalization** In some species, the cold temperature requirement for flowering. The term is derived from the word for "spring."

**very low–fluence responses VLFRs** See VLFR.

**vesicles** In cell biology, a small, spherical, membrane-enclosed compartment that functions in the intracellular transport of dissolved substances or membranes from one part of the cell to another. Vesicles typically are derived from components of the endomembrane system, such as the ER, Golgi, or plasma membrane, by a process of budding.

**vesicular-arbuscular mycorrhizal fungi** A nondense mycelium within the root itself and extending into the soil; some hyphae penetrate individual cells of the root cortex as well as extending through the regions between cells.

**vesicular shuttle model** Model for Golgi membrane development in which the *cis*, *medial*, and *trans*

cisternae are stable structures. Cargo movement is via vesicles.

**vessel** A stack of two or more vessel elements.

**vessel elements** Nonliving water-conducting cells with perforated end walls found only in angiosperms and a small group of gymnosperms.

**violaxanthin** The $C_{40}$ xanthophyll (oxygenated carotenoid) that is an intermediate in the pathway for biosynthesis of abscisic acid.

**viscoelastic properties** Properties that are intermediate between those of a solid and those of a liquid and combine viscous and elastic behavior.

**vitamin B6** See pyridoxal phosphate.

**vivipary** The precocious germination of seeds in the fruit while still attached to the plant.

**VLFR (very low fluence response)** Phytochrome responses whose magnitude is proportional to very low fluence (1 to 100 nmol m$^{-2}$).

**volatiles** Plant secondary metabolites that evaporate at ambient temperatures. May be produced and released in response to insect herbivore damage.

**wall creep** In isolated primary cell walls, the time-dependent irreversible extension of cell wall polymers due to their slippage relative to one another.

**wall extensibility (m)** During primary cell wall expansion, the coefficient that relates growth rate to the turgor pressure that is in excess of the yield threshold.

**water deficit** Any water content of a plant cell or tissue below its water content when fully hydrated.

**water potential, $\Psi_w$** Water potential is a measure of the free energy associated with water per unit volume (J m$^{-3}$). These units are equivalent to pressure units such as pascals. $\Psi_w$ is a function of the solute potential, the pressure potential, and the potential due to gravity: $\Psi_w = \Psi_s + \Psi_p + \Psi_g$. The term $\Psi_g$ is often ignored because it is negligible for heights under five meters.

**water potential difference, $\Delta\Psi_w$** The water potential difference across the plasma membrane (expressed in megapascals).

**wavelength (λ)** A unit of measurement for characterization of light energy. The distance between successive wave crests. In the visible spectrum, corresponds to a color.

**waxes** Complex mixtures of extremely hydrophobic lipids synthesized and secreted by epidermal cells and that make up the protective cuticle that reduces water loss from exposed plant tissues. Waxes are mostly alkanes and alcohols of 25- to 35-carbon atoms.

**wetlands** Land that is saturated with water on a regular basis.

**whorl** Pertaining to the concentric pattern of a set of organs that are initiated around the flanks of the meristem.

**whorled phyllotaxy** Several leaves or floral organs of a given type are attached to the stem at the same node.

**wilting** Plant loss of rigidity, leading to a flaccid state, due to turgor pressure falling to zero.

**witches' broom** An abnormal growth of shoots caused by infection with various pathogens.

**wound callose** Callose deposited in the sieve pores of damaged sieve elements that seals them off from surrounding intact tissue. As sieve elements recover, the callose disappears from pores.

**xanthophyll cycle** The interconvertion of violaxanthin and zeaxanthin via the intermediate antheraxanthin. Zeaxanthin accumulates under conditions of excess energy.

**xanthophylls** Carotenoids involved in nonphotochemical quenching. The xanthophyll zeaxanthin is associated with the quenched state of photosystem II, and violaxanthin is associated with the unquenched state.

**xanthoxin** An intermediate in the biosynthetic pathway for abscissic acid (ABA) and a neutral growth inhibitor that has physiological properties similar to those of ABA.

**xylan** A polymer of the five-carbon sugar xylose.

**xylem** The vascular tissue that transports water and ions from the root to the other parts of the plant.

**xylem loading** The process whereby ions exit the symplast and enter the conducting cells of the xylem.

**xyloglucan** A hemicellulose with a backbone of 1→4-linked β-D-glucose residues and short side chains that contain xylose, galactose and sometimes fucose. It is the most abundant hemicellulose in the primary walls of most plants (in grasses it is present, but less abundant).

**xyloglucan endotransglucosylase (XET)** *See* xyloglucan endotransglucosylase/hydrolases (XHTs).

**xyloglucan endotransglucosylase/ hydrolases (XHTs)** A large family of enzymes, which includes xyloglucan endotransglucosylase (XET), having the ability to cut the backbone of a xyloglucan in the cell wall and join one end of the cut chain with the free end of an acceptor xyloglucan.

**yellow cameleon** A calcium indicator protein whose expression in transgenic plants permits monitoring of changes in cellular calcium.

**yield threshold (Y)** The minimum value for turgor pressure at which measurable extension of the cell wall begins.

**yielding** Long-term irreversible stretching that is characteristic of growing (expanding) cell walls. Nearly lacking in nonexpanding walls.

**yielding properties of the cell wall** The capacity of the cell wall to loosen and irreversibly stretch in different ways in response to different internal and external factors.

**Z ("zigzag") scheme** Arrangement of the reaction center and antenna complexes of photosystem II and photosystem I, and the electron transport chain that links them by their midpoint redox potential. The resulting alignment is shaped like a letter "Z."

**zeatin** A naturally occurring cytokinin that stimulates mature plant cells to divide when added to a culture medium along with an auxin. Chemically, *trans*-6-(4-hydroxy-3-methylbut-2-enylamino) purine. *See trans*-zeatin.

**zeatin ribosides** Zeatin with a ribose attached to the amino purine moiety. The main cytokinin in the xylem exudate.

**zeaxanthin** A carotenoid implicated as a blue-light photoreceptor. A component of the xanthophyll cycle of chloroplasts, which protects from excess excitation energy.

**zeitgebers** Environmental signals such as light-to-dark or dark-to-light transitions that synchronize the endogenous oscillator to a 24-hour periodicity.

**zonation** Regional cytological differences in cell division in the shoot apical meristems of seed plants.

**zygote** The single-celled product of the union of an egg and a sperm.

**zygotic** Pertaining to the zygote.

# Author Index

# Subject Index

Entries in *italic* type indicate that the information will be found in an illustration. The designation "n" following a page number indicates the information is found in a footnote.

overview, 20–21
in photorespiration, 210, *211*
speed of reorientation, 420
stromules, 22
structure, 172–173
Chlorosis, 114, 115, 116
Cholesterol, 707
Cholodny, Nicolai, 567
Cholodny–Went model, 567
Chromatin, 36
histone modifications and, 50
structure, 9, *10*
30-nm Chromatin fiber, 9
Chromatin remodeling, 743
*Chromatium*, 352
Chromophores, 499, 500, *501*
Chromoplasts, 21
Chromosomes
linear, 43–44
metaphase, *10, 26, 27*
structural landmarks, 36–37
structure, 9, *10*
Chronic photoinhibition, 254
Chronobiology, 730
Chrysanthemum, 728, 736
*Chrysanthemum*, 373
*Chrysanthemum x morifolium*, 728, 734
Cinnabar moth, 383
Cinnamic acid, 375
*trans*-Cinnamic acid, 376
Circadian clock
light regulation of, 511
photoperiodism and, 736, *737*
regulation of ethylene production, 652
transcriptional negative feedback in, 510–511
CIRCADIAN CLOCK-ASSOCIATED 1 (CCA1), 510, 511
Circadian rhythms
characteristic features, 730–732
examples, 730
fitness and, 511–512
gene expression and, 511
overview, 509–510
phase shifting, *731, 732*
phytochromes and cryptochromes entrain, 732
temperature compensation, 731–732
*Cis*-acting sequences, 48
*Cis*-elements, 50
Cisternae, 10, 12
Citrate. *See* Citric acid
Citrate synthase, *316,* 317
Citric acid
as a chelator, 112–113
citric acid cycle, 316, 317
mitochondrial transport, *323*
in root acidification of the soil, 363
Citric acid cycle
biosynthetic pathways and, *328*
discovery, 315

energy yield, *322*
in mitochondria, 315–316
overview, 307
regulation, *327*
steps in, 316–317
unique features in plants, 317
Citrulline, *279,* 358
*Citrus*, 464, *728*
*Citrus aurantifolia, 152*
*Citrus aurantium, 762*
Citrus fruit, degreening, 667
*Citrus limonia, 152*
*Citrus paradisi, 152*
CK2 protein kinase, 511
*CKI1* gene, 629
CKI1 protein, 629
4-Cl-Iaa, *548*
Cladodes, 261
*clam* mutation, *38*
Class III HD-Zip proteins, 483
Clathrin, *15,* 16
Clathrin-coated pits, *16*
Clausius, R. J., A1-5
*CLAVATA (CLV)* genes, 478–479
Clay soils, 86
categorization, 119–120
Climacteric, 329
Climate change, flowering and, 748
Clock genes, 510–511
Clock hypothesis, 736, *737*
*Clostridium,* 352
Clover, *353,* 358
*Clusia,* 223
CLV1/CLV2 complex, 478, 479
*CLV1* gene, 478
*CLV2* gene, 478
*CLV3* gene, *477,* 478
CLV3 protein, 479
$CO_2$ compensation point, 260
*CO* gene, 737–738, 746, 748
CO protein, 737–738, 739
Coat-imposed dormancy, 686
Coat proteins, 15
Cobalt
inhibition of ethylene synthesis, 653
in plant nutrition, 110
Cocaine, *382, 383*
Cocklebur, 348, 735
Coconut milk, 623
Codeine, *382, 383*
Coenzyme A (CoA), 317
Coenzymes, 327, A1-15
Cofactors, A1-15
Cohesins, 26
Cohesion–tension theory, 93–94
COI1, 388, 412
Coincidence model, 737–739
Colchicine, 41
*Colchicum autumnale,* 41
Cold, responses to, 773–775
Coleoptiles
auxin-induced elongation, 563, 565

gravity perception in, 569
phototropism, *522, 523, 546, 547*
Coleus, 288, 764
*Coleus,* 764
*Coleus blumei,* 288
Collard greens, *41*
Collenchyma cells, *5*
Colorado River, 761
Columella, 458, 569–570, *571*
Columella cells, 419
Columella initials, 470, *471*
Combinatorial model, 727
*Commelina communis,* 379, *526,* 682
Common bean
as a halophyte, *761*
rhizobial symbiont, *353*
root nodules, *352*
transported form of nitrogen in, 358
Companion cells, *5*
functions of, 276
P-proteins and, 275
in pressure-flow translocation, *282*
structure, *273*
types of, 276, *277*
Compatible solutes, 770
Competence, 728–730
vernalization and, 742
Competition, for sunlight, 246–247
Competitive inhibition, A1-18
Complementation, 55
Complex I, 319, 320, 326
Complex II, 319
Complex III, 319
Complex IV, 319–320
Complex V, 321, *323*
Compound fertilizers, 118
Concentration gradient, diffusion and, 72
Condensed tannins, *376, 380, 381*
Condensing enzyme, 334
Conifers
defined, 2
monoterpenes in, 373
pit pairs, 90, *91, 92*
root absorption of mineral nutrients, 123
xanthophyll cycle, 252
Coniferyl, 378
Coniine, *382, 383*
*Conium maculatum,* 383
Conjugated auxin, 550
*CONSTANS (CO)* gene, 737–738, 746, 748
CONSTANS (CO) protein, 737–738, 739
Constitutive defenses, 386
*CONSTITUTIVE PHOTOMORPHOGENESIS (COP)* gene, 508
Contact angle, 70
Contranslational insertion, 13